Analysis of Biological Development

Klaus Kalthoff

The University of Texas at Austin

Boston Burr Ridge, IL Dubuque, IA Madison, WI New York San Francisco St. Louis
Bangkok Bogotá Caracas Kuala Lumpur Lisbon London Madrid Mexico City
Milan Montreal New Delhi Santiago Seoul Singapore Sydney Taipei Toronto

McGraw-Hill Higher Education

*A Division of The **McGraw-Hill** Companies*

ANALYSIS OF BIOLOGICAL DEVELOPMENT, SECOND EDITION

Published by McGraw-Hill, an imprint of The McGraw-Hill Companies, Inc., 1221 Avenue of the Americas, New York, NY 10020.

This book is printed on recycled, acid-free paper containing 10% postconsumer waste.

1 2 3 4 5 6 7 8 9 0 QPH/QPH 0 9 8 7 6 5 4 3 2 1 0

ISBN 0-07-092037-0
ISBN 0-07-118078-8 (ISE)

Vice president and editor-in-chief: *Kevin T. Kane*
Publisher: *James M. Smith*
Senior developmental editor: *Jean Sims Fornango*
Project manager: *Joyce M. Berendes*
Media technology producer: *Lori A. Welsh*
Production supervisor: *Kara Kudronowicz*
Coordinator of freelance design: *David W. Hash*
Cover/interior designer: *Maureen McCutcheon*
Cover images: © Oxford Scientific Films (*Drosophila melanogaster*); © Visuals Unlimited (*Xenopus laevis*)
Senior photo research coordinator: *Carrie K. Burger*
Photo research: *Toni Michaels*
Senior supplement coordinator: *Candy M. Kuster*
Compositor: *GAC—Indianapolis*
Typeface: *10/12 Palatino*
Printer: *Quebecor Printing Book Group/Hawkins,TN*

The credits section for this book begins on page C-1 and is considered an extension of the copyright page.

Library of Congress Cataloging-in-Publication Data

Kalthoff, Klaus.
Analysis of biological development / Klaus Kalthoff.— 2nd ed.
p. cm.
ISBN 0-07-092037-0 — ISBN 0-07-118078-8 (International ed.)
1. Developmental biology. I. Title.

QH491 .K35 2001
571.8—dc21 00-061670
CIP

INTERNATIONAL EDITION ISBN 0-07-118078-8

www.mhhe.com

The cover epitomizes the evolutionary conservation of genetic hierarchies that control basic processes in development. In frogs, the dorsoventral body pattern is controlled by gene products including xolloid (Xld), which inhibits chordin (Chd), which in turn inhibits bone morphogenetic protein 4 (BMP-4). Chordin promotes the formation of dorsal structures including the central nervous system (purple color in cartoon of frog embryo) whereas BMP-4 specifies ventral structures, such as the heart and blood vessels (orange color). In flies, the dorsoventral body pattern is controlled by orthologous gene products including tolloid (Tld), which inhibits short gastrulation (Sog), which in turn inhibits decapentaplegic (Dpp). Sog promotes the formation of the central nervous system, which in flies is located ventrally, while Dpp specifies dorsal structures, such as the heart. Remarkably, Sog mRNA from flies, when injected into frog embryos, has the same effects as frog Chd mRNA, that is, the induction of ectopic dorsal structures. Conversely, frog chordin mRNA ventralizes fly embryos. See section 23.4.

part one

From Gametogenesis to Histogenesis

part two

Control of Gene Expression in Development

part three

Selected Topics in Developmental Biology

To My Teachers, Colleagues, and Students

basic principles

method boxes

key strategies

This text is written for undergraduates and beginning graduate students. Previous course work in cell biology and genetics are needed to make full use of this text. The text is complemented by my web site (www.esb.utexas.edu/kalthoff/bio349), which I use for my own teaching, and which contains regular updates, exam questions, links to movies of interest to developmental biologists, and much more. Independently, McGraw-Hill maintains a website related to this text at www.mhhe.com/kalthoff.

Developmental Biology as an Analytical Process

My central idea in writing the second as well as the first edition has been to present developmental biology as an ongoing process of enquiry. This emphasis gives the aspiring researcher a taste of things to come, and it introduces those who just seek an education to the ways in which developmental biologists gain their knowledge. In Chapter 1, I discuss the surgical removal and transplantation of embryonic parts as *experimental strategies* that have defined embryology as a discipline in its own right. In a preview of genetic analysis in the same chapter, I introduce null alleles and gain-of-function alleles as alternate ways of removing a defined part from a system or allowing it to function at a time or place where it is normally absent. In Chapter 15, where genetic and molecular analysis begin in earnest, I present gene overexpression and dominant interference as equivalents to the mutant alleles. Such discussions of strategies are marked by headings with alternate colors in both the table of contents and the body of the text. Commonly used *methods* are set aside in boxes for easy reference. Throughout the text, I present *key experiments* in more detail than usual, and I conclude these parts with a few questions that test the students' understanding of the analytical process. The answers to these questions will be on my web site.

Principles of Development Provide Structure and Continuity

Next to the emphasis on the analytical process, I have found it important to bring out a few *general principles of development*. For instance, many steps in development rely on overlapping mechanisms that complement or reinforce each other. Spemann referred to this as the principle of double insurance; I introduce this principle in the context of fertilization and then take it up again in chapters on induction and genetic control. Other principles, including default programs and reciprocal interaction, are treated in a similar fashion. I hope that these recurring principles, which are also marked by alternate colors, will provide a sense of continuity and structure in the plethora of details and new results that can easily clutter the mind.

Modern Tools Applied to Basic Questions Foster a Sense of Excitement

What I have tried most to preserve in this second edition is the excitement in the field that has come with the new opportunities to bring the tools of genetic and molecular analysis to bear on the basic questions that have defined the discipline since its inception. Correspondingly, the subdivision of the text into *three major parts* is still the same.

- **Part One** emphasizes classical methods of analysis and covers the series of embryonic stages from gametogenesis to histogenesis. Interspersed are basic conceptual topics such as nuclear totipotency, cell determination, cytoplasmic localization, induction, and morphogenesis.
- **Part Two** introduces the genetic and molecular analysis of development, beginning with a chapter on the use of mutants, DNA cloning, and transgenic organisms. Subsequent chapters explain the concept of differential gene expression with the goal of understanding how the genomic information is used to build a three-dimensional organism that unfolds in time. The chapter on paragenetic information, a topic that was somewhat offbeat in the first edition but has since been recognized by the award of the Nobel prize to Stanley Prusiner, is still providing a counterpoint to the emphasis on genetic information.
- **Part Three** freely combines classical and modern types of analysis and should be the most enjoyable portion of the book. It illustrates how the application of new research tools has led to a better understanding of long-standing issues in development. Near the beginning of this part there is a chapter on the classical analysis of pattern formation, perhaps the most central topic in developmental biology. The chapter on the genetic analysis of patterning in *Drosophila*, which already was the longest one in the first edition, has grown even more, but the award of the Nobel prize to Christiane Nüsslein-Volhard, Eric Wieschaus, and Edward Lewis, gave me the justification to indulge here. Chapters centered on *Mus musculus*, *Arabidopsis thaliana*, and *Caenorhabditis elegans* follow, as do sex determination, hormonal control, and growth.

New to this Edition

- A separate chapter on the evolutionary theory of senescence and some of its proximate causes, topics that have matured to the point where I think they merit discussion in a textbook, is now provided.
- The chapters on cell adhesion and the extracellular matrix, formerly in Part Three, have been combined and moved to Part One, where they fit in more naturally with gastrulation and organogenesis.
- Both text and illustrations are thoroughly updated. About half of all references are new, reflecting the rapid progress being made in the field. A similar fraction of line drawings are new or revised substantially.
- Full color has been introduced to improve the clarity and educational impact of the illustrations.

Acknowledgments

I am grateful to some special people who provided inspiration and support for writing this new edition. I still feel connected to my mentor, Klaus Sander, who long ago introduced me to the history and culture of developmental biology. At The University of Texas at Austin, I enjoy the company of a diverse but congenial group of colleagues who, through countless journal club sessions and individual conversations, have broadened my outlook on developmental biology and its practitioners. My wife, Karin, has again been a source of support and balance throughout the project.

It has been a great pleasure to work with Gwen Gage and Kristina Schlegel, two expert computer illustrators who converted my rough drafts into pieces of art. Ondine Cleaver was again part of the illustration team, this time as advisor and coordinator. The contributions of the McGraw-Hill team are also appreciated including those of publisher, Jim Smith, developmental editors, Deborah Allen and Jean Fornango, and the production team headed by Joyce Berendes.

The following reviewers read draft chapters and provided valuable criticisms.

Stephanie Aamodt (Louisiana State University, Shreveport)
Robert C. Angerer (University of Rochester)
Karl Aufderheide ((Texas A&M University, College Station)
Bruce Babiarz (Rutgers—The State University of New Jersey)
Michael Bender (University of Georgia)
Karen Bennet (University of Missouri)
Edward Berzin (Maimonides Medical Center, Brooklyn, New York)
Antonie W. Blackler (Cornell University)
Seth Blair (University of Wisconsin)
Hans Bode (University of California, Irvine)
Bradley Bowden (Alfred University)
Bruce Bowerman (University of Oregon, Eugene)
John L. Bowman (University of California, Davis)
Maureen Brandon (Idaho State University)
Marianne Bronner-Fraser (California Institute of Technology)
Daniel Brower (University of Arizona, Tucson)
Carole Browne (Wake Forest University)
Rudolf Brun (Texas Christian University)
Susan Bryant (University of California, Irvine)
James Bull (University of Texas, Austin)
David Capco (Arizona State University)
Thomas W. Cline (University of California, Berkeley)
Karen Crawford (St. Mary's College of Maryland)
Yolanda Cruz (Oberlin College)
Mark Dansker (University of Texas, Austin)
Heather Dawes (University of California, Berkeley)
Marie DiBerardino (Medical College of Pennsylvania)
Alyce DeMarais (University of Puget Sound)
Thomas Drysdale (University of Texas, Austin)
Susanne Dyby (U.S. Department of Agriculture, Gainesville, Florida)
Elizabeth Eldon (University of Notre Dame)
Richard Elinson (University of Toronto)
William Elmer (Emory University)
Dennis Englin (Masters College)
David Epel (Stanford University)
Carol Erickson (University of California, Davis)
Susan G. Ernst (Tufts University)
Charles Ettensohn (Carnegie-Mellon University)
Kathy Foltz (University of California, Santa Barbara)
Joseph Frankel (University of Iowa)
Gary Freeman (University of Texas, Austin)
John Gerhart (University of California, Berkeley)
Michael (Goldman (San Francisco State University)
Paul B. Green (Stanford University)
Edwin P. Groot (Miami University, Ohio)
Ernst Hafen (University of Zurich)
Jeffrey Hardin (University of Wisconsin, Madison)
Rosalind Herlands (Richard Stockton College)
Ira Herskowitz (University of California, San Francisco)
Erwin Huebner (University of Manitoba)
Nicholas Hole (University of Durham)
Becky A. Houck (University of Portland)
Laurie Item (Purdue University)
Herbert Jäckle (Max Planck Institute for Biophysical Chemistry, Göttingen, Germany)
Marcelo Jacobs-Lorena (Case Western Reserve University)
Antone G. Jacobson (University of Texas, Austin)
Laurinda Jaffe (University of Connecticut at Farmington)
Andrew D. Johnson (Florida State University)
Raymond Keller (University of Virginia, Charlottesville)
Gregory M. Kelly (University of Western Ontario)
Kenneth Kemphues (Cornell University)
Michael Kessel (Max Planck Institute for Biophysical Chemistry, Göttingen, Germany)
Chris Kintner (Salk Institute, San Diego)
Mark Kirkpatrick (University of Texas, Austin)
David Knecht (University of Connecticut at Storrs)
Ruth Lehmann (New York University, School of Medicine)
Sally J. Leevers (Ludwig Institute for Cancer Research)
Wallace LeStourgeon (Vanderbilt University)
Larry Liddle (Long Island University)
Jeanne Lust (St. John's University)
Hong Ma (Pennsylvania State University)

Vicki Martin (University of North Carolina)
Patrick H. Masson (University of Wisconsin)
James Mauseth (University of Texas, Austin)
Jeffery B. McCallum (East Carolina University)
Cathy McElwain (Loyola Marymount University)
Andrew McMahon (Harvard University)
Judy Medoff (St. Louis University)
Douglas Melton (Harvard University)
Marco Milan (European Molecular Biology Laboratory)
John Morrill (New College)
Susan Mosier (University of Nebraska)
Ken Muneoka (Tulane University)
Gerald L. Murison (Florida International University)
Diana G. Myles (University of California, Davis)
Jeanette Natzle (University of California, Davis)
Raymond Neubauer (University of Texas, Austin)
Richard Nuccitelli (University of California, Davis)
Deborah O'Dell (Mary Washington University)
Nipam Patel (Howard Hughes Medical Institute)
Gail R. Patt (Boston University)
Jane Petschek (Miami University of Ohio)
Mitchell Price (Pennsylvania State University)
JoAnn Render (University of Illinois)
Deborah Ricker (York College of Pennsylvania)
Lynn M. Riddiford (University of Washington)
Karel Rogers (Grand Valley State University)
Joel H. Rothman (University of California, Santa Barbara)
Stanley Roux (University of Texas, Austin)
Klaus Sander (University of Freiburg, Germany)
Amy Sater (University of Houston)
Helmut Sauer (Texas A & M University)
Gary Schoenwolf (University of Utah)
Anne M. Schneiderman (Cornell University)
Trudi Schüpbach (Princeton University)
Susan Singer (Carleton College)
Hazel L. Sive (Massachusetts Institute of Technology)
Mary Lee Sparling (California State University, Northridge)
David L. Stocum (Purdue University)
William S. Talbot (University of Oregon)
William H. Telfer (University of Pennsylvania)
William Terrell (University of Texas, Austin)
Kathryn Tosney (University of Michigan)
Akif Uzman (University of Houson)
Gunnar Valdimarsson (University of Manitoba)
Peter Vize (University of Texas, Austin)
Walter Walthall (Georgia State University)
Stanley Wang (University of Texas, Austin)
Detlef Weigel (California Institute of Technology)
William Wood (University of Colorado)
Gregory A. Wray (State University of New York at Stony Brook)
Paul Wright (Western Carolina University)
Phillip Zinsmeister (Oglethorpe University)

I thank all these individuals for their most valuable contributions.

Klaus Kalthoff

Welcome to the most modern introduction to this exciting and dynamic field. Developmental biology is presented with enthusiasm as an experimental science, as an ongoing process of enquiry. Carefully structured and clearly written, this text helps students develop a broad perspective on the major themes and supporting experimental details.

part three

Selected Topics in Developmental Biology

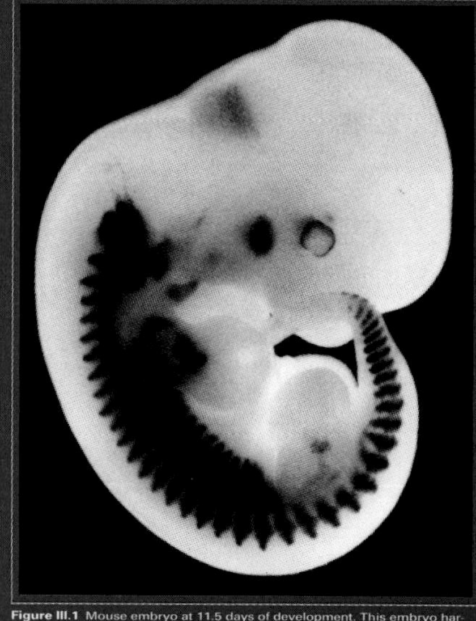

Figure III.1 Mouse embryo at 11.5 days of development. This embryo harbors a transgene consisting of the transcribed region of the bacterial β-galactosidase gene and the regulatory region of the mouse *myogenin* gene. The dark stain, caused by β-galactosidase, indicates that the transgene is expressed with the same pattern as the normal *myogenin* gene—that is, in the myotomal regions of the somites and the limb buds. The *myogenin* gene is necessary and sufficient for the differentiation of skeletal muscle tissue.

IN Part Three of this text, we will discuss a series of topics that are at once old and new. Although extensively investigated by early embryologists, these topics are still providing new frontiers for researchers. Interest in these questions has been revitalized by the arrival of new research methods such as advanced microscopy and genetic analysis, as well as DNA cloning and sequencing. Bringing these modern tools to bear on the long-standing problems of the discipline is what generates much of the current excitement in developmental biology. ● Two of the topics at the core of developmental biology are cell differentiation and pattern formation. How can cells that originate from the same rudiment produce different tissues such as bone and muscle? And how are different pathways of cell differentiation orchestrated in time and space, so that a well-formed organ like the human hand results instead of a randomly aggregated mass of different tissues? These questions will be discussed in general terms and specifically with regard to four of the organisms that are amenable to genetic analysis: the fruit fly *Drosophila*, the house mouse *Mus musculus*, the mouse ear cress *Arabidopsis thaliana*, and the roundworm *Caenorhabditis elegans*. ● The topics of sex determination, hormonal control, growth, and senescence are included in Part Three because their discussion relies heavily on genetic and molecular analysis. In particular, we will consider the following questions: How do genes or the environment determine sex? How do hormones regulate sexual development and metamorphosis? How are cell division and growth rates controlled in different tissues and in different body regions? What can we learn from tumors about the genes that control normal growth? What are the roles of programmed cell death and senescence in development? ● Developmental functions are controlled by autonomous programs within cells, by close-range interactions among neighboring cells, and by hormones and other long-distance signals. Integrating all these controls requires complex networks of genetic interactions. The analysis of these signaling pathways has become a major effort in contemporary developmental

499

Superior Organization

The subject is divided into three parts—Part One introduces classical methods of analysis and the stages of embryonic development from *gametogenesis* to *histogenesis*. Part Two covers the genetic and molecular analysis of development. Part Three illustrates how the application of new research tools has led to a better understanding of long-standing issues in development. Each section begins with an overview statement that provides the proper context for the coming chapters.

chapter 21

Pattern Formation and Embryonic Fields

Figure 21.1 Peacock displaying a dazzling array of eyespots on his tail feathers. The eyespots exemplify the ability of organisms to form patterns, harmonious arrays of different elements. In this case, the pattern arises through the synthesis of dark pigments in a circular territory of cells, and the synthesis of lighter pigments in drop-shaped territories of cells surrounding the dark spot. How are the signals that determine cells to make dark and light pigments coordinated in space so that a well-shaped eyespot results, rather than a salt-and-pepper mixture of dark and light cells?

PATTERN formation is one of the central topics in developmental biology. ***Patterns*** are harmonious arrays of different elements, such as the array of five fingers on a human hand. Patterns can be seen at different levels of organization. For instance, the overall body pattern of birds features a head, a body, a pair of wings, a pair of legs, and a tail. Each of these organs shows a typical pattern of its own, including feathers of different sizes, shapes, and colors. Each feather, in turn, is composed of a shaft with barbs and barbules. These consist of dead, pigmented epidermal cells, often arranged in eye-catching patterns (Fig. 21.1). Cells, too, have internal patterns, but in this chapter we will focus on patterns consisting of cells or larger elements.

Many patterns are formed when different types of cells mature in a coordinated fashion. For example, the eyespot pattern in a peacock feather results from the arrangement of cells containing different pigments in nested rings, with the darkest pigment occupying the center. The development of such patterns raises two questions. First, how do seemingly equal precursor cells give rise to different types of mature cells? These are the processes of *cell determination* and *cell differentiation* discussed in previous chapters. Second, how are multiple pathways of cell determination coordinated so that an orderly array is formed rather than a random mixture of cell types? In other words, how does spatial coordination occur? The answer to the second question may hold the key to the first: The same signals that bring about spatial coordination also control, at least in part, the determination of individual cell types.

Even though patterns become most obvious after cell differentiation, the process of pattern formation begins much earlier. Consider the development of the vertebrate limb. An early limb bud consists of an ectodermal cover and a mesenchymal core. The latter gives rise to muscles and to the cartilage models of bones. However, before the core cells become histologically different, they exhibit regular patterns of cell division and cell condensation. Cell division takes place more rapidly in a progress zone near the tip of the limb. Cell condensations are areas of higher cell density, which develop into either cartilage or muscle. The cartilage condensations proceed from proximal regions to distal and synthesize the extracellular matrix characteristic of cartilage (Fig. 21.2)

In addition to patterns based on cell division and condensation, there are patterns of *apoptosis*, or programmed cell death, especially in appendages with separated digits, such as the human hand or the foot of a chicken. The cells between the developing digits of the limb bud release a genetic program that kills the cells in a noninflammatory and economical way. As the dead cells are consumed by macrophages, the digits become separated. Thus, the processes of cell division, cell differentiation, and apoptosis are all involved in generating spatial patterns. As each pattern is created and refined, the visible complexity of the organism increases.

In many cases, the cellular territories in which patterns are formed have the properties of ***fields***. A field is defined as a group of cells that can form a certain structure, such as a limb. A key attribute of fields is that they can be enlarged or reduced in size and still give rise to the same complete and normally proportioned structure, except that the latter is now formed on a larger or smaller scale. This implies that fields are capable of *regulation*, an attribute discussed in Section 6.6. Cells in fields can adjust their state of determination according to signals received from other cells.

In Chapters 21 through 25, we will examine concepts and experimental strategies researchers have used in the study of pattern formation. In the present chapter, we will discuss mostly classic experiments in which cell isolation and transplantation have served as the principal methods. These experiments have provided the conceptual framework for the study of pattern formation. In Chapters 22 through 25, the emphasis will be on the genetic and molecular tools that have helped investigators to probe directly into the patterning mechanisms of fruit flies, vertebrates, flowers, and roundworms.

21.1 Regulation and the Field Concept

Most of the cells that engage in pattern formation are still capable of regulation: Their *potency* is greater than their *fate*. As an example, let us consider the cells that form the forelimb in the salamander *Ambystoma maculatum*. A *fate map* for the future forelimb at an advanced embryonic stage is shown in Figure 21.3. A circular area of somatic lateral mesoderm and overlying epidermis is fated to form the free limb. The free limb area is surrounded by a ring of cells that normally form the shoulder girdle and the peribrachial flank from which the free limb extends. Together, these areas are called the ***limb disc***.

Big Picture

Each chapter begins with an introduction that orients the student to the issues within the chapter.

method 8.1

In situ Hybridization

A key method in the genetic and molecular analysis of development is known as ***in situ hybridization*** (Lat. *in situ*, "in its natural position"). It is based on the fact that single-stranded DNA and RNA molecules form stable hybrids if their nucleotide sequences are complementary. The method is used most frequently to determine the location of a specific mRNA in wholemounts or sections of fixed tissue. For this purpose, one synthesizes a complementary strand of labeled DNA or RNA (see Fig. 15.11). This labeled probe is incubated with the fixed specimen to allow hybrid formation to occur. Unspecifically adhering probe is then washed off under conditions that preserve nucleic acid hybrids.

Commonly used labels are radioisotopes, which are detected by *autoradiography* (see Method 3.1). In this case, the hybridization and washing steps are followed by the application of a layer of photographic film emulsion (in the dark). The dried film layer is exposed to the radioactive signal for an appropriate period of time before it is developed and fixed like a photographic film. When viewed under the microscope, the silver traces generated by the radioactivity appear as dark grains in transmitted light and as white grains in reflected light (see Figs. 8.10 and 8.26).

More recently, most investigators have switched to chemically labeled DNA probes for in situ hybridization. In a modified nucleotide commonly used for this purpose, the methyl group of thymine is replaced with the plant steroid digoxygenin, which does not interfere with DNA synthesis or base pairing. The digoxygenin in turn is recognized by an antibody that carries a visible label. Thus, in situ hybridization with a chemically labeled probe works like *immunostaining* (see Method 4.1), except that the primary antibody is replaced with the chemically labeled hybridization probe. This type of probe has improved resolution, avoids the time-consuming exposure of photographic emulsion, and does not require the safety precautions associated with radioisotopes.

mRNA, are provided with an engineered oskar mRNA that mislocalizes to the anterior pole, then these eggs form polar granules and functional pole cells at both the anterior and the posterior pole. Embryos developing from such eggs show an abnormal body pattern in which head and thorax are replaced with a mirror-image duplication of abdominal segments (see Fig. 22.3).

The localization of oskar mRNA has been explored by *in situ hybridization*, a method that reveals the location of specific RNA or DNA sequences in histological sections and wholemounts of embryos (see Method 8.1). Transcribed in the nurse cells of each ovarian egg chamber, oskar mRNA is rapidly transferred via cytoplasmic bridges to the adjacent oocyte, where it forms a transient anterior accumulation. Subsequently, oskar mRNA moves all the way from the anterior to the posterior pole, where it forms a sharp localization (Fig. 8.10). This astounding movement depends on an array of polarized microtubules: *colchicine*, a drug that interferes with microtubule assembly, blocks oskar mRNA movement, and mutations that change microtubule polarity in oocytes cause characteristic mislocalizations of oskar mRNA (Theurkauf et al., 1993; Theurkauf, 1994; see Section 22.3). The normal movement of oskar mRNA also depends on the activity of additional genes including *staufen*$^+$. Staufen protein binds to oskar mRNA, apparently linking it to a kinesin *motor protein* moving to the plus end of microtubules (Clark et al., 1994). Once arrived at the posterior egg pole, oskar mRNA is translated into protein, which interacts with staufen protein and yet another protein encoded by the *vasa*$^+$ gene (Breitwieser et al., 1996; Rongo et al., 1995). These interactions apparently anchor oskar mRNA to cytoskeletal elements at the posterior pole and form a ribonucleoprotein complex that traps additional components to form mature polar granules. Trapping of the additional components seems to be facilitated by cytoplasmic streaming later during oogenesis (Gutzeit, 1986).

The molecular nature of the germ cell determinants that become sequestered in polar granules, and the

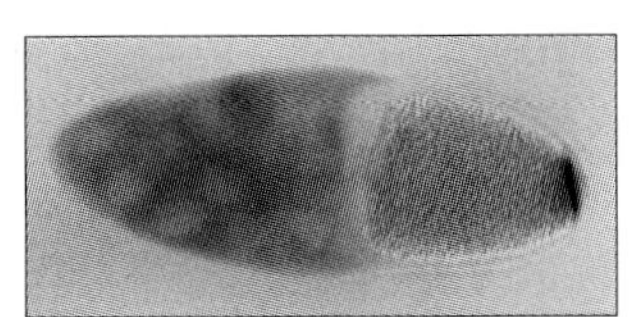

Figure 8.10 Localization of oskar mRNA during *Drosophila* oogenesis. Along with other maternal mRNAs, oskar mRNA is transcribed in the nurse cells of each ovarian egg chamber (to the left), from where it is rapidly transferred via cytoplasmic bridges to the adjacent oocyte. The part of the oocyte that is adjacent to the nurse cells will become the anterior pole region of the egg. From here, oskar mRNA moves all the way to the posterior pole, where it forms a sharp localization (dark band, made visible by in situ hybridization; see Method 8.1). This astounding movement depends on several other gene products as well as an array of polarized microtubules. Once arrived at the posterior egg pole, oskar mRNA is translated into protein, which interacts with other proteins in forming so-called polar granules, which are anchored to cytoskeletal elements and serve as repositories for germ cell and general posterior cytoplasmic determinants.

Modern Tools

Commonly used techniques are presented in *method boxes* and referred to throughout the text.

An Experimental Science

Key Experiments are presented in detail and highlighted by shading within the narrative. Each concludes with several thought questions that test the students' understanding of the experimental process. Students learn to think analytically and develop an appreciation for the connection between experiments and scientific models. (The answers to these questions will be on the McGraw-Hill, as well as the author's website.)

instead of arginine in position 171 are more prone to developing scrapie (Westaway et al., 1994). Conceivably, the mutations that predispose their carriers to prion diseases weaken the normal secondary structure of PrP^C and facilitate its refolding into PrP^{Sc}.

ONE prediction derived from the *"protein only" hypothesis* is that organisms without prion-precursor protein should be resistant to prion diseases. It became practical to test this hypothesis when it was discovered that transgenic mice without functional prion-precursor (PrP^C) genes either develop normally or show only slight neurological symptoms (Büeler et al., 1992; Collinge et al., 1994). Such "knockout" mice were inoculated, by injection into a cerebral hemisphere, with brain tissue from mice with scrapie (Büeler et al., 1993; Prusiner et al., 1993). The injected mice were observed for behavioral signs of scrapie, including abnormal gait, feet clasping when lifted, disorientation, and depression. At intervals, one of the injected mice was sacrificed and the brain examined histologically for the vacuolization and tissue damage that are characteristic of spongiform encephalopathy.

Homozygous mutant mice lacking both genes for prion-precursor protein remained symptom-free for more than 500 days (Fig. 19.12). In contrast, when mice carrying the wild-type alleles for prion-precursor protein were subjected to the same treatment they developed scrapie symptoms within less than 165 days. Heterozygous mice with one normal prion-precursor gene developed dysfunctions between 400 and 465 days after inoculation. Also, brain homogenates from inoculated knockout mice failed to cause scrapie in normal mice, whereas brain homogenates from inoculated wild-type mice transferred the disease.

Questions

1. The experiment described above proves that prion-precursor protein is *necessary* for contracting and transmitting scrapie. What kind of experiment would prove that the transition from prion-precursor protein to prion protein (PrP^C to PrP^{Sc}) by itself—without a chance for viral contamination—is *sufficient* to cause scrapie?
2. Suppose you would graft brain tissue from mice carrying wild-type prion genes into inoculated knockout mice. Which tissue(s) would you expect to develop the signs of spongiform encephalopathy: the grafted tissue, the host brain, or both? Remember that the prion-precursor protein is tethered to the plasma membrane.
3. Assuming that the biological functions of prion-precursor protein are equivalent in mice, sheep, and cattle, would it be possible to generate sheep and cattle whose meat could be eaten by humans without any risk of contracting spongiform encephalopathy?

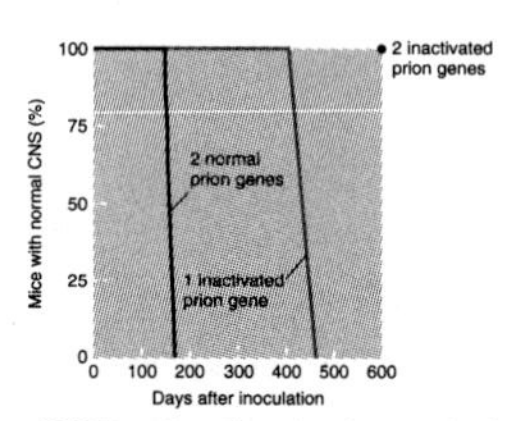

Figure 19.12 Dependence of scrapie on the expression of the gene for prion-precursor protein. Scrapie agent from diseased mice was transferred to healthy mice with two genes for normal prion-precursor protein, and to healthy transgenic mice in which one prion gene or both had been inactivated. The normal mice began to show neurological dysfunction within less than 165 days, and the mice with one functional prion-precursor gene developed the same symptoms after 400 to 465 days. In contrast, the mice without functional prion-precursor genes were symptom-free after more than 500 days.

In summary, the available data indicate that the prion protein exists in different three-dimensional conformations, in the normal and harmless PrP^C conformation and in one or more pathogenic PrP^{Sc} conformations. According to the "protein only" hypothesis proposed by Prusiner, and accepted by most others in the field, the PrP^{Sc} conformations by themselves are necessary and sufficient to cause scrapie diseases by directing the conversion of PrP^C into more PrP^{Sc} and thus disturbing the normal metabolism of prion protein. When Stanley B. Prusiner was awarded the Nobel prize in Physiology or Medicine for his pioneering work on prions in 1997, the Nobel Assembly of Stockholm broke with the tradition of considering only achievements that have won universal acceptance. Instead, the Assembly honored the main architect of an unorthodox theory of general importance, which is still disputed by some experts, on how much information cells can pass on in the form of protein structure, independently of DNA or RNA.

TUBULIN DIMERS ASSEMBLE INTO DIFFERENT ARRAYS OF MICROTUBULES

The tubulin polypeptides synthesized in eukaryotic cells form heterodimers, consisting of one α- and one β-tubulin. These dimers assemble into the cylinders known as *microtubules* (see Section 2.3). Depending on cell type and phase in the cell cycle, microtubules form different arrays (see Fig. 2.5). In mitotic cells, many short and straight microtubules radiate out from two *centrosomes* forming the asters at the spindle poles. During interphase, a loose network of long and wavy microtubules, often forming a tight meshwork around the

14.5 The Principle of Reciprocal Interaction

When two singers prepare to perform a duet in an opera, they will usually begin by practicing their parts separately but eventually rehearse together so that they can learn to pick up their cues from each other and adjust to each other's nuances. Similarly, many organs are constituted from more than one rudiment, which originate separately before they grow together and exchange inductive signals that coordinate their later phases of development. The eye, for example, originates from two major components: the *optic vesicle,* which gives rise to the optic cup and then the retina, and the *lens placode,* which forms the lens (see Fig. 1.16). One of the signals that contribute to lens induction comes from the optic vesicle. As a result, the lens placode is centered over the optic vesicle, and the developing lens fits the size of the underlying optic cup. However, there is not just one unidirectional signal from the optic vesicle that induces the overlying ectoderm to form a lens placode. Rather, the two rudiments exchange signals back and forth. This was shown by *heterospecific transplantations,* in which epidermis from a large-eyed donor species was transplanted over the optic vesicle of a small-eyed host species. The lens rudiment formed by the graft initially was too big for the host's optic cup, but as the chimeric eye grew, the lens slowed down in its growth while the optic cup grew larger than normal. The resulting eye was well-proportioned and intermediate in size between normal donor and host eyes (see Section 28.2).

The example described above illustrates the ***principle of reciprocal interactions,*** that is, the ability of parts of a developing organ to exchange signals that coordinate their development in space and time.

The development of the metanephros is another case illustrating the same principle (Saxén, 1987; Vize, 1997). As a first step, *metanephrogenic mesenchyme* induces the adjacent segment of the *Wolffian duct* to form the *ureteric bud.* As the bud enters the mesenchyme, it begins to branch (Fig. 14.26). The signal causing the branching seems to be a peptide growth factor that is secreted by the mesenchyme and acts on a tyrosine kinase receptor present on the bud epithelium (Robertson and Mason, 1997). In turn, the branching bud epithelium keeps adjacent metanephrogenic mesenchyme cells alive and causes them to condense and form nephric tubules (Fig. 14.27). Continued interaction between epithelium and mesenchyme causes the ureter to branch seven or eight times, generating the hierarchical system of excretory tubules. Simultaneously, the mesenchyme forms nearly 1 million nephrons in a human kidney (H. W. Smith, 1951). In addition to nephric tubules, the mesenchyme generates a supporting connective tissue known as *stroma.* The role of the stroma in kidney development is not well characterized, but seems required for the nephrons' further differentiation.

The transformation of metanephrogenic mesenchymal cells into tubular epithelium involves major changes in the cell adhesion, substrate adhesion, and extracellular matrix (ECM) composition. The formation of epithelial *zonula adherens* junctions (see Fig. 11.20) is associated with the synthesis *of E-cadherin* as usually. *Syndecan,* an adhesive proteoglycan (see Fig. 11.17), is synthesized heavily during the condensation of metanephric mesenchyme around the branches of the ureteric bud. Here, syndecan appears not only to enhance cell adhesion but also the proliferation of mesenchymal cells, presumably through syndecan's ability to sequester growth factors (Vaino et al., 1992). Some typical components of mesenchymal ECM, such as *fibronectin* and *collagen types I and III,* disappear while basement membrane components, including *collagen type IV, laminin,* and *nidogen,* are being synthesized cooperatively by mesenchymal and epithelial cells (Ekblom et al., 1994). Some or all of these activities may be coordinated by Pax-2, a gene-regulatory protein synthesized in both cell types (Torres et al., 1995).

Reciprocal interactions between epithelia and mesenchyme are observed in the development not only of

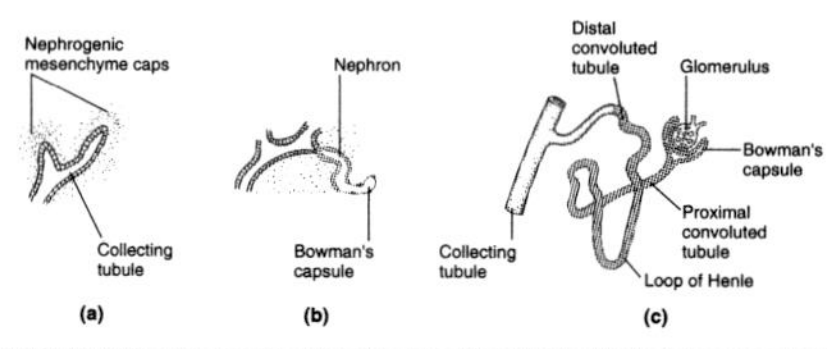

Figure 14.27 Nephron development in the metanephros. The tips of the collecting tubules induce caps of nephrogenic mesenchyme to form nephrons. The nephrons establish open communication with the collecting ducts, enabling urine to flow from Bowman's capsule to the renal pelvis.

Simplicity Among Details

Basic Principles showcase major phenomena and mechanisms of development and help students identify core unifying ideas. These principles are highlighted in color in the Table of Contents for easy reference.

Experimental Design

Key Strategies used by developmental biologists in designing experiments are discussed explicitly. For example, the strategy presented in chapter four is "A Bioassay is a Powerful Strategy to Reveal a Biologically Active Component."

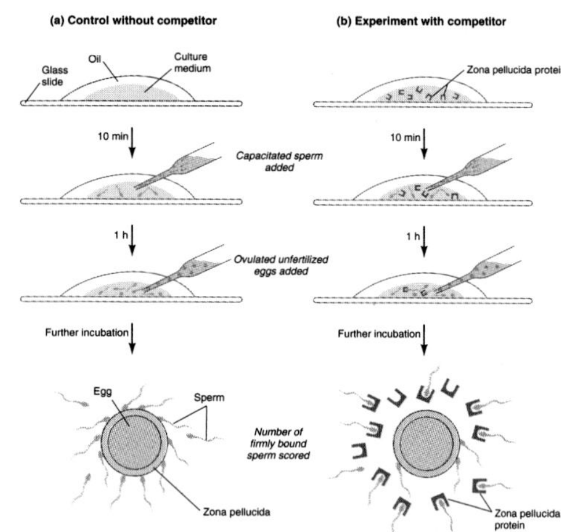

Figure 4.11 Bioassay to identify the mouse zona pellucida component to which the sperm adheres. **(a)** Control experiment without competitor. A drop of culture medium is kept on a glass slide under oil to prevent desiccation. Capacitated sperm are added to the drop from a glass micropipette. After 1 h of incubation, unfertilized ovulated eggs are added. After further incubation, the eggs are washed to remove loosely attached sperm. The average number of firmly bound sperm per egg is scored. **(b)** Same experiment except that zona proteins, or moieties thereof, are added to the culture medium. If the added components bind to sperm in solution, they effectively compete with the eggs for binding sites on the sperm, thus reducing the number of sperm bound to the eggs. For results see Figure 4.12.

(Wassarman, 1987, 1990). The zonae of other mammals, including humans, consist of similar glycoproteins.

IN order to identify the zona protein, and the exact moiety thereof, to which mouse sperm bind, Bleil and Wassarman (1980) used the bioassay described above (Fig. 4.11). As part of establishing this bioassay, they had already shown that total zona proteins from unfertilized eggs inhibit the binding of intact unfertilized eggs to capacitated sperm. In contrast, the same glycoproteins from 2-cell embryos did not have this inhibitory effect. Apparently, zona glycoproteins from unfertilized eggs were competing with intact unfertilized eggs for binding sites on the sperm. In 2-cell embryos, the zona glycoproteins seemed to be modified so that sperm could no longer bind to them.

Once this bioassay was established, each zona glycoprotein was purified and tested individually in the same way. Only ZP3 inhibited sperm-egg adhesion. The other zona proteins did not compete with unfertilized eggs for binding sites on the sperm (Fig. 4.12). Evidently, ZP3 is the only zona glycoprotein that binds to mouse sperm during the initial sperm-egg adhesion.

Molecular analysis of ZP3 showed that its backbone is a single polypeptide of about 400 amino acids (Wassarman, 1990). Extending from the polypeptide are many ***oligosac-***

Klaus Kalthoff is Professor and Chairman of the Section for Molecular Cell and Developmental Biology at The University of Texas at Austin. Born and raised in Germany, he studied biology at the universities of Erlangen, Hamburg, and Freiburg. He received his Ph.D. degree (Dr. rer. nat.) in 1971 from the University of Freiburg, where he completed his dissertation on pattern formation in insect embryos under the direction of Klaus Sander. Subsequent work done by Kalthoff and coworkers, also at Freiburg, showed that eggs of midges contain ribonucleoprotein particles acting as anterior determinants: they are localized near the anterior pole of the egg and direct the formation of anterior body parts. Kalthoff also discovered that damage from ultraviolet light to certain insect eggs is reversible in a catalyzed reaction that depends on light of longer wavelength. He was awarded the Prize of the Scientific Society of Freiburg in 1975.

Kalthoff moved to Austin in 1978, where he and his coworkers characterized the RNA moiety of the anterior determinants in *Chironomus samoensis* as small, polyadenylated, and cytoplasmic. He is author or co-author of numerous articles in scientific books and journals, including *Development, Developmental Biology, Nature, Photochemistry* and *Photobiology*, and *Proceedings of the National Academy of Sciences of the U.S.A.* He is an Honorable Guest Member of the Arthropodan Embryology Society of Japan. Kalthoff teaches courses in developmental biology and human biology. He won the College of Natural Science Teaching Excellence Award in 2000.

second edition

Analysis of Biological Development

part one

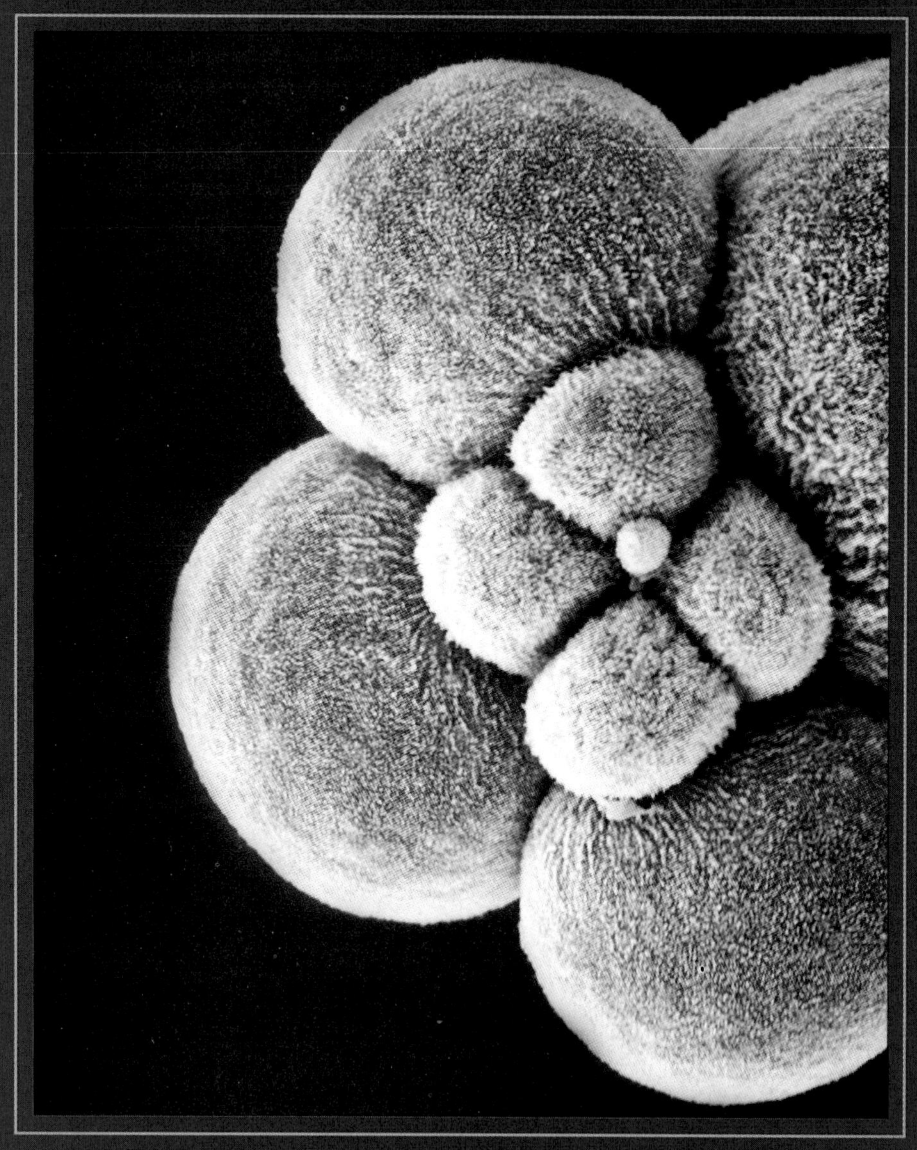

Figure I.1 Scanning electron micrograph of a snail embryo (*Ilyanassa obsoleta*) at the 8-cell stage. (The tiny cell at the center, a polar body, is not counted.) The four smaller cells, or micromeres, are rotated clockwise relative to their larger sister cells, or macromeres. The macromere at the top right is larger than the other three macromeres because it has incorporated a special mass of cytoplasm. The asymmetrical localization of this cytoplasm endows the largest macromere and its descendants with specific developmental capabilities.

From Gametogenesis to Histogenesis

IN Part One of this text, we will follow the development of several organisms through the early stages of life, a period called embryogenesis. Beginning with the production of egg and sperm, we will examine fertilization, the subsequent cleavage of the large egg cell into smaller embryonic cells, the dramatic movements that generate multiple cell layers and form organ rudiments, and finally the differentiation of functioning tissues. Developmental biologists traditionally focus on embryogenesis because it is during this stage that the basic body plan emerges and that later events in development are programmed to a large extent. Therefore, we will concentrate on this period first and explore some aspects of postembryonic development later in Part Three. ● As we examine the embryonic stages, we will draw examples from most of the species that have been the favorite research subjects of developmental biologists. These include frogs, chickens, mice, and humans among the vertebrates, and sea urchins, snails, and flies among the invertebrates. Some recent additions to this list, such as the plant *Arabidopsis thaliana* and the roundworm *Caenorhabditis elegans,* will be introduced in Part Three. ● Interwoven through Part One of the text will be discussions of important principles, theories, and research strategies. For example, the principles of cell determination by cytoplasmic localization and by induction will be discussed in several places. The theory of genomic equivalence—that each cell in an organism has a full complement of the organism's genetic information rather than just the information necessary for that cell—will be presented in the context of how different types of cells develop from a single fertilized egg cell. Analysis of embryonic movements in terms of cellular behavior will emerge as a major research strategy. ● At the end of Part One, readers should be familiar with the stages of embryonic development, some of the organisms used in developmental research, and the basic lines of reasoning and experimentation that have defined developmental biology as a classical discipline.

Analysis of Development

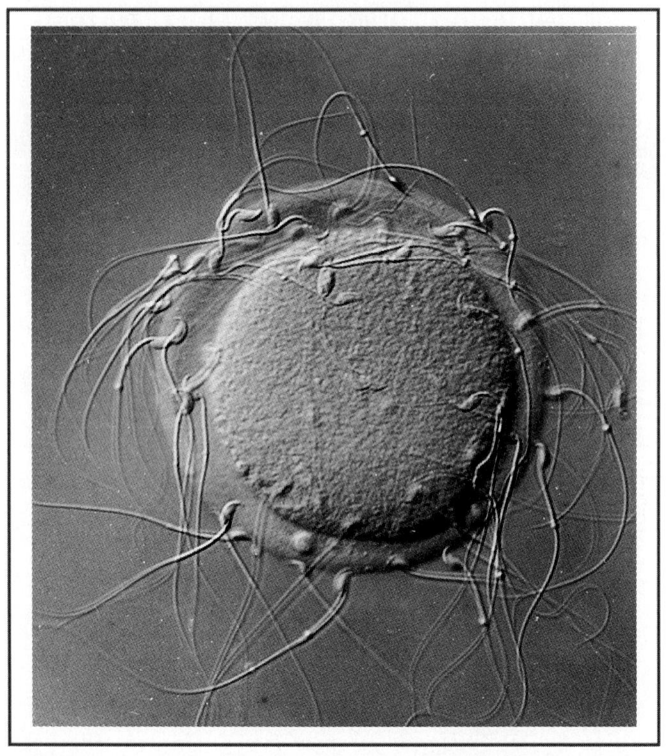

Figure 1.1 Photograph of a mouse egg being fertilized in a test tube. The egg is surrounded by a thick, transparent coat, called the zona pellucida. Many sperm adhere to the zona but none has fused with the egg yet. Under natural conditions, fewer sperm reach the site of fertilization in the oviduct.

IN this opening chapter, we will explore some of the central observations, questions, and experimental strategies that have defined developmental biology as a discipline. One central observation is that organisms form new structures from older ones as they proceed through their *life cycles*. Most individuals begin life as a fertilized egg (Fig. 1.1) and proceed through different forms of organization to adulthood. Adults, in turn, produce eggs and sperm, thus starting another life cycle with their offspring.

As biologists traced the life cycles of various species, it became clear that most embryonic cells have predictable fates, meaning that corresponding cells in different embryos of the same species will give rise to similar sets of structures. The first scientists to study how cells acquire their fates called themselves *experimental embryologists* and established a set of classical investigative strategies, which included the isolation, removal, and transplantation of embryonic parts. The general idea behind these strategies was to learn from the results of controlled interference, or, in the words of a leading embryologist, to maneuver embryos into more and more embarrassing situations, so that eventually they would have to reveal how they work.

While these classical strategies dominated embryology during the first half of the twentieth century, a few investigators began to examine the effects of mutations as an alternative way of analyzing development. Instead of applying fine surgical instruments to embryos, they took advantage of the molecular surgery that mutagenesis does on genes.

Both classical and genetic analyses of development were boosted enormously by advances in cellular and molecular biology, which transformed embryology into a new discipline called *developmental biology*. As the scope of investigations widened and deepened, developmental biologists sought to understand organismic development in terms of cellular behaviors, and cellular behaviors in terms of molecular events. This type of reductionist analysis has recently been complemented by the realization that molecular events in turn depend on events occurring in the entire embryo, such as the dramatic movements of cellular layers that generate the organ rudiments, which entail new juxtapositions for cells to exchange signals. Today, developmental biologists employ strategies of embryology, genetics, cell biology, and molecular biology in exciting combinations that yield an ever better understanding of how organisms proceed through their life cycles.

1.1 The Principle of Epigenesis

Watching the development of an embryo inspires us with a sense of awe, especially if we observe it in a timelapse movie. Figure 1.2 shows a few stages in the early development of a mouse embryo. Upon fertilization, the egg divides into two cells, then into four, and so on. A few days later, the developing embryo consists of about 60 cells arranged in a characteristic configuration known as the *blastocyst* (Gk. *blastos*, "germ"; *kystos*, "bag"). At this stage, the mammalian blastocyst implants itself into the wall of the maternal uterus, where cells keep multiplying while undergoing dramatic movements that shape the embryo into ever new configurations, which become more and more mouselike.

Our sense of awe is heightened when we recognize that development of an embryo does not necessarily follow a straight progression toward the familiar adult form. The human embryo shown in Figure 1.3 is 6 weeks old. On the one hand it looks sufficiently childlike to be recognized as a human being, yet on the other hand it has structures and proportions of its own, which make it appear unlike a human child or adult. The eye and forebrain loom large at the anterior end, while the remainder of the brain looks strangely incomplete. The five digits at the end of each rudimentary limb leave no

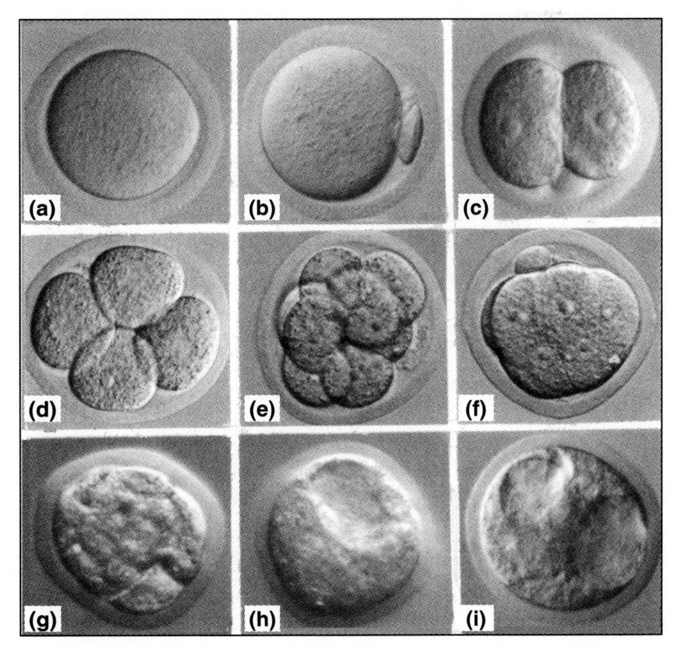

Figure 1.2 Mouse development prior to implantation. **(a)** Mature oocyte. **(b)** Fertilized egg. **(c)** 2-cell stage embryo. **(d)** 4-cell stage. **(e)** Early 8-cell stage. **(f)** Late 8-cell stage, after the embryonic cells have become more cohesive. **(g)** 16-cell stage. **(h)** 32-cell stage; a fluid-filled cavity has formed. The two cavities present in **(i)** will merge later.

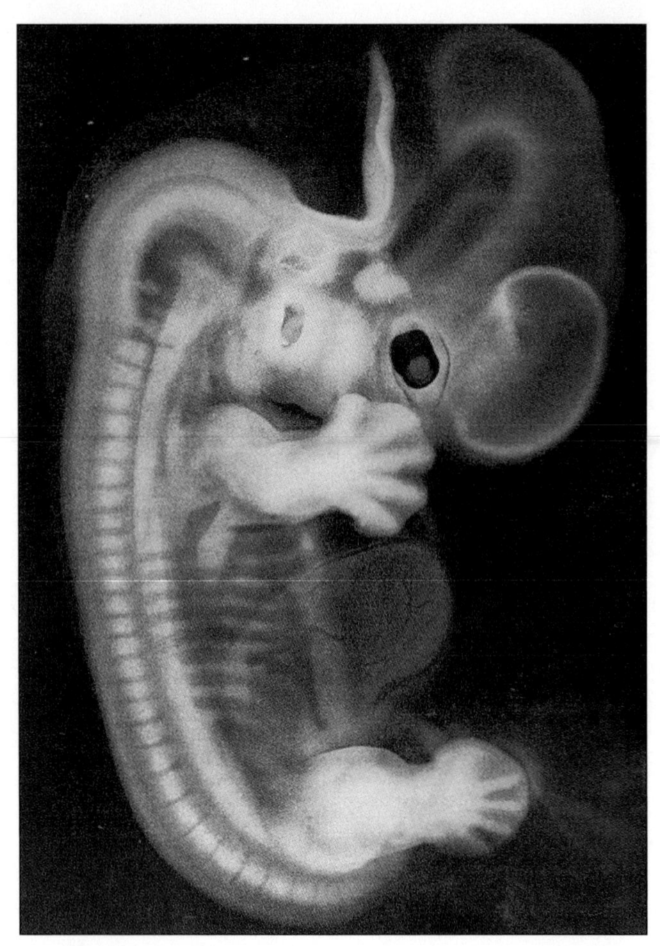

Figure 1.3 Photograph of a 6-week human embryo. Note the relatively large brain and comparatively short arm and leg rudiments.

doubt that these will be arms and legs, but the future limbs are proportionally too short. The spine and trunk muscles show a regularity that will later be lost, and the presence of a tail is an uncanny reminder that we humans have evolved from tailed ancestors.

The differences between successive stages of development are even more dramatic in animals with postembryonic forms called *larvae,* such as the tadpoles of frogs or the caterpillars of butterflies, which look only remotely like the adults they will become.

Thus, simple observation tells us that a single cell, the fertilized egg, has the ability to develop into a complex adult with a species-specific form and organization, but that it proceeds along this path with certain detours, giving each stage its own morphological characteristics. This type of progress differs greatly from the construction of a house, where the foundation is already laid according to the final size and major installations of every room. There are two fundamental differences between organismic development and house construction: First, organisms have to live and function at every stage of their development, whereas houses are typically uninhabited during construction. Second, house construction begins with a blueprint—that is, a detailed depiction of the final product—and progress toward the goal described in the blueprint is monitored by an architect or superintendent. In terms of this comparison, an embryo behaves like a pile of bricks that can replicate themselves, coat themselves with mortar, and assemble themselves into a house guided only by their own shapes and the attractive forces between them. The embryo performs this remarkable feat using instructions encoded in its genes and following a few spatial clues laid down in its cytoplasm. The key to understanding how the embryo works is to realize that its cells do *not* receive complete sets of instructions that direct them to their final destinations and tell them exactly what to do there. Rather, the embryo builds itself a step at a time. As each cell is born, it is given some preliminary directions and the molecular equivalent of a two-way radio that allows cells to communicate with one another and to coordinate their activities as development unfolds.

The successive generation of new structures from preexisting ones is called ***epigenesis****(Gk. *epi,* "upon"; *genesis,* "production"). A fundamental principle in development, epigenesis was first recognized by Aristotle (384–322 B.C.), who observed the development of chickens and other animals whose embryos are large enough to see with the naked eye (Fig. 1.4). Although Aristotle generally enjoyed widespread acceptance as an authority, the principle of epigenesis was rejected by subsequent generations of scholars. As late as the seventeenth century, nearly all biologists favored the opposite theory of ***preformation,*** according to which development simply means growth. Pursuant to this theory, each individual is fully formed within a germ cell and merely increases in size during development. The preformationists were divided into two camps, the "ovists" and the "spermists," who bitterly opposed each other. The ovists proposed that a female's ovaries contained miniature versions of her offspring, which in turn would hold miniatures of the following generation in their ovaries, and so forth. The seminal fluid contributed by the male was thought to merely activate the growth process. In contrast, the spermists proposed that sperm contained tiny but completely formed offspring, and that the role of the egg was simply to nourish the sperm. Distinguished scientists claimed to have seen miniature human figures curled up in the heads of sperm (Fig. 1.5). We must remember that the microscopes of the time were primitive and did not show any cellular detail, so it was tempting to fill the observational void with preconceived notions.

*Technical terms are printed in ***boldface italics*** at the point where they are fully defined. This is usually where the term first appears. Elsewhere, plain *italics* are used on occasion to remind readers that this is a defined term. If one does not remember its meaning, one can look it up in the index at the end of the book. Plain italics are also used for emphasis and species names, but the context should make clear when italics signal a technical term. References printed in italics refer to review articles of an introductory nature.

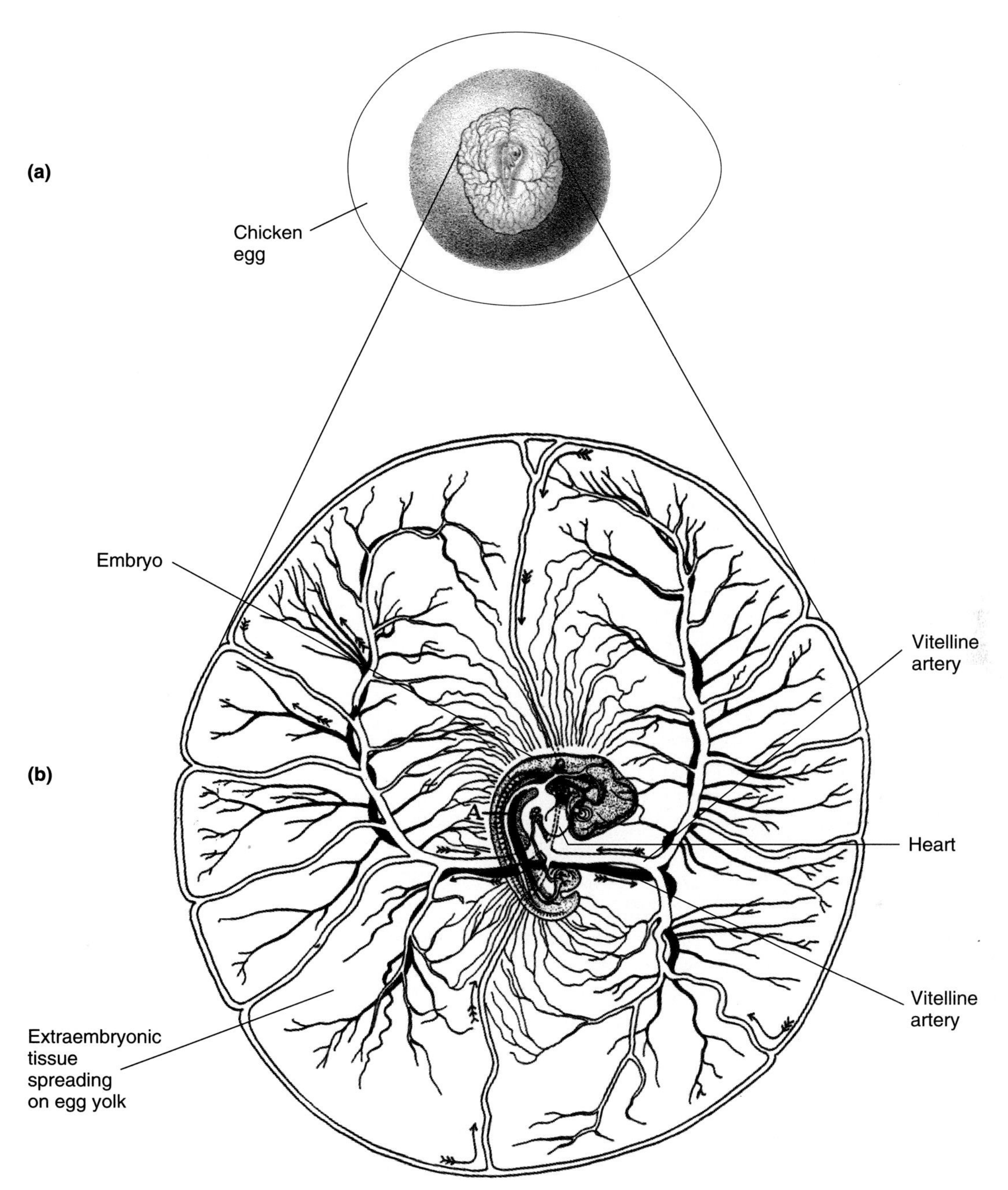

Figure 1.4 4-day chicken embryo. The heart pumps blood not only into the embryo but also into a system of vitelline arteries (black, outbound arrows), which branch on the surface of the egg yolk. Here, the blood picks up nutrients stored in the egg yolk and also oxygen, which diffuses through the porous eggshell and the egg white. The blood returns through vitelline veins (white, inbound arrows) and mixes with embryonic blood before entering the heart again.

With better microscopes and clearer reasoning, the champions of epigenesis came to prevail during the eighteenth century. The German anatomist Kaspar Friedrich Wolff (1733–1794) pointed out that if adults were preformed in the egg or the sperm, so that development meant only growing in size, then embryos would have to look like small adults. However, extending Aristotle's earlier observations, Wolff showed that the blood vessels, gut, and kidneys of a chicken embryo look quite different from the corresponding organs in an adult chicken (see Fig. 1.4). Unlike other anatomists before him, Wolff recognized the significance of stage-specific structures and began to interpret them as *temporary* adaptations for nourishment, respiration, and waste removal that are replaced or transformed later in development. As more scientists discovered differences between embryos, juvenile stages, and adults, preformationism faded away.

Epigenesis is now recognized as the guiding principle of development, but the underlying cellular and molecular mechanisms are only partly understood. We do know that the nuclei of both egg and sperm contain

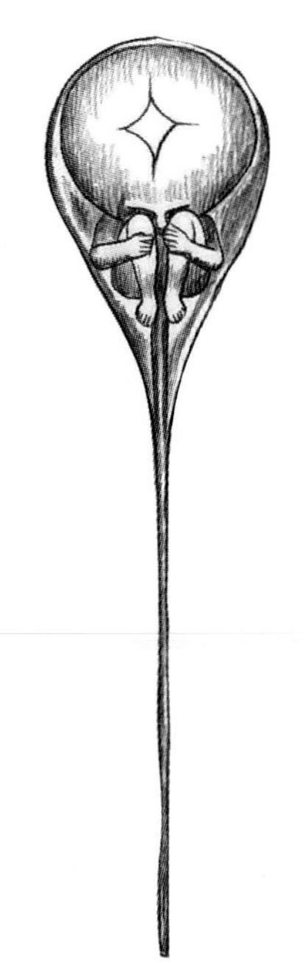

Figure 1.5 Miniature human ("homunculus") preformed in the sperm, as depicted by Niklaas Hartsoeker in 1694.

genetic instructions encoded in the nucleotide sequence of their DNA. The egg cytoplasm contains the initial molecular building blocks for the embryo, and more will be made from egg yolk according to the embryo's genetic instructions. In most animals, the egg cytoplasm also contains some localized components that are distributed in a coarse but regular order. These components become segregated into different embryonic cells so that, for instance, the cytoplasm of anterior cells differs from that of posterior cells. The differing cytoplasmic environments activate different combinations of genes in the nuclei of different groups of cells, so that the embryonic cells form a three-dimensional pattern according to the gene activities they exhibit. The ensuing interactions between different embryonic cells then generate an ever more complex network of cells and organs.

1.2 Developmental Periods and Stages in the Life Cycle

A basic phenomenon encountered by investigators of development is that all higher organisms exist in ***life cycles.*** Animals including humans begin their lives as fertilized eggs, which develop through embryonic and juvenile stages into adults. The adults produce eggs and sperm, and their fertilized eggs will start new life cycles while the adults eventually die. Thus, the first task for a developmental biologist who wants to investigate a new species is to observe it in its natural habitat, to culture it in the laboratory if possible, and to describe its entire life cycle.

Figure 1.6 shows the life cycle of the South African clawed frog, *Xenopus laevis.* Its embryos have become a favorite research subject of developmental biologists because they develop quickly and *Xenopus* females can be induced to lay eggs year-round in the laboratory. The embryos are also large enough to do transplantation experiments, and operated embryos as well as isolated embryonic parts are easy to culture in simple salt solutions.

Animal life cycles are conveniently subdivided into three major ***periods***: embryogenesis, postembryonic development, and adulthood. Each of these periods is further subdivided into ***stages.*** Such stages have been defined, described, and illustrated meticulously in *normal tables* for those animal species that are being studied extensively by developmental biologists. Referring to such normal tables helps greatly in obtaining reproducible results and in communicating effectively with other scientists.

EMBRYONIC DEVELOPMENT BEGINS WITH FERTILIZATION AND ENDS WITH THE COMPLETION OF HISTOGENESIS

The term ***embryo*** is generally used to describe the developing individual from fertilization through the formation of differentiated tissues. This period of development, called ***embryogenesis,*** is subdivided into the stages of fertilization, cleavage, gastrulation, organogenesis, and histogenesis.

Fertilization is the union of egg and sperm. The ***egg*** is an extremely large cell, loaded with nutrients to support development until the new organism can obtain nutrients from other sources. Most eggs have at least one polarity axis, known as the *animal-vegetal* axis. The ***animal pole*** is the pole closest to the egg nucleus, while the opposite pole is called the ***vegetal pole.*** In the case of *Xenopus,* the animal hemisphere is also pigmented, whereas the vegetal hemisphere is whitish with large amounts of yolk (Fig. 1.6). The ***sperm*** has the function of finding and fertilizing an egg. It contains a condensed nucleus and a flagellum to propel itself. After fertilization by a sperm, an egg is called a ***zygote.*** Fertilization triggers or accelerates many metabolic reactions, notably the synthesis of DNA and protein, which are required in large amounts for the subsequent stages of development.

During ***cleavage,*** all zygotes undergo a series of divisions that result in the formation of progressively smaller cells called ***blastomeres*** (Gk. *blastos,* "germ"; *meros,* "part"). For instance, a newt egg is about 1 mm in diameter, whereas at the end of cleavage an average blastomere measures about 50 μm across (Fig. 1.7). In this respect, cleavage differs from later stages, when cells after each division grow back to the size of their mother cell before dividing again. Cell growth requires the uptake of external nutrients, which are not available to most embryos. (Mammalian embryos, which implant themselves early into the maternal uterus, are excep-

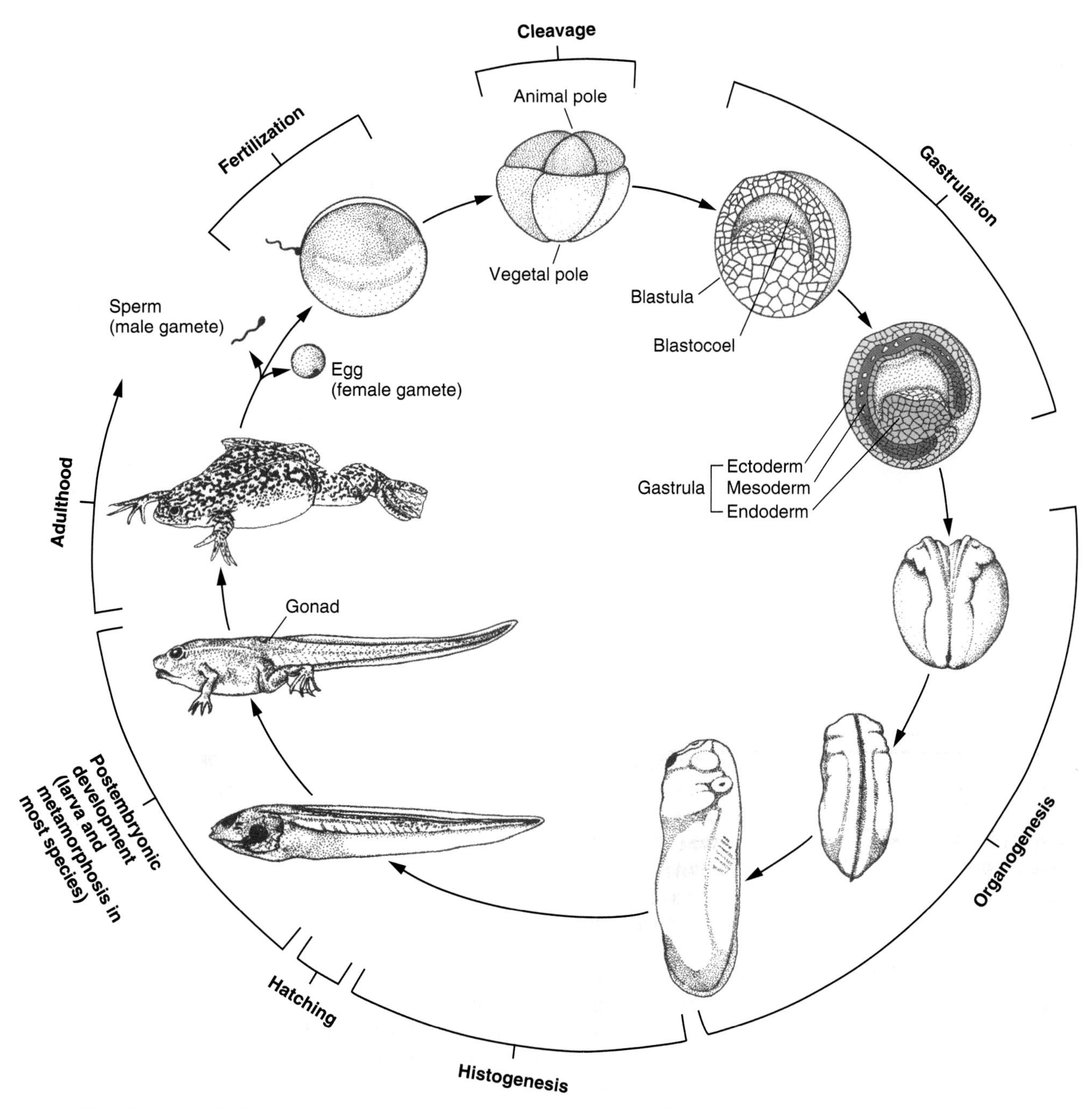

Figure 1.6 The life cycle of a frog. The embryonic period is subdivided into five stages: fertilization, cleavage, gastrulation, organogenesis, and histogenesis. The postembryonic period, in most frog species, begins with the hatching of a larva (called a tadpole) from the eggshell and ends with metamorphosis. The adult period begins with the production of eggs or sperm (gametogenesis) and ends with death.

tional in this respect.) By the end of cleavage, the embryos of frogs and many other animals consist of a hollow sphere of blastomeres surrounding a fluid-filled cavity called the ***blastocoel*** (Gk. *blastos,* "germ"; *koiloma,* "cavity"), and an entire embryo at this stage is known as a ***blastula*** (pl., *blastulae*).

The ensuing stage of ***gastrulation*** in all embryos involves a complex series of ***morphogenetic movements*** (Gk. *morphe,* "form"; *genesis,* "production"), in which cells rearrange and migrate while entire sheets of cells spread, bend, and fold. An embryo in the process of gastrulation is called a ***gastrula*** (pl., *gastrulae*). At the end of gastrulation the embryo consists of three concentric layers of cells called ***germ layers*** (Fig. 1.8). How the three germ layers are established varies considerably among groups of animals. Nevertheless, the resulting configuration of three germ layers and the derivatives formed later by each layer are remarkably consistent among animals of different phyla. The outermost layer is known as the ***ectoderm*** (Gk. *ectos,* "outside"; *derma,* "skin"); its major derivatives are the nervous system and the outer layer of the skin, or *epidermis.* The innermost germ layer, called the ***endoderm*** (Gk. *endon,* "within"), forms the inner lining of the digestive tract and its appendages. The

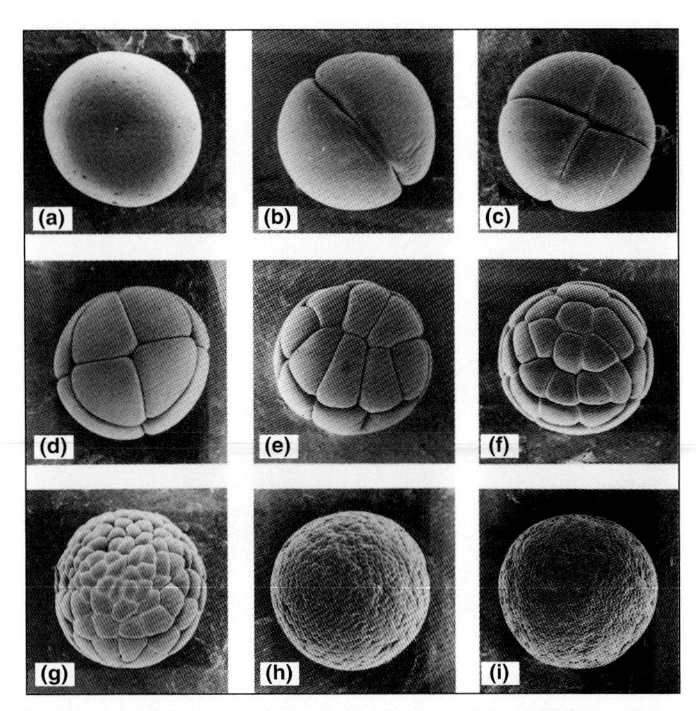

Figure 1.7 Early development of a newt (*Pleurodeles waltl*) embryo as seen in scanning electron micrographs. **(a)** Fertilized egg. **(b)** 2-cell stage. **(c)** 4-cell stage. **(d)** 8-cell stage. **(e)** 16-cell stage. **(f)** 32-cell stage. **(g)** Early blastula. **(h)** Midblastula. **(i)** Late blastula. Note the decreasing blastomere size.

intermediate germ layer, or ***mesoderm*** (Gk. *mesos*, "middle"), gives rise to a wide range of organs including bone and other skeletal structures, muscle, heart, blood vessels and blood cells, kidneys, and reproductive organs.

During the next stage of embryogenesis, the germ layers undergo further morphogenetic movements and interact with one another to form the rudiments of the larval organs. This process is known as ***organogenesis,*** and upon its completion the embryo shows the basic body plan of the animal group to which it belongs. One of the most spectacular events in vertebrate organogenesis is the formation of the rudiment of the central nervous system (Fig. 1.8). This process is known as *neurulation,* and the embryo is called a *neurula* during this stage. Ectodermal cells on the dorsal (back) side of the embryo elongate perpendicularly to the surface, forming a raised plate called the *neural plate.* The margins of the neural plate bend together and eventually fuse along the dorsal midline. The neural plate thus closes into the *neural tube,* which will ultimately become brain and spinal cord. Meanwhile the endoderm forms the gut rudiment, or *archenteron.* In between, the mesoderm divides into subunits that will give rise to backbone, dorsal musculature, kidneys, heart, and other structures. Toward the end of organogenesis, the embryo has become a complex animal, having the rudiments of most organs in place and typically exemplifying the basic body plan of its class or phylum.

Organogenesis blends into the subsequent stage of ***histogenesis*** (Gk. *histos*, "tissue"), during which cells—which by now have reached their final location—acquire their functional specialization. As tissues mature, most cells acquire special functions beyond their basic energy metabolism, protein synthesis, and other "housekeeping" functions. The special functions of cells are usually based on certain structural features. For instance, red blood cells, which function in gas exchange, contain massive amounts of hemoglobin, which binds oxygen and carbon dioxide, and have a doughnut-like shape, which provides a large surface for gas exchange while still allowing red blood cells to squeeze through narrow capillaries. Thus, histogenesis is associated with the emergence of many types of specialized cells, a process called ***cell differentiation.***

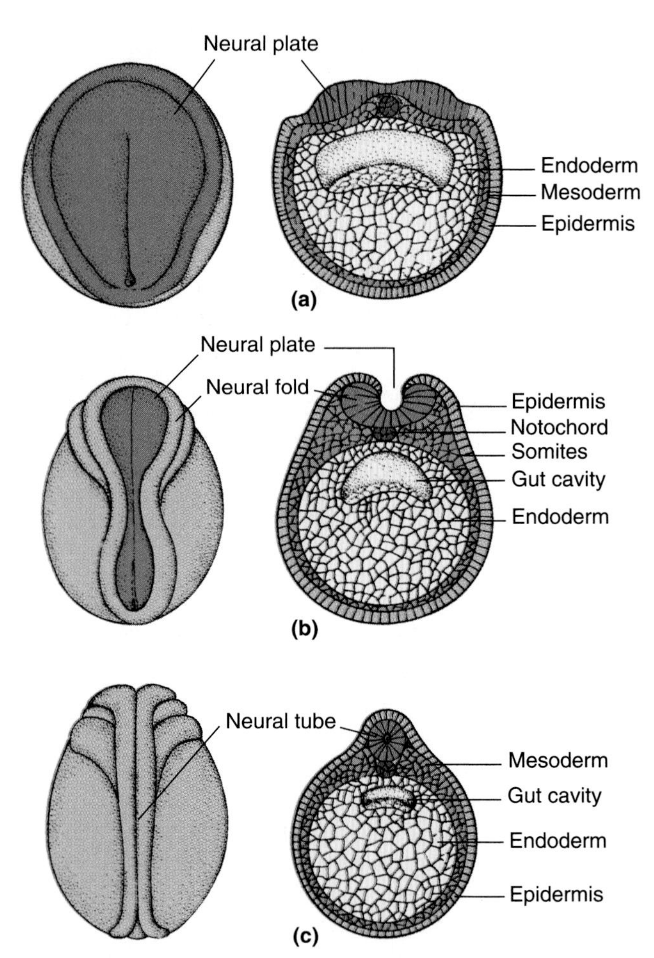

Figure 1.8 Frog embryos during neurulation. Left column, dorsal views; right column, transverse sections. After gastrulation, the embryo consists of three germ layers, namely ectoderm (outer layer), mesoderm (middle layer), and endoderm (inner layer). The dorsal ectodermal region forms the neural plate, which folds into the neural tube, later giving rise to the brain and spinal cord. The remainder of the ectoderm mainly forms the outer skin layer, the epidermis. The endoderm forms the gut. The mesodermal cells in between form a dorsal rod, the notochord, flanked by segmental units called somites, and other organ rudiments. **(a)** Early neurula. **(b)** Middle neurula. **(c)** Late neurula.

POSTEMBRYONIC DEVELOPMENT CAN BE DIRECT OR INDIRECT

The ***postembryonic*** period of development lasts from the end of embryogenesis to the beginning of adulthood. In most animal species, the postembryonic period begins when the developing animal hatches from its egg envelopes and begins to take up nutrients from outside. In mammals and other species that nourish their embryos through a placenta, however, the beginning of the postembryonic period is better defined as the period following organogenesis. The developing human, for example, is called an ***embryo*** during the first 8 weeks of gestation, and a ***fetus*** from the ninth week until birth.

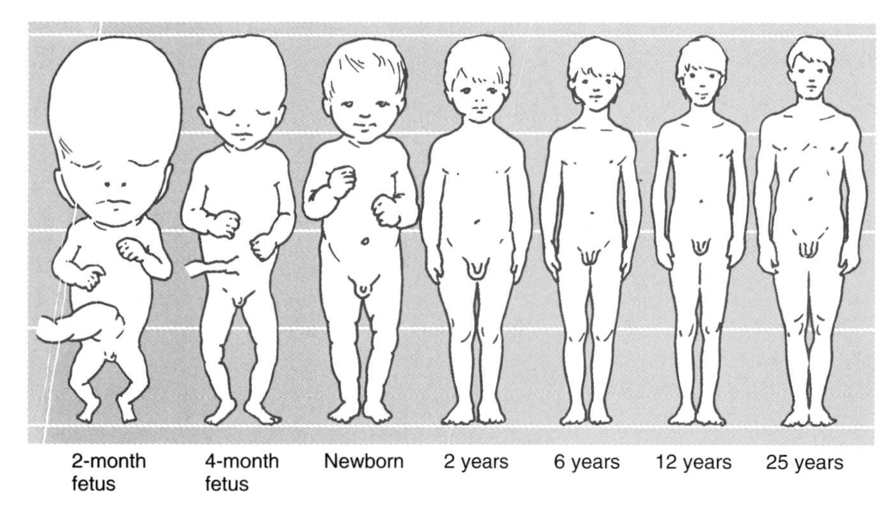

Figure 1.9 Changing proportions of the human body during fetal development and after birth. Note the disproportionately large head in the fetus at 2 months and the more rapid growth of trunk, arms, and legs thereafter.

During the postembryonic period, some animals look like miniature adults, except that some parts of the body are further advanced in growth than others and that the gonads and other sexual characteristics have not yet matured. Such animals undergo ***direct development*** and are called *fetuses, juveniles,* or more specific names, during various stages of their postembryonic period. Most vertebrates, including mammals, develop this way. In the human fetus and infant, for instance, the body parts are those of an adult, but the head is disproportionately large compared with the trunk, arms, and legs, which grow rapidly after birth (Fig. 1.9).

In other groups of animals, the postembryonic stages bear little resemblance to adults. Familiar examples include the tadpoles of frogs, the caterpillars of moths and butterflies, and the maggots of flies. Such animals undergo ***indirect development*** and are called ***larvae*** (sing., *larva*) during their postembryonic period. A typical larva is mobile, feeds, and gains substantially in body mass (Fig. 1.10). However, the larva differs in form and lifestyle from the adult and often has a different name; for example, the adult frog lives on the land and snares insects for food whereas the tadpole lives in the water and typically grazes algae from plants and rocks. Correspondingly, the tadpole has some of the organs found in the adult frog, such as eyes, brain, and heart, but also stage-specific organs that will disappear later, including gills and a finned tail. Conversely, some adult organs, including legs and lungs, are still missing from the tadpole. The transition from the larval to the adult period entails a dramatic sequence of events known as ***metamorphosis.***

Figure 1.10 Photograph showing silkworm larvae after successive molts. Feeding on mulberry leaves, a larva grows about 5000-fold in weight.

ADULTHOOD BEGINS WITH THE ONSET OF REPRODUCTION AND ENDS WITH DEATH

The adult period begins with the production of mature eggs and sperm and ends with death. Mating and parenting are the key issues during adult life. Access to mates may involve competition over territory and the establishment of dominance hierarchies; in some species, females choose males displaying certain appealing characteristics. After mating, adult animals spend varying amounts of time and energy caring for their offspring.

Maintaining a body requires energy, which living organisms take up either directly from sunlight or as chemical energy from food. However, individual organisms cannot sustain themselves indefinitely. Over time, energy-harvesting and maintenance mechanisms accumulate defects, which eventually overpower the organism's capabilities for repair and adjustment. As a result, all individuals that are not killed by accidents or predation eventually undergo a process of deterioration called ***senescence*** until eventually they die. The economy of this process may be apparent in the analogous situation of car maintenance. If one wants to maximize the number of miles driven per dollar spent on a car, it is generally best to take good care of the car for several years. However, when the car reaches the stage when several expensive repairs come due it is usually more economical to buy a new car. Similarly, organisms have evolved to maintain themselves until they have raised enough young who will keep the species going when they die.

The length of the adult life span before senescence differs widely in accord with the reproductive strategy of a species. Some insects do not even feed as adults; they mate, produce a large clutch of offspring, and die within a day, leaving the young to fend for themselves. At the other end of the scale, large mammals may survive several decades as adults while carefully raising a few young. Many species fall somewhere in between.

1.3 Classical Analytical Strategies in Developmental Biology

Describing the life cycles of organisms was originally part of an effort to fully describe their life history, anatomy, and behavior. A major goal of such work was to establish a taxonomic system of species, families, orders, and so on, in which, for example, a bottom-dwelling sea urchin and its free-swimming larva would be recognized as belonging to the same species. Since evolution became a widely accepted theory, characteristics of development were also taken into account in efforts to devise a "natural" taxonomic system, which would reflect the course of evolution. For example, modern taxonomists place tunicates and sea squirts along with vertebrates into the same phylum, the *chordates.* The reasons for doing so are not obvious at the adult stage, when tunicates and sea squirts look more like molluscs than like vertebrates. However, the similarities are obvious at the larval stage. A sea squirt larva is quite similar to a frog tadpole; shared characteristics include a dorsal nervous system and a *notochord*—that is, a cartilagenous rod later replaced with a backbone in vertebrates.

Late in the nineteenth century, a new generation of biologists emerged who would no longer consider the study of development as ancillary to taxonomy and evolution. Instead, they ventured into the *causal analysis* of development. With hand-made glass needles and hair loops, they removed and transplanted parts of frog and sea urchin embryos to see what the parts would do in isolation and how they would interact with cells that were not their normal neighbors. They focused on the embryonic period because this is when the basic body plan emerges and many later steps in development are programmed to a large extent. By defining a new set of terms based on the results of their experiments, these scientists established a new scientific discipline, which they called ***experimental embryology.***

EMBRYONIC CELLS HAVE PREDICTABLE FATES IN DEVELOPMENT

The first questions to present itself in the analysis of development is whether embryonic cells have predictable *fates.* One way of testing this is to label a particular blastomere in several embryos; from the same species at the same stage with a dye that does not interfere with normal development. Then we can observe whether the labeled blastomere in each embryo gives rise to the same embryonic structure—say, a patch of belly epidermis—or whether it contributes to different structures, such as brain in one embryo and gut in another.

Many experiments of this kind have shown that sooner or later embryonic cells do acquire predictable fates. However, species vary in the timing and the precision of fate-acquisition. A *fate map* indicating the fates of *Xenopus* blastomeres at the 32-cell stage is shown in Figure 1.11. While most of the blastomeres are fairly consistent in their fates, some are more variable. Blastomere

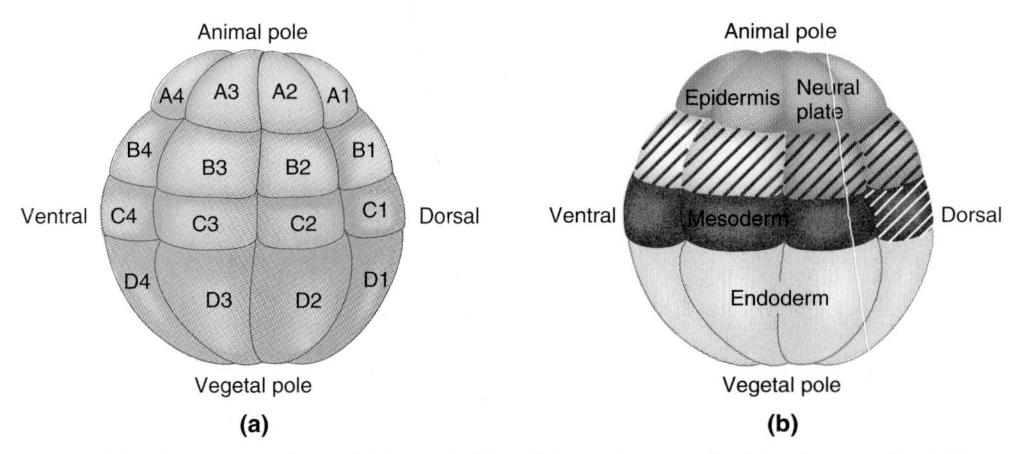

Figure 1.11 Fate map of the *Xenopus* embryo. **(a)** Lateral view of the embryo at the 32-cell stage. Each blastomere is identified by a letter and a number. **(b)** Fate map indicating which blastomeres will give rise to endoderm, mesoderm, and the major ectodermal derivatives: neural plate and epidermis. Blastomeres B1 and B2 contribute to neural plate and mesoderm, B3 and B4 to epidermis and mesoderm, and C1 to endoderm and mesoderm.

C1, for example, contributes, on average, 68% of its volume to endoderm (the innermost germ layer) and 25% of its volume to mesoderm (middle germ layer), with considerable variation among individual embryos (Dale and Slack, 1987; Moody, 1987).

In some species, fate maps can be drawn with great precision from the beginning of cleavage. These species include the roundworm *Caenorhabditis elegans* (*C. elegans*), which has become a favorite organism among developmental biologists (see Section 15.2 and Chapter 25). Part of the tiny worm's attraction stems from the fact that the number of nonreproductive cells is constant. Each adult male consists of exactly 1031 nonreproductive cells and about a thousand sperm, and each adult hermaphrodite (the equivalent of a female in this species) consists of exactly 959 nonreproductive cells plus about 2000 eggs and sperm. What is more, any given cell originates through the same lineage in each individual of the species, so that the fate of each cell at each stage of development is known with unparalleled precision. As an example, Figure 1.12 depicts part of the cell lineage that gives rise to the male's tail.

On the opposite end of the precision spectrum, there are species for which no fate map can be drawn at all for their early embryonic stages because the fate of their early blastomeres depends on environmental cues that do not act until development has already begun. For instance, in a chicken embryo, the anteroposterior (head-to-tail) axis is determined by the direction of gravity and by the way the egg rotates as it passes down the uterus (Kochav and Eyal-Giladi, 1971). During this time, cleavage is well under way, but the fate of each blastomere is still undetermined. Only after the chicken "egg" has been laid, at which time it is really an embryo consisting of thousands of cells, can fate maps be drawn.

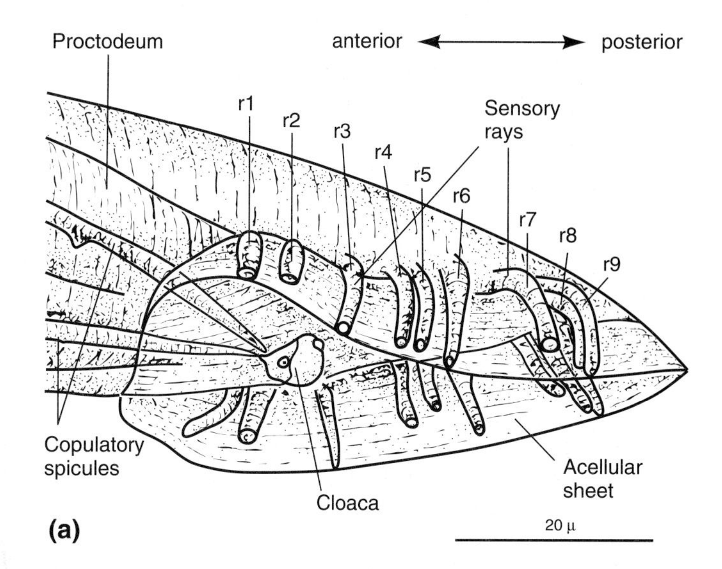

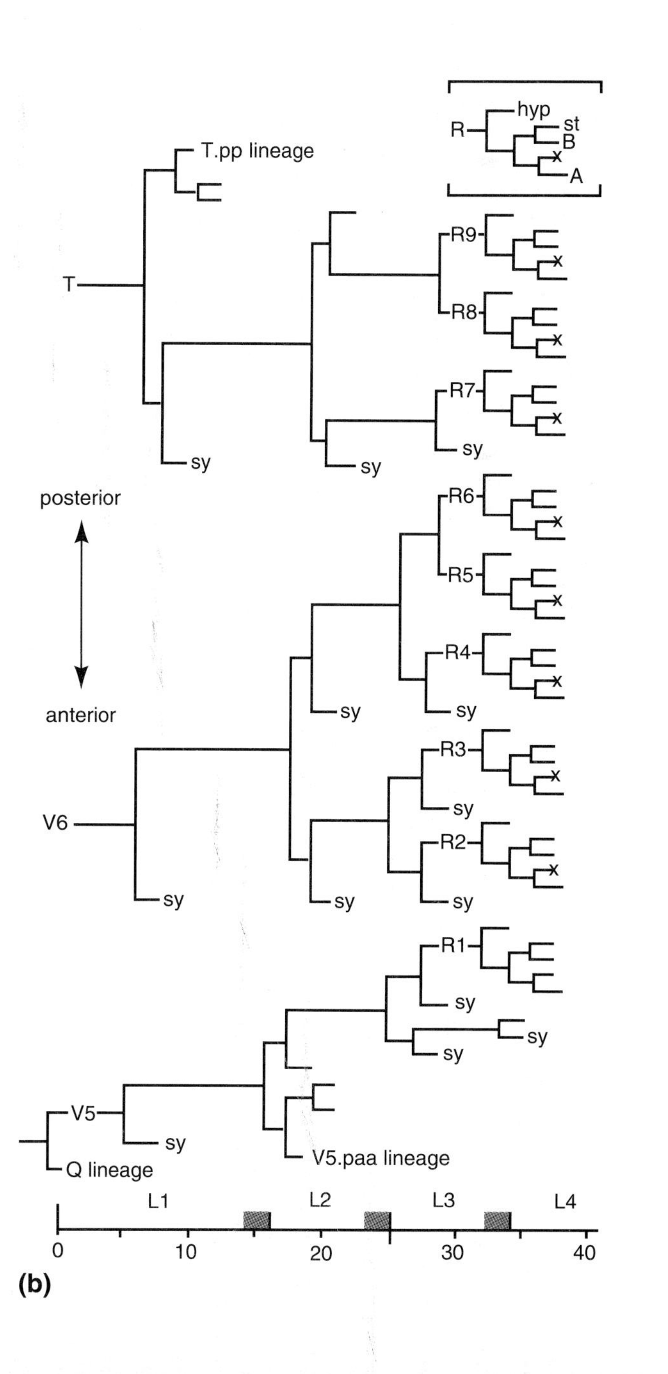

Figure 1.12 Cell lineage of the sensory rays in the tail of the roundworm *C. elegans*. **(a)** The tail of the adult male serves as a mechanosensory organ during copulation; it consists of a fan-shaped acellular sheet supported by nine pairs of rays (R1–R9). Located at the base of the tail is the cloaca, the common opening of gut and reproductive system. Two retractable spicules can protrude from the male's cloaca and are inserted into the hermaphrodite's vulva during copulation to aid in the transfer of sperm. **(b)** Cell lineages giving rise to the rays in the male's tail. The horizontal time axis at the bottom is calibrated in hours after hatching of the first larval stage (L1) from the eggshell. There are four larval stages followed by an adult stage (not shown), punctuated by molts. On each side of the body, three blast cells (V5, V6, and T) are set aside during embryonic development. During the larval stages, they undergo five rounds of standardized divisions, producing mostly skin, or syncytial hypodermis (sy), and nine ray precursor cells, R1 through R9 ray precursor cell undergoes three additional divisions (see bracket at top of part a), producing another hypodermis cell and a ray cell group consisting of two neurons (A and B) and one structural cell (st). The cells marked x undergo programmed death (apoptosis).

ANALYSIS OF DEVELOPMENT REQUIRES A STRATEGY OF CONTROLLED INTERFERENCE

Having seen that the fates of blastomeres sooner or later become predictable, embryologists then wondered how cells actually acquire their fates. Do all cells of an organism have the same genetic information? Are blastomeres endowed with different types of cytoplasm that control how their genetic information is being utilized? Do cells exchange signals that coordinate their development so that different fates are acquired in harmonious patterns?

To answer these questions investigators use strategies of ***controlled interference*** with development. First they identify the parameters (genetic information, cytoplasmic components, interactions with other cells, temperature, etc.) that affect the developmental process. Then they change one parameter at a time and observe any subsequent deviations from the normal course of development. From appropriate experiments of this kind they can conclude whether, say, cell A needs to be in contact with cell B in order for A to develop normally.

The credit for the first embryological experiment goes to the German biologist Wilhelm Roux. In 1888, he published a study on whether isolated blastomeres would develop in the same ways as they do in the intact organism. His way of isolating one blastomere of a frog embryo at the 2-cell stage was to disable its sister blastomere by stabbing it with a hot needle. The surviving blastomere frequently developed according to its fate: It produced a lateral half embryo (Fig.1.13).

A few years later, Oskar Hertwig and other German biologists completely separated the first two amphibian blastomeres from each other by constricting, or ***ligating,*** embryos with a hair loop along the first cleavage plane (Fig. 1.14). The separated blastomeres often developed into twin embryos that were half-sized but normally proportioned and grew up to be viable tadpoles. In a few cases, though, only one of the isolated blastomeres developed properly whereas its sister blastomere formed just a ball of undifferentiated cells. Close inspection revealed that the orientation of the first cleavage plane was critical. Usually, the first cleavage plane bisects a natural landmark known as the *gray crescent*, which is visible on the outside of the zygote and marks the future dorsal (back) side of the embryo (Fig. 1.15). In these cases, each blastomere would inherit part of the gray crescent material and would give rise to a complete embryo. However, if the orientation of the first cleavage plane was such that all of the gray crescent was segregated into one blastomere, then upon isolation only this blastomere would form a viable embryo while its sister blastomere, which did not receive any gray crescent material, developed into a disorganized cell mass.

The results of these experiments suggest several conclusions. First, each of the first two blastomeres of a frog embryo retains a full complement of genetic information; this is shown by the fact that normally each of two blastomeres, after gentle and complete isolation, can form a viable larva. Second, the kind of cytoplasm that a blastomere inherits during cleavage may be critical; blastomeres that do not receive some portion of the gray crescent (or any associated material) fail to develop normally. Third, adjacent blastomeres seem to interact with each other, since even the remnants of an incapacitated blastomere restrict the intact sister blastomere from forming a whole embryo to forming a half-embryo.

ISOLATION, REMOVAL, AND TRANSPLANTATION OF EMBRYONIC PARTS ARE KEY STRATEGIES OF EMBRYOLOGISTS

The tools of the early embryologists were simple: wax molds or glass strips with wells to hold embryos from various animal species, glass needles, hair loops and eyelashes used as surgical knives, and glass pipettes pulled from narrow tubes to transfer embryos or parts thereof. The delicate operations carried out by these pioneers required skill and practice, and work could be frustrating as many embryos would die from infections because antibiotics and fungicides were not yet available. Embryos that survived were observed, sketched at

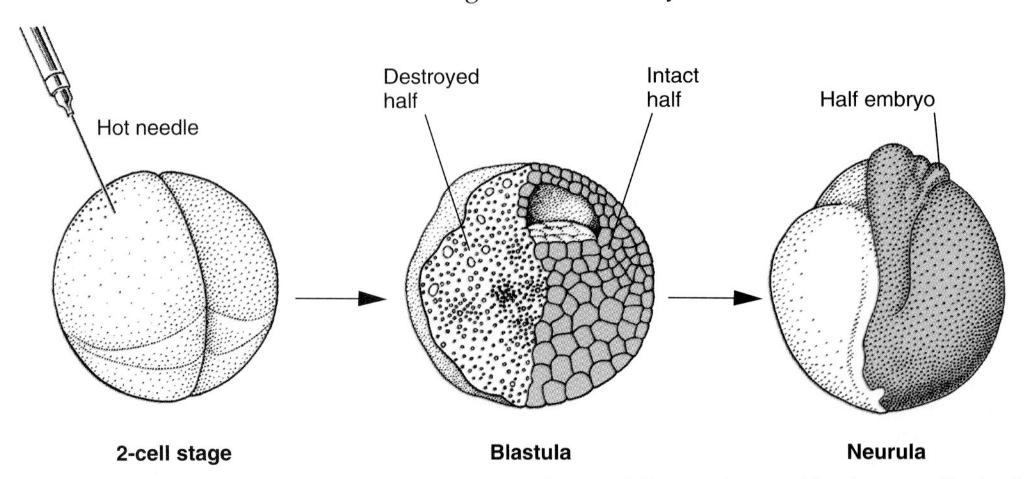

Figure 1.13 Partial development after disabling one blastomere of a 2-cell frog embryo with a hot needle. In this experiment, carried out by Wilhelm Roux in 1888, the intact blastomere developed into a half embryo.

intervals, and eventually fixed, sectioned, and stained for microscopic inspection. With these simple tools, or their modern replacements, embryologists pursue mostly three experimental strategies: isolation, removal, or transplantation.

In an ***isolation*** experiment, a part of an embryo is cultured by itself in order to disconnect it from its interactions with the remainder of the embryo. The culture medium is formulated to provide a life-sustaining environment for the isolated part, or *isolate.* However, the isolate does not have access to the developmental signals that it would have received in its normal location in the embryo. The isolate is monitored to find out whether or not it will develop according to fate. Any difference would indicate that for normal development the isolated part depends on signals received from elsewhere in the embryo. To identify the sources of these signals, one would proceed to coculture the isolate with various other parts of the embryo.

In the strategy of ***removal,*** an embryonic part is removed as is done for the purpose of isolation, but the focus is now on the remainder of the embryo. Will it continue to develop normally? If not, the removed part must be providing some signal or other function required for normal development of the remainder. Again, coculture of the removed part and the remainder may be a useful follow-up experiment: It may tell whether the function of the removed part can be provided through the medium without physical contact.

The third major strategy employed by embryologists is ***transplantation.*** Usually, an embryonic part is removed from a donor and implanted into a *different place* in a recipient of the same age. This version of the transplantation strategy is called ***heterotopic*** transplantation or *ectopic* transplantation (Gk. *heteros,* "other"; *ektos,*

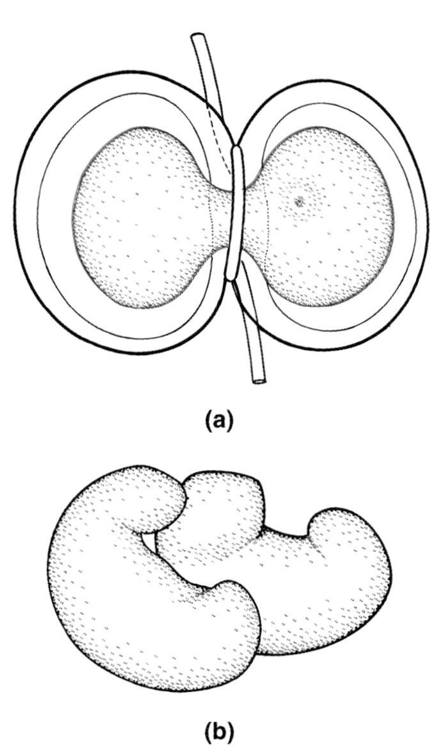

Figure 1.14 Development of twin embryos after complete separation of newt blastomeres. **(a)** The blastomeres were separated by ligation of a 2-cell embryo along the first cleavage furrow with a hair loop. **(b)** In many cases, both blastomeres developed into half-sized but normally proportioned embryos.

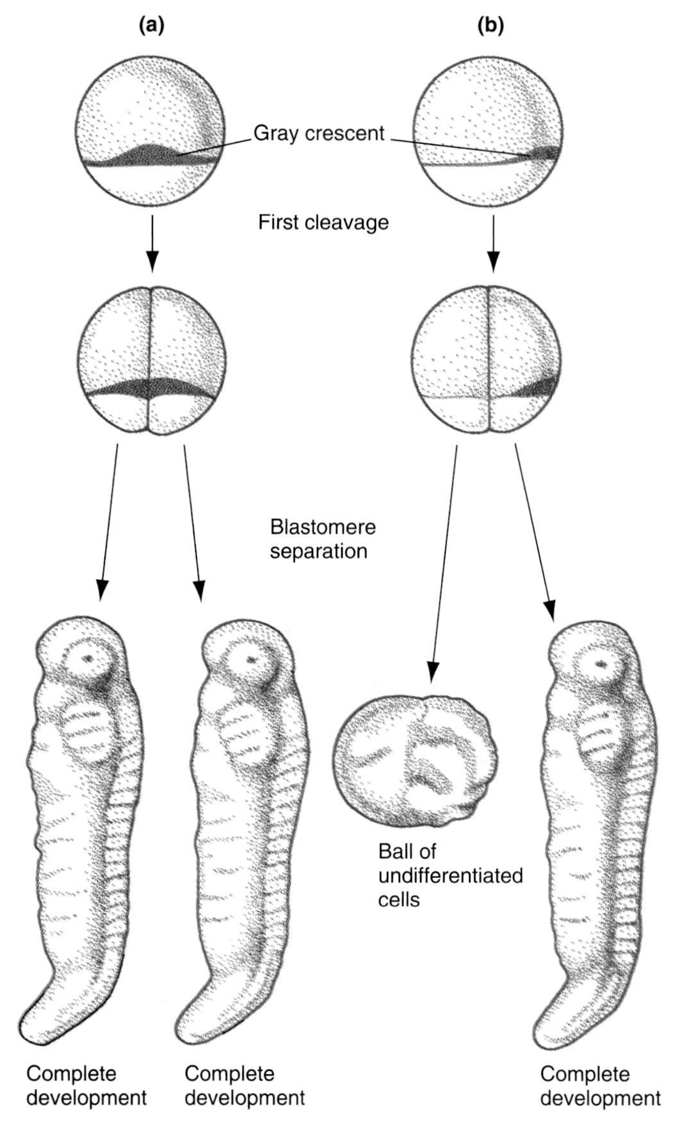

Figure 1.15 Dependence of twinning in amphibian embryos on the orientation of the first cleavage plane. The first two blastomeres were separated at the 2-cell stage as shown in Figure 1.14a. In many amphibian species, the future dorsal side is presaged in the fertilized egg by a natural structure known as the gray crescent. **(a)** In most embryos, the first cleavage plane bisected the gray crescent, creating a right blastomere and a left blastomere. In these cases, blastomere separation was followed by complete development of both blastomeres. **(b)** Cleavage furrows that did not divide the gray crescent were less frequent and were followed by complete development of the dorsal blastomere containing the gray crescent. The ventral blastomere formed only a round piece of unorganized tissue.

"outside"; *topos,* "place") because the position of the graft in the host is the varied experimental parameter. In another version of the transplantation strategy, an embryonic part is implanted into the same place from which it was taken but into a recipient of a *different* age. This is known as ***heterochronic*** transplantation (Gk. *chronos,* "time") because the age difference between graft and host is the varied parameter. Finally, transplantations may be *homospecific* (donor and host belong to the same species) or ***heterospecific*** (donor and host are from different species). The latter type of transplantation is sometimes used to find out whether signals exchanged between embryonic parts have been conserved during evolution. More often, donor and host are chosen from closely related species that differ in a visible marker, such as pigmentation, in order to be certain which structures in the graft/host combination are derived from the graft and which are contributed by the host. The following experiment will illustrate the power of heterotopic transplantation.

The vertebrate eye originates from two rudiments, the ***optic vesicle*** and the ***lens placode.*** The optic vesicle bulges out from the side of the brain and acquires a cup-shaped depression, the *optic cup,* which will form the retina of the eye. The lens placode is a raised portion of ectoderm that pinches off to the inside as a lens vesicle and then proceeds to form the lens of the eye (Fig. 1.16). The great precision with which the lens placode is centered over the optic vesicle, and the perfect synchrony in the formation of the two parts, suggests that the optic vesicle may send out a signal that causes the lens placode to form in exactly the right place. To test this hypothesis, the German embryologist Hans Spemann (1901) removed the optic vesicle from one side of a frog tadpole's brain while leaving the other side undisturbed as a control (Fig. 1.17). A few days later, a normal eye had formed on the control side whereas no eye parts could be detected on the operated side. This result showed that the optic vesicle was somehow *necessary* for lens formation. Without an optic vesicle there was no lens, at least not in *Rana fusca,* the species that Spemann used.

Generally, in order to test whether an event A is ***necessary*** for another event, B, to occur, one has to inhibit A and determine whether B now fails to occur. If so, A is necessary for B. In contrast, in order to establish that A is ***sufficient*** for B, one has to make A happen at a time or place where it normally does not. If A is still followed by B, then A is sufficient for B.

In order to test whether the optic vesicle is *sufficient* to induce lens formation, the American embryologist Warren H. Lewis (1904), working with the American frog *Rana palustris,* brought the optic vesicle in contact with foreign epidermis. He accomplished this in two ways, by transplanting the vesicle under flank epidermis or by replacing the epidermis over the optic vesicle with flank epidermis. In both cases, typical lenses formed, apparently from flank epidermis.

More recent work with marked transplants showed that the optic vesicle contributes only the last step to a series of interactions between ectoderm and underlying tissues that gradually *biases* the ectoderm toward lens

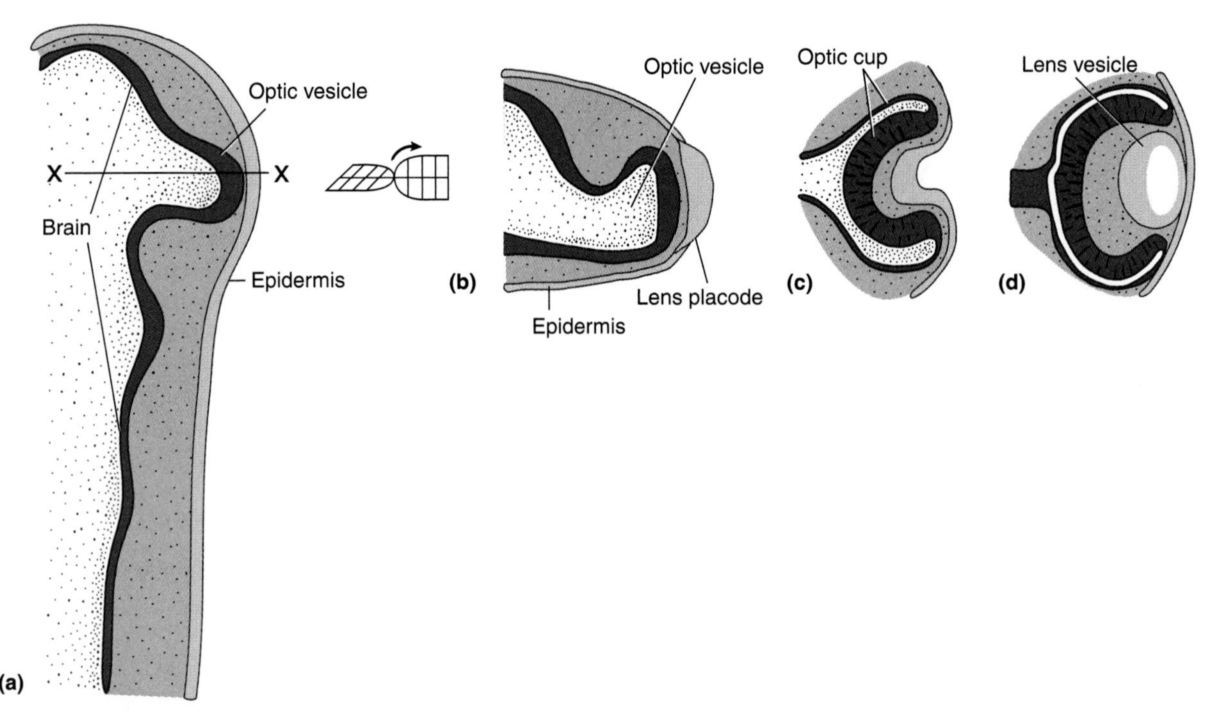

Figure 1.16 Development of the vertebrate eye. **(a)** Dorsal aspect of the embryo. The optic vesicle grows out from the brain and contacts the overlying layer of ectoderm, most of which will form epidermis. **(b–d)** Transverse sections, at level X–X indicated in part a, of subsequent stages showing the formation of the lens placode, optic cup, and lens vesicle.

formation (Jacobson, 1966; Grainger et al., 1988). It also turned out that even closely related species may differ in the relative contributions made by these various interactions to the overall result of lens formation.

The power of the strategies used by experimental embryologists attracted some of the brightest minds among biologists at the beginning of the twentieth century. The entire biological world took note in 1924, when Spemann and his graduate student Hilde Mangold published their famous "organizer" experiment. It showed that a specific piece from the dorsal side of an amphibian embryo, upon transplantation to the ventral side, could induce surrounding host tissue to form a secondary embryo. For this impressive experiment, and for his leading role in establishing the new discipline of experimental embryology, Spemann was awarded the Nobel prize in 1935. The scientific questions and experimental strategies that he and his contemporaries promoted are known today as *classical embryology*.

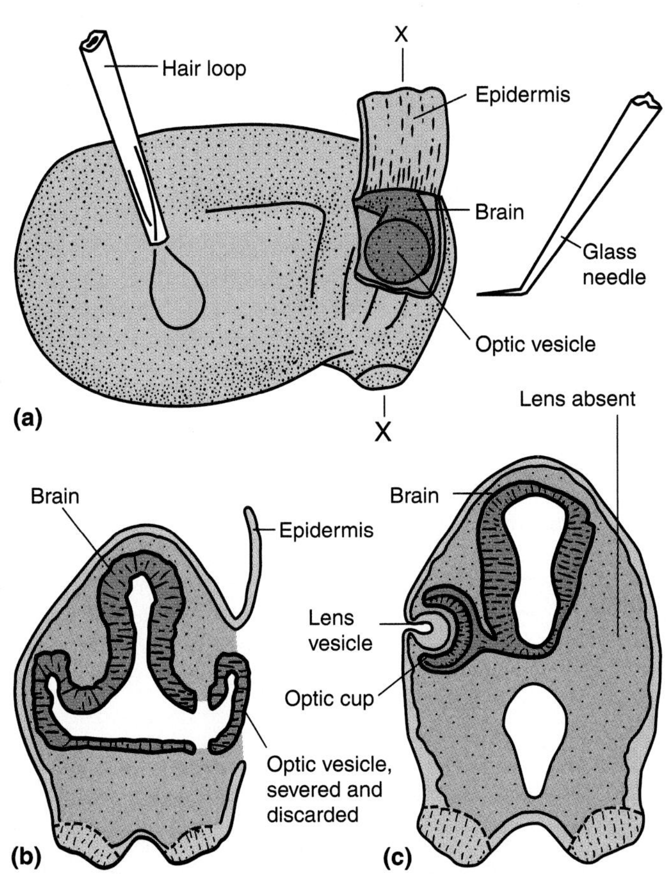

Figure 1.17 Removal of the right optic vesicle to test whether it is necessary for forming the lens of the eye. **(a)** Right side view of frog embryo. The ectodermal layer over the optic vesicle is cut with a glass needle and folded back with a hair loop to expose the optic vesicle. **(b)** Transverse section at level X–X indicated in part a. Optic vesicle is severed and removed. **(c)** Subsequently, no lens can be detected on the operated side while a normal lens develops on the unoperated side.

1.4 Genetic Analysis of Development

Given the experimental skill and intellectual acumen of Spemann and other early embryologists, it may seem surprising that most of them ignored an avenue for the analysis of development that became available in their time and attracted a lot of attention in its own right: genetics. Perhaps the embryologists felt that their discipline was still young and needed some time apart to develop its own concepts before forging an alliance with another new discipline.

All the instructions necessary for directing an organism through its entire life cycle is at one time contained in a single cell, the zygote. These instructions direct the physical form and, to a large extent, the behavior of an organism. Today we know that most of the information for heritable traits is encoded in deoxyribonucleic acid (DNA). The functional units of this information are called ***genes,*** and one full complement of DNA in an organism is called its ***genome.*** The genetic makeup of an organism is called its ***genotype.*** An organism's visible characteristics constitute its ***phenotype.***

Genetics research has traditionally been concerned with studying the transmission of genes from generation to generation, with mapping genes on chromosomes, and with describing their phenotypic effects in morphological and biochemical methods. The raw materials for such studies are ***mutants***—that is, variant stocks showing abnormal heritable phenotypes. Most abnormal phenotypes are caused by one or more ***mutations,*** or heritable changes in the genomic DNA, which occur as a result of errors in DNA replication, chromosomal exchanges during cell division, cosmic radiation, and chemical mutagenesis. Alternative forms of one gene are called ***alleles.*** The allele of each gene that predominates in natural populations is called the ***wild-type allele***; it is symbolized by a superscripted plus sign following the name of the gene. For instance, *shiverer*$^+$ refers to the wild-type allele of a particular mouse gene, whereas *shiverer* refers to most other alleles of the same gene.

In order to use mutants as analytical tools, it is critical to distinguish three types of mutant alleles. A ***null allele*** (Ger. *null,* "nothing") is characterized by the complete loss of a gene or its function. It is symbolized by a superscript minus sign following the name of the gene. For example, *hairy*$^-$ refers to the null allele of the *hairy* gene. In a ***loss-of-function allele,*** some but not all of a gene's function is lost. Most mutant alleles are loss-of-function alleles because random alterations of a gene usually leave some of its function intact. In ***gain-of-function alleles,*** a gene becomes active at a time when, or in a place where, it is normally silent. A gain-of-function allele typically differs from the wild-type allele in part of its *regulatory region,* which decides when and where a gene is on or off.

The name of a gene is often misleading because it refers to a mutant phenotype rather than the gene's normal function. For instance, the *white*$^+$ gene of the fruit fly *Drosophila melanogaster* is named after the white eye color of mutant flies that are defective in the *white* gene. The normal function of *white*$^+$ is to encode a transport protein that is involved in synthesizing the red eye pigment of normal flies. In the absence of this protein, the eyes of the fly lack the normal pigment and appear white. This system of naming genes after a defect caused by improper or lacking gene function, and not according to the normal gene function, can be confusing to the beginner. With the benefit of hindsight, it would have been more descriptive to name the *white*$^+$ gene *transport*$^+$ instead. However, scientists name a gene as soon as they isolate a mutant stock, which is normally long before they unravel the normal function of the wild-type allele.

For the analysis of development, mutant alleles can be used in much the same way as the embryological strategies of removal and transplantation. Null alleles and loss-of-function alleles of a gene are the equivalent of a removal experiment in that both are ways of studying a function by observing the consequences of its loss. Gain-of-function alleles are much like transplantation experiments in that both are ways of observing the consequences of a known function in the wrong time or place. However, only a few classical embryologists were farsighted and adventurous enough to harness the power of genetics for the analysis of development. To understand this, one needs to recall that the connection between genotype and phenotype was merely conceptual, and that the molecular mechanisms connecting the two were unknown through the first half of the twentieth century.

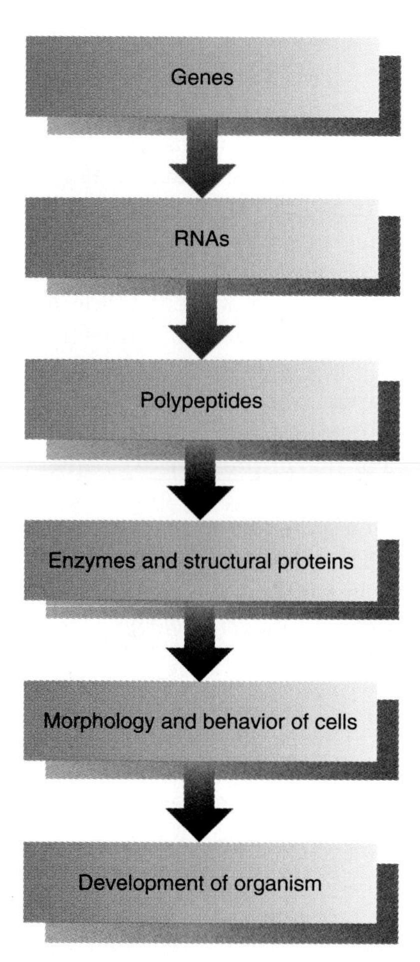

Figure 1.18 Genetic control of development. The colored arrows represent transcription and translation under the strict control of an individual's genes. The black arrows represent a strong but not exclusive genetic control of subsequent processes.

Today it is clear that genes control the production and properties of the very molecules on which development is based (Fig. 1.18). Because genes are transcribed into ribonucleic acid (RNA) and translated into polypeptides, the nucleotide sequence of the genomic DNA strictly determines the sequence of amino acids in polypeptides. Intermolecular forces between the amino acids of a polypeptide determine much of its three-dimensional configuration and its assembly into composite proteins, although preexisting molecules and cellular organelles may act as templates in the assembly of additional molecules. The properties of assembled structural proteins and enzymes determine much of the form and behavior of cells, and thus the development of the entire organism.

Initially a zygote's development is guided by its mother's genes (Fig. 1.19). RNA and proteins encoded by ***maternal effect genes***—or *maternal genes,* for short—are synthesized and deposited in the egg during oogenesis. The presence of these maternally provided products sustains the developing organism while its own DNA is replicating at a very rapid rate and is therefore only minimally transcribed. At a later stage, which depends on species, the direction of development is taken over by the embryonic genes, or ***zygotic genes,*** which are inherited from *both* father and mother.

The role of maternal gene products in an embryo's development is twofold. Some of these products, such as DNA polymerase and tubulin messenger RNA, are simply stockpiled to promote rapid cleavage and are used equally by all embryonic cells. Other maternal gene products are distributed unevenly in the egg, so that blastomeres inherit them in different quantities. At least some of these products act as *cytoplasmic determinants*—that is, they direct the development of certain embryonic regions in distinct ways. If such localized cytoplasmic determinants, or their distributions, are altered by mutations in the encoding maternal genes, the developing embryos may show severely disturbed body patterns.

For instance, females of the fruit fly *Drosophila melanogaster*, in which both copies of the *bicoid* gene are mutated, give rise to strikingly abnormal embryos (Frohnhöfer and Nüsslein-Volhard, 1986). These embryos have no head or thorax; instead, they have an extended abdomen, with a duplication of the last

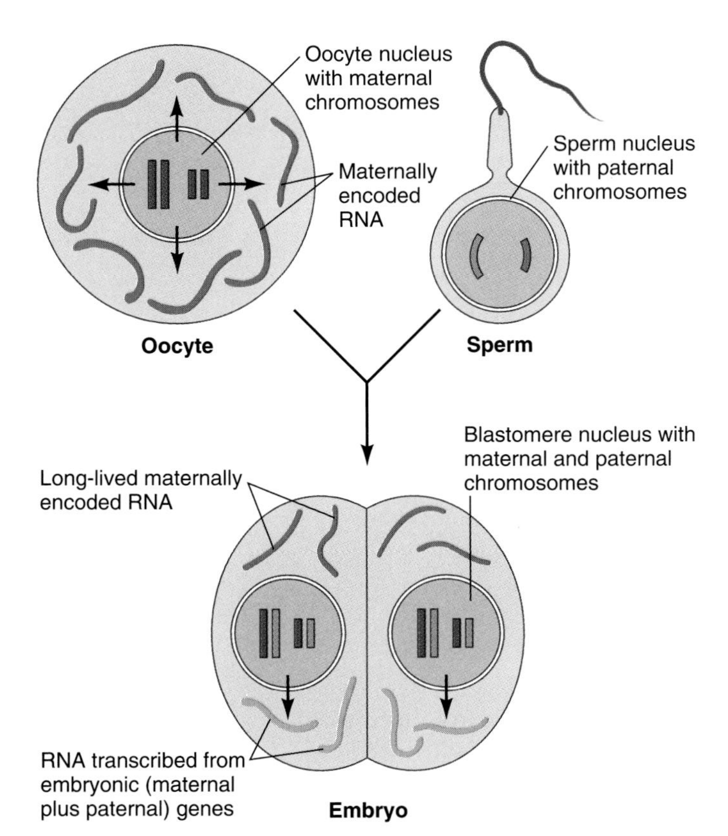

Figure 1.19 Two types of gene expression during embryonic development. Long-lived RNA (red) encoded by maternal genes accumulates during oogenesis. These transcripts typically support embryonic development during early cleavage. RNA (orange) transcribed from zygotic (maternal *and* paternal) genes begins to accumulate during cleavage, with the exact time of onset varying among species. (Through most of oogenesis, the oocyte is diploid, meaning that it has two homologous copies of each chromosome. Before fertilization, the oocyte completes a set of two specialized divisions called meiosis, in the course of which the fertilizable egg, or ovum, retains only one chromosome from each homologous pair. The ovum is then haploid, like the sperm, which has already passed meiosis. Fertilization restores the diploid state in the embryo.)

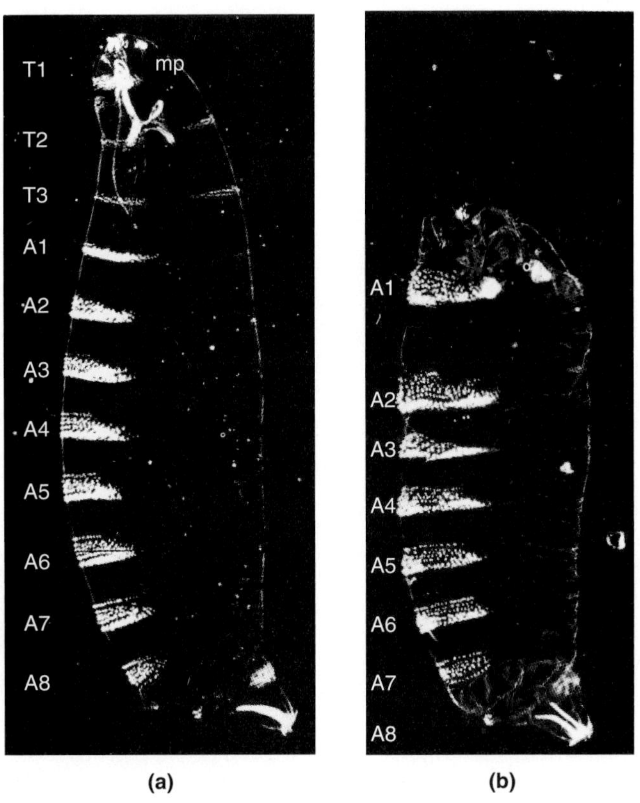

Figure 1.20 *Drosophila* larvae derived from **(a)** a normal wild-type (*bicoid*$^+$) mother and **(b)** a *bicoid*$^-$ mutant mother. The photographs show cuticle (skin) preparations of fixed and cleared specimens under dark-field illumination. The bright bands on the ventral side (to the left) are denticle belts, which the larvae use for traction during crawling. The belts are located near the anterior margins of thoracic (T1–T3) and abdominal (A1–A8) segments. The head contains mouthparts (mp) that are tucked inside the thorax. The tip of the abdomen is marked by anal plates and tracheal openings. The larva from the *bicoid*$^-$ mother is lacking head and thorax and shows a partial duplication of the abdominal tip at the anterior end.

abdominal element, the telson, attached to the anterior end of the abdomen in reverse polarity (Fig. 1.20). This phenotype develops independently of the *bicoid* allele inherited from the offspring's father. This is so because the fatal flaw in the embryo's development has already occurred while the egg was developing in its mother. Females without a functional *bicoid*$^+$ allele fail to deposit bicoid mRNA near the anterior egg pole of the eggs they produce. This mRNA is necessary to synthesize bicoid protein, which in turn is required to activate certain embryonic genes involved in head and thorax formation.

Mutations in zygotic genes can also have profound effects on the embryonic body pattern. For example, in *Drosophila* adults with mutations in the *Ultrabithorax* gene, the third thoracic segment (T3) looks like another copy of the second thoracic segment (T2). Normally, T2 carries a pair of wings while T3 carries a pair of balancer organs known as *halteres* (Fig. 1.21a). In the *Ultrabithorax* mutant, as part of transformation of T3 into T2, the halteres have been replaced with another pair of wings, thus creating a four-winged fly (Fig. 1.21c). Mutations that transform certain body regions into the likeness of another body region are known as ***homeotic mutations,*** and the genes that have this kind of mutant alleles are known as ***homeotic genes.***

Genetic analysis has shown that the *Ultrabithorax*$^+$ gene promotes the morphology characteristic of T3 and that this gene is normally active in T3 but not in T2. The mutant *Ultrabithorax* phenotype results from a failure to properly activate the gene in T3. In the absence of the normal *Ultrabithorax*$^+$ gene products, T3 assumes the morphological characteristics of T2 as a default program.

While the *bicoid* and *Ultrabithorax* phenotypes described above are caused by loss-of-function alleles of

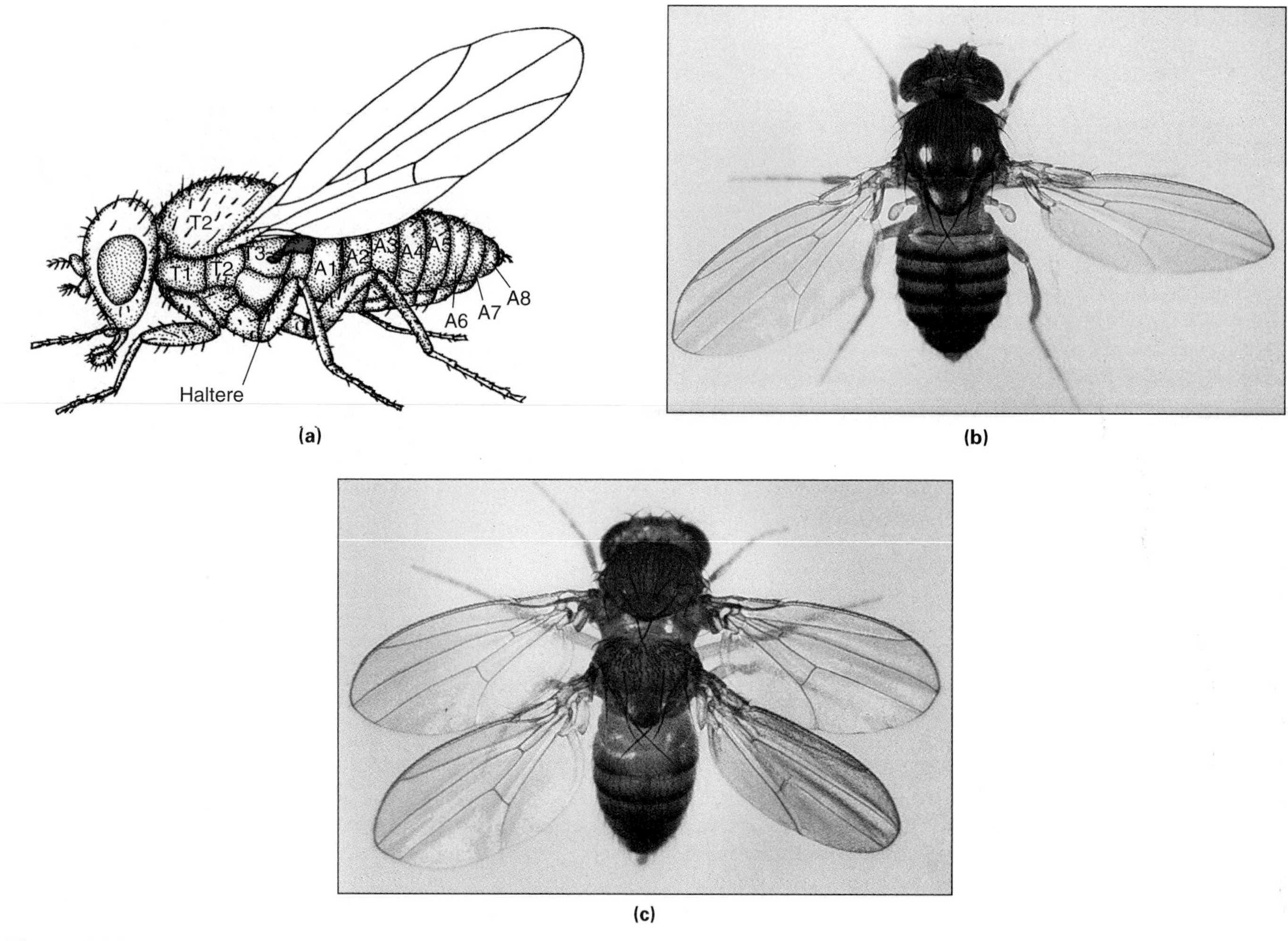

Figure 1.21 Abnormal body pattern caused by the loss of expression of a zygotic gene in the fruit fly *Drosophila melanogaster.* **(a)** Drawing of a wild-type fly. The second and third thoracic segments (T2, T3) differ in size and appendages. T2 carries a pair of wings; T3 carries a pair of smaller balancer organs called halteres (color). **(b)** Photograph of a wild-type fly. **(c)** Four-winged fly caused by mutations in regulatory regions of the *Ultrabithorax* gene. In the mutant phenotype, T3 (including halteres) is replaced by a duplication of T2 (including wings).

their respective genes, gain-of-function alleles may produce similarly dramatic phenotypes. For instance, the *Antennapedia*$^+$ gene is normally expressed in the thorax but not in the head. Faulty expression of this gene in the head results in the replacement of antennae with legs (Fig. 1.22).

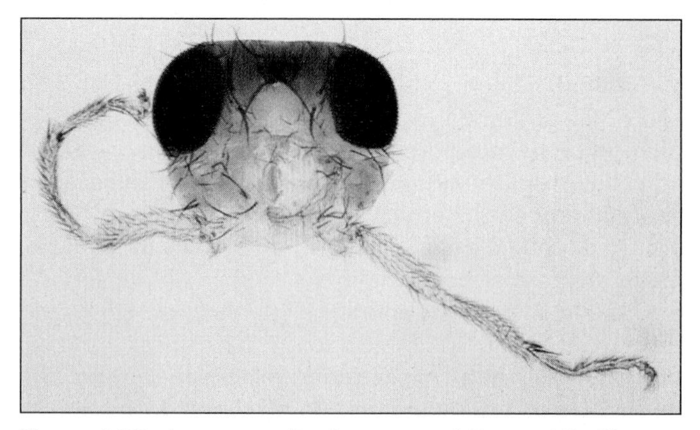

Figure 1.22 *Antennapedia* phenotype of *Drosophila.* The *Antennapedia*$^+$ gene normally is expressed in the thorax. Certain gain-of-function alleles are also expressed in the head, causing the antennae to be replaced by legs.

Homeotic genes including *Ultrabithorax*$^+$ and *Antennapedia*$^+$ were systematically analyzed in a life's work by Edward Lewis (1978). The *bicoid*$^+$ gene was characterized for the most part in the laboratory of Christiane Nüsslein-Volhard. She, her collaborator Eric Wieschaus, and others applied the method of *saturation mutagenesis screening* (see Section 15.2) in attempts to identify virtually all genes involved in building the embryonic body pattern of *Drosophila* embryos. The pioneering work of these investigators has snowballed into a worldwide effort involving hundreds of laboratories, and as a result of this collective effort, the genetic control of *Drosophila* development is now well understood. When Lewis, Nüsslein-Volhard, and Wieschaus received the Nobel prize in 1995, the award was a recognition and inspiration to all those investigators who use genetic tools for the analysis of development.

Powerful and elegant as genetic analysis of development can be, it is limited to those few organisms for which a large number of mutants are being maintained in research laboratories and stock centers. These genetically accessible organisms are the house mouse *Mus musculus,* the zebrafish *Danio rerio,* the fruit fly *Drosophila melanogaster,* the roundworm *Caenorhabditis elegans,* and the mouse ear cress *Arabidopsis thaliana.* However, investigators of other organisms use two ways of mitigating the lack of mutants. First, it turns out that many genes controlling basic developmental pathways have been exceedingly well conserved in evolution, so that once these genes are identified in a genetically accessible organism they are often found in other organisms as well. Second, certain advanced molecular methods allow investigators to mimic the effects of mutations very closely. These methods rely on DNA cloning and associated techniques, which will be discussed in the following section.

The discipline that began under the name of *experimental embryology* is called ***developmental biology*** today. This change in name reflects a broadening in scope from the study of mostly vertebrate embryos to investigations of all kinds of organisms, including in particular those that are amenable to genetic analysis. Perhaps more important was a deepening in the level of understanding that was sought. Developmental biologists set out to understand organisms in terms of cellular behaviors, and cellular behaviors in terms of molecular mechanism, which ultimately meant networks of genetic control.

1.5 Reductionist and Synthetic Analyses of Development

The seeking of ever deeper layers of understanding was driven to a large extent by technological progress. The introduction of electron microscopy made it possible to resolve not only the cells of an embryo but also cell organelles, and eventually molecules. Along with electron microscopes came centrifuges, electrophoretic chambers, and a host of other equipment to separate cells and their components. The new tools allowed investigators to see cellular details that no one had seen before, and they realized that the study of cellular anatomy and behavior could provide a better understanding of development. Because isolated cells kept in culture dishes are easier to observe and to manipulate than cells tucked away deep in an embryo, many developmental biologists began to investigate cells ***in vitro*** (Lat. "in glass"), rather than ***in vivo*** (Lat. "in the living organism").

Another technological breakthrough that has revolutionized the analysis of development has been the advent of DNA cloning, the ability to multiply any DNA segment indefinitely by splicing it into a viral or other type of DNA that multiplies naturally. This ability has boosted the power of genetic analysis enormously because the nucleotide sequence of any cloned gene can be determined, and this information often provides clues to the biochemical function of the protein encoded by the gene. Knowing both the biochemical function of certain genes from DNA sequencing and the biological function of the same genes from mutant analysis has often yielded a clear understanding of how these genes control a developmental process. This type of analysis will be greatly facilitated in the future since the genomes of *C. elegans* and *Drosophila melanogaster* have been completely sequenced.

The advances in cellular and molecular biology have promoted a ***reductionist analysis*** of development. *Analysis* is the separation of a whole into its component parts and the study of how the features of the parts determine the properties of the whole and vice versa. Most types of analysis tend to be ***reductionist,*** because they try to explain a complex system in terms of its elements, which are at a lower level of organization.

Explaining the properties of an organism in terms of the properties of its cells is an example of a reductionist analysis. This line of work began with studies on the role of cells and their organelles in heredity and in key developmental processes such as the localization of cytoplasmic components in eggs (E. B. Wilson, 1925). Later on, understanding complex morphogenetic movements such as gastrulation or neurulation as composites of simple, standardized cellular movements has been a major goal (Trinkaus, 1984). For example, the process of neurulation in amphibians (see Figure 1.8) has been analyzed in terms of coordinated cell movements and changes in cell shape.

Another level of reductionism is the attempt to explain cellular behaviors in terms of gene activities. The *principle of differential gene expression,* first developed for the metabolism of bacteria, was soon extended to the development of higher organisms (E. H. Davidson, 1986). Whereas all the cells of an organism have the same complement of genetic information, they nevertheless acquire different structures and functions by *expressing* different sets of genes. The spotlight of analysis then is on the networks of control formed by regulatory gene regions and certain proteins that interact with these regions. For instance, the bicoid protein, the absence of which causes the lack of anterior body parts (see Figure 1.20), activates embryonic genes required for head and thorax formation.

A potential problem with reductionist analysis is that the separation of a system's parts prevents any interactions between the parts; recall that this is the very reason for which classical embryologists used the strategy of isolation. Thus, the properties of cells or molecules studied in vitro may very well differ from their properties in vivo. Even worse, in the course of prolonged cell culture there may be an artificial selection process favoring cells that are well adapted to the conditions of laboratory culture but have less and less in common with normal cells in vivo.

To avoid the pitfalls of reductionist analysis, investigators must be aware of the limitations of their methods and of misleading artifacts that can result from conditions in vitro. One safeguard is the application of two independent methods, such as electron microscopy and biochemistry, to clarify the same point. Another safeguard is to use data obtained in vitro for deriving hypotheses that can be tested in vivo.

As the reductionist analysis proceeded by breaking down embryos into cells and cells into molecules, it became clear that some molecular events depend on activities that only an intact embryo can perform (Fig. 1.23). For example, lens formation in the eye requires that the appropriate portion of ectoderm receive a series of signals from adjacent endoderm, mesoderm, and optic vesicle. However, the required juxtapositions of future lens with these tissues must be brought about in morphogenetic movements that involve the entire embryo (see Figs. 1.8 and 1.16). Thus, the synthesis of *crystallins*, the proteins that characterize the lens of the eye, depends on signals generated by movements of the entire embryo. At this point, the continuous search for underlying causes comes full circle and turns into a ***synthetic analysis***, which considers the interdependence of molecular, cellular, and organismic events. Awareness of this interdependence has led to unexpected discoveries. For example, it turned out that the pattern of proteins synthesized by a cell may depend on what shape it is being held in.

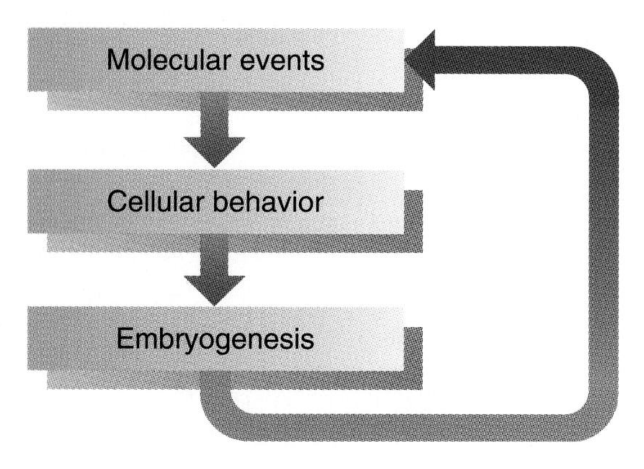

Figure 1.23 Synthetic analysis of development takes into account that molecular events, cellular behavior, and the development of the entire embryo control each other in a cyclic fashion.

Thus, modern developmental biologists can choose the best from two strategies. On the one hand, they can harness the powers of reductionist analysis, which reduces complexity and enhances the researcher's ability to change one experimental parameter at a time. On the other hand, they can take a synthetic approach and explore how events at a low level of organization are affected by activities that can be performed only at a higher level of organization.

Looking back over the past century, it is clear that developmental biology has become a mature discipline with a well-chosen set of basic questions and investigative strategies. A parallel maturation of genetics as a biological discipline, combined with great advances in methods of cellular and molecular biology, have vastly expanded the set of tools available to the modern developmental biologist. The opportunities for bringing powerful new tools to bear on a set of classical questions are generating a sense of excitement, and the prospects for attaining both a reductionist and a synthetic understanding of development are very bright indeed.

The Role of Cells in Development

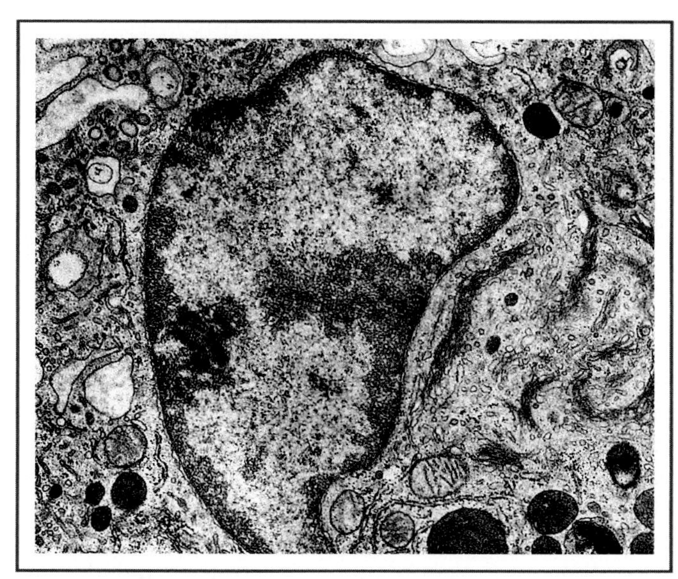

Figure 2.1 Transmission electron micrograph of a macrophage presenting the most common components of animal cells. The large organelle in the center—the nucleus—contains the chromatin, a meshwork consisting mostly of DNA and protein. The nucleus is surrounded by the nuclear envelope, a double membrane with openings that allow RNA and regulatory proteins to exit and enter selectively. The cytoplasm outside the nucleus includes various smaller organelles. The oval organelles with riblike projections, such as the one visible in an indentation of the nuclear envelope, are mitochondria, the power generators of the cell. The dark bodies are lysosomes, membranous bags filled with enzymes that digest materials taken up from outside the cell. Weaving between the mitochondria and lysosomes are flat membranous sacs dotted with ribosomes on the outside; they form the endoplasmic reticulum, or ER, a site of protein synthesis and modification. Proteins synthesized in the ER are further modified in curved stacks of membranous sacs known as the Golgi apparatus, visible to the right of the nucleus. The plasma membrane surrounding the cell often forms finger-like projections called microvilli.

CELLS are the smallest units of matter having all the attributes of living things. They have a distinct form and maintain a high degree of internal organization (Fig. 2.1). They take up energy and exchange materials with the environment. Most cells have a full complement of genetic information, and they reproduce by dividing into daughter cells. Moreover, cells communicate with each other by signals and receptors, and they have an elaborate system of internal messengers to process the information received from outside.

Cells are also the smallest units that can spontaneously aggregate into organisms. If a sponge is forced through a fine-meshed cloth, the surviving cells can reaggregate into a living sponge (H. W. Wilson, 1907). Likewise, isolated blastomeres from sea urchin embryos at the 16-cell stage can reassemble into almost normal-looking larvae (Freeman, 1988). Similarly, viable mice can be formed by combining separated blastomeres from different embryos (Markert and Petters, 1978). Such reconstitution of an organism has never been observed with units smaller than cells, such as organelles or molecules.

To developmental biologists, cells are the natural units from which organisms are made. Many processes in development have been analyzed in terms of cellular behavior. Growth, for example, can be studied in terms of cell division, and gastrulation can be studied in terms of cell movements and changes in shape. To prepare for this type of analysis, we will review some of the basic properties of cells in this chapter. Our review will be limited to eukaryotic cells (Gk. *eu*, "good," "normal"), which make up the organisms studied by most developmental biologists. Eukaryotic cells have a distinct nucleus and cytoplasmic organelles such as mitochondria and an endoplasmic reticulum. We will focus on those cellular properties that are especially relevant to development, including the cytoskeleton, the cell cycle, the cell membrane, cell movement, intercellular junctions, and cell signaling.

2.1 The Principle of Cellular Continuity

Today every student of biology learns that all organisms consist of cells. This well-known fact was first proposed as a theory in 1838 by Matthias Schleiden and a year later by Theodor Schwann. During the decades that followed, investigators found that cells arise *only* through the division of existing cells. This was an important discovery, because earlier scientists thought that cells might also form spontaneously from noncellular materials. However, this has never been found to occur under the conditions that now prevail on Earth, although cells probably evolved from simpler aggregates of organic matter under different environmental conditions during the early history of our planet. The early insights into the continuity of cells were summed up succinctly by Rudolf Virchow in 1858 in a famous aphorism, *omnis cellula e cellula*, meaning that all cells arise from cells.

Another important advance in cell theory was the discovery that eggs and sperm are specialized cells, and that they arise from less conspicuous cells in the ovary or testis. In 1841, Rudolf Albert Koelliker showed that spermatozoa are special cells originating in the testis and are not, as others had suggested, parasitic animals in the seminal fluid. Likewise, the egg was recognized as a single cell by Karl Gegenbauer in 1861. A few years later, in 1875, Oskar Hertwig observed two nuclei in fertilized sea urchin eggs, one derived from the sperm and the other from the egg.

Together, these discoveries established the ***principle of cellular continuity***: that all organisms have evolved through an uninterrupted series of cell divisions, punctuated by gamete formation and fusion. This means that the cells in our bodies today are the temporary ends of an *unbroken chain of cells*, extending back through our parents, grandparents, and earlier ancestors to Cro-Magnons, nonhuman primates, primitive mammals, reptiles, amphibians, fish, and invertebrates, and—ultimately—to a primordial unicellular organism that lived billions of years ago.

Modern biology began when the cell theory was combined with observations on chromosomal behavior. In 1883, Eduard van Beneden noted that the nuclei of conjugating gametes contribute equal numbers of chromosomes at fertilization. Chromosomal behavior during gamete formation and fertilization provided a physical basis for Mendel's laws of inheritance. At the beginning of the twentieth century it became clear that heredity is based on the continuity of cell division and on the faithful replication and distribution of chromosomes. The analysis of cell organelles other than the nucleus soon followed, and in 1896 the discoveries of cell biologists were summarized in Edmund B. Wilson's groundbreaking book *The Cell in Development and Heredity.*

2.2 The Cell and Its Organelles

Eukaryotic cells have several ***organelles***—that is, distinct units that have a particular function, such as the cell nucleus to contain the cell's DNA (Fig. 2.1). In contrast, *prokaryotic* cells, such as bacteria, have a much simpler organization without a separate nucleus.

The largest organelle in many cells, the ***nucleus,*** is bounded by a double membrane, the *nuclear envelope.* On the inside of its inner membrane, the nuclear envelope is lined with the *nuclear lamina,* a layer of fibrous proteins called *lamins.* The lamins provide attachment sites for the ***chromosomes*** (Gk. *chroma,* "color"; *soma,* "body"; referring to the deep staining of chromosomes in dividing cells). Chromosomes are threads of long DNA molecules associated with various structural and regulatory proteins. Except during cell divisions, chromosomes form a loose meshwork lodged mostly against the nuclear lamina (Fig. 2.1). Much like a train station, the cell nucleus seems to be a chaotic place at first glance, packed with a jumble of molecules transcribing genomic DNA into RNA precursors and then processing these precursors into functional RNAs. But, just as careful observation of a train station reveals an orderly flow of passengers to and from platforms at particular times, research is beginning to uncover some order in the flow of molecules from transcription sites to processing sites and then to the gates that connect the nucleus with the surrounding cytoplasm. These gates, known as *nuclear pore complexes,* are highly structured openings in the nuclear envelope (Fig. 2.2). The pore complexes allow small molecules to freely enter and leave the nucleus, but they selectively control the passage of larger molecules and particles. As a result, the nucleus and the surrounding cytoplasm are different cellular compartments adapted to different biological functions.

Outside the nuclear envelope is the cell ***cytoplasm*** (Gk. *cytos,* "cell"; *plasma,* a thing formed), which contains most other cell organelles embedded in a gelatinous matrix. While the main function of the nucleus is to manufacture RNAs, the synthesis of proteins occurs in the cytoplasm. When messenger RNAs (mRNAs) enter the cytoplasm through the nuclear pore complexes, most of them associate with many ribosomes to form ***polysomes,*** which synthesize proteins. Some of the newly synthesized proteins stay in the cytoplasmic matrix, where they function in the cell's own metabolism. Other proteins are destined to be released from the cell as signal molecules or as building blocks for the *extracellular matrix.* While these proteins are still in the cell cytoplasm, a signal sequence at their front end guides them into a labyrinth of membranous sacs called the ***endoplasmic reticulum,*** or ***ER*** (Lat. *reticulum,* "little net"). Because the beginning of the protein enters the ER while the rest of the protein is still being synthesized, the outside of the ER becomes studded with polysomes, giving it a rough appearance in electron micrographs, for which this part of the ER is called the (*rough ER*). Inside the rough ER, proteins are modified by the addition of hydroxyl groups, oligosaccharides, and other moieties.

Parts of the ER that are not associated with polysomes are known as the *smooth ER;* it serves in the synthesis of phospholipids for cellular membranes and other water-insoluble compounds. Certain regions of the ER are specialized for the sequestration of calcium ions, which are of critical importance to the regulation of many cellular activities.

Membranous *transport vesicles* shuttle the modified proteins from the rough ER to the ***Golgi apparatus,*** a stack of membranous sacs, where the proteins undergo further modifications (Fig. 2.2). Finally, proteins leave the Golgi apparatus in ***secretory vesicles,*** some of which move to the cell surface. Here the vesicular membranes fuse with the plasma membrane so that the vesicular

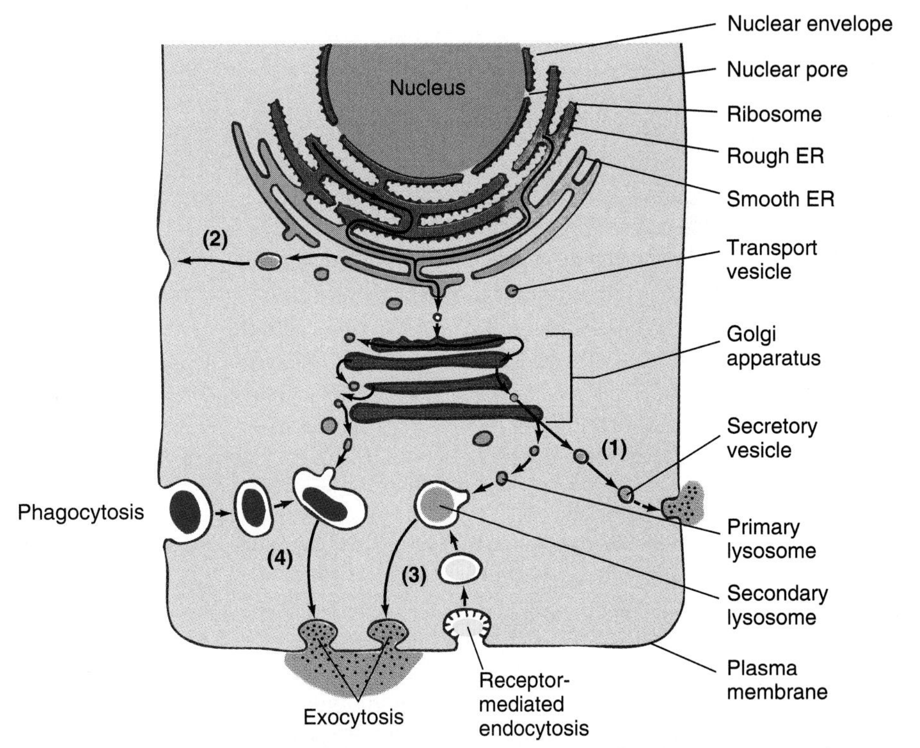

Figure 2.2 Functions of the endoplasmic reticulum (ER), Golgi apparatus, and associated vesicles. In the first pathway shown (**1**), proteins begin to be synthesized in the rough ER (studded with ribosomes) and are transported via the Golgi apparatus and secretory vesicles into the extracellular space, where the protein may serve, for instance, as an extracellular matrix component. Another pathway (**2**), beginning in the smooth ER (no ribosomes), is used simply to add more lipid bilayer to the plasma membrane. A third pathway (**3**), again beginning in the rough ER, is used to store digestive enzymes in lysosomes, which may serve in endocytosis. Finally (**4**), lysosomes may serve in the phagocytosis of larger pieces of foreign matter.

contents are shed outside the cell, a process known as ***exocytosis*** (Gk. *exo,* "outside"; *cytos,* "cell"). It allows cells to build their own environment by releasing fibrous and gelatinous materials collectively known as the ***extracellular matrix.*** Instead of being exported from the cell in secretory vesicles, certain digestive enzymes leave the Golgi apparatus in *lysosomes.* Lysosomes fuse inside the cell with other membranous vesicles that function in the uptake of external materials into the cell, a process known as ***endocytosis*** (Gk. *endon,* "within"). More specifically, the uptake of fluid in small vesicles is called *pinocytosis* (Gk. *pinein,* "to drink"), whereas the uptake of microorganisms or other firm particles in large vesicles is known as *phagocytosis* (Gk. *phagein,* "to eat").

The energy required to drive the synthesis of RNA and proteins, and to carry out most other cellular processes, is derived from small organelles known as ***mitochondria*** (sing., *mitochondrion*). They consist of two membranes, the outer one facing the cytoplasm and the inner one folded repeatedly. Proteins associated with the inner membrane reduce oxygen to water, generating in the process *adenosine triphosphate* (*ATP*), an energy-rich molecule that donates its energy to hundreds of molecular reactions.

Finally, each eukaryotic cell has a ***cytoskeleton*** consisting of interconnected threads, fibers, and lattices that give the cells its shape, internal organization, and ability to move. We will take a closer look at this system in the following section.

2.3 Cell Shape and the Cytoskeleton

CELLS CHANGE THEIR EXTERNAL SHAPE AS WELL AS THEIR INTERNAL ORDER

Cells have the capacity to either stabilize or change their overall shape and their internal organization. Morphogenetic processes such as *gastrulation* and *neurulation,* which shape the whole embryo, are partly the result of coordinated changes in cell shape. For instance, amphibian ectodermal cells assume very different shapes depending on whether they develop into epidermis or into neural plate (see Fig. 1.8). Prospective epidermal cells become ***squamous,*** which means flat or scalelike, whereas prospective neural plate cells become ***columnar***—that is, tall like a column (Fig. 2.3). As the neural

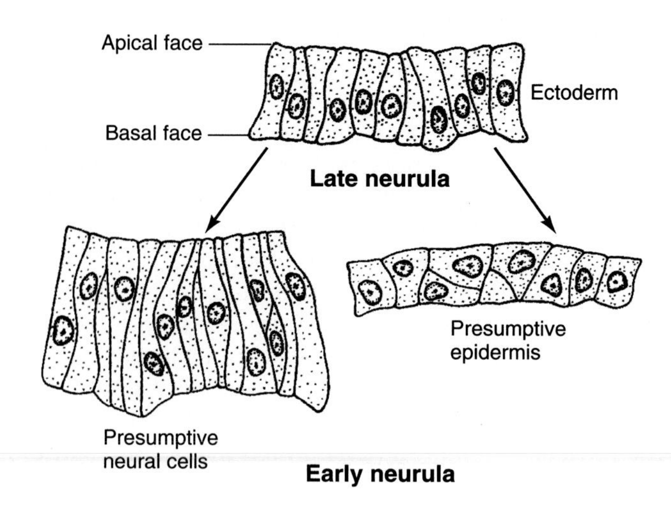

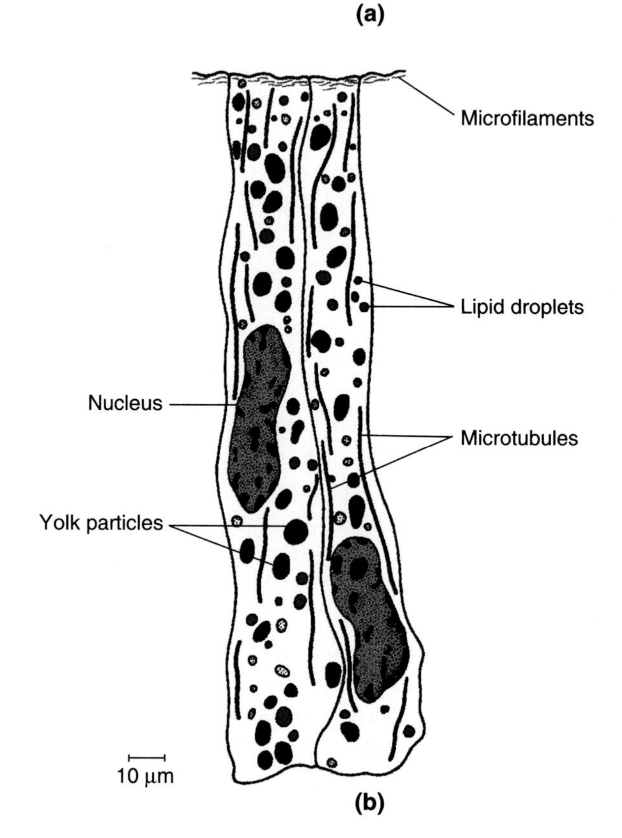

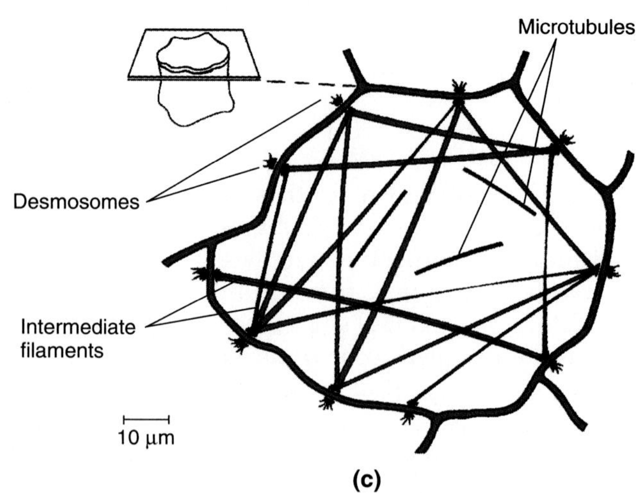

Figure 2.3 Shape changes in the ectoderm cells of a newt embryo during neurulation. **(a)** At the end of gastrulation, the ectoderm is an epithelium of short columnar cells. As neurulation proceeds, the future neural cells form the neural plate of tall columnar cells. Meanwhile, the future epidermal cells become much flatter. **(b)** Two neural plate cells. Numerous microtubules are aligned parallel to the long axis of the cells. Microfilaments are arranged in a bundle encircling the apical end of the cell like purse strings. **(c)** Flattening epidermis cell. Beneath the apical face, bundles of intermediate filaments span the cytoplasm between areas of cell contact. The few microtubules in these cells seem randomly oriented.

table 2.1 Cytoskeletal Components

	Microtubules	Microfilaments	Intermediate Filaments
Fiber diameter (nm)	24	7	8–10
Polypeptide subunit	Tubulin	Actin	Keratins; neurofilaments; vimentins; nuclear lamins
Unpolymerized form	Globular, dimer	Globular, monomer	Helical and globular domains, dimer
Polymerization inhibitors	Colchicine; nocodazole; low temperature; high pressure; antibodies	Cytochalasins; regulatory proteins; antibodies	Antibodies
Polymerization promoters	Taxol	Phalloidin	—

plate gives rise to the neural tube, the cells change shape further through constriction of their *apical* surfaces—that is, the surfaces that originally face outward. This *apical constriction* contributes to the bending of the neural plate into a tube.

In addition to a distinct outward shape, cells also maintain a high degree of inner order. Eggs, for instance, distribute certain cytoplasmic components unevenly, enriching some areas with specific mRNAs, pigment granules, or other organelles. Eggs and other cells can generate such asymmetrical arrangements in their cytoplasm by oriented transport. The fact that cells maintain these asymmetries against the force of diffusion suggests that there are "scaffolds" to hold cellular components in place.

How do cells change or maintain their shape? How do they conduct oriented transport and maintain asymmetries within the cytoplasm? Of particular importance for these activities is the *cytoskeleton*. Improved methods of light and electron microscopy have revealed three types of cytoskeletal elements that crisscross the cytoplasm (Table 2.1). Two types are present in all eukaryotic cells: *microtubules* and *microfilaments*, which have characteristic diameters of 24 nm and 7 nm, respectively. A third, more heterogeneous class of cytoskeletal fibers found in specialized cell types are intermediate in diameter (typically 8 to 10 nm) and are therefore called *intermediate filaments*. We will discuss each type of filament along with their roles in different cell activities.

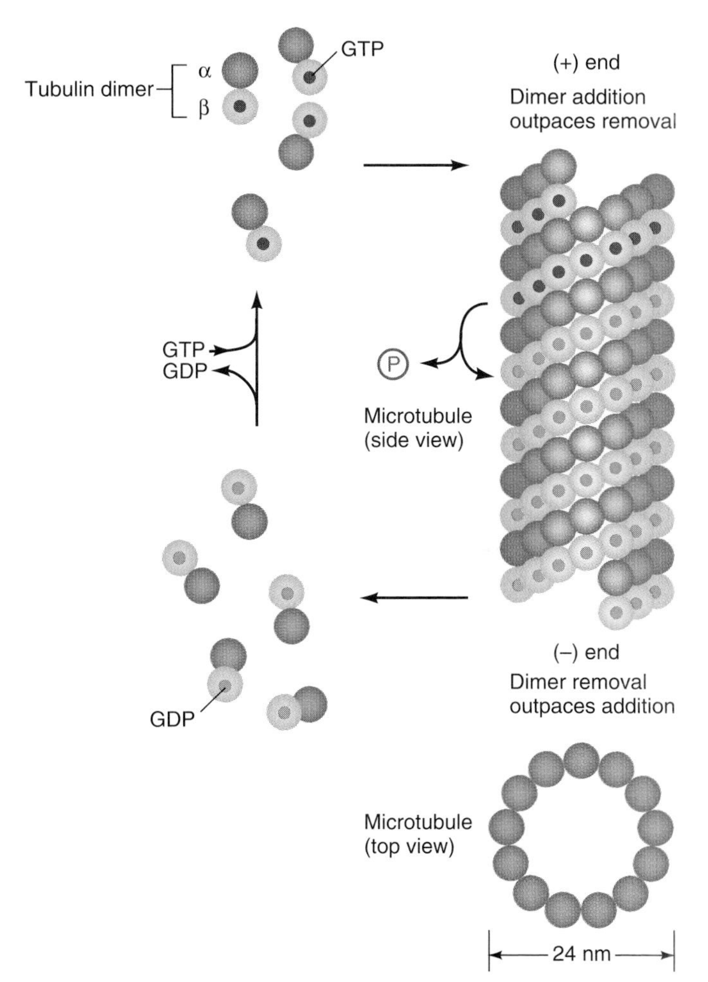

Figure 2.4 Microtubule assembly. The building blocks of microtubules are dimers consisting of two different tubulin polypeptides, usually α-tubulin and β-tubulin. Dimers are constantly added to microtubules and removed again. Dimer addition outpaces removal at one end of the microtubule, called the (+) end, whereas the converse is true at the other, called the (−) end. The addition of a dimer depends on the association of its β-tubulin with guanosine triphosphate (GTP). However, soon after addition to a microtubule of a tubulin dimer, its GTP is hydrolyzed to guanosine diphosphate (GDP). GDP-containing dimers disassemble rapidly and need to exchange their GDP for GTP before they can participate in microtubule assembly again.

MICROTUBULES MAINTAIN CELL SHAPE AND MEDIATE INTRACELLULAR TRANSPORT

In electron micrographs, ***microtubules*** look like hollow rods about 24 nm in diameter and up to 500 μm in length. Microtubules self-assemble from subunits, each of which is a dimer consisting of two different globular polypeptides called ***tubulins*** (Fig. 2.4). The tubulins are a family of very similar polypeptides encoded by a corresponding family of genes. The most common type of dimer is composed of one polypeptide from a subfamily called α-tubulins (alpha tubulins) and another polypeptide

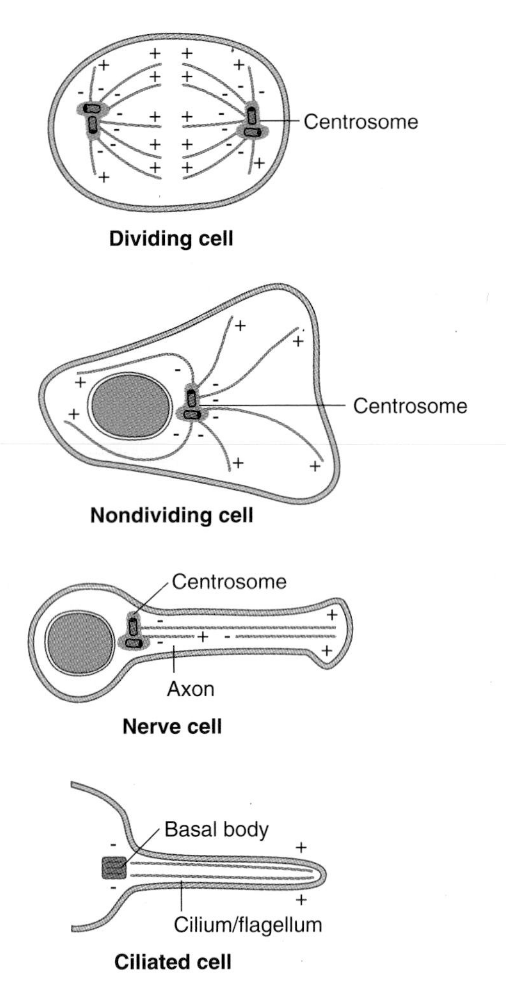

Figure 2.5 Microtubular arrays in various eukaryotic cells at different stages of the cell cycle. Most microtubules originate from a microtubule organizing center (MTOC). The two most common MTOCs are centrosomes and basal bodies. The position and properties of centrosomes change during the cell cycle. A plus sign indicates the growing end of a microtubule. A minus sign indicates the end that is shrinking unless it is anchored in an MTOC.

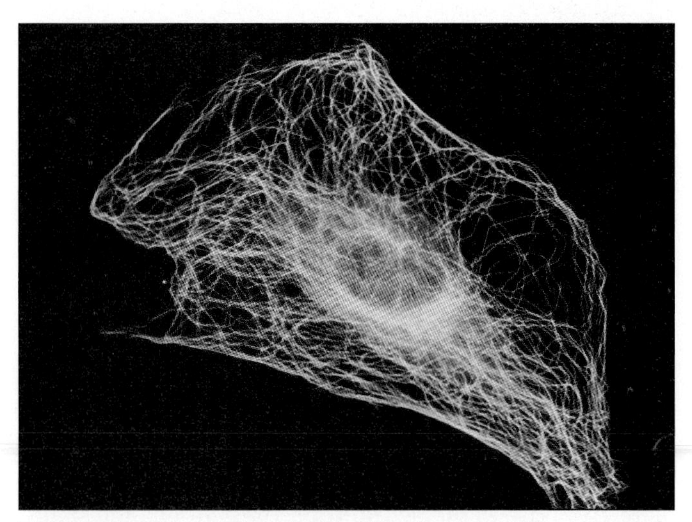

Figure 2.6 Microtubules in a cultured mouse 3T3 cell during interphase. The cells were fixed, and microtubules were made fluorescent by immunostaining (see Method 4.1). The microtubules radiate out from the centrosome, which lies just outside the nucleus.

from a subfamily called β-tubulins (beta tubulins); dimers composed of other tubulins are found in special cells, such as spermatozoa. Variants of a polypeptide that are encoded by similar genes, or derived from one gene by different processing steps, are called *isoforms.*

The assembly of microtubules is reversible and reveals an inherent polarity. At one end of the microtubule, designated the (+) end, assembly outpaces disassembly. The converse is true at the other end, called the (−) end, unless the microtubules are embedded in a ***microtubule organizing center (MTOC),*** which inhibits the disassembly process. The (+) end is characterized by *dynamic instability,* as it depolymerizes and repolymerizes continually, depending on whether the β-tubulin of a dimer is associated with guanosine triphosphate (GTP) or guanosine diphosphate (GDP). GTP tubulin readily polymerizes, but soon thereafter, its GTP is hydrolyzed to GDP, which tends to depolymerize rapidly.

The most common MTOCs are *centrosomes* and *basal bodies* (Fig. 2.5). In most animal cells, a ***centrosome*** consists of two small cylindrical bodies, called ***centrioles,*** which are positioned at right angles to each other and embedded in an amorphous substance (Glover et al., 1993). Each centriole consists of nine microtubule triplets. In dividing cells, there are two centrosomes forming the poles of the *mitotic spindle,* a system of short and straight microtubules that segregates chromosome copies into the two daughter cells. In nondividing cells there is only one centrosome, which is typically located close to the nucleus. The microtubules radiating out from these single centrosomes are long and wavy, and they tend to position the centrosome and the associated nucleus near the center of the cell (Fig. 2.6).

A particular type of microtubular array, known as the ***axoneme,*** forms the core of all the hairlike cellular extensions known as *cilia* (sing., *cilium*) and flagella (sing., *flagellum*). Eukaryotic flagella and cilia are very similar: Flagella are generally longer, and there are usually one or a few per cell. In all axonemes, microtubules are arranged in a universal pattern of nine doublets and two singlets (Fig. 2.7). Each singlet consists of 13 protofilaments (longitudinal rows of tubulin dimers) and is similar to a cytoplasmic microtubule. A doublet consists of two joined subfibers, designated A and B. In cross section, subfiber A contains the normal circular arrangement of 13 protofilaments. Subfiber B is sickle-shaped in cross section and consists of only 10 or 11 protofilaments. The doublets are associated with dynein and other proteins that generate a ciliary beat.

Axonemes grow out of ***basal bodies,*** cylindrical structures consisting of nine microtubule triplets, which are arranged with a slight tilt like the blades of a turbine (Fig. 2.8). Basal bodies are similar if not identical to the

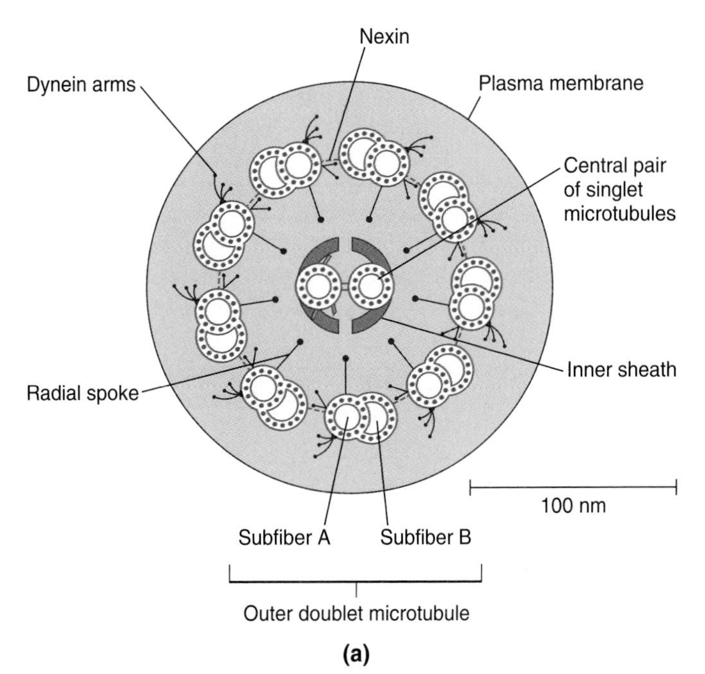

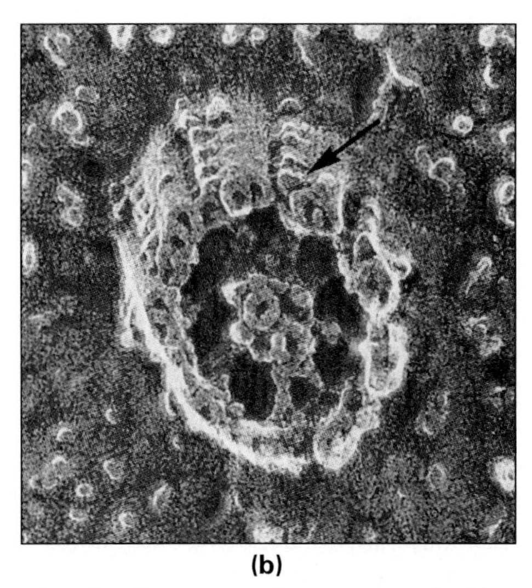

Figure 2.7 Axoneme structure shown in cross section. **(a)** Schematic diagram showing the universal arrangement of two singlet and nine microtubule doublet fibers. Each doublet fiber is composed of one circular subfiber (subfiber A) and a sickle-shaped subfiber (subfiber B). Associated with each A subfiber are "arms" of dynein molecules, which interact with the adjacent subfiber to generate the ciliary beat. Holding adjacent microtubule doublets together are proteins called nexins. Radial spoke proteins extend from each subfiber A to the center of the axoneme. Sheath projections extend from the central singlet microtubules. The entire cilium is bounded by the cell's plasma membrane. **(b)** Electron micrograph of an axoneme of *Chlamydomonas,* a flagellate protozoon. Arrow points to dynein arm.

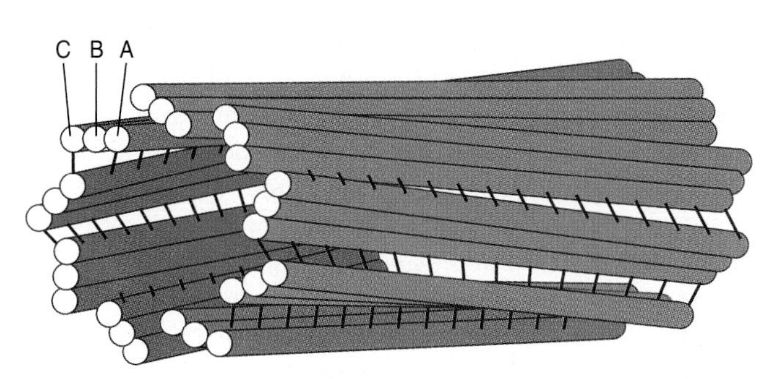

Figure 2.8 Schematic diagram of a basal body. The overall cylindrical structure is about 0.4 μm long and 0.2 μm wide. For the most part, it consists of nine microtubule triplets. The inner two subfibers of each triplet (designated A and B) continue into the corresponding subfibers of axoneme doublets (see Fig. 2.7). The outer subfiber terminates in a transition zone between the basal body and the axoneme shaft. The two central singlets of the axoneme also begin in this transition zone. Other proteins (black lines) link the triplets together.

centrioles that are embedded in centrosomes as discussed earlier. Indeed, some algae convert the basal bodies of their two flagella into centrioles when they divide. Nevertheless, one basal body anchors one complete axoneme, whereas one centrosome nucleates a variable number of individual microtubules.

The equilibrium between microtubule assembly and breakdown can be shifted through normal cellular regulation and by experimental means (Table 2.1; see Fig. 2.4). The assembly of microtubules is promoted by the availability of GTP. Once assembled, microtubules are stabilized by a wide variety of ***microtubule-associated proteins (MAPs),*** which also mediate interaction between microtubules and other cellular components. Low temperature as well as high pressure promote the diassembly of microtubules. The drugs *colchicine* and *nocodazole* bind to tubulin dimers and prevent their assembly into microtubules. This leads to the disappearance of microtubules, because their spontaneous disassembly is not blocked at the same time. A different type of drug, *taxol,* has the reverse effect. It binds tightly to microtubules, thus inhibiting their disassembly.

In addition to their role in maintaining cell shape, microtubules provide the tracks for ***motor proteins*** that transport organelles and molecules across the cell using energy from the hydrolysis of ATP. For instance, in the axons of *neurons* (nerve cells), microtubules convey an extensive traffic of materials in both directions between the cell body and the axonal terminus. The best-known families of motor proteins are *dyneins* and *kinesins* (Hirokawa, 1998). ***Dyneins*** are large proteins that move along microtubules toward the (−) end (Fig. 2.9). In the cytoplasmic matrix, they can carry various cargo molecules, whereas in axonemes, they pull microtubule doublets relative to their neighbors, thus creating the beat of cilia and flagella (see Fig. 2.7). ***Kinesins*** are similar

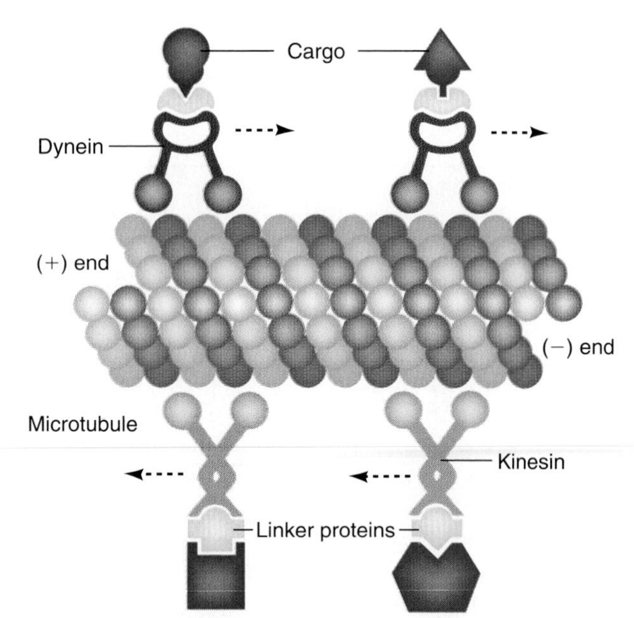

Figure 2.9 Motor proteins moving along microtubules. Dyneins move toward the (−) end whereas most kinesins move toward the (+) end. Both families of motor proteins have different members and presumably adapter proteins specialized to carry different molecules or organelles as cargos.

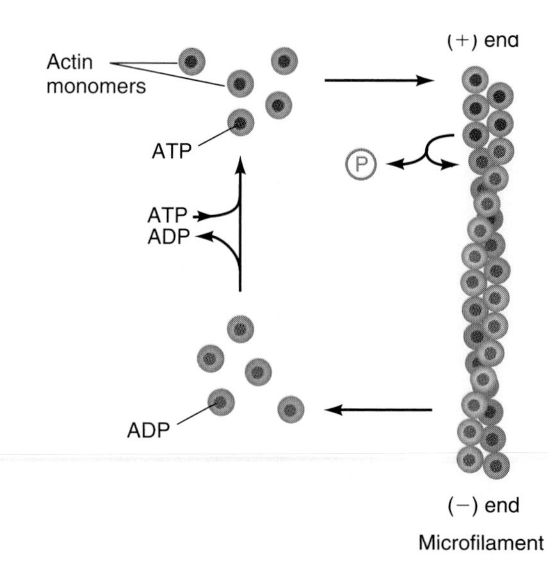

Figure 2.10 Reversible assembly of microfilaments from globular actin. The end of the microfilament where assembly outpaces disassembly is designated the (+) end. The assembly process is promoted by the association with adenosine triphosphate (ATP), in analogy to the role of GTP in the assembly of microtubules (see Fig. 2.4) However, soon after the addition of an actin monomer, its ATP is hydrolyzed to adenosine diphosphate, which dissociates more readily from the microfilament (shown here at the minus end).

motor molecules except that most of them move toward the (+) end of the microtubule. Motor proteins and their cargos are presumably linked by *adapter proteins.*

MICROFILAMENTS GENERATE CONTRACTING FORCES AND STABILIZE THE CELL SURFACE

A second class of cytoskeletal elements, the ***microfilaments,*** appear under the electron microscope as straight fibers about 7 nm in diameter. Microfilaments self-assemble as double-helical fibers from globular polypeptides called ***actins*** (Fig. 2.10). Actins, which may constitute as much as 10% of the total cellular protein, are some of the best-conserved proteins in evolution. Like tubulins, actins are a group of isoforms encoded by a family of very similar genes. Some types of actin are very abundant in muscle fibers, where they are part of the contractile system. Other types of actin form the microfilaments of nonmuscle cells.

The assembly of microfilaments is similar to that of microtubules in several respects. In vivo as well as in vitro, microfilaments self-assemble from actin in the presence of proper nucleation centers, potassium and magnesium ions, and a nucleotide triphosphate, ATP in the case of microfilaments. Also like microtubules, microfilaments are polarized, with assembly outpacing disassembly at the (+) end.

The equilibrium between assembled microfilaments and free actin can be modified experimentally by various drugs, again in parallel to microtubules (see Table 2.1). A family of drugs called *cytochalasins* bind to the (+) ends of microfilaments, preventing further elongation. *Phalloidin,* a poison made by the mushroom *Amanita phalloides,* has the opposite effect: It binds tightly along the side of microfilaments and inhibits their disassembly. At proper concentrations, these drugs interfere specifically with microfilaments and not with microtubules or intermediate filaments. Thus, cytochalasin blocks microfilament-driven cellular activities such as locomotion and the cleavage of blastomeres without inhibiting microtubule-dependent activities such as flagellum or mitotic spindle formation.

As cells change their shapes and move, they undergo extensive remodeling of their actin cytoskeleton. This phenomenon is mediated by various proteins that associate with either monomeric actin or with microfilaments (*Welch et al., 1997*). Such remodeling is often triggered by external stimuli, and the signaling pathways to actin-associated proteins are currently under active investigation (*Welch, 1999*; Maekawa et al., 1999).

A striking example of the rapid assembly of microfilaments from actin monomers occurs during the formation of the *acrosomal process,* a projection formed at the tip of many invertebrate spermatozoa when they approach an egg (see Section 4.2). The projection is driven by the assembly of microfilaments inside the acrosomal process. In sea cucumber sperm, the acrosomal process is extruded at a speed of 10 μm per second, penetrating the egg envelopes like a harpoon and bringing the sperm and egg plasma membranes into contact (Fig. 2.11).

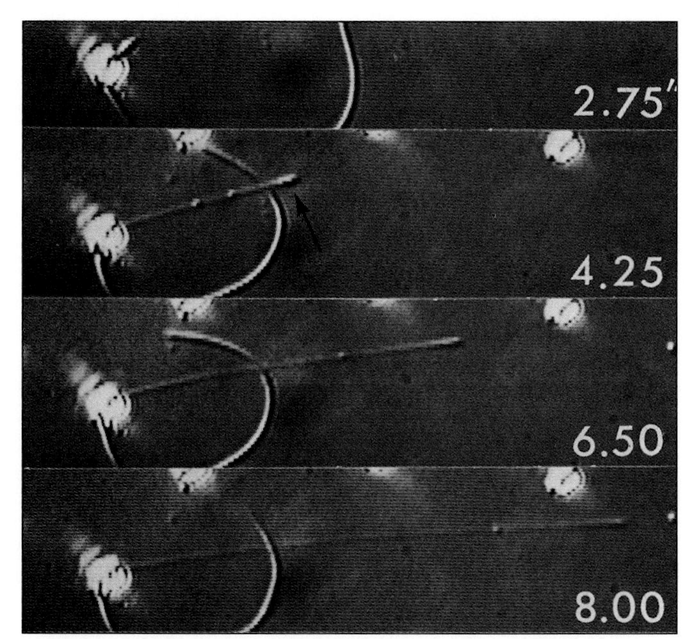

Figure 2.11 Acrosomal process formation in a sea cucumber sperm. The light micrographs were taken at intervals indicated in seconds to the lower right. The sperm head is to the left. The arc to the right of the sperm head is a portion of the sperm tail. The acrosomal process (arrow) is formed when the sperm contacts the jelly layer surrounding the egg. The signaling pathway triggered by the contact promotes the rapid self-assembly of microfilaments, which in turn causes the acrosomal process to extend.

Microfilaments in turn assemble into bundles and networks under the control of further actin-binding proteins. *α-actinin* and *fimbrin* cross-link microfilaments into parallel bundles, whereas *filamin* cross-ties microfilaments into meshworks. The various conformations of microfilaments reflect their diverse roles in the cell. For example, the widely spaced parallel bundles of microfilaments generated by α-actinin interact with myosin in the generation of contracting forces. This has been shown most clearly in muscle cells, which are specialized for this function. Almost certainly the same type of interaction occurs in nonmuscle cells during *cytokinesis* when the microfilaments that make up the contractile ring constrict the bridge between the dividing cells. This conclusion is supported by the inhibitory effects of both cytochalasin and myosin antibodies on the progress of cytokinesis. Microfilaments also form a contractile ring beneath the apical surface of neural plate cells, where they contribute to coordinated changes in cell shape (see Fig. 2.3).

As part of the cytoskeleton, microfilaments give dynamic structural support to the cell and its extensions. In the outer layer of cytoplasm beneath the plasma membrane, microfilaments form a meshwork that resists deformation and supports the plasma membrane, which by itself is fluid and has no mechanical strength. The reinforcement of the surface cytoplasm with microfilaments is especially prominent in large cells like eggs, which have a more resilient surface cytoplasm, known as the ***cortex*** (Lat. "rind"). Other roles performed by microfilaments in cell locomotion and in the localization of cytoplasmic components will be discussed in Sections 2.6, 8.5, and 8.9.

INTERMEDIATE FILAMENTS VARY AMONG DIFFERENT CELL TYPES

The third class of cytoskeletal elements is a diverse group of fibers collectively called ***intermediate filaments.*** Under the microscope, they are visible as arrays of straight or curved fibers typically 8 to 10 nm in diameter. There are several types of intermediate filaments, composed of different proteins including keratins, vimentin-like proteins, and neurofilament proteins (see Table 2.1). Unlike the globular actins and tubulins, the intermediate filament proteins themselves are long fibrous molecules capable of forming nonpolarized polymers. Intermediate filaments also vary more from species to species than tubulin and actin. Although intermediate filaments consist of repetitive subunits, they do not seem to undergo the rapid assembly and disassembly that characterize microtubules and microfilaments. Specific inhibitors or promoters of intermediate filaments have not been found; however, their assembly can be disrupted by injecting antibodies against their constituent proteins.

The functions of intermediate filaments are still being investigated (Fuchs and Cleveland, 1998). The filaments' variation in distribution and molecular nature suggests that different intermediate filaments may have different functions. One role of keratins seems to be stabilizing tightly knit sheets of cells known as *epithelia.* Epithelial cells are traversed by a wide variety of keratins, which connect at certain cellular junctions (see Fig. 2.23c). The transcellular network of fibers generated by this arrangement gives the epithelium tensile strength. Mutations that weaken this structural framework increase the risk of cell rupture, causing a variety of disorders in humans.

Massive networks of keratin filaments have also been found in *Xenopus* oocytes (Fig. 2.12). Experiments with cytochalasin B and nocodazole indicate that the organization of these keratin networks depends on intact microfilaments and microtubules (Gard et al., 1997). In starfish oocytes, keratin filaments are associated with a signal transduction protein, suggesting that a signal-dependent process may be facilitated by the keratin filaments (Chiba et al., 1995).

The lamin filaments mentioned earlier as attachment sites for chromosomes inside the nuclear envelope are also counted among the intermediate filaments. In contrast to the cytoplasmic intermediate filaments, which are fairly stable, the lamin filaments are broken down and reassembled along with the nuclear envelope during each cell division.

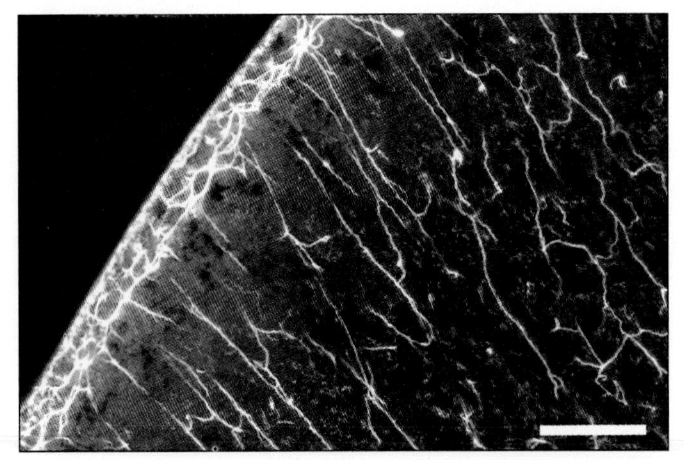

Figure 2.12 Keratin filaments in a *Xenopus* oocyte made visible by means of a fluorescent antibody staining technique. The filaments form a dense network in the outer layer of cytoplasm, or cortex, of the oocyte. In the inner cytoplasm, most keratin filaments are oriented radially while some are oriented transversely. Scale bar: 25 micrometers.

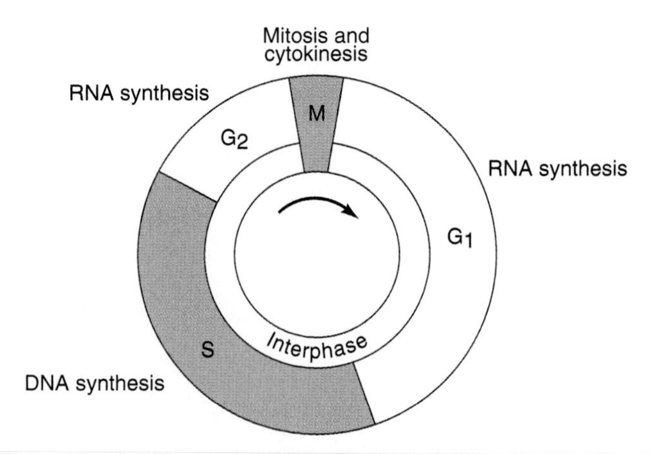

Figure 2.13 Schematic representation of the cell cycle. The relative lengths of the four phases vary among cell types, with G_1 exhibiting the greatest variation.

2.4 The Cell Cycle and Its Control

For an organism to develop and reproduce, its cells must increase in number. They do so in a cyclic process of growth and division known as the ***cell cycle*** (Fig. 2.13). A seemingly quiescent growth period known as ***interphase*** alternates with the dramatic process of cell division, called ***M phase.*** The major events that occur during M phase are mitosis and cytokinesis. ***Mitosis*** (Gk. *mitos,* "thread," alluding to the appearance of chromosomes during this phase, during which they are so condensed that they are visible under the light microscope) designates the phase in which the cell nucleus divides. ***Cytokinesis*** (Gk. *kytos,* "cell"; *kinesis,* "movement") refers to the division of the cell cytoplasm. During interphase, most chromosomes are too thin to be visible under the light microscope. This is the working phase for the cell nucleus, the period when DNA and RNA are synthesized. The replication of DNA is limited to an interval within the interphase called the ***S phase*** ("S" for synthesis of DNA). The synthesis of RNA occurs during the interval preceding S phase, known as the ***G_1 phase*** ("G" refers to a gap in DNA synthesis), and during the interval following S phase, known as the ***G_2 phase.*** Thus, a typical interphase is made up of successive G_1, S, and G_2 phases. Generally, a cell grows during interphase to the predivision size of its parent.

Nondividing cells may enter a modified G_1 state known as the ***G_0 state.*** In this state, a cell does not synthesize the proteins needed for DNA replication and mitosis. In multicellular organisms, many cells enter the G_0 state as part of their final differentiation, and most of these cells never divide again.

CHROMOSOMES ARE DUPLICATED DURING S PHASE

Nuclear DNA, the carrier of genetic information, is assembled in ***chromatin,*** which contains histones and other chromosomal proteins in addition to DNA. In the natural segments of chromatin, called ***chromosomes,*** continuous DNA molecules extend from one end to the other. The number and form of chromosomes are characteristic of each animal species. Before S phase, each chromosome contains one long DNA molecule. During S phase, the chromosomal DNA is faithfully replicated to preserve the genetic information; in fact, cells have elaborate enzymatic mechanisms for editing out errors that occur during replication. (The few errors that escape correction are mutations, which are inherited thereafter.) Following DNA replication, the chromatin structure of each of the twin DNA strands is restored by the addition of new histones and other proteins. Thereafter, each chromosome consists of twin ***chromatids*** held together at a region called the ***centromere.*** DNA synthesis and chromatid separation are delayed at the centromere; this region plays a key role in mitosis.

CHROMOSOME DUPLICATES ARE SPLIT BETWEEN DAUGHTER CELLS DURING MITOSIS

The events that occur during mitosis are of fundamental importance. Each chromosome splits into its two component chromatids, and an elaborate sequence of mechanisms ensures that each daughter cell will inherit one chromatid from each chromosome. In the daughter cells, each single chromatid will constitute a chromosome until after the following S phase, when each chromosome will again consist of two chromatids. Mitosis is divided into five stages, known as ***prophase, prometaphase, metaphase, anaphase,*** and ***telophase.*** A stage-by-stage account of the mitotic events accompanies Figure 2.14.

The nucleus changes dramatically during the cell cycle. The ***nuclear envelope,*** which surrounds the nucleus

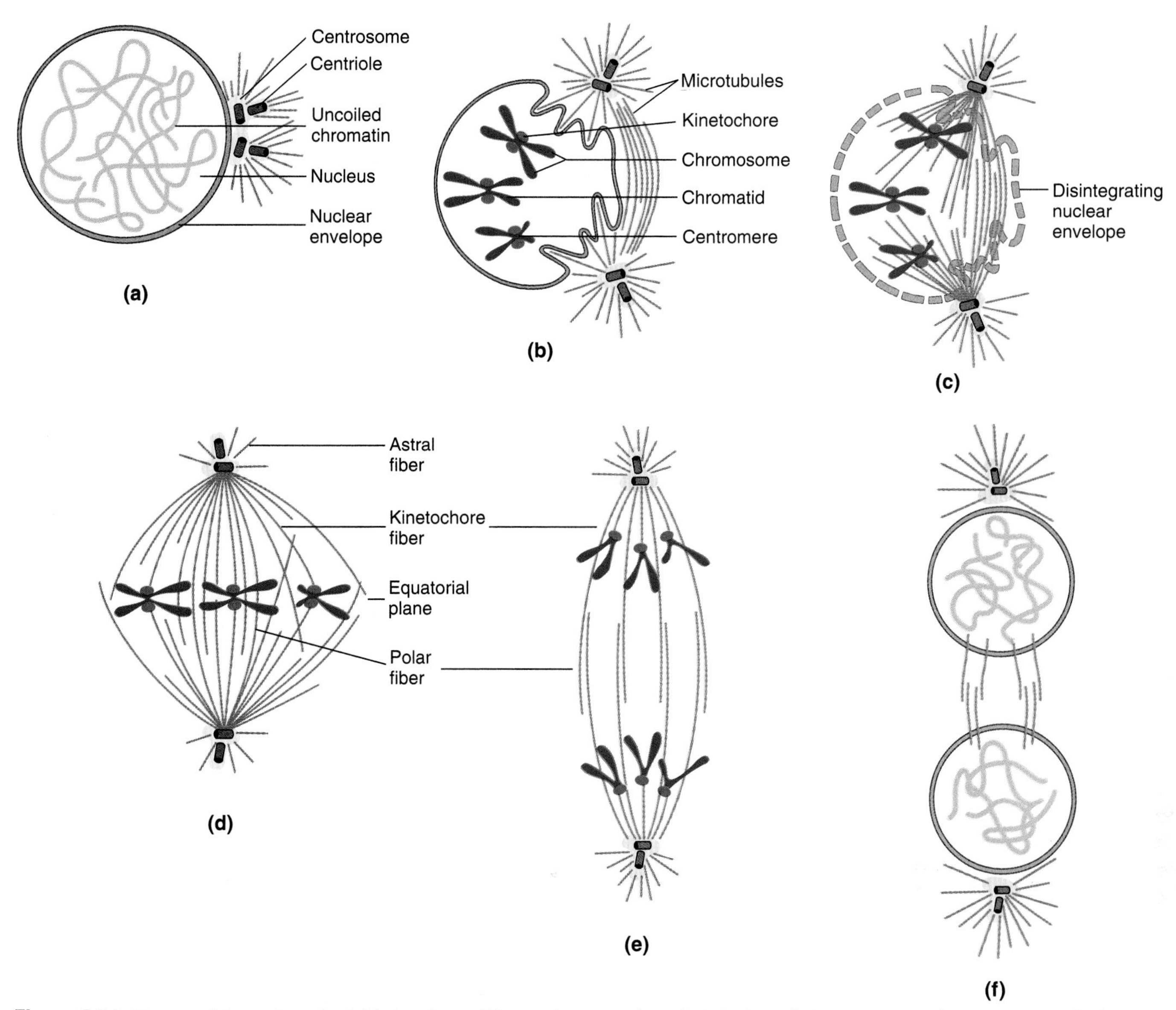

Figure 2.14 Stages of the cell cycle. **(a) Interphase.** The nuclear envelope is intact, and chromosomes form an entangled mesh. The genomic DNA and an extranuclear organelle, the centrosome, are both replicated. Each centrosome consists of two cylindrical bodies, called centrioles, and a surrounding matrix. **(b) Prophase.** Chromosomes condense into threads that are visible under the light microscope. Each chromosome consists of two chromatids, held together at the centromere, where each chromatid is associated with a kinetochore. The centrosomes move apart and begin to organize bundles of microtubules called spindle fibers. **(c) Prometaphase.** The centrosomes have moved to nearly opposite sides of the nucleus. The nuclear envelope disintegrates into small, membranous vesicles. This allows the spindle fibers to interact with the kinetochores. **(d) Metaphase.** The chromosomes are aligned in the metaphase plate midway between the centrosomes and perpendicular to the spindle axis. There are three classes of spindle fibers: polar fibers projecting from one centrosome toward the other centrosome; astral fibers radiating out from the centrosomes; and kinetochore fibers extending between a centrosome and a kinetochore. The kinetochore fibers pull the two kinetochores of each chromosome to opposite centrosomes, but the two chromatids are still held together at the centromere. **(e) Anaphase.** The chromatids separate, each becoming an independent chromosome. Each chromosome moves toward the centrosome to which it is connected by kinetochore fibers. The kinetochore fibers shorten while the polar fibers elongate and interact to push the spindle poles apart. **(f) Telophase.** Kinetochore fibers disappear. A new nuclear envelope forms around each group of chromosomes. Chromosomes uncoil.

during interphase, breaks up into small vesicles during prometaphase. At the end of mitosis, the fragments reunite to form new envelopes around each of the two daughter nuclei. Chromatids undergo an immense amount of coiling during prophase so that they do not become entangled when they segregate. The coiled chromatids are visible as short threads under the light microscope. The two joined chromatids of each chromosome form a distinct V or X shape, depending on whether the centromere is at the tip or near the middle of the chromosome. Two disc-shaped structures called ***kinetochores,*** each aligned with one chromatid, form next to the centromere. The kinetochores interact with certain fibers of the mitotic spindle.

The key organelle during mitosis is the ***mitotic spindle,*** which is visible from prometaphase to anaphase. The mitotic spindle extends between two poles, which are marked by *centrosomes* (not to be confused with *centromeres*). The plane midway between the spindle poles and perpendicular to the axis connecting them is called the ***equatorial plane.*** The most conspicuous elements of the spindle—the spindle fibers—consist of bundles of closely associated *microtubules.* The spindle fibers can be subdivided into three classes. ***Astral fibers*** radiate outward from the two centrosomes, keeping the latter in the central part of the cell. ***Polar fibers*** project from one centrosome toward the equatorial plane and beyond, where they interact with their counterparts from the opposite centrosome. ***Kinetochore fibers*** project from one centrosome and interact with one kinetochore of a chromosome. Since the minus ends of the microtubules are buried in the centrosomes, the free ends of the spindle fibers—those that interact with one another and with the kinetochores—are the plus ends. This polarity of the spindle fibers is the basis for the assembly of the chromosomes halfway between the spindle poles and for the eventual separation of the chromatids.

Because the two kinetochores of a chromosome face in opposite directions, each will interact with one pole of the spindle. During prometaphase, the chromosome thus becomes the object of a tug-of-war as each kinetochore is pulled by the kinetochore fibers to which it is attached. These forces reach an equilibrium when opposite kinetochore fibers have the same length, so that all chromosomes move to the equatorial plane, where collectively they form the ***metaphase plate.*** Each chromosome is then oriented with one of its chromatids facing one spindle pole and its sister chromatid facing the opposite spindle pole. This orientation ensures that every chromosome, when it divides, will contribute one of its chromatids to each of the daughter nuclei. At the same time, the bipolar spindle is maintained by a balance of opposing forces. They are generated by *kinesin* motor proteins, which link opposing polar microtubules and move toward one of the free ends, thus stemming the centrosomes apart, and *dynein* motor proteins, which have the opposite effect (Sharp et al., 1999).

When the centromere splits at the beginning of anaphase, the two chromatids become independent chromosomes and move to opposite spindle poles. The chromosomal movement during anaphase is rapid and is driven by two forces. First, each chromosome is pulled toward a spindle pole by kinetochore-associated proteins that break down microtubules at the plus end and/or act as motor proteins moving toward the minus end. During this time, the centrosomes that form the spindle poles may be anchored to the cell cortex via astral fibers. Second, *kinesin* motor proteins moving toward the plus ends of polar microtubules continue to stabilize the spindle while pushing its poles apart.

THE CELL CYTOPLASM IS DIVIDED DURING CYTOKINESIS

Mitosis is followed by cytokinesis, when the cytoplasm of a cell is divided between its daughter cells. Cytokinesis in animal cells begins during anaphase with the formation of a furrow in the plasma membrane, which is located in the same plane previously occupied by the metaphase plate. This furrow is often formed by the contraction of a circular bundle of actin and myosin filaments known as the ***contractile ring*** (Fig. 2.15). The

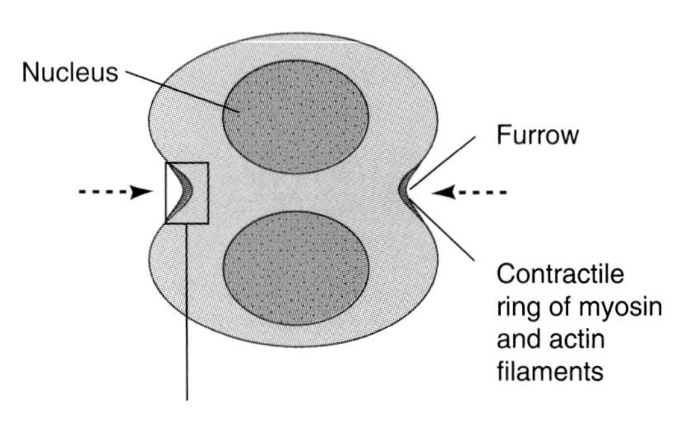

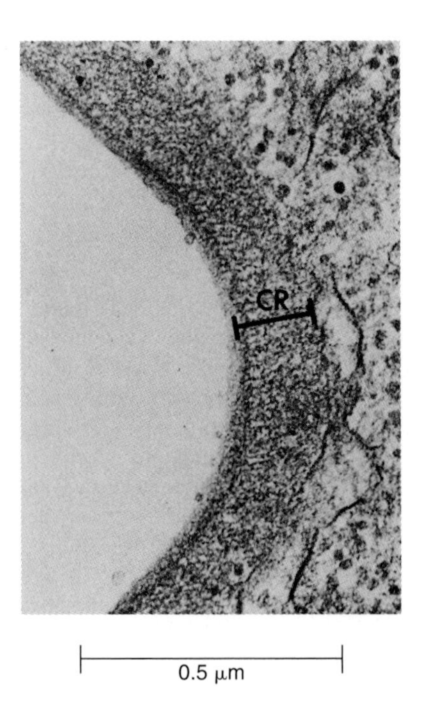

Figure 2.15 Contractile ring (CR) of a cell during cytokinesis. After mitotic telophase, a cortical ring of microfilaments and myosin fibers forms in the same plane that was occupied by the metaphase plate during mitosis. Contraction of the ring (arrows) pinches the two daughter cells apart. (*From H. W. Beams and R. G. Kessel. [1976] Cytokinesis: A comparative study of cytoplasmic division in animal cells,* Am. Sci. *64:279–290.*)

filaments of the contractile ring are anchored to proteins embedded in the cell plasma membrane, so that constriction of the ring causes the membrane to furrow. The force exerted by the contractile ring is most likely generated by sliding of the actin and myosin filaments past one another, a mechanism similar to the action of these proteins in muscle contraction. As the ring constricts, the two daughter cells are pinched apart. In some types of cells, cytokinesis is incomplete, so that a cytoplasmic bridge persists between the daughter cells (see Section 3.5).

The mechanism of cytokinesis in plants differs from that in animals. Instead of being pinched apart by the contraction of actin and myosin filaments, plant cells undergo cytokinesis by building a new cell wall from precursor materials contained in vesicles derived from the *Golgi apparatus*. These vesicles fuse to form a disclike structure called the *cell plate* while the vesicle contents assemble into pectin, hemicellulose, and other components of the primary cell wall. The orientation of the cell plate, like that of the metaphase plate, is perpendicular to the mitotic spindle axis.

THE CYCLIC ACTIVITY OF A PROTEIN COMPLEX CONTROLS THE CELL CYCLE

The cells in multicellular organisms divide at different rates. Embryonic cells may divide every 10 minutes, some epidermal cells go through a complete cell cycle within a day, and most adult neurons have stopped dividing altogether. The differences in cell cycle times are due mainly to variations in the length of the G_1 phase.

The progress of cells through the cell cycle is regulated by enzymes known as ***cyclin-dependent kinases.*** These enzymes have ***kinase*** activity, which means that they phosphorylate (add phosphate groups to) other proteins. This function is of great importance in cells because phosphorylation changes the three-dimensional structure and thus the biological activity of proteins. There are many types of kinases with different properties. The activity of cyclin-dependent kinases requires association with a cyclin. The ***cyclins*** are a family of proteins so named because they undergo a cycle of synthesis and destruction during each cell cycle. For instance, cyclin B accumulates during interphase and is abruptly degraded after metaphase. The first cyclin-dependent kinase to be discovered, CDK1, was characterized in fission yeast as the product of the *cdc2*$^{+}$ gene, but very similar proteins occur in all eukaryotic cells (Nurse, 1990). Most eukaryotic cells have multiple cyclin-dependent kinases and multiple cyclins, which in different combinations regulate progress through the successive phases of the cell cycle (Nigg, 1995).

The transition from G_2 phase to M phase is induced by a protein complex known as ***M-phase promoting factor (MPF),*** which consists of CDK1 and cyclin B (Fig. 2.16). (Historically, MPF was first described in the context of *meiosis* as a *maturation promoting factor* (see Section 3.5). Only later was it found that MPF stimulates mitotic cell divisions as well.) To form active MPF, CDK1 must be modified by phosphorylation of a particular threonine residue, Thr^{161}, and dephosphorylation of a tyrosine residue, Tyr^{15}. In addition, CDK1 must associate with *cyclin B,* which is also activated by phosphorylation. Directly or indirectly, the kinase activity of MPF is thought to cause the events that occur during prophase, including chromosome condensation, the breakdown of the nuclear envelope, and the assembly of the mitotic or meiotic spindle. Right after MPF has reached its peak activity at metaphase, cyclin B is rapidly degraded by complexes of ***proteases***—that is, enzymes that break down proteins, while the modification steps that activate CDK1 protein are reversed. Thus MPF becomes inactive and the cell proceeds to the next interphase.

Cyclic MPF activation by CDK1 modification and association with cyclin B is a basic cell cycle control mechanism that has been extremely well conserved in evolution. The CDK1 proteins from yeast and humans are so similar that they can functionally replace each other, and cyclins from clams and sea urchins function in frog oocytes (Draetta et al., 1987). However, the control of the cell cycle is modified in various ways at different stages of development (Whitaker and Patel, 1990; Murray, 1992).

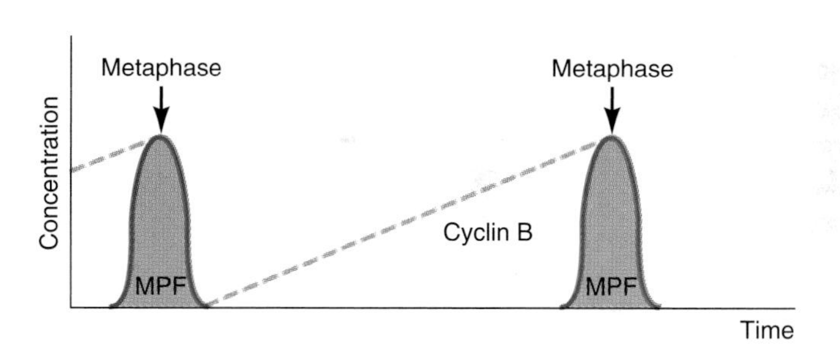

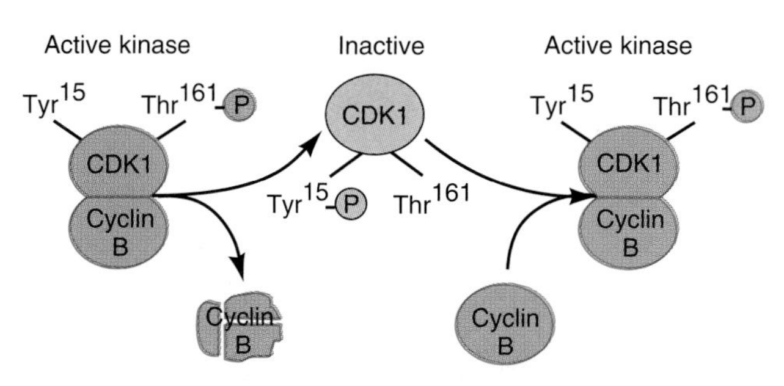

Figure 2.16 Cyclic generation and destruction of M-phase promoting factor (MPF) during the cell cycle. At the beginning of each M phase, CDK1 protein is modified and becomes associated with another protein, cyclin B. Specifically, CDK1 is phosphorylated at a particular threonine residue, Thr^{161}, and dephosphorylated at a tyrosine residue, Tyr^{15}. Cyclin B is synthesized throughout the cell cycle and rapidly destroyed by specific proteases after each metaphase. During its association with CDK1, cyclin B is also phosphorylated. After proper modifications and association, the CDK1–cyclin B complex (MPF) is an active kinase.

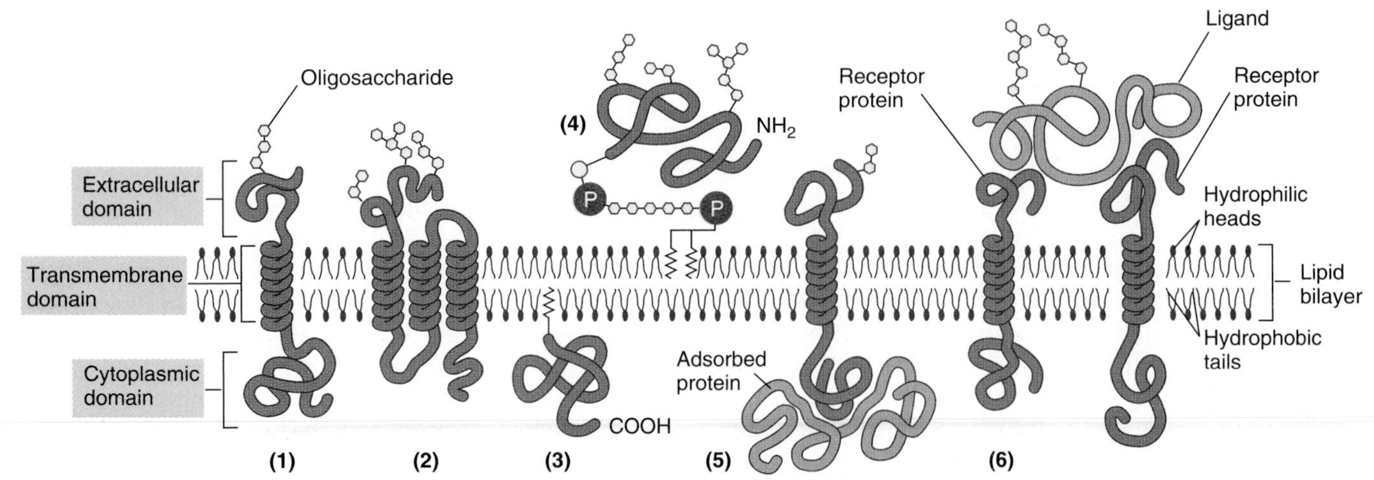

Figure 2.17 Model of a plasma membrane. Phospholipid molecules form a bilayer with their hydrophobic tails squeezed together, and with the hydrophilic heads of one layer facing the extracellular environment and those of the other layer facing the cytoplasm. Most membrane proteins are transmembrane proteins with an extracellular domain, one or more transmembrane domains, and a cytoplasmic domain **(1, 2)**. These proteins are held in the lipid bilayer due to the hydrophobic nature of their transmembrane domains. Other proteins are anchored in the lipid bilayer by a lipid covalently linked to the protein, either directly **(3)** or via a complex sugar **(4)**. Some molecules are held to the membrane only by noncovalent adsorption to other membrane proteins **(5, 6)**. The extracellular domains of many proteins have oligosaccharides (short chains of sugar residues) linked to them.

For instance, the entry of embryonic cells into M phase is regulated by the enzyme that dephosphorylates Tyr^{15} in CDK1. Mature cells generally have other checkpoints: DNA damage blocks the onset of DNA replication, unreplicated DNA blocks the cell cycle before entry into mitosis, and kinetochores that are not attached to the mitotic spindle block entry into anaphase.

2.5 Cell Membranes

The entire cell is surrounded by a thin membrane known as the ***plasma membrane,*** and many cell organelles are surrounded by similar *cellular membranes.* Their main function is to act as barriers to the diffusion of water-soluble molecules. In particular, the plasma membrane restricts the traffic, in and out of the cell, of molecules such as nucleic acids, proteins, and many small molecules and ions. This allows the cell to retain most biologically important molecules at concentrations different from those in their environment. For example, the concentrations of most enzymes is much higher inside of cells than outside, whereas the concentration of calcium ion, a very important intracellular signal molecule, is usually much higher in the extracellular environment than inside of cells. Likewise, the membranes surrounding some cellular organelles create subcellular compartments with specific mixtures of critical molecules.

The segregating function of cellular membranes is reflected in their basic molecular architecture. For the most part, cellular membranes consist of a *lipid bilayer.* The most common membrane lipids are phospholipids, polar molecules with hydrophilic (water-soluble) "heads" and hydrophobic (water-repellent) "tails." In an aqueous medium, phospholipids self-assemble into a double layer of molecules, with the hydrophobic tails facing each other (Fig. 2.17). The hydrophobic tails thus form an oily barrier to water-soluble molecules, preventing their free diffusion across cell membranes. Cellular membranes are like soap bubbles: They are effective barriers to diffusion, but they are fluid and have minimal mechanical strength. The relative stiffness of the cell surface is due not to its plasma membrane but to a superficial layer of cytoplasm called the *cortex,* which is reinforced with microfilaments as described earlier.

Like soap bubbles, cellular membranes easily fuse with one another to form new configurations. This property facilitates the breakdown and reconstitution of membranes during many developmental processes, such as the breakdown and reformation of the nuclear envelope during mitosis, or the formation of secretory vesicles, which pinch off from the Golgi apparatus and move to the plasma cell surface, where they fuse with the plasma membrane to discharge their contents into the extracellular space. The ease with which cell membranes break up and fuse depends on properties of the lipid bilayer, including the phospholipid:cholesterol ratio as well as fatty acid chain length and saturation.

Of course being a barrier is not the only function of cellular membranes. They also facilitate a *controlled* exchange of water-soluble molecules, they are involved in the exchange of signals between cells, and they mediate the adhesion of cells to other cells and to extracellular materials. These functions are carried out by various

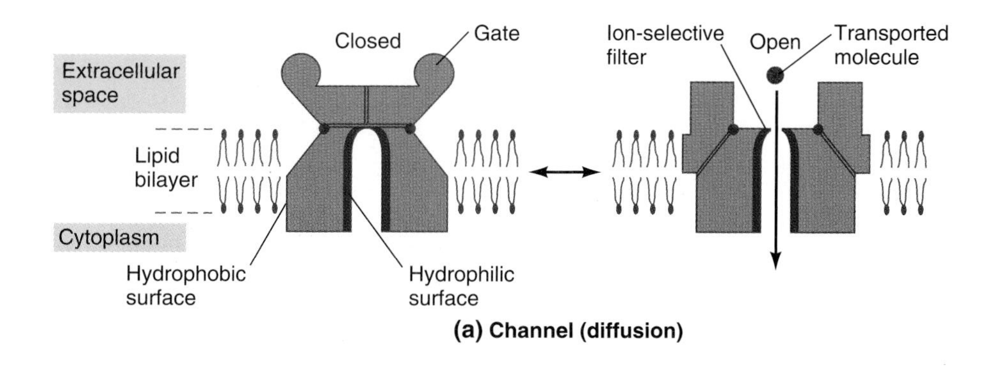

(a) Channel (diffusion)

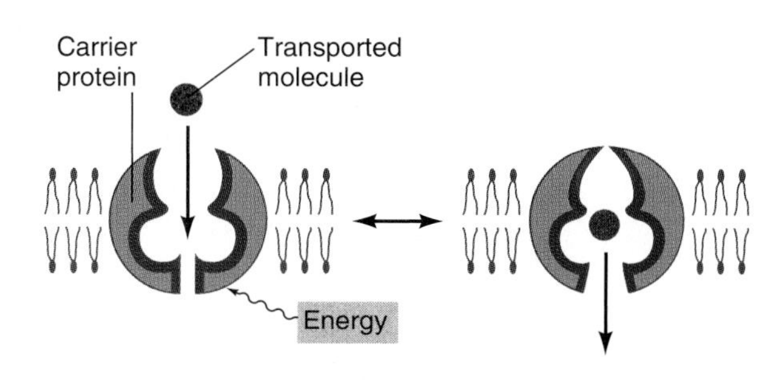

(b) Carrier (active transport)

Figure 2.18 Membrane transport proteins act as **(a)** channels for diffusion or as **(b)** carriers for active transport. Hydrophobic amino acid domains of these proteins are embedded in the lipid bilayer. Hydrophilic amino acid domains are positioned on the inside of the transporter protein, where they allow water-soluble molecules to pass.

membrane proteins. According to the widely accepted ***fluid mosaic model*** of membrane structure, membrane proteins are confined to the lipid bilayer yet they are free to float laterally within it. Some membrane proteins have hydrophobic domains that are enclosed by the tails of the membrane lipids and hydrophilic domains that extend into the cytoplasm or out into the extracellular environment (Fig. 2.17). Other proteins are attached to the membrane by covalent linkage to lipids or by weak interactions with other membrane proteins.

Membrane transport proteins serve as carriers or channels for specific ions or molecules, such as sodium ions or certain amino acids, to cross the plasma membrane. There are two major classes of membrane transport proteins: channel proteins and carrier proteins (Fig. 2.18). ***Channel proteins*** are charged, hydrophilic molecules forming water-filled pores that extend across the lipid bilayer. When these pores are open they allow ions of appropriate size and charge to pass through by diffusion. Such channels open and close in response to changes in electrical potential across the membrane, to mechanical stimuli, or to other signals. In contrast, ***carrier proteins*** bind the molecule to be transported and undergo an energy-dependent conformational change in order to move the molecule and release it on the other side of the membrane.

Membrane receptor proteins may function in the process of *endocytosis* (see Fig. 2.2) or start an intracellular signaling chain. Receptor proteins consist of an *extracellular domain,* a *transmembrane domain,* and a *cytoplasmic domain.* The extracellular domain binds specifically to a matching molecule or ***ligand*** (Lat. *ligare,* "to bind"). As a result of this interaction, the cytoplasmic domain undergoes a conformational change that modifies its physical properties and/or biochemical activity. In the case of ***receptor-mediated endocytosis,*** a particular large molecule binds as a ligand to its matching receptor. The loaded receptors are collected in a depression in the plasma membrane, which is then pinched off and internalized. This process allows cells to take up a specific kind of molecule in large quantities. For instance, oocytes accumulate large amounts of yolk proteins from maternal blood through receptor-mediated endocytosis. The role of receptor proteins in intracellular signaling will be discussed at the end of this chapter.

Cell adhesion molecules mediate the adhesion of cells to other cells and to extracellular materials. These molecules therefore play key roles in the assembly of cells into the shape and structure that characterize each species (Gumbiner, 1996). Like other membrane proteins, cell adhesion molecules would float laterally within the lipid bilayer of the plasma membrane if they were not anchored to some cytoplasmic structure, much like a buoy would drift on the water surface if it were not anchored to the bottom. Likewise, most cell adhesion molecules are anchored to cytoskeletal elements. The *cadherin* family of cell adhesion molecules, for example, has extracellular domains binding to the cadherins of adjacent cells while its cytoplasmic domains are anchored to microfilaments (Fig. 2.19). The anchorage to microfilaments is indirect and is mediated by linker proteins including *catenins*, which are also involved in cellular signaling (see Fig. 9.21). The connection between cell adhesion molecules and the cytoskeleton establishes firm links between the cytoskeletons of adjacent cells and between the cytoskeleton of a cell and the extracellular matrix. These links allow cells not only to adhere but also to exert traction for movement, as we will discuss next.

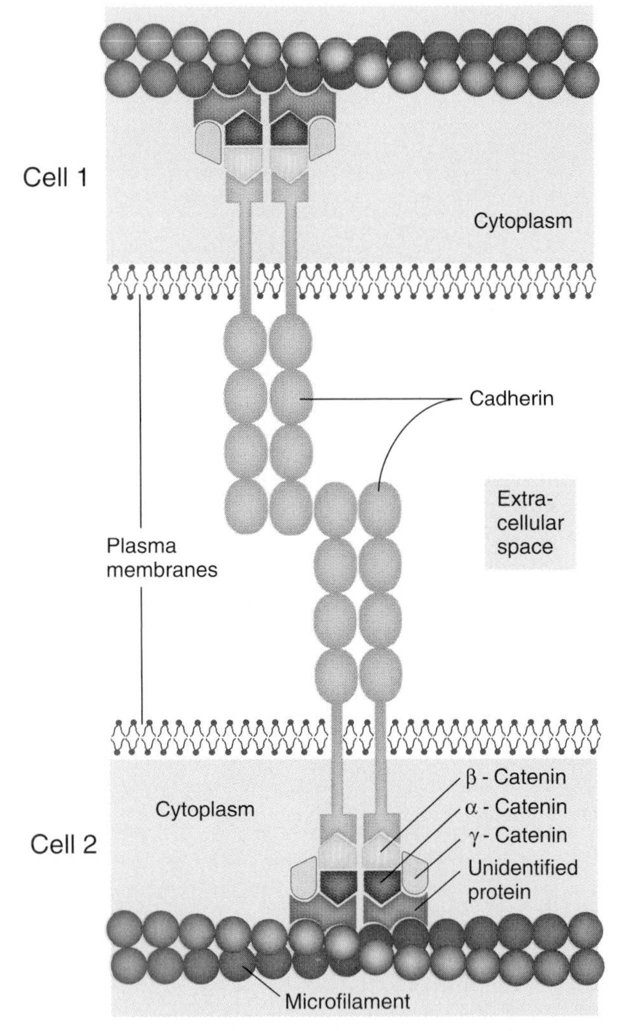

Figure 2.19 Association of cell adhesion molecules with the cytoskeleton via linker proteins. The cell adhesion molecule shown here is a cadherin, which forms dimers contacting opposite cadherin dimers near their tips in a zipperlike arrangement. The cadherin dimers are connected to microfilaments via linker proteins including α-catenin, β-catenin, and γ-catenin. The arrangement diagrammed here is thought to be common in the *zonula adherens* near the apical face of epithelial cells (see Fig. 2.23).

2.6 Cellular Movement

An important characteristic of cells during development is motility. This is dramatically demonstrated by *primordial germ cells,* which typically undergo an arduous journey from their site of origin to the gonad rudiment. Long migrations are also characteristic of *neural crest cells,* which originate from the crests of the *neural folds* of vertebrate embryos (see Fig. 1.8) and migrate to a wide range of destinations, where they form neurons, pigment cells, and many other derivatives.

Cell locomotion has been studied extensively using cultured mammalian ***fibroblasts,*** cells that adhere in vivo to collagen fibrils and remain fairly stationary except during wound healing. In vitro, they adhere to and move around on the bottom of the culture dish. The ***leading edge*** of a fibroblast in motion advances forward while the ***trailing edge*** drags behind. Fibroblasts move by expanding at the leading edge and contracting at the trailing edge (Fig. 2.20). The leading edge continually forms flat extensions of cortical cytoplasm called ***lamellipodia*** (sing., *lamellipodium;* Lat. *lamella,* "little plate," and Gk. *podos,* "foot"), some of which adhere to the substratum while previous points of adhesion near the trailing edge are severed.

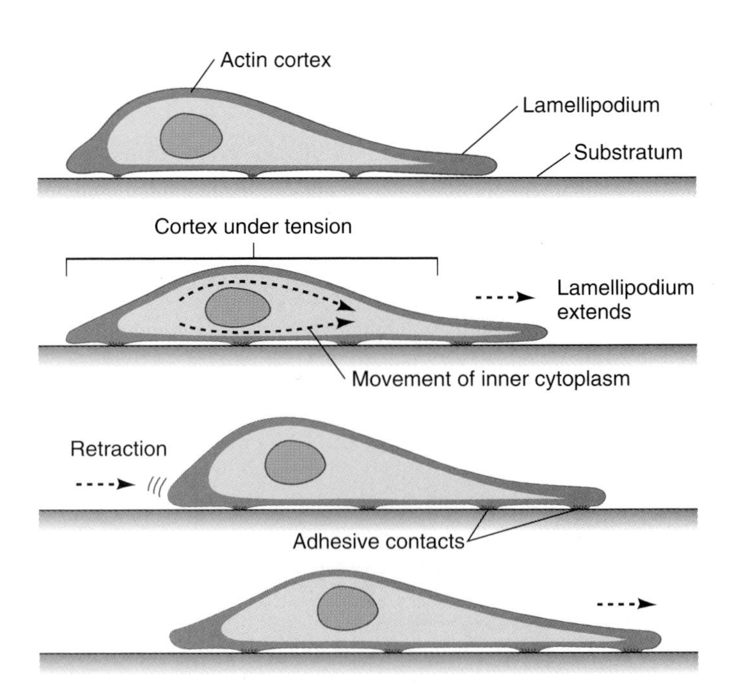

Figure 2.20 Movement of a fibroblast in culture. The cell extends a lamellipodium at the leading edge, makes new points of contact with the substratum, and severs older points of contact as the trailing edge contracts.

Another favorite subject for cell movement studies is the neuron—in particular, the growth cone at the tip of the growing axon (Fig. 2.21). The growth cone is rich in microfilaments, while the axon itself contains microtubules for reinforcement and transport. The growth cone forms lamellipodia and also extends spikelike processes called ***filopodia*** (Lat. *filum,* "thread"). Filopodia can grow to more than 50 μm in length, and they contain bundles of microfilaments. They are constantly extending and retreating, as if exploring their environment, and they are able to transmit environmental information to their parent growth cone (Davenport et al., 1993). For example, neurons respond to growth factors released by their target organs (see Section 28.4).

Exactly how cells turn internally generated forces into a translocation is still under investigation (Lauffenburger and Horwitz, 1996; Mitchison and Cremer, 1996). Microfilaments attached to membrane proteins (see Fig. 2.19) must play a major role, because *cytochalasin* stops cell movement. Indeed, new microfilaments are formed at the leading edge as a cell moves. It also seems significant that during lamellipodium formation the cortical cytoplasm undergoes cycles of relaxation and stiffening. The cytoplasm becomes more fluid when a lamellipodium originally forms and more gellike after the lamellipodium has made contact with the substratum. The relaxation may be caused by microfilament-severing proteins such as *gelsolin.* The stiffening may be accomplished by cross-linking proteins such as *α-actinin.* In concert with microfilament assembly, contraction, and disassembly, cell adhesion molecules must be deployed at the leading edge as it advances and somehow must be recycled from the trailing edge (Bretscher, 1996).

2.7 Cell Junctions in Epithelia and Mesenchyme

Embryonic tissues are broadly classified as either *epithelial* or *mesenchymal* (Fig. 2.22). An ***epithelium*** is a sheet of contiguous cells that rest on a layer of *extracellular matrix* called a ***basement membrane*** (not to be confused with the *plasma membrane* surrounding the cytoplasm). Epithelial cells are always polarized: The cell surface facing the basement membrane is the ***basal surface,*** and the opposite cell surface facing the outside world or certain

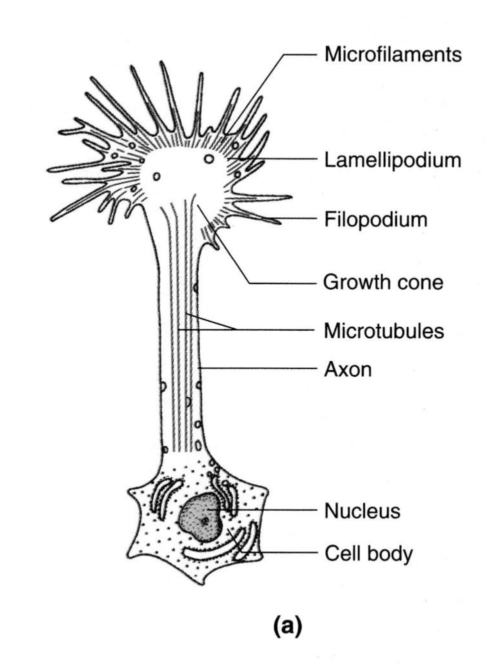

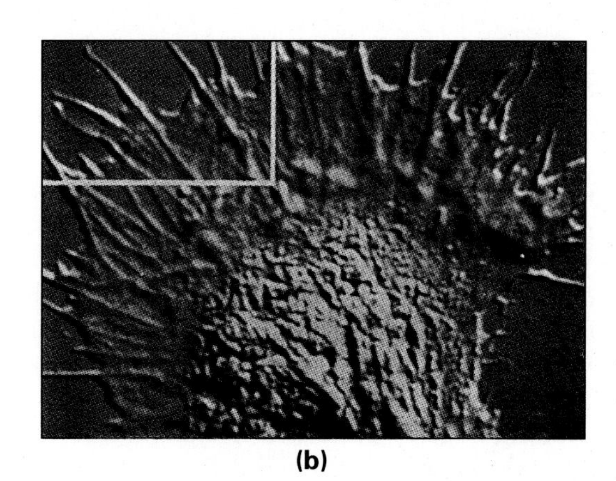

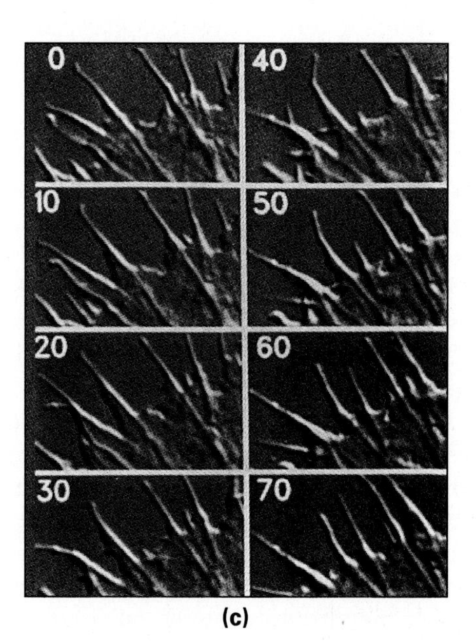

Figure 2.21 Growth cone of an axon. **(a)** Schematic drawing of a neuron with a growing axon. The diameter of the axon is exaggerated relative to the size of the cell body. **(b)** Video-enhanced image of a growth cone. Spikelike filopodia, reinforced with microfilament bundles, extend from the crescent-shaped lamellipodium. The lamellipodium excludes the granular organelles and microtubules present in the axon and the adjacent part of the growth cone. **(c)** A series of images taken every 10 s of the boxed region in part b. Note the constant protrusion and retraction of filopodia.

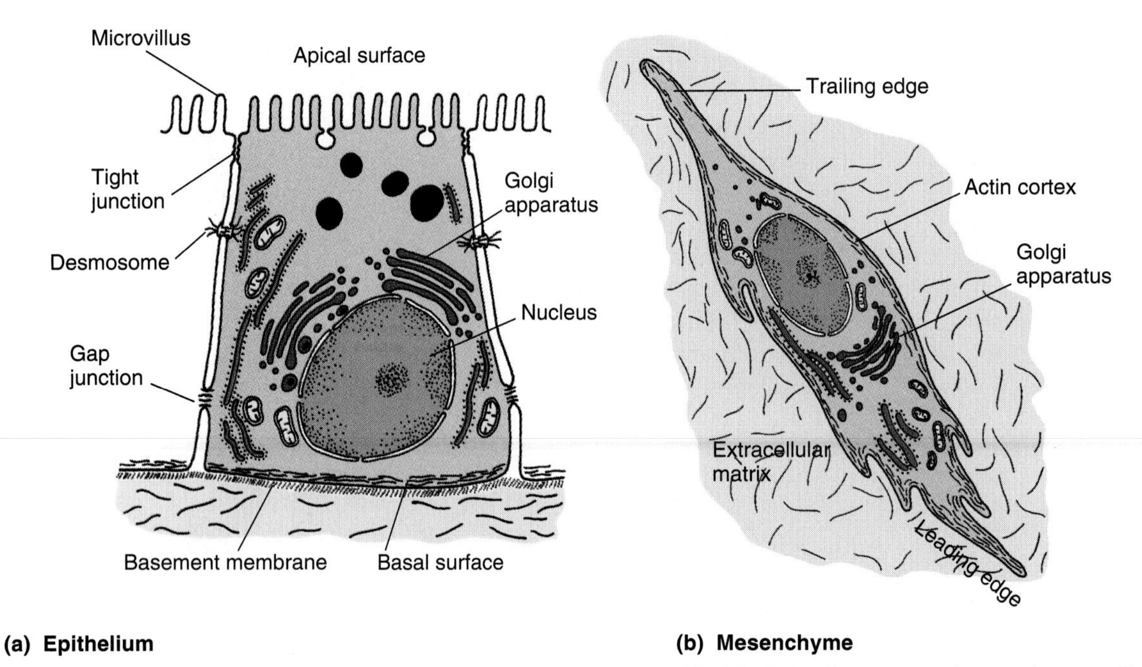

Figure 2.22 Differences between epithelial cells and mesenchymal cells. **(a)** Epithelial cells rest on a layer of extracellular matrix known as basement membrane. They are polarized, with the basal cell surface adjacent to the basement membrane and the apical cell surface facing the outside world or a body cavity. Epithelial cells are closely connected to their neighbors with tight junctions, gap junctions, and desmosomes. **(b)** Mesenchymal cells are only loosely connected to other cells or are surrounded entirely by extracellular matrix. When in motion, mesenchymal cells show a leading edge and a trailing edge.

body cavities is the ***apical surface.*** The apical surfaces of many epithelia have small, fingerlike projections, called ***microvilli,*** which increase the surface area available for endocytosis and exocytosis. In contrast, the cells of a ***mesenchyme*** have only sparse contacts with their neighbors or are surrounded on all sides by extracellular material. If they are migratory they show a leading edge and a trailing edge as discussed earlier.

Epithelial and mesenchymal cells are connected by different types of junctions. The common classes of intercellular junctions are *tight junctions,* which are formed between epithelial cells, as well as *anchoring junctions* and *gap junctions,* which occur between epithelial cells as well as between mesenchymal cells.

Tight junctions connect the adjacent membranes of epithelial cells near their apical surfaces (Fig. 2.23 a, b). Strings of membrane proteins run continuously around the apical circumference of each cell in an epithelium. A tight junction is a connection like a ziplock between a string of junctional proteins in one cell membrane and the corresponding proteins in the membrane of an adjacent cell. Tight junctions have two major functions. First, they form effective seals: There is no intercellular space left open for water-soluble molecules to diffuse between the epithelial cells. Instead, molecules entering or leaving the organism must be gated through transport proteins in the plasma membranes. Second, tight junctions prevent membrane proteins from floating between apical and lateral cell surfaces. Thus, cells can maintain different sets of membrane proteins on these surfaces so that they can interact differently with the external environment and with the organism's own intercellular space. By virtue of these two functions, tight junctions between epithelial cells permit an organism to control its internal environment.

Anchoring junctions enable epithelia to act as robust sheets by anchoring the cytoskeletal elements of each cell to those of its neighbors. There are three different types: *adherent junctions, desmosomes,* and *hemidesmosomes.* ***Adherent junctions*** connect microfilament systems of adjacent cells. In epithelia, they typically form a continuous *adhesion belt,* or *zonula adherens,* right below the tight junctions (Fig. 2.23a). Here the membranes are held together by cell adhesion molecules called *cadherins* (Gumbiner, 1996). These molecules traverse the cell membrane and are connected via linker proteins to rings of microfilaments in the cell cytoplasm (see Fig. 2.19). This linkage is critical for holding cells together when sheets of cells change shape. Another group of anchoring junctions, which is common in invertebrates, are the *septate junctions.* They connect microfilament systems of cells, as adherent junctions do, but septate junctions have a distinct appearance under the electron microscope: They are held together by junctional proteins in parallel rows arranged at a regular distance, which give sections of septate junctions a characteristic ladderlike appearance.

A different type of anchoring junction is the ***desmosome,*** which connects the intermediate filament systems of adjacent cells in vertebrates. At the desmosomal core, the adjacent plasma membranes are connected by a thin

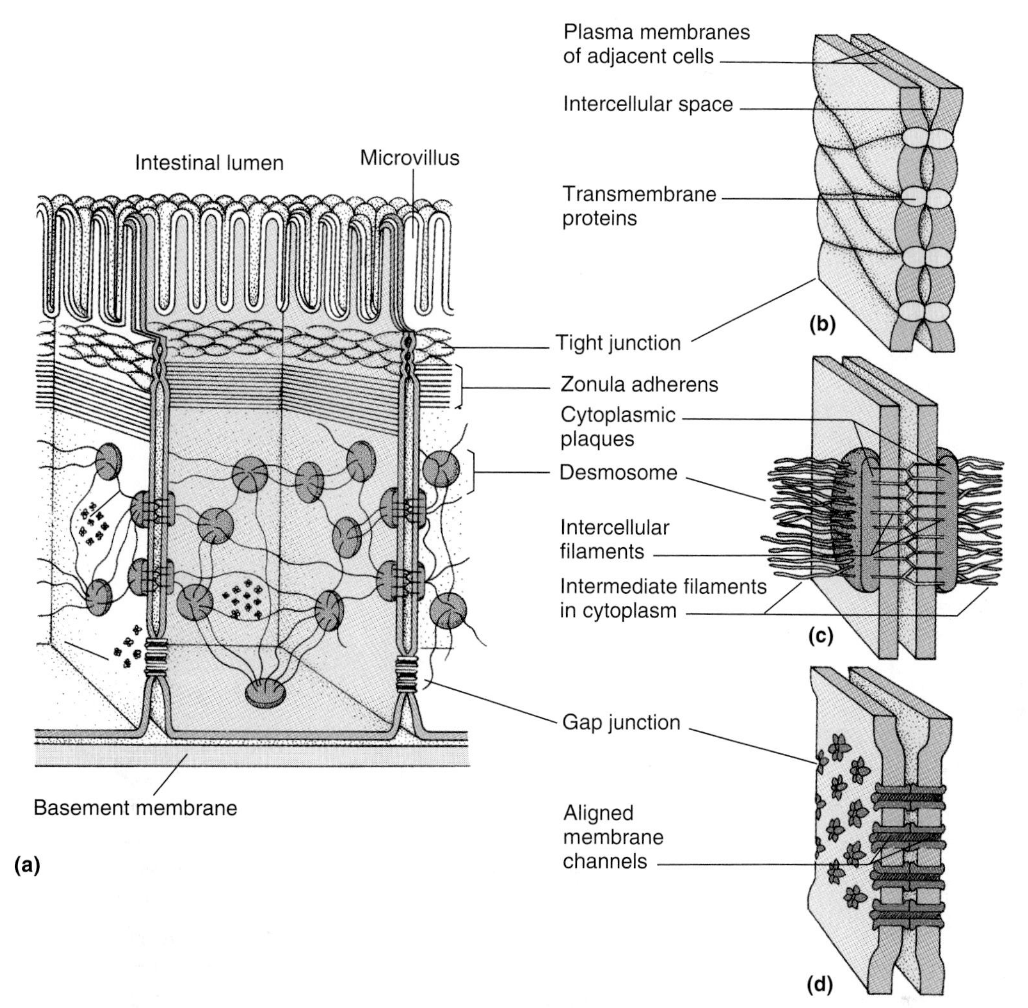

Figure 2.23 Three types of intercellular junctions in epithelial cells. **(a)** Schematic diagram showing the inner epithelium of the small intestine in a mammal; the microvilli on top are projections of the apical cell surface. The basal cell surface rests on the basement membrane, a sheet of extracellular material. **(b)** A tight junction is composed of adjacent rows of membrane proteins, which hold two adjacent membranes together like a ziplock. **(c)** In the zonula adherens beneath the tight junction, microfilament bundles running beneath the plasma membrane are linked to cell adhesion molecules (see Fig. 2.19). Desmosomes, a type of anchoring junction, are characterized by of a pair of cytoplasmic plaques, just inside the respective plasma membranes of a pair of adjacent cells. The core between the cell membranes is filled with fibrous material. Desmosomes are connected across the cytoplasm by intermediate filaments. **(d)** A gap junction is an array of cylindrical channels consisting of connexin proteins. Channels of adjacent plasma membranes are aligned to allow the transmission of small molecules.

layer of glycoprotein cement and fibrous material (Fig. 2.23c). The fibrous material seems to be connected to plaques of electron-dense material located on the cytoplasmic face of the plasma membrane on each side of the core. These plaques in turn serve as anchors for keratin and other cytoskeletal filaments that traverse the inside of the cell to end in another desmosome. The desmosomes thus link the intermediate filaments of adjacent cells to form a transcellular network that provides strength and elasticity to an epithelium. ***Hemidesmosomes,*** or half-desmosomes, connect the basal surfaces of epithelial cells to the underlying basement membrane.

Gap junctions, the third class of intercellular junctions, are arrays of miniature channels between cells (Fig. 2.23d). Each channel consists of two connecting cylinders, each formed by *connexin* proteins embedded in the abutting cell membranes (Kumar and Gilula, 1996). Cells joined by gap junctions are electrically coupled and can exchange molecules of up to about 1000 daltons. Thus, gap junctions serve as conduits for chemical signals traveling between cells by diffusion. Significantly, cells can modulate the degree of coupling with their neighbors. Even slight increases in the intracellular concentration of free calcium ions or slight changes in pH can trigger a decrease in gap junction permeability. These effects are thought to be mediated by changes in the conformation of the connexins that constitute the two cylinders of a gap junction.

In most early embryos, at least some of the blastomeres are connected by gap junctions. This was revealed by an experiment in which hydrophilic dye molecules were injected into one cell and the dye was observed diffusing into the neighboring cells. Gap junctions often appear at a stage when, according to experimental evidence, the blastomeres begin to communicate with each other. When antibodies to connexins are injected into specific blastomeres of frog or mouse embryos, the injected cells continue to divide but do not undergo normal development. Developmental defects arise among the descendants of the injected cell, indicating that some function of the gap junctions—presumably, the passage of signal molecules between embryonic blastomeres—is necessary for normal development (Kumar and Gilula, 1996).

2.8 Cellular Signaling

Throughout the life cycle of an organism, cells communicate by chemical signals. Eggs respond to sperm, vegetal cells modify the development of their animal neighbor cells, and organs grow in response to growth hormones. These signals are transduced along pathways that have been exceedingly well conserved in evolution. Elucidation of these pathways, and the complex networks they form, has become a major topic in contemporary developmental biology (Hunter, 2000). Some of these pathways will be introduced here to illustrate their general nature; others will be presented in specific developmental contexts.

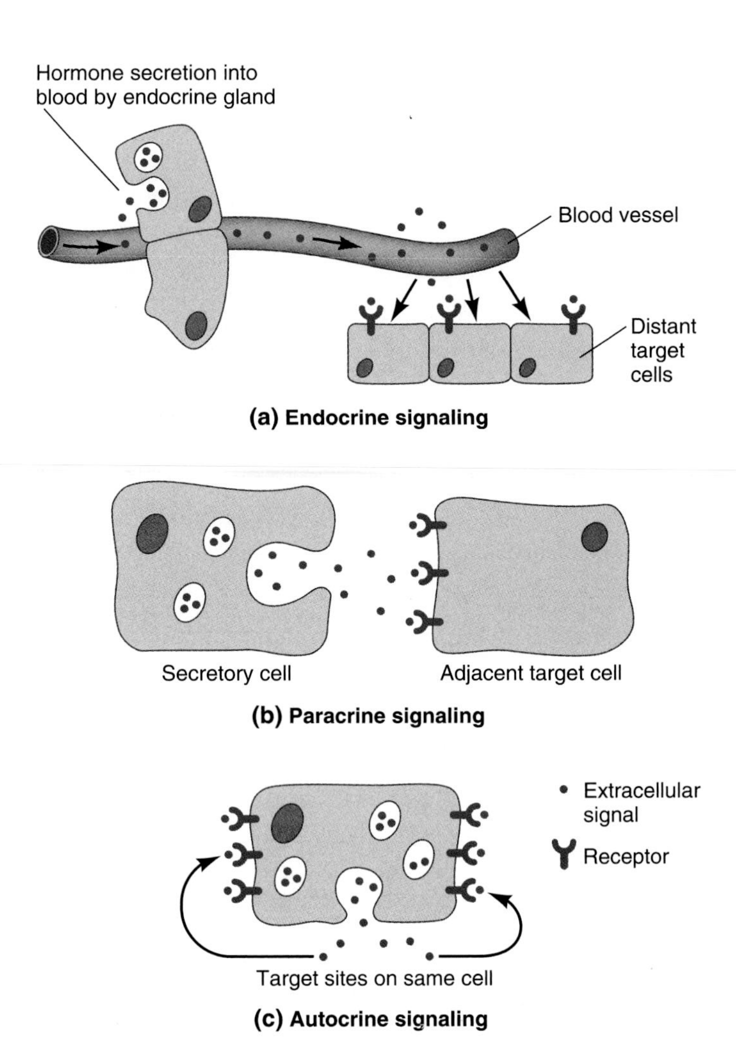

Figure 2.24 General schemes for cell-to-cell signaling using chemicals. The chemical signals may travel over longer or shorter distances, ranging from meters in endocrine signaling **(a)** to micrometers in paracrine **(b)** and autocrine signaling **(c)**.

INTERCELLULAR SIGNALS VARY WITH RESPECT TO DISTANCE, SPEED OF ACTION, AND COMPLEXITY

Animal cells respond to chemical signals coming from distant glands, from neighboring cells, and even from themselves (Fig. 2.24). In ***endocrine signaling,*** cells respond to a hormone released from a specific gland and distributed by blood or other body fluids. The sex hormones of vertebrates and the molting hormones of insects are well-known examples. In ***paracrine signaling,*** the signaling cell affects only nearby cells. In an amphibian blastula, for instance, vegetal cells signal to adjacent animal cells to form mesoderm. In ***autocrine signaling,*** cells respond to chemical signals that they themselves release. Several growth factors serve as both autocrine and paracrine signals by stimulating the secreting cells as well as neighboring cells to divide and/or to embark on certain pathways of development.

Most of the chemical signals exchanged between cells, especially in vertebrates, are either *steroids* or *polypeptides*. These two groups act through characteristic mechanisms that differ in speed of action (Fig. 2.25). Steroids may act on receptors located in the plasma membrane, but as a rule, they diffuse through the plasma membranes of their target cells. Inside the cytoplasm or the cell nucleus they bind to receptor proteins, forming a hormone-receptor complex. In the nucleus, the complex interacts with a small number of target genes whose transcription is specifically enhanced or inhibited. Since RNA transcripts need to be synthesized and processed in the nucleus and functional mRNAs have to be released into the cytoplasm before they can be translated into new proteins, the response to a steroid hormone generally takes hours or even days.

In contrast, polypeptides do not cross plasma membranes because they are large and water-soluble rather than lipid-soluble. Instead they act as *ligands,* binding specifically to receptor proteins in the plasma membrane of their target cells. The loaded receptors then trigger a signal transduction cascade, which usually causes the phosphorylation of one or more target proteins that are *already present* in the cell cytoplasm. The target protein may be involved in the control of translation (see Section 18.2), in which case the cell's response would be very fast. Of course, if the activated protein acts by controlling the transcription of genes, then the

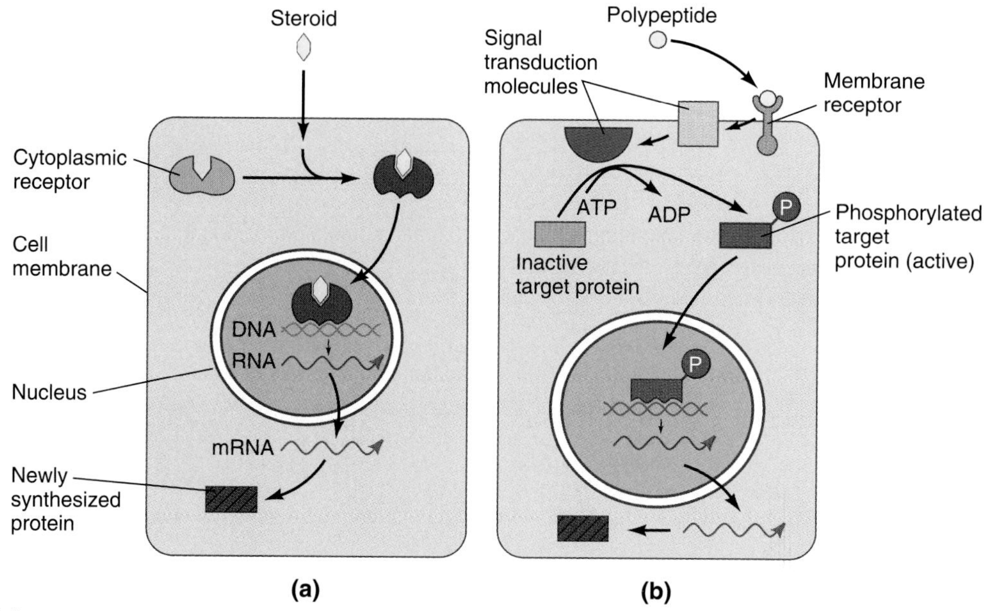

Figure 2.25 Cellular responses to external steroid and polypeptide signals. **(a)** A steroid typically crosses the plasma membrane and binds to a cytoplasmic or nuclear receptor protein in the target cell. The hormone-receptor complex interacts with specific genes to activate or inhibit their transcription into RNA. The transcripts are processed and released as mRNAs into the cytoplasm, where new proteins will be synthesized. **(b)** A polypeptide binds to a receptor protein in the plasma membrane of the target cell. The loaded receptor interacts with a set of signal transduction molecules to activate target proteins already present in the cytoplasm. This activation is often mediated through the transfer of a phosphate group (P) from adenosine triphosphate (ATP) to the inactive target protein. If the phosphorylation of a target protein is sufficient for the target cell to respond to the external signal, then this response is fast. However, if the phosphorylated protein is a gene regulatory protein, then the response will be slower because it will require RNA synthesis, processing, and translation into an effector protein.

cellular response to a polypeptide may be as time-consuming as the response to a steroid.

Cellular responses to polypeptide signals also tend to be more complex than responses to steroid signals because the signal transduction pathway from the plasma membrane to the cell nucleus involves more elements. The remainder of this section will focus on the transduction of polypeptide signals whereas responses to steroid hormones will be discussed in Chapter 27.

MEMBRANE RECEPTORS INITIATE DIFFERENT SIGNALING PATHWAYS

Different types of membrane receptors trigger different cellular signal transduction mechanisms. In most cases, the final result is a change in the three-dimensional conformation of a certain target protein and thus in its biological activity. If a target protein is a *transcription factor*, it will in turn activate or inhibit a group of genes. However, these end results are achieved through a variety of intermediate steps. Some receptors have cytoplasmic domains that act directly as *protein kinases*, transferring phosphate groups to other proteins and thus changing their conformation and activity. Other receptors are channel proteins that open in response to the binding of the ligand. The altered ion concentration inside the cytoplasm will in turn change the activity of certain proteins.

Most membrane receptors cooperate with other proteins that act as signal transducers. A large family of common signal transducers is called ***G proteins,*** because they bind guanosine triphosphate, or GTP (Neer, 1995). More specifically, the term is used for signal-transducing membrane proteins that consist of three polypeptides (Fig. 2.26). G proteins activate or inhibit a third group of membrane proteins that act as enzymes or ion channels. These proteins in turn generate small *intracellular* signal molecules known as ***second messengers.*** While some second messengers, such as *calcium ion* or *cyclic adenosine monophosphate,* diffuse through the cytoplasm, others, like *diacylglycerol,* remain bound to the plasma membrane.

ADENYLATE CYCLASE GENERATES cAMP AS A SECOND MESSENGER

One of the most common second messengers, ***cyclic adenosine monophosphate* (*cAMP*),** is made with the help of the enzyme adenylate cyclase. This enzyme floats in the plasma membrane and is activated (or inhibited) by various receptors via G proteins (Fig. 2.26). This mechanism introduces several important features into the signaling process. First, the extracellular signal is greatly amplified. Since the receptor and the G proteins diffuse freely in the plasma membrane, a single loaded receptor can stimulate many G protein and adenylate cyclase molecules, and each adenylate cyclase molecule will in turn synthesize many cAMP molecules. Second, different signals can be integrated when their different membrane receptors and associated G proteins act on the same set of adenylate cyclase molecules. Third, the G protein functions only as long as a GTP molecule is bound to it. Since the GTP is hydrolyzed in the course of the G protein's function, cells can dampen an overly strong response by reducing the supply of GTP.

The cAMP molecules generated in response to an external signal perform their second messenger function in different ways. One common pathway is through cAMP-dependent protein kinases, which phosphorylate other proteins, thus regulating their biological activity.

PHOSPHOLIPASE C-β GENERATES DIACYLGLYCEROL, INOSITOL TRISPHOSPHATE, AND CALCIUM IONS AS SECOND MESSENGERS

Receptor proteins and G proteins may act on other membrane components besides adenylate cyclase. ***Phosphatidylinositol bisphosphate*** **(PIP_2)** is one of several inositol phospholipids found in the cytoplasmic layer of the plasma membrane. ***Phospholipase C-β*** is a lipid-cleaving enzyme floating in the same membrane layer (Fig. 2.27). If stimulated by a G protein, phospholipase C-β cleaves PIP_2 into two components: ***diacylglycerol*** **(*DAG*)** and ***inositol trisphosphate*** **(IP_3)**. The water-insoluble DAG product remains in the plasma membrane, while the water-soluble IP_3 product diffuses into the cell cytoplasm. Here IP_3 releases stored calcium ions from certain parts of the endoplasmic reticulum. The liberated ions cooperate with DAG in activating another enzyme, ***protein kinase C.*** In its inactive state, protein kinase C is a cytoplasmic protein. In the presence of calcium ions, it inserts into the plasma membrane, where it is activated by DAG. The active protein kinase C then phosphorylates various target proteins for different cellular responses including changes in RNA and protein synthesis.

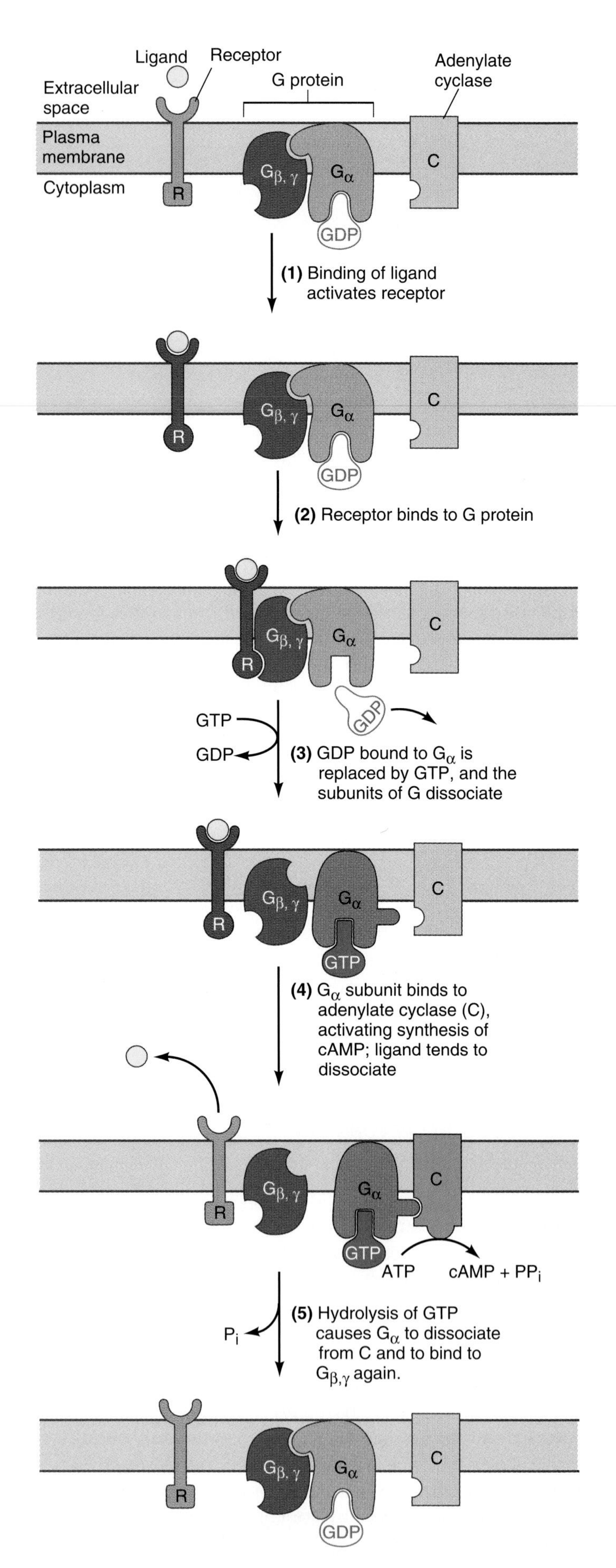

Figure 2.26 Signal transduction by a G protein from a plasma membrane receptor to a membrane-bound enzyme, exemplified here as an adenylate cyclase. The G protein consists of three peptides, $G_α$, $G_β$, and $G_γ$. Upon ligand binding, the cytoplasmic domain of the receptor changes so that it will interact with the G protein. This allows the α subunit to dissociate from the βγ subunits and to exchange its bound guanosine diphosphate (GDP) for guanosine triphosphate (GTP). In this short-lived active form, the α subunit has an exposed binding site for adenylate cyclase, which is activated. However, the α unit also hydrolyses its bound GTP to GDP, whereupon it dissociates again from the adenylate cyclase and returns to the βγ subunits, thus regenerating a conformation of G that can be activated by the receptor-ligand complex.

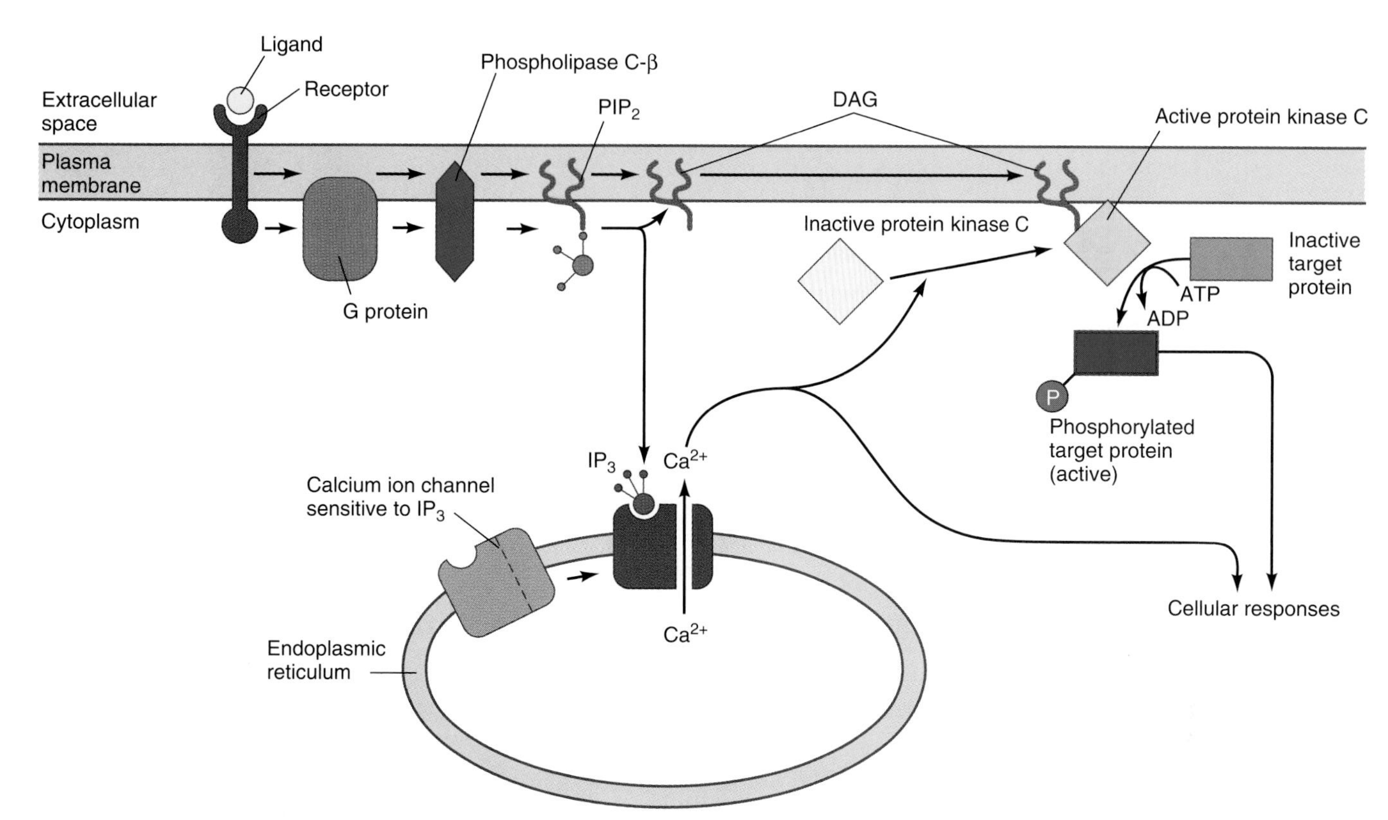

Figure 2.27 Second messengers in the phosphatidylinositol bisphosphate (PIP_2) pathway. Ligand binding activates G protein, which in turn activates phospholipase C-β. This enzyme cleaves PIP_2 into diacylglycerol (DAG) and inositol trisphosphate (IP_3). The latter diffuses through the cytoplasm, where it releases stored calcium ion (Ca^{2+}) from the endoplasmic reticulum. The released calcium ion allows inactive cytoplasmic protein kinase C to insert into the plasma membrane, where the protein kinase C is activated by DAG. The active protein kinase C then phosphorylates target proteins, which mediate a wide range of cellular responses. The released calcium ion may trigger additional cellular responses.

In addition to helping with the activation of protein kinase C, calcium ions have many more second messenger functions (Berridge et al., 1998). Most of these are carried out by calcium-dependent proteins, such as ***calmodulin.*** In its active form, with four calcium ions bound to it, calmodulin activates a wide variety of other proteins.

THE RTK-Ras-MAPK PATHWAY ACTIVATES TRANSCRIPTION FACTORS

Many signal transduction pathways go all the way from an active plasma membrane receptor to the regulation of a group of genes. One of these pathways begins with a large family of membrane receptors known as ***receptor tyrosine kinases (RTKs),*** so called because their cytoplasmic domains phosphorylate tyrosine residues of other proteins. The ligands of RTKs are signal proteins that cause a wide range of developmental effects, from cell division to the acquisition of various cell fates. Investigators studying these phenomena in mammals, flies, and roundworms in the 1980s were amazed to see that their research was converging on very similar signal transduction pathways!

A hallmark of RTKs and their ligands is that the ligands bind to two RTK molecules, thus causing them to form a unit consisting of two parts, or *dimer.* In the dimer conformation, the two RTKs cross-phosphorylate several tyrosine residues in their cytoplasmic domains (Fig. 2.28). The latter are recognized by a variety of so-called ***adapter proteins,*** which either activate or inhibit a member of a family of membrane-bound monomeric GTP-ases known as the ***Ras proteins*** (compare with the trimeric G protein shown in Fig. 2.26). Ras in turn triggers a cascade of protein kinases that convert the short-lived activation of Ras into a longer-lived signal. While many kinases can be involved in such cascades, one group known as the ***mitogen-activated protein kinases (MAPKs)*** or *extracellular-signal-regulated kinases* (*ERKs*) seems to play an especially prominent role. These kinases are turned on by a wide range of cell proliferation- and differentiation-inducing signals, some of which bind to RTKs while others act through G-proteins (see Fig. 2.26) or through protein kinase C (Fig. 2.28). A characteristic feature of MAPKs is that their activation requires the phosphorylation of both a threonine and a tyrosine, which are separated in the target protein by another amino acid. The specific enzyme that catalyses

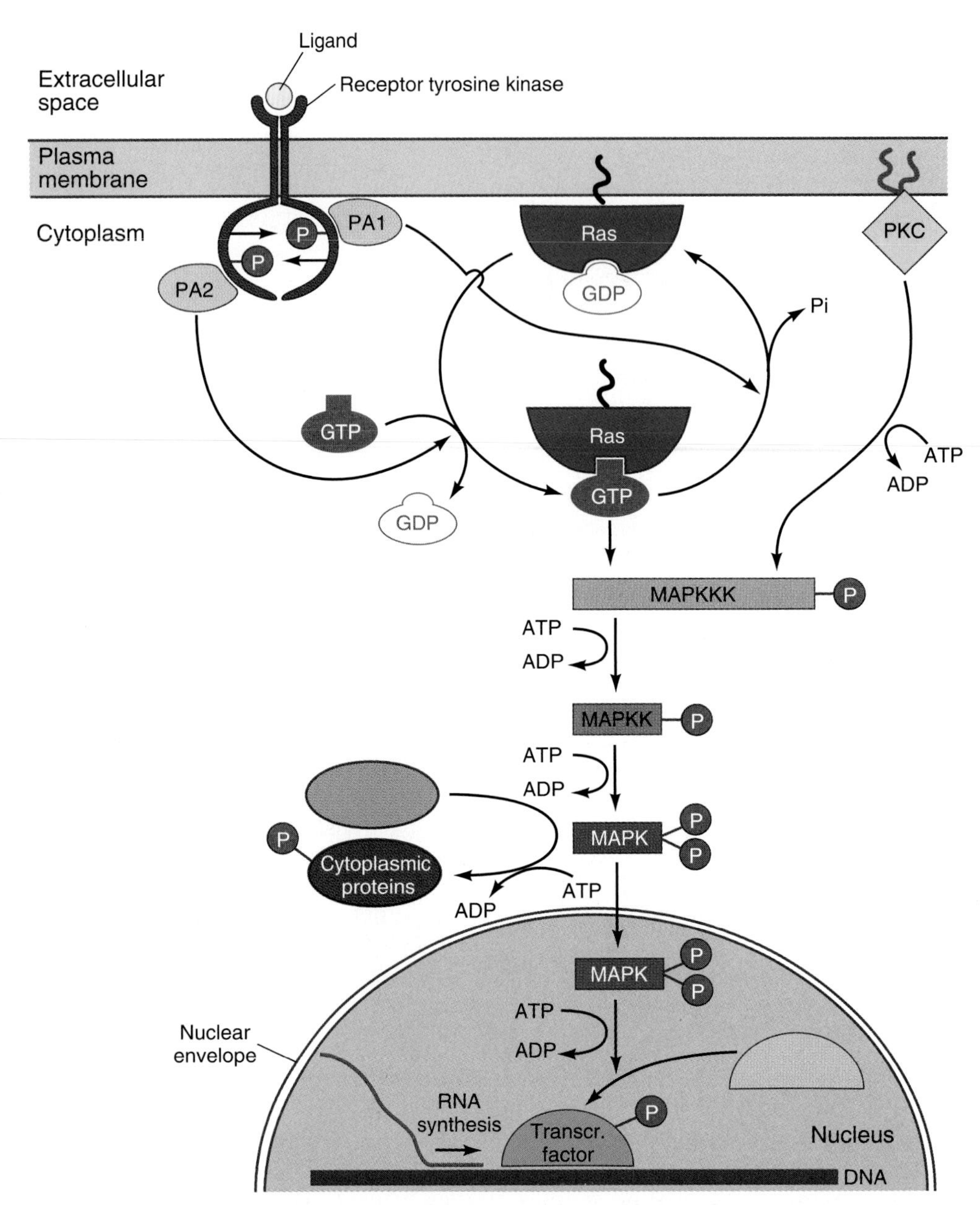

Figure 2.28 RTK-Ras-MAPK pathway. The ligand, a signal protein, causes dimerization and cross-phosphorylation of two receptor tyrosine kinase (RTK) molecules. Adapter proteins (PA1, PA2) inhibit or activate Ras protein by hydrolyzing its GTP or by allowing the exchange of GDP for GTP, respectively. Ras activates a mitogen-activated protein kinase-kinase-kinase (MAPKKK), which in turn activates a MAP-kinase-kinase (MAPKK) and then a MAP kinase (MAPK). The latter finally activates cytoplasmic proteins as well as nuclear transcription factors. MAPKKKs can also be stimulated by protein kinase C, an enzyme activated by G protein–associated receptors (see Figs. 2.26 and 2.27).

both these phosphorylations is known as *MAP-kinase-kinase* (*MAPKK*). MAPKK itself is activated by a phosphorylation catalysed by a *MAP-kinase-kinase-kinase* (*MAPKKK*), which in turn is stimulated by the binding of activated Ras.

SIGNAL TRANSDUCTION PATHWAYS ARE LINKED WITH ONE ANOTHER AND WITH CELL ADHESION

The cellular signal transduction pathways are connected in many ways. For example, the RTK-Ras-MAPK pathway can be activated by G protein–linked receptors as discussed earlier. Protein kinase C generated via G protein and phospholipase C-β may activate MAPKKK or similar seronine/threonine kinases, thus feeding into the MAPKKK-MAPKK-MAPK pathway (Fig. 2.28). Conversely, one of the intracellular signaling proteins that bind to activated RTKs is ***phospholipase C-γ***, which cleaves PIP_2 and causes the same downstream effects as phospholipase C-β. There is also a family of ***signal transducers and activators of transcription*** (***STATs***) that combine the abilities to recognize the phosphorylated cytoplasmic domains of RTKs and to activate gene transcription in one molecule, thus forging a direct pathway from the cell surface to the nucleus (Darnell, 1997).

Even more intriguing is the cross-linking of signal transduction pathways with cell adhesion in at least two ways. First, when cells interact through their cell adhesion molecules, these molecules may nudge adjacent RTKs into close proximity so that they start the same cross-phosphorylation that is triggered when a ligand causes RTKs to dimerize. Second, the same β-catenin molecule that is involved in anchoring cell adhesion molecules to microfilaments (see Fig. 2.19) is also active in signal transduction (*Gumbiner, 1995;* Fagotto and Gumbiner, 1996). Thus, the amount of β-catenin tied up by cell adhesion will affect the pool of β-catenin available for signal transduction.

The existence of signal transduction pathways and their cross-talking capabilities are very puzzling. Any one of these pathways transduces a multitude of signals: the second messengers Ca^{2+} and ATP control a plethora of cellular functions; heterotrimeric G proteins associate with receptors for signals ranging from neurotransmitters to mechanical stimuli; RTKs bind dozens of different ligands, each triggering a specific response even though they are using the same chain of kinases for signal transduction. This seems like a thousand people wanting to communicate over half a dozen telephone wires. Worse still, some of these wires are crossed, and some do double duty for transmitting telefax or computer data. How can this possibly work?

Part of the answer lies in the fact that each type of cell has only a limited set of signal receptors. If by experimental trickery one provides a cell with receptors that it does not normally have, the result can indeed be confusion. For example, if one injects frog eggs with mRNA for neurotransmitters, they will translate these mRNAs and insert the receptors in their plasma membrane. If these eggs are then bathed in a solution containing neurotransmitters, the eggs behave as if they had been fertilized and begin to cleave (Kline et al., 1988). Apparently, their sperm receptors utilize the same G proteins for signal transduction that are known to be used by neurotransmitter receptors. It follows that much of the possible confusion in intracellular signaling is prevented by limiting the number of receptors that each type of cell has at any given time. Similarly, certain components of the signaling pathways that connect receptors with their target genes are present in some cells but not in others (*Tan and Kim, 1999*). Many signal-transducing components act combinatorially, requiring the simultaneous presence of multiple components, which due to their modular nature can confer much specificity (Hunter, 2000). There is also evidence that cell types differ in which target genes are primed to respond to the gene regulatory proteins at the ends of signal transduction pathways.

Finally, we need to keep in mind that signal transduction pathways have evolved over billions of years, and that evolving organisms cannot close down for rewiring. Thus, one can only expect to find a signal transduction system that works, not necessarily one that is as streamlined and parsimonious as possible.

SUMMARY

Most cells share the general abilities to divide, to maintain or change shape and inner organization, to take up or release materials, to move, to form junctions, to become polarized, and to exchange signals with one another. Recent discoveries have greatly improved our understanding of the cell organelles and molecules involved in these activities, although many questions still need to be answered. However, the current knowledge already allows us to analyze developmental processes in terms of cellular behavior—an approach that we will employ repeatedly in Part One of this text.

Gametogenesis

Figure 3.1 Frontispiece of William Harvey's book on the generation of animals (Latin edition, 1651). The picture shows Zeus releasing many kinds of animals from an egg. The writing on the egg held by Zeus says, "All living beings come from an egg."

MOST metazoa procreate sexually through reproductive cells called ***gametes*** (Gk. *gamein*, "to marry"). Gametes always exist in two types, usually in the familiar forms of eggs and sperm. Appropriate mating behaviors of the adults work together with cellular interactions between the gametes to promote fertilization (see Chapter 4).

Gametogenesis, the formation of gametes, occurs in organs known as ***gonads,*** or more specifically as ***ovaries*** in females and ***testes*** (sing., *testis*) in males. Gametes differ from other cells in their number of chromosomes. Typical nonreproductive cells are ***diploid***: They carry two sets of chromosomes, one set derived from the male parent and the other from the female parent. Gametes are ***haploid***: They carry only one complete set of chromosomes. The reduction in number of chromosomes occurs as gamete precursor cells undergo two specialized cell divisions called *meiosis.* At the time of fertilization, when an egg and a sperm unite, their chromosomes combine so that the resulting zygote is diploid.

Although male and female gametes are both haploid, they differ from each other in all other respects. The male gamete, called a sperm or spermatozoon (pl., *spermatozoa*), is highly specialized for its sole function of finding and fertilizing an egg. The typical spermatozoon is a tiny cell consisting of a nucleus with a set of highly condensed chromosomes, a flagellum to propel itself, and mitochondria to provide chemical energy for flagellar movement. At its tip, the spermatozoon has certain recognition molecules that bind to an egg from the same species. Moreover, enzymes stored in the tip of the sperm help it penetrate the egg envelopes.

The female gamete, called the egg or ovum (pl., *ova*), is typically the largest cell made by a species. Besides providing genetic information, the egg's function is to sustain the development of the new individual until it can obtain nourishment from another source. Accordingly, eggs are filled with high-energy nutrients, such as glycoproteins, carbohydrates, and lipids. Eggs also contain building materials that can be rapidly assembled after fertilization. For instance, eggs store histones and tubulin, which are used during cleavage to make chromatin and microtubules, respectively. In most animals, the egg also provides spatial information through uneven distributions of cytoplasmic components, which later determine one or two of the embryo's body axes.

We will begin this chapter with a few historical comments on the discovery of the mammalian egg. Next we will examine the concept of the germ line as a lineage of potentially immortal cells, which is segregated from somatic cells early in development and eventually produces eggs or sperm. A discussion of meiosis will include the different timing of this process in males and females. Sperm development will be covered briefly, with emphasis on microscopic structure. Oogenesis will be described more extensively, focusing in particular on the processes by which the egg is supplied with RNA molecules and storage proteins. We will also discuss the process of egg maturation, when the egg is prepared for fertilization. Finally, we will explore some of the properties of egg envelopes, which protect the fragile egg cell but at the same time allow gas exchange to occur.

3.1 The Discovery of the Mammalian Egg

Humans have always been keenly interested in their own procreation as well as in the propagation of cultivated plants and animals. In most cultures, it has long been known that both males and females of a species contribute to the generation of offspring. It has also been recognized that semen is the male contribution to reproduction in both humans and other animals. Until relatively recently, the female contribution was less clear. Aristotle, although he had observed chicken embryos developing inside their eggshells, had no idea that the chicken egg was a single cell. Much less could he imagine a mammalian egg, which is invisible to the naked eye. He believed that mammalian development began from coagulated menstrual flow, a widely held idea in his time.

A major contribution to the eventual recognition of the mammalian egg came from William Harvey (1578–1657), who is also known for his discovery of blood circulation. As a physician to King Charles I of England, Harvey had access to the royal deer herd. Dissections of females during successive stages of their annual breeding cycle convinced him that mammalian embryos do not originate from coagulated menstrual flow. Although he could not see mammalian eggs, which are only about 0.1 mm in diameter, he postulated that every animal, including the human, originates from an egg (Fig. 3.1).

The Dutch biologist Regnier de Graaf (1641–1673) described the ovarian follicles that were later named after him, but he mistook the entire follicle for the egg.

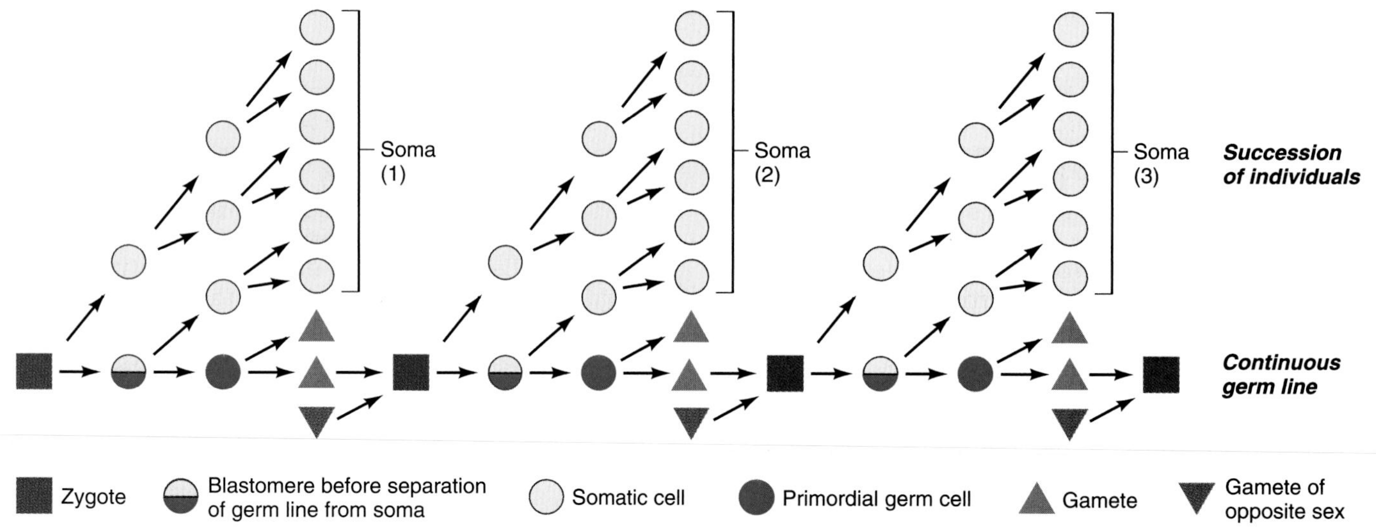

Figure 3.2 The germ line concept according to August Weismann (1834–1914).

Eventually, with improved light microscopy, the mammalian egg was identified as a small vesicle within the Graafian follicle. The German embryologist Karl Ernst von Baer (1792–1876) made this discovery in 1827. However, the cellular nature of the mammalian egg was not generally recognized until late in the nineteenth century.

3.2 The Germ Line Concept and the Dual Origin of Gonads

With the establishment of the cell theory, it became clear that all organisms consist of cells, that gametes are specialized cells, and that chromosomes are the carriers of genetic information. The far-reaching implications of these discoveries for heredity, development, and evolution were recognized by August Weismann (1834–1914). He developed the concept of the ***germ line,*** which is defined as the lineage of cells from which gametes arise (Fig. 3.2). The various non-germ line cells in an organism are collectively ***somatic cells,*** or *soma* for short. The zygote and those blastomeres that give rise to both gametes and somatic cells are included in the germ line. Germ line cells have a chance to live on forever in their descendants, whereas all somatic cells are bound to die. Thus the germ line cells can be viewed as a continuous lineage that weaves from generation to generation. From this perspective, each individual is a temporary caretaker entrusted with perpetuating his or her germ line.

Cells that will produce exclusively gametes are called ***primordial germ cells.*** In many animals, the primordial germ cells can be distinguished from somatic cells by their size, shape, division schedule, motility, and stainability for certain marker proteins (Nieuwkoop and Sutasurya, 1979, 1981; *Dixon, 1994*). In many groups of animals, primordial germ cells are determined early during embryonic development by localized components in the egg cytoplasm (see Section 8.3). In mice, however, the development of primordial germ cells depends on signals received from other cells (Tam and Zhou, 1996; Lawson et al., 1999). Cells receiving this signal maintain the expression of a particular gene, *Oct-4*$^+$, which seems to conserve cells in an undifferentiated state (Pesce et al., 1998).

Historically, the germ line concept has played an important role in defining those cells that pass on genetic information to the next generation. Most of Weismann's contemporaries thought that *all* parts of the parental bodies contributed to the characteristics of their offspring. This view was compatible with the Lamarckian idea that acquired traits are heritable. In contrast, the germ line concept conflicts with Lamarckism, because it holds that only germ line cells, not somatic cells, transfer genetic information to gametes. Today, the germ line concept is still relevant in the contexts of genetic transformation and senescence (see Sections 15.6 and 29.2).

In most animals, the primordial germ cells associate with somatic cells to form the gonads (ovaries or testes), in which the gametes develop. Typically, the somatic part of the gonad is derived from the *mesoderm,* the intermediate germ layer that forms during gastrulation. In many organisms, the primordial germ cells reach the mesodermal gonad rudiment after considerable migration. For instance, in a 3-week-old human embryo, the primordial germ cells are located in an appendage of the embryonic gut known as the *yolk sac* (Fig. 3.3). From there, the cells migrate around the hindgut, up the mesentery suspending the hindgut, and into the mesodermal gonad rudiment. In reptiles and birds, the primordial germ cells travel through the bloodstream, and when they reach the vicinity of the mesodermal gonad rudiment they exit the blood vessel and enter the gonad.

The primordial germ cells arrive at the mesodermal gonad rudiment during the *sexually indifferent* stage of development, which in humans lasts through the sixth week of gestation. The primordial germ cells proliferate

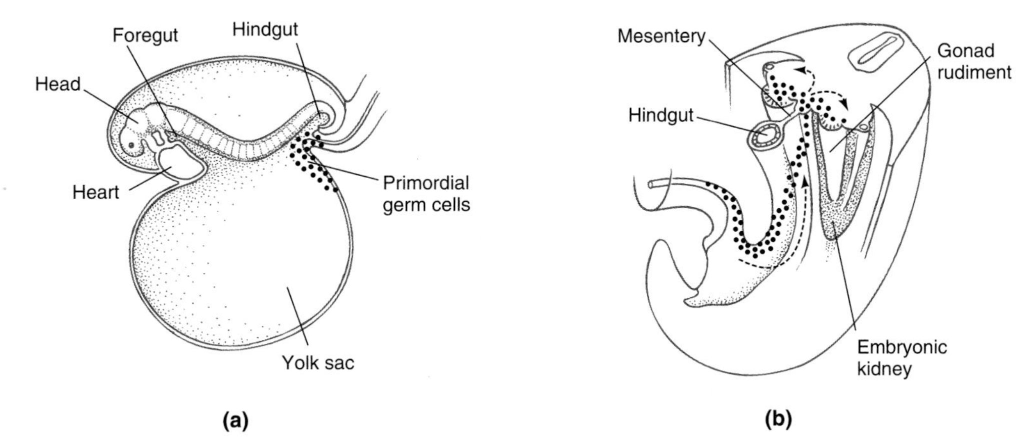

Figure 3.3 The migration of primordial germ cells in the human embryo. **(a)** Schematic drawing of a 3-week-old embryo in median section. **(b)** Posterior half of a 6-week-old embryo in ventrolateral aspect, with abdominal walls cut away. The primordial germ cells appear among the endodermal cells in the rear wall of the yolk sac, an extension of the gut. The primordial germ cells migrate around the wall of the hindgut and up the dorsal mesentery, which suspends the gut in the abdominal cavity. Eventually, they invade the mesodermal gonad rudiment, located on top of the embryonic kidney.

within the gonad via mitosis and then usually enter a quiescent period. Thereafter, they are distributed in specific parts of the gonad, and the male and female germ cells acquire different cytological characteristics; at this stage, the germ cells are called gonia (Gk. *gone,* "seed")—***spermatogonia*** in the male and ***oogonia*** in the female.

3.3 Meiosis

During gametogenesis, germ cells divide to reduce the number of their chromosomes by half, from a *diploid* to a *haploid* set. For example, a human's diploid cells contain 46 chromosomes, 23 derived from the father and 23 from the mother, whereas human gametes carry only 23 chromosomes derived randomly from either father or mother. This reduction in the number of chromosomes occurs in two successive cell divisions known as meiotic divisions I and II, or ***meiosis*** (Gk. *meioun,* "to diminish"). In diploid cells, each maternal chromosome has a paternal ***homolog*** (Gk. *homos,* "same"; *logos,* "word" or "sense"), and vice versa. Corresponding genes on homologous chromosomes are *allelic*—that is, they encode the same type of product, possibly with variations introduced by mutation. (Only the ***sex chromosomes,*** designated X and Y in mammals, are for the most part nonhomologous.) Thus, diploid cells have two identical or allelic copies of most genes. The hallmark of meiosis is the separation of homologous chromosome pairs and the formation of haploid daughter cells containing only one chromosome from each pair.

HOMOLOGOUS CHROMOSOMES ARE SEPARATED DURING MEIOSIS

Meiosis I is preceded by a round of DNA replication that results in two identical chromatids for each chromosome. Because meiosis entails two successive divisions without intervening DNA replication, the entire process yields four haploid cells. In the first meiotic division, pairs of homologous chromosomes are segregated between two daughter cells. The second meiotic division separates the two chromatids of each chromosome and creates four haploid cells. We will use the same names for the stages of meiosis that we have also used for mitosis—prophase, metaphase, and so on (see Section 2.4).

During meiotic prophase I, individual chromosomes become visible under the microscope as they coil and shorten. At this stage, the process resembles mitotic prophase. Subsequently, two fundamental differences between mitosis and meiosis become apparent. First, in meiosis, the twin chromatids of each chromosome are held together along their entire length by axial proteins (Fig. 3.4a). Second, homologous chromosomes join together in pairs, which are connected by a central element consisting of longitudinal and transverse proteins (Sym et al., 1993). This coupling, which is called ***synapsis,*** is characteristic of meiotic prophase I and is not observed in mitosis.

The configuration of four chromatids that results from synapsis is known as the ***synaptonemal complex.*** Its formation facilitates the genetically important process of ***crossing over,*** in which homologous chromosomes exchange parts of their chromatids (Fig. 3.4a). Crossing over begins with the formation of a *recombination nodule,* which is believed to contain the enzymes necessary for cutting and splicing chromosomal DNA (Prescott, 1988). The chromatids are severed at corresponding positions and rejoined crosswise. However, crossing over events do not become visible until metaphase I, when the synaptonemal complex disintegrates and the homologous chromosomes drift apart.

During metaphase I, the two chromatids of each chromosome are clearly recognizable and are seen to be crossed at points called ***chiasmata*** (sing., *chiasma*). Crossing over leaves most chromatids as composites of segments from maternal and paternal homologs. Within

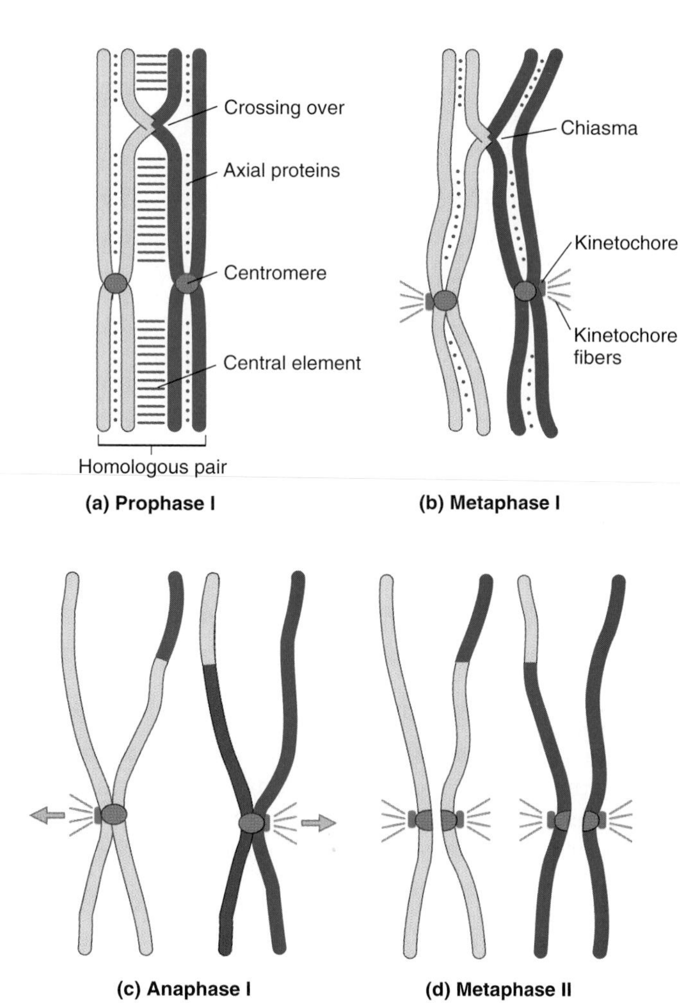

Figure 3.4 Meiosis. Schematic diagrams show the behavior of one pair of homologous chromosomes at various stages. **(a)** Prophase I. A germ cell entering meiosis is diploid, containing two homologs of each chromosome, one from the male parent and one from the female parent. Each homolog consists of twin chromatids connected by axial proteins (dots). Each pair of homologues is held together by a central element with transverse filamentous proteins (short lines), forming the synaptonemal complex. Crossing over events lead to the mismatch of chromatid segments. **(b)** Metaphase I. Each centromere sends kinetochore fibers to only one spindle pole, while the centromere of the homologous chromosome sends kinetochore fibers to the opposite pole. As a result, the homologous chromosomes are segregated into two daughter cells, while the twin chromatids of each chromosome remain joined. The crossing over points, called chiasmata, become visible when the central element breaks up and the homologs move apart. **(c)** Anaphase I. The axial connections between twin chromatids disappear. Chiasmata are resolved by exchange of homologous chromatid segments. **(d)** Metaphase II. Each centromere sends kinetochore fibers to both spindle poles, and the twin chromatids of each homolog are separated, as in mitosis (see Fig. 2.14d).

each chromosome, the two chromatids are still held together at the centromere, which stays intact throughout meiosis I. At this time spindles form, as they do during mitosis. However, during meiosis I, the homologous chromosomes are paired in such a way that all kinetochore fibers extending from one chromosome are oriented toward *one* spindle pole, and all kinetochore fibers extending from the homologous chromosome are oriented toward the *opposite* spindle pole (Fig. 3.4b).

During anaphase I, the homologous chromosomes are pulled apart at the centromeres, and the chromatid tangles at the chiasmata are resolved (Fig. 3.4c). During telophase I, two daughter nuclei are formed. Each nucleus contains one complete set of duplicated chromosomes, but due to crossing over events, most chromosomes now consist of nonidentical chromatids.

During meiosis II, each centromere now sends kinetochore fibers to both spindle poles. As a result, the chromatids of each chromosome are separated as in mitosis (Fig. 3.4d). The final step in meiosis II is the formation of four haploid cells. These cells are genetically nonequivalent because homologous genes are allelic but not necessarily identical. At some point during meiosis, the female germ cell is released from the ovary, an event called ***ovulation.*** In amphibians and most mammals, including humans, ovulation and fertilization occur at metaphase II.

Different terms are used to describe the germ cells at different points in meiosis (Fig. 3.5). When a gonial cell is about to undergo its first meiotic division, it is called a ***primary oocyte*** or ***primary spermatocyte.*** Following meiosis I, the germ cells are called ***secondary oocytes*** or ***secondary spermatocytes.*** After completion of meiosis, the male germ cells are referred to as ***spermatids,*** and the female germ cells are called *eggs*. However, the term "egg" is used loosely from ovulation, when in most species the female germ cell is technically an oocyte, until cleavage, when it is technically an embryo. While one primary spermatocyte gives rise to four spermatids and eventually to four functional spermatozoa, a primary oocyte produces only one large egg and three small sister cells called *polar bodies,* which have no known function and degenerate.

THE TIMING OF MEIOSIS DIFFERS BETWEEN MALES AND FEMALES

In humans and other mammals, the timing of meiosis is markedly different between females and males (Fig. 3.6). In the male germ line, cycles of meiosis begin with adulthood. In the human male, spermatogonia enter meiosis in waves from the onset of puberty throughout life. Immediately after completion of meiosis, which takes a few weeks, spermatids mature into spermatozoa (see Section 3.4). A man produces sperm in this fashion until the end of his life. A young man's ejaculation releases 100 to 500 million sperm.

In contrast, the first meiotic division of oogenesis usually starts long before the female reaches sexual maturity. An extended meiotic prophase I allows animals that produce large eggs to synthesize and accumulate molecular building materials for the embryo. Mammals,

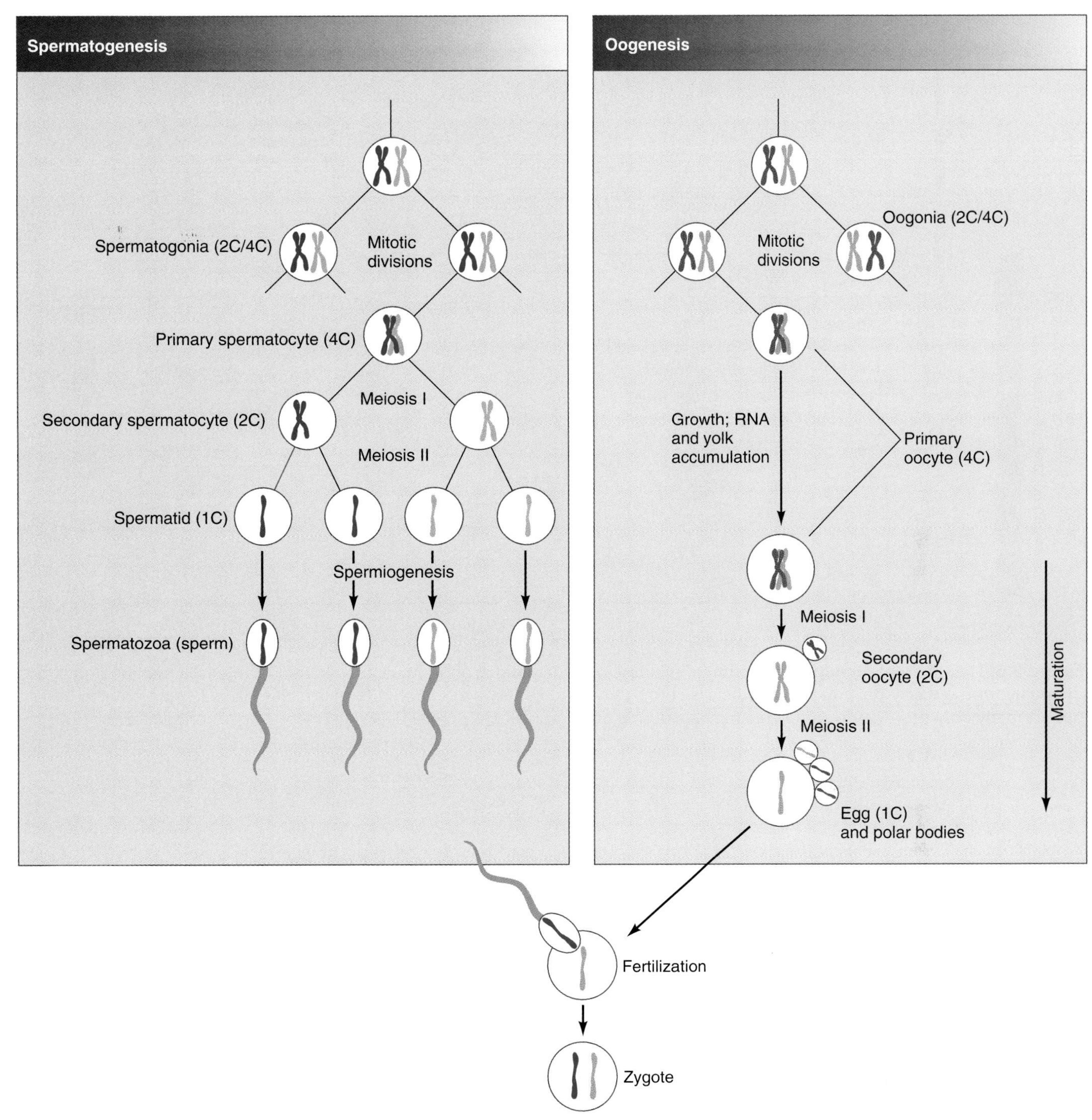

Figure 3.5 Comparison of spermatogenesis and oogenesis. Primordial germ cells divide mitotically, producing spermatogonia in males and oogonia in females. These cells are diploid, containing two or four genomic complements (2C or 4C), depending on their stage in the mitotic cycle. Before the gonia enter meiosis, their DNA replicates. They are then called primary spermatocytes or oocytes. After the first meiotic division, they contain two genomes (2C) and are called secondary spermatocytes or oocytes. After the second meiotic division, they are haploid (1C) spermatids or eggs. Note that the two rounds of meiosis produce four haploid spermatids, each of which develops into a spermatozoon, but only one egg. The egg's three small sister cells, known as polar bodies, have no known function and degenerate. Often the first polar body does not divide, so that only a total of two polar bodies is formed. Depending upon the species, eggs are fertilized at various stages of meiosis (see Fig. 3.18).

which produce unusually small eggs, nevertheless follow the same pattern. In the human female, meiosis begins before birth and is not completed until after fertilization (Fig. 3.6b). Some oocytes are therefore arrested in meiotic prophase I for as long as 50 years.

Prolonged arrest seems to affect the separation of homologous chromosomes or chromatids during meiotic anaphases, so that an egg may receive both homologs or none. If such eggs are fertilized and develop, they will show irregular numbers of chromosomes, or *aneuploidies*.

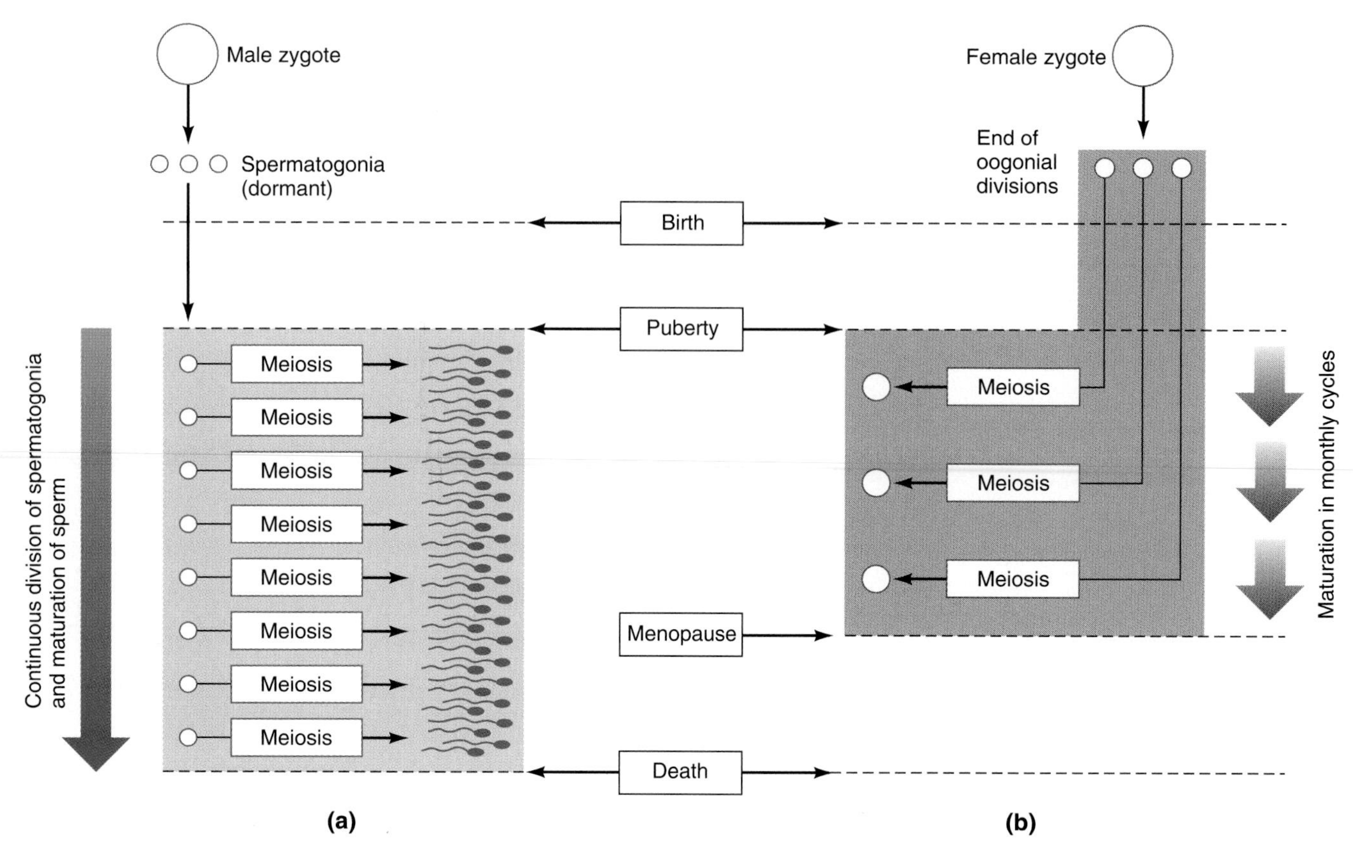

Figure 3.6 The different timing and duration of meiosis in human males and females. **(a)** In males, meioses are initiated continuously from puberty on, and each meiosis is completed within a few weeks. **(b)** In females, all meioses begin prior to birth, but normally only one oocyte completes meiosis during each reproductive cycle between puberty and menopause.

Aneuploid embryos are usually aborted spontaneously, but some may come to term. A fairly common human aneuploidy is *trisomy 21,* meaning the presence of three instead of two copies of chromosome 21. Individuals with this condition show *Down syndrome,* which includes short stature, round face, epicanthic fold (skin covering the inner corner of the eye), and mild to severe mental retardation. The frequency of children born with Down syndrome is less than 0.2% for mothers under age 30 but increases to more than 3% for mothers over 45 (Sutton, 1988).

MEIOSIS PROMOTES GENETIC VARIATION, HELPS TO ESTABLISH HOMOZYGOUS MUTANT ALLELES, AND ELIMINATES BAD GENES

Meiosis is inherently hazardous. Forming a synaptonemal complex before segregating homologous chromosomes may lead to *aneuploidies* as mentioned earlier. The cutting and splicing of genomic DNA during crossing over may cause the deletion or duplication of chromosomal segments. Indeed, the entire process of sexual reproduction seems wasteful and dangerous. Building large bodies that die while tiny fertilized eggs live on and have to grow again seems to be an extravagant way of propagating a species. Why don't most organisms procreate asexually, by forming stolons or buds?

The adaptive value of sexual reproduction has long been a matter of intense debate (Maynard-Smith, 1989; Michod and Levin, 1987). Most theories emphasize that sexual reproduction generates more genetic diversity than asexual reproduction and thereby facilitates adaptation and evolution. The root cause of genetic diversity are ***mutations,*** heritable changes to genetic information encoded in DNA. Such mutations occur randomly as a result of cosmic radiation, chemical mutagenesis, and the limited accuracy of DNA replication. Such mutations are passed on from the originally affected cell to all its daughter cells. If a mutation occurs in an organism with asexual reproduction, then the stolons or buds by which this organism propagates may consist of a mosaic of mutated and nonmutated cells, and it may take a long time until a new organism originates that consists exclusively of mutated cells. In contrast, if a mutation occurs in a sexually reproducing organism, and the mutated cell contributes to the formation of eggs or sperm, then any offspring derived from such gametes will instantly carry the mutation in *all* of its cells. Thus, the seemingly wasteful process of growing a whole new organism from a single fertilized egg in each generation ensures that new mutations appear in the entire mutant organism and are reliably passed on through the germ line.

However, sexual reproduction does more than forcing new genetic alleles through the bottlenecks of unicellular gametes. The process of meiosis also mixes and matches new and old genetic alleles in virtually unlimited numbers of combinations. Two events during meiosis are especially important here. First, meiotic anaphase I produces nuclei that may contain any combination of maternally and paternally derived chromosomes. During this phase, a given nucleus receiving, say, the maternal homologue of one chromosome may receive either the maternal or the paternal homologue of another chromosome and so forth. This phenomenon, known as ***independent assortment,*** makes it possible for one human, with 23 homologous chromosome pairs, to generate 2^{23} (about 8 million) *types* of gametes with different combinations of chromosomes. The second meiotic event that promotes genetic variation is *crossing over,* which results in new combinations of maternally and paternally derived genes, as described earlier.

The contributions of independent assortment and crossing over to genetic variation culminate at fertilization, when an egg with one unique combination of maternal genes unites with a sperm containing a different assortment of paternal genes. Because the activities of genes affect one another, a nearly limitless number of overall gene activity patterns are possible.

Independent assortment, crossing over, and the combination of maternal and paternal genes at fertilization are all associated with sexual reproduction and do not occur during asexual reproduction. This link, and the prevalence of sexual reproduction despite its high costs, suggest that creating genetic diversity as the raw material for natural selection is indeed beneficial. However, there are additional benefits derived from sexual reproduction.

Another advantage that sexually reproducing organisms have is the opportunity to perpetuate genetic mutations that have adaptive value for their species. To understand this point, one needs to remember that diploid cells have two homologous copies, or alleles, of each gene. If the two alleles are identical, the individual is said to be ***homozygous*** for this allele. If they are not identical, the individual is ***heterozygous.*** When a mutation occurs, the possibility that a matching mutation will spontaneously arise in a homologous gene is remote. Therefore, mutant alleles in diploid cells of asexually reproducing organisms virtually always exist in heterozygous combinations with wild-type alleles. Since the presence of one copy of the wild-type allele per cell is generally sufficient to support normal function, mutations tend to be ineffective in diploid cells and rarely produce a selectable phenotype. However, the ***segregation*** of homologous chromosomes during meiosis provides a mechanism by which mutant alleles [can quickly become homozygous] through inbreeding (Kirkpatrick and Jenkins, 1989).

Finally, meiosis provides a mechanism for the elimination of "bad," or nonadaptive, genes (Kondrashov, 1988). This is critical because the vast majority of mutations are deleterious. Meiosis provides two opportunities for losing chromosomal segments that are compromised by bad genes. First, each gamete inherits only half of the parental chromosomes. Second, crossing over swaps homologous chromosome segments, thus providing an opportunity for combining segments that are free of bad genes.

In summary, meiosis promotes genetic variation and provides mechanisms for establishing mutant alleles in homozygous condition as well as for eliminating nonadaptive genes. Thus, in comparison to asexual reproduction, sexual reproduction generates a much broader basis for natural selection.

3.4 Spermatogenesis

The entire process of sperm formation, beginning with spermatogonia and resulting in mature *spermatozoa,* is referred to as ***spermatogenesis.*** Spermatogenesis has been studied most extensively in the mammalian testis (Fawcett, 1975; Eddy and O'Brien, 1994). Mammalian spermatogenesis will therefore be presented first and then contrasted with spermatogenesis in other animals.

MALE GERM CELLS DEVELOP IN SEMINIFEROUS TUBULES

The mammalian testis is divided into portions called *lobules,* which contain tiny convoluted tubes, the ***seminiferous tubules*** (Lat. *semen,* "seed"; *ferre,* "to bear"), which are the functional units for spermatogenesis (Fig. 3.7). Inside a sheath of connective tissue cells, the developing germ cells are arranged in a distinct order. Cells in the earliest stages—that is, the spermatogonia—are located at the periphery; those in more advanced stages are placed successively toward the interior. Finally, mature spermatozoa are given off into the lumen (central open space) of the tubule.

Close examination reveals that the maturing sperm form groups of cells that are connected by cytoplasmic bridges and develop in strict synchrony (Fig. 3.8). The bridges, which result from incomplete cytokinesis after mitotic as well as meiotic divisions, allow the passage of large molecules and even organelles between connected cells. The development of all the sperm cells of a group can therefore be directed by gene products arising from both sets of parental chromosomes. This is particularly important in the case of genes located on the sex chromosomes, which for the most part are nonhomologous. Without cytoplasmic bridges, one half of the secondary

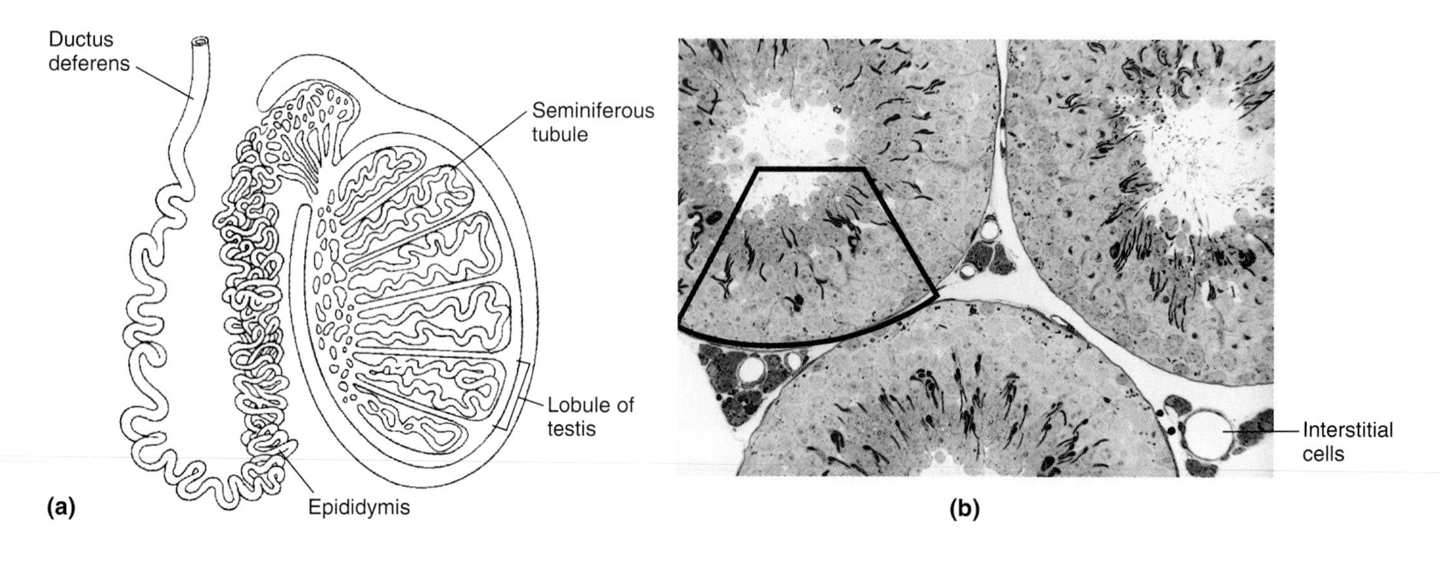

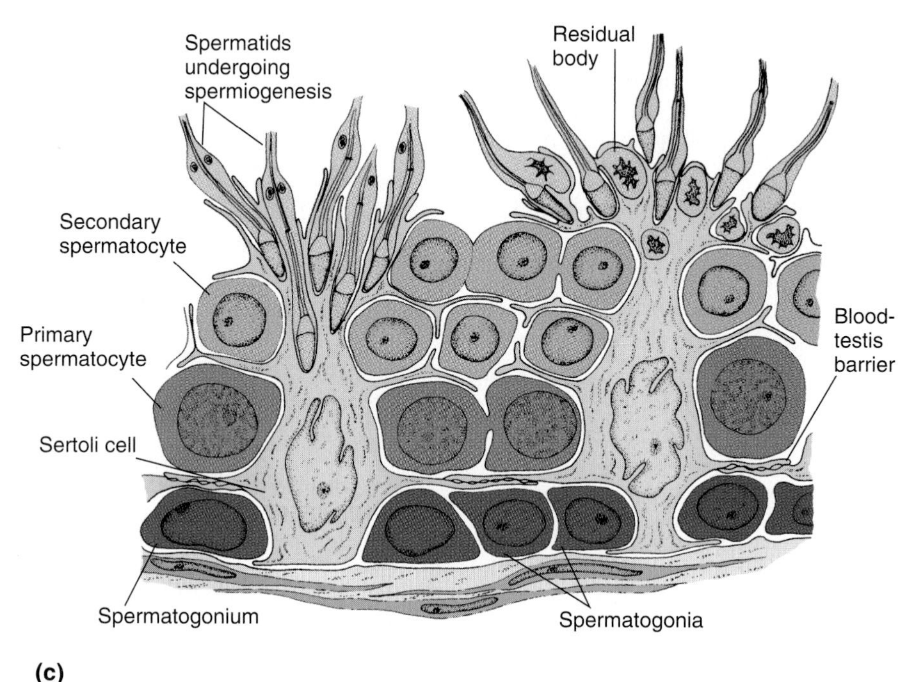

Figure 3.7 Seminiferous tubules in the mammalian testis. **(a)** Schematic diagram of a testis. Each testis is divided into lobules containing extensively coiled seminiferous tubules. These tubules are connected via a network of collecting ducts and the epididymis to the ductus deferens, which joins the urethra. **(b)** Light micrograph showing partial cross sections of seminiferous tubules and clusters of interstitial cells (Leydig cells) in the spaces between them. **(c)** Drawing of a segment (indicated by frame part b of a seminiferous tubule, showing Sertoli cells in close contact with germ cells at different stages of spermatogenesis. Refer to Figure 3.8 for origin of the residual body.

spermatocytes and their descendants would completely lack gene products from the Y chromosome, and the other half would lack gene products from the X chromosome.

In addition to germ line cells at different stages of maturation, the seminiferous tubules contain prominent somatic cells known as ***Sertoli cells.*** These large, columnar cells span the tubule wall from the connective tissue sheath to the lumen. Sertoli cells play many important roles in the testis, both structural and functional (Sharpe, 1993; Griswold, 1995). By forming extensions connected by *tight junctions,* the Sertoli cells create a *blood-testis barrier,* which divides the seminiferous tubule into two compartments (Fig. 3.7c). The exterior compartment contains the spermatogonia, whereas the interior compartment contains spermatocytes, spermatids, and sperm. The advancing germ line cells are gated through the blood-testis barrier as new junctional complexes are formed exteriorly while older ones disintegrate interiorly. Because the interior compartment is sequestered from blood and lymph fluid, its chemical composition is controlled locally by the secretions of Sertoli cells as well as meiotic and postmeiotic germ cells.

The germ line cells are lodged in deep recesses of the Sertoli cells, and they exchange close-range signals affecting each other's development. Before puberty, the association with Sertoli cells inhibits the proliferation of spermatogonia and prevents their entry into meiosis. After puberty, Sertoli cells stimulate spermatogonial divisions and provide nutrients and signaling to the developing sperm (Meehan et al., 2000; Meng et al., 2000). In addition, the Sertoli cells send and receive long-range hormonal signals. During fetal development, Sertoli cells release *anti-Müllerian duct hormone,* which inhibits the development of female reproductive ducts (see Section 27.2). Throughout development, Sertoli cells are involved in a hormonal feedback loop with the pituitary gland, and they also respond to testosterone produced by somatic *interstitial cells* located between the seminiferous tubules.

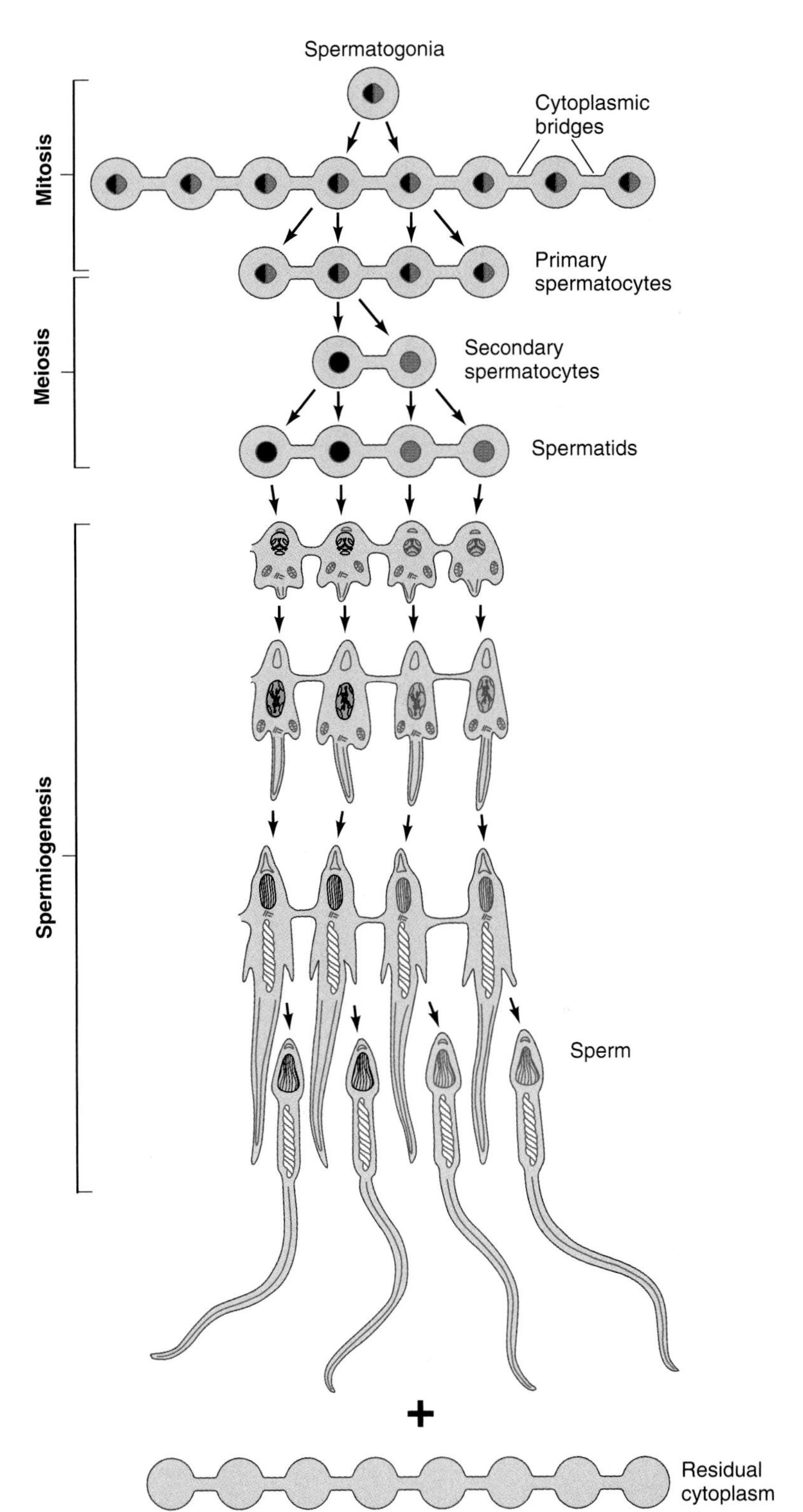

Figure 3.8 The development of mammalian sperm as groups of interconnected cells. The final mitotic divisions of spermatogonia, and the subsequent meiotic divisions, are followed by incomplete cytokineses, resulting in groups of germ cells connected by cytoplasmic bridges. The nuclei are shaded to show the presence of an X chromosome (black), a Y chromosome (color), or both sex chromosomes. The cytoplasmic bridges allow the germ cells to exchange gene products of maternal and paternal origin. During the subsequent process of spermiogenesis, individual sperm are separated from the residual cytoplasmic body of the entire group.

SPERMIOGENESIS

At the conclusion of meiosis, each primary spermatocyte has formed four haploid spermatids, which are inconspicuous, round cells. In most animal species, spermatids mature into uniquely shaped spermatozoa. The maturation process that transforms the spermatid into a spermatozoon is called ***spermiogenesis.***

The major steps of spermiogenesis are summarized in Figure 3.9. Vesicles pinched off from the spermatid's Golgi apparatus coalesce into a membranous organelle called the acrosomal vesicle, or ***acrosome*** (Gk. *akron,* "tip"; *soma,* "body"). It forms a cap over the nucleus, and the nucleus rotates so that the acrosome points away from the lumen of the seminiferous tubule. At the opposite pole of the spermatid, a centriole pair becomes the basis for the growth of an array of microtubules that supports a *flagellum* extending into the lumen of the tubule. The cell's mitochondria

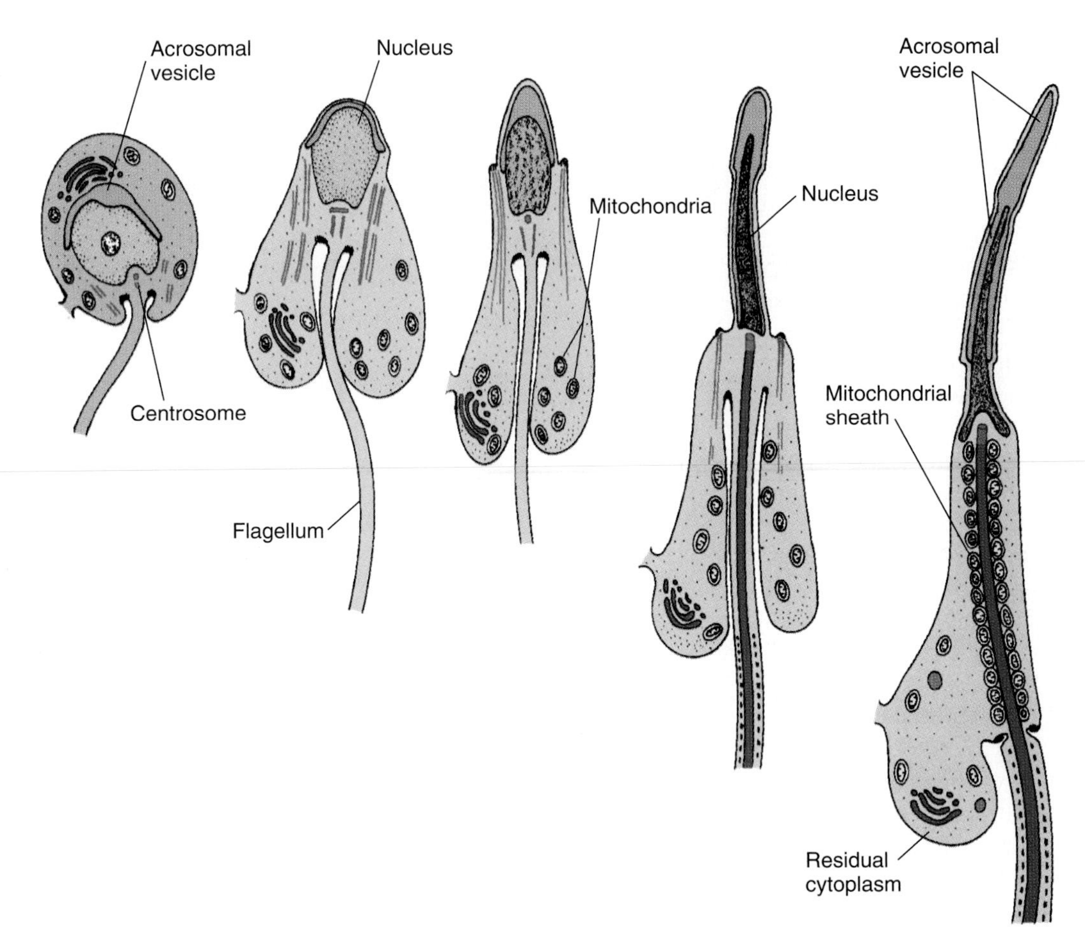

Figure 3.9 Successive stages in guinea pig spermiogenesis. The dramatic morphological changes include the formation of a cap-shaped acrosomal vesicle, condensation of the nucleus, outgrowth of a flagellum, and arrangement of a sheath of mitochondria around the base of the flagellum.

aggregate around the base of the flagellum. The nucleus condenses dramatically, as chromosomal histones are replaced with other proteins called *protamines.* Eventually the spermatozoa become individual cells by pinching off a ***residual body,*** which consists of the bulk of the cytoplasm—including the cytoplasmic bridges—from an entire group of spermatids.

The mature mammalian sperm is highly specialized for its function—the delivery of the male genome to the egg (Fig. 3.10). The nucleus and acrosome form the ***head*** of the sperm, which is joined by a neck to the remainder of the sperm, called the ***tail.*** The nucleus, of course, carries the genetic information of the male parent. The acrosome contains enzymes for penetrating the cells and envelopes surrounding the egg (see Section 4.2). The proximal part of the tail, called the ***midpiece,*** contains the mitochondria, placed strategically at the base of the propulsive organ, called a ***flagellum,*** which needs most of the chemical energy provided by the mitochondria. Like any other cell, the spermatozoon is surrounded by a plasma membrane, which contains special proteins that allow the sperm to respond to chemoattractants and to interact with the egg. These membrane proteins assemble in different regions of the head and the tail, and the differentiation of these domains continues as the spermatozoa leave the testis and traverse the epididymis (Phelps et al., 1990; Cowan and Myles, 1993).

Most nonmammalian spermatozoa generally resemble their mammalian counterparts, but on closer inspection show some differences. The sperm of marine and freshwater invertebrates, which are shed into the water, tend to have a less elaborate tail structure and a shorter midpiece. Many invertebrate sperm have a cup-shaped depression in the anterior face of the nucleus, which contains globular actin. At fertilization, the actin polymerizes and supports the formation of a fingerlike process that plays a role in sperm-egg adhesion (see Section 4.2). The sperm of some animal species have no resemblance at all to typical sperm. For instance, the sperm of some crayfish and crabs are star-shaped, and the sperm of roundworms are large cells that contain much cytoplasm, have no flagella, and move like amoebas.

SPERMATOGONIA BEHAVE AS STEM CELLS

In the male germ line of most animals, primordial germ cells show a pattern of cell division that defines a general class of cells called ***stem cells,*** which are defined by three

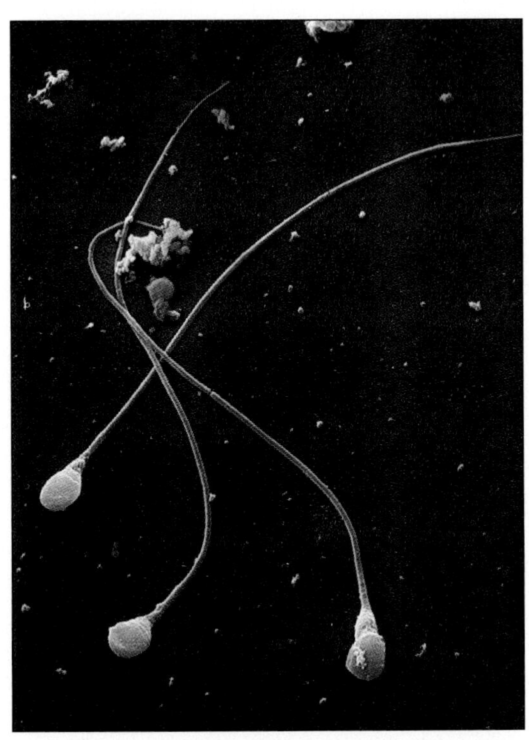

Figure 3.10 Scanning electron micrograph of human spermatozoa.

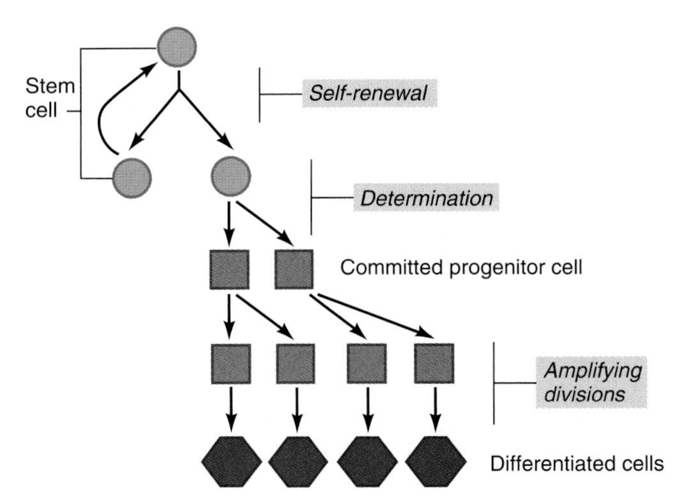

Figure 3.11 Stem cells are undifferentiated and have a large capacity for self-renewal, which often lasts throughout the life of the organism. Stem cells also give rise to committed progenitor cells, which undergo amplifying divisions before giving rise to differentiated cells.

criteria (Fig. 3.11; Hall and Watt, 1989; Morrison et al., 1997). First, stem cells are *undifferentiated*—that is, not yet adapted to a specialized function in the organism. Second, stem cells have the capacity for ***self-renewal***—that is, the ability to generate more stem cells through mitotic divisions. Typically, the ability to self-renew lasts for the life of the organism. Third, stem cells give rise to ***committed progenitor cells***—that is, cells that are still undifferentiated but already restricted in their potential to form certain differentiated cells. The committed progenitor cells go through further ***amplifying divisions*** before they become differentiated cells. In the case of the male germ line, a committed progenitor cell is a spermatogonium that undergoes only a limited number of mitotic divisions before its descendants will enter meiosis.

There are many other kinds of stem cells besides spermatogonia. For example, the external layer of our skin, the *epidermis,* undergoes a rapid turnover. In this case, committed progenitor cells are born from stem cells located in the deepest epidermal layers. Here they undergo amplifying divisions before they become differentiated as they move to the surface of the skin, where they are eventually sloughed off. Similarly, blood cells are relatively short-lived and are continuously being replenished from a small number of stem cells, which lie well protected in the bone marrow (see Section 20.4). These stem cells are called *pluripotent* because they give rise to more than one type of committed progenitor cell and eventually to several types of differentiated cells including red blood cells and various lymphocytes.

Being a stem cell lineage is a property of male germ cells that is not always shared by female germ cells. While the primordial germ cells in insect ovaries are stem cells, as will be discussed in the following section, the oogonia of mammals are produced in limited numbers, and only for a short period of time. As outlined in Section 3.3, all the oocytes that a female mammal will produce in her life are formed long before she reaches sexual maturity. From then on, oocytes are continuously lost through a genetically controlled process known as *programmed cell death* or *apoptosis* (see Section 25.4; Morita and Tilly, 1999). In human females, the depletion of the one-time endowment of oocytes occurs long before the end of the average life span. The steep decrease in estrogen synthesis caused by the disappearance of ovarian follicles also brings an end to a woman's menstrual cycles, an event known as *menopause.*

3.5 Oogenesis

Eggs differ from sperm in many respects. The basic functions of the egg are to contribute its own genome, to incorporate one sperm at fertilization, and to provide molecular building blocks, structural information, and protective envelopes for the developing embryo. These functions are anticipated in the provisions made during the egg's development in the ovary, a process called ***oogenesis.*** As we have already seen, meiosis starts much earlier in the female germ line and takes much longer to complete. In further contrast to the male germ line, the meiotic divisions of an oogonium produce only one large gamete, the egg; the tiny haploid sister cells of the egg, known as ***polar bodies,*** have no apparent function in later development (see Fig. 3.5).

Autoradiography

Tracing the synthesis and transport of molecules requires labeling the molecules of interest. Biological molecules can be traced effectively if they are synthesized in the presence of a radioactive precursor—a procedure termed ***radiolabeling.*** For instance, proteins are radiolabeled by adding radioactive amino acids, such as [^{3}H]leucine or [^{35}S]methionine, to the medium of a cell culture or by injecting them into an experimental animal. Cells take up the labeled amino acids and incorporate them, along with their own unlabeled amino acids, into newly synthesized proteins.

A sensitive method of tracing radiolabeled molecules is known as ***autoradiography*** (Lat. *radius*, "ray"; Gk. *autos*, "self"; *graphein*, "to write"). It is based on the fact that electrons from radioactive decay can be recorded on photographic film just like light. Exposure of the film to the radiolabel must be carried out in the dark . When the film is developed, areas of radioactivity show up as accumulations of silver grains.

Autoradiography can be used to reveal size, location, and other properties of radiolabeled molecules. For example, the size of radiolabeled proteins can be determined by extracting *all* proteins from the cells of interest and subjecting them to gel electrophoresis. The resulting gel will contain bands of proteins separated according to size. The gel is placed on a sheet of photographic film and kept in the dark for a suitable exposure time (Fig. 3.m1). After development, dark bands on the film reveal the location of the radiolabeled proteins in the gel. The location of an unknown protein in a gel relative to marker proteins of known size reveals the unknown protein's size.

All proteins, radioactive or not, can also be detected by stains, provided that enough molecules are present to bind visible amounts of stain. If a protein band is stainable but emits no radioactive signal, then the protein was synthesized before the radiolabeled precursor became available to the cells. If a protein band gives off a radioactive signal but is not stainable, then the protein is newly synthesized but has not accumulated in stainable amounts.

To identify the location of radiolabeled proteins in a tissue, the appropriate tissue is fixed and embedded in wax or plastic so that it can be cut into thin sections

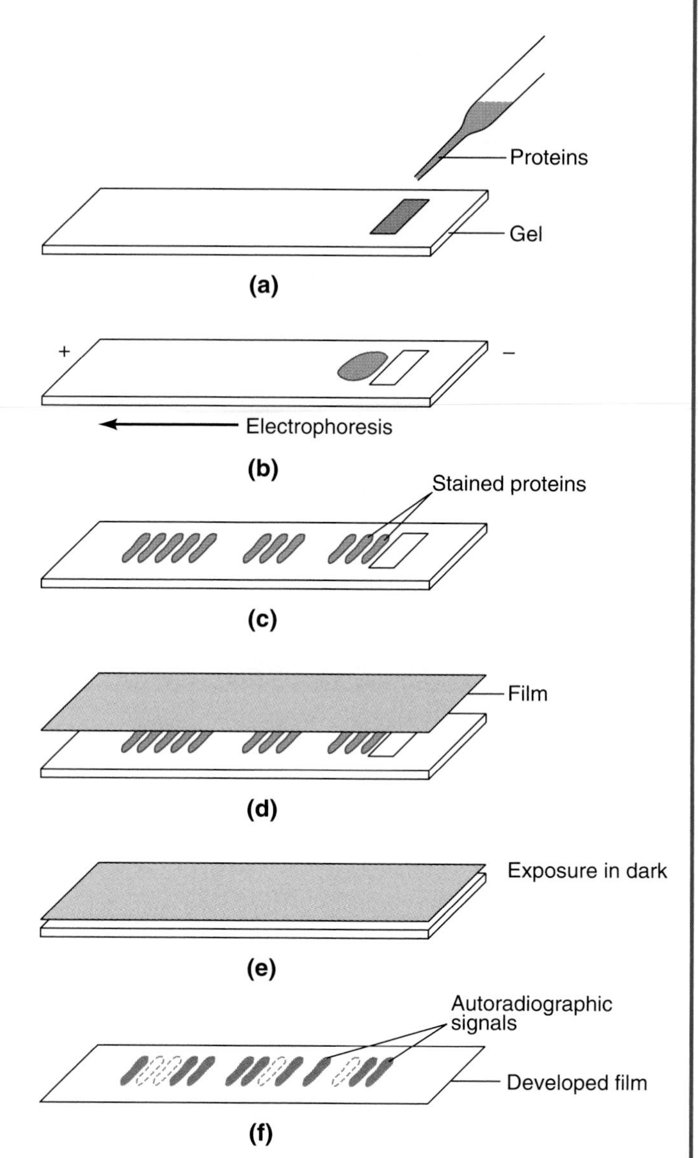

Figure 3.m1 Autoradiography to determine the size of newly synthesized proteins. Proteins are synthesized in vivo in the presence of a radioactive amino acid. **(a, b)** All proteins (labeled and unlabeled) are pipetted into the well of a gel and separated by electrophoresis. **(c)** Sufficiently abundant proteins are visible as stained bands. **(d, e)** To detect newly synthesized proteins, a photographic film is placed on the gel, exposed in the dark, and finally developed. **(f)** Dark bands on the processed film reveal the position of radiolabeled proteins in the gel. These bands may or may not coincide with stains produced by abundant proteins.

Eggs are storehouses filled with maternally provided building materials—mostly RNAs and proteins—for the developing embryo. These materials are incorporated into the oocyte during its meiotic arrest. After this growth phase, the oocyte is endowed with protective envelopes and prepared for its transition from the ovary to another environment. These processes will be described in the following sections.

OOCYTES ARE SUPPLIED WITH LARGE AMOUNTS OF RNA

During its arrest in meiotic prophase I, the oocyte nucleus in many species increases significantly in volume and becomes very active in RNA synthesis. This stage is well suited for RNA synthesis because four chromatids, and thus four sets of genes, are available for transcrip-

(Fig. 3.m2). The sections are placed on microscope slides and rinsed to remove any unincorporated amino acids—but not the labeled proteins—from the tissue. The rinsed sections are then covered with a layer of photographic emulsion in the dark. After appropriate exposure, the emulsion is developed, and the sections are viewed under the microscope. Areas of the tissue that are marked with silver grains from the photographic emulsion are the sites where the radiolabeled proteins accumulated.

Autoradiography is suitable also for determining the size and location of RNA, DNA, and other biomolecules that have been radiolabeled with appropriate precursors.

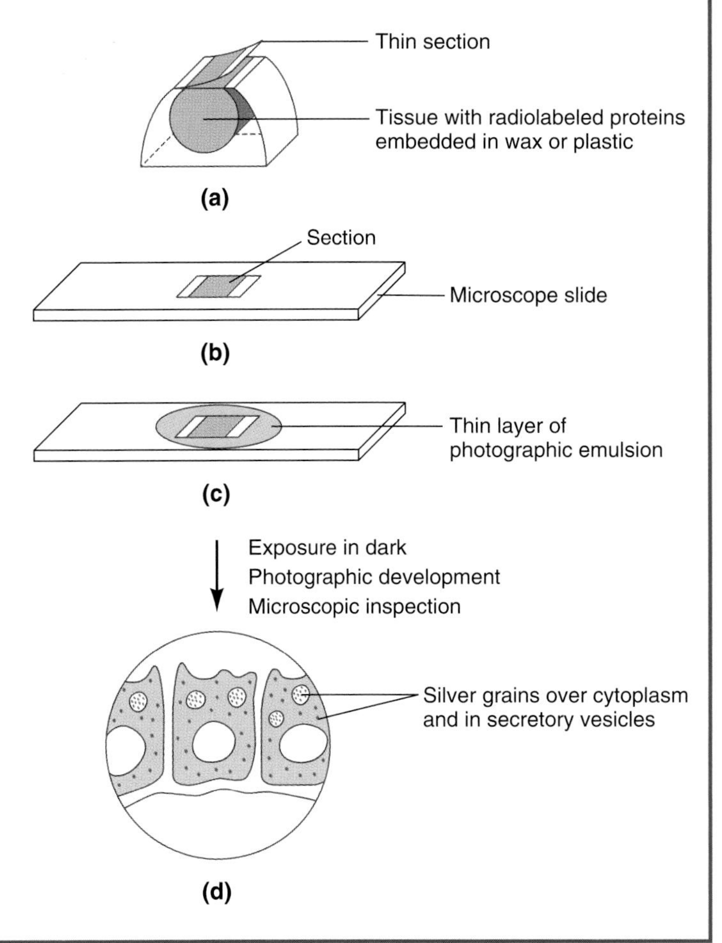

Figure 3.m2 Autoradiography to determine the locations where radiolabeled proteins accumulate in intact tissue. Proteins are synthesized in vivo in the presence of a radioactive amino acid. **(a)** Fixed and embedded tissue is sectioned. **(b)** Section is placed on a microscope slide and rinsed to remove unincorporated amino acids. **(c)** Photographic emulsion is pipetted onto the section and allowed to spread as a thin layer. **(d)** After exposure in the dark, the emulsion, still adhering to the tissue, is developed, and the whole preparation is viewed under the microscope. In transmitted light, dark silver grains in the emulsion indicate where radiolabeled proteins have accumulated in the tissue. In reflected light, the same silver grains are bright and more easily visible, as in Fig. 5.33a.

tion. The large nucleus of the primary oocyte is known as the ***germinal vesicle.*** In frogs and many other animals, the chromosomes in the germinal vesicle assume a characteristic configuration that appears when the synaptonemal complex dissolves. The two chromatids of each chromosome stay aligned with each other but become very elongated. Condensed chromosomal regions alternate with uncoiled chromatid loops extending symmetrically from the main axis of the chromosome (Fig. 3.12). Because the shape of these chromosomes reminded early researchers of the brushes used to clean the glass cylinders of kerosene lamps, they called them ***lampbrush chromosomes.*** The lampbrush configuration permits an extremely high level of transcription and supports the active transcription of many genes simultaneously. Both features are necessary to achieve the enormous amount and the sequence complexity of RNA stored in animal eggs (E. H. Davidson, 1986).

In keeping with the vast size of the oocyte relative to somatic cells, most cytoplasmic organelles are much more abundant in oocytes. For instance, a fully grown *Xenopus* oocyte contains 200,000 times as many ribosomes as an average somatic cell (Laskey, 1974). Transcribing the necessary amount of ribosomal RNA (rRNA) is a daunting task. Even somatic cells have serial, or tandem, repeats of several hundred copies of rRNA genes per haploid genome to satisfy their needs for rRNA (Lewin, 1980). In an apparent adaptation to the extremely high demand on ribosome synthesis in oocytes, these tandem repeats are again *selectively* multiplied in the germinal vesicles of amphibians and some other animal species (D. D. Brown and I. B. Dawid, 1968). This remarkable case of *selective gene amplification* will be described more fully in Section 7.4.

Another strategy for providing the oocyte with large amounts of all types of RNA is seen in certain insects, including flies, bees, butterflies, and beetles. Only some of their germ line cells develop into oocytes while the others become ***nurse cells.*** Similar to male germ line cells, one oocyte and several nurse cells develop as a clone of cells connected by ***cytoplasmic bridges,*** which result from incomplete cytokinesis. Instead of forming gametes, the nurse cells assume a helper role for oocytes to which they are connected (Telfer, 1975). The nurse cells become ***polyploid*** by undergoing multiple rounds of chromosome replication without mitosis. The nurse cell nuclei in a *Drosophila* ovary, for instance, contain about 1000 copies of each chromosome, so that the nurse cells can rapidly synthesize large supplies of RNA for the oocyte. During early and middle stages of oogenesis, the

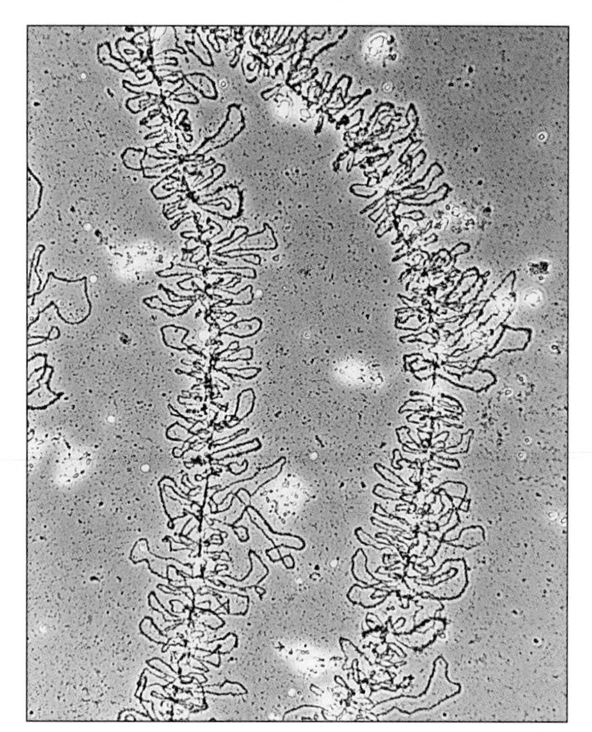

Figure 3.12 Lampbrush chromosomes in a newt (*Notophthalmus*) oocyte. This light micrograph shows loops of chromatin extending from a chromosomal axis. Closer analysis reveals that each chromosome consists of two chromatids with aligned sections that form the chromosome axis and symmetrical loops extending from the axis.

transfer of mRNAs from the nurse cells to the oocyte depends on a polarized system of microtubules that extends through the cytoplasmic bridges (Theurkauf, 1994). During late oogenesis, a microfilament-driven contraction of the nurse cells causes a massive stream of total cytoplasm into the oocyte (Mahajan-Miklos and Cooley, 1994).

The formation of the nurse cell–oocyte complex in *Drosophila* has been reconstructed from ovaries dissected at different stages of development (R. C. King, 1970). Each ovary consists of about 16 spindle-shaped tubes called ***ovarioles*** (Fig. 3.13). Oogonial stem cells located at the very tip of the ovariole produce committed progenitor cells that would normally be called oogonia. They are referred to as ***cystoblasts,*** however, because only some of their descendants form oocytes. Each cystoblast divides four times to produce a clone of 16 interconnected cells called ***cystocytes.*** Because of the successive order of incomplete divisions, only two of the cystocytes have four cytoplasmic bridges to other cystocytes (Fig. 3.14). These two cystocytes enter meiosis, but only one persists and becomes an oocyte; all of the remaining 15 cystocytes become polyploid nurse cells. The whole cluster of 16 germ line cells is surrounded by somatic cells that develop into ***follicle cells,*** which mediate the uptake of yolk proteins and synthesize eggshell proteins. The entire arrangement of one oocyte, the connected 15 nurse cells, and the surrounding follicle cells is called an ***egg chamber.*** Similar egg chambers are found in many groups of insects, including those orders that comprise the greatest numbers of species.

THE transfer of RNA from nurse cells to the oocyte in fly egg chambers was demonstrated by Bier (1963) using the technique of *autoradiography* (described in Method 3.1). To trace the movement of newly synthesized RNA, Bier injected female houseflies with a radiolabeled RNA precursor, [^{3}H]cytidine or [^{3}H]uridine. Because these nucleotides are turned over rapidly during cell metabolism, they are available for RNA synthesis for only a few hours. In contrast, most of the RNA synthesized during this period is much longer-lived. Thus, a few hours after injection, little radiolabel is left in the nucleotide pool, so that any RNA synthesized then will be labeled only weakly. In contrast, most of the RNA labeled shortly after injection will still be intact and strongly labeled. This type of labeling, called ***pulse labeling,*** allows investigators to monitor the fate of a cohort of long-lived macromolecules that were synthesized during a short labeling pulse (see also Method 17.2).

The flies were allowed to survive for different intervals after injection before their ovaries were removed and prepared for autoradiography. In ovaries removed 1 h after injection, much labeled RNA was located in the nuclei of nurse cells and follicle cells, indicating that these nuclei are the major sites of RNA synthesis and accumulation (Fig. 3.15). In ovaries removed 5 h after injection, only small amounts of labeled RNA were present in nuclei. Instead, the bulk of the label was present in the cytoplasm of nurse cells and follicle cells. Most intriguing was the presence of labeled material in the oocyte cytoplasm near the cytoplasmic connections with adjacent nurse cells. These observations show that large amounts of RNA are synthesized in the nurse cells and transferred into the oocyte cytoplasm.

Questions

1. In autoradiographs like the one shown in Figure 3.15a, the radiolabel over the oocyte nucleus is always spurious, even when the nucleus is clearly in the plane of section. What does this mean in terms of RNA synthesis in the oocyte?
2. The method of autoradiography registers the presence of the radioisotope used, regardless of the molecule in which it is incorporated. The radioisotope used in this experiment was tritium (^{3}H), which has the chemical properties of hydrogen and therefore enters the same metabolic pathways as hydrogen. The investigators therefore needed a way of testing whether the radioactivity they saw on their autoradiographs was present in RNA, rather than having been metabolized into other hydrogen-containing molecules. What do you think they may have done? (*Hint*: Their histological

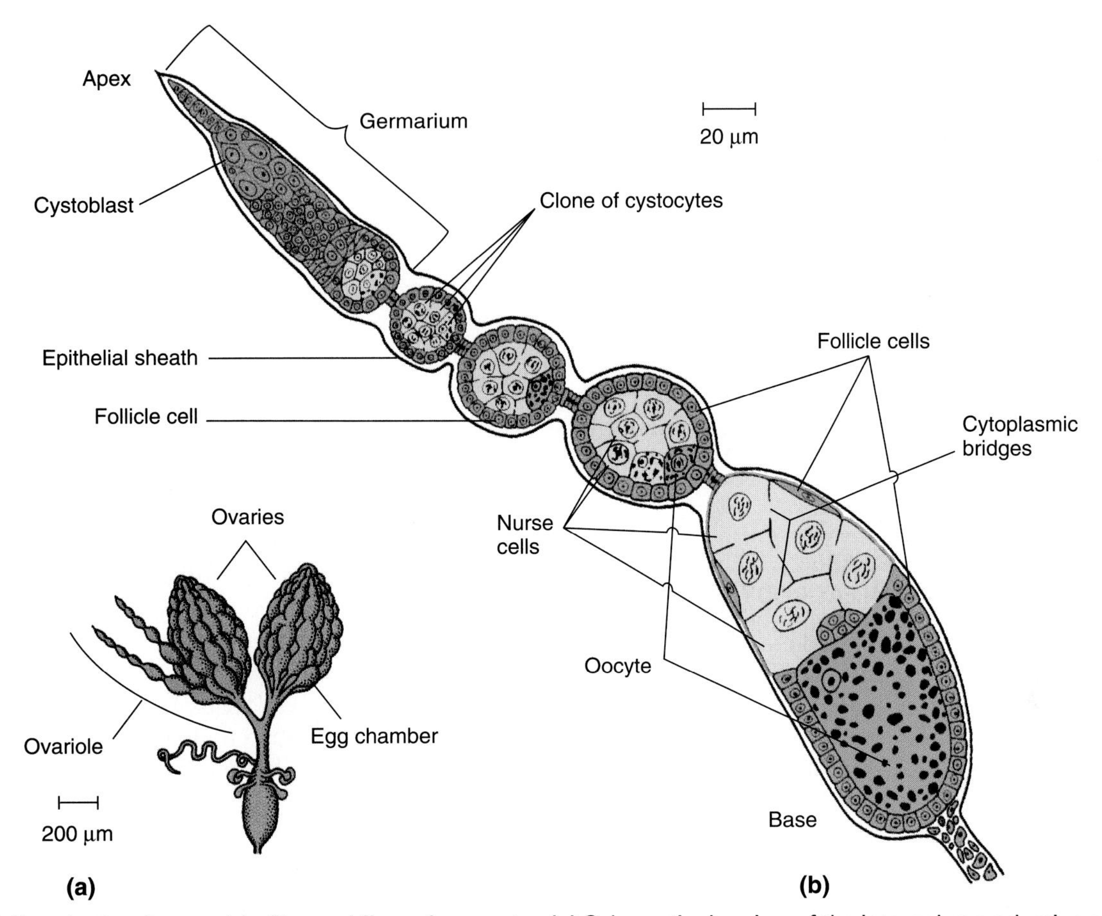

Figure 3.13 Oocyte development in *Drosophila melanogaster.* **(a)** Schematic drawing of the internal reproductive organs of an adult female. Each of the two ovaries consists of about 16 spindle-shaped ovarioles. **(b)** A single ovariole with several egg chambers. The most fully developed egg chamber, at the base of the ovariole, contains a large oocyte undergoing vitellogenesis, several large nurse cells, and an epithelium of follicle cells surrounding the nurse cell–oocyte complex. In the mid-range of the ovariole, the egg chambers are in earlier stages of development. The germarium near the apex of the ovariole contains germ cells called cystoblasts, which will give rise to oocytes and nurse cells, as well as somatic cells that will form follicle cells.

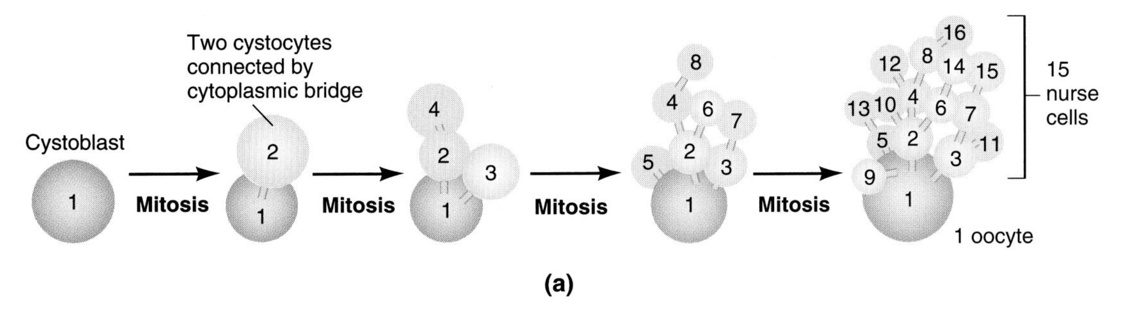

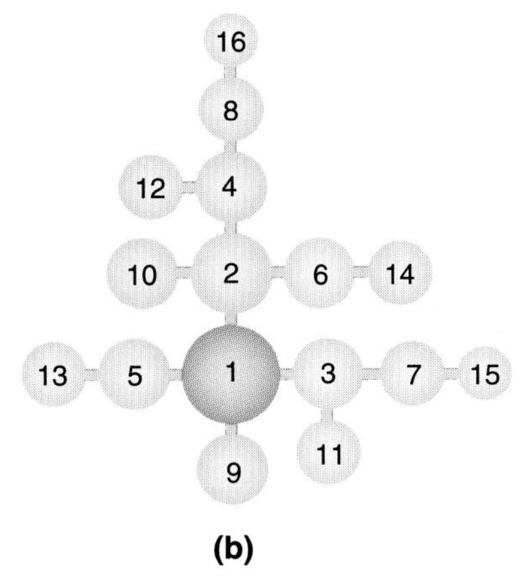

Figure 3.14 Development of the nurse cell–oocyte complex in the *Drosophila* ovary. The cystoblasts correspond to the oogonia of other species. **(a)** A cystoblast undergoes four mitotic divisions with incomplete cytokinesis, resulting in 16 interconnected cystocytes. One of two cystocytes with four cytoplasmic bridges becomes the oocyte. The other 15 cystocytes develop as nurse cells. Instead of undergoing meiosis, the nurse cells become highly polyploid and synthesize large quantities of macromolecules and cell organelles, which are transferred to the oocyte through the cytoplasmic bridges. **(b)** Spread-out view of the nurse cell–oocyte complex showing all cytoplasmic bridges.

sections were prepared in a way that leaves the tissue accessible to water-soluble agents.)

3. Do the autoradiographs show what type(s) of RNA are transferred from nurse cells to oocyte? If not, what additional experiment could the investigators do to find out? See Method 3.1.

The experiments of Bier and coworkers illustrate that insect ovaries with nurse cells are highly efficient in providing eggs quickly with large amounts of RNA. Indeed, a well-fed fly female during her period of maximum egg production will lay eggs equivalent to one-third of her body weight per day. This feat, however, requires the rapid synthesis and transfer not only of RNA but also of yolk proteins, as we will see next.

YOLK PROTEINS FOR OOCYTES ARE SYNTHESIZED IN THE LIVER OR FAT BODY

Animal eggs contain proteins, lipids, and glycogen to nourish the developing embryo. These materials, which are collectively called ***yolk,*** accumulate in the egg cytoplasm. Water-soluble yolk components and some insoluble ones are stored inside membranous vesicles; water-insoluble yolk components, such as lipids, also exist as free droplets. The amounts of yolk are minimal in mammalian eggs, which have to sustain embryogenesis only to a very immature stage. In contrast, the large eggs of birds, reptiles, and sharks have copious amounts of yolk to support embryonic development to advanced stages. In the course of embryogenesis, yolk is broken down into small molecules, such as amino acids, which are then used by embryonic cells in their synthetic processes. By the end of embryonic development, the yolk is used up, or its remnants are enclosed in the larval gut.

Like RNA, yolk accumulates in the oocyte during meiotic arrest and contributes to the oocyte's enormous growth. The process of yolk synthesis is generally similar to RNA synthesis in higher insects in that the oocyte has other cells working for it. In particular, this principle is observed in the synthesis of the major yolk proteins, called ***vitellins,*** which are formed from precursor proteins called ***vitellogenins*** (Lat. *vitellus,* "yolk"; Gr. *gennan,* "to produce"). In vertebrates vitellogenin synthesis occurs in the liver. In insects the major production site is the fat body (the insect equivalent of the liver), also in many insect species the follicle cells also contribute. Newly formed vitellogenins are released into the blood of vertebrates, or the *hemolymph* (blood equivalent) of invertebrates, from where they are taken into the oocyte by *receptor-mediated endocytosis* (see Sections 2.2 and 2.5). Once in the oocyte, the vitellogenins are modified to reduce their water solubility and, hence, increase their stability. Finally, they are packed as vitellins in a crystallike arrangement inside membranous yolk bodies. The entire process from vitellogenin synthesis to the sequestration of vitellins in the oocyte is termed ***vitellogenesis.*** Vitellogenesis has been studied in particular detail in insects and amphibians.

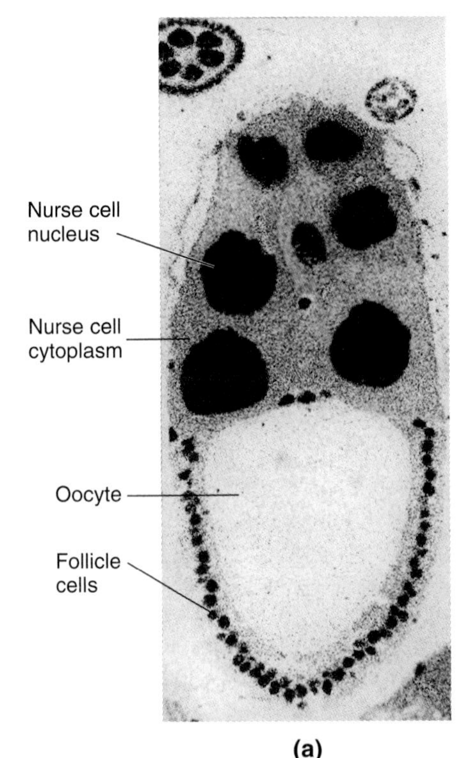

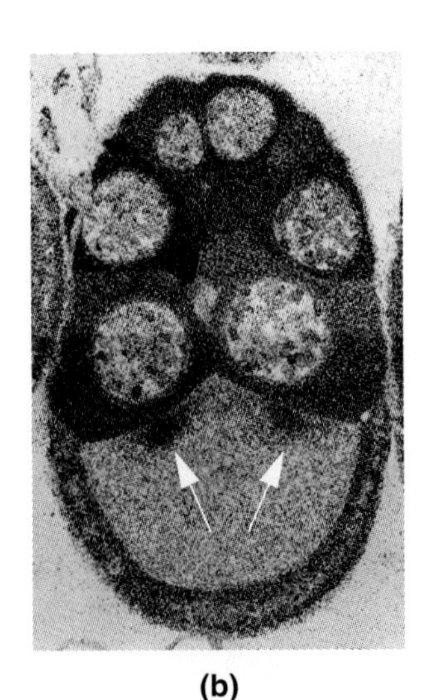

Figure 3.15 RNA synthesis in ovarian egg chambers of the housefly, *Musca domestica.* Newly synthesized RNA has been radiolabeled by injecting [^{3}H]cytidine into the abdomen. **(a)** Autoradiograph of an egg chamber fixed 1 h after the labeled ribonucleotide was injected. The nuclei of nurse cells and follicle cells are heavily labeled, indicating that they have synthesized large amounts of RNA. Some label is also seen in the cytoplasm of these cells. The oocyte is virtually free of label. **(b)** Somewhat younger egg chamber fixed 5 h after injection. The nurse cell nuclei are only weakly labeled, indicating that most of the radioactive cytidine has been used up. The labeled RNA synthesized previously has been transported to the nurse cell cytoplasm and passes into the oocyte through cytoplasmic bridges (arrows).

ONE of the milestone experiments in the study of insect vitellogenesis was done by Telfer (1965). When he analyzed the proteins in hemolymph of male and female moths, he discovered a female-specific protein. He also found that the concentration of this protein decreased in the hemolymph when yolk formation began in the ovaries, suggesting that the protein may be destroyed or sequestered during vitellogenesis. Since fluid collected from eggs contained the protein in concentrations 20 times higher than that in the hemolymph of females, Telfer devised a test to show whether the protein was taken up from the hemolymph or synthesized within the eggs. Hemolymph was transfused from one moth species, *Hyalophora cecropia*, to another, *Antherea polyphemus*. Following transfusion, female-specific protein from *Hyalophora* accumulated in the recipient's oocytes until its concentration was 20 times greater in the eggs than in the hemolymph. Other proteins that were transfused in the same manner did not accumulate in the recipient's eggs. Telfer's data demonstrated that certain female-specific hemolymph proteins are vitellogenins—proteins that are selectively taken up and stored by oocytes. Later experiments showed that the selectivity of the uptake depends on vitellogenin-specific receptors in the egg plasma membrane (Hagedorn and Kunkel, 1979).

Questions

1. The process of receptor-mediated endocytosis was not well understood when Telfer did the experiment described above. Which of his observation were strongly indicative of the involvement of a specific receptor in the uptake of the female-specific protein into oocytes?
2. Working to Telfer's advantage was the fact that vitellogenin from *Hyalophora cecropia* matches the vitellogenin receptor of *Antherea polyphemus* well enough to be taken up into the oocyte. If this had not been the case, what alternate experimental design could Telfer have used?
3. Given the advantages of synthesizing yolk proteins in a large organ outside the oocyte, what may be the adaptive value of using a *specific* family of storage proteins as opposed to withdrawing a fraction of *all* hemolymph proteins into oocytes?

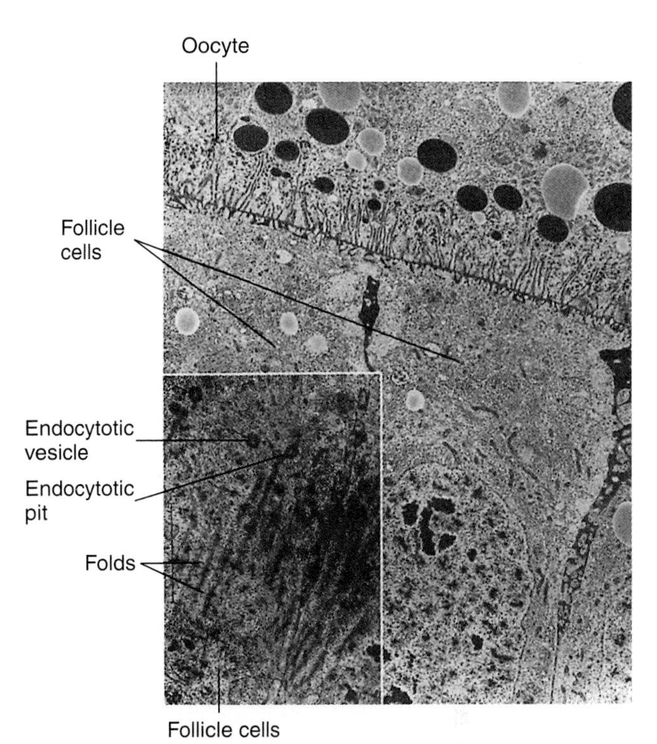

Figure 3.16 Vitellogenesis in a moth egg chamber. The oocyte surface is enlarged by deep folds with spaces between them containing electron-dense material (vitellogenin). The same material is also present between follicle cells. Endocytotic pits form mostly at the bases of these spaces (labeled in inset). These pits are pinched off as endocytotic vesicles, which are processed into membraneous yolk particles containing vitellins.

The amount of vitellins taken up by the oocyte during endocytosis is very impressive. This process is aided by folds and fingerlike extensions of the plasma membrane called *microvilli*, which increase the surface area available for endocytosis severalfold (Fig. 3.16). Even so, endocytosis would use up the entire plasma membrane of a *Hyalophora* oocyte in less than a minute if the membrane were not recycled (Telfer et al., 1982). In addition, the follicle cells facilitate the uptake of vitellogenin into the oocyte. Whereas most of the time the follicle cells form a tightly sealed epithelium, passageways open up between them during vitellogenesis. The follicle cells themselves create these passageways by secreting extracellular material that keeps their plasma membranes apart and allows hemolymph to pass between the follicle cells toward the oocyte plasma membrane (see Fig. 3.16).

The hormonal control of vitellogenesis is very similar in frogs and insects (Fig. 3.17). In both groups of animals, reproduction tends to be seasonal, regulated by environmental cues that are processed by the brain. In amphibians, the brain stimulates the hypothalamus to secrete *gonadotropin-releasing hormones.* They cause the pituitary gland to release *gonadotropins.* These hormones in turn stimulate the ovarian follicle cells to produce estrogen, which causes the liver to synthesize vitellogenins. In many insects the hormonal control of vitellogenesis follows a similar pattern. The *corpora allata,* a pair of endocrine glands attached to the insect brain, are analogous in function to the pituitary gland and release *juvenile hormone.* In dipterans, the juvenile hormone stimulates the ovarian follicle cells to produce a steroid hormone, *ecdysone.* This hormone, like estrogen in amphibians, stimulates both the fat body and the follicle cells themselves to produce vitellogenin. In other insects, juvenile hormone stimulates the fat body directly to synthesize vitellogenin.

In insects that feed on the blood of vertebrates, vitellogenesis is linked with their feeding. Some of these

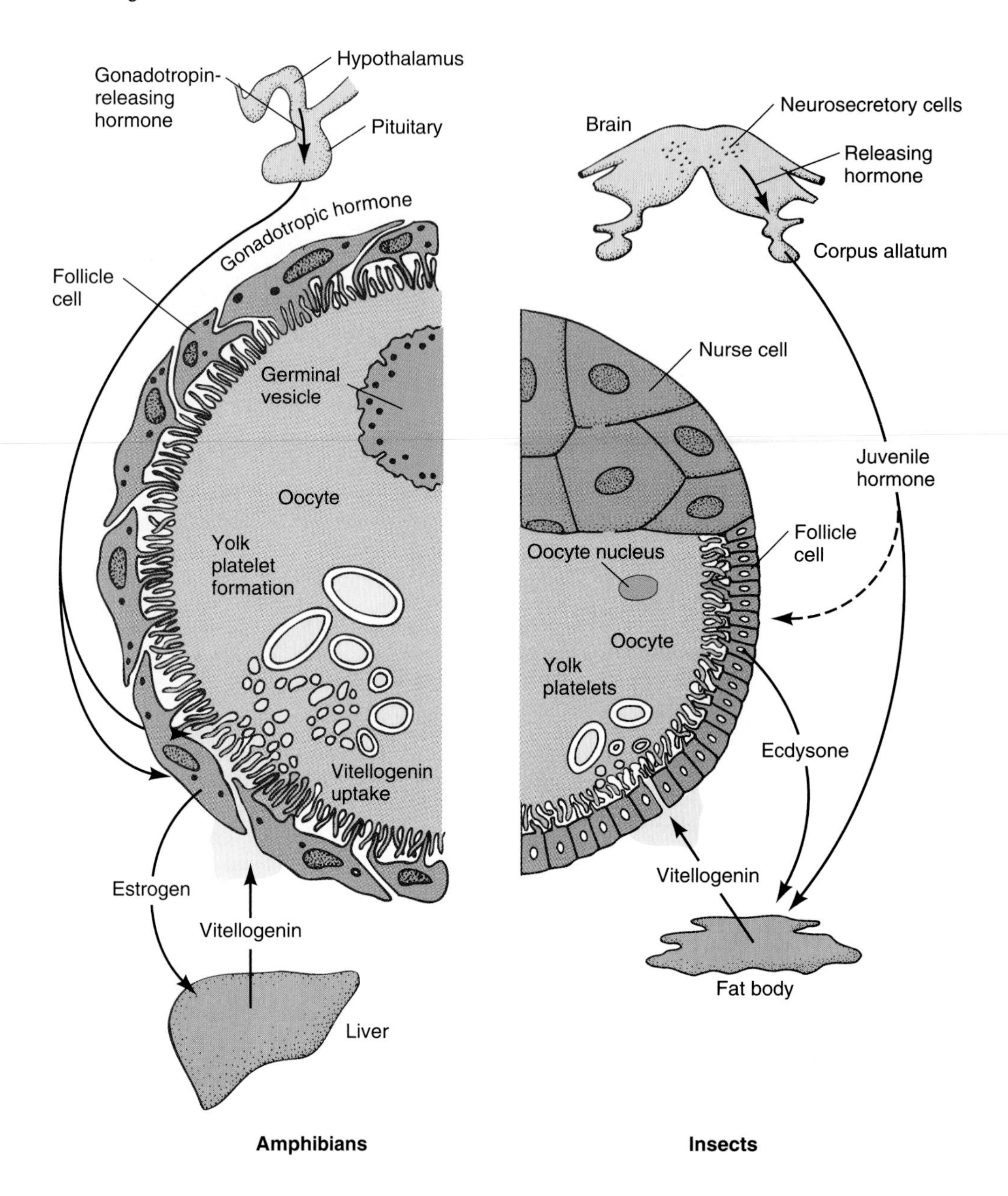

Figure 3.17 Comparison of the hormonal control of vitellogenesis in amphibians and insects. In both groups of animals, a part of the brain (hypothalamus or neurosecretory cells) stimulates a gland attached to the brain (pituitary gland or corpus allatum). In amphibians, the pituitary releases gonadotropic hormone, which stimulates the ovarian follicle cells to secrete estrogen, which in turn causes the liver to synthesize vitellogenin. In insects, the corpora allata release juvenile hormone (JH). In some species, JH stimulates the follicle cells to secrete ecdysone, which in turn causes the fat body to synthesize vitellogenin. In other species, JH acts directly on the fat body.

species, including bedbugs and lice, were actually the first insects in which vitellogenesis was studied. It was found that soon after a blood meal, breakdown products of the ingested blood appear in the insects' hemolymph and then in their oocytes (Wigglesworth, 1943).

THE OVARIAN ANATOMY IMPOSES POLARITY ON THE OOCYTE

Providing the oocyte with RNA and protein is a major task, but there is more to oogenesis. The oocytes of most animals also receive signals from surrounding cells in

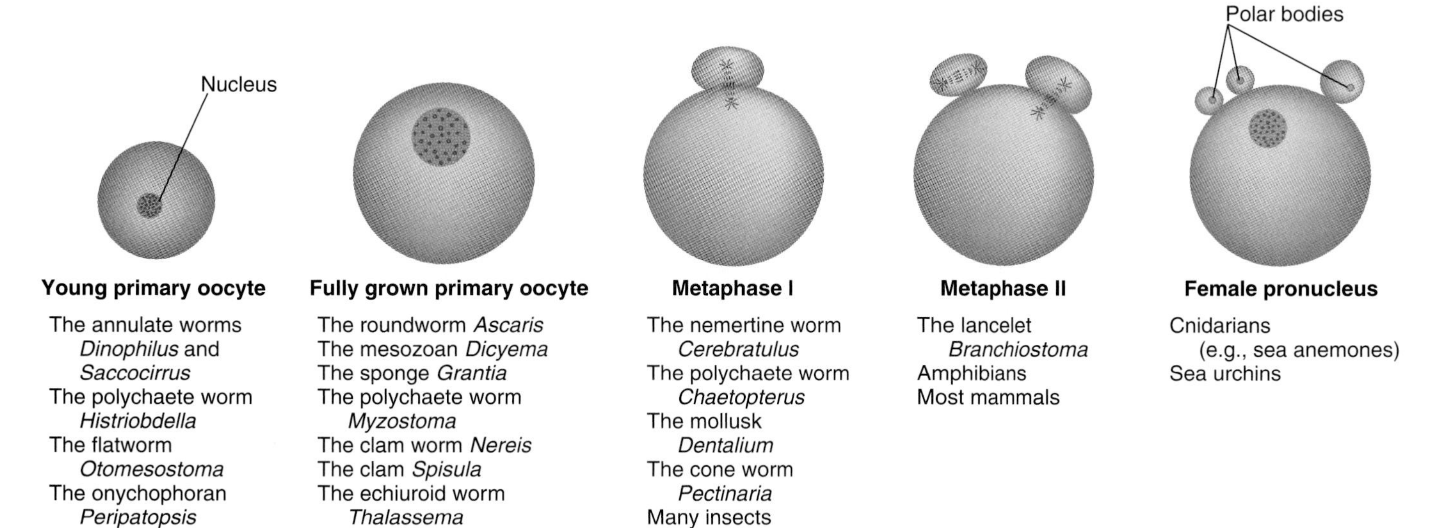

Figure 3.18 Different stages of egg maturation at the time of sperm entry in different animals.

the ovary that cause the cytoplasm to be asymmetrically distributed. In many species, the oocyte acquires an ***animal-vegetal polarity.*** The *animal pole,* defined by its proximity to the asymmetrically located oocyte nucleus, is marked by the extrusion of the polar bodies. The *vegetal pole* is located opposite to the animal pole and is often the site of greatest yolk accumulation. In other organisms, oocytes acquire a polarity that foretells the future anteroposterior axis of the embryo.

In the more highly evolved insects, the oocyte is associated with a cluster of nurse cells, which are invariably located on one side of the oocyte, facing the *germarium* at the tip of the ovariole (see Fig. 3.13). This side becomes the *anterior* pole of the egg, where the head of the embryo will be formed. The chain of events leading from the arrangement of the nurse cell–oocyte complex in the egg chamber to the anteroposterior body pattern of the embryo has been elucidated in part by genetic analysis. Certain *Drosophila* mutants with nurse cells on both ends of the oocyte form embryos with two heads! Mutations that interfere with the system of polarized microtubules in oocytes and nurse cells, and thus disturb the oriented transport of certain mRNAs, also cause the formation of embryos with abnormal anteroposterior body patterns (see Sections 8.4 and 22.3).

MATURATION PROCESSES PREPARE THE OOCYTE FOR OVULATION AND FERTILIZATION

A fully grown oocyte contains the maternal genetic information and all the cytoplasmic components necessary to begin the formation of an embryo. Yet the oocytes of most animals at this stage are not ready for fertilization. The chromatin has to be condensed before meiosis can advance. The permeability of the plasma membrane to small molecules and ions must be adjusted so that the cell can leave the ovary and function in other environments. The plasma membrane of the oocyte has to be provided with receptors for contact with sperm.

The processes that prepare the grown oocyte for fertilization are collectively called ***oocyte maturation.*** In most animals, oocyte maturation begins with the end of meiotic arrest in prophase I. Early maturation events include chromosome condensation and the disintegration of the nuclear envelope, a process known as ***germinal vesicle breakdown.*** Maturation also entails *ovulation,* the release of the oocyte from the ovary. After this event, the female gamete is called an *egg.*

Eggs of different animal species are ovulated and fertilized at different stages of meiosis (Fig. 3.18). At one extreme, eggs are ovulated and fertilized while the *germinal vesicle* is still intact. This pattern is shown by various worms, including *Ascaris* (a roundworm) and *Nereis* (a polychaete worm). At the other extreme, the eggs of some animals have completed both meiotic divisions by the time of fertilization. This pattern is shown by sea urchins and coelenterates. The eggs of most animals fall somewhere between these extremes. We will focus on the pattern seen in most vertebrates, including humans. In these species, maturation begins with the release of the oocyte from its arrest at prophase I and ends with a second arrest, at metaphase II.

The control of oocyte maturation has been studied most extensively in frogs, in particular, *Rana pipiens* and *Xenopus laevis* (Ferrel, 1999). Like vitellogenesis, oocyte maturation is controlled by hormonal interactions of hypothalamus, pituitary gland, and follicle cells (Fig. 3.19; Wasserman and Smith, 1978). Stimulated by releasing hormones from the hypothalamus, the pituitary gland produces gonadotropic hormones, which in turn prompt the follicle cells to secrete *progesterone.* Progesterone

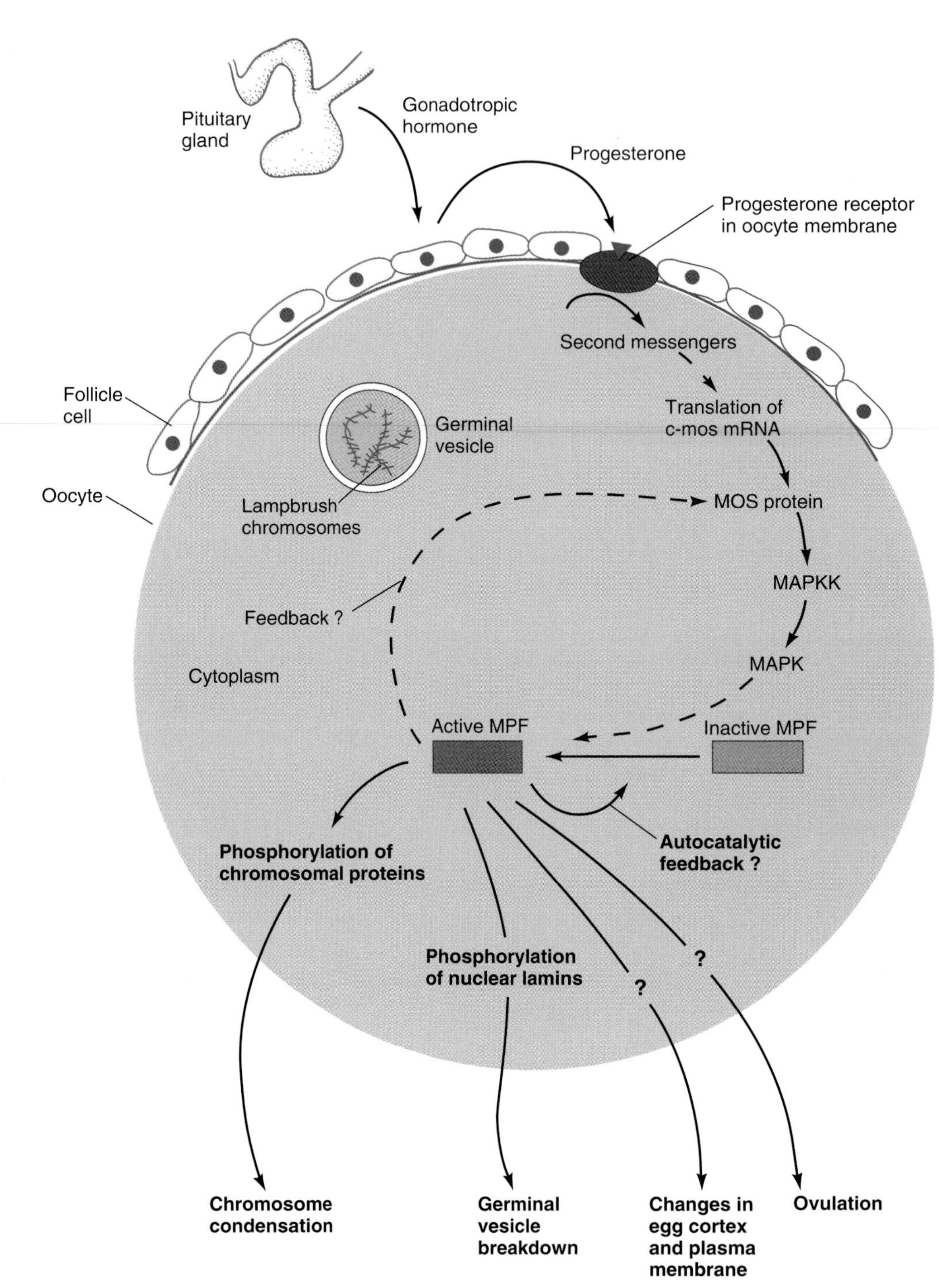

Figure 3.19 A model of oocyte maturation in amphibians. The binding of progesterone to the oocyte plasma membrane triggers second messenger events, which cause the translation of c-mos mRNA stored in the oocyte cytoplasm. The mos protein forms a positive feedback loop with M-phase promoting factor (MPF) through MAPKK and MAPK kinases (see Fig. 2.28). Through its protein kinase activity, MPF causes the phosphorylation of several other proteins, which trigger multiple maturation processes including germinal vesicle breakdown, chromosome condensation, cortical changes, and ovulation. Broken lines indicate indirect effects.

binds to a membrane receptor protein on the oocyte surface (Blondeau and Baulieu, 1984). Thus, unlike common steroid receptors, this receptor does not reside in the cytoplasm or nucleus and does not seem to act as a transcription factor. Oocyte maturation begins within a few hours after progesterone release.

To analyze the effect of progesterone on oocyte maturation, Masui and Markert (1971) transplanted cytoplasm from progesterone-treated oocytes at various stages of maturation into immature oocytes. The researchers found that the cytoplasm from the maturing oocytes induced maturation in the immature oocytes as well. They concluded that the progesterone-treated oocytes contained a *maturation promoting factor.* Subsequent work showed that this factor is identical to *M-phase promoting factor* (*MPF*), which plays a central role

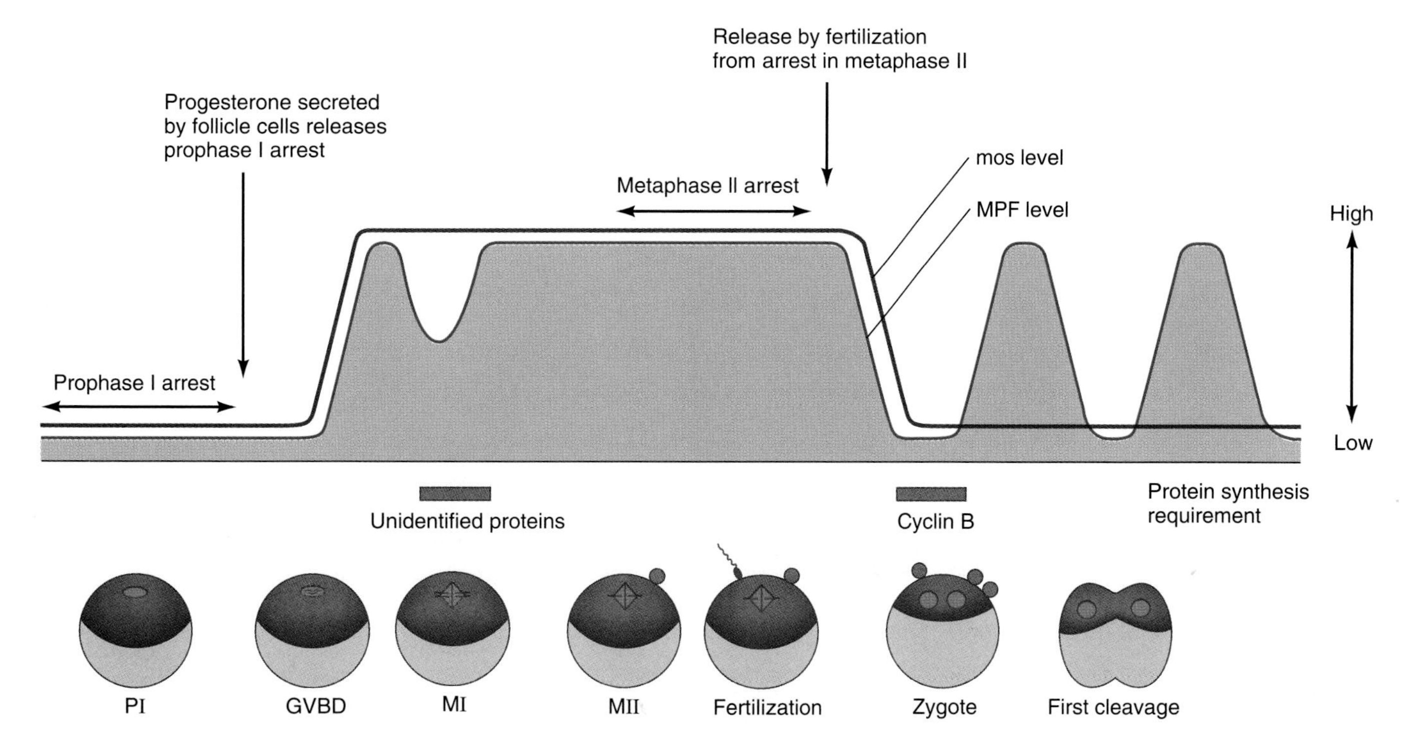

Figure 3.20 Control of *Xenopus* oocyte maturation by mos protein and M-phase promoting factor (MPF). Successive stages of development are shown along the abscissa. GVBD = germinal vesicle breakdown. The black curve represents the level of mos protein, and the colored area indicates the level of MPF. The two vertical arrows mark the stages when the oocyte's progress through the cell cycle depends on external signals.

in controlling the mitotic cycle of somatic cells (see Fig. 2.16). Thus, the release of the oocyte from meiotic arrest in prophase I is equivalent to the transition from G_2 phase to mitosis in a somatic cell.

How does progesterone promote MPF activity at the beginning of oocyte maturation? In fully grown oocytes, the binding of the hormone to its membrane receptor triggers a second messenger pathway involving a decrease in cAMP concentration (Taylor and Smith, 1987; L. D. Smith, 1989). This pathway initiates the translation of several maternal mRNAs. At least one of the resulting proteins is necessary for oocyte maturation because a translation inhibitor, *cycloheximide,* blocks germinal vesicle breakdown if administered right after progesterone treatment.

The key protein required for oocyte maturation in *Xenopus* is the *mos protein;* it is encoded by a gene that causes cancer if deregulated and is therefore known as the c-mos *proto-oncogene* (see Section 28.7). The mos mRNA is present throughout oogenesis and early embryogenesis, but the mos protein is detectable only during oocyte maturation and is destroyed upon fertilization (Fig. 3.20). However brief in its presence, the mos protein is both necessary and sufficient for releasing the oocyte from its dormancy (Gebauer and Richter, 1997; Sagata, 1997). No germinal vesicle breakdown or other signs of oocyte maturation are observed if immature oocytes are injected with *antisense oligodeoxynucleotides* that hybridize to mos mRNA, thus rendering it untranslatable and vulnerable to enzymatic breakdown. Conversely, precocious injection of mos protein activates MPF and triggers oocyte maturation. Thus, mos protein is not only *necessary* but also *sufficient* for completion of first meiosis.

How mos protein promotes MPF activity is still being investigated, but mos is a *protein kinase,* capable of phosphorylating itself as well as the MAPKK protein kinase (Resing et al., 1995). The MAPKK and MAPK kinases (see Fig. 2.28) seem to be involved in a feedback loop that stabilizes mos protein and activates MPF (see Fig. 3.19).

Once a certain amount of MPF activity is present, a burst of phosphorylation activity occurs, and oocyte maturation becomes independent of protein synthesis. MPF and other protein kinases act on a range of target proteins, bringing about the multiple cellular changes that characterize oocyte maturation (see Fig. 3.19). Most prominent among these changes is germinal vesicle breakdown, possibly caused by phosphorylation of nuclear *lamins,* and chromosome condensation caused by phosphorylation of chromosomal histone proteins. Other target proteins of MPF are thought to trigger ovulation and the changes in egg plasma membrane and cortical cytoplasm that occur during maturation.

A second function of mos protein is to inhibit DNA replication by prematurely reactivating MPF between the two meiotic divisions. As shown in Figure 3.20, MPF decreases somewhat at metaphase I but then increases again during anaphase I and remains high thereafter. This is in contrast to the changes in MPF that occur during the mitotic cell cycle, when MPF activity decreases after metaphase and remains low until the onset of the next mitosis. The premature reactivation of MPF at the end of meiosis I, which is necessary to inhibit DNA replication, requires mos and other, as yet unidentified, proteins (Furuno et al., 1994).

After its release from arrest at prophase I, the oocyte proceeds with meiosis through metaphase II, when it becomes arrested again. This second arrest is caused by a ***cytostatic factor* (*CSF*)** that was discovered, like MPF, by Masui and Markert (1971) in cytoplasmic transplantation experiments. The researchers injected cytoplasm from unfertilized frog eggs into blastomeres of cleaving embryos and found that the injected blastomeres were arrested in their next metaphase. No such arrest occurred when cytoplasm from fertilized eggs was transplanted. These experiments were extended by Sagata et al. (1989), who found that cleavage arrest by cytoplasm from unfertilized eggs did not occur if the cytoplasm was first depleted of mos protein. Thus, these investigators had shown that mos protein was at least a component of CSF. Additional work revealed that *continued high MPF activity* keeps cells arrested in metaphase II. In order for a cell in either mitosis or meiosis to proceed beyond metaphase, MPF activity must drop to a low level. These results raise the question why mos protein does not stop maturation at metaphase I. One possible answer is that mos protein needs to interact with other, as yet unidentified, proteins synthesized at the end of meiosis I to generate active CSF (Minshull, 1993).

The arrest of the egg in metaphase II is normally broken by fertilization, which is followed by a rapid decrease of both MPF and mos protein (Fig. 3.20). One of the responses to the entry of a sperm is a rise in the concentration of free calcium ions (Ca^{2+}) in the egg's cytoplasm (see Section 4.4). Ca^{2+}-dependent protein activities may cause the destruction of both MPF and the c-mos protein, thus allowing the fertilized egg to exit metaphase II and to begin the mitotic cycles of cleavage.

In the mammalian ovary, the oocytes are closely associated with somatic cells, in structures known as ***follicles.*** An immature follicle consists of one oocyte surrounded by several layers of small somatic cells called ***granulosa cells*** (Fig. 3.21). The oocyte synthesizes

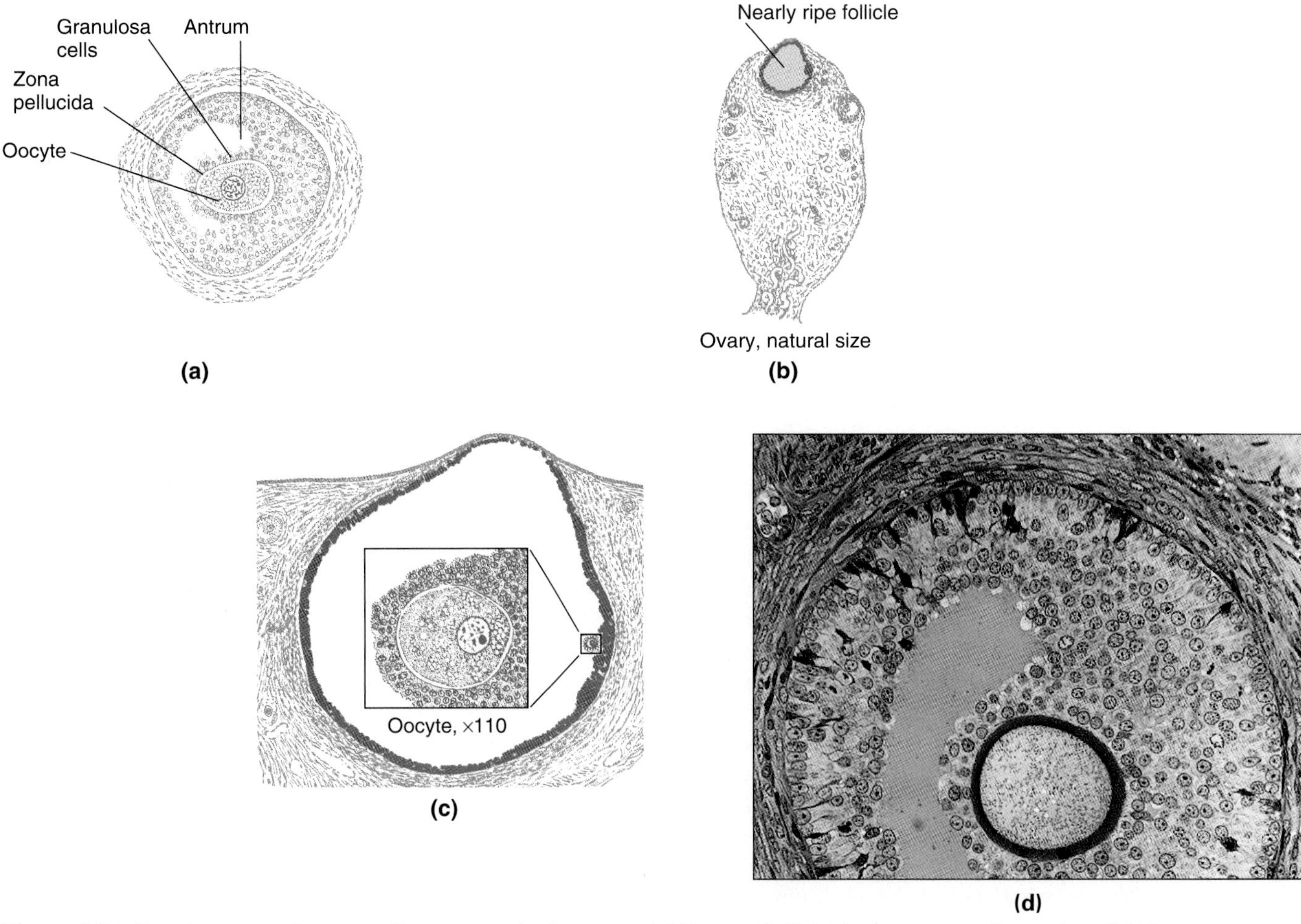

Figure 3.21 Development of mammalian oocytes in the ovary. **(a)** Human follicle during antrum formation. **(b)** Human ovary shown in actual size. **(c)** Mature human follicle; inset shows oocyte. **(d)** Photomicrograph of a follicle from a monkey ovary during the beginning of antrum formation. The oocyte is surrounded by the zona pellucida (dark) and by granulosa (follicle) cells.

glycoproteins that are deposited around it as a translucent layer known as the ***zona pellucida*** (Lat. *zona*, "girdle"; *pellucida*, "translucent"). The zona is traversed by microvilli that form gap junctions between the oocyte and the granulosa cells (Fig. 3.22). A mature follicle is often called a *Graafian* follicle, after its discoverer, Regnier de Graaf. It contains a fluid-filled space known as the *antrum*, which is large compared with the oocyte. The antrum is located near the periphery of the ovary. Toward the end of the maturation process, the follicular antrum bursts open and the oocyte, still surrounded by granulosa cells, is released.

The timing of oocyte maturation and ovulation in mammals varies. In some species, ovulation is prompted by seasonal cues. In others, such as mink and rabbits, ovulation is triggered by the act of mating. In primates, including humans, ovulation comes about as part of a monthly reproductive cycle. The underlying control mechanisms are based on signal exchanges between the oocyte, the surrounding granulosa cells, and the pituitary gland—much like the maturation of sperm is controlled by interactions between themselves, Sertoli cells, and the pituitary. Hormones controlling mammalian oocyte maturation include gonadotropin-releasing hormones, produced by the hypothalamus; follicle-stimulating hormone and luteinizing hormone, produced by the pituitary gland; and estrogen, produced by the granulosa cells. Also involved are cAMP and possibly other small molecules exchanged through the gap junctions between oocyte and surrounding granulosa cells. These interactions are regulated in such a way that only one or a few oocytes mature at a time.

In controlling maturation of the mammalian oocyte, mos protein is again a major player. Oocytes from female mice without functional *c-mos* genes initiate maturation and proceed through meiosis I normally except for some delay. However, they are not arrested in metaphase II and instead proceed into cleavage without being fertilized, indicating that the main function of mos in mammals is to act as CSF (Colledge et al., 1994; Hashimoto et al., 1994).

EGGS ARE PROTECTED BY ELABORATE ENVELOPES

The egg and the early embryo are vulnerable stages in the life cycle of many animals. A predator could hardly find a more nutritious meal. Desiccation and mechanical stress are also hazardous to eggs deposited in the open air. For protection, eggs have different types of envelopes. The plasma membrane of the egg cell is covered by a glycoprotein layer, called the *zona pellucida* in mammals and generally referred to as the ***vitelline envelope***, which plays an important role in fertilization (see Section 4.2 and 4.3). Most eggs have additional protective coats, some of which also serve as nutrients for the embryos or hatchlings. The egg coats are produced either before ovulation by the oocyte or follicle cells, or after ovulation by other cells. A jellylike coat surrounds many eggs spawned in water, such as sea urchin or frog eggs. In contrast, many types of eggs deposited on land,

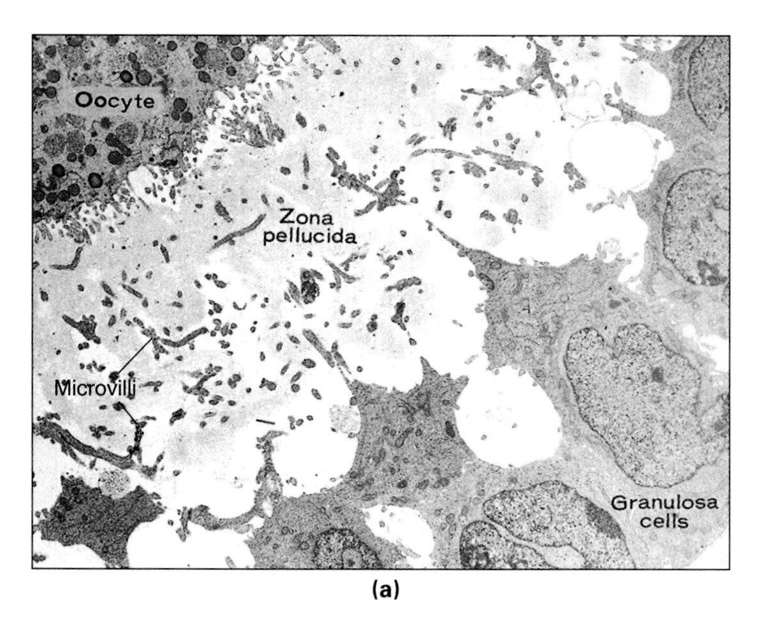

(a)

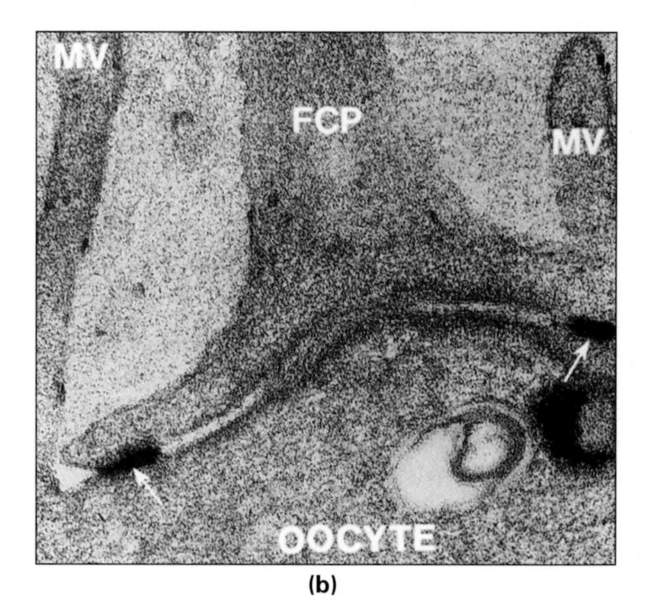

(b)

Figure 3.22 Mammalian oocytes and granulosa (follicle) cells. **(a)** Transmission electron micrograph showing part of an ovulated oocyte (upper left) and some follicle cells, now called granulosa cells (lower right), with the zona pellucida in between. Microvilli extend from both the oocyte and the granulosa cells into the amorphous substance of the zona pellucida. The microvilli look disconnected because they run into and out of the plane of section, but they actually form connections between the oocyte and the granulosa cells. **(b)** A granulosa cell at its point of contact with the oocyte. The dense patches (arrows) are gap junctions stained with lanthanum. FCP = follicle cell process; MV = oocyte microvillus.

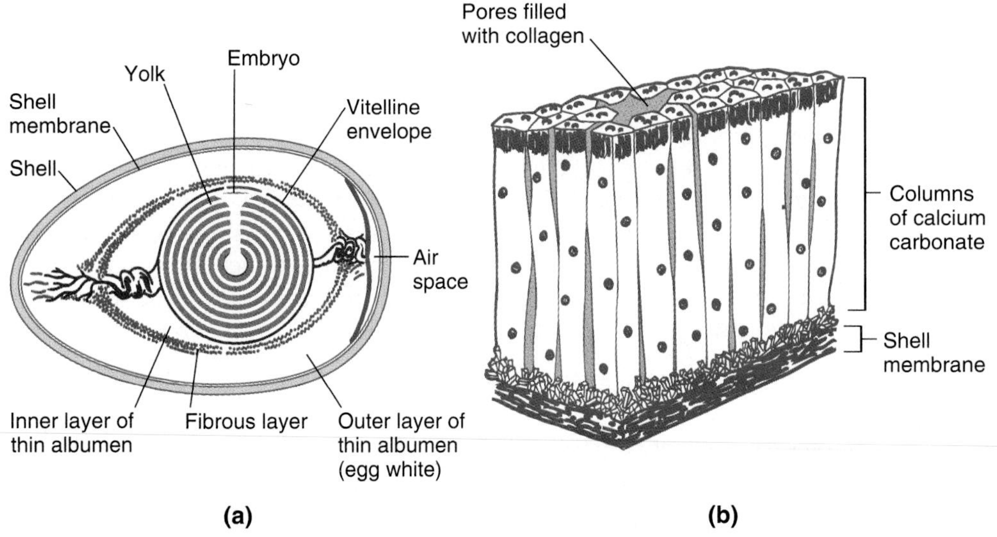

Figure 3.23 Chicken egg coats. **(a)** Drawing of a longitudinal section of a chicken egg. All egg coats outside the vitelline envelope are laid down after ovulation. **(b)** Drawing of a small segment of a bird's eggshell. The outermost shell consists mainly of calcium carbonate cylinders interspersed with collagen-filled pores to facilitate gas exchange.

such as those of reptiles and birds, have hard shells. For instance, the yolky chicken egg is surrounded initially by a fragile vitelline envelope (Fig. 3.23). As the egg moves down the oviduct, it is invested with several layers of "egg white," consisting of ovalbumin and related proteins. Then two layers of matted keratin fibers, called the *shell membranes,* are added. (These are the membranes found inside the hard outer shell of a boiled egg.) The outermost shell, which is added after fertilization, consists mostly of calcium carbonate with collagen-filled pores, an arrangement that regulates both water retention and respiration.

A very elaborate type of egg envelope, called a ***chorion,*** surrounds the insect egg. Insect species that lay their eggs in water tend to produce thin, transparent chorions including an area that functions as a physical gill, or plastron, repelling water and maintaining an air-filled space. Insect species that deposit eggs in the open air make thick chorions, which must allow respiration and at the same time limit the loss of water. As an apparent adaptation to these conflicting demands, most chorions provide an inner air space that is prevented from collapsing by chorionic columns (Fig. 3.24). In some species, the inner air space is vented to the outside by channels called *aeropyles* (Gk. *pyle,* "gate"), which can be short ducts or long, chimneylike structures (Hinton, 1970). An opening in the chorion, known as the *micropyle,* allows the sperm to enter the egg at the anterior pole.

The complex architecture that allows this compromise is based on a sophisticated program of protein synthesis. For instance, silk moth chorions are made up of more than 50 different proteins (Kafatos et al., 1977). These proteins are synthesized and laid down in a carefully orchestrated process. If released at the appropriate time, the proteins diffuse to their destinations. Subsequently, covalent bonds form between different proteins to anchor them in their final patterns.

Vitellogenesis and chorion formation are remarkable processes, in terms of both quantity and complexity. In insects, both activities are carried out by the ovarian follicle cells. These cells switch rapidly from helping with the synthesis of bulk vitellogenins to making large amounts of different chorion proteins. The genes for chorion proteins are selectively amplified in ovarian follicle cells after they have completed vitellogenesis but before they begin chorion synthesis (see Section 7.4). In this respect, chorion genes in insect follicle cells behave like the ribosomal RNA genes in amphibian oocytes.

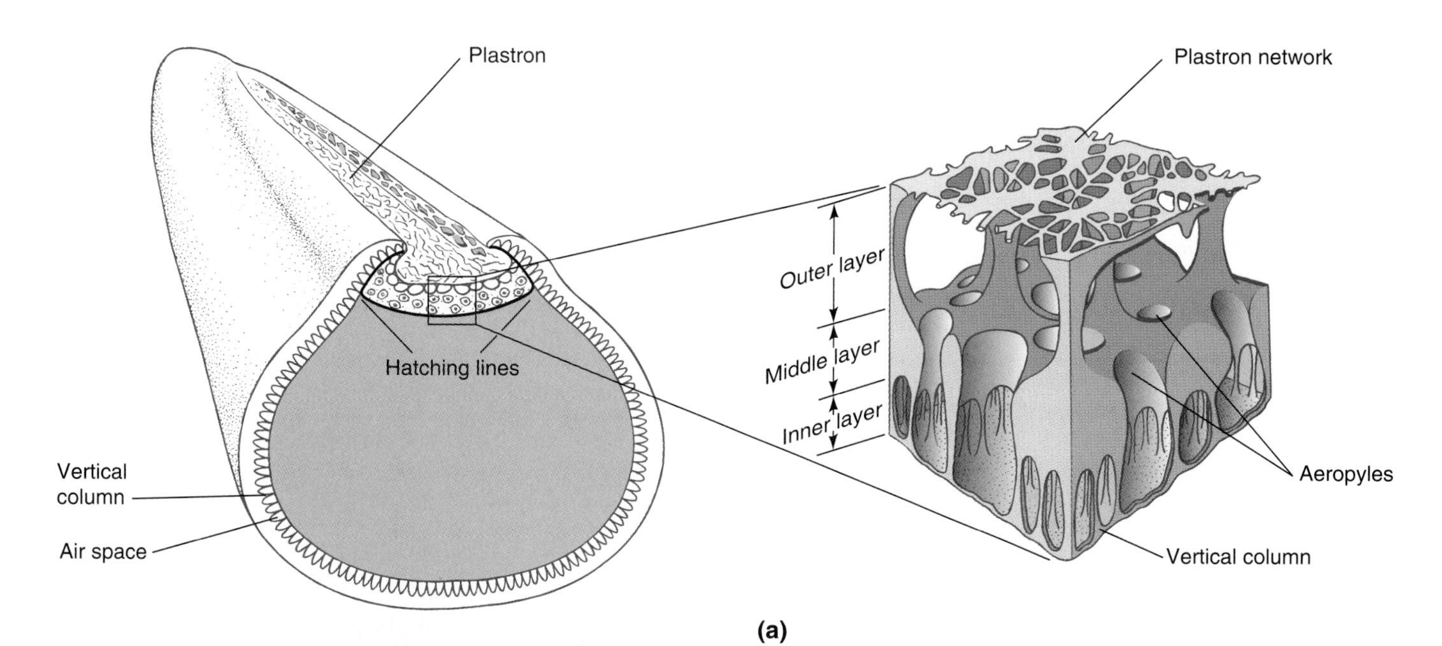

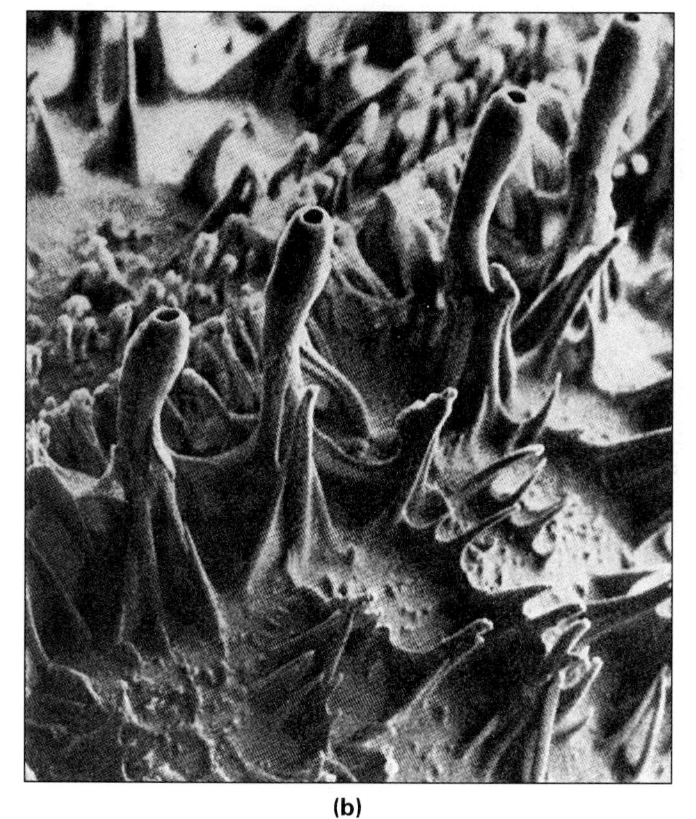

Figure 3.24 Insect egg chorion. **(a)** Drawing of the chorion of the blowfly *Calliphora erythrocephala.* The plastron (physical gill) structure is formed along a median strip lying between two hatching lines. A section of the plastron is enlarged to show an inner layer of thin, vertical columns, an outer layer with larger columns and spaces, a layer of chimneylike connections in between, and a superficial network on the outside. **(b)** Scanning electron micrograph of the chorion of a bug (*Piezodorus lituratus*) showing tall aeropyles that rise from its surface like chimneys.

SUMMARY

Animals of most species have a germ line, defined as the lineage of cells that gives rise to gametes—that is, eggs or sperm. Primordial germ cells segregate from somatic cells early in development and often migrate to the mesodermal gonad rudiments during embryogenesis. In the gonads, the primordial germ cells give rise to gametes, through a process known as gametogenesis. It entails meiosis—that is, two cell divisions during which homologous chromosomes separate and the dividing diploid cells give rise to haploid daughter cells.

In males, primordial germ cells develop into spermatogonia, some of which act as stem cells throughout life. At the stages in which they undergo meiotic division, the male germ cells are called primary and secondary spermatocytes. The haploid cells resulting from meiosis are called spermatids. Each primary spermatocyte ultimately gives rise to four spermatids. In most animal species, spermatids undergo a dramatic change in morphology called spermiogenesis, during which the nucleus becomes highly condensed while an acrosome is formed at the sperm's tip and a flagellum at the other end. After this transition, the male germ cells are called spermatozoa or sperm. In mammals, the mitotic divisions of committed spermatogonia and the ensuing meiotic divisions are followed by incomplete cytokinesis, so that spermatids form groups of interconnected cells. As these spermatids mature into spermatozoa, they pinch off from a clonal residual body of cytoplasm.

In females, primordial germ cells develop into oogonia. During the meiotic divisions, female germ cells are known as primary and secondary oocytes. The meiotic divisions begin in the ovaries at an early stage—in mammals, before birth. Primary oocytes are arrested for an extended period of time in meiotic prophase I. During this phase, they accumulate large amounts of RNA, functional proteins, and storage proteins. Other cells often help to synthesize these materials. Most oocytes also acquire one or two axes of polarity during this time. The meiotic divisions in female germ cells are characterized by unequal cytokinesis, generating one large egg cell and three tiny sister cells that do not develop further. The final phase of oocyte development is called maturation. It is initiated by gonadotropic hormones that act on somatic cells in the ovary, which then trigger a cascade of events in the oocyte that allow it to complete meiosis. Oocyte maturation also includes release from the ovary, an event called ovulation.

While typical sperm are tiny cells specialized to deliver their nucleus to an egg, eggs provide not only genetic information but also the necessary RNA and proteins as well as structural information for the development of the embryo. Thus, most eggs are very large cells surrounded by various envelopes that protect them against mechanical stress and loss of water.

Fertilization

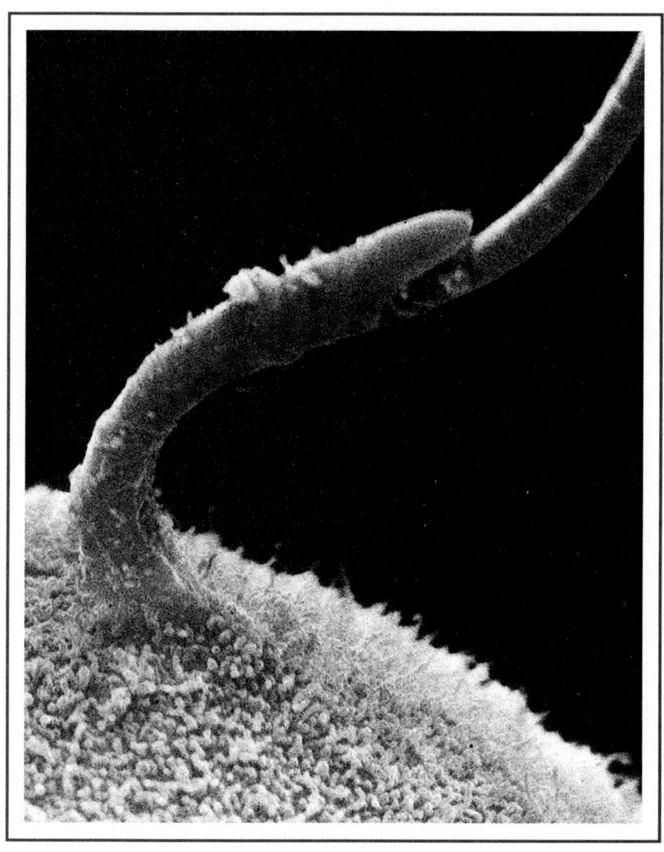

Figure 4.1 Fertilization in the rat. This scanning electron micrograph shows an egg from which the egg envelope, or zona pellucida, has been removed to make the fertilizing sperm better visible. The sperm's head has been engulfed by the egg; the midpiece and tail are still outside. The egg surface is covered with fingerlike projections called microvilli, which enlarge the egg surface and provide a reservoir of plasma membrane to be used during cleavage.

MOST higher organisms reproduce sexually by forming two types of haploid gametes, eggs and sperm. Both are highly specialized and short-lived unless they unite to form a newly developing organism, through a process called fertilization (Fig. 4.1). Fertilization begins with the approach of the sperm to the egg and ends with the formation of the diploid zygote. The entire process can be broken down into four major steps:

- The sperm's approach to the egg
- The sperm's penetration of egg envelopes with or without adhesion to the vitelline envelope
- Plasma membrane contact and fusion
- Egg activation

Many mechanisms have evolved in animals to ensure that gametes meet at the appropriate time and place and under the right conditions to achieve fertilization. Thus, fertilization follows a series of well-choreographed interactions between males and females, and between sperm and eggs. Parental mating behavior brings eggs and sperm close together before the sperm are released and begin to swim with their flagella. In some organisms, chemical attractants produced by the egg itself or by ovarian cells guide the sperm in their final approach to the egg.

Sperm that encounter an egg face a series of obstacles because the envelopes that protect the egg are also formidable barriers to sperm entry. Several solutions to this dilemma have evolved. Some of the hardest egg coats, such as the calciferous shells of bird eggs, are laid down only after sperm have been admitted to the egg. The eggs of other animals, including fish and insects, have tiny holes called *micropyles* in their shells through which sperm can enter. In most species, however, the sperm must penetrate one or more egg envelopes before the plasma membranes of the two gametes can fuse (Fig. 4.2). Of particular importance is the *vitelline envelope,* a coat of proteins laid down by the oocyte before ovulation. In mammals, the vitelline envelope is a thick, transparent layer of glycoproteins called the *zona pellucida.*

In many species, a critical step in fertilization is sperm-egg adhesion, the binding of sperm to the vitelline envelope. The underlying molecular mechanisms are species-specific: Sperm adhere more stably to eggs from the same species than to eggs from another species. This species-specificity is particularly important for aquatic organisms, whose eggs may be exposed to sperm from many other species.

After adhesion to an egg, sperm penetrate the vitelline envelope using enzymes produced in the *acrosome.* When the first sperm finally reaches the egg cell proper, the plasma membranes of the two gametes fuse, forming a cytoplasmic bridge between them. The bridge widens and allows the sperm head to enter the egg. The two haploid nuclei derived from the egg and the sperm unite so that both sets of chromosomes are passed on together during cleavage divisions.

Plasma membrane contact or gamete fusion triggers a chain of events collectively called *egg activation.* These events include resumption of the cell cycle in eggs that have been arrested since oocyte maturation. In many species, activation also entails measures to protect the fertilized egg against the entry of additional sperm, which would disrupt mitoses.

Much of our knowledge about fertilization stems from observations and experiments in vitro, which are easily carried out with animals that shed their gametes into water. Therefore sea urchins, marine worms, fish, and amphibians have traditionally been the organisms chosen for research on fertilization. More recent studies have focused on mammalian fertilization, with the goal of developing better infertility treatments and new methods of contraception. Since research on fertilization in mammals is now beginning to rival the depth and sophistication of work on aquatic organisms, this chapter will focus on fertilization in both sea urchins and mice.

4.1 Interactions before Sperm-Egg Adhesion

Parental mating behavior ensures that eggs and sperm are delivered close to each other. Further success of the fertilization process now depends on interactions between the gametes themselves. In many species, eggs or associated cells release chemical attractants to which sperm respond by swimming actively toward them. In mammals and other animals as well, sperm must undergo a biochemical conditioning process before they are capable of fertilizing an egg. The processes of chemical attraction and sperm conditioning may in fact be related. At least in mammals, there are indications that only those sperm that are at the peak of their fertilizing capability respond to a chemoattractant released from the ovary at ovulation.

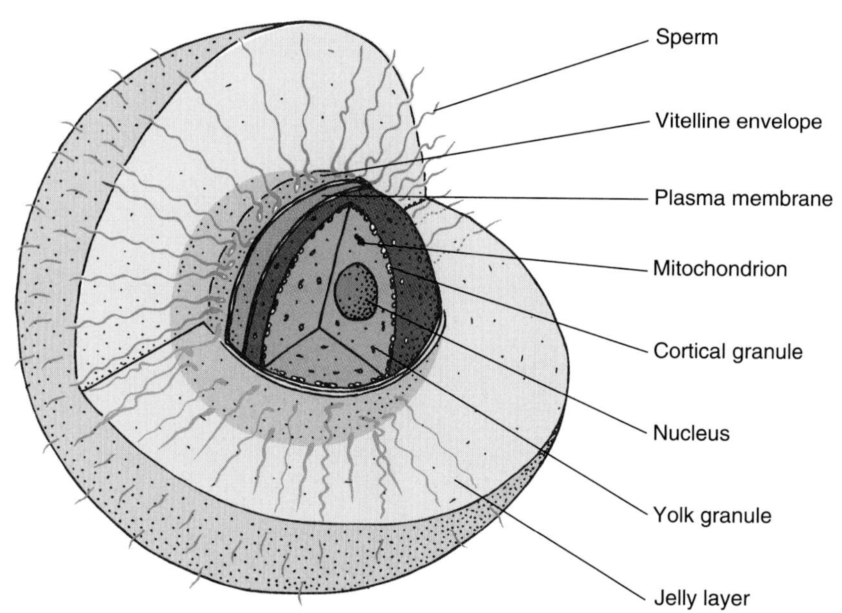

Figure 4.2 Sea urchin egg, drawn to show the egg coats that a sperm must penetrate before reaching the egg cell proper. The sperm first encounters a thick jelly layer, which it penetrates with digestive enzymes from its acrosome. Next the sperm binds to receptors at the level of the vitelline envelope and then pierces the vitelline envelope. Now the plasma membranes of sperm and egg make contact and fuse. Just beneath the egg plasma membrane are thousands of membranous vesicles called cortical granules. They will undergo exocytosis after sperm-egg fusion and shed their contents into the perivitelline space between egg and vitelline envelope, thus modifying the vitelline envelope so that no additional sperm can fertilize the egg. The diameter of the egg cell is about 75 μm.

SOME SPERM MUST UNDERGO CAPACITATION BEFORE THEY CAN FERTILIZE EGGS

Early attempts to fertilize mammalian eggs in vitro were unsuccessful because experimenters used freshly ejaculated sperm, which are unable to penetrate the *zona pellucida.* Sperm normally acquire this capability in the course of a few hours inside the female genital tract, through a process called ***capacitation*** (C. R. Austin, 1952). Capacitation also takes place in vitro when sperm are incubated with fluid from the uterus or oviduct. Prior to capacitation, mammalian sperm are in a state of low activity, apparently saving their energy and responsiveness until they are most likely to encounter an egg. A capacitated sperm is metabolically more active and beats its flagellum rapidly (Gwatkin, 1977). Similar behavior has also been observed in the sperm of some invertebrates (Wikramanayake et al., 1992).

The molecular mechanisms of capacitation are still under investigation. The sperm plasma membrane undergoes changes in lipid composition and other alterations that will facilitate fertilization (Yanagimachi, 1994). Other capacitation events may be reorganizations in the *cytoskeleton* at the sperm tip or the removal of "coating factors" that mask egg-binding sites at the sperm surface.

MANY SPERM ARE ATTRACTED TO EGGS BY CHEMICAL SIGNALS

Although parental mating behavior brings egg and sperm together, most sperm complete their journey to the egg by swimming. In mammals, sperm is ejaculated into the vagina or directly into the uterus. From here, the sperm must travel to the site of fertilization in the upper oviduct. Muscular actions of the uterus and oviducts contribute significantly to the movement of the sperm. Nevertheless, of the vast number of sperm (100 million to 500 million in humans) deposited in the vagina, only a few hundred reach the fertilization site. This extreme reduction in number is very intriguing and does not seem to be fully explained as a means of selecting against abnormal sperm.

Sperm movement is guided by chemical attractants produced by the egg or by ovarian cells, at least in some species. Oriented movement in response to an external chemical signal is called ***chemotaxis.*** Chemotactic behavior in sperm has been observed in various groups of animals, such as hydrozoans, molluscs, echinoderms, urochordates, and mammals, including humans (R. Miller, 1985; Eisenbach and Ralt, 1992). In the hydrozoan *Campanularia,* eggs develop in an ovary-like structure called a *gonangium,* which has a funnel-shaped opening through which sperm must enter to reach the eggs. Richard Miller (1966) showed that the funnel produces an attractant that activates sperm and causes them to turn toward the funnel (Fig. 4.3). This attractant is species-specific; thus, funnel extract from *C. calceofera* attracts sperm from the same species but not from a related species, *C. flexuosa.* In another hydrozoan, *Orthopyxis caliculata,* the attractant is released by the eggs instead of the gonangium (R. Miller, 1978). Its release is timed precisely: The sperm of this hydrozoan are attracted by eggs that have completed their second meiotic division, not by less mature eggs or oocytes.

Several sperm attractants have been isolated from sea urchins. One, called ***resact,*** is found in the jelly layer surrounding the egg of *Arbacia punctulata* (G. E. Ward et al., 1985). Resact is a small peptide consisting of 14 amino acids. At very low concentrations, resact attracts sperm in a species-specific way. However, when sperm are pretreated with high concentrations of resact they lose their chemotactic ability. This result suggests that the sperm surface may have specific receptors that become saturated at a certain resact concentration. In fact, such receptors have been identified in the plasma membrane of *Arbacia* sperm (Bentley et al., 1986). How these

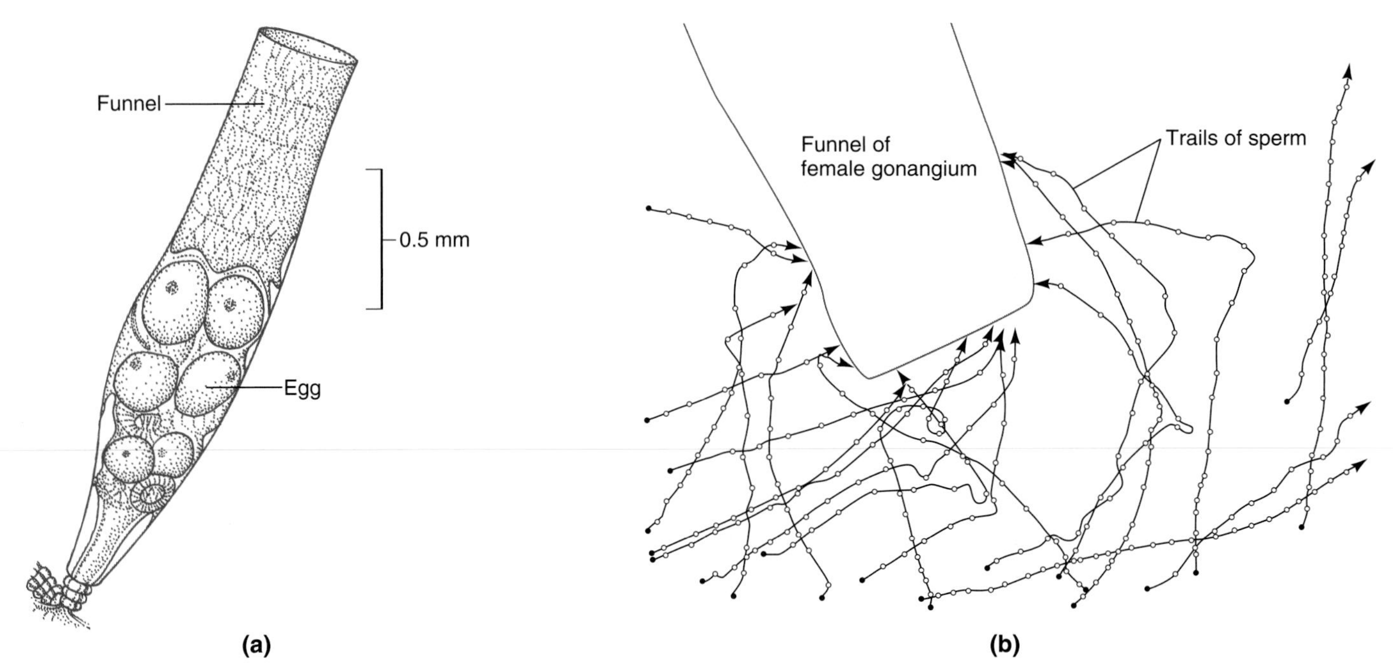

Figure 4.3 Chemotaxis in the hydrozoan *Campanularia flexuosa.* **(a)** Drawing of a female reproductive structure, the gonangium, from a colony of animals. The gonangium has a large funnel leading to the eggs within. **(b)** Plot of 21 sperm trails in the vicinity of the gonangium. The solid circles mark the start of each trail, and the open circles give the position of each sperm at 0.45-s intervals. Most trails lead to the funnel. The increasing distances between the circles indicate that the sperm accelerate as they draw closer to the funnel.

receptors allow sperm to find the direction of increasing chemoattractant concentration—that is, the way to the attractant's source, is still unresolved.

Preliminary evidence for sperm chemoattractants in mammals came from reports that mammalian sperm accumulate in the oviduct *below* the fertilization site and remain relatively motionless for hours until ovulation has occurred. Then the sperm become motile again, and a small number of them find their way to the fertilization site. These observations were corroborated by in vitro experiments indicating that human sperm respond chemotactically to a substance present in the follicular fluid released during ovulation (Ralt et al., 1991). However, the chemical nature of the putative attractant is not yet known, nor is it clear whether the signal is released by the egg itself or by other ovarian cells. Curiously, only a fraction of all sperm appear to be receptive to the attractant at any particular time (Cohen-Dayag et al., 1994). Thus, the putative chemoattractant in follicular fluid may selectively attract sperm that are at the peak of their ability to fertilize (*Eisenbach and Tur-Kaspa, 1999*). According to this hypothesis, the egg encounters only the selected sperm upon ovulation; the immature and overmature sperm are excluded from the competition.

4.2 Fertilization in Sea Urchins

Sea urchins have been classical animals for studies on fertilization (Epel, 1977). One advantage of these organisms is the copious amount of gametes they produce. During each breeding season, a sea urchin female spawns 400 million eggs, while a male releases 100 billion sperm. Producing so many gametes seems necessary because they quickly disperse in the ocean. A second advantage is that sea urchin fertilization occurs naturally in seawater and can therefore be easily studied in vitro. Early investigators may also have liked sea urchins because they provided a reason for them to spend their summers at marine stations, where they would meet biologists from other laboratories to trade new ideas and laboratory tricks.

The process of sea urchin fertilization can be broken down into five steps as shown in Figure 4.4. The sperm approaches by *chemotaxis* (step 1) and encounters the egg's protective coats. Upon contact with the egg jelly, the acrosome at the tip of the sperm releases its contents by exocytosis, in a process known as the ***acrosome reaction*** (step 2). Acrosomal enzymes digest a hole into the egg jelly while a slender projection called the *acrosomal process* rapidly forms at the tip of the sperm. Next the acrosomal process encounters the *vitelline envelope,* to which it adheres species-specifically, a step called ***sperm-egg adhesion*** (step 3). A hole lysed into the vitelline envelope finally gives the sperm access to the egg cell proper. The plasma membranes of egg and sperm touch, a step referred to as ***plasma membrane contact*** (step 4). It is followed quickly by ***gamete fusion*** (step 5), the event that most narrowly defines fertilization: The plasma

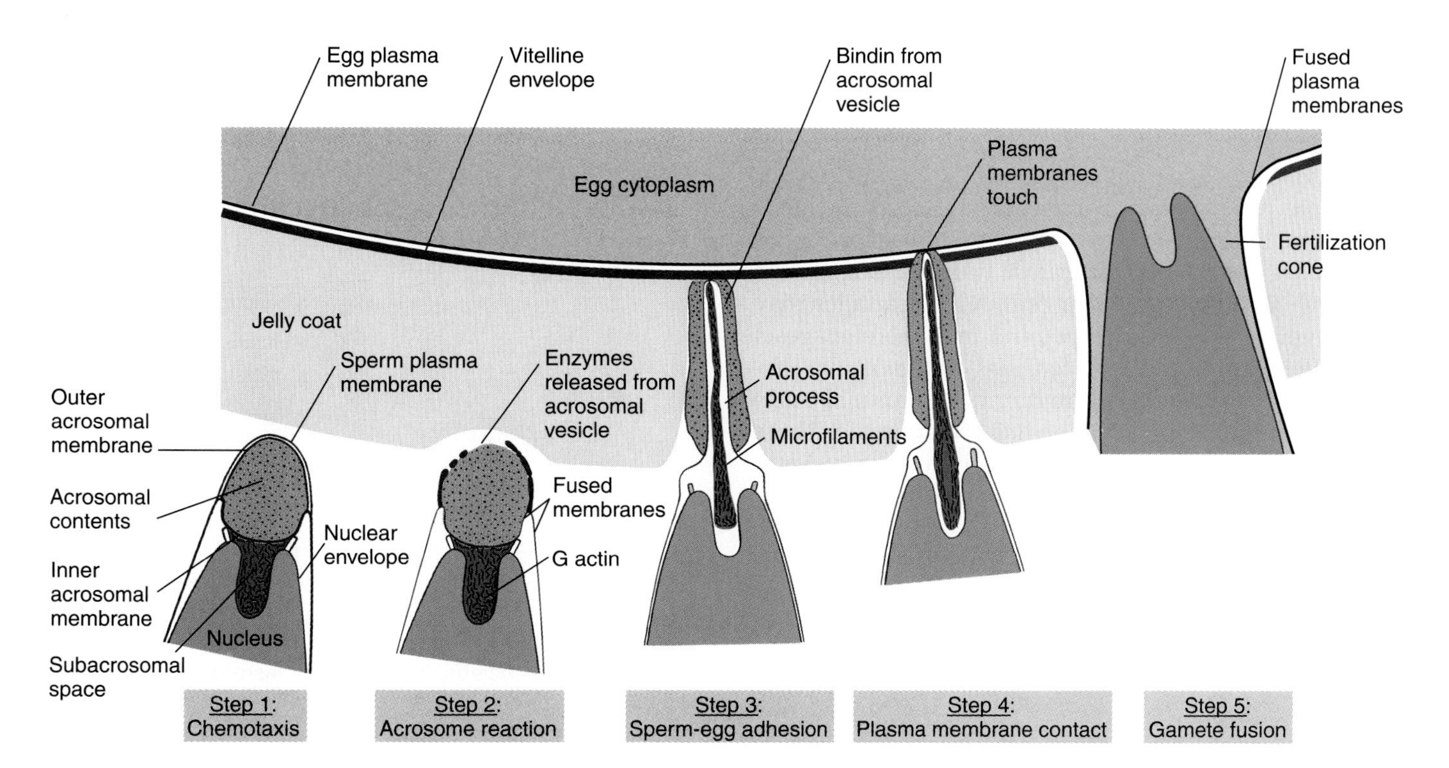

Figure 4.4 Overall process of sea urchin fertilization broken down into five steps. **Step 1:** Chemotactic approach of the sperm. The entire sperm is surrounded by a plasma membrane. The acrosome at the tip of the sperm is filled with contents including lytic enzymes and bindin. Its membrane can be regionally subdivided into the outer acrosomal membrane and the inner acrosomal membrane. The subacrosomal space between acrosome and nucleus is filled with globular actin (G actin). **Step 2:** Acrosome reaction. Upon contact with the egg's jelly coat, the acrosome undergoes exocytosis. The outer acrosomal membrane and the adjacent part of sperm plasma membrane vesiculate; the remainder of sperm plasma membrane and the inner acrosomal membrane fuse along the line indicated. Lytic enzymes released from the acrosome dissolve the jelly coat. The acrosomal process forms rapidly as globular actin from the subacrosomal space self-assembles into microfilaments (see step 3). The acrosomal process is covered with a membrane that originally served as the inner acrosomal membrane. **Step 3:** Sperm-egg adhesion. Bindin from the acrosome now sticking to the acrosomal process adheres species-specifically to the vitelline envelope. **Step 4:** Plasma membrane contact. After lysis of the vitelline envelope, the plasma membranes of sperm and egg touch, an event that may trigger the cascade of events collectively known as egg activation. **Step 5:** Gamete fusion. The plasma membranes of egg and sperm fuse, thus forming one membrane covering both gametes. The egg forms a cone of cytoplasm, called the fertilization cone, which actively engulfs the sperm nucleus. Note that the jelly layer is shown thinner, and the sperm larger, than proportional to the egg size.

membranes of egg and sperm fuse with each other to form a plasma membrane that surrounds both cells. Plasma membrane contact or gamete fusion trigger a cascade of events in the egg collectively called *egg activation,* which jumpstarts the egg on its way into embryonic development. One of the first visible signs of egg activation is the formation of the *fertilization cone,* a mound of cytoplasm in which the egg engulfs the sperm nucleus.

SEA URCHIN SPERM UNDERGO THE ACROSOME REACTION BEFORE THEY ADHERE TO THE VITELLINE ENVELOPE

In sea urchins, the *acrosome reaction* occurs when the sperm contacts certain components in the egg's jelly coat. The critical egg jelly components are generally glycoproteins (Ward and Kopf, 1993; S. H. Keller and Vacquier, 1994) although in *Strongylocentrotus purpuratus* it seems to be a fucose sulfate polymer without any associated peptide that triggers the acrosome reaction (Vacquier and Moy, 1997). The critical egg jelly component seems to bind to a receptor in the sperm plasma membrane, which releases intracellular signals, including a transient rise in calcium ion (Ca^{2+}) concentration. Interestingly, the same signals are also involved in coordinating the responses of the egg to fertilization (see Section 4.4).

The acrosome is the membraneous organelle at the tip of the sperm that contains enzymes and other components used by the sperm to adhere to the egg and to lyse its way through the egg's protective coats. The portion of the acrosomal membrane that faces the sperm plasma membrane is known as the ***outer acrosomal membrane.*** The remainder, called the ***inner acrosomal membrane,*** faces the *subacrosomal space* located between the acrosome and the sperm nucleus (Fig. 4.4). The subacrosomal space contains globular actin, which will

Immunostaining

The ***immunostaining*** procedure relies on blood serum proteins known as *immunoglobulins*, or ***antibodies,*** which are produced by B lymphocytes of vertebrates. Each clone of lymphocytes makes one type of antibody, directed against a specific *epitope* (binding site) on an antigen (a foreign molecule or cell). A given antigen may have several different epitopes and may therefore react with different antibodies. ***Polyclonal antibodies*** are prepared from blood serum of immunized mammals, into which an antigenic preparation has been injected. These antibodies are produced by multiple lymphocyte clones and directed against several epitopes of the antigen. In contrast, ***monoclonal antibodies*** are the products of single lymphocyte clones and directed against a single epitope (see Method 11.1).

Scientists use immunostaining to locate antigens at their normal sites of occurrence. If polyclonal antibodies are used, the antigen of interest is injected into a laboratory mammal, often a rabbit. Blood is taken from the rabbit at intervals, and the serum is tested for the presence of antibodies against the antigen. ***Preimmune serum*** prepared from the same rabbit before the first antigen injection is kept as a control. When the antibody concentration in the rabbit's blood is sufficiently high, a larger blood sample is taken, and the immunoglobulin fraction (the serum proteins containing the antibodies of interest) is prepared. This preparation is the ***primary antibody,*** which binds selectively to the antigen of interest.

One way of making the primary antibody visible is to conjugate it to a fluorescent molecule or some other label that can be detected under the microscope. Most researchers use a ***secondary antibody,*** prepared in a different mammalian species and directed against the primary antibody as an antigen. (Secondary antibodies avoid potentially detrimental effects of conjugated labels on the specificity of the primary antibody. Because one primary antibody molecule is bound by several secondary antibodies the latter also amplify the signal. Commercially available secondary antibodies are made in a goat or a swine and are directed against antigenic sites found in all rabbit immunoglobulins.) The secondary antibody is conjugated to a fluorescent dye, such as *fluorescein,* or to a stable enzyme, for example peroxidase, which generates a dark precipitate from two colorless reagents, diaminobenzidine and hydrogen peroxide.

For immunostaining, the primary and secondary antibodies are added in sequence to the antigen (Fig. 4.m1). The specimen containing the antigen is preserved with a fixative that leaves the antigenic sites exposed and fairly undisturbed so that they will bind to the primary antibody. Unless the specimen is transparent or can be cleared, it must be embedded and sectioned for microscopic viewing. On each section, a drop of primary antibody is pipetted and allowed to bind to the antigen. Unbound primary antibody is washed off before the secondary antibody is added and allowed to bind to the primary antibody. Unbound secondary antibody is washed off, so that only specifically bound antibody remains on the section. Depending on the conjugate attached to the secondary antibody, further reagents are added to make the conjugate visible.

When viewed under the microscope, the fluorescent stain or visible precipitate shows where the secondary antibody is located. Assuming that secondary antibody is bound *only* to primary antibody, and that primary antibody is bound *only* to the antigen of interest, the conjugate reveals where the antigen of interest is located in the sectioned specimen. These assumptions need to be confirmed by appropriate controls using specimens without antigen or with immunoglobulin prepared from preimmune serum in lieu of primary antibody.

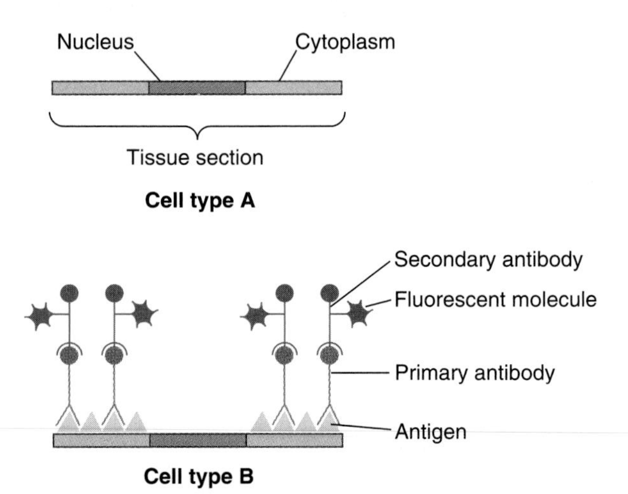

Figure 4.m1 Immunostaining is used to reveal the position of a known antigen in sections of fixed tissue. In this hypothetical example, the question is whether the antigen of interest is present in cell type A or B, and whether the antigen is located in the nucleus or in the cytoplasm. Primary antibody directed against the antigen (triangles) is allowed to interact with the sectioned tissue. Secondary antibody directed against the primary antibody is covalently linked with a fluorescent molecule or some other conjugate. The distribution of the conjugate over the section reveals that the antigen is present here in the cytoplasm of cell type B.

rapidly self-assemble into microfilaments once the acrosome reaction gets started.

The two major results of the acrosome reaction in sea urchin sperm are the *exocytosis* of the acrosomal contents and the formation of the *acrosomal process*. For exocytosis, the outer acrosomal membrane and the overlying portion of the sperm plasma membrane break up into small vesicles, thus releasing the acrosomal contents (Fig. 4.4). Acrosomal enzymes digest, or lyse, a hole through the egg jelly so that the sperm gains access to the vitelline envelope. Simultaneously, a long extension called the ***acrosomal process*** forms at the tip of the sperm. It is covered by what was originally the inner acrosomal membrane and has now become part of the sperm plasma membrane. The rapid extension of the acrosomal process is driven by the *self-assembly* of actin stored in the subacrosomal space (see Figs. 2.11 and 4.4).

Sea urchin sperm that have penetrated the egg jelly encounter the egg's vitelline envelope. If egg and sperm are from the same species, they adhere to each other stably. This step of *sperm-egg adhesion* is relatively *species-specific*; if sperm and egg are from different species they will adhere less stably or not at all. (Heterospecific fertilizations between closely related sea urchin species have been achieved in the laboratory, but they require about a million times higher sperm concentration than homospecific fertilizations.) The biological importance of this species-specificity is easy to see: Any gamete fertilized by a gamete from another species will not give rise to fertile offspring and therefore will not live on. It is also thought that the molecules mediating sperm-egg adhesion play critical roles in the evolution of new species (Vacquier, 1998).

Although other steps of fertilization, from mating behavior to gamete fusion, are also more or less species-specific, sperm-egg adhesion is particularly critical in animals that shed their gametes in the open water, where the chance of encountering gametes from other species is relatively high.

SEA URCHIN SPERM ADHERE TO EGGS WITH AN ACROSOMAL PROTEIN CALLED BINDIN

A sea urchin spermatozoon adheres to the egg by its acrosomal process. Investigators therefore searched for any molecules that would occur on the outside of the acrosomal process and stick to the vitelline envelope of eggs from the same species but not—or less so—to the vitelline envelope of eggs from other species.

The critical acrosomal component was first prepared from the sperm of the sea urchin *Strongylocentrotus purpuratus* (Vacquier and Moy, 1977; Vacquier, 1980). The sperm were suspended in a solution that dissolved the plasma membrane and acrosomal membrane but preserved the acrosomal contents as a more or less intact granule. The granules were separated from other sperm components and analyzed by gel electrophoresis. The researchers observed one major band, a protein with an apparent molecular weight of 30,500, which they termed ***bindin.*** Two experiments indicated that bindin is the adhesive material by which sea urchin sperm adhere to eggs.

TO test whether bindin was coating the acrosomal process as expected, Moy and Vacquier (1979) used a very versatile procedure known as *immunostaining* (Method 4.1). It revealed a thick coat of bindin covering the acrosomal process of acrosome-reacted sperm (Fig. 4.5). Control sperm, which had not undergone the acrosome reaction, were not stained by the procedure. In newly fertilized eggs, bindin was found at the site where acrosomal processes adhered. Thus bindin was present exactly when and where it should be as a mediator of sperm-egg adhesion.

To test whether bindin was adhering to eggs in a species-specific manner, Glabe and Lennarz (1979) designed a *competitive cell aggregation experiment.* They mixed equal numbers of dejellied *Strongylocentrotus purpuratus* and *Arbacia punctulata* eggs with bindin from either species in culture dishes. (Bindin is poorly soluble in water without detergent and forms particles in seawater.) The dishes were kept on a rotary shaker for a few minutes and then inspected under the microscope. Each bindin preparation agglutinated mostly eggs from its own species (Fig. 4.6). In order to test whether the bindin particles were directly holding the eggs together, the researchers labeled bindin with a fluorescent dye, fluorescein. Agglutinated eggs showed fluorescent particles precisely at the spots where they were stuck together. Thus bindins adhere specifically to eggs from the same species. Bindins from different sea urchin species are similar but not identical in size and amino acid composition, in accord with the species-specificity of their function (Glabe and Clark, 1991).

Questions

1. Why were the contents of the acrosomal vesicle a logical place to start the search for a bindinlike component?
2. If bindin from species A would agglutinate eggs from species A and S equally well, what should have been the relative frequencies of A-A, A-S, and S-S egg associations?

The experiments described above show that bindin is present exactly when and where a sperm adhesive is expected to act, and that bindin can account for the species-specificity of sperm-egg adhesion. Following sperm-egg adhesion, sperm penetrate the vitelline envelope with lytic enzymes stored in the acrosome (Colwin and Colwin, 1960). The sperm plasma membrane then comes in contact with the egg plasma membrane. This *plasma membrane contact* may trigger a series of events collectively called *egg activation*, which we will discuss in Section 4.4.

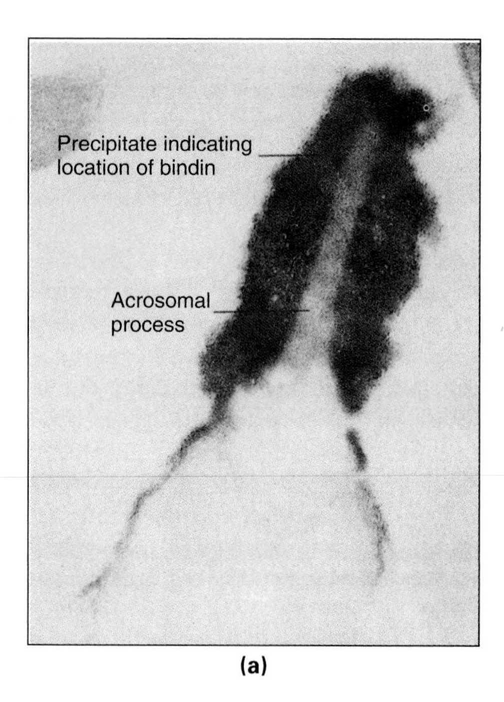

(a)

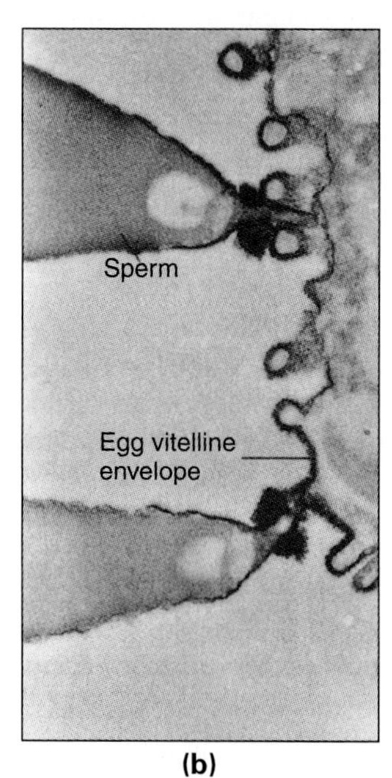

(b)

Figure 4.5 Localization of bindin on the acrosomal process of sea urchin sperm after the acrosomal reaction **(a)** and after binding to the egg at the level of the vitelline envelope **(b)**. The bindin location was revealed by immunostaining thin sections for viewing under the electron microscope. Primary rabbit antibodies were prepared against sea urchin bindin, and secondary swine antibodies against rabbit IgG were conjugated with peroxidase. When furnished with the appropriate substrates, the enzyme produced an electron-dense precipitate, which revealed the location of bindin.

The specific receptor(s) for bindin on the egg surface still need to be identified. It is not clear whether a single receptor molecule is sufficient to dock the sperm, trigger plasma membrane fusion, and initiate egg activation. Any candidate molecule should meet the following criteria: It is expected to be linked to the egg plasma membrane or to the vitelline envelope. It should also bind to sperm and its bindin species-specifically. Moreover, it should inhibit fertilization by competing with eggs for binding sites on sperm. Finally, antibodies against the candidate molecule should block fertilization.

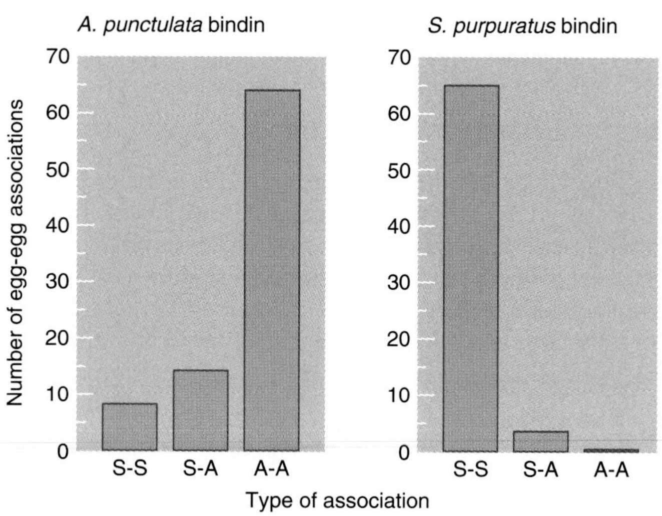

Figure 4.6 Species-specific agglutination of sea urchin eggs by bindin. A suspension containing equal numbers of eggs of *Strongylocentrotus purpuratus* (*S*) and *Arbacia punctulata* (*A*) was mixed with small amounts of bindin particles from one of the species and gently shaken for 2 to 5 min. The resulting egg-egg associations were scored on the basis of different egg pigmentations as *S-S, A-S, or A-A*. Each bindin agglutinated mostly eggs from its own species.

GAMETE FUSION LEADS TO THE FORMATION OF A FERTILIZATION CONE

After plasma membrane contact has occurred, egg and sperm proceed with *gamete fusion,* the event that lies at the heart of fertilization. In sea urchins, gamete fusion always occurs at the tip of the acrosomal process and often involves an egg microvillus (Fig. 4.7). In some species, the part of the egg membrane that fuses with the sperm is restricted to the area near the egg nucleus, which contains special glycoproteins in the plasma membrane (Freeman and Miller, 1982; Freeman, 1996).

At the site of gamete fusion, the fertilized egg forms a protrusion called the ***fertilization cone.*** The cone engulfs the sperm as it sinks into the egg (Fig. 4.8). This process involves a movement of egg cytoplasm into the region surrounding the sperm nucleus (Longo, 1989). The fertilization cone often grows for several minutes after gamete fusion and then regresses. The formation of the cone is associated with the polymerization of cortical actin into microfilaments; *cytochalasin,* a drug that interferes with microfilament assembly, inhibits cone formation. It is thought that the fertilization cone and microfilaments facilitate sperm entry, but the mechanisms involved are still unclear.

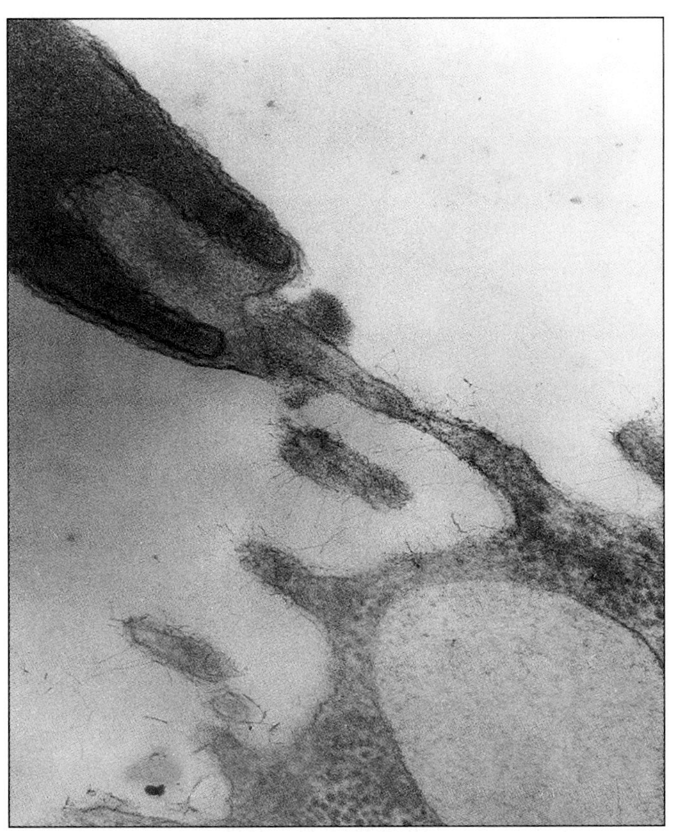

Figure 4.7 Gamete fusion during sea urchin fertilization. The electron micrograph shows a cytoplasmic connection between the acrosomal process of the sperm (top left) and a microvillus from the egg.

4.3 Fertilization in Mammals

Another group of animals frequently used for studies on fertilization are various nonhuman mammals (Yanagimachi, 1994). Mice and other laboratory mammals are often used as models for humans, who may benefit greatly from fertilization research for the purposes of overcoming infertility or preventing unwanted pregnancies. Investigators of mammalian fertilization have to contend with the fact that eggs are relatively difficult to obtain. Also, for in vitro studies, the fluids of the female reproductive tract must be mimicked by appropriately formulated media. In addition, for any observations on in vitro fertilized eggs beyond the blastocyst stage the embryos have to be implanted into the uterus of a female. However, researchers working with mice have the advantage of being able to harness the power of genetics, and this advantage will become more valuable as research in this field evolves toward an analysis of the key molecules involved (*Allen and Green, 1997;* Snell and White, 1996).

MOUSE SPERM UNDERGO THE ACROSOME REACTION AFTER THEY ADHERE TO THE ZONA PELLUCIDA

The first obstacle encountered by a mammalian sperm approaching a newly ovulated egg is the layer of *granulosa* cells embedded in a loose extracellular matrix (see Figs. 3.21 and 3.22). A sperm plasma membrane protein known as PH20 has an enzymatic activity that digests the major matrix component, hyaluronic acid. This action, along with active burrowing movements, allows the sperm to make its way to the zona pellucida (Lin et al., 1994).

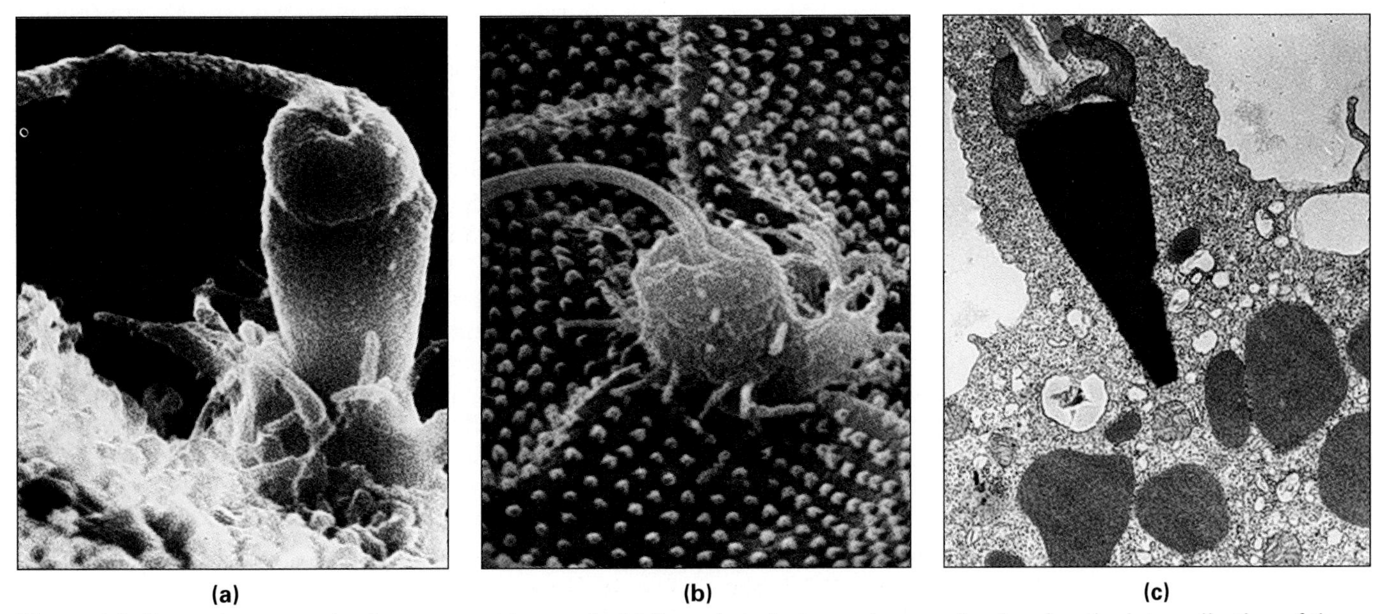

(a) (b) (c)

Figure 4.8 Sperm incorporation into sea urchin egg. **(a, b)** Scanning electron micrographs showing the internalization of the sperm head into the egg. **(c)** Transmission electron micrograph showing the sperm head engulfed by a cytoplasmic mound known as the fertilization cone.

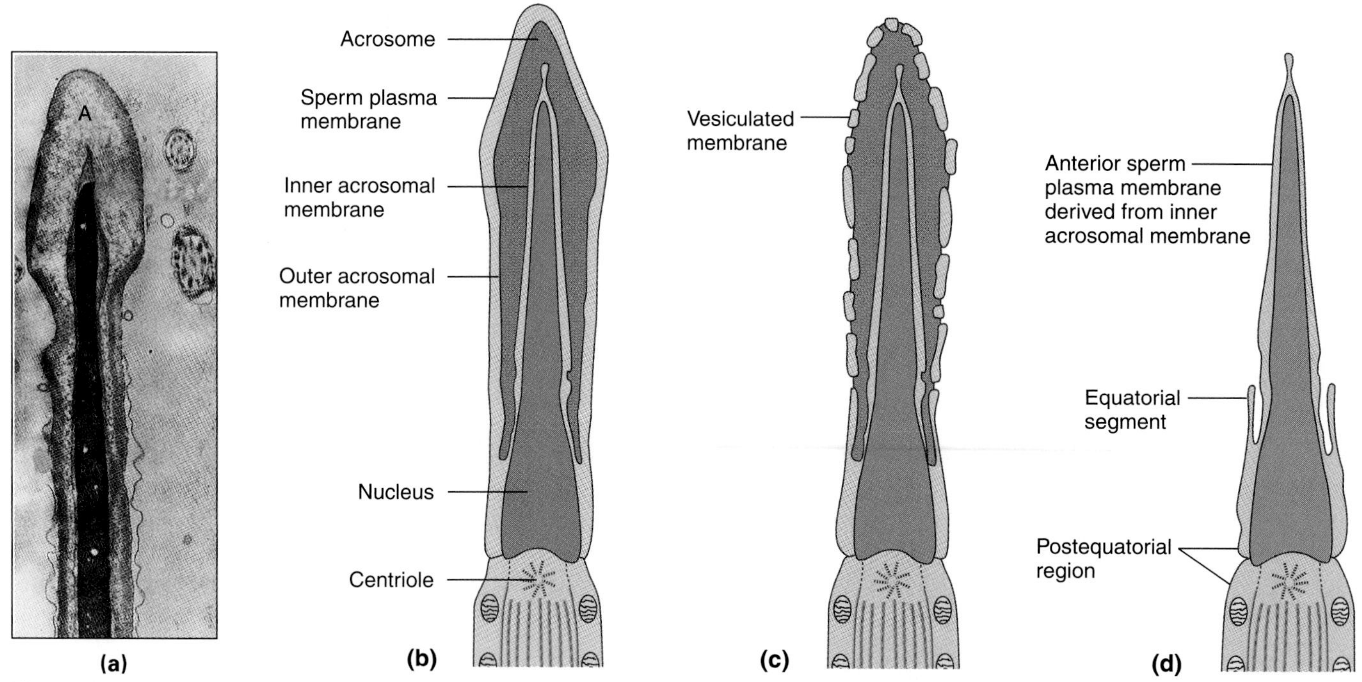

Figure 4.9 Acrosome reaction in mammals. **(a)** Transmission electron micrograph of a hamster sperm before the acrosome reaction. A = acrosome. **(b)** Drawing of sperm head before acrosome reaction, with sperm plasma membrane and acrosomal membrane intact. **(c)** During the acrosome reaction, the sperm plasma membrane and the outer acrosomal membrane fuse at many places, forming numerous vesicles. **(d)** After the acrosome reaction, the sperm plasma membrane consists of three regions: the anterior region derived from part of the inner acrosomal membrane, an equatorial segment, which has not vesiculated, and the postequatorial region.

The timing of the acrosome reaction may vary among mammalian species, and it is not always known with certainty due to the technical problems associated with studying mammalian fertilization in vivo. In rabbits, the acrosome reaction seems to be triggered when the sperm burrows through the *granulosa cells*. In mice, all available evidence indicates that the acrosome reaction is elicited when the sperm adheres to the *zona pellucida*, the mammalian equivalent of the vitelline envelope of sea urchins.

The acrosome in most mammals forms an extended cap over the anterior portion of the nucleus (Fig. 4.9). The *outer acrosomal membrane* lies underneath the plasma membrane while the *inner acrosomal membrane* lies close to the nuclear envelope. During the acrosome reaction, the outer acrosomal membrane and the sperm plasma membrane fuse at many points, breaking up the two membranes into many small vesicles. These vesicles remain connected to one another by a matrixlike component in the acrosomal contents until eventually the matrix is dissolved and the vesicles dissociate. The fusion of the outer acrosomal and plasma membranes, however extensive, stops short of the posterior edge of the acrosome, leaving a collarlike fold of membrane, the *equatorial segment,* to encircle the sperm head. As a result, the sperm plasma membrane now consists of three distinct regions: an anterior region derived from the inner acrosomal membrane, the equatorial segment, and the postequatorial region (Fig. 4.9d). In contrast to sea urchins, mammalian sperm also have less subacrosomal material and do not form an acrosomal process.

The sperm of mice and many other mammals undergo the acrosome reaction after adhering to the zona pellucida with the side of the anterior region of their heads (Fig. 4.10). Completion of the acrosome reaction depends on an increase in intracellular Ca^{2+} concentration (Shirikawa and Miyazaki, 1999), as is the case for sea urchins (see Section 4.2). Having undergone the acrosome reaction, the fertilizing mammalian sperm forms a hole—or widens a preexisting open area—in the zona pellucida by an as yet unknown mechanism. Thereafter, the sperm jettisons the acrosomal matrix with the membranous vesicles covering its head, and wriggles its way into the perivitelline space between the zona and the egg plasma membrane.

Once inside the perivitelline space, the sperm contacts the egg plasma membrane with the equatorial or the postequatorial region of their plasma membrane. Although both of these regions are exposed in acrosome-intact sperm, the acrosome reaction is nevertheless required for gamete fusion: Eggs from which the zona pellucida has been removed experimentally fuse only with acrosome-reacted sperm, not with acrosome-intact sperm. These observations suggest that the acrosome reaction modifies proteins located in the equatorial or postequatorial region, or that these regions somehow

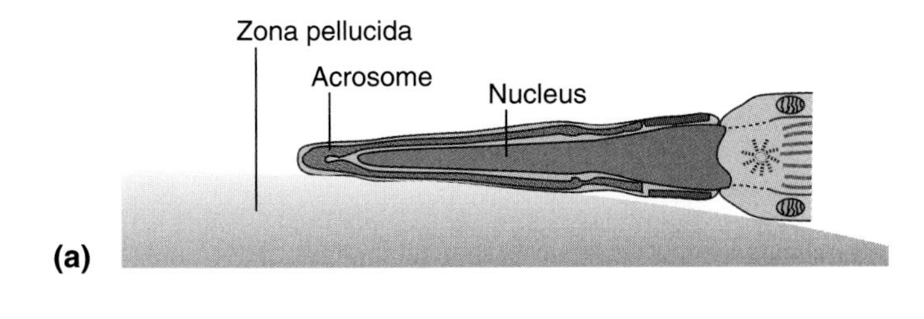

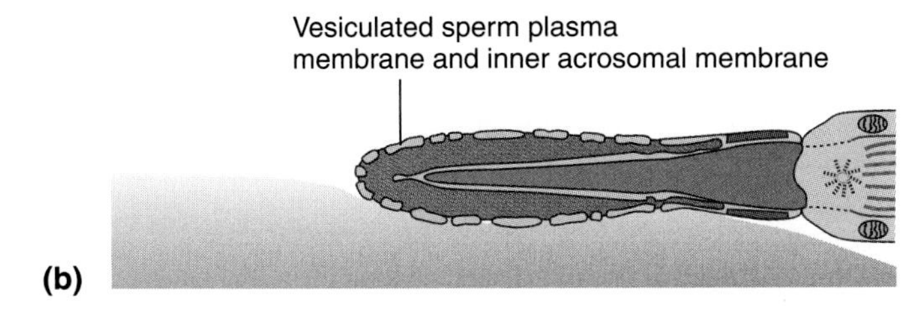

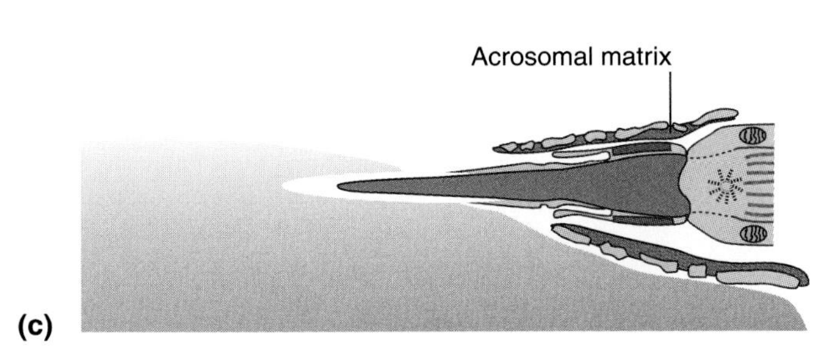

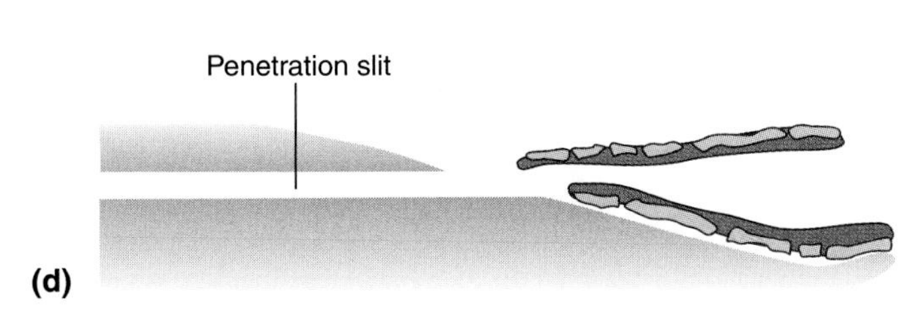

Figure 4.10 Mammalian sperm **(a)** adhering to the zona pellucida, **(b, c)** undergoing the acrosome reaction, and **(c, d)** breaching the zona pellucida on its way to the egg cell proper. The sperm overcomes the forces of adhesion between its own plasma membrane and the zona by leaving behind the shell of membranous vesicles and acrosomal matrix generated during the acrosome reaction.

acquire new molecules from elsewhere as part of the acrosome reaction.

The nature of the molecules involved in the contact and fusion of mammalian egg and sperm, and the mechanisms of their deployment, are currently under active investigation (*Allen and Green, 1997*). Particular interest has focused on a family of plasma membrane proteins with extracellular domains that have prometalloprotease activity and function in cell adhesion. A member of this family, known as PH30 or *fertilin,* accumulates in the plasma membrane of mammalian sperm. Mutant mice lacking fertilin have sperm that are deficient in migration from the uterus into the oviduct, in adhesion to the egg zona pellucida, in plasma membrane contact, and in sperm-egg fusion (Cho et al., 1998). A similar sperm protein is also required for fertilization in *Xenopus* (Shilling et al., 1997, 1998).

A BIOASSAY IS A POWERFUL STRATEGY TO REVEAL A BIOLOGICALLY ACTIVE COMPONENT

Sperm-egg adhesion in mice, as in sea urchins, is mediated by specific molecules on the surfaces of egg and sperm. In sea urchins, *bindin* has been established as the adhesive molecule on the part of the sperm, whereas the bindin receptor(s) of the egg are still being investigated as discussed earlier. The converse is the current situation in mammals: While the adhesive molecules on the side of the sperm is still a matter of debate, their counterpart on the zona pellucida is well characterized, at least in mice.

The sperm-binding component in mouse zona was identified by means of a ***bioassay***—that is, a strategy used to screen various cellular fractions or molecules for their ability to elicit a biological response that is clearly defined and readily scored. Such bioassays are generally very powerful, as the following example will demonstrate. In this instance, the investigators sought to identify the mouse zona pellucida component(s) to which mouse sperm naturally adhere.

To establish such a bioassay, Bleil and Wassarman (1980) fertilized mouse eggs under conditions in which the sperm were in limited supply (Fig. 4.11). Before fertilization, the sperm were preincubated in medium containing zona pellucida glycoproteins. Unfertilized eggs were then added to the cultures to allow sperm-egg adhesion. After gently pipetting the eggs to remove loosely "attached" sperm, the sperm firmly "bound" to each egg were counted. The average number of sperm bound after preincubation in pure medium was normalized to 100%. This percentage decreased dramatically when the sperm were preincubated with zona pellucida glycoproteins from unfertilized eggs.

The bioassay outlined above was designed to test zona glycoproteins and their components for the ability to compete with intact eggs for sperm adhesion. The most effective competitor(s) in this assay would thus be identified as the most likely candidate(s) for being the zona component to which sperm naturally adhere. Other bioassays to be discussed in Sections 8.3 and 9.5 have been designed to identify cytoplasmic components that are segregated unevenly during cleavage and impart distinct patterns of gene expression on blastomeres.

MOUSE SPERM ADHERE TO A SPECIFIC ZONA PELLUCIDA PROTEIN

The zona pellucida is synthesized by the growing oocyte as a thick but porous structure. Its major constituents are three glycoproteins (designated ZP1, ZP2, and ZP3), which assemble into long, interconnected filaments

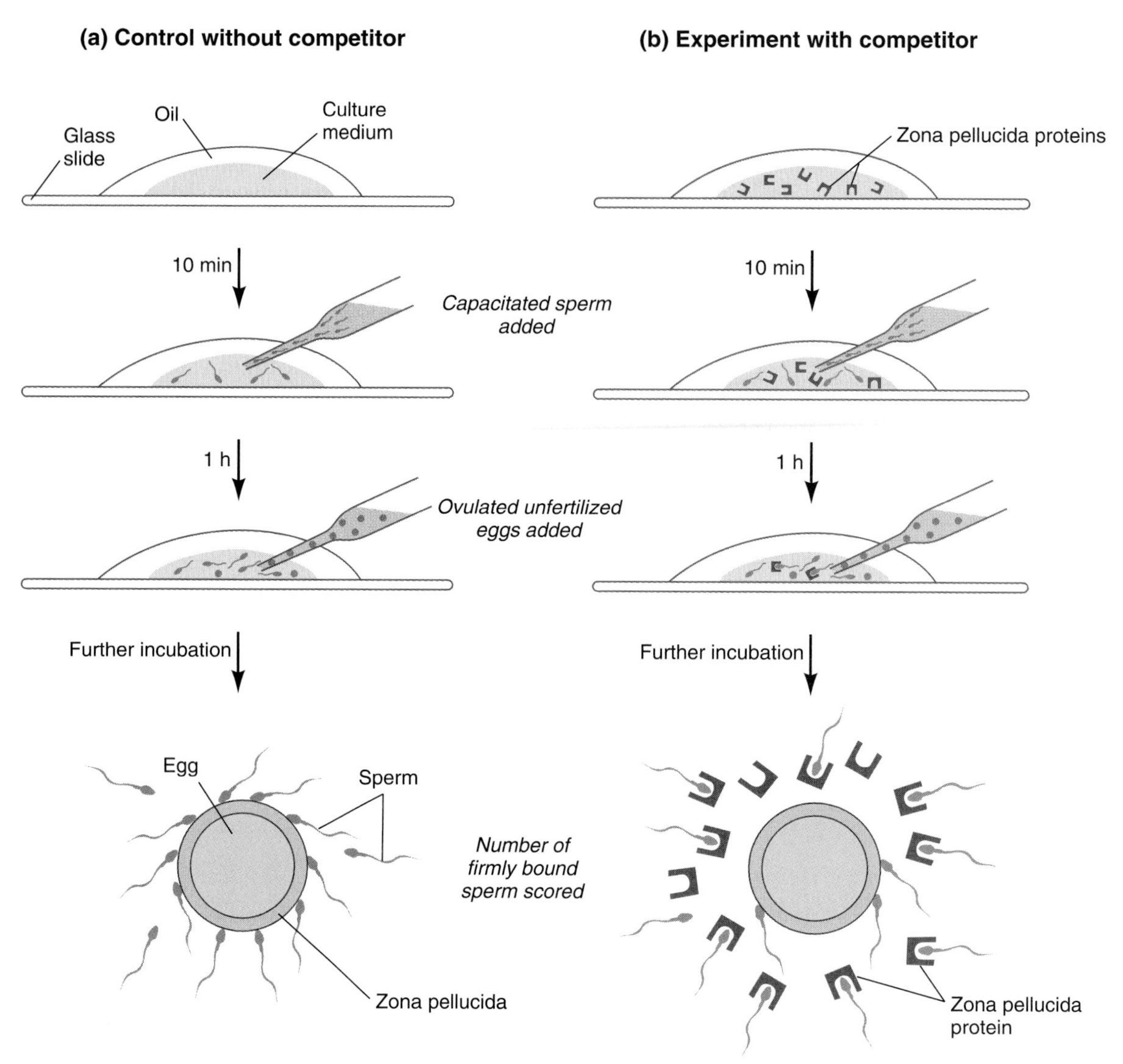

Figure 4.11 Bioassay to identify the mouse zona pellucida component to which the sperm adheres. **(a)** Control experiment without competitor. A drop of culture medium is kept on a glass slide under oil to prevent desiccation. Capacitated sperm are added to the drop from a glass micropipette. After 1 h of incubation, unfertilized ovulated eggs are added. After further incubation, the eggs are washed to remove loosely attached sperm. The average number of firmly bound sperm per egg is scored. **(b)** Same experiment except that zona proteins, or moieties thereof, are added to the culture medium. If the added components bind to sperm in solution, they effectively compete with the eggs for binding sites on the sperm, thus reducing the number of sperm bound to the eggs. For results see Figure 4.12.

(Wassarman, 1987, 1990). The zonae of other mammals, including humans, consist of similar glycoproteins.

IN order to identify the zona protein, and the exact moiety thereof, to which mouse sperm bind, Bleil and Wassarman (1980) used the bioassay described above (Fig. 4.11). As part of establishing this bioassay, they had already shown that total zona proteins from unfertilized eggs inhibit the binding of intact unfertilized eggs to capacitated sperm. In contrast, the same glycoproteins from 2-cell embryos did not have this inhibitory effect. Apparently, zona glycoproteins from unfertilized eggs were competing with intact unfertilized eggs for binding sites on the sperm. In 2-cell embryos, the zona glycoproteins seemed to be modified so that sperm could no longer bind to them.

Once this bioassay was established, each zona glycoprotein was purified and tested individually in the same way. Only ZP3 inhibited sperm-egg adhesion. The other zona proteins did not compete with unfertilized eggs for binding sites on the sperm (Fig. 4.12). Evidently, ZP3 is the only zona glycoprotein that binds to mouse sperm during the initial sperm-egg adhesion.

Molecular analysis of ZP3 showed that its backbone is a single polypeptide of about 400 amino acids (Wassarman, 1990). Extending from the polypeptide are many ***oligosac-***

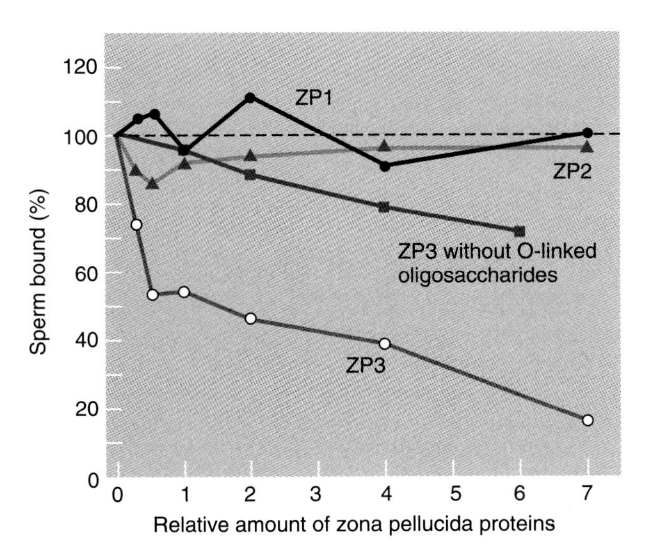

Figure 4.12 Competition of mouse ZP3 with intact eggs for sperm binding sites. Sperm were preincubated with purified proteins (ZP1, ZP2, ZP3) from mouse zonae pellucidae before unfertilized test eggs were added (see Fig. 4.11). The eggs were scored later for the number of firmly bound sperm. The hatched line (100% level) represents the average number of sperm bound per egg in control experiments without zona pellucida components added to the culture medium. Sperm preincubated with ZP1 or ZP2 bound to the test eggs at near-control levels, but sperm preincubated with ZP3 showed a reduced ability to bind. The inhibitory action of ZP3 depended largely on its O-linked oligosaccharides. (Apparent fertilization rates greater than 100% are an artifact of the method used for data processing.)

charides—that is, short chains of connected sugars. Some of these are ***O-linked oligosaccharides,*** which means that they are linked to an oxygen atom on either one of two amino acids: serine or threonine. To reveal which of the ZP3 components function in sperm adhesion, ZP3 was pretreated with various agents to destroy specific parts of the molecule. The pretreated ZP3 was then tested again as a competitor in the bioassay. Removal of the oligosaccharides destroyed the competitive function of ZP3, whereas cutting the polypeptide enzymatically into short pieces did not have this effect. Specifically, removing the O-linked oligosaccharides with a mild alkali left ZP3 an unfit competitor (Florman and Wassarman, 1985). Conversely, the O-linked oligosaccharides alone competed as well as intact ZP3. The exact nature of the moiety within the O-linked oligosaccharide that is critical for sperm adhesion still needs to be established. It also unclear whether the ZP3 polypeptide plays a direct role in adhesion or if it only determines the "presentation" of the O-linked oligosaccharides (Wassarman and Litscher, 1995).

Questions

1. In the bioassay described above, why was it important that the sperm be present in limited supply rather than in excess?
2. In the course of preparing zona proteins from eggs and 2-cell embryos, it became apparent that zona pellucida from embryos is much harder to dissolve in various aqueous media. What does this difference in solubility suggest?

Additional lines of evidence confirmed that the ZP3 glycoprotein is the critical molecule for sperm-egg adhesion in mice (Bleil and Wassarman, 1986). First, purified and radiolabeled ZP3 from *unfertilized eggs* binds only to the head of the sperm, not to its midpiece or tail. This is consistent with the observation that sperm bind by the head to the egg zona pellucida. Second, purified ZP3 from the *embryonic* zona pellucida does not bind to sperm, confirming the earlier observation that zona pellucida proteins from 2-cell embryos are inactive in the competitive binding assay.

The mouse sperm membrane protein(s) that bind specifically to ZP3 are still a matter of debate (Snell and White, 1996). Perhaps the leading candidate so far is an enzyme, *N-acetylglucosamine galactosyltransferase* (*GalTase*), which is located at the tip of acrosome-intact sperm (Shur, 1989; D. J. Miller et al., 1992; Gong et al., 1995). The biochemical function of GalTase is to transfer galactose to an oligosaccharide chain ending with *N*-acetylglucosamine. However, since galactose is not present in the oviduct, the enzyme is not able to complete its function and so remains stuck to the terminal *N*-acetylglucosamines of the O-linked oligosaccharides of ZP3. There is also evidence that other sperm proteins adhere to ZP3 (Bleil and Wassarman, 1990; Leyton et al., 1992, 1995), and it seems possible that two or more adhesion molecules could act synergistically in sperm-egg adhesion.

ANTIBODIES TO SPERM-EGG ADHESION PROTEINS CAN ACT AS CONTRACEPTIVES

Research on fertilization has included the development of medical applications designed either to help infertile couples to have children or to help fertile couples to avoid unwanted children. Past research into sperm capacitation has greatly improved the success of in vitro fertilization; knowledge of the hormonal control of ovulation has provided the basis for the development of the most commonly used oral contraceptive today.

A recently developed method for *intracytoplasmic sperm injection* (*ICSI*) inserts a single sperm directly into the cytoplasm of a mature oocyte. It has already helped men with very low sperm counts or high proportions of defective sperm to have children. This method avoids the rigorous sperm selection process that occurs during normal conception. The question whether the condition that has left the father infertile is heritable, and if so, how his child may be affected, is usually left unanswered. So far, it appears that ICSI does not significantly change the risk of spontaneous abortion, stillbirth, or malformation at birth (Bonduelle et al., 1999). However, many more ICSI children must be compared to normally conceived controls before any changes in the frequency of rare diseases become statistically detectable.

On the other hand, the overall growth of the human population on Earth is making effective contraceptive methods the only sustainable way of avoiding both large-scale famines and environmental degradation. Contraceptive vaccines may become viable alternatives to traditional contraceptives where the latter are impractical or deemed undesirable (Burks et al., 1995; Talwar and Raghupathy, 1994). Since sperm-egg adhesion and plasma membrane contact depend on a few specific molecules, it may be possible to block these processes using antibodies that do not interfere with other vital functions. Such antibodies might be elicited by injecting suitable antigens, such as the zona proteins or sperm proteins that mediate sperm adhesion.

Immunization with zona proteins from hamster makes female mice infertile, but after the antibody concentration decreases, the mice are capable of bearing normal young again (Gwatkin, 1977). Similarly, female mice immunized with mouse ZP3 produce antibodies that coat the zonae pellucidae of their developing oocytes. These mice are also infertile as long as the antibody concentration remains high (Millar et al., 1989). Unfortunately, the injected mice also developed ovarian diseases.

Sperm proteins are another target for contraceptive antibodies. Clinical data indicate that human infertility may be caused by anti-sperm antibodies in either the male or the female partner. (Normally, such an autoimmune reaction in males is prevented by the *blood-testis barrier* described in Section 3.4.) In guinea pigs immunized with a guinea pig sperm protein, the vaccination was fully contraceptive in both males and females, and the effects were reversible (Primakoff et al., 1988). None of the existing procedures is effective and safe for human application yet, but the results obtained so far are encouraging enough to continue related investigations (Feng et al., 1999).

4.4 Egg Activation

Fertilization triggers ***egg activation,*** a series of events that stimulates the quiescent egg to reenter the cell cycle and begin development (Table 4.1; Fig. 4.13; Epel, 1997). The activation cascade entails intracellular signaling mechanisms that coordinate the numerous steps involved. The signals include in particular ***calcium ion (Ca^{2+}) transients***—that is, short rises or oscillations in Ca^{2+} concentration, and the activation of protein kinase C. Activation also accelerates the egg's metabolism including DNA replication and protein synthesis in preparation for fast-paced cleavage divisions. Another important part of egg activation in many species is the prevention of multiple fertilizations (see Section 4.5.)

table 4.1 Sequence of Events in Sea Urchin Fertilization

Event	Time after Plasma Membrane Contact[a]
Fertilization potential appears (fast block to polyspermy)	Before 3 s
Plasma membrane fusion	Before 15 s[b]
Inositol trisphosphate and diacyclglycerol produced	Before 15 s[b]
Intracellular calcium release	40–120 s
Cortical reaction (slow block to polyspermy)	40–100 s
Sperm entry	1–2 min[b]
Activation of NAD kinase	1–2 min
Increase in O_2 consumption	1–3 min
Na^+/H^+ exchange, increase in pH	1–5 min
Sperm chromatin decondensation	2–10 min[c]
Sperm nucleus migration to egg center	2–10 min[c]
Activation of protein synthesis	After 5 min
Fusion of pronuclei	20 min[d]
Initiation of DNA synthesis	20–40 min
First mitotic prophase	60–80 min
First cleavage	85–95 min

[a] The times listed are estimates based on data from *Lytechinus pictus* kept at 16–18°C (Whitaker and Steinhardt, 1985), except as noted.
[b] Laurinda A. Jaffe (personal communication).
[c] Based on data from *Clypeaster japonicus* (Hamaguchi and Hiramoto, 1980).
[d] Based on data from *Strongylocentrotus purpuratus* (Epel, 1977).

EGG ACTIVATION MAY BE TRIGGERED BY DIFFERENT SIGNALING MECHANISMS

How does fertilization trigger egg activation? There are two likely scenarios (Whitaker and Swann, 1993). First, egg activation could be triggered by *gamete fusion, when sperm introduce the activating component into the egg.* Second, egg activation could be triggered earlier by *plasma membrane contact, when the sperm activates a receptor on the egg surface.*

According to the first scenario, a component introduced with the sperm cytoplasm into the egg cytoplasm, or with the sperm plasma membrane into the egg plasma membrane, triggers egg activation. In support of this hypothesis, egg activation responses were found to be triggered by cytoplasmic proteins isolated from sperm and injected into eggs of a marine worm (Stricker, 1997), a sea urchin (Galione et al., 1997), and a mouse (Sette et al., 1997). The activation responses so triggered always included Ca^{2+} oscillations, apparently released from Ca^{2+} stores in the *endoplasmic reticulum.*

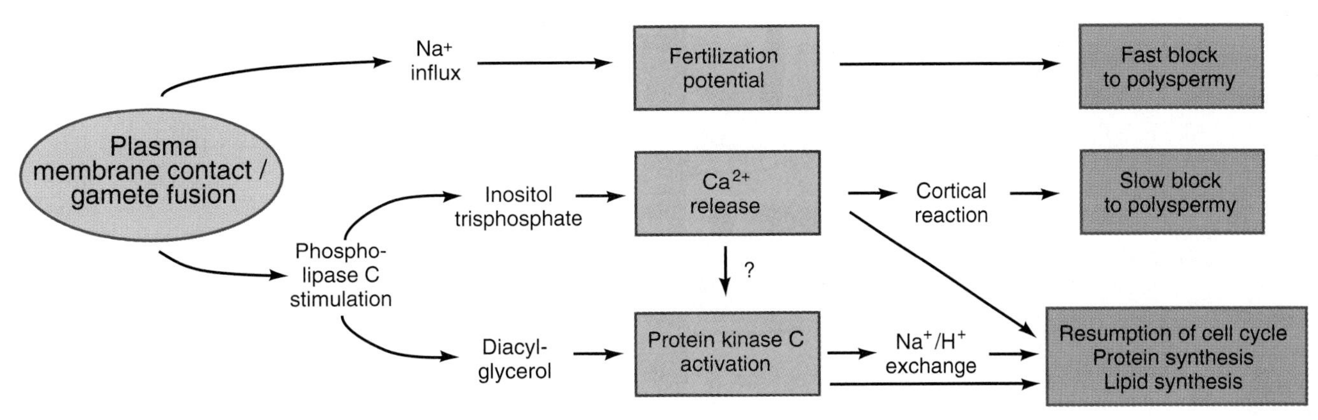

Figure 4.13 Flow chart of key steps in sea urchin egg activation. See also Table 4.1.

According to the second scenario, contact with sperm plasma membrane activates a sperm receptor on the egg surface, located presumably in the egg plasma membrane. (Note that these hypothetical sperm receptors in the egg plasma membrane are thought to be different from molecules like ZP3, which mediate sperm adhesion to the zona pellucida or other vitelline envelope.) The search for receptors that may function in egg activation have been guided by the observation that the activation process requires four well-known intracellular messengers: *inositol trisphosphate (IP_3), diacylglycerol (DAG), protein kinase C (PKC), and calcium ion (Ca^{2+})*. A key enzyme in the release of these messengers is the enzyme *phospholipase C (PLC)*, as diagrammed in Figure 2.27. PLC hydrolyses phosphatidylinositol bisphosphate (PIP_2) into IP_3, which releases Ca^{2+} from cytoplasmic storage sites, and DAG, which activates PKC.

IP_3 and a receptor to which it binds have been found in newly fertilized eggs of many species (Parys et al., 1994). Two modes of generating IP_3 involving different types of PLC have been tested. First, *G proteins*, which may be activated by the hypothetical sperm receptor in the egg plasma membrane, activate *phospholipase C-β (PLC-β)*, which cleaves membrane-bound PIP_2. Second, a *receptor tyrosine kinase*, or *RTK* (see Fig. 2.28), which may be stimulated by sperm-egg contact, or a cytoplasmic tyrosine kinase (TK) introduced by the sperm, activates *phospholipase C-γ (PLC-γ)*, which also cleaves PIP_2 and causes the same downstream events as PLC-β.

Several experiments indicate that both the G protein/PLC-β and the TK/PLC-γ pathways may be involved. For example, stimulation of G proteins by receptors not normally present in eggs is *sufficient* to trigger activation of frog and mouse eggs (Kline et al., 1988; G. D. Moore et al., 1993). The G protein/PLC-β pathway is also *necessary* for hamster egg activation (Miyazaki, 1988). In starfish eggs, activation of PLC-γ is necessary for egg activation, but the G protein/PLC-β pathway works as well (Shilling et al., 1994; Carroll et al., 1997). In *Xenopus* eggs, stimulation of an RTK that activates PLC-γ is *sufficient* to trigger complete activation (Yim et al., 1994). However, neither PLC-β nor PLC-γ is *necessary* for activating *Xenopus* eggs, suggesting that yet another pathway for making IP_3 is functioning in these eggs (Runft et al., 1999).

Taken together, the available data indicate that Ca^{2+} transients are a universal link in the chain of egg activation signals, and that IP_3 is commonly used to release Ca^{2+} from intracellular stores. The mechanisms for generating IP_3 differ among organisms and may be initiated by plasma membrane contact and/or gamete fusion. This diversity again demonstrates that critical biological processes such as egg activation are often supported by two or more molecular mechanisms, which may have evolved to act alternatively or synergistically, depending upon the species.

A TEMPORARY RISE IN CA^{2+} CONCENTRATION IS FOLLOWED BY ACTIVATION OF PROTEIN KINASE C

Ca^{2+} transients are naturally triggered by fertilization and have been found to be a regular feature of egg activation in all species studied so far (Stricker, 1999). In large eggs, the increased concentration of free Ca^{2+} spreads like a wave, beginning at the site of sperm entry (Fig. 4.14). The wavelike propagation seems to follow the release of Ca^{2+} from intracellular stores in response to either IP_3 or Ca^{2+} itself. The released Ca^{2+}, along with the DAG generated simultaneously with IP_3, activates PKC (see Fig. 2.27). Both Ca^{2+} and PKC are key regulators of the egg activation process (see Fig. 4.13).

To test whether Ca^{2+} transients are *sufficient* for later activation events to occur, investigators have added Ca^{2+} to unfertilized eggs. This was done by microinjecting Ca^{2+} into unfertilized eggs (Hamaguchi and Hiramoto, 1981), or by keeping unfertilized eggs in a Ca^{2+}-containing medium and making them permeable to Ca^{2+} with drugs called *ionophores* (Steinhardt and Epel, 1974). Either treatment triggers several downstream activation steps, showing that Ca^{2+} is sufficient for these steps to occur.

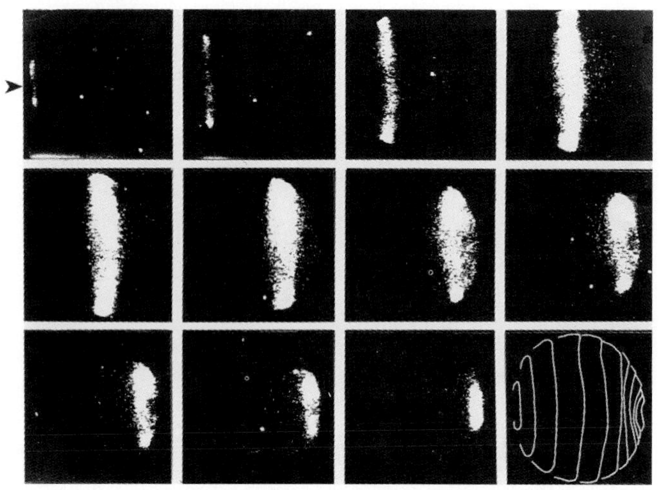

Figure 4.14 Ca^{2+} wave in a fertilized fish egg (*Oryzias latipes*). The egg has been injected with aequorin, a protein that emits light when it binds to Ca^{2+}ions. The photographs were taken at 10-s intervals to show the wave of Ca^{2+} that begins near the sperm entry point (arrowhead at the top left panel) and passes across the egg. The bottom right panel outlines the leading edges of the light front seen in the other 11 panels.

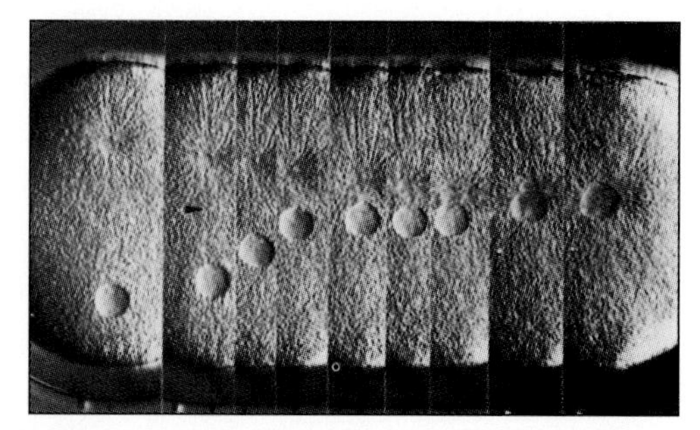

Figure 4.15 Fusion of pronuclei in the fertilized egg of the sea urchin *Clypeaster japonicus.* This series of photographs, which covers a time span of 800 s, shows the male pronucleus surrounded by an aster of microtubules. The pronuclei move toward each other and fuse.

To test whether Ca^{2+} transients are *necessary* for all subsequent activation events to occur, researchers have removed Ca^{2+} from sea urchin eggs before fertilization. This was achieved by injecting eggs with molecules that bind up Ca^{2+} (Zucker and Steinhart, 1978). The procedure rendered eggs metabolically inactive. However, mammalian eggs did undergo several activation steps even in the absence of a Ca^{2+} transient when PKC was stimulated by other means, and they failed to complete meiosis when PKC was inhibited (Gallicano et al., 1993, 1997a). It appears that groups of animals may differ in their dependence on PKC in their egg activation cascades (Gallicano et al., 1997b).

EGG ACTIVATION TRIGGERS THE COMPLETION OF MEIOSIS AND THE FUSION OF THE GAMETES' HAPLOID GENOMES

A very important aspect of egg activation is the resumption of the cell cycle, which occurs in three major steps. First, in most species, fertilization releases the egg from its second meiotic block, allowing the formation of a haploid egg nucleus. Second, the haploid egg nucleus and its male counterpart, the sperm nucleus, interact to form the diploid genome of the embryo. Third, egg activation initiates the rapid mitotic cycles that characterize embryonic cleavage.

The unfertilized eggs of amphibians and most mammals are arrested during the second metaphase of meiosis. The arrest is caused by a sustained high level of *M-phase promoting factor* (*MPF*). The high MPF level in turn depends on a high concentration of mos protein, and presumably another protein, during oocyte maturation (see Fig. 3.20). This situation changes when Ca^{2+} is released as part of the egg activation process. Ca^{2+} activates *calmodulin,* a Ca^{2+}-dependent protein, which in turn activates a calmodulin-dependent kinase, designated *CaM* K_{II}. The active CaM K_{II} promotes two effects, which seem to occur independently of each other (Lorca et al., 1993). First, the high level of MPF that has kept the egg arrested in metaphase II is reduced so that the cell cycle can continue. Second, the mos protein, the *cytostatic factor* that has maintained the MPF level high in the unfertilized egg, is degraded. In addition, the release of Ca^{2+} inhibits *MAPK,* a step that is necessary and sufficient for initiating DNA synthesis (Carroll et al., 2000).

Completion of meiosis renders a haploid egg nucleus, the ***female pronucleus.*** Its male counterpart derives from the sperm nucleus, the envelope of which disintegrates into small vesicles, thus exposing the sperm chromatin to the egg cytoplasm (G. R. Green and E. L. Poccia, 1985). The sperm nucleus then swells as its chromatin decondenses, and its envelope reconstitutes. At this time, the former sperm nucleus is called the ***male pronucleus.*** Meanwhile, the centrosome introduced by the sperm has organized an aster of microtubules, the *sperm aster* (Fig. 4.15). As soon as sperm aster microtubules contact the female pronucleus, *motor proteins* begin to move the pronuclei toward each other.

The fusion of a male and a female gamete concludes the process of fertilization by forming a single cell, the zygote. Both gametes contribute corresponding sets of chromosomes to the zygote. Nearly all of the zygote cytoplasm and its organelles are, of course, derived from the egg. However, the *centrosome* in most species, including humans, is lost during oogenesis and is restored to the zygote from the centriole pair located at the base of the sperm's flagellum. Upon fertilization, this centriole pair recruits egg components to rebuild a centrosome, which replicates and organizes the first mitotic spindle in the zygote (Schatten, 1994; Simerly 1995). The flagellum and the mitochondria of the sperm enter the egg but usually disintegrate, although in some species

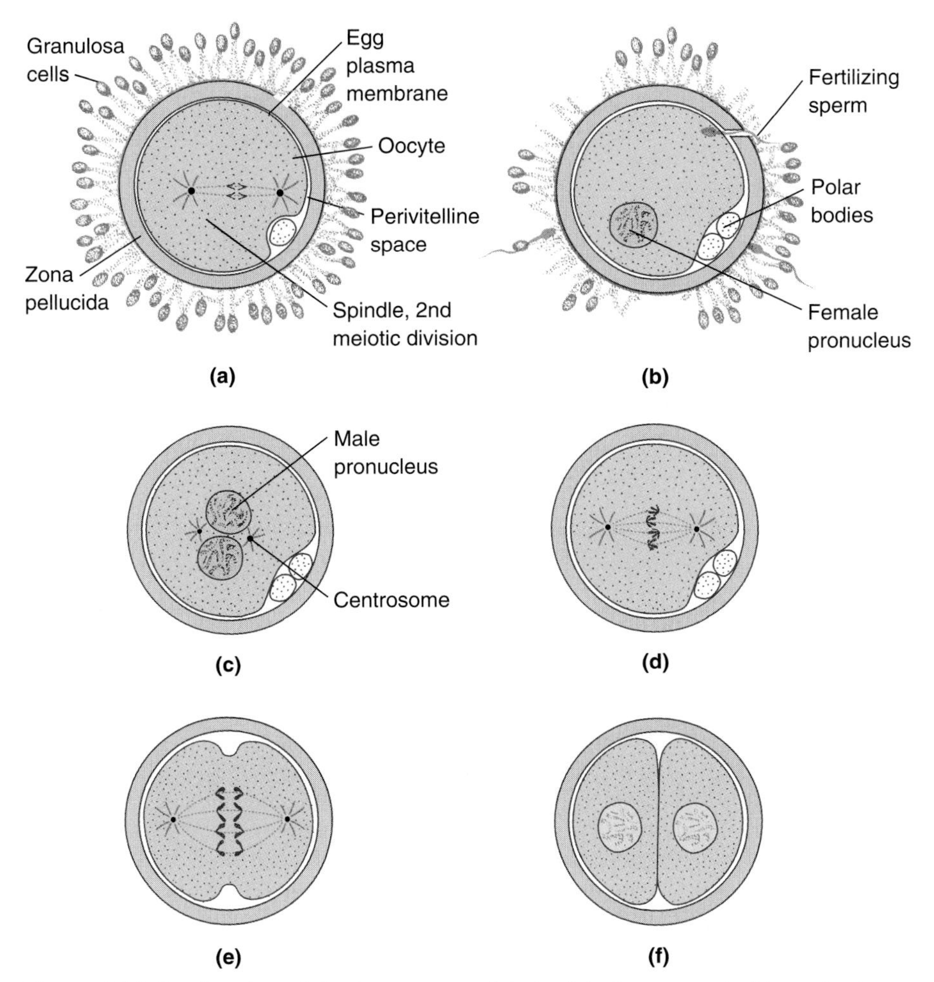

Figure 4.16 Fertilization of the human egg. **(a)** Ovulated egg arrested in metaphase II. **(b)** One sperm has entered the egg, which has completed meiosis II and formed a female pronucleus. The second polar body has been given off. **(c)** The sperm nucleus decondenses and forms the male pronucleus. **(d)** The nuclear envelopes of both pronuclei disintegrate, and their chromosomes are incorporated into a common mitotic spindle. **(e)** Anaphase of first embryonic cleavage. **(f)** Embryo at the 2-cell stage.

the sperm mitochondria are conserved during early cleavage (Giles et al., 1980; Zouros et al., 1992; Sutovsky et al., 1996). The plasma membrane of the zygote consists mostly of the egg membrane, with a small contribution by the sperm plasma membrane.

The process by which a diploid nucleus forms from the male and female pronuclei varies among groups of animals. In sea urchins and their relatives, the envelopes of the pronuclei fuse in a process called ***syngamy*** and form a common nuclear envelope around the maternal and paternal chromosomes. However, other animals, including most mammals, skip this step. Instead, the male and female pronuclei retain separate nuclear envelopes until just prior to mitosis. Then, both nuclear envelopes break up, and the maternal and paternal chromosomes align within the same mitotic spindle (Fig. 4.16). Diploid nuclei therefore do not originate until after the first mitosis in these animals.

ACTIVATION ACCELERATES THE EGG'S METABOLISM IN PREPARATION FOR CLEAVAGE

While the unfertilized egg is generally a sleepy cell, fertilization is the wake-up call to a stage of fast-paced action: *cleavage*. Cleavage entails the rapid replication of DNA, including the synthesis of its building blocks, the nucleotides. In addition, histones and other chromosomal proteins need to be synthesized in quantity. The same holds for cyclins, cyclin-dependent kinases, and other proteins involved in regulating the cell cycle. In most species, the mRNAs for the synthesis of early embryonic proteins have been stored during oogenesis. Thus, egg activation entails a massive recruitment of stored maternal mRNA into polysomes for rapid translation. Besides proteins, the embryo needs to synthesize large amounts of plasma membrane to cover the additional cell surfaces generated by cleavage. A key step in accelerating membrane formation is the activation of an enzyme, NAD^+ kinase, which phosphorylates NAD^+ to $NADP^+$, a coenzyme in lipid biosynthesis (Epel et al., 1981).

The start signal for many of these metabolic steps is a surge of intracellular Ca^{2+} concentration as discussed earlier. An additional signal, at least in sea urchin eggs, is a sudden *rise in intracellular pH,* which is caused by plasma membrane proteins that pump hydrogen ions (H^+) out of the egg by exchanging them for sodium ions (Na^+) from the surrounding medium (see Fig. 4.13). Procedures that prevent this increase in pH inhibit DNA and protein synthesis, whereas artificially raising the pH of unfertilized eggs boosts protein synthesis (Winkler, 1988).

4.5 Blocks to Polyspermy

Eggs of many species have thousands of adhesion sites on the vitelline envelope, presumably to ensure that fertilization can occur wherever a lone sperm may land on the egg. However, the excess of adhesion sites is potentially troublesome: it could lead to fertilization of a single egg by more than one sperm. Such a condition, known as ***polyspermy,*** would be disastrous in many species. The fertilization of a sea urchin egg by two sperm, for example, results in a zygote with three haploid sets of chromosomes. Also, the two centrosomes introduced by the two sperm set up a mitotic spindle with four poles. The ensuing cleavage produces blastomeres with irregular numbers of chromosomes (see Fig. 15.3). Such embryos usually die early (Boveri, 1907).

Multiple mechanisms have evolved to prevent either polyspermy itself or its consequences. Many species, including some insects, most salamanders, reptiles, and birds, are naturally polyspermic, but extra sperm are somehow inactivated in the eggs of these animals. In other species, polyspermy itself is prevented, typically by two independent mechanisms. One mechanism, known as the ***fast block to polyspermy,*** is rapid but temporary. The other mechanism, called the ***slow block to polyspermy,*** takes time to get under way but is permanent.

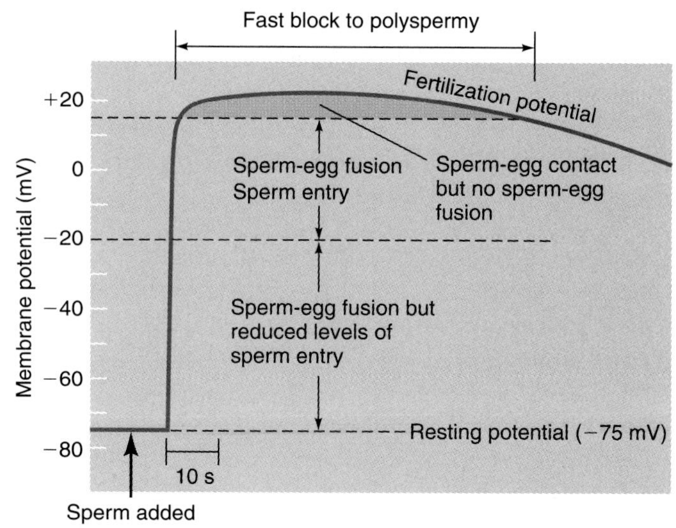

Figure 4.17 Resting potential and fertilization potential in the egg of the sea urchin *Strongylocentrotus purpuratus.* The resting potential is negative inside the cytoplasm (−75 mV). The first sperm-egg contact triggers a fertilization potential, which rises to near +20 mV for about 1 min. Whereas the resting potential permits membrane fusion, sperm entry requires a shift to a membrane potential between −20 and +17 mV. During this shift, which takes about 1 s (scale bar represents 10 s), a fused sperm begins to enter. A potential above +17 mV allows the first sperm to complete its entry, but any additional sperm make only membrane contact and do not proceed with sperm-egg fusion; thus the fertilization potential acts as a fast block to polyspermy. This block lasts for about 1 min, after which the membrane potential returns to intermediate values that allow all fertilization events to occur.

THE FERTILIZATION POTENTIAL SERVES AS A FAST BLOCK TO POLYSPERMY

When sea urchin eggs are fertilized in the laboratory with excess sperm, they rarely become polyspermic. As soon as one sperm contacts the egg, conditions change to prevent the plasma membrane from fusing with additional sperm. Some inhibitory signal must spread around the entire egg surface from the site of first plasma membrane contact. This mechanism is effective even when sperm concentrations are high enough for many sperm-egg collisions per second. Given an egg diameter of 0.1 mm, diffusion of a chemical signal would be much too slow for a response time of less than 1 s. Only an electrical signal can travel fast enough.

By inserting an electrode into an egg and placing another electrode on the outside, one can measure an electric potential across the plasma membrane. This is known as the ***resting potential,*** which is generated—as it is in neurons—by concentration differences in potassium (K^+) and sodium (Na^+) ions inside and outside the egg. In unfertilized sea urchin eggs, the resting potential is about −75 mV (L.A. Jaffe, 1976; Nuccitelli and Grey, 1984). Upon contact with a sperm, the membrane potential changes temporarily to a ***fertilization potential*** of near +20 mV (see Fig. 4.13; Fig. 4.17). The membrane potential remains positive for about 1 min and then gradually returns to the resting potential.

The fertilization potential is caused by increased membrane permeability to certain ions, a change that represents the first step in the egg activation cascade. The nature of the ion channels that generate the fertilization potential varies among animal groups (L. A. Jaffe and Gould, 1985; Elinson, 1986). In sea urchin eggs, the fertilization potential is caused by an increase in the membrane's permeability to sodium ions (Na^+), temporarily allowing these positively charged ions to enter the egg. In frog eggs, the fertilization potential results from an increase in the membrane's permeability to chloride ions (Cl^-), allowing these negatively charged ions to leave the egg.

The fertilization potential serves as a fast block to polyspermy in sea urchin eggs. This was shown by using electric current to maintain, or "clamp," the membrane potential of sea urchin eggs at different levels and then exposing these eggs to sperm (L. A. Jaffe, 1976). At a constant membrane potential of +20 mV or more, sperm adhered, but membrane fusion did not occur. The molecular basis for the observed voltage dependence of membrane fusion is still under investigation.

In addition to controlling membrane fusion, the membrane potential also affects fertilization cone formation and sperm entry, at least in sea urchins (Lynn et al., 1988). If the egg's membrane potential is held between −70 and −20 mV, fertilization cones begin to form. However, the cones do not develop fully, and sperm entry usually fails (Fig. 4.17). At membrane potentials between −20 and +17 mV, membrane contact is followed by membrane fusion, full fertilization cone formation, and sperm entry. At membrane potentials above +17 mV, sperm contact elicits no further response from the egg.

Evidently, a sperm can initiate fertilization at −70 mV, since this is the resting potential, but for sperm entry to proceed, the membrane potential must shift to a more positive level. Only during the fleeting moment (about 1 s) while the membrane potential rises from −70 to +17 mV is a sperm admitted to the egg. Once membrane fusion has occurred, entry of the *first* sperm can continue at +20 mV, but additional sperm cannot enter because they fail to fuse with the egg. However, the electrical block to polyspermy is only temporary, because the intermediate potential that allows fertilization is restored after about 1 min.

THE CORTICAL REACTION CAUSES A SLOW BLOCK TO POLYSPERMY

Another key event in egg activation is the ***cortical reaction,*** also known as the ***zona reaction*** in mammals (see Fig. 4.13). The cortical reaction is the exocytosis of ***cortical granules,*** which are membrane-bound vesicles derived from the Golgi apparatus and deployed beneath the egg plasma membrane during oocyte maturation (Laidlaw and Wessel, 1994). An unfertilized mouse egg has about 4000 cortical granules, while a sea urchin egg has 15,000 or so. During egg activation, the cortical granules undergo *exocytosis,* releasing their contents into the ***perivitelline space*** between the plasma membrane and the vitelline envelope (Fig. 4.18). In sea urchins, the cortical reaction is followed by the regulated exocytosis of additional vesicles that leads to the sequential construction of an extracellular layer of materials surrounding the fertilized egg (Matese et al., 1997).

The cortical reaction begins soon after sperm-egg contact and is completed in one or two minutes, depending on egg size. In mammals, the cortical reaction seems to be triggered by protein kinase C (PKC), which in turn is activated by diacylglycerol, and possibly, Ca^{2+} (see Fig. 2.27; Gallicano et al., 1993; Olds et al., 1995). In sea urchins, the cortical reaction appears to be triggered by a more direct effect of Ca^{2+}. Both a calcium wave and cortical granule exocytosis sweep across the egg in a wave initiated at the point of sperm entry, with exocytosis following calcium release after about 6 s (Matese and Clay, 1998). The exocytosis of cortical granules—similar to the exocytosis of the acrosomal vesicle—involves interactions between so-called SNARE proteins located on the cytoplasmic faces of the exocytosing vesicles and their target plasma membranes (Rothman, 1994; Conner et al., 1997).

In sea urchin eggs, the components released during the cortical reaction have three major effects (Fig. 4.18). First, proteases cleave the proteins that tether the vitelline envelope to the plasma membrane. At the same time, the cortical granules shed complex polysaccharides known as *glycosaminoglycans,* which attract water into the perivitelline space. The resulting gelatinous layer, known as the ***hyaline layer,*** lifts the vitelline envelope off the egg plasma membrane. Second, *peroxidases* harden the vitelline envelope by cross-linking adjacent proteins. The hardened envelope is then called the ***fertilization envelope.*** Third, enzymes modify sperm receptors and other components of the vitelline envelope so that sperm no longer adhere to the egg surface.

The cortical reaction causes a slow block to polyspermy. Its protective action may be required because the egg underneath the fertilization envelope is still fertilizable: If such an egg is stripped of its fertilization envelope and hyaline layer, then additional sperm will enter. However, the fertilization envelope permanently inhibits the penetration of more sperm and even causes previously adhering sperm to fall off (Fig. 4.19).

The *zona reaction* in mammalian eggs is similar to the cortical reaction in sea urchin eggs and has very similar results. Certain enzymes released from the cortical granules cross-link the zona proteins, thus making the zona impermeable to sperm. Other enzymes modify the ZP3 glycoprotein, so that ZP3 no longer binds to sperm or elicits the acrosome reaction.

4.6 The Principle of Overlapping Mechanisms

Many steps in development rely on two or more mechanisms that complement or reinforce each other. We call this the ***principle of overlapping mechanisms.*** The eminent embryologist Hans Spemann (1938) referred to it as the "principle of double assurance "or the "synergetic principle of development." The degree of overlap between different mechanisms that support the same biological process differs from case to case. In this chapter, we have come across overlapping mechanisms for sperm-egg adhesion, for the penetration of the zona pellucida, for triggering egg activation, and for preventing polyspermy. With regard to the last case, we saw that many animal eggs have two blocks to polyspermy: a fast block relying on an electric membrane potential, and a slow block relying on the formation of the fertilization envelope. The latter event in turn relies on three mechanisms—namely, the cleavage of the tethering proteins and formation of the hyaline layer, together lifting the vitelline envelope off the egg plasma membrane; the hardening of the vitelline envelope, which makes it impermeable to additional sperm; and the release of sperm-binding molecules from the vitelline envelope.

The fast block to polyspermy holds up about as long as it takes for the slow block to become effective (see Table 4.1). This is an example of minimal overlap between different mechanisms that serve the same biological function. In contrast, the three inhibitory mechanisms that constitute the slow block appear to be redundant. In theory, there should be no need for the vitelline envelope to harden or to lift off the plasma membrane when the sperm-binding molecules have been modified or shed. There are two possible explanations for the fact that these seemingly redundant mechanisms still exist. First, each mechanism might serve as a fail-safe mechanism to back up the other two. If each mechanism by itself eliminated only 9 out of 10 extra sperm, then the three mechanisms together would eliminate 999 of 1000 extra sperm. This would be adequate

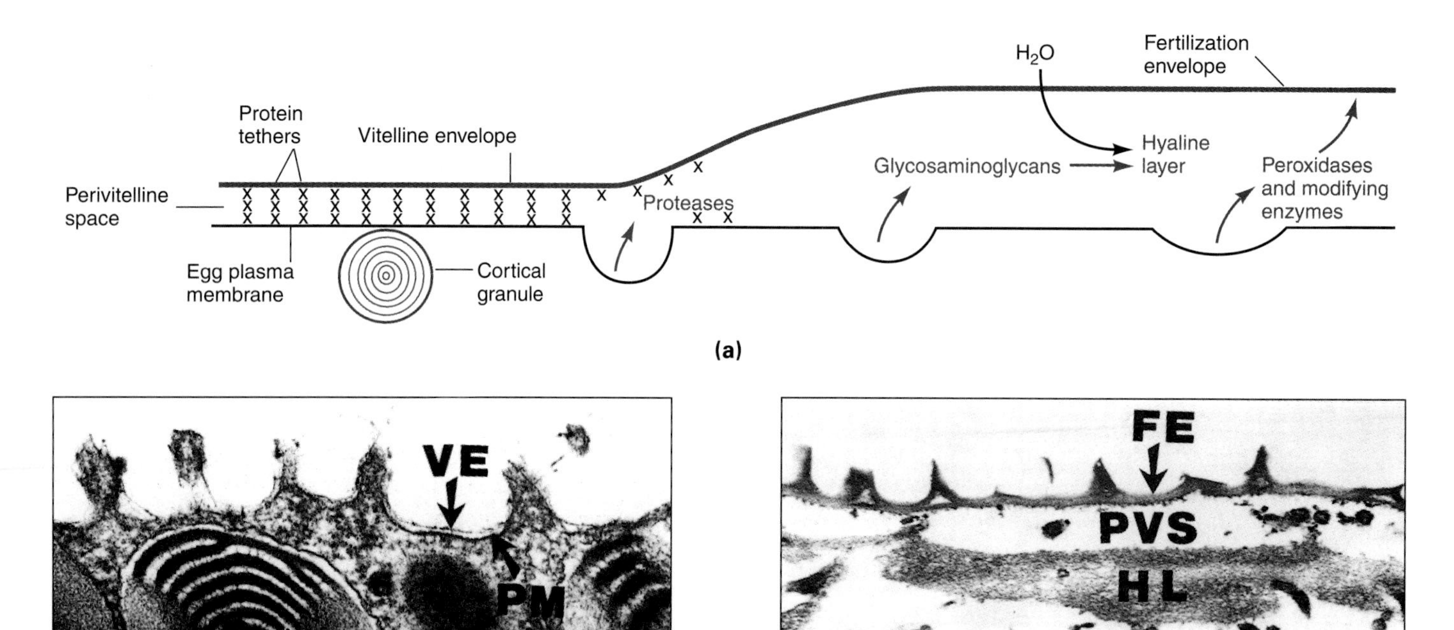

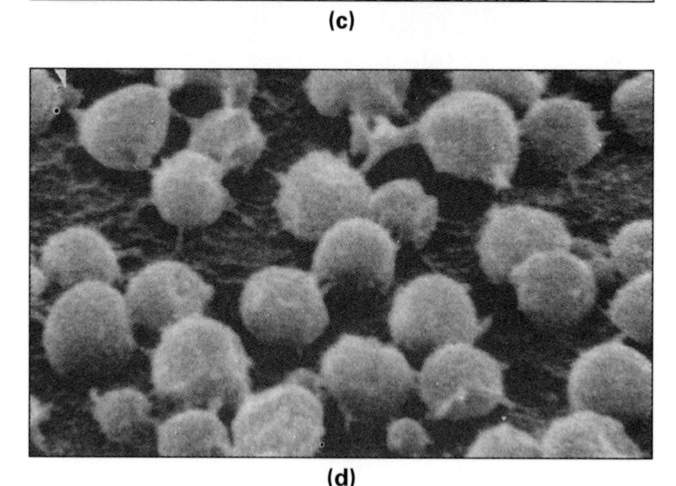

Figure 4.18 Cortical reaction in the sea urchin egg. **(a)** As part of egg activation, the cortical granules beneath the plasma membrane undergo exocytosis, releasing into the perivitelline space several components that support three major processes. First, proteases cleave the proteins that tether the vitelline envelope to the plasma membrane. At the same time, glycosaminoglycans also released from the cortical granules attract water into the perivitelline space and form a hyaline layer that lifts the vitelline envelope off the egg plasma membrane. Second, peroxidases harden the vitelline envelope by cross-linking adjacent proteins. The hardened envelope is then called the fertilization envelope. Third, other enzymes modify vitelline envelope components so that sperm no longer adhere to the egg surface. **(b)** Transmission electron micrograph showing the cortex of an unfertilized sea urchin egg (*Strongylocentrotus purpuratus*). The vitelline envelope (VE) is closely applied to the plasma membrane (PM) and follows the contours of the microvilli. The cortical granules contain lamellar and amorphous (asterisk) components. **(c)** Ten minutes after fertilization, the fertilization envelope (FE) is separated from the egg by a hyaline layer (HL), which builds up in the perivitelline space (PVS). **(d)** Scanning electron micrograph showing the inner aspect of the plasma membrane of an unfertilized egg. The cortical granules (round bodies) are still intact.

protection even for an egg that is exposed to an excess of sperm. Another explanation for the persistence of seemingly redundant mechanisms for one biological function is that each mechanism may have one or more additional functions. For instance, the hardened fertilization envelope also provides mechanical protection, and the hyaline layer also forms a firm coat around the blastula cells, which prevents these cells from leaving the embryo. Thus, what seem to be redundant mechanisms for one function may actually be part of a larger network of mechanisms serving several functions.

Overlapping mechanisms may be crucial for the organism, but they can also be a challenge to researchers. If a given biological function is supported by multiple mechanisms, it is very difficult to analyze any one of them. One way of testing whether a certain mechanism is *necessary* for a given function is to block the mechanism (by mutation, drugs, or other means) and to observe whether the function fails. This test may not reveal a necessary mechanism among a group of overlapping mechanisms, because each blocked mechanism is backed up, at least to an extent, by others that are still working. In this situation, investigators have to interfere with several mechanisms simultaneously before the biological function of interest fails and yields to analysis.

(a)

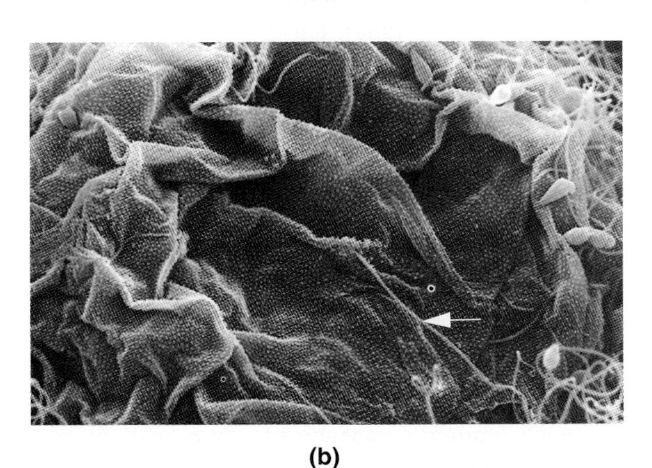

(b)

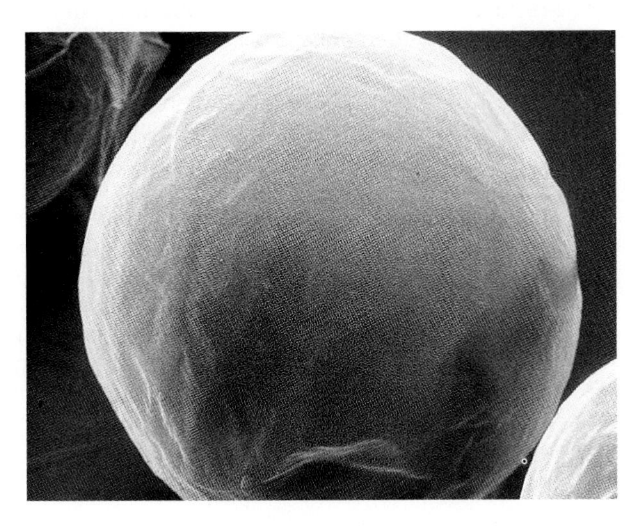

(c)

4.7 Parthenogenesis

Some animals can reproduce by ***parthenogenesis*** (Gk. *parthenos*, "virgin"; *genesis*, "origin"), which is the development of viable offspring from unfertilized eggs. In aphids and other insects, parthenogenesis alternates with sexual reproduction, a reproductive strategy known as ***facultative parthenogenesis.*** Females reproduce parthenogenetically during the summer when food is abundant. At the end of the season, a generation of males and females develops and reproduces sexually. The fertilized eggs from this generation overwinter and develop into females during the next spring. In other insects, including the honeybee, parthenogenesis is coupled with sex determination. Fertilized eggs give rise to diploid females, while unfertilized eggs develop into haploid males called *drones.* Some species, mostly invertebrates but also some lizards, reproduce by ***obligatory parthenogenesis.*** Their populations consist entirely of females.

Parthenogenetic species are faced with three problems. First, they need to compensate for the reduction in number of chromosomes that normally occurs during meiosis. They can do this through several mechanisms, such as skipping one meiotic division or fusing two haploid nuclei after meiosis. The second problem in parthenogenetic species is egg activation. The silver salamander, *Ambystoma platineum,* solves this problem by enlisting the help of males from another species. There are no *A. platineum* males, but the females mate with males of a closely related salamander, *A. jeffersonianum.* The sperm from these males only activate the eggs of *A. platineum* and do not contribute their genome to the offspring (Uzzell, 1964). Other parthenogenetic species use different stimuli to activate their eggs. In parasitic wasps, unfertilized eggs are activated by friction or distortion when they pass through the narrow ovipositor (Went and Krause, 1973). The third problem

Figure 4.19 Removal of excess sperm after the cortical reaction in the egg of the sea urchin *Strongylocentrotus purpuratus.* **(a)** Scanning electron micrograph of an egg 15 s after the addition of sperm. **(b)** Thirty seconds after the addition of sperm. The arrow marks the tail of the fertilizing sperm. The progress of the cortical reaction is indicated by the zone of sperm detachment surrounding the fertilizing sperm. The wrinkles in the fertilization envelope are an artifact of fixation. **(c)** Three minutes after the addition of sperm. The fertilization envelope has hardened and no longer wrinkles during fixation. All sperm have detached.

in parthenogenesis is the lack of a centrosome, which is normally introduced by the sperm and serves to organize the first mitotic spindle. However, a centrosome may also be contributed by the egg, or formed from smaller components in the egg cytoplasm, or replaced functionally by other microtubule organizing centers.

In the laboratory, a variety of physical and chemical treatments have been used to activate eggs. One of the most commonly used methods is simply to prick the eggs with a needle, preferably one that has been dipped in blood or other tissues. This treatment may trigger an early event in the activation cascade, such as the Ca^{2+} transient, which then elicits the subsequent events. Impurities on the pricking needle may also serve as nucleating centers for spindle formation. For development to advanced stages, the diploid chromosome number must usually be restored. However, adults have been obtained, with varying success, through artificial activation of unfertilized eggs from sea urchins, starfish, silk moths, fishes, and frogs (Beatty, 1967).

Natural parthenogenesis is unknown in mammals. Unfertilized mouse eggs activated in a series of experiments did not develop beyond day 11, halfway through their gestation. By transplanting pronuclei between fertilized mouse eggs, one can construct zygotes with two female pronuclei and no male pronucleus (Table 4.2). Such zygotes develop into bimaternal embryos, which cease development at about the same time as parthenogenetic embryos. Zygotes with two male pronuclei and no female pronucleus develop into bipaternal embryos; they also die midway through gestation but show different defects from the bimaternal embryos. Control zygotes with one male and one female pronucleus, generated by the same transplantation technique, can develop normally (McGrath and Solter, 1984). These results indicate that in mammals both a female pronucleus and a male pronucleus are needed for embryonic development. Certain mammalian genes can be activated only in the female pronucleus, whereas other genes must be introduced through the male germ line in order to be expressed (Surani et al., 1986; see also Section 16.6).

table 4.2 Pronuclear Transplantation Experiments with Fertilized Mouse Eggs

Class of Reconstructed Zygotes	Operation	Number of Successful Transplants	Number of Progeny Surviving
Bimaternal	♀ ♂ ♂ ♀	339	0
Bipaternal	♀ ♂ ♀ ♂	328	0
Control	♀ ♂ ♀ ♂	348	18

Source: McGrath and Solter (1984) and Gilbert (1991). Used by permission.

SUMMARY

Fertilization is the union of two haploid gametes—one egg and one sperm—to form a diploid zygote from which a new individual develops. In addition to properly timed parental mating behavior, some animals use chemical attractants and capacitation mechanisms to ensure that the egg and the sperm are ready for fertilization when they encounter each other. In most species, sperm penetrate several egg coats before reaching the egg plasma membrane. Of particular importance is the vitelline envelope—in mammals called the zona pellucida—which is a coat of proteins laid down by the oocyte before ovulation. The interaction of the sperm with the vitelline envelope or another egg coat triggers the acrosome reaction, in which the sperm releases the contents of its acrosome. Enzymes from the acrosome help the sperm to lyse its way to the egg cell proper. Sperm adhere to the egg surface in a species-specific way, by means of matching molecules on the sperm head and the vitelline envelope of an egg from the same species.

The contact or fusion of the egg and sperm plasma membranes triggers a cascade of egg responses collectively called egg activation. It includes sperm entry, transient rises in calcium ion (Ca^{2+}) concentration and pH, acceleration of many metabolic processes, and resumption of the cell cycle. In many species, egg activation also entails changes at the egg surface that prevent polyspermy. The eggs of such species typically use a fast but temporary electrical block before a slower, permanent block sets in. The slow block results from the cortical reaction, in which cortical granules located beneath the egg plasma membrane release their contents by exocytosis. The cortical reaction alters the vitelline envelope in several ways that make it impermeable to sperm.

Some animals can reproduce by parthenogenesis, the development of viable offspring from unfertilized eggs. These species use several mechanisms to avoid or compensate for the reduction in number of chromosomes that occurs during meiosis and to trigger egg activation.

Cleavage

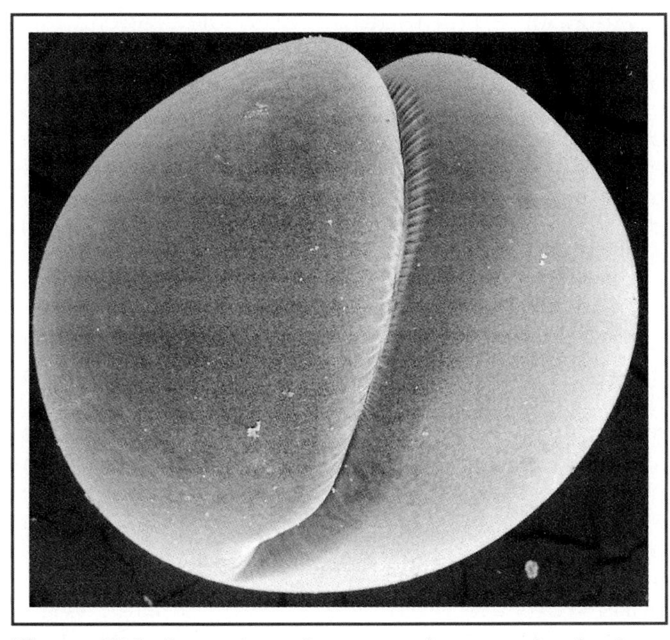

Figure 5.1 Scanning electron micrograph of a frog egg during its first cleavage. The first cleavage furrow forms at the end of the first mitosis. The furrow originates at the animal pole (facing up in this picture), where the asters of the mitotic spindle are closest to the egg surface. The furrow extends toward the vegetal pole, pinching the egg until it is divided into two blastomeres. The mechanical force to cut through the egg is generated by a contractile ring of actin and myosin filaments. These filaments assemble in the cytoplasm at the base of the furrow and are connected by linker proteins to the plasma membrane. The small folds visible near the base of the cleavage furrow reveal the stress that contraction of the ring imposes on the plasma membrane. (*From Beams H. W. and Kessel R. G. [1976] Cytokinesis: A comparative study of cytoplasmic division in animal cells,* Am. Sci. ***64****:279–290.*)

IN all metazoa, fertilization is followed by a series of rapid mitotic divisions. The zygote divides first into two blastomeres, then into four, and so forth, in exponential progression (Figs. 5.1 and 5.2). This embryonic stage is known as cleavage; its functions in the metazoan life cycle can be summarized as follows:

- Generating a large number of cells
- Generating many copies of the genome
- Segregating cytoplasmic components into different blastomeres
- Increasing the nucleocytoplasmic ratio

First, cleavage generates a large number of cells that can move relative to each other and undergo gastrulation and organogenesis. (By analogy, if you wanted to build a house from one large boulder, you would first have to cut the boulder into manageable stones.) Second, the mitotic divisions that take place during cleavage generate many copies of the zygotic genome. This allows cells to express distinct subsets of the complete genetic information and embark on different pathways of development. Third, cleavage may segregate unevenly distributed cytoplasmic components into different blastomeres, thus creating diverse cytoplasmic environments that will elicit different gene activities. Finally, cleavage increases the value of a critical cellular parameter, the ***nucleocytoplasmic ratio***; it is defined as the ratio of nuclear volume to cytoplasmic volume. Increasing the nucleocytoplasmic ratio is imperative for the turnover of RNA and proteins, which, like most other molecules in a cell, have short life spans; some proteins last only for minutes. Thus, just to sustain itself a cell must constantly synthesize RNA and proteins. The smaller the nucleocytoplasmic ratio, the more difficult it becomes for a cell to transcribe enough RNA to replace the protein losses from a large cytoplasm.

Animal eggs are inordinately large cells with very small nucleocytoplasmic ratios. Many animal species have eggs big enough to sustain the development of advanced larvae that can move and feed. However, the extremely unfavorable nucleocytoplasmic ratio of large eggs renders them highly perishable. For this reason, it would seem adaptive to increase the nucleocytoplasmic ratio and pass through this perilous stage of development as quickly as possible (Bier, 1964). Accordingly, cleavage divisions differ from later cell divisions in two respects. First, blastomeres do not grow between cleavage divisions as more mature cells do (Fig. 5.2). Second, cleavage proceeds more rapidly than later cell divisions; typical cleavage cycles are completed in less than an hour, whereas later cell cycles last several hours to many days.

Another advantage of rapid cleavage might be to reduce the risk of predation during an immobile and defenseless phase of the life cycle. In accord with this hypothesis, cleavage proceeds at a very leisurely pace in mammalian embryos, which are well protected in the maternal uterus. Mammalian eggs are also comparatively small, so the need to increase the nucleocytoplasmic ratio is less urgent.

While increasing the nucleocytoplasmic ratio is a universal aspect of cleavage, animal forms differ in their cleavage patterns—that is, the relative sizes of blastomeres and their configurations. In the first half of this chapter, we will study the cleavage patterns of various animals, including sea urchins, frogs, chickens, mammals, and insects. Some animals have strictly standardized cleavage patterns that each embryo faithfully replicates. The embryos of other species cleave less precisely, and any deviations from the norm are corrected later. While the former group relies on precision, the latter follows a strategy of stepwise approximation.

In the second half of this chapter, we will discuss some mechanical aspects of cleavage. What controls the positioning and orientation of mitotic spindles, and thus, the cleavage pattern? How is the cell cycle regulated first during cleavage and later during mature cell divisions? How does the embryo know when its cells have become small enough?

5.1 Yolk Distribution and Cleavage Pattern

The amount and distribution of yolk in an egg have a major impact on its cleavage pattern (Fig. 5.3 and Table 5.1). Eggs that have a small amount of evenly distributed yolk in the cytoplasm are called ***isolecithal*** (Gk. *isos,* "equal"; *lekithos,* "yolk"). Most echinoderms, molluscs, ascidians, and mammals produce this type of egg. (In this context, it is customary to speak of the cleavage of "eggs," although technically they are zygotes or embryos.) Eggs with a moderate amount of yolk present mostly in the vegetal hemisphere are called ***mesolecithal*** (Gk. *mesos,* "middle"); amphibians have mesolecithal eggs. Eggs with a large amount of yolk filling the entire egg except for a small area near the animal pole are called ***telolecithal*** (Gk. *telos,* "end"); most fish, reptiles, and birds have telolecithal eggs. Eggs in which the yolk

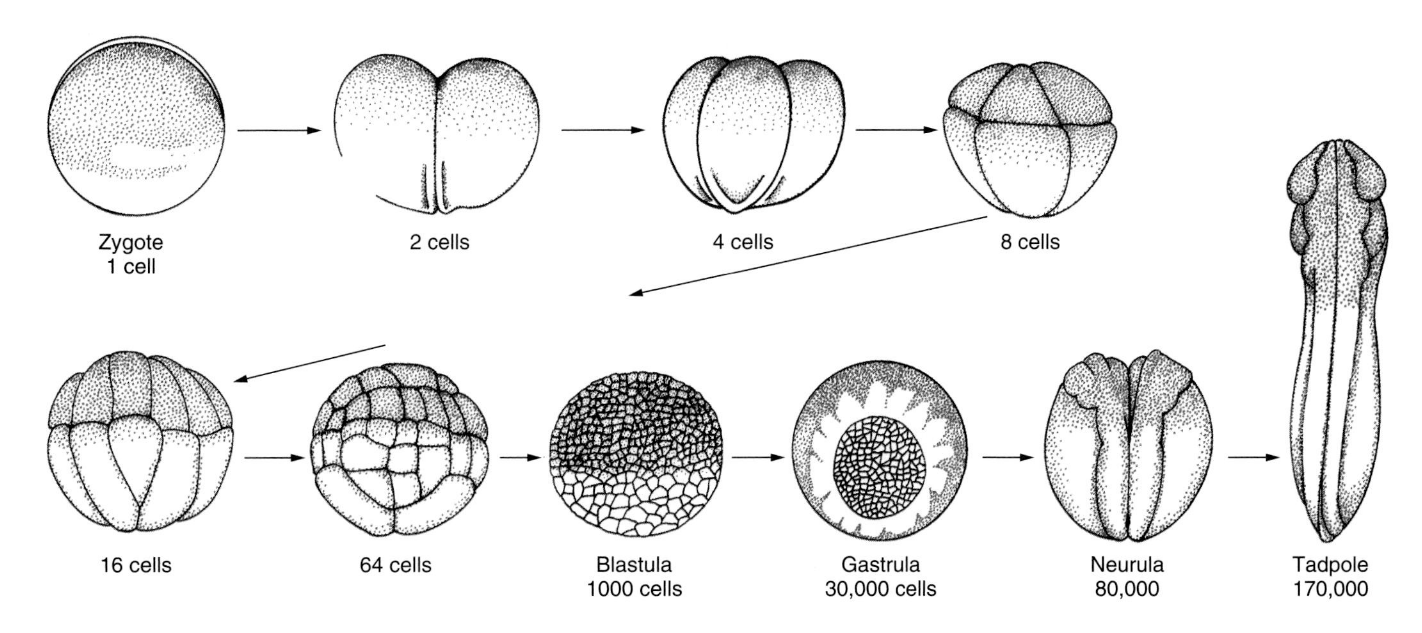

Figure 5.2 Development of the South African clawed frog, *Xenopus laevis,* from the fertilized egg to the tail bud stage. During this period, the total volume of the embryo remains nearly constant while the number of cells increases by several orders of magnitude.

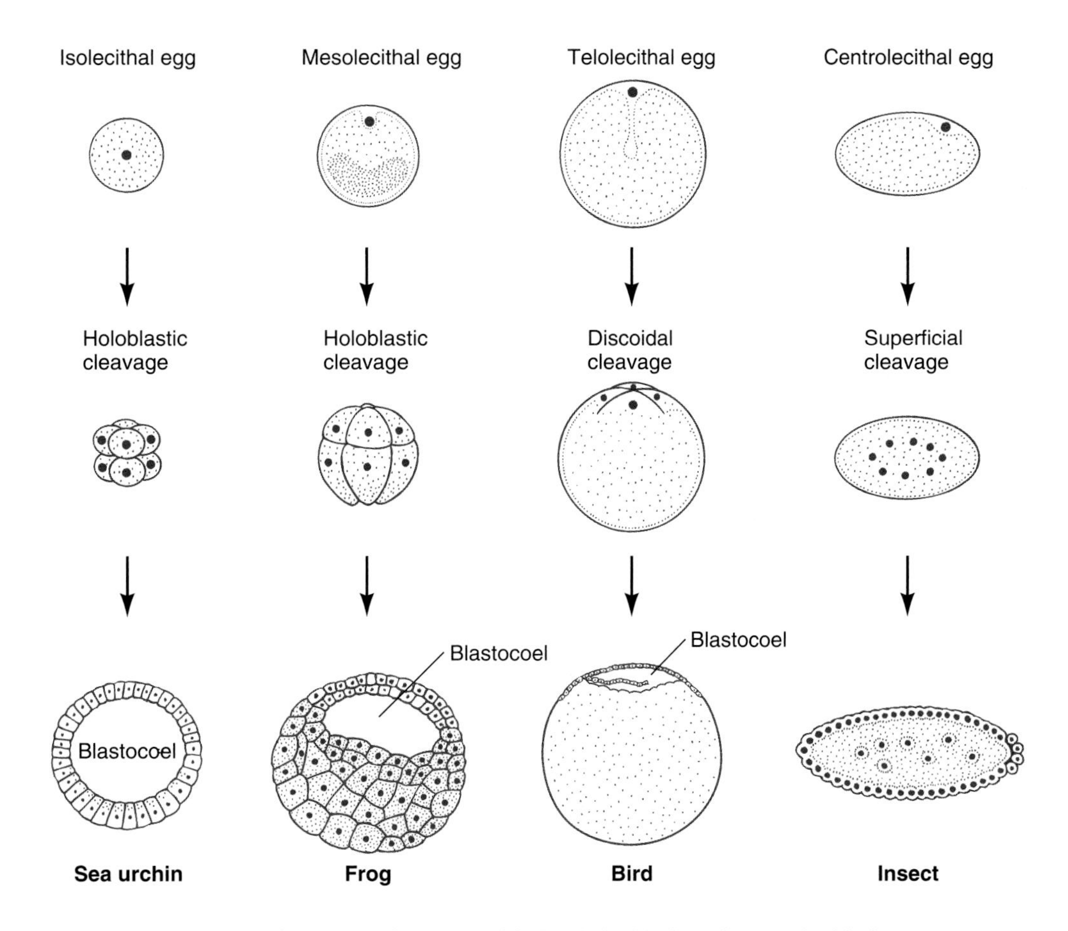

Figure 5.3 Typical cleavage patterns of isolecithal, mesolecithal, telolecithal, and centrolecithal eggs.

table 5.1 Cleavage Patterns in Relation to Yolk Distribution in Eggs*

Type of Egg (Based on Yolk Distribution)	Type of Cleavage	Representative Pattern of Cleavage	Blastula	Animal Groups
Isolecithal (little yolk, evenly distributed)	Holoblastic (egg cell completely cleaved)	Radial Bilateral Spiral Rotational	Spherical single layer of blastomeres surrounding large, fluid-filled blastocoel	Echinoderms, ascidians, molluscs, annelids, mammals
Mesolecithal (moderate amount of yolk, mostly in vegetal hemisphere)	Holoblastic	Radial	Sphere of multiple cell layers enclosing an eccentric, fluid-filled blastocoel	Amphibians, some fishes
Telolecithal (large amount of yolk, except for blastodisc at animal pole)	Meroblastic (egg cell incompletely cleaved)	Discoidal (blastomeres form disc on uncleaved yolk)	Cellular disc consisting of epiblast and hypoblast, with flat space in between	Most fishes, birds, reptiles
Centrolecithal (yolk concentrated in center of egg)	Meroblastic	Superficial (blastomeres form at the egg surface)	Ovoid consisting of single cell layer (blastoderm) surrounding central yolk	Insects and other arthropods

*See also Figure 5.3.
After B. M. Carlson (1988).

is concentrated in the central cytoplasm are called ***centrolecithal.*** The eggs of virtually all insects and many other arthropods belong to this type.

Isolecithal and mesolecithal eggs undergo a type of cleavage known as ***holoblastic cleavage*** (Gk. *holos,* "whole"; *blastos,* "germ"). This term indicates that the entire egg is cleaved during cytokinesis. Eggs that are characterized by partial cytokinesis are said to undergo ***meroblastic cleavage*** (Gk. *meros,* "part"). There are two major types of meroblastic cleavage. The first, ***discoidal cleavage,*** is restricted to a small disc of yolk-free cytoplasm at the animal pole; a large amount of yolk-rich cytoplasm at the vegetal pole remains uncleaved. Discoidal cleavage is characteristic of telolecithal eggs. In the second type of meroblastic cleavage, called ***superficial cleavage,*** cytokinesis is limited to a surface layer of clear cytoplasm while the yolk-rich inner cytoplasm remains uncleaved. Superficial cleavage is characteristic of centrolecithal eggs.

All large eggs have meroblastic cleavage, presumably because large masses of yolk interfere with cytokinesis. However, meroblastic cleavage is not limited to large eggs. Virtually all insect eggs, large or small, cleave superficially. Thus, if a phylogenetic group includes species with eggs of different sizes, then the meroblastic cleavage pattern, which is adapted to large eggs, may carry over to the species with small eggs as part of the overall developmental program that has evolved in the group.

Various groups of animals have particular blastomere arrangements or symmetries that gives their early embryos a unique appearance. Some of these patterns will be discussed in the following section.

5.2 Cleavage Patterns of Representative Animals

This section will describe the cleavage patterns of sea urchins, amphibians, snails, ascidians, mammals, fishes, birds, and insects.

SEA URCHINS HAVE ISOLECITHAL EGGS AND UNDERGO HOLOBLASTIC CLEAVAGE

Most sea urchin eggs are isolecithal and undergo holoblastic cleavage. As discussed in Section 4.4, fertilization in sea urchins leads to a union of pronuclei. Cleavage begins with the assembly of the first mitotic spindle, which is oriented perpendicular to the *animal-vegetal axis* (Fig. 5.4). After mitosis, cytokinesis occurs in the ***cleavage plane,*** which is perpendicular to the axis of the mitotic spindle.

The first cleavage plane passes through the animal and vegetal poles, creating two blastomeres of equal size. This type of cleavage is called ***meridional,*** because the cleavage furrow runs through the poles like a meridian on a globe. For the second mitosis, spindles form simultaneously in each of the two blastomeres. Their axes are still perpendicular to the animal-vegetal axis, and they are also perpendicular to the first mitotic spindle. Thus

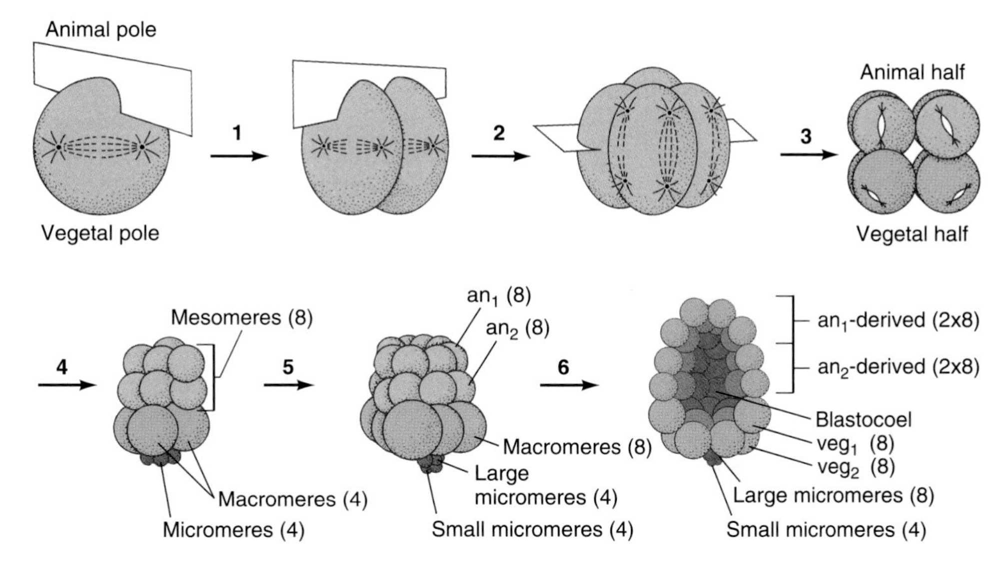

Figure 5.4 Cleavage of the sea urchin egg. Cleavages 1 and 2 are meridional, passing through the animal-vegetal axis. Cleavage 3 is equatorial and perpendicular to the animal-vegetal axis. Cleavage 4 features oblique spindle orientations. The animal blastomeres cleave nearly equally, producing eight mesomeres, whereas the vegetal blastomeres cleave unequally, producing four macromeres and four micromeres. Cleavage 5 produces two animal tiers (an_1 and an_2) of eight mesomeres each, one vegetal tier of eight macromeres, four large micromeres, and four small micromeres. The stage after the sixth cleavage is cut open to reveal the blastocoel. There are now two vegetal tiers (veg_1 and veg_2) of macromeres. The small micromeres do not divide during cleavage 6.

the second cleavage, which results in four blastomeres of equal size, is again meridional. For the third mitosis, the spindles are parallel to the animal-vegetal axis. The subsequent cleavage is called ***equatorial,*** because together the four cleavage furrows circle the embryo like the equator on a globe. This cleavage separates the embryo into four animal and four vegetal blastomeres.

The fourth cleavage in sea urchins shows a unique pattern (Summers et al., 1993). The animal blastomeres cleave in a slightly oblique orientation, generating two tiers of blastomeres that are somewhat offset (Fig. 5.4). The spindles in the vegetal blastomeres are displaced and tilted toward the vegetal pole. At cytokinesis, the vegetal blastomeres divide into four large cells, called ***macromeres,*** and four much smaller cells, called ***micromeres,*** which are pinched off at the vegetal pole from their larger sister cells. The eight animal blastomeres are intermediate in size and are known as ***mesomeres.*** The cleavage of the animal blastomeres features ***equal cytokinesis,*** because it produces daughter cells of nearly equal size, whereas the cleavage of the vegetal blastomeres shows ***unequal cytokinesis,*** since it gives rise to daughter cells of unequal size.

During the fifth cleavage, the mesomeres divide equally and meridionally, giving rise to two tiers of eight animal blastomeres each. The macromeres divide similarly to produce a tier of eight vegetal blastomeres. The micromeres divide unequally to produce four large and four small micromeres. At the sixth cleavage, mesomeres and macromeres cleave equatorially. The large micromeres also divide, but the small micromeres skip this cleavage, so that the resulting embryo consists of 60 cells. The cleavage pattern of sea urchin eggs is referred to as ***radial cleavage,*** because the blastomeres are arranged in radial symmetry around the animal-vegetal axis.

As cleavage progresses, the blastomeres continue to adhere to the *hyaline layer* formed as part of the *cortical reaction,* and the blastomeres also adhere to one another near their apical surfaces. However, the blastomeres separate at their inner surfaces, thus allowing the formation of the central cavity that is filled with a viscous fluid and known as the *blastocoel.*

AMPHIBIANS HAVE MESOLECITHAL EGGS BUT STILL CLEAVE HOLOBLASTICALLY

Cleavage in most amphibian embryos is holoblastic and radially symmetric, as in sea urchins. However, amphibian eggs contain much more yolk than sea urchin eggs, especially in the vegetal hemisphere, and are therefore classified as mesolecithal. The first cleavage furrow is meridional. It begins to form at the animal pole and progresses toward the vegetal pole, where it meets with resistance from the yolk-rich cytoplasm (see Fig. 5.1). In eggs of the Mexican axolotl, the furrow elongates at a rate of about 1 mm/min in the animal hemisphere but slows down to about 0.02 mm/min as it nears the vegetal pole (Hara, 1977). While the first furrow is still knifing through the vegetal hemisphere, the second furrow has already started at the animal pole, with its plane running perpendicular to the first.

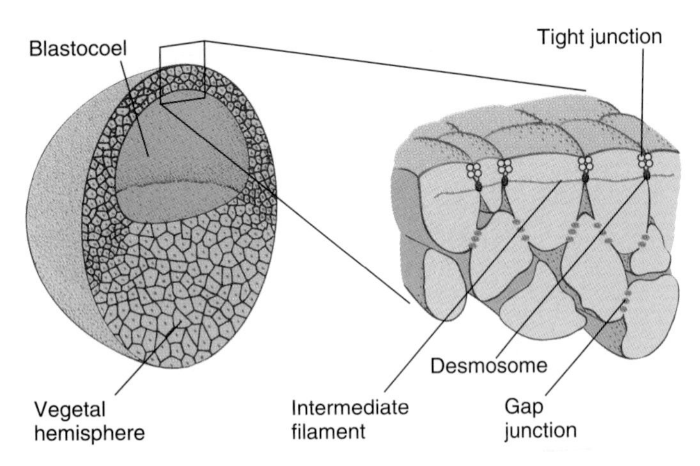

Figure 5.5 Frog blastula. The wall of the blastula is several cell layers thick. There are more layers—and the blastomeres are larger—in the vegetal hemisphere, so that the blastocoel is displaced toward the animal pole. The outermost cells are connected by tight junctions, which create a seal isolating the interior of the embryo from the external medium. These cells also form desmosomes, which connect their intermediate filament systems. Gap junctions provide channels of communication between all blastomeres.

The third cleavage of the amphibian embryo is equatorial. However, in contrast to the corresponding cleavage in the sea urchin, the mitotic spindles of the amphibian egg are displaced toward the animal pole, resulting in an unequal cleavage that separates four small animal blastomeres from four larger vegetal blastomeres (see Fig. 5.2). The fourth cleavage is meridional and the fifth equatorial. Because holoblastically cleaving embryos during the 16-cell and the 32-cell stages outwardly resemble mulberries, these stages are called ***morula*** stages (Lat. *morum,* "mulberry").

As cleavage of the amphibian embryo proceeds, a central blastocoel is formed much as in the sea urchin embryo. However, since the vegetal blastomeres are bigger and contain more yolk than the animal blastomeres, the blastocoel is displaced toward the animal pole. From the 128-cell stage on, the amphibian embryo is considered a *blastula.* In most amphibian species, the blastula develops into a multilayered *epithelium,* in which the outermost cells become *polarized.* Near the *apical* surface of the epithelium, *tight junctions* seal off the inside of the embryo from the external environment (Fig. 5.5; see also Fig. 2.23). The apical cells also form *desmosomes,* which connect the intermediate filament systems of adjacent cells and provide tensile strength to the epithelium. In addition, all adjacent cells in the blastula form *gap junctions,* channels of communication through which small molecules can pass.

The formation of tight junctions allows the embryo to segregate membrane proteins that face the external environment from other membrane proteins that face the blastocoel. This segregation allows blastomeres to secrete molecules toward the blastocoel that build up osmotic pressure.

SNAILS HAVE ISOLECITHAL EGGS AND FOLLOW A SPIRAL CLEAVAGE PATTERN

Spiral cleavage is seen in several groups of worms and in most molluscs, including snails. In this cleavage pattern, certain traits that are weakly expressed in other embryos with radial cleavage are so exaggerated that a distinctly new pattern results (Freeman, 1983). The first two cleavage planes run parallel to the animal-vegetal axis of the zygote, dividing it into four blastomeres. In contrast to sea urchins, the first four blastomeres in snails form a tight arrangement resembling a tetrahedron (compare Figs. 5.4 and 5.6). This tendency to minimize the outer surface area is also observed during later cleavages and makes the embryo look like a cluster of soap bubbles.

During subsequent cleavages, the four large blastomeres (macromeres) bud off quartets of smaller blastomeres (micromeres) that accumulate at the animal pole. However, these micromeres are not aligned with their sister macromeres. The first micromere quartet, given off during the third cleavage, is rotated clockwise when viewed from the animal pole of the embryo (see Figs. I.1 and 5.6). The rotated pattern, which results from the oblique orientation of the mitotic spindles relative to the animal-vegetal axis, gave the spiral cleavage pattern its name. During the fourth cleavage, the spindle orientation changes by about 90° so that the macromeres give off a second micromere quartet that is rotated counterclockwise. At the same time, the first micromere quartet divides in a similar fashion. During the following cleavage, the orientation of the spindles again shifts by 90°, and so forth.

THE ASCIDIAN CLEAVAGE PATTERN IS BILATERALLY SYMMETRICAL

Ascidians, such as the sea squirts, are relatives of the vertebrates, with which they share many features during the embryonic and larval stages. In addition to animal-vegetal polarity, ascidian eggs have an anteroposterior polarity caused by the asymmetrical distribution of several cytoplasmic components. The resulting bilateral symmetry is preserved throughout the cleavage of the zygote (Fig. 5.7). The first cleavage plane passes through the animal-vegetal axis and divides the cytoplasm evenly between the first two blastomeres. This cleavage plane corresponds to the future median plane, which separates the right and left halves of the embryo. The second cleavage is parallel to the animal-vegetal axis but slightly displaced toward the posterior. It separates two large anterior blastomeres from two smaller posterior blastomeres, which inherit cytoplasm of a different composition. This asymmetry is critical to the differential

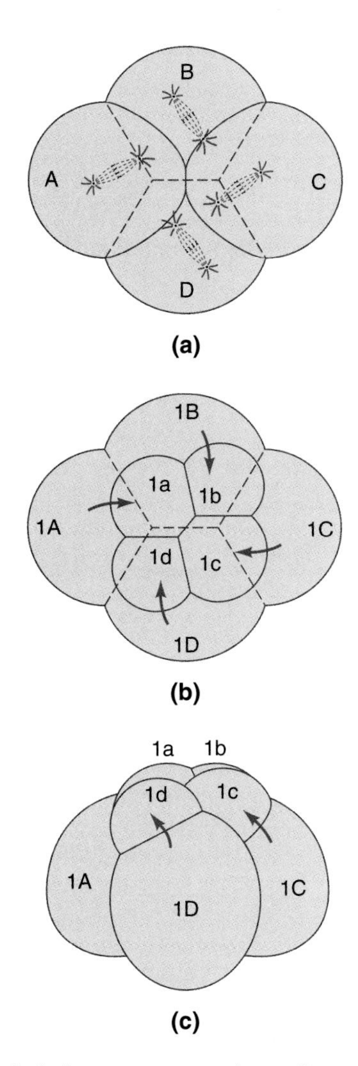

Figure 5.6 Spiral cleavage pattern in molluscs. **(a)** 4-cell stage seen from the animal pole. The first four blastomeres, called macromeres because of their large size, are designated clockwise A, B, C, and D. Blastomeres B and D are located opposite each other and touch at the vegetal pole. (In some species, one of these blastomeres is longer than the other and is then designated the D blastomere.) Blastomeres A and C, also opposite each other, touch at the animal pole. Note the spiral arrangement of the mitotic spindles in preparation for the third cleavage. **(b)** 8-cell stage seen from the animal pole. Arrows connect each micromere with the macromere from which it has budded off. See Figure I.1 on p. 2 for scanning electron micrograph. **(c)** Lateral view of the 8-cell stage.

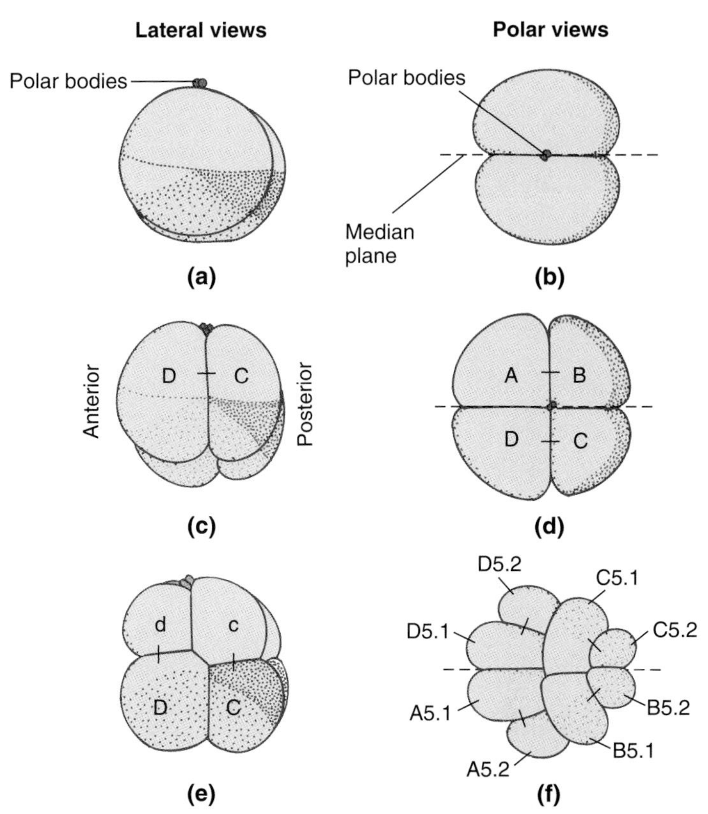

Figure 5.7 Bilaterally symmetrical cleavage pattern in ascidians. **(a)** 2-cell stage in lateral view, anterior to the left, animal pole marked by polar bodies. **(b)** Same stage viewed from animal pole. **(c, d)** 4-cell stage in lateral and animal pole views, respectively. **(e)** 8-cell stage in lateral view. **(f)** 16-cell stage seen from vegetal pole. Note that the embryo has only one plane of symmetry.

development of anterior and posterior blastomeres (see Section 8.5). At this point the cleavage pattern has only one plane of symmetry, the median plane, and is therefore described as *bilaterally symmetrical.* During subsequent cleavages, differences in cell size and shape enhance the bilateral symmetry of the embryo.

MAMMALIAN EGGS SHOW ROTATIONAL CLEAVAGE

The mammalian egg cleaves during its journey down the oviduct, which takes several days in humans (Fig. 5.8). With cell cycles of about 12 hours, cleavage of mammalian eggs is extremely slow. Mammalian cleavage is also unusual in that it is *asynchronous.* In other words, blastomeres do not divide at the same time. An embryo therefore does not always proceed regularly from two to four to eight blastomeres, and so on.

The first cleavage in mammals is meridional, passing through the animal-vegetal axis as in other eggs that undergo holoblastic cleavage (Fig. 5.9). However, the second cleavage is described as a ***rotational cleavage,*** because the two blastomeres divide in different planes: One blastomere divides meridionally, while the other divides equatorially. This results in a crosswise arrangement of cleavage furrows and blastomeres that is characteristic of most mammals, including humans. However, in a given strain of laboratory rabbits, only about half of the embryos may show rotational cleavage. In the rest, the second cleavage occurs more or less simultaneously in both blastomeres and with the same meridional orientation. Both types of embryos develop similarly thereafter, indicating that the orientation of the second cleavage is not critical, at least in rabbits.

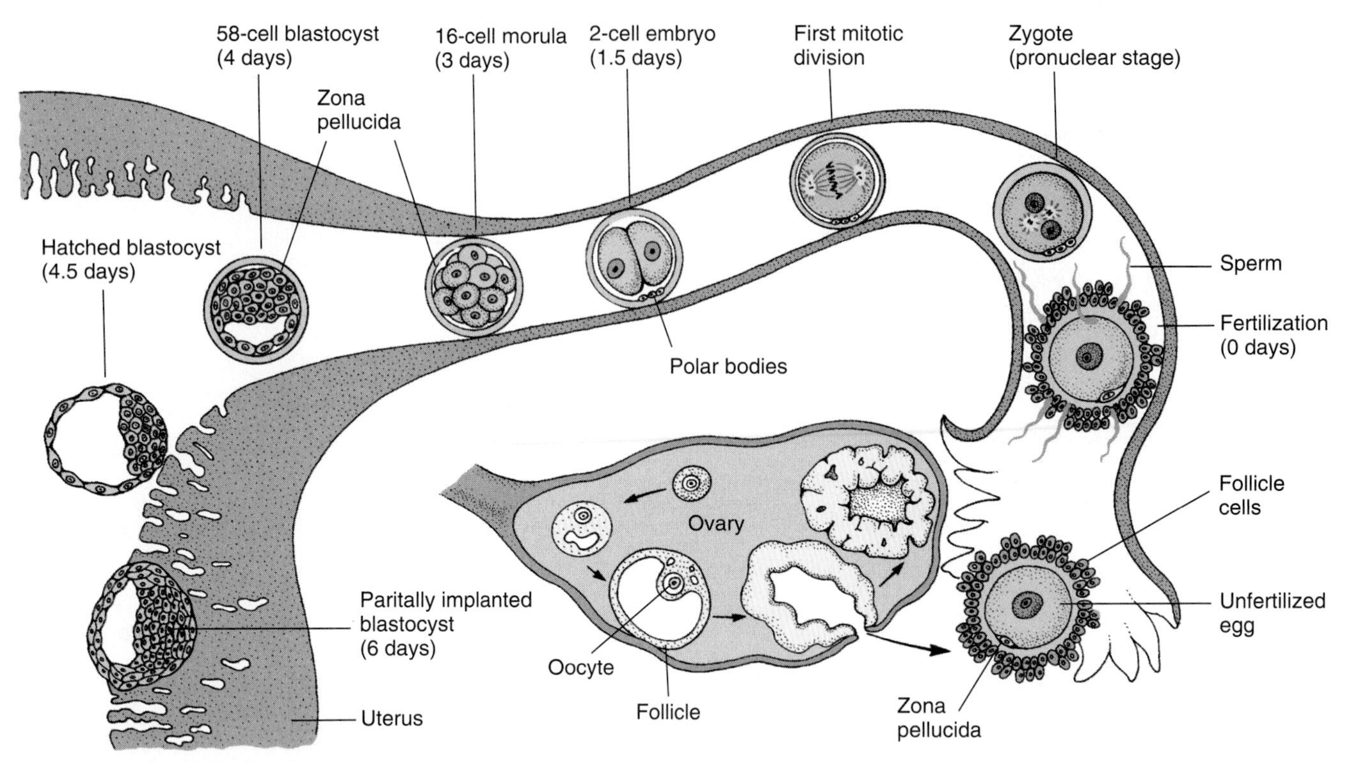

Figure 5.8 First week of human development. Fertilization occurs in the upper third of the oviduct. The zygote undergoes cleavage as it travels down the oviduct. About 4.5 days after fertilization, the embryo hatches from the zona pellucida and becomes implanted in the inner layer of the uterus.

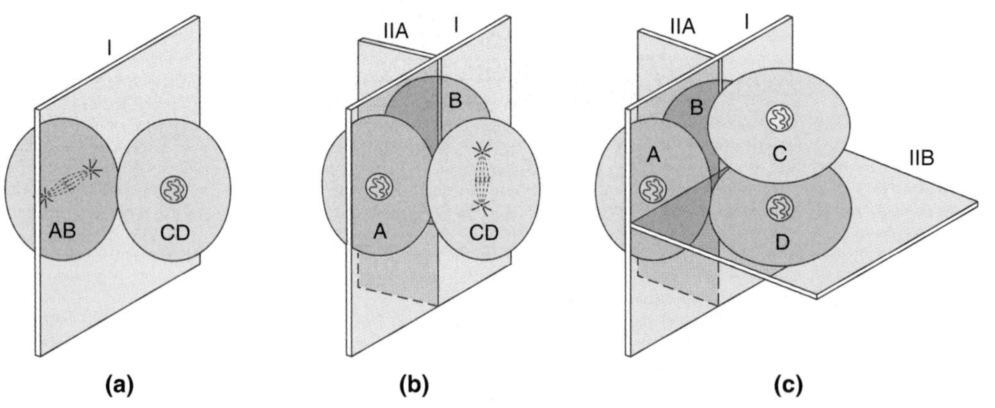

Figure 5.9 Rotational cleavage in mammals. **(a)** The first cleavage is on a meridional cleavage plane (I) passing through the animal-vegetal axis. The resulting two blastomeres are designated AB and CD. **(b)** The AB blastomere divides first, again on a meridional cleavage plane (IIA). The CD blastomere, which lags behind, elongates parallel to the furrow between A and B. Its mitotic spindle is oriented parallel to the long axis of the cell. **(c)** The CD blastomere divides on a cleavage plane (IIB) that is perpendicular to both I and IIA.

At the 8-cell stage, the mammalian embryo undergoes a striking change known as ***compaction.*** At first, the blastomeres form a loose arrangement, touching their neighbors only at small areas of their surfaces. After compaction, the blastomeres adhere tightly, maximizing the areas of contact and forming a solid ball of cells (Fig. 5.10). During compaction, each of the eight blastomeres undergoes a *polarization* process like the one that occurs in the outermost cells of amphibian blastulae (see Fig. 5.5). Each blastomere has an external surface that faces the outside environment as the apical face of an epithelial cell does. The other surfaces are internal and adjacent to other blastomeres. During compaction, *tight junctions* develop beneath the external surfaces, and *gap junctions* form between the internal surfaces. Polarization and the formation of tight junctions allow

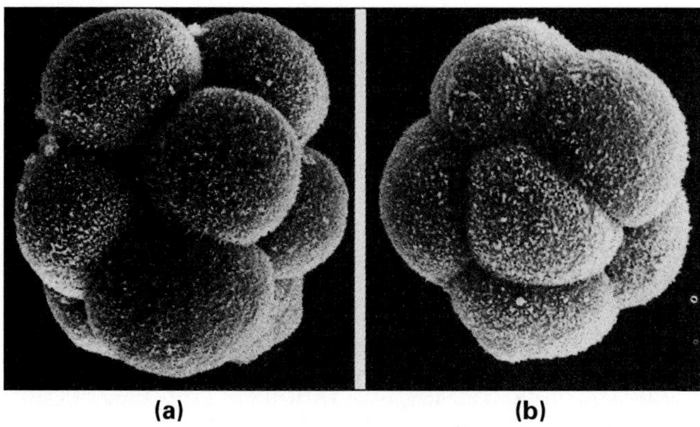

Figure 5.10 Scanning electron micrographs of mouse embryos at the 8-cell stage **(a)** before compaction and **(b)** after compaction. The rough appearance of the blastomere surfaces is caused by numerous microvilli.

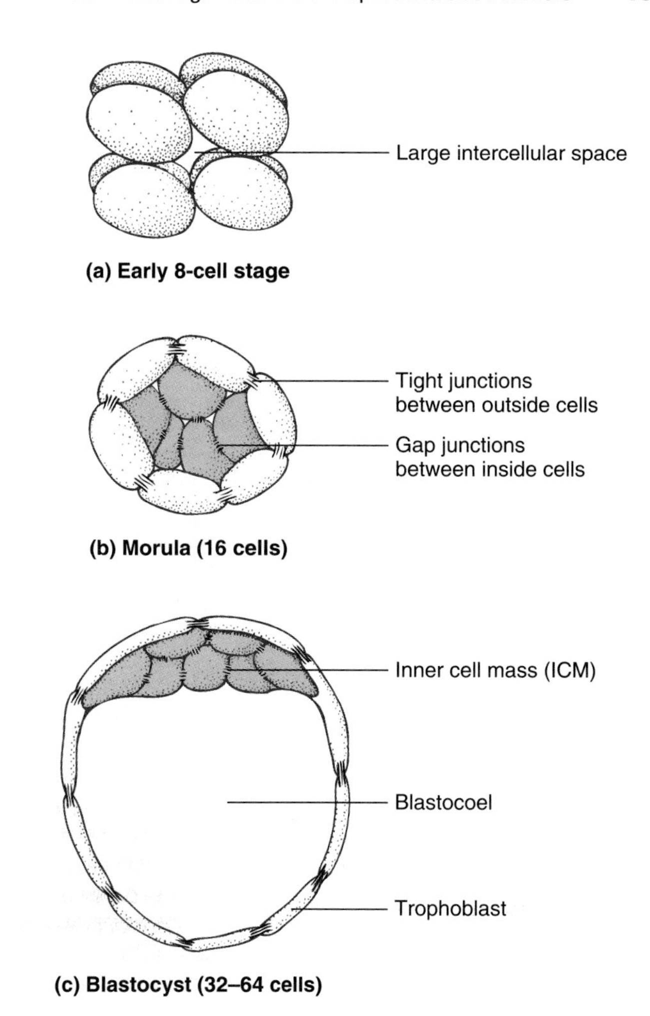

Figure 5.11 Compaction, cell junction formation, and cavitation in the mammalian embryo. **(a)** 8-cell stage before compaction. **(b)** At the morula stage, the embryo consists of 9 to 14 outside cells and 2 to 7 inside cells. The outside cells are connected by tight junctions to form a seal between the inside of the embryo and the external environment. The outside cells take up fluid and nutrients from the environment by endocytosis and give off fluid to the inside by exocytosis, causing cavitation (i.e., formation of the blastocoel). **(c)** The blastocyst consists of the inner cell mass (ICM), which gives rise to the embryo proper, and the trophoblast, which plays a role in the hatching of the embryo from the zona pellucida and in the embryo's implantation in the uterus.

the blastomeres to create an inner embryonic environment that is different from the outside environment (Rodriguez-Boulan and Nelson, 1989). Similar processes occur in other types of embryos, but compaction is an especially dramatic event in mammals.

The morula stage in mammals begins when the embryo consists of 16 blastomeres. In humans this occurs 3 to 4 days after fertilization, when the embryo passes from the oviduct into the uterus. A 16-cell morula consists of 9 to 14 polarized ***outside blastomeres*** surrounding 2 to 7 nonpolarized ***inside blastomeres*** (Fig. 5.11). As both groups of cells continue dividing, the outside blastomeres begin to pump fluid from the uterus into the embryo, taking up uterine fluid through their apical plasma membranes by means of endocytosis and giving it off again, in a somewhat altered composition, by exocytosis from their lateral and basal plasma membranes. This process, which results in the formation of fluid-filled cavities inside the embryo, is known as ***cavitation.*** These cavities merge and form the blastocoel.

Between the 32- and 64-cell stages, the mammalian embryo acquires the *blastocyst* configuration, which is the hallmark of early mammalian development. At the blastocyst stage the embryo consists of two groups of cells (Fig. 5.11). The outer layer of cells is called the ***trophoblast*** while the inner group of cells, known as the ***inner cell mass (ICM),*** forms a small mound eccentrically attached to the inside of the trophoblast. The trophoblast (Gk. *trophe,* "nourishment") participates in forming the placenta—that is, the disc-shaped structure in the uterus through which the fetus obtains oxygen and nutrients from its mother. The inner cell mass gives rise to the embryo proper.

Cells from the ICM of mammalian including human embryos can be cultured in vitro. Under suitable conditions, the cells remain undifferentiated, multiply indefinitely, and are then called ***embryonic stem cells (ES cells).*** ES cells are of great scientific and medical interest. As indicated by their ability to multiply indefinitely, they escape a mechanism that limits the number of mitoses for most mammalian cells (see Section 29.6). Mouse ES cells can also be manipulated genetically and then added back to the ICM of embryos, where they will contribute to all developing structures including the *germ line.* In addition, ES cells have an enormous potential as a primordial cell type from which any kind of cell or tissue may be grown in culture. This would be a great way of testing our current understanding of cell determination and differentiation while providing physicians with a new way of replacing cells and tissues lost to disease or injury (see Section 20.6).

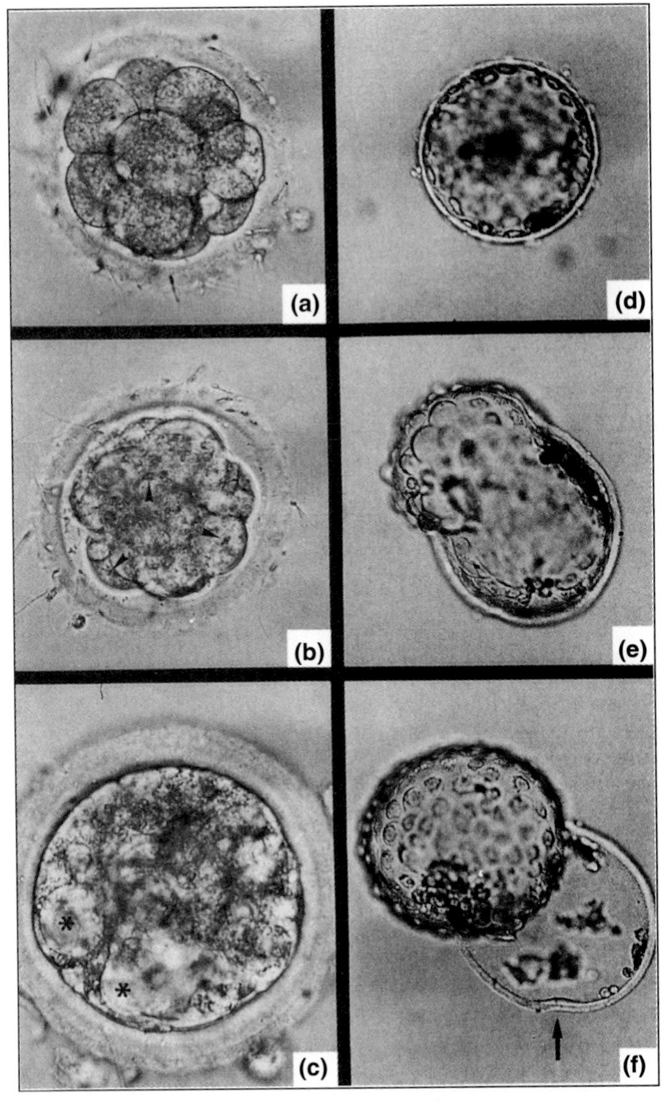

Figure 5.12 Hatching of a rhesus monkey blastocyst from the zona pellucida. **(a)** 16-cell embryo. **(b)** Morula after compaction. **(c)** Blastocyst with unfused blastocoel cavities (asterisks). **(d)** Fully formed blastocyst still inside the zona pellucida. **(e)** Blastocyst begins to hatch from zona. **(f)** Fully hatched blastocyst.

As the mammalian embryo moves down the oviduct, the zona pellucida prevents premature implantation there. Only after floating in the uterus for a day or two does the blastocyst hatch from the zona. The act of hatching involves the local digestion of the zona by an enzyme produced in a patch of trophoblast cells situated opposite the inner cell mass (Fig. 5.12). This location minimizes the risk of collateral enzymatic damage to the embryo (Perona and Wassarman, 1986). Having escaped from the zona, the blastocyst begins the process of ***implantation,*** which in humans occurs about 1 week after fertilization.

Intricate chemical communication between the embryo and the mother allows implantation to occur and enables the embryo to reside in the uterus and grow in its protective environment. When trophoblast cells come into contact with uterine cells, they proliferate and form two layers. The inner layer, called the ***cytotrophoblast,*** remains cellular. The outer layer is called the ***syncytiotrophoblast,*** because cells fuse to form a multinuclear mass of cytoplasm, or *syncytium.* Syncytiotrophoblast cells produce another enzyme that digests a small hole in the uterus, which the embryo penetrates. The syncytiotrophoblast erodes small maternal blood vessels, and the emanating maternal blood nourishes the embryo by diffusion. This is the beginning of placenta formation, which becomes more elaborate with the development of the heart, blood vessels, and umbilical cord (see Section 14.7). Meanwhile, trophoblast cells produce a hormone, known as *chorionic gonadotropin,* which, in humans and other primates, interrupts the menstrual cycle so that the uterus will not discharge the embryo. Other signals from the trophoblast cells regulate the maternal immune response so that the embryo will not be rejected like a tissue graft.

EGGS WITH VARIABLE CLEAVAGE SHOW REGULATIVE DEVELOPMENT

Some groups of animals show ***invariant cleavage,*** meaning that all embryos of a species cleave in exactly the same way, generating a constant number of cells in a stereotypical arrangement. Ascidian embryos (see Fig. 5.7) fit into this category, as do roundworm embryos, which will be described in Chapter 25. Most of the animals discussed so far, however, have ***variable cleavage.*** Their embryos all show the same general cleavage pattern, but they may differ in the exact number of cells and their arrangement.

THE allocation of cells to the *inner cell mass* (*ICM*) and the *trophoblast* of mouse embryos has been studied in detail. From direct observation it appears that much or all of the trophoblast originates from the outside cells at the morula stage (16 cells). Because only outside cells of embryos are capable of *endocytosis,* Fleming (1987) was able to label the outside cells selectively by "feeding" them tiny fluorescent latex beads. Using this technique, he found that the ratio of outside to inside cells in different embryos ranged from 9:7 to 14:2. When the labeled embryos were allowed to develop to the blastocyst stage (32 to 64 cells), the trophoblast proved to be derived mostly from morula outside cells. The origin of the ICM cells in the blastocysts was more varied. On average, about 75% of the ICM cells were unlabeled—that is, descended from cells that had been inside cells at the morula stage. The remainder of the ICM cells were found to be labeled with latex beads and therefore must have been derived from outside morula cells.

The contribution of labeled outside morula cells to the ICM was highest in embryos that had the fewest inside cells at the morula stage (Fig. 5.13). In blastocysts derived from morulae with only two inside cells, more than 65% of

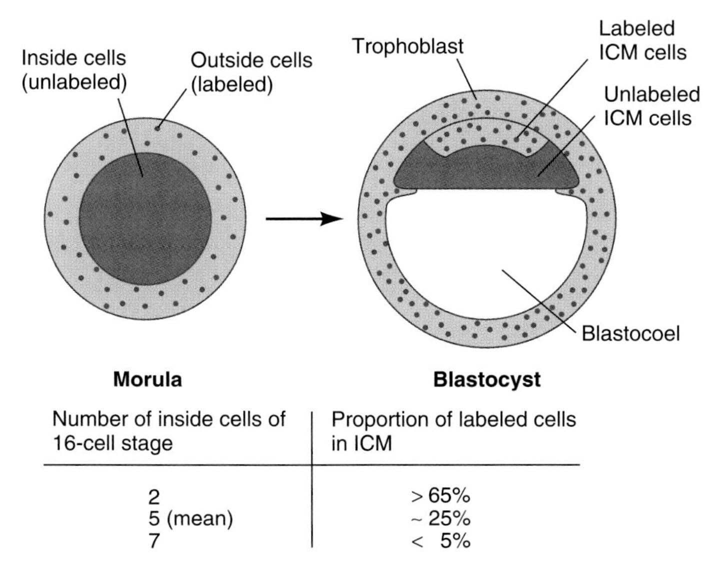

Number of inside cells of 16-cell stage	Proportion of labeled cells in ICM
2	> 65%
5 (mean)	~ 25%
7	< 5%

Figure 5.13 Stepwise formation of the inner cell mass (ICM) in mammalian embryos. Most of the ICM cells are derived from those cells that are in an inside position at the morula stage. Thus, after selectively labeling cells on the outside of a morula, most ICM cells of the developing blastocyst are unlabeled. However, in embryos that have few inside morula cells, additional ICM cells are generated by differential cleavage of outside morula cells.

the ICM cells were labeled. However, in blastocysts that developed from morulae with seven inside cells, fewer than 5% of the ICM cells were labeled. Any additional ICM cell seems to be generated by differential cleavage of an outside morula cell into one trophoblast cell and one ICM cell. This was indicated by the position of the labeled ICM cells, which were always found near the trophoblast, not near the blastocyst cavity.

Questions

1. Which general behavior of cells is utilized when researchers "feed" miniature plastic beads to them?
2. What does the stepwise formation of the inner cell mass imply for the likelihood with which a fate map predicts the fate of a given blastomere?

These results of Fleming show that the ICM, which gives rise to the embryo proper, originates by a process of ***stepwise approximation.*** The first step occurs during the fourth cleavage, when some blastomeres cleave parallel to their outer surface, thus giving off the first inside cells of the morula. Depending on the number of inside cells generated during this first step, a variable number of outside cells produce additional ICM cells by further cleavages parallel to their outer surface. In other words, the mouse embryo forms its ICM through stepwise approximation rather than by a precise initial allocation of cells. This is part of a more general phenomenon known as *regulative development*, which will be discussed more fully in Section 6.6.

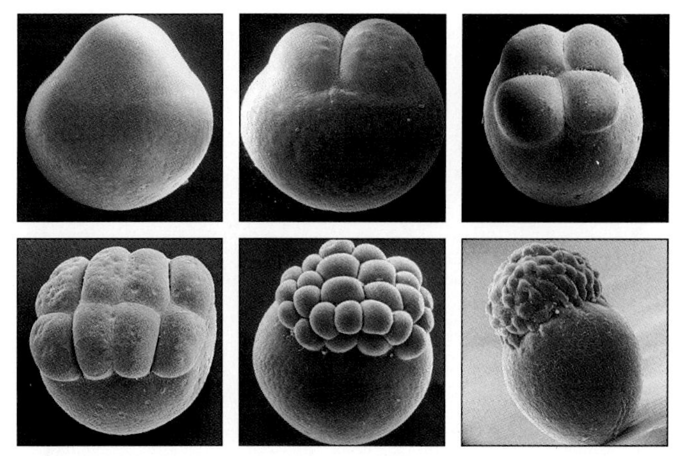

Figure 5.14 Discoidal cleavage in the zebrafish as seen in scanning electron micrographs. The cleavage furrows begin at the animal pole, but only the blastodisc cleaves, not the yolk-rich vegetal portion of the egg. (*From Beams H. W. and Kessel R. G. [1976] Cytokinesis: A comparative study of cytoplasmic division in animal cells,* Am. Sci. ***64****:279–290.*)

BIRDS, REPTILES, AND MANY FISHES HAVE TELOLECITHAL EGGS AND UNDERGO DISCOIDAL CLEAVAGE

As explained earlier, the amount and distribution of yolk in animal eggs are correlated with their cleavage patterns. In contrast to the holoblastic cleavage patterns discussed so far, in which the entire egg cell is cleaved during cytokinesis, meroblastic cleavage patterns leave a large yolk-rich portion of the egg uncleaved. There are two types of meroblastic cleavage, discoidal and superficial.

Discoidal cleavage is typical of fish, reptiles, and birds. The cleavage pattern of the zebrafish *Danio rerio,* which has relatively small eggs, is not far removed from the holoblastic pattern of amphibians (Fig. 5.14; Langeland and Kimmel, 1997). In the unfertilized zebrafish egg, the cytoplasm is distributed as a thin layer around the central yolk mass. Upon fertilization, cytoplasm streams to the animal pole, forming a mound called the ***blastodisc.*** The first cleavage cuts the blastodisc in half, with the cleavage plane being vertical, or perpendicular to the surface. The cleavage furrow begins to form at the animal pole, as in amphibian eggs, but instead of cutting all the way through the egg, it stops at the yolk. The second cleavage is again vertical and perpendicular to the first cleavage, and the resulting four blastomeres remain continuous with the yolk below and with the cytoplasmic layer at their outer margins. The next three cleavages continue making shallow perpendicular cuts in the blastodisc in a regular pattern. The sixth cleavage is the first to be horizontal, or parallel to the surface, producing two tiers of cells. The upper layer, along with the marginal cells of the lower layer, form the ***enveloping layer.*** The nonmarginal cells of the lower layer are called the ***deep cells.*** Continued cleavage builds up a mound of a

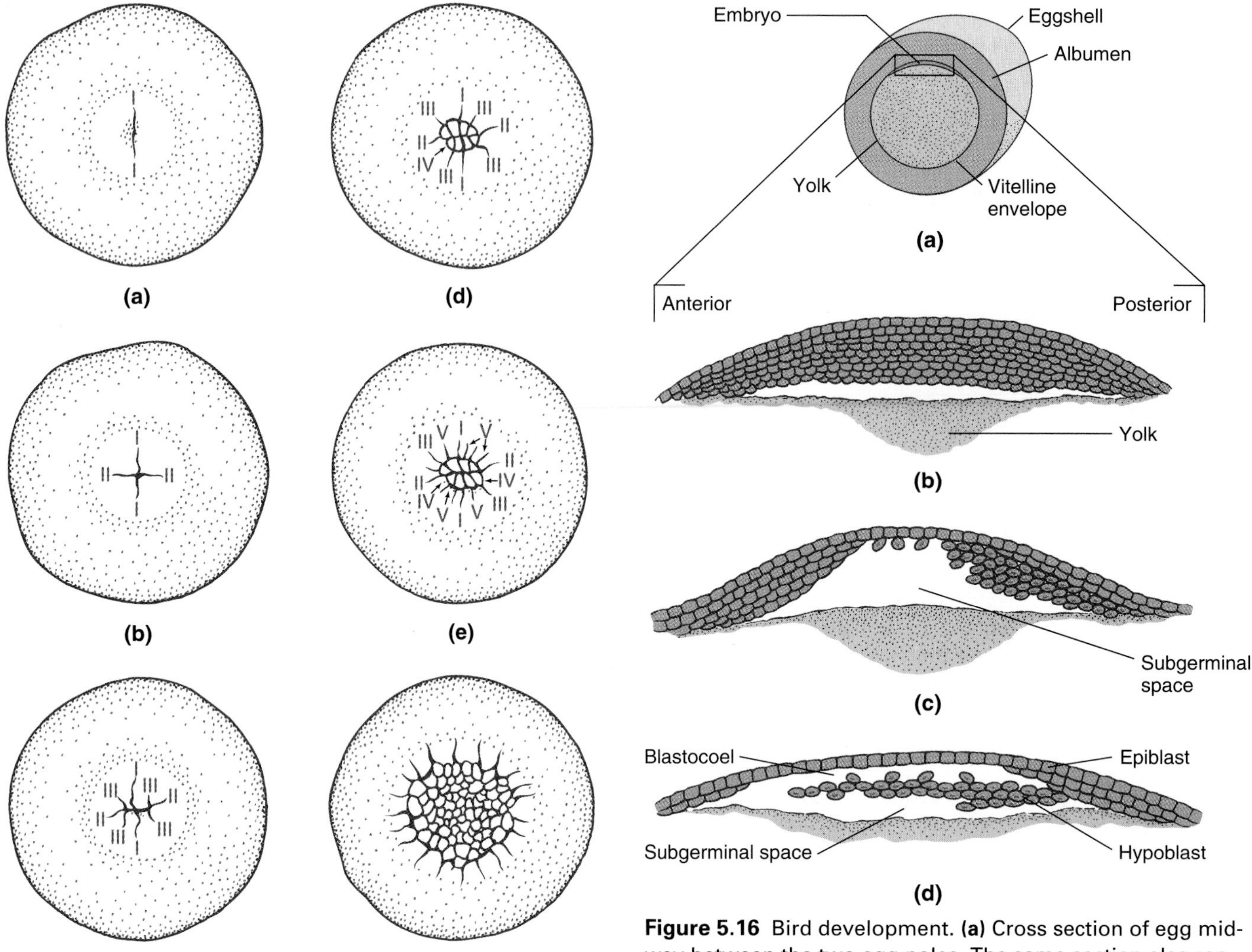

Figure 5.15 Cleavage in a pigeon's egg viewed from the animal pole. The drawings represent the blastodisc atop the uncleaved portion of the egg. Eggshells and albumen have been omitted. Roman numerals indicate the order in which the cleavage furrows appear.

Figure 5.16 Bird development. **(a)** Cross section of egg midway between the two egg poles. The same section also represents the median plane of the developing embryo. **(b)** Blastoderm stage. **(c, d)** Hypoblast formation. The separation of the hypoblast cells from the epiblast begins at the future posterior pole of the embryo.

few thousand cells, consisting of one enveloping layer and many layers of deep cells, perched atop an uncleaved mass of yolk.

In the large eggs of reptiles and birds, discoidal cleavage occurs in a similar fashion, except that the egg cell is much larger in relation to the blastodisc (Fig. 5.15). When cleavage has progressed to the point at which the cleavage planes become irregular and the number of blastomeres considerable, the term ***blastoderm*** is applied to the entire group of cells. (The fertilized "egg" laid by a hen actually contains an embryo at the blastoderm stage consisting of some 60,000 cells.) Between the blastoderm and the uncleaved yolk, a fluid-filled space, called the ***subgerminal space,*** is formed.

The next developmental step, shown in Figure 5.16, is the formation of two blastoderm layers, an upper layer, called the ***epiblast*** (Gk. *epi,* "upon"; *blastos,* "germ") and a lower layer, the ***hypoblast*** (Gk. *hypo,* "under"). The separation of the hypoblast and the epiblast is brought about by the detachment of interior groups of blastoderm cells from the overlying cells and by the formation of a solid lower shelf of cells near the posterior pole of the future embryo (Eyal-Giladi, 1984). The hypoblast is smaller than the epiblast, and so the two layers touch inside the margin of the subgerminal space. The cavity between the epiblast and the hypoblast, which forms at the expense of the subgerminal space, is the *blastocoel.* The epiblast gives rise to the embryo proper, while the hypoblast forms the extraembryonic endoderm that later surrounds the yolk (see Sections 10.4 and 14.7).

INSECTS HAVE CENTROLECITHAL EGGS AND UNDERGO SUPERFICIAL CLEAVAGE

In most insect eggs, cleavage is limited to a superficial layer of yolk-free cytoplasm called ***periplasm.*** The yolk-rich central cytoplasm, known as ***endoplasm,*** remains uncleaved. For insect embryos, the term "cleavage" is something of a misnomer, because cytokinesis is delayed until many rounds of mitosis have occurred. The division of the zygote nucleus starts deep in the endoplasm (Fig. 5.17). The multiplying daughter nuclei, with surrounding jackets of cytoplasm, move gradually toward the periplasm. Here they undergo further rounds of mitosis but are still not surrounded by plasma membranes. A few nuclei stay behind and become *vitellophages* (Gk. "yolk eaters"), which regulate the breakdown of yolk components.

Cleavage in *Drosophila* exemplifies the *superficial cleavage* pattern of insects (Fullilove and Jacobson, 1971; F. R. Turner and A. P. Mahowald, 1976; Foe and Alberts, 1983). The entire cleavage stage is subdivided into ***nuclear cycles,*** each cycle extending from the beginning of interphase to the end of M phase (Fig. 5.17). During the first eight cycles, all nuclei are hidden in the yolk-rich endoplasm. At the beginning of cycle 9, a few nuclei have reached the periplasm at the posterior pole. Here they become enclosed by plasma membranes to form the primordial germ cells, which are known as pole cells. At the beginning of cycle 10, most nuclei have arrived in the periplasm, each lodging beneath a small cytoplasmic mound that protrudes into the perivitelline space (Figs. 5.18 and 5.19). Cycles 10 through 13 are also known as the ***preblastoderm stage,*** during which most nuclei are located at the egg surface but still contained in a common cytoplasmic layer. In the preblastoderm embryo, the thickness of the yolk-free periplasm increases as the yolk components move closer to the center. Meanwhile, the pole cells divide with regular cytokinesis, although with a slower cell cycle than the somatic nuclei.

At the beginning of cycle 14, about 5000 nuclei are crowded into the periplasm of a *Drosophila* embryo, forming a dense hexagonal array while the somatic nuclei are still not enclosed by plasma membranes. This stage is also known as the ***syncytial blastoderm stage,*** which is a misnomer because a *syncytium* (Gk. *syn,* "together"; *kytos,* "cell") is a multinucleate body of cytoplasm that has originated by the fusion of individual, membrane-bound cells. A reverse process of cellularization begins as plasma membrane furrows cut in between the nuclei of the so-called syncytial blastoderm. At the same time, the nuclei elongate as they become enclosed by cages of microtubules (Fig. 5.19). The membranes deepen simultaneously between all nuclei in the syncytium, thus cutting a honeycomb pattern into the egg surface. Having cut deeper than the nuclei, the furrows broaden at their bases, gradually constricting the cytoplasmic connection between cells and endoplasm. This

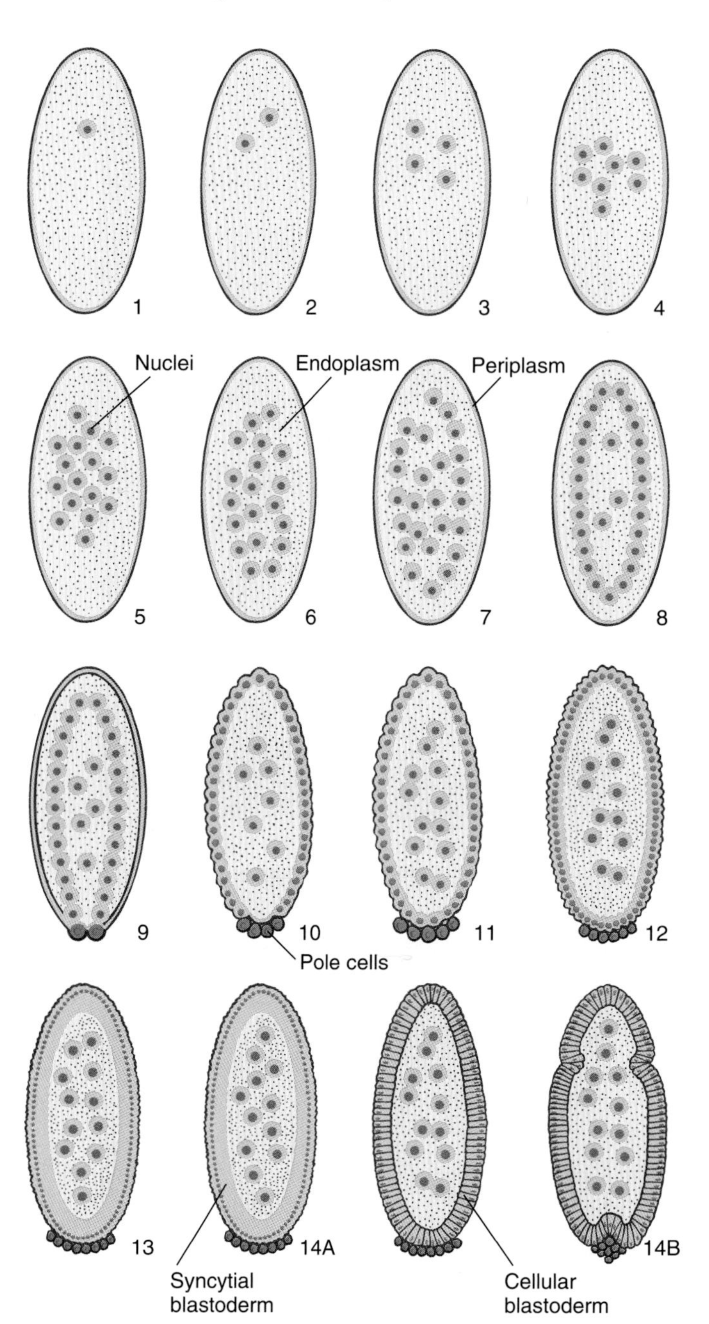

Figure 5.17 Superficial cleavage of the *Drosophila* embryo. In each diagram, the anterior pole is at the top. The number at the bottom of each diagram indicates the nuclear cycle. A cycle begins with the start of interphase and ends with the conclusion of M phase. Cycle 1 extends from fertilization through the first interphase and the first mitosis. All embryos are shown during interphase. The pole cells are pinched off at the posterior pole during cycle 10. The remainder of the embryo develops as a multinucleate syncytium through cycle 13, at the end of which thousands of nuclei have moved into a superficial layer of yolk-free cytoplasm called the periplasm. A few nuclei remain in the yolky endoplasm, where they are involved in metabolizing yolk. A single layer of somatic cells, the blastoderm, is generated during cycle 14 as folds of plasma membrane cut into the periplasm between the nuclei (see Fig. 5.19). Gastrulation movements begin late during cycle 14, at stage 14B.

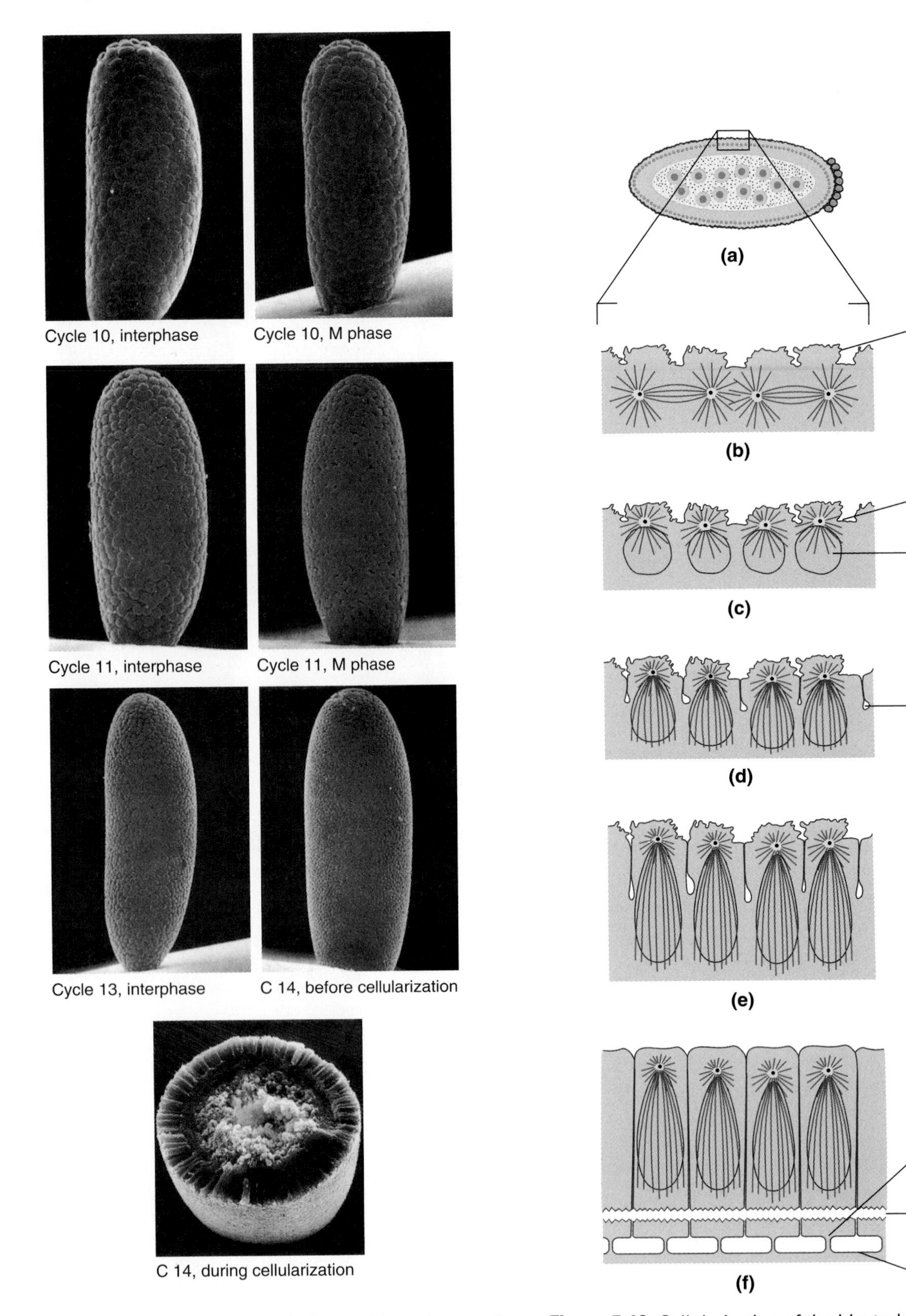

Figure 5.18 *Drosophila* embryos during preblastoderm and blastoderm stages shown in scanning electron micrographs with the posterior pole up. The bulges in the surface of the embryos are caps of cytoplasm, stiffened by microtubules and microfilaments, that lie over the nuclei near the surface. Notice that the bulges become increasingly smaller and more numerous with each cycle.

Figure 5.19 Cellularization of the blastoderm in *Drosophila*. **(a, b)** 13th mitosis (late cycle 13, see Fig. 5.17). **(c–f)** Cellularization during cycle 14. Cell membranes develop from furrows that surround each nucleus by folding in from the egg plasma membrane. Cytoplasmic stalks between blastoderm cells and uncleaved yolk-rich endoplasm persist until they are severed at the onset of gastrulation.

stage of nearly complete cellularization is referred to as the ***cellular blastoderm.*** It corresponds to the blastula stage of other embryos, although there is no fluid-filled blastocoel: The center of the egg is still filled with yolk-rich endoplasm.

5.3 Spatial Control of Cleavage: Positioning and Orientation of Mitotic Spindles

In the remainder of this chapter, we will consider the cellular mechanisms that bring about the cleavage patterns of embryos. This section will cover the spatial control of cleavage, in particular, the positioning and orientation of mitotic spindles. The next section will deal with the timing of cleavage divisions.

ACTIN AND MYOSIN FORM THE CONTRACTILE RING IN CYTOKINESIS

The cleavage pattern of an embryo, the relative size and spatial arrangement of its blastomeres, is controlled most directly by the positioning of ***cleavage furrows.*** In small eggs that undergo holoblastic cleavage, cleavage furrows constrict like tightening belts around the entire cell. In eggs that undergo meroblastic cleavage, cleavage furrows begin as infoldings of the egg plasma membrane at the animal pole and advance like cutting knives until they reach the yolk mass, which remains uncleaved. We will focus here on holoblastic cleavage.

The fourth cleavage of a sea urchin embryo generates a characteristic pattern of eight mesomeres, four macromeres, and four micromeres (see Fig. 5.4). The size difference between macromeres and micromeres, and the crowding of the micromeres into the interstitial space between the macromeres at the vegetal pole, are both caused by the positioning of the cleavage furrow that separates micromeres from macromeres. The asymmetric and oblique positioning of this furrow is conserved in a blastomere isolated at the 8-cell stage—that is, before the fourth cleavage (Fig. 5.20). What is this furrow made of, and how is it positioned?

Under the electron microscope, cleavage furrows show a thin, dense layer beneath the plasma membrane (see Fig. 2.15). This layer contains a bundle of filaments, called the ***contractile ring,*** which is oriented parallel to the plasma membrane and parallel to the cleavage plane (Schroeder, 1972). The force exerted by the contractile ring is strong enough to bend fine glass needles inserted into cleaving sea urchin embryos (Rappaport, 1967). Immunostaining (see Method 4.1) reveals that the contractile ring contains both actin microfilaments and myosin (Fig. 5.20). Most likely, the ring generates force by a musclelike sliding of actin and myosin filaments. The actin filaments must be anchored to the inside of the plasma membrane, so that shortening of the contractile ring constricts the membrane like an old-fashioned purse when its strings are pulled.

After cleavage, the contractile ring disappears as quickly as it assembled. In fact, the filaments seem to be dismantled even before cytokinesis is complete, because the contractile ring does not become thicker as it constricts. The contractile ring exemplifies how quickly cytoskeletal components can be assembled for a particular purpose and then disassembled until needed again, perhaps for a different function.

THE MITOTIC SPINDLE AXIS DETERMINES THE ORIENTATION OF THE CLEAVAGE PLANE

To approach the question of how cleavage furrows are positioned, we must first consider the connection between cytokinesis and mitosis. As we will see, the

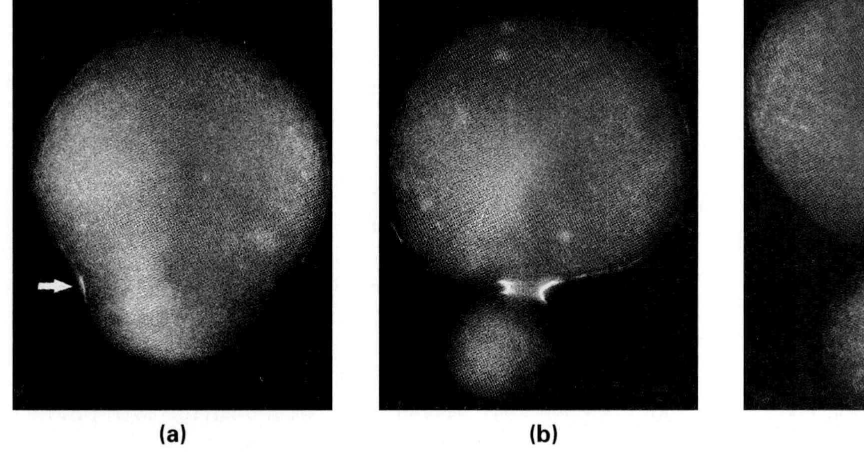

Figure 5.20 Localization of contractile rings in sea urchin blastomeres during fourth cleavage. Photographs show the unequal cleavage of an isolated vegetal blastomere into a macromere (top) and a micromere. The blastomeres were immunostained for myosin. Stainability for myosin begins at telophase **(a)** and lasts until the end of cytokinesis **(b, c)**.

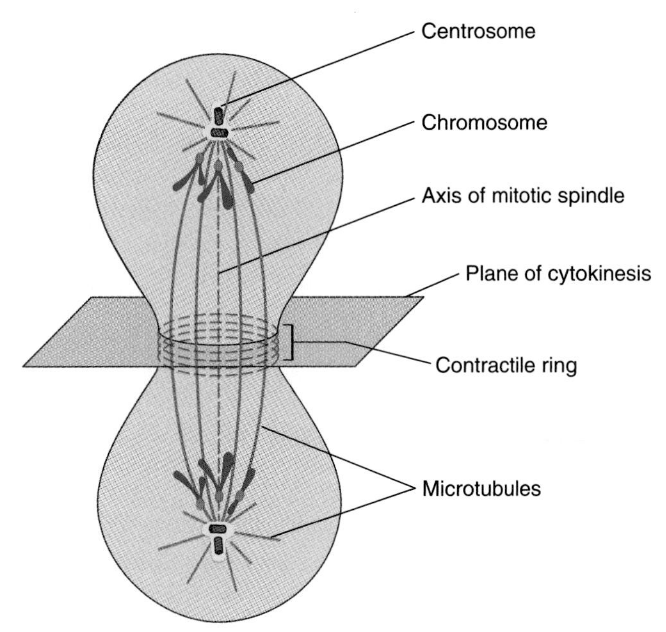

Figure 5.21 Cell division. The plane of cytokinesis develops perpendicular to the axis of the mitotic spindle.

positioning of the cleavage furrow is determined by interactions between components of the mitotic spindle and neighboring areas of egg cortex.

Mitosis is usually followed by cytokinesis. Exceptions to this rule include embryos with superficial cleavage, in which many rounds of mitosis occur without cytokinesis (see Fig. 5.17). A similar situation can be created in the laboratory with a procedure called ***cleavage arrest,*** which is often carried out by inhibiting cytokinesis with *cytochalasin.* For instance, ascidian embryos treated in this way proceed with their cell cycles and even with cellular differentiation in the absence of cytokinesis (Whittaker, 1973). Conversely, cytokinesis has been observed in sea urchin eggs from which the nuclei have been removed so that they cannot undergo mitosis (Harvey, 1956).

In most cases, however, mitosis not only sets the timing of cytokinesis but also determines the orientation of the cleavage plane (Strome, 1993). Invariably, cytokinesis occurs in a plane perpendicular to the axis of the mitotic spindle (Fig. 5.21). This indicates that either the spindle orientation itself determines the cleavage plane or each is determined independently by a third event preceding mitosis. The latter alternative has been ruled out by several investigators performing similar experiments on a variety of embryos. By squeezing blastomeres between two glass slides, they forced the mitotic spindles out of their normal orientation and into an orientation parallel to the plates. The plane of the subsequent cytokinesis was always perpendicular to the new orientation of the spindle. Thus, the spindle orientation itself must control the orientation of the contractile ring.

A role for the mitotic apparatus in cleavage plane orientation has also been indicated by comparative observations on cleavage in isolecithal and telolecithal eggs. When the mitotic spindle is centered, as it typically is in an isolecithal egg, the cleavage furrow forms simultaneously all around the circumference. If the mitotic spindle is displaced toward the animal pole, as it is in mesolecithal eggs, the furrow first appears at the animal pole and then cuts toward the vegetal pole (see Fig. 5.1). If the mitotic spindle is very eccentric, as it is in telolecithal eggs, the furrow forms only near the mitotic apparatus (see Figs. 5.14 and 5.15). Thus it seems that proximity between the egg cortex and one or more components of the mitotic apparatus is required for furrowing to occur.

WHICH parts of the mitotic apparatus are sufficient for the development of a normal-looking furrow? To answer this question, Rappaport (1974) deformed a sand dollar zygote with a fine glass tool, forcing it into a doughnut shape (Fig. 5.22a). As a result, the mitotic apparatus was displaced to one side, and the first cytokinesis was blocked by the glass tool. As a result, a horseshoe-shaped cell with two nuclei was formed (Fig. 5.22b). Next, a mitotic apparatus, complete with spindle fibers and chromosomes between asters at opposite spindle poles, was set up in each "leg" of the horseshoe. In the horseshoe bend, two asters belonging to different spindles were facing each other but had no chromosomes between them. Under these conditions, three cleavage furrows formed, generating four nucleated cells (Fig. 5.22 c, d). Two of these furrows were positioned normally—that is, along the corresponding metaphase plates where the mitotic spindles had been. The third cleavage furrow appeared in the horseshoe bend although no chromosomes had been in this location. Nevertheless, this furrow looked normal and appeared at the same time as the other two. Rappaport concluded that the presence of two asters—rather than a complete mitotic spindle—in the vicinity of the egg cortex can be sufficient to initiate cleavage furrow formation.

To further explore the role of the egg cortex in positioning the cleavage furrow, Rappaport and Rappaport (1994) forced sand dollar zygotes into conical shapes by sucking them into appropriately molded glass pipettes (Fig. 5.23). When the first mitotic spindle developed parallel to the cone axis, then the spindle pole closer to the vertex (tip) of the cone was also closer to the cortex of the zygote. Under these circumstances the cleavage furrow developed perpendicularly to the spindle axis, but instead of being equidistant to both spindle poles, the furrow was closer to the vertical pole. Such a shift of the furrow toward one spindle pole was never observed in control experiments, in which sand dollar zygotes cleaved after forcing them into elongated but cylindrical shapes. These results show that the cleavage furrow is not always equidistant from the two spindle poles. The fact that it usually forms in this position must be ascribed to the normally symmetrical relation between the spindle asters and the nearby cell surfaces. However, if this relation is asymmetrical, then the cleavage furrow is shifted toward the aster that is closer to the cell surface.

Questions

1. In the light of the experiments described above, how would you explain the fact that the cleavage furrows of the first two cleavages in frog eggs always begin to form at the animal pole (see Fig. 5.1)?
2. In which kind of embryo, illustrated in Section 5.2, does the majority of all cleavage furrows arise between asters that belong to different spindles?

Rappaport's experiments with sand dollar eggs show that interactions between two asters and the egg cortex are sufficient to form a cleavage furrow. In an equivalent experiment, Raff and Glover (1989) injected *Drosophila* embryos with *aphidicolin,* a drug that inhibits mitosis and nuclear migration but not the replication and movements of centrosomes. Remarkably, centrosomes and associated asters of microtubules that migrate to the posterior pole of such embryos initiate the formation of pole cells without nuclei. Thus, the position of the cleavage furrow is clearly *not* determined by any components of the metaphase plate, but by some interactions between spindle asters and cell cortex. The nature of these interactions is still unclear, but mutations that interfere with the positioning of cleavage furrows should provide a basis for further analysis.

MECHANICAL CONSTRAINTS MAY ORIENT MITOTIC SPINDLES

Given the overriding importance of asters, which are normally present as parts of mitotic spindles, for the positioning of cleavage furrows, investigations into the control of embryonic cleavage patterns have focused on the orientation of mitotic spindles and the positioning of centrosomes as spindle organizers.

The simplest forces that orient mitotic spindles are mechanical constraints. This is apparent from experiments described earlier, in which mitotic spindles were reoriented by squeezing embryos between glass plates. Other investigators have spun eggs in centrifuges to alter the shape of the entire cell and in particular the layer of yolk-free cytoplasm in which mitotic spindles could move. Invariably, the mitotic spindles formed in such cells were oriented with their axes parallel to the longest available dimension.

In order to test whether constraints on spindle orientation play a role in normal development, Meshcheryakov (1978) systematically altered the extent of contact between snail blastomeres during cleavage by adjusting the concentration of calcium ions (Ca^{2+}) in the culture medium. (Most likely, the Ca^{2+} concentration was exerting its effects through calcium-dependent cell adhesion molecules, *cadherins,* which will be discussed in Section 11.2.) When there was extensive contact between the blastomeres at the 2-cell stage, their shape was hemispherical (embryo A in Fig. 5.24). Under these circumstances, all spindles were aligned parallel to the contact area, which provided for the greatest available cell diameter. As the area of contact

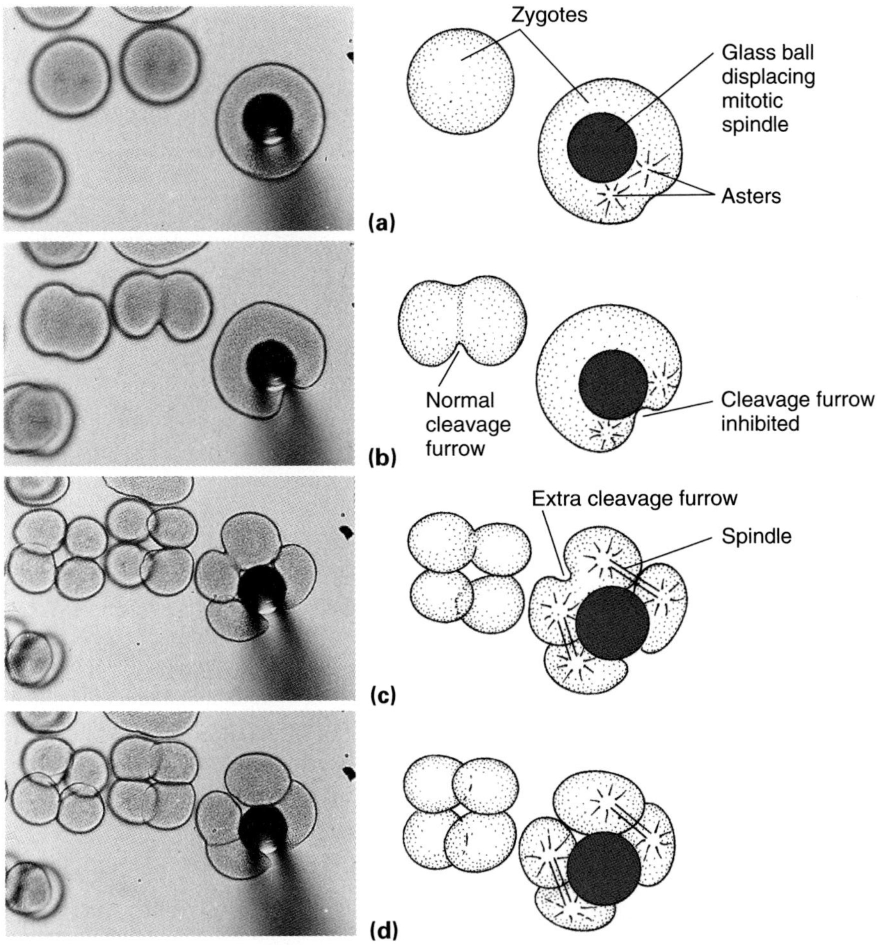

Figure 5.22 Formation of a new cleavage furrow between juxtaposed asters. The diagrams to the right are interpretations of the photographs shown to the left. **(a)** A glass ball forced into a sand dollar zygote creates a doughnut-shaped cell with an eccentric mitotic spindle. **(b)** The glass ball impedes the cleavage furrow, causing the formation of a horseshoe-shaped cell with two nuclei. **(c, d)** At the next cleavage, an extra furrow forms between adjacent asters, although no mitotic spindle was there before.

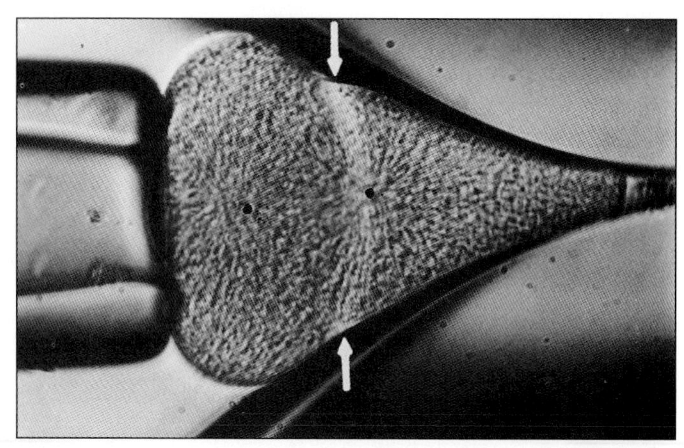

Figure 5.23 Cleavage of a sand dollar egg forced into a conical shape by sucking it into an appropriately molded glass pipette. In the egg shown here, the first mitotic spindle developed parallel to the cone axis (spindle poles marked with black dots). Under these conditions, the spindle pole located near the tip of the cone was closer to the egg cortex than the spindle pole located near the base of the cone. The cleavage furrow (arrows) then formed perpendicularly to the spindle axis, but instead of being equidistant to both spindle poles, the furrow was closer to the pole near the tip.

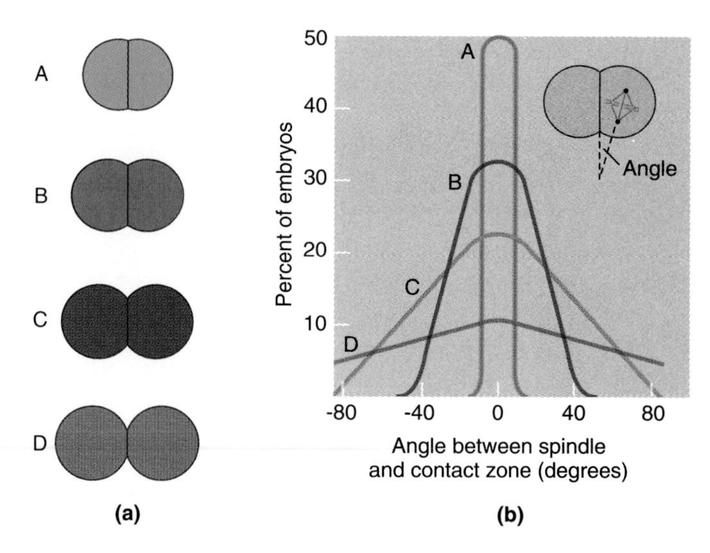

Figure 5.24 Dependence of mitotic spindle orientation on the extent of cell contact between snail blastomeres. **(a)** 2-cell embryos A through D show different degrees of contact reduction between blastomeres. **(b)** Graph showing distributions of the angle between the plane of blastomere contact and the spindle axes during second mitosis.

was reduced and the blastomeres became more spherical, the orientation of the mitotic spindles began to vary until eventually it was almost random (embryos B, C, and D in Fig. 5.24). Apparently, the orientation of the mitotic spindles was constrained under conditions of extensive cell contact.

CENTROSOMES ORGANIZE MITOTIC SPINDLES IN REGULAR WAYS DURING CLEAVAGE

While mechanical constraints may clearly be effective in small cells, where the length of the mitotic spindle can only be accommodated in certain dimensions, such constraints are not apparent in large eggs and blastomeres. In these cases, then, one needs to look for other forces that may position and orient mitotic spindles in certain ways. Since spindles are organized by *centrosomes*, these organelles have become a logical focus of attention.

The cleavage patterns of sea urchins (see Fig. 5.4) and amphibians (see Fig. 5.2) are based on a common pattern of centrosome duplication and migration. The fertilized egg contains one centrosome, which usually has been introduced by the sperm and lies to the side of the zygote nucleus (Fig. 5.25). Before the first mitosis, the centrosome replicates. Next, the daughter centrosomes move to opposite sides of the nucleus, in quarter-circular motions perpendicular to the animal-vegetal axis. Thus, the first mitotic spindle is also oriented perpendicular to the animal-vegetal axis, and the first cleavage furrow will form in a meridional plane that passes through the animal-vegetal axis. In each of the resulting two blastomeres, the centrosome divides again, and the daughter centrosomes move to opposite sides of the nuclei, in motions that are perpendicular to both the animal-vegetal axis and the centrosomal movements of the previous cycle. The second cleavage furrow will therefore be meridional and at right angles with the first furrow. In each of the resulting four blastomeres, the centrosome divides again, and the daughter centrosomes move along trajectories that are perpendicular to those of the preceding two cycles. The third cleavage furrow will therefore be perpendicular to the first two furrows—that is, equatorial.

This basic pattern of centrosome duplication and migration, which generates a succession of cleavage planes at right angles to each other, is found in the early embryos of many animal species but also in other proliferating cells. The mechanism that generates this pattern has been referred to as a ***cleavage clock*** because it can keep going even if mitosis and cytokinesis have been delayed experimentally (Hörstadius, 1973; Cather et al., 1986). Depending on the length of the delay, a 2-cell embryo may then show the cleavage pattern of a 4-cell embryo, or a pattern in between two normal cleavage patterns. The molecular processes underlying the cleavage clock are still being investigated (Rhyu and Knoblich, 1995; White and Strome, 1996).

SPECIFIC SITES IN THE EGG CORTEX ATTRACT AND ANCHOR CENTROSOMES

In many embryonic cleavages, cytokinesis is unequal, giving rise to daughter cells that differ in size. Examples include the fourth cleavage in sea urchins, where vege-

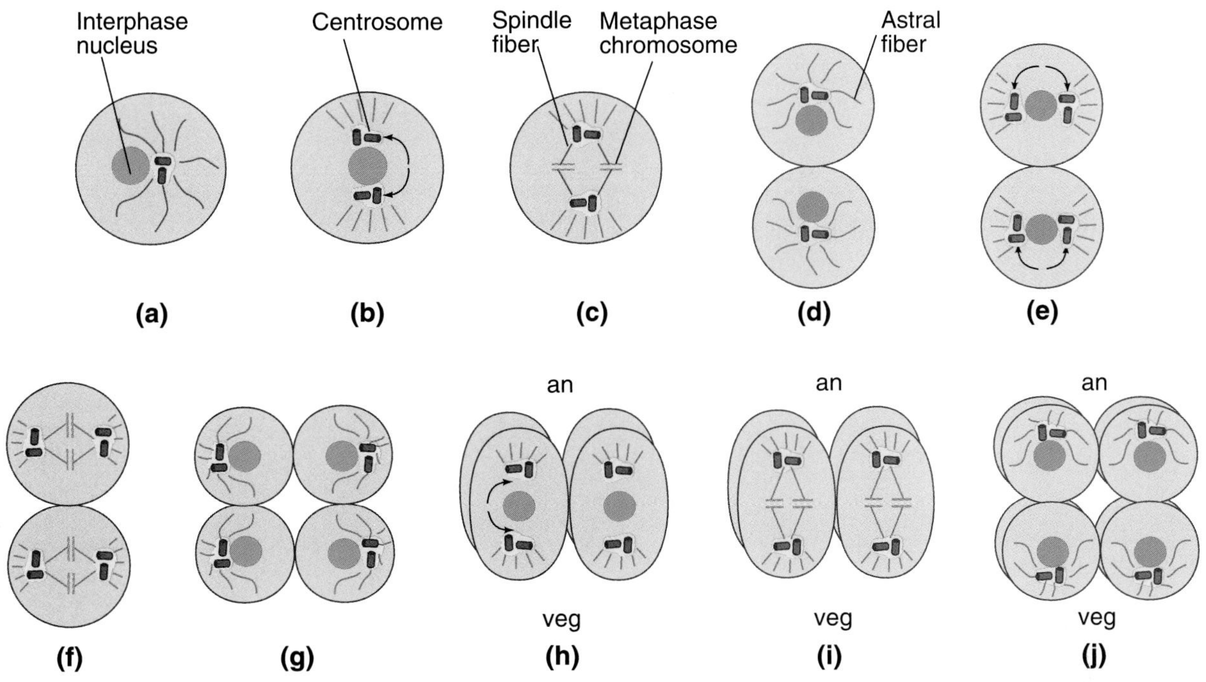

Figure 5.25 Centromere replication and movements during early cleavage. **(a–g)** Views from the animal pole; **(h–j)** lateral views, animal pole up. Centrosomes replicate and move to opposite poles of the nucleus during each cell cycle, thus generating a sequence of perpendicular spindle orientations and corresponding cleavage furrows.

tal blastomeres generate macromeres and micromeres (see Fig. 5.4), as well as the meiotic divisions in the female germ line, which produce one large oocyte and three small polar bodies (see Fig. 3.5). In both cases, the spindle *position* is eccentric (off the center of the dividing cell), and its *axis orientation* is perpendicular to the cellular cortex. Axis orientation is critical here: if the spindle axis were oriented parallel rather than perpendicular to the cortex, the resulting cleavage would be equal and possibly meroblastic instead of unequal and holoblastic (Fig. 5.26). The importance of spindle orientation relative to the cortex suggests that in the case of perpendicular orientation one spindle pole may be attracted and held by the adjacent cortical zone.

The mechanism of spindle pole attraction has been investigated in several species including the ascidian *Halocynthia roretzi* (Hibino et al., 1998; Nishikata et al., 1999). The two posterior vegetal blastomeres cleave asymmetrically, creating unequal pairs of daughter blastomeres. (The left posterior vegetal blastomere, designated B4.1, and its descendants are labeled in Figures 5.27 and 5.28. The counterpart of B4.1 on the right- hand side, designated C4.1 but not labeled here, behaves in a mirror-image fashion; see Fig. 5.7.) During the fourth cleavage, B4.1 divides into B5.1 and B5.2 (Fig. 5.27a, b). During the fifth cleavage, B5.2 again cleaves unequally into B6.3 and B6.4 (Fig. 5.27c). The orientations of the mitotic spindles change during these cleavage events, as indicated by the straight lines in Figure 5.27a–c.

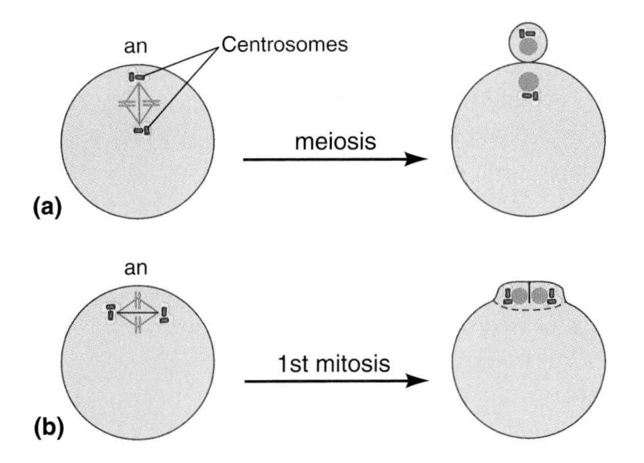

Figure 5.26 Importance of the axis orientation of a spindle positioned eccentrically near the animal pole (an). **(a)** If the spindle axis (between centrosomes) is perpendicular to the adjacent egg cortex, as in meiosis, the ensuing cleavage will be unequal, generating one large and one small cell. **(b)** If the spindle axis is parallel to the adjacent egg cortex, as during first mitosis, the ensuing cleavage will be equal and, depending on egg size, holoblastic or meroblastic.

If B4.1 is isolated and divided into an anterior and a posterior fragment, then the anterior fragment divides equally while the posterior one divides unequally (Fig. 5.27d). This means that a prerequisite for unequal cleavage is present in the posterior but not in the anterior

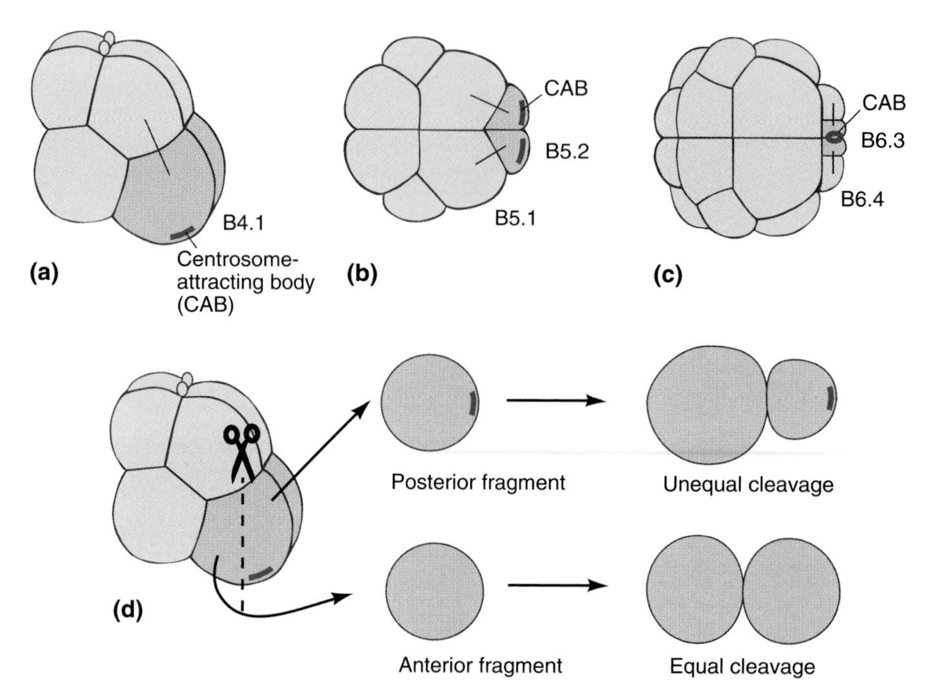

Figure 5.27 Asymmetrical division of posterior vegetal blastomeres in the ascidian *Halocynthia roretzi.* **(a)** 8-cell stage in lateral view. **(b)** 16-cell stage in vegetal view. Unequally dividing cells are characterized by a centrosome-attracting body (CAB). Unequal cleavage of B4.1 has produced two daughter cells of different sizes, B5.1 and B5.2. **(c)** 32-cell stage in vegetal view. Unequal cleavage of B5.2 has produced two daughter cells of different sizes, B6.3 and B6.4. A straight line connecting two cells indicates the spindle orientation during the mitosis that led to the formation of the two cells. **(d)** If B4.1 is isolated and divided into anterior and posterior fragments, the posterior fragment, which contains the CAB, divides unequally whereas the anterior fragment, which has no CAB, divides equally.

fragment. The nature of this prerequisite has been studied in embryos that were rendered transparent by extraction with detergent (Fig. 5.28). In equally dividing blastomeres *astral fibers* radiate symmetrically from both centrosomes, whereas in unequally dividing blastomeres the astral fibers originating from one centrosome form an unusual bundle (arrows in Fig. 5.28a). This bundle of microtubules connects the centrosome to a cortical organelle called a ***centrosome-attracting body,*** or ***CAB.*** The position of the CAB changes in concert with the position of the centrosome to which it is linked. In Figure 5.28, the CAB is intermediate between a posterior position, which it occupied during the fourth mitosis, and a median position, which it will occupy during the fifth mitosis. As the microtubular bundle shortens, the mitotic spindle is pulled toward the CAB, so that the subsequent cytokinesis will divide B5.2 into a small median daughter cell containing the CAB (B6.3) and a larger lateral daughter cell without CAB (B6.4).

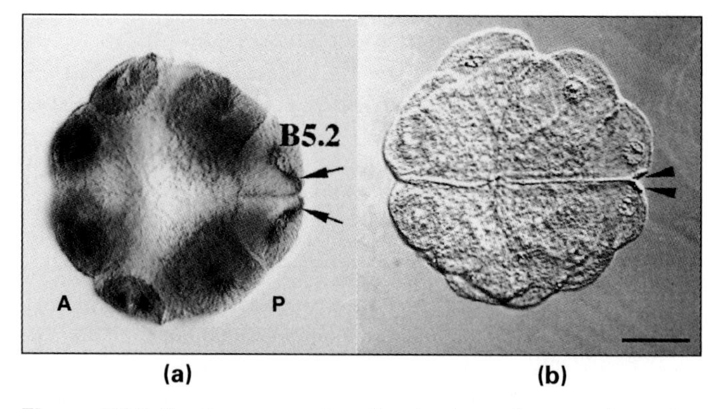

Figure 5.28 Centrosome-attracting body and eccentric position of nucleus prior to asymmetrical cleavage. Both photographs show *Halocynthia roretzi* embryos at the 16-cell stage in vegetal view, as diagrammed in Figure 5.27b. After extraction with detergent to make them transparent, the embryos were **(a)** immunostained with anti-α-tubulin or **(b)** viewed with differential interference contrast (Nomarski optics). The blastomeres are in prophase of the fifth mitosis, as indicated by the astral fibers radiating out from two centrosomes on opposite sides of each nucleus. In each of the B5.2 blastomeres the posteromedial centromere is connected by an unusual bundle of microtubules (arrow) to a cortical organelle known as centromere-attracting body, or CAB (arrowhead). As the bundle contracts, the connected centromere—and the associated nucleus—move toward the CAB. At the same time, the CABs move from a posterior position (shown in Fig. 5.27b) to a median position (shown in Fig. 5.27c). The ensuing cleavage will be unequal; the daughter cell with the CAB will be the smaller one. Scale bar: 50 μm.

The eccentric anchorage of *meiotic* spindles has been studied in maturing oocytes of a polychaete worm, *Chaetopterus.* Before the first polar body forms, the germinal vesicle moves toward the animal pole, where a meiotic spindle is formed with one spindle pole being anchored to the egg cortex and the spindle axis perpendicular to the oocyte surface. Using a fine glass needle mounted on a micromanipulator, Lutz et al. (1988)

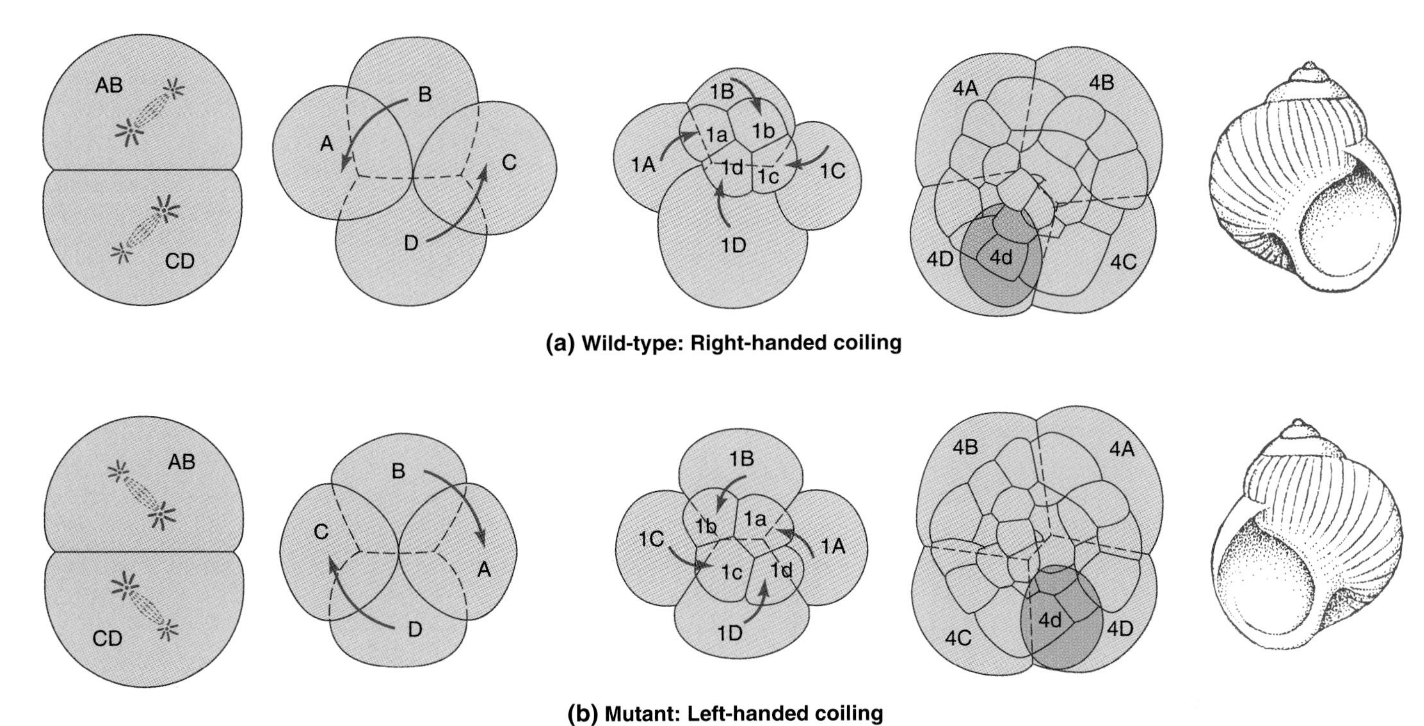

Figure 5.29 Genetic control of embryonic cleavage pattern and shell coiling in the snail *Lymnaea peregra.* **(a)** Wild-type showing a right-handed (dextral) coil of shell. **(b)** Mutant allele showing a left-handed (sinistral) coil of shell. The direction of shell coiling depends on the position of the 4d blastomere, which gives rise to the shell gland. The position of the 4d blastomere can be traced back to the orientation of the previous cleavages, which mirror each other in dextral and sinistral embryos. The dextral cleavage depends on the activity of a single gene during oogenesis (see Fig. 5.30).

reached into the oocyte and dislodged the spindle from its normal position. If the spindle was tugged toward the cell interior, the cell surface adjacent to the outer spindle pole dimpled inward. As the spindle was pulled further inward, the dimple suddenly receded, indicating the rupture of a mechanical link between the egg cortex and the outer spindle pole. When released from the glass needle, the spindle spontaneously returned to its original attachment site. If removed too far from the cortex, however, the spindle remained stationary until pushed closer to its attachment site. Spindles that were inverted after detachment reattached to the former cortical site but with the pole that had formerly faced the cytoplasm. Spindle poles pushed against other cortical regions, however, did not attach.

Together, the observations described above demonstrate the existence of localized cortical attachment sites in oocytes, eggs, and blastomeres of certain species. These sites are connected via astral fibers to one centrosome of a meiotic or mitotic spindle. As these fibers shorten, the spindle is pulled into an eccentric position, which in turn leads to an unequal cell division. The attachment sites are generated at the appropriate times and places during development. Recent studies with yeast cells, which divide asymmetrically, have identified a specific protein associated with the free ends of microtubules that interacts with a specific cortical protein (Lee et al., 2000).

MATERNAL GENE PRODUCTS MAY ORIENT MITOTIC SPINDLES

The factors that control the positioning and orientation of mitotic spindles must reside in the cytoplasm of the oocyte or egg. Any mutations that modify these processes should therefore be of the *maternal effect* type, meaning that the embryonic cleavage pattern should depend on the maternal but not the paternal genotype. This kind of mutation was indeed found in the freshwater snail *Lymnaea peregra.* The mutant cleavage pattern was discovered because it does not interfere with survival but has a dramatic effect on the coiling orientation, or *chirality,* of the snail's shell, which is readily observed.

The shell of *Lymnaea* normally shows ***dextral coiling,*** meaning that if one looks down on the top of the shell, its coil winds toward the opening in a clockwise or right-handed (dextral) spiral. However, in some broods of *Lymnaea,* the shell winds in the opposite direction, showing ***sinistral coiling.*** Crampton (1894) observed that the key difference between embryos from dextral and sinistral broods is the orientation of their mitotic spindles during the second cleavage (Fig. 5.29). As a result of an abnormal spindle orientation during second cleavage in sinistral broods, the A and C blastomeres in the sinistral embryos are swapped relative to dextral embryos, and all subsequent cleavages in sinistral embryos are therefore mirror images of those in dextral

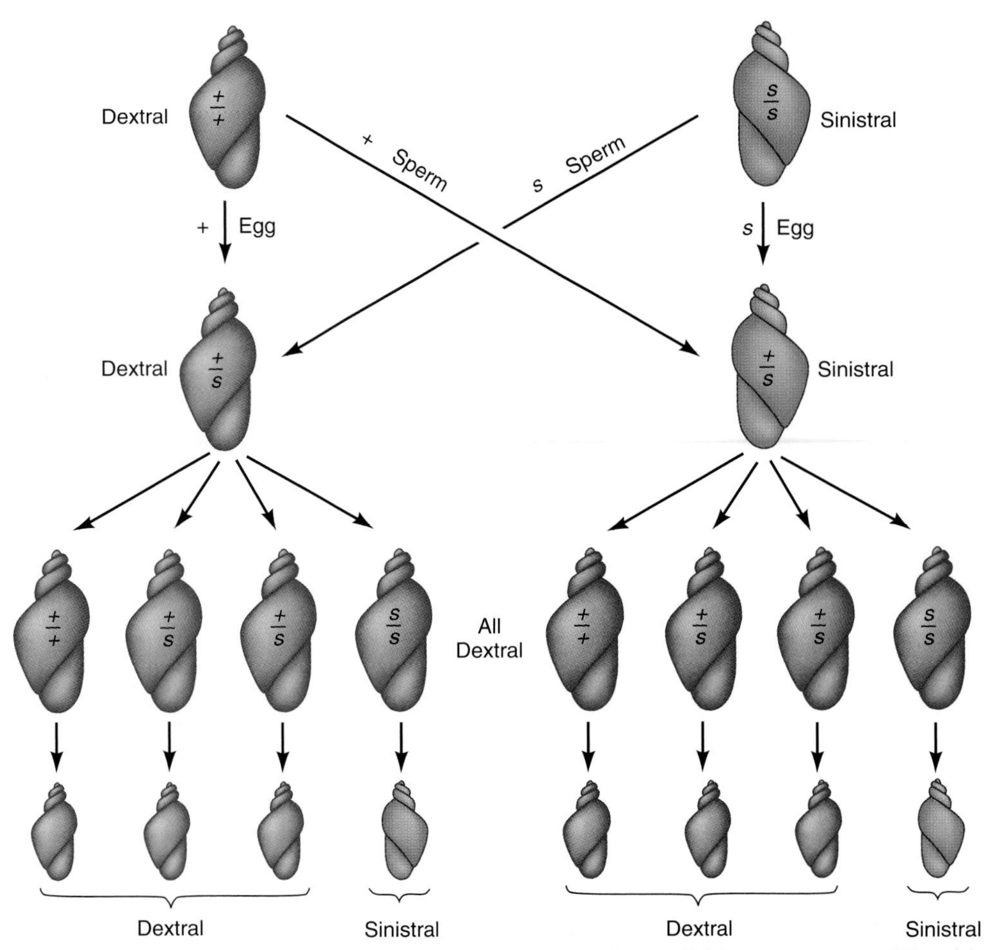

Figure 5.30 Maternal inheritance of direction of shell coiling (chirality) in the snail *Limnaea peregra.* The chirality is controlled by a single gene with two known alleles. The wild-type allele (+) promotes right-handed (dextral) coiling and is dominant over the mutant (*s*) allele, which promotes left-handed (sinstral) coiling. *Lymnaea,* like many snails, is hermaphroditic, meaning each individual produces eggs as well as sperm. If a +/+ individual mates with an *s/s* individual, then the *genotype* of their offspring will always be +/*s* regardless of which individual contributed the egg. However, the *phenotype* of their offspring depends on the genotype of the parent that contributed the egg. Eggs from +/+ parents develop into dextral adults whereas eggs from *s/s* parents develop into sinistral adults. Further crossing of these offspring among themselves shows that an individual will develop into a dextral adult regardless of its own genotype as long as its egg-contributing parent had at least one wild-type allele of the chirality gene.

embryos. Both shell chirality and the direction of embryonic cleavage are controlled by one gene. The *wild-type allele* (+) of this gene, which produces the dextral phenotype, is dominant over the mutant allele (*s*), which results in the sinistral phenotype. Most important, the sinistral form is a maternal effect mutant, as expected (Fig. 5.30). Thus, eggs derived from *s*/*s* individuals give rise to sinistral embryos, regardless of the allele introduced by the sperm, whereas all eggs from +/*s* or +/+ individuals become dextral embryos even if fertilized by *s* sperm. Therefore, the orientation of the mitotic spindles during cleavage must be controlled by components laid down in the egg cytoplasm during oogenesis.

To investigate this phenomenon further, Freeman and Lundelius (1982) transplanted cytoplasm between dextral and sinistral zygotes. Cytoplasm from dextral donors made recipients from sinistral broods cleave dextrally, whereas the converse transplantation had no effect. These results, together with the dominance of the genetic allele producing the dextral phenotype, indicate that the dextral pattern requires a certain gene product in the egg cytoplasm. In the absence of this product, the zygote switches to a default program, which results in sinistral cleavage. The nature of the product of the chirality gene is still unknown, but it is unlikely to be associated with the egg cortex because it readily moves into a glass needle. However, the transplantable gene product may be involved in the subsequent formation of a cortical centrosome attraction site.

In summary, the cleavage pattern of embryos is determined by interactions between the egg or blastomere cortex and the asters of mitotic spindles. A key role in the positioning and orientation of mitotic spindles is being played by the centrosomes, which undergo a regular cycle of replication and movements that generates a succession of perpendicular spindle orientations and corresponding cleavage furrows. However, spindle positioning and orientation may be modified by spatial constraints and by specific cortical sites in eggs and blastomeres that can attract and hold nuclei in eccentric positions. The positioning of such centrosome attracting sites may be influenced by maternally encoded cytoplasmic factors in the egg.

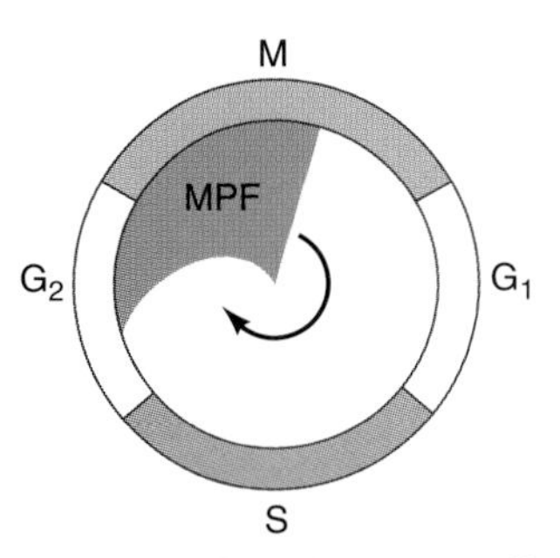

(a) Late embryonic and mature cell cycle

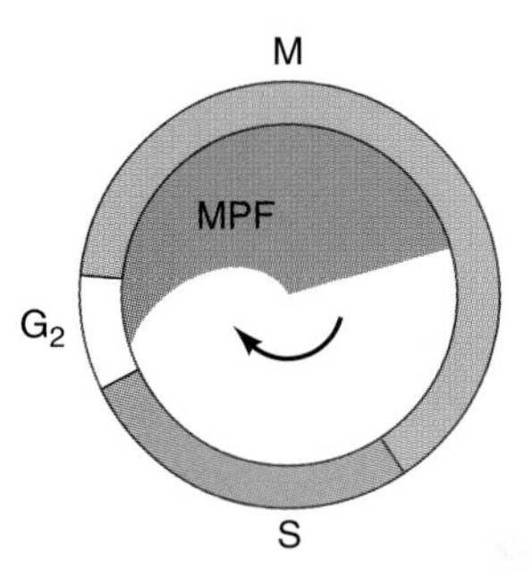

(b) Early embryonic cell cycle

Figure 5.31 Cell cycle in early embryonic and mature cells. **(a)** In mature cells, the cell cycle consists of mitosis (M), postmitotic gap phase (G_1), DNA synthesis (S), and premitotic gap phase (G_2). Synthesis of RNA is limited to the gap phases. **(b)** In early embryonic cells, there is no G_1 phase, because S phase begins at the end of M phase. S phase is accelerated, and G_2 phase is short or nonexistent. Both embryonic and mature cell cycles are controlled by M-phase promoting factor (MPF). MPF activity increases gradually during mitotic prophase and metaphase and decreases rapidly thereafter.

5.4 The Timing of Cleavage Divisions

The fast pace of embryonic cleavage is possible because the egg contains all necessary proteins, or at least the maternal mRNAs from which these proteins can be rapidly translated. These maternal reserves allow blastomeres to speed through several cycles of mitosis without having to transcribe the genes, or even translate the mRNAs, required for proteins such as DNA polymerase, histones, and tubulin. Even in those embryos that do transcribe their own genomes during cleavage, the contribution of the RNA synthesized in the few nuclei present is small compared to the large maternal supplies. As the number of embryonic nuclei increases, maternal supplies are depleted or actively destroyed, and embryos come to rely on their own RNA synthesis. When the *nucleocytoplasmic ratio* reaches a certain level, the pace of cell divisions slows down, and cells begin to grow between mitoses as the necessary nutrients are mobilized from yolk reservoirs or are obtained from outside. Both gene expression and cell division now occur in spatial patterns that are controlled by regional signals.

THE CELL CYCLE SLOWS DOWN DURING THE MIDBLASTULA TRANSITION (MBT)

Unlike mature eukaryotic cells, which have division cycles lasting from several hours to many days, an early *Drosophila* embryo undergoes a new round of mitosis every 9 min. Several features allow for a very rapid mitotic cycle in early embryos (Fig. 5.31). For instance, the S phase is very short because there are an exceptionally large number of initiation sites for DNA replication. In addition, cleaving blastomeres skip entire phases of the cell cycle. In mature cells, the cell cycle consists of mitosis (M phase), a postmitotic gap phase (G_1 phase), DNA synthesis (S phase), and a premitotic gap phase (G_2 phase). In contrast, the blastomeres of most early embryos have no G_1 phase; their DNA begins to replicate during the last stage (telophase) of mitosis. In some species including sea urchins, blastomeres have a short G_2 phase, during which a small amount of RNA is synthesized. In other species, such as *Drosophila* and *Xenopus*, early blastomeres skip the G_2 phase as well and synthesize no measurable amounts of RNA.

Toward the end of the cleavage stage, the cell cycle slows down as gap phases are added or elongated. Around the same time, many embryos begin to synthesize substantial amounts of RNA. In *Xenopus*, for instance, 12 rapid synchronous cleavages are followed by a period of slower, asynchronous cleavages. At the same time, blastomeres become motile and active in RNA synthesis. Together, these changes make up the ***midblastula transition***, or ***MBT*** (Newport and Kirschner, 1982a). In other species, however, the changes that occur simultaneously in *Xenopus* are more dissociated and spread out over several mitotic cycles. In *Drosophila*, for example, 10 rapid synchronous cell cycles precede three somewhat slower and nearly synchronous cell cycles, which in turn are followed by much longer, asynchronous cycles (Fig. 5.32). During the 11th cell cycle, nuclear RNA synthesis becomes detectable (Edgar and Schubiger, 1986). During the 14th cycle, the G_2 phase is extended, and large amounts of RNA are transcribed. Other animals, including mammals and sea urchins, undergo RNA synthesis throughout cleavage and thus have no midblastula transition in the originally defined sense.

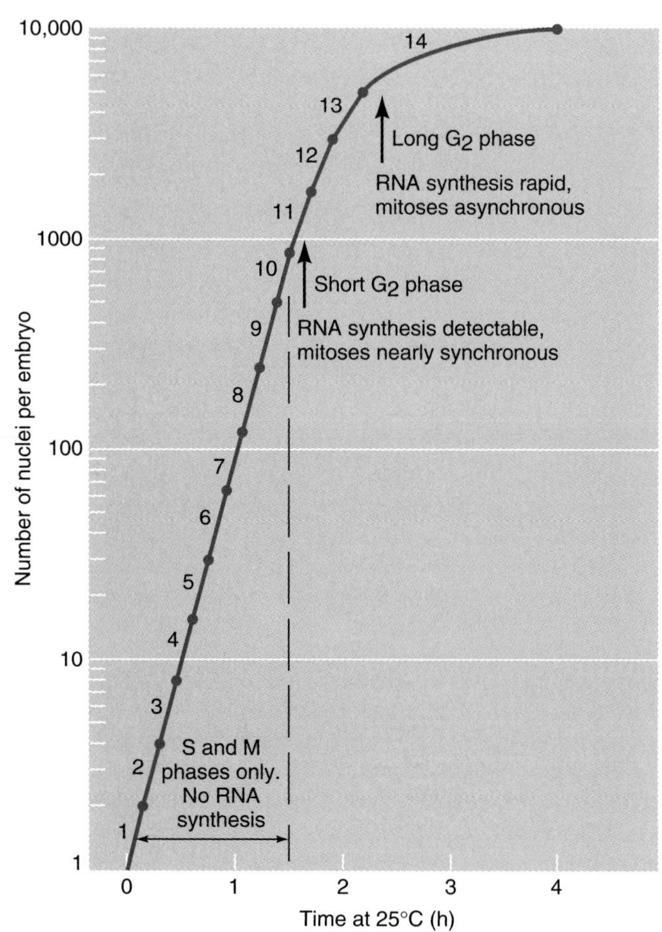

Figure 5.32 Nuclear divisions during early *Drosophila* embryogenesis. The numbers on the curve indicate the cell cycles (see Fig. 5.17). The first 10 cell cycles are synchronous and last 9 min each; little or no nuclear RNA is synthesized. The cell cycle during this period consists only of M and S phases. As the cell cycle slows down somewhat after cycle 10, a short G_2 phase is added, and nuclear RNA synthesis becomes detectable; mitoses are still nearly synchronous. After cell cycle 13, the G_2 phase becomes much longer, the rate of RNA synthesis increases dramatically, and cell divisions become asynchronous.

The following section focuses on cell cycle length and the nucleocytoplasmic ratio, the aspects of the midblastula transition that are most directly related to the cleavage process.

STAGE- AND REGION-SPECIFIC GENE ACTIVITIES MODULATE THE BASIC CELL CYCLE AFTER MBT

As discussed in Section 2.4, the cell cycle is regulated by a biochemical oscillator in the cytoplasm that generates *M-phase promoting factor* (*MPF*). This oscillator involves the periodic synthesis of *cyclins* and the concomitant activation of *cyclin-dependent kinases* (see Fig. 2.16). Genetic and molecular analysis have shown how this basic cell cycle is modified during *Drosophila* embryogenesis (Edgar et al., 1994a,b; Edgar and Lehner, 1996). During the first seven cell cycles, CDK1 and cyclin are continuously present and show little variation in abundance or activity, suggesting that these cycles do not require the typical MPF oscillation. During cycles 8 to 13 a cyclic degradation of cyclins leads to increasing oscillations of MPF activity. Mutants deficient in cyclin mRNAs show cell cycle delays, indicating that cyclin accumulation becomes rate-limiting during this period. During cell cycles 14 to 16, the protein encoded by the *string*$^+$ gene is necessary for initiating M phase (Edgar and O'Farrell, 1989, 1990). By modifying the supply of string protein, the *Drosophila* embryo now switches from rapid overall cleavage to regional patterns of mitoses at much longer intervals. Because the expression of the *string*$^+$ gene is controlled by the products of patterning genes, the *string*$^+$ gene becomes a link between regional patterning signals and cell division.

Throughout the first 13 cell cycles in *Drosophila*, the string protein is translated from an ample supply of maternal mRNA, and the mitoses that are part of these cycles occur rapidly and synchronously throughout the embryo. During the 14th cell cycle, the maternal string mRNA is degraded along with other maternal mRNAs, so that all subsequent mitoses depend on transcription of embryonic *string*$^+$ genes. These latter mitoses are synchronous within distinct regional domains, but the timing of mitosis varies from one domain to the next (Foe, 1989). In each domain, the onset of mitosis is forecast precisely by the transcription of the embryonic *string*$^+$ gene during G_2 phase (Fig. 5.33; Edgar and O'Farrell, 1989). At least some mitotic domains also delineate distinct cell fates (Cambridge et al., 1997).

The string protein acts on the basic cell cycle oscillator by activating one MPF component, the cyclin-dependent kinase CDK1 (Fig. 5.34). Specifically, string protein acts as a phosphatase, removing an inhibitory phosphate group from the tyrosine^{15} residue of CDK1 (Dunphy and Kumagai, 1991; Gautier et al., 1991). Proteins encoded by genes similar to *string*$^+$ have also been found in *Caenorhabditis elegans*, frogs, and mammals, including humans (Sadhu et al., 1990). These data indicate that the activation of CDK1 by a stringlike protein is a common way of controlling the cyclic generation of MPF.

THE NUCLEOCYTOPLASMIC RATIO MAY TRIGGER MBT ACCORDING TO A TITRATION MODEL

The detailed work on the role of string and similar proteins in the midblastula transition (MBT) has rekindled earlier speculations on how MBT might be triggered by the increase in the nucleocytoplasmic ratio during cleavage.

The MBT in amphibians begins when the nucleocytoplasmic ratio reaches a certain level. In haploid embryos, which start at half the normal nucleocytoplasmic ratio, the MBT is delayed by one cycle, whereas in

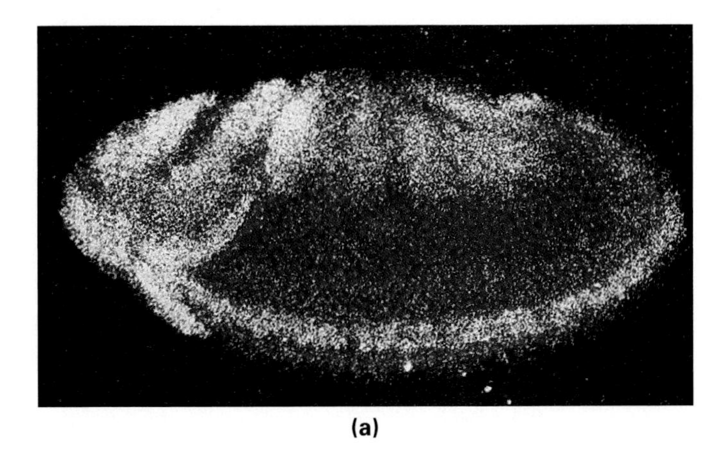

(a)

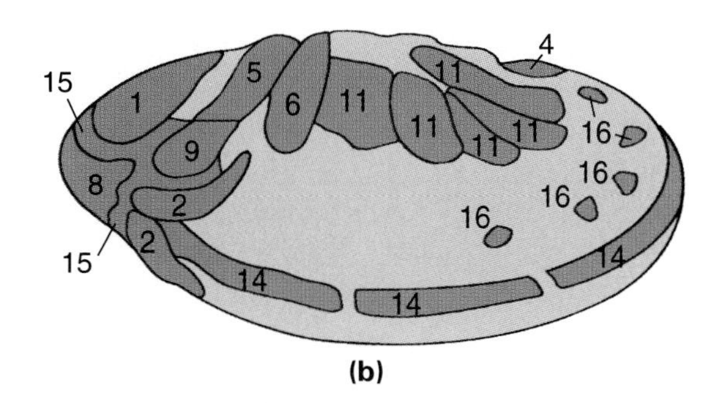

(b)

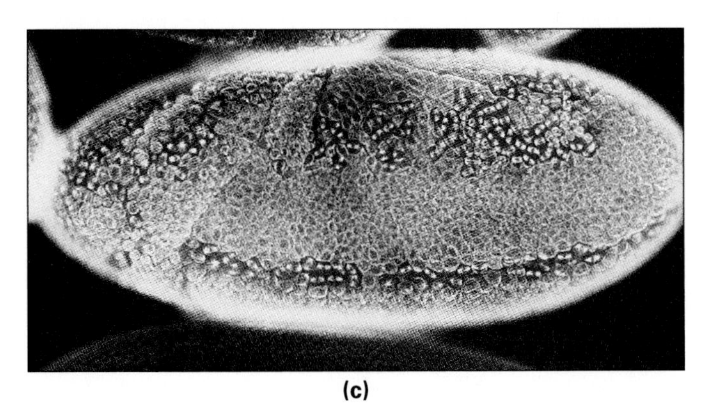

(c)

Figure 5.33 Correlation of mitotic patterns with *string*$^+$ gene expression in the *Drosophila* embryo. **(a)** Embryo late in cycle 14 (early gastrula). The white grains represent a labeled probe that binds specifically to string mRNA (see Method 8.1 for technique). The larger faint dots are nuclei stained with a fluorescent dye. **(b)** Tracing of the embryo shown in part a, showing the correlation of *string*$^+$ expression (shaded) with domains of cells undergoing mitosis (outlined and numbered areas). The numbers were assigned by Foe (1989) to each mitotic domain and indicate the sequence in which mitosis is initiated in each domain. (Some domains straddle the dorsal midline and cannot be seen in this ventrolateral aspect.) The embryos shown in parts a and c have initiated mitosis in domains 1 through 6, where *string*$^+$ expression is most intense. **(c)** Embryo photographed at a slightly later stage after immunostaining with antitubulin antibodies (see Method 4.1). This procedure imparts a bright stain to mitotic spindles. Embryos prepared in this fashion were used to map the mitotic domains shown in part b.

polyploid embryos, the MBT is accelerated. Experimental elimination of three-quarters of the egg volume by constriction induces the MBT two cell cycles early. Injection of DNA also hastens the MBT, and the amount of DNA needed to trigger the MBT is equal to the amount of nuclear DNA present after 12 mitoses (Newport and Kirschner, 1982b). A similar control seems to operate in *Drosophila* embryos. Again, haploid embryos undergo an extra mitotic cycle before MBT. As expected, experimental manipulations that delay the increase of the nucleocytoplasmic ratio also delay the maternal-zygotic transition in *string*$^+$ expression (Yasuda et al., 1992). Conversely, increasing the nucleocytoplasmic ratio by ligation (tying off some of the cytoplasm) can result in the omission of a mitotic cycle (Edgar et al., 1986). In zebrafish, the midblastula transition occurs after 10 cell cycles, again controlled by the nucleocytoplasmic ratio (Kane and Kimmel, 1993). In sea urchins, the nucleocytoplasmic ratio varies among the macromeres and micromeres that are generated by the fourth cleavage. The macromeres, which have the smallest nucleocytoplasmic ratio, undergo 10 additional cleavage divisions, whereas the micromeres, which have the greatest nucleocytoplasmic ratio, undergo only six additional divisions (Masuda and Sato, 1984).

The general significance of the nucleocytoplasmic ratio and the apparent universality of the basic cell cycle oscillator invite speculation that there may be a single mechanism by which the nucleocytoplasmic ratio is "read" and transmitted to the cell cycle oscillator. Many researchers have proposed a *titration model* to explain how cells might sense the nucleocytoplasmic ratio (Fig. 5.35). According to this model, a critical substance in the egg cytoplasm is bound (titrated) by DNA or some other nuclear component that increases exponentially during cleavage. When all of the critical substance is used up, the cell cycle slows down, and other events associated with MBT occur.

On the basis of their work with *Drosophila*, Edgar and Datar (1996) proposed the following sequence of events as a special version of the general titration model. First, nuclear proliferation causes a shortage of maternal factors required for cell cycling, most likely cyclins. Second, the resulting slowdown of the cell cycle allows the transcription of embryonic genes. Third, some of the new embryonic gene products destroy many maternal mRNAs. Fourth, the loss of phosphatase activity caused by the destruction of maternal string mRNA allows inhibitory phosphorylation of cyclin-dependent kinase 1 to persist, causing cell cycle arrest unless embryonic string mRNA is substituted. Fifth, cell division thus comes under the local control of embryonic genes such as *string*$^+$. It appears that this model will be of general utility although some aspects of MBT in *Xenopus* seem to be controlled by a maternal timing mechanism that is independent of the nucleocytoplasmic ratio (Clute and Masui, 1995; Hartley et al., 1997).

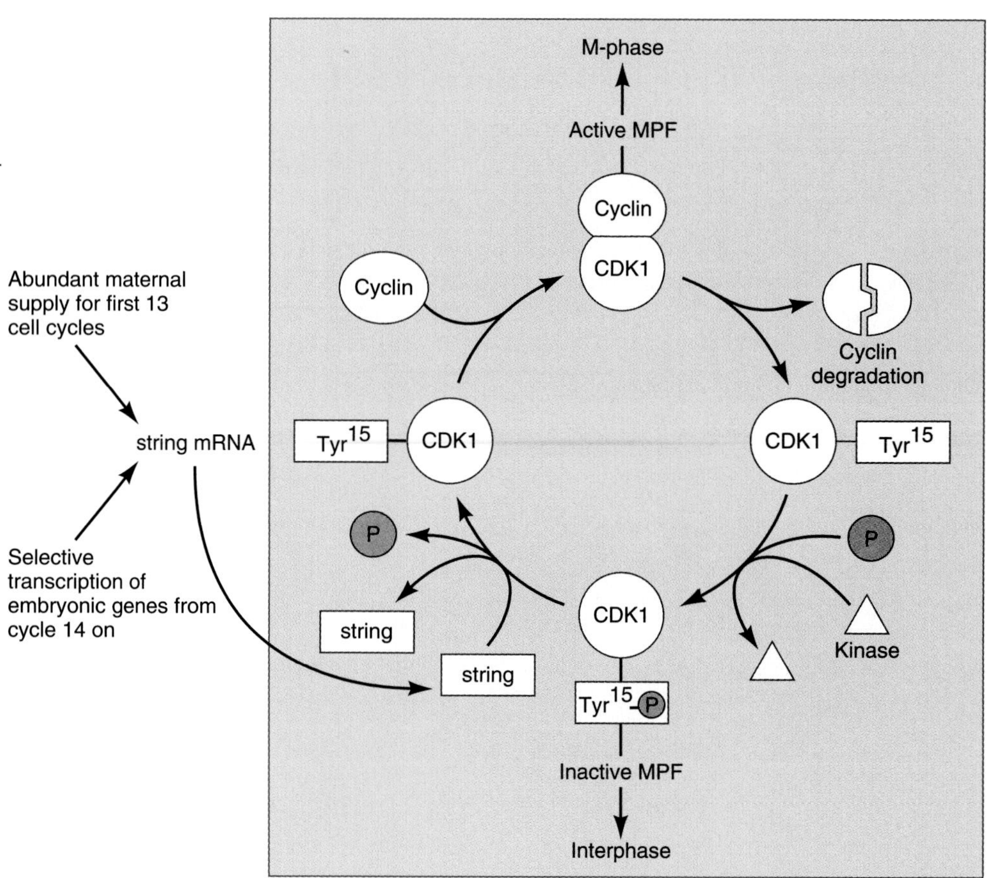

Figure 5.34 Control of the basic cell cycle oscillator by the string protein in *Drosophila.* In the basic oscillator, a cyclin-dependent kinase (CDK1) combines with periodically synthesized cyclin protein to form active M-phase promoting factor (MPF). When cyclin is degraded after metaphase, MPF becomes inactive and allows the remainder of M phase and the following interphase to proceed. The tyrosine^{15} residue of the dissociated CDK1 is phosphorylated by a kinase. (Other modifications of CDK1 and cyclin also occur but are not shown.) The tyrosine^{15} phosphorylation is reversed by the string protein, a phosphatase, before the CDK1 protein can form active MPF again. This basic oscillator runs freely during cleavage while a large supply of maternal string mRNA is available. After 13 mitotic cycles, the maternal string mRNA is destroyed, at which point mitosis becomes dependent on the selective transcription of embryonic *string*$^{+}$ genes.

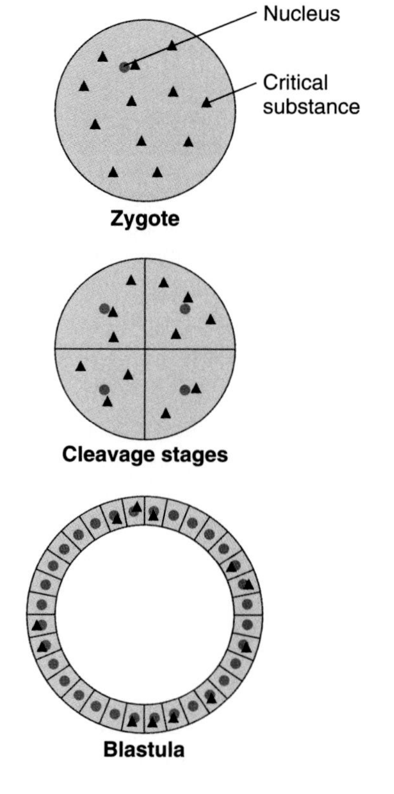

Figure 5.35 Titration model for the change in cell cycle control that occurs at the end of cleavage. A critical cytoplasmic substance (triangles) is bound (titrated) by genomic DNA (circles). As the amount of nuclear DNA increases exponentially during cleavage, all of the critical substance will eventually be bound. In the absence of critical substance, the changes associated with midblastula transition occur.

SUMMARY

The first divisions of the zygote are called cleavage divisions, and the resulting cells are called blastomeres. Unlike dividing mature cells, blastomeres do not grow back to their original size between cleavages. Correspondingly, cleavage divisions are usually fast, with cell cycles that have no G_1 phase and short S and G_2 phases.

The cleavage patterns of different animal groups are correlated with the amount and distribution of yolk in their eggs. The cleavages are termed holoblastic or meroblastic, depending on whether the egg cells are cleaved totally or partially. Holoblastic cleavage patterns include radial cleavage as seen in sea urchins and amphibians, bilateral cleavage as seen in ascidians, rotational cleavage as seen in mammals, and spiral cleavage as seen in molluscs and other groups of animals. Meroblastic cleavage is called discoidal when it is limited to a disc of cytoplasm at the animal pole, as in many fish, reptiles, and birds. In contrast, meroistic cleavage is superficial when it leads to a layer of cells surrounding a central mass of yolk, as in insects.

Embryonic cleavage, like mature cell division, involves mitosis and cytokinesis. Invariably, the plane of cytokinesis is perpendicular to the spindle axis of the preceding mitosis. A regular cycle of centrosome duplications and movements generates a succession of perpendicular spindle orientations and corresponding cleavage furrows. However, spindle positioning and orientation may be modified by spatial constraints and by specific cortical sites in eggs and blastomeres that can attract and hold spindles in eccentric positions. The positioning of such centrosome-attracting sites may be influenced by maternally encoded cytoplasmic factors in the egg.

The timing of cleavage is controlled by the same biochemical oscillator that governs meiosis and the division of mature somatic cells. This basic oscillator relies on the cyclic synthesis, activation, and breakdown of two proteins that together form M-phase promoting factor (MPF). In early embryos, this basic oscillator is free-running, because abundant maternal supplies of the accessory components required to maintain the oscillator are present. At the end of cleavage, the maternal supplies are degraded, and the cell cycle becomes dependent on the transcription of embryonic genes. This so-called midblastula transition is a distinct event in some species; its timing is controlled at least in part by the nucleocytoplasmic ratio.

Cell Fate, Potency, and Determination

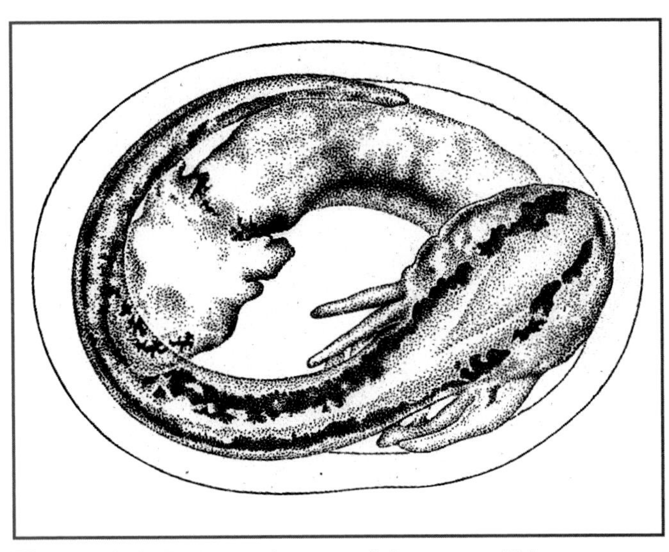

Figure 6.1 Twin embryos of the newt *Triton taeniatus* obtained after the first two blastomeres were separated by ligation with a hair loop.

THIS chapter will deal with three basic concepts in development: fate, potency, and determination of embryonic cells. The concept of cell *fate* is simple: A cell's fate is what the cell or its descendants will become in the course of normal development. Fate can be ascertained by labeling a cell with a dye that is passed on to its daughter cells and then finding out later which parts of the embryo contain the dye. The *potency* of a cell is a concept based on the observation that in many species embryonic cells can, up to a time, depart from their fate. If isolated early from their neighbors or transplanted into a foreign region of the embryo, they may display capabilities that they would not have revealed in normal development.

A dramatic example of the ability of embryonic cells to adjust to different circumstances is the development of twins. This was first observed by Driesch (1892), who found, much to his surprise, that each blastomere from a 2-cell or 4-cell sea urchin embryo could form a complete larva by itself. Other investigators, including Spemann (1938), obtained similar results with amphibian embryos (Fig. 6.1). In fact, the monozygotic ("identical") twins of humans arise by spontaneous splitting of the embryo during early development. Apparently, embryonic cells communicate with each other; they sense whether their normal neighbors are present and adjust their development if some or all of them are missing.

This striking ability of cells to adjust their development in response to signals from their neighbors diminishes over time. Most cells are eventually committed to a particular developmental pathway, from which they rarely depart. This process of gradual commitment is called cell *determination,* and its analysis is one of the principal goals of developmental biology.

In this chapter, we will define cell fate, potency, and determination and illustrate these concepts using insect, amphibian, and mammalian embryos as examples. Next we will discuss some properties of the determined state, in particular, its stability. Finally, we will revisit the patterns of invariant and variable cleavage, redefining them in terms of determination as mosaic versus regulative development. It will become apparent that both strategies are at work in all embryos, although to varying extents depending upon the species.

The concepts of fate, potency, and determination apply to single cells as well as to regions consisting of multiple cells. We will therefore use the terms "cell" and "region" in this chapter as convenient, and for the most part, interchangeably.

6.1 Fate Mapping

The ***fate*** of a cell is the sum of all structures that the cell or its descendants will form at a later stage of *normal* development. A ***fate map*** is a diagram of an organism at a specific stage of development, indicating the fate of its component cells or regions at a later stage. Fate maps are essential tools in most embryological experiments. Only if we know what becomes of cells in the course of normal development can we find out how their behavior is altered by grafting, isolation, or other experimental treatments. Thus, developmental biologists have established fate maps for all the species they study intensely.

Figure 6.2 shows the fate map of the frog *Xenopus laevis* at the blastula stage, as depicted by Kimelman et al., (1992). The map indicates which regions of the embryo will form the three *germ layers* and their major derivatives. Embryonic regions with a distinct fate are called ***primordia*** (sing., *primoridum*) or ***rudiments.*** Thus, those cells on both sides of the embryo that will form the heart are called the *heart primordia,* or *heart rudiments,* or *prospective heart.* Fate maps change over time because cells multiply and move relative to each other. For this reason, fate maps usually indicate fates at a closely

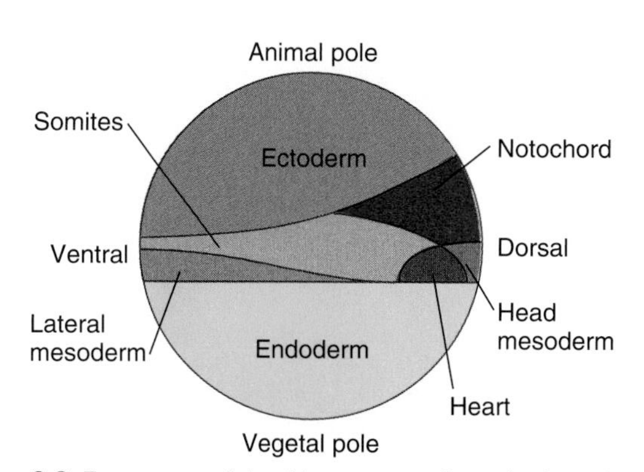

Figure 6.2 Fate map of the *Xenopus* embryo before the onset of gastrulation. The ectoderm arises from animal cells, while the endoderm originates in part from vegetal cells. An intervening zone, called the marginal zone (red and pink colors), gives rise to mesoderm (deep cells) and contributes to endoderm (superficial cells). The map also indicates the positions of major mesodermal derivatives: notochord, head mesoderm, heart, somites (which give rise to muscle, among other structures), and lateral mesoderm. During gastrulation, mesoderm and endoderm move inside the embryo (see Figs. 6.10, 6.11, and Section 10.3).

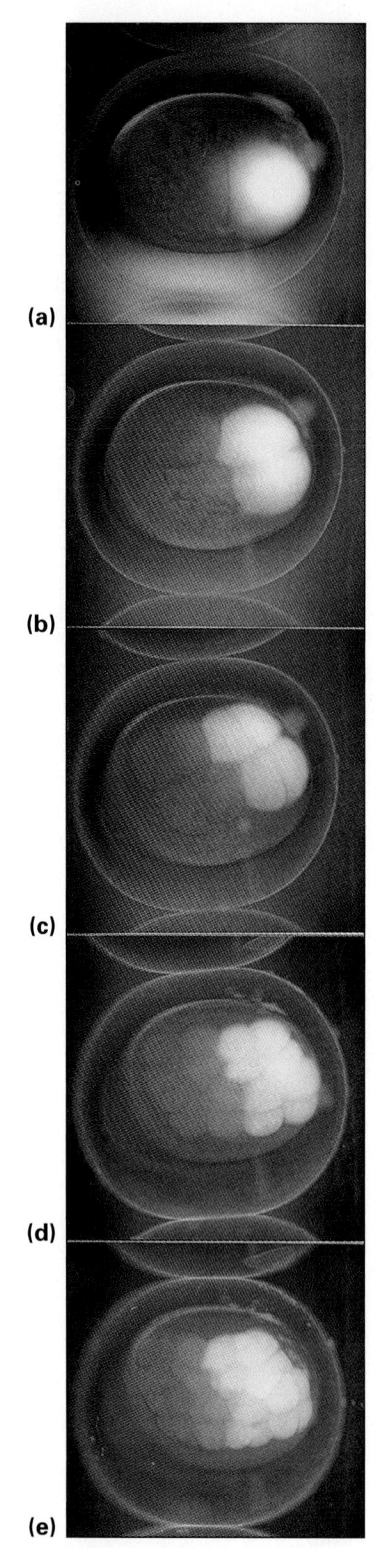

Figure 6.3 Injection of a fluorescent macromolecule as a label for fate mapping. This photograph shows a zebrafish embryo injected at the 2-cell stage with a fluorescent dye conjugated to dextran. **(a)** Embryo immediately after injection. The uninjected cell is seen in faint outline to the left. **(b–e)** Same embryo at the 4-, 8-, 16-, and 32-cell stages, respectively. At each stage, only half of the cells are labeled, indicating that the dye is passed on clonally but does not leak into other cells. Similar dye injections into single cells are still feasible at later stages when cells have become

following stage. A series of fate maps at consecutive stages shows the progression of different cells or regions through longer periods of development.

Different ways of establishing fate maps have been used over time. Some maps have been constructed from observations on living embryos under the microscope, augmented with histological sections and time-lapse motion pictures. For the mapping of large or opaque embryos, it is necessary to label specific cells. A label used for this purpose must not spread to neighboring cells, has to be readily detectable at later stages, and should not disturb normal development.

One commonly used labeling method is the injection of fluorescent dyes conjugated to large, metabolically inert carrier molecules that will not cross cell membranes or pass through gap junctions (Weisblat et al., 1980; Strehlow and Gilbert, 1993). When such molecules are microinjected into one or more cells, all descendants of the injected cell(s) are labeled distinctly (Fig. 6.3). Alternatively, stable mRNAs encoding fluorescent proteins may be microinjected to obtain a very long-lasting label (Zernicka-Goetz et al., 1996). Instead of labeling an embryonic region selectively, it is sometimes more convenient to label a donor in its entirety and transplant the region of interest to the corresponding position in an unlabeled recipient.

In embryos with *variable cleavage,* the accuracy of fate maps is limited by two phenomena. First, the size and position of primordia may vary from one embryo to the next, so that cells in the same topographical position contribute to different primordia in different embryos. Second, the descendants of a cell do not always stay together as a coherent group, but instead may mix with other cells. This is evident as groups of labeled cells are invaded and broken up by unlabeled cells as development proceeds. The extent of these limitations varies widely between species. For many fish, bird, and mammalian species it seems impossible to obtain fate maps before gastrulation (Helde et al., 1994). In *Xenopus,* there is a slow but progressive mixing of cells throughout development (Wetts and Fraser, 1989). For species with *invariant cleavage,* such as the roundworm *Caenorhabditis elegans,* fate maps tend to be very precise at every stage of development (see Chapter 25).

6.2 The Strategy of Clonal Analysis

An important experimental strategy is ***clonal analysis,*** meaning the isolation or labeling of a cell clone and its subsequent analysis. A ***cell clone*** consists of all surviving descendants of a single cell, the ***founder cell*** of the clone. To make a clone visible, it is necessary to label its founder cell in such a way that the label is passed on to all descendants of the founder cell but not to other cells.

There are two common ways to label a founder cell. In the first, the founder cell is selected by the experimenter and microinjected with a dye. This type of clonal analysis is equivalent to fate mapping at the single cell level. The second way depends on the availability of suitable genetic alleles. A cell displaying a distinct phenotype, such as an abnormal pigmentation, can then be generated by X-ray–induced *somatic crossover* (see Method 6.1). This procedure avoids the trauma of microinjection and generates a genetic label, which does not fade or become diluted over time. Unfortunately, X-rays cannot be focused on small target areas, so this technique is not suitable for fate mapping. Rather the resulting clones are positioned randomly. Also, because somatic crossover is a rare event, investigators must irradiate thousands of embryos or larvae and screen all these individuals at a later stage in order to find a small number of specimens with labeled clones. A more efficient way of generating genetically label clones relies on the random excision of certain segments of exogenous DNA (see Method 6.1).

The analysis of cell clones can provide valuable information about the development of an embryonic region. The outline of a clone reveals the degree of cell mixing and any preferred orientation of cell division or locomotion during development. In addition, the decreasing size of clones labeled at successive stages can be used to monitor the growth kinetics of an organ. Clone sizes can also be used to estimate the number of cells in an organ primordium at a certain stage. If a founder cell labeled at a particular stage produces clones covering a fraction $1/n$ of the organ, then the organ primordium presumably consisted of $n/2$ cells at the time of labeling. (The factor ½ accounts for the fact that after crossover, a cell divides into two daughters, only one of which is homozygously mutant). This calculation assumes that all cells of a primordium divide at the same rate, although exceptions to this rule occur both naturally and as a result of the X-irradiation used to induce somatic crossover.

Clonal analysis has led to the unexpected discovery, in insect epidermis, of *restriction lines* that are not crossed by clones labeled after certain stages in development (Garcia-Bellido, 1975). When investigators analyzed *Drosophila* clones labeled at the larval stage, they found that none of them crossed certain boundaries in the adult epidermis. For instance, clones did not extend from the underside to the top of the wing. And surprisingly, labeled clones in all individuals "respected" the same straight line running down the middle of the wing although this line did not coincide with any visible anatomical structure (Fig. 6.4). In fact, the boundary line was completely invisible unless demarcated by a labeled clone. In this case, the edge of the clone along the boundary line was straight, whereas everywhere else the edge of the clone was ragged. The boundary line was especially well demarcated by labeled clones that had been engineered by a genetic trick to grow faster than the surrounding cells. Despite their enormous size, all these clones stopped and formed straight edges at the same boundary lines. Areas of insect epidermis that

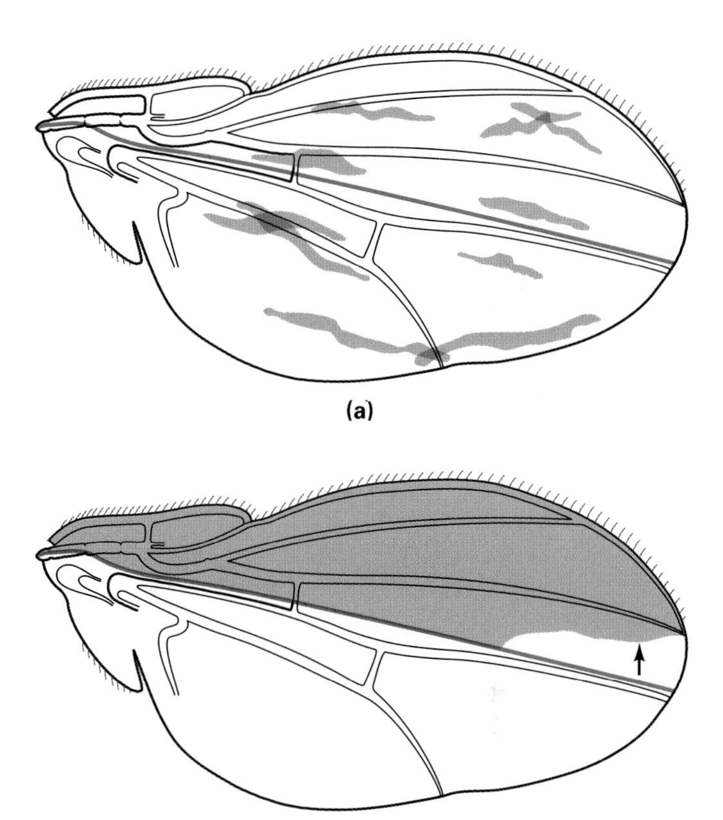

Figure 6.4 Clones and compartments in the wing of *Drosophila*. **(a)** Composite drawing in which clones (color shading) generated in the wings of different flies are superimposed. Each clone was founded by a homozygous mutant cell generated randomly by X-ray–induced somatic crossover (see Method 6.1). Note that the clones have jagged edges and that some of them overlap. Although many more clones were generated than shown here, none of them transgressed the compartment boundaries located at the wing margin and along an invisible line running down the middle of the wing (red color). **(b)** The boundaries of the anterior wing compartment (at top in drawing) are demarcated especially well by a large clone that was genetically engineered to grow faster than all other cells in the fly. Note that the wing is still normally sized and shaped. Also note that the large clone has smooth edges where it runs up against a compartment boundary, but a jagged edge where it abuts other cells within the same compartment (arrow).

are defined by such clonal restriction lines are called ***compartments.***

A compartment is never filled entirely by one clone but rather is made up of several clones, which together can be called a ***polyclone*** (Crick and Lawrence, 1975). Polyclones have several regulatory properties that clones lack. Labeled clones generated in different individuals, when superimposed on the same drawing, do *not* fit each other like a jigsaw puzzle; they overlap and differ in size and shape (Fig. 6.4a). However, polyclones *always* occupy a compartment that has the same distinct shape and position within all individuals of a species. If one member of the polyclone is given a genetic growth

method 6.1

Labeling Cells by Somatic Crossover

A common way of labeling cell clones in *Drosophila* and other species for which suitable mutants are available has been X-ray–induced ***somatic crossover.*** Such crossovers occur between homologous chromatids in somatic cells (Fig. 6.m1) as they do in germ line cells during meiosis (see Fig. 3.4). Normally, somatic crossovers are very rare because homologous chromosomes in somatic cells are not aligned as they are in meiotic prophase I. However, exposure to X-rays increases the frequency of somatic crossovers.

Figure 6.m1 represents a cell heterozygous for the mutant allele *multiple wing hairs (mwh)*. Homozygous (*mwh/mwh*) wing cells exhibit several irregular wing hairs as opposed to the one straight hair present on wild-type (+/+) and heterozygous (+/*mwh*) cells. In cells between S phase and mitotic anaphase, X-rays may induce crossovers between chromatids of homologous chromosomes. If such a cell is heterozygous for a recessive allele such as +/*mwh*, then there is a 50% chance that the next cell division will produce one wild-type (+/+) daughter cell and one homozygous mutant (*mwh/mwh/*) daughter cell. The latter will become the founder cell of a clone of *mwh/mwh* cells having multiple wing hairs (Fig. 6.m2).

An advanced technique for generating genetically labeled founder cells relies on the introduction of engineered *transgenes* into the genome of host organisms (see Section 15.6) and on the ability of a yeast enzyme, *flp* recombinase, to catalyze the removal of DNA segments located between certain target sites (Struhl and Basler, 1993). These "flp-out" events occur randomly, as do X-ray–induced somatic crossovers, but with higher frequency. An application of the flp-out technique is shown in Figs. 22.49 and 22.50.

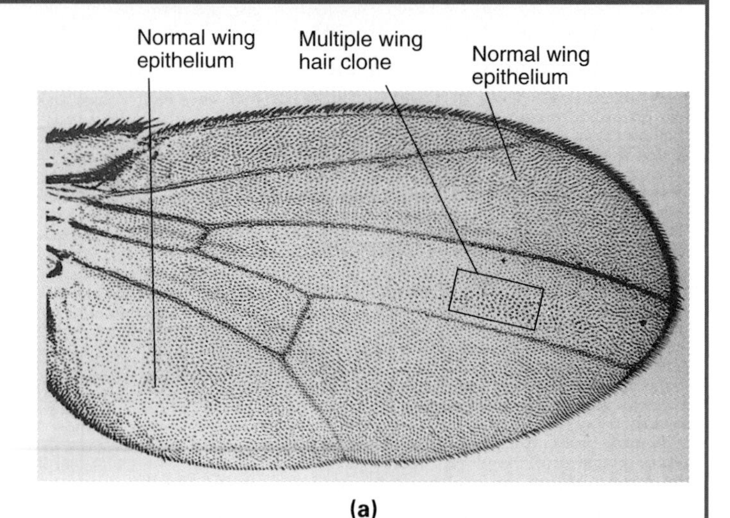

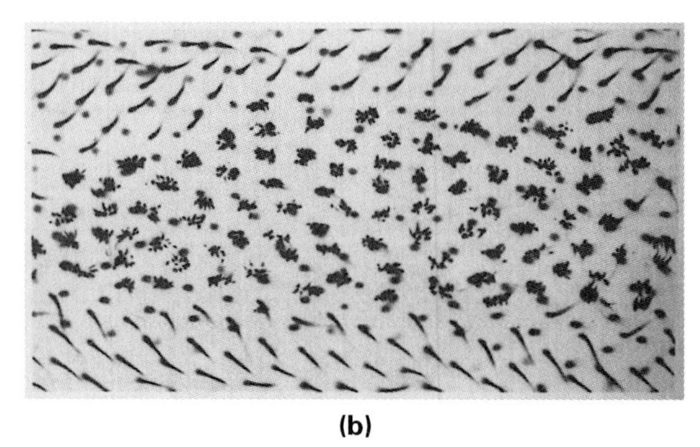

Figure 6.m2 *Drosophila* wing with a *multiple wing hair* clone. **(a)** Photomicrograph in which clone is boxed. **(b)** Clone shown at higher magnification.

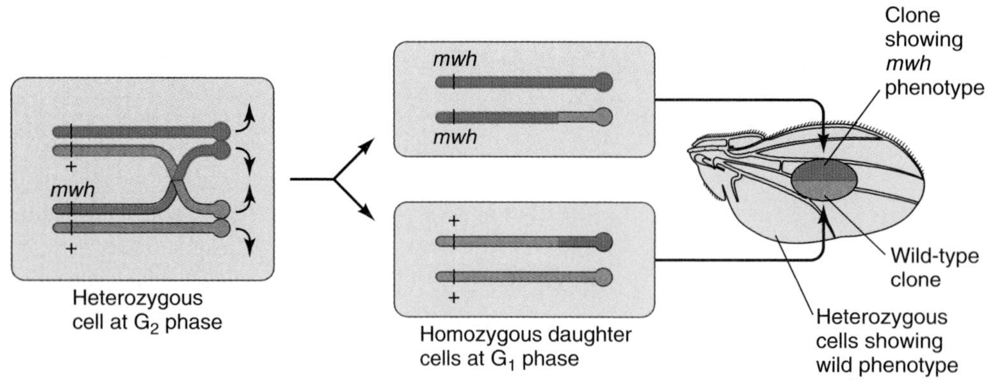

Figure 6.m1 Generation of genetically labeled cell clones by somatic crossover in *Drosophila.* To induce the crossover, individuals heterozygous for a suitable mutant allele such as multiple wing hair (*mwh*) are X-irradiated. In a cell at G_2 phase, a crossover may occur between chromatids of homologous chromosomes. Depending on the segregation of the chromatids during mitotic anaphase (curved arrows), there is a 50% chance that one daughter cell becomes homozygous for the wild-type allele (+) while its sister cell becomes homozygous for the mutant allele. This latter cell becomes the founder cell of a clone expressing the *mwh* mutant phenotype. Its sister cell produces a clone of homozygous wild-type cells, which are phenotypically indistinguishable from the heterozygous "background" cells.

advantage so that it outgrows all the other clones, the compartment is still normally sized and shaped (Fig. 6.4b). Under these circumstances, the remaining clones of the compartment—not of the entire wing—occupy a smaller area than they normally would. Thus, polyclones—rather than single clones or the entire wing—seem to be the units of size and shape regulation (see Section 28.8).

Moreover, polyclones seem to be units of gene regulation. In *Drosophila*, there is evidence that each compartment is characterized by a unique combination of active and inactive selector genes, which characterize each compartment like a binary zip code. Differences in selector gene activity seem to keep the boundaries between compartments straight (see Section 22.7).

6.3 Potency of Embryonic Cells

An embryonic cell during early development usually has a larger range of capabilities than its actual fate, especially in species with variable cleavage. This wider range of abilities is known as the ***potency*** of a cell, meaning the total of all the structures a cell or its descendants can form if placed in the appropriate environment (Slack, 1991). Such definitions that hinge on the outcome of experiments are known as ***operational definitions.*** It was through the use of such definitions that embryology established itself as an independent scientific discipline early during the twentieth century.

The two operations used to assess potency are *isolation* and *heterotopic transplantation* (Gk. *heteros,* "other"; *topos,* "place"). An isolation experiment tests the potency of a cell or region when it has been removed from the influence of other parts of the organism. Cells are explanted from an embryo and kept in vitro in a medium that nourishes them but supposedly does not enhance or inhibit the development of any particular structure. Heterotopic transplantation tests the potency of a cell or region when it is influenced by cells other than its normal neighbors. Cells are explanted from a donor and reimplanted into different regions of a *recipient,* or *host,* organism. In practice, the distinction between isolation and transplantation is not always clear: Tissues are sometimes transplanted into body cavities of a host, where they are nourished by the host's body fluid but do not necessarily receive any instructive signals from the host environment.

As an illustration, if cells from the prospective neural plate of an amphibian embryo are isolated prior to gastrulation, they form epidermis. If they are transplanted elsewhere in the same embryo or another embryo, the graft forms nose sensory epithelium, eye lens, inner ear, or various mesodermal structures, depending on the site of implantation. Thus, on the basis of just a few tests, it can be shown that the potency of prospective neural plate cells prior to gastrulation includes epidermis, nose, lens, ear, and various mesodermal structures as well as neural plate. Note that the potency of a region always includes its fate. If cells or regions can form more structures than their fate, they are called ***pluripotent.*** If they can give rise to a complete individual, they are called ***totipotent.***

On the basis of a large number of isolation and grafting experiments, one could draw a ***potency map*** of an embryo. A potency map would look like a fate map except that each area would have multiple labels. Thus, the area labeled simply "neural plate" on the fate map would be labeled "neural plate, epidermis, nose, eye, ear structures, or mesoderm" on the potency map.

A problem with the operational definition of potency is that the outcome of isolation and grafting experiments depends very much on technical detail. This is best illustrated by some of the earliest embryological studies, in which frog blastomeres were isolated at the 2-cell stage. Depending on the conditions of isolation, such blastomeres gave rise to complete embryos, half-embryos, or undifferentiated lumps of tissue (see Figs. 1.13–1.15).

The results not only of isolation, but also of transplantation, may depend on the number of cells that are transplanted. Single cells or small groups tend to depart from their fate and blend in with their new environment. In contrast, larger groups of equivalent cells are more likely to hold on to their original fate, a phenomenon that Gurdon et al. (1993) termed the ***community effect.*** (By analogy, a single family of German immigrants to the United States is likely to adapt, learn English well, buy American cars, let the children play baseball, and the like, whereas a large group of such immigrants will conserve much of their culture, including their language, choirs, soccer games, etc.) The community effect of cells may depend on intercellular signaling to activate and maintain the expression of certain genes. Because single cells are deprived of such signals, they may be unable to develop according to fate upon isolation or transplantation. Therefore, single cells or small groups may show an apparent lack of determination in isolation and transplantation tests while larger groups of the same cells seem to be determined.

Since the first investigators did not always recognize how much the outcome of operational tests for cell potency depended on the technique used, their different results sometimes led to spirited controversies with philosophical overtones. Roux (1885) believed that embryos were like machines in which each part had a strictly limited function. He found this idea confirmed by the development of a half-embryo from a frog blastomere that survived after its sister blastomere had been killed at the 2-cell stage (see Fig. 1.13). Another pioneer of developmental biology, Driesch (1909), adopted a radically different view. When he observed the development of whole sea urchin larvae from isolated blastomeres, he was so amazed by this result that he later postulated the existence of a goal-directed force (entelechy) in living organisms. With time, the effects of different experimental techniques have become better understood, but even contemporary researchers need to remember that operational criteria may depend on technical detail.

6.4 Determination of Embryonic Cells

In most animal species, the potency of an early embryonic blastomere will exceed its fate. However, as development progresses, the range of structures that a cell may form becomes more limited. When the potency of a cell is restricted to its fate, the cell is said to be ***determined.*** Thus, if the fate of a cell is to form heart, and if after various isolation and transplantation experiments it still forms heart, then this cell is said to be determined for heart development. The process by which fate and potency of a cell become identical is called ***determination.***

CELL DETERMINATION IS DISCOVERED THROUGH OPERATIONAL CRITERIA

To decide whether a cell or region is determined, one has to compare its fate with its potency. If a fate map is already available for the organism and developmental stage of interest, only the potency needs to be assessed. Isolation and heterotopic transplantation then serve as *operational criteria* for *embryonic determination.* According to these criteria, an embryonic cell or region is said to be determined when it develops according to fate upon isolation as well as heterotopic transplantation. Since finding the determined state of cells or regions involves assessing their potency, the difficulties with potency tests also affect the assessment of determination.

In addition to isolation and transplantation, there is a third criterion for determination, which is based on clonal analysis. If a labeled clone encompasses two different structures, such as leg and wing in a fly (Fig. 6.5), then the founder cell's fate, and hence its potency, was not limited to either leg or wing. This means that the founder cell was also not determined to form either one of these structures. Note that the *lack* of clonal restriction is used here as a *negative criterion* for determination. This strategy cannot show positively that a cell was determined because clonal analysis reveals the fate of a founder cell but not its potency. In short, the absence of clonal restriction to one type of structure reveals a lack of determination, whereas the presence of clonal restriction shows only a restriction in fate.

The three operational criteria for assessing cell determination, and the problems with these criteria, are summarized in Table 6.1. The application of these criteria will be illustrated as we discuss determination in insects, amphibians, and mammals.

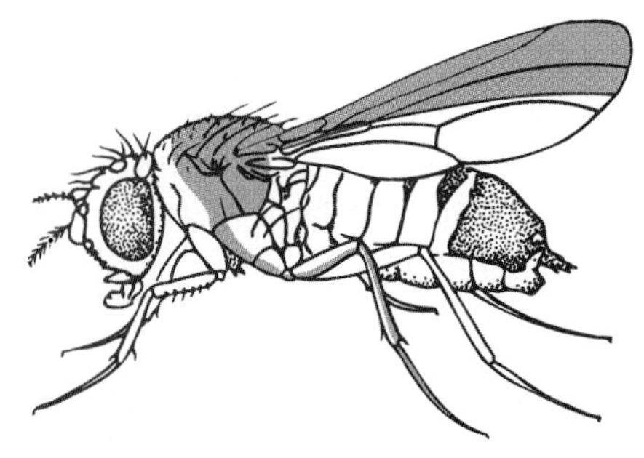

Figure 6.5 Clonal analysis as a test for nondetermination of the clonal founder cell. The diagram shows the outline of a *Drosophila* adult fly with a labeled clone (color shading) extending over most of the anterior compartment of the second thoracic segment, including wing and second leg. The clone was generated by X-ray–induced somatic crossover at the cellular blastoderm stage (see Method 6.1). The clone cells were homozygous for a cuticular marker gene and for a gene that confers a growth advantage. The fact that the clone contributes to both leg and wing shows that neither the potency nor the fate of the clone founder cell was restricted to either leg or wing development.

DROSOPHILA BLASTODERM CELLS ARE DETERMINED TO FORM STRUCTURES WITHIN A SINGLE SEGMENT

Cell determination in insect embryos has been tested by the strategies of clonal analysis, isolation, and heterotopic transplantation. The following experiments with *Drosophila* embryos were designed to reveal whether cells at the blastoderm stage are already determined to form all or parts of specific segments.

X-ray–induced somatic crossover (see Method 6.1) was applied to *Drosophila* embryos at the cellular blastoderm stage (Wieschaus and Gehring, 1976). All labeled clones were restricted to a single segment. However, within a segment some clones overlapped dorsal and ventral structures such as wings and second legs. The

table 6.1 Operational Criteria for Cell Determination

Operation	Potential Problems
Heterotopic transplantation	Dependence of cell potency on number of transplanted cells and contact between them Limited number of transplantation sites that can be tested
Isolation	Dependence of cell potency on number of isolated cells and contact between them Additional dependence of cell potency on method of isolation and culture medium
Clonal analysis	Only negative conclusion possible

possibility that the overlap was an appearance caused by two separately induced clones was rendered unlikely by the low frequency of clones in either wings or legs. Among 664 flies irradiated at the blastoderm stage and scored in the adult thoracic region, only 13 clones were found in second legs and 25 in wings. The probability of finding *one* fly with two *independently generated* clones in wing and second leg on the same side of the body was therefore $(13/664) \times (25/664) \times (664/2) = 0.24$. Actually, the investigators obtained *seven* flies with clones in both wings and second legs on the same side. The probability of obtaining this result by chance was $0.24^7 = 0.00005$. Therefore, in virtually all of these cases, a single labeled clone must have extended over both leg and wing. In an extension of these experiments, marked clones endowed with a genetic growth advantage were found to fill almost entirely either the anterior or the posterior *compartment* of the mesothorax, including wing and leg, on one side of the body (Fig. 6.5). The results show the blastoderm cells of *Drosophila* are not yet determined to form only dorsal or only ventral structures within a segment.

TO find out whether blastoderm cells were restricted in their potency to a single segment, Simcox and Sang (1983) transplanted groups of about six cells between *Drosophila* embryos at the cellular blastoderm stage (see Figs. 5.17–5.19; Fig. 6.6). Both donors and recipients were raised to adulthood. While the donors were genetically wild-type, the recipients carried several mutant alleles that gave the adults such phenotypic features as yellow cuticles and forked bristles. Wild-type donor tissue could therefore be distinguished from mutant host tissue anywhere in the adult epidermis. The donors were examined for defects while the recipients were scanned for patches of donor tissue. As expected, defects in donors were found in those areas that, according to known fate maps, might have been damaged by the removal of cells. Whenever matched pairs of donors and recipients survived, the transplants in the recipients had formed the same structures that were missing in the donors. Thus, after *heterotopic transplantation,* the donor cells still expressed fates corresponding to their origin. For instance, when cells from the prospective fifth abdominal segment were transplanted to the head region of a host at the blastoderm stage, the donor as an adult lacked part of the fifth abdominal segment while the recipient's head contained the same structures showing the genetic labels of the donor. These results strongly suggest that the potencies of blastoderm cells are restricted to one segment. In contrast, corresponding experiments at the syncytial blastoderm stage, before cellularization, showed that transplanted nuclei could still alter their fates according to their new positions. This was shown by the formation of

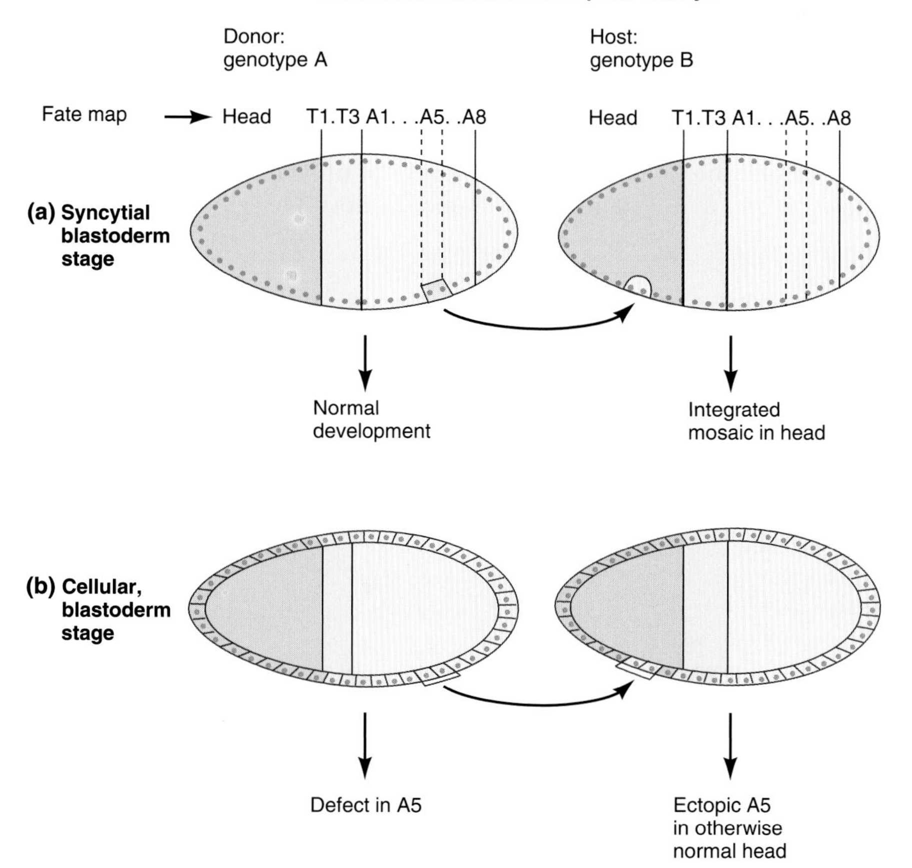

Figure 6.6 Transplantation test for cell determination in *Drosophila* embryos at the syncytial and cellular blastoderm stages. Genetic markers were used to distinguish adult structures formed by donor cells from those formed by host cells. Fate maps indicate the approximate positions of nuclei or cells that form head, thoracic segments (T1–T3), and abdominal segments (A1–A8). **(a)** At the syncytial blastoderm stage, nuclei and surrounding cytoplasm were grafted from a donor region fated to form A5 to the prospective head region of the host. The graft contributed to a normal head consisting of both donor and host cells (integrated mosaic), indicating that the transplanted nuclei had not been determined yet. **(b)** In a similar experiment, cells were transplanted at the cellular blastoderm stage. After metamorphosis, the donor was defective in A5, while the host had a head containing A5 structures carrying the genetic markers of the donor.

integrated mosaics—that is, normally composed structures consisting in part of genetically marked (recipient) and wild-type (donor) cells (Kauffman, 1980).

Questions

1. Why is *Drosophila melanogaster* a favorable organism for the experiment described here?
2. Under the conditions of this experiment, a significant restriction in the potency of blastoderm cells occurred around the time of cellularization. Can you think of a process at the cellular level that becomes restricted as a result of cellularization and may be related to cell potency?

Together, the foregoing experiments show that *Drosophila* blastoderm cells are determined for a single segment but not for dorsal or ventral structures within one segment. The restriction of potency to one segment occurs around the time of cellularization.

DROSOPHILA CELLS ARE BORN WITH A BIAS THAT IS HONED BY SUBSEQUENT INTERACTIONS

A fate map of the *Drosophila* embryo at an early gastrula stage shows that, within the prospective thorax and abdomen, dorsal ectoderm forms only epidermis whereas ventral ectoderm forms both epidermal and neural cells (Fig. 6.7). The dorsal area is therefore known as the ***dorsal epidermal anlage* (DEA)** while the ventral region is termed the ***ventral neurogenic region* (*VNR*).** To explore the process of cell determination in DEA and VNR, investigators have used the criteria of clonal analysis, isolation, and transplantation. They found that cell determination can be a stepwise process, and that full determination may be preceded by a state of preliminary determination that is described as ***bias.***

To isolate single cells, Lüer and Technau (1992) introduced fine beveled glass capillaries into *Drosophila* embryos at the early gastrula stage. The cells were taken from a strip extending around the middle of the embryo from the ventral to the dorsal midline (hatched area in Fig. 6.7). The isolated cells were kept for a day in culture chambers, where they formed clones of several cells, which were classified as neurons or epidermal cells based on their division pattern, morphology, and synthesis of specific proteins. Three types of clones were found in a characteristic spatial distribution: The majority of DEA cells formed pure epidermal clones, most VNR cells formed pure neural clones, and several cells from both DEA and VNR formed mixed clones (Fig. 6.8). The researchers concluded that DEA cells are biased to form epidermal cells while VNR cells have a preference to form neurons.

The bias of isolated DEA cells to form epidermis is in accord with their fate: *All* DEA cells in a normal embryo form epidermis. In contrast, the tendency of isolated VNR cells to form mostly neurons was not expected on the basis of their fate: Prospective neurons and epidermal cells are intermingled in the VNR, with *less than half* of them fated to be neurons and the rest bound to become epidermal cells. The comparison between the fate of VNR cells and their development in isolation suggests that most of them are "born" with a bias to form neurons but that many of them are "persuaded" by signals received later on to form epidermis instead. The following transplantation experiments further illustrate this point.

Drosophila eggs injected with dyes and other lineage tracers were allowed to develop until the early gastrula stage, when they were used as donors of labeled cells to be transplanted to unlabeled hosts at the same stage (Stüttem and Campos-Ortega, 1991). The types of

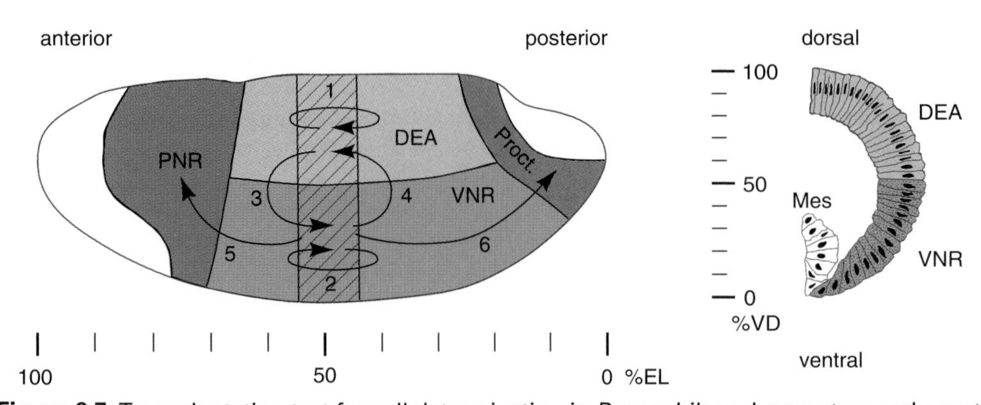

Figure 6.7 Transplantation test for cell determination in *Drosophila* embryos at an early gastrula stage. The diagrams show embryos in a lateral view (left) and in transverse section (right). The embryo's longitudinal axis is subdivided in % of total egg length (%EL), with 0% representing the posterior pole and 100% representing the anterior pole. The ventrodorsal dimension of the embryo is subdivided in % of the circumference (%VD), with 0%VD representing the ventral midline and 100%VD representing the dorsal midline. The dorsal epidermal anlage (DEA), ventral neurogenic region (VNR), procephalic neurogenic region (PNR), prospective proctodeum (Proct.), and mesoderm (Mes) are labeled. The hatched strip indicates the area from which cells were taken in the isolation and transplantation experiments described in the text. The numbered arrows refer to experiments listed in Table 6.2.

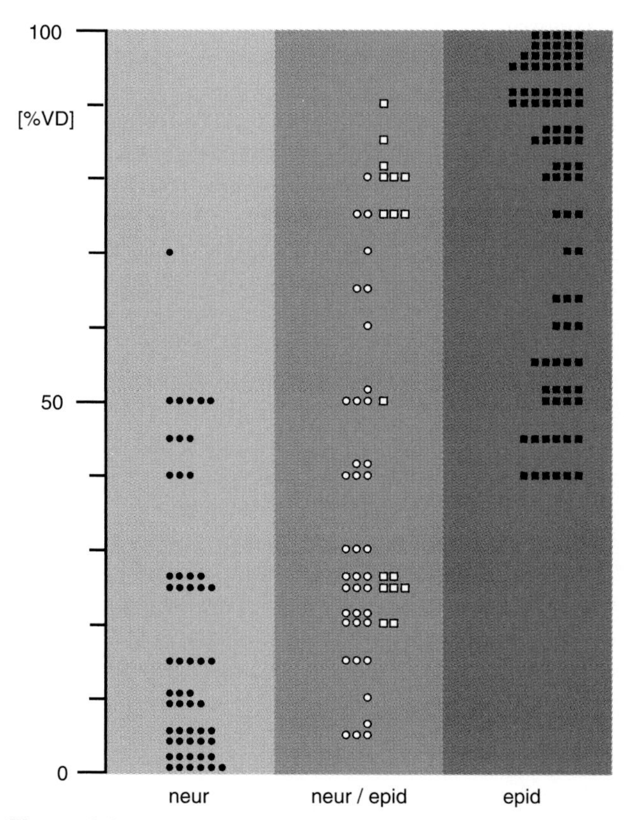

Figure 6.8 Potencies of isolated cells from *Drosophila* embryos at an early gastrula stage. Each cell was isolated during the early gastrula stage from an area between 45 and 55%EL, and between 0 and 100%DV (hatched area in Fig. 6.7), and then kept in culture for about a day. The cells gave rise to neural (neur), epidermal (epid), and mixed (neur/epid) clones with frequencies depending on the dorsoventral position from which the clone founder cell had been taken. The figure distinguishes between two types of mixed clones: open squares represent small clones (no more than 8 cells); open circles represent larger clones.

transplantations are indicated by numbered arrows in Figure 6.7, and their results summarized in Table 6.2. Single cells transplanted *homotopically* (Gk. *homos*, "same"; *topos*, "place") from DEA into DEA formed mostly epidermal clones and no neurons, and cells transplanted from VNR into VNR formed mostly pure neural or epidermal clones and some mixed clones. These results simply confirm that the transplantation procedure itself does not interfere with the normal process of fate acquisition. *Heterotopic* transplantations from DEA into VNR revealed a strong *community effect*. Nearly all transplanted *groups of cells* stuck to their original fate of forming epidermis, whereas many transplanted *single cells* adjusted to their new location, giving rise to neural or mixed clones. The development of heterotopically transplanted single VNR cells depended largely on their new location: In DEA, they formed neural and epidermal clones in proportions similar to those observed after homotopic transplantation into VNR. However, if transplanted into the *procephalic neuroectoderm* (*PNR*, which forms brain) or into the *proctodeum* (*Proct.*, which forms the ectodermal endpiece of the gut), VNR cells formed mostly neural and virtually no epidermal clones. The results indicate that the original bias of VNR cells to form neurons is counteracted more strongly in the DEA and VNR than in the PNR and proctodeum.

Extending these studies, Udolph et al. (1995) took advantage of the fact that cell clones developing in the *ventral VNR* (close to the ventral midline) differ morphologically from clones developing in the *lateral VNR*. Cells transplanted at the early gastrula stage from lateral VNR to ventral VNR positions developed in accord with their new position. In contrast, cells transplanted from ventral to lateral VNR migrated back ventrally and produced

table 6.2 Cell Transplantations between *Drosophila* Embryos at an Early Gastrula Stage

No. in Fig. 6.7	Type of Transplantation	No. of Cells per Transplantation	No. of Clones Analyzed	No. (%) of Neural Clones	No. (%) of Epid. Clones	No. (%) of N./E. Clones	No. (%) of Other Clones
1	DEA into DEA	1	52	0	40 (77)	0	12 (23)
2	VNR into VNR	1	189	84 (44)	73 (39)	26 (14)	6 (3)
3a	DEA into VNR	1	119	28 (23)	72 (61)	19 (16)	0
3b	DEA into VNR	3 or 4	97	5 (5)	88 (91)	4 (4)	0
3c	DEA into VNR	10	109	1 (1)	101 (93)	3 (3)	4 (4)
4	VNR into DEA	1	62	25 (40)	37 (60)	0	0
5	VNR into PNR	1	80	74 (93)	0	0	6 (7)
6	VNR into Proct.	1	61	49 (80)	1 (2)	0	11 (18)

Epid. = epidermal
N./E. = mixed neural and epidermal
PNR = procephalic neuroectoderm
Proct. = proctodeum
VNR = ventral neurogenic region
Source: Stüttem and Campos-Ortega (1991).

clones consistent with their origin. The researchers concluded that signals emanating from near the ventral midline determine VNR cells to assume both ventral positions and ventral fates.

Taken together, the isolation and transplantation experiments with *Drosophila* early gastrula cells reveal a common way for cells to acquire their fates: They are "born" with a *bias,* and this bias is either confirmed or modified through cell interactions later in development. A common way for cells to acquire a bias is through uneven distributions of cytoplasmic or membrane-bound activities. In the *Drosophila* embryo, the biases of the VNR and the DEA can be traced back to a gene-regulatory protein that accumulates in the nuclei of ventral but not dorsal blastoderm cells (see Section 22.6).

The ways in which cells modify their biases during cell determination can be divided into three classes. One class generates the *community effect,* the exchange among *equivalent* cells of signals that will *stabilize the same determined state for all of them,* as discussed earlier in this chapter.

The opposite effect is known as ***lateral inhibition.*** Here, *equivalent* cells *compete with one another to attain a preferred fate,* and a cell that has somehow gained an advantage in this competition will inhibit its neighbors from attaining the same fate. In the VNR of the *Drosophila* embryo, the preferred fate is to become a ***neuroblast*** (neuron percursor cell) rather than an epidermal cell. Of critical importance in this process are two classes of gene products. The first class are gene regulatory proteins named *basic helix-loop-helix (bHLH) proteins* after their molecular characteristics (see Sections 16.4 and 20.5). Representatives of this class are the Achete and scute proteins, which are synthesized in clusters of VNR cells that have the *potential* to become neuroblasts. The second class of critical proteins in neuroblast development are a plasma membrane receptor and its matching ligands. The receptor protein is encoded by the *Notch*$^+$ gene. Its most common ligand, also a membrane protein, is made from the *Delta*$^+$ gene. Notch and Delta mediate the *lateral inhibition* process that decides which of the cells in a cluster of *potential* neuroblasts will actually acquire this *fate* (Fig. 6.9). A cell that somehow gets the edge in producing Delta protein activates its neighbors' Notch receptors more strongly than its own receptors are stimulated. The signal cascade released by the activated receptor inhibits a set of genes that would otherwise determine a cell to become a neuroblast and also activate the *Delta*$^+$ gene. Thus, the dominant cells maintains its advantage in the production of Delta and attains the primary fate of becoming a neuroblast while inhibiting its neighbors from doing the same, thus forcing them into the secondary fate of making epidermis.

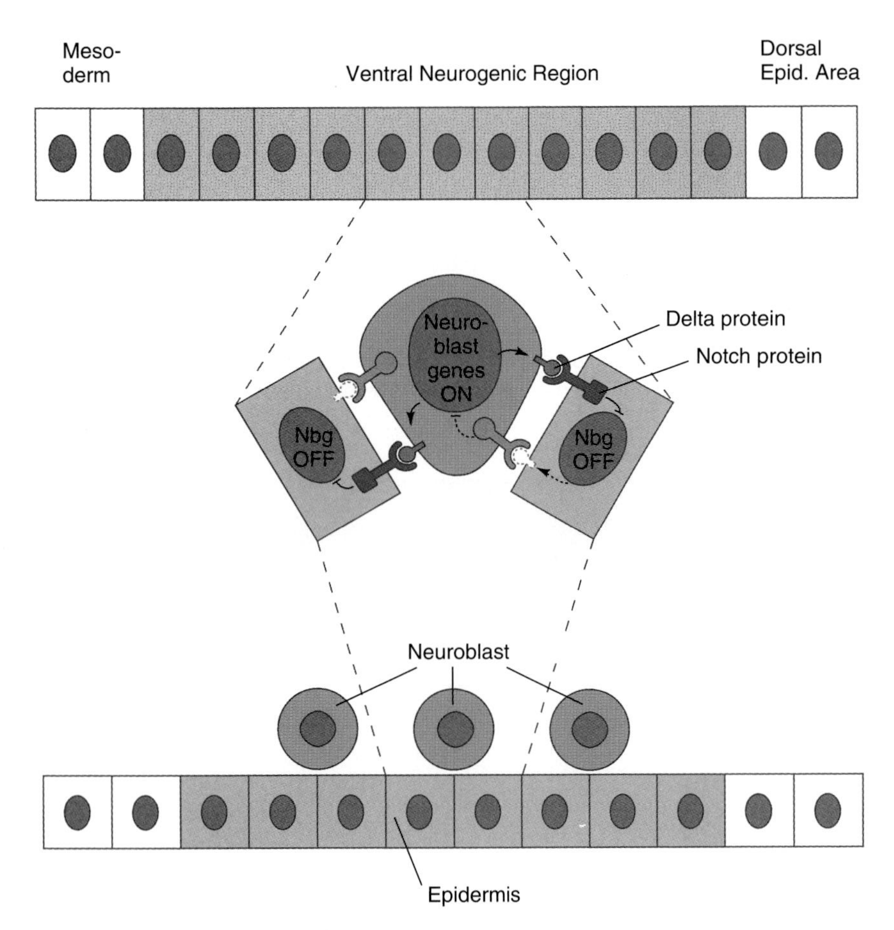

Figure 6.9 Lateral inhibition by Delta and Notch signaling. In the ventral neurogenic region (VNR) of *Drosophila* embryos, cells are biased to form neuroblasts (neuron precursor cells) or epidermal cells, but all cells are initially equivalent. All cells produce a receptor protein encoded by the *Notch*$^+$ gene. The receptor is activated by a ligand that is also a membrane protein and encoded by the *Delta*$^+$ gene. A cell that produces more Delta protein than its neighbors activates its neighbors' receptors more strongly. The signal cascade released by the activated receptor inhibits a set of genes that would otherwise determine a cell to become a neuroblast (neuroblast genes, Nbg). These cells are thus inhibited from becoming neuroblasts and form epidermis instead. Because neuroblast gene activity is also required for Delta protein synthesis, prospective epidermal cells continue to synthesize little or no Delta protein. The cell synthesizing more Delta protein thus retains its advantage and keeps its neuroblast genes active. As part of its neuroblast development, it leaves the VNR epithelium and moves to the basal side of the future epidermis.

The Notch/Delta mechanism of lateral inhibition is a very general one: In *Drosophila,* it is involved in the development not only of the central nervous system but also bristles, eyes, wings, ovaries (Artavanis-Tsakonas et al., 1995; 1999; Greenwald, 1998). The Notch/Delta mechanism of lateral inhibition has also been conserved exceedingly well in evolution: Corresponding proteins mediate lateral inhibition in the determination of various cell types in vertebrates and in the roundworm *C. elegans,* as we will see in future chapters.

The third way in which cells modify their biases in the process of determination is known as *embryonic induction.* Unlike the community effect and lateral inhibition, embryonic induction is defined as an interaction between *nonequivalent* cells. As a result of the interaction, at least one of the partners in the boundary region changes its determined state, as we will discuss more fully in Chapters 8, 9, and 12.

PROSPECTIVE NEURAL PLATE CELLS OF AMPHIBIANS ARE DETERMINED DURING GASTRULATION

The neural plate rudiment of an amphibian blastula is *fated but not yet determined* to become neural plate. The same is still true at the early gastrula stage, as shown by heterotopic transplantations carried out by Spemann (1938). The gastrulation movement begins at a depression called the ***blastopore,*** which is located below the egg equator near the area that was marked by the *gray crescent* in the zygote (see Fig. 1.15; Fig. 6.10). Here the vegetal cells buckle into the blastula, forming a curved groove that later extends into a circle. When the blastopore has just formed, the embryo is said to be at the ***early gastrula*** stage. During gastrulation, the animal and equatorial regions expand toward the blastopore rim, where the equatorial cells disappear to the inside. When the gastrulation process is complete, the embryo is called a ***late gastrula.*** At this stage, the embryo consists of an outer layer, called *ectoderm,* and an inner layer consisting of *mesoderm* and *endoderm.* The endoderm cells form the embryonic gut, or *archenteron.* The major derivatives of the ectoderm are the *epidermis* and the *neural plate.* The neural plate originates from the dorsal ectoderm between the blastopore and the animal pole. The neural plate first forms as an elevated area on the dorsal side of the embryo. Subsequently, the neural plate closes into a tube that is internalized and gives rise to brain and spinal cord.

To test the prospective neural plate region for its determined state, Spemann transplanted prospective neural plate from an early donor gastrula to a prospective epidermis region of an early host

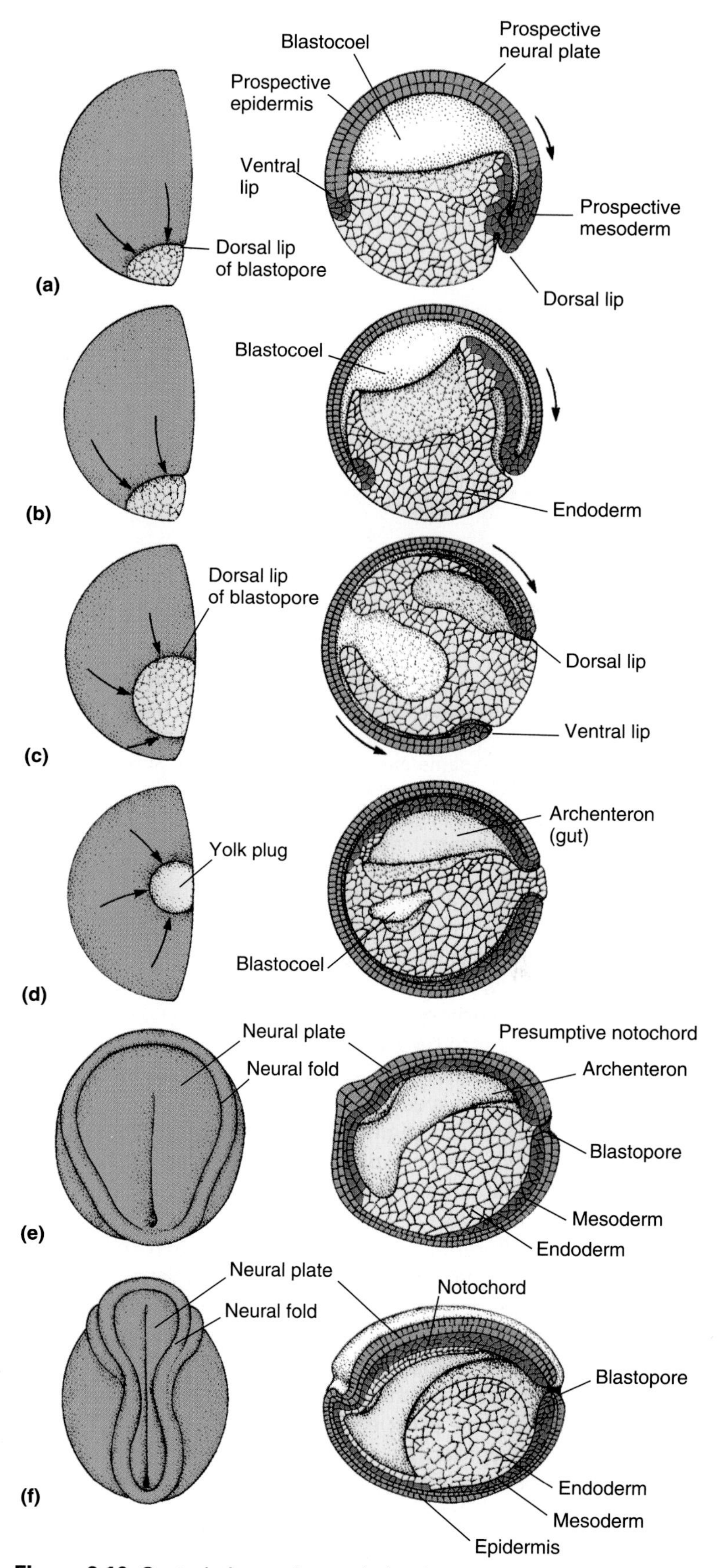

Figure 6.10 Gastrulation and neurulation in a typical amphibian embryo. **(a–d)** Gastrula stages. **(e, f)** Neurula stages. Left column shows surface views from a posterodorsal **(a–d)** or dorsal **(e, f)** angle. Right column shows right halves viewed from the left side. Gastrulation in *Xenopus,* described in Chapter 10, is somewhat different.

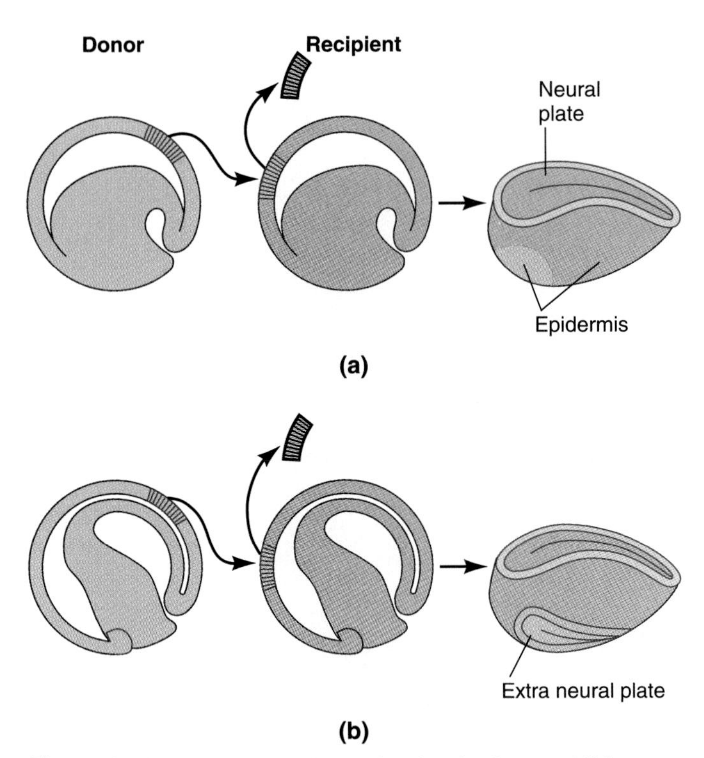

Figure 6.11 Neural plate determination in the amphibian embryo. (**a**) Prospective neural plate, if transplanted to the ventral portion at an early gastrula stage, forms epidermis (see Fig. 6.12). (**b**) If the same experiment is carried out at a late gastrula stage, the transplant forms (part of) an additional neural plate. These results show that the prospective neural plate region is determined during gastrulation.

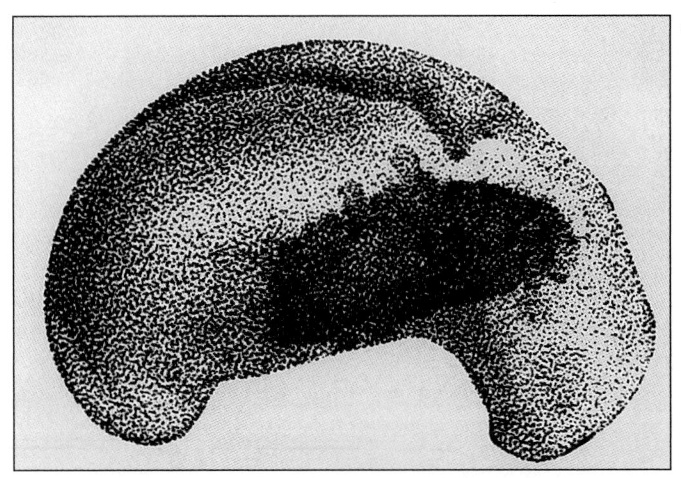

Figure 6.12 *Triton cristatus* embryo (lightly pigmented) with a graft from a *T. taeniatus* embryo (darkly pigmented). The graft, taken from the prospective neural plate of an early gastrula and grafted to prospective lateral epidermis of the host as shown in Figure 6.11a contributed to the epidermis of the host.

gastrula (Fig. 6.11a). The transplant developed in accord with its new surroundings and formed part of the host's epidermis. Similarly, prospective epidermis transplanted into the neural plate primordium contributed to the host's neural plate. In order to secure a reliable and lasting distinction between the transplant and the host, Spemann used two different newt species, the lightly pigmented *Triton cristatus* and the darkly pigmented *Triton taeniatus* (Fig. 6.12). The results showed clearly that, by the criterion of heterotopic transplantation, the prospective neural plate is not determined at the early gastrula stage.

An entirely different result was obtained when the same experiment was carried out at the *late* gastrula stage. This time, prospective neural plate developed into brain or spinal cord wherever it was placed in the embryo (Fig. 6.11b). Likewise, prospective epidermis had lost its ability to form structures of the nervous system. If transplanted into the prospective neural plate, it nevertheless formed epidermis. Spemann concluded that both the prospective epidermis and the prospective neural plate become determined during gastrulation.

The same conclusions were reached by Holtfreter, a coworker of Spemann's, from a corresponding series of isolation experiments. Holtfreter (1938) excised different parts from newt and frog gastrulae and cultured them in vitro in a saline solution. Isolated *regions* of prospective epidermis from early gastrulae as well as late gastrulae developed into epidermis and gland cells. (Isolated *cells* were later found to behave differently; see Section 12.5.) The same structures were derived from prospective neural plate isolated before gastrulation. After gastrulation, isolated prospective neural plate formed neural structures, including cells normally found in the brain or eye. These results confirmed the conclusion reached from Spemann's transplantation experiments that the prospective neural plate region is determined during gastrulation.

MOUSE EMBRYONIC CELLS ARE NOT DETERMINED UNTIL THE BLASTOCYST STAGE

Isolation and transplantation experiments suggest that mammalian blastomeres are pluripotent, if not totipotent, up to the 8-cell stage. To follow the development of isolated blastomeres, S. Kelly (1977) dissociated mouse embryos at the 4-cell or 8-cell stage by gently suctioning them into and out of a micropipette. Because single mouse blastomeres at this stage do not have enough mass to develop into complete blastocysts, she combined the isolated "donor" blastomeres with "carrier" blastomeres of a different genotype (Fig. 6.13). This method generates ***chimeras***—that is, organisms composed of cells with different genotypes. The chimeric mouse embryos were kept in vitro to form blastocysts before they were implanted into foster mothers and raised to midpregnancy or to term. Because of different genetic labels in the donor and the carrier, it was possible to decide which organs of the developing mice had been formed by the isolated donor blastomeres. In several cases three blastomeres obtained from the same 4-cell embryo each contributed to live-born mice, with donor characters found in skin, germ cells, and other organs. Together, the results show that most, if not all, mouse blastomeres up to the 8-cell stage are pluripotent.

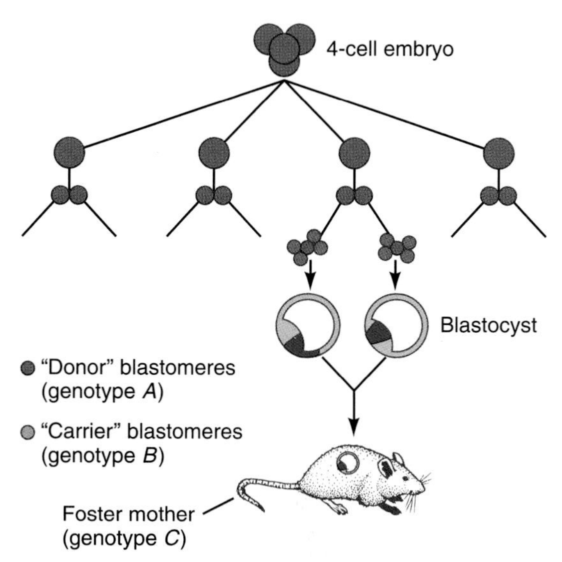

Figure 6.13 Test of isolated mouse blastomeres for totipotency. "Donor" embryos of genotype *A* were dissociated into single blastomeres at the 4-cell stage. Each blastomere cleaved again to reach the equivalent of the 8-cell stage. The pairs of 8-cell blastomeres were separated, and each was combined with four 8-cell–stage "carrier" blastomeres of genotype *B*. The composite embryos were cultured in vitro until the blastocyst stage, when they were implanted into the uterus of a foster mother. The developing offspring were of genotype *A*, genotype *B*, or composites of both genotypes.

At the 16-cell stage, some mouse blastomeres are still pluripotent while in others the process of determination is clearly progressing. At this stage, normal embryos consist two cell types. *Outside cells* are located at the surface of the embryo and are polarized with tight junctions near the apical face. *Inside cells* are located away from the surface, are nonpolarized, and connected by gap junctions. The inside cells are fated to form the *inner cell mass*, while most of the outside cells contribute to the *trophoblast* (see Section 5.2). To investigate the *potencies* of inside and outside cells at the morula stage, Ziomek and Johnson (1982) combined the strategies of heterotopic transplantation and clonal analysis (Fig. 6.14). They dissociated 16-cell embryos into single cells and classified the cells as outside or inside cells on the basis of morphological criteria such as the larger size and apical microvilli of the outside cells. After staining with a fluorescent dye, single stained cells were aggregated with 15 age-matched, unstained host blastomeres to reconstitute the equivalent of a morula. This reconstitution was done in two steps. First, six blastomeres were incubated for 1 h to form the core of the aggregate; then 10 more cells were added to form the *shell* of the aggregate. The aggregates were kept in vitro for 24 h, by which time they had formed blastocysts.

The results of this experiment are summarized in Table 6.3. When a stained outside cell was placed back in its "native" shell position, it contributed to the trophoblast in 13 of 14 cases. Similar results were obtained in 16 of 18 cases when an outside cell was placed in an

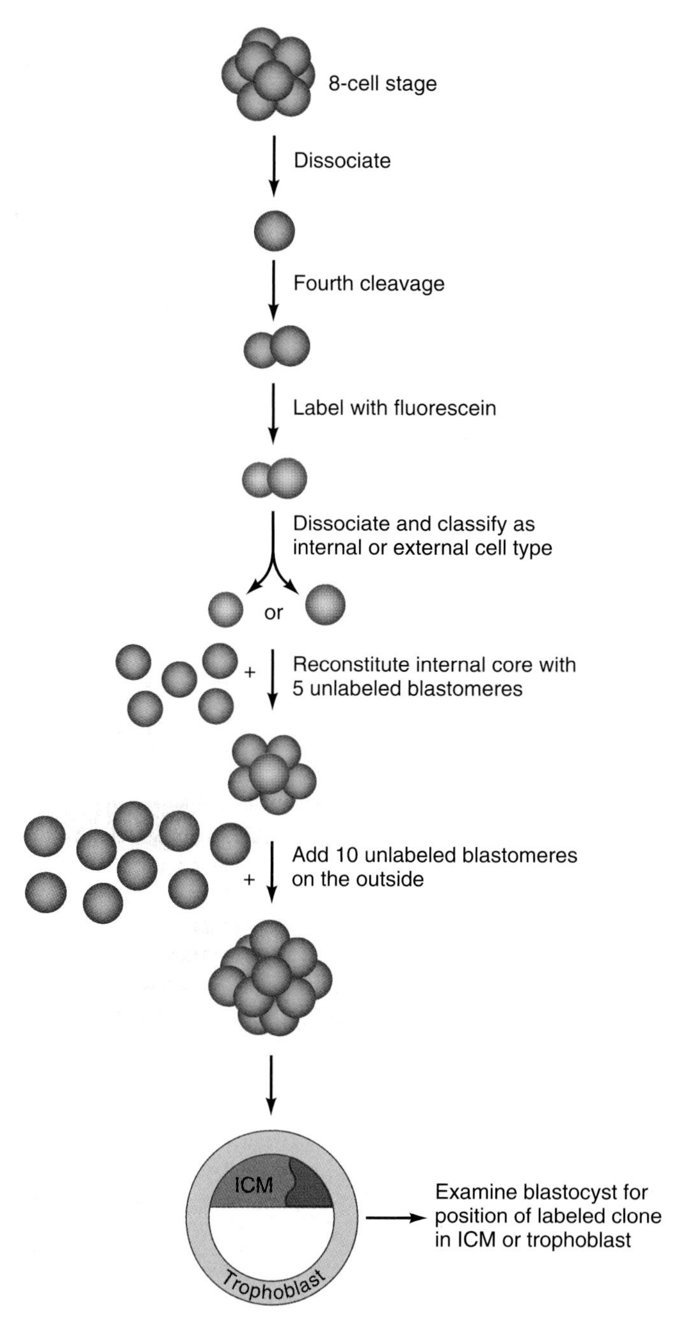

Figure 6.14 Experimental procedure used to test whether isolated 16-cell mouse blastomeres are determined to form inner cell mass or trophoblast. At the 16-cell stage, blastomeres are identified as outside cells (polarized) or inside cells (nonpolarized). Labeled and unlabeled cells are reaggregated, with a single labeled cell placed either into the internal cluster (core) or the external layer (shell) of the assembly. The fate of the descendants of the labeled cell is recorded at the blastocyst stage (see Table 6.3).

"alien" position in the core of the aggregate. Thus, most of the outside cells developed according to fate regardless of their newly acquired position in the host. In the exceptional cases, outside cells gave rise to contiguous clones that extended from the trophoblast to the inner cell mass. When a stained inside cell was placed back in its native core position, it contributed to the inner cell

table 6.3 Fate of Isolated Labeled Mouse $\frac{1}{16}$ Blastomeres in Reconstructed 16-Cell Aggregates

Type of Aggregate and Position of Labeled Cell (see Fig. 6.14)	Number of Blastocysts Analyzed	Number (%) of Blastocysts with Fluorescent Progeny in:		
		Tr* Only	ICM† Only	ICM + Tr
Labeled external $\frac{1}{16}$ cell				
Native, shell position	14	13 (93)	0	1 (7)
Alien, core position	18	16 (89)	0	2 (11)
Labeled internal $\frac{1}{16}$ cell				
Native, inside position	19	3 (17)	16 (84)	0
Alien, outside position	13	7 (54)	4 (31)	2 (15)

*Tr = trophoblast.

†ICM = inner cell mass.

(From Ziomek and Johnson [1982]. *Used by permission.)*

mass in 16 of 19 cases; stained parts appeared in the trophoblast in the other 3 cases. However, when an inside cell was placed in an alien shell position, the results were quite varied. Only in 4 of 13 cases did the donor cells develop according to fate, contributing to inner cell mass only. In the majority of cases, the transplanted inside cells contributed to trophoblast only or to both trophoblast and inner cell mass. These cells then developed in accord with their new positions in their hosts and not according to their fates in the donor embryos.

The experiments reviewed so far indicate that neither the outside nor the inside blastomeres of the mouse morula are strictly determined. Each type reveals a wider potency after heterotopic transplantation. Likewise, the observation of overlapping clones between trophoblast and inner cell mass indicates a genuine pluripotency of at least some blastomeres at the 16-cell stage. This pluripotency does not seem to be an artifact caused by heterotopic transplantation, because some of the stained cells gave rise to overlapping clones even if they were returned to their native positions. By comparison, the outside morula cells show a stronger bias toward developing in accord with their original fate than the inside morula cells. However, even this stronger bias can be overridden if the number of inside cells is too small to form a sufficient inner cell mass in the blastocyst (see Section 5.2).

The conclusions reached from the foregoing transplantation experiments and clonal analysis were confirmed and extended to the blastocyst stage by isolation experiments (Handyside, 1978; Rossant and Lis, 1979). These experiments focused on the ability of entire *inner cell masses* (*ICMs*) to form trophoblast structures or even complete embryos. Together, the results of transplantation and isolation experiments with mouse embryos indicate that blastomeres are totipotent through the 8-cell stage and perhaps longer. Inside and outside cells from morula stages are biased but not fully determined to form ICM and trophoblast, respectively. Even at the early blastocyst stage, entire ICMs are pluripotent, if not totipotent. Thus, mammalian cells show a very prolonged process of gradual determination.

6.5 Properties of the Determined State

In this section, we will explore some of the mechanisms by which cells are determined, and the connections between cell determination and other embryonic processes including pattern formation and cell division. We will also discuss a set of experiments that illuminates the stability of the determined state by pushing it to its limits.

CELL DETERMINATION OCCURS AS PART OF EMBRYONIC PATTERN FORMATION

In the larva and the adult, differentiated cells such as neurons or muscle cells are arranged in *spatial patterns* that enable the proper function of each organ and the entire organism. Patterns of differentiated cells—for instance, those on the wings of butterflies or in the plumage of birds—are often elaborate and appealing to the human eye. We therefore tend to associate body patterns with differentiated cells. However, it is important to realize that organized body patterns unfold from the earliest stages of development and *precede* the appearance of differentiated cell types. A coarse distribution of cytoplasmic determinants in the egg of many species sets up the initial pattern of animal and vegetal blastomeres, and, in some instances, the different between dorsal and ventral cells. Subsequent inductive interactions generate patterns of tissue and organ *rudiments*, which in turn set the stage for further inductive interactions.

Cells are usually determined long before they attain their mature differentiated states. In other words, cell determination is part of a long diversification process that precedes cell differentiation. It must be kept in mind, though, that cell determination is assessed by operational criteria, and that morphologically and biochemically diversified cells are not necessarily determined. For example, mammalian embryos at the morula stage consist of outside and inside cells. Of these two cell types, the outside cells can be distinguished by morphological criteria and their ability to pump fluid from outside to inside. Yet the transplantation experiments described in this chapter show that outside cells can still acquire the fate of inside cells and vice versa.

DETERMINATION IS A STEPWISE PROCESS OF INSTRUCTION AND COMMITMENT

The operational criteria described in this chapter may create the impression that determination is essentially a negative event. It seems to amount to a loss of potency, or a commitment to follow one path of development to the exclusion of others. However, it is important to realize that the process of fate acquisition often entails an element of instruction. For instance, the prospective neural plate region in the amphibian blastula and early gastrula will form epidermis when isolated as an epithelial sheet. Thus, forming epidermis would be the pathway taken in the absence of further instruction. In order to develop in accord with its fate, the prospective neural plate needs additional signals, which it receives in its normal location in the embryo during gastrulation.

There are several mechanisms by which embryonic cells are nudged toward their fates. In most animals, certain components in the egg cytoplasm are distributed unevenly and partitioned into distinct blastomeres during cleavage. Such components are called *localized cytoplasmic determinants*; they cause blastomeres to be determined, or at least biased, to form certain cell lineages or major body regions. For example, the posterior pole plasm of many insect eggs contains localized germ cell determinants. Cells that contain this type of cytoplasm are determined to form primordial germ cells (see Section 8.3).

Subsequent steps of cell determination may occur by signaling among equivalent cells or between nonequivalent cells. Among equivalent cells, exchanged signals may either stabilize all cells in the same determined state (community effect) or take the form of competition to attain a preferred state of determination (lateral inhibition). A common type of interaction between nonequivalent cells, known as *embryonic induction,* creates new types of cells in the boundary region between the interacting partners (see Chapters 8, 9, and 12). During morphogenetic movements in gastrulation and organogenesis, cells acquire several new neighbors and impart inductive signals to one another. As an example, prospective neural plate in amphibian gastrulae are induced in part by mesodermal tissue that moves into the blastocoel and contacts the prospective neural plate from underneath (see Figs. 6.10 and 6.11). Thus, the final body pattern results from a ***hierarchy of determinative events*** in which each step is based on the previous steps.

As an example of a hierarchy of determinative events, let us again consider an amphibian embryo. During cleavage, animal cells become different from vegetal cells due to localized cytoplasmic determinants. Next, the animal blastomeres closest to their vegetal neighbors are induced to form an intervening *marginal zone.* The three blastula regions resulting from this induction already differ in their potencies as indicated in part by isolation experiments (Fig. 6.15). Isolated regions from the animal cap develop into spheres of epidermis. Isolates

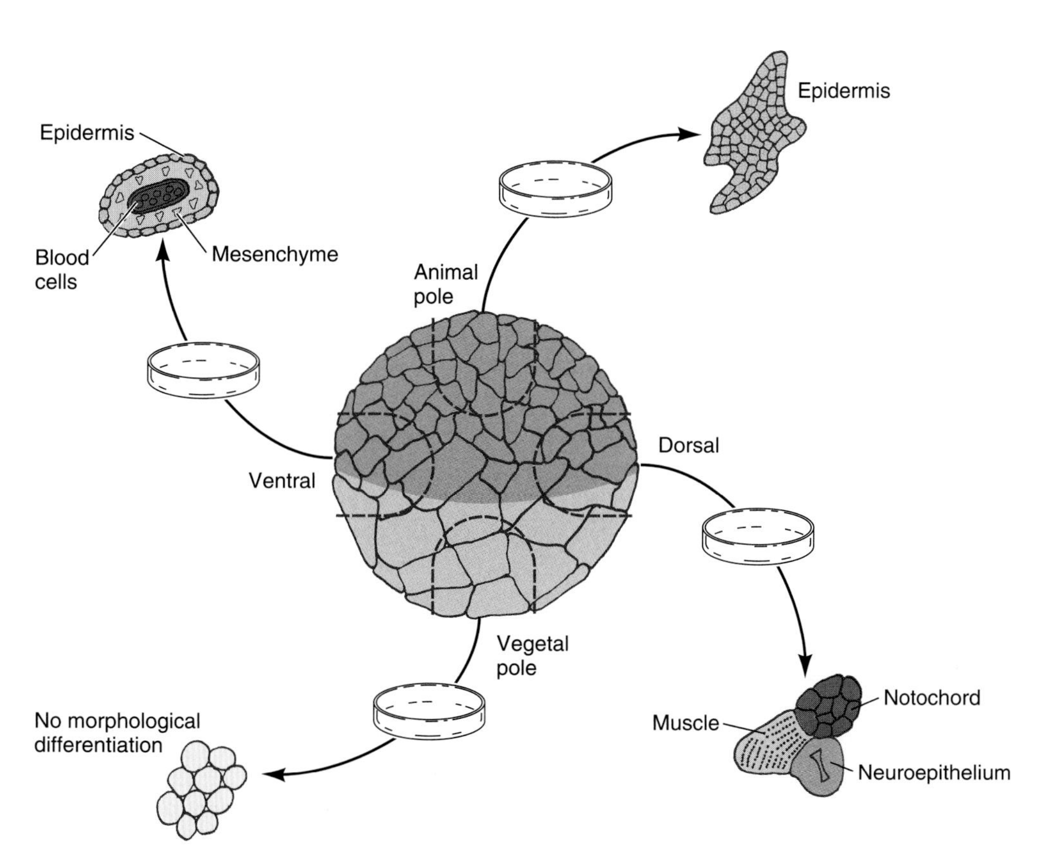

Figure 6.15 Development of isolated amphibian blastula regions in vitro.

from the marginal zone form *notochord* (an embryonic skeletal element), muscle, and mesenchyme. Vegetal isolates develop into an undifferentiated, gutlike tissue. Upon gastrulation the animal cells will form the outer germ layer called *ectoderm*. Within the ectoderm, the next determinative step is the induction of the neural plate by adjacent mesoderm, leaving the remaining animal cells in a still general state of determination for epidermis. Within the epidermal area, further determinative steps will lead to distinctions between epidermis proper and epidermal derivatives such as glands and sensory epithelia for nose and ear. A similar series of determinative steps are followed in the neural plate and in the other two germ layers.

Figure 6.16 is a simplified diagram showing some of the major determinative steps during amphibian development. In this diagram, each branching point is thought to be a determinative step, when one set of cells receives an instructive signal while other cells do not. An extended series of determinative events eventually commits cells to form a particular embryonic region and/or a particular type of differentiated cell.

In organisms that are amenable to genetic analysis, cell determination is now being analyzed in terms of gene regulation, with particular attention to "selector" genes, which in turn control the expression of other genes. In the *Drosophila* embryo, the ectoderm is patterned into segments that each have spatial polarity axes and individual identities (see Chapter 22). These patterning genes control a wide range of target genes including *achete*$^+$ and *scute*$^+$, which define clusters of potential neuroblasts. Within each cluster, the cell with the greatest *Delta*$^+$ activity actually acquires the neuroblast fate (see Fig. 6.9). In a series of asymmetrical divisions, which are controlled by further gene activities, the neuroblast gives rise to neurons (see Section 8.6). The gene activities that mediate the stepwise process of cell determination are stabilized by feedback loops of mutual control, which become more difficult to break as a cell proceeds toward its determined state. Thus, genetic and molecular analysis now offer a molecular understanding of the classical observations that the potential of a cell is restricted in a stepwise process and that with each step along the way a reversal of the process becomes less likely.

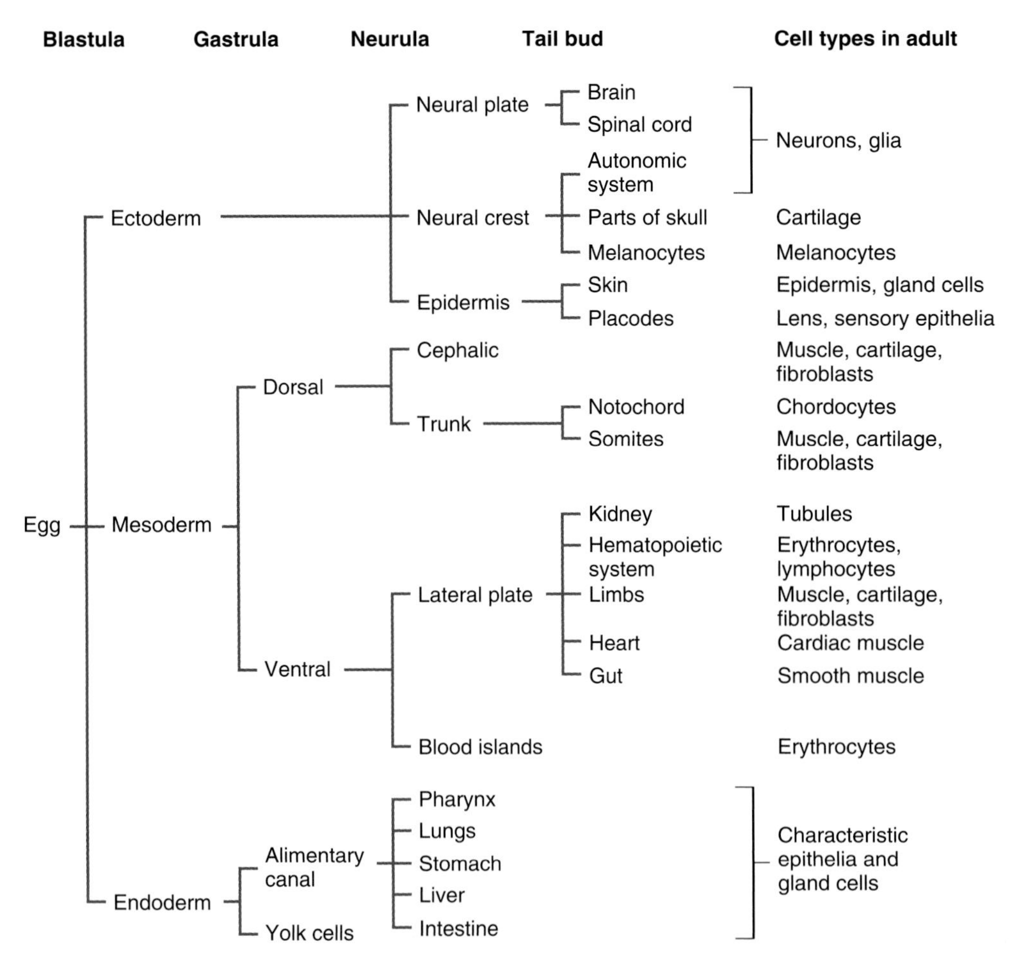

Figure 6.16 Series of determinative steps in amphibian development. Organismic development is represented as a hierarchy of determinative events, beginning with localized cytoplasmic determinants in the egg and continuing with inductive interactions. At the tail bud stage, the embryo shows the typical vertebrate body plan and has formed the major organ rudiments. Several further developmental decisions will in most cases be taken before the cells differentiate into the terminal cell types listed on the right-hand side. Note that some cell types, such as cartilage and gland cells, arise from more than one lineage.

THE DETERMINED STATE IS (ALMOST) STABLY PASSED ON DURING MITOSIS

The determined state is passed on from each cell to its descendants, at least under conditions of normal development. The mitotic progeny of cells determined to form cartilage, for example, are likewise determined to form cartilage. This stability is remarkable, and its underlying molecular mechanisms are still being investigated. In order to test the endurance of determination mechanisms, determined cells have been subjected to conditions of excessive proliferation. The most extensive test of this kind was carried out on those cells in *Drosophila* larvae that later form the epidermis of the adult.

The life cycle of *Drosophila* as well as other highly evolved insects includes a dramatic *metamorphosis*. The larva first enters an immobile and transient stage, during which the organism is called a *pupa,* and then reaches the adult stage, at which it is known as an *imago* (see Section 27.4). During this type of metamorphosis, most larval organs are resorbed while adult structures are built from groups of progenitor cells in the larva. The entire epidermis of the adult head and thorax, including legs, wings, and other appendages, is formed from pockets of larval epithelia called ***imaginal discs*** (Fig. 6.17). The imaginal discs are attached to the inside of the larval epidermis but have no function in larval life. However, imaginal discs grow dramatically as their cells undergo many rounds of mitosis. Each disc is constituted from an initial group of 20 to 50 cells but increases to tens of thousands of cells at the end of the larval period. Under the influence of the hormones that control metamorphosis, the folded discs evert and form a cuticle with hairs, bristles, and other markings. For instance, a pair of wing discs forms the wings and the dorsal parts of the second thoracic segment to which the wings are attached, and three pairs of leg discs form the legs and the ventral parts of the three thoracic segments to which the legs are attached (Fig. 6.18). Most imaginal discs are paired. The only unpaired disc is the genital imaginal disc, which gives rise to the external genitalia and anal structures of the adult.

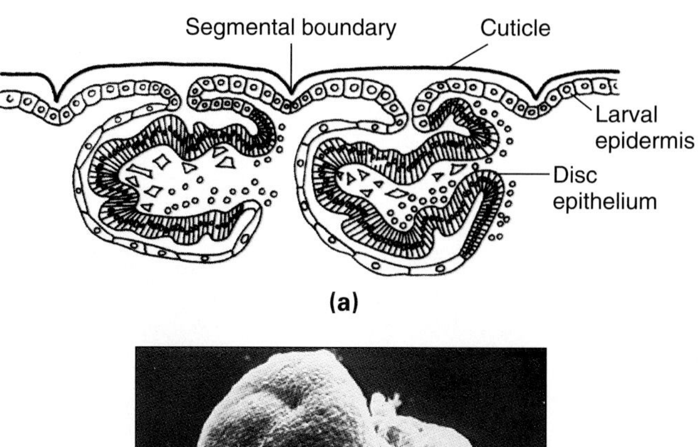

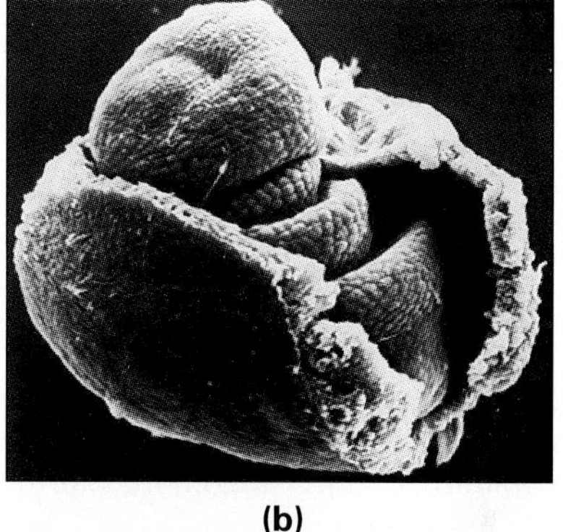

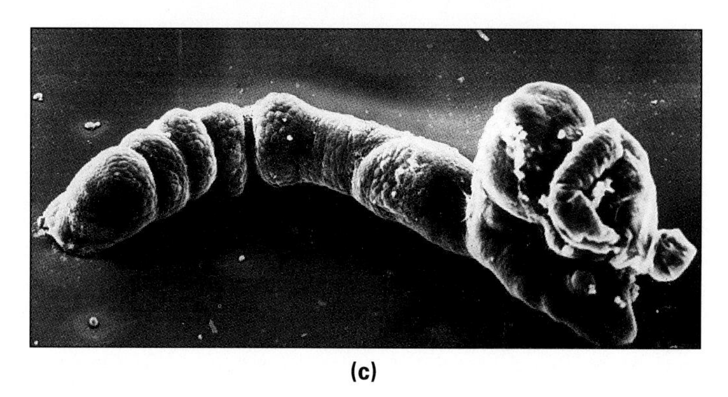

Figure 6.17 Eversion of imaginal discs during insect metamorphosis. **(a)** Paramedian section of the wing imaginal discs of an ant. Note the continuity of the disc epithelium with the larval epidermis. **(b)** Scanning electron micrograph showing a leg disc before eversion. This is the same stage shown in part a. To make the disc epithelium better visible, part of the overlying cellular membrane has been removed. **(c)** Leg disc after eversion.

The determined state of imaginal discs has been tested by transplanting them from donor larvae to the body cavity of host larvae. When the host larva goes through metamorphosis, its hormones act on the implanted disc, and an extra adult structure forms in the metamorphosing fly. When this extra structure is removed from its host and inspected under the microscope, its morphology always corresponds to the fate of the transplant in the donor, independently of its location in the host. Thus, according to the criterion of heterotopic transplantation, each imaginal disc is determined to form the imaginal structure for which it is fated. In fact, upon transplanting disc fragments, it turns out that each part of an imaginal disc is already determined to form a particular portion of the appropriate imaginal structure. Thus, certain regions of a male genital disc form the equivalent of a penis, others the sperm pump, and still others the hindgut.

IN order to test how extra rounds of mitosis would affect the stability of the determined state, Hadorn (1963) developed the serial transplantation technique shown in Figure 6.19. Here, an imaginal disc is first implanted into the body cavity of an *adult* host. Under these conditions, the disc bypasses the combination of hormones present in fully grown larvae and pupae that trigger metamorphosis; the adult fly's body fluid merely serves as a convenient culture medium in which the disc tissue can proliferate. Since the culture period is limited by the life span of the host, the

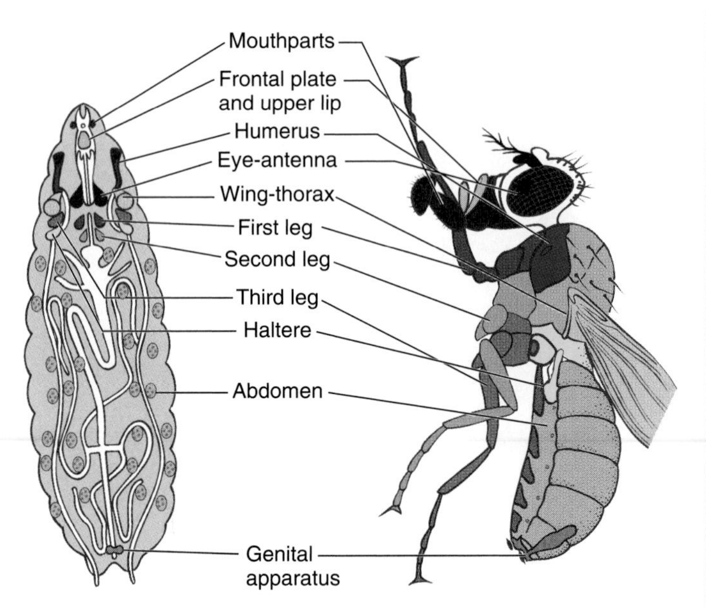

Figure 6.18 Imaginal discs of *Drosophila*. The diagram shows the position of the disc in the larva and their adult derivatives. The epidermis of the abdomen is formed from nests of isolated cells (purple).

implant must be transferred to a new adult host every few weeks. During the transfers, the implants can be cut and the fragments reimplanted into two or more new adult hosts. At the same time, the determined state of a disc may be checked by transferring part of it back into a *larva*, which the test implant is challenged to metamorphose along with its larval host.

Using the serial transplantation technique, Hadorn (1968) studied the determined state of imaginal discs that had been allowed to proliferate for several years. One extended series began with the implantation of a genital disc. During the first few transfer generations, the test implants formed only genital and anal structures, indicating that they had maintained their original determination (Fig. 6.20). Indeed, genital and anal structures were obtained for a total of 55 transfer generations. During this time, the imaginal disc tissue proliferated and passed on its determination for genital and anal structures to at least some of its descendants. However, from the eighth transfer generation on, head and leg structures appeared in addition to the genital and anal structures. The imaginal disc tissue in which these switches had occurred continued to form head and leg structures during subsequent test implantations. It behaved as if it were now determined to form head or leg. Each switch to a new determined state was termed a ***transdetermination.*** Further transdeterminations, those to wing and other thoracic structures, occurred during transfers 13 and 15. The thorax represented a ground state to which many implants had switched by the 100th transfer.

How can we be sure that transdetermination really represents a switch from one determined state to another, such as from antenna to wing? Could this be a false appearance generated by the *first* determination of previously unrecognized and *undetermined* reserve cells? To address this question, Gehring (1968) induced *somatic crossover* in antennal discs before they were subjected to serial transplantation. In some of the adult tissue showing transdetermination to wing, he observed *single clones* extending from antennal to wing structures. This showed that the founder cell of the labeled clone was indeed determined for antenna, because it passed this determination on to daughter cells before transdetermination to wing took place.

Questions

1. Would the outcome of Hadorn's experiment have been different if he had begun this serial transplantation experiment with a wing imaginal disc instead of a genital imaginal disc?
2. Do you see any similarities between the transdeterminations observed by Hadorn and the *homeotic mutations* discussed in Section 1.4?
3. How would you test whether transdetermination can occur in groups of cells rather than in single cells?

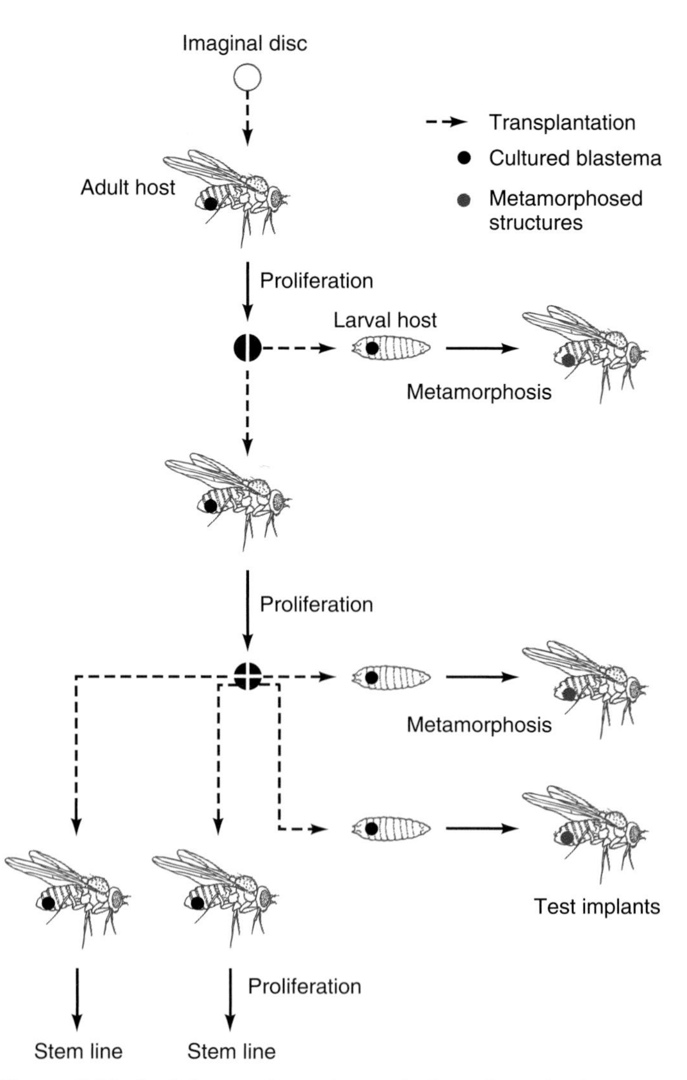

Figure 6.19 Serial transplantation technique for culturing imaginal discs in vivo.

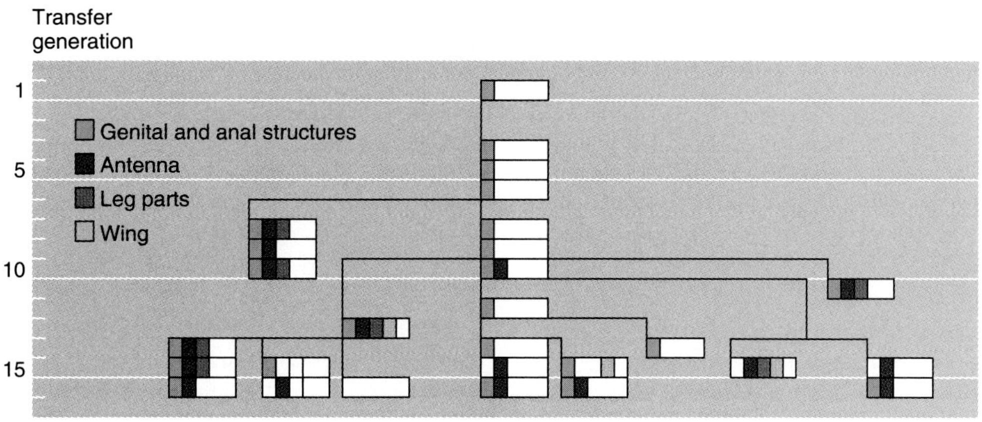

Figure 6.20 Transdetermination in imaginal discs of *Drosophila*. This chart summarizes the results of serial transplantations using the technique illustrated in Figure 6.19. The original transplant was taken from the genital disc, whose fate is to form sex organs and anal structures. The transplanted cells and their descendants formed only genital and anal structures during seven transfer generations (indicated on scale to the left). However, some cells transdetermined to antennal parts and leg structures by the eighth transfer, and soon thereafter to wing. The transdetermined states were passed on stably until another transdetermination occurred. The color codes in each rectangle indicate the imaginal structures found in a given stem line after the number of transfer generations indicated. Branch points show when a stem line was split.

Hadorn's experiments showed that the original determination of imaginal disc cells is stable and passed on faithfully as the cells divide. Only after excessive proliferation in culture did the determined state change in distinct steps, called *transdeterminations.*

Some of the transdeterminations observed by Hadorn and coworkers resemble the dramatic phenotypes caused by *homeotic mutations*. For example, the transdetermination from haltere to wing resembles the *Ultrabithorax* phenotype shown in Figure 1.21. Likewise, the transdetermination from antenna to leg mimicks the *Antennapedia* phenotype shown in Figure 1.22. However, transdeterminations differ from genetic mutations (heritable changes caused by alterations in genomic DNA) in several respects: First, no changes in the nucleotide sequence of homeotic genes have been found in transdetermined imaginal discs. Second, transdeterminations occur much more frequently than genetic mutations. Third, most transdeterminations are much more readily reversible than genetic mutations (Fig. 6.21). Fourth, transdeterminations have been found to occur simultaneously in two or more adjacent cells. While mutations are changes in the *genetic information itself,* transdeterminations result from changes in the *expression patterns* of certain homeotic genes. Nevertheless, the changed expression patterns are also passed on during mitotic divisions.

The molecular mechanisms underlying determination and transdetermination are still being investigated. We will discuss in Section 16.6 how interactions between genes and certain regulatory proteins generate *bistable control circuits* that can flip-flop between two states depending on the concentration of the regulatory proteins. It may be that the transdeterminations occur when such regulatory proteins are diluted as a result of many extra mitoses.

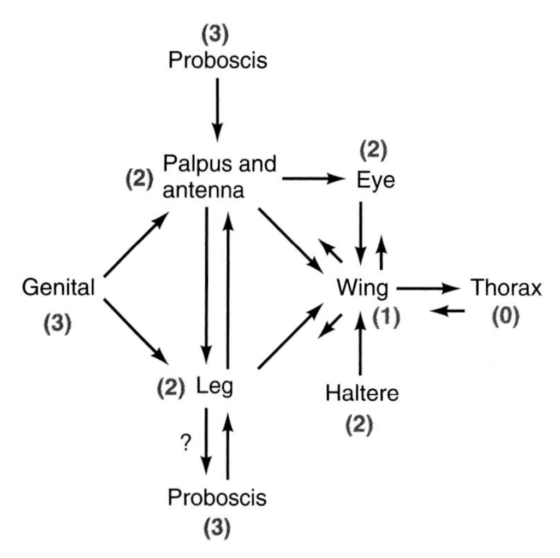

Figure 6.21 Flow diagram of transdetermination steps in *Drosophila*. The lengths of the arrows represent the relative frequencies of transdetermination. The numbers represent the minimum number of transdetermination steps needed before the disc can reach the determined state of thorax, which seems to be the ground state.

6.6 Regulation in Development

As mentioned in the previous chapter, embryos have different ways of orchestrating the development of each blastomere so that a complete and harmonious organism results. While embryos with *invariant cleavage* rely on utmost precision, embryos with *variable cleavage* use a process of stepwise approximation to correct any imbalances in cell numbers or arrangements that may have occurred. These two strategies can now be redefined in terms of cell determination as mosaic and regulative development. An embryo or embryonic region is said to undergo ***mosaic development*** when its potency map is identical to its fate map, or, in other words, when all its cells are determined. Under the criterion of isolation, an embryo or embryonic region shows mosaic development when it can be cut into pieces and each piece develops in the same way as it would in normal development. In contrast, an embryo or embryonic region is said to undergo ***regulative development*** when the potencies of its cells are greater than their fates, or, in other words, when its cells are not yet determined. The ability of a cell or of an embryonic region to deviate from its fate and to respond to varying circumstances is called ***regulation.***

These terms have been used to characterize, for example, nematode and ascidian embryos as "mosaic embryos" while amphibian and sea urchin embryos were classified as "regulative embryos." However, this dichotomy is misleading. As explained in this chapter, determination proceeds in steps, and most organisms will eventually undergo mosaic development. The question is: *How early* does the transition from regulative to mosaic development occur? The answer depends not only on the type of animal but also on the embryonic region under consideration. In a given embryo, development in some regions may be mosaic while in others it is still regulative. Most regions of an amphibian early gastrula, for example, are still capable of regulation, while the *dorsal blastopore lip,* located just above the blastopore, is already determined (see Figs. 6.11 and 12.17).

A common type of regulative development is ***twinning***—that is, the development of two complete individuals from an embryo that has spontaneously divided into two parts or has been split by an experimenter. Embryos of many species retain this capability during cleavage and later stages (Chan et al., 2000). Human twins arising through the division of one embryo are known as identical or ***monozygotic twins.*** (Fraternal, or ***dizygotic, twins*** result from two separately fertilized eggs.) Monozygotic twins can originate through division of the embryo at several stages of development (Fig. 6.22). The earliest possible separation is at the 2-cell

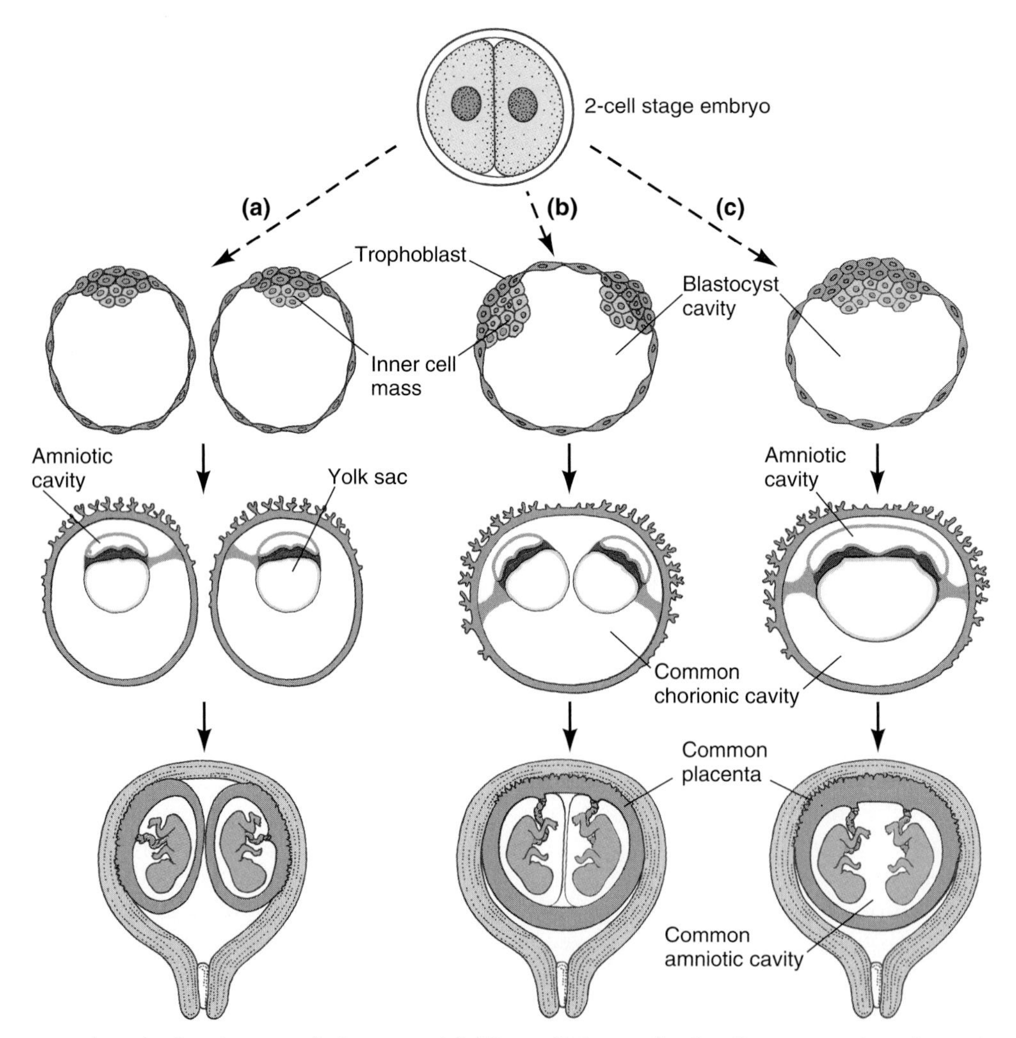

Figure 6.22 Monozygotic twin development in humans. **(a)** After splitting at the 2-cell stage, each embryo develops its own placenta, chorionic cavity, and amniotic cavity. **(b)** After complete splitting of the inner cell mass, the two embryos have a common placenta and chorionic cavity, but separate amniotic cavities. **(c)** After incomplete or late splitting, the embryos share a common placenta, chorionic cavity, and amniotic cavity.

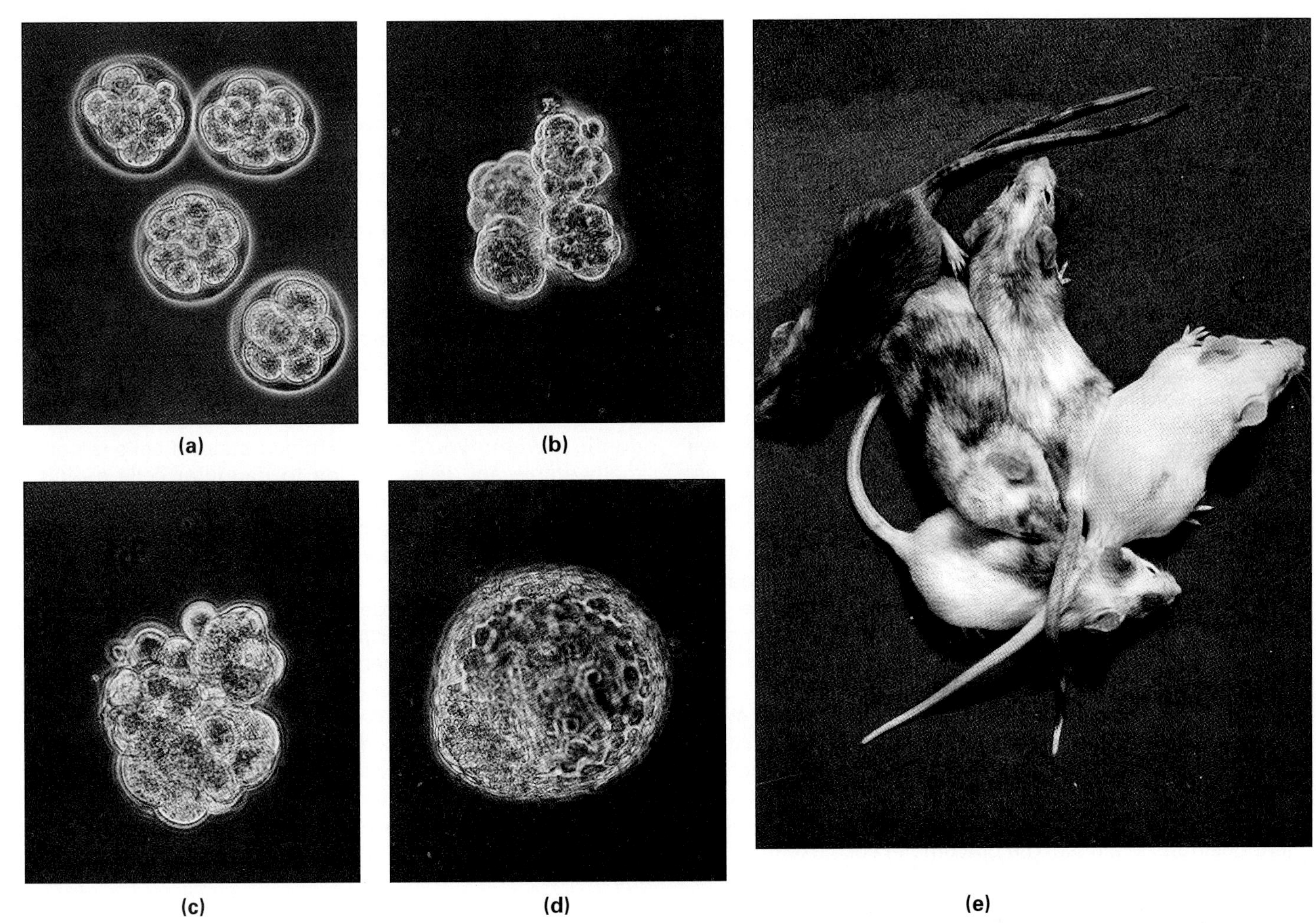

Figure 6.23 Regulation in mammalian embryogenesis. **(a–d)** Photographs showing the aggregation of four mouse embryos to form a single, large blastocyst. The embryos are removed from the zona pellucida at the 8-cell stage and placed into contact. They form an aggregated blastocyst, which is implanted into the uterus of a foster mother and brought to term. **(e)** Mouse chimeras made by combining embryos from strains with different pigmentation.

stage, resulting in two complete blastocysts that develop and implant separately. In this case, each twin will develop its own placenta and amniotic cavity. In most cases of human twinning, however, the inner cell mass of the embryo seems to divide during the early blastocyst stage. The resulting embryos have separate amniotic cavities but a common placenta. In rare cases the separation occurs at later embryonic stages, resulting in two partners with a common amniotic cavity and placenta. Incomplete separation at a late embryonic stage leads to the formation of conjoined twins. Depending on the nature of the bridge between them, they can sometimes be separated surgically.

The converse of twinning is the experimental ***fusion*** of two or more embryos resulting in the development of one embryo. Such fusion has been carried out extensively with mouse embryos of different genetic constitution (Mintz, 1965). The embryos stick to each other readily and form an unusually large blastocyst, which then implants and develops into a *chimeric* individual of normal size (Fig. 6.23).

Twinning and fusion dramatically demonstrate regulative development, but they are not the only examples of this phenomenon. In embryos of many animal species, when cells are removed, their neighbors often produce additional descendants that take over the role of the lost cell. Conversely, the addition of extra cells is compensated by fewer mitoses of the neighbors, which thus give up some of their normal fates to the extra cells. Such adjustments allow regulative embryos to compensate losses from injuries and to correct imbalances that may have resulted from earlier steps in development.

SUMMARY

It is important to distinguish the fate of embryonic cells from their potency and determination. A cell's fate is what becomes of it in the course of normal development. A fate map is a diagram of an embryo indicating which structures each region will form at a later stage. The most common method for establishing fate maps is the injection of dyes that do not impair development.

A technique relevant to the analysis of both fate and determination is the marking of clones. A cell clone consists of all surviving descendants of a single cell, the founder cell of the clone. A clone can be labeled by injecting a single cell with a dye that is passed on to daughter cells but is not transferred to neighboring cells. In organisms with suitable mutants, clones can also be labeled by X-ray–induced somatic crossover and other genetic techniques. However, such clones cannot be used for fate mapping because they occur randomly.

In the adult epidermis of *Drosophila*, labeled clones do not cross certain boundary lines. Such boundaries define compartments. Each compartment is filled by several clones, which collectively form a polyclone. Polyclones have the same size and place in each individual, whereas individual clones do not.

Most embryonic cells are able to form a range of structures that is wider than their fate. The total of all structures that a cell can form, given the appropriate environment, is the potency of the cell. The potency of cells is tested by isolation in culture media and by transplantation to different positions in a host. The outcome of such experiments, however, depends very much on experimental parameters, including the way cells are isolated and the number of cells that are isolated or transplanted together. Embryonic cells undergo a stepwise process, called determination, in which their potency becomes limited to their fate. To assess the determined state of a cell or of an embryonic region, its fate must be compared with its potency as revealed by isolation or heterotopic transplantation. The nondetermination of cells can be tested by clonal analysis, because each clone reveals a sample of those structures that its founder cell was still able to form.

The process of cell determination entails a loss of potency as well as the receipt of specific signals from localized cytoplasmic components or from other cells. The determination of each cell therefore depends on its position in the embryo. The determined state of cells is normally stable and is passed on to daughter cells during mitotic divisions. Under conditions of excessive proliferation, however, the determined state of cells may change—an event that presumably reflects alterations in control circuits of gene activity.

Cells differ in the stage at which they become determined. Within a given organism, some cells become determined earlier than others. In certain groups of animals, such as roundworms and ascidians, determination tends to occur early. These animals are said to show mosaic development as opposed to regulative development, which is characterized by late determination as observed in sea urchins and vertebrates. One characteristic of regulative embryos is the capacity for twinning—that is, the formation of small but complete individuals from isolated blastomeres.

Genomic Equivalence and the Cytoplasmic Environment

Figure 7.1 Carrot plant grown in an Erlenmeyer flask from a single differentiated root cell. Plant cells from which the rigid cell walls have been removed can be grown in culture. If treated with the appropriate nutrients, growth hormones, and light conditions, individual cells may give rise to complete plants. This ability demonstrates that differentiated plant cells retain a full complement of genetic information.

EMBRYONIC cells acquire distinct features as they undergo differentiation to become muscle fibers, neurons, and other kinds of cells. Even before cells become overtly different, they are usually determined to follow certain pathways of differentiation at the exclusion of others. Cell determination and cell differentiation occur through the synthesis of different proteins, which can ultimately be traced to differential gene activity.

Do all cells retain a full complement of genetic information and use it selectively, or is cell differentiation based on the selective loss and/or multiplication of certain genes? To answer this question, researchers have analyzed the ability of cells to change differentiated states in normal development and under conditions of regeneration. Others have compared chromosomes as well as specific segments of nuclear DNA from different tissues. As a particularly sensitive assay for completeness of genetic information, researchers have tested differentiated cells or their nuclei for *totipotency,* as indicated by their ability to promote the development of complete new organisms (Fig. 7.1). While differentiated plant cells turned out to be clearly totipotent, similar experiments with animal cells have not been successful. However, nuclear transfer experiments show that nuclei from differentiated animal cells can support the development of complete larvae or even adults. In particular, it has been possible to clone sheep by fusing enucleated eggs with embryonic cells or adult epithelial cells. However, only few of these experiments have been successful. Most of them have failed, apparently because nuclei from differentiated cells are ill-prepared to reenter the rapid mitotic cycle that characterizes embryonic cleavage.

On balance, it appears to be the rule that all cells of an organism have complete and equivalent sets of genetic information but use this information selectively. This raises the question, What determines which genes a given cell will express? Toward the end of this chapter, we will see that gene expression in a cell nucleus is controlled by its cytoplasmic environment.

7.1 Theories of Cell Differentiation

As cells undergo determination and differentiation, diverse sets of genes are activated to synthesize specific combinations of proteins. Since the late nineteenth century three hypotheses have been put forward to account for these differences. This section describes the three hypotheses, and subsequent sections will present some of the evidence on which the current thinking is based.

According to one hypothesis, first proposed by August Weismann (1834–1914) and Wilhelm Roux (1850–1924), cells become different by the *differential loss of genetic material.* Weismann thought that each cell would inherit only the particular fraction of the entire genetic material that it required to develop in accord with its fate (Fig. 7.2). He proposed that the genetic information present in the arm would differ from that in the leg, and the information in the upper arm would dif-

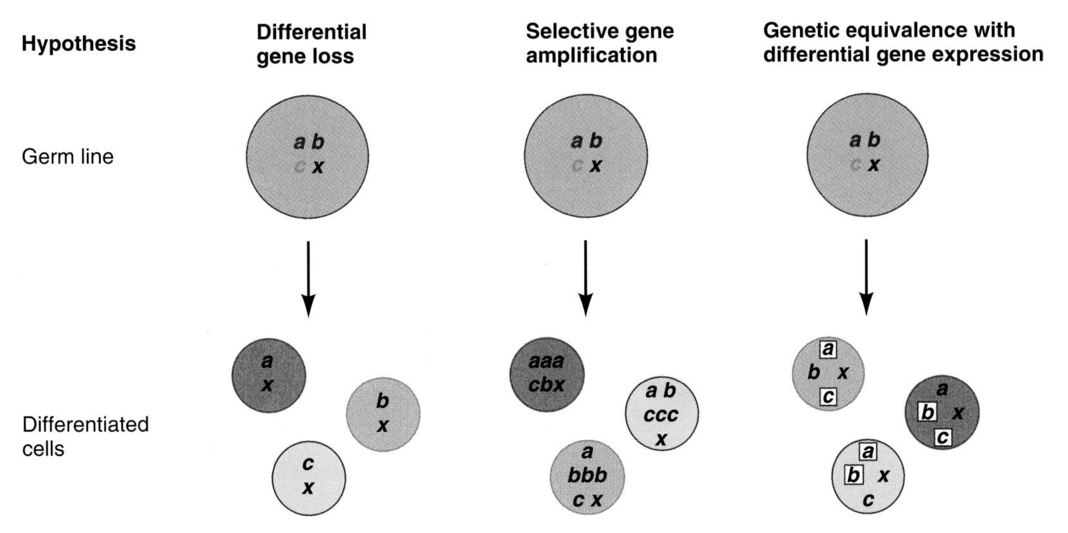

Figure 7.2 Three hypotheses of cell differentiation. Colored letters and color shading indicate gene activity. White boxes indicate gene is inactive. *x* = housekeeping genes; *a, b, c,* = tissue-specific genes.

fer from that in the lower arm, and so forth. This view seemed to be supported by observations of chromosomal losses during cleavage in certain animal species. For instance, in the nematode *Ascaris megalocephala*, early cleavage divisions are associated with the loss of chromosome fragments, a phenomenon termed ***chromatin diminution.*** Chromatin diminution occurs in all somatic cells of *Ascaris*; only the germ line cells retain the full complement of chromatin (Boveri, 1887; Müller et al., 1996). However, the lost chromatin appears to be similar if not identical in all kinds of somatic cells, so that only differences between germ line and soma, and not differences among somatic cells, might be ascribed to chromatin diminution.

Other studies have found that, rather than eliminating parts of their chromatin from somatic cell nuclei, a few types of cells undergo ***selective gene amplification.*** They replicate specifically those segments of nuclear DNA that they will utilize most intensely (Fig. 7.2). Frog oocytes, as we will see later in this chapter, have thousands of extra nucleoli, which contain the genes for large ribosomal RNAs. The selective amplification of these genes makes it possible to supply the oocyte's huge mass of cytoplasm with enough ribosomes.

A third hypothesis about cell differentiation is based on the theory of ***genomic equivalence*** and the *principle of differential gene expression*. Genomic equivalence means that all cells in an organism have complete and equivalent sets of genetic information, without losses, alterations, or selective amplifications. Differential gene expression means that cells utilize different subsets of their entire genetic information (Fig. 7.2). According to these concepts, cells use their genetic information the way different people read the newspaper. Everybody buys the same complete edition of the paper, and almost everybody will scan the front page news and the weather forecast, but beyond these sections of general interest, preferences differ. Some readers focus on national politics and sports, while others prefer local news or the comics. Similarly, it appears that most cells in most organisms have the same genetic information but use it differently. As we will see, the evidence of genetic equivalence in plants is direct and positive. For animals, the data is less compelling, but they have allowed scientists to build a good case.

7.2 Observations on Cells

Some of the observations relevant to the question of genomic equivalence have been made on intact cells, both in normal development and under experimental conditions.

CELLS MAY CARRY OUT DIFFERENT FUNCTIONS AT DIFFERENT TIMES

The hypotheses of cell differentiation by gene loss or selective gene amplification generally imply that mature cells are specialized for *one* function, and that once they have undergone differentiation, they do not "retool" to carry out any other functions. While this model applies to most cell types, some cells are known to undergo radical changes in structure and function as part of their *normal* development. This type of change is called ***sequential polymorphism*** (Gk. *poly-*, "many"; *morphe*, "form") and can be observed in several tissues (Kafatos, 1976). The ovarian follicle cells of insects are one example. During vitellogenesis, the follicle cells form an epithelium of tall, columnar cells (see Fig. 3.16; Fig. 7.3a). Channels between the follicle cells give blood proteins access to the oocyte surface. The follicle cells actively create these passageways by synthesizing extracellular material through which the blood proteins can diffuse. Depending on the species, the follicle cells may also

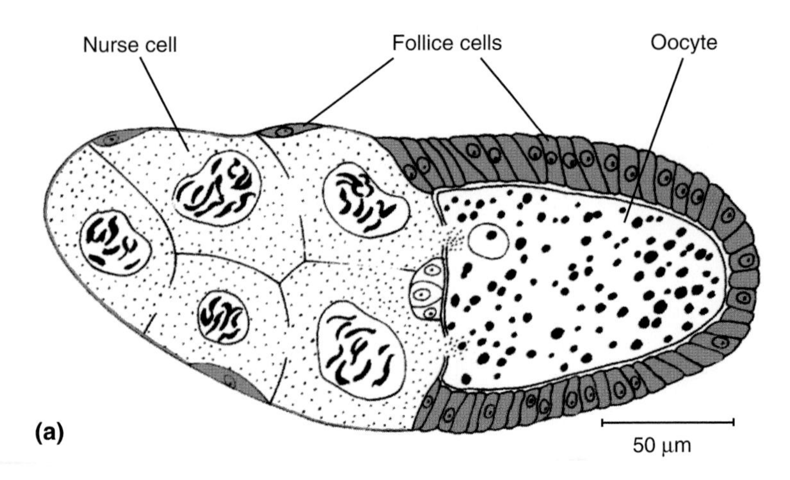

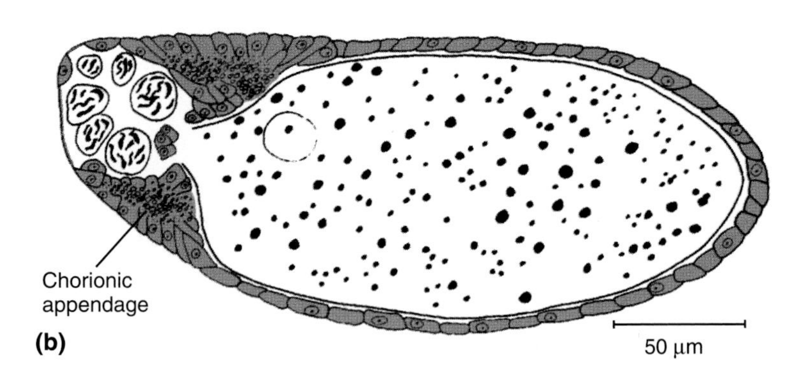

Figure 7.3 Sequential polymorphism of insect ovarian follicle cells. **(a)** Egg chamber of *Drosophila melanogaster* during vitellogenesis. Follicle cells (color) surrounding the oocyte are columnar and aid in vitellogenin synthesis and uptake. Follicle cells surrounding nurse cells are squamous. **(b)** Egg chamber at a late stage of oogenesis. Follicle cells surrounding the oocyte are now squamous and synthesize chorion proteins. Follicle cells surrounding the collapsed nurse cells are now forming the chorionic appendages.

synthesize vitellogenin. At the end of vitellogenesis, both the shape and protein synthesis of the follicle cells change dramatically. The cells assume a flat shape, with no openings between them, and they begin to synthesize large amounts of chorion proteins while the synthesis of other proteins declines sharply (Fig. 7.3b). Interspersed between the two differentiated states of the follicle cells is a period of "retooling," during which the follicle cells go through several rounds of DNA replication. After this period, at least in *Drosophila*, the chromosomal regions containing the genes for chorion proteins are also selectively amplified.

CELLS CHANGE THEIR DIFFERENTIATED STATE DURING REGENERATION

The ability of mature cells to convert to a different cell type is not only seen in normal development, but has also been observed in regeneration experiments; in this context the cells are said to undergo ***transdifferentiation.*** A striking example is the regeneration of a surgically removed eye lens from iris cells in the salamander (Fig. 7.4). This process involves a series of events in the iris cells including shape changes, loss of pigment, DNA replication and mitosis, ribosomal RNA synthesis, shaping of a new lens, and synthesis of lens-specific proteins (T. Yamada, 1967). These events are *not* a recapitulation of the normal steps of lens development: Normal lens formation occurs in ectodermal cells that would otherwise form epidermis, whereas the iris—along with the retina—arises from an outpocketing of the brain rudiment (see Fig. 1.16). Rather, lens regeneration exemplifies the ability of at least some cells to transdifferentiate from one cell type into another.

Other examples of transdifferentiation can be observed during the regeneration of lizard tails or amphibian limbs. When a salamander limb is amputated, epidermal cells first spread over the cut surface and form an ***apical epidermal cap.*** Underneath the cap, all tissues undergo a dramatic process of ***de-differentiation,*** in which the extracellular materials of bone, cartilage, and connective tissues dissolve and release the embedded cells. Multinucleated muscle fibers fragment into mononucleated cells. Cells from all these sources become histologically indistinguishable and form a cone of mesenchymal cells called a ***regeneration blastema.*** After a period of further proliferation, the blastema cells redifferentiate and reconstruct the missing limb.

Are the new bones formed only by former bone cells, and the new muscles only by former muscle fibers, or can blastema cells give rise to cell types that their precursors have never formed? To study this question, limbs were X-irradiated before amputation to prevent the resident cells from dividing. Unirradiated cartilage was transplanted so that it would serve as the only available source for forming a regeneration blastema. The limbs regenerated under these conditions were stunted and poorly shaped, but they did contain cartilage, dermis, and muscle, provided that the cartilage transplant came from a young donor (Wallace et al., 1974). Likewise, a muscle transplanted in a corresponding experiment regenerated all limb tissues including cartilage (Namenwirth, 1974). These and other data indicate that differentiated cells, if challenged by amputation, can dedifferentiate, proliferate, and then differentiate into new types of cells.

Both sequential polymorphism and transdifferentiation show that at least some cells have the ability to differentiate one way and then differentiate again in another way. In terms of gene activity, sequential polymorphism and transdifferentiation involve repressing one set of genes and activating a different set. In the case of the *Drosophila* follicle cells, the genes for chorion proteins must be present but dormant during vitellogene-

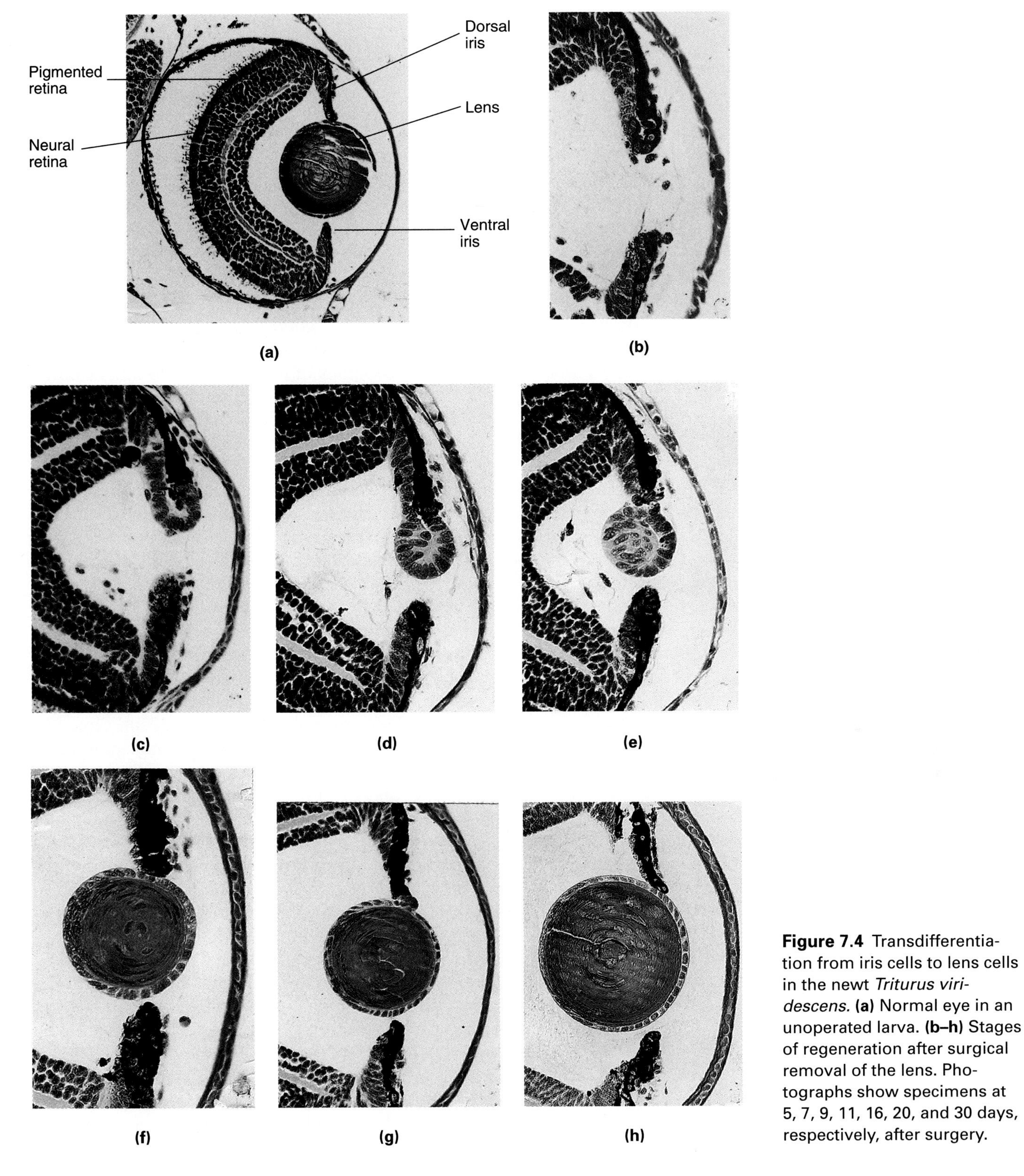

Figure 7.4 Transdifferentiation from iris cells to lens cells in the newt *Triturus viridescens.* **(a)** Normal eye in an unoperated larva. **(b–h)** Stages of regeneration after surgical removal of the lens. Photographs show specimens at 5, 7, 9, 11, 16, 20, and 30 days, respectively, after surgery.

sis, and the genes expressed during vitellogenesis are probably retained but inactive during chorion formation. Likewise, the iris cells regenerating a lens stop synthesizing pigment and eventually make large amounts of *crystallins,* the proteins characteristic of eye lenses. Clearly, the genes for lens crystallins must be present but not expressed while the progenitor cells of a regenerated lens are still functioning as iris cells. And presumably, the genes for the enzymes necessary to make the iris pigments are still available but silent in the regenerated lens cells.

7.3 Observations on Chromosomes

The elaborate process of mitosis ensures that chromosomes are faithfully replicated and delivered in complete sets to two daughter nuclei (see Fig. 2.14). Consequently, there should be no loss or gain of chromosomes, and all cells in an organism should have the same complement of genetic information. This expectation is generally met when chromosomes from different tissues of the same organism are viewed under the microscope.

DIFFERENT CELLS FROM THE SAME INDIVIDUAL SHOW THE SAME SETS OF CHROMOSOMES

Cells prepared from different organs of the same individual usually show the same numbers of chromosomes, with the same sizes, shapes, and staining patterns (Tijo and Puck, 1958). Only some cell types show a regular decrease or increase in the number of *entire chromosome sets.* While most eukaryotic cells are *diploid* (having two sets of chromosomes), gametes are *haploid* (having one chromosome set), and some specialized somatic cells are *polyploid* (have more than two chromosome sets). Gametes retain one full complement of chromosomes as is evident from their ability to form a new organism. Cells become polyploid by replicating their chromosomes and separating the chromatids without entering mitosis. Polyploidy is generally found in large, postmitotic cells that synthesize a few proteins in large quantities. For such cells, it seems adaptive to amplify the number of copies of those genes that they express abundantly. However, since all chromosomes are multiplied entirely, and by the same factor, there is no selective amplification or loss of genes in polyploid cells.

Most chromosomes are small, and during mitosis, when they are discernible under the microscope, their DNA is tightly coiled and supercoiled. Under these circumstances, microscopic inspection would reveal losses or duplications only of large chromosomal segments that contain many genes. Better resolution is obtained with a particular type of giant chromosomes, which are readily visible during interphase, and in which the loss or duplication of individual genes can be detected under the microscope. This type of chromosome will be discussed next.

POLYTENE CHROMOSOMES SHOW THE SAME BANDING PATTERN IN DIFFERENT TISSUES

A special form of polyploidy is polyteny. ***Polytene chromosomes*** originate through several rounds of chromosome replication without subsequent mitosis. The chromatids stick together and remain in register so that a polytene chromosome looks much like a cable composed of many wires (Fig. 7.5). In addition, pairs of homologous chromosomes are aligned side by side within the same polytene chromosome. Thus, in place of each pair of chromosomes in a normal cell, polytene cells have one polytene chromosome consisting of many chromatids. Polytene chromosomes occur in trophoblast cells of mammalian embryos as well as in specialized insect cells. The best-known examples are the salivary gland cells of *Drosophila* larvae. Here, each chromosome has gone through 10 rounds of replication, resulting in 1024 ($= 2^{10}$) chromatids. Since the paternal and maternal homologes are paired within a polytene chromosome, a mature salivary gland chromosome consists of 2048 chromatids. Because chromosomes can convert from the polytene to the polyploid state and vice versa, the structure of a polytene chromosome seems to be similar to that of a normal chromosome (Ribbert, 1979).

Although polytene cells are postmitotic, with many of their genes actively transcribed, they are nevertheless, easy to see under a light microscope. If stained for DNA, they show a distinct ***banding pattern,*** in which dense *bands* alternate with weakly stained *interbands* (Fig. 7.5). The bands are regions of relatively dense DNA coiling; their higher concentrations of DNA account for their more intense staining. Among insects with polytene chromosomes, each species has a distinct banding pattern, whereas different individuals of a given species have the same banding pattern. It is therefore possible to draw detailed ***chromosome maps*** for each of these species. Such maps can be used to identify species that are difficult to determine on the basis of morphological criteria. In *Drosophila melanogaster,* chromosome maps have been used to locate mutations that are associated with small deletions, duplications, or inversions of chromosome sections. The overall number of bands in *Drosophila* polytene chromosomes is about 5000, whereas the total number of genes in the *Drosophila* is estimated at 13,600 genes (Adams et al., 2000). Plus, one band and interband in a polytene chromosome comprise about 2.7 genes on average.

Since the loss or duplication of single bands is visible in polytene chromosomes, these chromosomes allow researchers to study the problem of genomic equivalence at much higher resolution. Beermann (1952) investigated the banding patterns of polytene chromosomes from different larval tissues of the midge *Chironomus tentans* (Fig. 7.6). He found that these chromosomes looked identical in different nuclei from the same tissue of a given individual. Chromosomes from different tissues or developmental stages differed in diameter, reflecting varying numbers of DNA replication rounds. However, the *banding patterns* of chromosomes from different tissues looked remarkably similar. Thus, whatever the molecular mechanism generating the banding pattern may be, it obviously produces very similar results in different tissues, suggesting that the chromosomal DNA sequences are also very similar or identical between tissues.

While the banding pattern of polytene chromosomes is generally constant for a given species, these chromosomes show variable local decondensations called ***puffs.*** Puffs appear and disappear in stage-dependent and tissue-specific patterns (Fig. 7.6; see also Fig. 27.15). The puffed areas of polytene chromosomes are areas of RNA

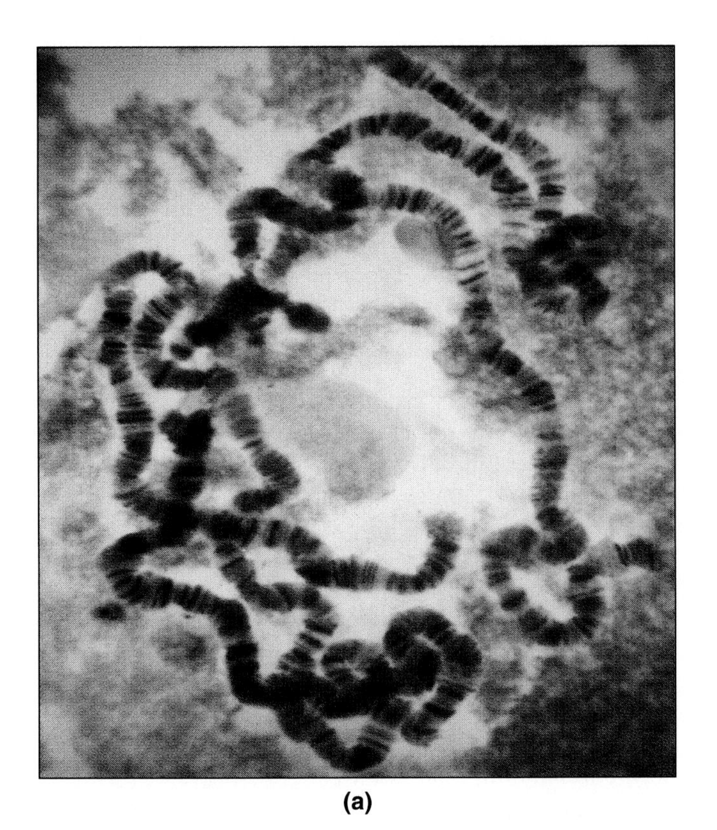

(a)

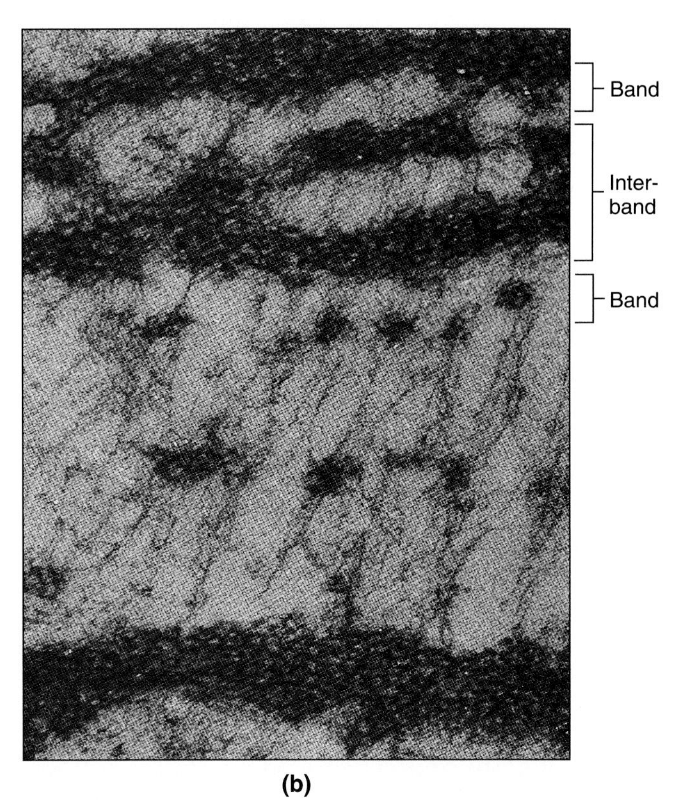

(b)

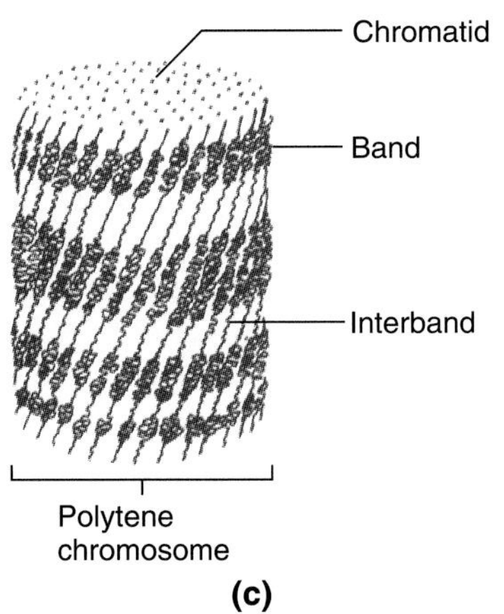

(c)

Figure 7.5 Polytene chromosomes from a *Drosophila* salivary gland cell. **(a)** Light micrograph of a complete chromosome set. The chromosomes were squashed between a microscope slide and a cover slip for better viewing. The centromeres of all chromosomes aggregate and form one chromocenter (arrow). The five long arms extending from the chromocenter are the X chromosome and the left and right arms of chromosomes 2 and 3. The short arm is chromosome 4. Each polytene chromosome represents a pair of homologous chromosomes, each of which has been replicated 10 times to yield $2^{10} = 1024$ chromatids. Because the polytene chromosomes are in interphase, they are less coiled and therefore much longer than normal metaphase chromosomes. Each chromosome shows bands of high DNA concentration between interbands of lower DNA concentration. **(b)** Transmission electron micrograph of a thin longitudinal section of a polytene chromosome, showing bands of highly coiled chromatin alternating with straight chromatin strands in the interbands. **(c)** Schematic diagram showing how a polytene chromosome is composed of many chromatids.

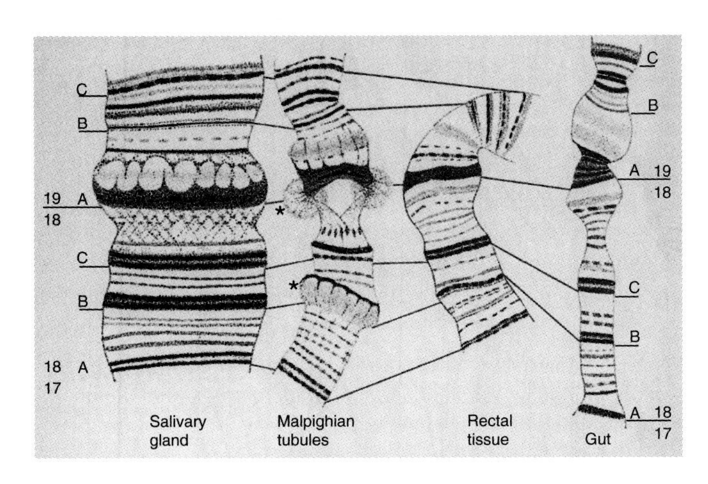

Figure 7.6 Drawings of polytene chromosome segments from different organs of a midge, *Chironomus tentans.* The degree of polyteny, indicated by the thickness of the chromosomes, varies among tissues. The puffed regions (asterisks) are also variable. However, the banding pattern of the chromosomes is remarkably constant among the tissues.

accumulation, probably caused by RNA synthesis (see Fig. 16.16). The variability of the puffing pattern therefore reflects stage-dependent and tissue-specific gene activity. Thus, inspection of polytene chromosomes supports not only the theory of genomic equivalence but also the principle of differential gene expression.

CHROMOSOME ELIMINATION IS ASSOCIATED WITH THE DICHOTOMY BETWEEN GERM LINE AND SOMATIC CELLS

Although chromosomes generally occur in the same number and morphology across all tissues of an individual, there are some cases in which germ line cells and somatic cells of the same organism differ dramatically in genetic constitution (Beams and Kessel, 1974). One well-known example is the chromatin diminution in the somatic cells of certain nematodes mentioned earlier. Several insect species undergo a similar process called ***chromosome elimination,*** in which whole chromosomes rather than chromosomal fragments are eliminated (Fig. 7.7). It occurs spontaneously during certain cleavage divisions, except in primordial germ cells. The protection of the primordial germ cells from chromosome elimination depends on a cytoplasmic component, presumably RNA, that is sensitive to ultraviolet light and localized near the posterior pole of the egg (Von Brunn and Kalthoff, 1983). If this RNA is removed or inactivated, chromosome elimination occurs in both somatic and germ line cells, and adults developing from treated embryos are sterile (Bantock, 1970).

The nature of germ line–specific DNA has been investigated in several roundworm species (Müller et al., 1996). Much of this DNA consists of repetitive sequences of unknown function. One identified gene encodes a germ line–specific ribosomal protein, designated RpS19G. It differs from a related protein, RpS19S, which occurs in both somatic and germ line cells. Thus, chromatin diminution generates different types of ribosomes in germ line versus somatic cells. It is thought that germ line–specific genes may generally be required for gametogenesis.

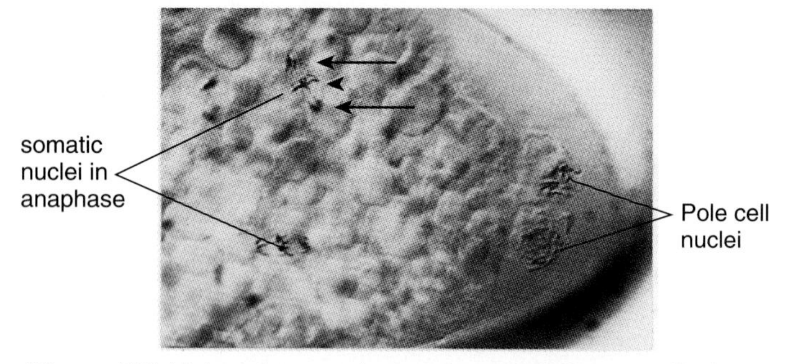

Figure 7.7 Light micrograph showing chromosome elimination in an insect embryo, *Smittia* sp. When somatic nuclei undergo mitosis in the yolk cytoplasm, some chromosomes (arrows) move normally to the spindle poles during anaphase. However, other chromosomes (arrowhead) are left behind in the metaphase plate, where they later disintegrate. Only the nuclei of pole cells (primordial germ cells) are protected from chromosome elimination and retain a full complement of chromosomes.

Chromosome elimination is relatively rare. The chromosomes that are retained only in germ cells seem to contain genes required for gametogenesis and are obviously dispensable for all somatic functions. It also appears that the same chromosomes are eliminated from all somatic cells. Thus, chromosome elimination is correlated with the dichotomy between germ line and somatic cells; comparable differences in number or identity of chromosomes have not been found among somatic cells.

7.4 Molecular Data on Genomic Equivalence

With the techniques of molecular biology, it is possible to clone any segment of DNA and to quantify the number of copies present in any type of cell (see Section 15.3). These techniques can reveal deletions or additions of DNA segments that are too small to detect by microscopic inspection of polytene chromosomes. Applications of these techniques have confirmed the rule of genetic equivalence, at least for most cell types. Normally, genes that are abundantly expressed in certain cell types are not amplified. For instance, there are no extra copies of globin genes in red blood cells and no extra copies of ovalbumin genes in chicken oviduct cells (P. R. Harrison et al., 1974; Sullivan et al., 1973). Randomly chosen *Drosophila* genes are present in different tissues in the same copy number, and in DNA segments of identical length (M. Levine et al., 1981).

Notable exceptions include the genes encoding ribosomal RNA (rRNA) which are selectively amplified in the oocytes of frogs and other animals (Fig. 7.8). These genes are repeated in tandem, forming a distinct region called a ***nucleolus organizer,*** which is normally present in two homologous chromosomes per diploid set. When a nucleolus organizer is actively transcribed, it forms a densely staining organelle called a ***nucleolus.*** Although two nucleoli can provide most somatic cells with enough rRNA, making all the ribosomes for an oocyte in this way would take several years. Instead, oocytes of amphibians and other animals accelerate the synthesis of large rRNAs by selective amplification of the nucleolus organizer (Brown and Dawid, 1968). As the genes in the amplified organizers become active in rRNA synthesis, they form extrachromosomal nucleoli—about 3000 per *Xenopus* oocyte. The extra nucleoli, along with the normal tandem repetition of rRNA genes in each nucleolus, ensure that large oocytes can accumulate enough ribosomes in a short period of time.

Similarly, the *Drosophila* genes encoding the proteins from which the egg envelope or *chorion* is made are

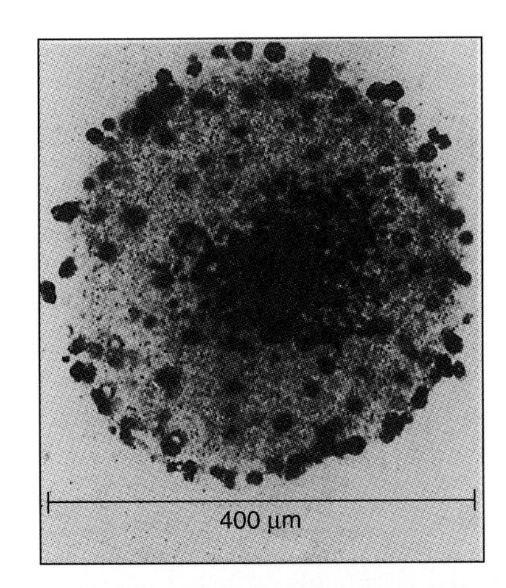

(a)

(b)

1 μm

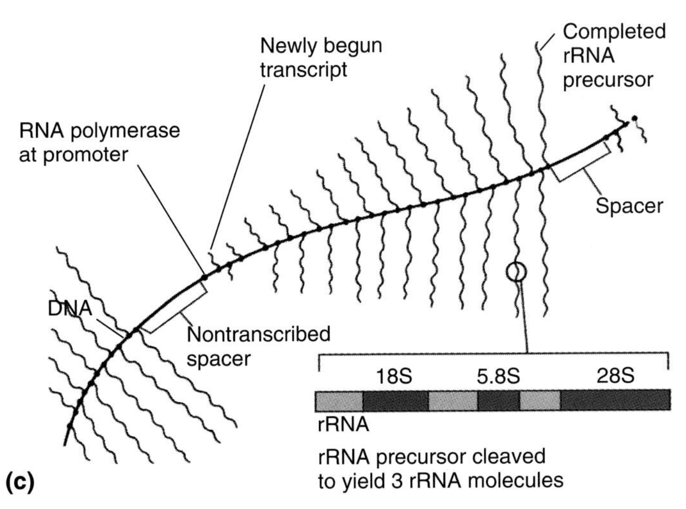

(c)

Figure 7.8 Selective amplification of genes for large ribosomal RNAs (rRNAs) in frog oocytes. **(a)** Light micrograph of an isolated nucleus (germinal vesicle) of a frog oocyte. Each dark dot represents a stained nucleolus. A normal diploid cell has only two nucleoli located on a pair of homologous chromosomes, but a *Xenopus* oocyte has about 3000 extra nucleoli that are not attached to chromosomes. **(b)** Electron micrograph of a portion of one nucleolus. The genomic DNA in each nucleolus contains several hundred copies of a gene from which a large rRNA precursor molecule is transcribed. The RNA molecules become longer as transcription proceeds, giving each transcribed gene the appearance of a fir tree. **(c)** Interpretive drawing of part b. The completed RNA transcripts are cleaved to yield three rRNA molecules having sedimentation values of 5.8S, 18S, and 28S.

multiplied by normal polyploidy and by selective amplification (Spradling and Mahowald, 1980; Spradling, 1981). The selective amplification occurs when the genomic DNA containing the chorion genes undergoes several rounds of extra replication. The selective amplification of chorion genes takes place immediately before chorion synthesis and is limited to the ovarian follicle cells, which synthesize chorion proteins (see Fig. 7.3).

Selective amplification is also common among genes encoding drug-detoxifying enzymes. This has been seen in cultured cells exposed to various drugs and in insect populations that have become resistant to pesticides (Mouches et al., 1990). Apparently, eukaryotic cells have the general potential to amplify specific genes. However, such amplifications seem to be maintained only in the presence of strong selection pressure.

A different kind of deviation from the rule of genomic equivalence is observed in maturing *B lymphocytes,* the mammalian blood cells that produce *antibodies,* or *immunoglobulins,* which are directed against foreign molecules, or *antigens.* A typical immunoglobulin is a Y-shaped molecule consisting of two large polypeptides, called *heavy chains,* and two smaller polypeptides, or *light chains* (Fig. 7.9a). At the tips of the forked portion of the Y are two identical *antigen-binding sites.* Those portions of the light chains and heavy chains that together form the antigen binding sites are called the *variable regions* because they come in a virtually unlimited variety of conformations, each binding tightly to a specific antigen. The remaining regions of the light and heavy chains are known as the *constant regions* because there are only a few types of them. The constant regions of the heavy chains form the tail of the Y, also known as the *effector* site of the molecule, which interacts with other blood serum proteins or with the lymphocyte's plasma membrane.

Mammals can produce a wider variety of antibody molecules than they have genes. How can this be accomplished? It turned out that each lymphocyte clone undergoes a different combination of DNA excisions and rearrangements, thus generating the enormous variety of genetic composites that encode the variable regions of light and heavy chains. We will focus here on the process that generates the composites for the light chains. For its discovery, Nobumichi Hozumi and Susumu Tonegawa (1976) were awarded the 1987 Nobel prize in physiology or medicine. Each light chain composite is spliced together from a long stretch of DNA containing about 300 V segments, four J segments, and a C segment, all separated by intervening sequences that are not represented in the final light chain polypeptide (Fig. 7.9b). By eliminating a large stretch of genomic DNA, one V segment and one J segment are joined. They are transcribed along with the C segment, and the resulting pre-mRNA is processed further to juxtapose the selected J segment with the C segment. This process allows the generation of more than a thousand different light chains. Similarly, heavy chains can be made with

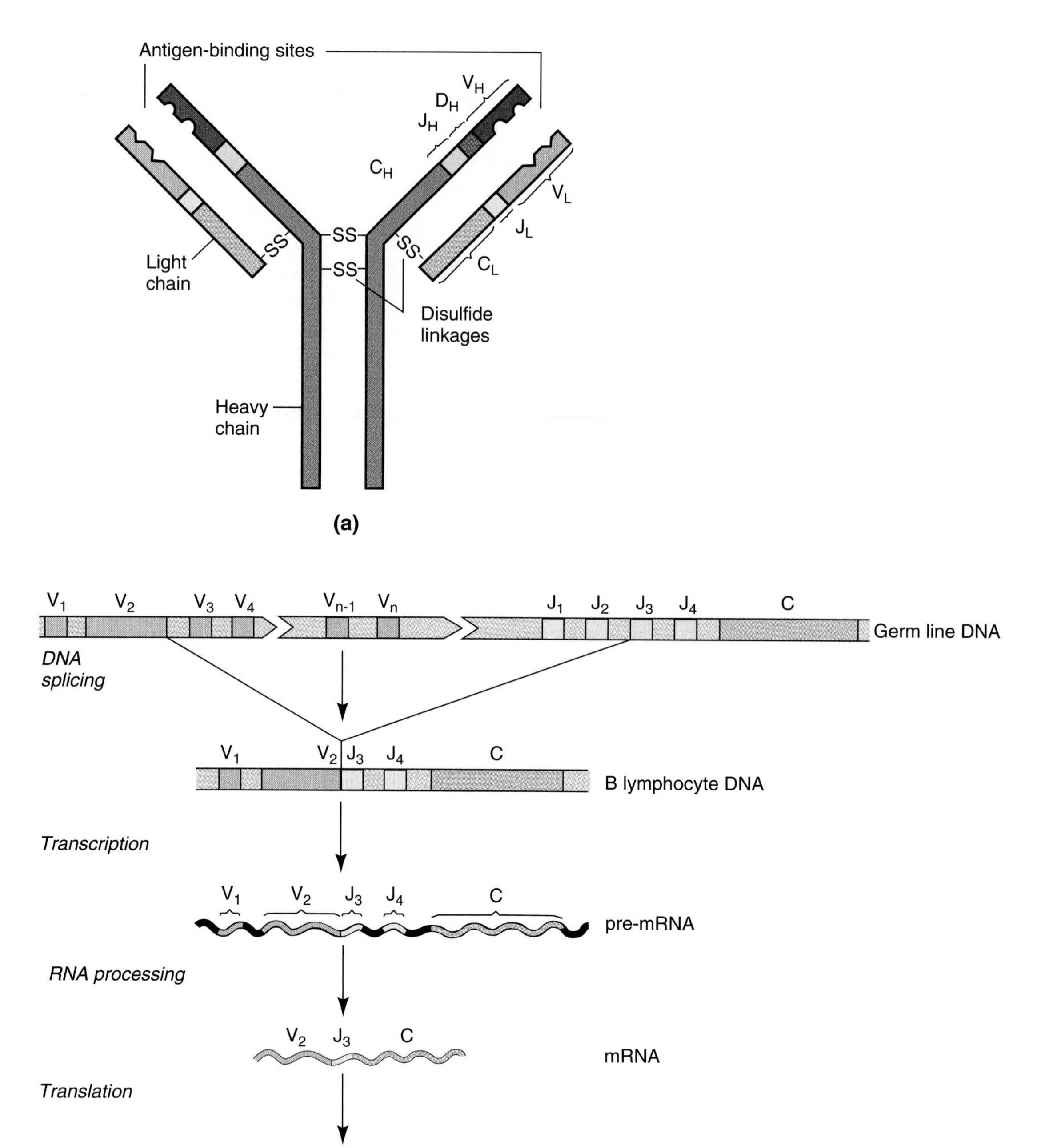

Figure 7.9 Irreversible loss of genomic DNA during lymphocyte maturation. **(a)** A typical immunoglobulin molecule consists of four polypeptides, so-called chains. Two identical light chains and two identical heavy chains are held together by disulfide linkages and weak intermolecular forces. Each chain consists of a constant region (C_H and C_L) and a variable region ($J_HD_HV_H$ and J_LV_L). Two identical antigen-binding sites are formed by the variable regions of both chains. At the opposite end of the molecule, the constant regions of the heavy chains form the effector site, which interacts with other serum molecules and with cell membranes. **(b)** Maturation and expression of the genetic elements for both the light and heavy chain involve losses of DNA and pre-mRNA segments to generate the variable regions. The genomic DNA encoding the light chain contains groups of about 300 V_L segments, four J_L segments, and a C element. During lymphocyte maturation, one V and one J segment are spliced together by eliminating the entire stretch of intervening DNA. The transcript from the spliced DNA undergoes more trimming and splicing until the final mRNA contains only those segments that are represented in the light chain. Similar processes generate an even greater variety of heavy chain variable regions.

more than 20,000 variable regions and different constant regions to different effector sites. The resulting combinatorial diversity of 20 million variable regions is boosted further by imprecise joining of segments and by frequent mutations.

On balance, the instances of selective gene amplification and losses of genomic DNA seem to be exceptions rather than the rule. Nevertheless, the apparent exceptions have had a disquieting effect because the question of how cells become different in development is so fundamental that one would like to have as much certainty as possible. The matter is further complicated by the possibility of subtle but irreversible changes to genomic DNA that would leave genes physically present but would nevertheless amount to a functional loss. Investigators have therefore sought to devise *functional* tests for the completeness of genetic information.

7.5 Totipotency of Differentiated Plant Cells

A very sensitive functional test for the completeness and intactness of a differentiated cell's genetic information is to test the cell for *totipotency*—that is, the ability to regenerate a whole new organism. If a differentiated cell can be shown to be totipotent, this does not rule out selective gene amplification, but it does prove that no essential part of the cell's genetic information was missing. It is important to realize, though, that this experiment is only conclusive if it succeeds. If differentiated cells do not support the development of a complete organism, many interpretations are possible. Genetic information may have been lost as an intrinsic part of cell differentiation, as in the case of lymphocyte maturation. Alternatively, differentiated cells may be less effective than germ line cells in repairing *random damage* to their DNA, which is caused inevitably by cosmic radiation, chemical mutagens, and errors in DNA replication. Other possibilities are that differentiated cells may no longer be able to replicate their DNA as rapidly as required by the fast pace of embryonic cleavage, or that there may be an upper limit to the number of mitotic divisions that a somatic cell may undergo.

Plant cells have passed the totipotency test with flying colors. Related techniques were developed in experiments originally designed to study the metabolism of plant cells stimulated to rapid growth (Steward, 1970). Small pieces of carrot root phloem (vascular tissue) were cultured in rotating flasks in a medium containing coconut milk, which normally serves to nourish the developing coconut embryo. Under these conditions the carrot tissue grew in unorganized masses called ***calluses.*** Studying cell behavior in these cultures, Steward and coworkers observed other groups of cells called ***embryoids,*** which resembled normal stages in the embryonic development of carrots. Presumably, the embryoids had arisen from carrot cells that had rubbed off the calluses. Under suitable culture conditions the embryoids grew into complete carrot plants (see Fig. 7.1).

In a similar experiment, Vasil and Hildebrandt (1965) generated whole tobacco plants from stem pith cells. These experimenters took care to isolate single cells sloughed off from calluses and kept them in microcultures. The cells grew into new calluses, which were placed on a semisolid culture medium with plant growth factors. The calluses formed plantlets with roots, shoots, and leaves, which, after transfer to soil, matured into flowering plants. Corresponding results were obtained with isolated leaf cells from potato plants (Fig. 7.10). These observations, made under well-controlled conditions, show that single cells from various plant tissues retain all the genetic information to build a whole new plant. In addition, the result shows that such cells can revert to rapid growth and to sending and receiving all the signals that are necessary for forming a complete and well-proportioned organism.

Currently, the regeneration of whole plants from single cells in the laboratory often starts with the preparation of ***protoplasts*** (Shepard, 1982). These are naked cells, from which the cell walls have been removed enzymatically (Fig. 7.10). Protoplasts, which grow and divide in culture, have several properties that make them useful in applied research aimed at crop improvement. First, they can be used to grow a virtually unlimited number of genetically identical plants. Second, protoplasts from different plant strains or even different species can be fused to create hybrid protoplasts. Plants grown from such hybrids combine traits of both parent plants. This method of making hybrids is faster than traditional genetic crossing, and it may be the only way of obtaining crosses between unrelated plants. Third, protoplasts can be genetically manipulated with cloned DNA that confers traits such as resistance to cold or pesticides. After these genetic manipulations, it is often necessary to screen thousands of hybrids or transformants before a favorable variety is found. Many of these screens can be done more quickly and economically with protoplasts or small calluses in the laboratory than with whole plants in the field.

7.6 Totipotency of Nuclei from Embryonic Animal Cells

Attempts to grow complete *animals* from isolated *animal cells* have met with very limited success. In many species, including sea urchins, frogs, and mammals, it is possible to isolate blastomeres at the 2-cell or 4-cell stage and raise them to normally proportioned larvae or even fertile adults. However, blastomeres isolated from later embryonic stages do not normally have this ability. Somehow, animal cells seem to lose their totipotency during cleavage. One possible explanation, proposed by

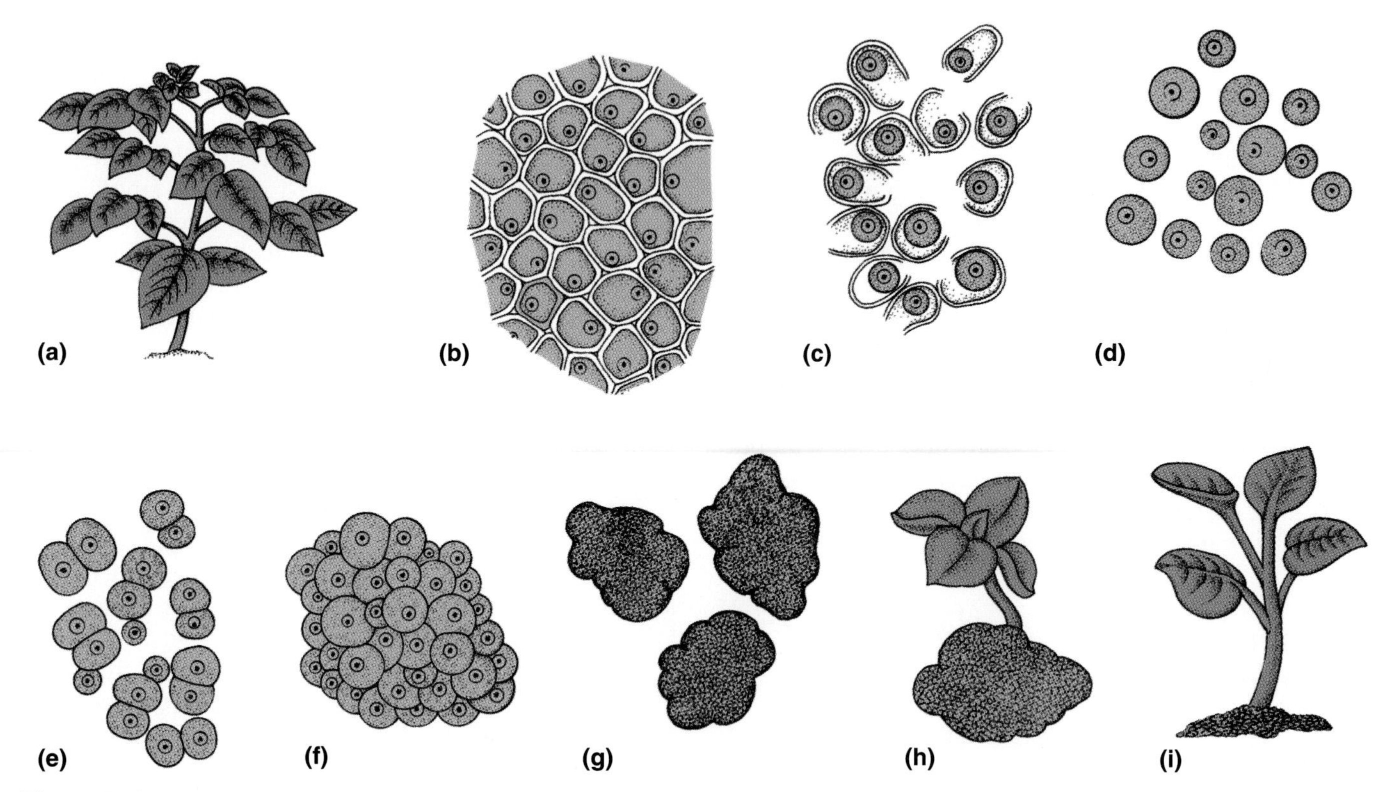

Figure 7.10 Regeneration of a complete potato plant from a single cell. **(a–c)** A leaf from a young potato plant is placed in an enzyme solution to remove the cell walls. **(d, e)** The isolated protoplasts are transferred to a culture medium, in which they divide and grow. **(f, g)** Each protoplast gives rise to a clump of undifferentiated cells, called a callus. **(h)** Callus cells differentiate to form a primordial shoot. **(i)** The shoot forms a plantlet that can be transferred into soil.

Weismann, is that animal cells lose part of their genetic information in the course of development. Alternatively, it may be that animal cells do preserve a full complement of genetic information in their nuclei, but that the properties of animal cytoplasm are too restrictive to allow proper expression of all the genetic information that is still present. To test the latter possibility, experimenters have sought to bring *nuclei* from advanced cells back into the cytoplasmic environment of an egg. If a transplanted nucleus is able to promote the development of a complete organism, then it must have retained a full complement of genetic information. This would rule out Weismann's hypothesis that somatic cells retain only those genes that are required for their specific functions.

NEWT BLASTOMERES DEVELOP NORMALLY AFTER DELAYED NUCLEATION

Spemann, at the University of Freiburg in Germany, where Weismann had worked before him, proceeded to test Weismann's hypothesis by a simple but ingenious experiment. With a baby hair tied into a loop, he ligated a newt zygote shortly after fertilization. The ligature was tight enough to confine the zygote nucleus to one egg half but sufficiently loose to maintain a cytoplasmic bridge between the nucleate and the anucleate egg halves (Fig. 7.11). While the nucleate half cleaved at a normal pace, the anucleate half remained undivided until eventually a nucleus slipped through the ligature. At this point, Spemann pulled the hair loop tight to completely separate the two embryonic halves. The newly nucleated half now contained a nucleus that had already gone through several rounds of mitosis. According to Weismann's hypothesis, this nucleus should have lost some portion of its genetic information. However, both halves developed into complete tadpoles, provided that the ligature had been in the median plane of the egg. The only difference between the twin embryos was that one lagged behind in its development because of the delayed nucleation. The result showed clearly that newt nuclei are still pluripotent after several rounds of mitosis. (Presumably, the nuclei are even totipotent, but to prove this, it would have been necessary to raise the newts to adulthood.) Spemann (1938) suggested that nuclei from more advanced cells should be tested in similar ways, but he did not develop the technical means to do so in his time.

NUCLEI FROM EMBRYONIC CELLS ARE STILL TOTIPOTENT

A straightforward way of testing the nucleus of any cell for totipotency is to transplant it into an egg of the same species. Of course, the recipient egg's own nucleus must be removed so that any subsequent nuclear activity can be ascribed with certainty to the transplanted nucleus.

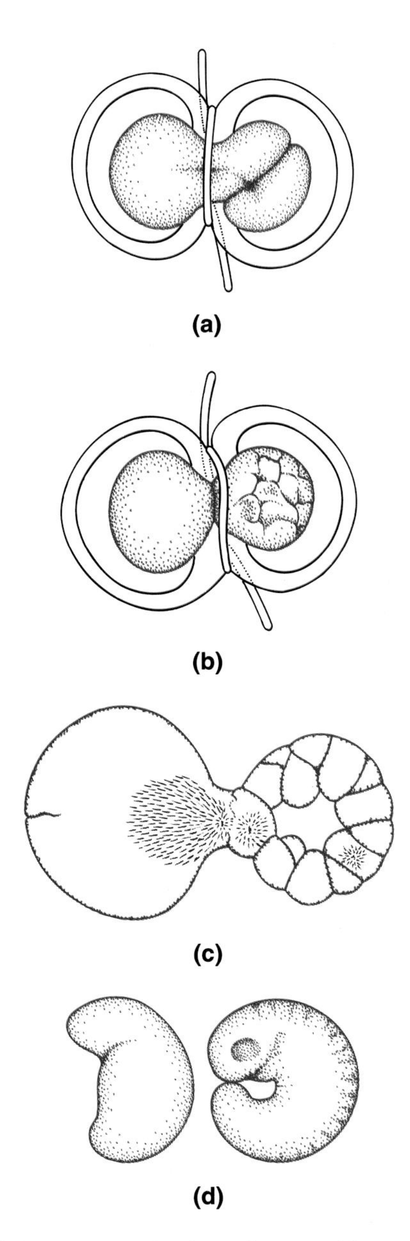

Figure 7.11 Delayed nucleation of a newt blastomere. **(a)** A fertilized egg was ligated, with the nucleus retained in one half. The ligature was incomplete, so that the halves were connected by a cytoplasmic bridge. Because the bridge was initially too narrow for a nucleus to pass through, cleavage was restricted to the nucleated half. **(b)** Several cleavages later, a nucleus slipped through the ligature. Now the other half egg was nucleated by a descendant of the zygote nucleus that had resulted from several mitoses. **(c)** Histological section of a stage similar to part b. **(d)** The ligature was closed completely, so that the two egg halves were fully separated. Each developed into a normally proportioned embryo, although the late-nucleated half lagged behind.

The recipient egg must also be artificially activated in the absence of the activation trigger normally provided by the sperm.

A technique that meets these requirements was developed by Briggs and King (1952). These investigators chose to work on the leopard frog, *Rana pipiens.* The

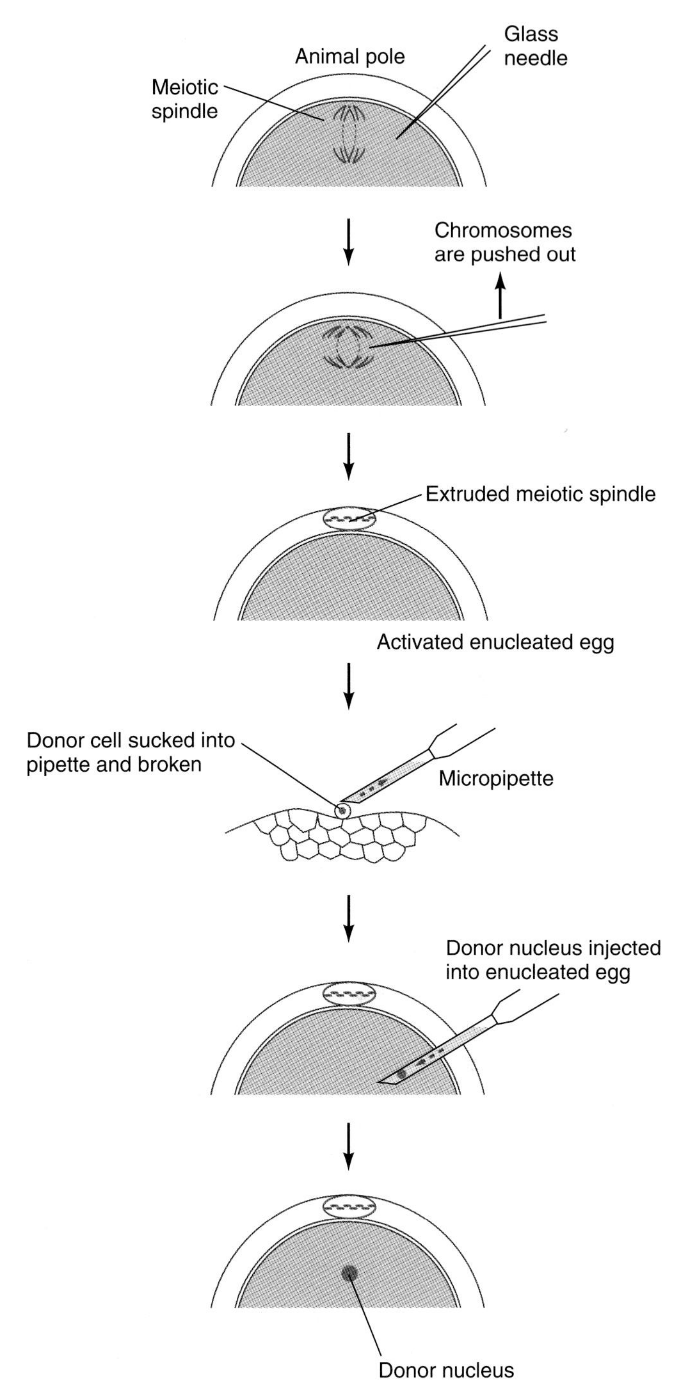

Figure 7.12 Technique used by Briggs and King (1952) for transferring nuclei from blastula cells into enucleated eggs of the leopard frog, *Rana pipiens.* See text. The relative size of the meiotic spindle is exaggerated.

recipient eggs were activated by pricking with a glass needle, and the egg nucleus was pushed out through the puncture (Fig. 7.12). Nuclei were taken from isolated embryonic donor cells. These cells were ruptured by being sucked into a glass pipette with a bore smaller than the cell but large enough to accommodate the nucleus. The aspirated nuclei and surrounding cytoplasm were then injected into enucleated host eggs. This technique and related ones developed later on have become known as *nuclear transplantation* or ***nuclear transfer.***

The first experiments were carried out with nuclei from donor cells at the blastula stage, and the results were encouraging: About 32% of all transferred nuclei promoted the development of their host eggs into complete blastulae; of these about 60% developed into swimming tadpoles. Other investigators extended these experiments and showed that many tadpoles derived from blastula nuclei developed into fertile adults (DiBerardino et al., 1984). Moreover, the technique employed in this experiment was so versatile that cells from later stages and from any tissue could now be used as donors of nuclei.

Nuclear transfers in *Drosophila* have yielded results similar to those in frogs. Transferred nuclei from donors at the early gastrula stage still promoted the development of fertile adults (Illmensee, 1973). Taken together, the transfer experiments described so far indicate that the cell nuclei of animal embryos up to the blastula or the early gastrula stage remain totipotent.

7.7 Pluripotency of Nuclei from Differentiated Animal Cells

The totipotency of nuclei from mature or differentiated animal cells has been more difficult to ascertain than expected. This section describes the modest results obtained as well as the difficulties encountered.

NUCLEI FROM OLDER DONOR CELLS SHOW A DECREASING ABILITY TO PROMOTE THE DEVELOPMENT OF A NEW ORGANISM

The nuclear transfer experiments that worked so well with blastula cells were extended to older cells. However, nuclei from *Rana* embryos beyond the gastrulation stage showed an ever-decreasing ability to promote normal development in host eggs (Fig. 7.13). Nuclei from donors at the tail bud stage were no longer capable of supporting the development of new tadpoles (McKinnell, 1978). Similar results were obtained with other amphibian species (Briggs, 1977; DiBerardino et al., 1984).

The results from these experiments showed that most nuclei from larval frog cells lose their ability to promote the development of an entire organism. However, it is uncertain which nuclear capacity diminishes with age. One possibility is that parts of the genetic information have been lost, as proposed by Weismann, or rendered unusable. If this were the only reason for the diminishing potency of animal nuclei, then the nuclei of *germ cells,* which are known to have a full complement of genetic information, should perform much better in transfer tests than somatic cell nuclei. However, transferred nuclei from spermatogonia gave the same low yields of viable embryos and larvae (DiBerardino and Hoffner, 1971). It appears that nuclei from mature animal cells have undergone another change that causes their diminished function after transfer into an egg.

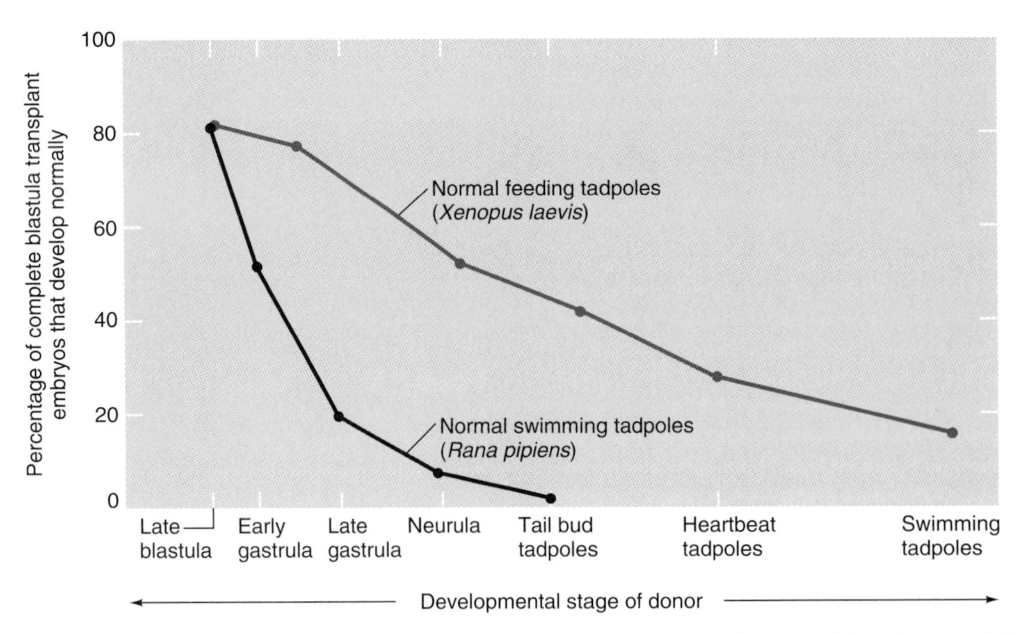

Figure 7.13 Success rate of nuclear transfers. The abscissa shows the developmental stage of the donors of the nuclei. Only those transfers that formed a complete blastula were counted. The percentage of complete blastulae developing into normal tadpoles was plotted on the ordinate. The success rate decreased steeply for *Rana pipiens,* in which nuclei from donors older than the tail bud stage did not support normal development. With *Xenopus laevis,* there was still a decline with age, but a small fraction of nuclei from swimming tadpoles supported the development of complete feeding tadpoles.

NUCLEI FROM MATURE CELLS ARE UNPREPARED FOR THE FAST MITOTIC CYCLES OF EARLY FROG EMBRYOS

To promote the development of a new organism after transfer into an egg, a nucleus must participate in the rapid mitotic cycles that characterize embryonic cleavage. Failure to do so would result in incomplete DNA replication and abnormal mitoses. It is quite possible that transferred nuclei lose parts of their genomic DNA when they are rushed into a fast series of mitoses for which they are no longer prepared.

This interpretation is supported by the following lines of evidence (Gurdon, 1968; DiBerardino et al., 1984). Both the S phase and the entire cell cycle are very brief during cleavage (see Section 5.4). In normally fertilized frog eggs, the S phase of the mitotic cycle lasts only a few minutes (Fig. 7.14). In contrast, when nuclei from mature cells are transferred to eggs, they replicate their DNA much more slowly. Nevertheless, the cyclic synthesis and destruction of *M-phase promoting factor* (*MPF*) that controls embryonic cleavage proceeds at its normal pace. Unlike mature cells, embryonic cells have no control mechanism that would halt mitosis until DNA replication is complete. And indeed, nuclear-transfer embryos frequently show chromosome breakages, which are apparently caused by precocious mitoses. Thus, abnormal development of these embryos most likely does result from the loss of genetic information, but this loss appears to be an experimental artifact and not an original attribute of nuclei in mature cells.

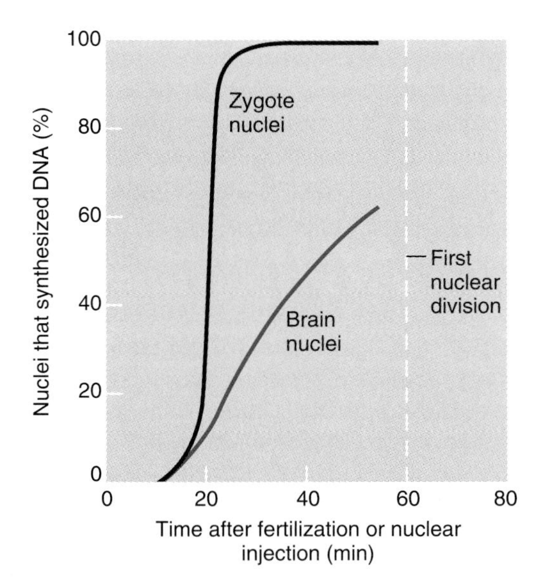

Figure 7.14 DNA replication of *Xenopus* eggs. The black curve represents a plot of DNA synthesis in normal zygote nuclei. The colored curve traces DNA synthesis in brain nuclei transferred into enucleated eggs. Note that many of the transfer nuclei had not completed DNA synthesis at the time of first mitosis.

SOME DIFFERENTIATED FROG CELLS CONTAIN HIGHLY PLURIPOTENT NUCLEI

A series of nuclear transfer experiments similar to those described so far was started by Gurdon (1962). He used a different frog species, *Xenopus laevis,* and a different technique to inactivate the host egg nuclei. Most importantly, he employed a *serial nuclear transfer technique,* in which embryos that developed from eggs with transferred nuclei served in turn as donors for nuclei. This way, Gurdon was able to salvage partially developing embryos while selecting for those nuclei that had successfully adapted to a rapid cell division cycle. Like others before him, Gurdon observed a steady decrease in the development capacity of the nuclei with advancing donor age, but the decline occurred less rapidly in his experiments (see Fig. 7.13). In particular, nuclei from tadpole intestinal cells, in 10 of 726 transfers, promoted the development of feeding tadpoles of normal appearance, and two of the tadpoles developed into fertile adult frogs (Fig. 7.15). Gurdon concluded that at least some of the transferred intestinal nuclei were highly pluripotent while those that gave rise to complete adults were totipotent.

These conclusions were criticized on the grounds that the totipotent nuclei might have accidentally been taken from migrating primordial germ cells or other

(a)

(b))

Figure 7.15 Normal frog and nuclear transfer frog (*Xenopus laevis*). **(a)** Normal frog raised in the laboratory from an egg that had been fertilized in the usual way by a sperm. **(b)** Nuclear transfer frog raised from an egg cell from which the nucleus had been removed and into which the nucleus of an intestine cell had been transferred. The frog is normal in all respects, indicating that the transferred nucleus had a full complement of genetic information.

undetermined cells. To counter this criticism, Gurdon and coworkers transferred nuclei from cultured cells derived from adult *Xenopus* skin and other tissues (Laskey and Gurdon, 1970; Gurdon et al., 1975; Reeves and Laskey, 1975). Of the cultured skin cells, 99.9% contained large amounts of keratin and were therefore considered to be terminally differentiated. Nuclei from skin, kidney, and lung cells gave rise to swimming tadpoles with functional muscles and nerves, a heartbeat and blood circulation, eyes with lenses, and other types of differentiated cells. However, despite the benefits derived from the serial transfer technique, all tadpoles died before reaching the feeding stage (Fig. 7.16). Similar results were obtained using as donor cells lymphocytes from adult *Xenopus* (Wabl et al., 1975) or erythrocytes from adult *Rana* (DiBerardino and Hoffner, 1983).

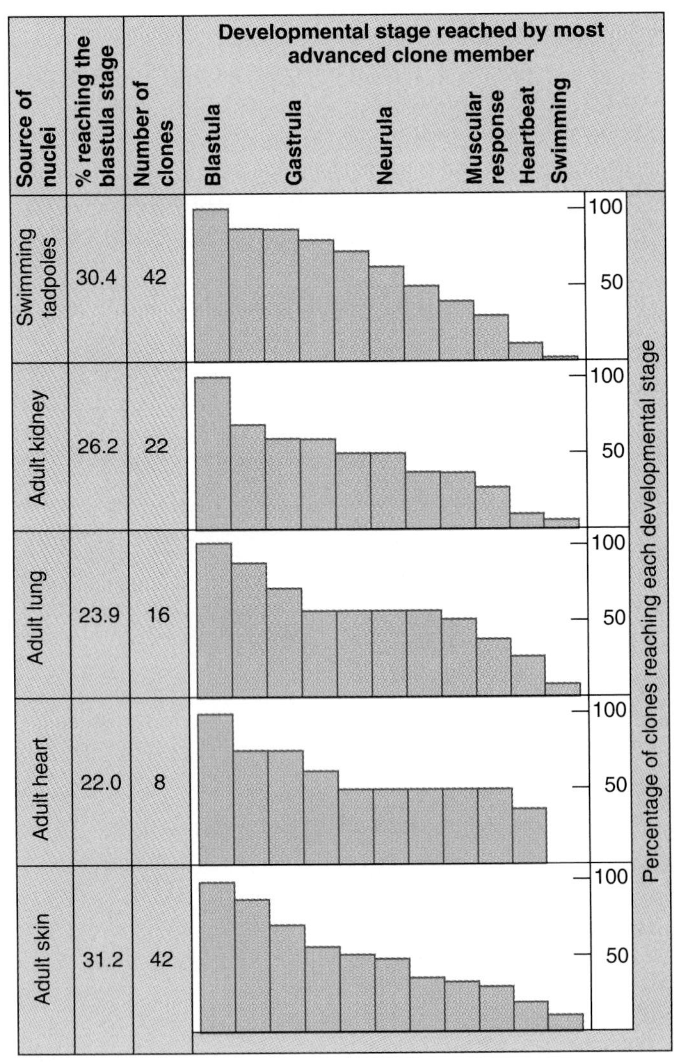

Figure 7.16 Results of transferring nuclei from cultured *Xenopus* cells into eggs. Nuclei were obtained from cell cultures prepared from tadpoles and four adult tissues. For each cell type, the percentage of nuclear transfers forming a partial or complete blastula is recorded in the second column. Several nuclei from each blastula were transferred into fresh enucleated eggs to obtain the number of clones indicated in the third column. The resulting embryos were monitored, and the developmental stage reached by the most advanced embryo of each clone was tallied.

Taken together, the transfers of frog nuclei from differentiated cells to enucleated eggs show that at least small fractions of such cells still have highly pluripotent nuclei. The low yields of successful transfers can plausibly be ascribed to a mismatch between the slow DNA replication in the transferred nuclei and the fast mitotic oscillator in the egg cytoplasm.

The failure of the tadpoles derived from differentiated cell nuclei to undergo complete development seems hard to explain on the basis of incomplete DNA replication. The latter tends to cause large chromosomal defects, which are likely to disrupt development at early stages. However, over the lifetime of an organism, all cells accumulate mutations resulting from errors in DNA replication, cosmic radiation, and chemical mutagens. Most of these mutations will be base substitutions, which are likely to have subtle effects. If somatic nuclei from adult differentiated cells are transferred into enucleated eggs and escape large-scale genetic damage, they may support development until the more subtle damage accumulated in the donor organisms is compounded by additional damage sustained in the recipient. This would explain why nuclear transfers from embryonic cells have been more successful than those with nuclei from adult cells. However, the DNA damage accumulation hypothesis raises the question why such accumulation is not happening in the *germ lines* of organisms, which have existed for hundreds of millions of years. One would have to postulate, then, that the repair of DNA damage, or selection against damaged cells, is much more efficient in the germ line than in somatic cells. Indeed, germ line cells generally divide slowly, which leaves repair mechanisms more time to act. In humans, mutations accumulate much faster in somatic cells as compared to germ line cells (see Section 29.2).

MAMMALS CAN BE CLONED BY FUSING FETAL OR ADULT CELLS WITH ENUCLEATED EGGS

The nuclear transfer technique has been applied to several mammalian species, with a prospect for major advances in basic research as well as in agriculture and medicine (*Fulka et al., 1998*).

AN extensive line of experiments has been carried out with sheep (Campbell et al., 1996; Wilmut et al., 1997). The experimental technique used (Fig. 7.17a) differed in two respects from the technique developed for frogs (see Fig. 7.12). First, instead of microinjecting donor cell nuclei, the experimenters made a wide slit into the *zona pellucida*, removed the egg's polar body and meiotic spindle, and then placed the entire donor cell next to the enucleated egg. Cell and egg were fused by means of an electric shock,

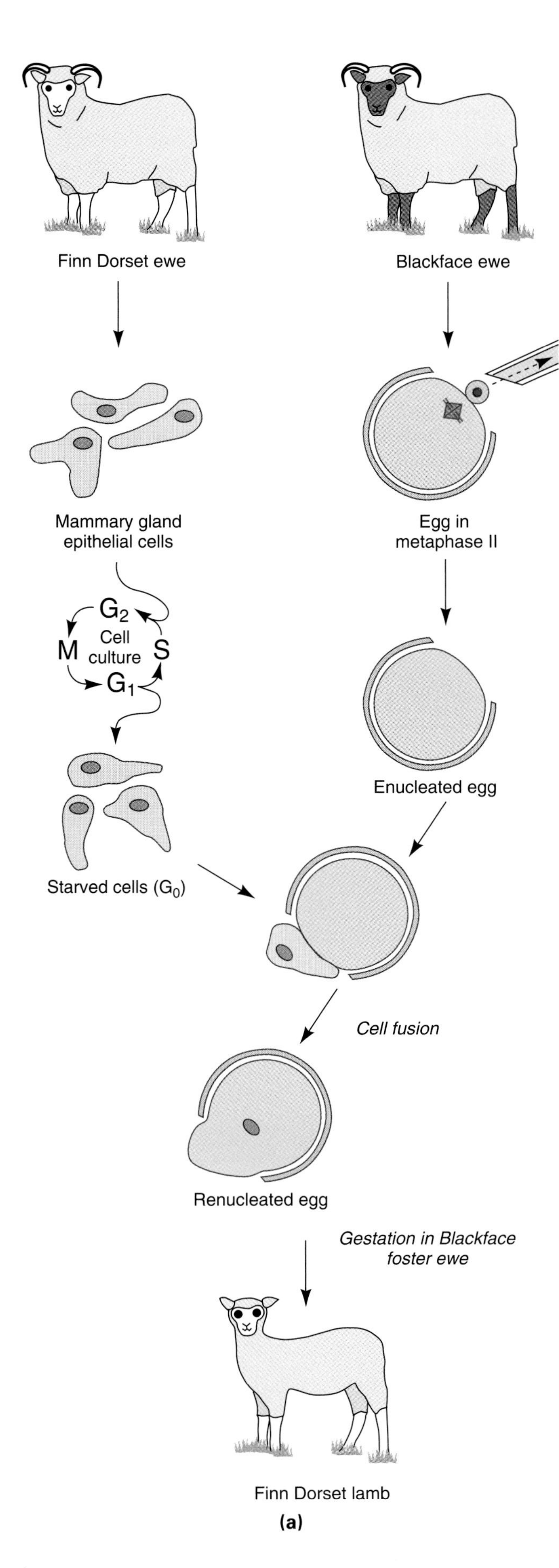

(b)

Figure 7.17 Sheep cloning. **(a)** Technique: Mammary epithelial cells derived from a Finn Dorset ewe were starved in culture so that they entered a quiescent state (G_0). Mature oocytes arrested in metaphase II were derived from a Blackface ewe. Through a slit made into the zona pellucida, the first polar body and the metaphase plate were removed. The enucleated oocyte was fused by electroshock with a cultured epithelial cell. The renucleated oocyte was cultured until the blastocyst stage and then implanted into the uterus of a Blackface foster ewe. One healthy lamb named Dolly was born showing the phenotypic characteristics and molecular markers of the Finn Dorset race. **(b)** Photograph of Dolly and her foster mother.

which also activated the egg. Second, the researchers paid much attention to the stage of the cell cycle in which the nuclear donor cells were. Starved culture cells that had left the cell cycle to enter the quiescent G_0 state were better donors than cells that were kept cycling in a rich medium. The G_0 state (a modified G_1 state; see Section 2.4) appears to facilitate the reprogramming of the nucleus for embryonic development. After cell fusion, the renucleated eggs were cultured in ligated sheep oviducts or in a chemically defined medium until they reached the blastocyst stage. At this point, the embryos were inspected, and the normal-looking ones were implanted into the uteri of foster ewes.

It was critical in these experiments to ensure that any lambs born did not result from host eggs that had not been enucleated properly or from unnoticed matings of foster ewes. Eggs and foster ewes were therefore chosen from a sheep race (Scottish Blackface) that differed from the races (Finn Dorset and others) used to provide the donor cells for nuclei. Even though the racial differences were clearly visible, DNA tests were done to prove independently that the lambs born did indeed belong to the race of the nuclear donor cells.

From more than 800 donor cells fused with enucleated eggs, eight live lambs were born, with losses occurring at various stages of the procedure (Table 7.1). One lamb, named "Dolly," was derived from a line of cultured epithelial cells derived from the mammary gland of a 6-year-old ewe (Fig. 7.17b). The other live-born lambs were derived from cells taken from a 26-day-old fetus and a 9-day-old embryo. These experiments have become known to the general public as "cloning" because an organism that develops from a renucleated egg is a genetically identical copy, or "clone," of the organism that donated the nucleus.

Questions

1. The high failure rate of nuclear transfer experiment in frogs has been ascribed, for the most part, to the inability of nuclei from differentiated cells to readjust to the fast pace of DNA synthesis during cleavage divisions. Could this adjustment be easier in mammals?
2. The cell culture from which the donor cell for the cloning of Dolly was derived has been established from tissue of a pregnant ewe. Under these circumstances, could Dolly have been cloned inadvertently from an embryonic or fetal cell rather than an adult cell?

The experiments described above show that at least some mammalian cells retain a totipotent nucleus. This conclusion seems safe for embryonic and fetal cells, which served as donors in additional experiments with sheep (Schnieke et al., 1997), cattle (Cibelli et al., 1998; Kato et al., 1998), and rhesus monkeys (Meng et al., 1997). The generation of Dolly from the nucleus of an *adult cell* has come under scrutiny because she was the only successful case obtained from 277 fused oocyte-donor cell couplets (Table 7.1). The cell culture from which the donor nucleus for Dolly was taken might have been contaminated with fetal cells. Such contamination might have occurred naturally (see Question 2 above) or through human error. To allay this concern, Ashworth et al. (1998) and Signer et al. (1998) compared nuclear DNA sequences from Dolly's blood, from the cell culture that had been the source of the donor cell nucleus for Dolly, and from the frozen tissue that had been the source of this donor cell culture. If Dolly had been cloned inadvertently from a contaminating fetal cell, her DNA should have shown the molecular characteristics not only of the fetus' mother, who donated the mammary tissue, but also of the fetus' father. However, the tests showed that the DNA samples were identical, without admixtures from another genome. These results prove beyond reasonable doubt that Dolly has been cloned from an adult cell.

Similar experiments were carried out with mice by Wakayama et al. (1998). These researchers transferred nuclei with fine glass needles, using a technique similar to the one originally devised for frogs. They took care to minimize the amount of cytoplasm transferred because the cytoplasm of the donor cells might contain factors that restrict the developmental potential of nuclei. They also allowed the transferred nuclei to adjust to the egg cytoplasm for several hours before activating the eggs. As donor cells for nuclei they used *neurons, Sertoli cells,* and ovarian *granulosa cells* (see Fig. 3.21). The nuclear transfers from granulosa cells resulted in several healthy adult females, while none of the nuclear transfers from Sertoli cells or neurons was successful. However, the yield of successful nuclear transfers from granulosa cell was small. From 2468 injected oocytes, the investigators recovered and implanted 1385 embryos into foster mothers. From these, 31 live mice were born, of which 22 developed into fertile adults. In subsequent experiments, male mice were cloned from tail tip cells (Wakayama and Yanagimachi, 1999). These results are of great significance because they confirm that nuclei from adult, differentiated mammalian cells introduced into enucleated oocytes can support full development. Also, because of the well-established genetics and short

table 7.1 Development of Cloned Sheep

Cell Type	No. (%) of Fused Couplets	No. (%) Recovered from Oviduct	No. Cultured In Vitro	No. (%) of Morulae/ Blastocysts	No. of Morulae/ Blastocysts Transferred	No. (%) of Pregnancies/ Recipients	No. (%) of Live lambs Born*
Mammary epithelium	277 (63.8)	247 (89.2)	0	29 (11.7)	29	1/13 (7.7)	1 (3.4)
Fetal fibroblast	172 (84.7)	124 (86.7)	—	34 (27.4)	34	4/10 (40.0)	2 (5.9)
		—	24	13 (54.2)	6	1/6 (16.6)	1 (16.6)†
Embryo-derived	385 (82.8)	231 (85.3)	—	90 (39.0)	72	14/27 (51.8)	4 (5.6)
		—	92	36 (39.0)	15	1/5 (20.0)	0

*As a proportion of morulae or blastocysts transferred

†This lamb died within a few minutes from birth.

Data from Wilmut et al. (1997).

gestation period of mice, further basic research on mammalian cloning will not proceed much faster than it could with large mammals.

NUCLEAR TRANSFER EXPERIMENTS WITH MAMMALS HAVE IMPORTANT APPLICATIONS IN SCIENCE, AGRICULTURE, AND MEDICINE

Nuclear transfer experiments with mammals will have important applications in the biological sciences, in agriculture, and in medicine. So far, we have discussed these experiments here as a way of testing whether differentiated cells retain a full complement of genetic information. In this context, it is still of great interest why most nuclear transfer experiments from adult donor cells have failed. Are the reasons technical, so that they will disappear as experimentors hone their skills? Or do the failures indicate a regular loss of genes in differentiated cells, as proposed by Weismann? Or are there other biological reasons for the large numbers of failed nuclear transfers?

Nuclear transfer experiments may provide new insights into DNA repair mechanisms and their control. There is an incessant trickle of damage to genomic DNA from errors made during DNA replication and from oxidants and other reactive chemicals. Most of the damage is removed by DNA repair mechanisms and through the immunological killing of cells displaying abnormal phenotypes. This surveillance should be nearly perfect in *germ line* cells, with just enough DNA modifications left to sustain an evolutionary process of mutation and natural selection. In somatic cells, one may expect that a compromise is struck between the costs of repair and the loss of vitality resulting from unrepaired damage. Little is known about the molecular mechanisms that gauge the cost and accuracy of DNA repair in germ line versus somatic cells, and nuclear transfer experiments should be useful in related studies.

Also, mammals have a set of *imprinted* genes, some of which undergo long-term inactivation during spermatogenesis but remain active in eggs. The converse occurs for other imprinted genes, so that a zygote contains some pairs of homologous genes of which only the paternal allele is expressed, and other homologs of which only maternal allele is expressed (see Section 16.6). The imprinted alleles remain inactive through most of development and are reactivated only in the primordial germ cells of the next generation. Similarly, female mammals have one inactive X chromosome through most of their life cycle (see Section 26.3). If nuclear transfer were to interfere with these normal gene inactivations, then this would be an opportunity to learn more about the underlying molecular mechanisms.

Another fascinating topic in basic research is *cellular senescence* (see Section 29.6). Somatic mammalian cells can undergo only a limited number of divisions. The apparent reason for the limit to cell proliferation is that the ends of chromosomal DNA, known as *telomeres*, become a little shorter with each round of DNA replication. Normal somatic cells either cease to divide or sustain chromosomal damage if they keep dividing after their telomeres are used up (Lundblad and Wright, 1996). Only germ line cells, stem cells, and most tumor cells escape this limitation. Since Dolly was cloned from a cell derived from a 6-year-old donor, one might expect that Dolly's telomere length is that of a so much older sheep, and this is the case (Shiels et al., 1999). So far, the premature senescence of Dolly's telomeres does not seem to interfere with her health. Indeed, she has given birth to healthy lambs. Measurements in sheep and mice (Herrera et al., 1999; Rudolph et al., 1999) suggest that the average telomere length in these species may be sufficient to sustain several lifetimes. However, telomere length may limit the usefulness of adult somatic cells for repeated nuclear transfers.

Notwithstanding their significance to basic research, most of the recent nuclear transfer experiments with large mammals have been fueled by commercial interests. Ranchers who wish to build a herd of domestic animals with a set of specific traits may soon be able to select the individual that comes closest to their ideal prototype and clone this specimen exactly. This way they can avoid the uncertainties of conventional breeding, although a herd without any genetic diversity will be very vulnerable to damage from parasites. Cloning of mammals will also help scientists to test the effects of drugs or other experimental regimens without the confounding variable of genetic diversity. This will allow them to obtain statistically significant data with smaller numbers of animals tested.

Health benefits to human patients, and profits to pharmaceutical companies, will come from genetically engineered domestic animals. Human therapeutic proteins like insulin or blood-clotting factors are difficult and expensive to make with current technologies. However, any human gene can be expressed in an animal host if inserted into an appropriate chromosomal region (see Sections 15.3 and 16.3). A *fusion gene* consisting of the *regulatory region* of a sheep milk protein gene and the *transcribed region* of human coagulation factor IX has already been inserted into the genome of fetal sheep fibroblasts. The cloned fibroblasts containing the fusion gene have been used as donors for nuclear transfer into oocytes, from which three healthy lambs were born (Schnieke et al., 1997). Lactating ewes raised from such lambs are expected to produce copious amounts of human coagulation factor IX with their milk.

WILL HUMANS BE CLONED IN THE FUTURE?

Methods that work with large mammals, including rhesus monkeys, are likely to work with humans. Nuclear transfer technology may be appealing to couples with specific reproductive handicaps that cannot be overcome with other methods. For example, mutations in mitochondrial DNA can cause blindness and other devastating afflictions, and an embryo conceived by a woman with such mitochondria will not grow up to be a healthy child. However, a cell from such an embryo, if fused with an enucleated egg donated by a woman with normal mitochondria, will have the nuclear genes of its biological parents plus enough intact mitochondria to develop normally. Parents who lost a child in an accident or crime may want to have the victim cloned from salvaged cells. Couples with one infertile partner may wish to clone a child from the other partner. For the time being, the grim losses incurred with the existing nuclear transfer procedure, and the unresolved basic biological questions about cellular senescence and the accumulation of genetic damage in somatic cells, makes this a clearly unethical option for humans.

If nuclear transfer ever became safe for human use, it could be used for ***gene therapy.*** This prospect is very attractive because at the heart of many human diseases lies a genetic defect. The basic strategy for a *germ line gene therapy* would involve adding a correct copy of the defective gene to cultured human cells, cloning of a successfully corrected cell, and then using the cloned cells as donors for nuclear transfer into oocytes. This procedure could open the possibility of raising a healthy child from a genetically treated embryonic cell, and it would eliminate the deficient gene from the child's *germ line.* However, as discussed above, this is not currently an option, and it may never become one for biological reasons. All currently approved clinical trials for human gene therapy are limited to *somatic* cells, so that even successfully treated patients will pass on their defective genes to any children they may have.

A worrisome aspect of human germ line therapy is that it may be used not only to repair clearly identified genetic deficiencies but also for attempts to improve upon human nature by adding genes for extra strength, intelligence, beauty, or whatever may seem desirable at the time. Such attempts have already been made with animals. The performance of mice in memory tests has been improved significantly by an added gene that encodes a receptor for a particular neurotransmitter (Tang et al., 1999). In another study, extra growth hormone genes were added to the genome of pigs so they would produce leaner pork. The results of the latter experiment have been very sobering: The genetically "enhanced" pigs, in comparison to their unmodified litter mates, grew faster and were leaner, but they also suffered from numerous diseases including gastric ulcers, arthritis, cardiomegaly, dermatitis, and renal disease (Pursel, 1989). It seems so difficult to predict the effects of adding another gene to the complex regulatory machinery of the existing genome that such attempts should never be made with humans.

Taken together, the observations reviewed in this section suggests that, as a rule, the nuclei of at least some differentiated animal cells retain a complete and functional set of genetic information. Consequently, genetic equivalence has become a widely accepted theory. The known cases of chromosome loss, selective gene amplification, and DNA rearrangements are considered exceptions to the rule. The high failure rates of nuclear transfer experiments using differentiated cells as donors may result from causes other than loss of genetic information. The most important implication of genomic equivalence is that cells must become different by using selected subsets of their total genetic information. Consequently, one would expect a multitude of regulatory signals to select the combination of gene activities appropriate for each cell type. Indeed, there are whole networks of genes whose function is to activate subordinate genes at the right time and in the right groups of cells (see Chapters 16 and 20 through 25).

7.8 Control of Nuclear Activities by the Cytoplasmic Environment

Having accepted genomic equivalence as a reasonable basis for further analysis, we want to know how various signals cause cells to express different portions of their genetic information. This is one of the focal questions in modern developmental biology. In the remainder of this chapter, we will explore two experimental situations in which the control of gene activities by the cytoplasmic environment has been directly demonstrated.

GENE EXPRESSION CHANGES UPON TRANSPLANTATION OF NUCLEI TO NEW CYTOPLASMIC ENVIRONMENTS

To analyze the effects of different cytoplasmic environment on nuclear activity, Gurdon (1968) injected nuclei from one type of cell into the cytoplasm of another type of cell. He monitored three basic types of nuclear activity—namely, DNA synthesis, RNA synthesis, and chromosome condensation in preparation for meiosis (Fig. 7.18). The test nuclei came either from *Xenopus* adult brain cells, which synthesize RNA but no DNA, or from early blastula cells, which replicate DNA but synthesize virtually no RNA. As host cells, Gurdon used growing oocytes, which synthesize RNA but no DNA; or mature oocytes, in which the nucleus is arrested in meiotic metaphase II; or activated eggs, the nuclei of which synthesize DNA but no RNA. In every combination, the transplanted nucleus adjusted its activity to the new cytoplasmic environment. Blastula nuclei injected into growing oocytes stopped synthesizing DNA and entered a phase of continuous RNA synthesis. Conversely, brain cell nuclei injected into activated eggs stopped RNA synthesis and began DNA synthesis. When the same nuclei were injected into mature oocytes, they synthesized neither DNA nor RNA but underwent chromosome condensation and formed a meiotic spindle.

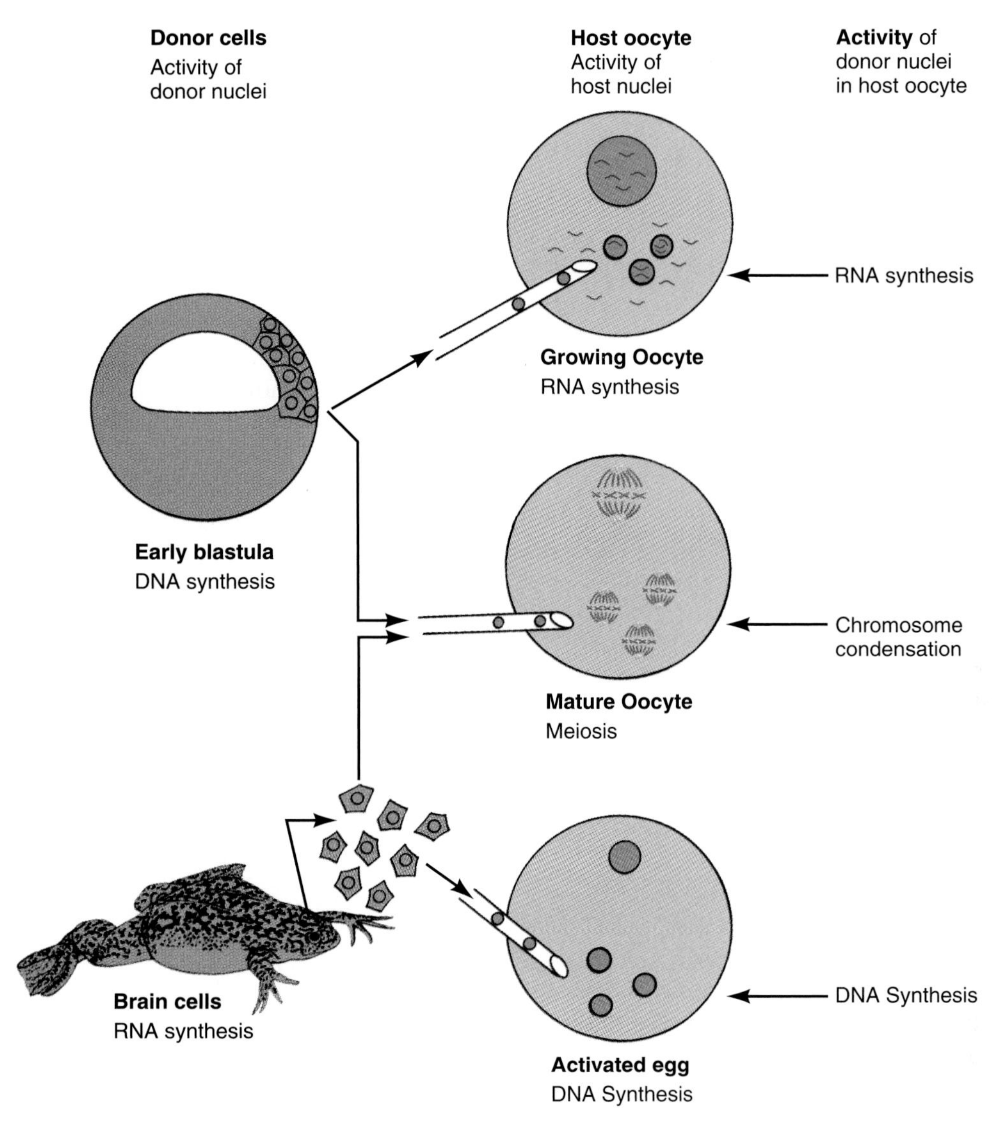

Figure 7.18 Transplantation experiments with nuclei from *Xenopus* showing that nuclear activities are controlled by the cytoplasmic environment.

TAKING this experiment a step further, De Robertis and Gurdon (1977) injected nuclei from kidney cells into growing oocytes and assayed for the reprogramming of the nuclei in terms of individual gene products. Would the transplanted nuclei continue to express the same set of genes in oocytes as they had done in kidney cells, or would they stop expressing those genes? Would the transplanted kidney cell nuclei begin to express genes that were normally active in oocytes only? The experimenters identified individual polypeptides (proteins), which are the products of single genes; the proteins were labeled with radioactive amino acids and separated by ***two-dimensional gel electrophoresis.*** This technique involves two rounds of electrophoresis through polyacrylamide gel, during which proteins are separated according to isoelectric point in one dimension and according to molecular weight in the other dimension. The resulting patterns of radioactive spots in the gel are made visible by *autoradiography* (see Method 3.1). This method detects several hundred of the most abundantly synthesized proteins in a given type of cell.

The authors first analyzed and compared the abundant proteins synthesized normally in *Xenopus* oocytes and kidney cells (Fig. 7.19). They found that most proteins were common to both kinds of cells; these are the "housekeeping" proteins that cells need as metabolic enzymes, for instance, or to assemble their cytoskeletons and their ribosomes. However, a few proteins occurred in oocytes but not in kidney cells or vice versa. These were termed *oocyte-specific proteins* or *kidney-specific proteins,* respectively.

Next, for their nuclear transplantation experiments, De Robertis and Gurdon injected nuclei from kidney cells of the frog *Xenopus laevis* into oocytes of the newt *Pleurodeles waltl* (Fig. 7.20). The host oocytes were cultured for several days, during which time newly synthesized mRNA accumulated in the cytoplasm. Then radioactive amino acids were added to the culture medium to label any polypeptides synthesized afterward. After a labeling period of 6 h, oocyte

Kidney Cells

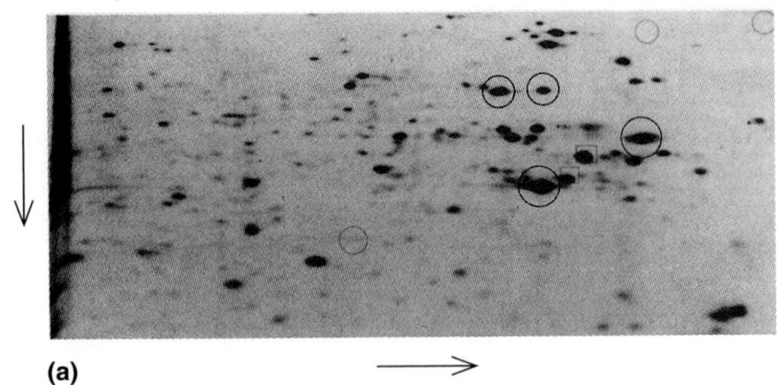

(a)

Oocytes

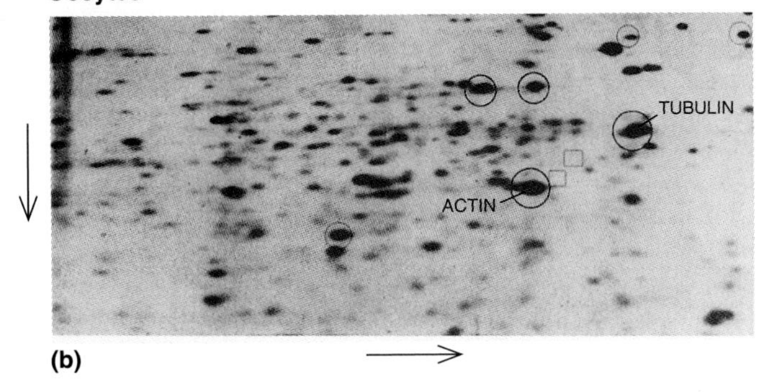

(b)

Figure 7.19 Comparison of *Xenopus* proteins synthesized by cultured kidney cells **(a)** and oocytes **(b)**. Proteins were radioactively labeled in vivo before they were extracted and separated by two-dimensional gel electrophoresis arrows. The proteins are seen as dark spots on an X-ray film exposed to the gel. Most proteins, including actin and tubulin, are made by both cell types. However, some proteins are oocyte-specific (empty circles in part a) while others are kidney-specific (empty squares in part b).

proteins were analyzed by two-dimensional gel electrophoresis. Among the radiolabeled proteins, at least seven were recognized as encoded by the *Xenopus* genome. They included three housekeeping proteins and four oocyte-specific proteins but no kidney-specific proteins. This meant that the oocyte environment had reprogrammed the kidney nuclei to express housekeeping genes and some oocyte-specific genes, but had inhibited the expression of kidney-specific genes.

Questions

1. Why did the investigators inject *Xenopus* kidney nuclei into *Pleurodeles* oocytes rather than into *Xenopus* oocytes?
2. What do the results show about the evolution of the signals that control gene expression?

The results of De Robertis and Gurdon show that cytoplasmic factors can control nuclear behavior not only generally with regard to DNA versus RNA synthesis, but also specifically in terms of activating certain genes while inactivating others.

CELL FUSION EXPOSES NUCLEI TO NEW CYSTOPLASMIC SIGNALS

A similar reprogramming of nuclei by their cytoplasmic environment was observed in cell fusion experiments. Cells can fuse to form a combined cell with two or more nuclei (Fig. 7.21). This happens normally in fertilization and during development of skeletal muscle (see Figs. 4.4 and 14.23). Cells that do not normally fuse can be stimulated to do so in vitro by exposing them to electric pulses, inactivated viruses, or polyethylene glycol, each of which alters plasma membranes so that they tend to fuse with one another. A hybrid cell with two or more nuclei from different species is called a ***heterokaryon*** (Gk. *heteros*, "other"; *karyon*, "nucleus"). Heterokaryons are capable of synthesizing DNA, RNA, and proteins and show rapid mixing of the cytoplasmic components and membrane proteins from their constituent cells.

Although heterokaryons may live for several weeks under favorable conditions, their continued survival depends on the formation of daughter cells containing a single hybrid nucleus. In binucleate heterokaryons, simultaneous mitosis of both nuclei sometimes leads to the formation of a single spindle, and subsequent cytokinesis produces two ***hybrid cells,*** each with one nucleus containing the chromosomes of both species (Fig. 7.21). For instance, a heterokaryon derived from one diploid rat cell (containing 2×21 chromosomes) and one diploid mouse cell (containing 2×20 chromosomes) may divide into two hybrid cells, each containing a nucleus with 42 rat plus 40 mouse chromosomes. After prolonged culture, such cells tend to become polyploid for some chromosomes while losing others.

Heterokaryons and hybrid cells have been used in a variety of studies. For research on nucleocytoplasmic interactions, a particularly useful heterokaryon was generated from chicken erythrocytes and human tumor cells known as HeLa cells (Harris, 1974). These two cell types display very different levels of DNA and RNA synthesis. HeLa cells, as actively proliferating cells, synthesize both DNA and RNA at high levels. By contrast, chicken erythrocytes are terminally differentiated, nondividing cells that retain a nucleus but synthesize no DNA and very little RNA. Upon cell fusion, the previously dormant erythrocyte nuclei are dramatically reactivated. Their volume increases 20- to 30-fold, and they resume the synthesis of both DNA and RNA, as shown by culturing the heterokaryons in the presence of radiolabeled thymidine or uridine and subsequent autoradiography (see Method 3.1). The results demonstrate that an almost completely inactive nucleus can resume DNA and RNA synthesis in response to cytoplasmic factors from a cell type with an active nucleus. In addition, the data show that the activating components must have been conserved remarkably well during evolution, because chicken nuclei respond to signals from human cytoplasm, although birds and mammals diverged more than 200 million years ago.

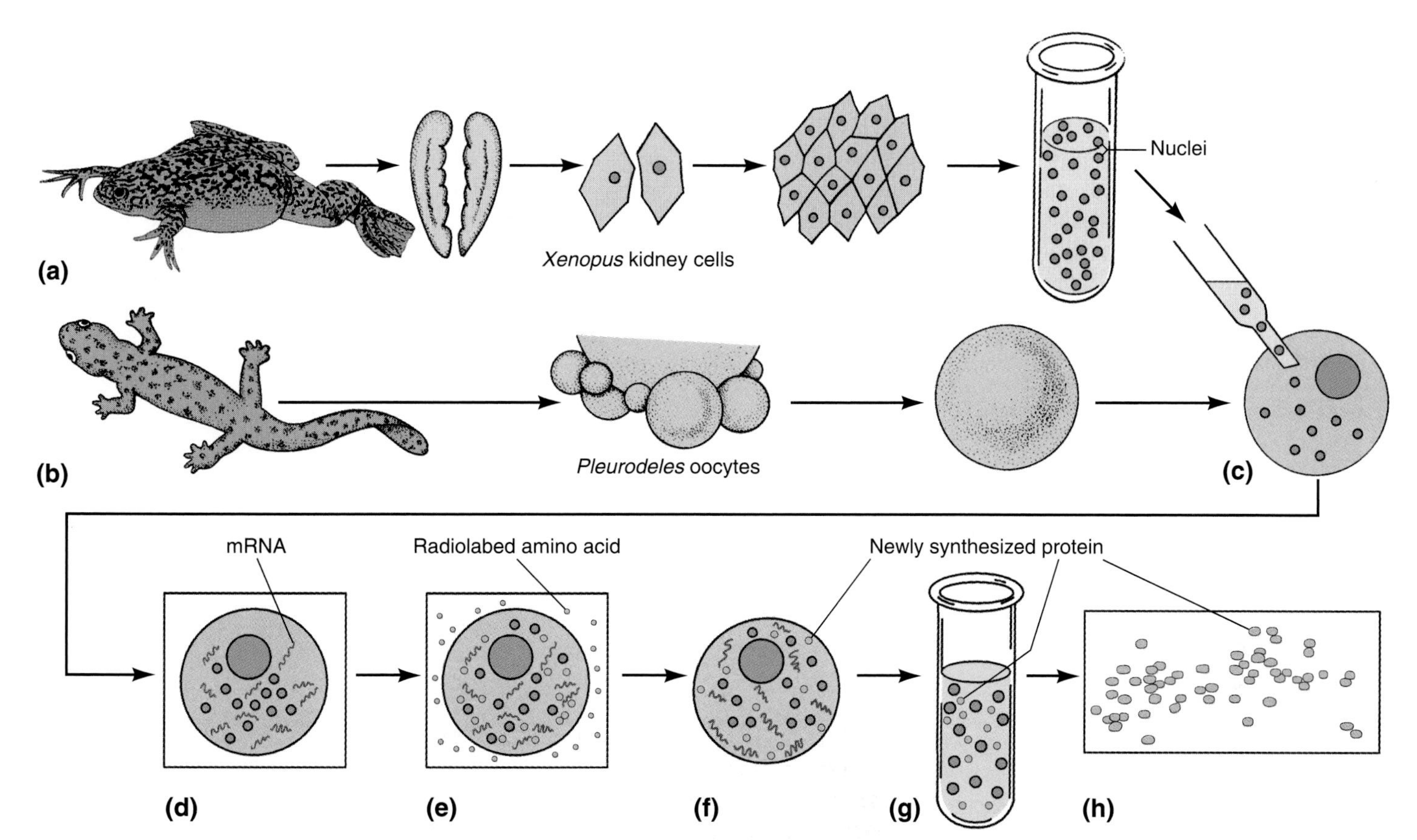

Figure 7.20 Nuclear transplantation experiment showing that oocyte cytoplasm can reprogram nuclei from kidney cells to express oocyte-specific proteins. **(a)** Kidney cells from the frog *Xenopus laevis* were used to prepare nuclei. **(b)** Oocytes were removed from the newt *Pleurodeles waltl.* **(c)** The oocytes were injected with the kidney nuclei and cultured for three days. **(d)** Messenger RNAs accumulated in the cytoplasm. **(e)** Radiolabeled amino acids were added to the medium. **(f–h)** Proteins synthesized during the following 6 h were extracted and separated by two-dimensional gel electrophoresis. Analysis of X-ray film exposed to the gel revealed proteins characteristic of *Xenopus* oocytes but none of the kidney-specific proteins.

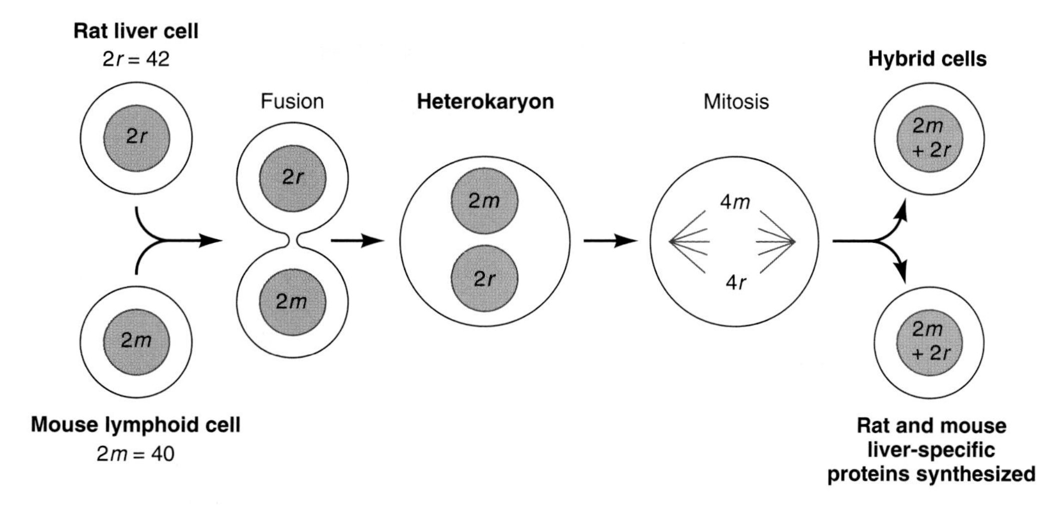

Figure 7.21 Somatic cells from different species, such as rat and mouse, can be fused by membrane-altering agents, including Sendai virus or polyethylene glycol. In the resulting heterokaryons, the nuclei of the two original cells stay separated. Simultaneous mitosis of the nuclei often generates two hybrid cells, each containing the chromosomes from both original cells. The mixed cytoplasm of heterokaryons and hybrid cells may activate previously silent genes, as shown by the synthesis of mouse liver proteins after fusion of rat liver cells with mouse lymphoid cells.

In a similar experiment, J. E. Brown and M. C. Weiss (1975) showed that nuclei from lymphoid cells were able to direct the synthesis of liver-specific enzymes after exposure to liver cell cytoplasm. By fusing rat liver tumor cells with mouse lymphoid cells, the investigators constructed hybrid cells containing rat chromosomes as well as mouse chromosomes (Fig. 7.21). The hybrid cells divided and secreted various proteins into the culture medium. Among these were liver-specific proteins such as alcohol dehydrogenase, aldolase, and tyrosine aminotransferase. Most important, the synthesis of these proteins was directed by genes from *both* species, rat and mouse. This was shown by electrophoretic mobility and other molecular characteristics that distinguish the mouse proteins from their rat counterparts. Since the mouse lymphoid cells, prior to fusion, had not expressed their genes for the liver-specific enzymes, these genes must have been activated by factors in the hybrid cell cytoplasm. These results show again that at least some of these gene-activating factors have been conserved between species during evolution.

SUMMARY

One of the basic questions in development is whether or not differentiated cells are still genetically equivalent. Transdifferentiation and sequential polymorphism indicate that some cells can differentiate, revert to a dedifferentiated state, and then differentiate in another way. Observations on chromosome number and morphology indicate that most cells inherit complete and unaltered sets of chromosomes. Cases of chromosome loss or selective gene amplification seem to be exceptions to the rule. Molecular investigations also indicate that most cells do not undergo loss, alteration, or selective amplification of their genomic DNA as part of normal differentiation.

As a biological test for genetic equivalence, single cells or nuclei have been examined for totipotency—that is, their ability to promote the development of a whole organism. Experiments with several types of differentiated plant cells indicate that they are clearly totipotent. When nuclei from differentiated animal cells were tested, some were shown to be highly pluripotent and very few possibly totipotent. The large number of cases in which nuclei from differentiated animal cells failed to support the full development of an organism may be ascribed to mutation accumulation in somatic cells, to the inability of mature cells to readjust to the fast pace of mitosis during embryonic cleavage, and possibly to an upper limit to the number of mitotic divisions that somatic cells can undergo.

Together, the available data strongly suggest that, as a rule, the cells of a given organism have complete and equivalent sets of genetic information. Although the evidence of genomic equivalence of mature animal cells is less than compelling, most biologists have adopted genetic equivalence as a well-founded theory. It implies that cells express different parts of their genetic information as they specialize for various roles in the organism. This principle of differential gene expression will be discussed more fully in subsequent chapters.

Gene activities are controlled by the cytoplasmic environment, as indicated by nuclear transplantation and cell fusion experiments. New cytoplasmic signals can silence previously active genes and activate previously silent genes. This type of control operates continuously during embryonic development, generating an ever-increasing variety of cells with different gene activities. Certain cytoplasmic components are partitioned into different blastomeres as the zygote cleaves, creating diverse cytoplasmic environments that initiate a primary pattern of embryonic cells that have distinct programs for gene expression. Later during development, interactions between blastomeres release new signals, which then determine additional groups of cells to activate new sets of genes.

Localized Cytoplasmic Determinants

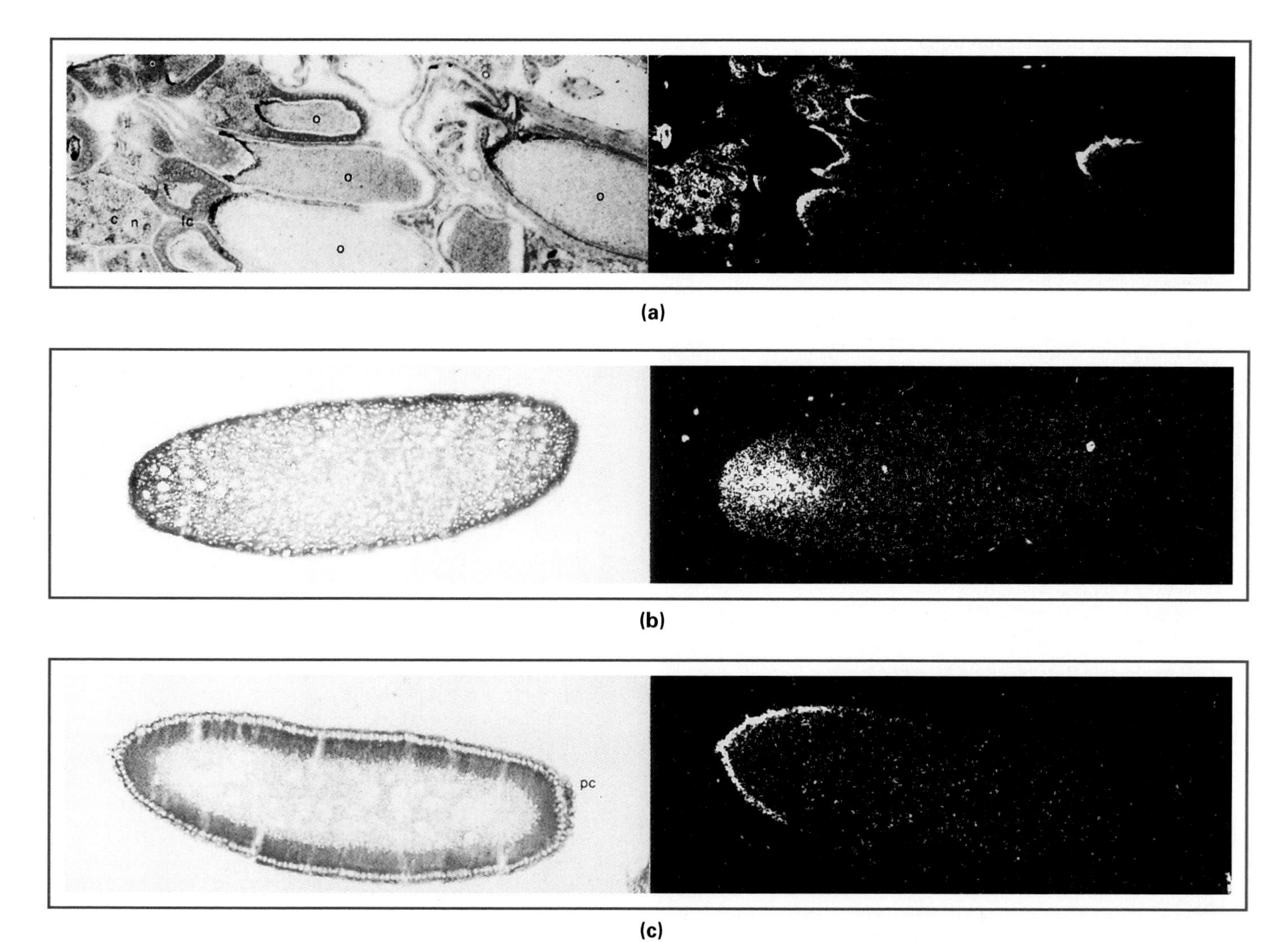

Figure 8.1 Localization of bicoid mRNA in *Drosophila* oocytes and embryos. All photographs show histological sections treated by *in situ hybridization* (see Method 8.1) with a cloned DNA probe that binds specifically to bicoid mRNA present in the sectioned tissue. Radioactive label in the probe is made visible by *autoradiography* (see Method 3.1), generating the black grains seen to the left under bright-field illumination, and the white grains seen in the same specimens to the right under dark-field illumination. All sections are oriented with the anterior pole to the left. Bicoid mRNA accumulates in the nurse cell cytoplasm and along the anterior margin of the oocytes **(a)**, in a deep anterior cone of cytoplasm during cleavage **(b)**, and in the anterior periplasm at the syncytial blastoderm stage **(c)**. c = nurse cell cytoplasm; fc = follicle cell; n = nurse cell nucleus; o = oocyte; pc = pole cells.

SINCE the beginning of the twentieth century, embryologists have proposed that certain components in the egg cytoplasm are segregated into different blastomeres, in which they determine different pathways of development (Conklin, 1905). The importance of this concept has been widely recognized, although the molecular nature of localized components remained enigmatic for a long time (E. B. Wilson, 1925; E. H. Davidson, 1986). With new genetic and experimental tools, investigators have now characterized several cytoplasmic determinants with respect to their molecular nature, localization mechanism, and other properties. These advances have rekindled interest in cytoplasmic determinants, so that today they are again a favorite topic of developmental biologists.

The concept of cytoplasmic localization is based on both experimental and genetic evidence. Experimental embryologists found that developing body patterns could be changed dramatically by removing or inactivating certain portions of egg cytoplasm, or by transplanting such cytoplasms to abnormal locations. Geneticists noticed that the phenotypes of certain maternal effect mutants show similarly dramatic regional abnormalities; for example, offspring of *bicoid* mutant *Drosophila* females lack anterior body parts (see Fig. 1.20). Closer analysis revealed that such phenotypes are due to the absence, or defective condition, of certain mRNAs that would normally be transcribed from maternal genes and localized in specific areas of egg cytoplasm (Fig. 8.1).

In this chapter, we will first discuss cytoplasmic localization as a basic principle of embryonic development. Using the germ cell determinants in insect eggs as an example, we will define rescue and heterotopic transplantation as operational criteria to prove the existence of localized cytoplasmic determinants. Additional examples of such determinants in various eggs and late embryonic cells will reveal differences in the timing and cellular mechanism of localization. Bioassays will be discussed as a means of characterizing the molecular nature, evolutionary conservation, and other properties of localized determinants. Finally, localized cytoplasmic determinants will be used to introduce the principle of default programs in development.

8.1 The Principle of Cytoplasmic Localization

The ***principle of cytoplasmic localization*** is simple and applies to most types of animals. Eggs have unevenly distributed cytoplasmic components, which are partitioned among different blastomeres as the egg cleaves (Fig. 8.2). The dissimilar cytoplasmic environments elicit distinct gene activities in the blastomeres, thus initiating different pathways of development. Localized cytoplasmic components that affect the determined state and eventual fate of the blastomeres containing them are called ***localized cytoplasmic determinants***—or *localized determinants, cytoplasmic determinants,* or *localizations,* for short. Some cytoplasmic determinants are associated with visible markers, such as pigmented yolk particles or

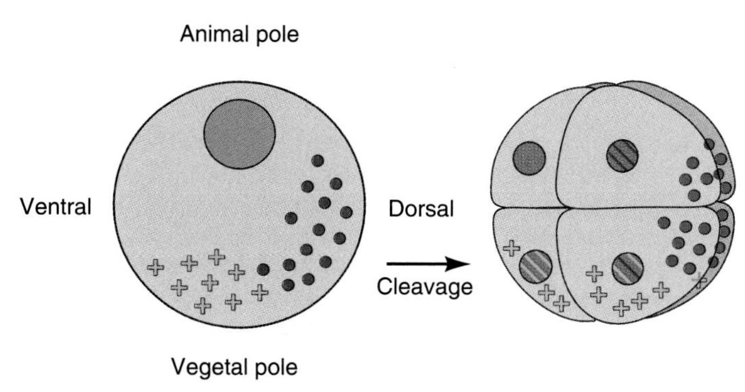

Figure 8.2 Principle of cytoplasmic localization. Most animal eggs contain unevenly distributed cytoplasmic components along their animal-vegetal axis (yellow crosses) and perhaps a second axis, such as the dorsoventral axis (red dots). These components are allocated to distinct blastomeres as the egg cleaves. In the resulting blastomeres, the segregated components generate different cytoplasmic environments, which elicit different patterns of gene activity (colored stripes). These differences in gene activity initiate different pathways of cell determination.

other organelles, that directly reveal their localization. Other cytoplasmic determinants are devoid of such convenient markers. Conversely, some visible cytoplasmic components are localized but have no effect on cell determination. For instance, the dark pigment in the cortex of amphibian eggs is limited to the animal hemisphere, but albino mutants lacking this localized pigment develop otherwise normally. To prove that a cytoplasmic component is a localized determinant, one has to demonstrate that the material of interest is unevenly distributed *and* that it affects the fate of the cells containing it.

Experimental evidence for localized cytoplasmic determinants started with the simple observation that, in most eggs, yolk and other components are distributed unevenly along the animal-vegetal axis. When an egg is cut along this axis and each half fertilized separately, both halves often develop into complete larvae (Fig. 8.3). When the same operation is carried out except with the egg cut perpendicular to the animal-vegetal axis, both halves develop abnormally. These results show that animal as well as vegetal cytoplasm contains localized components and that both sets of components are necessary for normal development.

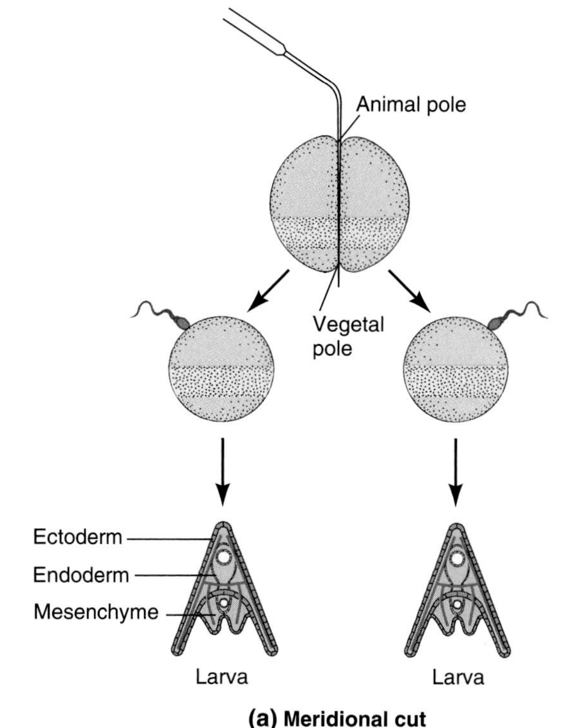

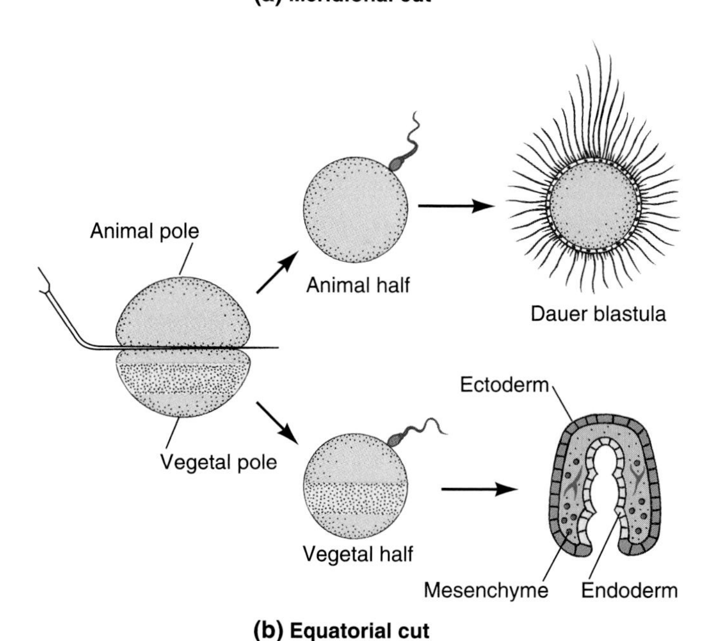

Figure 8.3 Development of sea urchin eggs cut in halves and fertilized. **(a)** After a meridional cut (parallel to the animal-vegetal axis), each half develops into a half-sized but normally proportioned larva. **(b)** After an equatorial cut (perpendicular to the animal-vegetal axis), the animal half develops into a dauer blastula (a ciliated blastula that does not gastrulate). The vegetal half forms a malformed embryo.

8.2 Polar Lobe Formation as a Means of Cytoplasmic Localization

Dramatic cases of cytoplasmic localization are observed in many animals with spiral cleavage, including several mollusc and annelid species. The first embryonic cleavages involve the cyclic appearance, as the vegetal pole, of a cytoplasmic protrusion known as the ***polar lobe*** (Fig. 8.4). The lobe formed during the first mitotic cleavage is referred to as the *first polar lobe,* although in some species, earlier polar lobes are observed during meiotic divisions. As the first two blastomeres become separated by cytokinesis, the polar lobe constricts until it is connected by only a thin cytoplasmic stalk to the rest of

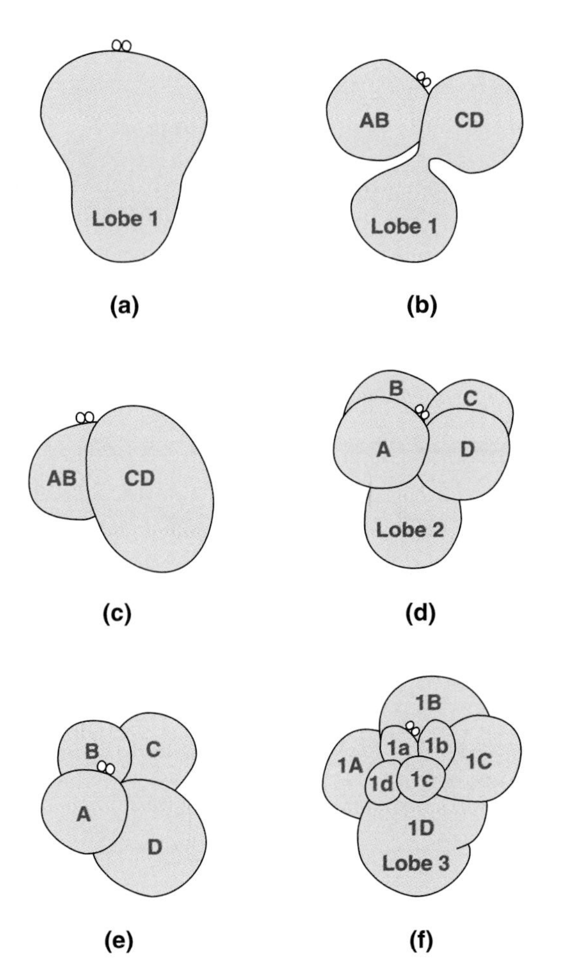

Figure 8.4 Polar lobe formation during early cleavages in the snail *Ilyanassa obsoleta.* **(a–c)** The first lobe forms at the vegetal pole of the zygote before the first cleavage and is thereafter resorbed into one blastomere designated CD. **(d, e)** The second lobe forms at the vegetal pole of the CD blastomere before the second cleavage and is thereafter resorbed into one blastomere designated D. **(f)** The third polar lobe is formed incompletely and resorbed before D divides.

the embryo (Fig. 8.5). This stage, when the polar lobe looks like a third blastomere, is called the ***trefoil stage.*** After completion of the first cytokinesis, the polar lobe is resorbed into just one of the two blastomeres, which is designated the CD blastomere. Before the second cleavage, the CD blastomere forms another polar lobe, which is selectively funneled back into one of its daughter cells, the D blastomere. As a result, the D blastomere becomes larger than blastomeres A, B, and C.

As cleavage continues, the A, B, C, and D blastomeres give off successive quartets of small blastomeres called *micromeres* (see Figs. I.1, 5.6 and 8.4f). The first quartet originates when blastomere A divides into macromere 1A and micromere 1a, blastomere B divides into macromere 1B and micromere 1b, and so forth. Similarly, the second quartet is given off when blastomere 1A divides into macromere 2A and micromere 2a, blastomere 1B divides into macromere 2B and micromere 2b, and so on. Of particular interest within the fourth quartet is the 4d micromere (see Fig. 5.29). Proper formation of these cells is critical to the formation of a normal larva.

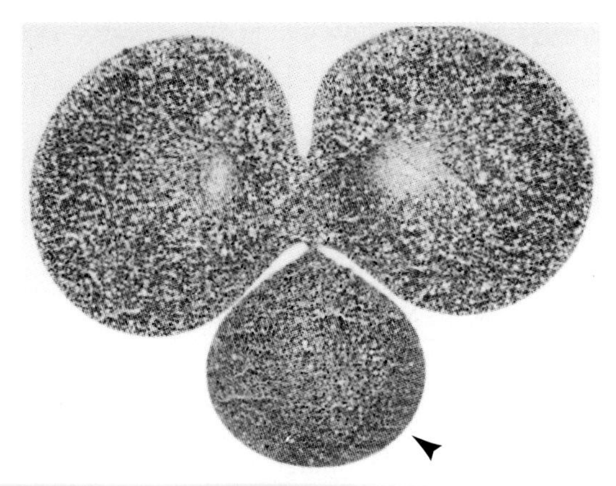

Figure 8.5 Photomicrograph of a mollusk embryo (*Dentalium*) at the trefoil stage. The arrowhead points to the polar lobe, which is connected by a thin cytoplasmic bridge to the dividing zygote.

To explore the significance of polar lobe formation, investigators removed the lobe at various cleavage stages. The most detailed investigations were carried out with embryos of the scaphopod *Dentalium* (E. B. Wilson, 1904) and the mud snail *Ilyanassa obsoleta* (Clement, 1962). Cutting off the polar lobe at the trefoil stage had dramatic consequences for subsequent development. The larger size and special cleavage schedule of the D blastomere were no longer observed, no mesodermal founder cells were formed, and the operated embryos cleaved in a radially symmetrical way. The developing larvae were severely disorganized and almost devoid of inner structures (Fig. 8.6). The same type of abnormal development was observed in larvae raised from AB blastomere pairs and other combinations lacking the D blastomere. In contrast, CD blastomere pairs and other combinations containing D developed rather normally. Similar results were obtained with other molluscs (Verdonk and Cather, 1983).

The general conclusion from these experiments is that the polar lobe contains cytoplasmic components, most of which are funneled selectively into the D blastomere and are, directly or indirectly, required for forming the missing structures. The mechanism of setting aside the polar lobe during the first two cleavages ensures its selective allocation to the D blastomere and its descendants.

8.3 Germ Cell Determinants in Insect Eggs

In many insect embryos, the primordial germ cells form at the posterior pole and are known as *pole cells.* They are easy to distinguish from somatic cells because they form

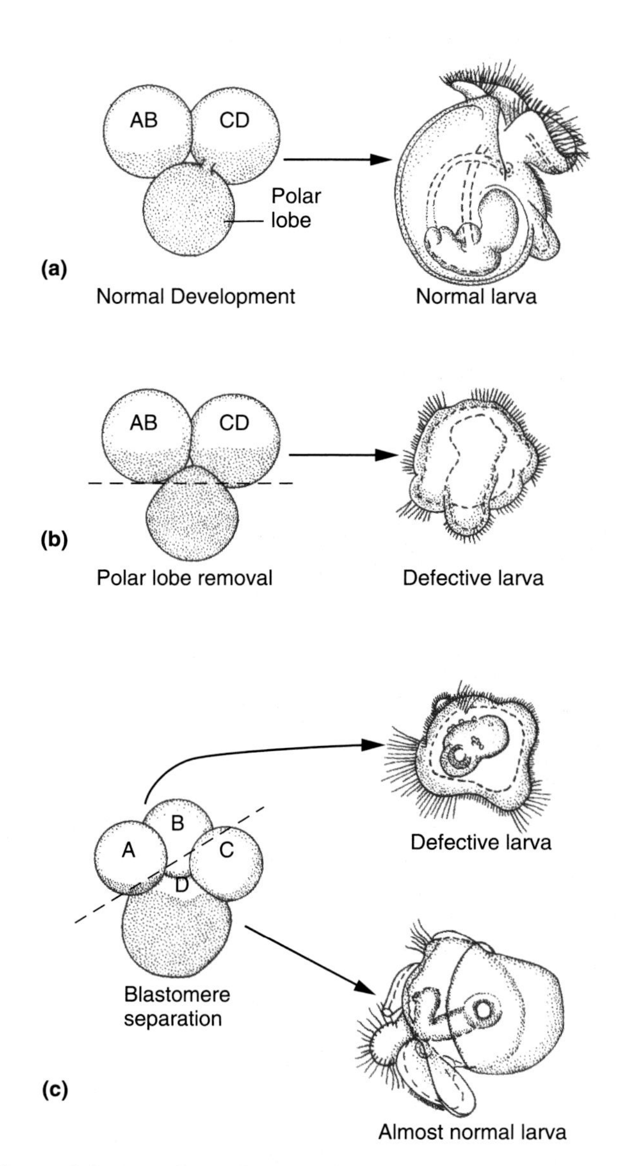

Figure 8.6 The effect of polar lobe removal on the development of *Ilyanassa.* **(a)** Development of a normal larva. **(b)** A highly defective larva is formed after polar lobe removal at the trefoil stage. **(c)** Blastomere separation at the 4-cell stage. The CD pair gives rise to an almost normal larva, while the AB pair forms a larva with internal structures severely disorganized or missing.

earlier than the *blastoderm* cells, are larger, and are round rather than columnar in shape (Fig. 8.7). The cytoplasm that is enclosed in pole cells is free of yolk and contains fibrous granules, called ***polar granules,*** that stain for RNA. In transmission electron micrographs, polar granules appear dark and are sometimes associated with polysomes—that is, aggregates of mRNA and ribosomes that synthesize protein (Fig. 8.7d). Similar granules are associated with germ cell determinants in the eggs of frogs and nematodes (see Section 25.2).

RESCUE EXPERIMENTS CAN RESTORE DEFECTIVE EMBRYOS TO NORMAL DEVELOPMENT

To explore the function of polar granules, investigators have removed or damaged them mechanically, by heat, or by irradiation. When the posterior pole of a *Drosophila* embryo is irradiated with ultraviolet (UV) light, no pole cells form, and the developing adults have gonads without gametes. This demonstrates that some components of the posterior cytoplasm are necessary for pole cell formation. The ability of UV-irradiated embryos to form pole cells and normal gonads can be restored, at least partially, by injecting them with posterior pole plasm from unirradiated donors (Fig. 8.8; Okada et al., 1975). This experiment shows that the rescuing components, those necessary for pole cell formation, are transplantable, but it does not prove that these components are localized. They could have been nonlocalized organelles, such as ribosomes or mitochondria, that are necessary for the survival of any cell.

In order to prove that the rescuing components are localized, Okada and coworkers performed the following control. They took cytoplasm from the anterior pole and injected it in the same fashion into the posterior pole region of UV-irradiated embryos. No rescue was observed: The recipients were unable to form pole cells. Together, the experiment and the control show that components required for pole cell formation are present in the posterior but not in the anterior pole of *Drosophila* eggs. Because these components are localized and change the fate of cells from somatic to pole cells, they act as pole cell determinants. Since the fate of pole cells is to form gametes, these determinants are usually called ***germ cell determinants.***

Experiments like the one described above, in which a defective embryo is restored to normal development, are known as ***rescue experiments.*** Instead of using embryos first rendered defective by experimental treatments like UV irradiation, the rescue strategy can also be used in conjunction with maternal effect mutants. In these studies, recipient eggs defective in localized cytoplasmic determinants are derived from mutant females. For example, eggs from females that are mutant for the *bicoid* gene do not accumulate bicoid mRNA near the anterior pole like eggs from wild-type females (see Fig. 8.1). Such eggs develop into larvae in which head and thorax are missing (see Fig. 1.20). However, these embryos can be restored to normal development by injecting anterior cytoplasm from normal embryos or bicoid mRNA synthesized in vitro.

To appreciate the logical structure of rescue experiments, it is important to realize that the *rescue proper* is already the second step in the experiment. The first step, the *removal* of a cytoplasmic component from the embryo, is designed to test whether the removed component is *necessary* for a certain event to occur. In the case of germ cell determination, posterior pole plasm

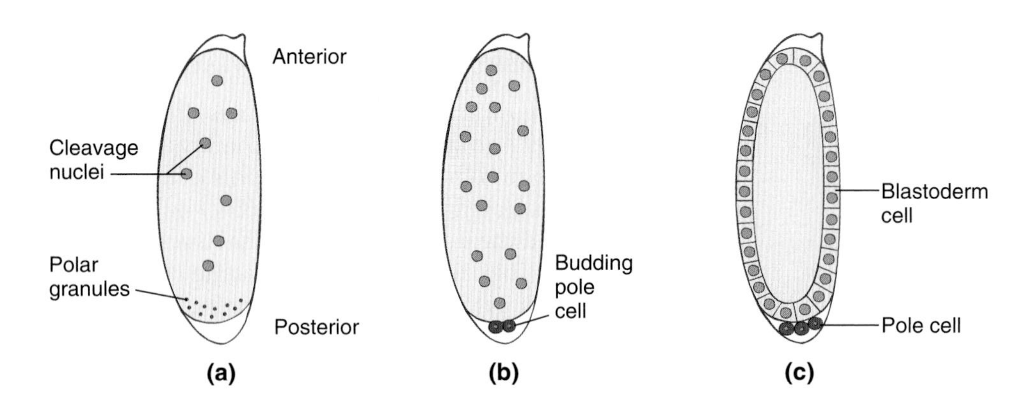

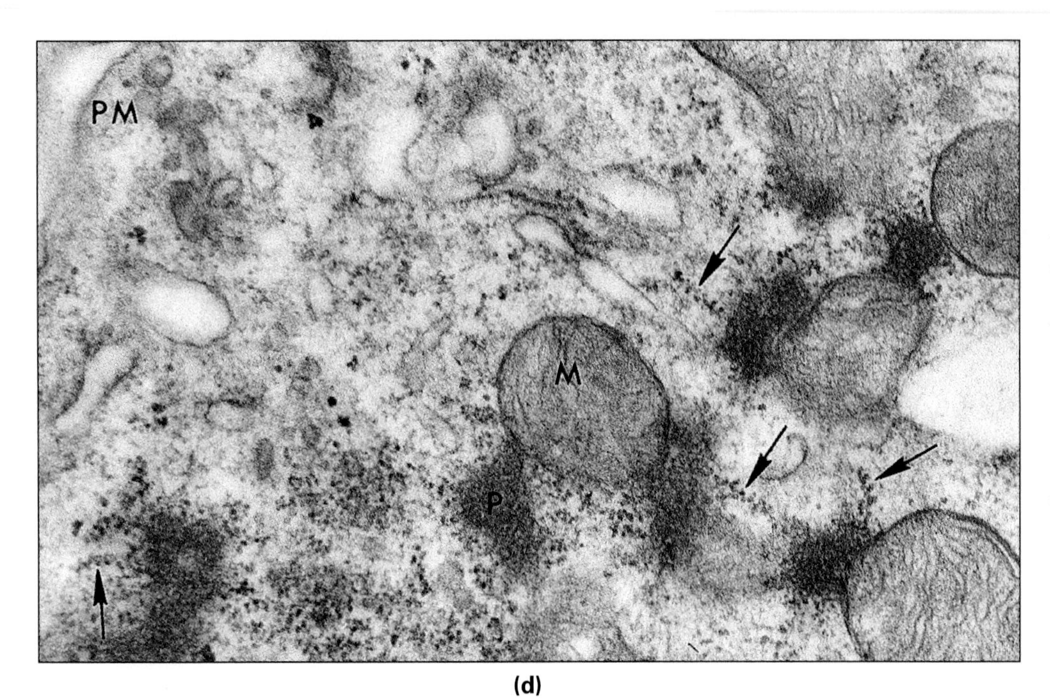

Figure 8.7 Polar granules and pole cell formation in the *Drosophila* embryo. **(a)** Schematic outline of an embryo during early cleavage, indicating the position of polar granules near the posterior pole. **(b)** As the pole cells form, they include the cytoplasm containing polar granules. **(c)** Embryo at the cellular blastoderm stage. Pole cells differ in size and shape from blastoderm cells. **(d)** Transmission electron micrograph showing polar plasm of a *Drosophila* embryo 15 min. after fertilization. The polar granules (P) are attached to mitochondria (M) and are associated with helical structures resembling polysomes (arrows). PM = egg plasma membrane.

was shown to be necessary. The rescue experiment is used as an extension to show that the rescuing activity is *localized*. In addition, the rescue step serves as a *control* demonstrating that the defect observed after the first step was caused by the absence of the removed component and not by some unintended side effect of the removal operation.

HETEROTOPIC TRANSPLANTATION TESTS CYTOPLASMIC DETERMINANTS FOR ACTIVITY IN ABNORMAL LOCATIONS

To test whether a component is *sufficient* for a certain event to occur, one has to add the component at a time or place where it is not normally present. In the case of germ cell determinants, this was accomplished by *heterotopic transplantation*—that is, transplantation to a place these determinants do not normally occur.

USING this strategy, Illmensee and Mahowald (1974) showed that posterior pole plasm from *Drosophila* eggs caused *ectopic* pole cell formation, meaning the formation of pole cells that were out of place. The first step in their experiment was the microinjection of posterior pole plasm from a donor embryo during early cleavage to the anterior pole of a primary host at the same stage of development (Fig. 8.9). As a result, the host formed cells at the anterior pole that resembled pole cells in their size, shape, and ultrastructure. Specifically, the modified anterior cells incorporated the polar granules transferred with the posterior pole plasm and formed electron-dense nuclear bodies that are characteristic of pole cells.

In order to test whether these ectopic pole cells would be able to function as primordial germ cells, the cells would need to find their way into a developing gonad. However, while posterior pole cells are swept to the vicinity of the gonad rudiment by morphogenetic movements associated with hindgut formation, no such helpful movements occur anteriorly. Therefore, the researchers transplanted the ectopic pole cells from the anterior pole of the primary host to the posterior pole of a secondary host embryo. The secondary host's gonads then became populated by both its

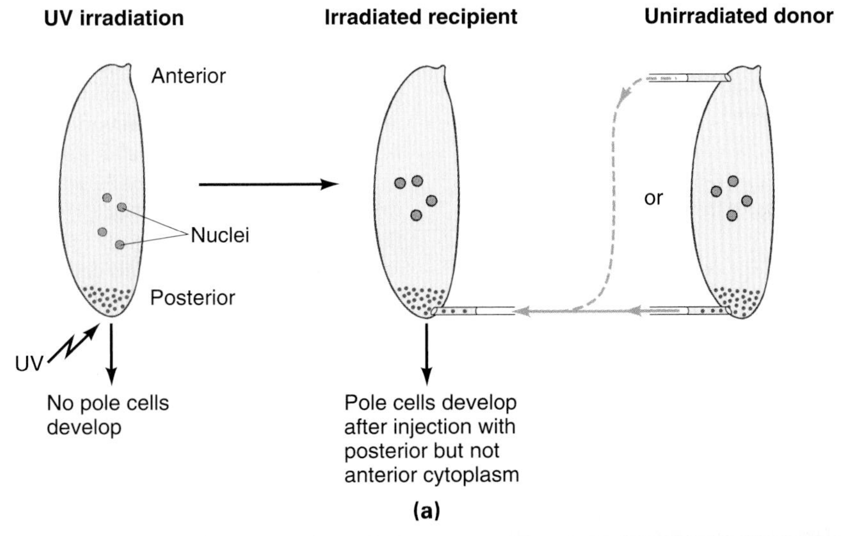

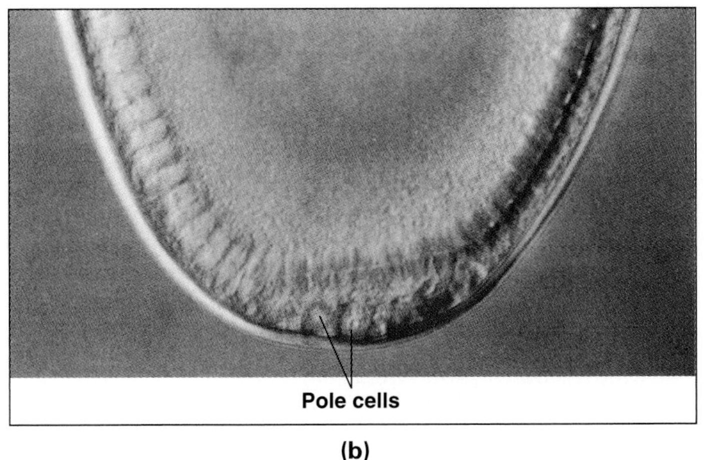

(b)

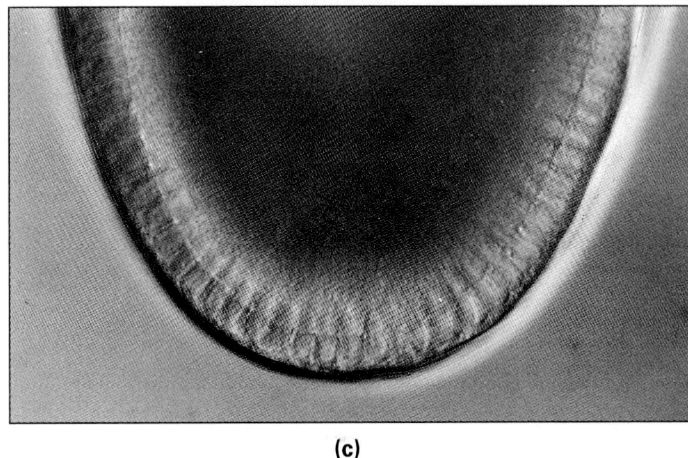

(c)

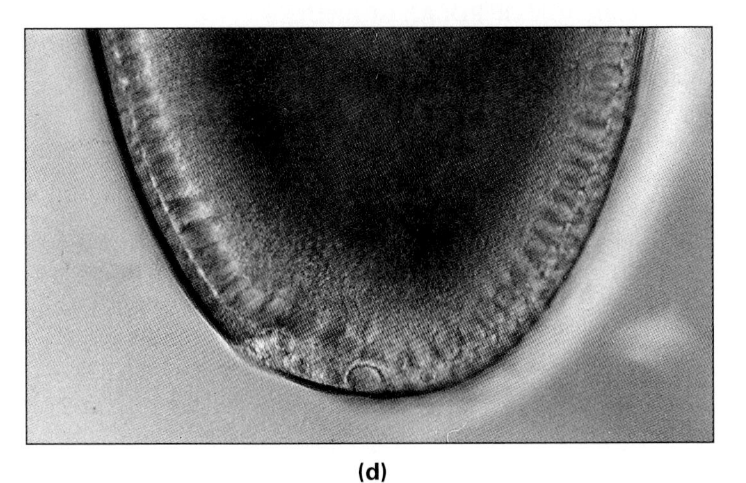

(d)

Figure 8.8 Rescue of *Drosophila* embryos sterilized by UV irradiation by subsequent injection with cytoplasm from an unirradiated donor. **(a)** Experimental technique. The cytoplasm was transferred with a fine, beveled glass needle. Only posterior cytoplasm restored the ability of the irradiated recipient to form pole cells. Light micrographs showing **(b)** normal embryo with pole cells, **(c)** UV-irradiated embryo without pole cells, and **(d)** rescued embryo with pole cell.

own pole cells and the transferred ectopic pole cells from the primary host. To distinguish between the gametes produced by the two different sets of pole cells, the investigators used recessive genetic markers. (Such markers are phenotypically expressed only if all copies of the gene in a cell are mutated.) The transplanted cells carried the nuclei of the primary host, which was homozygous for *multiple wing hair* (*mwh*) and *ebony* (*e*), whereas the secondary host was homozygous for, *yellow* (*y*), *white* (*w*), and *singed* (sn^3). When adult flies developing from secondary host embryos were mated with partners of the same genotype, most of their offspring showed the *y w* sn^3 phenotype. However, 4 of 92 matings also produced offspring showing the wild phenotype, which therefore had to be heterozygous for each of the recessive genetic markers used. Such offspring could arise only with the participation of gametes derived

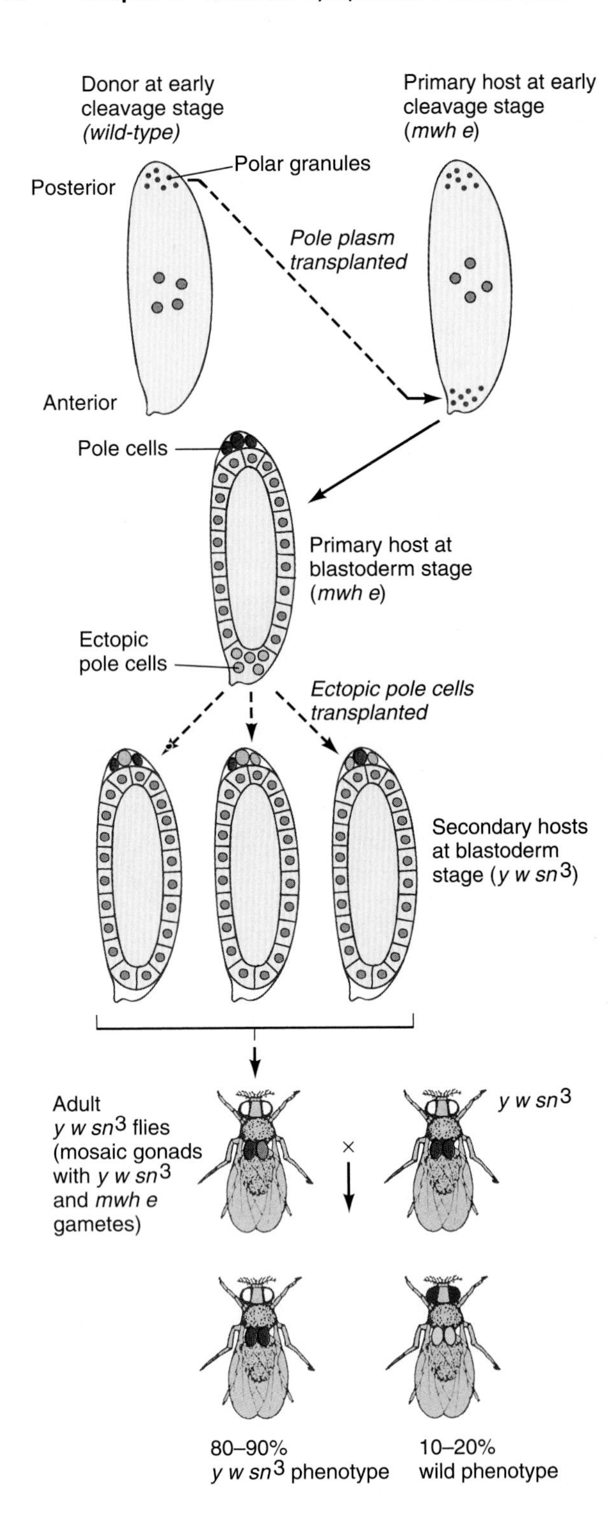

Figure 8.9 Heterotopic transplantation experiment showing the localization of germ cell determinants at the posterior pole of *Drosophila* eggs. Posterior pole plasm from a wild-type donor was transplanted to the anterior pole of a primary host at the same early cleavage stage. As a result, cells with the appearance of pole cells (ectopic pole cells) developed at the injection site. These cells were transplanted to the posterior pole of a secondary host, from which they could reach the secondary host's gonad. To distinguish any gametes formed by the transplanted cells from those formed by the secondary host's own pole cells, the transplanted cells were marked by the mutations *multiple wing hair* (*mwh*) and *ebony* (*e*), whereas the secondary host carried a different set of mutant alleles, *yellow* (*y*), *white* (*w*), and *singed* (sn^3). Secondary hosts grown to adulthood and outcrossed with *y w* sn^3 partners produced some phenotypically wild offspring, indicating that some of their gametes were derived from the transplanted pole cells.

from the transplanted pole cells. These results therefore showed that the posterior pole plasm transferred from the donor had caused the ectopic formation of functional primordial germ cells in the first host.

Questions

1. Which control experiments need to be done in order to test whether the ectopic pole cells are formed in response to a component localized in posterior cytoplasm or to the act of injecting *any* egg cytoplasm?
2. Which control experiments need to be done in order to test whether the ability to form additional gametes is restricted to ectopic pole cells, or whether this ability is acquired by any transplanted embryonic cell?

The experiments described above prove that a cytoplasmic component localized to the posterior pole of *Drosophila* eggs is not only necessary but also sufficient to cause the formation of primordial germ cells.

LOCALIZATION OF POLAR GRANULES REQUIRES RNA TRANSPORT ALONG MICROTUBULES

Genetic analysis of germ cell determination in *Drosophila* has revealed the involvement of several maternal effect genes in the process. The products of most of these genes are required to assemble the *polar granules,* those RNA-containing organelles that originally drew investigators' attention to the posterior pole plasm of insects. These granules act as repositories not only for germ cell determinants as discussed here, but also for other localized cytoplasmic determinants that are required for the formation of abdominal segments (see Section 22.3).

The first polar granule component to become localized to the posterior pole of *Drosophila* eggs is mRNA transcribed from the maternal $oskar^+$ gene (Micklem, 1995; St Johnston, 1995). Oskar mRNA is both necessary and sufficient for polar granule formation. Embryos derived from $oskar^-$ mothers lack polar granules, fail to form pole cells, and give rise to embryos with missing abdominal segments. However, such embryos can be restored to fertility and a normal body pattern by injection of posterior cytoplasm from donor eggs derived from $oskar^+$ mothers (Lehmann and Nüsslein-Volhard, 1986). Conversely, if eggs, in addition to their normal oskar

method 8.1

In situ Hybridization

A key method in the genetic and molecular analysis of development is known as ***in situ hybridization*** (Lat. *in situ,* "in its natural position"). It is based on the fact that single-stranded DNA and RNA molecules form stable hybrids if their nucleotide sequences are complementary. The method is used most frequently to determine the location of a specific mRNA in wholemounts or sections of fixed tissue. For this purpose, one synthesizes a complementary strand of labeled DNA or RNA (see Fig. 15.11). This labeled probe is incubated with the fixed specimen to allow hybrid formation to occur. Unspecifically adhering probe is then washed off under conditions that preserve nucleic acid hybrids.

Commonly used labels are radioisotopes, which are detected by *autoradiography* (see Method 3.1). In this case, the hybridization and washing steps are followed by the application of a layer of photographic film emulsion (in the dark). The dried film layer is exposed to the radioactive signal for an appropriate period of time before it is developed and fixed like a photographic film. When viewed under the microscope, the silver traces generated by the radioactivity appear as dark grains in transmitted light and as white grains in reflected light (see Figs. 8.10 and 8.26).

More recently, most investigators have switched to chemically labeled DNA probes for in situ hybridization. In a modified nucleotide commonly used for this purpose, the methyl group of thymine is replaced with the plant steroid digoxygenin, which does not interfere with DNA synthesis or base pairing. The digoxygenin in turn is recognized by an antibody that carries a visible label. Thus, in situ hybridization with a chemically labeled probe works like *immunostaining* (see Method 4.1), except that the primary antibody is replaced with the chemically labeled hybridization probe. This type of probe has improved resolution, avoids the time-consuming exposure of photographic emulsion, and does not require the safety precautions associated with radioisotopes.

mRNA, are provided with an engineered oskar mRNA that mislocalizes to the anterior pole, then these eggs form polar granules and functional pole cells at both the anterior and the posterior pole. Embryos developing from such eggs show an abnormal body pattern in which head and thorax are replaced with a mirror-image duplication of abdominal segments (see Fig. 22.3).

The localization of oskar mRNA has been explored by *in situ hybridization,* a method that reveals the location of specific RNA or DNA sequences in histological sections and wholemounts of embryos (see Method 8.1). Transcribed in the nurse cells of each ovarian egg chamber, oskar mRNA is rapidly transferred via cytoplasmic bridges to the adjacent oocyte, where it forms a transient anterior accumulation. Subsequently, oskar mRNA moves all the way from the anterior to the posterior pole, where it forms a sharp localization (Fig. 8.10). This astounding movement depends on an array of polarized microtubules: *colchicine,* a drug that interferes with microtubule assembly, blocks oskar mRNA movement, and mutations that change microtubule polarity in oocytes cause characteristic mislocalizations of oskar mRNA (Theurkauf et al., 1993; Theurkauf, 1994; see Section 22.3). The normal movement of oskar mRNA also depends on the activity of additional genes including *staufen*$^+$. Staufen protein binds to oskar mRNA, apparently linking it to a kinesin *motor protein* moving to the plus end of microtubules (Clark et al., 1994). Once arrived at the posterior egg pole, oskar mRNA is translated into protein, which interacts with staufen protein and yet another protein encoded by the *vasa*$^+$ gene (Breitwieser et al., 1996; Rongo et al., 1995). These interactions apparently anchor oskar mRNA to cytoskeletal elements at the posterior pole and form a ribonucleoprotein complex that traps additional components to form mature polar granules. Trapping of the additional components seems to be facilitated by cytoplasmic streaming later during oogenesis (Gutzeit, 1986).

The molecular nature of the germ cell determinants that become sequestered in polar granules, and the

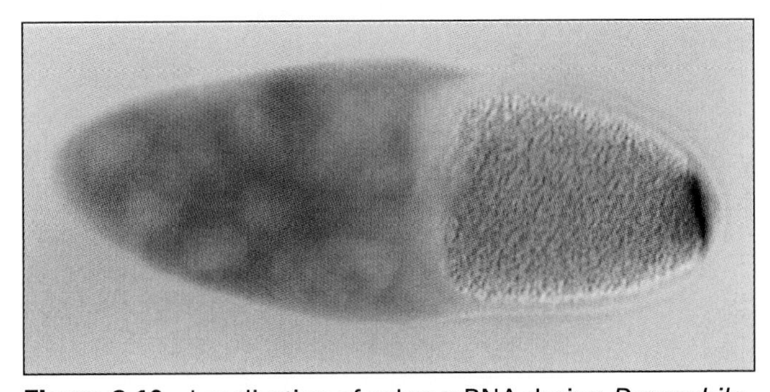

Figure 8.10 Localization of oskar mRNA during *Drosophila* oogenesis. Along with other maternal mRNAs, oskar mRNA is transcribed in the nurse cells of each ovarian egg chamber (to the left), from where it is rapidly transferred via cytoplasmic bridges to the adjacent oocyte. The part of the oocyte that is adjacent to the nurse cells will become the anterior pole region of the egg. From here, oskar mRNA moves all the way to the posterior pole, where it forms a sharp localization (dark band, made visible by in situ hybridization; see Method 8.1). This astounding movement depends on several other gene products as well as an array of polarized microtubules. Once arrived at the posterior egg pole, oskar mRNA is translated into protein, which interacts with other proteins in forming so-called polar granules, which are anchored to cytoskeletal elements and serve as repositories for germ cell and general posterior cytoplasmic determinants.

mechanisms by which these molecules promote the characteristic morphology of pole cells and their ability to form gametes, are still under investigation. One polar granule component, the mRNA transcribed from the *germ cell–less*$^+$ gene, is necessary for the development of pole cells but not of somatic abdominal structures; it encodes a pole cell–specific protein associated with nuclear pores (Jongens et al., 1994; S. E. Robertson et al., 1999). Another polar granule component is the large ribosomal RNA of mitochondria, which can rescue pole cell formation in UV-irradiated embryos (Kobayashi and Okada, 1989; Kobayashi et al., 1993). The molecular mechanism of this rescue is not understood.

8.4 Bicoid mRNA in *Drosophila* Eggs

Ligation and transplantation experiments with various insect eggs have been interpreted in terms of a model postulating cytoplasmic determinants localized near the anterior and posterior egg poles (Sander 1975, 1976). These determinants are thought to specify, in an overall way, the orderly arrangement of anterior and posterior body segments. The *global* effects of these determinants on the overall body pattern differ markedly from the *local* effects of germ cell determinants on one specific cell lineage, as discussed in the previous section. Genetic analysis of development in *Drosophila* has confirmed the existence of anterior and posterior determinants and has allowed investigators to characterize them with great precision (see Section 22.3). We will focus here on one anterior determinant, the mRNA transcribed from the *bicoid*$^+$ gene.

Embryos derived from females lacking bicoid function ("bicoid embryos") show a dramatically abnormal body pattern: Their head and thorax are missing, and their expanded abdomen carries at its anterior end a mirror-image duplication of the posteriormost abdominal element, the *telson* (see Fig. 1.20). Similar embryos can be produced by pricking wild-type *Drosophila* eggs with a fine needle, so that a small amount of anterior cytoplasm leaks out before the wound seals. Conversely, bicoid embryos can be partially rescued by injecting them anteriorly with anterior cytoplasm from wild-type donors (Frohnhöfer and Nüsslein-Volhard, 1986) Moreover, if anterior cytoplasm is injected laterally into bicoid embryos then the recipients will form head structures flanked by thoracic structures near the site of injection (Fig. 8.11). The simplest interpretation of these data is

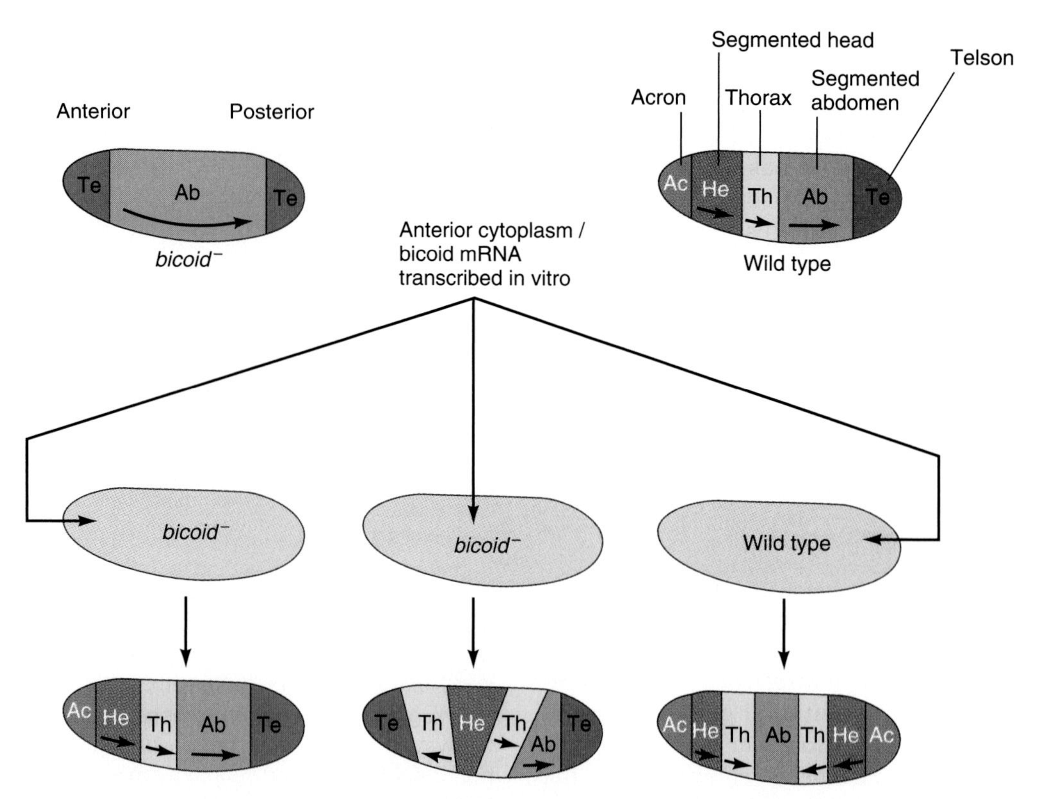

Figure 8.11 Rescue and heterotopic transplantation reveal bicoid mRNA as an anterior determinant in *Drosophila* eggs. Anterior pole to the left, dorsal side up. Top row: Fate maps of embryos derived from *bicoid*$^-$ mutant females and wild-type females. Middle: Outline of experiments. Anterior cytoplasm from donor eggs derived from wild-type females, or bicoid mRNA transcribed in vitro, was microinjected as indicated by arrows. Bottom row: Fate maps of embryos developing from recipient eggs. Anterior injection into embryos derived from *bicoid*$^-$ mutant females restored the normal body pattern. Lateral or posterior injection caused the formation of ectopic head structures flanked by thoracic structures. Arrows inside body regions indicate anteroposterior polarity.

that the *bicoid*$^+$ gene encodes a maternally synthesized product that acts as a localized anterior determinant.

When the *bicoid*$^+$ gene was cloned it became possible to synthesize labeled probes that would hybridize in situ to bicoid mRNA (see Method 8.1). These probes revealed a striking localization of bicoid mRNA near the anterior pole of the egg (see Fig. 8.1). Moreover, the rescue and heterotopic transplantation experiments described above yielded the same results when bicoid mRNA was injected instead of anterior cytoplasm from wild-type donors (Driever at al., 1990). Thus, based on the criteria of rescue and heterotopic transplantation, bicoid mRNA acts as a localized anterior determinant in *Drosophila* eggs.

How does bicoid mRNA become localized to the anterior pole? Like polar granule components, bicoid mRNA is synthesized in the nurse cells of each egg chamber and transferred to the oocyte. This movement depends in part on a system of microtubules that extends from the nurse cells through the cytoplasmic bridges into the oocyte. In addition, nurse cell contractions toward the end of oogenesis squeeze the entire nurse cell cytoplasm into the oocyte. As a result, bicoid mRNA first accumulates in an anterior cap of cortical cytoplasm and then becomes localized to a deep cone of cytoplasm behind the anterior pole.

The localization process depends on ***cis-acting*** elements (Lat. *cis*, "on this side," meaning parts of the molecule of interest itself) and ***trans-acting*** elements (Lat. *trans*, "on the other side," meaning other molecules or parts thereof). If as an analogy to localizing a cytoplasmic determinant one takes delivering a letter, then the zip code and street address on the envelope would be cis-acting elements, while the mail sorting system and the delivery person would be the trans-acting elements.

Cis-acting elements have been mapped to the bicoid mRNA's *trailer region*, also known as the *3′ untranslated region* (*3′UTR*). This is a standard segment of mRNA that is not translated into protein but has important functions in controlling the mRNA's longevity, translatability, and localization (see Chapter 17). If certain segments are deleted from the bicoid 3′UTR, then the remaining mRNA is not localized and diffuses all over the oocyte (Macdonald et al., 1993; Ferrandon et al., 1994; Macdonald and Kerr, 1998). Conversely, if these segments from bicoid mRNA are added to another mRNA, such as oskar mRNA, then this composite mRNA becomes localized just like bicoid mRNA (see Section 22.3). Thus, the 3′UTR segments identified in these experiments are necessary and sufficient for the localization of bicoid mRNA.

Trans-acting elements for bicoid mRNA localization include microtubules and some maternally synthesized proteins. If ovaries are cultured in the presence of microtubule inhibitors, such as colchicine, bicoid mRNA does not become localized. Also, in eggs derived from females deficient in genes including *exuperantia*$^+$, *swallow*$^+$, and *staufen*$^+$, bicoid diffuses posteriorly, and embryos developing from such eggs show anterior defects similar to bicoid embryos (see Section 22.3). It appears that bicoid mRNAs can form dimers or multimers via their 3′UTRs, which interact with staufen protein to form ribonucleoprotein particles for transport and localization (Ferrandon et al., 1997). In addition, bicoid mRNA localization seems to depend on other cytoskeletal structures for anchorage (Ding et al., 1993).

8.5 Myoplasm in Ascidian Eggs

Ascidian embryos have been favorite objects of developmental biologists since the beginning of the twentieth century (Satoh, 1994; Satoh and Jeffery, 1995; Jeffery and Swalla, 1997). Commonly known as sea squirts, ascidians live as filter feeders attached to rocks and shells in shallow seawater. Because their embryos develop rapidly, consist of relatively few cells, and show *invariant* cleavage, it is relatively easy to identify specific blastomeres and to trace cell lineages. Another interesting feature of ascidians is that phylogenetically they are closely related to vertebrates. This is not obvious at all from adult ascidians, but is readily apparent from their tadpole larvae, which have a *notochord*, a *dorsal nerve chord*, an *endodermal strand*, and *muscle cells*, all in the same relative positions as the *notochord*, *neural tube*, *archenteron*, and *somites* of a vertebrate embryo (compare Figs. 1.8 and 8.12).

Perhaps the greatest attraction of ascidians are various pigmented inclusions in their egg cytoplasm. Early investigators have described as many as five differently colored types of cytoplasm in ascidian eggs. Most important, Conklin (1905) reported that differently pigmented cytoplasms are segregated into different embryonic cells, which then proceed to form distinct tissues or organs. This observation suggested that the colored pigments

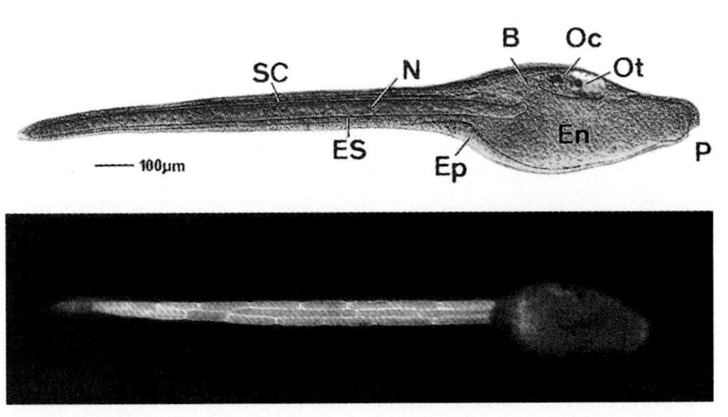

Figure 8.12 Photographs of ascidian tadpoles (*Halocynthia roretzi*). **(a)** Transmitted light with differential-interference-contrast showing all major tissues of the larva. **(b)** Fluorescent stain used to highlight tail muscle cells. B = brain; En = endoderm; Ep = epidermis; ES = endodermal strand; N = notochord; Oc = ocellus; Ot = otolith; P = palps; SC = spinal cord.

might be associated with cytoplasmic determinants. Aside from the function of these determinants, the process of their localization has become of interest in itself because it occurs in a dramatic set of movements.

CYTOPLASMIC COMPONENTS OF ASCIDIAN EGGS ARE SEGREGATED UPON FERTILIZATION

Adult ascidians shed their eggs as primary oocytes still arrested in meiotic prophase I. These oocytes are radially symmetrical relative to the animal-vegetal axis, and several organelles in the egg cortex are distributed with an animal-vegetal polarity (Sardet et al., 1992). Oocyte maturation is triggered by sperm entry, which may occur anywhere on the egg surface. In newly fertilized eggs of *Styela partita,* a layer of cytoplasm containing yellow pigment granules lies just beneath the plasma membrane (Fig. 8.13a). This yellow cytoplasm is called ***myoplasm*** (Gk. *mys,* "muscle"), because it is segregated into cells that later form most of the tail muscle. A different mass of clear cytoplasm, derived from the contents of the germinal vesicle, is located in the animal half of the egg. This material is called ***ectoplasm,*** because the cells inheriting it will form mostly ectodermal structures. The remainder of the egg is filled with gray, yolk-rich cytoplasm that will be enclosed by the cells forming the endodermal strand.

As part of egg activation, the contents of *Styela* eggs become dramatically rearranged, with the result that different types of cytoplasm become distinctly localized (Fig. 8.13b–d). The rearrangements, which are referred to as ***ooplasmic segregation,*** can be divided into two phases (Jeffrey, 1984). The first phase occurs while the egg is still completing its meiotic divisions. The myoplasm streams down the egg periphery toward the vegetal pole, where it accumulates as a yellow cap. In the wake of the myoplasm, islets of ectoplasm also flow to the vegetal pole and form a clear layer above the yellow cap. As both myoplasm and ectoplasm are streaming into the vegetal half of the egg, the gray cytoplasm is displaced toward the

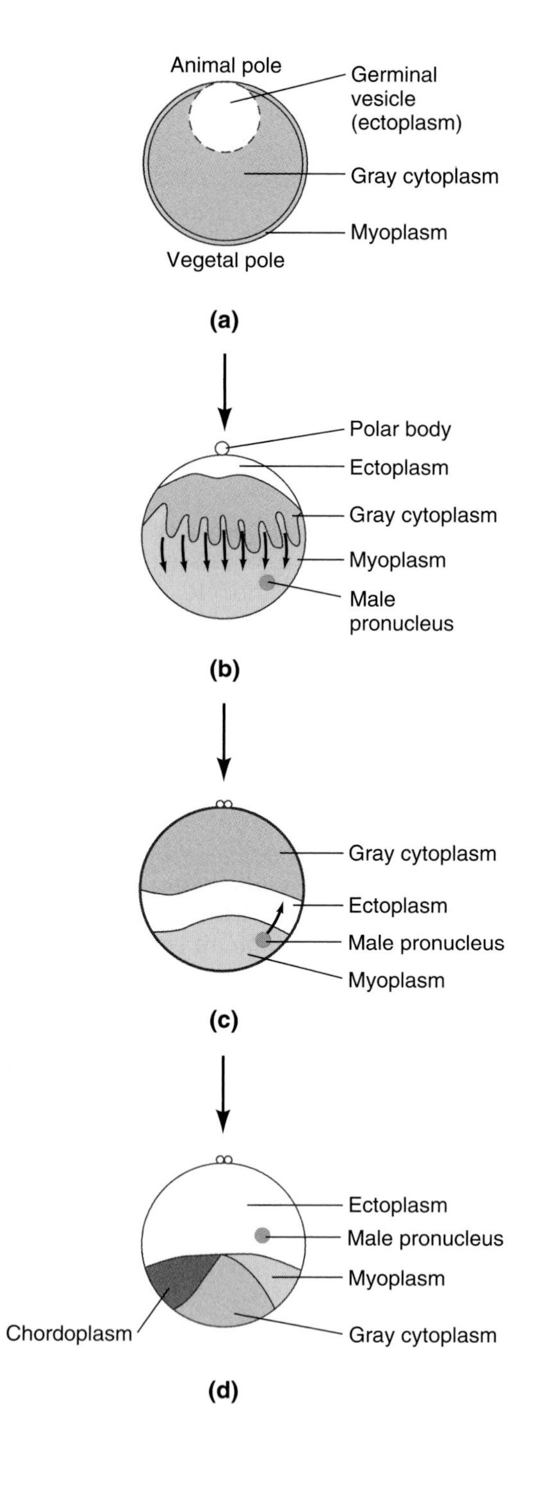

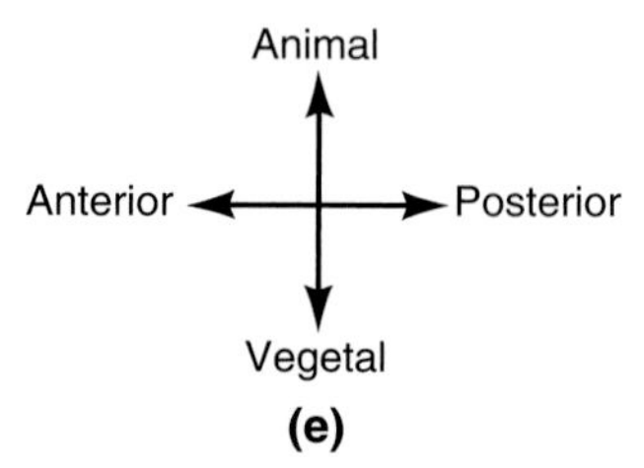

Figure 8.13 Ooplasmic segregation in *Styela partita.* **(a)** Unfertilized egg: the entire cortical cytoplasm contains yellow pigment granules (myoplasm). The egg has only one polarity axis, the animal-vegetal axis. **(b, c)** Fertilized egg during the first phase of ooplasmic segregation. Myoplasm streams down the egg periphery and accumulates as a yellow cap at the vegetal pole. In the wake of the myoplasm, clear cytoplasm (ectoplasm) and gray yolky cytoplasm (endoplasm) also become stratified. **(d)** Second phase of ooplasmic segregation. The male pronucleus migrates along the egg periphery toward the animal pole. Both myoplasm and ectoplasm move with the male pronucleus. While the ectoplasm returns to the animal hemisphere, the myoplasm is left behind to form the yellow crescent below the egg equator. An area of dark gray chordoplasm forms opposite the yellow crescent. **(e)** The fertilized egg has now acquired a second polarity axis, the anteroposterior axis, with the yellow crescent marking the posterior pole.

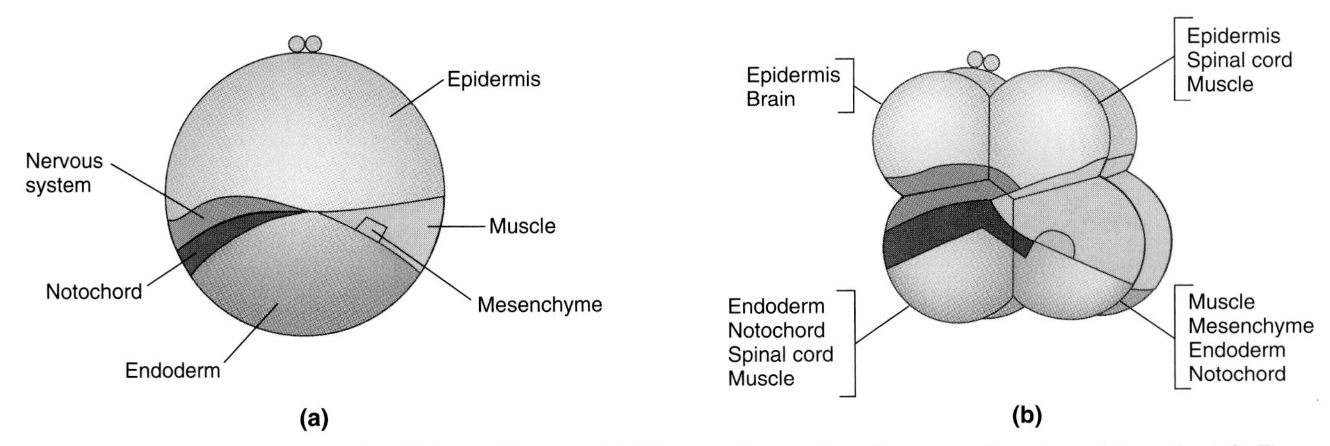

Figure 8.14 Fate maps of the ascidian *Halocynthia roretzi.* **(a)** Zygote after ooplasmic segregation viewed from the left. The map indicates the germ layers and derivatives into which various cytoplasmic regions will be partitioned. **(b)** 8-cell embryo with the major tissues derived from each blastomere.

animal pole. At the end of the first phase, the gray cytoplasm, ectoplasm, and myoplasm are stratified perpendicular to the *animal-vegetal axis,* with the myoplasm occupying the vegetalmost position.

The second phase of ooplasmic segregation begins after the meiotic divisions have been completed. The sperm nucleus has then become the male pronucleus and is located near the vegetal pole, either because the sperm entered there or because the sperm nucleus was swept there along with the myoplasm. The male pronucleus forms an aster of microtubules and migrates along the periphery of the egg in the direction of the animal pole. Both myoplasm and ectoplasm move along with the male pronucleus. At a position below the egg equator, the myoplasm is left behind and forms a ***yellow crescent,*** which marks the future posterior end of the embryo. After the yellow crescent has formed, the ectoplasm returns to the animal pole along with the male pronucleus. At the same time, a major portion of the gray cytoplasm returns to the vegetal hemisphere. A fourth cytoplasmic region, the ***chordoplasm,*** forms opposite the yellow crescent and is later included into the blastomeres that give rise to the *notochord.*

At the close of the ooplasmic segregation, the ascidian egg has two axes of polarity. One is the *animal-vegetal axis,* with the vegetal pole predicting the site of *blastopore* formation and the future *dorsal side* of the embryo. In addition, the ascidian egg has acquired an *anteroposterior axis,* with the posterior pole marked by the yellow crescent. The animal-vegetal and anteroposterior axes define the *median plane*—that is, the plane separating the right and left halves of the embryo.

The bilateral symmetry of the ascidian egg is reflected by the cleavage pattern of the embryo, with the first cleavage plane separating the right and left halves of the embryo (see Fig. 5.7). Using the egg's natural pigments and additional injected tracers, investigators have been able to establish the fates of embryonic cells and indeed to trace some of the cell lineages back to the cytoplasmic pigments that are segregated into these cells. Fate maps of a zygote and an 8-cell embryo are shown in Figure 8.14. We will focus on the myoplasm, which is segregated into those cells that will form most of the tadpole's tail muscle cells.

MYOPLASM IS NECESSARY AND SUFFICIENT FOR TAIL MUSCLE FORMATION

The yellow crescent of myoplasm has received much attention because its localization is so dramatic, and because it is segregated into a cell lineage that will form most of the larval tail muscle (Nishida, 1987). The fact that every myoplasm-including cell in an ascidian embryo develops into a muscle cell suggests that myoplasm may contain a cytoplasmic determinant that determines either muscle cell development or, more generally, tail development. Rescue and heterotopic transplantation experiments indicate that this is in fact the case.

Nishida (1992) devised the equivalent of a heterotopic transplantation experiment by fusing blastomeres that do not normally form muscle cells with cytoplasmic fragments containing myoplasm. Anterior animal blastomeres at the 8-cell stage are fated to form epidermis and brain (Fig. 8.14). When cultured in isolation, these blastomeres developed into epidermis but never expressed any muscle cell markers. However, the same blastomeres became capable of forming muscle cells upon fusion with enucleate cytoplasmic fragments containing myoplasm. Such fragments were cut with fine glass needles from eggs at different stages of ooplasmic segregation and from blastomeres of 8-cell embryos (Fig. 8.15). Before the *complementary* fragments were discarded, Nishida tested them for the presence of a nucleus by monitoring their ability to divide and by staining for nuclear DNA to make certain that no nucleus had slipped inadvertently into the fragments used for fusion. Fusion of test blastomeres with cytoplasmic fragments was promoted by polyethylene glycol and electric current. The fusion products were allowed to develop until control embryos hatched and then tested

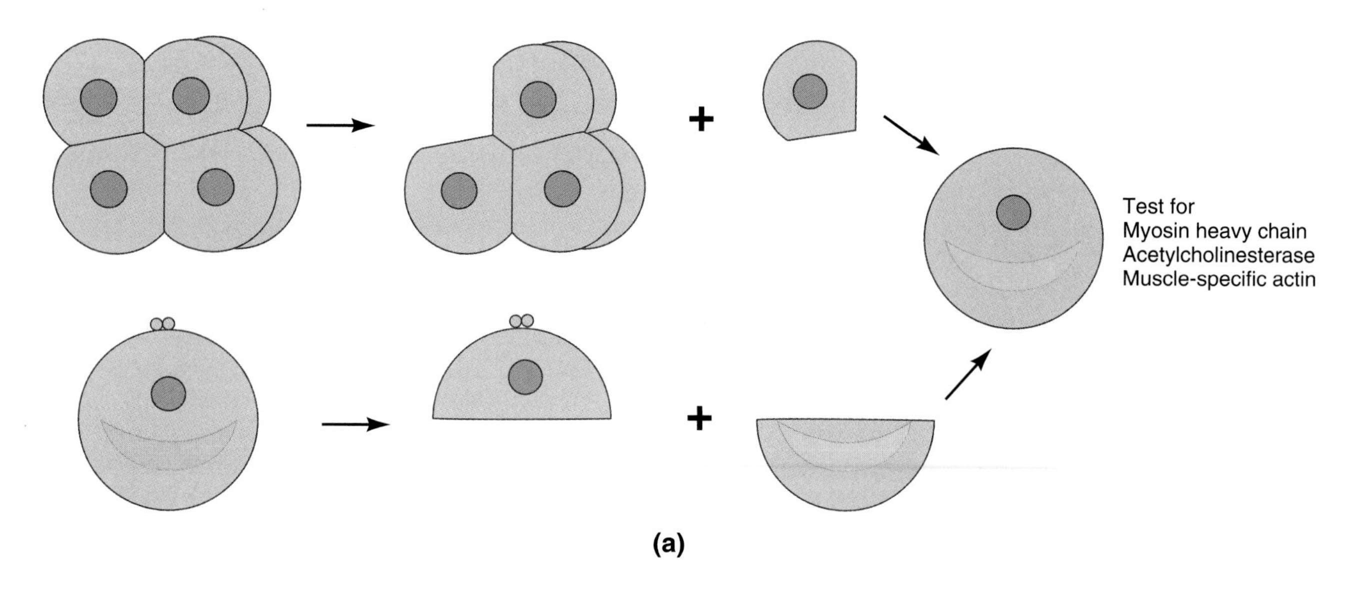

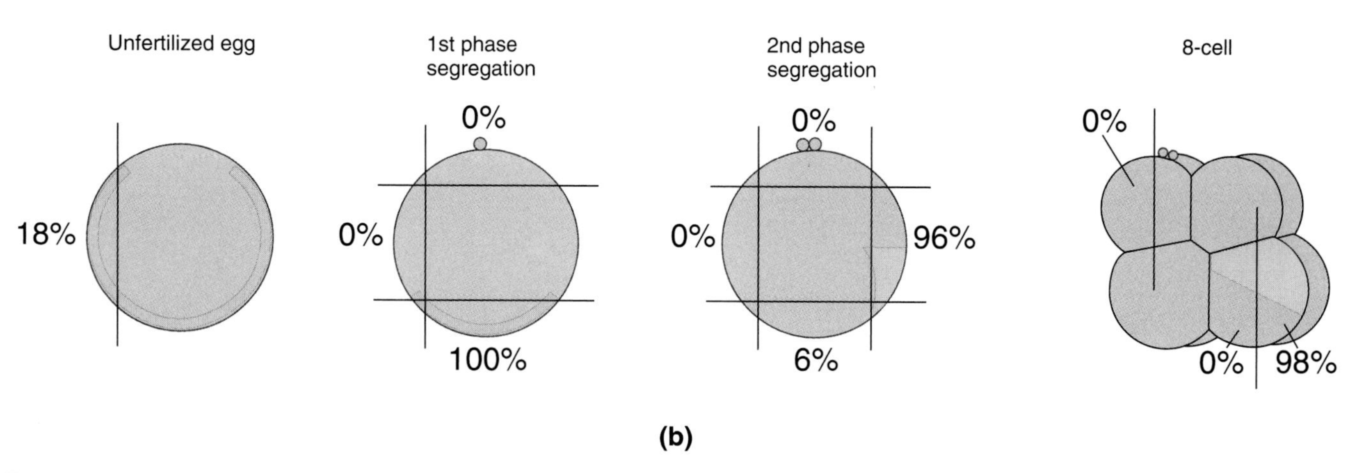

Figure 8.15 Experiments demonstrating colocalization of muscle cell determinants with myoplasm. **(a)** Experimental procedure. Isolated anterior animal blastomeres were fused with various enucleate cytoplasmic fragments cut from eggs or from blastomeres of 8-cell embryos. Fusion products were allowed to develop and then tested for the presence of three muscle-specific marker proteins. **(b)** Results. Percentages next to egg fragments indicate the proportion of anterior animal blastomeres forming muscle-specific proteins after fusion with the respective egg fragment. Fusion with any egg cortical region before ooplasmic segregation gave a low yield of muscle cells. After segregation, fusion with non-myoplasm fragments did not lead to muscle cell formation, whereas fusion with myoplasm-containing cytoplasmic fragments consistently resulted in muscle cell development.

for the presence of three muscle-specific marker proteins: myosin, muscle-specific actin, and acetylcholinesterase (an enzyme involved in the innervation of muscle cells by motor neurons). These marker proteins were synthesized by descendants of anterior animal blastomeres that had been fused with cytoplasmic fragments containing myoplasm, but not by descendants of blastomeres fused with various other cytoplasmic fragments.

The results of Nishida demonstrate that myoplasm co-localizes with a determinant that is sufficient for muscle cell development. However, myoplasm may also act as a general posterior determinant before cleavage. This is indicated by additional microsurgical experiments in which myoplasm was removed after the second phase of ooplasmic segregation (Nishida, 1994). The developing embryos not only lacked muscle cells but generally showed no signs of anteroposterior polarity. Instead, cells with features normally limited to the anterior pole region are formed all around the embryo, suggesting that anterior structures develop by default. Even more impressive, heterotopic transplantation of myoplasm to the anterior pole of eggs lacking myoplasm in its normal posterior location produced tadpoles with reversed anteroposterior polarity. These results indicate that myoplasm acts first as a *regional* determinant, specifying generally the development of the posterior elements of the body pattern. By the 8-cell stage, when myoplasm has been segregated into the posterior vegetal blastomeres, its regional activity may have subsided or become segregated into other blastomeres, so that henceforth myoplasm will serve as a *lineage-specific* determinant for tail muscle cells.

MYOPLASM SEGREGATION INVOLVES A PLASMA MEMBRANE LAMINA

The close association of muscle cell—or posterior—determinants with visible pigment granules in ascidian myoplasm provides an opportunity to study the cellular mechanism of a dramatic localization process by microscopic inspection of intact eggs. By extracting *Styela plicata* eggs with detergents, Jeffery and Meier (1983) removed much of the cytoplasmic matrix and made the cytoskeletal elements of the egg cortex better visible. They found the yellow pigment granules of the myoplasm to be associated with a *filamentous lattice*, presumably consisting of intermediate filaments, which in turn is connected to a *plasma membrane lamina* (*PML*) that is underlying the egg plasma membrane and appears to be rich in actin *microfilaments*. In unfertililzed eggs, the PML is present in the entire cortex. When the yellow granules recede to the vegetal pole after fertilization, so does the PML (Fig. 8.16). The authors propose that contraction of the PML provides the driving force for the movement of the myoplasm to the vegetal pole. They also suggest that PML contraction displaces the gray cytoplasm into its temporary position near the animal pole. This would explain the bulge of cytoplasm at the animal pole that widens considerably before the spherical shape of the egg is restored.

The role of cytoskeletal elements in ooplasmic segregation was also tested by application of appropriate inhibitors. The first phase of segregation, but not the second, was inhibited by *cytochalasin B* (Sawada and Schatten, 1988). This indicates again that microfilaments must function during the first phase as concluded from the observations of Jeffery and Meier. Conversely, the microtubule inhibitors *colcemid* and *nocodazole* inhibit the second phase of ooplasmic segregation but not the first. Sawada and Schatten concluded that the aster of microtubules originating near the male pronucleus is

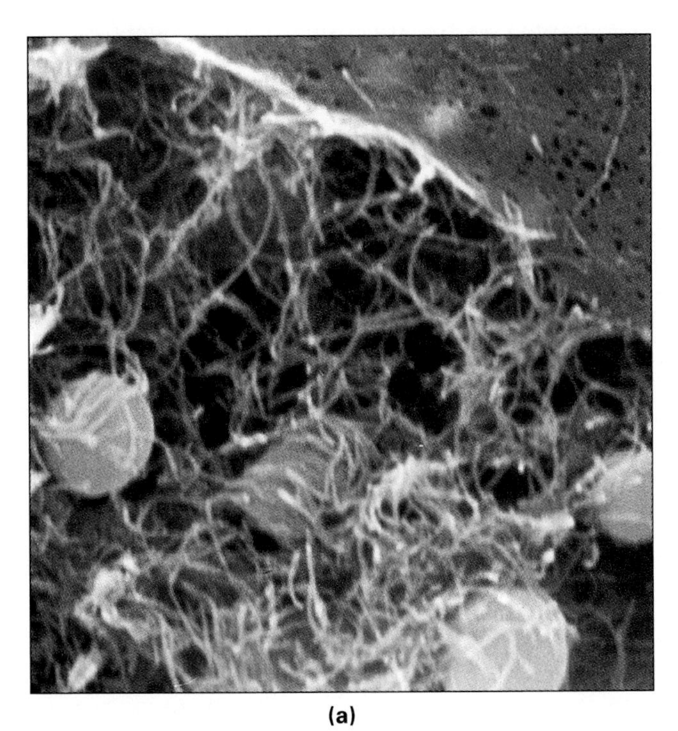

(a)

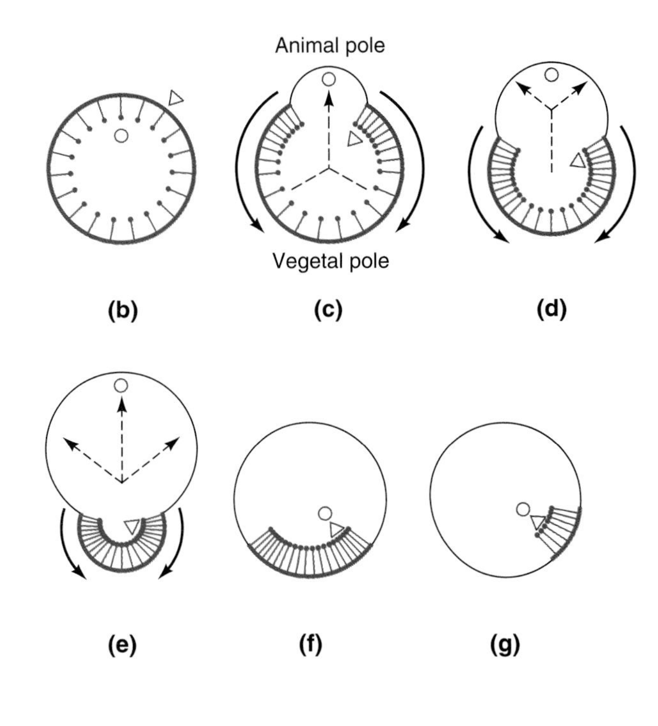

Figure 8.16 Association of pigment granules with a filamentous lattice and a plasma membrane lamina in the myoplasm of an ascidian, *Styela plicata.* **(a)** The scanning electron micrograph shows an egg that was extracted with detergent to make cytoskeletal structures more visible. A plasma membrane lamina (top) and deep filamentous lattice (below) are viewed from the side. Note the association of the lattice filaments with both the plasma membrane lamina and round pigment granules (lower left). **(b–g)** Model for ooplasmic segregation. In each diagram, the thick, colored part of the egg contour represents the part of the plasma membrane with plasma membrane lamina underneath, and the thin, black part of the egg contour represents the part of the plasma membrane without the plasma membrane lamina. The structures attached to the inside of the plasma membrane lamina, including the deep filamentous lattice and pigment granules, are represented by radial lines and dots. **(b)** Unfertilized egg. **(c–f)** First phase of ooplasmic segregation. **(g)** Second phase of segregation; most of the myoplasm shifts posteriorly (see also Fig. 8.13). Direction of ooplasmic movement: arrows with solid lines, myoplasm; arrows with broken lines, gray cytoplasm. Black circle, oocyte nucleus/female pronucleus. Black triangle, sperm nucleus/male pronucleus.

necessary for the second segregation phase. They proposed that during the first phase, microfilament action sweeps the sperm nucleus and the myoplasm together near the vegetal pole. Then, as the microtubules of the sperm aster assemble, the growth and movement of the aster toward the animal pole might serve to bring the myoplasm into its final crescent position.

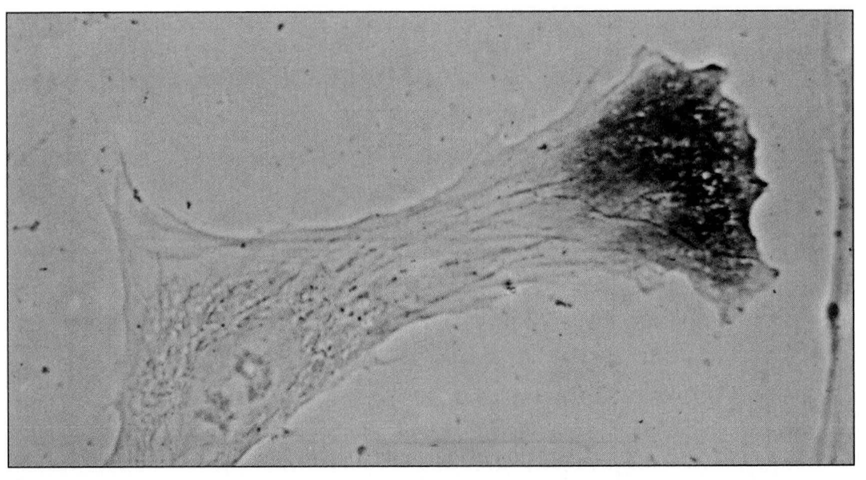

Figure 8.17 Localization of β-actin mRNA to the leading edge (upper right) of cultured chicken embryonic fibroblasts. The β-actin mRNA was detected by in situ hybridization (see Method 8.1) with a DNA probe that binds specifically to this mRNA and can in turn be detected by immunostaining.

MYOPLASM IS ASSOCIATED WITH LOCALIZED MATERNAL RNAs

The molecular nature of the muscle cell—or posterior—determinants in ascidian myoplasm is still under investigation. Preliminary evidence suggests that RNA molecules may be involved, as in other instances of localized cytoplasmic determinants. UV irradiation of myoplasm at a wavelength that inactivates nucleic acids has similar effects as surgical removal of myoplasm—that is, loss of posterior structures and radialization of the embryo (Jeffery, 1990). A polyadenylated but apparently noncoding RNA was found to be associated with the cytoskeleton of the myoplasm in *Styela clava* (Swalla and Jeffery, 1995). In *Ciona savignyi*, the maternal transcript of the *posterior end mark*$^{+}$ gene co-localizes approximately with myoplasm, and microinjection of this transcript into fertilized eggs causes the development of tadpoles in which posterior structures are expanded at the cost of anterior ones (Yoshida et al., 1996).

Taken together, the experiments described in this section indicate that the myoplasm of ascidian eggs contains maternal RNAs as well as pigmented granules that associate with cytoskeletal elements. Some of these components act as cytoplasmic determinants for the cell lineage that gives rise to most of the larval tail muscles. Other components of the myoplasm may act more generally as posterior determinants.

8.6 Cytoplasmic Localization at Advanced Embryonic Stages

Cytoplasmic localization has been investigated mostly in eggs because eggs are especially large cells. However, such localizations are found in cells at more advanced stages of development as well (Jan and Jan, 2000). Some of these cases are simply uneven distributions of molecules to distinct functional domains of mature, nondividing cells. In other cases, localizations occur in dividing cells, setting up asymmetrical cell divisions that generate daughter cells with different determined states.

Uneven distributions of mRNAs have been observed in *the leading edges* of mobile cells. For example, β-actin and the mRNA encoding it are restricted to the leading edge of cultured chicken embryonic fibroblasts (Fig. 8.17; Lawrence and Singer, 1986). Apparently, β-actin is better adapted to the functional requirements of the leading edge than other actin isoforms, which occur elsewhere in the cell. The cis-acting element for the spatial restriction of β-actin mRNA has been mapped to a nucleotide sequence of its 3′UTR—as in the case of bicoid mRNA discussed earlier. If the localization sequence of β-actin mRNA is inactivated by oligonucleotides that hybridize to it, then β-actin and its mRNA are still present in the fibroblast but no longer restricted to the leading edge. As a consequence, the edge collapses, and the cell loses its polarization (Kislauskis et al., 1994). These observations clearly demonstrate that the spatial restriction of an mRNA can be necessary for maintaining the polarity of a differentiated cell.

Dividing cells with cytoplasmic localizations that generate pairs of different daughter cells have been investigated in *Drosophila* (Hirata et al., 1995; Knoblich et al., 1995; Kraut et al., 1996; Spana and Doe, 1995). The central nervous system in the *Drosophila* embryo develops from cells called *neuroblasts,* which segregate from epidermal cells by *lateral inhibition* using the Notch/Delta signaling (see Fig. 6.9). Neuroblasts divide repeatedly and asymmetrically, each time giving rise to another neuroblast and a *ganglion mother cell* (*GMC*). The GMC divides further and eventually forms neurons. Similar asymmetrical divisions occur in the peripheral nervous system and elsewhere in *Drosophila.*

The asymmetrical division of neuroblasts is associated with the synthesis and unequal distribution of a protein, prospero, which occurs in a cell cycle–dependent manner (Fig 8.18). Prospero has been shown to be a *transcription factor* required for determination and differentiation of neurons in the nervous system of *Drosophila* (Knoblich,

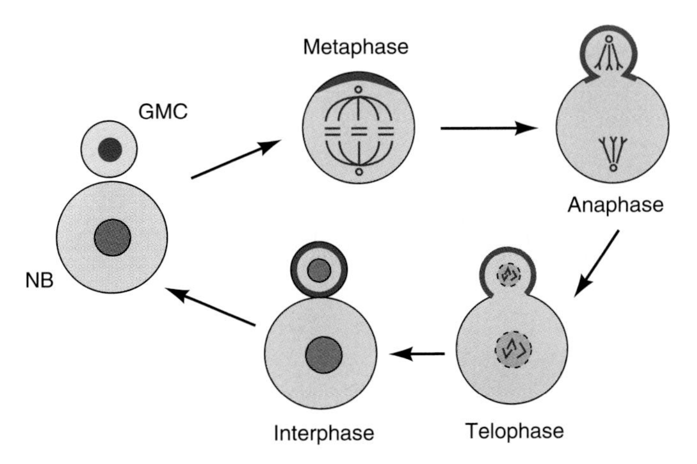

Figure 8.18 Cell cycle–dependent localization of prospero protein (color) in *Drosophila* neuroblasts. NB = neuroblast; GMC = ganglion mother cell.

1997; Jan and Jan, 1998). During interphase, prospero is faintly visible in the cytoplasm of the neuroblast. When the neuroblast enters mitosis, prospero accumulates in a cortical crescent overlying the centrosome on the neuroblast's basal side, where the GMC will bud off. During anaphase, when a bud appears, prospero moves into the bud. By the end of cytokinesis, prospero has been partitioned completely into the cortex of the GMC, with no prospero left in the neuroblast. During the following interphase, the neuroblast begins another round of prospero synthesis. Meanwhile in the GMC, prospero moves from the cortex to the nucleus, where it inhibits genes involved in cell proliferation and activates genes that promote neuronal differentiation (L. Li and H. Vaessin, 2000).

Another protein, numb, is synthesized in neuroblasts and segregated into GMCs just like prospero, except that numb does not move into the GMC nucleus and remains associated with the plasma membrane. Like prospero, numb is required for the asymmetrical division of neuroblasts in the developing nervous system (M. Guo et al., 1996). In the absence of numb, a neuroblast divides symmetrically into two neuroblasts. If numb is overexpressed, a neuroblast divides symmetrically into ganglion mother cells. Apparently, numb interferes with Notch signaling, so that cells with high numb activity are refractory to the *Notch/Delta mechanism* of *lateral inhibition* and advance toward becoming a neuron (see Fig. 6.9).

A third protein, inscuteable, is localized in the same cells that localize numb and prospero (Chia et al., 1997). However, inscuteable is localized to the neuroblast's apical pole, opposite to numb and prospero. During each mitotic cycle, inscuteable becomes localized before numb and prospero do. In null alleles of *inscuteable,* neuroblasts synthesize numb and prospero but do not segregate either protein asymmetrically. Embryos with *inscuteable* null alleles are also abnormal in the orientation of their mitotic spindles. In contrast to wild-type embryos, where prospective *epidermal* cells orient their mitotic spindles parallel to embryo's surface, and *neuroblasts* orient their mitotic spindles perpendicular to the surface, *inscuteable* embryos have spindles parallel to the embryo's surface in *all ectodermal* cells. Thus, inscuteable is required for both proper mitotic spindle orientation and asymmetrical segregation of numb and prospero. The localization of inscuteable in turn depends on the *bazooka*$^+$ gene, which encodes a protein that seems to provide an asymmetrical cue for anchoring the inscuteable protein to the apical cytocortex (Wodarz et al., 1999).

8.7 Bioassays for Localized Cytoplasmic Determinants

Heterotopic transplantations, as well as rescue experiments, can be used as *bioassays* to further characterize any cytoplasmic determinant of interest with regard to its stage of localization, species specificity, and molecular nature. The following examples will illustrate this strategy.

The heterotopic transplantation of posterior pole plasm in *Drosophila* (see Fig. 8.9) was used to define the *developmental stage* at which the germ cell determinants become localized at the posterior pole. To this end, the transplanted cytoplasm was taken from the posterior poles of eggs and oocytes at progressively earlier stages (Illmensee et al., 1976). In each case, the cytoplasm was injected near the anterior pole of a genetically marked host embryo. The ectopic pole cells that developed were examined for their morphological characteristics and for their ability to give rise to gametes, as described previously. Functional ectopic pole cells were obtained with posterior cytoplasm from oocytes as early as the stage at which vitellogenesis is complete and the nurse cells have just squeezed their contents into the oocyte. In donor oocytes at even earlier stages, polar granules had accumulated in the posterior cytoplasm, but this cytoplasm did not cause ectopic pole cell formation. Thus, it appears that the polar granules visible in younger oocytes are not loaded yet with germ cell determinants.

In a similar set of experiments, Mahowald et al. (1976) explored whether localized germ cell determinants from one species were compatible with the responding system of another species. To this end, these researchers transplanted posterior pole plasm from *Drosophila immigrans* donor embryos to the anterior pole region of *Drosophila melanogaster* recipients. As a result, the recipients developed hybrid pole cells containing a *D. melanogaster* nucleus and *D. immigrans* cytoplasm. When these cells were transplanted back to the posterior pole of a genetically marked host, they gave rise to functional gametes. The results show that germ cell determinants have been conserved well enough during the

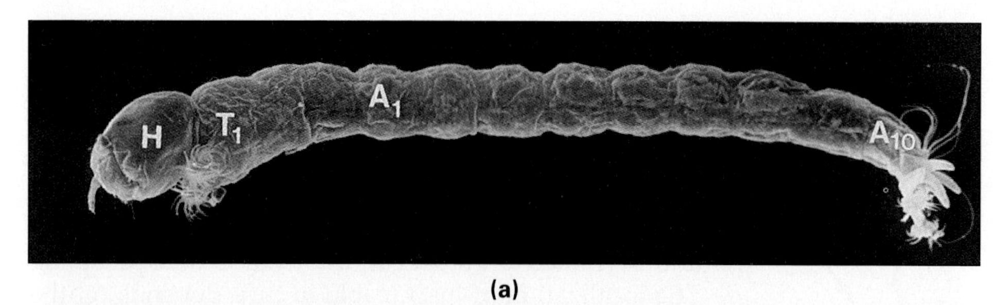

(a)

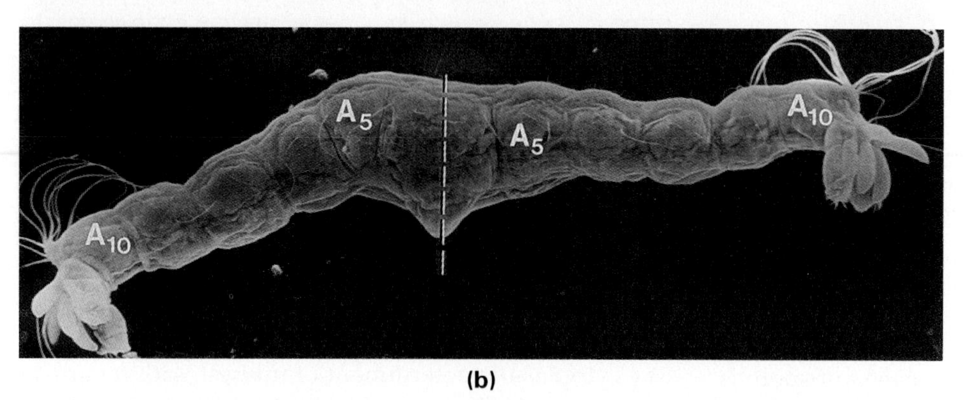

(b)

Figure 8.19 Normal larva of *Chironomus samoensis* **(a)** and double-abdomen larva obtained after anterior UV irradiation of the early embryo **(b)**. Development of the anterior body pattern depends on the function of small polyadenylated RNA molecules near the anterior pole of the embryo. If these anterior determinants are inactivated or displaced, those cells that would normally give rise to head and thorax form a mirror-image duplication of the abdomen instead. H = head; T1 = first thoracic segment; A1, A5, A10 = first, fifth, and last abdominal segments.

process of evolution to be recognized by cells of a closely related species.

Rescue experiments with eggs of a midge, *Chironomus samoensis,* have been used as a bioassay to identify the molecular nature of a localized anterior determinant that specifies the anterior part of the body pattern. Eggs of *Chironomus* and a related species, *Smittia,* form strikingly abnormal body patterns in response to various experimental manipulations. In the ***double-abdomen*** pattern, anterior segments including head and thorax are replaced by a mirror-image duplication of abdominal segments (Fig. 8.19; Percy et al., 1986). This pattern is formed after exposure of the anterior pole region of wild-type embryos to ultraviolet light or RNase, an enzyme that specifically degrades RNA (Yajima, 1964; Kandler-Singer and Kalthoff, 1976). Embryos programmed for double-abdomen development are restored to normal development by the microinjection of cytoplasm from unirradiated wild-type donors (Fig. 8.20). A rescue bioassay based on this result confirmed that the rescuing activity is present in anterior, but not posterior, egg cytoplasm of *Chironomus* and closely related species. The same rescuing activity was found in total egg RNA, and in RNA containing a poly(A) segment, a characteristic of mRNA. When different-size classes of RNA were tested in the bioassay, the strongest rescue activity was associated with the fraction containing RNA molecules between 250 and 600 nucleotides in length (Elbetieha and Kalthoff, 1988).

8.8 The Principle of Default Programs

The determination of embryonic cells by cytoplasmic localizations illustrates the ***principle of default programs,*** which can be observed at many branchpoints of developmental pathways. At each of these points, cells acquire one fate if they receive a specific signal from a cytoplasmic localization, from a neighboring cell, or in the form of a hormone. In the absence of such a signal, cells fall back on a program of development that does not depend on receiving the additional signal. Such programs, known as *default programs,* are very common.

The anterior determinants of *Chironomus* and *Smittia* are necessary and sufficient for head and thorax formation. In the absence of anterior determinants, eggs develop into *double-abdomen* embryos (Fig. 8.20). Double abdomens look the same, whether the anterior determinants have been removed by centrifugation or inactivated by means of enzymes or ultraviolet light. Thus, formation of a double-abdomen embryo is a default program followed by the egg in the absence of anterior determinants.

To test whether forming the anterior half of a double abdomen requires any interactions with the posterior half of the egg, Ritter (1976) performed a combined UV irradiation and ligation experiment (Fig. 8.21). Transverse ligation separated an anterior embryonic fragment, which produced a head, from a posterior fragment producing a set of abdominal segments. However, when the anterior pole region was UV-irradiated after ligation, the anterior fragment formed, instead of a head, another set of abdominal segments with reversed polarity. Therefore, everything required for abdomen development must be present in both anterior and posterior halves of the egg. In other words, formation of an abdomen is the default program of the anterior egg half in the absence of anterior determinants. This result raises the question of whether the posterior egg half, if provided with anterior determinants, will form head and thorax; as we will see, it does.

In addition to normal embryos and double-abdomen specimens, the eggs of *Chironomus* and *Smittia* can form

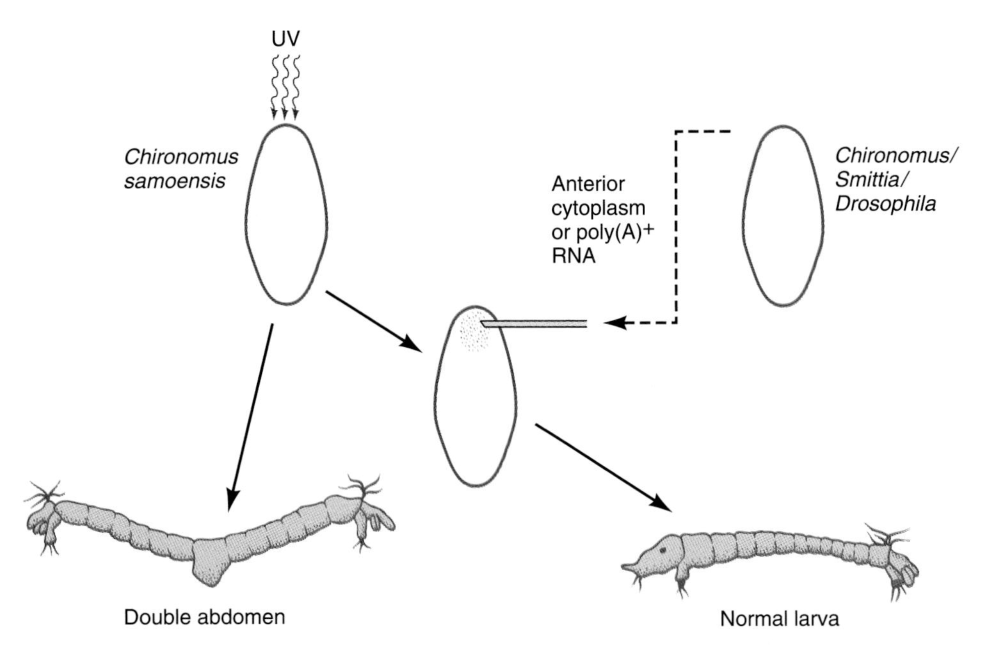

Figure 8.20 Rescue bioassay for anterior determinants in eggs of *Chironomus samoensis.* Eggs are programmed for double-abdomen development by anterior UV irradiation. Embryos can be restored to normal development by injecting anterior cytoplasm or poly(A)-containing RNA from unirradiated donor eggs. Anterior determinant activity is measured as an increase in the percentage of normal embryos among the surviving embryos.

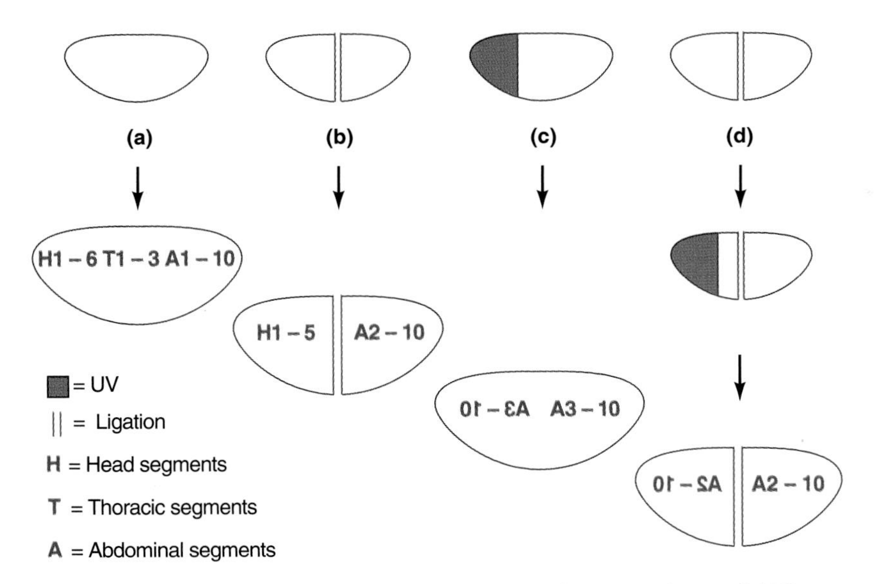

Figure 8.21 Combined ligation and UV irradiation experiment with *Smittia* embryos. **(a)** The normal segment pattern comprises the head, 3 thoracic segments, and 10 abdominal segments. **(b)** Upon transverse ligation, the anterior egg fragment develops an incomplete head, and the posterior fragment forms a partial abdomen. **(c)** UV irradiation of the anterior pole region causes the formation of double-abdomen embryos, similar to the one shown in Figure 8.19b. **(d)** After ligation and subsequent UV irradiation of the anterior pole region, the anterior fragment forms an abdomen with reversed polarity, indicating that all components necessary for abdomen formation are present in the anterior egg half.

two additional types of abnormal body pattern, called double-cephalon embryos and inverted embryos (Fig. 8.22). ***Double-cephalon*** embryos have mirror-image duplications of the head but are missing thorax and abdomen. ***Inverted embryos*** look normal, but the abdomen develops at the anterior end while the head originates posteriorly. Consequently, the pole cells formed at the posterior pole end up in the head. Both types of abnormal body pattern are obtained after centrifugation of eggs, an operation that apparently causes a variable displacement of anterior determinants (Rau and Kalthoff, 1980; Kalthoff et al., 1982; Yajima, 1983). In

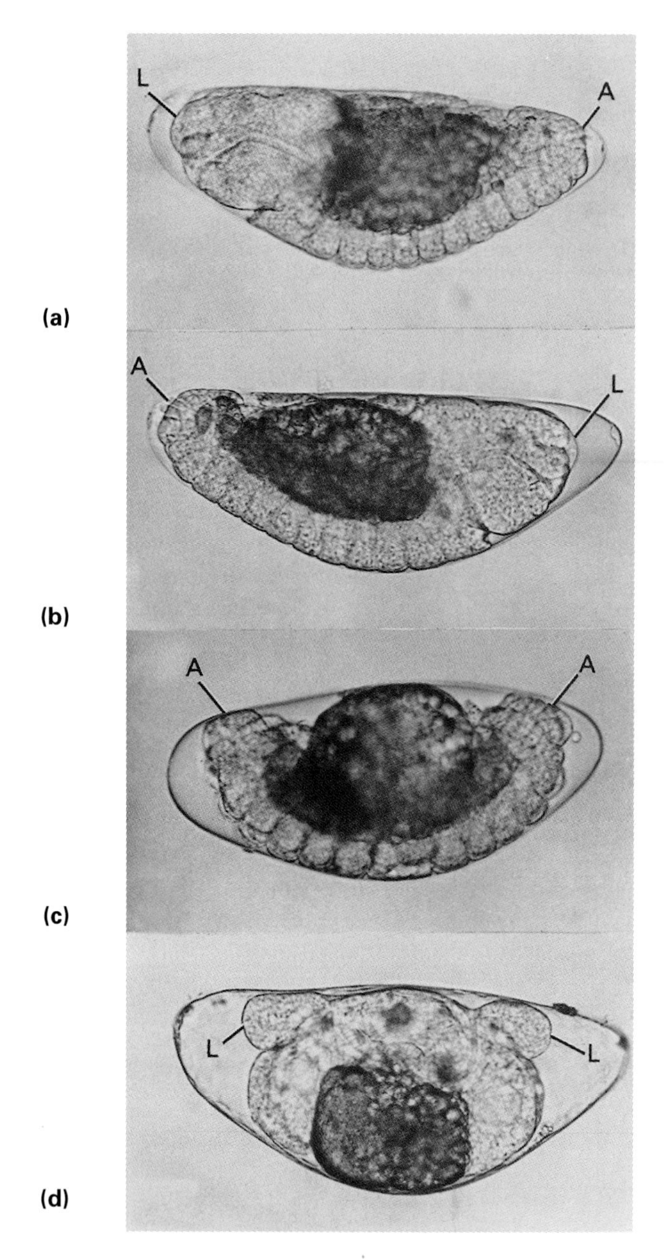

Figure 8.22 Four basic body patterns observed in embryos of *Smittia* sp. and related midges. In these photomicrographs, the anterior pole as marked by the micropyle is to the left, the ventral side facing down. **(a)** In the normal body pattern, the head is marked by the labrum (L) while the abdominal end is tipped with anal papillae (A). **(b)** Inverted embryos develop with their anteroposterior axis completely reversed, except for the pole cells, which end up in the head. **(c)** Double-abdomen embryos show a mirror-image duplication of abdominal segments in the absence of head and thorax. **(d)** In double-cephalon embryos, cephalic segments are duplicated while thorax and abdomen are missing. All four body patterns were obtained after centrifugation during cleavage, a treatment that apparently displaces or inactivates localized anterior determinants.

accord with this interpretation, anterior UV irradiation after centrifugation enhances the yield of double-abdomen and inverted embryos, presumably by inactivating anterior determinants that have remained anterior in the egg. Conversely, posterior UV irradiation of centrifuged eggs enhances the yield of double-abdomen and normal embryos, apparently by inactivating anterior determinants that have moved toward the posterior during centrifugation.

The data obtained with *Chironomus* and *Smittia* indicate that both anterior and posterior egg halves can give rise to either anterior or posterior body parts. In the presence of anterior determinants, each egg half will form the anterior part of the body pattern. If these determinants are weakened by mutation or experimental means, then each egg half falls back independently on the default program of forming the posterior part of the body pattern.

The principle of default programs is also observed in other cases of cytoplasmic localization. In the absence of germ cell determinants, somatic cells are formed. For instance, UV irradiation of *Drosophila* eggs on the posterior pole prevents the formation of pole cells. Instead, the posterior pole is covered with the same type of blastoderm cells that are formed elsewhere (Fig. 8.8c). There are also default programs in the developmental pathways controlled by embryonic induction or homeotic genes, in sex determination, and in hormonal control, as we will discuss at the end of this chapter and in others.

8.9 Properties of Localized Cytoplasmic Determinants

Having discussed some well-investigated cases of cytoplasmic localization, we will now summarize the salient features of cytoplasmic determinants in a comparative fashion (Table 8.1).

CYTOPLASMIC DETERMINANTS CONTROL CERTAIN CELL LINEAGES OR ENTIRE BODY REGIONS

Localized cytoplasmic determinants may be *lineage-specific*—that is, controlling the fate of certain cells and their descendants. The germ cell determinants of insects clearly fall into this category. If germ cell determinants are eliminated by mutation or UV irradiation, developing animals are often normal in their anatomy and behavior except for their empty gonads. Lineage-specific determinants seem to be passed on clonally but not to affect neighboring cells.

Other cytoplasmic determinants are *region-specific*, meaning they control the development of an entire body region. Examples are bicoid mRNA in *Drosophila* and the anterior determinants in *Chironomus* and *Smittia*. The hallmark of these determinants is that they control major portions of an entire body pattern rather than specific cell lineages. Region-specific determinants seem to

table 8.1 Properties of Localized Cytoplasmic Determinants

Species/ Taxonomic Group or Cell Type	Determined Cell Lineage (C) or Body Region (R)	Activating (A) or Inhibiting (I) Determinant	Molecular Nature of Determinant	Visible Marker	Structures Involved in Localization	Timing of Localization	Illustration (Fig. No.)
Molluscs	Organs depending on D cell lineage (C and R)	A and I	Unknown	—	Polar lobe	Cleavage	8.4–8.6
Sabellaria/ polychaete worm	Apical tuft (C or R?)	A and I	Unknown	—	Polar lobe	Cleavage	8.23–8.24
Many insects	Primordial germ cells (C)	?	mRNAs	Polar granules	Microtubules	Oogenesis	8.7–8.10
Drosophila/ insect	Anterior body region (R)	A	Bicoid m mRNA	—	Microtubules	Oogenesis	8.1, 8.11
Chironomus and Smittia/ insects	Anterior body region (R)	A	Short, polyadenylated RNA	—	—	Oogenesis	8.19–8.22
Drosophila/ neuroblast	Ganglion mother cell	A	Proteins	—	Mitotic spindle	Late embryo	8.18
Ascidians	Muscle and entire tail region (C and R)	A	RNAs	Pigment granules	Plasma membrane lamina, fil. lattice	Egg activation	8.13–8.16
Frogs	Primordial germ cells (C)	?	RNA?	Polar granules	Mitochondrial cloud	Oogenesis	8.25
Frogs	Endoderm, mesoderm (R)	A	VegT, Vg1 mRNA	—	Endoplasmic reticulum	Late oogenesis	8.25–8.26
Fibroblast/ chicken	Leading edge	A	β-actin mRNA	—	—	Late embryo	8.17

generate long-range signals that affect cells regardless of their lineage. In the case of bicoid mRNA, the long-range signal generated is bicoid protein, as we will discuss in Sections 16.4 and 22.3.

In some cases, the distinction between lineage-specific and region-specific determinants is not so clear-cut. The polar lobe material in molluscs, for instance, affects primarily the D cell lineage. However, some descendants of D interact with neighboring cells from other lineages, so that loss of polar lobe material affects, for example, the development of certain C lineage cells, albeit indirectly. Also, the myoplasm of ascidians seems to act as a region-specific determinant after ooplasmic segregation and as a lineage-specific determinant from the 8-cell stage on, as discussed earlier.

LOCALIZED CYTOPLASMIC DETERMINANTS MAY BE ACTIVATING OR INHIBITORY

Most localized determinants seem to function by *activation,* because their presence causes the development of a particular structure that would otherwise not be formed. In other words, an activating determinant endows a cell or an embryonic region with a capability that otherwise would not be within its potential. However, some cytoplasmic determinants act through *inhibition,* by preventing cells from forming certain embryonic parts. Those cells may still be able to form more than one structure, but the inhibitory determinant restricts their potential without adding a new capability.

Both activating and inhibitory determinants are present in polar lobes of the polychaete worm *Sabellaria cementarium,* a species that shows spiral cleavage with polar lobe formation as diagrammed in Figure 8.4. Polychaete larvae have an apical tuft of long cilia and an equatorial band of shorter cilia (Fig. 8.23). No apical tuft will form if the polar lobe that protrudes before the first cleavage is removed (Render, 1983). This result indicates that the first polar lobe contains cytoplasmic components required for apical tuft formation. However, if the first cleavage is allowed to occur normally before the polar lobe is removed prior to the second cleavage, then isolated C and D blastomeres each form an apical tuft while A and B blastomeres do not. If instead the second cleavage is modified by addition of a detergent to the culture medium, so that both C and D blastomeres receive polar lobe material, no apical tuft is formed. These results indicate that the second polar lobe contains

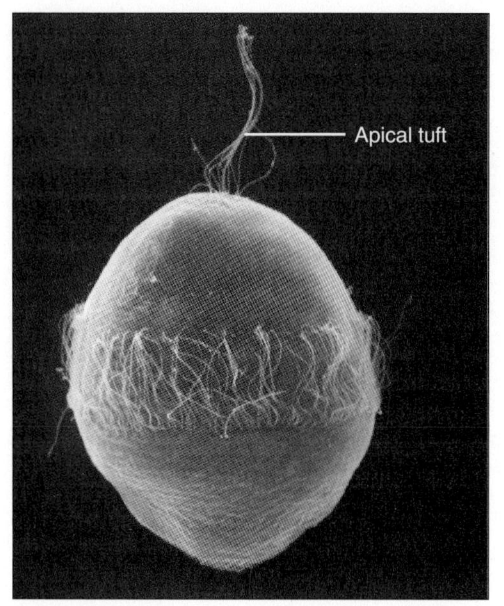

Figure 8.23 Scanning electron micrograph of *Sabellaria* larva showing apical tuft.

an inhibitor of apical tuft formation. When the first two cleavages are allowed to occur normally, any combination of cells containing the C blastomere but not the D blastomere, such as ABC, will form an apical tuft. Conversely, cell combinations containing the D blastomere but not the C blastomere, such as ABD, do not form apical tufts. Apparently, the inhibitor present in the second polar lobe is shunted selectively into the D blastomere during second cleavage.

Together, the results show that the first polar lobe contains both activating and inhibitory determinants of apical tuft formation (Fig. 8.24). Both determinants are shunted into the CD blastomere after the first cleavage, but only the activating determinants remain there; the inhibitory determinants move back into the second polar lobe. These determinants are shunted selectively into the D blastomere after completion of the second cleavage. As a result, the C blastomere inherits only activating determinants, while the D blastomere inherits activating as well as inhibitory determinants. The inhibitory ones prevail in the D blastomere.

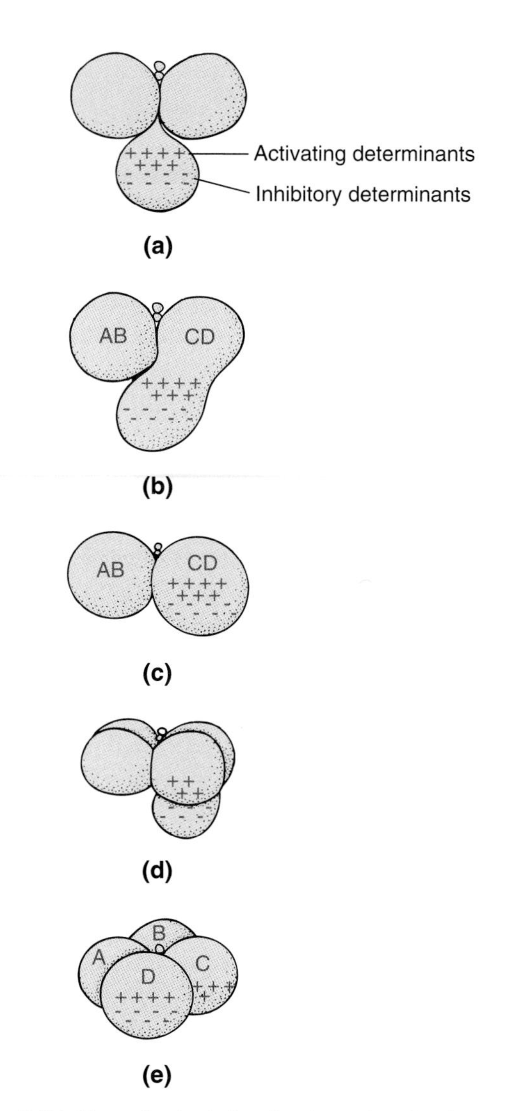

Figure 8.24 Hypothetical distribution of activating (+) and inhibitory (−) determinants for apical tuft formation in the polar lobes of *Sabellaria.* **(a)** First polar lobe at trefoil stage. **(b, c)** Fusion of the polar lobe with the CD blastomere. **(d)** Second polar lobe containing only the inhibitory determinants for tuft formation. **(e)** 4-cell stage after fusion of second polar lobe with D blastomere. Both C and D blastomeres now contain activating tuft determinants, whereas the D blastomere also contains inhibitory determinants.

CYTOPLASMIC DETERMINANTS MAY OR MAY NOT BE ASSOCIATED WITH VISIBLE MARKERS

Several cytoplasmic determinants are associated with distinct organelles, such as the polar lobe of molluscs, the yellow pigment granules in the myoplasm of ascidians, and the polar granules in the posterior pole plasm of *Drosophila* and other insects. In most cases, the association of a determinant with a visible marker may aid, at least temporarily, in the localization of a determinant. In other cases, the association with a visible organelle may be fortuitous: The determinant might function as well without the marker.

Polar granules are visible in germ line cells throughout the life cycle of *Drosophila* (Mahowald, 1971). The association of polar granules with germ cell determinants, as measured by rescue or heterotopic transplantation, lasts for a limited period of time from late oogenesis until pole cell formation. During this period, polar granules are associated with mRNAs and proteins, such as oskar mRNA or staufen protein, that are known to be necessary for the formation of pole cells, the abdomen, or both. During the same time, helical structures resembling polysomes—that is, mRNA in the process of translation—are sometimes found next to polar granules (see Fig. 8.7d). Taken together, these observations suggest that polar granules serve as temporary anchor sites for mRNAs acting as germ cell determinants. (The same polar granules are also repositories for determinants specifying the posterior body pattern. See Section 22.3.) Organelles resembling the polar granules in *Drosophila* are also associated with germ cell determinants in amphibians and other organisms (Wakahara, 1990).

Other cytoplasmic determinants become localized without the benefit of visible repositories or other markers. For example, bicoid mRNA and the anterior determinants found in chironomids are not associated with any unique visible structures. In these cases, the active molecules seem to be anchored to ubiquitous cytoskeletal structures.

MOST LOCALIZED CYTOPLASMIC DETERMINANTS ARE MATERNAL mRNAS

The molecular nature of cytoplasmic determinants, and the ways they control differential gene activity, have been of particular interest. In those cases where the nature of a localized determinant *in eggs* is known it is a maternally transcribed RNA. This is in contrast to embryonic *neuroblasts* in *Drosophila,* where the molecules segregated into ganglion mother cells are proteins (see Section 8.6). The difference may be related to the fact that RNAs are generally more stable than proteins and that localizations last longer in eggs than in more advanced cells. It would also seem more economical to transport and localize a few mRNA molecules, which can be translated many times, than to transport and localize many protein molecules. Perhaps most importantly, a combination of mRNA localization and translational control can prevent the occurrence of protein in the wrong places.

The localization of mRNAs depends on cis-acting and trans-acting elements, as discussed earlier. The cis-acting elements are often located within the mRNA's 3'UTR, as shown for several mRNAs in oocytes of *Drosophila* (St Johnston, 1995) and *Xenopus* (see further below). These elements tend to include complementary RNA sequences that fold back and hybridize, forming hairpin loops and other secondary structures that are recognized by RNA-binding proteins. These proteins in turn anchor the bound RNAs to the cytoskeleton or to the endoplasmic reticulum. While mRNAs are in the process of being localized their translation is generally inhibited. This ensures that their translational products are not spilled en route. (For the same reason, a waiter carrying a bottle of wine to a table will take care to keep it upright or corked.)

For example, oskar mRNA, which is localized to the posterior pole of *Drosophila* oocytes as discussed earlier, is translationally inhibited by an associated protein while en route to its localization site, and this inhibition is necessary for proper pole cell formation (Kim-Ha et al., 1995). Other localized mRNAs in *Drosophila* oocytes, which control the specification of the anteroposterior body pattern, are also not translated until egg activation (see Section 22.3).

Eventually, the mRNAs that serve as localized cytoplasmic determinants are translated into proteins. The life span and diffusion of the protein products tend to be limited so that the concentration of these proteins is greatest in the vicinity of their mRNA templates. Some of the protein products bind to nuclear DNA and control the activity of embryonic genes. Others interact with different mRNAs, affecting their translation and stability. Still other proteins translated from localized mRNAs are signal molecules that act locally in the cell or origin or bind to receptors on the surface of other cells. In conclusion, mRNAs, and perhaps regulatory RNAs that control the translation of mRNAs, have several properties that make them useful as localized cytoplasmic determinants.

CYTOPLASMIC LOCALIZATION OCCURS AT DIFFERENT STAGES OF DEVELOPMENT USING VARIOUS CELLULAR MECHANISMS

While the molecular nature of localized determinants appears to be well conserved, the timing of localization and the cellular mechanisms utilized in the process are quite variable. In most animals, cytoplasmic localization occurs during oogenesis. In *Drosophila,* the location of the nurse cells in the egg chamber imposes an anteroposterior polarity to the developing oocyte. This polarity includes a system of oriented microtubules, which are instrumental in localizing mRNAs to the anterior and posterior poles. For comparison, we will now consider RNA localization in the *Xenopus* oocyte, in which the animal-vegetal polarity develops without apparent cues from the ovarian anatomy. To round out our picture, we will revisit or briefly mention a few cases in which localization takes place after oogenesis, as part of egg activation or during cleavage.

The animal-vegetal polarity of frog oocytes is reflected in the uneven distribution of pigment, yolk, and several cytoplasmic components. At the vegetal pole, many amphibian eggs contain *polar granules* similar to those found in many insect eggs. They stain intensely for RNA, are associated with mitochondria, and become incorporated into primordial germ cells during cleavage. Amphibian oocytes and eggs also contain other RNAs and proteins that are localized to the animal pole or the vegetal pole but are not associated with visible markers (Jäckle and Eagleson, 1980; Dreyer at al., 1982; Rebagliati et al., 1985). We will focus here on the localization of RNAs to the vegetal pole.

An early sign of animal-vegetal polarity in immature frog oocytes is the ***mitochondrial cloud,*** located between the germinal vesicle and the vegetal pole and containing mitochondria, endoplasmic reticulum, and ribonucleoprotein particles (Fig. 8.25; Heasman et al., 1984). The mitochondrial cloud is involved in an early pathway of

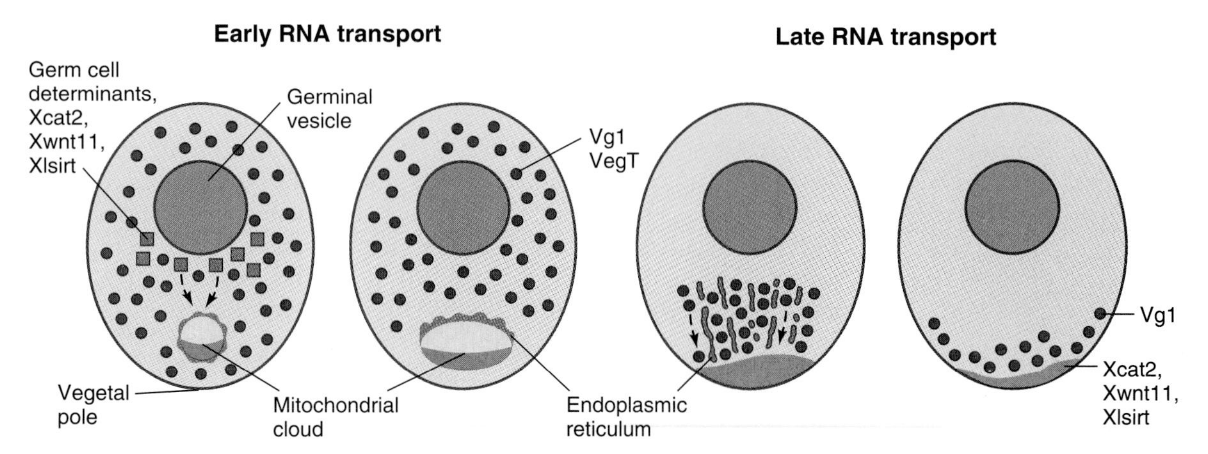

Figure 8.25 Two pathways of RNA transport and localization in *Xenopus* oocytes. RNAs using the early pathway, including Xcat2, Xwnt11, Xlsirt, and germ cell determinants, transfer from the germinal vesicle to the mitochondrial cloud. The cloud and its RNA cargo move to the vegetal pole, where the cloud breaks up and the associated RNAs become anchored to a small cortical area at the vegetal pole. RNAs using the late pathway, such as Vg1 and vegT mRNA, are evenly distributed in the oocyte cytoplasm before they associate with a cone of endoplasmic reticulum (ER) between the nucleus and the vegetal cortex. After traveling with the ER, Vg1 is eventually anchored in a broad cap of vegetal cytoplasm (see Fig. 8.26).

localization, which is taken by several RNAs, including two mRNAs, Xcat2 and Xwnt11, and a nontranslatable transcript, Xlsirt (Kloc and Etkin, 1995; Etkin, 1997; *M. L. King et al., 1999*). These RNAs are synthesized in the *germinal vesicle,* from where they seem to be trapped in specific regions within the mitochondrial cloud. Before the onset of vitellogenesis, the cloud and associated RNAs move toward the vegetal pole. Here the cloud fragments into islands located in a small region of cortical cytoplasm that is segregated into primordial germ cells. The association of these RNAs with the mitochondrial cloud, and their movement to the vegetal pole seem to be independent of microfilaments and microtubules, but their anchorage to the cortical cytoplasm is sensitive to *cytochlasin B,* a microfilament inhibitor.

A late pathway of localization is taken by two mRNAs, Vg1 and VegT. While Vg1 may be involved in establishing dorsal mesoderm in the *Xenopus* embryo (see Section 9.7), VegT seems to be instrumental in determining mesoderm and endoderm (see Section 9.3). Vg1 is still uniformly distributed in the oocyte while Xcat2, and Xlsirt are already riding the mitochondrial cloud to the vegetal pole. Subsequently, Vg1 accumulates around the nucleus before it moves to the vegetal pole. As a vehicle, Vg1 uses a cone of *endoplasmic reticulum (ER)* derived from the mitochondrial cloud and extending between the germinal vesicle and the vegetal pole. Eventually, Vg1 mRNA comes to occupy a broad cap at the vegetal pole (Fig. 8.26; Melton, 1987).

The cis-acting element for the transport of Vg1 mRNA is again localized to its 3′UTR (Mowry and Melton, 1992). Upon removal of this nucleotide sequence, Vg1 mRNA is no longer localized. Conversely, if this sequence from Vg1 mRNA is added to an unrelated mRNA such as globin mRNA, then the globin-Vg1 hybrid is localized to the vegetal pole just as the Vg1 mRNA is. Trans-acting elements include one or more RNA-binding proteins, which seem to anchor Vg1 mRNA to the ER (Mowry, 1996; Deshler et al., 1997). The transport of Vg1 mRNA also depends on microtubules, which presumably move the ER fragments to which Vg1 is bound. In addition, microfilaments are necessary to anchor the mRNA in the cortical cytoplasm around the vegetal pole (Yisraeli et al., 1990). Moreover, the anchorage of Vg1 near the vegetal pole depends on the previous localization of Xlsirt RNA; inactivation of this transcript with complementary oligonucleotides causes the release of anchored Vg1 mRNA (Kloc and Etkin, 1994).

While most cytoplasmic localizations occur during oogenesis, some happen as part of egg activation and others during cleavage. We have discussed the best-investigated example of cytoplasmic localization during egg activation, the ooplasmic segregation process in ascidians (Figs. 8.13 to 8.16). Localization during cleavage is especially apparent in molluscs that have a polar lobe (Figs. 8.4 to 8.6). Another case of localization during cleavage has been described in ctenophore *Mnemiopsis leidyi.* The larvae of these marine animals have cells with so-called comb plates for swimming and photocytes for generating light. These cell fates are determined by cytoplasmic localizations that depend on the proper orientation of mitotic spindles during early cleavage (Freeman, 1979). Taken together, the evidence discussed here indicates that a variety of cellular mechanisms, acting at different stages of development, have evolved to localize cytoplasmic determinants.

While cytoplasmic localization is nearly universal, it is certainly not the only way of cell determination. There are more than a hundred different cell types in an organism, but there are not nearly as many localized cytoplasmic determinants in any egg. It follows that, in addition to cytoplasmic localization, there must be other ways for cells to acquire different fates. An interaction between nonequivalent cells that changes the fate of at least one partner is called an *embryonic induction.* Their already existing differences may result from an earlier determinative event involving cytoplasmic localization. Inductive interactions in development often begin where cytoplasmic localizations leave off. For instance, the vegetal cortical cytoplasm of *Xenopus* eggs contains several localized RNAs (see Fig. 8.25) and other molecules, some of which determine cells that act as inducers later (see Section 9.3).

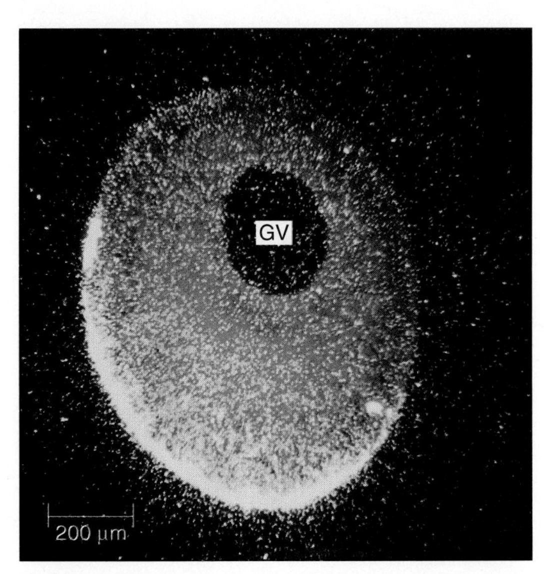

Figure 8.26 Localization of Vg1 mRNA near the vegetal pole of a *Xenopus* oocyte, as shown by in situ hybridization (see Method 8.1). Histological sections of the fixed oocyte were incubated with a radiolabeled RNA probe that forms stable hybrids with Vg1 mRNA. The radioactivity generates silver grains in a coat of film emulsion spread over the section. The silver grains appear as white dots under the dark-field illumination used for photography here. Control sections hybridized with probes binding to other mRNAs showed an even coverage with silver grains. GV = germinal vesicle.

SUMMARY

Most eggs have unevenly distributed components, called localized cytoplasmic determinants, that affect the fates of the blastomeres to which they are allocated. Such determinants become localized during oogenesis, upon fertilization, or during cleavage. As they are segregated into different blastomeres, they create different environments for their nuclei and cause differential gene expression.

Some determinants control the development of particular cell lineages, while others direct the formation of entire body regions encompassing many different cell types. In either case, cells that do not receive certain cytoplasmic determinants usually acquire alternate fates by default. Most determinants have an activating effect, although some are inhibitory.

Two strategies are generally used to identify localized cytoplasmic determinants. In one of these, known as the rescue strategy, a mutant or experimentally manipulated embryo is restored to normal development by transplantation of cytoplasm from a normal donor. If the rescuing activity is restricted to a specific area in the donor, then it follows that one or more components in the transferred cytoplasm are localized and necessary for the development of the parts that would otherwise be missing or abnormal.

Another strategy for demonstrating the involvement of a localized cytoplasmic determinant in embryonic development is heterotopic transplantation. In this procedure, cytoplasm containing a presumed determinant is transplanted from its normal location in the donor to a different region in the recipient. The same structure that would have formed at the donor site where the cytoplasm was removed may then be formed in the recipient at the site of transplantation. In this case, it can be concluded that one or more components of the transplanted cytoplasm are sufficient to cause the formation of the structure of interest. Both rescue and heterotopic transplantation can be used as bioassays to characterize the molecular nature and other properties of cytoplasmic determinants.

Maternally synthesized mRNAs seem to be the most common molecules to act as localized cytoplasmic determinants in eggs, whereas unevenly distributed proteins may control asymmetrical cell divisions of late embryonic cells. During the localization process, mRNAs acting as determinants may be associated via linker proteins with microtubules, the endoplasmic reticulum, or other cell organelles. Polar granules and yolk particles serve as temporary repositories and visible markers for some localized determinants. In other cases, localized determinants seem to be anchored to cytoskeletal components without visible markers.

Axis Formation and Mesoderm Induction

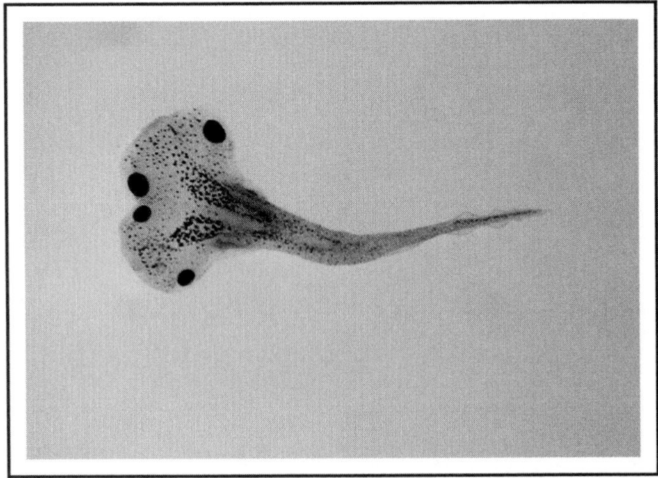

Figure 9.1 Duplication of dorsal organs in *Xenopus* embryos after ectopic injection of Xwnt-8 mRNA, which causes stabilization of β-catenin. This result and others demonstrate that β-catenin plays a central role in establishing the dorsal side of the embryo.

MOST animals have three body axes oriented at right angles to one another: anteroposterior, dorsoventral, and left-right. Some animals, and most plants, have just one axis. For instance, adult sea urchins have only the oral-aboral axis, which connects the oral pole (the mouth) with the aboral pole (near the anus). Generally, an axis is defined by two opposite poles, like the hardened tip and the notched end of an arrow. However, unlike an arrow whose shaft between the tip and the end is rather monotonous, a typical axis in an organism has various structures lined up from pole to pole in a specific order. For example, the bony elements along the anteroposterior axis of a mammal include the skull, the neck, the shoulder girdle with forelimbs, the thoracic vertebrae with ribs, and so forth, in this order.

How are axes specified during development? In most cases, it is useful to break this question down into two. First, how are the *poles* of an axis, such as the anterior and posterior poles, determined? Second, how is the *orderly array—or pattern—of structures* along the axis specified so that, for example, the thoracic ribs will follow behind the shoulder girdle and not the other way around?

Several ways of specifying embryonic axes have evolved (Goldstein and Freeman, 1997). A common way of determining one pole of an embryonic axis is by cytoplasmic localization—specifically, by the kind of cytoplasmic determinant that we have classified in Section 8.9 as region-specific as opposed to lineage-specific. As an example of such a determinant in the *Drosophila* oocyte, we have already introduced bicoid mRNA as defining the anterior pole. Later on, we will discuss how bicoid mRNA is translated into a protein with a concentration gradient that specifies the proper sequence of body elements along the anteroposterior axis.

For other embryonic axes, the poles are determined by external cues such as the point of sperm entry, gravity, or incident light. In amphibians, for example, sperm entry triggers movements in the egg cytoplasm that bias the side of sperm entry toward becoming the ventral side of the embryo. Interference with these cytoplasmic rearrangements or the subsequent signaling cascade can cause the loss or duplication of the dorsoventral axis (Fig. 9.1). The elaboration of the dorsoventral body pattern involves a series of *inductive interactions* by which cells influence the development of their neighbors. These interactions increase the complexity of the developing body pattern. From a simple axis with two poles, cellular interactions known as *embryonic induction* generate a complex pattern with several elements in a distinct order.

In this chapter, we will explore axis formation mainly in two organisms: a brown alga of the genus *Fucus* and the frog *Xenopus laevis*. *Fucus* has only one polar axis, which is induced by light and other environmental factors. In *Xenopus,* three body axes (anteroposterior, dorsoventral, and left-right) form in distinct ways. Our discussion will focus on the dorsoventral axis, which originates from cytoplasmic rearrangements triggered by sperm entry. Further establishment of the dorsoventral axis is linked to the development of the intermediate germ layer, the mesoderm, which arises from inductive interactions between animal and vegetal blastomeres.

Because induction plays a key role in stabilizing and elaborating embryonic body axes, the principle of induction will be introduced in this chapter, along with the criteria by which investigators prove the occurrence of inductive interactions between cells. Toward the end of the chapter, we will outline a molecular mechanism by which mesoderm induction and dorsoventral axis formation may act together in establishing Spemann's organizer.

9.1 Body Axes and Planes

Most metazoa have three ***body axes:*** an ***anteroposterior axis*** (Lat. *ante,* "before"; *post,* "after"), a ***dorsoventral axis*** (Lat. *dorsum,* "back"; *venter,* "belly"), and a ***left-right asymmetry*** (Fig. 9.2a). This nomenclature works well for most metazoa, for which the anterior pole comes first in normal forward locomotion and at the same time is the head end. In humans, because of their bipedalism, it is the ventral side rather than the head end that comes first in normal locomotion. Medical terminology therefore uses superior-inferior (top-to-bottom) or craniocaudal (head-to-tail) in lieu of anteroposterior. An additional axis, used in humans and nonhumans mostly for designating parts of a limb, is the ***proximodistal axis.*** *Proximal* (Lat. *proximus,* "next," "closest") describes the portion close to the point of attachment, to the center of the body, or to another point of reference. *Distal* (Lat. *distare,* "stand apart") refers to a part that is away (distant) from the point of reference. For example, the elbow is proximal to the wrist and distal to the shoulder. We also describe views of the body

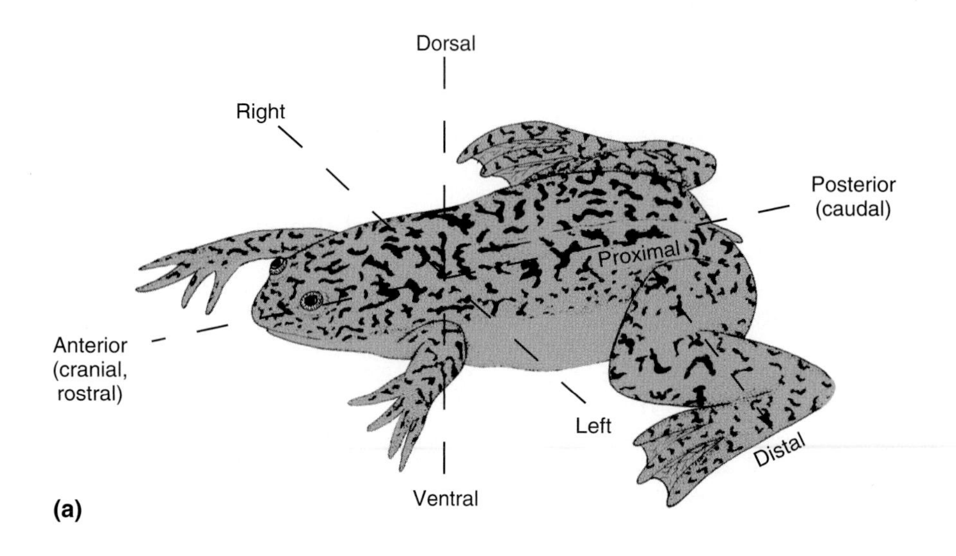

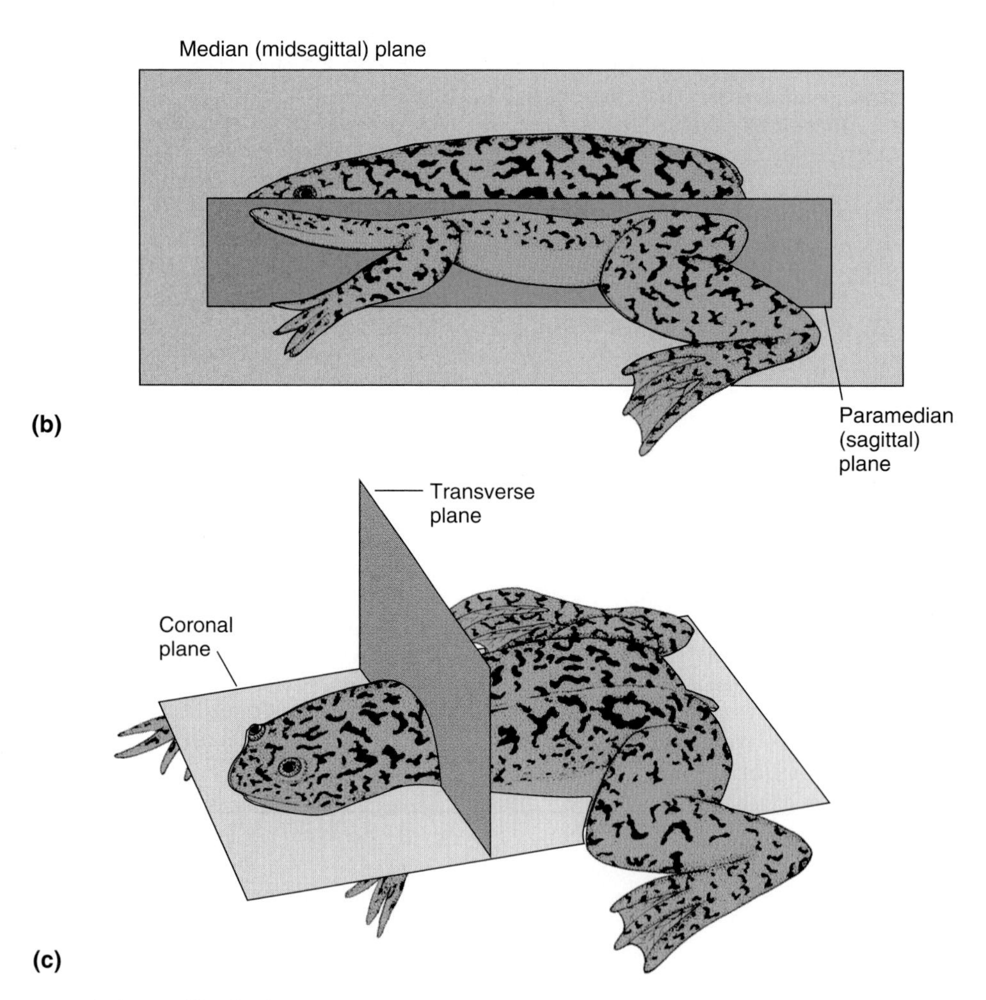

Figure 9.2 Body axes and planes. See text.

with regard to the poles of axes. For instance, a dorsal view is a view from the dorsal pole, and a lateral aspect is a view from the right or left side.

Sections of the body are named according to specific planes (Fig 9.2b, c). The ***median*** (or ***midsagittal***) ***plane*** divides the left and right sides of the body and is a plane of symmetry for most organs. A ***paramedian*** (or ***sagittal***) ***plane*** is any plane parallel to the median plane but displaced to the left or right side. The median plane and all paramedian planes are perpendicular to the left-right axis. Any plane perpendicular to the anteroposterior axis is a ***transverse plane,*** or, colloquially, a *cross section.*

Any plane perpendicular to the dorsoventral axis is a ***coronal plane,*** also known as a *frontal plane* in human anatomy.

9.2 Generation of Rhizoid-Thallus Axis in *Fucus*

The genus *Fucus* comprises large brown algae that are commonly seen in intertidal zones of North America and Europe. These algae have been favorite subjects for developmental studies because their organization is relatively simple (Fig. 9.3). They are also easy to keep and manipulate in the laboratory. In particular, large numbers of zygotes can be treated so that they develop in synchrony and with their polarity axes oriented the same way.

The egg of *Fucus* is a perfectly spherical cell with no apparent polarity. After fertilization, the egg forms a fertilization membrane, which is later modified into a cell wall like those that generally surround plant cells. At about 12 h after fertilization (at 15°C), the zygote bulges at one pole and becomes pear-shaped, in a process called ***germination*** (Fig. 9.4). About one day after fertilization, the zygote undergoes its first cell division, separating a small, pointed cell that includes the germination bulge from a larger, round cell. The small, pointed cell gives rise to the ***rhizoid,*** which anchors the plant to a rock or similar substrate. The large, round cell will form the leafy bulk of the plant called the ***thallus.*** The emerging axis between a thallus pole and a rhizoid pole is the *Fucus* plant's only polar axis. The embryonic development of many higher plants resembles that of *Fucus,* at least up to the 8-cell stage (see Section 24.1).

How is the rhizoid-thallus axis of *Fucus* established? In the absence of any orienting cues from the environment, the rhizoid pole forms at the site of sperm entry: The entering sperm somehow breaks the spherical symmetry of the egg, and the resulting bias prevails until the rhizoid pole becomes fixed many hours later (Hable and Kropf, 2000). However, the sperm's polarizing effect is easily overruled by various environmental cues. The best investigated of these signals, and presumably the natural orienting cue, is the direction of the incident light. The thallus pole forms on the side of the egg facing the light, and the rhizoid pole forms on the opposite side. This behavior seems highly adaptive because it maximizes the chance for the embryo to become anchored instead of being swept ashore or out to sea.

Under experimental conditions, the *Fucus* zygote can be polarized by light between 4 and 10 h after fertilization. During this period, the light-induced polarity remains as labile as the sperm-induced polarity: Another light pulse from a different direction will induce a new

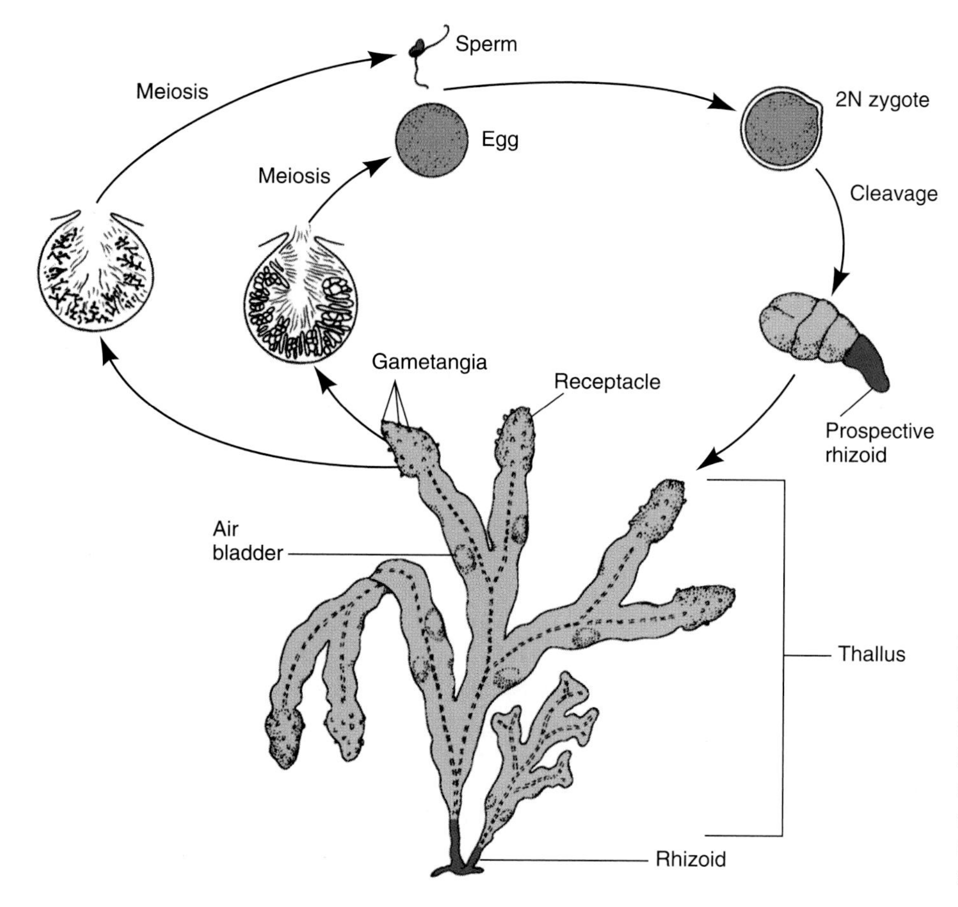

Figure 9.3 Life cycle of the brown alga *Fucus.* The adult plant consists of a rhizoid, which attaches the plant to a rock, and a large thallus, which is the photosynthetic organ. The receptacles at the tips of the thallus branches have openings leading to sex organs called gametangia, which produce eggs or sperm. Cleavage of the zygote is asymmetrical, separating prospective thallus cells from prospective rhizoid.

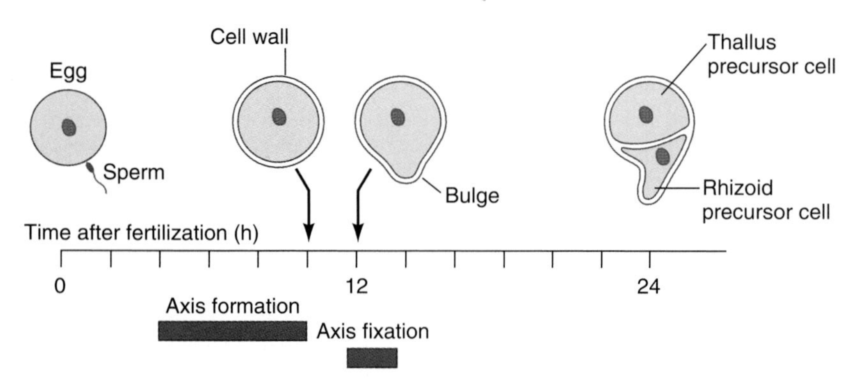

Figure 9.4 Axis formation and axis fixation in *Fucus.* Between 4 and 10 h after fertilization, the zygote appears spherically symmetrical but can be reversibly polarized by light and other environmental factors. Early during axis formation, Ca^{2+} channels accumulate near the future rhizoid pole, and the zygotes become electrically polarized. At the end of the axis formation period, microfilaments and Golgi-derived granules (F granules) accumulate near the future rhizoid pole. Between 10 and 12 h, the polarity of the zygote becomes fixed according to the last polarizing signal received. At about 12 h, a bulge appears at the future rhizoid pole. At about 24 h, the first cell division separates the rhizoid precursor cell from the thallus precursor cell.

polarity axis. However, between 10 and 12 h of development, the axis is fixed so that new orienting signals can no longer change it. We therefore distinguish a period of ***axis formation,*** when the thallus-rhizoid axis is set up in a preliminary way, from a period of ***axis fixation,*** when the axis is irreversibly established (Fig. 9.4).

The process of axis formation in *Fucus* zygotes was first analyzed by Lund (1922), who observed that zygotes grown in an electric field germinate toward the anode. Later, L. F. Jaffe (1966) found that in the absence of an external field, the germinating *Fucus* zygote polarizes electrically by itself, with the rhizoid end becoming negative. Similar observations were made on the zygote of a related genus of algae, *Pelvetia.* To measure the small voltages across the germinating *Fucus* zygote, Jaffe placed about 200 fertilized eggs in a glass capillary and induced axis formation by shining light from one end of the tube. This effectively connected the zygotes in series, thallus to rhizoid, so that the small electric potentials across the individual cells added up to a measurable potential across the entire tube. No voltage developed until 12 h after fertilization, when the zygotes began to germinate. However, as germination proceeded, there was a parallel rise in voltage across the tube; the end toward which the rhizoids formed became increasingly negative. No such voltage was observed in a control tube with zygotes that were illuminated from all sides and germinated randomly in all directions.

These measurements were refined with a vibrating electrode that can detect very small electric currents. Using this instrument, Nuccitelli and Jaffe (1974) found a small inward current at the prospective rhizoid site; this meant that either positively charged ions were entering or negatively charged ions were leaving the zygote at the prospective rhizoid site. These weak currents were detected before germination and often as early as 5 or 6 h after fertilization. In these cases, when the direction of the unilateral light was changed, the inward current shifted to the new prospective rhizoid position (Nuccitelli, 1978). Thus, the electric current through the zygote begins during axis formation and continues through axis fixation.

Further experiments were directed at identifying the carrier of the current. Initial observations suggested that the current may result from a flux of calcium ions (Ca^{2+}), which are present at much higher concentration in seawater than in living cells. Experiments with a calcium ionophore, an agent that makes plasma membranes permeable for Ca^{2+}, indicated that the rhizoid pole forms near the site of Ca^{2+} influx, which is then the site of highest Ca^{2+} concentration inside the cell. Further measurements showed that Ca^{2+} indeed enters germinating zygotes at the prospective rhizoid pole and leaves everywhere else (Robinson and Jaffe, 1975). In accord with these observations, a probe known to stain Ca^{2+} channels selectively labels the prospective germination site as early as 6 to 8 h after fertilization (Shaw and Quatrano, 1996). This localization of what seem to be Ca^{2+} channels is reversible: It can be changed—until the end of the axis formation phase—by changing the direction of the incident light.

Taken together, the experiments of Lund, Jaffe, and subsequent workers show that *Fucus* and *Pelvetia* zygotes drive an electric current, most likely carried by Ca^{2+} ions, through themselves as part of the axis formation process. The nature of the pigment that interacts with light in photopolarization, and the subsequent molecular events that orient the electric current through the zygote, are under active investigation.

Axis formation under the influence of unilateral light depends on microfilaments (Kropf, 1992). When *cytochalasin B,* a drug known to interfere with microfilament formation, is added to seawater with *Fucus* or *Pelvetia* zygotes during the axis formation period, photopolarization is inhibited. If the drug is subsequently removed, the rhizoids grow out in random directions. In *Pelvetia* zygotes polarized by light, a patch of microfilaments at the future rhizoid pole is an early—and reversible—sign of axis formation (Alessa and Kropf, 1999).

The process of *axis fixation* requires the participation of the *cell wall* (Quatrano and Shaw, 1997). Naked *protoplasts* isolated from *Fucus* zygotes retain their ability to repolarize in response to external light beyond the normal period of axis formation until they are allowed to form a cell wall again (Kropf et al., 1988). Normally, the zygote forms a cell wall within minutes after fertiliza-

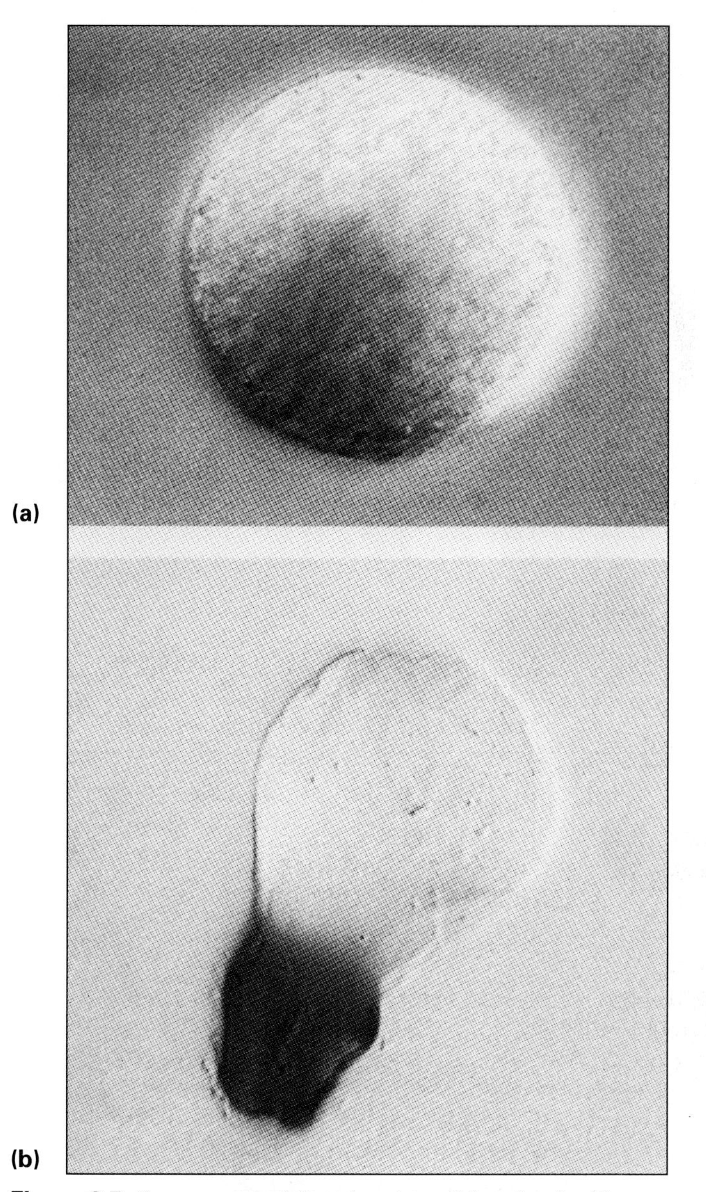

Figure 9.5 *Fucus* zygote **(a)** and embryo **(b)** stained with toluidine blue O to reveal the position of sulfated polysaccharides. A highly sulfated fucoidan designated F2 is locally incorporated into the cell wall before an asymmetry becomes morphologically apparent. The F2-containing part of the cell wall will cover the rhizoid cell of the embryo after the first cleavage division.

tion. The initial cell wall, which remains radially symmetrical throughout axis formation, consists of alginate, cellulose, and three fucoidans. One of the fucoidans, designated F2, becomes highly sulfated in specialized Golgi-derived vesicles, known as *F granules,* which can be stained specifically with toluidine blue O. Upon F2 sulfation, the F granules move toward the future rhizoid site, where F2 is incorporated in the cell wall (Fig. 9.5). The movement of F granules to the target site that is already identified by Ca^{2+} channels depends on microfilaments, suggesting that F2 deposition may be the step in axis fixation that is prevented by cytochalasin B.

To test the role of F2 incorporation into the cell wall, Shaw and Quatrano (1996) used an inhibitor, brefeldin A (BFA), which interferes with F granule exocytosis but not with the accumulation of Ca^{2+} channels or microfilaments. Much like the absence of a cell wall, the presence of BFA conserves the zygotes' ability to repolarize beyond its normal time until the BFA is washed out. Thus, F2 or other factors contained in F granules are required for and deposited at the time of axis fixation. BFA treatment does not interfere with the first cell division. When BFA is washed out after the first cleavage, the embryos form rhizoids belatedly. However, in contrast to normal embryos, where the first cleavage plane forms perpendicular to the budding rhizoid, the belated rhizoids of BFA-treated embryos have no apparent relation to the first cleavage plane. Indeed, some of these embryos form two rhizoids, apparently after the first cleavage has split the target site defined by Ca^{2+} channels and microfilaments. These data indicate that normal axis fixation entails an orientation of the first mitotic spindle parallel to the thallus-rhizoid axis. In contrast, BFA-treated cells seem to divide with a cleavage plane oriented randomly to the formed but not fixed thallus-rhizoid axis.

The current model of axis fixation in *Fucus* is diagrammed in Figure 9.6. Preceding steps of axis formation have culminated in a labile accumulation of Ca^{2+} channels, Ca^{2+}, and microfilaments at the prospective rhizoid pole. These factors promote the exocytosis of F granules and local incorporation of their contents in the cell wall. The locally modified cell wall signals back to the cell, providing orienting cues for the positioning of the first mitotic spindle and possibly other aspects of cell polarization.

In summary, studies on *Fucus* and *Pelvetia* have shown that the thallus-rhizoid axis is first formed in a preliminary way, oriented either by the site of sperm entry or by environmental cues. After a labile period, the axis is fixed by the local exocytosis of Golgi-derived secretory granules at the future rhizoid pole. A similar two-step process directs the development of the dorsoventral body axis of the *Xenopus* embryo, as will be discussed later on in this chapter.

9.3 Determination of the Animal-Vegetal Axis in Amphibians

Amphibian development involves the formation and fixation of three embryonic axes. The *anteroposterior* axis develops during gastrulation from the *animal-vegetal axis,* which originates during oogenesis. The *dorsoventral* axis is formed after fertilization and fixed before the first cleavage. The *left-right* asymmetry is not overt until the heart and other internal organs become asymmetrical during embryogenesis. However, molecular cues to left-right asymmetry have been detected during the blastula and early gastrula stages.

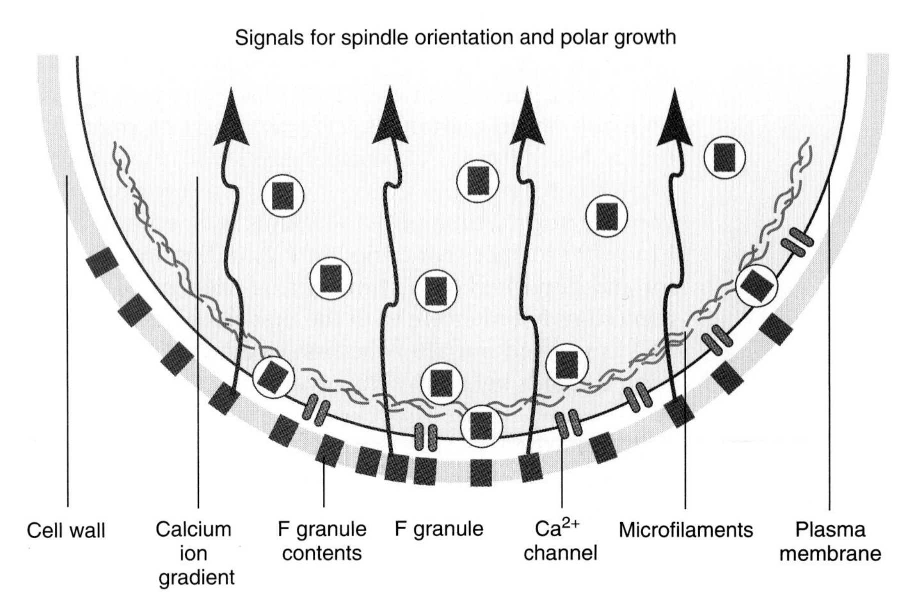

Figure 9.6 Model of axis fixation in the *Fucus* zygote. Accumulations of microfilaments, Ca^{2+} channels, and free Ca^{2+} ions target the future rhizoid pole in a labile way at the end of the axis formation period (see Fig. 9.4). Axis fixation occurs when Golgi-derived F granules move toward the prospective rhizoid pole and undergo exocytosis, so that F2 fucoidan and possibly other granule contents become incorporated into the cell wall. The locally modified cell wall is thought to signal back (arrows) to the cell, providing cues for the orientation of the first mitotic spindle and possibly other aspects of axis fixation.

Pigmentation of the animal cortex is not necessary for development of the normal body pattern; albino strains of Xenopus lacking the dark pigment develop otherwise normally, at least in the laboratory. In the frogs' natural habitat, where clutches of eggs float at the surface of ponds, the dark pigmentation of the animal hemisphere does confer advantages. First, it absorbs light, converting it to heat that will accelerate development. Second, the dark pigment of the animal half and the whitish color of the vegetal half both provide camouflage. Because amphibian eggs can rotate within the fertilization envelope, the vegetal half will always face down due to its greater buoyant density. The whitish color of the vegetal half is hard to see for predators in the water looking up against the bright sky. At the same time, the eggs' dark side is facing up, making them hard to spot for predators from land or air.

THE ANIMAL-VEGETAL AXIS ORIGINATES BY ORIENTED TRANSPORT DURING OOGENESIS

The animal-vegetal axis of the amphibian egg is revealed during oogenesis by the proximity of the *germinal vesicle* to the animal pole and by the accumulation of the *mitochondrial cloud* between the germinal vesicle and the vegetal pole (see Fig. 8.25). This structural polarity is part of a transport system that localizes several maternal RNAs to the vegetal pole, as discussed in Section 8.9. Other maternal RNAs are localized to the animal pole (Rebagliati et al., 1985; *M. L. King et al., 1999*).

The animal-vegetal polarity is also reflected in an accumulation of a yolk protein, *vitellin,* which is stored in the oocyte in crystallike *yolk platelets.* During the storage process, there is an overall transport of yolk protein toward the vegetal hemisphere, which eventually contains about 70% of the total vitellin in the oocyte. The same transport sets up a gradient in yolk platelet size. The largest platelets, having a diameter of 10 to 15 μm in *Xenopus,* are all located in the vegetal part of the egg. The animal region contains smaller platelets, about 2 to 4 μm in diameter; and intermediate sizes are found between the animal and vegetal pole regions.

Toward the end of oogenesis, the cortical cytoplasm in the animal hemisphere of amphibian eggs becomes darkly pigmented. There is no such pigment in the vegetal hemisphere, which therefore has a whitish appearance derived from densely packed yolk platelets.

THE ANIMAL-VEGETAL POLARITY DETERMINES THE SPATIAL ORDER OF THE GERM LAYERS

The animal-vegetal polarity in amphibian eggs is reflected in the spatial order of the germ layer rudiments at the blastula stage. Most of the animal half of the blastula forms ectoderm, and most of the vegetal half forms endoderm, while the intermediate zone gives rise to mesoderm (see Fig. 6.2). During gastrulation, mesoderm and endoderm turn inward while the ectoderm expands to cover the entire embryo (see Fig. 6.10). The portion of the expanded ectoderm that stays close to the animal pole will form anterior structures, such as brain, sense organs, and head epidermis. The opposite portion of the expanded ectoderm will give rise to posterior structures, such as spinal cord and the epidermis of trunk and tail. Thus, the animal-vegetal polarity of the blastula translates approximately into the anteroposterior polarity of the ectoderm. For mesoderm and endoderm, the relationship between animal-vegetal and anteroposterior polarity is complicated by gastrulation movements (see Section 10.3).

The determination of animal and vegetal blastomeres is controlled by cytoplasmic factors that show uneven distributions along the animal-vegetal axis (see Fig. 8.25). One of these determinants, a maternally encoded mRNA designated VegT, is localized to the vegetal pole region (J. Zhang and King, 1996; Stennard et al., 1999). Depletion of VegT leads to dramatic shifts in the blastula fate map (Fig. 9.7; J. Zhang et al., 1998). Endo-

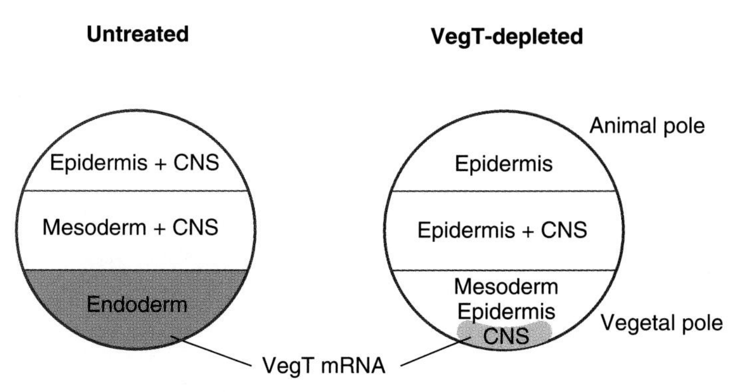

Figure 9.7 Effects of VegT mRNA depletion on the formation of germ layers in *Xenopus.* VegT mRNA (color) is localized to the vegetal pole region. Depletion of this mRNA leads to dramatic changes in the fate map at the blastula stage. In normal embryos, the vegetal region forms endoderm, the animal region forms epidermis and central nervous system (CNS), and the marginal zone in between forms mesoderm and CNS. In embryos depleted of VegT mRNA, the vegetal region forms mesoderm, epidermis, and CNS. The animal regions forms only epidermis, and the marginal zone forms epidermis and CNS.

derm is replaced with mesoderm and ectoderm, mesoderm is replaced with ectoderm, and ectoderm forms only epidermis and no nervous system. These results indicate that VegT protein is required for the proper formation of endoderm and mesoderm. Since VegT shows the molecular characteristics of a transcription factor, these results suggest that the involvement of VegT in germ layer formation does not begin until *midblastula transition.* However, other steps required for the generation of mesoderm seem to begin earlier, as indicated by the following observations.

Fate maps of *Xenopus* at the 32-cell stage have been prepared by several authors (see Fig. 1.11; Dale and Slack, 1987; Moody, 1987; Vodicka and Gerhart, 1995). The animal tier of blastomeres will form ectodermal structures, primarily epidermis. The second and third tiers of blastomeres will contribute to mesodermal structures, including the notochord, somites, lateral plates, and blood cells. The vegetal tier of blastomeres will form endodermal structures, mostly gut. If instead of fate one maps the *potency* of isolated tiers, a major difference emerges (Fig. 9.8). While the animal tier and the vegetal tier develop according to fate, the intermediate tiers do not: Instead of contributing to mesodermal structures they form mostly ectodermal derivatives (Smith, 1989). If the descendants of these cells are isolated at the 128-cell stage they form small amounts of mesodermal structures. The proportion of mesoderm increases as the cells are isolated later until eventually, at the late blastula stage, isolated intermediate cells develop in accord with their fate. These observations show that mesodermal cells are determined progressively during the blastula stage.

Because the cells with mesodermal fates arise between prospective ectoderm and endoderm, one possible mechanism for prospective mesodermal cells to acquire their fates would be by inductive interaction between animal and vegetal cells.

VEGETAL BLASTOMERES INDUCE THEIR ANIMAL NEIGHBORS TO FORM MESODERM

The origin of mesodermal cells by inductive interactions between vegetal and animal blastomeres was demonstrated by Nieuwkoop (1969a) with embryos of the axolotl *Ambystoma mexicanum,* and by Sudarwati and Nieuwkoop (1971) with *Xenopus* embryos. The investigators isolated and combined different parts in mid to late blastula stages, kept them in tissue culture, and

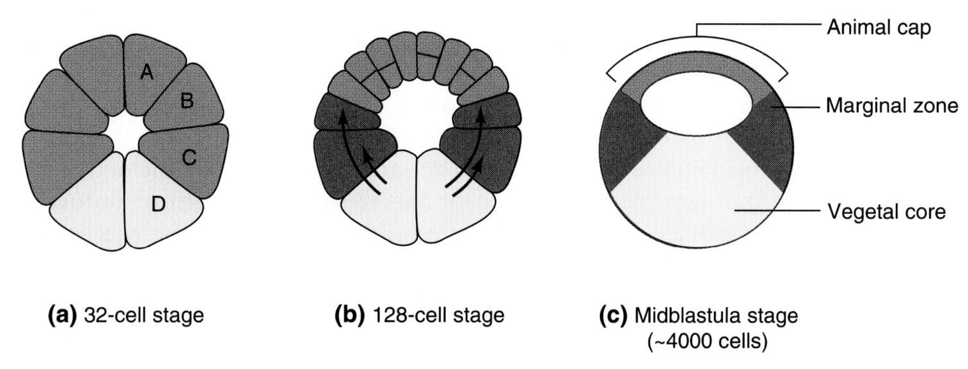

Figure 9.8 Development of isolated blastomere tiers in *Xenopus.* **(a)** At the 32-cell stage, an isolated animal tier (A) and an isolated vegetal tier (D) will develop according to fate, with A forming ectoderm and D forming endoderm. However, isolated intermediate tiers (B and C) do not develop according to fate. Instead of contributing to mesoderm, as they do in normal development (see Fig. 1.11), they will form mostly ectoderm if isolated at this early stage. **(b)** In contrast, at the 128-cell stage, descendants of tier C contribute to mesoderm both in normal development and upon isolation. Thus, an additional determined state (red color) has originated among the descendants of C. Further experiments (see Fig. 9.9) demonstrate that this change results from inductive signals (arrows) from vegetal blastomeres to their equatorial neighbors. **(c)** Continued induction establishes a girdle of prospective mesoderm cells known as the marginal zone.

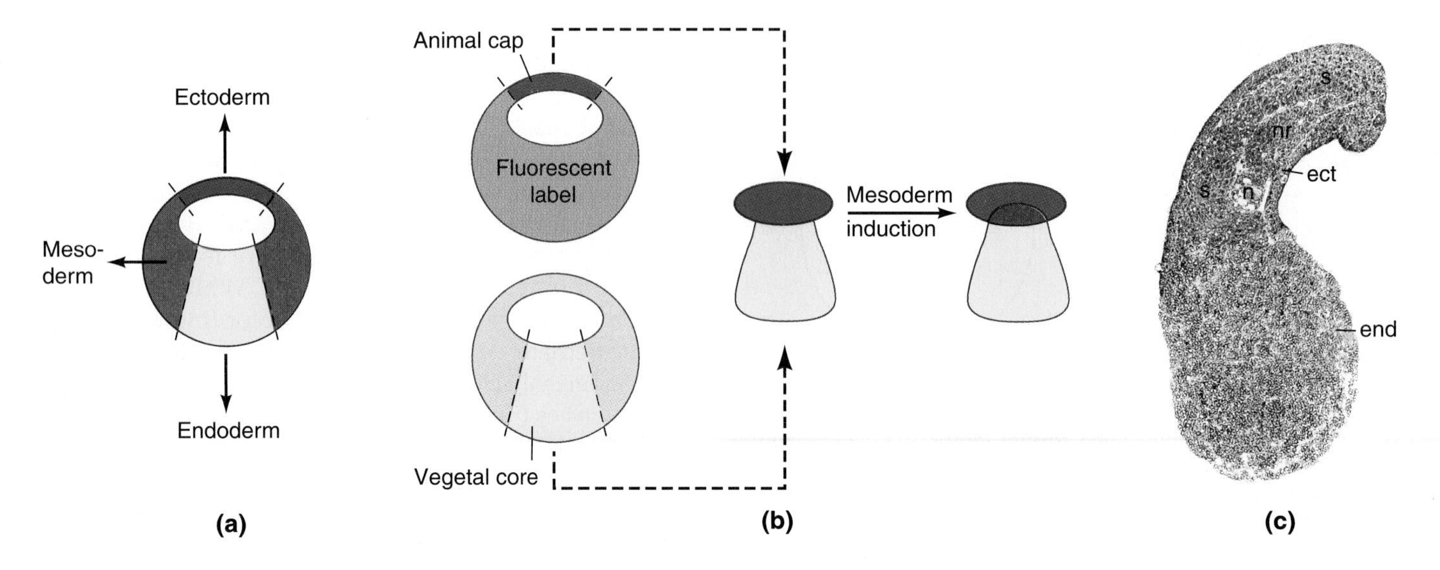

Figure 9.9 Mesoderm induction in the *Xenopus* blastula. **(a)** Potencies of isolated blastula areas in tissue culture: caps of animal cells give rise to ectodermal tissue (ciliated epidermis), while cores of vegetal cells form endodermal tissue (large, yolk-laden cells). The marginal cells, constituting a circular band between the animal cap and the vegetal core, form mostly mesodermal structures including the notochord, muscle, mesenchyme, and blood cells. The marginal zone is cut larger here than the fate map would require (see Fig. 9.8c), and the animal cap and vegetal core are cut correspondingly smaller, in order to ensure that the latter do not contain any prospective mesoderm. **(b)** Experiment demonstrating mesoderm induction: if animal caps from a fluorescently labeled donor are cultured in contact with unlabeled vegetal cores, the animal cells near the vegetal core are found to form mesodermal structures. **(c)** Histological section showing induced mesoderm, including notochord (n) and muscle tissue (s), between endodermal (end) and ectodermal (ect and nr) structures.

analyzed the resulting structures by microscopy. Figure 9.9 summarizes their findings. Isolated *vegetal core* pieces persisted as groups of large, yolk-laden cells similar to those forming the floor of the embryonic gut during normal development. Isolated *animal* regions, so-called *animal caps,* contracted into spheres of ciliated epidermis. Isolated intermediate zone elongated considerably in culture and formed mostly mesodermal structures, including a notochord, muscle precursors, kidney tubules, and blood cells.

The most revealing result was obtained when isolated vegetal cores were combined with isolated animal caps. Under such conditions, mesodermal structures were formed, including a notochord, muscle tissue, and embryonic kidney. While neither animal caps nor vegetal cores by themselves gave rise to mesoderm, mesoderm formed when the two were cultured in juxtaposition. In other words, mesoderm formed as the result of an interaction between animal and vegetal cells. This conclusion also explained the progressive formation of mesodermal structures by intermediate tiers of blastomeres isolated at successive stages as described earlier.

An interaction between *nonequivalent* cells that changes the fate of at least one partner is called an ***embryonic induction.*** Induction implies that the responding and inducing cells are already different before they interact. Their already existing differences may result from an earlier determinative event such as cytoplasmic localization.

The intermediate zone in amphibian blastulae that lies between animal and vegetal cells and becomes induced to form mesoderm is referred to as the ***marginal zone.*** In order to determine whether the marginal zone is derived from animal cells or vegetal cells, researchers have used labels or combined tissues from different species with different pigments. Dale and coworkers (1985) labeled animal caps with a fluorescent dye and combined them with unlabeled vegetal cores. They found that all clearly identifiable mesodermal structures were labeled and therefore derived from animal caps. The same experiments indicated that approximately half of the tissue derived from induced animal caps was mesodermal while the rest was ectodermal.

Several investigators have carried out similar experiments using molecular markers instead of histological analysis as a measure of mesoderm induction. For instance, Gurdon and coworkers (1985) monitored the synthesis of the mRNA for actin, an abundant protein in embryonic muscle. They found that this mRNA was not synthesized in isolated animal caps or vegetal cores, but was synthesized in combinations of animal caps with vegetal cores and in isolated marginal zones. Thus, the pattern of actin mRNA synthesis parallels the formation of the histological structures characteristic of mesoderm.

In summary, the vegetal cells of amphibian blastulae induce their animal neighbor cells to form mesoderm. The response to the inductive signal is limited to the marginal zone, where animal and vegetal cells are close to each other. The cells of the animal cap, which are separated from vegetal cells by many other animal cells and by the blastocoel, maintain their determination to form ectodermal derivatives. Presumably, the range of the inductive signals is limited by slow transport or rapid breakdown. Such properties would allow different region-specific signals to spread side by side.

9.4 The Principle of Induction

The determination of cells by embryonic induction is a very pervasive principle of development (Jacobson and Sater, 1988). Inductive interactions may occur between single cells or groups of cells. The cells that undergo a change in their determined state are called ***responding cells,*** and the cells that cause this change are called ***inducing cells.*** As a result of early inductive interactions, inducing and responding cells often form adjacent precursor cells or rudiments, as in the case of the amphibian germ layers discussed in the previous section. Inductive interactions that occur later, during *organogenesis,* typically generate complementary parts of the same organ. For instance, the rudiment of the retina of the eye—the *optic vesicle*—induces the formation of the rudiment of the lens—the lens placode (see Fig. 1.17).

Several advantages of induction as a developmental mechanism are apparent. First, induction increases complexity. In the amphibian blastula, the number of determined cell states at the 32-cell stage is two: ectoderm and endoderm (see Fig. 9.8a). After mesoderm induction, the number of determined states has increased to three: ectoderm, mesoderm, and endoderm (see Fig. 9.8b). As we will discuss later on, the combination of mesoderm-inducing and dorsalizing signals first generate two mesodermal states, dorsal and ventral, and then further interactions between dorsal and ventral mesoderm generate additional types of mesoderm in between. In other words—whereas the poles of an embryonic axis are typically determined by cytoplasmic localization, induction is the primary mechanism of generating arrays of different structures along these axes. Second, the proximity of inducing and responding cells is an effective guarantee that the structures for which they are fated will arise next to each other and in matching sizes. In the case of the eye, lens induction by the prospective retina prevents the lens from forming off-center, or in a size that does not fit the retina. Finally, because induction occurs typically at close range, many inductive interactions can go on at the same time. For example, key parts of nose, eye, and ear are all induced simultaneously by different parts of the brain in vertebrates.

Induction was already recognized as a basic principle by the earliest experimental embryologists. At the beginning of the twentieth century, the concept of induction was buttressed with rigorous *operational criteria* by the work of Hans Spemann in Germany and Warren Lewis in America (Hamburger, 1988). These criteria for demonstrating inductive interaction are *rescue* and *heterotopic transplantation,* the same criteria also used for proving the action of cytoplasmic determinants, except that cells or tissues are transplanted instead of cytoplasm. However, this is not the only parallel between cytoplasmic localization and induction. Both are also based on the *principle of default programs.* The absence of a particular localized determinant or inductive signal does not result in chaos or cell death; instead, cells have an alternative pathway of development to fall back on if they do not receive the particular signal. For instance, animal cells of a frog blastula that receive no mesoderm-inducing signal form ectodermal derivatives. Also, like cytoplasmic determinants, inductive signals may *activate or suppress* certain cell fates. For instance, the formation of the lens of the eye is promoted by inductive signals from pharyngeal endoderm, heart mesoderm, and the prospective retina, but inhibited by signals from *neural crest* cells, which arise at the interface between neural plate and prospective epidermis (Jacobson, 1966; Grainger et al., 1988).

Inductive interactions occur during certain sensitive phases of development. The ability of the responding tissue to react to an inducer by changing its determined state is called its ***competence.*** The period of competence typically begins some time before the normal inductive interaction occurs and ends thereafter. Likewise, the ability of an inducing tissue to affect a responding tissue is limited to a window of time before and after the normal inductive interaction. The periods of inductive ability and responsive competence can be revealed by *heterochronic transplantation*—that is, by combining an inducer of a given age with younger or older responsive tissues, and vice versa. Use of this strategy revealed, for instance, that animal caps of *Xenopus* embryos lose their competence to respond to the mesoderm-inducing signal at the early gastrula stage (Dale et al., 1985).

9.5 Determination of the Dorsoventral Axis in Amphibians

The second polarity axis of the amphibian embryo, the *dorsoventral axis,* arises from cytoplasmic rearrangements that occur as part of egg activation, not unlike the ooplasmic segregation movements of ascidians described in Section 8.5. These rearrangements occur during the first cell cycle, between egg activation and first cleavage. Similar to the thallus-rhizoid polarity of brown algae, the dorsoventral polarity of amphibians is reversible for a short period before it becomes fixed.

DEEP CYTOPLASM UNDERGOES REGULAR MOVEMENTS DURING EGG ACTIVATION

Several observations suggest that the proper formation of the dorsoventral axis may depend on ***deep cytoplasmic movements*** that occur during egg activation in a regular relation to the point of sperm entry. For instance, the dorsal blastopore lip, which is the site of beginning gastrulation and marks the future dorsal side of the embryo, always appears where cells containing large yolk platelets abut cells with small yolk platelets, suggesting that the intervening cytoplasm with intermediate platelets has been displaced (Pasteels, 1964).

To analyze the deep cytoplasmic movements in *Xenopus* eggs, Danilchik and Denegre (1991) *pulse-labeled* layers of yolk platelets by injecting females with a fluorescent dye that binds to vitellogenin. Eggs subsequently laid by such females were either fertilized or activated by electric current. Amphibian sperm enter the egg anywhere in the pigmented animal hemisphere, and the *sperm entry point* is visible for a short while as a concentration of pigment. This point was marked permanently with a dye spot applied to the surface of the fertilized eggs. Because the blastopore usually develops opposite the sperm entry point, the dye spot could be used to predict the future dorsal side of the embryo with some reliability.

Eggs prepared in this way were allowed to develop for increasing period of time before they were fixed and sectioned for microscopy. This procedure revealed complex movements of deep cytoplasm. During the first cell cycle following fertilization, such movements produced a *swirl of cytoplasmic layers* on the future dorsal side of the embryo (Fig. 9.10). This swirl consisted of alternating labeled and unlabeled layers of cytoplasm that had not been in contact previously. The swirl occurred in fertilized eggs as well as in eggs activated by electric current, although in the latter case the swirl's orientation was unpredictable.

How may these swirls be involved in setting up dorsoventral polarity? Conceivably, the same cytoplasmic rearrangements that mix labeled and unlabeled layers of yolk also bring together two previously segregated partners of a chemical reaction. For instance, a kinase or protease may be combined with its substrate protein, which may then be phosphorylated or cleaved so that it can snap into its biologically active conformation. Such an activated protein could in turn control the translation of a maternal mRNA or modify the assembly of a cytoskeletal component. Because cytoplasmic rearrangements were observed throughout the egg, factors segregated into animal as well as vegetal blastomeres might be affected by them.

CYTOPLASMIC REARRANGEMENTS FOLLOWING FERTILIZATION INVOLVE CORTICAL ROTATION

The best-investigated aspect of the cytoplasmic rearrangements in activated amphibian eggs in known as

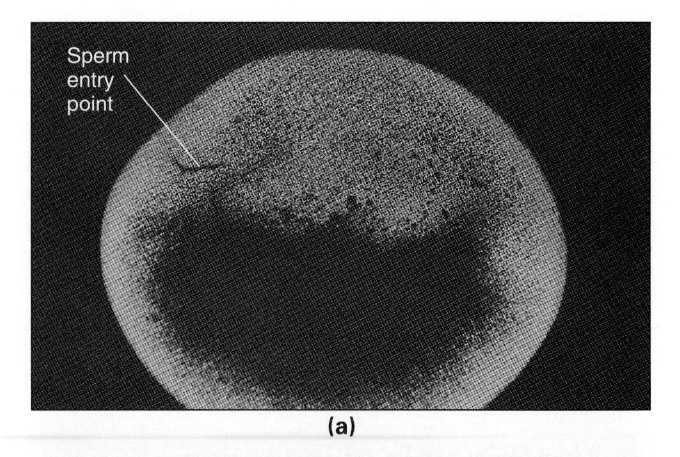

(a)

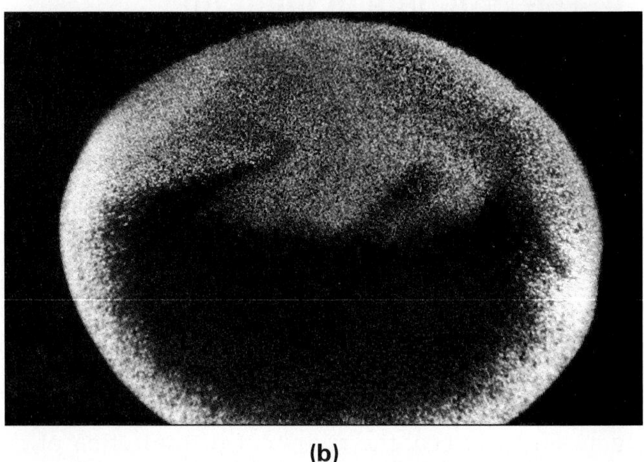

(b)

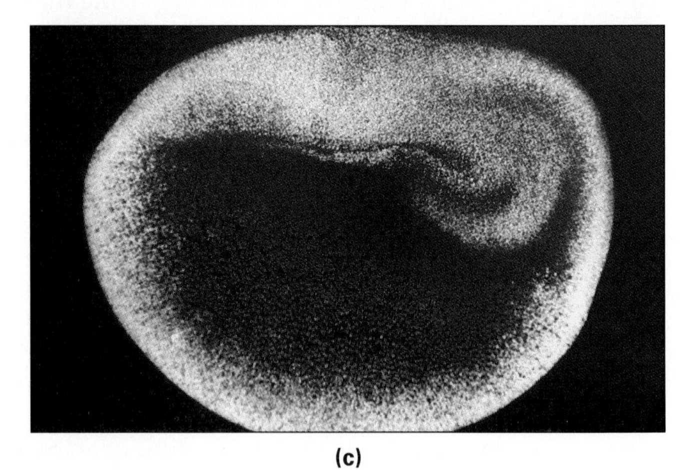

(c)

Figure 9.10 Rearrangement of cytoplasm in *Xenopus* embryos during the first cell cycle. *Xenopus* females were injected prior to spawning with a fluorescent dye that binds to vitellogenin. Yolk platelets that formed after the injection appear white in these photomicrographs. Eggs were fertilized, and embryos were fixed for histological sectioning at various times during the first cell cycle. All sections are oriented with the animal pole up and the sperm entry point to the left. Thus, the future dorsal side is to the right. **(a)** 30 min after fertilization. Labeled cytoplasm in the central animal hemisphere has shifted slightly in the dorsal direction. **(b)** 45 min after fertilization. Central animal cytoplasm has shifted further dorsally. **(c)** 90 min after fertilization (first mitosis). Shifted animal cytoplasm has generated a swirl, consisting of alternating labeled and unlabeled cytoplasm, in the dorsal region.

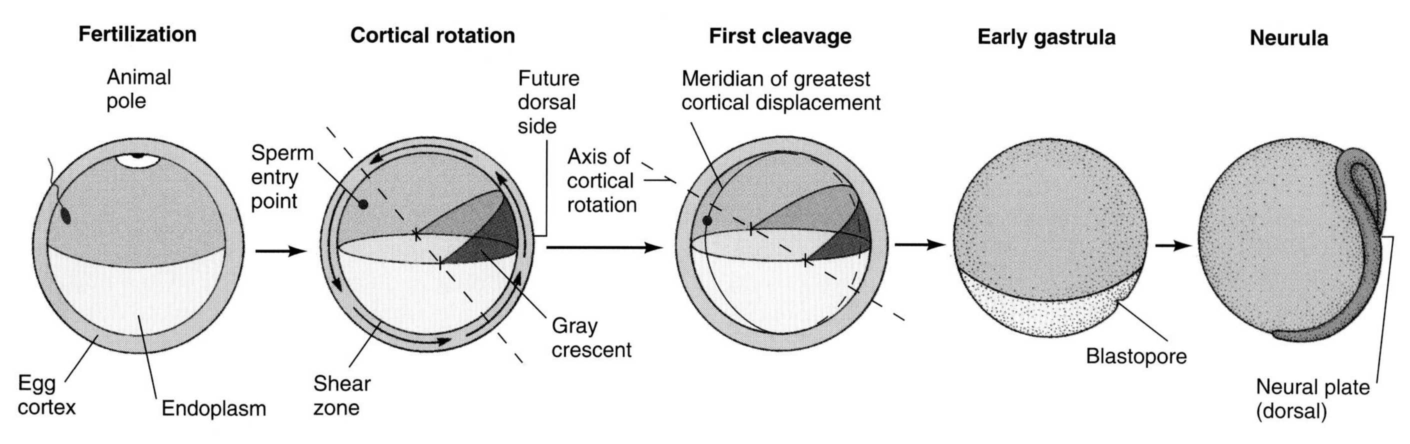

Figure 9.11 Relationship between fertilization, cortical rotation, gray crescent formation, and establishment of the dorsoventral axis in amphibian eggs; animal pole at the top, sperm entry point oriented to the left. The cortex of the egg is shown as a shell surrounding the deeper endoplasm. (The thickness of the cortex is exaggerated in this drawing; in reality, it accounts for about 1.5% of the egg radius; see Fig. 9.12b.) The cortical rotation (arrows) pivots on an axis (dashed line) that is perpendicular to the animal-vegetal axis. A shear zone marks the boundary between the rotating cortex and the endoplasmic core. In many amphibian species, a gray crescent forms near the egg equator and usually opposite the sperm entry point (see Fig. 9.13). The gray crescent marks the future dorsal side of the embryo, where the blastopore will originate and the neural plate will form later. The meridian that bisects the gray crescent is the line of greatest cortical displacement. This meridian is located within the future median plane, which separates the right and left halves of the body.

cortical rotation. A thin outer shell of cytoplasm, called the *cortex,* rotates relative to the massive inner core of the egg, called the *endoplasm* (Fig. 9.11; Gerhart et al., 1989). Cortical rotation pivots around an axis that is perpendicular to the animal-vegetal axis, and the rotation displaces the entire cortex relative to the core by an arc of about 30°. The most conspicuous result of cortical rotation—although mechanistically not the most important one—is the formation of a ***gray crescent*** at the egg surface near the equator. The gray crescent is best visible in those amphibian species that have most of their animal pigment associated with the cortex and some pigment with the endoplasm. Because the direction of cortical rotation in the area of gray crescent formation is toward the animal pole, the rotation causes the replacement of heavily pigmented animal cortex with clear vegetal cortex, making the sparse pigment of the exposed endoplasm appear gray (Fig. 9.11; see also Fig. 9.13).

Unless the egg is perturbed by further manipulations, the crescent predicts *exactly* where the blastopore will originate during gastrulation, and thus, where the dorsal organ rudiments will form. The widest part of the crescent is centered over the meridian of greatest cortical displacement toward the animal pole, and the plane of this meridian will become the embryo's *median plane.* As a rule, the gray crescent develops opposite the sperm entry point, and even when it does not, the gray crescent is a reliable predictor of the future dorsal side (Danilchik and Black, 1988). Similarly, the first cleavage furrow often bisects the gray crescent, so that each of the first two blastomeres will form a lateral half of the embryo. Again, in cases when the first cleavage furrow does not bisect the gray crescent, it is the gray crescent and not the furrow that predicts the median plane of the embryo.

For an experimental analysis of the cellular mechanism that drives cortical rotation, investigators have applied various inhibitors of cytoskeletal components. Agents that depolymerize microtubules, such as colchicine, nocodazole, cold shock, and hydrostatic pressure, all inhibit cortical rotation. Moreover, embryos that fail to undergo cortical rotation are ventralized, forming ventral structures such as gut tissue and blood cells but no dorsal structures such as notochord or central nervous system (Vincent and Gerhart, 1987). Irradiation of the vegetal surface of fertilized eggs with ultraviolet (UV) light also inhibits cortical rotation and causes ventralization (Malacinski et al., 1977). This result is significant because, unlike chemical microtubule inhibitors, UV light penetrates only a few micrometers into the cytoplasm and thus defines the location of its target. Indeed, other investigators (Elinson and Rowning, 1988; Houliston, 1994) have found an array of microtubules located just inside the vegetal cortex, present only during cortical rotation, and oriented with their plus ends toward the site of gray crescent formation.

The following microscopic analysis was designed to ascertain exactly where the microtubular array is located relative to the shear zone between the rotating cortex and core endoplasm. Another goal was to reveal exactly when the microtubular array was present relative to the timing and velocity of cortical rotation.

TO answer these questions, Larabell and coworkers (1996) used a *confocal microscope*—that is, a light microscope that "cuts optical sections" by producing a sharp image of the object in the focal plane, even if there are relatively thick layers of tissue above and below the focal plane. Their microscope was also *inverted,* meaning that the front lens was positioned *below* the object, so that they could observe the vegetal pole region of living eggs held in their normal position. To immobilize *Xenopus* eggs, they cemented the fertilization envelope on the thin glass cover slip that separated the egg from the front lens. They also removed osmotically the fluid from the perivitelline space between the fertilization envelope and the plasma membrane so that the egg was not stuck to the inside of the fertilization envelope. Under these circumstances, of course, the egg cortex could no longer rotate around the endoplasm. Instead, the endoplasm rotated in the opposite direction, away from the site of the gray crescent formation, and usually, toward the point of sperm entry.

Focusing first on the egg's plasma membrane and then progressively deeper into the egg, the investigators could not detect any moving cytoplasmic components within the first 4 μm of cortical cytoplasm. Between 4 and 14 μm into the vegetal cytoplasm, they observed yolk platelets moving with a regular timing and speed (Fig. 9.12). On a normalized time (NT) scale from 0.0 (fertilization) to 1.0 (first

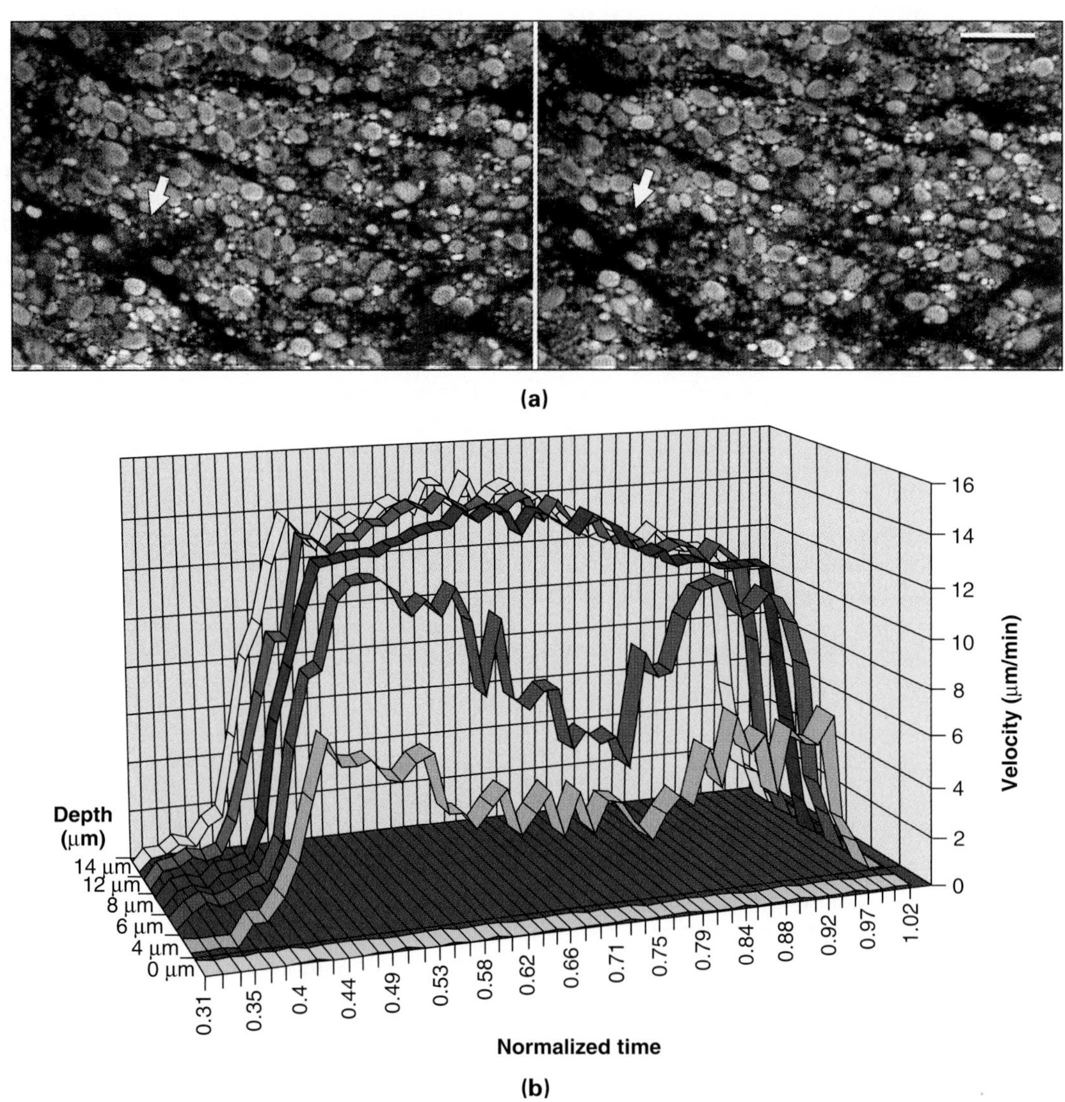

Figure 9.12 Cortical rotation in a living *Xenopus* egg. **(a)** Photographs taken with an inverted confocal microscope show optical sections at a depth of 8 μm inside the plasma membrane in the vegetal pole region. The pebble-shaped bodies are stained yolk platelets; the dark grooves between them result from the displacement of yolk platelets by microtubule bundles, as revealed by a different staining not shown here. The two frames were taken 3.1 min apart during the first cell cycle. During this interval, yolk bodies and microtubules traveled together over a distance of 24 μm (scale bar: 30 μm); note the same "X"-shaped configuration (white arrows) of crossed microtubule bundles with associated yolk platelets in both frames. **(b)** Timing and velocity of cortical rotation at different depths below the plasma membrane in the vegetal pole region of a living *Xenopus* egg. Data were obtained by videorecording as shown in part a. The normalized time (NT) axis covers most of the interval between fertilization (NT = 0.0) and first cleavage (NT = 1.0). No movement was detected immediately below the plasma membrane (0 μm). Progressive faster rotation was seen at depths of 4, 6, and 8 μm inside the plasma membrane, whereas beyond 8 μm no significant further increase in velocity was noted. Cortical rotation began slowly around 0.3 NT, accelerated sharply between 0.4 and 0.5 NT, plateaued between 0.5 and 0.8 NT, and ceased abruptly thereafter. Similar measurements on other *Xenopus* eggs showed similar kinetics, with an average plateau velocity of 11 μm/min.

cleavage), movements began slowly around 0.3 NT, accelerated between 0.4 and 0.5 NT, plateaued until 0.9 NT, and then decelerated sharply. The average plateau speed depended on the depth below the plasma membrane at which the measurements were taken, increasing from 4 μm per minute at 4 μm depth to 11 μm per minute at 8 μm and deeper levels.

Bundles of oriented microtubules were found precisely within the shear zone between 4 and 8 μm depth in which the velocity of the cortical rotation increases steeply. These microtubular bundles moved in lockstep with associated yolk platelets, and the overall orientation of the microtubules was in the direction of the movement. The simplest interpretation of these data is that *motor proteins* associated with the immobilized egg cortex drive the microtubules lying outermost in the shear zone, while motor proteins associated with these microtubules drive the microtubules lying to the inside of them, imparting greater speed on them than they have themselves, and so forth. This would be like people walking while carrying over their heads a long plank on which more people walk in the same direction carrying another plank over their heads, and so on.

Questions

1. If the investigators had not removed the perivitelline fluid from the eggs observed, at which level relative to the plasma membrane should they have seen the fastest movements of cytoplasmic components?
2. Remarkably, Larabell and her colleagues did not find oriented microtubules until 0.4 NT, which is well after the onset of slow yolk platelet movements, and bundled microtubules were first observed between 0.5 and 0.55 NT, when cortical rotation has already reached plateau speed. How do you interpret these observations?

The observations of Larabell and colleagues on the one hand confirm that the vegetal array of microtubules is located in the shear zone—that is, exactly where one would expect it to be for a critical role in cortical rotation. On the other hand, they raise the question of whether the microtubule bundles is orienting the cortical rotation or whether the rotation is orienting them.

MICROTUBULES MOVE CYTOPLASMIC COMPONENTS DORSALLY BEYOND CORTICAL DISPLACEMENT

During their analysis of cortical rotation by confocal microscopy, Rowning and coworkers (1997) made a surprising discovery. They employed the same technique for observing live eggs mounted on an inverted confocal microscope that was also used to collect the data shown in Figure 9.12. However, instead of staining yolk platelets or microtubules, they treated eggs with $DiOC_6$, a lipophilic dye that stains membrane-bound organelles such as mitochondria and endoplasmic reticulum. Much to their surprise, they found that about 10% of the stained organelles moved rapidly from the vegetal pole toward the dorsal side of the embryo! The fast-moving organelles traveled in the shear zone of cortical rotation, between 4 and 8 μm inside the plasma membrane, where an array of microtubule bundles is present during the plateau phase of cortical rotation as described earlier. However, these organelles moved at a velocity of 35 to 50 μm/min relative to the endoplasmic core, whereas the speed of cortical rotation is only about 10 μm/min. The rapid organelle movements showed the repeated, unidirectional saltations that are characteristic of motor-driven movements along microtubules, and no such movements were observed in eggs treated with microtubule inhibitors.

Because the stained organelles left the microscopic field of view rather quickly, they could not be traced over distances greater than 40 μm. The investigators therefore injected small fluorescent plastic beads that were known from other experiments to be transported toward the plus ends of microtubules, presumably via association with motor proteins. After injection into the vegetal shear zone before cortical rotation, most of the beads were located in the endoplasmic core and moved relative to the cortex at about 10 μm/min, the velocity of cortical rotation. But again, about 10% of the beads moved three to four times as fast toward the dorsal side of the embryo. Like the stained organelles observed before, these beads traveled in the shear zone of cortical rotation. Also, whereas the cortical rotation covers an angle of about 30°, the rapidly moving organelles and beads covered an angular distance of approximately 60° (Fig. 9.13).

The data reviewed here show that, during cortical rotation in *Xenopus*, cytoplasmic components located near the vegetal pole are transported dorsally along microtubules in the shear zone about twice as far as the maximum displacement of the rotating cortex. Thus, once the microtubules in the shear zone have become oriented as part of cortical rotation, they serve as tracks for transporting cytoplasmic components. On these tracks, these components travel faster and move further dorsally than they could have done within the rotating cortex itself.

While the molecular events that determine the dorsal pole is still unknown, a probable link in the signal chain is *β-catenin*, a protein involved in anchoring microfilaments to plasma membrane proteins as well as in intracellular signaling (see Section 9.7). In the current context, it is significant that β-catenin accumulates in the nuclei of dorsal vegetal blastomeres at the blastula stage. Also, β-catenin can act as part of a transcription factor that activates genes characteristic of the early gastrula region that forms dorsal mesoderm. Conceivably, the rapid movements in the shear zone of the dorsal cortex contribute to the restriction of β-catenin to the nuclei of dorsal vegetal blastomeres.

A DORSALIZING ACTIVITY MOVES FROM THE VEGETAL POLE TO THE DORSAL SIDE DURING CORTICAL ROTATION

Before Rowning and coworkers (1997) observed the movements of visible organelles, Yuge and coworkers

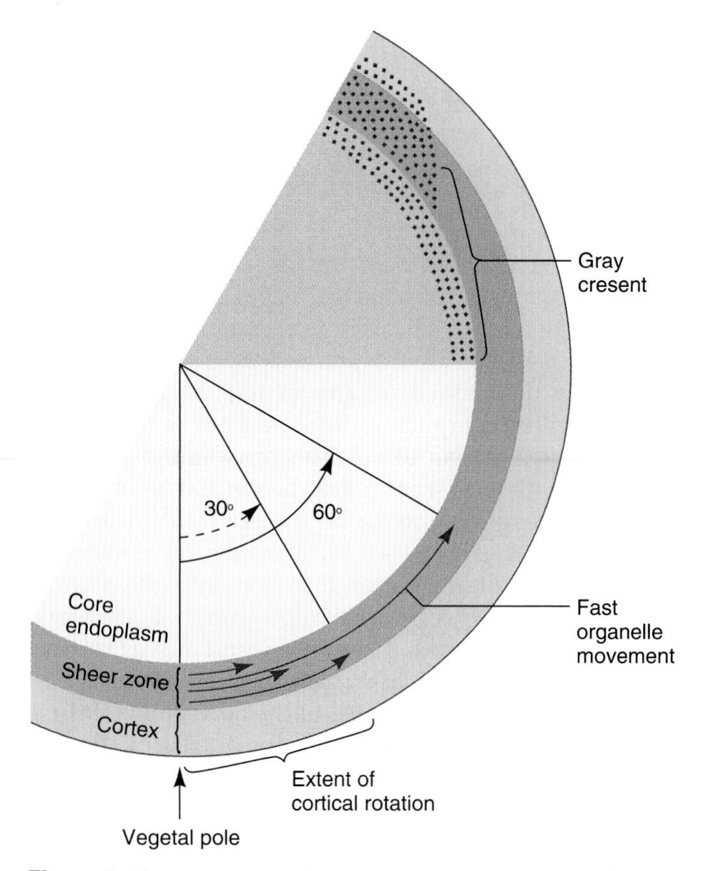

Figure 9.13 Rapid organelle transport from vegetal pole to dorsal side in *Xenopus* eggs. During cortical rotation, membrane-bound organelles move rapidly from the vegetal pole toward the dorsal side. The fast-moving organelles travel in the shear zone of cortical rotation, where an array of microtubule bundles is located. These organelles move more than three times faster than the speed of the cortex relative to the core endoplasm. The rapid movements show saltations characteristic of motor-driven movements along microtubules. The same kind of movement is shown by plastic beads known from other experiments to be transported toward the plus ends of microtubules. Whereas the cortical rotation covers only an angle of about 30°, the rapidly moving organelles and beads cover an angular distance of approximately 60°. The origin of the gray crescent can be explained by the distribution of black pigment, which is present in the animal hemisphere and seems to be associated primarily with the cortex and the shear zone. As the cortex rotates, most of the pigment moves away, exposing a small amount of pigment associated with the core endoplasm, which then appears gray. Note that the thickness of the cortex (about 10 μm) is exaggerated relative to the egg diameter (about 650 μm).

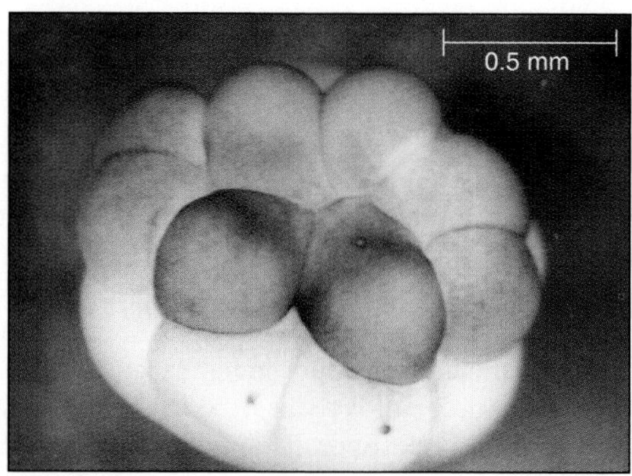

Figure 9.14 Bioassay for dorsalizing activity of microinjected cytoplasm or molecules. A *Xenopus* embryo at the 16-cell stage is stripped of its fertilization envelope and oriented with the animal pole up. The material to be tested is injected into the equatorial area of the two ventral vegetal blastomeres (injection sites marked with dots). If the transplanted material has dorsalizing activity, the host embryo forms an extra set of dorsal organs including a head with brain and sense organs, spinal cord, vertebrae, and dorsal trunk musculature. The embryo shown in Figure 9.1 was generated using this bioassay.

(1990) had traced the movements of the *dorsalizing activity*—that is, the activity of any material that promotes the formation of dorsal organs. For this purpose, they established a *bioassay* in which the materials to be tested were microinjected into the ventral vegetal blastomeres of *Xenopus* embryos at the *16-cell stage* (Fig. 9.14). When the researchers injected cytoplasm from *dorsal* vegetal blastomeres, 42% of the surviving recipients formed a secondary set of dorsal organs. In contrast, injection of cytoplasm from *ventral* vegetal cells into the same recipient area never caused secondary dorsal organ formation, and most recipients developed into normal tadpoles.

This bioassay was subsequently used to test cytoplasm from various regions of *eggs* at different stages for their dorsal organ–promoting activity (Fujisue et al., 1993; Holowacz and Elinson, 1993). Cortical (superficial) cytoplasm taken from the vegetal pole of eggs before cortical rotation, when tested in the bioassay, caused the formation of a secondary set of dorsal organs. In contrast, cortical cytoplasm from other egg regions or deep vegetal cytoplasm did not induce dorsal organ formation.

The dorsalizing activity present in the vegetal cortex of eggs disappeared from the vegetal pole area during the latter half of the first cell cycle—that is, between cortical rotation and first cleavage. During the same time interval, cytoplasm from the *dorsal subequatorial* region acquired dorsal organ–promoting activity, which had not been present there during the first half of the first cell cycle. The activity then remained present in the dorsal subequatorial region through the 16-cell stage.

When donor eggs were UV-irradiated after fertilization to block cortical rotation, the cytoplasmic activity that promoted dorsal organ formation remained at the vegetal pole and did not shift to a dorsal subequatorial position. This result shows that the normal shift of the activity depends on cortical rotation or some associated event, and that UV irradiation of the vegetal hemisphere after fertilization interferes with the transport—not the function—of the dorsalizing activity.

The results described above were extended by Holowacz and Elinson (1995), who injected cortical cytoplasm from newly fertilized eggs into animal blastomeres of UV-ventralized embryos at the 32-cell stage. These cortex-enriched animal blastomeres gave rise to dorsal mesodermal structures if two conditions were met. First, the injected cortical cytoplasm had to come from the vegetal rather than the animal pole region. Second, the vegetal–cortex-enriched animal blastomeres had to develop in contact with their normal neighbors through the blastula period. If vegetal–cortex-enriched animal caps were isolated at an early blastoderm stage, then they produced only epidermis. This result shows that vegetal cortex causes dorsal mesoderm formation not by itself but in conjunction with signals from vegetal or equatorial cells. Presumably, these are the same signals that determine endoderm and mesoderm.

Using a modified bioassay for dorsalizing activity, Kageura (1997) injected thin patches of egg cortex peeled from donor eggs rather than cytoplasm aspirated into a glass needle. The cortical patches were 4 to 8 μm thick, including the outermost layer that moves during cortical rotation and the underlying shear zone containing the mitochondrial array described earlier. Kageura confirmed the results of the earlier cytoplasmic transplantations, in particular that the dorsalizing activity is originally present in vegetal cortex and shifting to dorsal equatorial cortex during cortical rotation. While the transplanted cortex was much smaller in volume than the cytoplasm transplanted earlier, the cortical transplants were at least as active, strongly suggesting that all of the dorsalizing activity is indeed localized in the cortex. By transplanting active cortex to various positions, Kageura also showed that the transplants worked most reliably and induced the most complete secondary embryos if injected into the equatorial zone of the recipient—that is, into the cytoplasm later allocated to marginal blastomeres. Again, this result indicates that the dorsalizing cytoplasm does not cause dorsal organ formation by itself but in conjunction with mesoderm-inducing signals.

To test whether vegetal cortical cytoplasm is not only sufficient but also necessary to induce dorsal mesoderm, Sakai (1996) removed it at various times during the first cell cycle. He found that dorsal organ formation was consistently inhibited if vegetal cytoplasm was removed before 0.5 NT—that is, before cortical rotation. No defects were observed when the same operation was carried out after cortical rotation, at 0.8 NT or later. After early removal, embryos could be rescued by reinjecting the removed vegetal cytoplasm but not by injecting other cytoplasm.

Taken together, the results reviewed in this section show that an activity promoting dorsal organ formation is initially present near the vegetal pole and then shifts to a dorsal subequatorial position during cortical rotation. This activity acts in concert with other signals that remain centered around the vegetal pole and specify the formation of endoderm and mesoderm. It is in the area of overlap between dorsalizing and vegetal signals where dorsal organ formation is initiated.

DORSAL VEGETAL AND EQUATORIAL BLASTOMERES RESCUE VENTRALIZED EMBRYOS

The cytoplasmic activity that promotes dorsal organ formation is normally allocated to the dorsal blastomeres and can be transplanted with these blastomeres at the 32-cell and 64-cell stages. This was observed by Gimlich and Gerhart (1984), who transplanted various blastomeres from normal donor embryos to recipients in which the dorsoventral axis had been abolished by UV irradiation (Fig. 9.15). As described earlier, UV irradiation of the vegetal hemisphere of uncleaved eggs prevents the displacement of dorsalizing activity from the vegetal pole to the dorsal side that normally occurs during cortical rotation. The resulting embryos are radially symmetrical around the animal-vegetal axis and lack dorsal organs. The investigators used such radially ventralized embryos as recipients for blastomeres from normal donors. The question was which, if any, of the grafted cells would restore dorsoventral polarity to the recipients.

GIMLICH and Gerhart were guided by earlier observations of Nieuwkoop (1969b), who had rotated the animal hemisphere of amphibian blastulae relative to the vegetal hemisphere and observed that the vegetal hemisphere determined the dorsoventral polarity of the reconstituted embryos. Gimlich and Gerhart therefore expected that vegetal blastomeres would be critical to the establishment of dorsoventral polarity. To test this hypothesis, they transplanted vegetal *cells* from normal to radially ventralized embryos at the 32- or 64-cell stage. Within the vegetal tier of cells, they considered the quadrant centered over the meridian of the sperm entry point to be the ventralmost quadrant (blastomeres D4 and D4′ in Fig. 9.16c); the opposite quadrant (blastomeres D1 and D1′) they considered the dorsalmost quadrant, and the quadrants in between they called lateral quadrants. In each experiment, the researchers removed the cells of an entire quadrant from the vegetal tier of a radially ventralized recipient and replaced them with the corresponding cells of a normal donor. Because of variations in the cleavage pattern, the actual number of blastomeres transplanted varied from 1 to 3. The grafts healed into place within 1 h, and the recipients were allowed to develop until normal control embryos reached the tadpole stage. Then the recipients as well as the UV-irradiated control embryos, which had received no grafts, were scored for completeness of their dorsal organs.

Almost all of the UV-irradiated control embryos were lacking dorsal structures such as a notochord, a brain and spinal cord, and the rudiments of eyes and ears. With grafted vegetal blastomeres from the *dorsalmost* quadrant (D1 and D1′), the recipients were substantially rescued,

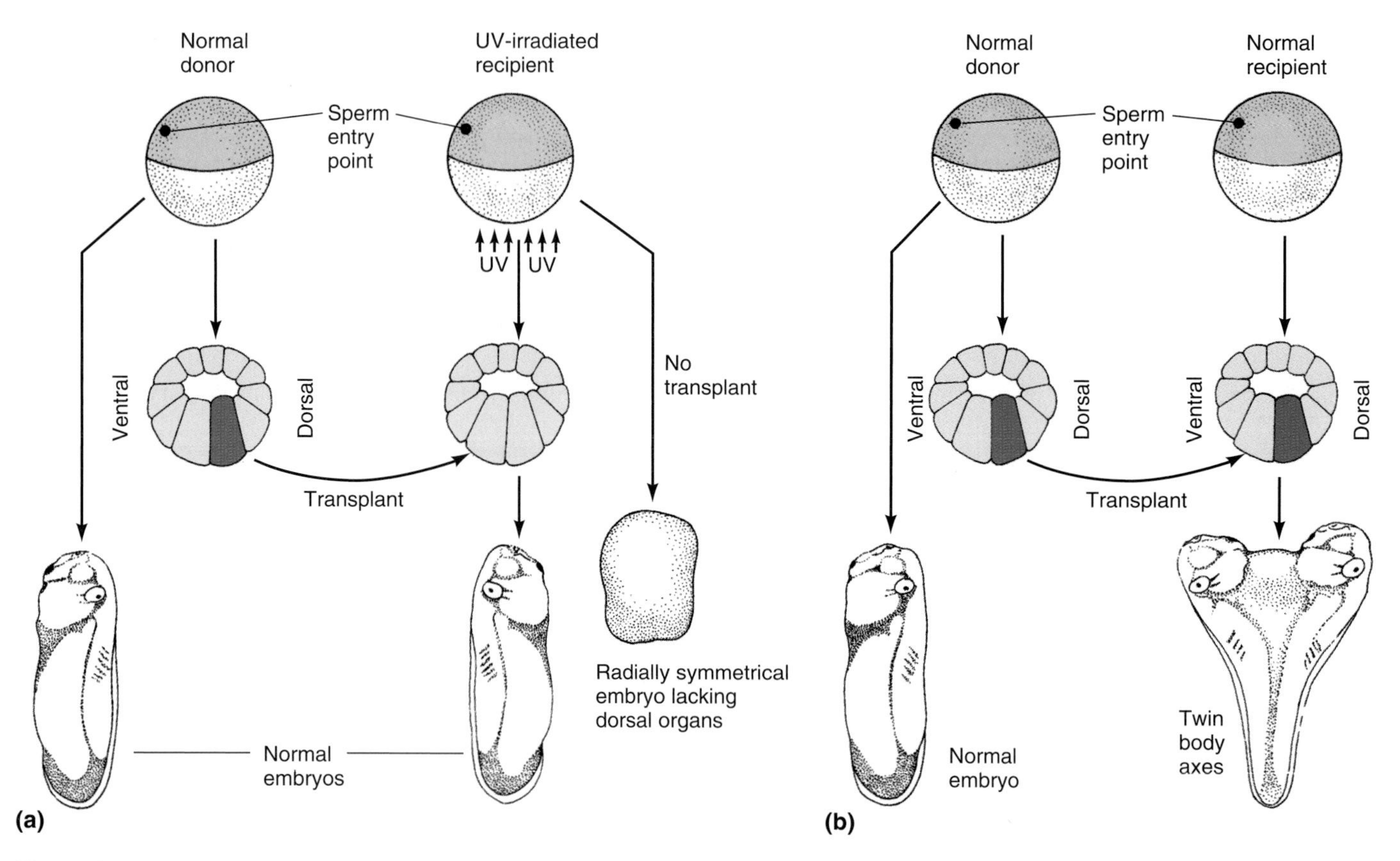

Figure 9.15 Transplantation of dorsal vegetal blastomeres in *Xenopus* embryos. **(a)** Rescue experiment. The dorsoventral axis of recipients was abolished by UV irradiation of the vegetal hemisphere of fertilized eggs. Eggs so treated give rise to radially symmetric embryos lacking all dorsal organs such as the brain, eyes, ears, spinal cord, and notochord. The radially ventralized embryos were restored to normal development by transplanted dorsal vegetal blastomeres from normal donors. **(b)** Heterotopic transplantation. Transplantation of dorsal vegetal blastomeres to the ventral side of normal recipients resulted in the development of embryos with two sets of dorsal organs.

with many of them showing complete sets of dorsal organs (Fig. 9.15a). In contrast, similar grafts of lateral (e.g., D2 and D3) and ventralmost (D4 and D4′) vegetal cells had small if any rescuing effects. Likewise, heterotopic transplantation of dorsal vegetal blastomeres to the ventral side of normal recipients caused the formation of an additional set of dorsal organs (Fig. 9.15b). The investigators concluded that within the vegetal tier of blastomeres the ability to establish dorsoventral polarity is concentrated for the most part in the dorsalmost quadrant (labeled D1/D1′ in Fig. 9.16).

In follow-up experiments, Gimlich (1986) and Kageura (1990) transplanted blastomeres from a wider range of locations. Some of the blastomeres transplanted in these experiments, in particular B1/B1′ and C1/C1′, are fated to form notochord and dorsal muscles (Bauer et al., 1994; Vodicka and Gerhart, 1995). The results of the follow-up resembled those of the original experiment. Dorsal blastomeres, but not their ventral counterparts, rescued the irradiated hosts' ability to form dorsal organs. At the 32-cell stage, the greatest rescuing activity was found in blastomeres C1 and C1′, and nearly as much in D1/D1′ and in B1/B1′. Lower levels of activity were present in A1/A1′, C2/C2′, and D2/D2′ (Fig. 9.16). Thus the dorsalizing activity ranges over a broad patch of mostly dorsal and vegetal-to-equatorial blastomeres, with the highest activity levels in those blastomeres that either induce or form dorsal mesoderm.

Questions

1. The fate of the D1/D1′ blastomeres in normal embryos is to form endoderm. Therefore, the blastomeres transplanted in the experiment of Gimlich and Gerhart most likely did not *form* the dorsal mesodermal structures in the rescued embryos but rather *induce* them. How would you test this hypothesis?
2. How do you interpret the blastomere transplantation experiments of Gimlich, Gerhart, and Kageura in the light of the subsequent cytoplasmic transplantation experiments of Yuge, Fujisue, Holowacz, and their colleagues?

Taken together, the results discussed in this section show that the dorsal side of the *Xenopus* embryo is determined by events associated with cytoplasmic rearrangements following egg activation. An activity that promotes the formation of dorsal organs is first present in cortical cytoplasm near the vegetal pole and shifts

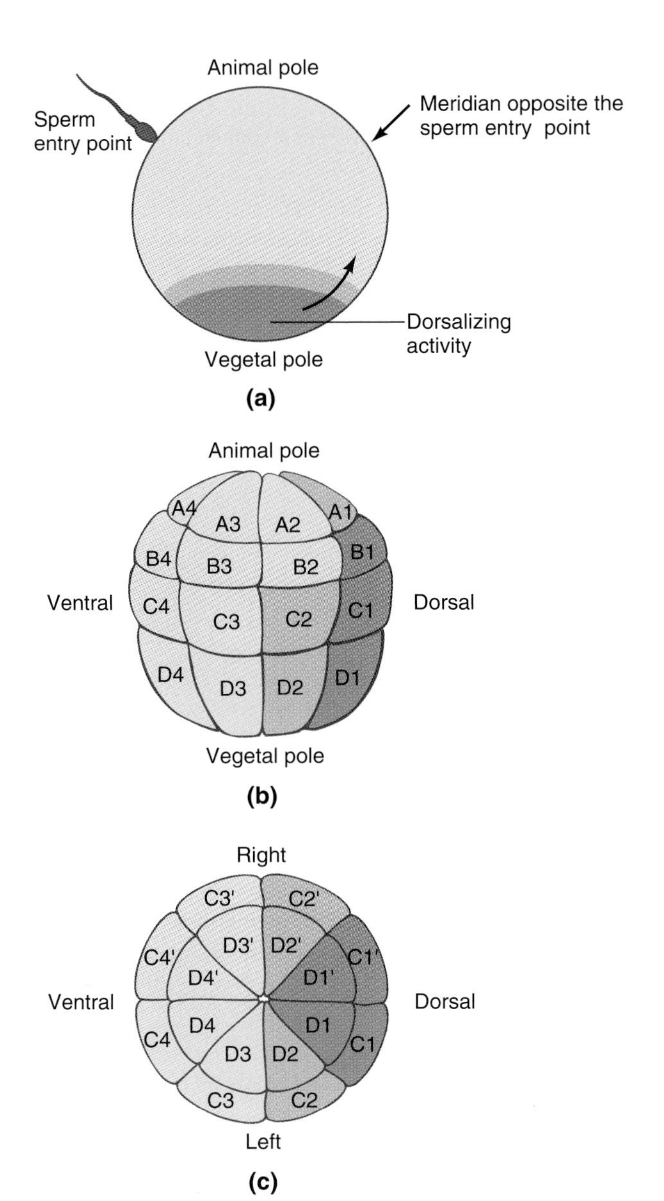

Figure 9.16 Dorsalizing activity (color) in the *Xenopus* embryo, revealed by the transplantation of cytoplasm or blastomeres as shown in Figures 9.14 and 9.15. Dorsalizing activity first occurs in vegetal cytoplasm and then shifts to a dorsal subequatorial region, where it is enclosed in dorsal vegetal and equatorial blastomeres. At the 32-cell stage, the dorsalizing activity is strongest in blastomeres B1/B1′, C1/C1′, and D1/D1′, and somewhat weaker in neighboring blastomeres. **(a)** Lateral view of the egg. **(b)** Lateral view of a 32-cell embryo. **(c)** Vegetal view of a 32-cell embryo.

dorsally late during the first cell cycle. The shift depends on microtubules, presumably for the transport of the activity from the vegetal pole to a dorsal subequatorial position. The dorsal blastomeres to which the activity is allocated during cleavage acquire the ability to rescue embryos ventralized by UV irradiation and to induce ectopic dorsal organs. Some of these blastomeres (C, C1′, B1, B1′) form dorsal mesoderm themselves while others (D1, D1′) induce dorsal blastomeres.

9.6 Effect of Dorsoventral Polarity on Mesoderm Induction in *Xenopus*

The establishment of dorsoventral polarity and its effect on mesoderm induction in amphibian embryos have attracted much attention for a particular reason. Dorsal mesoderm at the early gastrula stage occupies the dorsal lip of the blastopore. This material, beyond its own fate of forming notochord, induces lateral mesoderm to give rise to its appropriate structures and induces overlying ectoderm to form brain and spinal cord. These organizing powers of the dorsal blastopore lip came to light in a landmark experiment by Spemann and Mangold (1924a, b), in which dorsal blastopore lip was transplanted heterotopically and induced the formation of a secondary set of dorsal organs (see Section 12.3). Hence, the dorsal blastopore lip is also known as ***Spemann's organizer*** or simply the organizer. Finding out how the organizer acquires its special properties has been a longstanding quest of developmental biologists. While our knowledge is still incomplete, advances have been made. Having discussed mesoderm induction and the establishment of dorsoventral polarity in amphibian eggs, we will now discuss how these two processes may interact to generate Spemann's organizer.

MESODERM IS INDUCED WITH A RUDIMENTARY DORSOVENTRAL PATTERN

Vegetal blastomeres at the 32-cell stage show a *dorsoventral bias* in their ability to induce a secondary axis upon heterotopic transplantation. Dorsal and dorsolateral vegetal blastomeres (D1, D1′, D2, and D2′ in Fig. 9.16) have this capacity whereas their ventrolateral and ventral counterparts do not. The secondary embryos induced by D1 and D1′ are also longer and more complete than those induced by D2 and D2′ (Kageura, 1990). Does this bias affect only the likelihood with which mesoderm is induced and its overall mass, or is there also a dorsoventral polarity with regard to the specific types of mesodermal organs that are induced?

To answer this question, Dale and Slack (1987b) studied the mesoderm-inducing capacities of *isolated* vegetal blastomeres by combining them with animal caps as shown in Fig. 9.17. The animal cells were labeled with fluorescent dye to distinguish their descendants from those of the vegetal cells, which were not labeled. The researchers found that dorsal vegetal blastomeres induced mostly dorsal mesodermal derivatives, in particular notochord and muscle, some intermediate mesoderm such as kidney, but no ventral mesodermal structures. In contrast, ventral and lateral vegetal blastomeres induced the formation of ventral mesodermal derivatives, in particular blood cells and mesenchyme, along with many lateral and a few dorsal mesodermal structures. These data confirm that dorsal vegetal blastomeres differ from their lateral and ventral

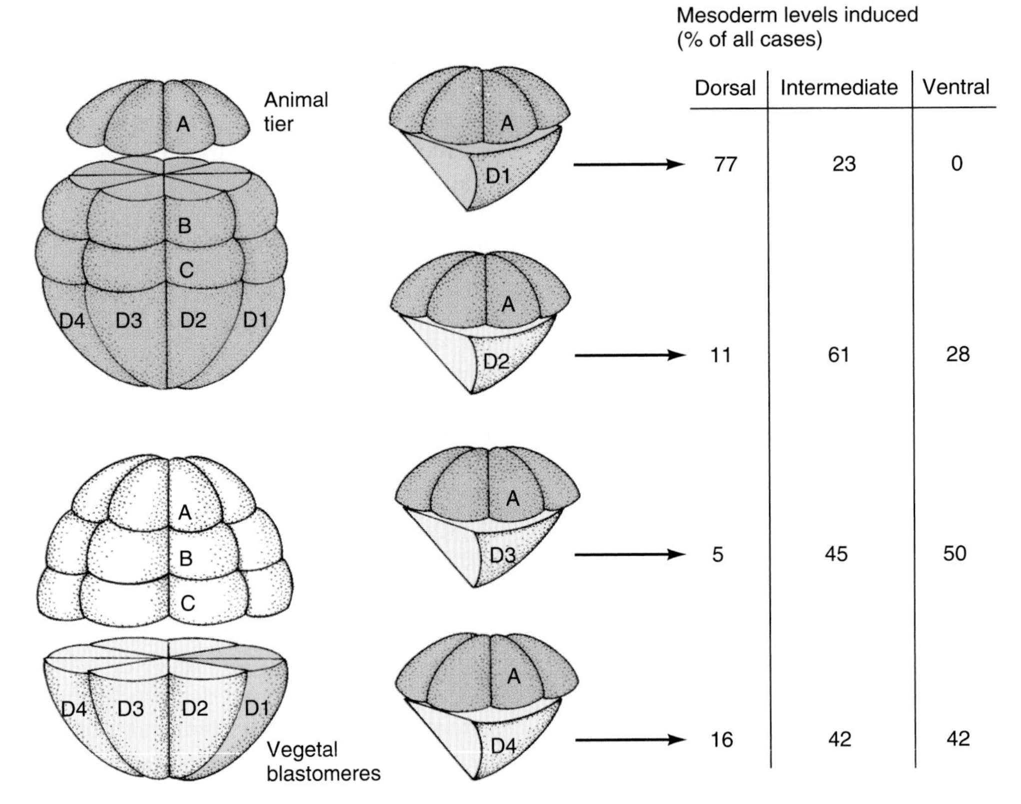

Figure 9.17 Regional specificity of mesoderm induction in *Xenopus.* Animal blastomere tiers (A) were combined with single vegetal blastomeres (D1 to D4) from 32-cell embryos. The animal tiers were labeled with fluorescent dye to distinguish their descendants from those of the vegetal blastomeres. After culture in vitro, the resulting tissues were fixed and sectioned. The mesodermal structures formed by the labeled animal tiers were classified as dorsal mesoderm, intermediate mesoderm, or ventral mesoderm. Dorsal vegetal blastomeres induced primarily dorsal mesoderm, while lateral and ventral vegetal blastomeres induced intermediate and ventral mesoderm in similar proportions.

neighbors in the kinds of mesodermal structures they *induce* most frequently.

Since the deep cytoplasmic movements shown in Figure 9.10 affect animal as well as vegetal cytoplasm, and because there are no definitive data on how far dorsalizing determinants are displaced from the vegetal pole, it is also possible that animal cells already have a dorsoventral bias in their *response* to the mesoderm-inducing signals from their vegetal neighbors. Indeed, the data compiled in Figure 9.17 show that animal caps formed some dorsal mesodermal structures even in combination with ventral vegetal blastomeres, a result that may be attributed to dorsal bias of some of the animal cap cells. Also, in combinations of animal and vegetal blastula pieces as shown in Figure 9.9, the most dorsal mesodermal structure—notochord—is induced only if the animal piece comes from the dorsal rather than from the ventral side of the embryo (Sutasurya and Nieuwkoop, 1974). In addition, treatment of dorsal and ventral halves of animal caps with proteins that represent or mimic mesoderm-inducing signals induces more dorsal mesoderm in the dorsal animal cap halves (Sokol and Melton, 1991; Kimelman and Maas, 1992).

In summary, the dorsoventral polarity of the amphibian embryo imposes a bias on both inducing and responding cells in mesoderm induction (Fig. 9.18). Since blastomeres become biased during cleavage and early blastula stages—that is, before midblastula transition MBT)—the molecular mechanisms creating the biases should not require transcription of the embryonic genome. Thus, the signals involved are likely to depend on the use and modification of maternally supplied molecules. Whatever the chain of dorsalizing signals may be, a critical link seems to depend on cortical rotation because eggs treated with UV light or microtubule inhibitors do not establish a dorsal side.

DORSAL MARGINAL CELLS INDUCE AN ARRAY OF MESODERMAL ORGAN RUDIMENTS

Within the newly induced mesoderm, the dorsoventral pattern begins as a simple difference between dorsal mesodermal cells, also known as Spemann's organizer, versus other mesodermal cells that form mostly ventral structures. The organizer then acts in two ways: First, it induces ectoderm to form brain and spinal cord. Second, it induces the adjacent portions of ventral mesoderm to form intermediate mesoderm. The first action, known as *neural induction,* will be discussed in Section 12.3 along with the organogenesis of the central nervous system.

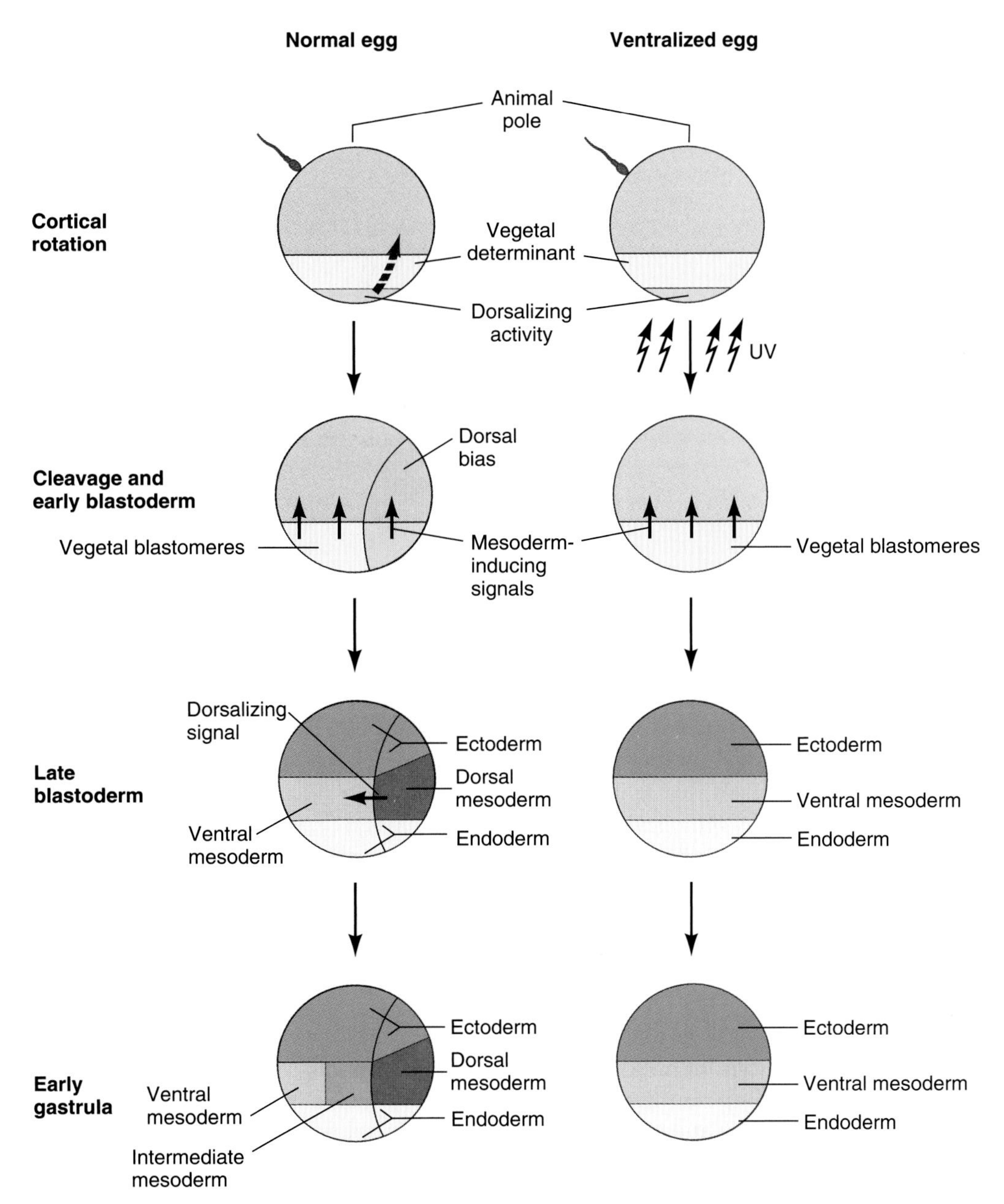

Figure 9.18 Model of germ layer formation and dorsalization in amphibian development. In newly laid eggs, both vegetal determinants and a dorsalizing activity are localized in the vegetal pole area. During cortical rotation, the dorsalizing activity moves toward the future dorsal side while the vegetal determinants stay in place. The latter become segregated into the vegetal blastomeres, which give rise to endoderm. The blastomeres inheriting the dorsalizing activity develop a bias for dorsal development. During cleavage, vegetal blastomeres send mesoderm-inducing signals to their equatorial neighbors. These respond by making dorsal mesoderm if they and their inducers are so biased; otherwise the equatorial cells respond by making ventral mesoderm. After midblastula transition, dorsal mesoderm sends dorsalizing signals to adjacent mesoderm, which in response changes its determination to intermediate mesoderm. In eggs that have been ventralized by UV irradiation or by other means interfering with cortical rotation, the dorsalizing activity remains stuck near the vegetal pole. As a result, none of the mesodermal blastomeres acquires a dorsal bias, so that all mesoderm becomes ventral.

The second organizer action, which generates intermediate mesodermal structures, will be discussed here to conclude the topic of dorsoventral axis formation.

In the experiment shown in Figure 9.17, the mesodermal structures induced by *ventral* vegetal blastomeres did not differ significantly from those induced by *lateral* vegetal blastomeres. The structures induced in this experiment therefore do not accurately reflect the sequence of mesodermal organ rudiments observed later during development, when we can distinguish the notochord, somites, nephrotomes, and lateral plates (Fig. 9.19). The *notochord* is a precursor of part of the spinal column. *Somites* give rise to other parts of the vertebral column, to skeletal muscle, and to the deeper parts of

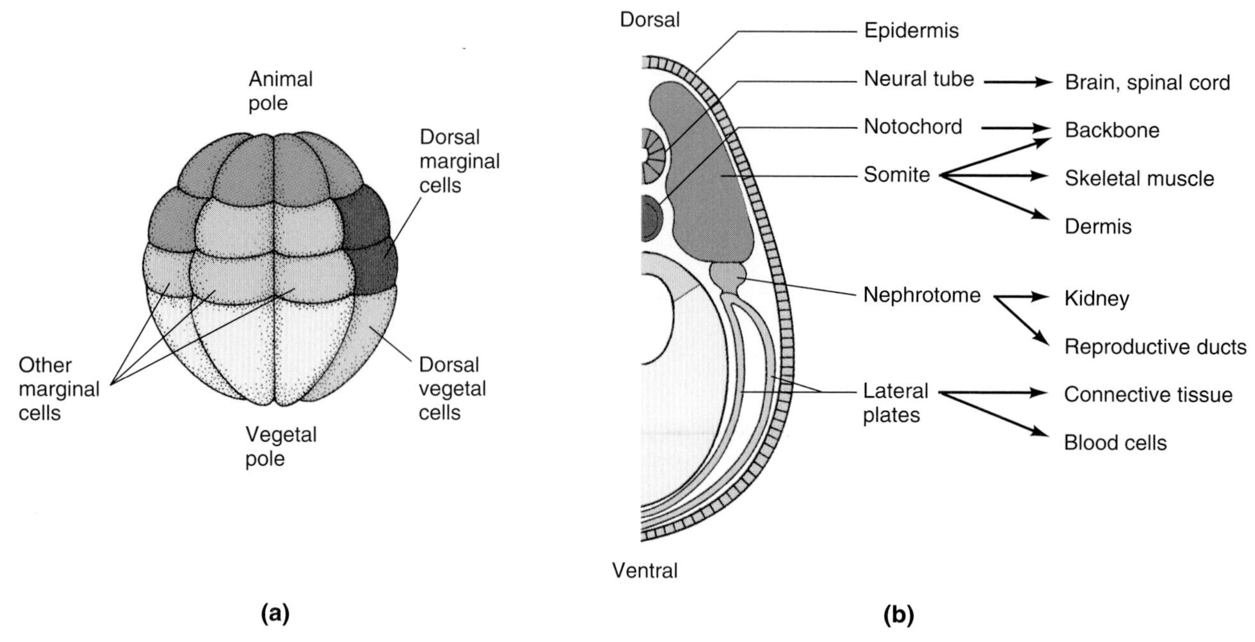

Figure 9.19 Development of the dorsoventral mesoderm pattern in the *Xenopus* embryo. **(a)** Side view at the 32-cell stage. Isolation and induction experiments (Fig. 9.17) reveal only two mesodermal states, one in the dorsal marginal cells and the other in all other marginal cells. **(b)** Transverse section at an advanced embryonic stage. The mesoderm contains a complex dorsoventral pattern of different elements, including notochord, somites, nephrotomes, and lateral plates.

the skin. *Nephrotomes,* also known as *intermediate mesoderm,* generate the embryonic kidneys. *Lateral plates* form the smooth muscle of internal organs, the connective tissues of trunk and limbs, and the circulatory system, including blood cells. How does this complex mesodermal pattern arise from the simpler pattern of potencies induced in dorsal marginal cells versus lateral and ventral marginal cells?

To analyze the further development of mesoderm in *Xenopus,* Dale and Slack (1987b) compared the *fate* and *potency* of prospective mesoderm cells at early blastula stages. To study mesodermal cell fate, they labeled marginal cells with a fluorescent dye and examined the labeled tissues later at a tadpole stage. To explore the potency of the same cells, they isolated them and let them develop in tissue culture. Comparing the results of the two procedures, they found that isolated dorsal and ventral marginal cells developed according to fate. In contrast, isolated *lateral* marginal cells developed more ventral structures than expected from the fate map. In particular, the lateral isolates produced substantial numbers of blood cells, which according to the fate map were to be expected mostly of ventral marginal cells. The investigators concluded that in the intact embryo the lateral marginal cells receive a signal that diverts them from ventral to lateral fates.

On the basis of earlier data, the researchers postulated that lateral marginal cells in the intact embryo receive an inductive signal from their dorsal neighbors. To test their hypothesis, the combined labeled ventral marginal parts with unlabeled dorsal or dorsolateral marginal parts (Fig. 9.20). After culture in combination, most of the ventral marginal parts were dorsalized, forming large amounts of muscle instead of blood. Only in a few cases, where the unlabeled dorsolateral part itself had formed ventral structures, did its labeled counterpart do the same. Conversely, the unlabeled dorsal and dorsolateral parts were not ventralized by the labeled ventral parts.

These results confirm that signals released from dorsal marginal cells change the development of their lateral neighbors toward intermediate mesodermal fates (see Fig. 9.18). These inductive interactions occur *after midblastula transition,* during late blastula and early gastrula stages. In UV-irradiated eggs without dorsoventral polarity, because no dorsal mesoderm is induced, only ventral mesodermal structures are formed.

9.7 Molecular Mechanisms of Dorsoventral Axis Formation and Mesoderm Induction

Given the intense interest of developmental biologists in how Spemann's organizer originates and how it works, it has been a major goal for many laboratories to understand mesoderm induction and the establishment of dorsoventral polarity in amphibians in molecular terms. What is the nature of the factor(s) that are segregated into dorsal blastomeres and impart a dorsal bias on them? Which molecules act as inducing signals to persuade marginal cells to form mesoderm instead of

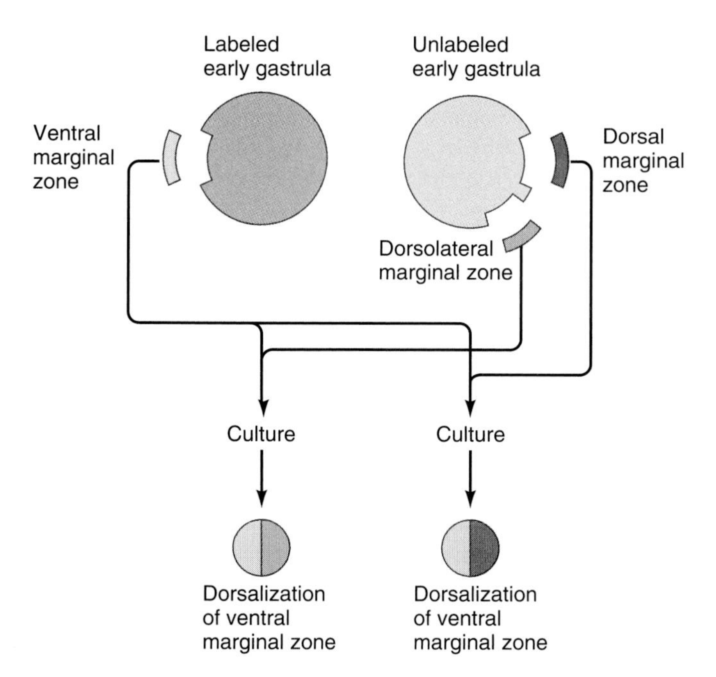

Figure 9.20 Dorsalization of the ventral marginal zone by dorsal or dorsolateral zone of the early amphibian gastrula. A piece of ventral marginal zone from a fluorescently labeled gastrula was combined with a piece of dorsal or dorsolateral marginal zone from an unlabeled donor. The mesodermal structures formed by each zone were analyzed after culture. The ventral zones developed more dorsal structures than they would have in isolation or in the intact embryo. The development of their dorsal or dorsolateral partners was not changed by the coculture.

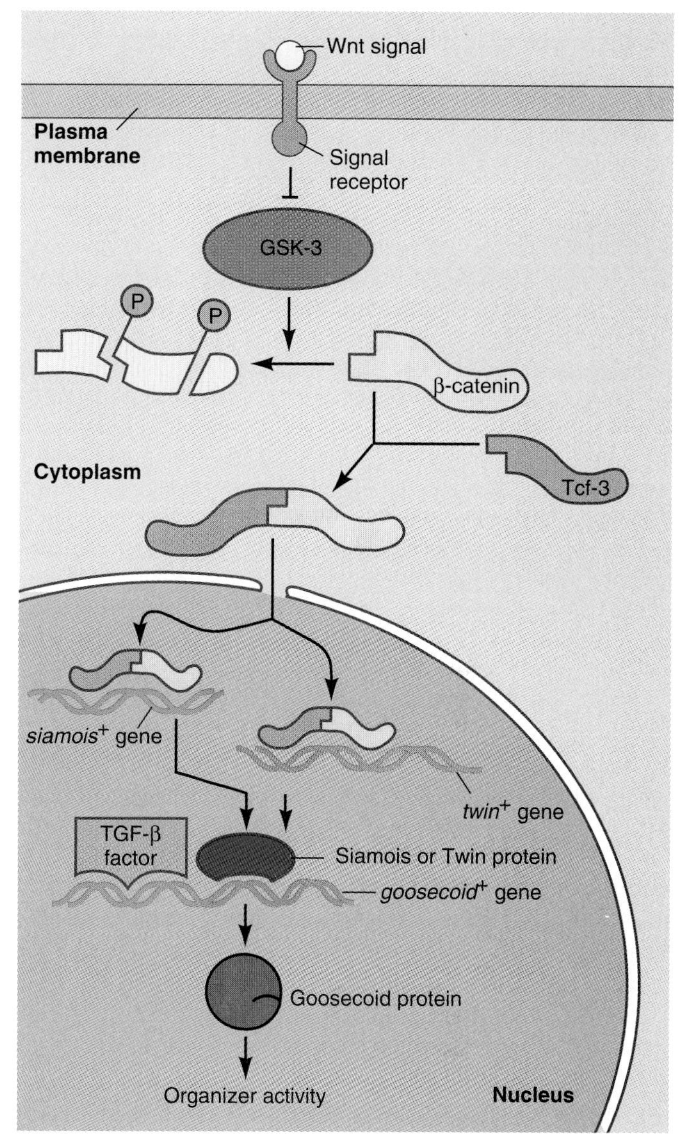

Figure 9.21 Proposed molecular mechanism for establishing Spemann's organizer. A key component is β-catenin, a signal molecule that can combine with another peptide, designated Tcf-3, to form a transcription factor. The stability of β-catenin is limited by GSK-3 kinase, which phosphorylates β-catenin, thus promoting its proteolytic breakdown. GSK-3 in turn is inhibited by Wnt signaling. Undegraded β-catenin associates with Tcf-3 and accumulates in the nucleus, where it is stable. At the blastula stage, β-catenin accumulates especially in dorsal nuclei (see. Fig. 9.22). The β-catenin/Tcf-3 complex activates the *siamois*$^{+}$ and *twin*$^{+}$ genes in the dorsal mesoderm that will act as Spemann's organizer. Siamois and twin proteins, along with a mesoderm-inducing signal of the TGF-β family, activate *goosecoid*$^{+}$, another gene specifically expressed in the organizer.

ectoderm? A candidate molecule for either of these functions must fulfill the following minimum requirements. First, the molecule must be present in the embryo at the required concentration and in the predicted region. It must be available as a maternally supplied component if it is proposed to act before MBT, and it should be synthesized from the embryonic genome if it is thought to act after MBT. Third, blocking the action of the molecule in vivo must interfere with its proposed biological function. Fourth, it must be active in rescue and/or heterotopic transplantation experiments with whole embryos. Corresponding requirements must be met for the receptor and other downstream signal components through which the molecule acts.

β-CATENIN MAY SPECIFY DORSOVENTRAL POLARITY

A likely candidate for establishing the dorsal pole of the embryo, as mentioned earlier, is *β-catenin* (Haesman, 1997; Moon and Kimelman, 1998). It was originally discovered in the context of cell adhesion, as a molecule that helps to anchor plasma membrane molecules known as *cadherins* to the cytoskeleton. Only later it became apparent that the non-membrane-associated, or cytoplasmic, fraction of β-catenin is involved in a signaling pathway known as the ***Wnt pathway.*** The pathway is named after an evolutionary highly conserved family of secreted signal proteins designated Wnt in vertebrates. They act on matching receptors in the plasma membrane of responding cells, which in turn inactivate a protein kinase designated GSK-3 (Fig. 9.21). GSK-3

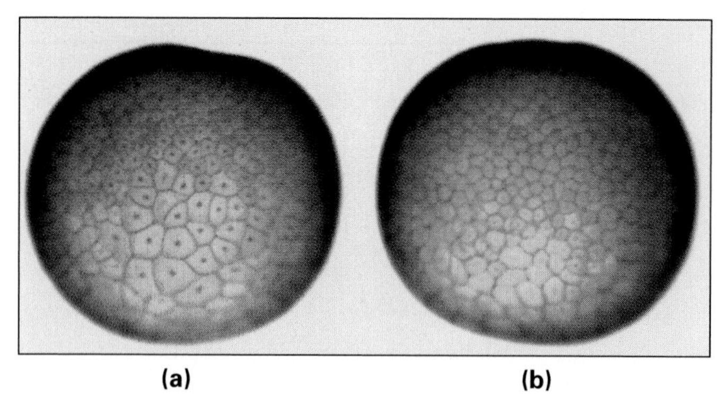

Figure 9.22 Nuclear localization of β-catenin in *Xenopus* embryos at the blastula stage. Embryos are photographed after immunostaining (see Method 4.1) with a polyclonal antibody against β-catenin. **(a)** Dorsal aspect of blastula with β-catenin accumulated in nuclei. **(b)** Ventral aspect of blastula without β-catenin in nuclei. In both photographs, cells are stained along the plasma membranes, where β-catenin is associated with cell adhesion molecules known as cadherins.

seems to hasten the proteolytic breakdown of β-catenin by phosphorylating a critical group of its serine and theronine residues (Aberle, 1997). Cytoplasmic β-catenin that is not degraded associates with a transcription factor designated Tcf-3 and accumulates in nuclei, where the complex is stable. After MBT, the complex activates the *siamois*$^+$ and *twin*$^+$ genes, which are closely associated with organizer activity (Brannon et al., 1997; Fan and Sokol, 1997; Laurent et al., 1997).

β-catenin occurs in *Xenopus* embryos at the right place and time (Larabell et al., 1997). During cortical rotation, it was found in association with oriented microtubules that move cytoplasmic components from the vegetal pole region to the dorsal side of the embryo (see Section 9.5). Significantly, UV irradiation of fertilized eggs under conditions blocking cortical rotation will trap, in the vegetal pole region, β-catenin and its downstream signaling steps such as the activation of *siamois*$^+$ and *twin*$^+$ genes. By the blastoderm stage and possibly earlier, β-catenin becomes enriched in the nuclei of dorsal but not ventral blastomeres (Fig. 9.22; Schneider et al., 1996). In experimentally ventralized frog embryos the dorsoventral pattern of β-catenin nuclear staining is abolished. Conversely, embryos hyperdorsalized by injection of Xwnt8 mRNA exhibit an enhanced nuclear accumulation of β-catenin.

Removal of β-catenin abolishes the dorsoventral polarity and creates radially ventralized embryos. This was first demonstrated by an experiment that depleted *Xenopus* eggs of the maternal mRNA encoding β-catenin (Heasman et al., 1994). The depletion was achieved by injecting eggs with *oligodeoxynucleotides* that hybridize in vivo to critical sequences within β-catenin mRNA. The resulting DNA/RNA hybrid was cleaved by an endogenous RNA-degrading enzyme, *Rnase H*, so that the maternal supply of β-catenin mRNA in the egg was destroyed and β-catenin protein depleted because naturally degraded protein could no longer be replaced. Embryos developing from such eggs were radially symmetrical and ventralized just like embryos in which cortical rotation has been prevented by UV-irradiation. The same ventralized appearance developed in embryos in which cytoplasmic β-catenin was diverted by excess amounts of cadherin, the cell adhesion molecule that is anchored to microfilaments by β-catenin.

Embryos that have been ventralized by UV irradiation or by excess cadherin can be rescued by microinjection of β-catenin mRNA (Heasman et al., 1994; Guger and Gumbiner, 1995). Finally, ectopic application of β-catenin or inhibition of GSK-3 leads to the formation of a second set of dorsal organs (see Fig. 9.1; Funayama et al., 1995). Similar axis duplications had been observed earlier after injection of lithium ion (Li^+) into ventral vegetal and equatorial blastomeres (Kao et al., 1986; Busa and Gimlich, 1989), and Li^+ has now been shown to boost β-catenin by inhibiting GSK-3 (Klein and Melton, 1996; Hedgepeth et al., 1997).

The evidence summarized above indicates that β-catenin meets all the criteria for being a critical link in the signaling chain that determines the dorsal pole of the dorsoventral axis in *Xenopus*. Less extensive data indicate similar roles for β-catenin in establishing the dorsoventral axis in zebrafish (Schneider et al., 1996), and in the animal-vegetal axis in sea urchins (Wikramanayake et al., 1998). A major unresolved question is how the nuclear accumulation of β-catenin becomes limited to the dorsal side of the *Xenopus* embryo. One possibility is that the microtubule-dependent transport of β-catenin from the vegetal pole region to the dorsal side during cortical rotation is sufficient. Alternatively, the membranous vesicles that undergo the same translocation may interact with the β-catenin pathway—for example by reducing the activity of GSK-3. Indeed, inhibition of GSK-3 on the ventral side of *Xenopus* embryos causes the formation of an ectopic dorsal axis whereas overexpression of GSK-3 on the dorsal side has a ventralizing effect (He et al., 1995; Yost et al., 1996). Endogenous molecular mechanisms are present for reducing the activity of GSK-3 either by inhibiting its synthesis or by reducing its activity (Dominguez and Green, 2000).

SEVERAL GROWTH FACTORS HAVE MESODERM-INDUCING ACTIVITY

The manifestation of β-catenin as a dorsalizing signal depends on mesoderm induction, as discussed earlier. To explore whether mesoderm induction requires direct cell contact, Grunz and Tacke (1986) modified the experiment shown in Figure 9.9 by placing a filter with very fine pores (diameter 0.4 μm) between vegetal and animal blastula cells. They reasoned that the filter would interfere with induction if direct cell contact was required. However, mesoderm was induced across the filter. The reacting ectoderm formed mostly ventral mesodermal structures, but a few animal caps also formed dorsal mesoderm. Electron-microscopic inspection of the filters failed to reveal any cell outgrowths in the filter pores, indicating that transmission of diffusible molecules is sufficient for mesoderm induction to occur.

To further analyze the nature of the diffusible signals that induce mesoderm formation, animal caps from blastulae were cultured in media containing various cell extracts or molecular fractions. In response, blastula ectoderm was often induced to form mesoderm, as indicated by strong elongation of the animal caps, by histological features characteristic of mesodermal tissues, and by the synthesis of mesoderm-specific mRNAs or proteins. These observations revealed the mesoderm-inducing capacity of several ***growth factors,*** secreted or membrane-bound peptides that activate matching plasma membrane receptors present on neighboring cells (Kimelman et al., 1992; Sive, 1993; J. C. Smith, 1989). In particular, several peptides of the ***fibroblast growth factor* (*FGF*)** family and the ***transforming growth factor-β* (*TGF-β*)** family can act as mesoderm-inducing signals.

Basic fibroblast growth factor (*bFGF*) meets several criteria of a natural signal that would be necessary for mesoderm induction. It is present in the embryo at the correct time and in sufficient concentrations, as are receptors for FGF (Gillespie et al., 1989). Suppressing the function of FGF receptor eliminates molecular markers of muscle development in vitro and interferes with gastrulation and subsequent trunk mesoderm development in vivo (Amaya et al., 1991, 1993).

Several members of the TGF-β family have also been considered as mesoderm inducers. The VegT protein exerts its strong effect on endoderm and mesoderm formation (see Fig. 9.7) through stimulating the synthesis of secreteal proteins of the TGF-β family (Clements et al., 1999; Agius et al., 2000). Inactivation of certain TGF-β receptors prevents mesoderm induction, but interpretation of this result has been hampered by the fact that these receptors are bound by several members of the TGF-β family (Hemmati-Brivanlou and Melton, 1992; Kessler and Melton, 1994). Some members of the TGF-β family, including Vg1 and activin, can promote dorsalization (Sokol and Melton, 1992; Thomsen and Melton, 1993), suggesting that they may act in conjunction with the β-catenin pathway. Another member of the TGF-β family, designated BMP-4, promotes the formation of ventral mesoderm at the expense of dorsal mesoderm (Graff et al., 1994; Hawley et al., 1995; Schmidt et al., 1995). BMP-4 signaling becomes critical after midblastula transition and will be discussed in the context of neural induction (see Section 12.5).

TGF-β MEMBERS AND β-CATENIN MAY ACT COMBINATORIALLY IN INDUCING SPEMANN'S ORGANIZER

According to the simplest hypothesis explaining the origin of Spemann's organizer, the β-catenin pathway acts combinatorially with signals of the FGF and TGF-β families (Kimelman et al., 1992; Moon and Kimelman, 1998). Evidence clearly supporting this hypothesis was described by Watanabe and colleagues (1995), who analyzed the activation of *goosecoid*$^+$, a gene specifically expressed in the organizer. Like other genes, *goosecoid*$^+$ has a regulatory region with several *response elements*—that is, nucleotide sequences that interact with proteins encoded by other genes. One of these elements activates transcription in response to signals of the TGF-β family while another response element interacts with proteins encoded by the *siamois*$^+$ and *twin*$^+$ genes, which in turn are activated directly by the β-catenin/Tcf complex as discussed earlier (see Fig. 9.21). Thus, an organizer-specific gene is activated by a combination of TGF-β and β-catenin signaling.

9.8 Determination of Left-Right Asymmetry

As one can easily test with a pair of hands or gloves, they can be aligned at most with two, but never with all three, of their axes. Any two objects with this property differ in their *chirality,* or **handedness,** and every three-dimensional object with three polarity axes—molecules, gloves, cars—exist or could be made in two versions that differ only in their handedness. These two versions are called right-handed (R) or left-handed (L) by some natural or arbitrarily defined relationship to our hands. Molecules that differ only in their handedness are known as *stereoisomeres* and have different physical and physiological properties. For example, L-ascorbic acid, also known as vitamin C, protects humans against scurvy whereas its stereoisomer, D-ascorbic acid, does not have this effect.

Most animals are *bilaterally symmetrical* relative to the *median plane,* which is defined by the anteroposterior and dorsoventral axes (Fig. 9.2). The lateral halves of bilaterally symmetrical animals also have a mediolateral polarity because their medial and lateral structures, such as the spinal column and ribs of a vertebrate, are different. (By contrast, a rectangular piece of cake, with a crust on one side and icing on one surface, would have exactly one median plane along which it could be cut into two equal halves, but these halves would not necessarily have a mediolateral polarity.) The left and right sides of bilaterally symmetrical animals differ in their handedness because their mediolateral axes point in opposite directions.

In many so-called bilaterally symmetrical organisms, some of the inner organs, such as stomach and liver, are positioned asymmetrically. This phenomenon is known as ***left-right asymmetry.*** It might be argued that internal organs simply fold themselves asymmetrically because this is the most economic way of packing growing viscera into body cavities of limited volume. This benefit, however, would also be achieved by *random asymmetry,* with asymmetrical organs positioned randomly one way or another. Instead, most left-right asymmetries are oriented the same way in nearly all members of a species. In most humans, for instance, the stomach normally curves to the left while most of the liver and the entire spleen are located on the right. In other words, most organisms show ***regular asymmetry***—that is, a left-right asymmetry that is oriented consistently.

How does regular left-right asymmetry arise? This question is not only interesting to scientists who are curious about basic natural phenomena but is also medically relevant. Disturbances in the development of left-right asymmetry may result in *situs inversus* (complete reversal of the regular left-right asymmetry in all organs), *heterotaxis* (only some organs reversed), or *isomerisms* (normally asymmetrical organs are duplicated or missing). Among these, only the rare cases of complete situs inversus are not associated with any health problems, whereas the other abnormalities tend to be serious or lethal.

In many species, the orientation of the right-left asymmetry relative to the other two body axes is robust enough to withstand experimental manipulation. Two studies, one carried out with frogs and the other with sea urchins, will illustrate this point. The dorsoventral axis in *Xenopus* can be abolished by UV irradiation of the vegetal pole region before cortical rotation, as discussed in Section 9.5. After irradiation with weak UV doses, embryos develop with reduced dorsal and anterior structures, and such embryos also show left-right reversal of heart looping (Danos and Jost, 1995). The frequency of left-right reversals is correlated with the severity of losses in anterior and dorsal structures.

When the dorsal and ventral halves of a sea urchin embryo are separated at the 16-cell stage, both halves develop into normally proportioned larvae. Moreover, fate-mapping experiments show that the ventral half retains both its animal-vegetal axis and its dorsoventral axis. The dorsal half also maintains its animal-vegetal axis while frequently reversing its dorsoventral axis. In those cases where the dorsoventral axis reverses so does the left-right axis. This is indicated by the position of the rudiment that gives rise to the adult sea urchin, which invariably develops on the left side of the larva (McCain and McClay, 1994). Thus, embryos with a reversed dorsoventral axis nevertheless retain their normal handedness. Similar observations have been made with embryos of the roundworm *Caenorhabditis elegans* (Priess and Thomson, 1987; Wood, 1991).

The dependent specification of the left-right asymmetry relative to the other two body axes has guided the formulation of hypotheses on the origin of left-right asymmetry in development. Huxley and deBeer (1963) suggested that an electric current flowing from anterior to posterior would create a circular magnetic field oriented from right to left dorsally and from left to right ventrally. Brown and Wolpert (1990) pointed out that any stereoisomeric molecule oriented relative to the anteroposterior and dorsoventral axes would define a left-right polarity, which conceivably could be translated into a cellular and organismic asymmetry. Recent genetic and molecular data on left-right asymmetry in mice can be interpreted in terms of the latter model (Levin and Mercola, 1998; *Yost, 1999;* Capdevila et al., 2000).

In mice, several genes are known to affect left-right asymmetry. One of them, *situs inversus viscerum*$^{+}$ (*iv*$^{+}$) is known from a mutant where about half of the homozygotes show situs inversus (Hummel and Chapman, 1959). The *iv*$^{+}$ gene therefore seems to be involved in orienting the left-right asymmetry. In the absence of *iv*$^{+}$ function, asymmetry still develops, but its alignment with the other body axes is abolished.

The *iv*$^{+}$ gene and its protein product have been characterized (Supp et al., 1997, 1999). The iv protein, also known as left-right dynein (LRD), is a major part of multisubunit proteins known as *dyneins.* Dyneins are microtubule-associated proteins that can be classified as either axonemal or cytoplasmic. Axonemal dyneins mediate the sliding between adjacent microtubules, thus causing the oriented movement of flagella and cilia (see Fig. 2.7). Cytoplasmic dyneins are motor proteins that transport cellular cargo toward the minus ends of microtubules (see Fig. 2.9). The absence of ciliary dynein is thought to cause *Kartagener's syndrome* in humans, which includes sperm immobility and frequent bronchitis caused by failure to clear mucus out of the trachea and bronchi (Afzelius, 1976). It was therefore of great interest to observe that certain cilia in homozygous mutant *iv*$^{-}$/*iv*$^{-}$ mouse embryos are immobile (Supp et al., 1999). These cilia, known as *monocilia,* are located in a critical growth in area of the mouse embryo called *Hensen's node,* or the *node* for short.

The dynein molecules have a handedness. Those associated with the microtubules of monocilia curve clockwise as seen from the base of the cilia, and the ciliary rotation is in the same direction (Afzelius, 1999). The resulting flow of fluid on the surface of the node is toward the left in wild-type mouse embryos, and this flow is absent in mutant *iv*$^{-}$/ *iv*$^{-}$ mouse embryos (Okada, 1999). Similar observations were made on a mouse mutant affecting kif3B, a protein of the *kinesin* superfamily (Nonaka et al., 1998). Homozygous mutant embryos show randomized left-right asymmetry and die in midgestation. Electron microscopy reveals that the monocilia are missing from Hensen's node in kif3B^{-}/kif3B^{-}

mice, whereas they are immobile in as iv^-/iv^- embryos as discussed earlier. One interpretation of these findings in that the monocilia in the node area, through their oriented movement, generate an oriented flow of extracellular signal molecules to the left side of the embryo. Conceivably, this flow would cause the activation or inhibition of embryonic patterning genes specifically on the left side of the body.

Another mouse gene was named *inversion of embryonic turning*$^+$ (*inv*$^+$) because interruption of this gene causes a striking phenotype during early embryogenesis: Homozygous mutants make a counterclockwise turn inside the amniotic cavity, whereas heterozygous and wild-type embryos make a clockwise turn (Yokoyama et al., 1993). Later during development, nearly 100% of all iv^-/iv^- individuals develop situs inversus combined with an enlarged spleen and severely abnormal kidneys. The reversal of left-right asymmetry in *virtually all* iv^-/iv^- individuals is in marked contrast to occurrence of situs inversus in *about half* of the iv^-/iv^- individuals. The comparison indicates that the orientation of the left-right asymmetry is *randomized* by the lack of iv^+ but truly *reversed* by the lack of inv^+. The inv^+ gene encodes a novel protein most likely located in the cell cytoplasm (Mochizuki et al., 1998). The mechanism by which the inv protein may affect left-right asymmetry still needs to be elucidated.

Both iv^+ and inv^+ act through the *nodal*$^+$ gene, which encodes an intercellular signaling protein of the TGF-β family. The nodal protein is involved in determining the left-right asymmetry of mice, chicken, frogs, and zebrafish (Collignon et al., 1996; Levin et al., 1995; Lohr et al., 1997; Rebagliati et al., 1998). The protein is synthesized in the left lateral plate mesoderm, most of which contributes to bilaterally symmetrical structures while some of the adjacent mesoderm gives rise to asymmetrical structures. The expression pattern of *nodal*$^+$ in mice is altered in iv^- and inv^- mutants (Fig. 9.23; Lowe et al., 1996; Meno et al., 1996). Ectopic expression of *nodal*$^+$ on the right side in chicken randomizes the orientation of heart looping (Levi et al., 1997). In *Xenopus*, experimental manipulations that interfere with the development of dorsoanterior structures not only cause the reversal of heart looping, as discussed earlier, but also affect the lateralized expression of a *Xenopus* nodal-related gene, *Xnr-1*$^+$ (Lohr et al., 1997). It appears that the *nodal*$^+$ gene may be involved in establishing right-left asymmetry in many or all vertebrates even though the control of *nodal*$^+$ expression may differ among vertebrate classes (Yost, 1999).

Being a secreted protein, nodal is well suited for cellular interaction. However, since left-right asymmetry is likely to involve changes in the expression of many genes, one might expect gene regulatory proteins to be involved in the signaling chain. The *Pitx2*$^+$ gene encodes a transcriptional regulator of the bicoid family and is expressed on the left side of the lateral plate mesoderm, cardiac tube and gut in mouse chicken, and frog embryos (Ryan et al., 1998; Campione, 1998). Inhibition or ectopic expression of *Pitx2*$^+$ interferes with left-right asymmetry. The expression of *Pitx2*$^+$ depends on the activities of iv^+, inv^+, and *nodal*$^+$, indicating that these genes are upstream of *Pitx2*$^+$ in the genetic hierarchy that controls left-right asymmetry.

In summary, it is clear that polarity axes are formed in a variety of ways. Even in one species, *Xenopus*, the three body axes are established at different stages of development and by different mechanisms. Oriented transport of molecules and organelles during oogenesis plays a critical role in establishing the future anteroposterior axis. Dorsoventral axis formation involves major cytoplasmic rearrangements including cortical rotation after fertilization. The origin of left-right asymmetry is thought to rely on handed molecules that determine a left-right polarity if they are oriented with regard to the anteroposterior and dorsoventral axes. By orienting the movements of cellular organelles, such as cilia, the handedness of oriented molecules may cause the oriented transport of signals, which in turn may direct left-right asymmetry in gene expression and morphological development.

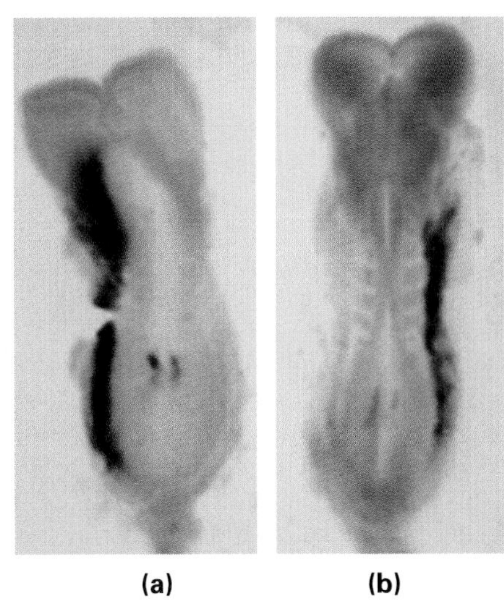

Figure 9.23 Lateralilzed expression of the *nodal*$^+$ gene and its dependence on the activity of *inv*$^+$. This photo shows two mouse embryos, oriented with the anterior end up and viewed dorsally, so that each embryo's left side is toward the left in the figure. The purple color is generated by a probe hybridized in situ (see Method 8.1) to nodal mRNA. Both embryos were obtained by crossing two heterozygous (*inv*/+) parents. Offspring (to the left in figure) carrying at least one wild-type allele (*inv*$^+$) express *nodal*$^+$ in lateral mesoderm on the left side of their body. In contrast, homozygous mutant *inv/inv* offspring (to the right in figure) express *nodal*$^+$ in the right half of their body.

SUMMARY

Most organisms acquire one or more body axes in the course of development. The brown alga *Fucus* has perfectly spherical eggs in which one axis develops between the future thallus and the future rhizoid. This axis may be determined by sperm entry or by environmental factors, such as light. A period of axis formation, when the thallus-rhizoid axis is set up in a preliminary way, is followed by a period of axis fixation, when the axis is irreversibly established. Axis formation by light results in a labile accumulation of Ca^{2+} channels, Ca^{2+}, and microfilaments at the prospective rhizoid pole. These factors promote local changes in the composition of the cell wall, which signals back to the cell, providing orienting cues for the positioning of the first mitotic spindle.

Metazoa are, for the most part, bilaterally symmetrical and have three body axes: anteroposterior, dorsoventral, and left-right. In *Xenopus,* the anteroposterior axis develops from the animal-vegetal axis, which originates during oogenesis. The dorsoventral axis is formed after fertilization and fixed before the first cleavage. Molecular cues to left-right asymmetry have been detected during the blastula and early gastrula stages.

The animal-vegetal polarity in amphibian eggs determines the spatial organization of the germ layer rudiments during the blastula stage. Most of the animal half of the blastula forms ectoderm, and most of the vegetal half forms endoderm, while a marginal zone in between gives rise to mesoderm. Mesoderm formation is based on embryonic induction, a pervasive principle of development. It is defined as an interaction between nonequivalent cells in which one partner changes the fate of a responding partner. In this case the endoderm induces the marginal cells, which would otherwise have formed ectoderm, to form mesoderm instead. Specific behaviors of the germ layers during gastrulation transform the original animal-vegetal polarity of the egg into the anteroposterior body pattern of the postgastrula embryo.

The dorsoventral axis in the amphibian embryo forms after fertilization. The point of sperm entry orients cytoplasmic rearrangements including a rotation of the egg cortex relative to the endoplasm. The meridian of greatest cortical displacement, opposite the sperm entry point, becomes the dorsal midline of the embryo. During cortical rotation, cytoplasmic components originally located near the vegetal pole are transported along microtubules to the prospective dorsal side of the embryo. Here, they interact with mesoderm-inducing factors in determining marginal cells to form dorsal mesodermal structures, in particular notochord. Prospective notochord, also known as Spemann's organizer, induces neighboring ectoderm to form neural tissue. The organizer also induces adjacent mesoderm to make lateral instead of ventral structures, thus generating the full spectrum of mesodermal organs along the dorsoventral axis. The molecular chain of events that leads to the formation of Spemann's organizer seems to include β-catenin, which acts as part of a transcription factor that becomes restricted to the nuclei of dorsal cells at the blastula stage.

The origin of left-right asymmetry in metazoa may be based on oriented stereoisomeric molecules that define a left-right polarity if they are oriented with regard to the anteroposterior and dorsoventral body axes. In mice and other vertebrates, several genes have lateralized expression patterns and control the development of left-right asymmetry. These lateralized gene activities may be triggered by signals carried with the flow, from right to left, of extracellular fluid covering Hensen's node, a critical growth area of the embryo. The flow appears to be driven by the oriented movement of certain cilia, which in turn seems to depend on the presence of oriented and handed dynein molecules.

Gastrulation

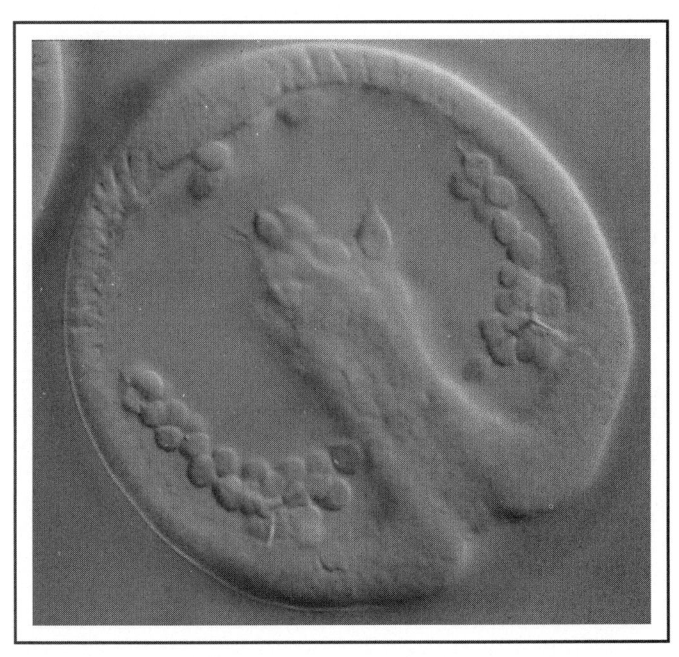

Figure 10.1 Live sea urchin embryo during gastrulation, photographed with differential interference contrast. The outer layer of cells is the ectoderm, which will produce epidermis and neurons. The columnar structure inside is the incipient inner germ layer, or endoderm. It will give rise to the inner lining of the gut. The cells in between constitute the middle germ layer, or mesoderm. Some mesodermal cells are arranged in a ring-shaped pattern around the base of the primitive gut and form two tri-radiate skeletal rudiments. Other mesodermal cells are given off at the tip of the primitive gut and send out filamentous extensions, which will pull the tip of the gut to the overlying ectoderm. At the point of contact between gut and ectoderm is where the mouth of the sea urchin larva will form. The original opening at the other end of the primitive gut will become the anus.

DURING cleavage, the fertilized egg is divided into many cells, which typically form a fluid-filled sphere, the *blastula.* During the following stages of development, the embryo profoundly rearranges its cells to form the body plan characteristic of its species. The first stage in this process is called *gastrulation* (Gk. *gaster,* "stomach"), because, among other things, it generates the rudiment of a digestive tract. Gastrulation transforms the spherical blastula into a more complex configuration of three *germ layers* (Figs. 10.1 and 10.2). The outer layer, which will be exposed to the external environment, is the *ectoderm.* It gives rise, for the most part, to the epidermis and the nervous system of the developing organism. The inner layer of cells is the *endoderm,* which forms the primitive gut, or ***archenteron*** (Gk. *arche,* "origin"; *enteron,* "gut"). Numerous structures in between develop from an intermediate layer called the *mesoderm.* The ectoderm and endoderm are *epithelia*—that is, closely packed sheets of cells connected by tight junctions and resting on a basement membrane (see Fig. 2.22). The mesoderm is sometimes epithelial, but in other cases it is *mesenchymal*—that is, forming a loose arrangement of cells surrounded by much extracellular material. An embryo in the process of gastrulation is called a *gastrula* (pl. *gastrulae*).

A common feature of early gastrulae is the ***blastopore,*** an indentation or groove through which cells move inside to form the endoderm and mesoderm. The fate of the blastopore is used as a basis for classifying animals into two groups. Those in which the blastopore develops into the mouth are known as ***protostomes*** (Gk. *proto-,* "first"; *stoma,* "mouth"). Animals in which the blastopore becomes the anus, and mouth formation is secondary, are called ***deuterostomes*** (Gk. *deutero-,* "second"). Deuterostomes include vertebrates and echinoderms (sea urchins and their relatives), whereas the protostomes include most other animals. Among the deuterostomes, as well as among the protostomes, there are further differences in regard to the cellular movements that establish the germ layers.

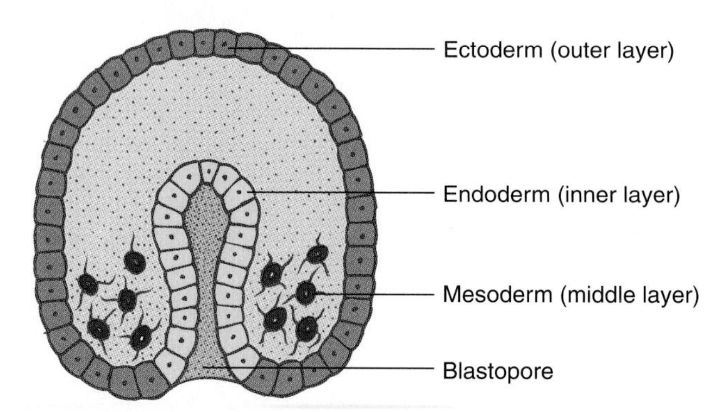

Figure 10.2 Schematic diagram of a gastrula. Formation of the primitive gut (archenteron) has begun at the blastopore. The embryo consists of three germ layers. The inner and outer layers, ectoderm and endoderm, are epithelia. The intermediate layer, mesoderm, sometimes consists of mesenchyme, that is, single cells surrounded by extracellular matrix.

In the analysis of gastrulation, researchers have identified recurring types of epithelial movements and cellular activities. Epithelia expand and buckle and so create folds and cavities, and cells have ways of changing their positions relative to each other. As we will see in this chapter, the complex gastrulation processes in various embryos can be understood as combinations of recurring elementary activities of epithelia and individual cells. In subsequent chapters, epithelial and cellular activities will in turn be analyzed in molecular terms including in particular those molecules that mediate cell adhesion. This reductionist approach was already adopted by the early embryologists and was boosted later with the introduction of advanced microscopy and molecular genetics. It has provided a better understanding not only of gastrulation but also of later events that generate the organ rudiments and shape the basic body plan of the organism. In this chapter, the emphasis will be on gastrulation in sea urchins and amphibians, which have been analyzed in incisive sets of experiments.

10.1 The Analysis of Morphogenesis

Gastrulation marks the onset of morphogenesis, when epithelia and individual cells undergo dramatic movements that generate the embryo's organ rudiments and in the process give rise to an overall body plan. These events are collectively called *morphogenetic movements* or ***morphogenesis*** (Gk. *morphe,* "form"; *genesis,* "creation").

During morphogenesis, individual cells move relative to each other, while entire epithelia spread, fold, bend, and roll dramatically. These movements bring about gastrulation and the subsequent stage of development, known as *organogenesis,* when subdivisions of the three germ layers cooperate to form the organ rudiments.

Morphogenetic movements are striking in their complexity and orderliness, as can be appreciated best in time-lapse movies (see www.esb.utexas.edu.kalthoff/bio 349). The behavior of cells during morphogenesis gives the impression of a grand ballet. Each "performance" creates a new organism of gradually increasing complexity. Once the organ rudiments are formed, the embryo has acquired its ***basic body plan;*** the appearance of an embryo at this stage is characteristic of its phylogenetic group. Any insect embryo at this stage begins to look unmistakably like an insect, and any vertebrate embryo at this stage can be recognized as such. The generation of a basic body plan by morphogenetic movements is fundamental to development and stunning to observe.

As we examine morphogenetic events more closely, we see how cells divide, wedge themselves between other cells, and even detach from epithelia and move into body cavities. Note that similar behaviors later in life may be fatal if performed by cancer cells. Thus, the harnessing of these cellular behaviors for the purposes of morphogenesis and their subsequent restriction represent a major regulatory achievement.

The analysis of morphogenetic movements presents researchers with formidable challenges. How can complex movements involving the entire embryo be broken down into smaller, more readily analyzable events? Which of the signals exchanged between cells are most important: the *physical* forces of cells pulling and pushing one another, or the *chemical* signals that are exchanged between cells and act on matching receptors? Does gene expression control cell shape, or does cell shape control gene activity? Some researchers have proposed that all cells in an embryo initially express the same genes until their physical interactions generate groups of cells with different mechanical states, which then secondarily cause these cells to express specific combinations of genes (R. Gordon and G. W. Brodland, 1987). However, most researchers today assume that cells are first programmed to express certain master genes, which in turn coordinate other genes with more limited functions. The latter class includes, for example, genes encoding cell adhesion molecules and the proteins that are involved in cell movements and shape changes. It is the latter approach that we will explore in this text.

MORPHOGENESIS INVOLVES TYPICAL EPITHELIAL MOVEMENTS

In the course of gastrulation and organogenesis, a small set of epithelial movements occurs in a large variety of combinations. Epithelia are defined as cellular layers that rest on a basement membrane and form tight junctions at the apical, or outer, surface (see Fig. 2.22). Some epithelia are *simple,* or consisting of a single sheet of cells, as in a sea urchin blastula (Fig. 10.2). Other epithelia are *stratified,* or consisting of multiple layers of cells, as in a frog blastula (see Fig. 5.5). In the latter case, only the outermost layer of cells is connected by tight junctions, whereas all layers tend to be connected by *gap junctions.*

Most morphogenetic movements of epithelia can be broken down into the following elements (Fig. 10.3):

Invagination: Local inward buckling of an epithelium to create a depression or cavity, much like the deformation sustained by a soft tennis ball poked with a finger.

Involution: Inward movement of an expanding epithelium around an edge.

Convergent extension: Elongation of an epithelium in one dimension while shortening in another dimension.

Epiboly: The spreading movement of an epithelium to envelop a yolk mass or deeper mass of cells.

Delamination: Splitting of one layer of cells into two parallel layers.

Passive movements: Movements in which one part of an epithelium is simply pushed or dragged along by another part of an epithelium.

Different combinations of these elements yield a wide variety of morphogenetic movements in different animals.

MORPHOGENESIS IS BASED ON A SMALL REPERTOIRE OF CELL ACTIVITIES

In addition to epithelia, embryos consist of *mesenchymes*—that is, aggregates of single cells surrounded by large amounts of extracellular material. Epithelia and mesenchymes are interconvertible by *epithelial-mesenchymal transitions,* which are a common phenomenon in embryonic development. The behavior of cells within epithelia and mesenchymes, and during epithelial-mesenchymal transitions, can be broken down into the following components (Fig. 10.4):

Migration: Movement of individual cells over a substratum of other cells or extracellular material.

Intercalation: Wedging of cells between their neighbors. *Lateral intercalations* within the same layer of cells and may cause convergent extension. *Radial intercalations* are perpendicular to the surface and between adjacent layers of cells; they may cause epiboly as the epithelium's surface increases while the number of its layers decreases.

Ingression: The transition of individual cells or small groups of cells from an epithelium into an embryonic cavity.

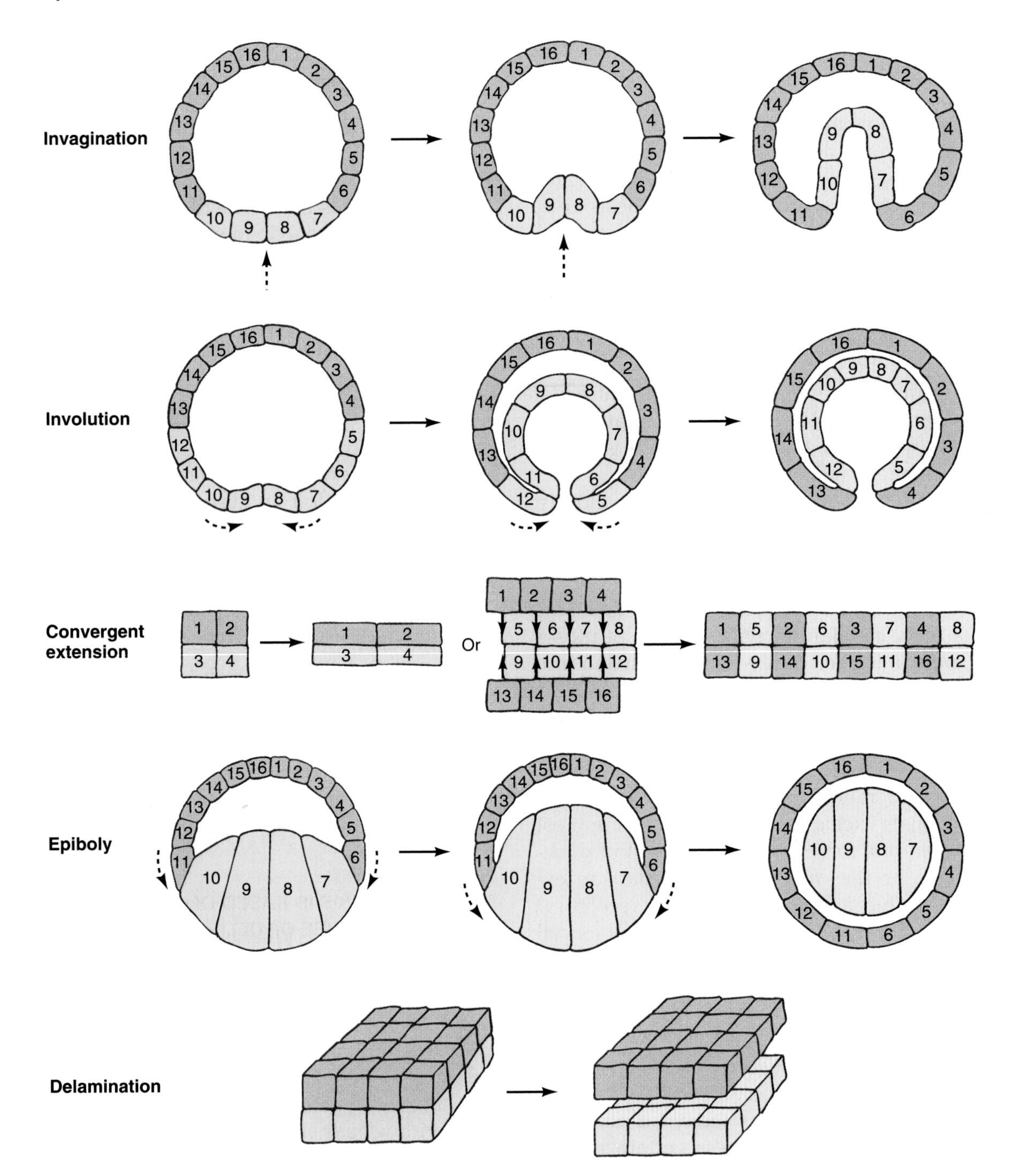

Figure 10.3 Schematic diagrams of typical epithelial movements. See text.

Shape changes: Coordinated changes in cell shape can cause cell layers to buckle or to undergo convergent extension.

Division: Cell division without growth, as in cleavage, increases the number of building blocks and the possibilities for fine morphogenetic movements in an embryo. In combination with cell growth, oriented cell divisions can change the size and shape of a cell layer.

Changes in adhesiveness: Cells adhere to one another, and to extracellular material, by means of certain membrane proteins. Changes in adhesiveness contribute to cell migration, ingression, delamination, and epiboly.

Programmed cell death: Cells can die in a programmed way. Such programmed cell death may carve out, or "sculpt," certain organs such as the digits of hands and feet.

The cellular behaviors just listed are thought to be regulated by hierarchies of gene activity. For instance, cyclins and cyclin-dependent kinases control the cell cycle as discussed earlier. A hierarchy of genes first studied in the roundworm *Caenorhabditis elegans* controls cell

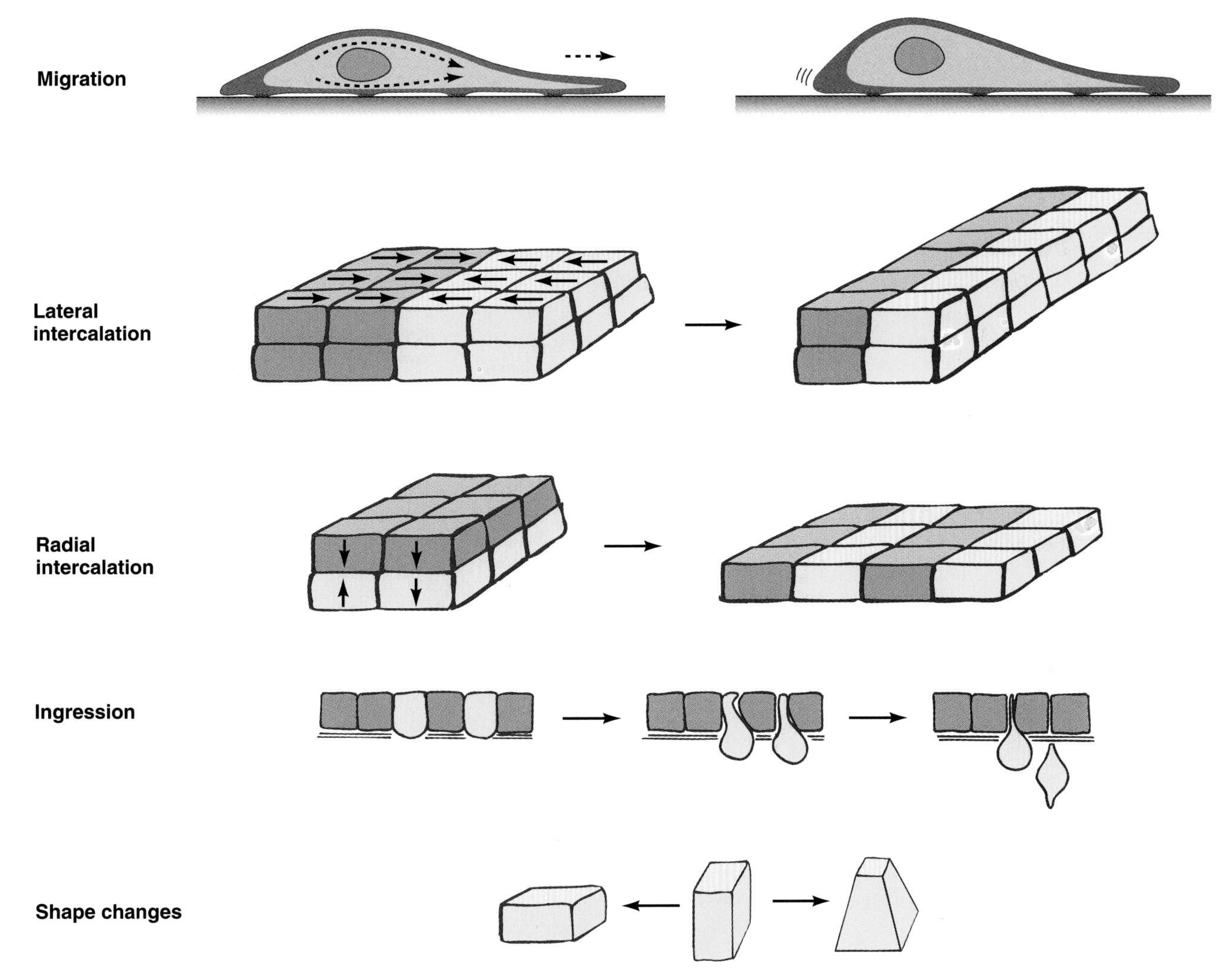

Figure 10.4 Schematic diagrams of cellular activities in morphogenesis. See text.

death (see Section 25.4). Similarly, the synthesis of plasma membrane proteins known as *cell adhesion molecules* and *substrate adhesion molecules* governs cell adhesion. These molecules are thought to play key roles in morphoregulatory cycles, in which gene activities control cellular behaviors, which regulate morphogenetic movements, which in turn feed back on gene expression (see Fig. 11.25).

10.2 Gastrulation in Sea Urchins

Sea urchins have long been favorite organisms for the study of gastrulation, because their transparency allows observation of the entire process in a living animal *(McClay et al., 1992).* The sea urchin blastula is a sphere of about 1000 cells surrounding a *blastocoel* filled with a gelatinous fluid. These cells are arranged as a single-layered epithelium with a *basement membrane* on the inside, and with cilia and a *hyaline layer* on the outside. The arrangement of cells in the free-swimming blastula, and the fates of these cells after gastrulation, reflect the animal-to-vegetal arrangement of blastomere tiers present at the 60-cell state (Figure 10.5; *Ettensohn, 1992;* Ruffins and Ettensohn, 1993, 1996). A patch of blastomeres with long and coarse cilia that make up the ***apical tuft*** are derived from the most animal tier of blastomeres. These blastomeres and the following animal and vegetal 1 tiers will form the larval ectoderm after gastrulation. The vegetal 2 tier will give rise to the endoderm and part of the mesoderm. The descendants of the large and small micromeres, located at the vegetal pole of the blastula, will form other portions of the mesoderm.

At the end of the blastula stage, the blastomeres in the vegetal area become tall and columnar, forming a thick, flat plate known as the ***vegetal plate.*** The first step in gastrulation is the *ingression* of a group of cells near the center of the vegetal plate. First these cells, which are descendants of the large micromeres, begin to pulsate and bulge on their basal surfaces while their apical surfaces constrict. Then they undergo a *change in adhesion:* They break the tight junctions that seal them to neighboring vegetal cells, and they withdraw their cilia and microvilli from the hyaline layer that covers the outside of the embryo. At the same time, they develop an affinity

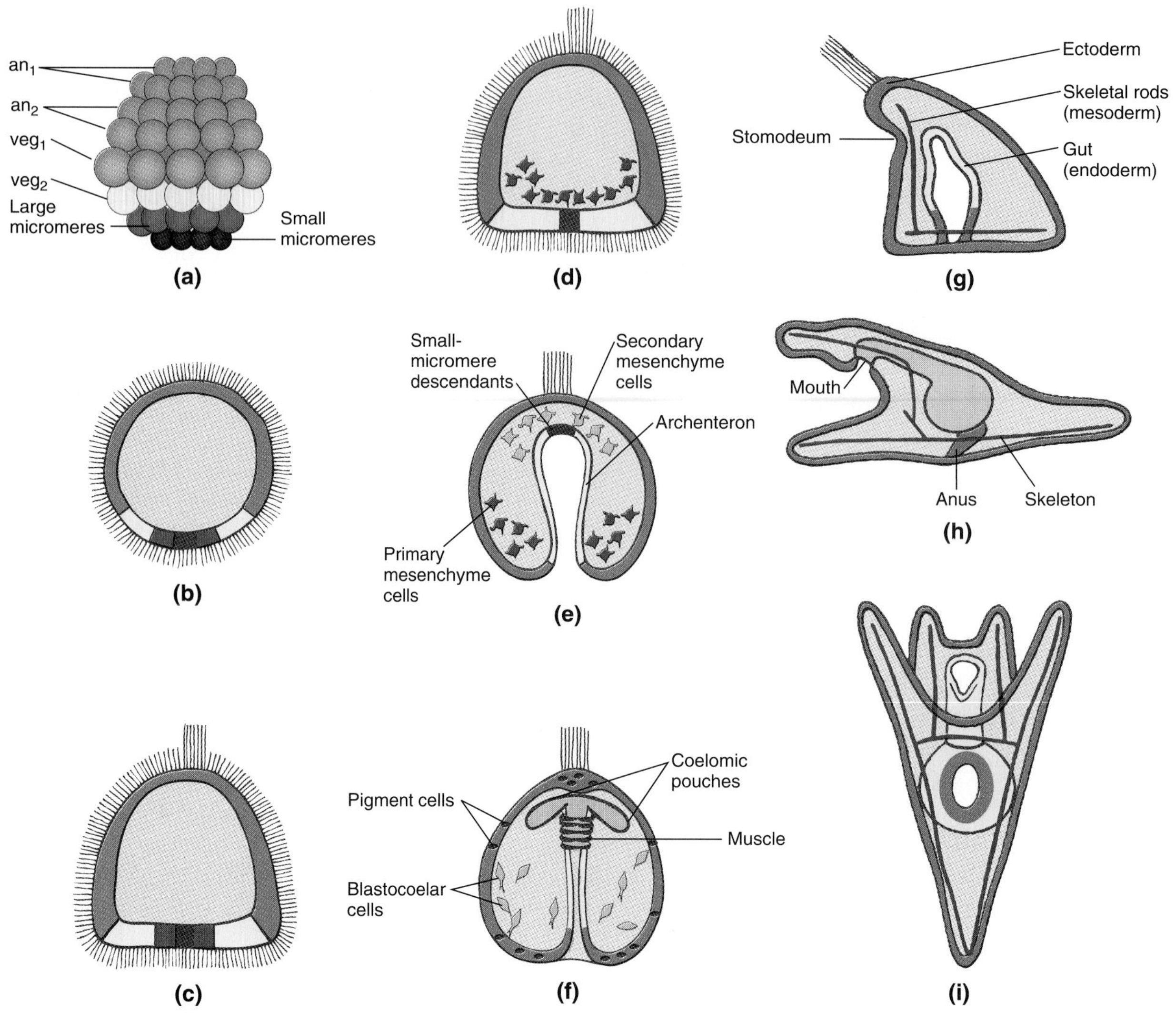

Figure 10.5 Sea urchin gastrulation. **(a)** Early blastula consisting of 60 cells. The top two tiers, each comprising eight blastomeres, are the animal 1 (an_1) tiers. The next two are the animal 2 (an_2) tiers. Following are one vegetal 1 (veg_1) tier, one vegetal 2 (veg_2) tier, eight large micromeres, and four small micromeres. **(b)** Midblastula. **(c)** Late blastula. The descendants of the vegetal 2 blastomeres, large micromeres, and small micromeres form a vegetal plate of columnar cells. **(d)** Early gastrula, showing ingression of primary mesenchyme cells, the descendants of the large micromeres. **(e)** Midgastrula with archenteron, showing ingression of secondary mesenchyme cells, which are derived from vegetal 2 blastomeres and the small micromeres. **(f)** Late gastrula, showing derivatives of the secondary mesenchyme cells and the small-micromere descendants; the latter contribute to the coelomic pouches that extend from the archenteron tip. Primary mesenchyme derivatives are omitted for clarity. **(g)** Prism-stage larva showing skeletal rods derived from primary mesenchyme cells. The archenteron bends toward an ectodermal area, which forms a depression, the stomodeum. The secondary mesenchyme cell derivatives have been omitted for clarity from this and the following diagrams. **(h, i)** Pluteus larva viewed from the side and from the vegetal pole. The archenteron tip and the stomodeum have joined to form the mouth, while the blastopore develops into the anus.

for the basement membrane on the inside of the blastula (Fink and McClay, 1985; Anstrom et al., 1987). As a result of these changes, the large-micromere descendants detach from the vegetal plate and ingress into the blastocoel as ***primary mesenchyme cells.*** After a period of migration on the basement membrane, the primary mesenchyme cells settle into a ringlike pattern on the vegetal side of the blastocoel. Here they fuse to form a *syncytium,* a multinucleated cytoplasmic body that remains anchored to the outer epithelium by filopodia. The syncytium secretes the skeleton of spicules supporting the arms of *pluteus larva,* which is characteristic of sea urchins.

As the primary mesenchyme cells begin to migrate, the remaining cells of the vegetal plate start to form the *archenteron* (see Figure 10.5e). Archenteron formation in sea urchins occurs in multiple phases. During the first phase, the blastopore forms as the vegetal plate *invaginates,* buckles inward, and extends about one-third into the blastocoel (see Fig. 10.1). During the second phase, the archenteron *elongates by convergent extension* until its tip almost reaches the animal pole. Toward the end of the elongation phase, the archenteron tip is also *pulled by filopodia* toward the ectodermal cells that will form the mouth. Finally, additional cells are recruited to the midgut and hindgut by *involution*.

The forces driving invagination are generated within the vegetal plate itself and are independent of the rest of the embryo. When A. R. Moore and A. S. Burt (1939) isolated vegetal halves of starfish embryos at the beginning of gastrulation, the vegetal parts invaginated on their own, and the rim of each vegetal plate rolled up and closed over the archenteron, forming a gastrula-shaped vesicle. Also, when an isolated vegetal plate was cut radially from the periphery to the center of an isolated vegetal plate, the cut edges sprang apart, indicating that the vegetal plate was under tension. These results indicated that the forces generated within the vegetal part of the embryo were sufficient to drive the first phase of archenteron formation.

The cellular and molecular mechanisms generating these forces remain to be established (Ettensohn, 1984a; L. A. Davidson et al., 1995). One of the current models, the *gel swelling model*, is built on the observation that the *hyaline layer* covering the vegetal plate cells consists of several laminae of extracellular material. If an inner lamina were to attract more water and thus swell more strongly than an outer lamina, then the entire hyaline layer and the adjacent vegetal plate would buckle like the bimetallic strip of a thermostat (Fig. 10.6). In accord with this model, the vegetal plate cells do secrete a hygroscopic (water-attracting) material prior to invagination, and treatments that disturb this process also interfere with invagination (Lane et al., 1993). According to another model, the *apical constriction model*, the apical surfaces of cells located at the center of the vegetal plate constrict, thus forcing the plate to buckle inward. This model is also used to explain the folding of the neural plate into the neural tube and will be discussed more fully in that context (see Section 12.2).

The second phase of archenteron formation is a dramatic case of convergent extension: The archenteron elongates from a short, squat cylinder to a long and slender one (Fig. 10.7). This approximately threefold increase in overall length cannot be explained by cell division and growth, since only 10 to 20% of the cells divide during this time. Involution does not seem to play a major role either, since only a few cells turn inside at the blastopore during this phase. Two coordinated cell behaviors could account for the convergent extension of the archenteron: cell *shape changes* and cell *intercalation* (Fig. 10.8).

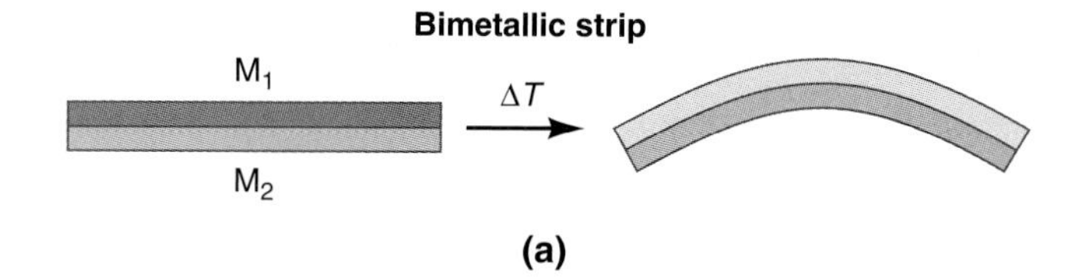

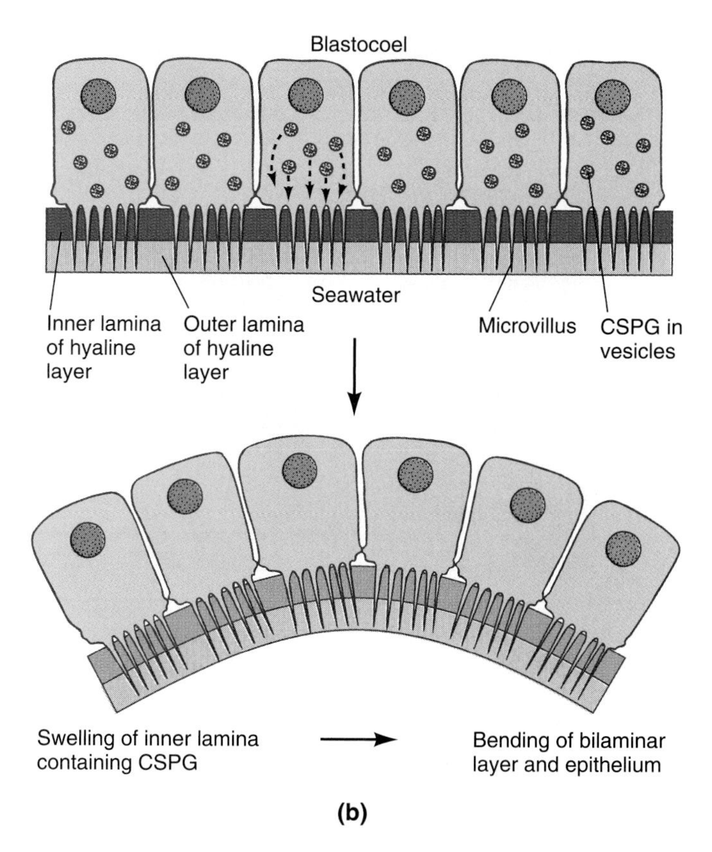

Figure 10.6 Gel-swelling model for vegetal plate invagination in sea urchin embryos by the secretion of a hygroscopic (water-attracting) molecule into the hyaline layer adjacent to the vegetal plate. **(a)** The bimetallic strip in a thermostat bends in response to temperature changes (ΔT) because one metal (M_1) has a greater thermal expansion coefficient than the other metal (M_2). **(b)** Vegetal plate of a sea urchin embryo. The hyaline layer covering the apical surface of the vegetal plate cells is shown for simplicity as consisting of one inner and one outer lamina. The cells contain secretory vesicles with chondroitin sulfate proteoglycan (CSPG). Secretion of CSPG (arrows) makes the inner lamina more hygroscopic, so that it swells more strongly than the outer lamina. Thus, the bilayered hyaline layer and the adjacent epithelium buckle.

MEASUREMENTS by Ettensohn (1985b) as well as Hardin and Cheng (1986) on two sea urchin species revealed that both types of cellular behavior occur during the second phase of archenteron formation. Archenteron cells grew longer, while their width and thickness decreased, indicating that a coordinated change in cell shape plays a role in archenteron elongation. At the same time, the *number* of cells along the length of the archenteron increased while the circumferential number of cells decreased from 16 to 10 in one species (*Strongylocentrotus purpuratus*) and from 18

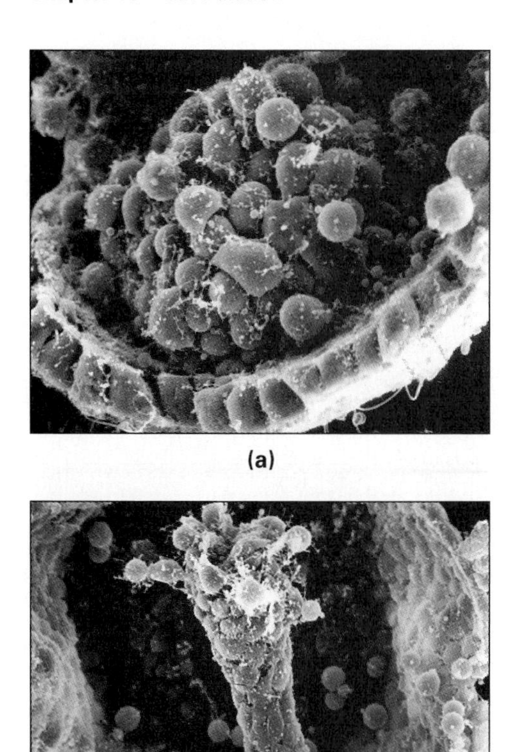

Figure 10.7 Archenteron elongation in the sea urchin *Lytechinus variegatus.* The scanning electron micrographs show the interior of the embryo after removal of the animal ectoderm. **(a)** Midgastrula. **(b)** Late gastrula.

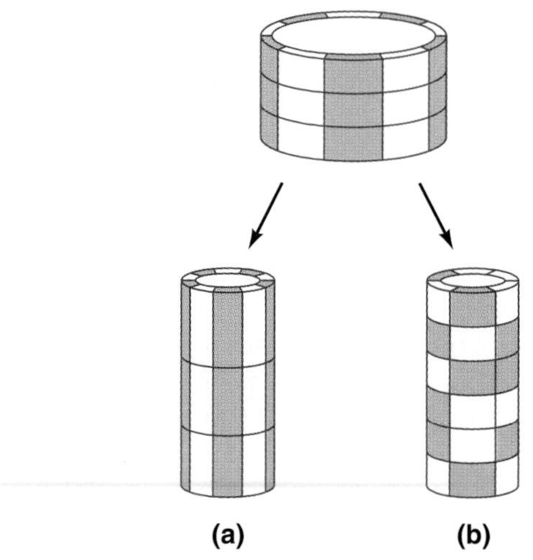

Figure 10.8 Two types of cellular behavior that could explain the convergent extension movement observed during the second phase of archenteron formation in sea urchins. In these schematic diagrams, the archenteron is represented as a cylinder. **(a)** Coordinated change of cell shape. Each cell may elongate while shrinking in width and possibly also in depth. **(b)** Cell intercalation may increase the number of cells longitudinally while reducing the number of cells in the cylinder's circumference.

to 7 in another species (*Lytechinus pictus*). This decrease could not have resulted from cell death, since no dying cells were observed and the overall mass of the archenteron did not change. This means that convergent extension of the archenteron must be associated with cell rearrangements.

Subsequent tracing of archenteron cells in a third species, *Eucidaris tribuloides,* by videomicroscopy showed directly that the archenteron cells *intercalate*—that is, wedge between their neighbors (Hardin, 1989). The intercalations are oriented in such a way that the *number* of cells increases in the longitudinal dimension of the archenteron (Fig. 10.9). The intercalating cells seem to displace their neighbors actively, beginning with the formation of small *lamellipodia* near the basal surface. The intercalating movements are strictly local and preserve the relative position of cells over the length of the archenteron. Thus, the cells that are located close to the center of the vegetal plate and invaginate first remain near the tip of the archenteron, and cells that are located at the periphery of the vegetal plate and invaginate last remain near the base of the archenteron. The signals that orient the cell intercalations perpendicular to the archenteron axis so that they result in an elongation are not yet known.

Questions

1. In Figure 10.8a, the convergent extension of the archenteron by means of coordinated cell shape changes is drawn in a way that involves a decrease in the archenteron's diameter. Could there be a convergent extension by coordinated cell shape change that leads to an elongation of the archenteron without a concomitant decrease in diameter?
2. Convergent extension, whether it is driven by coordinated cell shape change or by cell intercalation, is thought to be an *active* movement—that is, driven by forces acting within the archenteron itself. Could the elongation of the archenteron also be a *passive* movement, driven by forces outside the archenteron? How could this proposition be tested?

While archenteron elongation is still under way, another set of mesenchymal cells become motile at the archenteron tip (see Fig. 10.7b). These cells, called ***secondary mesenchyme cells,*** are derived from the vegetal 2 tier of blastomeres and the small micromeres present at the 60-cell stage (see Fig. 10.5a, e).

Still attached to the archenteron tip, the secondary mesenchyme cells initiate a third phase of archenteron elongation by extending filopodia through the blastocoel. With a diameter of about 1 μm, these filopodia are relatively thick and serve both sensory and mechanical functions. They first probe the inside of the blastocoel wall, attach at suitable sites, and then contract, further

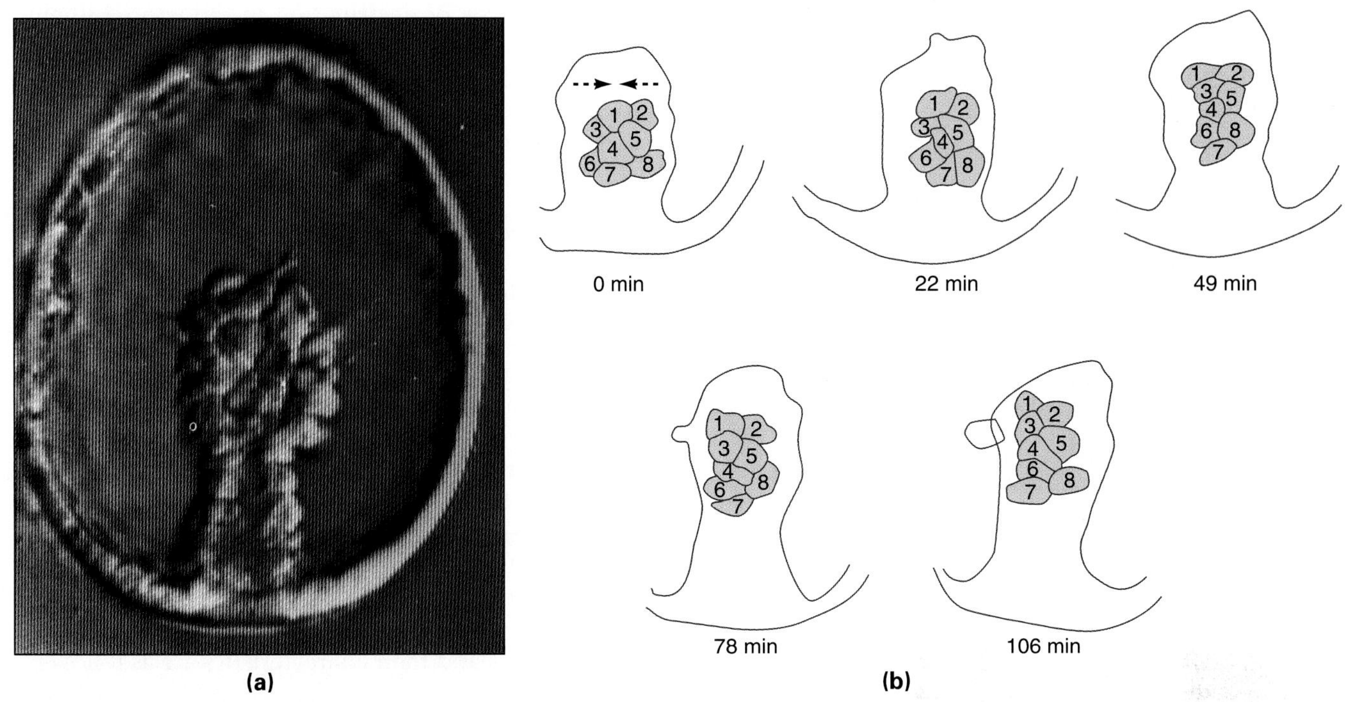

Figure 10.9 Cell rearrangements during gastrulation in the sea urchin *Eucidaris tribuloides.* **(a)** Midgastrula photographed from a video image. **(b)** Tracings made on the video monitor at the times indicated. Note that cell intercalations occurred in a circumferential direction (arrows), with cell 3 wedging between cells 1 and 4 and cell 6 wedging between 4 and 7. During the period of observation, the array of cells narrowed from about three cell diameters to two, while elongating from about three cell diameters to more than four. The length of the archenteron increased correspondingly.

extending the archenteron and orienting the tip toward its target site (Hardin and McClay, 1990). If secondary mesenchyme cells are killed with a laser beam, the elongation of the archenteron remains incomplete (Hardin, 1988). Thus, the shape change (elongation) of the archenteron cells is, at least in part, caused by the contraction of secondary mesenchyme cells.

The tip of the extending archenteron is met by the *stomodeum,* a depression in the oral surface of the ectoderm. Fusion of the stomodeum with the archenteron tip generates the mouth of the pluteus larva, while the blastopore becomes the anus. Thus, sea urchin gastrulation follows the pattern typical of deuterostomes.

As the archenteron tip approaches the stomodeum, the secondary mesenchyme cells form two clusters of cells on either side of the foregut. These cells form pouches of mesodermal epithelium, which displace the blastocoel and form a secondary body cavity called the ***coelom.*** Other descendants of the secondary mesenchyme cells form pigment cells, muscles, and blastocoelar cells (see Fig. 10.5f). While in most sea urchins there is little or no overlap in the *fates* of primary mesenchyme cells and secondary mesenchyme cells, the *potency* of secondary mesenchyme cells includes the major fate of primary mesenchyme cells—that is, skeleton formation. Ettensohn and McClay (1988) observed that secondary mesenchyme cells substitute for primary mesenchyme cells in skeleton formation if the number of primary mesenchyme cells is experimentally reduced. They found that after the surgical removal of different numbers of primary mesenchyme cells, a fraction of secondary mesenchyme cells converted to the primary mesenchyme cells' fate—that is, participated in skeleton formation. In fact, some sea urchins have only one burst of mesenchyme formation (Schroeder, 1981).

Toward the end of gastrulation, additional cells are recruited to the archenteron from the area surrounding the blastopore. In a movement of *involution,* cells derived from the veg1 tier of the late cleavage stage move over the blastopore lip and join the veg2-derived cells that invaginated earlier (see Fig. 10.5e, f). The involution movement occurs slowly, beginning when the archenteron has reached about half of its final length and extending until the stage when the embryo has the shape of a prism (see Fig. 10.5g). Perhaps due to the slow speed of this movement, the component of involution has long been overlooked in sea urchin gastrulation until it was recognized recently in several species (Martins et al., 1998; *Ettensohn, 1999*).

Many steps in sea urchin gastrulation require cellular interactions. For instance, primary mesenchyme cells (PMC) will form skeleton only if positioned near a ring of ectodermal cells close to the vegetal base. At the same time, PMC inhibit secondary mesenchyme cells (SMC) from making skeleton. SMC at the archenteron tip probe the blastocoel roof before pulling the archenteron tip to

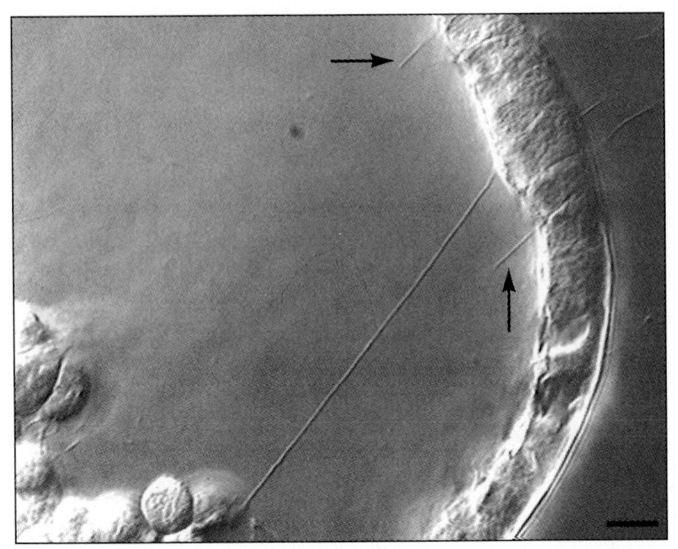

Figure 10.10 Thin filopodia extended by ectodermal cells (to the right) and primary mesenchyme cells (bottom left) in a sea urchin (*Lytechinus variegatus*) gastrula. These filopodia are 0.2 to 0.4 μm in diameter, and the one extended from the primary mesenchyme cell is about 80 μm long. These filopodia traverse the basement membrane of the ectodermal epithelium.

its target site where the mouth will form. The interacting cells contact one another by means of *thin filopodia,* which measure only 0.2 to 0.4 μm in diameter but extend to more than 80 μm in length (Fig. 10.10). The location and timing of filopodial contacts do not correlate with cell locomotion. Thus, in contrast to the thick filopodia mentioned earlier, the thin filopodia seem to be involved only in intercellular communication (Malinda et al., 1995; Miller et al., 1995). The nature of the signals transmitted remains to be elucidated.

In summary, sea urchin gastrulation combines several different types of morphogenetic movement. Mesoderm is formed by *ingression* of one or two waves of mesenchyme cells. Endoderm arises through *invagination* of the vegetal plate followed by *convergent extension* of the archenteron. The convergent extension movement is based on both cell intercalations and coordinated shape changes. The latter are caused in part by constrictions of secondary mesenchyme cells that pull the archenteron tip toward the prospective mouth region of the ectoderm. Toward the end of gastrulation, additional cells are recruited to the archenteron by involution.

10.3 Gastrulation in Amphibians

Gastrulation movements in amphibians are more complex than those in sea urchins, partly because the solid mass of cells in the vegetal half of the egg is an impediment to invagination. Also, the amphibian blastula is several cell layers thick, and in some areas the outer cell layers and the inner ones behave differently. Nevertheless, the analysis of gastrulation in terms of epithelial movements and cellular behaviors has been particularly rewarding with amphibian embryos because they are eminently suitable for transplantation experiments. Since much recent work has been done with embryos of the frog *Xenopus laevis,* our discussion will focus on this species.

DIFFERENT GASTRULA AREAS SHOW DISTINCT CELLULAR BEHAVIORS

Along the animal-vegetal axis of a *Xenopus* early gastrula, one can define six regions by their distinct cellular and epithelial behaviors (Gerhart and Keller, 1986). These regions are mapped in Figure 10.11, which also serves as a reference for much of the subsequent discussion. Each region of the early gastrula is initially symmetrical about the animal-vegetal axis. However, most cellular and epithelial movements begin first, and continue most extensively, on the dorsal side of the embryo. The six regions and their characteristics are as follows:

1. The ***animal cap,*** about three cell layers deep, is derived from the pigmented animal hemisphere of the egg. The animal cap expands by *epiboly* until it covers about half of the late gastrula surface (Fig. 10.11a, b).
2. Adjacent to the animal cap is the ***noninvoluting marginal zone,*** a wide girdle of cells about four to five layers deep. It is so named because it has certain cell behaviors in common with the adjacent *involuting marginal zone.* Both zones expand along the dorsal midline in the animal-vegetal direction while shrinking in other dimensions. This *convergent extension* movement begins midway during gastrulation and causes major distortions in the fate maps of gastrulae at successive stages of development (Fig. 10.11b, c). In addition, the noninvoluting marginal zone, like the animal cap, undergoes *epiboly.*
3. Following the noninvoluting marginal zone on the vegetal side is the ***involuting marginal zone.*** It forms the margin of the blastopore, which *involutes* during gastrulation while the noninvoluting marginal zone fills the surface area vacated by the involuting marginal zone. Therefore, the boundary between the involuting marginal zone and the noninvoluting marginal zone, called the *limit of involution,* reaches the edge of the blastopore by the end of gastrulation (Fig. 10.11c). Within the involuting marginal zone, we distinguish a superficial layer from a deep layer. The ***superficial layer*** of the involuting marginal zone forms the *archenteron roof*—that is, part of the lining of the future gut. Thus, the superficial layer of the involuting marginal zone stays at the surface of the embryo, even though it faces the cavity of the archenteron instead of the external environment (Fig. 10.11e, j–m). Meanwhile, the ***deep layer*** of the involuting marginal zone stays inside the embryo,

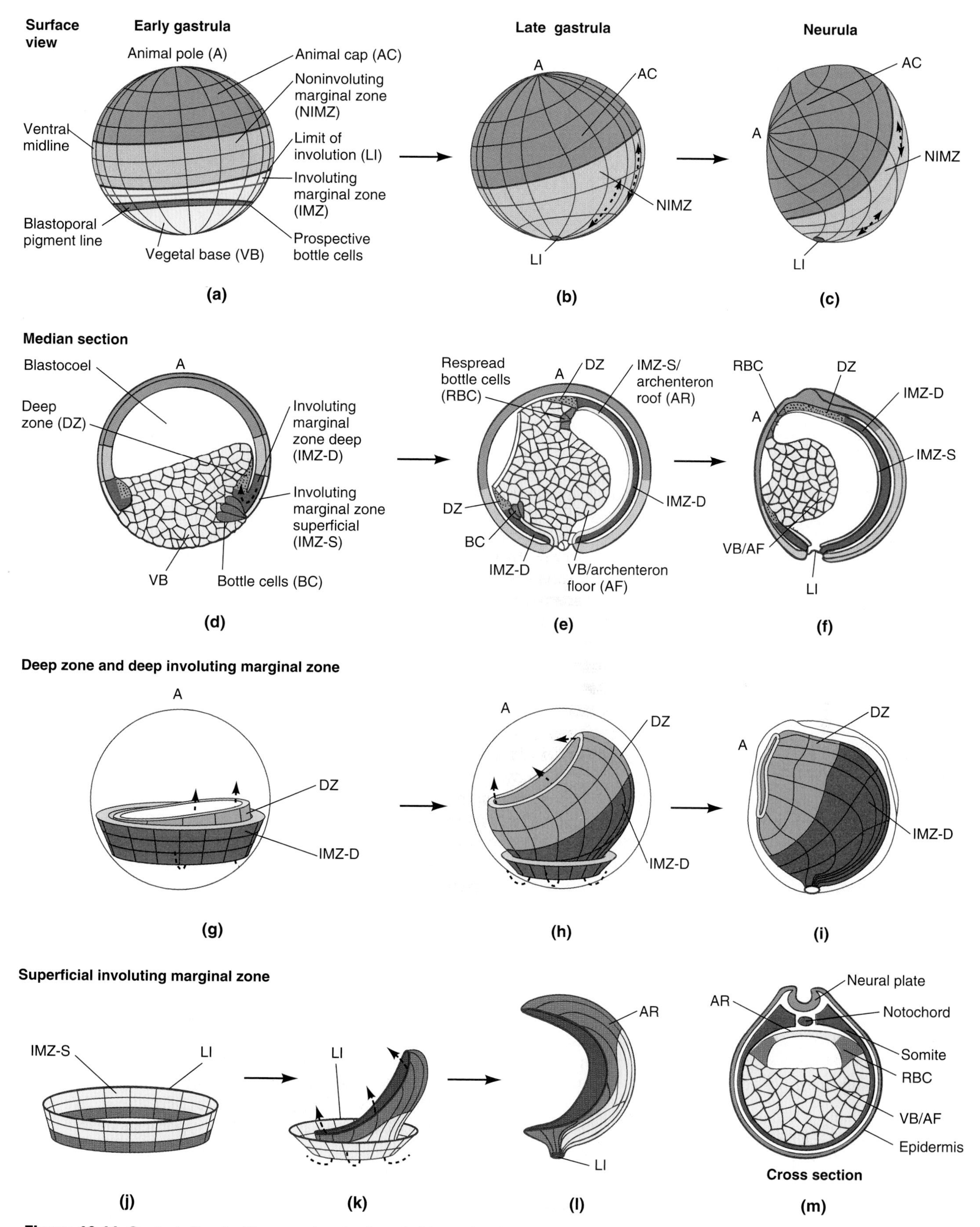

Figure 10.11 Gastrulation in *Xenopus laevis;* dorsal side to the right, animal pole up. Arrows indicate the direction of movement.

giving rise to much of the mesoderm (Fig 10.11e, g–i). The term "marginal zone" has been used in previous chapters as a short form for involuting marginal zone.

4. The ***deep zone*** is a ring of cells between the deep involuting marginal zone and the large vegetal cells (Fig. 10.11d–i). Deep zone cells migrate toward the animal pole on the basal surface of the involuting marginal zone, noninvoluting marginal zone, and animal cap, with the involuting marginal zone following behind. At the same time, the deep zone seems to drag the dorsal margin of yolky vegetal base toward the animal pole. Deep zone cells become head mesoderm and heart, while the deep involuting marginal zone cells form the trunk mesoderm.
5. At the vegetal border of the involuting marginal zone, a ring of superficial cells constrict at their apical surfaces while expanding at their bases, each cell thus acquiring a bottle shape (Fig. 10.11a, d–f); from this they were given the name ***bottle cells.*** As the apical constrictions cause the aggregation of cortical pigment granules, the bottle cells collectively generate a thin, dark ring called the *blastoporal pigment line.* The apical contractions begin at the dorsal midline, where the blastopore first becomes visible, and progress bilaterally over a period of 2 h until they reach the ventral midline.
6. The ***vegetal base*** is formed by large yolky cells, extending from the bottom of the blastocoel to the outer surface of the vegetal hemisphere (Fig. 10.11d–f). During gastrulation the vegetal base is *tilted* and *displaced ventrally,* so that it will form the *archenteron floor.* Note that the vegetal base is still contiguous with the respread bottle cells forming the archenteron sides and with the superficial involuting marginal zone forming the archenteron roof (Fig. 10.11m).

As the vegetal base is tilted inside, the involuting marginal zone involutes, the noninvoluting marginal zone and animal cap spread by epiboly, the limit of involution moves toward the vegetal pole, and the blastopore closes. In the course of *Xenopus* gastrulation, the ectoderm is formed by the animal cap and the noninvoluting marginal zone. The endoderm (archenteron) is composed of the vegetal base (floor), the respread bottle cells (sides), and the superficial layer of the involuting marginal zone (roof). The mesoderm is derived from the deep zone in the head region and from the deep layer of the involuting marginal zone in the trunk region.

BOTTLE CELLS GENERATE THE INITIAL DEPRESSION OF THE BLASTOPORE

Amphibian gastrulation becomes externally visible when the bottle cells acquire their characteristic shape by constricting at their apical surfaces and expanding at their bases. The bottle shape becomes especially prominent in the dorsal region of the embryo, where this activity begins, and where the bottle cells sink deep into the vegetal base (Fig. 10.12). These events initiate the formation of the *blastopore,* a curved depression that begins dorsally and expands laterally and ventrally (Fig. 10.13). The ledge of cells on the animal side of the blastopore is called the ***blastopore lip.***

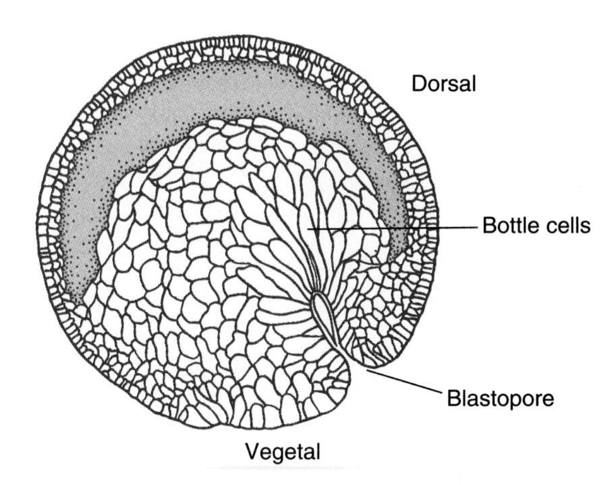

Figure 10.12 Drawing of early amphibian gastrula in median section. Bottle cells initiate the formation of the blastopore.

The bottle cells on the dorsal side remain in the deepest part of the blastopore and become the tip of the advancing archenteron (see Fig. 10.11d, e). It has therefore been thought that the bottle cells migrate inside the gastrula and pull the rest of the archenteron with them. To test this hypothesis, R. E. Keller (1981) surgically removed the bottle cells after blastopore formation. The embryos healed within 30 min of the operation, and virtually all of them gastrulated and developed normally except that the archenteron was truncated at the cephalic end. The results show that the bottle cells are *not necessary* for continued gastrulation in *Xenopus.* Their role seems to be limited to generating the initial depression of the blastopore. In accord with this conclusion, the bottle cells respread in the course of archenteron elongation to form the archenteron tip and lateral walls (see Fig. 10.11e, f, m).

As an alternative explanation of archenteron elongation, R. E. Keller (1981, 1986) proposed that the gastrulation process is driven to a large extent by *involution* and *convergent extension* of the deep involuting marginal zone. According to this hypothesis, the main forces driving archenteron elongation would be generated in prospective mesodermal cells, and the adjacent endodermal cells would simply be carried along. We will look at evidence supporting this hypothesis next.

DEEP MARGINAL ZONE CELLS ARE NECESSARY FOR INVOLUTION

In contrast to sea urchins, in which *invagination* is a major step in archenteron formation, amphibians gastrulate

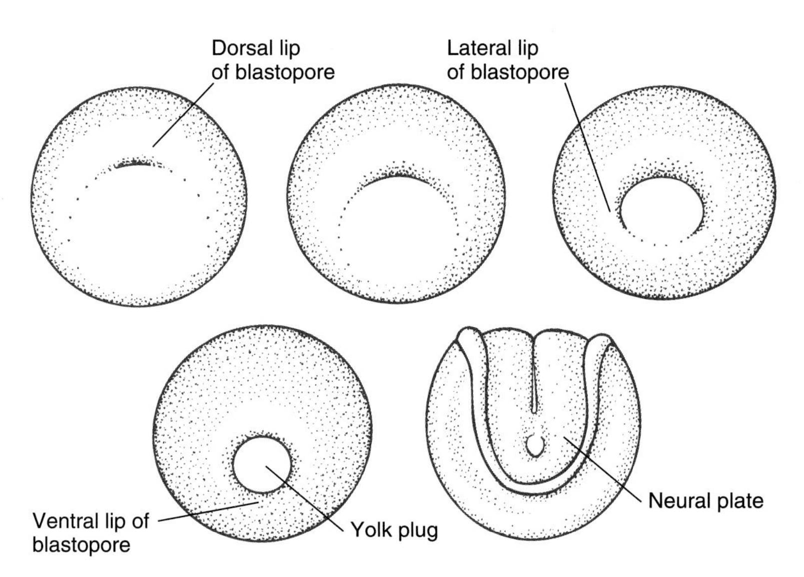

Figure 10.13 Drawings showing successive stages of amphibian blastopore formation in a vegetal view. Note that the blastopore surrounds the vegetal base as it first extends laterally and then closes into a circle. The part of the vegetal base that protrudes from behind into the blastopore is called the yolk plug. After the blastopore is closed, the dorsal surface of the embryo flattens and neurulation begins.

in large part by *involution.* As illustrated in Figure 10.11, the involuting marginal zone turns inward over the blastopore lip, and then back on itself. This was first demonstrated by Vogt (1929), using dyes that stained embryonic cells but did not interfere with normal development. Small agar pieces soaked with such dyes were pressed against newt embryos that were immobilized in wax (Fig. 10.14). When enough stain had been taken up by the embryos, they were released from their molds and observed in successive stages. During gastrulation, the stain marks disappeared into the blastopore. When the embryos were fixed and sectioned, the dye marks were found—in the original sequence—along the archenteron. Some of the marks had elongated dramatically, indicating that *convergent extension* had also occurred in those areas. From these observations, it was clear that cells *involuted* around the entire circumference of the blastopore, but most extensively on the dorsal side (see Fig. 10.11j–l). In addition, the blastopore lip enclosed the vegetal base by *epiboly,* constricting the blastopore to a small opening at the vegetal pole. The small protrusion of the vegetal base into the constricted blastopore is called the ***yolk plug,*** and the circular furrow between the yolk plug and the blastopore lip is called the ***blastopore groove.***

To determine whether involution requires any special cellular capabilities, R. E. Keller (1981) replaced involuting marginal zone (IMZ) layers with noninvoluting gastrula tissues. For instance, he excised the deep layer, the superficial layer, or both from the dorsal IMZ and replaced them with the corresponding layers of cells from animal cap (Fig. 10.15). The grafted patches were thus challenged to involute. Those animal patches replacing the deep layer of the IMZ balked at the blastopore lip and did not move in, while the host tissue on both sides of the graft involuted properly. In contrast, animal patches replacing only the *superficial layer* of the IMZ involuted and became part of the

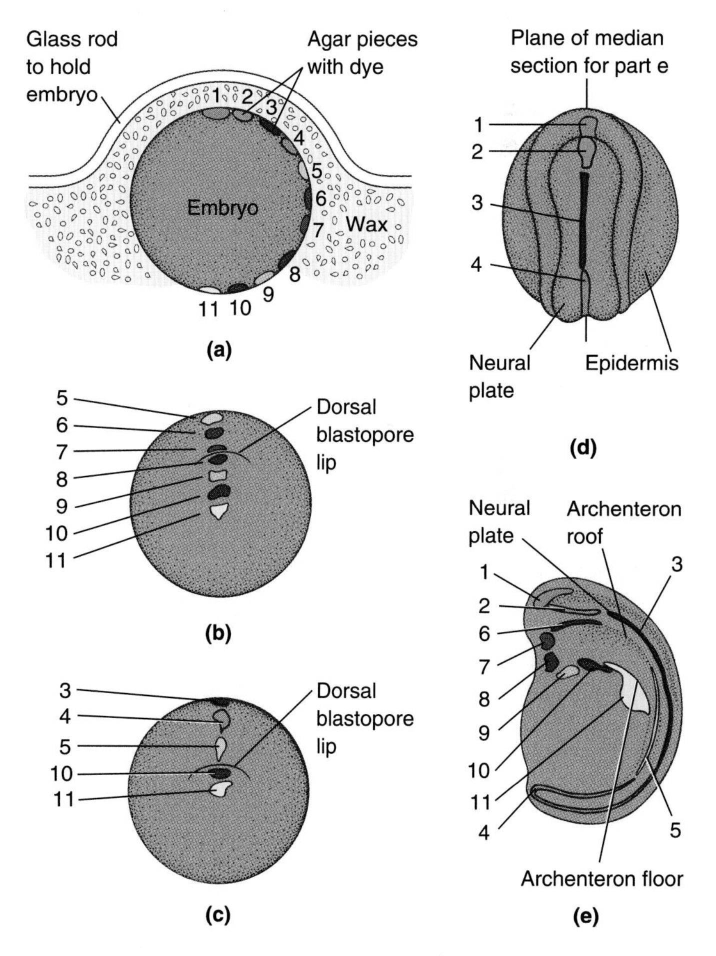

Figure 10.14 Involution and epiboly during gastrulation of the newt *Triturus.* **(a)** Vogt's method of placing dye marks on the surface of the embryo. **(b)** Early gastrula, surface view from vegetal pole. **(c)** Midgastrula, same view. Dye marks 7 and 6 have turned inside by involution. Dye marks 8 and 9 were covered as the dorsal blastopore lip moved toward the vegetal pole as a result of epiboly. **(d)** Embryo after neural plate formation, seen from a dorsal-anterior angle. Note elongation of marks 3 and 4. **(e)** Median section of embryo at tail bud stage showing the archenteron floor and roof. Note elongation of marks 3, 4, and 5. See also Fig. 10.11d–f.

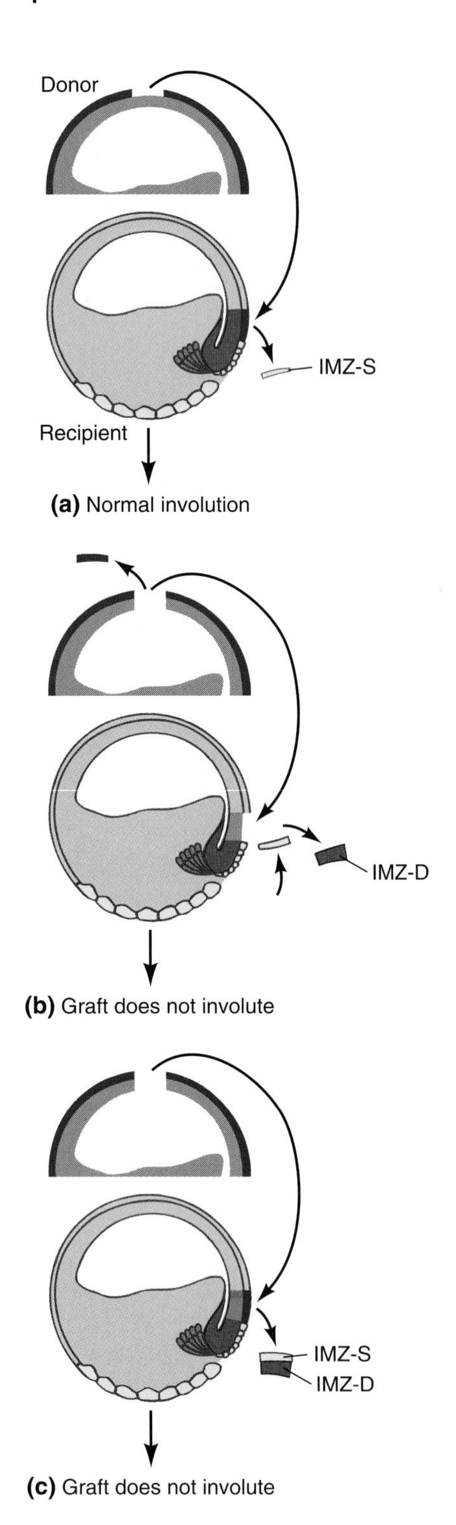

Figure 10.15 Analyzing the ability of different areas of *Xenopus* early gastrulae to undergo involution. Each drawing shows the animal cap of the donor and the entire recipient; animal pole up, dorsal side to right. **(a)** Replacing the superficial layer of the involuting marginal zone alone did not interfere with gastrulation. **(b)** Replacing the deep layer alone stopped the involution of the graft area. **(c)** Replacing both layers had the same result as replacing the deep layer alone.

archenteron roof. Control embryos in which patches of deep and superficial layers of IMZ had been excised and reinserted in the same place showed only minor defects. These results indicate that cells in the deep layer of the IMZ have unique properties that are necessary for involution.

DEEP ZONE CELLS AND INVOLUTING MARGINAL ZONE CELLS MIGRATE ON THE INSIDE OF THE BLASTOCOEL ROOF

During gastrulation, the deep zone cells migrate from their original position near the equator to the animal pole (see Fig. 10.11d–i). The deep involuting marginal zone cells follow suit: As they turn inside at the blastopore lip these cells become migratory as well (R. E. Keller, 1986). Microcinematography of cultured, open gastrulae shows waves of involuting mesodermal cells advancing across the inside of the blastocoel roof.

What role does cell migration along the blastocoel roof play in the overall process of gastrulation? In newts, migration of both deep zone cells and involuting marginal zone cells seems necessary for the internalization of the marginal zone (see Section 11.4). In frog embryos, however, the marginal zone is internalized even in the absence of migration. This was shown by Holtfreter (1933), who completely removed the blastocoel roof from the early gastrula stages of *Hyla* (Fig. 10.16). The result was stunning. Invagination and involution proceeded almost normally, although the deep zone and the involuting marginal zone had no blastocoel roof to migrate on. In addition, the blastopore constricted, and *convergent extension* of the involuted mesoderm occurred in a completely autonomous way. R. E. Keller and S. Jansa (1992) repeated Holtfreter's experiment with *Xenopus* and obtained similar results. It appears that in anurans, cell migration does not need to contribute to the forces that move the marginal zone inside. However, migration seems necessary for the deep zone mesoderm to spread out properly for future head organization. Furthermore, even subtle disturbances of the blastocoel roof, on which the mesodermal cells migrate, interfere with the right-left asymmetry of both heart and gut as described in Section 9.8.

CONVERGENT EXTENSION IS ESPECIALLY STRONG IN THE DORSAL MARGINAL ZONE

As gastrulation proceeds, amphibian fate maps undergo major distortions, especially in the dorsal area. The greatest shape changes occur during the second half of gastrulation, when the dorsal marginal zone undergoes *convergent extension.* Both involuting and noninvoluting marginal zones extend dramatically in the animal-vegetal direction while shrinking in circumference and depth. This movement displaces the animal pole, and to a smaller extent the vegetal pole, toward the ventral side

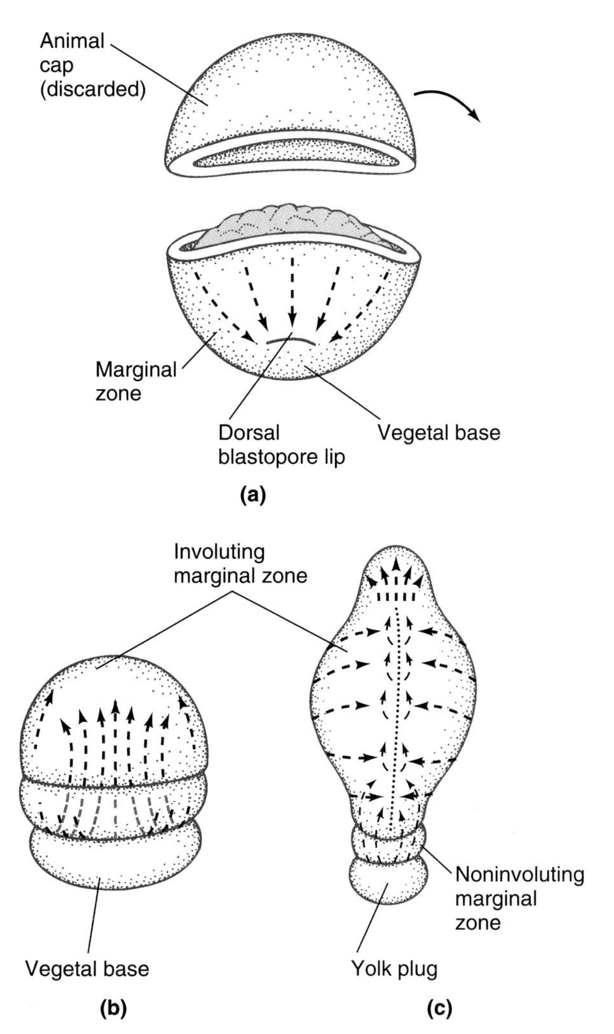

Figure 10.16 Gastrulation movements of *Hyla* embryos after removal of the blastocoel roof; embryos shown in dorsal view, animal pole up. **(a)** The animal hemisphere was removed completely. **(b)** The isolated vegetal hemisphere underwent invagination and involution. The epiboly movement, which normally surrounds the vegetal base, was disturbed. **(c)** The involuted marginal zone extended in the anteroposterior direction while shrinking laterally.

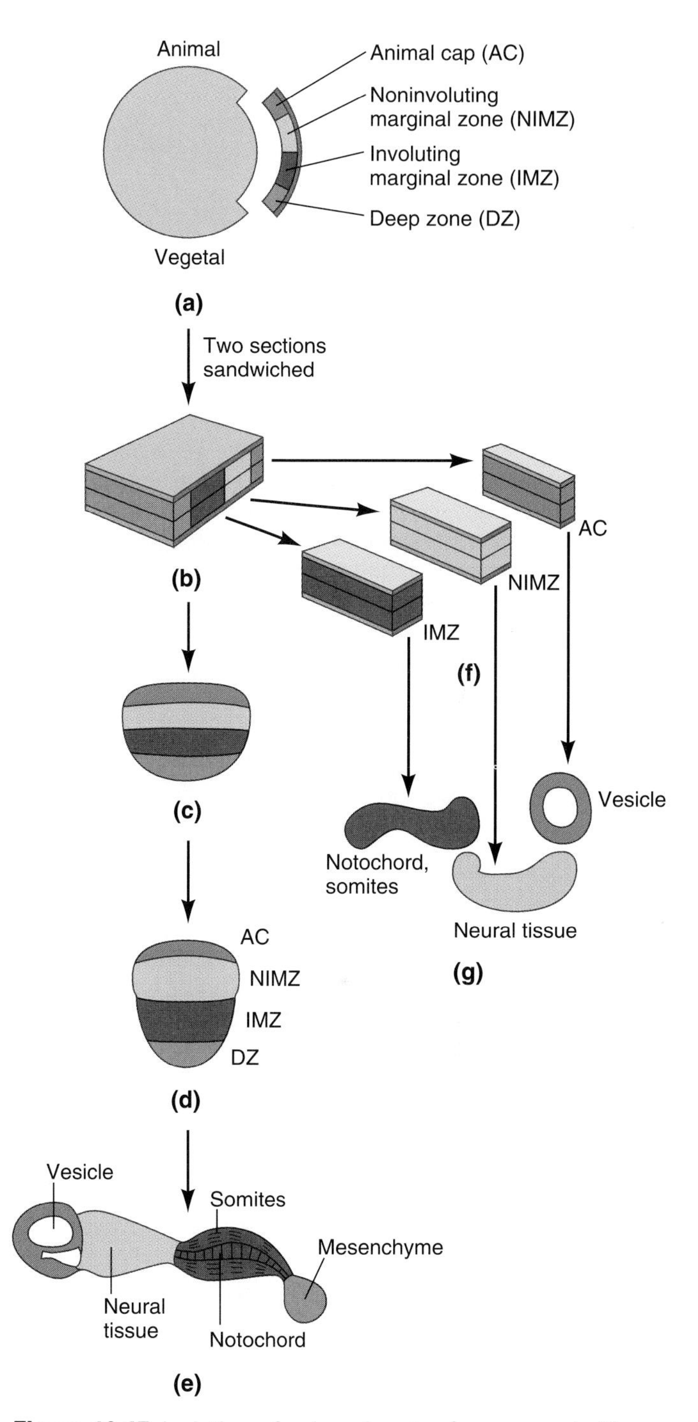

Figure 10.17 Isolation of a dorsal sector from an early *Xenopus* gastrula. **(a)** Embryo shown in lateral view, dorsal side to the right. The isolated sector included portions of the animal cap, noninvoluting marginal zone, involuting marginal zone, and deep zone. **(b)** Two such sectors were sandwiched with their inner surfaces together. **(c–e)** Upon culture in vitro, the "sandwich" extended markedly in the anteroposterior direction while shrinking laterally. In addition, each region differentiated according to its fate in the normal embryo, forming a fluid-filled vesicle, neural tissue, and notochord as well as somites and mesenchyme. **(f, g)** The same differentiations occurred when the regions were isolated after sandwiching and cultured separately.

of the embryo (see Fig. 10.11a–c). At the same time, the animal cap and noninvoluting marginal zone continue with epiboly until the *limit of involution* reaches the blastopore.

The autonomy of the convergent extension process was demonstrated by isolation experiments (R. E. Keller et al., 1985). Dorsal sectors from two early gastrulae were excised, sandwiched with their inner sides together, and kept in tissue culture (Fig. 10.17). The explants comprised parts of the animal cap, noninvoluting marginal zone, involuting marginal zone, and deep zone. After the edges of the sandwiches had healed together, no major changes occurred until the time at which untreated control embryos reached the midgastrula stage. Then the marginal zones began to narrow and lengthen while the animal cap and deep zone

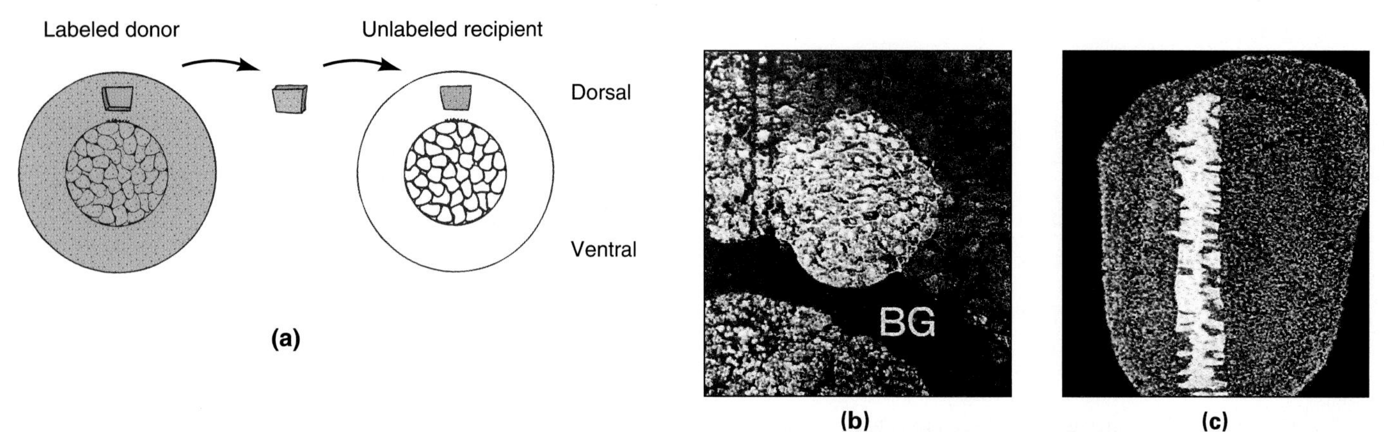

Figure 10.18 Convergent extension in *Xenopus* gastrula. **(a)** A fluorescently labeled donor (color) and an unlabeled recipient are shown in vegetal view, dorsal side up. A piece of dorsal involuting marginal zone, comprising both deep and superficial layers, was grafted. **(b)** Photomicrograph of a histological section showing the fluorescent graft after healing in. The graft was a compact patch, separated by the blastoporal groove (BG) from the yolk plug below. **(c)** In a specimen fixed and sectioned after neurulation, the graft had contributed to notochord and somites. The labeled cells had intercalated with unlabeled recipient cells and extended over almost the length of the embryo.

became spherical or knoblike. At the postgastrula stage, cell differentiation begins and, as far as recognizable, each zone developed according to fate. The animal cap cells formed fluid-filled epithelial vesicles while the noninvoluting marginal zone gave rise to an elongated mass of neural cells. The deep layer of the involuting marginal zone differentiated into a notochord flanked by somitic mesoderm. The deep zone developed into a ball of mesenchymal cells.

In our context of morphogenetic movements, the salient observation is that the marginal zone parts underwent convergent extension in isolation, showing that no forces or signals from other parts of the embryo were required. Similar explants of lateral or ventral sectors showed progressively weaker convergent extension. These results demonstrate that convergent extension is an inherent property of the marginal zone, present especially in the dorsal area.

HOW do the marginal cells bring about convergent extension? To answer this question, R. E. Keller and P. Tibbetts (1989) transplanted pieces of involuting marginal zone that were labeled with a fluorescent dye to an unlabeled host (Fig. 10.18a). The transplantations were *homotopic,* meaning between equivalent of the embryo, and *isochronic,* meaning between a donor and a recipient of the same age. Each labeled transplant healed in quickly as a continuous patch (Fig. 10.18b). At different intervals after transplantation, the recipients were fixed, sectioned, and studied by fluorescence microscopy. The most dramatic results were obtained after grafting patches of dorsal involuting marginal zone at the midgastrula stage. The dorsal marginal zone is fated to form notochord (backbone precursor) and somites (mesodermal spheres that will form dorsal skeletal muscles, among other things. See Section 14.2). The labeled graft cells intercalated extensively with unlabeled recipient cells. When the gastrulation movements were completed, the labeled cells had spread over the entire trunk region of the dorsal axis (Fig. 10.18c). In contrast, grafts forming head mesoderm showed little or no cell intercalation. These results indicate that the prospective notochord and somite areas undergo convergent extension by means of intercalation.

Questions

1. Do you expect the extensive cell intercalations shown in Figure 10.18c to occur mainly in the deep layer or in the superficial layer of the involuting marginal zone? How would you test your hypothesis experimentally?
2. Based on the results shown in Figure 10.11h, i, do you expect the extensive cell intercalations shown in Figure 10.18c to occur on the ventral side of the embryo as well?

By excising a dorsal piece of involuting marginal zone, keeping it in tissue culture, and recording the cellular movements in the deep layer by videomicroscopy, R. E. Keller and colleagues (1985) were able to observe the process of cell intercalation directly (Figure 10.19a). They found that cell intercalations occur in two directions. Early gastrulae show ***radial intercalation,*** meaning that cells intercalate *perpendicular* to the surface of the embryo (Fig. 10.19b). This behavior reduces the number of deep cell layers while increasing their surface area. Presumably due to constraints imposed by the overall shape of the embryo, the surface area increases only in the animal-vegetal direction, rather than in all directions as is observed in epiboly (discussed later in this section). Radial intercalation is followed by ***mediolateral intercalation:*** Now the deep cells intercalate *parallel* to the surface of the embryo, moving mostly from lateral to medial (Fig. 10.19d). This activity causes exten-

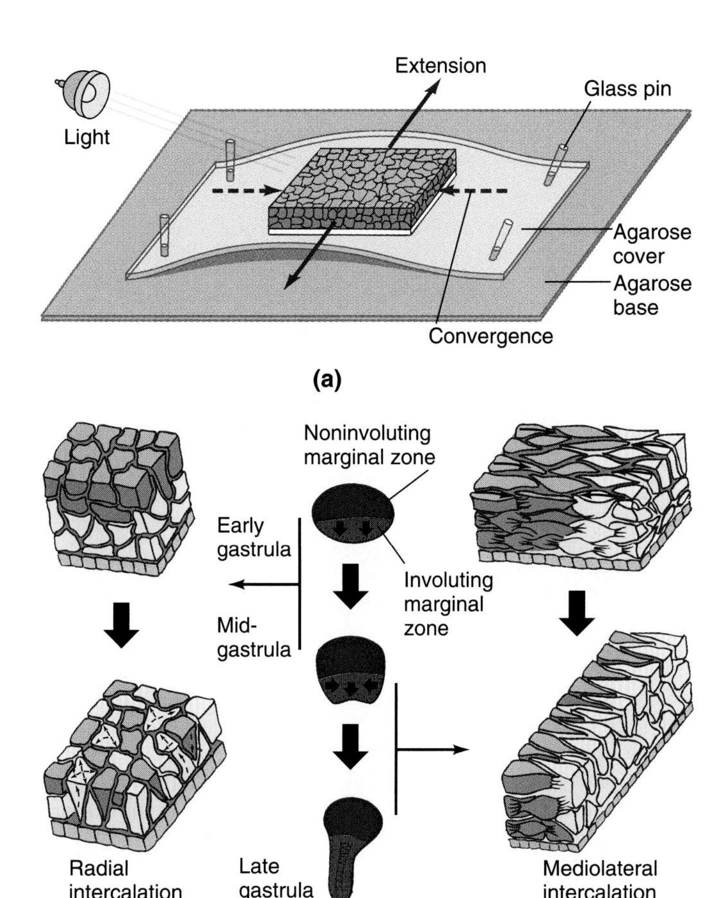

Figure 10.19 Direct observation of cell intercalation during gastrulation in *Xenopus.* **(a)** A dorsal sector comprising both involuting and noninvoluting marginal zone is cultured between two sheets of agarose. The inner, deepest cell layer is on top while the epithelial layer, which normally forms the outer surface of the embryo, is at the bottom. As in the intact embryo, the cultured tissue extends in the animal-vegetal direction (solid arrows) while converging in the mediolateral direction (dashed arrows). **(b)** Radial intercalation during early and midgastrulation. The thin arrows indicate radial cell intercalation (perpendicular to the surface), as well as the resulting increase in surface area and corresponding thinning of the tissue. **(c)** Overall changes in surface area and shape of the cultured tissue. Convergent extension is stronger in the involuting marginal zone. **(d)** Mediolateral intercalation during mid- and late gastrula stages. The black arrows show mediolateral cell intercalation (parallel to the surface), which causes convergent extension.

sion of the involuting marginal zone in the animal-vegetal direction and a proportional convergence in the mediolateral direction.

Recent studies are turning to the signals that control cell intercalation behavior. The mediolateral intercalations, which are the strongest motor of convergent extension, start about midway through gastrulation. Cellular protrusions, which so far had occurred randomly, are now becoming oriented medially and laterally. At the same time, cells become spindle-shaped, align in the mediolateral direction, and then intercalate mediolaterally to form a longer and narrower array (Fig. 10.19d). This *mediolateral intercalation behavior* occurs in a particular pattern, progressing from anterior to posterior and from lateral to medial (Shih and Keller, 1992; Domingo and Keller, 1995). The signals inducing mediolateral intercalation behavior seem to emanate from boundaries that develop between future notochord cells and prospective somite cells. A similar boundary effect is observed in the overlying neural plate (see Section 12.2).

ANIMAL CAP AND NONINVOLUTING MARGINAL ZONE UNDERGO EPIBOLY

While the involuting marginal zone and the vegetal base turn inside, the animal cap and—to a smaller extent—the noninvoluting marginal zone undergo the spreading movement called *epiboly.* Microcinematography reveals that the superficial cells in these regions divide, with their mitotic spindles oriented parallel to the surface. They also become thinner as gastrulation proceeds. At the same time, major cellular rearrangements occur in the deep layer underneath (R. E. Keller, 1978). Scanning electron micrographs show that the number of deep cell layers in the animal cap decreases from about two at the onset of gastrulation to one at the end of gastrulation (Fig 10.20; R. E. Keller, 1980). During the transition, the deep cells elongate temporarily and extend lamellipodia from one to another in a radial direction—that is, perpendicular to the surface of the embryo. After the radial intercalation is complete, the deep cells become flatter again. Thus, the *epiboly* of the animal cap and the noninvoluting marginal zone is associated with cell *division* and *cell flattening* in the superficial layer, and with *radial intercalation* in the deep layer. Whether these cellular behaviors actually drive the epiboly movement or whether they occur in response to forces generated elsewhere is undetermined.

HOW ARE PATTERNS OF CELL BEHAVIOR RELATED TO GENE EXPRESSION?

As discussed in previous parts of this section, amphibian gastrulation is composed of several cellular movements that complement one another. Interaction of bottle cells with the vegetal base produces a small *invagination* at the dorsal blastopore. The *involution* of a part of the marginal zone extends the archenteron and places the mesoderm between the archenteron (endoderm) and the ectoderm, which remains outside. *Migration* helps to spread out the cells of the deep zone and involuting marginal zone as they move to the animal pole. *Convergent extension* causes anteroposterior elongation of the embryo on the dorsal side. As the vegetal parts of the gastrula turn inside, they are enveloped by *epiboly* of the animal portions of the gastrula until the

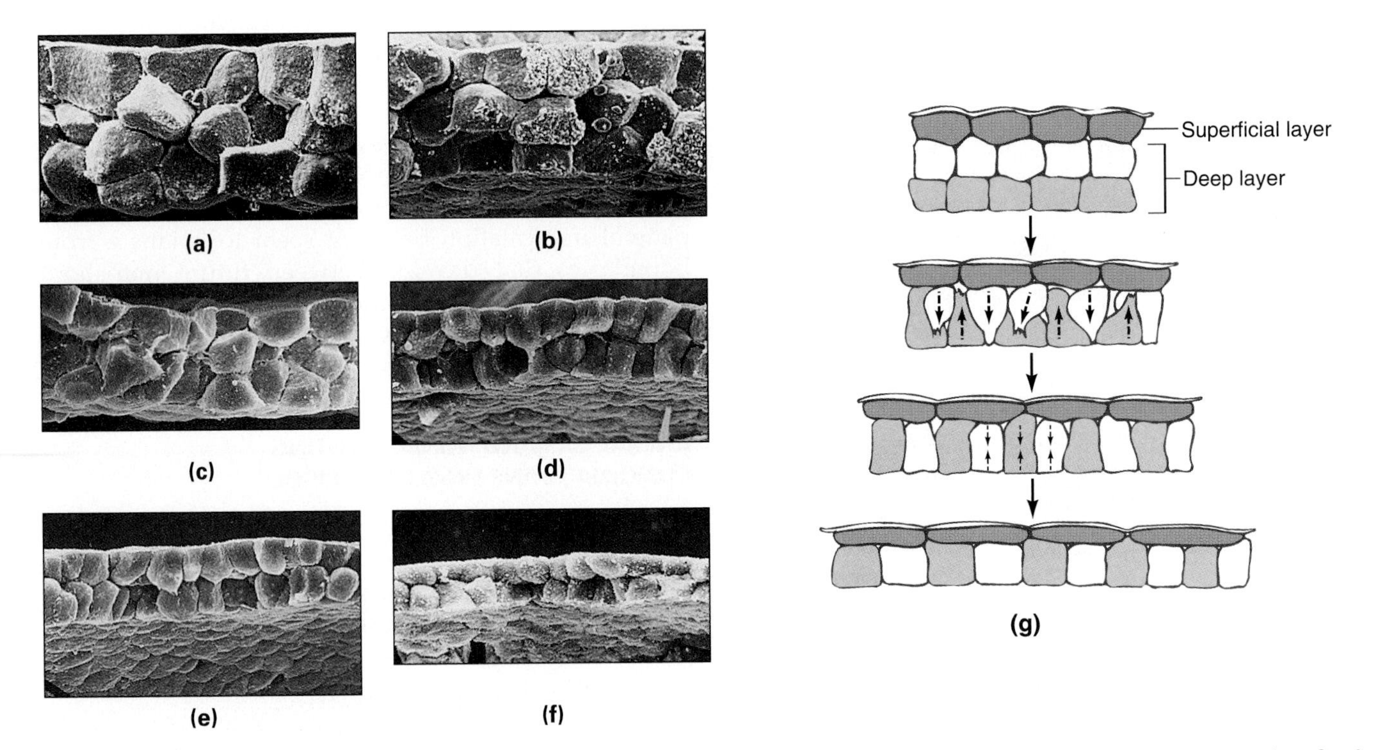

Figure 10.20 Cellular basis of epiboly in the animal cap of the *Xenopus* gastrula. **(a–f)** Scanning electron micrographs of animal cap cells at successive stages of gastrulation; superficial cell layer at the top. **(g)** Interpretative diagram showing the superficial layer (color) and the deep layer (gray and white). The superficial cells flatten considerably, and the deep cells undergo radial intercalation and some flattening (small arrows).

blastopore closes. How are all these movements orchestrated so that they produce a well-formed gastrula that will proceed with organogenesis to generate the basic body plan?

There are at least two general hypotheses that link gastrulation movements to the earlier processes of cell determination by localized cytoplasmic determinants and inductive interactions. First, one could assume that the cellular behaviors observed in each gastrula region (e.g., animal cap, dorsal marginal zone) are controlled locally and individually by the cytoplasmic localizations and inductive events that have occurred in that region. Thus, localized mRNAs in animal cap cells might cause the descendants of these cells to undergo epiboly, while the specific combination of inductive signals received in the dorsal marginal zone might promote convergent extension, and so forth. According to the second hypothesis, one region of the early gastrula could be determined to carry out its own cellular movements *and* to organize the movements of the other regions. Both hypotheses may be valid to an extent, according to the *principle of overlapping mechanisms*.

The most suitable candidate for an organizer role would be the dorsal blastopore lip, where gastrulation begins. Indeed, transplantation of this region to the ventral side of a host embryo leads to the formation of a second invagination site and eventually a second embryo (see Section 12.3). Because of the organizer capabilities of the dorsal blastopore lip, researchers have focused on several genes that are specifically expressed in this region and encode proteins that in turn regulate the activity of other genes. Such genes have indeed been found, including *goosecoid*$^+$ (Cho et al., 1991), *noggin*$^+$ (Smith and Harland, 1992), *Xnr3*$^+$ (Smith et al., 1995), and others (see Section 12.5).

Messenger RNA transcribed from one of these genes, *goosecoid*$^+$, has been tested for its effects on gastrulation movements (Niehrs et al., 1993). When goosecoid mRNA was injected into ventral marginal blastomeres (labeled C4 and C4′ in Fig. 9.16) the descendants of the injected cells acquired many properties of the normal dorsal blastopore lip. They formed notochord and other dorsal structures, and they recruited neighboring cells to contribute to the formation of a secondary embryonic axis. Moreover, the progeny of cells injected with goosecoid mRNA involuted earlier, and moved farther anterior, than control-injected cells. Thus, the ectopic goosecoid function promoted cell movements and cell fates normally displayed by dorsal mesoderm.

However, none of the organizer-specific genes mentioned above has an expression pattern that coincides neatly with a specific cellular behavior involved in gastrulation. Thus, it appears that the genetic control of cellular behaviors involved in gastrulation is more complex than being promoted by the activity of one particular gene.

10.4 Gastrulation in Fishes and Birds

As examples of superficial cleavage, we discussed the cleavage patterns of zebrafish and chicken (see Section 5.2). In both species, a disc of cells is formed at the animal pole of a large yolky egg that remains for the most part uncleaved. Fish and chicken embryos also look quite similar once the rudiments of their brains, dorsal muscles, and other organs have been formed. However, the intervening gastrulation movements that generate the germ layers are not alike. Even within the class of fishes, different egg sizes are associated with different gastrulation movements. We will maintain our focus on zebrafish and chicken as two species investigated in many laboratories.

ZEBRAFISH EMBRYOS DEVELOP FROM TWO CELL LAYERS MOSTLY BY CONVERGENT EXTENSION

The zebrafish embryo at the end of cleavage consists of a mound of cells perched atop of a ball of uncleaved yolk (see Fig. 5.14). The outermost cells, called the *enveloping layer* (*ENL*), cover several layers of *deep cells,* and both rest on top the ***yolk syncytial layer* (*YSL*),** which arises when cells in contact with the uncleaved yolk become confluent (Fig. 10.21a).

Gastrulation begins with *epiboly,* in a movement that has been likened to pulling a knitted ski cap over one's head (Langeland and Kimmel, 1997). The deep cells undergo *radial intercalation,* thus increasing the embryo's surface area while reducing its number of cellular layers. However, these intercalations may be just accommodating epiboly rather than driving it. The driving mechanism seems to reside in a system of microtubules that reach from the deep cells and the YSL into the yolk; UV light and chemical inhibitors of microtubules cause the epibolic movement to halt (Solnica-Krezel and Driever, 1994).

The next step in gastrulation is the formation of a *germ ring* around the entire margin of the embryo. Here, the deep cells form two layers, an outer *epiblast* and an inner *hypoblast.* Formation of the hypoblast is variously described as *involution* of a contiguous cell layer (Langeland and Kimmel, 1997) or as *ingression* of single cells (Driever, 1995; Trinkaus, 1996). In any event, the hypoblast originates at the germ ring and spreads underneath the epiblast toward the animal pole (Fig. 10.21b). The process begins on the future dorsal side of the embryo but spreads around the entire circumference of the embryo. While hypoblast and epiblast are being formed, the process of epiboly continues until the entire yolk is covered (Fig. 10.21c).

As epiblast and hypoblast are being formed, yet another gastrulation movement is superimposed on those that are already going on (Fig. 10.22). Cells move to the future dorsal midline, causing both epiblast and hypoblast to converge mediolaterally and to extend anteriorly (Driever, 1995). The initial result of this *convergent extension* is the formation of the ***embryonic shield,*** a triangular area with the dorsal portion of the germ ring as its base and a tip pointed toward the animal pole. As epiboly drives the base of the shield toward the vegetal pole, and extension pushes the tip of the shield toward the animal pole, a ridge of cells emerges that extends from the animal to the vegetal pole. The part between animal pole and equator will form head, and the part between the equator and the vegetal pole will give rise to trunk (see Fig. 10.21d). Indeed, the trunk portion soon thereafter extends beyond the vegetal pole to form a *tailbud.* These processes occur in both the epiblast and the hypoblast, with the epiblast forming ectoderm and the hypoblast forming endoderm and mesoderm.

CHICKEN EMBRYOS DEVELOP FROM ONE CELL LAYER MOSTLY BY INGRESSION

The chicken embryo at the end of cleavage is a disc of many cells, the *blastoderm,* located at the animal pole of a very large yolky cell. The central and peripheral portions of the blastoderm have different optical properties (Fig. 10.23). The central portion, which is separated from the yolk by the *subgerminal space,* is the ***area pellucida*** (Lat. *pellucidus,* "transparent"). The peripheral portion, which is in direct contact with the underlying yolk, is called the ***area opaca*** (Lat. *opacus,* "dark"). Similar blastulae are formed by many reptiles, fishes, and other animals that produce large, yolky eggs.

The chicken blastoderm consists of an upper layer, the *epiblast,* and a lower layer, the *hypoblast,* with a shallow cavity called *blastocoel* in between (see Fig. 5.16). However, the chicken blastocoel is not equivalent to the amphibian blastocoel, and the chicken hypoblast—unlike the zebrafish hypoblast and the frog vegetal base—does not form embryonic endoderm or mesoderm. Instead, the chicken hypoblast gives rise to primordial germ cells and to extraembryonic endoderm surrounding the yolk. *All three* germ layers of the chicken embryo proper are derived from the epiblast, so that a fate map of the chicken embryo prior to gastrulation is essentially a map of the epiblast.

Gastrulation in birds is initiated by extensive cell rearrangements within the posterior half of the epiblast. Cells move first to the midline and then anteriorly, generating a fountainlike flow (Fig. 10.23). As a result, most of the cells that were spread over the posterior half of the epiblast are now aggregated into a solid median ridge, the ***primitive streak,*** while the epiblast assumes the outline of a pear. As the primitive streak elongates, a furrow called the ***primitive groove*** forms between two ***primitive ridges*** along the dorsal midline. The primitive ridges terminate anteriorly in a thickening known as ***Hensen's node.*** The node contains a funnel-shaped depression, called the ***primitive pit,*** which marks the anterior end of the primitive groove.

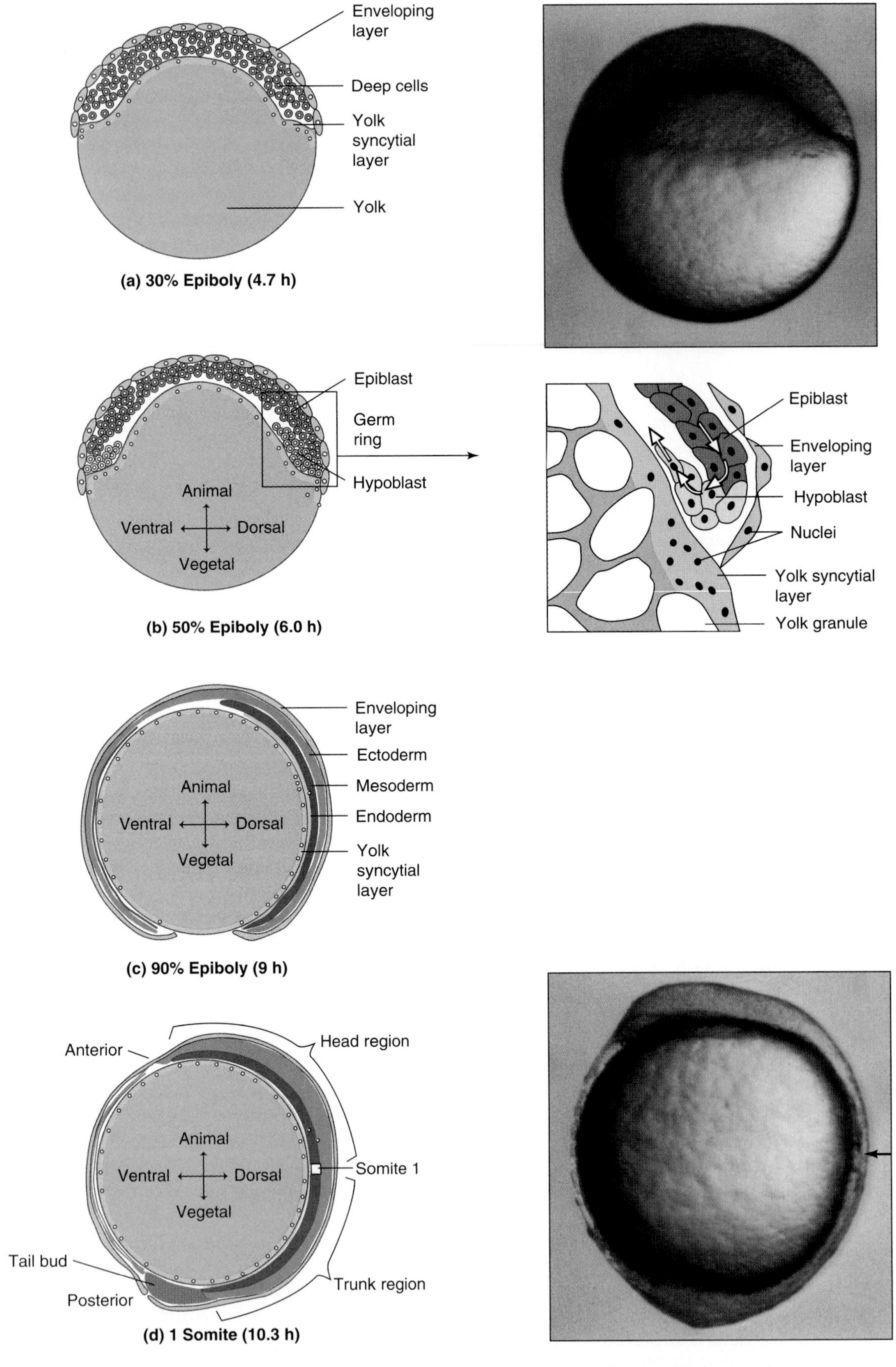

Figure 10.21 Gastrulation in zebrafish, shown in median sections. Times are in hours after fertilization at 28°C. Note progress of epiboly from **(a)** to **(d)**. The hypoblast is formed by involution and/or ingression **(b)**. The epiblast gives rise to ectoderm while the hypoblast forms mesoderm and endoderm **(b, c)**. The first somite (arrowhead) serves as a convenient landmark **(d)**.

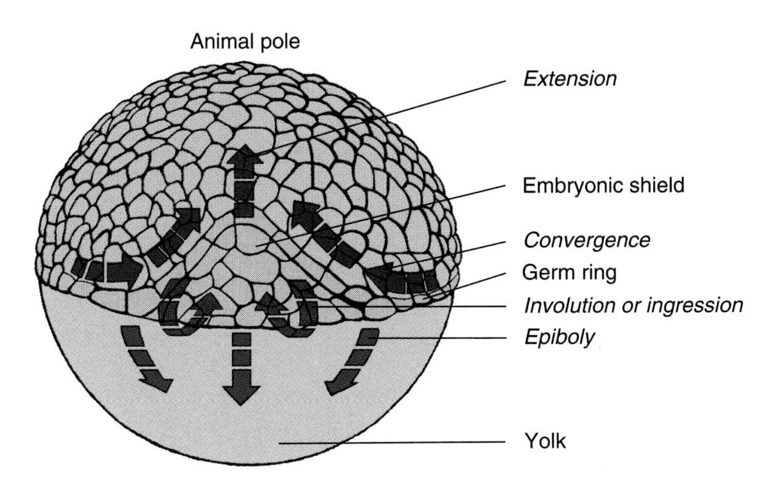

Figure 10.22 Superimposed morphogenetic movements (dashed arrows) during zebrafish gastrulation. Dorsal view. As epiboly spreads the embryo over the yolk, involution or ingression generates hypoblast and epiblast (see Fig. 10.21b). Convergent extension moves the bulk of both epiblast and hypoblast toward the future dorsal midline.

The primitive groove and primitive pit are the sites of major gastrulation events in birds. These are (1) the *involution* of epiblast cells over the edges of the primitive groove and primitive pit and (2) the subsequent *ingression* of the involuted cells into the blastocoel (Fig. 10.24). The cells that constitute the edges of the primitive groove and pit change constantly, as do those on the amphibian blastopore lip described earlier. The primitive groove and pit in birds are therefore the functional equivalents of the amphibian blastopore. As the epiblast cells enter the primitive groove, they undergo major changes. Their apical ends constrict while the basal ends expand, so that the cells take on a bottle shape, as seen in the amphibian blastula (compare Fig. 10.25 with Fig. 10.12). However, instead of moving in as a contiguous cell layer, avian epiblast cells break their junctions and ingress as single cells. Once inside the blastocoel, they flatten and migrate, an activity that is facilitated by the synthesis of certain extracellular matrix molecules. Because of its loose configuration, the bird mesoderm is first classified as a *mesenchyme*. Later, however, the loose and wandering mesenchymal cells will form epithelial spheres called *somites* and other contiguous structures (see Section 14.2).

Fate maps of avian embryos have been established by various methods (Rosenquist, 1966; Nicolet, 1971; Vakaet, 1984; Mittenthal and Jacobson, 1990; Selleck and Stern, 1991; Catala et al., 1996; Psychoyos and Stern, 1996; Schoenwolf, 1997). The maps indicate the position of the germ layers and some of their subdivisions during primitive streak formation (Fig. 10.26). Most of the ectodermal primordia are located anteriorly in the epiblast and do not ingress. The endodermal and mesodermal primordia are arranged in nested strips within the posterior half of the epiblast. Among the first cells to

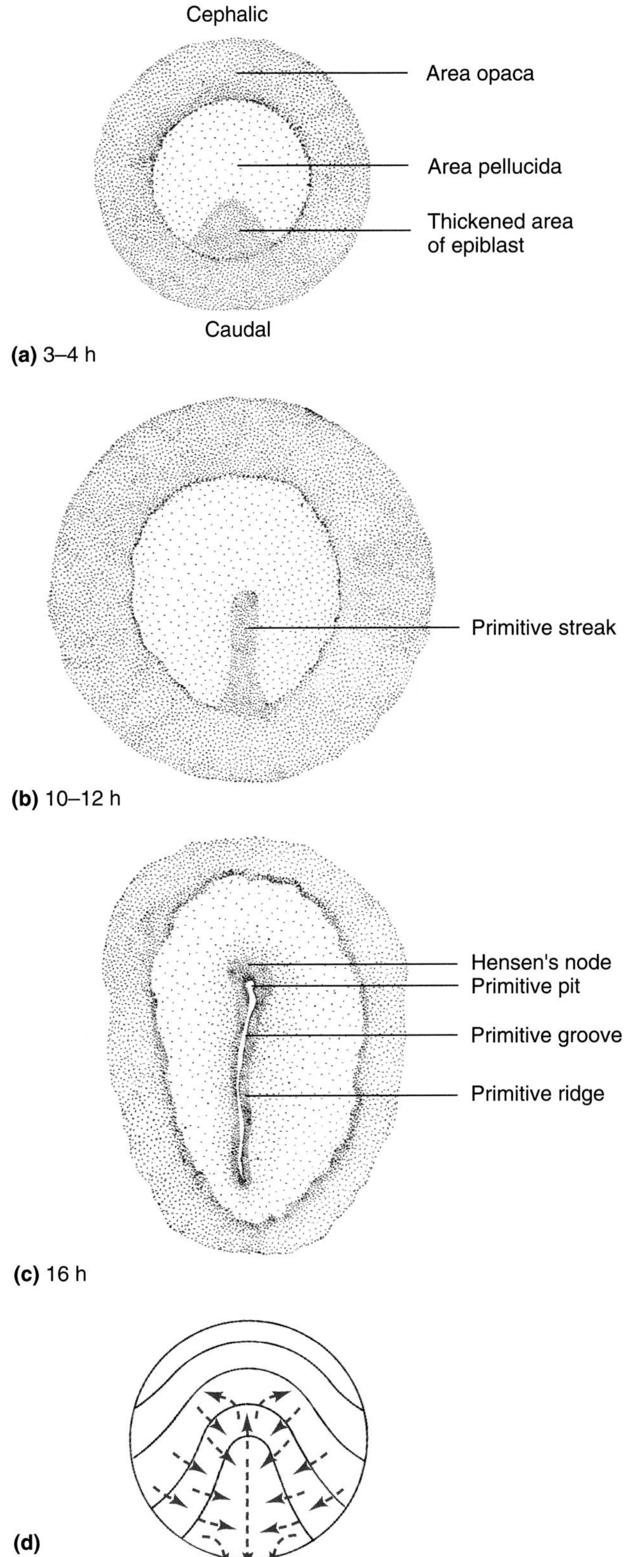

Figure 10.23 Primitive streak formation in the chicken embryo. All drawings show dorsal views of the epiblast. Most of the large uncleaved yolk has been omitted. **(a)** 3 to 4 h incubation. **(b)** 10 to 12 h incubation. **(c)** 16 h incubation. **(d)** Interpretative diagram of cell movements.

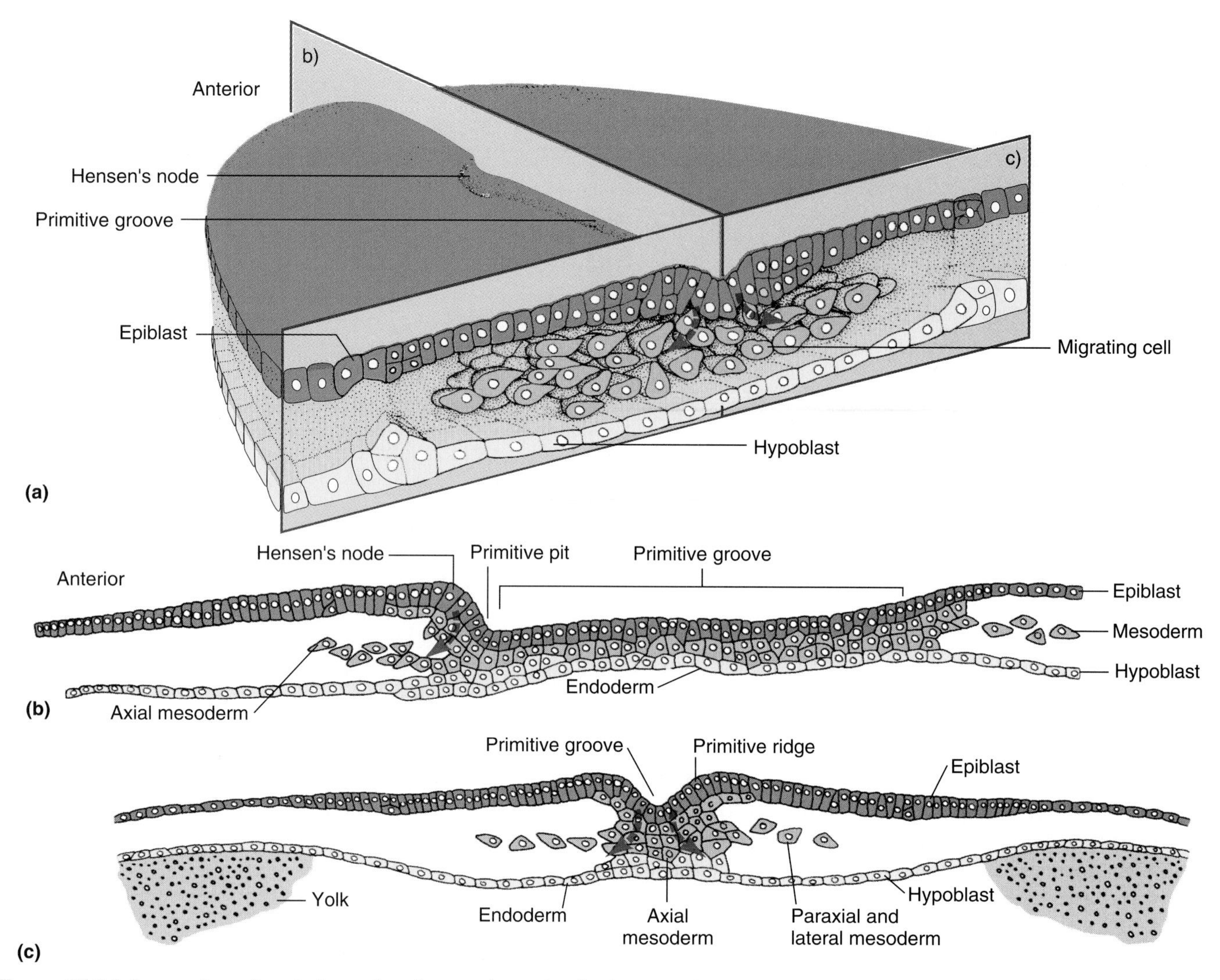

Figure 10.24 Ingression of endodermal and mesodermal cells through the primitive pit and primitive groove in bird embryos. **(a)** Three-dimensional view. The endodermal cells ingress first and displace the hypoblast. **(b)** Median section. Cells ingressing through the primitive pit form axial mesoderm (notochord) and paraxial mesoderm. **(c)** Transverse section. Cells ingressing through the primitive groove form axial, paraxial, intermediate, and lateral mesoderm.

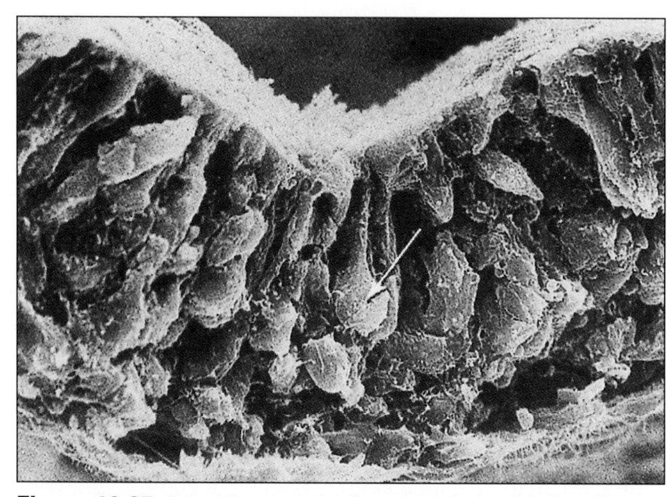

Figure 10.25 Primitive streak of a gastrulating bird embryo. The scanning electron micrograph shows a transversely fractured embryo. Cells entering the primitive groove become bottle-shaped (arrow) as they prepare to ingress into the blastocoel.

ingress through the primitive groove are the endodermal cells. They join the hypoblast cells and gradually displace them toward the margin.

While the ingression of endodermal cells is still under way, the primitive streak shortens, reversing its earlier advance, and gradually shifting Hensen's node to a more posterior position. As the node retreats, it leaves in its wake mesoderm cells ingressing from the anterior sector of the primitive pit (Figs. 10.24 and 10.27). Cells from the anterior sector of the node form the *axial mesoderm* along the dorsal midline, which will form the *notochord*. Cells from lateral portions of the node contribute to the *paraxial mesoderm,* which will give rise to the paired *somites* on both sides of the notochord. Curiously, only the medial somitic halves derive from cells that ingress through the node, whereas the lateral somitic halves come from cells that ingress from the streak posterior to the node (Selleck and Stern, 1991). The latter cells also establish the mesodermal primordia that come

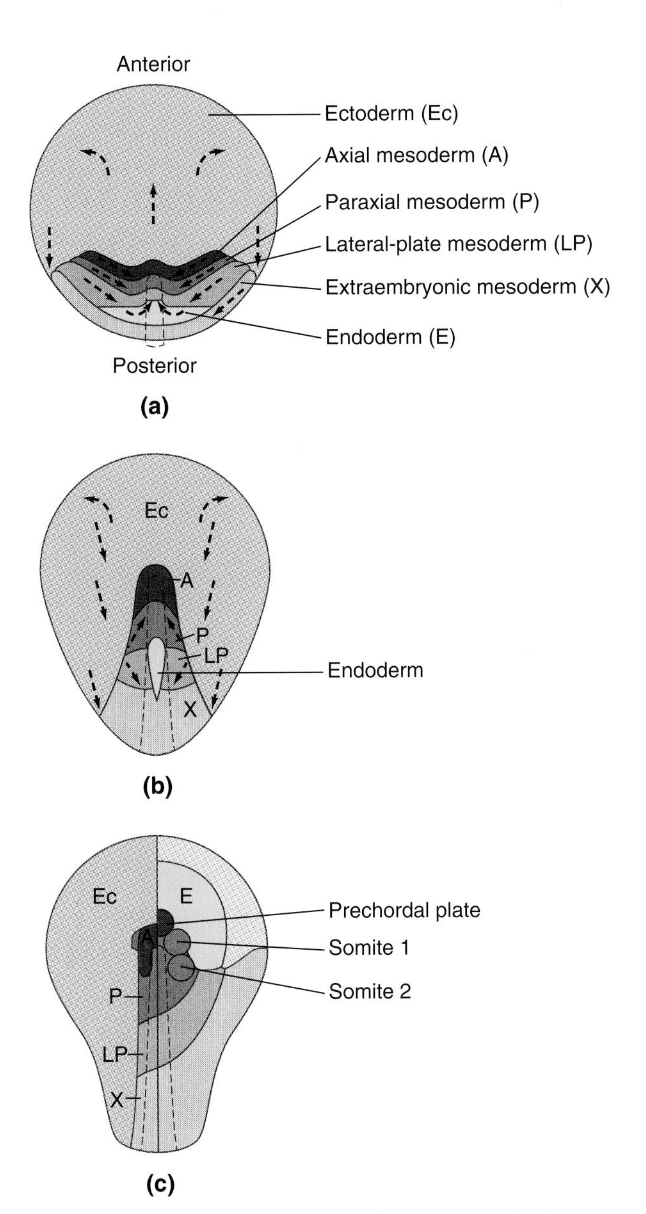

Figure 10.26 Fate maps of the chicken embryo during successive stages of development from early primitive streak formation **(a)** to maximum primitive streak length **(c)**. All maps show dorsal views of the epiblast. The area inside the hatched line represents the primitive streak. The epiblast is removed on the right side of part c to reveal the underlying endoderm and mesoderm.

to lie more laterally, including the *intermediate mesoderm*, the *lateral plates*, and the *extraembryonic mesoderm*.

Surprisingly, the same pool of cells that give rise to axial mesoderm also form the median strip of the *neural plate*, which will later give rise to the *floor plate* of the spinal cord (see Section 13.1). Moreover, the same pool of cells contributes to the dorsalmost part of the gut (Fig. 10.28). Thus, a streak of cells contributing to all three germ layers ingresses from Hensen's node and is laid down from anterior to posterior as the node regresses. Then this streak is subdivided into three horizontal layers: an upper layer, which becomes part of the neural ectoderm, a middle layer, which forms axial mesoderm, and a lower layer, which contributes to the endoderm.

After gastrulation, the remaining epiblast consists only of ectoderm, the hypoblast has been replaced by endoderm, and the mesenchymal cells spread between them. Thus, the result of the combined gastrulation movements is again an embryo consisting of three germ layers. However, in contrast to the organisms discussed earlier, in which the germ layers originate as concentric spheres or tubes, the germ layers of birds begin as a stack of three discs resting on a large mass of uncleaved yolk.

Once the germ layers have been established, they spread beyond the embryo proper to form *extraembryonic membranes* that surround the entire egg yolk. The ectodermal cells spread by *epiboly*, advancing as a circular front from the margin of the embryo to the vegetal pole. The leading cells adhere to the inside of the vitelline membrane, which surrounds the egg yolk. At the same time, hypoblast cells spread along the surface of the yolk mass while mesodermal cells expand in between. The extraembryonic membranes serve to break down the yolk into smaller molecules, which are delivered as nutrients to the embryo through blood vessels that form in the extraembryonic mesoderm and join the embryonic circulation. As the embryo grows, it lifts itself off the yolk, beginning with the formation of the *subcephalic space* under the head (see Fig. 10.27). This process involves closing the ventral side of the body, which has been open to the uncleaved yolk. As this ventral closure proceeds, first from the anterior and then from the posterior, the remaining yolk is restricted to a ventral appendage of the gut, the *yolk sac*. The originally disc-shaped rudiment has now closed into a tube-shaped embryo displaying the vertebrate body plan. Extraembryonic membranes protect the embryo against dehydration and mechanical stress and also play a role in the storage and exchange of waste products (see Section 14.7).

10.5 Gastrulation in Humans

Gastrulation has not been studied as extensively in mammals as in other animals. This is because mammalian embryos are very difficult to maintain in culture beyond the *blastocyst* stage, when they normally implant themselves into the uterus. After the blastocyst stage, mammalian embryos have different ways of establishing a ***bilaminar germ disc***, consisting of two cellular layers that correspond to the epiblast and hypoblast of chicken embryos. Our description of this period will focus on the development of primates, in particular humans (Langman, 1981; Larsen, 1993). Rodents and some other mammals develop differently during this period. The gastrulation events in mammals following the bilaminar germ disc stage seem to be similar to corresponding stages in birds.

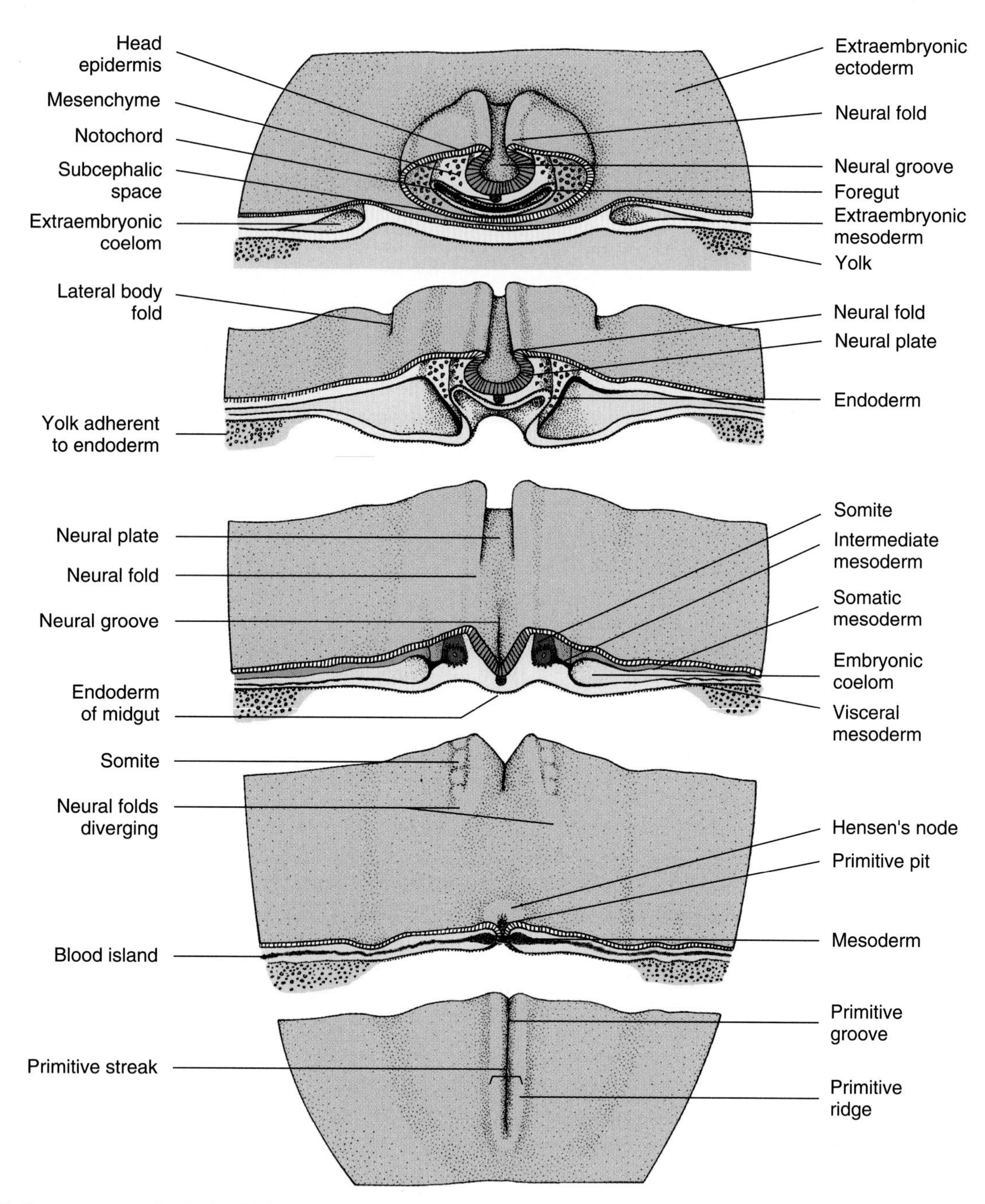

Figure 10.27 Stereogram of a 24-h chicken embryo. Note how much the anterior end of the embryo (top) is ahead of the posterior in development.

At the blastocyst stage, the primate embryo consists of an outer *trophoblast* surrounding a blastocoel and an eccentric *inner cell mass* (Fig. 10.29). The subsequent events leading to the establishment of embryonic germ layers and extraembryonic tissues are outlined in Fig. 10.30. A thin layer called the *hypoblast* delaminates from the inner cell mass of the blastocyst. After delamination of the hypoblast, the remainder of the inner cell mass is referred to as the *epiblast*. Subsequently, the epiblast delaminates another cell layer on the opposite side, the ***amniotic ectoderm*** (Fig. 10.31). The remainder of the epiblast is now referred to as the ***embryonic epiblast,*** and the fluid-filled space between the amniotic ectoderm and the embryonic epiblast is the ***amniotic cavity.*** Together, the embryonic epiblast and the underlying hypoblast form the ***bilaminar germ disc.*** This stage is comparable to the bird blastula before primitive streak formation. As in bird embryos, the hypoblast does not contribute to the embryo proper. Rather, hypoblast cells spread out to surround the blastocoel, thereafter called the ***primitive yolk sac,*** or *yolk sac* for short.

The bilaminar disc stage shown in Figure 10.31 is reached toward the end of the second week of human gestation. At this point, the disc is still very small (0.1 to 0.2 mm). The upper layer, the embryonic epiblast, will give rise to the embryo proper. The other structures will

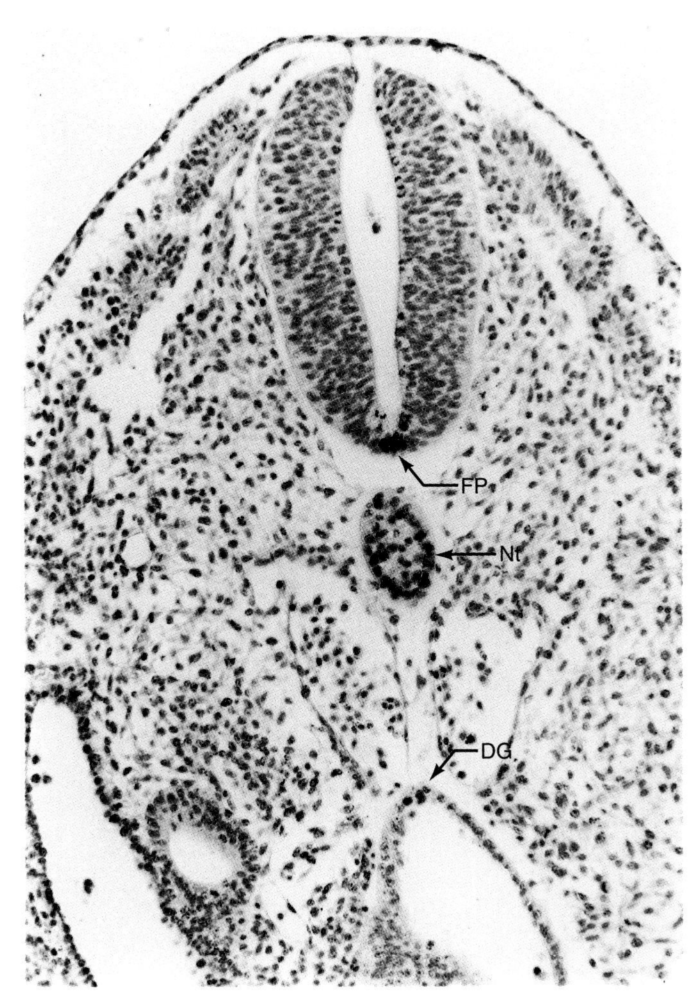

Figure 10.28 Transverse section of a chicken embryo to which a quail segment including Hensen's node and the underlying endoderm have been transplanted. The intensely staining quail heterochromatin marker reveals that the graft gave rise to floor plate (FP), notochord (Nt), and dorsal gut (DG). If only Hensen's node but not underlying endoderm was transplanted, the graft did not contribute to the gut.

(a)

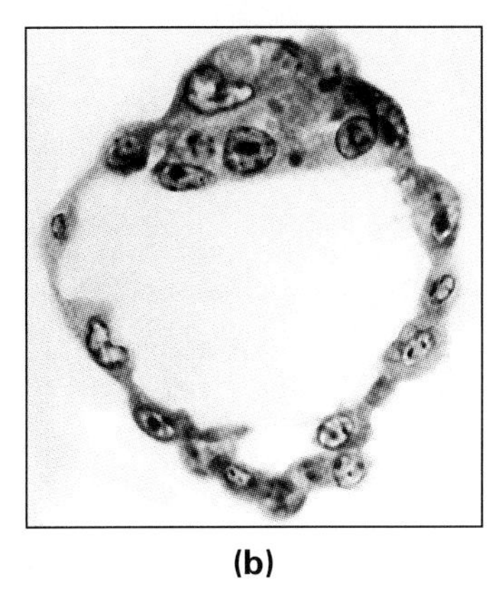

(b)

Figure 10.29 Blastocyst stage in primates. **(a)** Graphic reconstruction of a monkey blastocyst on day 9 after fertilization. The light areas of the inner cell mass represent hypoblast cells. **(b)** Photomicrograph of a sectioned human blastocyst about 5 days after fertilization.

form extraembryonic tissues. The yolk sac will become an appendage to the gut, as in bird embryos. The amniotic cavity will expand and surround the entire embryo as a fluid-filled cushion.

While the bilaminar germ disc is forming from the inner cell mass, the other portion of the blastocyst—the trophoblast—undergoes major changes as well. Its syncytial portion, the *syncytiotrophoblast,* forms internal spaces called ***lacunae*** and, at the same time, opens maternal blood vessels located in the uterus. This is particularly evident at the embryonic pole of the blastocyst, where maternal blood begins to fill the lacunae and to nourish the embryo by diffusion (see Fig. 14.39). In the cellular layer of the trophoblast, the *cytotrophoblast,* a new population of cells appears on the side facing the yolk sac. These cells form a loose meshwork that makes up much of the *extraembryonic mesoderm.* Large cavities that open up in this layer become confluent and fill with fluid. The fluid-filled space, referred to as the ***extraembryonic coelom,*** surrounds the entire yolk sac, the amniotic cavity, and the bilaminar germ disc in between.

The formation of the *primitive streak* in mammals begins, as in birds, in the posterior part of the embryonic epiblast. The cell movements causing primitive streak formation in mammals seem to be very similar to those observed in birds (compare Fig. 10.32 with Fig. 10.24). The embryonic endoderm and mesoderm originate by *involution* and *ingression* through the primitive streak. Epiblast cells ingressing through the primitive pit move anteriorly to form the prechordal plate and the notochord. Cells ingressing through the primitive groove spread laterally to form other mesodermal structures as well as the endoderm. At the end of gastrulation, the cells remaining in the epiblast have become the ectoderm, the hypoblast has been replaced with endodermal cells, and the embryonic mesoderm has spread in between. This stage, also known as the ***trilaminar germ disc,*** is reached toward the end of the third week of human gestation.

Although the embryos of placental mammals are very small and cleave holoblastically, their gastrulation movements are remarkably similar to those observed in

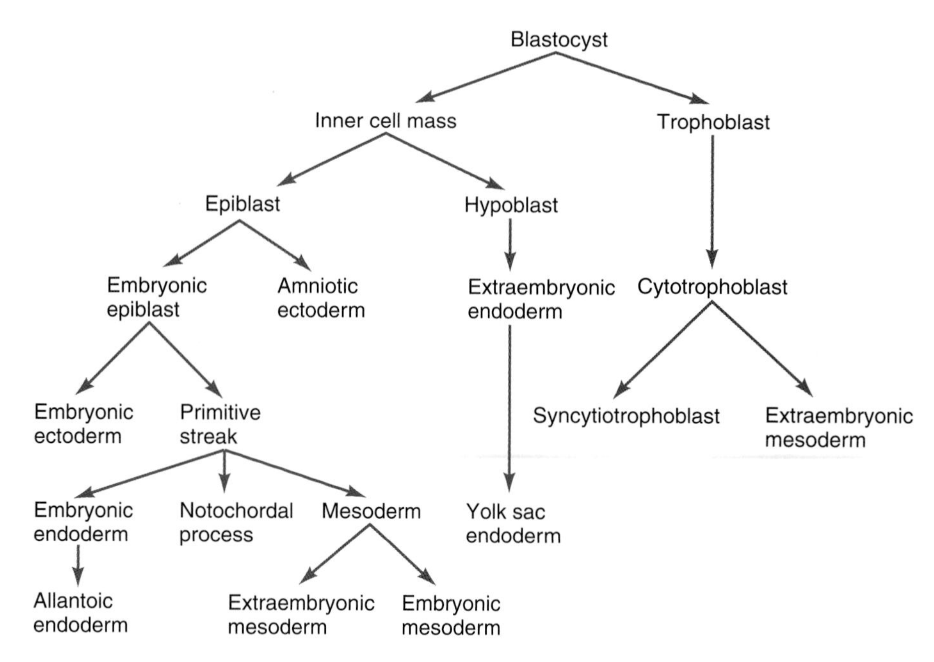

Figure 10.30 Flow diagram showing the origin of the germ layers and their derivatives in primates.

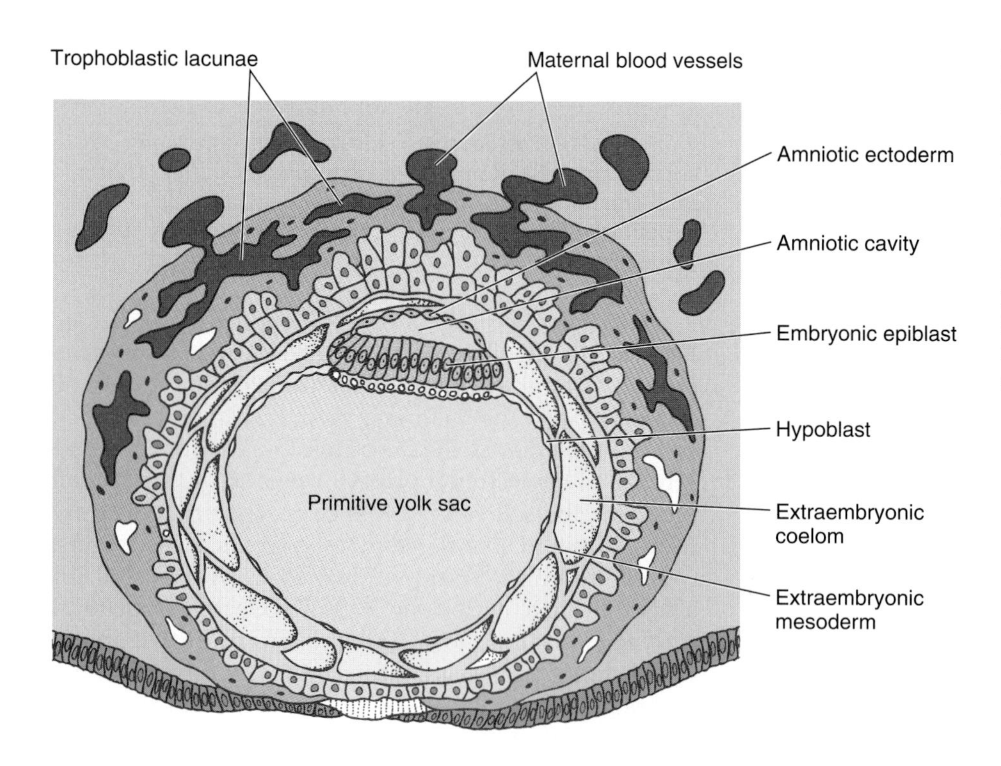

Figure 10.31 Drawing of a human blastocyst of approximately 12 days. The trophoblastic lacunae at the embryonic pole connect with maternal blood vessels. The embryo proper forms the bilaminar germ disc consisting of the epiblast and the hypoblast. Extraembryonic mesoderm proliferates and fills the space between the trophoblast on the outside and the embryo with yolk sac and amnion on the inside.

birds. The bilaminar germ disc of placental mammals behaves as if it were resting on a mass of yolk, although there is none. This is surprising from a functional point of view, because the small embryos of placental mammals could easily gastrulate by invagination or involution. Instead, gastrulation in placental mammals recapitulates a pattern established by their reptilian ancestors. Reptiles are also the ancestors of birds, and many reptiles gastrulate the same way as birds. When the placenta evolved in mammals, their eggs became small but their mode of gastrulation did not change correspondingly. Neither did they revert to the pattern of their more distant ancestors, the amphibians, nor did they jump forward to a radically new pattern, even though doing so would seem functionally more direct and economical. This is an example of the conservative way in which developmental processes have evolved (see Section 14.1).

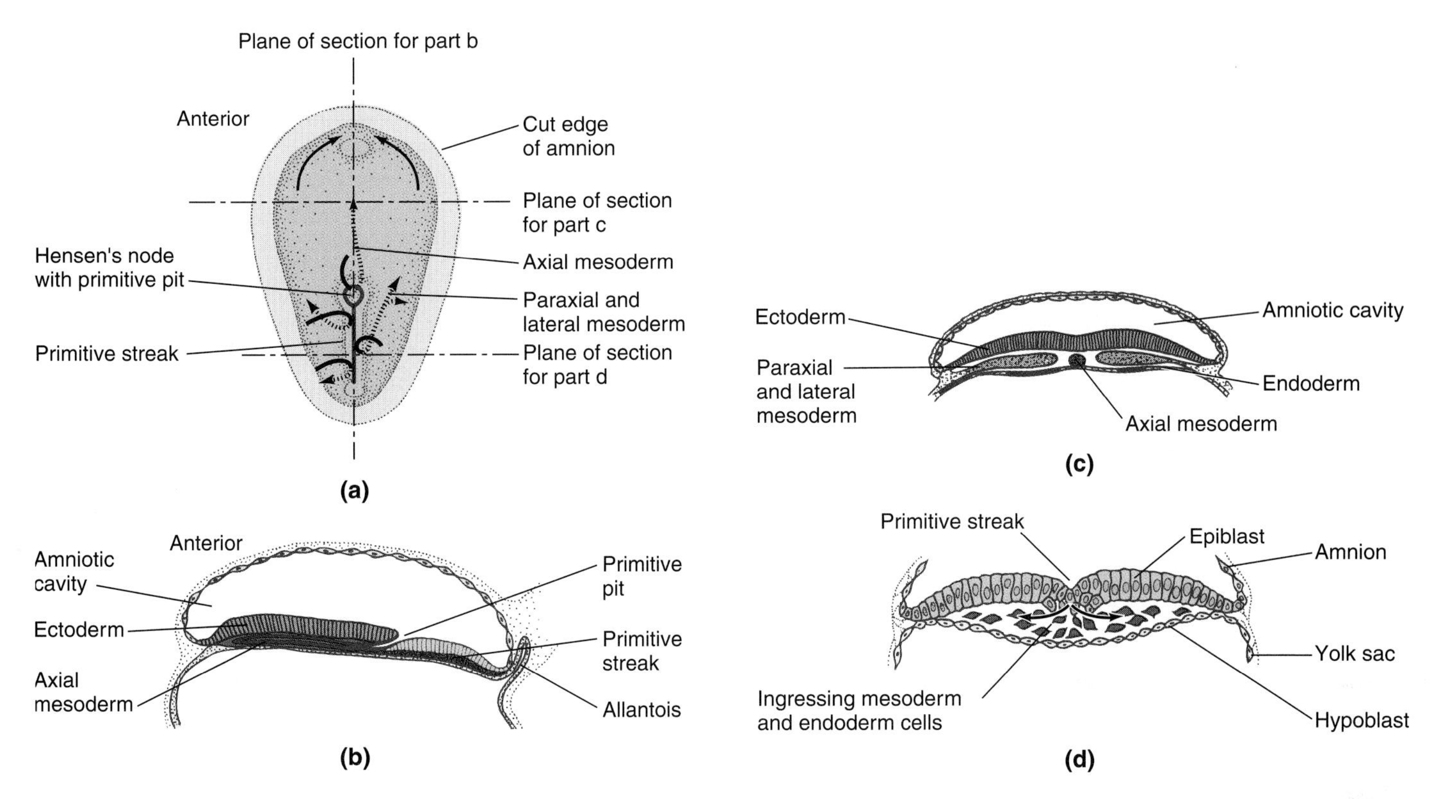

Figure 10.32 Gastrulation in the 16-day human embryo. **(a)** Surface view of the epiblast, with the amnion cut away. Endodermal and mesodermal cells converge toward the primitive streak (solid lines), ingress through the primitive pit and the primitive groove, and spread between the epiblast and the hypoblast (broken arrows). **(b)** Median section. **(c)** Transverse section anterior to Hensen's node. **(d)** Transverse section posterior to Hensen's node.

SUMMARY

During the period of gastrulation, the embryo is transformed from a simple sphere or disc to a more complex geometrical configuration of three germ layers. The outer layer, or ectoderm, gives rise to the epidermis and the nervous system. The inner layer, or endoderm, forms the inner lining of the intestine and its appendages. The numerous structures in between are derived from the middle layer, or mesoderm. Although these germ layers are formed universally in the animal kingdom, the cellular movements that establish them differ from one animal group to another.

Gastrulation, as well as the subsequent period of organogenesis, involves extensive morphogenetic movements. Through these movements the embryo molds itself into the three-dimensional form that characterizes its phylogenetic group.

The morphogenetic movements observed in different animals are combinations of elementary epithelial movements (invagination, involution, convergent extension, epiboly, and delamination) and cellular activities (migration, intercalation, ingression, shape changes, division, changes in adhesiveness, and programmed cell death). In another step of reductionist analysis, some of these cellular activities are now being studied in terms of specific gene activities.

Gastrulation in sea urchins involves the ingression of two waves of mesenchymal cells and the formation of the primitive gut, the archenteron. After an initial invagination phase, the archenteron undergoes convergent extension resulting from a combination of cell intercalation and shape changes. Finally, filopodia pull the archenteron tip toward the ectodermal site where the mouth will be formed, while additional cells are recruited to the anal end of the archenteron by involution.

During amphibian gastrulation, different areas of the blastula cooperate in carrying out several types of morphogenetic movements. First, bottle cells in the vegetal hemisphere form the blastopore. This is followed by involution of the marginal zone, which moves around the blastopore lip and turns inside. The involuting cells migrate along the inside of the blastocoel, pulling or at least guiding the cells that involute after them. The involution process generates the archenteron and places the mesoderm between the archenteron (endoderm) and the ectoderm, which remains outside. Convergent extension of the marginal zone causes anteroposterior elongation of the embryo on the dorsal side. As the vegetal parts of the gastrula turn inside, they are enveloped by the animal portions of the gastrula in a process of epiboly.

In the zebrafish, the embryonic disc engulfs the uncleaved yolk by epiboly while forming two layers, epiblast and hypoblast. In both layers, cells converge toward the future dorsal midline and forming an embryonic shield that extends anteroposteriorly. The epiblast forms the embryonic ectoderm while the hypoblast gives rise to endoderm and mesoderm.

In birds, gastrulation begins with convergent extension of epiblast cells to form the primitive streak along the embryo's posterior midline. A depression in the primitive streak, the primitive groove, is analogous in function to the blastopore in amphibians. Epiblast cells involute via the primitive groove before they ingress into the underlying blastocoel. Endodermal cells turn inside first and displace the underlying hypoblast. Mesodermal cells ingress thereafter and spread as a mesenchyme between the endoderm and the remainder of the epiblast, which remains outside to form the ectoderm. Once the germ layers are established, their extraembryonic portions spread by epiboly until the entire yolk is surrounded. Many reptiles gastrulate in the same way as birds.

Although placental mammals have small eggs and undergo holoblastic cleavage, their mode of gastrulation is patterned after that of their reptilian ancestors. In humans and other primates, the inner cell mass splits into the hypoblast, which forms the yolk sac, and the epiblast, which gives rise to the embryo proper. As observed in birds, gastrulation begins with primitive streak formation in the epiblast and ends with a trilaminar germ disc consisting of the endoderm adjacent to the yolk sac, the ectoderm facing the amniotic cavity, and the mesoderm in between.

Cell Adhesion and Morphogenesis

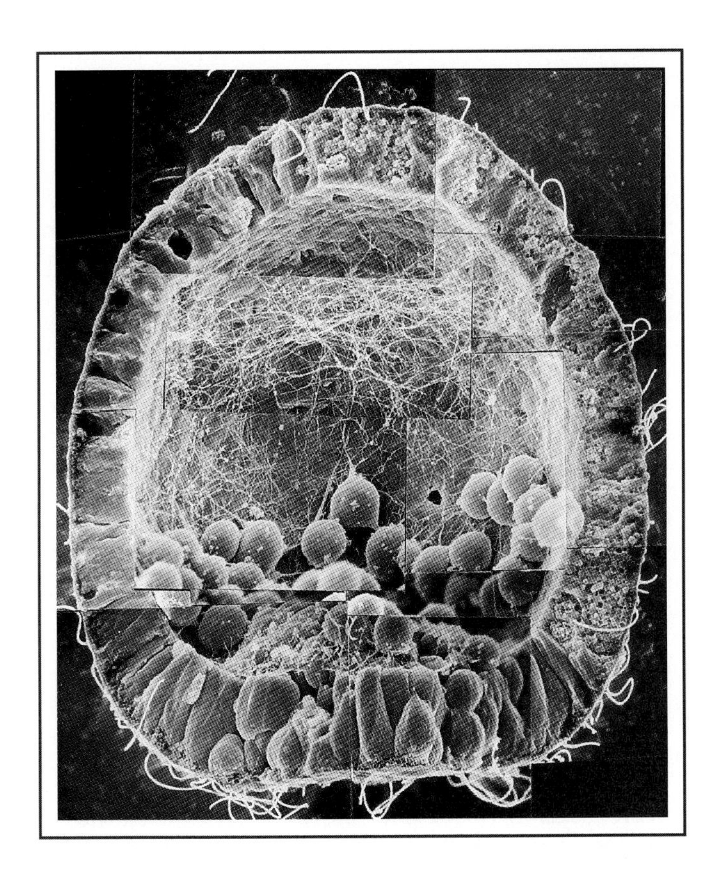

Figure 11.1 Scanning electron micrograph showing the inside of an early sea urchin gastrula. The round cells near the bottom of the embryonic cavity (blastocoel) are the primary mesenchyme cells. They were part of the outer epithelium before they lost adhesiveness to their neighbors and moved into the blastocoel. Changes in cell adhesiveness coupled with cell motility are driving forces in morphogenesis.

THE morphogenetic movements of gastrulation generate three germ layers, which will interact to form the organ rudiments and bring forth the overall body plan of the taxonomic group to which the embryo belongs. This is an astounding process, and watching it has driven a respected embryologist to the point of postulating a goal-oriented metaphysical force (Driesch, 1894, 1909). Staying within the realm of science, one has to assume that each embryo carries the instructions for its morphogenetic movements in its genetic information and in the three-dimensional configurations of its localized cytoplasmic determinants. How detailed are these instructions? Do they include some sort of zip code that would tell every cell where it is located at the beginning of morphogenesis and which of many trajectories it should use to reach its destination?

One way of getting an answer to this question would be to scramble an embryo by mixing up its cells. This is exactly what some classical embryologists did, mostly with sea urchin and amphibian embryos—that is, embryos with variable cleavage and late determination. Such embryos were dissociated into cells, mixed and reaggregated. The cells then underwent seemingly random movements but adhered to one another selectively. As a result, they separated into different layers, forming ***embryoids*** that were somewhat similar to normal embryos. Such embryoids are formed reproducibly, indicating that the spatial order generated does *not* depend on the positions in which cells ended up after mixing and reaggregation. Rather, cells behave to an extent like immiscible fluids. Small drops of oil in water coalesce until they have formed one big drop, whereas a drop of gasoline spreads on the surface of water until it forms a layer that is only one molecule thick. Thus, observations on reaggregating cells suggest that their genetic instructions may impart on them only general properties of mobility and adhesion, and that when cells behave according to these properties, an embryolike structure emerges. Cell adhesion has therefore come to be recognized as a driving force in morphogenesis.

The adhesiveness between cells changes in the course of development. Embryos consist of closely knit epithelia and loose mesenchyme (see Fig. 2.22). However, these distinct conformations wax and wane as development proceeds. Epithelial cells become mesenchymal in the sea urchin embryo at the beginning of gastrulation, when cells from the vegetal plate ingress and form the primary mesenchyme cells (Fig. 11.1). Conversely, mesenchymal cells may band together and form epithelia as, for example, in chicken embryos, where individual cells ingress through the primitive streak and later form the somites and lateral plates of the mesoderm.

Even within an intact epithelium, cells may change neighbors. Morphogenetic movements such as convergent extension and epiboly are associated with cell intercalations, in which cells detach from some of their neighbors and insert themselves between others. Thus, the organized making and breaking of cell contacts is an integral part of morphogenetic movements.

Modern techniques involving the use of antibodies and cloned DNA have allowed investigators to isolate several types of ***cell adhesion molecules (CAMs),*** by which cells adhere to one another. The same techniques have revealed ***substrate adhesion molecules (SAMs),*** a wide range of molecules by which cells adhere to the *extracellular matrix* (*ECM*), in which cells are embedded. Thus, the SAMs include certain ECM molecules as well as matching receptor-like molecules in the plasma membrane of cells. As expected, CAMs and SAMs change in correlation with morphogenetic events. Indeed, researchers have been able to show that certain CAMS or SAMs are necessary for certain steps in gastrulation or morphogenesis to occur normally. Thus, we are beginning to understand morphogenetic movements in terms of the ability of cells to synthesize changing combinations of CAMs and SAMs.

In the course of this chapter, we will first summarize the classic experiments that have pointed out the central role of cell adhesion in embryonic development. Next we will examine the molecular characteristics of several CAMs, SAMs, and ECM molecules. Then we will explore the functions of these molecules in morphogenetic processes including sea urchin and amphibian gastrulation. Finally, we will discuss evidence that gene activity, cell adhesion, and morphogenesis control one another in morphoregulatory cycles.

11.1 Cell Aggregation Studies in Vitro

To analyze cell adhesion, investigators often disaggregate embryonic tissues into single cells. Depending on the species, disaggregation may be achieved mechanically by forcing the tissue through a fine-meshed sieve, by changing the pH of the medium, by removing calcium ions (Ca^{2+}) and magnesium ions (Mg^{2+}) from the medium, or by adding proteases to partially digest the bonds between the cells. After any of these treatments, cells are allowed to recover in a medium mimicking their normal embryonic environment before their behavior is studied.

CELLS FROM DIFFERENT TISSUES ADHERE TO ONE ANOTHER SELECTIVELY

The most dramatic morphogenetic movements in embryonic development occur during gastrulation and organogenesis. At those stages, do cells adhere to one another selectively? This question was addressed in studies using cells from amphibian embryos. Cells from different germ layers or regions typically differ in size, shape, or pigmentation, so that they can be distinguished with relative ease. Moreover, cells readily dissociate if the pH of the culture medium is raised briefly.

IN a classical experiment, Townes and Holtfreter (1955) dissociated neural plate cells and prospective epidermis cells from newt embryos and mixed them together (Fig. 11.2). The cells initially formed aggregates in which the two cell types were randomly mixed. However, as cells began to move within the aggregate, the neural plate cells disappeared to the inside while the prospective epidermal cells came to the surface. When axial mesoderm cells were added, they were sandwiched between the neural cells on the inside and the future epidermal cells on the outside. In similar experiments, a mixture of prospective endodermal cells and mesodermal cells sorted out so that mesoderm was formed inside and endoderm outside. However, when epidermal cells were added to the mixture, the reaggregated endoderm was surrounded by mesoderm, and the mesoderm in turn was enveloped by epidermis. Generally, the epithelia reconstituted by the cells showed the appro-

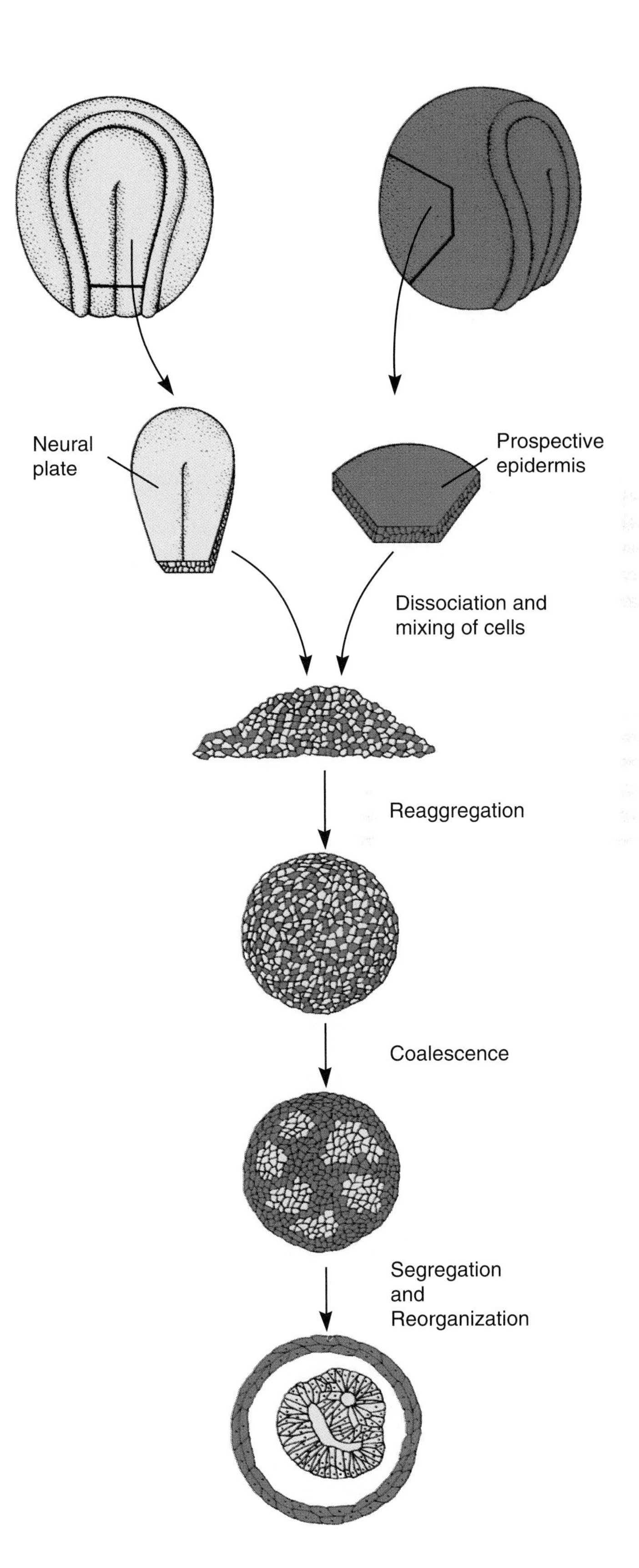

Figure 11.2 Segregation and reorganization of mixed cells from an amphibian neurula. Prospective brain and spinal cord from an unpigmented species and prospective epidermis from a pigmented species were excised and dissociated into single cells. When mixed together, the neural cells sorted out, forming small internal islands of tissue that coalesced and displaced the epidermal cells to the outside position. Both types of cells then reorganized into a semblance of the tissue that they would have normally formed, and the relative position of the two tissues was the same as in the embryo. The top three diagrams represent external views, the bottom three represent sections.

priate polarity, with the apical surface on the outside and the basal surface on the inside.

Questions

1. Why are amphibian embryos the most suitable vertebrate embryos for cell dissociation experiments? (Consider how amphibian and other vertebrate embryonic cells obtain their nutrients.)
2. Which of the results obtained by Townes and Holtfreter seems odd and not easily explained?

Two conclusions from these experiments are important. First, cells from different germ layers sort out from one another, thus revealing distinct adhesive properties. Second, the positions occupied by the cells in reaggregates reflect their relative positions in normal embryos. The behavior of reaggregating cells in vitro suggests that selective cell affinities help establish and maintain the spatial order of different tissues in the embryo.

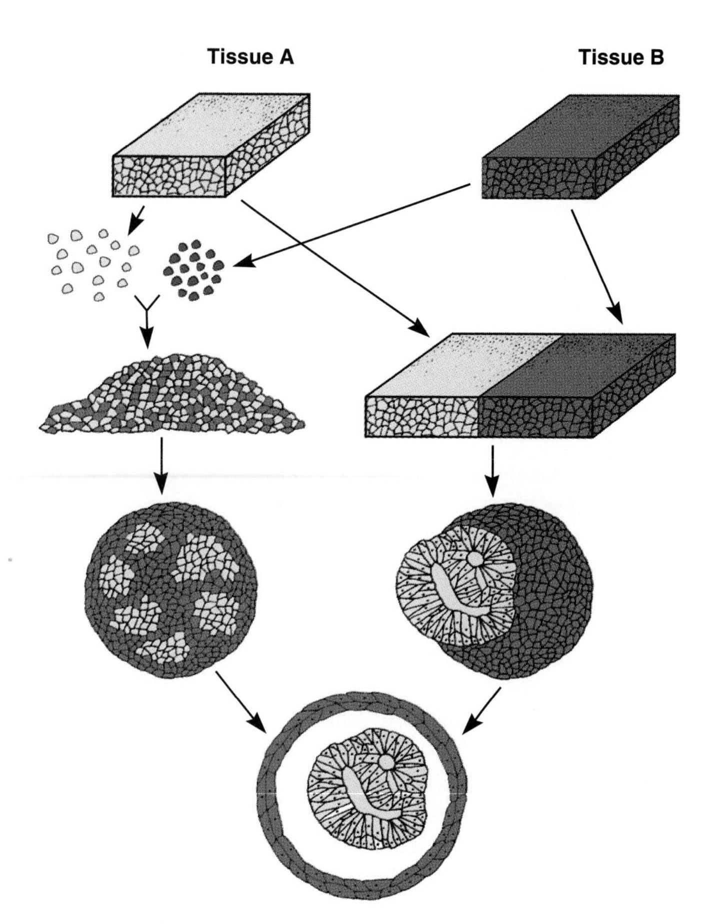

Figure 11.3 Cells reach the same final configuration on different trajectories. Two embryonic tissues, A and B, may be dissociated and then mixed and reaggregated (to the left) or simply pushed together (right). The cells in the mixed reaggregate sort out, with cells from tissue A forming small internal islands that coalesce into larger islands, displacing cells from tissue B to the outside. When pushed together as contiguous masses, B spreads over A to approach the same final configuration as by sorting out. Two immiscible fluids behave the same way.

TISSUES FORM HIERARCHIES OF ADHESIVENESS

Mixtures of different cell types sort out not only during gastrulation and organogenesis but also during later stages of development. When chicken cells from embryonic heart, liver, retina, cartilage, and other tissues are mixed, they segregate and arrange themselves in concentric layers. The aggregates mimic the normal relative positions of cells; for instance, muscle cells surround chondrocytes (cartilage cells). Even when the participating cells would not normally occur next to each other, they reaggregate in a reproducible order. For example, when pigmented retina cells and heart muscle cells from chicken embryos are mixed, the heart cells first form islets surrounded by retina cells. The islets of heart cells subsequently coalesce until eventually all heart cells are together on the inside; never do the heart cells envelope the retina cells.

Interestingly, two types of cells will reach the same final configuration regardless of their initial configuration. Whether cells are dissociated and mixed or pushed together as two blocks of pure tissue, they wind up as one cell type enveloping the other (Fig. 11.3).

The experiments described above show that embryonic cells can sort out from various unnatural arrangements and reconstitute a semblance of an embryo. This result is at variance with the idea that embryonic cells follow detailed assembly instructions encoded in their genome. Rather, gene activities seem to impose on cells certain rules of behavior that guide the cells' assembly toward a specific configuration with little concern for the pathways leading toward this end.

On the basis of their behavior in mixed aggregates, different types of cells can be ranked according to their ability to occupy the center position. Chondrocytes, heart cells, and liver cells are ranked in this order because in mixed aggregates heart cells are always enveloped by liver cells, and chondrocytes are enveloped by heart or liver cells (Fig. 11.4). This hierarchy, and the behavior of the cells during the process of sorting out, were interpreted by Steinberg (1963, 1970, 1998) in terms of quantitative differences in ***cell adhesion***—that is, the energy that must be expended to separate adhering cells. Steinberg proposed that cells in a mixed aggregate move randomly until the sum of all cell adhesions has reached a maximum. This is the state of minimum free energy, at which further cell rearrangements require the input of energy. This state is reached when the cell type with maximum adhesiveness has formed a core surrounded by concentric spheres of cells with progressively lower adhesiveness. This ***differential adhesion hypothesis*** is in accord with the general thermodynamic principle that closed systems naturally move toward a state of minimum free energy.

As a direct test of the differential adhesion hypothesis, Phillips and Steinberg (1969) centrifuged round cell aggregates and took the resistance of the aggregates to

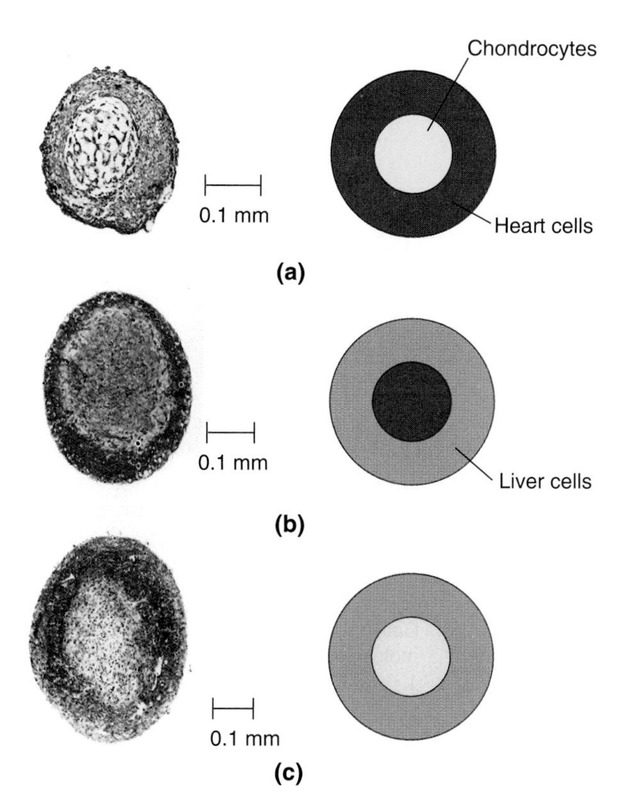

Figure 11.4 Sorting out and spatial ordering in mixed aggregates of different chicken cell types. Each photograph shows the segregation and reorganization of the two types of cells indicated in the diagram to the right. **(a)** Chondrocytes move to the center after mixing with heart cells. **(b)** Heart cells move to the center when mixed with liver cells. **(c)** Cartilage cells move to the center when mixed with liver cells. The three cell types form a hierarchical order with regard to their ability to occupy the center position.

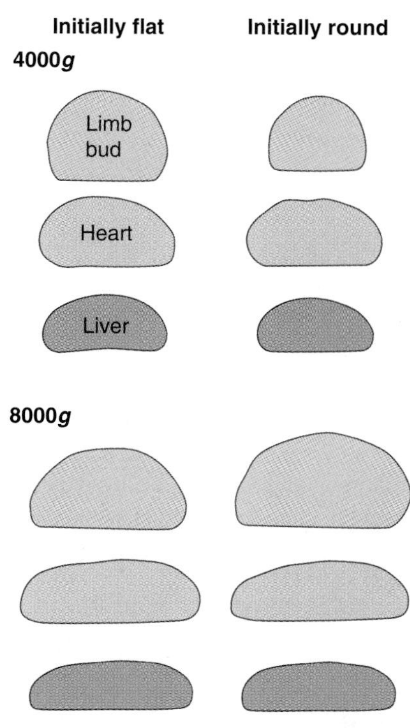

Figure 11.5 Profiles of centrifuged aggregates of different chicken cells. Starting from either round or flat (pancakelike) shapes, the aggregates assumed a shape somewhat like a hamburger bun. This shape represents a state of equilibrium at which the flattening centrifugal force just balances the forces of cell adhesion, which work toward imposing a spherical shape. The exact shape of the aggregates at equilibrium depended on the strength of the centrifugal force (4000*g* or 8000*g*) and on the cell type used. Limb bud cells (mostly chondrocytes) formed the most-rounded buns, indicating greatest adhesiveness, while liver cells were least adhesive and heart cells in between. This order of adhesiveness correlates with the ability of these cell types to occupy the central position in mixed aggregates (see Fig. 11.4).

deformation as a measure of the cells' adhesiveness. They chose centrifugation conditions that compressed originally round aggregates while allowing flat aggregates ("pancakes") of the same cells to assume a somewhat rounder shape (like that of a hamburger bun). They reasoned that the intermediate shape would represent an equilibrium at which the forces of cell adhesion, which work toward imposing a round shape, would balance the centrifugal force flattening the aggregate. As expected, the shapes of aggregates of different cell types at equilibrium reflected their behavior in aggregation experiments. Cells that tended to occupy the center of mixed aggregates—chondrocytes, in Steinberg's experiments—formed the roundest bun shape, indicating that they had the strongest adhesion (Fig. 11.5). Conversely, cells that tended to be displaced to the outside in mixed aggregates—the liver cells—formed the flattest buns, indicating the weakest adhesion. These results confirmed the hypothesis that the behavior of different cell types in mixed aggregates reflects quantitative differences in adhesiveness.

In an extension of this work, investigators used a thin quartz fiber to gently squeeze spherical cell aggregates against a microscope cover slip from below while observing the deformation of the sphere in the microscope (Foty et al., 1996; Davis et al., 1997). From the shape of the squeezed aggregate and the applied force, they calculated the *surface tension* as a quantitative measure of the aggregate's resistance against an increase of its surface. As predicted by the differential adhesion hypothesis, they found that any tissue that enveloped another tissue had the smaller surface tension of the two. Among five embryonic chicken tissues tested—limb bud, pigmented epithelium, heart, liver, and neural retina—the surface tension decreased in this order, and tissues with lower surface tension invariably enveloped tissues with higher surface tension.

The differential cell adhesion hypothesis does not predict the molecular mechanisms that cell adhesion is based on. Nor does it specify how many types of cell adhesion molecules may be involved. It just states that the *overall* strength of adhesion dictates the relative positions of two cell types in an experimental cell arrangement and, where applicable, in a normal embryo. Molecular studies described in the following sections

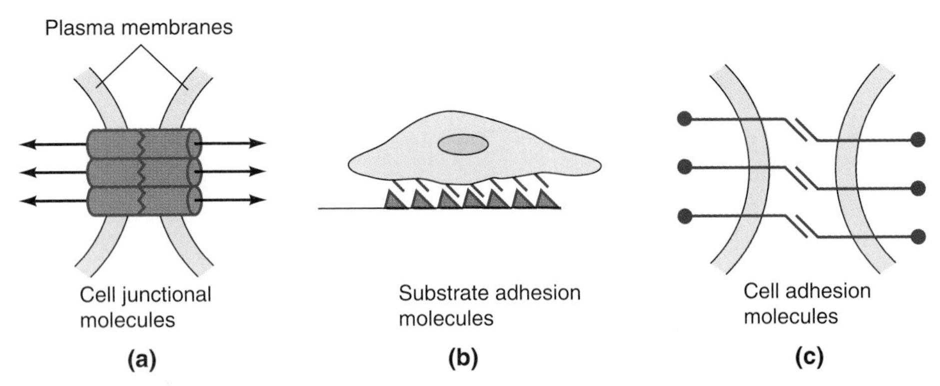

Figure 11.6 Three types of molecules (color) involved in connections between cells. **(a)** Cell junctional molecules form communication channels (gap junctions), seals between the intercellular space and the external environment (tight junctions), or stable adhesion sites (anchoring junctions). **(b)** Substrate adhesion molecules include extracellular molecules (triangles) and the corresponding cellular receptors (slanted lines) whose interaction causes cells to adhere to extracellular substrates. **(c)** Cell adhesion molecules (bent lines) are embedded in the cell membranes and link cells to each other quickly but weakly.

show that several types of cell adhesion molecules contribute to this phenomenon. Differences between these molecules allow cells to adhere to one another more selectively than would be possible with a single type of adhesion molecule, and they also link cell adhesion to cell differentiation.

11.2 Cell Adhesion Molecules

Cells adhere to other cells and to extracellular materials by means of *cell junctions, substrate adhesion molecules,* and *cell adhesion molecules* (Fig. 11.6). Cell junctions are large molecular structures that form slowly and are generally strong and durable (see Section 2.7). Substrate adhesion molecules, which will be discussed later in this chapter, are a group made up of components of the extracellular matrix and their matching receptors in cell membranes. ***Cell adhesion molecules* (*CAMs*)** are single molecules traversing cell plasma membranes, by which cells adhere to one another quickly, selectively, and relatively weakly.

A typical CAM is a glycoprotein with three domains: a large *extracellular domain,* a *transmembrane domain,* and a *cytoplasmic domain* (Fig. 11.7). By means of its extracellular domain, one CAM binds to another CAM. ***Homotypic binding*** is the binding that occurs between CAMs of the same type, whereas ***heterotypic binding*** is that between different types of CAMs. The binding specificity depends on both the protein and the carbohydrate moieties of CAMs. The transmembrane domain links a CAM to the plasma membrane by hydrophobic interactions. The cytoplasmic domain is often connected, via *linker proteins,* to microfilaments or other cytoskeletal structures. This connection is important because the phospholipid bilayer of the plasma membrane is fluid and offers no resistance to lateral floating. Without linkage to the cytoskeleton, a cell adhesion molecule would behave like an unanchored buoy. Such is the case for some CAMs

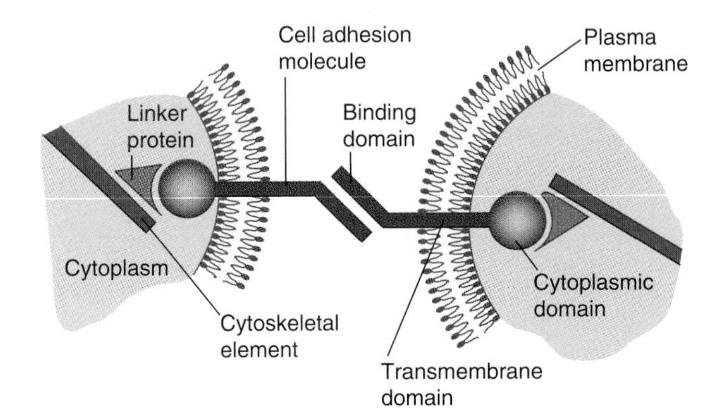

Figure 11.7 Schematic representation of two cell adhesion molecules (CAMs) binding to each other. A typical CAM consists of an extracellular binding domain, a transmembrane domain, and a cytoplasmic domain, which is connected, via one or more linker proteins, to the cytoskeleton.

that consist only of an extracellular domain bound covalently to a modified membrane phospholipid.

To obtain purified CAMs, researchers have made ingenious use of antibodies (Method 11.1). Purified CAMs have in turn allowed investigators to clone and sequence the genes and mRNAs encoding CAMs. The amino acid sequence of CAMs as predicted from the nucleotide sequence of their mRNAs reveals the molecular properties of CAMs in much detail. The cloning of CAM genes has also set the stage for exciting new research on how their expression is regulated as part of the overall processes of cell determination and morphogenesis. On the basis of their molecular characteristics, CAMs can be subdivided into three families: immunoglobulin-like CAMs, cadherins, and lectins.

IMMUNOGLOBULIN-LIKE CAMs MAY ALLOW OR HINDER CELL ADHESION

The first-known cell adhesion molecule was isolated from the neural retina of chickens and named the *neural*

Isolating Cell Adhesion Molecules and Their Genes with Antibodies

Antibodies, or *immunoglobulins,* are vertebrate blood serum proteins that are produced by B lymphocytes. *Polyclonal antibodies* are produced by injecting a preparation of the antigen into a laboratory mammal, typically a rabbit (see Method 4.1). *Monoclonal antibodies* are the products of single lymphocyte clones, which are obtained by fusing lymphocytes from immunized mice with tumor cells. The fusion cells, called *hybridoma cells,* are capable of unlimited mitosis, like tumor cells, but also retain the characteristic function of a normal lymphocyte: to generate a clone of cells that produce the same type of antibody against a specific epitope, or antigen-binding site (Köhler and Milstein, 1975). From a single immunized mouse, one can obtain many hybridoma cell clones that can be screened on the basis of the antibodies they produce. Even if a mouse is immunized with a mixture of many antigenic molecules, each hybridoma cell clone produces only one type of antibody.

To obtain polyclonal antibodies against CAMs, plasma membranes from the cell type of interest are injected into a rabbit. The antibodies obtained are tested for the ability to inhibit aggregation of the cells. (For this test, the antibodies must be cleaved into fragments—known as *Fab fragments* or *univalent antibodies*—having only one antigen-binding site. Normal antibodies are bivalent, having two binding sites that agglutinate the cells to which they bind, as shown in Fig. 7.9.) The inhibition of cell aggregation by Fab fragments is then used as a screening procedure to identify one or more CAMs. Cell surface proteins from the cells of interest are separated by electrophoresis and tested individually for the ability to *overcome* the inhibition of cell aggregation by the antibody (Fig. 11.m1). A protein that has this effect must compete with a required CAM on the cells for the binding sites of antibodies and must therefore be the same CAM.

Monoclonal antibodies against cell surface glycoproteins can also be tested for the ability to inhibit cell aggregation. An antibody that does so is again most likely directed against a CAM. Such a monoclonal antibody can be used to isolate the CAM itself from the mixed proteins of the cell membranes that were used as an antigenic preparation. For this purpose, the antibody is immobilized on a chromatography column. The antigenic preparation is passed over the column under conditions that allow the immobilized antibody to bind to its antigen, the CAM, while all other components of the preparation flow through. A different medium is then passed over the column to release the CAM from the antibody. Either of these strategies yields a purified CAM and an antibody directed against it. These preparations can be used in turn to clone the gene encoding the CAM.

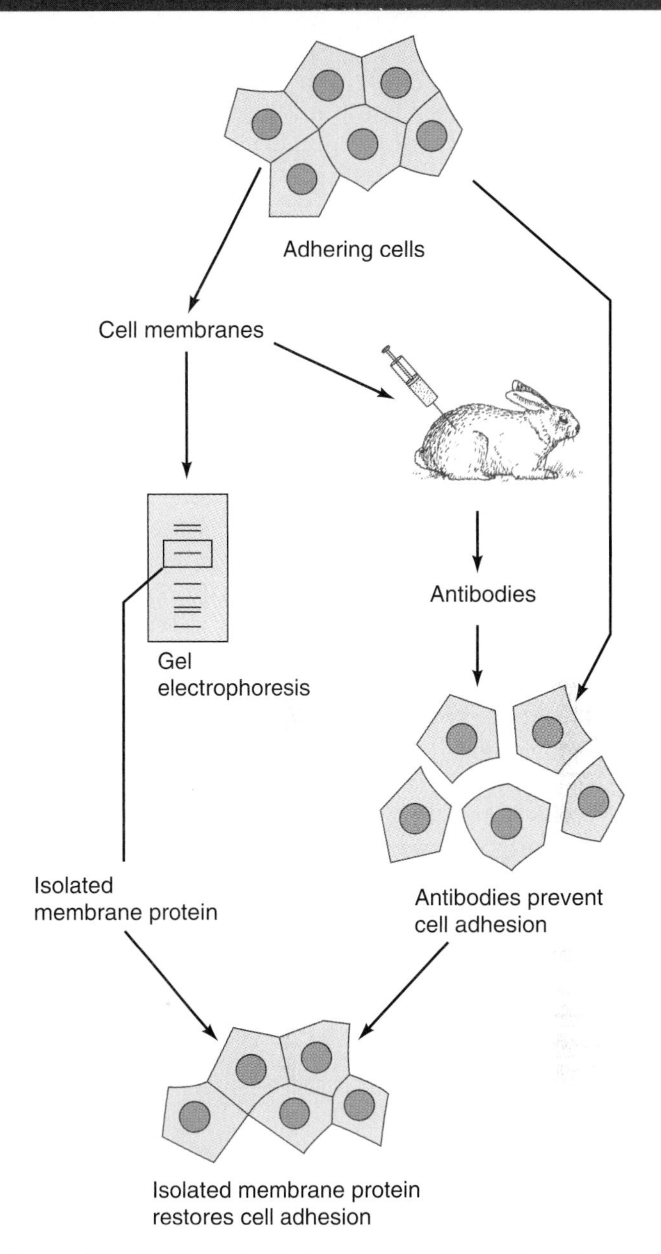

Figure 11.m1 Use of polyclonal antibodies to identify cell adhesion molecules (CAMs). Membranes from cells that adhere in culture are injected into a rabbit. Univalent antibodies prepared from the rabbit's blood serum prevent cell adhesion, indicating that the rabbit is producing antibodies against one or more CAMs. Cell membrane proteins isolated by electrophoresis are then tested individually for the ability to overcome the rabbit antibodies' inhibition of cell adhesion. A protein that has this ability must be able to compete with a CAM on the cell surfaces for the antigen-binding sites of antibodies and most likely will be the CAM itself.

cell adhesion molecule, or *N-CAM* (Brackenbury et al., 1977; Thiery et al., 1977; Edelman, 1983). N-CAM function is *necessary* for neural development. Antibodies to N-CAM interfere with the development of chicken retina in culture, with the establishment of proper connections between the optic nerve and the brain in *Xenopus,* and with other neural processes (Buskirk et al., 1980; S. E. Fraser et al., 1988). N-CAM is also *sufficient* for cell adhesion. This has been shown by the use of ***transgenes,*** engineered genes supplied to cells or organisms in such a way that they are expressed in addition to the host's resident genes (see Section 15.4). When tumor cells that did not express any resident genes for cell adhesion molecules and did not adhere to one another were supplied with an active transgene for N-CAM, they became adhesive (Mege et al., 1988).

Molecular analysis of N-CAM and related proteins has revealed the following features (Cunningham et al., 1987; Rutishauser et al., 1988). The hallmark of these proteins is the presence, in the extracellular domain, of repetitive domains that are also found in immunoglobulins (Fig. 11.8). These domains are thought to fold into complementary structures that allow homotypic binding. The similarities between immunoglobulin-like CAMs and immunoglobulins indicate that both evolved from a common precursor. Because insects do not have immunoglobulins but do have immunoglobulin-like CAMs, it is likely that the vertebrate immunoglobulins evolved from immunoglobulin-like CAMs (Grenningloh et al., 1990).

The extracellular domain of immunoglobulin-like CAMs begins with the end of the protein that is in general terminology referred to as the ***N-terminus*** for its free amino group. The opposite end of the protein, generally known as the ***C-terminus*** for its free carboxyl group, typically marks the end of the cytoplasmic domain in immunoglobulin-like CAMs.

The best-investigated family of immunoglobulin-like CAMS includes the one discovered first, N-CAM. An intriguing feature of N-CAM synthesis is the complexity of posttranscriptional and posttranslational modifications that generate more than 100 different proteins from a single gene. Proteins derived from the same gene are called ***isoforms.*** In the brain, the three most abundant N-CAM isoforms have molecular weights of approximately 180, 140 and 120 kd (Fig. 11.8). The two larger isoforms are transmembrane proteins that differ in the length of their cytoplasmic domain. The C-terminus of the third isoform is linked covalently to a glycosylphosphatidylinositol (GPI) molecule in the outer phospholipid leaf of the plasma membrane (He et al., 1986).

The large extracellular domain of all N-CAMs contains several carbohydrate moieties including *polysialic acid (PSA),* a negatively charged polysaccharide (Rutishauser et al., 1988). Three long PSA chains are attached to carbohydrate moieties in the region of the fifth immunoglobulin domain (Fig. 11.8). These chains vary in length during development and account for 10 to 30%

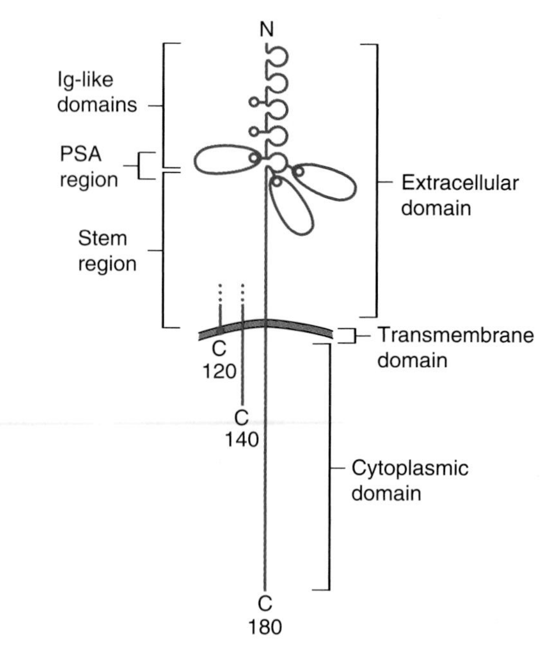

Figure 11.8 Schematic representation of N-CAM, a typical immunoglobulin-like cell adhesion molecule. The major isoforms of N-CAM (with molecules measuring 180, 140, and 120 kd) have identical extracellular domains. The two long isoforms have transmembrane and cytoplasmic domains; the shortest isoform is linked to the outer phospholipid layer of the plasma membrane. The large extracellular domain has several carbohydrate chains (small circles) and five immunoglobulin-like domains (loops). Three polysialic acid (PSA) chains (depicted as clubs) are attached in the area of the fifth immunoglobulin domain. N and C refer to the N-terminus (free amino group) and C-terminus (free carboxyl group) of the N-CAM's polypeptide chain.

of the total weight of the N-CAM. Because of their negative charge, PSA molecules repel one another and thus prevent cell adhesion. In this sense, N-CAM functions as a cell adhesion molecule or as an anti-cell adhesion molecule, depending on PSA content. Generally, embryos have high-PSA forms of N-CAM, which facilitate cellular movements, whereas adult tissues express low-PSA forms, which promote more stable cell adhesion (Rieger et al., 1985).

Since N-CAMs were discovered, researchers have characterized several other CAMs that have multiple immunoglobulin-like domains. Members of this superfamily include *Ng-CAM* and *L1* in vertebrates and *neuroglian* and *fasciclins* in arthropods (Grenningloh et al., 1990).

CADHERINS MEDIATE CA^{2+}-DEPENDENT CELL ADHESION

The ***cadherins*** are a family of CAMs defined by their dependence on calcium ions (Takeichi, 1977). In the absence of Ca^{2+}, they undergo a conformational change that renders them susceptible to breakdown by proteases. The biological significance of this Ca^{2+}

dependence is not known; other CAMs are not Ca^{2+}-dependent. Cells in virtually all tissues express some members of the cadherin family, making them the most prevalent CAMs in vertebrates (Takeichi, 1991, *1995*). Cadherins seem to work primarily by homotypic binding, as the immunoglobulin-like CAMs do.

Four main cadherin subfamilies are named for the tissues in which they were first found: *E-cadherin* (epithelial cadherin, also known as *uvomorulin*), *P-cadherin* (placental cadherin), *N-cadherin* (neural cadherin), and *L-CAM* (liver cell adhesion molecule). However, additional types of cadherins are still being identified. All cadherins have the same basic structure (see Fig. 2.19). They are transmembrane glycoproteins of approximately 125 kd, with the cytoplasmic domain ending in the C-terminus of the polypeptide chain, as is the case for immunoglobulin-like CAMs. To determine which domains in the cadherin molecule are responsible for its binding specificity, Nose and coworkers (1990) constructed chimeric molecules consisting of E-cadherin and P-cadherin domains. They found that the binding specificity resides within the first 113 amino acids of the extracellular domain. Most of these amino acids are conserved between different cadherins, and changing just a few of the variable ones alters the specificity of cadherin binding.

Cadherins are necessary for cell adhesion: Antibodies against cadherins interfere with cell adhesion. Similarly, the inactivation of maternal cadherin mRNA in frog oocytes by injection of antisense oligonucleotides reduces the adhesion of blastomeres in embryos that develop from such oocytes (Heasman et al., 1994).

Cadherins are also *sufficient* to confer adhesiveness on cells that would otherwise not adhere. This has been demonstrated in genetically manipulated mouse cells that normally do not express cell adhesion molecules or adhere to one another (Nagafuchi et al., 1987). However, such cells aggregate in a Ca^{2+}-dependent way and form compact colonies if supplied with a *transgene* encoding E-cadherin.

Experiments with transgenes have confirmed the *differential cell adhesion hypothesis* discussed in Section 11.1. When cells expressing P-cadherin transgenes at different levels are mixed, the cells synthesizing more cadherin are enveloped by the cells synthesizing less (Steinberg and Takeichi, 1994). Thus, differences in the expression levels of cadherin can mediate—at least in part—the ability of different cell types to organize themselves in concentric spheres, which was demonstrated so strikingly by the cell reaggregation experiments described earlier. With regard to the molecular basis of cell adhesion, experiments with transgenes have shown that cell adhesion may depend not only on the overall amount of cell adhesion molecules present but also on how they are matched. When cultured cells known as L cells, which do not normally synthesize E-cadherin, are mixed with various other cells, they associate most strongly with mesenchymal cells, which also do not synthesize E-cadherin. However, if the same L cells are supplied with a transgene encoding E-cadherin, they prefer to associate with epithelial cells, which naturally express E-cadherin (Nose et al., 1988).

Experiments with cadherins have also improved our understanding of the cytoplasmic domain of cell adhesion molecules. In particular, these experiments have corroborated the function of *catenins* as linker proteins between cadherins and the cytoskeleton (see Fig. 2.19). By providing cells with cadherin transgenes that were missing certain portions of the cytoplasmic domain, it has been shown that a region close to the C-terminus is required for cadherin interaction with catenins. Cadherins lacking this region insert into the plasma membrane but do not associate with catenins and do not accumulate near cell junctions like normal cadherins do (Nagafuchi and Takeichi, 1988; S. Schneider et al., 1993).

The functional significance of cadherin linkage to catenins in embryonic development was demonstrated by a competitive inhibition experiment. Kintner (1992) injected mRNA encoding an N-cadherin without a functional extracellular domain into one blastomere of *Xenopus* embryos at the 4-cell stage. The truncated N-cadherin transcribed from this mRNA was expected to bind to catenins and possibly other cytoplasmic linker proteins without contributing to cell adhesion. The excess of nonfunctional cadherin should therefore displace its functional counterpart from linker proteins, resulting in reduced cell adhesion. Indeed, neural tissue derived from the injected blastomere was severely disturbed (Fig. 11.9). Additional deletions from the cytoplasmic domain of the truncated cadherin showed that its anti-adhesion effect depended on the region involved in catenin binding. These results show that cell adhesion by cadherins depends on cadherin interaction with cytoplasmic linker proteins such as catenins.

LECTINS BIND HETEROTYPICALLY TO SUGAR RESIDUES

A third group of CAMs mediates *heterotypic binding* between *oligosaccharides* and matching receptor proteins known as *lectins*. The ***oligosaccharides,*** short chains of various sugar residues, extend from transmembrane proteins or are anchored to phospholipids in the plasma membrane (see Fig. 2.17). The oligosaccharides occur in great diversity, providing a molecular basis for specific binding to receptor proteins known as ***lectins.*** Lectins, like other CAMs, have large extracellular domains, transmembrane domains, and cytoplasmic domains. The heterotypic binding between oligosaccharides and lectins is weak, as is the case for the homotypic binding between other CAMs.

One group of lectins, known as ***selectins,*** function in an inflammatory response that recruits certain white blood cells to inflamed tissue areas (Lasky, 1992). Chemical signals released by the inflamed tissue cause endothelial cells in nearby blood vessels to insert selectins into the plasma membrane facing the blood vessel's

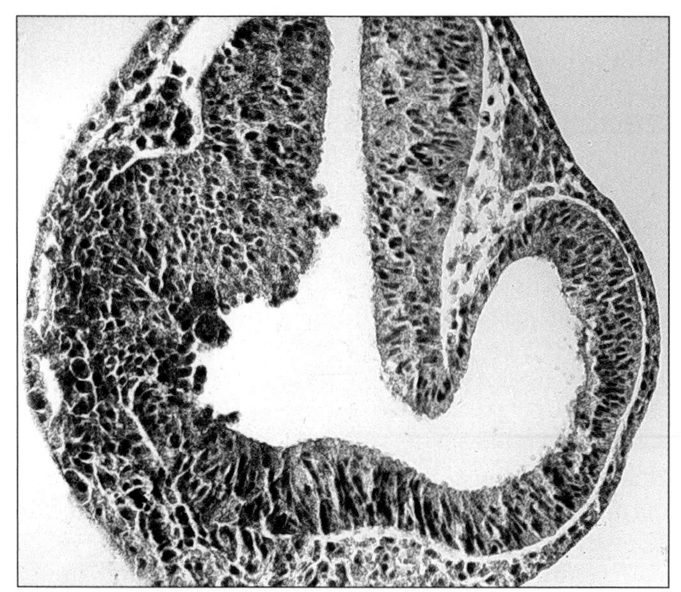

Figure 11.9 The role of cytoplasmic linker proteins in cadherin-mediated cell adhesion. This photograph shows a transverse section of the brain of a *Xenopus* embryo in which one blastomere at the 4-cell state was injected with mRNA encoding a truncated N-cadherin that had no functional extracellular domain. In the right half, brain development is normal; the section passes through the forming eye vesicle (see Fig. 1.16). A tracer coinjected with the mRNA reveals that the left half is derived from the injected blastomere. Here the neural tissue is severely disturbed, with detaching cells entering the brain ventricle. No such disturbances were observed in embryos injected with mRNAs encoding truncated cadherins with additional deletions in the cytoplasmic domain.

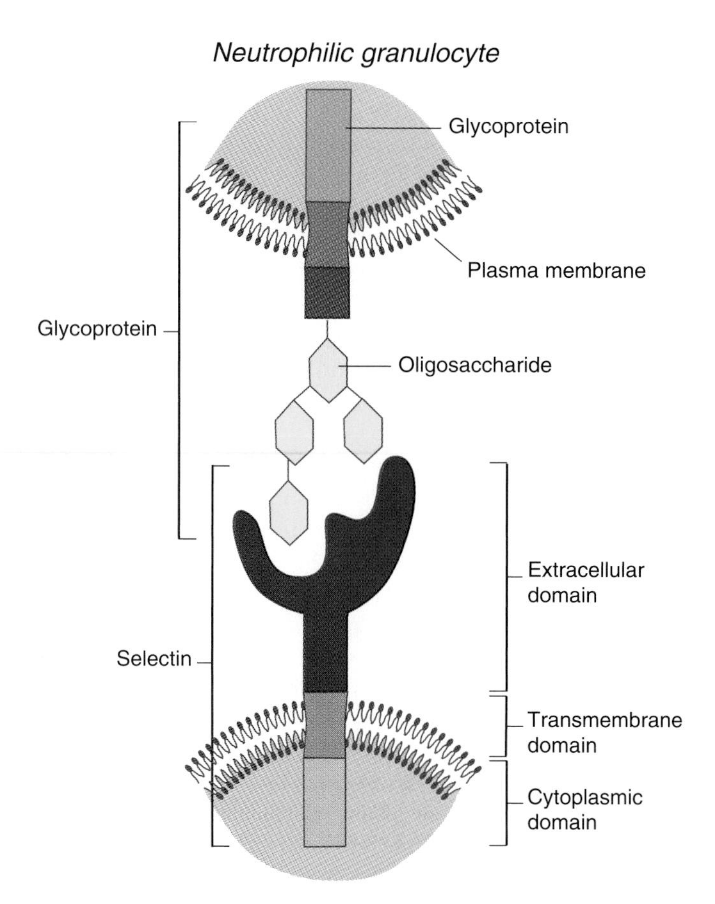

Figure 11.10 Heterotypic binding of a lectin to an oligosaccharide. In the example shown here, the lectin is a selectin molecule present on the apical surface of an endothelial cell lining the inside of a blood vessel. Its ligand is the oligosaccharide moiety of a glycoprotein exposed on the surface of a white blood cell, a neutrophilic granulocyte. Like most other cell adhesion molecules, lectins have an extracellular domain, and transmembrane domain, and a cytoplasmic domain.

lumen. These selectins bind specifically to oligosaccharides present on the surface of white blood cells known as *neutrophilic granulocytes*, or *neutrophils* (Fig. 11.10). The resulting adhesion slows down the circulating neutrophils like tennis balls rolling onto a patch of Velcro. Once slowed down, the neutrophils engage in further interactions that eventually allow them to crawl between the endothelial cells and into the inflamed tissue.

CAMs related to lectins are ***glycosyltransferases,*** enzymes that transfer monosaccharides from activated donor molecules to the oligosaccharides of glycoproteins. Known primarily from their role in the *endoplasmic reticulum* and *Golgi apparatus,* these enzymes are also found on cell surfaces. Here they can act as CAMs when they bind to one of their substrates, the oligosaccharide moiety of a glycoprotein in the plasma membrane of an adjacent cell. This binding will last as long as the enzyme's other substrate, an activated donor of a monosaccharide, is unavailable. If such a donor is added to adhering cells in vitro, the glycosyltransferase performs its enzymatic function, and then the cells detach. Whether this reaction normally occurs in vivo is now known with certainty; it would seem to be an elegant way of making cell adhesion reversible.

Most attention has focused on the membrane localization and function of ***galactosyltransferases*** **(*GalTases*),** a subfamily of glycosyltransferases that transfer galactose from a nucleotide donor (such as uridinediphosphate galactose, or UDPGal) to an oligosaccharide. Like most other CAMs, GalTase is a transmembrane protein, although in contrast to CAMs and cadherins, the cytoplasmic domain of GalTase begins with the amino terminus and its extracellular domain ends with the carboxyl terminus (Shaper et al., 1988). One mouse GalTase gene encodes two mRNAs, which as translated into two different proteins. One isoform, 399 amino acids long, seems to be targeted to the plasma membrane, where it functions as a CAM (Shur, 1991; D. J. Miller et al., 1992; see also Section 4.3). The other isoform, which is 13 amino acids shorter, is targeted to the Golgi apparatus, where it functions in protein glycosylation (Lopez et al., 1991).

Most of the CAMs described in this section mediate a selective adhesiveness of relatively low strength. The weak binding allows cells to easily join and dissociate again in morphogenetic movements. In addition to the three families of CAMs described here, there are other molecules that have similar properties. Some *growth factors* are also membrane-bound, as are their receptors. Expression of these molecules from transgenes can mediate selective cell adhesion, at least in vitro. Thus, some intercellular signal molecules also function in cell adhesion.

11.3 ECM Molecules and Their Cellular Receptors

Organisms do not consist entirely of packed cells. The interstices between cells are filled with materials collectively called the *extracellular matrix (ECM)*. The ECM consists of two components: an amorphous ground substance and a meshwork of fibers. The ground substance attracts water by osmotic pressure and forms a gel, whereas the fibers reinforce the ground substance and resist its tendency to expand. Because the ECM components are linked together by multiple binding domains, they form a matrix, which serves a wide variety of functions. In bones and teeth, the ECM hardens by calcification. In tendons, the ECM accumulates though fibers that maximize tensile strength. In the cornea, the ECM forms transparent layers.

Because the ECM is most plentiful in connective tissues, it has traditionally been thought of as an inert scaffolding that determines primarily the physical properties of a tissue. However, it is now clear that the ECM is very interactive (J. C. Adams and F. M. Watt, 1993; Hynes, 1994; Gumbiner, 1996). The ECM influences cell division, shape, movement, and differentiation. ECM components bind to certain cellular receptors, which in turn affect the cell's gene activity. The interactive nature of the ECM is of particular interest in embryos, where the ECM helps to direct cellular and epithelial movements.

The principal embryonic tissues, mesenchyme and epithelia, have characteristic types of ECM (see Fig. 2.22). Mesenchymal cells are surrounded on all sides by interstitial spaces filled with ECM. Beneath epithelia, the ECM takes the form of a dense sheet called a *basement membrane,* which separates the basal surface of the epithelium from the underlying mesenchyme (Fig. 11.11).

The ECM is composed of a variety of molecules that are released by exocytosis from the producing cells into the extracellular space, where the molecules *self-assemble* (see Fig. 18.20). Thus, cells create their own extracellular environment by synthesizing the ECM building blocks and by maintaining appropriate conditions for their self-assembly. Many molecules found in the ECM have been very well conserved during evolution. For instance, collagen fibers are found not only in vertebrates but also in sea urchins, roundworms, and many other invertebrates.

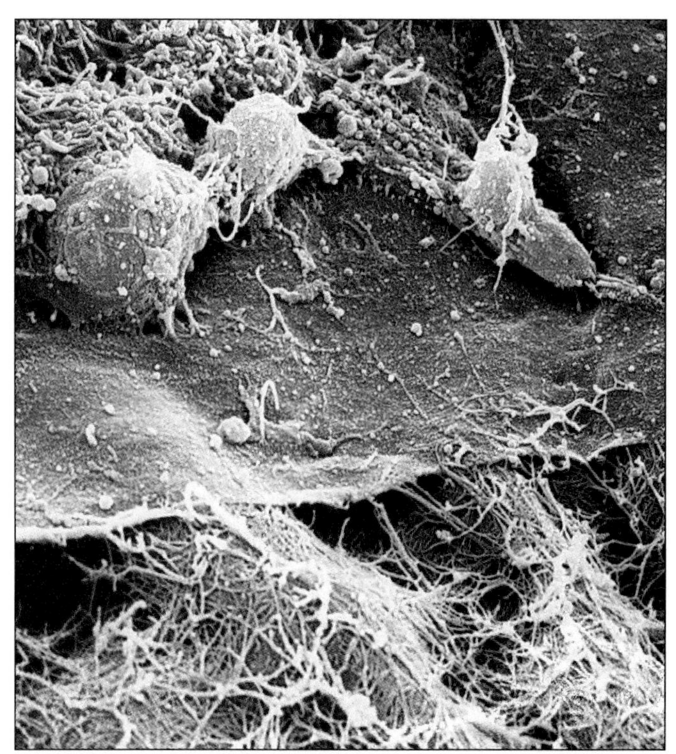

Figure 11.11 Scanning electron micrograph of the cornea of a chicken embryo. The specimen was torn to expose the interface between the outer embryonic epithelium (top) and the underlying mesenchyme (embedded in fibrous material, bottom). Epithelial and mesenchymal cells are separated by a dense sheet of extracellular matrix, called a basement membrane (center). The interstices between the mesenchymal cells and, to a lesser extent, between the epithelial cells are filled with a different kind of extracellular matrix. Its fibrous components, mostly collagen, are well preserved in the fixation process that precedes electron microscopy. The amorphous components of the matrix—mostly water and large polysaccharides—tend to be washed out or to collapse around the fibers during fixation.

In this section, we will summarize the molecular properties of the most common substrate adhesion molecules: the glycosaminoglycans and proteoglycans that form the gellike component of the ECM, the fibrous glycoproteins that make up the meshwork of the ECM, and the receptors in the cell surface that mediate cell adhesion to ECM components.

GLYCOSAMINOGLYCANS AND PROTEOGLYCANS FORM AN AMORPHOUS, HYDROPHILIC GROUND SUBSTANCE

Glycosaminoglycans are long, unbranched polysaccharide chains composed of repeating disaccharide units (Wight et al., 1991). One sugar in each disaccharide is an amino sugar (*N*-acetylglucosamine or *N*-acetylgalactosamine), and the other sugar is a uronic acid

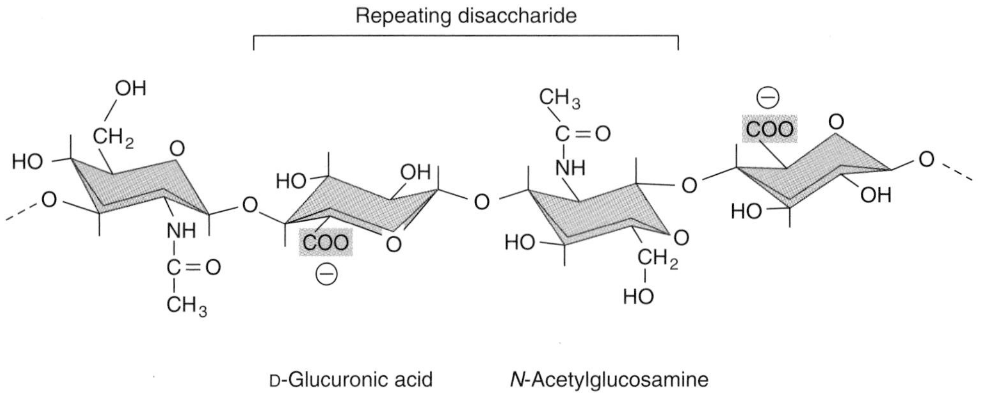

Repeating disaccharide

D-Glucuronic acid N-Acetylglucosamine

Figure 11.12 Portion of hyaluronic acid, a glycosaminoglycan. Hyaluronic acid consists of thousands of repeating disaccharides, each composed of D-glucuronic acid and *N*-acetylglucosamine. Note the negatively charged carboxyl groups. In glycosaminoglycans with sulfate groups, the density of negative charges is even higher.

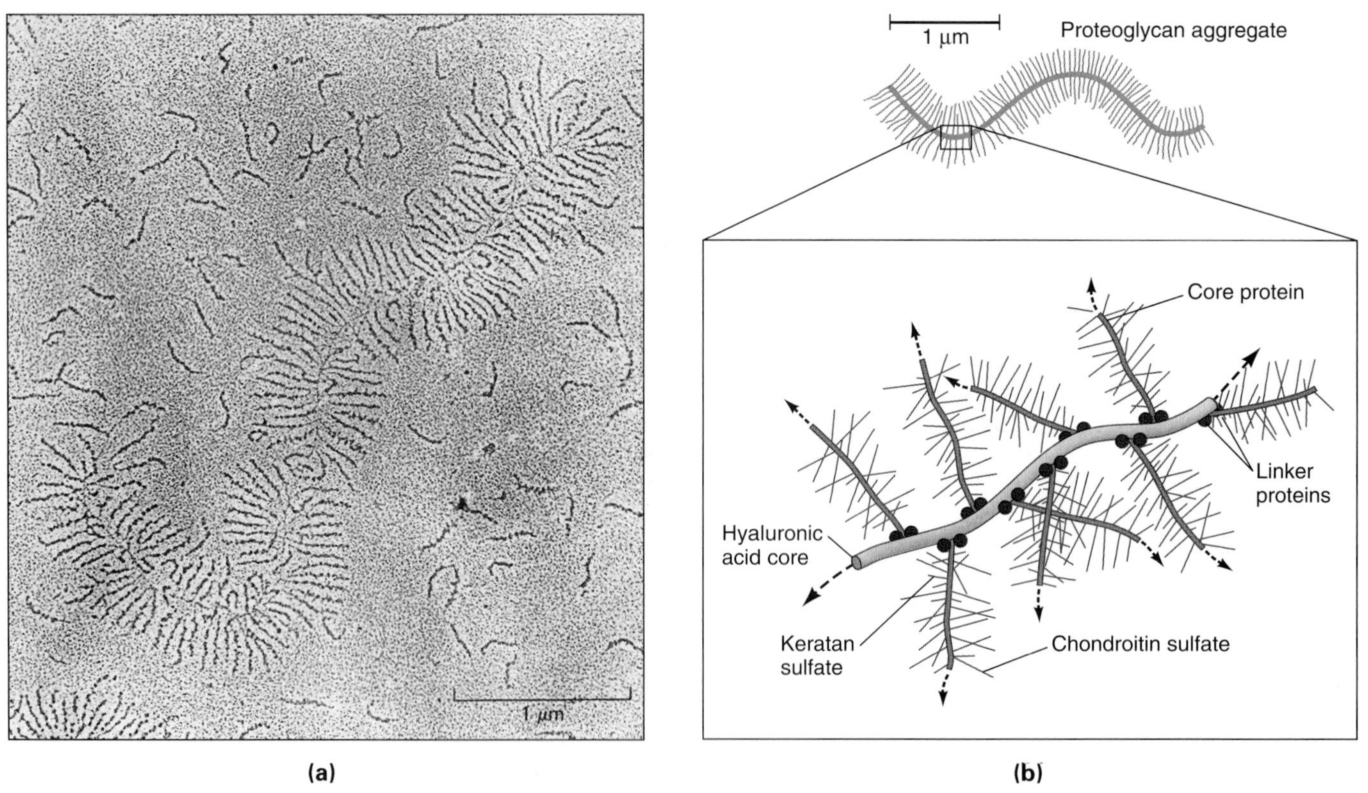

Figure 11.13 Large proteoglycan aggregate from fetal bovine cartilage. **(a)** Electron micrograph of molecules contrasted with platinum. **(b)** Interpretative drawing of the large aggregate shown in the micrograph. Two types of glycosaminoglycans, keratan sulfate and chondroitin sulfate, are linked to core proteins, which in turn are linked to a hyaluronic acid core.

(Fig. 11.12). Different glycosaminoglycans are distinguished by the nature of their sugar monomers, the type of linkage between them, and the number and location of added sulfate groups. The most abundant glycosaminoglycans are *hyaluronic acid, chondroitin sulfate, dermatan sulfate, heparin, heparan sulfate,* and *keratan sulfate.* Most glycosaminoglycans are covalently linked to core proteins, which in turn form the side chains of a long glycosaminoglycan (Fig. 11.13). The resulting large aggregates, called ***proteoglycans,*** may have molecular weights of 10^8 or more. An exception is *hyaluronic acid,* which is just one large glycosaminoglycan.

Because of their mutually repelling negative charges, glycosaminoglycans do not fold tightly but form extended random coils. They attract cations such as the sodium ion (Na^+), which in turn attract large clouds of water. As a result, glycosaminoglycans occupy much space and resist compression. Their most common function seems to be filling space while allowing cell migration and the diffusion of water-soluble molecules. This is the case particularly for hyaluronic acid, which is most abundant in embryos and in regenerating and healing tissue.

In addition to their mechanical function, glycosaminoglycans and proteoglycans are involved in intercellular signaling. This role is based in part on their ability to bind signal proteins acting as *growth factors* or *morphogens* (Reichsman, 1996; Binari, 1997). This mechanism allows the tethering of intercellular signal molecules in ECM domains containing appropriate glycosaminoglycans or proteoglycans. Through the use of enzymes that cut the tethers, sequestered signal molecules can be selective released. Recent genetic studies also show that enzymes modifying the carbohydrate moieties of ECM components can have major effects on intercellular signaling (Sen et al., 1998; X. Lin et al., 1999).

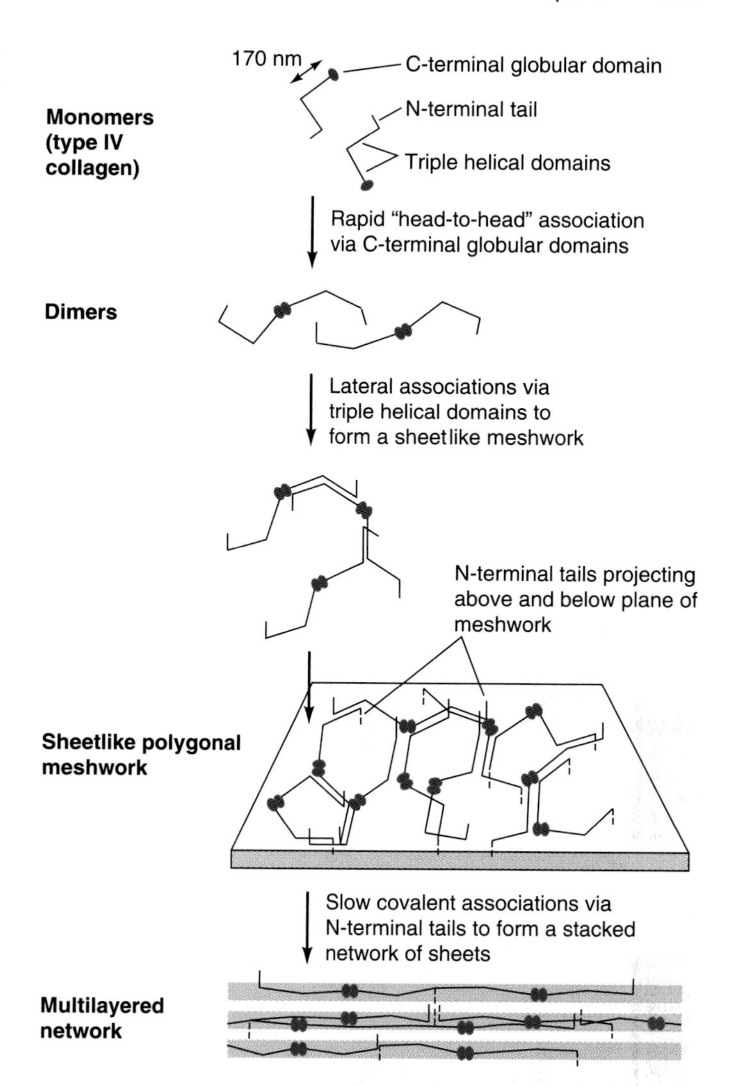

Figure 11.14 Model of type IV collagen self-assembly into a multilayered network as found in basement membranes. The model is based on the electron-microscopic analysis of supramolecular structures that self-assemble in vitro.

FIBROUS GLYCOPROTEINS MAKE UP THE DYNAMIC MESHWORK OF THE ECM

The fibrous components of ECM consist mostly of ***glycoproteins***—that is, proteins with attached oligosaccharides. In contrast to *proteoglycans,* the sugar moieties of glycoproteins are more diverse and account for less than half of the total molecular mass.

Collagens A family of stiff glycoproteins that are found in all metazoa are known as ***collagens*** (Linsenmayer, 1991). Their critical importance is underlined by the lethality of several mutations in collagen genes of mammals and other organisms. In mammals, collagens are the most abundant proteins, accounting for 25% of the animal's total protein.

Many organisms have several collagen genes, each encoding a particular polypeptide. Moreover, collagen pre-mRNAs and polypeptides are often processed in multiple ways, so that even more varieties of collagen molecules are generated. All collagen molecules are secreted from cells as *procollagens*—that is, as triple helices with globular ends. However, the procollagen types differ in their processing and self-assembly in the extracellular space. In the processing of most procollagens, including types I through III, the globular ends are cleaved off to generate rod-shaped collagen molecules, which then self-assemble into long, cablelike fibrils (see Fig. 18.20). Such fibrils have great tensile strength and are abundant in bones, tendons, the dermal layer of the skin, and other connective tissues. In contrast, type IV procollagen retains its nonhelical ends and self-assembles into multilayered networks, which form the core of basement membranes (Fig. 11.14).

Fibronectins Several ECM glycoproteins have specific binding domains for cells and other ECM components (K. M. Yamada, 1991). Their multiple binding sites enable these glycoproteins to cross-link other ECM elements and cells into three-dimensional meshworks. Best characterized among such interactive glycoproteins is ***fibronectin,*** which plays a major role in cell migration. The fibronectin molecule is a dimer of two long polypeptides linked by disulfide bonds near their carboxyl termini (Fig. 11.15). Each polypeptide is folded into several globular domains connected by flexible chains. The globular domains have binding sites for cells and for collagen, heparin, and other ECM components. Fibronectins form extended fibrils that facilitate the migration of embryonic cells as discussed later in this chapter.

The cell-binding activity of fibronectin depends primarily on a tripeptide sequence known as the ***RGD sequence*** (arginine-glycine-aspartate), which is recognized by a family of cell surface receptors called *integrins.* In solution, synthetic peptides that contain the RGD sequence compete with fibronectin for cell receptors and so inhibit cell adhesion to fibronectin. However, cell adhesion to the RGD sequence in fibronectin is only part of

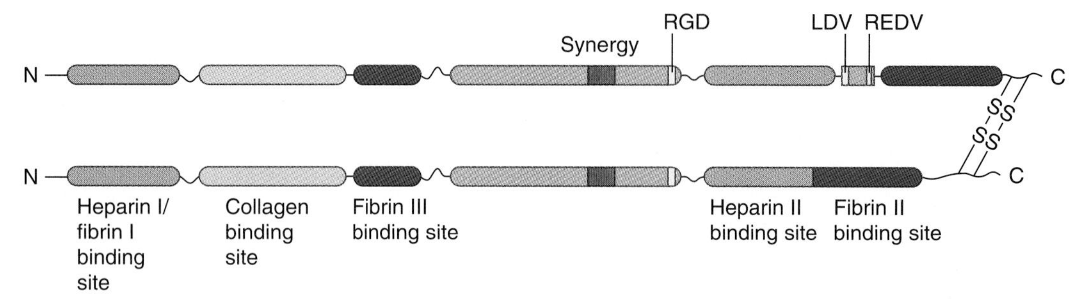

Figure 11.15 Schematic diagram of a fibronectin molecule. The large glycoprotein is a dimer of two similar polypeptides, linked by a pair of disulfide bonds near their C-termini. Each polypeptide is divided into several domains that bond to other ECM components or to cells. The cell-binding domains (labeled at the top) include the arginine-glycine-aspartate (RGD) sequence, the synergy region that acts together with the RGD sequence, and an alternatively spliced region (rectangle) with a leucine-aspartate-valine (LDV) and an arginine-glutamine-aspartate-valine (REDV) sequence. The major ECM components bound by each domain are labeled only at the bottom.

a more complex set of interactions between cells and the ECM. The RGD sequence occurs not only in fibronectin but also in other ECM components, and domains of the fibronectin molecule other than the RGD sequence contribute to cell binding.

Laminins Another family of large interactive glycoproteins, called ***laminins,*** are especially plentiful in basement membranes. Laminins consist of three polypeptides arranged in the shape of a cross and held together by disulfide bridges. As in the case of collagen, the component polypeptides of laminin are encoded by small families of different genes, and resulting combinations are tissue-specific. Like fibronectin, laminins have several domains that bind to other ECM components or cells. Some of these domains bind to type IV collagen; another, to heparin. There are at least two cell-binding regions in the center of the molecule and several additional sites in the long arm. Laminins promote adhesion of many cell types and in particular the extension of *neurites* from nerve cells. In basement membranes, laminins interact with collagen and epithelial cells on one side while fibronectin interacts with collagen and mesenchymal cells on the other.

INTEGRINS MEDIATE CELL ADHESION TO ECM MOLECULES

The complexity of the ECM is mirrored by a heterogeneous array of cell surface proteins that interact with ECM components. We will focus on two types of such molecules: *integrins* and *proteoglycans.*

Integrins are a family of transmembrane receptor glycoproteins, each consisting of two peptides, or chains, one α-chain and one β-chain (Ruoslahti, 1991; Hynes, 1994; Fig 11.16). The basic integrin structure has been highly conserved in evolution and is very similar in vertebrates and invertebrates. In mammals, at least 40 α-chains and 8 β-chains allow for the formation of a wide variety of dimeric integrins. Each chain is encoded by a different gene, and alternative splicing of pre-mRNAs allows for additional variation. Integrins can be in active and inactive states, but the nature of the activation step is still unclear. The affinity and binding specificity between integrins and their ligands depends on the presence of magnesium or calcium ions, for which the α-chain has multiple binding sites.

The extracellular domains of the α- and β-chains combine to form a binding site for RGD or other specific amino acid sequences. Because these sequences occur not only in fibronectin but also in other ECM molecules, most integrins bind to multiple ligands. Conversely, most ECM molecules have several binding sites for different integrins. These overlapping binding specificities allow cells to respond to various combinations of molecules in their extracellular environment.

The cytoplasmic domains of one or both integrin chains are connected, via linker proteins such as *talin* and *α-actinin,* to microfilament bundles inside the cell. Wang and coworkers (1993) demonstrated the connection between integrins and cytoskeletal elements directly by applying a magnetic twisting device to cell surfaces. Twisting of integrins caused a force-dependent stiffening response of the cells. This response was abolished by cytoskeletal inhibitors such as *cytochalasin D,* and no response was observed when the twisting device was applied to nonadhesion receptors. Thus, integrins link a cell's cytoskeleton to the ECM just as cell adhesion molecules connect the cytoskeletons of two cells. Moreover, the linkage of integrins to the cytoskeleton allows cells to control the arrangement to fibronectins on their surface in ways that promote or inhibit interactions with specific ECM components.

Another type of molecule that links cells to the ECM are ***cell surface proteoglycans.*** Their core proteins typically have a cytoplasmic domain, a transmembrane domain, and an extracellular domain, just as most cell adhesion molecules have. One of the best-investigated cell surface proteoglycans, ***syndecan,*** is diagrammed in Figure 11.17. Its extracellular domain comprises different glycosaminoglycans that interact with collagens, fibronectin, and possibly other ECM components

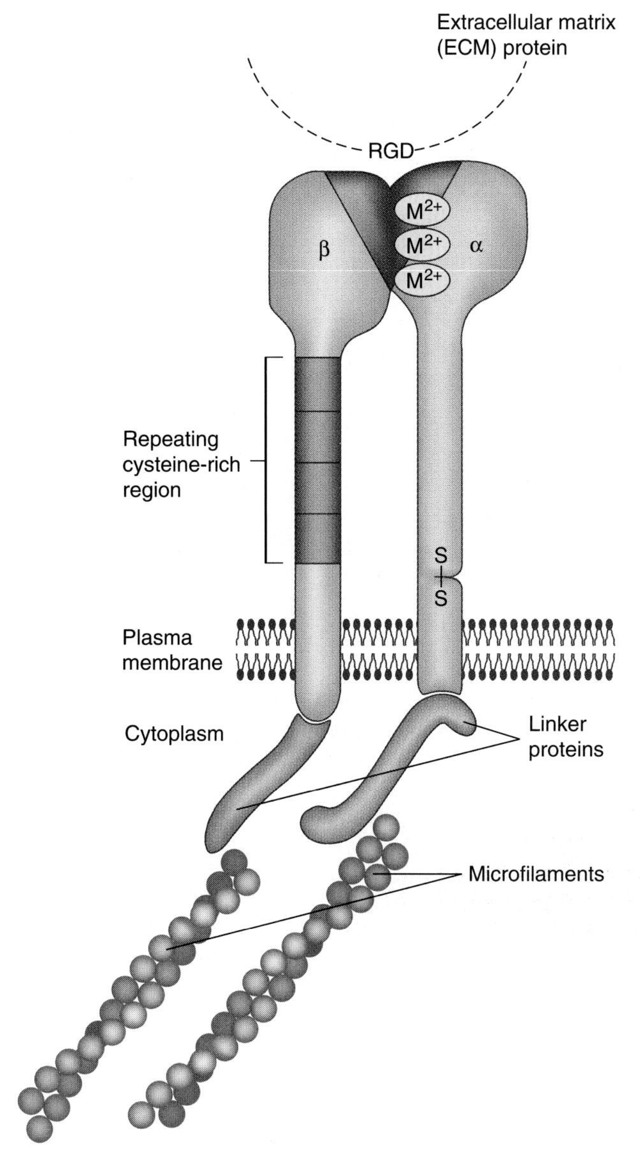

Figure 11.16 Structural features of integrins. The integrins are receptor molecules embedded in cell plasma membranes. Each integrin consists of two glycosylated polypeptides, the α-chain and the β-chain, which are held together noncovalently. The α-chain has been processed posttranslationally into two disulfide-linked fragments. The α-chain has several binding sites for divalent cations (labeled M^{2+}), which stabilize the connection to the β-chain. The β-chain has repetitive cysteine-rich sequences. Both chains cross the plasma membrane. Together, their extracellular domains form a binding domain for ligands such as the arginine-glycine-aspartate (RGD) sequence (see Fig. 11.15). The cytoplasmic domains of either the β-chain or both chains are connected to microfilaments by linker proteins, including talin and α-actinin.

(Bernfield and Sanderson, 1990). Close to the plasma membrane, the core protein has a protease-sensitive site that, when cleaved, releases the extracellular domain. The cytoplasmic domains of syndecan and several other cell surface proteoglycans may be connected with microfilaments, especially when their extracellular domains are cross-linked with other ECM components.

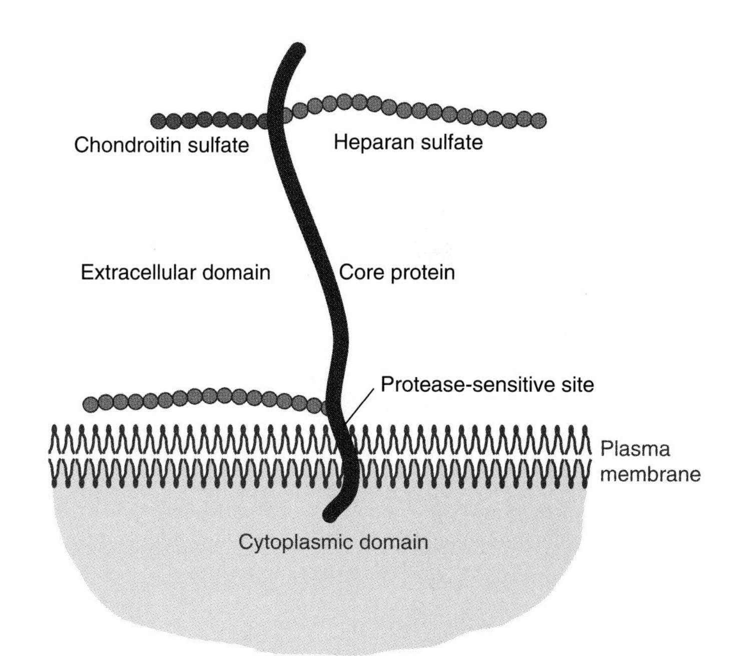

Figure 11.17 Schematic drawing of syndecan, a cell surface proteoglycan. The core protein has an extracellular domain, a transmembrane domain, and a cytoplasmic domain. The extracellular domain bears chondroitin sulfate and heparan sulfate chains, which bind to other ECM components. A protease-sensitive site is right outside the plasma membrane; cleavage at this site releases the entire extracellular domain. The cytoplasmic domain associates with microfilaments when the extracellular domain is cross-linked with other ECM molecules.

In addition to their mechanical function of linking the cytoskeleton to extracellular matrix molecules, integrins and cell surface proteoglycans also act as signal-transducing molecules (Hynes, 1994; Clark and Brugge, 1995; Gumbiner, 1996). Signal transfer from the ECM into cells is indicated by the dependence of the differentiation of certain cells on the functions of specific integrins. For example, osteocyte differentiation can be prevented by antibodies to the integrin β_1 subunit.

11.4 The Role of Cell and Substrate Adhesion Molecules in Morphogenesis

As we explore the role of CAMs in developing embryos, we must remember that morphogenesis is a complex phenomenon involving cell movements and cell shape changes, as well as alterations in the ECM composition. We will focus here on the involvement of the CAMs and SAMs described in the previous section. Are their expression patterns consistent with the view that they are the critical agents of cell adhesion? Do perturbations of CAMs and SAMs have the expected effects on morphogenetic movements? What are the roles of CAMs and

SAMs in cell development? How is their synthesis regulated, and do they have regulatory functions themselves?

CAM EXPRESSION IS CORRELATED WITH CELL FATES

The accumulation of CAMs is developmentally regulated, as shown by *in situ hybridization* or *immunostaining* (see Methods 4.1 and 8.1). CAMs are present in certain embryonic areas, and at specific stages, and their expression patterns are dynamic. Cells with different fates often express different CAMs. For instance, in a chicken embryo before gastrulation, the epiblast cells show an intricate pattern of N-CAM and E-cadherin expression (Edelman, 1984). The highest concentration of N-CAM is found in prospective neural plate, while lower concentrations are found in future notochord, somites, and lateral mesoderm (Fig. 11.18). Prospective urinary tract cells express both N-CAM and E-cadherin. All CAM-expressing areas are surrounded by endoderm and prospective epidermis, which express E-cadherin.

Many CAM expression patterns change during *induction* events, suggesting that the accumulation of specific CAMs may be an early step in the determination of embryonic tissues. In *Xenopus,* neural induction triggers N-cadherin and N-CAM synthesis in prospective neural epithelium (Kintner and Melton, 1987; Detrick et al., 1990). N-CAM is also synthesized when animal cap and vegetal base from *Xenopus* blastulae are cultured together, a situation that leads first to mesoderm induction and then to neural induction (see Figs. 9.9 and 12.17). In control experiments in which animal cap and vegetal base are cultured separately, no neural tissue forms and no N-CAM is synthesized. Thus, N-CAM synthesis is a reliable marker of neural induction.

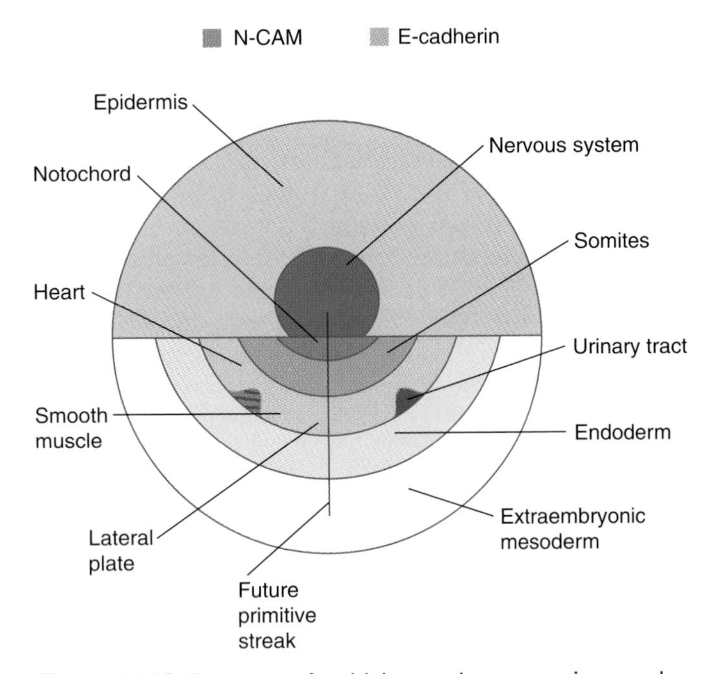

Figure 11.18 Fate map of a chicken embryo superimposed on an expression map of two CAMs: N-CAM (color) and E-cadherin (gray). The map shows a dorsal view of the epiblast before primitive streak formation. The vertical line indicates the position of the future primitive streak. The labels indicate the fates of different epiblast areas. Most areas express either E-cadherin or N-CAM. Prospective urinary tract expresses both CAMs, while prospective smooth muscle and extraembryonic mesoderm express neither CAM.

CELL ADHESIVENESS CHANGES DURING SEA URCHIN GASTRULATION

If cell adhesion plays a significant role in morphogenesis, then changes in cell adhesion as measured in vitro should correlate with the morphogenetic movements observed in vivo. To look for such correlations, researchers isolated embryonic cells from specific regions and at defined stages, measured their adhesiveness in culture, and tested whether regional or temporal changes in adhesiveness correlated with changing cellular behaviors in the intact embryo. Some of the most dramatic morphogenetic movements occur during gastrulation. Changes in cell adhesiveness during this period have been observed in several organisms, including fishes and sea urchins.

In fish and in other groups that undergo discoidal cleavage, the pregastrula embryo forms a disc-shaped *blastoderm* consisting of several cell layers on top of the uncleaved yolk. Initially, the blastoderm cells are nearly round and do not seem to adhere to one another very strongly (see Fig. 5.14). However, as cleavage continues, the blastoderm forms an outer epithelial layer, called the *enveloping layer,* and multiple layers of deep cells (see Fig. 10.21). While the deep cells are only loosely connected, the cells of the enveloping layer adhere very tightly as an epithelium. During gastrulation, both the enveloping layer and the deep layer spread over the uncleaved yolk reservoir, in a movement known as *epiboly.* During this movement, some cells of the enveloping layer become extremely flat, apparently in a passive response to stretching caused by radial intercalation of the deep cells. If this is correct, the stretching would indicate a very strong adhesion between the enveloping cells. Alternative, the flattening of the enveloping cells could result from increased adhesion to the underlying yolk syncytial layer.

To decide between these two hypotheses, Trinkaus (1963) examined whether isolated blastoderm cells flattened in culture. He dissociated blastoderm cells of the killifish *Fundulus heteroclitus* and observed their behavior in a culture medium under the microscope. Pregastrula cells remained spheroid for 6 h after culturing. In contrast, many gastrula cells flattened extensively within an hour or two on the glass bottom of the culture chamber. Similar flattening occurred when cells from pregastrula embryos were kept in prolonged culture until control intact embryos underwent epiboly. Thus, the flattening of isolated cells in vitro was correlated with the onset of epiboly in vivo. Trinkaus concluded that the gastrula cells were not flattened passively by stretching. Instead, he ascribed the flattening to an increase in the cells' adhesiveness to glass and to their natural substratum in the intact embryo. Furthermore, he proposed that the flattening of individual cells may actively contribute to the morphogenetic movement of epiboly.

Another study correlating morphogenetic movements with apparent changes in cell adhesiveness was carried out by Fink and McClay (1985) on the primary mesenchyme cells of sea urchin embryos. These cells undergo ingression at the beginning of gastrulation, when they leave the *vegetal plate* epithelium and move into the *blastocoel* (see Section 10.2). Just before the primary mesenchyme cells ingress into the blastocoel, their apical parts become very elongated as if the cells were still adhering to the *hyaline layer* surrounding the embryo before letting go (Fig. 11.19). If this ingression movement were based on changes in cell adhesion, the primary mesenchyme cells should lose adhesiveness to vegetal plate cells and to the hyaline layer while gaining adhesiveness to the *basement membrane* underneath the vegetal plate and to the extracellular material in the blastocoel. These predictions were tested in the following experiment.

TO assess the adhesiveness of primary mesenchyme cells to neighboring cells and to extracellular materials, Fink and McClay (1985) used a quantitative assay designed to measure the force necessary to separate radiolabeled test cells from various layers to which they adhered. The test cells were allowed to settle briefly in small dishes whose bottoms were covered with a layer of cultured cells or with a coat of isolated extracellular material. After the test cells had attached, the dishes were centrifuged with the test cells facing outward so that the centrifugal force dislodged them from the layer on which they had settled. By running the centrifuge at different speeds, the researchers determined the force necessary to dislodge half of the test cells.

The assay was applied to various types of cells isolated from sea urchin embryos. For example, micromeres were isolated by placing 16-cell embryos in calcium-free seawater, in which cells lose their adhesiveness. From the resulting mixture of blastomeres, the micromeres were isolated based on their size and buoyant density. Each type of test cell was allowed to settle on a layer of cultured gastrula cells, on a layer of isolated basement membrane, or on a layer of hyalin (a protein from the hyaline layer). The measurement results are compiled in Table 11.1.

Micromeres adhered to hyalin, but in their descendants, the migratory mesenchyme cells, adhesiveness decreased to less than 1% of the original force. In contrast, gastrula ectoderm and endoderm cells showed no such decrease during the same time interval. Micromeres also adhered to gastrula cell layers, but their adhesiveness was lost around

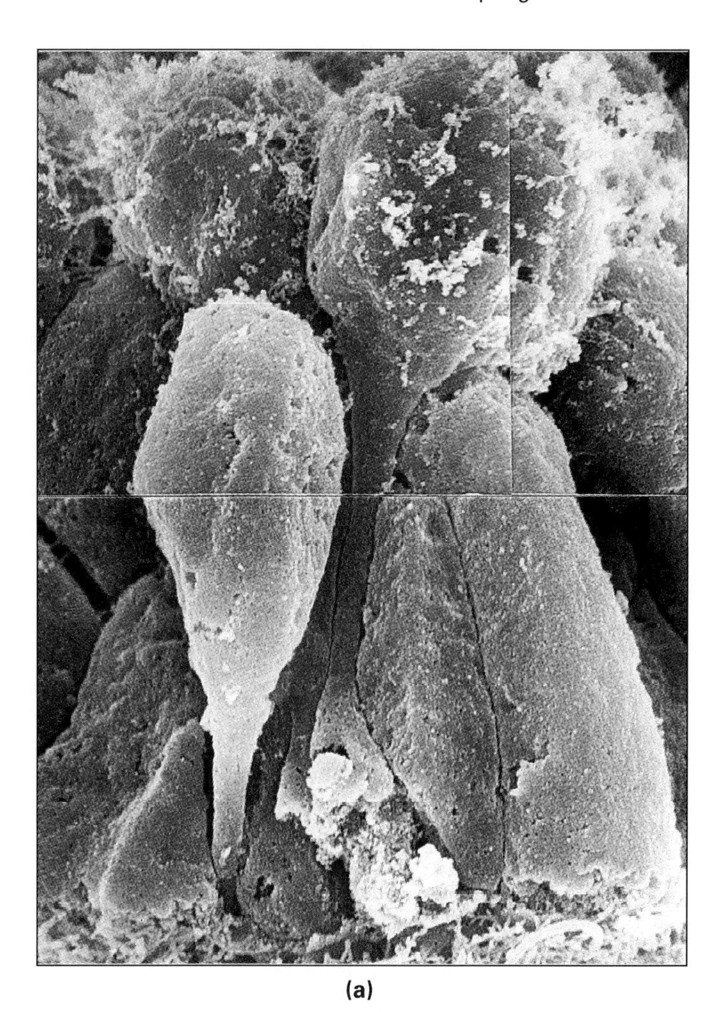

(a)

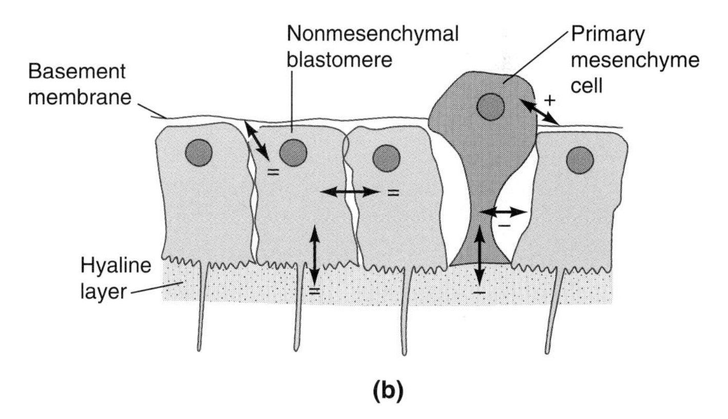

(b)

Figure 11.19 Ingression of primary mesenchyme cells during sea urchin gastrulation. **(a)** Scanning electron micrograph shows the vegetal plate viewed from inside the blastocoel. The basement membrane located on the basal cell surfaces appears as flocculent material at the top of the micrograph. An ingressing mesenchyme cell is still connected with a stalk to the hyaline layer on the outside of the embryo, shown at the bottom of the photograph. **(b)** Diagram showing a strip of vegetal plate cells in lateral view. The blastocoel, into which the primary mesenchyme cells move, is oriented to the top. The hyaline layer surrounding the embryo is at the bottom. The symbols reflect the changes in cell adhesiveness compiled in Table 11.1.

table 11.1 Binding Strength of Sea Urchin Micromeres, Primary Mesenchyme Cells, and Control Cells to Other Gastrula Cells, Basement Membrane, and Hyaline Layer

Test Cells	Force (μdyn) Required to Dislodge Test Cells From: Hyalin	Gastrula Cells	Basement Membrane
16-cell-stage micromeres	58.0	68.0	0.5
Migratory-stage mesenchyme cells	0.1	0.1	15.0
Gastrula ectoderm and endoderm cells	50.0	50.0	0.5

Binding strength is calculated from the centrifugal force that removes 50% of the test cells from the substrate.

From Fink and McClay (1985). Used with permission.

the time when the primary mesenchyme cells ingressed in vivo. Conversely, micromeres adhered very weakly to basement membrane, but this adhesiveness increased about 30-fold in migratory mesenchyme cells.

Questions

1. The force required to dislodge the test cells increased with the length of time that the test cells were allowed to settle on their substratum (cultured cells or ECM layer). The investigators decided to keep the settle-down time constant at 10 min. Why would this seem to be an appropriate interval, given the different types of connections formed between cells?
2. The particular adhesiveness of wandering mesenchyme cells to fibronectin suggests that the latter may be necessary for normal mesenchyme cell migration. How could this hypothesis be tested? How could researchers find out whether fibronectin is associated with the basement membrane on which newly ingressed primary mesenchyme cells migrate?
3. If cells that move relative to one another are generally less adhesive than stationary cells, which other cells in sea urchin embryos should then lose adhesiveness during gastrulation?

The experiments of Fink and McClay (1985) show that the ingression of primary mesenchyme cells in vivo is accompanied by dramatic changes in cell adhesiveness (Fig. 11.19b). As the cells ingress, they lose adhesiveness to other cells and to hyalin while gaining adhesiveness to basement membrane and presumably to other extracellular materials in the blastocoel. These data confirm the hypothesis that changes in cell adhesion contribute to the ingression process.

CAMS FACILITATE THE FORMATION OF CELL JUNCTIONS

The forces holding CAMs together are weak. Like the hooks and loops of Velcro, they are designed for quick attachment and detachment, and their strength is based on large numbers. Thus, the main function of CAMs seems to be to provide reversible adhesion that allows for cell migration and intercalation.

More stable connections between cells are provided by *gap junctions, anchoring junctions,* and *tight junctions* (see Section 2.7). These junctions are large, complex structures, designed not only to hold cells together but also to form channels of communication and tightly seal off the internal intracellular spaces of an organism from the external environment. Such stable junctions take time to establish, and before this process is complete, cells typically adhere to each other by means of CAMs. Two examples will illustrate this point.

A dramatic event in early mammalian development is *compaction,* when blastomeres become polarized and form tight junctions near their apical surfaces (see Section 5.2). Before compaction, mouse embryos express E-cadherin. Immunostaining for E-cadherin is especially strong where the blastomeres touch one another (Johnson et al., 1986; Vestweber et al., 1987). When raised in media with antibodies to E-cadherin, embryos fail to undergo compaction; no tight junctions form, and their blastomeres do not become polarized. Similar effects are noticeable after incubation with antibodies against GalTase (Bayna et al., 1988). It appears that both CAMs are required for the formation of tight junctions and that tight junctions in turn are necessary for blastomeres to become polarized, so that they acquire an apical surface with certain membrane proteins facing the outside and different surfaces with other membrane proteins facing the inside.

Polarized cells connected by tight junctions are a hallmark of epithelia. Parallel to the tight junctions, there is often an *adherent junction,* characterized by rings of microfilaments that circle beneath the plasma membrane of each cell like a purse string (see Fig. 2.3). Across the cell membrane from the microfilament rings, the intercellular space stains positive for E-cadherin (Fig. 11.20). Thus, adherent junctions appear to be a special form of general cell adhesion based on E-cadherin. Indeed, adherent junctions form rapidly, and dependent on E-cadherin, when isolated epithelial cells are allowed to make contact again (Gumbiner, 1990).

Adherent junctions change during sea urchin gastrulation (Miller and McClay, 1997a, b). During ingression

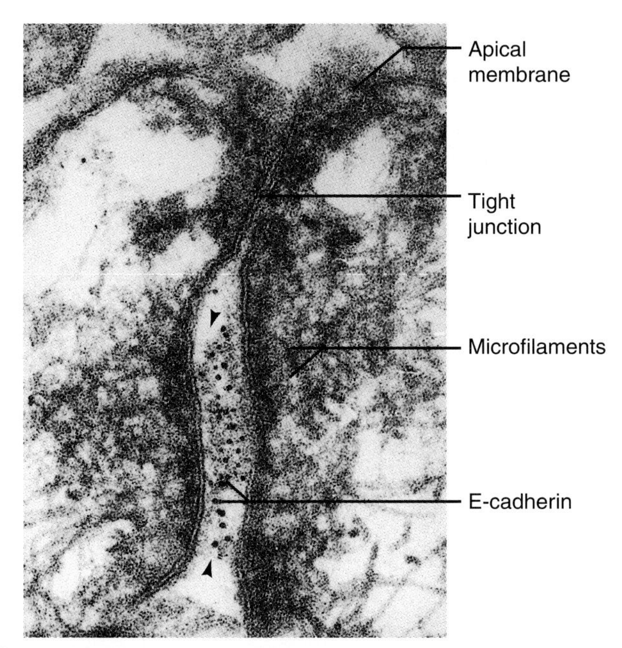

Figure 11.20 Presence of E-cadherin in the zonula adherens of mouse intestinal epithelium. In this transmission electron micrograph, the apical membrane is at the top. Two abutting epithelial cells are linked by a tight junction. Beneath the tight junction is the zonula adherens, characterized by rings of microfilaments circling the inside of each cell like a purse string (refer to Fig. 2.23). The extracellular space across the cell membranes from the microfilament rings is filled with E-cadherin (between arrowheads). This is shown here by immunostaining using a primary antibody against E-cadherin and a secondary antibody that is conjugated with fine gold particles. These particles absorb electrons and thus generate black dots in electron micrographs.

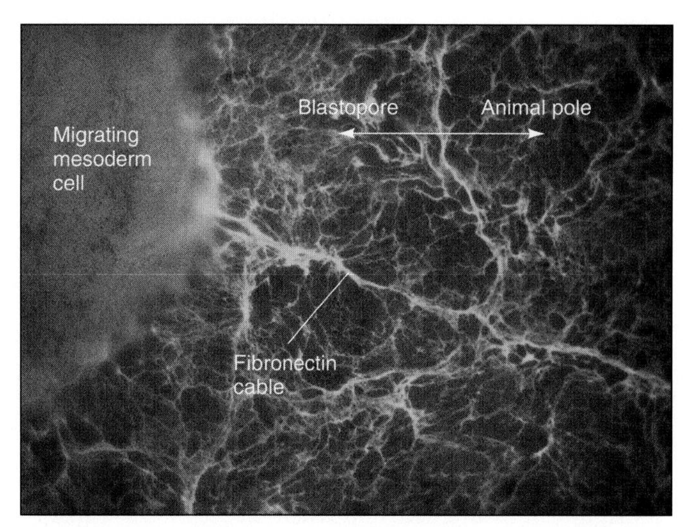

Figure 11.21 Migrating mesoderm cell (large body at upper left) inside an early gastrula of a salamander (*Ambystoma maculatum*). The fluorescent light micrograph shows the basal (inner) surface of cells forming the blastocoel roof. The fibrous ECM cables produced by these cells are immunostained (see Method 4.1) for fibronectin. Statistically, the preferred orientation of the fibers is along the blastopore–-animal pole axis (double-headed arrow).

of primary mesenchyme cells, immunostainability for cadherin is rapidly lost from adherent junctions while increasing in intracellular organelles. These observations suggest that primary mesenchyme ingression involves controlled endocytosis of cadherin molecule. The ingression of both primary and secondary mesenchyme cells also coincides with a dramatic loss of β-catenin from adherens junctions. Similarly, archenteron cells display a significant decrease in the level of adherent junction-associated β-catenin during *convergent extension.* These data are consistent with a role of β-catenin in regulating cell adhesion during multiple morphogenetic movements during sea urchin gastrulation.

FIBROUS ECM COMPONENTS PROVIDE CONTACT GUIDANCE TO CELLS

Many morphogenetic processes in embryos involve oriented cell migration. During amphibian gastrulation, for example, mesodermal cells of the deep zone and the involuting marginal zone migrate on the basal surface of the ectoderm (blastocoel roof) toward the animal pole (see Fig. 10.11). ECM fibrils found on this surface are scarce at blastula stage, but become more numerous in early gastrulae and continue to accumulate during gastrulation (Fig. 11.21). These fibrils have a tendency to become oriented along meridional lines between the dorsal blastopore and the animal pole (K. E. Johnson et al., 1990). Thus, the preferred orientation of the fibrils is along the route taken by mesodermal cells during gastrulation. A possible cause of the fibril alignment is the attachment of the fibrils to integrins or other fibronectin receptors in cell plasma membranes. As the dorsal ectodermal cells undergo convergent extension, the integrins are rearranged correspondingly, so that any attached fibrils should become oriented along the blastopore–animal pole axis.

The filopodia of the mesoderm cells that migrate on the inside of the blastocoel roof are attached to the extracellular fibrils between the dorsal blastopore and the animal pole (Nakatsuji et al., 1982). This behavior suggests that the fibrils may orient the cell migration. The phenomenon of cells being guided by fibril orientation or similar mechanical clues, which has been reported by several authors, is known as ***contact guidance*** (Weiss, 1958).

TO test the role of contact guidance in amphibian gastrulation, Nakatsuji and Johnson (1983) developed culture conditions under which gastrula cells could be observed under the microscope but still behaved as they would in whole embryos. The researchers transferred ECM from the inside of a blastocoel roof to a glass cover slip by isolating an ectoderm piece and placing it basal surface down on the glass surface. After 5 h of culture, they marked the margins of the ectoderm piece and the blastopore–animal pole axis on the cover slip before they flushed off the tissue. As a result of this procedure, the glass surface was now coated, or *conditioned*, with ECM residue. When mesodermal cells from gastrulae were placed on the conditioned area, the cells quickly attached and formed lamellipodia. The researchers recorded the subsequent cell movements by videomicroscopy and then fixed the cover slips for examination with a scanning electron microscope. They compared the recorded cell movements with the orientations of ECM fibrils seen under the microscope. Indeed, the cell migrations tended to follow the local fibril orientation: In areas where the cell trails were unoriented, the fibrils were not significantly aligned; and where the cell trails were oriented, so were the fibrils.

In an extension of this experiment, De-Li Shi and colleagues (1989) conditioned the surfaces of plastic culture dishes with ECM deposits from large rectangular explants of blastocoel roofs from early newt gastrulae (*Pleurodeles waltl*). The perimeter of each explant used for conditioning, and its animal pole–blastopore axis, were marked on the dish (Fig. 11.22). A smaller explant comprising marginal zone and adjacent ectoderm was then placed in the middle of the conditioned area with its animal pole–blastopore axis perpendicular to that of the ECM deposit. During the following culture period, mesoderm cells from the second explant migrated on the ECM deposited by the first explant. In 53 out of 78 such trials, the new mesoderm cells migrated toward the animal pole of the ECM deposit. In 2 cases, they migrated toward the blastopore pole of the ECM, and in 15 cases the cells moved in both directions. No migration was observed in the remaining 8 cases.

Questions

1. Could the orientation of the migrating mesoderm cells observed in the latter experiment have been an artifact introduced by the way in which the ECM coat of the blastocoel roof was transferred to the glass cover slips?
2. What controls can you think of to settle the artifact problem raised in question 1?

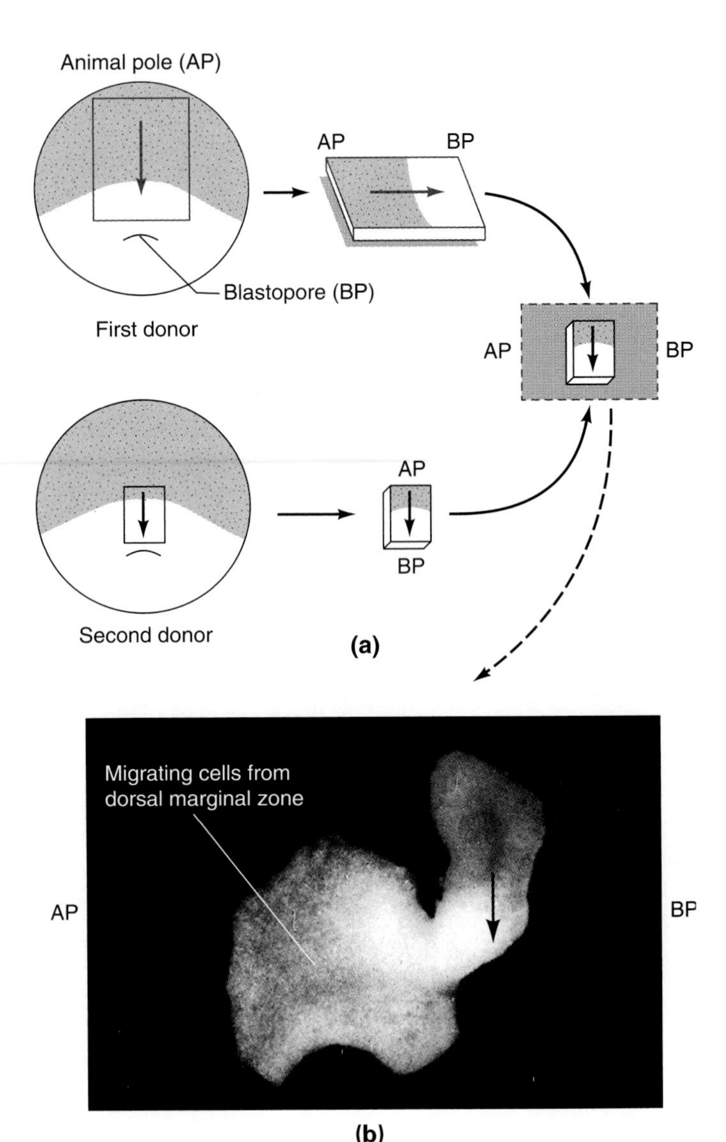

Figure 11.22 ECM components deposited by the blastocoel roof orient migrating mesoderm cells in the newt *Pleurodeles waltl.* **(a)** The bottom of a plastic culture dish was conditioned with a large explanted section of blastocoel roof left basal surface down in the dish for 2 h. The perimeter of the explant and its animal pole (AP)–blastopore (BP) axis (colored arrow) were marked on the dish. When the explant was flushed off, it left behind a deposit of ECM material. A smaller explant of dorsal marginal zone was placed in the center of the conditioned area with its animal pole-blastopore axis (black arrow) perpendicular to that of the ECM deposit. **(b)** Photograph of the area represented by the colored rectangle in the diagram. During the 24-h period in which the explant was cultured, mesodermal cells from the dorsal marginal zone migrated out. In most cases, these cells moved toward the animal pole of the ECM deposit.

The experiments just described demonstrate that the direction in which mesoderm cells from *Pleurodeles* gastrulae migrate is biased by contact with ectodermal ECM components. The orienting clue seems to come from a preferred orientation of fibronectin (FN) fibrils along the blastopore–animal pole axis. How the mesodermal cells migrating along these fibrils distinguish the opposing directions toward the animal pole and toward the blastopore is not understood. Perhaps the FN molecules are all oriented with their C-termini pointing in one direction, or they are assembled into fibronectin fibrils in a way that imparts directionality. In vivo, contact guidance by ECM fibrils also acts in concert with additional signals. The same signals that generate the endoderm-mesoderm-ectoderm pattern may also provide cues to migrating cells. In addition, amphibian mesoderm cells and many other cells demonstrate ***contact inhibition:*** When two cells collide, they both stop and then move in different directions. Statistically, contact inhibition should cause mesoderm cells to move away from the crowded blastopore region. Thus, contact inhibition by other cells and contact guidance by ECM fibrils together may direct the cells toward the animal pole.

NEWT GASTRULATION REQUIRES FIBRONECTIN ON THE INNER SURFACE OF THE BLASTOCOEL ROOF

To test whether contact guidance of migrating cells by ECM materials is necessary for amphibian gastrulation in vivo, Boucaut and coworkers (1984) carried out grafting experiments with gastrulae of the newt *Pleurodeles waltl.* The investigators cut out patches of dorsal ectoderm from early gastrulae and reinserted them inside out (Fig. 11.23). Consequently, during gastrulation, migrating mesoderm cells faced the outer ectodermal surface instead of the inner ectodermal surface, which is their normal substratum. In late gastrulae, mesodermal cells were found migrating on the normally oriented ectoderm but not on the inverted patches. This result showed that proper mesodermal cell adhesion and migration require a component that is present on the inner ectodermal surface but not on the outer one. Most likely, the critical components were ECM fibers.

The ECM fibers and integrins on the inner ectodermal surface include FN and FN receptors, as revealed by *immunostaining* (see Method 4.1). To test specifically for the involvement of FN in mesodermal cell migration, Boucaut et al. (1984) applied probes that disrupt cell-FN interaction. First, they injected *univalent antibodies* against FN into the blastocoels of early newt gastrulae (Fig. 11.24). This treatment completely upset the gastrulation process. Few if any mesodermal cells moved onto the inner ectodermal surface, and the ectoderm was thrown into deep folds. Presumably, the failure of the mesoderm cells to migrate on the blastocoel roof blocked the normal involution of the involuting marginal zone, which therefore made no room for the epiboly of the noninvoluting marginal zone and the animal cap. Control gastrulae injected with immunoglobulins from preimmune serum developed normally. Thus, gastrulation was not inhibited by some unspecific immunoglobulin activity but by the specific binding of antibodies to FN.

Anti-FN antibody injected into the blastocoel might inhibit gastrulation by blocking the interaction of the FN *RGD sequence* with its receptors on the migrating mesoderm cells. To test this hypothesis, Boucaut and coworkers (1985) injected synthetic peptides containing the RGD sequence into the blastocoels of early gastrulae. The effects were very similar to those obtained with anti-FN antibodies. Control embryos injected with a different peptide representing the collagen-binding domain of FN developed normally. The researchers concluded that the injected RGD sequences blocked the FN receptors of the migrating mesoderm cells, thus preventing them from interacting with FN. Confirming this conclusion, the researchers found that univalent antibodies to the *FN receptor* inhibited gastrulation in the same way (Darribè et al., 1988). Together, the observations described in this section indicate that the adhesion and/or

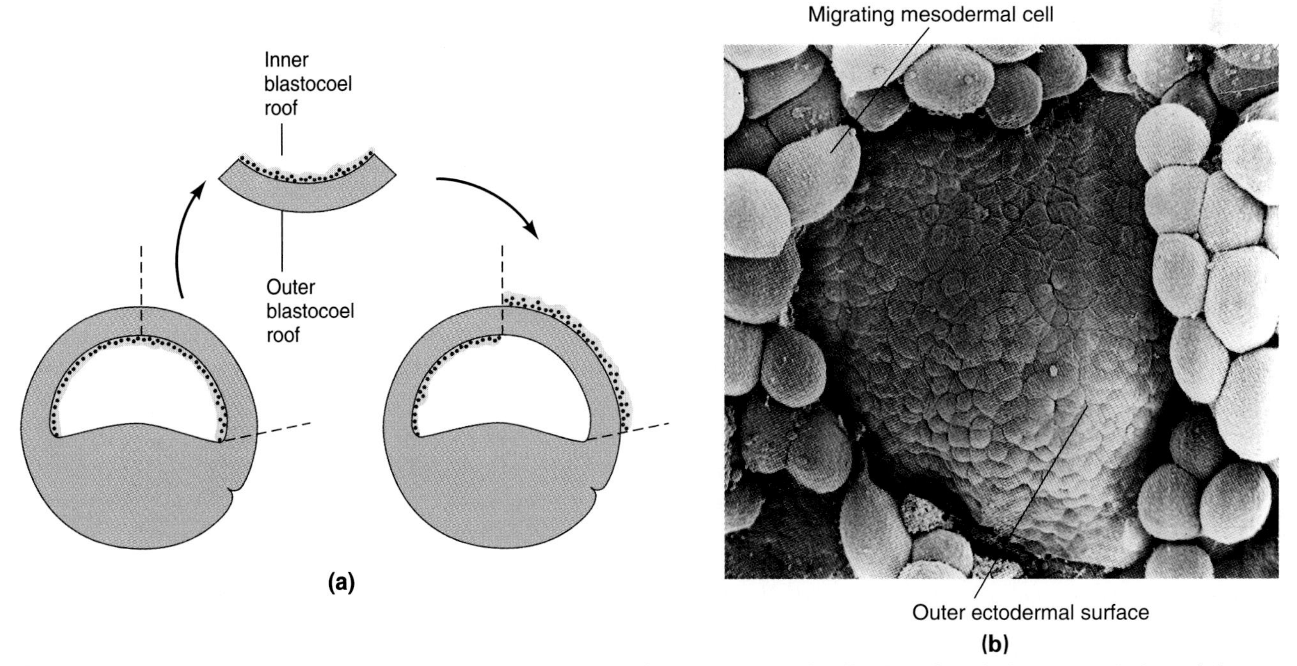

Figure 11.23 Role of the inner surface of the blastocoel roof in mesodermal cell migration during gastrulation of the newt *Pleurodeles waltl.* **(a)** Grafting procedure. A segment of dorsal blastocoel roof was excised and reinserted with the inner surface turned outside. **(b)** Scanning electron micrograph showing the inside of a gastrula 3 h after grafting. The migrating mesodermal cells adhered to the inside of the blastocoel roof except for the graft. Apparently, the roof's outer surface did not provide suitable attachment sites.

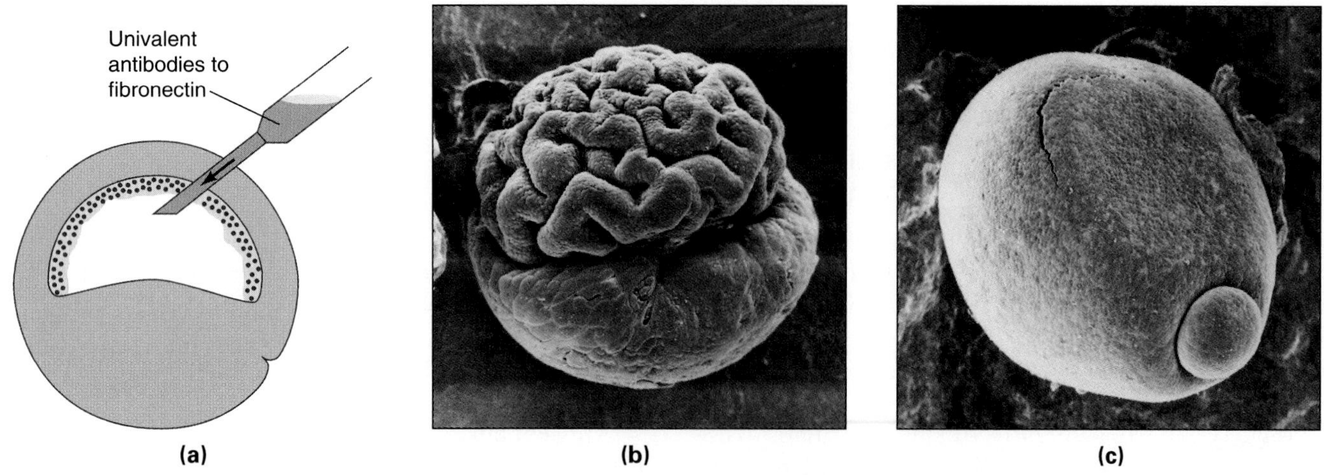

Figure 11.24 Dependence of gastrulation in the newt *Pleurodeles waltl* on cell-fibronectin interaction. **(a)** The blastocoel was injected at the early gastrula stage with univalent antibodies to fibronectin (see Method 11.1). **(b)** The normal gastrulation movements did not occur, the blastocoel roof was thrown into convoluted folds, and the entire vegetal hemisphere remained at the surface of the embryo. Similar effects were observed after injection of peptides containing the RGD sequence and after injection of univalent antibodies to fibronectin receptor. **(c)** Control embryo (blastopore to the lower right), injected with univalent immunoglobulins from preimmune serum. Such embryos developed normally.

migration of mesodermal cells during *Pleurodeles* gastrulation requires interactions between the mesodermal cells and FN fibrils on the inside of the blastoderm roof. The fibrils may provide adhesion sites and/or contact guidance to migrating mesoderm cells.

Similar experiments were carried out with *Xenopus* gastrulae (Winklbauer and Nagel, 1991; Winklbauer and Keller, 1996). Involuted head mesoderm and preinvolution mesoderm cells migrated on the inside of isolated blastocoel roof and on plastic surfaces coated with ECM from the blastocoel roof. Addition of RGD-containing peptide or univalent anti-FN antibodies reduced but did not abolish the adhesion of mesoderm cells to their substrate. However, these agents inhibited cell migration. In particular, the interaction with FN is necessary for the mesoderm cells to form lamelliform protrusions. Nevertheless, gastrulation of whole embryos was only slightly disturbed by the presence of the FN inhibitors. The latter result is in accord with observations showing that mesoderm involution and convergent extension proceed nearly normally even if the blastocoel roof has been removed (Keller and Jansa, 1992; see also Fig. 10.16). Thus, it appears that the extent to which gastrulation relies on convergent extension or cell migration differs between amphibian species. Again illustrating the *principle of overlapping mechanisms,* the epithelial and cellular movements that drive gastrulation complement and can partially replace one another.

Xenopus gastrulation also depends on C-cadherin (Lee and Gumbiner, 1995). If mRNA encoding a truncated C-cadherin is injected into the prospective dorsal involuting marginal zone (DIMZ), gastrulation remains incomplete, and the blastopore does not close. Similar injections elsewhere in the embryo have no effect, indicating that the DIMZ is particularly sensitive to this type of disturbance. The interference with involution must be caused by the displacement of functional cadherin by the truncated cadherin—rather than by an unspecific toxicity of the injected material—because normal development can be restored by additional injection of mRNA encoding complete C-cadherin. The critical need for C-cadherin may be for the cellular rearrangements involved in involution and/or convergent extension. Alternatively or in addition, tight adherence may be required to stiffen the dorsal involuting marginal zone for the process of convergent extension.

11.5 Morphoregulatory Roles of Cell and Substrate Adhesion Molecules

Over the last two chapters, we have sought to understand morphogenetic movements in terms of cellular behaviors, cellular behaviors in terms of adhesion molecules, and these molecules as the results of gene activities. This *reductionist analysis* has been widely successful; it holds the promise for us to eventually understand in molecular terms why, for example, the shapes of ears and noses are heritable. However, the view of gene activities as first causes and morphogenetic movements as their ultimate effects has recently been complemented by the realization that morphogenesis in turn affects gene activity.

Cell adhesion molecules (CAMs) and substrate adhesion molecules (SAMs) have been termed ***morphoregulatory molecules*** (Edelman, 1988) to emphasize the central role in a *cyclic process* of gene regulation and morphogenesis (Fig. 11.25). Other key players in this process are known as ***selector genes,*** master genes that control various target genes including those encoding

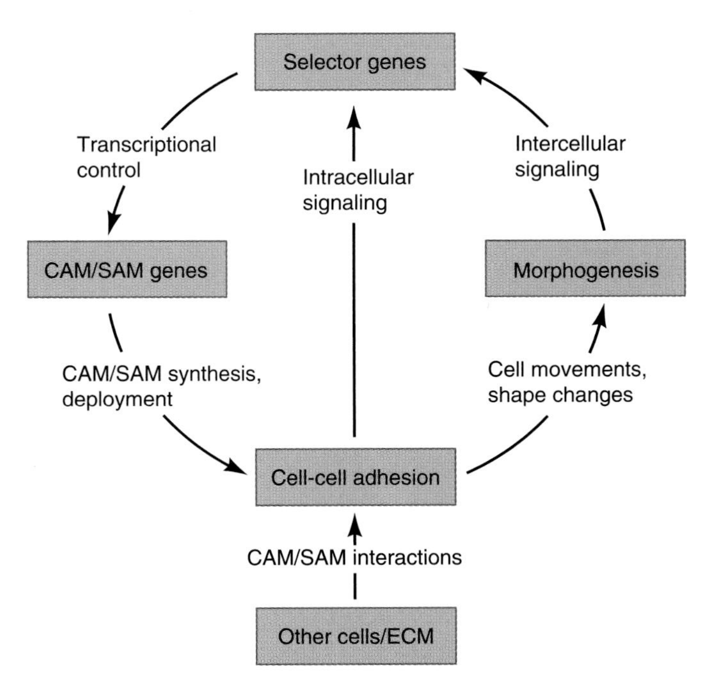

Figure 11.25 Model showing the possible role of CAMs and SAMs in morphoregulatory cycles. Selector genes control the expression of CAM/SAM genes. The synthesis and deployment of CAMs/SAMs modulates cell-cell adhesion, which is also affected by the CAMs of other cells and ECM materials. Altered cell-cell or cell-substrate adhesion causes changes in cell shape and cell movements, two major cell attributes that affect morphogenesis. Morphogenetic movements in turn generate new cell contacts, which set the stage for the exchange of new intercellular signals. In addition, altered cell shapes and contacts affect intracellular signaling through multiple mechanisms. Both intercellular and intracellular signals feed back into the expression of selector genes, thus completing the morphoregulatory cycle.

CAMs and SAMs. (Other target genes of selector genes are thought to regulate cell division, cell differentiation, cell death, and other morphogenetically important processes.) The synthesis of CAMs and SAMs mediates cell-cell and cell-substrate adhesion, which in turn control morphogenetic movements. Completing the morphoregulatory cycle, morphogenetic movements feed back on selector gene activity. So far in this chapter, we have covered the part of the cycle that extends from CAM and SAM genes to morphogenesis. In the remainder of this chapter, we will review evidence in support of the rest of the cycle.

CAM AND SAM GENES ARE CONTROLLED BY SELECTOR GENES

A well-known selector gene in *Drosophila* is *Ultrabithorax*$^{+}$ (*Ubx*$^{+}$). Genetic analysis has shown that *Ubx*$^{+}$ promotes the morphological characteristics of the third thoracic segment (T3) and that *Ubx*$^{+}$ is normally active in T3 but not in the second thoracic segment (T2). In adult flies with complete loss of *Ubx* function, T3 looks like another copy of T2, showing in particular a replacement of the *halteres*, the balancer organs that are characteristic of the wild-type T3, with a second pair of wings (Fig. 1.21).

Wings, halteres, and the thoracic structures to which they are hinged, are derived from *imaginal discs*, epithelial bags of cells that grow during larval development and evert during metamorphosis (see Figs. 6.17 and 6.18). These imaginal discs can be removed, manipulated experimentally, and reimplanted into larvae. The implants then undergo metamorphosis along with their hosts, giving rise to adult structures tucked into the host's body cavity.

TO test the effect of *Ubx*$^{+}$ on cell adhesion, Garcia-Bellido and Lewis (1976) removed imaginal discs from larvae carrying either the wild-type allele (+) of *Ubx* or one of two loss-of-function alleles known as *bithorax* (*bx*) and *postbithorax* (*pbx*). In *bx* homozygotes, the anterior *compartment* of T3 is replaced with the anterior compartment of T2, so that the anterior half of the haltere is replaced with the anterior half of a wing. In *pbx* homozygotes, the posterior compartment of T3 is replaced with the posterior compartment of T2 (see insets in Fig. 11.26).

From fully grown wild-type larvae, the investigators removed the dorsal T2 discs, which give rise to wings. These discs were dissociated into single cells and mixed with cells obtained similarly from dorsal T3 discs, which give rise to halteres or combined wing and haltere compartments depending on *Ubx* allelism. The mixed cells were reaggregated by spinning them down in a centrifuge tube, and pieces of the mixed reaggregate were cultured for a few days in the abdomen of adult flies. The grown reaggregates were then removed from the adult hosts, cut into smaller pieces, and implanted into larvae. When the larval hosts metamorphosed, the implants formed vesicles of cuticular structures within their hosts. These vesicles were inspected for the type of adult structures they represented and with regard to their origin from T2 discs and/or T3 discs. To distinguish the derivatives of the two discs, the investigators obtained them from fly strains carrying different sets of genetic marker alleles that confer distinct cuticular colors, bristle shapes, and wing hair arrangements.

With regard to origin, the adult structures derived from the reaggregated disc cells occurred in two forms, monotypic vesicles and integrated mosaics. The *monotypic vesicles* were composed of cells showing only one set of cuticular marker alleles indicating their origin from either T2 or T3 discs, but not from both. The *integrated mosaics* were contiguous wing areas composed of cells showing one set of cuticular markers intermingled with cells showing the other set of cuticular markers; the integrated mosaics therefore were structures to which both T2 and T3 discs had jointly contributed.

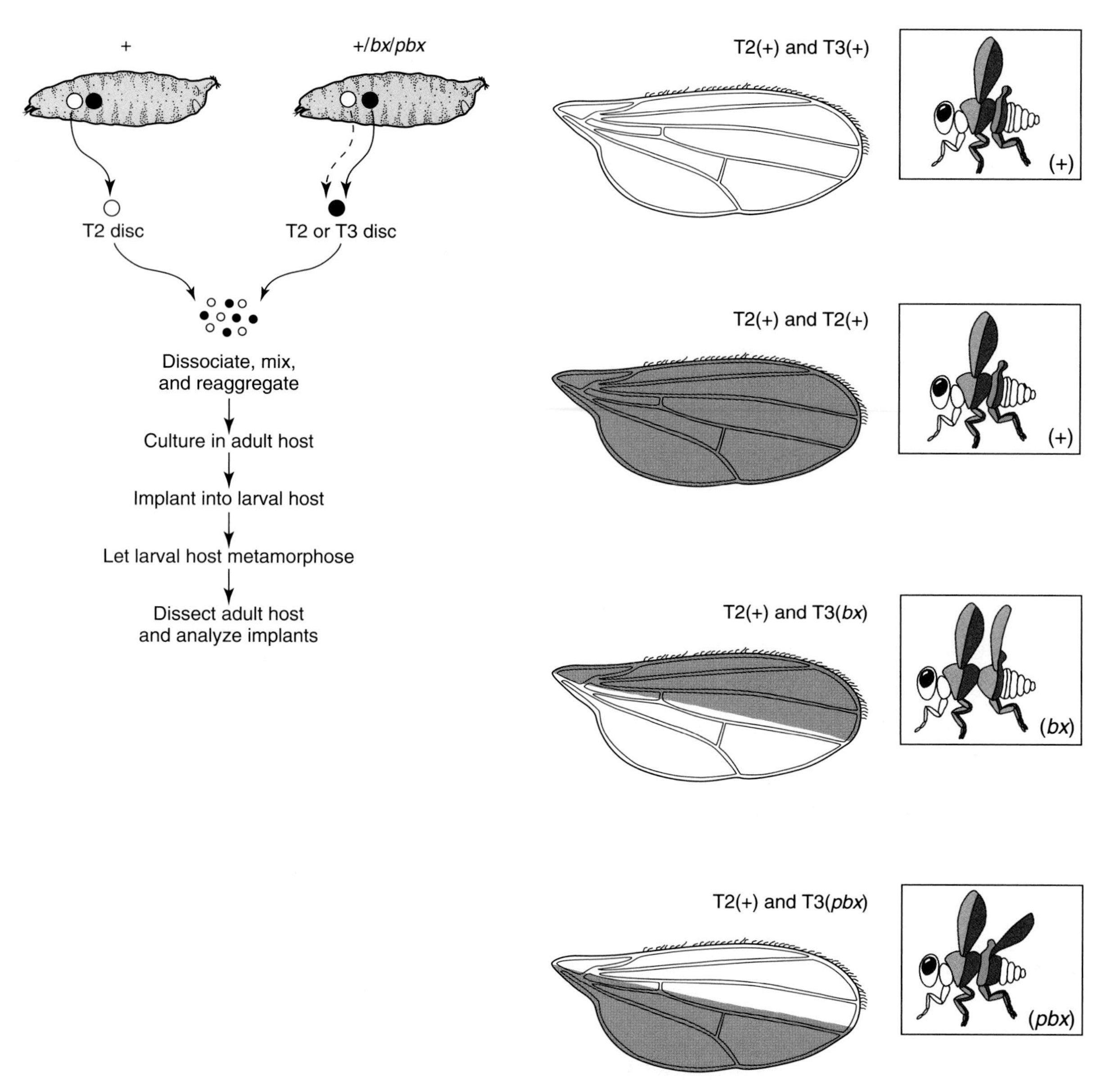

Figure 11.26 Control of cell affinity by the selector gene *Untrabithorax* (*Ubx*). Cells from two dorsal imaginal discs of *Drosophila* larvae were dissociated into single cells, mixed, and reaggregated. *Ubx* activity (pink/red color in inset fly cartoons) was restricted through the use of various *Ubx* alleles: wild-type (+; *Ubx* active in entire T3 but not in T2), *bithorax* (*bx; Ubx* active only in the posterior compartment of T3), and *postbithorax* (*pbx; Ubx* active only in the anterior compartment of T3). The two contributing discs also carried different sets of marker alleles that would allow the investigators to trace any cuticular structures formed later in the experiments to one of the two discs. The mixed cell aggregates were cultured in adult hosts and then forced through metamorphosis by implanting them into metamorphosing larvae. The adult implants showed integrated mosaics, composed of cells from both contributing discs, only from those wing areas (blue color) in which *Ubx* was inactive in both contributing discs. Cells in which *Ubx* was active did not form integrated mosaics with cells in which *Ubx* was inactive.

The wing structures that occurred as integrated mosaics depended on the *Ubx* allele present in the T3 discs (Fig. 11.26). With few exceptions, cells from wild-type T2 discs and wild-type T3 discs did not form integrated mosaics. The result from these experiments contrasted with the results from controls, in which cells from T2 discs mixed with cells from other T2 discs had produced integrated mosaics representing every region of the wing. Thus, cells derived from different wild-type T2 discs formed integrated mosaics, whereas cells derived from wild-type T2 and T3 discs did not. However, cells derived from T2 discs did form integrated mosaics with cells from T3 discs that were obtained from larvae with mutant *Ubx* alleles. In particular, such mosaics representing *anterior* wing structures were formed by mixed aggregates from wild-type T2 and *bx* T3 cells. Similarly, mosaics representing *posterior* wing structures were formed by mixed aggregates from wild-type T2 and *pbx* T3 cells.

Questions

1. The investigators injected the reaggregated cells into adult flies before transferring them into larvae. Could they have simplified their experiments by leaving the reaggregates in adults for a longer time or by injecting the reaggregates straight into larvae?
2. In the mutant *Contrabithorax,* the *Ubx* gene is active in the anterior T2, and anterior wings are replaced with anterior halteres. Do you expect cells from dorsal T2 discs of this mutant to form integrated mosaics with dorsal T3 discs from wild-type flies? If so, which body parts (wings, haltere, etc.) should be represented in the integrated mosaics?

The experiments of Garcia-Bellido and Lewis demonstrate that *Ubx* alleles have dramatic effects on adhesion between cells derived from dorsal T2 and T3 imaginal disc cells. Mixed reaggregates of cells from these two discs will form integrated mosaics representing those wing regions in which Ubx^{+} is either *off in both* discs or *on in both* discs. However, if Ubx^{+} is *on in one group of disc cells and off in the other,* then the cells segregate and form separate adult structures. The simplest interpretation of these observations is that the target genes controlled by Ubx^{+} include genes for CAMs or SAMs.

CELL-CELL ADHESION AND CELL-SUBSTRATUM INTERACTIONS AFFECT GENE ACTIVITY

The most novel aspect in the cyclic, or synthetic, investigation of morphogenesis is the realization that morphogenetic movements can feed back on gene expression in at least two ways (see Fig. 11.25). First, morphogenetic movements may turn cells that used to be far apart in the embryo into next-door neighbors, thus providing closer proximity for the new neighbors to interact. For instance, after the sea urchin archenteron has undergone convergent extension, the cells at its tip have reached a position from where they can send inductive signals to the oral ectoderm that will cause mouth formation.

In addition, cell and substrate adhesion molecules are themselves involved in cellular signaling. Just as our hands allow us to hold on to external objects while at the same time sensing their temperature, smoothness, and so on, so do cell and substrate adhesion molecules have mechanical as well as signal transduction functions. The following examples illustrate how the gene activities of cells, as indicated by their morphological appearance and/or proteins synthesized, change in response to cell-cell or cell-ECM interactions.

A striking example of cells changing their morphological appearance in response to cell adhesion was reported for PC 12 cells, rat tumor cells that resemble *chromaffin* cells (glandular cells normally found in the adrenal medulla). With nerve growth factor and other stimuli, these cells can be converted to a neuronal phenotype with extended *neurites* (Doherty et al., 1991). PC 12 cells express both N-CAM and N-cadherin. When these cells are cultured on a layer of human 3T3 cells, which do not normally express N-CAM or N-cadherin, the PC 12 cells maintain their chromaffin cell appearance and do not convert to the neuronal phenotype (Fig. 11.27). However, if cultured on 3T3 cells are genetically transformed with N-CAM or N-cadherin transgenes, adherent PC 12 cells do convert to the neuronal phenotype. The conversion can be blocked by antibodies directed against N-CAM (or N-cadherin) and by drugs known to interfere with Ca^{2+} channels. These observations demonstrate that the PC 12 cells adhere homotypically to the transformed 3T3 cells, using whatever CAM the 3T3 cells are deploying. This interaction seems to trigger a Ca^{2+}-dependent change in intracellular signaling that promotes the conversion to the neuronal phenotype.

A case of cells changing their protein synthesis in response to their ECM environment was reported for the *cornea* of the eye in chicken embryos. The cornea is a multilayered structure consisting of an outer epithelium, an inner endothelium, and an intervening tissue called *stroma.* The stroma contains fibroblasts in a collagen-rich ECM. A series of experiments by Hay (1981) and her coworkers has shown that cornea formation is induced by the underlying lens and depends on signals emanating from the *lens capsule,* a thick basement membrane surrounding the lens. Isolated corneal epithelium placed on a lens capsule in vitro produces primary stroma, while corneal epithelium placed on plastic does not. When the lens capsule is replaced by collagen-containing substrata, the corneal epithelium still forms stroma, but noncollagenous substrata do not induce stroma formation (Fig. 11.28). Additional experiments suggested the cells' response may be mediated by integrins.

Similar observations have been made with other organisms and cells. In cultured vertebrate cells, both cell shape and the types of protein synthesized may depend on whether the cells are suspended in gels or kept on flat surfaces. For example, rabbit chondrocytes (cartilage-forming cells) are round and synthesize large amounts of type II collagen and chondroitin sulfate in vivo. They retain the same shape and synthesize these same proteins in large quantities when suspended in an agarose gel (Benya and Shaffer, 1982). However, when kept on flat surfaces in monolayer cultures, the cells are spindle-shaped like fibroblasts and synthesize mostly type I collagen and low levels of proteoglycans. Thus, these cells can be driven back and forth between two states of cell shape and gene expression depending on the substrate used for cell culture.

The observations on chicken cornea and rabbit chondrocytes lend themselves to three different interpretations, each of which may be valid under certain

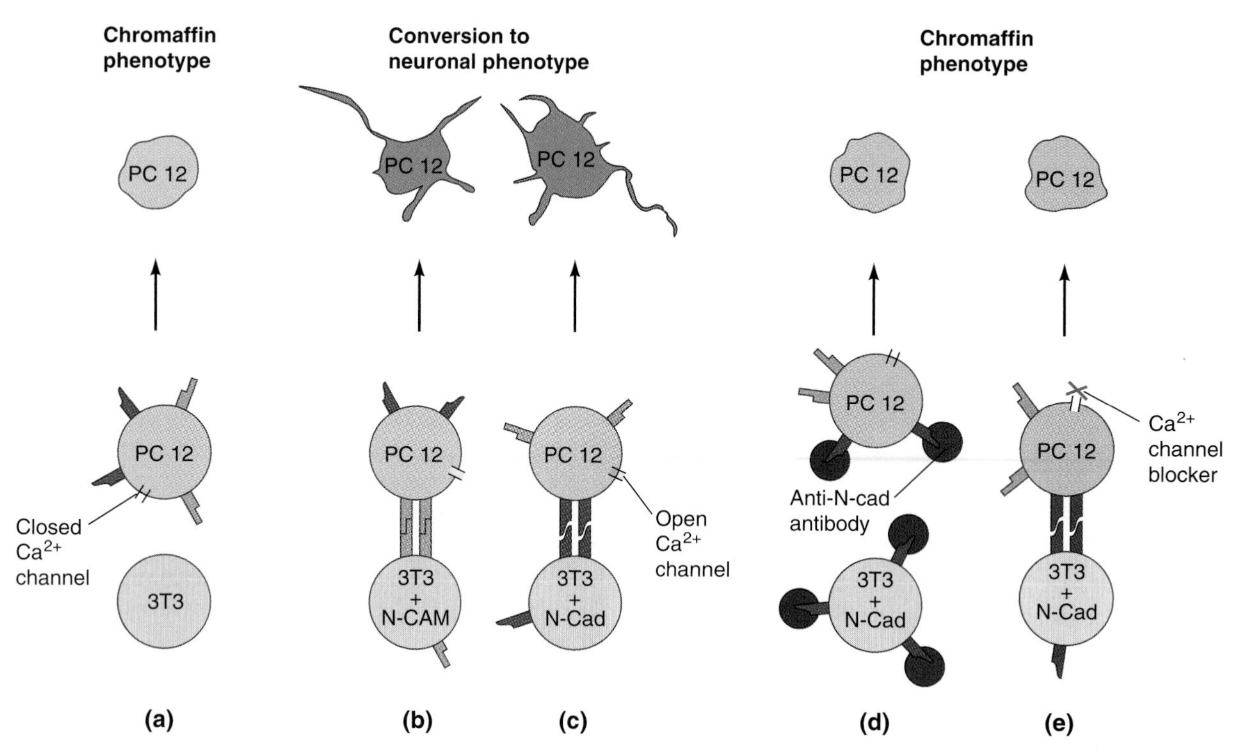

Figure 11.27 Phenotypical conversion of cells triggered by homotypic interactions of CAMs. A rat tumor cell line, PC 12, is grown on a layer of human cells, 3T3. The PC 12 cells synthesize both N-CAM and N-cadherin, whereas the 3T3 cells by themselves do not make either of these CAMs. **(a)** PC 12 cells grown on nontransformed 3T3 cells resemble chromaffin cells, which are normally found in the adrenal medulla. **(b, c)** If PC 12 cells are grown on 3T3 cells transfected with cDNA encoding N-CAM or N-cadherin, the PC 12 cells are converted to a neuronal phenotype characterized by neurites. **(d, e)** The conversion is inhibited by antibodies to the CAM present on the 3T3 cells and by drugs that block Ca^{2+} channels. Apparently, the interaction of PC 12 cells with N-CAM or N-cadherin on the transfected 3T3 cells causes a Ca^{2+}-dependent change in intracellular signaling that promotes the phenotypical conversion.

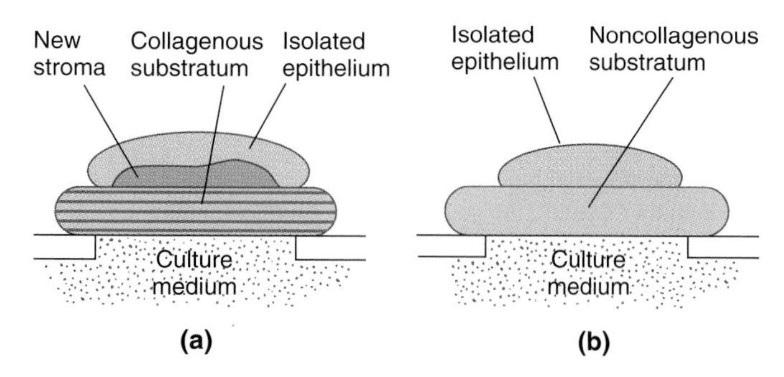

Figure 11.28 Effect of collagen on stroma formation by corneal epithelium in vitro. **(a)** Isolated corneal epithelium placed on substrata containing collagen is induced to form stroma as it does in vivo. **(b)** In contrast, corneal epithelium on noncollagenous substrata does not form stroma.

conditions in vivo. First, ECM binding may affect cell shape and gene expression independently of each other. Second, ECM binding may directly affect gene expression, which in turn may affect cell shape. Third, ECM binding—or the substrate used for cell culture—may directly influence cell shape, which in turn may affect gene expression. The third possibility is supported by the fact that integrins link the ECM with microfilaments and by the following experiments, which show that the microfilament system can regulate gene expression.

In a study on the connection between the state of the cytoskeleton and gene activity, Zanetti and Solursh (1984) monitored the effects of cytoskeletal inhibitors on *chondrogenesis* (cartilage formation) in cultured chicken limb bud cells. To detect chondrogenesis, the researchers used immunostaining of type II collagen, and alcian blue staining of acidic glycosaminoglycans. Without cytoskeletal inhibitors, the cells spread out on plastic culture dishes, developed numerous microtubules and actin bundles, and did not initiate chondrogenesis. When *cytochalasin D*, a microfilament inhibitor, was added to the medium, the cells became round, lost their actin cables, and began chondrogenesis. In contrast to microfilament inhibitors, agents that disrupt microtubules had no apparent effect. These results indicate that the state of the microfilament system affects gene activities involved in the synthesis of type II collagen and glycosaminoglycans.

The molecular mechanisms by which cell adhesion or the state of a cell's microfilament system might influence gene expression are still being investigated. Some molecules, such as β-catenins, function in *both* cell adhesion *and* intracellular signaling, and they are more available for signaling as adhesion diminishes and vice versa (Fagotto et al., 1996). Similarly, integrins not only function in substrate adhesion but also regulate the phosphatidylinositol cycle, sodium/hydrogen ion exchange pumps, and specific protein phosphorylations (Hynes, 1994; Clark and Brugge, 1995; Gumbiner, 1996). Likewise, certain *receptor tyrosine kinases* (*RTKs*) are bound by collagen molecules but respond through the same canonical Ras-MAPK pathway as the RTKs that are bound by growth factors and similar molecules commonly associated with intercellular signaling (see. Fig. 2.28). Such collagen-activated RTKs may act synergistically with integrins (Schlessinger, 1997). In addition, mechanical tensions generated by microfilaments between different ECM adhesion sites may change the permeability of stretch-activated membrane channels, which in turn affects gene expression.

SUMMARY

When cells from different embryonic tissues are mixed, they segregate and form concentric arrangements in which the most adhesive cells occupy the central position. Morphogenetic movements such as epiboly and ingression are associated with measurable changes in cell adhesiveness.

Cell adhesion molecules (CAMs) are plasma membrane proteins by which cells adhere quickly, selectively, and relatively weakly to other cells. Adhesion by means of CAMs seems to precede the formation of more elaborate and stable junctions between cells. A typical CAM has three domains: a large extracellular binding domain that adheres to another CAM, a transmembrane domain, and a cytoplasmic domain. The cytoplasmic domain is connected to the cytoskeleton by linker proteins including β-catenin.

There are three major families of CAMs. One family, which includes N-CAM, is immunoglobulin-like; its polysialic acid chains, which are attached to the extracellular domain, prevent or permit homotypic binding, depending on their quantity. The second family, called cadherins, includes the most prevalent CAMs. They are Ca^{2+}-dependent and also bind homotypically. A third family of CAMs, the lectins and glycosyltransferases, bind heterotypically to oligosaccharide moieties of glycoproteins. CAMs are very well conserved between invertebrates and vertebrates.

The expression patterns of CAMs are correlated with cell fates. They change dynamically, especially as cells shift from epithelial to mesenchymal organization or vice versa. CAM expression also changes during inductive events, indicating that CAM expression is closely associated with cell determination.

Cells also adhere to certain components of the extracellular matrix (ECM), and the molecules involved are collectively called substrate adhesion molecules (SAMs). The ECM consists of a meshwork of fibers embedded in an amorphous ground substance. The ground substance is composed of glycosaminoglycans and proteoglycans, which hold water by osmotic pressure. Fibrous ECM components, such as fibronectin, collagens, and laminins, are glycoproteins that reinforce the ground substance and resist its expansive forces. Because the ECM components are linked to one another by multiple binding regions, they form a stable, multifunctional matrix.

Cells are linked to ECM components mostly by heterodimeric receptor proteins called integrins. The α and β subunits of integrins have extracellular domains, which bind to certain ECM sites; transmembrane domains; and cytoplasmic domains linked to microfilaments. ECM molecules and integrins have been highly conserved during evolution.

Cells and ECM interact in many ways. Cells synthesize stage-dependent and region-specific ECM components in accord with prevailing cell activities. Cells also orient ECM fibers by means of their integrins, and conversely, ECM components guide cell movements by adhesion and contact guidance.

Both CAMs and SAMs are thought to play key roles in morphoregulatory cycles in which gene expression controls morphogenetic events and vice versa. The stage-dependent and region-specific distributions of CAMs and SAMs promote morphogenetic movements that generate new intracellular and intercellular signals. These signals may alter the activity of certain selector genes, which in turn regulate CAM and SAM genes.

Neurulation and Axis Induction

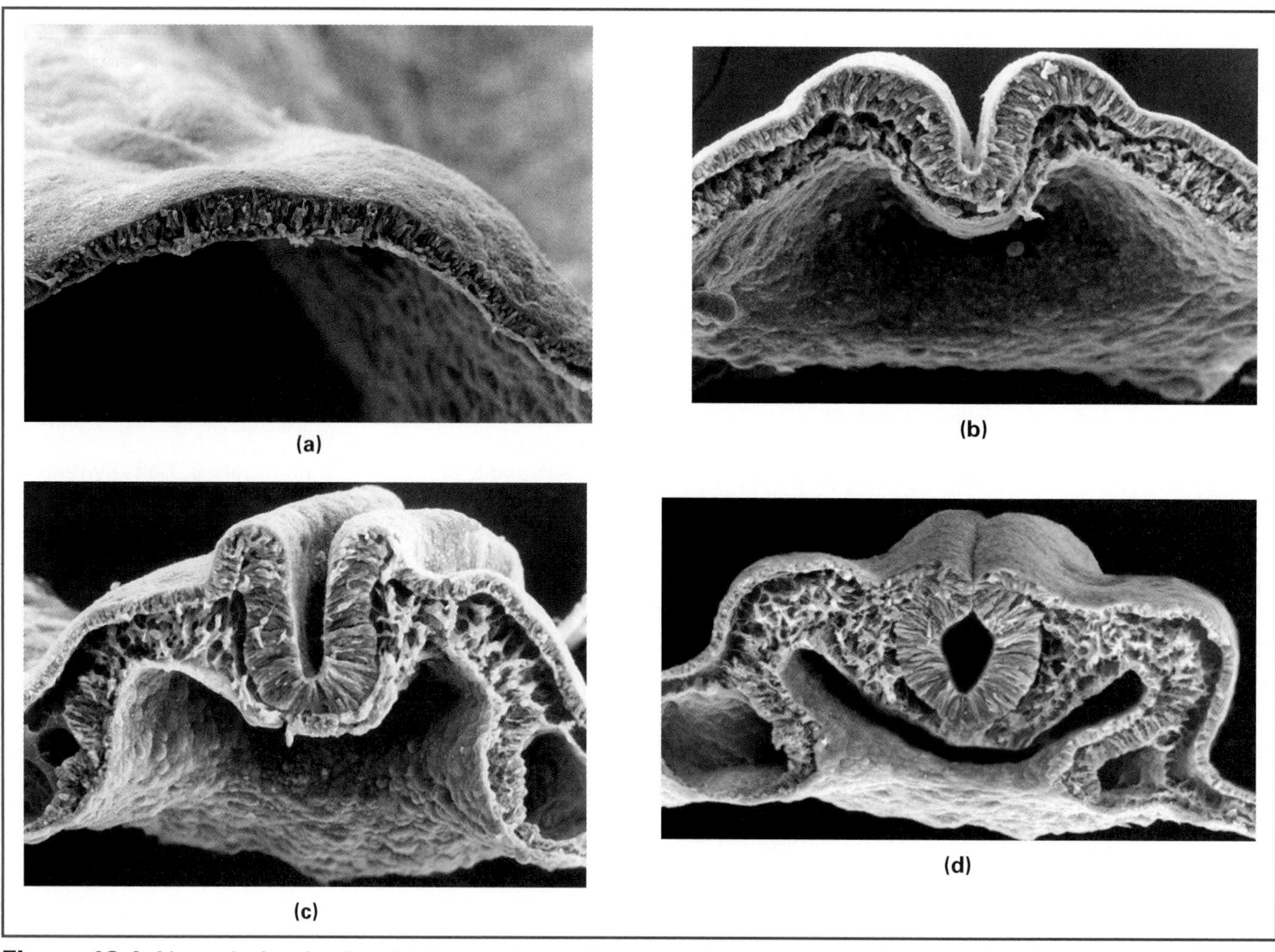

Figure 12.1 Neurulation in the chicken embryo. The scanning electron micrographs show transversely fractured embryos at successive stages of development. **(a)** Fracture made anterior to Hensen's node and the primitive streak, through which endodermal and mesodermal cells have ingressed. The ectodermal cells on both sides of the dorsal midline (top) become columnar and form a neural plate. **(b)** The neural plate cells have formed a neural groove. **(c)** Raised neural folds have formed at the boundaries between the neural plate and the adjacent epidermal ectoderm. The folds will grow together and close the neural plate into the neural tube, which will give rise to brain and spinal cord. The two halves of epidermal ectoderm will also seal together and form a contiguous sheet of epidermis overlying the neural tube.

WITH the completion of gastrulation, the embryo has taken a major developmental step. The germ layers are now arranged according to their ultimate positions in the body. During the subsequent period of organ formation, or *organogenesis,* the germ layers interact to form the organ rudiments. In vertebrates, the most conspicuous part of organogenesis is *neurulation,* the beginning of brain and spinal cord formation. Both organs originate from the same dorsal rudiment, the neural plate, which closes to form the neural tube (Fig. 12.1). Neurulation, like many other events in organogenesis, begins right after gastrulation. The end of organogenesis is less well defined, because organogenesis blends into a long period of tissue and cell differentiation during which organ rudiments are transformed into functioning organs. Overall, organogenesis in the human is considered to be essentially completed after 6 to 8 weeks of development.

As organ rudiments take shape in the appropriate positions, a basic body plan emerges that is characteristic not only of a particular species but also of the entire phylogenetic group to which it belongs. For instance, a 5-week-old human embryo (Fig. 12.2a) has a head with brain, eye, and ear rudiments, and a dorsally segmented trunk with tail and limb buds. Inside the embryo, the rudiments of gut, heart, lung, kidney, and most other organs have formed. The shapes of these organ rudiments and their positions relative to one another are characteristic of mammals and vertebrates in general.

The shaping of organ rudiments involves many of the same cellular behaviors that drive gastrulation. In Chapter 10, we have analyzed complex gastrulation processes in terms of a small number of simpler epithelial movements and cellular activities. In this chapter, we will see how neurulation can be treated similarly as a composite of the same elementary epithelial and cellular behaviors. The appeal of this reductionist analysis is the prospect of understanding a wide variety of complex morphogenetic movements in terms of a small number of simple cellular activities.

Many events in organogenesis are controlled by inductive interactions. A landmark experiment by Spemann and Mangold (1924) has identified the dorsal blastopore lip in the amphibian gastrula as the inducer of the embryonic axis, meaning the entire set of dorsal organ rudiments including neural tube, notochord, somites, and embryonic kidney. If transplanted to the ventral ectoderm of early gastrula hosts, blastopore lip induces the surrounding host tissue to cooperate in the formation of an entire axis, including most conspicuously a secondary neural plate. This experiment, for which Spemann received the 1935 Nobel prize in Physiology or Medicine, is widely seen as the epitome of classical experimental embryology.

Because of its accessibility and its central importance, neurulation is the best-studied example of organogenesis. Consequently, this chapter will focus on the morphogenetic movements that bring about neurulation. Other examples of organogenesis will be included in Chapters 13 and 14. Similarly, in discussing mechanisms of axis induction, we will emphasize the part of neural induction, that is, the

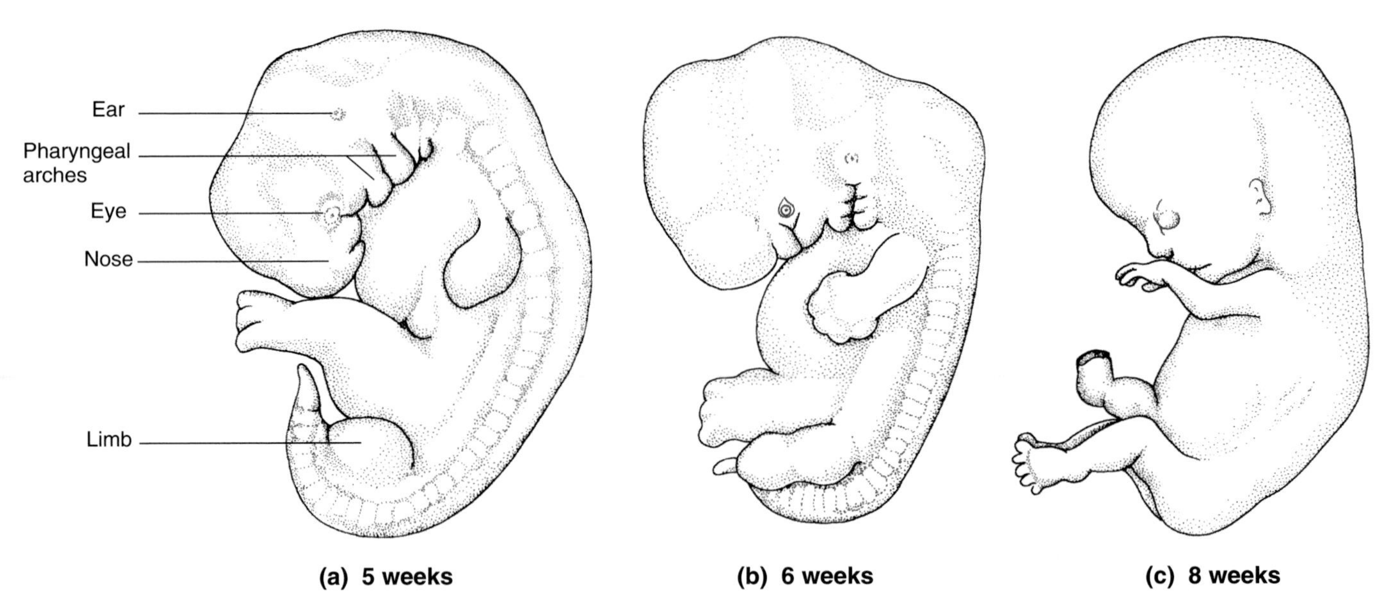

Figure 12.2 Drawings of human embryos during organogenesis, showing the development of nose, eye, ear, and limbs: **(a)** at 5 weeks, **(b)** at 6 weeks, **(c)** at 8 weeks.

induction of neurulation. Spemann's discovery was followed by a rush of investigations into the molecular basis of axis induction. Although the quest was originally unsuccessful, it has been taken up again with an armamentarium of new research tools. Toward the end of this chapter, we will review how modern investigators are using DNA cloning and related techniques to define axis induction in molecular terms.

12.1 Neurulation as an Example of Organogenesis

Neurulation is a dramatic sequence of morphogenetic events generating the rudiment of the central nervous system. Neurulation is first evident when an area of dorsal ectoderm is transformed into a plate of tall cells, the ***neural plate.*** In most vertebrates, the neural plate subsequently closes into a hollow tube, the ***neural tube,*** which gives rise to the brain and the spinal cord (Fig. 12.3). In typical fishes, however, the neural plate first forms a solid rod, which subsequently becomes a hollow neural tube. An embryo in the process of neurulation is referred to as a ***neurula*** (pl., *neurulae*), just as a gastrulating embryo is called a *gastrula.* Neurulation occurs in similar ways in most vertebrates, but closer analysis has also revealed some differences between classes of vertebrates, such as amphibia and birds.

NEURULATION IS OF SCIENTIFIC AND MEDICAL INTEREST

Neurulation is of major scientific interest because the size and accessibility of the neural plate facilitate experimental manipulation. In addition, a better understanding of neurulation will have significant medical benefits. One of the most common congenital malformations in humans is known as ***spina bifida*** (Lat. "divided spine"). It involves defects or delays in the closure of the neural tube and abnormal developments of the bone, muscle, and skin surrounding the brain and spinal cord. The mildest form of spina bifida, called *spina bifida occulta* (Lat. *occultus,* "obscure," "concealed"), results from failure of the two arches of a vertebra to fuse dorsally (Fig. 12.4). As many as 10% of all people have this defect, which causes no pain or neurological disorder (K. L. Moore, 1982). The only overt sign of its presence may be a dimple in the skin or a small tuft of hair over the affected area. If more than one or two vertebrae are involved, the spinal cord bulges out dorsally, and a cyst covered with skin forms on the outside. This congenital malformation, known as *spina bifida cystica,* occurs about once in every 1000 births. It is associated with neurological disorders, the severity of which depends on the extent to which nerve tissue protrudes into the cyst.

Failure of the cephalic part of the neural tube to close leads to ***anencephaly*** (Gk. *a*(*n*)-, "not"; *enkephalos,* "brain") associated with ***acrania*** (lack of the vault of the skull, from Gk. *kranion,* "skull") and severe spina bifida. The frequency of anencephaly varies greatly among

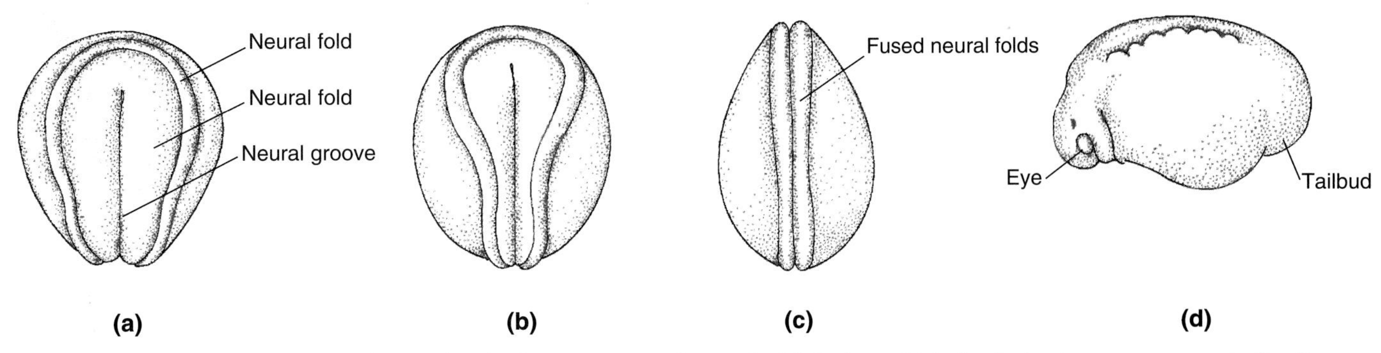

Figure 12.3 Neurulation in the salamander *Ambystoma maculatum.* **(a, b)** Early and late keyhole stages, so named after the outline of the neural plate in dorsal view. **(c)** Closed neural tube, dorsal view. **(d)** Early tail bud stage, lateral view.

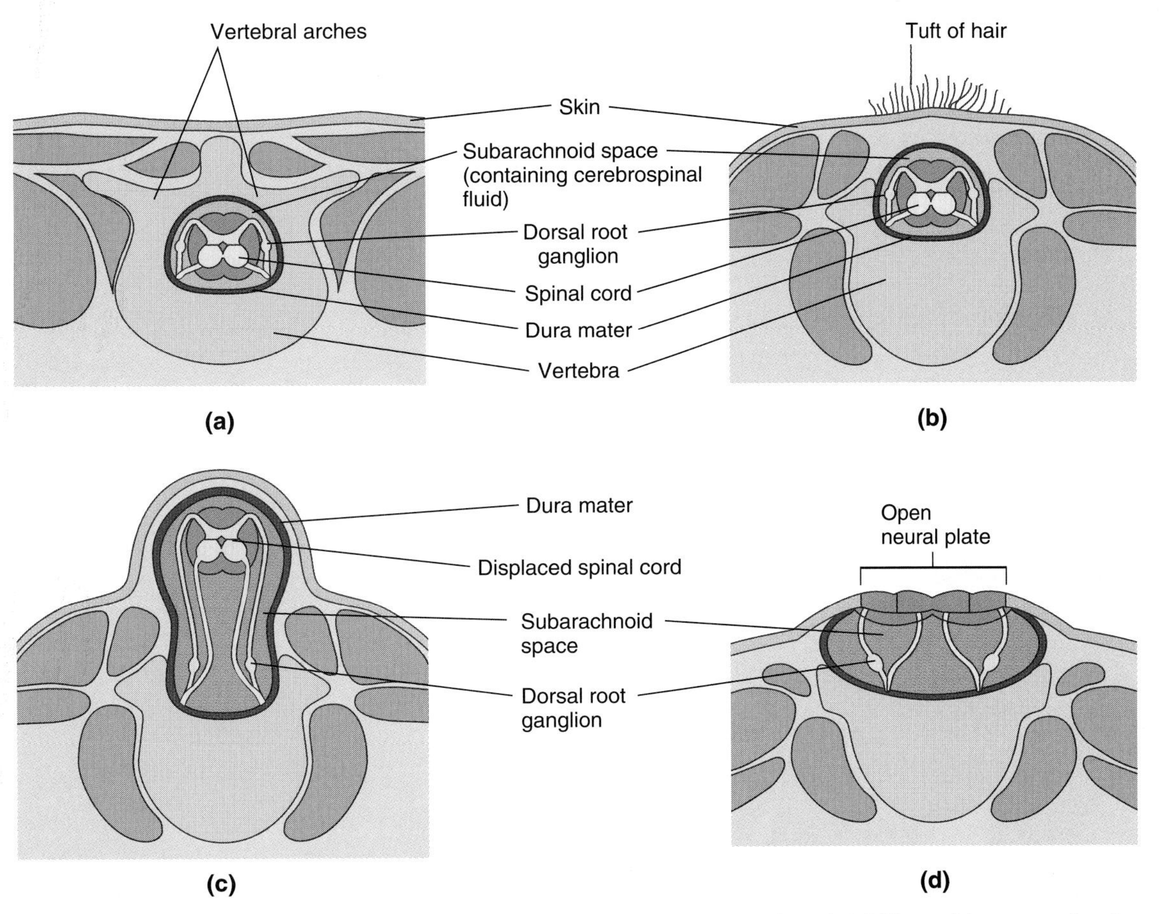

Figure 12.4 Degrees of spina bifida and associated malformations of the spinal cord. **(a)** Normal human spine in transverse section. **(b)** Spina bifida occulta, a failure of the vertebral arches to close over the spinal cord. This is a common defect of the fifth lumbar and/or first sacral vertebra, causing no clinical symptoms. **(c)** Spina bifida cystica with protruding meninges and displaced spinal cord. **(d)** Spina bifida with open neural plate. The *dura mater* is one of three membranes that form a fluid-filled cushion between the spinal cord and the vertebrae.

human populations, ranging from 0.1 to 6.7 per 1000 births (Shulman, 1974). Infants with anencephaly are stillborn or die shortly after birth. Multidisciplinary studies have shown that increasing the amount of folic acid in the maternal diet lowers the frequency of anencephaly and spina bifida in human newborns. Folic acid is a vitamin that is metabolized into coenzymes that play important roles in several biosynthetic processes.

NEURULATION IN AMPHIBIANS OCCURS IN TWO PHASES

In amphibian embryos, the morphogenetic process of neurulation can be conveniently divided into two phases. The first phase is the formation of the *neural plate*. This phase concludes with the ***keyhole stage,*** so named for the peculiar outline of the neural plate at that stage (see Fig. 12.3). The main event of the second phase is the closure of the neural plate into the *neural tube.*

The first phase of neurulation begins with a change in the behavior of the ***neural ectoderm*** cells, which occupy the dorsal surface of the embryo. These cells have moved posteriorly, toward the blastopore, as part of *epiboly* during gastrulation. At the beginning of neurulation, they begin to move toward the dorsal midline and the anterior. Simultaneously, the cells become *columnar* in shape and form a raised plate, the *neural plate,* on the dorsal side of the embryo (Fig. 12.5). A depression called the ***neural groove*** develops along the midline of the neural plate (see Fig. 12.3). At the same time, ridges of cells called ***neural folds*** arise along the boundary between the neural plate and the surrounding epidermis. Once the neural folds have emerged, the neural plate extends anteroposteriorly and shrinks laterally, especially in the posterior. As a result, the surface area of the neural plate decreases, and it assumes the characteristic keyhole shape. The anterior part of the keyhole region gives rise to the brain while the posterior part forms the spinal cord.

Unlike the neural ectoderm cells, the remaining ectodermal cells assume a *squamous,* or flat, shape (Fig. 12.5). Because these cells are fated to form epidermis, they are called the ***epidermal ectoderm.***

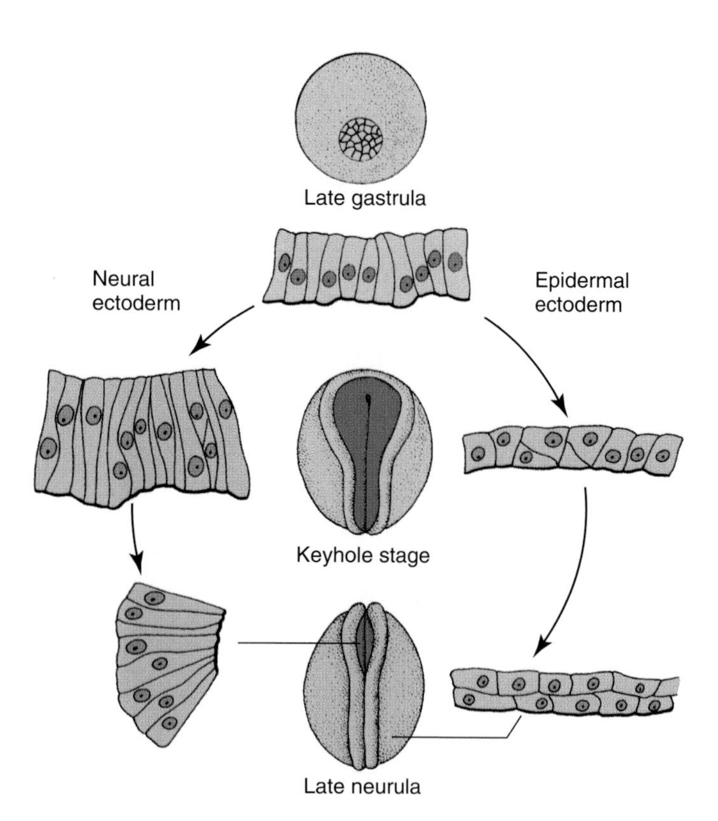

Figure 12.5 Changes in shape of ectodermal cells during neurulation in the newt *Taricha torosa.* The neural ectoderm cells become columnar, while the epidermal ectoderm cells become squamous. (See also Figure 2.3.)

The second phase of neurulation, which begins after the keyhole stage, is short and dramatic. The neural plate undergoes a spurt of anteroposterior extension and simultaneously curls up so that the neural folds meet along the dorsal midline, thus closing the neural plate into the neural tube. At the same time, the adjacent sheets of epidermal ectoderm fuse above the neural tube along the dorsal midline. As the neural tube separates from the future epidermis, the intervening cells generate a lineage of their own. These cells were originally located at the crest of the neural folds and are therefore called *neural crest cells.* They will migrate to different positions all over the body and give rise to a wide variety of differentiated cells (see Section 13.2).

NEURAL TUBES OF BIRD EMBRYOS HAVE HINGE REGIONS

In chicken and other bird embryos, the neural plate forms in the wake of Hensen's node as the latter regresses from anterior to posterior (see Fig. 10.24; Smith and Schoenwolf, 1997). Thus, at each stage of neurulation, the anterior of the embryo is ahead of the posterior. The lateral parts of the neural plate are derived from the epiblast. In contrast, at least some of the cells that form the median strip of the neural plate belong to the same population of cells that also form notochord as well as dorsal gut and are derived from Hensen's node (see Fig. 10.28).

Shortly after the neural plate is formed, it undergoes convergent extension, becoming longer, narrower, and thicker in the process (Fig. 12.6; see also Fig. 12.1). One type of cellular behavior contributing to convergent extension of the neural plate is *coordinated shape change.* The neural plate cells become columnar, except for those median cells that are derived from Hensen's node, which become wedge-shaped. These latter cells, along with the underlying notochord, form the so-called *median hinge point,* a crease that extends the entire length of the neural plate. Two similar creases, known as *dorsolateral hinge points,* extend along the sides of the anterior plate. All three hinge points seem to facilitate the bending and closure of the anterior neural plate into the tube that will form the brain. The posterior neural plate, which later will form the spinal cord, behaves differently. It closes like a book, using only the median hinge point. In contrast to the prospective brain, which has a wide lumen, the future spinal cord surrounds only a narrow vertical slit.

Another cellular behavior contributing to the convergent extension of the neural plate is *mediolateral cell intercalation,* which increases the number of cells in the anteroposterior dimension while reducing the number of cells in the lateral dimension of the plate. In addition, the neuroepithelial cells undergo two or three rounds of division with a preferred *anteroposterior spindle orientation.* Because cell division in bird embryos is associated with growth, oriented mitoses translate into overall

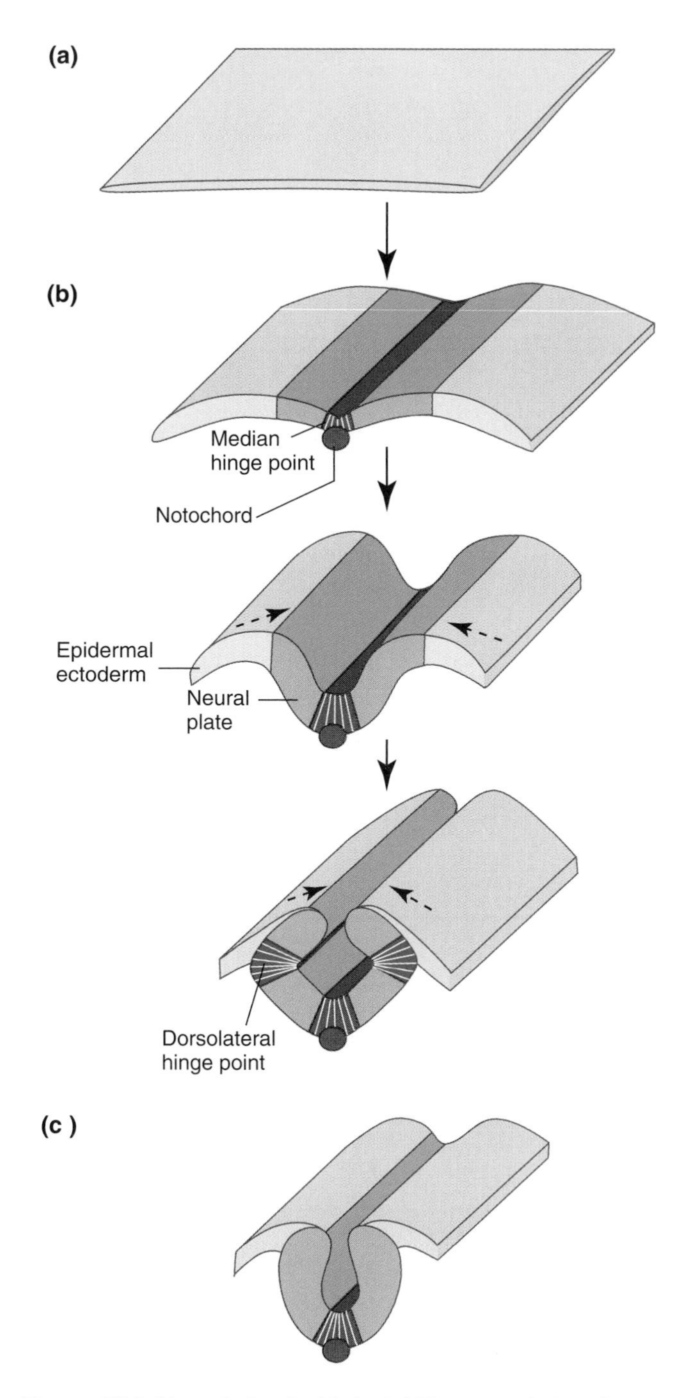

Figure 12.6 Neurulation in birds. **(a)** The neural plate forms as its cells become columnar. **(b)** Morphogenetic movements in the future brain region. A strip of neural plate cells contacting the notochord form the so-called median hinge point; these cells are wedge-shaped and not as tall as the other neural plate cells. The neural plate is creased along the median hinge point and forms raised neural folds where neural plate and epidermal ectoderm are juxtaposed. Additional strips of wedge-shaped neural plate cells form at so-called dorsolateral hinge points, so that the V-shaped cross section of the neural tube becomes diamond-shaped. Thus, the neural folds bend together and eventually close into a tube with a wide lumen. **(c)** In the region of the future spinal cord, no dorsolateral hinge points are formed. The two halves of the neural plate pivot together until they leave only a narrow slitlike lumen between them. (See Fig. 12.1 for electron micrographs.)

elongation in one dimension. Finally, the *epidermal ectoderm* cells located to the sides of the neural plate divide with a preferred *mediolateral spindle orientation*. The resulting lateral expansion of the future epidermis may help to push the neural plate closed.

IN HUMANS, NEURAL TUBE CLOSURE BEGINS IN THE NECK REGION

Neurulation in humans proceeds generally as in birds, except that the anterior part of the neural plate, which gives rise to the brain, closes with some delay (Fig. 12.7). The shaping of the neural plate is primarily associated with coordinated shape changes and cell intercalations. The *neural groove* again marks the *median hinge point*, where the neural plate is attached to the underlying notochord. Although the formation of the neural plate proceeds from anterior to posterior as in chicken, the neural tube closes first in the neck region of the human embryo. Apparently, the bulk of the anterior neural plate is a mechanical impediment to the growing together of the neural folds. From the neck region, the closure proceeds anteriorly and posteriorly. Unclosed residual openings, known as *anterior neuropore* and *posterior neuropore*, remain as temporary connections with the *amniotic cavity* until they close a few days later.

IN FISHES, THE NEURAL TUBE ORIGINATES AS A SOLID ROD

In zebrafish and most other teleosts (bony fish), the neural plate forms on the dorsal side of the epiblast (see Fig. 10.21). As in other vertebrates, the neural plate cells are distinguished by their columnar shape. However, instead of curling up and closing into a tube, cells on both sides of the median hinge point grow together with their apical faces, forming a longitudinal structure called the *neural keel* because in cross section it looks somewhat like a boat (Fig. 12.8; Langeland and Kimmel, 1997). The keel rounds up into a cylindrical rod before the juxtaposed halves of the neural plate separate again, thus creating a lumen. This type of neurulation, in which lumen formation occurs as a secondary step, is known as *secondary neurulation*, as opposed to *primary neurulation*, where the lumen is formed immediately. Primary neurulation is characteristic of amphibians, reptiles, birds, and mammals, as discussed earlier in this section, but only for the head and trunk regions. In their embryonic tail regions, the land-dwelling vertebrates also show secondary neurulation.

12.2 Mechanisms of Neurulation in Amphibians

Neurulation has occupied embryologists for more than 100 years (reviewed by R. Gordon, 1985; Jacobson, 1994; Schoenwolf and Smith, 1990; Smith and Schoenwolf,

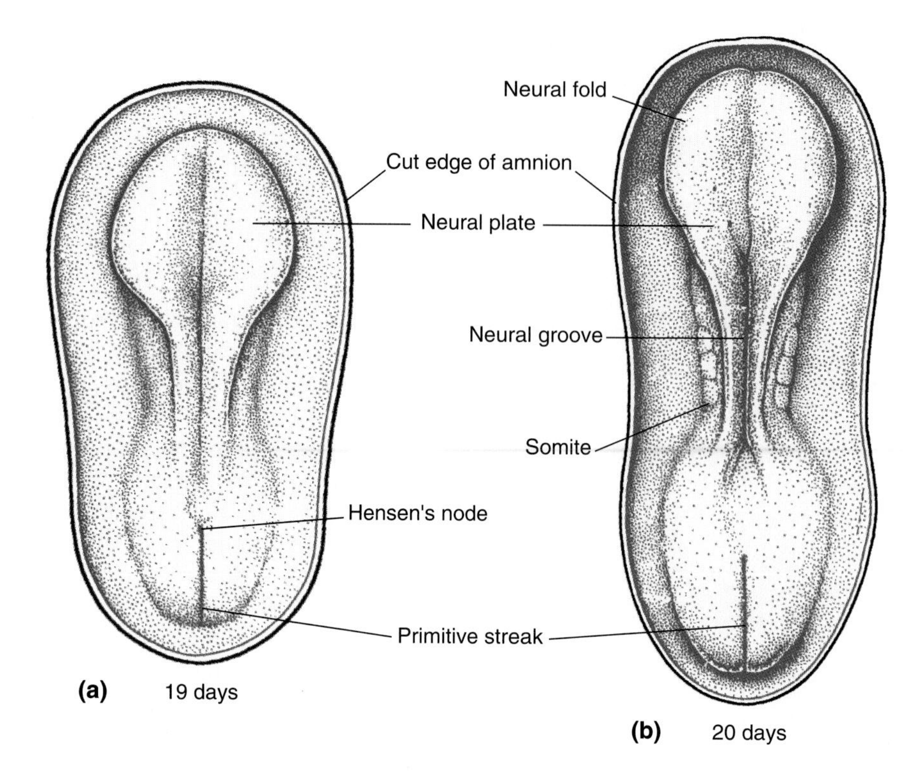

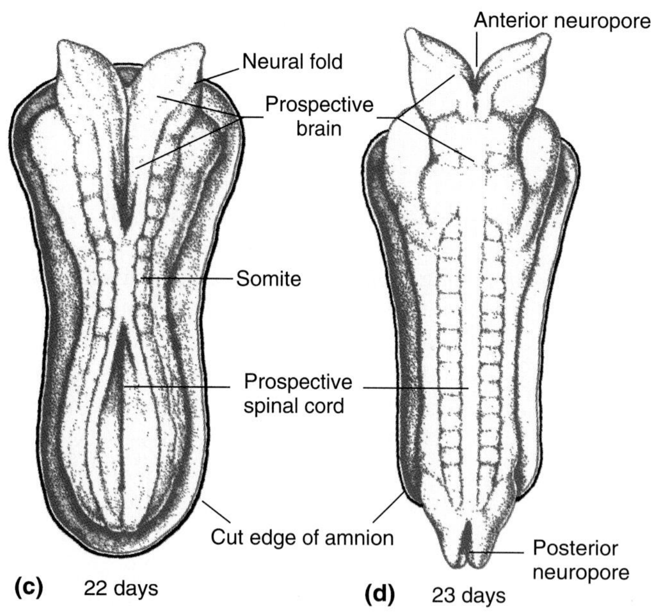

Figure 12.7 Neurulation in humans shown in dorsal views, with the amnion cut away. **(a, b)** The neural plate is formed in the wake of the regressing Hensen's node. The anterior part of the neural plate, which will give rise to brain, is much bulkier than the posterior part, which will form spinal cord. **(c, d)** Closure of the neural fold into the neural tube begins in the neck region of the embryo and proceeds to the anterior and posterior. The anterior and posterior neuropore form temporary openings connecting the lumen of the neural tube with the amniotic cavity.

1997). The analysis of neurulation in amphibians is simplified by the fact that their nervous system does not grow during embryogenesis (Jacobson, 1978). Furthermore, cell division is slow during amphibian neurulation, and in the absence of cell growth, the morphogenetic effects of mitosis are minor. Most amphibians, including newts, have the added advantage that the neural plate cells form a single layer, so that each cell has an apical face that can be seen from the outside. (The frog *Xenopus laevis*, a species that is preferred by many modern researchers because it reproduces year-round in the laboratory, belongs to the minority of

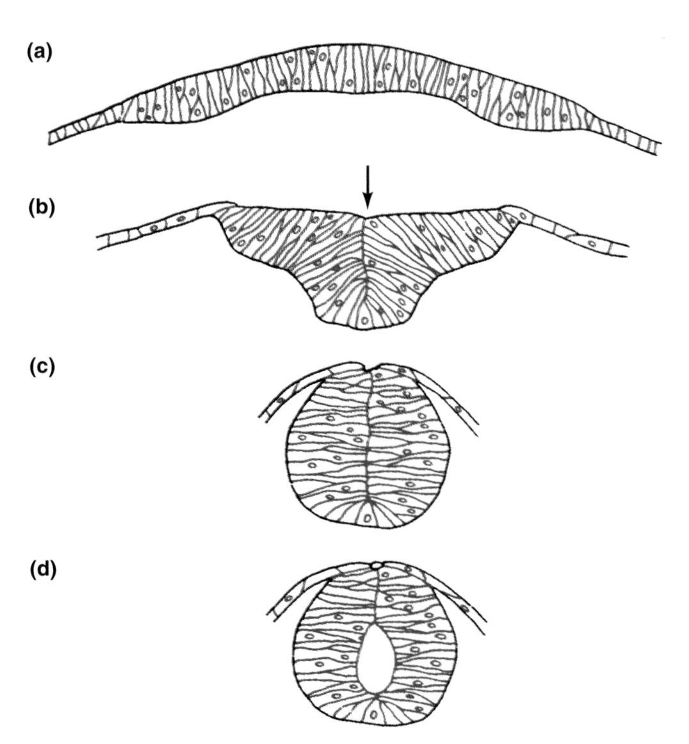

Figure 12.8 Neurulation in zebrafish. **(a)** The neural plate forms as an epithelium of distinctly columnar cells. **(b)** The neural plate infolds at the midline (arrow), juxtaposing the lateral halves of the neural plate with their apical faces. The resulting formation is known as the neural keel. **(c)** The keel continues to infold and rounds up into the cylindrical neural rod. **(d)** The juxtaposed halves of the neural plate separate again, forming a fluid-filled lumen.

amphibians with a neural plate consisting of more than one cell layer.)

NEURULATION DEPENDS ON TISSUES ADJACENT TO THE NEURAL PLATE

In the analysis of any epithelial movement, it is important to assess whether the movement is *autonomous*—that is, relying only on forces generated by the epithelium itself, or whether it depends on forces created elsewhere. In an early study of neurulation, W. His (1874) proposed that the epidermis on both sides of the neural plate expands actively and thereby compresses the neural plate, making it buckle and close into a tube. This notion of lateral compression was dismissed, at least for newt embryos, by Jacobson and Gordon (1976), who made incisions in the epidermis surrounding the neural plate. They found that the cuts gaped immediately and widely regardless of their orientation. These results indicate that the epidermis is under considerable tension in every direction and that it cannot be pushing the neural folds together when the neural plate closes into a tube.

Nevertheless, the epidermal ectoderm surrounding the neural plate contributes to neurulation in amphibians, although not by way of pushing. This was revealed by experiments in which neural plates from axolotl (salamander) embryos were isolated with or without a rim of epidermis (Jacobson and Moury, 1995). The cuts were made sufficiently deep so that the mesoderm and endoderm that are directly underlying the neural plate were included in the explants. Most of the explants that included a rim of epidermis closed completely into a tube. In contrast, the explants without an epidermal rim closed only in the caudal region, which is fated to form spinal cord, but remained open in the cranial region, which is fated to form brain.

Another tissue outside the neural plate that is required for complete neurulation is the *axial mesoderm*, or future *notochord* (see Fig. 6.10). In contrast to other mesoderm, the axial mesoderm adheres tightly to the overlying neural plate. The development of neural plates without axial mesoderm depends on the time or isolation. If isolated at the beginning of neurulation, the mesodermless neural plates fail to elongate and do not assume the proper keyhole shape (Jacobson and Gordon, 1976). Control neural plates isolated at the same stage with axial mesoderm attached do form keyhole-shaped plates. Similar experiments carried out at later stages have somewhat different effects. Neural plates isolated without notochords at the keyhole stage elongate considerably, although not as much as control neural plates isolated with notochords (Jacobson, 1985).

Taken together, these results indicate that the neural plate cells must interact with adjacent epidermal ectoderm and with underlying axial mesoderm for neurulation to proceed normally. As we will see shortly, the contact between neural plate cells and adjoining cells seems necessary for certain neural plate cell behaviors, which in turn generate the mechanical forces within the neural plate that drive neurulation.

NEURAL PLATE CELLS UNDERGO COLUMNARIZATION

In the course of neurulation, the cells of the neural plate undergo major shape changes while retaining the same volume. During the phase preceding the keyhole stage, the neural plate cells elongate perpendicular to the surface of the embryo while their apical and basal surfaces shrink (see Fig. 12.5). This process, called ***columnarization,*** is commonly observed in epithelial areas at the beginning of morphogenetic movements.

The columnarization of epithelial cells is generally correlated with the alignment of their microtubules. At the beginning of neurulation, neural plate cell microtubules are oriented parallel to the axis of cell elongation, whereas previously they were scattered throughout the cells in a seemingly random manner (see Fig. 2.3). Microtubules appear to be necessary for elongation, since the process is stopped by colchicine, an inhibitor of microtubule polymerization (Burnside, 1973). It is not clear whether microtubules actually cause the

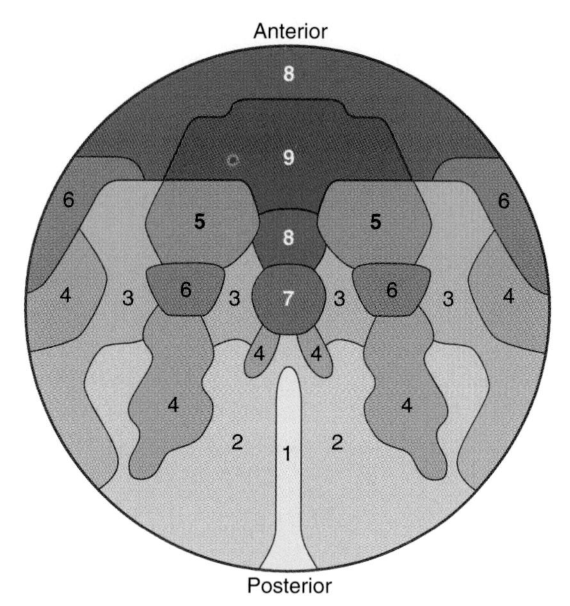

Figure 12.9 Pattern of cell elongation in the neural plate of the newt *Taricha torosa*. The drawing shows a dorsal aspect of the neural plate. The amount by which each cell elongated (perpendicular to the plane of the paper) is indicated here by color intensity, with highest intensity representing the greatest elongation.

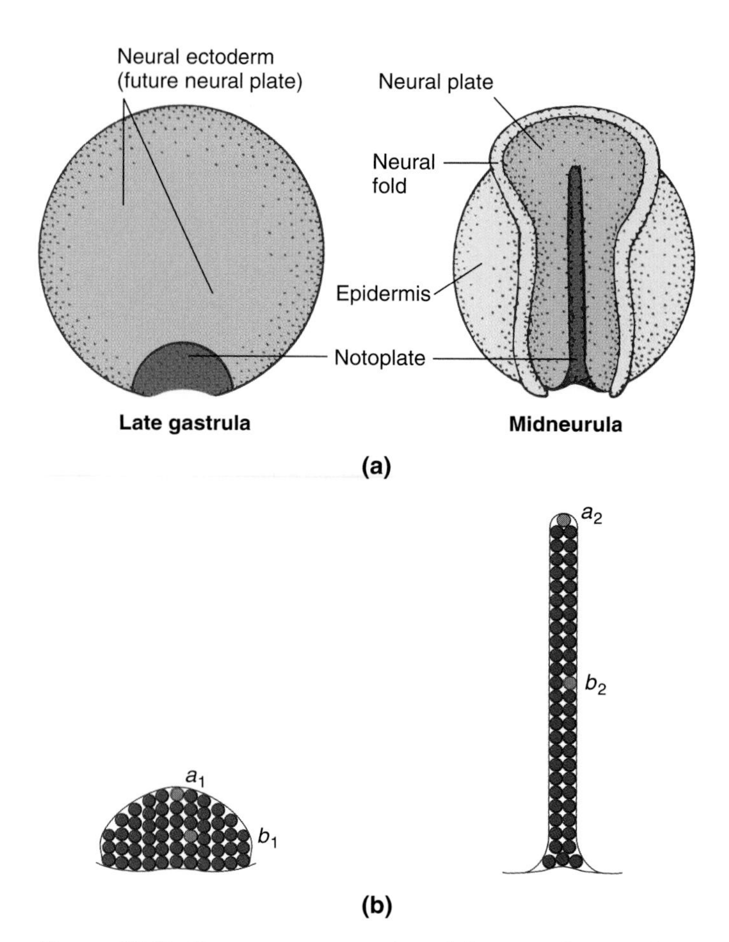

Figure 12.10 Convergent extension of the notoplate (region of the neural plate overlying the notochord) in the newt *Taricha torosa*. **(a)** Dorsal views of entire embryos. **(b)** Cell rearrangement in the notoplate. The two highlighted cells move from a_1 to a_2 and from b_1 to b_2. Thus, different sections of the notoplate extend proportionally.

columnarization of cells by orienting their cytoplasmic transport or whether they simply stabilize the elongated cell shape after it has been generated independently.

The degree of columnarization varies among different regions of the neural plate. The longest cells occur in a crescent-shaped anterior area, and the shortest cells, along the posterior midline (Fig. 12.9). When Jacobson (1981) transplanted pieces of neural plate from an area of greater elongation to an area of lesser elongation and vice versa, he found that that transplanted cells elongated according to the region from which they were taken. Holtfreter (1946) observed that isolated neural plate cells completed their normal elongation in vitro. Therefore, by the criteria of transplantation and isolation, cells in each neural plate area are *determined* to elongate by a certain amount.

The columnarization of individual neural plate cells adds up to an overall reduction in the surface area of the entire neural plate. However, the plate's anteroposterior extension and its transformation to the keyhole shape cannot be explained by columnarization alone.

INTERCALATION OF NEURAL PLATE CELLS CAUSES CONVERGENT EXTENSION

In addition to columnarization, the neural plate is shaped by *convergent extension*. Using time-lapse cinematography, Burnside and Jacobson (1968) surveyed the movements of neural plate cells in the California newt, *Taricha torosa*, from the end of gastrulation to the keyhole stage (Fig. 12.10a). Because this newt has a salt-and-pepper pigmentation, they were able to trace the movements of individual cells. In a frame-by-frame analysis, they mapped the movements of cells at the intersections of a superimposed coordinate grid. The pathways turned out to be consistent from one embryo to another, and the overall cell movements were toward the midline and the anterior. The most conspicuous anteroposterior extension of the grid occurred in the ***notoplate,*** the narrow median portion of the neural plate overlying the notochord (Fig. 12.10b).

Analysis of cellular movements in the notoplate revealed extensive *cell intercalation*. At the late gastrula stage, the notoplate cells occupy a semicircular region at the posterior margin of the neural plate (Fig. 12.10). As lateral notoplate cells wedge between medial cells, the notoplate converges laterally and extends anteroposteriorly. At the keyhole stage, the notoplate has been repacked into a narrow strip extending along most of the midline. Convergent extension also occurs elsewhere in the posterior region of the neural plate, although to a lesser degree.

The cell behaviors during convergent extension of the posterior neural plate in newts were tracked in time lapse movies of normal embryos (Jacobson et al., 1986). These observations indicated that the intercalation movements of neural plate cells are random except that cells do not cross the notoplate/neural plate boundaries. Cells contacting a boundary from either side are trapped at the boundary and stay there for hours. These cell behaviors resemble those observed during convergent extension of dorsal mesoderm during *Xenopus* gastrulation (see Section 10.3). In both cases, certain boundaries between cellular domains are not crossed by intercalating cells. Rather, cells bumping up against these boundaries become stuck, presumably by inhibition of an activity necessary for extending lamellipods. As more and more cells are lined up along the boundaries, the overall result will be convergent extension.

As an analogy, one could think of a schoolyard with children running around randomly. If one would stretch a rope across the yard and institute a rule requiring each child who bumps into the rope to hold on to it, then all children would soon be lined up in a double row on both sides of the rope. If, in addition, one would allow for the rope to be longer than the yard, and for the children to spread out along the rope until they have the same average distance from one another as they had when they were running freely, then the result would be a dramatic case of convergent extension.

In the late gastrula, dorsal mesodermal cells "stay put" along boundaries that extend between *axial mesoderm*, or future notochord, and *paraxial mesoderm*, or future somites (see Figs. 9.19 and 10.19). In the neural plate of the mid- and late neurula, similar boundaries extend between the notoplate and the lateral portions (Fig. 12.10). Because the notoplate is located right above the axial mesoderm, the boundaries in the neural plate are parallel and right on top of the boundaries in the dorsal mesoderm.

To experimentally test whether the notoplate/neural plate boundaries are necessary for neural plate elongation, Jacobson (1991) made microsurgical cuts along the notoplate of newt embryos at a midneurula stage. If a cut was made along one side only then the part of the neural plate with the notoplate elongated while the part without the notoplate did not (Fig. 12.11a). If cuts were made along both sides of the notoplate then the lateral parts did not change significantly in length while the notoplate itself shrank (Fig. 12.11b). As a control experiment, the neural plate was cut right down the midline to produce right and left halves, each with some notoplate. Under these circumstances, both halves elongated somewhat.

The results of Jacobson and coworkers indicate that the notoplate/neural plate boundaries are necessary for the convergent extension that occurs in the posterior part of the neural plate. The trapping of cells at the boundaries seems to be the principal mechanism of convergent extension in the neural plate. The same mechanism seems to act in dorsal mesoderm along the boundaries between axial and paraxial mesoderm. However, in dorsal mesoderm, the boundaries also appear to be a source of a signal that *orients* cell intercalations in a mediolateral direction, as discussed in Section 10.3. In this case, oriented cell intercalation and cell trapping at the boundaries both promote convergent extension, illustrating again the *principle of overlapping mechanisms.*

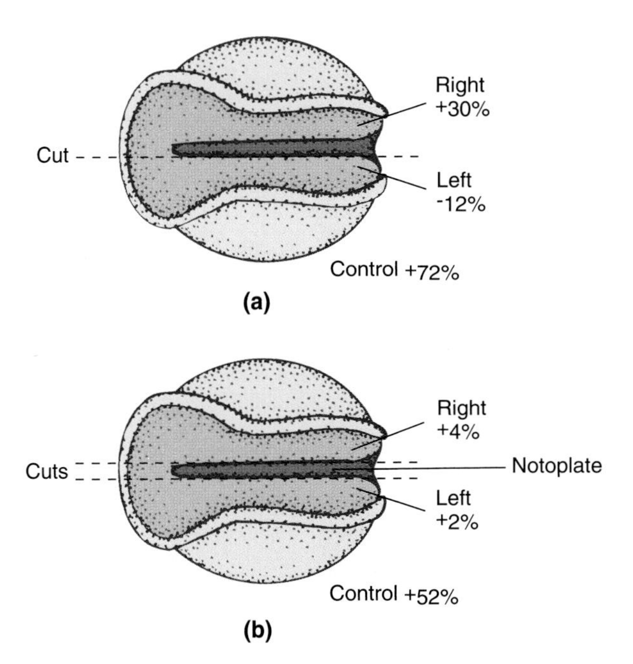

Figure 12.11 Importance of boundary regions between notoplate and adjoining neural plate as revealed by microsurgical cuts at the midneurula stage. **(a)** Neural plate severed along the left boundary of the notoplate. **(b)** Neural plate severed along both boundaries of the notoplate. The partial neural plates were measured before the cuts were made and when untreated control embryos had closed their neural plates into tubes. Percentages indicate the relative increases or decreases in the lengths of the cut edges during this period and are averaged from 12 cases each.

BOTH COLUMNARIZATION AND CELL INTERCALATION CONTRIBUTE TO GENERATING THE KEYHOLE SHAPE

So far, we have discussed two types of cellular behavior involved in neural plate formation: columnarization and mediolateral intercalation. Their relative importance in bringing about the keyhole shape of the neural plate could be assessed if each could be selectively inhibited. However, since both behaviors are presumably driven by microfilament action, inhibitors like cytochalasin B are not useful for isolating each behavior.

In cases like this, computer simulations can be helpful. Jacobson and Gordon (1976) simulated both columnarization and intercalation in a comprehensive computer program, which closely approximated the generation of the keyhole shape in the newt embryo (Fig. 12.12a, b). By

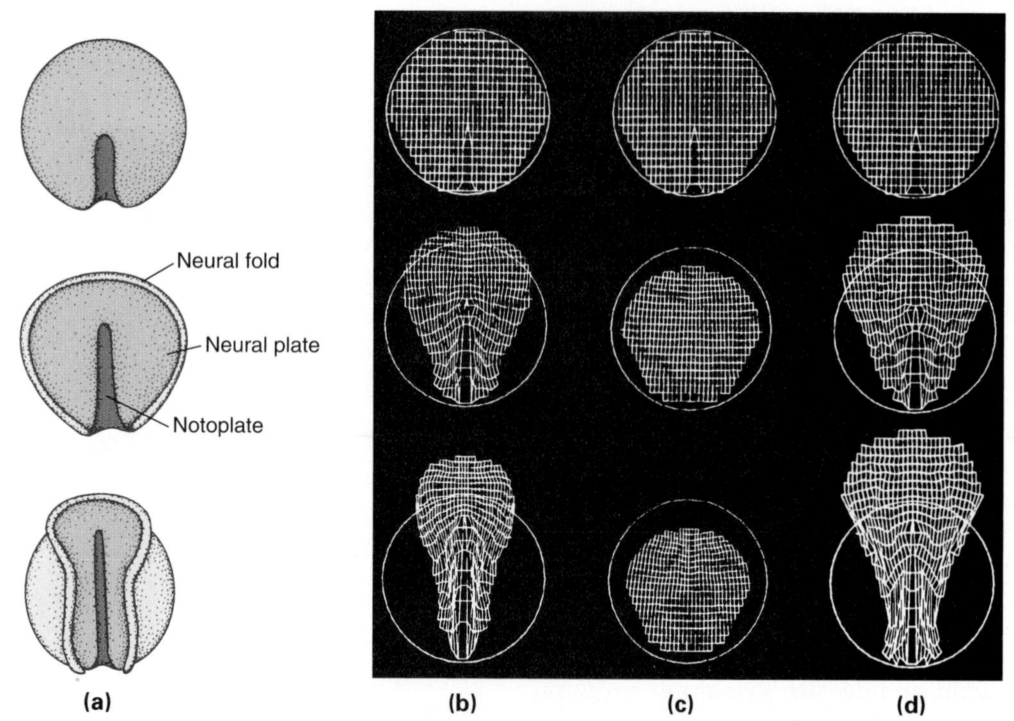

Figure 12.12 Computer modeling of neural plate formation in the newt *Taricha torosa.* **(a)** Sequence of diagrams of the developing neural plate, drawn from time-lapse frames. **(b)** Computer simulation incorporating two forces: columnarization, which is especially strong in anterior neural plate cells, and convergent extension, which is most prevalent in posterior neural plate cells. The resulting transformation of a coordinate grid placed over the neural ectoderm resembled closely the formation of the keyhole-shaped neural plate in the embryo. **(c)** Turning off the convergent extension part of the computer program resulted in a poor simulation. **(d)** Turning off the columnarization part resulted in a fair simulation, but the surface of the anterior neural plate was too large.

omitting the intercalation part of the program, they obtained a poor simulation that featured a reduction in surface area but no anteroposterior extension, much like the behavior of the neural plate isolated from the underlying mesoderm (Fig. 12.12c). Conversely, turning off the columnarization part of the computer program produced a good simulation of the anteroposterior extension and keyhole formation, but the surface of the simulated neural plate was too large (Fig. 12.12d). The investigators concluded that both mediolateral intercalation and columnarization are necessary to shape the neural plate properly, but that intercalation makes the greater contribution to anteroposterior extension and generation of the keyhole shape.

NEURAL TUBE CLOSURE IS ASSOCIATED WITH APICAL CONSTRICTION, RAPID ANTEROPOSTERIOR EXTENSION, AND CELL CRAWLING

During the second phase of neurulation, when the neural plate closes into a tube, the apical cell surfaces continue the shrinking process that began with columnarization in the first phase (see Fig. 12.5). This process of ***wedging*** gives the neural plate cells a shape that allows the plate to curl into a tube. (Fig. 12.13). Burnside (1971) suggested that this shape change may be caused by ***apical constriction***—that is, the constriction of a band of microfilaments arranged like a purse string beneath the apical surface (see Fig. 2.3). In accord with this hypothesis, the bundle of microfilaments thickens during apical constriction, suggesting that the constriction might be caused by interdigitation of the microfilaments. Of course, apical constriction by itself does not generate a wedge-shaped cell but rather a pyramidal or cone-shaped cell. However, wedge-shaped cells may well be generated by a combination of apical constriction and anteroposterior elongation of the neural plate, as suggested by the computer simulations discussed earlier.

Indeed, closure of the neural tube coincides with a spurt of rapid anteroposterior extension. To quantify this observation, Jacobson and Gordon (1976) excised neural plates and neural tubes from newts, laid them flat on agar plates, and measured their length under a microscope with a scale built into the eyepiece. Plotting overall length at successive stages, they found that the extension rate changed abruptly and was *10 times faster during neural tube closure* than before and after closure (Fig. 12.14). Similarly, as neurulation in chickens proceeds from anterior to posterior, a wave of rapid extension accompanies tube closure (Jacobson, 1981). Thus, it appears that a combination of apical constriction and anteroposterior extension may bring about neural tube closure. Computer models indicate that constricting the

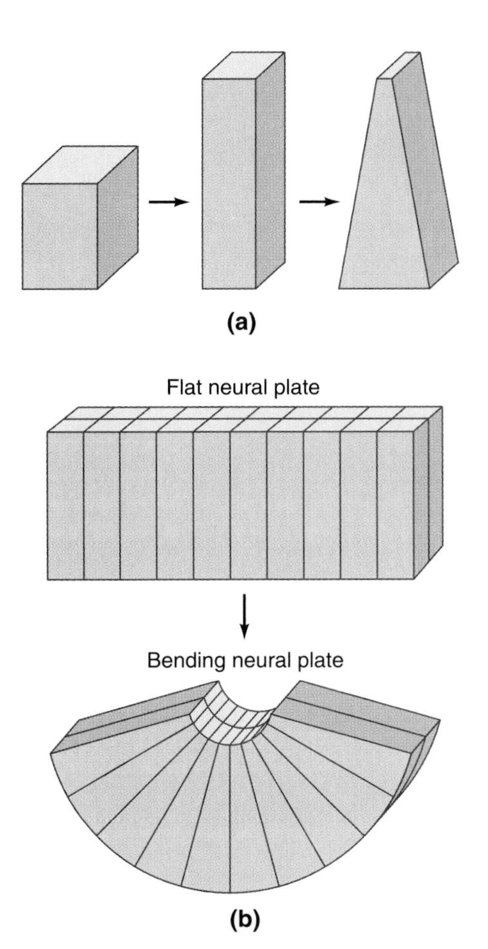

Figure 12.13 Cell wedging during neural tube closure. **(a)** Shape changes shown for a single neural plate cell: columnarization during the first phase of neurulation, and wedging during the second phase. **(b)** Wedging during the second phase of neurulation, shown for two transverse rows of cells.

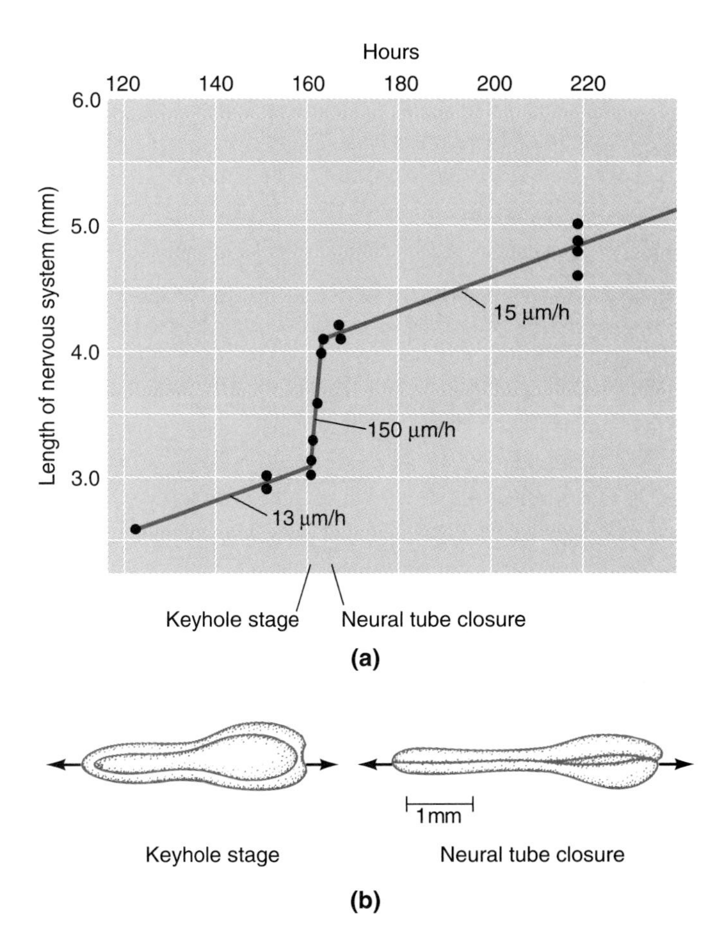

Figure 12.14 Abrupt change in the rate of elongation observed during the closure of the neural tube in the newt *Taricha torosa.* **(a)** Neural plates or tubes were excised and their lengths measured. The rate of elongation, indicated by the slope of the curve, increased 10-fold between the keyhole stage and the neural tube closure, when it returned to the old rate. **(b)** Neural plate and neural tube are drawn to the same scale to show the degree of extension between the two stages marked on the abscissa in part a.

apical surface of neural plate cells combined with anteroposterior extension of the neural plate can account for many salient features of neurulation, including the generation of the keyhole shape, neural folds, hinge formation, and neural tube closure (Clausi and Brodland, 1993). Nevertheless, a third mechanism may be involved.

At the neural folds, where the columnar neural plate cells are juxtaposed to the squamous epidermal ectoderm cells, it appears that the neural plate cells attempt to crawl under the epidermal cells (Fig. 12.15). The resulting traction on the basal face of the epidermal cells could generate an uplifting movement that may contribute to neural fold formation and eventually to neural tube closure. This notion is supported by the results of experiments discussed earlier, in which neural plates were isolated with and without margins of adjacent epidermis. If epidermal margins were present, the isolated plates rolled into complete tubes and elongated normally. However, if no epidermis was included, only a U shape was achieved, and the neural plates did not elongate as much as they normally do.

Together, the observations discussed in this section suggest that neural tube closure is based on at least three types of cellular behavior: apical constriction, intercalation, and crawling. Again, the *principle of overlapping mechanisms* applies. All three cellular behaviors normally work together to bring about neural tube closure. If one of them is disturbed, as by eliminating a prerequisite for cell crawling by removing the epidermis adjacent to the neural tube, the other two cell behaviors produce a partial but less-than-perfect result.

12.3 The Role of Induction in Axis Formation

The neural tube is an organ rudiment that is always formed in association with an underlying notochord, laterally adjacent somites, and other mesodermal structures (see Fig. 9.19). This set of mostly dorsal organ rudiments in vertebrates is collectively called the ***embryonic axis.*** (The term "axis" meaning this set of organ rudiments

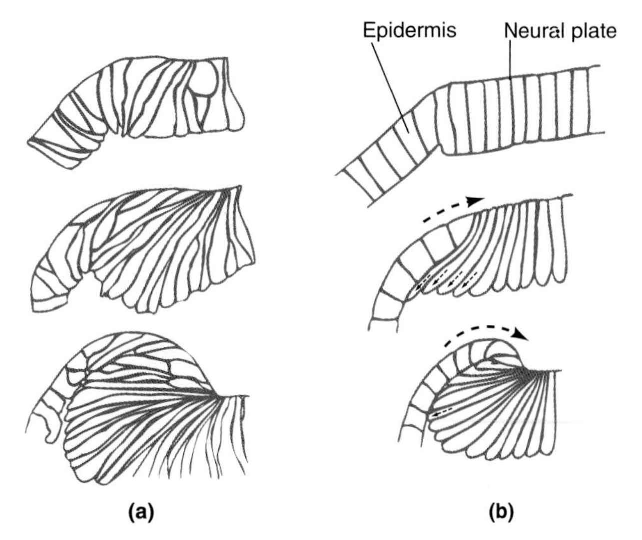

Figure 12.15 Cell crawling at the epidermis–neural plate boundary. **(a)** Enlarged tracings show cells from transverse sections of newt neurulae at successive stages. **(b)** Interpretative diagrams highlight major cell movements (arrows). The neural plate cells seem to crawl with their basal ends under the adjacent epidermis cells, thereby attenuating their apical surfaces. The combination of basal crawling and apical constriction appears to generate a movement that lifts up the neural folds and rolls them toward the dorsal midline.

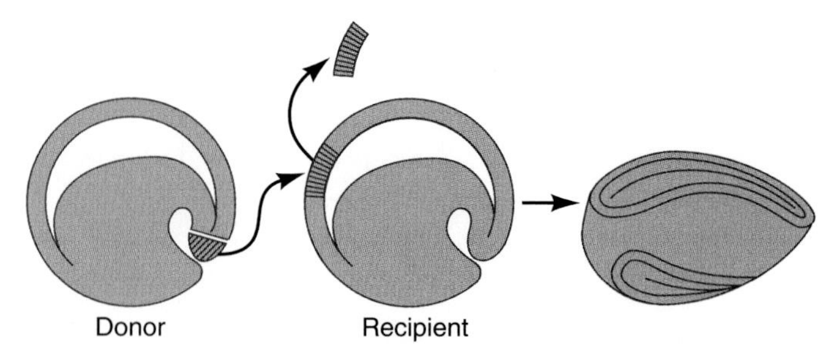

Figure 12.16 Spemann's early version of the organizer experiment. A dorsal blastopore lip from a newt early gastrula was transplanted to the ventral ectoderm of a recipient at the same state of development. The recipient formed two neural plates and later two embryonic axes. However, since the donor and the recipient were of the same species, their contributions to the secondary axis could not be distinguished with certainty.

should not be confused with the use of the same term to mean a line of orientation, such as the anteroposterior axis of an embryo.) The stereotypical arrangement of different organ rudiments in the embryonic axis (in the former sense) suggests that their development is coordinated by inductive events collectively known as ***axis induction.*** Indeed, the famous organizer experiment of Spemann and Mangold (1924) showed that all axis organs in newts are either formed or induced by the *dorsal blastopore lip* of the early gastrula.

THE DORSAL BLASTOPORE LIP ORGANIZES THE FORMATION OF AN ENTIRE EMBRYO

The organizer experiment was the culmination of Spemann's long quest to understand how the cells that form the embryonic axis organs are determined. From twinning experiments with newt eggs, Spemann already knew that the ability to form an axis depends on the presence of cytoplasm from the dorsal half of the egg (see Fig. 1.15). From transplantation experiments before and after gastrulation, he also knew that the neural ectoderm changes its state of determination during gastrulation (see Fig. 6.11).

Studying the progress of determination in different regions of the newt embryo, Spemann transplanted the upper blastoporal lip of early gastrula donors to the prospective flank epidermis of host embryos at the same stage of development (Fig. 12.16). Most of the host embryos developed two neural plates. When three specimens were fixed and sectioned after neural tube closure, two of them were found to have in their flanks an additional embryonic axis, including a neural tube, notochord, and somites. This experiment was first done in 1916, at a time when fate maps and trajectories of gastrulation movements in amphibians were not yet available. In the absence of such basic information, Spemann thought that the *entire* additional axis had been formed by the transplant—the neural tube from the superficial layer of the blastopore lip, which he presumed was ectodermal, and the mesodermal organs from the deep layer. He did not realize at the time that much or all of the blastopore lip was to involute and that the involuting marginal zone (IMZ) of newt gastrulae would give rise to mesoderm only. (This is in contrast to *Xenopus*, where the deep IMZ layers form mesoderm while the superficial IMZ layer contributes to endoderm, as shown in Fig. 10.11.)

When the experiment was resumed in 1921, Spemann had learned that the dorsal blastopore lip of an *early* gastrula does not include an ectodermal layer. He had also developed a *heterospecific transplantation* technique, in which he used donor and host embryos from different species, so that he could distinguish graft-derived tissues from host tissues on the basis of their different pigmentation. Having become the head of the Zoology Department at the University of Freiburg in Germany, Spemann assigned the task of repeating the transplantation of the blastopore lip to one of his graduate students, Hilde Proescholdt. Her experiment became the crowning achievement of Spemann's efforts to understand the determination of axial organs in vertebrate embryos. It is published under the names of Spemann and Mangold (1924) because Hilde had married a junior colleague of Spemann's, Otto Mangold, and adopted his name. (The paper has been reviewed in English by Spemann, 1938, and Hamburger, 1988.)

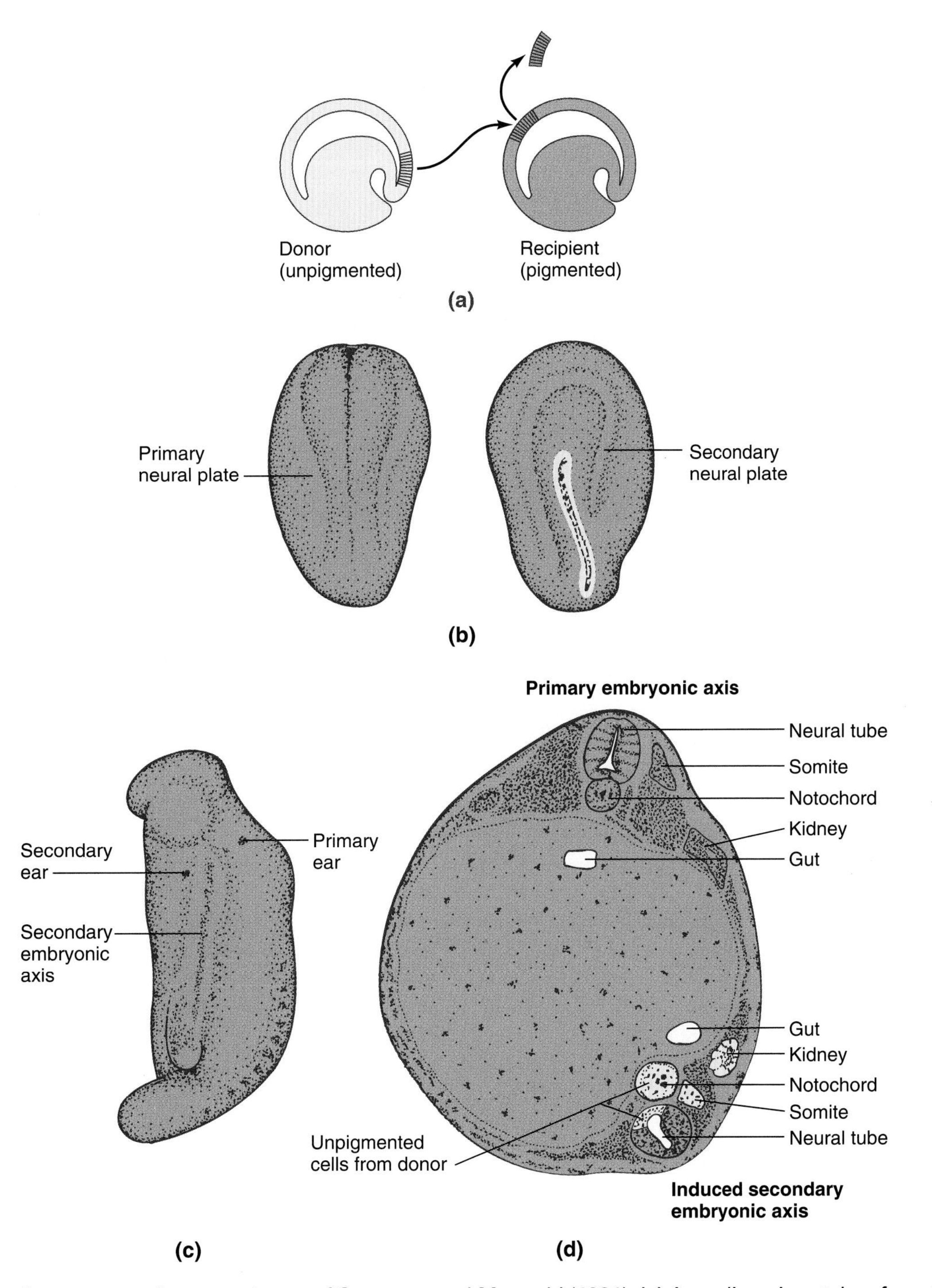

Figure 12.17 The famous organizer experiment of Spemann and Mangold (1924). **(a)** A median piece taken from just above the dorsal blastopore lip of a gastrula of *Triturus cristatus* (an unpigmented newt species, shown here in beige color) was transplanted to the ventral ectoderm of a gastrula of *Triturus taeniatus* (a pigmented species, shown here in green color). **(b)** At the neurula stage, the recipient formed a pigmented secondary neural plate with a streak of unpigmented tissue in the notoplate area. **(c)** The recipient proceeded to form a secondary embryonic axis. **(d)** Transverse section of the recipient shows the structures derived from the unpigmented transplant and the pigmented recipient.

PROESCHOLDT transplanted a median piece from the dorsal blastopore lip. In terms of the gastrula regions defined in Section 10.3, the transplant consisted of *involuting marginal zone* and, depending on the developmental stage of the donor, a smaller or greater portion of *noninvoluting marginal zone* (see Fig. 10.11). The transplant was taken from an (unpigmented) *Triturus cristatus* gastrula and implanted in prospective ventral epidermis of a (pigmented) *Triturus taeniatus* gastrula (Fig. 12.17). The graft performed the same morphogenetic movements it would have carried out normally: it involuted and extended into a long strip underneath and within the host ectoderm. As in Spemann's previous experiments, the host embryo developed an additional embryonic axis at the implantation site. However, because the host tissue was pigmented while the transplant was not, it was now unmistakable that much of the additional axis had been formed by the host.

The blastopore transplantation experiments were laborious because, due to the lack of sterile techniques, many embryos died before they could be analyzed. Five successful cases were described in detail. In the most completely developed case, the donor had been an advanced gastrula. In the host, the transplant involuted almost completely. Two days after the operation, when the embryo had developed to the neurula stage, a second neural plate complete with neural folds was clearly visible on the flank. Most of the neural plate was pigmented and therefore derived from the host. Only a median strip, called the *notoplate* in current terminology, was unpigmented and thus derived from the transplant. A day later both the original host embryo and the secondary embryo had advanced to the tail bud stage. The secondary embryo was shorter and lacked the anterior head parts. Fixation and sectioning showed that the secondary embryo had a complete set of axial organs including a neural tube, a notochord, somites, gut, and embryonic kidneys.

The outstanding feature of the secondary embryo was that it was an almost complete and normally proportioned embryo and also a chimera consisting of both host and graft tissues. The neural tube consisted almost entirely of host tissue, except for the notoplate. The notochord was entirely unpigmented—that is, derived from the transplant. The somites were composed of transplant and host tissue, while the other axial structures were mostly formed by the host.

Questions

1. Could the lack of a head in the secondary embryo shown in Figure 12.17 be related to the fact that the graft had come from a late gastrula donor? Explain.
2. The notoplate of the secondary embryo shown in Figure 12.17 is obviously derived from the graft. Can this be reconciled with the view that the involuting marginal zone does not contribute to ectoderm? If so, which experimental parameters may decide whether the notoplate of the secondary embryo is contributed by the host or by the graft?

Three major conclusions can be drawn from the organizer experiment. First, the grafted dorsal blastopore lip *developed according to its fate:* It underwent the normal morphogenetic movements and proceeded to form, for the most part, notochord. Second, the graft *dorsalized the host's mesoderm:* Tissue that would normally have formed hypodermis, blood, or other ventral mesodermal structures instead contributed to secondary somites and kidneys. Third, the *graft acted as a neural inducer.* It stimulated the host ectoderm to contribute most or all of a secondary neural plate, which closed into a neural tube running parallel to the graft-derived notochord. In short, the transplanted blastopore lip developed in accord with its own fate and marshaled the host tissues in such a way as to produce one integrated embryo. To emphasize this remarkable capability of the dorsal blastopore lip, Spemann called it the ***organizer.***

Axis induction in birds and mammals appears to occur much as it does in amphibians. As described in Section 10.4, the primitive groove and pit in birds correspond to the blastopore in amphibians, and Hensen's node is the equivalent of the dorsal blastopore lip. If Hensen's node from a duck donor is grafted beneath the epiblast of a chick host, it induces a secondary axis including neural tube, notochord, and somites (Fig. 12.18). Corresponding experiments with mice have yielded similar results (Beddington, 1994).

DOES THE ORGANIZER HAVE "STRUCTURE"?

The next question addressed by Spemann was to what extent the polarity, handedness, and spatial order of organ rudiments in a secondary (experimentally induced) embryo depended on the "structure" of the transplanted organizer. Conversely, how much of the seemingly organizing effect of the dorsal blastopore lip was really contributed by the host, which of course had its own polarity axes and biases with regard to cell determination? The answer to this question turned out to be complex (Spemann, 1931, 1938).

On the one hand, the transplanted blastopore lips gastrulated in accord with their own animal-vegetal orientation, but depending on the site of implantation, the transplants' gastrulation movements were more or less overpowered by those of the host. Eventually, almost all surviving secondary embryos were parallel to their primary counterparts, and the ear rudiments of the two embryos tended to be at the same level (see Fig. 12.17). These results indicate that a transplanted dorsal blastopore lip does not act unilaterally on an amorphous host area. Rather, the "organizer" interacts with host tissues that have a great deal of organization to themselves. On the other hand, transplanted blastopore lips do impose their handedness on the induced embryo. This was shown by Goerttler (reviewed by Hamburger, 1988), who replaced a lateral half of one blastopore lip with the contralateral half of another blastopore lip, thus creating, for example, a transplant consisting of two adjacent left halves. This resulted in the formation of two parallel left neural folds.

Another way in which Spemann tested the need for organizer structure was to destroy it; he transplanted blastopore lips that had been minced with glass needles or squeezed between glass slides. Nevertheless, secondary embryos formed, although the results were more variable than those obtained with live dorsal blastopore lips, indicating that organizer structure was not necessary but helped. Similar results were obtained

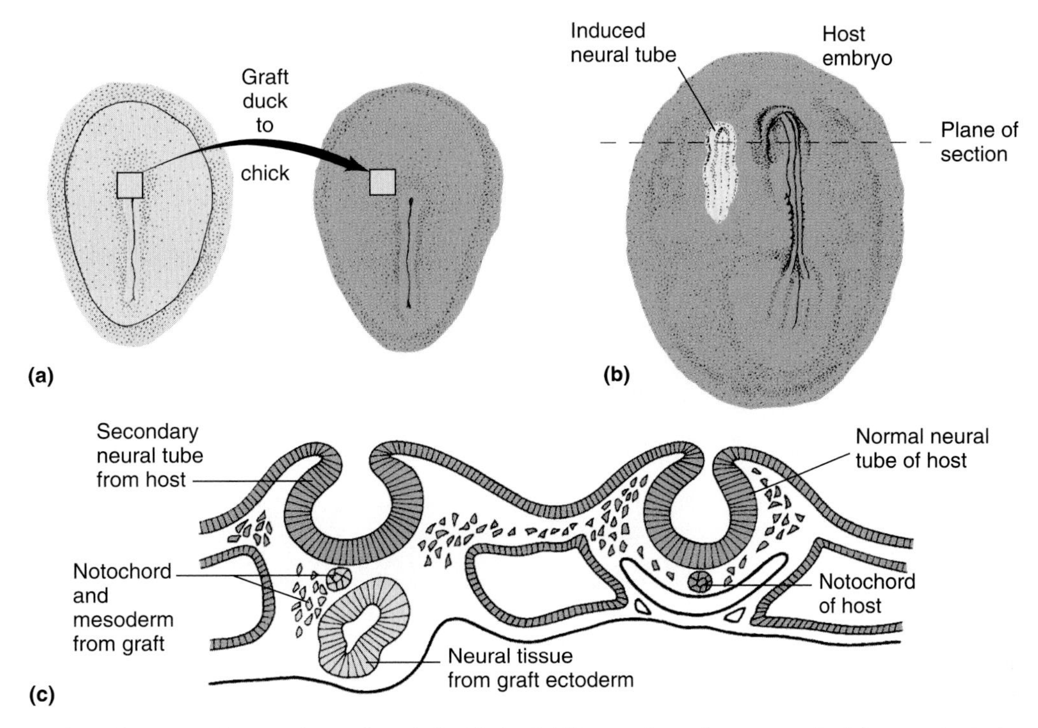

Figure 12.18 Induction of a secondary embryonic axis in birds. **(a)** Hensen's node was grafted from a duck donor to a chicken recipient. **(b)** The recipient developed an accessory neural tube. **(c)** Histological sections revealed that the graft itself formed a notochord and a neural tube underneath. In addition, the graft induced the formation of a neural tube from host tissue.

with inducers killed by heat, drying, freezing, or immersion in alcohol (Bautzmann et al., 1932; Spemann, 1938).

These results reveal a complex situation, which is commonly found between inducing and responding tissues. To an extent, inductive signals are *instructive* as they impart specific signals to the responding tissue. However, to a varying extent, the specificity of induction relies on the responding tissue's own polarities and biases, which have originated from previous inductive interactions and/or from cytoplasmic localization. In this respect, inductive signals merely trigger or release a set of prepatterned events. We will revisit this issue in Section 12.5, where we will discuss some of the molecular mechanisms involved in axis induction.

AXIS INDUCTION SHOWS REGIONAL SPECIFICITY

The instructive role of organizer tissue was also revealed by experiments triggered by the observation that the secondary embryos were usually missing either anterior or posterior parts. As possible causes of this incompleteness and its variability, Spemann and coworkers considered several experimental parameters. Some of these, including the exact place of the transplant in the host as well as the orientation of the transplant, had only minor effects, and henceforth most investigators used a simplified method of transplanting potential organizer tissues. Instead of painstakingly patching the graft into the host's ventral ectoderm, they simply inserted the graft into the host's blastocoel through a slit in the blastocoel roof (Figs. 12.19 and 12.20). During gastrulation, the host's vegetal base would then push the graft against the ventral ectoderm. This method of transplantation became known as the *insertion method*.

Parameters that did have major effects on the outcome of the organizer experiment included when and where the inducing tissue was isolated from the donor. In Proescholdt's best-developed embryo, the one shown in Figure 12.17, the graft had come from an *advanced* gastrula donor. In contrast, when Spemann (1931) grafted the blastopore lip from an *early* gastrula, the additional embryo had a complete head but no tail (see Fig. 12.19a). Thus, early-involuting mesoderm, which moves far anteriorly, tends to induce anterior axial structures, whereas late-involuting mesoderm, which remains in a more posterior position, is more likely to induce trunk and tail.

A regional specificity of the inducing mesoderm was also demonstrated by Otto Mangold (1933). He removed the neural plate from an early neurula to expose the underlying tissue, which had been the deep layer of the dorsal blastopore lip before it involuted. After involution, the same tissue is referred to as ***chordamesoderm,*** although its fate includes not only notochord but also adjacent somites. In Otto Mangold's experiment, the chordamesoderm was divided into four parts along the anteroposterior axis, and the isolated parts were inserted into the blastocoel of early host gastrulae

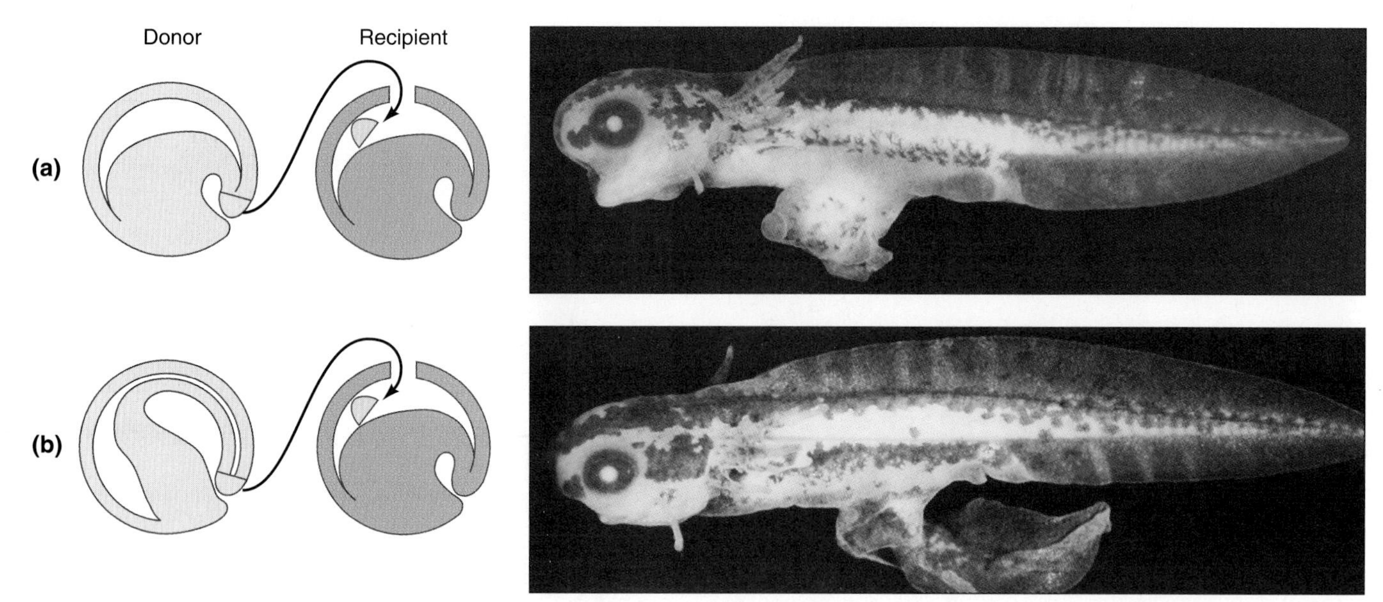

Figure 12.19 Stage dependence of the inductive capacity of the dorsal blastopore. **(a)** The dorsal blastopore lip from an early gastrula, upon insertion into the blastocoel of a host, induces the formation of a secondary head. **(b)** Inserting a late blastopore lip in the same way causes the formation of secondary trunk and tail.

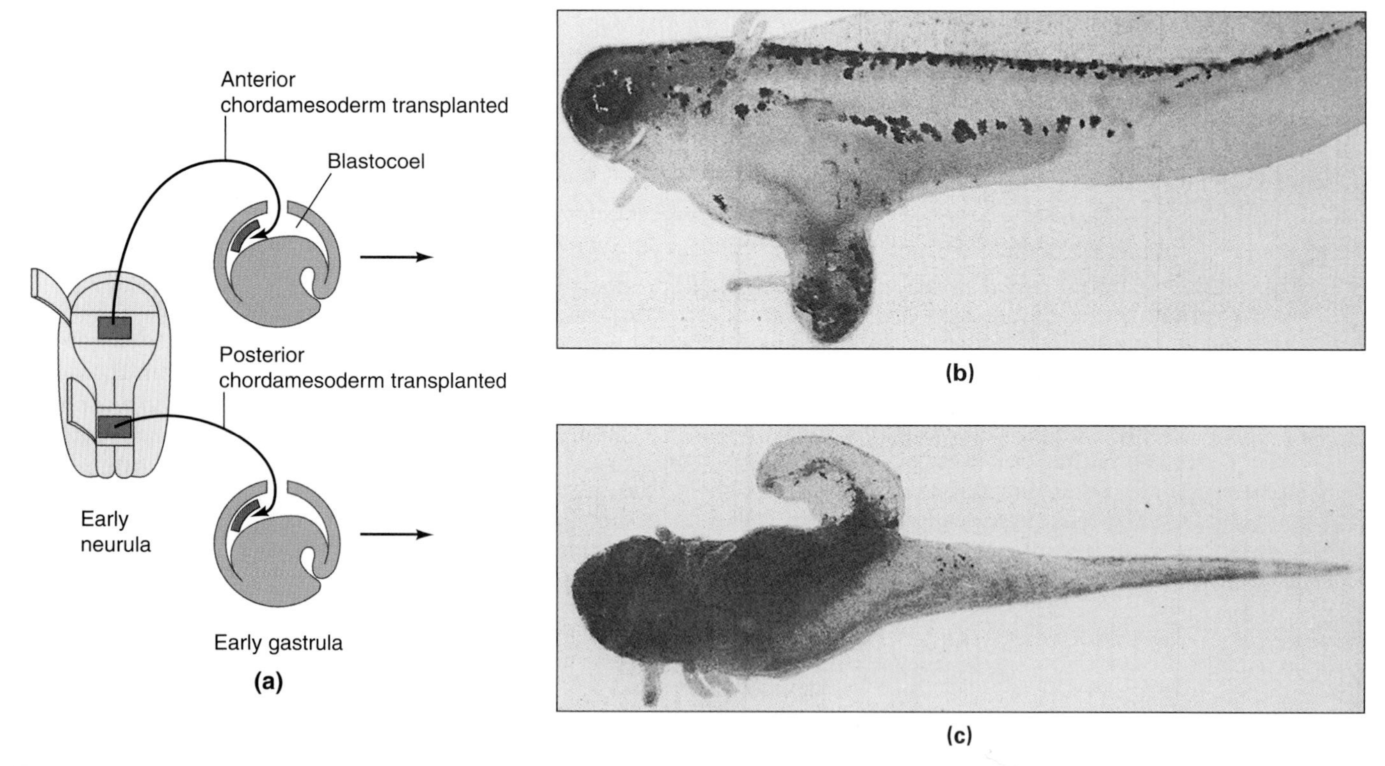

Figure 12.20 Regional specificity of neural induction. **(a)** Design of an experiment carried out by Otto Mangold (1933). He obtained different regions of chordamesoderm (red color) after removing the overlying neural plate from early neurulae and inserted the chordamesoderm pieces into blastocoels of early gastrulae. **(b)** If anterior chordamesoderm was inserted, the recipient formed a secondary head with balancers, forebrain, and eyes. **(c)** Posterior chordamesoderm inserted in the same way induced a secondary trunk and tail.

(Fig. 12.20). The secondary structures that protruded from the bellies of the hosts varied depending on which chordamesoderm part had been grafted.

The first chordamesoderm quarter induced mostly structures that lie in front of the brain in the embryo. This result was similar to the one obtained after the grafting of early dorsal blastopore lip, which forms anterior chordamesoderm after involution. The second chordamesoderm quarter induced a head with brain, nose, eye, and ear vesicles. The third quarter induced primarily hindbrain, spinal cord, and musculature, and the fourth quarter induced spinal cord, somites, kidney, and tail. There was considerable overlap among the structures induced by successive areas of chordamesoderm. However, the fourth-quarter graft had an effect similar to the one obtained by grafting *late* dorsal blastopore lip, which forms posterior chordamesoderm after involution. Clearly, blastopore lip removed at different stages of gastrulation, and the corresponding anteroposterior regions of chordamesoderm from late gastrulae, induce different sets of dorsal structures. This observation indicates that the organizer is a dynamic aggregate of tissues that will occupy different positions after gastrulation and have different inductive capabilities.

Indeed, more recent experiments indicate that the equivalents of Spemann's organizer in zebrafish and mouse embryos are split in components located at the *anterior and posterior* ends of the developing dorsal axis. In zebrafish gastrulae, the future notochord is located at the vegetal (posterior) margin of the *embryonic shield* and, upon transplantation, it shows organizer activity (Ho, 1992; Shih and Fraser, 1996). However, brain development also depends on a small population of anterior ectodermal cells. Removal of these cells causes perturbations in brain patterning that extend far beyond the fate of the removed cells, whereas heterotopic transplantation of the anterior cells into posterior neural plate causes surrounding host cells to express anterior marker genes (Houart et al., 1998). Both results indicate that these anterior ectodermal cells release inductive signals that affect anteroposterior patterning.

Similarly, in mouse embryos, heterotopic transplantation of Hensen's node causes ectopic axis formation as mentioned earlier. However, anterior embryonic endoderm and anterior extraembryonic tissues are also capable of changing the expression of marker genes along the anteroposterior axis (Tam and Steiner, 1999). Thus, it appears that in the course of vertebrate evolution different embryonic tissues have come to participate in organizing the anteroposterior pattern of dorsal organs.

12.4 Pathways of Neural Induction

The induction of an embryonic axis affects all three germ layers: The secondary embryo induced in the organizer experiment comprises organs derived from ectoderm as well as from mesoderm and endoderm (see Fig. 12.17). The events in the ectoderm, being most conspicuous, have attracted the greatest attention. Just as neurulation is the most salient part of axis formation, the most extensively studied aspect of axis induction is ***neural induction,*** the development of ectoderm into a neural tube under the influence of adjacent dorsal mesoderm. In this section, we will focus on neural induction, as Spemann and his coworkers did, before we return to the entire process of axis formation at the end of the chapter when we discuss some molecular aspects of axis induction.

THERE ARE TWO SIGNALING PATHWAYS—PLANAR AND VERTICAL—FOR NEURAL INDUCTION

Thinking very methodically about his experiments, Spemann considered two signaling pathways for neural induction (Fig. 12.21). First, in the blastula or early gastrula, when the organizer forms the dorsal *involuting marginal zone,* it might send ***planar signals*** through the plane of the ectoderm. At this stage, the prospective neural plate occupies a short and wide area that could easily be reached by a planar signal from the dorsal blastopore lip (R. Keller et al., 1992b). Second, after involution, when the organizer has formed the *chordamesoderm,* it might send ***vertical signals*** into the overlying ectoderm. At this later stage, the prospective neural plate has extended in anteroposterior direction and is more accessible to vertical signals from the chordamesoderm.

Originally, Spemann favored the notion of planar signals. However, the distinct inductive responses to different segments of chordamesoderm (see Fig. 12.20) were more readily interpreted by vertical signals. Additional results by Holtfreter (1933), another collaborator of Spemann's, were also indicative of vertical signals. Holtfreter stripped early axolotl gastrulae of their fertilization envelopes and kept them in a hypertonic salt solution. As a result, the gastrulation movements took a completely abnormal course called ***exogastrulation.*** The prospective endoderm and mesoderm, instead of turning inside the embryo, turned to the outside and left the ectoderm behind as an empty hull (Fig. 12.22). Under these circumstances, planar signals could still pass through a neck of tissue connecting the exogastrulated chordamesoderm with the ectodermal hull, while the possibility of vertical signals was excluded. In the developing exogastrulae, endodermal and mesodermal derivatives—including notochord, somites, and embryonic kidney—developed more or less normally. However, the ectodermal hull, deprived of its underlying mesoderm, failed to develop neural tissue that could have been identified on the basis of its *appearance in histological sections.* Thus, although the inducing tissue developed almost normally, and although planar inducing signals

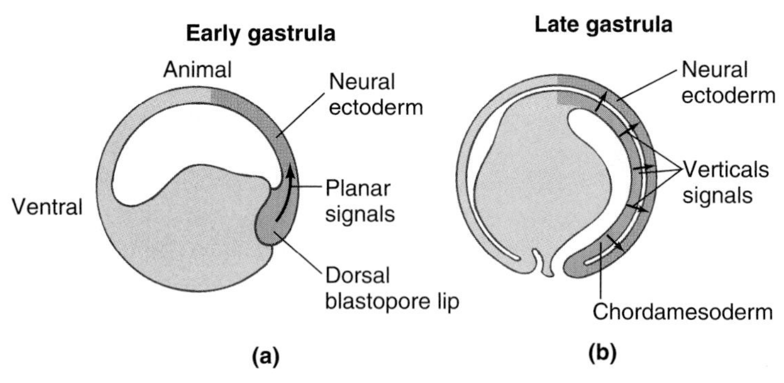

Figure 12.21 Planar and vertical signals in neural induction. **(a)** Early gastrula. The planar signals travel within the plane of the marginal zone from the dorsal blastopore lip to the adjacent neural ectoderm. **(b)** Late gastrula. The vertical signals pass from the same inducing tissue, now chordamesoderm, into the overlying neural ectoderm. The relative importance of the two sets of signals varies between species.

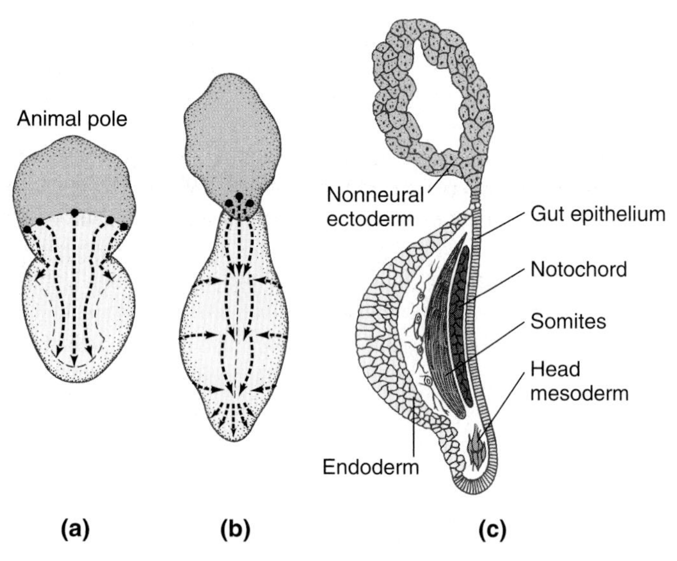

Figure 12.22 Exogastrulation in an axolotl embryo. **(a, b)** Marginal zone cells move (arrows) toward the blastopore region (constriction), but instead of involuting, these and the vegetal base cells are pushed outside, where the mesoderm nevertheless undergoes convergent extension. **(c)** Median section through exogastrula, showing an empty ectodermal hull separated by a constricted blastopore region from endodermal and mesodermal structures. The ectoderm forms no morphologically identifiable neural structures.

could have passed to the ectoderm, no neural response was detected.

From the results of these experiments with newt embryos, Spemann and his coworkers concluded that vertical signals were indispensable for neural induction, and that planar signals by themselves were not sufficient. Later investigators reached almost the reverse conclusion from *molecular data* obtained with *Xenopus* embryos.

PLANAR INDUCTION PLAYS A MAJOR ROLE IN *XENOPUS* EMBRYOS

Repeating the organizer experiment with *Xenopus* embryos confirmed the results of Spemann and Mangold (Gimlich and Cook, 1983). However, subsequent experiments using mostly molecular markers of neural development have shown that neural induction relies more on planar induction in *Xenopus* than it does in newts and other salamanders (Slack and Tannahill, 1992).

Studying neural induction in *Xenopus*, Kinter and Melton (1987) found that ectoderm cells express N-CAM, a neural *cell adhesion molecule* discussed in Section 11.2, as an early response to neural induction. They showed that in normal embryos, neural plate and neural tube express N-CAM while prospective epidermis does not. In in vitro experiments, isolated ectoderm did not synthesize N-CAM unless it was in contact with inducing mesoderm. Having established the usefulness of N-CAM as a marker for neural induction, the researchers examined the expression of N-CAM in exogastrulae. Unexpectedly, they found that exogastrulae synthesized nearly as much N-CAM as normal embryos, and that almost all the N-CAM synthesis occurred in the ectodermal portion of the exogastrula. Thus, *Xenopus* ectoderm did express a molecular marker for neural development under conditions that allowed planar but not vertical induction.

As an alternative test of *Xenopus* ectoderm development under conditions of planar induction only, Keller and Danilchik (1988) isolated patches comprising involuting and noninvoluting marginal zone from the dorsal side of early gastrulae (Fig. 12.23a, b). To keep these patches from curling up, they combined two of them with their deep zones face to face, creating a so-called Keller sandwich. This preparation allowed planar signals to travel between the involuting marginal zone (dorsal blastopore lip, organizer) and the noninvoluting marginal zone (the part that would normally form neural plate). Because the sandwich preparation did not allow involution to occur, prospective chordamesoderm never came to underlie neuroectoderm, and hence no vertical signals could pass between them. Despite these constraints, other morphogenetic movements and cell differentiations in the sandwich resembled those in intact embryos (Fig. 12.23c). In particular, the noninvoluting marginal zone formed a neural plate, which developed into a mass of cells that resembled multipolar neurons and was later shown to stain with a neuron-specific molecular probe.

On the basis of Otto Mangold's experiments with newts (see Fig. 12.20), it had been assumed that the *anteroposterior pattern* of different brain regions and the spinal cord relied on vertical signals from the chordamesoderm. However, the new results obtained with

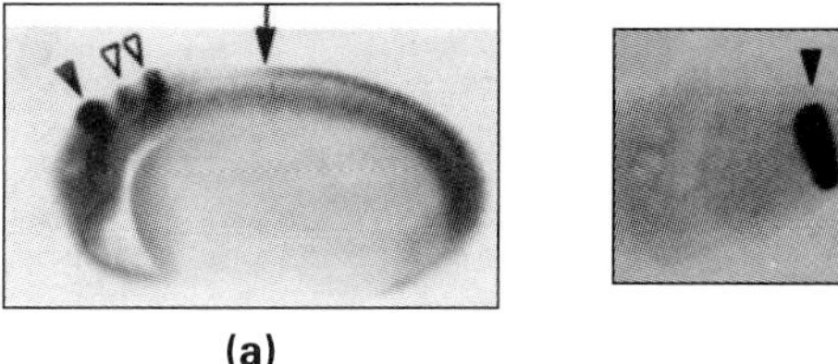

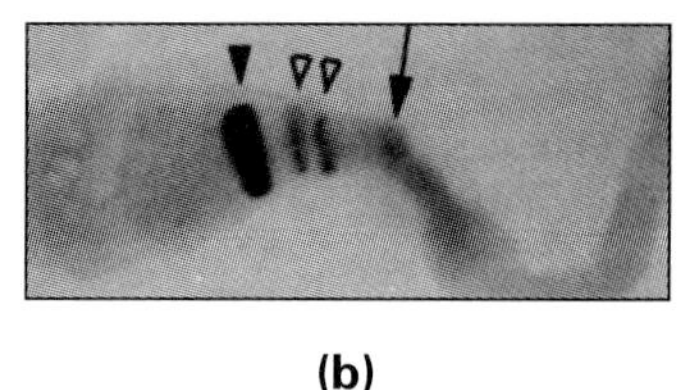

Figure 12.24 Expession of neural marker genes in *Xenopus* whole embryos and sandwiches of dorsal gastrula tissue. **(a)** Whole embryo, anterior to the left, dorsal side up. The dark areas in the brain and spinal cord are mRNAs labeled by in situ hybridization (see Method 8.1). The anteriormost band (filled arrowhead) represents mRNA transcribed from the *engrailed-2*$^{+}$ gene, which is expressed selectively at the boundary between midbrain and hindbrain. The doublet of bands (open arrowheads) represents Krox-20 mRNA, which is synthesized in two hindbrain segments. The dark streak (to the right of the arrow) represents XlHbox6 mRNA, which accumulates throughout the spinal cord. **(b)** Tissue developed from a sandwich prepared as shown in Figure 12.23, animal pole to the left. The narrowest region (to the right of the arrow) is prospective spinal cord, the flared region to the left is prospective brain, and the curved region to the right is prospective mesoderm. Note that contact between the neural region and the mesoderm is strictly planar. Nevertheless, the marker mRNAs are synthesized in the same pattern as in the intact embryo.

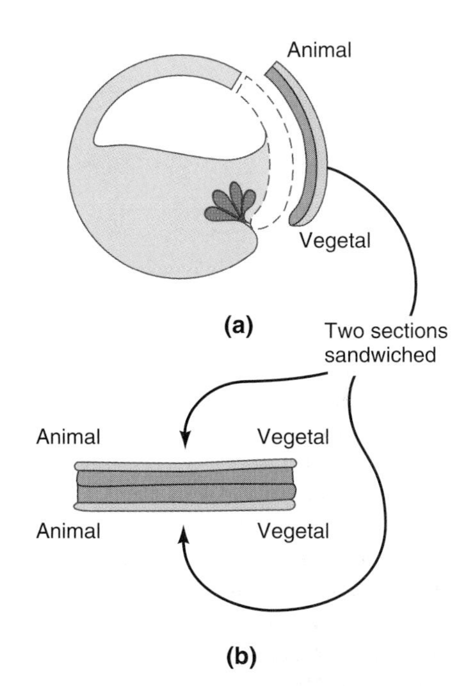

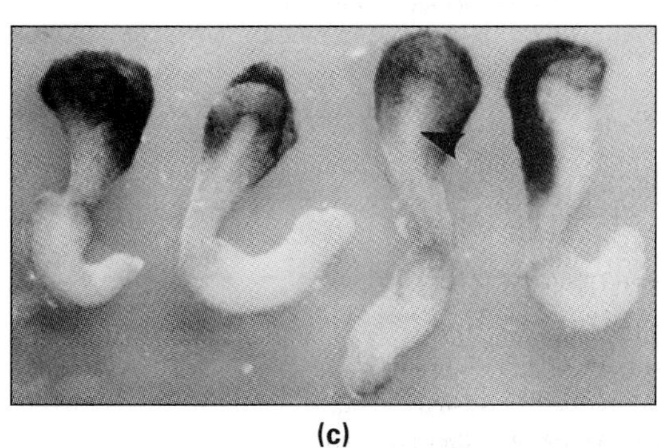

Figure 12.23 Test for neural induction in *Xenopus* by planar signals only. **(a, b)** Two dorsal flaps of tissue comprising involuting marginal zone, noninvoluting marginal zone, and animal cap were cut from two early gastrulae and stuck together with their inner cell layers face to face and with their animal-vegetal axes aligned. **(c)** After a few hours of culture, the involuting marginal zone (below arrowhead) formed notochord and somites while the noninvoluting marginal zone (above arrowhead) formed neural tissue.

Xenopus raised the question whether this pattern could also be generated by planar induction alone.

TO explore this possibility, Doniach et al. (1992) used *in situ hybridization* (see Method 8.1) to detect the mRNA products of certain marker genes that are expressed only in specific regions of the central nervous system (CNS). These methods made it possible to identify the products of three marker genes (*engrailed-2*$^{+}$, *Krox-20*$^{+}$, and *XlHbox6*$^{+}$) as a distinct series of transverse stripes in the developing midbrain, hindbrain, and spinal cord of intact embryos (Fig. 12.24a). To see whether this pattern would originate in the absence of vertical induction signals, the investigators prepared Keller sandwiches, which allowed only the exchange of planar signals between the dorsal involuting marginal zone and prospective neuroectoderm (see Fig. 12.23). In culture, these isolates underwent convergent extension and neural differentiation. When the developed tissues were tested for expression of region-specific marker genes, they indeed showed the same pattern of transverse stripes as the normal embryos (Fig. 12.24b). The investigators concluded that a pattern closely resembling the anteroposterior pattern of the normal brain and spinal cord can be induced by planar signals alone.

In a similar experiment, Zimmerman and coworkers (1993) monitored the expression of another neural marker gene, *XASH-3*$^{+}$, in normal *Xenopus* embryos and in Keller sandwiches. They found that the normal expression pattern of the gene in early neurulae includes two longitudinal stripes on both sides of the midline. These stripes are also expressed in cultured Keller sandwiches, indicating that some mediolateral patterning of the neural plate can occur independently of vertical neural induction.

Questions

1. The results shown in Figure 12.24 were obtained with *in situ hybridization* probes to three different mRNAs. Can several such probes be applied simultaneously to the same tissue section or embryo?
2. As an alternative to in situ hybridization, what other method could the investigators have used to compare patterns of region-specific gene expression in embryos versus Keller sandwiches?

3. The stunning results reported by Doniach and coworkers could be ascribed to an unnoticed planar migration of mesoderm cells from deep involuting marginal zone into the adjacent noninvoluting marginal zone and a subsequent response of the "invaded" noninvoluting marginal region to vertical signals from the putative migratory cells. How could the investigators have tested (and they did!) whether such a planar cell migration did indeed occur?

The observations discussed so far indicate that neural induction in amphibians depends on at least two types of signal, planar and vertical. The planar signals are sent within the outer plane of the early gastrula from the dorsal blastopore lip (organizer) to the adjacent dorsal ectoderm. The vertical signals are sent in the late gastrula from the same organizer, then called the chordamesoderm, to the overlying dorsal ectoderm. The classical results indicate that in newt embryos vertical signals play a dominant role whereas planar signals by themselves are insufficient. In contrast, more recent experiments with *Xenopus laevis* show that in this species planar signals alone cause neural differentiation, including the proper anteroposterior pattern of gene expression in midbrain, hindbrain, and spinal cord. However, forebrain structures seem to require vertical inductive signals as well. These results attest to the specificity of planar neural induction in *Xenopus,* but they also point out the necessity of vertical inductive signals, in particular for forebrain development.

Why do newts and *Xenopus* differ in relative importance of planar and vertical neural induction signals? This difference may be related to other differences that also distinguish these two forms. First, *Xenopus* mesoderm is located deep in the gastrula (see Fig. 10.11), whereas newt mesoderm includes the surface layer. *Xenopus* development is also very rapid, among the fastest of the amphibians. Species that develop rapidly often evolve in a way that emphasizes the use of developmental signals exchanged during earlier stages; thus, planar neural induction, which occurs before vertical induction, plays a greater role in *Xenopus* than it does in newts. Indeed, the experiments described next indicate that even earlier events in *Xenopus* embryos prepare, or bias, the dorsal ectoderm to respond to the subsequent neural induction signals.

NEURAL INDUCTION IS A MULTISTEP PROCESS

Whether ectoderm develops into neural plate or epidermis depends not only on the proximity of the *inducing* tissue but also on the preparedness of the *responding* tissue. (As explained in Section 9.4, the preparedness of a responding tissue to a specific inductive signal is called the competence, or bias, of the responding tissue.) At least in *Xenopus,* the dorsal ectoderm is more competent than the ventral ectoderm to respond to the neural induction signals that emanate from the dorsal mesoderm.

Differences between dorsal ectoderm, which will form neural plate, and ventral ectoderm, which will form epidermis, can be traced back to the 8-blastomere stage in *Xenopus.* This was shown in experiments using as a molecular marker *Epi 1,* a cell surface antigen that is present in future epidermal cells but not in prospective neural cells (London et al., 1988). The investigators isolated blastomeres and blastula regions of *Xenopus* embryos and kept them in culture until control embryos had reached the midneurula stage. The descendants of the isolated cells were then fixed for *immunostaining* (see Method 4.1) with an antibody against Epi 1. Descendants of ventral animal cells isolated as early as the 8-blastomere stage expressed the Epi 1 antigen more strongly and more consistently than the descendants of dorsal animal cells.

In subsequent experiments, Savage and Phillips (1989) used the same immunostaining procedure to monitor the *inhibitory* effect of neural inducers on Epi 1 expression. Indeed, ventral ectoderm kept in edge-to-edge contact with dorsal blastopore lip did not express Epi 1. Similarly, chordamesoderm sandwiched between ventral ectoderm layers inhibited Epi 1 expression.

The results of monitoring Epi 1 expression indicate that ventral animal blastomeres are already biased toward forming epidermis but that this early predisposition can be overridden by neural inducers. Conversely, the early bias of dorsal animal cells is not sufficient for them to form neural plate: they still express Epi 1 at low levels, and they form epidermis when explanted as a group during the blastula stage. However, subsequent neural induction signals completely inhibit Epi 1 expression and determine the cells for neural development. In short, a stepwise inhibition of Epi 1 expression in dorsal ectodermal cells is paralleled by a stepwise determination toward neural plate formation.

In a complementary experiment, Sharpe and coworkers (1987) monitored the synthesis of two marker mRNAs that characterize neural development: one encoding N-CAM and the other transcribed from *XlHbox6*$^{+}$, a gene expressed specifically in posterior neural cells of late gastrulae. In response to inductive signals from chordamesoderm, both markers were synthesized in larger amounts in dorsal ectoderm than in ventral ectoderm. Control pieces of dorsal and ventral ectoderm that had been kept without chordamesoderm did not accumulate either marker mRNA. The greater *competence* of the dorsal ectoderm—that is, its better ability to respond to inductive signals from chordamesoderm, must result from signals that were received earlier by dorsal ectoderm but not by ventral ectoderm.

Taken together, the experiments discussed in this chapter and in Chapter 9 support the view that neural induction is a multistep process (Fig. 12.25). The

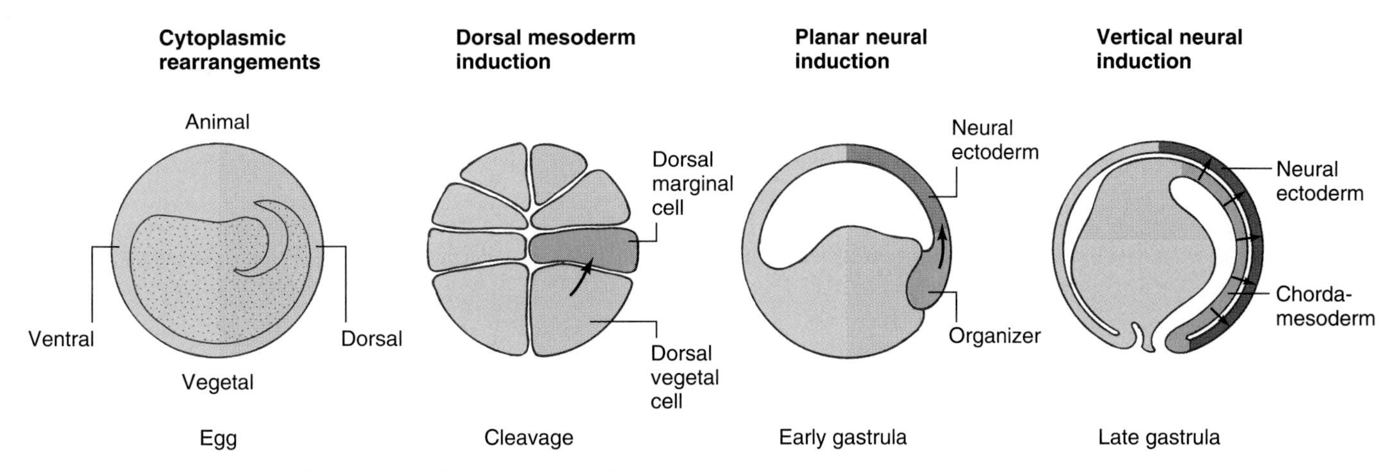

Figure 12.25 Neural induction as a multistep process. See text.

earliest dorsoventral bias is introduced by cytoplasmic movements in the fertilized egg during the first cleavage cycle. This bias is evident from the reduced expression of Epi 1 in dorsal animal blastomeres and from the enhanced response of dorsal animal caps at the blastula stage to mimics of mesoderm inducers (see Section 9.6). During cleavage, only dorsal vegetal blastomeres are able to induce animal cells to form dorsal mesoderm. The dorsal mesoderm cells give rise to the dorsal blastopore lip (Spemann's organizer), which sends planar neural induction signals at the early gastrula stage. After the dorsal blastopore lip has involuted and formed the chordamesoderm, it augments neural induction by sending vertical signals. The latter seem to be especially important for inducing anterior brain structures; these structures develop from the anterior neuroectoderm, which is located most distant from the source of the planar signal.

In the series of events that lead to neural induction, individual steps can be prevented experimentally with the result of a reduced level of neural development. These observations are in accord with the *principle of overlapping mechanisms,* which pervades many developmental processes (see Chapters 4, 10, 18, and 22). This principle is especially evident in inductive interactions, not only in neural induction, but also in the induction of the nose, eye, ear, and heart (Jacobson, 1966; Jacobson and Sater, 1988). The overlapping effect of several mechanisms in neural induction may also help to explain how the differences between newts and *Xenopus* with regard to planar and vertical induction may have evolved. If a biological function is based on several mechanisms with overlapping effects, one mechanism may increase while another decreases in its relative importance without jeopardizing the overall effect. In this way, fast-developing species like *Xenopus* may have come to rely more strongly on the earlier events in the sequence leading to neural development.

12.5 Axis Induction by Disinhibition

The discovery in the 1930s that dead organizer tissue still functioned in neural induction triggered a flurry of investigations into the chemical nature of the inducing signal. However, the investigators' enthusiasm was soon dampened by the discovery of a plethora of extraneous tissues and diverse biochemical fractions that all caused neural induction. These frustrating results culminated in the discovery that neuralization may be induced by stimuli as unspecific as changes in the ionic composition or pH of the culture medium (Holtfreter, 1945). Spemann was chagrined by the appearance that his "organizer" could be mimicked by grotesquely unspecific stimuli, but at the same time he felt confirmed in his view—published 3 years before the famous organizer experiment—that part of the specificity of an inductive interaction is inherent to the responding tissue.

After the advent of DNA cloning techniques, the quest to identify the molecular basis of neural induction was taken up again with new vigor. Investigators have identified individual genes that are expressed selectively in Spemann's organizer and encode proteins that mimic the organizer's effects. A gene product that effectively antagonizes the organizer actions has also been identified. From these investigations, we not only begin to understand neural induction in molecular terms but also come to appreciate again that the neural induction signals affect other aspects of dorsal axis formation as well.

DORSAL DEVELOPMENT OCCURS AS A DEFAULT PROGRAM IN *XENOPUS*

The specificity that resides in the responding partner of an inductive interaction may have two causes, which could act singly or in combination. First, as discussed earlier, embryonic inductions tends to occur in multiple steps. In the case of neural induction, neural ectoderm may be primed so well by events preceding its interaction with chordamesoderm that many types of stimuli may give it the final push to form neural tissue. Second, the inductive signal may act by ***disinhibition***—that is, it may interfere with an *inhibitory signal* that normally blocks a *default program*. In the case of axis induction, this would mean that vertebrate embryonic cells after midblastula transition develop dorsal organ rudiments unless they are told differently (Fig. 12.26; Hemmati-Brivanlou and Melton, 1997). The latter may happen by a signal that inhibits dorsal development and promotes ventral development instead. Axis induction may interfere with this inhibitory signal, thus releasing—by disinhibition—the default program of dorsal development. Such interference with an inhibitory signal could well be accomplished by various, not necessarily specific, stimuli.

The view of neural development as a default program has been supported by work with *Xenopus* embryos. Several investigators found that dissociated cells from *ventral* blastocoel roof developed into neurons. This was a remarkable result because the fate of these cells in intact embryos would have been to form epidermis (see Fig. 6.2), and because the isolated cells had no contact with prospective dorsal blastopore lip. Godsave and Slack (1991) found that their results depended on the number of cell per culture well: With more than 30 cells per well, most cells became epidermal, whereas, with less than 30 cells per well, neural development predominated. The researchers concluded that ventral blastocoel roof cells, as part of their normal development into epidermis, produce some inhibitor of neural development, and that this inhibitor is diluted out in culture, with neural development ensuing by default.

The default model of neural—or generally dorsal—development implies that ventral development requires an inhibitory signal. A candidate molecule for this signal is ***bone morphogenetic protein 4* (*BMP-4*)**, so called because members of the BMP family were first isolated for their ability to promote bone development. The BMPs belong to the transforming growth factor β (TGF-β) superfamily of secreted proteins, which serve as intercellular signals in many inductive interactions (Hogan, 1996). To test whether BMP-4 may be the natural signal that promotes ventral development while inhibiting dorsal development, investigators used the same criteria discussed earlier in the context of mesoderm-inducing and dorsalizing signals (see Section 9.7).

Microinjection of BMP-4 mRNA into *Xenopus* embryos causes the development of excess ventral structures such as blood, mesenchyme, and epidermis, at the expense of dorsal organs such as notochord, muscle, and nervous system (Graff, 1996). These results show that BMP-4 is *sufficient* to promote not only epidermal over neural development, but also ventral over dorsal mesoderm formation. Thus, BMP-4 seems to an antagonist to Spemann's organizer not only in neural induction but also in other aspects of axis induction. To test whether BMP-4 is

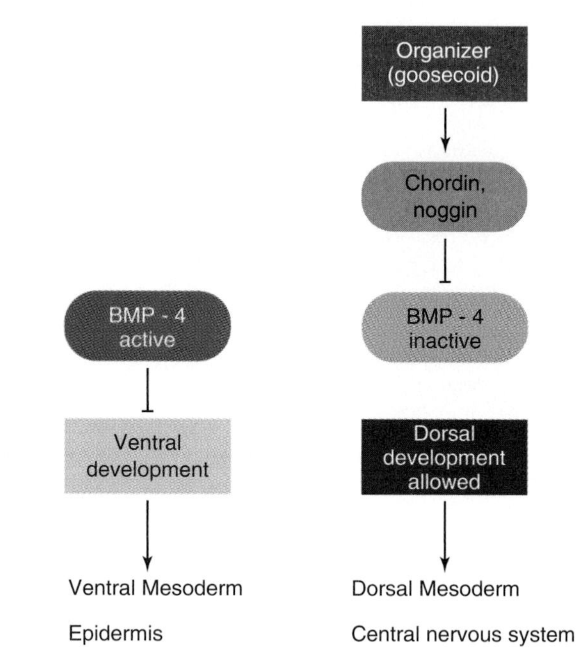

Figure 12.26 Model of axis induction by disinhibition. In *Xenopus,* dorsal development after midblastula transition seems to occur as a default program, in the absence of an inhibition. A likely candidate for being the natural inhibitor is bone morphogenetic protein 4 (BMP-4). Two secreted proteins, chordin and noggin, are synthesized specifically in Spemann's organizer. The synthesis of chordin and noggin is controlled, at least in part, by the transcription factor goosecoid. (The *goosecoid*$^{+}$ gene, in turn, is activated by a combination of maternal dorsalizing and mesoderm-inducing factors, as outlined in Fig. 9.21.) Chordin and noggin bind directly to BMP-4, thus interfering with its inhibitory function. As a result of the disinhibition, dorsal development occurs by default.

necessary for ventral development, investigators have used three different methods: selective elimination of BMP-4 mRNA, excess production of defective BMP-4 to outcompete functional BMP-4, and excess production of defective BMP-4 receptor to outcompete functional BMP-4 receptor. All three approaches produced similar results: Neural development occurred instead of epidermal development, and dorsal mesoderm replaced ventral mesoderm (Graff et al., 1994; Suzuki et al., 1994). In addition, BMP-4 and a suitable receptor occur at the appropriate time and place in normal embryos, as shown by in situ hybridization (see Method 8.1) of the corresponding mRNAs (Fainsod et al., 1994; Schmidt et al., 1995b). Thus, BMP-4 shows all the properties expected from a natural signal that promotes ventral development while inhibiting dorsal axis formation in *Xenopus.*

SPEMANN'S ORGANIZER INACTIVATES A VENTRALIZING SIGNAL

If BMP-4 is the natural inhibitor of dorsal axis formation, then Spemann's organizer must antagonize BMP-4. Indeed, two secreted proteins—***chordin*** and ***noggin***—are synthesized selectively in the organizer and inactivate BMP-4 signaling (Piccolo et al., 1996; Zimmerman et al.,

1996). Biochemical tests have shown that chordin and noggin directly bind BMP-4 and thereby prevent BMP-4 from activating its receptor (Fig. 12.26). These tests were prompted by the results of earlier investigations, which had shown that chordin and noggin have strong dorsalizing effects. For example, injection of noggin mRNA rescues *Xenopus* embryos ventralized by UV irradiation and, depending on dose, can cause excessive head development (Smith and Harland, 1992). Also, isolated ventral marginal zones from gastrulae kept in a culture medium containing noggin protein become dorsalized. This is indicated by the synthesis of muscle actin mRNA, which is abundant in skeletal muscle, a dorsal mesoderm derivative, but not in derivatives of lateral and ventral mesoderm (Smith et al., 1993). Similar results have been obtained with chordin (Sasai et al., 1994).

In one of the most rewarding recent discoveries, it was shown that the antagonistic action of BMP-4 versus chordin and noggin has been highly conserved in evolution. Corresponding proteins are also establishing the dorsoventral polarity in *Drosophila,* although in upside-down orientation relative to amphibians (see Section 23.4).

How are disinhibitors like chordin and noggin produced selectively in the organizer? In the context of dorsal mesoderm induction, we have discussed how a combination of maternal mesoderm-inducing and dorsalizing signals activate the *Xenopus goosecoid*$^+$ gene (see Fig. 9.21). This gene has attracted much attention because its expression is limited to Spemann's organizer. At the early gastrula stage, when a dorsal blastopore lip has formed, *goosecoid*$^+$ is transcribed in a patch of marginal cells occupying about a 60° arc centered over the dorsal blastopore (Blumberg et al., 1991). Specifically, goosecoid mRNA accumulates between the *midblastula transition* and the onset of neurulation in the deep layer of the involuting marginal zone, precisely where and when Spemann's organizer is thought to be active in mesoderm dorsalization and neural induction.

Because *goosecoid*$^+$ encodes a *transcription factor,* it has the capability to regulate several target genes, possibly including the genes for noggin and chordin. Indeed, it has been shown that ectopic microinjection of goosecoid mRNA activates the *chordin*$^+$ gene (Sasai et al., 1994). Thus, the simplest interpretation of the available data is that mesoderm-inducing and dorsalizing signals activate *goosecoid*$^+$, which in turn activates the genes encoding the disinhibiting protein(s) that cause neural development (Fig. 12.26). Further investigations are needed to test this hypothesis.

MAJOR QUESTIONS ARE STILL UNRESOLVED

Some fundamental questions about axial induction are still unresolved. One of them is how the complex dorsoventral body pattern is specified. With regard to ectoderm, only two major determined states—neural and epidermal—seem required for the next developmental step, that is, neurulation. By contrast, mesoderm development entails forming a dorsoventral sequence of several organ rudiments including notochord, somites, nephrotomes, and lateral plates (see Fig. 9.19). To account for the formation of this complex pattern, two research groups have proposed a model based on a graded distribution of BMP-4 receptor activity (Dosch et al., 1997; Jones and Smith, 1998). The highest receptor activity occurs ventrally where BMP-4 is produced but little or no noggin or chordin is found. The resulting high BMP-4 receptor activity would account for the formation of lateral plates, which later give rise to blood, cardiovascular system, and mesenchyme. Intermediate receptor activities are thought to occur laterally where BMP is still high but some chordin and noggin arrive by diffusion. This would lead to nephrotome and somite formation. The lowest receptor activity is assumed to occur dorsally where BMP-4 is not produced, leading to notochord formation.

Another challenge for the future will be to unravel the molecular mechanisms that specify the *anteroposterior pattern* of the embryonic axis. We have already discussed classical experiments by Spemann and Otto Mangold indicating that axis-inducing signals have an anteroposterior specificity (see Figs. 12.19 and 12.20). Following up on these results, Nieuwkoop (1952) and Toivonen and Saxén (1955) found that guinea pig bone marrow would induce anterior structures, such as forebrain, eye, and nose, whereas guinea pig liver would induce posterior structures including somites and tail fin. The most complete secondary embryos were induced by combinations of bone marrow and liver. These and other results have led to the proposal of a *two-signal model for axis induction:* The first signal is thought to induce generally dorsal and anterior axis elements, especially anterior brain, while the second signal induces more posterior and mesodermal structures, such as spinal cord and kidney. Chordin and noggin, as discussed above, induce dorsal and anterior structures ("noggin" is a slang word for "head"). Thus, chordin and noggin fit the role of the first signal in the two-signal model. The nature of the second signal is still under investigation (Doniach, 1995). One class of molecules that promotes the development of posterior neural structures are *retinoids* (Maden and Holder, 1992). Other lines of evidence point at a role for *basic fibroblast growth factor* (*bFGF*), the same molecule that also seems to act as a general mesoderm inducer as discussed in Section 9.7, but more work is needed to clarify how FGF works in either of these biological processes.

In summary, the combination of DNA cloning and related techniques with the classical strategies of isolation and transplantation has allowed researchers to work towards a molecular understanding of Spemann's organizer in *Xenopus.* This effort that is generating tremendous enthusiasm and is a major source of the current feeling that developmental biology has entered another golden age.

SUMMARY

After gastrulation, germ layers interact with one another to form the organ rudiments. During this period of organogenesis, a basic body plan emerges that is characteristic of the phylogenetic group to which an organism belongs. Organogenesis involves extensive morphogenetic movements. The process is particularly well studied in neurulation—that is, the formation of the central nervous system in vertebrates. During neurulation, a dorsal layer of ectodermal cells forms the neural plate, which closes into the neural tube and eventually generates the brain and spinal cord. Neurulation has been analyzed in terms of specific cell behaviors including coordinated shape changes, intercalation, and cell crawling, which cooperate as overlapping mechanisms. Neurulation is part of axis formation, meaning in this context the formation of an organized pattern of dorsal organ rudiments, including not only the neural tube but also notochord, somites, and embryonic kidneys.

Axis formation depends on inductive interactions, which are known as axis induction or—if focused on neurulation—as neural induction. A landmark experiment on axis induction was conducted by Spemann and Mangold (1924), who transplanted dorsal blastopore lip to the ventral ectoderm of newt embryos during the early gastrula stage and found that a secondary embryonic axis developed. Grafting of chordamesoderm, which arises from the dorsal blastopore lip during gastrulation, had the same effect. The grafted material (called the "organizer") developed in accord with its own fate and marshalled the surrounding host tissue to form a nearly complete and normally proportioned embryonic axis.

Modern work with *Xenopus* embryos has shown that neural induction is a multistep process. It begins with the rearrangement of cytoplasm in the fertilized egg and dorsal mesoderm induction in the blastula, which bias the prospective dorsal ectoderm for neural induction. During the gastrula stage, neural induction depends on both planar and vertical signals. The planar signals are sent from the dorsal blastopore lip within the outer plane of the embryo directly to the adjacent dorsal ectoderm. The vertical signals are sent in the late gastrula from the same inducer, which has then formed the chordamesoderm, to the overlying dorsal ectoderm. Presumably, the planar signals enhance the competence of the dorsal ectoderm to respond to the vertical signals received later.

Early attempts to identify the signal molecules involved in axis induction were frustrated by the finding that axis induction can be mimicked by a plethora of heterogeneous tissues and biochemical fractions. Recent work has revealed that axis induction occurs by disinhibition. Axis formation is inhibited by a ventralizing signal tentatively identified as bone morphogenetic protein 4 (BMP-4). Thus, axis formation occurs by default where BMP-4 signaling is inhibited. Two signal proteins, chordin and noggin, are synthesized specifically in the organizer and bind directly to BMP-4, thus preventing activation of the BMP-4 receptor.

Ectodermal Organs

Figure 13.1 Purkinje cell from the cerebellar cortex. The numerous dendritic branches provide an extensive area for thousands of synapses with other neurons.

DURING organogenesis, the embryo has formed organ rudiments, and the basic body pattern has been established. However, the organ rudiments cannot yet function: The eye rudiment cannot see, and the brain rudiment cannot serve as the main coordinating system of the body. To perform these tasks, cells must first acquire functional specialization. Photoreceptors in the retina must assemble certain pigments so that absorption of light triggers the release of neurotransmitters. Neurons in the nervous system must send out different types of processes to form connections with one another. This process of functional specialization is associated with further growth and occupies a relatively long time. In humans, cleavage takes only about 2 weeks, gastrulation another week, and organogenesis about 4 more weeks, whereas the subsequent growth and functional specialization occupy the last 7 months of gestation.

As a name for the overall process of functional maturation of embryonic organs, classical embryologists have coined the term *histogenesis* (Gk. *histos,* "web," or "tissue"; *genesis,* "origin"). This term places the emphasis on tissues, a level of organization between cells and organs. A tissue (Lat. *texere,* "to weave") consists of cells and extracellular materials that perform a particular set of functions. Typically, a tissue contains one or a few different cell types. For instance, nervous tissue consists mostly of neurons, glial cells, and extracellular material, functioning together in the selective transmission of signals. A typical neuron is characterized by numerous cytoplasmic extensions that provide the structural basis for connections with other neurons (Fig. 13.1). These extensions are readily discerned in microscopic sections and can therefore be used to monitor the functional maturation of nervous tissue. As an alternative way of monitoring the same process, scientists use the presence of characteristic molecules. In the case of neurons, a characteristic molecule would be a neurotransmitter, such as acetylcholine, or an enzyme involved in the metabolism of the neurotransmitter, such as acetylcholinesterase. Classical embryologists have used staining procedures that would identify such characteristic molecules more or less reliably. Modern developmental biologists prefer more specific techniques, such as *immunostaining* for characteristic proteins (see Method 4.1) or *in situ hybridization* to trace specific mRNAs (see Method 8.1).

In this chapter and the following one, we will examine the histogenesis of several vertebrate organs. In doing so, we will discuss examples of the four major tissue types, namely nervous, epithelial, muscular, and connective tissue. Because the discussion of organogenesis in Chapter 12 was limited mostly to the central nervous system, the early development of other organs will be included in Chapters 13 and 14 as needed.

The organization of the histogenesis chapters will be based on the three germ layers and their major subdivisions. Figure 13.2 is a flow chart relating these embryonic structures to adult tissues and organs. It reflects the important principle that the basic body plan develops through a hierarchy of determinative events, although the chart does not show the localizations and inductions involved in generating tissue diversity. The chart is also fairly general: It can be applied to all vertebrates. However, it is worth noting at the outset that so-called ectodermal and endodermal organs are not

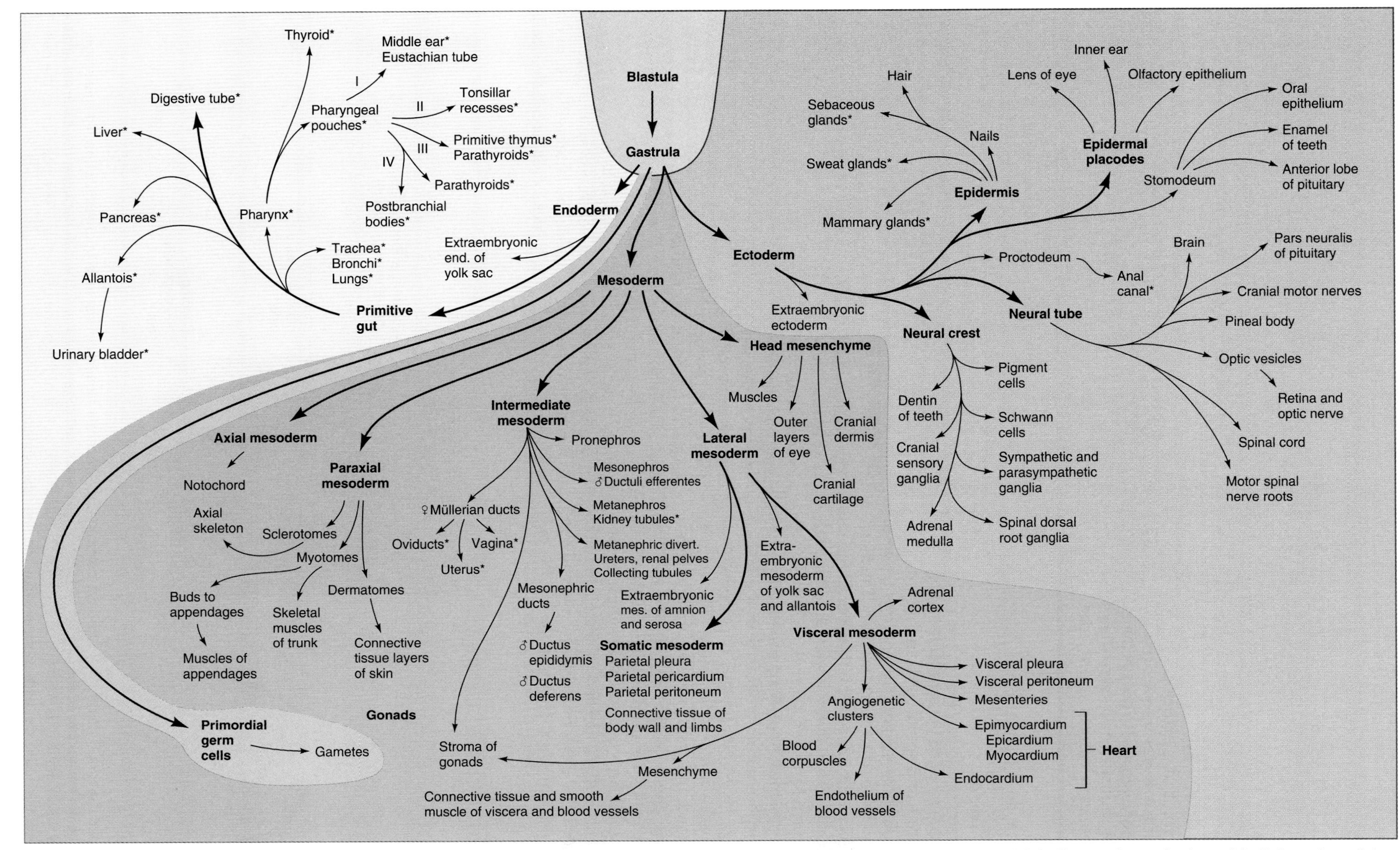

Figure 13.2 Flow chart showing the three germ layers, their subdivisions, and the organs or tissues derived from them. An asterisk indicates that only the epithelial portion of the marked organ is derived as indicated and that the organ has additional investments of mesodermal origin.

necessarily formed from one germ layer. In the majority of these organs, only the epithelial portions derive from ectoderm or endoderm, and the mesoderm makes substantial contributions. It is also important to realize that the same type of tissue may arise from different germ layers. Cartilage and muscle arise, for the most part, from mesoderm, but certain cranial cartilages and eye muscles are derived from ectoderm.

In this chapter, we will review the major derivatives of the ectodermal germ layer, namely the neural tube, the neural crest, the ectodermal placodes, and the epidermis. Neural tube development into the central nervous system will be considered in terms of overall morphology, cellular differentiation, and the molecular signals that specify dorsoventral patterning. The astounding neural crest cells will be explored in some detail because their extensive migrations and pluripotentiality have made them favorite objects for studying the roles of the extracellular matrix and diffusible growth factors in cell division, movement, and determination.

13.1 Neural Tube

The neural tube gives rise to the ***central nervous system (CNS)***—that is, brain and spinal cord, and it contributes to the ***peripheral nervous system,*** the nervous tissue located outside the skull and vertebral column. The development of the CNS is impressive by its sheer numbers. From neural tube closure in the embryo until after birth, the human organism forms an average of 250,000 neurons per minute until the CNS contains about 100 billion neurons. The morphological complexity of the CNS is equally daunting. However, tracing its development from the simple organization of the neural tube actually helps us understand and memorize the more intricate structures of later stages. In this section, we will first consider the basic cellular aspects of CNS development. Next we will examine how the relatively simple organization of the spinal cord is patterned by signals from the adjacent notochord and epidermis. Finally, we will survey how the more complex organization of various parts of the brain can be understood as modifications of the spinal cord organization.

NERVOUS TISSUE CONSISTS OF NEURONS AND GLIAL CELLS

The wall of the newly closed neural tube typically consists of one layer of ***neuroepithelial cells*** (Fig. 13.3). On the outside, the neural tube is covered by an *external limiting membrane* of extracellular material, which corresponds to the *basement membrane* generally found on the basal surface of an epithelium (see Fig. 2.23). The apical face of the neuroepithelium is directed toward the inner space, or *lumen,* of the neural tube. Here the cells form a seal of tight junctions as in all epithelia.

Once the neural tube has closed, the neuroepithelial cells behave as *stem cells*—that is, cells that have a virtually unlimited capacity for self-renewal and give rise to committed progenitor cells. The latter will in turn divide and produce certain types of differentiated cells

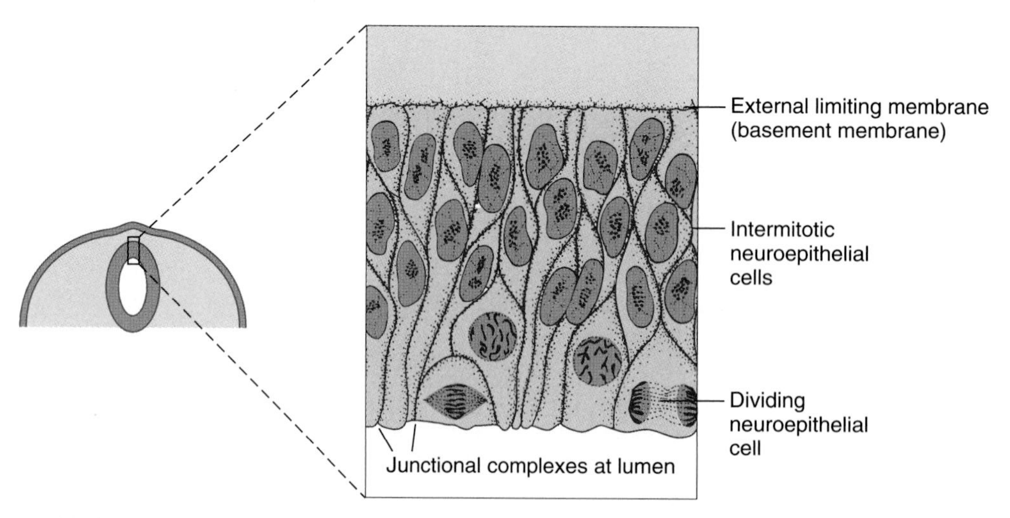

Figure 13.3 Neuroepithelium. The drawing shows a portion of the recently closed neural tube in cross section. At this stage, the neuroepithelial cells form a pseudostratified epithelium, with each cell extending the full width between the external limiting membrane and the lumen of the neural tube. (Most cells bend into or out of the plane of section.) The position of the cell nucleus varies during the cell cycle, with mitosis occurring near the lumen.

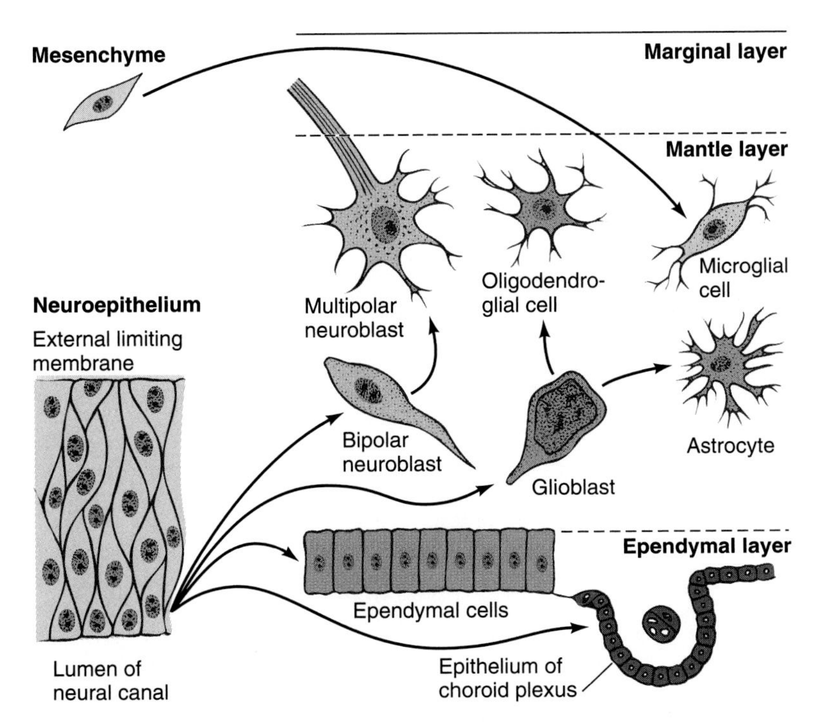

Figure 13.4 Development of the neuroepithelium. Its major derivatives are neuroblasts, which develop into neurons; and glioblasts, which give rise to two types of glial cells: oligodendrocytes (oligodendroglial cells) and astrocytes. Oligodendrocytes also derive from neural crest cells (not shown). Another type of glial cell, called microglia, is derived from mesenchymal cells. The remaining neuroepithelial cells later form the ependymal cells lining the lumen of the spinal cord and the brain.

(Fig. 13.4). Neuroepithelial stem cells produce committed progenitor cells known as ***neuroblasts*** (neuron precursor cells) and ***glioblasts*** (precursors of certain glial cells) (Fig. 13.4). Neuroepithelial stem cells persist in adult mammals, even though many species seem to form only small numbers of new neurons once the dramatic growth phase before and after birth is over. However, experiments with cultured neuroepithelial stem cells from embryonic and adult brain have led to the identification of certain growth factors that promote the formation of neurons, astrocytes, and oligodendrocytes (Johe et al., 1996; McKay, 1997). Attempts with laboratory animals are under way to release these growth factors from transgenic cells transplanted into the brain. Such experiments, if successful, hold great promise for the treatment of human neurodegenerative diseases.

Neuroblast development in vertebrates shows an astounding parallelism to the corresponding process in *Drosophila* (see Section 6.4). In *Xenopus* embryos at the beginning of neurulation, groups of cells that have the potential to form neuroblasts accumulate gene regulatory proteins known as neurogenin and Neuro D, which belong to the same class of gene bHLH proteins as the achete and scute proteins in *Drosophila* (Ma et al., 1996). Among each cluster of potential neuroblasts in the *Xenopus* neurula, the process of *lateral inhibition* selects the one cell that will actually become a neuroblast (Tanabe and Jessel, 1996). A key protein in this selection process is called X-Delta because its amino acid sequence is very similar to that of *Drosophila* Delta (Chitnis et al., 1995). Such Delta homologs have also been found in chicken and mice. Likewise, vertebrate embryos make homologs of the *Drosophila* Notch protein. In both vertebrates and *Drosophila,* overexpression of Delta or constitutively active forms of Notch inhibit the formation of neurons. Conversely, inactivation of Delta or inhibition of Notch signaling causes excessive neuron formation. Further extending the parallel to *Drosophila,* Notch/Delta signaling in vertebrates is not limited to the developing nervous system but seems to be a general mechanism for *lateral inhibition*.

Neuroblasts migrate toward the external limiting membrane, where they form a progressively thicker layer of nondividing cells, the ***mantle layer.*** A neuroblast matures by sprouting cytoplasmic extensions called ***neurites.*** The longest neurite is usually the ***axon,*** which transmits signals away from the body of the neuron to muscle cells or other neurons. ***Dendrites*** transmit signals toward the body of the neuron and are usually shorter than the axon.

After giving off large numbers of neuroblasts, the neuroepithelial cells produce a second type of committed progenitor cell, the *glioblast*. Glioblasts give rise to ***glial cells,*** which have supporting, insulating, and nutritive functions. There are about 10 times more glial cells than neurons in the vertebrate CNS. The two main types of glial cells derived from glioblasts are oligodendrocytes and astrocytes, although oligodendrocytes also derive from neural crest cells. ***Oligodendrocytes*** wrap around the axons of neurons repeatedly to create an insulating layer of multiple plasma membranes called the ***myelin sheath*** (Figs. 13.5 and 13.6). ***Astrocytes*** are associated with the *endothelial cells* of blood capillaries and form the ***blood-brain barrier,*** which restricts the passage of large molecules from the blood to the central nervous system. Additional glial cells, called ***microglia,*** are derived from mesenchymal cells. Tissues containing neurons and glial cells are classified as ***nervous tissue.***

As the developing neurons send their axons toward the outside, they create another layer, called the ***marginal layer,*** around the mantle layer (Fig. 13.7). Because of the myelination of the axons, the marginal layer takes on a whitish appearance and is then called the ***white matter*** of the spinal cord. This is in contrast to the ***gray matter,*** which derives from the mantle layer and contains the cell bodies of neurons as well as unmyelinated axons and dendrites.

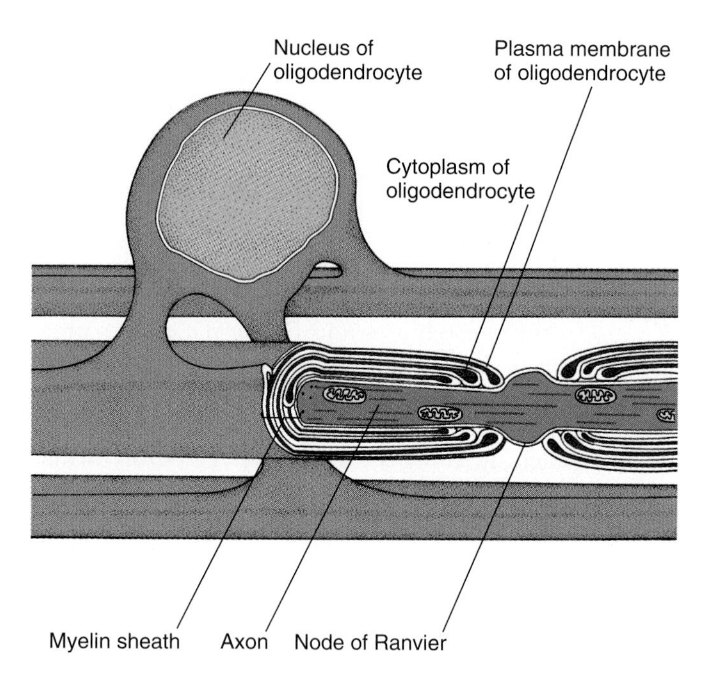

Figure 13.5 Origin of the myelin sheath surrounding vertebrate axons. Oligodendrocytes in the central nervous system wrap around the axons of neurons, surrounding the axon with multiple layers of oligodendrocyte plasma membrane. The compacted layers are called a myelin sheath. Each oligodendrocyte covers only a segment of an axon, leaving uncovered small intervening sections of axon, known as nodes of Ranvier. Similar myelin sheaths are formed by Schwann cells in the peripheral nervous system.

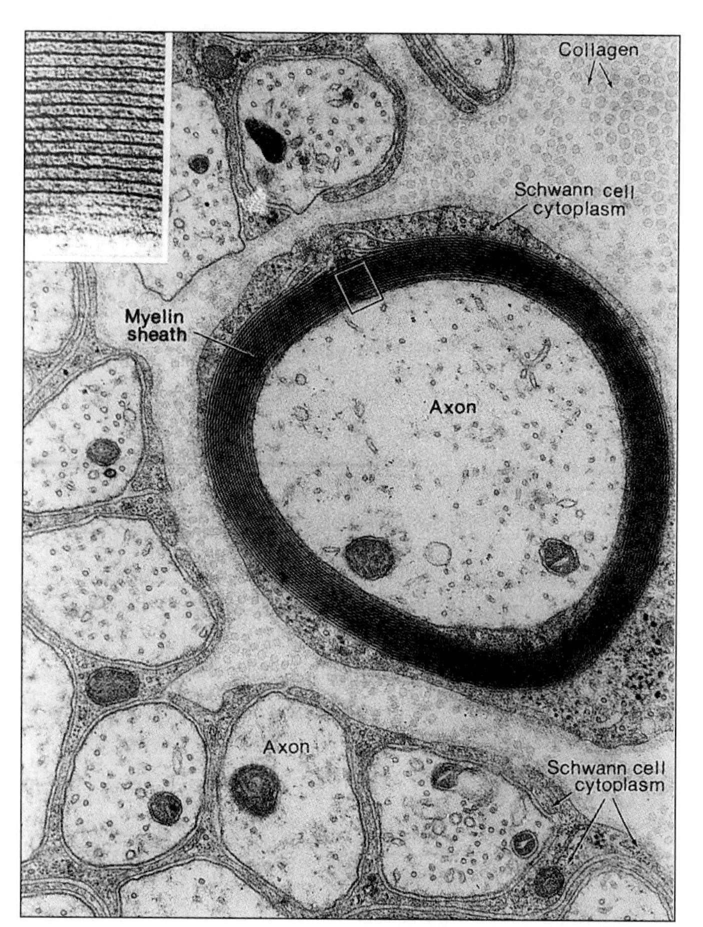

Figure 13.6 Electron micrograph of a small cross section of rat sciatic nerve. A nerve contains many axons. Most are embedded in the cytoplasm of certain glial cells called Schwann cells. Long axons are often surrounded by a myelin sheath, which consists of multiple layers of Schwann cell plasma membranes (inset).

THE SPINAL CORD IS PATTERNED BY SIGNALS FROM ADJACENT TISSUES

All neural functions, from simple reflexes to sophisticated cognitive processes, rely on appropriate neuronal circuits. The assembly of such circuits begins during embryonic development with the generation of certain classes of neurons in distinct locations. We will focus here on the dorsoventral pattern of cell types that emerges during neurulation and histogenesis of the spinal cord.

As more neuroblasts and glioblasts are added to the mantle layer, they form dorsal and ventral ridges of gray matter along each side of the neuroepithelium. The dorsal ridges are called ***alar plates,*** later referred to as ***dorsal columns*** or ***afferent columns*** (Lat. *afferre,* "to carry toward"). The latter designation alludes to the fact that interneurons in the afferent columns transmit sensory impulses arriving from the skin, muscles, tendons, and internal organs. The ventral ridges of gray matter are called ***basal plates,*** later referred to as ***ventral columns*** or ***efferent columns*** because their motor neurons send signals to muscles and glands (Lat. *efferre,* "to carry away"). The most dorsal portion of the neuroepithelium between the two alar plates is called the ***roof plate.*** Correspondingly, the most ventral portion of the neuroepithelium is known as the ***floor plate.*** Cells in the roof and floor plates are glial. In particular, floor plate cells promote and orient the growth of *commissural axons* that originate from neurons in a dorsal column and end on neurons in the ventral column of the opposite side (Fig. 13.7b).

The gray matter of the spinal cord thus acquires, during its organogenesis and histogenesis, a dorsoventral pattern of distinct cell types, including—from dorsal to ventral—roof plate, commissural interneurons, different types of ventral interneurons, motorneurons, and floor plate. How does this pattern arise? We have already discussed that *chordamesoderm* induces the neural plate, and that the two stay in close physical contact during neurulation (see Sections 12.2 and 12.3). We will see now that the chordamesoderm's derivative, the notochord, is also involved in patterning the ventral half of the neural tube.

TO explore the patterning effects of the notochord on the neural tube, Placzek et al. (1990) grafted notochord from chicken embryos along the side of closing neural plates (Fig. 13.8). In response, the adjacent part of the neural tube formed a wedge-shaped area resembling a floor plate. Both

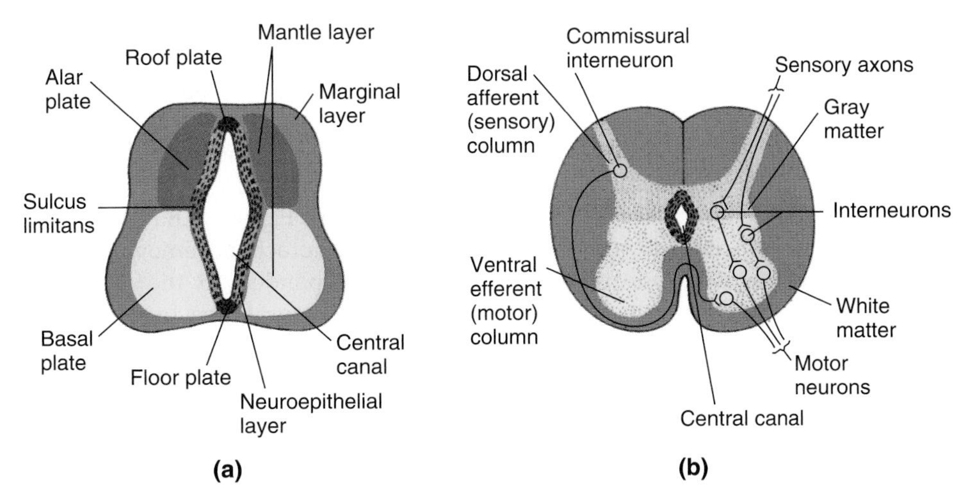

Figure 13.7 Development of the spinal cord in humans. **(a)** At 6 weeks. The central canal is surrounded by the neuroepithelium, which produces neuroblasts and glioblasts. Neuroblasts form the mantle layer, later called the gray matter, which will contain the cell bodies of neurons. The mantle layer is divided into dorsolateral alar plates and ventrolateral basal plates. Two medial derivatives of the neuroepithelium, known as floor plate and roof plate, are glial rather than neuronal. Outside the mantle layer is the marginal layer, which consists of sprouting neurites. This layer, which is later called the white matter, will contain the myelinated axons of neurons. **(b)** At 9 weeks, commissural neurons, interneurons, and motor neurons have formed in a characteristic dorsoventral pattern. Sensory axons arrive from dorsal root ganglia (see Fig. 13.20); motor neurons send axons to skeletal muscles.

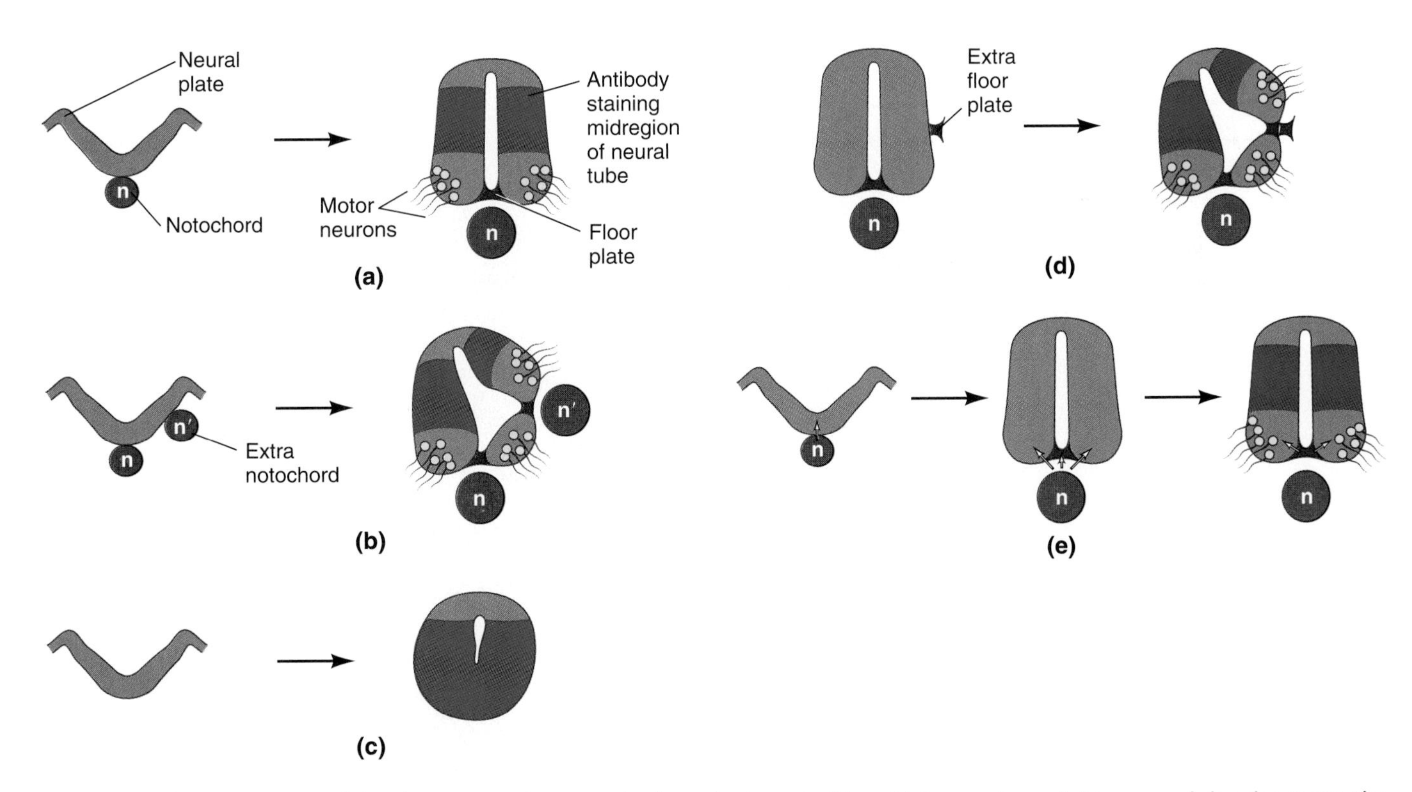

Figure 13.8 Establishment of the dorsoventral pattern in the spinal cord of the chicken embryo. **(a)** In normal development, the notochord is adjacent to open the neural plate but later becomes separated from the closed neural tube. The floor plate develops adjacent to the notochord, and motor neurons develop on both sides of the floor plate. Immunostaining reveals antigens specific to a wide transverse band within the neural tube (teal color). **(b)** Grafting an extra notochord (n′) alongside the closing neural plate results in the formation of another floor plate flanked by efferent neurons. The immunostainable antigen disappears near the grafted notochord. **(c)** Removing the notochord results in the absence of both floor plate and motor neurons, and the immunostainable antigen appears in the entire ventral region. **(d)** Grafting an extra floor plate alongside the closed neural tube has effects much like the operation shown in part b. **(e)** One hypothesis explaining these results is that notochord induces adjacent neural tube to form floor plate, and that induced floor plate releases an equivalent signal, which—depending on concentration—induces more floor plate and different types of neurons. (See also Fig. 13.10.)

the primary and the secondary floor plate were flanked by efferent columns sending out bundles of axons. The extra floor plate also promoted and guided the outgrowth of commissural fibers in vivo and in vitro. Conversely, removal of the notochord from the neural tube left the ventral neural tube cells unable to guide commissural fibers and resulted in failure to develop efferent columns. These results show that the presence of notochord is both sufficient and necessary for the development of floor plate and efferent columns, but they can be interpreted two ways: a signal released by the notochord could act as a *morphogen,* inducing both floor plate and efferent columns directly and in a concentration-dependent manner, with a high level of signal close to the notochord resulting in floor plate formation, and a lower level of signal further away from the notochord inducing efferent columns. Alternatively, there could be a cascade of inductions, with notochord inducing floor plate and floor plate in turn inducing motor columns. To further explore these possibilities, the researchers transplanted another *floor plate*—rather than a notochord—along the side of a closing neural plate. Again, the transplant caused the development of additional floor plate flanked by efferent columns. This result is compatible with the cascade model and with a version of the gradient model that postulates the release of equivalent signals by both notochord and floor plate. The latter hypothesis receives further support from molecular data discussed further below.

In an extension of this study, Yamada et al. (1991) used *immunostaining* (see Method 4.1) with antibodies directed against antigens in the floor plate, or antigens expressed in various horizontal bands of neural tube cells such as the one shaded in Figure 13.8. The immunostaining patterns shifted after notochord removal or transplantation, and the shifts corresponded to the morphological changes observed in the neural tube. In a similar investigation, Goulding and coworkers (1993) monitored transcription of a known regulatory gene, *Pax-3*$^+$, by *in situ hybridization* (see Method 8.1). Notochord removal or transplantation shifted the expression pattern of *Pax-3*$^+$ in ways suggesting that a signal from the notochord inhibits expression of this gene in the ventral portion of the spinal cord (Fig. 13.9; Goulding et al., 1993).

The best candidate molecule for a patterning signal released by notochord and floor plate is encoded by the *Sonic hedgehog*$^+$ *(Shh*$^+$*)* gene, so named after its *Drosophila* homolog *hedgehog*$^+$, a known patterning gene (see Sections 22.4 and 22.7). Shh is synthesized at the right time and place: During neurulation, *Shh*$^+$ is expressed in chordamesoderm and its derivative, the notochord. After neural tube closure, when the notochord becomes separated from

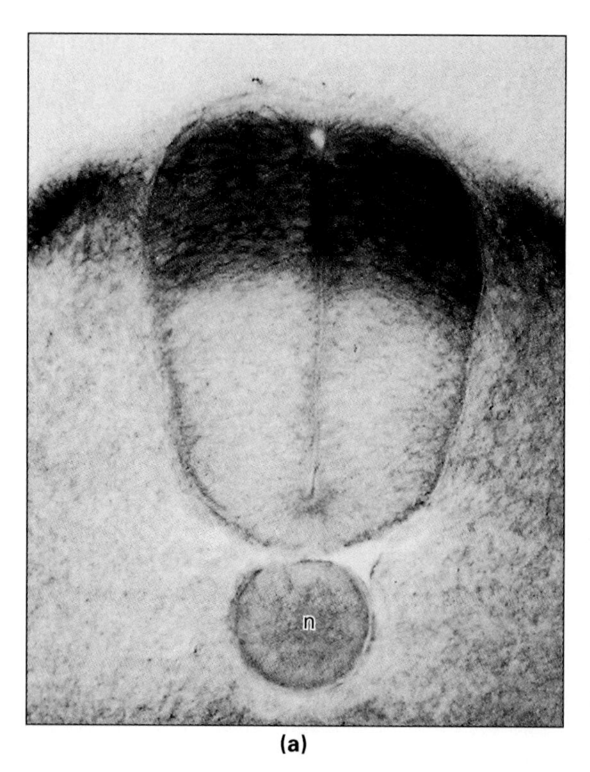

(a)

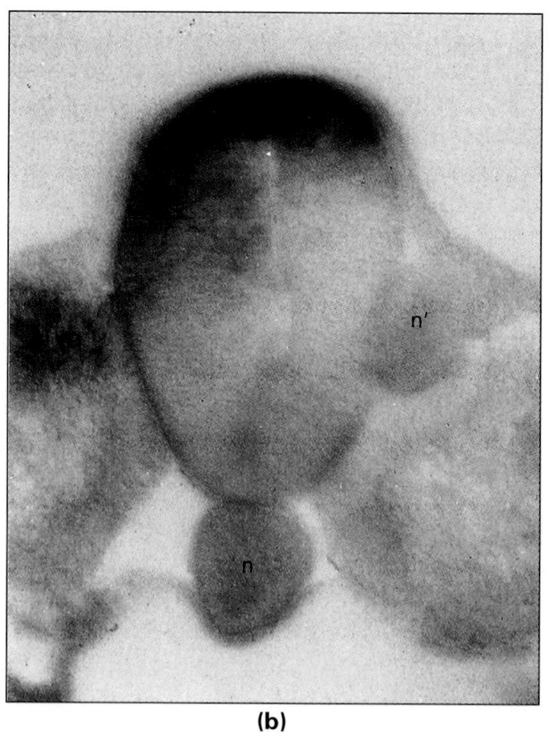

(b)

Figure 13.9 Inhibitory effect of the notochord on the expression of a regulatory gene (*Pax-3*$^+$) in the chicken embryo. The photos show cross sections of chicken embryos treated by in situ hybridization (see Method 8.1) to make Pax-3 mRNA visible as a dark stain. **(a)** Normal embryo. Pax-3 mRNA accumulates in the dorsal part of the neural tube, opposite the notochord (n). **(b)** Embryo that received an extra notochord (n') transplanted laterally under the open neural plate. A few hours later, when the neural plate had closed to form a tube, the extra notochord inhibited the expression of the *Pax-3*$^+$ gene in its vicinity. (See also Fig. 13.8.)

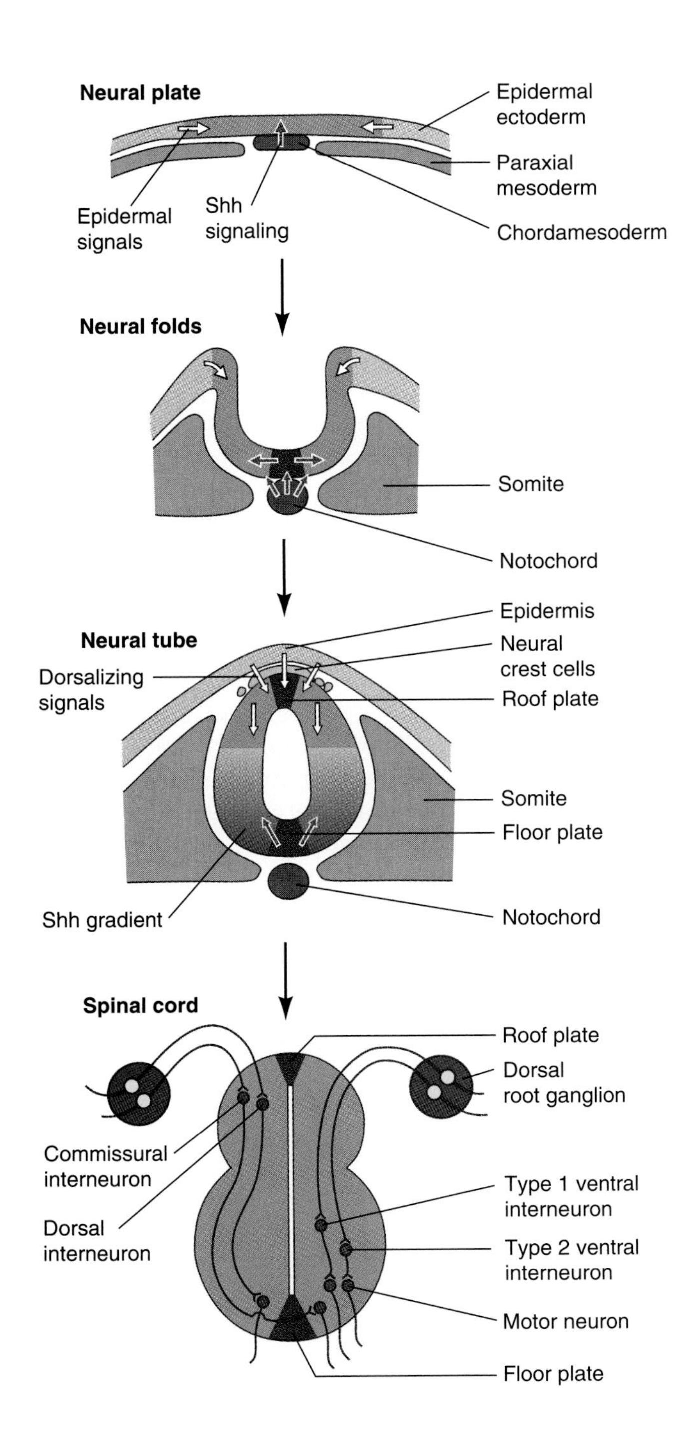

Figure 13.10 Dorsoventral patterning of the spinal cord by inductive signals from adjacent tissues. At the neural plate stage (top), Sonic hedgehog (Shh) peptide released from closely underlying chordamesoderm induces floor plate formation. Meanwhile, signals produced in epidermal ectoderm dorsalize laterally adjacent strips of neural plate. At the neural folds stage, notochord derived from the chordamesoderm as well as the induced floor plate release Shh while dorsalizing signals from the epidermal ectoderm continue. As the neural tube closes, it gives rise to the roof plate and to neural crest cells. On the ventral side, the notochord has detached while the floor plate continues to release Shh, setting up a gradient of Shh concentration. This gradient, along with various dorsalizing signals and dynamic changes in the competence of neural tube cells to respond, seems to specify the dorsoventral pattern of motorneurons and various types of interneurons in the spinal cord (bottom).

the overlying neural tissue, Shh is also produced by the floor plate (Fig. 13.10; Roelink et al., 1994; J. C. Smith, 1994; Tanabe and Jessel, 1996). Shh is necessary and sufficient for floor plate induction: Mice lacking *Shh*$^+$ function do not form floor plates, and the ability of transplanted notochords to induce ectopic floor plates is lost in the presence of antibodies directed against Shh protein (Chiang et al., 1996; Ericson et al., 1996). More detailed experiments revealed two critical periods of Shh signaling: an early period during which neural plate cells are ventralized, and a later period when Shh drives the differentiation of ventralized precursors into different types of neurons. Shh is a secreted glycoprotein, and its inductive activity resides in an amino-terminal peptide released by autoproteolytic cleavage (Marti et al., 1995). Interestingly, the active Shh fragment seems to act synergistically with fragments of an extracellular matrix protein, *vitronectin* (Pons and Marti, 2000).

The active Shh peptide is diffusible, so that its concentration decreases gradually with increasing distance from its producers, notochord and floor plate. It is therefore possible that Shh acts as a *morphogen,* inducing different types of neuronal structures in a concentration-dependent manner. To test this hypothesis, investigators cultured neural tube explants from chicken embryos in vitro in the presence of Shh at various concentrations (Roelink et al., 1995; Ericson et al., 1997a, b). These experiments were done during the later period of Shh signaling when Shh induces different types of neurons. In order to quantify the number of floor plate cells, motor neurons, and interneurons induced, spinal cord sections were immunostained for gene regulatory proteins characteristic of each cell type. The results were as predicted by the Shh morphogen hypothesis (Fig. 13.11). Each cell type was induced at certain concentration range, and the concentrations yielding maximum induction for each cell type decreased with increasing distance of the cell from the floor plate in vivo.

Questions

1. Morphogen models imply that the concentration of a diffusible molecule decreases with increasing distance from its source. What is implied about the turnover of the morphogen?
2. Morphogen models also imply that morphogens ultimately control differential gene expression. Since Shh is a secreted glycoprotein, it will not cross the plasma membrane of its target cells. How then can it control the target cells' gene activity?
3. While notochord and floor plate are closely attached during neurulation, the notochord subsequently separates from the neural tube. Does this separation affect the usefulness of the notochord as a source of a morphogen for the ventral portion of the neural tube?

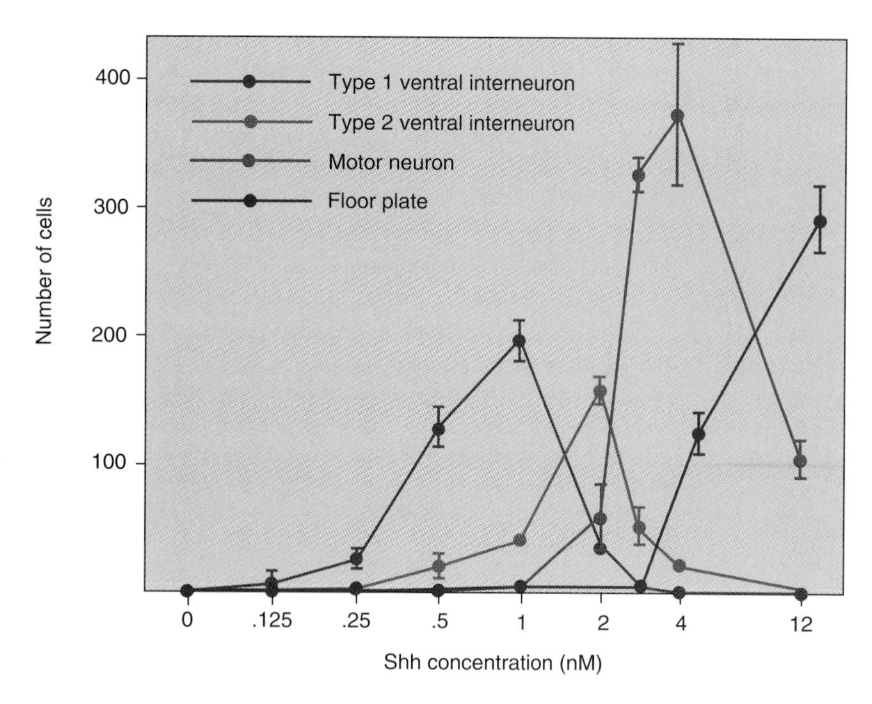

Figure 13.11 Induction of distinct types of neurons by different concentrations of Sonic hedgehog (Shh). Neural tube explants from chicken embryos were cultured in vitro in the presence of Shh at various concentrations and then immunostained for proteins characteristic of floor plate, motor neurons, or ventral interneurons. Labeled cells were counted in "optical sections" with a confocal microscope. Each cell type was induced within a certain range of Shh concentrations, and the concentrations yielding maximum induction for each cell type decreased with increasing distance of the cell from the floor plate in vivo (see Fig. 13.10).

The simplest interpretation of the available data is that Shh controls several regulatory genes such as *Pax-3*$^{+}$ in a concentration-dependent manner, and that these regulatory genes in turn regulate further genes that bring about the differentiation of floor plate cells, motor neurons, and different interneurons. However, it still needs to be resolved how Shh signaling interfaces with the basic neuronal cell determination by Notch/Delta signaling.

While the ventral portion of the neural tube is patterned by signals from notochord and floor plate, patterning of the dorsal neural tube depends on signals from adjacent epidermal ectoderm (Liem et al., 1995, 1997). Isolated neural plates acquire their characteristic patterns of gene expression only in contact with adjacent future epidermis. After neural tube closure, several bone morphogenetic proteins (BMPs) are synthesized in epidermal domains covering the dorsal midline, in the roof plate, and in adjacent portions of the neural tube (see Fig. 13.10; Barth et al., 1999). They seem to induce different subsets of dorsal interneurons and at the same time to inhibit the differentiation of ventral type neurons, just like Shh antagonizes genes necessary for dorsal neurons. Members of the Wnt family of signaling molecules are also expressed in the dorsal part of the neural tube and have been shown to promote the formation of the most dorsal neural structure, the *neural crest* (Saint-Jeannet et al., 1997; LaBonne and Bronner-Fraser, 1998).

In summary, the dorsoventral pattern of different cell types in the spinal cord is specified by antagonistic signals derived from adjacent tissues. Ventral signals are originally released from chordamesoderm and subsequently from notochord and floor plate. Dorsal signals originate first from epidermal ectoderm adjacent to the neural plate and later on from roof plate and dorsal spinal cord. These signals direct the specification of certain neuronal types in a dorsoventral pattern.

As the spinal cord completes its histogenesis, both white and gray matter increase in thickness. Mitotic activity in the neuroepithelium decreases. The neuroepithelial cells remaining in the adult spinal cord are called ***ependymal cells;*** they surround the lumen of the spinal cord, now called the ***central canal.*** On the outside of the spinal cord, mesenchymal cells form three covering epithelia, the *meninges.* The mature spinal cord consists, from the inside out, of the central canal, the surrounding ependymal cells, the gray matter, the white matter, and meninges.

THE BASIC ORGANIZATION OF THE SPINAL CORD IS MODIFIED IN THE BRAIN

The brain develops from the cranial part of the neural tube, and the initial steps in the development of brain and spinal cord are quite similar. In particular, the same molecular signals that set up the dorsoventral pattern of the spinal cord are also active in the brain (Ericson et al., 1995). However, the brain becomes more complex as the central canal dilates and forms fluid-filled spaces, called ***ventricles,*** which are surrounded by *brain regions* that develop regionally different structures and functions. Also, the relatively simple organization of the spinal cord into a mantle layer (gray matter) on the inside and a marginal layer (white matter) on the outside becomes more refined in the brain. Some clusters of mantle cells move out into the marginal layer and form islands of gray matter called ***nuclei*** that assume distinct functions. Other mantle cells move all the way to the periphery so that gray matter comes to surround white matter as the multilayered ***brain cortex.***

The development of the major adult brain regions is outlined in Figures 13.12 and 13.13. In a 4-week human embryo, three bulges appear in the cephalic end of the neural tube: ***prosencephalon, mesencephalon,*** and ***rhombencephalon.*** A week later the prosencephalon has

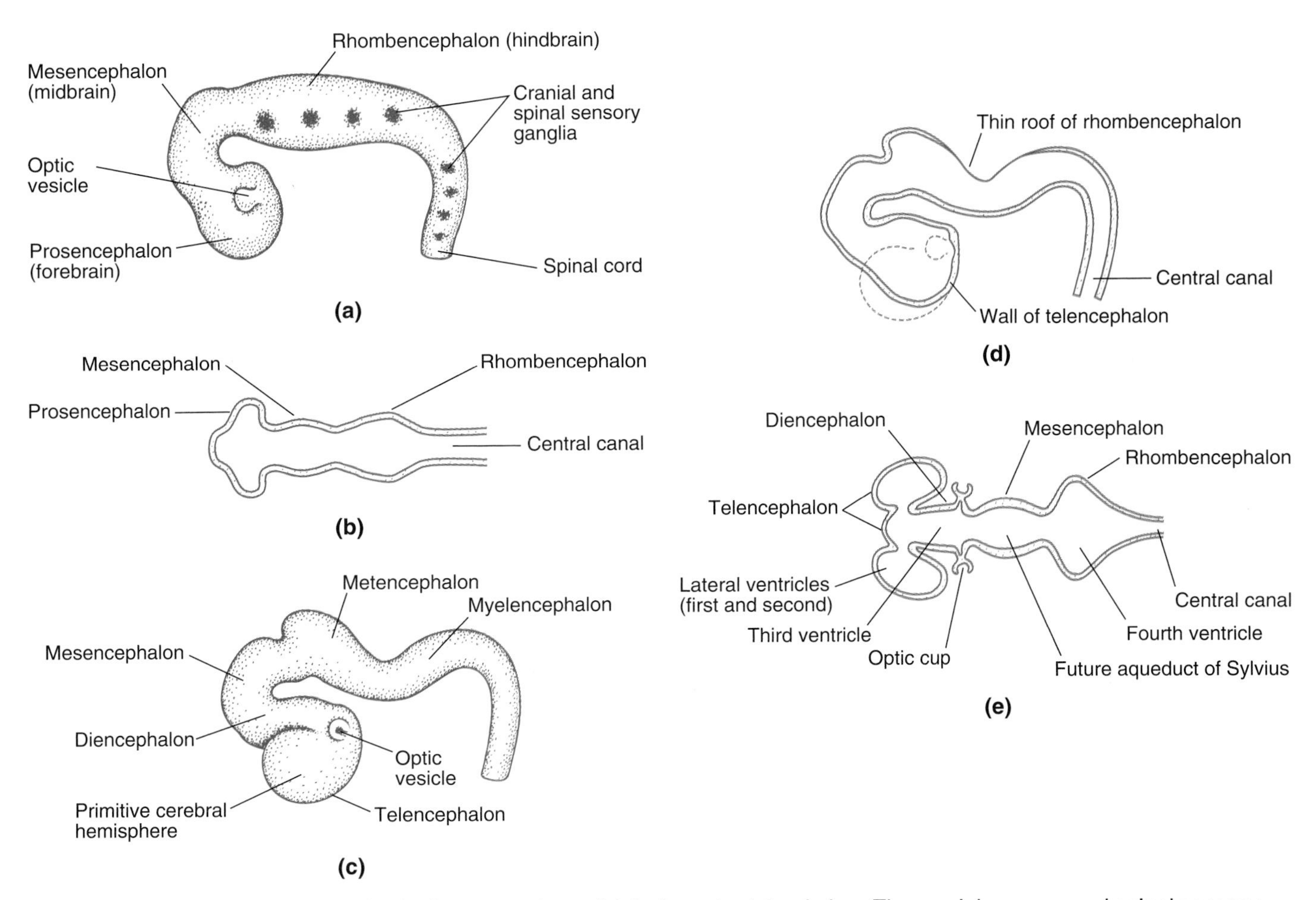

Figure 13.12 Brain development in the human embryo. **(a)** At 4 weeks, lateral view. The cranial sensory and spinal sensory (dorsal root) ganglia are derived from the neural crest (see also Fig. 13.27). **(b)** At 4 weeks, stretched-out brain in frontal section to show dilated portions of the central canal. **(c)** At 5 weeks, lateral view. **(d)** At 5 weeks, median section passing between the two cerebral hemispheres. **(e)** At 5 weeks, stretched-out brain in frontal section to show the brain ventricles.

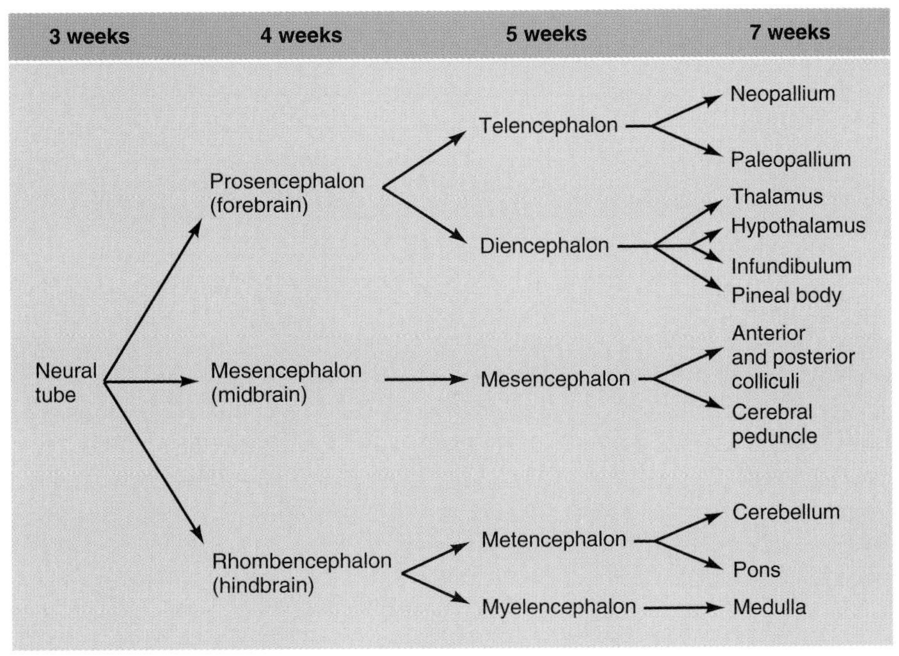

Figure 13.13 Flow chart of the subdivisions of the human brain at subsequent stages of development.

divided into the telencephalon and diencephalon. Anteriormost, the ***telencephalon*** consists of a median portion and two lateral outpocketings, the future hemispheres of the ***cerebrum*** (Lat. "brain"). Inside the major brain areas, the central canal is expanded into a larger spaces called *ventricles;* they are numbered as follows. Ventricles I and II are located inside the two cerebral hemispheres. The adjacent ***diencephalon*** contains ventricle III, which also extends into the two optic vesicles (see Fig. 1.16). The mesencephalon has a fluid-filled lumen called the *aqueduct of Sylvius,* which connects ventricle III with the more posteriorly located ventricle IV. The rhombencephalon subdivides into an anterior and dorsal portion, the ***metencephalon,*** and a posterior and ventral portion, the ***myelencephalon.*** The latter surrounds ventricle IV, which

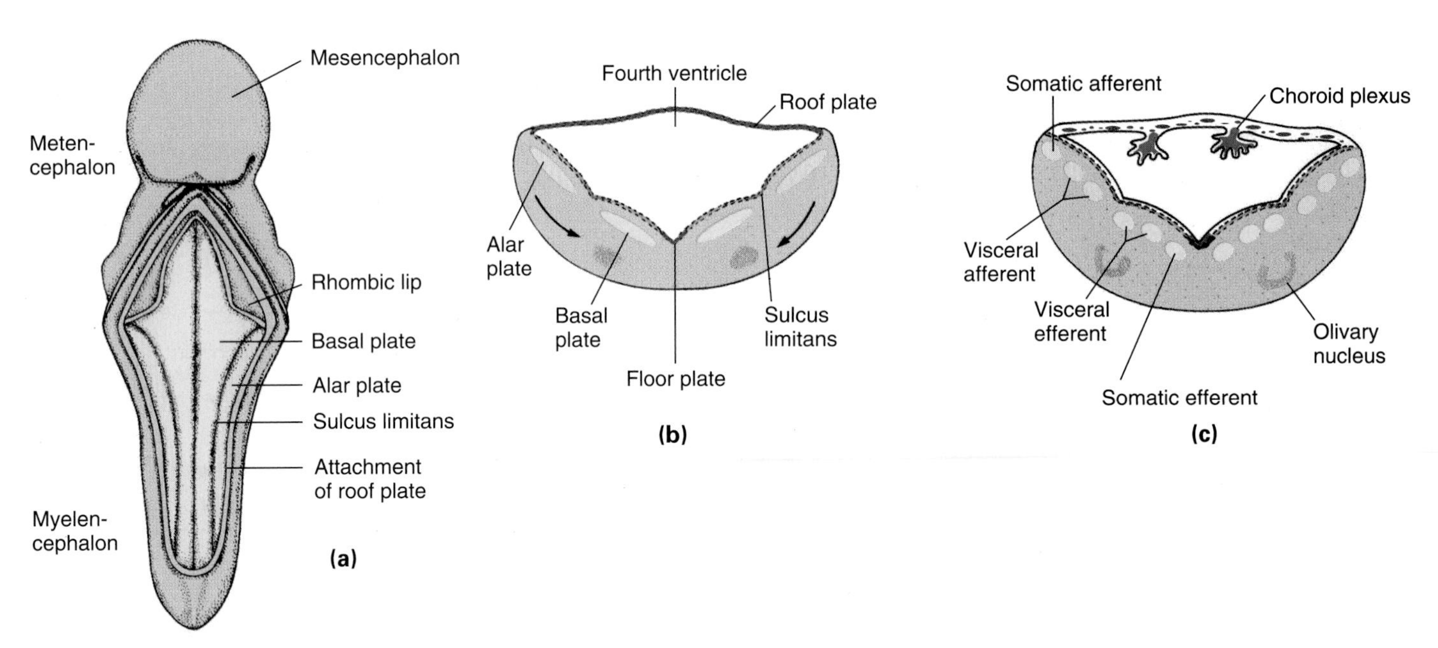

Figure 13.14 Development of the medulla oblongata. **(a)** Dorsal view of the rhombencephalon in the 6-week-old human embryo. The roof of the fourth ventricle has been cut away to expose the alar plates and basal plates, which are separated by the sulcus limitans. **(b, c)** Transverse sections showing the development of the choroid plexus of the fourth ventricle. Compare the position of the alar and basal plates here and in the spinal cord as shown in Figure 13.7. The olivary nucleus moves from its original position in the alar plate into the ventral white matter, from where it will project sensory input to the cerebellum.

blends posteriorly into the central canal of the spinal cord.

In the 7-week human embryo, the subdivisions of the brain have developed further. We will briefly characterize the brain regions at this stage, beginning caudally, where the organization of the brain is relatively simple.

Medulla The myelencephalon becomes the ***medulla oblongata*** (Lat. *medulla,* "innermost part," here referring to the spinal cord inside the spine; *oblongata,* "extended"), a fairly descriptive term, since this part of the brain retains the greatest similarity to the spinal cord (Fig. 13.14). The medulla controls the reflexes of the neck, throat, and tongue, just as the spinal cord mediates the reflexes of the trunk and appendages. However, the medulla differs from the spinal cord in that its lateral walls are arranged as if they had been rotated around an imaginary axis in the floor plate, a movement similar to opening a book. In the process, the neural canal expands into the fourth ventricle. As in the spinal cord, the basal plates of the medulla oblongata contain efferent neurons, and the alar plates contain afferent interneurons. The roof plate of the fourth ventricle consists of a layer of ependymal cells, which are in direct contact with the *meninges.* These membranes, along with the blood vessels contained in the meninges, form a ***choroid plexus,*** which releases a blood filtrate, ***cerebrospinal fluid,*** into the fourth ventricle. Similar plexuses are also formed in the other ventricles.

Cerebellum The metencephalon gives rise to the cerebellum and the pons. The ***cerebellum*** (Lat. "little brain") develops dorsally; it received its name because its cortex is folded and consists of gray matter like that of the cerebrum. The cerebellum arises from extensions of the alar plates that form the ***cerebellar plate*** anterior to the fourth ventricle (Fig. 13.15). Initially, the cerebellar plate consists of a neuroepithelium, a mantle layer, and a marginal layer (Fig. 13.16). Cells formed by the neuroepithelium migrate into the marginal layer to form the ***external granular layer.*** Cells given off later by the neuroepithelium include the ***Purkinje cells,*** which form one axon plus an enormous number of dendrites (see Fig. 13.1). One Purkinje neuron may form thousands of synapses (connections) with other neurons. Cells from the granular layer and Purkinje cells eventually form the cortex of the cerebellum, which functions as a coordination center for posture and movement.

Pons Opposite the cerebellum, on the ventral side of the metencephalon, the ***pons*** is formed. It serves as a pathway for nerve fibers between the spinal cord and the cerebellum as well as the cerebral hemispheres.

Mesencephalon The mesencephalon is similar in its morphology to the spinal cord. The marginal layer enlarges ventrally to accommodate nerve fibers connecting the cerebral cortex with the pons and spinal cord. The alar plates in the dorsal mesencephalon initially form two ridges, which become subdivided by transverse grooves into four elevations, or ***colliculi*** (Lat. *colliculus,* "little hill"). In mammals, the colliculi are relatively small relay stations for visual and auditory reflexes, known as anterior colliculi and posterior colliculi, re-

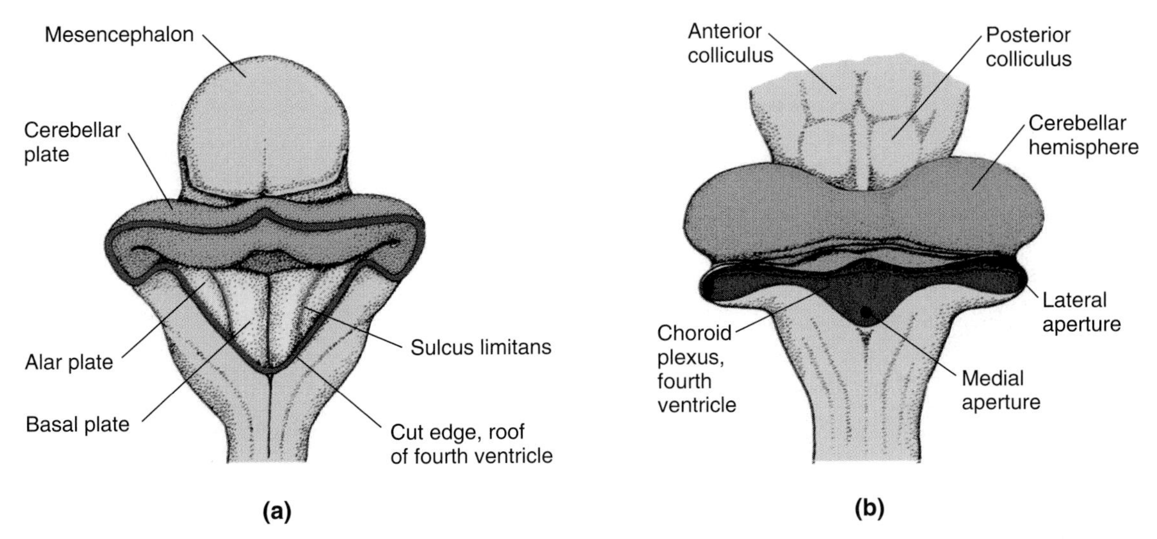

Figure 13.15 Dorsal views of the metencephalon and mesencephalon. **(a)** 8-week-old human embryo. The roof of the fourth ventricle has been cut away to expose the floor of the ventricle. **(b)** 4-month-old human fetus. The choroid plexus covering the fourth ventricle has acquired three apertures through which cerebrospinal fluid drains into the space between the meninges. The cerebellum develops from the cerebellar plate. The anterior and posterior colliculi have formed in the mesencephalon.

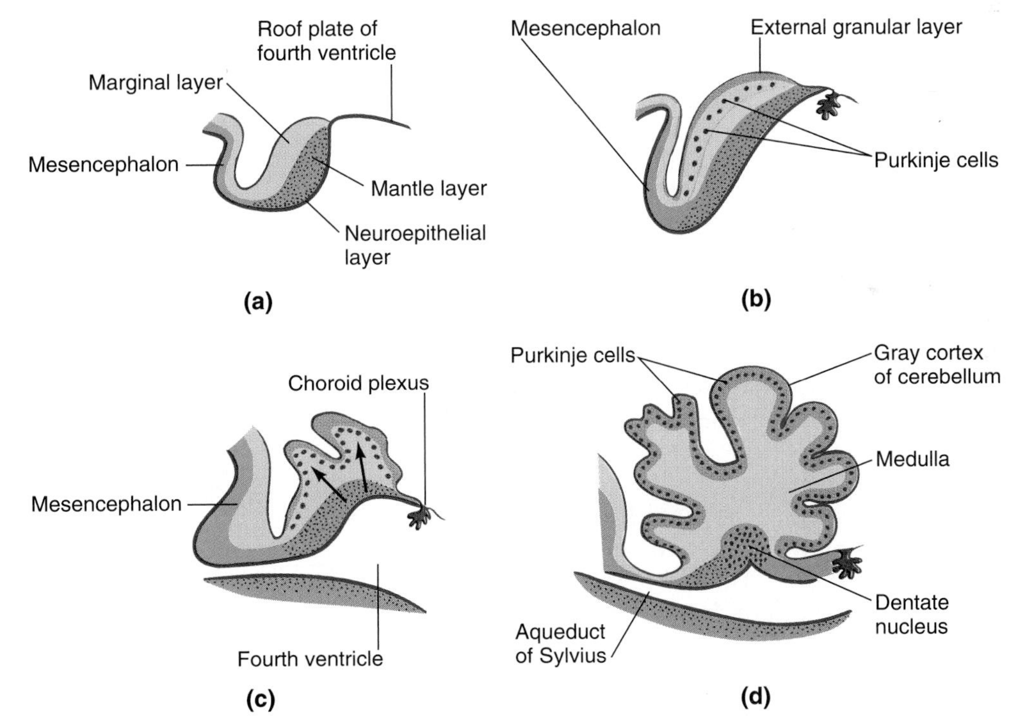

Figure 13.16 Development of the human cerebellum shown in median section at successive stages: **(a)** 8 weeks; **(b)** 12 weeks; **(c)** 13 weeks; **(d)** 15 weeks. The cells of the external granular layer and Purkinje cells are given off successively by the neuroepithelium. Both groups of cells move to the surface of the marginal layer, where they form the cerebellar cortex. The dentate nucleus is one of the deep cerebellar nuclei.

spectively (see Fig. 13.15). In nonmammalian vertebrates, the anterior colliculi are called the ***optic lobes;*** they are much larger and contain the visual projection areas that in mammals are located in the cerebral hemispheres.

Diencephalon The diencephalon seems to consist of two alar plates, delimiting a narrow third ventricle, while lacking basal plates (Fig. 13.17). The dorsal roof of the third ventricle is again formed by a choroid plexus. Located caudally to the plexus is the ***pineal gland,*** which through its hormonal secretions generates the circadian rhythm and, in some vertebrates, modulates the yearly reproductive cycle. A lateral groove on the inside of each alar plate separates the dorsal ***thalamus*** from the ventral ***hypothalamus.*** The thalamus has been dubbed the

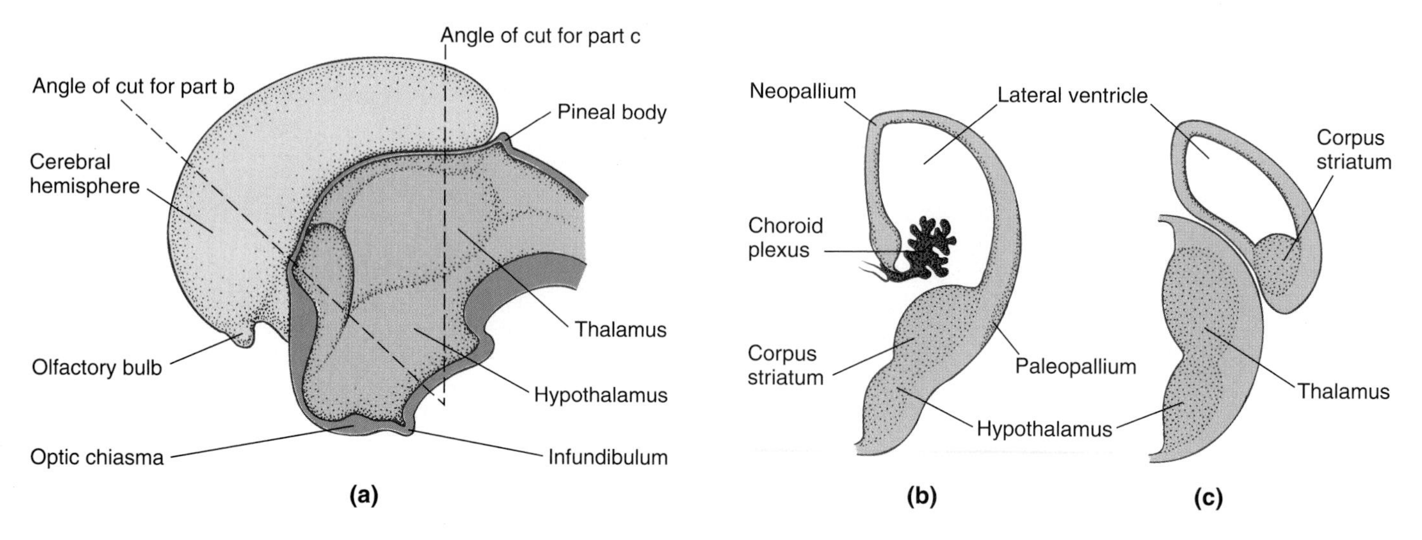

Figure 13.17 Diencephalon and telencephalon of the 8-week-old human embryo. **(a)** Medial view of the right half, from which the left half has been cut away. **(b, c)** Transverse sections at different angles as indicated in part a.

"antechamber of the cerebrum," because it serves as a gateway for sensory fibers passing from the spinal cord and the brain stem to the cerebral hemispheres. In the hypothalamus, several *nuclei* act as regulatory centers for visceral functions, including sleep, digestion, and body temperature, as well as aggression and other emotional behaviors. A prominent landmark at the bottom of the hypothalamus is the ***optic chiasma*** (Gk. *khiasma*, "cross"), an incomplete crossing of the *optic nerves*. Right behind the optic chiasma there is a median extension of the hypothalamus, the ***infundibulum*** (Lat. "funnel"), which forms the posterior lobe of the ***pituitary gland***, or *hypophysis*. The anterior lobe of the pituitary gland originates from ***Rathke's pouch***, an invagination from the stomodeum, the ectodermal epithelium lining the anterior oral cavity (Fig. 13.18).

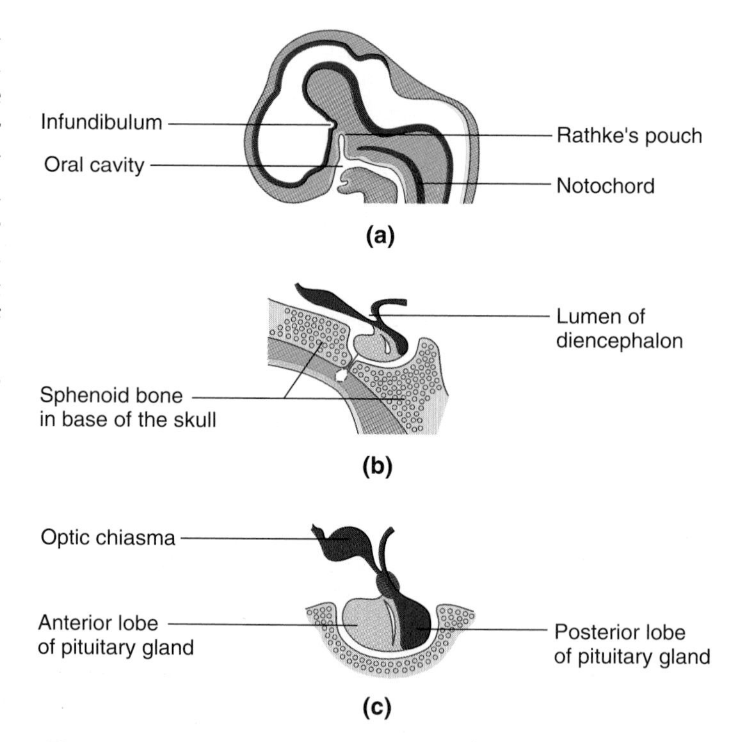

Figure 13.18 Development of the pituitary gland from Rathke's pouch, a median invagination from the stomodeum, the ectodermal lining of the oral cavity, and the infundibulum, an extension of the floor of the diencephalon.

Telencephalon The telencephalon, the anteriormost part of the brain, consists mainly of the two cerebral hemispheres, each surrounding a ventricle with a choroid plexus (see Fig. 13.17). Each of these ventricles opens into the third ventricle through large interventricular openings. To an extent, the development of the cerebral hemispheres reflects their phylogeny. A region that differentiates early in mammals and is relatively large in more primitive vertebrates is known as the ***paleopallium*** (Gk. *palaios*, "old"; Lat. *pallium*, "mantle"). It is located anteriorly and laterally, extends into the *olfactory bulb*, and is associated with the sense of smell. The phylogenetically newest part of the cerebrum, the ***neopallium***, is located dorsally and develops relatively late but then grows at a rapid rate. In humans, the neopallium eventually occupies about 90% of the cerebral hemispheres and, together with the cerebellum, covers most other parts of the brain.

In the adult, the right and left hemispheres are connected by several *commissures*, which cross the midline (Fig. 13.19). The first crossing axons to appear form the *anterior commissure*, which connects the olfactory bulb and related brain areas of one hemisphere to those of the other hemisphere. The most extensive commissure is the ***corpus callosum*** (Lat. "hard body"). It is formed gradually after the 10th week of human development and connects the nonolfactory portions of the cerebral hemispheres as they extend over the diencephalon and mesencephalon. Corpus callosum formation begins

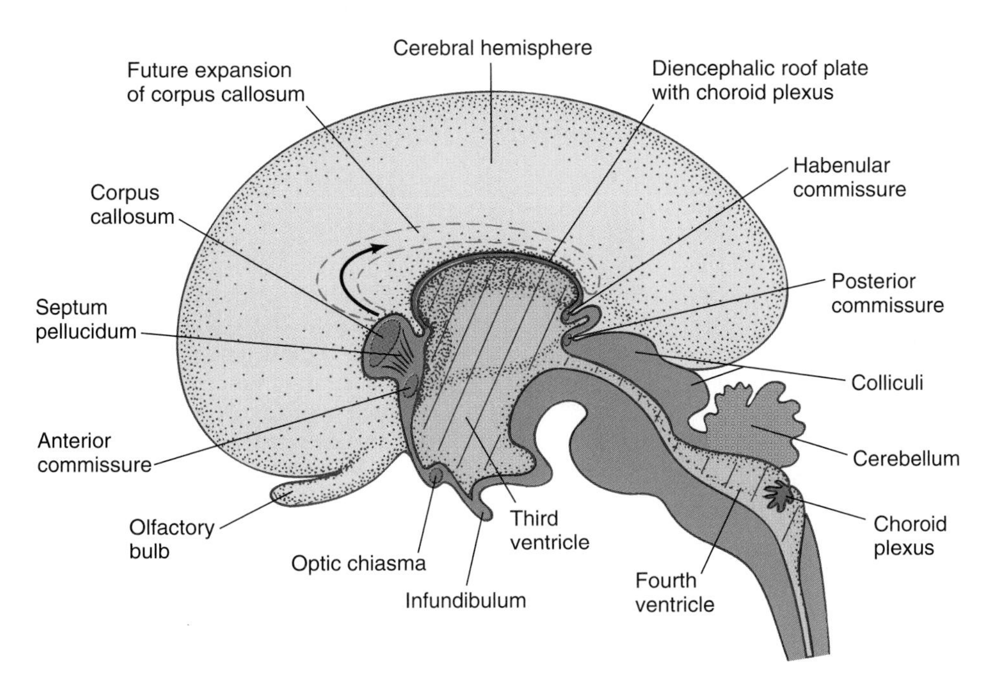

Figure 13.19 Brain of a 4-month human fetus. The brain is cut in the median plane, and the right half is shown from a medial view. The cut regions include several commissures (gray shading), bundles of axons that connect corresponding portions on the right and left sides. The arrow and broken lines indicate the future expansion of the largest commissure, the corpus callosum.

anteriorly above the *septum pellucidum,* the thin medial layer that separates the two lateral ventricles. From there, the corpus callosum extends posteriorly, arching over the thin roof of the diencephalon.

THE PERIPHERAL NERVOUS SYSTEM IS OF DIVERSE ORIGIN

The ***peripheral nervous system* (*PNS*)** includes all the nervous tissue outside the central nervous system. In part, the peripheral nervous system consists of the long axons that grow out of efferent neurons located in the central nervous system. In addition, the PNS includes many ***ganglia*** (sing., *ganglion,* a group of neurons) that originate from the neural crest and the ectodermal placodes (discussed later in this chapter). Most peripheral nerve fibers are myelinated in the same way as the axons in the white matter of the central nervous system (see Figs. 13.5 and 13.6). The myelin layers in the peripheral nervous system are built up by ***Schwann cells,*** which correspond to the oligodendroglial cells in the central nervous system. A typical peripheral ***nerve*** contains thousands of axons and/or dendrites, often bundled in groups and surrounded by connective tissue and fat cells derived from mesenchyme.

Efferent (motor) axons leave the spinal cord in segmental bundles called ***ventral roots*** (Fig. 13.20). Afferent (sensory) axons entering the spinal cord are bundled in segmental ***dorsal roots.*** These axons originate from neurons located in the ***dorsal root ganglia,*** which are derived from neural crest cells as we will discuss in the next section. The neurons of the dorsal root ganglia receive input via their long dendrites from sensory cells in the skin and elsewhere.

Dorsal and ventral roots unite to form ***spinal nerves,*** which leave the spine through canals formed between successive vertebrae. Outside the spine, each spinal nerve gives off a ***dorsal ramus*** (Lat. *ramus,* "branch"; pl., *rami*), supplying its respective segment of skin and muscles with both efferent and afferent axons. Each spinal nerve also sends out two ***communicating rami*** as connections to visceral ganglia, which derive from the neural crest. The remainder of a spinal nerve continues as a large ***ventral ramus,*** providing both motor and sensory supply to the ventral regions of the body and—in the case of some spinal nerves—the appendages.

There is a segmental pattern in the skin innervation by the sensory components of the spinal nerves that largely parallels the segmental pattern of the spinal nerves themselves (Fig. 13.21). However, the segmentation of the vertebrate body does not originate in the nervous system itself but is imposed by the segmental arrangement of the *paraxial mesoderm* (see Section 14.2).

Cranial nerves are associated with the brain much in the same way spinal nerves are associated with the spinal cord. However, the cranial nerves do not emerge at regularly spaced intervals, nor do they have similar functions. The union of dorsal and ventral roots characteristic of spinal nerves is not maintained in the brain. Several cranial nerves, including those innervating the eye and tongue muscles, have mostly efferent functions like the ventral roots of the spinal nerves. Other cranial nerves are strictly sensory, including the *olfactory tract* and the *optic nerve,* which are really connections within the telencephalon and the diencephalon, respectively. The *statoacoustic nerve* is an exclusively sensory nerve entering the myelencephalon. Still other cranial nerves—like the communicating rami given off by the

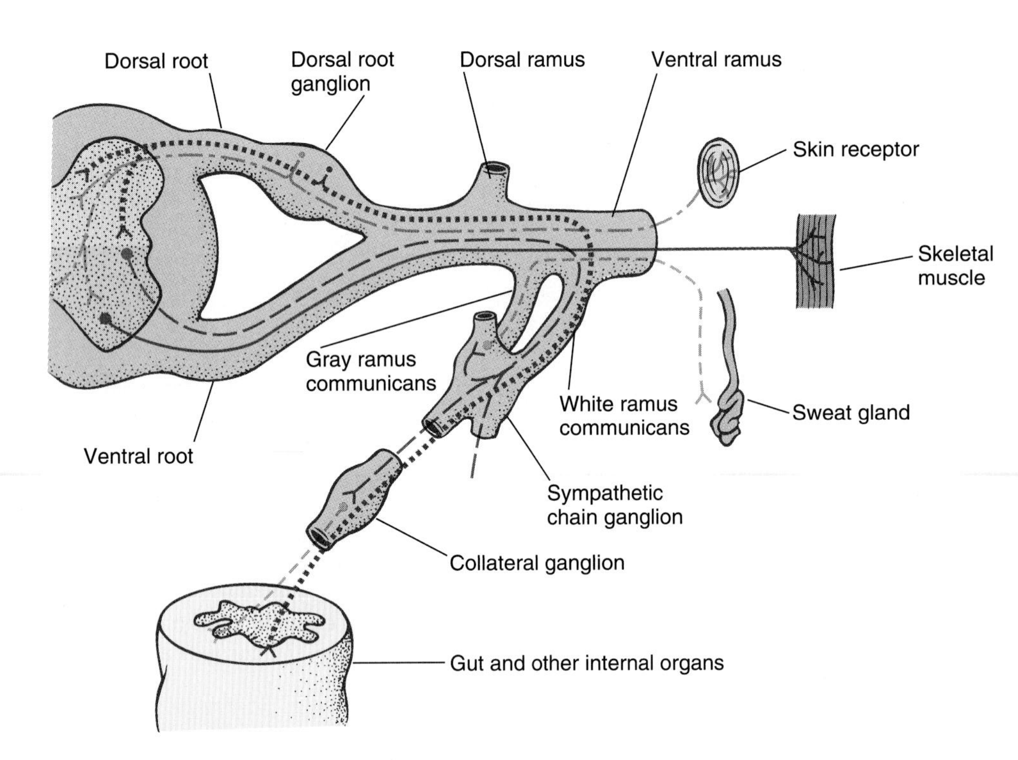

Figure 13.20 Diagram of a spinal nerve with its dorsal (sensory) and ventral (motor) roots. The chain ganglia of the sympathetic nervous system run parallel to the spinal cord outside the vertebral column. They are connected to each spinal nerve by two communicating rami. The nerve fibers are classified as somatic sensory (dashed/dotted blue), somatic motor (solid green), visceral sensory (dotted purple), and visceral motor (dashed red).

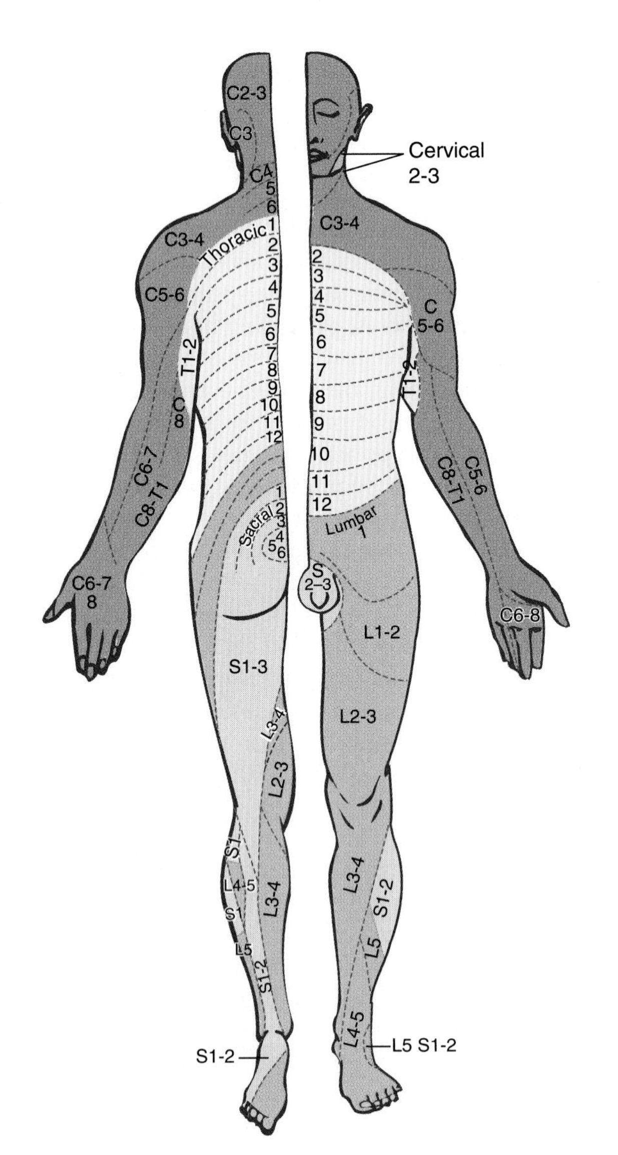

spinal nerves—serve mainly as connections to visceral ganglia. In contrast to the dorsal root ganglia of the spinal nerves, which originate from neural crest cells only, the sensory cranial nerves arise from both the neural crest and ectodermal placodes (see Section 13.3).

13.2 Neural Crest

Neural crest (NC) cells are found only in vertebrates; they originate during the process of neurulation. The name "neural crest cells" is indicative of their topographical position at the crests of the neural folds (Fig. 13.22). A neural fold comprises the margins of the neural plate and the adjoining epidermis. As neurulation proceeds, the folds rise and bend toward the dorsal midline of the embryo. When the folds fuse, epidermis joins epidermis, and neural plate joins neural plate to form the neural tube. During this process, cells that were originally located at the crest of the folds come to lie between the epidermis and the underlying neural tube and somites (Fig. 13.23). This location is called the ***migration staging area*** because it is from here that NC cells will embark on different migration routes to a wide range of destinations.

NC cells have held the attention of developmental biologists mainly because of their extensive migrations and because they give rise to a bewildering variety of cell types, including cartilage elements of the head, pig-

Figure 13.21 Segmental distribution of the skin areas innervated by the sensory components of the spinal nerves. C = cervical, T = thoracic, L = lumbar, S = sacral.

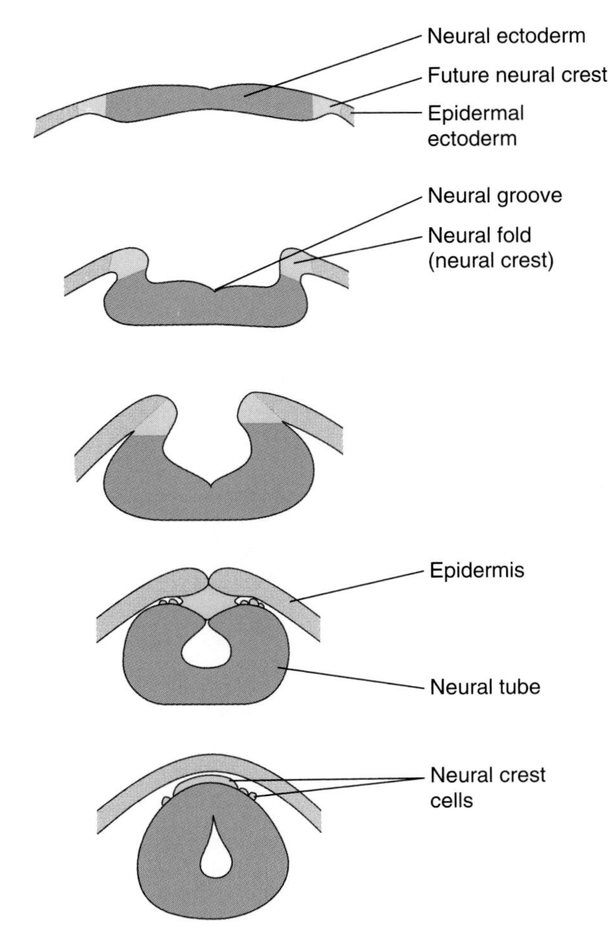

Figure 13.22 Successive stages of amphibian neurulation diagrammed in transverse sections. Each of the drawings shows the neural ectoderm, the epidermal ectoderm, and the neural crest. When the neural folds fuse, they close the neural tube under a continuous layer of epidermis. The cells originally located at the crests of the neural folds, known as neural crest cells, then come to lie on top of the neural tube, from where they migrate to different locations and have a wide variety of fates.

ment cells in the epidermis, various types of neurons, hormone-producing gland cells, and smooth muscle cells of the cardiovascular system. Because of these features, NC cells have long been favorite objects for studying cell migration and cell determination (B. K. Hall and S. Hörstadius, 1988; Selleck et al., 1993; Le Douarin et al., 1994; Anderson et al., 1997).

In this section, we will explore the embryonic origin of NC cells, their main routes of migration, and their major fates. Returning to the question of cell determination, we will examine to what extent NC cells are "born" with certain restrictions on their potency and how much their fate depends on the environment in which they settle. We will also discuss how the strategy of *heterochronic transplantations* has been used to reveal dynamic changes in the competence of neural crest cells and in the signals they receive. Finally, we will explore the role of extracellular matrix molecules and growth factors in guiding and determining NC cells.

NC CELLS ARISE AT THE BOUNDARY BETWEEN NEURAL PLATE AND EPIDERMIS

Do NC cells arise from neural plate cells, or epidermal ectoderm cells, or both? How do NC cells acquire the properties that distinguish them from the neighboring cells of the neural plate and epidermis? NC cells could possibly form where the vertical component of neural induction by *chordamesoderm* is at an intermediate level of strength, too strong to allow epidermis formation but too weak to induce neural plate formation. Alternatively, the NC could arise as a result of the juxtaposition of the neural plate and the adjacent epidermis.

To test the juxtaposition hypothesis, Moury and Jacobson (1989, 1990) swapped patches of neural plate and epidermal ectoderm (epidermis) between embryos of a salamander, *Ambystoma mexicanum* (Fig. 13.24). Soon after a graft healed into its new position, folds arose around the edges. In contrast, no fold formation was observed in control experiments when epidermal patches were transplanted into epidermis, and neural plate patches into neural plate. Also, the type of mesoderm underlying the transplants did not seem to affect the results. Thus, it was the juxtaposition of neural plate and epidermis, rather than wound healing or inductive influences from the mesoderm, that caused neural fold formation.

In follow-up experiments, the investigators replaced patches of neural fold in albino recipients with patches of either epidermis or neural plate from pigmented donors. By tracing the fates of the pigmented donor cells in the albino hosts, they found that both types of transplants gave rise to NC derivatives. In particular, transplanted neural plate gave rise to pigment cells, while transplanted epidermis gave rise to dorsal root ganglia. Similar results were obtained in experiments with *Xenopus* and chicken embryos (Mancilla and Mayor, 1996; Selleck and Bronner-Fraser, 1995). Together, these studies show that NC cells can be induced by local interactions between juxtaposed neural plate and epidermis. They suggest that, in normal development, the NC cells also arise from such interactions. In addition, the results show that both partners in this interaction can contribute to the NC cell pool.

NC CELLS HAVE DIFFERENT MIGRATION ROUTES AND A WIDE RANGE OF FATES

Early in their development, NC cells undergo a dramatic transition when they lose their epithelial connections and become migratory. In chicken embryos, this epitheliomesenchymal transition is associated with the activation of a regulatory gene, *slug*$^+$ (Nieto et al., 1994). The same gene is also expressed in the primitive streak, as cells ingress to form mesoderm and endoderm (see Section 10.4). When slug mRNA is degraded with antisense oligodeoxynucleotides and RNase H, embryos are deficient in the epitheliomesenchymal transitions that

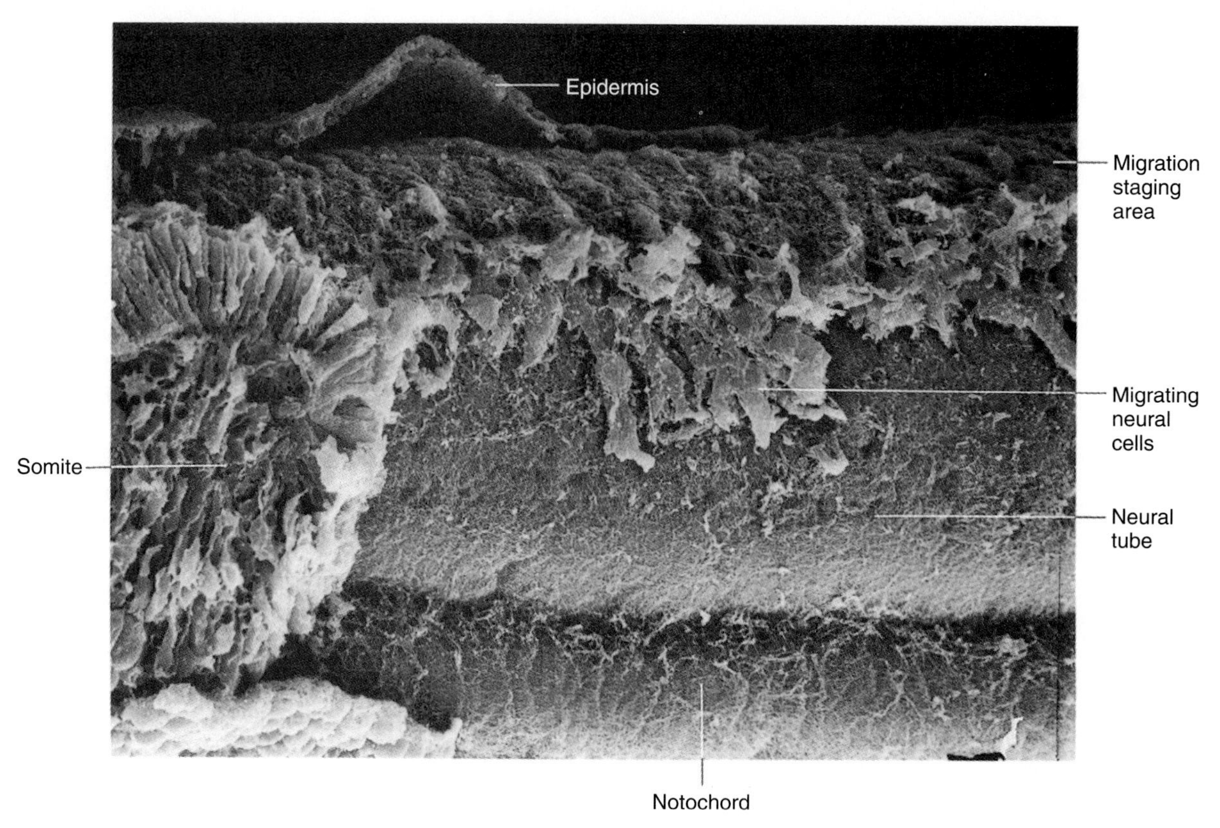

Figure 13.23 Scanning electron micrograph showing a lateral view of a chicken embryo from which much of the epidermis has been removed. The neural crest cells are leaving the migration staging area located between neural tube, epidermis, and somites.

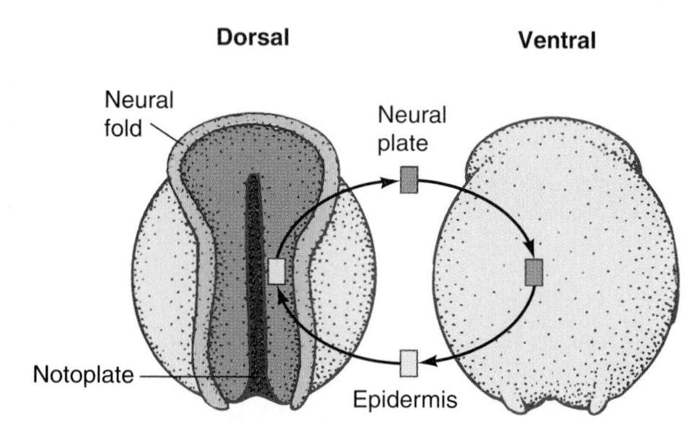

Figure 13.24 Boundary effect between neural plate and epidermis in the salamander neurula. In order to study the conditions of neural fold and neural crest cell formation, patches of neural plate and ventral epidermis were transplanted between salamander embryos (*Ambystoma mexicanum*). The grafts of neural plate were cut so as to avoid both the neural folds and the notoplate. In order to trace cells derived from the graft, the donor embryo was pigmented while the host was from an albino strain.

normally occur in primitive streak and NC. Because slug is a gene-regulatory protein, it may control many other genes involved in cell migration. For example, NC cells express different cadherins before and after the beginning of their migration (Nakagawa and Takeichi, 1995). The expression of *slug*$^+$ also causes the dissociation of *desmosomes,* anchoring junctions that connect the intermediate filaments of adjacent cells in epithelia (see Fig. 2.23; Savagner et al., 1997). Further analysis of the genes that control the activation of slug may provide a better understanding not only of NC and primitive streak development but also of the mechanisms of cellular migration.

Once NC cells have left the migration staging area, they take different pathways that lead them to a wide range of destinations and fates. Investigators have used different techniques to identify both the migratory pathways and the ultimate fates of NC cells. Weston (1963) and Johnston (1966) radiolabeled DNA by culturing chicken embryos in the presence of [^{3}H]thymidine. Segments of labeled neural tube, including NC, were then grafted *homotopically* into unlabeled embryos (Fig. 13.25). The hosts were allowed to develop for several days before microscopic sections were prepared for autoradiography (see Method 3.1). Silver grains generated by radioactivity revealed the positions to which the NC cells had migrated.

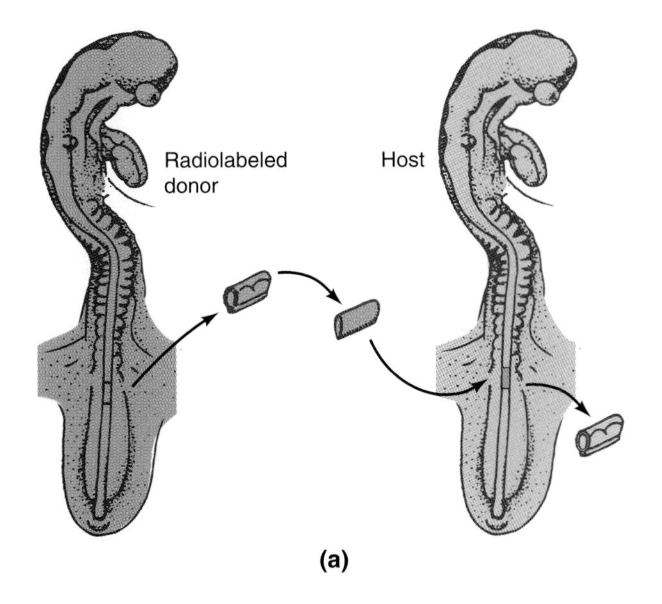

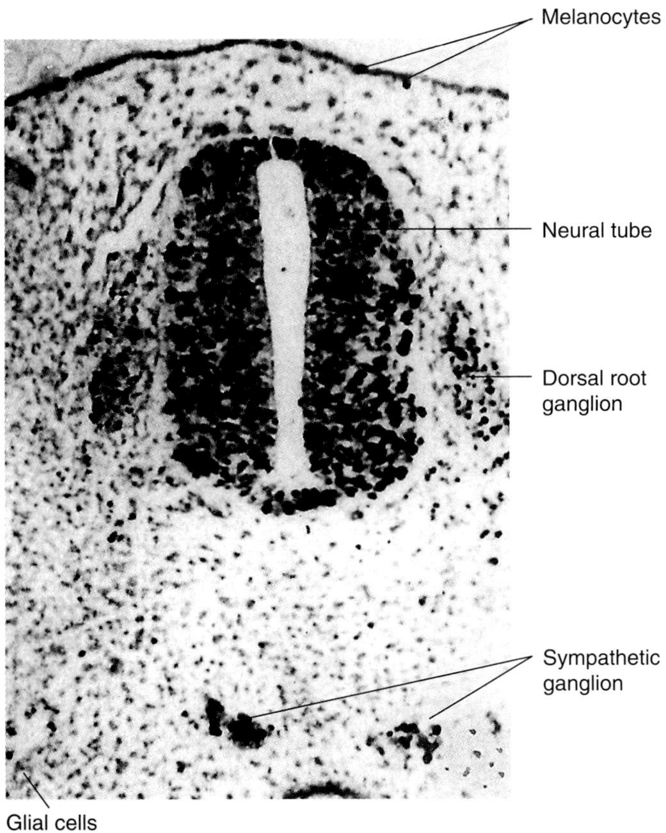

Figure 13.25 Fate-mapping technique for neural crest cells. **(a)** A piece of dorsal axis is removed from a donor labeled with [^{3}H]thymidine. The neural tube and adjacent neural crest are cleaned of other tissues and implanted in a host from which the corresponding piece has been removed. The descendants of the labeled cells are traced in the host by autoradiography(see Method 3.1). The neural tube cells remain stationary while the neural crest cells migrate extensively. **(b)** Autoradiograph showing locations of the transplanted neural tube and the descendants of neural crest cells that have formed melanocytes, dorsal root ganglia, sympathetic ganglia, and glial cells.

A genetic label, which has the advantage of not being diluted during cell divisions, was introduced by Le Douarin (1969, 1986). She found that embryonic parts can be transplanted between quail and chickens, and that both animals develop and hatch normally. Because quail nuclei have intensely staining *heterochromatin*, they can readily be distinguished from chicken nuclei in histological sections (Fig. 13.26). More recently, the use of fluorescent dyes and immunostaining (see Method 4.1) has further clarified the migration routes of NC cells (G. C. Tucker et al., 1984; Bronner-Fraser and Fraser, 1991). The combined use of these techniques has led to the following picture of NC development.

In the trunk (cervical, thoracic, and lumbosacral regions), NC cells leave by two major pathways (Fig. 13.27). Cells taking a ***dorsolateral pathway*** enter the skin and develop into melanocytes or ***xanthophores*** (yellow pigment cells). Cells that take the ***ventral pathway*** migrate to a variety of destinations. Some settle close to the neural tube and form the neurons and glial cells of the *dorsal root ganglia* (see Fig. 13.20). Other NC cells form most of the *visceral nervous system*, including the ***autonomic nervous system,*** which innervates all internal organs. The autonomic nervous system consists of ***sympathetic ganglia,*** derived from NC cells in the cervical, thoracic, and lumbar regions, and ***parasympathetic ganglia,*** derived from cephalic and pelvic NC cells. Sympathetic ganglia are centralized, forming two chains on the ventral side of the vertebral column, whereas parasympathetic ganglia are dispersed, being located near their target organs. Sympathetic neurons promote the activities of the lungs, heart, and other internal organs that support the fight-or-flight response. Parasympathetic neurons innervate the same organs antagonistically to promote rest and recreation.

Still other NC cells of the trunk form *Schwann cells*, which myelinate the peripheral nerves. Additional NC cells participate in forming the ***dorsal fin*** in amphibians. Finally, cells from the NC form the hormone-secreting ***adrenal medulla.*** Under stress, the adrenal medulla releases epinephrine and norepinephrine into the blood, thus enhancing the activity of the sympathetic neurons, which use the same hormones as transmitters.

In the head, so-called *cranial NC cells* form pigment cells, sensory cranial ganglia, parasympathetic ganglia, and hormone-producing cells, in ways that largely parallel the NC of the trunk. In addition, cranial NC cells give rise to bones and other connective tissues in the head (Le Douarin et al., 1994). Moreover, cranial NC cells contribute to certain structures within the eyes, ears, and teeth.

In an area overlapping the head and trunk regions, so-called ***cardiac NC cells*** give rise to melanocytes, neurons, cartilage and other connective tissues. In addition, cardiac NC cells contribute to the connective tissues and smooth muscle cells of the large blood vessels that emerge from the heart, including in particular the

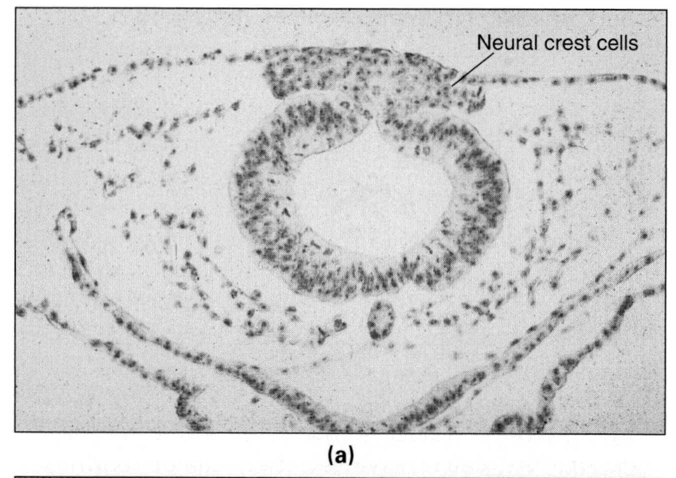

(a)

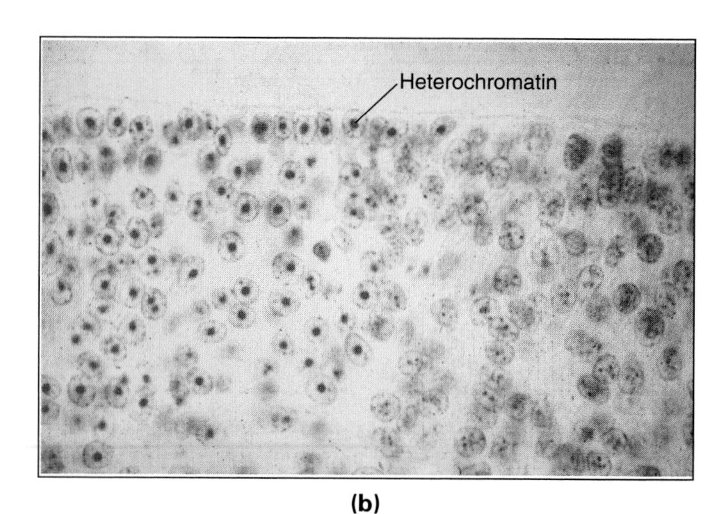

(b)

(c)

Figure 13.26 Photographs showing the use of quail-chicken chimeras for fate mapping. **(a)** Transverse section of the midbrain of a quail embryo. **(b)** Section of chicken embryo (right part) with a quail graft (left part). The quail cells are marked with deeply staining heterochromatin. **(c)** Chickens that have received grafts of quail neural tube and neural crest display normal behavior until their immune system matures and the graft is rejected. The pigment in their wing feathers is caused by melanocytes derived from quail neural crest cells.

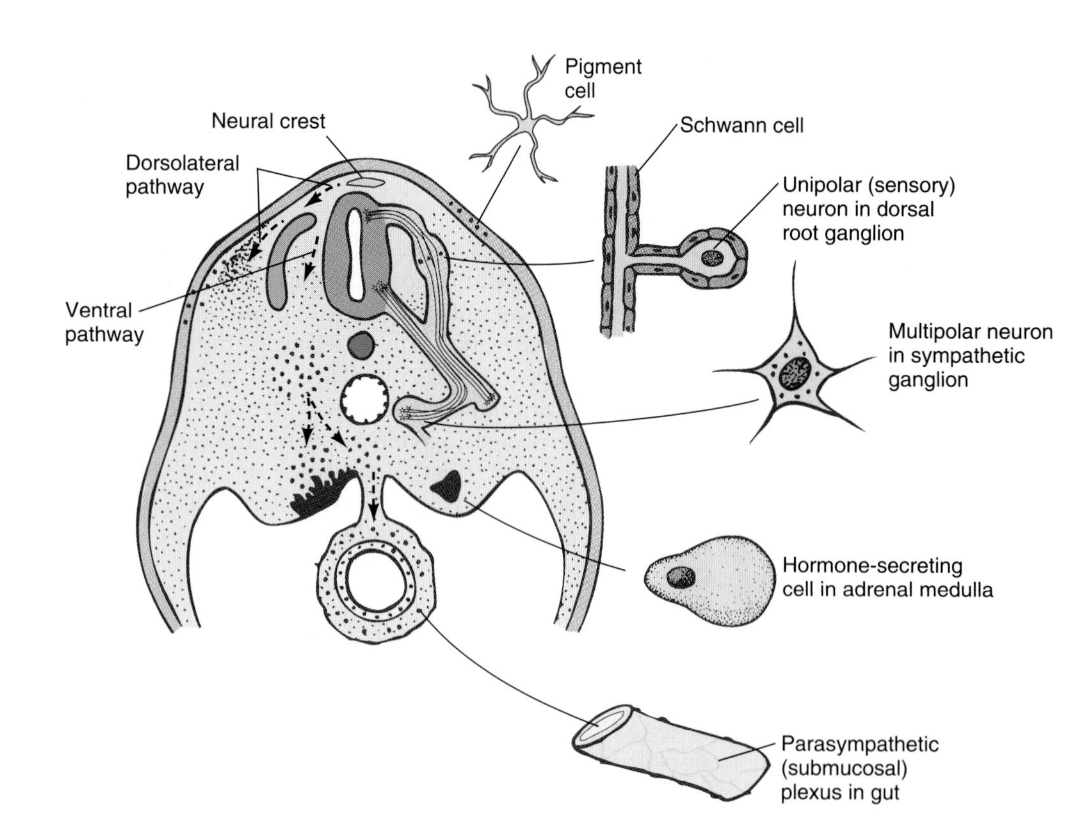

Figure 13.27 Development of neural crest cells in the trunk region. Schematic cross section of a vertebrate embryo showing the main pathways of migration (left side) and the major fates (right side).

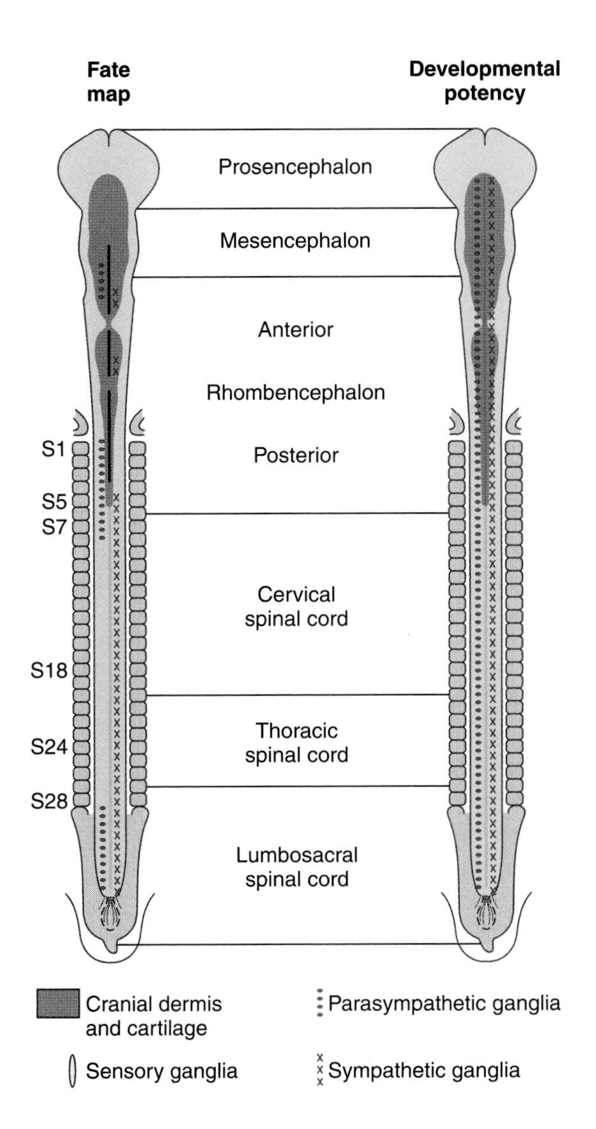

Figure 13.28 Fate map and developmental potency map of the chicken neural crest. Upon grafting to appropriate sites, neural crest cells from any level of the neural axis give rise to a wide range of different cell types. However, the potential of forming cranial cartilage and other connective tissues is limited to the head area down to the fifth somite (S5).

septum that separates the *aorta* from the *pulmonary artery* (Kirby, 1989; see also Section 14.6). Many of the structures formed by cranial and cardiac NC cells, as well as the dorsal fins in amphibians, arise from cells called ***ectomesenchymal cells*** because of their ectodermal origin and mesenchymal configuration. These cells do not fit the general pattern of the germ layers, because bone and blood vessels are generally derived from mesoderm rather than from ectoderm (see Section 14.6).

THE STRATEGY OF CLONAL ANALYSIS SHOWS THAT NC CELLS ARE A HETEROGENEOUS POPULATION OF PLURIPOTENT CELLS

How are NC cells determined? Is their potential limited to their fate when they leave the migration staging area? If not, at which points along their pathways, and in response to which signals, does determination occur? Transplantation experiments indicate that the potential to form cranial cartilage is limited to the NC of the head region (Fig. 13.28). No cartilage is formed when this section of the NC is removed or replaced with NC from the trunk (B. K. Hall and S. Hörstadius, 1988). Similarly, the potential to contribute to cardiovascular structures seems to be limited to the cardial NC (Kirby, 1989). However, the potential of cranial NC cells is not limited to their fate. When transplanted to the trunk region, they contribute to dorsal root ganglia, sympathetic ganglia, adrenal medulla, and Schwann cells (Schweizer et al., 1983).

While the potency to form certain head and cardiovascular structures seems restricted to cephalic and cardial NC, respectively, the potency to form all other NC cell derivatives is present along the entire anteroposteterior axis (Fig. 13.28). This conclusion is supported by extensive transplantation experiments. For instance, the posterior rhombencephalon and the thoracic region in chicken embryos differ in the types of visceral ganglia they normally give rise to. Rhombencephalic crest cells form parasympathetic neurons, which use acetylcholine as a transmitter, whereas thoracic NC cells produce sympathetic neurons, which use norepinephrine as a transmitter. However, upon reciprocal transplantation, each type of NC cell gives rise to the type of visceral ganglion that corresponds to its *new location* (Le Douarin, 1986). Therefore, the NC cells in each region have the potential of forming both types of autonomous neurons, although different sets of enzymes are required for the synthesis of the two transmitters. In fact, premigratory NC cells from head and trunk contain the enzymes to make both acetylcholine and norepinephrine (Kahn et al., 1980).

Because the transplantation experiments just described involved *potentially heterogeneous populations* of NC cells, the results are open to different interpretations (Fig. 13.29). According to one view, the ***pluripotency hypothesis,*** each individual cell has the potential to form *many* or *all* NC derivatives. This implies that the cells are not determined to form any particular NC derivative until they receive environmental signals. According to a radically different view, the ***selection hypothesis,*** any segment of NC contains a mixed population of already determined cells, each of which has only one possible fate. This implies that environmental signals allow only appropriately determined cells to *enter* certain pathways, and to further *divide* and *differentiate.* These two hypotheses represent two extremes. The following observations, however, suggest that a realistic picture should emerge somewhere in between.

A powerful strategy for testing whether individual cells are pluripotent is *clonal analysis,* which can be carried out in vitro or in vivo (see Sections 6.2 and 6.4).

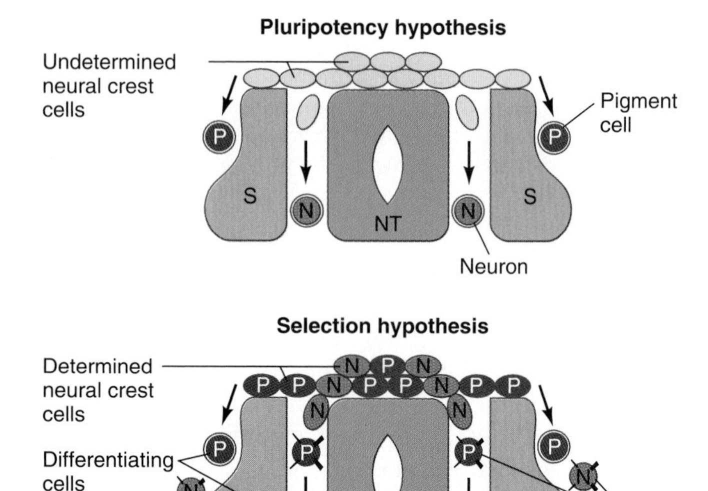

Figure 13.29 Alternative models of neural crest cell determination. The neural crest cell population is depicted by the ovals dorsal to the neural tube (NT) and somites (S). Empty ovals denote undetermined cells, whereas ovals labeled P or N represent cells determined to form pigment cells or neurons, respectively. For simplicity, the diagrams show only these two determined states, and only the two major pathways of neural crest cell migration. Double-contoured circles symbolize the expression of specific neural crest phenotypes at their eventual destination. Crossed circles represent cells that either do not arrive at a proper destination or die.

In vitro, single cells are cultured in a medium that supports their survival and proliferation without providing any signals that are known to restrict the potency of the cloned cells in a particular way. When the single cells have grown into clones they are analyzed for the types of differentiated cells they comprise, as indicated by cell morphology and *immunostaining* for characteristic proteins. If a clone consists of more than one cell type then the clone's founder cell must have been pluripotent.

Cohen and Konigsberg (1975) established such cultures from trunk NC cells of quail. A segment of neural tube with adhering NC was kept briefly in tissue culture until the NC cells had migrated away from the neural tube. When the NC cells adhered to the bottom of the culture dish, the neural tube was removed. NC cells were then dissociated and plated at low density so that individual cells could form new colonies. Three types of clones were obtained in this manner: all pigmented cells, all unpigmented cells, or some pigmented and some unpigmented cells within the same clone. The pigmented cells were melanocytes, and the unpigmented cells were identified as sympathetic neurons or adrenal medulla cells because they contained norepinephrine (Sieber-Blum and Cohen, 1980).

In corresponding experiments on cephalic NC cells from chicken and quail, Baroffio and coworkers (1991) analysed more than 500 clones and found that some of them comprised several cell types, such as cartilage, cholinergic neurons, adrenergic cells, glial cells, and melanocytes, whereas other clones contained only one or two cellular phenotypes. However, not all theoretical combinations of phenotypes were actually observed. For instance, the abilities to form cartilage and melanocytes were found to segregate early, while the ability to form glial cells was retained by all cells that were capable of forming two or more phenotypes. These observations suggest that many NC cells are originally pluripotent and go through a stepwise process of becoming restricted to their fates.

For clonal analysis in vivo, single cells are labeled, and the descendants of the labeled cell are traced at a later stage of development to establish their fates. If the descendants of one cell acquire different fates then the conclusion is again that the clone founder cell was pluripotent. Using this type of clonal analysis, Bronner-Fraser and Fraser (1991) microinjected chicken embryo NC cells with a fluorescent dye. The labeled cells divided and migrated in an apparently normal fashion, passing on the dye to their daughter cells at each division. After two days of development in vivo, the descendants of the labeled cells were classified according to their location and morphology (Fig. 13.30). Clones derived from *premigratory* NC cells included as many as four different cell types: dorsal root neurons, sympathetic neurons, Schwann cells, and either adrenal medulla or pigment cells. Many clones derived from *migratory* NC cells still contained more than one cell type. Similar results were obtained by Frank and Sanes (1991), who labeled NC cells with a recombinant virus, a technique that allowed them to follow the progeny of the labeled cells for a longer time, so that the possibility of selective cell death at the target site could be excluded.

While the results described above support the pluripotency hypothesis, evidence in favor of the selection hypothesis has also been obtained. For instance, in axolotl embryos, premigratory NC already contains distinct clusters of cells that appear to be prospective *xanthophores* (Epperlein and Löfberg, 1984). Also, certain cell culture media enhance the development of some NC subpopulations while inhibiting others (Le Douarin, 1986). These results support the selection hypothesis by showing that the developing NC contains subpopulations of cells with different biases and differentiation requirements. Taken together, the available evidence indicates that the determination of NC cells involves both the selection of subpopulations and the restriction of individual cell potentials (Le Douarin et al., 1994; Selleck et al., 1993). The experiments described next indicate that the bias in a subpopulation of premigratory chicken NC cells to form melanocytes depends on the timing of their detachment from the neural tube.

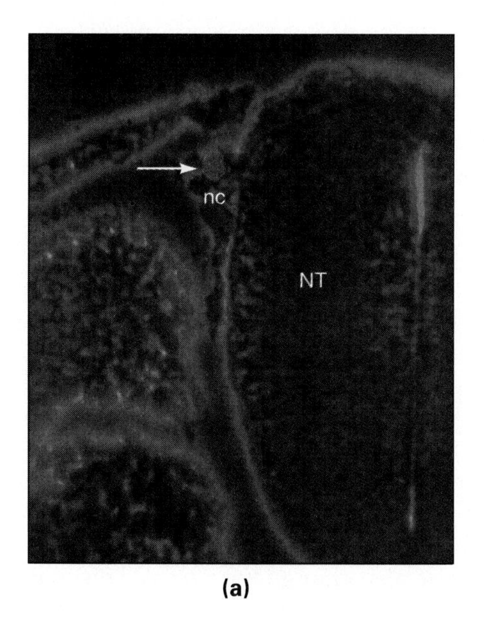

(a)

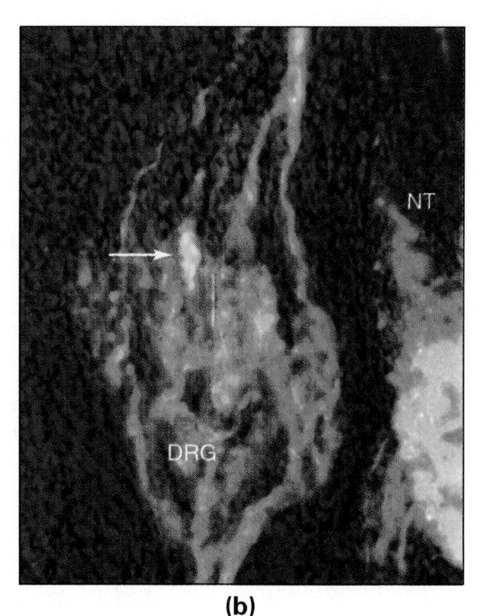

(b)

Figure 13.30 Pluripotency of migrating neural crest cells in chicken embryos. **(a)** Individual neural crest cells (nc) migrating away from the neural tube (NT) were microinjected with a red fluorescent dye; the descendants of the injected cell formed a clone of labeled cells. **(b)** Two days later embryos were fixed, sectioned, and immunostained with a primary antibody against neurofilaments and a secondary antibody conjugated with a green fluorescent dye. A cell (arrow) stained with both dyes was found in a dorsal root ganglion (DRG). Other neural crest cells from the same clone were located in a sympathetic ganglion and in a ventral root. Thus, the founder cell of the clone was not restricted to a single fate.

THE STRATEGIES OF HETEROTOPIC AND HETEROCHRONIC TRANSPLANTATION REVEAL SPATIAL AND TEMPORAL RESTRICTIONS TO NC CELL MIGRATION

Cell determination typically occurs in a stepwise process involving multiple signals (see Section 6.5). What may be the signals that determine and select NC cells? Before closure of the neural tube, NC cells are subject to the same signals that specify the anteroposterior and dorsoventral patterning of the central nervous system. As NC cells become migratory, their pathways affect their determination and/or selection as the cells encounter environments with various combinations of signals that change in space and time.

As NC cells leave the migration staging area, they organize into fairly distinct streams in which they embark on their migratory pathways. It appears that the formation of these streams involves the exchange of signals between the NC cells. Time lapse recordings of NC cell movements in live chicken embryos have shown that NC cells contact one another through long *filopodia* and may switch over to a neighboring stream (Kulesa and Fraser, 2000). Also, ablation of groups of premigratory NC cells is followed by chaotic movements of the neighboring NC cells (Kulesa et al., 2000). These interactions between NC cells have only recently come into view, and the nature of the signals they exchange still needs to be elucidated. In addition, longstanding investigations from several laboratories have revealed that NC cells follow signals they receive from other types of cells and from the extracellular matrix.

The fact that NC cells follow defined migration routes indicates that some external signals attract them to these routes and/or deter them from migrating elsewhere. For instance, trunk NC cells avoid a space around the notochord that is about 85 μm wide. *Heterotopic transplantations* of extra notochord segments indicate that the "forbidden" zone is established by an inhibitory signal from the notochord and that this signal is associated with a chondroitin sulfate–containing glycoprotein (Pettway et al., 1990).

Trunk NC cells taking the ventral pathway (see Fig. 13.27) pass through the *anterior* halves of somites while avoiding *posterior* somite halves. A similar restriction to the anterior half of each segment is seen in the motor axons growing out of the neural tube. To determine which tissue imposes this pattern, researchers reversed the anteroposterior axis of either the neural tube, to which the NC cells are originally attached, or the paraxial mesoderm bands from which the somites arise (Keynes and Stern, 1984; Bronner-Fraser and Stern, 1991). The results showed that the restriction of NC cell migration is imposed by the somitic mesoderm. The molecular mechanism of the restriction involves *Eph receptor tyrosine kinases* and their membrane-bound ligands, the *ephrins*, which play a key role in the formation of many segmental structures in vertebrates (Holder and Klein, 1999). In chicken embryos, EphB3 receptor localizes to the anterior sclerotome, including migrating neural crest cells, while ephrin-B1 and related ligands accumulate in the posterior sclerotome (Krull et al., 1997; H. U. Wang and D. J. Anderson, 1997). The addition of excess soluble ephrin-B1 disrupts the segmental pattern of NC cell migration, allowing these cells to enter the posterior halves of sclerotomes. The results suggest that in normal development ephrins cause a repulsive guidance of neural crest cell migration and motor axon outgrowth.

In addition to spatial restrictions on NC cell migration, there are also changes in time that can affect the use of a migration pathway. Trunk NC cells of chicken, for example, enter the ventral pathway first and the dorsolateral pathway only with a delay of about 24 h. The use of the dorsolateral pathway depends on the developmental history of the NC cells that use the pathway. This was shown by the strategy of *heterochronic transplantation*—that is, transplantations in which the graft and the host are at different stages of development.

TO investigate the possibility that NC cells acquire the ability to use the dorsolateral pathway with time, Erickson and Goins (1995) excised neural tubes and adherent NC cells from quail embryos in which NC cell migration had not begun yet. After removing epidermis, notochord, and somites with enzymes and fine tungsten needles, the clean neural tubes were incubated at 37°C in sterile salt solution supplemented with fetal calf serum and chicken embryo extract to mimic the NC cells' normal environment. Under these conditions, the NC cells began to migrate onto the bottoms of the plastic cultures dishes within a few hours. By 18 h of incubation, the NC cells also formed clusters on top of the neural tubes. Both the migratory cells and the clusters, aged from 12 to 96 h, were grafted to the migration staging area of chicken embryos at various stages of NC cell migration. In order to be able to trace the grafted NC cells in their host embryos, the cells were incubated with a fluorescent label prior to grafting.

When "old" NC cells (aged 24 h or longer) were transplanted to "young" host areas, where NC cells were just beginning to use the ventral pathway, graft cells utilized both dorsolateral and ventral pathways immediately while host cells waited 24 h before entering the dorsolateral pathway, as they normally do. Conversely, when "young" NC cells (aged 12 h) were transplanted into "old" host areas, where the resident NC cells were already using both pathways, no graft cells entered the dorsolateral pathway. In control experiments, when young NC cells were transplanted into young host areas, or when old NC cells are transplanted into old host areas, the graft NC cells behaved like their host counterparts.

In subsequent experiments, Reedy and coworkers (1998) found that an enzyme involved in the synthesis of black pigment, *tyrosinase-related protein-1* (*TRP-1*), is already synthesized in some NC cells before they reach the migration stage area. They also observed that all NC cells using the dorsolateral pathway contained TRP-1, whereas none of the cells using the ventral pathway stained positive for the enzyme. In addition, they found that NC cells migrating away early from cultured neural tubes did not contain TRP-1, whereas NC cells detaching late from neural tubes contained increasing amounts of TRP-1.

Questions

1. According to the experiments described here, NC cells that detach from the neural tube late and enter the dorsolateral pathway are *biased* to become pigment cells. However, the data do not show whether these NC cells are actually *determined* to form pigment cells. Alternatively, these cells could still be pluripotent, and their beginning differentiation as pigment cells might still be reversible. Ideally, which experimental strategy should be used to decide this question?
2. What kind of changes might occur in the late-detaching NC cells that make the dorsolateral pathway attractive to them or at least allows them to utilize it?

Taken together, these results described above indicate that NC cells detaching early from the neural tube are biased against differentiation as melanocytes and enter the ventral pathway. In contrast, NC cells detaching late from the neural tube begin their differentiation as melanocytes before they reach the migration stage area and take the dorsal pathway.

In addition to the changing bias of neural crest cells, the pathways they use also change with time. This was shown by another series of heterochronic transplantations using the *white* mutant of the axolotl *Ambystoma mexicanum.* In homozygous mutants, the pigmentation is restricted to a bilateral strip of epidermis at the level of the spine, which is close to the NC cell migration stage area in the embryo (Fig. 13.31). Transplantation of subepidermal ECM material from wild-type to mutant

(a)

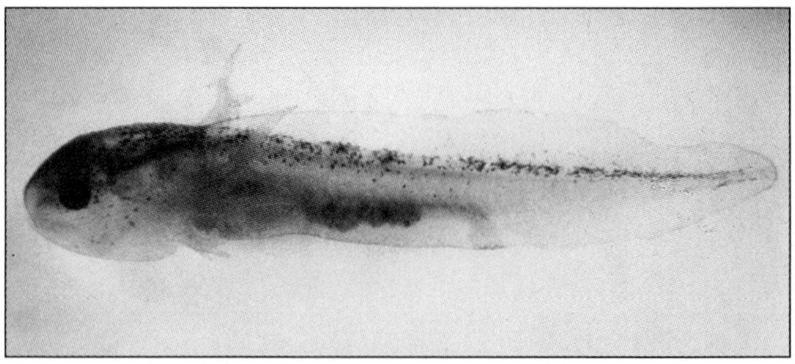

(b)

Figure 13.31 Swimming tadpoles of the axolotl *Ambystoma mexicanum.* **(a)** Wild phenotype. Pigment cells are spread over the entire body. The black pigment cells (melanocytes) are most conspicuous, while the yellow pigment cells (xanthophores) are in the lighter areas in between. **(b)** Homozygote for a mutant allele of the white gene, which is required for the proper spreading of pigment cell precursors. Note that the pigment cells of the larva are restricted to an irregular stripe at the level of the spine.

embryos locally restores normal pigmentation, indicating that the *white*$^+$ function is required for the synthesis or deployment of an ECM component required for the normal distribution of pigment cell precursors (Löfberg et al., 1989). As expected, ECM taken from mutant embryos *during* the time of pigment cell migration fails to support normal pigmentation in wild-type or mutant hosts. Surprisingly, however, ECM taken from mutant embryos *after* the time of pigment cell migration does restore normal pigmentation. Apparently, then, the rescuing ECM component is not totally lacking from the *white* mutant but becomes available too late to be utilized by the pigment cell precursors. Work aimed at the nature of this maturation process is in progress (Perris, 1997).

Mutants like *white*, which shift the relative timing of developmental events, are known as ***heterochronic mutants.*** Such mutants are the basis of ***heterochrony,*** the change of developmental timing in the course of evolution. A readily apparent kind of heterochrony is the retention of larval features by adults, such as the persistence of gills—typically a larval amphibian organ—in sexually mature axolotls. In this case, heterochrony is best explained by a changes in the response of various organs to hormones that control metamorphosis (see Section 27.5). Other types of heterochrony represent changes in the timing of growth. For example, the extraordinary relative size of the human brain can be ascribed to the retention of a fetal pattern of growth beyond the time of birth. Heterochronic mutations that affect periods of dramatic change in development, such as morphogenetic movements, rapid growth, or metamorphosis, may cascade into major phenotypic effects and thus have profound effects on evolution (Gould, 1977; Raff and Wray, 1989).

EXTRACELLULAR MATRIX AFFECTS THE DETERMINATION OF NC CELLS

Electron microscopic observations indicate that migrating neural crest cells extend *filopodia* that align with fibrils in the extracellular matrix (ECM) and form intimate contacts with other ECM components (Perris, 1997). Some of these contacts may provide signals that contribute to the determination of NC cells to form specific derivatives. This hypothesis is supported by the following experiment, in which regional ECM samples were obtained from living salamander embryos and tested in vitro for their effects on premigratory NC cells.

Perris and coworkers (1988) prepared miniature sheets of nitrocellulose to be used as microcarriers for ECM materials. The microcarriers were implanted into axolotl embryos, where they were left to adsorb ECM materials. The implantation sites were along the main migration routes of trunk NC cells: along the dorsolateral pathway, where prospective pigment cells migrate; and along the ventral pathway, where the dorsal root ganglia arise (Fig. 13.32). After they had adsorbed ECM materials for several hours, the microcarriers were removed from the embryos and transferred to small tissue culture dishes, where they were to serve as a substrate for culturing NC cells. Premigratory NC cells were detached from the dorsal side of spinal cords and deposited on the ECM-covered microcarriers.

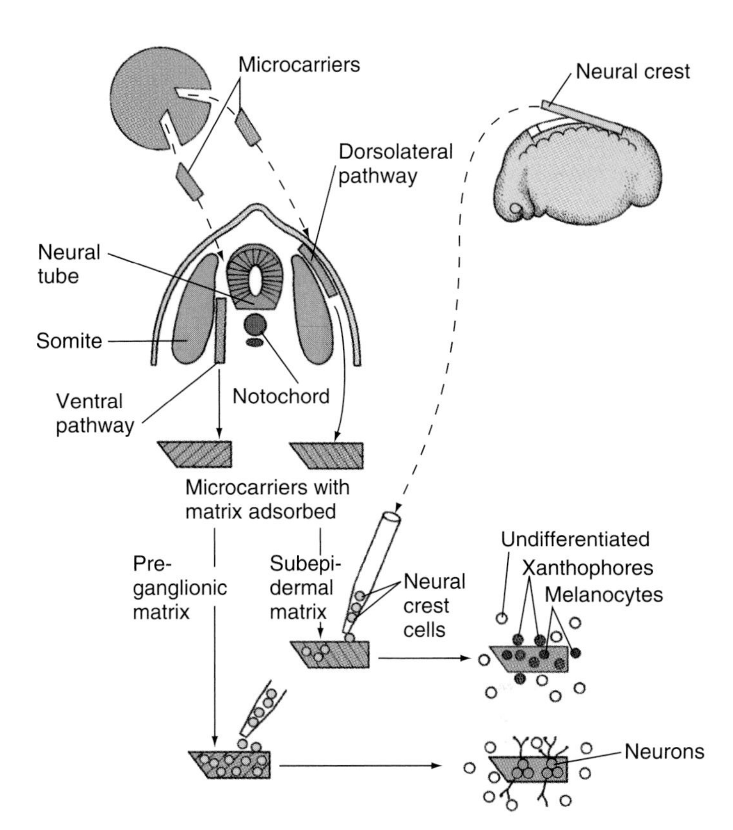

Figure 13.32 Determination of premigratory neural crest cells by region-specific ECM components. Microcarriers (approximate size 400 × 150 × 4 μm) made from sterile nitrocellulose were implanted into the trunk region of axolotl embryos. The implantation sites were along the two main migration routes of neural crest cells: along the dorsolateral pathway in the subepidermal space, and along the ventral pathway at the site where the dorsal root ganglia form. The carriers were left in the embryos to adsorb ECM materials for 10 to 12 h. Just before the onset of neural crest cell migration, the carriers were recovered from the embryos and transferred to small cell culture dishes. Premigratory neural crest cells removed from neural tubes were deposited on the ECM-covered microcarriers and incubated for up to 5 days. Neural crest cells cultured on microcarriers from the subepidermal space (dorsolateral pathway) developed into xanthophores or melanocytes. In contrast, premigratory neural crest cells cultured on microcarriers from the dorsal root ganglion area (ventral pathway) developed into neurons. Neural crest cells cultured without microcarrier contact remained undifferentiated.

After several days in culture, the NC cell derivatives were classified as pigment cells or neurons according to their morphological appearance and to the presence of pigments and marker proteins. NC cells cultured on microcarriers coated with subepidermal ECM developed into pigment cells and showed no signs of developing as neurons. Some of the pigment cells had moved away from their microcarriers during the culture period, indicating that temporary contact with subepidermal ECM was sufficient for subsequent differentiation as pigment cells.

Cells cultured on microcarriers coated with ECM from the ventral pathway developed into neurons and never showed the characteristics of pigment cells. Control NC cells grown in the same culture dishes but having no contact with the ECM-covered microcarriers did not express any of the traits indicative of either pigment cell or neuron development.

The results obtained by the experimenters indicate that groups of premigratory NC cells cultured in vitro can give rise to either pigment cells or nerve cells. The type of cell that develops depends on the source of the ECM with which the developing NC cells are in contact. Even temporary contact with subepidermal ECM from the dorsolateral pathway promotes pigment cell development, whereas contact with ECM from the ventral pathway promotes neuron development.

REGION-SPECIFIC GROWTH FACTORS ARE INVOLVED IN NC CELL DETERMINATION

ECM components can interact with integrins and other cell membrane proteins in ways that change gene expression (see Section 11.5). In addition, the ECM can serve as a repository for *growth factors*—that is, secreted or membrane-bound polypeptides that stimulate cells to divide and/or to differentiate in certain ways. Most NC derivatives are temporarily dependent on growth factors. This dependence confers a powerful combination of features—active division and invasiveness—to NC cells during embryonic development while minimizing the risk of cancer at later stages when the growth factors are no longer synthesized. Evidence for the involvement of growth factors in NC development has come from mutant analysis and from studies on cultured NC cells (Anderson et al., 1997; Perris, 1997; Wehrle-Haller and Weston, 1997).

Several mutations in mice affect subsets of NC derivatives. For example, melanocyte precursors depend on the so-called *Steel factor* for proliferation, and they respond to it by *positive chemotaxis.* The Steel factor is a polypeptide produced by developing skin prior to the onset of melanocyte precursor dispersal (Wehrle-Haller and Weston, 1995). *Steel* heterozygotes have a diminished population of melanocytes, which gives these mice a steel-gray coat color or, depending on genetic background, spots of white fur. The Steel factor's receptor, a tyrosine kinase known as c-kit, is present on melanocyte precursors; loss-of-function alleles of *c-kit* are known as *White-spotting* because heterozygotes have spots of white fur, again indicating a lack of melanocytes. The same condition in humans is known as piebaldism (Spritz et al., 1992).

Another set of growth factors, known as *endothelins,* and their receptors are required for the development of two NC derivatives, namely melanocytes and certain parasympathetic neurons associated with the gut. Mice deficient in the endothelin-3 ligand or its receptor, known as EDNRB, show areas of unpigmented fur, similar to *White-spotting,* and also a distention of the large intestine, or *colon* (Baynash et al., 1994). The latter symptom results from the lack of those parasympathetic neurons that activate the smooth muscles of the colon, which normally generate the peristaltic movements that expel feces. Similar conditions in humans, some caused by mutations in *EDNRB,* are known as Hirschsprung's disease (Puffenberger et al., 1994). Patients with this syndrome show irregular pigmentation of iris and skin as well as chronic and severe constipation.

Certain growth factors of the TGF-β superfamily can promote the development of NC cells cultured in vitro into distinct NC derivatives. When individual premigratory NC cells are cultured without growth factors they tend to form mixed clones comprising neurons, glial cells, pigment cells, and other NC derivatives, as discussed earlier. However, when the same NC cells are cultured in the presence of BMP-2, then about 50% of the clones consist of neurons only, 25% of the clones consist of smooth muscle cells, and the remaining clones consist of neurons and smooth muscle cells (Shah et al., 1996). In contrast, when the same NC cells are cultured in the presence of TGF-β_1, virtually all clones develop almost completely into smooth muscle cells. With either growth factor, virtually all single cells form large clones, indicating that these growth factors act by restricting the potential of the cultured NC cells rather than by selecting predetermined cells.

While the in vitro experiments outlined above only show that BMP-2 and TGF-β_1 *can* restrict the potential of NC cells, additional observations suggest that the same growth factors actually play this role in normal development. In mammals, BMP-2 is normally synthesized in organ rudiments near which autonomic neurons develop, such as heart, dorsal aorta, and lung. BMP-2 accumulation in the wall of the dorsal aorta is followed by the synthesis, in adjacent cells, of mammalian achete-scute homolog 1 (MASH1), a gene regulatory protein required for the development of visceral neurons (Guillemot, 1993). These and other observations strongly suggest that NC cells begin to express *MASH1*$^+$, and develop into visceral neurons, when they approach peripheral tissues that release BMP2. Future investigations are likely to address the following questions. First, what are the target genes of MASH1 that bring about the characteristics of autonomic neurons? Second, which additional signals determine prospective autonomic neurons to differentiate into sympathetic versus parasympathetic neurons?

13.3 Ectodermal Placodes

Several areas of ectoderm in the head region are induced by underlying parts of the brain to form *placodes*—that is, patches of columnar epithelium in a more squamous background (Fig. 13.33). The *ectodermal placodes* of vertebrates emerge in two rows: a lower row of epibranchial placodes, and an upper row of dorsolateral placodes. The epibranchial placodes, together with the neural crest of the head region, form the sensory ganglia of cranial nerves. The dorsolateral placodes also contribute to the cranial sensory ganglia and, in addition, form parts of the ear, eye, and nose. Because of their ability to form sense organs, neurons, and cranial cartilage, and because of their development in the vicinity of the central nervous system, the ectodermal placodes and the neural crest have many properties in common.

THE OTIC PLACODE FORMS THE INNER EAR

The *otic placode,* which will form the inner ear, is the first ectodermal placode to develop. In chicken and amphibian embryos, the otic placode is induced by underlying mesoderm and rhombencephalon. In human embryos, the otic placode appears during the third week on both sides of the rhombencephalon. During the fourth week, the placode invaginates to form the otic pit, which is subsequently pinched off as the *otic vesicle* (Fig. 13.34). On its median surface, the vesicle gives off a group of cells that develop into the *statoacoustic ganglion.*

Soon the otic vesicle begins to expand, pushing aside the surrounding mesenchyme cells. The expansion is unequal: The vesicle bulges out in some places and constricts in others until it assumes a complicated shape aptly called the *labyrinth* (Fig. 13.35). Some parts of the labyrinth consist of squamous epithelium while others form areas of columnar cells. These develop into sensory epithelia, which receive mechanical stimuli caused by sound, gravity, and body movements, and transmit them to the neurons of the statoacoustic ganglion.

The part of the labyrinth that is involved in the perception of sound in higher vertebrates is appropriately

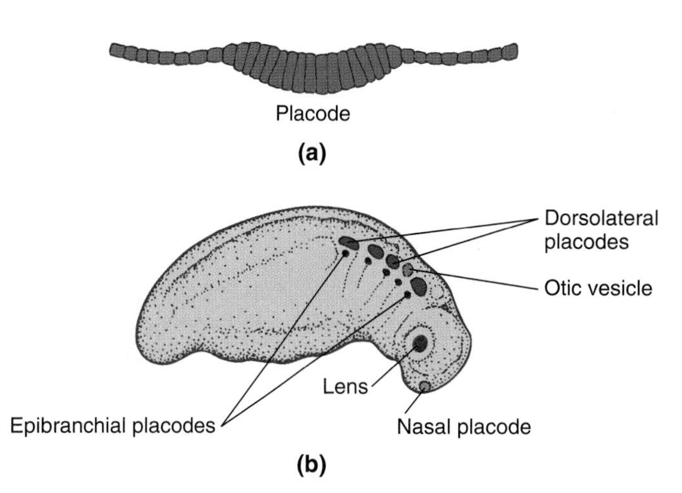

Figure 13.33 **(a)** Schematic diagram of a placode in an epithelium. **(b)** Epidermal placodes in the head of a salamander embryo.

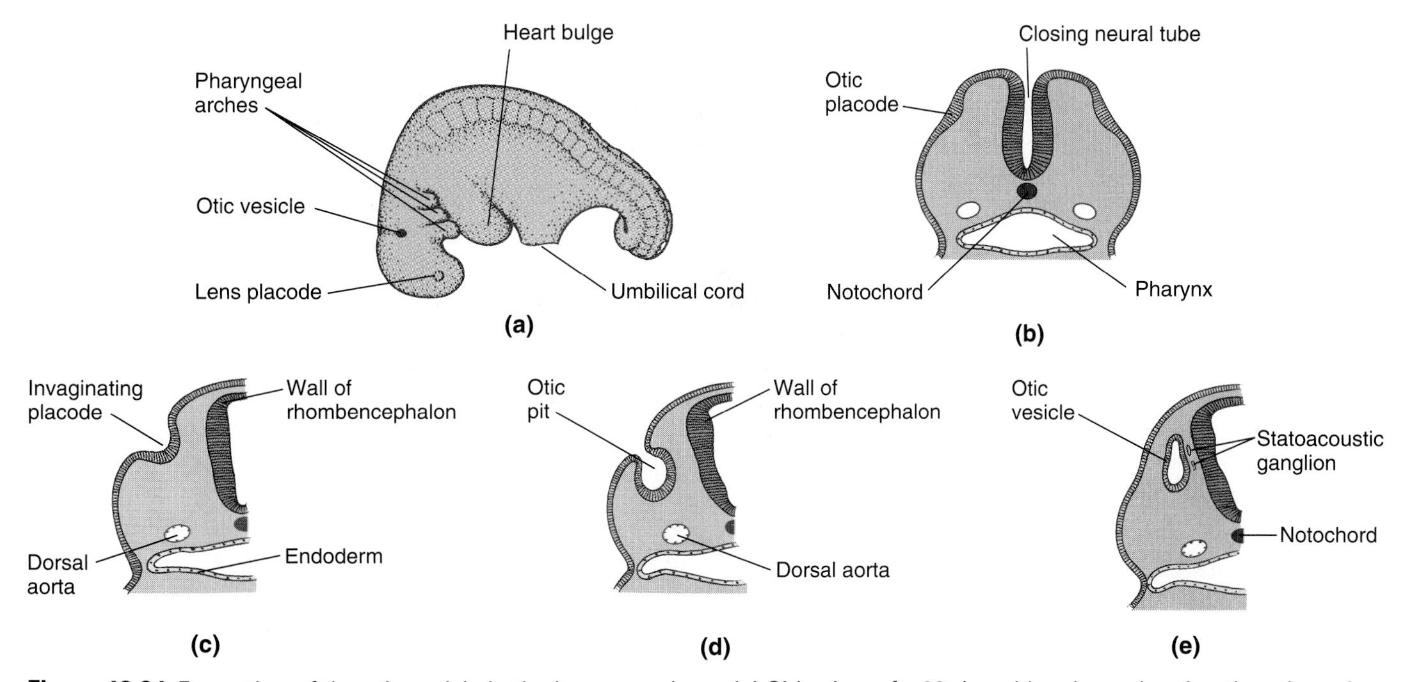

Figure 13.34 Formation of the otic vesicle in the human embryo. **(a)** Side view of a 28-day-old embryo showing the otic vesicle. **(b–e)** Schematic transverse sections through the region of the rhombencephalon showing the formation of the otic vesicle. Note the appearance of the statoacoustic ganglion. **(b)** 22 days; **(c)** 24 days; **(d)** 27 days; **(e)** 31 days.

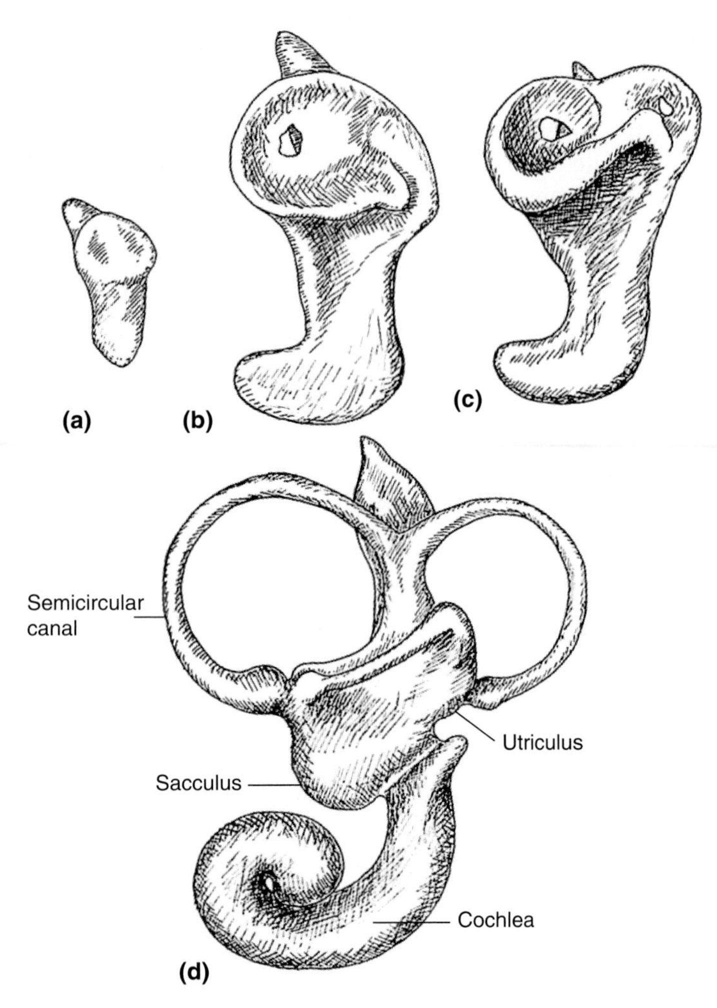

Figure 13.35 Development of the labyrinth from the otic vesicle in the human embryo. The drawings show lateral views of the left labyrinth.

called the ***cochlea*** (Gk. *kochlos*, "snail"). Developing in close association with surrounding tissues, the cochlea forms elaborate auxiliary structures that make even the most gentle sounds audible. Sounds are transmitted to the inner ear via a membranous window by tiny bones located in the *middle ear* (see Section 14.1). Another part of the labyrinth, the ***semicircular canals,*** registers the inertia of internal fluid to rotatory accelerations of the body in the three dimensions of space. The remainder of the inner ear registers the directions of gravity and linear accelerations. As the labyrinth develops, it is surrounded by mesenchymal cells, which produce a snugly fitting cartilage capsule. In most vertebrates, this capsule is later replaced with solid bone.

THE LENS PLACODE DEVELOPS TOGETHER WITH THE RETINA

The *lens placode* is formed in the head ectoderm as the result of a series of inductive interactions with pharyngeal endoderm, heart mesoderm, neural crest cells, and the *optic vesicle* (see Fig. 1.16; Jacobson, 1966; R. M. Grainger et al., 1988). The lens placode first invaginates and then pinches off as the *lens vesicle* (Fig. 13.36). Soon thereafter, the cells of the proximal layer of the vesicle elongate toward the distal layer, gradually filling the lumen of the vesicle. The elongated cells undergo a series of specializations including the synthesis of large amounts of proteins called ***lens crystallins*** (Piatigorsky, 1981). The differentiated cells, known as ***lens fibers,*** do not divide. Instead, the lens continues to grow as dividing outer epithelial cells are transformed into inner lens fibers.

While the lens placode invaginates to form the lens vesicle, the *optic vesicle* invaginates to form the *optic cup*. The invagination of the optic cup involves both the distal hemisphere and part of the ventral surface, where a groove called the ***choroid fissure*** forms. This fissure accommodates the ***hyaloid artery,*** which supplies blood to the interior of the eye. When the choroid fissure closes around the hyaloid artery, the opening of the optic cup acquires the round shape that characterizes the final derivative of this opening, the ***pupil.***

The outer layer of the optic cup develops into the ***pigmented retina,*** while the inner lining of the optic cup forms the ***neural retina.*** The cells of the neural retina divide much like the neuroepithelium in other areas of the brain. The innermost cells facing the pigmented retina form the light-receptive sensory cells known as ***rods*** and ***cones*** (Fig. 13.37). Adjacent to this layer of photoreceptors are two layers of neurons. The axons of these neurons converge toward the *optic stalk,* which connects the optic cup with the diencephalon. Once the optic stalk is filled with axons, it is called the ***optic nerve.***

The sequence of cell layers in the neural retina is contrary to what one might expect from a functional point of view. Light entering the eye has to pass through several layers of axons and neurons before reaching the rods and cones. Also, the light-sensitive tips of these photoreceptors are oriented away from the light instead of facing it. Functionally paradoxical as it may appear, this arrangement is in keeping with its developmental origin. The apical surface of the neural plate, from which the retina ultimately derives, faces the outside world (Fig. 13.38). A simple sensory cell in the epidermis would have the same orientation, with its apical (sensitive) surface facing out. But when the neural plate closes to form the neural tube, the apical surface comes to face inside toward the neural canal or the ventricles of the brain. Because the optic cup is an extension of the brain, its apical surface now faces inside, as do the rods and cones. Later, when the neural retina divides to form several cell layers, it gives off neuroblasts away from the lumen, as the neuroepithelium does elsewhere in the brain and spinal cord (see Fig. 13.4).

NASAL PLACODES FORM OLFACTORY SENSORY EPITHELIA

Development of the ***nasal placodes,*** located at the anterior tip of the embryo, is induced by the underlying endoderm and telencephalon. In humans, the nasal

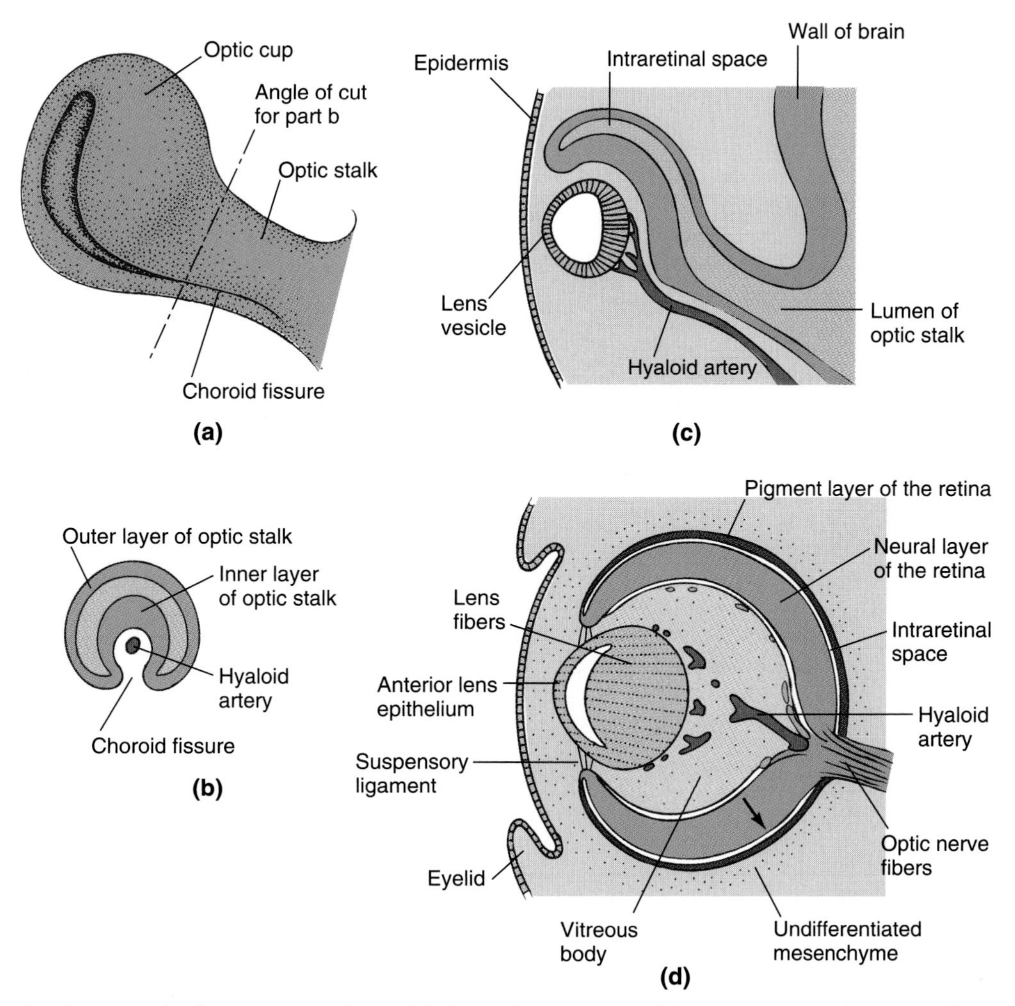

Figure 13.36 Eye development in the human embryo. **(a)** Ventrolateral view of the optic cup and optic stalk at 6 weeks. Note the choroid fissure accommodating the hyaloid artery on the ventral face. **(b)** Section as indicated in part a. **(c)** Section through the plane of the choroid fissure. **(d)** Section from a 7-week embryo. The outer layer of the optic cup has formed the pigment layer of the retina, and the inner layer of the optic cup has formed the neural layer of the retina. The tip of the arrow indicates the position of the light receptors. Nerve fibers emerging at the base of the neural layer (rear end of arrow) converge toward the optic nerve.

placodes appear at the end of the fourth week. During the fifth week, two fast-growing ridges, a ***lateral nasal swelling*** and a ***medial nasal swelling,*** surround each placode, which then forms the floor of a depression called the ***nasal pit*** (Fig. 13.39). During the following 2 weeks, the ***maxillary swellings,*** which will form most of the upper jaw, grow and push the nasal pits medially until the medial nasal swellings fuse. Thus, the upper lip is formed from parts of the maxillary and medial nasal swellings. The fusion of the latter extends also to a deeper level of the face, where they form part of the upper jaw and the ***primary palate.***

Meanwhile, the nasal pits have deepened until only a thin ***oronasal membrane*** separates each pit from the oral cavity (Fig. 13.40). Ruptures in the oronasal membranes then create openings, the ***primitive choanae,*** between the oral cavity and the nasal pits, which are then called ***nasal chambers.*** Elongation of the nasal chambers, combined with the formation of a ***secondary palate,*** displaces the choanae toward the *pharynx,* at which point they are called ***secondary choanae.*** The original epithelium of the nasal pit is now lining the "roof" of the nasal cavity, where it forms the ***olfactory epithelium.*** Its sensory cells send their axons directly into ***olfactory bulbs,*** stemlike extensions of the telencephalon.

13.4 Epidermis

The largest derivative of the ectoderm is the ***epidermis,*** which forms the outer layer of the skin. The epidermis represents a type of tissue known as *epithelium.* It is characterized by the close apposition of cells, which are sealed together by *tight junctions* at the apical surface (see Fig. 2.23). Initially only one cell layer thick, the epidermis soon divides into a temporary outer layer, the ***periderm,*** and a permanent inner layer, the *basal layer,* or ***germinative layer.*** The cells in the germinative layer behave as *stem cells;* they divide actively, thus renewing the stem cell population and creating committed progenitor

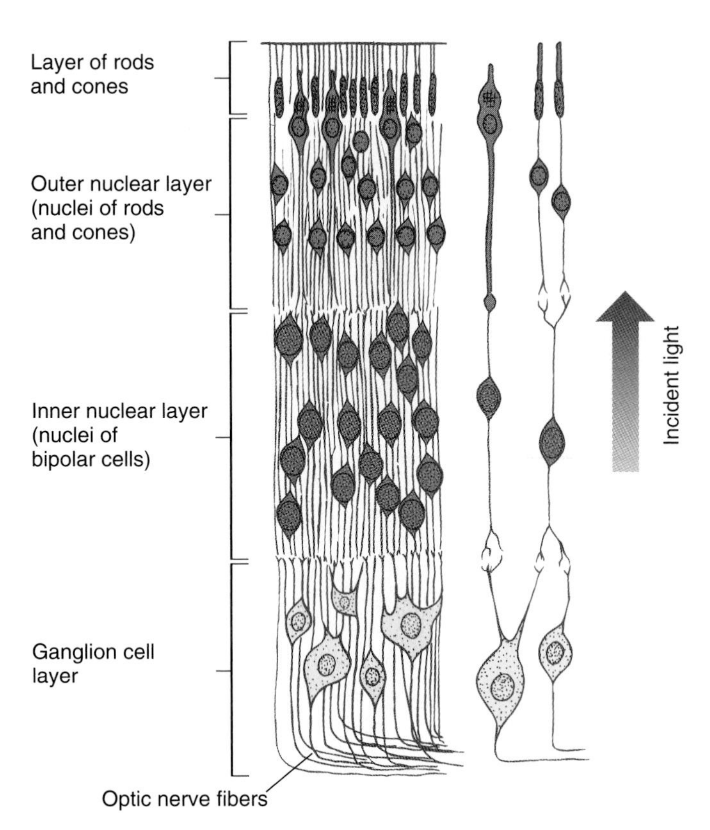

Figure 13.37 Schematic drawing of the neural retina in a 25-week human fetus.

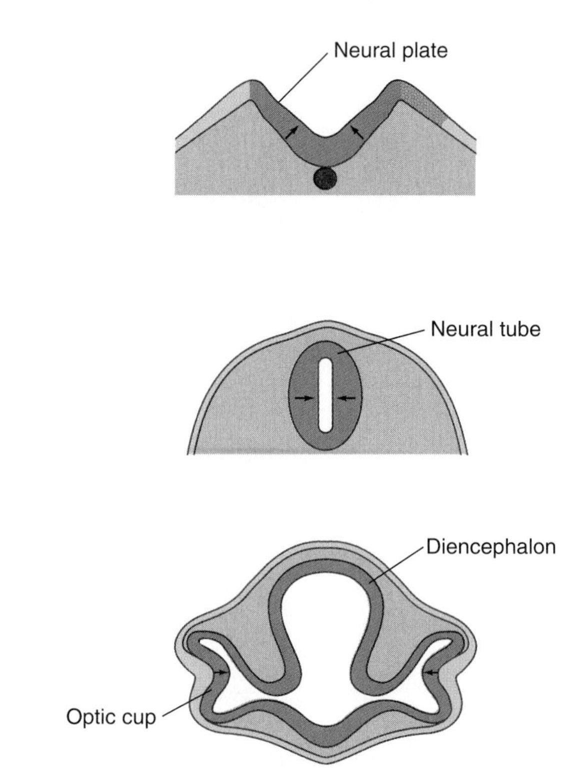

Figure 13.38 Schematic diagrams tracing the orientation of the sensory cells in the neural retina to the apical-basal polarity of the neural plate. The arrow tips indicate the apical (sensory) surfaces of the cells.

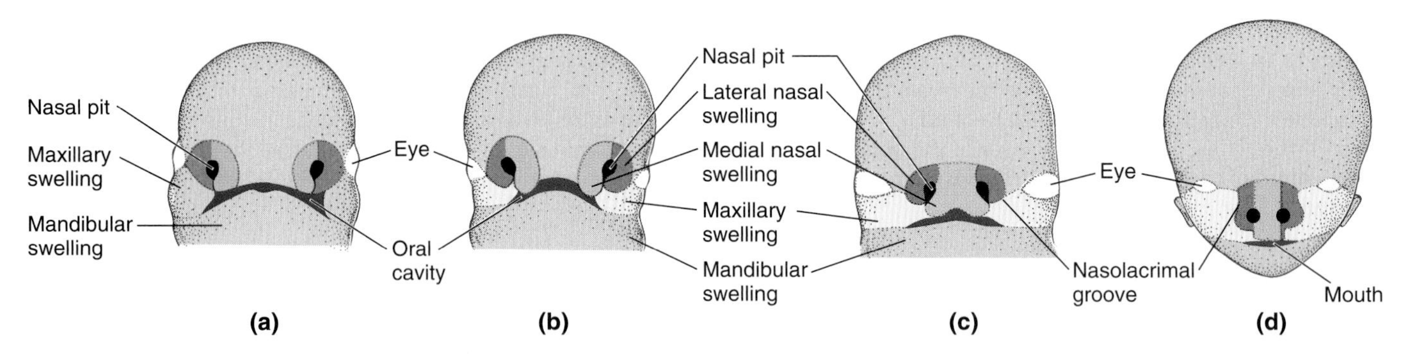

Figure 13.39 Nose development in the human embryo as seen in frontal aspects of the face. The nasal placodes are surrounded by the nasal swellings to form the nasal pits. The lateral nasal swelling is separated by the nasolacrimal groove from the maxillary swelling, which forms the upper jaw. The upper lip is formed by the two medial nasal swellings and the two maxillary swellings. **(a)** 5 weeks; **(b)** 6 weeks; **(c)** 7 weeks; **(d)** 10 weeks.

cells that differentiate into epidermal cells (Jones et al., 1995). The committed progenitor cells are pushed into the upper layers as they mature (Fig. 13.41); those that lose contact with the underlying dermis form the ***spinous layer.*** As part of their differentiation, the epidermal cells synthesize massive amounts of *keratin,* a protein that accumulates in granules. Cells at this stage cease to divide; they form the ***granular layer*** of the epidermis. When the cells are filled with keratin, they die and form a tough outer layer called the ***horny layer.*** Throughout life, the outermost cells of the horny layer are sloughed off and replaced by deeper cells, which ultimately originate from the dividing cells in the basal and spinous layers. In adult human skin, a cell born in the basal layer takes about 7 weeks to travel to the surface (Halprin, 1972).

The epidermis is supported by a mesenchymal component of the skin, the ***dermis.*** The dermis induces the overlying epidermis to form various derivatives and appendages, which include—depending on the body region and the class of vertebrates—scales, feathers, hair, and different glands. As examples, we will briefly con-

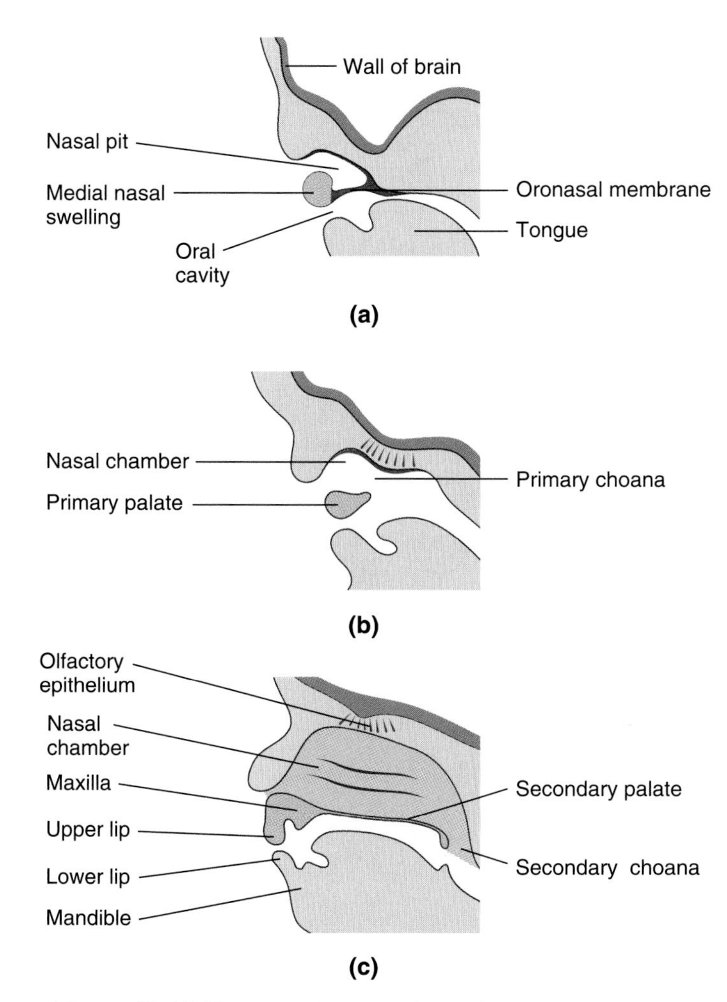

Figure 13.40 Nose development in the human embryo as seen in paramedian sections through one of the two nasal pits/chambers. **(a)** At 6 weeks. The nasal pit is separated from the oral cavity by the oronasal membrane. **(b)** At 7 weeks. The oronasal membrane has broken down, so that the nasal chamber is connected with the oral cavity by an opening, the primary choana. **(c)** At 9 weeks. After the development of the secondary palate, the nasal cavity is connected to the pharynx by a secondary choana.

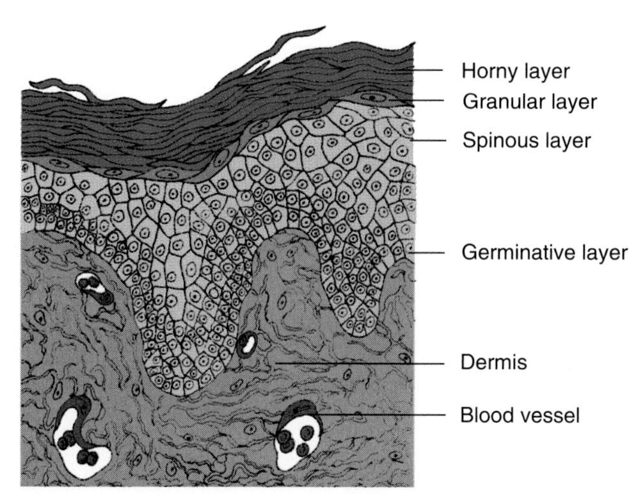

Figure 13.41 Section through the skin of the human shoulder. Epidermal cells originate by mitosis in the germinative or the spinous layer and are pushed via the granular and horny layers to the surface as they mature.

sider the development of hair and mammary glands, both characteristic of mammals.

The formation of a hair begins with a proliferation of epidermal cells, a ***hair bud,*** penetrating into the underlying dermis (Fig. 13.42). The growth of the hair bud is induced by an aggregation of dermal mesenchyme cells, which form a ***hair papilla*** as they are engulfed by the base of the hair bud. The entire organ rudiment is then called a ***hair follicle.*** As blood vessels and nerve endings develop in the papilla at the base, the rudiment of a gland forms on the side. Soon the core cells of the hair follicle are keratinized and form the ***hair shaft,*** which is eventually pushed outside. Further proliferation of the epidermal cells at the base of the shaft causes a continuous growth of the hair. The peripheral cells of the follicle form the ***epidermal hair sheath,*** which is surrounded by

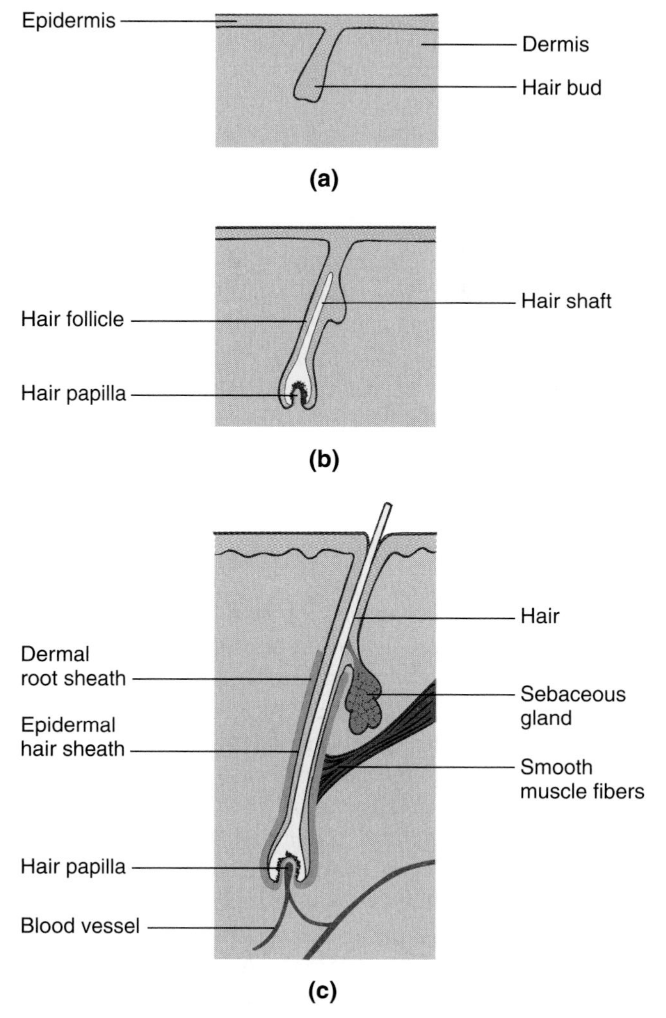

Figure 13.42 Hair development in the human: **(a)** 4 months; **(b)** 6 months; **(c)** newborn.

a ***dermal root sheath*** derived from surrounding mesenchyme. A little smooth muscle, also formed by mesenchyme, usually attaches to the dermal root sheath. Contraction of this muscle tilts the follicle perpendicular to the skin surface so that the hair "stands on end." The epidermal hair sheath usually buds out to form a ***sebaceous gland,*** which gives off an oily substance into the hair follicle, from where it reaches the surface of the skin.

The development of feathers begins in a similar fashion. However, their epidermal sheaths show intricate patterns of local proliferation and programmed cell death, which generate the more complicated structures of feathers.

Epidermal gland development, like hair and feather development, begins with epidermal proliferation resulting from inductive interaction with the underlying mesenchyme. The mammalian epidermis forms several types of glands, including sebaceous glands, sweat glands, scent glands, and mammary glands. The latter develop from a pair of bandlike epidermal thickenings, the ***mammary ridges*** (Fig. 13.43). Depending on the species, one or more segments of the ridge persist on each side. In each of these locations, solid epithelial cords sprouting lateral buds penetrate the underlying mesenchyme. At birth the buds have developed into small glands, and the cords have formed hollow ***lactiferous ducts*** (Lat. *lac,* "milk"; *ferre,* "to bear"). The lactiferous ducts at first open into a small pit, which is later transformed into a nipple by proliferation of the underlying mesenchyme.

In a 7-week-old human embryo, the mammary ridge extends from the armpit to the groin, as it does in other mammalian embryos. In normal development, only a small portion of the mammary ridge persists in the midthoracic region, giving rise to one pair of breasts in women or nipples in men. In some individuals, other segments of the mammary ridge fail to degenerate, so that accessory nipples or breasts are formed. This condition makes one wonder why the human mammary ridge is not limited to the midthoracic region to begin with.

What seems to be an error-prone detour can be explained against the background of phylogeny. The primitive mammals from which primates, including humans, evolved were small creatures, much like tree shrews, that presumably nursed litters of several young. An extended mammary ridge that gave rise to multiple mammary glands would appear to be adaptive for these species. For humans and other primates, which nurse only one or two young at a time, two breasts are sufficient and may, in fact, be more adaptive. Nevertheless, human development recapitulates the embryonic stage of the more primitive mammals. It is noteworthy that *normal* primate development recapitulates only the *embryonic* stage of more primitive mammals with extended mammary ridges, not the adult stage with multiple

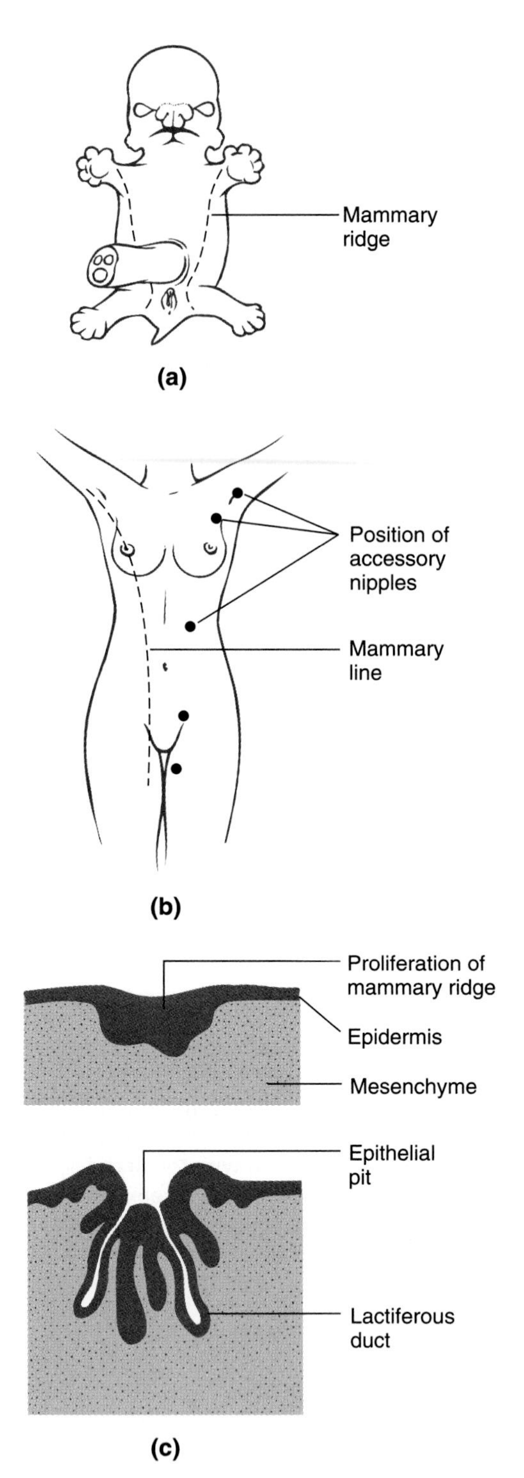

Figure 13.43 Mammary gland development. **(a)** Generalized mammalian embryo. Dashed lines indicate the normal position of the mammary ridge (milk line) on each side. **(b)** Accessory nipples or breasts in the adult human result from an abnormal development of the embryonic mammary ridge outside the thoracic region. **(c)** Sections of the developing mammary gland in the human during the third and the eighth month.

breasts or nipples. The *occasional* persistence of a primitive *adult* feature, such as multiple mammary glands, in an evolved species is an abnormality known as ***atavism.***

Presumably, it took fewer steps of mutation and selection to keep the extended mammary ridge and then let most of it degenerate than to develop a new way of inducing just one mammary gland in the thorax. This principle of preserving and modifying old developmental patterns, rather than replacing them, is pervasive in vertebrate development. We will encounter it frequently as we turn to endodermal and mesodermal derivatives in Chapter 14.

SUMMARY

The development of organ rudiments is followed by an extended period of histogenesis, during which tissues mature to their functional state. Surveys of histogenesis often use the germ layers as an organizing principle, starting from the major derivatives of each layer and tracing their development in turn. The major derivatives of the ectoderm are the neural tube, the neural crest (NC), ectodermal placodes, and the epidermis.

The newly closed neural tube consists of a layer of neuroepithelial cells, which give rise to large numbers of neuroblasts and glioblasts. These develop into neurons and glial cells, the cell types that characterize nervous tissue. The neural tube gives rise to the central nervous system. Its anterior portion forms the brain, and its posterior portion forms the spinal cord. Both show anteroposterior and dorsoventral patterns in their overall morphological features and in the types of neural and glial cells present. The signals specifying the dorsoventral pattern originate from adjacent tissues during neural plate closure and are released later on from the dorsalmost and ventralmost structures of the neural tube. Sonic hedgehog peptide acts as a ventral morphogen while bone morphogenetic proteins and Wnt proteins act as dorsalizing signals.

The brain is divided into five major regions, called (from cranial to caudal) telencephalon, diencephalon, mesencephalon, metencephalon, and myelencephalon. Axons grow out from motor neurons located in the central nervous system and leave via cranial or spinal nerves to become part of the peripheral nervous system. In addition, the peripheral nervous system comprises sensory and visceral ganglia derived from NC cells and ectodermal placodes.

NC cells, which are unique to vertebrates, are located at the crests of the neural folds during neurulation. During neural tube closure, NC cells come to lie on top of the neural tube, from where they migrate along different pathways. Some NC cells move dorsolaterally to enter the skin and form pigment cells. Other NC cells migrate ventrally and form dorsal root ganglia, visceral ganglia, Schwann cells, and adrenal medulla. In the head and heart regions, NC cells also form ectomesenchyme contributing to cartilage, other connective tissues, muscle, teeth, heart, and blood vessels.

The determination of NC cells for these various fates appears to be a stepwise process involving the selection of appropriate subpopulations as well as the restriction of cellular pluripotency. The strategy of heterochronic transplantation has revealed dynamic changes both in the migratory competence of NC cells and in the types of signals they receive. Components of the extracellular matrix (ECM) can facilitate or inhibit NC cell migration. There is also evidence for the involvement of ECM components and diffusible growth factors in NC cell determination.

Formation of ectodermal placodes is induced in head epidermis by interactions with underlying tissues. Most of these placodes contribute to cranial sensory ganglia together with the cranial neural crest. Three ectodermal placodes form the lens of the eye and the sensory epithelia of the nose and ear. The remainder of the ectoderm gives rise to the epidermis. Together with the underlying mesenchyme, it forms the skin. The epidermis is a stratified epithelium consisting of several layers. Stem cell divisions occurring in the basal layers push older cells outward into the upper layers, where they fill with keratin, die, and are eventually sloughed off. Interactions between the epidermis and the underlying mesenchyme induce the formation of various epidermal derivatives, including hair, feathers, scales, and different types of glands.

Endodermal and Mesodermal Organs

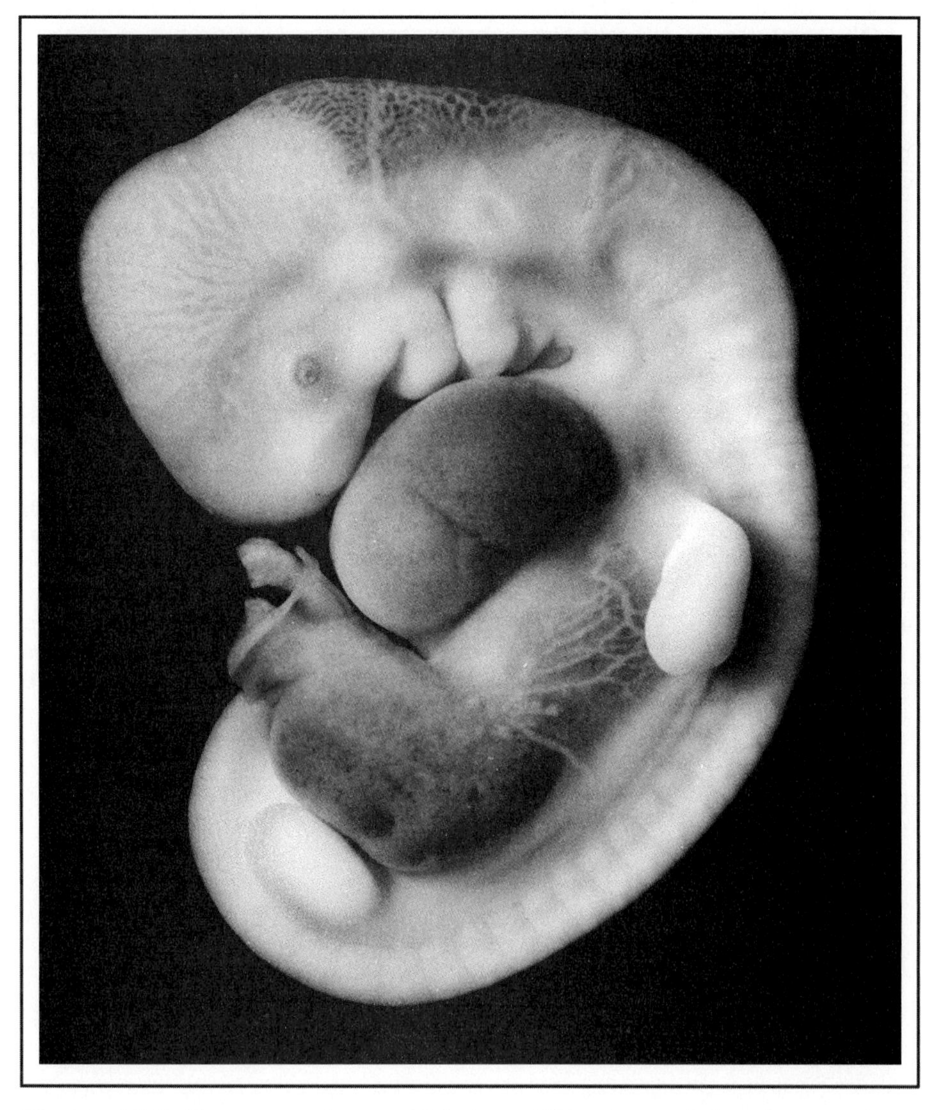

Figure 14.1 Human embryo at 31 days. This stage of development represents the phylotypic stage characteristic of all vertebrates. The deep grooves in the neck region are pharyngeal clefts, which in fishes and amphibian larvae would form gill slits. The bulge below the pharyngeal clefts contains the heart rudiment. The lateral protrusions are the limb buds.

CONTINUING the discussion of histogenesis begun in Chapter 13, we will now consider the derivatives of endoderm and mesoderm in vertebrates. The endodermal derivatives consist of the inner lining of the digestive tract and its derivatives, such as lung and liver. Mesodermal derivatives include a wide range of organs including bones, muscles, urogenital organs, and the circulatory system. The major tissue types to be considered here are connective tissue and muscle tissue. The chapter will also describe the extraembryonic membranes, on which embryos rely for their nutrition, respiration, and waste removal.

The material presented here will provide opportunities to discuss two general topics of great interest. The first is the occurrence of a phylotypic stage in development, at which all species of a phylogenetic group show an uncanny similarity. This stage is reached after organogenesis, when most organ rudiments are present in a basic body plan (Fig. 14.1). The second general topic is the principle of reciprocal interaction. Many organs originate from two embryonic rudiments—usually one epithelium and one mesenchyme—derived from different germ layers. As the organs develop, the rudiments exchange signals with each other, which ensure that the partners develop harmoniously. These interactions, as well as other signals exchanged during organogenesis, are now being analyzed in molecular terms, a development that is attracting many new investigators to this traditional field of work.

14.1 Endodermal Derivatives

The fate of the embryonic endoderm is to form the *inner epithelium* of the gut and its derivatives, whereas the connective tissues, blood vessels, and muscles of the gut and associated organs are contributed by the mesoderm. The spatial configuration of the endoderm in the vertebrate gastrula varies, as described in Chapter 10. In amphibians, the endoderm begins as the innermost sphere of the gastrula, the primitive gut or *archenteron.* In most other vertebrate embryos, the endoderm originally forms a disc adjacent to the yolk or to a cavity called the *yolk sac.* As these embryos develop, their flanks bend together ventrally, closing the endoderm to create the ***gut*** (Fig. 14.2). This process is referred to as ***lateral folding.*** It is enhanced by ***craniocaudal flexion*** of the embryo, driven mainly by the rapid extension of the neural plate (Fig. 14.3). In the course of these movements, the originally wide opening between the gut and the yolk sac is reduced to a narrow passage called the ***vitelline duct.***

The cranial and caudal ends of the digestive tube are closed temporarily by the ***buccopharyngeal membrane*** and the ***cloacal membrane,*** respectively.

The rudiments of certain appendages to the gut are convenient landmarks for scientists to subdivide the digestive tube into sections (Fig. 14.4). A small ventral outpocketing formed near the cranial end of the digestive tube is the rudiment of the ***trachea,*** or windpipe, which will convey air to and from the lungs; the section of the digestive tube between the buccopharyngeal membrane and the tracheal rudiment is called the ***pharynx.*** The

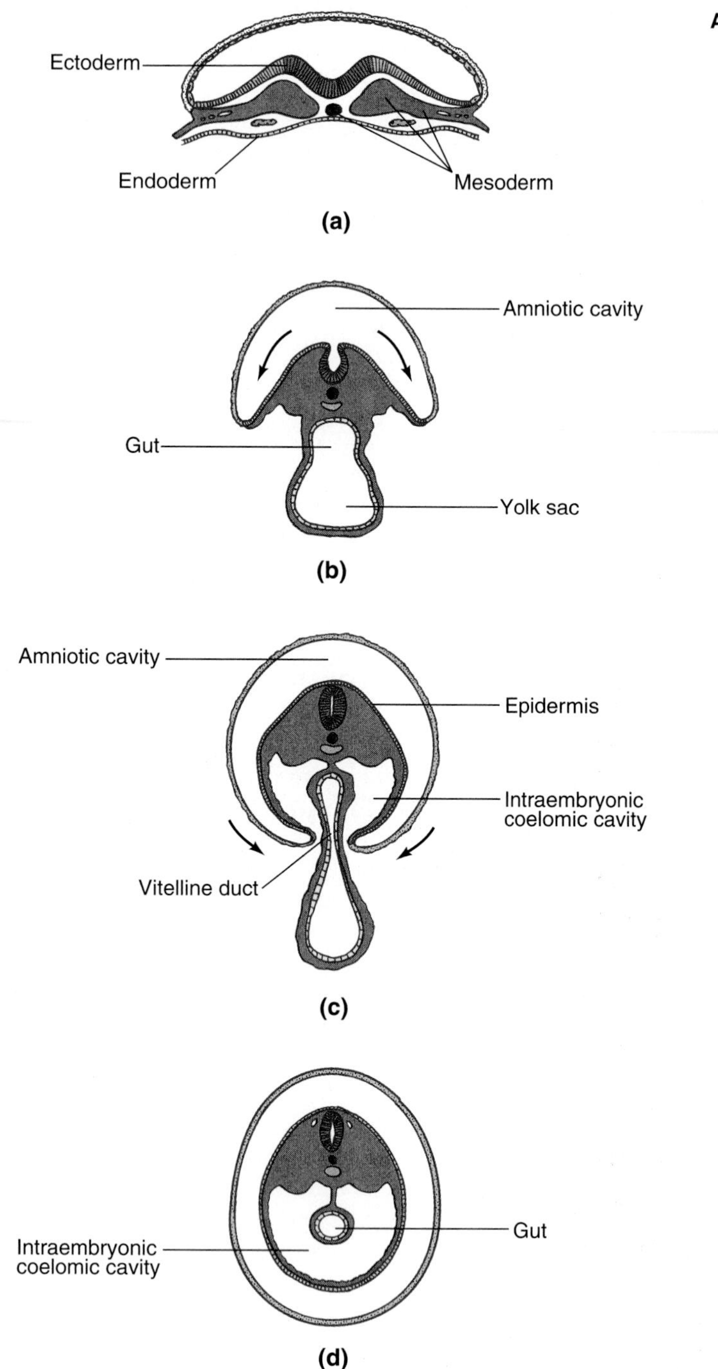

Figure 14.2 Lateral folding in the human embryo. **(a)** 19 days; **(b)** 20 days; **(c)** 21 days. Section through the midgut region shows the vitelline duct connecting the gut with the yolk sac. Lateral folding (arrows) closes the body cavity. **(d)** 21 days. Section through the hindgut region shows closed abdominal wall.

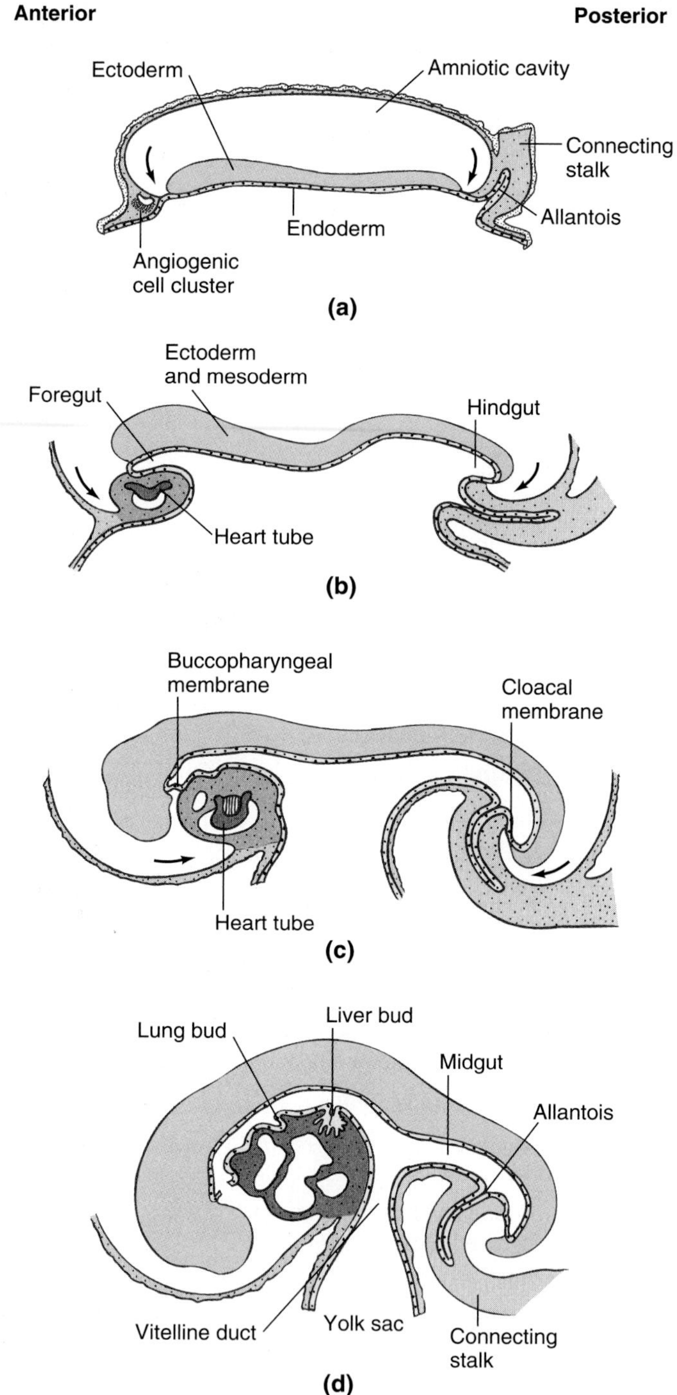

Figure 14.3 Craniocaudal flexion in the human embryo: **(a)** 18 days; **(b)** 21 days; **(c)** 24 days; **(d)** 30 days. Angiogenic clusters give rise to blood cells, blood vessels, and heart. The embryonic gut is closed anteriorly by the buccopharyngeal membrane and posteriorly by the cloacal membrane. The yolk sac and the allantois are ventral extensions of the gut.

next section, the ***foregut,*** extends from the trachea to the rudiments of the ***liver*** and the ***pancreas.*** From there to the cloacal membrane extend the ***midgut*** and the ***hindgut,*** which are not clearly distinguished in the embryo.

THE EMBRYONIC PHARYNX CONTAINS A SERIES OF ARCHES

The pharynx is crucial to the development of the neck region in vertebrates. An outstanding feature is a series

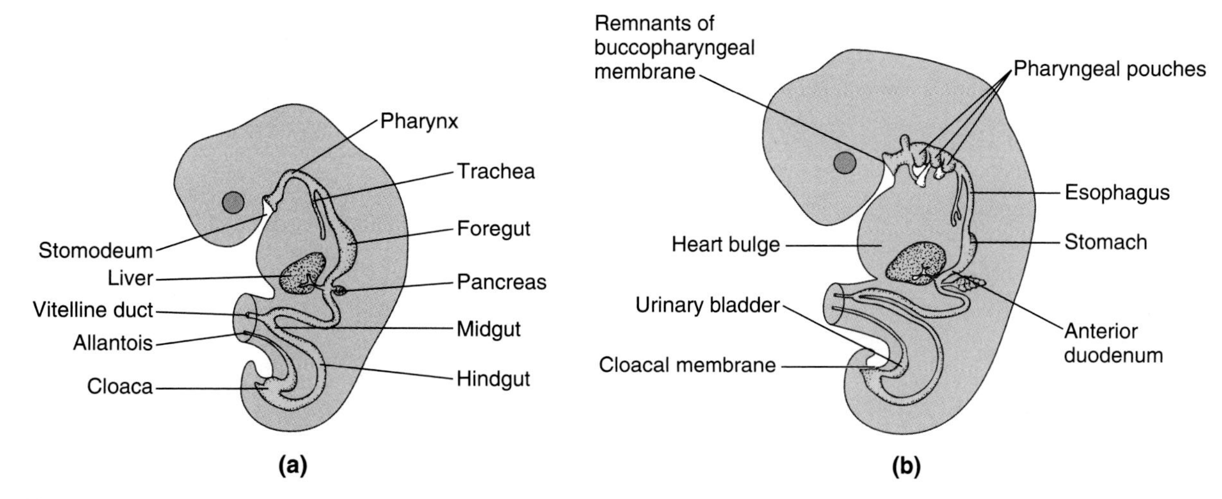

Figure 14.4 Endoderm of the human embryo: **(a)** 4 weeks; **(b)** 5 weeks. Note the development of the pharyngeal pouches, a characteristic of all vertebrate embryos.

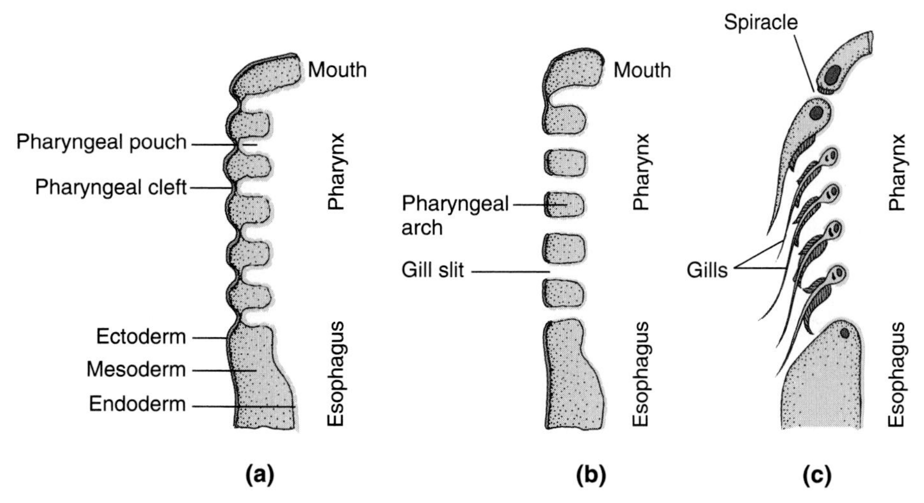

Figure 14.5 Schematic drawings showing the left half of the pharyngeal region in frontal section (see Fig. 9.2). **(a, b)** Generalized vertebrate embryo at successive stages. The pharyngeal pouches formed by the endoderm induce the formation of pharyngeal clefts in the ectoderm. Clefts and pouches fuse to form gill slits separated by pharyngeal arches. **(c)** Adult shark. The first gill slit has developed into a circular opening known as the spiracle, and gills have developed on the posterior pharyngeal arches.

of ***pharyngeal pouches***, which bulge out laterally from the pharyngeal endoderm and displace surrounding mesenchyme (Fig. 14.5). Where a pharyngeal pouch approaches the overlying ectoderm, it induces the formation of a ***pharyngeal cleft*** (see Figs. 14.1 and 14.5). The columns of neck tissue that remain between successive pharyngeal pouches and clefts are known as ***pharyngeal arches***. In primitive vertebrates and their chordate relatives, the pouches and clefts fuse to form slits through which water can stream from the pharynx to the outside. The vertebrate ancestors probably used this arrangement to strain food particles from a continuous flow of water moving through the pharynx (Torrey and Feduccia, 1991). Later, the pharyngeal arches became the basis for the formation of gills and jaws in primitive fish. In land-dwelling vertebrates, the respiratory function of the gills has been taken over by lungs, and the pharyngeal pouches and clefts have been put to other uses.

All vertebrate embryos form several pairs of pharyngeal arches, each containing precartilage cells of neural crest origin, premuscle mesenchyme, a blood vessel, and a cranial nerve. The cartilages that form in the first pharyngeal arch are known as the ***palatoquadrate cartilage*** and the ***mandibular cartilage*** (Fig. 14.6; Table 14.1). In fishes, their proximal portions persist as the *quadrate* and *articular* bones. They form the *primary jaw joint*, as discussed later. The second pharyngeal arch contains two embryonic cartilages, the ***hyomandibular***

table 14.1 Cartilage Elements in Pharyngeal Arches and Their Derivatives

		Adult Cartilages or Bones		
Pharyngeal Arch Number	**Embryonic Cartilage Elements**	**Fishes**	**Amphibians, Reptiles, Birds**	**Mammals**
I	Palatoquadrate	Quadrate	Quadrate	Incus
	Mandibular	Articular	Articular	Malleus Meckel's cartilage
II	Hyomandibular	Hyomandibular	Columella	Stapes, styloid process
	Hyal	Hyal	Part of hyoid	Part of hyoid
III	(Several elements)	Gill bar	Part of hyoid	Part of hyoid
IV–VI	(Several elements)	Gill bars	Laryngeal and tracheal skeleton	Laryngeal and tracheal skeleton

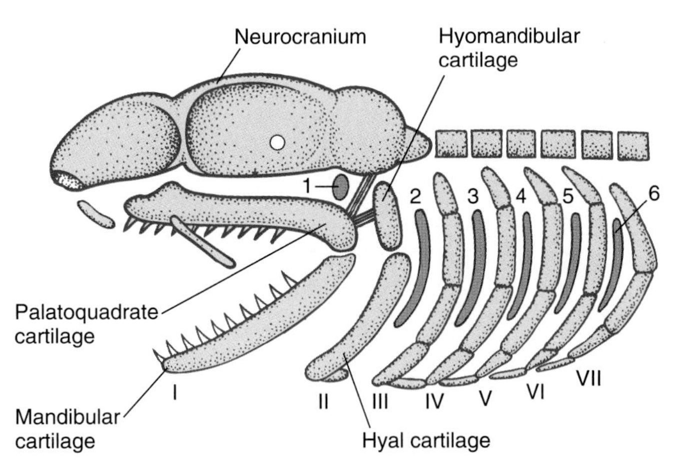

Figure 14.6 Schematic drawing of a developing shark skull in lateral view. In the neurocranium, which contains the brain, the nasal region is anteriormost (to the left); it is followed (in order) by the eye region (with an opening for the optic nerve), the otic region, and the occipital region. The cartilage elements of the pharyngeal arches (Roman numerals) are separated by gill slits (Arabic numerals), the first of which has formed the spiracle. The major cartilage elements of the first arch, the palatoquadrate and the mandibular, form the upper and lower jaw elements, respectively. The cartilage elements of the second pharyngeal arch are the hyomandibular and the hyal. Note that the palatoquadrate is only loosely attached to the neurocranium. A more stable connection is formed in bony fish by the hyomandibular.

and the ***hyal***. In fishes, the hyomandibular cartilage develops into a stout bone that connects the jaw to the braincase.

During mammalian development, the primary jaws are replaced with a second generation of jaw bones, the ***dentary*** for the lower jaw, and the ***maxillary*** bone and the ***squamosal*** (a portion of the temporal) bone for the upper jaw (Fig. 14.7). However, the vestiges of the palatoquadrate and mandibular cartilages persist as two tiny middle ear ossicles, the ***incus*** (Lat. "anvil") and ***malleus*** (Lat. "hammer"), respectively. The malleus attaches to the eardrum and transfers its vibrations to the incus. The incus, in turn, is linked to a third middle ear ossicle, the ***stapes*** (Lat. "stirrup"), which is homologous to the hyomandibular bone of fishes. Together, the three middle ear ossicles of mammals act as levers to increase the pressure of the sound waves as they travel from the eardrum to the inner ear. The other cartilage of the second pharyngeal arch, the *hyal*, forms part of the mammalian ***hyoid*** bone (Gk. *hyoides*, "U-shaped"), which supports the base of the tongue. The third and following pharyngeal arches also contribute to the hyoid bone and to the voice box, or ***larynx***.

The pharyngeal pouches and clefts still develop in mammalian embryos as they do in primitive vertebrates, but the pharyngeal endoderm and the adjoining epidermis are not perforated to form openings. The pouches and clefts, no longer needed for gill slits, assume a variety of other functions (Fig. 14.8). The first pharyngeal pouch, located between the first and second pharyngeal arches, gives rise to the middle ear, which contains the sound-transmitting ossicles already described. The original connection of the pouch to the pharynx persists as the ***pharyngotympanic tube***, or *Eustachian tube*, which equalizes the pressure between the middle ear and the outside. The first pharyngeal cleft gives rise to the outer ear. Where the first pouch meets the first cleft, the ***tympanic membrane*** (eardrum) is formed. The endoderm of the second pouch proliferates, penetrating the adjoining mesenchyme and forming the *palatine tonsil*. Part of the pouch remains and is still found as a groove in the adult tonsil. The third pharyngeal pouch forms the *thymus gland* and the lower portion of the *parathyroid gland*. Both glands lose their connection with the pharynx and move caudally and medially to their adult positions. Similarly, the fourth pharyngeal pouch gives rise to the upper portion of the gland, which also associates with the thyroid gland. The

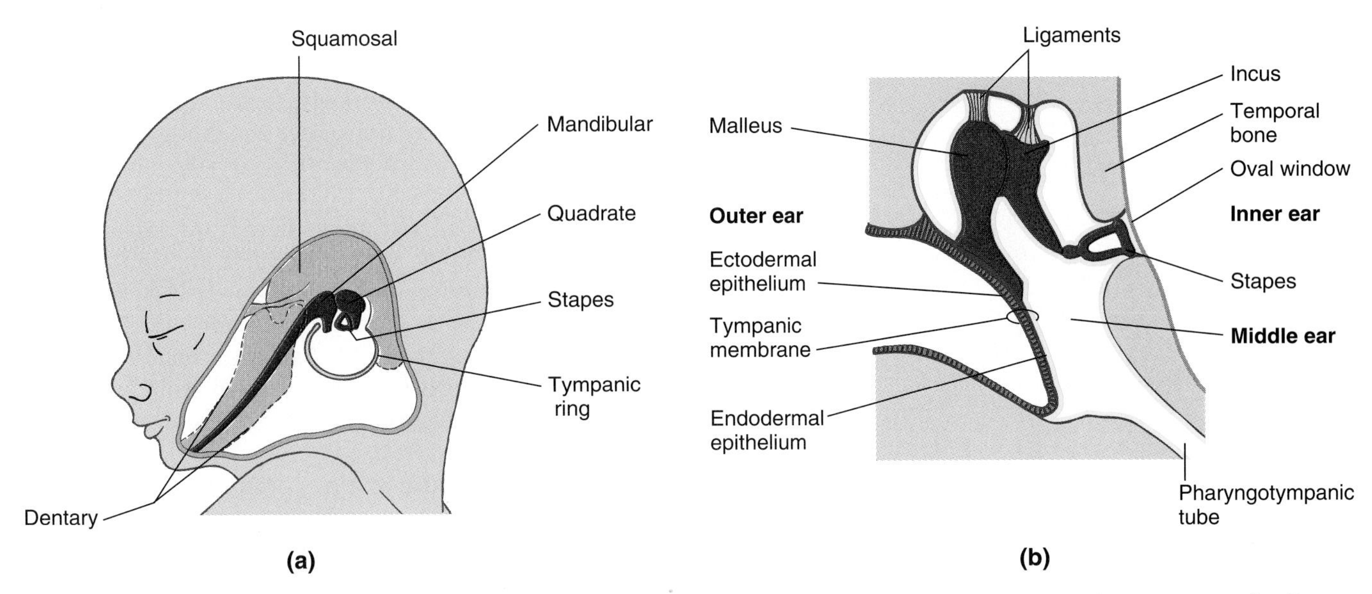

Figure 14.7 Development of the jaws and the middle ear ossicles in mammals. **(a)** Detail of a human fetus at 10 weeks. Two jaw joints are visible: The primary jaw joint is formed by the mandibular cartilage and the quadrate cartilage; the latter is the proximal part of the palatoquadrate, which originally supports the upper jaw. The secondary jaw joint is formed by the squamosal and the dentary. While the secondary jaw joint becomes the definitive one, vestiges of the primary jaw elements are transformed into middle ear ossicles. The malleus, which descends from the proximal part of the mandibular cartilage, will attach to the eardrum. The incus, which develops from the quadrate, will articulate with the malleus as it did in the primary jaw joint. The stapes, derived from the hyomandibular cartilage, will connect the incus with the inner ear. The squamosal and tympanic ring, along with other rudiments, will form the temporal bone of the adult skull. **(b)** Schematic drawing of the middle ear in the human adult. The malleus transmits vibrations from the eardrum via the incus to the stapes, the foot plate of which is in contact with the cochlea (see Fig. 13.35).

second to fourth pharyngeal clefts disappear during later development.

EMBRYOS PASS THROUGH A PHYLOTYPIC STAGE AFTER ORGANOGENESIS

The derivatives of the pharyngeal arches illustrate the indirect course by which many mammalian organs develop. The German embryologist C. B. Reichert, who discovered the origin of the middle ear ossicles in mammals in 1837, could hardly believe what he saw. Why would these tiny ear bones develop from cartilages that look like the rudiments of jaws? Why do mammalian embryos have deep grooves in the neck that look like the gill slits of fish? Answers to these questions had to await the study of the fossil record on vertebrate evolution (Romer, 1976; Colbert, 1980; S. J. Gould, 1977, 1990; McKinney and McNamara, 1991). We will briefly summarize the findings in this area and their implications for development before we return to our discussion of endodermal derivatives.

Fishes have an inner ear with which they perceive gravity and their own body movements. The same organ can provide a sense of hearing, provided that there is a device to transmit sound waves from the environment. One device of this kind evolved from a bone element located in the vicinity of the inner ear. The *hyomandibular bone* of fish, derived from the dorsal part of the second pharyngeal arch, is a stout bone that anchors the jaws to the braincase. The same hyomandibular bone of some ancestral fishes apparently also played a role in hearing, as indicated by the insertion of the bone into the ear region of the braincase (Clack, 1989). As land-dwelling vertebrates evolved, their upper jaws became attached directly to the braincase, so that the hyomandibular bone was no longer necessary as a brace. This bone then became more slender, in accord with its secondary function in hearing. Amphibians, reptiles, and birds still have a single middle ear ossicle, the ***columella***, which is derived from the hyomandibular cartilage (see Table 14.1).

A similar shift of functions occurred during the evolution of the mammalian jawbones. The ***primary jaw joint***, still found in all vertebrates except mammals, is formed by the ***quadrate*** and ***articulare***, two bones that develop from the *palatoquadrate cartilage* and the *mandibular cartilage*, respectively, of the first pharyngeal arch. During mammalian evolution, a ***secondary jaw joint*** between the *dentary bone* (lower jaw) and the *squamosal*, a part of the temporal bone, was added to the primary jaw joint. Mammal-like reptiles that lived during the Jurassic period (about 180 million years ago) actually used both the primary and the secondary jaw joints as adults. But the existence of a secondary jaw joint eventually relieved the bones of the primary jaw joint of their original function and allowed them to

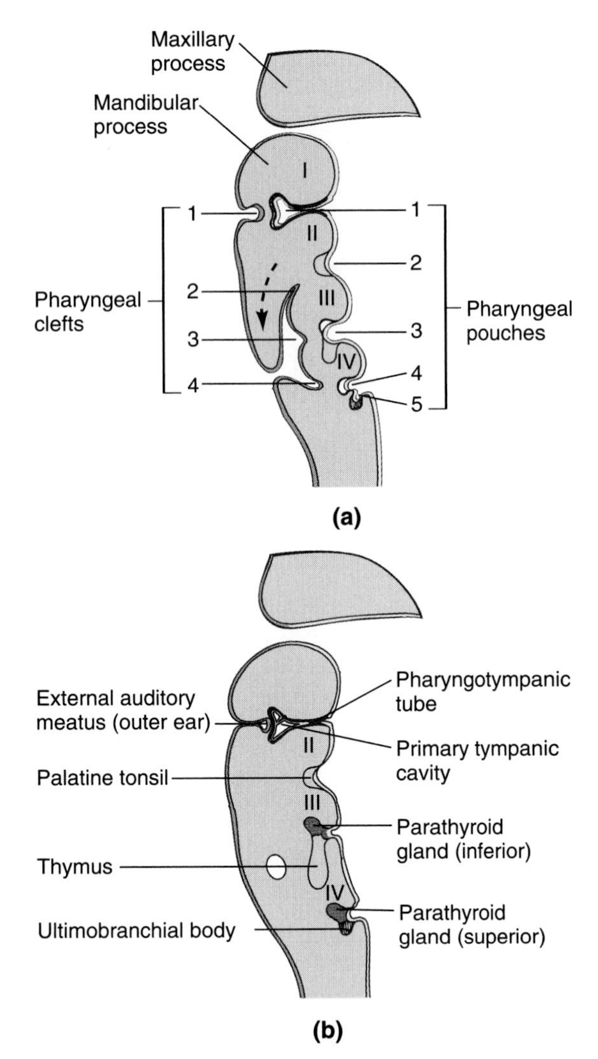

Figure 14.8 Fates of the pharyngeal pouches and clefts in the human embryo, drawn in frontal sections: **(a)** 5 weeks; **(b)** 6 weeks. Roman numerals designate pharyngeal arches; Arabic numerals designate pharyngeal pouches and clefts. The first cleft gives rise to the outer ear, while the other clefts are overgrown. The first pouch develops into the middle ear and the pharyngotympanic tube. The other pouches give rise to palatine tonsils and several endocrine glands.

evolve into two additional middle ear ossicles, the *malleus* and the *incus*.

It is a striking phenomenon that the evolution of jawbones into middle ear ossicles is not only shown in the fossil record but is also reenacted in the development of each mammal today. Every mammal, including the human, retraces in its normal fetal development the evolution of the middle ear ossicles from jawbones (see Fig. 14.7). This process exemplifies a phenomenon—termed ***recapitulation***—that is very pervasive in vertebrate development. It appears that nature often prefers the modification of what it has over the construction of novelties.

It is important to realize that advanced vertebrates recapitulate the *embryonic rather than the adult* stages of their phylogenetic ancestors. In *normal* human development, it is the embryonic mammary ridge, not the adult condition of multiple mammary glands, that is recapitulated (see Fig. 13.43). Similarly, the development of the middle ear ossicles recapitulates the embryonic stages, not the adult stages, of the primary jaw joint and the hyomandibular bone. A false impression that adult conditions are recapitulated might stem from the fact that in primitive vertebrates, adult structures are similar to embryonic structures, whereas in advanced vertebrates, the adult stage is farther removed from the embryonic stage. For instance, the gills of an adult fish remain similar to the pharyngeal arches of a fish embryo, but the neck of an adult mammal is very different from its embryonic rudiment.

Comparative studies on vertebrate development were first carried out in a systematic way by Kare Ernst von Baer (1828), who discovered the notochord, the mammalian egg, and the human egg (see Section 3.1). He pointed out that all vertebrate embryos look very similar after organogenesis, when the notochord, the rudiments of brain, sense organs, heart, and gut, are all present in a spatial arrangement known as the ***basic body plan*** (Fig. 14.9). This phenomenon is not limited to vertebrates: The embryos of insects and other arthropods are also very similar after organogenesis, at a stage known as the *segmented germ band* (see Figs. 22.2 and 22.53; Sander, 1983). Such a stage that is shared by all species of a phylum—or a subphylum, as in the case of the vertebrates—is called the ***phylotypic stage***.

Reportedly, von Baer's way of convincing his students that vertebrates have a phylotypic stage was to show them jars with preserved embryos after organogenesis and to pretend that the labels identifying the species of each specimen had come off. He then challenged his audience to help him reconstruct which of the embryos was a fish, amphibian, reptile, and so on. Hardly anyone was able to do this. Modern scholars can make such distinctions based on differences in growth patterns and heterochronies in the formation of many organ rudiments (Richardson, 1995). However, differences between vertebrates at the phylotypic stage are much more subtle than differences at earlier stages, when egg size, mode of cleavage, and gastrulation movements differ markedly among the embryos of fishes, amphibians, reptiles, birds, and mammals. Likewise, during advanced embryonic and fetal stages, class-specific characteristics such as the beak and feathers of a bird become readily apparent.

Why the phylotypic stage has been so resistant to evolutionary change has been a much-debated question. The most common speculation is that the genes involved in organogenesis are typically functioning in multiple control circuits so that mutations in these genes are inevitably lethal.

The developmental processes that occur after the phylotypic stage have changed more readily during evolution. The tendency of these processes to recapitu-

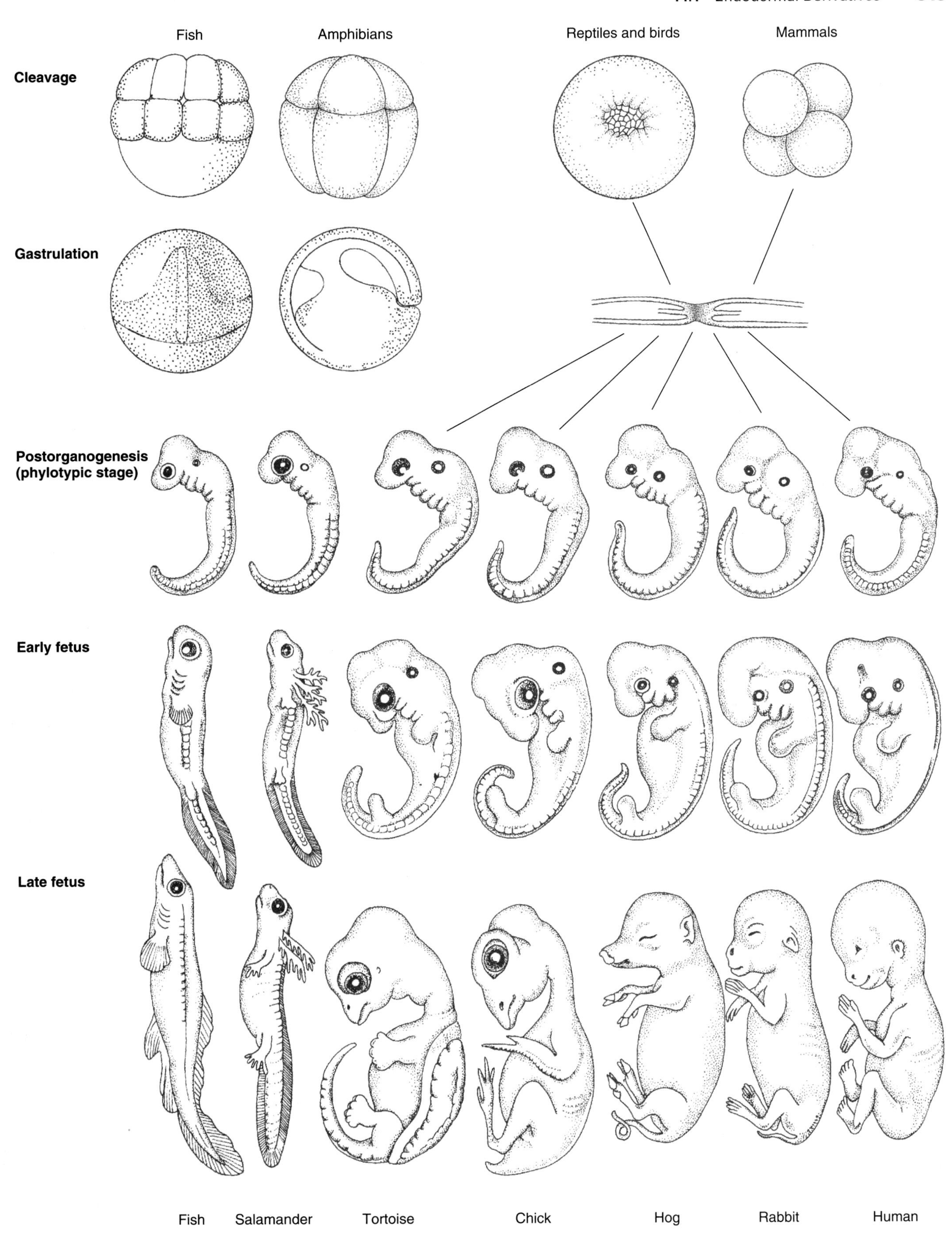

Figure 14.9 The phylotypic stage in vertebrates. Early during development, vertebrates differ with regard to cleavage and gastrulation patterns. After organogenesis—at the phylotypic stage—embryos from different vertebrate classes are very similar. After the phylotypic stage, embryos and fetuses become recognizable first as members of their class, then their order, and so on, and finally their species.

late phylogenetically old patterns suggests that the evolutionary changes were caused by gene activities that modify existing regulatory networks rather than creating entirely new ones. This view is supported by the occurrence of ***atavisms,*** that is, *occasional* reappearances of phylogenetically older morphological traits in *adults.* Examples of atavisms in humans include multiple nipples or breasts (see Fig. 13.43) and a divided uterus that are reminiscent of the uteri of more primitive mammals. In horses and mules, occasional supernumerary hoofs resemble the three-toed feet of ancestral horses that lived 15 million years ago. Such atavisms demonstrate vividly that phylogenetically old patterns of development are still present in evolved species but are normally suppressed or modified.

The genes that control the suppression or modification are best known in insects. The *Ultrabithorax*$^+$ gene of *Drosophila,* for example, modifies the development of the third thoracic segment so that it will differ from the second thoracic segment. Failure of the *Ultrabithorax*$^+$ gene to function properly leads to the development of a four-winged fly (see Fig. 1.21), a phenotype reminiscent of fossil ancestors of modern flies.

THE ENDODERM LINES THE INSIDE OF THE INTESTINE AND ITS APPENDAGES

Soon after growing out from the foregut, the tracheal rudiment forms two lateral buds. In the human, these buds divide again into three branches on the right side and two on the left (Fig. 14.10). These branches develop into the main ***bronchi,*** in advance of the development of the lobes of the lung. The bronchi divide many times into more and finer ***bronchioli.*** Eventually, the epithelial cells in the terminal sacs of the bronchioli change from a cuboidal to a *squamous* shape, forming the grapelike ***alveoli.*** The alveoli become intimately associated with blood capillaries so that a very large and thin interface is formed between the air in the alveoli and the blood in the capillaries. Only the inner epithelium of the trachea, bronchi, and lungs is derived from endoderm. The other tissues, including the cartilage reinforcements of the trachea and bronchi, the blood vessels supplying the lungs, and the connective tissue in which these structures are embedded, are of mesodermal origin.

The part of the foregut that is located caudal to the tracheal rudiment develops into the ***esophagus, stom-***

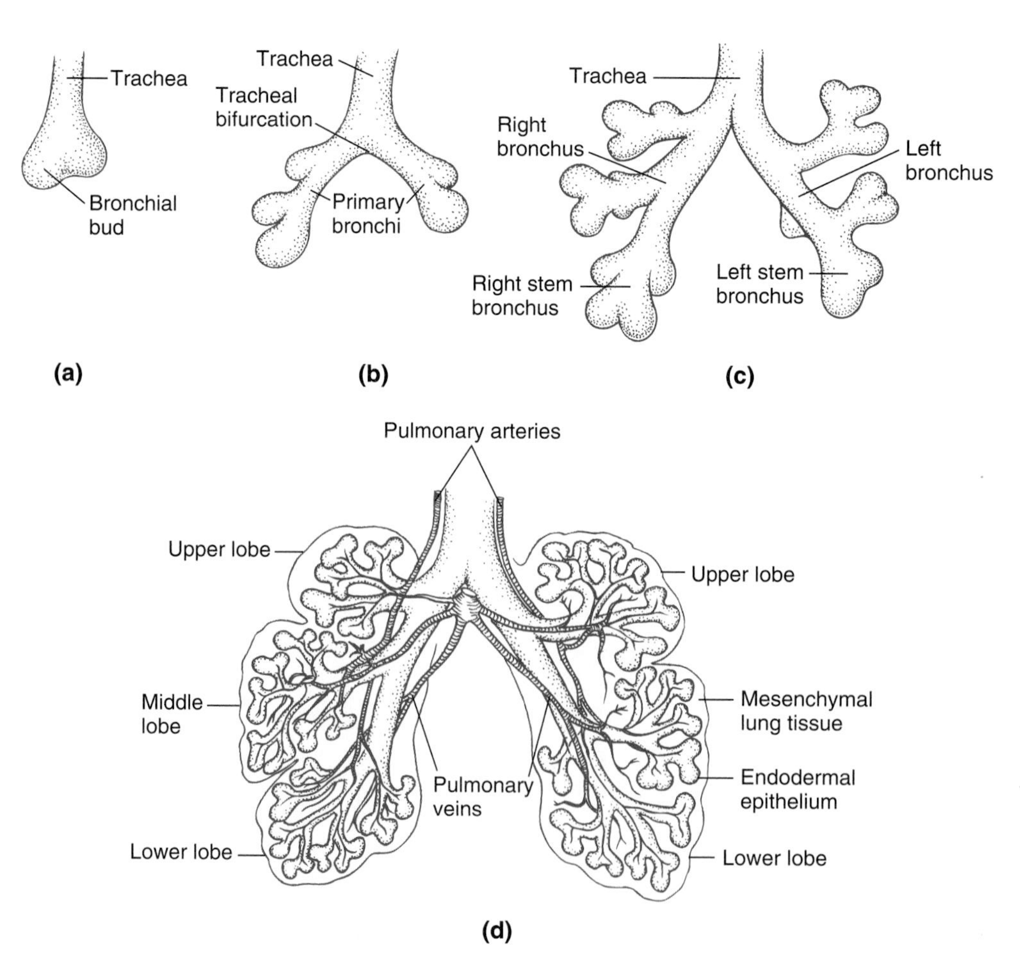

Figure 14.10 Development of the human lung. The diagrams show ventral aspects. The branching endodermal epithelium is surrounded by mesodermal tissues including blood vessels: **(a)** 28 days; **(b)** 33 days; **(c)** 39 days; **(d)** 50 days.

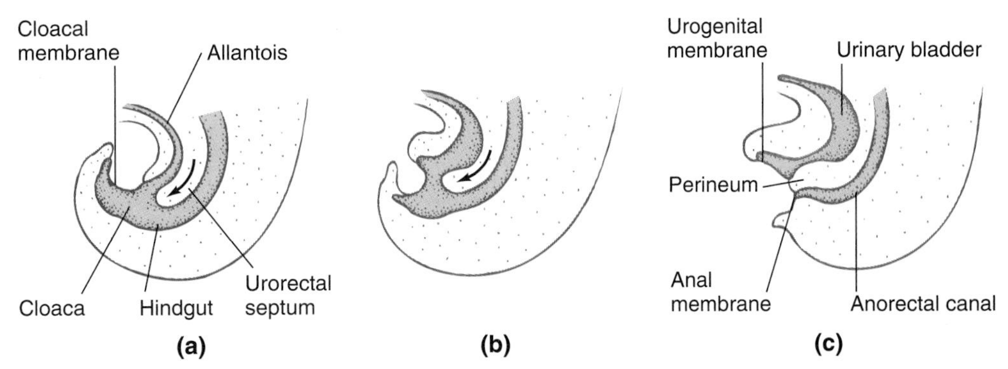

Figure 14.11 Development of the cloacal region in the human embryo: **(a)** 5 weeks; **(b)** 7 weeks; **(c)** 8 weeks. Arrows indicate the movement of the urorectal septum, which forms the perineum and divides the cloacal membrane into the anal membrane and the urogenital membrane.

ach, and the anterior part of the ***duodenum*** (see Fig. 14.4). The esophagus is a straight connection between the pharynx and the stomach. The stomach, already discernible in the 4-week human embryo, is the most muscular and expandable portion of the digestive tract. The remainder of the embryonic foregut, between the stomach and the liver, is the rudiment of the anterior duodenum. The adjoining midgut, beginning with the posterior duodenum, forms the ***small intestine,*** in which the epithelium is specialized so as to maximize the area available for taking up nutrients. Mesodermal layers surrounding the epithelium provide the blood vessels and smooth muscle layers involved in digestion. The embryonic midgut also forms a small part of the ***colon,*** the remainder of which is derived from the hindgut.

The hindgut gives rise to the greater part of the *colon,* the ***rectum,*** and the ***anal canal*** of the adult. In addition, the endoderm of the hindgut forms the inner lining of the bladder and urethra. The latter structures arise from the ***allantois,*** an appendage of the embryonic hindgut, and the embryonic ***cloaca*** (Fig. 14.11). In most vertebrates, the cloaca serves as a common opening for the gut, the bladder, and the reproductive system. In mammals, the cloaca is partitioned by the development of a transverse mesenchymal ridge, the ***perineum,*** which divides the *cloacal membrane* into the ***urogenital membrane*** and the ***anal membrane.*** The anal membrane is displaced to the interior by an ectodermal depression, the ***proctodeum,*** before it ruptures to open a passage between the rectum and the outside.

14.2 Axial and Paraxial Mesoderm

The two germ layers we have looked at so far in this text, the ectoderm and the endoderm, give rise mostly to *epithelia,* including the epidermis of the skin and the inner epithelium of the gut. The intervening germ layer, the mesoderm, forms both epithelia and large amounts of *mesenchyme.* Mesenchyme consists of isolated cells surrounded by large amounts of extracellular material, as opposed to epithelia, which are sheets of tightly joined cells (see Fig. 2.22). Mesenchymes are typically derived from the mesodermal germ layer, but may also arise from ectoderm, as discussed in Sections 13.2 and 13.3 in connection with the *neural crest cells* and some *ectodermal placodes.*

The overall shape of the mesoderm in the gastrula varies with the class of vertebrates. In typical amphibians, the germ layers are arranged like three nested tubes. In other vertebrates the mesoderm arises as a flat layer between an ectodermal roof and an endodermal floor. Regardless of whether the mesodermal germ layer is tubelike or flat, it becomes subdivided in a pattern that is characteristic of all vertebrate embryos (Fig. 14.12); its major subdivisions are the *axial mesoderm,* the *paraxial mesoderm,* the *intermediate mesoderm,* and the *lateral plates.*

AXIAL MESODERM FORMS THE PRECHORDAL PLATE AND THE NOTOCHORD

The ***axial mesoderm*** is located along the dorsal midline. In amphibians, it assumes its position mostly by convergent extension (see Figs. 10.11 and 10.18). In mammals, birds, and some reptiles, the axial mesoderm ingresses through *Hensen's node* as the *primitive streak* retreats from anterior to posterior (see Figs. 10.24 and 10.32). The axial mesoderm gives rise to the ***prechordal plate*** in the anterior head and to the ***notochord*** in the posterior head, neck, trunk, and tail (Fig. 14.13). The prechordal plate is a source of mesenchyme, which, together with *ectomesenchyme* derived from neural crest cells, forms cranial cartilage. The notochord (Lat. *notum,* "back"; Gk. *chorde,* "cord") is a dorsal rod of cartilagelike connective tissue. The notochord persists in the adult stages of small marine creatures making up the subphyla Urochordata and Cephalochordata, which, together with

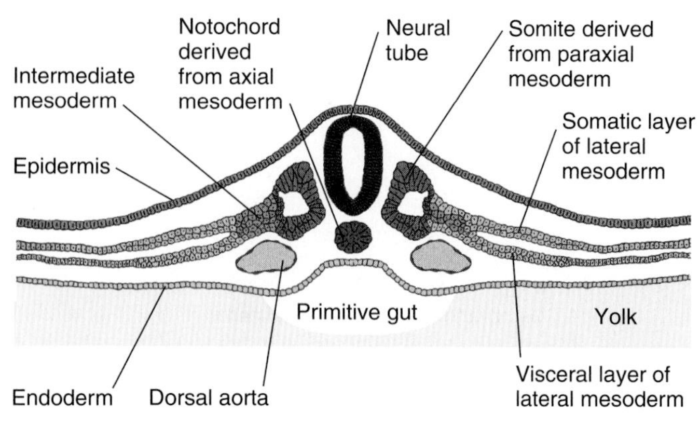

(a) Section of normal chick embryo

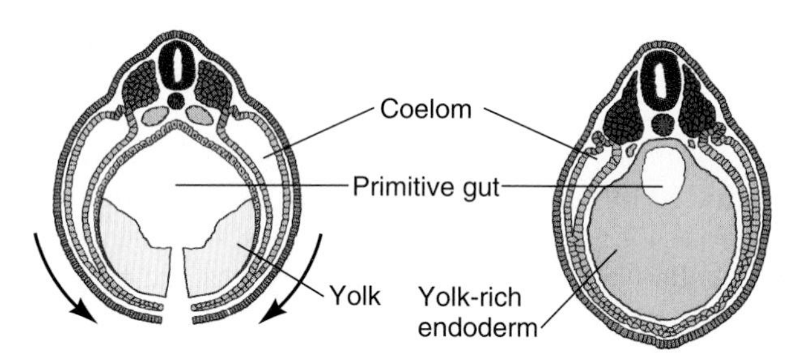

(b) Chick embryo removed from yolk, edges pulled together

(c) Section of frog embryo

Figure 14.12 Major subdivisions of the mesoderm in bird and amphibian embryos shown in transverse sections. **(a)** Normal chick embryo showing mesoderm and mesoderm-derived structures. **(b)** Chick embryo with yolk removed for easier comparison with frog embryo. **(c)** Frog embryo. Note the close resemblance between the bent chick embryo and the frog embryo at the same stage of development.

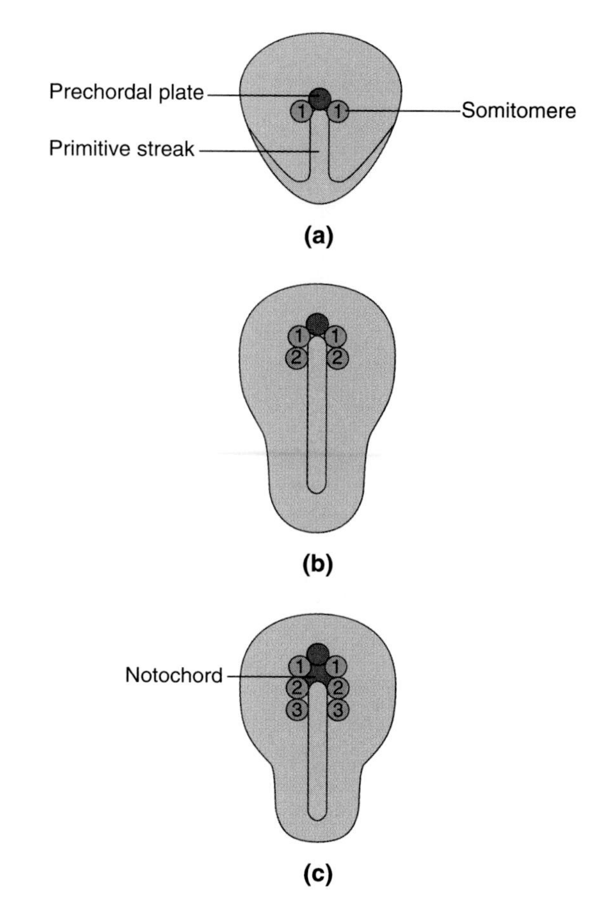

Figure 14.13 Development of the axial mesoderm (prechordal plate and notochord) and paraxial mesoderm (somitomeres). The diagrams show bird embryos during **(a)** primitive streak elongation, **(b)** maximal elongation of the streak, and **(c)** streak regression. The epiblast has been removed to reveal the underlying mesoderm, but the position of the primitive streak is represented. The prechordal plate is the most anterior axial mesoderm. The somitomeres are numbered in the sequence in which they form.

the vertebrates, constitute the phylum Chordata. In vertebrates, much of the notochord is replaced by the vertebral column. In mammals, including humans, small portions of notochord persist as a pulpy, elastic tissue inside the ***intervertebral discs*** located between vertebrae.

PARAXIAL MESODERM FORMS PRESOMITIC PLATES AND SOMITES

The ***paraxial mesoderm*** (Gk. *para,* "beside") originates on both sides of the axial mesoderm (Tam et al., 2000). About the same time the neural tube closes, the paraxial mesoderm subdivides into segments called ***somites.*** If the mesoderm is formed with Hensen's node regressing from anterior to posterior, the somites appear in the same sequence. The final number of somites formed is characteristic of a species.

Before the somites become distinct under the light microscope, the paraxial mesoderm forms two ridges of mesenchymal cells that are not overtly segmented, the ***presomitic plates.*** In chickens, the presomitic plates reach from the last-formed somite to Hensen's node (Fig. 14.14a). As the node regresses, the presomitic plates extend posteriorly while giving off overt somites anteriorly.

Close inspection of chicken presomitic plates with the scanning electron microscope reveals that they consist of mesenchymal cells arranged in whorls termed ***somitomeres*** (Fig. 14.14b; Meier, 1979). The size, position, and morphology of the somitomeres indicate that they are precursors of somites. The most mature somitomeres, closest to the emerging somites, are clearly distinct by the radial orientation of their cells, even though they are contiguous at their cranial and caudal borders. The least mature somitomeres, barely discernible in stereoscopic pairs of scanning electron micrographs, appear posteriorly on both sides of Hensen's node. As the node regresses, leaving in its wake the axial mesoderm, the left and right rows of somitomeres remain separated by the developing notochord.

IN search for evidence that somitomeres are the precursors of somites, Packard and Meier (1983) isolated presomitic

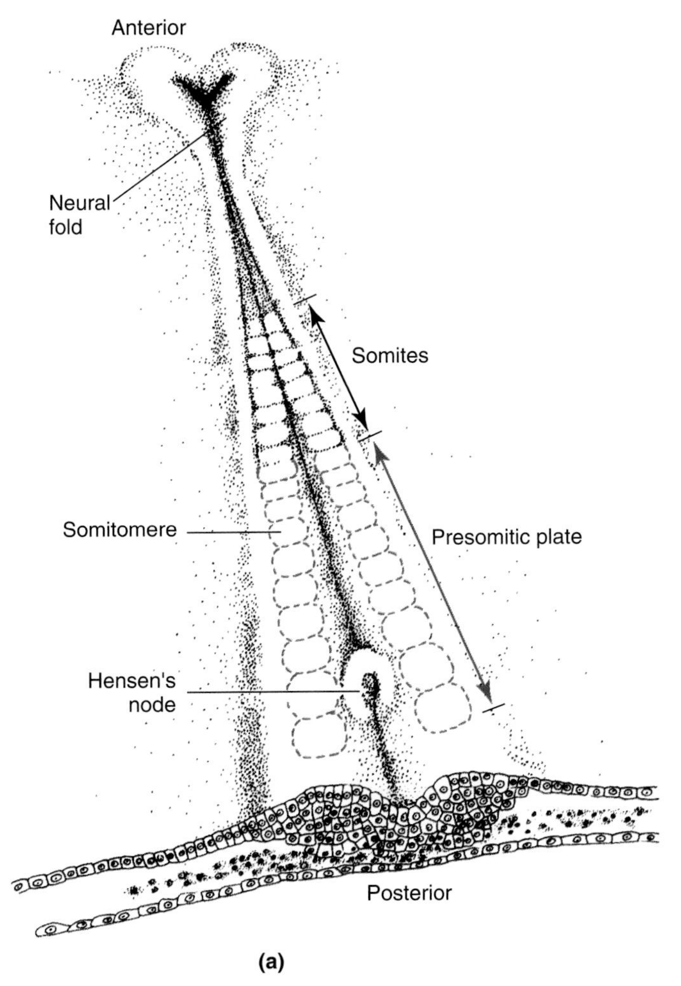

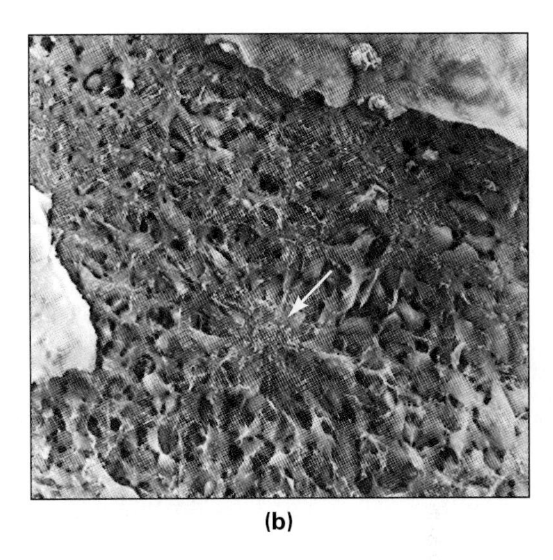

Figure 14.14 Somite development in the chicken embryo. **(a)** Survey diagram showing the anterior end of a chicken embryo sectioned posterior to Hensen's node. The dashed outlines represent somitomeres in the presomitic plates. The somites anterior to the presomitic plates are discernible under the light microscope. Note how the anterior end of the embryo is ahead of the posterior with respect to neurulation and somite formation. **(b)** Scanning electron micrograph of a somitomere in the chicken embryo. The "hub" in the center (arrow) is formed by a cluster of cell extensions (microvilli).

plates from chicken or quail embryos. Hensen's node was excluded from the isolates in order to prevent further presomitic plate formation during experimentation and culture. In a first series of experiments, the investigators fixed the right and left presomitic plates simultaneously for electron microscopy and counted the number of somitomeres formed. As expected, two presomitic plates from one embryo usually formed the same number of somites; the small discrepancies that occurred could be ascribed to slightly oblique cuts made during isolation.

In another series of experiments, the investigators fixed the left half immediately for scanning electron microscopy while the right half was allowed to develop in tissue culture for 5 to 8 h before fixation. During culture in vitro, the right halves continued to form somites of normal size and shape at the cranial end. However, the combined number of somites and somitomeres in the right half virtually equaled the number of somitomeres in the matching left half. Because the first series had shown that matching left and right halves produced the same numbers of somitomeres, the results of the second series indicated that for each somitomere lost, a somite had been gained. Thus, the somitomeres must be the precursors of somites.

Questions

1. How does the configuration of paraxial cells change as somitomeres turn into somites?
2. How do you expect the transition from somitomeres to somites to be reflected in the synthesis of cell adhesion molecules?

Since somitomeres have been identified in mammals, reptiles, amphibians, and teleosts, it is believed that somitomeres are the first visible signs of segmentation in all vertebrates (Jacobson, 1988). Later expressions of segmentation such as nerves or blood vessels are determined by the pattern of somitomeres and somites (Jacobson, 1998).

What are the cellular and molecular mechanisms that subdivide the presomitic plates into somitomeres? What keeps the somitomeres on the right and left sides of the body neatly in register? These questions have puzzled embryologists for a long time, and recent progress has rekindled the interest in the subject (Tajbakhsh and Spöhrle, 1998; Tam et al., 1998). Investigators who tracked the accumulation of certain mRNAs by *in situ hybridization* (see Method 8.1) found that they are synthesized with periodicities corresponding to the somitomeric pattern. For example, *Eph receptor tyrosine kinases* and their membrane-bound ligands, the *ephrins,* are expressed in a periodic pattern in the presomitic mesoderm, and interference with the synthesis of their mRNAs disturbs somite boundary formation and myotome development (Durbin et al., 1998). One of the targets of Eph activity seems to be the *Notch signaling pathway*—the same pathway discussed earlier in the context of neural development in *Drosophila* as well as in vertebrates (see Sections 6.4 and 13.1). *Notch*$^+$ is expressed in somitomeric patterns in a wide range of vertebrates (Fig. 14.15). If this pathway is interrupted, the subdivision of paraxial mesoderm into somites is severely disturbed even though histogenesis of somitic derivatives, such as vertebrae and skeletal muscle, generally proceeds (Conlon et al., 1995; Jen et al., 1997; Evrard et al., 1998).

Figure 14.15 Expression pattern of *X-Delta-2*$^+$ in presomitic plates of *Xenopus* embryos at the tailbud stage. Dorsolateral view, anterior to the left. Transient accumulations of X-Delta-2 mRNA have been made visible by in situ hybridization (see Method 8.1). The transcripts are made in a striped pattern located in the paraxial mesoderm prior to the appearance of somites. The patterned synthesis of X-Delta-2 mRNA progresses as a wave from anterior to posterior and in each location precedes the appearance of somites. The stripes have a spacing of about ten cells, the length of a prospective somite. These observations suggest that X-Delta-2, via its receptor X-Notch-1, plays a role in somite formation.

Another type of molecule that accumulates with a somitomeric periodicity in presomitic plates of chicken is encoded by *c-hairy1*$^+$ (Palmeirim et al., 1997). This gene was isolated based on its sequence similarity to the *Drosophila* gene *hairy*$^+$, which is involved in generating the segmented germ band of *Drosophila* and other insects (see Section 22.4). C-hairy mRNA accumulates in a cyclic pattern with the length of a somitomere and the periodicity of 90 min, the time that elapses between the appearance of consecutive somites. Subsequently, c-hairy mRNA is present continuously in the posterior half of each somite.

In chickens and many other vertebrates, a somitomere develops into a somite by changing from a solid ball of mesenchymal cells into a hollow ball of epithelial cells. During this transition, the cells become polarized, expressing N-cadherin at their subapical surfaces and forming a *zonula adherens* (Fig. 14.16). The apical surfaces of the somitic cells are oriented toward the cavity while a basement membrane is laid down around the outer surface.

Different parts of each somite give rise to different structures. We will focus here on those differences that appear in a *transverse section* of a somite, in a plane that is perpendicular to the anteroposterior axis and has a dorsoventral and a mediolateral dimension (see Fig. 9.2). To establish transverse *fate maps* of somites, researchers have transplanted stained portions of somites or have relied on the deeply staining heterochromatin

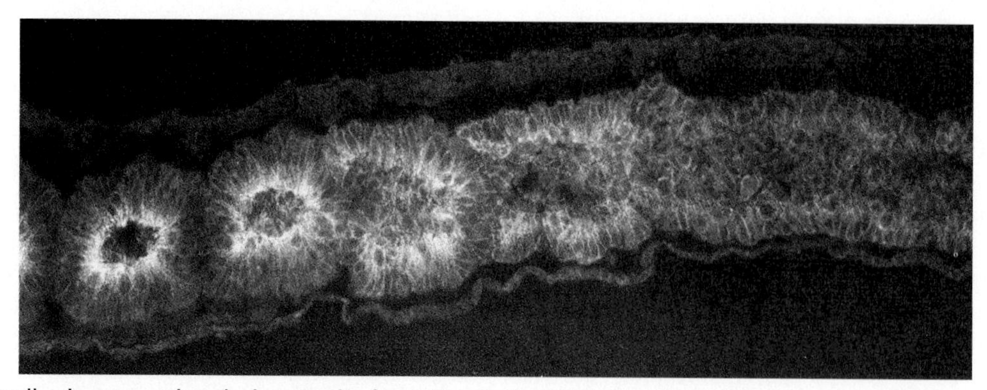

Figure 14.16 N-cadherin expression during somite formation in the chicken embryo. The photograph shows a paramedian section at the somite level, with anterior to the left. The somites are formed as Hensen's node regresses from anterior to posterior (see Fig. 14.14). Somite formation is therefore more advanced in the anterior of the embryo. The formation of somites as epithelial spheres correlates with the accumulation of N-cadherin, revealed by immunostaining (see Method 4.1) with a fluorescent secondary antibody.

present in quail cell nuclei (see Fig. 13.26). After replacing parts of chicken somites with the labeled or marked grafts, they were able to trace these grafts and establish their fates.

Cells from a somite's ventromedial portion become mesenchymal again and are known as the ***sclerotome*** (Gk. *skleros*, "hard"; *tome*, "cut", referring to the sliced appearance of the somitic derivatives). Sclerotome cells undergo a burst of mitotic activity and migrate away from the remainder of the somite to surround the notochord and the neural tube (Fig. 14.17). Secreting large amounts of extracellular material, they form cartilage elements that replace the notochord and are later transformed into vertebrae. In the thoracic region, sclerotome cells also form ribs.

After the sclerotomal cells have become mesenchymal, the epithelial remainder of the somite forms a transitional structure called the *dermomyotome*. A lip formed at its dorsomedial margin forms the ***epaxial myotome***, while a lip at the ventrolateral margin forms the ***hypaxial myotome*** (Gk. *mys*, "muscle"; *epi*, "upon"; *hypo*, "below"). Epaxial and hypaxial myotomes together, possibly with contributions from other cells, form a contiguous sheet of cells, the *myotome*. Although temporarily united, the epaxial and hypaxial myotomes have different origins and fates, at least in chickens (Ordahl and Le Douarin, 1992). Epaxial myotome, like other medial somitic structures, is derived from cells that have ingressed through Hensen's node during gastrulation. In contrast, hypaxial myotome, like other lateral somitic structures, is derived from cells that have ingressed through more posterior portions of the primitive streak (see Fig. 10.24). Epaxial myotome cells stay together in the myotome and give rise to dorsal trunk muscles, whereas hypaxial myotome cells detach from the myotome and migrate ventrolaterally to form limb muscles and ventral trunk muscles (Fig. 14.18).

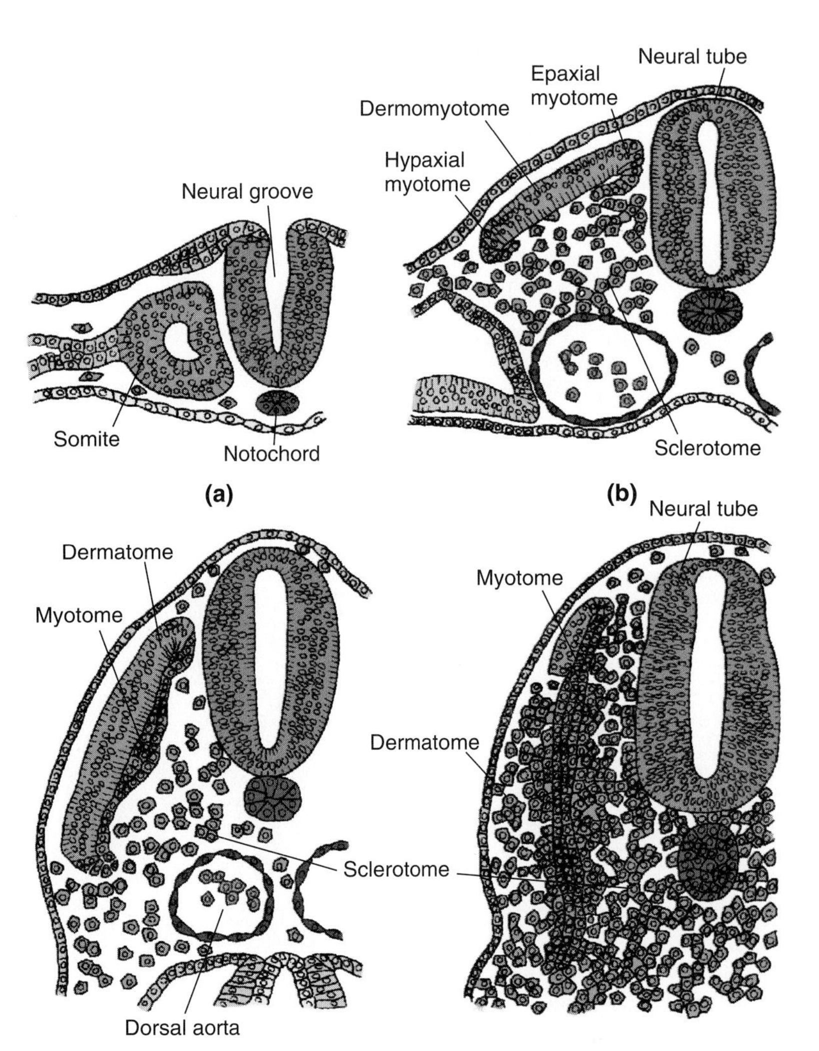

Figure 14.17 Somite development in the human embryo at successive stages. **(a)** The somite forms an epithelial sac around a small cavity. **(b)** The ventral and medial cells of the somite lose their epithelial connections and migrate toward the notochord and around the neural tube. These cells, which form the cartilage precursors of vertebrae, are called the sclerotome; the remainder of the somite is then called the dermomyotome. Lips forming at the dorsomedial and ventrolateral margins of the dermomyotome form the epaxial and hypaxial myotomes, respectively. **(c)** The epaxial and hypaxial myotome, possible with contributions from other cells, form a contiguous layer of cells, the myotome. The remainder of the dermomyotome is now called the dermatome. **(d)** The dermatome cells become mesenchymal and spread out under the epidermis to form the dermis.

After myotome formation, the remainder of the dermomyotome is called the ***dermatome*** because it will give rise to the *dermis* (Gk. *dermatos*, "skin"), the mesoderm-derived part of the skin. Like sclerotome cells, dermatome cells become mesenchymal and migrate. However, instead of surrounding notochord and neural tube, dermatome cells come to underly the epidermis.

SOMITES ARE PATTERNED BY SIGNALS FROM SURROUNDING ORGAN RUDIMENTS

How do the various parts of a somite acquire their specific fates? Specifically, at which stage are somitic parts *determined* to develop in accord with their fates, and what are the key signals in these determination events?

In order to test the determined state of somitic regions, their *potency* has been assessed in *heterotopic transplantations*. (Remember that the fate maps of somites were established by *homotopic transplantations*.) These experiments have indicated that the cells of a young somite are largely undetermined. For example, if the medial half of a young somite is used to replace the lateral half of a somite, then the graft develops according to its new location, contributing to limb muscles rather than dorsal trunk muscles (Ordahl and Le Douarin, 1992). Conversely, a lateral somitic half from a young somite transplanted into a medial position will form medial derivatives including dorsal trunk muscles and

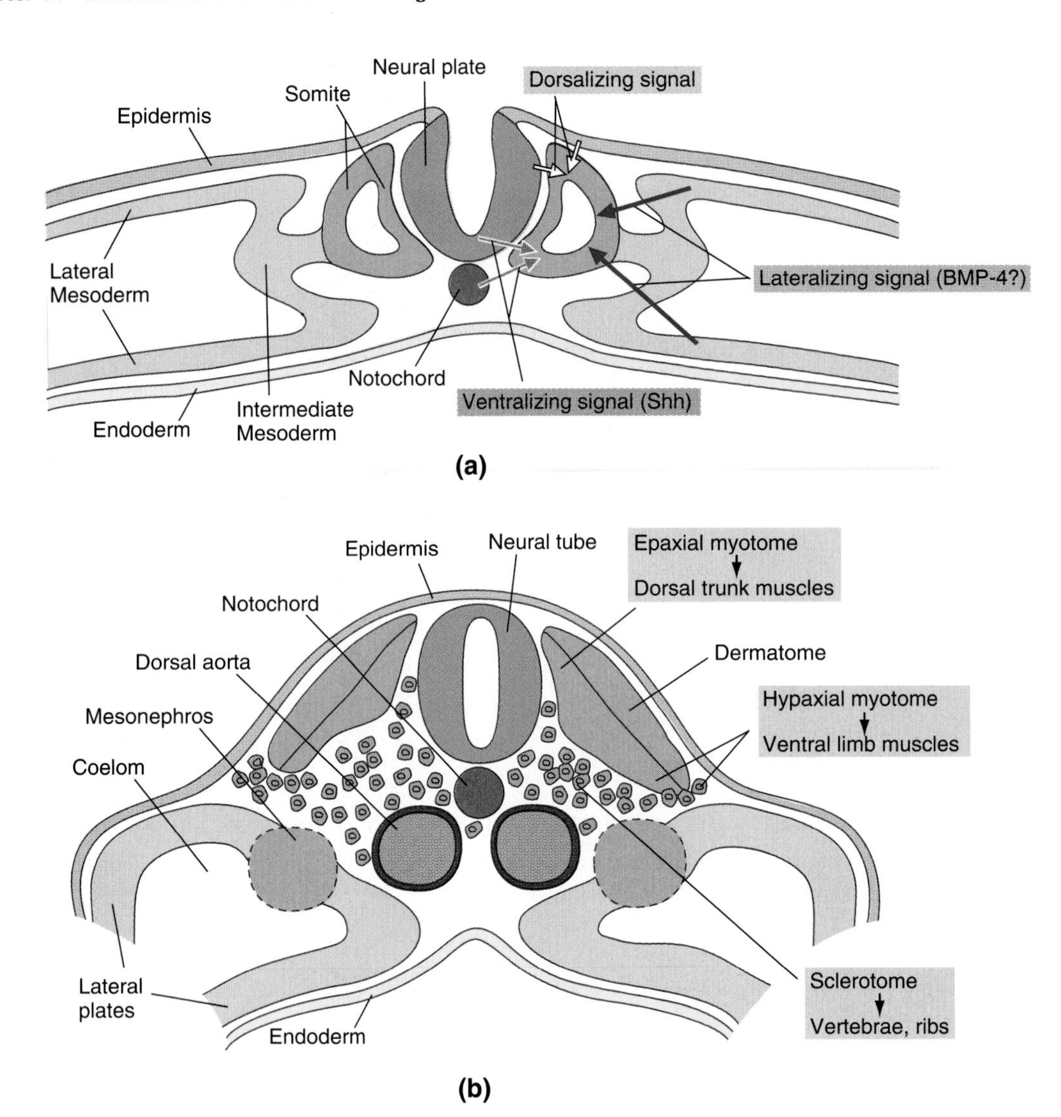

Figure 14.18 Model of transverse pattern specification in somites by signals from surrounding organ rudiments. Part **(a)** shows the known signals, and part **(b)** identifies the major elements of the somitic pattern with highlighted labels. A ventralizing signal from notochord and floor plate, most likely Sonic hedgehog (Shh) protein, determines sclerotome, which forms vertebrae and ribs. Dorsalizing signals from dorsal neural tube and epidermis determine primarily epaxial myotome, which develops into dorsal trunk muscles. A lateralizing signal, originating from lateral plate mesoderm and possibly carried by bone morphogenetic protein 4 (BMP-4), determines primarily hypaxial myotome, which gives rise to skeletal muscles of the ventral trunk and limbs. All three signals interact, specifying the transverse pattern of the somite through their relative concentrations.

sclerotome. However, when the same experiments are carried out except that older somitic halves are transplanted, then the transplants are biased to develop corresponding to their original location, which means in accord with their fate. These and other results indicate that the different parts of a somite are determined during its early development. The same results also show that the determination of distinct somitic parts may depend on adjacent organ primordia.

A role for notochord and neural tube in the patterning of somites was revealed by experiments in which both neural tube and notochord were removed from chicken embryos. This led to the disintegration of somites and subsequent lack of dorsal trunk muscles while the limb muscles developed normally (Teillet and Le Douarin, 1983; Rong et al., 1992). These results show that notochord and/or neural tube are necessary for somite integrity and for the development of dorsal trunk muscles. However, it is not clear from these results whether the epaxial myotome cells—the precursors of dorsal trunk musculature—do not survive without notochord and/or neural tube or whether they change fate and perhaps join the hypaxial myotome cells in forming limb or ventral body muscles. To distinguish between these possibilities, Ordahl and Le Douarin (1992) removed notochord and neural tube from chicken embryos and also replaced medial somitic halves on one side of the operated area with medial somitic halves from quail. All somites that were not adjacent to notochord/neural tube disintegrated, as they had done in the earlier experiments. In addition, the researchers did not find any surviving quail cells in the host chicken, suggesting that medial somitic cells undergo cell death if they are deprived of the normally adjacent notochord/neural tube.

In subsequent experiments, many researchers have tested dorsal ectoderm, neural tube, and notochord for their roles in specifying the dorsoventral pattern of developing somites (Dietrich et al., 1997, and literature quoted therein). In the case of the neural tube, ventral and dorsal parts were tested separately and by dorsoventral inversion of whole neural tubes. If there were indications that some parts were sending redundant signals, these parts were tested singly and in combination. All signaling parts were tested by removal and by heterotopic transplantation. For many of these microsurgical studies there are corresponding genetic analyses using loss-of-function alleles in lieu of surgical removals or overexpression of genes in lieu of heterotopic transplantations. In similar experiments, other investigators have tested the effect of lateral mesoderm on the mediolateral pattern of developing somites (Pourquié et al., 1995). The results of these investigations are summarized in Figure 14.18.

The transverse pattern of the somite is controlled by signals from three directions: Signals from the notochord and the floor plate of the neural tube ventralize (and somewhat medialize) the somite. Signals from dorsal epidermis and the roof plate dorsalize the somite. Signals from lateral plate lateralize the somite. The ventromedial signal from notochord and floor plate acts in a dose-dependent fashion, with high doses inducing sclerotome and lower doses inducing myotome. This signal is carried by Sonic hedgehog (Shh) peptide, the same molecule that acts as a ventralizing morphogen in the neural tube as discussed in Section 13.1. Shh is both sufficient and necessary to activate key regulatory genes for sclerotome and myotome, as shown by experiments using Shh-coated beads as well as anti-sense inhibition of Shh mRNA (Borycki et al., 1998). The dorsalizing signals from roof plate and dorsal epidermis are antagonistic to the ventromedial signal. The relative levels of dorsal and ventromedial signal control the identity assumed by somitic cells. Candidate molecules for the dorsal signal include members of the Wnt family (Marcelle et al., 1997; Capdevila et al., 1998; Kispert et al., 1998). An inhibitory signal from the lateral plate is defining the ventral border of hypaxial myotome. A candidate molecule for this signal is bone morphogenetic protein 4 (BMP-4), the same molecule discussed earlier as an antagonist of Spemann's organizer (see Section 12.5). Extending the parallelism with the organizer, the effect of BMP-4 on the hypaxial myotome is in turn antagonized by factors including noggin, which is made by dorsomedial cells of the dermomyotome (Tonegawa et al., 1997; Reshef et al., 1998).

14.3 Connective Tissue and Skeletal Muscle

Once the transverse pattern of somites is specified, its components begin two of the most spectacular processes in histogenesis, the formation of connective tissue and muscle tissue.

CONNECTIVE TISSUE CONTAINS LARGE AMOUNTS OF EXTRACELLULAR MATRIX

Several types of vertebrate tissue, including cartilage, bone, tendons, and adipose tissue, are collectively called ***connective tissue.*** These tissues are characterized by large amounts of *extracellular matrix,* the specific composition of which determines the mechanical properties of the tissue. The extracellular matrix consists of two types of molecules, which are linked together covalently in a meshwork (see Section 11.3). Molecules of one type, called *proteoglycans,* attract water and form a gelatinous mass. The other molecules are proteins that assemble into fibrils reinforcing the proteoglycans and limiting their expansive properties.

All extracellular matrix components are synthesized locally by connective tissue cells. The most common type of connective tissue cell is the *fibroblast* (Fig. 14.19). Fibroblasts are migratory cells that are prevalent in embryonic mesenchyme and play an important part in wound healing. In vitro they are convertible into other types of connective tissue cells, such as ***osteoblasts*** (bone-forming cells), ***chondrocytes*** (cartilage cells), and ***adipocytes*** (fat cells). In addition, fibroblasts can form smooth muscle cells, as we will discuss later in this chapter. Genetically manipulated fibroblasts can also form skeletal muscle (see Section 20.5). For the remainder of this section, we will consider the formation of two abundant connective tissues: cartilage and bone.

Cartilage tissue originates from mesenchyme in several regions of the embryo, including the skull, vertebrae, and limbs. The chondrocytes secrete large amounts of extracellular matrix, which changes in composition as the cartilage matures. Cartilage grows on the inside when chondrocytes divide and secrete more matrix, and grows on the outside by converting adjacent fibroblasts into chondrocytes. As the amount of extracellular material increases, the chondrocytes become more widely separated (Fig. 14.20). Typical cartilage contains no blood vessels; the chondrocytes are sustained instead by diffusion of materials through the extracellular matrix.

The formation of bone tissue involves different types of cells. *Osteoblasts* lay down the extracellular bone matrix and eventually become entrapped in it as ***osteocytes.*** The extracellular matrix produced by osteoblasts has a tendency to calcify instead of taking up water as cartilage does. Another part of bone development is bone removal, a function carried out by multinucleate

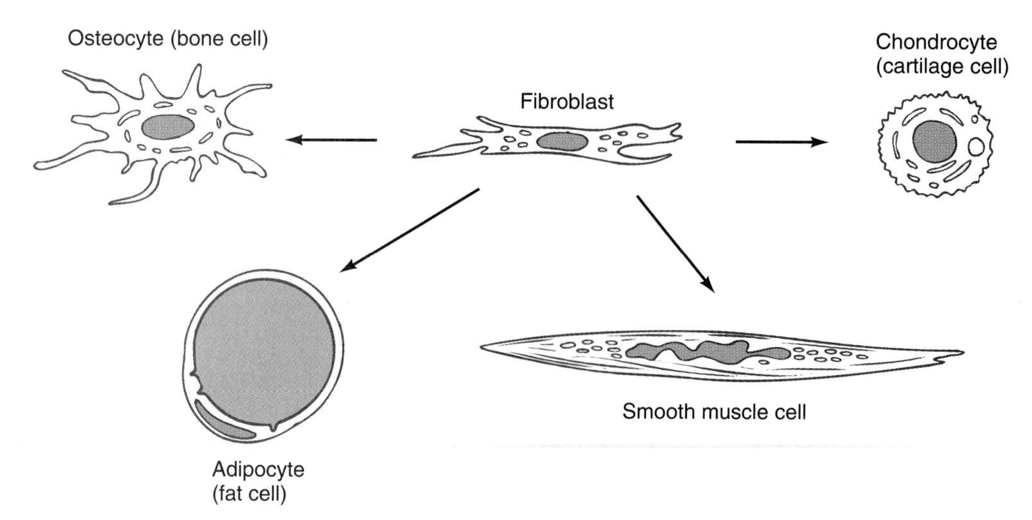

Figure 14.19 Vertebrate cells that can be derived from fibroblasts in vitro. For simplicity, only one type of fibroblast is shown, although there may actually be several types, perhaps with restricted potentials.

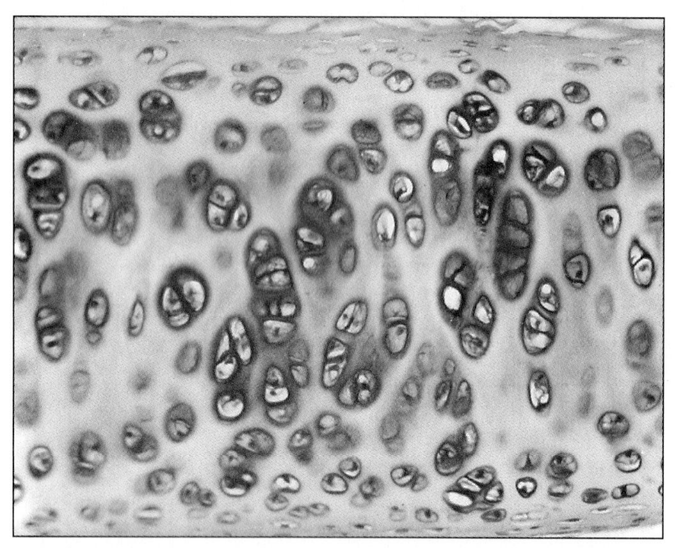

Figure 14.20 Cartilage from growing human bone shown in a light micrograph of a fixed and stained section. The cartilage cells, or chondrocytes, are embedded in abundant extracellular material. The round or oval cavities in the extracellular material are called lacunae. A living chondrocyte completely fills its lacuna, but the fixation process causes chondrocytes to shrink.

cells called ***osteoclasts.*** They belong to the family of blood cells, as do the bone marrow cells.

The growth and maintenance of bone requires a delicate balance between the deposition of new matrix and the degradation of previously deposited material. Typically, a healthy mammal destroys and regenerates 5 to 10% of its bone every year. This constant remodeling makes it possible for bone to respond to external pressure or tension with the formation of properly oriented beams or ridges of extracellular material known as ***trabeculae*** (Lat. *trabecula,* "little beam"). If bone degradation outpaces the deposition of new bone material, bones become fragile, a condition known as ***osteoporosis.***

Bone tissue originates in two ways. The flat bones of the skull and face are formed directly from mesenchyme in a process called ***membranous ossification*** (Fig. 14.21). It begins with the assembly of type I collagen, a fibrous extracellular matrix component (see Section 11.3). Osteoblasts line up along the fibrils and synthesize a matrix that hardens from deposition of calcium phosphate crystals. The resulting *trabeculae* eventually interconnect to form a meshwork of bone, with marrow filling the intervening spaces.

For most other bones, including the vertebrae and in the long bones of limbs, chondrocytes first form a cartilage model. Subsequently, the cartilage matrix is replaced in a process known as ***endochondrial ossification.*** The chondrocytes swell and die, leaving large cavities. Osteoclasts and cells that form blood vessels invade the cavities and erode the residual cartilage, while osteoblasts deposit bone matrix, which becomes mineralized from the deposition of calcium phosphate crystals. In the long bones of newborn vertebrates, cartilage is formed at the ends of the bones, called ***epiphyses,*** while mineralization occurs in the central part of the bone, the ***diaphysis*** (Fig. 14.22). In juvenile vertebrates, the terminal ends of the epiphyses also mineralize while subterminal portions, the ***epiphyseal discs,*** remain cartilaginous until longitudinal growth ceases. In adults, only thin layers of cartilage remain at those surface areas of bones that form joints.

SKELETAL MUSCLE FIBERS ARISE THROUGH CELL FUSION

Muscle cells, like many other differentiated cells, have taken a general cellular feature to a high degree of specialization. A contractile system involving the proteins actin and myosin is present in most animal cells, and families of actin and myosin genes encode different

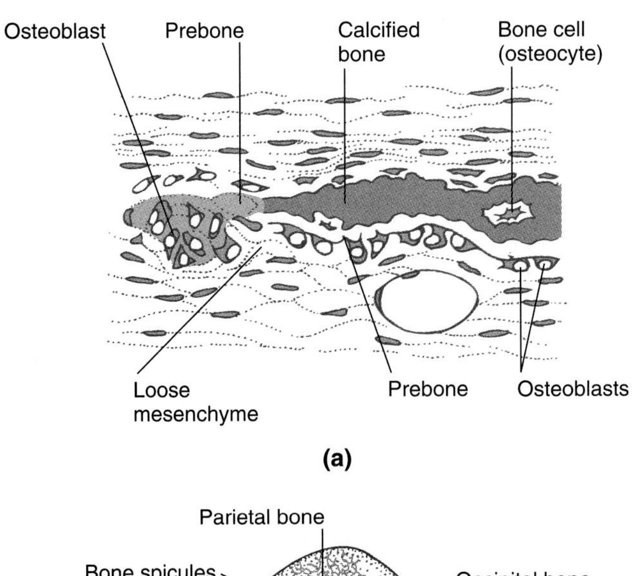

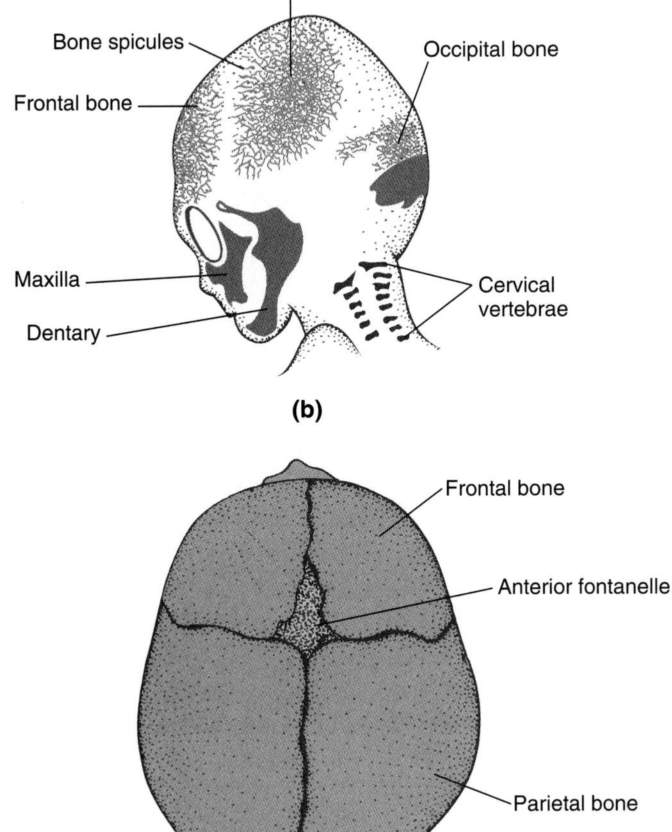

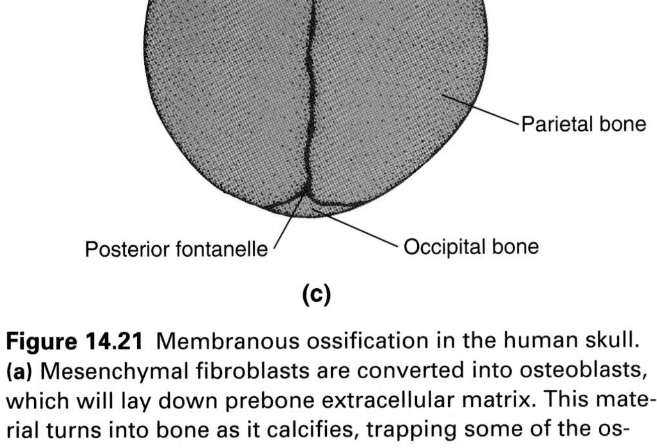

Figure 14.21 Membranous ossification in the human skull. **(a)** Mesenchymal fibroblasts are converted into osteoblasts, which will lay down prebone extracellular matrix. This material turns into bone as it calcifies, trapping some of the osteoblasts as osteocytes. **(b)** Bones of the skull in a 3-month-old human embryo. The webbed areas represent bone spicules that originate by membraneous ossification and grow together to form the flat bones of the skull. **(c)** Skull of a newborn seen from above. Note the open areas (fontanelles) where the flat bones are incompletely closed.

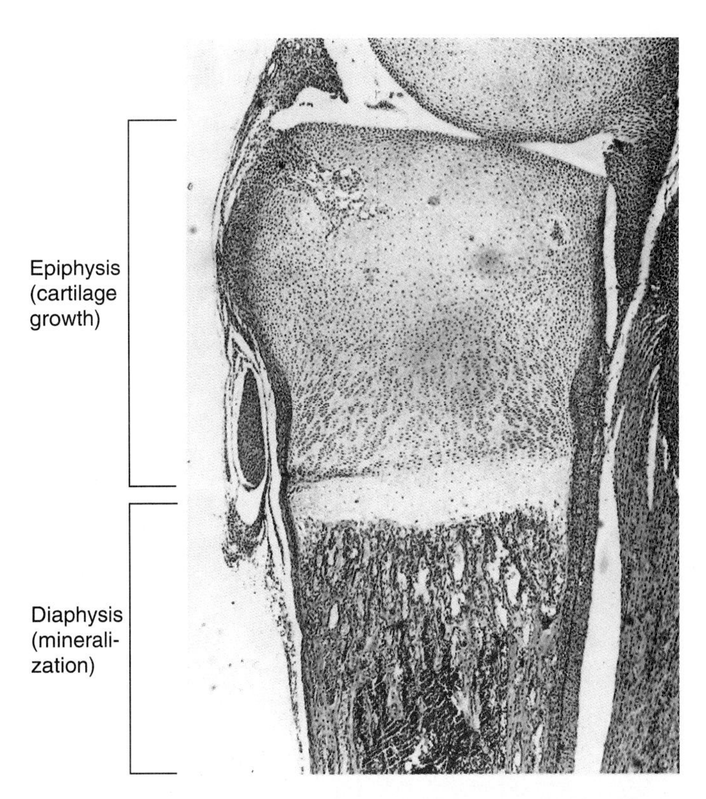

Figure 14.22 Bone formation by endochondrial ossification in the newborn rat. The light micrograph shows the proximal (upper) end of the tibia (shin bone) in longitudinal section. In the cartilaginous ends of the bone (epiphyses), chondrocytes proliferate and grow in size. In the middle part of the bone (diaphysis), extracellular matrix laid down by the chondrocytes hardens by deposition of calcium crystals (mineralization).

versions of these proteins that are optimized for different functions. However, muscle cells have developed especially large and organized filament systems composed of certain actins and myosins along with other molecules that control muscle contraction. *Skeletal* muscle tissue develops from the *myotome* portions of the somites (see Figs. 14.17 and 14.18). Other types of muscle tissue, such as *smooth muscle* and *cardiac muscle*, derive from lateral plates.

The myotome cells that form skeletal muscle are called ***myoblasts.*** After a period of proliferation, myoblasts fuse with one another to form multinucleate ***myotubes*** (Fig. 14.23). Myotubes develop into ***muscle fibers*** by recruiting additional myoblasts and by assembling a characteristic arrangement of actin and myosin fibrils. Muscle fibers are the building blocks of skeletal muscle; they are about 0.1 mm in diameter and can be half a meter long in humans. Muscle fibers are contractile, have a characteristic striated appearance, and are surrounded by a basement membrane.

The process of skeletal muscle fiber formation, called ***myogenesis,*** has been studied in vitro and in vivo. In vitro, myoblasts migrate and form chains before they fuse (Nameroff and Munar, 1976). Chain formation is

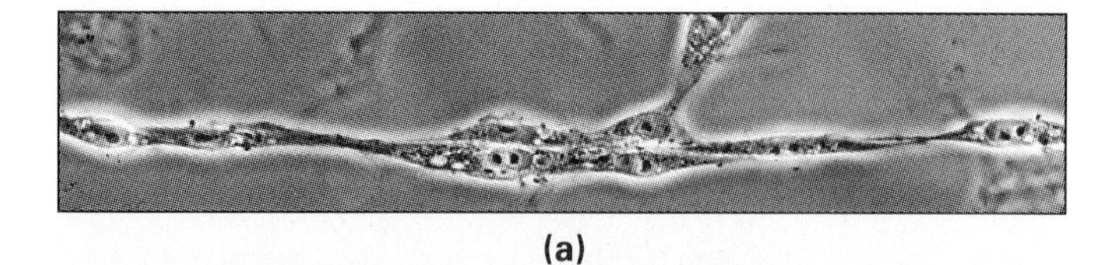

(a)

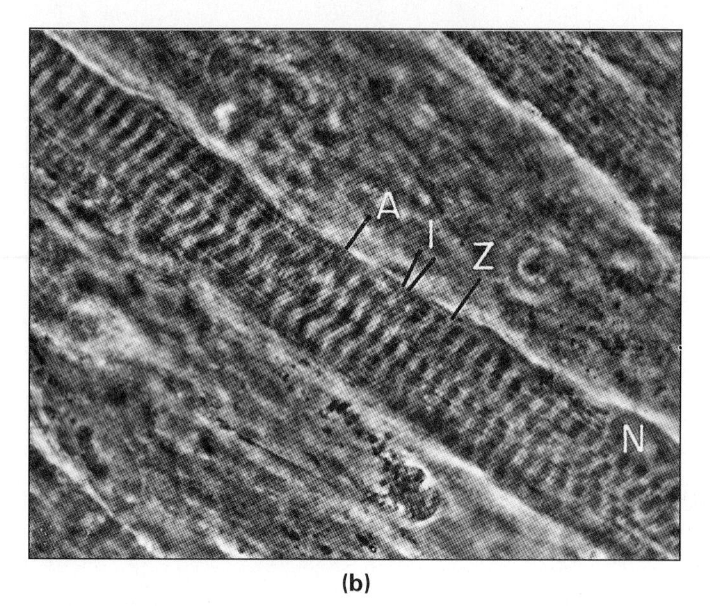

(b)

Figure 14.23 Myogenesis of embryonic cells in tissue culture. **(a)** After 2 days of culture, myoblasts line up to fuse and form myotubes. **(b)** After 12 days of culture, the cytoplasm of myofibrils shows the characteristic cross striation (A, I, and Z bands) of skeletal muscle fibers. The position of the nucleus (N) at the periphery is also typical of advanced differentiation.

sufficient to stop further mitoses of myoblasts. Actual cell fusion seems to require some recognition event between myoblasts, since they do not fuse with other cell types. The molecular control of myogenesis has been examined in great detail and will be discussed as an example of cell differentiation in Section 20.5.

The adult number of muscle fibers is attained early in development, before birth in mammals. The subsequent increase in muscle mass is achieved mostly through the enlargement of existing muscle fibers. They grow in length by recruiting more myoblasts at their ends. Growth in girth occurs in response to training and depends mainly on an increase in the number of actin and myosin fibrils per muscle fiber. However, a few myoblasts persist in contact with the mature muscle fiber and inside its basal lamina. If a muscle is damaged, these myoblasts can be reactivated to proliferate, and their descendants can fuse to form new muscle fibers.

14.4 Intermediate Mesoderm

The *intermediate mesoderm* is located dorsolaterally between the somites and the lateral plates (see Fig. 14.12). It gives rise to the kidneys and most of the reproductive structures (Saxén, 1987). The principal function of the kidneys is to eliminate waste, in particular nitrogen-containing waste such as urea. This is achieved in part by *excretion,* using specific *membrane transport proteins* that take up specific waste products from blood serum and deliver them to the kidney's duct system. Alternatively, or in addition, kidneys work by combining *filtration* with *selective reabsorption.* Here, kidneys first generate an ultrafiltrate containing all components of blood serum except for proteins. From this ultrafiltrate, sugars, amino acids, and other molecules of value are reabsorbed into the bloodstream, again through the use of specific transport proteins. Filtration depends on high blood pressure and an adequate supply of water but has the advantage of eliminating *any* small and water-soluble molecule unless it is specifically recognized by the transport proteins that reabsorb known valuables.

Embryologists distinguish three types of vertebrate kidney (Fig. 14.24), which are referred to as the *pronephros* (Gk. *pro,* "before"; *nephros,* "kidney"), the *mesonephros* (Gk. *mesos,* "middle"), and the *metanephros* (Gk. *meta,* "after"). The pronephros is formed in the neck region of all vertebrate embryos but persist to adulthood only in a few fishes. The mesonephros is formed over much of the length of the trunk in the embryos of almost all vertebrates; it persists to adulthood in most fishes and amphibians. The metanephros originates most posteriorly and becomes the definitive kidney of reptiles, birds, and mammals.

The ***pronephros*** is the most primitive type of vertebrate kidney (Fig. 14.25). Here, a blood filtration unit is usually present but separated from the reabsorption unit by a portion of *coelom,* the fluid-filled body cavity located between the two layers of lateral mesoderm (see Fig. 14.24 a, d). The filtration unit projects into the coelom and is supplied with blood by an arteriole branching off the dorsal aorta. Such a unit is called a *glomerulus* (Lat. *glomerulus,* "little ball") if it is one body segment long and a *glomus* if it occupies two or more body segments. The blood ultrafiltrate released into the coelom is swept by ciliary beat into a funnel called the *nephrostome.* The latter is the beginning of the reabsorption unit, called the *pronephric tubule;* it is surrounded by a blood sinus, where valuable molecules are reabsorbed from the tubular fluid and returned to the bloodstream. The fluid that remains in the tubule is collected by the ***pronephric duct,*** which links the pronephros to the *cloaca.* Pronephros and pronephric duct arise from consecutive portions of the intermediate mesoderm (Fig. 14.25). The tip of the duct grows posteriorly in a movement that involves active migration of duct cells guided by cues from the overlying epidermis (Drawbridge et al., 1995).

The ***mesonephros*** develops after the pronephros and caudal to it (see Fig. 14.24b, c). The mesonephros consists of many segmental units called ***nephrons.*** In a typical nephron, a *nephric tubule* begins blindly in a

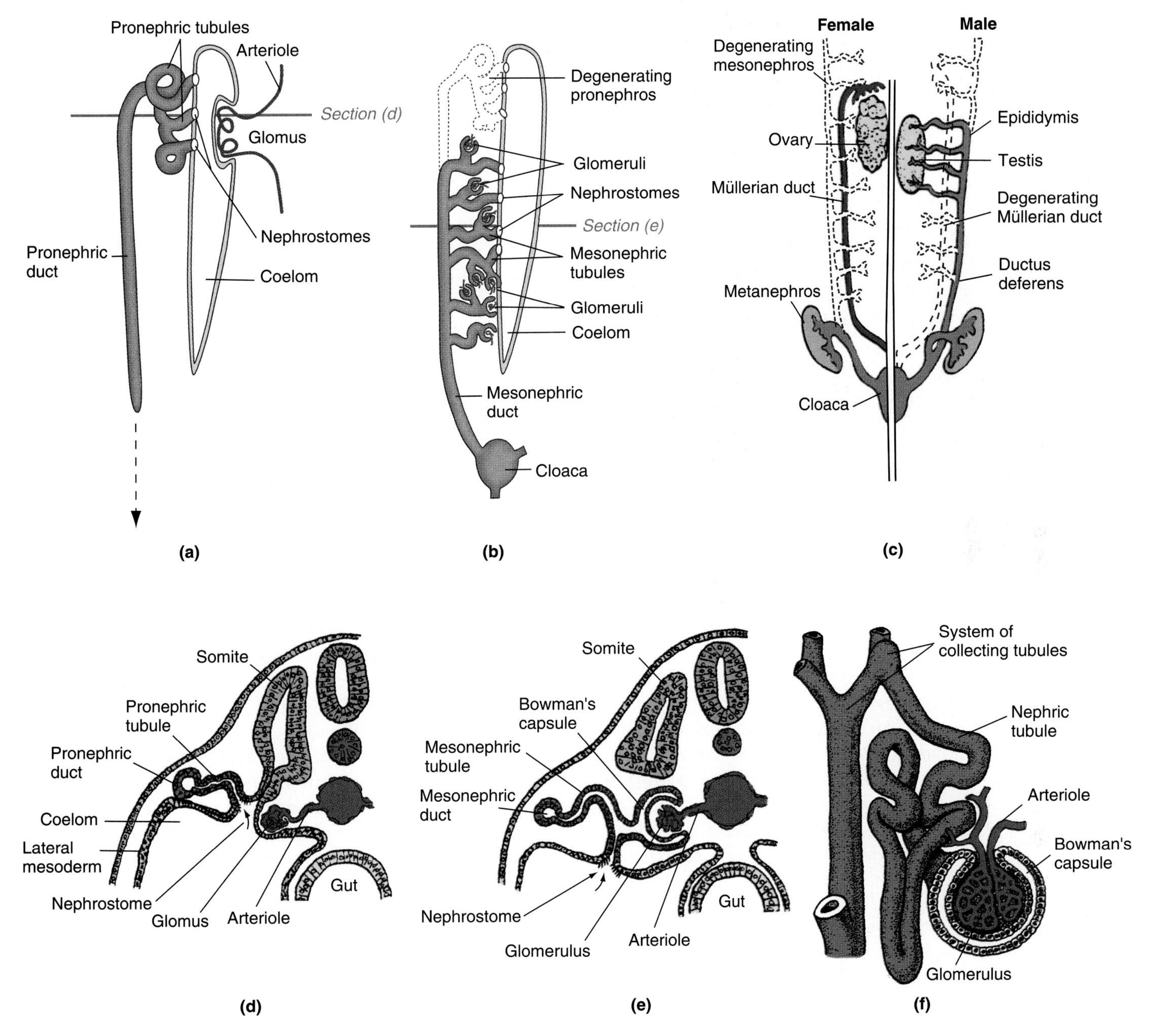

Figure 14.24 Three types of kidney in vertebrates. **(a, d)** Pronephros. An ultrafiltrate of blood supplied by an arteriole is released into the coelom. Coelomic fluid is driven by ciliary beat into nephrostomes, funnel-shaped openings of pronephric tubules. After reabsorption of known valuable molecules, coelomic fluid is collected from the pronephric tubules into the pronephric duct. The tip of the pronephric duct is shown here growing toward the cloaca. **(b, e)** Mesonephros. Anterior nephrons are connected by nephrostomes with the coelom, as in the pronephros. In addition, all nephrons begin blindly with a Bowman's capsule surrounding a glomerulus—that is, a convoluted blood capillary. **(c, f)** Male and female urogenital systems. In the male, mesonephric tubules and mesonephric duct are converted into epididymis and ductus deferens (or Wolffian duct). In the female, the mesonephros degenerates, and the parallel Müllerian duct forms much of the female reproductive tract. The metanephros develops—similarly in both sexes—from an outgrowth of the mesonephric duct and posterior intermediate mesoderm.

cup-shaped depression known as *Bowman's capsule.* Lodged inside this capsule, without an intervening coelom, is a *glomerulus,* a convolute of small capillaries fed by an arteriole. Depending on species and stage of development, the anterior tubules of a mesonephros—in addition to originating from a Bowman's capsule—may be connected to the coelom by branches with nephrostomes, similar to pronephric tubules. The tubular portion of each nephron is surrounded by a plexus of fine capillaries, which return reabsorbed molecules into a

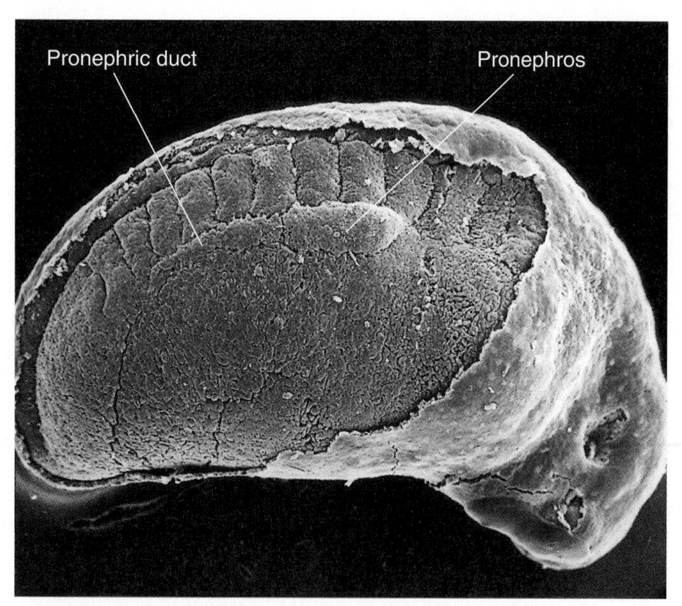

Figure 14.25 Pronephros of a salamander embryo. For this scanning electron micrograph, the epidermis has been removed to expose the underlying mesoderm. Intermediate mesoderm gives rise to the pronephros and the pronephric duct.

system of veins. Animals with mesonephric kidneys living in environments that withdraw much water from their bodies, such as deserts or oceans, rely more on tubular excretion than on filtration and reabsorption. The *mesonephric tubules* hook up to the pronephric duct, which is then called the *mesonephric duct* or ***Wolffian duct***. In the human embryo, the mesonephros forms an elongated organ on the inside of the dorsal body wall (see Fig. 3.3). The mesodermal portion of the gonad rudiment, the *genital ridge*, develops medially and ventrally to the mesonephros.

The ***metanephros*** consists of a very large number of nephrons, which are similar to their mesonephric counterparts (see Fig. 14.24c, f). However, they are arranged in a way that allows the recovery of almost all water from the blood filtrate by a countercurrent exchange mechanism. The metanephros arises from two embryonic rudiments. One rudiment is the ***ureteric bud***, a small outgrowth of the mesonephric duct near the cloaca, which gives rise to the ***ureter***, the ***renal pelvis*** and its branches, or ***calyces***, and ***collecting tubules*** (Figs. 14.24c and 14.26). The other rudiment is the caudal portion of the intermediate mesoderm, called the ***metanephrogenic mesenchyme***, which forms *nephrons*. The coordinate development of ureteric bud and metanephrogenic mesenchyme illustrates the *principle of reciprocal interactions*, which will be discussed in the following section.

As the metanephros develops, the mesonephros degenerates, except that some of the mesonephric tubules and the Wolffian duct contribute to the male reproductive system. The ***Müllerian duct***, originally formed parallel to the Wolffian duct in both sexes, degenerates in males but gives rise to parts of the female reproductive system (see Section 27.2).

The more evolved vertebrates go through the pronephric and mesonephric kidney forms before they develop and maintain metanephric kidneys. This recapitulation of phylogenetically old embryonic characters, like other recapitulations discussed earlier in this chapter, may be taken as yet another indication that it can be easier to modify existing sets of genetic control circuits than to generate new ones. However, the temporary conservation of the older kidney types may have provided functional benefits as well (Vize et al., 1997). The needs to remove waste and control osmotic pressure arise early in development, whereas the metanephros takes much longer time to develop than pronephros and mesonephros. The older kidney forms, with their open connections to the coelom and cilia-driven fluid movement, also function at the low blood pressures that prevail in early embryos, whereas the metanephros depends on higher blood pressure. Thus, pronephros and mesonephros allow embryos of the more evolved vertebrates to survive long enough for the more complex and efficient metanephric kidneys to develop.

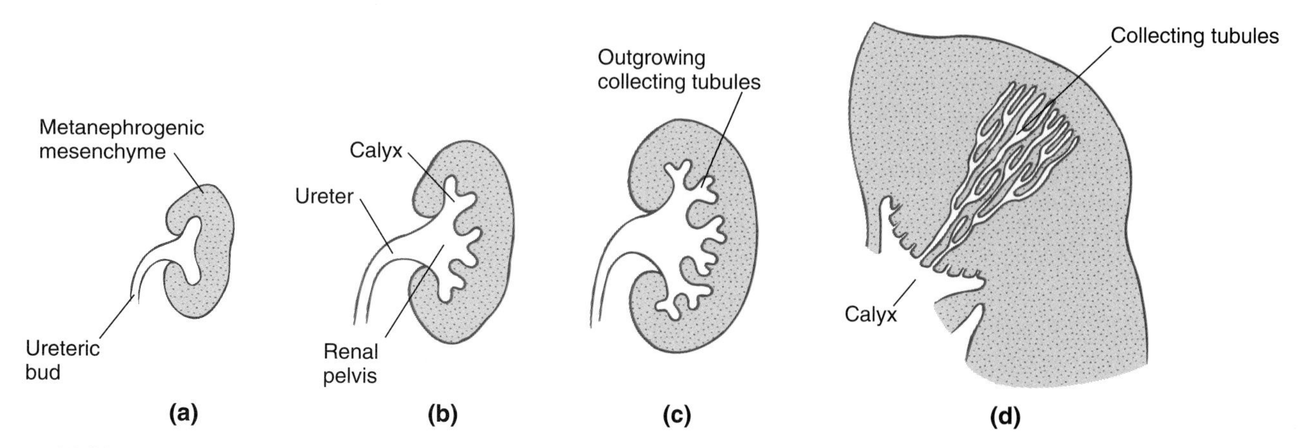

Figure 14.26 Metanephros development in the human embryo: **(a–c)** 5–7 weeks; **(d)** newborn. The ureteric bud gives rise to the structures that drain urine from the kidney: ureter, renal pelvis, calyces, and collecting tubules.

14.5 The Principle of Reciprocal Interaction

When two singers prepare to perform a duet in an opera, they will usually begin by practicing their parts separately but eventually rehearse together so that they can learn to pick up their cues from each other and adjust to each other's nuances. Similarly, many organs are constituted from more than one rudiment, which originate separately before they grow together and exchange inductive signals that coordinate their later phases of development. The eye, for example, originates from two major components: the *optic vesicle,* which gives rise to the optic cup and then the retina, and the *lens placode,* which forms the lens (see Fig. 1.16). One of the signals that contribute to lens induction comes from the optic vesicle. As a result, the lens placode is centered over the optic vesicle, and the developing lens fits the size of the underlying optic cup. However, there is not just one unidirectional signal from the optic vesicle that induces the overlying ectoderm to form a lens placode. Rather, the two rudiments exchange signals back and forth. This was shown by *heterospecific transplantations,* in which epidermis from a large-eyed donor species was transplanted over the optic vesicle of a small-eyed host species. The lens rudiment formed by the graft initially was too big for the host's optic cup, but as the chimeric eye grew, the lens slowed down in its growth while the optic cup grew larger than normal. The resulting eye was well-proportioned and intermediate in size between normal donor and host eyes (see Section 28.2).

The example described above illustrates the ***principle of reciprocal interactions,*** that is, the ability of parts of a developing organ to exchange signals that coordinate their development in space and time.

The development of the metanephros is another case illustrating the same principle (Saxén, 1987; Vize, 1997). As a first step, *metanephrogenic mesenchyme* induces the adjacent segment of the *Wolffian duct* to form the *ureteric bud.* As the bud enters the mesenchyme, it begins to branch (Fig. 14.26). The signal causing the branching seems to be a peptide growth factor that is secreted by the mesenchyme and acts on a tyrosine kinase receptor present on the bud epithelium (Robertson and Mason, 1997). In turn, the branching bud epithelium keeps adjacent metanephrogenic mesenchyme cells alive and causes them to condense and form nephric tubules (Fig. 14.27). Continued interaction between epithelium and mesenchyme causes the ureter to branch seven or eight times, generating the hierarchical system of excretory tubules. Simultaneously, the mesenchyme forms nearly 1 million nephrons in a human kidney (H. W. Smith, 1951). In addition to nephric tubules, the mesenchyme generates a supporting connective tissue known as *stroma.* The role of the stroma in kidney development is not well characterized, but seems required for the nephrons' further differentiation.

The transformation of metanephrogenic mesenchymal cells into tubular epithelium involves major changes in the cell adhesion, substrate adhesion, and extracellular matrix (ECM) composition. The formation of epithelial *zonula adherens* junctions (see Fig. 11.20) is associated with the synthesis *of E-cadherin* as usually. *Syndecan,* an adhesive proteoglycan (see Fig. 11.17), is synthesized heavily during the condensation of metanephric mesenchyme around the branches of the ureteric bud. Here, syndecan appears not only to enhance cell adhesion but also the proliferation of mesenchymal cells, presumably through syndecan's ability to sequester growth factors (Vaino et al., 1992). Some typical components of mesenchymal ECM, such as *fibronectin* and *collagen types I and III,* disappear while basement membrane components, including *collagen type IV, laminin,* and *nidogen,* are being synthesized cooperatively by mesenchymal and epithelial cells (Ekblom et al., 1994). Some or all of these activities may be coordinated by Pax-2, a gene-regulatory protein synthesized in both cell types (Torres et al., 1995).

Reciprocal interactions between epithelia and mesenchyme are observed in the development not only of

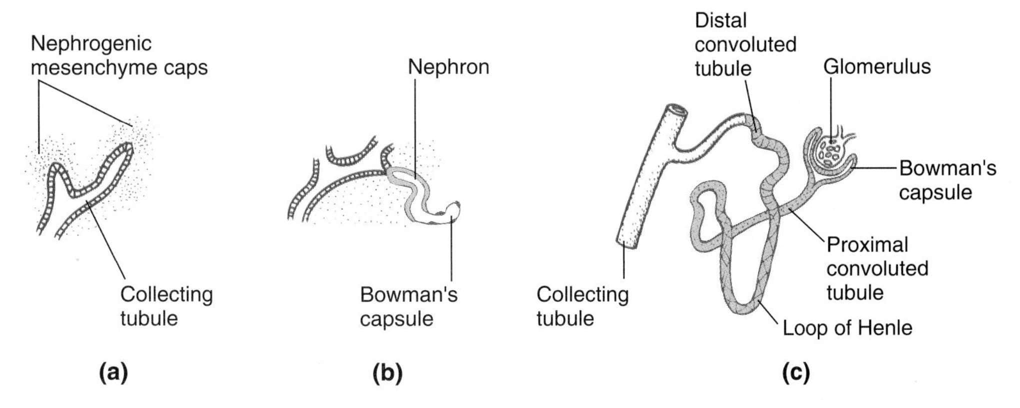

Figure 14.27 Nephron development in the metanephros. The tips of the collecting tubules induce caps of nephrogenic mesenchyme to form nephrons. The nephrons establish open communication with the collecting ducts, enabling urine to flow from Bowman's capsule to the renal pelvis.

eye and kidney but also lung and many glands. Other reciprocal interactions, again between epithelium and mesenchyme as partners, will be discussed in the context of limb formation (see Section 23.5).

14.6 Lateral Plates

The lateral mesoderm occupies a lateral or ventral position in the body (see Fig. 14.12). Very early in development, it splits into two layers known as the lateral plates. The ***somatic layer*** develops in association with the ectoderm, while the ***visceral layer*** develops in association with the internal organs. The space between the two layers is called the embryonic *coelom*. Thus, the lateral plates form the lubricated membranes that line the body cavities derived from the coelom. A major derivative of the visceral layer is the cardiovascular system. In addition, the visceral layer gives rise to smooth muscle and connective tissues surrounding the blood vessels, intestines, and other internal organs. The somatic layer, on the other hand, contributes the connective tissues of the body wall and the limbs.

THE LATERAL PLATES SURROUND THE COELOMIC CAVITIES

The subdivision of the coelom is a complicated process and will only be summarized here. Initially, the coelom is divided into two lateral compartments separated by a median partition, the *primary mesentery*, which consists of the visceral layers from each side. As it is formed, the primary mesentery attaches the embryonic gut to the dorsal and ventral midlines. However, the ventral part of the mesentery disappears soon thereafter; what is left is called the ***dorsal mesentery*** (Fig. 14.28a). Next, a ventral anterior portion of the coelom is closed off as a small ***pericardial cavity*** surrounding the heart (Figs. 14.28b–d). The membranes of the pericardial cavity,

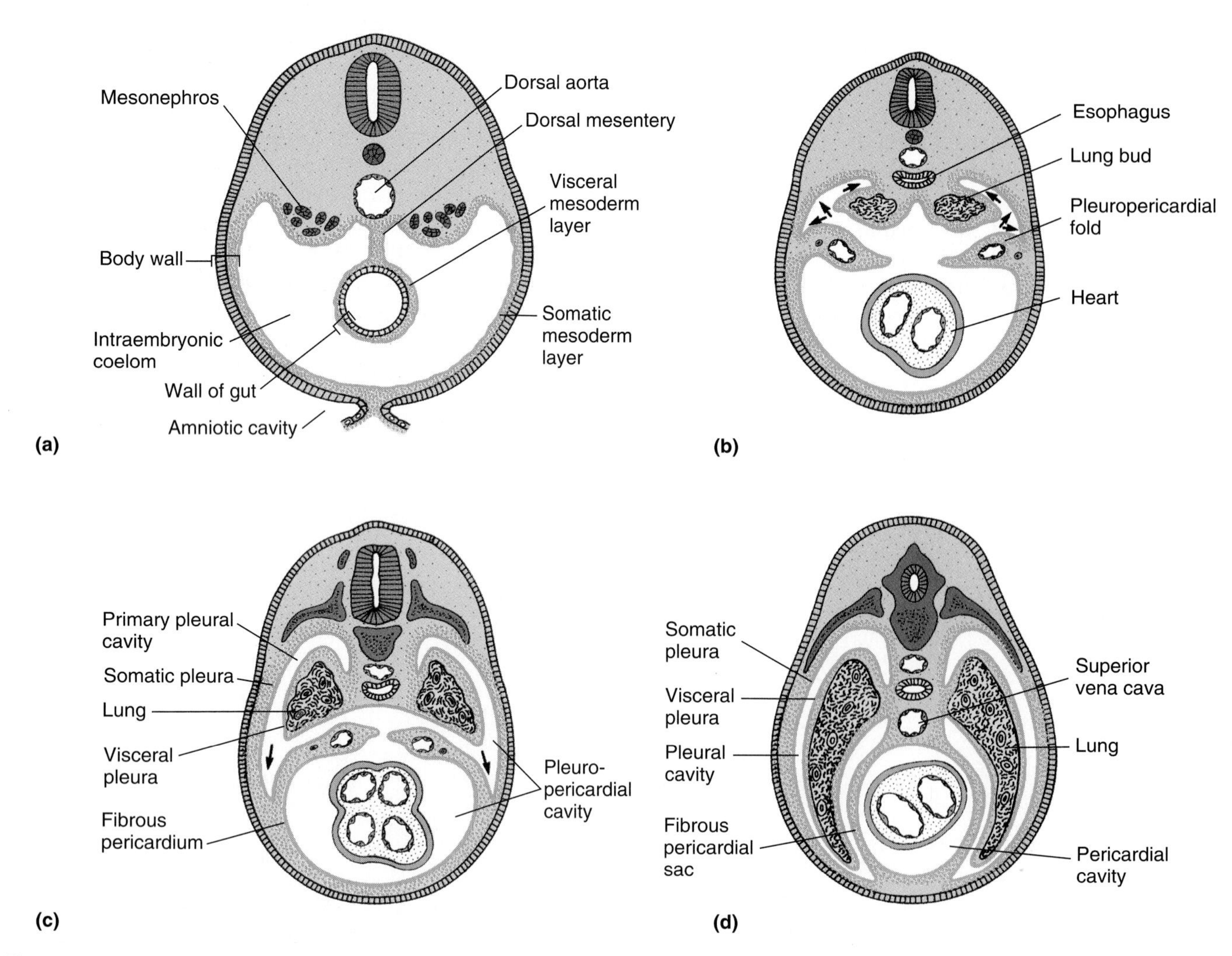

Figure 14.28 Coelomic cavities in the human embryo. **(a)** Transverse section through the abdominal region at the end of the fourth week. The ventral part of the primary mesentery has disappeared, leaving the dorsal mesentery. **(b)** Transverse section through the heart region at the end of the fifth week. The pleuropericardial folds begin to close off the pericardial cavity from the pleural cavity. **(c, d)** During subsequent stages, the growing lungs displace the pleuropericardial folds, and a separate pericardial sac is formed.

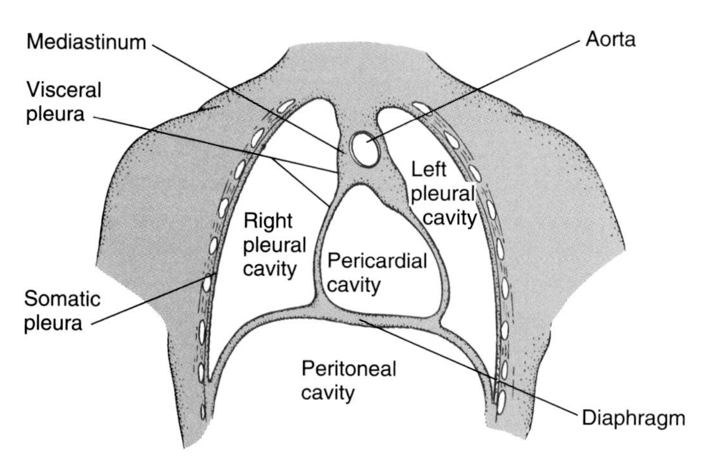

Figure 14.29 Coelomic cavities in the human adult. The abdominal (peritoneal) cavity is separated by the diaphragm from the thoracic cavities. In the thorax, two pleural cavities containing the lungs are separated from each other by the mediastinum, a mass of connective tissue that contains major blood vessels and the pericardial cavity with the heart.

together called the ***pericardium,*** consist of a visceral layer applied to the heart and a somatic layer originally applied to the body wall. Later, the somatic layer is displaced from the body wall, so that the heart comes to lie in a separate pericardial sac.

In land-dwelling vertebrates, the lungs grow out from the foregut and around the heart. The part of the coelomic cavity surrounding each lung is called a ***pleural cavity*** (Gk. *pleura,* "side"), again with the visceral layer applied to the lung and the somatic layer lining the body wall (Fig. 14.29). The pleural cavities are separated from each other and from the pericardial cavity. In mammals, these cavities are also walled off by the ***diaphragm*** from the ***peritoneal cavity*** (abdominal cavity), which contains the stomach, the remainder of the digestive tract, and other internal organs. The dome-shaped diaphragm arises from several embryonic rudiments, including skeletal muscle derived from myotomes.

THE CARDIOVASCULAR SYSTEM DEVELOPS FROM VARIOUS MESODERMAL PRECURSORS

The ***cardiovascular system,*** consisting of heart, blood vessels, and blood cells, begins to function very early in development when diffusion alone can no longer satisfy the metabolic needs of the embryo. In the chicken embryo, the heart rudiment begins to beat during the second day of incubation; the human heart rudiment begins to beat around the 23rd day of gestation.

Two principal mechanisms are distinguished during formation of the cardiovascular system; the nomenclature used to describe these mechanisms, unfortunately, is somewhat confusing. The first mechanism generates the earliest blood vessels in the embryo and is known as ***vasculogenesis*** (Lat. *vasculum,* "small vessel"; Gk. *gennan,* "to produce"). The key cells involved in this mechanism are called ***angioblasts*** (Gk. *angeion,* "vessel"; *blastos,* "germ," in the sense of "embryo"). The second mechanism operates at later stages of development and is called ***angiogenesis.*** Here, existing blood vessels sprout and send capillary branches into adjacent tissue (see Fig. 20.6).

Angioblasts originate mainly from two different embryonic regions, namely paraxial mesoderm and visceral lateral plate (Pardanaud et al., 1996). While some embryonic blood vessels take shape where the angioblasts have originated, other blood vessels are formed elsewhere by angioblasts that have undergone considerable migration. In either case, angioblasts join into a tight, squamous epithelium, which is referred to as the ***endothelium*** (Gk. *endon,* "within") because it forms not only blood capillaries but also the innermost lining of arteries and veins. In these larger blood vessels, the endothelium is surrounded by additional layers of smooth muscle and connective tissue.

An interesting variant of vasculogenesis involves *blood islands* or *angiogenic clusters,* which originate in particular in the visceral layer of the lateral plates. Within these clusters, the central cells give rise to blood cells, while the peripheral cells behave like angioblasts and form endothelium (Fig. 14.30). The close vicinity of angioblast and blood cells precursors (hematopoietic cells) in angiogenic clusters has led to the hypothesis that these two cell types may have a common progenitor, dubbed a *hemangioblast* (see Section 20.4).

The mechanisms that control the patterning of the cardiovascular system—what kind of blood vessels are formed and where—are not well understood. Both vasculogenesis and angiogenesis depend on *vascular endothelial growth factor* (*VEGF*), a protein that promotes the division and differentiation of angioblasts and even seems to attract them chemotactically, as shown by experiments described below. In mouse embryos deficient for one (out of two known) *VEGF* genes, both angiogenesis and blood island formation are abnormal, indicating a close dependence of both processes on the amounts of VEGF protein available (Carmeliet et al., 1996; Ferrara et al., 1996). A receptor for VEGF, known as flk-1, is specifically present on angioblasts, and mice deficient for this receptor die from the lack of vasculogenesis and blood island formation (Shalaby et al., 1995).

TO examine the signals that control formation of the dorsal aorta in *Xenopus* embryos, Cleaver and Krieg (1998) studied the expression patterns of VEGF and flk-1. Using in situ hybridization (see Method 8.1) of labeled probes to embryos before dorsal aorta formation, they found that VEGF mRNA is present at the highest concentration in the *hypochord,* a transient structure located just ventral of the notochord in fish and amphibian embryos (Fig. 14.31a, b).

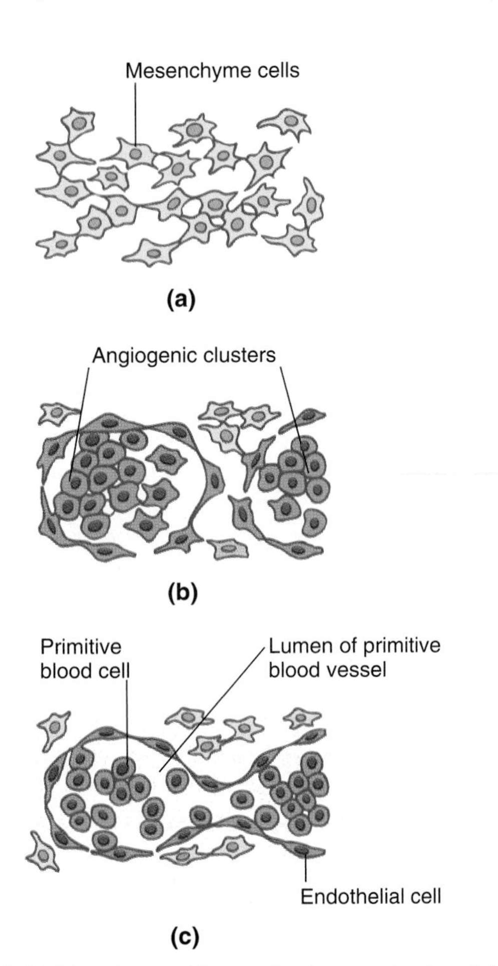

Figure 14.30 Blood vessel formation in vertebrates: **(a)** undifferentiated mesenchymal cells in the visceral layer of the lateral mesoderm; **(b)** angiogenic clusters; **(c)** primitive capillary. Note that the mesenchymal cells differentiate into either blood cells or endothelial cells.

The position in which the dorsal aorta will form is located immediately ventral to the hypchord. Shortly after the dorsal aorta has been formed, the VEGF transcripts disappear from the hypochord, and subsequently the hypochord disintegrates.

Cells expressing flk-1, the receptor of VEGF, are first detected many cell diameters away from the hypochord in lateral mesoderm (Fig. 14.31c). This is the location in which a pair of prominent embryonic veins, the *posterior cardinal veins,* will be formed. At subsequent stages, cells with high concentrations of flk-1 mRNA are observed between the positions of the future cardinal veins and the dorsal aorta (Fig. 14.31d) and then in the aorta itself once it is formed (Fig. 14.31e, f). These observations suggest that the angioblasts forming the posterior cardinal veins and those forming the dorsal aorta originate together dorsolaterally. While the former stay in position, the latter seem to migrate medially before they form the aorta.

To test whether the cells that form the aorta indeed originate from lateral plate mesoderm, the investigators removed segments of lateral plate from both sides of embryos and allowed them to develop to a stage when dorsal aorta formation is normally complete. Indeed, the operated embryos showed a normal aorta anterior and posterior to the operated segment, whereas the aorta was missing from the operated segment. Thus, lateral mesoderm is necessary for aorta formation, most likely as a source of angioblasts.

Next the researchers explored the mechanism that guides the migrating angioblasts to the site of aorta formation. Knowing that VEGF is necessary for vasculogenesis, they considered the hypothesis that VEGF serves as a chemoattractant to some of the angioblasts. A testable prediction derived from this hypothesis was that angioblasts should be diverted from their normal pathway of migration by an ectopic source of VEGF. Indeed, when cells with transgenic DNA encoding VEGF were implanted ventrally into embryos, angioblast were diverted from their normal migration route and trapped in the VEGF-producing implant.

Questions

1. To directly observe whether lateral mesoderm cells actually migrate to the site of dorsal aorta formation, the investigators did an additional experiment. What do you think the experiment may have been?
2. What may cause some angioblasts to migrate medially and form aorta while other angioblasts stay behind and form cardinal veins?

The results of Cleaver and Krieg strongly suggest that the hypochord functions as a source of VEGF, which serves as a chemoattractant to angioblasts that originate in lateral mesoderm and migrate medially, where they form the endothelium of the dorsal aorta. Other angioblasts, however, remain near their site of origin, where they form the posterior cardinal veins.

CARDIOVASCULAR SYSTEM DEVELOPMENT RECAPITULATES THE PHYLOTYPIC STAGE

The formation of organ rudiments is typically controlled by a *series* of determinative events, and the heart is no exception. In a fate map of *Xenopus* at the early gastrula stage, the heart appears as a bilateral pair of rudiments in the dorsolateral mesoderm (see Fig. 6.2). Deletion and transplantation experiments indicate that early steps of heart determination are part of the dorsalizing effect of the dorsal blastopore lip (Sater and Jacobson, 1990a). During gastrulation, involution carries the heart rudiments forward into the neck region, where they seem to receive further inductive signals from pharyngeal endoderm (Sater and Jacobson, 1990b). Further development of the heart has been analyzed especially in chicken embryos, where bilateral heart rudiments originate in similar anterolateral positions between visceral lateral mesoderm and pharyngeal endoderm.

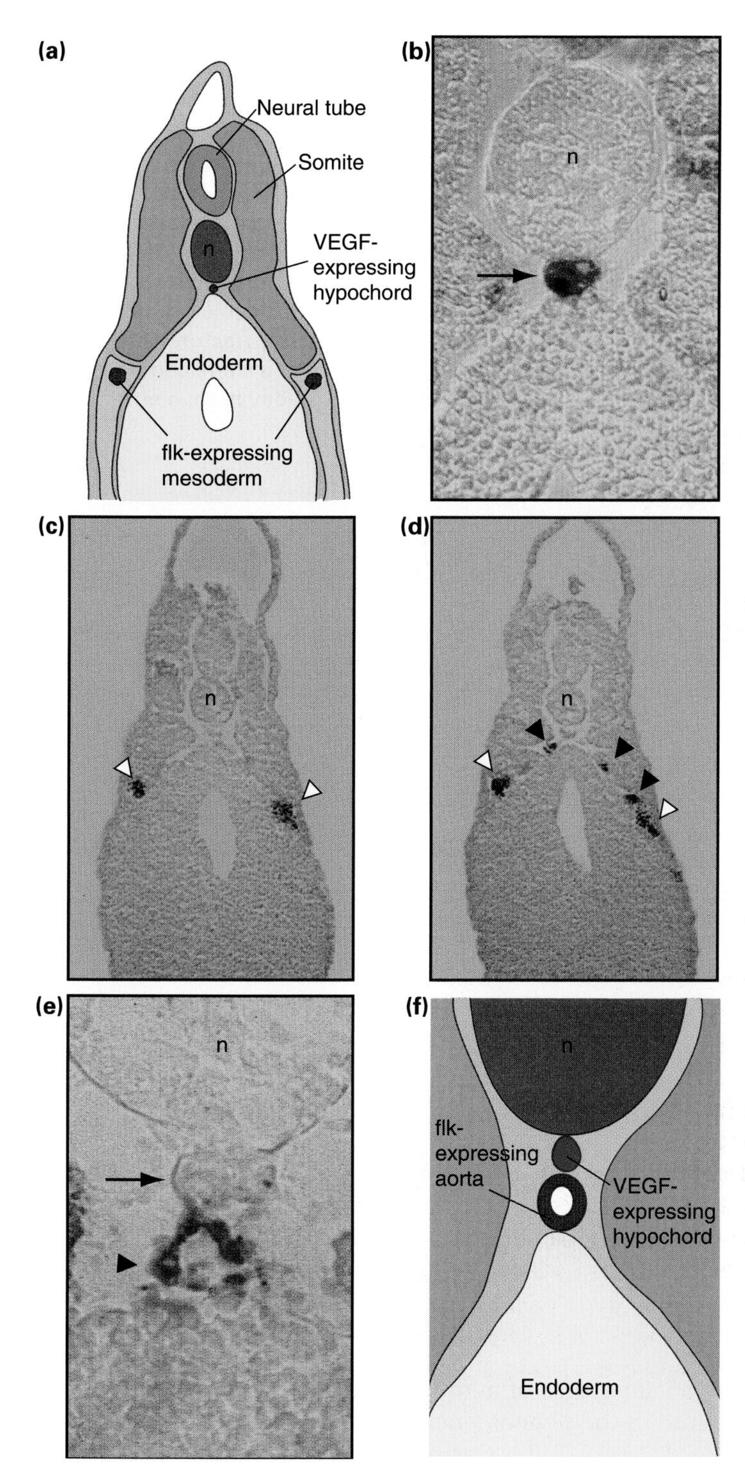

Figure 14.31 Apparent chemoattraction of angioblasts to the site of dorsal aorta formation in *Xenopus* embryos. The stages of development are numbered here according to the normal table of Nieuwkoop and Faber (1994). The dorsal aorta is formed around stage 34. **(a)** Schematic drawing of an embryo at stage 30 showing tissues synthesizing vascular epithelial growth factor (VEGF) and other cells synthesizing VEGF receptor, flk-1. **(b)** Light micrograph of a transverse section of an embryo at stage 30 after in situ hybridization (see Method 8.1) with a probe hybridizing to VEGF mRNA. The highest concentration of VEGF mRNA is found in the hypochord (arrow) located beneath the notochord (n). **(c)** Transverse section of an embryo at stage 28 after in situ hybridization with a probe hybridizing to flk-1 mRNA, which is specifically synthesized in angioblasts. Labeled cells (open arrowheads) are located where posterior cardinal veins will be formed later, around stage 34. **(d)** Same as part c but at stage 33. Some of the cells accumulating flk-1 mRNA (solid arrowheads) seem to migrate medially from the position of future cardianal vein formation (open arrowheads). **(e)** Section of embryo at stage 35, when the endothelium of the dorsal aorta (solid arrowhead) is formed beneath the hypochord (arrow). **(f)** Interpretative diagram of part e.

Fusion of the bilateral heart primordia gives rise to a single ***heart tube*** located ventrally beneath the foregut (Fig. 14.32). The heart tube, as well as its paired precursors, consists of two layers of visceral mesoderm separated by extracellular matrix. The innermost layer, the ***endocardium,*** is continuous with the endothelium of the adjoining blood vessels. The endocardium is surrounded by ***cardiac jelly,*** a layer of extracellular matrix rich in *hyaluronic acid,* which facilitates cell movements during subsequent remodeling of the endocardium. Surrounding the cardiac jelly is the ***myocardium,*** the portion of visceral mesoderm that gives rise to the heart muscle, or ***cardiac muscle.*** On the outside of the cardiac muscle, other mesoderm cells form a slippery membrane that represents the visceral layer of the pericardium.

The embryonic heart tube is subdivided into four consecutive chambers, named—from posterior to anterior, and in the direction of the blood flow—***sinus venosus, atrium, ventricle,*** and ***truncus arteriosus*** (Fig. 14.33). This primitive arrangement is fairly well conserved in adult sharks and bony fishes (Fig. 14.34a). In these animals, blood from the truncus arteriosus enters the ***ventral aorta.*** The ventral aorta gives off paired arteries called ***aortic arches,*** which are located inside the *pharyngeal arches* described earlier in this chapter (see Fig. 14.5). Here the deoxygenated blood arriving from the body is pumped with maximum pressure through the narrow capillaries of the gills. From the gills the blood is collected in paired ***dorsal aortae,*** which unite caudally and supply the large arteries of the body with oxygenated blood.

In land-dwelling vertebrates, which do not need blood vessels to supply gills, the aortic arches are nevertheless present during the *phylotypic stage.* Because this embryonic circulatory system is modified little during further development of fishes, there is an uncanny resemblance between the circulatory systems of a human embryo and adult fish (Fig. 14.34). In contrast, the circulatory system changes greatly from the phylotypic stage

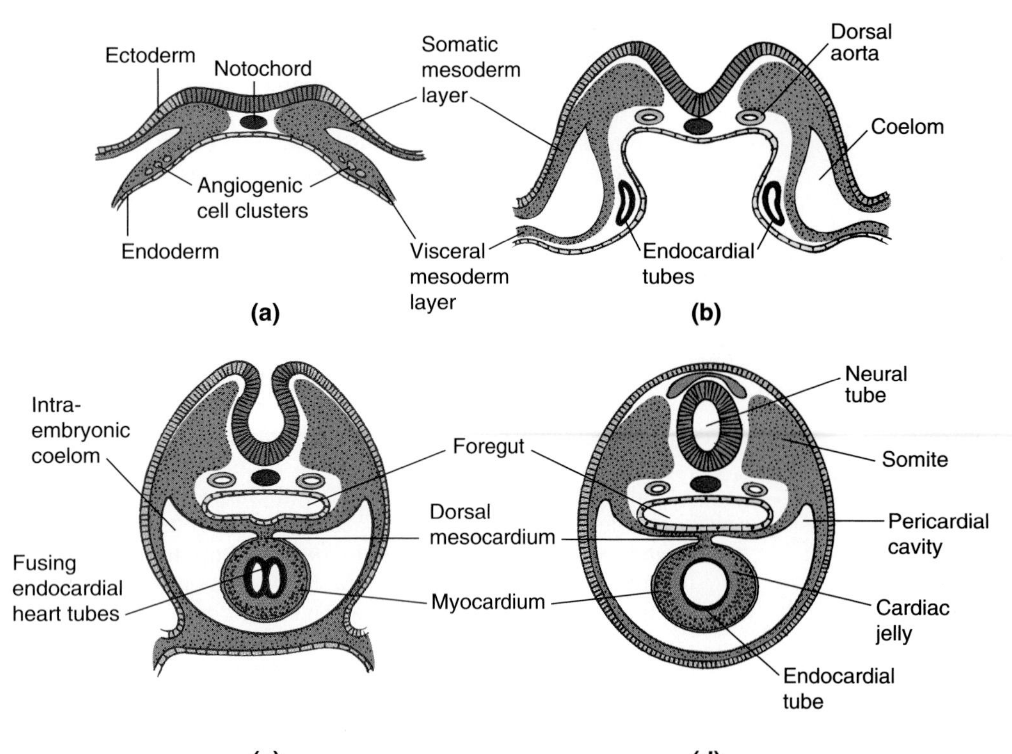

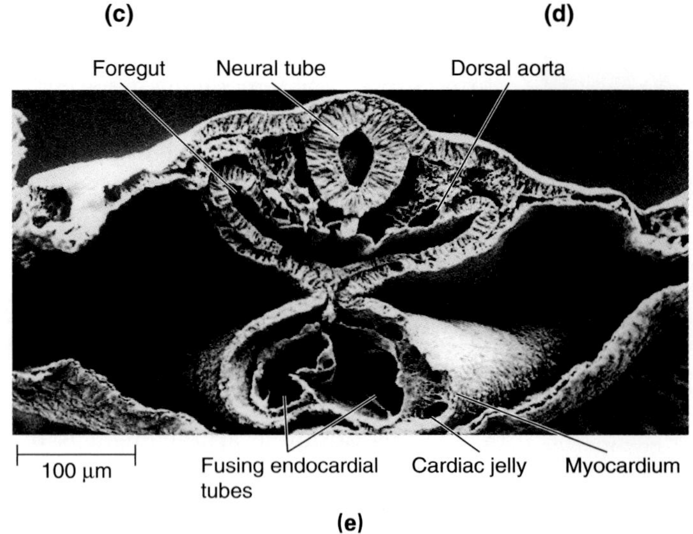

Figure 14.32 Formation of the heart tube by fusion of paired endocardial tubes and subsequent investment with cardiac jelly and myocardium. **(a–d)** Schematic cross sections of human embryos aged 17 days, 18 days, 21 days, and 22 days, respectively. **(e)** Scanning electron micrograph of a transversely fractured chicken embryo at a stage comparable to the 21-day human embryo.

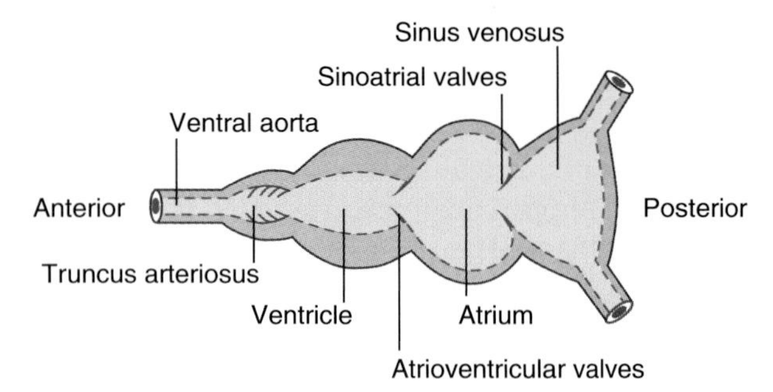

Figure 14.33 Schematic diagram of the primitive vertebrate heart.

to later stages of mammalian development, again illustrating the phenomenon of *recapitulation.*

Most of the aortic arches in mammals acquire secondary fates (Fig. 14.35). The first two aortic arches, located in pharyngeal arches I and II, are modified early and form minor arteries in the head. The third aortic arch forms the large carotid *arteries.* The fourth arch forms part of the aorta on the left side and a segment of the right *subclavian artery* on the right. The fifth arch is often underdeveloped or missing altogether. The fate of the sixth arch is particularly interesting, because it gives off a branch toward the developing lung on each side. This branch, together with the proximal part of the sixth aortic arch, gives rise to the *pulmonary artery* on each side. The distal part of the sixth arch on the left side

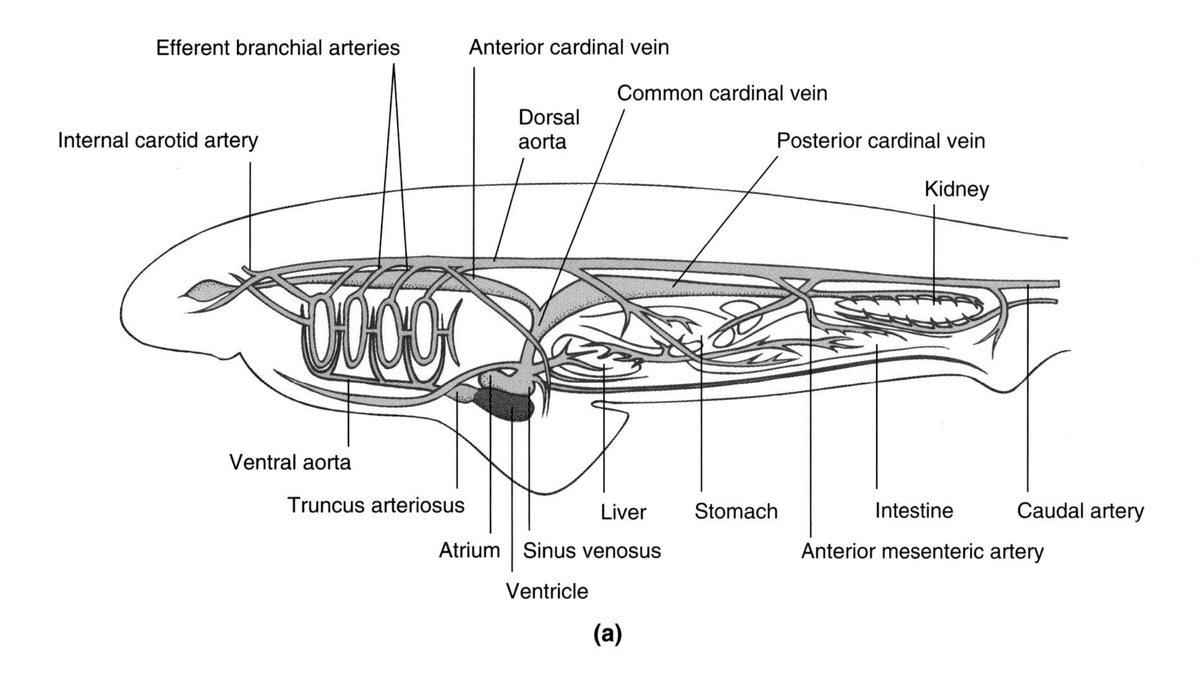

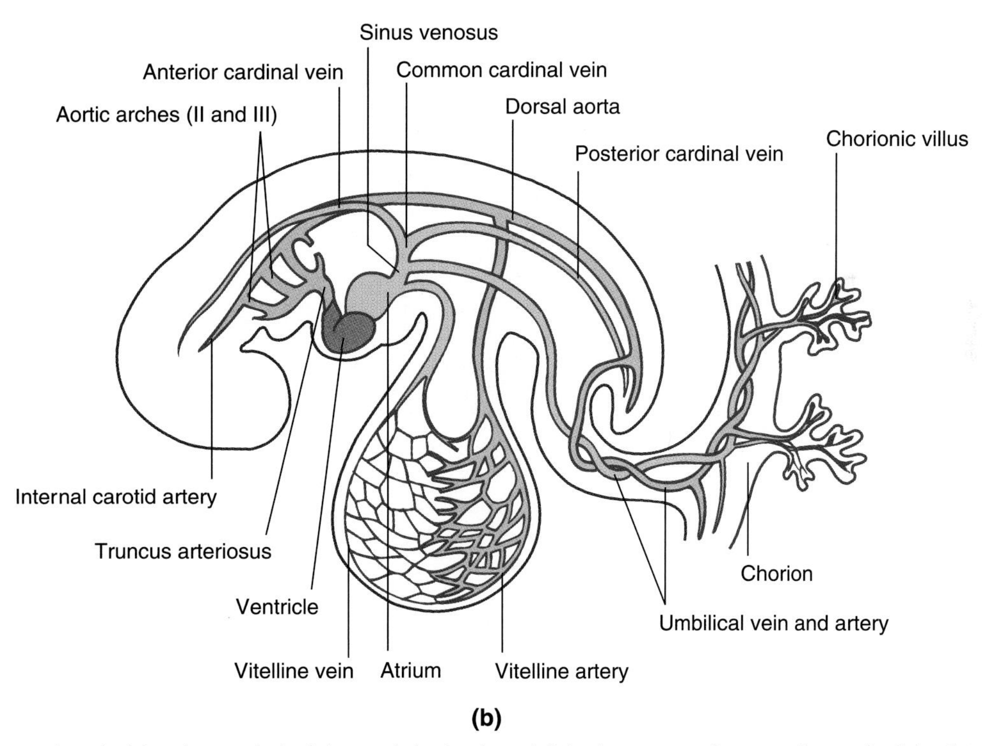

Figure 14.34 Heart and main blood vessels in **(a)** an adult shark and **(b)** a human embryo at the end of the fourth week. Only the blood vessels on the left side are shown.

forms the ***ductus arteriosus*** (not to be confused with the *truncus arteriosus*). The ductus arteriosus, which connects the left pulmonary artery with the aorta, shunts blood past the lung before birth and constricts soon thereafter into a ligament, the *ligamentum arteriosum* (Fig. 14.35c).

Also recapitulated during mammalian development is the evolution of the *pulmonary circulation*, which drives blood through the lungs and is separate from the *systemic circulation* supplying the rest of the body. The embryonic heart tube, which undergoes little further modification in fishes, undergoes a dramatic series of additional morphogenetic movements in mammals. As part of these movements, an *interatrial septum* develops between the right half of the atrium, which receives blood from the body, and the left half, which receives blood from the lungs (Fig. 14.35c). Similarly, the ventricle is divided into right and left halves by the growth of

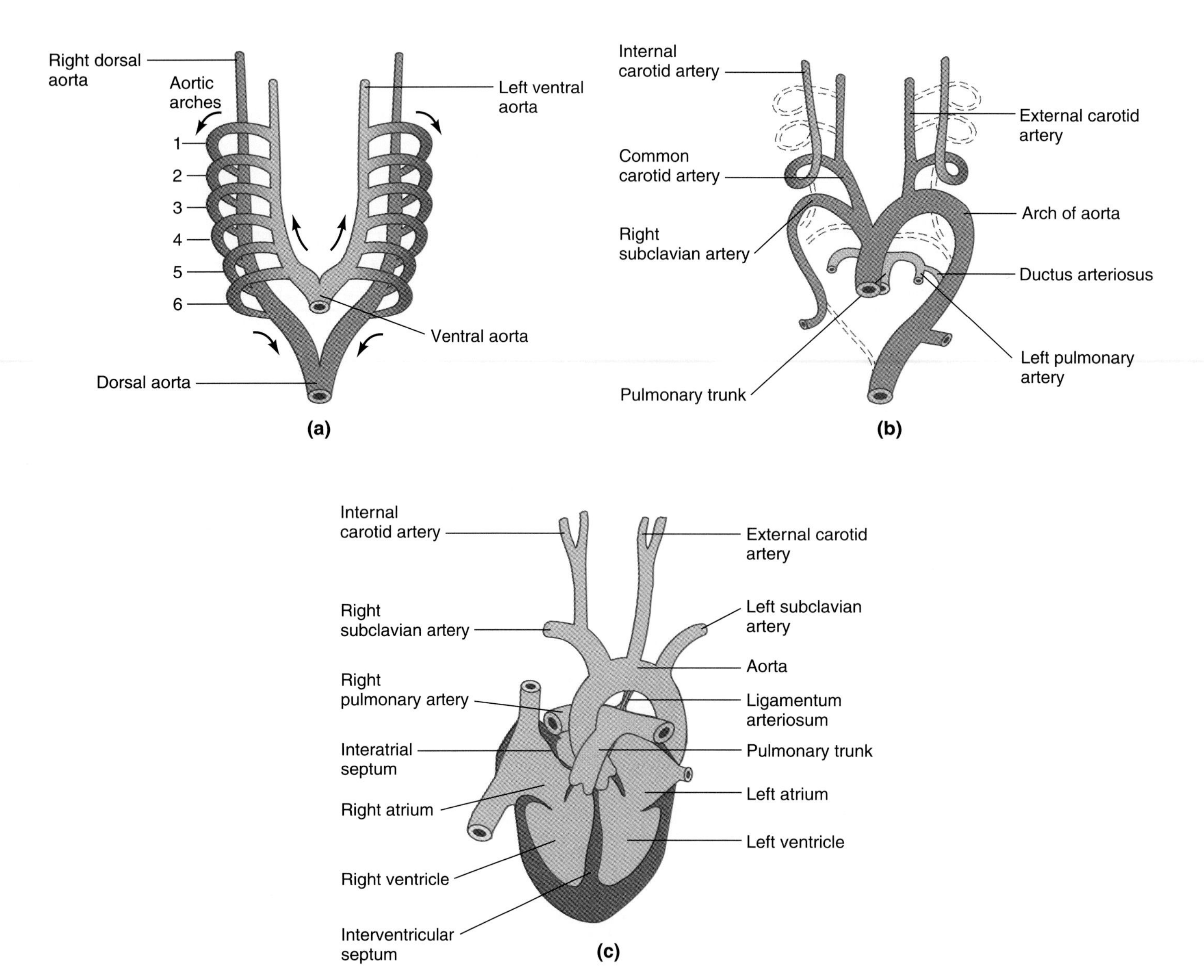

Figure 14.35 Fate of the aortic arches in mammalian development. All diagrams show ventral aspects. **(a)** Generalized early embryonic pattern. Arrows indicate the blood flow from the ventral aorta through the aortic arches to the dorsal aortae. **(b)** Degeneration of arches 1, 2, and 5 and transformation of arches 3, 4, and 6. **(c)** Heart and major arteries in the human adult. An interatrial septum has separated the right atrium from the left atrium. Similarly, the ventricle has been partitioned by an interventricular septum. The truncus arteriosus has been divided into the aorta, which emerges from the left ventricle; and a pulmonary trunk, which emerges from the right ventricle. A vestige of the ductus arteriosus, which shunts blood past the lungs in the embryo, collapses into the ligamentum arteriosum.

an *interventricular septum.* At the same time, the atrioventricular canal between the old atrium and ventricle is remodeled so that the new atria and ventricles on each side have separate connections. Moreover, the truncus arteriosus splits into the aorta and the pulmonary trunk. The ***aorta*** connects with the left ventricle and pumps blood into the body, while the ***pulmonary trunk*** carries blood from the right ventricle to the pulmonary arteries. The split of the truncus arteriosus is accomplished by ridges on the inside, which grow together in a spiral pattern foreshadowing the way in which the pulmonary trunk and the aorta later twist around each other.

CARDIAC MUSCLE AND SMOOTH MUSCLE CONSIST OF SINGLE CELLS

In contrast to skeletal muscle, which is composed of multinucleate muscle fibers (see Fig. 14.23), ***cardiac muscle*** consists of individual cells. However, both skeletal and cardiac muscle are striated because of the alignment of their actin and myosin fibrils.

A third type of muscle tissue, which under the microscope does not have the striated appearance of skeletal and cardiac muscle, is known as ***smooth muscle.*** It consists of single cells that are derived from mesoderm or ectomesenchyme. Smooth muscle tissue derived from

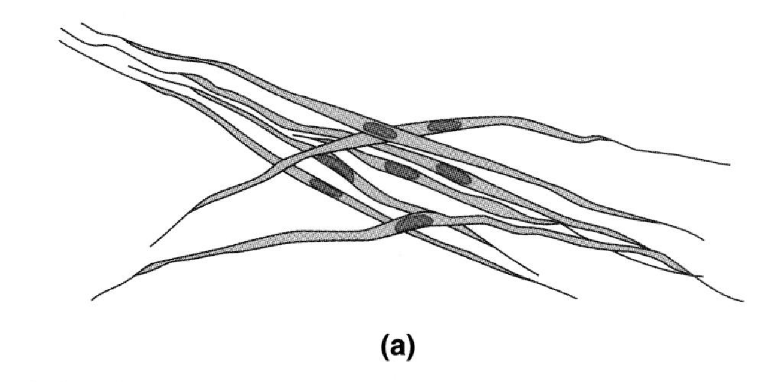

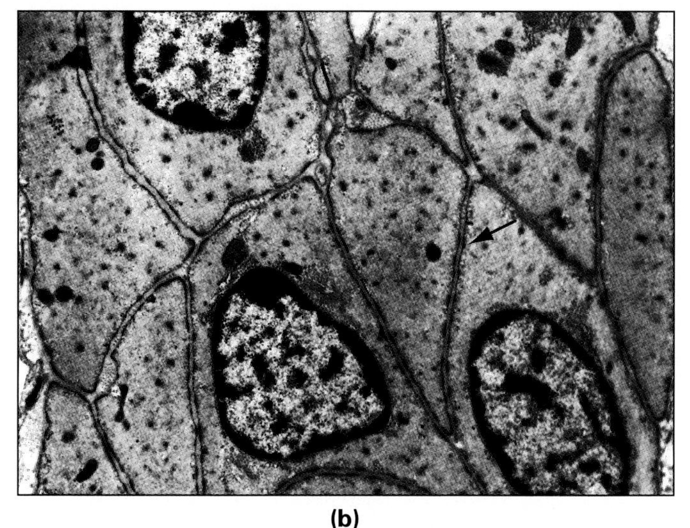

Figure 14.36 Smooth muscle cells. **(a)** Isolated smooth muscle cells from the stomach wall of a cat. **(b)** Transmission electron micrograph of smooth muscle in transverse section. The dense areas (arrows) around the inside of the plasma membrane are sites of insertion of contractile filaments.

the visceral layer of lateral mesoderm is found in the digestive tract and its appendages, such as glands and respiratory passages. Smooth muscle in blood vessels controls vessel diameter and thus the blood pressure and the rate of blood flow. However, smooth muscle also occurs in the urinary and genital ducts, notably in the oviducts and uterus. Mammalian skin contains minute smooth muscle cells that can erect the hairs. In the eye, smooth muscle is responsible for lens accommodation and regulating the diameter of the pupil.

A smooth muscle cell is long and spindle-shaped, with a single nucleus positioned about halfway along its length (Fig. 14.36). When smooth muscle cells form bundles or sheets, the cells are offset so that the thick central portion of one cell is juxtaposed to the thin end of another. Smooth muscle represents a primitive type of muscle cell, in its similarity to nonmuscle cells. Smooth muscle is not striated, because its contractile filaments are not arranged in the strictly ordered pattern observed in skeletal and cardiac muscle. Instead, the fibers are attached obliquely to the plasma membrane at disclike junctions that connect smooth muscle cells.

14.7 Extraembryonic Membranes

To fish and amphibian embryos, the water in which they develop affords easily accessible food, protection against desiccation and mechanical shock, and a vast repository for metabolic waste. Since reptiles first began to lay eggs on dry land, they and their avian and mammalian descendants have had to compensate for the losses incurred by leaving the water. Birds and most reptiles develop inside hard eggshells, and most mammals inside the maternal uterus. In these three classes of vertebrates, membranes outside the embryo proper—collectively called ***extraembryonic membranes***—provide for nourishment, protection, respiration, and excretion. One of these membranes, directly surrounding the embryo, is known as the ***amnion.*** Because they share this important feature, reptiles, birds, and mammals have become known by the collective term ***amniota.***

The amniota have four principal extraembryonic membranes (Fig. 14.37). Two of them, the *amnion* and *chorion,* originate from raised folds of a double membrane called ***somatopleure*** because it consists of an ectodermal layer and the adjacent somatic layer of lateral plate mesoderm. The two other extraembryonic membranes, the *yolk sac* and the *allantois,* develop essentially as parts of the gut. Accordingly, these membranes consist of an endodermal layer with the adjacent visceral layer of lateral plate mesoderm, together called ***splanchnopleure.*** The terms "somatopleure" and "splanchnopleure" are also used for the corresponding double layers where they occur inside the embryo.

THE AMNION AND CHORION ARE FORMED BY LAYERS OF ECTODERM AND MESODERM

The amnion and chorion of reptiles and birds originate when folds of the somatopleure arise on all sides of the embryo and fuse dorsally above (Fig. 14.37c, d). The embryo is then covered by two layers of somatopleure. The inner layer, with the ectoderm facing the embryo, is the *amnion.* The outer layer, with the ectoderm facing the eggshell or uterus, is called the ***chorion*** or *serosa.* The space between amnion and chorion is the *extraembryonic coelom,* which is continuous with the embryonic coelom. The cavity enclosed by the amnion, the *amniotic cavity,* is filled with fluid so that the embryo floats in its own private pool. In mammals, amnion and chorion arise differently but assume the same spatial relation to each other and to the embryo.

THE YOLK SAC AND ALLANTOIS ARE FORMED BY LAYERS OF ENDODERM AND MESODERM

In reptile and bird embryos, both splanchnopleure and somatopleure grow out over and around the uncleaved egg yolk. Where these membranes extend beyond the region of embryo formation, they are considered

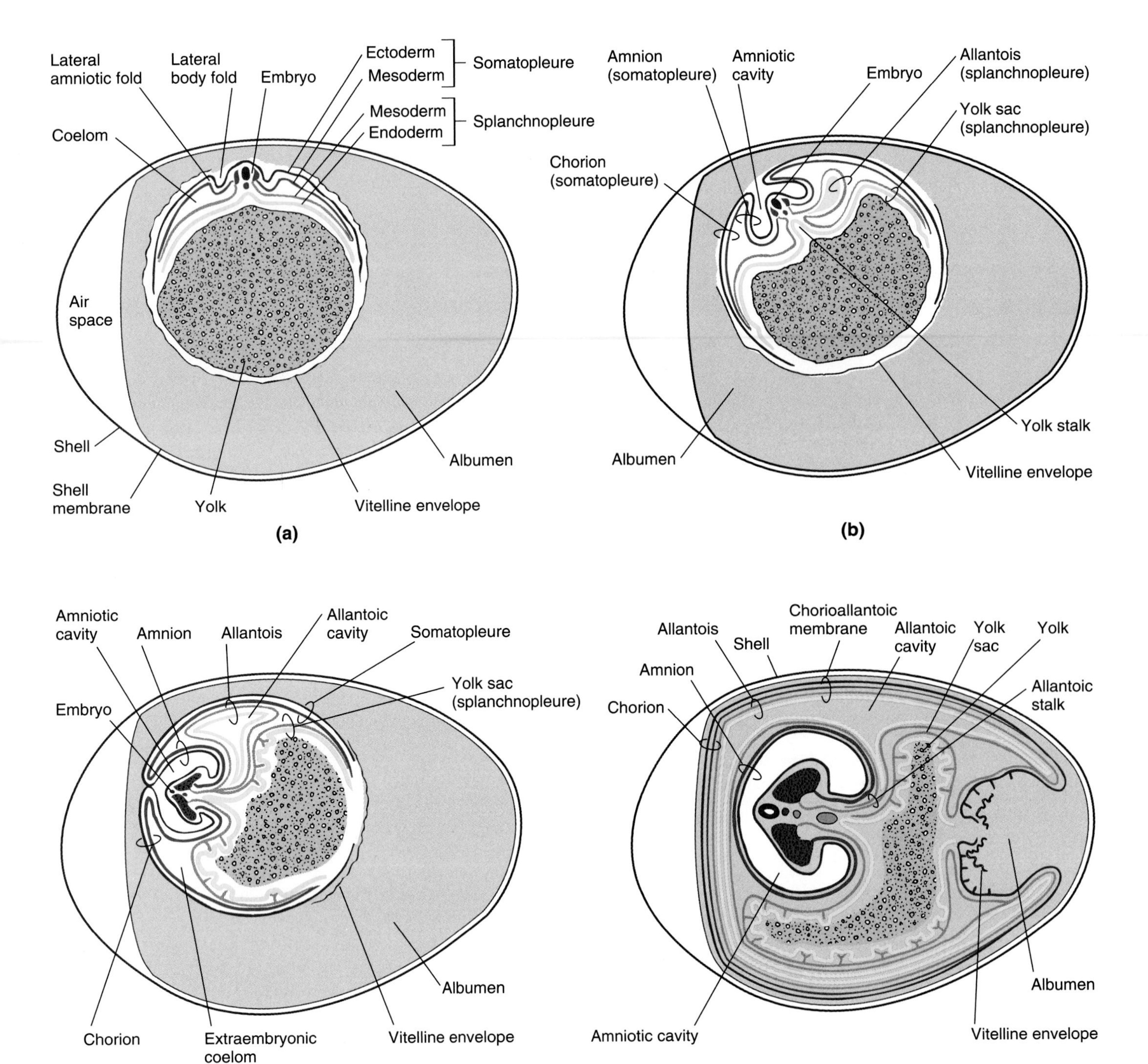

Figure 14.37 Extraembryonic membranes of the chicken. Each diagram represents a transverse section through the embryo. After **(a)** 2 days of incubation; **(b)** 3 days; **(c)** 5 days; **(d)** 14 days.

extraembryonic. When the splanchnopleure has overgrown the egg yolk, it is called the *yolk sac.* The yolk sac turns into a large appendage of the midgut when the hindgut and foregut close to form tubes (see Figs. 14.2 and 14.3). Since the embryonic and the extraembryonic portions of the splanchnopleure are continuous, and because the formation of angiogenic clusters occurs simultaneously in both portions, the developing blood vessels carry the digested yolk components directly to the embryo. A yolk sac is found in all vertebrates except amphibians, where each embryonic cell has its own yolk supply.

The *allantois* develops as an evagination of the hindgut splanchnopleure. In reptiles and birds, the allantois spreads into the extraembryonic coelom until it surrounds the entire embryo (Fig. 14.37c, d). The somatic mesoderm of the chorion, together with adjacent visceral mesoderm of the allantois, forms the ***chorioal-***

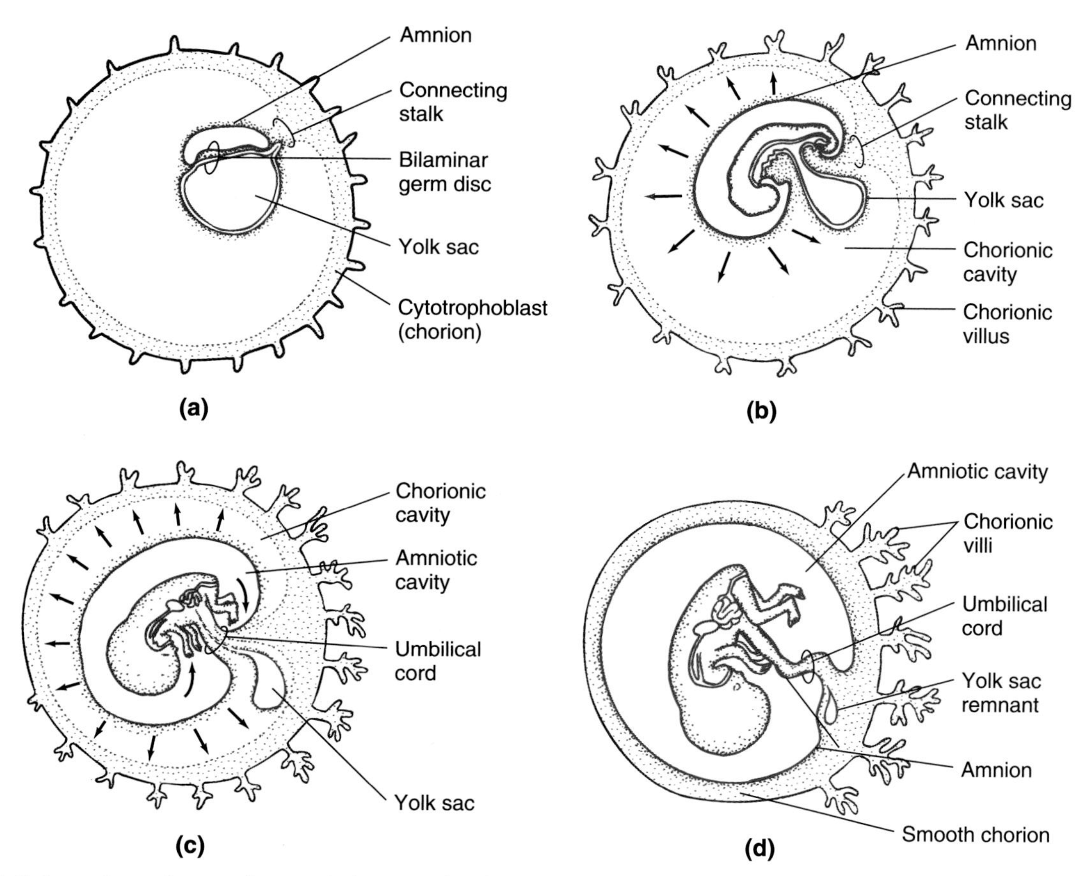

Figure 14.38 Extraembryonic membranes in human development: **(a)** at 3 weeks; **(b)** at 4 weeks; **(c)** at 10 weeks; **(d)** at 20 weeks. The connecting stalk develops into the umbilical cord. The amniotic cavity expands (arrows) until it completely fills the chorionic cavity and envelops the umbilical cord plus the remnant of the yolk sac. The chorionic villi near the umbilical cord branch and form the embryonic portion of the placenta. The other villi disappear.

lantoic membrane. As blood vessels develop therein, the blood circulates directly under the eggshell, providing for an effective gas exchange with the outside world. In addition to its function as an embryonic lung, the allantois serves as a repository for metabolic waste products, especially uric acid. In some groups of mammals, the allantois retains some of these functions while contributing to the placenta.

THE MAMMALIAN PLACENTA IS FORMED FROM THE EMBRYONIC TROPHOBLAST AND THE UTERINE ENDOMETRIUM

Mammals form the same extraembryonic structures as do reptiles and birds, but certain modifications occur, apparently as adaptations to intrauterine development. Although mammalian eggs contain no yolk, a yolk sac is formed nevertheless, providing another example of *recapitulation.* However, the yolk sac of mammals retains its function as the site of primordial germ cell formation (see Fig. 3.3).

During early mammalian development, the blastocyst implants itself in the inner lining of the uterus, the ***endometrium.*** The cells of the *syncytiotrophoblast* erode the endometrium, including its blood vessels, creating blood-filled spaces called *lacunae* between syncytiotrophoblast and uterine tissues. In the human, this occurs at the end of the second week when the embryo is a bilaminar germ disc (Figs. 10.31 and 14.38). The ***connecting stalk,*** which connects the embryo to the *cytotrophoblast,* will develop into the umbilical cord. The cytotrophoblast is now called the ***chorion,*** since its position relative to the amnion is analogous to the chorion in reptiles and birds (see Figs. 14.37c and 14.38). The cavity surrounded by the chorion is the ***chorionic cavity.***

The chorion sends out fingerlike ***villi,*** which invade the lacunae formed by the syncytiotrophoblast. The formation of *angiogenic clusters* spreads to the yolk sac and the allantois, which reach down the connecting stalk and into the chorionic villi. From the fourth week on, two ***umbilical arteries*** and one ***umbilical vein*** in the stalk connect the embryonic cardiovascular system with the chorionic villi (Fig. 14.39). The villi, in turn, are surrounded by maternal blood, which gives off nutrients and oxygen to the embryonic blood and takes up waste products from it. However, all molecules transferred between embryonic and maternal blood need to diffuse through the tissues of the villi.

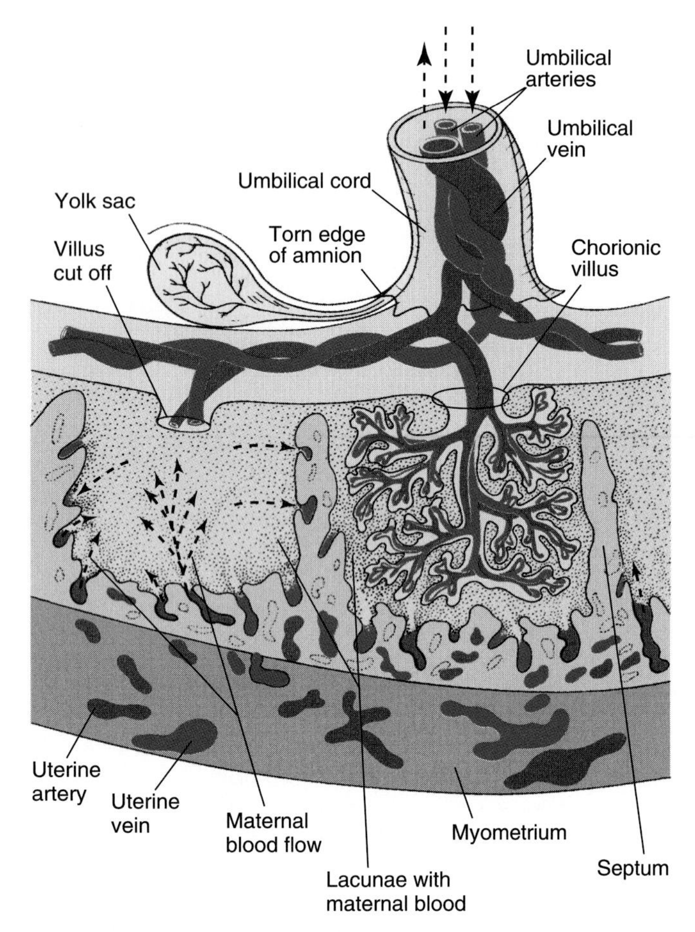

Figure 14.39 The human placenta. The embryonic tissues are labeled along the top of the drawing and the maternal tissues along the bottom. Note that the chorionic villi float in maternal blood; one villus has been removed in the drawing to show the maternal blood flow from uterine arteries into the lacunae and back into uterine veins.

As pregnancy advances, the majority of the chorionic villi disappear (Fig. 14.38c, d). The villi near the connecting stalk grow and form branches, anchoring themselves firmly in the endometrium. The villous portion of the fetal chorion and the associated part of the uterine endometrium make up the organ known as the ***placenta*** (Fig. 14.39). The placenta connects the fetus with the uterus, mediating the exchange of nutrients, oxygen, hormones, and waste products throughout pregnancy. As the fetus and the amniotic cavity enlarge, the amnion fuses with the inside of the chorion, so the chorionic cavity disappears. At the same time, the growing fetus and the amniotic cavity distend the covering area of the endometrium, thus displacing most of the uterine cavity.

SUMMARY

The embryonic endoderm forms the inner linings of the gut and its appendages, such as lungs, liver, and pancreatic gland. Of particular interest in vertebrates are the pharyngeal pouches, which bulge out from the endoderm in the neck region and induce the formation of matching pharyngeal clefts in the overlying ectoderm. Where the pharyngeal pouches and clefts fuse, they create slits connecting the pharynx with the outside world. The tissue between the slits forms pharyngeal arches, which are the sites of gill formation in fishes and amphibian larvae. In more advanced vertebrates, pharyngeal clefts and pouches give rise to other structures, including parts of the ear and various glands.

The recapitulation of pharyngeal pouches and clefts in the development of land-living vertebrates is part of the evolutionary conservation of the phylotypic stage, which is reached after organogenesis. The primitive species of a phylum remain close to the phylotypic stage in their adult organization, whereas the advanced species of a phylum usually undergo major changes in their development from the phylotypic stage to later stages.

Mesoderm originates partly in epithelial form but also forms large amounts of mesenchyme, an embryonic tissue that is characterized by a loose aggregation of cells surrounded by extracellular matrix. The connective tissues of adult vertebrates, including bone, cartilage, tendons, and adipose tissue, are generally derived from mesenchyme.

The mesoderm of vertebrate embryos consists of axial mesoderm, paraxial mesoderm, intermediate mesoderm, and lateral plates. The axial mesoderm gives rise to the prechordal plate and the notochord. The paraxial mesoderm forms presomitic plates, which become subdivided first into somitomeres and then into somites. Identified signals from adjacent organ rudiments pattern the somites to form vertebrae, dermis, and two groups of skeletal muscle.

The intermediate mesoderm forms the kidneys and some reproductive structures. Depending on taxonomic group and stage of development, the kidneys are present in one of three forms of organization: pronephros, mesonephros, and metanephros.

The metanephros develops from two rudiments, the ureteric bud epithelium and the metanephrogenic mesenchyme. This development exemplifies the principle of reciprocal interactions, the ability of organ rudiments to coordinate their development in space and time through the mutual exchange of signals.

The lateral plates consist of a somatic layer and a visceral layer. Together, they form the inner linings of the coelom and its derivatives, the pericardium, the pleural cavities, and the peritoneal cavity. The visceral layer also produces smooth muscle and the entire cardiovascular system, whereas the somatic layer contributes the body wall and the limb buds.

In addition to the embryo proper, developing vertebrates form several extraembryonic membranes on which they depend for their protection, nutrition, respiration, and waste removal. A yolk sac is formed by all vertebrates except most amphibians. The other extraembryonic membranes, known as amnion, chorion, and allantois, characterize reptiles, birds, and mammals. The chorion and uterus participate in the formation of the placenta in mammals.

part two

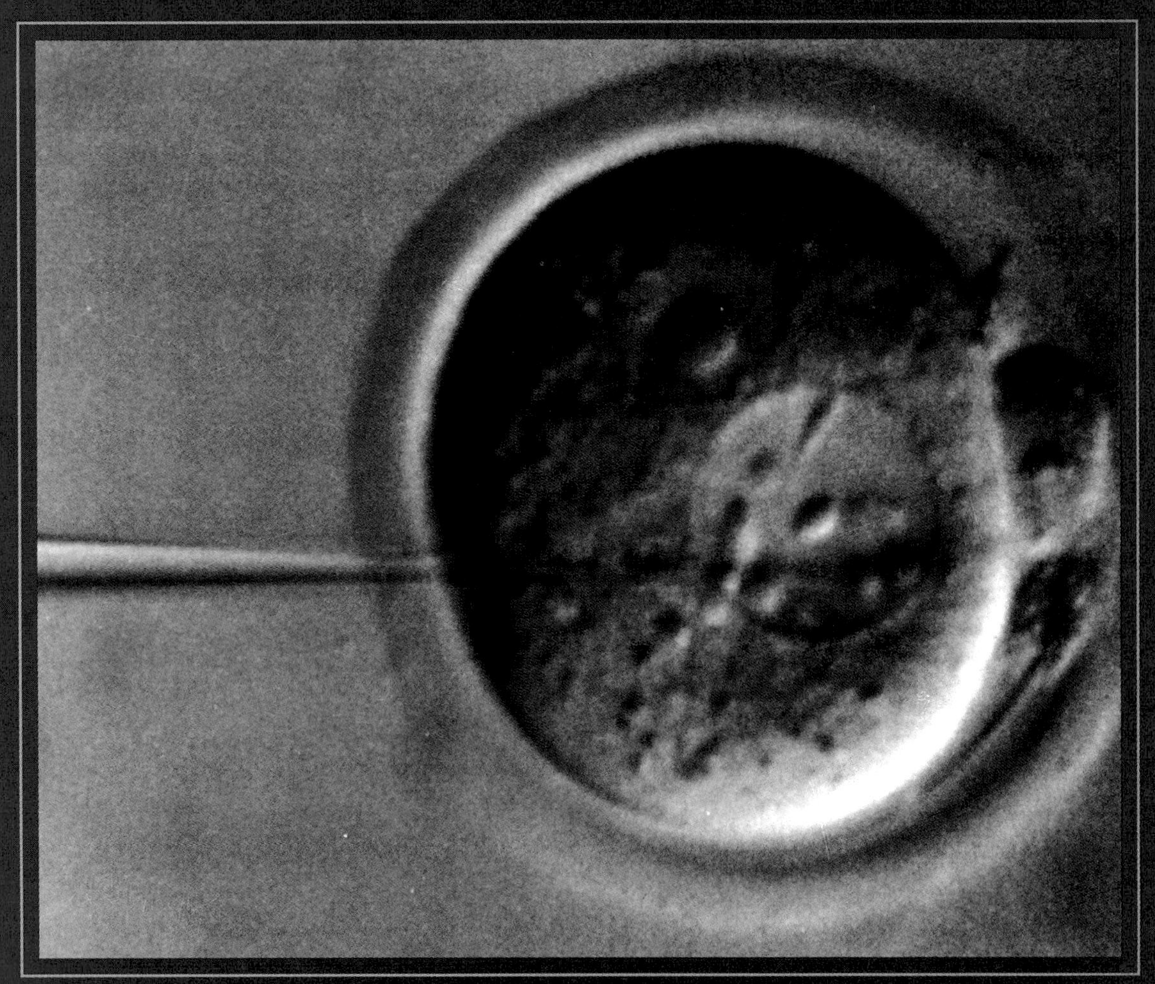

Figure II.1 Genetic transformation of animals. This photomicrograph shows the microinjection of cloned DNA into one pronucleus of a fertilized mouse egg. The DNA is integrated into the host DNA and then passed on to daughter cells and offspring as part of the mouse genome. Transgenic organisms are most valuable for the genetic analysis of development and for many applications in agriculture, medicine, and pharmacy.

Control of Gene Expression in Development

THIS part of the text will introduce the genetic and molecular analysis of development. We will examine how the genetic information encoded in linear DNA molecules is used to build a three-dimensional organism that unfolds in time. The advent of molecular biology in the 1960s provided the missing link between genes as hereditary units and the physiology of the cell. Work at this new frontier was boosted again in the 1970s by a set of modern techniques using cloned DNA. This term refers to the amplification of DNA in vitro or in bacterial host cells, so that any gene of interest can be produced in large amounts for detailed analysis of the gene itself and its products. The turn of the twentyfirst century then brought the complete genomic sequence information of several microorganisms as well as two higher organisms, the roundworm *Caenorhabditis elegans* and the fruit fly *Drosophila melanogaster*. The genomic sequences of several other organisms, including the human, will follow in short order. These data, and new techniques of comparing tenthousands of cloned DNA sequences at a time, will greatly enhance the scope and speed of genetic and molecular analysis. ● In Chapter 15, we will explore the use of *mutants,* organisms with altered genes, in the analysis of development. Many mutants display abnormal traits that provide clues to the normal functions of the affected genes. For instance, the additional pair of wings seen in *Ultrabithorax* mutants of *Drosophila* (see Fig. 1.21) indicates that the normal function of this gene is to somehow cause the formation of balancer organs (halteres) instead of wings. ● Investigators also make increasing use of *transgenic cells* and *transgenic organisms,* which result from insertion of an engineered gene into the nucleus of a host. The engineered gene, called a *transgene,* may be from the same species as the host organism or from a different species. If integrated into the genome of an egg or a germ line cell, a transgene is passed on to offspring like an ordinary mutation. The power of this technique (Fig. II.1) was demonstrated in a spectacular way when mice transformed with a rat growth hormone gene grew to twice the normal size, apparently as a result of excessive growth hormone synthesis. Today, genetic transformation has found a wide range of applications in basic research as well as in medicine, agriculture, and pharmacy. ● Chapters 16 through 18 examine the flow of information from genes to proteins in specific developmental events chosen as examples. The first step in gene expression, *transcription,* is the synthesis of RNA from a DNA template. The transcripts of most genes are *messenger RNAs (mRNAs),* which in turn serve as templates during protein synthesis, a step that is known as *translation.* In eukaryotes, there are many levels of gene regulation, reflecting the eukaryotic cell's advanced organization (Fig. II.2). The nuclear envelope separates transcription in the nucleus from translation in the cytoplasm. Nuclear pre-mRNA undergoes several processing steps before mRNA is released into the cytoplasm, where most mRNAs are immediately translated into proteins while some are stored as inactive messenger ribonucleoprotein (mRNP) particles to be translated later. Following translation, proteins undergo several modifications and often associate with other molecules to form multimolecular complexes or cell organelles. ● Many molecular complexes and even some cell organelles snap together in a process of *self-assembly,* driven only by forces between the participating molecules. Other cell organelles, as we will see in Chapter 19, do not have this capability. These organelles can be assembled only using preexisting organelles as templates. An assembly process that requires structural templates in addition to molecular building blocks is called *directed assembly.* As an example of directed assembly, we will study a

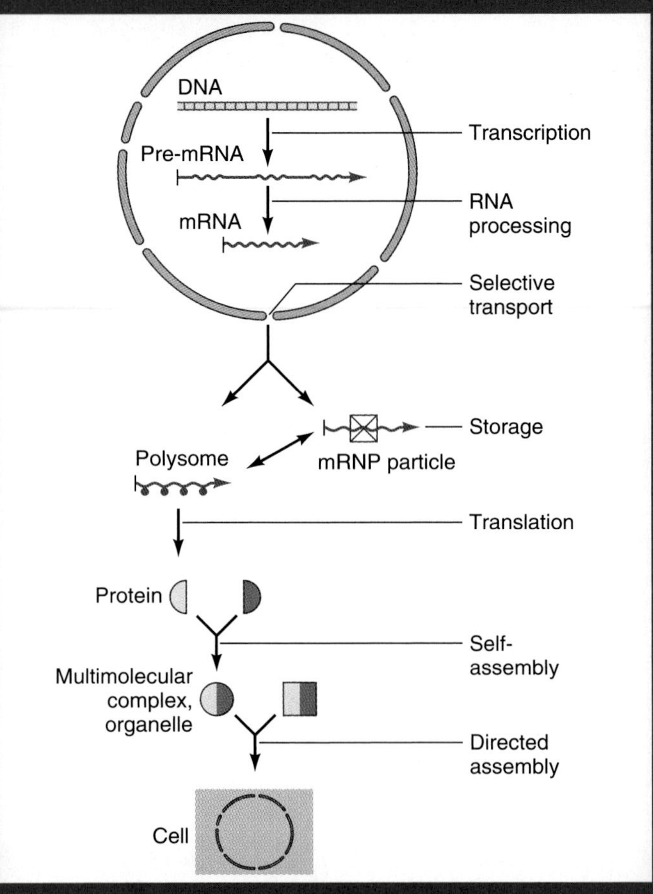

Figure II.2 Gene expression in eukaryotic cells. The genetic information encoded in nuclear DNA is transcribed into pre-mRNA, which is processed into mRNA. Some mRNAs are transported selectively from the nucleus into the cytoplasm. Cytoplasmic mRNA may be stored in messenger ribonucleoprotein (mRNP) particles or recruited into polysomes. The polysomes translate mRNA into protein. Proteins and other molecules self-assemble into composite molecules or cellular organelles. Organelles are directed by existing cellular structures to assemble into additional cellular structures.

unicellular organism, *Paramecium*. It has an intricately structured cell cortex, with cilia for swimming and sting organelles for defense. Occasional injuries may cause cortical irregularities that are not associated with any genetic changes. Nevertheless, these abnormalities are passed on for hundreds of generations as the animal multiplies by cell division. Apparently, some of the cortical structures, be they normal or abnormal, are duplicated by directed assembly. ● Structural information that is passed on during cell division but not encoded in DNA is called *paragenetic information*. Animal cells inherit both genetic and paragenetic information. ● During the heady years of beginning molecular biology, a leading researcher expressed his hope that within a few years one would be able to "compute a mouse." He thought that the application of genetic and molecular methods would lead to a complete understanding of development. We are still far from reaching this goal. So far, nobody has even "computed" a cell. However, geneticists and molecular biologists have made great forays, and their tools have proven to be exceedingly powerful.

chapter 15

The Use of Mutants and Transgenic Organisms in the Analysis of Development

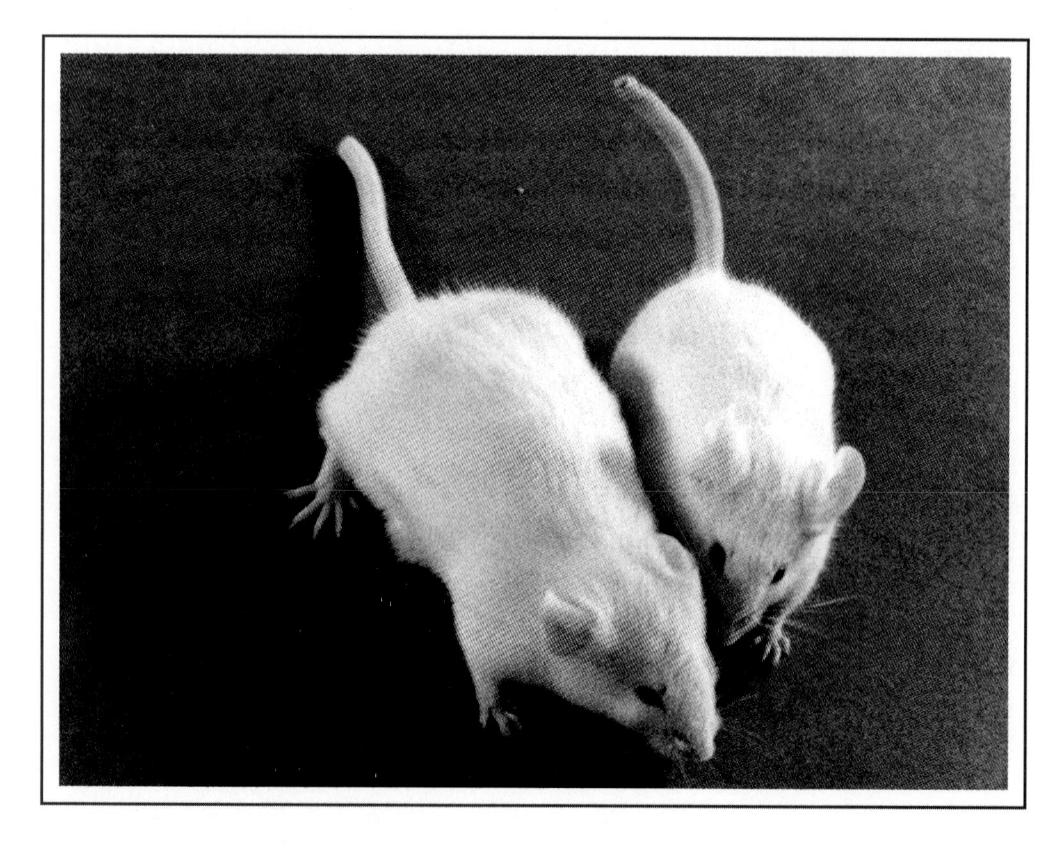

Figure 15.1 Gigantic mouse (left) raised from an egg injected with a cloned gene for rat growth hormone. The mouse grew to about twice the weight of his uninjected littermate (right).

ONE way of analyzing a complex system is to modify parts of it and then observe its performance. This *strategy of controlled interference* is especially instructive if the modifications affect single components instead of causing more widespread damage. In terms of an everyday analogy, one could learn what the alternator does in an automobile by removing it—or by cutting the alternator drive belt—and watching the electrical system fail after the battery has been drained. Dropping a boulder on the engine would cause general damage and would be much less informative.

Developmental biologists have used the strategy of controlled interference extensively, transplanting or removing cells or portions of cytoplasm and then analyzing the resulting defects (see Section 1.3). However, there are disadvantages to this type of controlled interference. Surgical procedures often cause unintended damage to embryos, and even if this problem can be minimized, the removal of whole cells or portions of cytoplasm is unlikely to provide much insight at the molecular level. In contrast, geneticists interfere with individual genes, each of which controls the synthesis of one or a few specific proteins in the organism.

The tools for genetic analysis are mutants, stocks of individuals in which one or more genes are altered. Some mutants show easily detectable abnormalities that can provide clues to the normal function of the affected gene. For example, *Drosophila* females without a functioning copy of the *bicoid* gene give rise to nonviable embryos without head and thorax (see Fig. 1.20). Genetic and molecular analysis has shown that *bicoid*$^+$ is a maternal effect gene whose transcript serves as a localized anterior determinant (see Section 8.4).

We will begin this chapter with a historical note on the strangely unproductive relationship that prevailed between geneticists and developmental biologists before molecular biology brought them together. Next, we will explore how mutants are used today in the genetic analysis of the development of five organisms: the fruit fly *Drosophila melanogaster;* the house mouse, *Mus musculus;* the roundworm *Caenorhabditis elegans;* the zebrafish *Danio rerio;* and the mouse ear cress *Arabidopsis thaliana.* For each species, particularly interesting mutants have been isolated and used to investigate developmental events. Subsequently, we will see how investigators analyze the functions of certain genes even in the absence of suitable mutants. One way of generating the functional equivalents of mutants is to clone a modified gene and insert it as a *transgene* into the genome of a cultured cell or an entire organism (Fig. 15.1). Alternatively, modified mRNAs transcribed in vitro from engineered DNA templates are injected into cells. This strategy of making the equivalents of mutant cells and organisms by molecular techniques is used widely today to investigate the biological effects of specific gene activities.

Although mutations and transgenes are generally finer tools than surgical scalpels, even the smallest genetic change may affect several observable traits, especially if the gene is expressed in different tissues (Fig. 15.2). This property of genes is known as ***pleiotropy*** (Gk. *pleion*, "more"; *tropos*, "direction"). Conversely, many phenotypic traits are ***polygenic,*** meaning that they are influenced by several genes. These genes may be unrelated, or they may have evolved through a process of gene duplication and mutation. In the latter case, researchers may have to interfere with all of the redundant or overlapping gene functions before a mutant phenotype can be observed.

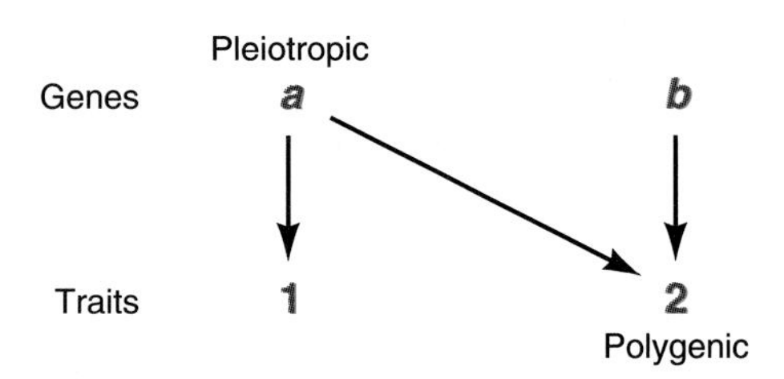

Figure 15.2 Relationships between genes (*a, b*) and observable traits (1, 2). Gene *a*, affecting more than one trait, is pleiotropic. Trait 2, affected by more than one gene, is polygenic.

15.1 The Historical Separation of Genetics from Developmental Biology

The science of genetics dates from 1865, when Gregor Mendel made public the results of his breeding experiments with peas. From these results, he inferred the laws of genetic segregation and independent assortment. However, Mendel's abstract laws were not fully accepted until after 1900, when several investigators identified chromosomes as the physical carriers of heritable traits.

A critical experiment to test the role of chromosomes in early development was performed by Boveri (1902). He added excess amounts of sperm to sea urchin eggs to obtain eggs fertilized by two sperm, called ***dispermic eggs*** (Fig. 15.3). These eggs had two centrosomes, one introduced by each sperm, and three haploid pronuclei, two from the sperm and one from the egg. Each chromosome replicated its DNA and formed two chromatids. The two centrosomes also replicated and set up four mitotic spindles, which competed for the three sets of chromosomes. During mitotic anaphase, each spindle pole attracted various combinations of chromatids from multiple spindles. Cytokinesis then generated four blastomeres simultaneously.

Boveri separated the blastomeres from one another and observed their development. From previous experiments it was known that each quarter blastomere derived from a normal egg that had been fertilized by one sperm would usually develop into a complete larva. In contrast, only a

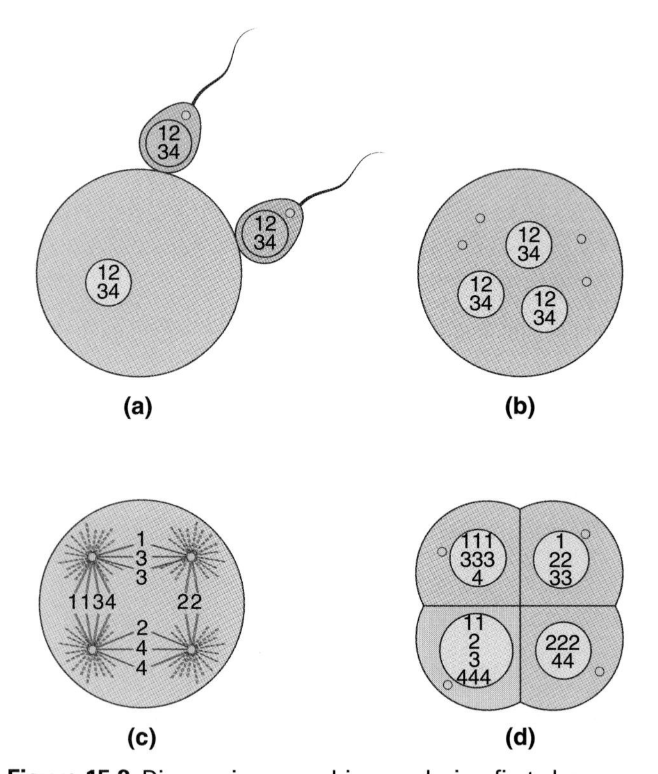

Figure 15.3 Dispermic sea urchin egg during first cleavage. **(a)** Egg being fertilized by two sperm. Each sperm introduces a male pronucleus and a centrosome. Each of the male pronuclei, as well as the female pronucleus in the egg, contains a haploid set of chromosomes, designated 1, 2, 3, and 4. **(b, c)** Both centrosomes replicate in the fertilized egg and set up four mitotic spindles, which attract random assortments of chromosomes. **(d)** The first cleavage generates four blastomeres simultaneously. In the case shown, only one blastomere (pink color) inherits at least one copy of each chromosome.

few isolated blastomeres from dispermic eggs developed into complete larvae. Most important, the other blastomeres stopped developing at different stages and with different signs of abnormality. It was also known from the development of unfertilized but artificially activated sea urchin eggs that a haploid but complete set of chromosomes was sufficient to support normal embryonic development. Most blastomeres had more than a haploid number of chromosomes, but apparently only a few had a complete set. Boveri concluded that *each chromosome was different* and had a specific effect on development.

Questions

1. Could Boveri have explained his results by the hypothesis that each blastomere simply needed a *minimal number* of chromosomes?
2. How are the disastrous consequences of multiple fertilizations avoided in normal development of sea urchins and many other species?

Geneticists were originally concerned with the *transmission* of heritable traits from one generation to the next. Only a few insightful pioneers recognized the value of genetics to the analysis of development (Horder et al., 1985). One of them was Hadorn (1948), who focused on the stages of developmental arrest in mutants carrying ***lethal factors,*** mutated genes that cause early death. These studies showed certain genes needed to be active at specific stages of development and in particular organs. In addition, the importance of *homeotic genes,* such as *Ultrabithorax*$^+$, for the development of the overall body pattern was recognized early (Goldschmidt, 1938). These genes were investigated systematically by Lewis (1978). However, most researchers in developmental biology remained uninterested in genetics, and vice versa. Until recently, the gap between genes and their manifestation in development seemed too wide to bridge.

The relationship between embryology and genetics began to change with the advent of molecular biology. DNA was revealed as the carrier molecule of genetic information (Avery et al., 1944), and its double-helical structure was clarified (Watson and Crick, 1953). DNA transcription into messenger RNA (mRNA) and other RNAs, and the translation of mRNA into protein, were recognized a few years later. The first model of gene regulation (Jacob and Monod, 1961) provided biologists with a concept of how genes could be turned on and off in living cells. The model proposed that the transcribed portion of a gene was associated with *regulatory DNA sequences,* to which *regulatory proteins* would bind to control the transcription of the gene. Since the regulatory proteins themselves were gene products, their genes in effect controlled the activity of the regulated genes. Thus, development came to be seen as a "program" encoded in DNA and enacted through networks of gene regulation.

15.2 Modern Genetic Analysis of Development

The primary tools for the genetic analysis of development are mutants in which genes that control key developmental steps are altered. In this section, we will discuss some general strategies for the generation and maintenance of mutants. We will also introduce five organisms that have been used extensively for genetic analysis, so that hundreds or thousands of mutant strains are available from stock centers and individual laboratories. These species are the fruit fly, *Drosophila melanogaster;* the roundworm, *Caenorhabditis elegans* (commonly abbreviated as *C. elegans*); the house mouse, *Mus musculus* (Wilkins, 1993), the zebrafish, *Danio rerio* (Mullins et al., 1994), and the mouse ear cress, *Arabidopsis thaliana.* While the mouse and *Drosophila* have been used for genetic work since the beginning of the twentieth century, the zebrafish and the mouse ear cress have come into wide use only recently. Some genetically relevant data on these species are compiled in Table 15.1. Among these data, the ***genome size*** refers to the number of DNA base pairs that comprise one haploid *genome.*

table 15.1 Genetic Characteristics of Representative Organisms

	Genome Size (millions of base pairs)	Non-repetitive DNA (% of total DNA)	Haploid Number of Chromosomes	Generation Time (from egg to fertile adult)
Escherichia coli	4.2	100	1	30 min
Arabidopsis thaliana	130	80	5	2 months
Caenorhabditis elegans	97	83	6	3 days
Drosophila melanogaster	180	70	4	3 weeks
Danio rerio	1700	?	25	2 months
Mus musculus	3000	60	20	2 months
Homo sapiens	3000	70	23	12–14 yr

A small genome size is advantageous for efforts to sequence the entire genome of an organism and for attempts to identify nearly *all* genes involved in a biological process of interest. The latter is done in *saturation mutagenesis screens,* in which a large number of mutant lines is generated and screened for abnormal phenotypes that seem relevant to the process of interest.

In order to appreciate the genetic analysis of development, one needs to be familiar with a few genetic terms and concepts. Some of these have already been introduced in Section 1.4, and others are compiled for easy reference in this chapter under Method 15.1. More comprehensive discussions of these concepts can be found in textbooks of genetics.

MUTANTS REVEAL THE HIDDEN "LOGIC" OF EMBRYONIC DEVELOPMENT

Mutant phenotypes can reveal the hidden logic of the genetic control of development. The *Drosophila* maternal effect mutant *torso (tor)* may serve as an example. Females *homozygous* for *loss-of-function* alleles (see Section 1.4) of *tor* produce nonviable embryos that are deficient in their anteriormost as well as their posteriormost body parts (Fig. 15.4; Schüpbach and Wieschaus, 1986a). In contrast, females with at least one *gain-of-function* allele of *tor* have offspring with enlarged terminal regions and little in between. Apparently, tor^{+} encodes a product required for the formation of the anteriormost as well as the posteriormost body parts. This was an unexpected conclusion because classical embryological experiments had revealed the existence of cytoplasmic determinants that are necessary for development of either anterior or posterior body parts, but had provided no evidence for a single maternal factor required for both terminal regions.

Similarly unexpected phenotypes were found for a group of genes aptly termed *pair-rule genes* because they affect the body pattern with a bisegmental periodicity. For instance, *Drosophila* embryos homozygous for *null alleles* of the *fushi tarazu* gene show half the normal number of segments, with each of the segments being longer than normal (Fig. 15.5). Indeed, *fushi tarazu* is Japanese for "not enough segments." Closer analysis reveals that the mutants lack portions of the body pattern containing every other segment boundary. This means that *fushi tarazu*$^{+}$ and several other pair-rule genes are active in regions with a *bisegmental periodicity.* Again, the

Figure 15.4 *Drosophila* larvae derived from **(a)** a normal wild-type mother and **(b)** a homozygous *torso*$^{-}$ *(tor*$^{-}$*)* mutant mother. The photographs show cuticle preparations of fixed and cleared specimens under dark-field illumination (see legend to Fig. 1.20 for more explanation). Larvae derived from *tor*$^{-}$ mothers are lacking terminal anterior and posterior parts.

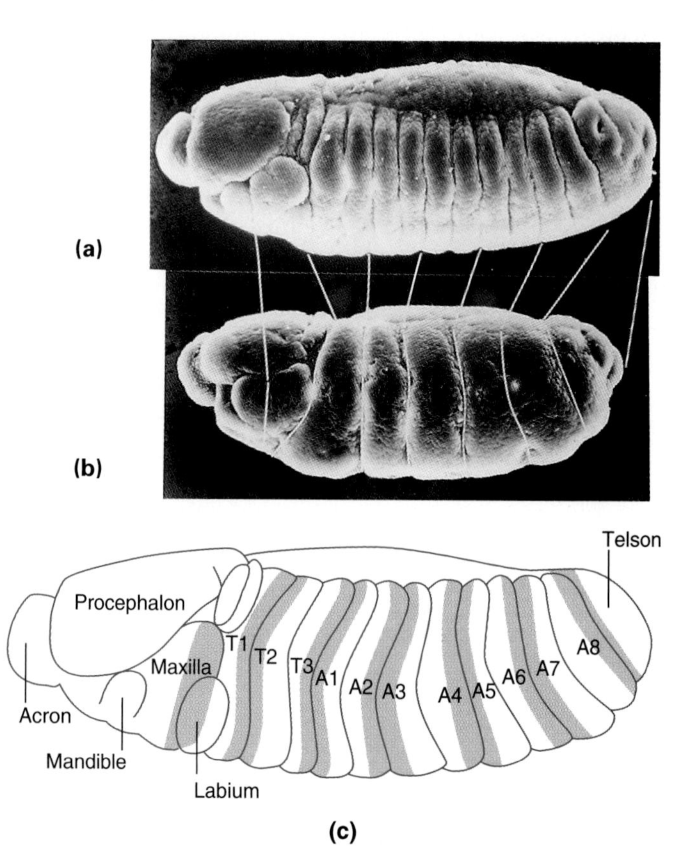

Figure 15.5 Phenotype of the *Drosophila* mutant *fushi tarazu*^{2}. **(a)** Scanning electron micrograph of a wild-type embryo in lateral view. **(b)** *Fushi tarazu*^{2} embryo at a corresponding stage of development. White lines between parts a and b connect homologous portions of the segmented germ band. **(c)** Diagram of a wild-type embryo. The colored areas are missing in the *fushi tarazu*^{2} embryo. These areas are the expression domains of *fushi tarazu*$^{+}$ during the blastoderm stage.

conclusion that *Drosophila* embryos at this stage of development consist of units that are two segments long was completely unexpected.

Additional information about the hierarchies of genes that control development can be derived from analyzing the phenotypes of *double mutants*—that is, strains that are mutant in two different (nonallelic) genes. The *tor* gene discussed earlier may again serve as an example. Females homozygous for loss-of-function alleles of another gene, *torsolike* (*tsl*) give rise to embryos with a phenotype much like those derived from tor^- females. Apparently, then, tor^+ and tsl^+ act within the same pathway of control. Does the torsolike gene product act upstream or downstream of the torso product in the signaling pathway that controls the anteroposterior body pattern? Double mutant females heterozygous for a gain-of-function allele of *tor* and homozygous for a loss-of-function allele of *tsl* produce offspring showing the same phenotype as offspring from females carrying only the *tor* gain-of-function allele. In technical terms, the gain-of-function allele of *tor* is ***epistatic*** (Gk. *epi*, "upon"; *statikos*, "standing") over the loss-of-function allele of *tsl*. The epistasis suggests that the torso product acts downstream of the torsolike product, because otherwise it would be difficult to explain how torso could gain function in the absence of torsolike. Indeed,

method 15.1

The Generation and Maintenance of Mutants

In diploid organisms, which carry two sets of chromosomes, the corresponding maternal and paternal genes and chromosomes are *homologs* of each other. Chromosomes that differ between males and females of the same species are called *sex chromosomes;* they are often symbolized by the letters X and Y. All other chromosomes are collectively called ***autosomes.*** Genes located on sex chromosomes are known as ***sex-linked*** (or ***X-linked*** or ***Y-linked***) genes.

Alternative forms of a gene are referred to as *alleles.* The allele of a gene most commonly found in natural populations is called the *wild-type allele;* it is marked by a superscript plus sign following the name or abbreviated name of the gene. Other alleles are referred to as variant or mutant alleles. The position on a chromosome occupied by a particular set of homologous alleles is called the ***locus*** for those alleles.

A diploid organism carrying the same allele on two homologous chromosomes is *homozygous* for the allele. An organism carrying different alleles of a gene is *heterozygous* for the gene. Sometimes, a diploid organism has only one copy of a gene because it is located on the X chromosome in an XY individual, or because the homologous gene has been lost. The organism is then described as ***hemizygous*** for the allele that is present. If two alleles *A* and *a* of a gene produce different phenotypes and the phenotype of the heterozygote *A*/*a* is the same as the phenotype of the homozygote *A*/*A*, then *A* is called the ***dominant*** allele and *a* the ***recessive*** allele. (Note that capital and lowercase letters are sometimes used to denote dominant and recessive alleles, respectively.) Most mutant alleles are recessive to the wild type.

Mutant alleles are classified according to the ways they change the function of the gene. Most mutant alleles cause some loss of normal gene function. Mutations that completely abolish a gene function produce *null alleles.* Their phenotypes are equivalent to the phenotypes of chromosomal deletions in which the gene is physically lost. Null alleles are often marked by a superscript minus sign. For instance, $hairy^-$ is the null allele of the *hairy* gene. Mutations that weaken the function of a gene produce *loss-of-function alleles.* A loss-of-function allele is usually recessive to its wild-type allele except when one copy of the wild-type allele is insufficient to produce the wild phenotype; such genes are called ***haploinsufficient,*** and their loss-of-function alleles can be dominant. Another type of mutation, which causes a gene to be activated in the wrong place or at the wrong time, generates a *gain-of-function allele.* This type of allele also tends to be dominant. Alleles are called recessive or dominant *lethal factors* if they cause premature death when homozygous or heterozygous, respectively.

The genotype of an organism, in principle, refers to all the genes of an individual, regardless of whether they are expressed. Likewise, the *phenotype* of an individual encompasses all its observable characteristics. The common *phenotype* in a natural population is the *wild phenotype.* In practice, the genotypes and phenotypes of organisms are usually compared selectively for the gene(s) under consideration.

While some mutants have been found by chance, most have been generated intentionally by means of chemical or physical mutagens. A widely used chemical mutagen, ethylmethane sulfonate (EMS), causes single-base substitutions in DNA, referred to as ***point mutations.*** Larger chromosomal rearrangements are caused by X-rays, which break the phosphodiester "backbone"

(continued)

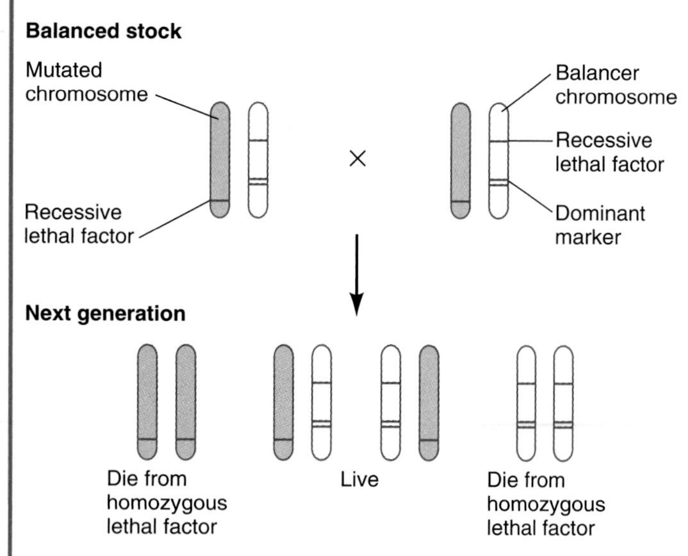

Figure 15.m1 Use of a balancer chromosome to maintain a homologous mutated chromosome carrying one or more lethal factors. Since the balancer chromosome carries a different lethal factor itself, both homozygotes die while the heterozygotes survive. Because inversions in the balancer chromosome prevent crossing over the genes on the mutated chromosome are permanently linked to each other.

of DNA. This procedure results in chromosomal ***deletions*** (missing segments), ***translocations*** (segments moved from one chromosome to another), or ***inversions*** (segments cut out and reinserted with reversed orientation). A special class of mutagens are mobile DNA elements, which cause deletions as well as ***insertions*** (see Section 15.5).

Of particular advantage to developmental biologists are ***temperature-sensitive* (*ts*)** mutants, which develop the wild phenotype when raised at a ***permissive temperature*** but express the mutant phenotype when raised at a ***restrictive temperature.*** Most ts mutants have a single amino acid substitution in the protein encoded by the mutated gene, a change that renders the protein unstable at restrictive temperatures. Stocks of ts mutants can be maintained conveniently at permissive temperatures, even if the mutant allele is lethal at the restrictive temperature. To study the effects of temperature-sensitive mutations on development, ts mutants are simply subjected to a restrictive temperature. Temperature shifts at regular intervals make it possible to identify the ***temperature-sensitive period* (*TSP*),** the period during which mutant individuals must be kept at a permissive temperature in order to develop the wild phenotype. The TSP reveals the stage at which the wild-type gene function is required for normal development. The TSP is also used in experiments to eliminate individuals with certain genotypes from a study population (see Fig. 15.m2).

The maintenance of non-ts mutants can be facilitated by the generation of ***balanced stocks.*** These involve the use of ***balancer chromosomes,*** which carry dominant marker mutations, recessive lethal factors, and multiple inversions, which suppress recombination (Fig. 15.m1). These chromosomes are used to maintain any homologous chromosome carrying another recessive lethal factor. In a strain with such a pair of chromosomes, both types of homozygotes die while the heterozygotes survive because each lethal factor on one of the two chromosomes is covered by the wild-type allele on its homolog.

subsequent molecular analysis has shown that *tor*$^+$ encodes a receptor tyrosine kinase, and that the *tor* gain-of-function allele produces a receptor that is active constitutively, with or without its ligand. The *tsl*$^+$ gene encodes a product that seems to be involved in the processing of the torso ligand (see Section 22.3).

DROSOPHILA MUTANTS ALLOW THE ANALYSIS OF A COMPLEX BODY PATTERN

One of the core topics that have defined developmental biology as a discipline has been *pattern formation* (to be discussed in Chapters 21–24). A *pattern* can be defined as a harmonious array of different elements arranged in distinct sequences and proportions. Patterns abound in living organisms and can be recognized at different levels of organization. For instance, the body of the fruit fly *Drosophila melanogaster* is subdivided into head, thorax, and abdomen. The head consists of a capsule carrying antennae, eyes, and brain, followed by three segments with mouthparts. The thorax is composed of three segments carrying legs, wings, and balancers. The abdomen has eight segments distinguishable by their pigmentation and genital appendages. Each of the segments and appendages in turn has its own array of distinct elements. For instance, the wing has a row of strong bristles at its anterior margin, a characteristic pattern of veins, and a thin hair growing on each cell of the wing epidermis (see Fig. 6.4).

Because of their rich and detailed body patterns, insects have been the favorite objects of many investigators of pattern formation (Sander, 1976). For the genetic analysis of pattern formation in insects, *Drosophila melanogaster* has become the preferred organism. There is an abundance of known genes affecting every aspect of the body pattern, ranging from *maternal effect genes* that affect the overall body pattern, such as *bicoid*$^+$ (see Fig. 1.20), to cell-autonomous genes, such as *multiple wing hairs*$^+$, which control the number and shape of hairs on single cells (see Fig. 6.m2).

Saturation Mutagenesis Screens

Saturation mutagenesis screens are of great importance to the genetic analysis of biological processes, because they help researchers identify *nearly all* genes involved in a biological process of interest. To this end, large numbers of individuals are mutagenized and mated with partners of suitable genotype to breed lines of offspring that are screened for abnormal phenotypes with regard to the trait of interest. Any trait that can be screened for, such as male fertility, having a normal body pattern, or being able to recognize colors, can be investigated in this fashion.

To obtain at least one mutant allele of almost every gene involved in the process of interest, most genes have to be "hit" several times. (As an analogy, one may think of rain beginning to fall on a cobblestone pavement. By the time at least one drop has fallen on *nearly every* stone, several drops have fallen on *most* stones.) A mathematical formula known as the Poisson distribution indicates that an average of five mutant alleles per gene is required to ensure that 99% of the genes have been mutated at least once. It is also desirable to have multiple mutant alleles of a gene because the resulting range of phenotypes provides a better basis for judging the function of the wild-type allele than a single mutant phenotype would.

The number of mutagenized lines that must be screened to achieve saturation depends on the number of genes per genome and on the efficiency of the mutagenizing treatment (number of mutations per treated genome). Thus, if the organism used has 5000 genes and the regimen employed for mutagenesis generates 0.7 mutations per treated genome, then the number of lines to be screened for 99% saturation is $5000 \times 5 : 0.7 = 36{,}000$. Whether it is feasible to screen this many lines depends on several parameters, including space and funds required to maintain thousands of lines, and the labor involved in breeding the mutagenized individuals and screening their offspring. Clearly, the preferred organisms for this type of analysis have small numbers of genes, are easy to breed and maintain, and tolerate high mutagen doses.

With multiple mutant alleles per gene expected, it is important to determine whether any two mutants that have similar phenotypes are *allelic*—that is, caused by mutations in the same gene. This is done by a *complementation test.* First, each line is crossed with wild-type individuals to determine whether the mutant allele is recessive to the wild-type allele, which is usually the case. If so, heterozygous individuals from two possibly allelic lines are crossed. If all the offspring show the wild phenotype, then each mutated gene from one mutant line is apparently complemented by the wild-type allele of the same gene from the other mutant line. Thus, the two lines are probably mutated in different genes. However, if a cross between two mutant lines produces offspring that do not show the wild phenotype, then the mutated genes fail to complement each other and are probably allelic.

The career of *Drosophila* as a laboratory animal began in Thomas H. Morgan's "fly room" at Columbia University, where he and a group of enthusiastic coworkers began genetic studies on *Drosophila* around 1910. Fruit flies are easy to keep in the laboratory, and they mate as single pairs. Most important, the flies survive well with both chemically and X-ray–induced mutations. An added advantage of *Drosophila* is the occurrence of large polytene chromosomes, especially in the salivary glands of larvae (see Section 7.5). Their characteristic banding pattern makes it easy to identify chromosomal deletions, duplications, and inversions. Correlations between such chromosomal rearrangements and mutant phenotypes have made it possible to establish *cytogenetic maps* of *Drosophila* chromosomes, which show the physical location of thousands of genes with great precision. Our knowledge about mutagenesis, the linear order of genes on a chromosome, genetic sex determination, and other genetic concepts arose mostly from the work of Morgan and his associates. In recognition of these achievements, Morgan was awarded the 1933 Nobel prize in Physiology or Medicine.

The use of modern genetic and molecular tools in the analysis of development has been exceedingly successful in *Drosophila*. In 1980, Nüsslein-Volhard and Wieschaus published the first part of a genetic analysis of pattern formation in the *Drosophila* embryo. In this and many following studies, the researchers asked how maternally and embryonically expressed genes act together in forming the body pattern seen in the hatched larva. This work, and the earlier investigation of homeotic mutants by Lewis (1978), were recognized in 1995 by the award of the Nobel prize in Physiology or Medicine to Lewis, Nüsslein-Volhard, and Wieschaus.

The genetic and molecular analysis of *Drosophila* development involved several *saturation mutagenesis screens* (see Method 15.2) generating multiple mutant alleles in hundreds of genes involved in embryonic pattern formation.

NÜSSLEIN-VOLHARD and coworkers (1984) based the following mutagenesis screen on the assumption that mutant *Drosophila larvae* with abnormal body patterns would

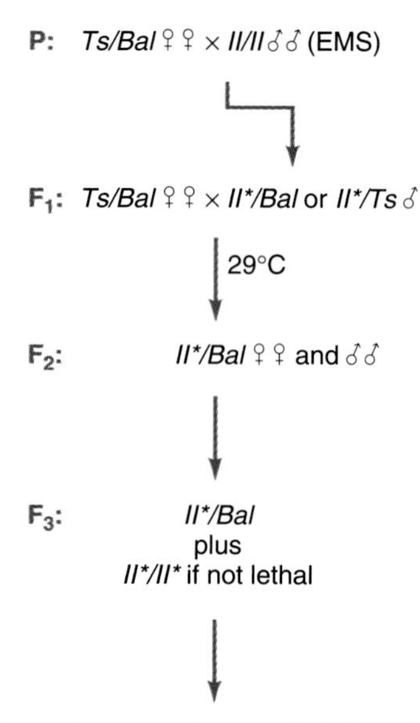

Figure 15.6 Crossing scheme used by Nüsslein-Volhard and coworkers (1984) to establish mutant *Drosophila* lines, each carrying a particular lethal mutant allele. Males homozygous for a chromosome II carrying a set of recessive marker genes (*II*) are mutagenized with ethylmethane sulfonate (EMS). They are mated (P generation) with females having specially engineered chromosomes II, one being a balancer chromosome (*Bal*) and the other carrying a dominant lethal but temperature-sensitive mutant allele (*Ts*). In the offspring, each individual carries a particular mutated (*II**) second chromosome. Individual males are mated again with *Bal/Ts* females (F_1 generation). Their offspring are kept at the restrictive temperature to kill all individuals with a *Ts* chromosome, so that the survivors (F_2 generation) must be heterozygous *II*/Bal.* Their offspring (F_3 generation) are *II*/Bal* (showing the *Bal* phenotype) or *II*/II** (not showing the *Bal* phenotype). Lines showing only the *Bal* phenotype are kept, since their *II** chromosome must carry a lethal factor.

not be able to hatch from the *chorion* (egg envelope), because larvae must be fully vigorous to accomplish this step. Therefore, they screened for embryonically expressed genes that could be mutated to lethality. Figure 15.6 shows the crossing scheme employed to identify such genes on the second chromosome. (Similar screens were carried out separately for mutations on the first and third chromosomes.) Males homozygous for a second chromosome carrying a set of recessive marker genes were fed ethylmethane sulfonate (EMS), a chemical mutagen that causes single base changes, so-called point mutations, in DNA. Thus, these males would produce mutated sperm. The overall goal was to recover *balanced stocks,* each with a unique mutated second chromosome present in each individual of that stock. To this end, the mutagenized males were mated with females having a pair of specially engineered second chromosomes. One was a *balancer chromosome,* while the other carried a dominant lethal but *temperature-sensitive* (*Ts*) allele. From the offspring (F_1), *individual* males were mated again with females of the mother's genotype, and the offspring (F_2) of each male were kept as a separate *line.* In each line, the researchers killed all individuals except those that carried the mutated chromosome over the balancer chromosome. This was accomplished by raising the temperature to a level (29°C) at which all carriers of the *Ts* allele would die. The surviving F_2 individuals of each line were crossed among each other to test the lethality of those of their F_3 offspring that were homozygous for the mutated chromosome. Only those lines were kept where *all* F_3 individuals showed the phenotype conferred by the balancer chromosome. These lines evidently carried a lethal factor on the second chromosome.

From about 10,000 F_1 crosses set up, 4580 lines were recovered as balanced stocks—each with a unique second chromosome carrying a homozygous lethal mutation. The 4580 lines mutated to general lethality were tested further to identify those lines having *embryonic* lethality. In these lines, about 25% of all eggs laid were expected not to hatch because the embryos were homozygous for a mutation interfering with embryogenesis. Eggs were inspected microscopically from the 2843 lines selected under this criterion. A total of 272 lines produced embryos that were clearly distinguishable from wild-type embryos in cuticle wholemounts of the type shown in Figure 15.4. *Complementation tests* among these lines showed that they represented 61 different genes. Thus, the average number of alleles per gene was about 4.5, indicating that the screen was close to saturation (see Method 15.2). The investigators concluded that the 61 genes identified represented the majority of the genes on the second chromosome that could be mutated to an embryonic lethal phenotype with a recognizable alteration in the larval cuticle. Similar estimates were obtained for the other chromosomes, except for the fourth, which is very small.

Questions

1. Why did the investigators mate *single* males in the F_1 generation whereas *groups* of males were mated in the P and F_2 generations?
2. Less than 10% of the lines showing embryonic lethality showed deviations from the normal body pattern that could be discerned in cuticle preparations. Why may the larvae of the other 90% of the lines have been unable to hatch?

The results of saturation mutagenesis screens carried out by Nüsslein-Volhard and coworkers (1984) suggest that a little more than 3 × 61, or approximately 200, embryonically expressed genes make a detectable contribution to the cuticular body pattern of the larva. This is a relatively small fraction, representing about 2% of all *Drosophila* genes. Some of these genes make dramatic contributions, as will be discussed in Chapter 22, while the effects of other genes are barely detectable.

CAENORHABDITIS ELEGANS MUTANTS UNCOVER GENE ACTIVITIES CONTROLLING CELL LINEAGES

Caenorhabditis elegans is a tiny, free-living nematode, measuring just over 1 mm in length and 70 μm in diameter (see Fig. 25.2). Adults of *C. elegans* are mostly self-fertilizing ***hermaphrodites,*** with gonads producing both sperm and oocytes. Males develop infrequently but hermaphrodites, when fertilizing their eggs, prefer sperm obtained from males over their own. In laboratory cultures, the worms can be handled with the same ease as bacteria. *C. elegans* completes its entire life cycle in just 3 days, and 100,000 individuals can live in a Petri dish. Because the worm is transparent, investigators can watch its development live under the microscope.

The tiny worm rose to scientific fame through the determined efforts of Brenner (1974) and his coworkers, who wanted to study the development of the nervous system in a simple animal that had a small repertoire of standardized behaviors. Nematodes seemed best suited to this project, because they are simple but "real" animals: they have skin, muscles, a gut, and a nervous system, and when touched with a brush, they react.

Most important, each nematode species has a fixed number of somatic cells, which always develop in the same lineages (see Section 1.3). In *C. elegans*, hermaphrodites have exactly 959 somatic cells and about 2000 gametes, while males have exactly 1031 somatic cells and about 1000 sperm. The anatomy of each developmental stage has been reconstructed, cell by cell, from serial sections viewed under the electron microscope (Sulston et al., 1983). Thus, *C. elegans* became the first animal with a complete "parts list" (13 pages long!) and detailed cell lineage maps from egg to adult (Wood, 1988a).

The cell constancy of *C. elegans* has made it possible to isolate mutants with abnormal lineages. These mutants provide unique opportunities for scientists to study those gene activities that control the normal sequence of cell divisions. For instance, mutations of the *unc-86* gene alter in equivalent ways the lineages of three different postembryonic neuroblasts (Figs. 1.12 and 15.7). In each of these mutant lineages, cells repeat division patterns normally displayed by their wild-type lineage precursors. The exact knowledge of all cell lineages in *C. elegans* has also allowed investigators to analyze inductive interactions between cells, the timing of cell divisions, and programmed cell death with unparalleled precision (see Chapter 25).

C. elegans is very amenable to genetic and molecular analysis (Brenner, 1974; Kemphues, 1989). The propagation of *C. elegans* by self-fertilization makes the isolation of homozygotes for recessive mutations very easy. A mutated hermaphrodite is simply allowed to reproduce for two generations to obtain homozygous mutants. Diploid cells have five pairs of *autosomes* plus one X chromosome in males and two X chromosomes in hermaphrodites (see Table 15.1). The genome size of

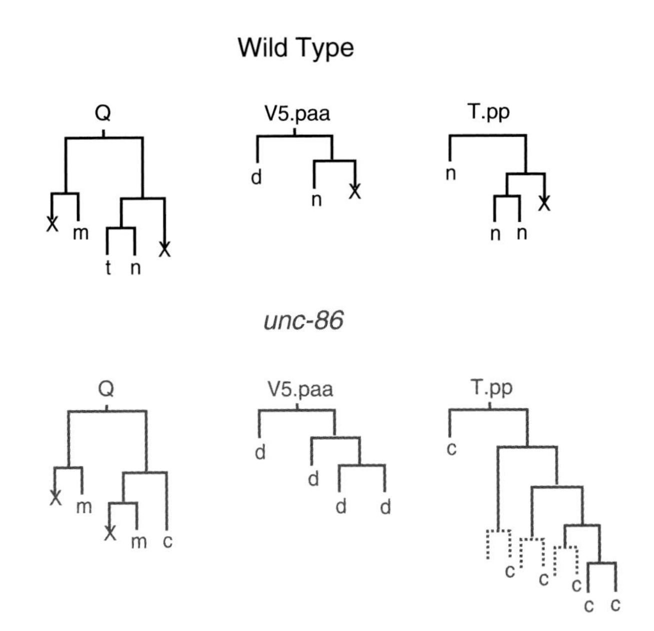

Figure 15.7 Lineages of three postembryonic neuroblasts (Q; V5.paa; and T.pp) in wild-type and *unc-86* mutant *C. elegans*. (The origin of these neuroblasts from blast cells can be gleaned from Figure 1.12b. T.pp is the *p*osterior daughter cell of the *p*osterior daughter cell of T. V5.paa is the *a*nterior daughter of the *a*nterior daughter of the *p*osterior daughter of V5.) c = neuron with compact nucleus; d = dopaminergic neuron; m = migrating neuron; n = neuron; t = microtubule cell; X = programmed cell death. In the mutant, posterior daughter cells (drawn to the right) tend to undergo extra divisions repeating the lineage pattern normally shown by the parental cell.

C. elegans, at less than 100 million base pairs, is one of the smallest in the animal kingdom.

Because of their small genome sizes, and due to the vision and determined efforts of several investigators, *C. elegans* and *Drosophila* have become the first multicellular organisms with a completely sequenced genome (*C. elegans* Sequencing Consortium, 1998; M. D. Adams et al., 2000). About 19,000 genes have been identified in the worm's genome, and about 14,000 in the fly's. Strategies to inactivate each of these genes in order to study the biological effects of its loss are being explored (*Plasterk, 1999*). Because many developmentally important genes have been well conserved in evolution, the genetic and molecular analysis of *C. elegans* and *Drosophila* will greatly accelerate the analysis of higher organisms as well (Kornberg and Krasnow, 2000). The conserved genes include in particular those involved in controlling growth and cell differentiation. Mutations in these genes are often involved in cancer and other human diseases (see Section 28.7). To explore the entire network of genetic controls involved in such diseases, a small and genetically well-known organism is almost always the best point to start. The DNA sequence data from *C. elegans* and *Drosophila* are therefore certain to become invaluable resources for basic as well as applied research.

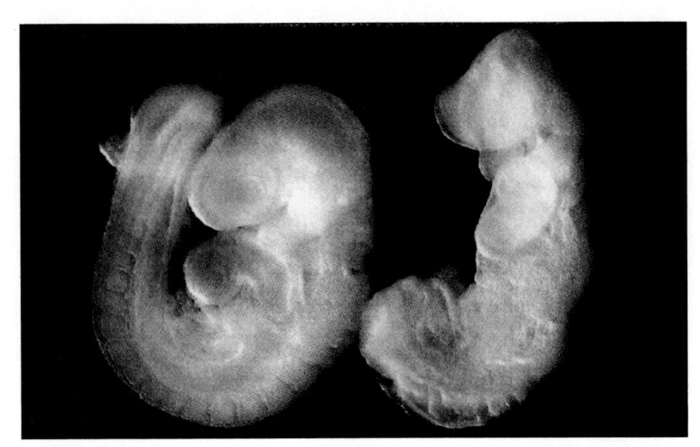

Figure 15.8 Phenotype of the mouse mutant *Brachyury* (*T*). A wild-type mouse embryo at 9.5 days of gestation is shown to the left. To the right is a *T/T* embryo of the same age; it has failed to make posterior notochord and somites.

GENETIC ANALYSIS OF THE MOUSE *MUS MUSCULUS* USES EMBRYONIC STEM CELLS

The mouse has been used for genetic experiments since the beginning of the twentieth century (Wilkins, 1993; I. J. Jackson, 1989). Several human genetic disorders have close parallels in the mouse. Human development is also similar to that of mice, except for the period between the blastocyst stage and gastrulation. Therefore, the mouse is often considered a model for human genetics and development.

Many mouse genes were discovered as spontaneous mutants. One of these was originally named *Brachyury* (Gk. *brachys*, "short"; *oura*, "tail") in reference to the short, kinky tails of heterozygotes (Dobrovolskaia-Zavovskaia, 1927). Homozygotes die before birth showing severe disturbance of posterior axial and paraxial mesoderm (Fig. 15.8). The *Brachyury* gene locus became later known as the *T-locus* due to its complex interactions with other genes in an adjacent chromosome segment known as the t complex (Bennett, 1975). T^+ is first expressed early in the entire embryonic mesoderm but later is restricted to chordamesoderm. Proper expression of T^+ is necessary for normal mesodermal cell movements during gastrulation and somite development (Wilson and Beddington, 1997). T^+ encodes a transcription factor, and its target genes appear to include cell adhesion molecules (Wilson et al., 1995).

Unfortunately, saturation mutagenesis screens are impractical with mice because of their relatively small litter size, long generation time, large genome size, and high maintenance costs (see Table 15.1). However, a technique to be discussed more fully in Section 15.5 makes it possible to generate mutants for any mouse gene provided the gene has been cloned. This technique relies on a type of cells specifically available in mammals and known as *embryonic stem cells* (*ES cells*). Being derived from the *inner cell mass* (*ICM*) of blastocysts (see Section 5.2), they multiply indefinitely in culture and lend themselves to genetic manipulation. When added back to the ICM of embryos, the genetically engineered ES cells will contribute to all embryonic structures including the germ line. By breeding such chimeric animals, one can obtain mice that are heterozygous or homozygous for any cloned gene.

In order to clone a mouse gene for which no mutant is available yet, a gene with sufficient sequence similarity first needs to be identified by mutant alleles and cloned from another species. Genes that have evolved from a common precursor gene are called *homologous genes*. One of the most remarkable results from the molecular analysis of development has been that homologous genes are often very well conserved in evolution, at least in their functional domains. This is not only intellectually rewarding but also of great practical importance. If a developmentally important gene has been cloned from one species, chances are that homologous genes can be found in other, often very distantly related, species. Molecular techniques that use a cloned segment from a gene of species A to probe for homologous genes in species B are discussed in Section 15.3.

THE ZEBRAFISH *DANIO RERIO* IS GENETICALLY TRACTABLE AND SUITABLE FOR CELL TRANSPLANTATION

The drawbacks of mouse genetics have kept developmental biologists looking for another vertebrate species that would be genetically more tractable but could still be considered a fair model for the human. The small zebrafish *Danio rerio* is emerging as the species of choice.

Danio can be handled in sufficiently large numbers to carry out saturation mutagenesis screens. These are facilitated by the fact that eggs fertilized with UV-irradiated sperm develop as haploid embryos. Thus, researchers can detect early effects of recessive mutant alleles without having to breed them to homozygosity. Two large mutagenesis screens for genes affecting embryonic development have already been carried out (Haffter et al., 1996; Driever et al., 1996). With average allele frequencies of 1.5 and 2.4, neither screen came close to saturation, but each has identified hundreds of genes affecting morphogenetic movements, the specification of embryonic axes, and the development of various organs.

In contrast to the intrauterine development of mice, the external development of fish makes continuous observation of normal and mutant embryos very easy. The large size and transparency of *Danio* embryos (Fig. 15.9) also facilitate classical types of analysis involving the removal and transplantation of cells.

The next major efforts will be to map the identified genes relative to the positions of known molecular markers on the 25 chromosomes of *Danio* and to prepare sets of overlapping segments of cloned DNA spanning each chromosome (Postlethwait and Talbot, 1997). With these prerequisites in place, it will be possible to

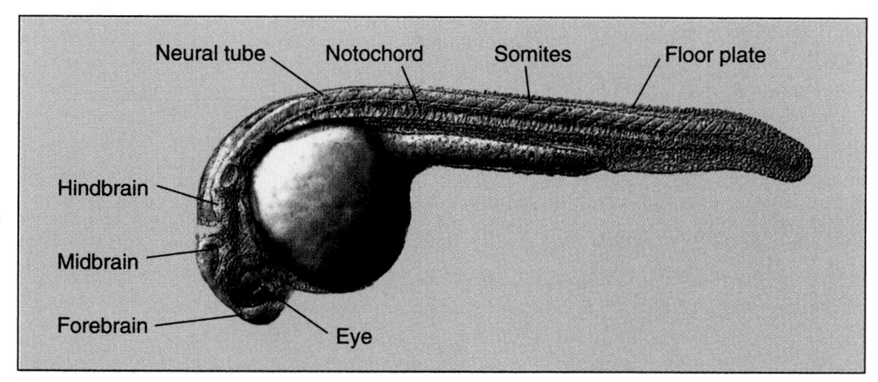

Figure 15.9 Zebrafish embryo after 29 h of development, photographed with Nomarski optics.

sequence the entire genome and to clone any gene of interest. At that point, *Danio* will be the only vertebrate in which any mutable gene can be found and analyzed by both embryological and molecular techniques.

GENETIC ANALYSIS REVEALS SIMILAR CONTROL CIRCUITS IN PLANT AND ANIMAL DEVELOPMENT

The great success of genetic and molecular studies on fly and mouse development has enticed several researchers to apply the same methods to entirely different systems—plants. Plant genetics has a long history, not only in agriculture but also in science: the Mendelian laws of genetics are based on Gregor Mendel's cross-pollinations of peas. The snapdragon *Antirrhinum majus* and the corn plant, *Zea mays*, have been favorite subjects of genetic research in the twentieth century. During recent years, the mouse ear cress *Arabidopsis thaliana* has emerged as a model plant with very favorable traits for genetic studies. It belongs to the flowering plants, or angiosperms, and among these it represents the dicotyledons—that is, the plants with two cotyledons, or seed leaves.

Arabidopsis is small enough to be grown in test tubes, so that large numbers can be kept in the laboratory. Its generation time is about 2 months, and each plant produces many seeds. It self-fertilizes, so that mutant alleles are easily made homozygous. Most important, its genome size (7×10^7 nucleotide pairs) is even smaller than those of *Caenorhabditis elegans* and *Drosophila* (see Table 15.1). These features combine to make saturation screens for certain classes of genes feasible and should allow investigators to clone any gene on the basis of its phenotype and map position.

The application of genetic and molecular tools to plant development has proved very rewarding. In particular, investigators have isolated mutants with specific abnormalities in embryonic pattern formation. Some of these mutants show striking similarities to patterning mutants of *Drosophila*. Homeotic genes in plants, like their animal counterparts, are characterized by transformations of certain organs into other organs that are normally formed elsewhere (Fig. 15.10). For instance, one of these mutants transforms stamens (the organs that produce pollen grains) into petals (the colored leaflike organs of flowers). Also like animal homeotic genes, plant homeotic genes have restricted expression domains and encode transcription factors. The parallels between plants and animals with regard to genetic control of pattern formation, and some of the differences between animal and plant development, will be discussed in Chapter 24.

15.3 DNA Cloning and Sequencing

While genetic analysis is a powerful strategy in itself, its application has been vastly expanded and accelerated by the advent of ***DNA cloning,*** techniques that allow investigators to produce billions of copies of any given segment of DNA. Since their invention in the 1970s, these techniques have revolutionized many areas of science, medicine, agriculture, and law. For

(a)

(b)

Figure 15.10 Wild-type and homeotic mutants of the mouse ear cress *Arabidopsis thaliana.* **(a)** Wild type with sepals (hidden), four white petals, six stamens, and a pistil. **(b)** An *agamous* mutant showing extra petals and sepals in the absence of stamens and a pistil.

developmental biologists, DNA cloning has provided the means to study *differential gene expression* at the molecular level. Cell differentiation and pattern formation can now be analyzed in terms of specific DNA sequences and regulatory proteins.

Any cloned gene lends itself to detailed analysis. ***DNA sequencing*** means determining the precise nucleotide sequence of a DNA segment. From the known nucleotide sequence and the genetic code, it is possible to predict the amino acid sequence of the protein encoded by the gene, which often provides clues to the function of the protein in vivo. For instance, the DNA binding regions of many transcription factors have certain "signature" sequences by which they can be identified with virtual certainty. The biochemical information on gene function inferred from sequence analysis is a powerful adjunct to the biological information on gene function derived from mutant analysis. For instance, the biochemical knowledge that *bicoid*$^+$ encodes a transcription factor, when combined with the biological information that the loss-of-function phenotypes of *bicoid* and another gene, *hunchback,* are very similar, has led to the hypothesis that bicoid protein may promote *hunchback*$^+$ transcription. The testing of this hypothesis, along with many related experiments, has led to a much deeper and more accurate understanding of embryonic development than had previously been possible (see Section 16.4 and Chapter 22).

Once a gene is cloned, its RNA and protein products can also be synthesized in vitro. In turn, it is possible to prepare specific molecular probes that allow investigators to trace these gene products in living cells or in fixed material. Moreover, cloned genes can be modified by removing or adding certain DNA segments, and by changing single base pairs at will. Again based on the strategy of controlled modification, the testing of engineered genes is a powerful method of learning how genes work.

DNA CAN BE REPLICATED, TRANSCRIBED, AND REVERSE-TRANSCRIBED IN VITRO

Techniques for purifying enzymes have made it possible to replicate DNA in vitro. Indeed, it turned out that of the many enzymes involved in the natural DNA replication process only DNA polymerase is required. This discovery has made it possible to devise a technique, known as ***polymerase chain reaction (PCR),*** that will amplify minuscule amounts of DNA into large amounts sufficient for all kinds of analysis (see Method 15.3).

Similarly, the transcription of DNA into RNA can be carried out in vitro using purified *RNA polymerases.* Conversely, the reverse transcription of RNA into DNA can also be done in vitro (Fig. 15.11). The key enzyme involved in this process is *reverse transcriptase,* an enzyme made naturally by RNA viruses. The reverse transcript of an RNA is known as ***complementary DNA (cDNA),*** because the transcript and its template are *complementary,* meaning that their 5′ to 3′ polarities are antiparallel, that each guanine in one strand is opposite to a cytosine in the other strand, and that likewise each adenine is opposite to thymine (or uracil).

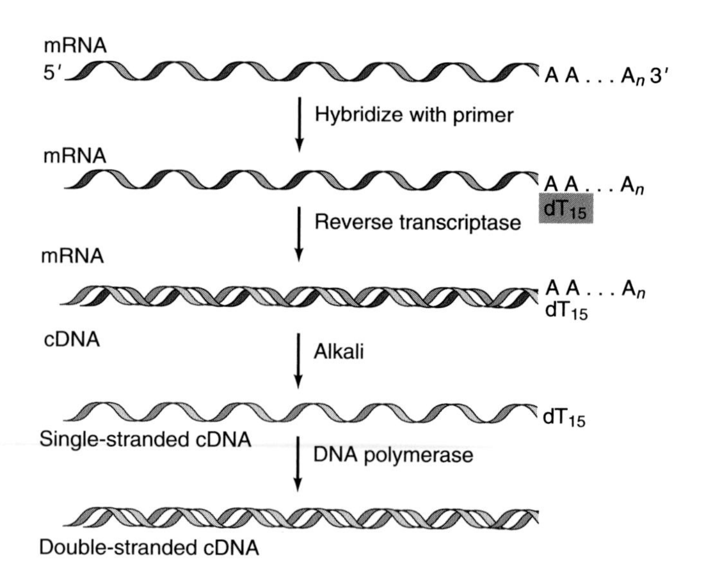

Figure 15.11 Synthesis of complementary DNA (cDNA) from messenger RNA (mRNA). Most mRNAs end with a series of adenosine residues (AA . . . A$_n$) called a poly-A tail. The poly-A tail hybridizes to a primer consisting of about 15 thymidine residues (dT$_{15}$). Reverse transcriptase, an enzyme encoded by retroviruses, transcribes the mRNA into single-stranded cDNA, starting from the primer. The cDNA is isolated by alkaline digestion of the mRNA. Finally, the single-stranded cDNA is made double-stranded with DNA polymerase.

Synthesizing DNA or RNA in vitro provides an opportunity to *label* them by adding to the reaction mixture a nucleotide that contains a radioisotope or antigenic nucleotide analog so that it can be traced later. A labeled transcript can be used as a ***probe***—that is, as a means to detect another molecule to which it binds specifically. This exceedingly useful property of nucleic acids depends on the ability of single-stranded nucleic acids to form a stable association with a complementary single-stranded nucleic acid.

NUCLEIC ACID HYBRIDIZATION ALLOWS THE DETECTION OF SPECIFIC NUCLEOTIDE SEQUENCES

When double-stranded DNA is heated in aqueous solution to 95°C the hydrogen bonds that hold the two strands together are disrupted, and the double helix dissociates into two single strands. This process of ***DNA denaturation*** is reversible. At temperatures around 60°C, single strands will readily rejoin into double strands in a process called *DNA renaturation* or ***DNA hybridization.*** However, double strands are stable only if their component single strands are complementary. Exactly how perfect the match of their sequences needs to be depends on the so-called ***stringency conditions***—that is, the temperature and ionic strength of the solution in which the DNA is dissolved. The sequence complementarity needs to be nearly perfect under *high stringency conditions* whereas

Polymerase Chain Reaction

One way of cloning DNA is cyclic replication in vitro, a process known as the *polymerase chain reaction* (*PCR*). The terminal nucleotide sequences of the DNA segment targeted for replication must be known (or guessed correctly), so that complementary oligonucleotides can be synthesized chemically. The DNA to be amplified is denatured by heat, and to each strand an oligonucleotide is hybridized at opposite ends of the targeted DNA segment (Fig. 15.m2). The oligonucleotides serve as primers for the replication of each strand by a DNA polymerase. Thus, two new DNA double helices are formed, each consisting of one original and one newly synthesized strand, and both being identical to each other and to the original double helix. The process is then repeated, with the first cycle yielding two DNA molecules, the second cycle 4 DNA molecules, and the nth cycle a clone of 2^n molecules. The key to keeping the reaction mixture cycling rapidly is the use of a DNA polymerase purified from heat-tolerant bacteria or algae, which withstand the temperature necessary for melting the double-stranded DNA molecules present at the beginning of each cycle.

The PCR is extremely sensitive and very fast. In theory, it can detect a single DNA molecule in a sample, and because each cycle requires only a few minutes, a billionfold amplification takes only half a day. The PCR's ability to detect minuscule amounts of DNA makes it the method of choice whenever the amount of material available for analysis is limited, as is often the case with mammalian embryos. PCR also holds great promise for other applications including forensic medicine and paleontology. Indeed, PCR has been used to clone and analyze the DNA from a small amount of Neanderthal bone (Krings et al., 1997). Trace amounts of RNA can also be analyzed after reverse-transcribing it into DNA.

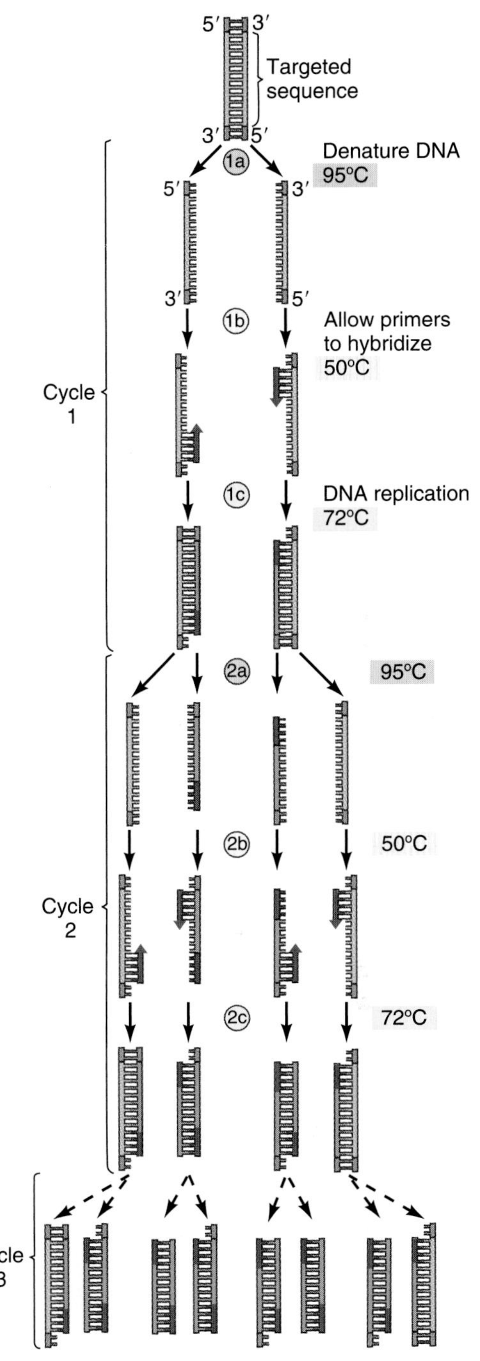

Figure 15.m2 Polymerase chain reaction (PCR). The PCR is a cyclic process that replicates every present copy of a targeted DNA segment during each cycle. A cycle consists of three steps. During step a, double-stranded DNA is separated into single strands at 95°C. In step b, chemically synthesized oligonucleotide primers are hybridized to the separated strands at 50°C. The DNA segment between the primers is targeted for replication. In step c, the targeted DNA is replicated at 72°C using DNA polymerase from heat-tolerant bacteria or algae.

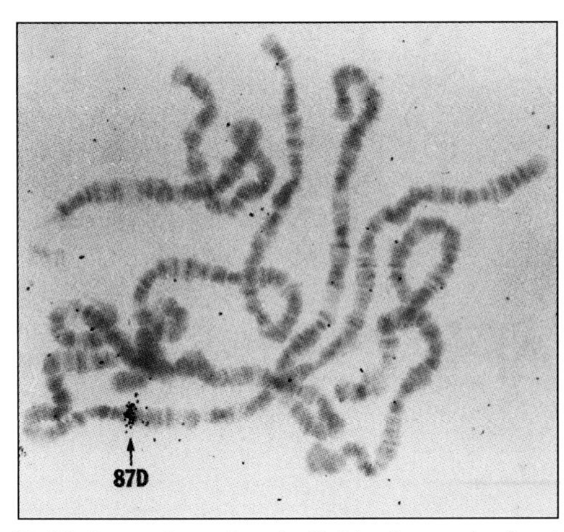

Figure 15.12 Detection of an individual gene by in situ hybridization (see Method 8.1) of a labeled probe (complementary DNA or RNA) to polytene chromosomes. In this case, the gene detected was the *rosy*$^+$ gene of *Drosophila melanogaster,* which is located on the third chromosome and has the position 87D on the cytogenetic map. The probe was radiolabeled and detected by autoradiography (see Method 3.1).

some mismatch is still compatible with hybridization under *low stringency conditions.* (The latter are used in particular to finds *homologs* of a given gene in the genomes of other species.) Thus, by gauging the stringency conditions, an experimenter can use DNA hybridization to detect nucleotide sequences that are more or less exactly complementary. Similar hybridization reactions occur between two complementary RNA strands, and between a single DNA strand and complementary RNA.

Many steps in DNA cloning and associated techniques rely on nucleic acid hybridization. Complementary nucleic acids will hybridize even if one partner is preserved in fixed cells or tissues. The powerful method of *in situ hybridization* (see Method 8.1) is so named (Lat. *situs,* "position") because it detects nucleic acids in their natural position. In a polytene chromosome, for example, any gene can easily be detected in its natural position on the chromosome (Fig. 15.12). In this case, the chromosomal DNA is denatured by heat so that it will hybridize with a single-stranded DNA or RNA *probe.* Such a detection of single genes is facilitated by the fact that polytene chromosomes consist of many identical chromatids bundled neatly in register so that many probe molecules can bind next to each other, making the hybridization signal stand out easily. Similarly, in situ hybridization with a labeled probe can be used to detect localized RNA sequences in cells, such as vg1 mRNA in amphibian oocytes (see Fig. 8.26), or the expression patterns of genes in multicellular organisms (see Fig. 22.1).

Complementary nucleic acids will also hybridize if one partner is immobilized on a nylon or nitrocellulose membrane. Experimenters take advantage of this fact in a method known as ***Northern blotting,*** in which RNA molecules are separated according to size by gel electrophoresis, blotted onto membrane, and hybridized with a labeled probe (Method 15.4). This method can be used, for instance, to find out whether a particular gene is transcribed in vivo, and if so, in which part(s) of an organism and at which stage(s) of development. One can also determine the size of an RNA molecule by comparing its position in a Northern blot to the positions of RNA markers of known size.

In a related technique, known as ***Southern blotting,*** DNA segments of various sizes are separated according to size and then processed similarly to RNA molecules in Northern blotting. The DNA segments are generated by cutting DNA with enzymes that recognize specific nucleotide sequences, as will be described shortly.

A simplified method of RNA blotting, known as ***RNA dot blotting,*** eliminates the electrophoresis step. Instead, the RNA mixture to be analyzed is simply applied to a small area of a filter and immobilized before the filter is immersed in the hybridization probe. This method makes it possible to screen many RNA samples simultaneously for the presence of a particular sequence, but it provides no information about the size of the RNA molecule(s) recognized by the probe.

DNA CAN BE CUT AND SPLICED ENZYMATICALLY

Many bacteria produce enzymes called ***restriction nucleases,*** which cut double-stranded DNA at specific ***restriction sites*** of four to eight nucleotide pairs. (These enzymes protect bacterial cells from viruses by degrading viral DNA. Bacteria protect their own DNA against degradation by methylating the restriction sites for their own restriction nucleases while leaving foreign DNA unmethylated.) More than 150 restriction nucleases purified from various bacteria are available commercially. DNA segments cut by restriction nucleases are called ***restriction fragments.*** The length of the DNA fragments cut by a given restriction nuclease varies widely, depending on the spacing of the enzyme's restriction site in the DNA being digested.

Many restriction nucleases produce staggered cuts, leaving short single-stranded DNA sequences at the two ends of each fragment. Such ends are called *cohesive ends,* or "sticky ends," because they hybridize to each other and to one end of any restriction fragment cut by the same enzyme—or by another enzyme producing the same ends. Because cohesive ends are short their hybrids are not very stable. However, hybridized cohesive ends can be joined covalently by an enzyme known as DNA ligase. Thus, through the use of restriction nucleases and DNA ligase, DNA fragments from different organisms can be spliced together to form ***recombinant DNA molecules.*** In particular, any DNA fragment of potential interest can be inserted into ***DNA vectors*** (Lat. *vector,* "carrier") that replicate naturally, such as plasmid or bacteriophage DNA. As the vector replicates, so does the insert (Method 15.5). Thus, the ability to make recombinant DNA has become the basis for another way of DNA cloning.

Northern Blotting and Southern Blotting

Frequently, it is of interest whether a mixture of RNA molecules, such as the total mRNA synthesized in a certain embryonic region, contains the transcript(s) of a particular gene, and if so, whether the size or relative amount of the transcript is stage-dependent. All this kind of information can be obtained by a method known as *Northern blotting*.

The first step in Northern blotting is to separate the RNA molecules present in the mixture according to size by gel electrophoresis (Fig. 15.m3). Typically, several samples obtained from various tissues or successive stages of development are electrophoresed in parallel. Next, the separated RNA molecules are transferred by capillary action (blotting) to a nylon membrane or nitrocellulose filter placed on the gel. This blotting step preserves exactly the size order in which the RNA molecules were present in the gel. Further treatment permanently links the RNA molecules to the membrane, so that they will not diffuse or be washed off during the subsequent steps. The membrane is then exposed to a solution containing a labeled probe that is complementary to the transcript of interest. This probe will bind selectively wherever it finds a matching RNA sequence on the membrane. The sites of probe binding are made visible by autoradiography (see Method 3.1) if the probe was radiolabeled, or by other means according to the nature of the label used in the probe.

A Northern blot may show one labeled band for each lane of electrophoresed RNA, or several bands, or none. The position of a single band is indicative of the size of the target RNA to which the probe has bound, and the size of the target RNA can be quantified by running labeled marker RNAs of known sizes in a parallel lane. The intensity of the band indicates the relative amount target RNA. Multiple labeled bands may reflect the processing of primary transcript into multiple mRNAs (see Section 17.2), or the probe may hybridize to transcripts from more than one gene. In the latter case, increasing the *stringency* of the hybridization conditions may cause all but one band to disappear. If a lane does not yield a labeled band, then the RNA sample analyzed may have been degraded or may not have contained a sequence hybridizing to the probe.

A similar method, named ***Southern blotting*** is used to analyze DNA rather than RNA (Southern, 1975). The DNA of interest is cut into *restriction fragments* with restriction nucleases before the fragments are separated by electrophoresis and treated further as described above. Southern blotting can be used, for example, to distinguish the wild-type allele of a gene from mutant allele that has been altered by a major deletion or insertion (see Fig. 15.20).

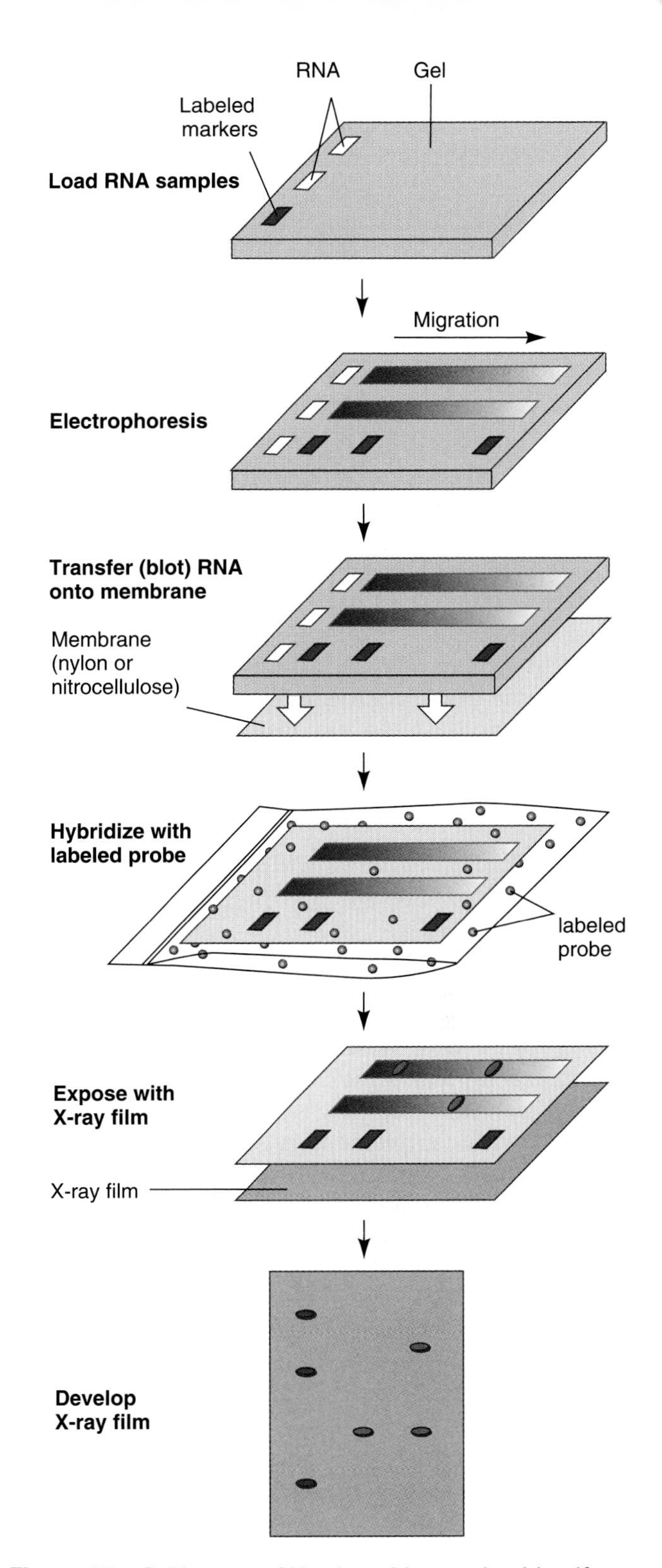

Figure 15.m3 Diagram of Northern blot used to identify a specific RNA sequence in a mixture of RNA molecules. See text.

Recombinant DNA Techniques

Genes can be cloned not only by PCR but also by *recombinant DNA* techniques. The DNA segment to be cloned is inserted into a DNA *vector* using restriction nucleases and DNA ligase as discussed in the main text. As the vector is replicated in its host bacteria or other cells, so is the insert. A commonly used class of vectors are *plasmids*—that is, extrachromosomal rings of bacterial DNA. Plasmids can be engineered to contain a *selectable marker,* such as a gene conferring resistance against an antibiotic to the host bacterium (Fig. 15.m4). In a medium containing this antibiotic, only those bacteria that contain the recombinant plasmid will grow. This trick greatly enhances the yield of recombinant plasmids obtained from a bacterial culture.

Recombinant DNA techniques can be used to prepare "libraries" of plasmid or bacteriophage clones, each containing a specific DNA fragment from the organism of interest. A genomic DNA library represents the complete genome of the organism; it is prepared by cutting the entire genomic DNA into restriction fragments and cloning them in a suitable vector. Another type of library, called a cDNA library, represents all the mRNA sequences present in a cell type or tissue of interest; it is prepared by reverse-transcribing mRNA into complementary DNA (cDNA) and cloning the cDNA molecules in suitable vectors.

Both genomic and cDNA libraries can be screened with specific probes to identify, among hundreds of thousands of clones, those containing the specific gene or cDNA of interest. Most screening methods are based on nucleic acid hybridization, the formation of double-stranded hybrids from single-stranded DNA or RNA molecules with complementary sequences. The protocols vary according to the vector and host cell chosen. Figure 15.m5 illustrates a commonly used procedure to screen a library prepared in a bacteriophage vector.

Recovering specific DNA sequences from a library requires suitable probes for screening. Such probes may represent a single sequence or multiple sequences. For instance, a single segment amplified by PCR may be used to screen a cDNA or genomic library to obtain a more complete sequence. Another strategy for single-sequence screening relies on the observation that many DNA sequences have been conserved so well during evolution that homologous sequences from different species form stable DNA hybrids under low-stringency conditions. Many cloned *Drosophila* genes have been used in this fashion to identify corresponding genes in vertebrates including humans.

Probes consisting of multiple sequences can be used for ***differential screening.*** For example, DNA sequences representing neural plate–specific mRNAs can be identified by screening a library first with a cDNA probe transcribed from mRNA of normal neurulae, and then with a control probe transcribed from mRNA of embryos ventralized by UV irradiation (see Section 9.5). Clones hybridizing with the first probe but not with the

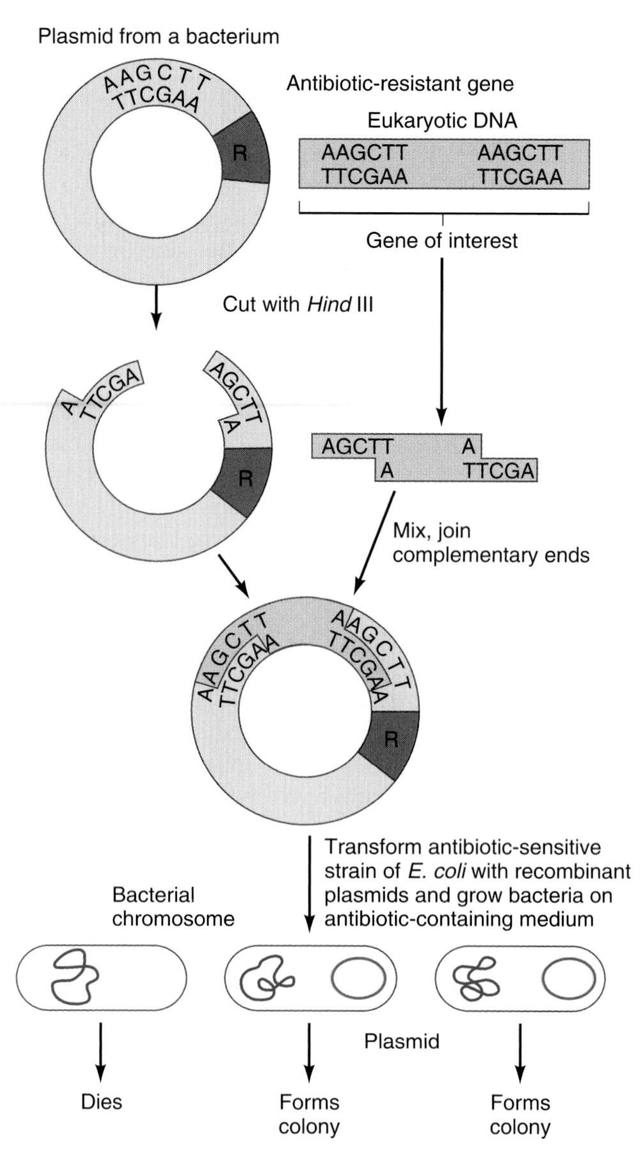

Figure 15.m4 Cloning of a eukaryotic gene in a plasmid vector. A eukaryotic DNA fragment containing the gene of interest is cut with a restriction nuclease such as *Hind* III and spliced into a plasmid vector cut with the same enzyme. The recombinant circular DNA is used to infect host bacteria. The plasmid vector also contains a gene conferring on its host bacteria resistance (R) to an antibiotic. When the host bacteria are plated on a medium containing this antibiotic only bacteria containing the R gene will form colonies. Individual colonies are tested for the presence of the gene of interest. A positive colony is used to grow a clone of bacteria large enough to obtain the desired amount of recombinant DNA.

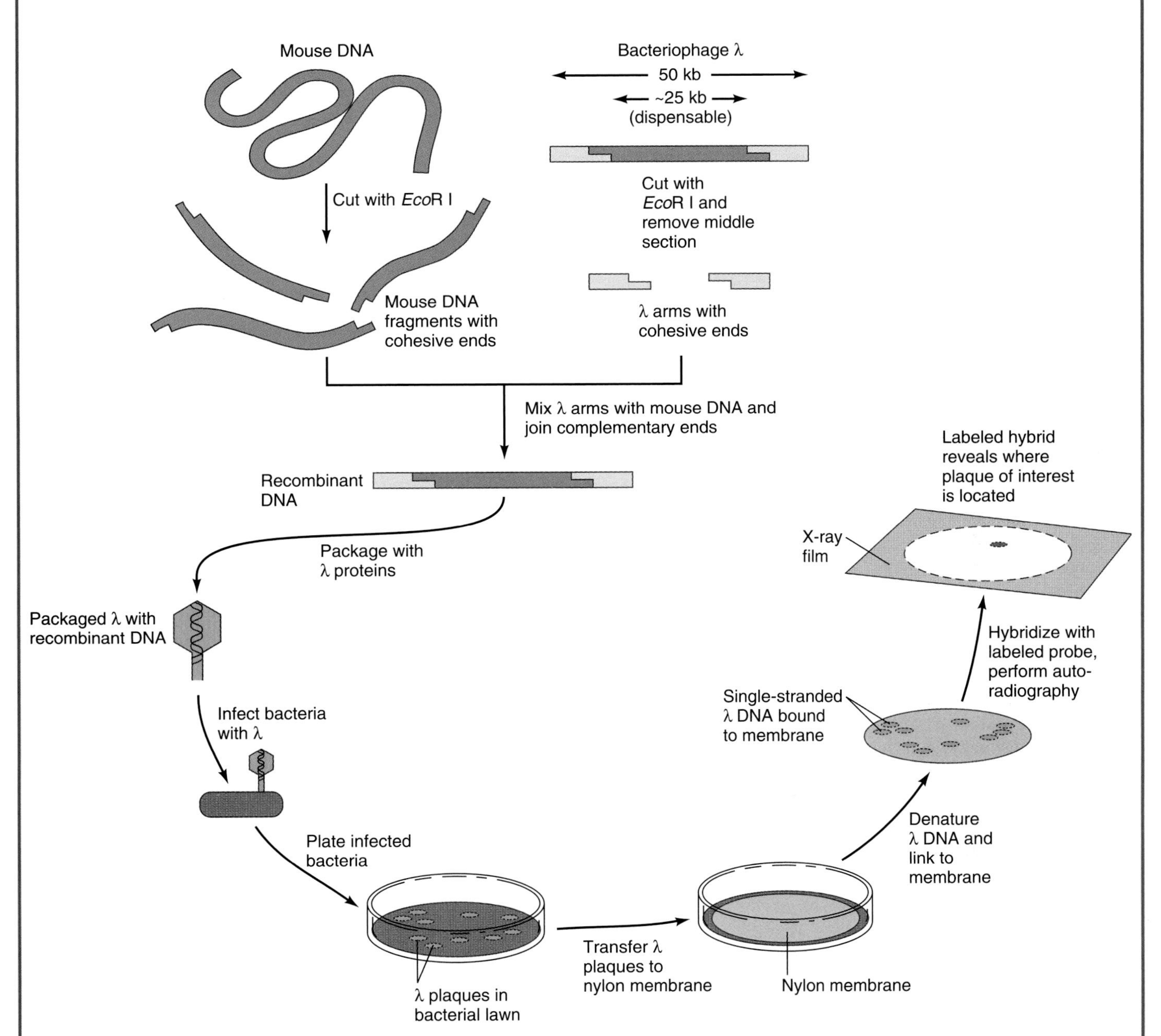

Figure 15.m5 Preparation and screening of a genomic or cDNA library prepared in bacteriophage (phage) DNA. A mixture of genomic DNA fragments or cDNAs is cut with *Eco*R I restriction nuclease or fitted with *Eco*R I ends. These DNA inserts are spliced into phage DNA, from which a dispensable central segment has been removed. The recombinant DNA is packaged in phage proteins, and the reconstituted phages are added to suitable host bacteria. These bacteria, some of which are now infected with a single phage, are spread on agar plates. As the bacteria form a lawn on the agar, each phage inside a bacterium multiplies, lyses the host cell, and reinfects surrounding cells. This leads to the formation of a plaque of lysed bacteria with millions of copies of cloned phage DNA. Each clone contains one specific DNA insert that has been replicated along with the vector DNA. To identify the clone that contains the DNA of interest, a replica of the agar plate is prepared by placing a nylon membrane on it. From each plaque, the membrane picks up some of the phage DNA, which is then made single-stranded and linked to the membrane. Next, the membrane is incubated with a radiolabeled DNA or RNA probe representing the gene or cDNA of interest. The probe hybridizes selectively to any complementary DNA sequences attached to the membrane. The position of labeled spots on the membrane, which is detected by autoradiography (see Method 3.1), reveals which of the plaques contain the insert of interest. These plaques are picked and amplified for testing and analysis.

second are likely to contain neural plate–specific genes. Likewise, cDNAs representing RNA sequences that are localized during oogenesis can be identified by differential screening of an egg cDNA library with probes representing mRNA extracted from different egg regions (Rebagliati et al., 1985).

DNA SEQUENCING MAY REVEAL THE BIOCHEMICAL ACTIVITY OF A GENE PRODUCT

The nucleotide sequence of a gene or mRNA represents exceedingly valuable information because it allows investigators to predict the amino acid sequence of the protein encoded by the gene. The amino acid sequence of a gene product, in turn, often reveals characteristic domains that are known to be associated with specific protein functions, such as DNA binding or protein phosphorylation. The amino acid sequences of more than 120,000 proteins are stored in computerized data bases, and more are being added regularly. Thus, a researcher who has cloned a novel gene can often obtain valuable clues to the biochemical function of the novel gene product by comparing its predicted amino acid sequence to the sequences of proteins with known functions.

Determining the nucleotide sequence of a cloned DNA segment has become a routine operation that can largely be carried out by robots. Indeed, the complete genomes of *C. elegans*, *Drosophila melanogaster*, as well as several bacteria and viruses have already been sequenced. The same information will become available in the near future for *Mus musculus* and the human.

The commonly used method for DNA sequencing relies on the use of dideoxyribonucleosidetriphosphates (ddNTPs), which can be incorporated along with the normal deoxyribonucleosidetriphosphates (dNTPs) by DNA polymerases into nascent DNA. However, the ddNTPs block the further growth of the new DNA strand. For the purpose of sequencing, the DNA segment of interest is replicated in vitro in a reaction mixture containing large amounts of all four dNTPs and a small amount of *one* ddNTP, for example ddGTP. The synthesis of the new strand will then terminate randomly wherever ddGTP is inserted instead of dGTP for the first time. The reaction mixture therefore produces a staggered series of new DNA segments, each terminating with a ddGTP in one of those positions that normally would be filled with a dGTP. After separation by electrophoresis, the position of each segment reveals its total number of nucleotides, and thus, the position of the terminal ddGTP. Replicating the same DNA segment in four different reaction mixtures, each with a different ddNTP added, will reveal the entire DNA sequence (Fig. 15.13).

Together, DNA cloning and the associated methods for the synthesis and modification of nucleic acids have provided a set of exceedingly powerful tools that complement the genetic analysis of development. In addition, DNA cloning and associated methods allow investigators to generate the equivalents of mutant cells and organisms in species in which traditional mutants are not available, such as *Xenopus laevis*.

15.4 Transfection and Genetic Transformation of Cells

Recombinant DNA techniques allow researchers to study gene expression by modifying various parts of a gene and introducing the modified gene, or its mRNA transcript, into cells.

Many types of cells take up foreign DNA by *endocytosis*. This process is rare in nature, but its frequency can be greatly enhanced by temperature shock, an electric field, or other means. A foreign gene in a host cell is called a ***transgene***. Typically, transgenes are introduced into cells along with a ***selectable marker***, a device that allows investigators to retrieve those cells that have taken up the transgene from a large number of unaffected cells. A commonly used selectable marker is a gene that enables cells to grow in the presence of a toxic agent, such as the antibiotic *neomycin* (Fig. 15.14).

Transgenes may persist in the host cell for some time without being integrated into the genome; this method of generating a temporarily transgenic cell is called ***transfection***. Under certain conditions, however, transfected cells integrate the transgene into their own genome, where it is replicated and passed on during cell divisions along with the host cell's own genes. This process of generating a stable genetic change by introducing an exogenous gene into the genome of a host cell or organism is referred to as ***genetic transformation***. It was first demonstrated in bacteria, in the same experiments that also established the role of DNA as the carrier of genetic information (Griffith, 1928; Avery et al., 1944). Researchers later extended this procedure to eukaryotic cells by showing that DNA from mouse tumor cells transformed normal fibroblasts into tumor cells.

A transgene that includes all its regulatory regions is expressed in the same way as its normal counterpart according to the host cell's control signals. Genetically transformed cells can therefore be used to find out where a gene product of interest accumulates (Fig. 15.15). For this purpose, one can modify the gene by adding to its transcribed region a nucleotide sequence encoding a ***green fluorescent protein* (*GFP*)**. Because the GFP sequence is inserted in a position where it is unlikely to interfere with the normal transport and turnover of the targeted protein, the fluorescent fusion protein will probably trace its unmodified counterpart. However, the principal use of transformed cells is to study the role of specific genes in cell determination and cell differentiation, as will be discussed next.

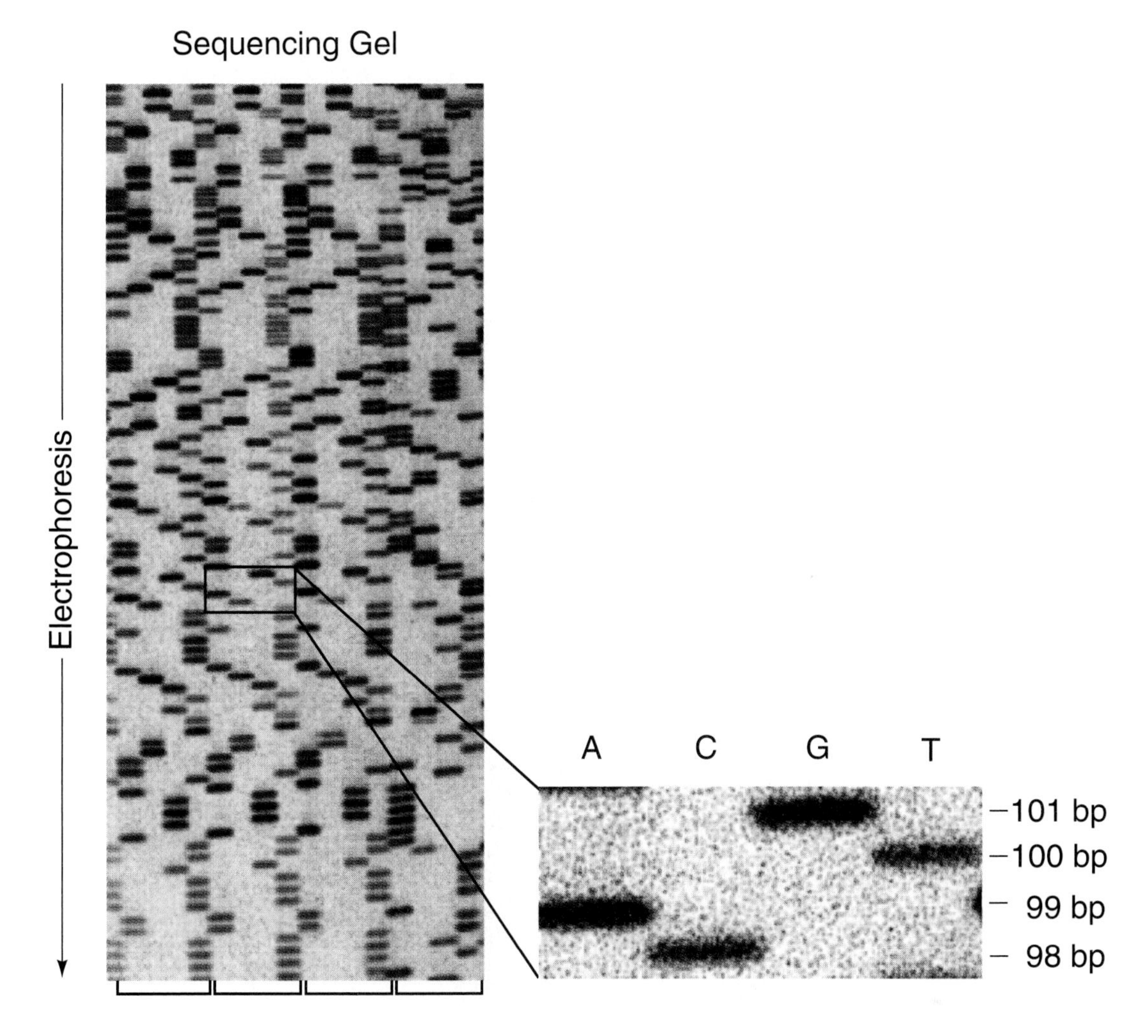

Figure 15.13 DNA sequencing gel. The DNA segment to be sequenced is replicated in vitro in a way that generates a staggered series of newly synthesized strands that are incomplete (see text). These strands are labeled by using a radioactive primer, which is extended by DNA polymerase from 5′ to 3′. The labeled strands are separated by gel electrophoresis, and made visible by autoradiography (see Method 3.1). This photograph shows the results of four sequencing reactions, one each for A, C, G, and T, repeated four times (brackets at bottom). The length of the newly synthesized strand decreases from the top (origin of the gel) to the bottom. Each G lane, for instance, shows all the strands that were terminated by the incorporation of a modified GTP. The length of each strand, and hence the position of the terminating GTP, can be determined with suitable markers. The inset shows newly synthesized strands terminating with G (101 bp long), T (100 bp long), A (99 bp long) and C (98 bp long). Thus, the sequence of the parental (complementary) strand, read from 5′ to 3′, is CATG.

15.5 The Strategies of Gene Overexpression, Dominant Interference, and Gene Knockout

To study how the activity of a specific gene affects the activity of other genes and the development of a cell lineage, transformed cells can be engineered as the equivalents of *gain-of-function*, *loss-of-function*, and *null* mutants.

To generate the equivalent of a *gain-of-function* mutant, researchers use the strategy of ***gene overexpression.*** Instead of waiting for mutagenesis to yield a gain-of-function allele of the gene of interest, they manipulate the expression of an extra copy of the gene's wild-type allele so that is expressed beyond its normal level, or under conditions where it would normally be silent. To this end, they may engineer a cloned gene so that it is expressed continuously or in response to external signals such as elevated temperature. For example, the *MyoD*$^+$ gene, which is specifically expressed in developing skeletal muscle, was cloned and provided with the regulatory region of an actin gene, so that the fusion gene would be expressed in any type of cell. When the composite gene was introduced into fibroblasts, which normally give rise to connective tissue, these cells instead formed skeletal muscle! This stunning result showed that MyoD protein is sufficient to determine some mesodermal cells to develop as myoblasts. Overexpression of *MyoD* also allowed investigators to analyze how this gene controls other muscle-specific genes (see Section 20.5).

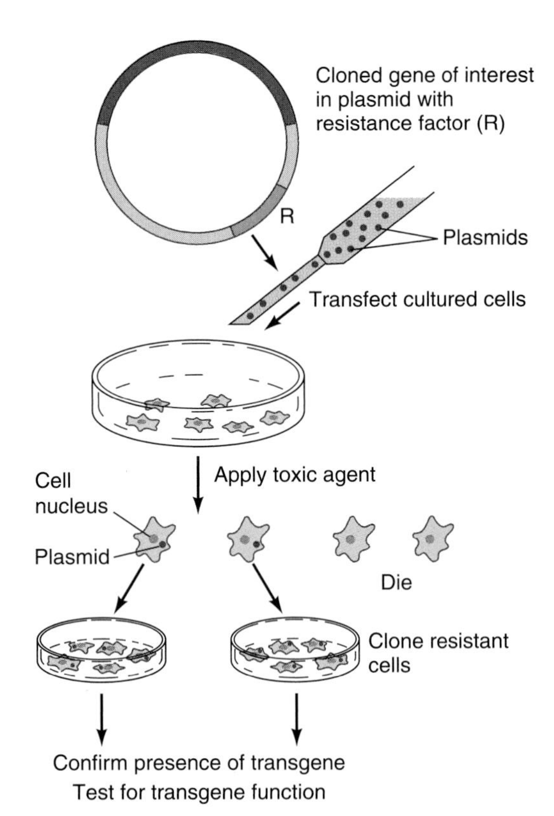

Figure 15.14 Cell transfection. The gene of interest (red) is cloned in a plasmid vector. The vector also contains a gene conferring resistance against a toxic agent, such as neomycin. The cloned DNA is added to cultured cells, some of which take up the DNA by endocytosis. Only these cells survive in the presence of the toxic agent. These cells are cloned individually, and the presence of the transgene is confirmed by Southern blotting (see Method 15.4). Transfected cells are then examined to determine whether the transgene functions as expected.

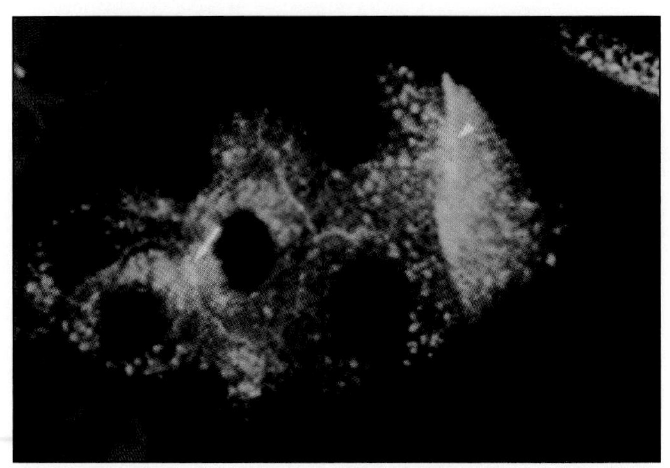

Figure 15.15 Use of green fluorescent protein (GFP) to study the distribution of exuperantia (exu), a protein involved in anteroposterior patterning of the *Drosophila* embryo (see Section 22.3). The photograph, made with a confocal microscope, shows an "optical section" of an egg chamber (see Fig. 3.14). The triangular cell to the right with the arrowhead label is the oocyte. The exu-GFP protein appears green. The egg chamber was also treated with rhodamine-conjugated phalloidin, which stains microfilaments red. The yellow structures, resulting from a superimposition of green and red stains, are ring canals surrounding the cytoplasmic bridges between two nurse cells (arrow), and between a nurse cell and the oocyte (arrowhead). The large black holes represent nurse cell nuclei. The green color is densest in the cytoplasmic bridges, and where the oocyte abuts the nurse cells, suggesting that exu protein may be synthesized in the nurse cells and transferred to the oocyte.

A *loss-of-function* mutant can be mimicked by transforming a cell with a "flipped" transgene in such a way that the normally untranscribed DNA strand serves as a template for RNA synthesis (Fig. 15.16a). The resulting RNA, called ***antisense RNA***, has a nucleotide sequence complementary to the normal mRNA. If present in excess, it prevents the normal mRNA from being translated, presumably by forming a hybrid. Another method for generating the equivalent of a loss-of-function mutant, known as ***dominant interference***, is to transform a cell with a transgene that encodes a defective polypeptide in excess quantity. The latter is often called a *dominant-negative* because it outcompetes its normal counterpart in interactions with other molecules, so that the normal gene function is sharply reduced (Fig. 15.16b).

The equivalent of a *null* mutant can be generated by transforming cells with a DNA segment that will insert itself into the targeted gene by *homologous recombination*. This process is based on the natural alignment of identical DNA sequences, which also causes the crossing over of homologous chromatid segments during meiosis (see Section 3.3) and in X-irradiated somatic cells (see Method 6.1). The insertion of the transgene is usually designed to completely block the expression of the targeted gene and is then referred to as *gene knockout*. Since gene knockouts are mostly done with transgenic mice it will be discussed in the following section.

The strategies discussed here have been applied to cells cultured in vitro as well as to embryonic cells in vivo. To deliver transgenes to embryonic cells, investigators have spliced the transgenes into the genomes of viruses that naturally insert their DNA into their host cells' DNA. For example, the effects of certain genes on generating the pattern of digits in the chicken wing have been studied by injecting a retrovirus containing the transgene into the wing bud (see Fig. 23.27).

Instead of transforming embryonic cells genetically, one can also microinject them with long-lived gene products. If a *Xenopus* blastomere is injected with a long-lived mRNA, then most or all of the injected blastomere's descendants will inherit some of the mRNA and translate it into protein. Depending on the quantity injected, this procedure amounts to a *gene overexpression*. For example, the *Xenopus* embryo shown in Figure 9.1

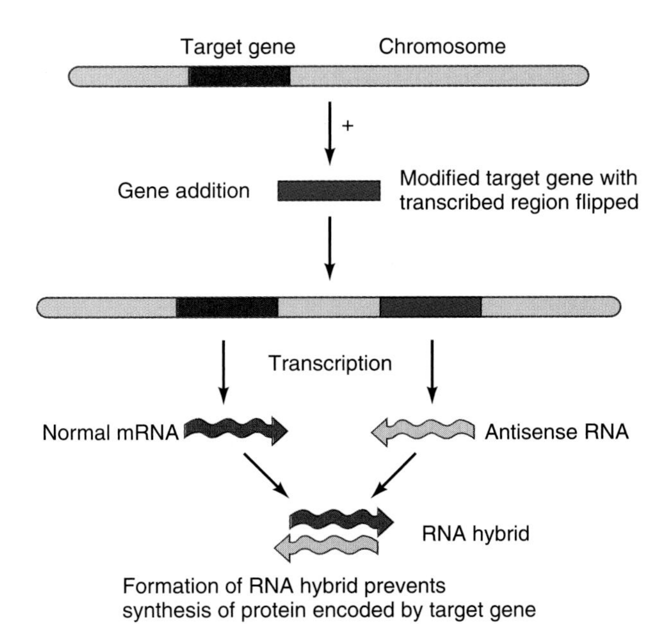

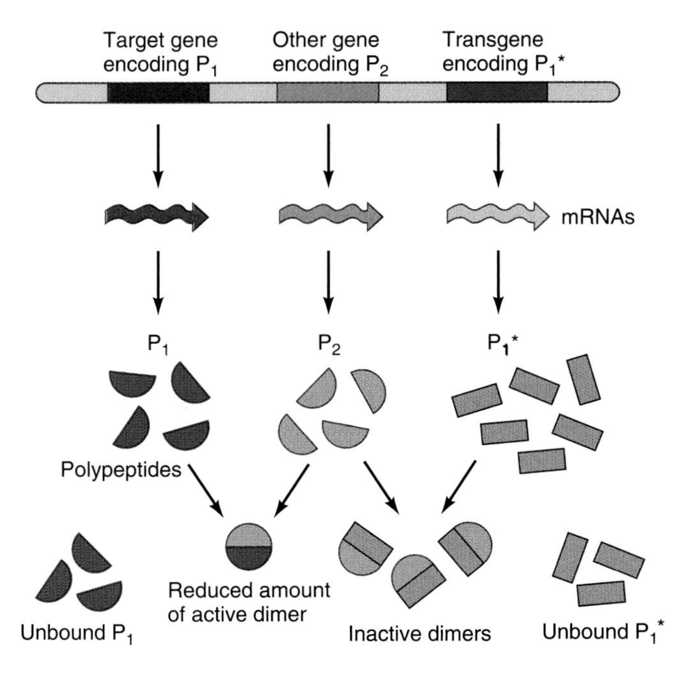

Figure 15.16 Two strategies for mimicking a loss-of-function mutation of a known target gene. **(a)** Antisense RNA strategy. Cells are transfected with a modified version of the target gene in which the transcribed region has been flipped so that transcribed RNA is complementary in sequence to the mRNA transcribed from the normally oriented target gene. This antisense RNA, if produced in excess quantity, inactivates most of the normal mRNA, presumably by forming a double-stranded hybrid that is not translated into protein. **(b)** Dominant interference strategy. The target gene encodes a polypeptide (P_1) that must combine with another polypeptide (P_2) to form a functional dimeric protein. Cells are transfected with a mutant allele of the target gene that encodes P_1*, an inactive version of P_1. An excess of P_1* will bind up most of P_2 in inactive dimers, thus diminishing the function of P_1.

was generated by injecting Xwnt8 mRNA into ventral vegetal blastomeres. (In order to be certain which cells inherited the injected mRNA, many investigators mix into the RNA solution a nontoxic dye, or an mRNA encoding a visible protein, such as GFP.)

Instead of injecting normal mRNA, one can also inject the RNA transcripts of engineered cDNAs. By generating a series of deletions, for example, one can map a certain biological function to a specific segment of an mRNA. Likewise, the strategy of *dominant interference* can be applied by injecting an engineered mRNA that is translated into a defective protein. Figure 11.9 illustrates the use of this strategy. Here, one blastomere of a *Xenopus* embryo at the 4-cell stage had been injected with mRNA encoding defective N-cadherin, and as a result, the portion of the brain derived from this blastomere developed abnormally.

While dominant interference generates the equivalent of loss-of-function alleles, the equivalent of null alleles can be generated with a new technique known as ***RNA interference*** **(*RNAi*).** Injection of double-stranded RNA specifically suppresses the expression of the gene(s) containing the corresponding DNA sequences but has no effect on the expression of other genes. Originally observed in *C. elegans* (Fire et al., 1998), this phenomenon has since been observed in a wide variety of plants, invertebrates, and vertebrates. In most cases, if not all, the suppression seems to be based on the post-transcriptional degradation of the native mRNA (Zamore et al., 2000).

15.6 Germ Line Transformation

While some questions about gene activity in cell determination and cell differentiation are best answered using cultured cells or defined cell lineages, other studies call for the analysis of ***transgenic organisms***—that is, organisms in which *all* cells are transformed with the same transgene(s). Making transgenic plants is relatively simple since complete, fertile plants can be regenerated from single differentiated cells (see Section 7.5). In contrast, the current methods for generating transgenic animals are based on genetic transformation of germ line cells. In addition, the technique of cloning animals by fusing somatic cells with enucleated oocytes (see Section 7.7) may become a way of generating transgenic animals from genetically engineered somatic cells.

Germ line transformation is a procedure in which an exogenous gene is stably integrated into the genome of *germ line* cells. So far, germ line transformations have been performed mostly with *Drosophila, C. elegans,* and *Mus musculus,* although some of the procedures used can be applied to other organisms as well. We will first explore the technique used in *Drosophila,* where gene insertions are aided by a naturally occurring mobile genetic element. Then we will examine germ line transformation in the mouse, the favorite animal for medical applications. In this context, we will see how germ line transformation can be used to interrupt and mutagenize known genes and to isolate previously unknown genes.

THE GENETIC TRANSFORMATION OF *DROSOPHILA* UTILIZES TRANSPOSABLE DNA ELEMENTS

Germ line transformation in *Drosophila* makes use of naturally occurring ***transposable DNA elements* (*transposons*)** that jump randomly into and out of genomic DNA. The existence of transposons was first proposed on the basis of genetic experiments with maize by Barbara McClintock (1952). For this discovery, she received the 1983 Nobel prize in Physiology or Medicine. Active transposons and inactive transposon residues are common in the genomes of mammals, including humans, where they amount to a substantial fraction of the entire genome.

The best-investigated transposons are the ***P elements*** of *Drosophila.* Molecular analysis has shown that P elements encode an enzyme, ***transposase,*** which is expressed selectively in the germ line (M. D. Adams et al., 1997). Recognizing specific nucleotide sequences on both ends of the P element, the enzyme cuts the transposon out of its place in the genome and inserts it elsewhere (Fig. 15.17; Beall and Rio, 1997). In addition to full-length P elements, there are *incomplete P elements,* which do not contain a functional gene but do include the recognition sites for transposase. These incomplete elements do not move spontaneously but are excised and reinserted if transposase is provided from another source. In fact, any piece of DNA flanked by P-element recognition sequences is treated by transposase as an incomplete P element. P elements are therefore convenient vehicles for inserting foreign DNA into *Drosophila* genomic DNA. This method, known as ***P-element transformation,*** has been used successfully to insert a wide variety of transgenes into the germ line cells of *Drosophila* embryos.

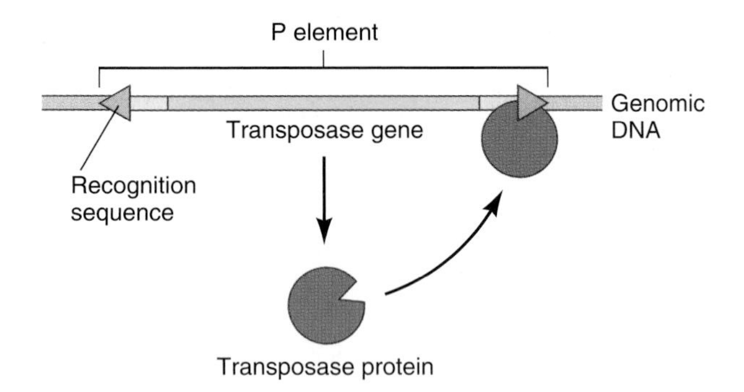

Figure 15.17 Structure of the *Drosophila* P element. A full-length P element has specific recognition sequences flanking a gene that encodes the enzyme transposase. The enzyme binds to the recognition sequences in the process of cutting out the P element from genomic DNA and reinserting it elsewhere. Full-length P elements are about 2.9 kilobases (kb) long. Short P elements have the recognition sequences, but the transposase gene is incomplete or missing.

The first P-element transformation in *Drosophila* had the overall design of a germ line therapy (Fig. 15.18). It was carried out on a mutant homozygous for a *null* allele of the *rosy* gene. This gene encodes xanthine dehydrogenase (XDH), an enzyme required to synthesize the red eye pigment. Rubin and Spradling (1982) inserted the corresponding wild-type allele (*rosy*$^+$) into a short P element and injected the construct into the primordial germ cells of early embryos. Some of the surviving flies produced offspring that had the P element along with the inserted *rosy*$^+$ allele stably integrated into their genomes. Therefore, these flies and their progeny were permanently cured of their genetic defect.

IN preparation for their transformation experiment, Rubin and Spradling constructed a circular P-element vector from a plasmid and a piece of *Drosophila* genomic DNA containing an incomplete P element (Fig. 15.19). The plasmid portion allowed the entire DNA molecule to be cloned in bacterial host cells, while the P element provided the recognition sequences for the transposase. Inserted between the recognition sequences was a fragment of *Drosophila* genomic DNA containing the *rosy*$^+$ gene. This insert was 8.1 kilobases (kb) long (1 kb = 1000 base pairs). The entire construct was designated *pry 1.*

The *pry 1* DNA was injected into the posterior pole region of *rosy* host embryos, which came from a *Drosophila* strain that was devoid of P elements. The injections were done before *pole cell* formation, so that the injected DNA would be included in the *germ line* of the developing embryos. Since neither the embryos nor the *pry 1* construct provided transposase activity, the investigators mixed the *pry 1* DNA with a small amount of a so-called helper plasmid containing a full-length P element. Upon entering the egg cytoplasm, the helper plasmid was to produce a burst of transposase activity that would excise the P element from *pry 1* and insert it into the pole cell genomic DNA.

Of more than 1000 injected embryos, 82 developed into fertile adults. Some or all of these were expected to contain the transgene in their germ line. To identify these germ line chimeras, the investigators mated each adult individually with rosy mutant partners. Their progeny were screened for red (wild-type) eye color indicating the presence of *rosy*$^+$ allele in the genome. About 25% of the matings pro-

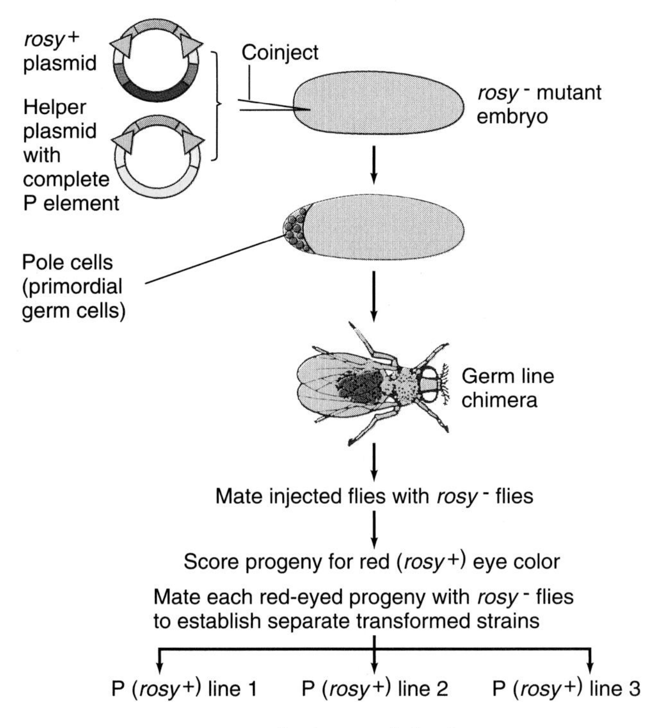

Figure 15.18 Gene therapy by P-element transformation of germ line cells, applied to a *Drosophila* mutant homozygous for a null allele (*rosy*$^-$) of the gene for xanthine dehydrogenase, an enzyme required for synthesizing the wild-type red eye pigment. The wild-type allele (*rosy*$^+$) of the gene was inserted between the inverted repeats of a P element and cloned in a plasmid vector. A complete P element containing the *transposase* gene was cloned in a separate helper plasmid. Both plasmids were injected into the posterior pole plasm of early embryos that were homozygous for *rosy*^{60}, a null allele. Some of the developing flies were germ line chimeras, with some of their gametes carrying the *rosy*$^+$ transgene. Repeated mating of these flies with *rosy*$^-$ partners resulted in stable lines of offspring that were permanently cured of the genetic deficiency. Because the *rosy*$^+$ transgene is embedded in a P element, the transgenic lines of flies are designated P(*rosy*$^+$).

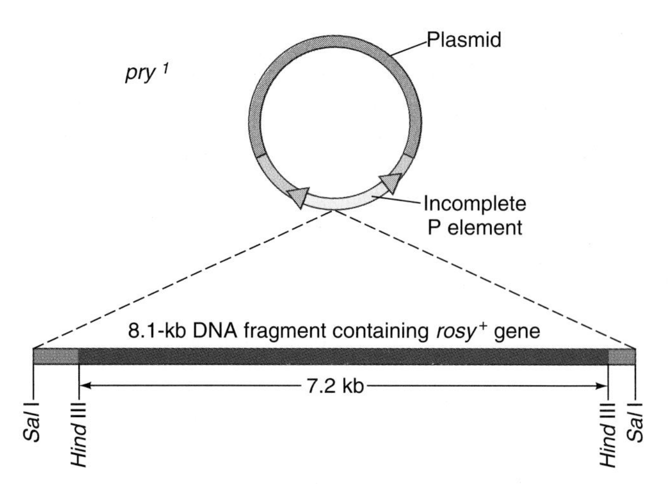

Figure 15.19 Construction of the recombinant vector *pry 1*, used by Rubin and Spradling (1982) to genetically transform the *Drosophila* mutant *rosy*$^-$. An 8.1-kb fragment of *Drosophila* genomic DNA containing the wild-type allele rosy$^+$ was isolated with *Sal* I restriction nuclease and cloned. The *Sal* I fragment was inserted into a vector (circle) consisting of an incomplete P element, small adjacent pieces of *Drosophila* DNA (pale blue), and a plasmid for replication. Within the *Sal* I fragment, a *Hind* III restriction fragment (teal) containing the *rosy*$^+$ gene was 7.2 kb long. The corresponding *Hind* III fragment from the *rosy*^{60} mutant was only 6.3 kb long (see Fig. 15.20).

duced offspring with red eyes. Each of these offspring was maintained as a ***transformant line,*** a separate strain of transgenic flies. Within each line, the red eye color was passed on as a stable, dominant marker.

To confirm that the red eye color in their transformation lines was due to the integration of the *rosy*$^+$ transgene into the host germ line, Rubin and Spradling (1982) used the *Southern blotting* technique (see Method 15.4). Genomic DNA from wild-type flies, from the *rosy* mutant strain that supplied the host embryos, and from each line of transformants was hybridized with a labeled probe representing the *rosy*$^+$ gene (Fig. 15.20). Wild-type DNA showed one labeled *Hind* III restriction fragment containing the entire *rosy*$^+$ gene. In DNA from the mutant rosy allele, the labeled probe hybridized with a shorter *Hind* III fragment because the mutation had caused a deletion in the *rosy*$^+$ gene. Both bands were labeled in the transformant lines, since they contained the resident mutant rosy allele and the *rosy*$^+$ transgene.

As controls, the investigators probed genomic DNA from some of the lines derived from injected embryos that had no red-eyed progeny. Their Southern blots showed the same labeling pattern as the uninjected *rosy* mutant. Thus, the inheritance of the red eye color correlated perfectly with the integration of the transgene in genomic DNA.

In a subsequent study, Spradling and Rubin (1983) examined whether the activity of a *rosy*$^+$ transgene depended on its site of integration into the host genome. They measured the specific activity of the rosy enzyme, xanthine dehydrogenase (XDH), in each of their transformant lines of flies. In most lines, the activity level was between 30 and 130% of the wild-type activity. For each transformant line, however, the activity level was stable from one generation to the next. When the investigators examined XDH activity in different tissues, they found high XDH levels in fat body and Malpighian tubules in both wild-type flies and transformants, and this activity distribution in transformants was found regardless of the chromosomal position of the transgene. Thus, the transgene was expressed with the

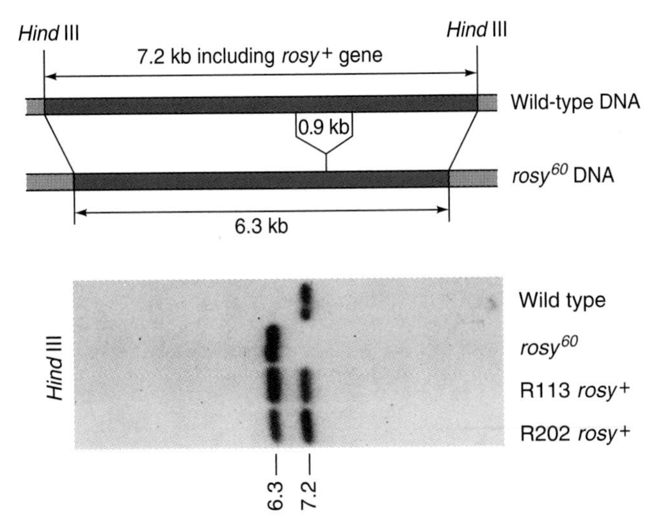

Figure 15.20 Analysis of genomic DNA from *rosy* transformant lines by Southern blotting (see Method 15.4). *Drosophila* genomic DNA fragments cut with *Hind* III were hybridized with a *rosy* probe. The probe hybridized to a 7.2-kb fragment of wild-type DNA, as expected from the known size of the *Hind* III fragment containing the *rosy*$^+$ gene. The *rosy* mutant, *rosy*^{60}, had a 0.9-kb deletion, which reduced the length of the corresponding *Hind* III fragment to 6.3 kb. This was exactly the length of the genomic fragment hybridized by the rosy probe in the Southern blot of *rosy*^{60} DNA. In the DNA from transformants designated R113 and R202, both fragments were hybridized by the probe: the 6.3 kb fragment from the resident *rosy*^{60} gene, and the 7.2 kb fragment from the *rosy*$^+$ transgene.

normal tissue specificity and at approximately normal levels regardless of the transgene integration site. These results indicated that the transgene contained not only the transcribed region of the *rosy*$^+$ gene but also its regulatory regions necessary for tissue-specific expression.

Questions

1. For their pioneering experiment, Rubin and Spradling chose the *rosy*$^+$ gene because *null* mutants are fully viable, and because only a few percent of the normal XDH level are necessary to produce the red eye color of wild-type flies. Why were these criteria important for choosing the first transgene?
2. The transposase activity required to insert the *rosy*$^+$ transgene into the host genome was provided by a separate "helper" plasmid coinjected with the *pry*^{1} plasmid. If the investigators had used instead a single plasmid containing both the *rosy*$^+$ gene and the transposase gene, what would have happened with their transformant lines?
3. How do you think the investigators established the chromosomal site of transgene integration for each of their transformant lines? *Hint:* One of the photographs shown in this chapter illustrates the key method used.

The results of many genetic transformation experiments in *Drosophila* show that P elements can be used effectively to insert exogenous genes into the germ line. The transgenes are stably integrated into the genome and passed on to future generations. Most transgenes, if their regulatory sequences are present, show normal expression patterns and provide a nearly normal level of gene activity that varies somewhat with the site of integration.

In later experiments, Spradling, Rubin, and many other investigators, used transgenes that conferred phenotypic effects more difficult to score than eye color. To facilitate screening the surviving hosts for transformants, researchers constructed new P-element vectors with visible or selectable marker genes. These markers were included between the recognition sequences of the P elements so that they were excised from the vector and reinserted along with the gene of interest. The *rosy*$^+$ gene is a convenient visible marker. Even more convenient are selectable markers such as the bacterial *neo*^{R} gene, which renders *Drosophila* larvae resistant to the drug neomycin (Steller and Pirrotta, 1985). When kept on food containing neomycin, only those larvae that express the *neo*^{R} gene survive. These individuals can be expected to carry both the *neo*^{R} marker and the gene of interest as transgenes.

MAMMALS CAN BE TRANSFORMED BY INJECTING TRANSGENES DIRECTLY INTO AN EGG PRONUCLEUS

Transgenic mammals are generated much as are transgenic fruit flies, but with three differences. First, the cloned transgene is injected directly into one of the two *pronuclei* of the fertilized egg (Fig. 15.21; see also Fig. II.1). This procedure is facilitated by the leisurely pace of events after fertilization in mammals. Second, the transgene is injected without the aid of a natural transposon. This method is therefore applicable to all mammals, whereas the P-element transformation is limited to *Drosophila melanogaster* and a few closely related species. Third, mammalian transgenes are injected as linear DNA fragments without a cloning vector, since plasmid or bacteriophage DNA inserted into the host genome may have undesirable effects. Under the conditions used in mammals, tandem arrays of many copies of the transgene are often integrated into the same site of the host genome. For the most part, these multiple integrations are stable in successive generations.

Injected mammalian eggs are allowed to develop in vitro until the blastocyst stage, when surviving embryos are implanted into foster mothers (Fig. 15.21). Their offspring are tested directly for transgene integration by *Southern blotting.* Many of the offspring that test positive have the transgene in all tissues, not just in some of the germ line cells as in *Drosophila.* Apparently, foreign DNA injected into a mammalian pronucleus is usually

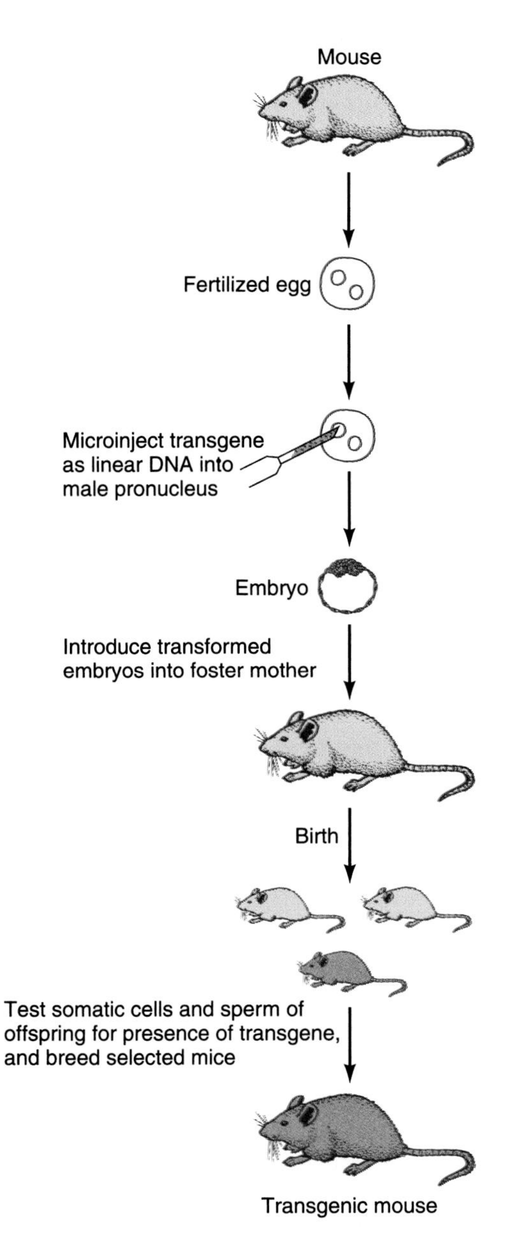

Figure 15.21 Technique for generating transgenic mice by microinjection of DNA into an egg. Fertilized eggs are removed from the oviduct of the donor mother. The cloned foreign gene is injected directly into a pronucleus (see Fig. II.1). Developing blastocysts are inserted into the oviducts of foster mothers. Offspring are screened for transgenes in their somatic cells by Southern blotting. Transgenic mice are also bred to test for the presence of the transgene in the germ line.

integrated either before the first genomic DNA replication or not at all. The site of transgene integration into the host chromosomes seems to be random, as in *Drosophila*. Most transgenes in mammals are expressed at a lower-than-normal level, although exact quantification is difficult because of the unknown number of transgene copies inserted. Some chromosomal sites seem to silence inserted transgenes so that they are not expressed. For these reasons, researchers routinely generate several lines of transformed mice for each transgene to compare its effects in different positions within the host genome.

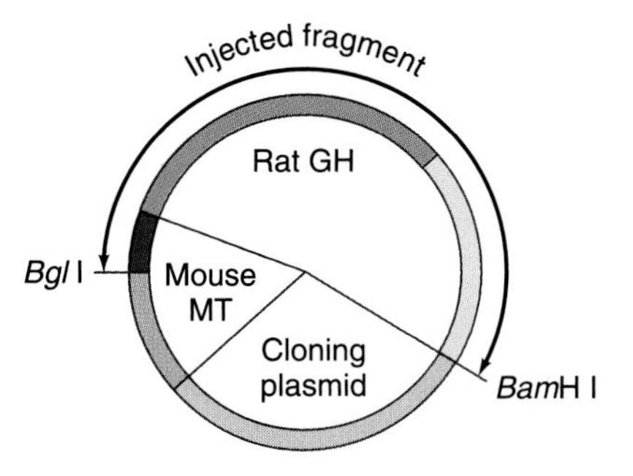

Figure 15.22 Mouse metallothionein (MT)/rat growth hormone (GH) fusion gene created by Palmiter et al. (1982). A segment of mouse genomic DNA including the regulatory region of the MT gene (teal) was spliced to a segment of rat genomic DNA including the transcribed region of the GH gene. The fusion gene was cloned in a plasmid vector. The DNA fragment to be injected was cut out with *Bgl* I and *Bam*H I restriction nucleases.

A striking demonstration of transgene activity in mice came from an experiment with a mouse metallothionein-rat growth hormone fusion gene (Palmiter et al., 1982). ***Metallothionein* (*MT*)** is a protein synthesized mostly in liver and kidney that removes ingested heavy metals such as cadmium or zinc. The investigators isolated the regulatory region that controls the expression of the MT gene. An important feature of this control region is that its activity is inducible by the presence of heavy metals. The investigators also cloned the rat gene for *growth hormone* (*GH*). Normally produced in the pituitary gland, GH enhances the growth of bones and muscle and reduces fat deposition. Using recombinant DNA techniques, the investigators spliced the mouse MT regulatory region to the transcribed region of the rat GH gene, thus creating a composite gene, or ***fusion gene*** (Fig. 15.22). They expected the fusion gene to be expressed primarily in liver and kidney, because the regulatory region of a gene, and not its transcribed region, controls its expression (see Section 16.3).

The MT-GH fusion gene was amplified in a plasmid vector, excised from the vector, and injected into pronuclei of mouse eggs. From 170 developing embryos implanted into foster mothers, 21 mice were born alive, and 7 of these were transgenic for MT-GH. When these seven mice were placed on a diet containing zinc sulfate ($ZnSO_4$) to activate the MT promoter activity, they underwent extraordinary growth. One of them reached twice the weight of the nontransgenic littermates that served as controls (Fig. 15.23; see also Fig. 15.1).

It remained to be tested whether the unusual growth of the transgenic mice was indeed caused by growth hormone encoded by the transgene or resulted from unexpected

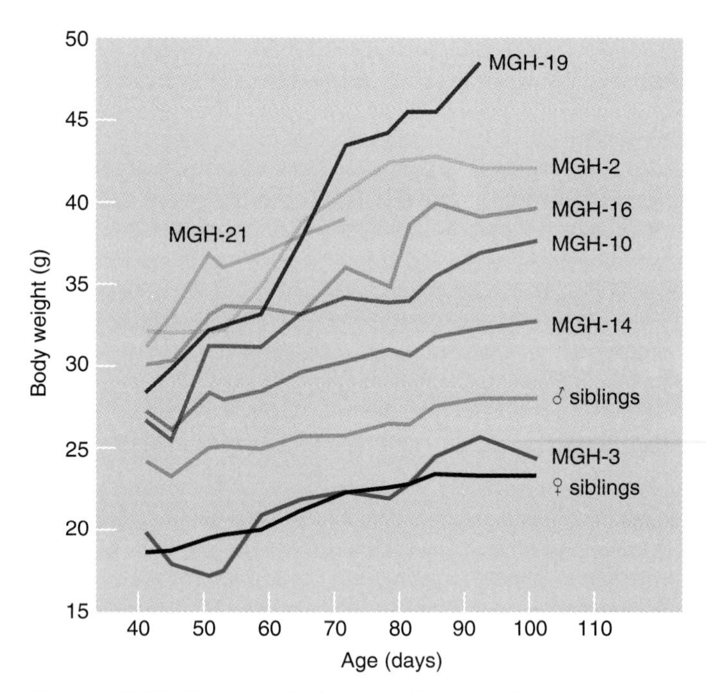

Figure 15.23 Growth of mice transformed with a mouse metallothionein (MT)/rat growth hormone (GH) fusion gene. At 33 days, the mice were weaned, and their drinking water was supplemented with zinc sulfate to stimulate transgene expression. The body weight of each transgenic mouse (designated MGH-2, MGH-3, etc.) was measured individually. For comparison, the average weight of the mice's nontransgenic siblings was also recorded.

activation of the resident mouse GH gene. As proof, the investigators showed that the liver RNA in the transgenic mice contained a 74-nucleotide segment present in transgene mRNA but not in the resident growth hormone mRNA or metallothionein mRNA.

An unexpected result was obtained later when the transgene expression patterns were examined (Swanson et al., 1985). In addition to the tissues normally expressing the MT gene, the investigators also found novel expression areas of the MT-GH fusion gene. It was expressed, for instance, in the hypothalamus and in distinct regions of the cerebral cortex, where neither MT nor GH is normally produced. These results were not readily explainable and suggested that there might be unexpected interactions between transgenes and neighboring resident genes in mammals.

Questions

1. How do you think the investigators established which of the mice derived from injected eggs were transgenic for the MT-GH fusion gene?
2. How could the investigators have tested their expectation that the MT-GH fusion gene would be expressed primarily in the livers and kidneys of the transgenic mice?

The experiments of Palmiter and coworkers confirmed and extended results of an earlier study by Gordon and colleagues (1980), who had shown that a viral gene injected into an egg pronucleus integrated stably into the mouse genome. Since then, mice and other mammals have been the subject of germ line transformations with a large variety of transgenes, many of them taken from species other than the host. Once integrated into the genome of the host's germ line, the transgenes are passed on to offspring in a fairly stable manner. Most transgenes introduced by pronuclear injection are introduced for the purpose of *overexpression*, that is, to generate the equivalent of a gain-of-function allele.

TRANSGENIC MICE CAN BE RAISED FROM TRANSFORMED EMBRYONIC STEM CELLS

An alternate way of generating transgenic mice takes advantage of two features of mammalian development. First, cells from the *inner cell mass* (*ICM*) of blastocysts can be cultured in vitro as *embryonic stem cells* (*ES cells*). Second, when ICM or ES cells from one mouse strain are injected into blastocysts from a genetically different mouse strain, the developing offspring is often *chimeric*—that is, composed of cells from both genotypes. These two features have been combined in a powerful technique for ***gene targeting***, also known as *gene knockout*. This technique essentially allows investigators to raise completely transgenic mice from transgenic ES cells.

The first step in the common form of gene targeting is to engineer a transgene that will insert itself into, and thus inactivate, the gene of interest in an ES cell (Fig. 15.24). The targeted insertion relies on the fact that DNA strands with identical nucleotide sequences tend to pair up in cells, as they do in *polytene chromosomes* and during *meiosis*. Such paired DNA segments facilitate ***homologous recombination***—that is, crossing-over events between homologous DNA sequences. The transgene is therefore engineered so that it will pair up with the target gene at two sites flanking a gene segment that is critical to the gene product's function. If homologous recombination occurs at both sites, the intervening DNA segments will be swapped between the transgene and the target gene. The transformed ES cell then carries one wild-type allele and one null allele of the targeted gene.

The DNA segment thus inserted into the target gene contains the *neo*^{R} gene, which has two effects on the transgenic ES cell. First, it completely inactivates the target gene by interrupting a critical part of its coding sequence. Second, it allows the transgenic cell to grow in the presence of neomycin while nontransformed cells are killed. This selection step is critical because homologous recombination is a rare event.

The second step in gene targeting turns single transgenic ES cells into strains of transgenic mice. To this end, each transgenic ES cell is grown into a cell clone, and several cells from a given clone are injected into a blastocyst from a mouse strain that differs in genetic

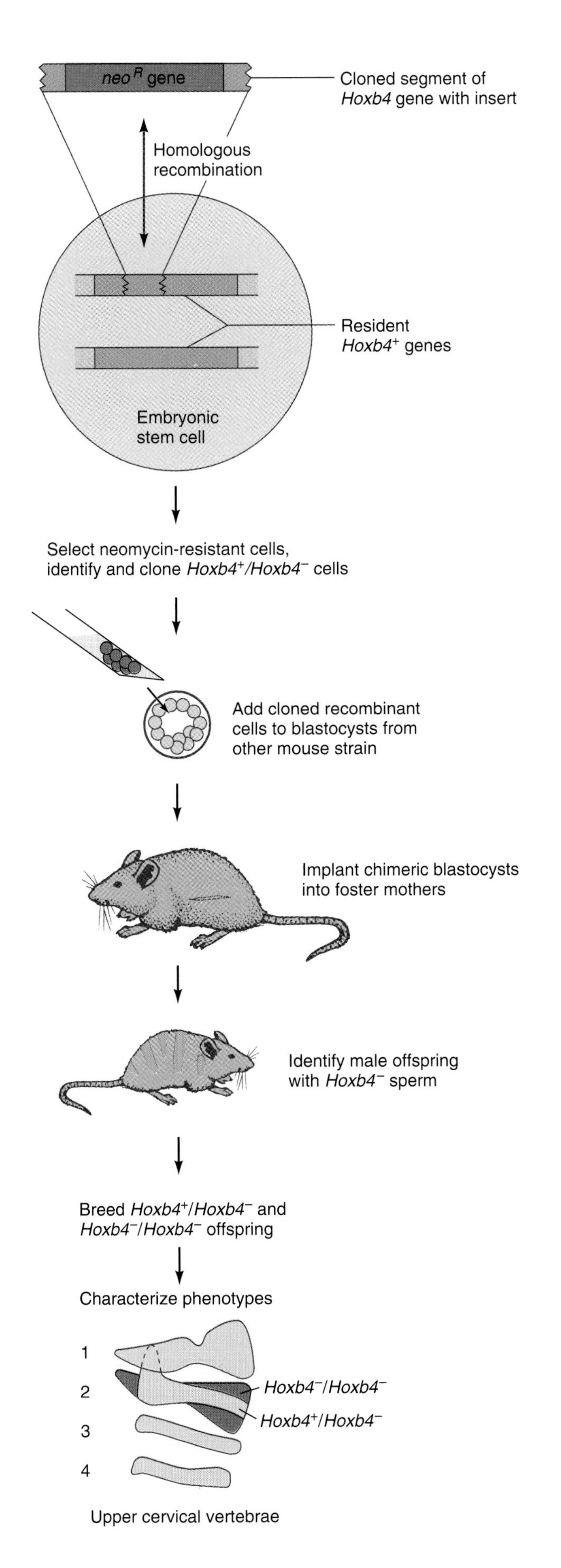

Figure 15.24 Gene knockout technique in mice. Cultured *embryonic stem cells* are transformed with a modified segment from the coding region of the gene of interest, for example the *Hoxb4* gene. The transgene carries as an insertion the *neo*^{R} gene, which destroys the normal function of the Hoxb4 protein while allowing transformed cells to grow in the presence of neomycin. The preferred sites of transgene recombination into the host genome will be the two resident alleles of *Hoxb4*$^+$ due to the sequence homology of the latter with the transgene. Stem cells with this insertion carry one wild-type and one null allele (*Hoxb4*$^-$/*Hoxb4*$^+$). Stem cells of this type (red circles) are injected into blastocysts from another mouse strain carrying different marker genes for fur color and other traits. The added stem cell can contribute to various cell lineages within the developing embryo, including the germ line. Males with sperm carrying the null allele (*Hoxb4*$^-$) are mated with wild-type females, and their heterozygous offspring (*Hoxb4*$^-$/*Hoxb4*$^+$) are mated with one another to generate homozygotes for the null allele (*Hoxb4*$^-$/*Hoxb4*$^-$). Typically, only the homozygotes show a mutant phenotype. In our example, the second cervical vertebra will be affected. The bottom diagram shows the outlines of the cervical vertebrae in wild-type and heterozygotes (pink) and in the homozygous (*Hoxb4*$^-$/*Hoxb4*$^-$) null mutant (red). The four cervical vertebrae shown are numbered, starting at the head. Dorsal is to the right. In the mutant, the second cervical vertebra has been transformed into the likeness of a first cervical vertebra.

markers from the strain that provided the ES cells. The injected ES cells tend to join the ICM of the host blastocyst and to contribute to all types of cell lineages, including the germ line of the developing chimeras (Bradley, 1987). Males with chimeric germ lines will produce some sperm that contain the null allele of the targeted gene. Such males are mated with wild-type females to obtain offspring that are heterozygous for the targeted gene. If these heterozygotes are crossed among themselves a quarter of their offspring will be homozygous for the null allele of the targeted gene. Usually only the homozygous mutants show an abnormal phenotype.

Gene targeting is used mostly to generate the equivalent of null mutants, and such "knockout mice" have been exceedingly valuable in establishing the biological functions of cloned genes. Similar procedures can be used to "knock in" gain-of-function alleles that will be expressed with an abnormal tissue and/or stage-dependence (Hanks et al. 1995).

A possible disadvantage of knock-out mice is that the presence of the *neo*^{R} gene may inactivate not only the targeted gene but also neighboring genes (Olson et al., 1996). Some investigators have overcome this problem by targeting the same gene in different mouse strains carrying the targeted gene in different genetic "neighborhoods." Only the phenotypic effects observed in all mouse strains can be ascribed to inactivation of the targeted gene.

An alternate method of generating knockout mice avoids the problem described above by physically removing the targeted gene. It is based on the ***Cre-loxP recombination system*** of the bacteriophage P1. Wherever the recombinase Cre encounters two loxP sequences in a DNA strand, it ligates the two sites while cutting out the DNA segment between them. The gene for Cre is introduced into one transgenic mouse strain while the target gene flanked by loxP sequences is introduced into another mouse strain by homologous recombination. Crossing the two strains results in offspring that will remove the target gene wherever Cre is expressed. This method allows investigators to target genes for selective removal by splicing Cre to a promoter that will activate Cre at a certain stage of development and/or only in specific tissues (Gu et al., 1994). Such *conditional knockouts* are required to study genes that function early and late in development and for which loss of the early function is lethal so that later functions cannot be analyzed by conventional gene targeting. Instead of removing the targeted gene, one can also engineer transgenes in such a way that excision of the element between the loxP sequences will generate a loss-of-function or a gain-of-function allele. In principle, this method allows investigators to generate any desired allelic series for misexpressing a gene when and where they want.

DNA INSERTION CAN BE USED FOR MUTAGENESIS AND PROMOTER TRAPPING

Transgenes that are not specifically designed for homologous recombination will usually insert themselves randomly into the host genome. If insertion occurs within a functional part of a gene, the activity of the host gene will be disrupted. The generation of mutants by random insertion of foreign DNA into a host genome is called ***insertional mutagenesis.*** It may occur as an unwanted side effect in a genetic transformation experiment. However, insertional mutagenesis also provides a *method* of generating new mutant strains.

Mutants generated by insertion of engineered segments of DNA greatly facilitate the isolation of the mutated gene. If the inserted DNA contains a selectable marker, such as a gene conferring resistance to an otherwise poisonous drug, only transformed individuals survive on an appropriate medium. If the insert contains a unique sequence that occurs nowhere else in the host genome, this sequence can be used as a tag to identify the mutated gene (see Method 15.5). A labeled probe hybridizing with the tag will identify the mutated gene in a genomic library prepared from the mutant strain. With this method, it is but one step from the genetic to the molecular analysis of a gene.

Variants of the insertional mutagenesis method are known as ***promoter trapping.*** *Promoters* and other regulatory gene elements will be discussed more fully in Section 16.3. These elements are not transcribed themselves; instead, they control the region- and stage-specific transcription of the coding regions in their vicinity. If the transcribed region of a transgene is inserted between the promoter and the transcribed region of a resident gene, then the effect is that of a cuckoo laying an egg in another bird's nest. The resident promoter "adopts" the transgene and transcribes it while the resident gene typically loses its function. If the transgene causes the formation of a readily detectable product—such as a blue stain—then the staining pattern in an embryo will shed light on the normal transcription pattern of the neighboring resident gene. Thus, for example, researchers can detect genes involved in neurulation by randomly inserting stain-producing transgenes and selecting those transgenic lines that show selective staining in the neural plate. An application of this technique is shown in Figure 23.1.

One of the genes used for promoter trapping in *Drosophila* encodes the yeast transcriptional activator, GAL4, which acts on an upstream activating site (UAS) present in the genome of yeast but not *Drosophila.* A wide array of fly strains is available in which a *GAL4* transgene is driven by one of many promoters in the *Drosophila* genome that have tissue-specific activity. For example, *GAL4* trapped next to the promoter of the *apterous*$^+$ gene is expressed on the dorsal but not on the ventral surfaces of the fly's wings. The synthesis of GAL4 protein itself is of little consequence because few if any promoters in the *Drosophila* genome recognize it. However, any additional transgene engineered to contain the yeast UAS sequence in its regulatory region will respond to GAL4 protein (Brand and Perrimon, 1993). An investigator who wants to get, for instance, green fluorescent protein (GFP) synthesized specifically in the dorsal wing blade can construct a transgene with the transcribed region for GFP driven by UAS. If flies transgenic for this construct are mated to flies that have *GAL4* driven by the *apterous* regulatory region, then 25% of the offspring will have green fluorescent dorsal wing blades.

TRANSGENIC ORGANISMS HAVE MANY USES IN BASIC AND APPLIED SCIENCE

DNA cloning and transgenic organisms have revolutionized many areas in basic and applied science. In basic biology, one important use is to confirm the identity of cloned genes. Many cloning strategies leave some uncertainty whether the cloned gene does in fact encode the protein of interest. In these cases, proof comes from genetically transforming the entire organism with the cloned gene and monitoring the gene's function in vivo. This test is especially convincing if the transgene restores the normal phenotype for the gene of interest in a *loss-of-function* mutant.

The most important application of transgenic organisms in basic research is in the analysis of gene-regulatory regions. As will be described in Section 16.3, genes have regulatory sequences controlling their expression. The location and function of each regulatory sequence can be revealed by removing DNA fragments from cloned genes. Testing the effects of such modifications in transgenic organisms is especially valuable for investigating regulatory sequences that respond to signals not easily mimicked in cell culture, such as region- or tissue-specific signals.

In addition to their invaluable role in basic research, transgenic organisms have many applications in medicine, agriculture, and pharmacy. For instance, many types of cancer are caused by genes called *oncogenes.* Most oncogenes are mutant alleles of normal cellular genes that encode growth factors, growth factor receptors, and intracellular signals (see Section 28.7). In many cases, an oncogene causes cancer only in combination with other mutations or environmental factors. While some of these combinations can be tested in cell culture, transgenic animals are unsurpassed in modeling the full range of factors contributing to development of cancer or other diseases in humans.

Experimental gene therapy of human somatic cells has begun for patients with certain genetic diseases that would otherwise be fatal (Bank, 1996; Friedmann, 1999; Mountain, 2000). In a commonly used type of therapy, the abnormal somatic cells are removed from the patient and transformed with a cloned wild-type allele of the defective gene. After testing in vitro, successfully transformed cells are reintroduced into the patient's body. Somatic gene therapy is effective only if the transformed cells reach their appropriate destination and last for some time. To work out these problems, experiments on genetically modified mice and other mammals that serve as models for humans with genetic diseases have been very useful.

Transgenes are also being used to improve agricultural plants and livestock. Many agricultural plants have been genetically engineered to improve crops or make them resistant to cold or to pests or—more problematically—to pesticides. Human pharmaceutic proteins, such as insulin and blood clotting factors, will soon be produced by transgenic sheep or other large mammals. These proteins will be safer, more effective, and less expensive than the homologous proteins obtained from unmodified animals so far. The transgenic animals will carry fusion genes combining the regulatory region of one of the animal's milk protein, such as casein, with the transcribed region the human therapeutic protein (Schnieke et al., 1997; Velander et al., 1997). Transgenic female mammals will then mass-produce the human pharmaceutic proteins in their milk. Other efforts are under way to genetically modify pigs and other human-sized mammals so that their organs can be transferred to human recipients without being rejected by the human immune system.

Not everything that can be done with modern DNA techniques is safe and morally acceptable. It will take a close cooperation between scientists and lawmakers to establish legal boundaries that will prevent misuse while still allowing basic research and work on beneficial applications to proceed.

SUMMARY

Life scientists frequently use the strategy of controlled modification. Developmental biologists remove or transplant cells in order to learn from the resulting abnormalities about the normal functions of the manipulated cells. Loss-of-function and gain-of-function alleles of genes are similarly used to infer the functions of the wild-type alleles.

In the fruit fly *Drosophila melanogaster*, and in other organisms amenable to genetic analysis, saturation mutagenesis screens have been used to identify virtually all genes involved in certain developmental processes, such as the formation of the embryonic body pattern. Saturating chromosomes with mutations is a strategy that enables researchers to estimate the total number of genes involved in any biological process for which an appropriate screening procedure is available. Mutant phenotypes often reveal in a striking way what types of decisions cells make during development. They also show which embryonic areas are affected by these decisions.

The power of genetic analysis has been boosted enormously by the advent of DNA cloning, DNA sequencing, and associated techniques. While mutant phenotypes are indicative of the biological function(s) of a gene, the gene's nucleotide sequence often reveals the biochemical activity of the gene's protein product. From the biological and biochemical functions of a gene, and from knowing its position in a hierarchy of other genes, one can often infer its role in the genetic control of development.

Even in the absence of mutants, recombinant DNA techniques often allow investigators to construct the equivalents of mutants in order to study the effects of gene overexpression, dominant interference, and gene inactivation. Some of these studies are done using transgenic cells—that is, cells that have taken up an exogenous gene, or transgene. Such cells are called *transformed* if the transgene is integrated in the host cell's genome. An integrated transgene is replicated during mitosis and expressed along with the resident alleles of the same gene in the cell. With large cells, such as animal eggs and blastomeres, gene overexpression and dominant interference can also be achieved by injecting excess amounts of mRNA transcribed in vitro.

Transgenic plants can be regenerated from genetically transformed protoplasts derived from various cell types.

Transgenic animals must be grown from genetically transformed germ line cells. An effective method for making transgenic *Drosophila* flies utilizes naturally mobile DNA segments known as P elements. Transgenic mice and other mammals can be generated either by injecting the cloned transgene into an egg pronucleus or by adding transformed embryonic stem cells to the inner cell mass of host blastocysts. Related techniques can be used for insertional mutagenesis and simultaneous tagging of the mutagenized gene.

Transgenic organisms have become invaluable for the genetic and molecular analysis of development. They are widely used to confirm the identity of cloned genes, and to delimit the positions and functions of gene regulatory sequences. In particular, they allow investigators to identify specific regulatory regions that control gene expression in response to localized cytoplasmic determinants, inductive interactions, and other signals. In addition to these uses in basic research, there are many applications in medicine, agriculture, and pharmacy.

Transcriptional Control

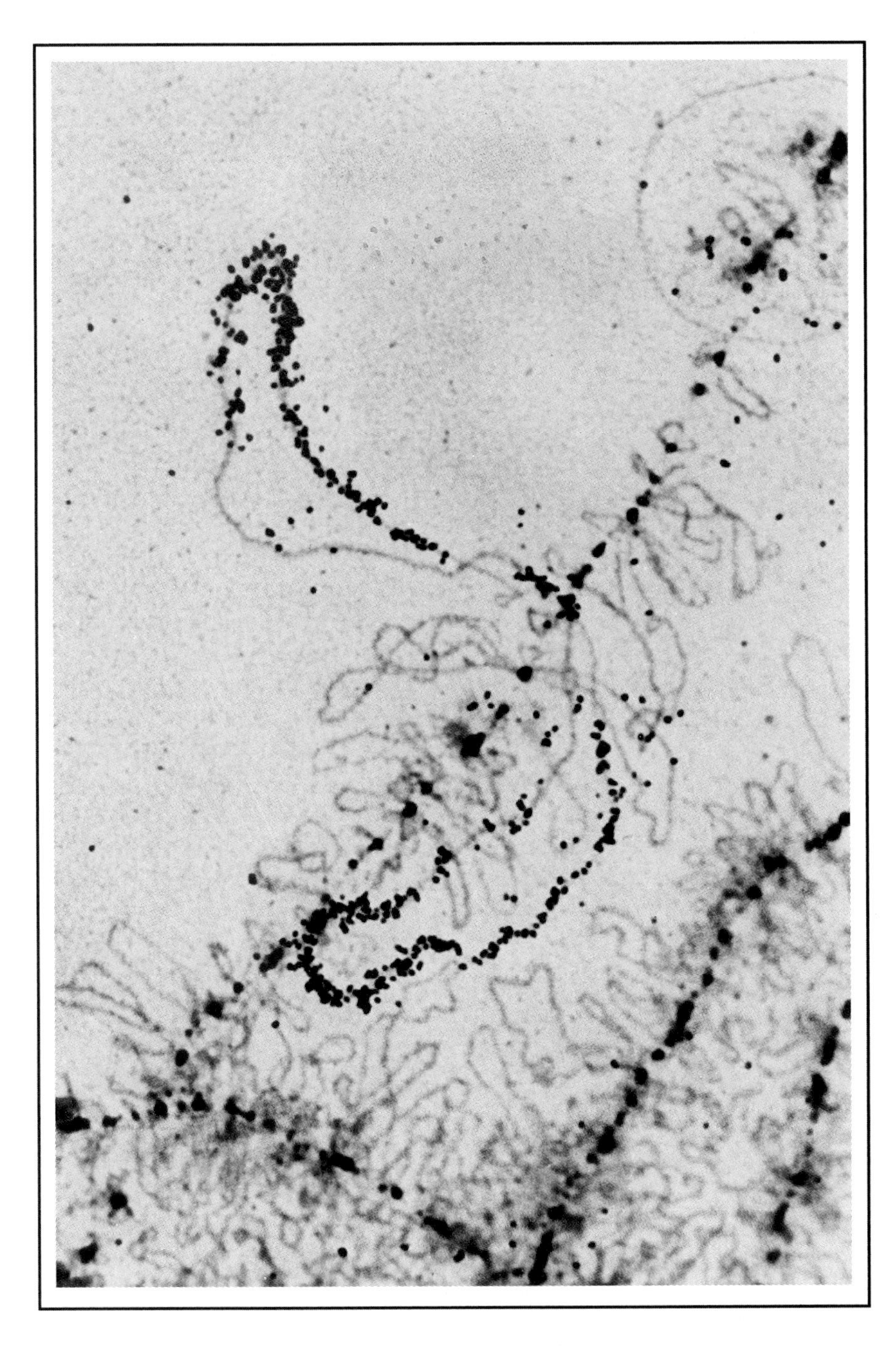

Figure 16.1 Transcription from a lampbrush chromosome in a newt oocyte. Each chromosome consists of an axis with paired loops extending from it. In situ hybridization (see Method 8.1) with a radiolabeled histone DNA probe identifies a pair of loops with histone genes in the process of transcription. The probe hybridizes to the transcripts while they are still attached to the genes. As the length of the transcript increases, so does the intensity of the label.

THE nuclear transplantation experiments and other observations discussed in Chapter 7 support the theory of genomic equivalence—that most differentiated cells retain a full and functional complement of the organism's genetic information. Nevertheless, cells become clearly different in their structure and function as development proceeds. At the molecular level, cells differ in the proteins they synthesize. This is revealed by comparing, among different types of cells, the proteins sufficiently abundant to be detected by two-dimensional electrophoresis (see Fig. 7.19). Analogous methods reveal that different types of cells also accumulate distinct combinations of mRNAs. It follows that different cells in an organism generally have the same genetic information but use, or express, different subsets of genes. In Chapters 16 through 18, we will examine various mechanisms of differential gene expression.

16.1 The Principle of Differential Gene Expression

The ***principle of differential gene expression*** is based on the well-founded theory of genomic equivalence and on the observation that cells synthesize gene products (mRNAs and proteins) in stage-dependent and region-specific ways. The principle states that cells in an organism develop differently not because they have different genetic information, but because they express different subsets of the same information. As suggested in Section 7.1, the principle of differential gene expression can be compared to the selective ways different people read different parts of the same newspaper. All cells in an organism express a large set of so-called *housekeeping genes*, or **common genes,** thought to encode all the proteins that cells need to maintain themselves. In addition, different cells express different ***cell-type-specific genes*** to support their particular functions as nerve cells or muscle cells, for example. However, the expression of cell-type-specific genes is often a matter of different quantities and combinations. Only a small percentage of proteins are expressed exclusively in one cell type.

Some cell-type-specific proteins are synthesized in large amounts by fully differentiated cells. For example, hemoglobin accounts for the bulk of all proteins in red blood cells. Other cell-type-specific proteins are present in only small quantities but nevertheless have important regulatory functions. Some of these proteins are synthesized early in development, before cells become visibly different from one another. Such proteins are of particular interest because they may be involved in the process of *cell determination*. In a *Drosophila* embryo, for instance, the fushi tarazu and evenskipped proteins are each synthesized in distinct stripes of blastoderm cells, at a time when these cells are morphologically indistinguishable from neighboring cells that do not express these genes (see Fig. 22.4). Genetic and molecular analyses have shown that *fushi tarazu* and *evenskipped* belong to a hierarchy of locally expressed genes that determine the embryonic body pattern (see Section 22.2).

Gene expression involves many steps from transcription to the assembly of functional multimeric proteins (see Fig. II.2). As the first step in this process, a cell *transcribes* a selected set of cell-type-specific genes in addition to the common genes (Fig. 16.1). The set of genes transcribed in each cell depends on its stage of development, tissue type, position in the body, and physiological conditions. This selective use of DNA as a template for RNA synthesis—as a result of any of these factors—is called ***transcriptional control.*** The concept of transcriptional control was first proposed by Jacob and Monod (1961), who found that bacteria transcribe certain groups of genes only when the enzymes encoded by those genes are needed to metabolize those nutrients that are actually present. Since then it has been shown that transcriptional control is the most prevalent type of gene regulation in bacteria as well as in eukaryotes.

In this chapter, we will review how transcriptional control works in development. The selectivity of transcription depends on *regulatory DNA sequences* associated with the transcribed regions of genes. These sequences interact with *regulatory proteins* that enhance or inhibit transcription. Some of these regulatory proteins are translated from localized maternal mRNAs and activate genes in certain areas of developing embryos. Other regulatory proteins are translated from zygotic mRNAs. Some regulatory proteins are themselves regulated by hormones and other signal molecules that cells use to communicate with one another.

In our study of transcriptional control, we must take into account that the genomic DNA of eukaryotes is tightly associated with structural proteins in chromosomes. We will examine how the availability of genomic DNA for transcription depends on local decondensation of the chromatin structure. Then, having familiarized ourselves with the molecular mechanism of transcriptional control, we will relate it to the process of *cell determination* in development and to the *principle of default programs,* both of which are based in part on transcriptional control.

16.2 Evidence for Transcriptional Control

Because transcription is the first step of gene expression, transcriptional control would seem to be the most economical way to achieve differential gene expression. Indeed, several lines of evidence indicate that transcriptional control is the prevalent mechanism for using genes selectively. The variability of *puffs in polytene chromosomes,* and the correlation between puffing and RNA synthesis (see Section 16.5), were early indications of differential gene expression. More recently, comparisons of *cDNA libraries* (see Method 15.5) from different sources has shown that certain mRNAs occur specifically in certain organs or at certain stages of development. Some mRNAs are found only in liver, not in kidney or brain.

Figure 16.2 Expression of the *Cwnt-8C*$^+$ gene in the hindbrain neuroepithelium of a chicken embryo. The accumulated Cwnt-8C mRNA is made visible by in situ hybridization (see Method 8.1). The gene is transiently expressed in a specific region of the developing hindbrain (arrow).

In embryos, several mRNAs are present only at the gastrula stage and not at other stages of development.

Another technique for revealing the stage- or region-specific occurrence of specific RNA sequences is *in situ hybridization* (see Method 8.1), which allows investigators to see the distribution of a particular RNA sequence directly in wholemounts or sections of embryos. Results obtained with a wide variety of organisms show that many mRNAs accumulate specifically in certain tissues or regions and at certain stages during development (Fig. 16.2). Of particular interest are those expression patterns that appear before embryonic cells become overtly differentiated. These patterns offer us a glimpse into gene activities that must be part of the determination and differentiation process itself.

A critical reader might object that the assays discussed so far are suggestive of transcriptional control but do not actually prove it. Indeed, both the screening of cDNA libraries and in situ hybridization detect all RNAs *present,* not just RNA that is *newly transcribed.* The RNA distributions observed with these techniques can also be biased by posttranscriptional events, including RNA transport and RNA degradation. To test whether transcription itself is being controlled, Derman and coworkers (1981) used a "nuclear run-on" procedure that is carried out with isolated nuclei. Newly synthesized RNA was radiolabeled during a brief period of transcription, but the subsequent steps of RNA modification and transport were curtailed. Samples of nuclear run-on RNA from mammalian liver, kidney, and brain were hybridized with various cloned cDNAs that were unlabeled. Some cDNAs, including actin cDNA and tubulin cDNA, hybridized with nuclear run-on probes from liver, kidney, and brain. This showed that actin and tubulin genes were transcribed in all three of these tissues. Other cDNAs hybridized with only one or two of the probes, demonstrating that those genes were transcribed selectively in these tissues. These experiments proved that the expression of some genes is controlled directly at the level of transcription.

16.3 DNA Sequences Controlling Transcription

Transcription requires enzymes known as ***RNA polymerases.*** In eukaryotes, there are three types of RNA polymerase, which synthesize different classes of RNA. We will focus on the activity of RNA polymerase II, which synthesizes mRNA precursors known as primary transcripts or ***pre-mRNA.*** In particular, we will examine how the activity of this enzyme is directed toward specific genes in different cells.

A typical eukaryotic gene is shown in Figure 16.3. It consists of a ***transcribed region*** and a ***regulatory region.*** The transcribed region begins with the first transcribed nucleotide of the gene, which is called the ***transcription start site*** and is given the number +1. The part of the regulatory region that is located close to, and for the most part upstream, of the start site is known as the ***promoter.*** An RNA polymerase molecule is positioned at the promoter by a large complex of gene regulatory proteins. Once positioned, the RNA polymerase moves on to nucleotides 2, 3, 4, and so forth, generating an RNA transcript of one DNA strand in the process. While the RNA polymerase moves relative to its DNA template, it may be well be stationary in relation to other cell structures. In other words, rather than racing down the DNA like a locomotive on railroad tracks, the RNA polymerase may reel its template and extrude newly synthesized RNA, which is immediately modified by other enzymes clustered together with RNA polymerase in a factory-like arrangement (Cook, 1999).

The transcribed region of the gene consists of *exons,* which will be represented in the final mRNA, and intervening *introns.* Both are transcribed into pre-mRNA, but the introns are removed during the subsequent processing of pre-mRNA into mRNA (see Section 17.2).

The regulatory DNA sequences located upstream of the transcription start site are assigned negative numbers, beginning with −1 directly upstream of +1. The function of regulatory DNA sequences is to accelerate or inhibit the binding of RNA polymerase to the transcription start site. This regulation is indirect. The regulatory DNA sequences are recognized and bound by proteins called ***transcription factors,*** which regulate the assembly of the large protein complex that positions and determines the activity of RNA polymerase.

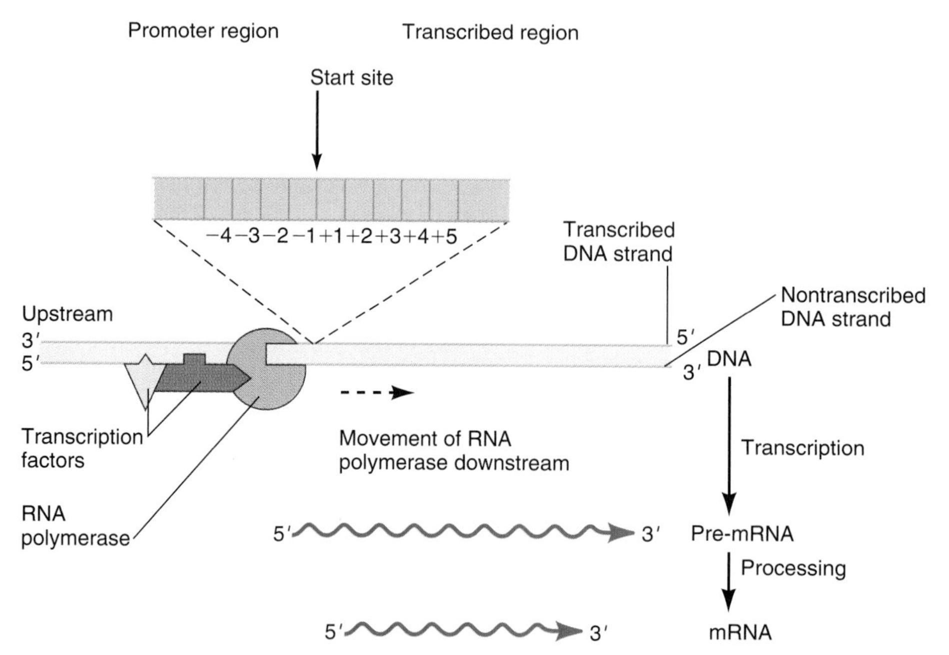

Figure 16.3 Promoter region and transcribed region of a eukaryotic gene. In the transcribed region, nucleotides are numbered +1, +2, +3, etc., with +1 marking the start site of transcription. The nucleotides upstream of the start site, numbered −1, −2, −3, etc., constitute the principal part of the gene's regulatory region, known as the promoter. Specific DNA sequences within the promoter interact with regulatory proteins called transcription factors, which position RNA polymerase at the start site and regulate its activity. The movement of the RNA polymerase is downstream, or from 3′ to 5′, on the transcribed DNA strand of the gene. The transcription start site is also called the 5′ end of the gene. It corresponds to the 5′ end of the nontranscribed DNA strand, which has the same polarity and nucleotide sequence as the RNA transcript. The primary transcript, called pre-mRNA, undergoes several processing steps to yield mRNA. The processing includes the removal of pre-mRNA segments that are not represented in the final mRNA.

The regulatory DNA sequences of a gene are *cis-acting elements* because they are part of the regulated DNA molecule itself. (Similarly, we have referred to the nucleotide sequences of mRNAs that are required for the cytoplasmic localization of the same mRNAs as cis-acting sequences; see Section 8.4.) According to the theory of genomic equivalence, the cis-acting elements of a gene are the same in all cells of an organism because they are part of the genome. In contrast, the transcription factors acting on a gene are *trans-acting elements* because they originate independently of their target genes. The combination of transcription factors acting on a given gene may vary from one cell type to another. In the remainder of this section and in the next section, we will examine first some properties of regulatory DNA sequences and then some features of transcription factors.

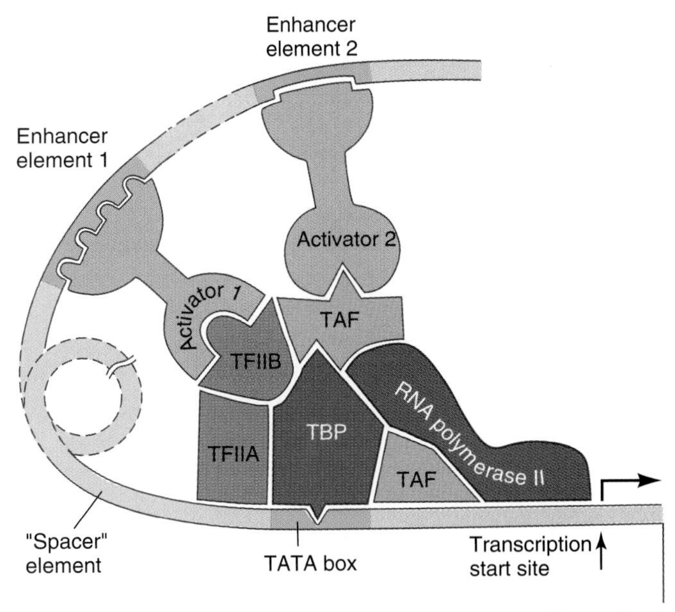

Figure 16.4 Simplified model of transcriptional control. An RNA polymerase II molecule is positioned upstream of the transcription start site by a set of general transcription factors (green colors), including the TATA-binding protein (TBP) and TBP-associated factors (TAFs), some of which are bound to the TATA box and other promoter elements. The general transcription factors are sufficient for a low, basal level of transcription. This basal level can be enhanced by transcriptional activators, which are bound specifically to certain enhancer elements and to matching domains of general transcription factors. The DNA "spacer" elements between promoter and enhancers are looped out, thus allowing assembly of the transcription initiation complex. The designations TFIIA and TFIIB refer to general transcription factors associated with RNA polymerase II.

PROMOTERS AND ENHANCERS ARE REGULATORY DNA REGIONS WITH DIFFERENT PROPERTIES

Two types of regulatory DNA regions—promoters and enhancers—are found in eukaryotic genes (Fig. 16.4). The *promoter* is necessary for transcription to occur even at a slow rate; some scientists refer to it as *core promoter* or *basal promoter* to emphasize this fact. Other researchers prefer the term *proximal promoter* because most of its parts are located immediately upstream of the transcription start site. Comparisons of promoter nucleotide sequences from different genes and species have revealed common cis-elements, or *consensus sequences*, that are shared by many different promoters. Such sequences are the recognition elements bound by the gene-regulatory proteins described later in this chapter.

The best-investigated consensus sequence, named the ***TATA box*** because it consists only of A/T base pairs, is usually located at −25 to −30. The conservation of the TATA box in life forms from yeast to humans underlines its fundamental importance (Hoffman et al., 1990). Another motif, located immediately upstream of the TATA box and having the consensus sequence G/C-G/C-G/A-C-G-C-C, has been termed the *IIB recognition element* (*BRE*) because it is bound by *transcription factor IIB*, to be discussed in Section 16.4 (Lagrange et al., 1998). A less well-conserved element, located at the transcription initiation site, is known as the *initiator element* (Weis and Reinberg, 1992). Finally, a conserved element with the sequence A/G-G-A/T-C-G-T-G is found especially in TATA-less promoters. Because of its position near +30 it is called *downstream promoter element* (Burke and Kadonaga, 1997).

Every gene needs a core promoter, but no two core promoters seem to be identical (Goodrich et al., 1996). Any one of the typical cis-elements may be missing from a given promoter. While most promoters have a TATA box, some are TATA-less. There is also some variance in the nucleotide sequence of each cis-element and in its exact placement relative to the transcription initiation site. Thus, there is considerable individuality to promoters, and their subtle differences may very well restrict the regulatory proteins to which they respond as well as the resulting rate of transcription. Indeed, some genes, such as the *hunchback*$^+$ gene of *Drosophila* discussed in the following section, use two core promoters in conjunction with alternate patterns of RNA processing. If a gene has two core promoters, they respond to different combinations of regulatory proteins, which in turn interact with other DNA elements to be discussed next.

In addition to promoters, eukaryotic genes have cis-regulatory sequences known as ***enhancers,*** which can *increase* or *inhibit* the activity of certain core promoters in their vicinity. Enhancers are distinguished from promoters in three ways. First, enhancers need promoters to work, while promoters will work by themselves, albeit slowly. Second, whereas promoter elements depend fairly closely on their positions relative to the transcription start site, most enhancers are effective even if they are moved thousands of base pairs upstream or downstream from their normal positions. Third, most enhancers are effective in reverse orientation whereas promoters are not.

Enhancers are modular, like promoters, and may comprise just a few cis-elements or many. Some of these elements are unique, while others occur repeatedly within the same enhancer. A given element may also be shared among different enhancers, providing a mechanism to coordinate expression of different genes.

In accord with the principle of genetic equivalence, the promoters and enhancers of a gene appear to be identical in all cells of a given organism. The stage- and organ-specific transcription of genes must therefore be mediated by *trans-acting* molecules that interact with promoters and enhancers.

REGULATORY DNA SEQUENCES ARE STUDIED WITH FUSION GENES

The regulatory regions of many eukaryotic genes are large and complex, often extending over more than 10 kb and containing dozens of binding sites for transcription factors (Arnone and Davidson, 1997). To facilitate the analysis of the regulatory region of a eukaryotic gene, researchers often replace its transcribed region with the transcribed region of a bacterial ***reporter gene***, a gene that encodes a product that is readily detected and measured. The composite gene, consisting of the regulatory region from the eukaryotic gene of interest and the transcribed region of the reporter gene, is called a ***fusion gene***. If a fusion gene is introduced into a eukaryotic cell, the gene's regulatory region responds to the host cell's transcription factors. Provided all regulatory cis-elements have been included, the reporter part of the fusion gene is transcribed with the same stage- and region-specificity as the normal gene of interest.

A commonly used reporter gene encodes the bacterial enzyme ***chloramphenicol acetyltransferase* (*CAT*)**. Its expression in cultured cells is easily measured by the acetylation of labeled chloramphenicol added to a cell homogenate (Fig. 16.5). Because CAT is not produced in normal eukaryotic cells, its synthesis in a transfected cell accurately reflects the activity of the reporter gene. Another bacterial reporter gene, designated ***lacZ***, encodes β-galactosidase. This enzyme produces a blue stain in the presence of a suitable substrate. The *lacZ* reporter gene is especially useful for studying region-specific gene expression because the cells expressing it stain dark blue against an unstained background (Fig. 16.6). Green fluorescent protein (GFP, see Section 15.4) works similar to lacZ, with the additional advantage that GFP can be detected in vivo whereas lacZ requires fixation of the material being studied. The synthesis and processing of GFP, as well as the proteins tagged with GFP, can therefore be traced by microcinematography (see Fig. 15.15).

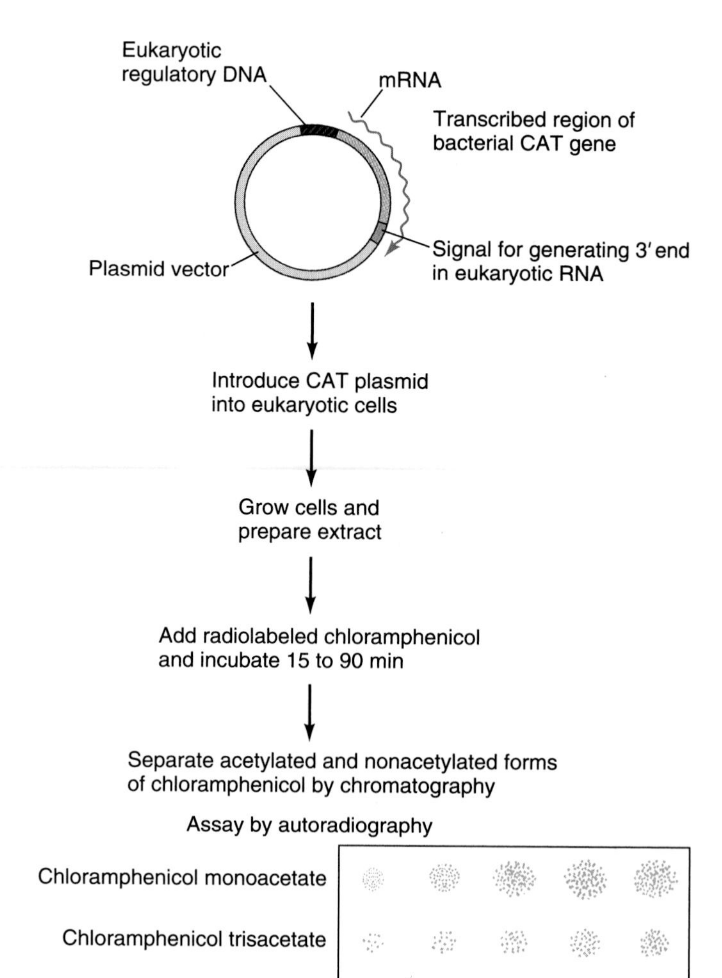

Figure 16.5 Use of a bacterial reporter gene to analyze the regulatory region of a eukaryotic gene. The transcribed region of the chloramphenicol acetyltransferase (CAT) reporter gene is fused with the regulatory region from the eukaryotic gene of interest. The activity of the fusion gene in transfected cells is measured by the ability of a cell extract to convert radiolabeled chloramphenicol to its acetylated forms (monoacetate and trisacetate). The proportion of acetylated chloramphenicol increases with time and with the amount of CAT present.

Fusion genes are frequently used for the techniques known as ***deletion mapping*** and ***point mutagenesis***. The regulatory DNA segment of interest is modified by deletions or single-base substitutions. The effects of these manipulations on the expression level of the fusion gene show which DNA sequences in a regulatory region are functionally important. For instance, deletion mapping has identified different regulatory elements in the *fushi tarazu*$^+$ gene of *Drosophila* (Hiromi and Gehring, 1987; Tsai and Gergen, 1995). One regulatory element, called the *zebra element*, is needed for expression in seven

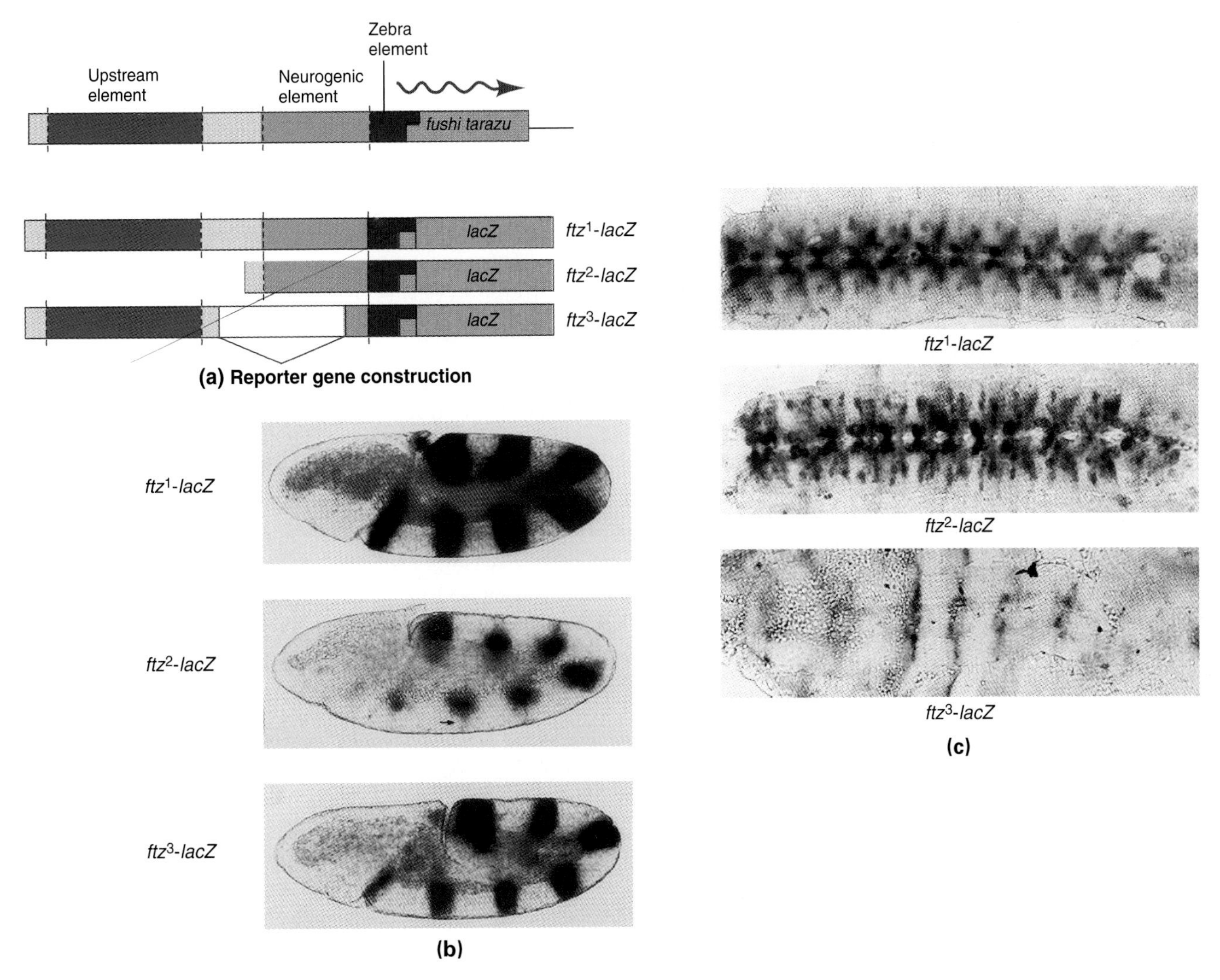

Figure 16.6 Deletion analysis of the regulatory DNA sequences of the *fushi tarazu*$^+$ (*ftz*$^+$) gene in *Drosophila* using the *lacZ* reporter gene. **(a)** Construction of fusion genes with different regulatory regions from the *ftz*$^+$ gene. The *ftz*^{1}-*lacZ* construct includes a complete *fushi tarazu*$^+$ regulatory region and the transcribed region of *lacZ*. In the *ftz*^{2}-*lacZ* construct, a large upstream element is deleted. The *ftz*^{3}-*lacZ* construct is missing the greater part of a different regulatory element, called the neurogenic element. **(b)** Expression pattern of the fusion genes in early embryos. The complete fusion gene (*ftz*^{1}-*lacZ*) is expressed in segmentlike areas (dark bands) of both the ectoderm and the mesoderm. Deletion of the neurogenic element (in the *ftz*^{3}-*lacZ* construct) does not change this pattern. However, deletion of the upstream element (in the *ftz*^{2}-*lacZ* construct) reduces the expression, especially in the ectoderm. **(c)** Expression pattern of the fusion gene after development of the central nervous system. Deletion of the upstream element does not change the normal expression pattern. However, deletion of the neurogenic element precludes expression in the central nervous system.

stripes in the early embryo (see Figs. 16.5 and 22.21). A second regulatory element, the *neurogenic element,* is necessary for later expression in the central nervous system. A third regulatory element, the *upstream element,* is bound by the fushi tarazu protein as part of a positive feedback loop.

Point mutagenesis is a laborious, but high-resolution, method of analyzing regulatory regions by changing one base pair at a time. Most single substitutions do not change the level of transcription from the reporter gene. However, mutations in certain base pairs cause sharp downturns or increases in the transcription rate. These base pairs are clustered in short, distinct sequences of about 4 to 15 nucleotides variously called ***boxes, sites, cis-elements,*** or ***motifs.*** Such sequences are the critical parts of regulatory DNA regions; most of them are located upstream of the transcription start site, but some have been found at the 3′ end or in introns.

16.4 Transcription Factors and Their Role in Development

As mentioned earlier, the binding of RNA polymerase to the promoter depends on regulatory proteins called *transcription factors.* A typical transcription factor has two functional domains: a ***DNA-binding domain***, which binds to a promoter or enhancer element, and an ***activation domain***, which interacts with other transcription factors and/or with RNA polymerase (see Fig. 16.4). Many transcription factor are *dimeric* proteins—that is, they consist of two polypeptides. In this case, the component polypeptides have distinct ***dimerization domains.*** The interactions of transcription factors with DNA elements, and with one another, create a large three-dimensional complex known as the ***transcription complex***, which enhances or inhibits the ability of polymerase to dock at the transcription start site. The quality of the fit between the combined transcription factors and the RNA polymerase also regulates the speed with which a new RNA polymerase molecule is seated at the promoter after the previous one has departed on its journey down the transcribed region.

Transcription factors connect genes in complex regulatory networks. Frequently, a given enhancer may bind more than one transcription factor, each of which ultimately has a different effect on the assembly or activity of the transcription complex. The net outcome then depends on the relative concentration of each competing factor. Conversely, a given transcription factor may interact with similar regulatory cis-elements in many genes, including its own encoding gene and genes encoding other transcription factors. Thus, a maze of interactions among different transcription factors, their encoding genes, their target sequences, and RNA polymerase form the molecular basis of transcriptional control. As key elements in these interactions, transcription factors are of particular interest for a molecular understanding of transcriptional control.

GENERAL TRANSCRIPTION FACTORS BIND TO ALL PROMOTERS

Some transcription factors, called ***general transcription factors*** or *basal transcription factors*, bind to the core promoters of most genes. About 20 general transcription factors have been identified that act together in different combinations to position an RNA polymerase molecule near the transcription start site and to initiate transcription (Buratowski, 1994). The first general transcription factors found to be associated with RNA polymerase II have been named TFIIA, TFIIB, and so forth, in the order of their discovery. TFIID subsequently turned out to consist of several proteins, a ***TATA box–binding protein* (*TBP*)** and at least eight ***TBP-associated factors,*** or ***TAFs*** (Goodrich et al., 1996). Certain TAFs interact not only with TBP but also with the *downstream promoter element* if present (Burke and Kadonaga, 1997).

Several of the general transcription factors interact with promoter elements. TBP, as its name suggests, binds to the TATA box. Nevertheless, a set of general transcription factors lacking TBP can nevertheless promote transcription (Wieczorek et al., 1998). Thus, the *principle of overlapping mechanisms* applies to general transcription factors, and different combinations of these factors may be assembled as a result of the somewhat variable composition and spacing of promoter elements discussed earlier.

The levels of general transcription factors present in an entire cell nucleus are probably similar among the cells of an organism. Therefore, these factors are likely to contribute little, if anything, to the stage-dependent or region-specific transcription of certain genes. Such selective regulatory effects depend on other transcription factors that stimulate or repress the activity of the general transcription factors. This interesting class of factors will be discussed next.

TRANSCRIPTIONAL ACTIVATORS AND REPRESSORS ASSOCIATE WITH RESTRICTED SETS OF GENES AND MAY OCCUR ONLY IN CERTAIN CELLS

A different class of transcription factors, called ***transciptional activators*** and ***transcriptional repressors***, act more selectively than their general counterparts (Table 16.1). Whereas general transcription factors bind strictly to *basal promoters* and associate with the majority of all genes, transcriptional activators and repressors bind to *upstream promoter elements* and *enhancers* of a restricted number of target genes. Most importantly, the concentrations of transcriptional activators and repressors present varies greatly from one cell type to another. As a result, the rate at which a given gene is transcribed may vary by several orders of magnitude between different cells. Thus, it is essentially the combined action of transcriptional activators and repressors that control differential gene expression.

How can a transcriptional activator bind to an enhancer located thousands of nucleotides away from the promoter and still affect the rate of transcription? It appears likely that the DNA "spacer" segments that separate enhancers from promoters, and enhancers from one another, loop out so that the associated activators and

table 16.1 Two Types of Transcription Factors

	General Transcription Factors	Transcriptional Activators and Repressors
Target genes	Most genes	Selected genes
Occurrence in	All cells	Certain cells
Recognition elements located in	Basal promoter	Enhancers
Examples	TBP, TFAs	bicoid protein, steroid receptors

repressors can contact the general transcription factors (see Fig. 16.4). Many transcriptional activators act on TAFs associated with TBP. Thus, TFIID (consisting of TBP and its associated TAFs) has come to be viewed as a central processing unit that integrates the modifying signals received from several enhancers simultaneously (Sauer et al., 1995a, b, 1996).

The salient aspect of transcriptional control is that a relatively small number of transcriptional activators and repressors can control, in combinatorial ways, the activity of many target genes (Tjian and Maniatis, 1994). Part of this combinatorial action occurs when transcriptional activators and repressors bind to enhancers with ***composite response elements***—that is, closely spaced or overlapping recognition sequences. Some composite response elements are bound *competitively* by activators or repressors that mutually displace each other (see Section 22.4). Other complex response elements are bound *simultaneously* by two or more activators/repressors in such a way that the combination, rather than any single factor, determines the overall effect on a general transcription factor, and eventually, RNA polymerase (Fig. 16.7; Starr et al., 1996). Other combinatorial actions, as mentioned above, occur when several TAFs together impinge on another transcription factor.

Several important questions about transcriptional control are unresolved, and fundamentally new discoveries are still being made. Recent studies have revealed a large heterogeneous class of proteins called ***transcriptional coregulators*** (*Lemon and Freedman, 1999;* Mannervik et al., 1999; Xu et al., 1999; Glass and Rosenfeld, 2000). Some act as bridges and integrating elements that link stage-dependent and tissue-specific transcriptional activators and inhibitors to the basal transcription complex. In this sense, they function similar to TAFs, but they act on parts of the general transcription complex other than TBP. Other transcriptional coregulators modify histones so that chromatin becomes more open or closed to the proteins of the basal transcription complex (see Section 16.5).

Another unresolved question is how the action of enhancers is normally restricted to their nearest promoters. In some cases, the restriction may simply be due to the fact that only the nearest promoter is located in an open section of chromatin. In other cases, novel DNA sequence elements termed ***insulators*** have been identified, which associate with specific proteins. Some insulators have the effect of allowing an enhancer to interact with promoters on one side but not the other (*A. C. Bell and G. Felsenfeld, 1999*). Other insulators protect promoters from the encroachment of mechanisms that generate a closed chromatin conformation. Conversely, certain ***promoter targeting sequences*** have evolved that facilitate the interaction of enhancers with promoters across intervening insulators (Dorsett, 1999; Zhou and Levine, 1999).

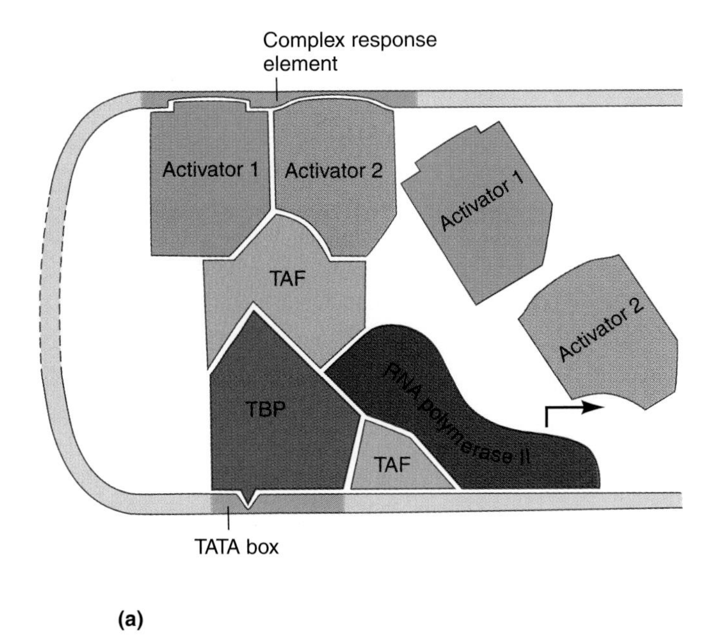

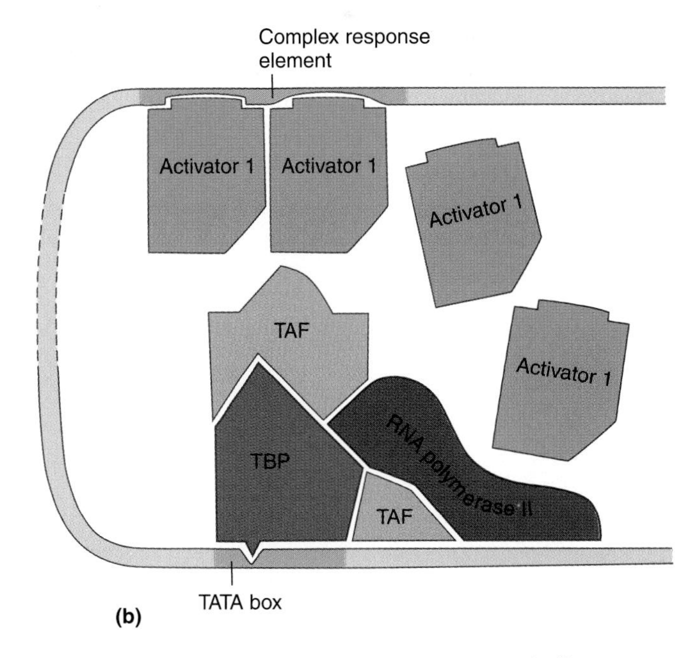

Figure 16.7 Combinatorial action of transcriptional activators. **(a)** Two transcriptional activators, both present at similar concentrations, are assumed to bind to matching DNA sequences located next to each other in a complex response element (CRE). Their combined activation domains then interact with a matching domain of a general transcription factor such as a TBP-associated factor (TAF), with the result of RNA polymerase activity. **(b)** If Activator 1 is present in excess over Activator 2, and if Activator 1 can bind—albeit poorly—to the response element normally occupied by Activator 2, then two adjacent Activator 1 molecules may occupy CRE. However, two Activator 1 molecules may not fit the same TAF domain as the Activator 1/Activator 2 combination did, and RNA polymerase may fail to become activated. Thus, even though Activator 1 is a transcriptional activator, increasing its concentration will not necessarily cause further transcriptional activation.

Given the enormous regulatory power of transcriptional activation and repression, it is not surprising that they have come to be recognized as the principal mechanisms by which cells acquire different states of determination and differentiation. In short, cells become different by synthesizing different combinations of transcriptional activators and repressors. Because the transcription factors themselves are gene products, their synthesis in turn is controlled by transcription factors. Thus, the cells of an organism acquire different fates by establishing different hierarchies, feedback loops, and mutual inhibitions of transcription factors and their encoding genes.

TRANSCRIPTIONAL ACTIVATORS HAVE HIGHLY CONSERVED DNA-BINDING DOMAINS

Transcriptional activators are grouped into families based on the structure of their functional domains, especially their *DNA-binding domains.* In addition, some transcription factors are themselves multimeric proteins, in which case their *multimerization domains* may aid in their classification.

One of the most common DNA-binding domains is known as the ***helix-turn-helix*** motif. It consists of two α helices connected by a short nonhelical stretch, which constitutes the turn. One of the helices, the so-called recognition helix, fits in the major groove of DNA. Its amino acid residues, which vary from one helix-turn-helix domain to another, interact with specific nucleotide sequences. A related DNA-binding domain is known as the ***homeodomain***, so named because it was first discovered in transcriptional activators encoded by *homeotic* genes such as *Ultrabithorax*$^+$. Another common DNA-binding domain in transcriptional activators is the ***zinc finger***, a sequence of about 30 amino acids folded around a zinc atom. Typically, two or more zinc fingers are found in a row, extending into neighboring sections of the major DNA groove.

The activation domains of transcription factors have been classified as acidic, glutamine-rich, and proline-rich, although the functional significance of this classification is still under investigation.

A common dimerization domain is the ***leucine zipper***, a leucine-rich strip that zips two polypeptides together into a Y-shaped dimer, with the two arms of the Y engaging the DNA like a clothes pin. Similar Y-shaped dimeric DNA-binding domains are formed by transcriptional activators containing the ***basic helix-loop-helix*** motif, not to be confused with the helix-turn-helix motif mentioned earlier, which acts as a monomer.

The functional domains of general transcription factors as well as transcriptional activators have been exceedingly well conserved in evolution. A similar conservation of functional protein domains has been observed in cell cycle proteins, cell adhesion molecules, and growth factors and their receptors. These similarities are of fundamental importance. They indicate that, in widely divergent organisms, similar sets of molecules control some of the key regulatory processes in development. Learning of the generality of these processes has been intellectually rewarding because general mechanisms are viewed as more noteworthy than species-specific ones. The same similarities have also been of great practical importance because the conserved protein domains reflect similar genomic DNA sequences, so that a cloned segment encoding a conserved functional domain in one species can be used as a probe to isolate genes encoding proteins with corresponding domains in other species.

THE BICOID PROTEIN ACTS AS A TRANSCRIPTIONAL ACTIVATOR ON THE *HUNCHBACK*$^+$ GENE IN THE *DROSOPHILA* EMBRYO

Transcriptional activators, conferring specific gene activities upon different cells, are very important regulators of embryogenesis. As an example, let us examine the bicoid protein in *Drosophila.* Embryos derived from mothers lacking a functional *bicoid*$^+$ gene have no head or thorax; they consist of an extended abdomen carrying at its front end a duplication of a posterior terminal structure known as a *telson* (see Fig. 1.20). In wild-type females, ovarian nurse cells synthesize bicoid mRNA and transfer it to the anterior pole of the oocyte. Following fertilization, the bicoid mRNA serves as a localized source of bicoid protein, which diffuses posteriorly, forming a concentration gradient with a maximum near the anterior pole of the embryo (Fig. 16.8).

The bicoid protein contains a *homeodomain,* providing a first indication that this protein may act as a transcription factor. A plausible target gene of bicoid protein is the *hunchback*$^+$ gene, which is one of the first genes transcribed in the *Drosophila* embryo. Embryos homozygous for loss-of-function alleles of *hunchback* lack posterior head parts and the entire thorax. The similarity between the *bicoid* and *hunchback* phenotypes suggests that much of the *bicoid* phenotype may be caused by a failure to activate the embryonic *hunchback*$^+$ gene. This hypothesis is confirmed by monitoring the synthesis of hunchback RNA or protein in embryos derived from wild-type versus *bicoid* mutant mothers (Schröder et al., 1988; Tautz, 1988).

In offspring from wild-type mothers, the *hunchback*$^+$ gene is expressed in the anterior half of embryos at preblastoderm stages (Fig. 16.9a). After formation of the cellular *blastoderm,* an additional *hunchback*$^+$ expression domain appears as a subterminal posterior band (Fig. 16.9b). In offspring from bicoid loss-of-function mothers, the large anterior *hunchback*$^+$ expression domain is missing while the posterior domain appears twice—in its normal posterior location and as an anterior duplication corresponding to the duplication of the *telson* (Fig. 16.9c). Genetic and molecular analyses show that tran-

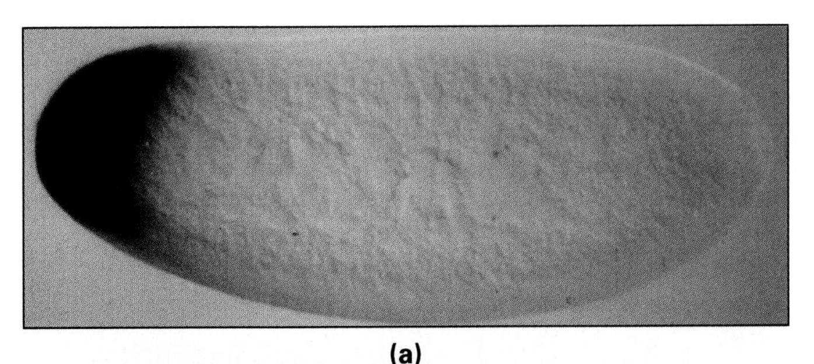

(a)

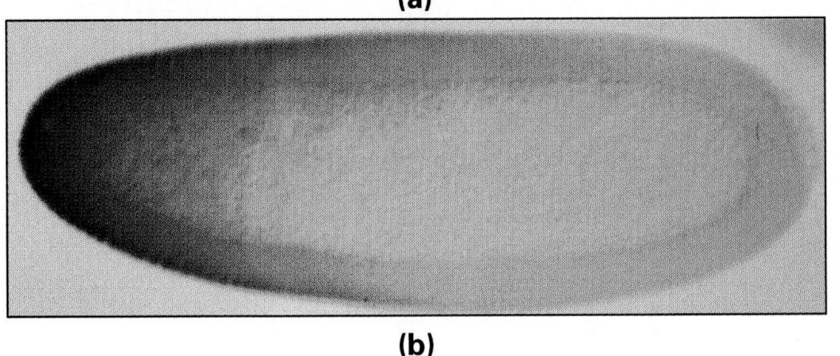

(b)

Figure 16.8 Regional control of embryonic gene expression by a maternally encoded transcriptional activator in *Drosophila melanogaster.* Anterior egg pole to the left. **(a)** Maternal bicoid mRNA, stained dark here by in situ hybridization (see Method 8.1), becomes localized near the anterior pole while the oocyte develops in the ovary. **(b)** Bicoid protein, made visible by immunostaining (see Method 4.1), is synthesized after egg deposition. The protein decays as it diffuses away from its mRNA, thus forming a concentration gradient. Note the accumulation of bicoid protein in the nuclei of the blastoderm (small circles near anterior surface of embryo). Above a threshold value that is normally surpassed in the anterior egg half, bicoid protein activates the embryonic *hunchback*$^+$ gene, as shown in Figure 16.9.

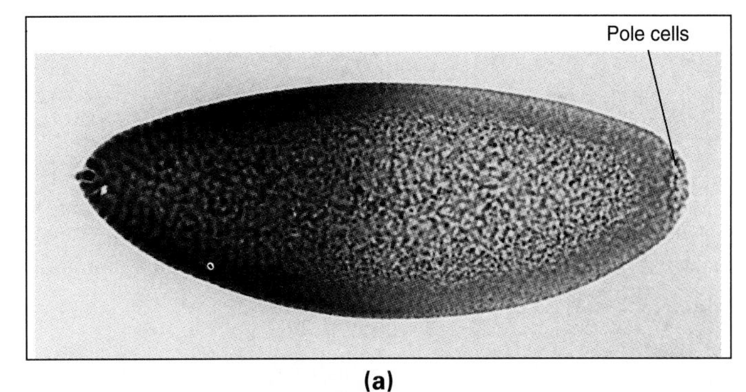

(a)

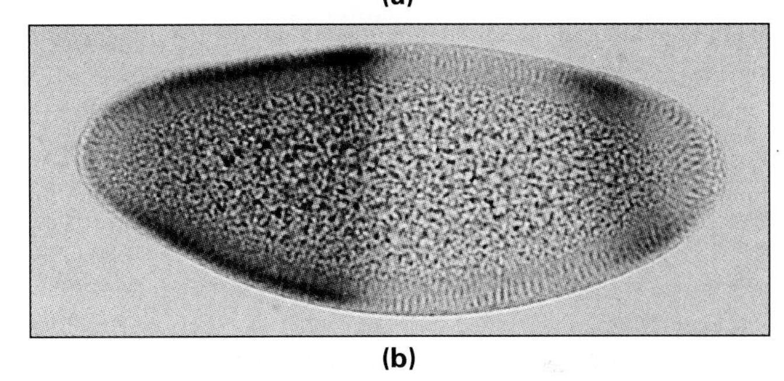

(b)

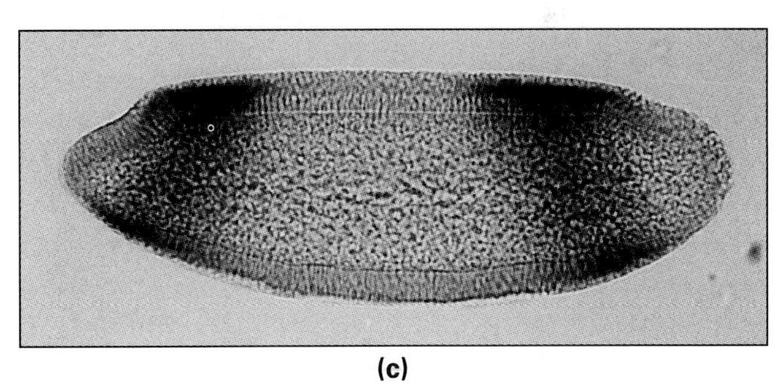

(c)

Figure 16.9 Expression of the *hunchback*$^+$ gene in *Drosophila* embryos shown by immunostaining. The posterior pole (to the right) is marked by pole cells. **(a, b)** Wild-type embryos. Before blastoderm formation **(a)**, hunchback protein is found only in the anterior half. At the cellular blastoderm stage **(b)**, an additional expression domain appears as a subterminal band near the posterior pole. **(c)** Embryo derived from a *bicoid* mutant mother. The anterior *hunchback*$^+$ expression domain is missing, while the posterior domain is duplicated anteriorly.

scripts in the normal posterior and anterior domains start from two different promoters, designated P_1 and P_2, respectively (Fig. 16.10). The transcript from promoter P_1 is made in the posterior expression domain at the cellular blastoderm stage (see Fig. 16.9b) and also maternally during oogenesis. In contrast, the transcript from promoter P_2 is made in the anterior expression domain at the cellular blastoderm stage. Taken together, these results suggest that the activity of promoter P_2 depends on bicoid protein, presumably acting as a transcriptional activator, while promoter P_1 can be activated without bicoid protein.

To test whether bicoid protein binds directly to enhancer regions associated with promoter P_2 of *hunchback*$^+$, Driever and Nüsslein-Volhard (1989) used a technique known as ***DNase footprinting.*** For this procedure the investigators cloned a *hunchback*$^+$ gene segment containing the P_2 promoter and the upstream region, which most likely contained any enhancers associated with P_2. This DNA was allowed to bind to bicoid protein in vitro and then was exposed to deoxyribonuclease I (DNase I), an enzyme that digests naked DNA but not DNA associated with protein. Any segments protected from DNase attack therefore had to be bound by bicoid protein. The most strongly protected DNA elements, termed A_1, A_2, and A_3, were located just upstream of the P_2 promoter (Fig. 16.10).

TO test the function of elements A_1 through A_3 in vivo, Driever and coworkers (1989a) spliced a DNA segment containing these elements and the P_2 promoter in front of the transcribed region of the *lacZ reporter gene.* This fusion gene was introduced into the germ line of host embryos by P-element transformation (see Section 15.6). The expression pattern of the transgene closely mimicked the anterior expression domain of the resident *hunchback*$^+$ gene as

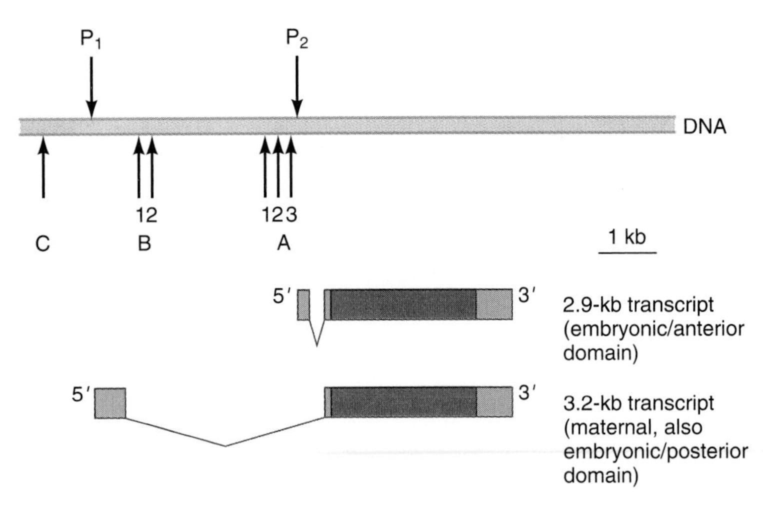

Figure 16.10 Map of the *hunchback*$^+$ gene showing the binding sites for bicoid protein. The *hunchback*$^+$ gene is transcribed from two promoters, P_1 and P_2. The mRNA originating from P_2 is 2.9 kb long and is expressed in the anterior half of the embryo. The mRNA originating from P_1 is 3.2 kb long and is expressed near the posterior pole of the embryo. (This mRNA is also transcribed maternally and deposited in the oocyte during oogenesis.) The translated regions (deep red) of the two mRNAs are identical. A, B, and C are enhancer regions protected from DNase I in the presence of bicoid protein. Most strongly protected are the elements A_1, A_2, and A_3, which are located next to the P_2 promoter.

expected (compare Figs. 16.9b and 16.11a). Upon successive deletion of elements A_1 to A_3, the expression of the fusion gene became progressively weaker, and the boundary of the expression domain became less sharp (Fig. 16.11b–d). Also, the boundary of the expression domain seemed to recede toward the anterior pole.

In further experiments, the DNA domain containing A_1 to A_3 revealed the characteristics of an enhancer: Its orientation could be reversed without any loss of activity, and it could be moved more than 1 kb from the promoter and still retain considerable activity.

Questions

1. Where would you expect the *hunchback-lacZ* fusion gene to be expressed if it were introduced into embryos derived from females lacking a functional *bicoid* gene?
2. Given that deletion of the A_1 enhancer element resulted in expression of the fusion gene in a smaller anterior domain, how could the gradient of bicoid protein in the *Drosophila* control the expression of several natural target genes with staggered expression domains?

The results just described show that the maternally encoded bicoid protein acts as a transcriptional activator on the P_2 promoter of the embryonic *hunchback*$^+$ gene. Above a certain threshold concentration of the bicoid gradient, blastoderm cells express *hunchback*$^+$ at high level. As we will discuss in Sections 22.3 and 22.4, bicoid protein enhances or inhibits the region-specific expression of several other embryonic genes.

THE ACTIVITY OF TRANSCRIPTION FACTORS THEMSELVES MAY BE REGULATED

Given the importance of transcriptional control, it is no surprise that several regulatory mechanisms have evolved that tie transcriptional control to environmental or endogenous signals. For example, all species seem to have a set of genes, called *heat shock genes,* that are transcribed in response to heat and other types of environmental stress. In vertebrates, the presence of toxic substances in the blood triggers the transcription of genes that encode detoxifying proteins such as the *metallothioneins,* which remove cadmium and other heavy metals from the blood serum. Researchers often use the regulatory regions of such genes in transgenes designed to be expressed conditionally, as described in Section 15.6.

Many transcription factors are synthesized in an inactive precursor form, so that their activation itself becomes a level of regulation. As mentioned earlier, some factors are composed of two or more polypeptides that form an active protein only upon multimerization. For other transcription factors, the regulated step is the transport from the cytoplasm, where they are synthesized, to the nucleus, where they function. For example, the dorsal protein, which has a similar function for the dorsoventral axis of the *Drosophila* embryo as bicoid has for the anteroposterior axis, is prevented from entering the nucleus by an associated regulatory protein, cactus (see Section 22.6). Still other transcription factors are activated by phosphorylation and/or dephosphorylation of certain amino acids (T. Hunter and M. Karin, 1992). The activity of the *protein kinases* and *protein phosphatases* catalyzing these steps may depend in turn on their own phosphorylation state, so that they form regulatory cascades that ultimately regulate the activity of a transcription factor (see Fig. 2.28). Finally, many transcription factors are activated by the binding of a *ligand*—that is, a small matching molecule such as a hormone. In relation to the ligand, such transcription factors are also called *receptors.*

Hormones are signal molecules that are carried by the bloodstream to all cells in the body. Yet each hormone elicits a response only in some cells, termed ***target cells,*** and only from a few genes, called ***target genes.*** For instance, the production of egg white proteins in the chicken oviduct is controlled by the sex hormones estrogen and progesterone. In the presence of one of these hormones, certain oviduct cells synthesize ovalbumin and other egg white proteins in large quantities. After 10 h of hormone stimulation, the ovalbumin level in these cells increases dramatically from trace amounts to more than 50% of all newly synthesized protein. The surge in ovalbumin synthesis results from a steep increase in the production of ovalbumin mRNA, indicat-

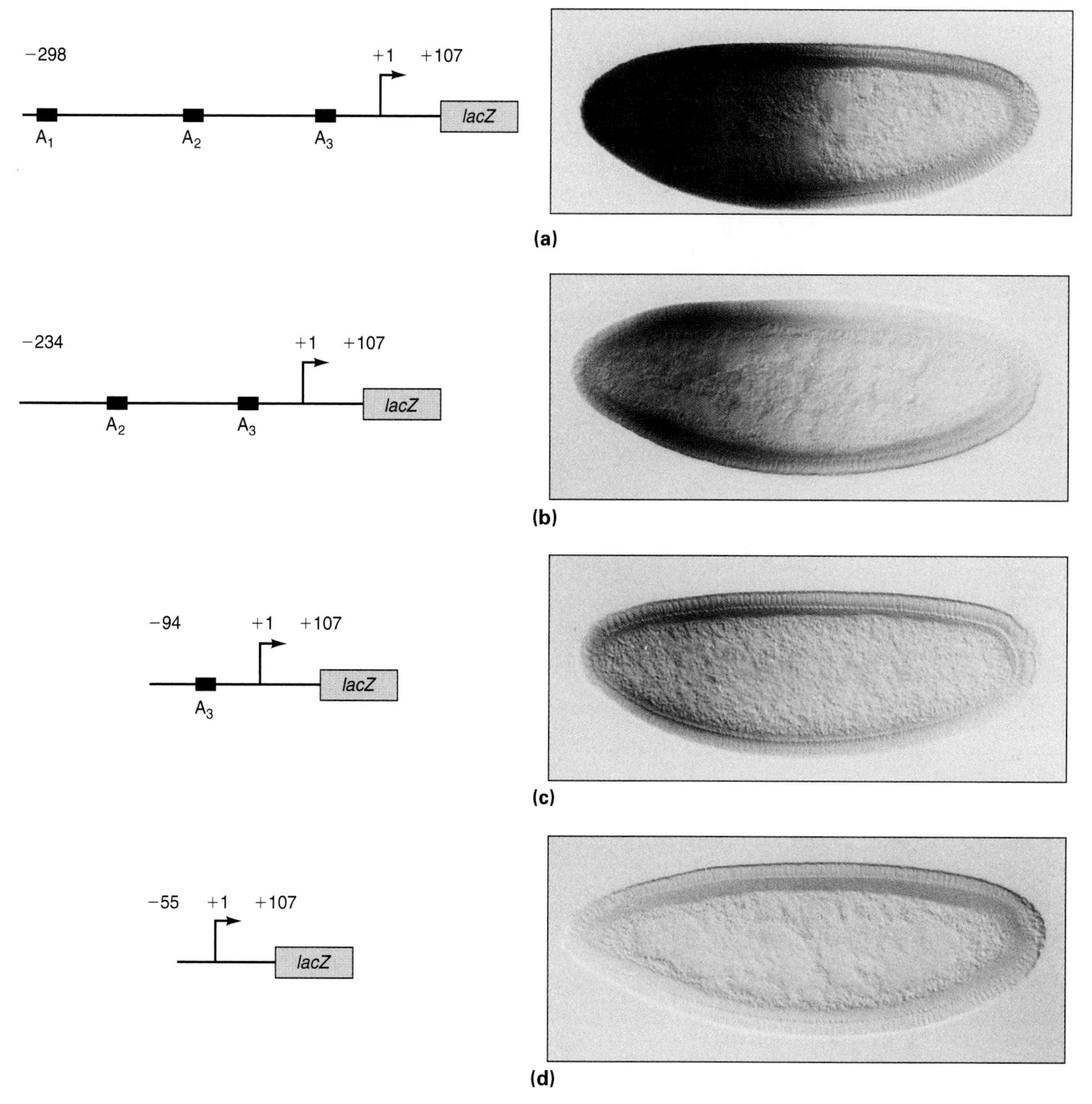

Figure 16.11 Analysis of *hunchback*$^+$ enhancer sequences by fusion genes consisting of the *hunchback* P_2 promoter and a variable number of enhancer elements (A_1 to A_3) spliced to the *lacZ* reporter gene. Each fusion gene is used to create transgenic *Drosophila* lines by P-element transformation. The diagrams show the *hunchback* enhancer sequences present in the respective fusion genes. The photomicrographs show embryos stained for *lacZ* expression. **(a)** When enhancer elements A_1 to A_3 all are present, the fusion gene is expressed in a pattern similar to that of the resident *hunchback*$^+$ gene in the anterior domain (compare with Fig. 16.9b). **(b–d)** As the enhancer elements are progressively removed, *lacZ* expression becomes progressively weaker, and the posterior boundary of the expression domain becomes less distinct.

ing transcriptional regulation of the ovalbumin gene (Fig. 16.12). The astounding specificity of the hormone action is due to the molecular mechanisms by which steroid hormones act on their target genes.

Steroid hormones are small lipid-soluble molecules. They can cross plasma membranes by simple diffusion and permeate all cells of an organism. *Target cells* are distinguished from non-target cells by the presence of a matching *receptor.* The receptors for some steroid hormones, including progesterone and testosterone, are typically located in the cell cytoplasm. Each of these receptors is composed of two receptor polypeptides and associated proteins (Fig. 16.13). Upon hormone binding, the receptor undergoes an activation process involving the release of an inhibitory protein. The hormone-receptor complex then moves into the nucleus and binds

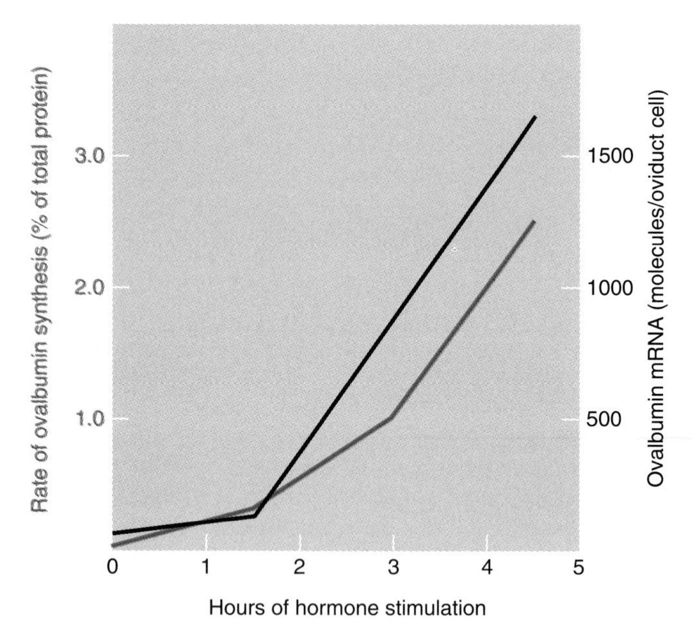

Figure 16.12 Ovalbumin synthesis in oviduct cells of chickens injected with progesterone. The hormone application caused sharp increases in the concentration of ovalbumin mRNA (black line) and the rate of ovalbumin synthesis (colored line).

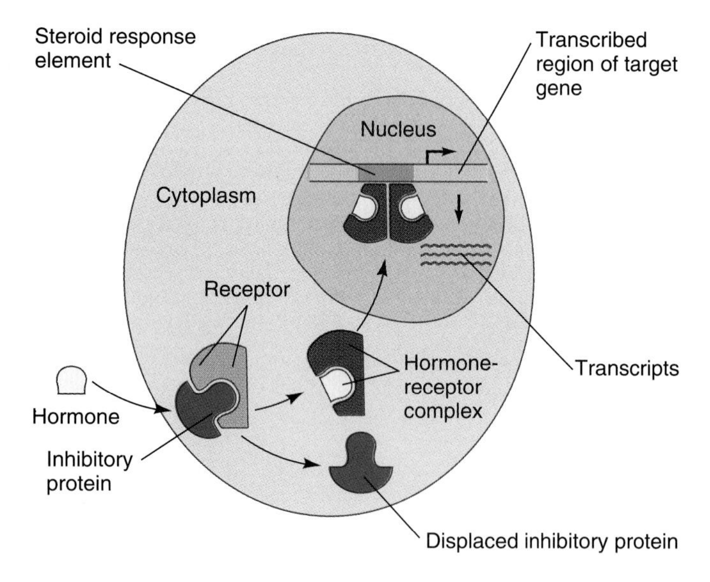

Figure 16.13 General model of gene activation by a group of steroid hormones including progesterone and testosterone. The hormone reaches all cells in an organism through the bloodstream and by diffusion. However, only some cells (called the target cells of the hormone) have a matching receptor protein to which the hormone can bind. The receptor is typically located in the cytoplasm, where it is associated with other proteins. Hormone binding displaces an inhibitory protein previously associated with the receptor. The hormone-receptor complex then enters the nucleus and binds to certain genes (called the target genes of the hormone) that have an enhancer with a matching steroid response element (SRE). The hormone-receptor complex binds to the SRE as a dimer. This event usually activates, but in some cases inhibits, transcription of the target gene.

to specific enhancer sequences known as *steroid response elements.* Such elements are the characteristic feature of the *target genes* that respond to a steroid hormone. The receptors for other steroid hormones, including estrogen and the insect molting hormone ecdysone, act similarly except that they are already located in the nucleus before they are bound by their respective ligands. Indeed, these receptors bind to their response elements and inhibit transcription of their target genes before the receptors are bound by their ligands and then act as transcriptional activators. For a typical steroid hormone receptor, there may be about 50 target genes.

The receptors for steroid hormones, vitamin D, and retinoic acid form a family of similar proteins, all of which act as transcription factors. The similarities within the family of steroid receptors suggest that their genes have evolved by duplication and subsequent sequence divergence.

16.5 Chromatin Structure and Transcription

So far in our discussion of transcriptional control, we have assumed that genomic DNA is readily accessible for interactions with RNA polymerase and transcription factors. However, the genomic DNA of eukaryotes is contained in *chromatin,* where the DNA is associated with histones and various nonhistone proteins. These associations, which change during the cell cycle and depend on the developmental stage of a cell, may greatly affect the accessibility of genes for transcription. As a rule, DNA in condensed chromatin is transcribed less actively than DNA in decondensed chromatin.

HETEROCHROMATIC CHROMOSOME REGIONS ARE NOT TRANSCRIBED

During the cell cycle, most chromosomes alternate between a highly condensed state and a more diffuse state. The condensed state during M phase allows the chromosomes to disentangle so that mitosis can occur. No RNA synthesis is observed during this phase. The diffuse state during interphase allows the chromosomal DNA to be transcribed. Chromosomes, or parts thereof, showing this cyclical change are called ***euchromatin*** (Gk. *eu,* "good", also "true"; *chroma,* "color", referring to the stainability of condensed chromatin). However, certain chromosome regions remain highly compacted even during interphase. The chromatin in these regions is called ***heterochromatin*** (Gk. *heteros,* "other"). Autoradiographs of cells incubated with [^{3}H]uridine show that little or no RNA is synthesized in heterochromatin (Fig. 16.14).

Some heterochromatic regions, called *constitutive heterochromatin,* are present in all individuals and cells of a species. Other chromosomal regions, known as *facultative*

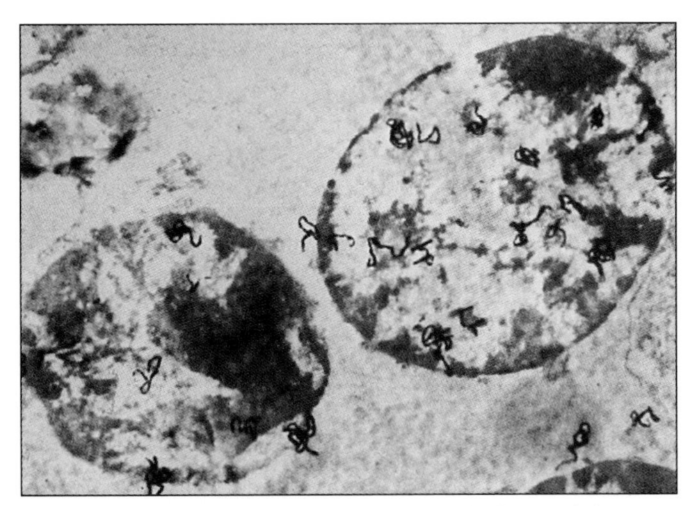

Figure 16.14 Correlation between transcription and decondensed chromatin status. This transmission electron micrograph shows a calf thymus nucleus after incubation with [^{3}H]uridine. The dark wiggly lines are silver grains (autoradiographic signals; see Method 3.1) indicating the presence of newly synthesized RNA. Note that these signals appear mostly over the decondensed (light) areas of chromatin, even though the DNA concentration is much higher in the condensed areas.

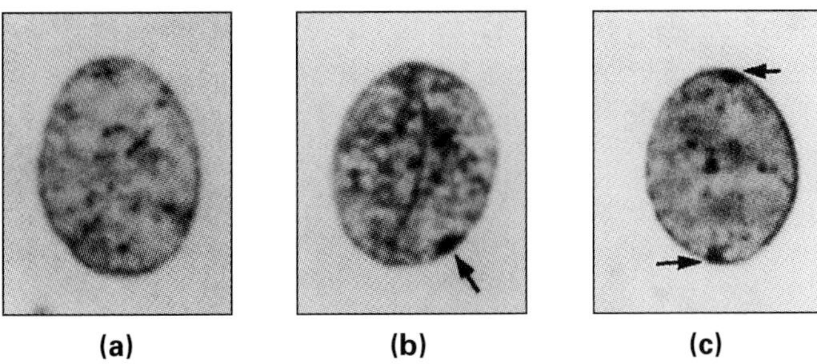

Figure 16.15 Barr bodies in nuclei of human oral epithelial cells stained with cresylecht violet. **(a)** Normal XY male, with no Barr body. **(b)** Normal XX female, with one Barr body. **(c)** Female with three X chromosomes and two Barr bodies. In each case, one X chromosome per nucleus is left active.

heterochromatin, are heterochromatic only in one sex or at certain stages of development. A prominent example of facultative heterochromatin can be found in the cells of normal female mammals, in each of which one of the two X chromosomes is heterochromatic. This heterochromatic X chromosome, called the ***Barr body,*** after its discoverer, is absent from normal males (Fig. 16.15).

Further analysis of Barr body formation in mammals with abnormal numbers of X chromosomes has shown that most cells inactivate all X chromosomes except for one. This is a means of *dosage compensation*—that is, of adjusting the expression of X-linked genes so that the same amounts of gene products are made in both sexes (see Section 26.3). The inactivation of supernumerary X chromosomes occurs during cleavage and appears to occur randomly in the cells of the *inner cell mass,* which give rise to the embryo proper. From then on, the active or inactive state of each X chromosome is passed on clonally.

PUFFS IN POLYTENE CHROMOSOMES ARE ACTIVELY TRANSCRIBED

In euchromatin during interphase, there are varying levels of decondensation. Some of these can be distinguished under the light microscope in *polytene chromosomes,* which consist of many chromatids aligned neatly in register (see Section 7.3). Polytene chromosomes have areas of decondensation known as ***puffs.*** Puffs respond to gene regulatory signals such as hormones, suggesting that they might represent sites of active transcription. Experiments in which cells with polytene chromosomes are *pulse-labeled* with [^{3}H]uridine confirm this (Pelling, 1959). Autoradiographs prepared from such cells show that puffs are areas of very active RNA accumulation (Fig. 16.16). These results confirm that decondensed chromosomal regions are transcribed more actively than condensed regions. However, the data do not show whether puffing is a consequence or a prerequisite of transcription.

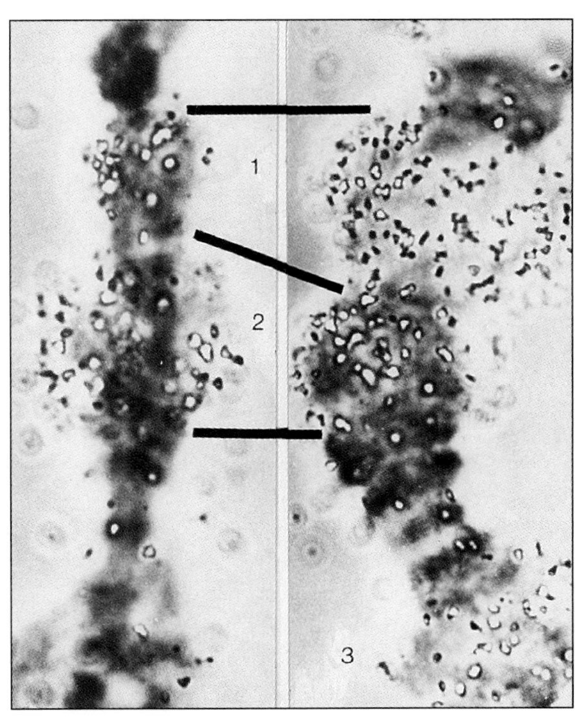

Figure 16.16 Correlation between puffing and RNA synthesis in polytene chromosomes. **(a)** Salivary gland chromosome from a normal larva of the midge *Chironomus tentans* after injection with a radiolabeled RNA precursor, [^{3}H]uridine. **(b)** Corresponding chromosome from larva also injected with the molting hormone, ecdysone. Note size increase in puff 1. The grains over the puffs were generated through autoradiography (see Method 3.1) and indicate newly synthesized RNA. The rate of RNA synthesis over puff 1 increases with the puff's size.

DNA IN TRANSCRIBED CHROMATIN IS SENSITIVE TO DNASE I DIGESTION

The general correlation between chromatin decondensation and transcription is also found at the molecular level. The basic packaging units of chromatin are known as ***nucleosomes*** (Fig. 16.17a). Each nucleosome consists of about 150 bp of DNA wrapped around a core of eight histone proteins, two molecules each of histones H2A, H2B, H3, and H4. Neighboring nucleosomes are connected by stretches of "linker" DNA, which is sometimes associated with a linker histone, H1. This ***beaded string configuration*** leaves the DNA fairly accessible to transcription factors and RNA polymerase. More condensed chromatin fibers measure 30 nm in diameter (Fig. 16.17b). These fibers are arranged in a tight helical configuration of six nucleosomes per turn, known as the ***solenoid configuration,*** which leaves DNA less accessible to transcription factors and RNA polymerase (Felsenfeld and McGhee, 1986).

Observations on a wide range of organisms indicate that nucleosome packing is indeed relaxed in transcribed chromatin regions (Felsenfeld, 1992). This correlation can be inferred from tests exposing chromatin to very low concentrations of DNase I. After a mild digestion of chromatin or whole nuclei with DNase I, DNA extracted from chromatin can be tested for the presence and integrity of any gene for which a labeled hybridization probe is available. Such tests show regularly that those genes that are most actively transcribed are also most vulnerable to DNase I attack. For example, globin genes are especially sensitive to DNase I in red blood cells but not in other cells. In chicken oviduct cells, the ovalbumin gene is DNase-sensitive, while the globin genes are DNase-resistant (Stalder et al., 1980). These results show that transcribed genes are in an *open*, or decondensed, state that leaves them more accessible to DNase, and, by implication, to transcription factors and RNA polymerase.

DNA regions sensitive to DNase I are often very large, spanning 100 kb or more. Within such a region, there are much smaller ***DNase I-hypersensitive sites.*** These are degraded by even lower concentrations of DNase I. In contrast to general *DNase sensitivity*, which reflects a decondensed state in which the chromatin is uncoiled but the *nucleosome* structure is left intact, *DNase I hypersensitivity* may result from a disruption of nucleosomes (C. C. Adams and J. L. Workman, 1993). The hypersensitive sites are of particular interest because they coincide with promoters and enhancers.

The observations described in this section support a two-step model of chromatin decondensation and transcriptional availability. In step 1, chromatin regions are

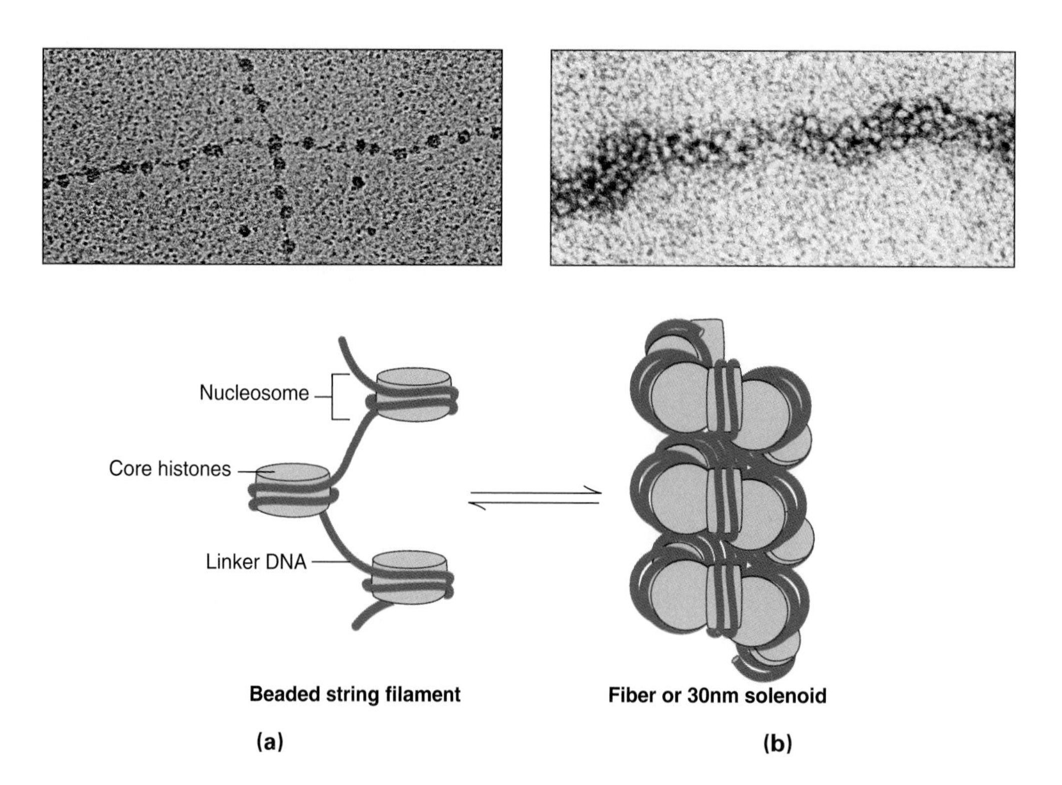

Figure 16.17 Nucleosome structure of eukaryotic chromatin. Each nucleosome consists of a core of eight histone proteins with about 150 bp of DNA wrapped around it in 1.8 turns. Nucleosomes are connected by stretches of linker DNA, which vary in length but measure about 60 bp on average. **(a)** Electron micrograph and schematic diagram of decondensed chromatin. In this beaded string configuration, the linker DNA is unprotected. **(b)** Condensed chromatin, forming a fiber measuring about 30 nm in diameter. The nucleosomes are packed tightly in a solenoid arrangement, in which the linker DNA is less exposed.

decondensed without disrupting the nucleosome structure. This step of "opening the chromatin" confers DNase sensitivity to large stretches of DNA and makes them fairly accessible to nucleases as well as transcription factors. This step may occur before transcription actually begins. In step 2, the nucleosome structure is disrupted as transcription factors and RNA polymerase bind to promoters and enhancers. This step confers DNase I hypersensitivity to the gene-regulatory regions.

TRANSCRIPTIONAL CONTROL DEPENDS ON HISTONE ACETYLATION AND DEACETYLATION

The eight core histones in nucleosomes are arranged in such a way that their amino termini are sticking out. This makes them accessible to modifying enzymes, such as ***acetyltransferases*** and ***deacetylases,*** which add or remove acetyl moieties, respectively. The amino acid residues modified by these enzymes are lysines, and addition of the acetyl group neutralizes the lysine's positive charge (Fig. 16.18). As a result, the histones bind DNA and certain proteins less tightly, so that the DNA becomes more accessible to transcription factors. The level of histone acetylation changes with the cell cycle and has profound effects on gene activity (Grunstein, 1997). For example, the inactive X chromosome in female mammals is distinguished by a lack of histone H4 acetylation, and H4 is generally less acetylated in heterochromatin than in euchromatin (Jeppesen and Turner, 1993; O'Neill and Turner, 1995).

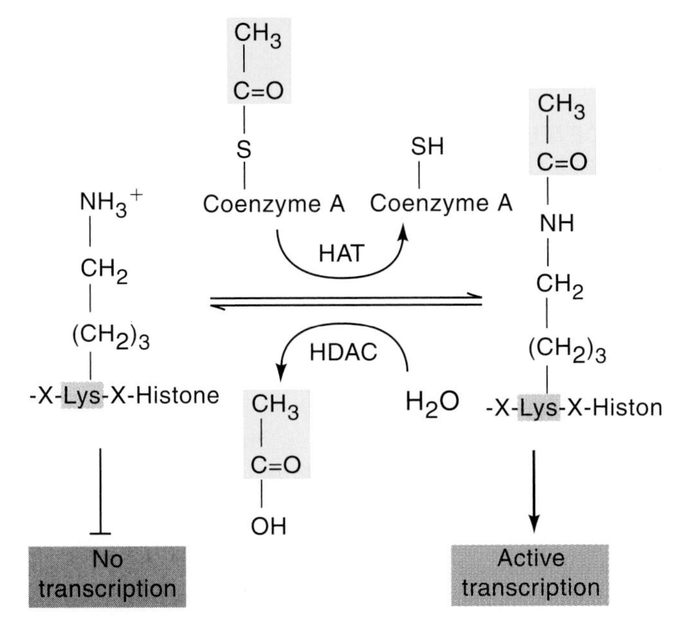

Figure 16.18 Histone acetyltransferase (HAT) and deacetylase (HDAC). Histone acetyltransferases transfer the acetyl moiety of donor coenzyme A to the amino group of internal lysine residues of histones. The reverse reaction is catalized by histone deacetylases. Introduction of the acetyl group to lysine neutralizes its positive charge.

The enzymatic activities that control the level of histone acetylation are inherent to several proteins involved in transcriptional control (Kuo and Allis, 1998; K. Struhl, 1998). Some histone acetyltransferase activity is associated with certain TAFs and should therefore act genome-wide. Other histone acetyltransferase and deacetylase activities are associated with the *transcriptional coregulators* mentioned in Section 16.4. One transcriptional coactivator, designated GCN5, is found in a wide range of organisms from yeast to humans. The most detailed studies on GCN5 have been done with yeast, where GCN5 is necessary for maximal expression of target genes encoding enzymes involved in the biosynthesis of certain amino acids. Yeast mutants with loss-of-function alleles of *GCN5* therefore do not grow well on media that do not contain these amino acids.

IN order to test whether GCN5 activates its target genes by acetylating the histones of associated nucleosomes, Kuo and coworkers (1998) used a yeast strain carrying a deletion (null allele) for *GCN5.* This strain could be "rescued"—that is, restored to normal growth and expression of GCN5 target genes, by transforming it with a cloned plasmid containing the wild-type allele, *GCN5*$^+$. Using this *rescue bioassay,* the investigators introduced mutations into the *GCN5* transgene and then assessed the effects of these mutations on the transgene's rescuing ability and on the histone acetyltransferase activity of GCN5 protein synthesized from the cloned plasmids in bacterial host cells. They found that the same mutations that interfered with the rescuing ability of the transgene also destroyed the histone acetylating activity of its protein product in vitro.

To confirm that GCN5 was acting as a histone acetyltransferase in vivo, the investigators overexpressed wild-type and mutant *GCN5* transgenes in yeast cells and then analyzed the total histones from these cells using a gel electrophoresis that separates histones according to acetylation state. They found that transgenes with gain-of-function mutations causing overproduction of wild-type GCN5 caused widespread histone hyperacetylation in vivo. In control experiments with mutant *GCN5* transgenes that did not rescue, overexpression did not lead to hyperacetylation. These results showed that GCN5 acts as a histone acetyl transferase in vivo, and that the same active domain of the GCN5 molecule that causes hyperacetylation is also necessary for rescue.

To detect the acetylation state of histones specifically associated with GCN5 target genes, the researchers used an antibody directed against histones acetylated in specific lysine positions targeted by GCN5. This antibody precipitated chromatin fragments with appropriately acetylated histones. DNA isolated from precipitated fragments was subsequently analyzed for the presence of target gene nucleotide sequences by *dot blotting.* The results showed that the promoters of GCN5 target genes were associated with hyperacetylated histones whenever a transgene encoding active GCN5 was used. Remarkably, only histones from nucleosomes located in or near the promoter were hyperacetylated, indicating that the GCN5-mediated histone acetylation is a very specific and locally confined event. Again, in control experiments with mutant transgenes that failed to rescue, there was also no histone hyperacetylation at the promoters of target genes.

Questions

1. How do you think the investigators assayed for the activation of GCN5 target genes?
2. The N-terminus of yeast histone H4 contains four highly conserved lysine residues that are reversibly acetylated in H4 molecules located near the promoters of genes that are activated in the presence of a particular nutrient, such as galactose. How do you expect the activation of these genes to change in mutants in which the H4 lysine residues are changed to arginine—an amino acid that conserves the positive charge of lysine but cannot be acetylated?

The experiments of Kuo and coworkers indicate that GNC5 acts as a promoter-specific histone acetyltransferase in vivo, and that this activity is necessary for the activation of certain GNC5 target genes. GNC5 seems to act as part of a multiprotein complex that binds to certain transcriptional activators and disrupts nucleosomes in its vicinity, thus facilitating the access of general transcription factors to nearby promoters while serving at the same time as a bridge between general transcription factors and their activators. However, other mechanisms of GNC5 action have not been ruled out. Because the methods for analyzing the effects of histone acetyltransferases and deacetylases in vivo have been developed only recently, it can be expected that much more will be learned in the near future about the wide range of proteins that provide these critical activities in more or less gene-specific ways.

16.6 Transcriptional Control and Cell Determination

Transcription factors are of great interest to developmental biologists because they can explain how thousands of different gene expression patterns can be generated by various combinations of a comparatively small number of regulatory proteins. Transcriptional control occurs at all stages of development, including the earliest steps of *cell determination*, that is, the process by which a cell's potency is narrowed down to its fate (see Section 6.4). For instance, the regional expression of the *fushi tarazu*$^+$ and *hunchback*$^+$ genes in the *Drosophila* embryo begins before the cellular blastoderm stage, when cell determination occurs. This timing suggests that the regional expression of certain genes is part of the cell determination process itself. In this section, we will examine the role of transcriptional control in cell determination.

The active or inactive states of genes may be passed on by different molecular mechanisms, which may vary among animal groups. We will now examine two proposed models. One is based on bistable control circuits, in which transcription factors act as signals. The other relies on DNA methylation.

COMBINATORIAL ACTION OF TRANSCRIPTION FACTORS EXPLAINS THE STEPWISE PROCESS OF CELL DETERMINATION

Cell determination typically occurs in a multistep process, during which cells receive instructive signals from localized cytoplasms and inductive interactions with neighboring cells. As these signals are received, originally pluripotent cells also become committed to their fate. Eventually, cells become differentiated—that is, overtly specialized in structure and function. This stepwise process of instruction and commitment is often represented as a tree with many branches, where each branch point represents a binary decision (see Fig. 6.16).

In terms of transcriptional control, cell differentiation means differential gene expression under the control of a specific combination of transcription factors. The process of cell determination may therefore be seen as a sequential accumulation of transcription factors until each of the factors required for a particular fate is in place. Can the stepwise acquisition of transcription factors also explain the commitment of cells—that is, their "stubborn" behavior in isolation and transplantation experiments, which led classical embryologists to coin the term "cell determination"?

CELL DETERMINATION MAY BE BASED ON BISTABLE CONTROL CIRCUITS OF SWITCH GENES

As a rule, the determined states of cells are passed on faithfully during cell division. This is shown most clearly in the *imaginal disc* (*ID*) cells of *Drosophila* larvae (see Section 6.5). During serial transplantations, the determined states of ID cells are passed on through many rounds of mitosis. Only rarely do cells undergo *transdetermination*—that is, a switch to another determined state. In those cases, the transdetermined state of a cell is then passed on to its descendants during subsequent mitoses.

A molecular understanding of determination and transdetermination was first reached by researchers working with the bacteriophage *lambda*. Two mutually exclusive stages in this phage's development are controlled by a pair of proteins that inhibit each other's synthesis. A generalized version of this regulatory system is shown in Figure 16.19. Because the system has two stable states, it is called a ***bistable control circuit***, and the genes that constitute the circuit are called ***switch genes***.

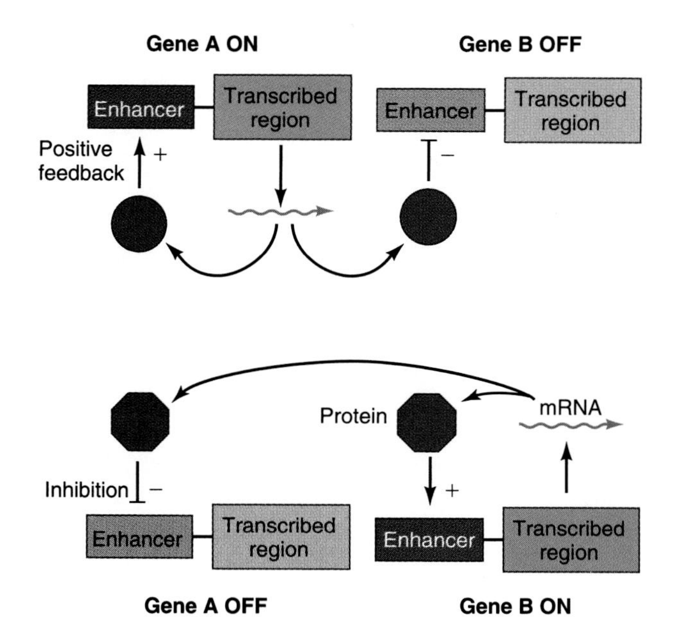

Figure 16.19 Bistable control circuit formed by two genes. Each gene encodes a transcription factor that activates the transcription of its own gene while inhibiting the other gene. The circuit has two stable states, with either gene A on and gene B off, or with gene B on and gene A off.

Kauffman (1973) has proposed that cell determination in *Drosophila* IDs is under control of five bistable control circuits formed by homeotic genes, such as *Antennapedia*$^+$ and *Ultrabithorax*$^+$. Designating the two stable states of each circuit as 1 and 0, he represented the determined states of ID cells with five-digit binary numbers (Fig. 16.20). Transdeterminations, in terms of this model, occur when one of the circuits switches from 0 to 1 or vice versa. This model predicts that transdetermination steps requiring changes in two or more circuits simultaneously should be rare, which is generally the case.

Another type of switch gene, exemplified by the *MyoD*$^+$gene of vertebrates, can switch fibroblasts from connective tissue to muscle fiber development (see Section 20.5).

DROSOPHILA HOMEOTIC GENES SHOW SWITCH GENE CHARACTERISTICS

The genetic analysis of *Drosophila* development has revealed the existence of homeotic genes and other *selector genes*, which control the development of whole body regions by regulating batteries of target genes involved in morphogenesis, such as genes promoting cell division and cell adhesion (see Sections 11.5 and 22.7). Do selector genes show the characteristics of switch genes in bistable control circuits—that is, positive feedback on their own expression and inhibition of other selector genes?

As examples, we will examine the homeotic genes *Ultrabithorax*$^+$ (*Ubx*$^+$) and *Antennapedia*$^+$ (*Antp*$^+$) genes. As

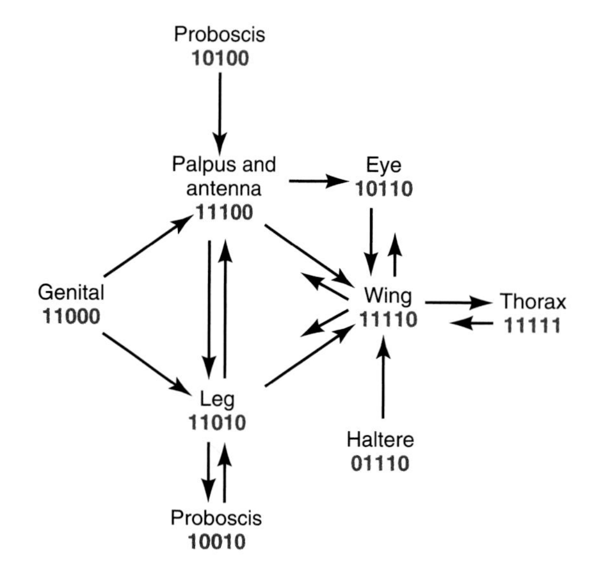

Figure 16.20 Model representing the determined state of *Drosophila* imaginal disc cells as controlled by five bistable control circuits of the type shown in Figure 16.19. The more stable of the two states is designated 1, the alternate state is designated 0. The arrows represent the complete set of actually observed observed transdeterminations (see Section 6.5), with the length of each arrow indicating their relative frequency. Each transdetermination, except for leg to antenna and vice versa, requires only a change in the state of one circuit.

described earlier, individuals carrying certain *Ubx* loss-of-function alleles develop as four-winged flies (see Fig. 1.21): The third thoracic segment (T3) develops as a repeat of the second thoracic segment (T2). Therefore, the normal activity of the *Ubx*$^+$ gene must be necessary for generating the morphological characteristics of T3 instead of T2. Similarly, the normal function of the *Antp*$^+$ gene is necessary for generating the characteristics of the second thoracic segment (T2). In gain-of-function alleles of *Antp*, the gene is ectopically expressed in the head, causing the development of legs instead of antennae (see Fig. 1.22). The following experiments have shown that the *Ubx*$^+$ gene encodes a protein that enhances the transcription of its own gene and inhibits the expression of the *Antp*$^+$ gene. Both interactions have been analyzed in whole embryos and in cultured cells.

HAFEN and coworkers (1984) studied the expression pattern of the *Antp*$^+$ gene using *in situ hybridization* (Fig. 16.21; see Method 8.1). Antennapedia mRNA accumulated strongly in an area overlapping the first and second thoracic segments. Posterior to this domain, *Antp*$^+$ was expressed only weakly in wild-type embryos. However, the expression domain of *Antp*$^+$ was extended dramatically in *Ubx* mutant embryos, which transcribed the *Antp*$^+$ gene throughout the thoracic region. These results indicate that the wild-type function of *Ubx*$^+$ is involved in restricting the expression domain of *Antp*$^+$.

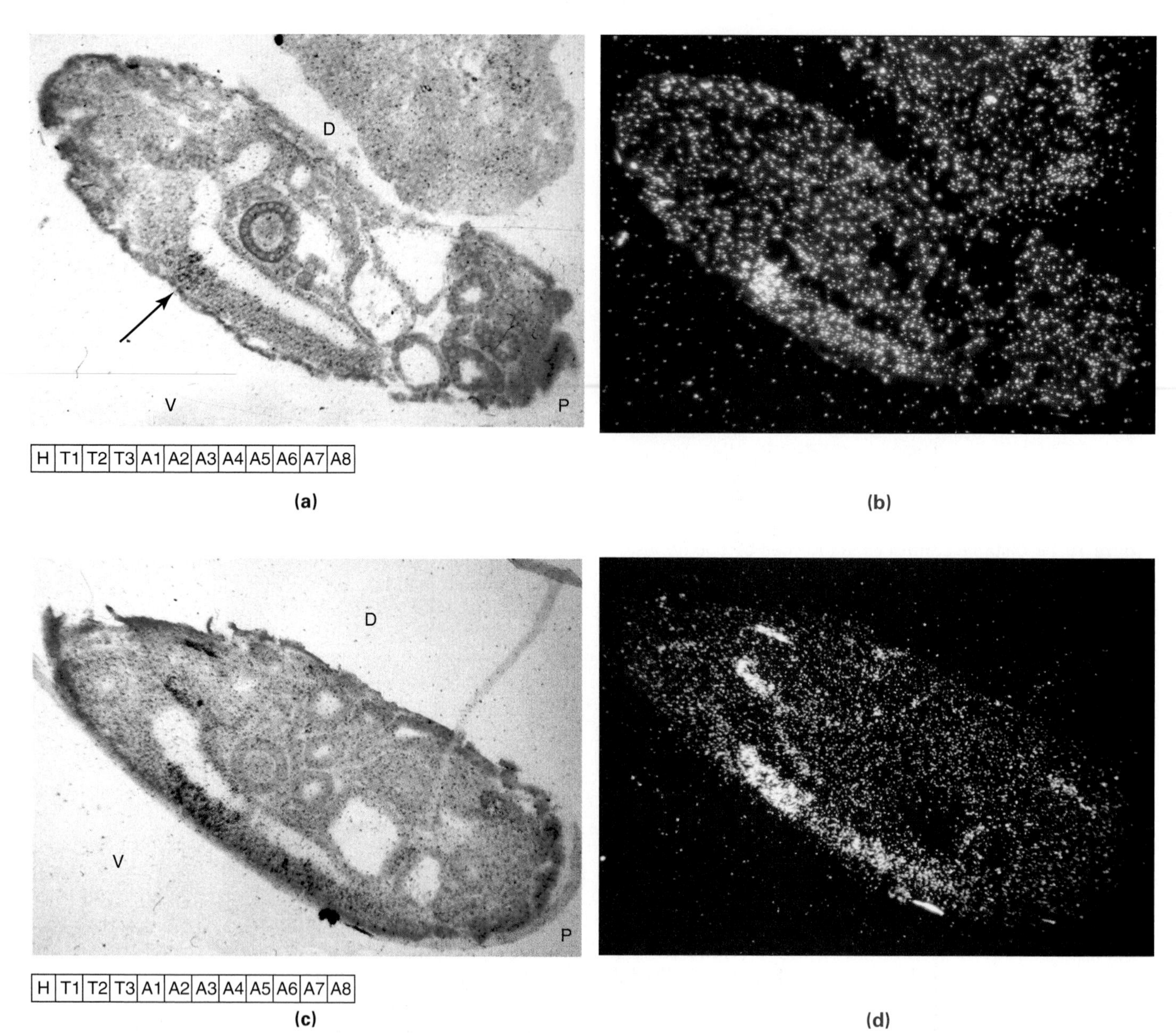

Figure 16.21 Expression of the *Antennapedia*⁺ (*Antp*⁺) gene in normal and mutant *Drosophila* embryos. The transcription of *Antp*⁺ is monitored by in situ hybridization with a radiolabeled cDNA probe (see Method 8.1). Photomicrographs of paramedian sections are shown in transmitted light (left) and in dark-field illumination to make the autoradiographic signal more visible (white grains, right). The diagrams show the embryonic segment pattern, with H representing the head, T1 to T3 the thoracic segments, and A1 to A8 the abdominal segments. **(a, b)** Normal embryo expressing the *Antp*⁺ gene in a region overlapping T2 (indicated by the arrow in part a). **(c, d)** Embryo mutant for *Ultrabithorax* (*Ubx*^{105}). This allele transforms segments T3 and A1 morphologically toward T2. Correspondingly, the expression domain of *Antp*⁺ is extended posteriorly. D = dorsal side of the embryo; P = posterior pole of the embryo; V = ventral side of the embryo.

Results obtained with cultured *Drosophila* cells, which can be manipulated more easily than entire embryos, confirmed the conclusions drawn from the preceding observations (Krasnow et al., 1989). Cells were cotransfected with two types of plasmids (Fig. 16.22). One plasmid, called a ***reporter plasmid***, contained the transcribed region of the bacterial *CAT* gene spliced to the promoter and an enhancer of a *Drosophila* gene. The other plasmid, called an ***effector plasmid***, encoded a *Drosophila* regulatory protein that would bind to the enhancer in the reporter plasmid. The cultured cells were transfected with both plasmids simultaneously for studying the impact of the protein encoded by the effector plasmid on the activity of the reporter plasmid.

In one experiment, Beachy and coworkers (1988) used a reporter plasmid containing regulatory sequences from the *Ubx*⁺ gene. The effector plasmid contained the transcribed region of the *Ubx*⁺ gene driven by an actin promoter. (The latter is known to be fairly active in virtually every type of

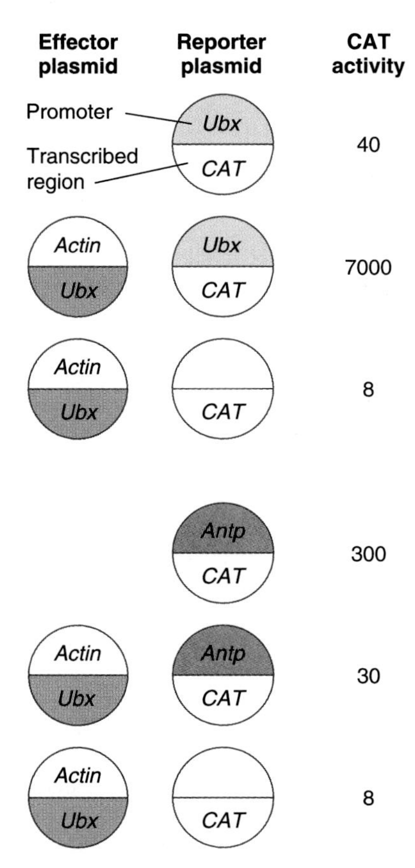

Figure 16.22 Enhancement of *Ultrabithorax*$^+$ (*Ubx*$^+$) gene expression by Ultrabithorax protein, and negative regulation of *Antennapedia*$^+$ (*Antp*$^+$) gene expression by the same protein. Cultured *Drosophila* cells are infected simultaneously with an effector plasmid and a reporter plasmid. The effector plasmid contains the promoter from the *actin*$^+$ gene and the transcribed region of the *Ubx*$^+$ gene. The reporter plasmid contains the promoter from either the *Ubx*$^+$ gene or the *Antp*$^+$ gene, or no promoter, and the transcribed region of the *CAT*$^+$ gene. The CAT activity is determined in relative units (see Fig. 16.5). While the activity of the *Ubx*$^+$ promoter is enhanced by the Ubx gene product, the *Antp*$^+$ promoter is inhibited by the Ubx gene product.

cell.) Transfection of host cells with the reporter plasmid alone produced only weak CAT activity. However, cotransfection with the effector plasmid caused an almost 200-fold increase in reporter gene activity. In a similar experiment, a reporter gene driven by *Antp* was inhibited by the expression of the *Ubx* effector gene. Thus, the presence of the *Ubx* gene product strongly enhanced the activity of the *Ubx* promoter while inhibiting the *Antp* promoter. In accord with these observations, the researchers found that a protein encoded by the *Ubx*$^+$ gene binds tightly to specific DNA sequences, presumably enhancer elements, near the *Ubx*$^+$ promoter and near one of the two *Antp*$^+$ promoters.

Questions

1. Using the in situ hybridization technique as illustrated in Figure 16.21, where would you expect to find *Antennapedia* mRNA accumulation in embryos from an *Antp* gain-of-function mutant in which the adults show antennae replaced with legs?
2. The results summarized in Figure 16.22 indicate that *Ubx*$^+$ gene expression in cultured cells feeds back positively on itself. Based on this information, where would you expect a transgene consisting of the *Ubx*$^+$ regulatory region and the *lacZ* transcribed region to be expressed in wild-type and in *Ubx*-deficient embryos?

The results described above show that *Ubx*$^+$ and *Antp*$^+$ do show some of the properties expected of switch genes in bistable control circuits. In particular, the Ultrabithorax protein promotes the activity of its own gene, *Ubx*$^+$, while inhibiting the *Antp*$^+$ gene. However, *Ubx*$^+$, *Antp*$^+$, and other *selector genes* in *Drosophila*, have large regulatory regions with many enhancer elements binding the proteins encoded by several other selector genes. Thus, instead of pairs of genes forming simple bistable control circuits, we find many selector genes and their products engaged in complex and overlapping regulatory networks. However, these networks can switch between two or more stable states, as indicated by the transdetermination experiments described in Section 6.5. The *principle of default programs*—that is, the existence of alternative developmental pathways that are taken in the absence of a signal—is often mediated through bistable genetic control circuits.

Many transdetermination steps, such as leg-to-antenna, or haltere-to-wing, mimic homeotic mutations. While transdeterminations reflect changes in selector gene expression, homeotic mutations occur in the selector genes themselves. Thus, it is apparent that cell determination involves control circuits formed by selector genes and the transcription factors they encode. The "stubbornness" of determined cells, the tenacity with which they hold to their fates in transplantation and isolation experiments, can be ascribed to the stability of these control circuits. This stability is obvious in the case of the simple bistable control circuit shown in Figure 16.19. A positive feedback loop stabilizing one gene's activity, combined with inhibition of the other gene, needs a major upset to switch from one stable state to the other. In particular, a dividing cell will pass on its stable state during cell division, simply because the transcription factor encoded by the active gene is passed on to both daughter cells. More complex control circuits seem to be stable under most circumstances, as indicated by the basic phenomenon of cell determination, but some are labile enough to flip into another stable state occasionally, as shown by transdeterminations.

DNA METHYLATION MAINTAINS PATTERNS OF GENE EXPRESSION

Another molecular basis for the inheritance of gene activity patterns, known as ***DNA methylation***, is observed in mammals and flowering plants (Bird, 1992). About

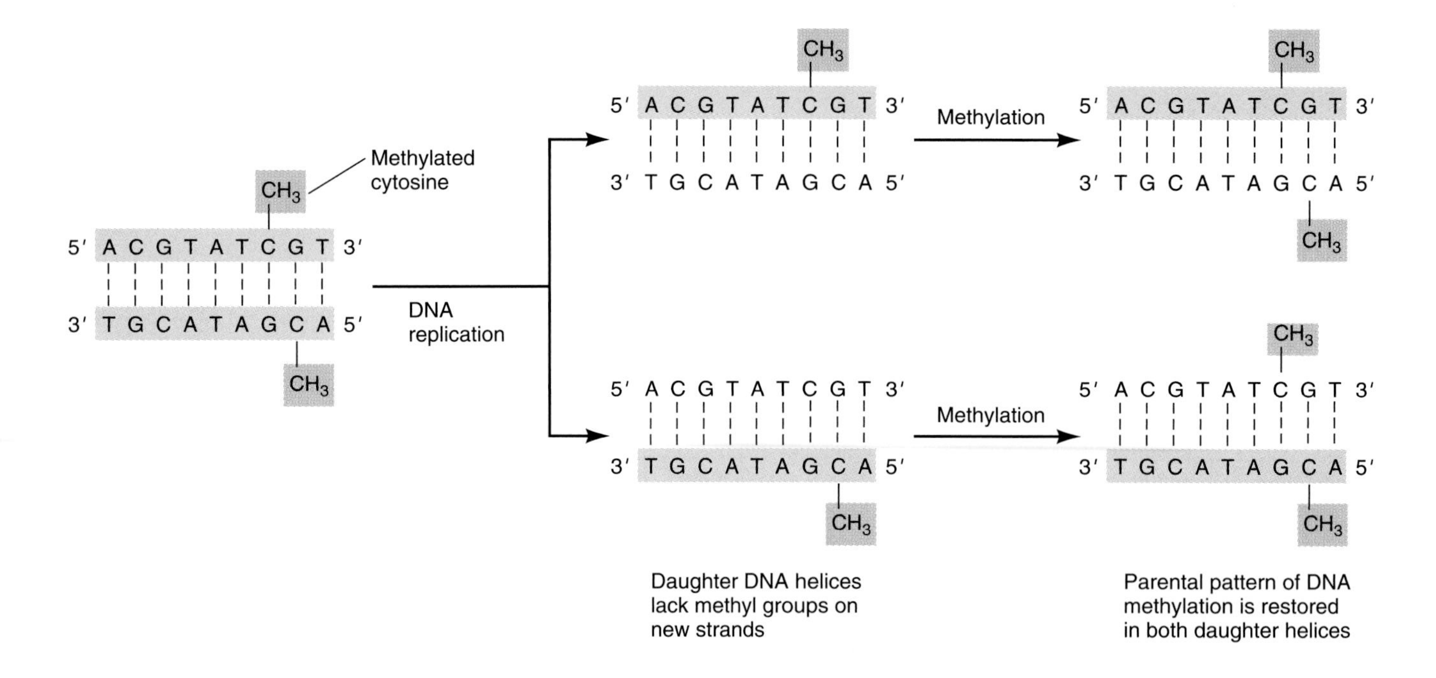

Figure 16.23 Maintenance of DNA methylation patterns. Once a pattern of cytosine methylation is established, it is propagated after each DNA replication by an enzyme, DNA (cytosine-5) methyltransferase, or Dnmt1. The enzyme recognizes unmethylated CpG sequences in a new DNA strand that are base-paired to a methylated CpG sequence in the old strand.

4% of the cytosine nucleotides in mammalian DNA are modified with a methyl group to form *5-methyl cytosine* (Yoder et al., 1997). This step is catalyzed by several enzymes, including ***DNA (cytosine-5) methyltransferase 1 (Dnmt1).*** Dnmt1 acts on previously methylated cytosine nucleotides that occur in the dinucleotide sequence CpG. Recognizing the half-methylated CpG sequences that exist after DNA replication, Dnmt1 methylates the newly synthesized CpG sequences, in effect propagating, or providing a cellular memory, of the preexisting methylation pattern (Fig. 16.23). Cytosine methylation also occurs *de novo,* as transgenes tend to become CpG methylated soon after they are taken up into mammalian cells, but the mechanism is unknown so far. Dnmt1 is critical for normal development in mice. *Knock-out mice* homozygous for a null allele of *Dnmt1* show about one third of the normal level of DNA methylation and die during gestation (E. Li et al., 1992).

In mammals and other organisms with cytosine methylation, CpG dinucleotides have been selectively depleted from genomic DNA through a conversion of methylated cytosine to thymidine via a slow natural deamination process (Antequera and Bird, 1993). The human genome has only 10% of the expected frequency of CpG dinucleotides, and about 75% of these are methylated. However, small regions of genomic DNA are not depleted of CpGs. These so-called ***CpG islands*** are unmethylated and are located in the promoter regions of almost half of all human genes (Antequera and Bird, 1999; *P. A. Jones, 1999*). Most of these are thought to be the *housekeeping genes* that are expressed in nearly all cell types.

The effect of CpG methylation on transcription is inhibitory. Several studies have shown that most genes—especially their promoter regions—are methylated when they are inactive and unmethylated when they are active. At least one mechanism by which CpG methylation inhibits transcription is through *histone deacetylation.* A protein known as MeCP2 binds to methylated CpG and is associated with proteins that function as histone deacetylases (Nan et al., 1998). Thus, it appears that methylated CpGs direct histone deacetylases to associated histones, with the result of rendering nearby transcription start sites inaccessible (*Ng and Bird, 1999*).

DNA methylation plays a key role in parent-specific gene inactivation, a phenomenon termed ***genomic imprinting.*** About 20 imprinted genes have been identified in mammals (Surani, 1998; Latham, 1999). Some undergo long-term inactivation during spermatogenesis but remain active in eggs. The converse occurs for other imprinted genes, so that a zygote contains some pairs of homologous genes of which only the paternal allele is expressed, and other genes of which only the maternal allele is expressed (Fig. 16.24). The imprinted states remain stable after fertilization in embryonic cells and throughout the subsequent development of somatic cells. Each of these cells has one active allele and one inactive allele of each imprinted gene. However, the imprinting is erased in the primordial germ cells of each new generation. (Around the same time, heterochro-

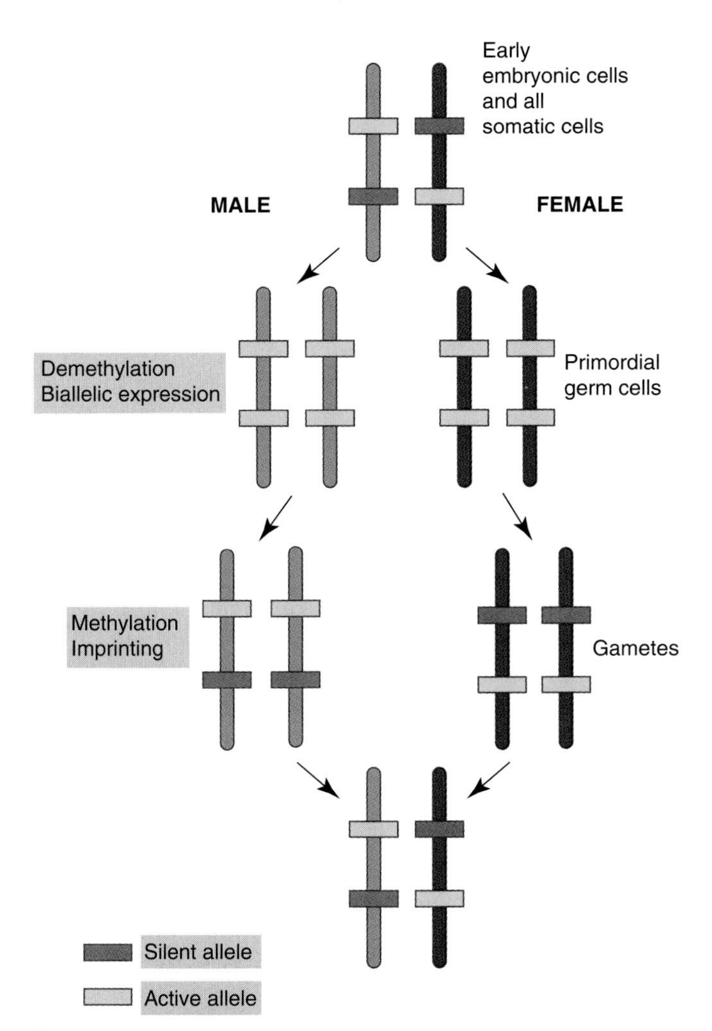

Figure 16.24 Genomic imprinting in mammalian embryos. Parental imprinting (silencing) persists through early embryonic development but is erased in primordial germ cells. The resulting gene activation is usually associated with DNA demethylation. New sex-specific imprints are generated during gametogenesis.

matic X chromosomes are reactivated in females; see Section 26.3.) Germ line cells then express both alleles of imprinted genes until during gametogenesis parents again imprint genes for the next generation.

Genomic imprinting was originally inferred from pronuclear transplantations and genetic studies. Mammalian eggs in which the female pronucleus has been replaced with another male pronucleus do not develop normally, and neither do eggs with two female pronuclei and no male pronucleus (see Section 4.7). Even humans with *uniparental disomy* (*UPD*), who—for just one chromosome—have two copies from one parent and no homolog from the other parent, while all other chromosomes are represented normally with one copy from each parent, show serious clinical symptoms (*W. P. Robinson, 2000*). Patients with maternal UPD 15 (two chromosomes 15 from mother, none from father) show *Prader-Willi syndrome* (*PWS*), which is characterized by severe obesity, gonadal dysfunction, short stature, small hands and feet, and mental retardation. Paternal UPD 15 causes *Angelman syndrome* (*AS*), which includes seizures, sleep disorder, severe mental retardation, and lack of speech. To avoid PWS, then, at least one gene on chromosome 15 must be paternally derived and cannot be replaced by an extra maternally derived copy, while the converse must be the case to avoid AS. As expected, certain paternal (but not maternal) deletions from chromosome 15 cause PWS, while maternal (but not paternal) deletions of *another part* of chromosome 15 cause AS.

Imprinting generates information that is passed on during cell division for much of the life cycle and affects the phenotype of offspring. However, since this information is erased and reset in the germ line of each generation, it cannot be encoded in the nucleotide sequence of DNA. This kind of information has therefore been called the ***epigenotype***. What is its molecular basis? What controls the activity of imprinted genes? Which parts of imprinted genes are involved in resetting the epigenotype in the germ line according to the sex of its carrier? These questions are currently being investigated using mainly two approaches. Human geneticists examine the various types of mutations that cause PWS or AS (Nicholls et al., 1998). Focusing on patients who are unable to reset the epigenotype from female-specific to male-specific or vice versa, they found that many of those patients carry microdeletions in certain regions of chromosome 15. Areas of overlap between these deletions should contain genetic elements that are critical to epigenotype resetting. Other investigators study mice because they allow the use of transgenes.

There is a general correlation in mice and other mammals that imprinted (inactive) genes are methylated, whereas their active homologs are not. For instance, mice express a tumor-promoting transgene (c-myc) in the heart if it is paternally derived but not at all if it is maternally derived (Swain et al., 1987). This pattern of expression correlates with the parentally controlled methylation state. The transgene is methylated during the gene's passage through the female germ line, but not the male germ line, and the female-specific methylation is erased in the primordial germ cells of the following generation. Some imprinted genes behave similarly, while in others it is the paternally derived allele that is transcriptionally inactive as well as methylated. However, the mouse gene for insulinlike growth factor-2 receptor (Igf2r) does not fit the rule that the methylated allele is inactive. In this case, the maternally derived allele is both methylated and expressed (Stöger et al., 1993).

Regardless of whether the active allele of an imprinted gene is methylated or not, a normal level of DNA methylation is necessary for differential expression of the maternal and paternal alleles of imprinted genes. In mouse embryos lacking the gene encoding DNA (cytosine-5) methyltransferase (Dnmt1), the expression patterns of imprinted genes including *Igf2r*$^+$ are disturbed (Li et al., 1993). However, it takes more than Dnmt1 to impart the normal methylation and activity pattern on imprinted genes. When embryonic stem cells lacking Dnmt1 were transformed with a transgene containing cDNA encoding Dnmt1, the normal levels of DNA methylation and activity were restored for nonimprinted genes but not for imprinted genes (Tucker et al., 1996). In order to restore normal transcription and methylation to imprinted genes, these genes had to pass through the germ line first. This result is consistent with the hypothesis that methyl transferase activities other than Dnmt1 are present during gametogenesis and are required for appropriate gene imprinting. This conclusion is also supported by the fact that the methylation pattern of imprinted genes is undisturbed by general waves of demethylation and remethylation that occur between cleavage and gastrulation (Razin and Shemer, 1995).

While the current picture of DNA methylation and gene imprinting is more complex than originally thought, the sophisticated methods for analysis that are now available should provide a better understanding in a not too distant future.

SUMMARY

The cells in an organism have thousands of proteins in common but are distinguished by a few hundred cell-type-specific proteins. Some of the latter, such as globin or ovalbumin, are synthesized in large quantities and reflect the terminal differentiation of the cells producing them. Other cell-type-specific proteins have gene-regulatory functions and are part of the cell determination process itself.

Transcription, the first step in gene expression, is the DNA-dependent synthesis of RNA by an enzyme, RNA polymerase. The control of transcription involves regulatory DNA sequences. One regulatory region, the promoter, is located immediately upstream of the transcription start site. Other regulatory regions, called enhancers, occur in various positions.

Transcription factors are gene-regulatory proteins that bind to specific nucleotide sequences in both promoters and enhancers. General transcription factors occur in all cells of an organism. They bind to the promoter and to each other, forming a transcription complex that positions the RNA polymerase at the transcription start site. Other transcription factors bind to enhancer elements of specific genes to increase or decrease their rates of transcription. These transcriptional activators and inhibitors occur preferentially in certain cells. In insect embryos at the syncytial blastoderm stage, some transcriptional activators are translated from localized maternal mRNAs and control the region-specific expression of embryonic genes. Other transcriptional regulators are in turn controlled by the binding of a hormone or a posttranslational modification.

Chromosome regions containing actively transcribed genes are less condensed than nontranscribed chromosome regions. The decondensation may occur in two steps, one encompassing the entire gene and the second affecting in particular the promoter and enhancer regions.

Transcriptional activators and inhibitors are the molecular basis of differential gene transcription. Some of these factors are encoded by switch genes, which, by their activity or nonactivity, can shift cells between two possible fates. These transcription factors link switch genes in control circuits of autocatalytic activation and mutual repression. Such control circuits have alternate stable states that are passed on during mitosis. In mammals, the methylation of inactive genes is an additional mechanism for stabilizing the inactive state of genes and propagating it during mitosis. This mechanism may account for the phenomenon of gene imprinting in mammalian embryos.

chapter 17

RNA Processing

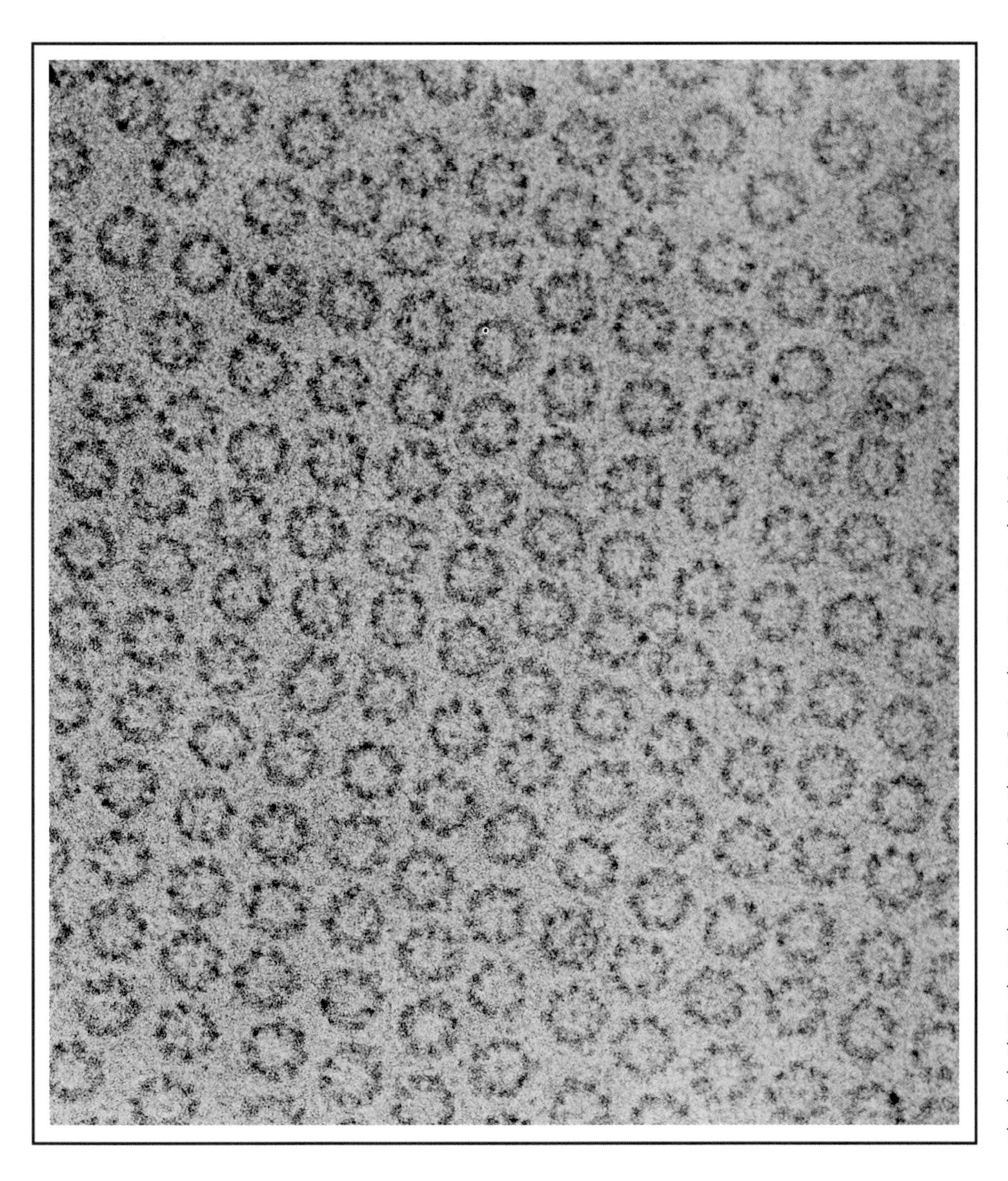

Figure 17.1 Electron micrograph of the nuclear envelope of a newt oocyte. The envelope contains numerous nuclear pore complexes, openings that allow the active transport of macromolecules between nucleus and cytoplasm. Each nuclear pore complex has an eightfold radial symmetry and a central plug, or transporter element. These and other features show that nuclear pore complexes are elaborate gates for regulated transport. The passage of messenger RNP particles through nuclear pore complexes is one of many controlled processing steps that mRNAs undergo between transcription and translation.

AS soon as a gene is transcribed into RNA, the transcript undergoes a series of modifications inside the nucleus. While transcription is still going on, the nascent transcript associates with proteins to form ***ribonucleoprotein particles*** (***RNP particles,*** or ***RNPs***). All subsequent modifications are made to RNPs, not to RNAs, although we will usually call them RNAs when we focus on the RNA moiety. While a few aspects of processing ribosomal RNA have been mentioned earlier (see Fig. 7.8), the following discussion will focus ***messenger RNPs*** (***mRNPs***)—that is, RNPs containing mRNA. The primary transcript that is the precursor molecule of an mRNA is known as ***pre-mRNA***. Most pre-mRNAs are cut and spliced and become extended at both ends before they pass from the nucleus into the cytoplasm. Their passage usually takes the form of active transport through gates in the nuclear envelope known as ***nuclear pore complexes*** (Fig. 17.1). These processing and gating steps are characteristic of eukaryotes, and all of them have become post-transcriptional levels of controlling gene expression.

One of the steps in pre-mRNA processing is reminiscent of the way movies are made. The total film footage shot on scene is typically much longer than what is eventually released to movie theaters. While some parts of the original footage are spliced together to create the final product, the remaining scenes are edited out and never shown. Likewise, while pre-mRNA is a faithful copy of the entire gene, only certain pre-mRNA segments make it into the mRNA product that eventually leaves the nucleus (Fig. 17.2). These parts of the pre-mRNA, and the encoding gene, are called expressed sequences, or ***exons.*** They are separated by intervening sequences, or ***introns***, which are cut out of pre-mRNA while successive exons are ligated—or spliced—together into one mRNA molecule.

The exon-intron structure, also known as *split-gene organization,* is found in most genes of all higher eukaryotes and came as a great surprise when it was discovered in 1976. Since then, it has also become clear that most of the enzymatic functions involved in cutting and splicing pre-mRNA into mRNA are carried out by RNA molecules rather than by proteins. Thus, RNA processing has opened a window into an ancient world in which RNAs were the dominant biological molecules, being not only the carriers of genetic information but also the preferred catalysts of biochemical reactions.

Many pre-mRNAs are spliced in two or more ways, yielding different mRNAs and often different proteins. The generation of two or more different mRNAs from the same type of pre-mRNA is called ***alternative splicing.*** Next to differential transcription, alternative splicing is the most important molecular mechanism by which cells assume different fates during development. Alternative splicing may be *constitutive,* meaning that all cells that produce a pre-mRNA process it into the same group of multiple mRNAs, which in turn are translated into a family of related proteins or *isoforms*. Of greater interest in the context of differential gene expression is *regulated alternative splicing,* where different types of cells use different splicing patterns to generate different mRNAs and eventually different proteins.

We will examine two examples of regulated alternative splicing to illustrate its importance in development. One example is sex differentiation in *Drosophila:* Whether a fly grows up to be male or female depends on alternative splicing of several gene products. The second example is a rat gene transcript that is spliced in two ways: One way is carried out in gland cells and yields the mRNA for a hormone, while the other way of splicing the same pre-mRNA occurs in neurons and yields the mRNA for a neuropeptide instead.

After splicing and further modifications, mRNPs are transported through the nuclear

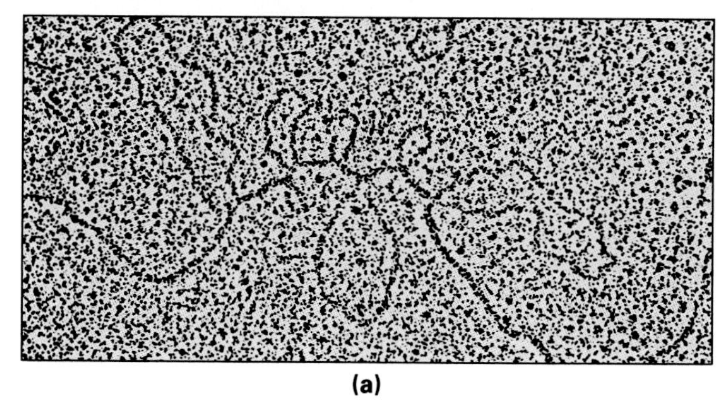

(a)

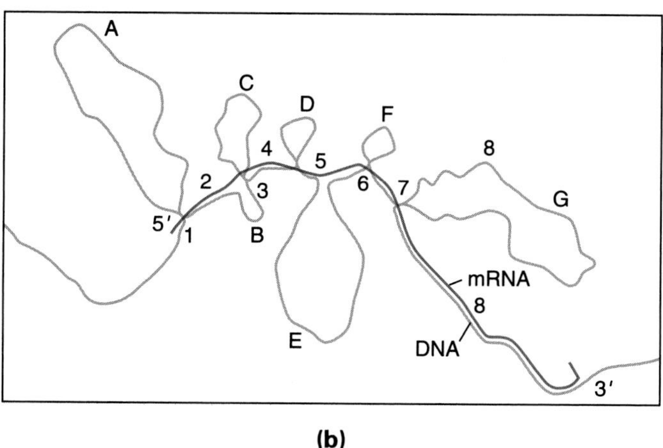

(b)

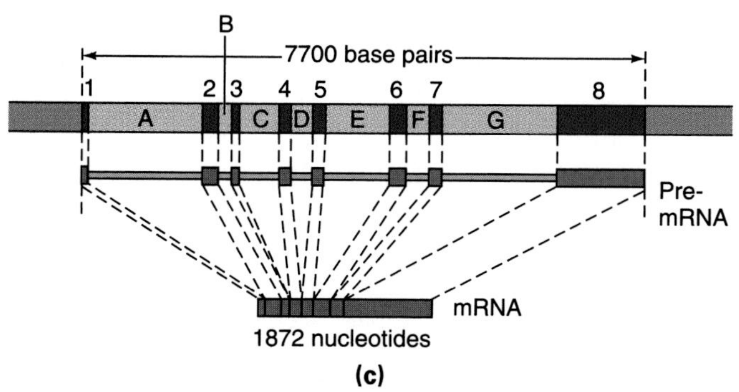

(c)

Figure 17.2 Exon-intron structure of eukaryotic genes. **(a)** Electron micrograph showing the transcribed strand of the chicken ovalbumin gene after hybridization to ovalbumin mRNA (magnification 130,000×). **(b)** Interpretative diagram indicating that only small sections of the genomic DNA (blue) hybridize with the mRNA (red), the molecule from which the ovalbumin protein is translated. These expressed DNA sections are known as exons. Between the exons, large intervening DNA segments, called introns, are looped out because they have no mRNA sequences to hybridize with. **(c)** Schematic representation of the ovalbumin gene showing eight exons (labeled 1–8) and seven introns (labeled A–G). The primary gene transcript (pre-mRNA) is 7700 bases long and contains both exons and introns. The mRNA processed from this pre-mRNA consists only of exons and is 1872 bases long.

pore complexes into the cytoplasm. There the association of mRNA with proteins and possibly other factors controls its lifetime and translatability. For instance, in the mammary glands of nursing mammalian females, the mRNA for casein, an abundant milk protein, is associated with certain proteins that protect it against degradation. As a result, a large number of casein mRNA molecules are available for milk production. In nonnursing females, however, casein mRNA is not associated with protective proteins and is rapidly destroyed.

In this chapter, we will first examine the processing of pre-mRNA into mRNA, with an emphasis on alternative splicing. Next we will consider the export of mature mRNPs from the nucleus, and finally the controlled degradation of mRNA in the cytoplasm. For each of these steps, we will discuss how it is used for differential gene expression in development.

17.1 Posttranscriptional Modifications of Pre-Messenger RNA

The major steps in the ***posttranscriptional modification*** of pre-mRNA are its association with proteins, the "capping" of the 5′ end, the "tailing" of the 3′ end, and the splicing of exons.

Association with Proteins Newly made pre-mRNA in eukaryotes immediately associates with proteins to form a series of closely spaced particles of about 20 nm diameter called heterogeneous nuclear ribonucleoproteins or *ribonucleosomes.* Each ribonucleosome consists of about 700 nucleotides (nt) of RNA folded through and around particles consisting of several proteins (Beyer and Osheim, 1990). This arrangement is similar to the packaging of DNA in nucleosomes. The function of ribonucleosomes is not well understood. Conceivably, they keep the nascent RNA single-stranded and accessible to processing proteins by preventing the formation of double-stranded regions.

Capping of the 5′ End While the pre-mRNA molecule is still being synthesized, its 5′ end is extended by the addition of a methylated guanosine triphosphate (7-methylguanylate), which is attached enzymatically to the pre-mRNA in an unusual 5′-to-5′ phosphotriester linkage (Fig. 17.3). The first two original (transcribed) nucleotides are methylated in their ribose moieties, and collectively these modifications of the 5′ end are called the ***cap*** of the mRNA. The cap protects the growing RNA transcript against degradation by nucleases that attack unprotected 5′ ends (see Section 17.4). The cap

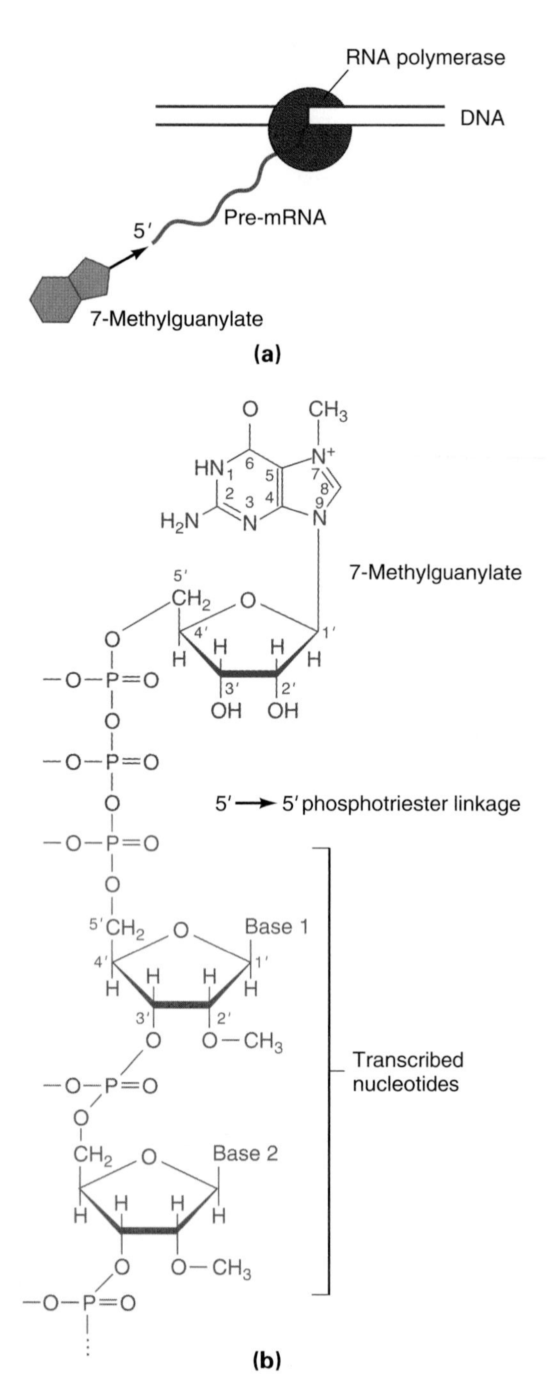

Figure 17.3 Cap structure at the 5′ end of eukaryotic pre-mRNA and mRNA. **(a)** A "capping" enzyme adds 7-methylguanylate to the 5′ end of the pre-mRNA while the rest of the pre-mRNA is still being transcribed. **(b)** The 7-methylguanylate is linked via a 5′-to-5′ phosphotriester linkage to the first transcribed nucleotide of the mRNA. The ribose moieties of the first one or two regular nucleotides are also variably methylated.

also promotes the translation of many mRNAs once they have been released into the cytoplasm.

Polyadenylation of the 3′ End The 3′ end of pre-mRNAs is not formed simply by termination of transcription. Rather, the RNA polymerase transcribes beyond the future end of the pre-mRNA molecule, which is then

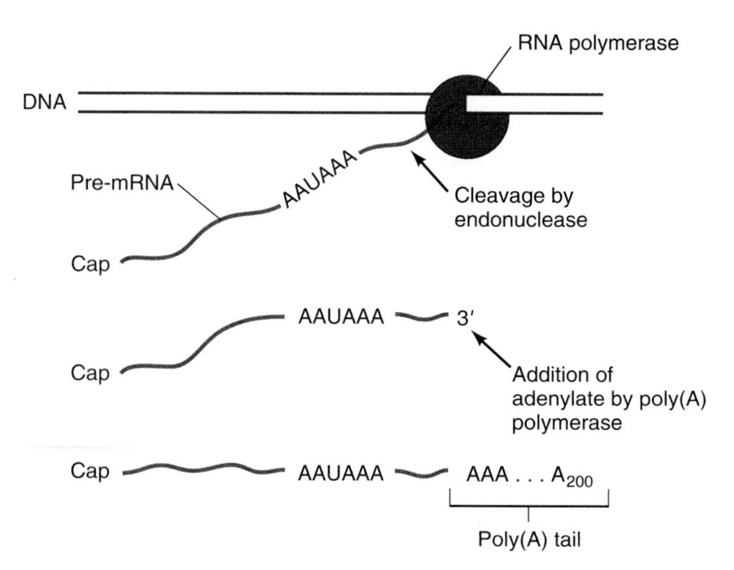

Figure 17.4 Generation of the poly(A) tail of eukaryotic mRNA. RNA polymerase transcribes past the site corresponding to the 3′ end of the pre-mRNA. An endonuclease cuts the pre-mRNA, guided by the polyadenylation signal AAUAAA. A poly(A) polymerase adds about 200 single adenylate residues, which make up the poly(A) tail.

generated through cleage by an enzyme complex. This complex recognizes a specific nucleotide sequence, AAUAAA, which act as a ***polyadenylation signal*** (Fig. 17.4). The pre-mRNA is cut downstream of the polyadenylation signal, which remains part of the mRNA 3′ UTR region (Colgan and Manley, 1997; W. Keller, 1995). The 3′ end of the mRNA is then extended by another enzyme, poly(A) polymerase, which adds adenylate residues. This activity generates a ***poly(A) tail,*** which in most vertebrate mRNAs is about 200 nucleotides long. The polyadenylation of mRNAs is a reversible, dynamic process that continues after the mRNA has been transported into the cytoplasm and plays an important role in the regulation of translation (see Section 18.2). For most mRNAs, the poly(A) tail is also necessary for nucleocytoplasmic transport and for protection against enzymatic degradation, as will be discussed in Sections 17.3 and 17.4.

Aside from its biological functions, the poly(A) tail of mRNAs is a gift of nature to molecular biologists. It allows them to isolate the relatively rare mRNAs from the much more abundant ribosomal and transfer RNAs. This is accomplished by hybridization of the poly(A) tails with poly(U) or poly(T) bound to chromatographic carrier materials.

Splicing of Exons Pre-mRNA associates with nuclear ribonucleoprotein particles known as ***small nuclear RNP particles* (*snRNPs*).** Each snRNP consists of one or two small RNA molecules and up to 10 different proteins. (The protein composition of snRNPs has been intensively investigated because humans with an autoimmune disorder, *lupus erythematosus,* produce antibodies against their own snRNPs.) Five different snRNPs carry out the majority of all exon splicing (Maniatis, 1991; Gall, 1991). A few more snRNPs are involved in splicing exons with unusual junction sites and with processing the 3′ ends of mRNAs that are not polyadenylated.

The basic steps of intron removal are summarized in Figure 17.5. Each intron has a ***splice donor site*** at its 5′

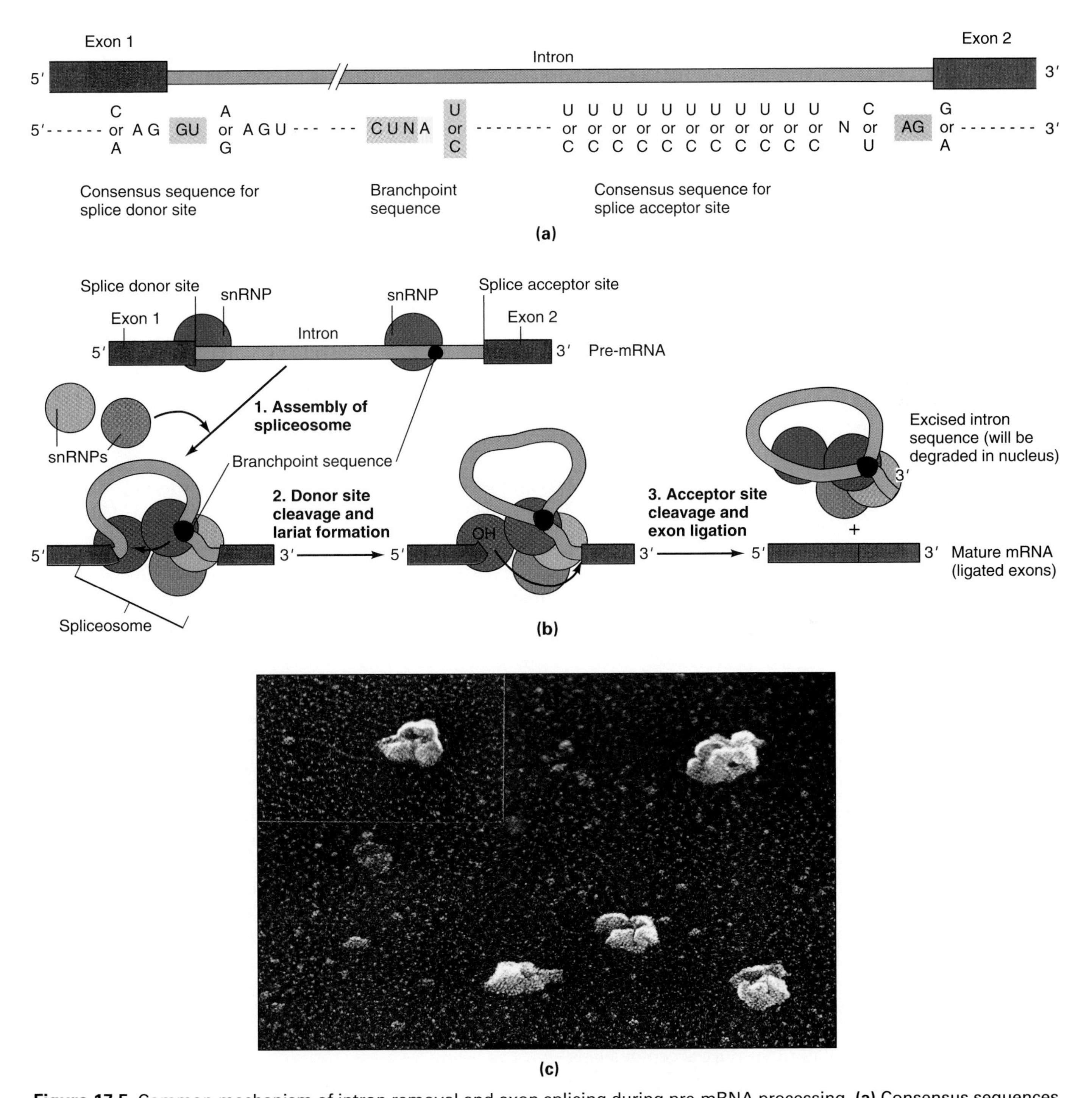

Figure 17.5 Common mechanism of intron removal and exon splicing during pre-mRNA processing. **(a)** Consensus sequences at the splice donor site (3′ end of exon 1, 5′ end of intron) and splice acceptor site (3′ end of intron, 5′ end of exon 2). The shaded GU and AG dinucleotides at these sites are nearly invariant. N stands for any of the four RNA nucleotides. The branchpoint sequence, located about 30 nucleotides upstream of the splice acceptor site, contains an A nucleotide that will attack the splice donor site and form the lariat as shown in part b. **(b)** Three major steps of intron removal and exon splicing. Step 1: spliceosome formation from several snRNPs (green colors) and proteins (not shown). Step 2: reaction of branchpoint with donor site, causing donor site cleavage and lariat formation. Step 3: acceptor site cleavage and exon ligation. **(c)** Electron micrograph of spliceosomes isolated by gel column chromatography, fixed with glutaraldehyde, mounted onto thin carbon film, and shadow-cast with tungsten.

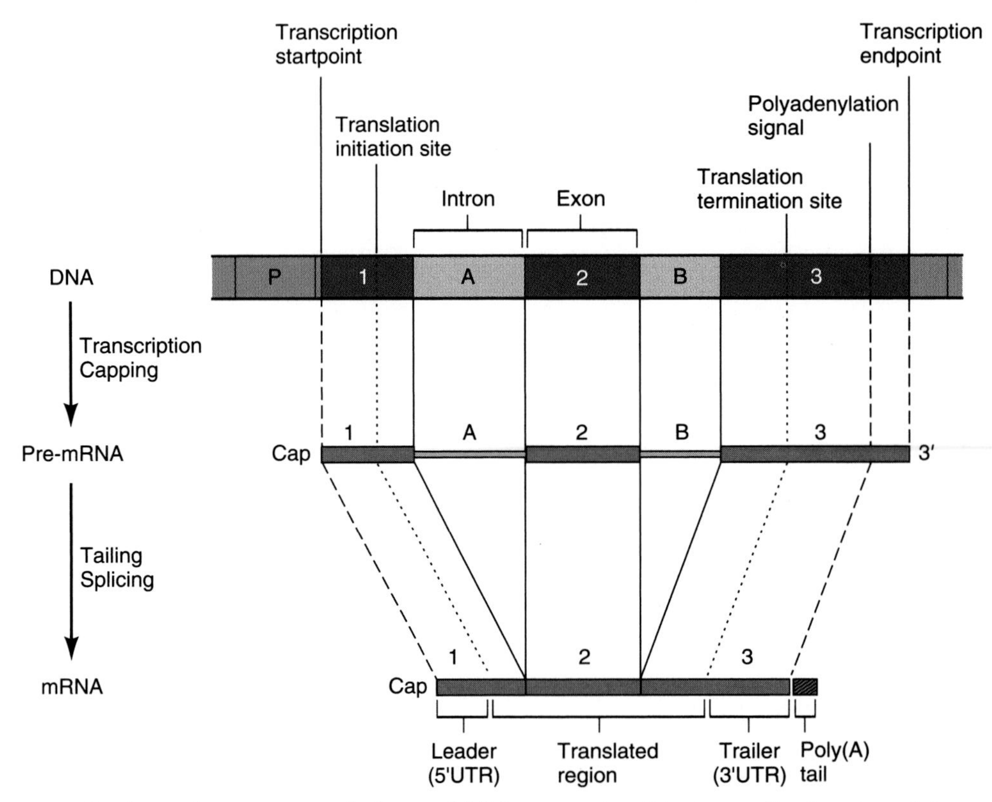

Figure 17.6 Summary of eukaryotic gene transcription and RNA processing. The transcribed region of the gene begins at the transcription start site and extends to a poorly defined transcription endpoint. The transcript (pre-mRNA) is capped at the 5′ end and cleaved behind the polyadenylation signal. The resulting 3′ end is extended into a poly(A) tail. The introns are removed and the exons (1, 2, 3) spliced together. The translated region of the mRNA extends from the translation initiation site to the translation termination site. The segment between the cap and the translation initiation site is called the 5′ untranslated region (5′ UTR) or leader sequence. Likewise, there is a 3′ untranslated region (3′ UTR), or trailer sequence, between the translation termination site and the beginning of the poly(A) tail.

end and a ***splice acceptor site*** at its 3′ end. These sites have consensus sequences that are very similar among the introns of vertebrates and many other groups of organisms. These consensus sequences are located, for the most part, at the ends of the intron but extend slightly into the adjoining exons. In the most common splicing mechanism, one type of snRNP binds to the splice donor site, and another binds to a ***branchpoint sequence*** close to the acceptor site. These two snRNPs, together with other snRNPs and dozens of different proteins, assemble into a composite organelle known as a ***spliceosome.*** It promotes the next step of the splicing reaction, which is an attack of the A nucleotide in the branchpoint sequence on the splice donor site, thus cleaving the splice donor site and forming a lariat by joining the free intron end to the branchpoint. Finally, the splice acceptor site is cleaved, the adjacent exons are joined together. The excised intron, in most cases, has no apparent function and is degraded. Exon splicing must be done with great precision because any deletion or addition of nucleotides would disrupt the reading frame of nucleotide triplets in the mRNA, which is translated into a particular amino acid sequence later.

Only RNA polymerase II transcripts have caps and poly(A) tails, and only RNA polymerase II has a functional domain that associates with enzymes involved in capping, splicing, and polyadenylation (Steinmetz, 1997). This association explains why pre-mRNA processing steps commonly occur while transcription is still in progress.

The processing of a pre-mRNA into an mRNA is summarized in Figure 17.6. The processed mRNA consists of a cap, a ***5′ untranslated region* (*5′ UTR*),** a translated region, or ***coding region,*** a ***3′ untranslated region* (*3′ UTR*),** and a poly(A) tail. The 5′ UTR and 3′UTR sequences of mRNAs are also known as *leader,* or *trailer,* regions, respectively. They are involved in several functions, including mRNA localization, regulation of lifetime, and translational control (see Sections 8.4, 17.4, and 18.2, respectively).

17.2 Control of Development by Alternative Splicing

Generally, the splicing machinery combines the splice donor and acceptor sites of the same intron, so that adjacent exons are joined. However, departures from this rule may occur if an intrinsically weak splice site competes with stronger splice sites in their vicinity, or if

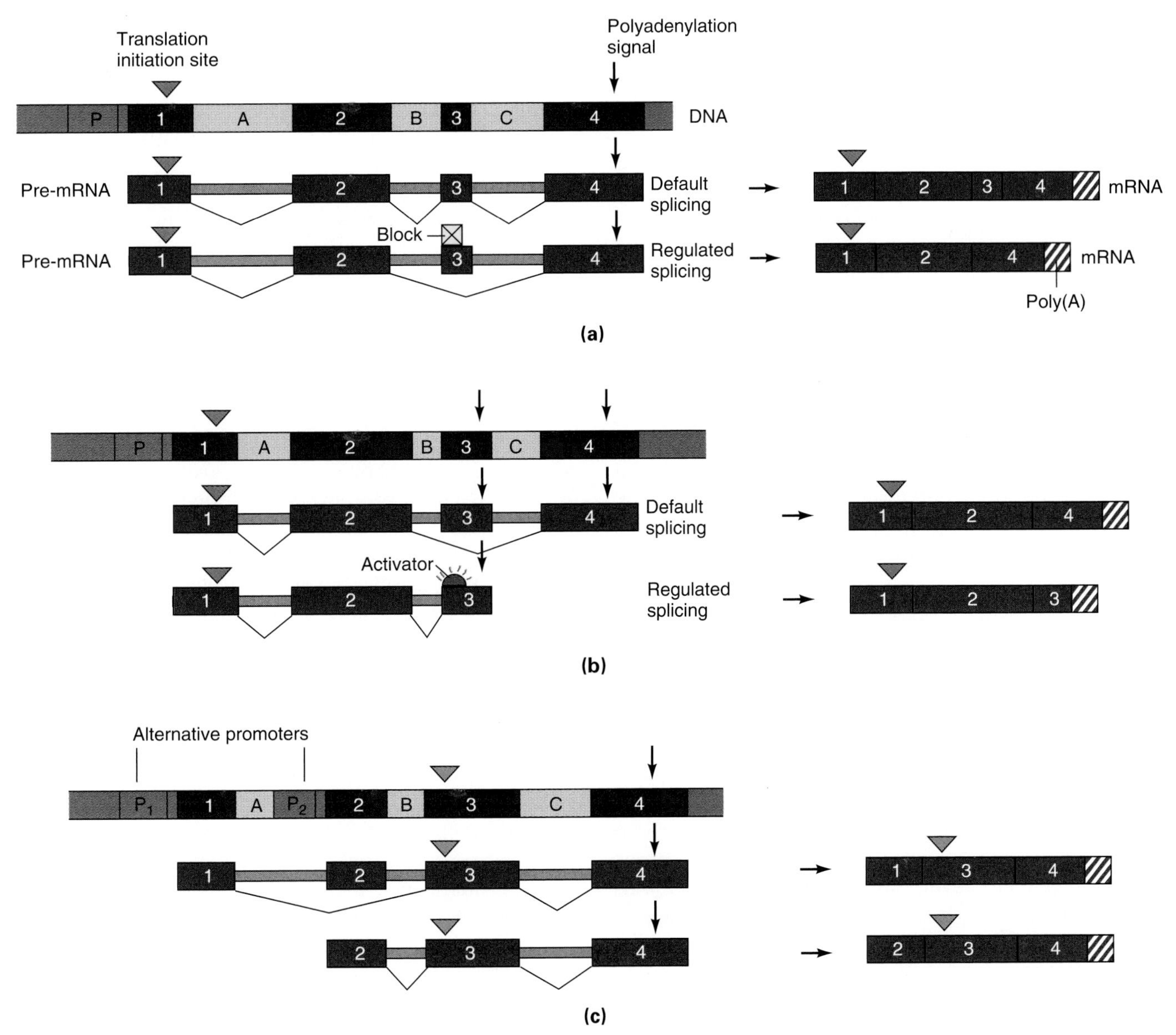

Figure 17.7 Common modes of alternative splicing. **(a)** Binding of a regulatory protein blocks a splice acceptor site (the 5′ end of exon 3, in this diagram), causing the alternative use of the next acceptor site downstream. The two mRNAs resulting from the regulated and the default splicing differ only in the presence versus absence of exon 3. They are transcribed from the same promoter and have the same transcription start site, translation initiation site (triangle), and polyadenylation signal (arrow). **(b)** An inherently poor splice acceptor site (again shown here at the 5′ end of exon 3) is skipped in the default program. However, binding of a regulatory protein can activate this acceptor site, causing its preferred use over the default acceptor site at the beginning of exon 4. In the case shown, exons 3 and 4 both have polyadenylation signals, so that the 3′ ends of the two mRNAs are cut and polyadenylated differently. **(c)** Alternatively processed pre-mRNAs may arise from the use of different promoters (P_1 and P_2). In the case shown, the alternative mRNAs have the same translated region but different 5′ UTRs.

regulatory proteins either block certain splice sites or make intrinsically weak splice sites stronger. The presence of regulatory proteins provides a mechanistic basis for *alternative splicing*. In such cases, there is a ***regulated splicing pattern,*** which depends on one or more regulatory proteins, and a ***default splicing pattern,*** which occurs in the absence of these regulatory proteins. Frequently, both patterns lead to functional proteins. Even if one of the proteins is nonfunctional, its lack of function often triggers an alternative developmental pathway, again illustrating the *principle of default programs* in development.

Three common modes of alternative splicing are illustrated in Figure 17.7. First, a splice acceptor site may be subject to *blockage* by a regulatory protein. The donor site of one exon will then be combined with the acceptor site of an exon further downstream. Second, an intrinsically poor splice acceptor site may be skipped in the default splicing pattern, whereas in the regulated pattern it is subject to *enhancement* by an activating protein. Third, pre-mRNAs initiated from different promoters may generate alternative splice *donor* sites that are combined with the same acceptor site.

Alternative splicing patterns usually generate mRNAs with different translated regions, which give rise to different proteins. However, some alternatively spliced mRNAs have the same translated regions and differ only in their 3′ UTR, or trailer, sequences. The latter may contain different regulatory sequences for mRNA localization and control of lifetime. Alternative promoters also allow organisms to synthesize the same protein under two sets of transcriptional control signals (Fig. 17.7c).

Alternative splicing patterns complicate the definitions of *exon* and *intron,* because what is an exon in one pattern may be part of an intron in another. Generally, though, the context clarifies any ambiguities. The following two examples will show how alternative splicing mediates sexual differentiation in fruit flies and tissue-specific gene expression in mammals.

A CASCADE OF ALTERNATIVE SPLICING STEPS CONTROLS SEX DEVELOPMENT IN *DROSOPHILA*

The majority of animal species have two sexes, ***males*** as defined by having testes, and ***females*** as defined by having ovaries. Most of these species have a genetic mechanism that determines their sex (see Section 26.2). In fruit flies as well as in mammals, normal females have two X chromosomes and two sets of *autosomes* (abbreviated "A") per diploid cell. Normal males have one X chromosome, one Y chromosome, and two sets of autosomes per diploid cell. However, this similarity between mammals and flies is deceptive. In mammals, the presence or absence of a specific gene on the Y chromosome decides whether an individual develops as a male or a female (see Section 26.3). In *Drosophila,* the Y chromosome is of secondary importance: It is not involved in sex determination and contains only a few genes necessary for spermatogenesis.

The primary sex-determining signal in *Drosophila* is the ***X:A ratio***—that is, the number of X chromosomes relative to sets of autosomes. Individuals with one X chromosome and two sets of autosomes per diploid cell have an X:A ratio of 0.5 and become males. Individuals with two X chromosomes and two sets of autosomes per diploid cell have an X:A ratio of 1.0 and become females. How *Drosophila* cells sense the X:A ratio is at least partially understood (see Section 26.5). The numerator of the X:A ratio is measured by the amounts of transcriptional activators encoded by X-linked genes, including *sisterless-a*$^+$ and *sisterless-b*$^+$ (Keyes et al., 1992; J. W. Erickson and T. W. Cline, 1993). These activators act on *Sex-lethal*$^+$, the master gene for sexual differentiation in *Drosophila.* The more X chromosomes there are, the more sisterless proteins are being made, and the more active *Sex-lethal*$^+$ becomes. The denominator of the X:A ratio is measured by the nuclear volume, which increases with the number of autosomes and dilutes the concentration of sisterless proteins. In addition, there seem to be specific proteins encoded by autosomal genes that antagonize the sisterless proteins.

The *Sex-lethal*$^+$ gene transduces the graded signal of the X:A ratio into an all-or-none activity. This transduction involves the two promoters of *Sex-lethal*$^+$, known as the ***establishment promoter*** **(P_e)** and the ***maintainance promoter*** **(P_m)**. The activity of P_e increases with the X:A ratio but is limited to a short period (cell cycles 12–14) ending with the cellular blastoderm stage (Barbash and Cline, 1995). The activity of P_m, which becomes active after P_e, is independent of the X:A ratio. However, the pre-mRNA transcribed from P_m can be spliced in two ways: in a *female-specific splicing pattern* and in a *male-specific splicing pattern.* Which one is chosen depends on the previous activity level of P_e and on a positive feedback loop of the Sex-lethal protein on the splicing of its own pre-mRNA (Fig. 17.8). Due to the feedback loop, which will be detailed later, a cell will produce either the full amount of functional Sex-lethal protein or none.

The presence or absence of functional Sex-lethal protein controls three key elements of sexual differentiation in *Drosophila* (Fig. 17.8). One aspect is the *sexual differentiation of the germ line*—that is, the formation of eggs versus sperm. The second aspect, less obvious but critical, is called *dosage compensation.* It ensures that males and females, even though they have different numbers of X chromosomes, still transcribe the same amounts of RNA per cell from X-linked genes (see Section 26.5). The third element of sexual differentiation controlled by Sex-lethal protein, and the focus of the following discussion, is known as *somatic sexual differentiation.* It encompasses all differences between the sexes other than ovaries versus testes. In *Drosophila,* such somatic sexual dimorphisms include copulatory organs, the pigmentation of the abdomen, and the "sex comb" at the male foreleg (Fig. 17.9).

The main regulatory genes controlling somatic sexual differentiation are the *Sex-lethal*$^+$ gene, the *transformer*$^+$ gene, and the *doublesex*$^+$ gene (Baker, 1989). They form a regulatory hierarchy, in which the protein encoded by each gene controls the splicing of the pre-mRNA transcribed from its subordinate gene (Fig. 17.10; MacDougall et al., 1995). A general characteristic of this hierarchy is that the regulated splicing patterns generate female-specific mRNAs, whereas the default splicing patterns yield male-specific or non-sex-specific mRNAs. The following paragraphs outline the functions of these genes and the alternative splicing patterns of their pre-mRNAs.

Alternative Splicing of Sex-lethal Pre-mRNA The control of *Sex-lethal*$^+$ expression by the X:A ratio is a two-step process involving the gene's two promoters, the *establishment promoter* (P_e) and the *maintenance promoter* (P_m). The first step is the activation of P_e by the sisterless

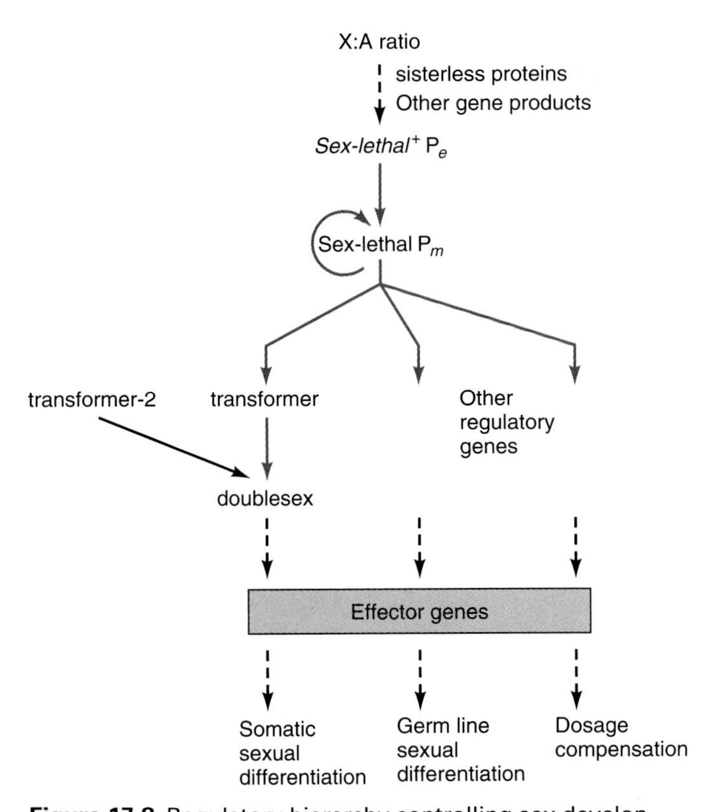

Figure 17.8 Regulatory hierarchy controlling sex development in *Drosophila.* The sex-determining signal is the X:A ratio. Its numerator is measured by the amounts of sisterless proteins and other X-linked gene products. The sisterless proteins activate the establishment promoter (P_e) of the *Sex-lethal*$^+$ gene. The Sex-lethal protein derived from P_e activity controls the female-specific splicing pattern of pre-mRNA transcribed later from the maintenance promoter (P_m) of the same gene. At an X:A ratio of 1.0, the start-up amount of Sex-lethal protein from P_e is sufficient to establish the positive feedback loop of Sex-lethal protein on the splicing of its own pre-mRNA. At an X:A ratio of 0.5, the initial amount of Sex-lethal protein from P_e is not sufficient to establish the feedback loop. By default, Sex-lethal pre-mRNA is then spliced in a male-specific pattern, which results in a truncated, inactive Sex-lethal protein. The Sex-lethal protein also controls the female-specific splicing pattern of transformer pre-mRNA. The transformer protein, together with transformer-2 protein, controls the splicing of doublesex pre-mRNA. The resulting two doublesex mRNAs, and their protein products, are sex-specific and control, via effector genes, the somatic sexual differentiation of the fly. In addition, the Sex-lethal protein controls other regulatory and effector genes that direct the sexual differentiation of the germ line and the transcription of the X-linked genes as a means of dosage compensation. The colored arrows represent control steps based on RNA splicing.

proteins, which act as transcriptional activators as mentioned earlier. The nuclear concentration of sisterless protein is higher, and hence P_e becomes more active, if the X:A ratio is 1.0 (as in XXAA individuals) rather than 0.5 (as in XYAA individuals). The second step in the control of *Sex-lethal*$^+$ expression begins at the cellular blastoderm stage, when P_e becomes inactive. At the same time, P_m becomes active, but the level of its activity is independent of the X:A ratio. In contrast to the pre-mRNA transcribed from P_e, the pre-mRNA transcribed from P_m can be spliced in two different ways, yielding either a *male-specific mRNA* or a *female-specific mRNA* (Bell et al., 1988). The male-specific mRNA includes exon 3, which is absent from the female-specific mRNA (Fig. 17.10). Exon 3 contains a ***stop codon***—that is, a nucleotide triplet terminating translation, so male-specific mRNA yields a truncated Sex-lethal protein that is biologically inactive. The female-specific mRNA, because it excludes exon 3, avoids this stop codon and gives rise to biologically active Sex-lethal protein.

Which splicing pattern is initiated depends on the availability of functional Sex-lethal protein, which directs the female-specific splicing pattern of Sex-lethal pre-mRNA. At an X:A ratio of 1.0, the strong activation of the P_e promoter by sisterless protein "jumpstarts" the synthesis of functional Sex-lethal protein, which then perpetuates the female-specific splicing of pre-mRNA and hence the synthesis of functional Sex-lethal protein. At an X:A ratio of 0.5, the start-up amount of Sex-lethal protein resulting from P_e is not sufficient to initiate the positive feedback loop of Sex-lethal protein on the splicing of its own pre-mRNA. Thus, the male-specific splicing pattern is initiated by default, and this pattern is continued because the resulting Sex-lethal protein is not functional.

The mechanism by which exon 3 is excluded from the Sex-lethal pre-mRNA transcribed from P_e is incompletely understood. The P_e-derived pre-mRNA has a unique first exon that seems cause the skipping of exon 3 without requiring any trans-acting factors (Horabin and Schedl, 1996; Zhu et al., 1997).

RNA-Binding Domain in Sex-lethal Protein Sequence analysis of the Sex-lethal protein has revealed an ***RNA-binding domain*** consisting of about 90 amino acids. This domain has been conserved during evolution of different RNA-binding proteins in life forms ranging from yeasts to humans, indicating that it fulfills critical functions in all organisms (Bandziulis et al., 1989). Generally, proteins with RNA-binding domains interact with specific RNA sequences; their interactions are similar to those of transcription factors with their DNA recognition motifs in promoters and enhancers. In the case of the Sex-lethal protein, the analogy with transcription factors extends even further. Just as several transcription factors act cooperatively in regulating promoter activity, the Sex-lethal protein interacts with other RNA-binding proteins in blocking the male-specific splice acceptor site of exon 3 in Sex-lethal pre-mRNA (Horabin and Schedl, 1993; J. Wang and L. R. Bell, 1994).

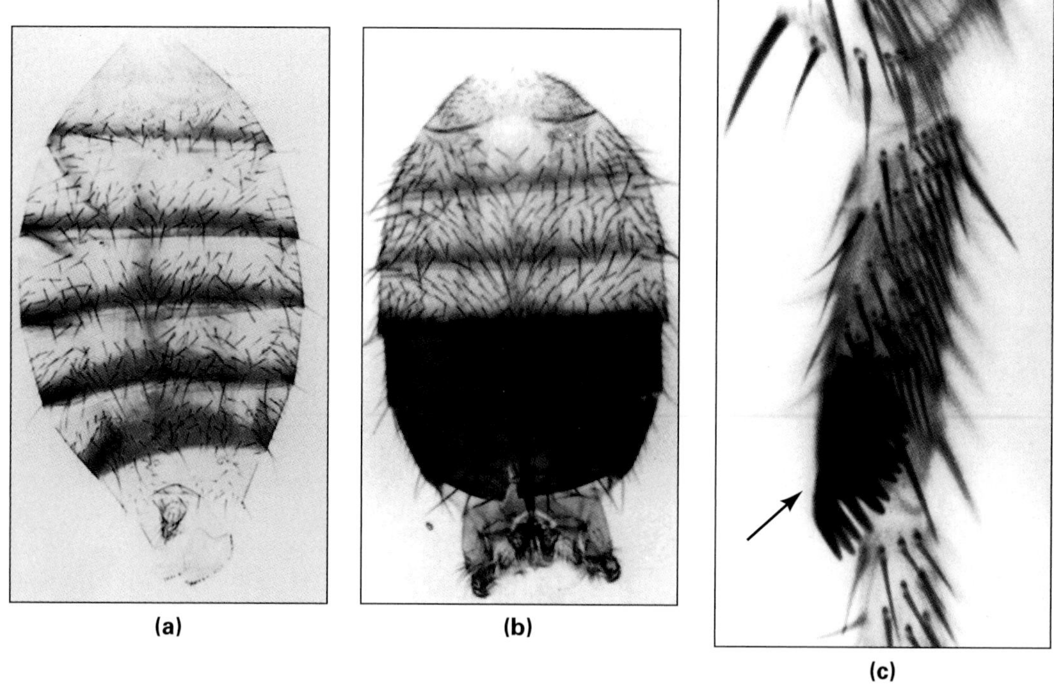

Figure 17.9 Somatic sexual differentiation in Drosophila. **(a)** Female abdomen. **(b)** Male abdomen. **(c)** Sex comb (arrow) on the male foreleg.

The *transformer*$^+$ Gene In addition to promoting the female-specific splicing pattern of its own pre-mRNA, the Sex-lethal protein has a similar effect on the alternative splicing of the transformer pre-mRNA. In both cases, Sex-lethal protein blocks the splice acceptor site of an exon that contains a premature stop codon.

In the alternative splicing of transformer pre-mRNA, the default pattern is called non-sex-specific because it occurs in both sexes (Fig. 17.10). The resulting mRNA includes exon 2, which contains a stop codon truncating the translated protein (Boggs et al., 1987). The regulated splicing pattern is female-specific. It exludes exon 2, leading to the synthesis of a functional transformer protein. The female-specific splicing pattern depends on functional Sex-lethal protein (Nagoshi et al., 1988). This is indicated by the transformer mRNAs produced in males and females carrying different alleles of *Sex-lethal*$^+$ (Fig. 17.11). The female-specific splicing pattern is used when the non-sex-specific splice acceptor site is blocked by the Sex-lethal protein (Valcárcel et al., 1993). The Sex-lethal protein is not very effective at this function; about half of the transformer pre-mRNA in females is wasted because it follows the non-sex-specific splicing pattern. However, enough functional transformer protein is apparently generated to ensure female development. The active transformer protein cooperates with the transformer-2 protein, which itself is not sex-specific, in regulating the splicing pattern of the next pre-mRNA in the genetic hierarchy. In genetic tests, the *transformer*$^+$ and *transformer-2*$^+$ genes behave similarly; loss-of-function alleles cause male somatic sexual development, irrespective of the X:A ratio.

The *doublesex*$^+$ Gene A target gene controlled cooperatively by the transformer and transformer-2 proteins is *doublesex*$^+$, a gene that encodes two functional mRNAs. The female-specific mRNA comprises exons 1 through 4 (see Fig. 17.10). The male-specific mRNA comprises exons 1 through 3, 5, and 6. As in the cases of Sex-lethal and transformer pre-mRNA, the male-specific splicing pattern of doublesex pre-mRNA is the default program depending only on the basic cellular splicing machinery. The female-specific splicing pattern requires the presence of both transformer and transformer-2 proteins.

In *doublesex* mutants with defects in exons 1 through 3, both XX and XY individuals develop as intersexes, showing mixed male and female genitalia as well as intermediate secondary sex characteristics. Defects in exon 4 cause bisexual development in females but leave males unaffected, while the converse is true for defects in exons 5 or 6. These observations indicate that the female-specific mRNA is translated into a regulatory protein that inhibits male development, while the converse holds for the male-specific mRNA.

The two doublesex mRNAs are generated by alternative splicing in conjunction with the use of different polyadenylation sites. The switch from the male-specific to the female-specific splicing pattern might be explained by either of two models (Fig. 17.12). According to the *blockage model*, the transformer and transformer-2

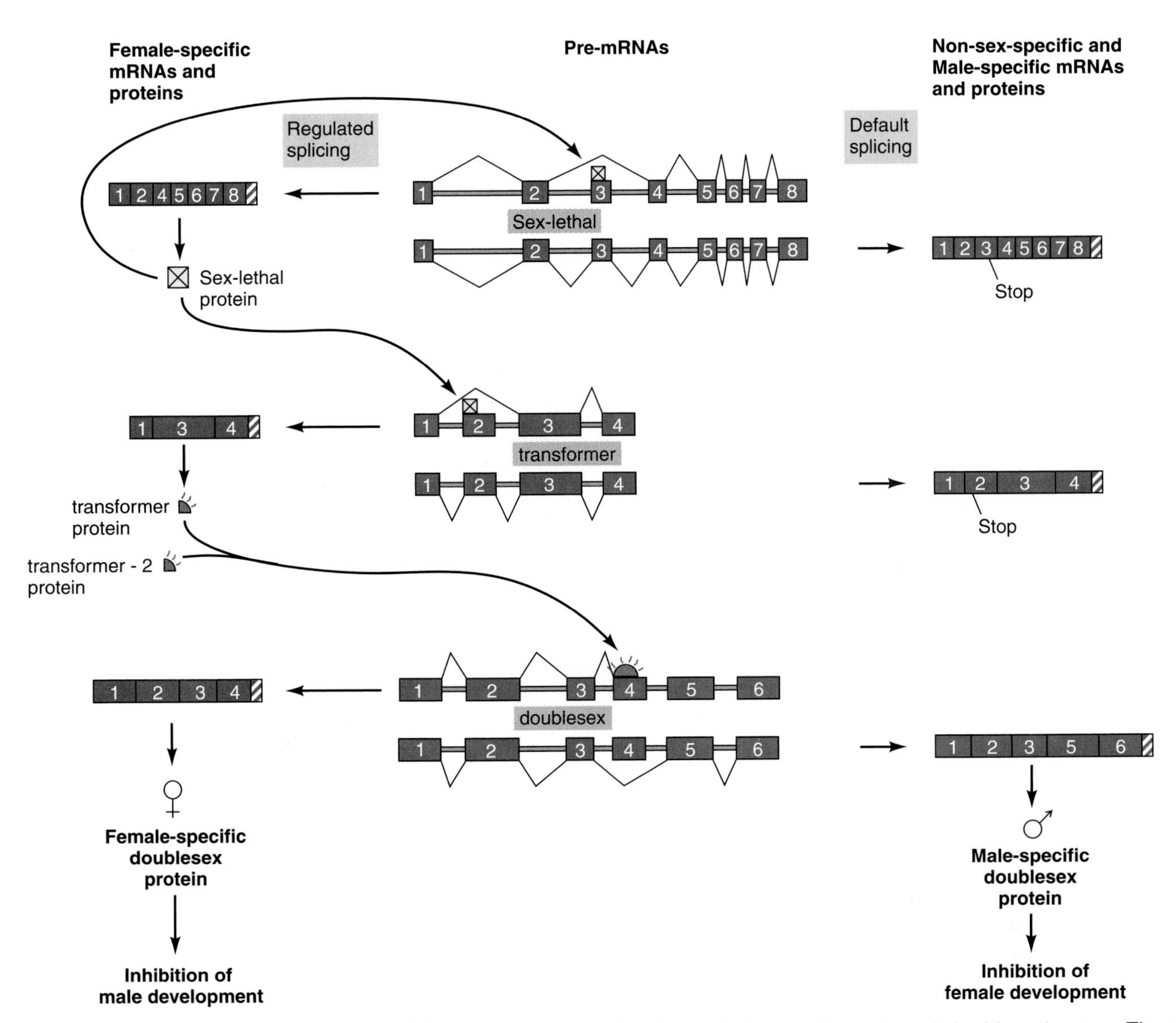

Figure 17.10 Alternative splicing patterns of the transcripts from the *Sex-lethal*$^+$, *transformer*$^+$, and *doublesex*$^+$ genes. The respective primary transcripts (pre-mRNAs) of these genes are identical in males and females (central portion of figure). However, they are spliced differently. The default patterns give rise to the mRNAs shown on the right-hand side. These mRNAs are male-specific in the cases of Sex-lethal and doublesex. The mRNA resulting from default splicing of transformer pre-mRNA is called non-sex-specific as it occurs in both males and females. The Sex-lethal and transformer default mRNAs give rise to truncated and inactive proteins because they contain early translation stop codons. Among the default mRNAs, only the male-specific doublesex mRNA encodes an active protein. The regulated splicing patterns produce the female-specific mRNAs shown on the left-hand side. All these mRNAs are translated into active proteins. The Sex-lethal protein promotes the female-specific splicing patterns of Sex-lethal pre-mRNA transcribed from the maintenance promoter as well as transformer pre-mRNA. The transformer protein, along with the transformer-2 protein, promotes the female-specific splicing pattern of doublesex pre-mRNA.

proteins block the splice acceptor site of exon 5 (the first of the male-specific exons). This model is in parallel to the action of Sex-lethal protein on the transformer pre-mRNA described previously. According to the *activation model,* the transformer and transformer-2 proteins make the acceptor site of exon 4 (the female-specific exon) more attractive to the splicing machinery. The following observations indicate that the activation model is correct.

THE activation model implies that the splice acceptor site of exon 4 is intrinsically poor, so that the cellular splicing machinery ignores this site unless the transformer and transformer-2 proteins are present. In order to test this implication, Burtis and Baker (1989) compared the splice acceptor site of exon 4 to that of exon 5 and other exons in *Drosophila.* Most sites showed the consensus sequence of $(C/U)_n$NCAG shown in Figure 17.5, with uninterrupted

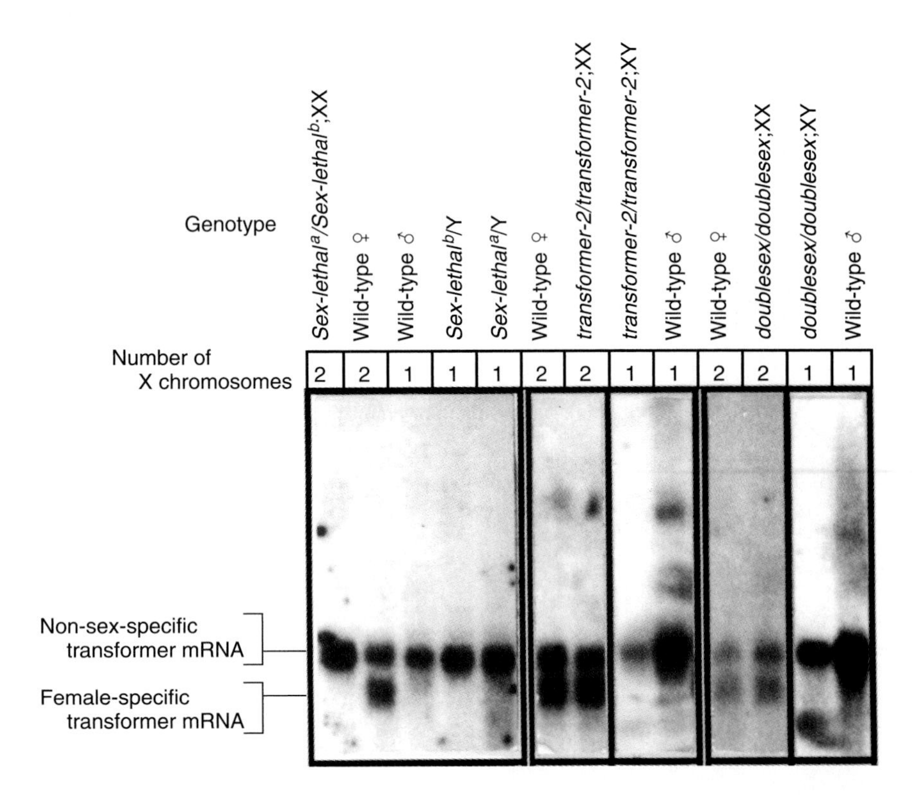

Figure 17.11 Splicing of transformer mRNAs in fruit flies with different mutations. The transformer female-specific and non-sex-specific mRNA were identified by Northern blotting (see Method 15.4). RNA samples from different fly stocks were separated by gel electrophoresis and probed with a labeled cDNA hybridizing with both female-specific and non-sex-specific mRNA. All flies made the non-sex-specific mRNA. The female-specific mRNA was produced only in flies with two X chromosomes (resulting in an X:A ratio of 1.0) and two wild-type alleles of *Sex-lethal*$^{+}$. (*Sex-lethal*^{a} and *Sex-lethal*^{b} are two different loss-of-function alleles.) Mutant alleles of *transformer-2* or *doublesex* did not affect the splicing pattern of transformer pre-mRNA. The data show that the splicing pattern of transformer pre-mRNA depends on the expression of *Sex-lethal*$^{+}$ but not of *transformer-2*$^{+}$ or *doublesex*$^{+}$.

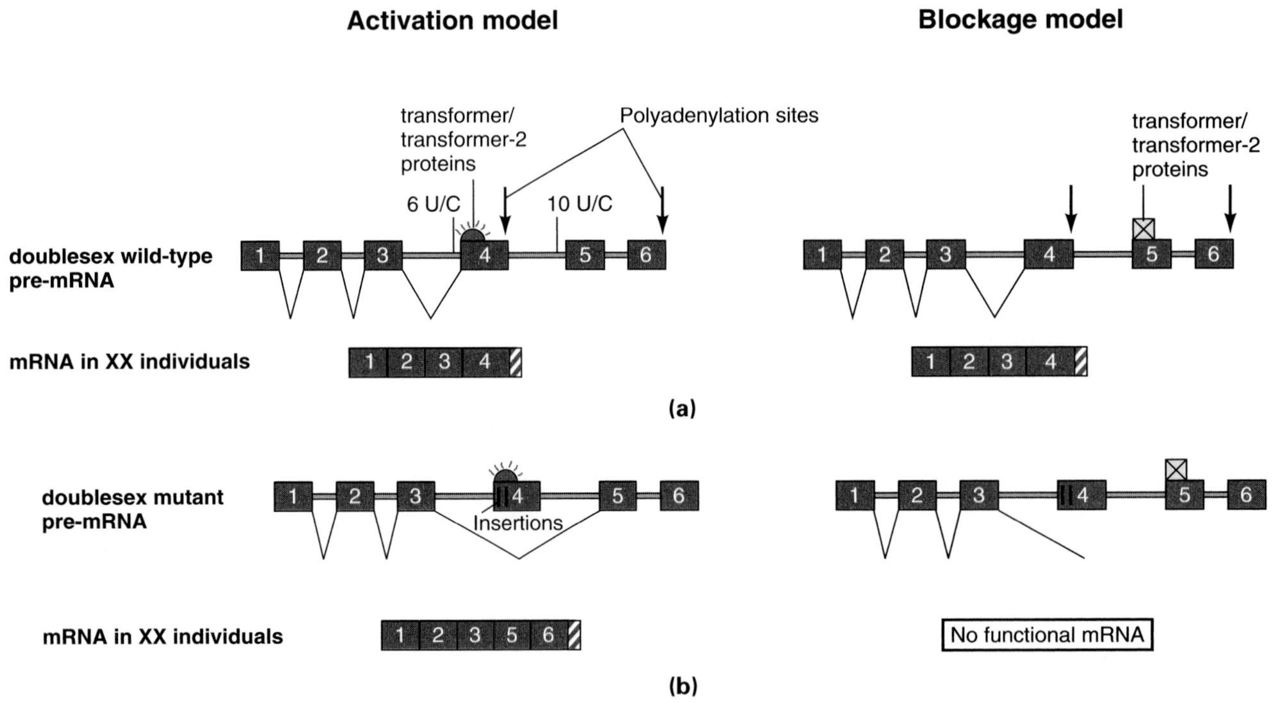

Figure 17.12 Activation model and blockage model for the splicing of doublesex mRNAs. The wild-type female-specific mRNA comprises exons 1 to 4. The male-specific mRNA comprises exons 1 through 3, 5, and 6 (see Fig. 17.10). The male-specific splicing pattern is the default pattern because the acceptor site for exon 5 has a stretch of ten U or C nucleotides as compared with only six in the acceptor site for exon 4. **(a)** For wild-type doublesex pre-mRNA, the activation model predicts that in XX individuals the transformer and transformer-2 proteins will associate with the acceptor site for exon 4 to make it more attractive to the splicing machinery. According to the blockage model, the transformer and transformer-2 proteins will block the acceptor site for exon 5, forcing the splicing machinery to use the acceptor site for exon 4 instead. **(b)** For *doublesex* mutants with DNA insertions or deletions in the splice acceptor site of exon 4, the activation model predicts that XX individuals will produce male-specific mRNA. In contrast, the blockage model predicts that XX individuals will produce no functional mRNA, because they cannot use either splice acceptor site. Experimental data support the activation model.

stretches of 9 to 11 pyrimidines (C or U). In contrast, the doublesex exon 4 acceptor site had only 6 pyrimidines, making it appear weaker than the other acceptor sites.

As a direct test of the blockage model, Nagoshi and Baker (1990) analyzed mutants in which they expected the acceptor site of exon 4 to be inactivated. Under this condition, the blockage model predicts that XX individuals should produce no functional doublesex mRNA, since the exon 4 splice acceptor site is unavailable due to mutational damage and the exon 5 splice acceptor site should be unavailable due to blockage by transformer and transformer-2 proteins (Fig. 17.12). In contrast, the activation model predicts that mutant XX individuals should produce male-specific mRNA because the splice acceptor site of exon 5 is the only site available. The actual results were as predicted by the activation model: XX flies from each of four mutant strains with deletions or insertions in the exon 4 splice acceptor site produced male-specific doublesex mRNA.

Questions

1. Figure 17.12 shows the splicing patterns for doublesex pre-mRNA in wild-type and mutant XX individuals as predicted on the basis of the activation model versus the blockage model. What are the corresponding splicing patterns predicted for XY individuals?
2. The splicing of doublesex pre-mRNA can be studied in cultured *Drosophila* cells transfected with plasmids containing an abbreviated *doublesex*$^+$ gene and cDNAs encoding transformer-2 mRNA and the female-specific transformer mRNA (Hoshijima et al., 1991). In this experimental system, each of the three transgenes can be modified and the effects of the modifications studied with relative ease. Based on the analysis of Burtis and Baker reviewed above, what splicing pattern would you expect if the short pyrimidine tract in the splice acceptor site of exon 4 were extended to ten or more pyrimidines?
3. The regulated splicing pattern of transformer pre-mRNA is "leaky," producing a lot of non-sex-specific mRNA along with female-specific mRNA (see Fig. 17.11). In contrast, the alternative splicing of doublesex pre-mRNA is very precise: There is little if any female-specific doublesex mRNA in males or vice versa. Why would you expect a strong selection for precision here?

In summary, development of sexual dimorphism in *Drosophila* depends on alternative splicing of a hierarchy of gene products. At each level of the regulatory cascade, the default splicing pattern leads to male development, whereas the female-specific pattern requires an active protein made at the same level or the next higher level. The Sex-lethal protein regulates the splicing of both its own pre-mRNA and of transformer pre-mRNA by blockage. In contrast, the transformer and transformer-2 proteins regulate the splicing of doublesex pre-mRNA by activation. Thus, both types of splicing regulation operate in the same cascade.

ALTERNATIVE SPLICING OF CALCITONIN AND NEUROPEPTIDE mRNA IS REGULATED BY BLOCKAGE OF THE CALCITONIN-SPECIFIC SPLICE ACCEPTOR SITE

Alternative splicing of the same pre-mRNA can also generate different mRNAs that have entirely unrelated functions. As an example, we will examine a rat gene that encodes two proteins: *calcitonin* and a *neuropeptide.* The hormone ***calcitonin*** is produced in the thyroid gland and lowers the level of calcium in the blood. The particular ***neuropeptide*** encoded by the gene is synthesized in certain neurons of the brain and in the pituitary gland. The two proteins are translated from *different* mRNAs, which are polyadenylated and spliced alternatively from the *same* type of pre-mRNA (Fig. 17.13). The calcitonin/neuropeptide gene and its primary transcript contain six exons and two polyadenylation sites. The first three exons (1, 2, and 3) are used in thyroid cells and neurons alike. Thyroid cells splice exon 3 to exon 4 and use polyadenylation signal poly(A)$_1$, which is located in exon 4. Neurons splice exon 3 to exon 5 and use polyadenylation signal poly(A)$_2$ present in exon 6. Thus, differential splicing and polyadenylation of the same

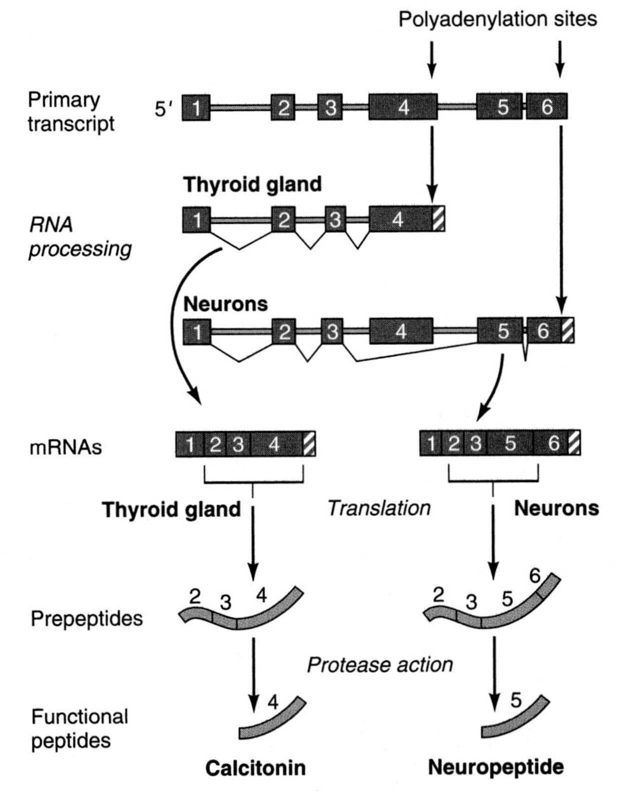

Figure 17.13 Alternative splicing and polyadenylation of the primary transcript from the calcitonin/neuropeptide gene. In thyroid gland cells, mRNA for calcitonin, a hormone, is formed as exon 3 is spliced to exon 4 and polyadenylation occurs at the site in exon 4. In certain neurons, mRNA for a neuropeptide is created by splicing exon 3 to exon 5 and polyadenylation at the site in exon 6. The proteins translated from both types of mRNA are modified by proteases that remove the domains encoded by exons 1, 2, 3, and 6.

pre-mRNA results in two different mRNAs: calcitonin mRNA in the thyroid gland and neuropeptide mRNA in certain neurons.

The two mRNAs are translated into the corresponding proteins, each in the appropriate cell type. Both these proteins are initially synthesized as inactive precursors. Functional proteins are generated by proteases that clip off the parts corresponding to exons 1 through 3. Thus, in their final active forms, the two proteins encoded by the same gene have no parts in common.

To analyze the alternative splicing of calcitonin/neuropeptide pre-mRNA, investigators have used two lines of cultured cells that reproduced the respective splicing patterns observed in the two rat tissues. Only with cultured cells was it practical to introduce suitable reporter genes with engineered splice acceptor sites and other alterations. A line of cancerous human epithelial cells, when transfected with the rat calcitonin/neuropeptide transcribed region under control of a suitable promoter, mimicked the splicing pattern of the rat thyroid gland. For simplicity, we designate these the C (calcitonin-producing) cells. Another cell line, derived from a mouse teratocarcinoma and containing the same transgene, imitated the splicing pattern of rat neurons. We designate the cells of this line the N (neuropeptide-producing) cells. To monitor the splicing pattern in each cell line, the investigators employed the *ribonuclease protection assay* described in Method 17.1. It showed that each cell line produced not only the correct final mRNA but also the expected intermediate splice products (Fig. 17.14).

Using these cell lines, Emeson and coworkers (1990) tested the hypothesis that the splice acceptor sites for exons 4 and 5 were critical to the alternative splicing. They transfected C cells and N cells with modified

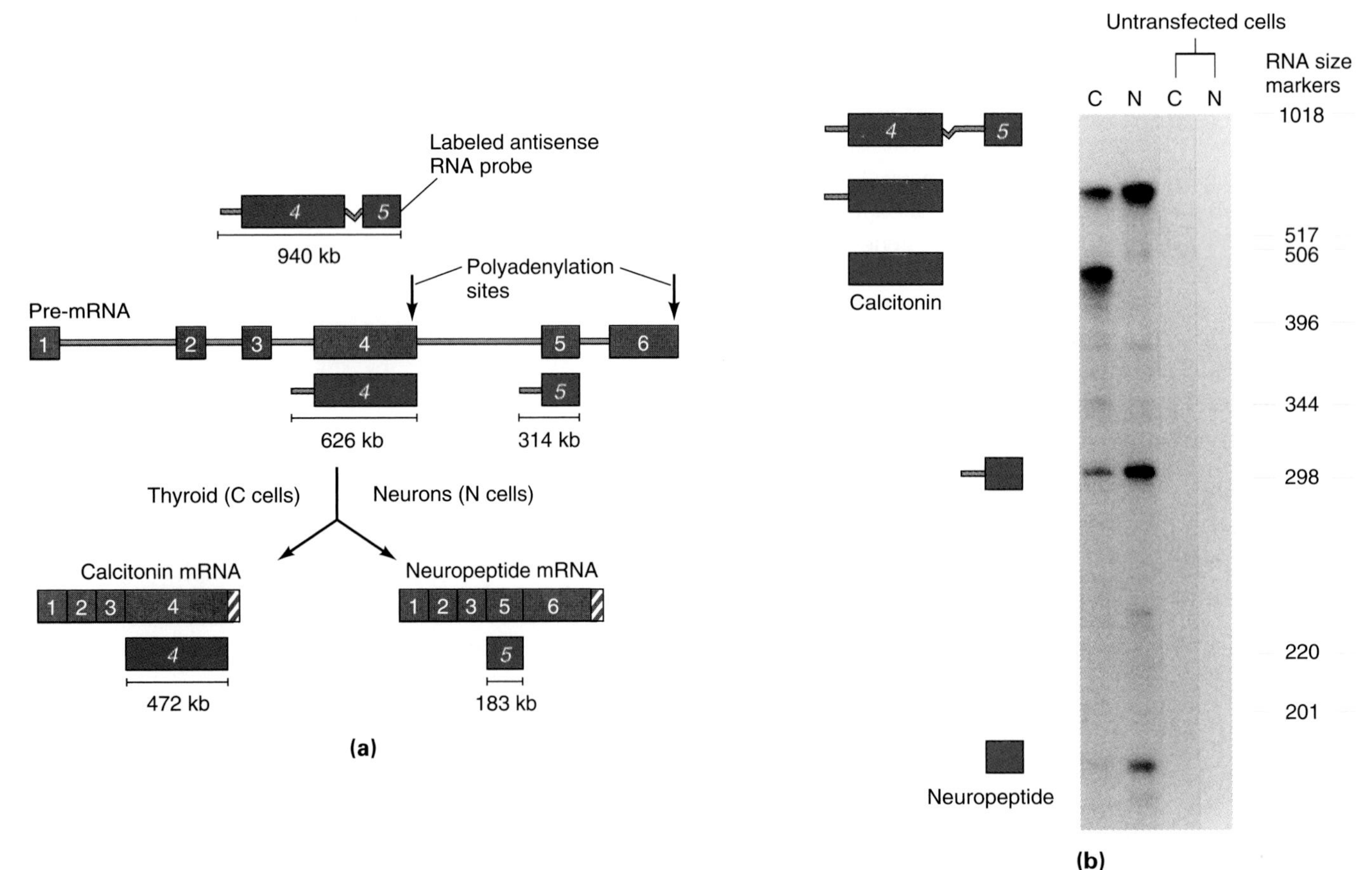

Figure 17.14 Differential splicing of calcitonin/neuropeptide pre-mRNA in thyroid glands and neurons was simulated in cultured cell lines designated C cells and N cells, respectively. The two mature mRNAs and their splicing intermediates were identified by RNase protection (see Method 17.1). **(a)** Schematic representation of a labeled antisense RNA probe, the calcitonin/neuropeptide pre-mRNA, and the two mRNAs for calcitonin and neuropeptide. The probe was an antisense RNA transcript from an engineered cDNA consisting of exons 4 and 5 along with their adjoining splice acceptor sites. For the pre-mRNA and each mRNA, the protected probe segments are indicated in blue color. **(b)** Results of a ribonuclease protection assay, using RNA from C cells and N cells to protect the probe shown in part a. RNA from untransfected cells protected no piece of the probe, indicating that these cells produced no complementary RNA sequences. RNA from C cells transfected with an actively transcribed calcitonin/neuropeptide transgene included the mature calcitonin mRNA and splicing intermediates of both mRNAs. RNA from N cells containing the same transgene included the mature neuropeptide mRNA and splicing intermediates of both mRNAs.

method 17.1

Ribonuclease Protection Assay

The ***ribonuclease protection assay*** is used to detect small quantities of a particular RNA sequence in a mixture of many RNAs. The assay is based on the fact that the enzyme ribonuclease A degrades single-stranded RNA but not double-stranded RNA (Fig. 17.m1). Therefore, the presence of the RNA sequence of interest can be tested by preparing a complementary RNA sequence and adding it as a labeled probe. If the sequence of interest is present in the mixture of RNAs, it will protect the probe from enzymatic degradation, so that the probe still moves as one intact piece during gel electrophoresis. If the RNA sequence of interest is not present in the mixture, then the probe is degraded by the enzyme and can no longer be detected.

As an example, Figure 17.14 shows a ribonuclease protection assay using as a labeled probe antisense RNA transcribed in vitro from a DNA template comprising exons 4 and 5 from the rat calcitonin/neuropeptide gene.

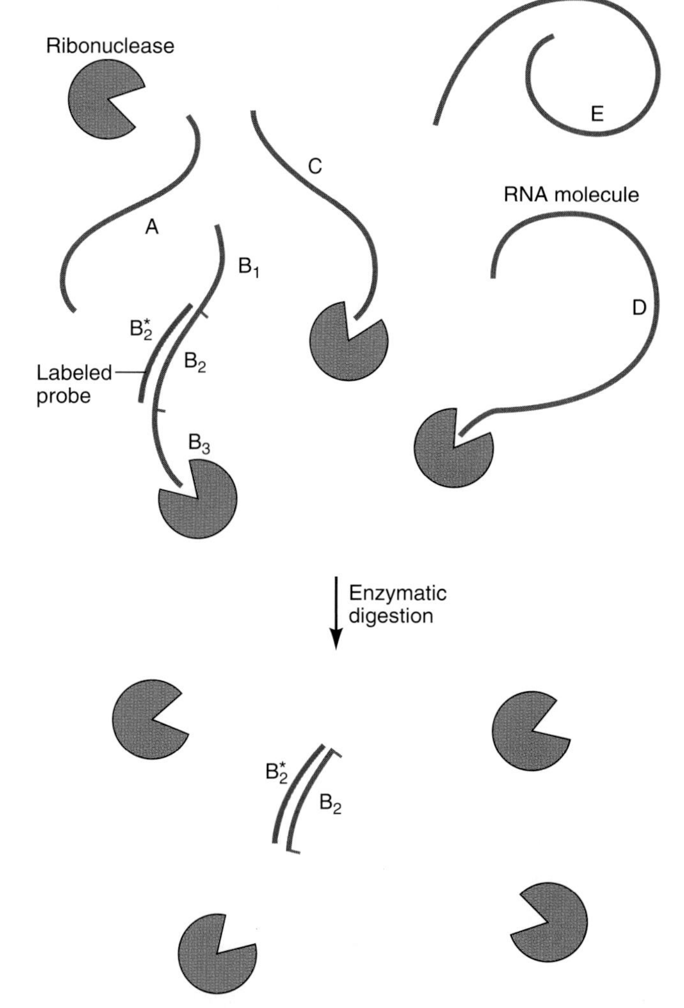

Figure 17.m1 Ribonuclease protection assay. A mixture of unlabeled RNA molecules (A through E) is tested for the presence of a particular sequence, B_2. B_2 is part of a larger molecule, $B_1 + B_2 + B_3$. To test for the presence of B_2, a complementary labeled probe, B_2^*, is added to the mixture and allowed to hybridize. Then a small amount of ribonuclease is added. This enzyme digests single-stranded but not double-stranded RNA. Thus, after a short period, all RNA is digested except for the B_2/B_2^* hybrid. Upon electrophoresis, the label associated with B_2^* still migrates in one piece. The protection of the labeled probe shows that B_2 was present in the mixture.

calcitonin/neuropeptide genes from which one of the two acceptor sites had been removed (Fig. 17.15). With a transgene lacking the acceptor site for exon 4, N cells produced neuropeptide mRNA as expected. However, C cells no longer produced calcitonin mRNA but spliced neuropeptide mRNA instead. With a transgene lacking the acceptor site for exon 5, C cells kept splicing calcitonin mRNA, whereas N cells produced neither neuropeptide mRNA nor calcitonin mRNA. The investigators concluded that N cells contained an agent that *blocked* the use of the calcitonin-specific acceptor. No corresponding blockage of the neuropeptide-specific acceptor seemed to exist in C cells, because they were able to use this site when the alternative site was deleted. Thus, the calcitonin-specific splicing pattern seems to be used by default.

The experiments described above indicate that the rat calcitonin/neuropeptide gene gives rise to two products by means of alternative splicing. The default splicing pattern is carried out in thyroid cells and gives rise to

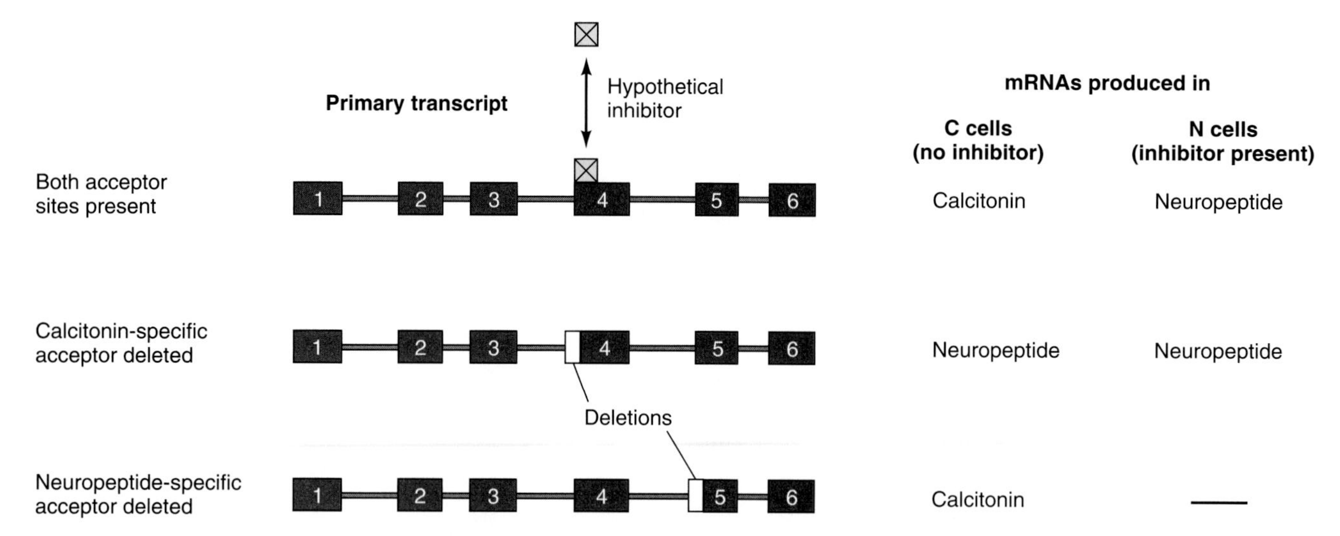

Figure 17.15 Analysis of calcitonin/neuropeptide pre-mRNA splicing by deletion of the alternatively used splice acceptor sites (for exons 4 and 5). With a transgene containing both acceptor sites, C cells produced calcitonin mRNA and N cells produced neuropeptide mRNA. With a transgene lacking the acceptor site for exon 4, both cell types produced neuropeptide mRNA. With a transgene lacking the acceptor site for exon 5, only C cells produced calcitonin mRNA while N cells produced neither mRNA. The results are best explained by assuming that the calcitonin mRNA is produced by default, and that N cells contain an agent that blocks the calcitonin-specific (default) splicing acceptor site for exon 4.

calcitonin mRNA. Neurons contain a component that blocks the calcitonin-specific splice acceptor site. Consequently, the splicing machinery is redirected to the alternative acceptor site and neuropeptide mRNA is generated instead.

17.3 Messenger RNP Transport from Nucleus to Cytoplasm

Once pre-mRNPs have been processed into mRNPs, most of them are released into the cytoplasm. Usually, the release occurs during the interphase of the cell cycle, when the nucleus is partitioned from the cytoplasm by the nuclear envelope. Under these conditions, molecules entering or leaving the nucleus must pass through the *nuclear pore complexes* (*NPCs*). Traffic through these pore complexes is regulated in both directions (Görlich and Mattaj, 1996; Nigg, 1997; Stutz and Rosbash, 1998; Nakielny and Dreyfuss, 1999). This regulation is yet another means of differential gene expression, and the underlying molecular mechanisms are actively being investigated.

In this section, we will first survey some observations on the properties of NPCs and some general data on nucleocytoplasmic transport. We will then present some quantitative measurements on the selective transport of specific RNPs.

NUCLEAR PORE COMPLEXES ARE CONTROLLED GATES FOR THE TRANSPORT OF RNPs TO THE CYTOPLASM

NPCs are flat, cylindrical openings that span the outer and inner membranes of the nuclear envelope (Fig. 17.16). Each NPC is composed of about 100 different proteins, collectively called ***nucleoporins.*** Many nucleoporins are present in groups of 8 or 16 copies and arranged in eightfold radial symmetry (see Fig. 17.1). Their molecular composition and ultrastructure have been highly conserved during the evolution of eukaryotes.

Microinjection experiments with labeled molecules of different sizes indicate that NPCs have two functions. On the one hand, they simply act like the openings of a sieve, allowing small molecules such as nucleotides or amino acids to diffuse freely into and out of the nucleus. The upper size limit for diffusible molecules depends on the Ca^{2+} concentration inside the nuclear envelope (Perez-Terzic et al., 1997). This space is connected to the *endoplasmic reticulum,* part of which serves as a Ca^{2+} reservoir (see Fig. 2.2). Molecules larger than 20 kd pass slowly, and spherical molecules larger than 60 kd do not cross the nuclear envelope by diffusion. On the other hand, NPCs are part of active transport systems that require energy, are saturable, and accept only specific molecules as cargo. In these active transport systems, certain nucleoporins interact with carrier proteins known as ***nuclear transport receptors*** to shuttle specific protein molecules and RNP particles into and out of the nucleus.

The import of proteins from the cytoplasm into the nucleus has been well characterized (Silver, 1991; Dingwall and Laskey, 1992). For example, chromosomal proteins, after being synthesized in the cytoplasm, are targeted to the nucleus by *nuclear localization signals.* These consist of one or more clusters of basic amino acid residues, which are recognized by shuttle proteins, which in turn interact with nucleoporins in hauling their cargo across the NPC. The import of many proteins is

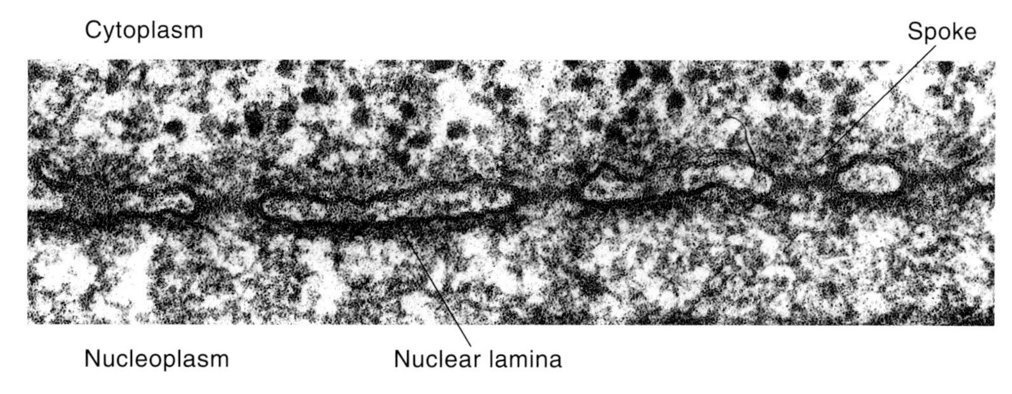

Figure 17.16 Nuclear lamina and nuclear pore complexes in a frog oocyte. This transmission electron micrograph shows the sectioned nuclear envelope with the outer membrane facing the cytoplasm at the top and the inner membrane facing the nucleoplasm at the bottom of the photograph. The dark, fibrous material adjacent to the inner membrane is the nuclear lamina. The openings in the nuclear envelope are nuclear pore complexes (see also Fig. 17.1). Radially oriented proteins inside the pore complex are known as spokes.

regulated by exposing or masking their nuclear localization signals, either through chemical modifications or by association with other proteins.

The export of mRNP particles from the nucleus to the cytoplasm is less well understood (Nakielny and Dreyfuss, 1997). Splicing signals and introns generally prevent mRNA export, possibly because they associate with spliceosomes, which render mRNPs too bulky for passing through the NPCs. The 5′ cap promotes export from the nucleus, as does the 3′ poly(A) tail. These and possibly other cis-acting signals are likely to act through RNA-binding proteins with nuclear export signals and/or nuclear retention signals that have antagonistic effects. Such RNA-binding proteins may interact directly with nucleoporins or with carrier proteins that in turn interact with nucleoporins.

EXPERIMENTS WITH CLONED cDNAs INDICATE DIFFERENTIAL mRNP RETENTION

Several investigators, working with various species, have used cloned cDNAs to monitor the release of mRNPs from the nucleus into the cytoplasm. Such cDNA probes are reverse-transcribed from cytoplasmic or polysomal mRNAs and cloned in a suitable vector. Because cDNAs represent individual mRNAs they are highly specific tracers. Two examples will illustrate this type of analysis.

One way of tracing the release of mRNA from the nucleus is to monitor its recruitment into *polysomes*, the cytoplasmic organelles of protein synthesis. When studying the recruitment of mRNA for histone H3 in sea urchin embryos, Wells and coworkers (1981) observed that it lagged about 90 min behind the recruitment of most other mRNAs (Fig. 17.17). A similar time interval passed between fertilization and the breakdown of the zygote nucleus before first cleavage. Thus, the data were compatible with the hypothesis that the mRNPs for histone H3, and possibly other histones, were retained selectively in the nucleus. This conclusion was confirmed by *in situ hybridization* (see Method 8.1)

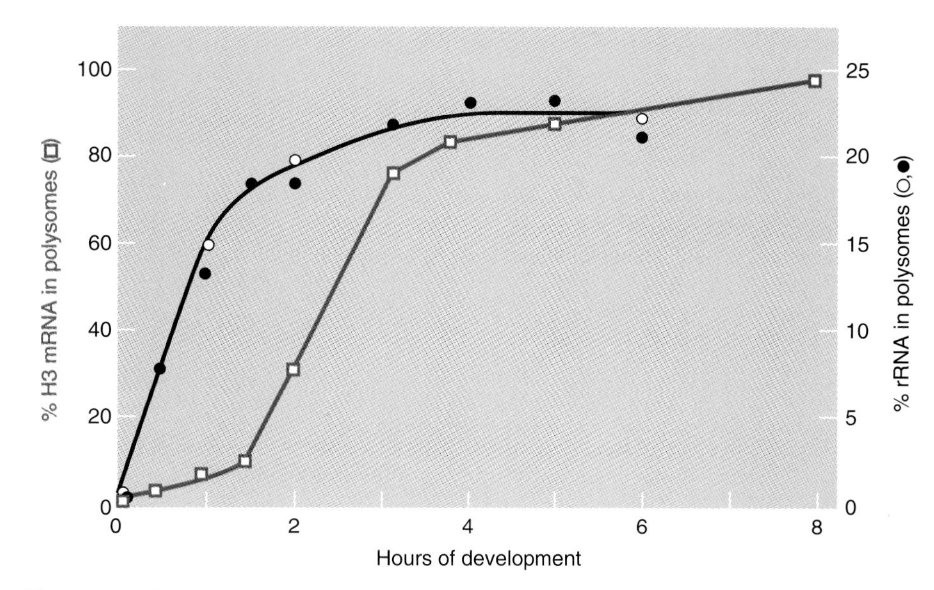

Figure 17.17 Delayed recruitment of histone H3 mRNA into polysomes after fertilization in sea urchin eggs. The fraction of H3 mRNA in polysomes (left ordinate, squares) was measured at intervals after fertilization (abscissa). The amounts of polysomal H3 mRNA and total H3 mRNA were determined by hybridization with a radiolabeled cDNA probe. For comparison, the fraction of ribosomal RNA in polysomes was measured in the same way (right ordinate, open circles) or by optical density (filled circles). This measurement reflected the overall amount of all mRNAs loaded into polysomes. The comparison showed that the overall assembly of polysomes reached a plateau within 90 min after fertilization. In contrast, histone H3 mRNA was loaded into polysomes mostly between 90 min and 3 h after fertilization. The cause for this specific delay is illustrated in Figure 17.18.

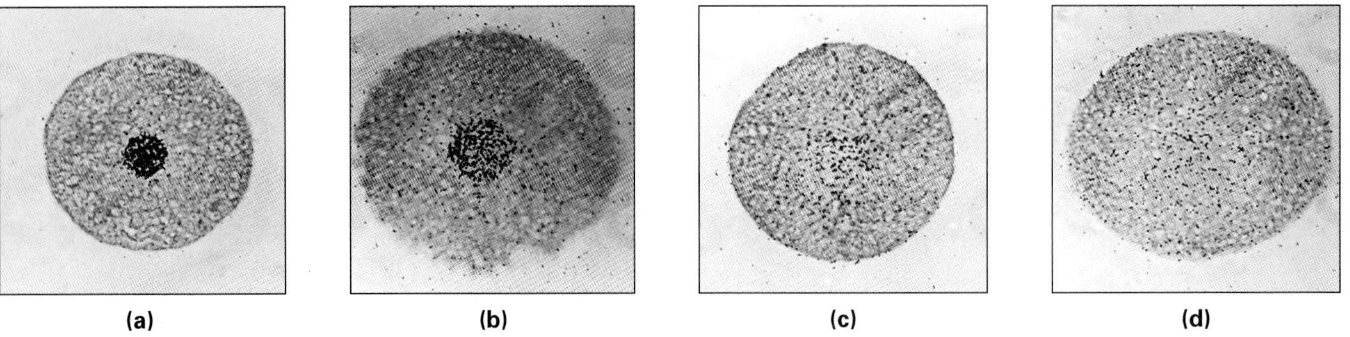

Figure 17.18 In situ hybridization (see Method 8.1) of a radiolabeled histone mRNA probe with sections of sea urchin eggs fixed at 70 min **(a)**, 80 min **(b, c)**, and 90 min **(d)** after fertilization. The intensely labeled area in the center of the egg is the zygote nucleus, which breaks down about 80 min after fertilization during the first mitosis. These data explain the delay, relative to other mRNAs, before histone mRNA is recruited into polysomes (see Fig. 17.17).

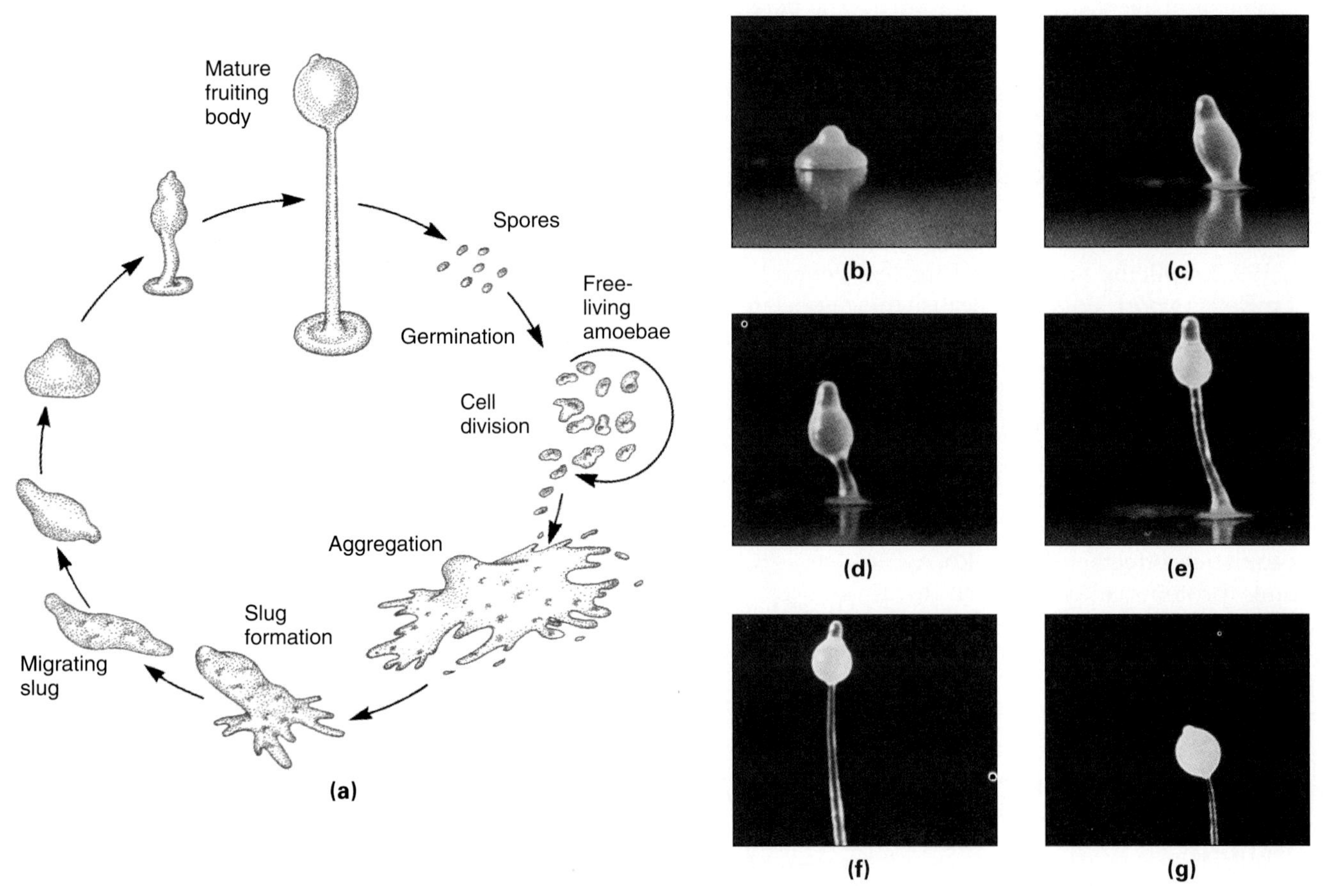

Figure 17.19 Life cycle of the slime mold *Dictyostelium discoideum.* **(a)** Individual amoebalike cells feed on bacteria and divide. When starved, the cells aggregate and form migrating slugs. The slugs form fruiting bodies, which release spores. The spores germinate when living conditions are favorable, and the life cycle begins anew. **(b–g)** Photographs showing successive stages in the development of the fruiting body.

with probes specific for histone mRNPs (DeLeon et al., 1983). These mRNPs are sequestered in the nucleus until first mitosis, when they were rapidly released into the cytoplasm (Fig. 17.18). This sequestration also occurs during the subsequent mitotic cycles and is specific to embryonic histone mRNPs, which are made only during cleavage. Their passive release during mitosis makes large quantities of histone mRNAs available exactly when they are needed.

The following study on the slime mold *Dictyostelium discoideum* shows that some mRNAs move quickly from the nucleus into the cytoplasm while others are released more slowly. In the course of their life cycle, *Dictyostelium* cells first grow and multiply as individual amoebae feeding on bacteria. When starved, the amoebae aggregate, differentiate into a fruiting body, and form spores that can survive adverse conditions for a long time (Fig. 17.19). Some mRNAs, called *common*

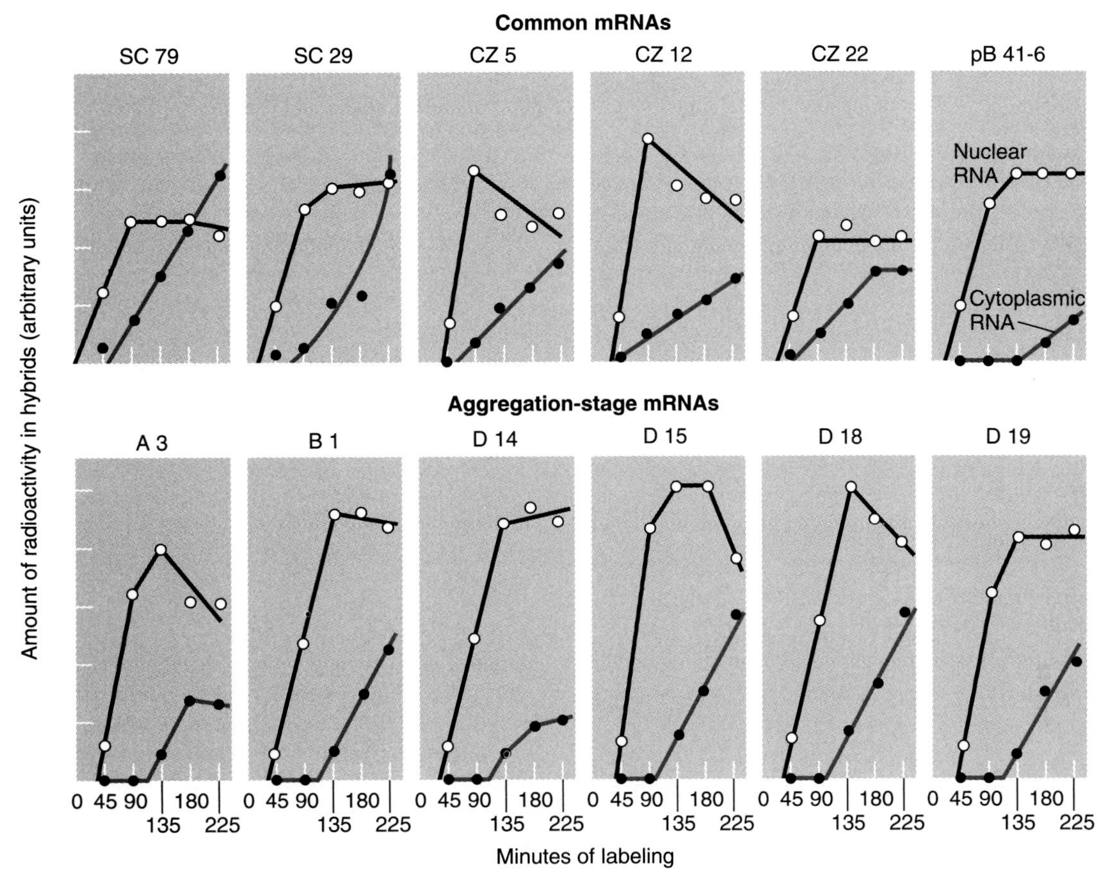

Figure 17.20 Differential release from the nucleus to the cytoplasm of specific mRNAs in the slime mold. In this experiment, all newly synthesized RNA was labeled in vivo, beginning at 14 h of development. During the labeling period, both common and aggregation-specific mRNAs were synthesized. After labeling periods ranging from 45 min to 225 min, cells were harvested to prepare nuclear and cytoplasmic RNA. Each fraction was hybridized separately with 12 cloned cDNAs known to represent either common mRNAs or aggregation-specific mRNAs. The cloned cDNAs were made single-stranded and immobilized on membranes. The amount of label binding to each cDNA indicated how much of the corresponding mRNA was present in the nucleus (open circles) and in the cytoplasm (filled circles). Every sequence was detected in the nucleus within 45 min. The common mRNAs also appeared in the cytoplasm within 90 min, except for one sequence (pB 41-6), which was also unusual in other respects. However, the aggregation-specific mRNAs were not detected in the cytoplasm until after 90 min. The results indicate a sequence specificity in the release of different mRNAs from the nucleus to the cytoplasm.

mRNAs, are present in the cytoplasm during all stages of development. Other mRNAs, called *developmentally regulated mRNAs,* are present only at certain stages.

To analyze the mechanisms involved in the regulation of mRNA transport in *Dictyostelium,* Mangiarotti and coworkers (1983) traced individual mRNA sequences known from previous studies to be either common or aggregation-specific. Hybridization of individual cDNA clones with *pulse-labeled* nuclear and cytoplasmic mRNA fractions revealed different patterns of processing and/or nucleocytoplasmic transport (Fig. 17.20). Common mRNAs remained in the nucleus for about 45 min before they appeared in the cytoplasm. In contrast, *aggregation-specific* mRNAs did not appear in the cytoplasm until after 90 min. Depending on the particular mRNA, between 20 and 60% of the nuclear transcripts were transported into the cytoplasm. A third type of RNA, expressed predominantly during cell differentiation, was degraded extensively within the nucleus; what remained of it was released into the cytoplasm after a delay of over 2 h.

The foregoing studies show that mRNP processing and/or transport from the nucleus into the cytoplasm can regulate the timing and quantity of gene expression. The molecular mechanisms underlying these control steps are currently under active investigation, as discussed earlier.

17.4 Messenger RNA Degradation

Once in the cytoplasm, most mRNPs are incorporated into polysomes to synthesize proteins. (The storage of some mRNPs in the form of inactive RNP particles will be discussed in Section 18.1.) The rates at which different proteins are synthesized in a cell are precisely regulated. The synthesis rate for a protein in a cell is determined to a large extent by the abundance of the encoding mRNA, which ranges approximately from 100 to

100,000 molecules per cell. The abundance in turn depends on the rates of mRNA *production* and *degradation*. The importance of the rate of production is obvious and has been discussed previously. However, for quick adjustment of the available mRNA, the rate of degradation is just as critical (Ross, 1989). As an analogy, consider the task of adjusting fluid in a tank to varying levels. To raise the level, one needs a controlled influx. To lower the fluid level, waiting for evaporation or leakage to do the job would be slow. A quicker adjustment requires a controlled efflux. This is precisely what controlled degradation of mRNA does in a cell.

THE HALF-LIFE OF MESSENGER RNAs IN CELLS IS REGULATED SELECTIVELY

The longevity of mRNA in cells is measured by its ***half-life***, which is defined as the time after which an original number of intact molecules has been reduced to one-half (Method 17.2). The half-life of eukaryotic mRNAs in active cells ranges from minutes to weeks.

The half-life of mRNAs is controlled in cells in adaptive and selective ways. A few examples follow.

Globin mRNA As bone marrow cells mature into red blood cells, they undergo several steps of differentiation, including mass production of hemoglobin and, in mammals, elimination of the cell nucleus. In the enucleate cells, called *reticulocytes*, no additional RNA can be synthesized. Nevertheless, the RNA composition continues to change by differential degradation. This is shown by comparing the proteins synthesized and the mRNAs present at different stages of development. While young reticulocytes synthesize globins as well as other proteins, mature reticulocytes synthesize globins almost exclusively. Analysis of the mRNAs shows that globin mRNA in rabbit reticulocytes has a half-life about four times as long as that of the mRNA for protein I, the next most abundant reticulocyte protein (Lodish and Small, 1976). Since differential mRNA production cannot occur in this case, the result must be caused by differential degradation. The adaptive value of this type of regulation seems to be that it maximizes the synthesis of globins by reducing the concurrent synthesis of other proteins.

Casein mRNA The mammary gland is a target tissue for several hormones that interact during pregnancy to cause growth and differentiation of the gland cells. In nursing females, the hormone *prolactin* stimulates lactation, the production of milk. This response can also be observed in cultured mammary gland tissue, using the synthesis of casein, an abundant milk protein, as a biochemical marker (Fig. 17.21). Upon prolactin stimulation, the abundance of casein mRNA increases from 50 to 2000 molecules per cell (Guyette et al., 1979). Most of this increase is caused by an increase in the half-life of

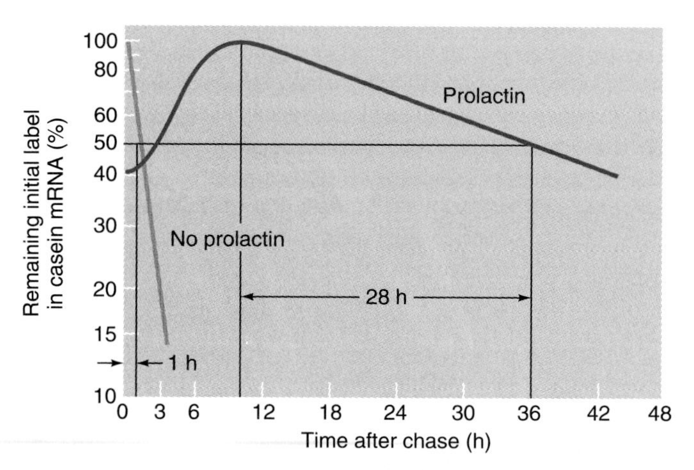

Figure 17.21 Stabilization of the mRNA for the mammalian milk protein casein in the presence of the pituitary hormone prolactin. In this experiment, breast tissue from lactating rats was placed in a medium without prolactin to dilute the endogenous hormone. One sample was then incubated with prolactin and pulse-labeled with uridine (see Method 17.2). Another sample was also pulse-labeled with uridine, but without prolactin. After 3 h, the labeled uridine was removed, and unlabeled uridine was added to chase the labeled uridine from the tissues. Samples taken at intervals thereafter were analyzed for casein mRNA by hybridization with cloned, unlabeled cDNA. The culture with prolactin accumulated labeled casein mRNA for several hours, presumably because the chase of labeled uridine was ineffective. The eventual decay of mRNA was much slower ($t_{1/2}$ = 28 h) in the presence of prolactin than without the hormone ($t_{1/2}$ = 1 h).

the casein mRNA, which is approximately 1 h without prolactin and 28 h with prolactin. This dramatic increase is highly selective because the half-life of other mRNAs changes little in response to the hormone treatment. Similarly, the presence of estrogen causes a substantial increase in the half-lives of vitellogenin mRNA in liver cells of frogs and birds, and of ovalbumin mRNA in hen oviduct cells (Shapiro et al., 1987).

Oncogene mRNA Many cancer-causing genes, called *oncogenes*, are mutant alleles of normal cellular genes, called *proto-oncogenes*, which are involved in the control of the normal cell cycle (see Section 28.7). For example, the c-*myc* gene is a proto-oncogene encoding a protein involved in the control of cell proliferation. However, certain mutant alleles of this gene cause lymph node cancer.

Comparing the c-*myc* proto-oncogene with its cancer-causing alleles (designated *myc*), Piechaczyk and coworkers (1985) observed that the mRNAs transcribed from *all* alleles had identical translated regions, so that the encoded proteins had to be the same. The mutated sequences were in the 5′ UTR regions of the mRNAs. In correlation with these changes, the mutant mRNAs had half-lives three to five times longer than wild-type

Pulse-Labeling of Molecules and Their Half-Life in Cells

Researchers can measure the half-life of a molecule in a cell by *pulse-labeling* the molecule during a short period of synthesis and then measuring the time until half of the labeled molecules have decayed. The following procedure is used to measure the half-life of an RNA sequence; however, the same principles apply to measuring the half-lives of proteins and other molecules in cells.

To pulse-label an RNA synthesized in cells, the investigator adds a labeled nucleotide to the culture medium for a short period of time. The labeled nucleotide is then "chased" by adding an excess of the same nucleotide in unlabeled form. The unlabeled nucleotide dilutes the labeled nucleotide so that the subsequent uptake of label into RNA is negligibly small. The cells are allowed to survive for various intervals before their RNA is extracted. An unlabeled cDNA representing the sequence of interest is hybridized with the labeled cell RNA. The amount of label driven into hybrids indicates how many copies of the RNA sequence of interest are present in the RNA preparation. This number decreases exponentially with time. The time after which the label in hybrids has decreased to half of its original value is the ***half-life*** (abbreviated $t_{1/2}$).

mRNA. The resulting overproduction of the c-myc protein appeared to be causing the cancers. Evidently, the half-life of a proto-oncogene mRNA must be strictly controlled.

This example illustrates not only that maintaining the correct half-life of an mRNA may be critical, but also that the regulation process may depend on sequences in the 5′ UTR region.

THE DEGRADATION OF mRNAs IN CELLS IS CONTROLLED BY PROTEINS BINDING TO SPECIFIC RNA MOTIFS

The cellular agents of mRNA degradation are enzymes called ***ribonucleases (RNases).*** The destruction of RNA molecules by RNases is regulated by *cis-acting signals* and *trans-acting factors.* The cis-acting signals are specific nucleotide sequences or secondary structures of the mRNA. The transacting factors are regulatory proteins that bind to the cis-acting signals. Some of the trans-acting proteins exert a protective effect by masking nuclease-sensitive sites that would otherwise be attacked by RNases. Other trans-acting signals exert a destabilizing effect by recruiting ribonucleases, thus hastening the destruction of the bound RNA. In either event, the specificity with which trans-acting proteins bind to cis-acting RNA motifs explains the selective control of mRNA lifetime.

The cases investigated so far have revealed several locations of cis-acting signals (Sachs, 1993). The importance of the 5′ UTR region in the c-myc mRNA has already been mentioned. In other cases, regulatory proteins bind to the poly(A) tail or to the 3′ UTR region. The poly(A) tail at the 3′ end prolongs the half-life of most mRNAs. The tail's stabilizing effect is mediated by a ***poly(A)-binding protein (PABP)*** that binds specifically and tightly to poly(A) segments. At least in yeast, the protective function of PABP also depends on its interaction with proteins that control the initiation of translation (Coller et al., 1998; see Section 18.2). This interaction must close the mRNA into a circle, which seems to protect it against ribonuclease attack.

The 3′ UTR affects the half-lives of many mRNAs, as shown by cell transfection experiments with cDNAs encoding chimeric mRNAs. For instance, globin mRNAs can be engineered with the 3′ UTRs from other mRNAs instead of their own (Fig. 17.22). If the 3′ UTR comes from a stable mRNA, the chimeric globin mRNA is stabilized; if the 3′ UTR comes from an unstable mRNA, the half-life of the chimeric globin mRNA is reduced. Instability is conferred in particular by sequences rich in A and U nucleotides (Shaw and Kamen, 1986). These results suggest that certain sequences of these two bases are recognized by regulatory proteins that bind to specific mRNAs and increase their susceptibility to ribonuclease.

Other regulatory proteins that bind to the 3′ UTR of mRNAs protect them from ribonuclease attack. Some of these proteins are themselves activated by a *ligand,* such as prolactin in the case of casein mRNA protection described earlier. Another protective protein binds to a so-called *iron-response element* in the 3′ UTR region of an mRNA encoding transferrin receptor, a receptor involved in regulating the uptake of iron into red blood cells (see Fig. 18.8). The regulation of RNA-binding proteins by ligands is similar to the regulation of transcription factors by ligands, as described in Section 16.4 for steroid hormone receptors.

How the various mechanisms involved in regulating mRNA degradation may be coordinated is still under investigation (Beelman and Parker, 1995). A key event, at least in yeast, is the enzymatic removal of the cap structure present at the 5′ end of mRNA (see Fig. 17.3). In wild-type cells, decapping triggers rapid mRNA degradation, whereas cells deficient for the decapping enzyme accumulate capped mRNA (Beelman et al., 1996).

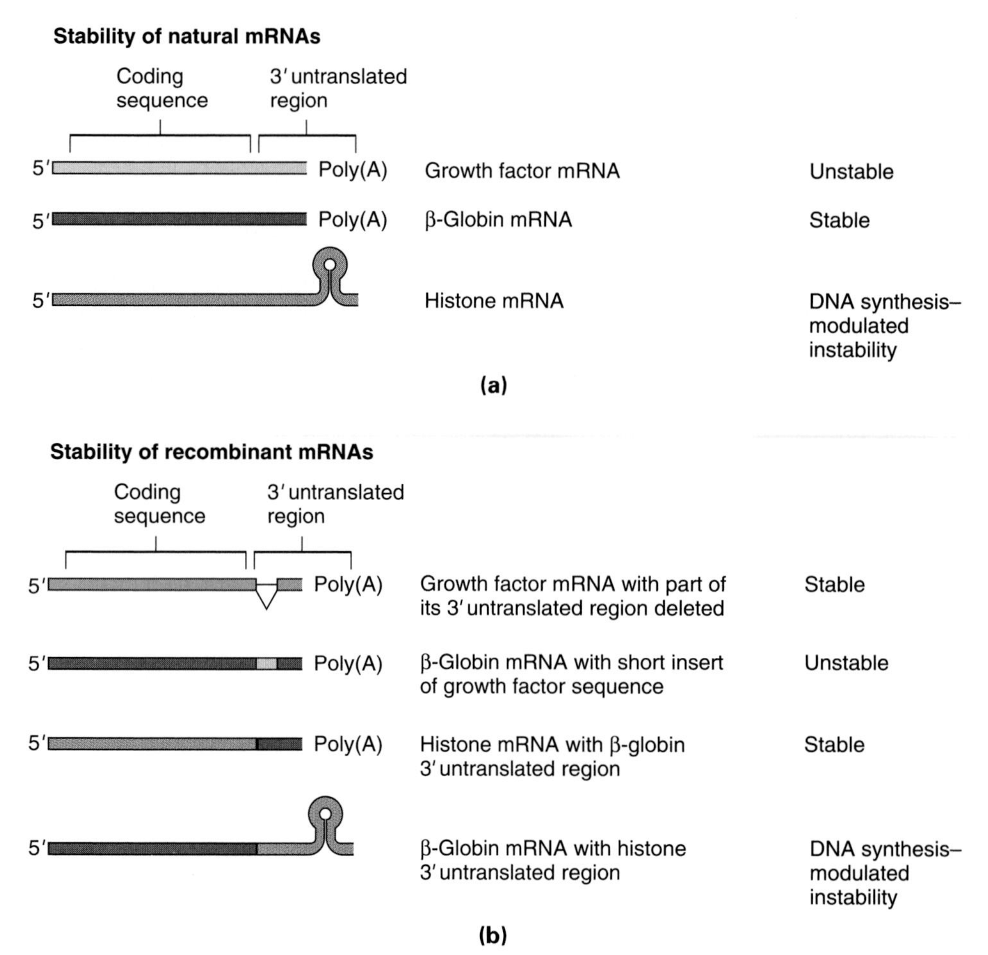

Figure 17.22 Importance of the 3′ untranslated region (3′ UTR) for mRNA stability. **(a)** Three naturally occurring mRNAs with different half-lives. Globin mRNA is stable, with $t_{1/2}$ = 10 h. Growth factor mRNA is unstable, with $t_{1/2}$ = 0.5 h. Histone mRNA, which is unusual in that its 3′ UTR ends with a stem-loop structure instead of a poly(A) tail, has a modulated instability: Its half-life is 1 h when cells are synthesizing DNA and 0.2 h during the rest of the cell cycle. **(b)** By cell transfection with engineered cDNA transgenes, it is possible to create recombinant mRNAs. Growth factor mRNA becomes stable when a certain part of its 3′ UTR region has been deleted. Grafting this 3′ UTR sequence from growth factor mRNA to globin mRNA makes the globin mRNA unstable. Swapping 3′ UTRs between histone mRNA and globin mRNA renders histone mRNA stable while conferring modulated instability on globin mRNA.

Some of the other mechanisms that regulate mRNA lifetime seem to act through the decapping step. Most importantly, PABP inhibits the decapping reaction, which explains the protective effect of the poly(A) tail on many mRNAs. Conversely, decapping is triggered by a surveillance mechanism that detects abnormal nucleotides causing premature translation termination (Hagan et al., 1995).

SUMMARY

Newly synthesized nuclear RNA undergoes several modifications, which typically begin while transcription is still going on. The processing of pre-mRNA into mRNA includes an association with proteins to form mRNP particles, the addition of a 7-methylguanylate cap at the 5′ end, the polyadenylation of the 3′ end, and the removal of introns. Capping and polyadenylation protect the mRNA against rapid degradation by ribonucleases. For intron removal, pre-mRNA associates with a family of small nuclear RNP particles (snRNPs), which interact to form organelles called spliceosomes. They remove introns from pre-mRNA and splice exons together to form mature mRNA. Most of these processing steps serve as points of regulation for differential gene expression.

During the splicing of two exons, the splice donor site at the 3′ end of an exon is usually combined with the splice acceptor site at the 5′ end of the following exon. However, the blocking of this splice acceptor site by a regulatory protein may cause the donor site to be combined with the acceptor site of an alternate exon. The splicing machinery may also deviate from its usual pattern if an intrinsically poor splice site is made more attractive by a regulatory protein. Finally, the presence of more than one promoter within the same gene may generate alternative donor sites cooperating with the same acceptor site. These regulatory processes allow cells to generate two or more different mRNAs from one pre-mRNA, an ability called alternative splicing.

Alternative splicing controls gene expression in a variety of organisms at different stages of development. In *Drosophila,* the development of sexual dimorphism depends on splicing patterns resulting in a regulatory cascade of gene products. At each level of the cascade, a default splicing pattern leads to male development; the female-specific pattern requires an active protein made at the same or a higher regulatory level. Alternative splicing can also generate entirely different proteins in different types of cells. For example, the pre-mRNA transcribed from one rat gene is spliced in two ways, giving rise to either the hormone calcitonin or a neuropeptide.

After pre-mRNPs have been processed into mRNPs, most of the mRNPs are transported into cytoplasm. Nucleocytoplasmic transport provides another opportunity for differential gene expression. This has been shown by monitoring the release of individual transcripts into the cytoplasm with cloned cDNA tracers. The ultrastructure of the nuclear pore complexes, and the restricted movements of macromolecules through nuclear pore complexes, indicate that the nucleocytoplasmic transport of mRNPs is selective and energy-dependent.

The half-life of eukaryotic mRNAs in cytoplasm ranges from minutes to weeks and is regulated in selective and adaptive ways. The agents of active mRNA degradation are cytoplasmic ribonucleases. These enzymes in turn are regulated by proteins that recognize certain nucleotide sequences in different regions of the mRNA.

Translational Control and Posttranslational Modifications

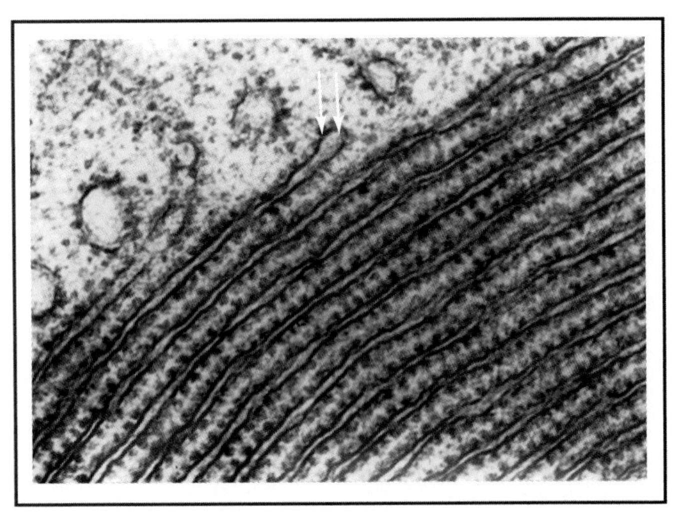

Figure 18.1 This transmission electron micrograph shows part of the cytoplasmic area from a glandular cell. The proteins secreted by the cell are synthesized and processed in the endoplasmic reticulum (ER), located to the lower right. The ER is a stack of membranes connected with the nuclear envelope (see Fig. 2.2). The cytoplasmic side of each membrane is studded with ribosomes (arrows), where polypeptides are first synthesized. Through translocator proteins in the membrane, the nascent polypeptide crosses into the inner space (lumen) of the ER (see Fig. 18.17). Here and in the adjacent Golgi apparatus, proteins receive hydroxyl groups, oligocaccharides, and other substitutions, before they are released from the cell.

CELLS owe their form and function primarily to proteins. Cytoskeletal proteins such as tubulins and actins support the cell's shape and its motility. Cell cycle proteins such as cyclins control cell divisions. Many proteins are enzymes that serve as the catalysts of cell metabolism. Some of the cell's key regulatory molecules are DNA-binding and RNA-binding proteins. Thus, an organism develops essentially through changes in the protein composition of its cells and/or regulation of protein activity. Indeed, cells expend a great amount of energy synthesizing and modifying proteins (Fig. 18.1).

The overall activity of a given type of protein in a cell depends on how many molecules of that protein are present and on the percentage of the molecules that are in the appropriate conformation. In many cases, the number of protein molecules present is controlled ultimately at the levels of transcription and RNA processing, as discussed in previous chapters. However, additional levels of control operate after mRNAs have been exported to the cytoplasm. Translational control mechanisms regulate the translation of mRNAs into polypeptides. Posttranslational mechanisms control modification and assembly of polypeptides to form active proteins.

Translational and posttranslational controls are based on a wide variety of mechanisms, which seem to have evolved to meet the specific needs of various cell types. Some mRNAs are made temporarily untranslatable by association with "masking" proteins or RNAs. The translation of other mRNAs is regulated by adjusting the length of their poly(A) tails. Still other control mechanisms target regulatory proteins that are required to initiate the translation process.

Organisms especially rely on translational and posttranslational control mechanisms when transcriptional control and RNA processing are not operational or would be too slow. Such conditions occur during certain stages of development including egg activation and spermatogenesis.

Animal oocytes are generally provided with large amounts of maternally synthesized RNA that carry them through early development. The translation of stored maternal mRNA is carefully regulated. Fully grown oocytes are generally quiescent and preserve their maternal endowment until fertilization activates the egg and boosts the overall rate of translation. Eggs also have the ability to translate certain maternal mRNAs selectively before oocyte maturation and to translate different mRNAs thereafter. These changes in protein synthesis occur while the egg nucleus is transcriptionally inactive in many species. Even in those species where transcription occurs in eggs and early embryos, the amount of RNA transcribed is often insignificant because of the extremely small *nucleocytoplasmic ratio.* Thus, transcriptional control and RNA processing cannot account for many of the changes in protein synthesis observed in early development.

Spermatogenesis is another developmental process in which translational control is critical. Most sperm undergo dramatic changes in their morphology, which require the synthesis of different proteins at different times. While these changes are taking place, the chromatin in the nucleus is packed very densely, making transcription impossible. Again, translational control is used to orchestrate protein synthesis during a stage when transcriptional control is not available.

Toward the end of this chapter, we will survey how newly synthesized proteins undergo posttranslational modifications before they assume their cellular functions. Some of these modifications direct proteins to their appropriate cellular compartments, such as the endoplasmic reticulum. Other modifications give proteins the three-dimensional conformation and chemical substitutions that are necessary for their biological functions. These modifications occur selectively in certain proteins and promote their particular functions and their ability to assemble into multimeric proteins and cell organelles.

Finally, we will examine how the longevity of proteins is controlled in cells. As with mRNA, we will see that the breakdown of proteins is an active enzymatic process targeting specific proteins in adaptive ways.

18.1 Formation of Polysomes and Nontranslated mRNP Particles

During *translation,* the nucleotide sequence of the mRNA directs the formation of a ***polypeptide***—that is, a chain of amino acid residues linked by peptide bonds. In each polypeptide, the first amino acid has a free amino group that marks the ***N-terminus,*** and the last amino acid has a free carboxyl group that marks the ***C-terminus.*** In this chapter, the term "polypeptide" will be used for the immediate product of translation, and the term "protein" for processed and assembled polypeptides. In contexts where this distinction is not important, the shorter term "protein" will be used for both polypeptides and proteins.

MOST mRNAs ARE IMMEDIATELY RECRUITED INTO POLYSOMES AND TRANSLATED

In most cell types, mRNAs released from the cell nucleus immediately associate with ribosomes to form polyribosomes, or ***polysomes***. Each ribosome of a polysome is engaged in translating the associated mRNA into a polypeptide. The process of translation consists of three phases: initiation, elongation, and termination. We will briefly review elongation and termination before we focus on initiation, the step that is most critical to translational control.

During ***elongation***, the ribosome travels along the mRNA, at each step adding another amino acid residue to the nascent polypeptide chain. This process depends on the ability of transfer RNAs (tRNAs) to load a specific amino acid and to recognize a corresponding nucleotide triplet (codon) in an mRNA. Translation also relies on the ability of ribosomes to hold the mRNA and two tRNAs in a specific configuration. The small ribosomal subunit has the binding site for mRNA while the large ribosomal subunits holds the two binding sites for the loaded tRNAs (Fig. 18.2a). The site that holds the tRNA with the nascent polypeptide chain is known as the ***peptidyl-tRNA-binding site***, or ***P site***. The site for the incoming tRNA with the next amino acid to be added is called the ***aminoacyl-tRNA-binding site***, or ***A site***. The A site and the P site are so close to each other that bound tRNAs have to use adjacent codons.

Elongation is a cyclical process that can be broken down into three steps. In step one, a loaded aminoacyl-tRNA binds to a vacant A site next to an occupied P site.

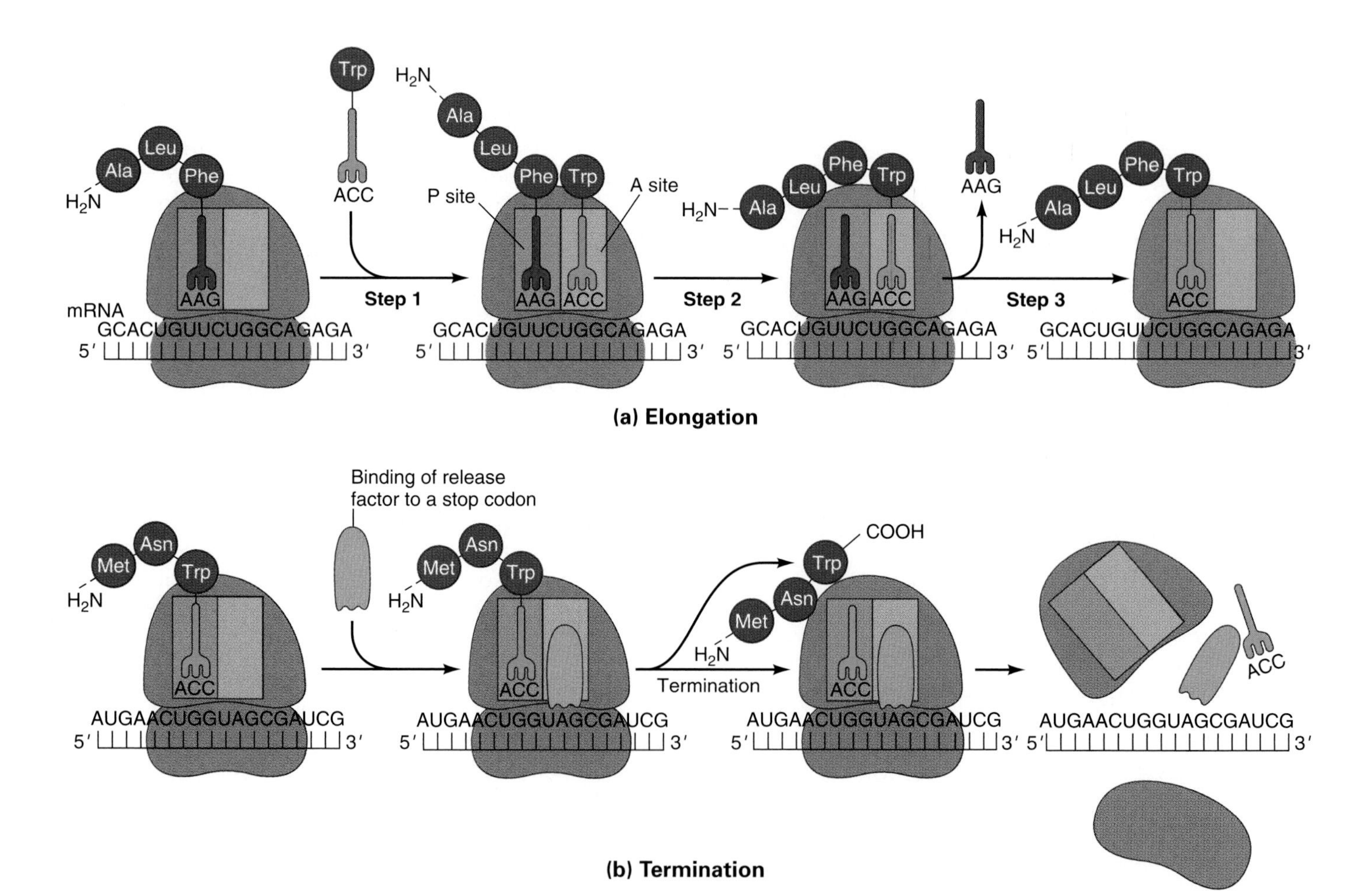

Figure 18.2 Elongation and termination phases of protein synthesis. These steps involve three binding sites of the ribosome: the site for the mRNA, the aminoacyl-tRNA binding site (A site), and the peptidyl-tRNA binding site (P site). **(a)** During elongation, a three-step cycle is repeated over and over. In step 1, a new tRNA loaded with an amino acid (aminoacyl-tRNA) binds to the A site of the ribosome, next to a P site occupied by the tRNA loaded with the nascent polypeptide. In step 2, the polypeptide is uncoupled from its tRNA and joined by a new peptide bond to the new amino acid held in the A site by its tRNA. In step 3, this tRNA moves from the A site to the P site, ejecting the old tRNA, while the ribosome moves one codon down on the mRNA. **(b)** Termination occurs when the A site is lined up with any of three stop codons: UAA, UAG, or UGA. Each stop codon binds to a release factor instead of an aminoacyl-tRNA. This causes translation to stop, the growing polypeptide to be released, and the ribosomal subunits to dissociate from the mRNA.

In step two, the C-terminus of the nascent polypeptide is uncoupled from its tRNA and joined by a peptide bond to the neighboring amino acid docked with its tRNA in the A site. In step 3, the unloaded tRNA leaves the P site, and the tRNA with the newly elongated polypeptide shifts from the A site to the P site, while the ribosome moves one codon downstream on the mRNA.

The final phase of translation, ***termination***, occurs when the ribosome encounters one of the three ***stop codons***—UAA, UAG, or UGA—none of which is recognized by a tRNA (Nakamura et al., 1996). Instead, proteins known as *release factors* bind to the stop codon and cause the release of the polypeptide from the ribosome (Fig. 18.2b). The vacant ribosome then dissociates into its two subunits, which are recycled.

The most important phase for the control of translation is ***initiation***, in which the mRNA and the first loaded tRNA are bound to a ribosome. This process involves the mRNA, the small ribosomal subunit, a special initiator tRNA loaded with the amino acid methionine (methionyl initiator tRNA, or Met-tRNA$_i$), and several catalytic proteins called ***eukaryotic initiation factors (eIFs).*** In the first initiation step, eIF4E and eIF4G bind to the cap of the mRNA. This may be followed by the unwinding of any stem-loop or other secondary structures in the 5′ UTR (Fig. 18.3a), which otherwise would

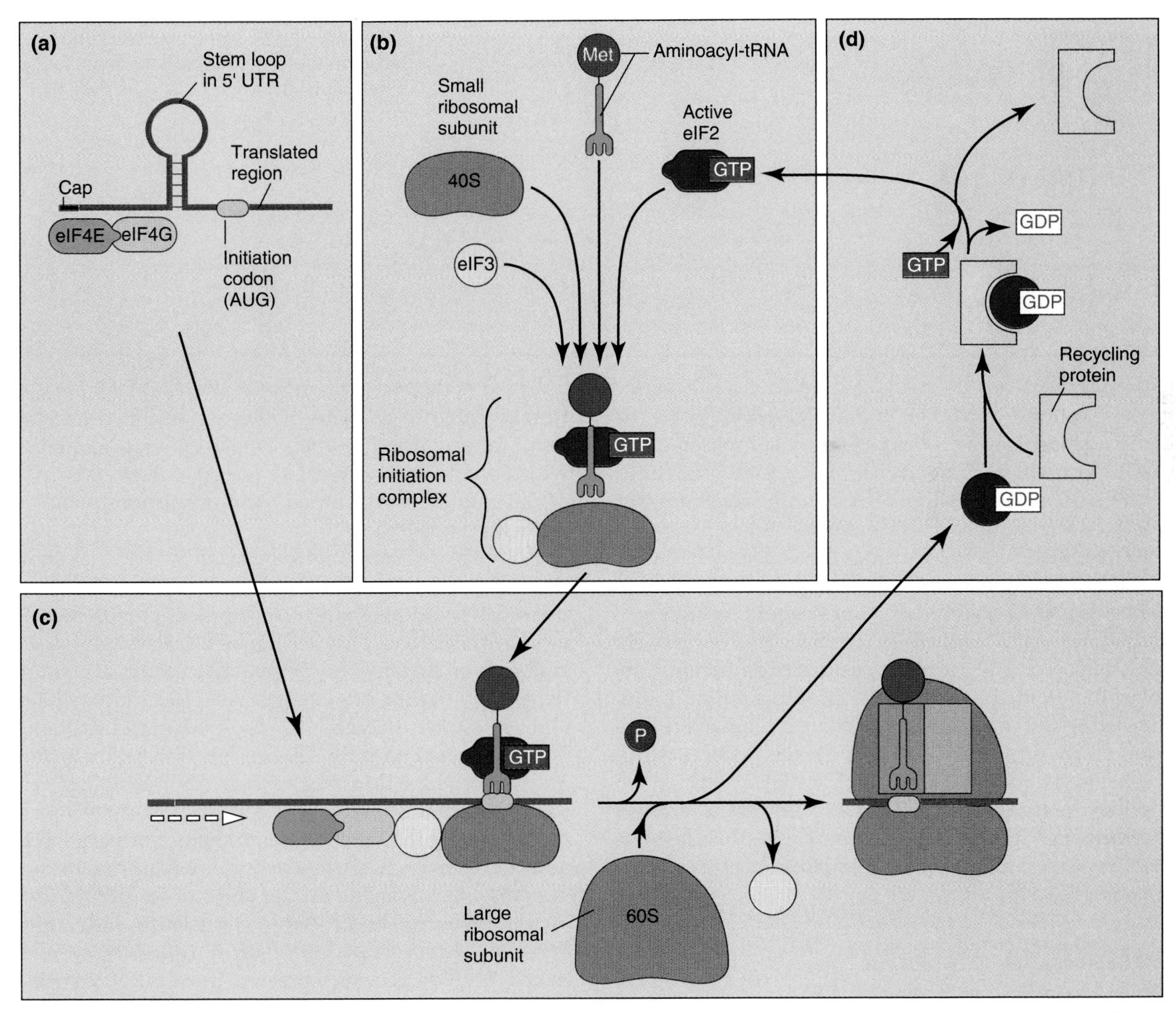

Figure 18.3 Simplified model of initiation of protein synthesis. **(a)** Several proteins collectively known as eukaryotic initiation factor 4F (eIF4F) associate with the cap structure of the mRNA. The 5′ UTR segment between the cap and the initiation codon (AUG) may contain stem-loop structures as detailed in Figure 18.8. **(b)** The small ribosomal subunit (40S) associates with methionyl initiator tRNA (Met-tRNA$_i$), and two initiation factors (eIF2 and eIF3) to form the ribosomal initiation complex. **(c)** The initiation factors associated with the cap interact with those of the initiation complex, thus recruiting the complex to the cap site. The initiation complex scans the mRNA until it finds the initiation codon. When the initiation complex is properly positioned, eIF3 is recycled. Likewise, eIF2 hydrolizes its GTP to GDP and is released in this inactive form. **(d)** Inactive eIF2 is bound by a recycling protein, which restores eIF2 to its active form by exchanging GDP for GTP.

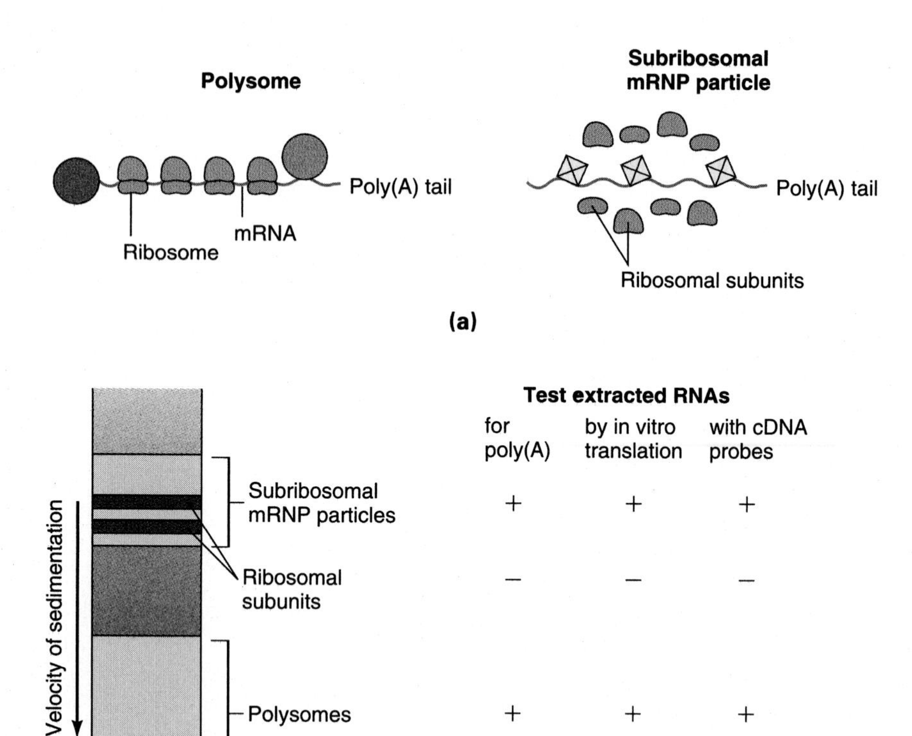

Figure 18.4 Analysis of messenger ribonucleoprotein (mRNP) particles. **(a)** Schematic diagram showing cytoplasmic mRNAs associated with two sets of proteins or other factors. One set of factors (green circles) makes mRNAs available for recruitment into polysomes. With other factors (crossed yellow boxes), they form temporarily untranslatable mRNP particles. **(b)** Centrifugation on a sucrose gradient separates polysomes, ribosomal subunits, and subribosomal mRNP particles (so named because some of them sediment behind the ribosomal subunits). Each fraction can be assayed for the presence of mRNA by extracting RNA from it and testing the RNA moiety for the presence of poly(A) tails, translatability in vitro, and hybridization with cloned cDNA probes. Based on these criteria, mRNA is present in both polysomes and subribosomal RNP particles.

block initiation. Next eIF4E and eIF4G recruit to the cap region a ***ribosomal initiation complex*** composed of the small ribosomal subunit, Met-tRNA$_i$, and two initiation factors, eIF2 and eIF3 (Fig. 18.3b, c). In its active form, eIF2 is associated with a GTP molecule as an energy carrier. The initiation complex slides down the mRNA until the Met-tRNA$_i$ is positioned over a ***translation initiation codon, AUG*** (Fig. 18.3c). After proper positioning of the initiation complex, eIF2 hydrolizes its GTP to GDP and is released. The properly positioned initiation complex is then joined by the large ribosomal subunit, thus completing the initiation step. Finally, the inactive eIF2 is bound by a recycling protein, which restores active eIF2 by exchanging its GDP for GTP (Fig. 18.3d).

When the ribosome vacates the initiation site, another ribosome can be assembled there. In this fashion, mRNAs associate with multiple ribosomes to form a polysome.

SOME mRNAs ARE STORED AS NONTRANSLATED mRNP PARTICLES

Some cell types, including gametes and early embryonic cells, can store mRNAs in the form of nontranslated messenger ribonucleoprotein (mRNP) particles, known as *postribosomal RNP particles* or ***subribosomal RNP particles***, because of the way they separate from polysomes and ribosomal subunits when they are centrifuged through sucrose gradients. Different RNP particles then separate according to their velocity of sedimentation, with polysomes sedimenting fastest, followed by ribosomal subunits and a broad band of subribosomal RNP particles (Fig. 18.4).

RNA isolated from subribosomal RNP particles shows three characteristics of mRNA. First, it is polyadenylated. Second, it hybridizes with cDNA probes complementary to known mRNAs. Third, it can be translated in ***cell-free translation systems***. Such systems contain all components required for protein synthesis—including ribosomal subunits, tRNAs, amino acids, and initiation factors—except for mRNAs. Cell-free translation systems are prepared by breaking up actively translating cells, such as immature red blood cells or wheat germ cells, and removing their nuclei, their membranous organelles, and their own mRNAs. Any protein synthesis in these systems then depends on the addition of mRNA from another source. The process of translating mRNA in a cell-free system is called ***in vitro translation***. When naked RNA molecules extracted from mRNP particles are added to cell-free translation systems, they stimulate protein synthesis. In contrast, whole subribosomal RNP particles stimulate little or no protein synthesis. Thus, subribosomal RNP particles appear to be a storage form of mRNA that is temporarily untranslatable.

For certain applications, some investigators prefer to use mature frog oocytes as "living test tubes" rather than truly cell-free translation systems. The mRNA to be tested is simply injected into the oocytes, which proceed to translate the exogenous mRNA along with their own. Naked RNA prepared from either polysomes or subribosomal RNP particles translates very well when injected into mature frog oocytes, again revealing the presence of mRNA in both fractions.

Several studies have shown that gametes and early embryos contain long-lived mRNAs that can shift from the subribosomal fraction, where they are not translated, to polysomes, where they are translated. What are the factors that either hold an mRNA molecule back in the subribosomal fraction or promote its recruitment into polysomes?

18.2 Mechanisms of Translational Control

Unlike transcriptional control, translational control is based on a wide variety of molecular mechanisms (Jackson and Wickens, 1997; Sachs et al., 1997; Wickens et al., 1997). Some of these are more suited to regulate the overall rate of protein synthesis, while others control—directly or indirectly—the relative translation rates of *specific* mRNAs.

CALCIUM IONS AND pH MAY REGULATE THE OVERALL RATE OF PROTEIN SYNTHESIS AT FERTILIZATION

Some translational control mechanisms do not involve the recognition of specific RNA sequences and are therefore nonselective with regard to the mRNAs they affect. For example, the *egg activation* process in many species entails an increase in the overall rate of protein synthesis (see Section 4.4). Two events that are necessary and sufficient for accelerating protein synthesis in sea urchin eggs are an increase in the concentration of free calcium ions (Ca^{2+}) and an elevation of the intracellular pH. These changes are thought to promote the shift of mRNAs from subribosomal RNPs to polysomes, and to accelerate elongation (Winkler, 1988). However, the molecular mechanisms linking these events are still being investigated.

PHOSPHORYLATION OF INITIATION FACTORS AND ASSOCIATED PROTEINS CONTROLS TRANSLATION

Regulating the activity of any of the translation initiation factors is an effective way of controlling protein synthesis. Because mRNAs may differ in their dependence on the same initiation factor, any change in initiation factor activity may affect the absolute and/or the relative rates of protein synthesis. The activity of translation initiation factors can be regulated by phosphorylation of specific amino acid residues, as is the case for many proteins. And just as many cellular signaling pathways involve kinases that phosphorylate transcription factors (see Fig. 2.28), other kinases regulate the activity of translation initiation factors.

A well-investigated example is the synthesis of hemoglobin, the most abundant protein in the red blood cells of vertebrates. The hemoglobin molecule consists of four polypeptides, two α-globins and two β-globins, each associated with a prosthetic (helper) group, heme, which mediates the respiratory exchange of oxygen and carbon dioxide. The heme molecule plays a key role in regulating the overall amount of hemoglobin per cell, and in establishing the correct heme:globin ratio. Heme regulates its own synthesis by means of a feedback loop that inhibits a key enzyme involved in heme synthesis. In addition, heme stimulates the translation of globin polypeptides from their mRNAs by inactivating a protein kinase that would otherwise phosphorylate eIF2 (Fig. 18.5; Fagard and London, 1981; Grace et al., 1984). Phosphorylated eIF2 is not active in translation initiations. Similar translational control mechanisms have been observed in other eukaryotic cells.

Phosphorylation of eIF4E, the cap-binding protein, seems to enhance its binding to the cap structure as well

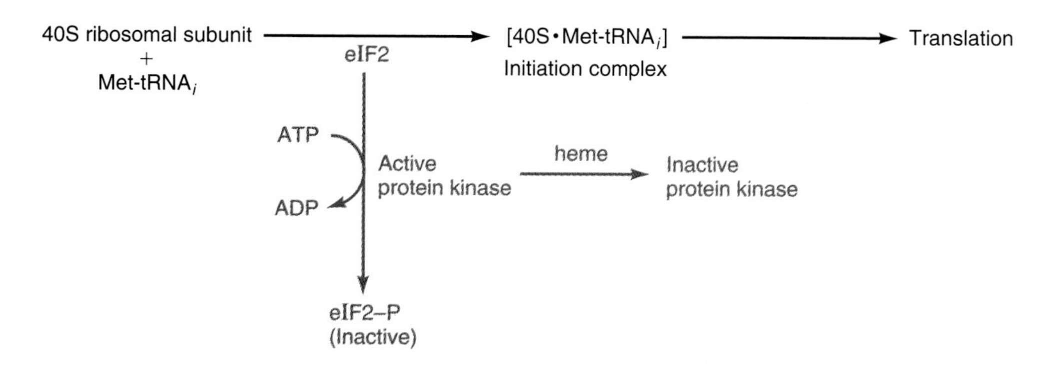

Figure 18.5 Translational control of globin synthesis by its prosthetic group, heme. Heme inhibits a protein kinase that would otherwise phosphorylate translation initiation factor eIF2. Phosphorylated eIF2 is not active as a translation initiation factor.

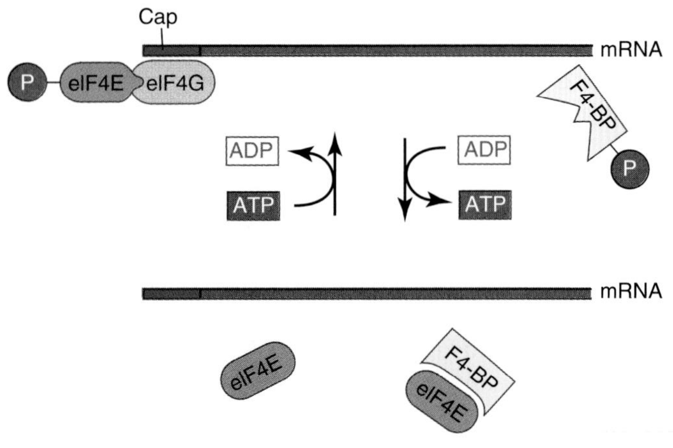

Figure 18.6 Translational control by phosphorylation of eukaryotic initiation factor eIF4E and its binding proteins (4E-BPs). Phosphorylation of eIF4E itself promotes its interactions with the cap structure and with another initiation factor, eIF4G. In addition, phosphorylation of 4E-BPs reduces their affinity for eIF4E, thus making more of it available for translation initiation.

as its interaction with eIF4G (Fig. 18.6). Certain viruses interfere with eIF4E phosphorylation as a strategy to reduce host cell mRNA translation. (The viral RNAs are less dependent on eIF4E because of special 5′ UTR sequences.)

In addition to *activity* of translation initiation factors, eukaryotic cells also control their *availability*. For example, availability of eIF4E is limited by so-called 4E-binding proteins (4E-BPs). In response to signals that promote cell growth, 4E-BPs are phosphorylated, which prevents their binding to eIF4E, thus making more eIF4E available for translation (Pause et al., 1994). Indeed, too much eIF4E can be deleterious: oversupplying the cap-binding protein by a transgene causes cultured cells to form tumors (Lazaris-Karatzas et al., 1990).

REGULATORY PROTEINS OR RNAs "MASK" CRITICAL SEQUENCES OF SPECIFIC mRNAs

As mentioned earlier, mRNA in eggs and other cells can be stored temporarily in the form of subribosomal RNP particles. There are two principal ways of accounting for this observation. First, the mRNAs stored in mRNP particles may be *untranslated by default*, simply because the cell's translational capacity is limited by a lack of initiation factors or other components required to recruit the mRNP particles into polysomes. Second, the mRNAs contained in mRNP particles may be *untranslatable due to association with proteins or other molecules* that actively inhibit their recruitment into polysomes. The latter hypothesis has come to be known as the ***masked mRNA hypothesis***.

According to the masked mRNA hypothesis, masking proteins should be present in cells that have major amounts of subribosomal RNP particles, such as oocytes, but not in adult tissues that do not have detectable amounts of subribosomal RNP particles. To test this prediction, Richter and Smith (1984) prepared subribosomal RNP particles from young *Xenopus* oocytes and isolated five major proteins from them. These proteins were mixed with naked oocyte mRNA to allow the reconstitution of oocyte mRNP particles (Fig. 18.7). These particles were injected into mature oocytes, which were used as "living test tubes" as discussed in Section 18.1. The re-

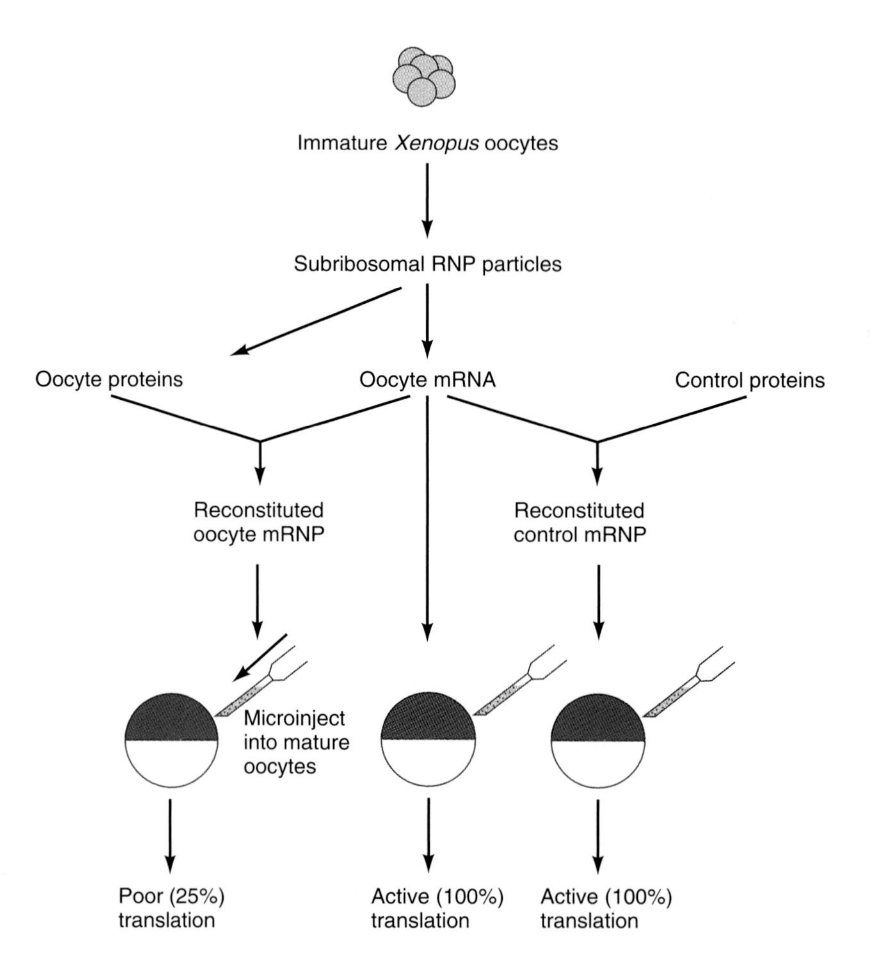

Figure 18.7 Demonstration that proteins inhibiting mRNA translation ("masking" proteins) are present in oocytes but not in adult tissues. Subribosomal RNP particles were prepared from immature *Xenopus* oocytes. From these RNP particles, five oocyte proteins were isolated. These proteins were combined with naked oocyte mRNA to reconstitute oocyte mRNP particles. These particles and two control preparations were injected into mature *Xenopus* oocytes, which served as "living test tubes" to measure the extra protein synthesis stimulated by the injectant. The reconsituted oocyte mRNPs translated poorly in comparison with two controls, namely naked oocyte mRNA and mRNP particles reconstituted from oocyte mRNA and proteins obtained from adult *Xenopus* tissues (liver, leg muscle, or heart).

constituted oocyte mRNP particles translated poorly as compared to naked oocyte mRNA, indicating that association with oocyte proteins inhibits the translation of oocyte mRNA. The reconstituted oocyte mRNP particles were also translated ineffectively when compared to the control mRNP particles reconstituted from oocyte mRNA and proteins from adult frog tissues. These results indicate that masking proteins inhibiting translation are present in frog oocytes but not in adult tissues.

Suitable locations for masking proteins to bind mRNA would seem to be in the 5′ UTRs, where they would impair binding of the ribosomal initiation complex to the mRNA. In accord with this expectation, the 5′ UTR regions of many mRNAs have stem-loop structures that are bound by regulatory proteins. The best-known example is found in the mRNA encoding *ferritin,* an intestinal cell protein that stores iron resorbed from food. The ferritin mRNA 5′ UTR has a stem loop known as the *iron response element* (*IRE*), which interacts with an *iron regulatory protein,* or *IRP* (Fig. 18.8). The activity of IRP depends on the concentration of iron ions (Fe^{2+}) in the blood serum. At high Fe^{2+} concentration, IRP is inactive. At low Fe^{2+} concentration, IRP binds to IRE, inhibiting translation of ferritin mRNA (Goossen et al., 1990).

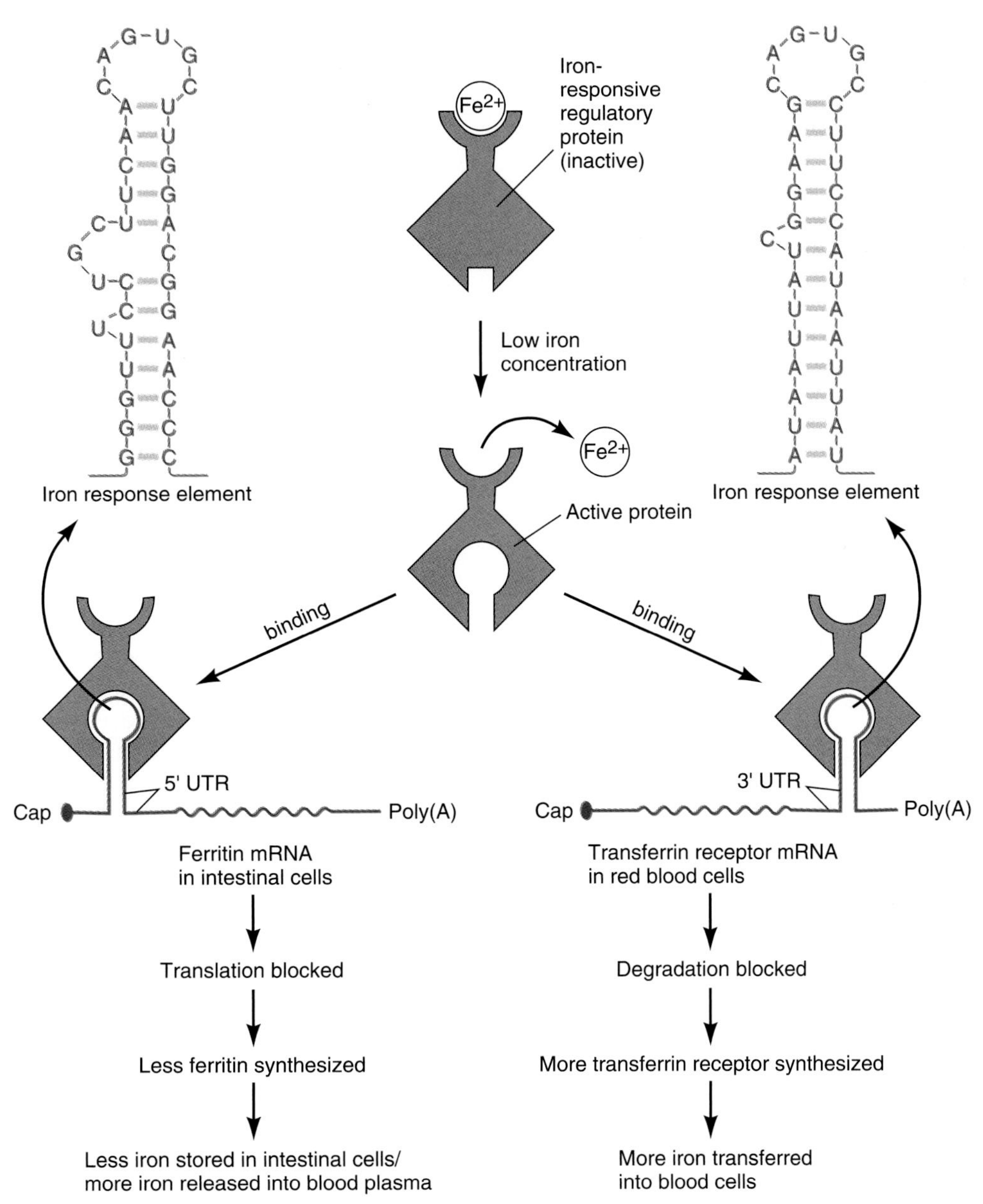

Figure 18.8 The translation of ferritin mRNA and the degradation of transferrin receptor mRNA are controlled by the same iron-responsive regulatory protein binding to nearly identical stem-loop structures called iron response elements, or IREs, in the two mRNAs. At low ionic iron concentration, the regulatory protein releases bound iron and becomes capable of binding to the IREs of both mRNAs. The IRE of ferritin mRNA is located in the 5′ UTR; binding of the iron-responsive protein inhibits translation of the mRNA. The IRE of transferrin receptor mRNA is located in the 3′ UTR; binding of the iron-responsive protein inhibits degradation of the mRNA. The overall physiological effect is an increase in the free iron concentration in the blood plasma and in the blood cells.

As an aside, the same IRP that binds to the IRE in the 5′ UTR of *ferritin* mRNA also binds to a similar IRE in the 3′ UTR of *transferrin receptor* mRNA. Transferrin is a blood serum protein that serves as a carrier for Fe^{2+}, and transferrin receptor on the surface of red blood cells mediates the uptake of transferrin with along with Fe^{2+} into red blood cells. The binding of IRP to the two IREs has synergistic effects on the concentration of iron in blood plasma and blood cells. At low iron concentration, the regulatory protein binds to both IREs. The resulting inhibition of ferritin mRNA translation means that intestinal epithelial cells synthesize less iron storage protein and therefore release more iron into blood serum. At the same time, red blood cells take up more iron by blocking the degradation of transferrin receptor mRNA.

Perhaps contrary to intuition, a binding site for a masking protein was found in the 3′ UTR region of the mRNA for fibroblast growth factor receptor, or FGFR, in *Xenopus* oocytes (Robbie et al., 1995). Fibroblast growth factor, or FGF, is an important embryonic signaling molecule thought to be involved in mesoderm induction and neural patterning (see Sections 9.7 and 12.5). Its receptor, FGFR, is translated from maternally supplied mRNA that is present but not translated in oocytes. However, translation begins upon egg activation. Premature translation is prevented by a *cis-acting* signal referred to as translation inhibition element (TIE). The TIE is about 180 nucleotides long and located immediately downstream of the translated region of FGFR mRNA. The TIE is necessary and sufficient for translation inhibition, as shown by deleting TIE from FGFR mRNA and by translational inhibition of engineered reporter mRNAs containing the TIE. The *transacting signals* binding to TIE include an oocyte protein of 43 kDa. The nature of this protein and how it inhibits translation remain to be elucidated.

In addition to proteins, complementary RNAs might be expected to function as "masks" that could render mRNAs unavailable for recruitment into polysomes. Indeed, the *lin-4⁺* gene of the roundworm *Caenorhabditis elegans* encodes two small inhibitory RNAs that bind specifically to the 3′ UTR region of the mRNA transcribed from another gene, *lin-14⁺* (see Section 25.3).

POLYADENYLATION AND DEADENYLATION CONTROL THE TRANSLATION OF SPECIFIC mRNAs

In the oocytes, eggs, and embryos of many species, there is a general correlation between the translational activity of an mRNA and the length of its poly(A) tail (Curtis et al., 1995a). In oocytes of the clam *Spisula solidissima*, Rosenthal and coworkers (1983) traced individual mRNAs for which cloned cDNAs were available. They found four mRNAs that were inactive in the oocyte but were recruited into polysomes after fertilization. All four of these mRNAs had very short poly(A) tails in the oocyte, which lengthened considerably after fertilization. Conversely, one mRNA was long-tailed and actively translated in the oocyte and became completely deadenylated and disappeared from polysomes after fertilization.

A general correlation between polyadenylation and translational activation is also found in *Xenopus* oocytes, although there are exceptions to this rule. The mRNAs synthesized during oogenesis can be grouped broadly into two classes (R. J. Jackson, 1993). One class is represented by actin and ribosomal protein mRNAs, which have long poly(A) tails and are actively translated during oogenesis. But shortly after the beginning of egg activation, their poly(A) tails are shortened, and their translation is reduced. Other mRNAs are stored in the oocyte as nontranslated mRNPs with short poly(A) tails; these mRNAs, including those for c-mos and cyclins, undergo polyadenylation and become translated either during egg activation or upon fertilization (Sheets et al., 1994).

Cytoplasmic polyadenylation, like the nuclear polyadenylation discussed in Section 17.1, requires the AAUAAA polydenylation signal. As an additional cis-acting signal, cytoplasmic polyadenylation requires uridine-rich sequences located close to AAUAAA and known as ***cytoplasmic polyadenylation elements (CPEs).*** Transacting signals for cytoplasmic polyadenylation include poly(A) polymerase, CPE-binding protein, and other proteins (Stebbins-Boaz et al., 1996).

TO examine whether polyadenylation is just correlated with translation or actually causing it, Sheets and coworkers (1995) experimentally manipulated the mRNA encoding the c-mos protein in *Xenopus*. This mRNA is synthesized during oocyte growth and translated during oocyte maturation, when it is necessary for *germinal vesicle breakdown* (*GVBD*) and other oocyte maturation steps to occur (see Section 3.5). Translation is associated with an elongation of the poly(A) tail as is the case for most mRNAs.

In order to test whether polyadenylation is *necessary* for c-mos mRNA translation and oocyte maturation, the investigators removed portions of the 3′ UTR region that contain the polyadenylation signals (AAUAAA and CPE). They accomplished these amputations by injecting fully grown oocytes with short antisense DNA oligonucleotides designed to hybridize with 3′ UTR sequences located upstream of the polyadenylation signals (Fig. 18.9). Endogenous RNaseH would then cut the double-stranded segment, thus removing the polyadenylation signals and any sequences downstream. The success of this chemical amputation, and the stability of the amputated c-mos mRNA, were verified by Northern blotting (see Method 15.4). To assay for oocyte maturation, the investigators scored the appearance, at the animal pole, of a white spot that signals GVBD. Most of the oocytes injected with either of the two antisense oligonucleotides did not mature. In contrast, most of the control oocytes injected with sense

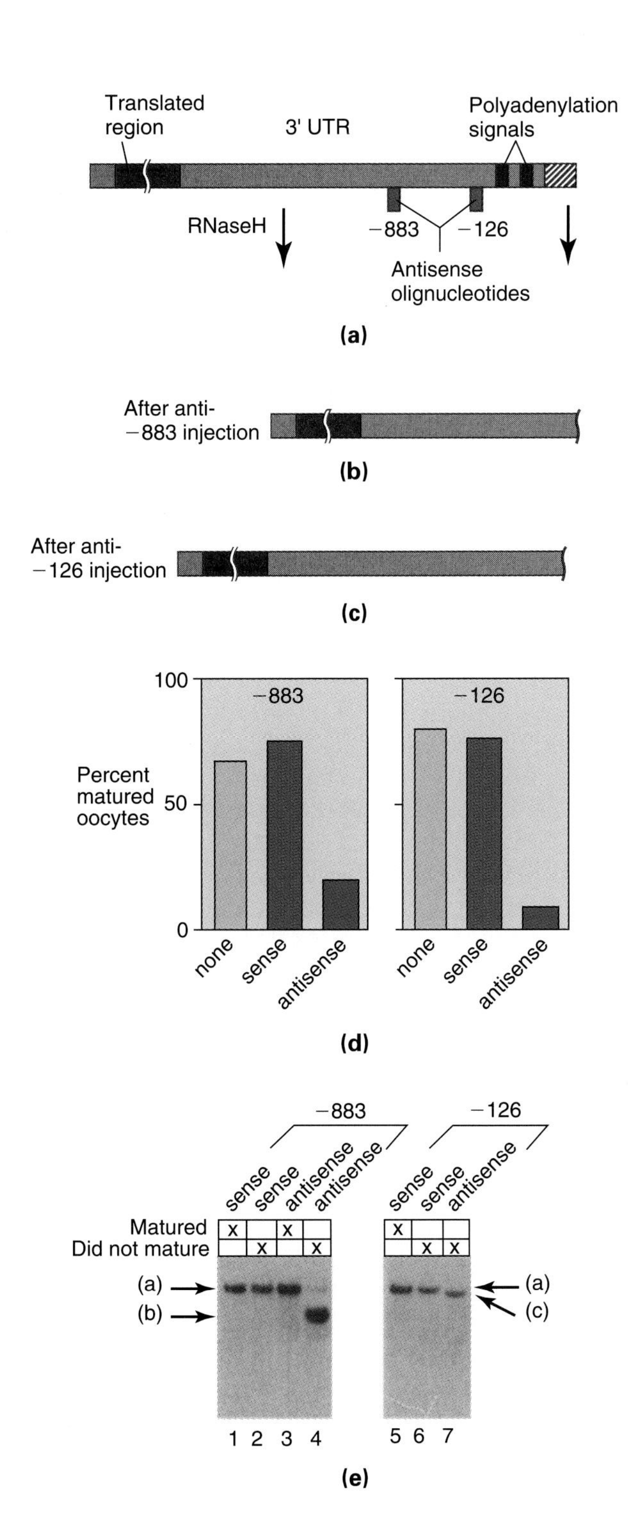

Figure 18.9 Inhibition of *Xenopus* oocyte maturation by removal of polyadenylation signals from the 3′ UTR of c-mos mRNA. **(a–c)** Experimental design. Fully grown oocytes were injected with short DNA antisense oligonucleotides hybridrizing to 3′ UTR regions located 126 or 883 nucleotides upstream of the poly(A) tail. These oligonucleotides were designated anti- −126 and anti- −883, respectively. The truncated mRNAs expected after RNaseH digestion of the hydrids are shown. **(d)** The percentage of oocytes that matured upon exposure to progesterone was dramatically reduced after injection with either antisense oligonucleotide, as compared to controls that were either uninjected or injected with sense oligonucleotides. **(e)** Northern blots (see Method 15.4) of oocyte mRNA probed with a labeled antisense RNA fragment hybridizing to c-mos mRNA upstream of the −883 site. In oocytes that did not mature after injection with antisense oligonucleotides, c-mos mRNA was shortened as predicted in parts b and c but was not significantly reduced in abundance.

oligonucleotides, which do not cause amputation of c-mos mRNA, did mature.

To see whether the failure of the oocytes injected with antisense oligonucleotides to mature was caused by the amputation of c-mos mRNA, rather than by some side effect of the antisense oligonucleotides, the investigators sought to reverse the amputation. To this end, they engineered a prosthetic 3′ UTR region that hybridized to the amputation site and contained the missing polyadenylation signals (Fig. 18.10). As expected, the prosthetic 3′ UTR rescued the oocytes' ability to mature. Together, these data indicate that removal of the 3′ end of c-mos mRNA, including the polyadenylation signals, prevents oocyte maturation. The most likely cause of this result is that the truncated c-mos mRNA is not translated into c-mos protein.

Questions

1. The few oocytes that matured in spite of injection with antisense oligonucleotide contained full-length c-mos mRNA (Fig. 18.9c, lane 3). Why is this result significant?
2. The investigators tested whether the polyadenylation elements near the 3′ end of c-mos mRNA can confer polyadenylation and translational activation to other mRNAs. In these experiments, they also showed that the polyadenylation signals of c-mos mRNA stimulate translational activation of the mRNA before GVBD. How do you think they accomplished these goals?

A dependence of translational activation on polyadenylation was also found by genetic analysis using *Drosophila* embryos (Sallés et al., 1994; Lieberfarb et al., 1996). Three mRNAs involved in embryonic axis formation, namely bicoid, Toll, and torso, are polyadenylated at the time of their translational activation. In embryos derived from females deficient for the genes *cortex* or *grauzone*, bicoid mRNAs have abnormally short poly(A) tails and fail to be translationally activated. These phenotypes can be rescued by injection of bicoid mRNA with a long poly(A) tail. Thus, *cortex* and *grauzone* seem to be part of a genetic network that controls polyadenylation as a means of activating bicoid mRNA.

The molecular mechanisms that trigger polyadenylation are still being explored. In *Xenopus* oocytes, the polyadenylation of c-mos mRNA occurs as part of the oocyte's maturation in response to progesterone (see Section 3.5). A critical link in the causal chain of events is the phosphorylation of a CPE-binding protein at a specific amino acid residue, serine 174 (Mendez et al., 2000).

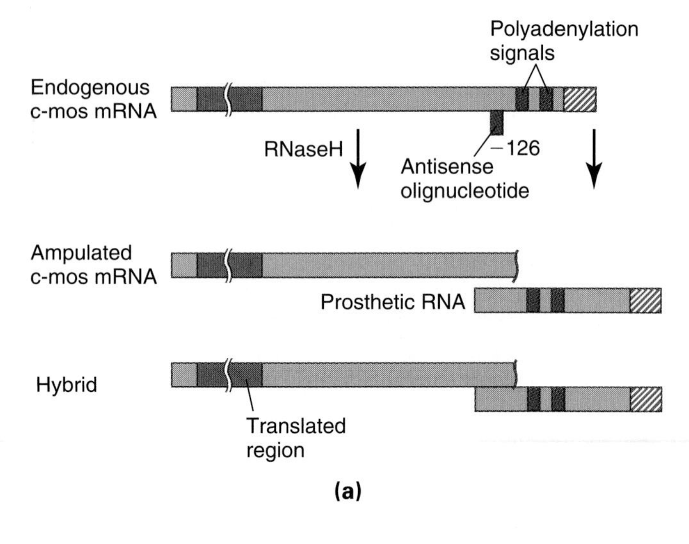

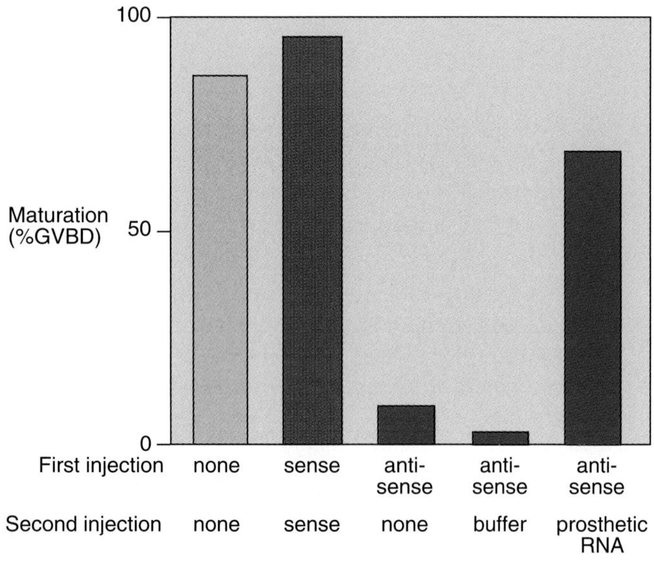

Figure 18.10 Rescue of *Xenopus* oocyte maturation by a prosthetic RNA segment containing polyadenylation signals. **(a)** Experimental strategy. The c-mos mRNA was amputated as explained under Figure 18.9. The 5′ end of the prosthetic RNA was designed to hybridize with a sequence of 54 complementary nucleotides at the 3′ end of the amputated mRNA. The 3′ end of the prosthetic RNA contained the polyadenylation signals (AAUAAA and CPE) of c-mos mRNA. **(b)** Results. Fully grown oocytes were first injected with the anti- −126 oligonucleotide, or with the corresponding sense oligonucleotide, or not at all. One hour later, the oocytes received a second injection of either buffer or prosthetic RNA. Following exposure to progesterone, the percentage of mature oocytes was tallied. The prosthetic RNA restored maturation to the majority of the oocytes previously injected with antisense oligonucleotides.

This step in turn may promote the recruitment of the enzymes that cleave pre-mRNA and polyadenylate the upstream cleavage product.

The molecular mechanisms that link polyadenylation with translational activation are also under investigation. As described in Section 17.4, the poly(A) tail of mRNAs is tightly associated with a *poly(A)-binding protein* (*PABP*), which protects mRNAs against premature degradation. The same protein seems also involved in translation initiation. At least in yeast, poly(A)-bound PABP interacts with the translation initiation factor eIF4G, and this interaction is thought to promote the recruitment of the *ribosomal initiation complex* to the mRNA (Tarun and Sachs, 1996). The interaction between PABP and eIF4G must join the 5′ UTR and 3′ UTR regions of an mRNA into a circle (Preiss and Hentze, 1999). Such an arrangement would seem to be adaptive in that cells would initiate the energy-consuming process of translation only for complete mRNAs. The same rationale—that cells circularize mRNAs as a test of their usefulness—would also apply to the observations that both intact cap structures and poly(A) tails tend to promote the export of mRNA from the nucleus and to protect mRNA from breakdown in the cytoplasm.

In summary, there is a wide variety of translational control mechanisms, many of which are also involved in regulating other steps of gene expression. General signals, such as the Ca^{2+} concentration, control the overall rate of protein synthesis, at least in sea urchins. The availability of initiation factors is another general control mechanism, but since mRNAs may differ in their dependence on a given initiation factor, its availability may affect some mRNAs more than others. Regulatory proteins binding to specific RNA sequences control the translation of individual mRNAs. Some of these proteins, binding to recognition sites in mRNA 5′ UTR regions, interfere directly with translation initiation. Other regulatory proteins bind to recognition sites in mRNA 3′ UTR regions and seem to affect translation via polyadenylation.

It is important to note that the translational control mechanisms described here may act in combination, in accord with the *principle of overlapping mechanisms.* For example, translational activation of a particular mRNA in mouse oocytes requires displacement of a masking protein from the 3′ UTR region, while at the same time elongation of the poly(A) tail is necessary to stabilize the mRNA (Stutz et al., 1998). Similarly, the activation of cyclin B1 mRNA during meiotic maturation in *Xenopus* requires both poly(A) tail elongation and the removal of a trans-acting repressor. Indeed, the binding site for the repressor overlaps the CPEs that are known to promote polyadenylation (Barkoff et al., 2000).

18.3 Translational Control in Oocytes, Eggs, and Embryos

Translational control is a common way of regulating gene expression in oocytes, eggs, and early embryos. As discussed in Section 3.5, large amounts of RNAs, including mRNAs, are deposited in growing oocytes. These RNAs are stored to support development during the phase of rapid cleavages, when most embryos do not transcribe enough RNA from their own genome. It is critical to conserve the maternal supplies during the time before fertilization. Indeed, all of the molecular mechanisms discussed in the previous section are utilized in keeping certain oocyte mRNAs in an untranslated state until embryonic development is imminent.

In this section, we will first discuss evidence that maternally encoded mRNA is translated during early embryogenesis. We will see that normal development and protein synthesis in early embryos do not require transcription of the embryonic genome. Next we will consider translational control as a *general* mechanism affecting the overall rate of protein synthesis in oocytes and eggs. Then we will examine data showing that translational control in eggs can be *sequence-specific*—that is, that it can inhibit the synthesis of specific proteins while allowing the synthesis of others. Finally, we examine cases where the translation of mRNAs is linked to their localization within the cell.

EARLY EMBRYOS USE mRNA SYNTHESIZED DURING OOGENESIS

Interspecies Hybrids The control of early embryonic development by maternally derived factors was first observed in embryos obtained by crossing different animal species. The resulting hybrid embryos were monitored for expression of maternal, paternal, or intermediate traits. Working with sea urchin hybrids, Driesch (1898) observed that the number of *primary mesenchyme cells* formed at the beginning of gastrulation was characteristic of the species that contributed the egg and was unaffected by the species contributing the sperm (Table 18.1). When reexamining the same hybrids at the pluteus stage, after the skeletal rods had formed, Driesch found characteristics of *both* parental species. These results showed that early embryonic traits were controlled only by maternal factors while traits expressed later were influenced by both parents.

table 18.1 Primary Mesenchyme Cells in Sea Urchin Hybrids

Egg		Sperm	Average Number of Primary Mesenchyme Cells*
Echinus	×	*Echinus*	55 ± 4
Spherechinus	×	*Spherechinus*	33 ± 4
Spherechinus	×	*Echinus*	35 ± 5
Strongylocentrotus	×	*Strongylocentrotus*	49 ± 3
Spherechinus	×	*Strongylocentrotus*	33 ± 3

* The mesenchyme cells of 15, 25, 47, 15, and 22 embryos were counted in the five samples, respectively. The average and range of the counts are given.

Source: Davidson (1976), after data of Driesch (1898). Used with permission.

Maternal Effect Mutants Control of early embryonic development by maternal gene products is also indicated by the effects of *maternal effect mutants*. By definition, the embryonic phenotype of such a mutant is affected only by the *maternal* genotype, and is completely independent of the paternal genotype. Examples include the gene controlling the orientation of mitotic spindles in snail embryos (see Section 5.3) and the *bicoid*$^{+}$ gene of *Drosophila* (see Sections 8.4 and 22.3). Genetic analysis of maternal effect mutants shows that the embryonic phenotype depends on *both of the two homologous alleles* of the gene present in the *mother*, not just on the *one* allele inherited by each *embryo*. For instance, all offspring of *bicoid/bicoid*$^{+}$ mothers develop normally, not just the 50% that inherit the wild-type allele. These observations indicate that maternal effect genes control embryonic phenotypes through maternally encoded RNAs and proteins, which are synthesized and stored during oogenesis.

Enucleated Eggs The presence of mRNA accumulated during oogenesis is also demonstrated by the development of egg fragments without nuclei. Such fragments can be obtained by centrifugation of sea urchin eggs in a sucrose solution (Fig. 18.11). When placed in hypotonic seawater to artificially activate them, they undergo several rounds of cytokinesis, forming an irregular blastula that hatches successfully (Harvey, 1940). Activated enucleated eggs synthesize proteins at a rate comparable with their nucleated counterparts (Denny and Tyler, 1964). These results indicate that the cytoplasm of sea urchin eggs contains maternally encoded mRNAs and other components for synthesizing proteins. These components can support early development without the presence of nuclear DNA.

Inhibition of mRNA Synthesis An alternative to physical enucleation is a biochemical experiment in which eggs are incubated with an inhibitor of transcription, such as actinomycin D. Using this strategy, Gross and Cousineau (1964) fertilized sea urchin eggs in water containing enough actinomycin D to shut down almost all RNA synthesis. Nevertheless, the eggs synthesized protein and developed morphologically like their untreated controls until the blastula stage. In contrast, similar experiments with inhibitors of protein synthesis caused immediate arrest of development. Thus, early embryos are absolutely dependent on protein synthesis, but many can do without RNA synthesis.

Midblastula Transition Revisited The point at which many embryos begin to replace maternal mRNA with their own occurs during the *midblastula transition* (*MBT*).

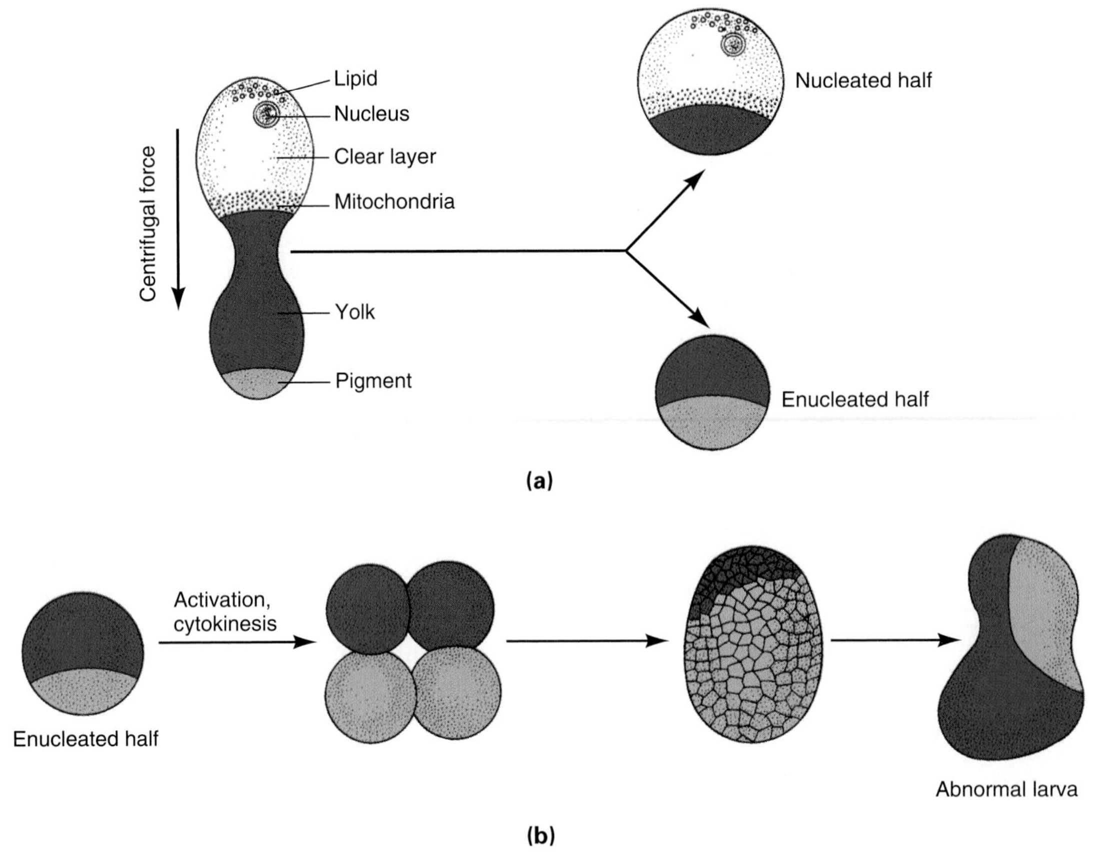

Figure 18.11 Development of enucleated egg halves. **(a)** When sea urchin eggs are centrifuged in sucrose solution, their components stratify under the influence of the centrifugal force. The half containing the light components of the cytoplasm is pulled to the top, while the other half, containing the heavy cytoplasmic components, is pulled to the bottom. The upper half retains the egg nucleus. **(b)** The enucleated lower half may nevertheless develop to stages resembling an abnormal blastula or larva.

Before MBT, embryos undergo very rapid mitoses, which leave little or no opportunity for RNA synthesis (see Section 5.4). Even if some RNA is synthesized, the amount produced by the few nuclei present in the embryo at this stage is generally insufficient to produce the proteins needed to support the relatively large masses of cytoplasm surrounding them. Thus, most or all of the proteins that are synthesized during this time relies on maternally supplied RNA. At MBT, when an embryo has produced many nuclei and mitosis has slowed down, the embryo can transcribe enough RNA from its own genomic DNA, and the maternal RNA supply is degraded.

SPECIFIC mRNAs SHIFT FROM SUBRIBOSOMAL mRNP PARTICLES TO POLYSOMES DURING OOCYTE MATURATION

Ovulation marks the transition from the ovarian to the embryonic environment, and egg activation means a shift from a quiescent state to rapid cleavage. The proteins required before and after these transitions are different, but they must all be translated from the same pool of maternal mRNAs. Given the limited storage capacity of an oocyte, and the high energetic costs of protein synthesis, it would seem adaptive if eggs could adjust both the overall rate of protein synthesis, and the kinds of proteins being made, to meet the requirements of different stages. The following studies focus on sequence-specific translational control in oocytes versus eggs and embryos.

THE synthesis of stage-specific proteins was directly observed in the surf clam *Spisula solidissima*. (In this species, oocytes are fertilized while still in meiotic prophase I. The events normally referred to as oocyte maturation, beginning with germinal vesicle breakdown, start 10 min after fertilization.) Rosenthal and coworkers (1980) compared proteins synthesized in unfertilized oocytes, in fertilized eggs during maturation, and in embryos after first cleavage. Each stage was incubated for 20 min with radioactive amino acids; this method labels proteins synthesized during the time that the radioactive precursor is present. Following the extraction and separation of all proteins by electrophoresis, the labeled proteins were identified by autoradiography (see Method 3.1). At least three proteins (designated X, Y, and Z) were synthesized prominently in

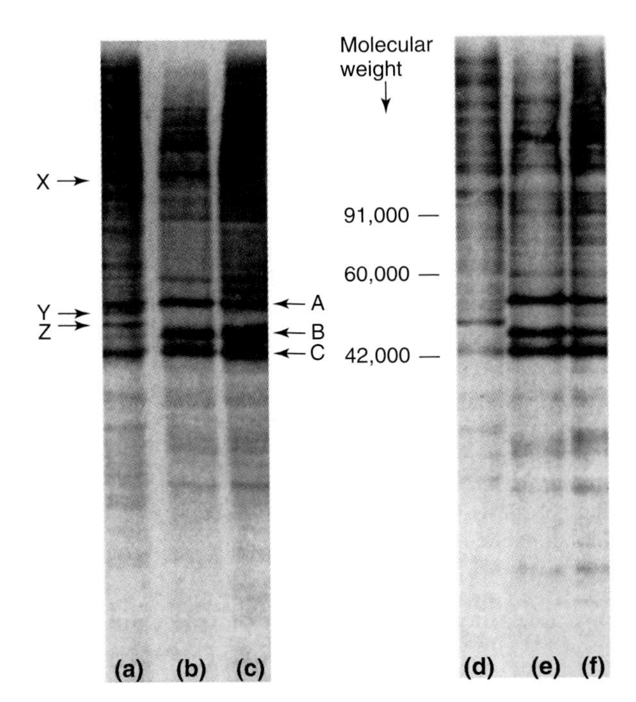

Figure 18.12 Sequence-specific translational control in the surf clam *Spisula solidissima.* The autoradiographs show proteins synthesized by oocytes, eggs, and embryos labeled for 20 min with [^{35}S]methionine (a–c) or [^{3}H]leucine (d–f). **(a, d)** Oocytes; **(b, e)** eggs labeled after germinal vesicle breakdown; **(c, f)** embryos labeled after first cleavage. The locations of molecular weight markers (91,000, 60,000, and 42,000 Daltons) are indicated. Proteins X, Y, and Z were synthesized specifically in oocytes but not in eggs and embryos. Conversely, proteins A, B, and C were synthesized in eggs and embryos but not in oocytes.

oocytes, but much less so in eggs or embryos (Fig. 18.12). Conversely, at least three other proteins (designated A, B, and C) were synthesized abundantly in eggs and embryos, but only sparsely in oocytes. Proteins A and B were later identified as *cyclins A* and *B,* two major proteins that control the cell cycle during cleavage. Protein C turned out to be the small subunit of ribonucleotide reductase, a major enzyme required for DNA synthesis during cleavage.

Thus, the pattern of proteins synthesized in *Spisula* eggs changed significantly after fertilization. In particular, there was a marked increase in the synthetic rate of some proteins needed specifically during cleavage. The investigators concluded that mRNAs must be recruited into polysomes in stage-specific ways.

To test this conclusion, the investigators prepared polysomal and subribosomal RNP fractions from *Spisula* oocytes, eggs, and embryos. From each preparation, RNA was extracted with phenol to remove associated proteins, and the naked RNA was translated in vitro. As expected, mRNAs encoding the proteins A, B, and C described previously were present in subribosomal RNP particles at the oocyte stage and in polysomes at the embryonic stage (Fig. 18.13). Moreover, the mRNA encoding protein X was present in oocyte polysomal fractions. Because of quantitative limitations, it was not clear whether the mRNAs encoding protein X were later degraded or stored in subribosomal mRNP particles.

Questions

1. Ascribing the synthesis of stage-specific proteins in *Spisula* as seen in Figure 18.12 to selective recruitment of mRNAs into polysomes implies that oocytes, eggs, and embryos contain identical inventories of mRNAs. What would be a direct way of testing this implication?
2. Do the results shown in Figures 18.12 and 18.13 show which of the translational control mechanisms discussed in Section 18.2 are involved in this case?

The experiments just described show that eggs can control the translation of stored mRNA by selective recruitment into polysomes. Similar shifts of mRNAs from subribosomal mRNP particles to polysomes during egg maturation have also been observed in other organisms. In a corresponding study on *Drosophila* oocytes and embryos, Mermod and coworkers (1980) focused on proteins that bind to DNA and are therefore presumed to have gene-regulatory functions. First, the researchers prepared polysomes and subribosomal mRNPs separately from oocytes and embryos. From each of the four fractions, they extracted RNA with phenol to remove any associated proteins. Next the naked mRNAs were translated in vitro, and the synthesized proteins were separated by two-dimensional gel electrophoresis (see Fig. 7.19). The results showed that at least three mRNAs encoding DNA-binding proteins shifted from subribosomal RNP particles to polysomes at some time between the oocyte stage and the embryo stage.

TRANSLATION OF SOME mRNAs DEPENDS ON THEIR CYTOPLASMIC LOCALIZATION

A particular class of maternally supplied mRNAs in oocytes and eggs are those that are localized to certain cytoplasmic regions and determine cell lineages or embryonic axes. For some of these mRNAs, localization and translational control are linked in intriguing ways. The transcripts of two *Drosophila* genes, *oskar*$^{+}$ and *nanos*$^{+}$, will serve as our examples.

Oskar mRNA is synthesized in ovarian nurse cells and transported across the oocyte to the posterior pole, where it forms a sharp localization (see Fig. 8.10). The protein translated from this mRNA is involved in specifying posterior embryonic structures and primordial germ cells. Throughout the localization process, oskar mRNA is translationally repressed, a mechanism that prevents the ectopic formation of germ cells and posterior structures. The *cis-acting signal* for this repression, located in the 3′ UTR region of oskar mRNA, has been named the *bruno response element* (*BRE*) because a *trans-acting factor* that binds to the BRE is a protein encoded by the *bruno*$^{+}$ gene (Kim-Ha et al., 1995). Both bruno

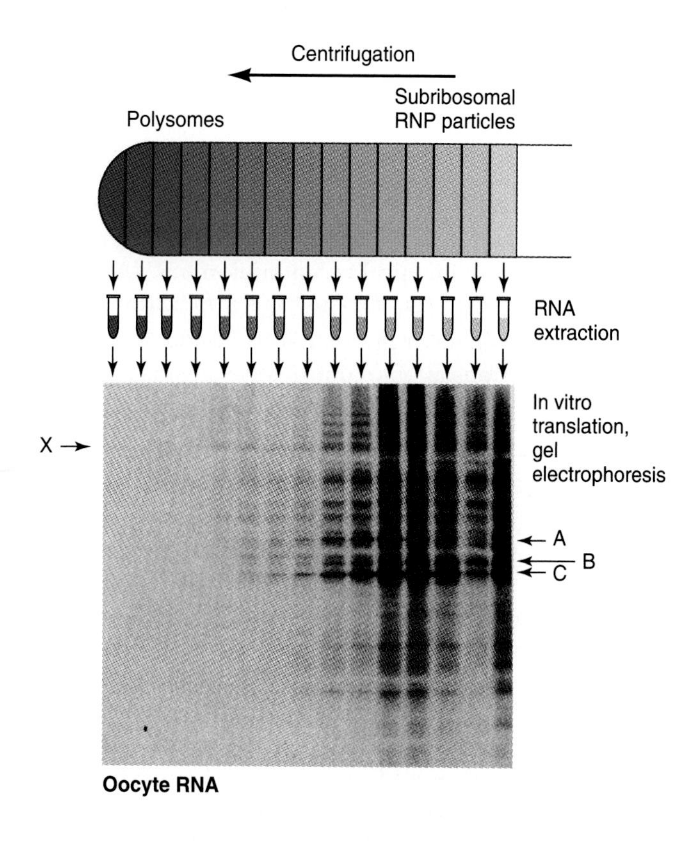

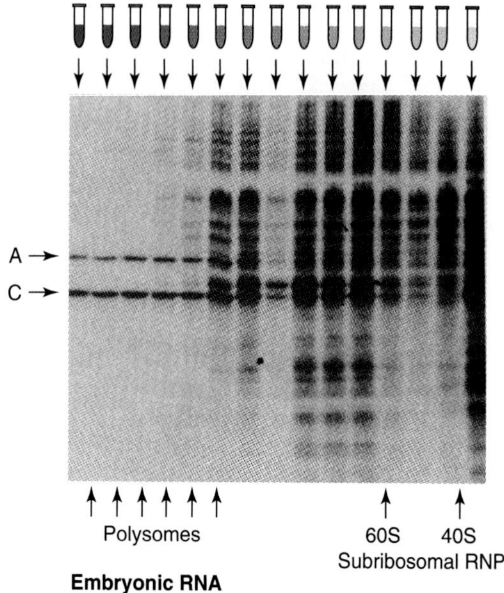

Figure 18.13 Selective shift of mRNAs from subribosomal RNP particles to polysomes in *Spisula*. Polysomes and subribosomal RNP particles were separated by gradient centrifugation. RNA was extracted from each gradient fraction and translated in vitro. The diagram on top shows the position of the gradient fraction from which the mRNA for each lane was collected. The photographs show newly synthesized proteins after gel electrophoresis and autoradiography. Proteins A, B, C, and X are the same as in Figure 18.12. The mRNAs encoding proteins A and C were present in subribosomal mRNP particles of oocytes and in polysomes of embryos. The mRNA encoding protein B underwent only a minor shift, and the causes for this different behavior were not clear. The mRNA encoding protein X was present in some of the oocyte polysomal fractions. However, the mRNA encoding this protein was not abundant enough to be detected at the embryo stage.

protein and the BRE are necessary for translational repression of oskar mRNA. Embryos derived from *bruno*$^-$ females synthesize oskar protein everywhere and form posterior structures ectopically. The same lethal effects occur in embryos that are derived from *oskar*$^-$ females and are supplied with a transgene encoding an oskar mRNA with defective BRE elements. BRE is also sufficient as a cis-element for translational inhibition; addition of BRE to heterologous mRNA renders them susceptible to translational repression in the ovary. Genetic and molecular analysis of *bruno*$^+$ indicates that it regulates multiple mRNAs involved in oogenesis and spermatogenesis (Webster et al., 1997).

Once oskar mRNA is properly localized at the posterior egg pole, its translational repression by bruno protein is somehow overcome. The molecular mechanism of this derepression is still being investigated. It may involve other gene products localized together with oskar mRNA in the electron-dense organelles at the posterior egg pole known as *polar granules* (see Section 8.3). Curiously, the derepression of oskar mRNA translation requires a cis-acting control element located in the 5′ UTR of oskar mRNA (Gunkel et al., 1998). This derepressor element interacts specifically with two proteins, one of which also binds to the BRE, the binding site of bruno protein in the 3′ UTR. These results again highlight the importance of interactions between proteins associated with the two ends of mRNA for translational activation.

A similar linkage of mRNA localization and translation has been found for nanos mRNA (Dahanukar and Wharton, 1996; Smibert et al., 1996; Bergsten and Gavis, 1999). Like oskar mRNA, nanos mRNA is part of the polar granules that assemble at the posterior pole of *Drosophila* oocytes and specify posterior embryonic structures. The nanos case is more extreme than that of oskar because the majority of nanos mRNA is never localized to the posterior pole but remains distributed throughout the cytoplasm. Thus, translational control of nanos mRNA is the major mechanism for generating a localized source of nanos protein in the embryo. Like oskar mRNA, nanos mRNA has a cis-acting *translational control element* (*TCE*) in its 3′ UTR region. However, the TCE overlaps with other 3′ UTR sequences that promote localization of nanos mRNA. Thus, translational derepression of nanos mRNA might occur not only by a polar granule component that antagonizes a repressor protein but also by simple competition of trans-acting repression and localization signals that have overlapping binding sites.

18.4 Translational Control during Spermiogenesis

Male gametes, after meiosis, undergo dramatic morphological changes, collectively called *spermiogenesis,* which prepare the developing sperm for their specific functions. These changes include formation of the acrosome, growth of the flagellum, and condensation of nuclear chromatin (see Section 3.4). These processes require carefully orchestrated synthesis of many proteins, some of them in large quantities. At the same time, meiosis and chromatin condensation impede RNA synthesis, thus excluding transcriptional control as an effective means of regulating gene expression. Therefore, spermiogenesis is another biological situation in which *sequence-specific translational control* would seem to be required.

PROTAMINE mRNA IS STORED IN SUBRIBOSOMAL RNP PARTICLES BEFORE TRANSLATION

Many studies on translational control during spermiogenesis in vertebrates have focused on the synthesis of *protamines,* abundant proteins that replace histones during chromatin condensation in many species. Early studies in this area, taking advantage of the abundant sperm produced by fish, have established that protamines are not synthesized until the *spermatid* stage, but that protamine mRNA is already transcribed at the *primary spermatocyte* stage and stored in subribosomal mRNP particles. The storage lasts between 15 and 30 days, during which time many other mRNAs are translated. It must be concluded that protamine mRNAs contain cis-acting signals that earmark them for storage.

Comparisons of protamine mRNAs from different vertebrates show a remarkable sequence similarity in the 3′ UTR region, although this region is not well conserved in the evolution of most other mRNAs. The unusual degree of sequence conservation of the protamine mRNA 3′ UTR region suggests that it has an important function. This has been confirmed by the experiments with transgenic mice described below.

TO study the role of the 3′ UTR region in the translation of mouse protamine 1 (mP1) mRNA, Braun and coworkers (1989) transformed mice with a matched pair of transgenes (Fig. 18.14). Both transgenes contained the regulatory region (promoter and enhancers) and the 5′ UTR sequence of an mP1 gene spliced to the translated region of a human reporter (hR) gene. The transgenes differed only in the sequences encoding the *3′ UTR* regions of their mRNAs: One encoded part of the mP1 3′ UTR sequence, while the other encoded the hR 3′ UTR region. *Northern blots* and *in situ hybridization* indicated that the transgenic mice transcribed both transgenes with virtually the same stage and tissue specificity as the resident mP1 gene (for the techniques, see Methods 8.1 and 15.4). In this respect, both transgene mRNAs could be considered valid substitutes for endogenous mP1 mRNA, which was first synthesized at the round-spermatid stage (day 25 after birth). At the elongating-spermatid stage (day 28), however, the

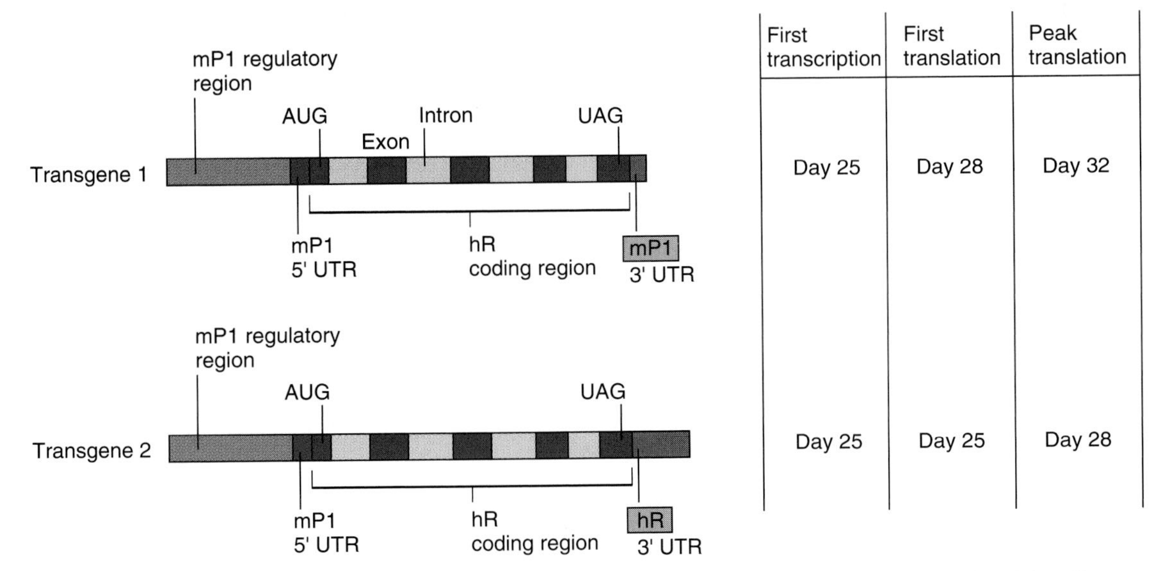

Figure 18.14 Role of the 3′ UTR region in the translational control of mouse protamine 1 (mP1) mRNA. The diagram shows a matching pair of transgenes used to genetically transform mice. Both transgenes contain the coding region of a human reporter (hR) gene, the 5′ UTR region of the mP1 gene, and 4.1 kb of 5′ regulatory region (promoter and enhancers) of the mP1 gene. The two transgenes differ in their 3′ UTR region: one 3′ UTR is from the mP1 gene and the other from the hR gene. Both transgenes are transcribed first on day 25 after birth. However, the mP1 3′ UTR causes a marked delay in mRNA translation. Protein synthesis from the mRNA with the mP1 3′ UTR begins on day 28 and peaks on day 32, whereas translation from the control mRNA with the hR 3′ UTR begins right after transcription on day 25 and peaks on day 28.

transgene mRNA with the hR 3′ UTR *disappeared,* while the transgene mRNA with the mP1 3′ UTR and the endogenous mP1 mRNA *persisted.* These results suggested that the mRNA with the hR 3′ UTR was less stable than the other two mRNAs, which had mP1 3′ UTRs.

The translation of endogenous mouse protamine mRNA as well as the two transgene mRNAs was monitored by *immunostaining* (see Method 4.1). In mice containing the transgene with the mP1 3′ UTR region, hR protein was first detected on day 28 (Fig. 18.14). This was the same time when endogenous mP1 protein was also first observed. However, in mice containing the transgene with the hR 3′ UTR region, hR protein was already present on day 25. Thus, although both transgenes were under the same transcriptional control, the mRNA with the mP1 3′ UTR was translated later than its counterpart with the hR 3′ UTR. The delayed translation had to be ascribed to the mP1 3′ UTR sequence. In essence, a short segment of mP1 3′ UTR sequence was sufficient to enhance the stability of a transgenic mRNA and to confer on it an mP-like translational regulation.

Questions

1. The first round of spermatogenesis in the mouse is synchronous. How does this fact facilitate collection of the data discussed above?
2. A transgene introduced into a diploid cell, such as a primordial germ cell, is usually integrated only into one chromosome of a homologous pair. Thus, after meiosis, the transgene will be present only in half of the gametes. Is this a complication for the experiments reviewed here?
3. How may the investigators proceed to delimit the exact sequence within the mP1 3′ UTR region that is sufficient for delaying mP1 mRNA translation?

Similar observations of delayed translation from stored mRNPs were also made for other proteins involved in spermiogenesis, including phosphoglycerate kinase (Gold et al., 1983) and acrosin (Nayernia et al., 1994).

The molecular mechanisms that control translation in developing sperm are still being investigated (Schäfer et al., 1995). Curiously, the translational activation of several sperm mRNAs is associated with shortening, rather than elongation, of the poly(A) tail (Kleene, 1989). The 3′ UTR region of protamine mRNAs seems to contain at least two cis-acting signals that are recognized by trans-acting proteins, one with inhibitory and the other with activating effects (Kwon and Hecht, 1993; Fajardo et al., 1997).

MESSENGER RNA MAY BE EARMARKED FOR STORAGE BY SEQUENCES IN ITS 5′ UTR OR 3′ UTR

In *Drosophila,* the most advanced stage of spermiogenesis at which RNA synthesis can be detected by *autoradiography* is the *primary spermatocyte* stage, which is right before the onset of meiosis (Gould-Somero and Holland, 1974). In contrast, the stage-specific synthesis of particular proteins can be monitored through advanced stages of spermiogenesis (Fig. 18.15; Schäfer et al., 1993). Genes regulated in this fashion include several members of the *Mst(3)CGP* family, which encode structural proteins in the sperm tail. Sequence comparisons among members of this family reveal a conserved sequence element within the mRNA 5′ UTR region, not the 3′ UTR region as in the case of mouse protamin mRNAs. The *translational control element* (*TCE*) in Mst(3)CPG mRNAs is 12 nucleotides long, has the consensus sequence ACATCNAAATTT, and is invariably located at position +28 to +39, counting from the transcription start site.

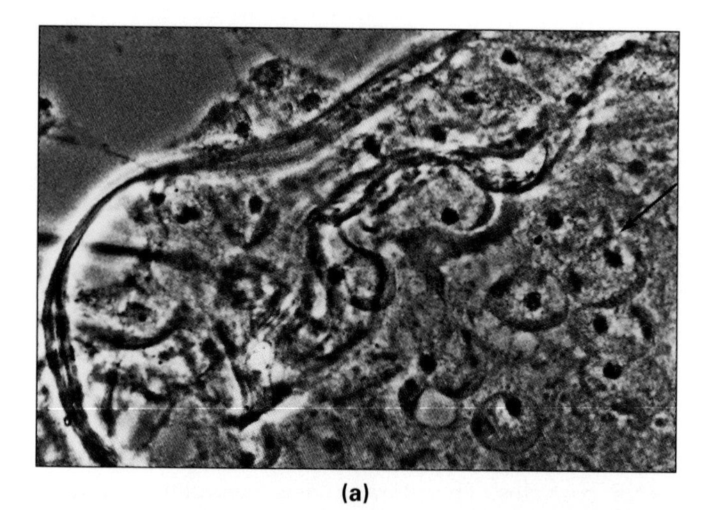

(a)

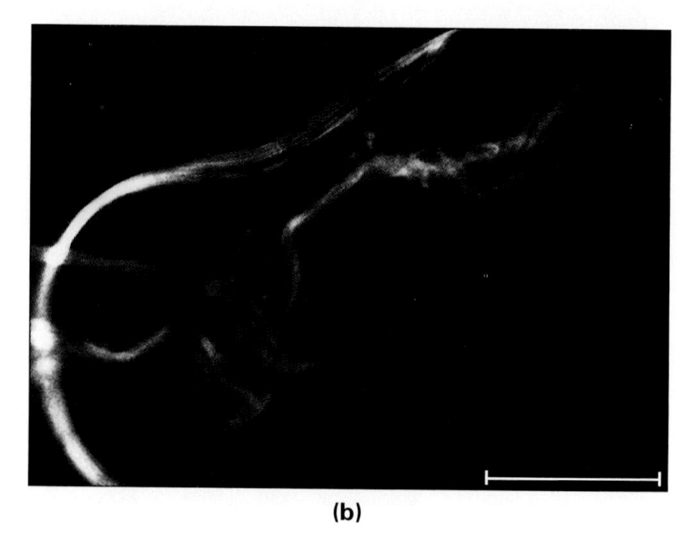

(b)

Figure 18.15 Translational control during spermiogenesis in *Drosophila.* **(a)** Photomicrograph (transmitted light, phase contrast) of a testis squash preparation showing spermatocytes (arrowhead) and elongated spermatids. **(b)** Photomicrograph (epifluorescence) of the same preparation after *immunostaining* (see Method 4.1) with a primary antibody against protein Mst98C, a structural component of the sperm tail. The protein is present in the elongated spermatids but not in the spermatocytes, although the mRNA encoding this protein is present at both stages. Scale bar: 50 μm.

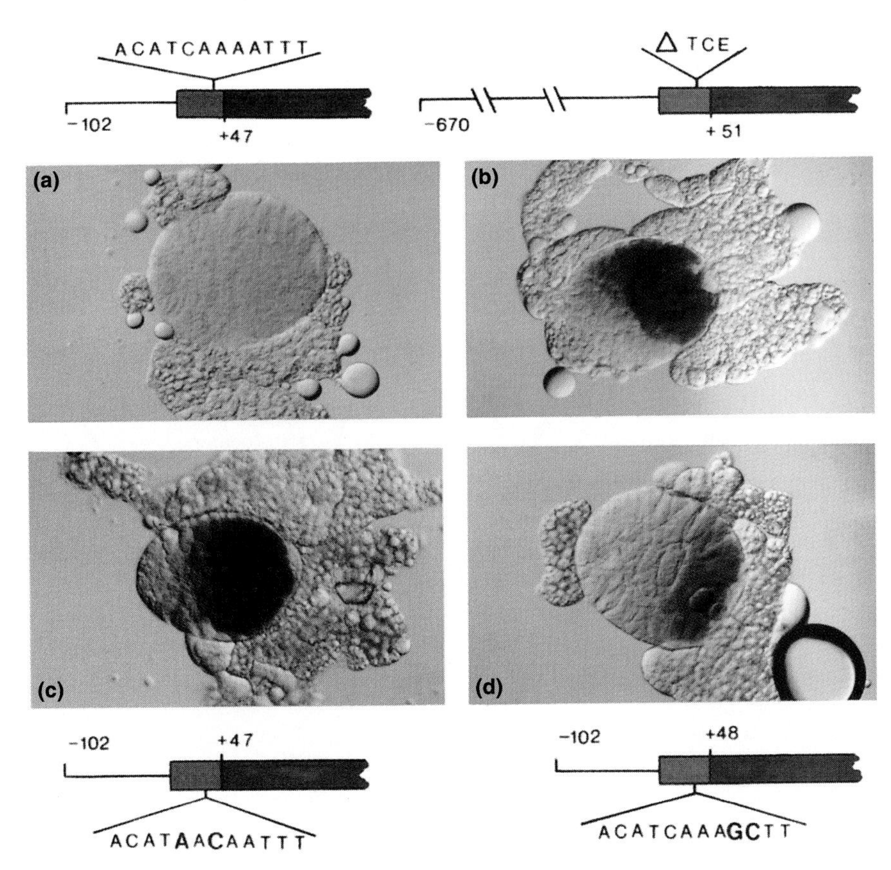

Figure 18.16 Role of a translational control element (TCE) in the expression of the *Mst87F*$^+$ gene, which encodes a structural protein in *Drosophila* sperm (similar to the Mst98C protein immunostained in Fig. 18.15). Intense blue color in the photographs indicates expression of the fusion genes diagrammed above each photo. Each transgene consists of a regulatory segment from the *Mst87F*$^+$ gene (transcribed sequences shown in red) and the coding region of the *lacZ* reporter gene (purple). The regulatory segment encompass nucleotides −102 through approximately +47, which includes the TCE. **(a)** The normal TCE (ACATCAAAATTT) inhibits the synthesis of lacZ protein. **(b)** Deletion of TCE permits the synthesis of lacZ protein, even if a larger portion of the transcriptional regulatory region of *Mst98C* is included in the transgene. **(c, d)** Point mutations in TCE abolish or weaken the translational inhibition of lacZ protein synthesis.

The involvement of the TCE in translational control was demonstrated in transgenic flies with reporter genes composed of a regulatory region of the *Mst87F* gene and the coding region of the bacterial *lacZ* gene. In *Drosophila,* development and spermatogenesis are correlated in such a way that only premeiotic germ cells are present in the gonads of third instar larvae. Therefore, properly controlled Mst87F mRNA is not translated in the testis of a third instar larva. This repression is still mediated by a gene fragment spanning from −102 to +47, which includes in particular the TCE (Fig. 18.16a). However, TCE deletion or mutation abolishes or weakens the inhibition of transgene expression (Fig. 18.16b–d). Thus, TCE is sufficient for translational inhibition of mRNAs in premeiotic male germ cells.

The molecular mechanism by which translation of the Mst(3)CPG mRNAs is regulated is still being investigated. It appears that different mechanisms have evolved to regulate mRNA translation during spermiogenesis, given that cis-acting signals have been found in the 3′ UTR region as well as in the 5′ UTR region.

18.5 Posttranslational Polypeptide Modifications

After translation, many polypeptides undergo further processing before they become functional proteins. First, polypeptides are directed to one of several cellular destinations. Second, many polypeptides are cleaved or receive certain chemical substitutions before attaining their biologically active form. Finally, polypeptides are degraded in a controlled fashion. These processing steps are regulated, at least in part, by certain amino acids, known as *signal sequences,* that are part of the polypeptide itself.

POLYPEPTIDES ARE DIRECTED TO DIFFERENT CELLULAR DESTINATIONS

The synthesis of all polypeptides begins in the ***cytosol***—that is, the cytoplasmic matrix that fills the spaces between the membranous organelles of a cell. Some polypeptides, including RNA-binding proteins,

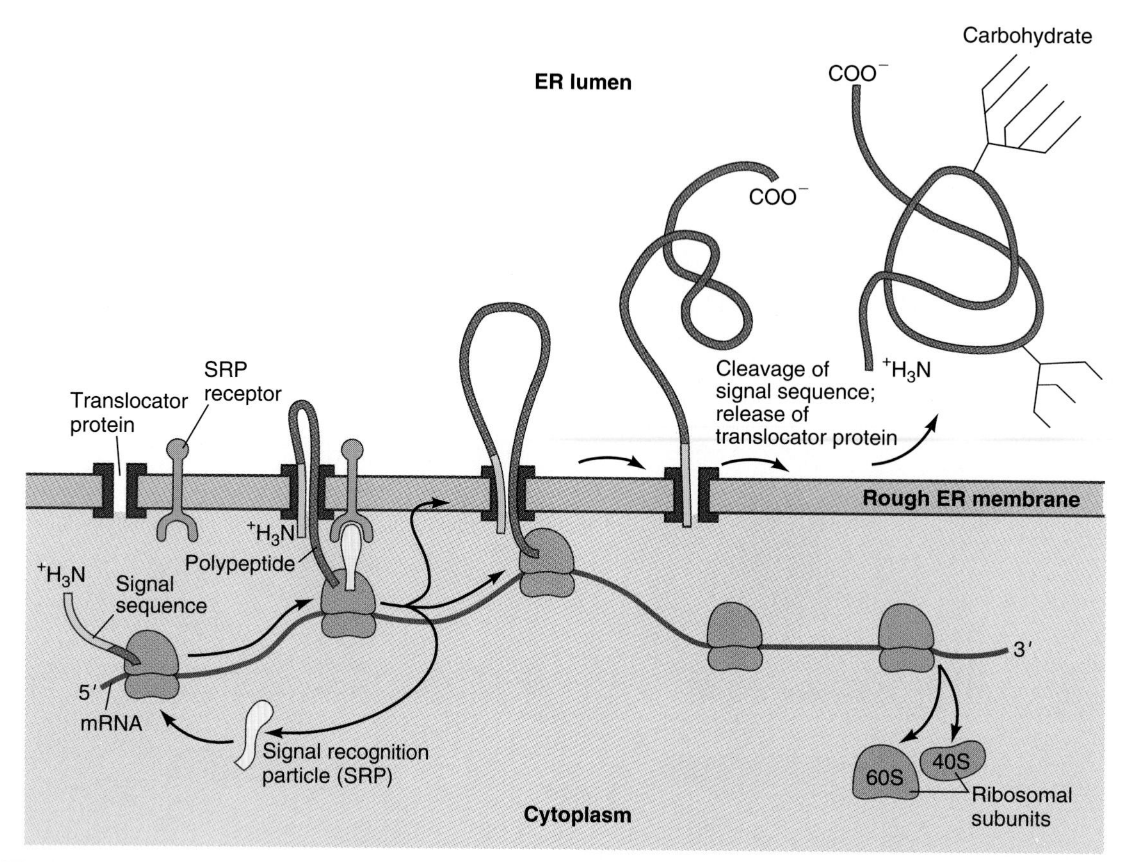

Figure 18.17 Model for the synthesis of a secreted polypeptide on the endoplasmic reticulum (ER). Polypeptides to be released into the ER have a signal sequence of about 30 amino acids at their N-terminus. When the signal sequence emerges from a ribosome, it combines with a signal recognition particle (SRP), which docks at an SRP receptor protein on the outside of the ER. The signal sequence inserts itself into a translocator protein located next to the SRP receptor in the ER membrane. As translation continues, the polypeptide is threaded into the ER lumen while the SRP particle is recycled. When the polypeptide's C-terminus has been released into the ER lumen, the signal sequence is cleaved off and degraded, and the ribosomal subunits are released and recycled. Once inside the ER, the polypeptide receives carbohydrates and other substitutions.

cytoskeletal proteins, and many enzymes, stay in the cytosol. Other polypeptides are synthesized with signal regions that target them to other cellular compartments.

A large class of polypeptides enter the *endoplasmic reticulum* (*ER*) while they are still being synthesized. The ER is a closed system of folded membranes connected with the nuclear envelope; the cytoplasmic face of much of the ER membrane is studded with polysomes (Fig. 18.1). From the ER, many polypeptides are given off in *transport vesicles* that are directed by a system of signal and docking proteins to the *Golgi apparatus*, a stack of closed membranous sacs. Inside the Golgi apparatus, polypeptides are modified, and many are then released in secretory vesicles. These vesicles are similarly targeted to the plasma membrane—where they release their contents via exocytosis—or to other cellular organelles.

Some of the targeting signals that regulate the protein traffic in cells take the form of a ***signal sequence,*** typically a continuous stretch of 15 to 60 amino acids located at the N-terminus, the end of the polypeptide that is synthesized first. Other targeting signals occur as *signal patches*, three-dimensional arrangements of amino acids that may be located at different positions within the polypeptide but come together when polypeptides fold into their active conformation. The targeting signals interact with different types of receptors that mediate the transport of the polypeptide across a membrane into another cellular compartment.

Polypeptides destined for the ER have a signal sequence of about 30 amino acids at their N-terminus. This sequence becomes bound to a ***signal recognition particle*** (***SRP***), a small cytoplasmatic RNP particle (Fig. 18.17). The SRP stops further translation until it docks at an ***SRP receptor,*** a protein in the ER membrane. The SRP receptor is associated with a ***translocator protein*** in the ER membrane, with which the signal sequence becomes aligned so that the polypeptide is threaded into the ER lumen. Translation now continues until the polypeptide's C-terminus has been released into the ER lumen. Then the signal sequence is clipped off and degraded, so that the amino terminus is no longer tethered to the translocator protein. Most polypeptides targeted to the

ER enter it completely. However, some polypeptides have a *stop transfer sequence* that keeps them embedded in the ER membrane; these polypeptides become plasma membrane proteins when ER vesicles bud off and fuse with the cell plasma membrane.

POLYPEPTIDES MAY UNDERGO SEVERAL POSTTRANSLATIONAL MODIFICATIONS

Many newly synthesized polypeptides are inactive until they have undergone further modifications, such as formation of disulfide bridges, limited proteolysis, folding of the linear polypeptide into a three-dimensional configuration, and addition of carbohydrates or other substitutions. Some of these modifications are necessary to stabilize the three-dimensional shape of the polypeptide. Others contribute to the formation of signal patches that direct the polypeptide further along its pathway in the cell. Still other modifications are required to form the active sites for the polypeptide's biological function. We will briefly survey these modifications and then consider, as an example, how they contribute to the synthesis of collagen.

Disulfide Bridges Disulfide bridges between two cysteine residues are critical to the stability of many proteins (Fig. 18.18). Disulfide bridges are generated enzymatically in the ER, which provides a more oxidizing environment than the cytosol. Hence, these bridges are commonly found in membrane proteins and in proteins that are secreted from the cell.

Limited Proteolysis Many polypeptides undergo enzymatic cleavages between specific amino acids, a process called ***limited proteolysis.*** This is known for several hormones and enzymes, which are synthesized as inactive prohormones or proenzymes. For instance, the hormone insulin is synthesized in pancreatic cells as a long polypeptide called *preproinsulin* (Fig. 18.18). Inside the ER, the signal sequence is removed and three disulfide bridges are formed. These disulfide bridges stabilize the shape of the molecule, which is now called *proinsulin.* A central domain is then removed in two proteolytic steps while the two lateral domains are held together by disulfide bridges. The resulting insulin is now in its biologically active form.

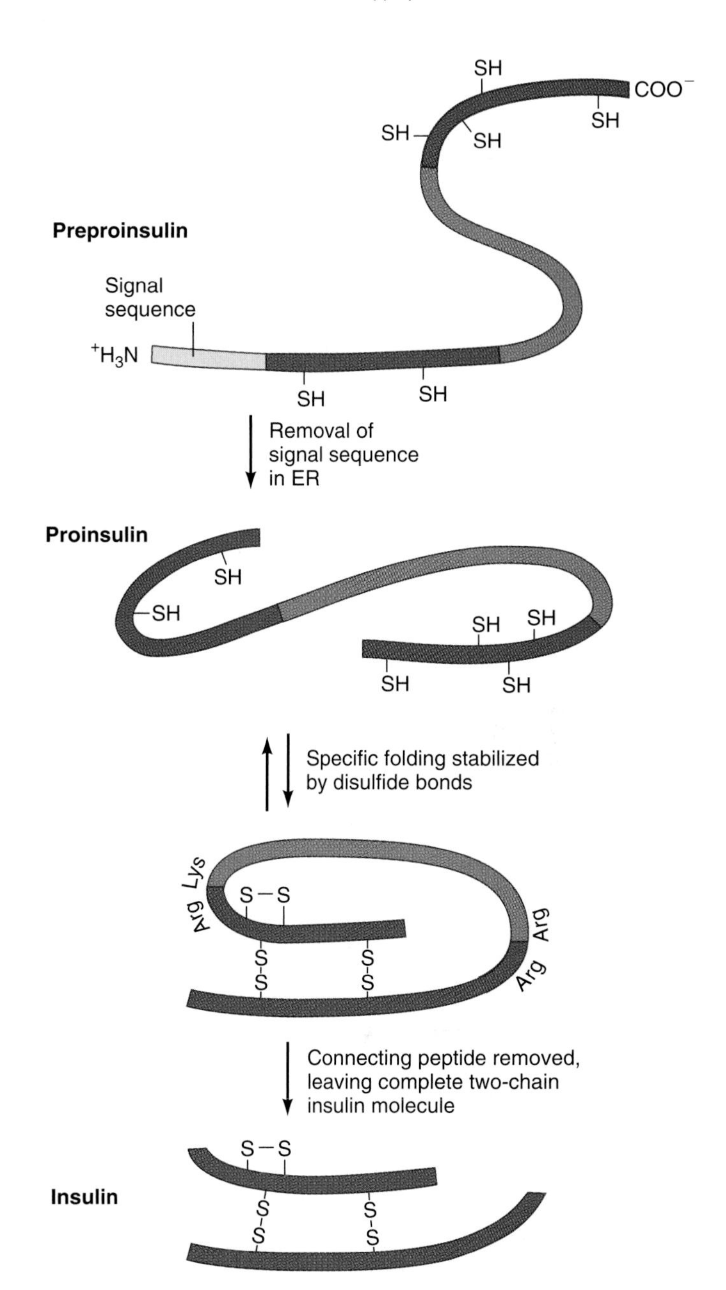

Figure 18.18 The hormone insulin is synthesized as a much larger precursor peptide, preproinsulin. Removal of the signal sequence in the ER generates proinsulin. The proinsulin folds into a specific shape that is determined by the position of the cysteine residues forming disulfide bridges. A central portion of the peptide is then removed by proteases recognizing the flanking arginine and lysine residues. Now the remaining insulin has acquired its biologically active form.

Folding All biologically active polypeptides have a characteristic conformation, which can be described at four levels (Fig. 18.19). The ***primary structure*** of a polypeptide is its linear sequence of amino acids; it is determined directly by the genetic code.

The term ***secondary structure*** refers to the folding of certain polypeptide sections into regular patterns such as α helices and β pleated sheets. These patterns are stabilized by hydrogen bonds between C=O and N–H groups. In most polypeptides, α helices and β pleated sheets are intermixed with a less regular pattern known as a *random coil.*

The ***tertiary structure*** of a polypeptide is formed by the folding of the helices, sheets, and coils into a three-dimensional shape. This shape is often stabilized by disulfide bridges and hydrophobic interactions. The secondary and tertiary structures dictate which amino acids are located at the surface of the polypeptide and therefore determine its chemical properties.

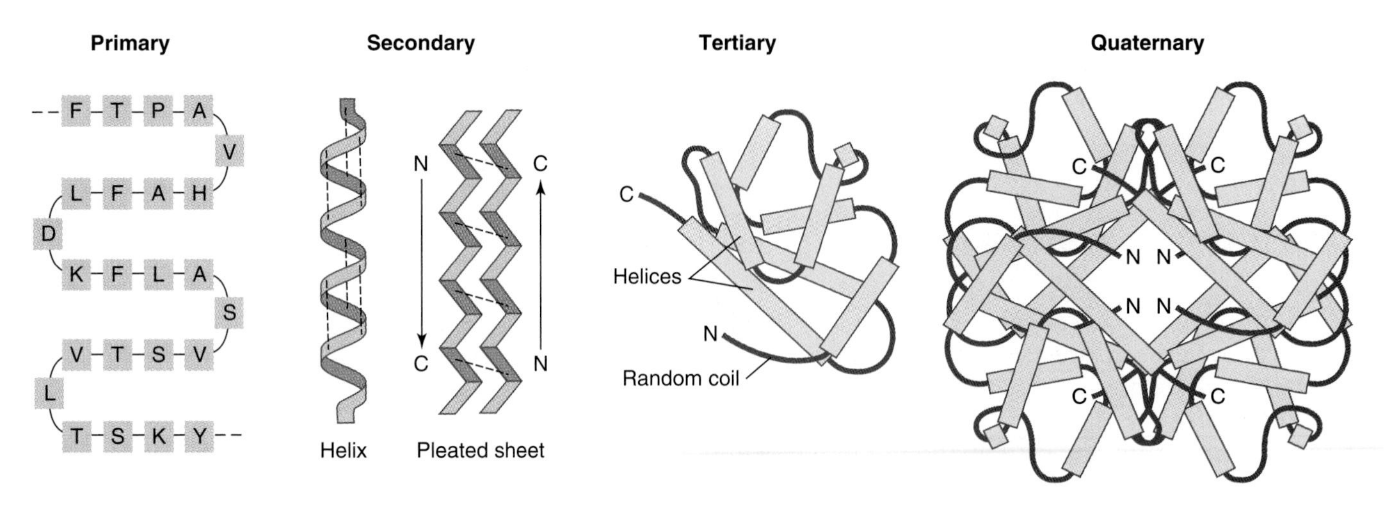

Figure 18.19 Primary, secondary, tertiary, and quaternary structure of proteins. Common secondary structures of proteins are the α helix and the β pleated sheet. The latter consists of two antiparallel peptide domains. Both secondary structures are stabilized by hydrogen bonds (hatched lines) between C=O and N–H groups of amino acid residues that are several positions away. The tertiary structure shown here consists of several α helices and random coil segments arranged in a specific three-dimensional conformation. The quaternary structure shown is made up of four polypeptides.

Finally, a protein may consist of two or more polypeptides arranged in a ***quaternary structure.*** The formation of tertiary and quaternary structures may bring into close proximity amino acids that are far apart according to the primary structure alone. Such grouping of amino acids may form active sites of a particular shape suitable for interacting with complementary sites on other molecules.

Many polypeptides seem to have just one secondary and one tertiary structure, which are dictated by the primary structure and formed under normal conditions in cells. In particular, some sequences of amino acids are conducive to α helix formation whereas other sequences favor the formation of β pleated sheets. In addition, the positions of cysteine residues limit the disulfide bridges that can be formed. However, some polypeptides exist in multiple conformations that can be converted into one another via unfolding and refolding (Fox et al., 1986).

The ability of some polypeptides to fold in more than one way seems to have prompted the evolution of a class of proteins referred to as ***molecular chaperones.*** They associate with newly translated polypeptides in ways that inhibit improper folding while translation is still going on, and direct the appropriate folding of the polypeptide once it is released from the ribosome (Frydman et al., 1994). Molecular chaperones include the class of *heat shock* proteins, which are synthesized by all cells in the presence of heat or under other stressful conditions. They seem to rescue unfolded polypeptides by restoring them to their active conformations and to hasten the degradation of polypeptides that are denatured beyond repair (Craig, 1993).

Oligomeric Proteins Many proteins consist of multiple *subunits* and are therefore called *oligomers* (Gk. *oligo-*, "few"; *meros*, "part"). Each subunit typically is a polypeptide encoded by a separate gene. The configuration of an oligomeric protein is also known as its *quaternary structure,* as discussed previously. For example, the quaternary structure shown in Figure 18.19 is that of adult hemoglobin, which is composed of four polypeptides, two α-globins and two β-globins. For another example, many transcription factors consist of two polypeptides (see Section 16.4 and Fig. 16.13).

The formation of oligomeric proteins from two or more polypeptides occurs spontaneously under the conditions prevailing in cells. The subunits are attracted to each other by hydrogen bonding, electric charges, and hydrophobic interactions. Some associations between subunits are later stabilized by disulfide bonds. Other oligomeric associations are reversed as part of an activation process. For instance, the binding of a steroid hormone to its receptor is associated with the release of an inhibitory protein from the receptor (see Fig. 16.13).

Phosphorylation, Glycosylation, and Other Substitutions The biological functions of many polypeptides in the cytosol are regulated by *protein kinases,* which add phosphate groups to certain amino acids, and by *phosphatases,* which remove phosphate groups. The resulting changes in the tertiary structure of the polypeptide increase or reduce its biological activity. Phosphorylation is reversible and serves to regulate the activity of enzymes, transcription factors, eukaryotic initiation factors, and many other proteins. Other reversible polypeptide modifications are methylation and acetylation. Polypeptides synthesized in the ER and passed through the Golgi apparatus may undergo further enzymatic modifications. Some enzymes add carbohydrates to certain amino acids, a process called *glycosylation.* Other enzymes add hydroxyl groups, a step known as *hydroxylation.* Still others link fatty acids to polypeptides, thus targeting them for anchorage in membranes.

Collagen Synthesis To illustrate the modification steps discussed above, let us consider the formation of collagen, the most abundant structural protein in vertebrates. After synthesis inside *fibroblasts,* collagen undergoes *exocytosis* to become part of the *extracellular matrix* (*ECM*). There are at least 12 types of collagen, each optimized for a different function. While some collagens provide strength to tendons, bones, and dermis, other collagens play an important role in cell-ECM interactions (see Sections 11.3, 11.5, and 14.3). The following description fits type I collagen, which forms the fibrous components of tendons and similar tissues.

As a future component of the extracellular matrix, the *preprocollagen* polypeptide is synthesized on the rough ER (Fig. 18.20). After the signal peptide has been clipped off, the remaining procollagen polypeptide is hydroxylated by different enzymes recognizing proline and lysine residues in specific contexts of other amino acids. Next, hydroxylysine residues are glycosylated by addition of galactose. In the next step, three procollagen polypeptides are aligned and disulfide bonds form among them. The linked polypeptides then form a triple helix over most of their length, resulting in a rod-shaped *procollagen protein* held together by disulfide bridges and hydrogen bonds. Only the ends of the procollagen protein remain globular. At this point, the protein is transported to the Golgi apparatus, where further glycosylation steps occur. After passage through secretory vesicles and exocytosis, the globular ends are clipped off. The remaining triple helical protein, now called a *collagen protein,* is about 300 nm long.

The collagen proteins spontaneously form staggered lateral associations called *collagen fibrils* (Fig. 18.20, part 12). The staggered association creates bands of aligned gaps, which stain more intensely in electron micrographs than the interbands (Fig. 18.21). Type I collagen fibrils average about 50 nm in diameter and can be up to several micrometers in length. The stability of these fibrils is greatly enhanced by the formation of additional disulfide bridges between adjacent collagen proteins. The fibroblasts that synthesize collagen also control the density and orientation in which the fibrils assemble.

PROTEIN DEGRADATION IS DIFFERENTIALLY CONTROLLED

Like most molecules in an organism, proteins are subject to ***turnover.*** This means that they are broken down and resynthesized at short intervals relative to the organism's life span. As discussed previously in the context of mRNA longevity, both the synthesis and breakdown of a gene product are effective means of controlling gene expression. In mammalian cells, the *half-lives* of proteins range from minutes to weeks. Proteins with short half-lives include cyclins, which are broken down within a few minutes after each M phase in the cell cycle (Murray, 1995). Likewise, most regulatory proteins have short half-lives, a property that makes it possible to limit their activity to short periods of time. A protein with a short half-life, if translated from a localized mRNA, will also form a concentration gradient within a cell, as in the case of bicoid (see Fig. 16.8).

Protein degradation is carried out by cellular enzymes called *proteases.* Many proteins are earmarked for degradation by the addition of a small protein, *ubiquitin.* The ubiquitin additions, carried out enzymatically and in a sequence-specific way, target proteins to *proteasomes,* particles consisting of multiple proteins that specialize in the rapid destruction of ubiquitinated proteins (Jentsch and Schlenker, 1995).

The selectivity of the ubiquitin addition process depends in part on signals provided by the structure of the protein to be broken down. For instance, incomplete or misfolded polypeptides are degraded rapidly. Another signal for rapid degradation seems to be provided by amino acid clusters that are rich in proline, glutamate, serine, and threonine. Such clusters are common in polypeptides with half-lives of less than 2 h but rare in long-lived polypeptides (Rogers et al., 1986).

In summary, cells are able to degrade proteins quickly and selectively in response to appropriate signals. The initiation of the breakdown process depends on the presence of certain amino acids in the polypeptide, much as the selective degradation of mRNAs depends on certain nucleotide sequences in their 5′ UTR or 3′ UTR regions.

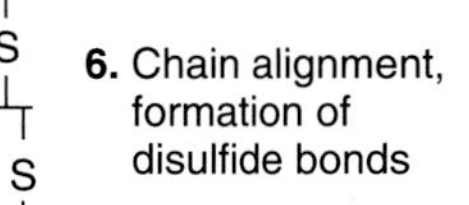

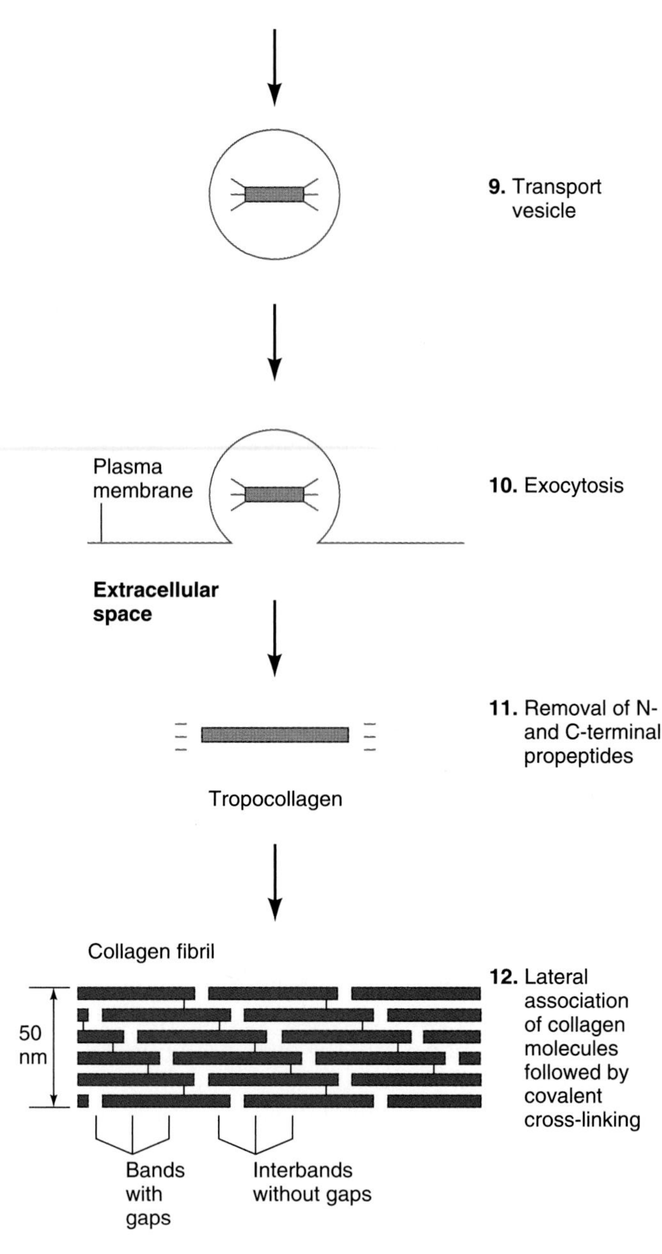

Figure 18.20 Summary of the major steps in collagen biosynthesis. Modifications of the collagen polypeptide include hydroxylation, glycosylation, disulfide bond formation, and limited proteolysis. These steps occur in a precise sequence in the rough ER, the Golgi complex, and the extracellular space. The modifications allow formation of a stable triple helix and also the staggered lateral association and covalent cross-linking of helices into fibrils of about 50 nm diameter. The staggered association of the collagen molecules lines up the gaps between consecutive molecules in transverse bands. Because these bands stain differently from the interbands, which have no gaps, fibrous collagen has a striated appearance in electron micrographs (see Fig. 18.21).

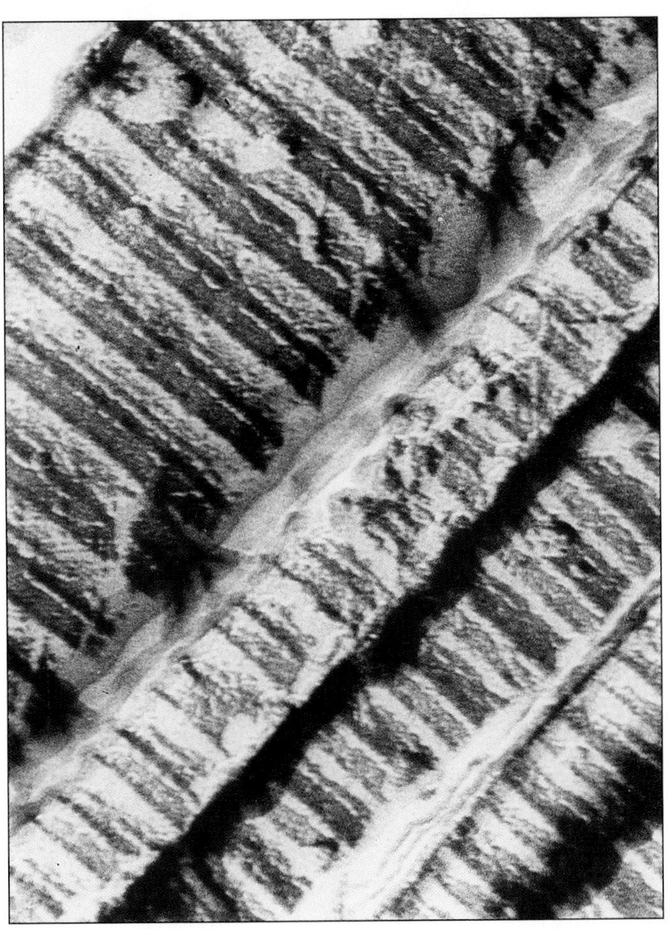

Figure 18.21 Electron micrograph of collagen fibrils from a tendon. The transverse banding reflects the fact that collagen molecules assemble in a staggered pattern but with the gaps between consecutive molecules at distinct levels (see Fig. 18.20).

SUMMARY

During translation, the nucleotide sequence of mRNA is used to order and join amino acids into a polypeptide. The initiation phase of translation, which involves several proteins called eukaryotic initiation factors, is most important for translational control. In cells, most mRNAs released from the nucleus are immediately recruited into polysomes and translated into polypeptides. However, certain cell types can store mRNA temporarily in the form of messenger ribonucleoprotein (mRNP) particles.

There are a wide variety of translational control mechanisms, including several that are sequence-specific. Some of these act through polyadenylation and deadenylation of mRNAs, with cis-acting signals located in the 3′ UTR region. Other sequence-specific controls involve the masking of critical mRNA regions by bound proteins or short regulatory RNAs.

Translational control is a prominent way of regulating gene expression in oocytes, eggs, and embryos. Many of the mRNAs stored in the oocyte are left untranslated until embryonic development begins. Oocyte maturation or fertilization are typically followed by an overall increase in protein synthesis. Also, the synthesis of some oocyte-specific proteins decreases while the synthesis of some cleavage-specific proteins increases. These changes, which occur in the absence of transcription, demonstrate that the translation of maternally supplied mRNA is controllable in eggs in a sequence-specific way. Similarly, the translation of mRNAs encoding protamine and other sperm-specific proteins is delayed selectively during spermiogenesis.

After translation, many polypeptides undergo further processing before they become functional proteins. The presence of specific amino acid sequences directs a polypeptide to the cytosol or into the endoplasmic reticulum (ER), and further into various cellular compartments. Many polypeptides are cleaved by limited proteolysis and form disulfide bridges before they assume their active three-dimensional configuration. Several cytosolic polypeptides are phosphorylated to modulate their activity. Polypeptides synthesized in the ER are often glycosylated by the addition of carbohydrates. These processing steps, and finally the controlled degradation of polypeptides, rely on signals provided by certain amino acid sequences or other structural aspects of the polypeptide.

Genetic and Paragenetic Information

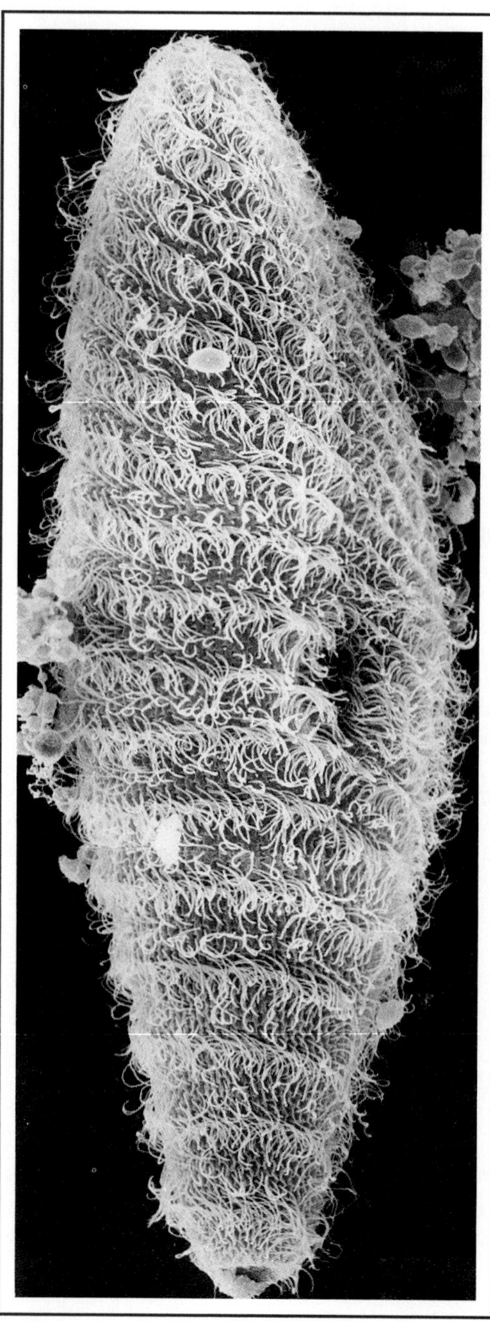

Figure 19.1 Scanning electron micrograph of *Paramecium,* a ciliate protozoon (anterior at the top, ventral to the right). This unicellular protist is about 170 μm long and 50 μm wide. It is covered with thousands of cilia, which propel it through the water by coordinated strokes. Modified cilia also drive food particles into an oral apparatus, the opening of which is visible on the ventral surface. *Paramecium* reproduces by cell division, and the pattern of the mouth, cilia, and other organelles is duplicated before each division. Studies on *Paramecium* and other ciliates have played an essential role in establishing the concept of paragenetic information.

IN the previous chapters, we have traced the expression of genetic information from genes to proteins. Yet, we are still far from understanding how a single cell is made. How do polypeptides reach their target sites within a cell? We have learned that signal sequences in polypeptides interact with signal recognition particles (SRPs) and that matching receptors guide the loaded SRPs to their destinations. But how are the receptors laid out so that they will guide the cellular protein traffic properly? This question has been difficult to answer, because no one has ever assembled a living cell in a test tube. No matter how carefully one breaks a cell down into its component molecules, the molecules will not reassemble into a cell. The only way for a cell to be generated is by division of a progenitor cell. This fundamental observation was summarized succinctly in Latin by the German pathologist Rudolf Virchow: *Omnis cellula e cellula* ("every cell comes from a cell"). Coined in 1858, this aphorism still holds true.

19.1 The Principle of Genetic and Paragenetic Information

Virchow's aphorism is restated, somewhat less concisely, in the *principle of cellular continuity* (see Section 2.1): All cells of currently living organisms are the temporary ends of *uninterrupted* cell lineages extending back through their ancestors' germ lines to primordial cells billions of years ago. Two types of information have been passed on through these lineages. One is *genetic information*, which is encoded in RNA or DNA. Equally important is the unbroken chain of *structural organization* that is passed on directly, without being encoded in genes, from each cell to its daughter cells. Because this organization is inherited in parallel to the genetic information, it is called ***paragenetic information*** (Gk. *para*, "near," "beside," "beyond"). Genetic and paragenetic information complement each other functionally but are inherited in different ways.

According to the ***principle of genetic and paragenetic information***, cells pass on both types of information to their daughter cells. As a result, organisms pass on both types of information to the next generation, and have done so throughout the course of evolution. Whereas the last four chapters dealt with the role of genes in development, this chapter will address the nature and inheritance of paragenetic information.

To define the role of paragenetic information, we will examine different modes of molecular assembly in cells. In the simplest types, *self-assembly* and *aided assembly*, molecules snap together without paragenetic information. By describing specific examples, we will explore the role of conformational changes and auxiliary proteins in these processes. Of particular interest are *seed* structures, which form nucleation centers for the assembly process. In self-assembly and aided assembly, seeds are not absolutely necessary, but they greatly accelerate the initiation phase.

Directed assembly is the most prevalent mode of molecular assembly in cells. Here, paragenetic information is required in the form of preexisting seed structures, which not only start the assembly process, but may also determine the type of product to be assembled.

Directed assembly occurs at the molecular as well as the cellular level. Preexisting proteins may propagate their three-dimensional structure by associating with newly synthesized proteins. The importance of this phenomenon has been recognized by the award of the 1997

Nobel prize in Physiology or Medicine to Stanley B. Prusiner for his pioneering work on prions, proteins that exist in normal, functional configurations and in pathogenic forms that cause dreadful diseases of the central nervous system.

At the cellular level, alternative seed structures direct the assembly of different products from the same pool of building blocks. Eukaryotic cells, for example, use the same tubulin molecules to assemble different arrays of microtubules in the cytoplasm as well as in the *axonemes* of cilia. Which of these microtubular arrays are assembled depends on the available seeds, or *organizing centers*. By controlling the replication and placement of these organizing centers, and by modifying their molecular composition, cells direct the assembly of various microtubular arrays according to cell type and phase of the cell cycle. The role of paragenetic information in directed assembly is most evident in ciliated protozoa (Fig. 19.1). The outer membrane and associated structures of these animals embody much paragenetic information. This becomes apparent when occasional irregularities in surface structures are propagated during cell division, even though the genetic information is completely unaltered.

In addition to *local* directed assembly, ciliates have independent mechanisms of *global patterning*. These utilize a type of paragenetic information that preserves the global features of the cell pattern—including the number and orientation of major organelles—during cell division and *encystment*. Like ciliated protozoa, metazoan cells also pass on features of global organization during cell division.

19.2 Self-Assembly

Self-assembly is the simplest way in which molecules combine and form larger structures: the molecules just snap together spontaneously. All the energy and information needed to form the final structure are contributed by its constituent molecules. No other molecules are required as scaffolds, catalysts, or energy sources. Only water and small ions must be present to provide a solution of suitable pH and ionic strength. Self-assembly processes are energetically favorable—that is, they release energy. Some examples of self-assembly have already been discussed in previous chapters. Hemoglobin self-assembles from two α-globin polypeptides, two β-globin polypeptides, and four heme molecules; even larger structures, such as collagen fibrils, self-assemble from collagen proteins. A simple in vitro test can be used for classifying an assembly process as self-assembly: If a structure can arise from its component molecules in a simple aqueous solution, it is self-assembling. Thus, self-assembly does not require paragenetic information.

SELF-ASSEMBLY IS UNDER TIGHT GENETIC CONTROL

Self-assembly can generate entire cell organelles. This process has been investigated in detail for bacterial ribosomes (Nomura, 1973). Like eukaryotic ribosomes, bacterial ribosomes consist of a large and a small subunit. Each subunit can be broken down into ribosomal RNA and proteins (Fig. 19.2). When these components are recombined in appropriate salt solutions in a test tube, they spontaneously reconstitute a functional ribosomal subunit. The same self-assembly process observed in vitro may also occur during the normal formation of ribosomal subunits in vivo.

The self-assembly of ribosomes is under tight genetic control. If ribosomal rRNA from one bacterial species is mixed with ribosomal proteins from a *closely related* species, functional particles still form. However, components from *distantly related* bacteria do not self-assemble. Apparently, ribosome self-assembly requires a close fit between certain RNA sequences and some protein domains, and the quality of the fit decreases as mutations accumulate in the course of evolution.

Stringent genetic control is also indicated by the effects of some point mutations that cause abnormal self-assembly. A dramatic example is the heritable human

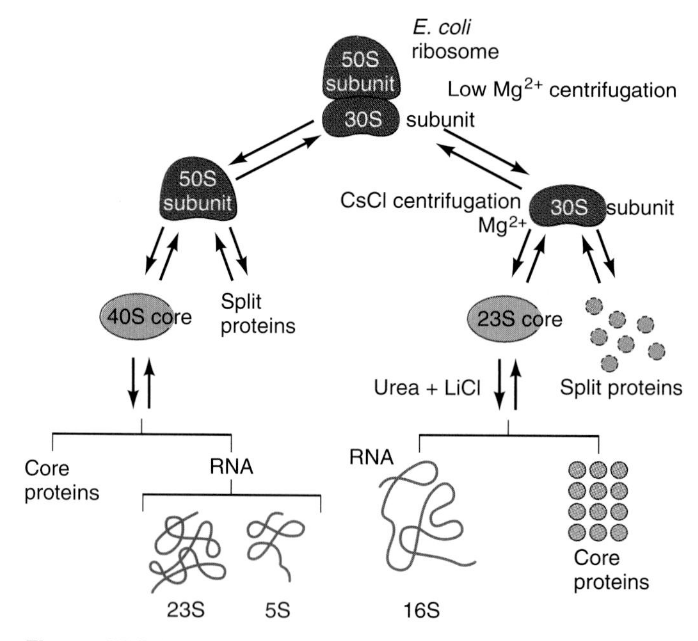

Figure 19.2 Dissociation and self-assembly of bacterial ribosomes. The ribosome of *Escherichia coli* consists of a large (50S) and a small (30S) subunit, which can be dissociated by centrifugation at low magnesium concentration. The small subunit is dissociated, by centrifugation in cesium chloride, into a 23S core and 7 split proteins. Finally, the core is dissociated by denaturing agents into 16S rRNA and 12 core proteins. Under suitable conditions, each step is reversible and complete ribosomes will self-assemble from their single components. Similar dissociation and reassembly steps are found with the large ribosomal subunit.

disease known as *sickle cell anemia.* It is caused by a single base pair change that results in the substitution of valine for glutamine in position 6 of the β-globin polypeptide. The mutant β-globin assumes a different *tertiary structure,* which in turn causes the mutant hemoglobin molecules to attach to each other to form long rods. These rods distort red blood cells, forcing them into an abnormal sickle shape (Fig. 19.3). The sickle cells are less elastic than normal red cells and therefore tend to clog small blood vessels, causing painful, life-threatening crises. Thus, a small change in the genetic information can affect not only the structure of a polypeptide but also the self-assembly properties of an oligomeric protein.

THE INITIATION OF SELF-ASSEMBLY IS ACCELERATED BY SEED STRUCTURES

Some products of self-assembly, such as microtubules and microfilaments, consist of many repeating units (see Section 2.3). The assembly of such polymers usually starts from ***seed structures*** that provide nucleation centers for polymerization. The importance of seed structures has been documented especially well in the self-assembly of a simple virus.

The *tobacco mosaic virus* (*TMV*) infects tobacco plants, causing their leaves to become mottled and wrinkled. This virus is rod-shaped, with a coat of proteins surrounding an RNA core in a helical pattern (Fig. 19.4). Each rod consists of one 6400-nucleotide RNA molecule and 2130 identical polypeptides that form the cylindrical coat. TMV can easily be dissociated into its components. The resulting mixture of RNA and proteins will spontaneously reassemble into a complete virus, and this reassembly will proceed in vitro under proper conditions of ionic strength and pH. The biological activity of self-assembled TMV can easily be tested by spreading it on tobacco plant leaves: Depending on assembly conditions, at least some of the TMV particles are infective.

Studies of TMV have provided much insight into the self-assembly process (*P. J. Butler and A. Klug, 1978*). In one of the earliest experiments, when TMV protein and RNA were mixed together in vitro, it was several hours before maximum yields of infective virus were reached (Fig. 19.5). Throughout the self-assembly process, the investigators took samples of the reaction mixture and analyzed them under the electron microscope. Initially, they saw only single coat proteins and aggregates of several coat proteins. Only after a few hours did they find a more advanced intermediate product in the reaction mixture: circular discs consisting of two layers of coat proteins. When such discs were purified and added to a fresh mixture of TMV RNA and protein, the results were dramatic: Complete virus particles formed within minutes. This outcome suggests that incomplete discs are unstable, and that therefore the formation of a

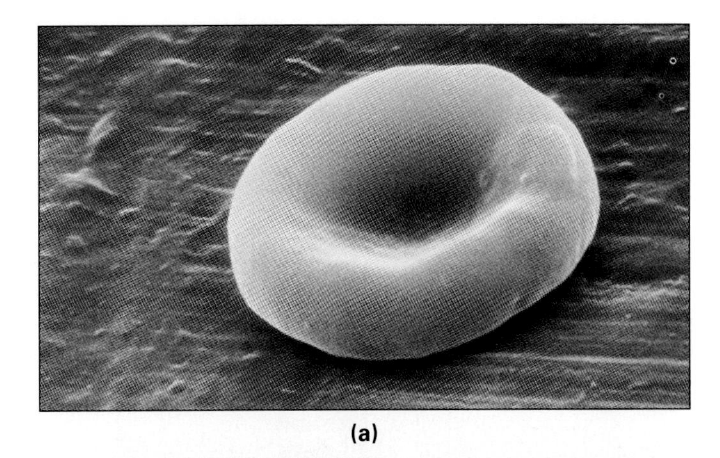

(a)

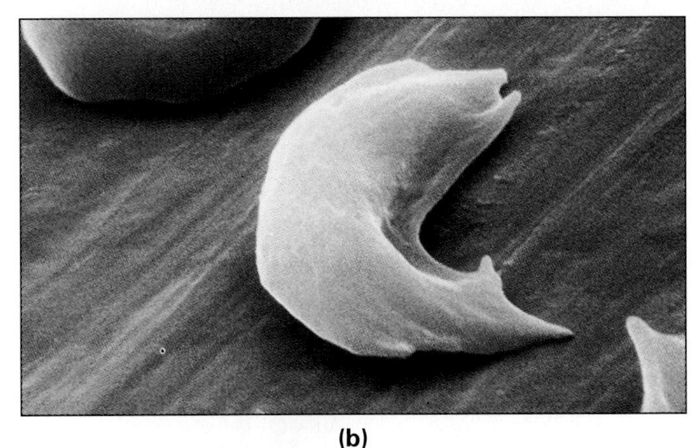

(b)

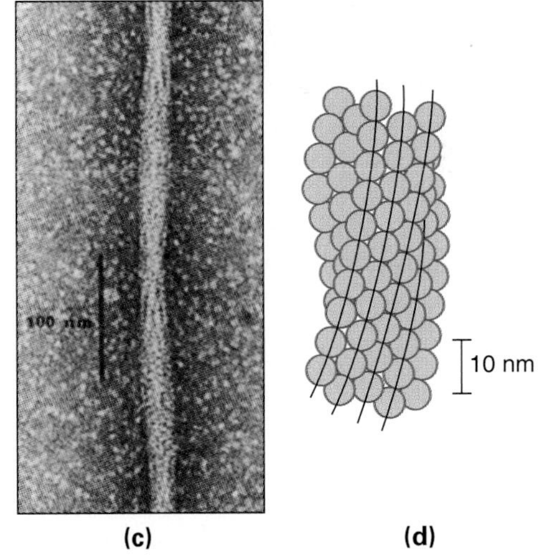

(c) (d)

Figure 19.3 Abnormal self-assembly of hemoglobin molecules in humans with sickle cell anemia. **(a)** Scanning electron micrograph of a normal red blood cell. **(b)** Sickled red blood cell. **(c)** Electron micrograph of fiber formed by hemoglobin S (hemoglobin tetramer with β-globin carrying the sickle mutation). **(d)** Model of self-assembled hemoglobin S. Each circle represents a complete hemoglobin S tetramer. Normal hemoglobin tetramers do not assemble in this fashion.

(a)

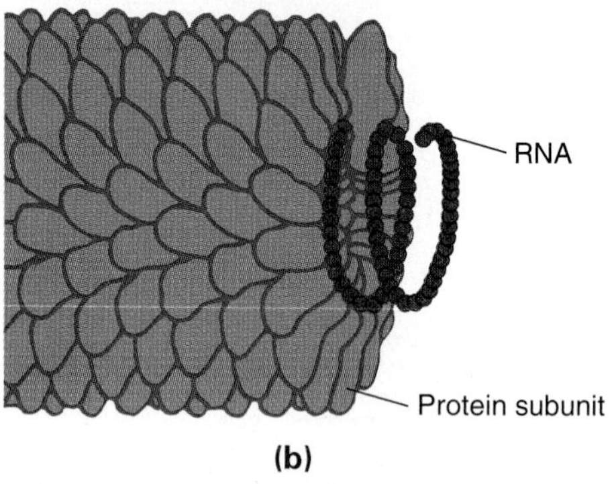

(b)

Figure 19.4 Tobacco mosaic virus (TMV). **(a)** Electron micrograph of the rod-shaped virus particles. **(b)** Schematic drawing of a portion of TMV with some of the coat proteins removed to expose the viral RNA. Note the spiral arrangement of the RNA and the repetitive coat proteins.

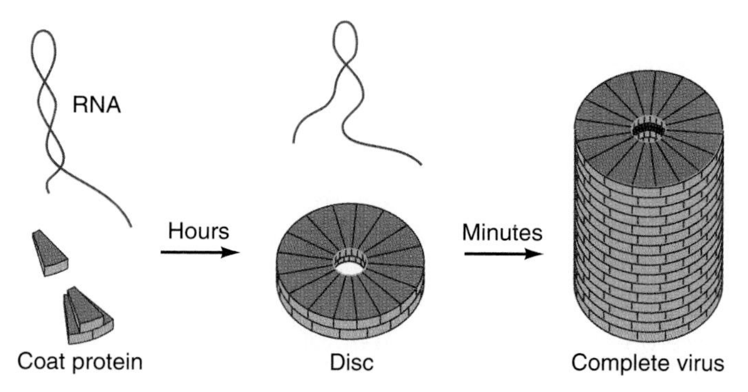

Figure 19.5 Self-assembly of TMV from its RNA and coat proteins. As an intermediate product, protein discs devoid of RNA are formed. Each disc consists of two circular layers of coat proteins, each layer containing 17 protein units. This number corresponds almost exactly to the number of coat proteins per helical turn in the fully assembled virus. The self-assembly of a disc from single coat proteins takes several hours. However, once a disc has formed and has associated with a specific segment of the viral RNA, the RNA-disc complex serves as a seed structure, facilitating the assembly of the complete virus within minutes.

complete disc is a slow and rate-limiting step. In contrast, a complete disc is a stable intermediate. Also, after one disc has associated with a specific segment of the viral RNA, this complex acts as a seed structure, to which further discs can be added in rapid succession.

THE CONFORMATION OF PROTEINS MAY CHANGE DURING SELF-ASSEMBLY

When molecules snap together during self-assembly, they release energy and become thermodynamically more stable. Participating proteins may change in tertiary or quaternary structure in the process. Each conformational change associated with a self-assembly step may in turn facilitate the addition of the next building block (Fig. 19.6). Such self-propagating waves of conformational changes are thought to play a role in many assembly processes.

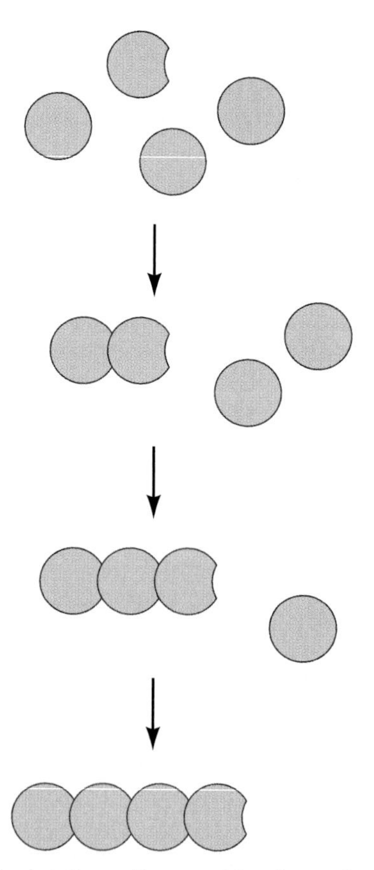

Figure 19.6 During the self-assembly of a molecular complex or an organelle, some proteins undergo a change in tertiary or quaternary structure (represented by the shallow depression in the circle). The new conformation of each added protein in turn facilitates the addition of the next protein. This self-propagating conformational change imposes a direction on the self-assembly process.

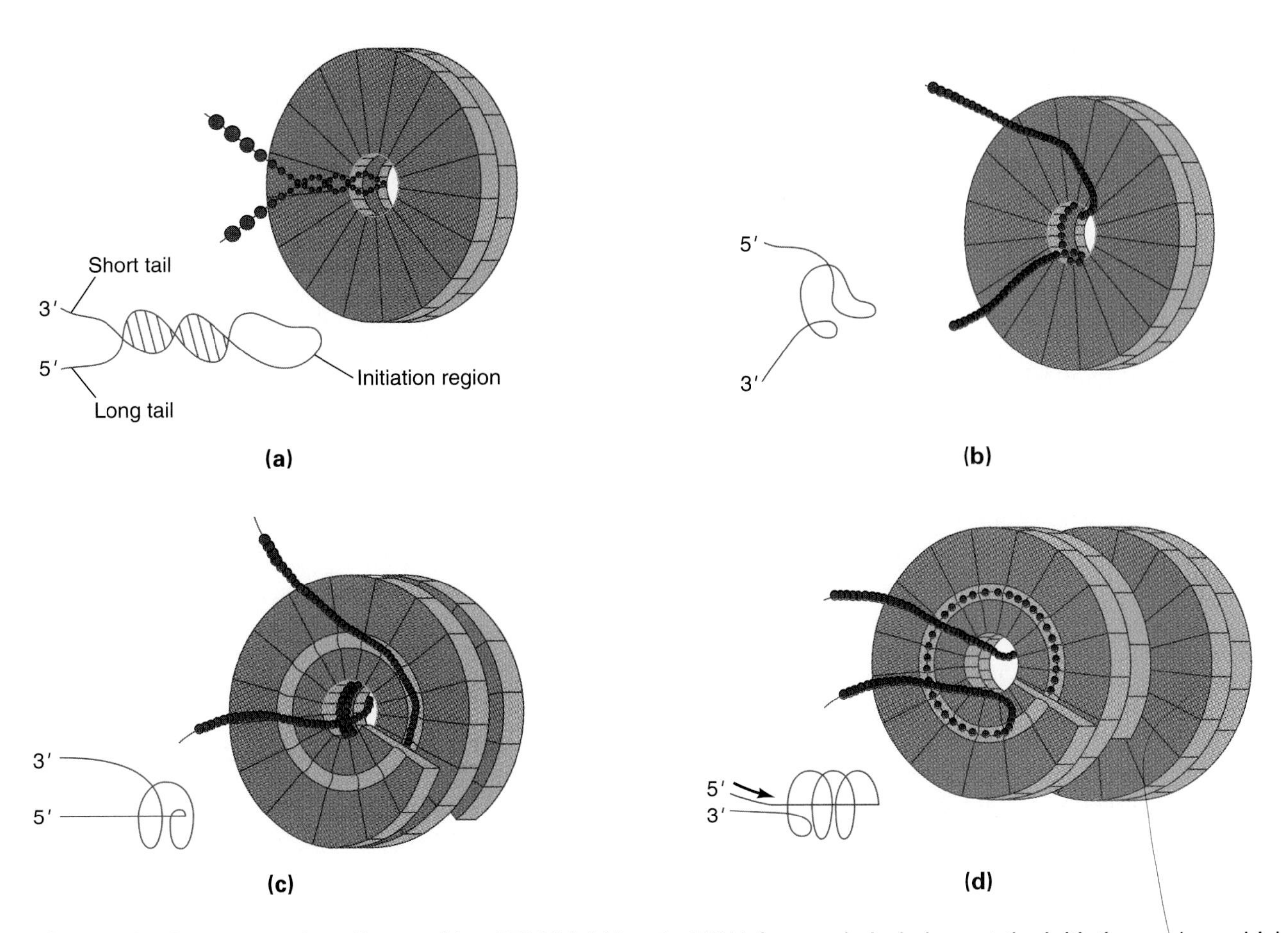

Figure 19.7 Nucleation process in self-assembly of TMV. **(a)** The viral RNA forms a hairpin loop at the initiation region, which is about 50 nucleotides long. This region is located about 1 kb away from the 3′ end (short tail) of the viral RNA, and about 5 kb away from the 5′ end (long tail). The loop inserts into a disc of coat proteins. **(b)** The RNA initiation region intercalates between the two layers of the disc, which form an open cleft on the inside. **(c, d)** The intercalation triggers a change in the conformation of the disc, which now assumes a helical shape like a lock washer. This change also closes the cleft between the protein layers and traps the RNA inside. In addition, the change activates a new ring of RNA binding sites on the outside surfaces of the lock washer. The lock washer–RNA complex serves as a nucleation site for the self-assembly of a complete virus by the rapid addition of further discs.

To illustrate this phenomenon, we will continue our exploration of TMV self-assembly. As already noted, TMV coat proteins self-assemble into two-layered discs. Such discs interact with a unique initiation region of the TMV RNA (Fig. 19.7). This region is about 50 nucleotides long and has a specific initiation sequence that binds stably to the disc of viral coat proteins. (RNAs lacking this sequence will not support TMV self-assembly.) The binding of the initiation sequence causes a conversion of the disc into two turns of a flat helix that looks like a lock washer. This conformational change traps the RNA initiation sequence between the lock washer proteins and generates new binding sites for RNA on the outside of the proteins. The resulting lock washer–RNA complex serves as a nucleation site for the rapid addition of further coat protein discs. As each new disc interacts with the viral RNA, it is converted to the lock washer shape and added on top of the growing helix. The resulting structure is increasingly stable, and further additions are no longer dependent on the specific nucleotide sequence of the viral RNA.

In summary, TMV self-assembly involves the initial formation of protein discs, the association of a disc with a specific segment of viral RNA, the resulting change of the original disc to a lock washer conformation, and a self-propagating wave of disc additions and changes to the lock washer conformation.

Together, these observations show that self-assembly is a powerful process that generates cell organelles and simple viruses just by permitting their parts to snap together. Many cellular structures self-assemble in vitro and presumably originate by self-assembly in vivo. These structures include the oligomeric proteins, collagen fibers, ribosomes, microtubules, and microfilaments already mentioned. In addition, simple cell membranes self-assemble in vitro.

The self-assembly of proteins depends on their *tertiary* and *quaternary* structure. These are determined, to a large extent, by the *primary* and *secondary* structure of their component polypeptides and are therefore under direct genetic control. Self-assembly processes are greatly accelerated in the presence of seed structures and are often associated with conformational changes in their building blocks. These properties help to restrict self-assembly processes to certain areas within cells and to control the direction in which they progress.

19.3 Aided Assembly

Notwithstanding its great powers, self-assembly is limited in the complexity of the structures it can produce. The limits of self-assembly are best illustrated by the morphogenesis of another virus, the T4 *bacteriophage* (Fig. 19.8; Wood, 1980). In order to multiply, the virus injects its DNA into a host bacterium. Once inside, the viral DNA commandeers the bacterium's synthetic machinery to replicate the viral DNA and to synthesize viral proteins. About 40 of these proteins associate with viral DNA to form new bacteriophages. The phage particles are synthesized and assembled inside the host bacterium, which is eventually lysed by the new phages which then move on to infect other hosts.

If the phage components are mixed together in aqueous solution, phage parts will assemble, but no complete phage will originate. Indeed, mutational analysis has shown that T4 assembly depends on at least six auxiliary proteins that are encoded by the phage genome but are not present in the mature phage. These proteins have auxiliary functions during the assembly process. Therefore, the assembly of the T4 bacteriophage does not fit the definition of self-assembly. Rather, it is classified as ***aided assembly*** to reflect its dependence on auxiliary components. Like self-assembly, however, aided assembly does not require paragenetic information.

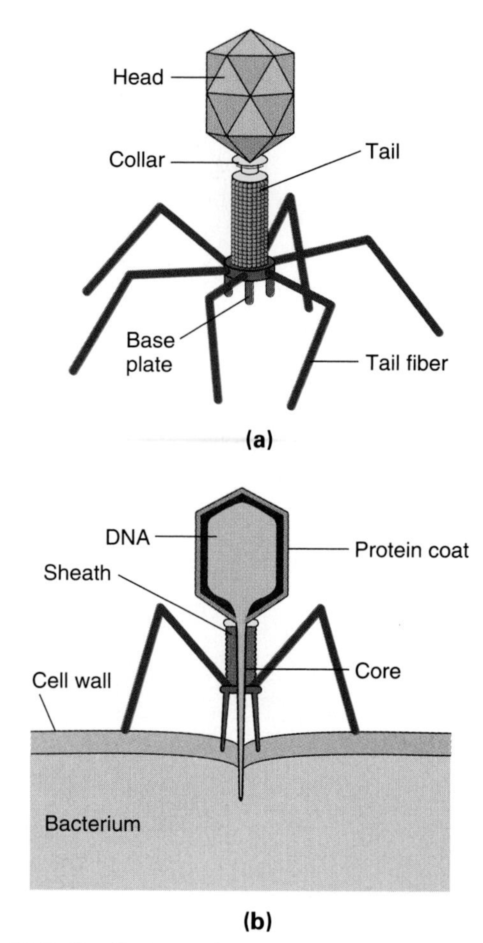

Figure 19.8 The T4 bacteriophage. **(a)** View of the phage with head, collar, tail, base plate, and tail fibers. **(b)** Schematic section showing how the phage infects a host bacterium. The phage binds to the bacterial surface, using receptor proteins at the tail fiber ends. The contact triggers a contraction of the tail, and the viral genome is injected into the bacterium. Inside the host, the phage DNA is replicated and transcribed, and the transcripts are translated into proteins.

BACTERIOPHAGE ASSEMBLY REQUIRES ACCESSORY PROTEINS AND OCCURS IN A STRICT SEQUENTIAL ORDER

Detailed studies of T4 assembly have revealed two features of general interest that are also important principles underlying the assembly of cellular organelles. The first is the dependence of assembly on auxiliary proteins, as already noted. The second interesting feature is the *strict sequential order* of phage assembly. Each assembly step depends on completion of the preceding steps. These features are emphasized in the following description of T4 assembly.

To study T4 assembly, investigators have isolated a large number of conditionally lethal mutants of the phage. Under permissive conditions, such as low temperature, these mutant phages assemble normally. Under restrictive conditions, such as high temperature, assembly stops. Each mutant is characterized by an abnormal or a missing protein and by a set of partially assembled phage particles inside the host bacteria. This type of analysis has provided a detailed picture of T4 assembly (Fig. 19.9). The major viral components (head, tail, and tail fibers) are assembled independently before they are finally mounted together. A defect that stops the assembly of one of the components does not affect the assembly of the other two.

The number of genes involved in phage assembly exceeds the number of coat proteins, indicating that the assembly of the coat proteins depends on auxiliary proteins. In particular, the tail fibers consist of only four proteins but require at least seven genes for their assembly. Likewise, three proteins necessary for making the base plate of the tail do not become part of the final structure. Moreover, the assembly of the head capsule requires prior formation of an inner scaffold or core that is later destroyed enzymatically. Finally, some of the head proteins are cleaved by limited proteolysis before

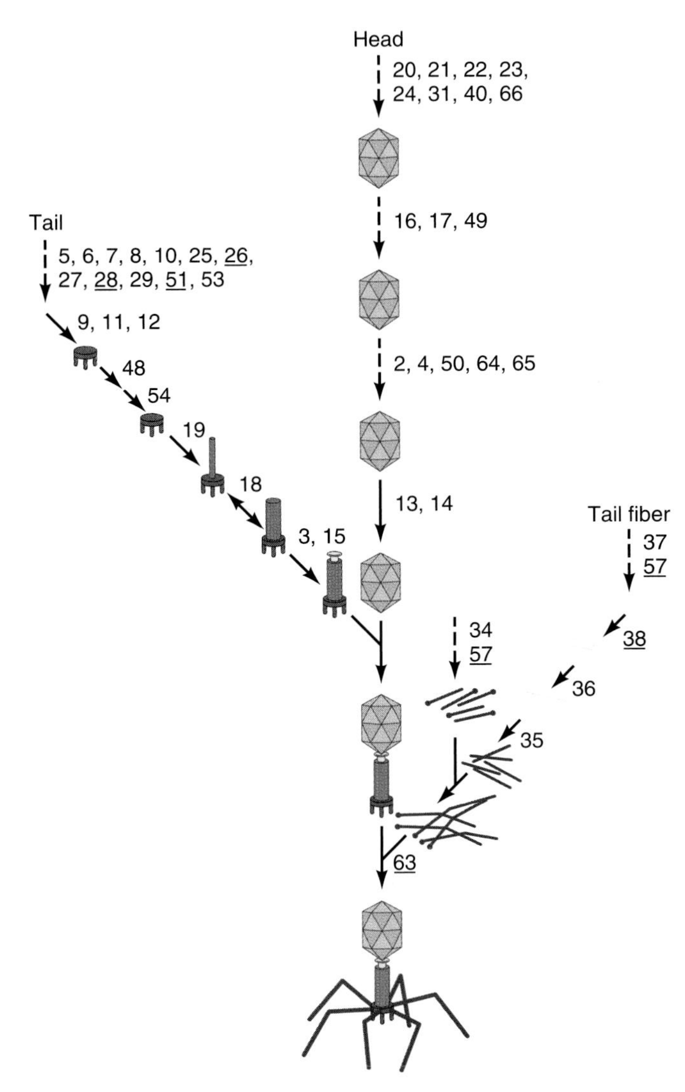

Figure 19.9 Flow chart for the aided assembly of the T4 bacteriophage. Each number represents a gene function revealed by mutant analysis. The underlined numbers refer to accessory proteins that are required for assembly but are not included in the final structure. The major components of the phage (head, tail, and tail fibers) are assembled independently before they are mounted together. The components are assembled in a strict sequential order.

assembly. The need for enzymes or scaffold proteins constitutes a clear departure from self-assembly and justifies the term *aided assembly.*

Components are assembled in a fixed sequence. For instance, the sheath of the tail is not assembled unless the base plate has been assembled first. The order in which parts are assembled cannot be attributed to the timing of protein synthesis, because all proteins are synthesized simultaneously. Rather, it seems that each assembly step is associated with a conformational change generating the seed structure that accelerates, or is a prerequisite to, the following step. This type of sequential assembly allows different building blocks to be present simultaneously without making "wrong" connections.

AIDED ASSEMBLY IS COMMON IN PROKARYOTIC AND EUKARYOTIC CELLS

The aided assembly of proteins into complex structures is a common occurrence in prokaryotic and eukaryotic cells. Indeed, some of the proteins that aid in the assembly of bacteriophage heads are encoded by their host bacteria. These proteins belong to a large group of proteins called *molecular chaperones* (Hemmingsen et al., 1988). Chaperone proteins are also found in mitochondria, in plant chloroplasts, and in the nuclei and cytoplasm of eukaryotic cells. Their main function, as discussed in Section 18.5, is to ensure the proper folding of polypeptides into their secondary and tertiary structures, so that they will assemble correctly into oligomeric proteins or other complex structures. For instance, the assembly of nucleosomes depends on an acidic nuclear protein, *nucleoplasmin,* which itself does not become part of the nucleosomes.

19.4 Directed Assembly

In the types of assembly discussed so far, we have assumed that a given building block, such as a T4 tail fiber protein, can serve in the assembly of *only one* structure, in this case a T4 bacteriophage. In many instances, however, the same pool of building blocks can participate in the assembly of *two or more* different structures. Which of the possible structures is assembled depends on what kind of nucleation center is present. A nucleation center may be a simple seed structure, as in self-assembly, or may be a more complex ***organizing center*** such as an entire centrosome. The type of assembly in which the presence of a particular seed or organizing center directs the assembly of one structure at the expense of another structure is called ***directed assembly.*** The term is also used if a seed or an organizing center directs the assembly of only one structure but is required for the assembly process rather than just accelerating it as in self-assembly. (This distinction may be difficult to make in practice, though.) Both types of directed assembly depend on paragenetic information in the form of seeds or organizing centers. Of all molecular assembly types, directed assembly is probably the most prevalent (Grimes and Aufderheide, 1991). The following examples will illustrate directed assembly at different levels of complexity.

ONE BACTERIAL PROTEIN CAN ASSEMBLE INTO TWO TYPES OF FLAGELLA

Many bacteria swim by means of a *flagellum* attached to a rotating "hook," which is driven by a "motor" that traverses the bacterial cell wall. A bacterial flagellum is built much more simply than a eukaryotic flagellum, consisting of a helical filament that assembles from one type of protein, ***flagellin.*** Flagella harvested from bacterial cells can be dissociated into flagellin monomers, and

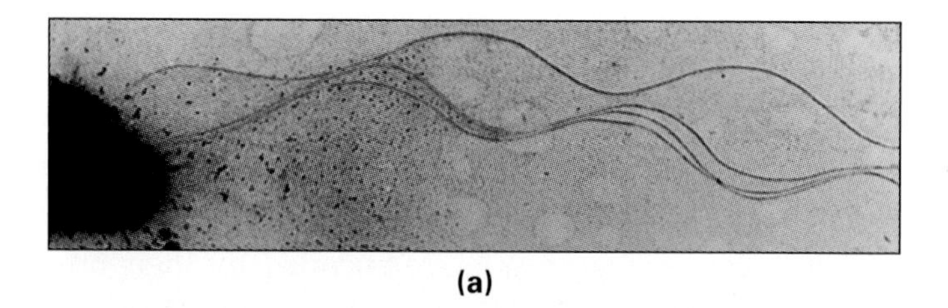

(a)

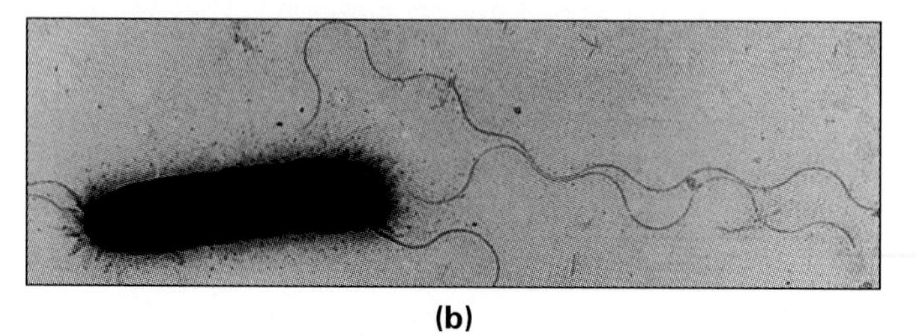

(b)

Figure 19.10 Electron micrographs of *Salmonella* bacterial cells. **(a)** Wild-type. **(b)** *curly* mutant with flagella of shorter wavelength.

the assembly of flagella from flagellin monomers can be studied in vitro. These studies show that flagella assemble from flagellin and that the shape of the assembled flagellum depends on the available seed structure, such as a hook protein or a segment of a flagellum.

Asakura and coworkers (1966) used different strains of the *Salmonella* bacterium to study flagellum assembly. One strain has wild-type flagella bent in a long helical pattern with a 2.5-μm wavelength (Fig. 19.10). A *curly* mutant has flagella with a much shorter wavelength of 1.1 μm. Instead of swimming in straight lines, the *curly* bacteria tumble and rotate about themselves. The mutant phenotype is caused by a single amino acid substitution in the flagellin monomer. The substitution restricts the ability of the monomer to assemble into a flagellum. Whereas the wild-type flagellin can assemble into either wild-type flagella or *curly* flagella, the mutant flagellin can assemble only into *curly* flagella. This was revealed by the following observations.

When either wild-type or curly flagellin was mixed with different seed structures, the phenotype of the reassembled flagella was generally determined by the nature of the flagellin monomer (Table 19.1). However, when wild-type flagellin monomers were mixed with moderately long (1.0-μm) *curly* flagellum segments as seeds, the result was surprising: The reassembled flagella had the *curly* phenotype. In contrast, the same wild-type monomers, when mixed with short (0.2-μm) *curly* flagellum segments or wild-type flagellum segments, assembled into wild-type flagella. The investigators also found that a mixture of wild-type and curly monomers assembled into *curly*-like flagella with any type of seed. Moreover, wild-type flagella eventually assumed the *curly* phenotype when incubated for a long time. These transformed flagella would revert to the wild-type conformation after treatment with low concentrations of pyrophosphate or adenosine triphosphate.

These results indicate that wild-type flagellin can reversibly change its tertiary structure between a state assembling into wild-type flagella and another state assembling into *curly*-like flagella. Which conformation the protein assumes depends on the chemical composition of the medium and on the *nature of the primer* that nucleates its assembly. With a wild-type primer, each added wild-type flagellin molecule seems to assume the wild-type configuration, thus propagating the assembly of monomers in this state. In contrast, a curly primer seems to start a self-propagating wave of monomer additions in the *curly*-like state, regardless of the flagellin type. Thus, the assembly of wild-type *Salmonella* flagellin is a simple case of *directed assembly*, although otherwise it has all the characteristics of a *self-assembly* process. Many cases of directed assembly also depend on accessory proteins and energy sources, as does aided assembly.

table 19.1 Directed Assembly of a Bacterial Flagellum

Seed Structure	Genotype of Flagellin Monomer	Phenotype of Flagellum
Wild-type bacterial "hook"	Wild type	Wild type
Wild-type fragment*	Wild type	Wild type
Wild-type fragment	*curly*	Curly
curly 0.2-μm fragment	Wild type	Wild type
curly 1.0-μm fragment	Wild type	Curly
curly fragment	*curly*	Curly
Any fragment	Wild type + *curly*	Curly

*"fragment" refers to a segment of a flagellum

PRION PROTEINS OCCUR IN NORMAL AND PATHOGENIC CONFORMATIONS

The directed assembly of flagellin has an uncanny parallel in the behavior of certain proteins involved in deadly diseases (Prusiner, 1997, 1998; Jackson et al., 1999). A condition known as *scrapie* in sheep and goats causes its victims to scratch and scrape, and eventually leads to debility and death. Histological examination of the victims' brains reveals a progressive vacuolization and disorganization of brain tissue, a condition called *spongiform encephalopathy*. Scrapie-like diseases in other mammals are known as bovine spongiform encephalopathy (BSE) and transmissible mink encephalopathy. Similar human conditions are known as *Creutzfeldt-Jakob disease* (*CJD*), *Gerstmann-Sträussler Scheinker* disease (*GSS*), and *fatal familial insomnia* (*FFI*). All of these can be propagated, within and between species, by transferring brain tissue from a diseased organism to a healthy one.

However, normal individuals can also become infected by eating the brains of diseased individuals. For instance, a BSE epidemic among cattle in England was traced to the use of apparently diseased sheep carcasses in cow feed. In turn, an epidemic of BSE in England was followed by the appearance of a new variant of CJD with characteristics indicating that these human patients most likely have contracted the disease by exposure to BSE (Collinge et al., 1996; Hill et al., 1997). Another form of human CJD known as *kuru* to the Fore people of Papua New Guinea is almost certainly contracted by ritual cannibalism involving eating of the brain.

Although the infective agents for scrapie diseases were originally thought to be viruses, no nucleic acid component has been found. Instead, Prusiner (1982) isolated from the brains of scrapie-infected hamsters a nonviral agent that was capable of transmitting scrapie. For this agent he coined the term ***prion*** (for *pro*teinaceous *in*fectious particle). He proposed that each mammalian species produces a normal cellular prion-precursor protein (PrP^C), and that the pathogenic prion represents an altered, scrapie-inducing form (PrP^{Sc}) of that species' normal prion-precursor protein. The PrP^C form is a secreted glycoprotein that is tethered to the plasma membrane of neurons and, to a lesser extent, of lymphocytes and other cells. The pathogenic PrP^{Sc} form appears to originate from the harmless PrP^C form through a posttranslational conversion, in which the PrP^{Sc} form—of the same species or another—acts as a seed (Fig. 19.11). The PrP^{Sc} form collects mostly in membranous vesicles inside the cells. The accumulation of these vesicles, or the formation of fibrilar aggregates by PrP^{Sc}, seems to cause the histological and behavioral symptoms of neurodegenerative disease.

Prion proteins have been found in all mammals investigated. In each species, the PrP^{Sc} form has a large hydrophobic core that is resistant to acids and proteases, in accord with epidemiological data showing that the agents causing BSE and CJD can be transmitted through the gastrointestinal tract when brains from diseased individuals are eaten. Prusiner and most others in the field propose that the PrP^{Sc} conformations by themselves are necessary and sufficient to cause scrapie diseases by directing the conversion of PrP^C into more PrP^{Sc} and thus disturbing the normal metabolism of prion protein. This ***"protein only" hypothesis*** is disputed by some investigators. They suspect the involvement of a viruslike undiscovered RNA or DNA component because scrapie agents occur in multiple "strains" that differ in their incubation time (time elapsed between inoculation and onset of neurodegenerative disease) and in the severity of histological degeneration they cause. Prusiner and followers counter that a given PrP^C molecule may be refolded into more than one version of PrP^{Sc}, each capable of replicating itself. Indeed, protease-resistant PrP^{Sc} fragments of two different sizes were obtained from the same transgenic PrP^C molecule, depending on the scrapie agent used for inoculation (Telling et al., 1996). Thus, alternative refolding and variable glycosylation may account for the generation and propagation of different prion strains (Collinge et al., 1996; Wadsworth et al., 1999).

Physicochemical studies have unraveled the three-dimensional structure of mouse and human PrP^C (Riek et al., 1996; Donne et al., 1997; G.S. Jackson et al., 1999). It consists of three α helices, a short β pleated sheet, and random coil, which together form a stable globular structure. The structure of PrP^{Sc} is different from that of PrP^C in having a higher content of β pleated sheet. This conformation is more resistant to proteinase K, and it is a direct precursor of fibrilar structures closely resembling those isolated from diseased brains.

The initial conversion of PrP^C to PrP^{Sc} is assumed to occur spontaneously, although very rarely, in wild-type prion-precursor proteins. However, the probability of spontaneous conversion appears to be greater in proteins encoded by certain mutant alleles of prion-precursor genes. The prion genes of human families with hereditary CJD or GSS have been screened for mutations. In each family, at least one prion-precursor gene was found to be mutated. In 10 families from 9 different countries, the mutations caused the same amino acid substitution: leucine for proline at residue 102, counted from the amino terminus (G. A. Carlson et al., 1991). Other investigators examined the prion-precursor genes of individuals with nonhereditary CJD; almost all these patients were found to be homozygous at amino acid residue 129 of the prion-precursor protein, a site where the majority of the normal population is polymorphic and heterozygous (M. S. Palmer et al., 1991). These data show that homozygous mutations at residue 129 predispose their carriers to CJD. Similarly, sheep homozygous for a prion-precursor allele that encodes glutamine

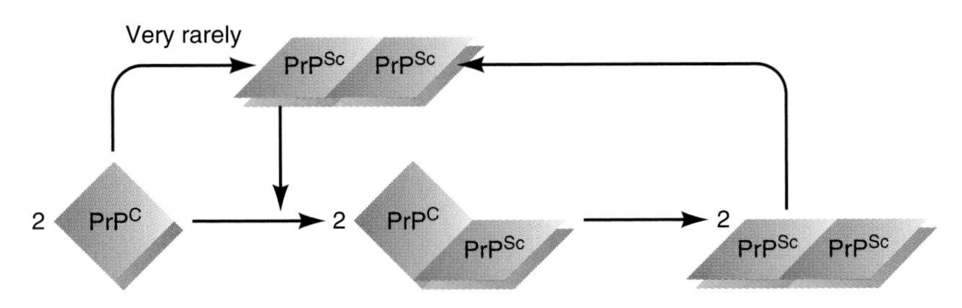

Figure 19.11 Simple model of the posttranslational modification of prion-precursor protein from the normal form (PrP^C) to the scrapie form (PrP^{Sc}). The modification is a conformational change that is thought to occur very rarely in wild-type prion-precursor protein but more frequently in certain mutant prion-precursor proteins. When present, PrP^{Sc} can associate with PrP^C and distort it to yield more PrP^{Sc}. More complex models involve additional proteins.

instead of arginine in position 171 are more prone to developing scrapie (Westaway et al., 1994). Conceivably, the mutations that predispose their carriers to prion diseases weaken the normal secondary structure of PrP^C and facilitate its refolding into PrP^{Sc}.

ONE prediction derived from the *"protein only" hypothesis* is that organisms without prion-precursor protein should be resistant to prion diseases. It became practical to test this hypothesis when it was discovered that transgenic mice without functional prion-precursor (PrP^C) genes either develop normally or show only slight neurological symptoms (Büeler et al., 1992; Collinge et al., 1994). Such "knockout" mice were inoculated, by injection into a cerebral hemisphere, with brain tissue from mice with scrapie (Büeler et al., 1993; Prusiner et al., 1993). The injected mice were observed for behavioral signs of scrapie, including abnormal gait, feet clasping when lifted, disorientation, and depression. At intervals, one of the injected mice was sacrificed and the brain examined histologically for the vacuolization and tissue damage that are characteristic of spongiform encephalopathy.

Homozygous mutant mice lacking both genes for prion-precursor protein remained symptom-free for more than 500 days (Fig. 19.12). In contrast, when mice carrying the wild-type alleles for prion-precursor protein were subjected to the same treatment they developed scrapie symptoms within less than 165 days. Heterozygous mice with one normal prion-precursor gene developed dysfunctions between 400 and 465 days after inoculation. Also, brain homogenates from inoculated knockout mice failed to cause scrapie in normal mice, whereas brain homogenates from inoculated wild-type mice transferred the disease.

Questions

1. The experiment described above proves that prion-precursor protein is *necessary* for contracting and transmitting scrapie. What kind of experiment would prove that the transition from prion-precursor protein to prion protein (PrP^C to PrP^{Sc}) by itself—without a chance for viral contamination—is *sufficient* to cause scrapie?
2. Suppose you would graft brain tissue from mice carrying wild-type prion genes into inoculated knockout mice. Which tissue(s) would you expect to develop the signs of spongiform encephalopathy: the grafted tissue, the host brain, or both? Remember that the prion-precursor protein is tethered to the plasma membrane.
3. Assuming that the biological functions of prion-precursor protein are equivalent in mice, sheep, and cattle, would it be possible to generate sheep and cattle whose meat could be eaten by humans without any risk of contracting spongiform encephalopathy?

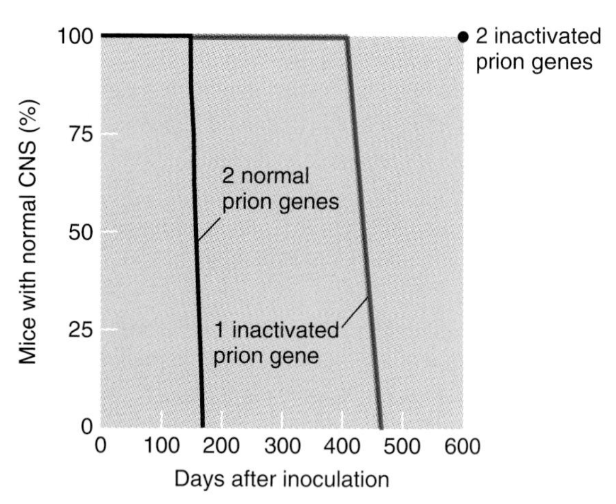

Figure 19.12 Dependence of scrapie on the expression of the gene for prion-precursor protein. Scrapie agent from diseased mice was transferred to healthy mice with two genes for normal prion-precursor protein, and to healthy transgenic mice in which one prion gene or both had been inactivated. The normal mice began to show neurological dysfunction within less than 165 days, and the mice with one functional prion-precursor gene developed the same symptoms after 400 to 465 days. In contrast, the mice without functional prion-precursor genes were symptom-free after more than 500 days.

In summary, the available data indicate that the prion protein exists in different three-dimensional conformations, in the normal and harmless PrP^C conformation and in one or more pathogenic PrP^{Sc} conformations. According to the "protein only" hypothesis proposed by Prusiner, and accepted by most others in the field, the PrP^{Sc} conformations by themselves are necessary and sufficient to cause scrapie diseases by directing the conversion of PrP^C into more PrP^{Sc} and thus disturbing the normal metabolism of prion protein. When Stanley B. Prusiner was awarded the Nobel prize in Physiology or Medicine for his pioneering work on prions in 1997, the Nobel Assembly of Stockholm broke with the tradition of considering only achievements that have won universal acceptance. Instead, the Assembly honored the main architect of an unorthodox theory of general importance, which is still disputed by some experts, on how much information cells can pass on in the form of protein structure, independently of DNA or RNA.

TUBULIN DIMERS ASSEMBLE INTO DIFFERENT ARRAYS OF MICROTUBULES

The tubulin polypeptides synthesized in eukaryotic cells form heterodimers, consisting of one α- and one β-tubulin. These dimers assemble into the cylinders known as *microtubules* (see Section 2.3). Depending on cell type and phase in the cell cycle, microtubules form different arrays (see Fig. 2.5). In mitotic cells, many short and straight microtubules radiate out from two *centrosomes* forming the asters at the spindle poles. During interphase, a loose network of long and wavy microtubules, often forming a tight meshwork around the

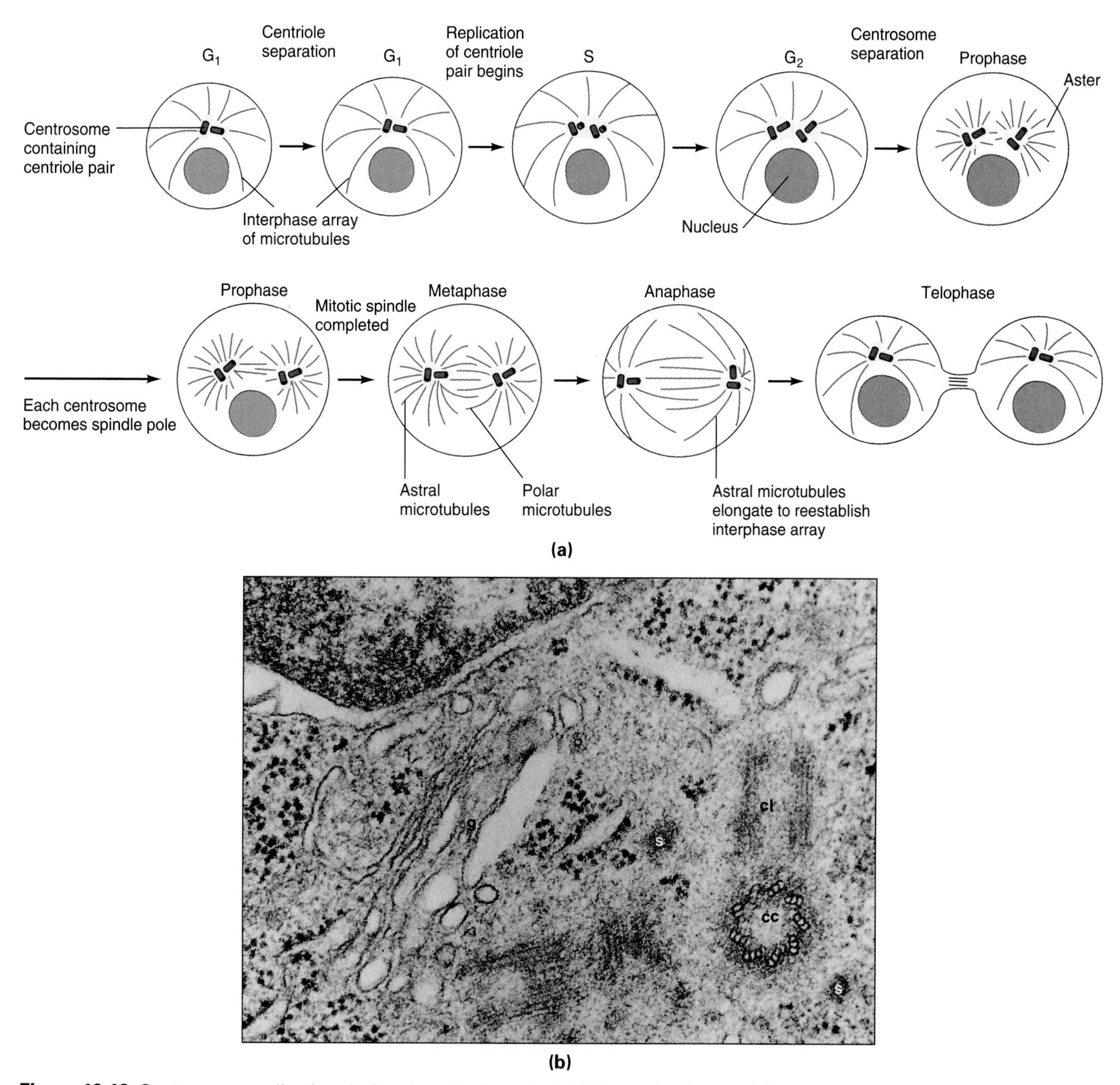

Figure 19.13 Centrosome replication during the mitotic cycle. **(a)** Schematic diagram. The centrosome consists of a pair of centrioles (gray cylinders) surrounded by amorphous material (yellow color). During interphase (G_1, S, G_2), the centrosome anchors a relatively small number of long microtubules. During S phase, each centriole directs the assembly of a daughter centriole oriented at right angles to its parent centriole. During prophase, the two centriole pairs split and move apart, forming two centrosomes. At the same time, the molecular composition of the amorphous material changes. Each centrosome then nucleates a larger number of shorter microtubules, forming an aster. After the nuclear envelope breaks down, the two asters set up the mitotic spindle. (Chromosomes are not shown during metaphase and anaphase.) During cytokinesis, each daughter cell inherits one centrosome to organize a new interphase array. **(b)** Transmission electron micrograph showing two pairs of centrioles at G_2 phase. One centriole of each pair appears in longitudinal section (cl) and the other in cross section (cc) because they are oriented at right angles to each other. g = Golgi apparatus; s = Satellite.

nucleus, extend from the single centrosome (Fig. 19.13). In neural plate cells, oriented microtubules support the columnar cell shape (see Fig. 2.3). In many protozoa and epithelial cells, microtubules form the axoneme of *cilia* and *flagella,* as we will see. In neurons, microtubules are part of the cytoskeletal lattice stabilizing the axon and providing a bidirectional transport system.

The different arrays of microtubules and axonemes assemble from the same cellular pool of tubulin dimers. Nevertheless, each type of cell displays a distinctive microtubular array rather than bits of everything. Indeed, the tubulin needed for the assembly of one array is often supplied by the disassembly of another array. For instance, during mitosis, the microtubular meshwork characteristic of interphase disappears while the mitotic spindle is built. These observations indicate that cells *direct* the assembly of microtubular arrays to specific sites and configurations depending on cell type and cell cycle stage.

Which type of microtubular array is assembled in a given cell at a particular time depends on available organizing structures and on other proteins that associate with microtubules. The organizing structures for microtubule assembly in cells are called *microtubule organizing centers (MTOCs)*. The assembly of each microtubule begins with the (−) end (see Section 2.3), which is stabilized by the MTOC. Construction of most cytoplasmic microtubule arrays is initiated by MTOCs known as *centrosomes*. Other MTOCs known as *basal bodies* initiate the assembly of the axonemes in cilia and flagella.

Microtubule-associated proteins act to stabilize microtubules against disassembly and to mediate their interactions with other cell components. By controlling the localization and molecular composition of these MTOCs, and by regulating the availability of microtubule-associated proteins, cells can direct the assembly of different microtubular arrays.

CENTRIOLES AND BASAL BODIES MULTIPLY LOCALLY AND BY DIRECTED ASSEMBLY

Having recognized the importance of *centrosomes* and *basal bodies* as MTOCs, we will now examine how these critical structures are replicated and positioned in cells, and how their composition changes during the cell cycle. Basal bodies will also figure prominently in our subsequent discussion of the directed assembly of larger cortical structures in ciliates.

A ***basal body*** is located at the base of each cilium beneath the plasma membrane. The basal body is required for cilium formation—cilia without basal bodies have not been found. Each basal body is a small cylinder consisting of nine microtubule triplets (see Fig. 2.8). Each triplet consists of three subfibers, designated A, B, and C. Subfibers A and B extend directly into the corresponding subfibers of the axoneme. In ciliate protozoa, new basal bodies originate at right angles to preexisting basal bodies. Later, the daughter basal body tilts up and assumes its definitive position parallel to the parent basal body and perpendicular to the cell surface.

The replication of *centrosomes* is coordinated with the cell cycle (Fig. 19.13). Most animal cells during interphase contain one centrosome near the nucleus. It consists of a pair of *centrioles* and some surrounding amorphous material. Each centriole is built like a basal body, except that it is not connected to an axoneme. The two centrioles are oriented at right angles to each other, forming an L-shaped configuration. The centrosomes of higher plants are devoid of centrioles, indicating that the functional part of centrosomes, at least in these types of cells, might be the amorphous material. This material is poorly characterized except that it associates with γ-tubulin as an apparent link to the (−) end of microtubules (Stearns and Kirschner, 1994).

During S phase, each centriole promotes the assembly of a daughter centriole in its vicinity. The daughter centriole is oriented perpendicular to the parent centriole. During mitotic prophase, the two daughter centrosomes move to opposite poles of the nucleus, where they direct the assembly of the mitotic spindle.

The molecular composition of both centrosomes and microtubule-associated proteins changes with the cell cycle. Each centrosome seems to be able to organize a maximum number of microtubules. This number increases considerably during mitotic prophase. The change may explain the transition from the interphase array of relatively few, long microtubules to the mitotic spindle containing many short microtubules.

The stereotypical arrangements of replicating centrioles and basal bodies indicate that each parent structure directs the assembly of its daughter. It has been a matter of considerable debate whether basal bodies and centrioles contain DNA or RNA. Nucleic acids may encode proteins that play special roles in the replication of basal bodies and centrioles. Conceivably, the local availability of special proteins and the *spatial order* of the preexisting MTOCs could promote the replication of the MTOCs.

Although basal bodies and centrioles usually originate by duplication of their respective parent structures, they can also arise from less conspicuous precursors. Some epithelial cells form hundreds of cilia on their apical surfaces. The basal bodies of these cilia arise simultaneously from electron-dense "satellites" formed in close association with the centrosome of the nonciliated precursor cell. Likewise, the eggs of many animal species lack conspicuous centrosomes and use the centrosome introduced by the sperm to set up their first mitotic spindle (Schatten, 1994). If such eggs are activated parthenogenetically, some nevertheless form a centrosome. Some precursor structures seem to exist in eggs and normally remain quiescent in the presence of a fully formed centrosome introduced by the sperm. Presumably, the directed assembly using the preexisting centrosome as a seed is faster than the assembly of a new centrosome from smaller precursors, and the latter pathway is triggered only in the absence of the former.

CILIATED PROTOZOA INHERIT ACCIDENTAL CORTICAL REARRANGEMENTS

A spectacular instance of directed assembly occurs in the ciliated protozoon *Paramecium tetraurelia*. To appreciate the experimental results obtained with this unicellular creature, we must first familiarize ourselves with its morphology and reproduction.

Dubbed the "slipper animal" because of its overall shape, *Paramecium* consists of one large cell (Fig. 19.1) and is about 0.2 mm long. The cell has an anteroposterior and a dorsoventral polarity. The anterior end leads during normal swimming, and the posterior end has a tuft of elongated cilia. An ***oral apparatus*** on the ventral side gathers food for phagocytosis, while two or more contractile vacuoles, which regulate the osmotic pressure, open on the dorsal surface. The right and left sides of the cell are defined by analogy to bilaterally symmetrical metazoa.

The surface layer, called the *cortex,* is more viscous and structured than the interior cytoplasm. The cortex consists of about 4000 *cortical units* arranged in longitudinal rows. Each unit has one or two cilia, each anchored in a basal body. A ***kinetodesmal fiber*** extends anteriorly from the right side of the basal body, and a ***parasomal sac*** is located to the right of the kinetodesmal fiber (Fig. 19.14). This arrangement, along with other cytoskeletal structures, gives each cortical unit an anteroposterior polarity and a right-left asymmetry. Normally, all cortical units in an individual are oriented the same way. This uniformity is functionally important because it provides for a uniform orientation of the cilia, which in turn is a prerequisite for the coordinated ciliary strokes that propel the animal.

Paramecium reproduces by a transverse division process called ***fission*** (Fig. 19.15). Each fission is preceded by both longitudinal growth and the duplication of each cell organelle. Likewise, the number of cortical units doubles during each growth cycle. This involves the generation of new basal bodies, a process that closely resembles the generation of new centrioles (see Fig. 19.13). Each new basal body arises *immediately anterior* to an existing basal body and at right angles to it. Once formed, the new basal body moves away from its parental basal body and tilts upward to make contact with the cell surface. Here it is positioned exactly in line with the other basal bodies of the same row. The other organelles are assembled around each new basal body, and the cortical unit elongates. The unit is then cross-partitioned, giving rise to two daughter units next to each other in the same row.

Independently of its reproduction by fission, *Paramecium* engages in a sexual process called ***conjugation*** (Fig. 19.16a), during which two conjugands adhere to each other with their oral apparatuses. Each conjugand forms two haploid nuclei and transfers one of them to its mate. Thereafter, in each conjugand, the transferred haploid nucleus and its stationary counterpart fuse to form a diploid zygote nucleus.

When two individuals separate after conjugation, they sometimes take portions of their mate's cortex with them. This "cortical picking" amounts to a transplantation of one or two partial rows of cortical units from one animal to another. Occasionally, the transplanted row heals in being rotated 180° in the plane of the cell surface (Fig. 19.16b). As a result, the power stroke of the cilia is

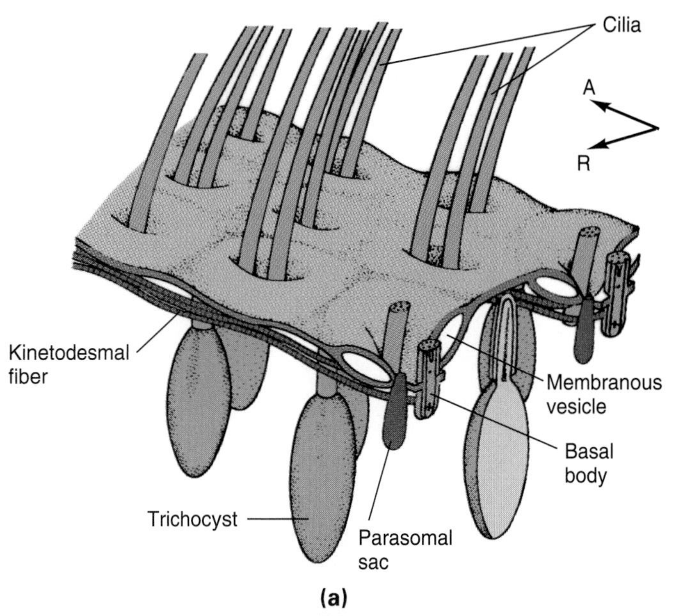

(a)

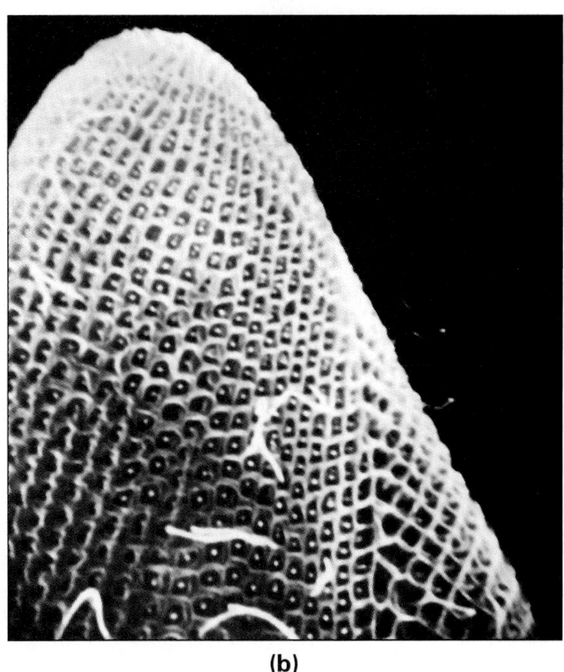

(b)

Figure 19.14 Organization of the cortex of *Paramecium* in repeated units. Cortical units are arranged in longitudinal rows. Organelles called trichocysts, which function in self-defense, are located between consecutive units in the same row. Each cortical unit is covered by a membranous vesicle. From a depression in each vesicle emerge one or two cilia, each anchored in a basal body. A striated band called a kinetodesmal fiber originates from each basal body and extends anteriorly past the basal bodies of the anteriorly adjacent units. To the right of the kinetodesmal fiber in each unit there is a blind sac called the parasomal sac. The arrangement of these structures gives each unit an anteroposterior and a right-left polarity, indicated by the intersecting arrows. (In describing ciliates, the same right-left convention is used as for the human body: In a ventral view, the cell's right is on the observer's left-hand side.) **(a)** Interpretative drawing; **(b)** scanning electron micrograph of the anterior ventral surface of a *Paramecium* from which most cilia have been removed to make the cortical units visible.

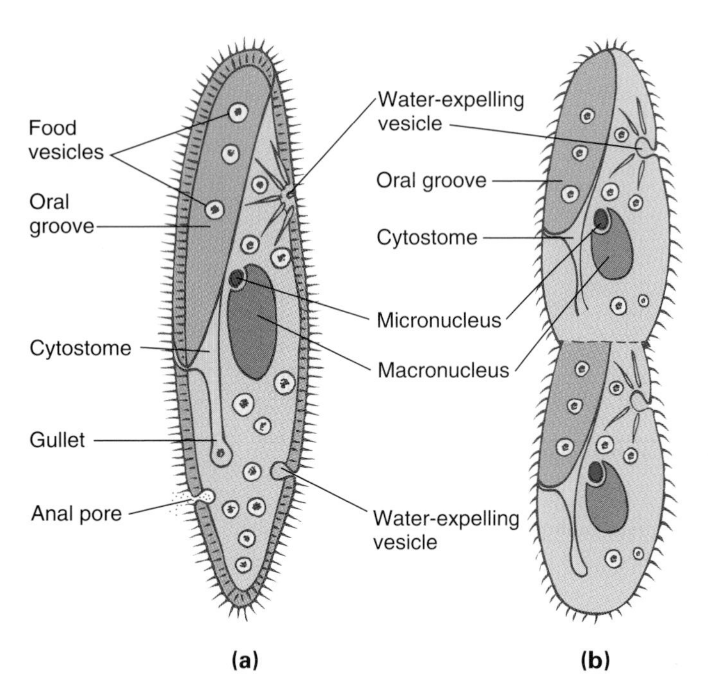

Figure 19.15 Organization and fission of *Paramecium caudatum.* **(a)** The ventral aspect shows the oral groove, through which cilia drive algae and other food particles into the cytostome (cell mouth), from where the food passes down a gullet. Food vesicles bud off the gullet and circulate through the cell until they are exocytosed at the anal pore. Two water-expelling vesicles regulate the osmotic pressure inside the cell. The genes of the polyploid macronucleus are actively transcribed, whereas the diploid micronucleus undergoes meiosis during conjugation. **(b)** *Paramecium* propagates by transverse fission. Before each fission, all cell organelles are duplicated.

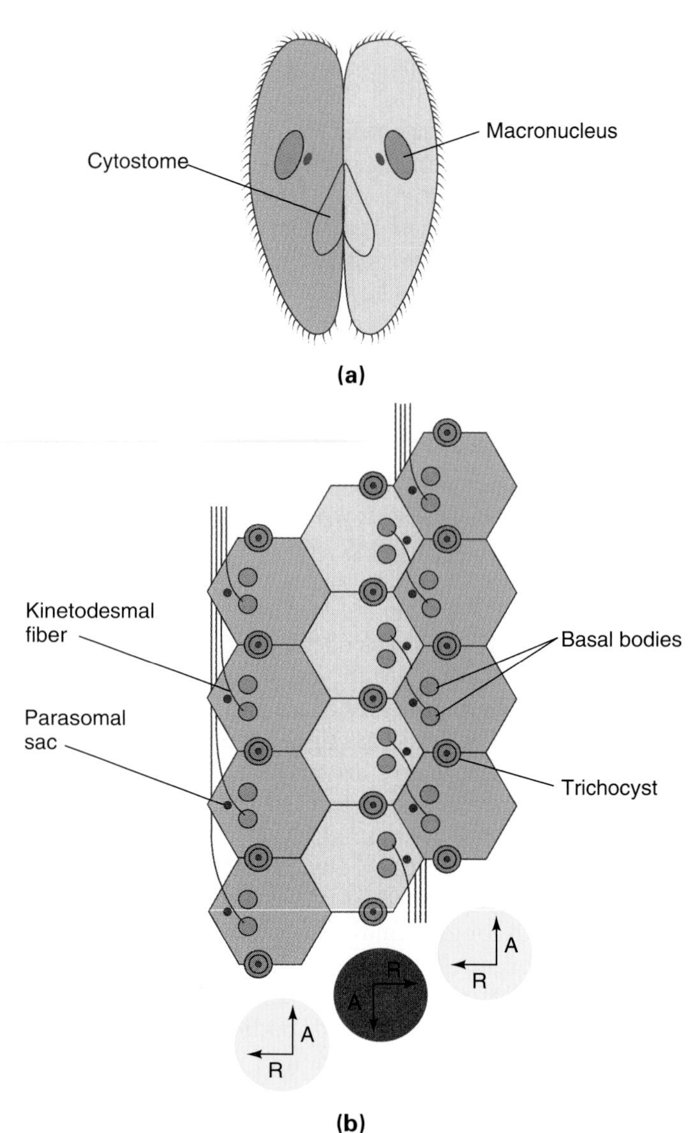

Figure 19.16 Conjugation and "cortical picking" in *Paramecium.* **(a)** Drawing of a conjugating pair before meiosis and exchange of haploid nuclei. When the partners separate after conjugation, one may take a piece of cortex from the other. **(b)** Transplanted row of cortical units observed after conjugation. This diagram shows part of two normally oriented rows of cortical units flanking a single row that is rotated 180°. A, anterior, R, right.

reversed in the transplanted row. This causes an abnormal swimming pattern, which is readily identifiable under the microscope. Cells displaying this feature can be isolated and propagated into a clone of cells.

Observation of *Paramecium* individuals with cortical transplants has provided the most striking example of directed assembly (Beisson and Sonneborn, 1965). They observed that the transplanted patches grew anteroposteriorly during fission cycles until they were as long as normal ciliary rows. In the inverted transplant, new basal bodies were formed with the correct orientation—that is, *anterior* to their "parent" basal body with regard to their cortical units and *posterior* to their "parent" basal body with regard to the entire cell. After extension of the transplant to a full row, the progeny kept inheriting complete inverted rows. Occasionally, transplanted rows were lost during fission, as revealed by the recovery of cells showing the normal swimming pattern. When such cells were eliminated from the culture, the original inversion could be maintained for hundreds of generations and remained unaffected by subsequent conjugations with normal cells. The experiment has been repeated, with the same results, by several investigators.

The experimental results indicate that the assembly of new cortical units in *Paramecium* is directed by the "local geography" of preexisting units in the same row. The molecular basis of this directed assembly is still unknown. It is noteworthy that normal and inverted cortical units can be duplicated side by side in adjacent rows. Thus, the cellular genome encodes the molecular building blocks for both normal and inverted cortical units without directing the assembly of one or the other. The genome does not even provide a bias, because conjugations between normal and altered mates have no effect on the propagation of inverted cortical units. Rather, there seems to be an unbroken chain of *paragenetic infor-*

mation that is passed on from one cortical unit to the progeny units in the same cortical row. This conclusion holds true whether or not the basal bodies in *Paramecium* contain DNA or RNA or no nucleic acid. The act of cortical picking (which starts the inversion of cortical rows) cannot possibly change the *nucleotide sequence* (*genetic information*) of the basal body DNA or RNA of dozens of cortical units in the same way. Of course, the cortical picking could invert the *spatial orientation* of any basal body DNA or RNA, as it does with all other molecules in a cortical unit. And the three-dimensional orientation of basal body DNA or RNA might be part of the local geography that perpetuates itself when cortical units are duplicated before fission. Thus, with or without basal body DNA or RNA, it is the *local orientation of supramolecular assembly* (*paragenetic information*) that is passed on from each cortical unit to its daughter units.

19.5 Global Patterning in Ciliates

In addition to the orientation of their small cortical units, ciliates have a *global body pattern* of major cell organelles (Frankel, 1989). For instance, *Pleurotricha lanceolata* (Fig. 19.17a) has an oral apparatus in the anterior left quadrant of its ventral side and a characteristic pattern of ***cirri*** (sing., *cirrus*)—that is, compound organelles made of clusters of hexagonally packed cilia. A single row of cirri is found on the left margin of the ventral side, a double row on the right margin, and a regular pattern of cirri in between. The dorsal surface is covered with a thin lawn of cilia. These structures confer distinct anteroposterior, dorsoventral, and right-left polarities on the cell.

The maintenance of the global cell pattern throughout the life cycle of ciliates is called ***global patterning.*** We will examine this process in *Pleurotricha* with three thoughts in mind. First, it will become apparent that part of the global patterning process relies on paragenetic information. Second, we will see that paragenetic global patterning preserves the overall body pattern but not necessarily all its details. Third, we will discuss a critical observation showing that paragenetic global patterning is independent of the local directed assembly of individual organelles.

THE GLOBAL PATTERN IN CILIATES IS INHERITED DURING FISSION

Each time a ciliate divides, the cell's global pattern is duplicated ahead of time. However, the mechanisms underlying the duplication of the global pattern appear to be different from the directed assembly processes discussed so far. For instance, in *Pleurotricha*, and several other ciliates, a second oral apparatus is formed far away from the original apparatus in the posterior half of the cell, so that after fission both daughter cells have an oral apparatus in the appropriate place (see Fig. 19.15). The new oral apparatus, along with other duplicated pattern elements, is positioned with great precision. This suggests that ciliates have some *long-range* control mechanisms that maintain their global patterns during fission.

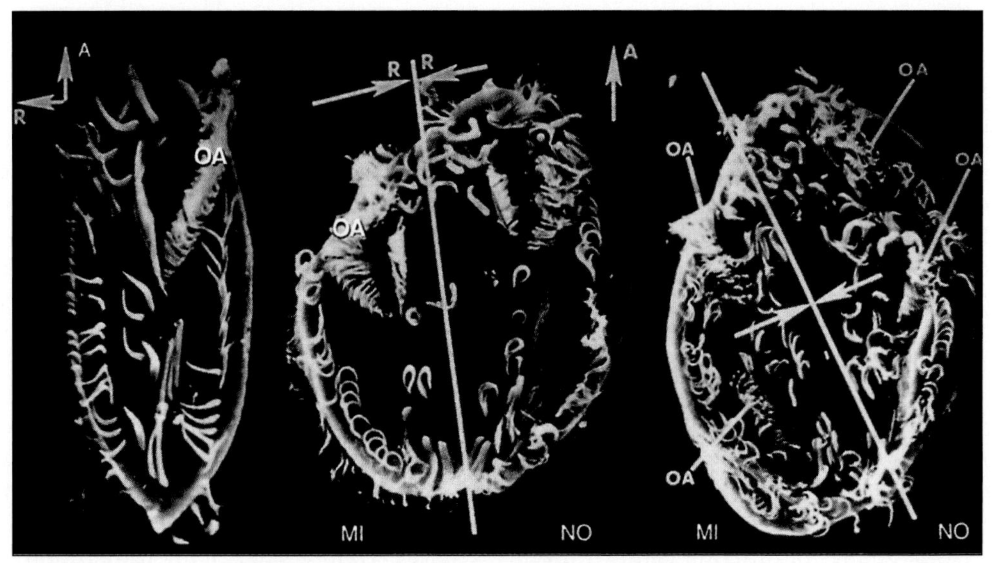

Figure 19.17 Scanning electron micrographs of the ciliate *Pleurotricha lanceolata.* **(a)** Ventral view of a normal cell. Arrows indicate the position of the anterior pole and the right-hand side. On the ventral surface are an oral apparatus (OA) and additional arrangements of packed cilia. The most conspicuous part of the oral apparatus is an oblique band of membranelles in the anterior left quadrant. Each membranelle consists of about 30 cilia packed in a rectangular array. In addition, clusters of cilia are arranged in one row to the left, in two rows to the right, and in a regular pattern in between. **(b)** Mirror-image doublet, fused at the right cell margins and showing duplications of the oral apparatus and the left ciliary row on both sides of the symmetry plane (white line). **(c)** Prefission mirror-image doublet, showing duplication of the oral apparatus and other ciliary pattern elements. NO = normal half of doublet; MI = mirror-image half of doublet.

These mechanisms seem to differ from the *close-range* mechanisms of directed assembly discussed so far, in which the newly assembled structures originate next to their seed structures or organization centers.

To analyze global patterning, investigators have generated abnormal cellular patterns by heat-shocking, by preventing cells from separating after conjugation, or by microsurgery (X. Shi et al., 1991). Incidentally, none of these treatments is known to cause any changes in DNA or RNA sequences. The most spectacular of the resulting pattern abnormalities is the ***mirror-image doublet*** cell. The doublets are like a cell and its mirror image, joined at the right margins, with both left margins turned laterally, anterior poles pointing in the same direction, and ventral structures facing the same way (Fig. 19.17b). The overall arrangement resembles the two hands of a person held side by side with both palms facing in the same direction. The global ciliary pattern of the left half corresponds to that of a normal singlet cell, while the right half takes on a mirror-image pattern. We will refer to the two parts of such a doublet cell as the *normal half* and the *mirror-image half.*

Before mirror-image doublets complete division, the components of the ciliary patterns are duplicated in both halves (Fig. 19.17c). Each half generates a second oral apparatus in its posterior region. Thus, doublets beget doublets. Occasionally doublets revert to singlets by resorption of the right (mirror-image) component. However, when such revertants are removed from a culture, the doublet pattern is stable for hundreds of generations. These observations show that the global cell pattern, whether normal or abnormal, is maintained during the cell cycle.

THE GLOBAL CELL PATTERN IS MAINTAINED DURING ENCYSTMENT

Pleurotricha and related ciliates can survive periods of starvation by ***encystment:*** They form a resting structure called a cyst, which is surrounded by a protective coat. When environmental conditions are favorable again, the cells undergo ***excystment:*** They emerge from the protective coat and resume a normal growth and division cycle. What happens with respect to global patterning during encystment and excystment is striking. All ciliary structures seem to disappear during encystment; no parts of cilia or basal bodies, or even microtubules, have been found in cysts by electron microscopy (Grimes and Hammersmith, 1980). However, these structures apparently reassemble from smaller organizing centers after excystment, perhaps in a process similar to the assembly of centrosomes in unfertilized eggs as discussed earlier. The disappearance of all ciliary structures during encystment means the loss of an overt global body pattern. However, the faithful reappearance of the same body pattern after excystment shows that the pattern is preserved by some other, as yet unknown type of paragenetic information.

Investigators have studied encystment to find out which aspects of *abnormal* body patterns are maintained in the absence of ciliary structures. Under varying environmental conditions, doublets encyst and excyst like normal cells. Surprisingly, and so far inexplicably, doublets always excyst as doublets, thus restoring the global body pattern that existed before encystment (Grimes, 1990). In contrast, local irregularities, such as some ventral cilia placed on the dorsal side of otherwise normal cells, are not reconstituted upon excystment. (Remember that similar irregularities are propagated faithfully during fission.) Apparently, the paragenetic information passed on through encystment is sufficient to preserve the *global* aspects of both normal and abnormal body patterns. However, the paragenetic information maintained through encystment does not suffice to preserve *local abnormalities* in the ciliary pattern.

GLOBAL PATTERNING IS INDEPENDENT OF LOCAL ASSEMBLY

The global pattern of *Pleurotricha* mirror-image doublets gives the impression of perfect mirror-image symmetry. For example, the oral apparatus of the normal half curves to the left, while that of the mirror-image half curves to the right (Fig. 19.17b). However, the impression of symmetry vanishes when we study the *local* organization of individual pattern elements. Each oral apparatus consists of many ***membranelles,*** each of which is typically made up of four rows of basal bodies with cilia (Fig. 19.18). These rows assemble from pairs of basal bodies that have a left-right *handedness,* which they impose on the membranelle. This handedness is the *same,* rather than *symmetrical,* in the two oral apparatuses of the doublet (Grimes et al., 1980). This is significant because it shows that the same local membranelle organization is compatible with *both* the normal global pattern and its mirror image. In this case, the global pattern and the local assembly of the pattern elements are not coordinated and, presumably, are determined by different types of paragenetic information.

A close look at the local assembly of oral apparatus membranelles will illustrate this important point. Each membranelle of the oral apparatus is composed of four rows of cilia. In a normal animal, and in the normal part of a doublet, the two posterior rows are the longest. The anteriormost row is composed of only three cilia at the anterior right corner of each membranelle. If this local assembly were coordinated with the overall pattern, then the membranelles in the two parts of the doublet would be mirror images of each other. However, this is not the case. In the mirror-image part of the cell, *both* the anteroposterior polarity and the right-left polarity are reversed. It is as though the membranelle had gone

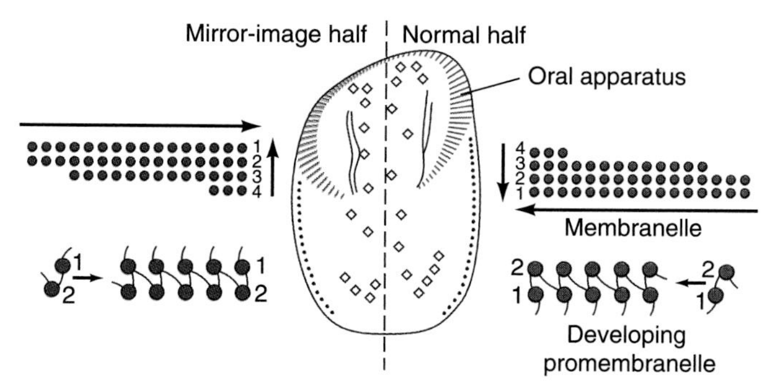

Figure 19.18 Discrepancy between the global pattern and the local assembly of oral apparatus membranelles in *Pleurotricha lanceolata.* The diagram in the center represents the ventral aspect of a mirror-image doublet, with each part having its own oral apparatus. The plane of symmetry is indicated by the dashed line. The colored lines around each oral apparatus represent membranelles. The top diagrams on each side represent mature membranelles, each consisting of four rows of basal bodies with cilia. The bottom diagrams show the assembly of the first two rows of a membranelle from dimers of basal bodies. Even though the overall arrangement of membranelles around the two oral apparatuses is approximately symmetrical, the individual membranelles in the two halves of the doublet are not mirror images of each other. Instead, the membranelles in the mirror-image half of the doublet are rotated 180° relative to the membranelles in the normal half. It appears that the basal body dimers can be assembled in only one configuration, so that true mirror-image assembly of a membranelle is not possible.

through a planar 180° rotation instead of developing as a mirror image. The local membranelle organization and the overall pattern of the oral apparatus are therefore incongruent. This inconsistency is also reflected in the abnormal function of the oral apparatus in the mirror-image half of the doublet. Instead of moving food particles into the oral apparatus, the membranelles move them out. Consequently, doublets feed only through the oral apparatus of the normal half, and an isolated mirror-image half of a doublet will starve to death.

The progress of oral apparatus formation suggests that membranelles can be assembled only in their normal configuration. This restriction seems to be imposed by the three-dimensional structure of basal body pairs. The development of true mirror images of membranelles would require right-handed and left-handed versions of basal body pairs. Since these do not exist, the anteroposterior axis has to be inverted along with the left-right axis. What is important for us to observe is that the rotated assembly pattern of the membranelles does not impose a rotated pattern on the entire oral apparatus. Therefore, the global pattern of the cell can be determined independently of the local membranelle assembly pattern.

Ciliates, then, employ more than one type of paragenetic information to maintain their body pattern through fission and encystment. Cortical units, membranelles, and other local pattern elements originate through local directed assembly. However, the global cellular pattern, including the orientation and position of the oral apparatus, is determined independently.

The physical basis of global patterning in ciliates is not yet known. The observations reviewed in this section suggest that it is not the close-range mechanisms that underly the cases of directed assembly discussed previously. Nor can genetic mutations explain the global pattern changes discussed here, for two reasons. First, the experimental manipulations that generate doublets are not mutagenic. Second, the doublet pattern is unaffected by conjugation with either normal cells or doublets. In corresponding experiments with *Paramecium,* Sonneborn (1963) showed that a doublet phenotype is not associated with nuclear genes or with components of internal cytoplasm that are exchanged during conjugation. It appears that the factors determining the global pattern are strictly associated with the cell cortex. However, the loss of cilia and basal bodies during encystment indicates that the pattern determinants can function independently of these overt structures.

19.6 Paragenetic Information in Metazoan Cells and Organisms

What do the lessons learned from ciliates tell us about paragenetic information in the cells of metazoans, such as fruit flies or mammals? Clearly, the cell surfaces of ciliates are highly specialized for multiple functions in swimming, self-defense, and nutrition. However, the cells of metazoa also have an internal organization that is critical to their survival and function. Are some aspects of this spatial organization of a cell passed on paragenetically, or is the entire organization lost at each cell division and reestablished through external cues, such as contact with other cells or extracellular materials? Both processes occur to varying degrees in different cells. We will focus here on paragenetic information.

In every cell, newly synthesized polypeptides have to be targeted to the appropriate cellular compartments. As discussed in Section 18.5, polypeptides have specific *signal sequences* or *signal patches,* which interact with matching cytosolic targeting particles and membranous *receptor proteins.* The interactions either initiate the transfer of the polypeptide across a cellular membrane or keep the polypeptide where it is (Fig. 19.19). For instance, proteins entering the endoplasmic reticulum (ER) have at their amino terminus a signal sequence that associates with a signal *recognition particle* and then with a *receptor protein* in the ER membrane (see Fig. 18.17). In addition, proteins enclosed in transport vesicles are

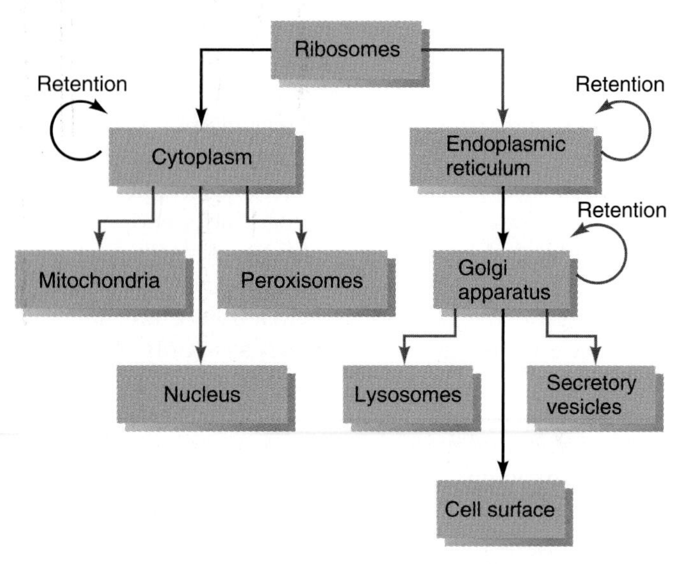

Figure 19.19 Simplified "road map" showing how polypeptides reach their cellular destinations. The signals that guide a polypeptide's movement along its route are contained in its amino acid sequence and in peptides on the outside of transport vesicles. These signals interact with matching receptors on the membranes of the appropriate cellular compartments. The journey of a polypeptide begins with its synthesis on a ribosome. At each station along its way, a decision is made whether the polypeptide is to be retained or transported further. A signal may be required either for retention or for moving on. (The signal-dependent pathways are shown in red color.) In the absence of the required signal, the polypeptide takes a default pathway (black). The signal sequences that determine the route of a polypeptide are encoded by the gene for the polypeptide. The molecular properties of the receptors with which the signal sequences interact are also encoded genetically. However, the distribution of these receptors in the cell is part of the preexisting cell structure. This structure is passed on, independently of DNA replication, during cell division.

guided to their destinations by proteins on the vesicle surfaces. These mechanisms direct the traffic of vesicles to mitochondria, to the nucleus, and toward other cellular destinations. Without the appropriate guidance of its polypeptide traffic, a cell would not be viable.

Targeting the correct destination depends on each polypeptide's own signal sequences and on matching sequences in targeting particles and receptors, all of which are controlled by the genes encoding these components. However, the *correct distribution* of receptors is not encoded in genes but is part of the preexisting spatial order in the cell. During mitosis, each new cell inherits a cytoplasmic organization that already has in place a complement of organelles with receptors for the signal sequences of new polypeptides. The correct positioning of the receptors presumably depends on the preexisting membrane structure and therefore could be classified as *directed assembly.*

Evidence for the direct propagation of cytoplasmic organization during cell division has also been obtained from observations on cytoskeletal elements in cultured fibroblasts (Albrecht-Buehler, 1977). During mitosis, these cells produce pairs of sister cells that look either identical or like mirror images of each other. The similarity is especially striking with respect to cell shape and microfilament organization (Fig. 19.20). Moreover, paired daughter cells leave behind mirror-image tracks as they move. Corresponding observations have been made on neuroblastoma cells (Solomon, 1981). These cells display a wide repertoire of morphological characteristics. Mitotic pairs of these cells are very similar with respect to cell shape and number of neurites as well as neurite length, thickness, and branching pattern. These observations show that cytoplasmic structures controlling cell morphology and cytoskeletal organization are passed on during cell division.

Finally, one can argue that spatial attributes of multicellular organisms, such as their anteroposterior and dorsoventral polarity, are passed on as paragenetic information throughout the life cycle. A fertilized egg contains all the genetic and paragenetic information necessary to carry the organism through all the subsequent stages in its life. Genetic information is encoded in the DNA of the nucleus, of the mitochondria, and possibly of other organelles. Paragenetic information is contained in several egg features, including localized *cytoplasmic determinants.*

How do animal eggs propagate the paragenetic information encoded in their cytoplasmic organization? In order to answer this question, we take the view that a fly (or other adult organism) is an egg's way of making another egg (Fig. 19.21). The more common view is that an egg is a fly's way of making another fly, but both views are equally valid since life cycles are cyclical.

Clearly, the egg organization is *not* passed on in its entirety to all blastomeres. Instead, different blastomeres inherit different cytoplasmic components and are thus determined to form different parts of the developing embryo. Inductive interactions between different cells, and morphogenetic movements guided by these events, generate the embryonic body pattern, which in turn gives rise to the adult body pattern. The paragenetic information that was originally localized in specific regions of the egg cytoplasm is now manifest in spatial relationships between the adult cells and organs. These relationships include the configuration of oocytes and other cells in the ovary.

The adult anatomy in turn regenerates the cytoplasmic organization in the next generation of eggs. As described in Section 3.5, *Drosophila* females have egg chambers, in which all nurse cells point toward the distal end of the ovariole and are connected to each other and to the oocyte by oriented microtubules. The organization of the nurse cell–oocyte complex causes bicoid

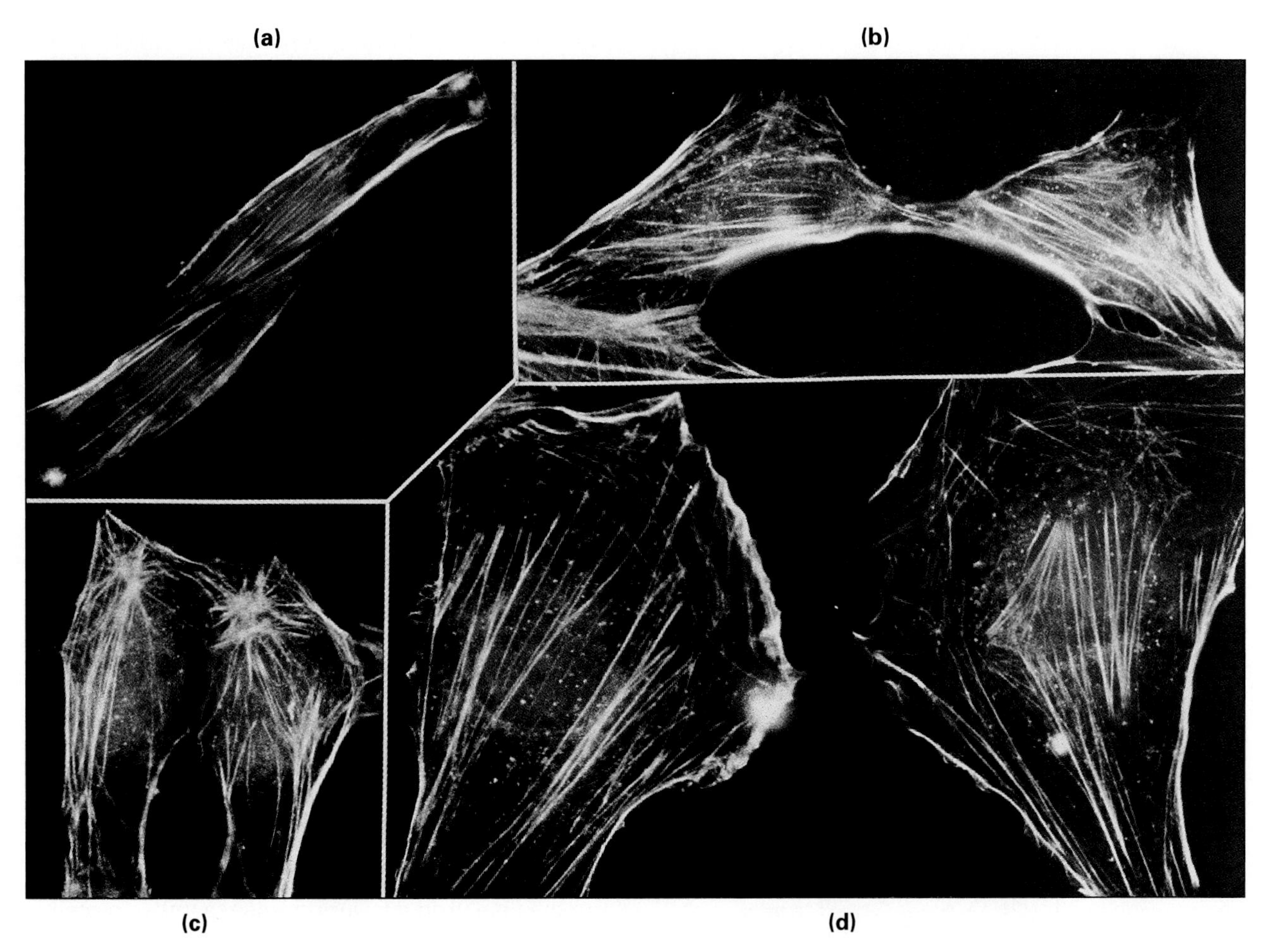

Figure 19.20 Microfilament bundles in mouse 3T3 fibroblasts immunostained for actin. Each photograph shows a pair of sister cells after mitosis. Note that the overall cell shape and the distribution of microfilaments are either identical **(a, c)** or mirror-symmetrical **(b, d)**.

mRNA to be localized near its point of entry into the oocyte (see Fig. 8.1). The site of bicoid mRNA localization in the oocyte then establishes its anterior pole of the embryo, and eventually the adult. Thus, eggs pass on the paragenetic information contained in their cytoplasmic organization by building adults that will recreate the same organization in their eggs.

In order to prove that cytoplasmic localization is necessary for an animal to complete its life cycle, it must be shown that disturbances of localization processes interfere with survival or reproduction. This has been shown by both genetic and experimental means. For instance, there are maternal effect mutants of *Drosophila* that transfer normal amounts of functional bicoid mRNA to the oocyte but fail to properly localize the mRNA at the entry point. Embryos developing from such eggs have abnormal body patterns and are not viable. Similarly, if anterior determinants in chironomid eggs are dislodged or inactivated by experimental means that do not affect nuclear DNA, the eggs nevertheless develop abnormal body patterns characterized by mirror-image duplications of the head or abdomen or by complete inversion of the anteroposterior polarity (see Fig. 8.22). Such embryos are inviable or infertile.

In summary, both genetic and paragenetic information are passed on throughout the life cycles of higher organisms, and both are necessary for survival. Some types of paragenetic information, such as cell shape and cytoskeletal organization, may be passed on directly during mitosis. Other types of paragenetic information, such as cytoplasmic localization in eggs, are propagated in a cyclical fashion by different, stage-specific means.

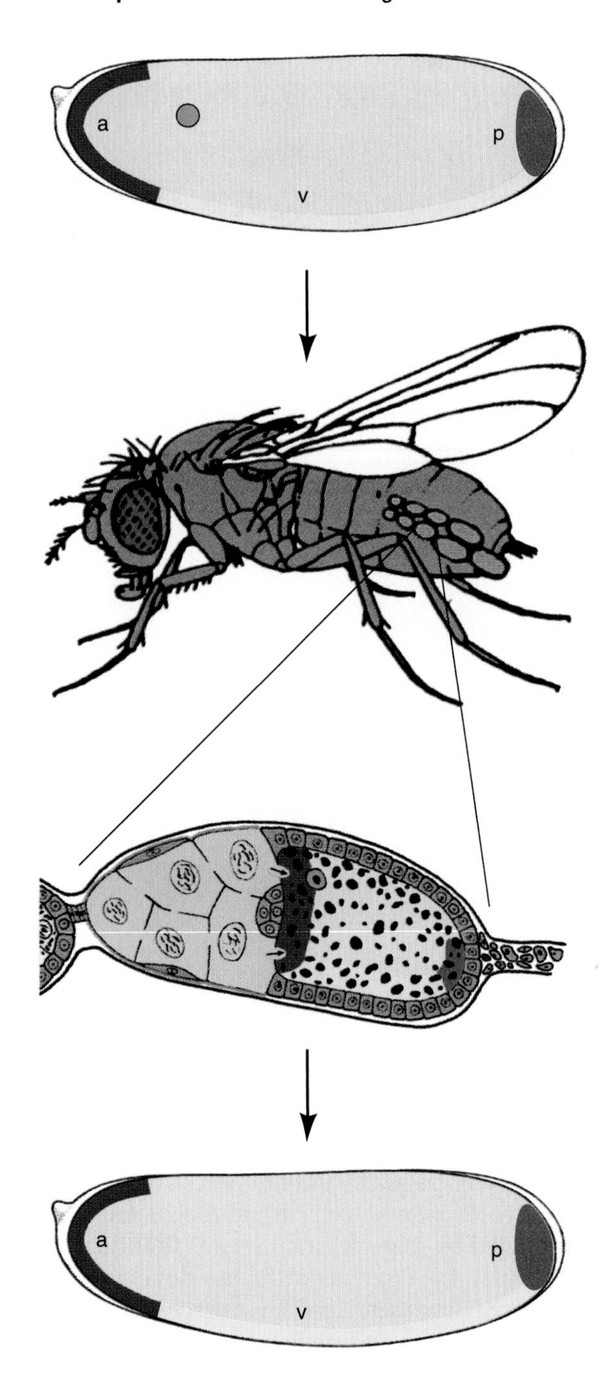

Figure 19.21 Propagation of cytoplasmic egg organization from one egg generation to the next. The adult fly is seen here as a device of the egg to make more eggs. The cytoplasmic egg organization encompasses anterior, posterior, and ventral cytoplasmic determinants. These are segregated into different blastomeres, which have inductive interactions with one another and undergo morphogenetic movements. The paragenetic information that was originally localized in specific regions of the egg cytoplasm is thus transformed into spatial relationships between the adult cells and organs. These relationships include the configuration of oocytes, nurse cells, and connecting microtubules in the ovary. The adult anatomy in turn regenerates the cytoplasmic organization in the next generation of eggs. *Drosophila* females have egg chambers, in which all nurse cells point toward the distal end of the ovariole. The organization of the nurse cell–oocyte complex causes bicoid mRNA and other anterior determinants (a) to be localized near the nurse cell end of the oocyte while posterior (p) determinants move to the opposite end of the oocyte. Ventral determinants (v) originate through an interaction of the oocyte nucleus, which is located dorsally, with adjacent follicle cells. Thus, eggs pass on the paragenetic information contained in their cytoplasmic organization by building adults that will recreate the same organization in their eggs.

SUMMARY

Cells originate only from preexisting cells. During division, cells inherit two types of information: genetic and paragenetic. Genetic information, encoded in DNA, is used in daughter cells to synthesize molecular building blocks for maintenance and growth. Paragenetic information is present in structural aspects of cellular organization and is not encoded in DNA.

Many biomolecules undergo self-assembly, a spontaneous association for which all the necessary energy and information are contributed by those components that form the final product. During self-assembly, proteins may change in conformation as they release energy and assume a more stable state. Self-assembly generates oligomeric proteins, small organelles such as ribosomes, and simple viruses. More complex organelles and viruses, on the other hand, are formed by aided assembly, a process that requires auxiliary components—such as catalysts, energy sources, or scaffolds—that will not become part of the assembled product. Self-assembly and aided assembly do not strictly depend on paragenetic information. However, their initial phases are greatly accelerated by the presence of seed structures.

The most prevalent assembly process in cells is directed assembly, a type of assembly that requires seeds or larger organizing centers. Alternate seeds often determine which one of multiple structures is assembled from a given pool of molecular building blocks. At the molecular level, certain proteins can impose their own three-dimensional conformation on other proteins with which they associate.

In ciliated protozoa, basal bodies and other cortical structures are duplicated by directed assembly before cells undergo fission. Local irregularities in the arrangement of the cortex are passed on indefinitely in this process. The inheritance of such cortical abnormalities is not associated with any genetic changes, but is propagated locally by the directed assembly of cortical units.

In addition to replicating local cortical structures, ciliates also propagate their global body pattern during fission. Under certain conditions, global patterning can be made to take place separately from the local assembly of ciliary structures. The mechanisms underlying global cell patterning appear to differ from those of locally directed assembly.

Metazoan cells pass on paragenetic information during cell division. Paragenetic patterns include the distribution of receptor proteins that target newly synthesized polypeptides to their destinations.

The eggs of most metazoa contain paragenetic information in the form of localized cytoplasmic determinants. The cytoplasmic organization of the egg is passed on through the embryonic and adult body pattern until the anatomy of the adult ovary regenerates the egg's cytoplasmic organization. The effects of genetically or experimentally induced changes in cytoplasmic localization show that both paragenetic and genetic information are necessary for survival and must be passed on properly throughout the life cycle.

part three

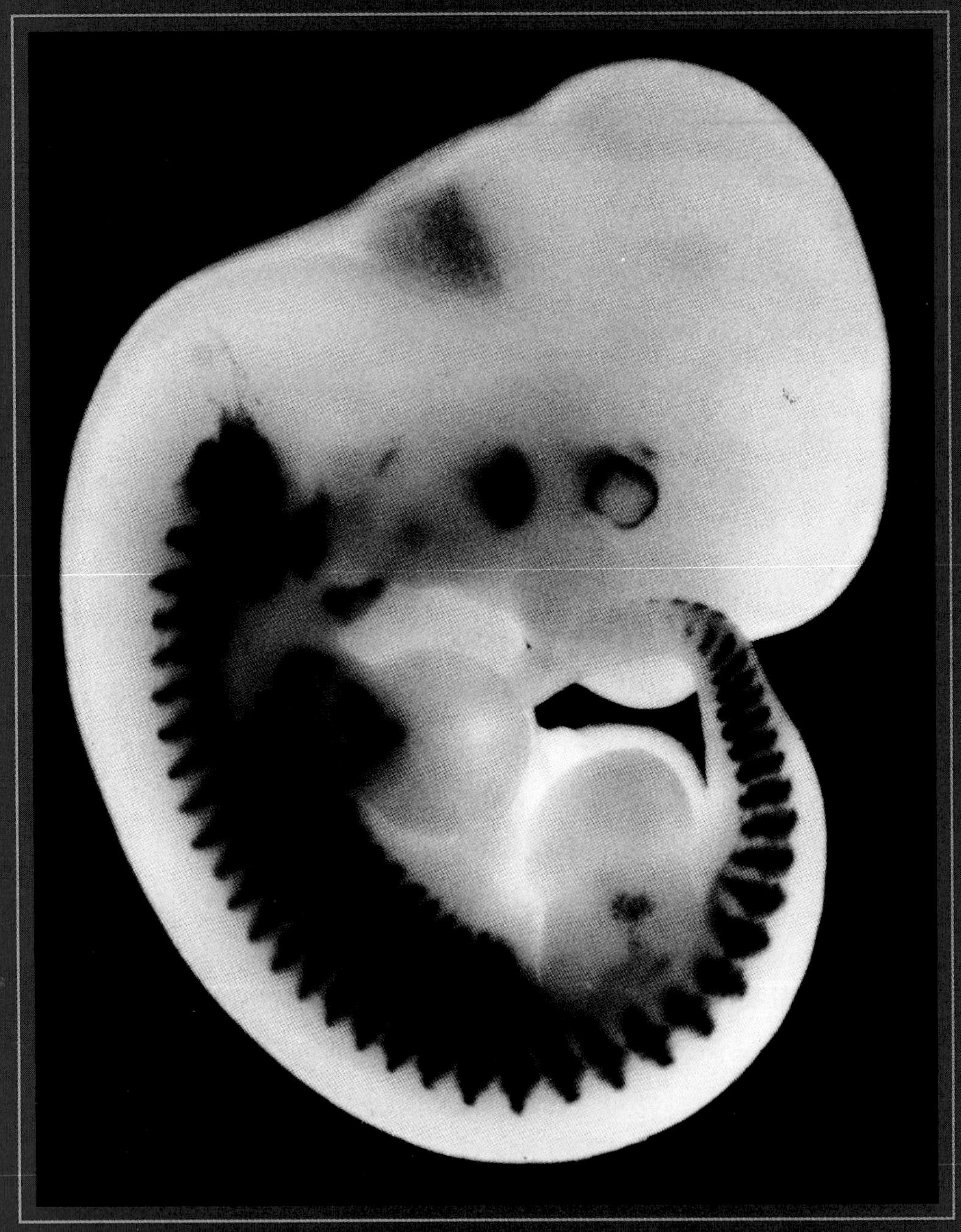

Figure III.1 Mouse embryo at 11.5 days of development. This embryo harbors a transgene consisting of the transcribed region of the bacterial β-galactosidase gene and the regulatory region of the mouse *myogenin*$^+$ gene. The dark stain, caused by β-galactosidase, indicates that the transgene is expressed with the same pattern as the normal *myogenin*$^+$ gene—that is, in the myotomal regions of the somites and the limb buds. The *myogenin*$^+$ gene is necessary and sufficient for the differentiation of skeletal muscle tissue.

Selected Topics in Developmental Biology

IN Part Three of this text, we will discuss a series of topics that are at once old and new. Although extensively investigated by early embryologists, these topics are still providing new frontiers for researchers. Interest in these questions has been revitalized by the arrival of new research methods such as advanced microscopy and genetic analysis, as well as DNA cloning and sequencing. Bringing these modern tools to bear on the long-standing problems of the discipline is what generates much of the current excitement in developmental biology. ● Two of the topics at the core of developmental biology are cell differentiation and pattern formation. How can cells that originate from the same rudiment produce different tissues such as bone and muscle? And how are different pathways of cell differentiation orchestrated in time and space, so that a well-formed organ like the human hand results instead of a randomly aggregated mass of different tissues? These questions will be discussed in general terms and specifically with regard to four of the organisms that are amenable to genetic analysis: the fruit fly *Drosophila,* the house mouse *Mus musculus,* the mouse ear cress *Arabidopsis thaliana,* and the roundworm *Caenorhabditis elegans.* ● The topics of sex determination, hormonal control, growth, and senescence are included in Part Three because their discussion relies heavily on genetic and molecular analysis. In particular, we will consider the following questions: How do genes or the environment determine sex? How do hormones regulate sexual development and metamorphosis? How are cell division and growth rates controlled in different tissues and in different body regions? What can we learn from tumors about the genes that control normal growth? What are the roles of programmed cell death and senescence in development? ● Developmental functions are controlled by autonomous programs within cells, by close-range interactions among neighboring cells, and by hormones and other long-distance signals. Integrating all these controls requires complex networks of genetic interactions. The analysis of these signaling pathways has become a major effort in contemporary developmental biology. More often than not, regulatory networks turn out to be amazingly complex, perhaps more so than would be required for the functions to be performed. However, such networks were not designed all at once; they have evolved. In this respect, they resemble European cathedrals whose construction began in medieval times. Through the ages, these churches have been expanded, renovated, and redecorated according to the needs and tastes of each time. The cumulative results of these changes are still functional and often marvelous buildings. However, some of their design features are more intricate than their current functions require and are best appreciated from a historical perspective.

chapter 20

Cell Differentiation

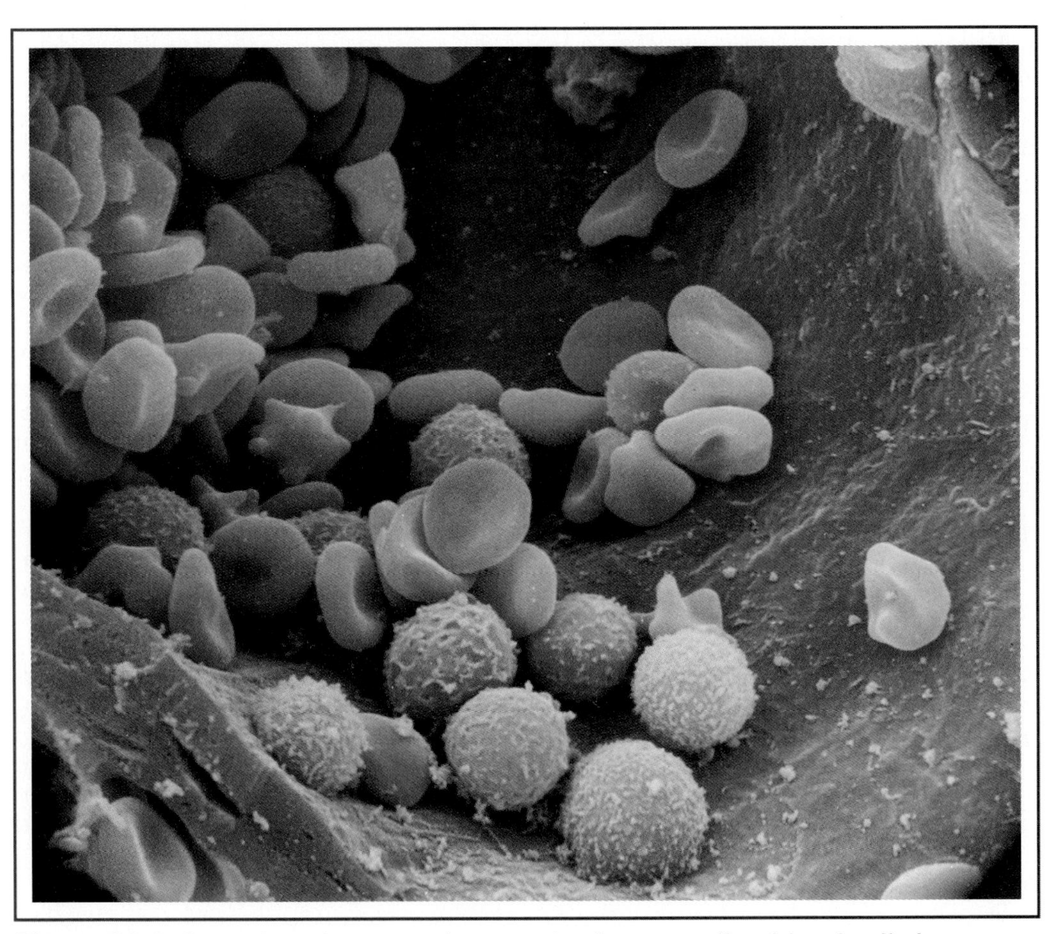

Figure 20.1 Scanning electron micrograph of mammalian blood cells in a small blood vessel. The doughnut-shaped cells are erythrocytes. The spherical cells in the foreground are leukocytes. All types of blood cells arise from one type of stem cell. *(From* Tissues and Organs: A Text Atlas of Scanning Electron Microscopy, *by Richard G. Kessel and Randy H. Kardon. W. H. Freeman & Co., 1979.)*

THE number of cells in organisms ranges from one to the tens of trillions. Regardless of the total number, each organism has a relatively small catalog of *cell types*. Each cell type is defined by its morphology and by characteristic molecules that can be detected with antibodies or other specific probes. Most cell type designations, such as neuron or muscle fiber, refer to mature and specialized cells that have reached the final stage of development in their lineage. The mature state of a cell is also called its *differentiated state* to emphasize the fact that organisms have different types of mature cells. The process by which a cell reaches this state is called *cell differentiation.* It initially occurs during histogenesis, when tissues assume their special functions in the larva or adult (see Chapters 13 and 14), and it continues throughout the life of an organism as worn-out cells are replaced by new ones. Recent work has revealed some of the molecular mechanisms underlying cell differentiation. They provide a basis for understanding two classical features of cell differentiation: the limited number of differentiated cell states, and their relative stability.

In this chapter, we will inspect two types of differentiating cell populations. In one type, which includes liver cells, differentiated cells can still divide. In the other type, cells no longer divide in the differentiated state. Instead, these cell populations renew themselves from a small pool of undifferentiated reserve cells known as *stem cells.* We will briefly examine both types of cell populations in small freshwater polyps. The main attraction of these organisms is their simple anatomy and the ease with which they can be manipulated. Next, we will consider the differentiation of mammalian blood cells, arguably the best-investigated stem cell population (Fig. 20.1). Many of the genes that regulate blood cell development have recently been cloned and characterized. In addition, we will explore the genetic control of muscle cell differentiation. This involves an interesting class of transcription factors that form heterodimers in which one monomer enhances or inhibits the other's activity. Finally, we will briefly consider the wide-ranging potency of stem cells and some prospects for their medical use.

20.1 The Principle of Cell Differentiation

Developmental biologists have long studied how various cell types reach their differentiated state. The general characteristics of this process are summarized in the ***principle of cell differentiation:*** Multicellular organisms develop a limited catalog of distinct cell types that are usually stable once they have matured. This principle applies to all multicellular organisms and is basic to development.

EACH ORGANISM HAS A LIMITED NUMBER OF CELL TYPES

The number of cell types in an organism is much smaller than its total number of cells. For instance, a hydrozoan like the freshwater polyp consists of tens of thousands of cells but has only seven cell types. Humans and other large vertebrates are made up of trillions of cells but only about 200 cell types. Familiar examples of vertebrate cell types include epithelial cells, nerve cells, blood cells, and muscle cells. Each of these cell types is subdivided further. Muscle cells, for example, include skeletal muscle fibers, cardiac muscle cells, smooth muscle cells, and myoepithelial cells. Skeletal muscle fibers, in turn, are subdivided into red fibers, white fibers, intermediate fibers, muscle spindles, and satellite cells. Some cells take many different shapes. In particular, neurons have various processes of different lengths and branching patterns, which make classification difficult and somewhat arbitrary.

Each cell type has a distinctive set of structural and functional characteristics. The *endothelial cells* that line the inside of the cardiovascular system form a very thin cellular layer surrounded by a basal lamina (Fig. 20.2). These cells are adapted to the efficient transport of material between the lumen of the blood vessel and the surrounding tissue, and they regulate the transit of white blood cells in and out of the blood. In contrast, the *photoreceptors* in the retina of the eye are specialized for responding to light (Fig. 20.3). Their light-sensitive outer segments—known as rods and cones according to their shape—contain different complexes of proteins with

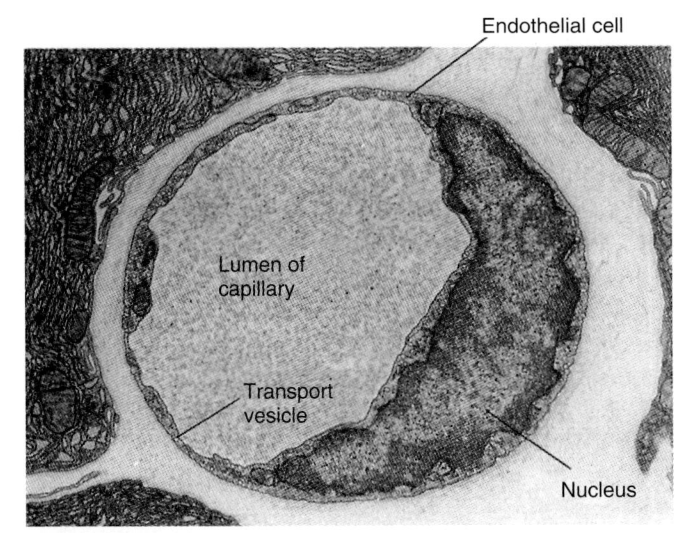

Figure 20.2 Endothelial cell forming a blood capillary. This electron micrograph shows a cross section of a single endothelial cell with joined margins to enclose the lumen of the capillary. The cell cytoplasm contains small vesicles that transport large molecules across the endothelium; the vesicles originate by endocytosis on one side and are given off by exocytosis on the other side.

Figure 20.3 The retina of the vertebrate eye consists of several layers of cells. The slender elements at the top—known as rods and cones—are the outer segments of a layer of photoreceptors. Stimulation of the rods and cones by light is transmitted through subsequent layers of interneurons and ganglion cells to the brain. (*From Gene Shih and Richard Kessel, Living Images. Science Books International, 1982.*)

visual pigments. There are many other distinct cell types, each with specific morphological and molecular features.

Mature cell types are discrete, with no intermediate forms. The names of some cell types, such as *myoepithelial cells*, suggest that the cells are intermediate between two distinct types. However, myoepithelial cells are branched cells containing myofibrils, which typically surround gland cells like a basket and help to empty them by constricting (Fig. 20.4). Thus, myoepithelial cells are muscle cells and not intermediate between muscle cells and epithelial cells. Intermediates, when found, are usually in some stage of a developmental sequence. Gradual changes take place during the development of each cell type, but in their mature states cells are usually of discrete types.

THE DIFFERENTIATED STATE IS GENERALLY STABLE

A differentiated cell is generally stable. Most differentiated cells do not transform into other cell types as part of normal development. Bone cells do not normally become muscle cells, and muscle cells do not turn into epithelial cells. Under experimental conditions, one can observe dramatic exceptions to this rule. For instance, an isolated differentiated plant cell may *regenerate* entire new plants with all the differentiated cell types characteristic of the species (see Section 7.5). The iris of a salamander's eye can regenerate a new lens, whose cells have an entirely different morphology and synthesize a new set of proteins (see Fig. 7.4). Typically, regeneration involves

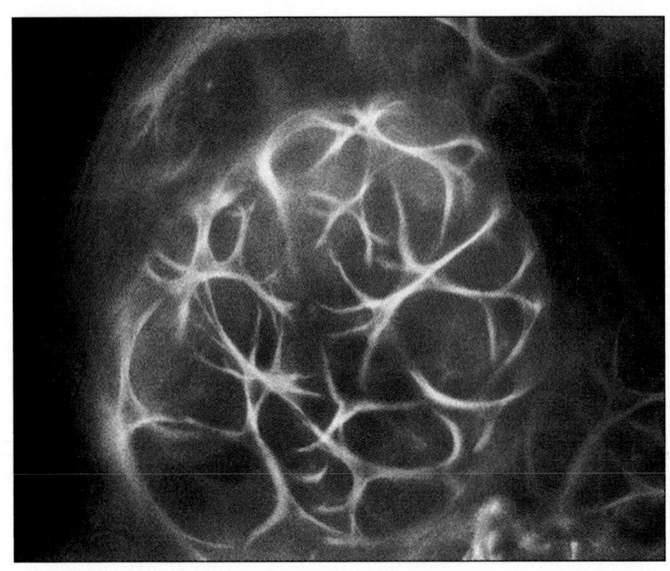

Figure 20.4 Myoepithelial cells in a mouse mammary gland, made visible by immunostaining (see Method 4.1) of their abundant actin filaments. Myoepithelial cells are a specific type of muscle cell that is associated with ectodermal epithelia such as glands. The "arms" of each myoepithelial cell embrace a few gland cells and serve to expel milk and other secretions.

phases of *dedifferentiation,* during which cells lose their differentiated character and proliferate before they differentiate again in new ways. Finally, some cells, such as the follicle cells of ovarian egg chambers in insects, show *sequential polymorphism* (see Fig. 7.3). These cells assume different structures and functions at successive stages of development. However, such cells represent exceptions rather than the rule.

THE SAME CELL TYPE MAY BE FORMED THROUGH DIFFERENT DEVELOPMENTAL PATHWAYS

Cells become different in a stepwise process, which may begin early in development. In many embryos, the animal blastomeres are smaller and contain less yolk than vegetal blastomeres. In mammalian embryos at the morula stage, the external cells are polar and form tight junctions, while the internal cells are apolar and connected by gap junctions. Some of these differences are directly visible under the microscope. Others, such as the synthesis of particular gene products, can be detected only with molecular probes. However, calling cells *differentiated* as soon as they become *distinguishable* would inflate the term "differentiated" to the point of rendering it useless. We will therefore use this term only in reference to mature cells.

The concept of cell types pertains to the final structure of cells, regardless of their location in the body. For instance, the same types of skeletal muscle cells are found in the head, trunk, arms, and legs. Thus, two cells may be of the same cell type but have different developmental histories. Cells may even reach the same differentiated state through different mechanisms of cell determination: the same type of ascidian tail muscle cell may arise from primary muscle cells containing myoplasm or from secondary muscle cells without myoplasm (see Section 8.5).

Many differentiated cells are highly specialized and are dependent on other cells for their support. For instance, the neurons in the central nervous system depend on *glial cells* for insulation. The entire nervous system is highly dependent on the respiratory and digestive systems for nourishment and gas exchange. Thus, maintaining the differentiated state of one cell type often depends on the differentiated state of other cells.

20.2 Cell Differentiation and Cell Division

The initial differentiation of many cell types occurs simultaneously during the embryonic period of histogenesis. However, cell differentiation is also an ongoing process that occurs throughout life. Since most cells have a shorter life span than the organism in which they are found, cells in most tissues are continually dying and being replaced. The rate of this turnover differs from tissue to tissue. The epithelial cells of the small intestine are renewed every few days, whereas the turnover of cells in the pancreatic gland takes a year or more. Many tissues in which turnover is normally slow can be stimulated to replace cells faster when the need arises.

During adult life, new differentiated cells are produced in one of two ways. First, differentiated cells can divide, thus producing a pair of daughter cells of the same type. Second, new cells can differentiate from a pool of undifferentiated reserve cells called stem cells.

SOME CELLS DIVIDE IN THE DIFFERENTIATED STATE

In the epithelia of many glands, differentiated cells turn over slowly by cell division and cell death. The liver, which in part functions as a large gland, renews itself this way. Normally, liver cells are long-lived and divide slowly. However, in response to food poisoning or injury, the surviving cells divide at a more rapid pace. A standard operation on rats to investigate this response has been *partial hepatectomy*, the surgical removal of two-thirds of the liver. With great regularity, the remaining one-third regenerates a nearly normal-sized liver within 2 weeks (*Michalopoulos and DeFrances, 1997*). This capability of the liver was already described in the Greek myth of Prometheus. When Prometheus stole the heavenly fire, Zeus chained him to a rock and sent an eagle to eat from his liver, which constantly renewed itself.

Like other organs, the liver contains several cell types (Fig. 20.5). The liver's principal cells are known as

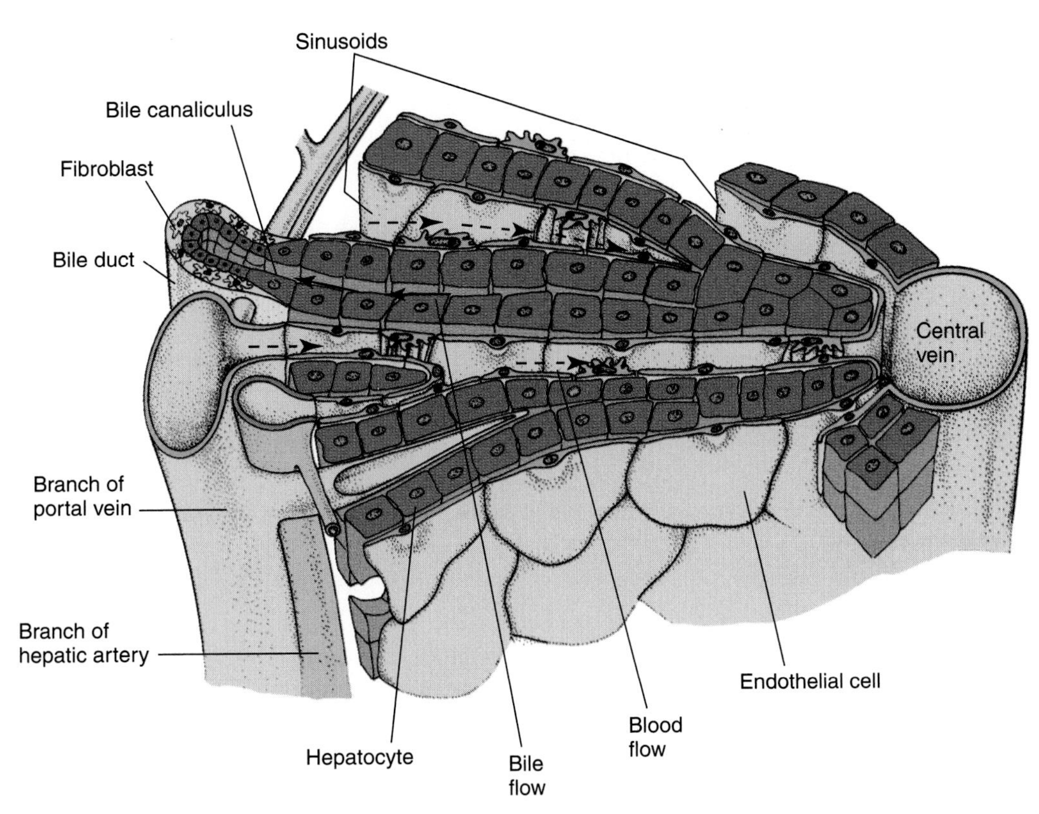

Figure 20.5 Structure of the liver. The hepatocytes (green) are arranged in plates between blood-filled sinusoids connecting arteries and veins. The sinusoids are lined with endothelial cells. Bile formed by the hepatocytes drains into bile ducts. Portions of liver tissue are held together by connective tissue.

hepatocytes. Their major function is to secrete bile salts, which solubilize dietary fat. In addition, hepatocytes are involved in a wide range of metabolic functions, including glucose and cholesterol metabolism as well as the synthesis and degradation of blood proteins. Other abundant liver cells are *endothelial cells,* which line the blood-filled *sinusoids* that connect liver arteries and veins. In addition, liver contains *fibroblasts,* which provide a framework of supporting connective tissue. Moreover, there are *macrophages,* which break down worn-out blood cells, and several other cell types.

For balanced liver regeneration, each cell type must divide according to the losses sustained. Food poisoning affects hepatocytes more strongly than the other liver cells, and hepatocytes respond most quickly and vigorously to the signals that trigger regeneration. Indeed, a single healthy rat hepatocyte can regenerate enough cells for 50 rat livers. However, if the liver is poisoned continually with alcohol or other drugs, the hepatocytes no longer divide fast enough, and the liver becomes clogged with connective tissue—a condition known as *cirrhosis.*

The signals that stimulate and coordinate the division of liver cells have been extensively investigated. To test whether these signals spread through the bloodstream, investigators joined the blood circulations of pairs of rats and then removed part or all of one rat's liver. Indeed, hepatectomy in one rat triggered extra growth of the other rat's intact liver. Molecular studies have revealed that liver regeneration involves a complex network of signals including *hepatocyte growth factor, epidermal growth factor, transforming growth factor,* and several other regulatory molecules.

Cells of another type that divide in the differentiated state are the *endothelial cells* of the circulatory system. During growth, wound healing, and regeneration, blood vessels supply nearly all tissues according to their physiological needs. New blood vessels always originate from the endothelial cells of existing blood vessels. These cells first form a solid sprout, which then hollows out to form a tube (Fig. 20.6). This process is known as *angiogenesis,* and the substances stimulating it are called *angiogenic factors.* These factors are released by the surrounding cells in response to oxygen deprivation or a wound. Several angiogenic factors have been isolated (*Risau, 1997; Yancopoulos et al., 1998*). Some of these, including *fibroblast growth factor* and *transforming growth factor,* also stimulate the proliferation of hepatocytes (see above) and act as inductive signals during embryogenesis (see Section 9.7). Other angiogenic factors, such as *vascular endothelial growth factor* (*VEGF*), seem more specifically involved in *angiogenesis* and *vasculogenesis* (see Section 14.6).

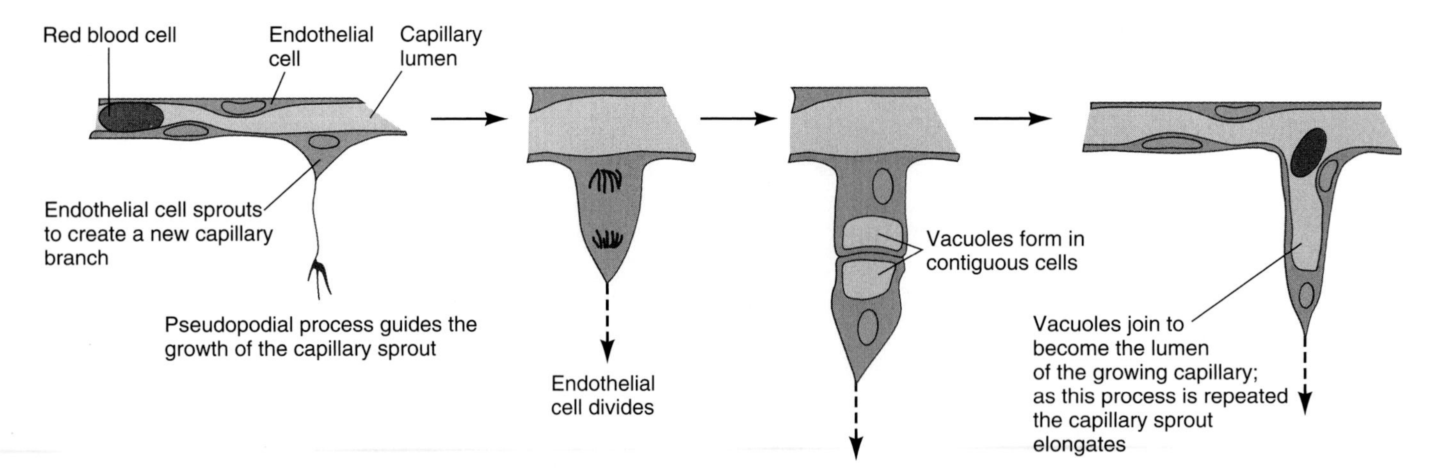

Figure 20.6 Schematic illustration of angiogenesis. A new blood capillary forms through the sprouting and mitosis of an endothelial cell in an existing capillary.

OTHER CELL POPULATIONS ARE RENEWED FROM STEM CELLS

Tissues in which differentiated cells no longer divide contain a type of reserve cells, known as *stem cells,* that remain capable of dividing. Stem cells have three main characteristics (see Fig. 3.11; *Morrison et al., 1997;* Fuchs and Segre, 2000; *Watt and Hogan, 2000*). First, they are *undifferentiated*—that is, not yet adapted to a specialized function in the organism. Second, stem cells have the capacity for *self-renewal*—that is, the ability to generate more stem cells through mitotic divisions. Typically, the *ability* to self-renew lasts for the life of the organism, even though the *actual pace* of stem cell divisions may be slow. Third, stem cells give rise to *committed progenitor cells*—that is, cells that are still undifferentiated but already restricted in their potential to the formation of certain differentiated cells. (Most investigators of stem cells use the term "commitment" as a synonym for *determination* as defined in Section 6.4.) The commitment of the progenitor cells is irreversible; their daughter cells do not normally become stem cells again. Typically, the committed progenitor cells go through further *amplifying divisions* before they become differentiated cells.

Renewal by stem cells is observed in a wide range of cell populations, including blood cells, epithelial cells, and spermatogonia in vertebrates. Typically, stem cells reside in safe and sequestered places, while their committed descendants move out to more exposed sites in the course of maturation. For instance, the epithelium lining the inner surface of the intestine consists of a single layer of cells. The epithelium covers the villi projecting into the lumen of the gut as well as the deep crypts that descend into the underlying connective tissue. The stem cells of the epithelium lie in the most protected positions near the bases of the crypts (Fig. 20.7). The committed progenitor cells, in the course of their amplifying divisions and subsequent differentiation, are pushed out of the crypt and up the villi, where they are exposed to abrasion from food and chemical attack from digestive enzymes. After a few days, they reach the tips of the villi, from which they are discarded. Similarly, epidermal cells are "born" in the deepest layer of the epidermis, and blood cells are born in bone marrow.

STEM CELLS MAY BE UNIPOTENT OR PLURIPOTENT

Some stem cells produce only one type of committed progenitor cell and are therefore called ***unipotent.*** Vertebrate stem cells forming the intestinal epithelium, the epidermis, or spermatogonia belong to this type. ***Pluripotent*** stem cells, on the other hand, give rise to several types of committed progenitor cells. Thus, when a pluripotent stem cell divides, the daughter cells can again be stem cells or any one of a small number of committed progenitor cell types; each progenitor cell then undergoes amplifying divisions and eventually produces a clone of differentiated cells. Most of the stem cells discussed in this chapter are pluripotent.

We will examine two pluripotent stem cell systems: the *interstitial* cells of freshwater polyps, and the *blood cells* of mammals.

20.3 Stem Cell Development in *Hydra*

Small freshwater polyps of the genus *Hydra* have been the subject of experimental studies since 1740, when this creature caught the eye of Abraham Trembley, a Swiss naturalist, philosopher, and tutor of children from affluent families (Lenhoff and Lenhoff, 1988). With only magnifying glasses and boars' hairs as tools, Trembley discovered that *Hydra* could reproduce asexually by budding, and that cut parts would regenerate whole polyps.

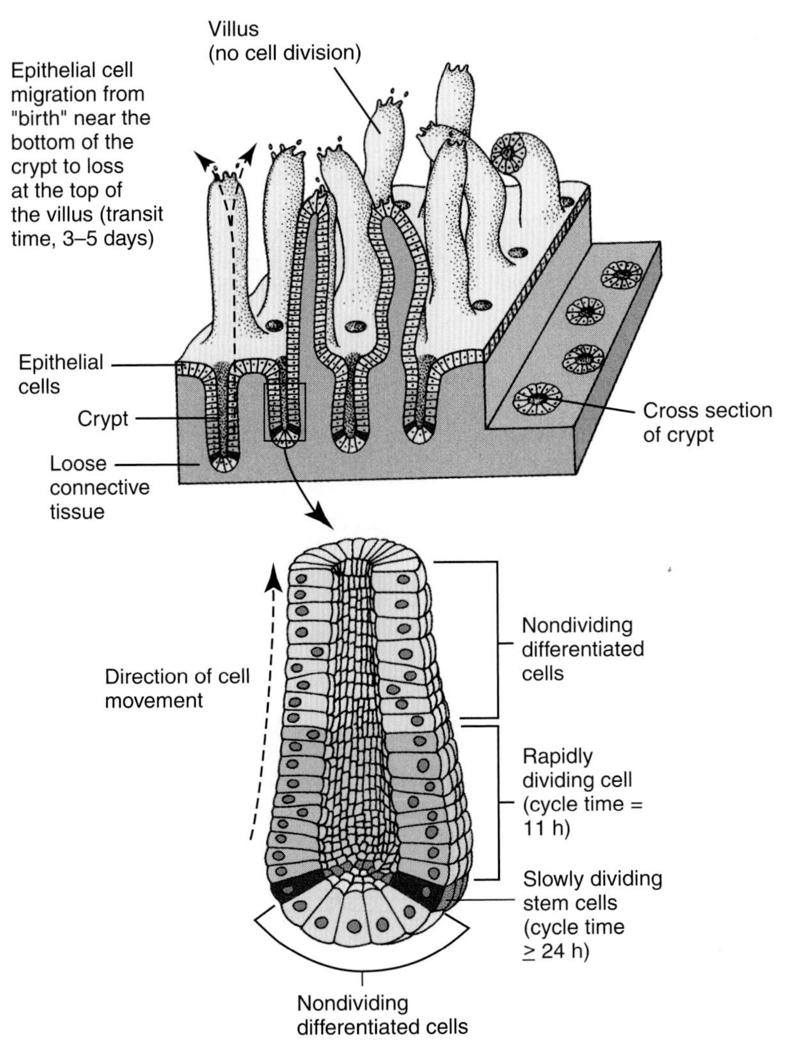

Figure 20.7 Turnover of epithelial cells in the small intestine. Stem cells located near the bases of crypts divide slowly and produce rapidly dividing committed progenitor cells. The cells differentiate as they move out of the crypt and onto the villi, where they are most exposed to digestive enzymes and soon die.

THE ORGANIZATION OF *HYDRA* IS RELATIVELY SIMPLE

Since Trembley's time, many zoologists have noted that *Hydra* offers several advantages for experimental work. First, *Hydra's* powers of regeneration are indeed enormous. (The polyp is named after the mythical serpent with multiple heads, any of which, if cut off, would grow back as two.) Even in a small segment of a *Hydra* polyp, cells will reorganize to create a small but complete individual. Second, with only a handful of somatic cell types, the organization of *Hydra* is relatively simple. Third, when an *adult* polyp is dissociated into single cells, the cells will reaggregate and form a polyp again. This capability, which in other animals is limited to *embryonic* cells, allows investigators to isolate *differentiated* cells and recombine them into entire organisms.

The body of *Hydra* is essentially a tube with one open end, called the ***hypostome,*** which serves as both mouth and anus (Fig. 20.8). The hypostome is surrounded by a ring of tentacles, which are used for defense and for catching prey, such as small crustaceans. The hypostome and the surrounding tentacles together are called the ***head*** of the hydra. At the opposite end of the tube there is a mucus-secreting ***foot*** with which the animal attaches to the substratum. Between the head and the foot extends the ***body column.*** Its central region, which is distended after a meal, is called the ***gastric region.***

Hydra normally reproduces asexually by budding, although under adverse circumstances it may reproduce sexually by means of eggs and sperm. In a well-fed individual, there is a continuous flow of cells from the middle of the animal toward the extremities. Most cells are born in the gastric region. From there, they move to the foot, to the buds, which are eventually pinched off, and to the tentacles, where most cells are lost at the tips.

The cells of *Hydra* are arranged in two epithelia: an outer epidermis, and an inner ***gastrodermis.*** These epithelia are separated by the ***mesoglea,*** a layer of extracellular material to which the two cellular layers adhere. Both cellular layers consist for the most part of ***epitheliomuscular cells.*** These cells form tightly coherent epithelial sheets. In addition, the epitheliomuscular cells have a broadened base with contractile filaments that allow the animal to elongate and bend, as it must in the pursuit of prey. Despite their muscular properties, the epitheliomuscular cells are usually called *epithelial cells* for short. In the gastrodermis, the epithelial cells can also absorb nutrients. In addition, both epidermis and gastrodermis contain nerve cells. Moreover, each epithelium contains cells to carry out its special functions. The gastrodermis contains gland cells, which secrete digestive enzymes; and the epidermis of the body column may hold gametes (eggs or sperm). The epidermis of the tentacles also contains numerous ***nematocytes*** (sting cells), which are equipped with elaborate devices to poison and hold prey.

Interspersed between the epithelial cells of both layers are *Hydra's* most versatile cells, the ***interstitial cells.*** There are large interstitial cells (about 10 μm in diameter) and small interstitial cells (about 5 μm). The large interstitial cells include *pluripotent stem cells* for nematocytes, gland cells, nerve cells, eggs, and sperm, as well as unipotent stem cells for eggs or sperm (Fig. 20.9). The small interstitial cells are thought to be mostly precursor cells of nematocytes and nerve cells.

In summary, *Hydra* consists of three cell lineages: the epithelial cells of the epidermis, the epithelial cells of the gastroderm, and the interstitial cell system. The two epithelial lineages generate one cell type each, both of

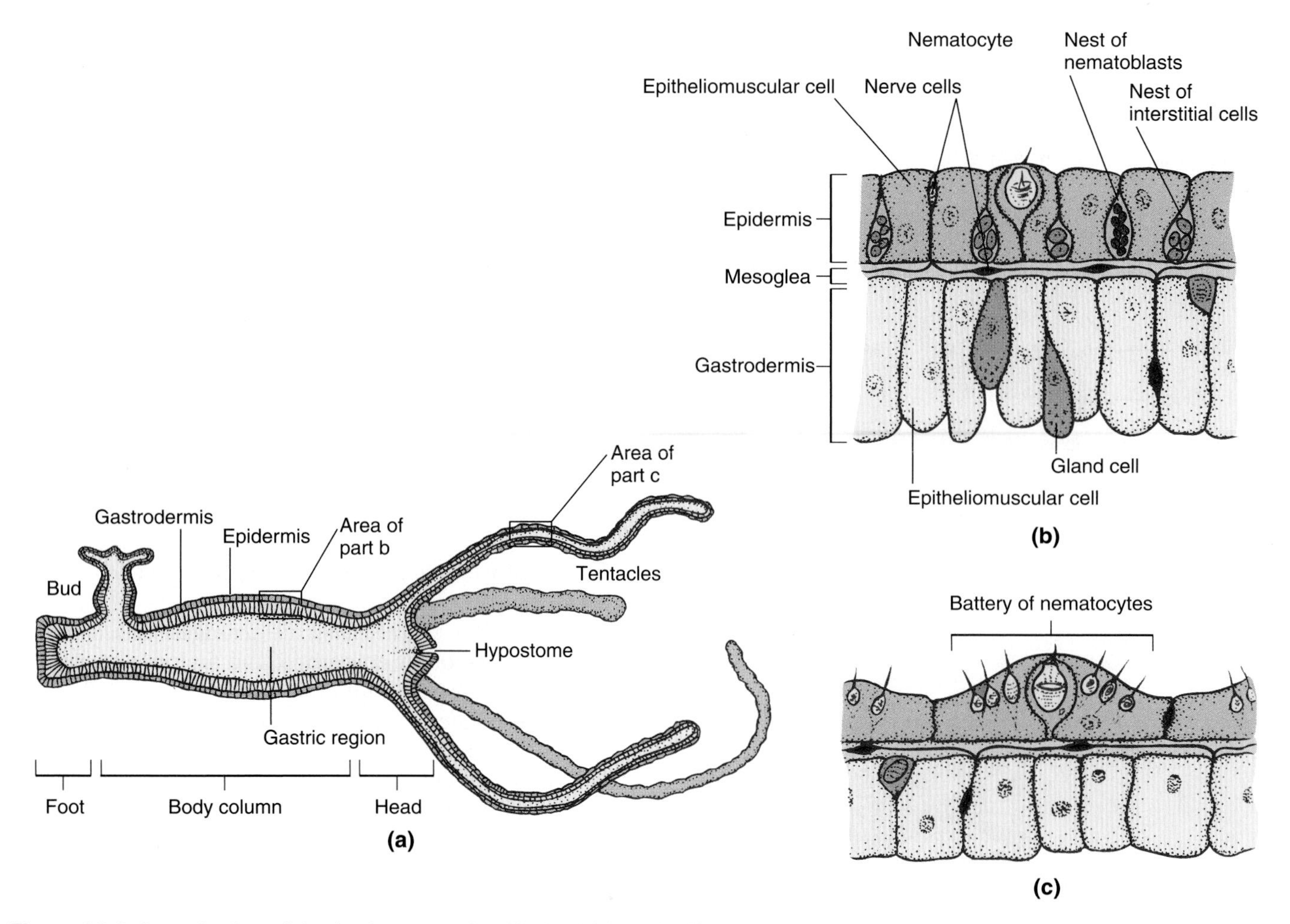

Figure 20.8 Organization of the freshwater polyp *Hydra.* **(a)** Longitudinal section of the entire animal. **(b)** Section of the gastric region. Two cellular layers, epidermis and gastrodermis, adhere to an acellular mesoglea. The gastrodermis, lining the inside of the animal, consists of epitheliomuscular cells and gland cells. The epidermis, on the outside, contains epitheliomuscular cells, nematocytes (sting cells), and developing nematocytes (nematoblasts). Nerve cells and interstitial cells (including stem cells) are found in both layers. **(c)** Section of a tentacle showing the numerous nematocytes in the epidermis.

which divide as differentiated cells. The interstitial cells form a pluripotent stem cell system that generates five cell types: eggs, sperm, nerve cells, nematocytes, and gland cells.

INTERSTITIAL CELLS CONTAIN PLURIPOTENT AND UNIPOTENT STEM CELLS

The interstitial cell lineage is more sensitive than the other two lineages to certain drugs including colchicine, hydroxyurea, and nitrogen mustard. Researchers have used these drugs to generate so-called epithelial animals that consist only of epidermal and gastrodermal epithelial cells with mesoglea in between (Bode et al., 1976; Marcum and Campbell, 1978). Such animals have swollen bodies because the absence of nerve cells leaves them unable to regulate their osmotic pressure (Fig. 20.10). They must be hand-fed because they cannot catch prey, but they reproduce by budding and can be maintained indefinitely as epithelial animals. Apparently, they form enough gland cells from gastrodermis for food digestion to occur. By using low doses of those drugs that kill only interstitial cells, researchers have also produced animals in which stem cells were reduced to unipotent stem cells that gave rise only to gametes (eggs or sperm, depending on whether the animal was female or male). Alternatively, investigators have combined cells from epithelial animals with small numbers of interstitial cells from normal donors to study the development of the transplanted interstitial cells in their hosts.

The presence of pluripotent stem cells among the interstitial cells of *Hydra* was revealed by David and Murphy (1977), who used the strategy of *clonal analysis* as illustrated in Figure 20.11. Cells from normal *Hydra vulgaris* were mixed with an excess of cells from nitrogen mustard–treated individuals of the same species. Under these circumstances, *single* interstitial cells contained in the normal cell samples formed clones separated by large numbers of epithelial cells from the treated animals. Such clones of cells were identified by toluidine blue, which stains interstitial cells more intensely than it

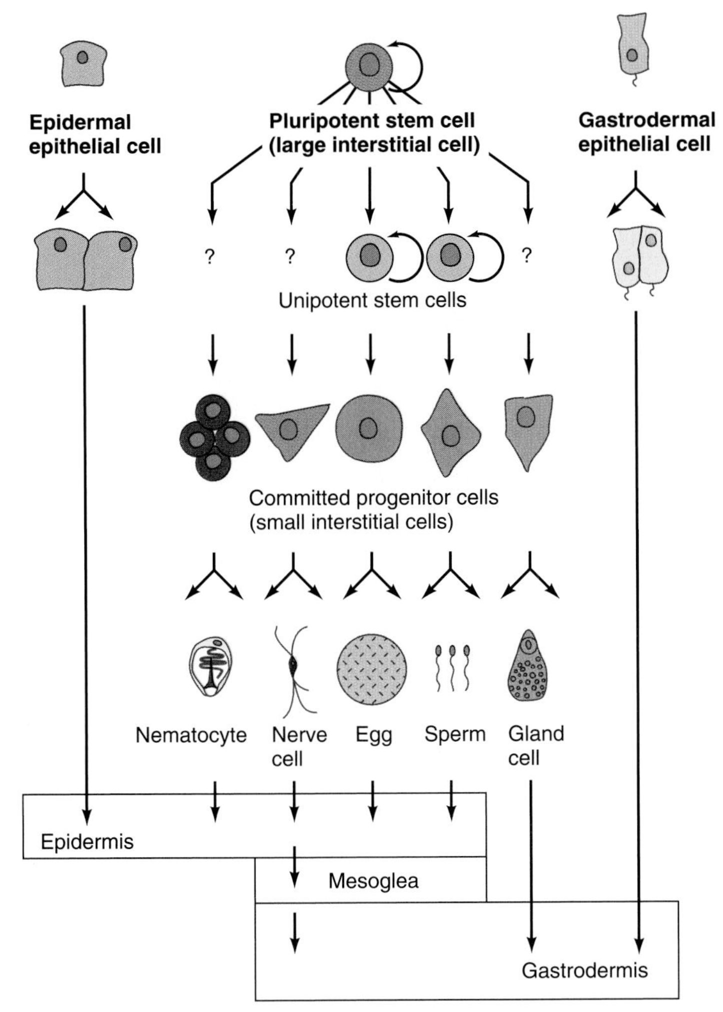

Figure 20.9 The three cell lineages of *Hydra.* Epidermal epithelial cells (top left) and gastrodermal epithelial cells (top right) divide as differentiated cells and form most of the epidermis and gastrodermis, with an acellular mesoglea in between (bottom). All other cells are formed by the interstitial cells, which include pluripotent and unipotent stem cells. The stem cells generate committed progenitor cells, which undergo amplifying divisions before they differentiate into nematocytes, nerve cells, gland cells, and, at least in some species, gametes.

stains epithelial cells. Of nine clones analyzed, each contained both developing nematocytes and nerve cells, indicating that they were derived from pluripotent stem cells.

In addition to pluripotent stem cells, at least some *Hydra* species also contain unipotent stem cells that are restricted to forming gametes. Such unipotent stem cells were first observed in *Hydra vulgaris* treated with colchicine or hydroxyurea. In most of the treated animals, the entire interstitial cell population was eliminated. However, a few animals were found that did not form nematocytes or nerve cells but occasionally produced either oocytes or sperm. In a similar experiment, Littlefield (1985, 1991) used *Hydra oligactis,* a species in which gamete formation is triggered by low temperature. From animals treated with marginal doses of hydroxyurea, she isolated lines of animals with low levels of interstitial cells (< 10% of normal). Such animals reproduced by budding and could be maintained for years without forming nerve cells or nematocytes. However, when exposed to low temperature, they reliably produced sperm or eggs (see Fig. 20.10). Those interstitial cells that gave rise to sperm also stained specifically with a monoclonal antibody (Littlefield et al., 1985). These results demonstrate that the interstitial cells of *Hydra oligactis* contain a subpopulation of unipotent stem cells that are restricted to forming either eggs or sperm. Corresponding experiments with *Hydra magnipapillata* revealed a similar subpopulation of interstitial cells that can form sperm but not nerve cells or nematocytes (Nishimiya-Fujisawa and Sugiyama, 1993). Presumably, these unipotent stem cells are derived from pluripotent stem cells in the interstitial cell population.

In summary, experiments with *Hydra* have shown that the interstitial cells include stem cells that give rise to all cell types of this organism except the epithelial cells of epidermis and gastrodermis. These pluripotent stem cells give rise to different types of committed progenitor cells and to unipotent stem cells for either eggs or sperm. Similar characteristics are found in other stem cell lineages, including the mammalian blood cells, which will be discussed next.

20.4 Growth and Differentiation of Blood Cells

Although separated by nearly a billion years of evolution, the interstitial cells of hydrozoans and the blood cells of mammals both show the basic characteristics of stem cell systems. The former are easier to manipulate experimentally, and they are more deeply involved in the growth and reproduction of the entire animal. The latter have more immediate relevance to human health. Fatal illnesses such as leukemia can be understood as disorders in the progression from stem cells to committed progenitor cells and differentiated cells. The availability of transgenic mice has also made it possible to take the study of the mammalian blood cell system to the molecular level.

The blood of vertebrates contains many types of cells with different functions, ranging from the transport of oxygen to the destruction of foreign cells (see Fig. 20.1). All mature blood cells are short-lived and must be replaced continuously in a process called ***hematopoiesis*** (Gk. *haimat-,* "blood"; *poiein,* "to make"). An adult human produces billions of blood cells each hour just to replace normal loss. All blood cells originate from pluripotent stem cells called ***hematopoietic stem cells***

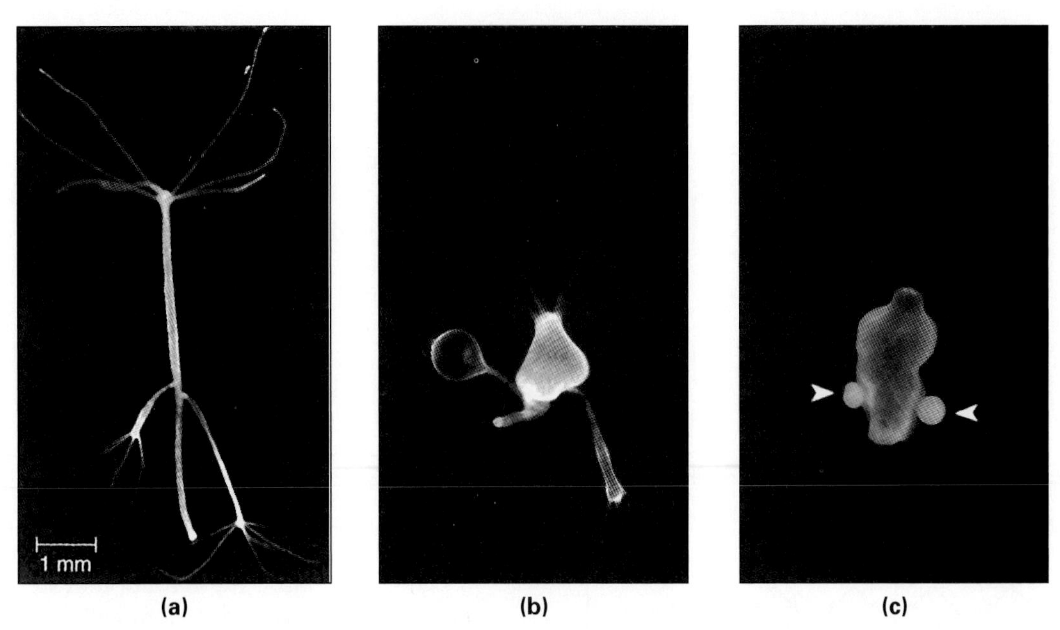

Figure 20.10 Normal and "epithelial" specimens of *Hydra oligactis.* **(a)** Normal individual with buds. **(b)** Animal from which all interstitial cells have been eliminated except for a few that are restricted to generating eggs. This specimen has the typical morphology of epithelial *Hydra:* a swollen body resulting from the inability to osmoregulate, and thin, translucent tentacles due to the absence of nematocytes. This individual has produced two buds. **(c)** Animal from the same clone as the specimen shown in part b. This animal has been kept at a low temperature to induce the formation of eggs (arrowheads).

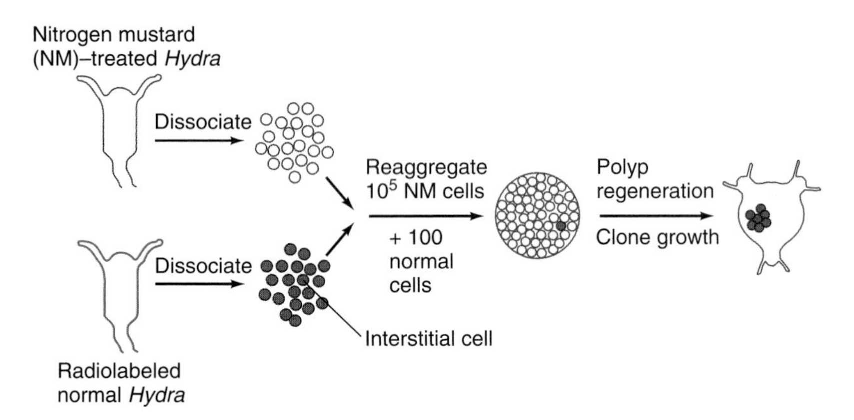

Figure 20.11 Clonal analysis of interstitial cells in *Hydra.* Normal animals and animals treated with nitrogen mustard (NM) to eliminate interstitial cells were dissociated into single cells. Large numbers of cells (about 100,000) from NM–treated animals were mixed with small numbers of cells (20–400) from radiolabeled but otherwise normal animals (red color). The cells were pelleted by centrifugation and allowed to regenerate polyps. Clones derived from interstitial cells in the regenerates were identified by their radiolabel and by staining with toluidine blue. The cell types found in individual clones were identified morphologically.

(Fig. 20.12). In mammalian embryos, they originate in the yolk sac as well as in tissues surrounding the aorta, mesonephros, and gonad (Medvinsky et al., 1993). From these locations, hematopoietic stem cells are carried by the blood stream to the fetal liver and from there to adult hematopoietic tissues including bone marrow, spleen, and lymph nodes (Metcalf and Moore, 1971).

A group of blood cells known as *lymphocytes* are involved in cellular immune reactions (*T cells*) and antibody production (*B cells*). They mature in the thymus gland, lymph nodes, and other so-called *lymphoid organs.* Accordingly, lymphocytes and their precursors are referred to as ***lymphoid cells,*** in contrast to all other blood cells, which are collectively called ***myeloid cells.*** More than 99% of the myeloid cells are red blood cells, or ***erythrocytes.*** They transport oxygen (O_2) and carbon dioxide (CO_2) and are confined to blood vessels. Other myeloid cells, as well as lymphoid cells, use blood vessels mainly for rapid transport. After crossing the endothelium of small blood vessels, they invade other tissues, where they combat infections and kill the body's own tumorous and senescent cells. In addition, blood contains *platelets,* cell fragments without nuclei that pinch off from large bone marrow cells called *megakaryocytes.* Platelets are involved in blood clotting and blood vessel repair. Other myeloid cells, known as *monocytes,* give rise to *macrophages,* cells that can engulf and destroy worn-out and abnormal cells. Another group of myeloid cells, called *neutrophilic granulocytes, basophilic granulocytes,* and *eosinophilic granulocytes,* are involved in ingestion of bacteria, wound healing, inflammation, and allergic reactions.

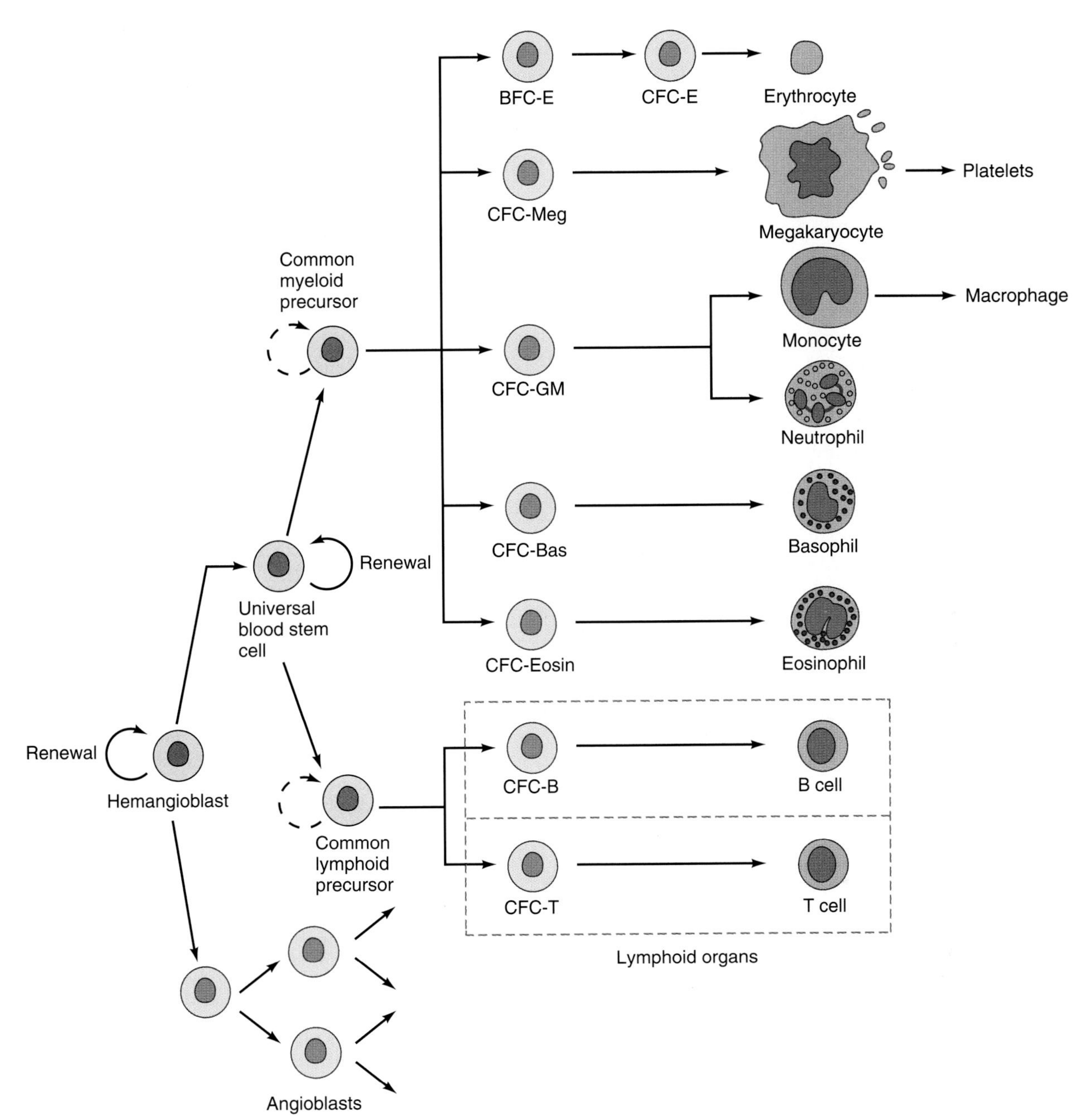

Figure 20.12 A tentative scheme of hematopoiesis. An embryonic stem cell-derived cell, the hemangioblast, gives rise to angioblasts, which form the endothelium of blood vessels, and universal blood stem cells. The latter divide for self-renewal (circular arrow) and to produce cells with more restricted potencies. The daughter cells are either common myeloid precursor cells or common lymphoid precursor cells. Both types of cell have a limited potential for self-renewal (dashed circular arrows) and produce different types of committed progenitors for myeloid cells and lymphoid cells, respectively. The committed progenitor cells in this system are referred to as CFCs (for colony-forming cells). They undergo amplifying divisions (not shown) under the control of specific regulatory proteins before they differentiate (see Table 20.1). Most CFCs produce only one type of differentiated cell. However, CFC-GM gives rise to both neutrophilic granulocytes and macrophages. The differentiated cells live for a few days or weeks and carry out specific functions.

The concentrations of different blood cells differ considerably. One ml of human blood typically contains 5 billion erythrocytes, but only 3 million lymphocytes. Moreover, the concentration of each blood cell type is regulated individually. Under conditions of low oxygen pressure, the concentration of erythrocytes increases selectively. In contrast, bacterial infections drive up the concentration of *neutrophilic granulocytes,* whereas infections with certain protozoa tend to cause an increase in *eosinophilic granulocytes.* Given the wide differences between blood cells in concentration and mechanisms that control them, it was a surprise to find that all blood cells, and the endothelial cells of blood vessels as well, are derived from one type of universal stem cell.

THE SAME UNIVERSAL STEM CELL FORMS ALL TYPES OF BLOOD CELLS AS WELL AS ENDOTHELIAL CELLS

Proving the existence of hematopoietic stem cells and describing some of their properties has been a difficult task (Dexter and Spooncer, 1987). All types of blood cells begin their development in the bone marrow and look very similar in their early stages. Thus, the existence of stem cells can only be inferred from *clonal analysis*—that is, by tracing the progeny of single hematopoietic cells.

The ***spleen colony assay*** pioneered by Till and McCulloch (1961) has been used by many researchers to assess the potency of bone marrow cells for forming different blood cell types. For this assay, an animal is exposed to a heavy dose of X-rays so that its hematopoietic cells can no longer form new blood cells. However, the animal is saved by a transfusion of bone marrow cells from an unirradiated and immunologically compatible donor (Fig. 20.13). The restoration of the hematopoietic system in the recipient can be observed in the blood and, in particular, in the spleen. Two weeks after transfusion, the spleen shows several distinct nodules, each containing a colony of new blood cells. The spherical shape of the nodules and their isolation from each other suggest that each is derived from a single precursor cell. Microscopic examination of such nodules shows that they contain erythrocytes, granulocytes, and other myeloid cells. Some colonies contain only one type of blood cell, while others contain several myeloid cell types. These observations suggest that each nodule originates either from a pluripotent myeloid stem cell or from a committed progenitor cell.

To conclusively test whether the nodules observed in spleen colony assays are indeed clones generated by single founder cells, Becker and coworkers (1963) genetically marked the marrow cells used for transfusion. For this purpose, the experimenters induced chromosome abnormalities by X-irradiating the donor marrow cells after they had been taken from the donor and before they were transfused to the recipient. Spleen nodules founded by such cells still contained different types of myeloid cells, but each cell in a given nodule had the same chromosome anomaly. Therefore, all marked blood cells in a nodule had to be derived from a single founder cell, which the researchers termed a ***colony-forming cell (CFC)***.

The survival of the mice that received transfusions of bone marrow cells showed that these cells could restore not only the myeloid but also the lymphoid cells of irradiated recipients. This raised the question whether there are universal blood stem cells that give rise to both myeloid and lymphoid cells.

(a) X-irradiation halts blood cell production; mouse would die without further treatment

(b) Inject bone marrow cells from healthy donor

(c) Mouse survives; 2 weeks after injection, many newly formed blood cells are in circulation

(d) Examination of spleen reveals large nodules on its surface

Each spleen nodule contains a clone of hematopoietic cells, descended from one of the injected bone marrow cells

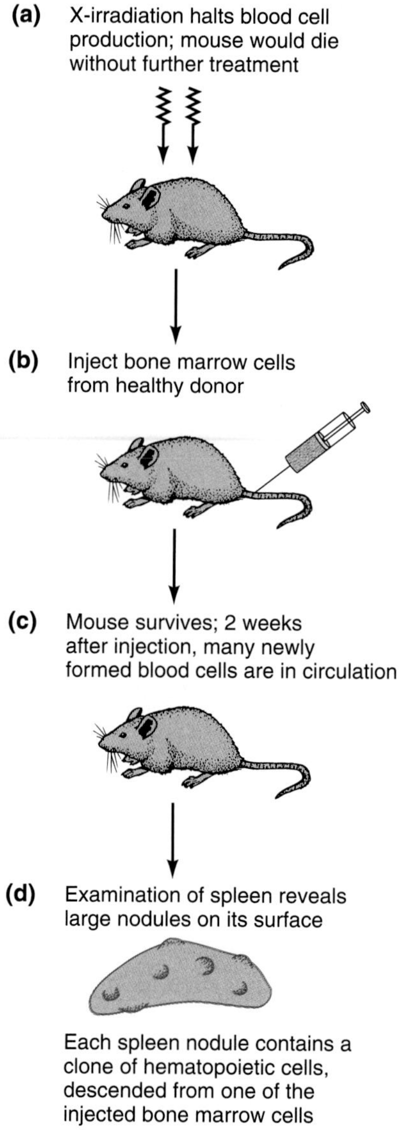

Figure 20.13 Spleen colony assay to assess the potency of injected bone marrow cells. **(a)** A mouse is heavily X-irradiated to prevent all resident hematopoietic stem cells from further division. **(b)** The mouse receives bone marrow cells from a healthy and immunologically compatible donor. **(c)** Two weeks after treatment, newly formed blood cells are in circulation. **(d)** The spleen of the reconstituted mouse has multiple nodules, each containing a clone of blood cells derived from a single injected bone marrow cell.

TO investigate whether universal blood stem cells exist, G. Keller and coworkers (1985) used a marker that allowed them to trace both myeloid and lymphoid cells in the same animal. These investigators infected the transfused marrow cells with a virus containing the *neo*^{R} gene as a selectable marker (Fig. 20.14). The virus inserted itself randomly into the genomic DNA of bone marrow cells, conferring on them two valuable properties. First, the infected cells were resistant to neomycin and could therefore be selected from uninfected cells by culture in a neomycin-containing medium. Second, because the virus was replicated in the

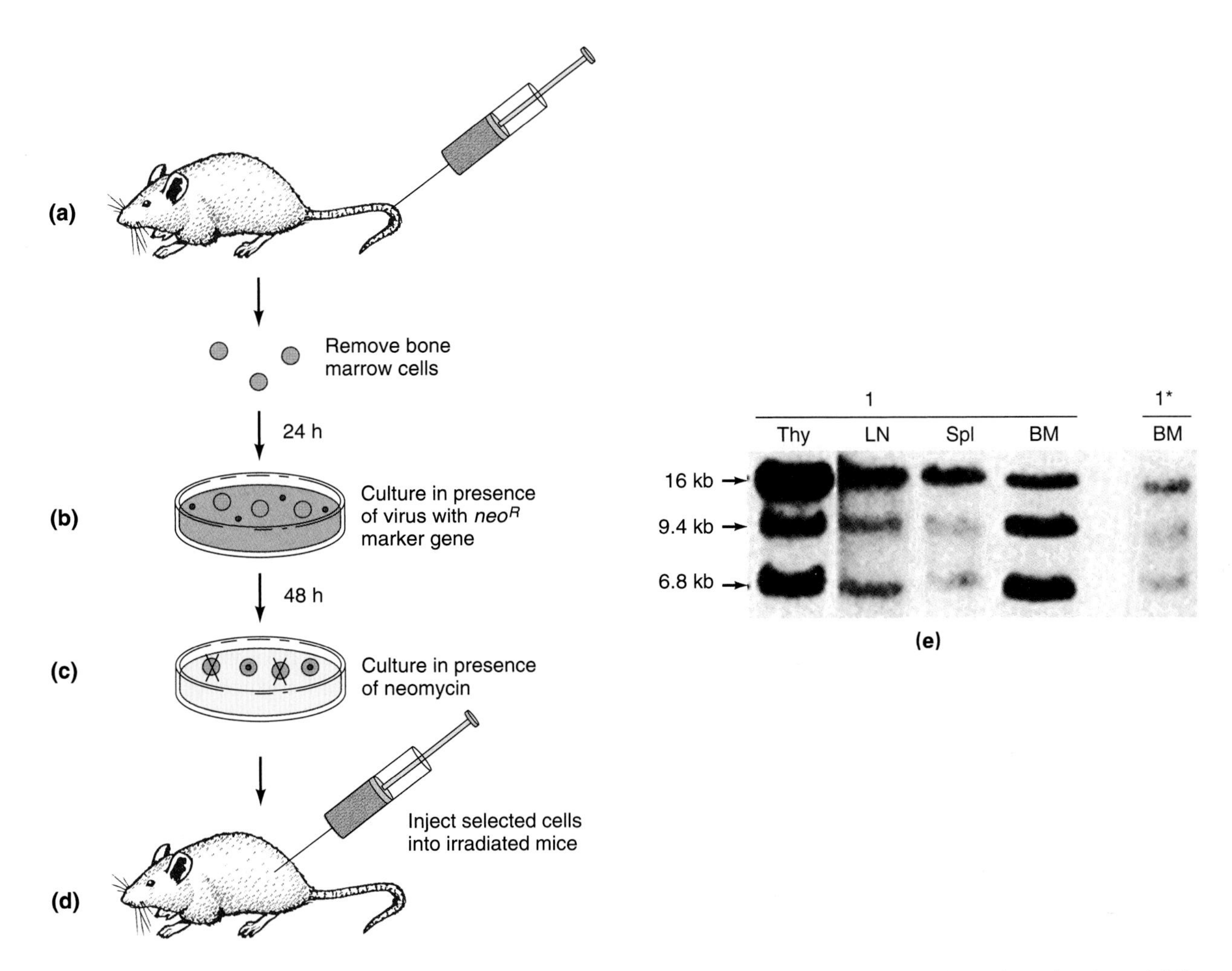

Figure 20.14 Tracing of blood cell clones founded by single bone marrow cells marked by the insertion of a virus. **(a)** Bone marrow cells are removed from a healthy mouse. **(b)** The cells are cultured in the presence of a virus (red dots) carrying the *neo*^{R} gene as a selectable marker. The virus inserts itself into the genomic DNA of some of the bone marrow cells, conferring on these resistance to neomycin and marking each cell with the virus insertion site. **(c)** The cells are cultured in the presence of neomycin to select the virus-infected cells. **(d)** The selected cells are injected into X-irradiated mice as described in Figure 20.13. **(e)** Genomic DNA is prepared from thymus (Thy), lymph node (LN), spleen (Spl), and bone marrow (BM) of the mouse with a reconstituted hematopoietic system (designated 1). Genomic DNA is also prepared from the bone marrow of another irradiated mouse (designated 1*), which has previously received bone marrow cells from mouse 1. DNA from each source is analyzed by Southern blotting (see Method 15.4), using as a probe the *neo*^{R} sequence. In each case, three DNA restriction fragments of 6.8 kb, 9.4 kb, and 16 kb are labeled. The results indicate that a maximum of three transplanted bone marrow cells reconstituted the entire hematopoietic system and that each of these three cells contributed to all hematopoietic tissues.

same genomic position whenever its host cell divided, each infected cell formed a clone carrying the insertion site of the virus as a unique marker. Cells containing the virus in the same genomic site therefore had to belong to the same clone. To characterize the virus insertion site in the blood cell DNA, the investigators used the *Southern blotting* technique (see Method 15.4). When they tested DNA prepared from different tissues of mice with reconstituted hematopoietic systems, they found that bone marrow, spleen, lymph nodes, and thymus gland cells did indeed contain the marker virus in DNA restriction fragments of the same lengths. These results proved the existence of universal blood stem cells.

Questions

1. What seem to be the advantages of genetically labeling cells with a virus over the older method of using X-rays to induce chromosomal abnormalities?
2. If the Southern blot in Figure 20.14e would show only one band per lane, and the band would be in the same position in each lane, what would this result indicate?
3. If the bands in Figure 20.14e would differ from lane to lane, what would this result indicate?

The reconstitution experiments just described show that a small number of cells, sometimes just one cell, can restore the myeloid and lymphoid blood cells of an entire mouse. Moreover, an irradiated mouse saved by a transfusion of bone marrow cells can in turn be used as a donor of bone marrow cells to save another irradiated mouse, and the experiment can be repeated at least one more time. To fully appreciate this result, one needs to keep in mind that the number of transplanted bone marrow cells in a reconstitution experiment is much smaller than the normal complement of bone marrow in a mouse. Thus, self-renewal displayed by those cells that become the founder cells of reconstituted hematopoietic systems in these experiments greatly exceeds the normal self-renewal of hematopoietic stem cells.

It has long been suspected that there is not only a universal hematopoietic stem cell but probably an even more general type of cell called a ***hemangioblast*** because it gives rise to both blood cells and *endothelial cells.* The precursors of blood cells and endothelial cells are closely associated and morphologically indistinguishable in *angiogenic clusters,* which originate in particular in the *visceral layer* of the *lateral plates.* Within these clusters, the central cells give rise to blood cells, while the peripheral cells behave like angioblasts and form endothelium (see Fig. 14.30). Also, *knockout* mice that are genetically deficient for vascular endothelial growth factor receptor-2 (VEGF-R2) lack both hematopoietic and angioblastic cells (Shalaby et al., 1995). Finally, it has been shown that mouse *embryonic stem (ES) cells* can be cultured in such a way that they will produce hematopoietic cells as well as adjacent cells of a different type that show morphological and molecular characteristics of endothelial cells (Choi et al., 1998). If ES cells carrying different genetic markers were mixed in these experiments, the hematopoietic and adjacent endothelial cells always carried the same marker, indicating that they were derived from the same founder cell, a hemangioblast.

The conclusions reached from reconstitution experiments in vivo have been confirmed by observations made in vitro (Metcalf, 1977). Since blood cells can be cultured in dishes with soft gels and defined media, it is possible to monitor the capacity of individual cells to divide and differentiate. Some cells behave in culture like pluripotent stem cells, forming successive colonies containing all myeloid cell types. Other isolated cells reveal more limited potencies, forming only two differentiated cell types, such as macrophages and neutrophilic granulocytes. Still other cells form just one type of blood cell. As the potencies of blood cells become more limited, their capacity for self-renewal also decreases, at least normally.

The results described so far, and many others, can be summarized as follows (see Fig. 20.12; Kondo et al., 1997; Akashi et al., 2000). Hemangioblasts give rise to endothelial cells and universal hematopoietic stem cells. The latter form ***common myeloid precursor cells*** and ***common lymphoid precursor cells.*** Each type of precursor cell still has a limited potential for self-renewal and produces several types of committed progenitor cells, which undergo a potentially large but limited number of amplifying divisions before terminal differentiation. These divisions will be described later in the context of erythrocyte differentiation.

PROGENITOR CELL COMMITMENT MAY DEPEND ON STABLE CONTROL CIRCUITS OF GENES AND TRANSCRIPTION FACTORS

What are the factors that regulate hematopoiesis? Since the most prominent derivative of isolated ventral mesoderm in vertebrate embryos is blood, determination of hematopoietic stem cells can be expected to occur as part of mesoderm induction and the general establishment of dorsoventral polarity (see Sections 9.6 and 14.6). Indeed, bone morphogenetic protein 4 (BMP-4), a secreted protein that generally promotes ventral development, induces in particular red blood cell formation (Huber et al., 1998). This effect is greatly enhanced by other signal proteins that induce mesoderm, such as activin or fibroblast growth factor (FGF). Thus, hemangioblasts or hematopoietic stem cells must be determined by a combination of ventralizing and mesoderm-inducing signals.

The subsequent steps of stem cell renewal and progenitor cell commitment are still enigmatic. Observations made in vitro suggest that there is an element of chance in the way an individual hematopoietic cell behaves (Whetton and Dexter, 1993). Cells isolated from apparently homogeneous populations produce colonies of remarkably different sizes and characters. Even sister cells cultured under identical conditions will often produce colonies with different numbers or types of cells. Thus, the transition from stem cells to a committed progenitor cell seems to involve random or very labile processes.

What does the transition from a pluripotent stem cell to a committed progenitor cell mean in terms of gene activity? This question has been addressed by amplification of mRNAs from single cells by the *polymerase chain reaction* (see Method 15.3). This has revealed the simultaneous presence of gene products characteristic of multiple types of differentiated blood cells, such as erythrocytes and granulocytes, in *stem cells* and *precursor pluripotent cells* (Hu et al., 1997). In contrast, *committed progenitor cells* express only the genes appropriate to their lineage. Presumably, the restriction of gene activity during lineage commitment involves a "shake-out" between competing genes and transcription factors until eventually they form stable control circuits, similar to the ones underlying determination and transdetermination of imaginal disc cells in *Drosophila* (see Section 16.6).

For a genetic analysis of progenitor cell commitment, some investigators have taken advantage of mutations associated with certain types of human leukemia, while others have generated transgenic mice or screened frog

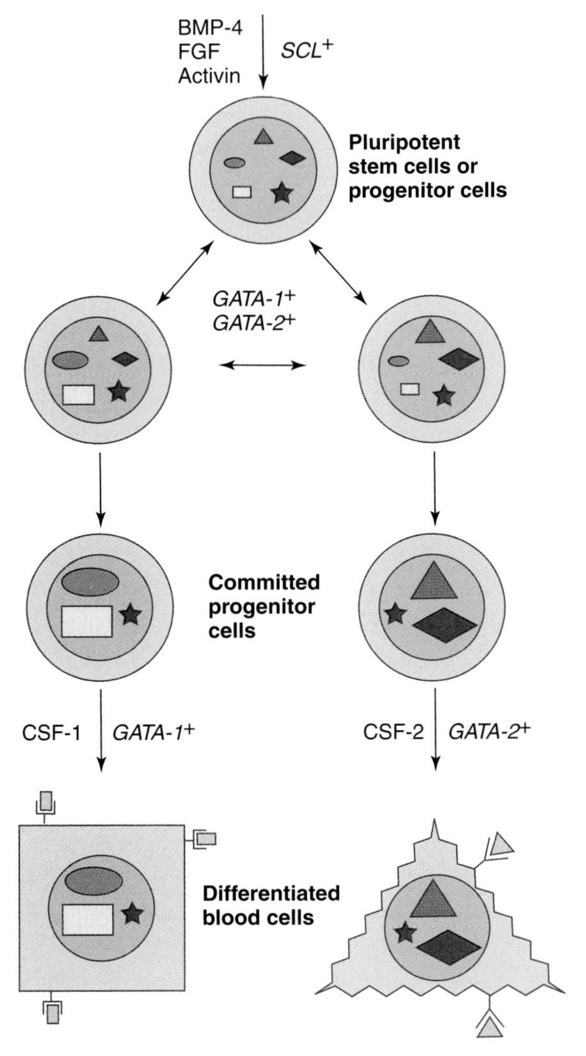

Figure 20.15 Model of progenitor cell commitment during hematopoiesis. In pluripotent stem cells and progenitor cells, multiple transcription factors (colored symbols within nuclei) are initially present at fluctuating but low levels. In committed progenitor cells, some transcription factors have become dominant while others have disappeared as the result of positive and negative feedback loops. In differentiated cells, the dominant transcription factors have activated genes that encode proteins conferring on cells their characteristic cell shape and membrane proteins. The latter include the receptors for growth factors, or colony stimulating factors, which blood cells need to survive. The genes *SCL*$^+$, *GATA-1*$^+$, and *GATA-2*$^+$ encode just three of the many transcription factors involved. Likewise, CSF-1 and CSF-2 represent larger, and partially overlapping, sets of colony stimulating factors (see Table 20.1).

cDNA libraries for genes involved in hematopoiesis (*Cross and Enver, 1997*). Greatly simplifying a complex picture, we will focus here on the roles of three genes—namely, *SCL*$^+$, *GATA-1*$^+$, and *GATA-2*$^+$, all of which encode transcription factors (Fig. 20.15).

The *SCL*$^+$ gene was discovered because many cases of human leukemia are associated with chromosomal translocations that juxtapose the transcribed region of the *SCL* gene with powerful promoters, thus causing SCL protein synthesis at higher than normal rates (Robb and Begley, *1997*). In mouse embryos, SCL protein is detectable very early in both endothelial and hematopoietic cells of the yolk sac (Kallianpur et al., 1994). In *knockout* mice lacking *SCL*$^+$ genes, blood vessels are formed, but blood cells are absent altogether (Shivdasani et al., 1995). In *chimeric* mice, stem cells containing only null alleles of *SCL* do not contribute to any hematopoietic cell lines (Porcher et al., 1996). However, the contributions of the null *SCL* stem cells to hematopoiesis are restored by an *SCL*$^+$ transgene. These results indicate that the requirement for *SCL*$^+$ gene activity begins at an early step of hematopoiesis, most likely the formation or self-renewal of hematopoietic stem cells.

Experiments with frogs have confirmed the key role of *SCL*$^+$ in early hematopoiesis and have also shown that *SCL*$^+$ is controlled by the ventralizing signal BMP-4 (Mead et al., 1998). In *Xenopus* embryos, SCL transcripts are first detected on the ventral side of early neurulae. SCL synthesis is induced by BMP-4, as shown by the lack of SCL transcripts in embryos with a *dominant negative* BMP-4 receptor. Ectopic synthesis of SCL causes ectopic erythropoiesis but does not otherwise disturb the normal dorsoventral polarity. Together, the results suggest that BMP-4 promotes the synthesis of SCL as part of ventral mesoderm specification and that SCL is both necessary and sufficient for the initiation of hematopoiesis. In addition, SCL suppresses the activation of muscle-specific genes by another transcription factor, myogenin (Hofmann and Cole, 1996; see Section 20.5). Thus, SCL may act as a general master regulator for hematopoiesis.

Other key elements in the genetic control of hematopoiesis are transcription factors named GATA proteins, after the GATA consensus element in the DNA sequences to which they bind. GATA-1 and GATA-2 are restricted, for the most part, to blood cells (Bockamp et al., 1994). GATA-1 is present in erythrocytes, megakaryocytes, and eosinophils as well as in their precursor cells. GATA-2 accumulates in the same cells as GATA-1 except the erythroid lineage. However, the presence of a protein in a cell does not prove that it is necessary or even functional. To determine in which blood cell lineages a given GATA factor is necessary, investigators have constructed *chimeric* mice with embryonic stem cells lacking particular GATA genes. Cells without a functioning *GATA-1* gene contribute to all kinds of tissue including some white blood cells but not to erythrocytes (Pevny et al., 1991). Follow-up experiments in vitro showed that the absence of GATA-1 blocks the development of the committed progenitor cells for erythrocytes (M. C. Simon et al., 1992). Corresponding experiments indicated that GATA-2 is necessary for the survival or self-renewal of lymphoid precursor cells.

The GATA proteins are involved in the decision of progenitor cell commitment versus self-renewal. For example, GATA-2 blocks erythroid differentiation and seems to promote the development of hematopoietic stem cells and committed progenitors (Briegel et al., 1993; Tsai et al., 1994). Conversely, GATA-1 promotes the development of erythrocytes, as indicated by morphology and hemoglobin content, while at the same time suppressing the genes for GATA-2 and other factors involved in cell proliferation (Briegel et al., 1996). Indeed, forced expression of a *GATA-1* transgene in macrophage precursors, which do not normally make GATA-1, reprograms these cells into erythroblasts and eosinophils (H. Kulessa, 1995).

The determination of lymphoid cells occurs in a similar fashion, with different sets of transcription factors and colony-stimulating factors involved (Rothenberg et al., 1999). T cell development is special in that most of it occurs apart from the other hematopoietic cells in the thymus. Both B and T cells develop in clones expressing unique antigen receptors on their plasma membranes. These receptors are encoded by genes assembled by mechanisms including DNA rearrangements and somatic mutagenesis (see Section 7.4). These DNA modifications are closely associated with the commitments of the B and T cell lineages.

In summary, the commitment of hematopoietic progenitors cells seems to emerge from a labile state that seems almost chaotic because of its simultaneous synthesis of proteins that are later made only in lineage-restricted ways. Feedback loops between transcription factors and their target genes seem to result in a "shake-out" process, from which a more limited set of gene activities emerges. The lineage-specific transcription factors that come to prevail during the commitment of a given progenitor cell are thought to activate the genes that are required in the respective cell lineage. The latter include the genes for receptors that allow blood cells to respond to certain growth factors, which will be discussed next.

BLOOD CELL DEVELOPMENT DEPENDS ON COLONY-STIMULATING FACTORS (CSFs)

The development of hematopoietic cells depends on secreted regulatory proteins. For instance, a hematopoietic stem cell inhibitor synthesized by bone marrow cells limits stem cell proliferation (G. J. Graham et al., 1990). The majority of the known regulatory proteins have stimulating effects and are known as *growth factors, cytokines,* or ***colony-stimulating factors* (*CSFs*).** Their presence is required for the survival, division, and differentiation of hematopoietic cells (Dexter and Spooncer, 1987; Metcalf, 1989; Whetton and Dexter, 1993; Orkin, 1996). In the absence of CSFs, the cells die, regardless of their stage of development. The CSFs are glycoproteins that bind to matching receptors in the plasma membranes of their target cells.

Some progenitor cells have more than one type of receptor and develop according to which CSF is present in higher concentration. For example, certain progenitor cells can form two types of leukocytes, namely, macrophages and neutrophilic granulocytes (see Fig. 20.12). Each of these cell types is enhanced by its own CSF. Thus, granulocyte CSF fosters granulocyte development while macrophage CSF favors macrophage development. There are also CSFs that stimulate the formation of both cell types. Thus, the overall rate of progenitor cell production, and the degree to which each type is amplified, depend on the overall amount and mixture of CSFs present.

Many CSFs have been cloned and characterized molecularly along with their receptors. Table 20.1 lists these factors along with their target cells. As can be seen from the table, most CSFs act on more than one target cell, and conversely, most target cells are stimulated by several CSFs. The overall result depends on the relative concentrations of different CSFs and their receptors. CSF concentrations may vary between local microenvironments created by surrounding tissues.

While most CSFs are diffusible, some are tethered to their producer cells, or to components of the extracellular matrix (ECM). In living bone marrow, hematopoietic cells develop amid a supporting tissue known as ***stroma***. When stroma and hematopoietic cells are cocultured in vitro, stroma cells show two interesting properties (Dexter, 1982). First, they support hematopoiesis even without diffusible CSFs in the medium. Second, this support requires direct contact between the stroma cells and the hematopoietic cells. Only stroma cells from bone marrow—not those from other tissues—support the growth of hematopoietic cells. These observations suggest that bone marrow stroma cells may deploy CSFs in their membranes or in ECM. Indeed, the growth factor encoded by the *Steel*$^+$ gene exists in two forms, one released and one membrane-bound, generated by alternative splicing of the pre-mRNA (Flanagan et al., 1991). The membrane-bound form is expressed in the stroma cells of bone marrow. Mice mutant in the membrane-bound but not the released form of Steel have severe hematopoietic defects, indicating that normal hematopoiesis depends on direct contact between hematopoietic cells and stroma cells. Other CSFs are tethered to ECM components of bone marrow stroma (M. Y. Gordon et al., 1987; R. Roberts et al., 1988).

The membrane-bound and ECM-tethered deployment of CSFs enables hematopoietic tissues to establish local microenvironments, or ***niches,*** each containing a unique mixture of immobilized CSFs. Becoming trapped in such niches appears to be indispensable for the development of hematopoietic stem cells. Stem cells that are deficient in β_1-integrin are unable to colonize the fetal

table 20.1 Hematopoietic Growth Factors and Their Target Cells

		Target Cells*										
Factor	Symbol	Erythroid BFC-EC	Cell CFC-E	Granulocyte Eosinophil	Neutrophil	Macrophage	Mast Cell	Megakaryocyte	Lymphocyte B	T	Pluripotent CFC-GM Cell	Stem Cell
Multipotential colony–stimulating factor, Interleukin-3	Multi-CSF IL-3	X		X	X	X	X	X	X	X	X	X
Granulocyte-macrophage colony–stimulating factor	GM-CSF	X		X	X	X		X			X	
Granulocyte colony–stimulating factor	G-CSF				X							
Macrophage colony–stimulating factor	M-CSF					X						
Mast-cell growth factor	MGF	X			X		X	X				X
Erythropoietin	Epo		X					X				
Interleukin-1	IL-1				X			X				X
Interleukin-2	IL-2								X	X		
Interleukin-4	IL-4						X		X	X		
Interleukin-5	IL-5			X					X			
Interleukin-6	IL-6				X	X		X	X	X	X	
Steel factor	Sl	X										X
Leukemia inhibitory factor	LIF					X						

*For brevity, most target cell listings refer to the differentiated cell state, such as T lymphocyte. However, it must be understood that the growth factors act on stem cells and committed progenitor cells, such as T-lymphocyte colony–forming cells (CFC-Ts).

Sources: Dexter and Spooncer (1987), © 1987 Annual Reviews, Inc.; Metcalf (1989), © 1989 Macmillan Magazines Limited; Copeland et al. (1990), © Cell Press, and others. Used with permission.

liver and fail to produce any blood cells, even though they are perfectly capable of forming differentiated blood cells when kept in culture with appropriate CSFs (Hirsch et al., 1996). Thus, it appears that the hematopoietic tissues have two major properties. First, they entrap hematopoietic stem cells during certain stages of their development. Second, they expose the entrapped stem cells and their descendants to a unique combination of CSFs, at least some of which seem to bind to extracellular matrix components of hematopoietic tissues.

ERYTHROCYTE DEVELOPMENT DEPENDS ON SUCCESSIVE EXPOSURE TO DIFFERENT CSFs

Erythrocytes are by far the most common blood cells. Highly specialized, they are packed with hemoglobin for the transport of O_2 and CO_2. Erythrocytes mature in the bone marrow from precursor cells known as ***erythroblasts.*** In mammals, maturation of an erythroblast entails the loss of the nucleus (Fig. 20.16). Once released into the bloodstream, erythrocytes have a life span of a few months before they are broken down in the liver and spleen.

The continuous formation of new erythrocytes occurs in two major steps that involve different CSFs (Fig. 20.17). Myeloid stem cells give off a type of committed progenitor cell called an ***erythrocyte burst–forming cell (BFC-E).*** The name derives from the fact that these cells can produce bursts of up to 5000 erythrocytes. The BFC-Es respond to a CSF called ***interleukin-3 (IL-3),*** which generally promotes the proliferation of hematopoietic stem cells and committed progenitor cells (Table 20.1). Being committed precursor cells, BFC-Es have only a limited capacity to divide and produce colonies that consist only of erythrocytes, even under culture conditions that enable other committed progenitor cells to form other types of differentiated blood cells. After about six divisions, the BFC-E descendants begin to respond to another CSF known as ***erythropoietin.*** These cells are called ***erythrocyte colony–forming cells (CFC-E).*** Erythropoietin is a hormone formed in the kidney; it binds to and activates a matching receptor on the cell surface of CFC-Es. The receptor in turn activates a protein kinase, which phosphorylates other proteins to eventually stimulate the mitotic division of CFC-Es (Witthuhn et al., 1993). Thus, in the presence of erythropoietin, a CFC-E divides

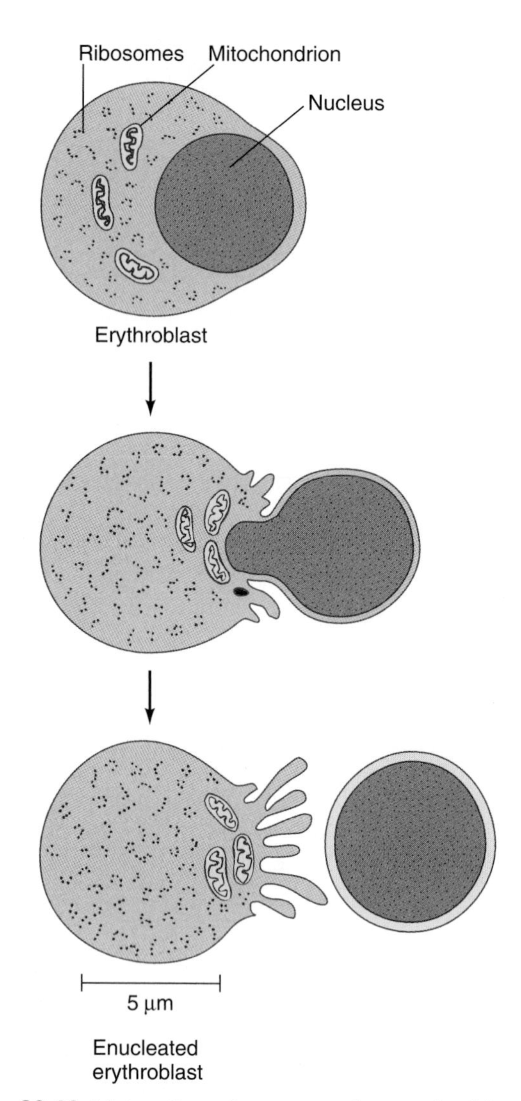

Figure 20.16 Maturation of a mammalian erythroblast. While an erythroblast is still in the bone marrow, its nucleus is pinched off and digested by macrophages. The enucleate erythroblast is released into the circulation.

five or six times, giving rise to about 30 to 60 erythroblasts, which then mature to erythrocytes.

The concentration of erythropoietin, which controls the number of final divisions, increases in response to any shortage of erythrocytes or lack of oxygen in the blood. In this fashion, the amount of erythrocytes is continuously adjusted to physiological conditions. For instance, with increasing height above sea level, the oxygen pressure falls and the red blood cell count increases in compensation. Similarly, local or systemic infections produce signals that lead to dramatic increases in the concentration of leukocytes.

THE ABUNDANCE OF BLOOD CELLS IS CONTROLLED BY CELL PROLIFERATION AND CELL DEATH

Just as the concentration of molecules in cells is controlled by the rates at which they are synthesized and degraded, the abundance of any cell type in an organism depends on the rates of cell proliferation and cell removal. In some stem cell systems, such as the epidermis and the gut epithelium, the differentiated cells are eventually sloughed off. In the blood, billions of cells are removed each day by a process known as *programmed cell death*, or *apoptosis*.

At the heart of apoptosis is a genetic program that commits cells to destroy themselves, *by default*, unless they are instructed otherwise (see Sections 25.4 and 28.4). One part of the genetic control of apoptosis is the integration of external signals that affect the survival of cells. In the case of blood, each cell type depends on a combination of CSFs to survive. In the absence of any of its vital CSFs, a blood cell will enter apoptosis. Another element of apoptosis control is the registration of internal signals that gauge the ability of a cell to repair damage to its molecules. The execution of apoptosis involves controlled changes in the molecules that a cell presents on its surface and by which it interacts with macrophages and other neighboring cells. In blood cells, such changes may involve certain phospholipids in the plasma membrane and specific sugar groups in membrane glycoproteins.

In summary, the molecular characterization of hematopoietic transcription factors and CSFs has greatly enhanced our understanding of blood cell differentiation. The molecular events underlying the commitment of stem cells to certain blood cell lineages are beginning to be understood, but much more needs to be learned. CSFs regulate not only the proliferation and maturation of different types of blood cells in response to changing physiological conditions but also their removal by apoptosis. While some CSFs are diffusible, others are bound locally to stroma cell surfaces or extracellular matrix.

20.5 Genetic Control of Muscle Cell Differentiation

In search of master genes that can reprogram the process of cell differentiation, investigators have genetically transformed cells of one type using genes associated with the differentiation of cells of another type. In the first successful experiments of this kind, connective tissue cells were converted into muscle cells by transfection with cDNA encoding MyoD, a muscle-specific protein (R. L. Davis et al., 1987). The transformed cells proceeded to undergo *myogenesis*—that is, they fused like *myoblasts* to form *myotubes* and proceeded to synthesize myosin protein and form skeletal *muscle fibers* (see Section 14.3). This was an astounding result, because fibroblasts do not normally form cardiac or skeletal muscle (see Fig. 14.19). Thus, the activity of a single gene, *MyoD*$^+$, could completely redirect a cell and force it onto a new pathway of differentiation.

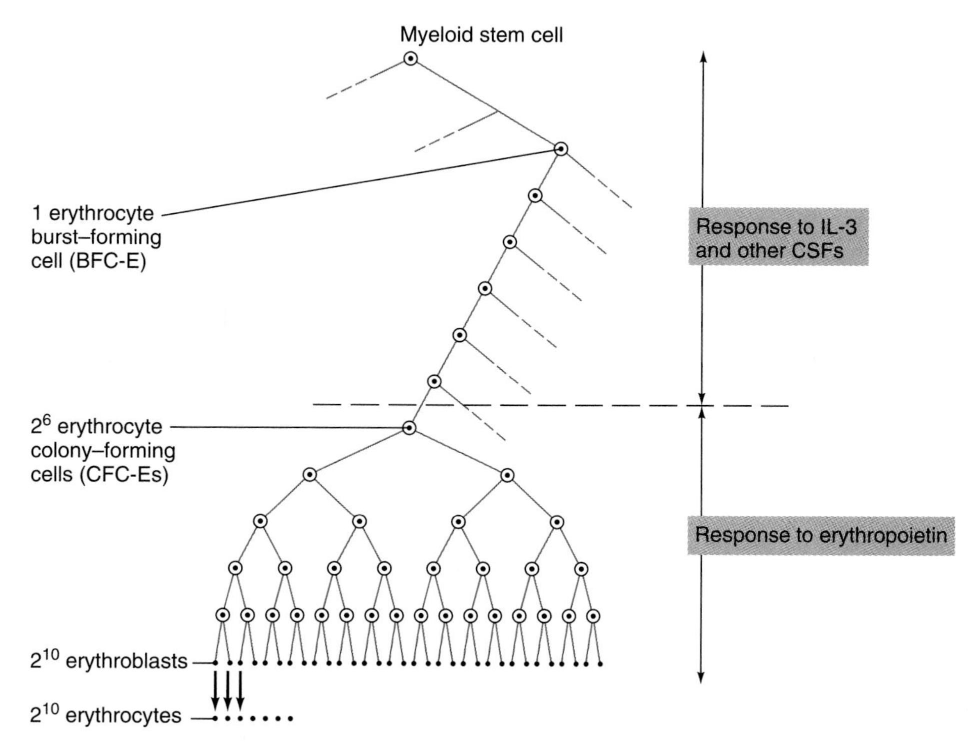

Figure 20.17 Flow diagram of erythropoiesis. Pluripotent myeloid stem cells give rise to erythrocyte burst–forming cells (BFC-Es), which undergo up to six amplifying divisions. The formation of BFC-Es and their initial divisions are stimulated by interleukin-3 (IL-3) and other hematopoietic growth factors but not by erythropoietin (see Table 20.1). BFC-E-derived cells that respond to erythropoietin are called erythrocyte colony–forming cells (CFC-Es). After further amplifying divisions, CFC-Es give rise to erythroblasts, which will differentiate into mature erythrocytes.

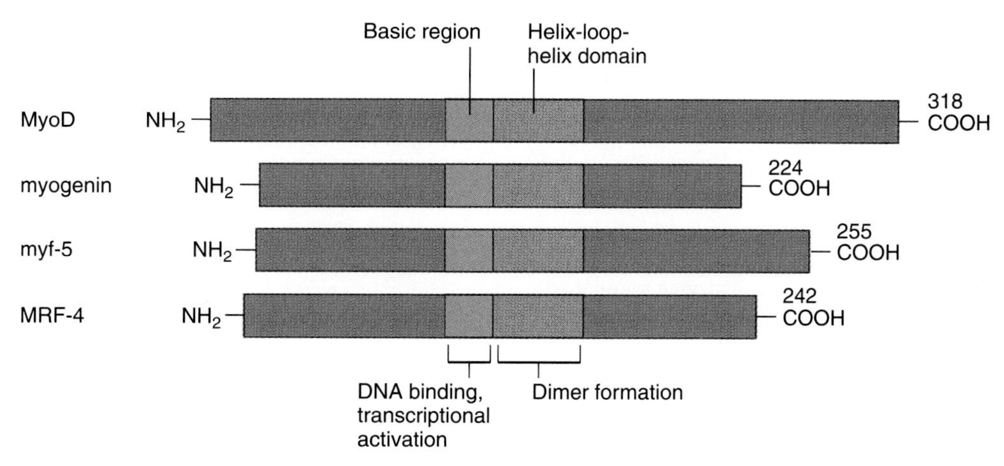

Figure 20.18 General structure of four mammalian myogenic bHLH proteins. The helix-loop-helix domain is required for dimer formation. The basic region is necessary for DNA binding and transcriptional activation. The number at the end of each box indicates the number of amino acid residues in the respective protein.

MYOGENESIS IS CONTROLLED BY A FAMILY OF MYOGENIC bHLH PROTEINS

Cell transformations similar to the one just described have revealed that $MyoD^+$ belongs to a family of genes that also includes $myogenin^+$, $myf\text{-}5^+$, and $MRF\text{-}4^+$ (Weintraub, 1993; Molkentin and Olson, 1996). Each of these genes can initiate muscle development in non-muscle cells. The proteins encoded by these genes are transcription factors that have about 80% sequence similarity in a domain of 70 amino acids known as the *basic helix-loop-helix* (*bHLH*) domain (Fig. 20.18). It contains a basic region (b), which acts as a weak base, and a helix-loop-helix (HLH) motif, where two α helices are connected by a peptide loop. The basic region is necessary for DNA binding and transcriptional activation, while the HLH motif is responsible for dimer formation. Because the MyoD, myogenin, myf-5, and MRF-4 proteins share the control of myogenesis and the bHLH domain as common features, they are called ***myogenic bHLH proteins.*** These proteins have been found in a wide range of animals, including vertebrates, sea urchins, *Drosophila,* and *Caenorhabditis elegans.* In all animals investigated so far, myogenic bHLH proteins are synthesized specifically during muscle development (see Fig. III.1).

The myogenic bHLH proteins share the bHLH motif with several other gene-regulatory proteins referred to as the ***bHLH superfamily.*** Only a few amino acid sequences in the basic region distinguish the myogenic bHLH proteins from other bHLH proteins and account for their specific role in myogenesis (Brennan et al., 1991). Nevertheless, the specific effects of the myogenic bHLH proteins depend on their cooperation with more common members of the bHLH superfamily. These includes the so-called E proteins, which bind to specific DNA sequences known as E boxes and are expressed in a wide range of tissues (Ellenberger et al., 1994). Another ubiquitous member of the bHLH superfamily is a protein called Id (for inhibitor of differentiation). It contains the HLH motif, which mediates dimer formation, but not the basic region, which binds to DNA. When Id protein dimerizes with a typical bHLH protein such as MyoD, then the heterodimer is inactive because a single basic region by itself is insufficient to bind to its recognition motif (Fig. 20.19).

Heterodimers involving myogenic bHLH proteins may be activating or inhibitory transcription factors (Lassar et al., 1991). An example of the inhibitory effects of dimerization is seen in the heterodimer formed by MyoD and Id. Id prevents MyoD from activating muscle-specific enzymes, such as kreatin kinase, and retards the formation of myotubes in muscle cell lines (Benezra et al., 1990; Jen et al., 1992). The production of Id protein declines when myoblasts begin their final differentiation, the formation of muscle fibers. This explains nicely the apparent paradox that myogenic bHLH proteins are present in muscle cell lineages long before myogenesis begins.

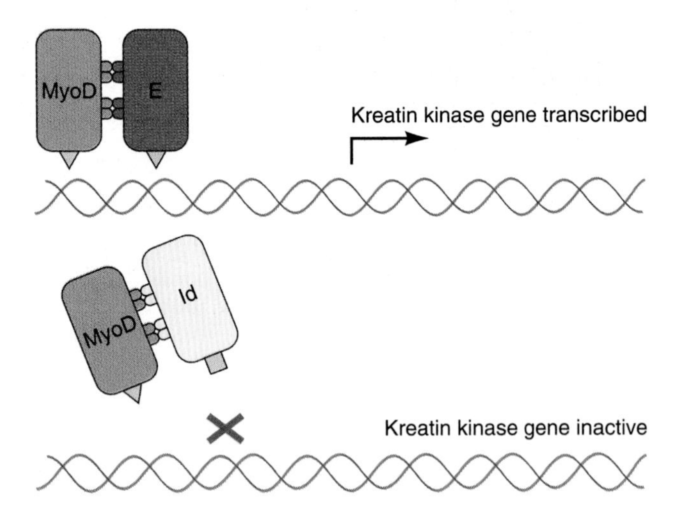

Figure 20.19 Gene regulation by heterodimers of basic helix-loop-helix proteins. Binding of the MyoD/E protein heterodimer enhances the transcription of the muscle kreatin kinase gene and other muscle-specific target genes. Dimerization of the MyoD protein with the Id protein precludes binding to the creatine kinase enhancer, so that the gene is not activated.

MYOGENIC bHLH PROTEINS HAVE PARTIALLY OVERLAPPING FUNCTIONS IN VIVO

Although much has been learned about the function of myogenic bHLH proteins in cultured cells, the importance of these proteins in vivo has been demonstrated most clearly in *knockout* mice (see Fig. 15.24) that are homozygous for *null alleles* of one or two myogenic bHLH genes. Mice lacking MyoD protein were viable and fertile but deficient in muscle regeneration as adults (Rudnicki et al., 1992; Megeny et al., 1996). Mice without myf-5 protein were unable to breathe and died immediately after birth, but histological examination of skeletal muscle revealed no defects. Apparently, the inability to breathe was caused by the lack of distal parts of the ribs (T. Braun et al., 1992). In contrast, transgenic mice lacking *both* MyoD and myf-5 protein died after birth from a complete absence of skeletal muscle (Rudnicki et al., 1993). Thus, neither MyoD nor myf-5 protein is strictly required for myogenesis, but at least one of the two must be present.

Close examination revealed that *MyoD*$^+$ and *myf-5*$^+$ are required in complementary portions of trunk musculature (Kablar et al., 1997). While mouse embryos without MyoD showed normal development of *epaxial* (dorsal trunk) muscles, their *hypaxial* (ventral trunk and limb) muscles were delayed by about 2.5 days. The converse was true for embryos lacking myf-5. These observations indicate that the epaxial and hypaxial myotome differ not only in their fate and in the signals that induce them (see Fig. 14.18), but also in the myogenic bHLH proteins that govern the beginning of their differentiation. Epaxial muscle determination depends on *myf-5*$^+$, whereas hypaxial muscle requires *MyoD*$^+$. However, these two cell lineages seem to communicate their presence to each other and respond to the absence of their counterparts by expanding beyond their normal realm.

While *MyoD*$^+$ and *myf-5*$^+$ are required early during myogenesis—that is, for *myoblast* formation—the *myogenin*$^+$ gene appears to act later, initiating the formation of *myotubes* (Fig. 20.20; Venuti et al., 1995). Experiments knocking out the *myogenin*$^+$ gene had disastrous effects on skeletal muscle development (Hasty et al., 1993; Nabeshima et al., 1993). Mice without a functional copy of this gene did not breathe and died at birth, apparently because their diaphragms consisted of disorganized cells and very few skeletal muscle fibers had developed. Similarly, other skeletal muscles were severely disorga-

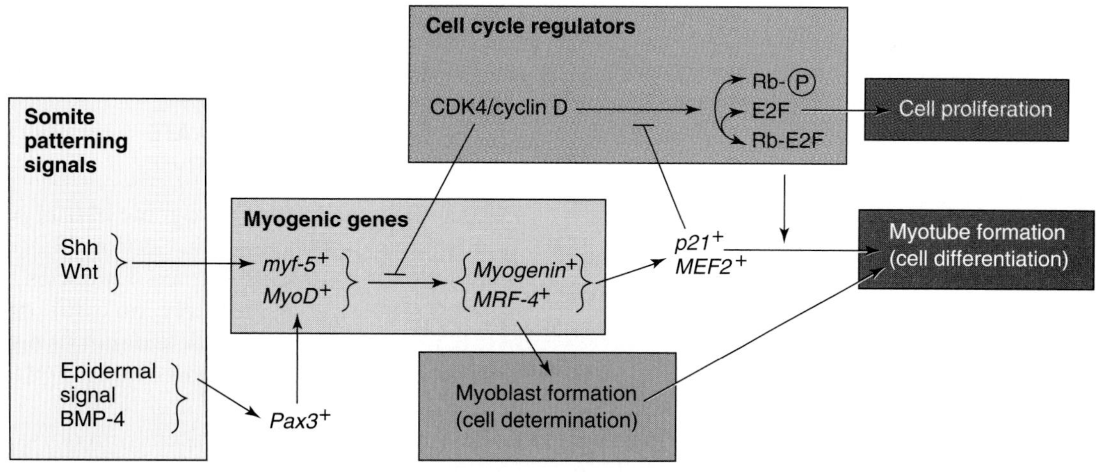

Figure 20.20 Model of interactions between myogenic bHLH proteins with cell cycle regulators. As part of somite patterning, *myf-5*$^+$ is activated by Sonic hedgehog (Shh) and Wnt signals, whereas *MyoD*$^+$ is regulated, via *Pax3*$^+$, by epidermal signals and bone morphogenetic protein 4 (BMP-4). *myf-5*$^+$ and *MyoD*$^+$ form a stable control circuit with *myogenin*$^+$ and *MRF-4*$^+$, which determines cells to develop as myoblasts. *myogenin*$^+$ and *MRF-4*$^+$ activate, through *p21*$^+$ and *MEF2*$^+$, the genes required for myotube differentiation. Some of the gene products involved in myogenesis interact by mutual inhibition with cell cycle regulators. Active CDK4 phosphorylates MyoD and retinoblastoma (Rb) protein. Phosphorylated MyoD is inactive, and phosphorylated Rb cannot sequester E2F, a transcription factor promoting cell proliferation. Conversely, p21 inactivates CDK4.

nized and reduced in mass. In contrast, the heart was formed and beat normally at birth, and other internal organs appeared normal as well. Apparently, the lack of myogenin interrupted the histogenesis of skeletal muscle, whereas cardiac and smooth muscle developed normally. Thus, myogenin is required specifically for skeletal muscle differentiation.

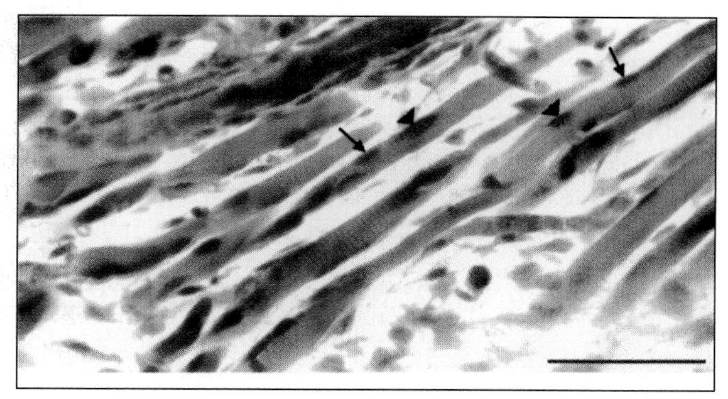

Figure 20.21 Role of myogenin in myogenesis. This photograph shows a sectioned skeletal muscle from a myogenin-null/wild-type chimeric mouse. The nuclei with blue stain (arrowheads) are derived from myogenin-null embryonic stem cells labeled with a *lacZ*$^+$ transgene. The pink nuclei (arrows) are derived from wild-type blastomeres. The striated muscle fibers contain both types of nuclei, sometimes with two or more stained nuclei in a row. These observations indicate that myogenin-null myoblasts are able to fuse with one another and with wild-type myoblasts.

TO examine more closely which steps of myogenesis depend on myogenin, Myer et al. (1997) generated chimeric mouse embryos composed in part of cells with *myogenin*$^+$ wild-type alleles and in part of cells that were homozygous for a null allele of *myogenin*. Unlike embryos consisting entirely of myogenin-null cells, at least some of the chimeras were expected to be viable and to provide a nearly normal environment in which the behavior of myogenin-null cells could be studied. In order to make the myogenin-null cell derivatives discernible from their wild-type counterparts, the null cells were labeled with a *lacZ*$^+$ transgene.

The embryos obtained fell in two overlapping classes. Class 1 embryos appeared normal, and less than 40% of their cells lacked myogenin. Class 2 embryos with greater proportions of myogenin-null cells tended to show spinal curvature and swelling similar to embryos consisting entirely of myogenin-null cells.

Histological sections of class 1 embryos revealed substantial skeletal muscle, with many muscle fibers stained partially for lacZ (Fig. 20.21). Generally, the blue staining caused by β-galactosidase activity was restricted to certain nuclei and surrounding jackets of cytoplasm. Such nuclei were sometimes interspersed with unstained nuclei, suggesting that mygenin-null myoblasts had fused with wild-type myoblasts. In other areas, several stained nuclei and cytoplasmic jackets were next to each other, suggesting that myogenin-null myoblasts had fused with one another. The authors concluded that the scarcity of skeletal muscle fibers in entire myogenin-null mice does not result from an intrinsic inability of myogenin-null cells to fuse but from some extracellular factor that decreases with the proportion of wild-type myoblasts in the fibers.

The chimeric muscle fibers did not synthesize the normal amount of muscle kreatin kinase, a muscle-specific protein. Indeed, the level of kreatin kinase in skeletal muscle decreased roughly in proportion to the amount of myogenin present. These data indicate that myogenin is a limiting factor to a muscle fiber's ability to form muscle-specific proteins.

Questions

1. How do you think the investigators constructed their chimeric embryos?
2. How could the investigators estimate the percentage of myogenin-null cells in a given tissue?
3. Assume the process that makes myoblasts fusion competent includes an *autocrine* feedback loop (see Fig. 2.24) consisting of a secreted signal and a matching receptor. Based on the data presented here, would you expect that myogenin is required for producing the signal or the receptor?

Additional experiments have shown that myogenin can to a large extent substitute for myf-5 in promoting myogenesis. If the coding region of *myf-5*$^+$ is replaced, by homologous recombination, with *myogenin* cDNA, then myogenin is synthesized with the same spatial and temporal pattern as myf-5. Such myogenin "knock-in" mice are viable and show normal rib formation, indicating that myogenin can functionally replace myf-5 in skeletal morphogenesis (Y. Wang et al., 1996). However, myogenin is a less than perfect substitute for myf-5 in myogenesis, which is quantitatively reduced in myogenin knock-ins that are deficient for both myf-5 and MyoD (Y. Wang and R. Jaenisch, 1997). Whether the functional differences between myf-5 and myogenin results from differences in their properties as gene regulators and/or from their different spatiotemporal patterns of synthesis remains to be seen.

The *MRF4*$^+$ gene is first expressed in embryonic myotomes and is up-regulated in differentiated muscle fibers, making it the predominant myogenic bHLH protein in adult muscle. Knockout mice lacking MRF4 were viable but showed a dramatic increase in myogenin synthesis, indicating that MRF4 is required for the down-regulation of myogenin synthesis that normally occurs after birth (W. Zhang et al., 1995). However, MRF4 can partially substitute for myogenin if it is expressed in the same spatiotemporal pattern as myogenin (Z. Zhu and J.B. Miller, 1997).

Together, the myogenic bHLH proteins and their genes form a complex regulatory network that acts as the hub in the genetic control of myogenesis. The discovery of this network has opened the way for a genetic and molecular understanding of cell differentiation. While some of the pathways within this network have been explored, a satisfactory understanding of their entire hierarchical relationships still needs to be achieved. In the meantime, researchers have also begun to analyze some of the signals that activate the network.

INDUCTIVE SIGNALS REGULATE MYOGENIC bHLH GENE ACTIVITY

How is the synthesis of myogenic bHLH proteins initiated in vivo? As discussed in Section 14.2 and earlier in this section, epaxial and hypaxial muscles are activated by different signals and rely on different myogenic bHLH proteins. Epaxial muscle depends on *myf-5*$^+$ gene activity and is induced by Sonic hedgehog (Shh) protein emanating from notochord and floor plate in combination with Wnt proteins released from dorsal neural tube and epidermis. In contrast, hypaxial muscle depends on *MyoD*$^+$ activity and is induced by secreted proteins from overlying epidermis while being inhibited by BMP-4 protein from lateral mesoderm. The simplest interpretation of these observations is that *myf-5*$^+$ is activated by a combination of Shh and Wnt while *MyoD*$^+$ is regulated by an epidermal signal and BMP-4 (see Fig. 20.20). However, there is evidence that a *homeobox* gene, *Pax-3*$^+$, is involved as an intermediary between the secreted signals and the myogenic genes *myf-5*$^+$ and *MyoD*$^+$ (Maroto, 1997; Tajbakhsh et al., 1997).

MYOGENIC bHLH PROTEINS INTERACT WITH CELL CYCLE REGULATORS

Many classical observations on cell populations that renew from stem cells have indicated that *cell proliferation* and *cell differentiation* exclude each other. The modern analysis of muscle differentiation has provided an opportunity to study the molecular mechanisms underlying this mutual exclusion.

The pacemakers of the cell cycle are *cyclin-dependent kinases (CDKs)*, kinases activated by association with proteins known as *cyclins* (see Section 2.4). Cyclin D accumulates specifically during the transition from G_1 to S phase and activates CDK4. In addition to promoting the cell cycle, CDK4/cyclin D inhibits the differentiation of skeletal muscle cells (see Fig. 20.20). This inhibition is caused directly and indirectly. The direct inhibition occurs by phosphorylating MyoD (Skapek et al., 1995). The indirect inhibition is mediated by proteins of the retinoblastoma (Rb) family, so called because the *Rb* gene was discovered through null alleles found in human patients with tumors of the retina (see Section 28.7). In its active (nonphosphorylated) state, Rb sequesters transcription factor E2F, which promotes cell proliferation. However, CDK4/cyclinD phosphorylates Rb, thus interfering with its ability to sequester E2F and allowing cell proliferation to proceed. In addition to sequestering E2F, the Rb-E2F complex promotes myogenesis (Shin et al., 1995). Thus, by phosphorylating both MyoD and Rb,

CDK4/cyclin D promotes cell proliferation while inhibiting muscle differentiation.

Conversely, the myogenic bHLH proteins not only promote myogenesis but also inhibit cell proliferation. The genetic cascade from myogenin to muscle differentiation includes protein p21, which is synthesized at low levels in proliferating myoblasts but up-regulated dramatically during normal muscle differentiation and upon experimental overexpression of *MyoD*. In turn, p21 inhibits CDKs and augments the synthesis of muscle-specific proteins such as myosin and muscle kreatin kinase (Halevy et al., 1995; Skapek et al., 1995).

Together, these studies reveal a mutual inhibition between myogenic bHLH factors and some of the proteins that regulate the cell cycle. While our knowledge of this mutual inhibition is far from being complete, the available data have provided a molecular understanding of the older observations that differentiating cells in stem cell populations withdraw from the cell cycle.

MYOGENIC bHLH PROTEINS COOPERATE WITH MEF2 FACTORS IN ACTIVATING MUSCLE-SPECIFIC TARGET GENES

The final differentiation of muscle requires the synthesis of muscle-specific proteins, such as myosin, muscle kreatin kinase, and desmin, a muscle-specific *intermediate filament* protein. The genes for these proteins are controlled jointly by the myogenic bHLH proteins and members of the MEF2 family of transcription factors (MEF stands for myocyte enhancer factor). Vertebrates have four *MEF2*$^+$ genes with overlapping expression domains, making it difficult to assess their individual functions. *Drosophila*, however, has only one such gene, *D-mef2*$^+$. Like its vertebrate homologs, D-mef2 protein is synthesized in developing skeletal, cardiac, and visceral muscle. Homozygous null alleles of *D-mef2* do not interfere with the deployment of myoblasts but with the synthesis of muscle-specific proteins, causing the absence of differentiated muscle and death (Raganayaku et al., 1995).

The enhancer regions of several muscle-specific protein have binding sites for both myogenic bHLH proteins and MEF2. For example, the mouse gene for desmin has an enhancer with an MEF2-binding site next to an E-box, the typical binding site for myogenic bHLH proteins (Kuisk et al., 1996). Mutations in the MEF2-binding site completely suppress desmin gene expression in heart and skeletal muscle, whereas mutations in the E-box interfere only with skeletal muscle development. Deletion mapping of the functional domains required for the cooperative action of MEF2 and myogenic bHLH protein indicate that the two proteins interact with each other during transcriptional activation (Molkentin et al., 1995; Ficket, 1996).

Myogenesis has become a paradigm for understanding the molecular mechanisms underlying cell differentiation. Classical studies have identified the embryonic origins of skeletal muscle and the surrounding tissues that induce myotome formation. Genetic and molecular work has revealed a network of myogenic bHLH proteins, which control the synthesis of muscle-specific proteins. The myogenic bHLH proteins have attracted particular attention because of their stunning ability to redirect non-muscle cell precursors into making muscle fibers. However, the existence of small networks of transcriptional regulators that stabilize their own activity and marshal the expression of batteries of target genes may turn out to be a general principle. The SCL and GATA factors play a similar role in hematopoiesis (see Section 20.4). Proteins encoded by the *achete*$^+$ and *scute*$^+$ genes in *Drosophila* (see Section 6.4), and by the *NeuroD*$^+$ and *neurogenin*$^+$ genes in vertebrates (Ma et al., 1996) have a corresponding function in neuronal development. Transfection of mouse embryonic carcinoma cells with cDNAs encoding neurogenic bHLH proteins and their putative dimerization partner E12 causes the differentiation of these cells as neurons (Farah et al., 2000). In another parallel to myogenesis, Id proteins prevents neuroblasts from premature differentiation (Lyden et al., 1999).

Networks of a few transcriptional regulators seem to acquire stable states through a "shake-out" process involving many competing transcription factors synthesized at low and fluctuating levels (see Section 20.4). The number of stable states that can be formed is limited by the number of competing transcription factors and by their regulatory properties. These limitations can explain, at least in principle, two classical features of cell differentiation: the finite number of mature cell types in each species and the stability of differentiated cell states.

20.6 Unexpected Potency of Stem Cells and Prospects for Medical Uses

For a long time, developmental biologists have assumed that the potency of stem cells would be limited to those differentiated cell types that occur naturally in the tissue in which the stem cells reside. That is, hematopoietic stem cells were expected to form only blood cells, epidermal stem cells only epidermis, and so on. However, experiments to test this assumption have revealed a much greater potency of stem cells than anticipated. Ironically, the most dramatic results were obtained with cells from an organ that was thought to retain few if any stem cells during adulthood, the mammalian central nervous system (CNS).

The CNS of mammals, in particular humans, regenerates poorly and has long been thought to be altogether incapable of forming new neurons in adults. However, Eriksson and coworkers (1998) were able to show that new neurons can originate in the adult human brain. The authors obtained human brain tissue from deceased cancer patients who had been treated with bromodeoxyuridine (BrdU), a thymidine analog that labels newly synthesized DNA. This treatment allowed them to identify new neurons as cells bearing the BrdU label and displaying neuron-specific morphology as well as marker proteins. Such neurons were indeed found in a brain area known as the *dentate gyrus* of the *hippocampus*. These results indicate that the human brain retains its ability to generate at least some neurons throughout life.

In similar experiments with mice, Johansson and coworkers (1999) found stem cells in the *ependymal layer*, an epithelium surrounding the fluid-filled central canal of the spinal cord and the ventricles of the brain. Thus, the ependymal layer of the adult occupies the same location as the *neuroepithelial cells* of the embryo, which give rise to the entire central nervous system during embryogenesis (see Fig. 13.4). The investigators were able to grow single ependymal cells into clones of undifferentiated cells, which, upon addition of growth factor–containing serum, synthesized marker proteins characteristic of *neurons* and two types of glia cells, namely *astrocytes* and *oligodendrocytes*. In vivo, some of the neurons migrated to an anterior portion of the brain known as the olfactory bulb. In the spinal chord, the ependymal stem cells divided very rarely, but the division rate increased dramatically after spinal cord injury. These observations show that at least some ependymal cells in mice are neural stem cells that divide rarely but do so in response to injury.

Even more spectacular, mouse neural stem cells can give rise to blood cells including both myeloid cells and lymphoid cell types (Fig. 20.22; Bjornson et al., 1999). This was shown by injecting neural stem cells from embryonic or adult forebrain into the bloodstream of mice whose own hematopoietic system had been nearly destroyed by X-irradiation (see Fig. 20.13). Because the donor mice were labeled with a *lacZ*$^+$ transgene, the investigators could use the *polymerase chain reaction* (see Method 15.3) to establish that the clones of newly formed blood cells in the spleens of recipient mice were indeed derived from the transferred brain cells. To exclude the possibility that the transferred brain cells were contaminated with blood cells, the investigators transferred clones derived from single adult neural stem cells in some of their experiments.

The experiment described above shows that neural stem cells have a much wider potency than originally thought. Indeed these cells, which normally produce ectodermal derivatives (neurons and glia), can be manipulated so that they will form a mesodermal derivative (blood). How this transition occurs is not known with certainty. Circumstantial evidence suggests that the transferred neural stem cells do not produce committed blood cell progenitors directly but form hematopoietic stem cells first (Fig. 20.22). The time interval between the

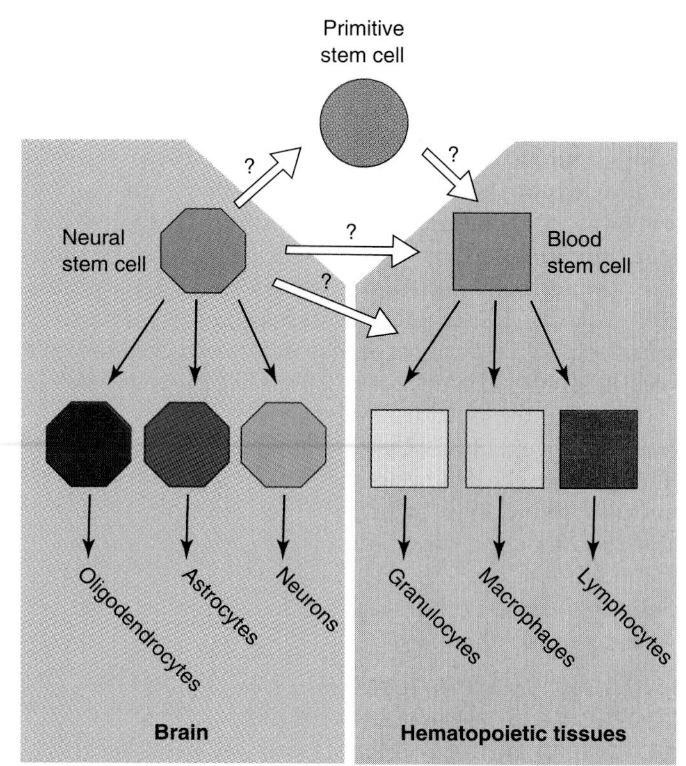

Figure 20.22 Possible ways in which neural and blood stem cells could be related to each other. Mouse neural stem cells injected into the bloodstream can give rise to differentiated blood cells as discussed in the main text. The blood cells may have been formed directly from the neural stem cells when the latter became integrated into hematopoietic tissues. Alternatively, the neural stem cells may have produced blood stem cells first or reverted to a more primitive type of stem cell, which in turn can form blood (and possibly other) stem cells.

transfer of neural stem cells and the appearance of the first blood cells was longer than the interval between the transfer of hematopoietic stem cells and the appearance of their first blood cell derivatives. Also, neural stem cells were never observed to produce blood cells in vitro.

Similar transplantation experiments revealed that stem cells from bone marrow can give rise to liver cells (Pedersen et al., 1999) as well as to adipocytes, chondrocytes, and osteocytes (Pittenger et al., 1999). These experiments provide dramatic confirmation of the *principles of genomic equivalence* and *differential gene expression* (see Section 7.1). More specifically, the results show that the potency of stem cells is not limited to the cell types of the tissues in which they reside. Little is known about the specific signals that bring about the dramatic changes of fate of the stem cells used in these experiments.

The enormous potential of stem cells to multiply and produce wide ranges of differentiated cell types is of great medical interest (*Pedersen, 1999;* Weissman, 2000). For example, the general procedure employed in the *spleen colony assay* (see Fig. 20.13) is also used by clinicians to reconstitute the hematopoietic systems of humans whose stem cells are deficient or have been incapacitated by cancer therapy (*Golde, 1991*). As the signals that direct the behavior of various stem cells become better understood, stem cells will become increasingly useful for cell replacement therapies of various human afflictions. The latter include neurodegenerative diseases like Alzheimer's and Parkinson's as well as losses of brain tissue resulting from stroke or injuries. Manipulation of pancreatic reserve cells may enable physicians to combat certain forms of diabetes. Stem cells that make chondrocytes may help to restore cartilage layers damaged by arthritis. Some of these treatments have already been successful with mutant mouse strains that mimic human diseases (Studer et al., 1998; Yandava et al., 1999). Attempts to grow relatively simple organs, such as skin or urinary bladders, from stem cells are also well under way (*Mooney and Mikos, 1999*). A problem inherent to cell and tissue replacement therapies is of course graft rejection by the recipient's immune system. One way of overcoming this problem is to genetically engineer the stem cells used for producing the graft in such a way that their descendants will not present the signals on their plasma membranes to which the host immune system reacts.

Of particular interest are human *embryonic stem (ES) cells*—that is, cells obtained from the *inner cell mass (ICM)* of embryos at the *blastocyst* stage and cultured in vitro (see Section 5.2). Under suitable conditions, human ES cells can divide indefinitely and give rise a wide range of differentiated cells (*Thomson and Marshall, 1998; Pedersen, 1999*). When transplanted under the skin of immunodeficient mice, ES cells give rise to disorganized tissues including derivatives from each of the three germ layers (Thomson et al., 1998). In addition to being virtually totipotent, human ES cells may also be clonable on demand and in a way that avoids immunological reaction, using the nuclear transfer technique discussed in Section 7.7.

In an ideal futuristic scenario, a person who suffered a heart attack could be treated as follows. A nucleus from one of the patient's cells is transplanted into an enucleated egg, which is allowed to grow in vitro to the blastocyst stage. ICM cells are isolated to obtain ES cells, and these are cultured with a suitable mix of signals to make them develop as heart muscle cells. These cells could replace the damaged part of the person's heart and would not be rejected because of their perfect immunological match with the recipient.

Research on human ES cells has been hampered by legislative restrictions in the United States and many European countries. Many people consider the harvesting of ICM cells from human blastocysts obtained by in vitro fertilization unethical, so that this procedure has been limited to privately funded research, which in turn tends to be tied up with proprietary interests and is therefore communicated less freely than government-supported research. The cloning of human blastocysts for the purpose of making immunologically compatible ES cells is likely to be even more controversial because it would be a major step toward cloning human babies. As is often the case with technological progress, the manipulation of human embryos can be used for beneficial and detrimental purposes. It is important that scientists keep educating the public about the progress being made, so that informed public discussions and legislative acts can determine what is acceptable and put appropriate regulations in place.

SUMMARY

Every species has a limited number of cell types, each with a distinct morphology and molecular makeup. The same cell type may appear in several body regions and may even develop from different embryonic cell lineages. Cells that have acquired the distinctive features of their type are called mature or differentiated cells. The differentiated state is generally stable: Differentiated cells do not normally turn into other cell types. Many differentiated cells are highly specialized and depend on other kinds of cells for support and maintenance.

Cell differentiation occurs initially during the embryonic period of histogenesis. In some differentiated tissues, including nerve tissue and skeletal muscle, there is normally little turnover or addition of cells after the embryonic period. In other tissues, cell differentiation continues throughout adulthood and cells are turned over many times during the life of the organism. The replacement of differentiated cells occurs in one of two ways. In certain tissues, including liver and blood vessel endothelium, differentiated cells are still capable of dividing. In others, including epidermis and blood, worn-out cells are replaced from a pool of undifferentiated stem cells. Stem cells retain their capacity for self-renewal throughout the life of the organism. As stem cells divide, they produce both new stem cells and committed progenitor cells. The latter typically go through several rounds of amplifying divisions before they differentiate. Stem cells are either unipotent or pluripotent—that is, they give rise to either one or several types of committed progenitor cells.

The freshwater polyp *Hydra* consists of three cell lineages. Two of these lineages are epithelial; their cells divide in the differentiated state. The third cell lineage, the interstitial cells, forms nematocytes, nerve cells, gland cells, and gametes. The interstitial cells renew from pluripotent stem cells and from unipotent stem cells for either eggs or sperm.

The blood cells of vertebrates are continuously renewed from a small pool of hematopoietic stem cells located mostly in bone marrow. A single pluripotent stem cell can form committed progenitors for all types of blood cells as well as the endothelial cells that line the inside of the heart and blood vessels. The commitment of hematopoietic progenitor cells seems to be based on a "shake-out" process between many competing transcription factors synthesized at low and fluctuating levels. The lineage-specific transcription factors that come to prevail during the commitment of a given progenitor cell are thought to activate the genes required in the respective cell lineage.

The survival, proliferation, and differentiation of blood cells depend on regulatory proteins, called colony-stimulating factors, which act on specific receptors in their target cells. Some colony-stimulating factors are diffusible, while others are bound to bone marrow stroma cells and extracellular matrix.

The process of skeletal muscle differentiation is known as myogenesis. It is controlled by a family of transcription factors called myogenic bHLH proteins, which have distinct but partially overlapping functions. They cooperate with other transcription factors in activating muscle-specific target genes. The myogenic bHLH proteins are also engaged in mutual inhibition with a different set of proteins that promote the cell cycle.

The potency of stem cells is not limited to the cell types that occur naturally in the tissues in which stem cells reside. For instance, neural stem cells can give rise to different types of blood cells. Nearly unlimited potency is observed in so-called embryonic stem cells—that is, cells taken from the inner cell mass of mammalian blastocysts and cultured in vitro. Both types of stem cell hold great promise for cell replacement therapies of human diseases.

chapter 21

Pattern Formation and Embryonic Fields

Figure 21.1 Peacock displaying a dazzling array of eyespots on his tail feathers. The eyespots exemplify the ability of organisms to form patterns, harmonious arrays of different elements. In this case, the pattern arises through the synthesis of dark pigments in a circular territory of cells, and the synthesis of lighter pigments in drop-shaped territories of cells surrounding the dark spot. How are the signals that determine cells to make dark and light pigments coordinated in space so that a well-shaped eyespot results, rather than a salt-and-pepper mixture of dark and light cells?

PATTERN formation is one of the central topics in developmental biology. ***Patterns*** are harmonious arrays of different elements, such as the array of five fingers on a human hand. Patterns can be seen at different levels of organization. For instance, the overall body pattern of birds features a head, a body, a pair of wings, a pair of legs, and a tail. Each of these organs shows a typical pattern of its own, including feathers of different sizes, shapes, and colors. Each feather, in turn, is composed of a shaft with barbs and barbules. These consist of dead, pigmented epidermal cells, often arranged in eye-catching patterns (Fig. 21.1). Cells, too, have internal patterns, but in this chapter we will focus on patterns consisting of cells or larger elements.

Many patterns are formed when different types of cells mature in a coordinated fashion. For example, the eyespot pattern in a peacock feather results from the arrangement of cells containing different pigments in nested rings or ovals, with the darkest pigments occupying the center. The development of such patterns raises two questions. First, how do seemingly equal precursor cells give rise to different types of mature cells? These are the processes of *cell determination* and *cell differentiation* discussed in previous chapters. Second, how are multiple pathways of cell determination coordinated so that an orderly array is formed rather than a random mixture of cell types? In other words, how does spatial coordination occur? The answer to the second question may hold the key to the first: The same signals that bring about spatial coordination also control, at least in part, the determination of individual cell types.

Even though patterns become most obvious after cell differentiation, the process of pattern formation begins much earlier. Consider the development of the vertebrate limb. An early limb bud consists of an ectodermal cover and a mesenchymal core. The latter gives rise to muscles and to the cartilage models of bones. However, before the core cells become histologically different, they exhibit regular patterns of cell division and cell condensation. Cell division takes place more rapidly in a progress zone near the tip of the limb. Cell condensations are areas of higher cell density, which develop into either cartilage or muscle. The cartilage condensations proceed from proximal regions to distal and synthesize the extracellular matrix characteristic of cartilage (Fig. 21.2)

In addition to patterns based on cell division and condensation, there are patterns of *apoptosis,* or programmed cell death, especially in appendages with separated digits, such as the human hand or the foot of a chicken. The cells between the developing digits of the limb bud release a genetic program that kills the cells in a noninflammatory and economical way. As the dead cells are consumed by macrophages, the digits become separated. Thus, the processes of cell division, cell differentiation, and apoptosis are all involved in generating spatial patterns. As each pattern is created and refined, the visible complexity of the organism increases.

In many cases, the cellular territories in which patterns are formed have the properties of ***fields.*** A field is defined as a group of cells that can form a certain structure, such as a limb. A key attribute of fields is that they can be enlarged or reduced in size and still give rise to the same complete and normally proportioned structure, except that the latter is now formed on a larger or smaller scale. This implies that fields are capable of *regulation,* an attribute discussed in Section 6.6. Cells in fields can adjust their state of determination according to signals received from other cells.

In Chapters 21 through 25, we will examine concepts and experimental strategies researchers have used in the study of pattern formation. In the present chapter, we will discuss mostly classic experiments in which cell isolation and transplantation have served as the principal methods. These experiments have provided the conceptual framework for the study of pattern formation. In Chapters 22 through 25, the emphasis will be on the genetic and molecular tools that have helped investigators to probe directly into the patterning mechanisms of fruit flies, vertebrates, flowers, and roundworms.

21.1 Regulation and the Field Concept

Most of the cells that engage in pattern formation are still capable of regulation: Their *potency* is greater than their *fate.* As an example, let us consider the cells that form the forelimb in the salamander *Ambystoma maculatum.* A *fate map* for the future forelimb at an advanced embryonic stage is shown in Figure 21.3. A circular area of somatic lateral mesoderm and overlying epidermis is fated to form the free limb. The free limb area is surrounded by a ring of cells that normally form the shoulder girdle and the peribrachial flank from which the free limb extends. Together, these areas are called the ***limb disc.***

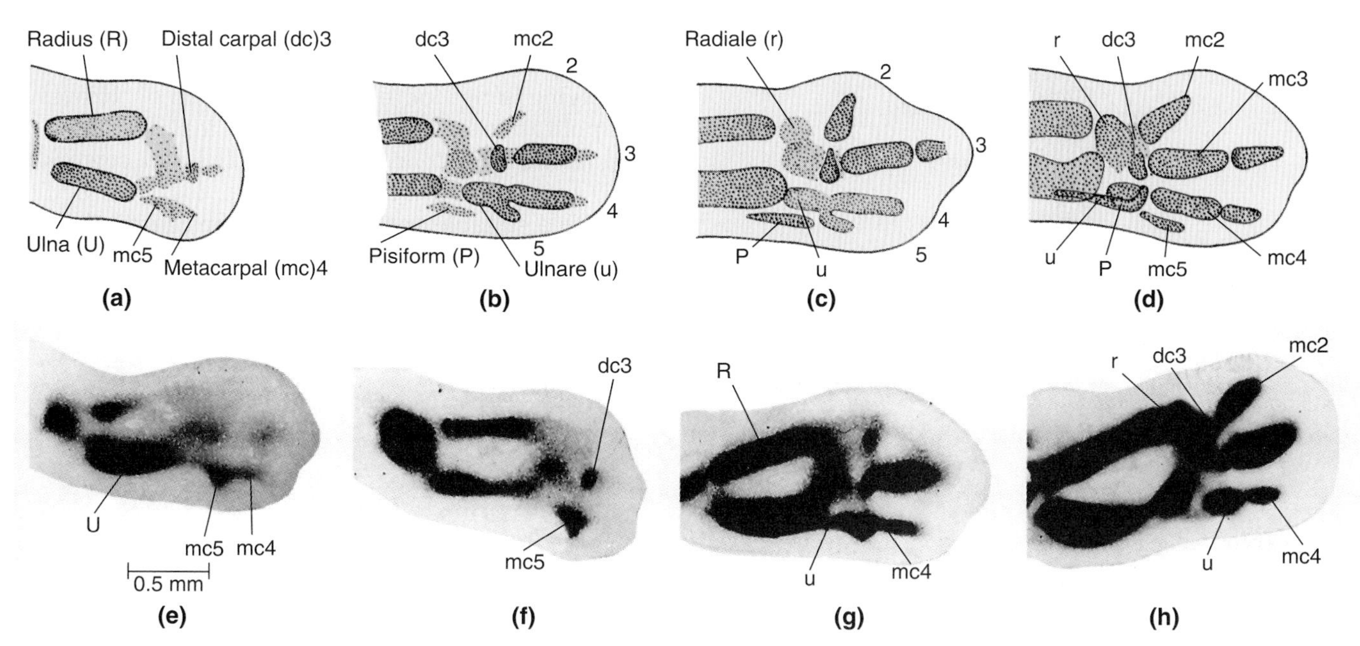

Figure 21.2 Pattern of precartilaginous cell condensations in the chick forelimb. **(a–d)** Drawings of precartilaginous cell condensations (stippled) and cartilaginous elements (stippled and outlined) in chicken wing buds at successive stages. The drawings were made from autoradiographs such as the ones shown in parts e through h. **(e–h)** Autoradiographs of sectioned chicken wing buds after labeling with $^{35}SO_4$. This label is incorporated into chondroitin sulfate, an extracellular matrix component characteristic of cartilage. Note that the intensity of the label progresses from proximal to distal with time.

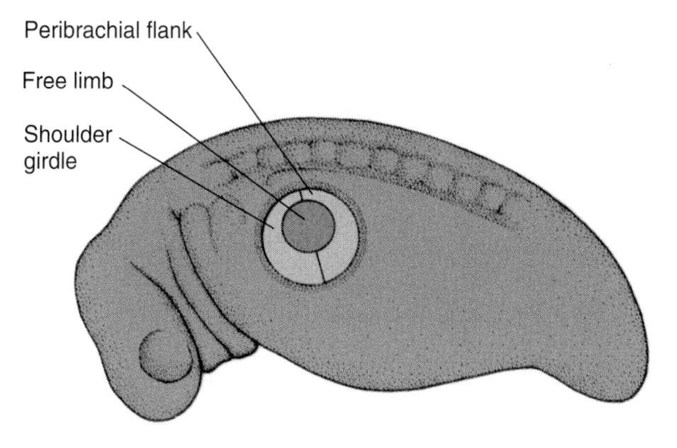

Figure 21.3 Forelimb field of the newt *Ambystoma maculatum.* The field is located in the somatic layer of the lateral mesoderm but also affects the development of the overlying ectoderm. The central area of the field contains the cells fated to form the free limb. These cells are surrounded by other cells that normally give rise to the shoulder girdle and peribrachial flank but will also form limb if the central cells are removed. A circle of cells outside this region does not normally contribute to the forelimb but may do so if all interior cells are removed.

A different picture arises when the developmental potencies of the limb disc and surrounding areas are tested by removal or transplantation. If the free limb area is removed, some of the cells that would normally contribute to shoulder girdle and peribrachial flank will deviate from their fate and form the free limb. If the entire limb disc is removed, an additional ring of cells surrounding the disc will form limb, shoulder girdle, and peribrachial flank, although with considerable delay. If the surrounding ring of embryonic cells is also removed, no limb will form. These results indicate that the potency of forming a complete limb is inherent in a larger group of cells than the area fated to form the limb. This larger area is called the ***limb field.***

If a limb disc is transplanted to another location in the embryonic lateral mesoderm, it will give rise to a supernumerary limb. If the limb disc is removed and split before being reimplanted, *two* supernumerary limbs will develop. Such an outcome was observed in a natural experiment that took place in California, Minnesota, and other U.S. states. In many natural ponds, frogs and salamanders showed supernumerary limbs (Fig. 21.4) and other limb deformities. Inspection revealed that parasitic flatworms (trematodes) were burrowing into developing limb buds. To test whether the worms were not only associated with the deformities but actually causing them, investigators inserted beads into the frogs' limb buds to mimic the mechanical effects of the cysts that the trematodes form (Sessions and Ruth, 1990). Another team, led by a student who had just received his Bachelor's degree, infected frogs with trematode larvae in the laboratory (P. T. J. Johnson et al., 1999). In either case, the resulting limb abnormalities closely resembled those observed in the wild. Presumably, the trematode activity can split frog limb fields into two or more so that they give rise to multiple limbs.

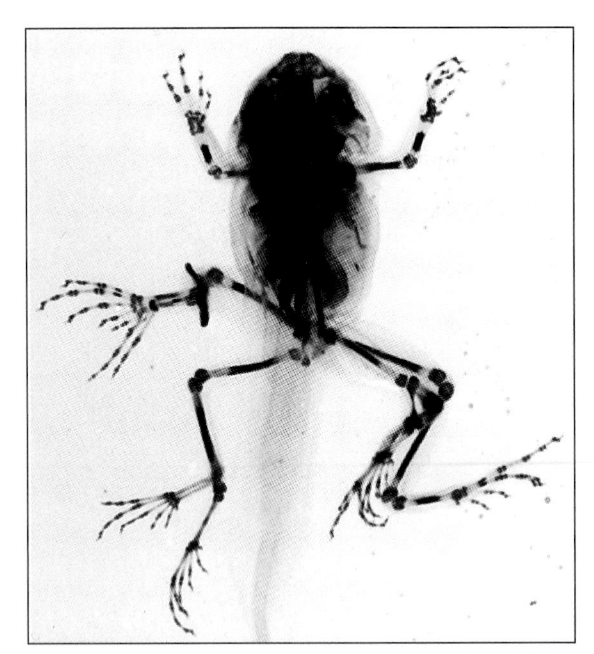

Figure 21.4 Adult frog (*Hyla regilla*) with multiple hindlimbs, from a pond inhabited by parasitic worms, which apparently partitioned the hindlimb buds of tadpoles so that the fragments formed limbs, or distal parts thereof, independently. The adult frog was cleared and stained to make cartilage and bone visible.

Similar regulative capabilities may be shown by entire embryos. Many animal embryos are capable of *twinning*. For instance, when a frog embryo is split at the 2-cell stage, or if human blastomeres separate spontaneously at the same stage, each blastomere will develop into a small but normally proportioned organism (see Figs. 6.1 and 6.22a). Conversely, when two or three mouse embryos are stuck together, they will also give rise to one normally proportioned mouse (see Fig. 6.23).

The German embryologist Hans Driesch (1894), who studied twinning in sea urchins, described the embryo as a *harmonisches äquipotentielles System* (harmonious equipotential system), meaning that isolated blastomeres had the same ability as an entire embryo to form a whole larva. The American embryologist Ross G. Harrison (1918) adopted the English translation of this phrase to characterize the same properties of the amphibian limb disc. Later investigators coined the shorter term *field* to describe any group of cells that cooperate in forming a complete and well-proportioned organ or embryo. The ability of fields to form the same pattern after experimental size reduction or enlargement is now referred to as ***size invariance.*** This term is often misunderstood: It does *not* mean that the size of a field is invariant. Rather, it means the pattern is invariant regardless of field size.

Classical embryologists recognized the importance of fields to the understanding of development. Weiss (1939), in his textbook *Principles of Development*, devoted 148 out of 573 pages to the topic of gradual determination and the field concept. Later on, fields fell out of fashion, presumably because they stood in the way of simplifying things for the purpose of analysis. However, the application of molecular probes may lead to a revival of the field concept. De Robertis and coworkers (1991) have shown that immunostaining (see Method 4.1) with a suitable antibody labels the cells in the pectoral fin bud of the zebra fish embryo as well as the cells of the fin disc before outgrowth (Fig. 21.5). The antibody in this experiment was directed against the antigenic protein's *homeodomain*, a specific DNA-binding domain that characterizes a family of transcription factors (see Section 23.2). The result indicates that all cells in the fin field share at least one transcription factor with a very similar DNA-binding domain. This observation and others have led to the notion of a field as a group of cells with coordinated gene activities.

Many territories in which pattern formation takes place exhibit the properties of fields. They have been found to do so in all embryos that are capable of regulation. For the embryo, fields are a marvelous way of compensating for growth irregularities and injuries. For the researcher, the field phenomenon is a formidable complication of the study of pattern formation. However, any serious model of pattern formation must deal with fields and their properties.

21.2 Characteristics of Pattern Formation

In this section, we will examine a few key experiments that establish some major points regarding pattern formation: Patterning depends on cellular interactions, on the genetic information of the interacting cells, and on their developmental history. The studies also indicate that at least some patterning signals are fairly general: They have been well conserved during evolution, and they are used in different parts of the body either simultaneously or successively.

PATTERN FORMATION DEPENDS UPON CELLULAR INTERACTIONS

In embryos with regulative capabilities, cells are determined gradually and, to a large extent, by interactions with other cells. In such embryos, *embryonic induction* is a prevalent patterning mechanism (see Sections 9.4, 9.6, 9.7 and 12.3–5). To investigate this process, classical embryologists performed transplantation experiments. By grafting tissues between two closely related species, such as a light-colored newt and a darkly pigmented newt, it was possible to ascertain which of the developing structures were formed by the graft and which by the host. In one experiment of this kind, Spemann and Schotté (1932) exchanged a piece of future belly epidermis with a piece of future mouth epidermis at the early

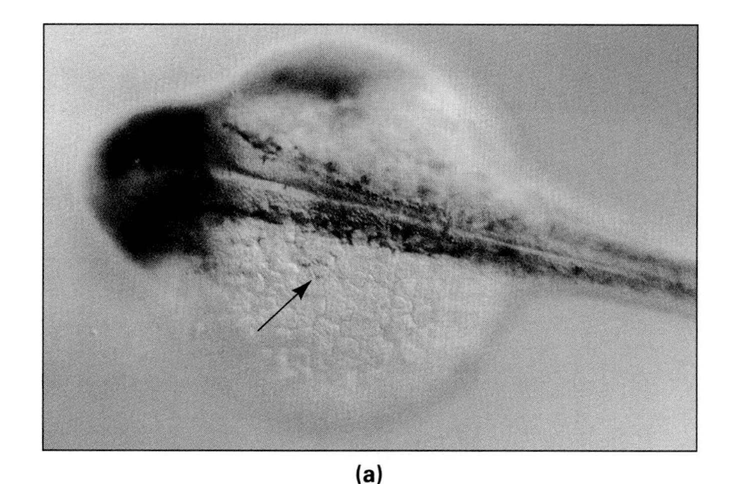

(a)

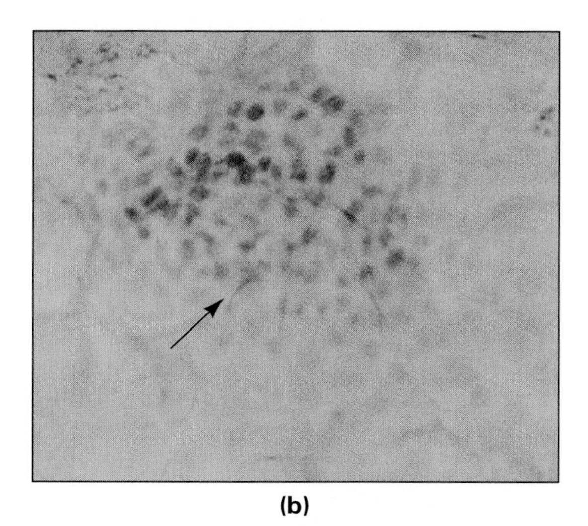

(b)

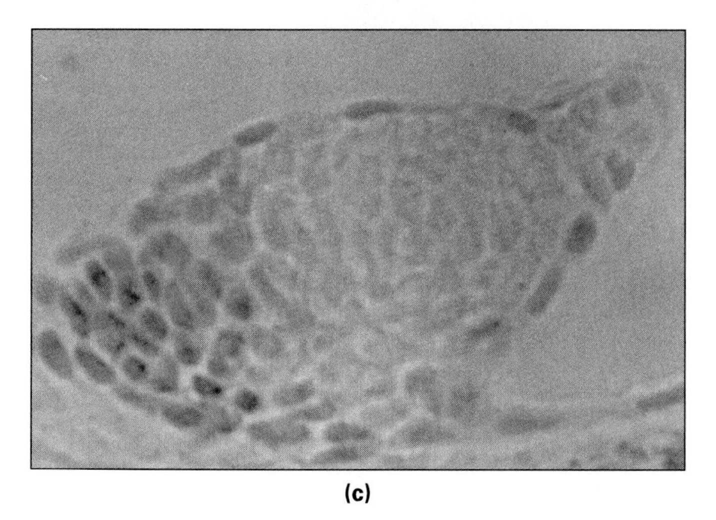

(c)

Figure 21.5 Fin field in the zebrafish embryo. **(a)** Zebrafish embryo immunostained (see Method 4.1) with a primary antibody against the homeodomain from *Xenopus* XlHbox 1 protein. The arrow points to a circular area containing the cells that will form the pectoral fin (fin field). **(b)** High-power magnification shows that the fin field cells express XlHbox 1 antigen. **(c)** Histological section indicating that the antigen is expressed in anterior and proximal mesoderm as well as in overlying epidermal ectoderm.

gastrula stage. Each transplanted piece developed according to its *new* position in the host: prospective belly epidermis now formed mouth structures, and vice versa. Because the adult structures formed by the transplants depended upon surrounding tissue, their development was classified as *dependent differentiation.*

Not all grafted pieces of early amphibian gastrula undergo dependent differentiation. In the famous organizer experiment of Spemann and Mangold (1924), dorsal blastopore lip of the early gastrula underwent *self-differentiation:* Upon grafting, it did not adapt to its new location but stuck to its original fate of forming notochord. Moreover, the graft induced the surrounding host tissue to cooperate with the graft in forming an additional set of dorsal embryonic organs (see Fig. 12.17). In other words, the transplanted organizer set up an additional field for dorsal organs.

These experiments, and many similar ones carried out with other species, demonstrate that pattern formation depends on cellular interactions. Cells in some embryonic regions are determined early, act as inducers on their surroundings, and undergo self-differentiation. Cells in other regions are determined later, respond to the inducing signals from other cells, and undergo dependent differentiation.

THE RESPONSE TO PATTERNING SIGNALS DEPENDS ON AVAILABLE GENES

In the experiments described above, grafts were exchanged between two species that were chosen to differ as little as possible in their development except for their pigmentation. In other experiments, the investigators asked how the results might be different if corresponding grafts were exchanged between distantly related species, such the salamander *Triturus taeniatus* and the frog *Rana esculenta.* These species belong to two separate subclasses of amphibians, urodeles (tailed amphibians) and anurans (tailless amphibians). For each species, Spemann and Schotté (1932) had shown that prospective belly epidermis, upon transplantation to the future mouth region, would undergo dependent differentiation—that is, contribute to mouthparts (Table 21.1). Next, the same investigators transplanted prospective belly ectoderm from a *salamander* gastrula to the prospective mouth ectoderm of a *frog* gastrula. Their question was whether the transplant still undergo dependent differentiation—that is, develop mouthparts. In other words, would the graft "understand" the patterning signals from the host? If so, would the mouthparts derived from the graft be teeth and balancers (characteristic of the salamander), or horn-covered jaws and a sucker (characteristic of the frog)?

The results of this experiment are summarized in Table 21.1 and Figure 21.6. Indeed, prospective belly epidermis from the salamander gastrula, when transplanted into the prospective mouth region of the frog,

table 21.1 Transplantation Experiment of Spemann and Schotté (1932) Involving Frog and Salamander Gastrulae

Donor Species of Prospective Belly Epidermis Graft	Structures Formed after Transplantation to Mouth Region of Frog	Salamander
Frog	Horned jaw, sucker	Horned jaw, sucker
Salamander	Teeth, balancers	Teeth, balancers

gave rise to mouthparts. And clearly, the structures formed by the graft were the balancers and teeth characteristic of the salamander. Conversely, prospective frog belly epidermis in the salamander mouth region gave rise to frog mouthparts. Thus, regardless of the phylogenetic distance between donor and host, the transplants underwent dependent differentiation, but the type of mouthparts formed was characteristic of the donor species.

The limitation of the mouth structures formed to the "repertoire" of the species that provides the mouth tissue could be ascribed to either genetic or nongenetic factors. Additional experiments to be described shortly will show that the response of cells to patterning signals depends critically on the cells' gene activity. Frog cells do not seem to have (or use) the genetic information for forming sala-

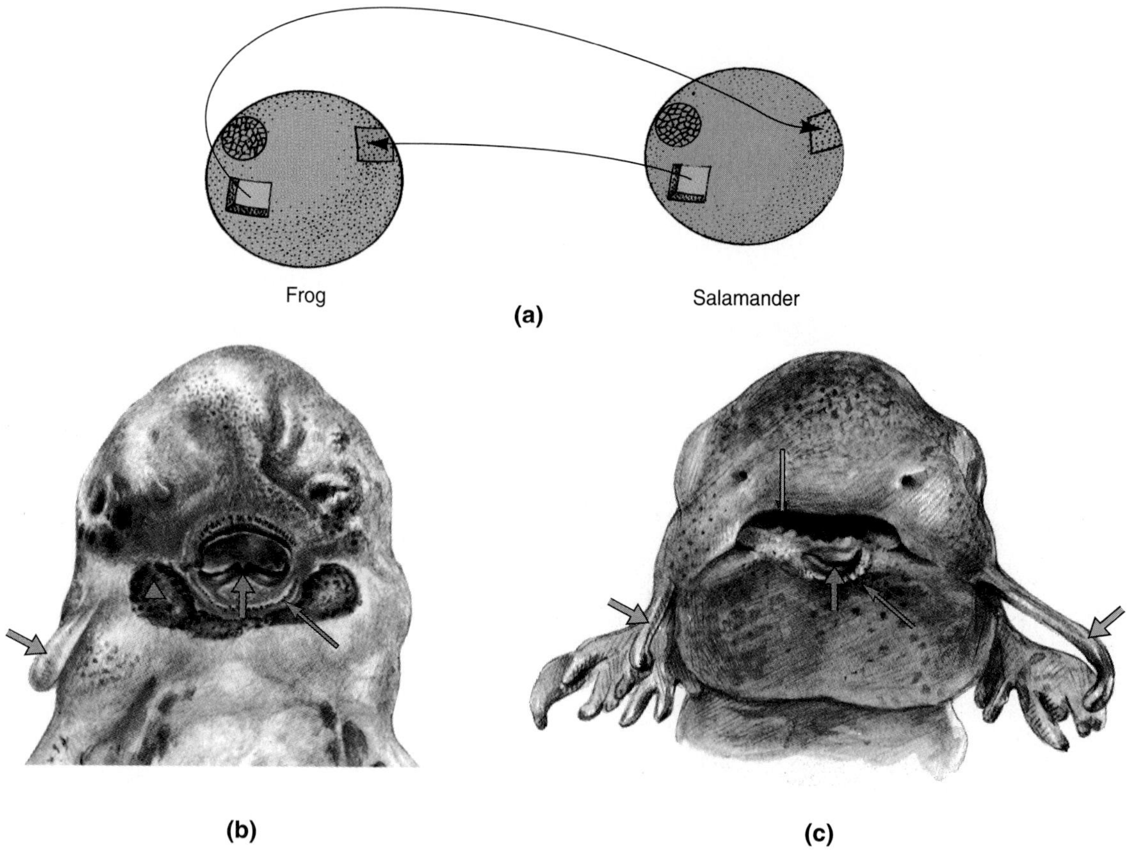

Figure 21.6 Dependence of inductive interactions on the genotype of the responding tissue. The diagram shows a reciprocal transplantation experiment between gastrulae of a frog (*Rana esculenta*) and a salamander (*Triton taeniatus*). The two species differ in their mouth structures. The frog has keratinized horn plates (short green arrow) and horn pins (long green arrows) as mouthparts and an adhesive organ called a sucker (green triangle) behind the mouth. The salamander has teeth on its jaws (long red arrow) and a pair of "balancer" organs (short red arrows) on both sides of the mouth. **(a)** Prospective belly epidermis from the salamander gastrula was transplanted to the prospective mouth region of the frog gastrula, and vice versa. Both transplants developed according to their new location, forming mouth structures that fit topographically with the mouth structures derived from host tissue. However, the mouth structures formed by the transplants had the morphological characteristics of the donor. **(b)** Head of a salamander larva that has formed its right balancer from its own (host) tissue. The mouth is derived from the transplant and is completely froglike, with horn plates and horn pins around the mouth, and a sucker behind the mouth. **(c)** Salamander host that has made an almost complete mouth with teeth from host tissue, while a small patch of grafted frog tissue has formed a small piece of horn plate with adjacent horn pins.

mander mouth structures, and vice versa. Nevertheless, the transplants from both species respond to apparently similar patterning signals that prompt them in a general way to form mouthparts—whatever mouthparts they have in their genetic repertoire.

THE RESPONSE TO PATTERNING SIGNALS DEPENDS ON DEVELOPMENTAL HISTORY

The response of a tissue to inductive signals is also limited by its developmental history. As we saw in previous chapters, induction is often a cumulative process involving multiple inducers. For instance, the lens of the eye is formed by head ectoderm in response to three inducing tissues. During gastrula stages, the head ectoderm that will eventually form the lens is underlain by the endoderm of the future pharynx. This endoderm is the first inducer of the lens. As gastrulation proceeds, mesoderm that will later form the heart comes to lie next to the future lens ectoderm. The heart mesoderm is the second inducer of the lens. Finally, as the optic vesicles bulge out from the embryonic brain, the future retina comes in contact with the overlying head ectoderm (see Fig. 1.16). The optic vesicle, then, is the third inducer of the lens, although it acts indirectly by displacing mesenchymal cells that would otherwise *inhibit* lens formation.

The response of the head ectoderm to the second and third lens inducers depends on its previous interactions with the first inducer. Jacobson (1966) studied this phenomenon quantitatively in a salamander, *Taricha torosa.* In one experiment, he removed the first inducer, the pharyngeal endoderm (Fig. 21.7). The earlier the removal occurred, the lower was the percentage of successful lens inductions. Similar results were obtained after removal of the second inducer, heart mesoderm. Thus, the early history of inductive interactions clearly primed the response of head ectoderm to the later inducers.

In most inductive interactions, the *competence* of a tissue to respond to an inductive stimulus depends on the developmental history of the responding tissue. In normal development, competence typically arises some time before the responding tissue is exposed to the new stimulus and lasts for a while thereafter.

PATTERNING SIGNALS HAVE A HIGH DEGREE OF GENERALITY

The most enthusiastic students of pattern formation have hoped to find that patterning signals are universal, that is, the same in all organisms, as is the genetic code. Indeed, some patterning signals do show a remarkable degree of phylogenetic conservation. The transplantation experiment summarized in Figure 21.6 and Table 21.1 demonstrates that a salamander graft responds to inductive signals provided by a frog host and vice versa. This is an astounding result, given that frogs and salamanders have evolved independently for more than 200 million years.

The patterning signals exchanged between cells are also general in the sense that they are used in different regions of the same embryo—either simultaneously, or successively at different stages of development. This is illustrated by the results of the following study.

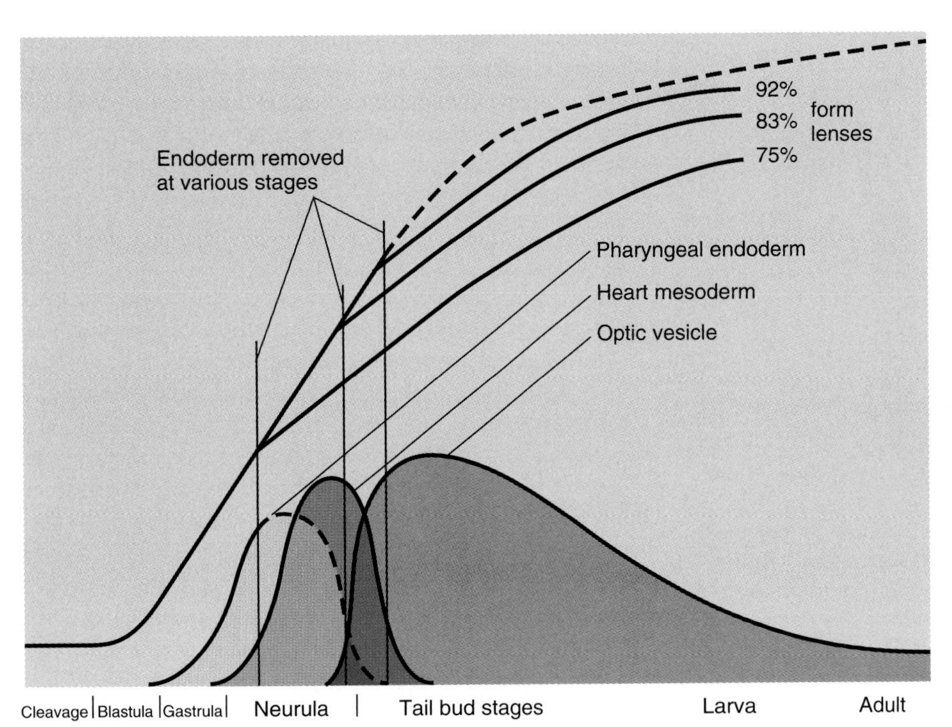

Figure 21.7 Dependence of inductive interactions on the developmental history of the responding tissue. The lens of the eye in the salamander *Taricha torosa* is induced in head ectoderm by successive interactions with pharyngeal endoderm (light color), heart mesoderm (dark color), and the optic vesicle (gray). The height of each curve (ordinate) represents the stage-dependent capacity to induce lens formation. Removal of the endoderm during the neurula and early tail bud stages (vertical lines) leaves some embryos unable to subsequently respond to inducers. The longer the endoderm is present, the greater the percentage of individuals that form lenses.

THE wild-type allele of the *Antennapedia*$^+$ gene in *Drosophila* is most active in the thoracic segments and plays a key role in leg development. The *Antennapedia*^{R} allele is characterized by a small chromosomal inversion that destabilizes the transcriptional control of the gene. This mutation causes sporadic activation of the gene in the antenna, where it is normally silent. The ectopic gene activity causes the formation of leg structures instead of antennal parts. Rarely is the entire antenna transformed into an entire leg. More commonly, parts of the antenna are replaced with parts of the second leg. When Postlethwait and Schneiderman (1971) examined a large number of these replacements under the microscope, they found that they were strictly position-specific (Fig. 21.8). The most distal part of the antenna, the *arista,* was always replaced with the last tarsal segment carrying the claws at the tip. The most proximal part of the antenna was always replaced by the most proximal leg structure, known as the *trochanter.* Intermediate parts of the antenna were replaced with intermediate leg parts.

Questions

1. Does the *Antennapedia* allele used in this study represent a loss-of-function or a gain-of-function allele?
2. Assume you make *Drosophila* transgenic for a fusion gene consisting of the coding region of *Antennapedia* spliced to the regulatory region of an actin gene. Assuming that the transgenic flies survive to adulthood, what would you expect their antennae to look like?

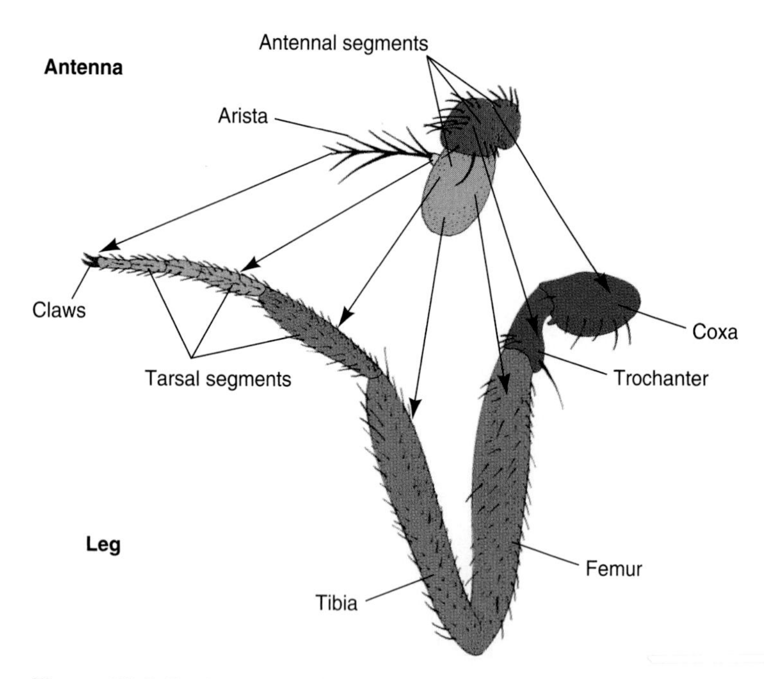

Figure 21.8 Position-specific replacement of antennal parts with corresponding parts of the second leg in the homeotic mutant *Antennapedia*^{R} of *Drosophila.* Sporadic expression of the deregulated gene in the antenna causes the transformation of antennal parts into leg parts. Each arrow points from the antennal area to the region of the leg that replaces it. Note that the replacements maintain the same proximodistal order.

The strict order of the homeotic transformations in the *Antennapedia*^{R} mutant suggests that leg and antenna respond to similar or identical patterning signals specifying the proximodistal order of their segments. The character of the structures (leg or antenna) formed in response to a signal depends on the local activity or inactivity of the *Antennapedia* gene. Cells in which the gene is inactive form antennal structures, while cells with an active *Antennapedia* gene form leg structures. This result confirms our interpretation of the transplantation experiments between frog and salamander embryos, in which we ascribed the species-specific response to patterning signals to the genetic limitations of the reacting cells.

In summary, pattern formation depends on intercellular signals. At least some of these signals have been well conserved during evolution. The response of cells to patterning signals depends on the cells' genetic repertoire and developmental history. These characteristics of pattern formation have shaped the concepts used in its analysis.

21.3 The Concept of Receiving and Interpreting Positional Value

When Hans Driesch discovered what is now termed an embryonic field, he immediately realized that a cell's fate depends on its *position* in the field. Driesch made this a salient point in his treatise *Analytische Theorie der organischen Entwicklung* ("Analytical Theory of Organic Development," 1894). As a cautionary explanation, however, he added that it was not position itself that determined the fates of cells but rather the *different signals* received by the cells according to their positions. In discussing this point, several biologists have used the analogy of a mountain, which supports a sequence of different vegetations, from tropical or deciduous forests at the bottom, through grasslands and coniferous trees, to mosses and lichens at the top (Sander, 1990). A similar pattern of change in plant communities is observed as one moves at low altitude from the equator to the poles of the earth. At first glance, it may seem that position itself (altitude or latitude) favors the growth of specific plants. However, further comparison and experimentation show that the factors of more direct importance are the temperature and the amount of precipitation at each position. Similarly, the analysis of embryonic pattern formation needs to proceed from the study of the effects of cell position to the study of molecular signals received in different positions.

The apparent effect of a cell's position on its fate was cast into a widely accepted set of terms by Wolpert (1969, *1978, 1989*), who proposed that pattern formation is a two-step process. First the cells of a territory or field receive positional values, and then they interpret these values according to their genetic information and developmental history. ***Positional value,*** in terms of this concept, is assigned to a cell in much the way geometric coordinates define a point in a grid. ***Interpretation*** is the general process whereby a positional value leads to a particular cell activity or differentiated state. If we heed Driesch's cautionary remark, then positional value is a preliminary parameter that must be understood in terms of the signals that cells receive from elsewhere. Similarly, interpretation must eventually be described as the conversion of these signals into gene activity.

Of critical importance in this concept are the cells that provide the ***reference points*** for the assignment of positional values. The reference points correspond to the axes of a coordinate grid. Experimental changes to the size of an embryonic field would mean the establishment of new reference points for positional information.

Wolpert has used the French flag as an analogy to emphasize the two components to biological pattern formation, the receipt of signals from other cells ("assignment of positional value") and the conversion of the signals into gene activity ("interpretation of positional value"). In the French flag analogy, the patterning process is assumed to occur in a rectangular territory of cells as shown in Figure 21.9. The positional value assigned to each cell specifies whether its X coordinate is within the first, second, or third portion of the flag territory. (The Y coordinate is unimportant in this case.) At first, the cells are still "embryonic," or unpigmented. Subsequently, the cells interpret their positional values by synthesizing blue, white, or red pigment. The interpretation, however, depends on the cells' genetic information, which is "French." Cells with "American" genes would respond to the same positional values but would interpret them by forming stars and stripes.

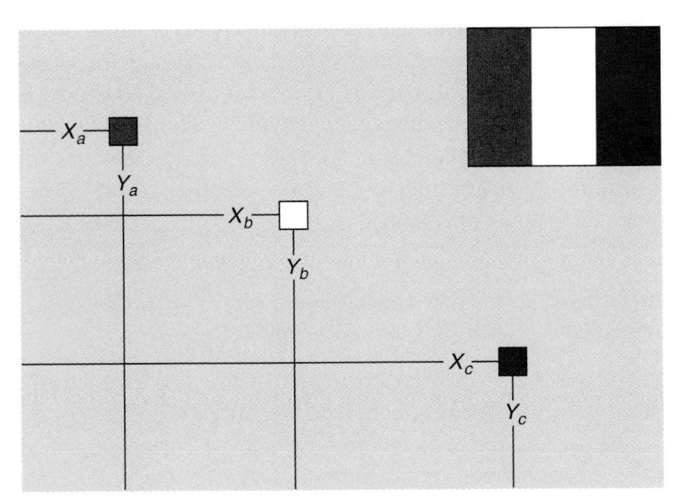

Figure 21.9 French flag model used by Wolpert (1978) to explain his concept of pattern formation. A two-dimensional territory of embryonic cells undergoes differentiation, forming cells with three kinds of pigment: blue, white, or red. The process is thought to require two steps. First, each cell is assigned a positional value (X_a/Y_a, X_b/Y_b, X_c/Y_c) with regard to the two axes of the flag. The Y coordinate is unimportant in this case. In the second step of the patterning process, cells interpret their positional values. Cells with low X values produce blue pigment; cells with intermediate X values produce white pigment; cells with high X values produce red pigment. As a result, the array of differentiated cells resembles the French flag (shown as inset in upper right corner).

The French flag analogy explains nicely the transplantation experiments discussed earlier. In the experiment of Spemann and Schotté, salamander belly ectoderm transplanted into the mouth region of a frog formed mouthparts, according to its new positional value (see Fig. 21.6). However, the rules for interpreting the new positional values were set by the genetic information in the salamander graft. Likewise, when antennal parts of *Antennapedia*^{R} mutant fruit flies were replaced by leg structures of corresponding positional value, the change in interpretation resulted from faulty activation of the *Antennapedia* gene (see Fig. 21.8).

21.4 Limb Regeneration and the Polar Coordinate Model

Of the many ways in which positional value might be conferred to cells, we will review two models that have received general recognition. In this section, we will explore the *polar coordinate model,* so named because it uses polar coordinates rather than the more familiar Cartesian ones. This model relies on *close-range interactions* between neighboring cells. In the following section, we will examine the *gradient model.* It is based on *long-range signals* that specify positional values to all cells within a field.

The polar coordinate model was conceived to explain strikingly similar results of transplantation experiments obtained with experimental systems as different as newt limbs, roach legs, and *Drosophila* imaginal discs (French et al., 1976; S. V. Bryant, 1977; S. V. Bryant et al., 1981). We will first introduce some of the newt limb experiments, which will make the special features of the polar coordinate model readily intelligible.

REGENERATION RESTORES THE ELEMENTS DISTAL TO THE CUT SURFACE

When a salamander limb or tail is amputated, epidermal cells spread over the wound surface and form an apical epidermal cap. Underneath the cap, all tissues undergo dedifferentiation and generate a mound of cells called a ***regeneration blastema*** (Fig. 21.10). After a period of further proliferation, blastema cells redifferentiate and restore the missing parts.

When an appendage has been severed, only the stump—not the severed part—is supplied with blood, and thus only the stump can survive. The parts regenerated by the stump are the missing parts, which are also

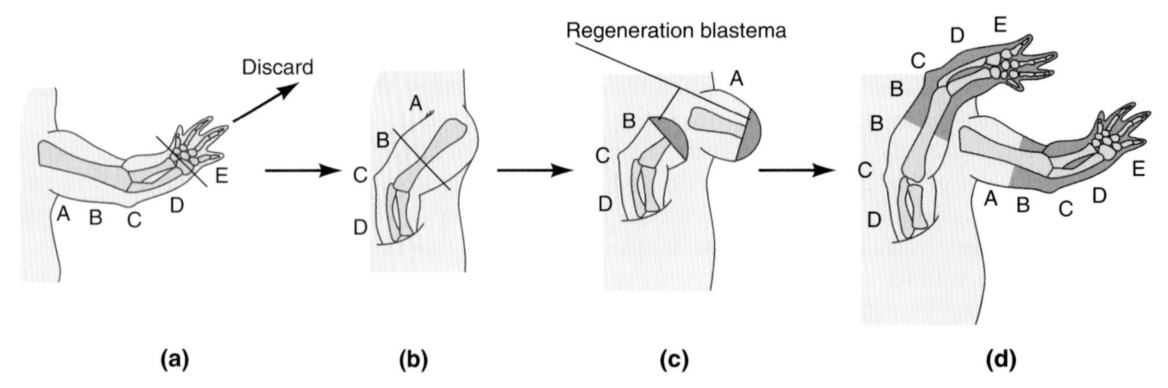

Figure 21.10 Distal transformation in a regenerating salamander limb. The capital letters mark different proximodistal levels of the forelimb. **(a)** The forelimb is amputated at the wrist, and the cut surface of the stump is inserted into the flank. **(b)** After healing, the inserted limb is cut again at the level of the humerus (upper foreleg). **(c)** The second cut has generated two different stumps: a proximal stump consisting of the shoulder and the proximal part of the upper foreleg, and a distal stump consisting of the lower foreleg and the distal part of the upper foreleg. However, both stumps end where the second cut was made, and both form regeneration blastemas. **(d)** Both stumps regenerate the same limb parts: the ones normally located distal to the cut surface.

the parts distal to the cut surface. What would the severed part do if it could be supplied with blood? Would the leg, if its life were sustained, regenerate a hip, or even an entire animal?

To answer this question, E. G. Butler (1955) cut off a foot of a salamander and stuck the open end of the limb into a slit in the flank, where it connected with the nervous and circulatory systems. After healing, he cut the inserted limb again at the level of the humerus (upper foreleg). This second cut generated two different stumps. The *proximal stump* consisted of the shoulder and the proximal part of the upper foreleg, whereas the *distal stump* consisted of the lower foreleg and the distal part of the upper foreleg. Both stumps ended where the second cut had been made, and both these surfaces formed regeneration blastemas. The proximal stump regenerated the missing parts as usually. The novel result was that the distal stump regenerated the same parts, so that most of the regenerate was a mirror image of the distal stump (Fig. 21.10). In other words, both stumps regenerated *all structures distal to the cut surface.* This result shows that the regenerated limb pattern is determined by the *proximodistal level of the cut* and not by other properties of the stump. This rule, which has also been observed in other regenerating organisms, is known as the ***rule of distal transformation***: It states that regeneration restores those elements that are normally present distal to the cut surface.

INTERCALARY REGENERATION RESTORES MISSING SEGMENTS BETWEEN UNLIKE PARTS

Regeneration is also observed in experiments that juxtapose tissues having unlike fates. The confrontation of nonneighbor cells triggers a process called ***intercalary regeneration,*** in which cells proliferate and restore the intervening parts. Differences in pigmentation can be used to discern which of the juxtaposed tissues gives rise to the regenerate. In salamanders, intercalary regeneration can be stimulated by grafting a blastema resulting from the amputation of a foot to the stump of an amputated leg (Fig. 21.11). The blastema heals onto the stump and then forms a foot as if it had remained in its original position at the ankle. In addition, cells from the stump divide and regenerate the intervening parts between the foot and the upper foreleg. This type of intercalary regeneration obeys the rule of distal transformation, since the intervening parts are distal to the cells from which they are derived. However, in similar experiments with insect legs, the rule of distal transformation is violated. Here, it is mostly the distal leg part that regenerates the intervening structures (Bohn, 1976).

LIMB REGENERATION RESEMBLES EMBRYONIC LIMB BUD DEVELOPMENT

A regenerating limb is similar to a normal embryonic limb bud in many respects (see Section 23.5). Both have a mesenchymal core covered by a layer of ectodermal cells. Removal of the ectodermal cover stops both limb regeneration and normal development. The distal part of the mesenchymal core is a zone of rapid mitotic divisions in a limb bud as well as in a regeneration blastema. If mitoses are blocked by X-irradiation, neither normal limb development nor regeneration will proceed.

What is most important for pattern formation is that limb buds and regenerating limbs respond to similar if not identical patterning signals (Muneoka and Bryant, 1982). This is indicated by the results of transplantation experiments with axolotl larvae. Hindlimbs develop more slowly than forelimbs in axolotl, so that tips of developing hindlimb buds and blastemas of regenerating

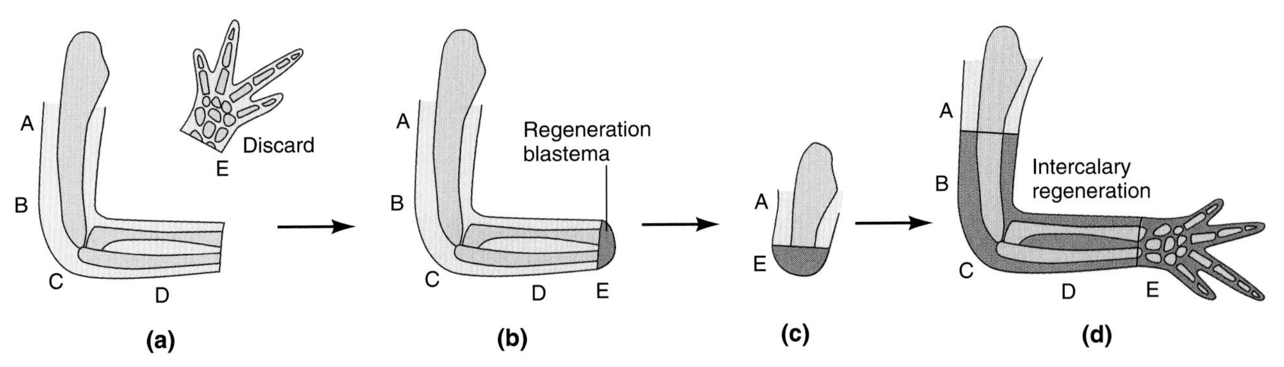

Figure 21.11 Intercalary regeneration in the salamander limb. The capital letters mark different proximodistal levels of the forelimb. **(a)** The limb is amputated at the wrist. **(b)** A regeneration blastema forms at the amputation site. **(c)** The regeneration blastema is grafted to the freshly severed stump of another limb amputated through the upper foreleg. This juxtaposes tissues from levels A and E. **(d)** The distal limb part is regenerated from the grafted blastema (level E), while the intervening levels (purple) are regenerated from stump tissue (level A).

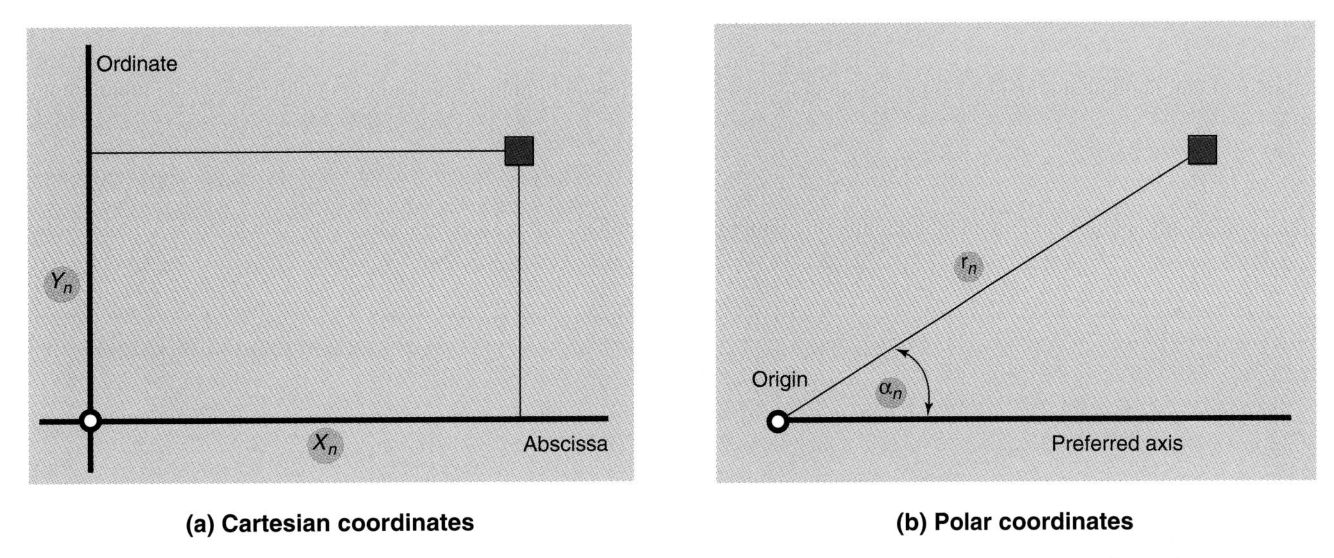

Figure 21.12 Two coordinate systems for specifying the position of a point in a plane. **(a)** In the familiar Cartesian system, two intersecting axes—known as *X* and *Y* axes, or as abscissa and ordinate—are used as reference system. Each point is defined by its distances, X_n and Y_n, from the two axes. **(b)** In the polar coordinate system, the reference system consists of a single point, the origin, and a preferred axis extending from the origin in a particular direction. Each point is defined by its distance from the origin (r_n) and its angular orientation (α_n) relative to the origin and the preferred axis.

forelimbs can be swapped between animals at similar stages of development. If the transplants are properly aligned, limb bud tips cooperate with stumps of regenerating limbs in forming new limbs. Likewise, regeneration blastemas cooperate with limb bud stumps. Most intriguingly, misalignments cause characteristic pattern abnormalities that are the same in limb buds, regenerating limbs, and combinations of both. These results indicate that the patterning mechanisms in limb buds and regenerating limbs are similar. Presumably, cells in regeneration blastemas produce the same signals and activate the same genes that were involved in the original development of the limb. This conclusion is in line with the generality of patterning signals discussed previously.

THE POLAR COORDINATE MODEL IS BASED ON TWO EMPIRICAL RULES

Many embryonic fields, such as the limb field shown in Figure 21.3, have a well-defined focal area surrounded by zones of decreasing potential that are not clearly demarcated. Such fields are not represented well by rectangular coordinates, which seem to fit a notepad or computer screen better than a limb bud or imaginal disc. In an alternative system of *polar coordinates*, any point is defined by its *distance* from a center, or origin, and by its *angular orientation* relative to a *preferred axis* (Fig. 21.12). This coordinate system is more suited to round or conical structures. For instance, all points with the same distance from the origin lie on a circle.

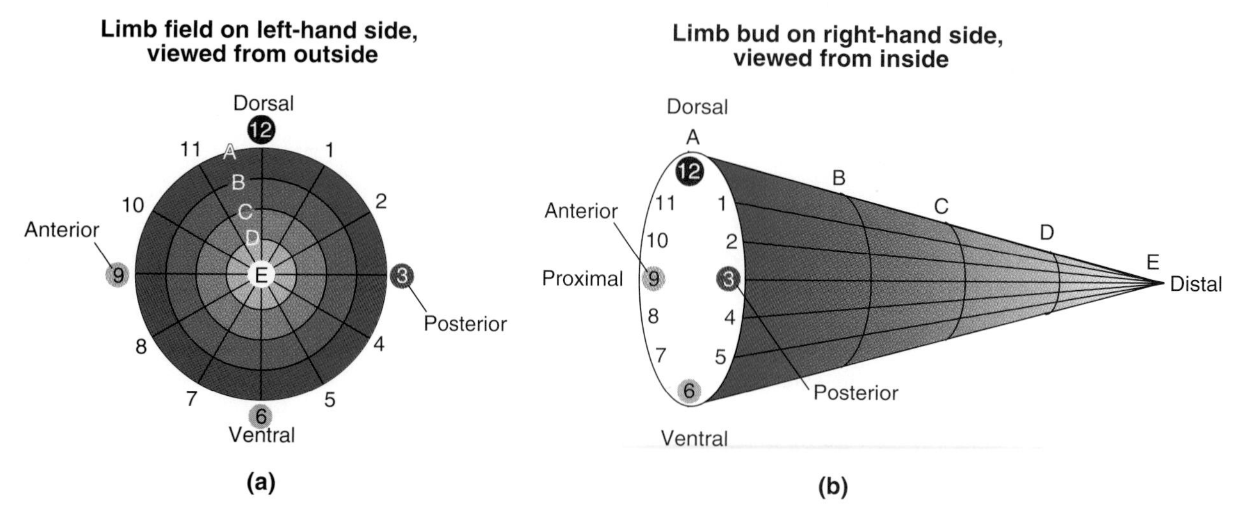

Figure 21.13 Positional values of cells in a limb field and a limb bud according to the polar coordinate model. **(a)** The flat limb field is modeled as a clock face. Each cell is assumed to have coordinates, or positional values, in the proximodistal direction (A–E) and in circumference (1–12). Positional value A belongs to the perimeter of the field, and positional value E belongs to its center. Positional value 12 marks the dorsal pole of the disc, and 6 represents the ventral pole. Positional value 9 is anterior, and 3 is posterior. Thus, part a represents a limb disc on the left-hand side of the embryo and *viewed from the outside.* **(b)** The three-dimensional limb bud is modeled as a cone. Now positional value A belongs to the proximal part (base) of the limb, and positional value E belongs to the distal part (tip). Thus, part b represents a limb bud on the right-hand side of the embryo, cut off at the base and *viewed from inside* to show positional values around the entire circumference.

In the ***polar coordinate model*** of embryonic pattern formation, positional values are assigned to cells as if they were located on the surface of a clock face or dart board (Fig. 21.13a). The ***circular coordinate*** of each cell (its angular orientation) is given by the numbers of the clock face. As additional conventions, the dorsal orientation is represented by 12, anterior is 9, ventral is 6, and posterior is 3. The ***proximodistal coordinate*** of each cell (its distance from the origin) is indicated by the letters A through E, with A at the perimeter of the disc and E at the center. The clock face is used to model flat structures, such as a limb disc or the epithelium of an insect imaginal disc. However, the clock face is easily transformed into a cone by pulling on the center of the clock. The resulting a cone is suited to modeling limb buds (Fig. 21.13b). The base of the cone represents the proximal end of the limb, and the tip represents the distal end.

Two simple, empirical rules govern the behavior of cells according to the polar coordinate model. The first is the ***shortest intercalation rule.*** It states that wherever cells with nonadjacent positional values confront one another, the incongruity triggers cell proliferation. The growth continues until cells with all intermediate positional values have been regenerated. If cells with different circumferential values confront each other, it is always the shorter of the two possible sets of intermediate values that is intercalated (Fig. 21.14). For instance, if cells with positional values 1 and 4 are grafted next to each other, cells with values 2 and 3 will be intercalated rather than cells with values 5 through 12.

The second rule of the polar coordinate model is the ***distalization rule.*** It states that cells can regenerate only cells with more distal positional values. In a limb amputated at level A, blastema cells can give rise only to cells with B values, which in turn can produce only cells with C values, and so forth (Fig. 21.15). This rule forces the blastema cells to gradually adopt more distal positional values until the limb is completely regenerated.

The polar coordinate model makes no explicit assumptions about the molecular mechanisms by which cells acquire positional values. However, since the model is based on interactions between neighboring cells, the implication is that positional values are set by signals exchanged between nearest neighbors. These could be chemical signals or a physical signal such as cell adhesiveness. Indeed, cells that form the distal part of a limb adhere to one another more strongly than cells that form the proximal part (Crawford and Stocum, 1988).

THE POLAR COORDINATE MODEL EXPLAINS SUPERNUMERARY LIMB FORMATION FROM MISALIGNED REGENERATES

The polar coordinate model explains not only the progress of normal limb regeneration but also the formation of supernumerary limbs, which frequently grow out from misaligned grafts. One such case, reported by S. V. Bryant and L. Iten (1976), is summarized in Figure 21.16. It shows that the model correctly predicts not only

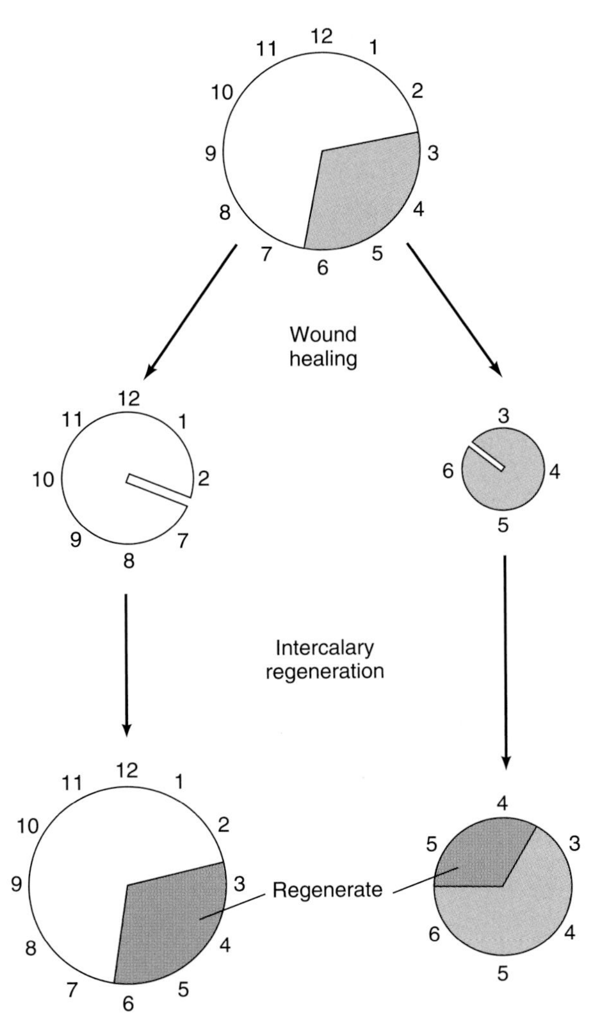

Figure 21.14 Shortest intercalation rule of the polar coordinate model. A small segment, of circumferential values 3 to 6, is cut out. Wound healing juxtaposes values 2 and 7 in the large piece as well as 3 and 6 in the small piece. Intercalary regeneration restores the missing values (color) along the shortest circumferential route. This process regenerates the full set of values in the large piece but causes duplication in the small piece.

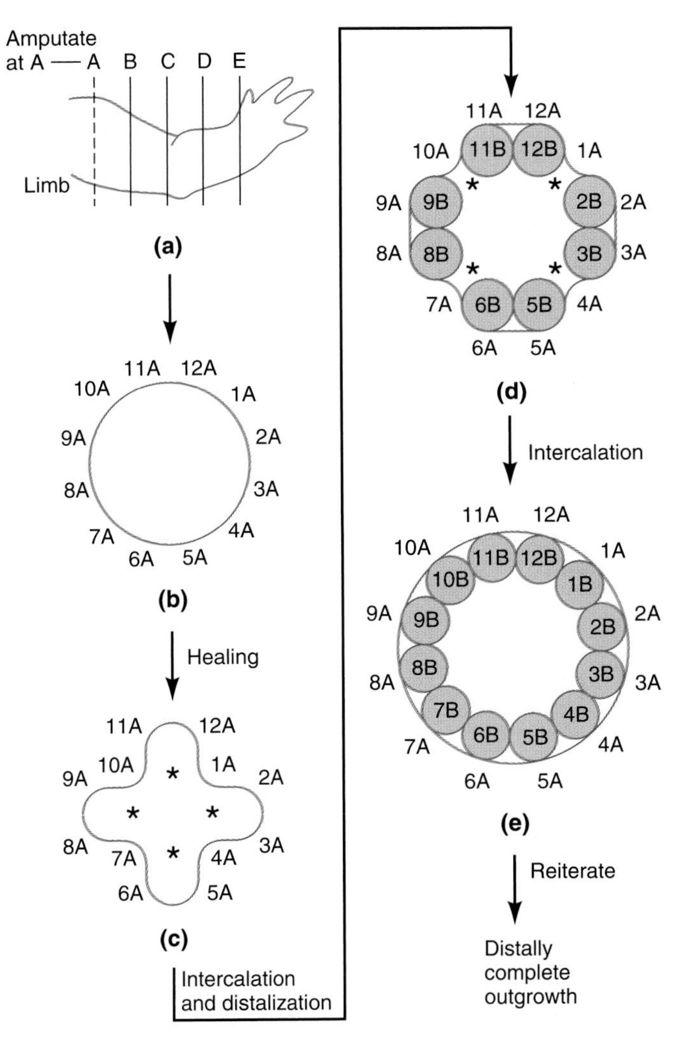

Figure 21.15 Distalization rule of the polar coordinate model. **(a)** Limb amputation removes levels B through E, leaving level A. **(b)** The circle represents the edge of the wound. **(c)** Healing involves the migration of dermal fibroblasts from the wound margin to the center. As a result, fibroblasts with different circumferential values confront one another (asterisks). **(d)** Intercalation restores the missing circumferential values according to the shortest intercalation rule. In addition, the distalization rule forces the new cells to adopt the next more distal value, B (color). Among the new cells, several circumferential values are missing, creating more confrontations of nonneighboring cells (asterisks). **(e)** Intercalary regeneration completes a continuous set of circumferential values at the B level. The process is reiterated until the most distal level (E) is completed.

the number and location of the supernumerary limbs but also their orientation and handedness. This remarkable accuracy validates the polar coordinate model and the general concept of positional value. However, the physical basis of positional value, in this case and in many others, remains to be elucidated.

The rules of shortest intercalation and distalization, on which the polar coordinate model is based, suggest that positional values of regenerated cells are determined by the positional values of their nearest neighbors. Such close-range interactions contrast with the long-range signals that are postulated by the gradient models discussed in the following section.

21.5 Morphogen Gradients as Mediators of Positional Value

The models proposed most frequently as a means of specifying positional values are known as gradient models (Neumann and Cohen, 1997b). A ***gradient*** may be set up by any property that varies continuously depending on location. As an example, a heat source generates a

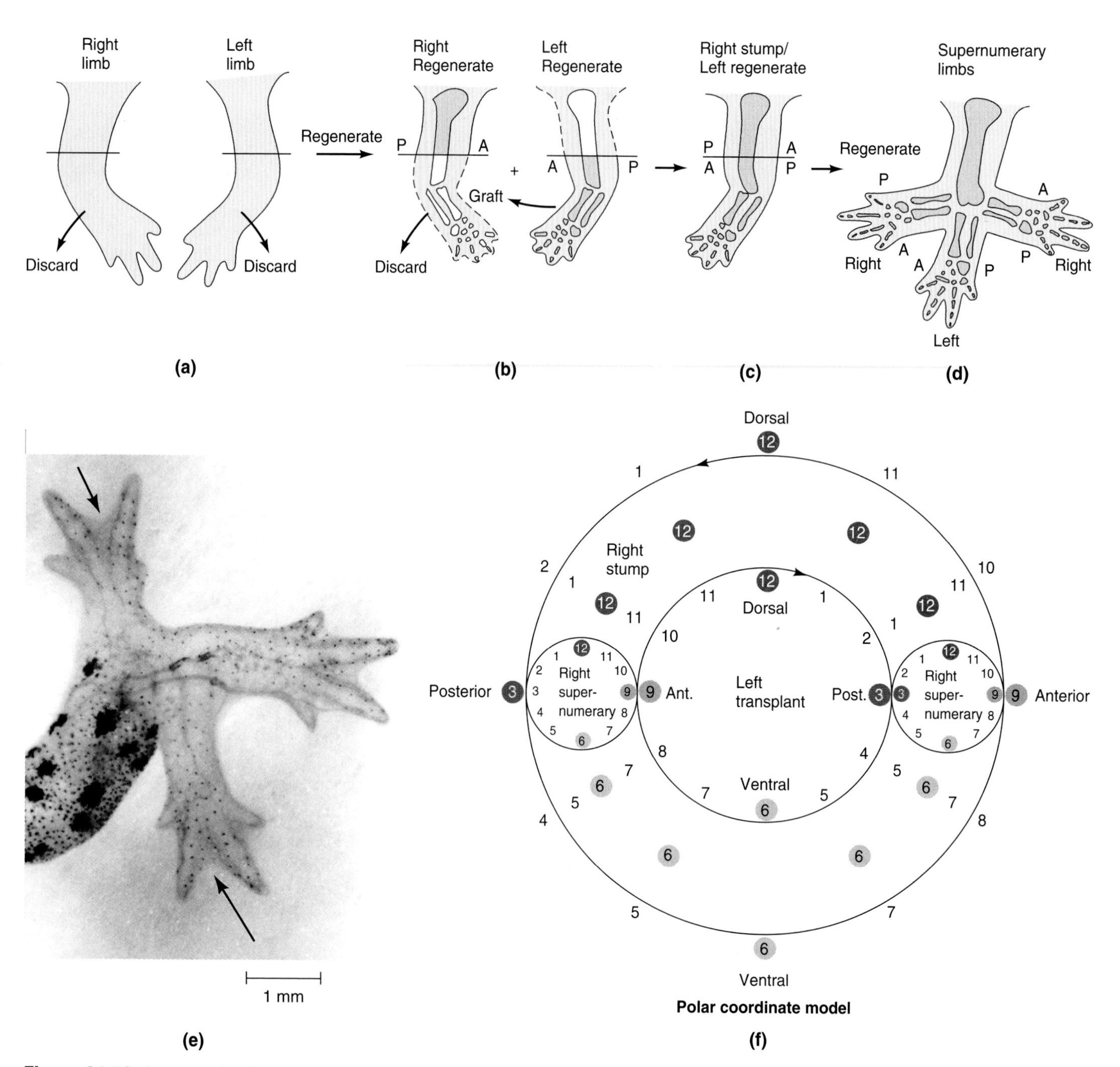

Figure 21.16 Outgrowth of supernumerary limbs from a misaligned regenerate in the salamander *Notophthalmus viridescens.* **(a)** Both hindlimbs are amputated through the femur and allowed to regenerate. **(b)** When the regenerates begin to form digits, the right regenerate is removed and the left regenerate is grafted to the right stump. **(c)** This transplantation juxtaposes anterior tissue of the stump and posterior tissue of the regenerate, and vice versa. **(d)** Two supernumerary limbs form at the sites of maximum disparity. **(e)** Photograph of the regenerated left hindlimb and two supernumerary right limbs (arrows). **(f)** Schematic cross section of the graft-host junction, viewed from the outside. The inner and outer circles represent the respective circumferences of the graft and the stump. The diameters of the graft and the stump are shown different for clarity only. The numbers are positional values according to the polar coordinate model (see Fig. 21.13). The values between the circles are generated according to the shortest intercalation rule. At the two sites of maximum incongruity, the shortest intercalation goes in either direction, depending on the neighboring intercalations. The resulting complete circles of positional values behave like regenerating limbs, causing the outgrowth of supernumeraries. The model predicts that the supernumeraries will have the same handedness and orientation as the stump. These predictions are confirmed by the results, as shown in part e.

gradient of decreasing temperature with increasing distance from the source. The gradients thought to guide pattern formation in embryos are usually associated with varying concentrations of signal molecules. A molecule that triggers several different developmental pathways depending on its local concentration or activity is called a ***morphogen*** (Gk. *morphe*, "form"; *genesis*, "production"). Morphogen gradients have been proposed to explain many instances of pattern formation in development.

A MORPHOGEN GRADIENT CAN SPECIFY A RANGE OF POSITIONAL VALUES AND POLARITY TO A FIELD

Most gradient models propose that a distinct group of cells act as the source of a morphogen (Fig. 21.17). The morphogen moves away from the source, by diffusion or active transport, until eventually it is destroyed—that is, metabolized or sequestered. The destruction site may be another special group of cells acting as a *sink*. In this case, the morphogen concentration will decrease *linearly* between the source and the sink. Alternatively, all cells outside the source may destroy the morphogen, in which case its concentration will decrease exponentially with distance from the source. Models based on morphogen gradients imply that cells respond to long-range signals from distant reference points, rather than short-range signals from their nearest neighbors.

Many gradient models assume that the morphogen *diffuses* from its source to the other cells in the field, crossing in and out of cells or using extracellular material as a conduit. However, it is also conceivable that a morphogen spreads by active transport, as long as its concentration or activity decreases with increasing distance from the source. A particular way for morphogens to spread within a field could be by means of thin *filopodia*, or *cytonemes*, extending between the source and the other cells in the field. Such processes were discovered in insect imaginal discs (Ramírez-Weber and Kornberg, 1999) and will be described in Section 22.8.

In addition to a range of positional values, a morphogen gradient also specifies the *polarity*, or the direction of maximal change in morphogen concentration. At any point in the gradient landscape, the polarity may therefore be symbolized by an arrow indicating the direction in which the morphogen concentration decreases most steeply. Polarity may be expressed, for example, by the orientation or asymmetrical location of organelles in a cell.

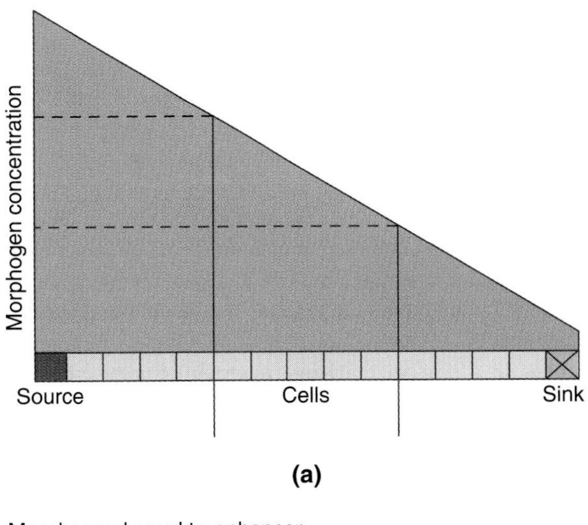

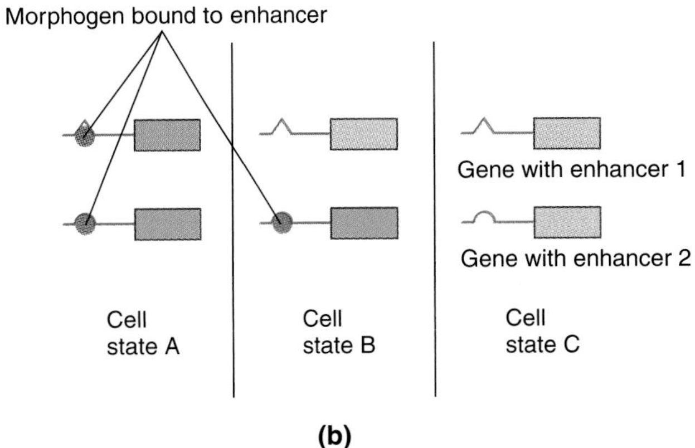

Figure 21.17 Model for specification of positional values by a morphogen gradient and interpretation of positional values by genes. **(a)** In a string of cells, the cell at one end is a morphogen source. The morphogen diffuses through the string until it is destroyed in the cell at the opposite end, which acts as a morphogen sink. At equilibrium, the morphogen concentration decreases linearly between the source and the sink. The local concentration then specifies a positional value to each cell in the string, with the source and sink serving as reference points. **(b)** Positional values can be interpreted by cells if different local morphogen concentrations cause different gene activities. It is assumed here that the morphogen is a transcription factor binding to two enhancers, 1 and 2, belonging to different target genes. Enhancer 1 has little affinity for the morphogen and is bound only at high concentration. Enhancer 2 has a higher affinity and is bound at high and intermediate morphogen concentrations. Neither enhancer is bound at low morphogen concentration. The threshold concentrations between these states define boundary lines at which the activity patterns of the target genes change. The resulting gene activity patterns define three cell states: State A has genes with enhancer 1 and with enhancer 2 active, state B has only genes with enhancer 2 active, and state C has neither of these genes active.

POSITIONAL VALUES SPECIFIED BY MORPHOGEN GRADIENTS MAY ELICIT DIFFERENTIAL GENE ACTIVITIES AND CELL BEHAVIORS

Cells can interpret positional values if some of their genes are activated or inhibited by the morphogen at concentrations within the range of the gradient. Each positional value then elicits a unique combination of active and inactive genes. If these genes were to control certain pathways of cell behavior and differentiation, each level of morphogen concentration would evoke a corresponding type of cell activity.

The interaction between a morphogen and its target genes can be direct if the morphogen is a transcription factor. Such a factor could activate or inhibit different sets of target genes depending on its local concentration and its affinity for each target gene enhancer (Fig. 21.17). For example, the bicoid protein forms an anteroposterior gradient in the *Drosophila* egg, activating several embryonic genes in a concentration-dependent way (see Figs. 16.8 and 22.10). However, transcription factors can act as morphogens only in embryos with delayed cellularization or intercellular bridges that allow large molecules to pass freely. In other systems, morphogens may be small cytoplasmic molecules that can pass through gap junctions, lipid-soluble molecules that diffuse across plasma membranes, or water-soluble molecules that are secreted by cells, diffuse through the extracellular matrix, and act on receptors in the plasma membrane of their target cells. In any case, if such molecules were to activate or inhibit transcription factors in a concentration-dependent way, then the model outlined in Figure 21.17 would apply as if the transcription factor itself were the morphogen.

Although the gradient model is encompassed by the more general *principle of induction*, gradients come with two specific connotations that are not fulfilled in all cases of induction. First, the gradient model requires a long-range signal acting across several cell diameters, whereas induction may be a close-range interaction between cells that are next neighbors. Second, the gradient model implies that the cells responding to the morphogen can adopt at least three different states of determination: one default state and two or more signal-dependent states. The latter are triggered when the morphogen concentration exceeds certain thresholds (Fig. 21.17). In contrast, the responding cells in an induction may have only two states of determination: default and induced.

MORPHOGEN GRADIENTS CONFER SIZE INVARIANCE ON EMBRYONIC FIELDS

Morphogen gradients have held the interest of developmental biologists throughout the twentieth century because they explain how several *qualitatively different cell states* (e.g., blue, white, and red cells, in the French flag model) are specified by *quantitative variations* of a *single parameter*, the morphogen concentration. They also offer a simple explanation for the polarized sequence with which the different elements of biological patterns may arise. Last but not least, morphogen gradients offer a plausible explanation for the *size invariance* that characterizes pattern formation in embryonic fields.

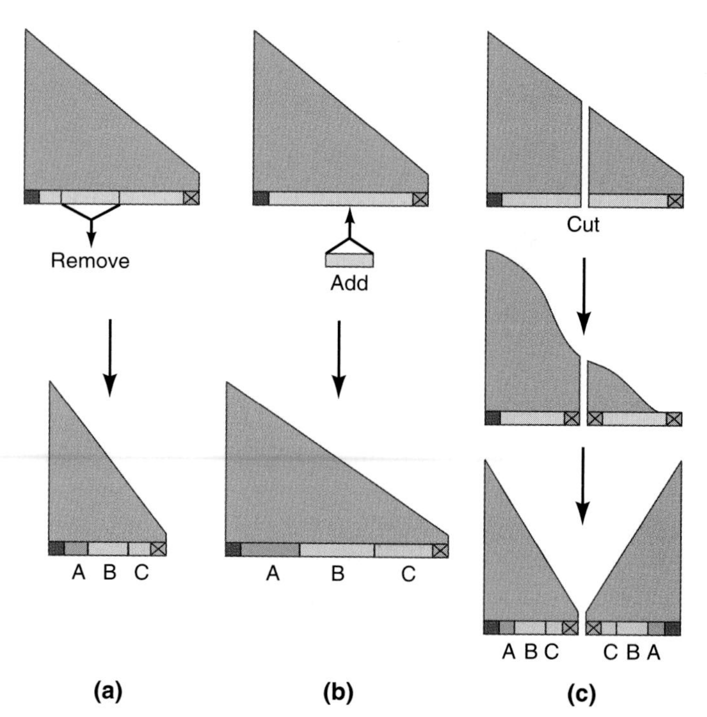

Figure 21.18 Size invariance of an embryonic field specified by a morphogen gradient. In a string of cells with a morphogen source at one end and a sink at the other end, a pattern of three cell states (A, B, and C) is specified, as explained in Figure 21.17. Note that the morphogen concentration at the sink is greater than zero. **(a, b)** Removal or addition of cells between source and sink makes the gradient steeper or gentler, with the resulting pattern of cell states compressed or spread out proportionally. **(c)** Cutting the cellular string in half is thought to cause the formation of new sinks at the cut ends. In addition, the old sink is assumed to turn into a new source as the local morphogen concentration falls to zero. Under these conditions, both halves form two compressed but complete patterns in mirror-image orientation.

To understand how size invariance is conferred on patterns specified by a gradient, consider a string of cells with a source at one end and a sink at the other. The gradient may specify a pattern of three cell states—A, B, and C—as modeled in Figure 21.17. Removing or adding cells somewhere between the source and the sink will simply cause the formation of a gradient with a steeper or gentler slope, and thus a pattern that is compressed or spread out accordingly (Fig. 21.18a, b).

What happens if the cellular string is cut in half, leaving one half without a sink and the other half without a source (Fig. 21.18c)? If the missing reference points are not restored, the half without a sink will generate only cell state A (because maximal levels of morphogen will accumulate everywhere) and the half without a source will generate only state C (because the sink will drain the entire string of the morphogen). To restore the missing reference points, the cells need additional mechanisms. For instance, each cell might be programmed to become a sink if it finds itself at the end of the string, except if the morphogen concentration is near zero, in which case the end cell will become a source. Under these additional rules, each half will form a new sink at the cut end. In addition, the old sink will turn into a new source when the morphogen concentration has fallen close to zero. Other rules would be necessary to explain how a normally proportioned pattern is formed when two or more strings of cells are fused together.

In the following, we will discuss two studies on morphogen gradients. The first uses cultured embryonic tissues exposed to a defined morphogen. The other uses the classical methods of ligation and transplantation on an insect embryo.

ACTIVIN CAN FORM A MORPHOGEN GRADIENT IN *XENOPUS* EMBRYOS

Mesoderm is induced in the *marginal zone* of the amphibian blastula by signals from vegetal blastomeres. The inductive effect has been demonstrated in a classical experiment, in which a vegetal core piece and a labeled animal cap, both isolated from midblastulae, were stuck together (Nieuwkoop, 1969a; see Fig. 9.9). In such conjugates, part of the animal cap formed mesodermal derivatives, whereas in the normal embryo the entire cap's fate would have been to form ectordermal structures. While the nature of the inducing signals is still unresolved, *activin* and other members of the transforming growth factor-beta (TGF-β) superfamily have emerged as strong candidates (see Section 9.7). Activin is a secreted protein that acts on other cells through receptors with tyrosine kinase activity (see Section 2.8).

In order to explore the morphogen properties of activin, Green and coworkers (1992) dispersed animal caps into single cells and cultured them in the presence of purified activin at different concentrations. RNA extracted from the aggregates was tested by RNase protection (see Method 17.1) for the presence of mRNAs characteristic of certain mesodermal derivatives. The results showed that staggered concentration ranges of activin induced the accumulation of specific mRNAs (Fig. 21.19). The lowest activin concentration led to the accumulation of Xhox3, XlHbox6, and *Xenopus* brachyury (Xbra) mRNAs, all synthesized in the posterior lateral mesoderm of normal embryos. In contrast, the highest activin concentration caused the accumulation of goosecoid mRNA, which is characteristic of Spemann's organizer region. These results are compatible with the hypothesis that activin forms a morphogen gradient, and that different genes are activated depending on local activin concentration. However, the experimental strategy used in this experiment could not test the basic assumption of the gradient model that different morphogen concentrations are generated as the morphogen diffuses away from its source and is degraded.

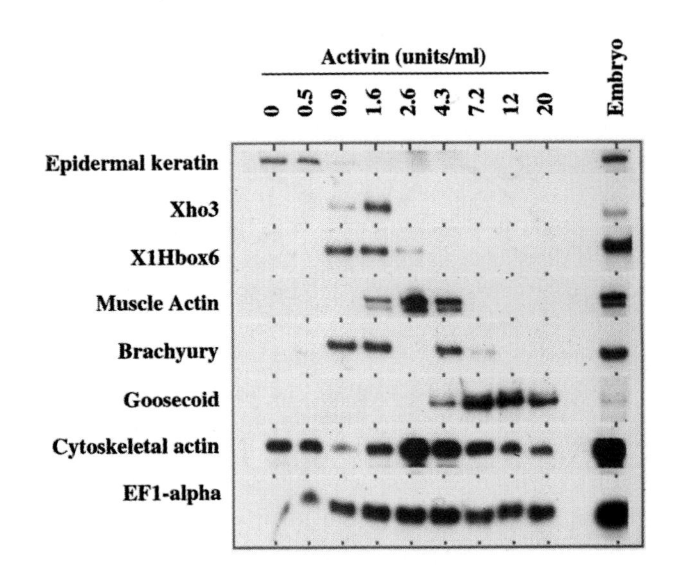

Figure 21.19 Induction of mesodermal marker genes by activin at different concentrations. *Xenopus* animal cap cells were dispersed into single cells and cultured in the presence of purified activin at different concentrations (top). Subsequently, the cells were washed, reaggregated, and cultured until control embryos from the same batch had reached the tailbud stage. RNA extracted from the aggregates was tested by RNase protection (see Method 17.1) for the presence of mRNAs characteristic of certain mesodermal derivatives. The mRNAs for cytoskeletal actin and the translation elongation factor EF1-alpha, which accumulate in virtually all cells, were used as controls to assure that similar amounts of total RNA were loaded in each vertical lane. At the lowest activin concentrations (0 and 0.5 units/ml), no mesodermal marker genes were activated. At the same time, the presence of epidermal keratin mRNA indicates differentiation of epidermal cells, in accord with the fate of the animal cap in normal embryos. Intermediate concentrations of activin led to the accumulation of Xhox3, XlHbox6, muscle actin, and *Xenopus* brachyury (Xbra) mRNAs, all synthesized in the posterior lateral mesoderm of normal embryos. The highest activin concentrations caused the accumulation of goosecoid mRNA, which is characteristic of Spemann's organizer region. These results are compatible with the hypothesis that activin forms a morphogen gradient, and that different genes are activated depending on local activin concentration.

IN order to test the spatial aspect of the gradient model more closely, Gurdon and coworkers (1994) performed a series of experiments using variations of Nieuwkoop's classical design. In the first set of tests, vegetal cores were prepared from an embryo that had been preloaded with varying amounts of activin mRNA at the 2-cell stage (Fig. 21.20a). The animal cap of the conjugate was labeled by injection of a red fluorescent macromolecule, rhodamine lysinated dextran (RLDx). The conjugates were allowed to develop for 5 to 6 hr until control embryos reached the early gastrula stage, at which time the conjugates were fixed and sectioned. The sections were hybridized in situ (see Method 8.1) for Xbra mRNA, which in normal embryos at this stage accumulates in the entire mesoderm. In the conjugates, Xbra mRNA accumulated in a layer of animal cells, and the location of this layer depended on the amount of activin mRNA injected (Fig. 21.21). With increasing amounts of activin mRNA in the vegetal core, the layer of animal cells synthesizing Xbra mRNA became located further away from the activin source. The authors coined the term *distance effect* to refer to this phenomenon.

The distance effect was generated in a second set of experiments in which the activin source consisted of cells that do not normally induce mesoderm, such as animal caps (see Fig. 21.20b). Thus, activin can induce $Xbra^+$ gene expression at a distance irrespective of the nature of the cells supplying the activin. Indeed, agarose beads loaded with purified activin protein, when sandwiched between two animal caps (see Fig. 21.20c), induced $Xbra^+$ gene expression with the same distance effect.

Third, the investigators analyzed the possible role of cell movement in the distance effect. According to the gradient model, Xbra mRNA is synthesized in response to a narrow range of activin concentrations, which is established at some distance from the source as the activin diffuses away and is broken down. The distance from the source at which activin reaches the "right" concentration for inducing Xbra mRNA synthesis would then depend on activin concentration at the source, diffusion parameters, and the half-life of activin in the tissue. Alternatively, the cells containing activin mRNA might induce only a few adjacent layers of cells to express the $Xbra^+$ gene and at the same time cause the responding cells to move away from the inducer, with the extent of the movement depending on the strength of the activin signal. To test this possibility, the researchers constructed *two layers of responding cells* labeled with different dyes (see Fig. 21.20d). Between the previously used layers they now inserted an additional layer of animal cap cells labeled with a green fluorescent dye, fluorescein lysinated dextran (FLDx). If cell movement was involved in generating the distance effect then the FLDx-labeled cells, since they were closest to the activin source, should have moved into the more distant layer of RLDx-labeled cells. In fact, no such movement was observed in conjugates fixed at the early gastrula stage. The distance effect could therefore not be ascribed to cell movement.

In a fourth set of experiments, the researchers addressed the question whether the inducing signal does indeed spread by diffusion, as implied in the gradient model. Alternatively, the signal might spread by a relay mechanism, in which each cell that receives the signal produces more of it in response. With such a relay mechanism, the concentration of the signal substance would not necessarily decrease with increasing distance from the original source. As a test of the relay hypothesis, the investigators inserted *nonresponding cells* between the inducing vegetal core and the responding animal cap (see Fig. 21.20e). As nonresponding cells they chose FLDx-labeled gut cells from tadpoles, since these cells did not induce $Xbra^+$ gene expression in animal caps nor did they express $Xbra^+$ themselves in response to mesoderm-inducing signals. Nevertheless, these incompetent cells successfully transmitted the $Xbra^+$-inducing signal from vegetal cells to competent animal cap cells. This result still left open the possibility that the gut cells responded to the activin signal from the inducer by synthesizing some other (nonactivin) inducer, which causes $Xbra^+$ gene expression in distant cells but not themselves. To test this possibility, the researchers treated the gut cells with *cycloheximide,* an inhibitor of protein synthesis. When such treated cells were inserted between inducing and responding cells, the latter expressed $Xbra^+$ while the inserted gut cells did not. Thus, the signal that induces $Xbra^+$ expression can be transmitted passively through nonresponding cells, either through the cells themselves or, more likely, around the cells through the extracellular matrix.

Finally, the investigators wanted to know which genes, if any, were expressed in the space between a strong activin source and responding tissue synthesizing Xbra mRNA. A good candidate was the *Xenopus goosecoid*$^+$ ($Xgscd^+$) gene, which is expressed specifically in the dorsal lip of normal gastrulae. So they prepared conjugates in which the inducing tissue was loaded with a large amount (20 pg or more) of activin mRNA (see Fig. 21.20f). After incubation, the fixed material was sectioned alternatively onto two slides, which were processed for in situ hybridization with Xbra or Xgscd antisense RNA probes. As previously, the responding tissue expressed $Xbra^+$ at a distance from the inducer. In addition, $Xgscd^+$ was expressed in the intervening space between the inducer and the cells expressing $Xbra^+$. This result indicated that activin induced dorsal marker gene expression at a high concentration and another mesodermal marker gene at a somewhat lower concentration, precisely as predicted by the gradient model. To measure the concentration-dependence of $Xgscd^+$ and $Xbra^+$ expression more precisely, the researchers prepared conjugates of two animal caps, each of which was loaded with activin mRNA and served as an inducer and responding tissue at the same time (see Fig. 21.20g). The amounts of Xbra and Xgscd mRNA synthesized were determined by RNase protection (see Method 17.1) and plotted versus the amount of activin mRNA loaded (Fig. 21.22). The result shows that the expression profiles of $Xgscd^+$ and $Xbra^+$ differ dramatically in their activin concentration dependence. While $Xbra^+$ expression peaked at 1.5 pg activin per animal cap and became negligible at higher concentrations, $Xgscd^+$ expression reached a broad maximum around 6.2 pg and was still significant at 50 pg.

Questions

1. Why was it necessary in this experiment to label the responding tissue (animal caps)?
2. Since the investigators wanted to test the inducing effect of activin, why did they supply activin mRNA, rather than protein, to the inducing tissue?
3. The authors were concerned that the distance effect might be a deceptive appearance caused by the movement of responding cells away from the inducer. To

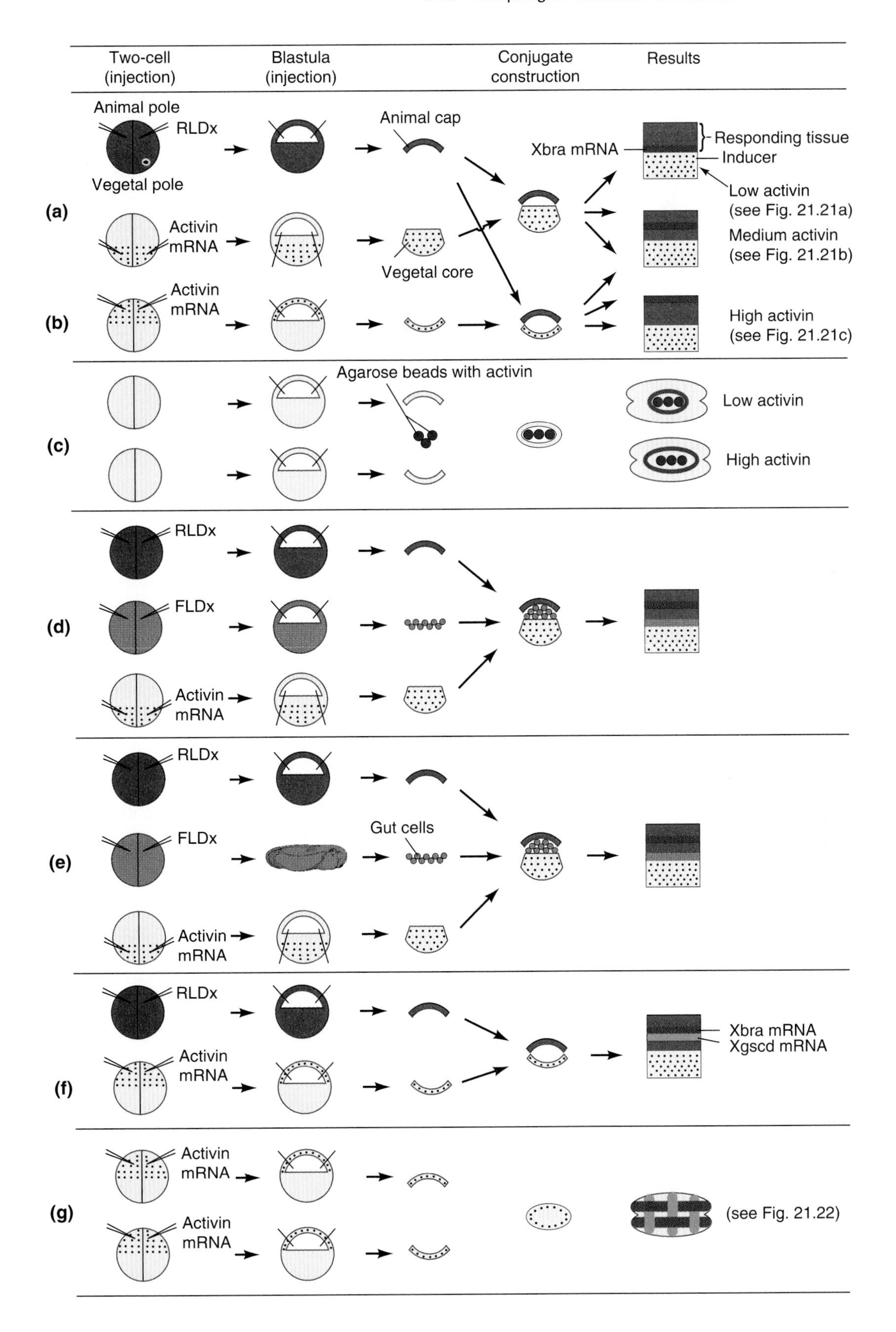

Figure 21.20 Design of experiments to test the hypothesis that activin acts as a morphogen gradient in *Xenopus* embryos. At the 2-cell stage (left-hand column), embryos were injected with activin mRNA, a red fluorescent lineage tracer (RLDx), or a green fluorescent lineage tracer (FLDx). At the blastula stage (next two columns), embryos were dissected to obtain animal caps and vegetal cores. The FLDx-labeled embryos were used to obtain dissociated cells from animal cap or tadpole gut. Conjugates were constructed as indicated in the following column and cultured for 5 to 6 h. Results are shown schematically here (right-hand column) and in more detail in Figures 21.21 and 21.22.

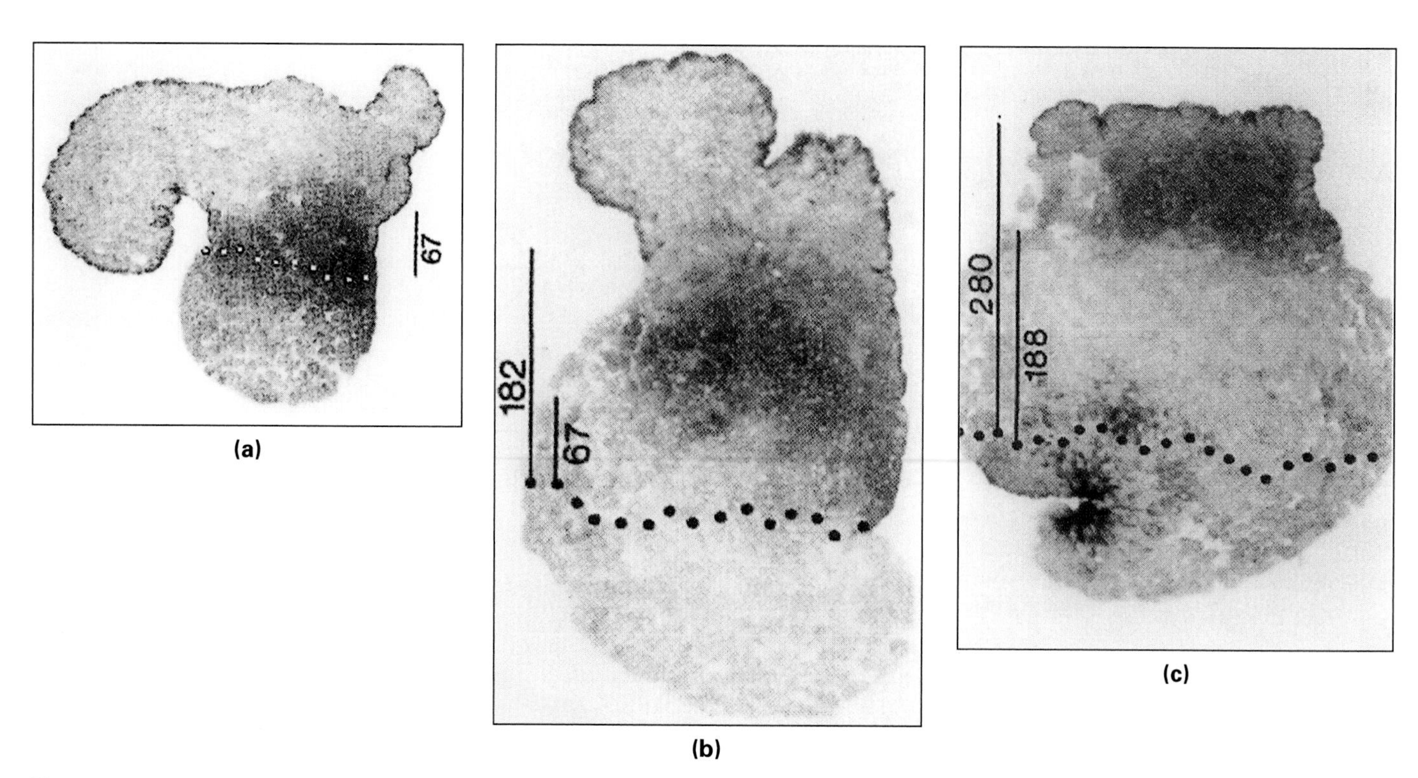

Figure 21.21 Distance effect in the induction of *Xbra*⁺ gene expression by activin. Animal caps were conjugated with vegetal cores as indicated in Figure 21.20a, cultured until controls reached the early gastrula stage, and then hybridized in situ with an Xbra antisense mRNA probe (see Method 8.1). The vegetal core had been preloaded with **(a)** 2 pg, **(b)** 7 pg, and **(c)** 20 pg of activin mRNA. The purple stain indicates accumulation of Xbra mRNA. The vertical lines denote minimum and maximum distances of Xbra mRNA accumulation from the demarcation line (dots) between inducing and responding tissue.

investigate this possibility, the investigators constructed two layers of potentially responding cells labeled with two different dyes. No movement of the green fluorescent cells, which were located next to the inducer, was observed. Did this test cover the possibility that such cell movement might be mimicked by the induction of rapid cell divisions in the layer immediately adjacent to the inducer, with concomitant loss of *Xbra*$^+$ gene expression in the daughter cells from this layer? If so, explain. If not, how could this possibility be tested independently?

Taken together, the results described above show that activin in *Xenopus* embryonic tissue can act as a diffusible morphogen, activating different target genes in a concentration-dependent manner.

In the experiments described above, conjugates were cultured routinely for 5 to 6 hr. In a follow-up study, Gurdon and coworkers (1995) examined what happens in a responding tissue while the morphogen concentration increases, as the gradient is being built up, or decreases, when the morphogen source ceases to produce. To this end, the investigators prepared animal cap sandwiches containing agarose beads loaded with activin (see Fig. 21.20c) and cultured these sandwiches for a total period of 2 or 4 hr. However, after various intervals during the total incubation period, the beads were removed or replaced by beads with different activin concentrations. At the end of the total incubation time, the sandwiches were processed for in situ hybridization with an Xbra antisense RNA probe. The results are summarized in Figure 21.23. As expected, the distance between the beads and the *Xbra*$^+$ expressing cells increased with time and with activin concentration in the beads. Importantly, the distance *increased* when weak (low activin) beads were replaced with strong (high activin) beads, but the distance *did not decrease* when strong beads were replaced with weak beads. This means the cells responded to the highest activin concentration to which they had been exposed for a sufficiently long time interval during the total incubation period. In other words, the response to varying activin concentrations was like a ratchet mechanism, which can only go further up but not down.

INSECT DEVELOPMENT HAS BEEN MODELED BY MORPHOGEN GRADIENTS AND LOCAL INTERACTIONS

Insect embryos have long been favorite objects of embryologists interested in pattern formation (Sander, 1975, 1976). The data from insect embryos have been interpreted in terms of various models, including one postulating a pair of antagonistic morphogenetic gradients.

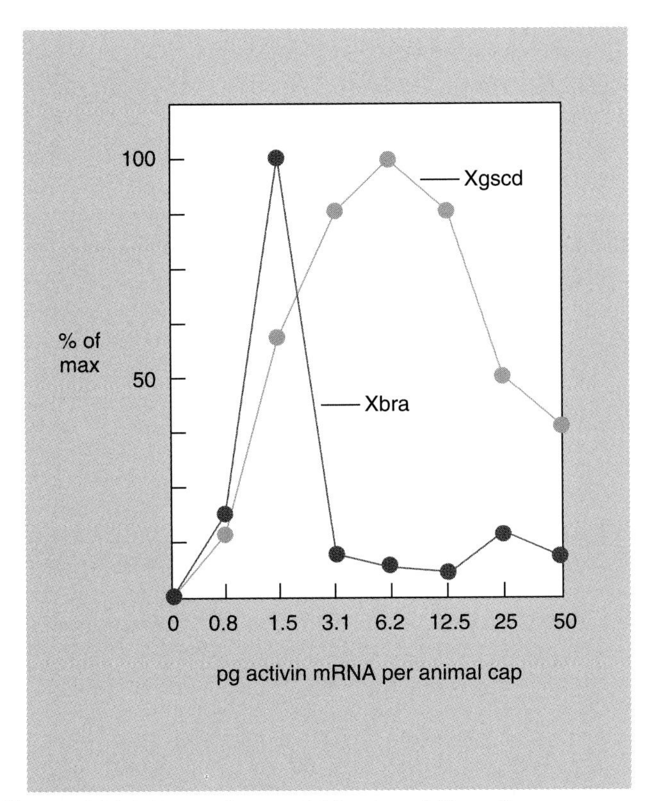

Figure 21.22 Dependence of *Xbra*$^+$ and *Xgscd*$^+$ gene expression on activin mRNA concentration. Sandwiches of animal caps preloaded with activin mRNA were prepared as indicated in Figure 21.20g, cultured for 5 to 6 h, and then assayed for Xbra and Xgscd mRNA by RNase protection (see Method 17.1). The two mRNAs accumulated at overlapping but distinct concentrations of activin.

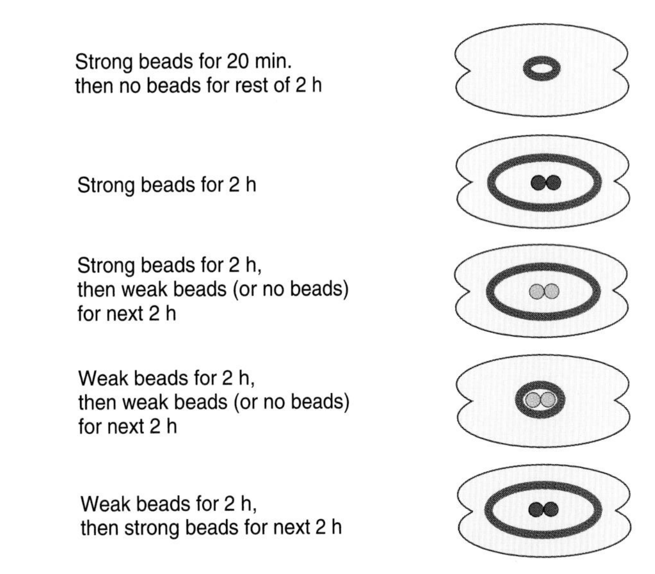

Figure 21.23 Ratchetlike progression of *Xbra*$^+$ gene expression in response to activin. Animal cap sandwiches were prepared with strong (15 nM) or weak (1.5 nM) activin beads. The distance between the beads and the *Xbra*$^+$ expressing cells increased with time and with activin concentration in the beads, in accord with the results shown in Figures 21.20a–c and 21.21. Importantly, the distance *increased* when weak beads were replaced with strong beads, but the distance *did not decrease* when strong beads were replaced with weak beads.

However, additional data indicate that the actual patterning process in the embryo must be more complex than predicted by a pure gradient model.

Experiments demonstrating the action of a posterior cytoplasmic determinant were carried out by Sander (1960) with eggs of the leafhopper *Euscelis plebejus* (Fig. 21.24). The eggs are limp and can be invaginated by poking with a blunt needle. Sander used this technique to transplant posterior cytoplasm anteriorly, using a cluster of symbiotic bacteria at the posterior pole as a convenient marker to monitor the position of the displaced material. He also separated anterior and posterior egg fragments by ligation with hair loops or by pinching the eggs between razor blades with blunted edges. The embryos that developed from such manipulated eggs showed clearly that the posterior cytoplasm plays a key role in specifying the anteroposterior body pattern of *Euscelis*.

Large anterior egg fragments, generated by ligation during cleavage stages, by themselves produced only head structures (Fig. 21.24c). However, such fragments became capable of forming more segments or even complete embryos if posterior cytoplasm was shifted before ligation so that it was included in the anterior fragment (Fig. 21.24e, f). Was this dramatic increase in potency caused by the posterior cytoplasm itself, or by a factor emanating from it? To answer this question, Sander shifted the posterior pole material anteriorly and left it there for several hours before placing a ligature in front of or behind the material (Fig. 21.24j, k). The potential of the anterior fragment increased all the same, indicating that a signal produced by the posterior cytoplasm played the key role. In posterior egg fragments, the combined transplantations and ligations sometimes caused the formation of abdomens with reversed polarity (Fig. 21.24g, j). In other cases, the same experiments produced double abdomens, consisting of two sets of thoracic and abdominal segments joined in mirror-image symmetry (Fig. 21.24h, k). Thus, the posteriorizing signal seemed to emanate from transplanted posterior cytoplasm and in some cases also from cytoplasm that had been left behind at the posterior pole.

The signal spreading from the posterior cytoplasm of *Euscelis* eggs has two important characteristics. First, it does not always cause anterior fragments to form complete embryos. In cases when it does not, the additional segments formed are not the terminal abdominal segments normally formed next to the posterior pole. Instead, middle segments are added *continuously* to the head segments that develop in anterior fragments. Second, the length of a given set of segments formed may vary by a factor of more than 2. (Compare the arrays labeled CDE in Fig. 21.24g, j, and k.) One can explain these

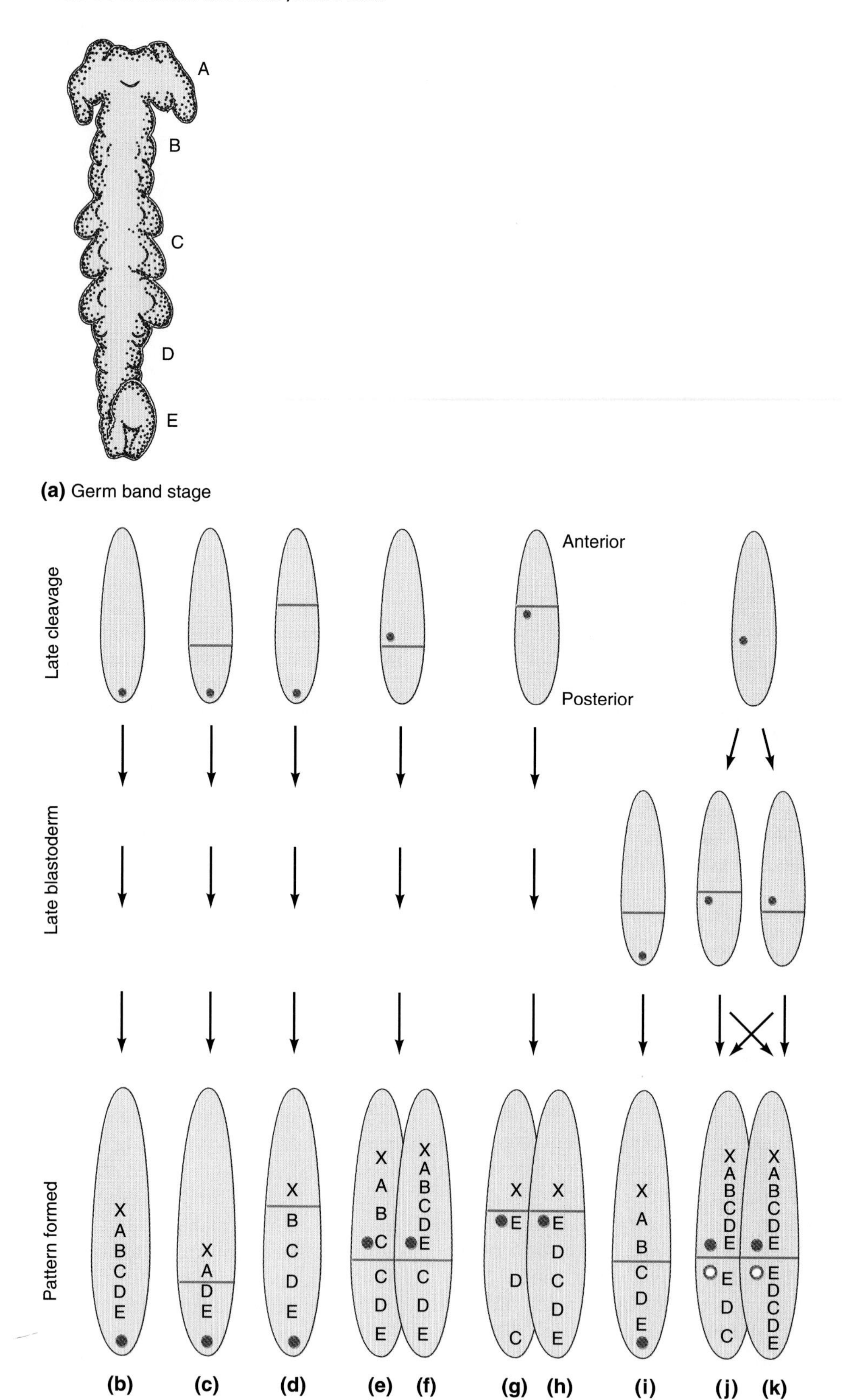

Figure 21.24 Ligation or combined ligation and translocation of posterior cytoplasm in leafhopper (*Euscelis plebejus*) eggs: anterior pole up, stages of operation indicated at left. **(a)** Embryo at the germ band stage, with capital letters assigned to body regions. A = procephalon; B = mouthparts; C = thoracic segments; D = abdominal segments; E = telson; X = extraembryonic membranes. **(b–k)** Different sets of experiments. The red line within the egg outline represents ligation at the respective level. The green dot represents the ball of symbiotic bacteria used to monitor the position of posterior cytoplasm. The open circles in parts j and k represent alternative positions of the symbiotic bacteria observed in some experiments. The pairs **(e, f)**, **(g, h)**, and **(j, k)** represent different results obtained in the same series. Capital letters represent the embryonic body patterns identified at the germ band stage shown in part a.

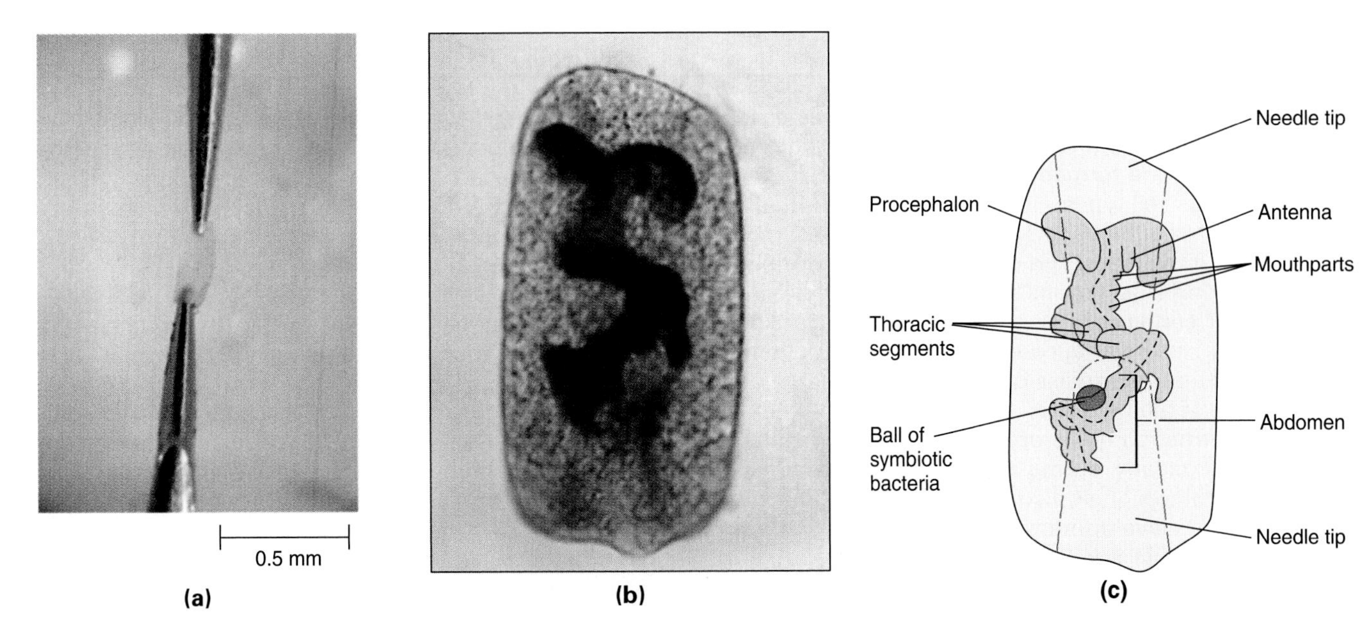

Figure 21.25 Development of a complete embryo in a drastically deformed leafhopper egg. **(a)** Photograph of egg being invaginated with blunt needle tips at both the anterior and the posterior pole. **(b)** Photograph of a fixed and stained embryo that developed in an egg that had been invaginated at both ends until the needle tips met. **(c)** Interpretative drawing of part b. The embryo has formed in a spiral around the juxtaposed needle tips (faint lines).

characteristics by assuming that the signal emanating from the posterior cytoplasm has the form of a gradient with the highest concentration near the ball of symbiotic bacteria.

In addition to a posterior morphogen, Sander (1975) proposed that a second morphogen is produced in a source near the anterior pole. He suggested that the segments of the *Euscelis* embryo were specified by the ratio of anterior to posterior morphogen concentration. He inferred the existence of the anterior morphogen partly from the effects of early ligation: Posterior egg fragments form partial embryos missing anterior parts. Specifically, these partial embryos have fewer segments than expected according to the fate map. Instead, the reduced sets of segments are stretched out farther than usual (Fig. 21.24b, d). Sander concluded that the posterior fragments failed to develop according to fate because they were cut off from the anterior morphogen source.

Sander's gradient model was challenged by Vogel (1978, 1982), who pushed in both ends of the *Euscelis* egg with two blunt needles until the presumed anterior and posterior morphogen sources were next to each other (Fig. 21.25). Under these conditions, the ratio of anterior to posterior morphogen should have been nearly uniform throughout the egg. Nevertheless, about half of the badly deformed eggs formed complete embryos, which were often twisted in a spiral around the two needle tips. Vogel concluded that the body pattern of the embryo must be specified by signals other than the ratio of freely diffusible morphogens.

Vogel (1982) also ligated *Euscelis* eggs. Confirming Sander's results, he found that anterior fragments formed partial embryos. They began anteriorly with the procephalon containing brain and eyes, and terminated posteriorly in mouthparts or thoracic segments. The character of the posteriormost segment depended on the length of the anterior fragment. Short fragments contained only procephalon, while longer fragments terminated in more posterior segments. When he tabulated his results, Vogel found that the number of segments *did not increase steadily* with the length of the anterior fragment. The embryos tended to have either no mouthparts at all or all three of them (mandible, maxilla, and labium). This discontinuity in segment number was at odds with the smooth decrease of the anterior to posterior morphogen ratio implied in Sander's model. Vogel suggested that the body pattern might be specified in several steps, with broad anterior, posterior, and middle areas determined initially by localized cytoplasmic factors. Local interactions between these areas, instead of long-range signals from the poles, would then specify smaller groups of segments and eventually individual segment characters.

The results obtained with insect and many other embryos indicate that the process of embryonic pattern formation is too complex to be described in terms of a single model (Held, 1992). It appears that patterning mechanisms have evolved into intricate networks that encompass morphogen gradients as well as close-range interactions between neighboring cells. To unravel processes of this complexity, genetic and molecular methods have proven to be very powerful, particularly in studies of *Drosophila.* This work has considerably advanced our understanding of embryonic pattern formation and has brought us to a new era in which modern tools are applied to classical topics. Some of these exciting advances will be discussed in the following chapters.

SUMMARY

Pattern formation is the process by which organisms form harmonious arrays of different elements. Many patterns in adult organisms are formed by the coordinated differentiation of various cell types. However, the formation of embryonic patterns often begins with distinct areas of mitotic division, cell movement, and cell death. Pattern formation often occurs in fields, groups of cells that cooperate in the formation of certain organs. The main characteristic of fields is their size invariance—that is, their ability to form the same pattern regardless of experimental reduction or enlargement of the field size.

Pattern formation relies on interactions between cells, as seen most clearly during induction. The response of cells to patterning signals also depends on their genetic information and developmental history. Some patterning signals are used in similar ways in different body parts and during several stages of development. Transplantations between species show that many of these signals have been conserved during evolution.

The fate of a cell depends upon its position within a territory of cells. This observation has led to the concept of positional values, which are analogous to the coordinates of a point in a geometric coordinate system. According to this concept, pattern formation is a two-step process. First, the cells within a territory receive positional values with regard to certain foci or boundaries that serve as reference points. Second, each cell interprets its positional value by engaging in activities that depend on its genetic makeup and developmental history.

The concept of positional value is implied in the polar coordinate model of limb development and regeneration. This model, based on the two rules of shortest intercalation and distalization, predicts the results of various transplantation experiments in great detail.

Although positional values may be specified by various means, they are most frequently associated with gradients of diffusible signal molecules called morphogens. Morphogens are thought to emanate from sources that serve as reference points for cellular territories. As a morphogen diffuses away from its source and is degraded, its concentration decreases. Different local concentration levels would then confer positional values on the cells within the territory. Interpretation of these values would be controlled by genes with enhancers that have different affinities for the morphogen itself or for a transcription factor generated in proportion to the morphogen concentration. Gradient models propose that more than two cell states can be controlled by the quantitative variation of one signal.

Gradient models offer a simple rationale for the sequence and polarity with which the different elements of a pattern are formed. With some additional assumptions, they can also explain the size invariance of embryonic fields. Because of these properties, morphogen gradients have been invoked to explain pattern formation in amphibian embryos, insect embryos, and many other developmental systems. However, many patterning events in development may be specified in part by combinations of morphogen gradients and close-range cell interactions.

chapter 22

Genetic and Molecular Analysis of Pattern Formation in the *Drosophila* Embryo

Figure 22.1 *Drosophila* embryos stained to reveal the expression domains of an embryonic patterning gene, *paired*$^{+}$, during different phases of development. An initial pattern of broad domains evolves into seven stripes as the embryo forms a superficial layer of cells. Later, a segmental pattern of 14 stripes appears. At midembryogenesis, *paired*$^{+}$ is expressed in specific regions of the head and central nervous system. The first critical function of the gene seems to occur during the period when seven stripes are visible: These stripes have a bisegmental periodicity, and every other segment is deleted in paired-deficient embryos.

THE analysis of pattern formation has been revolutionized by the introduction of genetic and molecular techniques. These powerful tools have been brought to bear in studies on several species—the fruit fly *Drosophila melanogaster,* the house mouse, the roundworm *Caenorhabditis elegans,* and two flowering plants—which will be covered in this and the following chapters. The zebrafish has also recently become the object of large mutagenesis screens; as the mutants collected in these screens become analyzed the zebrafish is beginning to rival the mouse as a vertebrate that is amenable to genetic analysis.

At present, genetic analysis of pattern formation has advanced furthest in the fruit fly *Drosophila melanogaster.* Two circumstances particularly favor study of this species: First, an extensive set of genetic tools is available, including balancer chromosomes and all kinds of mutants as well as transgenic strains (see Section 15.6). These tools have recently been complemented by the full genomic sequence (see Section 15.2). Second, *Drosophila* embryos are large enough to permit transplantation experiments. Indeed, *Drosophila* and other insect embryos have long been subjects of classical analysis (see Section 21.5; Counce, 1973; Sander, 1976).

In 1980, Nüsslein-Volhard and Wieschaus reported on a new approach to the analysis of insect development. Using large *mutagenesis screens,* they and others identified and characterized *Drosophila* genes involved in embryonic pattern formation. This endeavor was noteworthy in several respects. First, unlike most geneticists, these investigators did not restrict themselves to viable mutants but realized that dead embryos can be instructive. Second, the researchers aimed at identifying virtually all genes involved in the pattern formation process, because they sensed that only the analysis of entire control networks would lead to an adequate understanding of development. Finally, they were prepared to invest long and hard work into the screening of tens of thousands of mutant lines.

Although saturation mutagenesis screening is primarily a genetic strategy, its rewards were amplified tremendously by DNA cloning methods, which came into wide use soon after. These techniques allowed researchers to view the expression patterns of embryonic genes directly (Fig. 22.1), and to study the molecular nature of the proteins encoded by these genes. Thus, it became possible to see a newly discovered gene's biological function, as inferred from its mutant phenotype, in the light of the gene product's biochemical function, which could often be deduced from its amino acid sequence. These prospects generated tremendous enthusiasm, attracting dozens and then hundreds of laboratories to *Drosophila* research. Today, most of the gene activities that direct the development of *Drosophila* from an egg to a segmented embryo are basically understood. However, new and exciting discoveries are still being made, and many investigators are now turning to the gastrulation movements, organogenesis, and histogenesis of *Drosophila* as new frontiers for genetic and molecular analysis.

In 1995, the Nobel prize for Physiology or Medicine was awarded jointly to Christiane Nüsslein-Volhard, Eric Wieschaus, and one of the early pioneers of genetic analysis of *Drosophila* development, Edward Lewis. This award was made in recognition of a unique achievement, understanding how genes control the development of the *Drosophila* embryo. At the same time, this award was an inspiration to all those researchers who applied genetic and molecular tools to long-standing questions in development.

22.1 Review of *Drosophila* Oogenesis and Embryogenesis

To discuss the genetic control of development in *Drosophila,* we need to review some basic facts about oogenesis and embryogenesis. All the materials and information necessary to form a larva must be assembled in the egg. Three cell types within the *Drosophila* ovary are involved in this task: *oocytes, nurse cells,* and *follicle cells* (Fig. 22.2a, b). The oocyte is connected by cytoplasmic bridges to 15 sister cells, the nurse cells, which synthesize large amounts of RNA and cytoplasm for transfer to the oocyte. Both oocytes and nurse cells are germ line cells. The follicle cells originate independently of the germ line from embryonic mesoderm and surround the oocyte–nurse cell complex. In addition to playing a role in *vitellogenesis,* follicle cells help to establish the anteroposterior and dorsoventral polarity axes of the egg. The follicle cells also synthesize the *vitelline envelope* and the *chorion,* which form the eggshell including the respiratory appendages that stick out like rabbit ears dorsally and anteriorly from the eggshell.

Like other insect eggs, the *Drosophila* egg undergoes superficial cleavage (Fig. 22.2c–e; see also Figs. 5.17 through 5.19). At the beginning of the tenth nuclear cycle, the primordial germ cells—also known as *pole cells*—are formed at the posterior pole. During the 14th nuclear cycle, thousands of nuclei are lined up beneath the egg surface but are not yet separated by cell membranes. This stage is called the *syncytial blastoderm.* Infoldings of plasma membrane then separate the nuclei from each other, thus generating a monolayer of cells called the *cellular blastoderm.* Up to the syncytial blastoderm stage, the insect embryo is one large, multinucleate cell in which nuclei can communicate with one another directly using transcription factors and other large proteins as signal molecules.

Blastoderm formation is followed by dramatic morphogenetic movements, including gastrulation, which generate an embryonic rudiment known as the ***germ band.*** It consists of the three germ layers and essentially represents the ventral part of the future larva. The germ band elongates until, doubling back on itself, it has become twice as long as the egg—a stage known as the *extended germ band* (Fig. 22.2f). Constrictions divide the germ band into units called ***parasegments;*** they have the length of segments but are out of register with the definitive *segments* of the advanced embryo and larva (Fig. 22.2g). The epidermis of each parasegment consists of two *compartments,* which are separated by clonal restriction lines (see Section 6.2). The germ band shortens again so that its posterior tip returns to the posterior egg pole. During germ band shortening, the parasegmental boundaries disappear, and segmental boundaries develop. The latter are positioned so that the posterior compartment of one parasegment and the anterior compartment of the following parasegment together form one segment. Each definitive ***segment*** contains a pair of ganglia and paired portions of mesoderm. In a process called ***dorsal closure,*** the flanks of the embryo grow laterally and dorsally around the remaining yolk, generating a cylindrical embryo that resembles the future larva.

The body of *Drosophila* is subdivided into head, thorax, and abdomen. The head comprises a nonsegmental tip called the ***acron,*** two or more fused segments that together form the ***procephalon*** and contain most of the brain, and three segments that carry mouthparts: ***mandible, maxilla,*** and ***labium.*** The three thoracic segments are known as the ***prothorax, mesothorax,*** and ***metathorax.*** The abdomen, behind the thorax, has eight

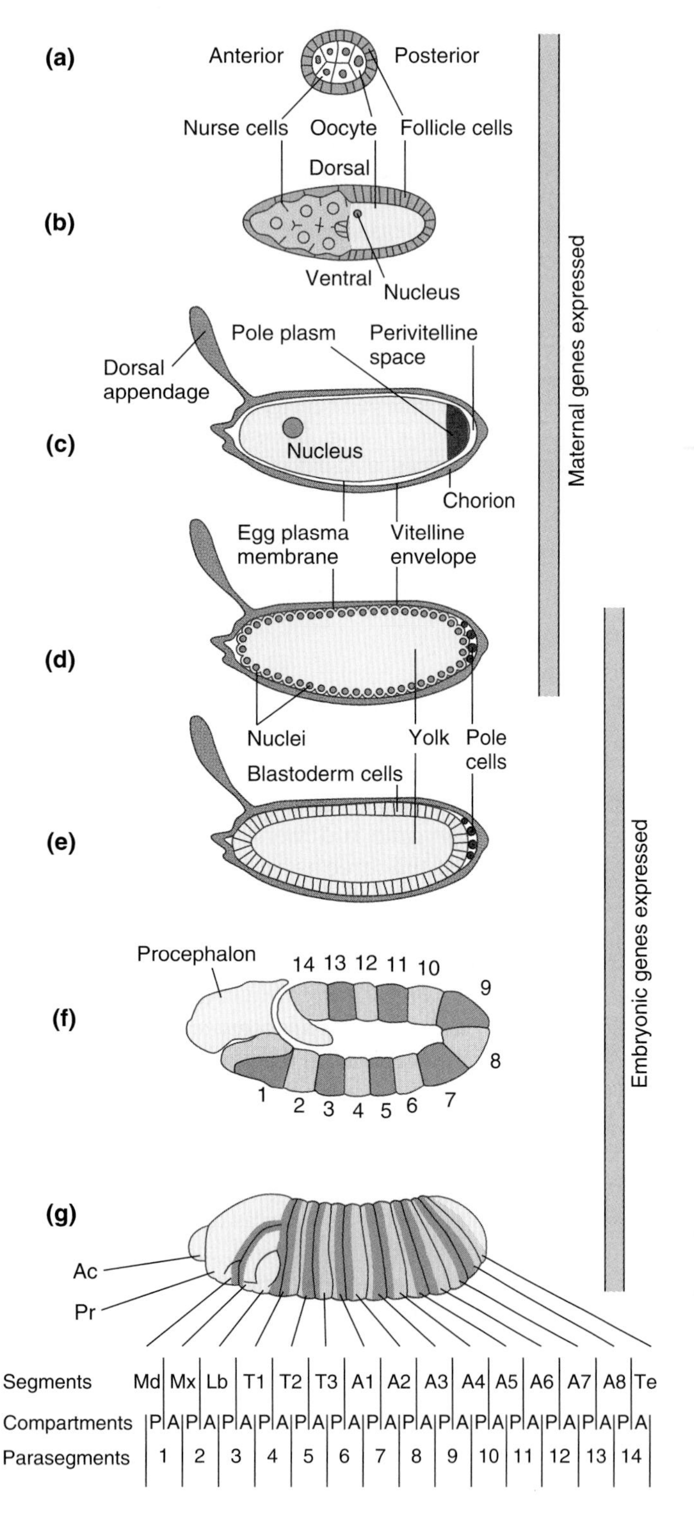

Figure 22.2 Schematic illustrations of oogenesis and embryogenesis in *Drosophila.* **(a)** Egg chamber at an early stage of oogenesis. The germ line–derived oocyte–nurse cell complex is surrounded by mesodermal follicle cells. The nurse cells are located anterior to the oocyte. **(b)** Egg chamber at an intermediate stage of oogenesis. The oocyte nucleus is displaced anteriorly and dorsally. **(c)** Newly laid egg. The egg cell is surrounded by two covers, the vitelline envelope and the chorion, both produced by the follicle cells. Near the anterior and posterior tips, the egg plasma membrane and the vitelline envelope are separated by a fluid-filled space, the perivitelline space. **(d)** Syncytial blastoderm stage. Pole cells have budded off at the posterior pole. Most nuclei have moved to the egg periphery but are not yet separated by plasma membranes. Transcription of the embryonic genome begins. **(e)** Cellular blastoderm stage. Membrane infoldings between the nuclei have created a monolayer of about 5000 uniform cells. **(f)** Extended germ band stage. The rudiment of the embryo proper, called the germ band, has extended posteriorly and doubled up until the posterior tip touches the head rudiment dorsally. The germ band is subdivided into parasegments numbered from 1 to 14. **(g)** Advanced embryo after germ band shortening and dorsal closure. The embryo is now subdivided into segments corresponding to the segments of the larva and adult. Each segment consists of an anterior compartment (A) and a posterior compartment (P), derived from two successive parasegments. Egg covers have been deleted in parts f and g for clarity. Ac = acron; Pr = procephalon; Md = mandible; Mx = maxilla; Lb = labium; T1 = prothorax; T2 = mesothorax; T3 = metathorax; A1 to A8 = abdominal segments; Te = telson.

visible segments and a nonsegmental end called the ***telson.***

All substances deposited in the egg have been synthesized during oogenesis under the control of the maternal genome. Maternal gene products support the *Drosophila* embryo through most of the cleavage stage. Starting with the 11th nuclear cycle, maternal mRNAs are broken down while transcription of the embryo's own genes is phased in. Mutations in either maternal or embryonic genes may cause abnormal pattern formation in the developing embryo.

22.2 Cascade of Developmental Gene Regulation

The epigenesis of the *Drosophila* embryo proceeds from a relatively uniform egg to a blastoderm stage in which there are two morphologically different cell types (pole cells and somatic cells), and then to an embryo with 16 different segments, each consisting of dozens of cell types. Underlying this increase in morphological complexity is a cascade of genetic control, which begins with a small number of maternal gene products and develops into an ever more complex network of embryonic genes and signaling molecules. To facilitate keeping track of the multitude of genes involved, Table 22.1 at the end of this chapter lists their names and main characteristics.

The cascade of gene regulation in *Drosophila* begins with a few localized products of *maternal effect genes* (Fig. 22.3). These products direct the segregation of pole cells and the activation of certain embryonic genes in specific domains of blastoderm cells. Some embryonic

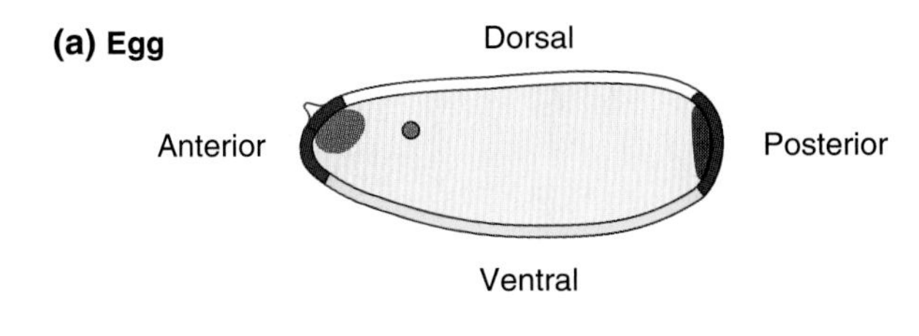

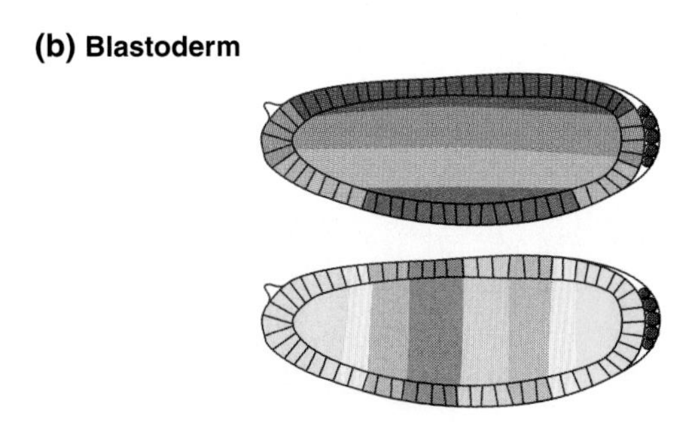

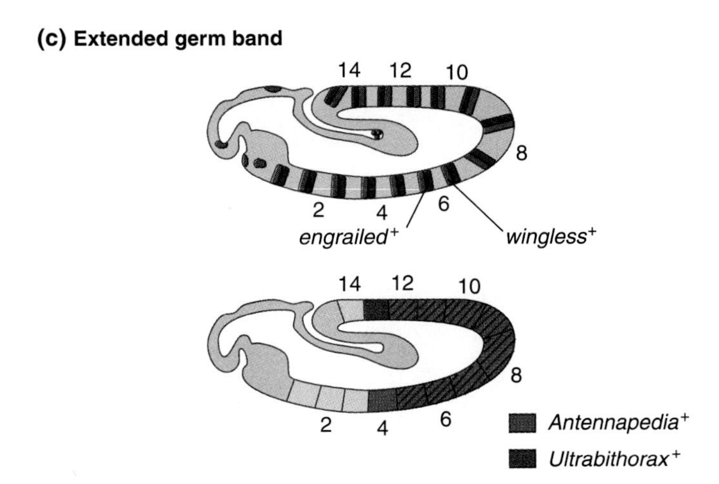

Figure 22.3 Increasing complexity of gene expression patterns during early embryogenesis in *Drosophila.* **(a)** The egg contains four localized products of maternal effect genes, which are deposited anteriorly, posteriorly, at both poles, or ventrally. They reside in part in the perivitelline space (between contours) and in part in the cytoplasm. **(b)** At the blastoderm stage, embryonic gene expression defines several longitudinal and transverse stripes without periodicity. **(c)** At the extended germ band stage, each parasegment is delineated anteriorly by a strip of cells expressing the *engrailed*$^+$ gene and posteriorly by cells expressing the *wingless*$^+$ gene. In addition, each parasegment is characterized by a unique combination of active homeotic genes, such as *Antennapedia*$^+$ in parasegments 4 through 12 and *Ultrabithorax*$^+$ in parasegments 5 through 13. Many more patterning genes are active at this stage but are not shown. The chorion has been deleted throughout. The vitelline envelope is also deleted in part c.

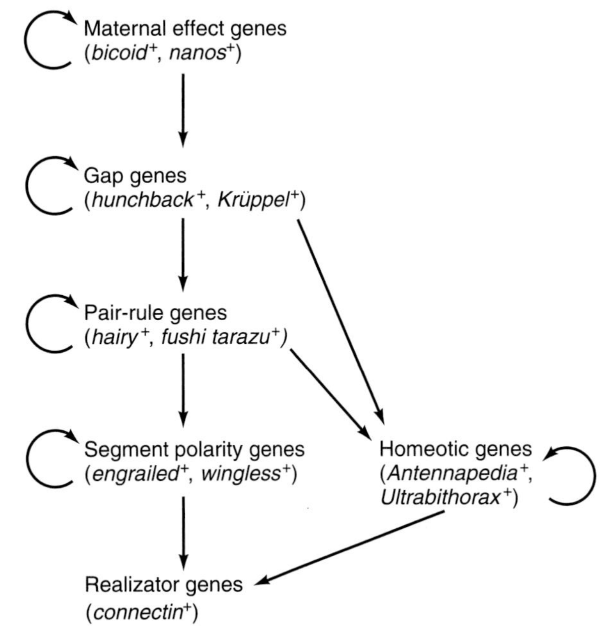

Figure 22.4 Cascade of gene regulation controlling the development of the anteroposterior body pattern in *Drosophila.* Under each group heading, one or two representative genes are named as examples. The straight arrows indicate the control of a gene of one group by genes of another group. The circular arrows symbolize interactions between genes of the same group.

genes are expressed in longitudinal domains that are parallel to the anteroposterior axis of the embryo. These longitudinal domains include a ventral strip of invaginating mesoderm, adjacent ventral neuroepidermis, and dorsal epidermis. Other embryonic genes are expressed in transverse bands of cells that are parallel to the dorsoventral axis (or circumference) of the embryo. These bands run parallel to the future segment boundaries on the blastoderm fate map but are much broader than one segment. By further genetic interactions, these broad transverse expression domains generate narrower expression domains, which eventually assume parasegmental periodicity.

The stepwise refinement of embryonic gene expression domains has been particularly well investigated for the anteroposterior pattern (Fig. 22.4). The first embryonic genes expressed in this class are called ***gap genes,*** because their loss-of-function phenotypes show gaps in the body pattern that are several segments wide. Combinations of gap gene activities control the expression of a second tier of embryonic genes, which are known as ***pair-rule genes,*** since their mutant alleles cause defects or abnormal development in every other segment. Combinations of pair-rule gene activities control the expression of a third tier of embryonic genes, called ***segment polarity genes,*** because their mutant alleles disturb the anteroposterior polarity of each segment. Collectively, gap genes, pair-rule genes, and segment polarity genes are called ***segmentation genes,*** because this entire class serves to subdivide the germ band into parasegments and later into segments. (Note that segment polarity genes and segmentation genes are *not synonymous.*)

Gap genes and pair-rule genes also control the expression of the *homeotic genes* already mentioned in previous chapters. Homeotic mutant phenotypes are characterized by striking transformations of certain body parts, such as the replacement of balancers with wings or antennae with legs (see Figs. 1.21 and 1.22). Finally, there must be a large class of ***realizator genes*** that are targets of the patterning genes and control morphological details such as the size of a segment and the position of its bristles.

The hierarchy depicted in Figure 22.4 is based mainly on three lines of evidence. First, comparing the effects of two mutations singly and in combination often reveals one gene as being *epistatic,* which usually means it is lower in the hierarchy (see Section 15.2). Second, the expression pattern of any cloned gene can be assayed by *immunostaining* or *in situ hybridization* (see Methods 4.1 and 8.1). By comparing the expression patterns of the same gene (gene A) in wild-type embryos and in embryos mutant for another gene (gene B), one can see directly whether gene B affects the expression of gene A. For instance, the abnormal expression pattern of a pair-rule gene (*hairy*$^+$) in a mutant for a gap gene (*knirps*) shows that *knirps*$^+$ directly or indirectly controls *hairy*$^+$ (Fig. 22.5). Third, specific binding of a protein encoded by one gene to the enhancer region of another gene suggests that the first gene *directly* controls the expression of its target gene. Indeed, most early patterning genes in *Drosophila* encode *transcription factors* controlling the expression of their target genes.

The timing of the gene activities shown in Figure 22.4 offers some clues to the overall functions of the respective genes. The maternal gene products last from oogenesis to early embryogenesis. The gap genes and the pair-rule genes are expressed for short periods of time from the preblastoderm or syncytial blastoderm to the extended germ band stage, and the mRNAs transcribed from these genes have short half-lives. In contrast, segment polarity and homeotic genes are expressed continuously from the late blastoderm stage on. These observations suggest that maternal, gap, and pair-rule genes play an important but temporary role: They establish the expression patterns of both segment polarity and homeotic genes. The latter have the lasting function of defining and maintaining the anteroposterior body pattern of the organism. These critical genes begin their activity at the same time that the potency of blastoderm cells is restricted to individual segments (see Fig. 6.4). The coincidence suggests that establishing the expression patterns of segment polarity genes and homeotic genes is part of the cell determination process itself. This hypothesis is in accord with models of cell determina-

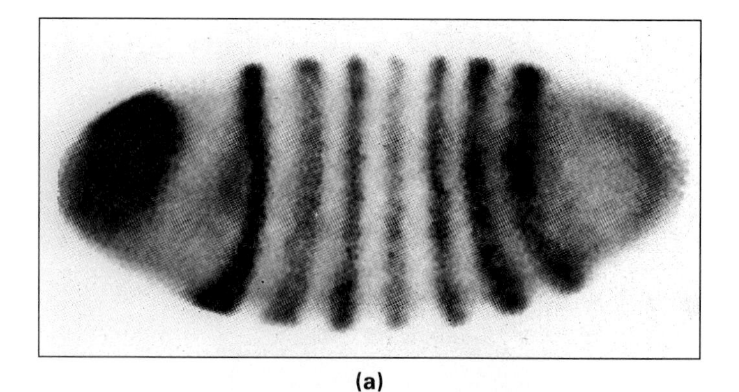

(a)

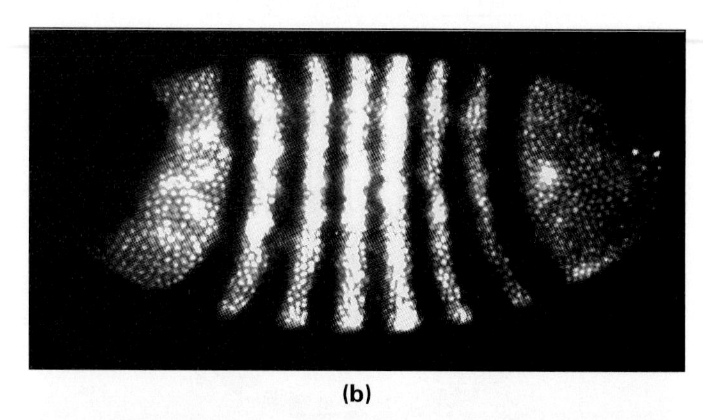

(b)

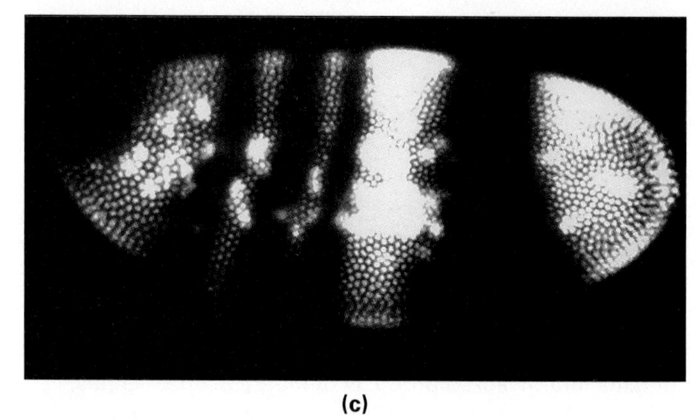

(c)

Figure 22.5 Establishing genetic hierarchies by comparing the expression pattern of the same gene in wild-type embryos and in embryos mutant for another gene. Here, expression of the pair-rule gene *hairy*$^+$ is monitored in wild-type and in *knirps*$^-$ mutant embryos at the blastoderm stage. Anterior pole is to the left. **(a)** Bright-field photograph of a wild-type embryo immunostained for hairy protein (see Method 4.1). The dark stain reveals the presence of hairy protein in an anterior dorsal patch and in seven stripes around the embryo. **(b)** Same embryo, now under fluorescent illumination. All nuclei appear bright from a fluorescent DNA strain, except in the cells expressing the *hairy*$^+$ gene, where the immunostain quenches the fluorescence. **(c)** Embryo homozygous for a *knirps* loss-of-function allele, also shown under fluorescent illumination. The *hairy*$^+$ gene is expressed normally in the anterior patch and the first three stripes. The other stripes have run together into one broad domain. Comparison of parts b and c shows that the normal function of the *knirps*$^+$ gene directly or indirectly affects the expression of the *hairy*$^+$ gene in the posterior four stripes.

tion based on bistable control circuits of switch genes (see Section 16.6).

In the remainder of this chapter, we will see how the cascade of embryonic patterning genes unfolds in time. First, we will examine the role of maternal effect genes, segmentation genes, and homeotic genes in forming the anteroposterior body pattern. This is the best-understood case of biological pattern formation to date. Next, we will look briefly at the genetic control of dorsoventral pattern formation. Occasionally, we will compare the results from genetic and molecular analysis with the older gradient model of embryonic pattern formation. Thereafter, we will discuss the compartment hypothesis, which links the activity of certain "selector genes" to *compartments* in the adult epidermis identified by clonal analysis. We will also see how compartment boundaries act as organizing centers for the outgrowth of legs and wings. Finally, we will consider which parts of the genetic hierarchy governing pattern formation in *Drosophila* have been found in other insects.

22.3 Maternal Genes Affecting the Anteroposterior Body Pattern

More than 20 maternal effect genes involved in anteroposterior pattern formation in *Drosophila* have been isolated, and these are probably the majority of all maternal genes with this function. Females mutant in any of these genes produce offspring lacking particular body regions while other regions are sometimes enlarged or duplicated. The number of mutant phenotypes is much smaller than the number of genes, suggesting that groups of maternal effect genes cooperate in generating certain parts of the embryo. Each group acts on specific gap genes, so that a group of maternal effect genes and their target gap genes form a ***genetic control system.***

There are three control systems, affecting respectively the anterior, the posterior, and the terminal parts of the anteroposterior body pattern (Figs. 22.6 and 22.7; *Nüsslein-Volhard, 1991*). Maternal effect mutations affecting the anterior system, such as *bicoid,* show reductions or losses of head and thoracic structures, and in some cases, their replacement with posterior structures. Corresponding mutations in the posterior system, such as *nanos,* cause the loss of abdominal segments, leading to very short embryos. Mutations in the terminal system, such as *torso,* eliminate the acron and telson as well as adjacent terminal segments. Each of the three systems can act independently of the others: even in double mutants, in which two systems are inactivated, the third system still operates and forms its part of the body pattern (Nüsslein-Volhard et al., 1987).

Although the anterior, posterior, and terminal control systems employ different molecular mechanisms, they share three basic features (Fig. 22.7). First, the activity of each system involves a *localized maternal signal* (see Fig. 22.3a). In the anterior and posterior systems, this signal is carried by mRNAs localized in the oocyte. In the terminal system, the follicle cells release a protein that seems to become localized in the vitelline envelope near the egg poles. Second, the localized signal triggers a mechanism that causes the *uneven distribution or activity of a maternally encoded transcription factor* in the early embryo. Third, the transcription factors set up *specific expression domains of embryonically expressed gap genes.*

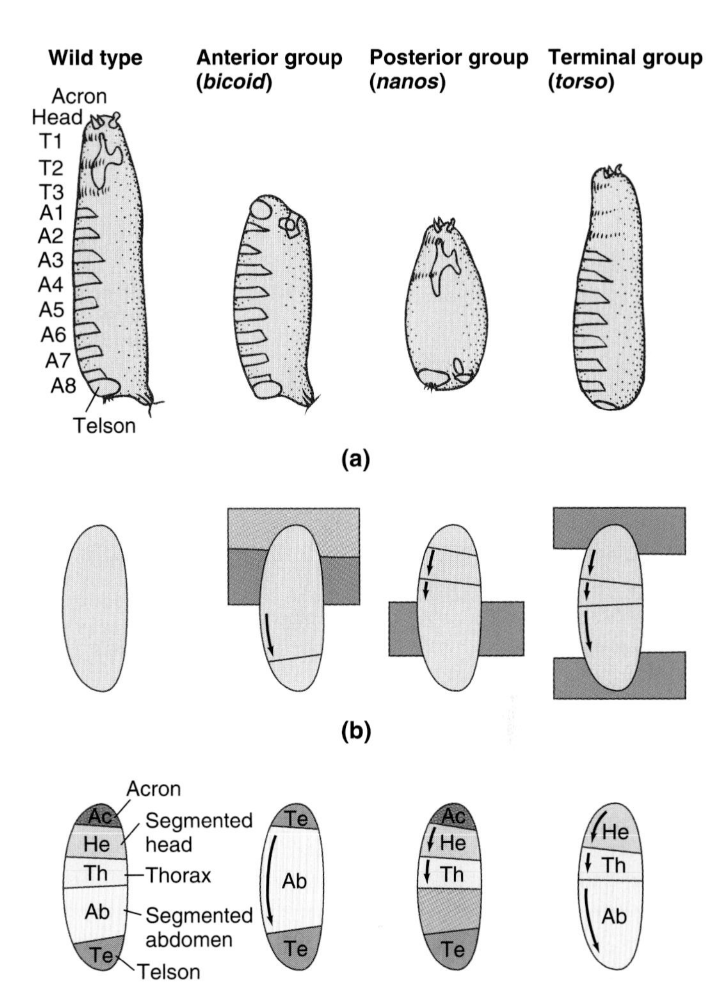

Figure 22.6 Three groups of maternal effect genes affect different regions of the embryonic body pattern. **(a)** For each group, a representative gene and its loss-of-function phenotype are indicated. **(b)** The rectangles indicate the embryonic parts on the blastoderm fate map that are not formed in mutant embryos. For instance, in bicoid embryos, head and thorax (shaded area) are missing and the acron area (lightly shaded) forms a telson. **(c)** The remaining areas are usually enlarged or transformed, so that the entire fate map is changed in the mutant embryos. Arrows indicate anteroposterior polarity.

The anterior and terminal transcription factors act as *morphogen gradients,* controlling two or more target genes in concentration-dependent ways. The posterior system, in contrast, acts by limiting the realm of an anterior signal. For the most part, the phenotypes resulting from abnormal shapes of these gradients are not caused

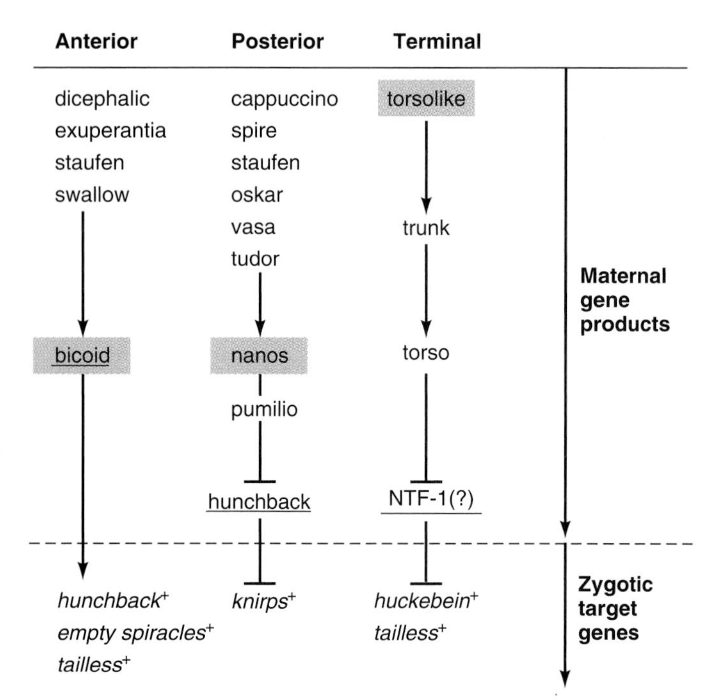

Figure 22.7 Three systems of genes and their products, each generating a part (anterior, posterior, or terminal) of the anteroposterior body pattern. Highlighted are three of the four egg components also shown in Figure 22.3a. Underlined gene products are unevenly distributed transcription factors. Arrowheads symbolize activation; perpendiculars symbolize inhibition.

by cell death or cell movement. Rather, the cells receive abnormal determination signals, so that the fate map of the entire embryo is changed.

THE ANTEROPOSTERIOR AND DORSOVENTRAL AXES ORIGINATE FROM THE SAME SIGNAL

The two polarity axes of the *Drosophila* embryo, anteroposterior and dorsoventral, are both generated during oogenesis, and both polarizations involve the same pair of genes, *gurken*$^+$ and *torpedo*$^+$ (Roth et al., 1995; Gonzales-Reyes et al., 1995; *Munn and Steward, 1995*). Gurken is a secreted protein showing sequence similarity with the vertebrate transforming growth factor-alpha (TGF-α). Torpedo is a receptor tyrosine kinase homologous to the vertebrate epidermal growth factor receptor (EGFR) and most likely is the receptor for gurken. Early during oogenesis, the oocyte sends out the gurken signal, and the follicle cells nearest to the nucleus respond by adopting a posterior identity (Fig. 22.8a, b). Later, the oocyte nucleus moves into an anterior and eccentric position in the oocyte, which then sends out the same gurken signal again. The follicle cells that are now closest to the oocyte nucleus are derived from a different lineage than the posterior follicle cells, and their response to the gurken signal is to take on a dorsal identity. We will focus here on the establishment of the anteroposterior axis and explore the dorsoventral axis in Section 22.6.

The earliest sign of anteroposterior polarity develop in the *germarium*, located at the tip of each ovariole (see Fig. 22.8a). In each developing egg chamber, 1 of 16 germ line cells becomes the oocyte, while the other 15 become nurse cells (see Figs. 3.13 and 3.14). In wild-type egg chambers, the oocyte invariably occupies the position away from the tip of the germarium, while the nurse cells form one cluster oriented toward the tip. The part of the oocyte facing the nurse cells, where nurse cell cytoplasm will enter the oocyte through cytoplasmic bridges, becomes the oocyte's anterior pole. The opposite pole, pointing away from the nurse cells, becomes the oocyte's posterior pole. Mutant alleles of *dicephalic* and the *spindle* group of genes lead to variable positioning of the oocyte within the egg chamber, such as an oocyte flanked by two opposite clusters of nurse cells (Fig. 22.9). Embryos developing embryos from such egg chambers show mirror-image duplications of head parts with no abdomen (Lohs-Schardin, 1982; Gonzales-Reyes et al., 1998).

Within the oocyte, the nucleus is originally located near the posterior pole. During this early stage, microtubules are oriented with their minus ends to the oocyte nucleus and their plus ends ending anteriorly in the oocyte or reaching through cytoplasmic bridges into the nurse cells (Theurkauf et al., 1993). This orientation favors the transfer of molecules and organelles from nurse cells to oocytes by *dynein* and similar minus-oriented motor proteins. Thus, the oocyte has an overt anteroposterior polarity at this stage. However, this early polarity does *not* develop cell-autonomously into the localization of cytoplasmic determinants, including bicoid and nanos mRNA. Rather, the polarized oocyte signals to its posteriorly adjacent follicle cells, causing them to assume a posterior identity and to signal this identity back to the oocyte.

It is the signaling step from the oocyte to the posterior follicle cells that requires *gurken*$^+$ and *torpedo*$^+$. Loss-of-function alleles in these and a few other genes interfere with the development of anteroposterior polarity from early oogenesis on. In particular, such mutations prevent the activation of a marker gene that is normally expressed in the posterior follicle cells adjacent to the oocyte. (The marker gene activity was discovered by *promoter trapping*, a strategy discussed in Section 15.6. A promoter trap line known as 998/12 has the *lacZ* reporter gene inserted next to the marker gene.) In *dicephalic* mutants, in which the oocyte position is variable, the marker gene can be activated in anterior and posterior follicle cells but not in midbody follicle cells (Fig. 22.9; Gonzales-Reyes et al., 1998). These results indicate that both anterior and posterior follicle cells are competent to assume a posterior identity, and that this step requires gurken/torpedo signaling from the oocyte.

After the posterior follicle cells have acquired their gurken/torpedo–dependent identity, they seem to send a signal back to the oocyte. This is inferred from the ob-

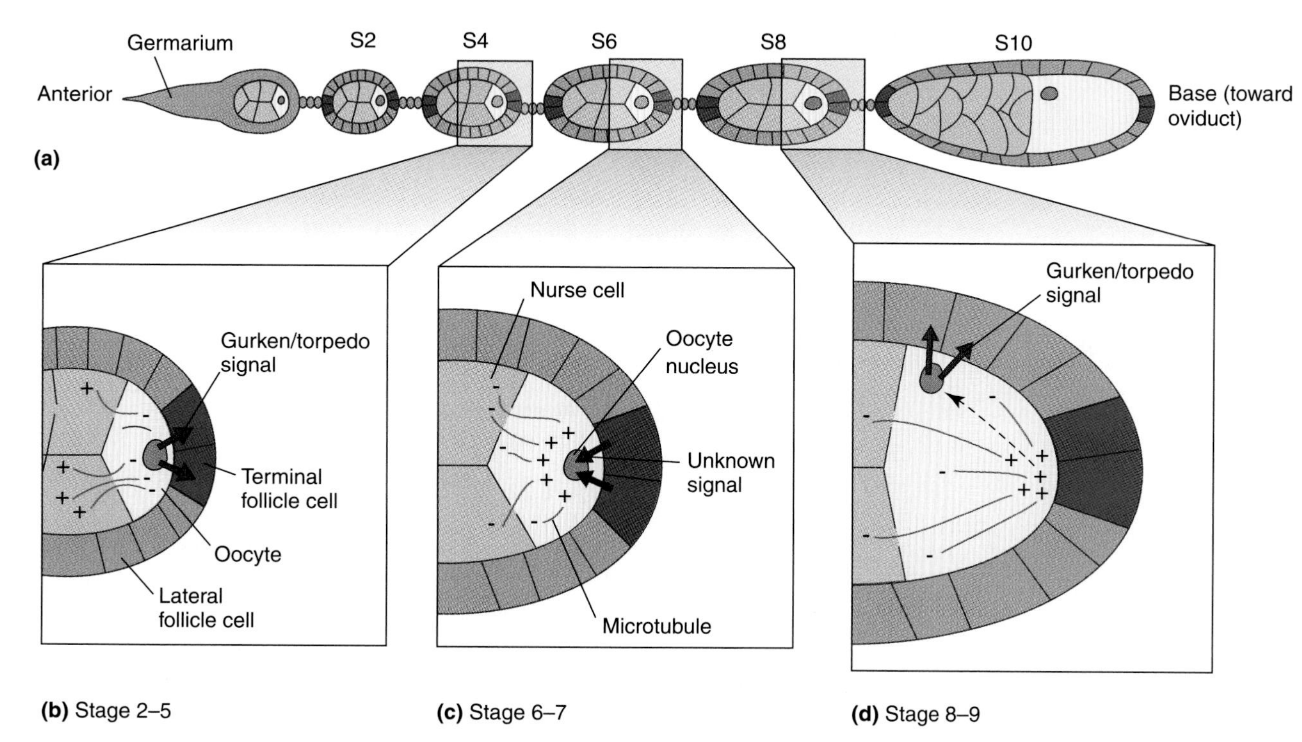

Figure 22.8 Establishment of the anteroposterior and dorsoventral polarity in the *Drosophila* oocyte. **(a)** Ovariole with germarium and egg chambers at stages S2, S4, S6, S8, and S10. (A mature oocyte is stage 14.) In each egg chamber, the oocyte (with blue nucleus) is positioned toward the base of the ovariole, where the ovariole opens into the oviduct. The nurse cells are clustered toward the germarium. **(b)** The anterior pole of the oocyte develops where it abuts the nurse cell cluster. Microtubules are oriented with their minus ends toward the oocyte nucleus, while their plus ends often reach through cytoplasmic bridges into adjacent nurse cells. The follicle cells adjacent to the anterior and posterior pole regions of the oocyte, called terminal follicle cells (teal color), differ in their cell lineage from the lateral, or main body, follicle cells (light green). At some time between stages 2 and 5, the oocyte nucleus sends a signal dependent on *gurken*$^+$ and *torpedo*$^+$ to adjacent terminal follicle cells, which as a result assume a posterior identity. **(c)** During stage 6 to 7, the posterior follicle cells send an as yet unidentified signal back to the oocyte, causing a repolarization of microtubules. **(d)** During stage 8 to 9, the oocyte nucleus moves to an anterior and eccentric position in the oocyte, and subsequently sends the same *gurken*$^+$ and *torpedo*$^+$ -dependent signal to adjacent main body follicle cells. As a result, these follicle cells assume a dorsal identity (see Fig. 22.37).

servation that now the oocyte undergoes a repolarization of microtubules, upon which kinesin and other plus-oriented motor protein will carry cargo into the oocyte and to its posterior pole. This repolarization was made directly visible through the use of a kinesin/β-galactosidase fusion protein, which changes its direction of movement from anteriorly to posteriorly during midoogenesis in wild-type egg chambers. In egg chambers from *dicephalic*, *spindle*, and *gurken* mutants, the repolarization of microtubules as well as the localization of bicoid and oskar mRNA are disturbed (Clark et al., 1994).

THE ANTERIOR GROUP GENERATES AN AUTONOMOUS SIGNAL

Once the microtubules of the egg chamber are repolarized and large amounts of cytoplasm are being transferred from the nurse cells to the oocytes, bicoid mRNA accumulates at the anterior face of the oocyte right behind the adjoining nurse cells. This accumulation depends on *cis-acting elements* located in an extended region of the bicoid mRNA's 3' UTR (see Section 8.4). Trans-acting elements required for localizing bicoid mRNA include microtubules and the products of at least three genes: *exuperantia*$^+$, *swallow*$^+$, and *staufen*$^+$ (Frohnhöfer and Nüsslein-Volhard, 1987; St. Johnston et al., 1989; Pokrywka and Stephenson, 1991). In eggs from females deficient in any one of these genes, bicoid mRNA diffuses posteriorly, and embryos developing from such eggs show anterior defects similar to embryos derived from *bicoid*$^-$ mothers. The staufen protein associates with the 3' UTR region of bicoid mRNA to form particles that move in a microtubule-dependent manner (Ferrandon et al., 1994).

The exuperantia protein also forms particles with bicoid mRNA. While some movements of these particles seem microtubule-dependent, their transport from nurse cells to oocytes via cytoplasmic bridges seems microtubule-independent (S. Wang and T. Hazelrigg, 1994; Macdonald and Kerr, 1997; Theurkauf and Hazelrigg, 1998).

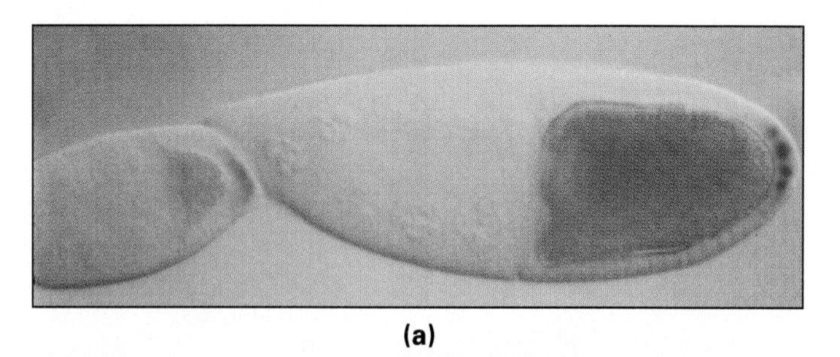

(a)

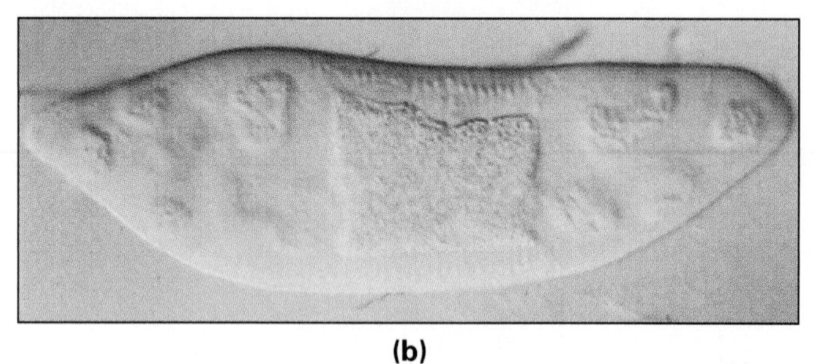

(b)

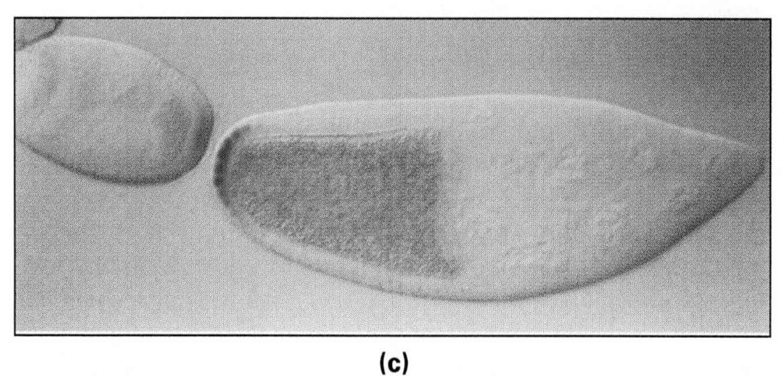

(c)

Figure 22.9 Egg chambers from *dicephalic* mutant females reveal that only terminal follicle cells can acquire the fate of posterior follicle cells. This fate is indicated here by the expression of a lacZ transgene in an enhancer trap line, known as 998/12. **(a)** Egg chamber in which the oocyte is positioned normally; the transgene is expressed by the posterior terminal follicle cells of the chamber. **(b)** Bipolar egg chamber: The transgene is not expressed. **(c)** Egg chamber in which the oocyte is positioned anteriorly, as determined by the overall polarity of the ovariole: The transgene is expressed by the anterior terminal follicle cells of the chamber.

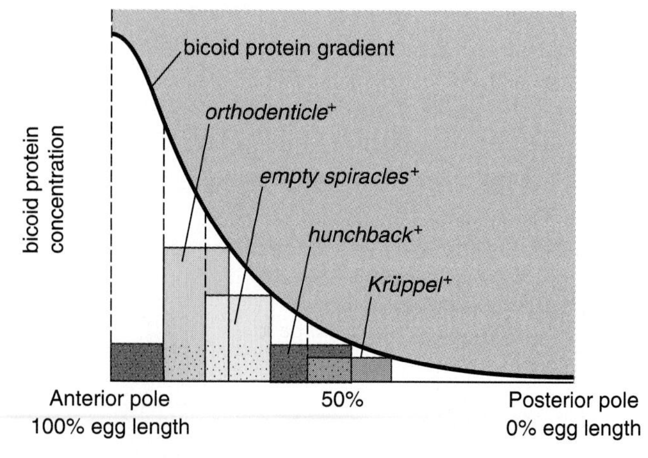

Figure 22.10 The bicoid protein as a morphogen. The bicoid protein concentration forms a gradient with a maximum near the anterior pole and an exponential decline toward the posterior. Depending on its local concentration, bicoid activates or represses several target genes including *orthodenticle*$^+$, *empty spiracles*$^+$, *hunchback*$^+$, and *Krüppel*$^+$. The threshold around which the target genes are activated or inhibited are shown here as vertical lines, although the expression domains of the target genes have graded boundaries as shown in Figures 22.18 and 22.25.

Translation of the localized bicoid mRNA begins after fertilization. While bicoid protein diffuses away from its mRNA source, it is degraded with a half-life of about 30 min (Driever and Nüsslein-Volhard, 1988). These processes result in a gradient of bicoid protein, which acts as a *morphogen*, generating several cell states characterized by different gene activities (Fig. 22.10; see also Fig. 16.8). Above a certain threshold concentration, the protein activates transcription of *Krüppel*$^+$, a gap gene. In normal embryos, the threshold for activating *Krüppel*$^+$ is reached near 40% egg length (EL), with 0% being defined as the posterior pole. At a higher concentration, normally reached around 50% egg length, bicoid protein also activates transcription of another gap gene, *hunchback*$^+$, as described in detail in Section 16.4. At high concentrations, reached beyond 58% egg length, hunchback protein inhibits transcription of *Krüppel*$^+$, thus limiting accumulation of Krüppel mRNA to a band around the egg equator. In addition, the local concentration of bicoid protein controls the expression *of empty spiracles*$^+$, a gene expressed in a band of blastoderm cells between 67 and 78% egg length (Walldorf and Gehring, 1992). Finally, bicoid concentrations normally reached between 75 and 92% egg length directly activate the *orthodenticle*$^+$ gene (Gao and Finkelstein, 1998). Thus, directly or indirectly, the bicoid gradient controls the expression of several embryonic genes. Historically, the bicoid protein has become the first defined molecule that seems to act according to the morphogen gradient model that has been on the minds of developmental biologists for decades (see Section 21.5).

The thresholds for target gene activation and inhibition shown in Figure 22.10 were observed in normal embryos derived from mothers with two copies of the *bicoid*$^+$ gene. In embryos from mothers with fewer or more *bicoid*$^+$ copies, the thresholds are shifted anteriorly or posteriorly. Nevertheless, such embryos develop into almost normal larvae and adults, apparently through correction mechanisms involving cell death (Namba et al., 1997).

In addition to the bicoid gradient, the *Drosophila* embryo contains a gradient of maternally encoded hunchback protein. (The *hunchback*$^+$ gene is expressed as a maternal effect gene and as an embryonic gap gene.) The maternal and embryonic hunchback mRNAs are transcribed from different promoters but encode the same protein (see Fig. 16.10). The concentration of maternal hunchback protein is high in the anterior third of the egg and then declines to the posterior. This profile is shaped indirectly by the product of *nanos*$^+$, a member of the posterior group described in the following section. When embryonic transcription begins, the maternal hunchback protein is replaced with larger amounts of embryonic hunchback protein, but the restriction of hunchback to the anterior half due to antagonism by nanos remains. The maternal hunchback and bicoid proteins have similar effects in controlling the expression of certain gap genes (G. Struhl et al., 1992; Simpson-Brose et al., 1994).

THE POSTERIOR GROUP ACTS BY DEREPRESSION

The posterior group contains more than ten maternal effect genes. Females deficient in any of these genes lay eggs that develop into embryos without an abdomen. The key component of the posterior group is the *nanos*$^+$ gene (Lehmann and Nüsslein-Volhard, 1991; C. Wang and R. Lehmann, 1991). As a counterpart to bicoid mRNA, nanos mRNA is partially localized to the *polar granules,* the electron-dense ribonucleoprotein particles that assemble near the posterior pole during oogenesis (see Fig. 8.7). Nanos mRNA associated with polar granules acts as a source for a gradient of nanos protein (Fig. 22.11a). In contrast to bicoid mRNA, the localization of nanos mRNA is imperfect, leaving much of it distributed through the egg cytoplasm. However, the nonlocalized nanos mRNA is translationally repressed, and this repression is overcome only by other polar granule components. Thus, the gradient of nanos protein is generated mostly by translational control and only in part by mRNA localization (see Section 18.3).

The *nanos*$^+$ function is required to allow the transcription of *knirps*$^+$, a gap gene expressed in a broad posterior band of blastoderm cells. However, in contrast to bicoid protein, the nanos protein is not a transcriptional activator. Instead, nanos acts by *derepression.* The presence of nanos protein blocks hunchback mRNA translation, and indirectly promotes the degradation of hunchback mRNA (Wharton and Struhl, 1991). Thus, nanos protein prevents the accumulation of maternal hunchback protein in the posterior half of the egg, where it would otherwise inhibit the expression of *knirps*$^+$ (see Fig. 22.7). Experimenters demonstrated this effect by engineering females with germ lines deficient in both hunchback and nanos (Hülskamp et al., 1989; Irish et al., 1989; G. Struhl, 1989). Despite the absence of both maternal gene products, eggs from these females developed into fertile flies.

The inhibition of hunchback function in the posterior half of the embryo by nanos protein depends on the *pumilio*$^+$ gene (Barker et al., 1992). It appears that pumilio protein binds to a specific recognition sequence in the 3′ UTR region of hunchback mRNA and recruits nanos protein to the binding site. The presence of both proteins at this site seems to promote the deadenylation of hunchback mRNA, thus inhibiting its translation and hastening its degradation (Murata and Wharton, 1995; Wreden et al., 1997).

The development of normal embryos from eggs lacking both maternal nanos and hunchback products reveals an apparent redundancy. It seems that the *embryonic* expression of the *hunchback*$^+$ gene in the anterior half of the embryo (see Fig. 16.9b) can substitute for the earlier synthesis of maternal hunchback protein. According to the *principle of overlapping mechanisms,* one could view the overlap as a safeguard that ensures normal development even if one control is not fully functional under certain environmental conditions or because of natural variation. This overlap may have allowed the genetic control of anteroposterior pattern formation to evolve in different ways (see Section 22.9 at the end of this chapter).

MANY POSTERIOR-GROUP GENES ARE ALSO REQUIRED FOR GERM LINE DEVELOPMENT

Aside from its role in embryonic pattern formation, nanos protein also functions in germ line development. The *polar granules,* which are critically involved in the derepression of nanos mRNA and the resulting posterior gene activity as discussed above, are subsequently sequestered in the pole cells—that is, the primordial germ cells. Pole cells lacking maternally provided nanos protein fail to slow down their mitotic cycle and do not inhibit the inappropriate expression of embryonic genes (Deshpande et al., 1999). Such pole cells also do not migrate to the gonads and thus do not become functional primordial germ cells (Kobayashi et al., 1996).

Moreover, embryonic expression of both *nanos*$^+$ and *pumilio*$^+$ is required for normal development of the female germ line (Forbes and Lehmann, 1998). In *pumilio*-deficient females, ovaries fail to maintain stem cells so that their ability to form *cystoblasts* is soon exhausted. In females without *nanos* function, the differentiation of the cystoblasts themselves is abnormal. The phenomenon that many genes have multiple functions is known as *pleiotropy* and is very common in development (see Fig. 15.2).

Pleiotropic function in posterior patterning and germ line development has also been observed for several other maternal effect genes including *cappuccino*$^+$, *oskar*$^+$, *spire*$^+$, *staufen*$^+$, *tudor*$^+$, and *vasa*$^+$. Eggs derived from females deficient in any of these genes show a lack of polar granules before displaying a lack of pole cells and developing abdominal defects. This combination of phenotypes raised the possibility that some or all of the genes mentioned above may be encoding building blocks for polar granules as repositories for posterior and germ cell determinants, rather than the determinants themselves (Lehmann and Rongo, 1993).

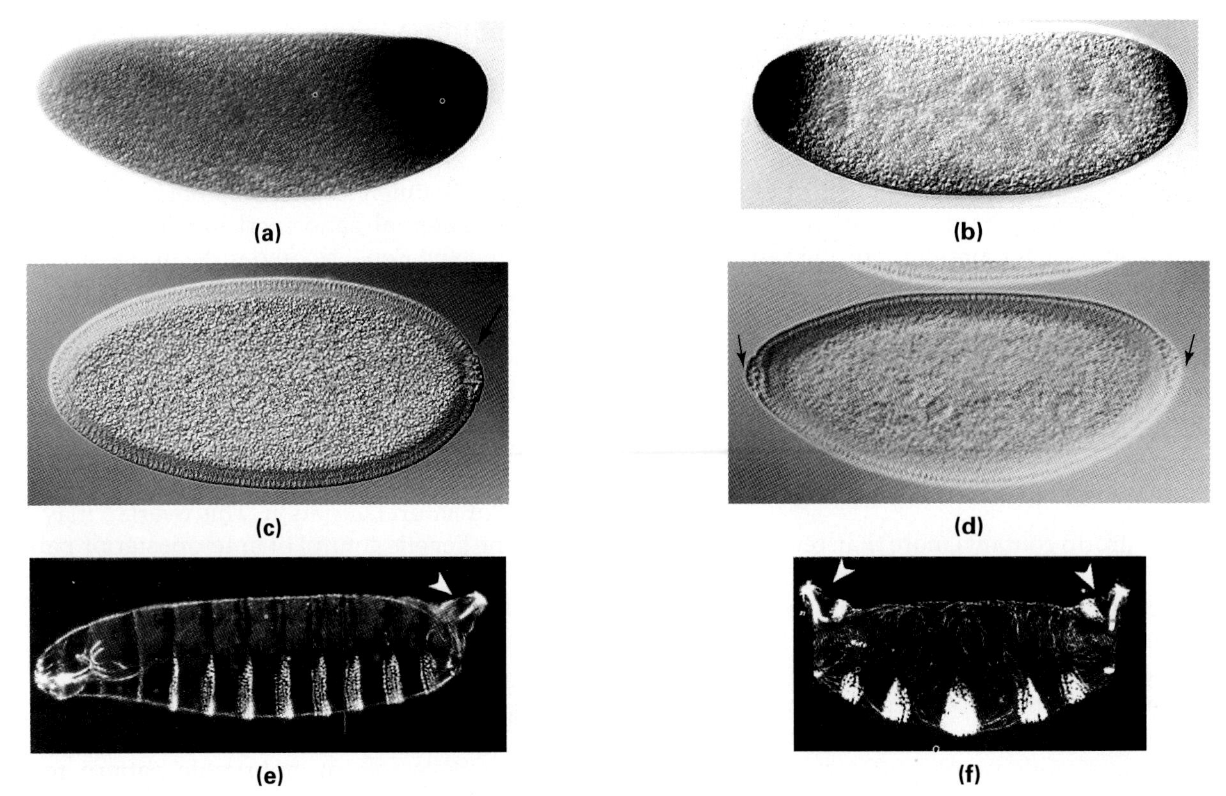

Figure 22.11 Effects of mislocalizing oskar mRNA to the anterior pole of *Drosophila* embryos; anterior pole to left. **(a, c, e)** Wild-type embryos; **(b, d, f)** embryos derived from mothers containing a transgene that encodes a modified oskar mRNA with the 3′ UTR region of bicoid mRNA. The latter causes mislocalization of the modified oskar RNA to the anterior pole. **(a)** Normal embryo during intravitelline cleavage immunostained for the nanos protein (see Method 4.1). **(b)** Embryo from transgenic mother, with normal nanos mRNA at the posterior and mislocalized nanos mRNA at the anterior pole (labeled by in situ hybridization; see Method 8.1). **(c)** Normal embryo at preblastoderm stage with pole cells (arrow) at the posterior pole. **(d)** Embryo from transgenic mother, with pole cells (arrows) at the anterior and posterior poles. **(e)** Normal larva, with head skeleton to the left and spiracles (arrowhead) at the posterior tip of the abdomen. **(f)** Embryo from transgenic mother, showing mirror-image duplication of posterior abdominal segments.

IN order to distinguish genes encoding genuine germ cell and posterior determinants from genes required for accessory functions, Ephrussi and Lehmann (1992) devised an experiment in which oskar mRNA was mislocalized to the anterior pole (Fig. 22.11). To this end, they cloned a cDNA encoding a hybrid mRNA consisting of the 5′ UTR and translated regions of oskar mRNA and the 3′ UTR of bicoid mRNA. Females containing this cDNA as a transgene produced eggs in which the modified oskar mRNA was localized to the anterior pole along with the normally supplied bicoid mRNA. The embryos developing from these eggs showed a stunning combination of phenotypes: ectopic pole cells at the anterior pole and a mirror-image duplication of the posterior abdomen. By transplanting the ectopic pole cells to the posterior pole of pole cell–deficient host embryos, the researchers confirmed that the ectopic pole cells were able to produce functional gametes.

Having set up this experimental system, the investigators proceeded to test which of the genes required for normal polar granule formation were also required for ectopic pole cell formation. The idea was that the mislocalized oskar mRNA should function independently of all genes that are merely required for the normal localization of oskar mRNA at the posterior pole. For these tests, the researchers generated females carrying the *oskar/bicoid* hybrid cDNA as a transgene but were deficient for one of the other genes required for normal pole cell formation. The results showed that two of these genes, *vasa*$^+$ and *tudor*$^+$, were required for ectopic pole cell formation. Three other genes, *cappuccino*$^+$, *spire*$^+$, and *staufen*$^+$, were dispensable for ectopic pole cell formation. These genes must therefore be involved in localizing oskar mRNA and assembling the polar granules at the posterior pole.

Questions

1. Why was it important to *replace* the oskar 3′ UTR with the bicoid 3′ UTR in the transgene construct, rather than simply *adding* the bicoid 3′ UTR to a complete oskar cDNA?
2. What phenotypes would you expect in embryos derived from females that carry the transgene but are lacking (a) *bicoid*$^+$, (b) *exuperantia*$^+$, or (c) *oskar*$^+$?

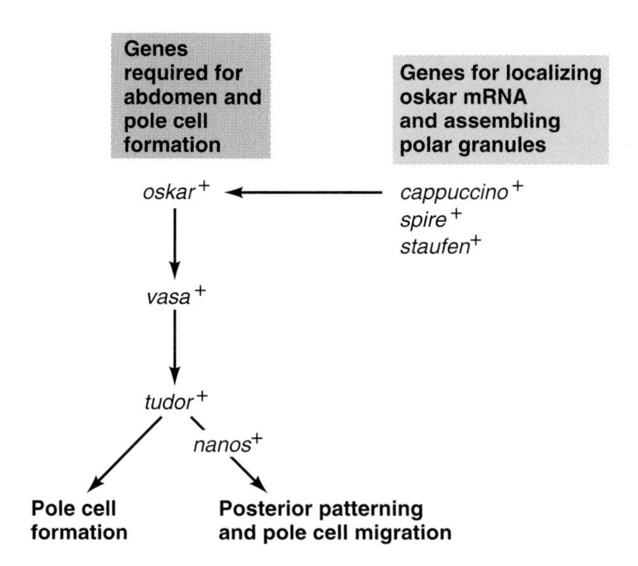

Figure 22.12 Hierarchy of genes involved in pole cell formation, in abdominal patterning, in pole cell migration, and in assembling polar granules as repositories for posterior and germ cell determinants.

The results of the experiments described above indicate that *vasa*$^+$ and *tudor*$^+$ are required for pole cell formation itself, presumably downstream of oskar (Fig. 22.12). In contrast, *cappuccino*$^+$, *spire*$^+$, and *staufen*$^+$ are required for localizing oskar and assembling polar granules at the posterior pole.

THE TERMINAL GROUP RELIES ON THE LOCAL ACTIVATION OF A RECEPTOR

The anteriormost and posteriormost embryonic regions, despite their physical separation, are affected by the same hierarchy of genes, the terminal system (see Fig. 22.7). The maternally expressed genes in this system include and *torso*$^+$, *torsolike*$^+$, and *trunk*$^+$. Embryos from females deficient in any of these gene activities show the same distinct phenotype; they are lacking the most anterior structures, including the acron and some mouthparts, as well as the most posterior structures, including the telson and the last abdominal segment (see Fig. 15.4).

The expression of *torsolike*$^+$ is restricted to a particular lineage of follicle cells that form two patches next to the anterior and posterior poles of the oocyte (Fig. 22.13a, b; Savant-Bhonsale and Montell, 1993). The torsolike protein has the characteristics of a modifying protein. In the egg and early embryo, the torsolike protein forms a symmetrical gradient with maximum concentrations at the egg poles and seems to be linked to the vitelline envelope (Martin et al., 1994).

While the activity of the *torsolike*$^+$ gene is critically required in the polar follicle cells, the activity of *trunk*$^+$ is required in the germ line (Schüpbach and Wieschaus, 1986b; Stevens, 1990). The trunk protein is a secreted protein that appears to be released into the perivitelline fluid between the vitelline envelope and the plasma membrane (Fig. 22.13c, d). Apparently, the localized torsolike protein processes the trunk protein, so that the modified trunk protein is present mostly in the pole regions of the perivitelline space.

The modified trunk protein seems to be the ligand of a receptor protein encoded by the *torso*$^+$ gene (Casanova et al., 1995). The receptor is synthesized in the egg and is incorporated into the entire egg plasma membrane (Sprenger et al., 1989). However, the receptor seems to be deployed in excess over its ligand (presumably the processed trunk protein), so that the ligand is trapped immediately at its site of origin—that is, in the polar regions of the embryo. Therefore, and most importantly, the receptor is activated only in the pole regions (Sprenger and Nüsslein-Volhard, 1992; Casanova and Struhl, 1993). A very similar mechanism for the local activation of a ubiquitous receptor by a localized ligand is also involved in setting up the dorsoventral polarity of the egg (see Section 22.6).

The torso protein is a receptor tyrosine kinase, which regulates the activity of transcription factors through a signaling pathway involving Ras and MAPK proteins (see Fig. 2.28). Eventually, this signaling cascade activates two gap genes, *huckebein*$^+$ and *tailless*$^+$ (Duffy and Perrimon, 1994). Both target genes are expressed in anterior and posterior pole regions, with *tailless*$^+$ extending further toward the middle of the embryo. The Ras protein and other elements in the signaling pathway show morphogen properties; high levels of Ras activate both *huckebein*$^+$ and *tailless*$^+$ whereas low levels activate only *tailless*$^+$ (Greenwood and Struhl, 1997; Ghiglione et al., 1999). The transcription factor(s) that are regulated by the torso signaling cascade and in turn control *huckebein*$^+$ and *tailless*$^+$ are not yet known with certainty. One candidate is the NTF-1 protein, which inhibits the *tailless*$^+$ gene by binding to a distinct response element in its promoter (Liaw et al., 1995). The simplest hypothesis involving NTF-1 would be that it acts as a transcriptional repressor and is inactivated by phosphorylation through the torso signaling pathway. Different levels of signaling activity could possibly translate into graded levels of NTF-1 inactivation, and consequently, target gene derepression.

In addition, the signal transduction cascade started by torso protein inactivates bicoid protein, thus downregulating the expression of its target genes (Ronchi et al., 1993).

In summary, the three groups of maternal effect genes control their embryonic target genes through different mechanisms. The anterior group generates localized bicoid mRNA, giving rise to a gradient of bicoid protein, which acts as a transcriptional activator for several genes including *hunchback*$^+$ and *Krüppel*$^+$. The posterior group produces the localized nanos signal, which by

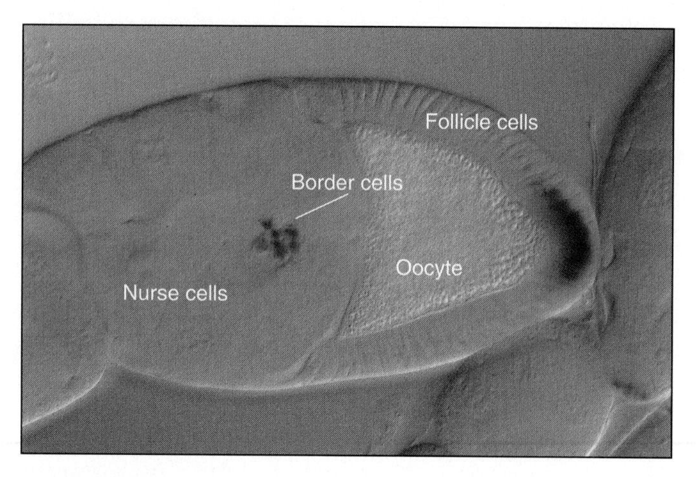

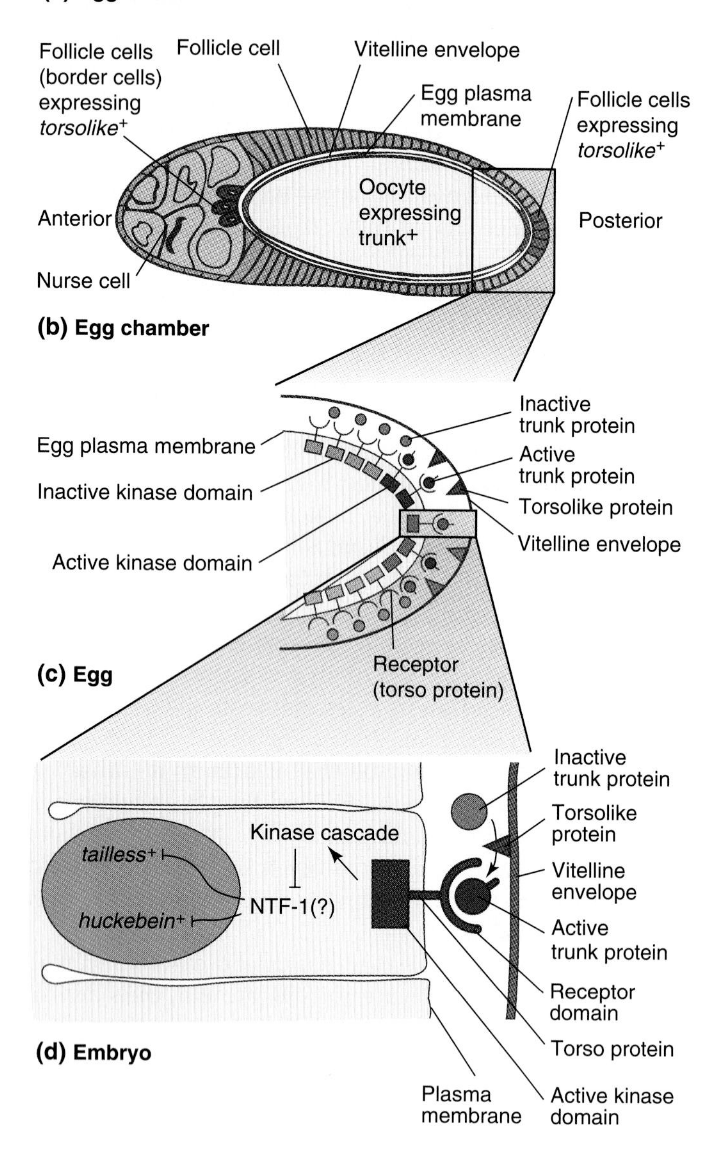

Figure 22.13 Model for the action of the terminal system of genes controlling the anteriormost and posteriormost parts of the body pattern in *Drosophila.* **(a)** Photograph of an egg chamber during midoogenesis. Nurse cells are to the left; the oocyte surrounded by follicle cells is to the right. The dark stain reveals the activity of a lacZ transgene inserted next to the promoter of a resident *torsolike*$^+$ gene. The transgene is expressed in the same region as the normal *torsolike*$^+$ gene: in posterior follicle cells and in the border cells, which will become anterior follicle cells. **(b)** Late during oogenesis, anterior and posterior patches of follicle cells express the *torsolike*$^+$ gene, which seems to encode a modifying protein. This protein becomes associated with the vitelline envelope near the anterior and posterior egg poles. **(c)** In the egg, the receptor tyrosine kinase encoded by the *torso*$^+$ gene is inserted into the entire plasma membrane. A signal precursor encoded by the *trunk*$^+$ gene is released everywhere from the egg into the perivitelline space. Where torsolike protein is present—that is, at the poles—the trunk precursor is converted into active trunk protein, which seems to act as a ligand to the torso receptor. **(d)** When blastoderm cells are formed in the embryo, the receptor kinase initiates a sequence of protein kinase reactions, which seem to inactivate a transcriptional inhibitor encoded by the *NTF-1*$^+$ gene. NTF-1 protein is thought to act as a morphogen, repressing embryonic target genes (*tailless*$^+$ and *huckebein*$^+$) depending on its local activity.

double negative control allows the embryonic *knirps*$^+$ gene to become active. The terminal group causes the localized expression of the *torsolike*$^+$ gene in polar patches of follicle cells. The torsolike protein appears to activate the *trunk*$^+$ gene product, which acts as a ligand, causing the local activation of an embryonic receptor encoded by the *torso*$^+$ gene. The torso receptor starts a signal transduction cascade thought to inactivate a transcriptional repressor of two target genes: *tailless*$^+$ and *huckebein*$^+$.

22.4 Segmentation Genes

The signals provided by the maternal gene products regulate a class of embryonic genes collectively called *segmentation genes* (Akam, 1987; Ingham, 1988). These genes control the initial subdivision of the embryo into *parasegments.* Many of these genes play additional roles later during development, but we will focus here on their early function of generating the parasegmental body pattern.

Three groups of segmentation genes are distinguished by their expression patterns during the blastoderm stage and by the phenotypes of their null alleles (Fig. 22.14). Typical *gap genes* are expressed in one or two broad transverse domains, and their null mutants are characterized by the loss of an entire series of segments. *Pair-rule* genes are transcribed in seven or eight regular transverse stripes. Their null phenotypes are defective in corresponding parts of every other parasegment, and

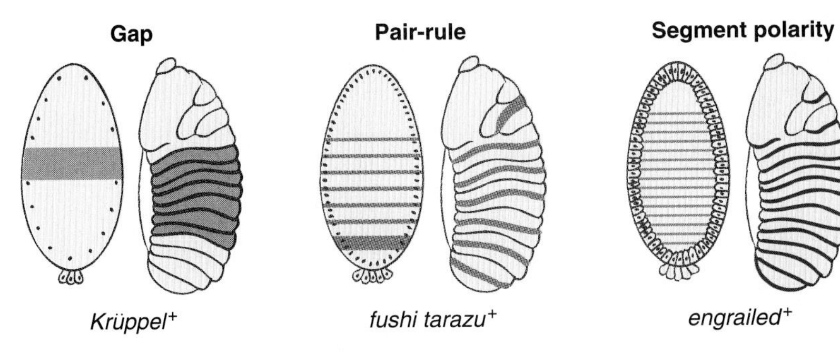

Figure 22.14 Three groups of segmentation genes are illustrated by their transcription patterns during late cleavage (color in left-hand diagram of each pair) and by those larval areas that are missing or transformed in loss-of-function phenotypes (color in right-hand diagrams of each pair). Transcription patterns are revealed by in situ hybridization (see Method 8.1). The gap gene *Krüppel*$^+$ is first expressed after the 12th nuclear division in a belt near the middle of the embryo. Null mutants lack the thorax and anterior abdominal segments. The pair-rule gene *fushi tarazu*$^+$ is transcribed in seven distinct stripes after the 13th nuclear division, before cellularization. In the null phenotype, alternate segment boundaries are deleted. The segment polarity gene *engrailed*$^+$ is transcribed in 14 narrow stripes at the cellular blastoderm stage. Mutants have a pattern duplication at the posterior margin of each segment.

the deficiencies are often coupled with duplications of the remaining parts. *Segment polarity* genes are expressed in 14 periodic stripes, and their null alleles cause deletions and duplications in each parasegment. The three groups form a genetic hierarchy, with gap genes regulating pair-rule genes, and these in turn controlling segment polarity genes (see Fig. 22.4). At each level of the hierarchy, there are also interactions among genes of the same group.

GAP GENES ARE CONTROLLED BY MATERNAL PRODUCTS AND BY INTERACTIONS AMONG THEMSELVES

Gap genes are the critical links between localized maternal gene products and all other embryonic genes involved in specifying the anteroposterior body pattern (Hülskamp and Tautz, 1991). We will discuss primarily four genes from this class: *hunchback*$^+$, *Krüppel*$^+$, *knirps*$^+$, and *tailless*$^+$ (Fig. 22.15). The *hunchback*$^+$ gene is expressed, for the most part, in

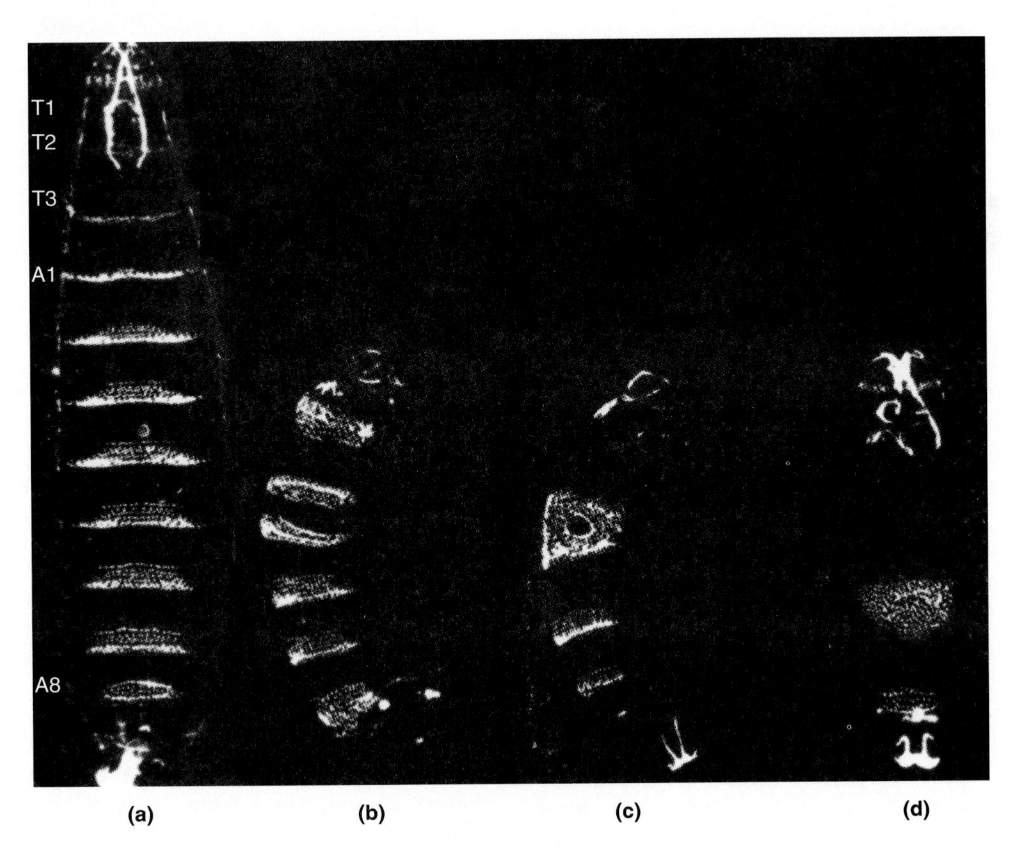

Figure 22.15 Phenotypes of gap mutants. The photographs show cuticle preparations under dark-field illumination. **(a)** Wild-type larva having eight abdominal segments (A1–A8), three thoracic segments (T1–T3), and an involuted head with mouth hooks (tucked inside T1 and T2). **(b)** *hunchback* embryo with thorax and head parts deleted. In this mutant allele, anterior abdominal segments are also replaced by mirror-image duplications of posterior abdominal segments. **(c)** *Krüppel* embryo missing thorax and A1 through A5 and having instead a mirror-image duplication of A6. **(d)** *knirps* embryo in which A2 through A6 are missing and A1 is fused to A7.

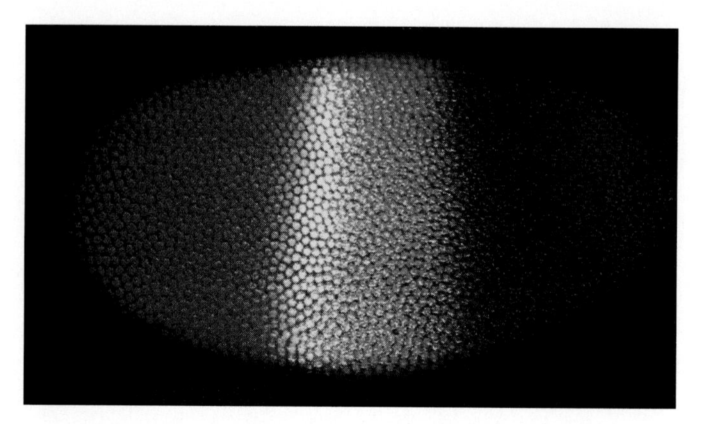

Figure 22.16 Expression of two gap genes, *hunchback*$^+$ (bright red) and *Krüppel*$^+$ (green), in a *Drosophila* embryo at the cellular blastoderm stage. Anterior to the left. The embryo was immunostained for each of the two gap gene proteins, and the two immunofluorescent images were superimposed by computer. Note the overlap between two expression domains (yellow) and the gradual decrease of expression toward the margins. The purple color to the right is from a DNA stain.

the anterior half of the blastoderm embryo, and null mutants lack the posteriormost head segment as well as all thoracic segments. The *Krüppel*$^+$ gene is expressed in a central domain (Fig. 22.16), and the null phenotype lacks the entire thorax and the first five abdominal segments. The *knirps*$^+$ gene is expressed in a posterior domain, and null mutants lack almost the entire abdomen. Finally, *tailless*$^+$ is expressed in the two terminal domains near the egg poles, and its null phenotype lacks anterior head parts as well as the eighth abdominal segment plus telson.

Within each domain, the gap protein concentration peaks near the center and falls off gradually toward the margins (Figs. 22.16 and 22.17a). If the expression domain includes an egg pole, the protein concentration is highest there. The gap domains overlap considerably at their margins, and their profiles change as development proceeds. For instance, embryonic *hunchback*$^+$ expression is observed in the anterior half of the embryo at the preblastoderm stage. At the blastoderm stage, a posterior domain is added, and the expression then recedes from both poles (see Fig. 16.9b). These dynamic changes reflect the multiplicity of control signals that affect gap gene expression.

The earliest gap gene expression domains are delineated at the preblastoderm stage by activating and inhibitory effects of maternal gene products (Fig. 22.17d, e and 22.18a). Expression of the *tailless*$^+$ gene is activated by the maternal *torso*$^+$ function in both pole regions of the embryo. In the anterior pole region, *tailless*$^+$ expression is also affected in a complex way by the maternal bicoid protein (Pignoni et al., 1992). Embryonic *hunchback*$^+$ gene expression is activated anteriorly by the bicoid protein and inhibited posteriorly by the nanos protein, as discussed earlier. The *Krüppel*$^+$ gene is acti-

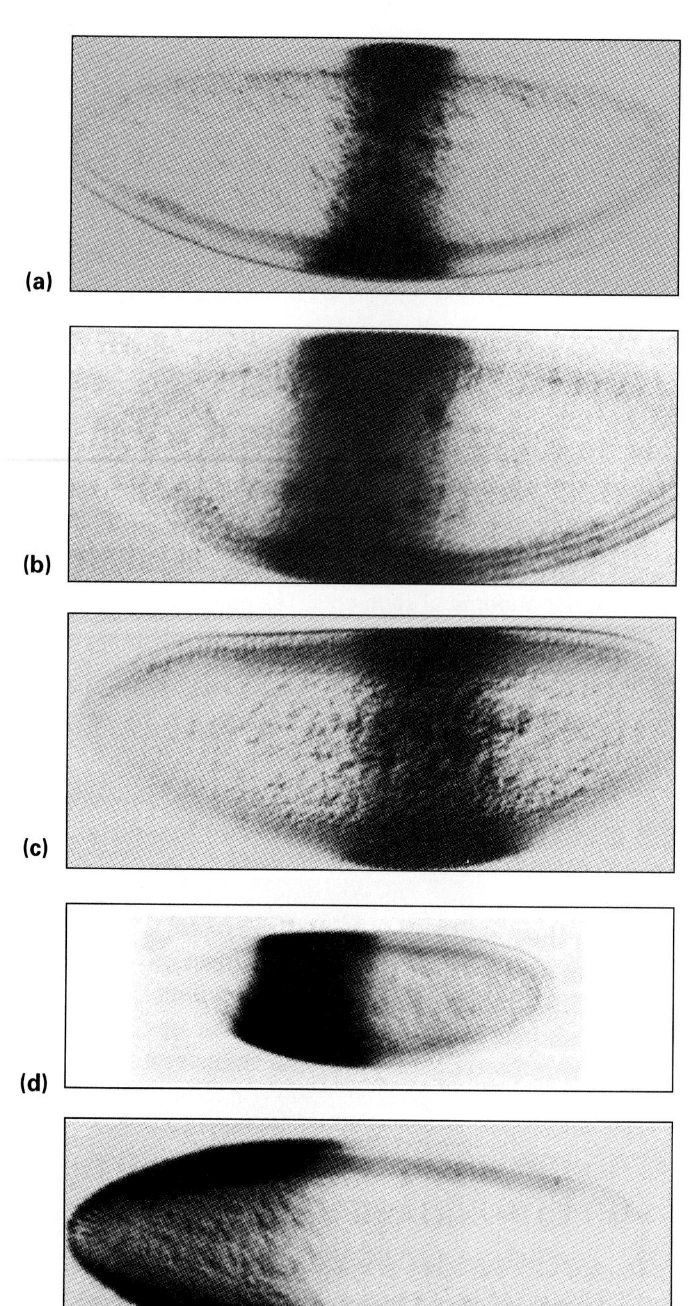

Figure 22.17 Regulation of *Krüppel*$^+$ gene expression by maternal and embryonic gene products (refer to Fig. 22.18). The expression of *Krüppel*$^+$ is assayed at the blastoderm stage by immunostaining (see Method 4.1). Anterior pole to the left. **(a)** Wild-type embryo. Krüppel protein is present in a central band of cells. **(b)** *hunchback*$^-$ mutant embryo. The *Krüppel*$^+$ expression domain is broadened and shifted anteriorly. **(c)** *tailless*$^-$ mutant embryo. The *Krüppel*$^+$ expression domain is slightly broadened in both directions. **(d)** Embryo derived from a *bicoid*$^-$ female. The *Krüppel*$^+$ domain is expanded far anteriorly, reflecting the lack of embryonic hunchback protein, which embryos from *bicoid*$^-$ mothers do not produce. **(e)** Embryo from a *bicoid*$^-$, *torsolike*$^-$ double mutant female. The *Krüppel*$^+$ domain is shifted and extends to the anterior pole, reflecting the absence of embryonic hunchback and tailless proteins as regulators.

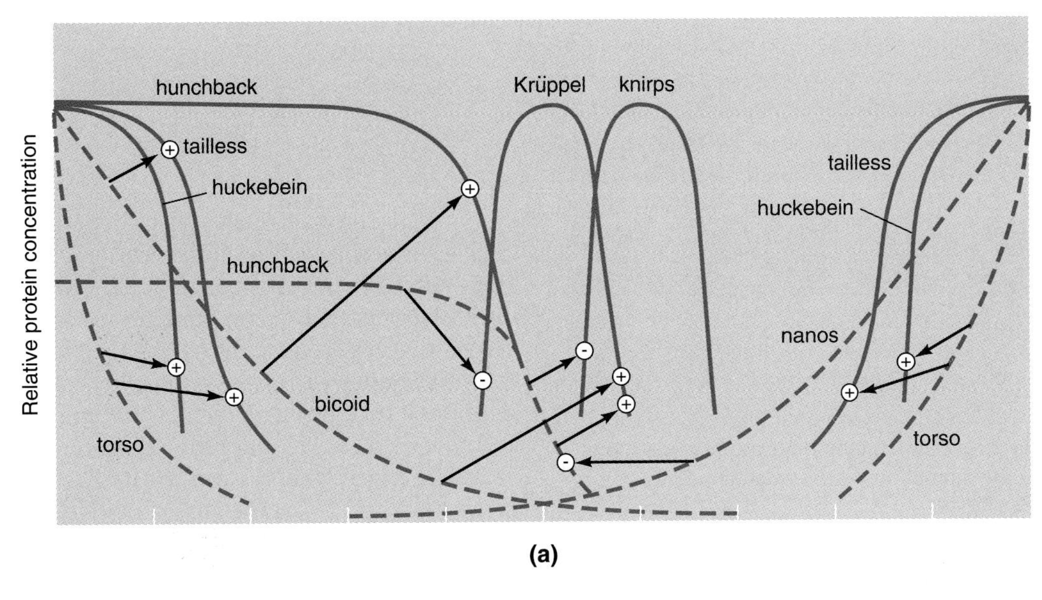

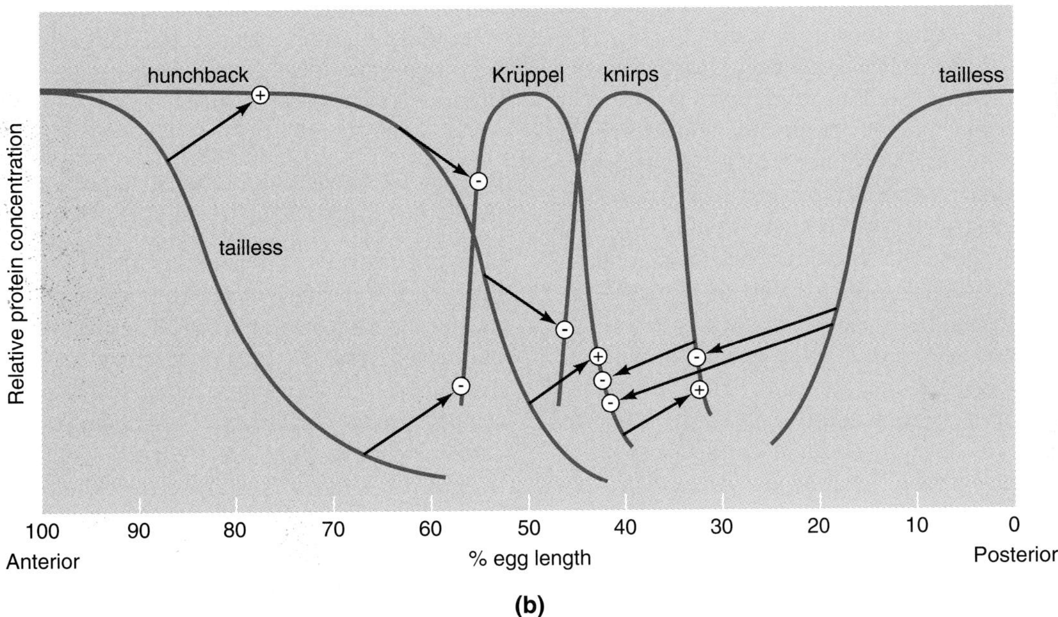

Figure 22.18 Simplified scheme of gap gene regulation. Some gap gene domains have been omitted. Several quantitative aspects have been estimated rather than measured. **(a)** Activating (+) and inhibitory (−) effects of maternal encoded proteins. Relative concentrations of maternal proteins are indicated by dashed lines; relative concentrations of gap gene proteins are shown as solid lines. The origins of the arrows indicate whether high, medium, or low levels of the respective protein are required for the regulatory effect. **(b)** Early embryonic interactions among gap genes and their proteins.

vated by bicoid protein, which sets the posterior boundary of *Krüppel*$^+$ expression (Fig. 22.17d). The *knirps*$^+$ gene seems to become active spontaneously, with its anterior boundary set through inhibition by hunchback protein.

In addition to being controlled by maternal effect gene products, gap genes also interact among themselves (Figs. 22.17b, c and 22.18b). Embryonically expressed hunchback protein has the same activating and inhibitory effects on *Krüppel*$^+$ and *knirps*$^+$ expression as the maternally expressed hunchback protein. Thus, in *hunchback*$^-$ mutant embryos, the *Krüppel*$^+$ expression domain is broadened and shifted anteriorly (see Fig. 22.17b). A similar shift of the *Krüppel*$^+$ expression domain in embryos without bicoid protein (see Fig. 22.17d) is presumably caused in part by the lack of *hunchback*$^+$ gene activity. Because the hunchback protein inhibits the *knirps*$^+$ gene more efficiently than the *Krüppel*$^+$ gene, *knirps*$^+$ is expressed more posteriorly. Likewise, in tailless$^-$ mutant embryos, the *Krüppel*$^+$ expression domain is broadened, indicating an inhibitory effect of the tailless protein on *Krüppel*$^+$ expression (Fig. 22.17c). Similarly, the expression domain of *knirps*$^+$ is limited by hunchback protein anteriorly and by tailless protein posteriorly. Both proteins seem to bind directly to enhancer regions of *knirps*$^+$ (Pankratz et al., 1992).

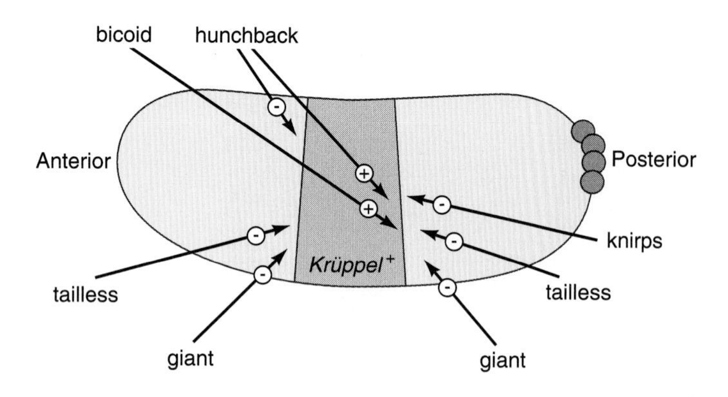

Figure 22.19 Regulation of the gap gene *Krüppel*+ by multiple maternal and embryonic proteins. The protein encoded by the gap gene *giant*+ is included here even if it does not appear in Figure 22.18.

In the regulation of the *Krüppel*+ gene, there is much overlap in both activating and inhibitory control functions (Fig. 22.19; Tautz, 1992). Four regulatory proteins, bicoid, hunchback, tailless, and knirps are involved in setting the anterior and the posterior boundaries of *Krüppel*+ expression, as mentioned earlier. In addition, the protein encoded by another gap gene, *giant*+, acts as a negative regulator of *Krüppel*+ (X. Wu et al., 1998). The lack of any one of these five regulatory proteins usually causes just a small effect, as illustrated in Figure 22.17. Only if two or three regulatory proteins are missing does the expression of *Krüppel*+ change substantially.

These observations illustrate again the *principle of overlapping mechanisms* discussed previously (see Sections 4.6, 12.2, 12.4, and 16.4). Each of the two explanations that have been proposed to explain the pervasiveness of overlapping mechanisms appears to be valid in the case of *Krüppel*+ regulation. First, the multiple controls may be *fail-safe mechanisms* for one another, making the control of *Krüppel*+ more reliable under a wide range of genetic and environmental circumstances. Second, most genes involved in regulating *Krüppel*+ are *pleiotropic*—that is, have additional functions. If one function of a pleiotropic gene causes a phenotype that is selected against, this critical function is likely to keep the gene "in business" also in its noncritical, or overlapping, functions.

All gap gene–regulating proteins have the characteristics of transcription factors. The binding of bicoid protein to *hunchback*+ enhancers has already been described in Section 16.4. Other *hunchback*+ enhancers are bound by the hunchback and Krüppel proteins, presumably mediating a positive feedback loop of the *hunchback*+ gene on its own expression and an inhibition of *hunchback*+ by *Krüppel*+ (Treisman and Desplan, 1989).

In the regulatory region of the *Krüppel*+ gene, a 730-bp segment contains at least 6 binding sites for bicoid protein, 10 binding sites for hunchback protein, 1 binding site for knirps protein, and 7 binding sites for tailless protein (Hoch et al., 1991, 1992). One of these regions, a 16-bp element bound by both bicoid and knirps proteins, was analyzed in detail. Cultured cells were transfected with plasmids containing the chloramphenicol acetyltransferase (CAT) reporter gene under control of the 16-bp element. Addition of another plasmid encoding bicoid protein caused a dose-dependent increase in CAT expression. When knirps protein was introduced along with bicoid protein, the bicoid-dependent increase in CAT expression was reduced by an amount depending on the dosage of knirps protein. These results indicate that genes can be regulated by activating and an inhibitory transcription factors that bind competitively and reversibly to overlapping enhancer elements.

In summary, genetic and molecular analysis of gap genes has shown that they are controlled by morphogen gradients of maternally encoded transcription factors. The proteins encoded by the gap genes are also transcription factors, which mediate additional regulatory interactions among gap genes. These multiple regulatory proteins delineate the expression of gap genes in broad transverse domains.

PAIR-RULE GENES ARE CONTROLLED BY GAP GENES AND BY OTHER PAIR-RULE GENES

Gap gene expression domains are broad and aperiodic; they do not correspond to the pattern of parasegments on the blastoderm fate map. The first embryonic genes expressed with a periodic pattern reminiscent of segmentation are the *pair-rule genes*. Three of them, *even-skipped*+, *hairy*+, and *runt*+, are called ***primary pair-rule genes***, because they are expressed first and are controlled directly by the gap genes; the others are called ***secondary pair-rule genes***, since they are controlled by the primary pair-rule genes. Of particular interest among the secondary pair-rule genes is *fushi tarazu*+, which is critical to the control of both segment polarity genes and homeotic genes. In null phenotypes of *fushi tarazu*, all even-numbered parasegments (parasegments 2, 4, etc.) are missing. Because each parasegment contains a future segment boundary, the mutant larvae have only half the normal number of segment boundaries (see Fig. 15.5).

The expression of pair-rule genes begins in fuzzy, broad, and irregular stripes during the formation of the cellular blastoderm. By the time cellularization is complete, pair-rule genes are expressed in seven or eight sharp, narrow, and evenly spaced stripes that encircle the entire embryo. A typical pair-rule stripe at this stage is as wide as three to four blastoderm cells or one prospective parasegment (Fig. 22.20). The stripes are separated by interstripes—also initially four blastoderm cells wide—where the gene is not expressed. A stripe and an interstripe together are as wide as two segments, so the expression pattern of each pair-rule gene has a ***bisegmental periodicity***. The expression patterns of *fushi*

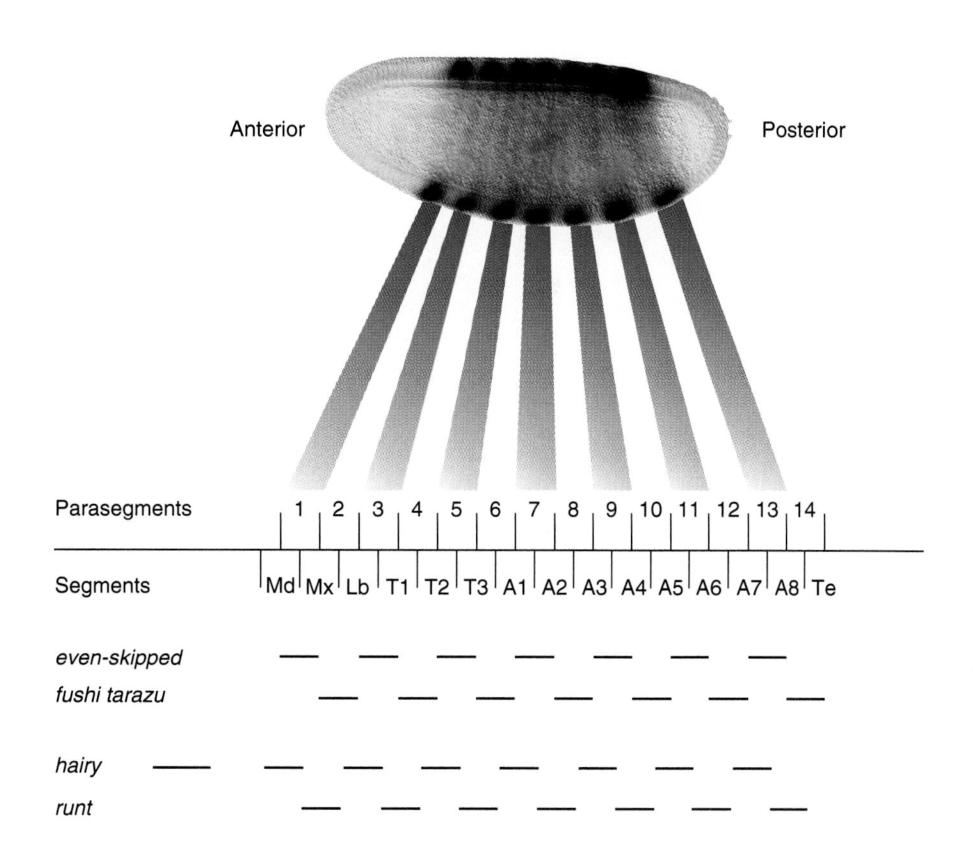

Figure 22.20 Expression of pair-rule genes in the *Drosophila* embryo. The photograph shows an embryo immunostained (see Method 4.1) for the protein encoded by the *even-skipped*+ gene. The embryo is at an advanced stage of blastoderm formation; the plasma membrane infoldings have passed the nuclei (see Fig. 5.19). The even-skipped protein has accumulated in seven regular stripes, which will give rise to the odd-numbered parasegments (1, 3, 5, etc.). When cellularization is complete, expression remains strong in the anterior two cell rows of each parasegment but decreases toward the posterior margin (see Fig. 22.21). The scales indicate the positions of segments and parasegments, with segment names abbreviated as in Figure 22.2. The lines at the bottom represent the initial expression domains of further pair-rule genes.

tarazu+ and *even-skipped*+ are in register with the parasegmental primordia, with *fushi tarazu*+ expressed in even-numbered parasegments and *even-skipped*+ in odd-numbered parasegments (P. A. Lawrence and Johnston, 1989). Closer analysis shows that the expression of both genes recedes from the posterior margins of their domains, while the anterior borders, each marking the anterior border of a parasegment, remain sharp (Fig. 22.21). The expression patterns of *hairy*+ and *runt*+ are also complementary to each other but are out of register with both parasegmental and segmental primordia.

How are the periodic (bisegmental) patterns of primary pair-rule genes generated from the aperiodic (irregular) pattern of gap gene domains? The gap gene proteins are transcription factors that activate and inhibit primary pair-rule genes so that they are expressed in periodic patterns of stripes. This control occurs in a piecemeal fashion because each gap gene affects only part of the entire stripe pattern of a pair-rule gene. As an example, we will consider the control of the *hairy*+ gene. Preliminary evidence for a piecemeal control of *hairy*+ came from expression studies on mutant embryos, which revealed that each gap gene affects only part of the entire *hairy*+ expression pattern (see Fig. 22.5; Carrol et al., 1988). Similarly, mutations in many enhancer regions of *hairy*+ affect only one or a few stripes rather than all of them (Fig. 22.22; Howard et al., 1988). These observations suggest that different combinations of gap gene proteins drive the expression of *hairy*+ in successive stripes by binding to different enhancers.

TO definitively test the notion of stripe-specific enhancers, Pankratz and coworkers (1990) isolated various *restriction fragments* from the upstream regulatory region of the *hairy*+ gene and spliced each fragment in front of a *lacZ reporter gene* (Fig. 22.23). Each of the resulting fusion genes was introduced into wild-type flies by *P-element transformation.* When embryos from transformant lines were assayed for *lacZ* expression, each fragment from the *hairy*+ regulatory region was found to enhance a characteristic subset of stripes. One fragment located about 8 kb upstream of the *hairy*+ promoter enhanced the formation of stripe 6; a neighboring fragment enhanced the formation of stripe 7.

(a)

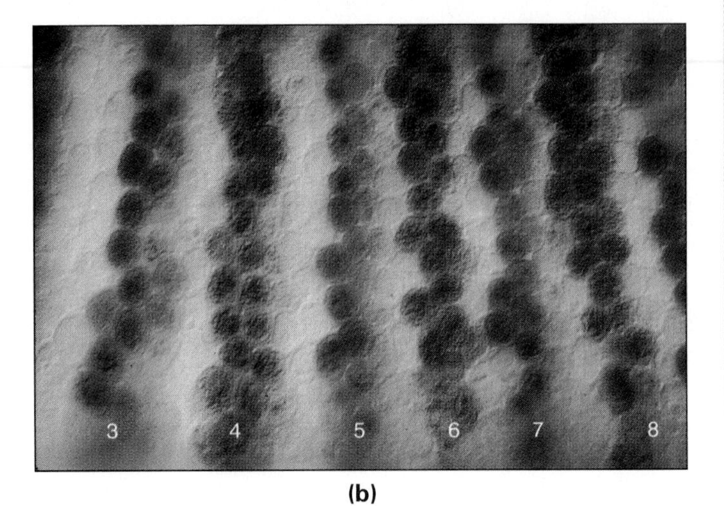

(b)

Figure 22.21 Expression of pair-rule genes in the *Drosophila* embryo. Anterior pole to the left, ventral side down. Embryos were immunostained for fushi tarazu protein (brown) and even-skipped protein (blue). Cells with the highest concentrations of these proteins will form the anterior margins of parasegments (numbered). Each parasegment at this stage is three to four cells wide. Each pair-rule gene is expressed with a bisegmental periodicity, *fushi tarazu*$^+$ in even-numbered parasegments and *even-skipped*$^+$ in odd-numbered parasegments. **(a)** Embryo at the cellular blastoderm stage. **(b)** Similar embryo at higher magnification.

Some fragments from the *hairy*$^+$ enhancer region drove the expression of more than one stripe, but each fragment enhanced the expression of a characteristic subpattern.

To find out which gene products interact with the specific enhancer fragments of the *hairy*$^+$ gene, the researchers introduced their fusion gene constructs into mutant flies that were deficient in various gap genes (Fig. 22.24). They found that loss of *Krüppel*$^+$ activity caused stripe 5 of *hairy*$^+$ to disappear and stripe 6 to expand anteriorly. Loss of *knirps*$^+$ activity caused *hairy*$^+$ stripe 6 to disappear and *hairy*$^+$ stripe 7 to expand anteriorly. They concluded that *Krüppel*$^+$ activity is required to express hairy$^+$ in stripe 5 and to set the anterior boundary of stripe 6 (Fig. 22.25). This conclusion implies that at a low concentration Krüppel protein interacts with one enhancer region to inhibit the expression of the *hairy*$^+$ gene in stripe 6, while at a higher concentration Krüppel protein interacts with another enhancer region to promote the expression of *hairy*$^+$ in stripe 5. Likewise, a low concentration of knirps protein inhibits *hairy*$^+$ expression in stripe 7, while a higher concentration of knirps protein enhances *hairy*$^+$ expression in stripe 6.

Questions

1. In the first part of their experiment (shown in Fig. 22.23), the researchers transformed wild-type flies. Wouldn't it have been better to transform a *hairy* null mutant in order to avoid a possible disturbance of the transgene expression by the resident *hairy*$^+$ gene?
2. How could the investigators have tested whether the Krüppel and knirps proteins that control *hairy*$^+$ expression in stripes 6 and 7 bind directly to enhancer elements RK and RR?
3. As an analogy to the control of *hairy*$^+$ gene expression, imagine that you drive across the country in your car, and that you listen to the same National Public Radio program (*hairy*$^+$ gene expression) broadcast on different frequencies by local radio stations (the eight hairy expression domains) as you pass on your way (from the anterior to the posterior egg pole). Instead of adjusting the tuner of your radio as you move from one station to the next, imagine that you have an array of fixed antennal circuits, with at least one circuit adjusted to the frequency of each station. In terms of this analogy, which part of the entire radio transmission system would correspond to (a) to the stripe-specific enhancers of the *hairy*$^+$ gene and (b) to the gap gene proteins that drive *hairy*$^+$ expression from these enhancers?

The experiments described above indicate that the expression of *hairy*$^+$ is controlled by local combinations of different gap gene proteins acting on stripe-specific enhancers. Other investigators identified gap gene products that control the expression of *hairy*$^+$ in the remaining stripes. They also observed that another primary pair-rule gene, *runt*$^+$, contributes to the even spacing of the hairy$^+$ stripes (Howard and Struhl, 1990; Riddihough and Ish-Horowicz, 1991). These studies have also shown that a given gap gene protein may enhance or inhibit transcription of *hairy*$^+$, depending on gap protein concentration and on what other transcription factors are bound simultaneously. Corresponding results have been reported for the two other primary pair-rule genes, *even-skipped*$^+$ (Stanojevíc et al. 1989, 1991; S. Small et al., 1996) and *runt*$^+$ (Klingler et al., 1996).

The piecemeal fashion in which the deceptively regular expression patterns of primary pair-rule genes are cobbled together by various combinations of gap gene proteins came as a surprise to some investigators who had envisioned more elegant ways of generating stripes by chemical oscillators. However, while the human minds prefers elegant simplicity, organisms often reflect

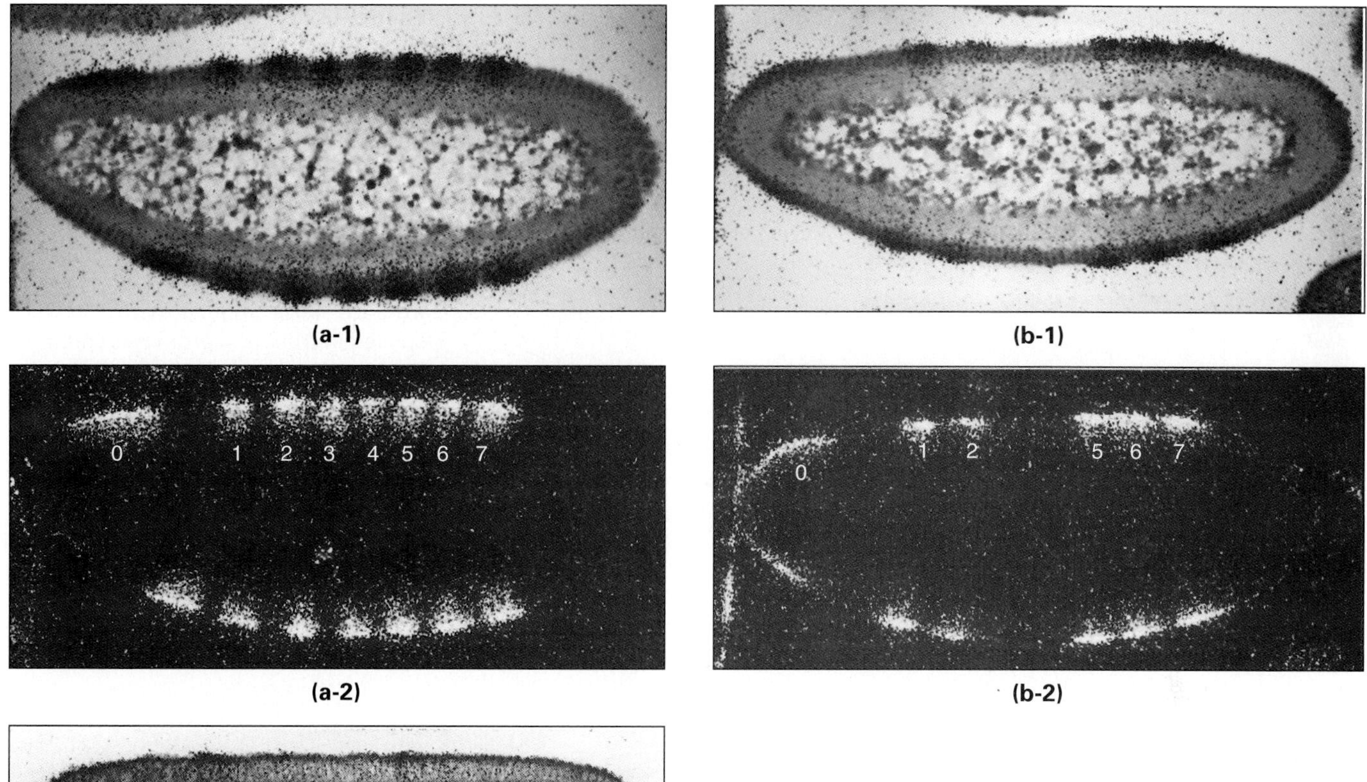

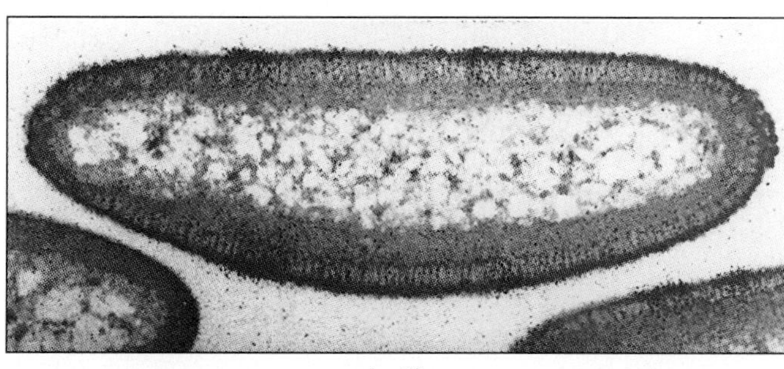

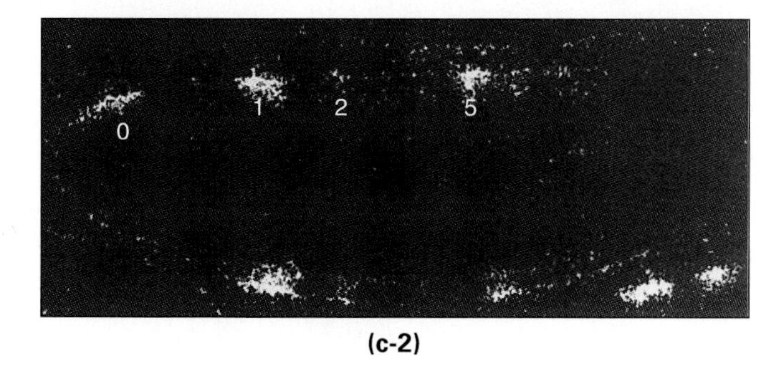

Figure 22.22 Effects of mutations in different regulatory regions of the *hairy*$^+$ gene. The panels show sections of blastoderm embryos hybridized in situ with a radiolabeled probe binding to hairy mRNA (see Method 8.1). Each specimen is photographed in bright-field illumination (top) and in dark-field illumination (bottom). **(a)** In the wild-type pattern there are eight expression domains: an anterior dorsal patch (0) and seven stripes (1–7). **(b)** In the mutant *hairy*^{m3} phenotype, stripes 3 and 4 are suppressed while stripes 5 through 7 are broadened. **(c)** In the *hairy*^{m7} phenotype, stripes 3, 4, 6, and 7 are suppressed while stripe 2 is quite weak.

the haphazard ways of evolution, which just builds on what already exists.

In addition to being controlled by gap genes, the primary pair-rule genes feed back on themselves and interact with one another to delineate their expression domains. Their interactions cause dynamic changes in the expression pattern. For instance, *even-skipped*$^+$ is initially expressed in seven broad stripes with poorly defined edges. Within 30 min, these stripes are sharpened and refined so that they are only two or three cells wide and show a marked concentration profile. The even-skipped protein concentration becomes highest in the anteriormost cells of each stripe and declines toward the posterior (see Fig. 22.21). Modulations of this kind, which are also observed for other pair-rule genes, are important for the control of segment polarity genes, as we will see.

The expression of *secondary pair-rule genes* is regulated mainly by primary pair-rule genes. At the preblastoderm stage, *fushi tarazu*$^+$ is inhibited by *even-skipped*$^+$, so that the two genes' striping patterns are complementary. At the blastoderm stage, *fushi tarazu*$^+$ is activated by *runt*$^+$ and inhibited by *hairy*$^+$, so that the *fushi tarazu*$^+$ stripes are then narrowed to the first one or two rows of

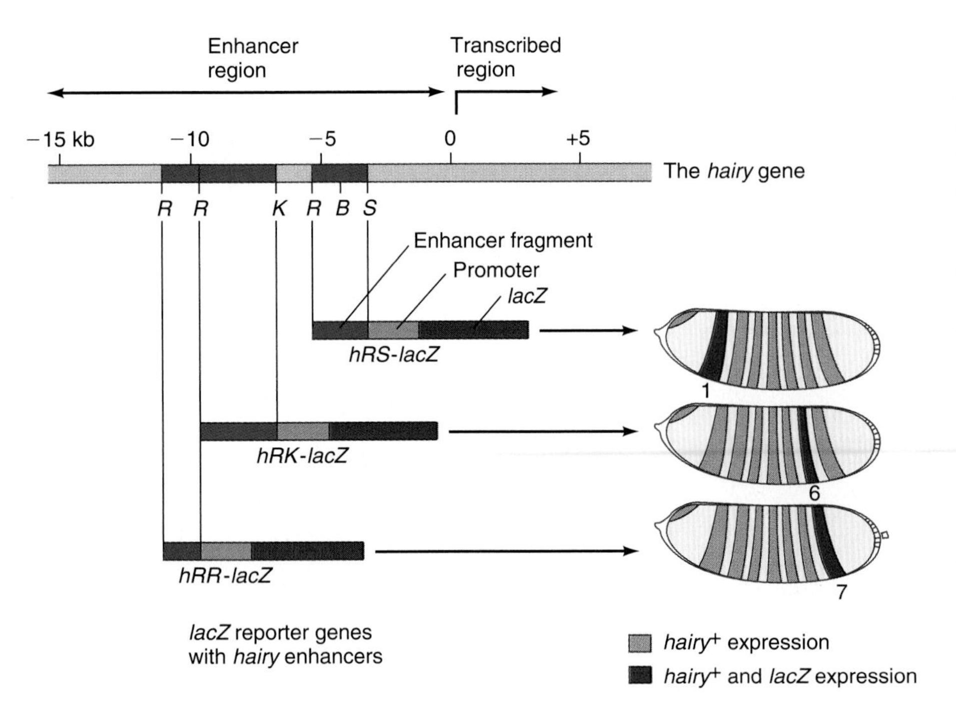

Figure 22.23 Expression patterns of *lacZ* fusion genes driven by specific enhancer fragments of the *hairy*+ gene. The top of the diagram shows the *hairy*+ gene with its transcribed region and a large regulatory region extending about 15 kilobases (kb) upstream. The letters represent restriction sites: B = *BamH* I; K = *Kpn* I; R = *EcoR* I; S = *Sal* I. Restriction fragments from the *hairy*+ regulatory region were spliced in front of a promoter and the *lacZ* reporter gene. These fusion genes are named according to their *hairy*+ restriction fragment. For instance, fusion gene *hRR-lacZ* contains the hairy fragment between the two *EcoR* I sites located about 10 kb upstream. The embryos to the right show which stripes of the normal *hairy*+ pattern are expressed by each fusion gene. See also Fig. 22.24.

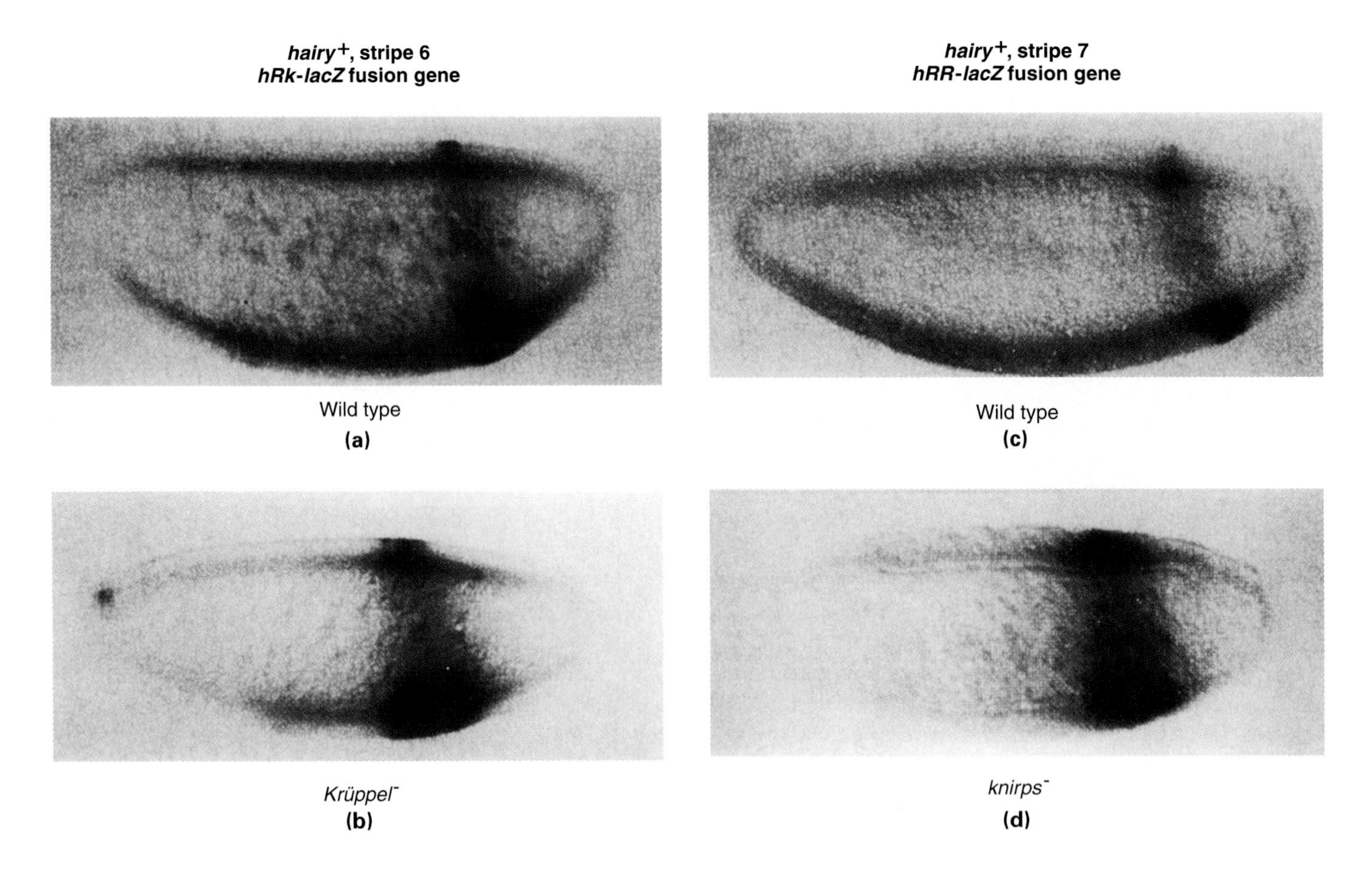

Figure 22.24 Control of individual *hairy*+ stripes by gap genes. **(a)** A stripe corresponding to stripe 6 of the normal *hairy*+ phenotype is expressed by fusion gene *hRK-lacZ* in a wild-type host embryo (see also Fig. 22.23). **(b)** In a *Krüppel*− mutant embryo transformed with *hRK-lacZ,* stripe 6 expands anteriorly. **(c)** A stripe corresponding to stripe 7 of the normal *hairy*+ phenotype is expressed by fusion gene *hRR-lacZ* in a wild-type host embryo. **(d)** In a *knirps*− mutant embryo transformed with *hRR-lacZ,* stripe 7 expands anteriorly.

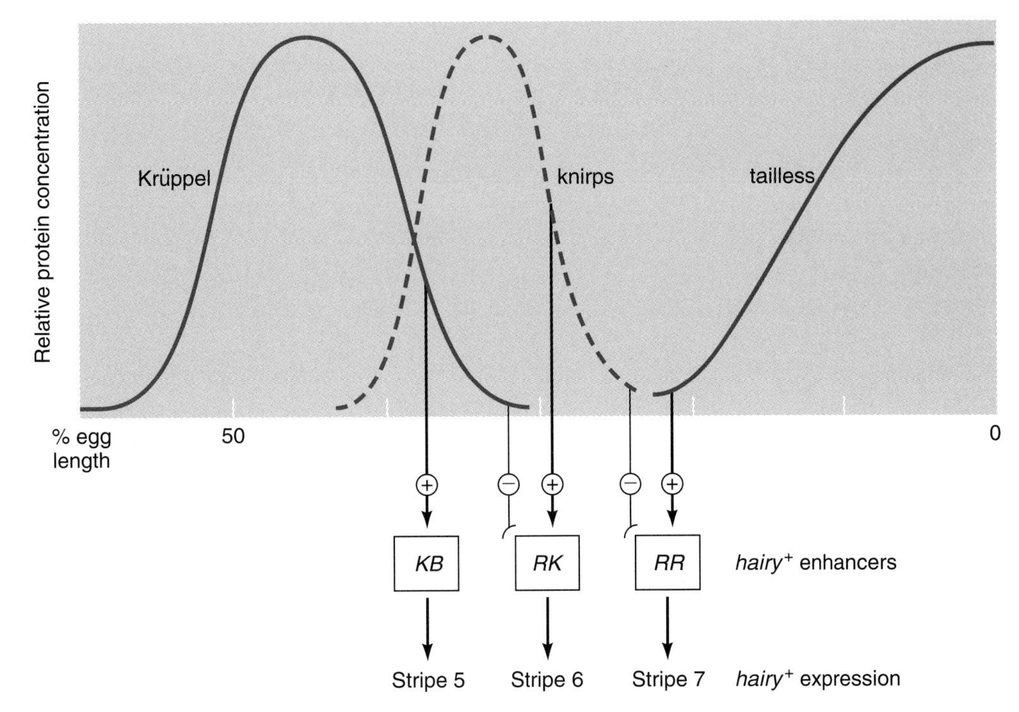

Figure 22.25 Model of gap gene control over the expression of a primary pair-rule gene. The gap gene proteins Krüppel, knirps, and tailless are expressed in broad, overlapping domains without periodicity. However, through unique combinations of positive and negative regulation, they control the expression of primary pair-rule genes in periodic stripes. The diagram summarizes regulatory effects on three enhancer regions of the *hairy*$^+$ gene, termed *KB, RK,* and *RR,* which control the expression of *hairy*$^+$ in stripes of 5, 6, and 7, respectively (see Fig. 22.24). 0% egg length is at the posterior pole.

cells within each even-numbered parasegment (see Fig. 22.21; C. Tsai and P. Gergen, 1995).

At the end of the blastoderm stage, the *Drosophila* embryo is encircled by the patterns of at least eight pair-rule genes, each expressed in seven or more stripes. These stripes are generally not in register, so each transverse row of blastoderm cells expresses a different combination of pair-rule genes (Fig. 22.26). The rudiment of each parasegment at the blastoderm stage is about four rows of cells wide. The pair-rule genes expressed in the anteriormost row of cells in a future odd-numbered parasegment include *hairy*$^+$, *even-skipped*$^+$, *paired*$^+$. The next row of cells to the posterior also expresses *hairy*$^+$ and *even-skipped*$^+$, but not *paired*$^+$.

Because of the *bisegmental* periodicity of the pair-rule patterns, corresponding rows in the rudiments of odd-numbered and even-numbered parasegments express different combinations of pair-rule genes. The anteriormost row of an even-numbered future parasegment expresses *fushi tarazu*$^+$ and *paired*$^+$, a combination that is quite different from the genes expressed in the anteriormost cells of a future odd-numbered parasegment, namely *paired*$^+$, *even-skipped*$^+$, and *hairy*$^+$.

The fireworks of pair-rule gene activity last only for a few hours. During that time, the genes produce different combinations of regulatory proteins that entrain the activity of genes that are expressed through the remaining life of the fly. These are the segment polarity genes, discussed below, and the homeotic genes, to be described thereafter.

SEGMENT POLARITY GENES DEFINE THE BOUNDARIES AND POLARITY OF PARASEGMENTS

Mutations in *segment polarity* genes cause deletions and duplications in corresponding portions of each segment. These deficiencies are often associated with a loss of anteroposterior polarity within the segment (Bejsovec and Wieschaus, 1993; Ingham and Martinez-Arias, 1992; Peifer and Bejsovec, 1992). This section will focus on three segment polarity genes: *engrailed*$^+$, *wingless*$^+$, and *hedgehog*$^+$.

At the blastoderm stage, each future parasegment consists of approximately four transverse rows of cells (Fig. 22.26). In each parasegment rudiment, the *engrailed*$^+$ gene is expressed in the anteriormost row and the *wingless*$^+$ gene in the posteriormost row. Researchers have analyzed the regulation of these genes by examining their expression patterns in embryos mutant in various pair-rule genes (Akam, 1987; Ingham et al., 1988; Mullen and DiNardo, 1995). For example, in *fushi tarazu*$^-$ mutants, *engrailed*$^+$ is not expressed in even-numbered parasegments, which normally express *fushi tarazu*$^+$. This result indicates that fushi tarazu is required, directly or indirectly, for *engrailed*$^+$ expression in even-numbered parasegments. Similarly, *paired*$^+$ is

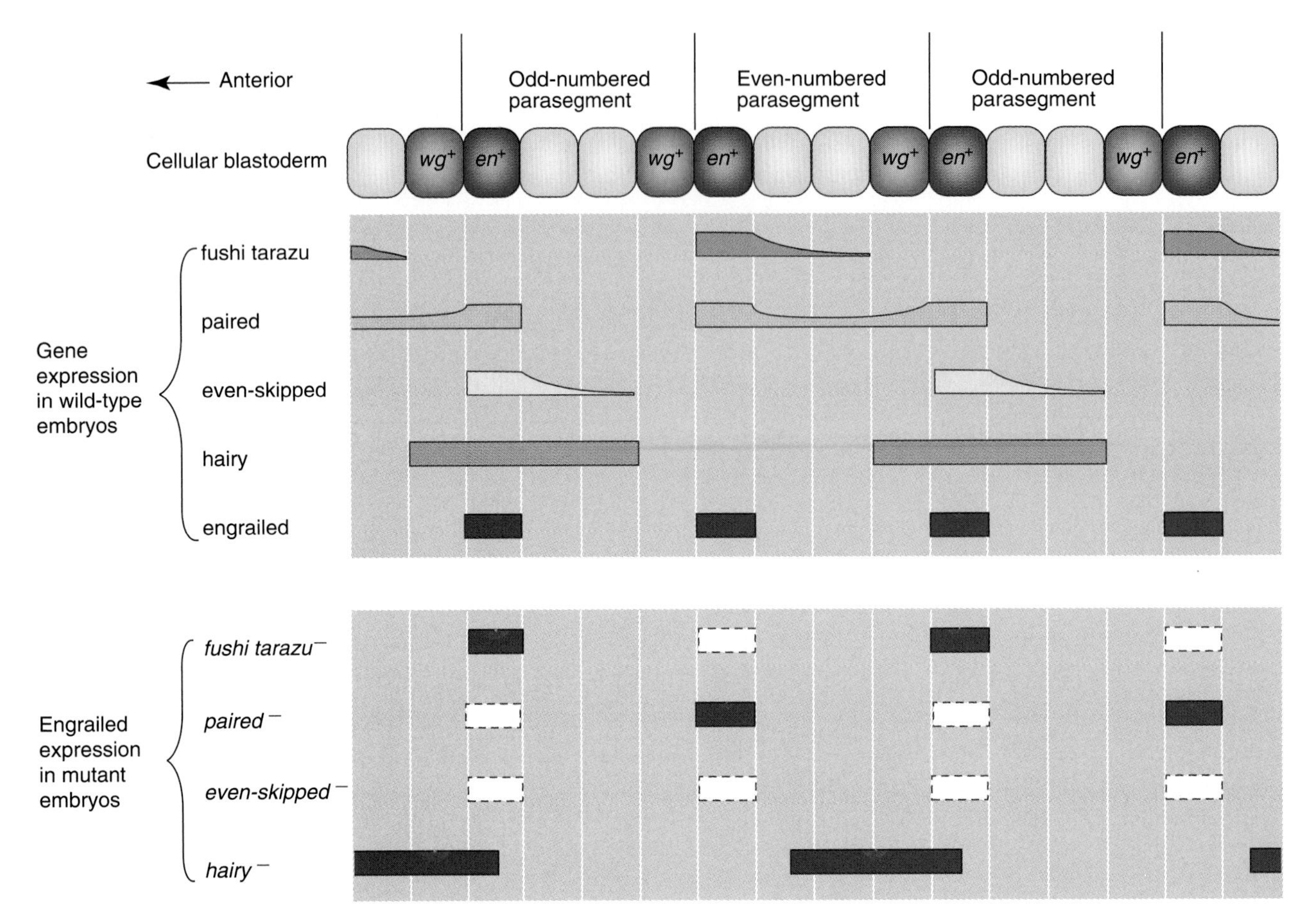

Figure 22.26 Expression domains of four pair-rule genes and one segment polarity gene at the cellular blastoderm stage. The colored rectangles at the top represent blastoderm cells. At this stage, the rudiment of each parasegment consists of about four transverse rows of cells. Each row expresses a unique combination of pair-rule genes, and these combinations differ between odd-numbered and even-numbered parasegment rudiments. Nevertheless, the expression patterns of segment polarity genes are identical in all parasegments. In particular, the anteriormost row expresses the *engrailed*$^+$ gene, and the posteriormost row expresses the *wingless*$^+$ gene. The changed expression domains of the *engrailed*$^+$ gene in embryos deficient for certain pair-rule genes indicates a complex regulation of *engrailed*$^+$ by pair-rule gene proteins (see text).

required for *engrailed*$^+$ expression in odd-numbered parasegments. The broadening of the stripes synthesizing engrailed protein in *hairy*$^-$ mutants indicates that hairy inhibits *engrailed*$^+$ expression, but the less-than perfect congruence between the broadened engrailed stripes and the expression domains of *hairy*$^+$ indicates that a combination of regulatory genes is involved. Also, the genes that control the expression of segment polarity genes change with time. The *engrailed*$^+$ gene, for instance, is originally controlled by pair-rule genes but later by *wingless*$^+$ and other genes, as we will see next.

While some segment polarity genes encode transcription factors, as do all gap genes and pair-rule genes, other segment polarity genes encode signal proteins and their receptors. The wingless product, for example, is a secreted protein acting on matching receptors in the plasma membranes of neighboring cells (Fig. 22.27). In particular, wingless protein released from the posteriormost cells in each parasegment acts on receptors on the anteriormost cells of the following parasegment. These receptors are encoded by the *frizzled*$^+$ and *frizzled-2*$^+$ genes and seem to be redundant (C.M. Chen and G. Struhl, 1999). Their activity keeps the *engrailed*$^+$ gene expressed after the blastoderm stage, when pair-rule genes are no longer active in segmentation. The engrailed protein is a transcription factor. One of its target genes is *hedgehog*$^+$, another segment polarity gene. It encodes a signal protein that acts on a receptor complex involving two proteins encoded by *patched*$^+$ and *smoothened*$^+$ (Y. Chen and G. Struhl, 1998). The activity of this complex, in turn, activates *wingless*$^+$, so that in effect *wingless*$^+$ and *engrailed*$^+$ stabilize each other's activity (P.A. Lawrence and G. Struhl, 1996). This mutual stabilization is critical for maintaining parasegmental boundaries. In loss-of-function alleles of either gene, the boundaries and polarity of parasegments are lost.

The molecular mechanisms by which *wingless*$^+$ and *engrailed*$^+$ mutually stabilize their expression in adjacent rows of cells reflect the loss of cytoplasmic continuity after completion of the cellular blastoderm in the

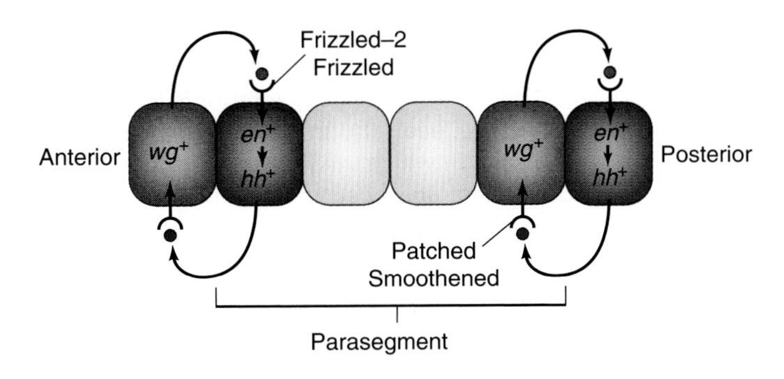

Figure 22.27 Interactions among segment polarity genes in the *Drosophila* embryo during the cellular blastoderm stage, gastrulation, and germ band extension. Within each parasegment, the anteriormost cells express *engrailed*$^+$ (*en*$^+$) and *hedgehog*$^+$ (*hh*$^+$), while the posteriormost cells express *wingless*$^+$ (*wg*$^+$). The engrailed protein, a transcription factor, enhances the expression of *hedgehog*$^+$ within the same cells. The hedgehog protein, a secreted protein, acts on a receptor complex involving proteins encoded by the *patched*$^+$ and *smoothened*$^+$ genes. The activity of this receptor complex, through other signaling proteins, maintains the expression of *wingless*$^+$ in anteriorly adjacent cells. (In posteriorly adjacent cells, the patched/smoothened complex and downstream signaling chain are incomplete and therefore inactive.) The *wingless*$^+$ gene, in turn, encodes itself a secreted signal protein, which acts on apparently redundant receptor proteins encoded by the *frizzled*$^+$ and *frizzled-2* genes. The activity of these receptors maintains the expression of *engrailed*$^+$ in neighboring cells in which other regulatory requirements for *engrailed*$^+$ expression are also present. In effect, the *engrailed*$^+$ and *wingless*$^+$ genes are mutually reinforcing their expression across parasegmental boundaries.

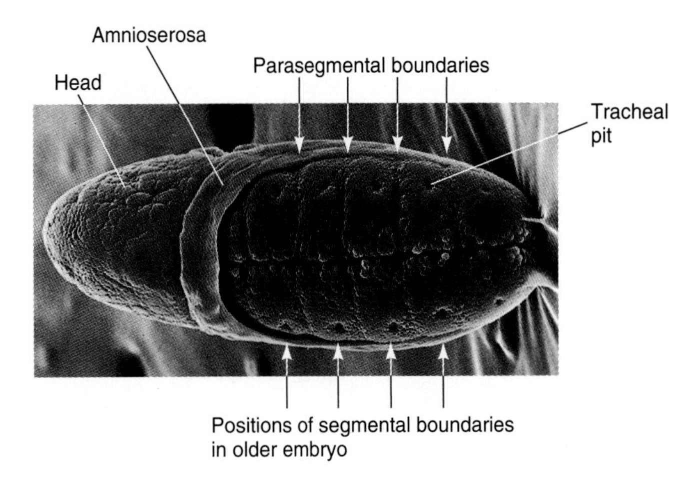

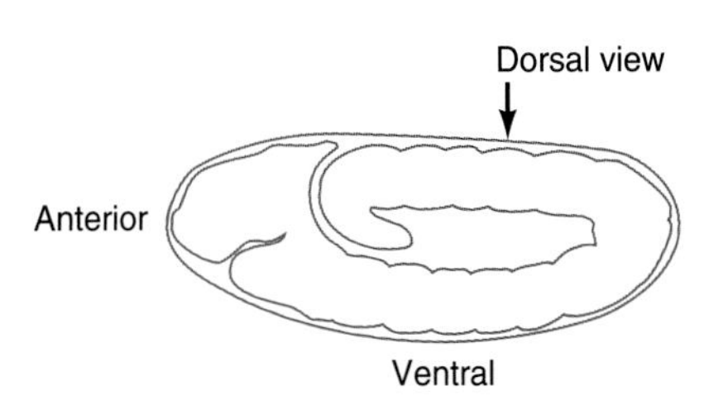

Figure 22.28 Parasegmental and segmental boundaries in the *Drosophila* embryo. The scanning electron micrograph shows a dorsal view of the elongated germ band; the diagram represents the same stage in lateral view. The abdomen is maximally extended and folded back on itself. The shallow grooves are parasegmental boundaries. In the middle of each parasegment there is a pair of tracheal pits, openings of the respiratory system. The parasegmental boundaries will soon disappear, and a new set of segmental boundaries will form where the tracheal pits are.

Drosophila embryo. During the preceding, syncytial blastoderm stage, transcription factors could freely diffuse and thus provide a very direct mechanism of control among nuclei and their surrounding jackets of cytoplasm. After cellularization, cells interact through signal peptides and their receptors.

MIDWAY THROUGH EMBRYOGENESIS, PARASEGMENTAL BOUNDARIES ARE REPLACED WITH SEGMENTAL BOUNDARIES

The first morphologically identifiable boundaries in the *Drosophila* embryo are shallow grooves in the epidermis that appear after gastrulation and germ band elongation (Fig. 22.28). These grooves separate *parasegments,* which are significant as modular units of gene expression in the early embryo. For example, the initial expression domains of *fushi tarazu*$^+$ and *even-skipped*$^+$ are parasegmental, and the stripes of *engrailed*$^+$ and *wingless*$^+$ expression mark the anterior and posterior margins of parasegments (see Figs. 22.3 and 22.26). Also, many of the homeotic genes to be discussed in the following section are initially expressed in parasegments.

A few hours after germ band elongation, the parasegmental boundaries are smoothed out and segmental boundaries are formed. The change in the register of segmentation can be observed conveniently in the position of the openings of the respiratory system, known as the *tracheal pits.* First located in the middle of each parasegment, the tracheal pits are later found in the grooves of the segmental boundaries. Thus, parasegmental and segmental boundaries have the same length but are out of register by about half a segment's length.

The segmental subdivision of the *Drosophila* larva carries over to the segmental subdivision of the adult (Fig. 22.29). The *imaginal discs,* which give rise to adult epidermis, originate after gastrulation when ectodermal cells segregate into those that will form larval epidermis and those that will form imaginal discs. Clonal analysis suggests that the earliest imaginal disc cells are already restricted in their fates to individual segments but not with regard to dorsal versus ventral organs, such as wings and legs (see Fig. 6.5). Although the imaginal discs form in register with segments, the older parasegmental

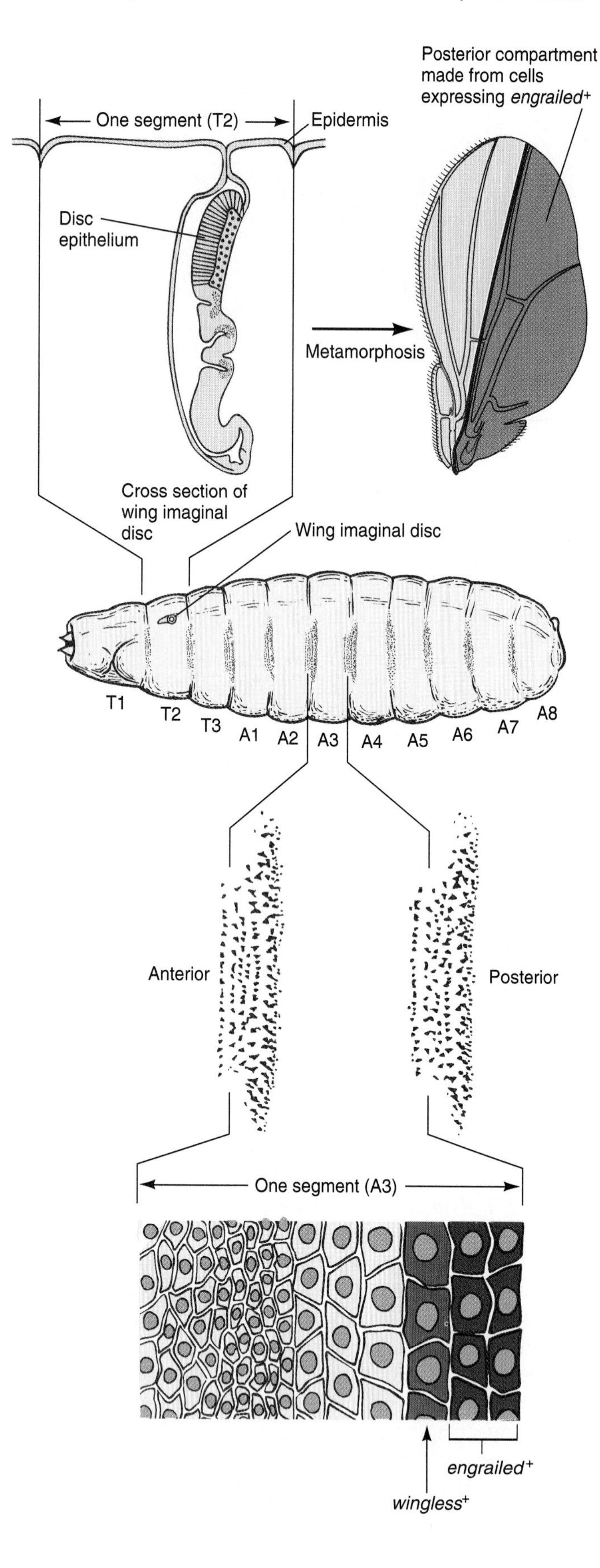

Figure 22.29 Development of epidermal structures and *engrailed*+ gene expression in *Drosophila.* At the center is a view of the ventral surface of the larva, anterior is to the left. On the anterior portion of each abdominal segment are several rows of denticles with which the larva crawls. The cells expressing *wingless*+ and *engrailed*+, which defined the parasegmental boundaries in the early embryo (see Figs. 22.26 and 2.27), are now within the posterior part of the segment; the posterior margin of the *engrailed*+ domain marks the posterior segment boundary. Attached to the larval epidermis are imaginal discs, epithelial bags that are everted during metamorphosis to form the adult epidermis. The adult parts are formed by the disc epithelium proper. Only the disc forming the left wing and associated part of the thorax is shown. (For other discs, see Figure 6.18.) The imaginal disc cells that express *engrailed*+ form the entire posterior compartment of the wing blade.

boundaries can still be traced later as clonal restriction lines and boundaries of gene expression domains. For instance, the imaginal disc cells that express *engrailed*+ form the posterior compartment of a wing or leg, as will be discussed later in this chapter.

In summary, the segmentation genes of *Drosophila* are arranged in a control hierarchy that proceeds from a set of aperiodic and simple expression domains to a periodic and detailed segmental pattern. In response to localized maternal gene products, the gap genes are activated in broad, overlapping domains. Various combinations of gap proteins generate individual stripes of primary pair-rule gene expression. Together, these stripes form regular patterns with bisegmental periodicity. The primary pair-rule genes control the secondary pair-rule genes, and together they regulate the segment polarity genes; these are expressed in narrow segmental stripes that initially delineate the anterior and posterior margins of parasegments. Midway through embryogenesis, the parasegmental boundaries are replaced with segmental boundaries that persist in the larva and adult. The segment polarity genes remain active in segments and imaginal discs, where they provide the basis for intrasegmental pattern formation (see Section 22.8).

22.5 Homeotic Genes

Unlike the segments of a centipede, which for the most part are alike, each segment in an insect differs from its neighbors in size, appendages, pigmentation, and other features. Such segment-specific traits are specified by homeotic genes. Their principal function is to specify regional differences in the anteroposterior body pattern (*P. A. Lawrence and Morata, 1994*).

Homeotic mutants of *Drosophila* and other insects have long been known for their dramatic phenotypes. The hallmark of these mutants is the transformation of one body part into another that is normally formed elsewhere. Familiar examples include *Ultrabithorax,* in

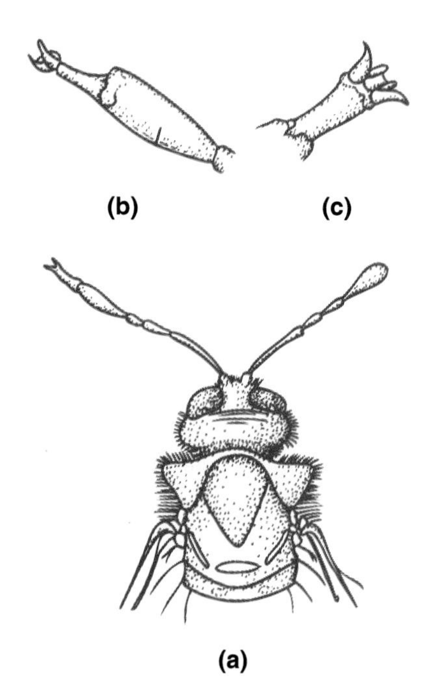

Figure 22.30 Homeotic transformation in a sawfly, *Cimbex axillaris.* **(a)** The anterior part of the fly is shown from a dorsal perspective. The distal portion of the left antenna developed as a foot. The right antenna is normal, ending in a club-shaped terminal joint. **(b)** The foot at the end of the left antenna seen from above. **(c)** The end of the left antenna seen from the front. The ectopic foot is smaller than a normal foot but is perfectly formed with claws and all other morphological characteristics.

which the metathorax is transformed into an extra mesothorax, thus producing a four-winged fly, and *Antennapedia,* in which antennae are replaced with legs (see Figs. 1.21 and 1.22). Such phenotypes are caused by loss-of-function alleles, as in the case of *Ultrabithorax,* as well as by gain-of-function alleles, as in the case of *Antennapedia.* Thus, either loss or gain of homeotic gene activity may trigger a switch to another developmental pathway, in accord with the *principle of default programs* already discussed in Sections 8.8, 16.6, and 17.2.

The first homeotic mutant was described by the entomologist G. Kraatz (1876), who found a sawfly of the species *Cimbex axillaris,* in which the distal part of the left antenna was replaced by a foot (Fig. 22.30). The English zoologist William Bateson (1894) discussed Kraatz' discovery and many related observations in a book on morphological discontinuities and their role in the origin of species. He termed replacements of antennal parts with leg parts and similar transformations "homoeotic." Biologists at his time were very interested in ***homologous organs***—that is, organs that have evolved from a common precursor. One of the criteria by which homologous organs are recognized is that they occupy corresponding positions in different animals. For instance, the balancer organs (halteres) on the metathorax of a fly are homologous to the second pair of wings in other insects such as dragonflies or bees. Organs are said to be ***serially homologous*** if they occupy corresponding positions in different segments of the same animal. Thus, fly balancers are serially homologous to fly wings, which occupy a corresponding position on the mesothorax. Since homeotic mutations cause transformations between homologous organs, changes in homeotic gene activities may have been key events in the evolution of insects and other arthropods (R. A. Raff and T. C. Kaufman, 1983).

HOMEOTIC GENES ARE EXPRESSED REGIONALLY AND SPECIFY SEGMENTAL CHARACTERISTICS

The segment-specific defects observed in homeotic mutants suggest that homeotic genes have regional patterns of expression. For instance, a null allele of *Antennapedia* transforms the second leg into an antenna and causes abnormalities in the other parts of the thorax, while leaving the head and abdomen unaffected (G. Struhl, 1981). It follows that the Antennapedia protein is, directly or indirectly, required for normal thorax development. As expected, the gene is strongly expressed in the thorax (see Fig. 16.21; Bermingham et al., 1990). Transcripts from both promoters are first detected during the blastoderm stage in cells that will form parasegments 4 through 6 (Fig. 22.31). At the extended germ band stage, transcripts are found in the epidermis and mesoderm of the thorax, mostly in parasegments 4 and 5, and in the central nervous system throughout the thorax and abdomen.

The *Antennapedia* alleles that gave the gene its name have the reverse effect of the null allele: They are gain-of-function alleles that transform antenna into leg (see Fig. 1.22). Such alleles are dominant over the wild type and are associated with chromosomal inversions. These inversions bring the transcribed region of *Antennapedia* under the control of promoters of other genes, causing inappropriate synthesis of Antennapedia protein in the head, where the gene is normally silent. This was demonstrated by Schneuwly et al. (1987), who spliced a heat shock promoter in front of an Antennapedia cDNA and introduced this construct into the *Drosophila* genome by P-element transformation (see Section 15.6). Heat activation of the transgene at certain larval stages caused the transformation of antennae into legs and the appearance of other thoracic structures in the head. This stunning result confirmed that *Antennapedia*$^+$ promotes leg development. Activation of this gene in the head region mimics a homeotic gene expression pattern that normally prevails in the thorax, and morphological development changes accordingly.

Expression of the *Ultrabithorax*$^+$ gene begins with a relatively simple pattern at the blastoderm stage in parasegments 6, 8, 10, and 12. Later, Ultrabithorax mRNAs are synthesized in intricate cellular mosaic

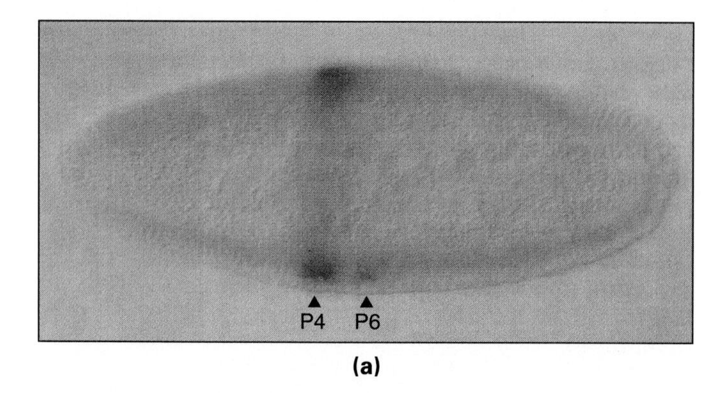

(a)

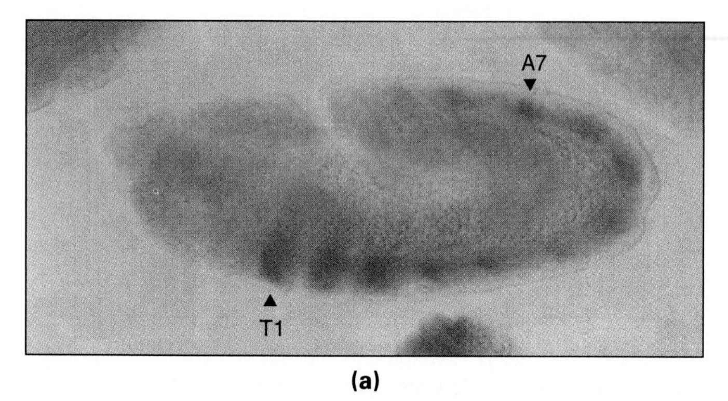

(a)

Figure 22.31 Transcription of the *Antennapedia*$^+$ gene in the *Drosophila* embryo shown by in situ hybridization (see Method 8.1). **(a)** Blastoderm stage. A suitable probe reveals transcription from promoter P_2 in the rudiments of parasegments 4 and 6. **(b)** Extended germ band stage: transcription from both promoters (P_1 and P_2) in the ectoderm and mesoderm of the thorax, and also in the nervous system of thorax and abdomen.

patterns in parasegments 5 through 12. These expression domains are controlled by several enhancer regions located upstream and downstream of the promoters, and even in introns of the transcribed region (Fig. 22.32a; Peifer et al., 1987; Irvine et al., 1991). Segments then differ in the particular mosaics of cells with certain combinations of active enhancers.

Mutations that interfere with particular enhancers of *Ultrabithorax* change the mosaic of gene expression in the metathorax and in abdominal segments, causing the development of morphological features normally found in the mesothorax. The most dramatic changes occur in the metathorax (Fig. 22.32b–e). For instance, mutant alleles inactivating two specific enhancers change the pattern of *Ultrabithorax* transcription in the anterior metathorax (aT3) so that it mimics the pattern of the anterior mesothorax (aT2). In flies surviving to adulthood, aT3 is replaced with an extra copy of aT2. Mutations in another *Ultrabithorax* enhancer change the expression pattern in the posterior metathorax (pT3), so that it resembles the pattern characteristic of the posterior mesothorax (pT2). Correspondingly, pT2 develops in place of pT3. A combination of these mutations generates the famous "four-winged fly" phenotype mentioned previously (see Fig. 1.21).

THE HOMEOTIC GENES OF *DROSOPHILA* ARE CLUSTERED IN TWO CHROMOSOMAL REGIONS

The homeotic genes of *Drosophila* are clustered in two regions of the third chromosome (Fig. 22.33a). One region, the ***Antennapedia* complex,** contains four homeotic genes: *labial*$^+$, *Deformed*$^+$, *Sex combs reduced*$^+$, and *Antennapedia*$^+$. These genes specify the morphological characteristics of anterior parasegments up to parasegment 5. The *Antennapedia* complex also contains nonhomeotic patterning genes including *bicoid*$^+$, *fushi tarazu*$^+$, and *zerknüllt*$^+$. The other chromosomal region, known as the ***bithorax* complex,** contains three homeotic genes: *Ultrabithorax*$^+$, *abdominal A*$^+$, and *Abdominal B*$^+$. These genes specify the characteristics of *posterior* parasegments, beginning with parasegment 6 and extending into the abdomen.

Strikingly, the physical order of these genes on the chromosome is the same as the anteroposterior order of their expression domains, especially with regard to the anterior boundary of each expression domain (Fig. 22.33b). The anterior region of each homeotic gene's expression domain usually gives rise to the body parts that are most strongly affected by loss-of-function alleles of the gene. The *labial*$^+$ gene, for example, is located at the 3′ end of the *Antennapedia* complex; the gene is expressed in the anterior head, and loss-of-function alleles affect the same area. In contrast, the *Antennapedia*$^+$ gene is located at the 5′ end of the *Antennapedia* complex, its expression domain begins in parasegment 4, and its loss-of-function alleles most strongly affect segments T1 and T2, to which parasegment 4 contributes. The *Abdominal B*$^+$ gene is located farthest toward the 5′ end in the *bithorax* complex; the gene is expressed in parasegments 10 to 14, and loss-of-function alleles affect the same region. Thus, the more 3′ a gene is located on the chromosome, the farther anterior reaches its expression domain, and the more anterior are the body regions affected by loss-of-function alleles. This ***colinearity rule*** was first noted by Lewis (1963, 1978).

The significance of the colinearity rule is still uncertain. At least in *Drosophila*, the homeotic genes do not need to be clustered the way they are for normal function. The separation of the *Antennapedia* and *bithorax* complexes in *Drosophila melanogaster* already represents a split condition of what is a single cluster of homeotic genes in most other species (see Sections 22.9 and 23.2). The same cluster is split differently in a related species, *Drosophila virilis* (Von Allmen, 1996), and the *bithorax* complex of *Drosophila melanogaster* can be split again without apparent loss of function (G. Struhl, 1984). However, while the cluster of homeotic genes seems to be falling apart in *Drosophila*, the same cluster has been preserved with astounding tenacity for more than 500 million years in other

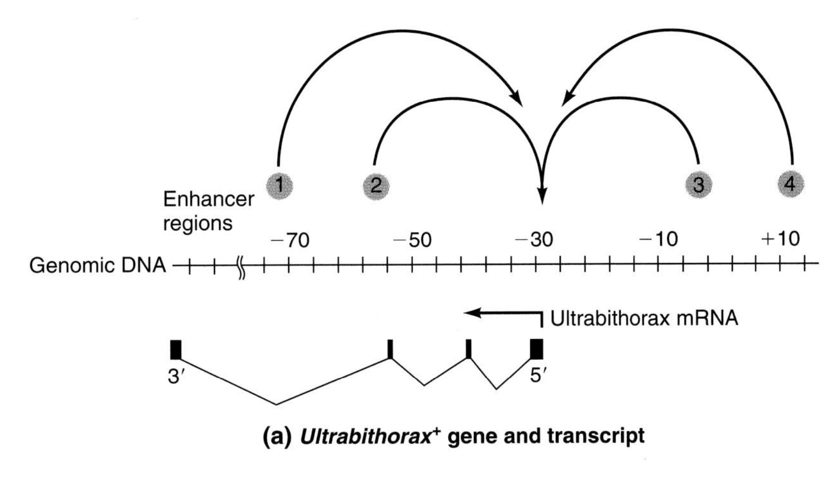

(a) ***Ultrabithorax*⁺ gene and transcript**

Adult phenotype

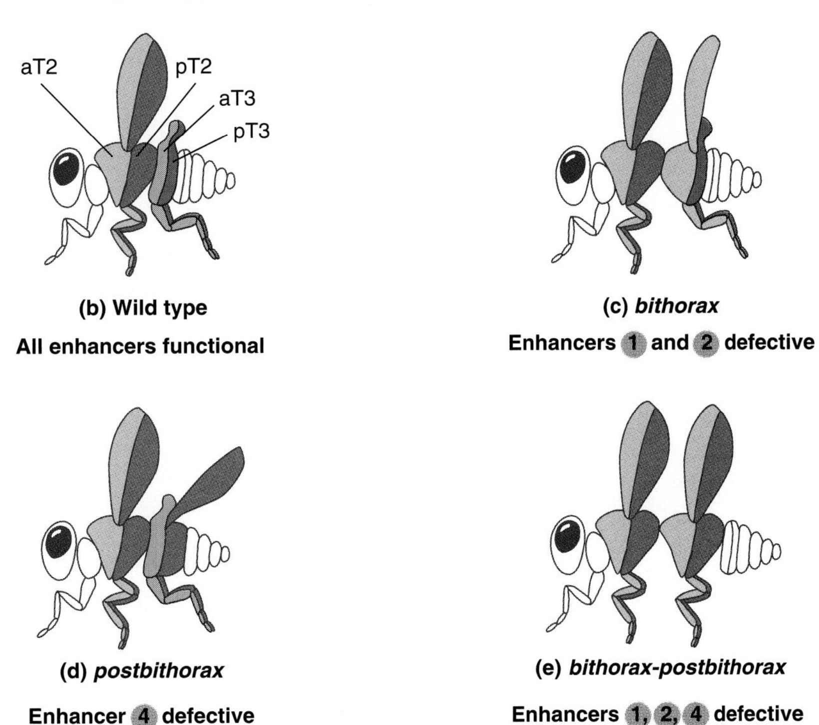

(b) Wild type
All enhancers functional

(c) *bithorax*
Enhancers 1 and 2 defective

(d) *postbithorax*
Enhancer 4 defective

(e) *bithorax-postbithorax*
Enhancers 1, 2, 4 defective

Figure 22.32 Locations of *Ultrabithorax*$^+$ enhancers and phenotypic effects of their inactivation. **(a)** Transcribed region of the gene and four enhancer regions (1–4). The numbers on the scale refer to kilobases (kb) of genomic DNA from an arbitrary reference point. The small blocks under the scale represent exons; the intervening brackets introns. The angular arrow indicates the beginning and direction of transcription; the curved arrows represent the actions of the enhancer regions. **(b–e)** The diagrams represent homeotic transformations in the adult. Thoracic segments T2 and T3 are shown as being composed of anterior and posterior compartments (aT2, pT2, etc.), in keeping with their origin from two parasegments. **(b)** Wild type. **(c)** Mutant alleles known as *bithorax* and *anterobithorax* show loss of function in enhancers 1 and 2. As a result, the aT3 compartment has the same lack of *Ultrabithorax*$^+$ expression and the same adult morphology as aT2. The loss of enhancer function also affects the abdominal segments, but the changes there are morphologically less dramatic. **(d)** Mutant alleles known as *postbithorax* have loss of function in enhancer 4. As a result, the pT3 compartment shows the same lack of *Ultrabithorax*$^+$ that normally characterizes pT2, with the same adult morphology resulting. **(e)** Triple mutants (*anterobithorax, bithorax,* and *postbithorax*) are lacking function in enhancers 1, 2, and 4. As a result, the entire segment T3 has the same lack of *Ultrabithorax*$^+$ expression, and the same morphology, as T2.

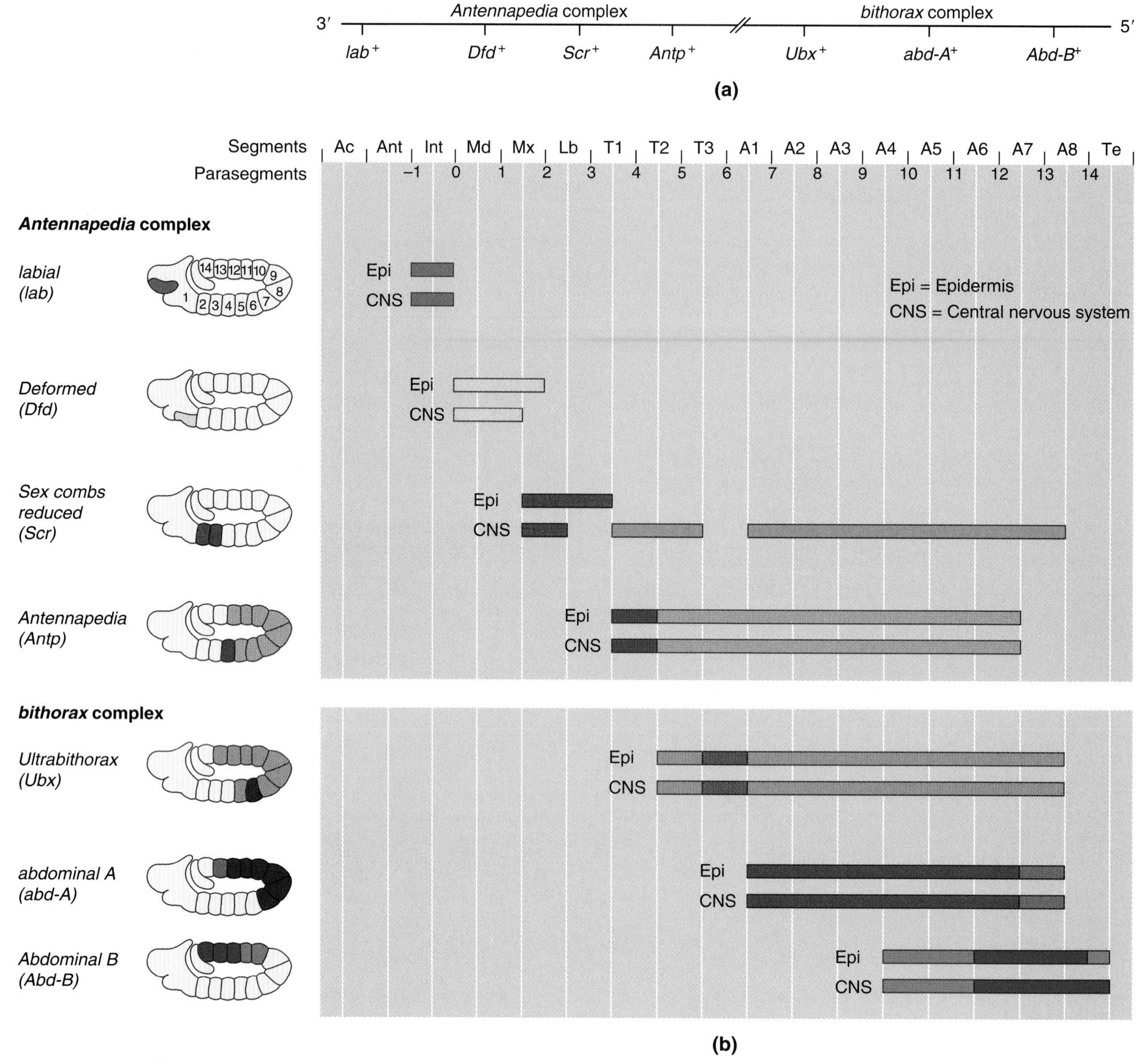

Figure 22.33 Chromosomal location and expression domains of homeotic genes in *Drosophila.* **(a)** The *Antennapedia* complex contains the *labial*$^+$, *Deformed*$^+$, *Sex combs reduced*$^+$, and *Antennapedia*$^+$ genes. The bithorax complex contains the *Ultrabithorax*$^+$, *abdominal A*$^+$, and *Abdominal B*$^+$ genes. **(b)** Domains of homeotic gene expression (mRNA and protein synthesis) in the embryonic epidermis and central nervous system. The darker color indicates the areas of strongest gene expression. Segments are labeled as in Figure 22.2, except that the procephalon is shown as an antennal (Ant) and an intercalary (Int) segment. In the diagrams to the left, epidermal expression is mapped relative to parasegmental boundaries. Note that the physical order of the genes from 3′ to 5′ on the chromosome correlates with the sequence of the anterior boundaries of the genes' expression domains in the embryo (colinearity rule).

animal phyla (see Section 23.2). In the latter, it appears that enhancer elements are shared between neighboring genes, so that the clustering has become a prerequisite of their normal function (*Mann, 1997*).

Because of the staggered expression patterns of homeotic genes, more homeotic genes tend to be active in posterior parasegments than in anterior parasegments (Fig. 22.33b). This rule allows two predictions about the phenotypes of homeotic mutants. First, a gain-of-function allele will often mimic a homeotic gene activity pattern that normally prevails more posteriorly and will therefore tend to transform structures into posterior serial homologues. The *Antennapedia* gain-of-function allele confirms this prediction. Second, a loss-of-function allele will generally mimic a homeotic gene activity pattern that normally prevails more anteriorly and will most likely transform structures into serial homologs that normally occur farther anterior. This is confirmed by the loss-of-function alleles of *Ultrabithorax*. Homeotic genes in vertebrates tend to follow the same rules (see Section 23.3).

HOMEOTIC GENES ARE PART OF A COMPLEX REGULATORY NETWORK

The molecular structure of homeotic genes suggests that they function in different contexts and that they are regulated by multiple signals. For instance, the *Antennapedia*$^+$ gene has two promoters and two polyadenylation sites (Kaufman et al., 1990). Although at least four different mRNAs are produced, they all encode the same protein. However, the two promoters and different 5′ UTR and 3′ UTR regions of the mRNAs confer different regulatory properties. Another striking feature of the *Antennapedia*$^+$ gene is the enormous size of some of its introns. At 57 kb, the largest intron is more than ten times longer than all exons together. It is estimated that the large introns add about an hour of lag time between the beginning and the end of the transcription of each pre-mRNA molecule, a feature that may be useful in regulation of the gene (Shermoen and O'Farrell, 1991). Similarly, the *Ultrabithorax*$^+$ gene is more than 300 kb long and produces several proteins by alternative splicing. Some of them are especially abundant during early embryogenesis, while others seem to function mostly in the organization of the central nervous system (Mann and Hogness, 1990).

Transcription of homeotic genes begins just before the cellular blastoderm stage. Initially, the homeotic genes are activated and repressed by gap genes (Akam, 1987; Ingham, 1988). For instance, the initial expression of *Antennapedia*$^+$ depends on both *hunchback*$^+$ and *Krüppel*$^+$ and is inhibited by *knirps*$^+$. The regulatory region of *Ultrabithorax*$^+$ contains target sequences for hunchback protein, while the regulatory region for *Abdominal-A*$^+$ has a binding site for Krüppel protein (Zhang et al., 1991; Shimell et al., 1994). The regulatory effects of gap genes are modulated by interactions of homeotic genes with pair-rule genes to generate parasegmental expression domains. In particular, expression of *fushi tarazu*$^+$ enhances the activation of *Antennapedia*$^+$ in parasegment 4 and the activation of *Ultrabithorax*$^+$ in parasegment 6.

The homeotic genes, like the segment polarity genes, remain active throughout development. Both types of genes are indispensable for maintaining the determined state of cells. When the segmentation genes that direct the initial expression of homeotic genes are no longer active, homeotic gene expression is maintained by two groups of genes that seem to modify the structure of chromatin so as to "lock in" their state of transcriptional activity. One group, known as the *Polycomb* genes, encodes proteins that inhibit homeotic gene expression, possibly by histone deacetylation and the resulting formation of *heterochromatin* (see Section 16.5; *Paro, 1993;* Lonie et al., 1994; *van Lohuizen, 1999*). At the same time, genes of the *trithorax* group maintain the activity of those homeotic genes that have already been turned on, apparently through the conservation of an active chromatin structure (*Pirrotta, 1998*).

Homeotic genes also maintain their expression domains by interactions among themselves. For instance, the Ultrabithorax proteins stimulate transcription of their own gene and inhibit transcription of *Antennapedia*$^+$ (see Figs. 16.21 and 16.22). Similarly, abdominal A protein inhibits the expression of *Ultrabithorax*$^+$. The *Deformed*$^+$ gene remains active through a positive feedback loop in which its own protein binds to its enhancer (Regulski et al., 1991).

In addition to being the subjects of complex genetic controls, homeotic genes also exert wide-ranging control functions themselves. Accordingly, all homeotic genes encode transcription factors. Some of their target genes encode again transcription factors, as discussed above, while other target genes encode proteins for intercellular signaling. For example, *Ultrabithorax*$^+$ and *abdominal A*$^+$ regulate the *wingless*$^+$ and *decapentaplegic*$^+$ genes, which encode secreted proteins with numerous functions in *Drosophila* development (see Sections 22.6 and 22.8).

Moreover, homeotic genes regulate genes for plasma membrane proteins that control cell adhesion. Using an antibody against Ultrabithorax protein to identify target genes it regulates, Gould and White (1992) isolated the *connectin*$^+$ gene. The connectin protein can mediate cell-cell adhesion, suggesting a direct link between homeotic gene function and cell-cell recognition. This link is of particular interest because it supports the compartment hypothesis discussed in Section 22.7.

In summary, homeotic genes are defined by the dramatic phenotypes of their mutant alleles, which cause substitutions of body parts by others that are normally found in more anterior or posterior positions. Homeotic genes encode transcription factors expressed in staggered domains and in intricate patterns within these domains. The combination of homeotic genes active in a segment controls its morphological identity. In *Drosophila*, the homeotic genes are clustered in two chromosomal regions, and the more 3′ a gene is located in a cluster, the more anterior begins its expression domain in the embryo. Clusters of homeotic genes have been conserved exceedingly well in evolution (see Section 23.2).

22.6 The Dorsoventral Body Pattern

Like the anteroposterior pattern, the dorsoventral pattern in *Drosophila* develops through the activity of maternal effect genes that establish a morphogen gradient (*Ray and Schüpbach, 1996*). The morphogen is a transcription factor encoded by the *dorsal*$^+$ gene. Embryos derived from mothers with null alleles of *dorsal* are lacking ventral and lateral structures, including the

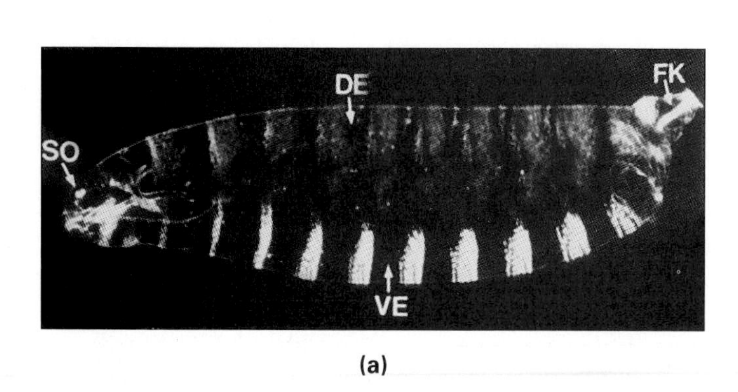

(a)

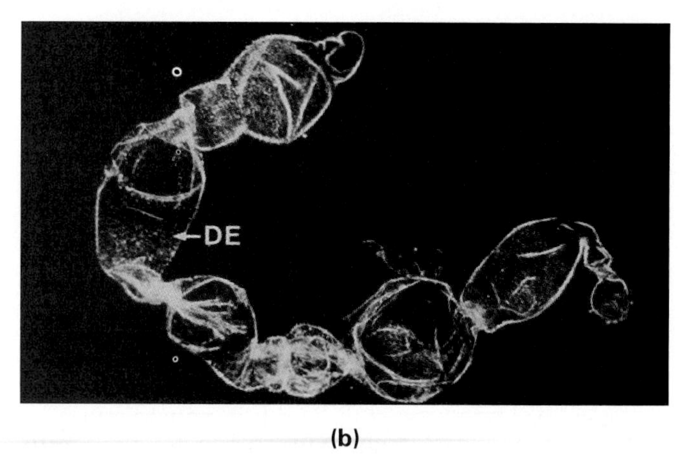

(b)

Figure 22.34 The dorsal phenotype. The photomicrographs show cuticle preparations under dark-field illumination. **(a)** Wild-type larva. Anterior end, marked by sense organ (SO), to the left; posterior end marked by Filzkörper (FK). Ventral side down. **(b)** Larva derived from a female homozygous for a dorsal null allele. The cuticle is a tube without dorsoventral polarity, having none of the ventral denticle belts that are prominent on the ventral epidermis (VE) of the wild type. The fine hair normally limited to the dorsal epidermis (DE) is seen around the circumference of the mutant larva.

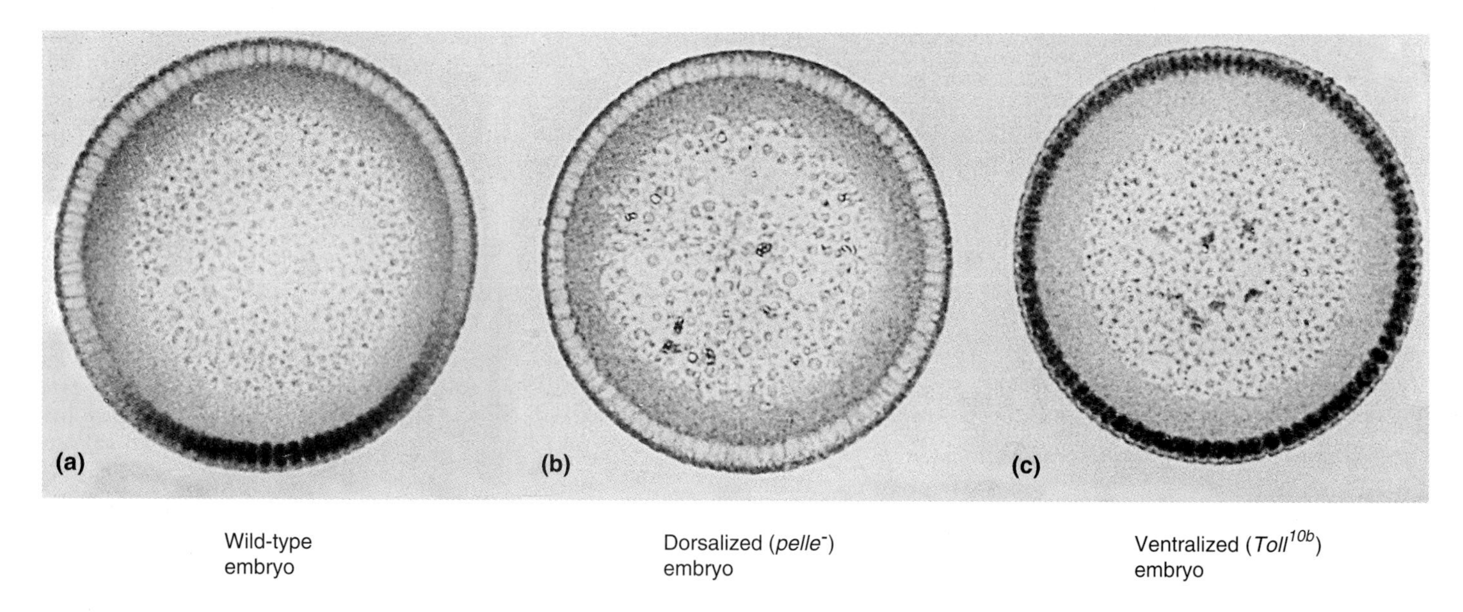

Wild-type embryo

Dorsalized (*pelle*$^-$) embryo

Ventralized (*Toll*^{10b}) embryo

Figure 22.35 Distribution of the dorsal protein in normal and mutant *Drosophila* embryos at the blastoderm stage. The photomicrographs show cross sections of embryos oriented with the ventral side down. The dorsal protein is detected by immunostaining (see Method 4.1). **(a)** Wild-type embryo. There is a gradient in the nuclear concentration of dorsal protein with a maximum around the ventral midline. **(b)** Dorsalized embryo. One of the genes (*pelle*) required for the deployment of dorsal protein is deficient, so that the dorsal state of the wild type is expanded around the embryo. **(c)** Ventralized embryo. The *Toll* gene, required for the proper formation of the dorsal protein gradient, is active independently of its normal spatial clues, so that dorsal protein accumulates around the entire circumference of the embryo.

mesoderm and central nervous system. Instead, the dorsalmost epidermal structure, a fleece of fine hair, develops around the entire circumference of these embryos (Fig. 22.34). As in the case of the anteroposterior pattern, the ventral and lateral pattern elements are missing not because their precursor cells died, but because cells are determined abnormally at the blastoderm stage. It follows that the normal function of the *dorsal*$^+$ gene is required to specify the ventral and lateral elements of the body pattern. (As explained in Section 1.4, the name given to a gene usually epitomizes a loss-of-function phenotype. Of course, this name is then contrary to the wild-type function of the gene.)

The dorsal protein forms a morphogen gradient with highest concentration in the ventral nuclei of the blastoderm embryo (Fig. 22.35a). This gradient fails to form properly in *dorsalized* embryos, in which no dorsal protein is synthesized, and in *ventralized* embryos, where synthesis or deployment is misregulated so that dorsal protein accumulates at high concentration in nuclei around the entire embryo (Fig. 22.35b, c). As a transcription factor, the dorsal protein activates or inhibits certain

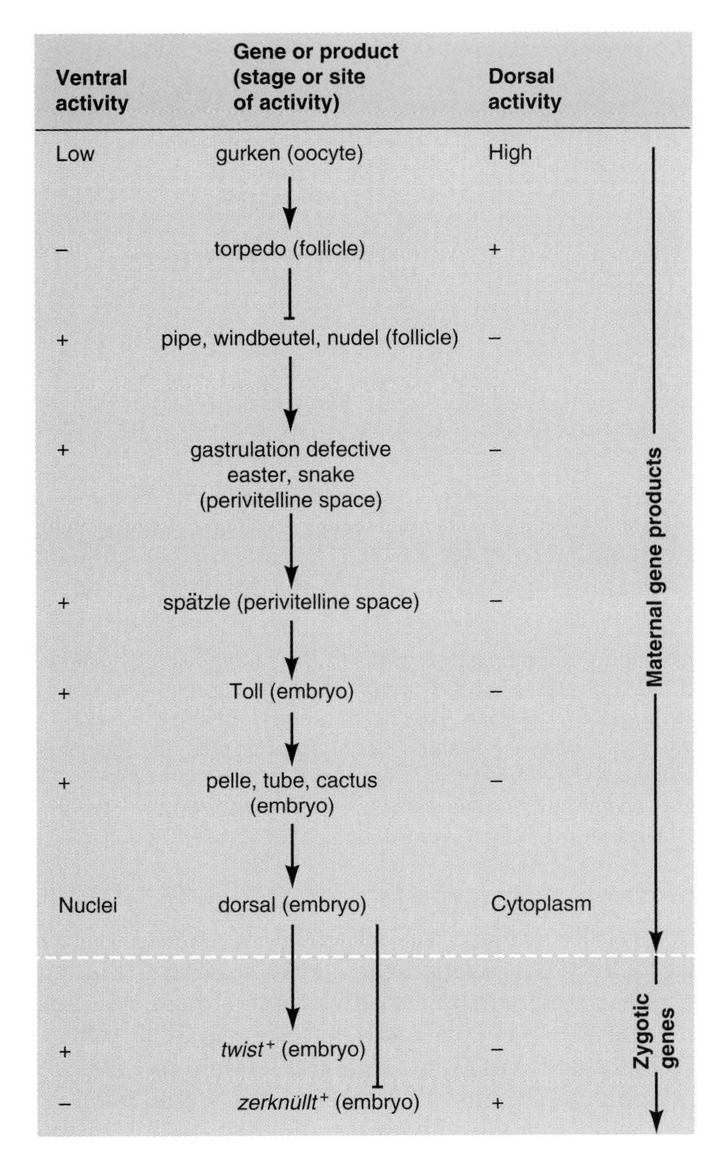

Figure 22.36 Cascade of maternal gene products and embryonic genes that establish dorsoventral polarity in the *Drosophila* embryo. The central column lists some of the key gene products and genes and indicates where they are expressed. The left and right columns indicate the level of activity of each gene product in the ventral and dorsal areas.

embryonic target genes in dorsoventral order (*Govind and Steward, 1991; Nüsslein-Volhard, 1991; K. V. Anderson, 1998*). The earliest expression domains of these genes at the blastoderm stage correlate approximately with certain cell fates during later development. For instance, the blastoderm cells on the ventral side, which acquire the highest concentration of dorsal protein in their nuclei, activate the *twist*$^+$ gene and proceed to form mesoderm.

The maternal gene products that eventually establish the gradient of dorsal protein interact in a complicated signal chain involving 14 proteins and weaving from the oocyte to the follicle cells and back to the embryo (Fig. 22.36). The first sign of dorsoventral polarity is the eccentric position of the oocyte nucleus near the dorsal surface of the oocyte (Fig. 22.37). Here, the oocyte nucleus sends a signal, and the follicle cells on the dorsal surface of the oocyte receive the signal most strongly because of the proximity of the nucleus (Roth et al., 1999). The signal depends on the *gurken*$^+$ and *torpedo*$^+$ genes, and is the same signal that earlier during oogenesis established the posterior identity of the egg chamber (see Fig. 22.8). Eggs laid by females homozygous for loss-of-function alleles of *gurken* or *torpedo* are also ventralized and have reduced dorsal structures (Schüpbach, 1987; Neuman-Silberberg and Schüpbach, 1994).

The activation of the torpedo receptor by its ligand has two major effects on the dorsal follicle cells. First, the signal from the receptor is processed further to pattern the dorsal structures of the chorion, including the respiratory appendages. Second, the activity of the torpedo protein inhibits another gene, *pipe*$^+$. This is indicated by the observation that *pipe*$^+$ is expressed uniformly around the ovarian follicle of females lacking the *gurken* gene function, whereas the transcription of *pipe*$^+$ is restricted to a ventral strip of ovarian follicle cells in wild-type females (Sen et al., 1998). Females deficient in *pipe* produce dorsalized offspring. The same phenotype is observed in offspring from females with loss-of-function alleles in *nudel* or *windbeutel*, two genes that are expressed in the entire ovarian follicle but are required for wild-type function only ventrally (Nilson and Schüpbach, 1998). The combined product of *pipe*$^+$, *nudel*$^+$, and *windbeutel*$^+$, known as the ***ventral polarizing activity***, is therefore limited to ventral follicle cells and the ventral aspect of the *vitelline envelope*, which is produced by the follicle cells and is the innermost layer of the protective covers surrounding the insect egg (Figs. 22.37 and 22.38; Stein et al., 1991; Stein and Nüsslein-Volhard, 1992).

To test whether *pipe*$^+$ can generate ventral polarizing activity ectopically, Sen and coworkers (1998) used females that (1) carried a transgene consisting of *pipe* cDNA spliced behind the yeast UAS promoter, (2) were otherwise homozygous mutant for *pipe*$^-$, and (3) expressed the yeast GAL-4 transcriptional activator in various follicular cell lineages (see Section 15.6). Progeny from such females formed their ventral sides next to the ovarian follicle cells in which the *pipe* transgene had been expressed. Thus, it is the *pipe*$^+$ function that normally orients the ventral polarizing activity in the ovarian egg chamber.

The molecular analysis of the ventral polarizing activity has begun with the cloning and sequencing of the three genes known to be involved in generating it. The *pipe*$^+$ gene encodes a glycosaminoglycan-modifying enzyme similar to heparan sulfate 2-O-sulfotransferase (Sen et al., 1998). The pipe protein is thought to be active in the Golgi apparatus of ventral follicle cells. *windbeutel*$^+$ encodes a protein that appears to reside in the endoplasmatic reticulum and is therefore an unlikely target of the pipe enzyme (Konsolaki and Schüpbach, 1998). *nudel*$^+$ encodes a secreted protein with protease activity and three sites for glycosaminoglycan addition, which

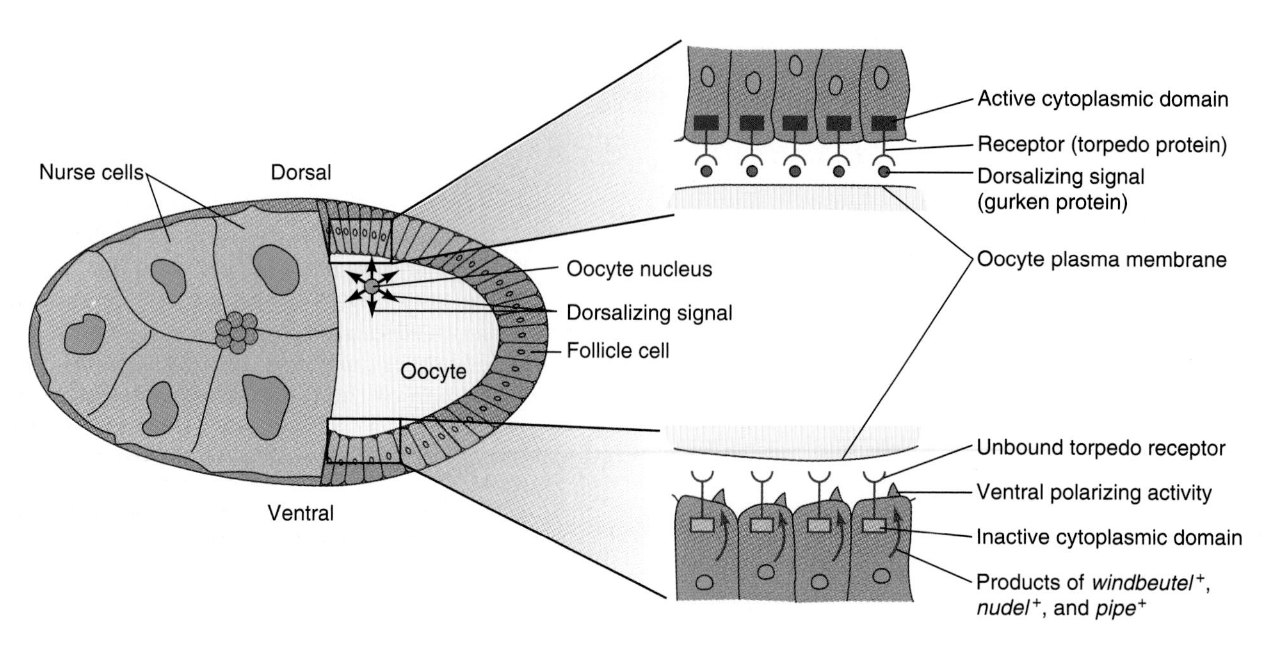

Figure 22.37 Origin of the dorsoventral polarity from the eccentric position of the oocyte nucleus in *Drosophila.* The diagram to the left shows a median section of an egg chamber; anterior to the left, dorsal side up. The *gurken*+ gene expressed in the oocyte nucleus generates a dorsalizing signal that is stronger toward the dorsal surface. The insets to the right show the dorsalizing signal as a ligand released dorsally, but not ventrally, into the space between the oocyte and follicle cells. The ligand binds to a receptor protein encoded by the *torpedo*+ gene in the follicle cells. In the ventral follicle cells, where the torpedo protein remains inactive, three genes, *windbeutel*+, *nudel*+, and *pipe*+, cooperate in generating a ventral polarizing activity. In the dorsal follicle cells, the activity of the torpedo protein prevents the generation of the ventral polarizing activity.

would seem to be likely targets of pipe (LeMosy et al., 1998). However, *genetic mosaic analysis* indicates that the nudel product is not required ventrally (Nilson and Schüpbach, 1998). Thus, the available evidence suggests that the pipe enzyme acts on an as yet unidentified component of the pathway that generates dorsoventral polarity. The modulation of glycosaminoglycans is a new and largely unexpected mechanism in embryonic axis formation.

During late stages of oogenesis, when the oocyte is surrounded by the vitelline envelope and the chorion, the ventral polarizing activity seems to become part of the vitelline envelope, presumably facing the perivitelline space (Fig. 22.38a, b). During embryonic cleavage, the precursor proteins for three proteases (encoded by the *gastrulation defective*+, *snake*+, and *easter*+ genes) and for a signal peptide (encoded by the *spätzle*+ gene) are released into the *perivitelline space.* Under the influence of the ventral polarizing activity, the proteases become locally active and cleave an active signal peptide from the spätzle precursor (C. L. Smith and R. DeLotto, 1994; Schneider et al., 1994). The active spätzle signal peptide in turn binds as a ligand to a receptor tyrosine kinase encoded by the *Toll*+ gene (D. S. Schneider et al., 1991; Morisato and Anderson, 1994). Although the Toll protein is present around the entire egg plasma membrane, it is activated mostly in the ventral area. Apparently, only a limited supply of the spätzle signal peptide is present on the ventral side and is immediately bound up by Toll receptor. (This activation mechanism is similar to the local activation of the torso receptor in the egg plasma membrane by the active trunk protein, which is present only near the egg poles. See Fig. 22.13.) Thus, Toll receptor is activated in a gradient around the egg periphery with a maximum around the ventral midline.

The Toll activity determines the distribution of the maternally encoded dorsal protein (Figs. 22.35a, c and 22.38c). The dorsal protein is initially retained in the egg cytoplasm by its association with the protein encoded by *cactus*+, another maternal effect gene expressed in the early embryo (Kidd, 1992). The cactus protein binds to a domain of the dorsal protein that is necessary for transport into the nucleus (Whalen and Steward, 1993). The Toll activity in the ventral region of the embryo, transduced by the products of *pelle*+ and *tube*+, releases the dorsal protein from the cactus protein (Drier et al., 1999; Grosshans et al., 1999). The dorsal protein then moves into nuclei, directed by its own nuclear localization sequence (Govind and Steward, 1991). Thus, the uneven distribution of Toll activity finally creates a gradient in the nuclear concentration of dorsal protein, with a maximum in the ventral nuclei.

The nuclear dorsal protein serves as a morphogen gradient, establishing a coarse dorsoventral pattern in the expression of several embryonic genes (*Rusch and Levine, 1996*). In wild-type embryos, *zerknüllt*+ is initially expressed in the dorsal half of the embryo but later restricted to a dorsal band five to six cells wide (Fig.

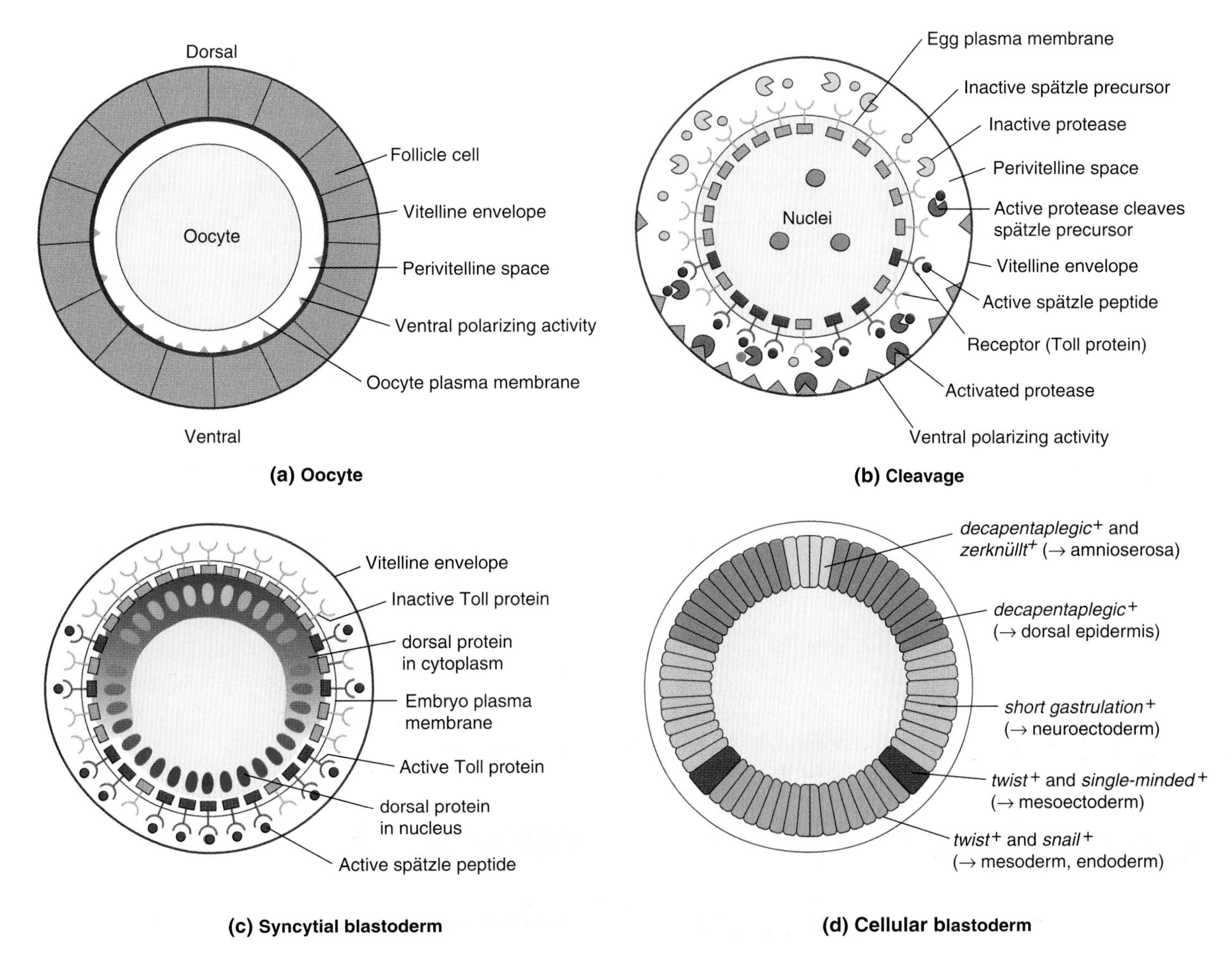

Figure 22.38 Development of the dorsoventral polarity in the *Drosophila* embryo. The schematic cross sections are oriented with the ventral side down. **(a)** Follicle cells synthesize the vitelline envelope surrounding the oocyte. The polarizing activity generated by the ventral follicle cells is present ventrally, but not dorsally, on the inside of the vitelline envelope (see Fig. 22.37). **(b)** After fertilization, precursor proteins for a cascade of proteases encoded by *gastrulation defective*$^+$, *snake*$^+$, and *easter*$^+$, and for a signal peptide encoded by *spätzle*$^+$, are all released into the perivitelline space between the vitelline envelope and the egg plasma membrane. The ventral polarizing activity presumably starts the protease cascade, which then cleaves an active spätzle peptide from its precursor. The spätzle peptide binds as a ligand to a receptor protein on the egg plasma membrane. The receptor protein, encoded by the *Toll*$^+$ gene, is deployed in excess around the entire plasma membrane of the embryo. However, because active spätzle ligand is produced mostly ventrally, Toll receptor is activated ventrally, to a lesser extent laterally, but not dorsally. **(c)** Dependent on the graded Toll activity, the maternally encoded dorsal protein is transported from the cytoplasm into the nuclei of the preblastoderm embryo. The Toll-dependent transport results in a gradient of nuclear dorsal protein with a maximum concentration ventrally. **(d)** Being a transcription factor and acting as a morphogen, the dorsal protein inhibits the embryonic genes *zerknüllt*$^+$ and *decapentaplegic*$^+$, limiting their expression to dorsal and dorsolateral cells. Laterally, dorsal protein activates the embryonic *short gastrulation*$^+$ gene. In ventral nuclei, the high dorsal protein concentration activates the embryonic *single-minded*$^+$, *twist*$^+$ and *snail*$^+$ genes. The expression domains of the embryonic genes coincide approximately with the major fates of these cells to become amnioserosa, dorsal epidermis, neuroectoderm, mesectoderm, and mesoderm or endoderm.

22.38d). These cells give rise to the amnioserosa, an extraembryonic membrane that temporarily closes the dorsal side of the embryo. Expression of *zerknüllt*$^+$ is inhibited by dorsal protein even at low levels, due to a cooperative action of dorsal with a non-DNA-binding corepressor (Dubnicoff et al., 1997). Similarly, the dorsal protein restricts expression of *decapentaplegic*$^+$ to the dorsal 40% of the embryo's circumference (Figs. 22.38d and 22.39a). The cells expressing *decapentaplegic*$^+$, but not *zerknüllt*$^+$, will form the dorsal epidermis.

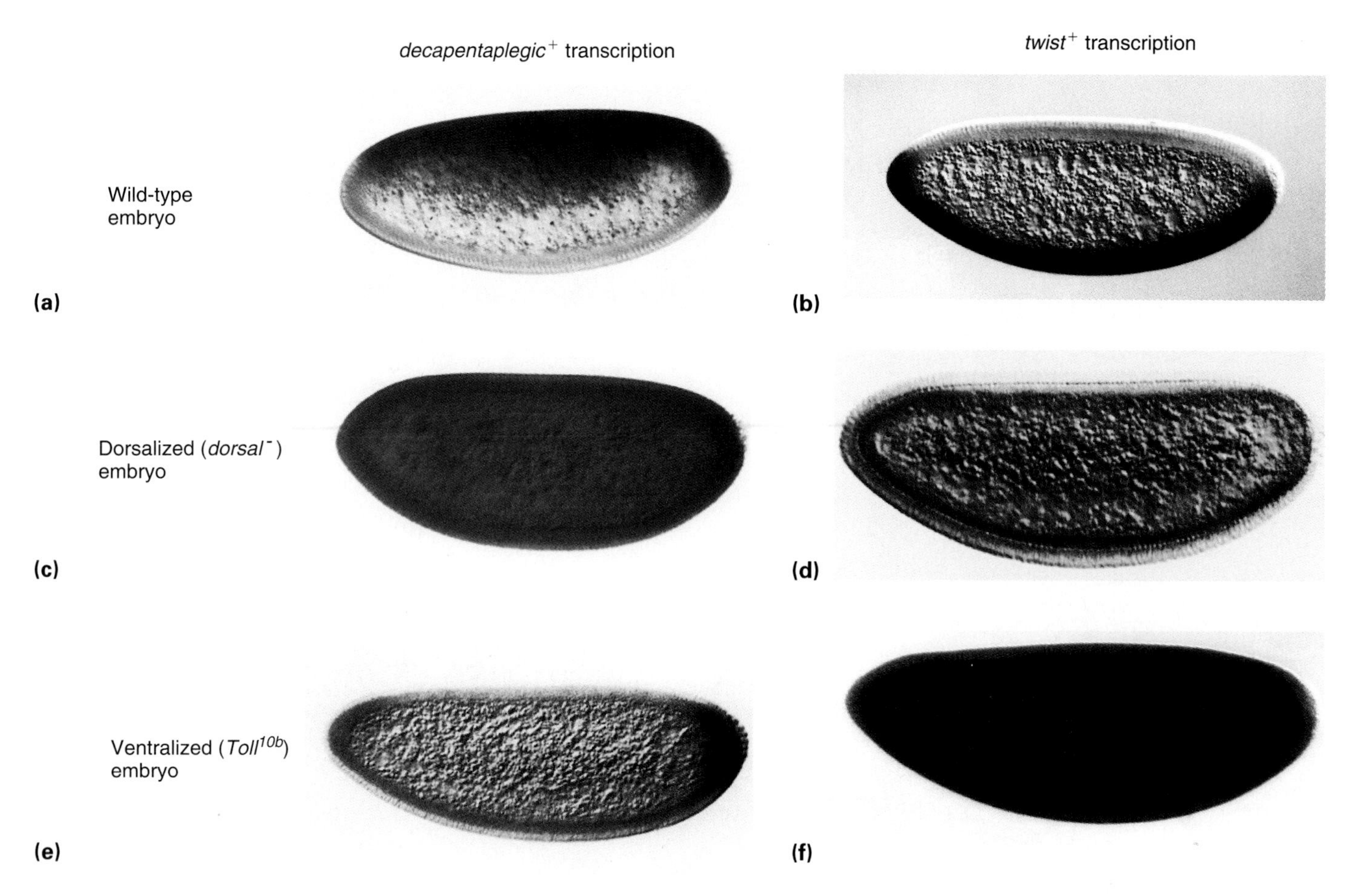

Figure 22.39 Expression patterns of the embryonic *decapentaplegic*$^+$ and *twist*$^+$ genes revealed by in situ hybridization (dark stain; see Method 8.1); dorsal is up and anterior pole to left. **(a, b)** In wild-type embryos, *decapentaplegic*$^+$ is transcribed in the dorsal 40% of the circumference and *twist*$^+$ in the ventral 20%. **(c, d)** In embryos dorsalized by a maternal *dorsal*$^-$ allele, *decapentaplegic*$^+$ is transcribed in all embryonic cells and *twist*$^+$ not at all. **(e, f)** In embryos ventralized by a maternal *Toll*^{10b} gain-of-function allele, *decapentaplegic*$^+$ is transcribed only in the polar regions and *twist*$^+$ over the entire embryo.

More laterally in the embryo, dorsal protein activates *short gastrulation*$^+$. This gene is expressed in two broad ventrolateral regions that will form the neuroectoderm. (*short gastrulation*$^+$ would also be expressed midventrally were it not for the inhibitory action of the *snail*$^+$ gene. See below.) At the high concentrations of dorsal protein reached on the ventral side of the embryo, the *twist*$^+$ gene is activated. Twist protein at high concentration cooperates with dorsal in activating the *snail*$^+$ gene in an expression domain that is a bit narrower than the *twist*$^+$ domain but has sharper boundaries. The ventral cells expressing *snail*$^+$ invaginate and form the mesoderm (Fig. 22.40). Low concentrations of twist and dorsal activate *single-minded*$^+$ in two ventrolateral stripes of cells known as mesectoderm, which form the neurons positioned along the ventral midline. (The endoderm, which gives rise to the midgut, is formed at the two ends of the embryo, which are not considered here.)

The orienting effect of Toll activity was first demonstrated by injecting cytoplasm from wild-type embryos into embryos derived from mutant females that could not synthesize Toll protein (K. V. Anderson and C. Nüsslein-Volhard, 1984; K. V. Anderson et al., 1985). In subsequent experiments, Roth (1993) delivered wild-type cytoplasm close to the dorsal, ventral, or lateral surface of the recipient at the syncytial blastoderm stage. Regardless of where the injection took place, the recipients formed a patch of cells around the injection site that behaved like ventral cells do in normal embryos. They accumulated dorsal protein in their nuclei with a graded concentration falling off toward the margin of the patch (Fig. 22.41). The same cells subsequently expressed the embryonic target gene *twist*$^+$, as ventral cells in normal embryos do (Figs. 22.39b and 22.42). The size of the dorsal-accumulating and *twist*$^+$-expressing cells increased with the amount of injected cytoplasm.

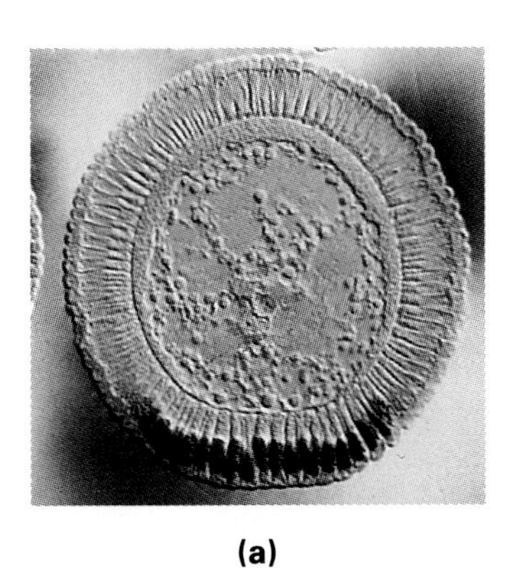
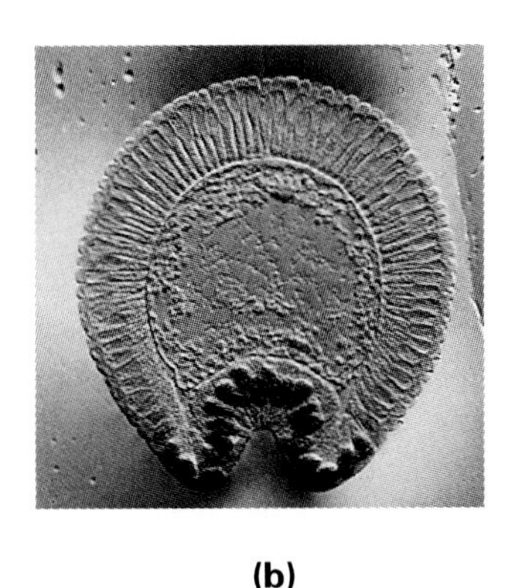
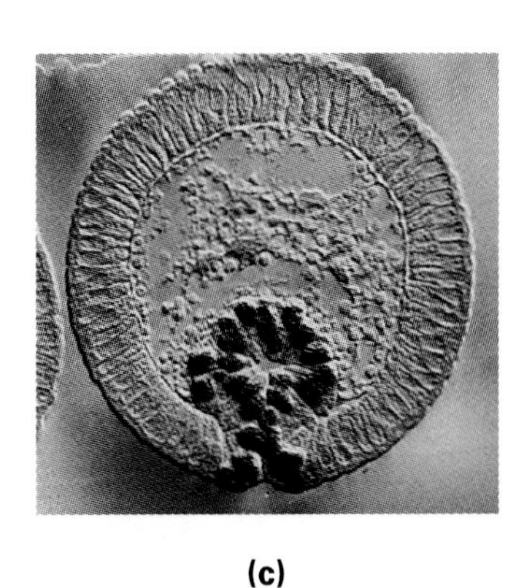

(a) (b) (c)

Figure 22.40 Expression of the *twist*$^+$ gene in the *Drosophila* embryo. The photomicrographs show transverse sections, ventral side down, immunostained with an antibody to twist protein (see Method 4.1). The twist protein, a transcription factor, accumulates in the nuclei of the producing cells. The nuclei of unstained cells appear as depressed ovals. **(a)** Beginning of gastrulation. The cells on the ventral side form a plate. Note the decrease of twist protein concentration toward the sides. **(b)** Invagination of the ventral plate forms the embryonic mesoderm. **(c)** Mesoderm invagination almost complete. The outermost cells stained for twist protein are mesectodermal cells, which will give rise to neurons.

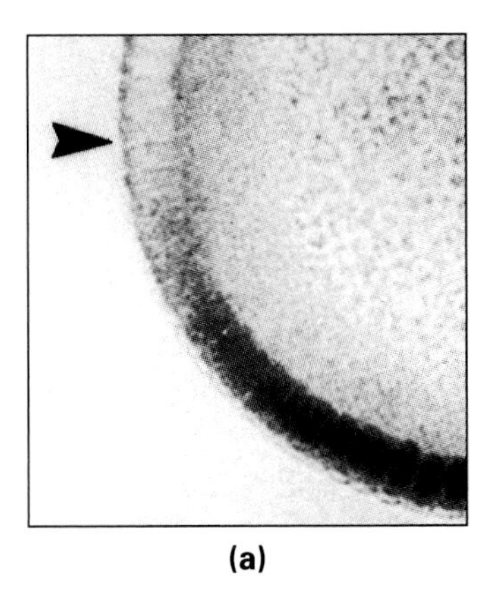
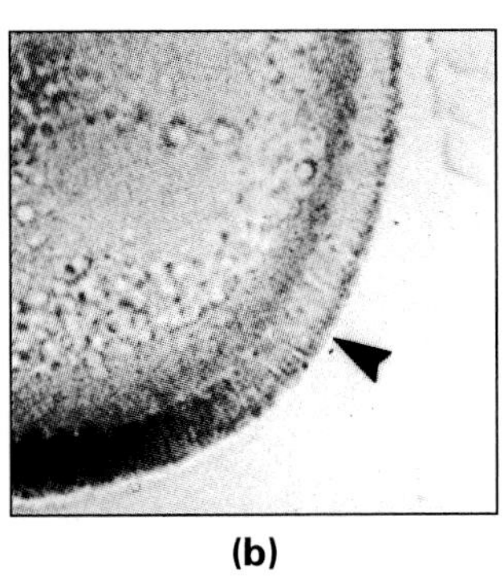

(a) (b)

Figure 22.41 Gradient of dorsal protein concentration in blastoderm cell nuclei shown in transverse sections. **(a)** Embryo derived from a wild-type mother and oriented with the ventral side down. **(b)** Embryo derived from a mother lacking a functional *Toll* gene, injected with wild-type egg cytoplasm on the dorsal side, and oriented with the injected side down. The arrowheads indicate the extention of the dorsal gradients. Both embryos were immunostained for dorsal protein. The transplanted cytoplasm causes a nuclear accumulation of dorsal protein around the site of injection. (see Method 4.1)

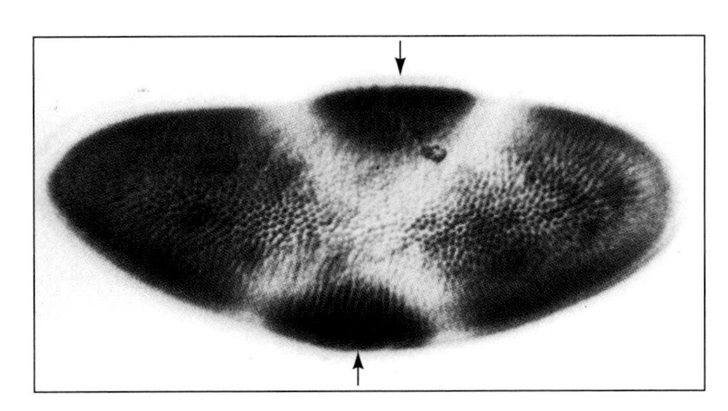

Figure 22.42 Effects of injected wild-type cytoplasm on an embryo derived from a mother lacking *Toll* gene activity. The embryo was injected dorsally and ventrally (arrows) at the syncytial blastoderm stage. At the cellular blastoderm stage, the cells surrounding the injection sites expressed the embryonic gene *twist*$^+$, as ventral cells in normal embryos do (Figs. 22.39b and 22.40). The size of the patch of *twist*$^+$-expressing cells increased with the amount of injected cytoplasm. The areas of ectopic *twist*$^+$ expression were surrounded by a ring of cells that expressed neither *twist*$^+$ nor another embryonic gene, *zerknüllt*$^+$. Outside this ring, the injected embryos did express *zerknüllt*$^+$, a gene that is expressed dorsally in normal embryos (see Fig. 22.38d) and over the entire surface of uninjected dorsalized embryos (see Fig. 22.39f).

Remarkably, the areas of ectopic *twist*$^+$ expression were surrounded by a ring of cells that expressed neither *twist*$^+$ nor *zerknüllt*$^+$. Outside this ring, the injected embryos expressed *zerknüllt*$^+$, a gene that is expressed over the entire surface of uninjected dorsalized embryos (see Figs. 22.39f and 22.42). Thus, the injected cytoplasm containing Toll activity generated concentric circles displaying the same sequence of embryonic gene expression (twist → other genes → zerknüllt) that normal embryos show in ventral to dorsal order.

These results support the hypothesis that local Toll activity orients the formation of a gradient of nuclear dorsal protein, which in turn serves as a morphogen, activating embryonic target genes in a specific sequence.

Questions

1. Given that the recipients were dorsalized due to the lack of local Toll activity, how could the investigator tell whether he was injecting a recipient embryo dorsally, ventrally, or laterally?
2. Which control could be done to ascertain that the local ventralization is caused by the Toll activity of the

injected cytoplasm and not by some stimulus associated with the injection procedure?

3. In the recipient embryos, the entire hierarchy of maternal effect genes down to the local generation of active spätzle ligand should work. Then why is the local ventralization occurring at the site of injection, rather than where the concentration of spätzle is highest?

The transcription patterns of embryonic dorsoventral genes change as expected in mutant embryos. If the low dorsal level of nuclear dorsal protein is extended around the embryo due to the lack of pelle, a protein required for the release of dorsal from cactus (see Fig. 22.35b), *twist*$^+$ is not expressed at all, while *decapentaplegic*$^+$ is transcribed over the entire embryo (see Fig. 22.39c, d). The converse happens in ventralized embryos, if a high level of nuclear dorsal protein is extended around the embryo by the maternal *Toll*^{10b} gain-of-function allele (see Fig. 22.35c). Here, *twist*$^+$ is transcribed all over the embryo, while *decapentaplegic*$^+$ is limited to the terminal regions (see Fig. 22.39e, f).

Just as the gap genes are first activated under the control of maternal determinants and then interact among themselves, so do the genes that establish the initial dorsoventral pattern. We will focus here on the *decapentaplegic*$^+$, *short gastrulation*$^+$, and *tolloid*$^+$ genes, all of which encode secreted proteins. Decapentaplegic (dpp) protein promotes the development of amnioserosa and epidermis while inhibiting neurogenesis. The short gastrulation (sog) protein acts on dpp, as indicated by the phenotypes of double mutants (Biehs et al., 1996). This inhibition is in turn counteracted by the product of the *tolloid*$^+$ gene, which is expressed on the dorsal side of the blastoderm embryo similar to *decapentaplegic*$^+$ (see Fig. 22.39a). Loss-of-function alleles of *tolloid*$^+$ and *decapentaplegic*$^+$ have very similar phenotypes (Ferguson and Anderson, 1992). Tolloid protein is a protease, and it cleaves sog, especially in the presence of dpp (Marqués et al., 1997). As a result of these interactions, dpp activity may form a morphogen gradient with the highest levels near the dorsal midline. The highest dpp activity level, along with expression of *zerknüllt*$^+$, seems to direct amnioserosa formation. Intermediate dpp activity levels, in dorsolateral blastoderm, still inhibit neurogenesis but promote dorsal epidermis. Low dpp activity levels in ventrolateral blastoderm allow the formations of both neurons and epidermal cells (see Section 6.4).

The *decapentaplegic*$^+$, *short gastrulation*$^+$, and *tolloid*$^+$ genes have been highly conserved during evolution. In vertebrates, the same genes are involved in establishing the dorsoventral polarity, although in upside-down orientation relative to insects (see Section 23.4).

The embryonic genes that are activated under control of the maternal dorsal protein are thought to control realizator genes that regulate cell division patterns, morphogenetic movements, and the eventual differentiation of the cells in their domains. In accord with this function, decapentaplegic is a secreted protein that appears to act as a morphogen, as discussed above, whereas the zerknüllt, twist, and snail proteins are all transcription factors.

In summary, the dorsoventral patterning system generates five major domains of gene expression and cell fates on both sides of the embryo (right and left). These domains are superimposed on the anteroposterior pattern of different parasegments determined by the segment polarity and homeotic genes. The anteroposterior and dorsoventral patterns are specified almost independently of each other; except for *gurken* and *torpedo*, mutations disrupting one pattern have no major effects on the other.

22.7 The Compartment Hypothesis

The expression patterns of segment polarity genes and homeotic genes in the *Drosophila* embryo create a patchwork of many domains, each filled with a few dozen to several hundred cells. How does this patchwork turn into an embryo, and later on into an adult? In Section 6.2, we have seen that the epidermis of insects is divided into *compartments* defined by clonal restriction lines. Remarkably, the cells from a given compartment stay together and do not mix with cells from other compartments, even though the formation and eversion of imaginal discs are associated with extensive morphogenetic movements (see Figs. 22.29 and 6.17). This cohesiveness raises the possibility that the morphological development as well as the selective cell adhesion within each compartment may be controlled by a unique combination of active patterning genes (*P.A. Lawrence and G. Struhl, 1996*).

COMPARTMENTS ARE CONTROLLED BY THE ACTIVITIES OF SPECIFIC SELECTOR GENES

Some compartment boundaries are anatomically prominent, such as segmental boundaries or the boundary between the upper and lower faces of a wing. Other compartment boundaries are anatomically inconspicuous. For instance, each body segment of *Drosophila* is subdivided into an anterior and a posterior compartment (see Fig. 22.2g). In the wing, this anteroposterior compartment boundary strikes across featureless terrain between the third and fourth veins, along a line that does not seem to have any anatomical or physiological significance (Fig. 22.43a).

Compartments in insect epidermis have remarkable properties. First, they are the basic units of growth and shape regulation. If one clone in a compartment is allowed by genetic trickery to grow larger than usual, the remaining clones of the same compartment become correspondingly smaller, and the compartment as a whole

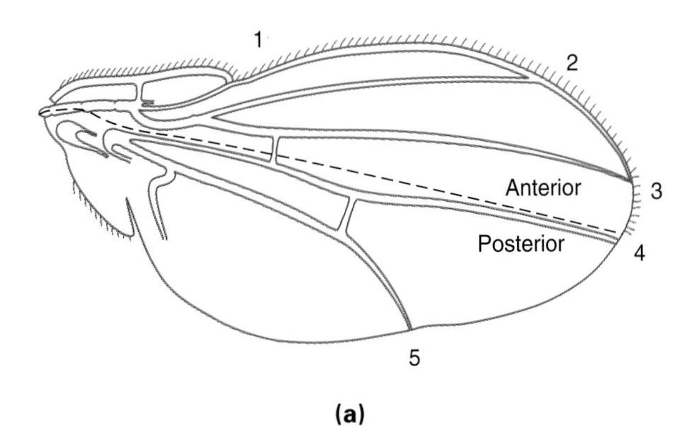

(a)

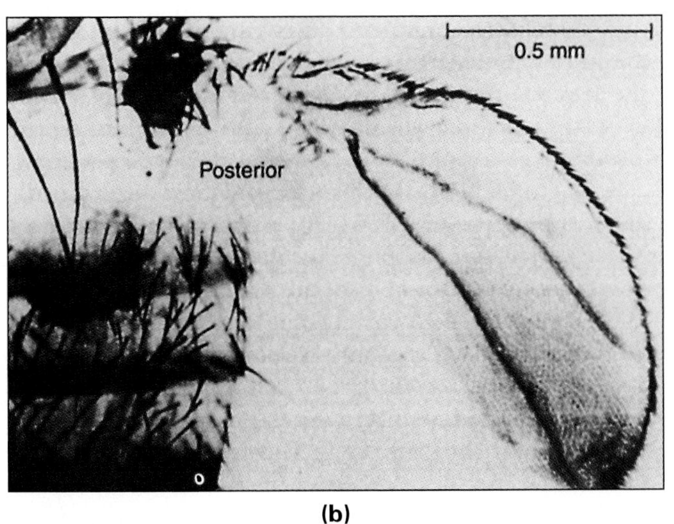

(b)

Figure 22.43 Coincidence of a compartment boundary, as defined by clonal restriction, with a functional domain of the *Ultrabithorax*$^+$ gene. **(a)** Schematic drawing of a *Drosophila* wing. Numbers indicate major wing veins. The compartment boundary (dashed line) between veins 3 and 4 is a line of clonal restriction between the anterior and posterior compartments. **(b)** Composite appendage of a mutant fly carrying the *bithorax* allele of the *Ultrabithorax*$^+$ gene. (See Fig. 22.32c for a schematic drawing of the phenotype.) The photomicrograph shows the anterior half of a wing attached to the posterior half of a haltere. Note that the posterior margin of the wing half coincides approximately with the compartment boundary shown in part a.

maintains its size and shape (see Fig. 6.4b). Second, adjacent compartments differ in their cell surface properties, and cells from different compartments do not intermingle. The result is that compartment boundaries are smooth, like the interface between oil and water.

Although compartments are originally defined by clonal restriction lines, Garcia-Bellido (1975) has linked compartments to patterning genes by the ***compartment hypothesis.*** He proposed that each compartment is characterized by a specific combination of active patterning genes, which he called ***selector genes.*** The selector genes are thought to control many ***realizator genes,*** which confer the different surface properties and other morphogenetic features that give each compartment its unique character. According to this hypothesis, selector genes have the following properties. First, the expression domains of selector genes coincide with certain compartments. Second, mutations in selector genes cause major changes in the morphology of compartments. Third, selector genes—directly or indirectly—affect cell surfaces. Fourth, the products of selector genes are gene-regulatory factors.

THE *ULTRABITHORAX*$^+$ EXPRESSION DOMAIN COINCIDES WITH A COMPARTMENT

The patterning genes that have the characteristics of selector genes include the homeotic gene *Ultrabithorax*$^+$, which is required for generating metathorax instead of mesothorax morphology. Flies carrying the *bithorax*^{3} mutant allele, in which one of the *Ultrabithorax*$^+$ enhancers is inactive, have an anterior mesothorax in place of the anterior metathorax (see Fig. 22.32). The boundary of this transformation closely follows the anteroposterior compartment boundary (Fig. 22.43b). Thus, a function of *Ultrabithorax*$^+$ that is driven by the mutated enhancer is necessary for normal development in the anterior but not in the posterior compartment of the metathorax.

Ultrabithorax$^+$ fulfills the criteria for a selector gene in terms of the compartment hypothesis. Loss of *Ultrabithorax*$^+$ function causes a major morphological transformations. It also changes the surface properties of cells as assayed by cell-mixing experiments (see Section 11.5). Finally, the Ultrabithorax protein is a transcription factor that is thought to regulate the activity of a large number of realizator genes. Similar experiments indicate that other homeotic genes function as selector genes as well.

ENGRAILED$^+$ CONTROLS THE ANTEROPOSTERIOR COMPARTMENT BOUNDARY

Another patterning gene with selector gene properties is the segment polarity gene *engrailed*$^+$. Flies homozygous for the mutant allele *engrailed*^{1} have wings with posterior margins that look like anterior wing margins. Other *engrailed* alleles are homozygous lethal, but heterozygous flies with homozygous mutant clones are viable. When such clones are generated in the anterior compartment of any segment, development is normal. However, similar clones in posterior compartments cause severe distortions (Kornberg, 1981). These results show that the *engrailed*$^+$ function is required for normal morphogenesis of the posterior—but not the anterior—compartment of each segment.

USING *clonal analysis,* Morata and Lawrence (1975) provided evidence that the anteroposterior compartment boundary within each segment is maintained by *engrailed*$^+$ expression in the posterior compartment and nonexpression in the anterior compartment. By X-ray–induced

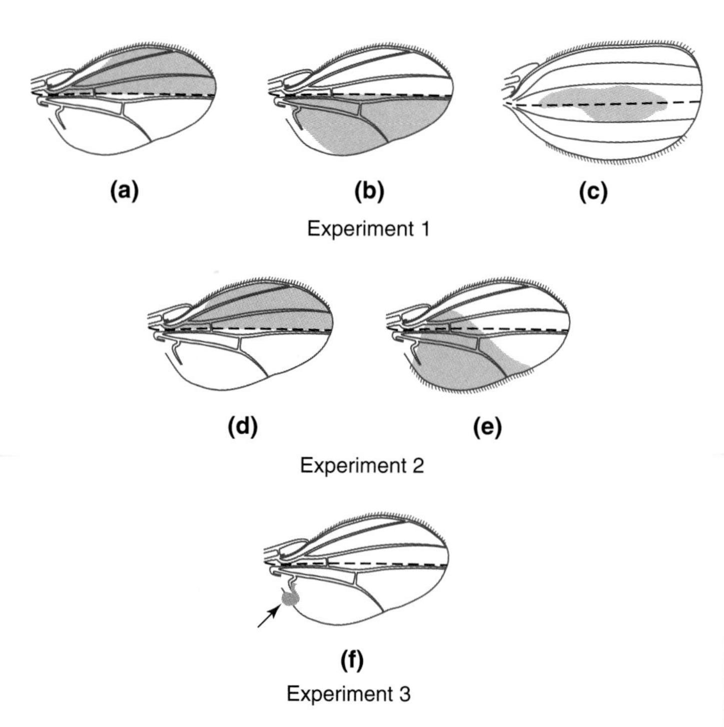

Figure 22.44 Dependence of the anteroposterior compartment boundary in the *Drosophila* wing on the function of the *engrailed*$^+$ gene. The diagrams summarize the results from three experiments. **Experiment 1:** The labeled clones (gray shading) were homozygous for a mutated cuticular marker gene, *multiple wing hair.* **(a, b)** In *engrailed*$^{1/+}$ flies, carrying one wild-type and one mutant allele of engrailed, such marked clones respected the anteroposterior compartment boundary (dashed lines), whether they were generated in the anterior compartment or in the posterior compartment. **(c)** Similar labeled clones were generated in homozygous *engrailed*$^{1/1}$ mutants, as indicated by the abnormally shaped wing. In this case, the clones crossed the boundary between the anterior half and the posterior half of the wing, indicating that there was no effective compartment boundary. **Experiment 2:** The labeled clones were homozygous for both *engrailed*^{1} and another mutated cuticular marker gene, *pawn*$^-$. (These two genes are located next to each other on the chromosome and are therefore rarely separated during somatic crossover.) **(d)** In flies that were *pawn*$^{-/+}$ and *engrailed*$^{1/+}$, homozygous mutant clones generated in the anterior compartment respected the compartment boundary and did not affect wing morphology. **(e)** Similar clones in the posterior compartment crossed into the anterior compartment and locally transformed the posterior wing margin into an anterior margin with sensory bristles. **(f) Experiment 3:** Clones of cells carrying two different loss-of-function alleles (*engrailed*$^{1/C2}$) were generated. These clones showed a more severe loss of function than homozygous *engrailed*$^{1/1}$ clones. If such clones originated in the anterior compartment, they respected the compartment boundary. When originating in the posterior compartment close to the boundary, the clones crossed into the anterior compartment as in part e. When originating deep in the posterior compartment, the mutant clone segregated as a vesicle (arrow) from the wing blade.

somatic crossover (see Method 6.1), they generated cell clones marked with a simple cuticular marker (Fig. 22.44a–c). When generated in flies with at least one wild-type allele of *engrailed*$^+$, such clones respected the anteroposterior compartment boundary in the wing. However, in flies homozygous for the *engrailed*^{1} allele, some of the marked clones straddled the boundary between the anterior and the posterior half of the wing. These results showed that the *engrailed*$^+$ function is required for maintaining the compartment boundary.

Another experiment provided evidence that in wild-type flies the *engrailed*$^+$ gene is active in the posterior compartment and inactive in the anterior compartment. Here, the labeled clones were homozygous for both the mutant *engrailed*^{1} allele and a cuticular marker gene. Such clones crossed from the posterior into the anterior compartment but not in the opposite direction (Fig. 22.44d, e). From these results, the researchers concluded that the *engrailed*$^+$ gene is normally expressed in the posterior but not in the anterior compartment. They proposed further that the *engrailed*^{1} allele causes a partial loss of function. (This assumption is supported by the *engrailed*^{1} phenotype, which shows an incomplete transformation of a posterior wing half into an anterior one.) Consistent with these proposals, Morata and Lawrence argued that an *engrailed*$^{1/1}$ clone originating in the posterior compartment has cell surface properties halfway between anterior cells (which do not express any *engrailed* allele) and posterior cells (which express the wild-type as well as the mutant allele). Such a clone therefore mixes with *engrailed*$^{1/+}$ cells in both compartments and crosses the boundary. In contrast, an *engrailed*$^{1/1}$ clone originating in the anterior compartment expresses zero *engrailed*^{1} activity because the gene is not active anteriorly. Its cells therefore do not mix with posterior cells expressing *engrailed*^{1} as well as *engrailed*$^+$ and do not cross the boundary.

Questions

1. In Figure 22.44, why is the wing outline in part c drawn differently from the outlines in parts a and b? Also, why has the labeled clone hitting the posterior wing margin a border of strong bristles in part e but not in part b?
2. In subsequent work with an engrailed allele (*engrailed*^{C2}) causing a more severe loss of function, the investigators generated (*engrailed*^{1}/*engrailed*^{C2}) heterozygous clones. If such clones originated in the posterior compartment but near the compartment border, they crossed into the anterior compartment, as seen in Figure 22.44e. However, when such a clone originates away from the compartment border, the margin of the clone constricted, and the narrowing ring cleanly pinched out the clone so that it separated from the wing blade as a vesicle (Fig. 22.44f). How do you explain these results?

The experiments just described indicate that the *engrailed*$^+$ gene directly or indirectly affects the surface properties of the cells expressing it. When two groups of cells, one expressing *engrailed*$^+$ and the other lacking it, contact each other, they generally remain separate and form a smooth boundary. Also, mutant alleles of the gene cause major morphological changes in the posterior wing compartment. In addition, the expression domain of *engrailed*$^+$ seems to coincide with the posterior wing compartment. Finally, the gene encodes a transcription factor. Because of these characteristics, *engrailed*$^+$ qualifies as a selector gene.

The compartment hypothesis places stringent requirements on the expression of selector genes. If selector gene activities are to specify the character of each compartment including its cell surface properties, the expression domains of selector genes should follow compartment boundaries *exactly*. Because compartments are established as early as the blastoderm stage, it follows that subsequently generated cell clones should fall entirely inside or outside a selector gene domain. In other words, such clones should not straddle the margins of selector gene expression domains. Studies designed to test this requirement revealed that it is indeed met, with few exceptions, for *engrailed*$^+$ as a selector gene (Blair, 1992; Vincent and O'Farrell, 1992).

Studies on *engrailed*$^+$ expression have also shown that compartment boundaries may be ragged when selector genes are first activated but straighten out with time. The boundary between the anterior and posterior wing compartments may again serve as an example. This boundary is revealed by clonal restriction as an astoundingly straight line in the adult wing (see Fig. 22.43a). In the early embryo, this line corresponds to the boundary between parasegments 4 and 5 (see Fig. 22.2g). At the cellular blastoderm stage, cells on the posterior side of this boundary express *evenskipped*$^+$, and subsequently *engrailed*$^+$ (see Figs. 22.21 and 22.26). At this time, the boundary is irregular and would not be recognizable without immunostaining for a marker gene product. Between the ragged state of the compartment boundary in the embryo, and its remarkably straight course in the adult wing, there is an intermediate stage in the wing imaginal disc. Here, the location of the anteroposterior compartment boundary is known from fate-mapping studies as well as from the expression of marker genes, such as *engrailed*$^+$, which is now active in the entire posterior compartment (Fig. 22.45a). The compartment boundary, while not a perfectly straight line, is now straighter than any neighboring line that one could draw between cells. Indeed, the boundary is now detectable, without immunostaining for a marker gene, as the smoothest line of demarcation between the anterior and posterior compartment (Fig. 22.45b).

The simplest interpretation for the gradual straightening of compartment boundaries is that the cells of adjacent compartments, under control of unlike selector gene activities, acquire different cell surface properties that cause them to minimize contacts across the boundary. This segregation process is likely to occur over time and to be facilitated by events that loosen the epithelial connections between cells, such as cell division or cell intercalation.

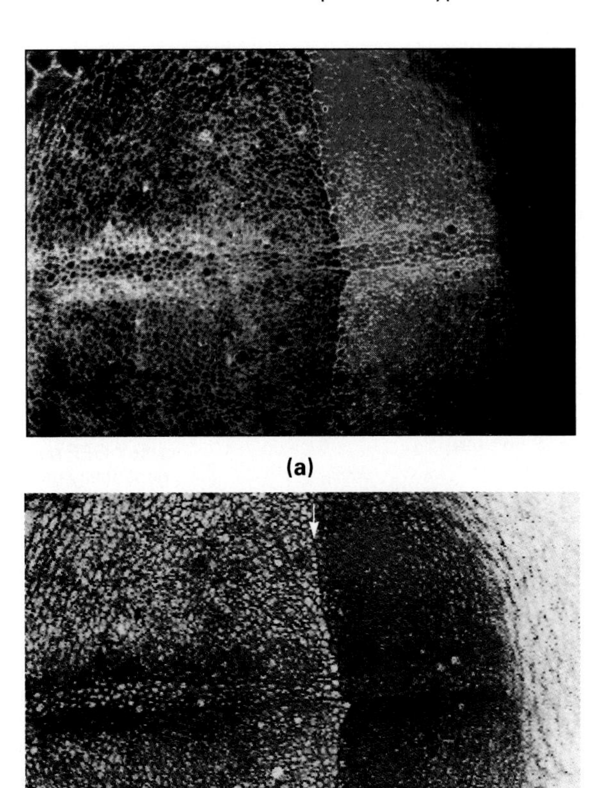

Figure 22.45 Cellular compartments in *Drosophila* imaginal discs. The two photos show the apical surface of the same wing imaginal disc, photographed with different types of illumination. Anterior compartment to the left. **(a)** The disc has been stained with a green fluorescent dye that outlines cell boundaries. In addition, the disc has been immunostained against engrailed protein with a red fluorescent dye. The anterior limit of *engrailed*$^+$ expression coincides with a relatively smooth line of cell boundaries, indicating that cells expressing *engrailed*$^+$ minimize their contacts with cells that do not express this gene. **(b)** Without immunostaining engrailed protein, in light that makes only cells visible, the same compartment boundary is discernible just based on its relatively straight course.

In the case of *engrailed*$^+$, the selector gene effect on cell surface properties is in part cell-autonomous and in part induced. Cells in the posterior compartment, under the influence of *engrailed*$^+$, release hedgehog (hh) protein (Fig. 22.46). Only cells that do not express *engrailed*$^+$ can respond to hh, apparently because *engrailed*$^+$ inhibits one or more genes involved in the hh response pathway (Zecca et al., 1995; *Ruiz i Altaba, 1997*). One component involved in this pathway is the smoothened

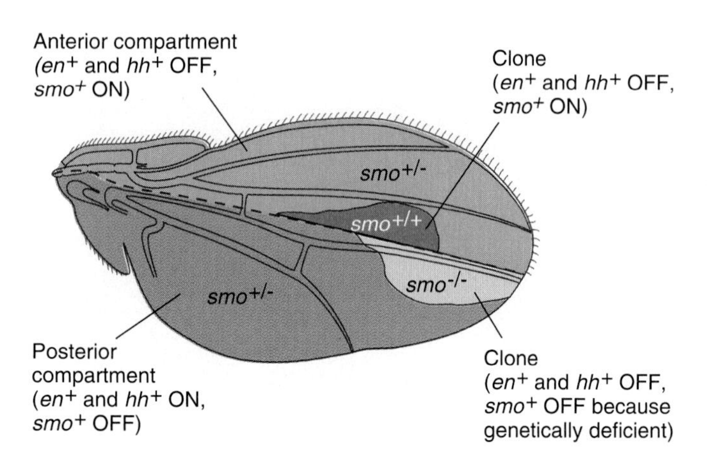

Figure 22.46 Role of the *smoothened*$^+$ (*smo*$^+$) gene in cell segregation along the boundary between anterior and posterior compartments in the *Drosophila* wing. The smoothened protein acts as a transducer in a signal chain across the anteroposterior compartment boundary (see Fig. 22.48). Twin spots of *smo*$^{+/+}$ and *smo*$^{-/-}$ were generated in a *smo*$^{+/-}$ background by somatic crossover. Expression of the *engrailed*$^+$ (*en*$^+$) gene, and consequent inactivity of *smo*$^+$, was monitored with an *en-lacZ* fusion gene. If a twin spot originated near the boundary in the anterior compartment, as indicated by inactivity of *en*$^+$, the *smo*$^{+/+}$ clone remained in the anterior compartment. By contrast, its *smo*$^{-/-}$ twin became part of the posterior compartment. Both clones had smooth edges along the compartment boundary but jagged edges elsewhere. Thus, the compartment boundary developed not between cells that did or did not express *en*$^+$, but rather between cells that did or did not make active smoothened protein.

protein, which is part of the hh receptor complex and limits the movement of hh (see Fig. 22.27; Y. Chen and G. Struhl, 1996; 1998).

Part of the signal chain downstream of smoothened must be changing the surface properties of cells so that they will segregate from cells that lack smoothened. This was shown by using the *somatic crossover* technique (see Method 6.1) on heterozygous *smoothened*$^{+/-}$ flies (Blair and Ralston, 1997; Rodriguez and Basler, 1997). This leads to the formation of "twin spots"—that is, marked sister clones located next to each other, one of which is *smoothened*$^{+/+}$ and the other *smoothened*$^{-/-}$ (Fig. 22.46). If such a twin spot originates near the compartment boundary in the anterior compartment, as indicated by the lack of *engrailed*$^+$ activity, then the two sister clones behave differently. The *smoothened*$^{+/+}$ clone stays in the anterior compartment, whereas its *smoothened*$^{-/-}$ twin becomes part of the posterior compartment and contributes to the straight boundary with the anterior compartment.

This result indicates that the cells of the anterior and posterior compartments do not segregate according to the inactivity versus activity of the *engrailed*$^+$ gene itself. Instead, they segregate according to the presence or absence of active smoothened protein and the resulting ability or inability to respond to hh. However, this latter distinction depends on *engrailed*$^+$ activity in two ways. First, *engrailed*$^+$ cell autonomously causes the synthesis of hh, while inhibiting *smoothened*$^+$ expression and thus a response to hh in the posterior compartment. Second, as the hh protein moves across the boundary, it induces the hh response, including the activity of smoothened protein, in anterior cells.

APTEROUS$^+$ CONTROLS THE DORSOVENTRAL COMPARTMENT BOUNDARY

The *apterous*$^+$ gene of *Drosophila* also has the properties of a selector gene. Apterous is derived from the Greek word for "wingless," and flies lacking *apterous*$^+$ activity fail to form wings and halteres. The *apterous*$^+$ gene is expressed dorsally, but not ventrally, in the imaginal discs for wings and halteres (Fig. 22.47; Blair et al., 1994). If cell clones that are deficient in *apterous* arise in the dorsal wing compartment near the dorsoventral compartment boundary, they are included in the ventral compartment, apparently because they adhere more strongly to cells without apterous protein than to cells with apterous protein. If *apterous*$^-$ clones originate deep within dorsal territory, they develop into a patch of adult wing tissue with a ventral morphology. They also form, at their junction with the surrounding dorsal cells, the type of bristles that are normally formed at the wing margin (Diaz-Benjumea and Cohen, 1993). In addition, such clones may extend out of the wing surface like an extra winglet. It appears that wing tissue grows out from the thorax wherever cells expressing *apterous*$^+$ are juxtaposed with cells that do not express this gene. This juxtaposition seems to trigger a complex set of gene activities including the activation of the *vestigial*$^+$ gene, which is expressed in the wing and haltere as opposed to the thorax (S. M. Cohen, 1996; Nagaraj et al., 1999).

The apterous protein also has the molecular characteristics of a transcription factor. It seems to control, or at least maintain, cell adhesiveness through the well-known class of substrate adhesion molecules known as *integrins* (*Bunch and Brower, 1993; N. H. Brown, 1993*). At least some of these integrins are expressed in position-specific way and are therefore known as PS integrins. Under the control of the *apterous*$^+$ gene, PS1 is expressed mainly in dorsal cells and PS2 primarily in ventral cells. However, it is not clear how much the PS integrins have to do with maintaining the dorsoventral compartment boundary.

In summary, selector genes are linked to epidermal compartments, which are defined by clonal restriction, through the compartment hypothesis. According to this hypothesis, compartments differ from one another in their unique combinations of selector genes including the *Ultrabithorax*$^+$, *engrailed*$^+$, and *apterous*$^+$ genes of *Drosophila*. They act in concert to control presumably

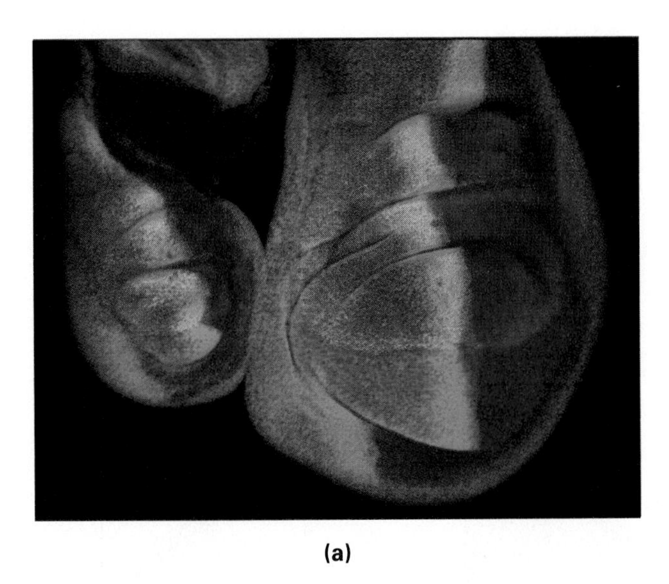

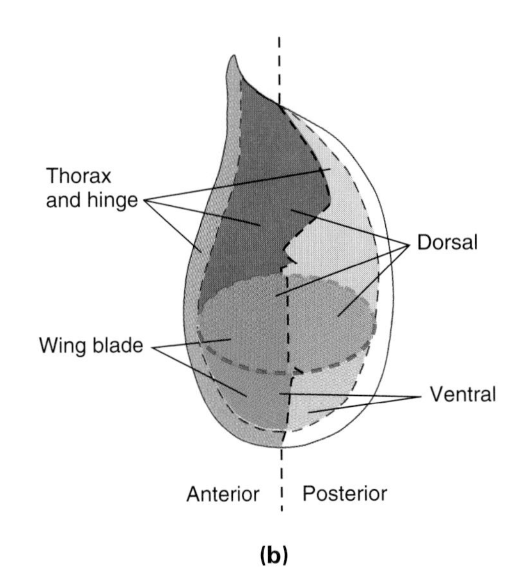

(a) (b)

Figure 22.47 Compartment-specific gene expression in *Drosophila* imaginal discs. **(a)** Imaginal discs for wing (right) and haltere (left) are shown in composite immunostaining patterns. *vestigial*$^+$ (blue) is expressed mostly in the appendages, as distinct from the thoracic segments to which they are hinged. *cubitus interruptus dominus*$^+$ (green) marks the anterior compartments. *apterous*$^+$ (red) is expressed only in the dorsal compartments. Overlap between red and green gives rise to yellow-orange; red and blue produce purple. **(b)** Fate map of the wing imaginal disc. The broken line running down centrally represents the anteroposterior compartment boundary.

large numbers of target genes, which direct development of the compartment-specific morphology as well as different cell surface properties. The latter generally prevent the mixing of cells from different compartments but may also promote adhesion between certain differentiated tissues.

22.8 Appendage Formation and Patterning

In the previous section, we have seen how compartments receive a genetic identity by having a unique set of active selector genes. However, these genes cannot account for any patterns *within* compartments, such as the veins within the anterior compartment of a wing or the rows of cuticular denticles in the anterior compartment of larval segments. In Wolpert's terminology, the role of selector genes as discussed so far has been in the *interpretation* of positional information, demonstrated dramatically by the local replacement of antennal parts with corresponding leg parts in certain *Antennapedia* mutants (see Fig. 21.8). In this instance, the same set of proximodistal values is interpreted in either of two ways, by forming leg or antenna, depending on the local activity or inactivity of the *Antennapedia*$^+$ gene.

In this section, we will see that selector genes are also involved in setting up morphogen gradients that *specify positional information* within compartments (*Blair, 1995; P.A. Lawrence and G. Struhl, 1996*). In essence, a compartment boundary can separate two cell populations, one capable of sending a short-range signal and the other capable of responding to it. This generates a strip of responding cells along the boundary, which becomes the source of a morphogen gradient that provides positional information to both adjoining compartments. In doing so, the compartment boundary combines two critical functions: It sets up a source for a morphogen and—because cells don't cross the compartment boundary, it maintains the morphogen source as a contiguous strip of cells despite the cell divisions and morphogenetic movements that occur during the development of imaginal discs.

COMPARTMENT INTERACTIONS ACROSS BOUNDARIES ORGANIZE PATTERN FORMATION WITHIN COMPARTMENTS

A series of investigations on the wing imaginal disc has led to the following hypothesis (see Figs. 22.27 and 22.48). The disc is divided into anterior and posterior compartments, depending on the inactivity or activity of the *engrailed*$^+$ gene, as discussed earlier. In the posterior compartment, *engrailed*$^+$ activates the synthesis of hedgehog (hh), a secreted protein. The latter diffuses for a short distance across the compartment boundary, where it activates a membrane receptor complex that includes the patched and smoothened proteins and maintains the activity of *wingless*$^+$. The same receptor complex also stabilizes a transcriptional activator encoded by the *cubitus interruptus*$^+$ gene in the anterior wing compartment (Hepker et al., 1997; Ruiz i Altaba,

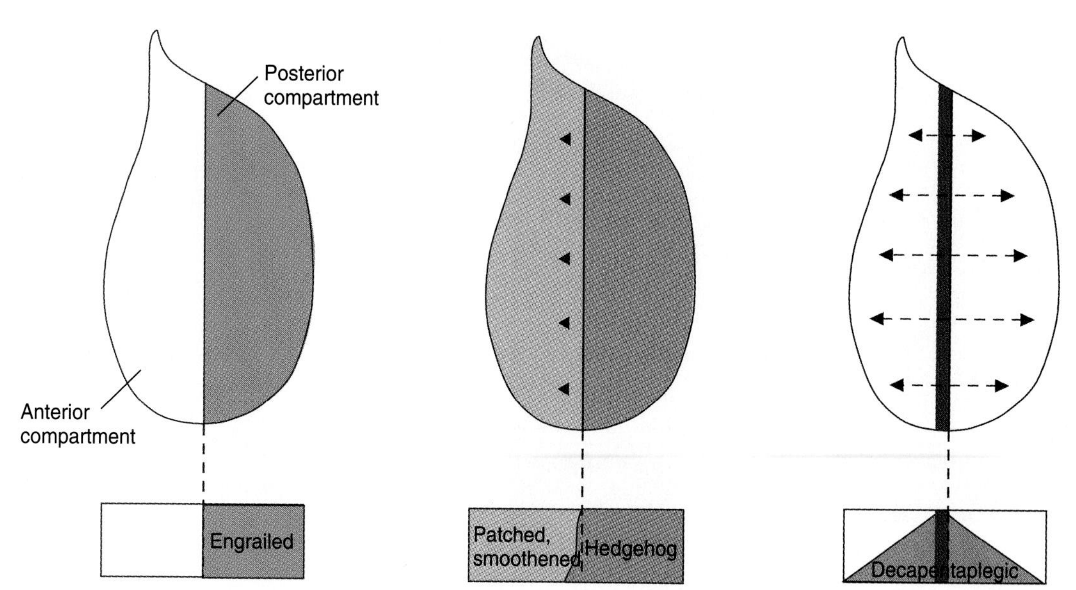

Figure 22.48 Model for the establishment of a morphogen gradient in the wing disc of *Drosophila.* The engrailed protein, a transcription factor, is synthesized in the posterior compartment. Here it causes the synthesis of hedgehog, a secreted protein, which moves a short distance into the anterior compartment. The cells of the anterior compartment, in contrast to those of the posterior compartment, express *patched*$^{+}$ and *smoothened*$^{+}$, which encode proteins acting as a receptor and signal transductor for hedgehog. Smoothened signaling results in the synthesis of decapentaplegic, another secreted protein. Because of the limited mobility of hedgehog, most of the decapentaplegic synthesis occurs in a narrow strip of cells just anterior to the compartment boundary. From here, decapentaplegic diffuses as it is degraded or sequestered, thus generating a concentration gradient.

1997; Q.T. Wang and R.A. Holmgren, 1999). One of the target genes of cubitus interruptus is *decapentaplegic*$^{+}$, which encodes a single protein of the TGF-β superfamily. (The same does not happen in the posterior compartment because *engrailed*$^{+}$ inhibits *cubitus interruptus*$^{+}$ and *smoothened*$^{+}$.) Thus, decapentaplegic (dpp) comes to be synthesized only in a narrow strip of cells located just anterior to the boundary (Fig. 22.48). The local concentration of dpp will therefore be highest in this strip and grade away towards the anterior and posterior wing margins as dpp diffuses while being sequestered or destroyed.

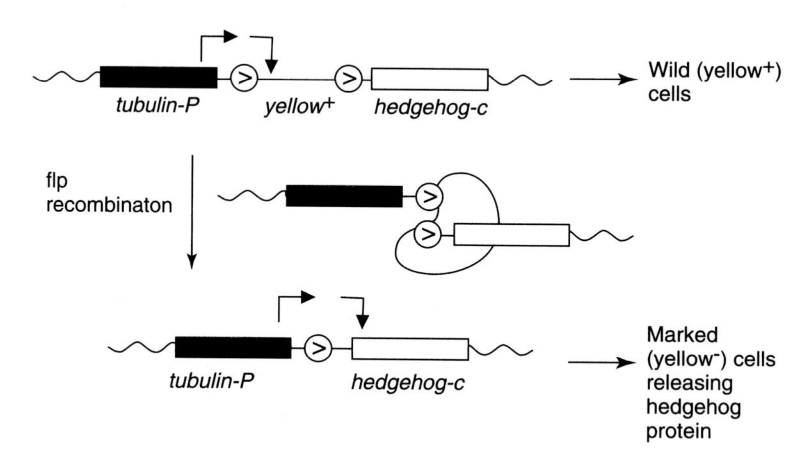

Figure 22.49 Use of the "flp-out" technique for generating a marked clone expressing *hedgehog*$^{+}$ constitutively. Flies were made transgenic for a construct consisting of three main elements: (1) a *tubulin*$^{+}$ promoter (*tubulin-P*), which drives the expression of any downstream gene in most cells; (2) the wild-type allele of the *yellow*$^{+}$ gene flanked by flp recombination targets (circled chevrons); and (3) the coding sequence of the *hedgehog*$^{+}$ gene (*hedgehog-c*). Without recombination, the promoter drives expression of *yellow*$^{+}$ but not of *hedgehog*$^{+}$, generating cells of wild (*yellow*$^{+}$) phenotype. A second transgene (not shown) contains the *flp*$^{+}$ gene under control of a heat shock promoter. Heat shocks will therefore induce flp-mediated recombination in a few cells, removing the *yellow*$^{+}$ gene and bringing *hedgehog*$^{+}$ under control of the tubulin promoter. Clones derived from these cells will be marked by the *yellow* mutant phenotype and synthesize hedgehog protein from the transgene.

IN order to test this hypothesis, Basler and Struhl (1994) engineered flies with randomly generated clones of cells expressing either a *hedgehog*$^{+}$ or a *decapentaplegic*$^{+}$ transgene. For this purpose, they used the "flp-out" technique based on the yeast recombinase flp, which cuts out DNA segments flanked by specific recognition sites known as flp recombination targets (FRTs). Figure 22.49 illustrates the use of this technique for generating a marked cell clone expressing the *hedgehog*$^{+}$ gene. The experimental fly embryos have two transgenes. One contains the yeast gene for flp driven by a heat shock promoter, which can be activated by a pulse of high ambient temperature. The other transgene contains *hedgehog*$^{+}$ cDNA that will ulti-

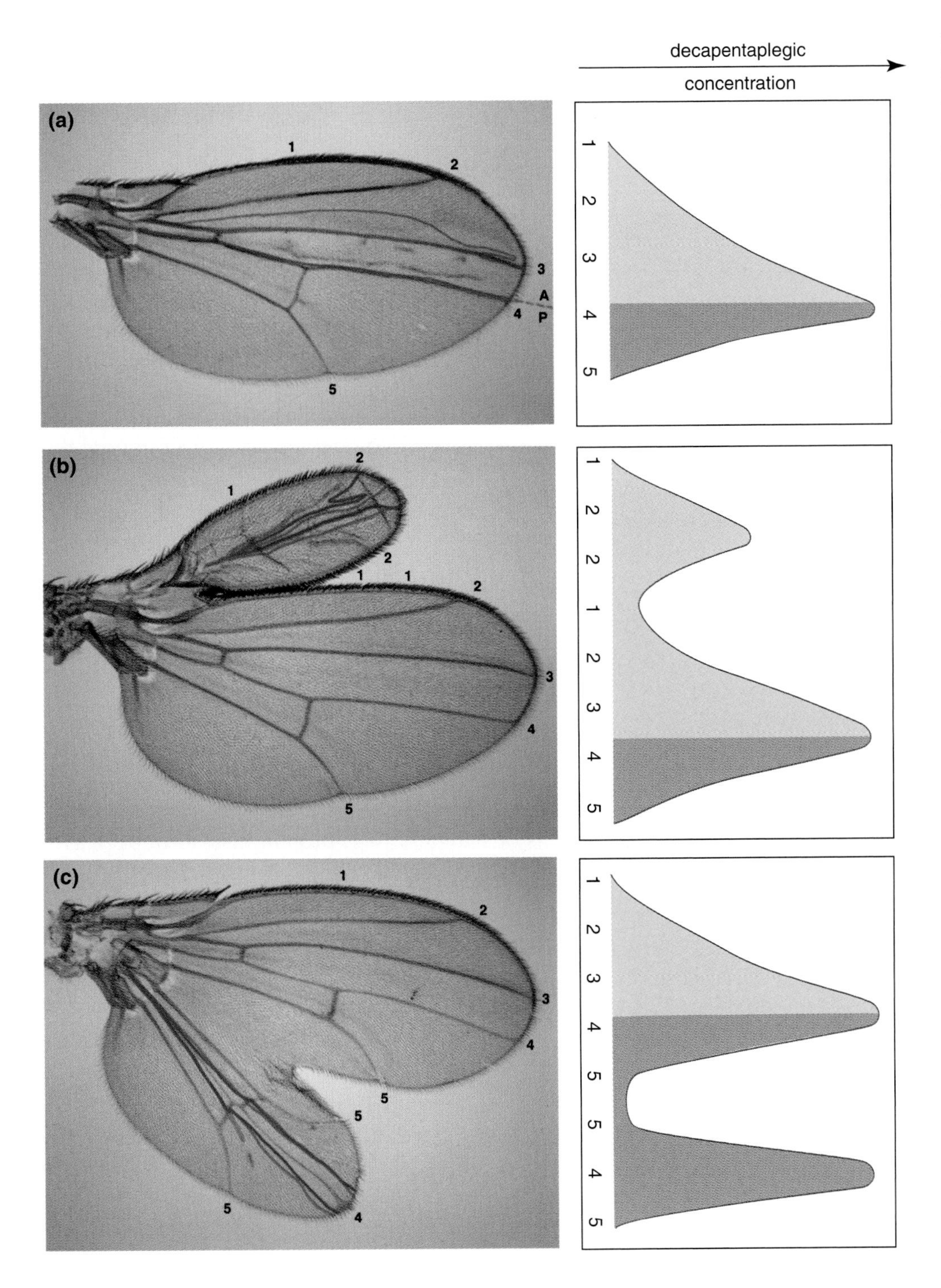

Figure 22.50 Reorganization of *Drosophila* wing patterns by ectopic expression of *decapentaplegic*$^+$ (*dpp*$^+$). Marked clones expressing *dpp*$^+$ were generated as shown in Figure 22.49. Clone borders are outlined in blue ventrally and in red dorsally. Veins are numbered along the wing margins. The dashed line marks the anteroposterior compartment boundary. **(a)** If a clone synthesizing dpp was formed in the anterior compartment next to the anteroposterior compartment boundary, where dpp synthesis occurs normally, only small disturbances of the wing pattern occurred. **(b)** If a dpp-synthesizing clone originated deep inside an anterior wing compartment, a mirror-image duplication of anterior wing structures was formed. **(c)** Clones producing dpp posteriorly led to accessory mirror-image duplications of posterior wing parts. The diagrams to the right show hypothetical gradients of dpp. The same dpp concentrations are thought to have different morphological effects, depending on different selector gene activities.

mately be driven by a constitutively active promoter (taken from a *tubulin*$^+$ gene). However, the *hedgehog*$^+$ gene and its promoter are initially separated by a cuticular marker gene (*yellow*$^+$) flanked by FRTs. The promoter will therefore drive the expression of *yellow*$^+$ rather than *hedgehog*$^+$, with no effect on wing patterning. If such embryos are heat-shocked, the flp recombinase will be activated randomly in single cells, where the DNA between the FRTs will be cut out. As a result, the cell will grow into a clone synthesizing hedgehog protein. This clone is also marked by the lack of yellow gene product, whereas all other cells express *yellow*$^+$ from the uncut transgene. Similarly, the researchers generated flies with marked clones synthesizing decapentaplegic protein ectopically.

The use of this technique had remarkable results on wing patterning (Zecca et al., 1995). If a clone synthesizing dpp was formed in the anterior compartment next to the compartment boundary—that is, where dpp synthesis occurs normally, only small disturbances of the wing pattern occurred (Fig. 22.50a). However, if clones synthesizing hh or dpp originated deep inside an anterior wing compartment, an extra mirror-image duplication of anterior wing structures was formed (Fig. 22.50b). Similar hh-synthesizing clones in posterior compartments did not cause dpp synthesis and had no morphological effect. In contrast, clones producing dpp posteriorly led to accessory mirror-image duplications of posterior wing parts (Fig. 22.50c). These observations are most readily interpreted by assuming that hh activates dpp synthesis in the anterior compartment cells, and that dpp acts as a morphogen in both compartments.

Questions

1. Why would the removal of the *yellow*$^+$ segment from the transgene by flp recombinase cause a phenotypic effect? Should not the resident gene copies of *yellow*$^+$ mask the loss of this gene from the transgenic construct?
2. The flp-out technique described above can be used similarly to generate marked cell clones that constitutively express *engrailed*$^+$. Based on the model diagrammed in Figure 22.48, what phenotypic effects would you expect if such clones form (a) in the posterior compartment and (b) deep in the anterior compartment?
3. By mitotic recombination, induced by X-rays (see Method 6.1) or flp, one can generate marked cell clones that are lacking endogenous *engrailed*$^+$ activity. Suppose you generate such clones in flies that contain a *lacZ* reporter gene driven (a) by a *hh* promoter or (b) by a *dpp* promoter. How should the expression of the *lacZ* reporter transgene change if clones without *engrailed*$^+$ activity form deep in the posterior wing compartment?

The results described above indicate that the normal anteroposterior pattern of the wing is specified by a morphogen gradient of dpp protein, which is synthesized in a narrow strip of cells just anterior to the compartment boundary and is degraded or sequestered as it diffuses anteriorly and posteriorly. How can such a symmetrical, or roof-shaped, gradient account for an asymmetrical anteroposterior wing pattern? The simplest interpretation is that activity of *engrailed*$^+$ in the posterior compartment and the absence of such activity in the anterior compartment lead to different interpretations of the same dpp levels. Thus, the lowest dpp level would result in a posterior wing margin when *engrailed*$^+$ is active, but the same low dpp level would result in an anterior wing margin when *engrailed*$^+$ is inactive.

The dorsoventral pattern of the wing disc seems to be specified in a similar manner. At the dorsoventral compartment boundary, dorsal cells expressing the *apterous*$^+$ gene are juxtaposed with ventral cells not expressing apterous (see Fig. 22.47). The apterous protein, a transcription factor, activates the expression of *fringe*$^+$, a gene that encodes a secreted protein and is necessary for wing formation (Irvine and Wieschaus, 1994). This is analogous to the activation of *hedgehog*$^+$ by the engrailed protein (see Fig. 22.48). Further extending the analogy, the generation of clones expressing *fringe*$^+$ in the ventral wing causes the outgrowth of an extra wing, as does the ectopic expression of *hedgehog*$^+$ or *decapentaplegic*$^+$ in the anterior compartment (Fig. 22.50). Finally, the counterpart to decapentaplegic protein in anteroposterior pattern formation seems to be the wingless protein in dorsoventral pattern formation. Wingless protein forms a double gradient of centered on the wing margin—that is, the boundary between the dorsal and ventral compartments domains (Zecca et al., 1996; Neumann and Cohen, 1997a). Wingless is synthesized in a narrow strip of wing imaginal disc cells straddling the compartment boundary, apparently triggered by the juxtaposition of cells that express *apterous*$^+$ and *fringe*$^+$ (dorsal compartment) with cells that don't (ventral compartment). From the strip of synthesis, wingless protein spreads out in concentration gradients over at least ten cells dorsally and ventrally (Fig. 22.51a). Depending on wingless concentration, the genes *Achete-scute*$^+$, *Distalless*$^+$, and *vestigial*$^+$ are expressed in nested domains (Fig. 22.51b). In the anterior compartment, the proneural genes *Achete*$^+$and *scute*$^+$ are expressed on both sides of the strip of wingless synthesis itself. The cells expressing these genes will form the sensory bristles that characterize the anterior wing margin.

The leg imaginal disc is patterned similarly to the wing disc. Again, the *engrailed*$^+$ gene is expressed in the posterior compartment, where it causes the release of hh protein. In the dorsal leg compartment, hh induces the synthesis of dpp in a strip of cells along the anteroposterior compartment boundary. The same happens in the ventral compartment except that here hh induces the synthesis of wingless (wg) protein instead of dpp. The dpp and wg strips meet where the anteroposterior and dorsoventral compartment boundaries cross at the center of leg disc. In a small area of cells around this junction, the *Distalless*$^+$ and *aristaless*$^+$ genes are activated, which are necessary for the growth of the disc along the proximodistal axis (Diaz-Benjumea et al., 1994; Lecuit and Cohen, 1997; J. Wu and S.M. Cohen, 1999).

DECAPENTAPLEGIC PROTEIN FORMS A MORPHOGEN GRADIENT IN THE WING IMAGINAL DISC

Additional studies have confirmed that dpp protein functions as a morphogen gradient, rather as the first link in a sequence of short-range inductions. Most of these studies have focused on two target genes of dpp, *spalt*$^+$ and *optomotor-blind*$^+$, both of which encode transcription factors that are required to form proper wing morphology (Lecuit et al., 1996; Nellen et al., 1996; Singer et al., 1997). These two genes are expressed in nested domains centered on the strip of dpp synthesizing cells along the anteroposterior compartment boundary in the wing imaginal disc (Fig. 22.52). Cells near the boundary express both *spalt*$^+$ and *optomotor-blind*$^+$, whereas cells further away from the boundary express only *optomotor-blind*$^+$. The effect of dpp is concentration-dependent, since overexpression of dpp in its normal domain causes broadening of the target gene expression domains and disc overgrowth. Conversely, cells that receive little dpp express *optomotor-blind*$^+$ but not *spalt*$^+$.

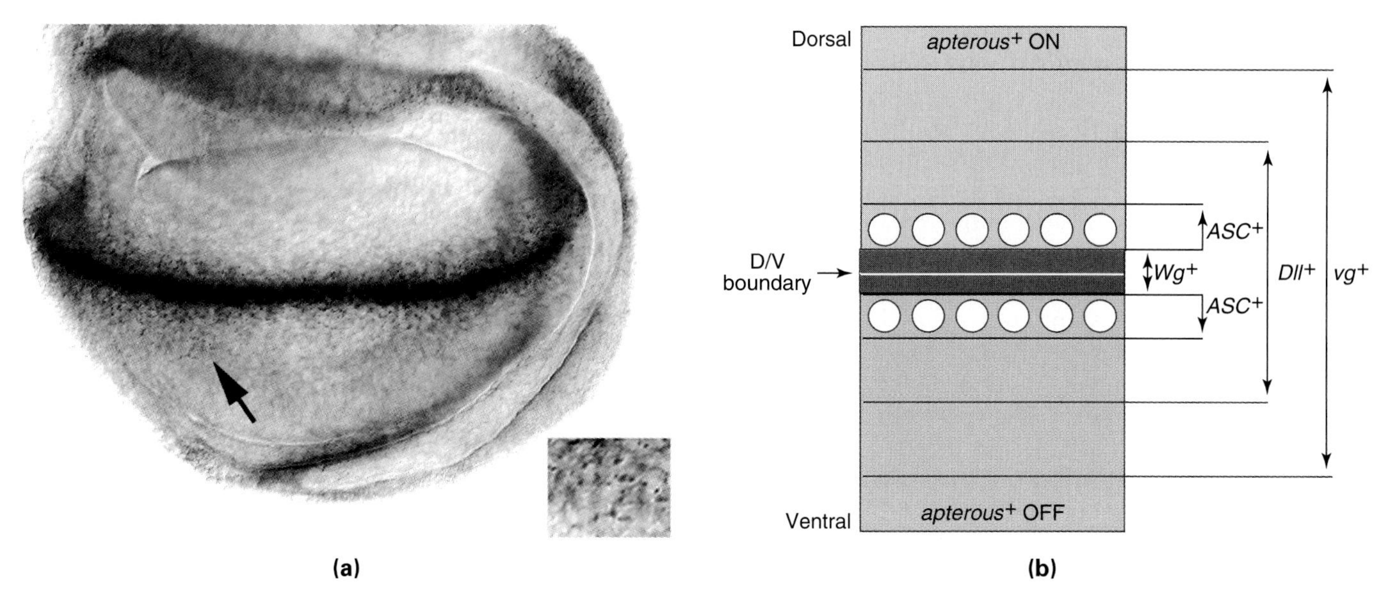

Figure 22.51 Long-range organizing activity of wingless protein in the dorsoventral pattern of the wing imaginal disc. **(a)** Wing imaginal disc from a *Drosophila* larva immunostained (see Method 4.1) for wingless protein. The protein is present in high concentration along a narrow strip of cells along the dorsoventral (DV) compartment boundary. Cells nearby show lower levels of wingless protein. The arrow indicates a distance of ten cell diameters from the DV boundary (see Fig. 22.47b). The inset shows an area near the tip of the arrow, where wingless protein can be seen in small vesicular structures. **(b)** Schematic drawing of part of the wing imaginal disc. The *apterous*$^+$ gene is expressed in the dorsal but not in the ventral compartment. *Wingless*$^+$ (*Wg*$^+$) is expressed in the cells adjacent to the compartment boundary on both sides. From the strip of synthesis, wingless protein spreads out in concentration gradients over at least ten cells dorsally and ventrally (see part a). Depending on wingless concentration, the genes *Achete-scute*$^+$ (*ASC*$^+$), *Distalless*$^+$ (*Dll*$^+$), and *vestigial*$^+$ (*vg*$^+$) are expressed in increasingly wider domains. However, *ASC*$^+$ is expressed only in the anterior compartment, and outside of the strip of *Wg*$^+$ activity. This expression domain is where the bristles (white circles) characterizing the anterior wing margin are formed. Thus, wingless may act as a morphogen, controlling the activity of multiple target genes by its local concentration.

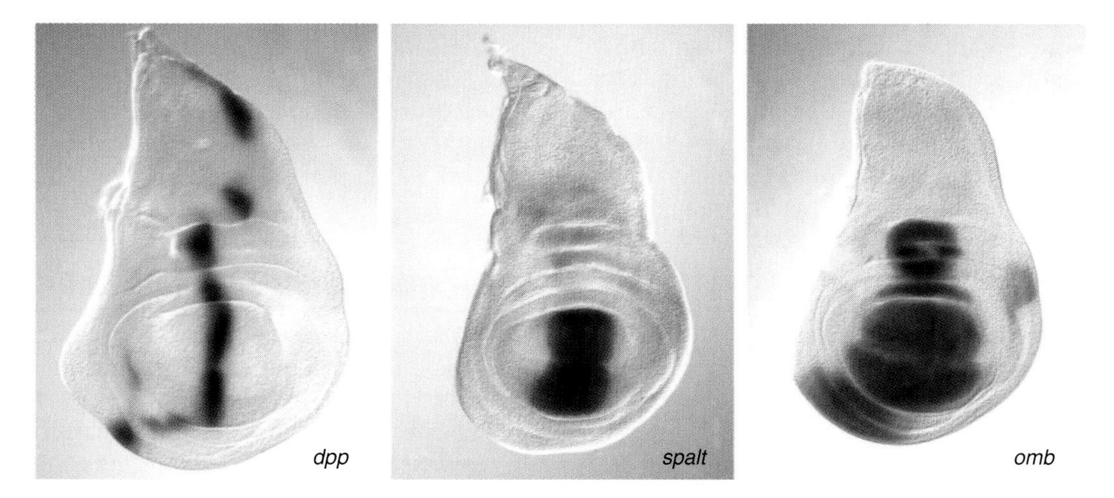

Figure 22.52 Expression patterns of *decapentaplegic*$^+$ (*dpp*$^+$), *spalt*$^+$, and *optomotor-blind*$^+$ (*omb*$^+$), as revealed by *lacZ* fusion genes (see Section 16.3), in the wing imaginal disc. *dpp*$^+$ is expressed along the anteroposterior compartment boundary (see Fig. 22.48). *omb*$^+$ and *spalt*$^+$ are expressed in nested domains centered on the strip of dpp synthesizing cells. Cells near the boundary express both *spalt*$^+$ and *omb*$^+$, whereas cells further away from the boundary express only *omb*$^+$.

The most compelling evidence for a long-range gradient—and against a sequential induction model—was obtained by generating cell clones with modified receptors for dpp. Clones with a constitutively active, ligand-independent dpp receptor expressed *spalt*$^+$ and *optomotor-blind*$^+$ ectopically in each cell of the clone but not in surrounding cells. This result shows that *spalt*$^+$ and *optomotor-blind*$^+$ are activated by dpp cell autonomously, not through sequential induction. If the latter were true, then the cell clone with the activated dpp receptors should have released the next signal in the inductive sequence, which should have caused *spalt*$^+$ and *optomotor-blind*$^+$ to be expressed outside the clone. However, this was not observed.

The sequential induction model would also predict that dpp receptor is required only close to the dpp source, and that other signals and receptors would take over at a distance from the dpp source. However, when cell clones deficient for dpp receptor originated away from the dpp source, dpp target genes were expressed outside the clones but not within the clones. This result proves that dpp receptor is required for activating target genes even at a distance from the dpp source. Moreover, clones with a reduced ability to receive the dpp signal showed shifts in cell fate, adopting more anterior fates when the clone was in the anterior compartment and more posterior fates when the clone was in the posterior compartment.

The genetic analysis of the dpp morphogen gradient described above has much similarity with the analysis of an activin gradient in an amphibian embryo by means of mRNA injection and transplantation discussed earlier (see Section 21.5). Both lines of work indicate that morphogen gradients can contribute to pattern formation in development, even though other mechanisms may be active at the same time.

IMAGINAL DISC CELL MAY COMMUNICATE THROUGH CYTONEMES

Although the influence of decapentaplegic (dpp) extends over the entire wing disc (Zecca et al., 1995), it is not clear how readily dpp will diffuse from its source near the anteroposterior compartment boundary over a distance of dozens of cell diameters to the edge of the disc. Indeed, some members of the TGF-β superfamily bind to extracellular material (Taipale and Keski-Oja, 1997). Pursuing this question, Ramírez-Weber and Kornberg (1999) screened transgenic fly strains in which green fluorescent protein (GFP) was expressed under control of promoter-trapped *GAL4* inserts (see Section 15.6). In one of these strains, wing imaginal discs showed strong green fluorescence in much of the disc except for a broad swaths along the anteroposterior compartment boundary.

Close inspection of living (unfixed) discs revealed very thin cytoplasmic extensions, which the investigators termed ***cytonemes*** (Gk. *kytos,* "cell"; *nema,* "thread"). At 0.2 μm diameter, the cytonemes were barely visible, and they did not survive any attempt at a histological fixation that would have allowed electron microscopic inspection. For comparison, the *thin filopodia* extending from the secondary mesenchyme cells to the blastocoel roof in sea urchins were 0.2–0.4 μm in diameter (see Fig. 10.10).

The cytonemes in *Drosophila* imaginal discs originated from anterior and posterior disc cells and were oriented toward the anteroposterior compartment boundary. To see whether they were growing in response to a chemoattractant emanating from the middle of the disc, the investigators cut out fragments from wing discs and cultured them together (Fig. 22.53a). The fragments used were cut from the anterior (A) or posterior (P) margin of the disc and from a central (C) area straddling the compartment boundary. When an A fragment was cultured next to a C fragment, the A cells extended long cytonemes toward C (Fig. 22.53b). Likewise, P cells extended cytonemes toward a C fragment, but few if any cytonemes were extended in A+A, P+P, or C+C combinations.

The C fragment needed at least one wild-type allele of the *hedgehog*$^+$ gene to induce the outgrowth of cytonemes. However, the hedgehog protein itself did not seem to be the chemoattractant because cultured cells with a transgenic cDNA for hedgehog protein did not elicit the formation of cytonemes from A or P fragments. In the same test, dpp protein, which is thought to act as a morphogen in wing imaginal discs as discussed earlier, also failed to elicit cytoneme formation. In contrast, cells with a transgenic cDNA for *Drosophila* fibroblast growth factor (dFGF) were as powerful in inducing cytonemes as C fragments from discs. However, *in situ hybridization* (see Method 8.1) showed that the mRNA for dFGF is distributed evenly in wing imaginal discs, making it unlikely that dFGF is the natural chemoattractant for cytonemes.

The investigators propose that the cytonemes are conduits for a signal that provides positional information to anterior and posterior imaginal disc cells relative to the central area around the compartment boundary. Such a signal could act as a morphogen as long as the signal itself—or a second messenger generated by the signal—decays with time or distance as it moves from the cytoneme's tip to its cell body. A morphogen gradient would then exist in thousands of cytonemes rather than in the extracellular space. However, direct evidence that cytonemes confer positional information still needs to be obtained.

22.9 Patterning Genes in Other Insects

Drosophila has become the species in which embryonic development is best understood in genetic and molecular terms. How general are the lessons learned from this embryo? To answer this question, many investigators are now using well-known *Drosophila* genes as molecular probes to screen DNA libraries from other species for similar genes or cDNAs (see Method 15.5). Genes with extensive sequence similarity are often referred to as *homologous genes, homologues,* or *homologs.* In strict usage, calling two genes or other structures homologous implies that both have evolved from the same ancestral structure. In a genetic context, the same terms refer to the corresponding pairs of genes and chromosomes inherited from both parents (see Section 3.3). In the context of development and evolution, it is often useful to distinguish two more specifically defined homologies. Genes of different species that have evolved from a common precursor gene through one or more speciation

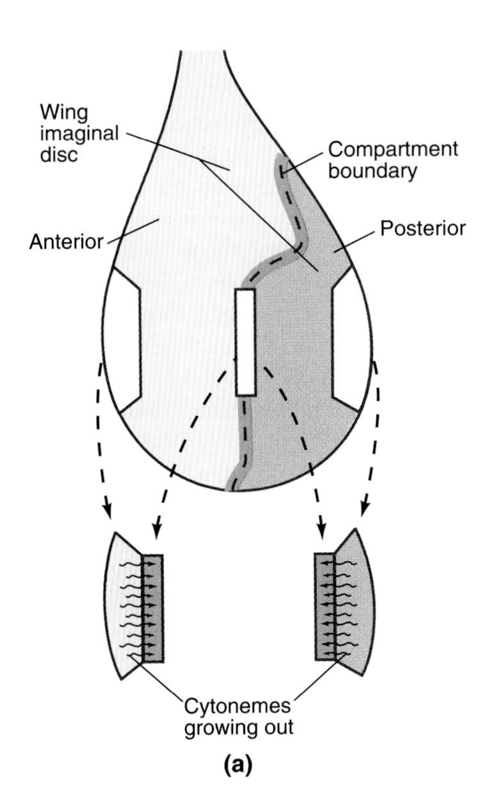

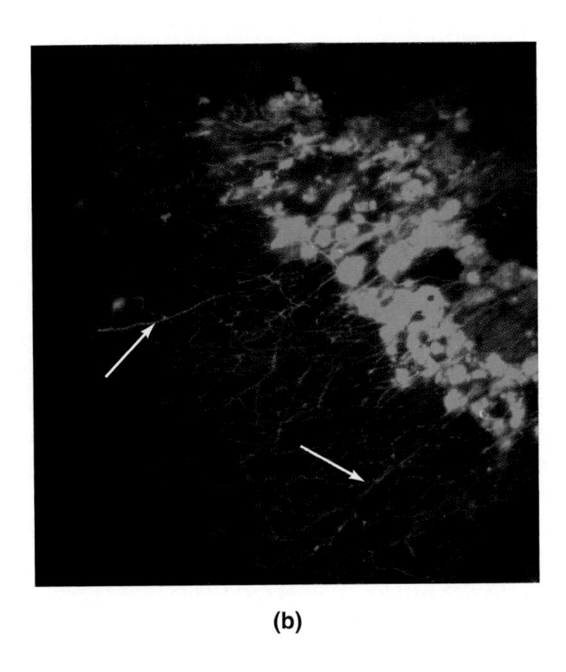

Figure 22.53 Cytonemes in imaginal discs of *Drosophila.* **(a)** Imaginal disc from a fully grown larva. (See Fig. 22.29 for a cross section of a disc.) The anterior and posterior compartments of the prospective wing and associated thorax are shown. Cells from the margins of the disc extend thin filamentous processes, called *cytonemes,* towards the compartment boundary. Explantation experiments were done to test whether the central part of the disc, straddling the compartment boundary, can induce marginal parts of the disc to extend cytonemes toward the central part. While no cytonemes were present during the first hour of coculture, numerous cytonemes grew out from the marginal parts toward the central part thereafter. **(b)** Photomicrograph showing marginal imaginal disc cells stained with green fluorescent protein. After being cultured next to a central part of the disc, the marginal cells sprout numerous cytonemes (arrows) in the direction of the central part.

events are called ***orthologous genes,*** or *orthologs* (Gk. *orthos,* "straight, upright"; *logos,* "word, meaning"). In contrast, genes that have originated by gene duplication *within the same chromosome* are called ***paralogous genes,*** or *paralogs* (Gk. *para,* "beside, near"). For example, the members of the actin gene family, encoding cytoskeletal actin, muscle actin, and so on, within the same species are paralogs. Even though orthologous and paralogous genes often have similar functions, this is not implied in the definition of orthology or paralogy. Rather, orthologous as well as paralogous genes are recognized by their sequence similarity.

Orthologs can usually be found between *Drosophila melanogaster* and its close relatives, such as the housefly *Musca domestica,* even though these two species have evolved separately for about 100 million years, or more than a billion generations. In particular, *bicoid*$^+$, *hunchback*$^+$, *Krüppel*$^+$, *knirps*$^+$, *tailless*$^+$, *hairy*$^+$, *engrailed*$^+$, and *Ultrabithorax*$^+$ are conserved with regard to their early embryonic expression patterns, while some of the later expression patterns have diverged (Sommer and Tautz, 1991). The key maternal gene of the posterior system in *Drosophila melanogaster, nanos*$^+$, is also well conserved with regard to early expression pattern. This was shown for a close relative, *Drosophila pseudoobscura,* for two more distantly related species from the same genus, *Drosophila virilis* and *Drosophila hydei,* and for a midge, *Chironomus samoensis* (Curtis et al., 1995b). The evolutionary conservation was less striking in a functional test, in which the nanos ortholog mRNAs from these species were injected into *Drosophila melanogaster* embryos derived from *nanos*$^-$ mothers. The ability of the injected mRNAs to restore abdominal segment formation to the recipients decreased markedly from *Drosophila melanogaster* to *Chironomus samoensis.*

In similar rescue experiments, anterior egg cytoplasm from various flies and midges was injected into *Drosophila melanogaster* embryos derived from *bicoid*$^-$ mothers (Schröder and Sander, 1993). Again, the ability to rescue decreased within the fly group and was no longer present in more primitive dipterans or in the honeybee. The localized anterior determinants in eggs of the midge *Chironomus samoensis* also differ functionally and molecularly from *Drosophila melanogaster* bicoid mRNA. Inactivation of the anterior determinants in *Chironomus* results in the *double-abdomen* phenotype, which features a mirror-image duplication of the abdomen, while the *bicoid*$^-$ phenotype tends to be highly asymmetrical. Mimicking the double-abdomen phenotype in *Drosophila* requires transplanting nanos activity to the anterior pole (Ephrussi and Lehmann, 1992). Moreover, rescue experiments indicate that the strongest anterior determinant activity in *Chironomus* eggs is associated with an mRNA size fraction that is much smaller than bicoid mRNA (Elbetieha and Kalthoff, 1988).

Of great interest was the search for orthologs to *Drosophila* patterning genes in insects that develop their *germ band* in a more gradual manner. *Drosophila* and other evolved insects are known as *long germ* types because the embryonic germ band is longer than the egg and comprises all segments when it is first formed. In contrast, grasshoppers and other primitive insects belong to the *short germ* type because their germ bands are short as compared to the egg and initially represent only the anterior part of the head, while subsequent segment rudiments are added gradually from anterior to posterior (Sander, 1976). Various intermediate forms between these two extremes are known as *semi-long germ* types.

The flour beetle *Tribolium castaneum,* which belongs to the semi-long germ types, has been analyzed genetically and molecularly. These studies have revealed a single cluster of homeotic genes, corresponding to the combined *Antennapedia* and *bithorax* complexes of *Drosophila* (Beeman et al., 1993). Comparisons of both DNA sequences and mutant phenotypes indicate that the biological functions of these genes have been largely conserved between the two species, which have evolved separately for more than 300 million years. For instance, *Drosophila* embryos homozygous for loss-of-function alleles of *abdominal A* show transformations of parasegments in the anterior abdomen toward parasegment 6 (consisting of the posterior compartment of T3 and the anterior compartment of A1). The *Tribolium* ortholog produces similar loss-of-function phenotypes. Similarly, the expression patterns of *even-skipped*$^+$ and *wingless*$^+$ in *Tribolium* resemble the patterns observed in *Drosophila* in most respects, even though the segments in *Tribolium* are formed sequentially (Patel et al., 1994; Nagy and Carrol, 1994). In addition, *Tribolium* has a *hunchback*$^+$ gene that has an expression domain similar to its *Drosophila* ortholog but seems differently regulated (Wolff et al., 1998).

The grasshopper *Schistocerca gregaria,* which belongs to the short-germ type, as well as other arthropods, have *engrailed*$^+$ orthologs with expression patterns very similar to the one observed in *Drosophila* (Patel, 1993). Accumulating in the posterior portion of each segment, engrailed protein appears sequentially as the grasshopper head and thorax is first formed, and then as abdominal segments are added sequentially from a proliferation zone near the posterior end (Fig. 22.54).

Although *even-skipped*$^+$ is of critical importance for the regulation of *engrailed*$^+$ in *Drosophila* (see Fig. 22.26),

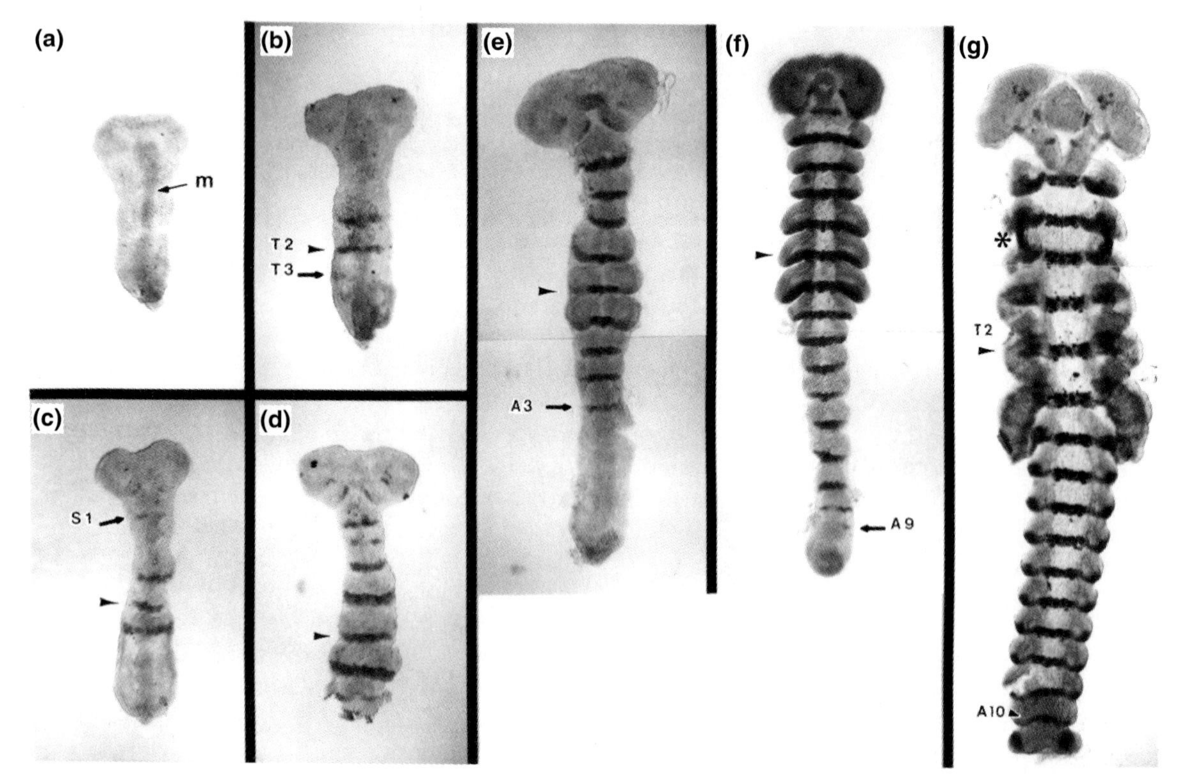

Figure 22.54 Expression of the *engrailed*$^+$ gene in the embryo of the grasshopper *Schistocerca gregaria.* Gene expression is shown by immunostaining using a monoclonal antibody to engrailed protein. The protein accumulates near the posterior margin of each segment. The photos are all at the same magnification, and anterior is always up. The position of the stripe belonging to the second thoracic segment is indicated by an arrowhead. **(a)** Early germ band primordium during mesoderm (m) formation. **(b)** Thoracic stripes are forming. **(c–f)** Additional stripes appear in head and abdomen as more segments are formed sequentially. **(g)** The final engrailed stripe has appeared in the 10th abdominal segment (A10).

the expression pattern of *even-skipped*$^+$ in *Schistocerca* differs greatly from the pattern observed in *Drosophila* (Patel et al., 1992). In the grasshopper embryo, *even-skipped*$^+$ activity never resolves into repetitive stripes, and even-skipped protein disappears before engrailed mRNA becomes detectable. Evidently, *even-skipped*$^+$ does not regulate the expression of *engrailed*$^+$ in *Schistocerca* as it does in *Drosophila*. Only the function of *even-skipped*$^+$ in generating the nervous system seems conserved between the fruit fly and the grasshopper. Likewise, the segmental expression pattern of *engrailed*$^+$ in the honeybee seems to be controlled without a *fushi tarazu*$^+$ ortholog (Walldorf et al., 1989). It appears that the overlapping control mechanisms of segmentation gene expression have allowed the regulatory hierarchy to evolve in different ways.

Of particular interest have also been studies on the development of the eyespot patterns that decorate the wings of many butterflies and seem to function in predator avoidance. These eyespots are formed by minute scales that are pigmented and secreted by the epidermal cells of the wing blades. The wing blades of butterflies develop from imaginal discs as in *Drosophila*. How is the determination of these cells coordinated so that they make scales with different pigments in concentric rings? To find out, Keys and coworkers (1999) cloned the butterfly *(Precis coenia)* orthologs of various *Drosophila* genes involved in wing patterning and studied their expression patterns in *Precis* imaginal discs by in situ hybridization (see Method 8.1). They discovered that the transcription patterns of *engrailed*$^+$, *hedgehog*$^+$, and two genes involved in the hedgehog signaling pathway (*patched*$^+$ and *cubitus interruptus*$^+$) were modified at the centers of developing eyespots. Thus, eyespots seems to have evolved through the redeployment of preexisting gene regulatory circuitry. Because such redeployments may lie at the heart of many evolutionary processes, the detailed knowledge of *Drosophila* development has become a favorite starting point for studies on the evolution of genetic control circuits.

Based on the small number of genes analyzed comparatively so far, it appears that the best-conserved patterning genes belong to the homeotic and segment polarity classes. These genes function not only in establishing the segmental body pattern, but also in maintaining it throughout development. Indeed, many of these genes have been conserved through most of the animal kingdom, as we will discuss in subsequent chapters.

table 22.1 Genes controlling pattern formation in *Drosophila* embryogenesis.

This table lists those patterning genes of *Drosophila* that are mentioned in this chapter. Expression patterns are usually given for transcription. If a protein is mentioned explicitly, then its pattern differs from that of the mRNA. Otherwise, the protein pattern can be assumed to follow the RNA pattern after a short time interval. Mutant phenotypes are usually those of homozygous loss-of-function alleles. Partial loss-of-function or gain-of-function alleles are mentioned explicitly. The molecular nature of a gene product is usually inferred from sequence comparisons with databases and in some cases supported by direct experiments. The developmental functions of a gene product are mainly inferred from mutant phenotype, molecular nature of gene product, and gene expression patterns in wild-type and mutant embryos.

Gene	Expression Pattern	Mutant Phenotype	Molecular Nature	Developmental Function
Maternal Anteroposterior Axis				
dicephalic (dic)	Maternal	Abnormal position of nurse cells relative to oocyte		Contributes to anteroposterior polarity of nurse cell-oocyte complex
gurken (grk)	Oocyte	Eggs ventralized with dorsal appendages reduced or missing	Similar to vertebrate transforming growth factor alpha (TGF-α), ligand for torpedo	Establishes first posterior and then dorsal identity of ovarian follicle cells
spindle (spn) (several loci)	Maternal germ line	Disturbed anteroposterior polarity of oocyte		Determination of prospective oocyte among cystocytes
torpedo (trp)	Ovarian follicle cells	Similar to *grk*	Receptor tyrosine kinase, receptor for grk	Same as grk

(continued on next page)

Gene	Expression Pattern	Mutant Phenotype	Molecular Nature	Developmental Function
Maternal Anterior Class				
bicoid (bcd)	mRNA transcribed in nurse cells and transported to anterior of oocyte. Translated during cleavage. Protein forms anteroposterior gradient.	Head and thorax replaced with mirror image of telson.	Transcriptional regulator (homeodomain and paired repeat)	Activates *hb* transcription; acts as morphogen on embryonic genes; expressed anteriorly
exuperantia (exu)	Testis and ovary	Reduction of head; defective gastrulation	RNA-binding protein	Needed to transport bcd mRNA from nurse cells to oocyte
hunchback (hb)	See Zygotic anteroposterior gap class			
staufen (stau)	Maternal germ line, embryonic central nervous system	Most abdominal segments missing, no pole cells, also head defects	Double-stranded RNA-binding protein	Required for bicoid mRNA localization and polar granule assembly
swallow (swa)	Maternal germ line	Reduction of head; cellularization defects	RNA-binding protein	Required for localization of bcd mRNA
Maternal Posterior Class				
cappuccino (capu)	Maternal	Abdominal segments A2–A7 missing; no pole cells	Putative transmembrane receptor	Required for polar granule assembly
nanos	Transcription in oocyte and nurse cells. Partial localization to polar granules. Only this localized mRNA is translated.	Absence of abdomen; pole cells normal	RNA-binding protein, translational repressor	Inhibits translation of maternal hb mRNA and hastens its destruction, hence allows activation of *kni* in posterior.
oskar (osk)	Maternal germ line	Similar to *capu*		Required for polar granule assembly, localizes nos mRNA
pumilio (pum)	Transcription maternal. Protein probably localized to posterior	Most of abdomen missing; pole cells normal	RNA-binding protein	Recruits nos protein to hb mRNA
spire (spir)	Maternal	Similar to *capu*	Actin-binding protein	Required for polar granule assembly
staufen (stau)	See maternal anterior system			
tudor (tud)	Male and female germ line	Similar to *capu*		Required for pole cell formation
vasa (vas)	Transcription in nurse cells and oocyte; mRNA uniform in early embryo. Protein localized in posterior of oocyte, egg, and in pole cells.	Similar to *capu*	RNA helicase	Required for pole cell formation
Maternal Terminal Class				
NTF-1			Transcriptional inhibitor	Inhibits transcription of *hkb and tll*

(continued on next page)

Gene	Expression Pattern	Mutant Phenotype	Molecular Nature	Developmental Function
torso (tor)	Transcription in nurse cells. Transport to oocyte, where protein distributed uniformly in plasma membrane.	Absence of labrum, reduction of head skeleton, absence of A8 and telson; gain of function; suppression of segmentation in thorax and abdomen	T[illegible]sine kinase receptor, likely receptor for active trunk protein	Activates *tll, hkb* by inhibiting *NTF-1*
torsolike (tsl)	Protein secreted by anterior and posterior ovarian follicle cells probably associates with vitelline envelope near egg poles	Similar to *tor* (loss of functon only)	Modifying protein	Generates active form of trunk near egg poles
Trunk (trk)	Maternal mRNA translated in early embryo, protein released into perivitelline space	Similar to *tor*	Secreted protein	Probably ligand of tor, after processing by tsl
Zygotic Anteroposterior Gap Class				
empty spiracles	Transcription during blastoderm 78–67% EL; later, in cells forming tracheal pits and spiracles	Defects in antennal segment, tracheal system	Transcriptional regulator (homeodomain)	Specifies procephalic segments
giant (gt)	Initial transcription at preblastoderm stage 82–60% and 33–0% EL; changes later	Defects in or absence of labrum, labium, and abdominal segments	Transcriptional regulator (Leucine zipper)	Activates *Scr;* represses *Kr*
huckebein (hkb)	Anterior and posterior ends of blastoderm	Defects in terminal body structures and gut	Transcriptional regulator (zinc finger domain)	Specifies terminal structures in concert with *tll*
hunchback (hb)	Maternal mRNA initially uniform; protein forms antero-posterior gradient. Embryonic transcription 100–50% EL and 20–10% EL. Posterior domain long-lasting.	Embryonic: deletion of labium, thorax, aA7/pA8 (PS 13). No pure maternal effect, but if mother and sperm are hb^{-}, then deletion of all mouthparts and thoracic segments and A1–A3, with mirror duplication of abdomen.	Transcriptional regulator (zinc finger domain)	Represses *kni;* represses *Kr* at high level and activates at low level; regulates *eve;* regulates *h, run, ftz;* activates *Antp* (with *ftz*); represses *Ubx*
knirps (kni)	Transcription starts at preblastoderm stage forming zone 45–30% EL. Ventral patch and narrow ring form in anterior by cellular blastoderm stage.	Replacement of A1–A7 by single A-type segment	Transcriptional regulator (zinc finger)	Represses *Kr;* regulates *h, run, eve, ftz;* represses *Abd-B* and *Antp*
Krüppel (Kr)	Transcription starts at preblastoderm stage in central zone, eventually expanding to 60–30% EL. Later also posterior patch and patch in head.	Purely embryonic. Head normal, deletion of thorax and A1–A5. Duplicated inverted A6.	Transcriptional regulator (zinc finger)	Activates *kni;* represses *gt;* regulates *eve;* regulates *h, run, ftz;* represses *Scr;* activates *Antp;* represses *Abd-B*
orthodenticle (otd)	Narrow band (75–92 %EL) in anterior blastoderm	Defects in eyes and anterior head parts	Transcriptional regulator (homeodomain, paired domain)	Activated by bcd and tor, otd—along with ems—specifies procephalic segments

(continued on next page)

Gene	Expression Pattern	Mutant Phenotype	Molecular Nature	Developmental Function
tailless (tll)	Transcription in termini of preblastoderm embryo	Head skeleton reduced but labrum present. Hindgut, Malpighian tubules absent.	Transcriptional regulator (steroid receptor family)	Activates *hb;* represses *Kr;* represses *kni;* activates *h*
Zygotic Pair-Rule Class				
even-skipped (eve)	Transcription starts at preblastoderm stage; 7 stripes corresponding to parasegments 1, 3, 5, etc., by cellularization stage. Fades during gastrulation.	Abolishes segmentation, giving lawn of denticles. Weak alleles delete odd parasegments.	Transcriptional regulator (homeodomain)	Represses *ftz, odd, run;* activates itself; activates *en;* represses *wg;* activates *Dfd*
fushi tarazu (ftz)	Transcription starts at preblastoderm stage; 7 stripes corresponding to parasegments 2, 4, 6, etc., by cellularization. Fades during germ band extension.	Deletes even parasegments	Transcriptional regulator (homeodomain)	Activates itself; activates *en;* activates *Antp;* activates *Ubx;* represses *wg*
hairy (h)	Transcription starts at preblastoderm stage; 7 stripes + dorsal head patch by time of cellularization. Fades during gastrulation.	Range from deletion of odd parasegments to formation of lawn of denticles. Also loss of labral tooth. Other alleles give extra hairs in adult.	Transcriptional regulator (helix-loop-helix domain)	Represses *run, ftz;* activates *eve*
paired (prd)	Transcription starts at preblastoderm stage; 8 stripes with dorsal head patch by cellularization. During germ band extension stripes 2–7 split and then fade.	Deletions about 1 segment wide starting in middle of denticle band affecting mainly odd abdominal segments	Transcriptional regulator (homeodomain)	Activates *wg, en*
runt (run)	Transcription starts at preblastoderm stage; 7 stripes. After cellularization, pattern becomes single-segment stripes, which persist through germ band extension.	Deletions more than 1 segment wide centered on T2, A1, A3, A5, A7, with mirror duplication of what is left	Transcriptional regulator (novel)	Represses *h, eve;* activates *ftz*
Zygotic Segment Polarity Class				
engrailed (en)	Transcription starts at blastoderm stage and produces 14 narrow stripes by extended germ band stage, marking the anterior of each parasegment. Expression in posterior compartments of larval imaginal discs.	Ventral cuticle a continuous lawn of denticles. Less severe alleles delete even-numbered parasegments	Transcriptional regulator (homeodomain)	Maintains *hh* and indirectly *wg* activity; selector gene defining posterior compartment
frizzled (fz)	Various imaginal discs	Variable orientation of wing hairs, irregular ommatidia in eye	Seven-pass transmembrane receptor	Receptor for wg

(continued on next page)

Gene	Expression Pattern	Mutant Phenotype	Molecular Nature	Developmental Function
frizzled-2 (*fz-2*)	Apparently redundant with *fz*			
hedgehog (*hh*)	Posterior compartments of larval segments and imaginal discs	Posterior compartment replaced with mirror image of anterior compartment	Secreted signaling molecule (TGF-β super family)	Maintains *wg* and indirectly *en;* defines source of segmental morphogen
patched (*ptc*)	In anterior compartments	Defects in segmentation and eye development	Multipass trans-membrane protein	Part of the receptor complex for hh
smoothened (*smo*)	In anterior compartments	Defects in various tissues including epidermis, wing, and eye	G protein-associated transmembrane receptor	Part of the receptor complex for hh; causes segregation of posterior from anterior compartment cells
wingless (*wg*)	Transcription starts at blastoderm stage and produces 14 narrow stripes by extended germ band stage, marking the posterior of each parasegment.	Ventral cuticle a continuous lawn of denticles, indicating deletion of 3/4 of each segmental unit	Secreted signaling molecule (Wnt family)	Maintains *en*
Homeotic Genes and Related Genes				
abdominal A (*abd-A*)	Transcribed in PS 7–13 of extended germ band	Transforms PS 7–9 to PS 6	Transcriptional regulator (homeodomain)	Represses *Ubx;* activates *wg, dpp;* specifies a.p. body pattern
Abdominal B (*Abd-B*)	Transcribed in PS 10–14 of extended germ band	Transforms PS 10–14 to PS 9	Transcriptional regulator (homeodomain)	Represses *Ubx;* specifies a.p. body pattern
Antennapedia (*Antp*)	Transcription starts in blastoderm; mainly PS 4, also in posterior parasegments. Thoracic imaginal discs.	Transforms PS 4, 5 to PS 3	Transcriptional regulator (homeodomain)	Specifies a.p. body pattern
Deformed (*Dfd*)	Transcription in preblastoderm stage in PS 1; later also PS 0. Eye-antennal disc.	Deletion of mandibular and maxillary segments	Transcriptional regulator (homeodomain)	Activates itself; specifies a.p. body pattern
labial (*lab*)	Transcription in extended germ band anterior to cephalic furrow and in posterior midgut	Deletes labial derivatives	Transcriptional regulator (homeodomain)	Specifies a.p. body pattern
Polycomb-group (*Pc-G*)		Multiple homeotic transformations		Maintains inhibition of inactive homeotic genes
Sex combs reduced (Scr)	Transcription in PS 2 in blastoderm; later PS 3 epidermis and mesoderm, and abdominal ganglia	Transforms PS 3 to PS 4 and PS 2 to PS 1	Transcriptional regulator (homeodomain)	Specifies a.p. body pattern
trithorax-group (*trx*-G)		Multiple homeotic transformations		Maintain expression of active homeotic genes

(continued on next page)

Gene	Expression Pattern	Mutant Phenotype	Molecular Nature	Developmental Function
Ultrabithorax (Ubx)	Transcription starts in cellular blastoderm, mainly in PS 6, but at lower levels in PS 5–13; more in even-numbered parasegments. Neuromeres; metathoracic discs.	Transforms PS 5, 6 to PS 4	Transcriptional regulator (homeodomain)	Represses *Scr;* represses *Antp;* activates itself; activates *wg, dpp;* selector gene
Maternal Dorsoventral System				
cactus (cac)	Maternal; action in embryo	Ventralizing	Genetic regulator	Inhibits entry of *dl* protein to nuclei
dorsal (dl)	Transcription in nurse cells; uniform mRNA in oocyte and egg; ventral-dorsal gradient of protein in preblastoderm nuclei	Embryos become tubes of dorsal epidermis. Some haploinsufficient alleles.	Genetic regulator	Activates *twi, sna;* represses *zen, dpp;* morphogen for dorsoventral pattern
easter (ea)	Maternal, released from embryo into perivitelline space	Similar to *dl*. Also weak ventralizing gain-of-function alleles.	Serine protease	Activation of spz (Tl ligand)
gastrulation defective (gd)	Maternal, released from embryo into perivitelline space	Similar to *dl*	Serine protease	Activation of spz (Tl ligand)
gurken (grk)	Maternal; needed in oocyte	Ventralizing; affects eggshell and embryo	Growth factor-like	Ligand for top protein; dorsalizing signal from oocyte
nudel (ndl)	Maternal; needed in follicle cells	Similar to *dl*	Protease activity	Contributes to ventral polarizing activity
pelle (pll)	Maternal, acts in embryo	Similar to *dl*	Protein kinase	Signal transduction from Tl to cac
pipe (pip)	Maternal; needed in follicle cells	Similar to *dl*	Glycosaminoglycan-modifying enzyme	Localized contribution to ventral polarizing activity
snake (snk)	Maternal, released from embryo into perivitelline space	Similar to *dl*	Serine protease	Activation of spz (Tl ligand)
spätzle (spz)	Maternal, released from embryo into perivitelline space	Similar to *dl*		Ligand for Tl protein
Toll (Tl)	Maternal; uniform distribution of protein in egg plasma membrane	Similar to *dl*. Gain-of-function alleles ventralize, producing denticle belts all around.	Receptor tyrosine kinase	Activated ventrally by spz; releases dl from cac, permits dl to enter nucleus
torpedo (top)	Maternal; needed in follicle cells	Similar to *grk*	Growth factor receptor	Receptor for grk; inhibits pip, thus orienting ventral polarizing activity.
tube (tub)	Maternal, acts in embryo	Similar to *dl*	Adapter protein	Signal transduction from Tl to cac
windbeutel (wbl)	Maternal; needed in follicle cells	Similar to *dl*	Enzyme associated with endoplasmic reticulum	Contributes to ventral polarizing activity.

(continued on next page)

Gene	Expression Pattern	Mutant Phenotype	Molecular Nature	Developmental Function
Zygotic Dorsoventral Class				
decapentaplegic (dpp)	Transcription in pre-blastoderm on dorsal side, curling around to ventral at poles. (see also Compartment-Related Genes)	Loss of amnioserosa and reduction of dorsal epidermis. Viable alleles produce multiple defects in imaginal disc derivatives.	Signaling molecule (TGF-β homologue)	Forms morphogen gradient with maximum dorsally. Promotes amnioserosa and epidermis; inhibits neurogenesis
short gastrulation (sog)	Protein synthesized blastoderm cells in ventral	Dorsalizes embryo	Plasma membrane protein	Anatagonizes dpp; specifies d.v. pattern
snail (sna)	Transcription in mid-ventral strip of prospective mesoderm	Loss of mesoderm	Transcriptional regulator (zinc finger domain)	Inhibits expression of other d.v. patterning genes in prospective mesoderm.
tolloid (tld)	Released from blastoderm cells into extracellular space	Ventralizes embryo	Metalloprotease	Cleaves extracellular domain of sog, especially in presence of dpp; specifies d.v. pattern
twist (twi)	Transcription in mid-ventral strip of prospective mesoderm and mesectoderm	Loss of mesoderm	Transcriptional regulator (basic helix-loop-helix domain)	Specifies dorsoventral pattern
zerknüllt (zen)	Transcription in preblastoderm stage on dorsal side	Loss of amnioserosa and optic lobe	Transcriptional regulator (homeodomain)	Promotes amnioserosa formation
Compartment-Related and Other Genes				
Achaete (Ac)	Proneural cells, prospective sensory bristles	Lack of sensory bristles and neurons	Transcriptional regulator (bHLH domain)	Many functions, incl. patterning around d.v. compartment boundary in wing
apterous (ap)	Dorsal compartments of wing and haltere discs	Wings and halteres missing	Transcriptional regulator (homeodomain)	Required for wing (haltere) outgrowth; selector gene
cubitus interruptus (ci)	Anterior compartment of each segment	Disturbed segmentation	Transcriptional regulator (zinc finger domain)	Activates *wg* and *dpp* in response to hh signaling
decapentaplegic (dpp)	In strip of cells anterior to a.p. compartment boundary (see also Zygotic Dorsoventral Class)	Abnormal wing patterns (see Fig. 22.50)	Secreted protein (TGF-β superfamily)	Morphogen for anteroposterior pattern of wing and haltere
Distalless (Dll)	Expressed around leg dorsoventral compartment boundary	Loss of legs and antennae	Transcriptional regulator (homeodomain)	Required for outgrowth of legs and antennae
fringe (fng)	Dorsal wing and haltere compartments	Loss of wings and halteres	Secreted protein	Juxtaposition of cells with and without fringe activates wg^{+}
hedgehog (hh)	see Zygotic Segment Polarity Class			
optomotor-blind (omb)	Prospective CNS and epidermal regions		Transcriptional regulator	Target of dpp in patterning of wing and abdominal segments

(continued on next page)

Gene	Expression Pattern	Mutant Phenotype	Molecular Nature	Developmental Function
scute (sc)	Similar to *Achaete*			
spalt (sal)	In endoderm and parts of wing imaginal disc		Transcriptional regulator (zinc finger domain)	Target of dpp in wing patterning
vestigial (vg)	In prospective appendages of wing and haltere imaginal discs	Loss of wings and halteres	Putative transcriptional regulator	Specification of wing and haltere as opposed to thorax

SUMMARY

Large mutagenesis screens have uncovered about 200 genes governing embryonic pattern formation in *Drosophila.* These genes form a regulatory cascade, which begins anew in each generation. During oogenesis, the products of a few maternally expressed genes are deposited in specific areas of the egg. These localized mRNAs and proteins generate uneven distributions of transcription factors, which in turn act on a small number of embryonic target genes. These first embryonic patterning genes are expressed in broad transverse stripes, which are the beginning of an anteroposterior body pattern, and in longitudinal stripes that form the rudiment of a dorsoventral body pattern.

The first embryonic genes involved in anteroposterior patterning are known as gap genes. Loss of function in a typical gap gene causes deletion of several contiguous segments. Gap gene expression domains differ in width and overlap to varying degrees, but none of them repeats periodically. Nevertheless, gap gene products direct the expression of a second tier of embryonic genes, the pair-rule genes, in periodic stripes. Each stripe and an adjacent interstripe together are as wide as two segments but are not necessarily congruent with prospective segments. Several pair-rule genes jointly control a third tier of embryonic genes, called segment polarity genes. Their expression domains delineate parasegments, metameric units of the early embryo that each correspond to the posterior compartment of a definitive segment plus the anterior compartment of the following segment.

Whereas the boundaries of parasegments are defined by the functions of segment polarity genes, the individual character of each segment is determined by a unique combination of homeotic genes. Homeotic genes are characterized by dramatic mutant phenotypes in which certain body parts are transformed into other parts normally formed elsewhere. Homeotic genes have staggered expression domains, and the combination of homeotic genes active in a segment controls its morphological identity. In *Drosophila,* the homeotic genes are clustered in two chromosomal regions, and the more 3′ a gene is located in a cluster, the more anterior begins its expression domain in the embryo.

The expression of segment polarity and homeotic genes begins at the blastoderm stage, when the fate of embryonic cells becomes restricted to their future segments. These genes are expressed throughout development and are the genes that maintain segmental boundaries and the individual characteristics of each segment.

The dorsoventral body pattern is specified in a similar manner. Maternal gene products set up an uneven distribution of a transcription factor that acts on a small number of embryonic target genes. The dorsoventral pattern is generated nearly independently of the anteroposterior pattern.

Some of the patterning genes identified in *Drosophila* have been linked by the compartment hypothesis to compartments—that is, areas of epidermis defined by clonal restriction lines. Such compartments appear to be the basic units of growth control and morphogenesis. According to the compartment hypothesis, each compartment is characterized by a specific combination of active selector genes that maintain its boundaries and generate its specific morphology by controlling presumably large numbers of realizator genes.

Imaginal discs are divided in anterior and posterior compartments. In the posterior compartment of the wing disc, the *engrailed*$^+$ gene activates the synthesis of hedgehog (hh), a secreted protein. The latter diffuses for a short distance across the compartment boundary, where it activates the gene for another secreted protein, decapentaplegic (dpp). The latter acts as a morphogen, having its highest concentration in a strip of cells anterior to the boundary and grading away toward the anterior and posterior wing margins.

The genetic and molecular analysis of pattern formation in *Drosophila* embryos has confirmed, to an extent, the classical gradient model of morphogens. At least three of the maternally encoded transcription factors—and several other gene products as well—control their target genes in a concentration-dependent way. However, most gradients in *Drosophila* govern only part of a body axis, and the number of target genes in each case is rather small. Local interactions between nuclei and the epigenetic generation of complex patterns from simpler precursors have been found to play a larger role than anticipated.

Some of the genetic hierarchies that govern pattern formation in *Drosophila* include overlapping regulatory mechanisms. Such mechanisms seem to be common among insects, and somewhat different systems of control have evolved from them in different insect groups. So far, it appears that the best-conserved patterning genes belong to the homeotic and segment polarity classes. These genes function not only in establishing the segmental body pattern, but also in maintaining it throughout development.

chapter 23

The Role of *Hox* Genes in Vertebrate Development

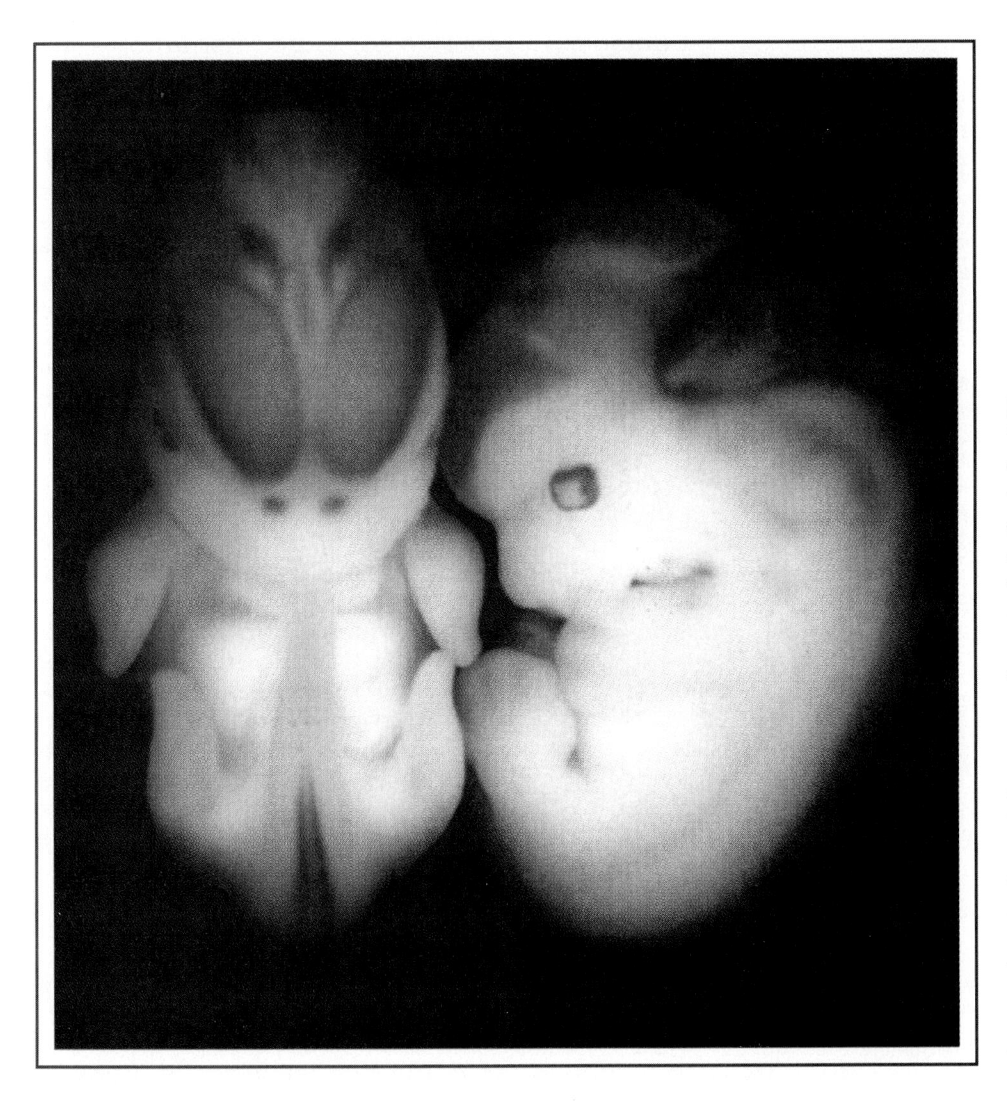

Figure 23.1 A method for identifying mouse genes involved in embryonic pattern formation relies on a transgene that produces a blue stain in fixed host tissues where the transgene has been active. The transgene has no promoter of its own, so that it reveals the activity pattern of a resident promoter located nearby in the host genome. The photograph shows transgenic fetuses with blue stain (dark in photograph) in the brain and eyes (top) as well as the spinal cord (visible in tail at bottom), indicating that the transgene is driven by a resident promoter that is active specifically in the central nervous system.

THE genetic and molecular analysis of pattern formation in *Drosophila* embryos has been a spectacular success. The application of these methods has yielded concrete and detailed answers to long-standing questions in development. But how general are the lessons learned from this fruit fly? As humans, we have a particular interest in the development of vertebrates, especially mammals. Naturally, we are eager to know how much of the insight gathered from flies applies to our closer phylogenetic relatives.

Work with *Drosophila* has shown that mutants are the most valuable tools for studying the functions of genes. Unfortunately, the most tractable vertebrates for experimentation—amphibians and birds—are unsuited for mutational analysis because of their large genomes and long generation times. The zebrafish has recently become the first vertebrate on which large-scale mutagenesis screens have been carried out (Rossant and Hopkins, 1992; Haffter et al., 1996; Driever et al., 1996). These efforts will pay rich dividends once the genes now identified by mutations have been mapped and cloned.

Genetically, the best-known vertebrate so far is the house mouse, *Mus musculus.* More than 2000 mouse genes have been mapped, and a growing number of these genes have been assigned to cloned DNA segments (M. C. Green, 1989; Wagner, 1990). This number is only about 5% of all mouse genes, but more mutants are constantly being generated by insertional mutagenesis and gene targeting (Fig. 23.1).

In this chapter, we will explore the use of genetic and molecular methods for the analysis of pattern formation in vertebrate embryos. Featured prominently in specifying the anteroposterior body pattern of flies and vertebrates is a set of homeobox-containing genes. After describing the evolution and expression patterns of these genes in mice, we will examine whether they act as homeotic selector genes like their counterparts in *Drosophila.*

The dorsoventral axis of vertebrates is controlled by a set of gene products that have exact homologs in *Drosophila.* Ironically, factors that are ventralizing in *Drosophila* are dorsalizing in vertebrates and vice versa, supporting the somewhat controversial view of Étienne Geoffroy Saint-Hilaire, a French naturalist, who proposed in 1822 that vertebrates and invertebrates shared the same basic body plan, except that one was upside down.

The study of vertebrate limb development, a long-standing interest of classical embryologists, has been rekindled through the use of transgenic mice and chickens to which transgenic cells have been grafted. These investigations are leading to a new understanding of limb formation in genetic and molecular terms.

Finally, limb development provides an opportunity to revisit the *principle of reciprocal interactions.* The mesenchymal core of a limb bud and its ectodermal cover exchange reciprocal signals through which the two components maintain each other and coordinate their contributions to limb development. The same principle is observed in many other organs that develop from mesenchymal and epithelial rudiments.

23.1 Strategies for Identifying Patterning Genes in Vertebrates

A variety of strategies have been used to identify patterning genes in mice (M. Kessel and P. Gruss, 1990). While traditional approaches proceed from analyzing mutant alleles to cloning the affected segments of genomic DNA, alternative approaches known as ***reverse genetics*** begin with the cloning of a gene and then employ molecular techniques to generate mutant alleles or their equivalents. The ultimate goal of both traditional and reverse genetics is to infer the biological role of a gene from its molecular features and from the phenotypes generated by mutant alleles.

GENES ARE CLONED ON THE BASIS OF THEIR CHROMOSOMAL MAP POSITION

The traditional strategy for identifying genes involved in a particular function starts with collecting mutant alleles that show related phenotypes. Such mutants are used to map a genetic locus based on the frequencies with which the mutant phenotypes of interest recombine with the phenotypes of known marker genes. With luck, a DNA segment close to the new locus of interest has already been cloned. In this case, one can "walk" from the previously cloned DNA segment to the new locus by cloning the intervening DNA in a series of large overlapping segments.

A mouse gene that was first discovered as a spontaneous mutant and later cloned on the basis of its mapped chromosomal position is *Brachyury,* also known as the *T-locus* (Dobrovolskaia-Zavovskaia, 1927; Herrmann et al., 1990). Proper expression of T^+ is necessary and sufficient for normal mesodermal cell movements during gastrulation and somite development (V. Wilson and R. S. Beddington, 1997). In situ hybridization (see Method 8.1) indicates that T^+ is first expressed early in the entire embryonic mesoderm but subsequently restricted to chordamesoderm. The Brachyury protein predicted on the basis of its cDNA sequence is a transcription factor, and its target genes appear to include cell adhesion molecules (V. Wilson et al., 1995).

PROMOTER TRAPPING IDENTIFIES PATTERNING GENES

In an alternative strategy, known as *promoter trapping,* researchers insert a bacterial reporter gene, such as *lacZ,* randomly into the mouse genome (see Section 15.6). If by chance the reporter gene "lands" next to a promoter that drives a patterning gene, then the reporter gene will be expressed in the transgenic mouse with the same tissue- and stage-specificity as the endogenous patterning gene would be in a normal mouse. For instance, if the blue stain caused by the *lacZ* gene product appears specifically in the neural tube, then *lacZ* would seem to be driven by the promoter of a gene involved in neurulation (Fig. 23.1). Because the reporter gene has separated the resident gene from its promoter, the resident gene typically loses some or all of its function. The reporter gene insertion then has created a loss-of-function allele, which can be bred to homozygosity to study its biological effects (Friedrich and Soriano, 1991). Finally, because the reporter gene is not a natural part of the mouse genome, the reporter gene also "tags" its insertion site for easy cloning.

VERTEBRATE GENES CAN BE ISOLATED WITH MOLECULAR PROBES FROM *DROSOPHILA* GENES

Another method for identifying previously unknown genes relies on the fact that complementary nucleic acids hybridize even if their nucleotide sequences do not match perfectly (see Section 15.3). Therefore, a gene from one species can be used as a hybridization probe to identify *orthologs*—that is, genes of similar nucleotide sequence in another organism. Under proper *stringency conditions,* a *Drosophila* probe will hybridize with any cloned segment of mouse genomic DNA or cDNA that contains a nearly complementary sequence. The amino acid sequence of the protein encoded by the *Drosophila* gene from which the probe was taken is then compared to the predicted amino acid sequence of its mouse ortholog. The more extensive the sequence similarities are, the greater is the chance that the two proteins have similar *biochemical* functions in the fly and the mouse. This strategy has allowed molecular biologists to "fish" for vertebrate genes that might be related to well-known genes from *Drosophila, Caenorhabditis elegans,* or yeasts.

Orthologs with similar *biochemical* functions do not necessarily have corresponding *biological* functions. The biological function of the mouse ortholog of a known *Drosophila* gene must therefore be established by studying the phenotypes of mice with mutated alleles of the cloned gene. But how does one decide whether the phenotype of an available mouse mutant is caused by an alteration in a gene that has been cloned as an ortholog of a *Drosophila* gene? One way of making this connection is to study the expression pattern of the cloned ortholog by *in situ hybridization* (see Method 8.1). This may lead to an informed guess on which of the known mutant mouse phenotypes might be generated by a gene with the expression pattern observed.

For example, the murine *Pax 1*$^+$ gene was obtained by screening a mouse DNA library with the *paired box,* a repetitive element contained in the *paired*$^+$ gene of *Drosophila* (M. Kessel and P. Gruss, 1990). In situ hybridization revealed that *Pax 1*$^+$ is expressed in intervertebral discs, the sternum (breastbone), and the thymus gland. The mouse mutant *undulated* is characterized by defective intervertebral discs, a coincidence suggesting that perhaps *undulated* may be a mutant allele of *Pax 1*$^+$. Sequence comparisons revealed that each mutant allele of *undulated* was also abnormal in its *Pax 1* DNA sequence and expressed *Pax 1* in an abnormal pattern (S. Dietrich and P. Gruss, 1995). These results proved beyond reasonable doubt that the *undulated*$^+$ gene and the cloned *Pax 1*$^+$ DNA are identical.

Another way of proving identity between a mutant allele and a cloned gene would be phenotypic rescue of mutant mice by introducing the cloned wild-type allele as a transgene. Yet another strategy for revealing the biological function of a cloned gene is to generate transgenic mice in which the gene of interest is "knocked out" by targeted recombination or overexpressed from an added transgene. We will discuss applications of these two techniques in Section 23.3.

23.2 Homeotic Genes and *Hox* Genes

Molecular analysis of *Drosophila* patterning genes revealed that many of them share a consensus sequence of 180 nucleotides (W. McGinnis et al., 1984; Scott and Weiner, 1984). This sequence was dubbed the ***homeobox*** because it was discovered in two homeotic genes, *Antennapedia*$^+$ and *Ultrabithorax*$^+$. Since then, homeoboxes have been found in all homeotic genes and many other patterning genes of *Drosophila.* As researchers began to use homeoboxes from *Drosophila* genes as probes to screen DNA libraries from other organisms, it became apparent that the homeobox has been highly conserved in evolution from yeasts through all plants and animals. A typical homeobox-containing gene, the *Antennapedia*$^+$ gene of *Drosophila,* is shown in Figure 23.2.

To avoid confusion, its important to distinguish genes defined by characteristic DNA sequences from genes defined by their biological effects. The term ***homeobox gene*** is generally used for genes that contain a homeobox sequence. In contrast, we will reserve the term *homeotic gene* for genes whose mutant alleles cause replacements of certain body parts with parts normally formed elsewhere. In *Drosophila,* all homeotic genes are also homeobox genes, but some homeobox genes (such as *fushi tarazu*$^+$ and *engrailed*$^+$) are not homeotic. Similarly, some but not all homeobox genes in vertebrates have phenotypic effects that can be interpreted as being homeotic (see Section 23.3).

THE HOMEODOMAIN IS A COOPERATIVE DNA-BINDING REGION

Every protein encoded by a homeobox gene contains a ***homeodomain***—that is, the segment of 60 amino acid residues that are encoded by the homeobox (Fig. 23.3). Sequence comparisons have revealed two important features of homeodomain-containing proteins. First, the homeodomains of the proteins are even more similar than the homeoboxes of the genes that encode them.

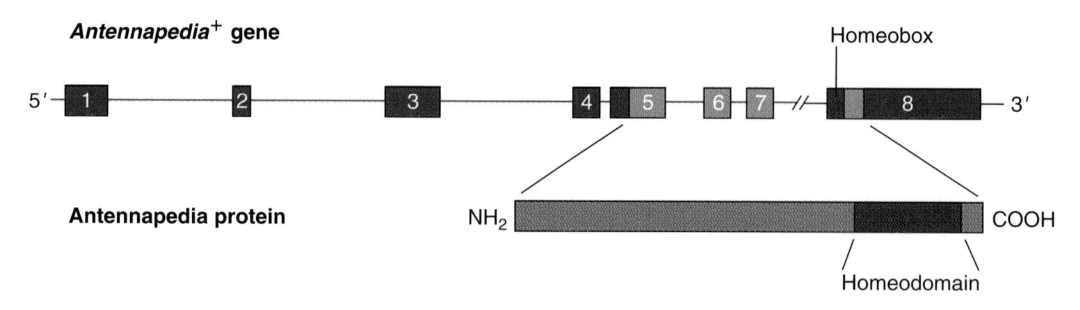

Figure 23.2 Structural organization of the *Antennapedia*$^+$ gene and the Antennapedia protein. Exons 1 through 8 are separated by introns, which are not drawn to scale. The homeobox, located in exon 8, is 180 nucleotides long. Within the Antennapedia protein, the homeodomain is located near the carboxyl terminus (COOH). The homeodomain encompasses 60 amino acids in a sequence that has been highly conserved during evolution.

Amino acid sequence of homeodomain

(Helix 1, Helix 2 and Helix 3 are shaded in the original figure.)

Species	Gene	Sequence
	Antp	RKRGRQTYTRYQTLELEKEFHFNRYLTRRRRIEIAHALCLTERQIKIWFQNRRMKWKKEN
Mm	*Hoxb9*	SRKK-CP--K----------L--M----D--H-V-RL-N-S---V---------M---M-
Mm	*Hoxa9*	TRKK-CP--KH---------L--M----D--Y-V-RL-N-----V---------M---I-
Mm	*Hoxc9*	TRKK-CP--K----------L--M----D--Y-V-RV-N-----V---------M---M-
Mm	*Hoxd9*	TRKK-CP--K----------L--M----D--Y-V-RI-N-----V---------M---MS
Dm	*Abd-B*	VRKK-KP-SKF---------L--A-VSKQK-W-L-RN-Q-----V---------N---NS
Bf	*Hox-9*	SRKK-CP---F---------LY-M----E--Y--SQHVN-----V---------M---MS
Mm	*Hoxb5*	G--A-TA------------------------------------S-----------------D-
Mm	*Hoxa5*	G--A-TA------------------------------------S-----------------D-
Dm	*Scr*	T--Q-TS---L----H
Bf	*Hox-5*	N--T-TA---
Mm	*Hoxb4*	P--S-TA---Q-V--------Y---------V--------S-----------------DH
Mm	*Hoxa4*	P--S-TA---Q-V-----------------------T---S---V-------------DH
Mm	*Hoxd4*	P--S-TA---Q-V-----------------------T---P-----------------DH
Dm	*Dfd*	P--Q-TA---H-I--------Y--------------T-V-S-----------------D-
Bf	*Hox-4*	T--S-TA---Q-V-----------------------S-G-------------------D-
Mm	*Hoxb3*	S--A-TA--SA-LV------------C-P--V-M-NL-N-S-------------Y---DQ
Mm	*Hoxa3*	S----TA---P-LV------------M-P--V-M-NL-N---------------Y---DQ
Mm	*Hoxd3*	S--A-TA--SA-LV-----------FV-P--VQM-NL-N-S-------------Y---DQ
Dm	*zenZ1*	L--S-TAF-SV-LV---N--KS-M--Y-T------QR-S-C---V---------F---DI
Bf	*Hox-3*	G--A-TA--SA-LV------------C-P--V-M-AM-N---------------Y----Q
Mm	*Hoxb1*	PRGL-TNF-TR-LT---------K--S----V---PT-E-N-T-V---------Q---RE
Mm	*Hoxa1*	PNAV-TNS-TK-LT---------K----AA-V---AS-Q-N-T-V---------Q---RE
Dm	*lab*	NNS--TNF-TK-LT--------------A------NT-Q-N-T-V---------Q---RV
Bf	*Hox-1*	PNN--TNF-TK-LT-------Y-K----A--V---A--N-N-T-V---------Q---RE
Dm	*Exd*	AR-K-RNFSKOASEI-NEY-YSHP-PSEEAKE-L-RKCGI-VSOVSN--G-K-IRY--NI
Homo	*Pbx1*	AR-K-RNFNKOASEI-NEY-YSHP-PSEEAKE-L-KKCGI-VSOVSN--G-K-IRY--NI

▲ LSN (arrow at the HP of the Exd/Pbx1 sequences)

Figure 23.3 Amino acid sequence comparisons of various homeodomains from house mouse (*Mus musculus, Mm*), fruit fly (*Drosophila melanogaster, Dm*), Amphioxus (*Branchiostoma floridae, Bf*), and human (*Homo sapiens*). The Antennapedia sequence of *Drosophila* shown at the top is used as the standard for all comparisons. Each of the letters in this sequence represents one amino acid. Dashes represent amino acids identical to the Antennapedia standard. The shaded areas within each homeodomain represent the segments that form the three α helices that characterize the homeodomain as it binds to DNA (see Fig. 23.4). Many homeodomains fall into natural groups based on the common deviations from the Antennapedia standard. These groups represent orthologous or pseudo-orthologous genes (see Fig. 23.6). The two homeodomains at the bottom are atypical. They represents the class of TALE (for Three Amino acid Loop Extension) proteins, so named for three extra amino acids—as compared to typical homeodomains—in the loop between helix 1 and helix 2.

(The genetic code is degenerate, meaning that two or more codons may encode the same amino acid.) Second, all homeodomain-containing proteins seem to be transcription factors (Scott et al., 1989; Gehring et al., 1994). Together, these observations suggest that the homeodomain has a critical and interactive function in this class of transcription factors, such as binding to specific DNA sequences. This hypothesis was confirmed by X-ray crystallography, which revealed the three-dimensional structure of the homeodomain as it binds to DNA (Gehring et al., 1994; Wolberger et al., 1991).

The homeodomain is characterized by three α helices (Fig. 23.4). Helix 3 is called the *recognition helix:* It is aligned with the major DNA groove, where it interacts with specific nucleotide sequences. This helix is also the best-conserved segment of all homeodomains. Helices 1 and 2 are at right angles to helix 3 and farther away from the DNA.

The evolutionary conservation of the homeodomain, in particular of helix 3, explains the "promiscuity" in the binding of isolated homeodomains to DNA in vitro. A given DNA nucleotide sequence will be bound by several homeodomains about equally well, and vice versa. These results have been very puzzling because the apparent lack of binding specificity makes it hard to account for the distinct morphological effects observed in each homeotic mutant of *Drosophila*. At least a partial resolution of this paradox has come from studies on a class of atypical homeodomain proteins that act as cofactors in the DNA binding of typical homeodomain

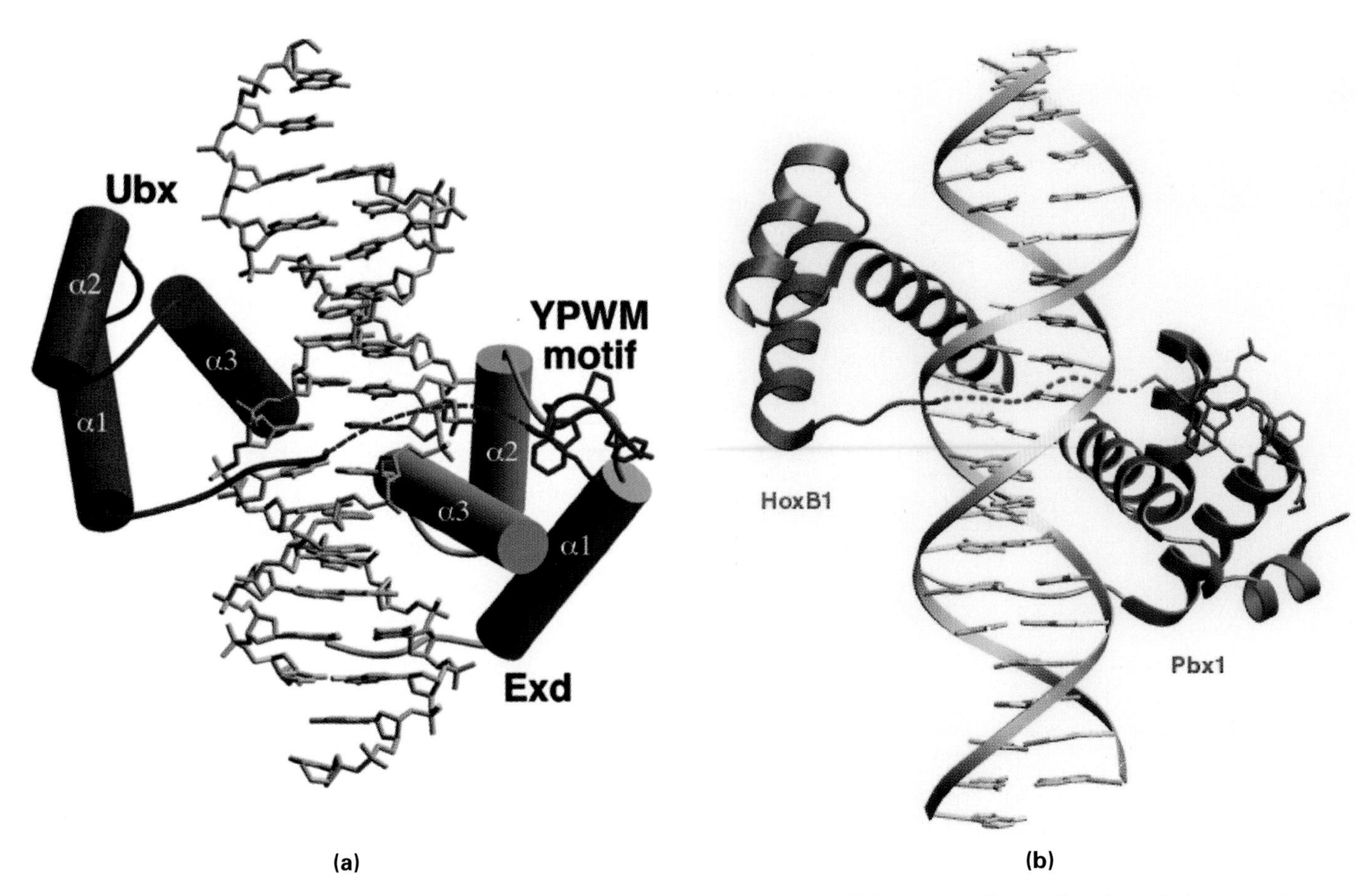

Figure 23.4 Models illustrating the binding of two homeodomains (red), each linked to a cofactor (blue), to their respective target DNAs (yellow). Both models are based on data obtained by X-ray crystallography. **(a)** *Drosophila* Ultrabithorax homeodomain with Extradenticle (Exd) cofactor. **(b)** Human HoxB1 homeodomain with Pbx1 cofactor. Note the similarity of the two models despite differences in rendition. The red and blue cylinders in part a, corresponding to the red and blue spirals in part b, represent the α helices formed by three segments within each homeodomain (see Fig. 23.3). The dashed red line in each model represents the linker between helix 1 and the YPWM sequence at the amino terminus of the homeoprotein. This sequence binds to a matching pocket of the cofactor. The helices marked α3 in part a are arranged in tandem, docking in the major groove of the DNA double helix on opposite sides of the molecule.

proteins such as Ultrabithorax. These cofactors are known as *TALE* (*Three Amino acid Loop Extension*) proteins because they have an extra three amino acids—as compared to typical homeodomains—in the loop between helix 1 and helix 2 (see Figs. 23.3 and 23.4).

The best-known member of the TALE class in *Drosophila* is encoded by the *extradenticle*$^+$ gene. Mutations in this gene cause multiple homeotic transformations even though the typical homeotic genes are intact and expressed normally, indicating that the latter genes are not regulated by *extradenticle*$^+$ (Peifer and Wieschaus, 1990). Instead, extradenticle protein (Exd) cooperates with typical homeotic proteins, such as Ultrabithorax (Ubx), in DNA binding (Passner et al., 1999). The two proteins bind on opposite sides of the DNA molecule, with their α3 helices close to each other and oriented in tandem (Fig. 23.4). Also, the amino terminus of Ubx reaches around the DNA double helix, inserting a conserved sequence of four amino acid residues (YPWM) into a matching pocket formed by the TALE region of Exd. It is thought that Exd binds first to its DNA target site, slightly distorting the DNA double helix and facilitating the binding of Ubx to its adjacent target site and to Exd. Under these conditions, the binding of Ubx depends not only on its match with the DNA target site but also on its match with Exd. Thus, Ubx binds more specifically to the target DNA-Exd complex than to the same target DNA alone.

In humans, a well-investigated TALE protein is known as Pbx1; it forms a heterodimer with the HoxB1 protein, a typical homeodomain protein (Piper et al., 1999). Again, the degree of conservation between the Ubx/Exd and HoxB1/Pbx1 heterodimers and the way they bind to each other and to their adjacent DNA target sites is astounding (see Figs. 23.3 and 23.4). About 70 TALE proteins are known from a wide variety of organisms ranging from fungi to various plants and animals. The TALE proteins are encoded by homeobox-containing genes located outside the homeotic gene clusters. Analysis of TALE protein sequences indicates that they form a distinct group that diverged long ago from the typical homeodomain proteins (Bürglin, 1997). The formation of heterodimers between these two classes of proteins, and their cooperative binding to adjacent DNA target sequences explains, at least in part, the specific morphological effects of homeotic mutations. However, it is still unclear how many target genes homeodomain proteins have, and whether they control these target genes directly or indirectly (*Mann, 1995; Graba et al., 1997; Mannervik, 1999*).

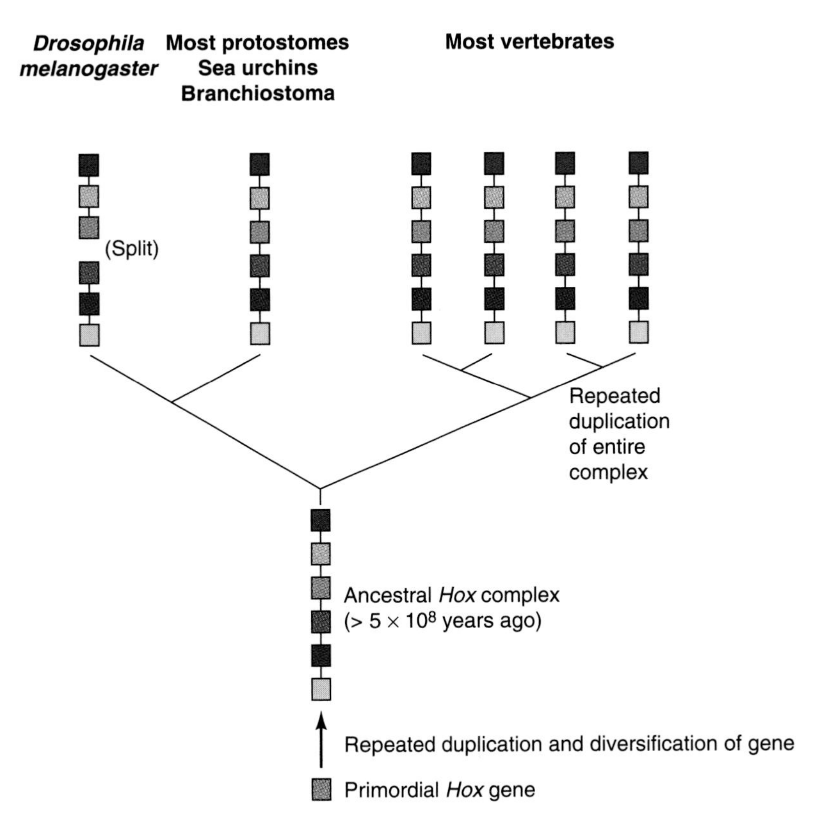

Figure 23.5 Simplified model of the evolution of the *Hox* complex. A primordial *Hox* gene was duplicated and diversified several times to generate a complex consisting of several *Hox* genes. This ancestral *Hox* complex must have existed in organisms that lived before the evolutionary separation of vertebrates and invertebrates. During vertebrate evolution, the entire complex was duplicated repeatedly, and four different chromosomes retained one copy each. During the evolution of *Drosophila,* the *Hox* complex was split into the *Antennapedia* complex (ANTP-C) and the bithorax complex (BX-C), both located on the same chromosome. Most invertebrates still have one *Hox* complex.

HOX GENES CAN BE DIVIDED IN PARALOGY AND ORTHOLOGY GROUPS

Throughout the animal kingdom, some homeobox genes are clustered in certain chromosomal regions while other homeobox genes occur outside these clusters. Examples of the latter group are *engrailed*$^{+}$ and the genes encoding the TALE proteins discussed above. We will now focus on the former group, the homeobox genes that occur in clusters. These genes have come to be known as ***Hox* genes,** and the clusters in which they occur are called ***Hox* complexes.**

One salient feature of *Hox* complexes is that they occur in all bilaterally symmetrical animals and have been exceedingly well conserved during evolution (Fig. 23.5; de Rosa et al., 1999). As discussed in Section 22.5, insects generally have one *Hox* complex, which in *Drosophila melanogaster* is split into the *Antennapedia complex* (*ANTP-C*) and the *bithorax complex* (*BX-C*). Similarly, all *protostomes* seem to have a single *Hox* complex. Among the *deuterostomes,* sea urchins also have one *Hox* complex, and the same is true for *Branchiostoma floridae,* representing the cephalochordates, a primitive sister group of the vertebrates (Garcia-Fernàndez and Holland, 1994). Mammals, including the mouse and the human, have four *Hox* complexes, and some fish species have six.

Another salient feature of *Hox* genes is that they can be arranged in *paralogy groups* and *orthology groups.* Figure 23.6 shows this arrangement for the *Hox* complexes of the fly *Drosophila melanogaster,* the mouse *Mus musculus,* and the lancelet *Branchiostoma floridae.* The ***paralogy groups*** simply are the *Hox* complexes—that is, the natural clusters of *Hox* genes on certain chromosomes. The order of *Hox* genes in a paralogy group is the cumulative result of those gene duplications that have occurred *within this chromosomal region* during evolution. The ***orthology groups*** are the vertical columns in Figure 23.6, and their existence is by no means trivial. The validity of the orthology groups is based on detailed sequence comparisons and on tests by heterospecific gene transfer.

Some of the sequencing data that form the basis for the orthology groupings shown in Figure 23.6 are shown in Figure 23.3. Here, the amino acid sequence of the Antennapedia homeodomain of *Drosophila melanogaster,* written out in the usual one-letter code, is used as a standard for comparisons with other homeodomains. For each of the other domains, an amino acid residue is shown as a dash if it is identical to the Antennapedia standard. Otherwise, the nature of the diverged amino acid residue is denoted by the appropriate letter. As it turns out, the deviations from the Antennapedia standard fall into natural groups characterized by similar sets of deviations. Therefore these groups represent genes with maximum sequence similarity. These genes fulfill the definition of *orthologs* if they belong to different species. If they belong to different *Hox* complexes of the same species, they are not technically orthologs, and we will refer to them as ***pseudo-orthologs.***

Remarkably, the genes of any given orthology group have corresponding positions on their respective

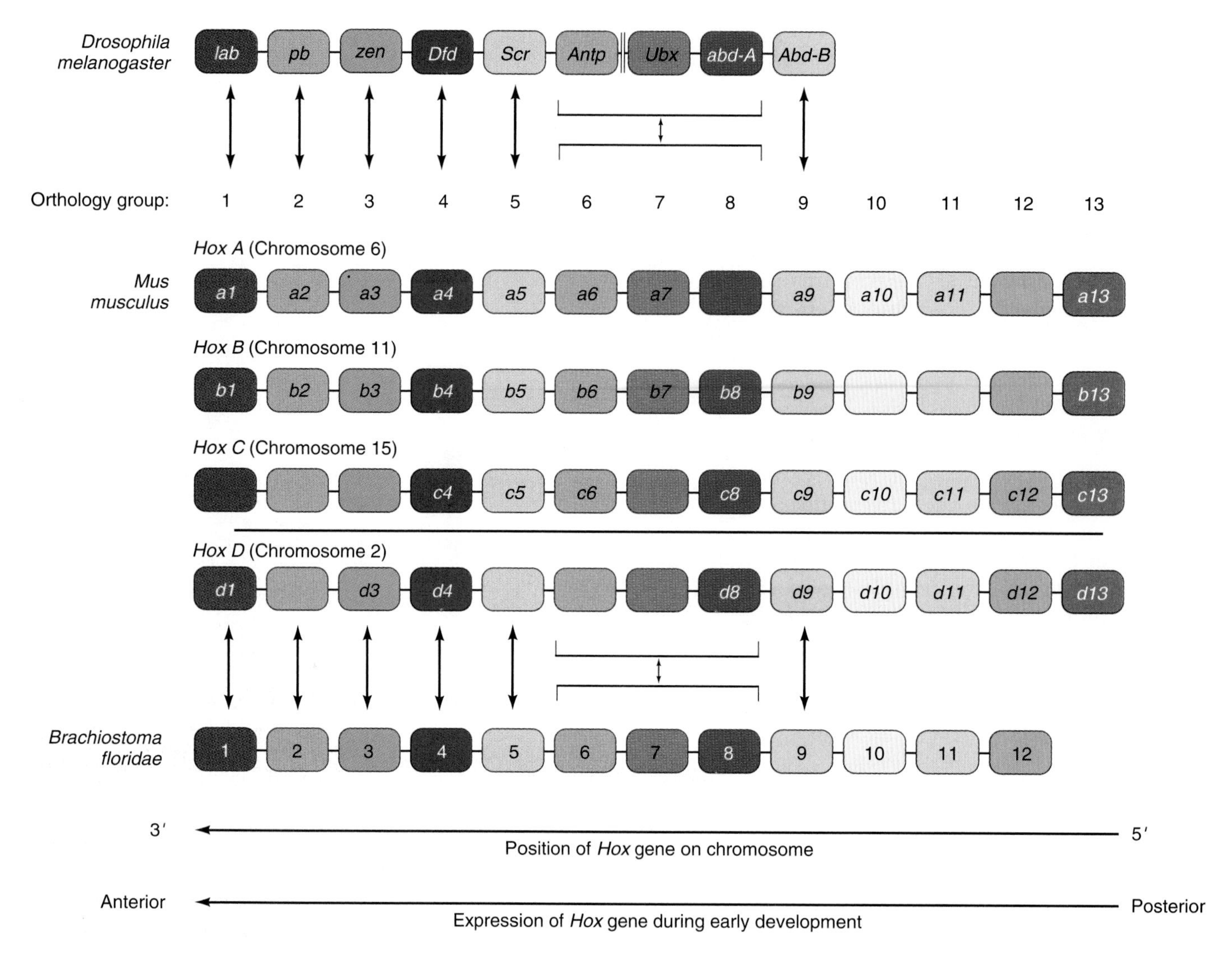

Figure 23.6 *Hox* genes in the fruit fly *Drosophila melanogaster,* the house mouse *Mus musculus,* and the cephalochordate *Branchiostoma floridae.* Each horizontal row represents a cluster of homeobox genes, or *Hox* complex, with genes shown in their physical order on the chromosome. Remarkably, the genes can be aligned so that the columns with vertical arrows represents orthology groups based on maximum sequence similarity among the homeodomains encoded (see Fig. 23.3). Orthology groups 1 through 5, and 9, are clearly distinct; the differences between the other vertical columns are less certain. The abbreviations for *Drosophila* genes refer to the homeotic genes illustrated in Figure 22.33 and to the *zerknüllt*$^+$ (*zen*$^+$) gene shown in Figure 22.38d. In the abbreviations for mouse genes, the letter refers to the *Hox* complex in which the gene is located, and the number indicates the orthology group. The horizontal arrows at the bottom indicate the direction in which the anterior boundary of the embryonic expression domain moves as one proceeds farther in the 3′ direction to the next gene (colinearity rule).

chromosomes. Thus, if the paralogy groups, or *Hox* clusters, are drawn as horizontal lines, then the orthology groups form 13 vertical columns. Orthology groups 1 through 5 (located toward the 3′ end of each paralogy group) and 9 are clearly distinct. The differences among groups 6 through 8, and those among groups 10 through 13, are less certain.

Since *Hox* genes can be arranged in paralogy and orthology groups, the following dual designation has been adopted for mouse and human *Hox* genes. A letter from A through D designates the *Hox* complex, and a number between 1 and 13 indicates the orthology group. The gene names are given as *Hoxa1*$^+$, *Hoxc4*$^+$, and so on, in mice, and as *HoxA1*$^+$, *HoxC4*$^+$, and so on, in humans (Scott, 1992).

The orthology groups based of homeodomain sequence comparisons were confirmed by generating transgenic organisms in which *Hox* genes were functionally replaced with orthologs from different species. For instance, flies were made transgenic for the mouse *Hoxb6*$^+$ gene, which was overexpressed under the control of a heat-inducible promoter (Malicki et al., 1990). In the flies' heads, where the resident *Antennapedia*$^+$ gene is silent, expression of the *Hoxb6*$^+$ transgene caused the same antenna-to-leg transformations also observed after overexpression of an *Antennapedia*$^+$ transgene (see Section 22.5). Similarly, the mouse *Hoxa5*$^+$ gene is functionally equivalent to the *Sex combs reduced*$^+$ gene in *Drosophila* (Zhao et al., 1993). Other experiments have shown that the human *HoxD4*$^+$ gene can substitute for the regulatory function of its *Drosophila* homolog, *Deformed*$^+$ (N. McGinnis et al., 1990). These results indicate that orthologous *Hox* genes from different species can functionally replace each other.

THE *HOX* COMPLEX IS HIGHLY CONSERVED IN EVOLUTION

Because homeoboxes are most similar among orthologous and pseudo-orthologous *Hox* genes, these must have become separated later in evolution than paralogous *Hox* genes. This conclusion leads to the following evolutionary scenario. All *Hox* genes found in different species today originated from one primordial *Hox* gene (see Fig. 23.5). This primordial gene has gone through a series of duplication and divergence steps within the same chromosome, thus generating the ***ancestral Hox complex*** (Lewis, 1978; A. Graham et al., 1989; Kappen et al., 1989; Murtha et al., 1991). This ancestral complex should have contained at least six genes, one for each of the orthology groups that can be clearly identified today (groups 1, 2, 3, 4, 5, and 9). The ancestral complex must have been present in animals that lived before protostomes (including insects) and deuterostomes (including vertebrates) began to diverge more than 500 million years ago. The entire ancestral *Hox* complex was duplicated repeatedly during the evolution of vertebrates, and four copies of the complex persist on different chromosomes in mice and humans. The single *Hox* complex present in protostomes is split in *Drosophila melanogaster*, by a chromosome break that probably occurred late in insect evolution, since a unified *Hox* complex is still present in other insects and the complex is broken up differently in *Drosophila virilis*. Orthology groups 6 through 8 and 10 through 13 may have originated the same way but may also contain genes that arose late in evolution by duplication within the same chromosome.

PSEUDO-ORTHOLOGOUS *HOX* GENES MAY ACT REDUNDANTLY, INDEPENDENTLY, OR SYNERGISTICALLY

Since pseudo-orthologous *Hox* genes show the greatest degree of sequence similarity, it has been expected that they would tend to have similar expression patterns and functions. This is indeed the case, often to the chagrin of investigators who find that targeted mutagenesis of a single *Hox* gene generates only a subtle, if any, mutant phenotype. And if a phenotype is observed, the affected area is often smaller than the expression domain of the respective gene.

Although many pseudo-orthologous genes have overlapping functions, some have diverged in evolution. For instance, knockout mice deficient in *Hoxa3* have disorders in structures derived from neural crest cells, whereas mice lacking *Hoxd3* are defective in derivatives of paraxial and lateral mesoderm (Chisaka and Capecchi, 1991; Condie and Capecchi, 1994).

Still other pairs of pseudo-orthologs function ***synergistically,*** meaning that their combined effect is *greater than the sum* of the two single gene effects. An example of such synergism was found between *Hoxa1*$^+$ and *Hoxb1*$^+$ (Studer et al., 1998; Gavalas et al., 1998; Rossel and Capecchi, 1999). *Hoxa1* mutations disrupt the organization of the hindbrain, whereas *Hoxb1* mutations affect the fate of neural crest cells. However, *Hoxa1/Hoxb1* double mutants show major additional defects including the complete loss of part of the hindbrain and the second pharyngeal arch. It appears that the simultaneous lack of both gene activities causes a failure in the determination of the missing parts followed by *apoptosis.* Similarly, mice lacking both *Hoxa9*$^+$ and *Hoxb9*$^+$ show more severe abnormalities of the ribs and sternum than predicted by simply adding the single mutant phenotypes (F. Chen and M.R. Capecchi, 1997).

The *Hox* complex is perhaps the most impressive example for the evolutionary conservation of genes that control development. As we will see in the following section, the *Hox* complex has been conserved not only in its molecular features but also in its fundamental function of specifying the anteroposterior body pattern. The *Hox* complex has therefore become a common object of developmental and evolutionary biologists who pursue the quest of tracing the evolution of body plans in terms of changing genetic control circuits.

23.3 The Role of *Hox* Genes in the Anteroposterior Body Pattern

Not only do *Drosophila* homeotic genes and mouse *Hox* genes share their chromosomal order and sequence information in critical domains, they also have similar functions in specifying the anteroposterior pattern (Wilkinson, 1993; Krumlauf, 1994). Most of the mouse *Hox* genes follow the same *colinearity rule* already discussed for *Drosophila* in Section 22.5: The physical order of *Hox* genes within each complex is related to the order of their expression domains along the anteroposterior axis of the embryo. In addition, some boundaries of mouse *Hox* gene expression are also clonal restriction lines, just as the expression domains of some *Drosophila* homeotic genes delineate compartment boundaries. Moreover, transgenic mice that mimic mutant alleles of certain *Hox* genes show phenotypes that can be interpreted as homeotic.

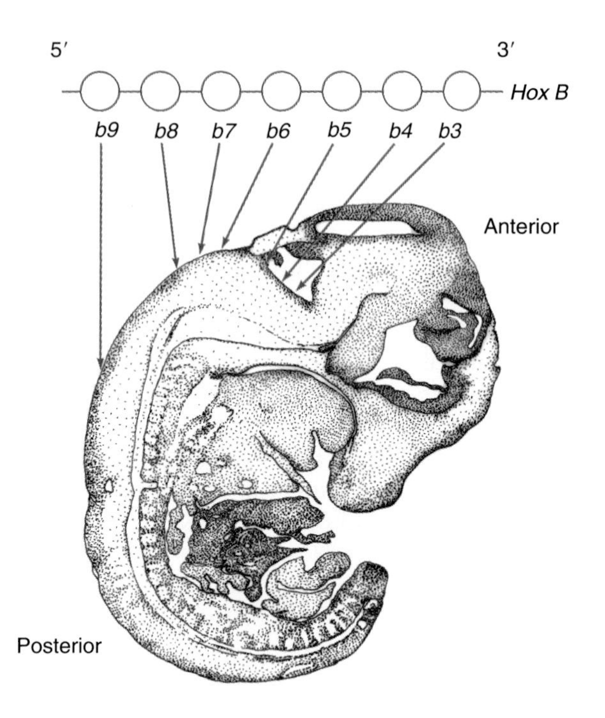

Figure 23.7 Anterior boundaries for the expression of *Hox B* genes in the mouse central nervous system, and the correlation of these boundaries with gene position in the *Hox B* complex. The upper diagram represents several genes from the *Hox B* complex. The lower diagram shows a 12.5-day embryo in median section, head to upper right. The arrows indicate the respective anterior limits of transcription of the genes, as revealed by in situ hybridization (see Method 8.1).

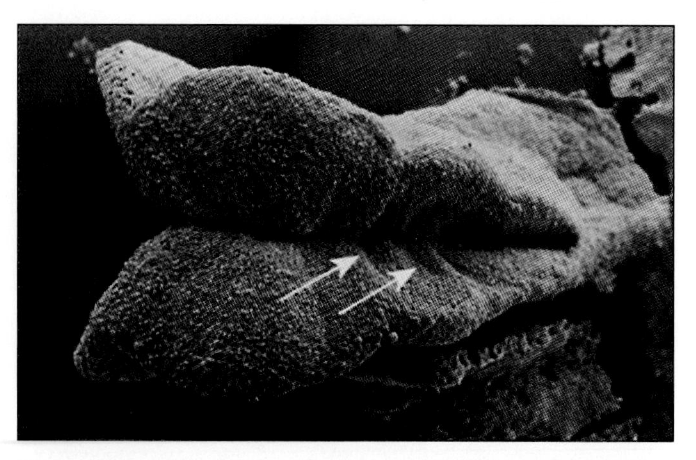

Figure 23.8 Rhombomeres in the developing mammalian brain. This scanning electron micrograph shows a mouse embryo during closure of the neural tube. The periodic ridges (arrows) in the bottom of the hindbrain are called rhombomeres.

THE ORDER OF *HOX* GENES ON THE CHROMOSOME IS COLINEAR WITH THEIR EXPRESSION IN THE EMBRYO

The colinearity rule was already noted for the homeotic genes of *Drosophila:* as one progresses from 3′ to 5′ through the *bithorax* complex and the *Antennapedia* complex, the anterior boundary of each gene's expression domain moves farther to the posterior (see Fig. 22.33). The same *colinearity rule* is observed for the expression of *Hox* genes in the mouse. The *Hox B* genes provide an example.

All genes of the mouse *Hox B* complex are transcribed in the central nervous system from neurulation until birth and beyond. The expression domain of each *Hox B* gene begins at a well-defined anterior boundary and extends posteriorly into the spinal cord, where the expression usually tapers off without a well-defined boundary. For genes near the 3′ end of the complex, such as *Hoxb3*$^+$, the anterior boundary maps in the hindbrain (Fig. 23.7). For the genes near the 5′ end of the complex, such as *Hoxb9*$^+$, the anterior boundary maps in the spinal cord (A. Graham et al., 1989; Wilkinson et al., 1989). The boundaries of *Hox B* gene expression domains in the hindbrain (*rhombencephalon*) were mapped with respect to a repeating pattern of bulges called ***rhombomeres*** (Fig. 23.8). In histological sections cut parallel to the axis of the hindbrain, the anterior boundaries of transcription were detected by in situ hybridization (Fig. 23.9). Clearly, the investigated genes have expression domains with anterior boundaries near the boundaries between rhombomeres. The results generally confirmed the colinearity rule with the exception of *Hoxb1*$^+$, a gene with a chromosomal location closer to 3′ than *Hoxb2*$^+$ but an anterior boundary behind that of *Hoxb2*$^+$. Corresponding results were obtained with genes from the other mouse *Hox* complexes. *Hox* genes belonging to the same pseudo-orthology group tend to have similar anterior boundaries in their expression domains.

In addition to rhombomeres, *Hox* genes are also expressed in embryonic mesoderm and in the neural crest cells that contribute to cranial nerves and pharyngeal arches. Since the neural crest cells tend to maintain the anteroposterior order in which they originate, the anterior limits of *Hox* gene expression in rhombomeres, cranial nerves, and pharyngeal arches tend to be similar (Fig. 23.10; Hunt et al., 1991b). Significantly, the expression of *Hox* genes in the mouse is reminiscent of a similar expression pattern of homeotic genes in *Drosophila* (compare with Fig. 22.33b). In both organisms, the *Hox* genes show a staggered expression pattern, with an increasing number of genes expressed posteriorly. Also in both organisms, the colinearity rule generally applies, with the expression of a *Hox* gene starting more posterior as the gene is located more toward the 5′ end of the chromosome.

What is the significance of this intriguing colinearity between the physical order of genes within a *Hox* complex and the anterior borders of their expression domains? Why has this correlation been conserved during hundreds of millions of years, instead of being interrupted by chromosome inversions or breaks?

In mice, the regulation of *Hox* genes may be facilitated by a process known as ***enhancer sharing.*** As dis-

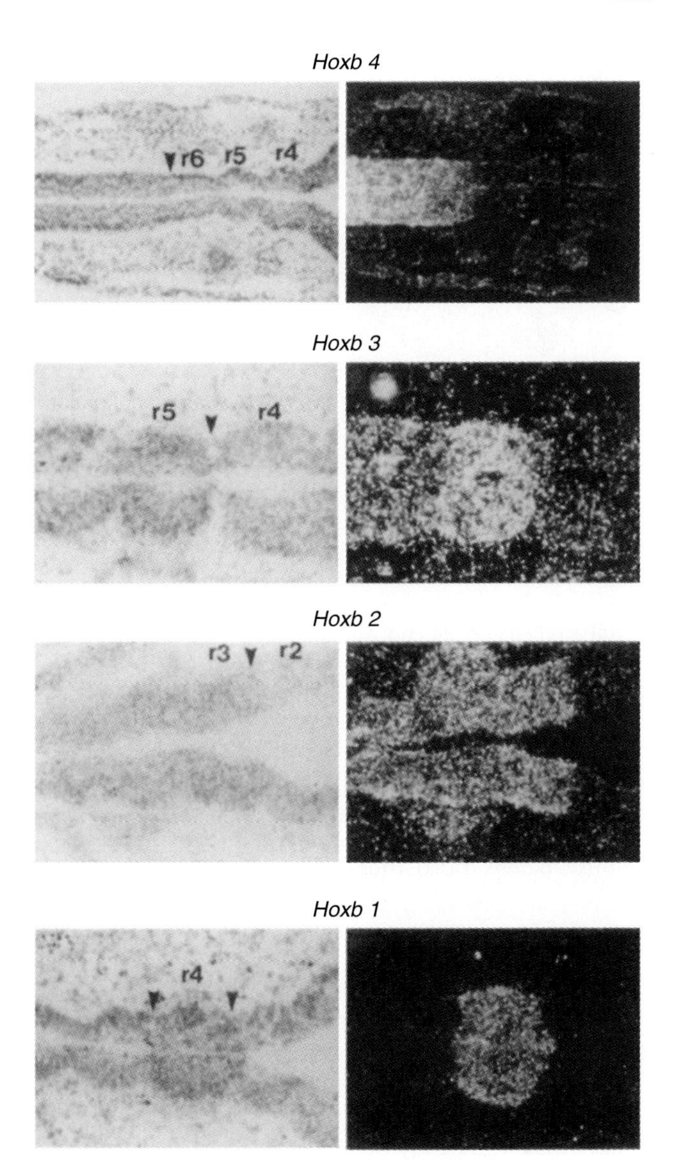

Figure 23.9 Transcription patterns of *Hox B* genes in the rhombencephalon (hindbrain) of a 9.5-day mouse embryo. Accumulated mRNA is made visible by in situ hybridization (see Method 8.1) using gene-specific probes. In each pair of photomicrographs, the left one was taken with bright-field illumination to show the sectioned tissue, and the right one was taken with dark-field illumination to show the labeled probes as white grains. The serial bulges (rhombomeres) of the hindbrain are numbered, from anterior to posterior, r1, r2, etc. Cranial ganglia VII and VIII, adjacent to r4, were used as landmarks. Arrowheads mark the anterior boundaries of the expression domains of four *Hox B* genes.

cussed in Sections 16.3 and 16.4, the gene-regulatory regions known as enhancers may act over large distances. Thus, enhancers may act on more than one promoter. Indeed, there is evidence for neighboring mouse *Hox* genes being linked into functional groups by shared enhancers. For example, expression of the $Hoxb3^+$ and $Hoxb4^+$ genes in a certain portion of the hindbrain is driven by a common enhancer (A. Gould et al., 1997). If enhancer sharing evolved to a point where it is neces-

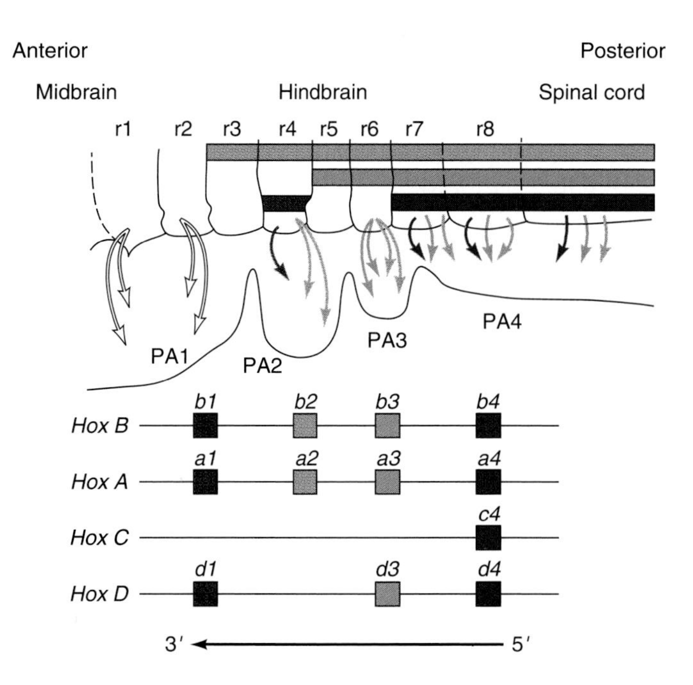

Figure 23.10 Summary of *Hox* gene expression in the head region of the mouse embryo. The top diagram shows a side view of the head region. The bottom diagram shows four pseudo-orthology groups (vertical columns) of *Hox* genes. Most groups are expressed posteriorly; progressively fewer, anteriorly. Expression domains of these pseudo-orthology groups include the spinal cord, hindbrain rhombomeres r3 to r8, and neural crest cells (arrows), which contribute to cranial ganglia and pharyngeal arches (PA1–PA4).

sary for generating critical *Hox* gene expression patterns, then mutations disrupting the *Hox* gene order would be selected against. Enhancer sharing therefore could account for both the colinearity rule and the uninterrupted nature of the *Hox* complexes in mice. This explanation is unlikely to hold for *Drosophila,* where insulating elements that functionally separate neighboring genes have been found in the *Antennapedia* and *bithorax* complexes (Mann, 1997). As expected in the absence of enhancer sharing, the *Hox* complex is breaking up in *Drosophila melanogaster* and its close relatives (see Sections 22.5 and 22.9). However, the genetic insulators observed in *Drosophila* seem to have evolved late, so that enhancer sharing may still account for the colinearity rule in most organisms.

BOUNDARIES OF *HOX* GENE EXPRESSION ARE ALSO CLONAL RESTRICTION LINES

The coincidence of some gene expression boundaries with rhombomere and segmental boundaries in mice parallels the properties of epidermal compartments in insects. However, insect compartments are defined by clonal restriction lines—that is, lines not transgressed by cell clones generated after a certain stage of development (see Sections 6.2 and 22.7). Are there similar clonal restriction lines between the rhombomeres of the vertebrate brain, or between groups of neural crest cells? This question is difficult to answer in regard to mice, because mammalian embryos past the implantation stage are difficult to manipulate and maintain. However, such experiments have been carried out in chickens, which are similar to mice in their modes of gastrulation and organogenesis. Chickens also have *Hox* genes with expression domains similar to those of mice. For instance, a chicken homolog of *Hoxb1*$^{+}$ is expressed in rhombomere 4, exactly like its mouse counterpart (Sundin and Eichele, 1990).

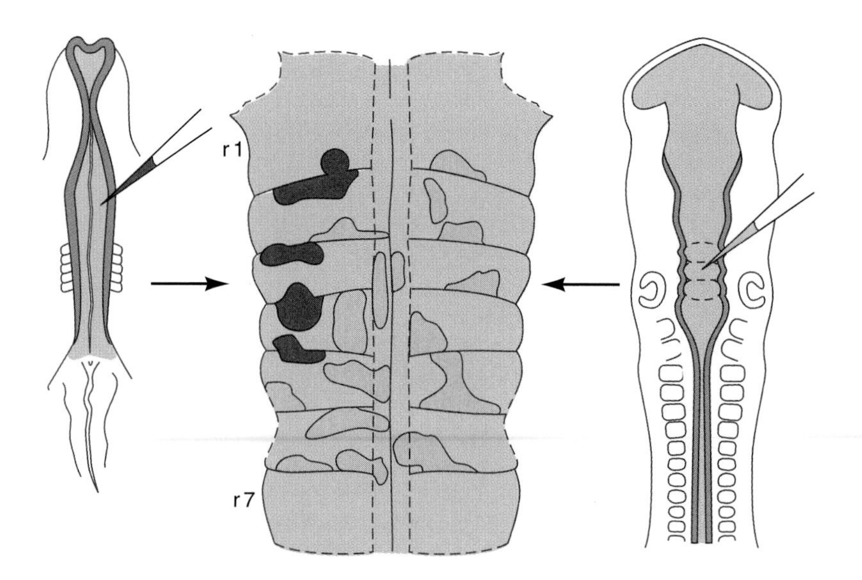

Figure 23.11 Clonal analysis of the rhombencephalon (hindbrain) development in the chick embryo. The center diagram shows the ventral aspect of a chicken hindbrain as if it had been cut open dorsally and spread out flat. The diagram on the left illustrates the marking of clone founder cells at the 5-somite stage before the appearance of rhombomeres. Of the resulting clones (red) some respect the rhombomere boundaries, while others overlap adjacent rhombomeres. The drawing on the right shows the result of the same procedure performed after the appearance of rhombomeres. Among these clones (pink), most respect the rhombomere boundaries.

In order to test chicken rhombomeres for clonal restriction lines, Fraser and coworkers (1990) injected a fluorescent dye into individual cells of living chicken embryos. After a period of development, clones formed by the descendants of the labeled cells were mapped to the hindbrain. The critical question was whether a marked clone would respect the boundaries between rhombomeres or transgress them. The answer depended on the stage when the clone founder cell was labeled, as in insects. When a founder cell was labeled before the appearance of morphological rhombomere boundaries, some of the resulting clones were found to overlap neighboring rhombomeres (Fig. 23.11). However, when founder cells were labeled after boundary appearance, most of the clones were restricted at the rhombomere boundaries (Birgbauer and Fraser, 1994). Thus, rhombomeres in the chicken brain seem to undergo a stage-dependent clonal restriction similar to that of epidermal compartments in insects.

To explore whether the clonal restriction lines in chicken rhombomeres were based on cell affinities, Guthrie and Lumsden (1991) transplanted rhombomere pieces. When parts from different rhombomeres were juxtaposed, they reestablished a boundary between them, especially if a part from an odd-numbered rhombomere was juxtaposed with a part from an even-numbered rhombomere. In contrast, parts from the same rhombomere joined without forming a boundary and showed extensive cell mixing (Guthrie et al., 1993). Thus, cells from different chicken rhombomeres minimized contact, whereas cells from the same rhombomere mixed freely, in parallel with cells from insect epidermal compartments.

The transplanted tissue parts also behaved autonomously—that is, were unaffected by their new environment—with regard to gene expression and cell differentiation. Rhombomere 4 tissue transplanted anteriorly into rhombomere 2 continued to express *Hoxb1*$^{+}$ and proceeded to form the nerve nuclei and cranial nerve roots characteristic of the rhombomere 4 level (Guthrie et al., 1992; Kuratani and Eichele, 1993). These observations show that certain brain regions are determined at the time they become clonally restricted units. In addition, the results suggest that the determined state of brain regions may be linked to the expression of their region-specific *Hox* genes.

The results described so far show that the homeotic genes of insects and the *Hox* genes of mice share several properties. First, their protein products contain homeodomains, which are characteristic of transcription factors. Second, these genes are clustered in *Hox* complexes that have been conserved exceedingly well during evolution. Third, these genes are expressed in staggered domains, and the anterior boundaries of their expression domains are colinear with the physical order of the genes within their complex. Fourth, some of the anterior expression boundaries coincide with compartment lines that seem to be based on different cell affinities.

In addition, homeotic genes in insects act as *selector genes* directing the *morphological* development of each compartment (see Section 22.7). This is proven by the striking phenotypes of homeotic mutants, which show characteristic transformations of one compartment into a duplication of another compartment, such as the replacement of halteres with wings or antennae with legs.

The following section describes attempts to generate equivalents of homeotic mutants in vertebrates by reverse genetics.

ELIMINATION OF *HOX* GENES MAY CAUSE TRANSFORMATIONS TOWARD MORE ANTERIOR FATES

To determine whether the *Hox* genes of mice and the corresponding genes of other vertebrates control the morphology of certain body regions, one needs to analyze mutants or their equivalents. Unfortunately, no spontaneous mutants for these genes have yet been found in vertebrates. Therefore, several groups of investigators have used previously cloned genes to obtain the equivalents of mutant phenotypes (*Krumlauf, 1994*).

One strategy used by researchers working with mice to analyze the biological function of cloned genes is known as *gene targeting*, or more colloquially, as *gene knockout* (see Fig. 15.24). First, they replace the gene of interest in an embryonic stem (ES) cell with an inactive allele of the same gene. Then, they use descendants of the transformed ES cell to create chimeric mice, in which the ES cells often contribute to the germ line. Finally, they breed males with transgenic sperm so that the altered gene will be passed on to all cells in some of their progeny.

In the example illustrated in Figure 15.24, mice in which both copies of the *Hoxb4*$^+$ gene were knocked out developed an abnormal second cervical vertebra that was morphologically similar to the first cervical vertebra. This type of anterior transformation is precisely the phenotype that one should expect from knocking out a *Hox* gene. Because of the staggered expression pattern of these genes, with more genes expressed posteriorly, knocking out one of them is likely to mimic an expression pattern normally found more anteriorly. Are there more examples of anterior transformations resulting from knocking out mouse *Hox* genes?

TO eliminate the *Hoxc8*$^+$ gene, Le Mouellic and coworkers (1990, 1992) transformed ES cells with a DNA construct containing the bacterial *lacZ* and *Neo*^{R} genes flanked by sequences from the transcribed region of *Hoxc8*$^+$. The flanking sequences were selected so that homologous recombination would replace a transcribed portion of the resident *Hoxc8*$^+$ gene with the *lacZ* and *Neo*^{R} genes. This replacement was designed to eliminate the function of the resident gene, to bring the *lacZ* gene under control of the resident *Hoxc8*$^+$ regulatory region, and to confer upon the transgenic cells the ability to grow in a medium containing the poisonous drug G418. Such cells were grown into clones, and tested to determine whether the transgene had become inserted into one of the two resident *Hoxc8*$^+$ genes.

Stem cell clones with the correctly inserted transgene were used to generate chimeric males producing sperm derived from the transformed ES cells. These males were mated with wild-type females to generate offspring heterozygous for the transgene (*Hoxc8*$^{+/-}$). Matings of heterozygotes then produced offspring homozygous for the transgene (*Hoxc8*$^{-/-}$). In both homozygotes and heterozygotes, the *lacZ* portion of the transgene was expressed under the control of the resident *Hoxc8*$^+$ promoter. As expected, *lacZ* was expressed more strongly in homozygotes, and the expression pattern of *lacZ* corresponded to the normal pattern of *Hoxc8*$^+$ expression.

In wild-type mice there is generally a pattern of 7 cervical vertebrae, 13 thoracic vertebrae with ribs, and 6 lumbar vertebrae before the hipbone (called *sacrum*). For easy reference, these vertebrae are numbered consecutively from 1 to 26 in Figure 23.12. In 12.5-day embryos, *Hoxc8*$^+$ expression begins with a sharp boundary behind the rudiment of vertebra 12, continues strongly until the future vertebra 22, and then decreases toward the posterior. Within this expression domain the researchers expected to find transformations toward more anterior fates.

One obvious transformation in the homozygous mutant (*Hoxc8*$^{-/-}$) mice affected vertebra 21. This is normally the first lumbar vertebra; in homozygotes, however, this vertebra had the morphological characteristics of a thoracic vertebra, with a pair of ribs attached to it (Fig. 23.13a). This vertebra was followed by 5 vertebrae without ribs, so that the total number of vertebrae before the sacrum was 26 as in wild-type mice. The development of a 14th vertebra with ribs must therefore be viewed as a homeotic transformation of the 21st vertebra from lumbar to thoracic and not as the development of an additional thoracic vertebra.

More transformations of a homeotic type were observed in the area of the breastbone (sternum). In wild-type animals, the first seven ribs on each side are attached to the sternum (Fig. 23.13b). The attachment sites of the first six ribs are separated by bony elements (sternebrae), while the sixth and seventh ribs are attached to the sternum at the same point. In homozygous mutants, an additional sternebra developed between the sixth and seventh ribs, and the eighth rib was fused with the sternum at the same point as the seventh rib (Fig. 23.13c). In addition, the 12th rib of homozygous mutants was transformed into a likeness of the 11th rib by the criteria of overall length and bone:cartilage ratio.

Questions

1. How would you test whether the *lacZ-Neo*^{R} transgene has indeed become inserted into the targeted location of the *Hoxc8*$^+$ gene?
2. Based on the data provided here, would you expect to find additional homeotic transformations in the area of the cervical vertebrae?
3. Homeotic transformations caused by homozygous null alleles in *Drosophila* tend to be complete and present in every surviving individual. In contrast, the *Hoxc8*$^{-/-}$ mice described here showed considerable variation. For instance, the ends of the eighth ribs were distant from

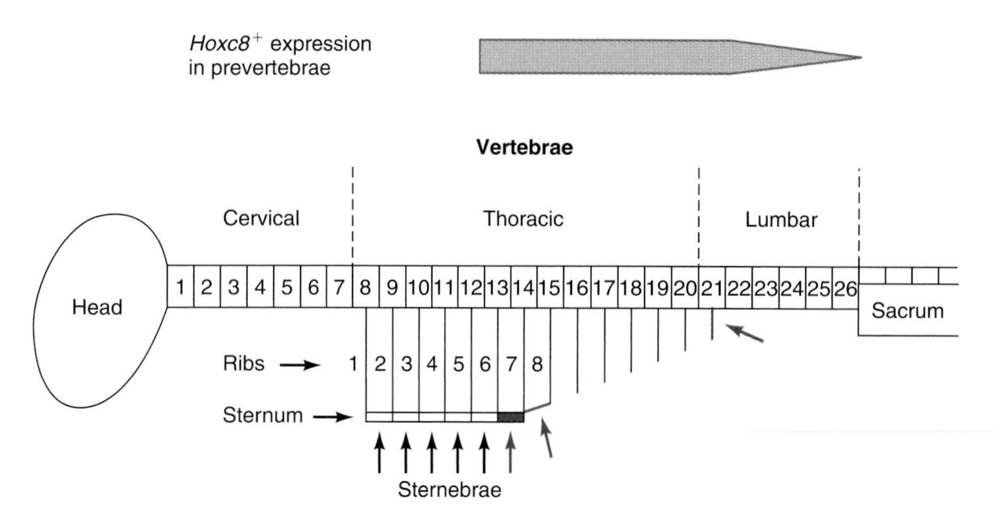

Figure 23.12 Homeotic transformations in mice homozygous for a null allele of *Hoxc8.* In wild-type 12.5-day embryos, *Hoxc8*+ expression begins with a sharp boundary behind prevertebra 12, continues strongly until prevertebra 22, and then decreases toward the posterior. This region develops abnormally in mutants homozygous for a null allele of this gene ($Hoxc8^{-/-}$). Wild-type mice have seven cervical vertebrae (vertebrae 1–7), 13 thoracic vertebrae (8–20) with ribs, and 6 lumbar vertebrae (21–26) without ribs. Homozygous mutants have the abnormalities drawn in color. Vertebra 21, normally the first lumbar vertebra, is transformed into a thoracic vertebra. The sternum (breastbone) has an extra element (sternebra) between the attachment sites of ribs 6 and 7. Rib 8, normally unattached, is attached to rib 7 or to the sternum. These abnormalities can be interpreted as transformations of each affected segment toward the morphological character of its anterior neighbor.

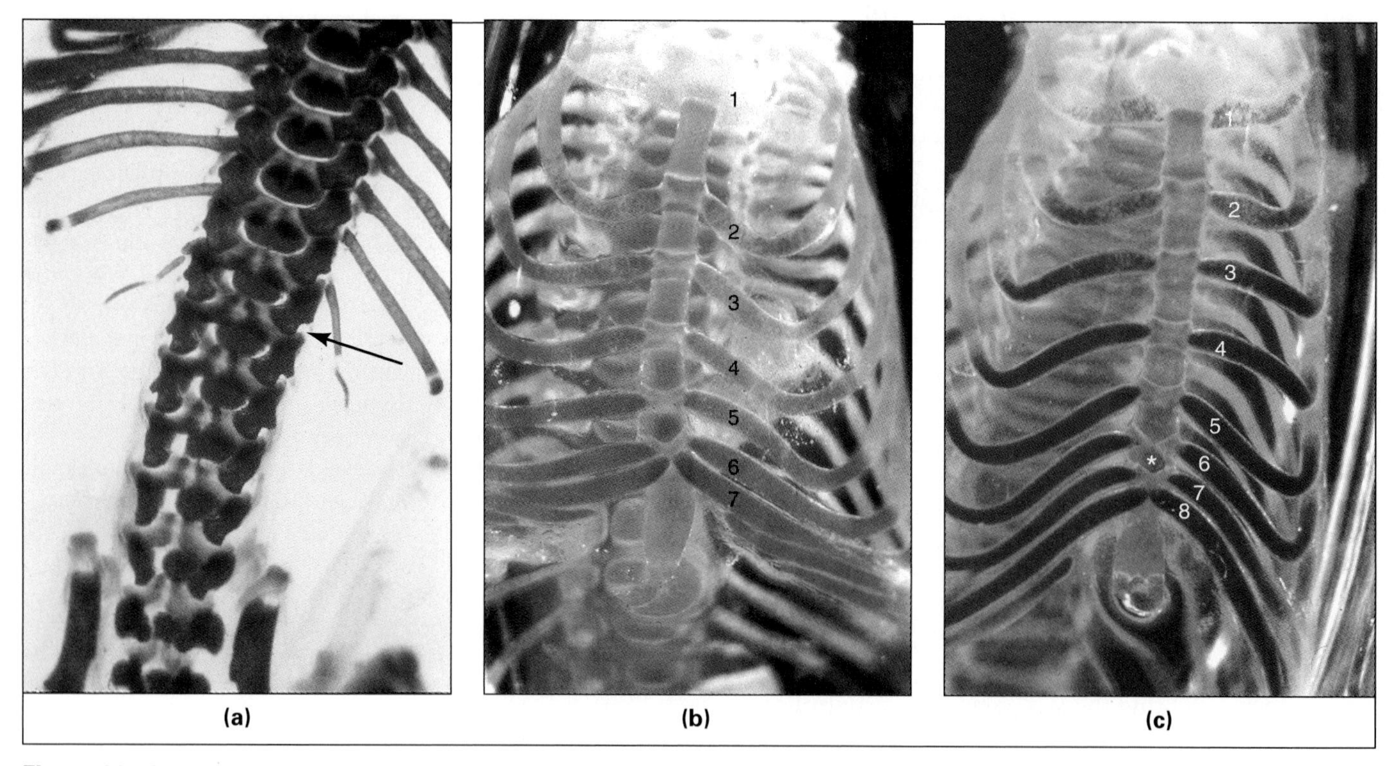

Figure 23.13 Homeotic transformations in mice homozygous for a null allele of a *Hox* gene (see Fig. 23.12). **(a)** Homozygous mutant ($Hoxc8^{-/-}$) having a pair of ribs attached to vertebra 21 (arrow). **(b)** Sternum region of the wild-type in ventral view. The first seven ribs (numbered 1 through 7) are attached to the sternum, with the attachment sites of the first six ribs separated by sternebrae. The eighth rib is floating. **(c)** Sternum region of a homozygous mutant. Eight ribs are attached to the sternum, and an extra sternebra (asterisk) has developed between ribs 6 and 7.

the sternum (as in wild-type mice) in 5 of 14 homozygous null individuals, fused to the seventh ribs in three, and attached to the sternum (as shown in Fig. 23.13c) in six. How do you explain the weaker expressivity of homeotic transformations in mice?

In their experiments, Le Mouellic and his coworkers found that elimination of $Hoxc8^+$ function in mice causes morphological transformations toward anterior fates in the body region where the gene is normally expressed. Anterior transformations were also obtained after targeted disruption of $Hoxd3^+$ (Condie and Capecchi, 1993). Likewise, knocking out the $Hoxa2^+$ gene was found to cause homeotic transformations of second to first pharyngeal arch elements (Rijli et al., 1993).

Not all knockout experiments with *Hox* genes, however, have led to anterior transformations. Inactivation of $Hoxa5^+$ or $Hoxa11^+$ caused posterior transformations, indicating that loss of function in either gene generates or mimics selector gene activities that normally prevail more posteriorly (Jeannottee et al., 1993; K. S. Small and S. Potter, 1993). Mice homozygous for null alleles of $Hoxa3^+$ (Chisaka and Capecchi, 1991) or $Hoxa1^+$ (Lufkin et al., 1991) had specific deficiencies and malformations but underwent no clear transformations toward anterior fates. These results can be interpreted in different ways. First, some *Hox* genes might not act as homeotic selector genes. Second, partial overlap of functions of *Hox* genes within a pseudo-orthology group might obscure the consequences of losing the activity of one group member.

Although much more needs to be learned about the functions of *Hox* genes, the experiments noted so far confirm and extend the parallelism between mouse *Hox* genes and *Drosophila* homeotic genes. The elimination of at least some *Hox* genes in mice causes transformations of body elements into anterior serial homologs. These transformations are similar to the phenotypes of most *Drosophila* mutants with loss-of-function alleles of homeotic genes, such as a transformation from metathorax to mesothorax (see Fig. 22.32). Conversely, *gain-of-function* alleles of certain homeotic genes in *Drosophila* cause mimicking of an activity pattern normally found more posteriorly. Such mutants, including the dominant *Antennapedia* alleles, result in replacements of structures with more posterior serial homologs, such as a transformation from antenna to leg. As we will see in the following, similar transformations occur in mice as well.

OVEREXPRESSION OF *HOX* GENES MAY CAUSE TRANSFORMATIONS TOWARD MORE POSTERIOR FATES

Many experiments in the functional analysis of *Hox* genes rely on ***overexpression***—that is, the expression of a transgene at stages or in areas where the resident pair of genes is silent. This strategy generates the equivalent of a gain-of-function allele.

The normal $Hoxa7^+$ gene is expressed in the spinal cord and in the mesoderm that forms the vertebrae and intervertebral discs. The anterior boundaries of these expression domains are in the lower neck, for the spinal cord, and in the thorax, for the mesoderm (Püschel et al., 1991). To cause overexpression of the $Hoxa7^+$ gene, Kessel and coworkers (1990) engineered a transgene containing the transcribed region of $Hoxa7^+$ under the control of an actin regulatory region. In mice transgenic for this construct, *Hoxa7* was expected to be expressed in the entire embryo, because actin promoters are active in all cells. The transgene's ectopic activity was expected to be critical in the head and the parts of the neck where the resident $Hoxa7^+$ genes are silent. If the $Hoxa7^+$ gene acted as a homeotic selector gene, then its overexpression should transform head and neck structures into structures normally found farther posterior. Such transformations were indeed observed.

The overt segmentation of vertebrates begins with the formation of *somites,* epithelial balls of paraxial mesoderm (see Section 14.2). Generally, the sclerotome portions of the somite form vertebral rudiments called *prevertebrae* (Fig. 23.14b). However, somite development is modified in the head and neck region (Jenkins, 1969; Romer, 1976; Verbout, 1985). The first somites, called *occipital somites,* do not generate vertebrae but form the ***occipital bones*** (Lat. *occiput,* "back of the head"). These bones surround the opening in the base of the skull, through which the spinal cord exits into the vertebral column. The fourth somite forms the rudiment of the ***proatlas,*** which develops into a separate vertebra in certain reptiles. In mammals, however, the proatlas rudiment contributes to the basioccipital bone and to the second vertebra.

The first seven vertebrae in mammals do not carry ribs. They are called ***cervical vertebrae*** (Lat. *cervix,* "neck"). The first two cervical vertebrae are specially adapted to allow the head to move relative to the vertebral column. The first vertebra is called the ***atlas,*** after the Greek mythological figure who supported the sky with his shoulders. The atlas lacks the vertebral body—that is, the weight-bearing, drum-shaped part of a standard vertebra (Fig. 23.14a, d). Instead, the atlas is held together ventrally by a slim *ventral arch.* What would have been the body of the atlas has become part of the second cervical vertebra, the ***axis.*** The axis has a central process called the ***dens*** (Lat. "tooth"), which develops from cells that would otherwise have formed the body of the atlas. This modification allows the atlas, along with the skull, to rotate around the dens.

In the experiment of M. Kessel and coworkers (1990), the atlas of transgenic mice overexpressing the *Hoxa7* transgene showed a striking abnormality: There was a vertebral body inside the ventral arch (Figs. 23.14e and 23.15b). Correspondingly, the axis of transgenic mice

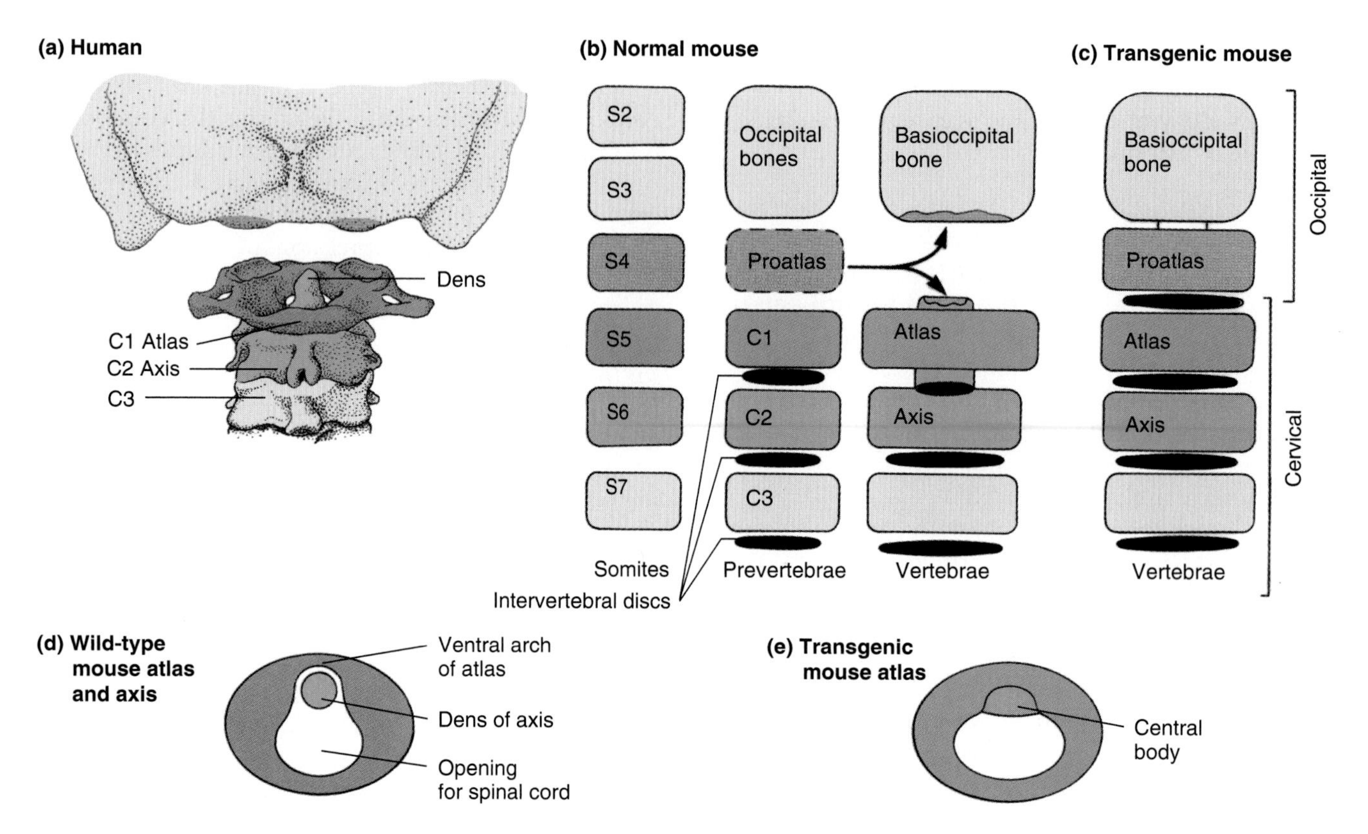

Figure 23.14 Development of cervical vertebrae in normal and transgenic mice overexpressing a *Hoxa7* transgene. **(a)** Dorsal view of part of the skull and anterior cervical vertebrae of a human. **(b)** Schematic diagrams showing normal development of six somites of the mouse (S2–S7, left column). They develop into occipital bones and two cervical prevertebrae (C1–C3), with a transient rudiment (proatlas) contributing to both occipital bones and cervical vertebrae. C1 and C2 form the anteriormost cervical vertebrae, the atlas and the axis. Part of C1, instead of forming the body of the atlas, gives rise to the dens of the axis. **(c)** Cervical vertebrae in postnatal transgenic mice overexpressing a *Hoxa7* transgene. The proatlas persists as a vertebra, separated from the atlas by an extra intervertebral disc (black). The atlas has a body, and the axis has no dens. **(d)** Anterior (cranial) view of the normal atlas and axis. The atlas has no central body. It rotates around the dens of the axis. Note the slim ventral arch of the atlas. **(e)** Anterior view of the atlas in mice expressing a *Hoxa7* transgene. The atlas has a central body like a more posterior cervical vertebra. The underlying axis (not shown) has no dens.

had no dens. Moreover, the small patch of intervertebral disc tissue that is normally embedded between the dens and the body of the axis was broadened into a full-sized intervertebral disc separating the atlas from the axis. In addition, transgenic mice had another intervertebral disc anterior to the atlas; and in front of this disc, a supernumerary vertebra corresponding to the proatlas was attached to the occipital bones (Fig. 23.14c). In sum, the characteristic adaptations of the atlas and axis had altogether disappeared, and the anterior cervical vertebrae had assumed the structure normally seen in more posterior cervical vertebrae.

The experiment just described shows that overexpression of the *Hoxa7* gene in mouse embryos changes the development of vertebrae in the head-neck zone so that they resemble more posterior vertebrae. Similar transformations were observed in mice with a transgene combining the transcribed region of *Hoxd4* with the regulatory region of *Hoxa1* (Lufkin et al., 1992). This fusion gene caused the transcription of Hoxd4 mRNA more anteriorly than its normal anterior boundary, which is at the level of the first cervical somites. This ectopic expression resulted in the expected posterior transformation: occipital bones were replaced with structures resembling cervical vertebrae. This result again confirms the general parallelism between mouse *Hox* genes and *Drosophila* homeotic genes: Overexpressions of both tend to cause morphological transformations toward more posterior fates.

The dramatic effects of misexpressing homeotic genes and *Hox* genes raise the question, How are these genes activated in exactly their appropriate domains during normal development? In Section 22.5, we discussed the relevant genetic control mechanisms in *Drosophila*. The following section will address the same question in mice.

ONLY SOME OF THE CONTROL SYSTEMS OF *HOX* GENE ACTIVITY ARE EVOLUTIONARILY CONSERVED

In the *Drosophila* embryo, *Hox* genes are controlled by gap genes and pair-rule genes, which in turn are regulated by concentration gradients of maternally encoded

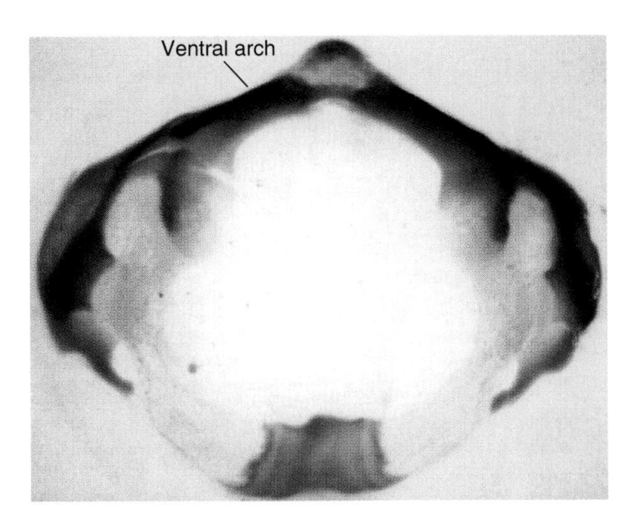

(a) Wild type

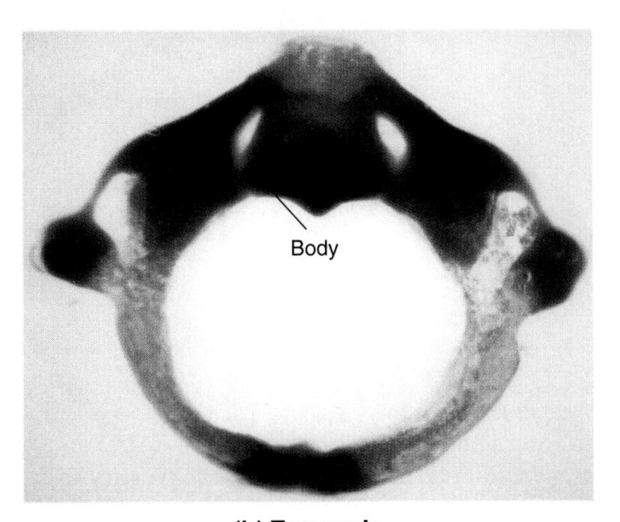

(b) Trangenic

Figure 23.15 The atlas bone of **(a)** a normal mouse and **(b)** a mouse overexpressing a *Hoxa7* transgene; refer to Figure 23.14d and e, respectively. The photographs show the vertebrae in anterior (crenial) view. Note the slim ventral arch of the normal atlas and the solid body of the transgenic atlas.

transcription factors. This type of gene regulation is facilitated by the superficial cleavage of insect eggs, which leaves large molecules free to diffuse during the first 14 cell cycles. The same mechanism is not found in mice; the holoblastic cleavage of mammals precludes it. However, gradients of transcription factor activity might be generated by a *ligand-dependent transcription factor* produced evenly in all cells and a ligand that is produced locally and diffuses across cell membranes.

A morphogen that would be compatible with holoblastic cleavage, and with the extensive cell migration and cell mixing that occur during vertebrate axis formation, may be provided by retinoic acid. Its receptors occur naturally in embryos and function as ligand-dependent transcription factors similar to steroid receptors (see Section 16.4). Retinoic acid and related molecules, such as vitamin A, have long been known as *teratogens* (Gk. *teras*, "monster," or severely malformed embryo). For instance, amputation of tails from frog tadpoles followed by application of retinyl palmitate to the wound may lead to the formation of extra hindlimbs (Maden, 1993). These properties have made retinoic acid a candidate for controlling *Hox* gene expression.

Retinoic acid fed to pregnant mice during day 7 of gestation causes transformations of cervical vertebrae similar to those observed after overexpression of *Hoxa7* (see Fig. 23.14). In addition, vertebra 7, which normally develops as a cervical vertebra, develops as a thoracic vertebra with ribs (M. Kessel and P. Gruss, 1991). These transformations of vertebrae toward more *posterior* fates correlate with an *anterior* shift in expression of the (resident) *Hoxa7*$^+$ genes during the development of prevertebrae. Similar effects of retinoic acid applied at 7.5 days of gestation are observed in the development of the hindbrain (H. Marshall et al., 1992; Conlon and Rossant, 1992). However, retinoic acid does not always cause posterior transformations. When administered at 8.5 days, its effects are opposite to those observed after application on day 7. Retinoic acid now causes anterior transformations in another region of the vertebral column, similar to those observed after elimination of the *Hoxc8*$^+$ gene.

The perplexing and sometimes paradoxical effects of retinoids are consistent with the general observation that a given transcription factor, such as a retinoic acid receptor, may have different effects on different target genes, effects that depend on cooperative or inhibitory actions of other transcription factors. It is apparent that retinoic acid *can* interfere with the process that activates *Hox* genes in their specific expression domains. Whether retinoids are *natural* activators of *Hox* genes—and, if so, how they act in the normal activation process—remains to be seen (*S. V. Bryant and D. M. Gardiner, 1992; Krumlauf, 1994*).

In contrast to the control system that *activates Hox* genes in their proper order, the genes that *maintain* the inactive state of *Hox* genes are well conserved between flies and mice. In particular, the *Drosophila Polycomb*$^+$ gene can be replaced functionally by its mouse ortholog, *M33*$^+$ as a transgene (J. Müller et al., 1995). Also, mice lacking the *bmi*$^+$ gene show posterior transformations along the entire body axis, just like *Drosophila* lacking the orthologous gene, *Posterior sex combs*$^+$, another member of the *Polycomb* group (van der Lugt et al., 1994). Mice lacking both *M33*$^+$ and *bmi*$^+$ die, for the most part, prenatally. Survivors show stronger skeletal alterations than each single mutant, with some abnormalities of the skull and clavicle (collar bone) detected exclusively in the double mutant (Bel et al., 1998).

In summary, the common function of *Hox* genes in flies, mice, and probably most other animals is to define anteroposterior spatial patterns of selector gene activities. The morphological development of a given embryonic area depends on its specific combination of active *Hox* genes, and mutants in these genes tend to show characteristic anterior or posterior homeotic transformations. At least some *Hox* genes are expressed in domains defined by clonal restriction lines, and their expression affects cell adhesion. The *Hox* genes encode transcription factors, which may control target genes affecting cell division, adhesion, death, migration, and other properties that are likely to be controlled by evolutionarily conserved genes. However, only a few of these target genes have been discovered, presumably because their mutant phenotypes are too subtle or too global to be selected in genetic screens for embryonic pattern formation.

23.4 The Dorsoventral Body Pattern

While *Hox* genes are involved in specifying the anteroposterior body pattern, the dorsoventral axis of vertebrates is determined by another set of genes. These genes are also very similar between vertebrates and invertebrates, although with an ironic twist.

A LOBSTER MAY BE VIEWED AS AN UPSIDE-DOWN RABBIT

In 1822, the French naturalist Étienne Geoffroy Saint-Hilaire provoked a heated discussion when he proposed that arthropods (such as insects and crustaceans) and vertebrates have a common body plan. Anecdote has it that this idea came to him while he was eating a lobster and was thus reminded that lobsters have most of their central nervous system on the ventral side while their heart is positioned dorsally. This is of course the reverse of the situation in vertebrates, where the central nervous system is located dorsally and the heart ventrally. Thus, a lobster could be viewed as an upside-down rabbit, or vice versa. The notion that most animals shared a common body plan had been an important part of Geoffroy's thinking throughout his career, in which he linked development and evolution by postulating that the major forms of life originated from one another through the occasional appearance of successful monsters.

Geoffroy's idea has recently received support from the discoveries of molecular biologists studying the genetic control systems that establish the dorsoventral polarity in developing embryos (*Holley and Ferguson, 1997*). As discussed in Section 22.6, this process in *Drosophila* involves the *decapentaplegic*$^+$, *short gastrulation*$^+$, and *tolloid*$^+$ genes, all of which encode secreted proteins. Decapentaplegic (dpp) protein is synthesized on the dorsal side of the blastoderm embryo, where it promotes the development of amnioserosa and dorsal epidermis while inhibiting neurogenesis (Fig. 23.16). Dpp is antagonized by the short gastrulation (sog) protein, which is synthesized ventrolaterally. The inhibitory action of sog is in turn counteracted by the tolloid (tld) protein, which is also synthesized dorsally. Tld has the molecular characteristics of a protease, and it cleaves

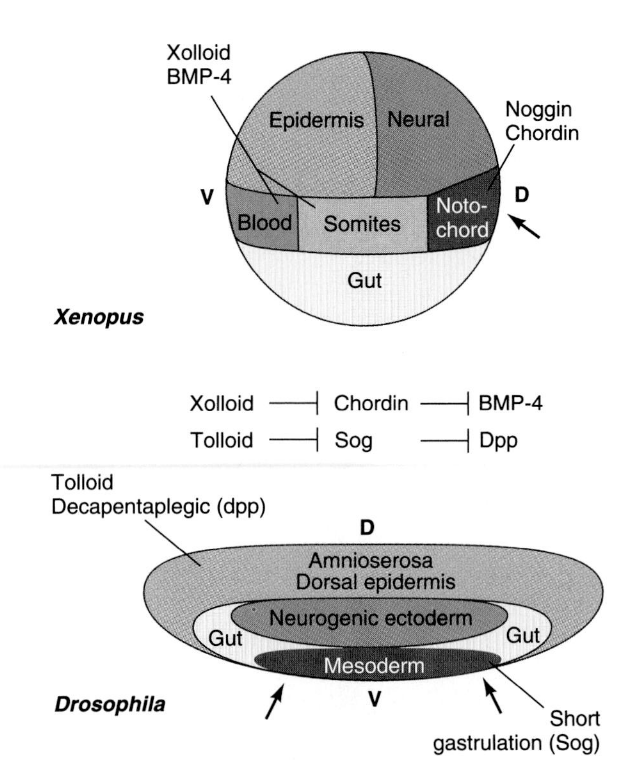

Figure 23.16 Fate maps of *Drosophila* and *Xenopus* embryos at the early gastrula stage superimposed with the expression patterns of genes establishing the dorsoventral axis.

sog, especially in the presence of dpp. As a result of these interactions, dpp activity seems to form a morphogen gradient with the highest levels near the dorsal midline and tapering off laterally.

In *Xenopus*, the development of dorsal organs, including neural tube and notochord, is promoted by chordin and noggin (see Section 12.5). These secreted proteins are synthesized specifically in the dorsal blastopore lip, also known as Spemann's organizer (Fig. 23.16). Both proteins dorsalize ventral mesoderm, induce ectoderm to form neural plate, and restore dorsal structures in embryos that have been ventralized by UV irradiation. Both proteins antagonize bone morphogenetic protein 4 (BMP-4), which is necessary and sufficient to promote ventral development of amphibian ectoderm and mesoderm. Chordin and noggin bind to BMP-4, thus interfering with the ability of BMP-4 to activate its receptor. As a result, the activity of the BMP-4 receptor may form a morphogen gradient with a maximum on the ventral side.

The molecules involved in establishing the dorsoventral axes of flies and frogs were discovered independently by different teams of investigators. It turned out, however, that dpp and BMP-4 are remarkably similar, both belonging to the family of transforming growth factor-β (TGF-β) factors. Likewise, sog and chordin share 29% sequence similarity (François and Bier, 1995). Given that dpp/BMP-4 and sog/chordin are orthologs by the criterion of sequence similarity, are their functions also homologous, albeit with a reversal of orientation in the ancestral line of either flies or frogs?

DORSOVENTRAL PATTERNING IS CONTROLLED BY THE SAME GENES IN FLIES AND FROGS

In order to test whether the dpp/BMP-4 and sog/chordin pairs of orthologs have similar functions, researchers in two laboratories swapped molecules between fly and frog embryos.

TO find out whether sog and chordin proteins could functionally replace each other, Holley and coworkers (1995) first confirmed that sog mRNA transcribed in vitro would function as expected in the species from which it was derived, that is, in *Drosophila*. In heterotopic transplantation experiments, they injected controlled amounts of sog mRNA dorsally into wild-type embryos at the syncytial blastoderm stage. Injection of 100 pg sog mRNA blocked the formation of the most dorsal structure, the amnioserosa (Fig. 23.17c, d). After injection of 200 pg, the resulting embryos showed ventral structures, such as denticle belts, on the dorsal side (Fig. 23.17a, b). Also after injection of 200 pg sog mRNA, many of the injected embryos developed ventral neurons on their dorsal sides (Fig. 23.17e, f). Control injections of mRNA encoding a truncated form of sog mRNA had no effect. Thus, ectopic localization of sog mRNA is sufficient to suppress dorsal structures and generate ventral structures in *Drosophila* embryos.

In another series of experiments, 100 pg sog mRNA were injected into *Drosophila* embryos derived from females homozygous for a null allele of *dorsal*, one of the maternal genes required for establishing the ventral side (see Section 22.6). Almost all of the injected embryos formed the dorsolateral structures associated with the opening of the respiratory organs known as tracheae. Thus, sog mRNA can ventralize the *Drosophila* embryo in the absence of the maternal gene product that normally defines the ventral side.

For the critical part of their experiments, the researchers injected sog mRNA into single ventral blastomeres of *Xenopus* embryos at the 4- to 16-cell stage. As a result, many of the injected embryos formed secondary blastopores and twin axes (Fig. 23.18a–c). Again, injections with control mRNAs had no effect. When injected into embryos that had been ventralized by UV irradiation (see Section 9.5), sog mRNA restored the formation of dorsal structures in a dose-dependent manner. While 25 pg sog mRNA caused a partial rescue, injection of 100 pg sog mRNA effected a virtually complete rescue (Fig. 23.18d–f). These experiments were carried out independently, and with similar results, by Schmidt and coworkers (1995a).

In an extension of their experiment, Holley and coworkers (1995) labeled the sog-injected *Xenopus* blastomere by coinjection of lacZ mRNA, which encodes β-galactosidase, an enzyme that generates a visible blue stain from a suitable substrate. Tracing the descendants of the injected cell revealed that they formed mostly notochord themselves and recruited neighboring cells to make somites and neural tissue (Fig. 23.18g–i). In summary, ventral blastomeres of *Xenopus* cells injected with sog mRNA from *Drosophila* gave rise to descendants that showed all the characteristics of Spemann's organizer.

For the converse experiment, the investigators injected chordin mRNA from *Xenopus* dorsally into *Drosophila*

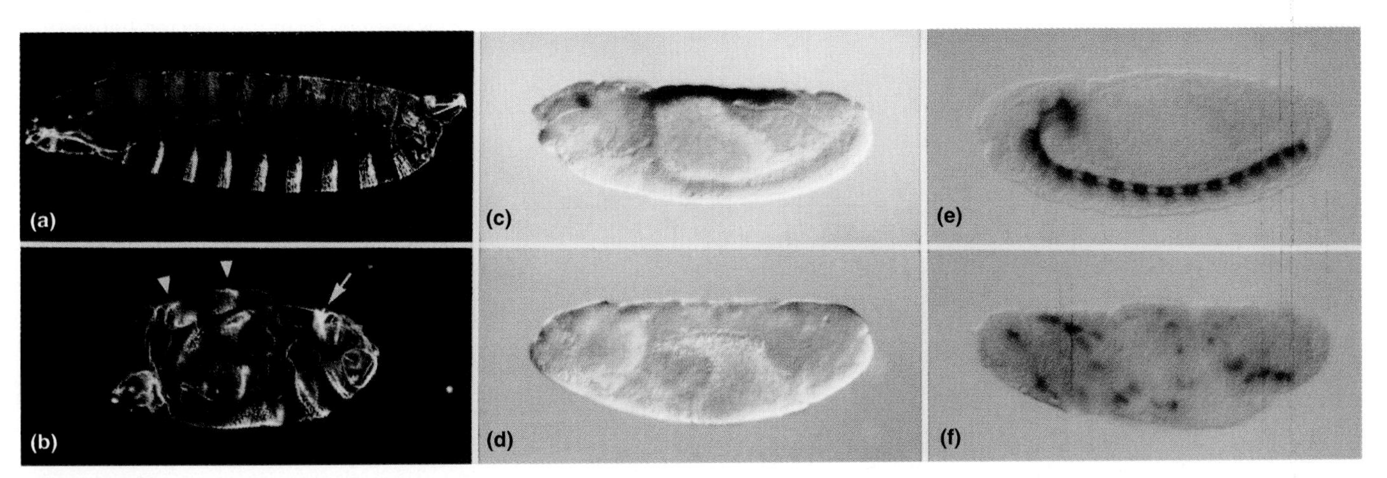

Figure 23.17 Ventralizing effects of short gastrulation (sog) mRNA in *Drosophila* embryos. **(a)** Cuticle of a normal wild-type embryo. Anterior to the left, ventral side down. **(b)** Embryo after dorsal injection of 200 pg sog mRNA. Ventral structures, such as denticle belts, are formed dorsally (arrowheads). **(c)** Wild-type embryo carrying a *lacZ* reporter gene expressed selectively in the amnioserosa, the dorsalmost structure of the embryo. **(d)** Similar embryo after injection of 100 pg sog mRNA. The reporter gene for amnioserosa is not expressed. **(e)** Normal wild-type embryo immunostained (see Method 4.1) for a protein characteristic of the central nervous system (CNS), located ventrally in the embryo. **(f)** Similar embryo after dorsal injection of 100 pg sog mRNA. The CNS is disrupted, with stained cells present dorsally as well as ventrally.

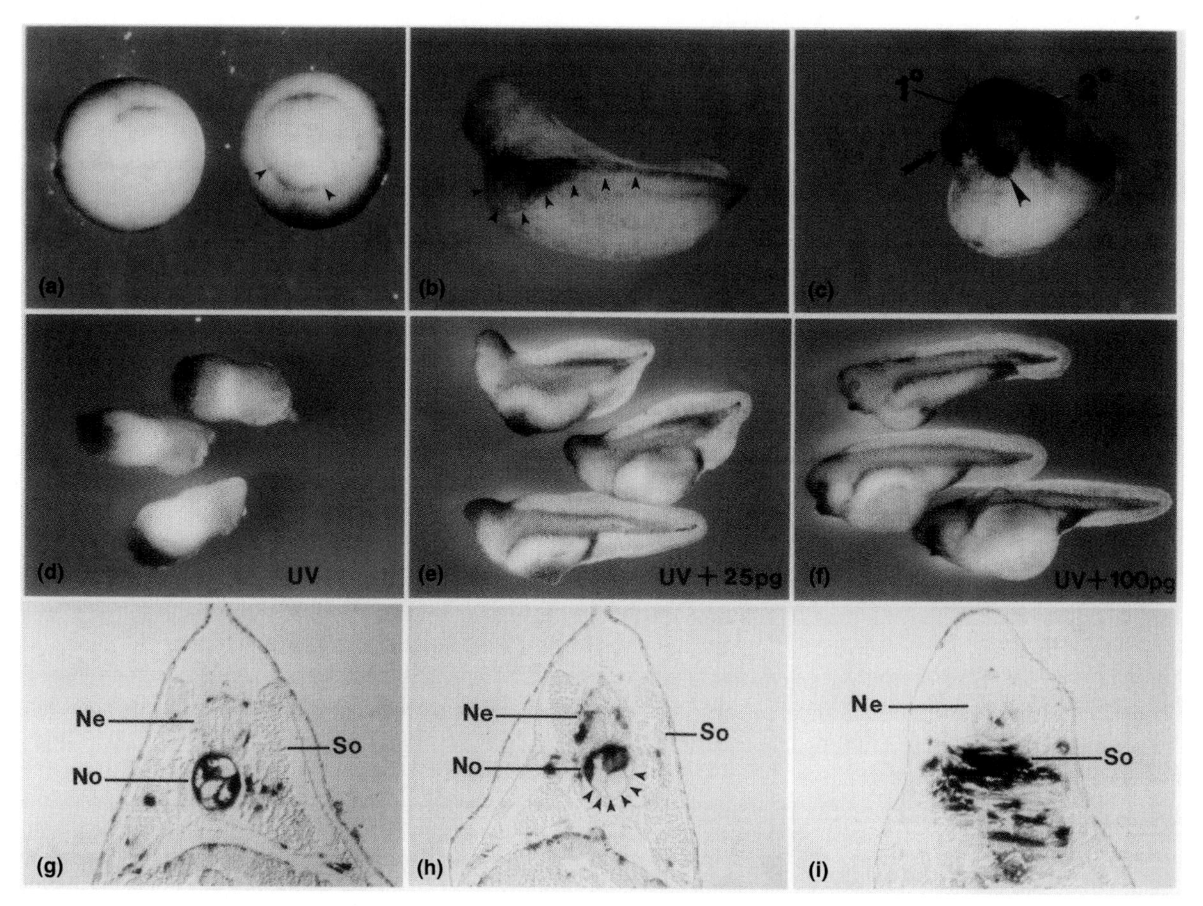

Figure 23.18 Dorsalizing effect of short gastrulation (sog) mRNA from *Drosophila* in *Xenopus* embryos. **(a)** The control embryo to the left is at the early gastrula stage, forming a blastopore on the dorsal side. The embryo to the right forms a secondary blastopore (arrowheads) after ventral injection of sog mRNA. **(b)** Injected embryo at a later stage showing secondary axis (arrowheads) induced by injected sog mRNA. **(c)** Head of an injected embryo with twin axes, displaying a secondary cement gland (arrowhead) in addition to the primary one (arrow). **(d)** Embryos completely ventralized by UV irradiation during the first cell cycle. **(e)** Partial rescue of ventralized embryos after injection of 25 pg sog mRNA. **(f)** Complete rescue of ventralized embryos after injection of 100 pg sog mRNA. **(g)** Histological section of a UV-irradiated embryo completely rescued after single-cell injection of sog mRNA at the 16-cell stage. (LacZ mRNA causing a blue stain was coinjected with the sog mRNA as a lineage tracer.) The descendants of the injected cell formed notochord (No) and a few somitic (So) cells. Neural tissue (Ne) was recruited from unlabeled cells. **(h)** Rescued embryo in which only part of the notochord is derived from the injected (labeled) cell. Arrowheads indicate unlabeled cells. **(i)** Section of an incompletely rescued embryo that lacks a notochord but has lineage label in somitic muscle cells (So), indicating that neural tube (Ne) can be induced in the absence of an organized notochord. See also cover illustration.

embryos, but with little effect. Since they suspected that the *signal sequences* at the amino terminus of chordin protein were not processed correctly by *Drosophila,* they engineered a mRNA that encoded a chimaeric protein consisting of the N-terminal signals of sog followed by the mature chordin protein. After injection of 200 pg of the chimaeric mRNA into the dorsal side of *Drosophila* embryos, most of the embryos failed to form amnioserosa and showed a weak ventralization of their cuticles. However, in contrast to injected sog mRNA, injected chordin chimaeric mRNA did not induce the ectopic formation of neurons. Overall, the wild-type *Drosophila* embryos injected with chimaeric chordin mRNA resembled the phenotypes produced by weak loss-of-function alleles of *decapentaplegic*$^+$. These data indicate that chordin mRNA has a ventralizing effect in *Drosophila,* although a much weaker one than sog.

Questions

1. The experiment described last shows that chordin mRNA is less effective in *Drosophila* than in *Xenopus.* This could be ascribed to a less than perfect match between *Xenopus* chordin and the remaining components of the dorsoventral system in *Drosophila.* Specifically, chordin and BMP-4 may both be intrinsically weaker than dpp and sog. Are there (quantitative) indications in the data reviewed above

that the latter, more specific hypothesis may be correct? Can you derive a testable prediction from this hypothesis?

2. How would you test whether dpp from *Drosophila* has a ventralizing effect in *Xenopus?*
3. Assuming that dpp does have a ventralizing effect in *Xenopus,* what kind of embryo would you expect to develop from an 8-cell embryo in which all four animal blastomeres have been injected with dpp and one vegetal blastomere with sog?

In additional experiments, Holley and coworkers (1996) used noggin mRNA from *Xenopus,* which—like chordin mRNA—is synthesized specifically in Spemann's organizer and has similar dorsalizing effects in amphibian embryos. When injected dorsally in *Drosophila* embryos, noggin mRNA not only blocked the development of amnioserosa and other dorsal structures, but also induced the formation of neurons. Thus, noggin mRNA displayed a more potent ventralizing effect than chordin in *Drosophila* embryos. Noggin mRNA also induced neuron formation in *Drosophila* embryos derived from females homozygous for a null allele of *dorsal.* This means noggin mRNA strongly ventralized *Drosophila* embryos in the absence of the endogenous maternal gene product that normally defines the ventral side.

The similarities between the dorsoventral patterning mechanisms of *Drosophila* and *Xenopus* extends even further (Marqués et al., 1997; Piccolo et al., 1997). As mentioned earlier, the inhibitory action of sog on dpp in *Drosophila* is counterbalanced by the tolloid (tld) protein, which is also synthesized dorsally. Tld has the molecular characteristics of a protease, and it cleaves sog, especially in the presence of dpp. The *Xenopus* ortholog of tld, Xolloid (Xld) has exactly corresponding properties: It encodes the same type of protease as tld and cleaves chordin (but not noggin) at specific sites. Moreover, it releases biologically active BMP-4 from BMP-4/chordin complexes. Most importantly, both tld and Xld inhibit the ability of sog or chordin to induce a secondary embryonic axis in *Xenopus* embryos. Thus, it is a set of at least three signal proteins—tld/Xld, sog/chordin and noggin, dpp/BMP-4—that are orthologous and set up the dorsoventral axis, although in reverse orientations, in flies versus frogs.

The conservation of the mechanism for setting up the dorsoventral axis also extends beyond *Drosophila* and *Xenopus.* In the zebrafish *Danio rerio,* two genes have been identified that seem to be orthologous to *BMP-4*$^{+}$ and *chordin*$^{+}$ (Hammerschmidt et al., 1996). The results reported above are a stunning demonstration that the genetic control mechanism for establishing the dorsoventral axis is very similar in vertebrates and invertebrates, just as the anteroposterior body pattern is specified by *Hox* genes in both groups of organisms.

The use of the same signaling pathway in two different developmental processes does not prove that the two processes are *homologous,* meaning that they have evolved from a common precursor. For instance, decapentaplegic protein seems to act as a morphogen in patterning processes that are clearly not homologous, such as the dorsoventral pattern of the entire embryo and the anteroposterior pattern of the wing in *Drosophila* (see Sections 22.6 and 22.8). However, in the case of the dorsoventral patterns in vertebrates and invertebrates, the detailed similarities between the two signaling pathways are striking, and the same body axis is involved in both groups of organisms. These commonalities have rekindled discussions of whether the dorsoventral axes of vertebrates and invertebrates are homologous, and when and how this axis may have become inverted in one of the groups. Some scenarios have been discussed, but definitive answers will require a better knowledge of the common ancestor of *protostomes* and *deuterostomes* than we have today (*Lacalli, 1996*).

23.5 Pattern Formation and *Hox* Gene Expression in Limb Buds

Vertebrate limbs develop relatively late during embryogenesis, when most other organs have already been established. In the human embryo, paddle-shaped limb buds emerge during the fifth week, and digits are formed thereafter (see Figs. 12.2 and 14.1). Limb buds have a core of mesenchyme cells and a covering layer of ectoderm. The ectodermal layer usually forms the ***apical ectodermal ridge* (*AER*)**, a ridge of columnar cells at the end of the limb bud running in anteroposterior direction (Fig. 23.19). The mesenchymal core originates in part from the somatic layer of the lateral plate and in part from myotomes.

To investigate the contributions from these two sources, Christ and coworkers (1977) removed a strip of newly formed somites from chicken embryos and replaced them with a corresponding strip of quail somites. (Quail cells can be distinguished from chicken cells by their brightly staining heterochromatin; see Fig. 13.26.) After incubation, the investigators found that the somitic cells formed muscle while somatic lateral plate mesoderm gave rise to cartilage and other connective tissues of the limb. Also, in the absence of somites, lateral mesoderm will form a muscleless limb with a normal skeleton (Kieny and Chevallier, 1979).

The limb bud is gradually shaped into a limb by differential growth, by programmed cell death, and through histogenesis. Cell divisions accompanied by cell growth generate the mass of the limb and its approximate shape. An area of actively dividing undifferentiated mesenchyme located beneath the AER is known as the ***progress zone.*** As the limb bud grows,

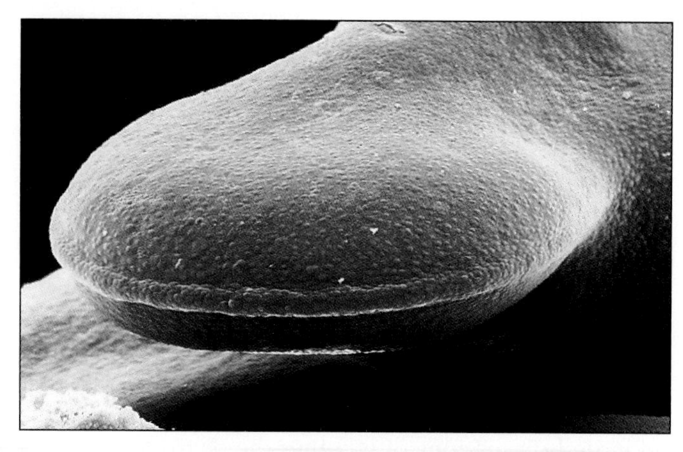

Figure 23.19 Limb bud of a hamster embryo as seen with the scanning electron microscope. Note the apical ectodermal ridge at the distal rim. Figure 14.1 shows an entire human embryo at a similar stage of development.

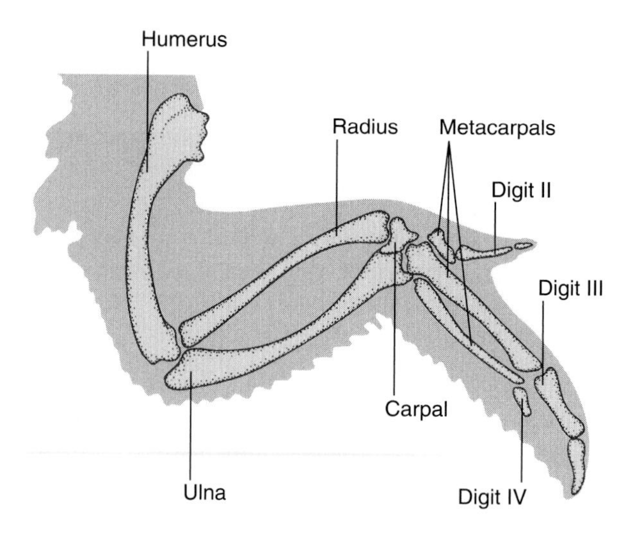

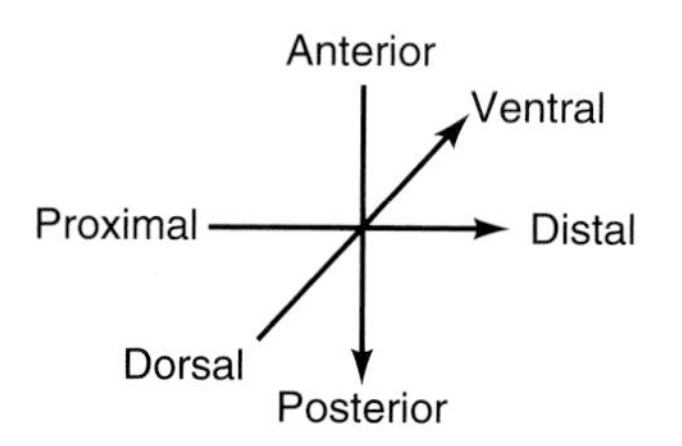

Figure 23.20 The three polarity axes of the chicken wing. The diagram shows a dorsal view of the right wing. The digits are numbered II, III, and IV, according to their homology with the digits of reptiles and mammals. Digit II is homologous to the index finger of the human hand; digit IV is homologous to the human ring finger. The wing has three

cells leave the progress zone and begin to differentiate. The limb bud's contour is refined by waves of *programmed cell death* sweeping along the edges. Cell death also causes erosion of the tissues between the digits, thus sculpting the fingers and toes. (In ducks and other web-footed vertebrates, interdigital cell death is limited to the distal margin of the limb.) While the external shape of the limb is still being established, mesenchyme condensations near the center of the limb form the cartilage models that foreshadow the development of bones (see Fig. 21.2). Other cells positioned more to the outside of the limb bud form myogenic condensations, the future muscles of the developing limb.

Vertebrate limbs have been favorite objects for investigators of pattern formation because the limb buds of salamanders and chicken are sufficiently large and accessible for transplantion. This classical method of investigation has recently been complemented by the transplantation of transgenic cells in chicken and by the study of transgenic mice. As a result, the 1990s have been years of great excitement and significant progress in understanding pattern formation in limbs (*R. L. Johnson and C. J. Tabin, 1997; Zeller and Duboule, 1997; Coates and Cohn, 1998*).

The first question that arises in the study of limb formation is about the positioning and regional character of *limb fields*—that is, the semiautonomous and self-regulating groups of cells that will give rise to limbs. Why do vertebrates have only two pairs of limbs, and how do hindlimbs become different from forelimbs? Other questions are about setting up the polarity axes of limbs (Fig. 23.20). How does each limb acquire its three polarity axes: anteroposterior, dorsoventral, and proximodistal? Further questions are about translating these polarities into patterns of different elements, such as the digits of a wing or hand. We will address these questions in turn, focussing on the anteroposterior pattern.

LIMB FIELD INITIATION IS AFFECTED BY FGF SIGNALING AND *HOX* GENE EXPRESSION

Given the late outgrowth of limb buds, it seems logical to assume that the initiation of limb fields should be controlled by signals from previously formed organ rudiments. Recent investigations have provided some clues about the molecular nature of the signals involved.

During normal limb initiation, fibroblast growth factor 10 (FGF-10) is synthesized in the mesenchyme, and FGF-8 is made in the ectoderm of the prospective limb. By applying FGF-soaked beads to lateral plate mesoderm between the prospective wings and legs of chicken embryos, additional limbs can be induced (Cohn et al., 1995, 1997). The ectopic wings are morphologically normal except that their anteroposterior polarity is reversed. Interestingly, FGF can induce either ectopic wings or ectopic legs according to the position at which it is applied. FGF-10 is not only sufficient but also necessary for the initiation of wing development (Sekine et al., 1999). Transgenic mice lacking FGF-10 show complete truncation of the fore- and hindlimbs. Analysis of marker gene expression indicates that the apical ectodermal ridge (AER) and the *zone of polarizing activity*

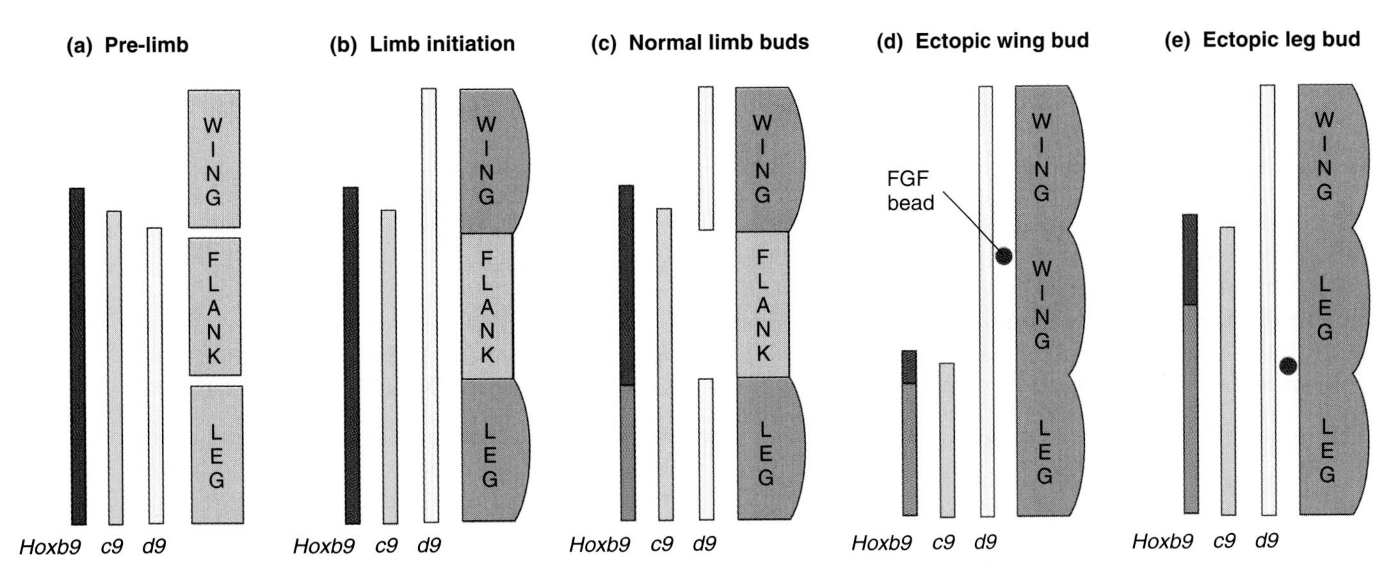

Figure 23.21 Limb initiation and expression of *Hox-9* genes in chicken lateral plate mesoderm. **(a)** Before limb initiation, *Hoxb9*⁺, *Hoxc9*⁺, and *Hoxd9*⁺ are expressed in lateral plate mesoderm with anterior boundaries staggered near the posterior boundary of the wing bud. **(b)** As limb budding is initiated, *Hoxd9*⁺ expression shifts anteriorly, where it marks the anterior boundary of the limb bud. **(c)** Later in limb development, *Hoxd9*⁺ switches off in the flank, and *Hoxb9*⁺ is reduced in the leg. Thus, the wing expresses *Hoxb9*⁺ and *Hoxc9*⁺ posteriorly, and *Hoxd9*⁺ throughout, while the flank expresses *Hoxb9*⁺ and *Hoxc9*⁺, and the leg expresses low levels of *Hoxb9*⁺ and high levels of *Hoxc9*⁺ and *Hoxd9*⁺. **(d)** Application of FGF to the anterior flank reprograms *Hox* gene expression in the lateral plate so that the wing pattern is recreated in the flank and an ectopic wing develops from flank cells. **(e)** Application of FGF to posterior flank results in the leg pattern of *Hox* gene expression being recreated in the flank, and an ectopic leg develops from the flank.

(*ZPA*) do not form. The same mutant mice are also deficient in the pulmonary branching of their lungs. Thus, FGF-10 may be necessary for a step in the epitheliomesenchymal interactions that is required for both limb and lung development.

The induction of ectopic limbs in chicken embryos by FGF-10 is accompanied by specific changes in the expression of *Hox* genes in lateral plate mesoderm. These changes mimic, in the flank, the expression patterns of *Hox9*⁺ pseudo-orthologs found at normal limb levels (Fig. 23.21). Observations on transgenic mice confirm that changes in *Hox* gene expression can cause changes in limb position and anteroposterior polarity. While loss of function of the *Hoxb5* gene causes the forelimb to shift anteriorly along the body axis, anterior extension of the *Hoxb8*⁺ expression domain leads to the formation of extra digits anteriorly in the forelimb (Charité et al., 1994; Rancourt et al., 1995). Thus, a combination of *Hox* gene activities and FGF signals may be involved in the overall positioning of limb buds.

T-BOX GENES CONTROL THE DIFFERENCE BETWEEN HINDLIMB AND FORELIMB

The hindlimbs and forelimbs of vertebrates are made from the same types of cells, and they are very similar in the overall pattern of bones and muscles. So what determines "legness" versus "handness" in a human, or "legness" versus "wingness" in a chicken? The answer to this question has eluded researchers for a long time. Recent studies have revealed three genes that are involved. One of them, designated *Pitx1*⁺, encodes a bicoid-related transcriptional activator and is expressed preferentially in the early hindlimb. The two other genes, *Tbx4*⁺ and *Tbx5*⁺, are members of a group of genes known as the T-box, or Tbx, family, which also encode transcription factors. While *Tbx4*⁺ is expressed in the early hindlimb *Tbx5*⁺ is expressed in the early forelimb.

Knockout mice (see Sections 15.5 and 15.6) lacking *Pitx1*⁺ have short, thin hindlimbs (Szeto et al., 1999). The bones in these limbs show several modifications that make them resemble the corresponding forelimb structures. Deletion of the *Pitx1*⁺ gene also results in decreased distal expression of the *Tbx4*⁺ gene in the hindlimb bud. These results indicate that *Pitx1*⁺ expression is necessary for normal hindlimb development and complete *Tbx4*⁺ activation. The same *Pitx1*⁺ gene–deleted mice also exhibit a cleft palate and other defects suggestive of insufficient cell proliferation.

In order to test whether *Pitx1*⁺ expression is not only necessary but also sufficient for hindlimb formation, investigators have used different ways of forcing the ectopic expression of this gene in chicken wing buds. Logan and Tabin (1999) inserted the *Pitx1*⁺ gene into a virus and infected the limb bud region of chicken embryos with the recombinant virus. This caused the ectopic activation of *Tbx4*$^{+}$, the member of T-box family that is normally expressed in the hindlimb only. Wing buds in which *Pitx1*$^{+}$ was misexpressed developed into limbs with some morphological characteristics of chicken legs. While normal wings bend posteriorly at the equivalent of the wrist, the infected chicken limbs grew straight, as do the corresponding bones of a chicken foot. Also, the digits of the infected wings were more toelike in their relative size and shape, and the wing muscle pattern was transformed to that of a leg. Similar results were reported by Takeuchi and coworkers (1999), who misexpressed *Tbx4*$^{+}$ in wing buds and *Tbx5*$^{+}$ in leg buds. Following this treatment, they observed almost complete transformations from wing to leg, and leg to wing, respectively.

The T-box genes, like many other patterning genes, have been well conserved in evolution. Mutations in the *TBX5*$^{+}$ gene, the human homolog of mouse *Tbx5*$^{+}$, cause severe defects in arm and heart development, a condition known as Holt-Oram syndrome (Basson et al., 1999). The zebrafish *Danio rerio* has three T-box genes, *Tbx2*$^{+}$, *Tbx4*$^{+}$, and *Tbx5*$^{+}$, with expression patterns that are remarkably similar to those of their tetrapod counterparts (Ruvinsky et al., 2000). In particular, expression of *Tbx4*$^{+}$ and *Tbx5*$^{+}$ is restricted to pelvic and pectoral fin buds, respectively. These results suggests that the last common ancestor of teleosts and tetrapods possessed the T-box genes, and that these genes had already acquired, and have subsequently maintained, their region-specific functions. This conclusion confirms that teleost pectoral and pelvic fins are homologous to tetrapod fore- and hindlimbs, respectively, a notion that has been held previously on the basis of comparative anatomy.

THE PROXIMODISTAL PATTERN IS FORMED IN THE PROGRESS ZONE

Once a limb bud has formed, its continued growth depends on the *apical ectodermal ridge* (*AER*). If the AER is removed the limb bud stops growing, and the resulting limb has distal truncations (Saunders, 1948; Summerbell, 1974). Conversely, ectopic transplantation of AER leads to the outgrowth of supernumerary limb buds. Supplying beads soaked in FGF-4 to AER-less limb buds restores both outgrowth and patterning and may allow the development of a complete limb (Niswander et al., 1993; Fallon et al., 1994). Together, these observations indicate that the AER induces the formation of the *progress zone*, the region of undifferentiated, actively dividing cells beneath the AER, and that the inducing signal may be FGF-4.

The level of truncation observed after excision of the AER depends on when the excision is made. Early removals lead to proximal truncations, whereas later removals allow for more distal outgrowth (Fig. 23.22). Cells that form the proximal parts of a limb differ from cells that form distal parts by the time they spend in the progress zone. Cells that exit this zone early form proximal limb parts, whereas cells leaving late form progressively more distal parts. How time spent in the progress zone translates into different cell determinations is not known.

SIGNALS FROM SOMITES AND LATERAL PLATE SPECIFY THE DORSOVENTRAL LIMB BUD AXIS

How does the limb bud acquire dorsoventral polarity so that, for example, the extensor muscles of the digits are formed on the dorsal side and the flexor muscles on the

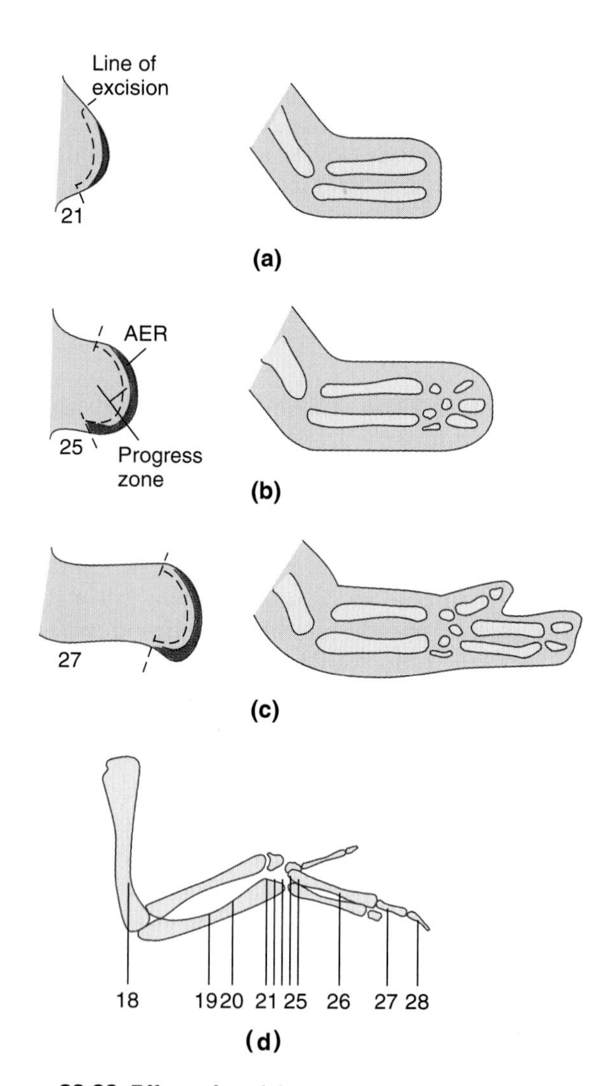

Figure 23.22 Effect of excising the apical ectodermal ridge (AER) from the chicken wing bud. The diagrams show dorsal views of the right wing bud and wing skeleton. **(a–c)** Excision at later stages of development is followed by truncation at more distal levels. **(d)** Summary diagram showing the level of truncation after AER excision at stages numbered 18 (earliest) to 28 (latest).

ventral side? As we will see, the underlying somites and lateral plate provide the orienting clues.

Fate-mapping studies using chick-quail chimeras have shown that, prior to limb bud outgrowth, its mesodermal portion is located in the medial part of the lateral mesoderm (Fig. 23.23a). At this stage, the entire overlying ectoderm is fated to form the AER. The prospective dorsal limb bud ectoderm is still overlying the somites and will be pulled over the bud mesenchyme as it grows out. Similarly, the future ventral bud ectoderm will be recruited from cells that are now overlying the lateral portion of the lateral plate.

The limb bud mesoderm is first induced by signals from adjacent somites, and in turn the bud mesoderm induces the overlying ectoderm to form AER. Subsequently, the limb bud mesoderm continues to receive dorsalizing signals from neighboring somites and ventralizing signals from adjacent lateral mesoderm. This was shown by heterotopic transplantations and by placing barriers between parts of the mesoderm (Michaud et al., 1997). For instance, limb fields wedged between two somites (by inserting them in the place of the neural tube) form a bidorsal limb bud. Dorsal and ventral limb bud ectoderm also receive inductive signals from somites and lateral portions of plate mesoderm, respectively (Fig. 23.23b). As the ectodermal sheets are pulled over the outgrowing limb bud mesenchyme, their dorsal and ventral determinations may feed back on the mesenchymal core, thus reinforcing its dorsoventral polarity.

Several important genes become active as the limb bud grows out. One key pair consists of *engrailed-1*$^+$ and *Radical fringe*$^+$, both cloned by using their *Drosophila* orthologs *engrailed*$^+$ and *fringe*$^+$, as probes. *Radical fringe*$^+$, like its *Drosophila* ortholog, encodes a secreted protein (see Section 22.8). Its synthesis begins in the dorsal ectoderm and future AER of the outgrowing limb bud

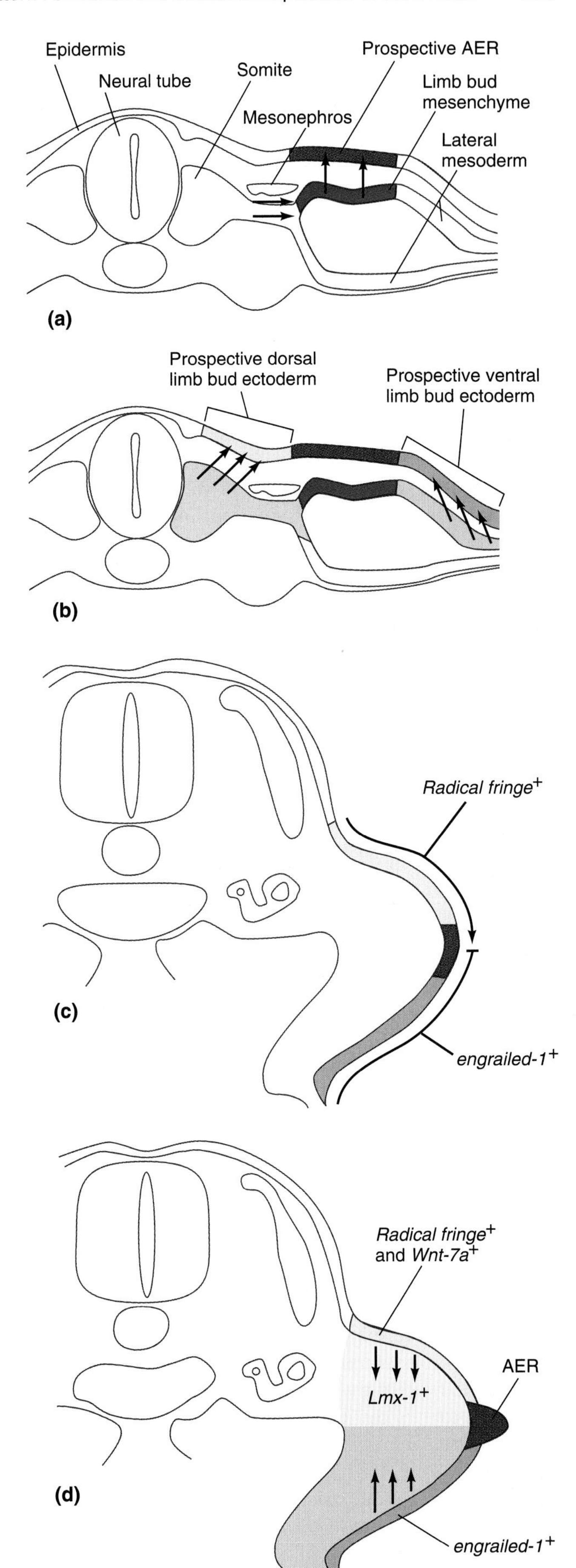

Figure 23.23 Specification of the dorsoventral polarity in the chicken limb bud. The diagrams show transverse sections of chicken embryos at early **(a, b)** and successively later **(c, d)** stages of development. **(a)** Somites and mesonephros induce the formation of limb bud mesenchyme, which in turn induces the overlying ectoderm to form the AER at the future dorsoventral interface. **(b)** Somites and lateral mesoderm continue to send dorsalizing and ventralizing signals to limb bud mesenchyme while also inducing overlying ectoderm to assume dorsal and ventral identities. **(c)** Activation of *engrailed-1*$^+$ in ventral ectoderm and *Radical fringe*$^+$ in dorsal ectoderm. *engrailed-1*$^+$ limits the expression of *Radical fringe*$^+$ to dorsal ectoderm, and the AER will form at the boundary between cells that express *Radical fringe*$^+$ and those that don't. **(d)** Expression of *Wnt-7a*$^+$ in the dorsal ectoderm, and the resulting activation of *Lmx-1*$^+$ in dorsal mesenchyme, impose dorsoventral polarity on the developing limb. Again, *engrailed-1*$^+$ limits the expression of *Wnt-7a*$^+$ to the dorsal side.

(Fig. 23.23c). Further extending the homology with *fringe*$^+$, ectopic but localized expression of *Radical fringe*$^+$ in ventral limb bud ectoderm leads to the formation of supernumerary, ventral AERs, whereas uniform expression suppresses AER formation (E. Laufer et al., 1997; Rodriguez-Esteban et al., 1997). Expression of *Radical fringe*$^+$ is restricted to dorsal ectoderm by *engrailed-1*$^+$, presumably through the activity of one or more intermediate genes. *Engrailed-1*$^+$ is expressed in ventral limb bud ectoderm up to the midline of the AER, and inactivation of *engrailed-1*$^+$ dorsalizes both ectoderm and mesenchyme of the limb bud. Conversely, ectopic dorsal expression inhibits *Radical fringe*$^+$, with the result of disrupting AER formation (Logan et al., 1997). Together, these observations indicate that the AER is formed where *Radical fringe*$^+$–expressing cells are juxtaposed to nonexpressing cells, and that *engrailed-1*$^+$ generates this juxtaposition by inhibiting *Radical fringe*$^+$.

Do the same genes that position the AER also determine the dorsoventral polarity of the limb? Signals from the ectoderm have been implicated in this process since inversion of the ectoderm's dorsoventral axis by surgical rotation causes the same inversion of the mesodermal limb parts (MacCabe et al., 1974). In addition to *engrailed-1*$^+$ and *Radical fringe*$^+$, two other genes are expressed differentially along the dorsoventral axis of the limb bud. *Wnt-7a*$^+$, an ortholog of the *Drosophila wingless*$^+$ gene, makes a secreted protein in the dorsal ectoderm. *Lmx-1*$^+$ encodes a homeodomain protein in the dorsal mesenchyme (Fig. 23.23d).

The functions of *Wnt-7a*$^+$ and *Lmx-1*$^+$ have been tested by ectopic expression in chick and by targeted disruption in mice (Parr and McMahon, 1995; Riddle et al., 1995; A. Vogel et al., 1995). Based on these investigations, Wnt-7a acts as a dorsal signal that activates the synthesis of Lmx-1 in dorsal mesenchyme. The expression of *Lmx-1*$^+$, as in the case of *Radical fringe*$^+$ discussed above, is restricted to dorsal mesenchyme by *engrailed-1*$^+$, presumably again through other genes. Thus, the positioning of the AER and the determination of dorsoventral polarity both involve *engrailed-1*$^+$ and *Radical fringe*$^+$, but each function may require its own set of intermediates.

A ZONE OF POLARIZING ACTIVITY DETERMINES THE ANTEROPOSTERIOR PATTERN

Transplantation experiments have revealed that the overall anteroposterior polarity axis is already specified in the limb field of the lateral plate mesoderm before the limb bud grows out (Zwilling, 1956). A key element in anteroposterior patterning process within the limb bud was also discovered by transplantation experiments. When Saunders and Gasseling (1968) removed a bit of tissue from the *posterior rim* of a chicken wing bud and inserted it into the *anterior rim* of a host wing bud they observed a stunning effect (Fig. 23.24). The normal sequence of digits in a chicken wing (II-III-IV) was changed to a mirror-image duplication (IV-III-II-III-IV). When the posterior tissue was implanted at the tip of the host wing bud rather than at the anterior rim, the resulting pattern of digits was II-III-IV-III-IV (from anterior to posterior). Heterospecific transplantations revealed that the extra digits were formed by cells of the host limb bud rather than by the transplant itself. The ability of the transplant to establish the posterior pole of an anteroposterior axis in its vicinity was termed ***polarizing activity.***

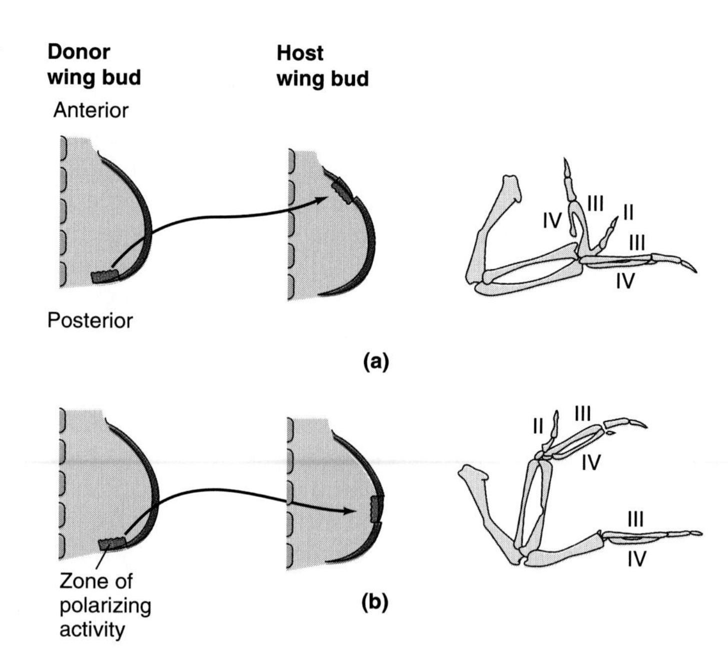

Figure 23.24 Effects of grafting tissue from the posterior margin of a chicken wing bud. The digits are numbered as in Figure 23.20. **(a)** Implantation of posterior tissue into the anterior margin of a host wing bud results in a mirror-image duplication of the digital pattern (IV-III-II-III-IV). **(b)** Implantation into the tip of a host wing bud results in a supernumerary ulna with additional digits III and IV. The tissue with this reorganizing activity has hence been known as the zone of polarizing activity (ZPA).

Other researchers repeated and extended these experiments, mapping the polarizing activities of different limb bud areas at different stages of development. In early limb buds, the activity is strongest at the posterior edge of the bud near its junction with the body wall (Fig. 23.25). In more advanced limb buds, the activity is greatest in a distal portion of the posterior edge. At each stage, the most active area is called the ***zone of polarizing activity* (ZPA).**

Polarizing activity was also found in the corresponding zones of reptilian and mammalian limb buds, indicating that the polarizing signal has been well conserved during evolution. Additional experiments revealed that polarizing activity is present in embryonic tissues other

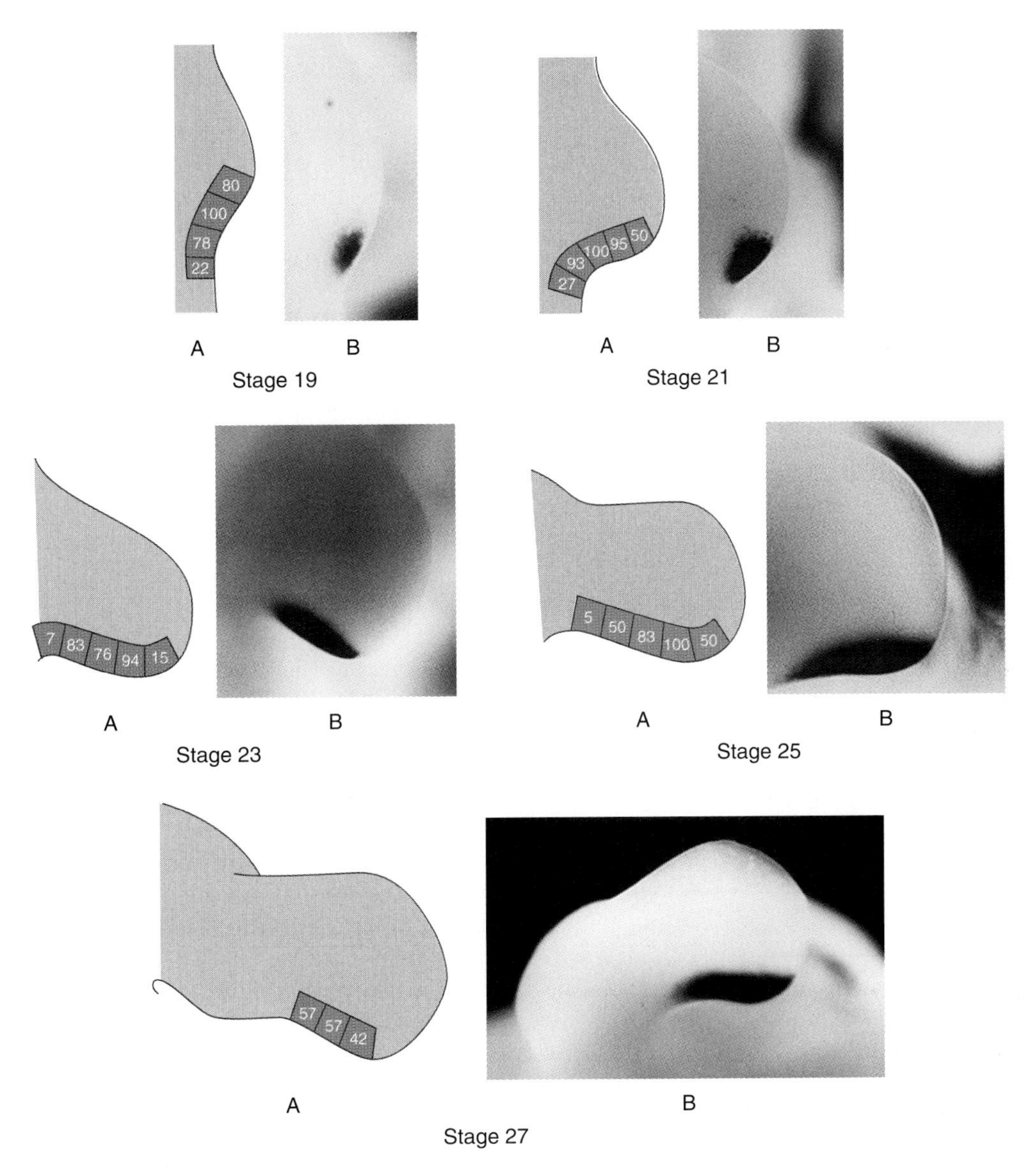

Figure 23.25 Polarizing activity and *Sonic hedgehog*$^+$ gene expression in chicken wing buds. The numbered stages of development are as defined by Hamburger and Hamilton (1951). For each stage, a diagram (A) and a photograph (B) of a limb bud are shown. In the diagrams, the number in each square indicates the relative strength (%) of the polarizing activity based on the number of extra digits formed in a standardized series of transplantations as diagrammed in Figure 23.24a (Honig and Summerbell, 1985). The photos show expression of the *Sonic hedgehog*$^+$ gene as revealed by in situ hybridization (see Method 8.1).

than limb buds, such as *Hensen's node* in chicken gastrulae, the *notochord*, the *floor plate* of the developing spinal cord, and the *genital tubercle.*

SONIC HEDGEHOG PROTEIN HAS POLARIZING ACTIVITY

Some of the tissues that showed ZPA activity in transplantation experiments, including the notochord, have since been found to secrete Sonic hedgehog protein (see Section 13.1 and 14.2). The *Sonic hedgehog*$^+$ gene was identified as a vertebrate homolog of the *Drosophila* segment polarity gene *hedgehog*$^+$ and named after a then-popular video game character. When tested in chicken limb buds at various stages, the accumulation of Sonic hedgehog mRNA, as revealed by *in situ hybridization,* coincided closely with the zone of polarizing activity, as assayed by transplantation experiments (Fig. 23.25; Riddle et al., 1993).

To test whether Sonic hedgehog protein actually establishes anteroposterior polarity in limb buds, Riddle and coworkers (1993) transfected cultured cells with *Sonic hedgehog*$^+$ gene and implanted a pellet of transfected cells anteriorly into limb buds. This operation caused duplications of the bone pattern in the wing similar to those obtained after transplanting posterior wing bud tissue. Thus, Sonic hedgehog signaling is sufficient to establish the anteroposterior pattern of the developing limb.

HOX GENE EXPRESSION MEDIATES ANTEROPOSTERIOR LIMB PATTERNING

The identification of *Hox* genes has provided a new basis for studying pattern formation in vertebrates. We have already noticed the role of Hox-9 pseudo-orthologs in positioning the limb buds along the anteroposterior body axis and in determining their character as forelimbs or hindlimbs (see Fig. 23.21). Another set of *Hox* genes, namely *Hoxd13*$^+$, *Hoxd12*$^+$, *Hoxd11*$^+$, *Hoxd10*$^+$, and *Hoxd9*$^+$, is expressed in a nested pattern within the limb (Fig. 23.26). The gene at the 3′ end of the cluster, *Hoxd9*$^+$, is expressed first at the posterior and distal margin, an area coinciding approximately with the zone of polarizing activity. The *Hoxd9*$^+$ expression domain expands anteriorly and proximally as the limb bud grows. Next, expression of *Hoxd10*$^+$ begins at the same posterior distal spot and spreads anteriorly and proximally, but not as far as that of *Hoxd9*$^+$. Genes are sequentially activated but each is successively more restricted in its spreading, so that the final expression domains form a nested set, with all genes expressed in the zone of polarizing activity.

To determine whether the *Hox D* genes are causally involved in establishing the anteroposterior pattern in developing limbs, the *Hox D* gene expression patterns had to be altered directly, and subsequent morphological developments had to be monitored. Such experiments were carried out in chicken embryos, where the

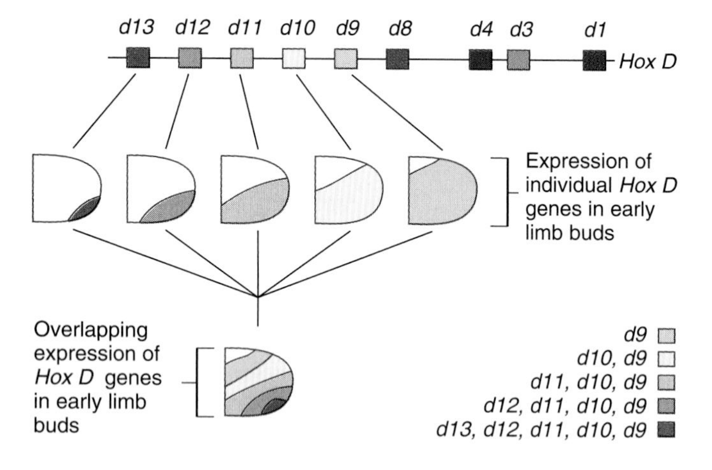

Figure 23.26 Expression of *Hox D* genes in the mouse limb bud. Genes *Hoxd9*$^+$ through *Hoxd13*$^+$ are expressed as a nested set, all domains beginning at the posterior distal margin and extending over staggered distances toward the anterior and proximal.

Hox D genes are expressed with the same patterns as in the mouse (Izpisúa-Belmonte et al., 1991; Nohno et al., 1991). Morgan and coworkers (1992) used a retrovirus vector to introduce a cloned mouse *Hoxd11* cDNA into the genome of embryonic chicken cells. After injection into an early limb bud, the virus spread locally and infected all cells of the bud as the embryo developed. Using in situ hybridization (see Method 8.1), the researchers ascertained that the transgene was expressed over the entire limb bud but did not affect the activity of resident chicken *Hox* genes.

Overexpression of mouse *Hoxd11*$^+$ in chicken limb buds generated an expression pattern of homeotic genes as diagrammed in Figure 23.27. In a normal chicken leg, the "big toe" (toe I) pointing backward on the foot develops from bud cells expressing chicken *Hoxd9*$^+$ and *Hoxd10*$^+$. In an infected leg bud, these cells express the same resident genes plus the *Hoxd11*$^+$ transgene. Such a combination of chicken *Hox* genes in a normal leg bud is associated with the formation of the "index" toe (toe II). It differs from toe I by being longer, having three bone elements instead of two, and by pointing forward. As expected, toe I from an infected leg bud showed the same morphological characteristics as toe II (see Fig. 23.28). A similar transformation, this one associated with the development of an extra digit, was observed in infected wing buds. These results show that *Hox* gene expression is directly involved in forming the pattern of digits in chicken limbs. Changes in the combination of *Hox* genes expressed are sufficient to cause corresponding changes in the digit pattern.

Extending these experiments, Riddle and coworkers (1993) transfected cultured chicken cells with *Sonic hedgehog* cDNA and implanted pellets of transfected cells anteriorly into chicken limb buds. As a result, the *Hox D* genes were expressed in mirror-image duplications of the pattern seen in normal limb buds. This result showed that Sonic hedgehog signaling is sufficient to generate the nested expression pattern of *Hox* genes in limb buds.

The experiments described above show that *Hox* genes play key roles in patterning the limbs. A polarizing signal associated with the expression of the *Sonic hedgehog*$^+$ gene is established at the posterior margin of the limb bud. This signal activates *Hox* genes in a nested set of expression domains, and the activities of these genes control the limb pattern to be formed. Since *Hox* genes encode transcription factors, they are expected to exert their morphogenetic effects by controlling various types of target genes.

Given the functional role of *Hox* genes in specifying digit characters, it may be significant that the expression domains of the *Hox D* genes in the limb bud define exactly five stripes with different combinations of *Hox D* gene activity (see Fig. 23.26). Since *Hox* genes and their functions have been exceedingly well conserved during evolution, the availability of five *Hox D* gene expression

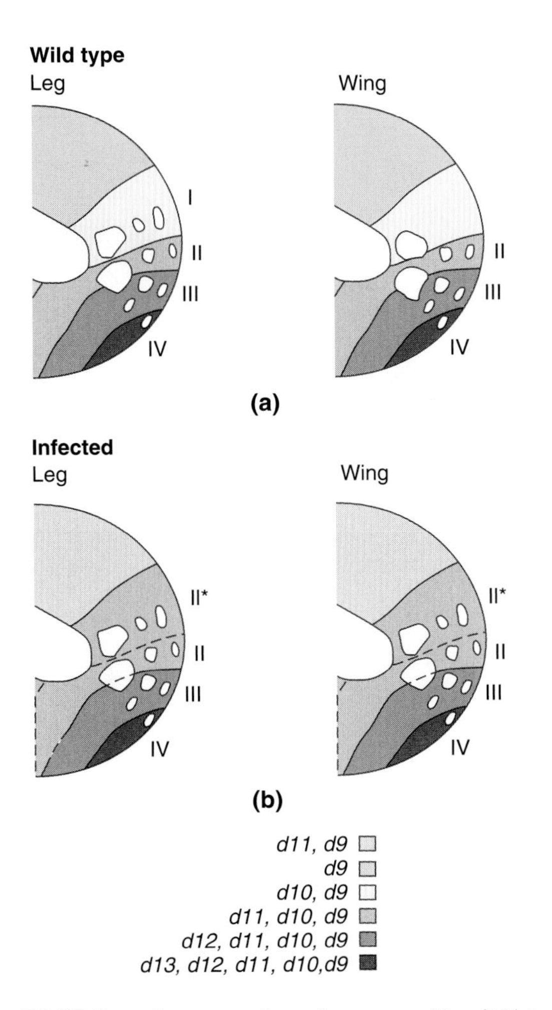

Figure 23.27 Ectopic expression of a mouse *Hoxd11*$^+$ transgene in chicken limb buds. **(a)** Normal expression patterns of chicken *Hox D* genes in the leg bud and the wing bud, superimposed on a fate map of the cartilage elements (white). The Roman numerals indicate prospective digits. **(b)** Altered expression patterns produced after limb bud infection with a virus containing the mouse *Hoxd11*$^+$ gene. (Mouse *Hoxd11*$^+$ and chicken *Hoxd11*$^+$ encode similar proteins and are not distinguished here.) Note that the leg bud region forming digit I expresses chicken *Hoxd9*$^+$ and chicken *Hoxd10*$^+$ in wild-type buds, whereas the same region in infected leg buds expresses *Hoxd11*$^+$ in addition. This is the same gene combination normally expressed in the region forming digit II, and correspondingly, digit I is transformed into an extra digit II (designated II*) in legs developing from infected buds. A corresponding change in the *Hox* gene expression pattern of wing buds is followed by the formation of an extra digit II (see Fig. 23.28).

domains may have acted as a constraint that has prevented the evolution of land-dwelling vertebrates with more than five different fingers or toes (Tabin, 1992). Mutant phenotypes with more than five fingers arise frequently, but an extra finger tends to have the same morphological features as one of its neighbors. In other words, the *number* of fingers is increased, but not the number of finger *characters*. A mere increase in number, however, without the potential for gene-regulatory and thus anatomical diversification, has apparently not been adaptive.

THE PRINCIPLE OF RECIPROCAL INTERACTION REVISITED

The analysis of limb morphogenesis reveals the *principle of reciprocal interaction* during development. As discussed earlier in the context of kidney formation (see Section 14.5), many interactions between tissues are not one-way communications but are more of a dialogue. Developmental cues are exchanged in both directions to balance and maintain the activities of all parts involved so that a harmoniously proportioned organ results.

Reciprocal interactions between the AER and the mesenchymal core of limb buds have been analyzed especially well in chicken embryos (see Hinchliffe and Johnson, 1980). First, limb bud mesenchyme induces the formation of the AER in the overlying ectoderm. Reuss and Saunders (1965) demonstrated this effect by transplanting wing bud mesenchyme that had been stripped of its ectodermal cover into the flank of a chicken embryo. Regenerating flank ectoderm soon healed over the wound. In many cases, an AER was induced over the graft, which then developed into an extra wing. The specific character of a developing limb is determined by the mesenchyme: Leg mesenchyme combined with wing ectoderm forms a leg, whereas wing mesenchyme combined with leg ectoderm forms a wing. The inductive capacity of the mesenchyme, and the competence of the ectoderm to respond to it, disappear after the stage when the AER is established in normal development.

Once established, the AER becomes necessary for the further proliferation of the limb bud mesenchyme. If the AER is removed, the mesenchyme ceases to proliferate, and the developing limb is truncated, as shown in Fig. 23.22. When two AERs are combined with a single mesenchymal core, the mesenchyme grows out under each AER. However, apart from promoting limb outgrowth, the action of the AER seems permissive rather than instructive. AERs can be exchanged between wing and leg buds without affecting the developing wing or leg pattern of the mesenchyme (Zwilling, 1955). Similarly, if the anteroposterior axis of the AER is reversed relative to the underlying mesenchyme, the developing limb pattern is not affected (Zwilling, 1956).

Conversely, the persistence of the AER depends on a maintenance factor from the mesenchyme. If limb mesenchyme is removed from a limb bud and replaced with nonlimb mesoderm, such as flank lateral plate or somites, the AER degenerates within 2 days. However,

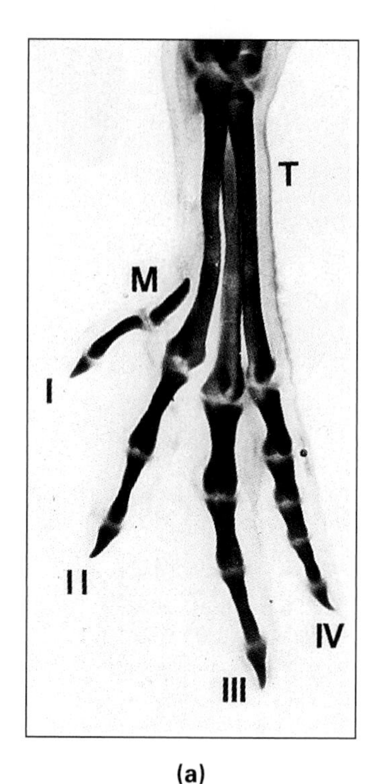

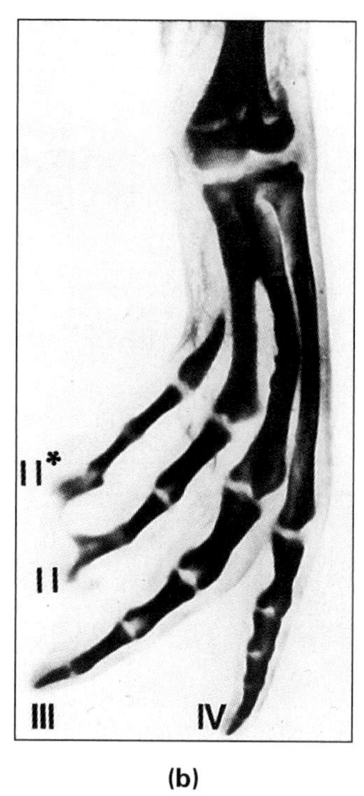

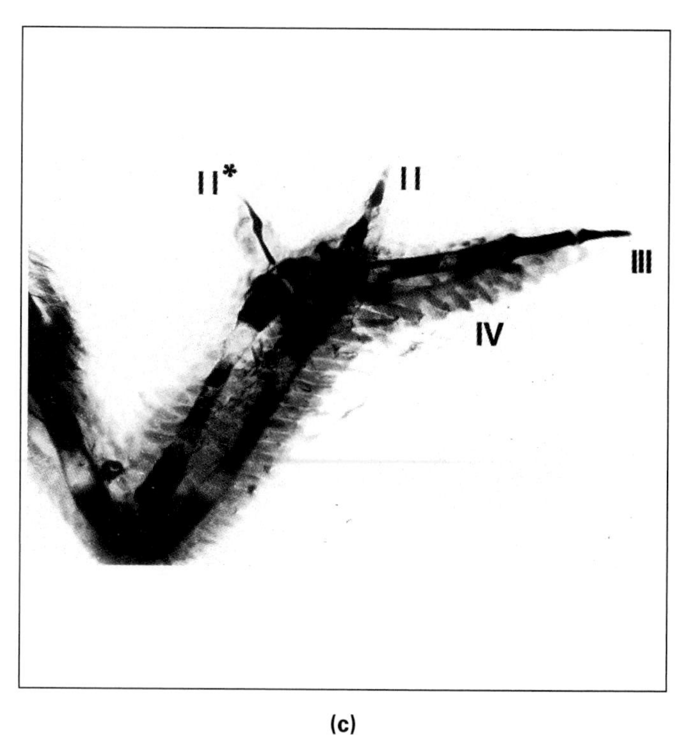

Figure 23.28 Homeotic transformations caused by ectopic expression of a mouse *Hoxd11*$^{+}$ transgene in chicken limb buds as diagrammed in Figure 23.27. **(a)** Skeleton of a wild-type leg. T = tarsometarsals; M = anterior metatarsal. Toe I, pointing backward, consists of two digital elements on a metatarsal (M). Toe II has three digital elements. **(b)** Leg derived from an infected leg bud expressing the transgene. Toe I is replaced with another toe (toe II*) resembling toe II. **(c)** Right wing derived from an infected wing bud expressing the mouse transgene. The wing has an extra digit (II*) anterior to digit II.

the AER survives if only a small piece of limb mesenchyme is added.

The principle of reciprocal interactions is also encountered in the development of the metanephros from the ureteric bud and the nephrogenic mesenchyme (see Section 14.5), as well as in the organogenesis and histogenesis of many other organs that develop from mesenchymal and epithelial components.

SUMMARY

Various strategies are being used to identify patterning genes in vertebrates. In the mouse, a small number of such genes have been found by mutant analysis and have subsequently been cloned. Additional mutants are generated by promoter or enhancer trapping. Alternative strategies—known as reverse genetics—begin with the cloning of genes that are likely to play key roles in pattern formation. One way of identifying such genes in vertebrates is based on their sequence similarity to well-characterized patterning genes from other organisms such as *Drosophila.* This strategy works well even if the sequence similarity is limited to certain motifs. Once the cloned genes are in hand, investigators proceed with molecular strategies to generate the equivalents of mutants. The goal of both genetics and reverse genetics is to characterize the biochemical and morphological functions of control genes in embryonic pattern formation.

All animals investigated so far have homeobox-containing genes, which encode transcription factors playing key roles in development. The protein domain encoded by a homeobox is known as a homeodomain; it serves as a DNA-binding region. Homeodomains have been conserved exceedingly well in evolution.

Subgroups of homeobox-containing genes are clustered in *Hox* complexes. Invertebrates have one *Hox* complex, while most vertebrate species have four complexes located on four different chromosomes. Each *Hox* complex contains about ten paralogous genes. Sequence comparisons indicate that all *Hox* complexes have evolved from the same ancestral *Hox* complex, which in turn has originated from a primordial *Hox* gene by duplication and diversification. The greatest sequence similarities are found between orthologous and pseudo-orthologous genes that are in corresponding positions within different *Hox* complexes.

A typical *Hox* gene is expressed during embryogenesis in an area with a sharp anterior boundary. The location of this boundary in the embryo and the chromosomal position of the gene are colinear: As one progresses from 3′ to 5′ through the *Hox* complex, the anterior boundary of the expression domain shifts posteriorly. Within the hindbrain of the chicken, the expression boundaries coincide with the limits of morphological units called rhombomeres. The rhombomeres are separated by clonal restriction lines somewhat similar to epidermal compartments in insects.

At least some mouse *Hox* genes act as homeotic selector genes, similar to the homeotic genes of *Drosophila.* Elimination of the *Hoxc8*$^+$ gene in mice causes transformations of lumbar and thoracic vertebrae as well as ribs toward more anterior fates. Conversely, overexpression of the *Hoxa7*$^+$ gene transforms anterior neck vertebrae so that they resemble more posterior vertebrae.

While *Hox* genes are involved mainly in specifying the anteroposterior body pattern, other highly conserved genes are determining the dorsoventral polarity. Three *Drosophila* genes, *decapentaplegic*$^+$, *short gastrulation*$^+$, and *tolloid*$^+$, have orthologs in *Xenopus* and other vertebrates that function in the same ways, except that ventralizing genes are dorsalizing in vertebrates and vice versa.

The development of vertebrate limbs, which has been investigated in many classical transplantation experiments, is also yielding to genetic and molecular analysis. For example, the zone of polarizing activity, which produces striking pattern duplications upon transplantation, is a source of Sonic hedgehog, a signal protein also involved in other patterning processes. A gradient of Sonic hedgehog protein seems to promote the synthesis of a nested set of *Hox* proteins, which in turn affect the anteroposterior pattern of digits formed in the limb. The establishment of the dorsoventral limb axis involves vertebrate orthologs of the *Drosophila* patterning genes *fringe*$^+$, *wingless*$^+$ *and engrailed*$^+$.

The development of vertebrate limbs also illustrates the principle of reciprocal interactions. The mesenchymal core of a limb bud induces the formation of an apical ectodermal ridge (AER) in the overlying ectoderm. Once established, the AER becomes necessary for the continued proliferation and proximodistal pattern formation in the mesenchymal core, which in turn produces maintenance factors for the AER. Similar reciprocal interactions have been found in many other organs that develop from mesenchymal and epithelial components.

Genetic and Molecular Analysis of Pattern Formation in Plants

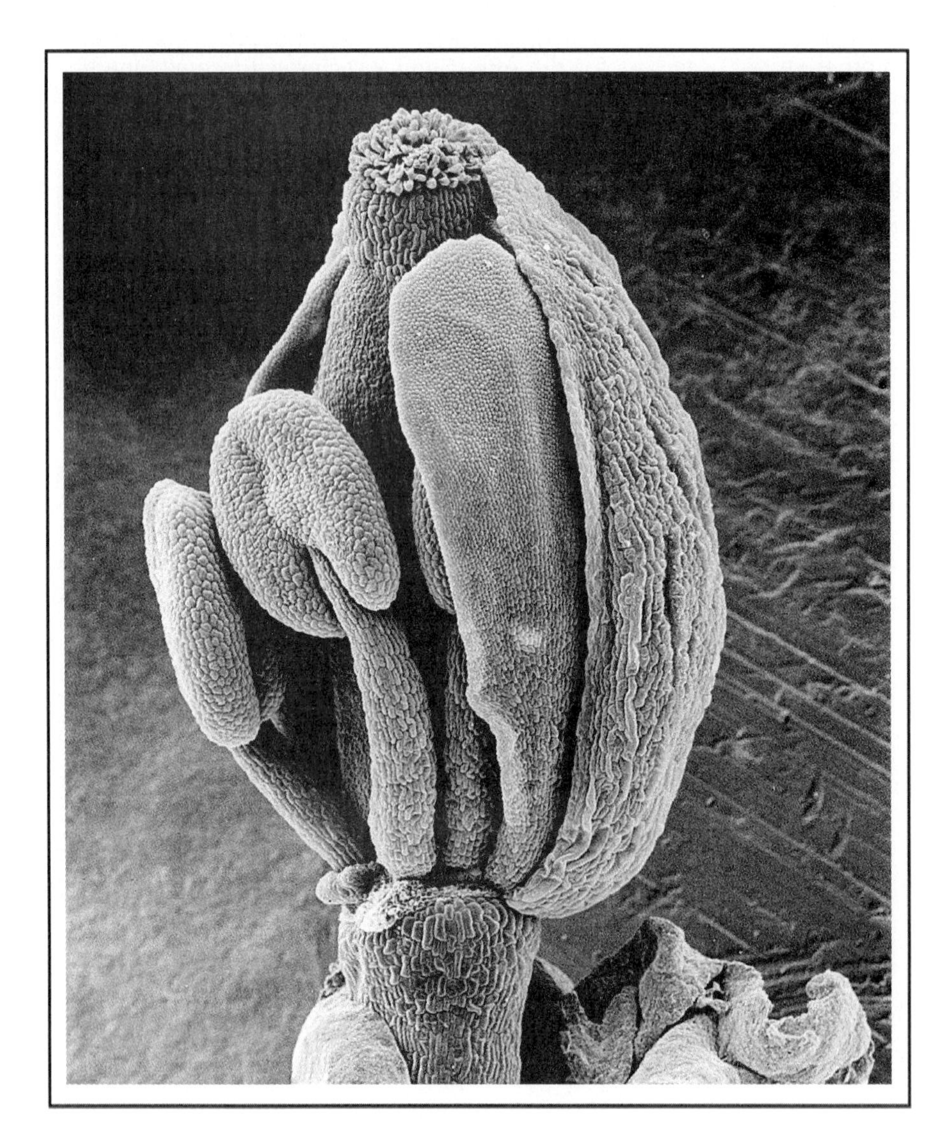

Figure 24.1 Nearly mature flower of the mouse ear cress *Arabidopsis thaliana.* For this scanning electron micrograph, parts of the flower have been removed to show the internal organs. The central flower organ, the pistil, is surrounded by six stamens, in which the pollen grains will be produced. The stamens are enveloped by four petals (one visible in front), which become large and white in the adult flower. The leaflike sepals (one visible to the right) on the outside protect the budding flower but are later outgrown by the other flower organs. *Arabidopsis thaliana* is used as a model plant because of its favorable traits for genetic studies.

TRADITIONALLY, plants have received less attention than animals at meetings of developmental biology societies and in textbooks of developmental biology. This bias may have been promoted by historical circumstances as well as by practical considerations, such as the shorter generation time of most animals, and the less confounding role of environmental factors in animal development. However, since the tools of genetic and molecular analysis apply equally well to plants and animals, many researchers have decided to study plant development. This trend was accelerated by the emergence of the mouse ear cress, *Arabidopsis thaliana,* as a genetically favorable model system (Fig. 24.1). It belongs to the flowering plants, or ***angiosperms***, and among these it represents the ***dicotyledons*** (or dicots, for short)—that is, the plants with two cotyledons, or seed leaves. *Arabidopsis* seems to be a fairly representative dicot, so the knowledge it yields is likely to apply to other dicots as well.

In this chapter, we will first review some basic aspects of angiosperm development. Next we will introduce the use of *saturation mutagenesis screens* in studying pattern formation in plant embryos. This will be followed by an overview of the genetic analysis of *floral induction*—that is, the transition from forming leaves to forming flowers.

A major topic will be the determination of flower organs, for which a genetic model has been proposed based on the phenotypes of homeotic mutants. Predictions derived from this model have been tested with genetic and molecular techniques, and examples of these studies will be presented. At the end of this chapter we will consider the relationship between homeotic genes and homeobox-containing genes.

24.1 Reproduction and Growth of Flowering Plants

To better appreciate the genetic and molecular analysis of pattern formation in plants, it will be useful to review some basic morphological aspects of angiosperm reproduction and growth. More information on these subjects may be found in the works of *Steeves and Sussex (1989), Goldberg et al. (1994), and Southworth (1996).*

SPORE-FORMING GENERATIONS ALTERNATE WITH GAMETE-FORMING GENERATIONS

In the life cycles of plants, a spore-forming generation alternates with a gamete-forming generation (Fig. 24.2). The spore-forming generation is diploid and is called the ***sporophyte*** (Gk. *phyton,* "plant"). In the grown sporophyte, haploid ***spores*** are produced by meiosis. The spores develop into haploid ***gametophytes,*** which in turn form *gametes.* Male and female gametes unite in

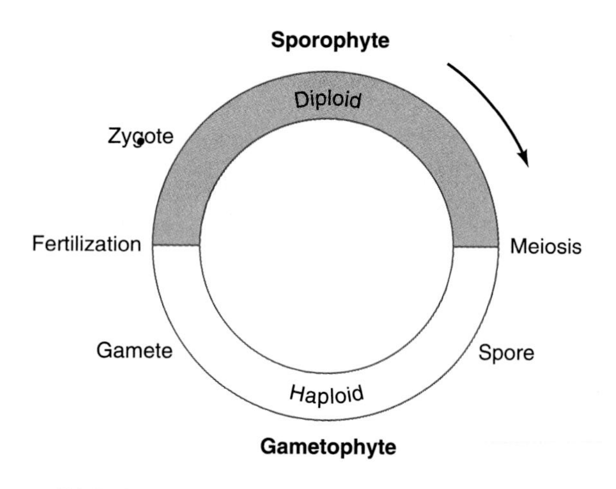

Figure 24.2 Generalized life cycle of a plant. In angiosperms, the gametophytes are reduced to microscopic organisms inside the flower.

fertilization and form a diploid zygote, which develops into the next sporophyte generation. In mosses and ferns, these alternating generations are quite distinct. In angiosperms, the sporophyte is all we see, but the male and female gametophytes are still present as tiny organisms within the flower.

The parts of the flower in which the spores and gametophytes develop are shown in Figure 24.3. Flowers produce two types of spores: microspores and megaspores. The microspores arise in flower organs called ***stamens,*** where they develop into tiny male gametophytes known as ***pollen grains.*** The megaspores develop inside another set of flower organs called ***carpels,*** which in many species fuse to form the central part of the flower, the ***pistil.*** Each carpel has one or more spore-forming tissues known as ***ovules.*** Each ovule contains a megaspore mother cell, which undergoes meiosis to produce one haploid megaspore. The megaspore divides mitotically and develops into a female gametophyte called the ***embryo sac.*** The embryo sac is a microscopic organism containing one egg cell and six other cells including a *central cell* with two haploid nuclei.

In the process of ***pollination,*** pollen grains are transferred by wind or by animals to the tip of the pistil. Here each pollen grain forms two sperm cells and a tube cell. The tube cell extends a long tip into an embryo sac, where one of the sperm cells fertilizes the egg to form the zygote. The other sperm cell unites with the central cell to form a triploid ***endosperm cell.***

PLANT EMBRYOS DEVELOP INSIDE THE FLOWER AND FRUIT

After fertilization, the embryo sac continues to develop within the ovule. The zygote divides and forms the *embryo.* The endosperm cell also divides and forms a temporary tissue, the ***endosperm,*** which is subsequently used to nourish the embryo.

The initial divisions of the plant zygote are analogous to the cleavage of animal eggs in that the daughter cells do not restore their original volume before dividing again. The first division of the *Arabidopsis* zygote is asymmetrical, generating a small apical cell and a large basal cell (Fig. 24.4). The apical cell gives rise to the embryo proper except for part of the root, which derives from the basal cell. The basal cell also forms the extraembryonic ***suspensor,*** which serves as a conduit for

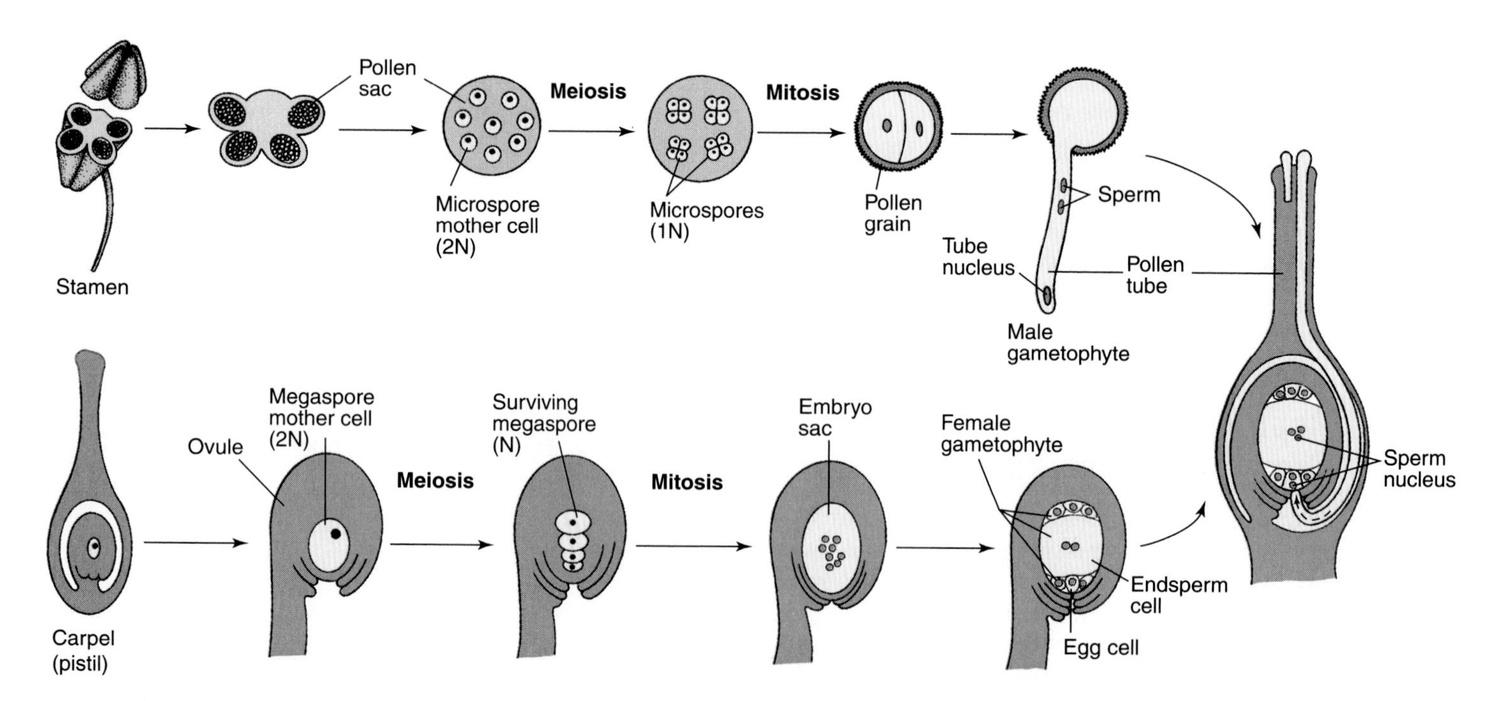

Figure 24.3 Spore and gametophyte formation in an angiosperm.

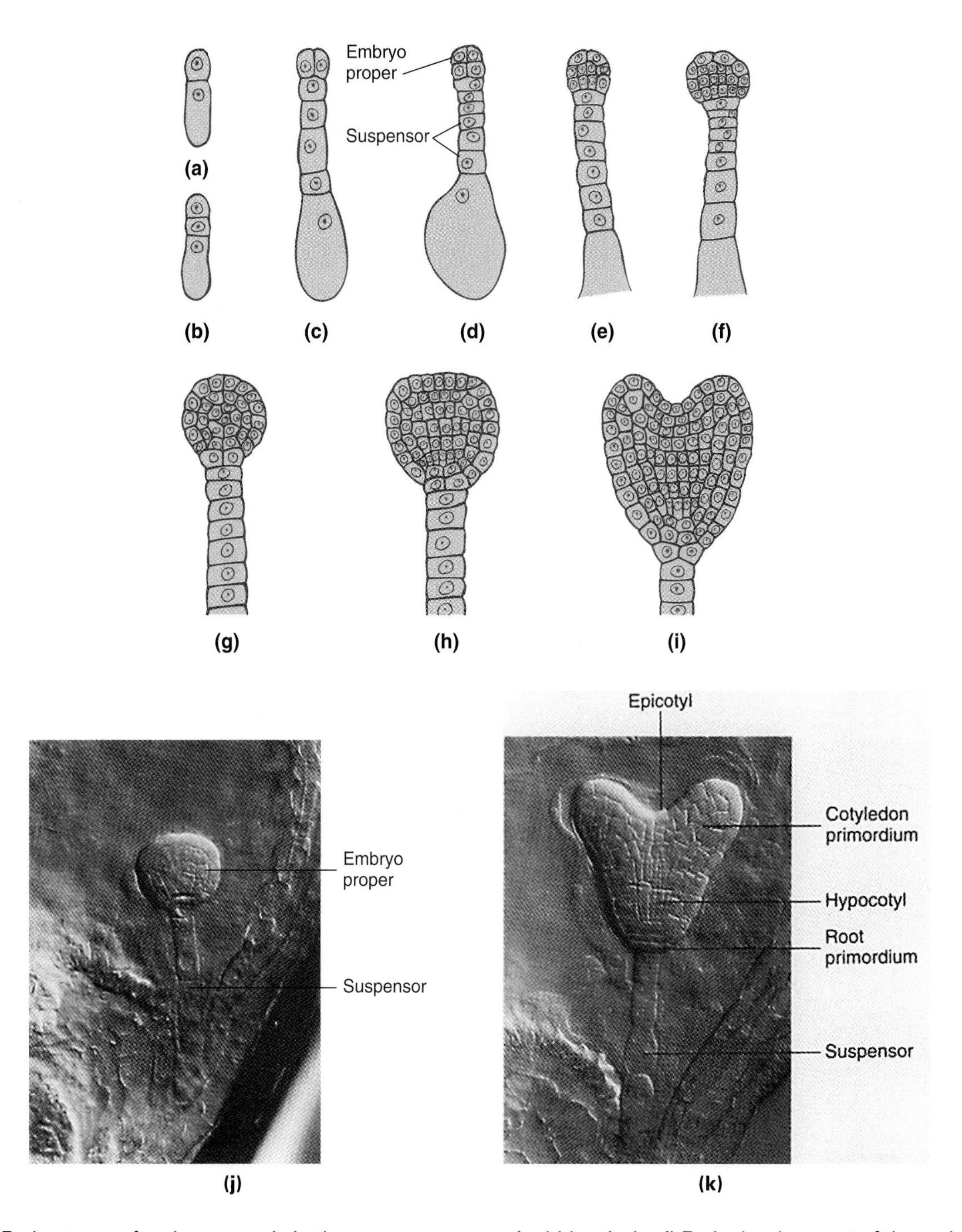

Figure 24.4 Early stages of embryogenesis in the mouse ear cress *Arabidopsis.* **(a–d)** Early development of the embryo and the suspensor. **(e)** At the 16-cell stage, the embryo proper consists of 8 outer cells and 8 inner cells. **(f–h)** Globular stages. **(i)** Heart stage. **(j, k)** Photographs of an early globular stage and the heart stage.

nutrients and anchors the growing embryo in the surrounding endosperm and ovule tissues. At the 16-cell stage, the embryo consists of 8 outer cells, which give rise to epidermis, and 8 inner cells, which give rise to deeper tissues. This stage also marks the beginning of the *globular stage,* during which the embryo proper is spherical. With the transition to the *heart stage,* the embryo establishes a bilateral symmetry by initiating cotyledon formation. When the embryo consists of about 100 cells, the primordia of root, shoot, vascular tissue, and ground tissue are discernible. Thus, many features of the basic plant pattern are already established at the heart stage.

As the embryo develops further, the ovule in which it resides is transformed into a ***seed*** (Fig. 24.5). The cotyledons, and the ***hypocotyl*** beneath the cotyledons, grow substantially using the endosperm and the suspensor as sources of nutrients. At the same time, embryonic cells continue to differentiate into epidermal cells, vascular cells, and storage cells. The storage cells of the cotyledons

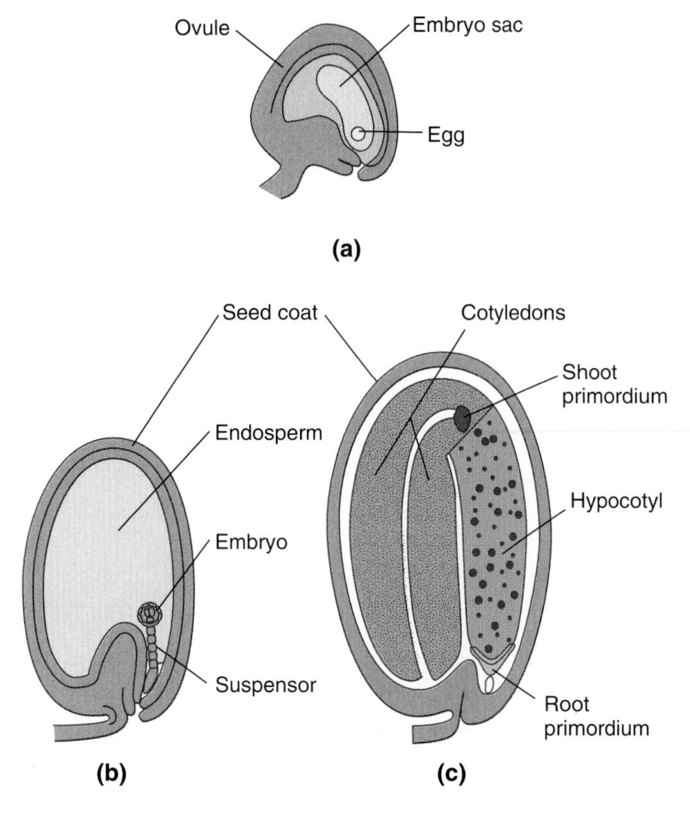

Figure 24.5 Seed development in *Arabidopsis.* **(a)** Ovule with embryo sac containing egg cell. **(b)** Immature seed with embryo at the globular stage, suspensor, and endosperm tissue. **(c)** Mature seed (0.5 mm long) with embryo. Visible are the two cotyledons, shoot primordium, hypocotyl, and root primordium. The endosperm and suspensor have been used up and are degenerating. The seed is surrounded by a tough coat.

and the hypocotyl accumulate large amounts of proteins and oils that will serve as nutrients later on. The shoot primordium between the two cotyledons and the root primordium at the basal tip of the hypocotyl remain quiescent until *germination.* The ovule is covered by integuments, which are formed by the sporophyte and develop into the tough coat of the seed. The surrounding pistil gives rise to the ***fruit.*** Typically, the seed now enters a period of dormancy until conditions are favorable for germination.

Plants that have only one cotyledon, such as grasses, cereals, and maize, develop similarly except that the seed retains much of its endosperm and the embryo is comparatively smaller. In all plants, the seed is a quiescent embryonic stage adapted for dispersal and survival outside the mother plant.

The embryo resumes its growth at ***germination,*** when the embryo sprouts from the seed. Germination depends on both internal and environmental factors. The food reserves for the early phase of germination are stored in the cotyledons or in the endosperm. The root usually emerges first to anchor the seedling and to take up water and dissolved minerals. The hypocotyl stretches before the shoot primordium begins to grow and form the shoot. The young seedling shows the same pattern that was already visible at the heart stage of the embryo (Fig. 24.6). From top to bottom we can distinguish four main elements: the shoot primordium, now called ***epicotyl;*** the *cotyledons;* the *hypocotyl;* and the *root primordium.* These elements constitute the ***apical-basal body pattern*** of the plant seedling. The seedling also has a ***radial pattern,*** which is most clearly seen in the hypocotyl and involves three major tissues: the outer epidermis, the centrally located vascular strands, and an intervening mass of ground tissue.

MERISTEMS AT THE TIPS OF ROOT AND SHOOT ALLOW PLANTS TO GROW CONTINUOUSLY

While most cells of a seedling proceed to differentiate for functions in photosynthesis, transport, and storage, certain nests of cells remain forever embryonic and

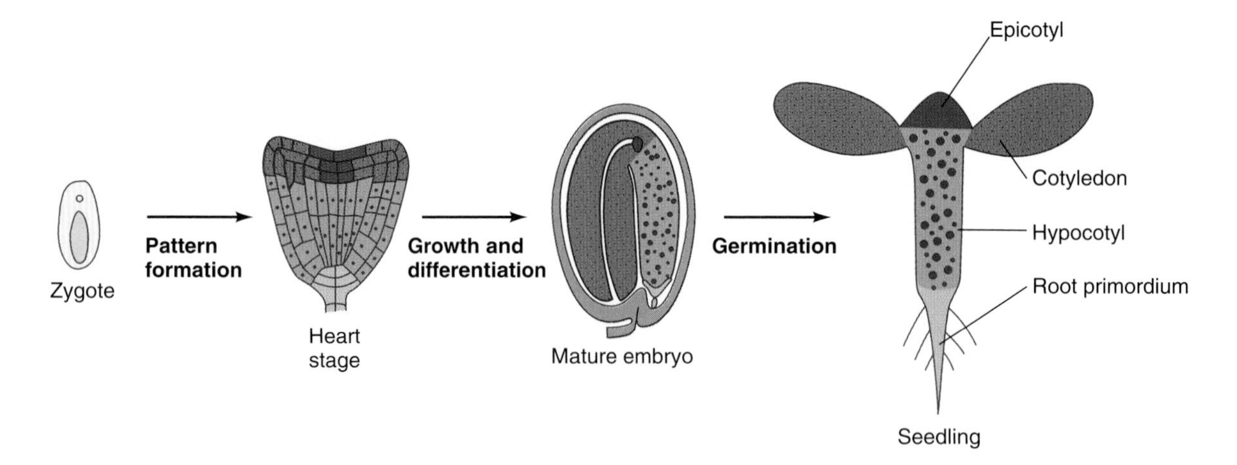

Figure 24.6 Pattern formation in the plant embryo. Four major elements (epicotyl, two cotyledons, hypocotyl, and root primordium) are laid down at the heart stage.

active in cell division. These nests of cells, called ***meristems,*** are located at the tips of the root and shoot primordia. Plant meristem cells are like animal *stem cells:* They are undifferentiated and can divide throughout the life of the organism. As they divide, they generate more stem cells while also giving rise to various progenitor cells of other tissues. Postembryonic plant development, which generates entire organ systems including the entire flowering part of a plant, depends largely on meristems. After germination, the root meristem forms the root of the mature plant while the shoot meristem gives rise to the stem, branches, leaves, and flowers. During each growing season, meristems add new sections of root, stem, and branches. Thus, unlike many animals, plants grow throughout their lives.

As the stem cells in the apical shoot meristem divide, they continuously displace daughter cells to the surrounding peripheral region, where they form the primordia of leaves and flowers. A balance between the creation of new cells and the departure of cells is required to maintain a functional meristem. Some of the mutations that disturb this balance, including *clavata,* cause the formation of large, club-shaped meristems and lead to the formation of extra floral organs. The *clavata1*$^{+}$ and *clavata3*$^{+}$ genes have been cloned and show the characteristics of a signal receptor kinase and its putative ligand, respectively (Fletcher et al., 1999). The clavata3 protein is synthesized in the superficial cell layer of the meristem, while the clavata1 protein is made in deeper parts of the meristem. Clavata1—the receptor—in turn associates with other proteins that either transmit its receptor activity or inhibit it (Trotochaud et al., 1999). Thus, the signals that control the balance between the addition of apical shoot meristem cells and their departure have begun to yield to genetic and molecular analysis.

At regular intervals, the meristem produces lateral swellings that develop into leaves, branches, or flowers. The points where leaf primordia arise from the stem are called ***nodes,*** and the length of stem between two nodes is called an ***internode.*** Internodes are short near the meristem and elongate later on. Because of its short internodes, the meristem is typically covered with leaf primordia (Fig. 24.7). The resulting compound structure, consisting of the meristem, leaf primordia, and short internodes, is called a ***bud.*** Roots grow similarly except that the root meristem is covered by a protective cap of nondividing cells that produce a lubricating substance.

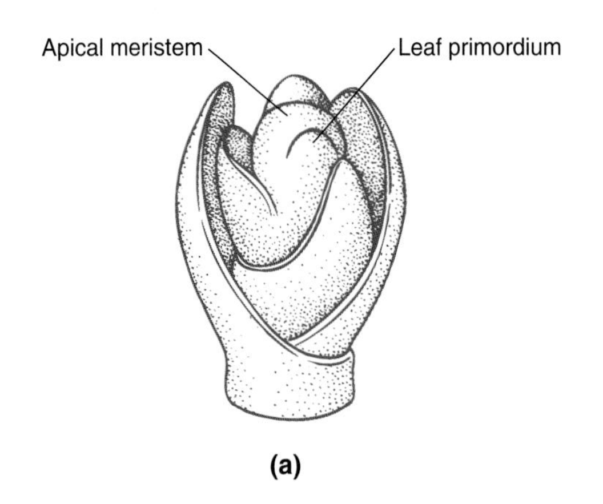

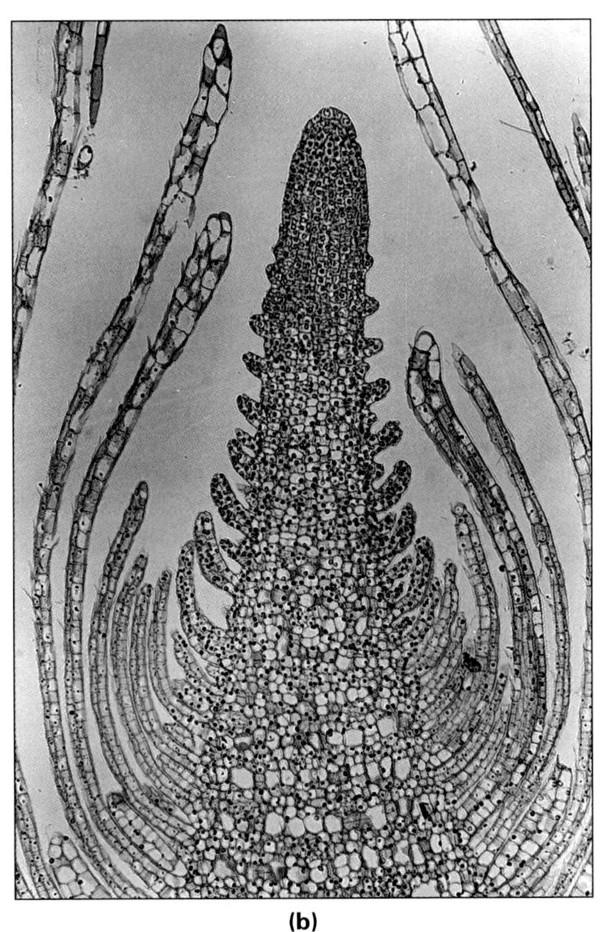

Figure 24.7 Growing bud of a marsh plant, *Elodea* sp. **(a)** Schematic diagram. **(b)** Photograph of a histological section. The apical meristem forms nodes with leaf primordia and short internodes. The growing leaf primordia curve upward and enclose the meristem in a bud.

The pattern in which an apical shoot meristem gives off the primordia of leaves, branches, and flowers generates much of the overall appearance of a plant. After germination of an *Arabidopsis* seed, the shoot meristem gives off a set of leaf primordia in a mostly spiral arrangement with short internodes (Fig. 24.8). Thus, the basal portion of the plant develops as a rosette. Subsequently, the pattern of cell division in the apical meristem changes to produce a few small leaves separated by long internodes. After this transient phase, the apical meristem produces floral primordia, still in a spiral pattern. Each floral primordium forms a flower or an entire branch with multiple flowers. Such a flowering branch is called a *secondary inflorescence,* or ***axillary inflorescence,*** in contrast to the flowering central shoot, which is then called the ***primary inflorescence.***

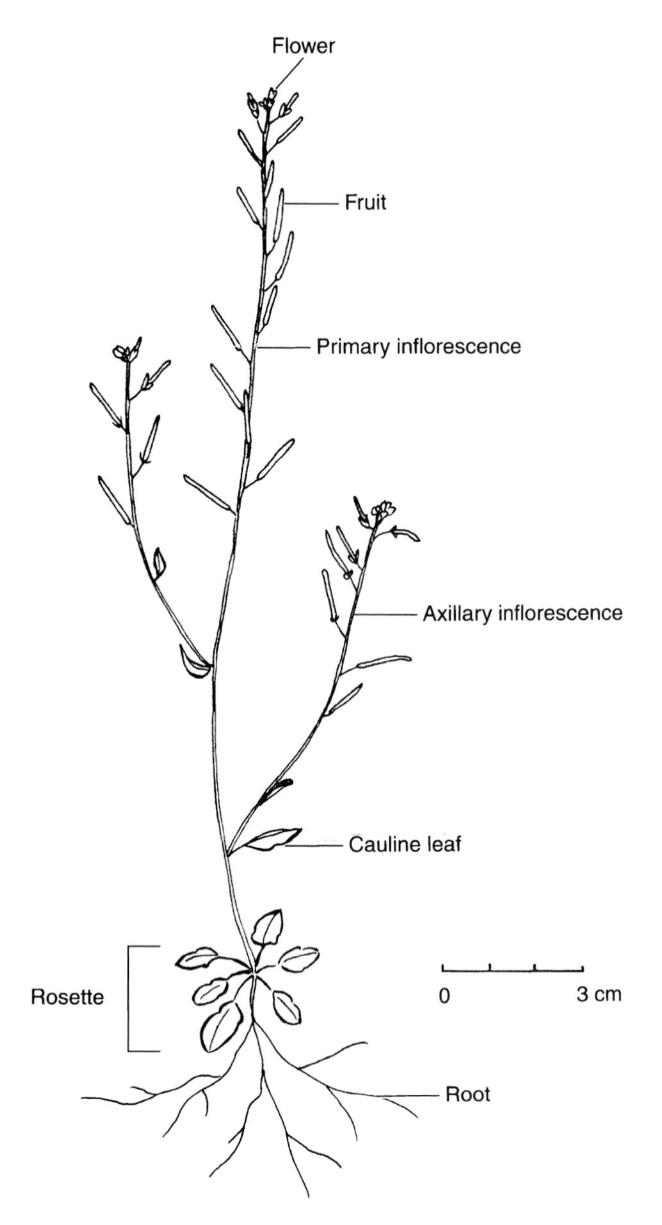

Figure 24.8 Drawing of the mouse ear cress *Arabidopsis thaliana,* showing the overall appearance of the plant. Initially, the apical meristem generates a basal rosette of leaves that are arranged in a mostly spiral pattern with short internodes. Later-formed leaves (cauline leaves) are separated by longer internodes. Upon floral induction, the apical meristem changes into the primary inflorescence meristem; it forms flowers instead of leaves, still in a spiral pattern. Each flower matures into an elongated fruit. Each of the cauline leaves contains an axillary inflorescence meristem that reiterates the development of the primary inflorescence.

In summary, plant development differs from animal development in several respects. The apical meristems allow plants to form new organs and to grow continuously throughout their lives. No germ line is set aside early during plant development: The flower organs that give rise to megaspores and microspores develop from the same meristem cells that have also produced the entire shoot. Correspondingly, many nonreproductive plant cells remain *totipotent,* retaining the capacity of forming an entirely new plant (see Section 7.5). Plants produce a smaller number of distinct cell types. Unlike animal cells, plant cells are surrounded by rigid cell walls, and except for pollen tubes, they do not move extensively.

Some of the differences between plant development and animal development are significant from a genetic point of view. The development of both male and female gametophytes from haploid spores requires gene expression and should therefore cause the elimination of many mutant alleles before fertilization can occur. The small size and maternal nourishment of the plant embryo make it less dependent than most animal embryos on maternally provided RNA. Instead, one might expect that comparatively more zygotic genes are activated early during plant embryogenesis.

24.2 Genetic Analysis of Pattern Formation in Plant Embryos

The great success of genetic and molecular studies on fly and mouse development has enticed several researchers to apply the same methods to plants. In addition to the traditional species used in plant genetics, including the snapdragon *Antirrhinum majus* and the corn plant *Zea mays,* the mouse ear cress *Arabidopsis thaliana* has emerged as a model plant with very favorable traits for genetic studies. Because of its small sporophyte size, small genome size, and other favorable traits for genetic analysis, *Arabidopsis* has become the equivalent of *Drosophila* among plant developmental biologists (see Section 15.2).

SIMILAR METHODS ARE USED FOR THE GENETIC ANALYSIS OF PLANT AND ANIMAL DEVELOPMENT

The methods of plant mutagenesis are essentially the same as those used for animals. Geneticists have employed X-rays as well as chemical mutagens. Researchers working with maize have taken advantage of transposons similar to the P elements of *Drosophila* (Chandlee, 1991; see Section 15.6). Many plants are also susceptible to genetic transformation using a naturally occurring bacterial plasmid as a vector (Azpiroz-Leehan and Feldmann, 1997). The bacterium *Agrobacterium tumefaciens* transfers a plasmid, called Ti (for tumor-inducing), to cells of its natural host plants. This step is followed by the integration of part of the plasmid, called *T-DNA* (for transferred DNA), into the plant genome. The T-DNA encodes products that induce crown gall tumors, and also make the plant produce certain amino acids for the bacterium.

The ability of T-DNA to stably integrate into its host's genome has made it a useful vector for transforming plant cells with engineered genes. For this purpose, the T-DNA has been disarmed so that it is still transferred but no longer induces tumors. Instead, it may contain genes that confer resistance to antibiotics or herbicides so that transformed cells can be selected in culture. Most important, any cloned genes included into the T-DNA will be introduced into the genome of host plant cells. The same system is also suitable for insertional mutagenesis, in which the inserted DNA is used to introduce a selectable marker and a tag to pull out the mutated gene from a genomic library (see Method 15.5).

Mutants have been used in several ways to study plant development (Meinke, 1991a, b). For example, genetic marking of single cells has been employed for clonal analysis just as in insects (see Method 6.1). Studies employing this technique indicate that shoot meristem cells do not always divide in the same manner or contribute to precisely the same portion of the adult plant (McDaniel and Poethig, 1988; Jegla and Sussex, 1989). Thus, cellular interactions must be involved in pattern formation, while cell lineage seems to play only a minor role in plant development. In addition, mutagenesis of plants has revealed a wide range of genes involved in embryogenesis as well as later development.

PATTERN FORMATION IN PLANT EMBRYOS INVOLVES 25 TO 50 SPECIFIC GENE FUNCTIONS

Encouraged by the successful genetic analysis of embryonic pattern formation in *Drosophila*, scientists set out on a similar analysis of plant embryogenesis (*Jürgens, 1995; Laux and Jürgens, 1997*). At the start of their work, they asked two questions. First, how many genes are involved in determining the basic body pattern of the plant embryo? Second, can the process of pattern formation be broken down into distinct events that are controlled by particular subsets of genes? To answer these questions, the researchers carried out a saturation mutagenesis screen with *Arabidopsis* seeds.

The screening procedure used by Jürgens and coworkers (1991) is outlined in Figure 24.9. Wild-type seeds are mutagenized with ethylmethylsulfonate (EMS) and grown into adult plants. The embryo contained in each seed has only two precursor cells that later give rise to spores. If one of these cells is mutated, it produces a clone of heterozygous meristem cells that, on average, populate the reproductive tissue in every other flower. Since each flower normally fertilizes itself, the seeds produced by one flower represent one mutagenized line. Seeds from each line are germinated and screened for abnormal development. Lines producing seedlings with distinct pattern abnormalities are

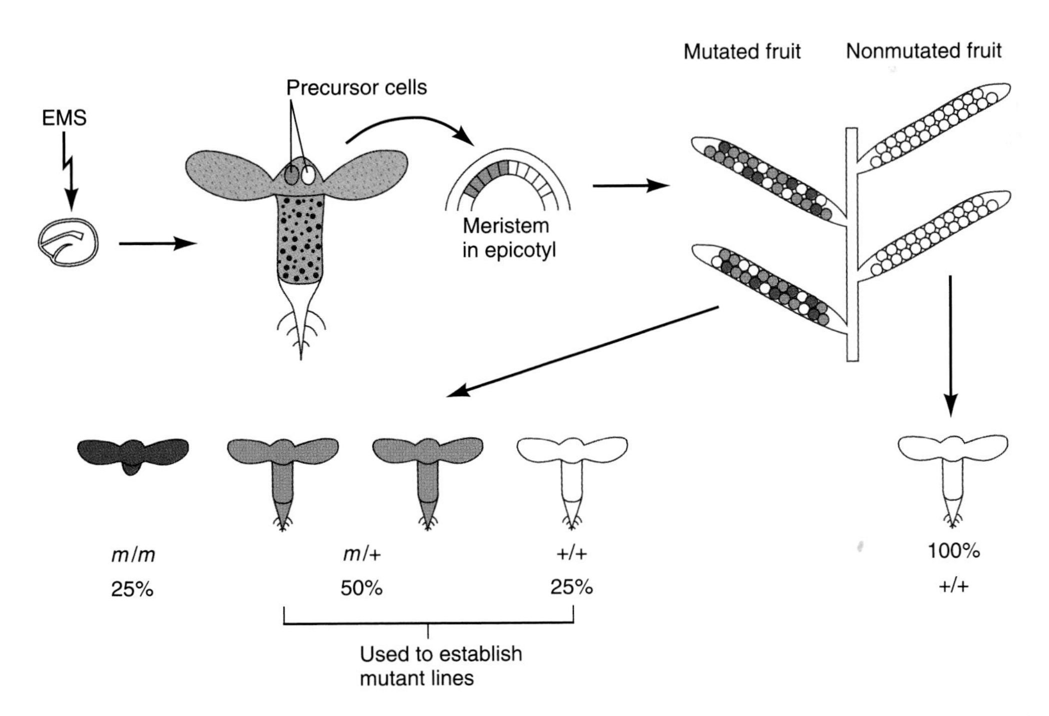

Figure 24.9 Mutagenesis screen to identify patterning genes in *Arabidopsis.* Seeds are exposed to the chemical mutagen ethylmethylsulfonate (EMS) and raised to adult plants. The embryo contained in each seed has only two precursor cells that later give rise to spores. One of these cells, on average, is mutagenized (light color) and produces a clone of heterozygous meristem cells that statistically populate the reproductive tissue in every flower. Since each flower normally fertilizes itself, about one-quarter of the seeds (bold color) from a mutated fruit will be homozygous for the mutant allele and give rise to an abnormal seedling (labeled *m/m*). Their normal-looking siblings (labeled *m/+* and *+/+*) are then used to establish a mutant line.

table 24.1 Mutations Affecting Body Organization in the *Arabidopsis* Embryo

Mutant Phenotype	Gene	Number of Alleles
Apical-basal–pattern defect		
Apical	*gurke*	9
Central	*fackel*	5
Basal	*monopteros*	11
	bodenlos	
Terminal	*gnom*	15
Radial-pattern defect	*keule*	9
	knolle	2
Shape-change defect	*fass*	12
	knopf	6
	mickey	8

From Mayer et al. (1991). Used with permission, © 1991 Macmillan Magazines Limited.

inspected and categorized, and their normal-looking siblings are used to breed the mutant lines.

Among 44,000 mutagenized lines screened, the researchers found 25,000 embryonic lethal mutants, which they did not analyze in detail. About 5000 lines produced seedlings with mutant phenotypes having abnormal pigmentation or abnormal morphology or both. The majority of the morphologically abnormal seedlings had reduced or misshapen cotyledons, and their roots tended to be short or less differentiated than normal. These phenotypes did not seem to provide any clues to patterning events that might be affected specifically, although this possibility could not be completely ruled out.

About 250 lines showed specific morphological abnormalities. These phenotypes were most likely the result of mutations in genes critical for distinct events during embryonic pattern formation. The investigators therefore classified these lines as putative patterning mutants. Lines with similar phenotypes were crossed in *complementation tests* to determine whether they were *allelic*—that is, representing different mutations of the same gene. For each of the first genes to be identified, nine mutant alleles were obtained on average (Table 24.1). The total number of patterning genes identified in this screen is expected to be about 25 to 50 when the analysis is complete.

GROUPS OF GENES CONTROL DISTINCT PATTERNING EVENTS

In the first sample of mutant phenotypes they studied, Jürgens and coworkers (1991) found mutations that affect three aspects of the body organization: the *apical-basal pattern*, the *radial pattern*, and the *shape* (Table 24.1; Mayer et al., 1991). The mutations affecting the apical-basal body pattern generate large deletions, which can be classified as *apical, basal, central,* and *terminal* (Fig. 24.10). Four mutant phenotypes illustrating such deletions are shown in Figure 24.11, along with several phenotypes having radial-pattern and shape-change defects.

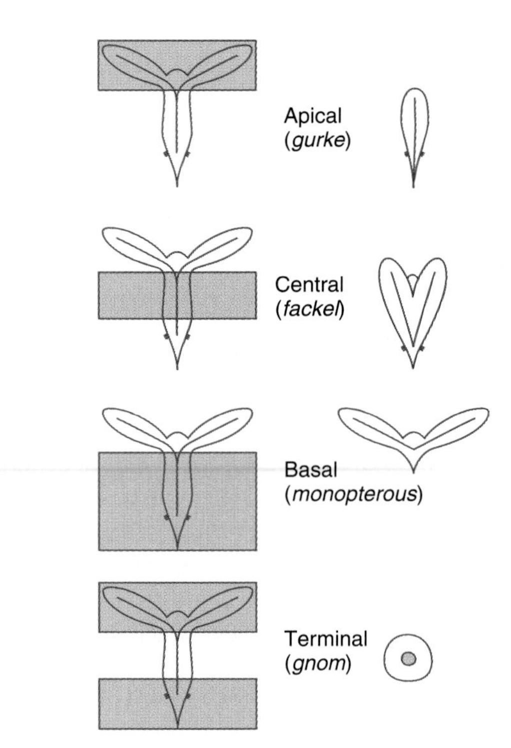

Figure 24.10 Apical-basal patterning mutants of *Arabidopsis.* Four types of pattern deletion, and representative mutant alleles causing them, are indicated. The wild-type seedling patterns with the deleted parts shaded are shown on the left, and the resulting mutant phenotypes are shown on the right. Note the complementarity between apical and basal deletions, as well as between central and terminal.

Two pairs of deletions—apical and basal, as well as central and terminal—add up to the entire apical-basal body pattern, suggesting that the corresponding gene activities may define boundaries along the axis. Each of these deletions can readily be detected in heart-stage embryos, indicating that the apical-basal pattern is formed before this stage.

In the *apical deletion* class, mutant alleles of the *gurke*$^+$ gene affect the shoot meristem and the cotyledons (Torres-Ruiz et al., 1996). In strong loss-of-function alleles, the epicotyl and cotyledons of the seedling are replaced with a disorganized mass of cells (Fig. 24.11b). The internal phenotype matches the external appearance: The vascular strands terminate apically without branching. Defects are first observed in the apical domain of the heart-stage embryo, indicating an early requirement for the wild-type gene product.

In the class of *basal deletions,* the phenotype of *monopteros* mutant seedlings is lacking both the root and

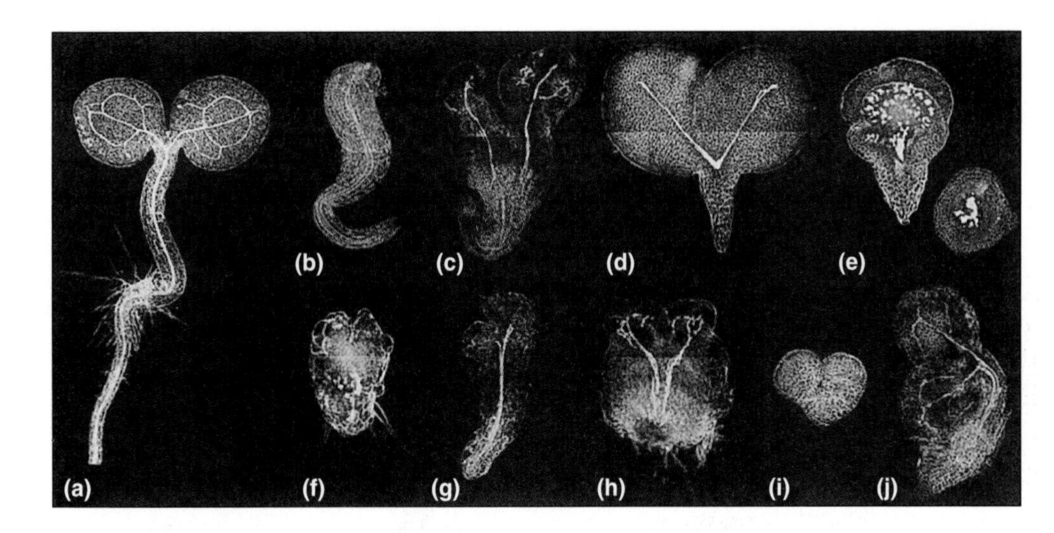

Figure 24.11 Phenotypes of mutant *Arabidopsis* seedlings: **(a)** wild-type; **(b)** apical deletion (*gurke*); **(c)** central deletion (*fackel*); **(d)** basal deletion (*monopteros*); **(e)** terminal deletion (*gnom*); **(f, g)** radial-pattern defects caused by mutations in *knolle* and *keule;* **(h, i, j)** shape-change mutants *fass, knopf,* and *mickey.*

the hypocotyl (Fig. 24.11d). This phenotype has been traced back to the early embryonic stage, when cells fail to establish the division patterns that would normally generate the basal body structures (Berleth and Jürgens, 1993). Seedlings of another mutant, *bodenlos,* are lacking the root meristem and, in strong alleles, also the hypocotyl (Hamann et al., 1999). *Bodenlos* embryos deviate from normal development as early as the second cleavage, when both embryonic cells divide abnormally. Apparently, *bodenlos*$^+$ as well as *monopteros*$^+$ are involved in the polar transport of auxin, a key growth hormone in plants, towards the root meristem (Hardtke and Berleth, 1998; Grebe et al., 2000).

The *central deletion* class is represented by *fackel* mutants; they lack the hypocotyl. The cotyledons appear directly attached to the root, and the vascular strands separate at the upper end of the root and diverge into the cotyledons (Fig. 24.11c). The defect becomes obvious during the midglobular stage (see Fig. 24.4g), when loss-of-function alleles do not form the elongated vascular precursor cells of the hypocotyl.

The *terminal class* of pattern deletions affects structures located at the two opposite ends of the seedling. Mutations in the *gnom* gene cause deletion of both the root and the cotyledons, leaving cone- or ball-shaped seedlings with no apparent apical-basal organization (Fig. 24.11e). Although vascular cells are present and well differentiated, they are not connected to form strands. The *gnom* phenotype has been traced back to the first division of the zygote, which generates two cells of nearly the same size instead of a small apical cell and a large basal cell (Mayer et al., 1993). The molecular characteristics of the gnom protein indicate that it may be involved in vesicle transport between the *endoplasmic reticulum* and the *Golgi apparatus* (Peyroche et al., 1996). If this were the case, the *gnom*$^+$ gene may be required to stabilize the apical-basal axis of the embryo through an exocytotic process similar to the stabilization of the rhizoid-thallus axis in the *Fucus* zygote (see Section 9.2).

Radial pattern defects are caused by mutations in two genes, *knolle* and *keule* (Fig. 24.11f, g). In *knolle* mutants, one cannot morphologically distinguish an outer epidermal cell layer from the internal layers (Lukowitz et al., 1996). Mutations in *keule* lead to an irregular appearance of epidermal cells. Both mutant phenotypes can already be recognized at the globular stage, when in the wild-type embryo an outer group of epidermal cells would surround a distinct group of inner cells.

Shape mutants, in contrast to apical-basal and radial mutants, do not affect specific pattern elements but rather cause grossly abnormal overall shapes. Seedlings with mutant alleles of the *fass* gene contain the normal pattern elements but are compressed in the apical-basal axis (Fig. 24.11h). Curiously, mutant alleles of *fass* show abnormal embryonic cleavage patterns but nevertheless give rise to a normal albeit compressed body pattern (Torres-Ruiz and Jürgens, 1994). This observation suggests that cell lineage is not critical in establishing the embryonic body pattern of plants. Instead, the body pattern may be established by interactions between the embryonic ells. Mutant *knopf* seedlings look small, round, and pale (Fig. 24.11i). Mutations in the *mickey* gene cause thickened, disc-shaped cotyledons, which appear disproportionately large relative to the short hypocotyl and root of the seedling (Fig. 24.11j).

Compared to the work done with *Drosophila,* the genetic analysis of pattern formation in plant embryos has just begun. However, the mutant lines of *Arabidopsis* that have been collected are raw material for further research, which will unfold rapidly. Researchers will next generate double mutants, because their phenotypes will provide clues to the genetic hierarchy that controls pattern formation. When some of the catalogued genes have been cloned, their expression patterns can be studied by *in situ hybridization* and *immunostaining* (see Methods 8.1 and 4.1, respectively). At the same time, sequencing data will help to characterize the molecular nature of the proteins encoded by these genes. Such results will allow investigators to piece together a molecular model of pattern formation in plant embryos. Indeed, work is already proceeding along these lines in the area of flower development, which will be described next.

24.3 Genetic Control of Flower Development

The seasonal cycle of angiosperms culminates with the development of flowers, which contain the reproductive organs of the plant. Flowers consist of four rings of organs called ***whorls.*** From outside to inside, the organs of the four whorls are known as sepals, petals, stamens, and carpels (Fig. 24.12). The ***sepals,*** forming the outermost whorl, are most similar to leaves. The ***petals,*** which form the next whorl, are often brightly colored and attract insects for pollination. Inside the petals is a whorl of *stamens,* in which pollen grains develop as described earlier. In the central whorl, formed by the carpels, megaspores develop inside ovules and form embryo sacs. The carpels fuse into the central organ of the flower, the *pistil,* which later gives rise to the fruit.

For centuries, naturalists and morphologists have recognized intuitively that all organs of the flower are modified leaves. As we will see, it takes only the activity of a few genes to transform leaves into flower organs.

FLORAL INDUCTION GENES CONTROL THE FORMATION OF AN INFLORESCENCE

After a period of vegetative growth, plants become capable of forming flowers. The process of flower development can be subdivided into three steps, each of which is affected by certain genes (Fig. 24.13; see also Weigel and Meyerowitz, 1993; H. Ma, 1994; Blazquez, 1997; Irish, 1999). The first step is ***floral induction.*** It is triggered by a combination of environmental and internal signals, including day length, temperature, the age of the plant, and its nutritional state, which are sensed by different parts of the plant. Floral induction results in the reorganization of the apical shoot meristem from a ***vegetative meristem*** producing leaves into an ***inflorescence meristem*** giving rise to flower primordia. It takes a few days for this florally induced state to be achieved, but thereafter it remains stable even if environmental conditions revert to those that promote vegetative growth (Steeves and Sussex, 1989). Floral induction is affected by several ***flowering genes;*** mutations in these genes accelerate or delay flowering (Koornneef et al.,

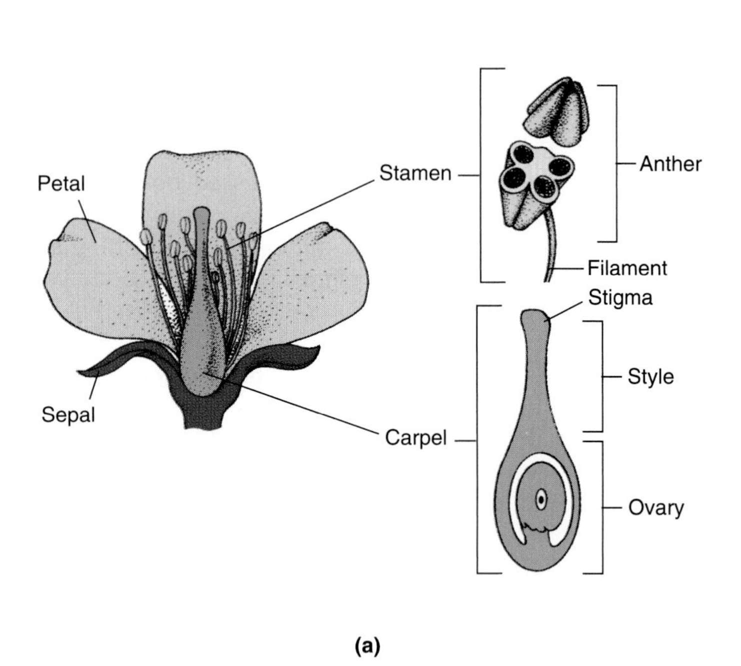

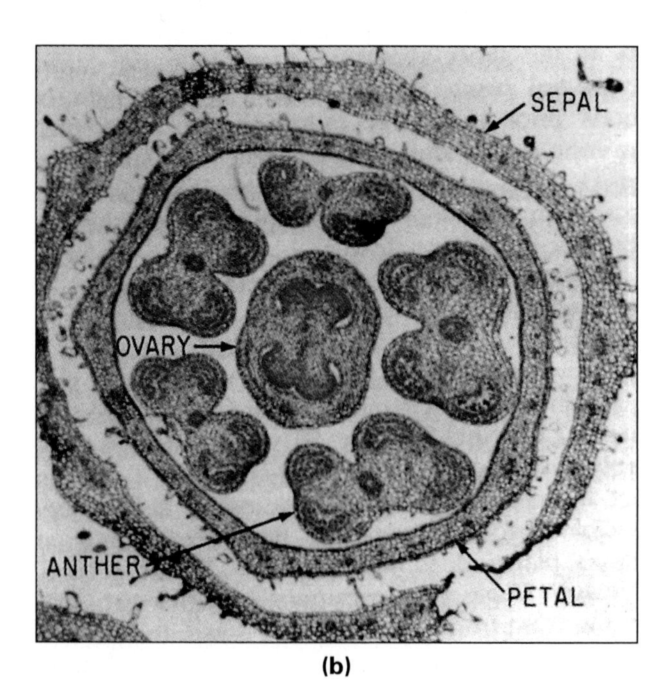

Figure 24.12 Flower structure. **(a)** Schematic lateral view of a generalized flower. **(b)** Photomicrograph of a histological section across a tobacco flower. A complete flower contains four whorls of organs: sepals, petals, stamens, and carpels. The leaflike sepals and the colored petals form the outer two whorls. These organs are often fused at their bases. Inside is a whorl of stamens, each consisting of a thin filament and a thicker organ called the anther, in which the pollen grains develop. The central organ of the flower, the pistil, is formed from a whorl of fused carpels. The pistil consists of a basal ovary containing the ovules; a top part called the stigma, to which the pollen grains attach; and a connecting tube, the style.

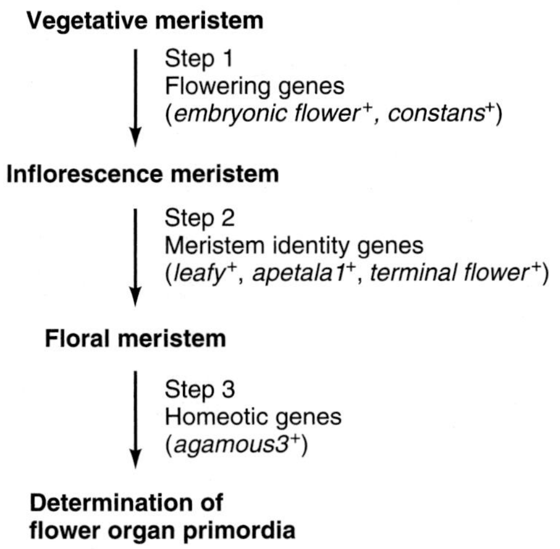

Vegetative meristem

Step 1
Flowering genes
(*embryonic flower*$^+$, *constans*$^+$)

Inflorescence meristem

Step 2
Meristem identity genes
(*leafy*$^+$, *apetala1*$^+$, *terminal flower*$^+$)

Floral meristem

Step 3
Homeotic genes
(*agamous3*$^+$)

Determination of flower organ primordia

Figure 24.13 Flower development can be broken down into three major steps, each depending on the functions of certain genes.

1998). For example, the *embryonic flower*$^+$ gene is required for the vegetative state of the apical shoot meristem; mutants with loss-of-function alleles of this gene form a single flower right after germination (Sung et al., 1992).

Other floral induction genes mediate the effects of environmental conditions, such as day length, on floral induction. Many plants, including *Arabidopsis,* flower readily in long days (LDs, 16 h of light), whereas floral induction is much delayed in short days (SD, 10 h of light). Floral induction under LD conditions is associated with the activation of *constans*$^+$, a gene that encodes a protein with the molecular characteristics of a transcription factor. Loss of *constans*$^+$ function delays flowering in LDs but not in SDs, indicating that *constans*$^+$ activity is necessary for rapid floral induction by LDs.

IN order to test whether expression of the *constans*$^+$ gene is sufficient for floral induction, R. Simon and coworkers (1996) constructed a transgene designed to produce constans mRNA independently of day length, and to permit the *activation* of translated constans protein at will. To this end, they spliced a constitutively active promoter in front of a cDNA encoding a fusion protein that combined the DNA-binding domain of constans with the ligand-binding domain of the rat glucocorticoid receptor. This transgene by itself did not change the behavior of *constans*$^-$ mutant plants. They kept flowering late even in LDs. This was expected because the ligand-binding domain of the glucocorticoid receptor needs to be loaded with a ligand for the fusion protein to pass into the nucleus (see Section 16.4). As soon as a suitable steroid ligand (dexamethasone) was provided to the transgenic plants through the soil, flowering was induced rapidly. This response to dexamethasone occurred at any stage of plant development. When dexamethasone was applied prematurely (less than 20 days after germination), the transgenic plants produced fewer leaves, shorter stems, and fewer flowers than untreated controls. This effect was observed in LDs as well as SDs, indicating that expression of *constans*$^+$ is sufficient to overcome the delay in flowering observed in wild-type plants kept in SDs.

Questions

1. In order to make their results conclusive, the investigators had to test whether dexamethasone was exerting its effects on plant flowering through the contans-glucocorticoid receptor fusion protein, as designed, or whether dexamethasone was acting independently in some unforeseen way. Which control experiment would settle this question?
2. The results of this experiment do not show whether expression of the *constans*$^+$ gene is controlled at the level of transcription. How could the authors have tested (and they did) whether this was the case?

The observations of R. Simon and coworkers (1996) and others indicate that a long photoperiod, through a light-responsive molecular mechanism, activates the *constans*$^+$ gene, which in turn controls target genes in ways that promote flowering. However, other pathways of floral induction are triggered by factors other than photoperiod and are independent of *constans*$^+$, again illustrating the *principle of overlapping mechanisms* in development.

MERISTEM IDENTITY GENES PROMOTE THE TRANSITION FROM FLORAL MERISTEMS TO INFLORESCENCE MERISTEMS

The second step in flower development is the actual formation of flower primordia, which arise through distinct patterns of cell division in the inflorescence meristem (Fig. 24.13). The pattern in which the flower primordia are generated varies among different groups of plants. In *Arabidopsis,* the flower primordia arise in a spiral pattern on the flanks of the inflorescence meristem (Fig. 24.14a). Correspondingly, the flowers and later the fruits of *Arabidopsis* are attached to the stem in a spiral (see Fig 24.8). Because young flower primordia show some meristem properties, they are also called ***floral meristems.*** The genes that specify the identity of floral meristems versus inflorescence meristems are called ***meristem identity genes.*** A representative of this group in *Arabidopsis* is the *leafy*$^+$ gene (Weigel et al., 1992). In leafy loss-of-function mutants, flowers are transformed into secondary inflorescence shoots. Conversely, plants that overexpress *leafy*$^+$ flower prematurely but still form leaves before they do. Thus, the normal *leafy*$^+$ activity promotes the transition from an inflorescence meristem to a floral meristem.

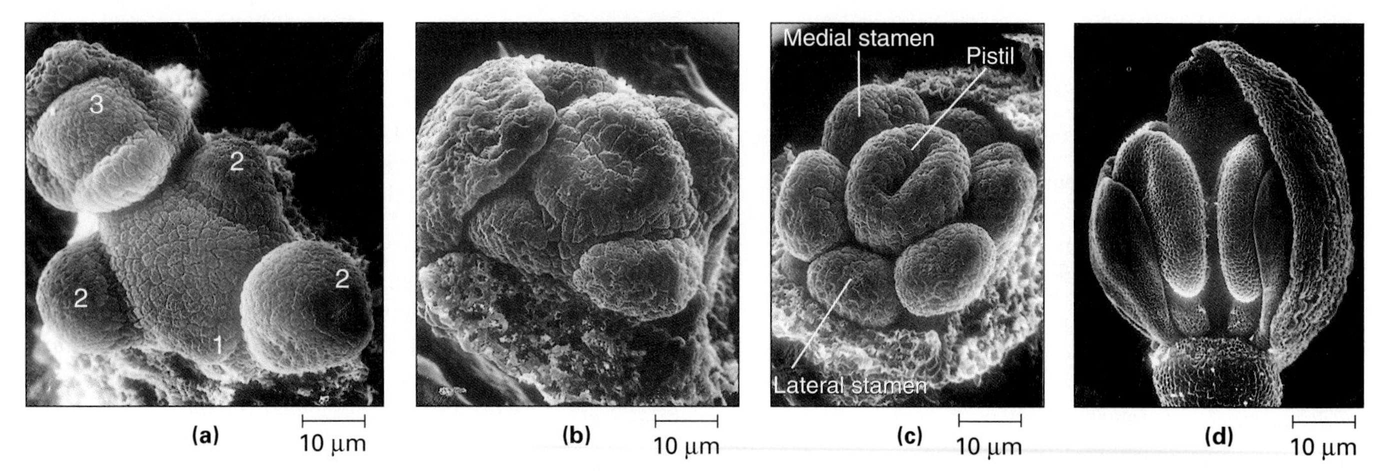

Figure 24.14 Scanning electron micrographs showing early flower development in *Arabidopsis.* **(a)** Inflorescence meristem of a plant. The apical part of the meristem is surrounded by flower primordia (also called floral meristems) numbered 1, 2, and 3 with increasing age. Sepal formation is beginning in the most advanced flower primordium (3). **(b)** Individual flower primordium at a later stage. Three sepals have been removed to show the underlying whorls. **(c)** Another developing flower with sepals and petals removed to reveal the primordia of two medial and four lateral stamens and two carpels, which have fused into one pistil or gynoecium. **(d)** Developing flower with two sepals removed to show the underlying petals, stamens, and pistil.

The $leafy^+$ gene acts together with other meristem identity genes, and a current model of their interactions is shown in Figure 24.15. The $apetala1^+$ gene and $leafy^+$ mutually activate each other, although each of them can be activated independently (Liljegren et al., 1999; R. Simon et al., 1996; D. Wagner et al., 1999). The $leafy^+$ and $apetala1^+$ genes together are in a state of mutual antagonism with the *terminal* $flower^+$ gene, which also appears to be activated by $constans^+$ (Ratcliffe et al., 1998, 1999). The null phenotype of *terminal* $flower^+$ is characterized by a flower at the tip of the inflorescence, indicating the replacement of the inflorescence meristem by a flower meristem. Conversely, plants overexpressing *terminal* $flower^+$ show an enlarged vegetative rosette of leaves and a highly branched inflorescence, similar to the loss-of-function phenotypes of $leafy^+$ and $apetala1^+$. In accord with these phenotypes, *terminal* $flower^+$ is normally expressed at the tip of the inflorescence meristem whereas $leafy^+$ and $apetala1^+$ are expressed laterally in the flower meristems. Therefore, the sharp transition from the production of leaves to the formation of the flowers, which occurs upon floral induction, is controlled by positive feedback interactions between $leafy^+$ and $apetala1^+$, together with negative interactions of these two genes with *terminal* $flower^+$.

The final stages of flower development are the determination of the organ primordia and their subsequent differentiation into organs appropriate for their positions. The determination of flower organ primordia is controlled by *homeotic genes,* which will be the focus of the following discussion. Several homeotic genes are activated by $leafy^+$ (Weigel et al., 1992), and leafy protein binds directly to the promoter of one homeotic gene, $agamous^+$ (Busch, 1999). Also $apetala1^+$, in addition to being a meristem identity gene, has a homeotic function.

24.4 The Role of Homeotic Genes in Flower Patterning

A particularly rewarding set of studies has focused on homeotic mutations, which change the morphological character of the flower organs formed in certain whorls (Coen and Meyerowitz, 1991; Meyerowitz et al., 1991; Weigel and Meyerowitz, 1994). In such mutants, the first steps of flower development are completed normally, so that all organ primordia form in the numbers and positions as they would in the wild types. However, the subsequent determination of flower organ primordia is abnormal, so certain flower organs are replaced with organs normally formed in a different whorl.

ARABIDOPSIS HAS HOMEOTIC GENES

In several mutants of *Arabidopsis,* certain flower organs are replaced by other organs that are normally formed elsewhere. This is the main characteristic of *homeotic mutants,* as we have discussed in the context of *Drosophila* and mouse development. The genes defined by the homeotic mutants of *Arabidopsis* are expressed in restricted domains, and they control in a combinatorial way the morphological development of these domains (Figs. 24.16 and 24.17). We therefore refer to these plant genes as *homeotic genes.*

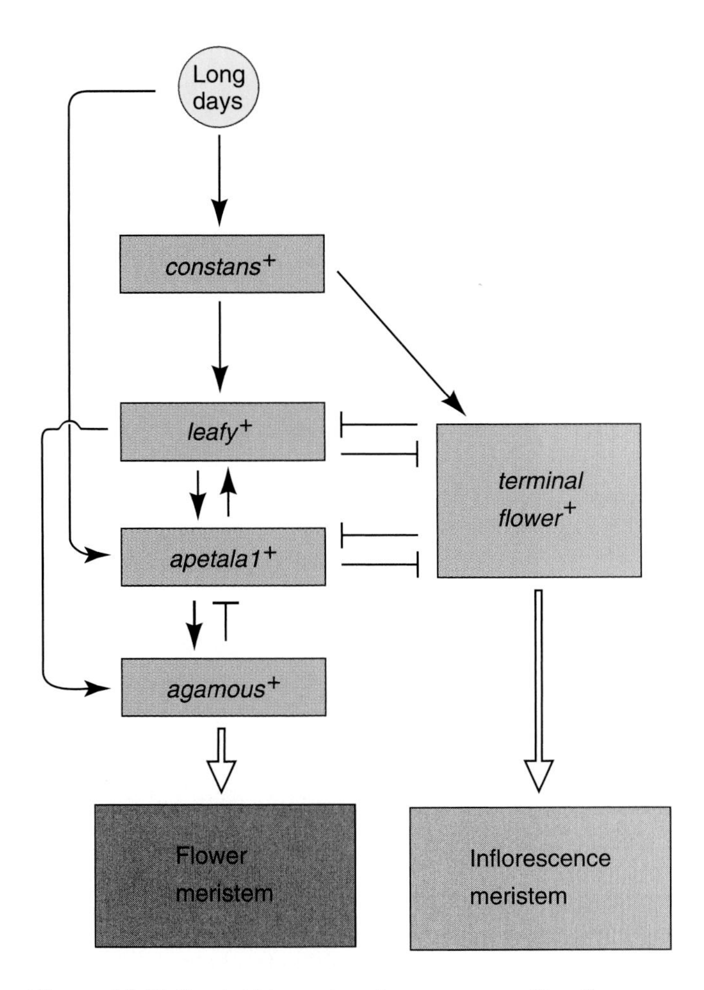

Figure 24.15 Partial hierarchy of genes controlling flower development in *Arabidopsis.* Long days lead to the activation of *constans*+, a flowering gene, which in turn activates *leafy*+, a meristem identity gene. *leafy*+ and another meristem identity gene, *apetala1*+, are mutually activating, but *apetala1*+ can also be activated independently of *constans*+. The *leafy*+/*apetala1*+ pair is in a state of mutual inhibition with a third meristem identity gene, *terminal flower*+, which also appears to be activated by *constans*+. The latter preserves the state of an inflorescence meristem whereas *leafy*+/*apetala1*+ promote the formation of flower meristems, a development that includes the activation of *agamous*+ and other homeotic genes.

All known homeotic mutations in *Arabidopsis* affect sets of two neighboring whorls: whorls 1 and 2 (normally forming sepals and petals), whorls 2 and 3 (normally petals and stamens), or whorls 3 and 4 (normally stamens and carpels). For each pair of whorls, one or two pertinent mutants will be described briefly here. More detailed descriptions and additional illustrations may be found in the accounts of Bowman and coworkers (1991) and Meyerowitz and coworkers (1991). Nearly all homeotic mutants are recessive, so that an abnormal phenotype develops only if both copies of the gene are mutated. Presumably these phenotypes result from a partial or complete loss of gene function.

Whorls 1 and 2 develop abnormally in mutants of the *apetala2* gene. In most alleles, including *apetala2-2,* the organs of whorl 1 (normally sepals) are carpels, leaves, or absent (Figs. 24.16 and 24.17d). The organs of whorl 2 (normally petals) are usually absent. In the *apetala2-1* allele, whorl 1 consists of leaves and whorl 2 consists of four stamens (or stamen-petal intermediates) in place of petals (Figs. 24.16 and 24.17e). Also required for the development of the development of sepal and petal identity is the *apetala1*+ gene, which also functions as a meristem identity gene as discussed earlier (Gustafson-Brown et al., 1994).

Whorls 2 and 3 are affected by mutations in two different genes, *apetala3* and *pistillata,* which cause similar phenotypes. (Despite their similar names, *apetala2* and *apetala3* are different genes with different functions.) Mutants homozygous for *apetala3* or *pistillata* have a normal first whorl of sepals, but the organs of whorl 2 develop as additional sepals rather than petals. In addition, the six organs of whorl 3 develop as carpels rather than stamens (Figs. 24.16 and 24.17c).

Whorls 3 and 4 are affected by mutations in the *agamous* gene. The six positions of whorl 3 are occupied by petals instead of stamens. Moreover, the cells that normally form whorl 4 develop into another flower consisting of one whorl of sepals and two whorls of petals (Figs. 24.16 and 24.17b). This process of forming flowers within flowers is reiterated several times.

In summary, each of the four homeotic genes described here affects two adjacent whorls of flower primordia while leaving the organ identity in the other two whorls nearly unaffected. The mutant phenotypes are characterized by transformations of flower organs into organs that normally form in different whorls. Most of these mutations affect only the morphological charter of the plant organs and not their number or positioning within the whorl.

THREE CLASSES OF HOMEOTIC GENES DETERMINE THE MORPHOLOGICAL CHARACTERS OF FOUR FLOWER WHORLS

On the basis of the homeotic phenotype just described, Bowman and coworkers (1989, 1991) have proposed the following model for the specification of the four organ characters in flowers. The homeotic genes affecting flower formation fall into three classes and are active in three overlapping rings (labeled A, B, and C in Fig. 24.18). The *apetala2*+ gene represents class A, which affects whorls 1 and 2. The *apetala3*+ and *pistillata*+ genes represent class B, affecting whorls 2 and 3. The *agamous*+ gene, which represents class C, affects whorls 3 and 4.

The critical postulate of the genetic ABC model is that the activity of these homeotic genes determines the

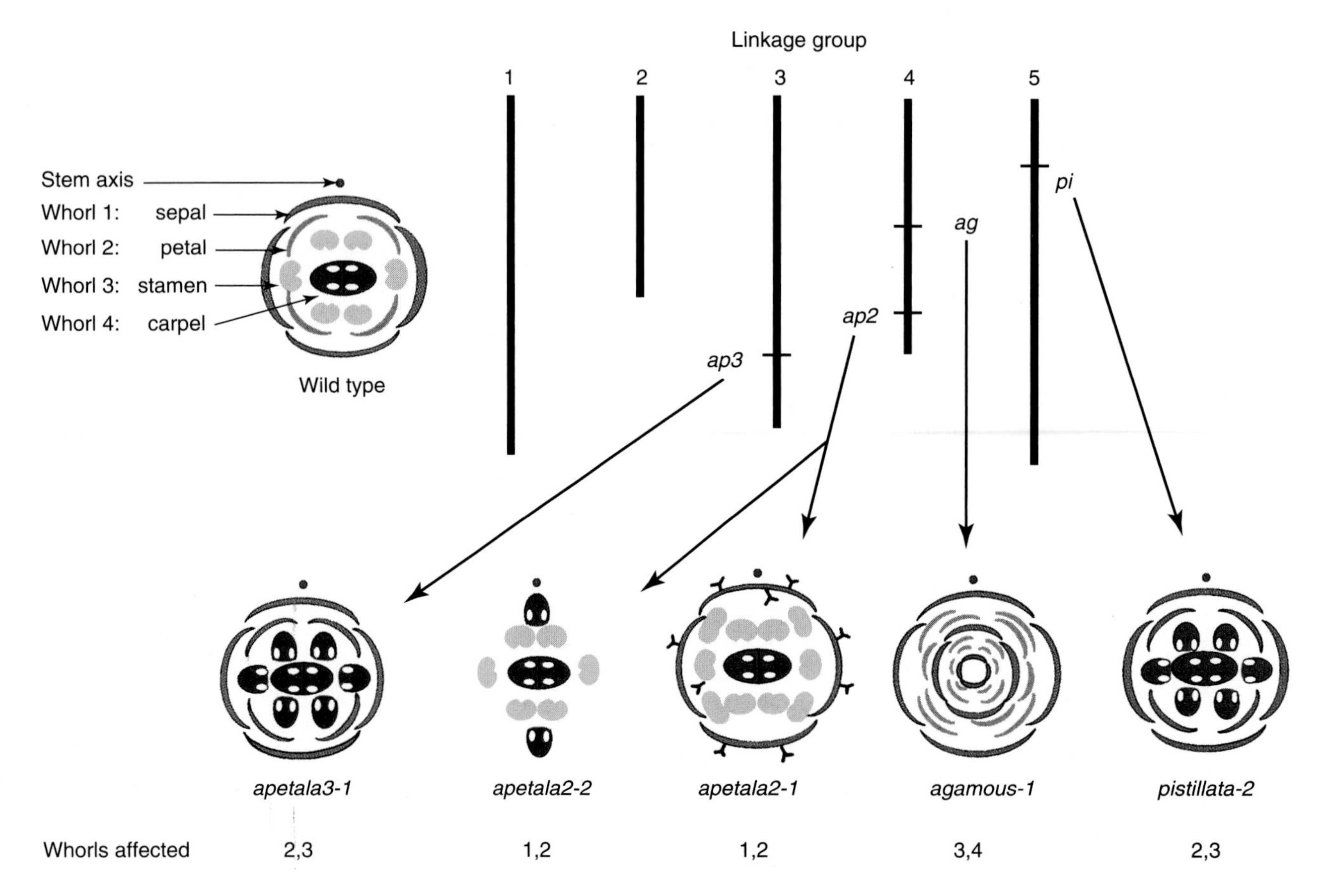

Figure 24.16 Homeotic mutants of *Arabidopsis thaliana.* The top diagrams show a schematic representation of the wild-type flower and the five linkage groups (chromosomes) of *Arabidopsis* with the map positions of four homeotic genes. The bottom diagrams represent the phenotypes of certain mutant alleles of those genes. In *apetala3-1,* petals are replaced with sepals, and stamens with carpels. In one *apetala2-2* phenotype, two sepals are replaced with carpels while the other sepals and all petals are missing. In *apetala2-1,* the organs of whorl 1 develop as leaves, and petals are replaced with stamens. In *agamous-1,* all six stamens are replaced with additional petals, and the carpels are replaced with another flower of the same mutant phenotype. The *pistillata-2* phenotype is very similar to that of *apetala3-1.*

character of the flower organs formed, regardless of where these genes are expressed. Thus, activity of

- *apetala2*$^+$ alone determines sepal development.
- *apetala2*$^+$, *apetala3*$^+$, and *pistillata*$^+$ together determine petal development.
- *agamous*$^+$, *apetala3*$^+$, and *pistillata*$^+$ together determine stamen development.
- *agamous*$^+$ alone determines carpel development.

In addition, the ABC model postulates that *apetala2*$^+$ directly or indirectly inhibits *agamous*$^+$ and vice versa. Moreover, *agamous*$^+$ is thought to *terminate* the development of additional flower organ primordia once the carpel primordia have been formed.

The determination of each flower organ character by its specific combination of gene activities is thought to occur even if the pattern of homeotic gene expression is changed by mutation. For instance, in a mutant without the *apetala2*$^+$ function, the *agamous*$^+$ gene is no longer inhibited in whorls 1 and 2. Consequently, whorl 1 now expresses *agamous*$^+$ and forms carpels, and whorl 2 expresses *agamous*$^+$, *apetala3*$^+$, and *pistillata*$^+$, causing stamen formation. In this fashion, the model explains the homeotic phenotypes described in this section.

The homeotic plant genes considered here, like their counterparts in animals, are thought to direct the morphological development of flower organs by controlling batteries of realizator genes. Accordingly, the homeotic genes are expected to become active before organ differentiation, and their products are expected to show the molecular characteristics of gene-regulatory proteins. Both these expectations are met in plants as they are in animals. Little is known about the hypothetical realizator genes, presumably because their mutant phenotypes are so general that they tend to go unselected in mutagenesis screens for patterning defects.

DOUBLE AND TRIPLE HOMEOTIC PHENOTYPES CONFIRM THE GENETIC ABC MODEL

To further test the validity of their ABC model, Bowman and coworkers (1989, 1991) generated double and triple

Figure 24.17 Wild-type and homeotic mutants of the mouse ear cress *Arabidopsis thaliana.* **(a)** Wild type with sepals (hidden), four petals, six stamens, and a pistil fused from two carpels. **(b)** *agamous* mutant showing extra petals and sepals in the absence of stamens and carpels. **(c)** *apetala3-1* mutant, with the outer two whorls (both sepals) removed to show the third whorl organs, which develop as carpels rather than stamens. **(d)** *apetala2-2* mutant showing first whorl carpels and absence of second whorl organs. **(e)** *apetala2-1* mutant with first whorl modified leaves and second whorl stamen-petal intermediates.

homeotic mutants of *Arabidopsis* and compared their phenotypes with the predictions of the model as shown in Figure 24.18. In mutants homozygous for loss-of-function alleles of *apetala2* and *apetala3* (or *apetala2* and *pistillata*), all whorls should express *agamous*$^+$, and no whorl should express both *apetala3*$^+$ and *pistillata*$^+$, so that all organ primordia should develop into carpels. This is indeed the phenotype of the double mutant (Fig. 24.19a).

Similarly, a double mutant for *apetala3* (or *pistillata*) and *agamous* retains only the *apetala2*$^+$ function, which should be expressed in all four whorls. Accordingly, all organ primordia should develop into sepals. Also, the loss of *agamous*$^+$ function should cause the formation of additional whorls. This is exactly what takes place in the actual phenotype (Fig. 24.19b).

The *apetala2, agamous* double mutant is the most interesting double mutant. According to the model, whorls 1 and 4 will not express any of the homeotic genes discussed here. Therefore, if at least one of these genes is necessary for developing a flower organ, then whorl-1 and whorl-4 primordia should develop unlike any of the organs normally seen in flowers. Indeed, these primordia develop as leaves, by many morphological criteria (Fig. 24.19c). The whorl-2 and whorl-3 organ primordia express the *apetala3*$^+$ and *pistillata*$^+$ genes. They have the potential to develop as petals or stamens, but the gene activities that normally decide between these two options are missing. Actually, the organs of whorls 2 and 3 develop as intermediates between petals and stamens. In addition, the loss of the *agamous*$^+$ function in whorl 4 again causes the formation of additional whorls.

Finally, the triple mutant combinations *apetala2, pistillata, agamous,* and *apetala2, apetala3, agamous* should reveal the ground state in every whorl and should not

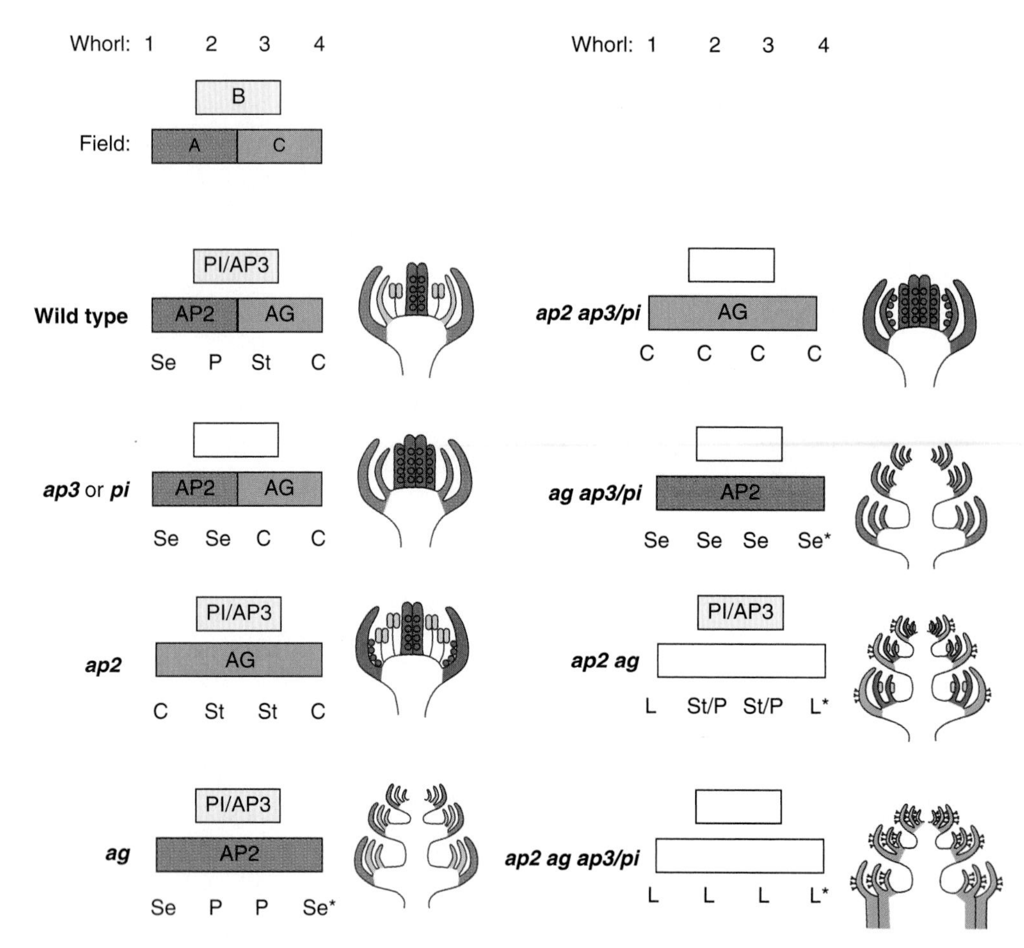

Figure 24.18 ABC model of whorl determination in *Arabidopsis* flowers. The four whorls of floral organs are determined by three sets of homeotic genes, expressed in three ring-shaped or circular fields (A, B, and C), which are represented as boxes in a schematic cross section of half a floral primordium (top left). In wild-type flowers, the *apetala2*$^+$ (*ap2*$^+$) gene is expressed in field A, and its product (AP2) determines the development of sepals (Se). The *pistillata*$^+$ (*pi*$^+$) and *apetala3*$^+$ (*ap3*$^+$) genes are expressed in field B, and their products (AP3 and PI), together with AP2, determine the development of petals (P). The *agamous*$^+$ (*ag*$^+$) gene is expressed in field C, and its product (AG), together with PI/AP3, determines the formation of stamens (St). By itself, AG determines the formation of carpels (C). The activities of *apetala2*$^+$ and *agamous*$^+$ inhibit each other, and *agamous*$^+$ also terminates the formation of additional flower organs after carpel formation. For single, double, and triple loss-of-function mutants (identified in boldface at left), the model predicts the distribution of gene products and correctly explains the phenotypes (schematic drawings at right). The asterisk is a reminder that loss of the *agamous*$^+$ function causes the formation of additional whorls inside whorl 4. L = leaves or carpel-like leaves.

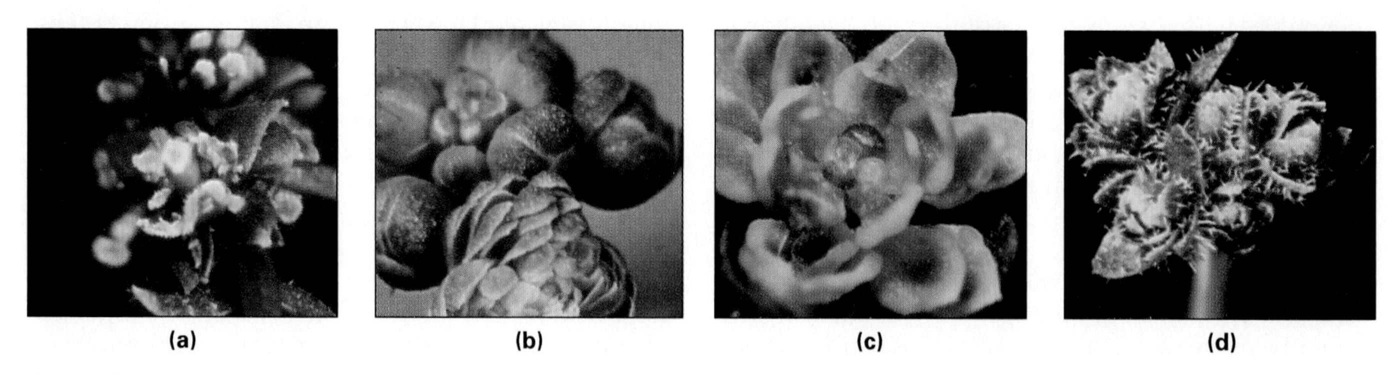

Figure 24.19 Phenotypes of double and triple homeotic mutants of *Arabidopsis.* **(a)** Flowers homozygous for both *apetala2-1* and *apetala3-1.* All organs are carpels or carpel-like leaves. **(b)** Flowers homozygous for both *apetala3-1* and *agamous.* All floral organs are sepals, and supernumerary whorls form. **(c)** Flowers homozygous for both *apetala2-1* and *agamous.* The organs of whorls 1 and 4 are leaves, the organs of whorls 2 and 3 are intermediate between petals and stamens, and additional whorls are formed. **(d)** Flowers homozygous for the three mutations *apetala2-1, pistillata,* and *agamous.* All floral organs develop as carpelloid leaves, and additional whorls form.

contain any floral organs. Indeed, all organs in the triple mutant flowers develop as carpelloid leaves, showing characteristics of both leaves and carpels (Fig. 24.19d).

In conclusion, the ABC model has proved very effective in explaining the morphological character of the flower organs in double and triple mutants for homeotic genes.

HOMEOTIC GENES ARE CONTROLLED BY REGULATOR GENES, MUTUAL INTERACTIONS, AND FEEDBACK LOOPS

While homeotic genes are thought to control batteries of realizator genes, they must themselves be under some sort of regulation to establish their expression domains. Some of this regulation is between the homeotic genes, as in the case of *apetala2*$^{+}$ and *agamous*$^{+}$, which inhibit each other. However, since *apetala2*$^{+}$ is expressed prior to *agamous*$^{+}$, the original limitation of *apetala2*$^{+}$ to whorls 1 and 2 still needs to be explained. Similarly, the restriction of *apetala3*$^{+}$ and *pistillata*$^{+}$ to whorls 2 and 3 calls for an explanation. Thus, plant homeotic genes, like their animal counterparts, are part of a genetic hierarchy but do not constitute its top tier.

One gene that defines the boundary of *apetala3*$^{+}$ and *pistillata*$^{+}$ expression between whorls 3 and 4 has been revealed by the *superman* mutant (Bowman et al., 1992). The phenotype of the recessive *superman-1* allele is an expansion of whorl 3 at the expense of whorl 4 (Figs. 24.20 and 24.21). Whorls 1, 2, and 3 contain the normal numbers of sepals, petals, and stamens, while the center whorl shows a partial or complete replacement of the pistil with additional stamens.

The superman phenotype suggests that the normal function of the *superman*$^{+}$ gene includes inhibiting the expression of *apetala3*$^{+}$ and *pistillata*$^{+}$ in whorl 4. Indeed, the accumulation of apetala3 mRNA, which is restricted to whorls 2 and 3 in the wild type, expands in *superman* flowers to include most of whorl 4. These observation were originally interpreted by postulating that *superman*$^{+}$ encodes a transcriptional repressor of *apetala3*$^{+}$. When the *superman*$^{+}$ gene was cloned the predicted protein actually showed the molecular characteristics of a transcription factor, but *in situ hybridization* (see Method 8.1) also revealed that superman mRNA is synthesized mainly in *whorl 3,* and that this mRNA is synthesized *after* apetala3 mRNA (H. Sakai et al., 1995). Thus, *superman*$^{+}$ seems to act in a more complex way to prevent the spreading of *apetala3*$^{+}$ and *pistillata*$^{+}$ expression to whorl 4.

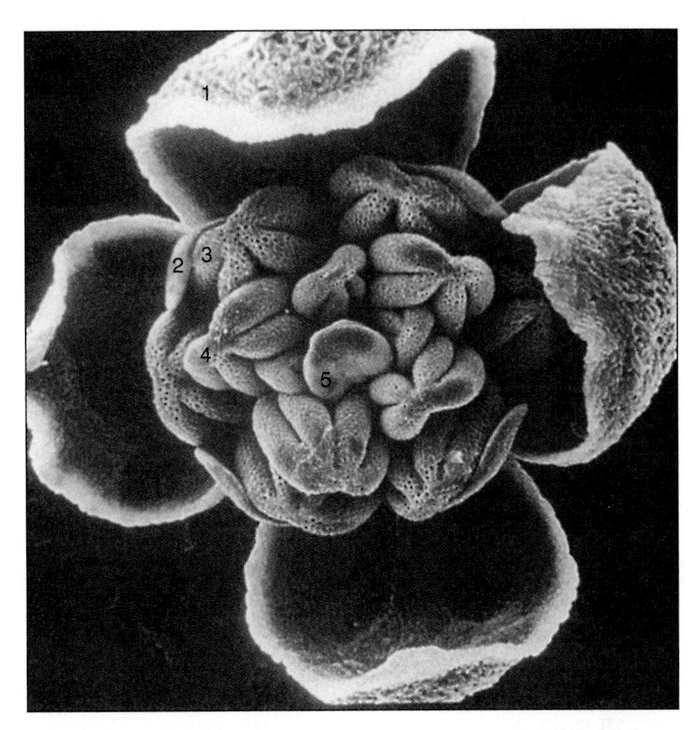

Figure 24.21 Scanning electron micrographs of developing *superman* flowers. This specimen has four sepals, four petals, and six stamens in their normal positions in the first three whorls (1–3). Eight additional stamens occupy two additional rings (4 and 5) in the center, from which carpels are missing.

Figure 24.20 Phenotype of the *superman* mutant of *Arabidopsis.* This flower shows four sepals, four petals, ten stamens, and a reduced pistil. See also Figure 24.21.

Another gene regulatory protein is encoded by the *curly leaf*$^{+}$ gene (Goodrich et al., 1997). Loss-of-function mutants show ectopic expression of *apetala3*$^{+}$ and *agamous*$^{+}$ in leaves and, apparently as a result, the leaf curling that gave the mutant its name. The cloning of *curly leaf*$^{+}$ led to the discovery of extended sequence similarities between the predicted curly leaf protein and a member of the Polycomb group of proteins, which in *Drosophila* maintains the inactive state of homeotic genes, presumably by modifying the chromatin structure (see Section 22.5). Indeed, *Drosophila* Polycomb protein, when labeled with *green fluorescent protein* and expressed from a transgene in tobacco plants, associates with specific chromatin regions and generates phenotypically altered leaves (Ingram et al., 1999). These startling discoveries reveal an astounding structural and functional similarity between the Polycomb proteins in *Drosophila* and their orthologs in plants.

Plant homeotic genes, like their animal counterparts, also control themselves by mutual interactions and positive feedback loops. For example, the *apetala3*$^{+}$ and *pistillata*$^{+}$ genes maintain each other's expression while each feeds back positively on its own expression (Goto and Meyerowitz, 1974; Jack et al., 1994).

Yet another level of control for the expression of plant homotic genes occurs at the posttranslational level. The apetala3, pistillata, and agamous proteins need to form dimers in order to be active. Among the possible combinations, only agamous homodimers and apetala3/pistillata heterodimers bind to DNA (Riechmann et al., 1996). Dimer formation may be necessary to either form or expose a localization signal required for the transport from the cytoplasm to the nucleus (McGonigle et al., 1996). The need for heterodimer formation between apetala3 and pistillata proteins also limits the activity of the corresponding genes to those flower parts that express both of them—that is, whorls 2 and 3.

OTHER PLANTS HAVE PATTERNING GENES SIMILAR TO THOSE OF *ARABIDOPSIS*

The insight derived from the genetic analysis of flower development in *Arabidopsis* probably applies to a wide range of plants. This is indicated by the isolation of very similar mutants in the snapdragon *Antirrhinum majus*. Within the dicotyledonous plants, *Arabidopsis* and *Antirrhinum* are not closely related, so characteristics shared by these species are likely to be common to the entire group.

Both *Arabidopsis* and *Antirrhinum* are well suited to genetic research (Coen and Meyerowitz, 1991). The advantages of *Arabidopsis* have already been explained. *Antirrhinum* is a much bigger plant, large enough to allow collection of floral tissues in quantity for biochemical analysis. Recent molecular studies indicate a high degree of DNA sequence conservation between the two species. It is therefore possible to combine the advantages of both species by using cloned genes from one species as probes to screen DNA libraries of the other species.

Morphologically, *Antirrhinum* differs from *Arabidopsis* in the number and shape of floral organs (Fig. 24.22). The snapdragon has five sepals and five petals; the dorsal (upper) stamen, however, is aborted during development. The petals are fused at the base, forming a tube that terminates in five lobes. The lowest lobe is larger, folds back to the outside, and is bilaterally symmetric, giving the flower much of its overall appearance. The two lateral and two dorsal petals are individually asymmetric, but the entire flower has bilateral symmetry. The pistil, like the one in *Arabidopsis*, consists of two united carpels.

The existence of transposons in the genome of *Antirrhinum* facilitates the isolation of mutants and cloning of

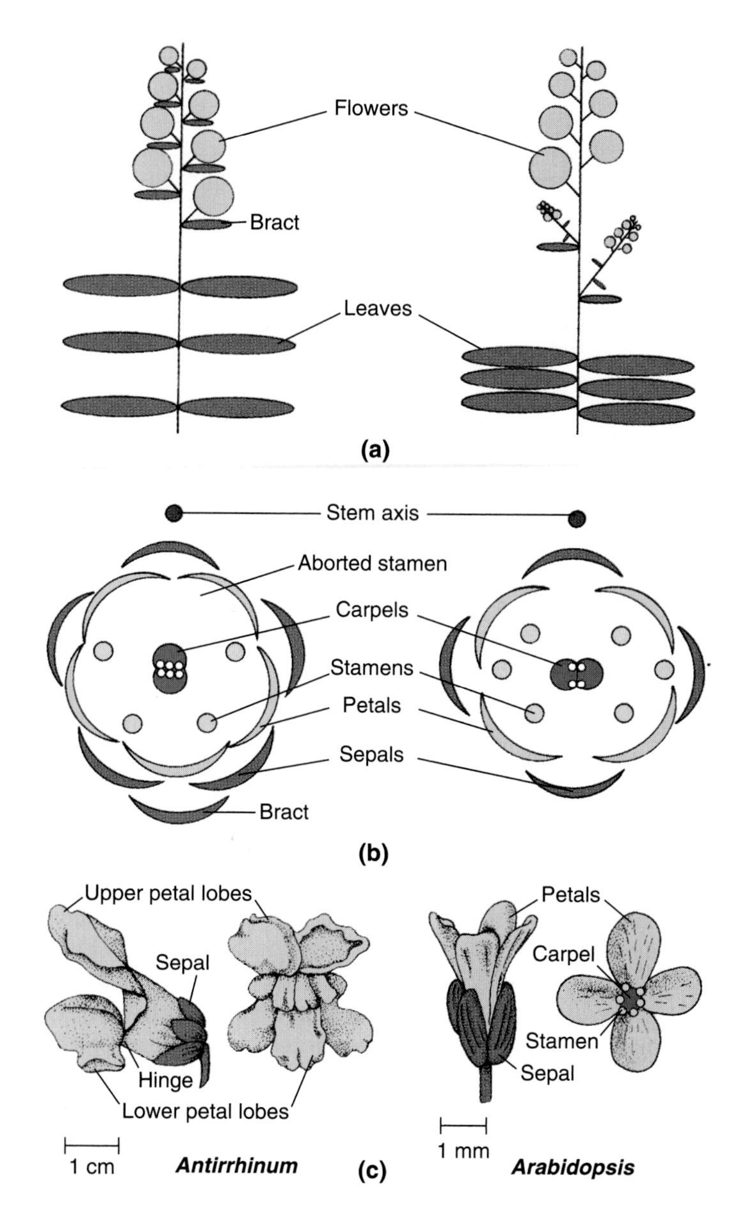

Figure 24.22 Comparison of the snapdragon *Antirrhinum majus* and the mouse ear cress *Arabidopsis thaliana*. **(a)** Schematic diagrams of the entire plants. **(b)** Floral diagrams indicating the numbers and positions of floral organs. **(c)** Drawings of flowers. Note that *Antirrhinum* flowers are about 10-fold larger in linear dimension and 1000-fold larger in mass than *Arabidopsis* flowers.

the corresponding genes, including a series of homeotic genes that produce phenotypes very similar to those of *Arabidopsis*. The similarity is especially noteworthy in view of the differences between the two species with respect to number and position of floral organs. This finding underlines the specific role of the homeotic genes in determining the *morphological character of entire whorls* of flower organs, not the number or specific variations of organs within a whorl. The parallel features of corresponding mutants are compiled in Table 24.2 and illus-

table 24.2 Phenotypes of Floral Homeotic Mutants in *Arabidopsis* and *Antirrhinum*

Mutant Class	Organ Type in Whorl				Mutants Obtained in:	
	1	2	3	4	*Antirrhinum*	*Arabidopsis*
—	Sepals	Petals	Stamens	Carpels	Wild type	Wild type
A	Carpels	Stamens	Stamens	Carpels	*ovulata* *macho*	*apetala2*
B	Sepals	Sepals	Carpels	Carpels	*deficiens* *globosa*	*apetala3* *pistillata*
C	Sepals	Petals	Petals	Sepals*	*pleniflora* *plena*	*agamous*

* Another *agamous* flower with sepals in the outer whorl (*Arabidopsis*) or extra whorls of petaloid structures (*Antirrhinum*).

After Drews et al. (1991b). Used with permission.

trated in Figure 24.23. The similarity between the homeotic genes of *Arabidopsis* and *Antirrhinum* is also observed at the molecular level, as will be discussed later.

ANTIRRHINUM HAS GENES CONTROLLING ORGAN VARIATIONS WITHIN THE SAME WHORL

The bilateral symmetry of *Antirrhinum* depends on genes that control organ variations within the same whorl. Mutations in the *cycloidea* and *dichotoma* genes generate more or less radially symmetrical flowers in which all petals and stamens resemble the ventral organs of the wildtype (Fig. 24.24). The stamen that is aborted in the wild-type flower also develops in *cycloidea* mutants, so that whorls 2 and 3 both have a fivefold radial symmetry. The *cycloidea*$^+$ and *dichotoma*$^+$ genes encode DNA binding proteins that are expressed in the dorsal parts of the flower (Luo et al., 1999). An ortholog of the *cycloidea*$^+$ gene, termed *Lcyc*$^+$, was found in the toad flax *Linaria vulgaris*, which has also bilaterally symmetrical flowers. A naturally occurring mutation in *Lcyc*$^+$ generates a radially symmetrical flower phenotype, which was originally described by Linnaeus more than 250 years ago. Curiously, this mutation consists of an extensive and heritable methylation of the *Lcyc*$^+$ gene (Cubas et al., 1999).

The radializing effect of *cycloidea* mutations on whorls 2 and 3 seems independent of the organs formed in these whorls. This can be seen in double mutants combining *cycloidea* with homeotic mutations affecting whorl 2 or 3. For instance, the *ovulata* mutation of *Antirrhinum*, like *apetala2-1* in *Arabidopsis*, transforms petals into stamens. In *cycloidea, ovulata* double mutants, whorls 2 and 3 both form five stamens in radial symmetry.

These observations show that in addition to the homeotic genes specifying the morphological characters of entire whorls, there are homeotic genes that specify the characters of different organs within a whorl. These two classes of homeotic genes seem to act both independently of each other and in combination.

24.5 Molecular Analysis of Homeotic Plant Genes

Although the ABC model of flower development explains mutant phenotypes well, it makes no predictions about the molecular mechanisms involved in the actions of the homeotic genes. It does not specify the nature of the signals to which the homeotic genes respond, nor does it predict how the homeotic genes act on their target genes. To study these mechanisms, researchers had to clone those genes.

PLANT HOMEOTIC GENES ENCODE TRANSCRIPTION FACTORS

The first homeotic gene of *Arabidopsis* to be cloned was the *agamous*$^+$ gene. The cloning strategy took advantage of a mutant allele, *agamous-2,* generated by insertional mutagenesis with T-DNA (Feldman et al., 1989). Using the T-DNA as a tag, Yanofsky and coworkers (1990) cloned the plant genomic DNA next to the insert and used this DNA as a probe to isolate a corresponding clone from a wild-type *Arabidopsis* genomic DNA library (see Method 15.5). This clone, when used to transform *agamous-2* mutant plants, restored the wild phenotype to the transformants. This proved that the cloned DNA did contain the *agamous*$^+$ gene.

Agamous mRNA is present only in flowers, as indicated by Northern blotting and in situ hybridization experiments (see Methods 15.4 and 8.1, respectively). Transcription begins with the formation of the first

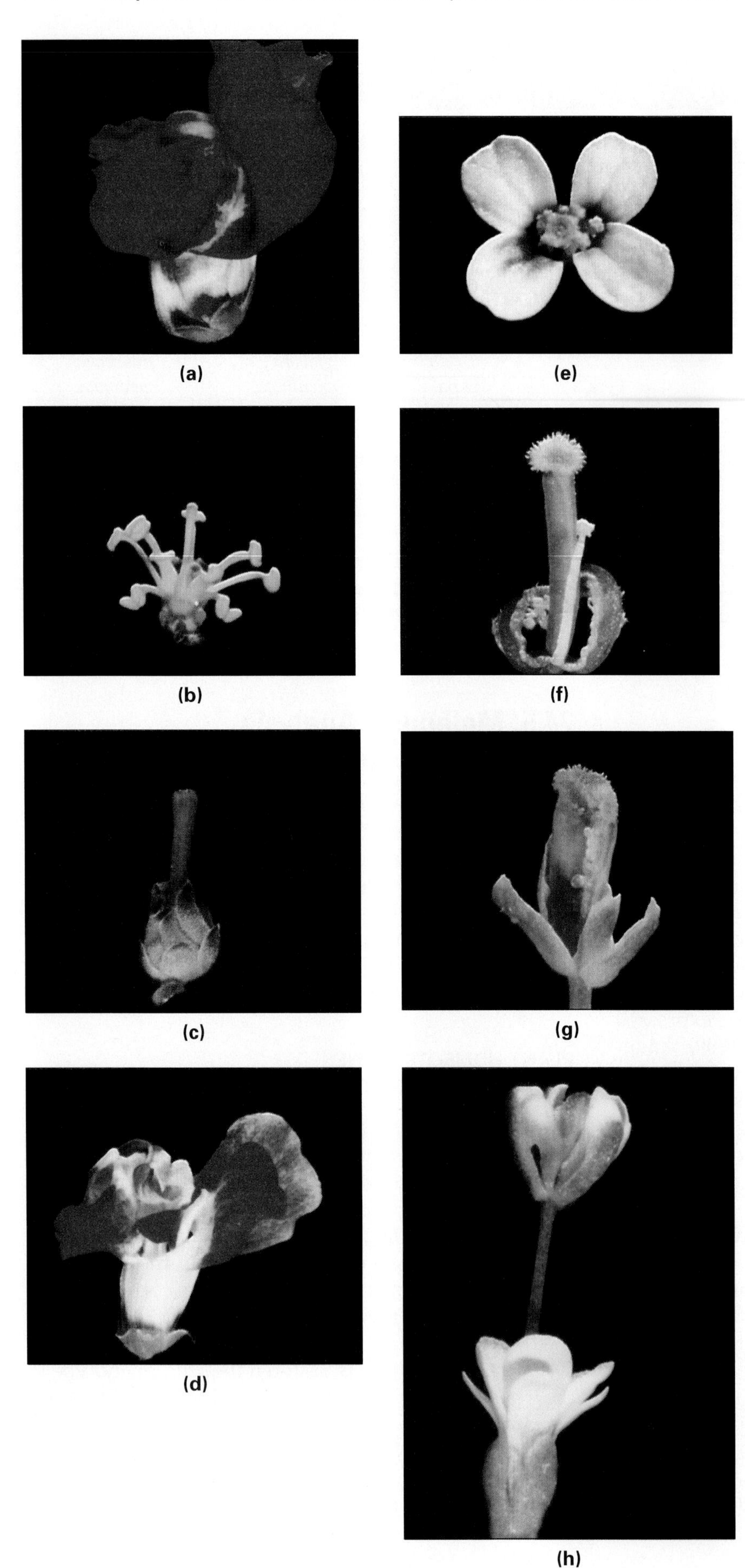

organ primordia and continues throughout flower development. In developing wild-type flowers, the *agamous*$^+$ gene is transcribed only in stamen and carpel primordia, not in sepal and petal primordia (Fig. 24.25a, b). Thus, expression of the *agamous*$^+$ gene is restricted to those primordia that are affected in loss-of-function mutant alleles, in accord with the ABC model (see Fig. 24.18).

Cloned agamous cDNA was used to test predictions derived from the ABC model. The complementary phenotypes generated by loss-of-function mutations in *agamous* and *apetala2* suggest that these two genes inhibit each other's expression. To test this hypothesis, Mizukami and Ma (1992) transformed *Arabidopsis* plants with a fusion gene consisting of a viral promoter and agamous cDNA. The fusion gene was expressed in all parts of the transgenic plants, which formed flowers resembling those of *apetala2* loss-of-function mutants. This result indicates that *agamous*$^+$ does indeed inhibit *apetala2*$^+$ function.

As another test of the ABC model, Drews and coworkers (1991a) examined whether *apetala2*$^+$ gene activity inhibits expression of the *agamous*$^+$ gene as postulated. If so, absence of *apetala2*$^+$ function should allow the *agamous*$^+$ gene to expand its expression domain to whorls 1 and 2. This expansion was indeed observed when a labeled probe for agamous mRNA was hybridized in situ with sections of *apetala2* flower primordia. Here the agamous mRNA had accumulated in all four whorls, instead of being restricted to whorls 3 and 4 (Fig. 24.25c, d). This result confirmed the prediction derived from the ABC model that *apetala2*$^+$ (directly or indirectly) controls *agamous*$^+$ expression. (Note that the *Arabidopsis* investigators used the same strat-

Figure 24.23 Parallel series of homeotic mutants of *Antirrhinum* (left) and *Arabidopsis* (right). **(a, e)** Wild type. **(b, f)** Class A mutants *ovulata* and *apetala2,* both with carpels in place of sepals and no petals. **(c, g)** Class B mutants *deficiens* and *pistillata,* both without petals and stamens. **(d, h)** Class C mutants *plena* and *agamous,* both with a reiterated sequence of sepal-petal-petal.

Figure 24.24 Flowers of *Antirrhinum* shown in face view. The wild type (left) is bilaterally symmetrical; the *cycloidea* mutant (right) is radially symmetrical.

egy employed by *Drosophila* researchers to establish a regulatory hierarchy among embryonic patterning genes. See Section 22.2.)

Molecular analysis of the *agamous*$^+$ gene revealed an extensive regulatory region with redundant *cis-acting elements* for repression by the apetala2 protein and for activation by the leafy protein (Bomblies et al., 1999; Busch et al., 1999). These and other studies, some of which will be discussed later, indicate that plant homeotic genes encode transcription factors like their animal counterparts. Extending this parallel, plant homeotic genes are expressed in restricted domains, and this restriction is mediated by interactions among themselves and by control genes such as *leafy*$^+$.

GENETIC AND MOLECULAR STUDIES REVEAL ORTHOLOGOUS GENES BETWEEN *ARABIDOPSIS* AND *ANTIRRHINUM*

In similar studies on the *deficiens*$^+$ gene of *Antirhinnum*, Schwarz-Sommer and coworkers (1990) observed that the *deficiens*$^+$ gene was transcribed most actively in those wild-type organ primordia that were transformed in the mutant phenotype: petals and stamens (Fig. 24.26). These results confirmed the expectation that homeotic plant genes in general encode regionally expressed transcription factors.

The *apetala3*$^+$ and *pistillata*$^+$ genes of *Arabidopsis* have orthologs in *Antirrhinum* known as *deficiens*$^+$ and *globosa*$^+$, respectively. Mutations in these genes transform petals into sepals and stamens into carpels (see Fig. 24.23c, g; Table 24.2). Their mRNAs show the strongest accumulation in the petal and stamen primordia of developing wild-type flowers of both species (Jack et al., 1992). The deficiens and globosa proteins are transcription factors and have the same need as their orthologs to form heterodimers in order to accumulate in the nucleus (Tröbner et al., 1992; Zachgo et al., 1995). Thus, the two pairs of genes produce equivalent mutant phenotypes, have corresponding expression patterns, and encode similar proteins.

Another orthologous pair of genes is the *leafy*$^+$ gene of *Arabidopsis* and the *floricaula*$^+$ gene of *Antirrhinum* (Coen et al., 1991; Weigel et al., 1992). Mutations in either gene cause the transformation of flowers into entire inflorescence shoots, although the phenotype is more severe in *Antirrhinum*. Both *leafy*$^+$ and *floricaula*$^+$ are expressed in floral meristems, and the proteins encoded by the two genes show 70% sequence identity. These observations indicate that not only the homeotic genes but also other floral patterning genes are conserved among dicotyledonous plants.

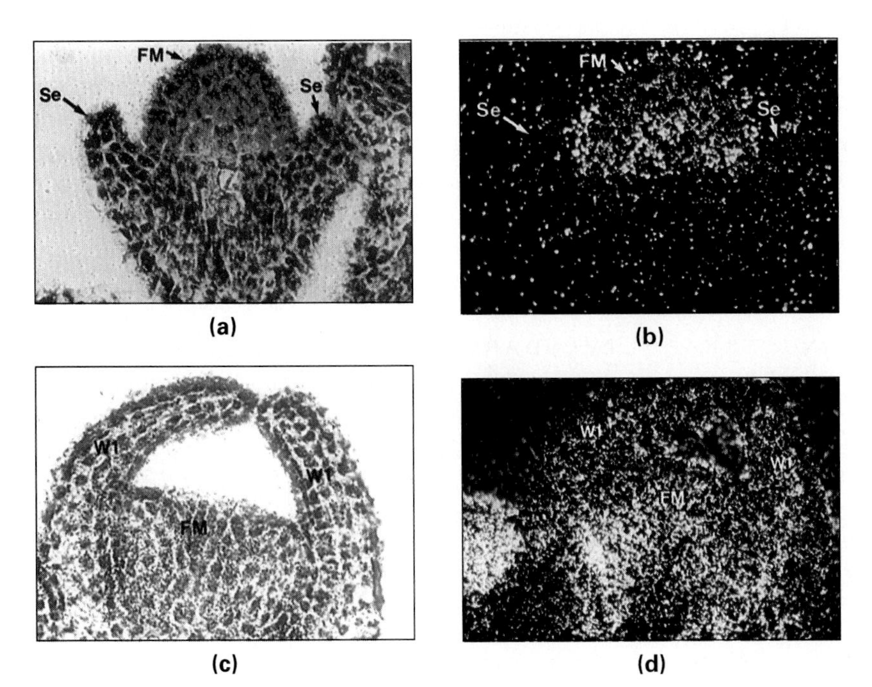

Figure 24.25 Distribution of agamous mRNA in *Arabidopsis* flower primordia. All panels show sections of flower primordia hybridized in situ with a radiolabeled agamous antisense RNA probe (see Method 8.1). The photographs in parts a and c were taken with bright-field illumination; radioactivity signals appear as red dots. Parts b and d show the same sections in dark-field illumination; radioactivity signals appear as white grains. **(a, b)** Wild-type flower primordium. The sepal primordia (Se) are unlabeled; the central floral meristem cells (FM) contain agamous mRNA. These are the cells from which the stamen and carpel primordia will arise. **(c, d)** *apetala2* mutant flower primordium at a corresponding stage. The organ primordia in whorl 1 (W1), which form carpels instead of sepals, contain agamous mRNA.

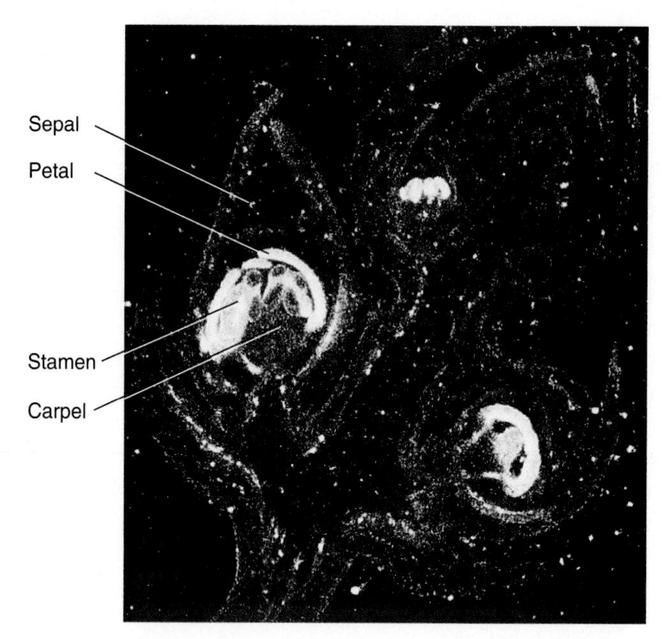

Figure 24.26 Spatial expression pattern of the *deficiens*⁺ gene in *Antirrhinum* shown by in situ hybridization (see Method 8.1). The tip of an inflorescence is shown in longitudinal section and with dark-field illumination. Most of the label (white grains) accumulates over the primordia of petals and stamens. The other parts of the inflorescence, including the primordia of sepals and carpels, are labeled less strongly.

THE KNOWN PLANT HOMEOTIC GENES HAVE A MADS BOX INSTEAD OF A HOMEOBOX

The known homeotic plant genes share the main characteristics of homeotic genes in *Drosophila* and the mouse: Mutations in these genes cause characteristic transformational phenotypes, their expression is restricted to certain organ primordia, and they encode transcription factors that have two functions—they connect homeotic genes in bistable control circuits and they presumably control many target genes that mediate morphological development. On the basis of these similarities, one might expect plant homeotic genes to share two other features with their counterparts in fruit flies and mice: the clustering in complexes and the *homeobox*. However, the plant homeotic genes cloned so far are scattered over several chromosomes (see Fig. 24.16). Also, the plant homeotic genes *do not* have a homeobox. Instead, all plant homeotic genes and several other plant patterning genes contain the ***MADS box,*** so called because it was first identified in the *MCM1*⁺ gene in yeast, in *agamous*⁺ of *Arabidopsis*, in *deficiens*⁺ of *Antirrhinum*, and in *SRF*⁺, a human serum response factor gene. The protein domain encoded by the MADS box has about the same length as the homeodomain, and it is thought to form similar α helices that bind to specific DNA recognition motifs, but there is no sequence similarity to the homeodomain (compare Figs. 23.3 and 24.27).

This is not to say that plants do not produce proteins with homeodomains (*R.W. Williams, 1998*). A gene encoding such a protein, *Knotted-1*⁺, was isolated from maize (Vollbrecht et al., 1991). The gene was identified by gain-of-function alleles that alter leaf development. Nests of cells in the mutant leaves fail to differentiate properly and continue to divide, throwing the leaf blades into folds and distorting their vein patterns. These observations, along with the phenotypes of loss-of-function alleles, indicate that the *Knotted-1*⁺ gene function keeps shoot meristem cells in an undifferentiated state or promotes the division of undifferentiated cells (Kerstetter et al., 1997).

On the basis of the homeotic genes and homeobox genes so far identified in plants, it appears that they may form two distinct groups. The main characteristic of *homeotic genes* is their *biological function:* They specify, by their activity or inactivity, a pattern of determined states along an embryonic axis. Apparently, this function can be carried out by transcription factors with a homeodomain and by other transcription factors with a different DNA-binding domain, such as the MADS box. The main characteristic of *homeobox genes* is a *biochemical feature:* They encode transcription factors with a homeodomain. In animals, most homeotic genes are homeobox genes that are arranged in highly conserved *Hox* clusters as discussed in (see Sections 22.5 and 23.2). Indeed, Slack and coworkers (1993) consider the *Hox* complex and its expression pattern in the embryo as the defining

```
AP3    RGKIQIKRIENQTNRQVTYSKRRNGLFKKAHELTVLCDARVSIIMFSSSNKLHEYISP
DEF    ------------------------------S------K------i--TQ---------  53/58
AG     ----E-------t-------Fc-------I---y--S------e-AL-V---rgR-y--snn  40/58
MCM1   -r--E--f---k-r-h--F---Kh-Im---f--S--tgTq-ILLVv-eTgIVytFsT-  26/58
SRF    -v--kMef-D-kir-yt-F---Kt-Im---y--St-tgTq-ILLVa-eTgHVytFaTr  19/58
```

Figure 24.27 Amino acid sequence comparison for the putative DNA-binding region of several proteins: apetala3 (AP3) of *Arabidopsis,* deficiens (DEF) of *Antirrhinum,* agamous (AG) of *Arabidopsis,* yeast MCM1 gene product, and human serum response factor (SRF). The top line is the AP3 sequence in single-letter code. In the sequences below, only amino acids diverging from the AP3 standard are shown, while identical amino acids are represented by a dash. Capitals represent amino acids that are similar to their counterparts in AP3; lowercase letters represent amino acids that are dissimilar to their AP3 counterparts. The numbers to the right indicate the fraction of amino acids a protein has in common with AP3.

character of the animal kingdom and propose that it be called the ***zootype.*** Plants clearly do not show this character. Here, homeotic genes and homeobox genes have evolved independently. The disjunction indicates that each of the two functions, generating patterns of determined states and producing transcription factors with homeodomains, was by itself sufficiently adaptive to be positively selected for.

It also appears that the homeotic genes of plants and animals have evolved independently. On the one hand, this is not surprising since the last common ancestor of animals and plant was a unicellular organism, which did not need genes for patterning multicellular arrays. On the other hand, it is remarkable that the biological characteristics of homeotic genes, including *default programs* and *bistable control circuits,* have evolved twice independently in so similar ways. It suggests that developmental mechanisms based on these principles are very adaptive indeed (*Meyerowitz, 1999*).

MANY QUESTIONS ABOUT PATTERN FORMATION IN FLOWERS ARE STILL UNANSWERED

Molecular analysis of homeotic plant genes has confirmed the ABC model, but many fundamental questions are still unanswered. Only a few genes that control homeotic plant genes are known, as is the case for the realizator genes presumed to be controlled by the homeotic genes. While plant homeotic genes are controlled at the transcriptional level and in turn encode transcription factors, many other control levels of gene expression seem to be involved that are just beginning to be explored.

The specific phenotypes caused by mutations in plant homeotic genes also contrast with the observation that their MADS domains all bind to similar DNA sequences (Riechmann and Meyerowitz, 1997), much in parallel to the "promiscuity" of homeodomain binding to DNA (see Section 23.2). In the case of the plant homeotic proteins, amino acid sequences outside the MADS domain may contribute to the DNA-binding specificity of the proteins.

An emerging topic impinging upon the action of plant homeotic genes is the existence of cytoplasmic connections, known as ***plasmodesmata*** (Gk. *plasma,* "a thing formed"; *desmos,* "band, connection"). These connections either persist between sister cells after cytokinesis or are formed secondarily between neighboring nonsister cells. Plasmodesmata serve as conduits for the transport of both RNA and proteins (*Lucas, 1995*). At least in some cases, the transport is polar, that is, restricted to certain adjoining cells and proceeding only in one direction. One such case is the transport of the class B gene products, deficiens and globosa, in the floral meristem of *Antirrhinum* (Perbal et al., 1996). This meristem, like most plant meristems, is composed of three layers. The outermost layer, designated L1, gives rise to epidermis, while the following layers, L2 and L3, produce the deeper plant tissues. The deficiens and globosa proteins move between these layers, and at least for deficiens protein, the movement is polar: It occurs only from the inner layers to the outer layer. There is little or no transport of the same gene products within any given layer. This restriction assures that the epidermal cells adopt the same general fate as the deeper layers while the boundaries between whorls are kept sharp. It will be interesting to see whether there are any shared functions and/or signaling mechanisms between plant plasmodesmata and the *cytonemes* found in *Drosophila* imaginal discs (see Section 22.8).

SUMMARY

The plant embryo already shows the basic pattern of the seedling that will emerge when the seed germinates: It contains an epicotyl, one or two cotyledons, a hypocotyl, and a root primordium. The epicotyl and the root primordium contain meristem cells, which retain their ability to divide throughout the life of the plant. Meristem cells are similar to the stem cells of animals, in that they give rise to the primordia of different plant organs while at the same time generating new meristem cells. Depending on internal and environmental factors, the apical shoot meristem produces the primordia of leaves, branches, or flowers.

Compared with animal development, plant development relies less on maternal RNA and more on early embryonic gene activity. Also, the overall appearance of a plant is generated primarily by the timing and pattern of meristem cell divisions; morphogenetic movements play a minor role. Moreover, plant have no germ line. Instead, differentiated plant cells remain totipotent.

The mouse ear cress *Arabidopsis thaliana* has emerged as a model for genetic and molecular studies of the development of angiosperms. A saturation mutagenesis screen for abnormal embryogenesis has yielded 250 mutant lines, which probably represent about 30 genes involved in embryonic pattern formation. In the mutant phenotypes, there are specific deletions along the apical-basal axis, defects in the radial stem pattern, or major shape changes. These observations show that specific steps of the patterning process are controlled by certain genes, which are now being characterized.

The process of flower development can be subdivided into three steps, each of which is affected by certain genes. The first step, called floral induction, results in the reorganization of the apical shoot meristem from a vegetative meristem producing leaves into an inflorescence meristem giving rise to flower primordia. The second step in flower development is the actual formation of flower primordia, which arise through distinct patterns of cell division in the inflorescence meristem.

In the third and final step of flower development, a set of homeotic genes controls the determination of flower organ primordia. These genes bear many similarities to their counterparts in fruit flies and mice: They encode transcription factors, their expression is restricted to certain domains, and their mutant phenotypes are characterized by homeotic transformations. For instance, in one type of homeotic mutant, the primordia that would form stamens in the wild type develop as petals instead. Plant homeotic genes are regulated by other patterning genes that control earlier steps in floral development. In addition, there are regulatory interactions among homeotic plant genes.

A simple genetic model explains most phenotypic traits of the known homeotic mutants and their combinations. The regulatory interactions implied in the model are confirmed by the available genetic and molecular data. The proteins encoded by the plant homeotic genes have a so-called MADS-box instead of a homeobox to encode the DNA-binding domain of their protein products. This observation suggests that the specification of developmental pathways by homeotic gene activities has evolved independently of the biochemical feature of the homeodomain, which is common to many transcription factors involved in the patterning of animals.

Arabidopsis and the snapdragon *Antirrhinum majus* have pairs of floral patterning genes that are not only orthologous in terms of similar DNA sequences but also share similar mutant phenotypes and expression domains. Thus, these genes have been well conserved during the evolution of dicotyledonous plants. In addition to those homeotic genes that control the organ character of entire whorls, *Antirrhinum* also has genes that control organ variations within the same whorl.

Experimental and Genetic Analysis of *Caenorhabditis elegans* Development

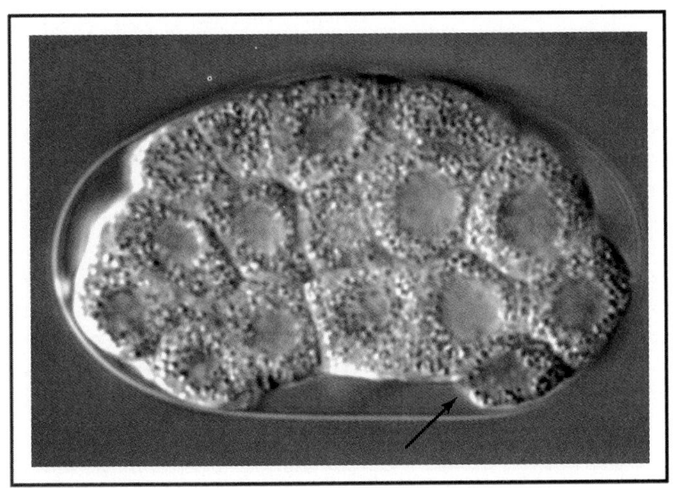

Figure 25.1 Living embryo of the roundworm *Caenorhabditis elegans* at the 28-cell stage; anterior pole to left, dorsal side at the top. The arrow marks the primordial germ cell. The other cells are somatic cells. The entire cell division pattern from zygote to adult is nearly invariant between individuals. The cell nuclei appear as depressions in this photograph because of the optical contrast enhancement used.

AFTER *Drosophila* and the house mouse, the soil nematode *Caenorhabditis elegans* is the third most popular animal subject for the genetic analysis of development. Living on agar plates or in liquid medium and feeding on bacteria, this tiny worm is easy to keep in the laboratory. Because *C. elegans* is transparent at all stages of its life, cell nuclei are readily observed in live specimens under the microscope (Fig. 25.1). And because an adult consists of only about 1000 somatic cells, the anatomy of *C. elegans* is comparatively simple.

The most characteristic attribute of *C. elegans,* which it shares with many other nematodes, is a striking invariance of development that makes individuals of the same species and sex exactly alike (Sulston et al., 1983). The number of cell divisions, as well as their timing and orientation, is strictly controlled. As a result, the number of cells per individual is constant, and it has been possible to trace the exact life history of each cell in this species.

Advantages of *C. elegans* for the genetic analysis of development include a short life cycle, and a small genome (Brenner, 1974). Moreover, the fact that most adults are self-fertilizing makes it easy to breed alleles to homozygosity. These advantages have been boosted enormously by the complete sequencing of the *C. elegans* genome (*C. elegans* Sequencing Consortium, 1998; see Section 15.2).

The roundworm's unique combination of advantages has allowed researchers to study some basic questions in development with unparalleled precision. Typically, they can carry out such studies on *C. elegans* at the level of individually known, single cells. We will illustrate this type of analysis by returning to two familiar topics: cytoplasmic localization and embryonic induction—and by looking at two new topics: the timing of developmental events and programmed cell death. In each case, mutant analysis and molecular techniques have confirmed and extended the conclusions reached from embryological studies.

25.1 Normal Development

HERMAPHRODITES AND MALES

Adult *C. elegans* worms are either hermaphroditic or male (Fig. 25.2). *Hermaphrodites* are defined as individuals that have both male and female gonads. In the case of *C. elegans,* hermaphrodites can fertilize their eggs either with their own sperm or with sperm obtained by copulating with males. The two-armed gonad of the hermaphrodite produces sperm during the last larval stage. These sperm are stored in ***spermathecae*** (sing., *spermatheca;* Gk. *theke,* "sheath"), receptacles located between each gonadal arm and the uterus. In the adult hermaphrodite, the gonads function as ovaries and produce oocytes. The mature eggs are fertilized as they pass the spermathecae on their way to the uterus. Fertilized eggs can develop completely independently of the mother but are usually retained in the uterus for a few cleavage divisions before they are born through a mid-ventral genital opening, the ***vulva.*** An adult *C. elegans* hermaphrodite consists of exactly 959 somatic cells and produces about 2000 gametes (Wood, 1988a).

The male of *C. elegans* has a one-armed gonad that produces only sperm, which are delivered to the vulva of the hermaphrodite via the copulatory organ at the male's posterior end. The adult male consists of exactly 1031 somatic cells and produces about 1000 sperm. Since hermaphrodites have two sex chromosomes and males only one, the self-fertilization of hermaphrodites produces almost exclusively more hermaphrodites; occasional males that develop after self-fertilization result from chromosomal nondisjunction during meiosis. However, sperm from a male outcompete sperm from a hermaphrodite at fertilization, so that hermaphrodites that have mated produce predominantly male-fertilized offspring, of which 50% are male and 50% hermaphroditic.

FERTILIZATION, CLEAVAGE, AND AXIS FORMATION

From fertilization to hatching, embryogenesis in *C. elegans* takes only 14 h (Fig. 25.3; Sulston et al., 1983). The anteroposterior polarity of the egg is established at fertilization, as the posterior pole develops next to the position of the male pronucleus after sperm entry. Usually, the sperm enters opposite to the position of the female

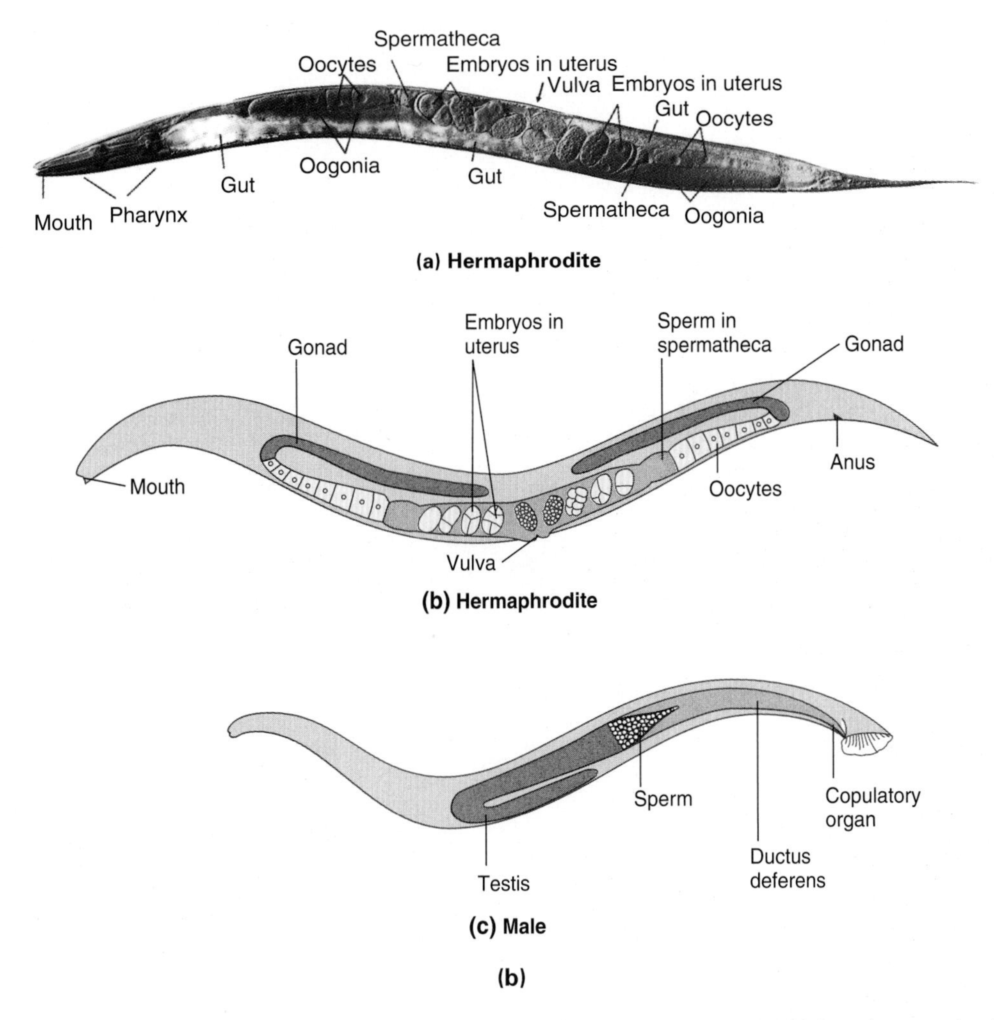

Figure 25.2 Adult morphology and reproductive organs of the roundworm *C. elegans.* **(a)** Light micrograph of a gravid hermaphrodite. The mouth, pharynx, and anterior gut are seen anteriorly (to the left). The two arms of the gonad (see part b) contain oogonia, which develop into mature oocytes and are fertilized with sperm stored in the spermatheca. The embryos undergo cleavage while still in the uterus. The vulva is a ventral opening for copulation and egg laying. **(b)** Schematic diagram of the adult hermaphrodite's reproductive organs. Each gonadal arm, now functioning as an ovary, is connected to a spermatheca, where eggs are fertilized by stored sperm derived either from the hermaphroditic gonad during the larval stage or obtained as an adult from copulating males. In the uterus, the fertilized eggs undergo cleavage before they are laid through the vulva. **(c)** Male reproductive organs. A single testis releases sperm via a ductus deferens to a copulatory organ at the posterior end.

pronucleus, but the sperm entry point may change under experimental conditions, and if it does, so does the posterior pole (Goldstein and Hird, 1996). Once fertilized, the egg forms a nearly impermeable outer shell. The female pronucleus resumes meiosis and gives off polar bodies at the anterior pole.

Upon completion of meiosis, the fertilized egg undergoes a dramatic reorganization. Internal cytoplasm streams posteriorly while cortical cytoplasm streams anteriorly. During this time, waves of cortical contraction sweep from posterior to anterior, eventually forming a temporary constriction known as the *pseudocleavage* furrow. At the same time, the female pronucleus moves posteriorly toward the male pronucleus until the two meet in the posterior egg half.

The early cleavages in *C. elegans* are asymmetrical and asynchronous. The first mitotic spindle forms and positions itself parallel to the anteroposterior axis and slightly posterior to the egg center. The pronuclei fuse, and the first mitosis and cleavage follow immediately, about 35 min after fertilization. Cytokinesis produces a larger anterior blastomere called AB and a smaller posterior blastomere called P_1. About 10 min later, the AB cell divides, producing an anterior daughter cell (ABa) and a posterior daughter cell (ABp). Shortly thereafter, P_1 divides into P_2 at the posterior pole and EMS opposite ABp. EMS defines the future ventral side of the embryo and ABp the future dorsal side. Shortly after the cleavage of AB, EMS divides into the E cell and the MS cell.

FOUNDER CELLS

Each of the early cleavages produces an anterior somatic cell (AB, EMS, C, and D) and a posterior germ line cell

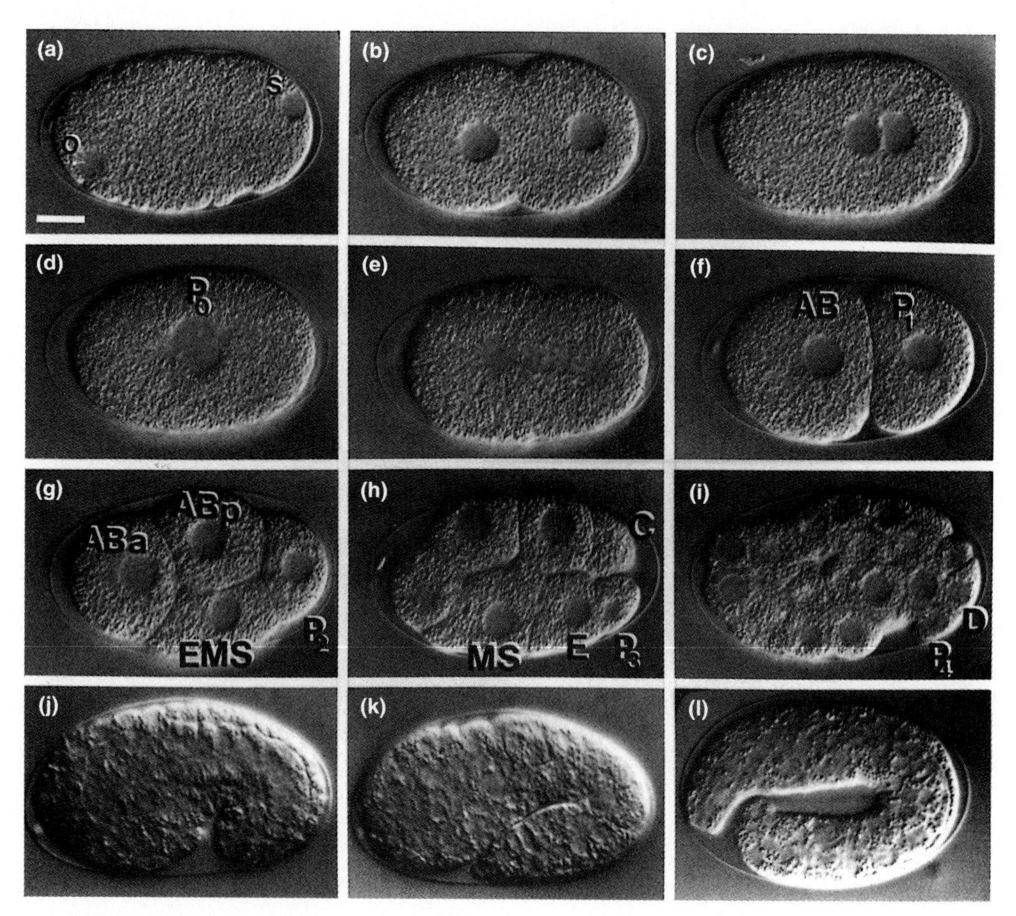

Figure 25.3 Cleavage of *C. elegans.* In each photograph, the anterior pole is oriented to the left. **(a)** About 30 min after fertilization, the female pronucleus (o) is visible anteriorly and the male pronucleus (s) posteriorly. **(b, c)** The pronuclei migrate until they meet in the posterior half of the egg, while the egg constricts temporarily in pseudocleavage. **(d)** The pronuclei rotate and fuse near the egg center. **(e)** The mitotic spindle orients parallel to the anteroposterior axis. **(f)** The first cleavage separates a large somatic cell (AB) from a smaller germ line cell (P_1). **(g)** After division of AB into ABa (anterior) and ABp (posterior), P_1 cleaves into a large somatic cell, EMS, and a smaller new germ line cell, P_2. **(h)** EMS divides into E, the gut precursor cell, and MS. P_2 also divides unequally into a somatic cell, C, and a smaller new germ line cell, P_3. **(i)** The somatic cells divide again. In the germ line, the last unequal cleavage generates the somatic founder cell D and the primordial germ cell P_4. Soon after, gastrulation begins with the immigration of Ea and Ep, the two daughter cells of E.

labeled P_1 after the first division, P_2 after the second division, and so forth until P_4 is formed as an exclusive germ line progenitor (Fig. 25.4). The resulting five cells (six after the division of EMS into E and MS) are called ***founder cells,*** because each of them develops in a characteristic way. The founder cells and their respective progeny continue to cleave at a rate that is roughly proportional to the founder cells' size and follows their order of origin. AB cleaves fastest, D slowest, and P_4—the primordial germ cell—cleaves only once more during embryogenesis.

Given the near-invariance of cell lineage in *C. elegans* development, one might expect each founder cell to give rise only to certain cell types and, conversely, all cells of a certain type to be derived from a single founder cell. However, lineage analysis has revealed that this is not generally true (Fig. 25.4). While intestinal cells and germ cells are indeed derived from single founder cells, this is not the case for hypodermis cells, neurons, and muscle. Thus, cells of similar type are not necessarily closely related, and closely related cells may turn into different types.

GASTRULATION, ORGANOGENESIS, AND HISTOGENESIS

Gastrulation begins at the 28-cell stage, when the 16 AB descendants lie anteriorly and laterally, the 4 MS derivatives lie ventrally, and the 4 C descendants lie posteriorly and dorsally. The two E derivatives (Ea and Ep) as well as D and P_4 lie ventrally and posteriorly (see Fig. 25.3i). At the start of gastrulation, the two E cells sink inward, followed by P_4 and the daughter cells of MS. As the entry zone widens, prospective pharynx cells and body muscle cells derived from the AB, C, and D lineages follow into the interior. Further divisions of the E cells and pharynx precursor cells generate the cells for the digestive tube. The body muscle cells are positioned

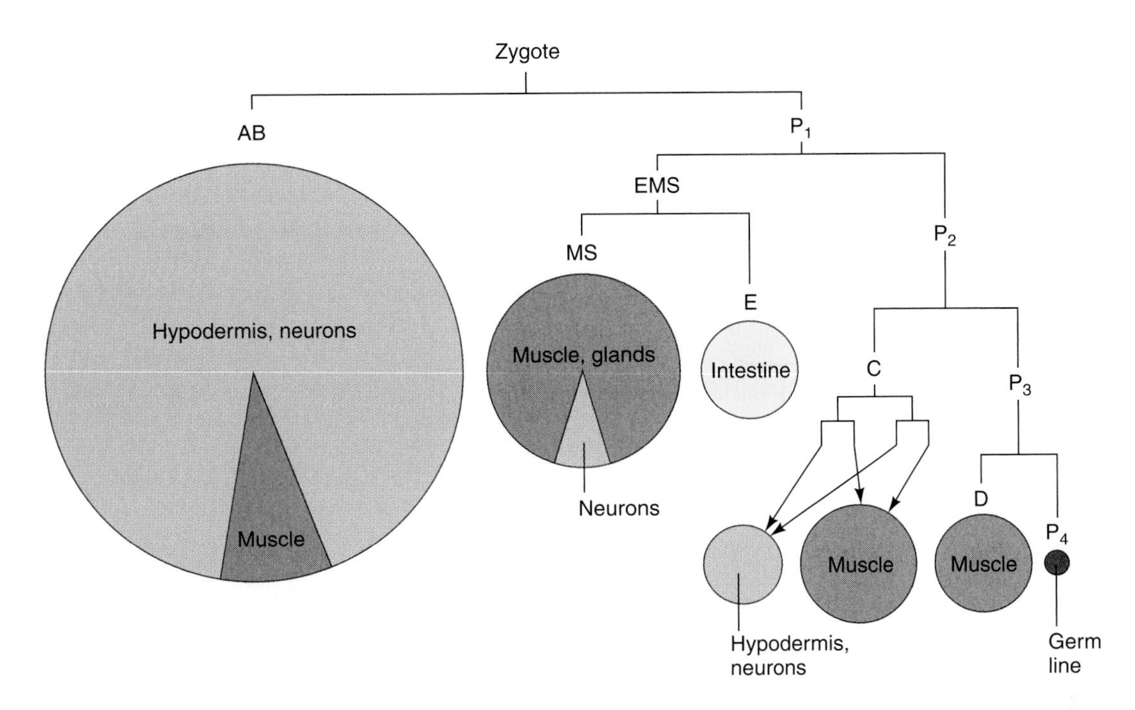

Figure 25.4 Generation of founder cells in *C. elegans,* and a summary of cell types derived from them. Areas of circles and sectors are proportional to numbers of cells. Founder cells E, D, and P_4 give rise to one cell type each, while all other founder cells have mixed progeny. Conversely, a given cell type, such as muscle, may originate from two or more founder cells.

between the gut and the outer cell layer, which forms the nervous system and the epidermis (called the hypodermis in roundworms).

Cell division and organogenesis continue until about 6 h after fertilization, when the hermaphrodite embryo consists of 558 cells and the male embryo of 560 cells. During the next 6 h, the embryo changes from a spheroid to an elongated shape as microfilaments and microtubules squeeze the embryo around its circumference (Priess and Hirsh, 1986). Subsequently, the worm's elongated shape is maintained by a collagenous cuticle that is secreted by the hypodermis. During the final 2 h of embryogenesis, the pharynx begins to pump and the eggshell is softened enzymatically so that the worm can hatch.

LARVAL DEVELOPMENT

After hatching, the larva takes 2 to 3 days to grow into an adult. During this period, the worm increases in length from 0.24 to 1.2 mm while molting and replacing its cuticle four times. The first molt is from the first larval stage (designated L1) to the second larval stage (L2), and the fourth molt is from L4 to the adult. From the time of hatching, the average life span is 18 days. However, under conditions of starvation or crowding, worms that would otherwise enter L3 can enter an alternative, longer-lived larval stage termed the ***dauer larva*** (Gk. *dauern,* "to last" or "persist"). Dauer larvae do not feed and do not move much except when attracted by food or water. They are resistant to desiccation and other harsh treatments and can survive for months. The principal signal for entry into the dauer stage is a pheromone, which is produced constitutively (Riddle and Albert, 1997). When the level of pheromone is high, indicating high population density, developing larvae will take the dauer, rather than the L3, route. When favorable conditions return, dauer larvae molt into L4's and then into adults.

During larval development, reproductive organs and other adult structures arise from a class of cells known as ***postembryonic blast cells.*** Figure 25.5 shows the

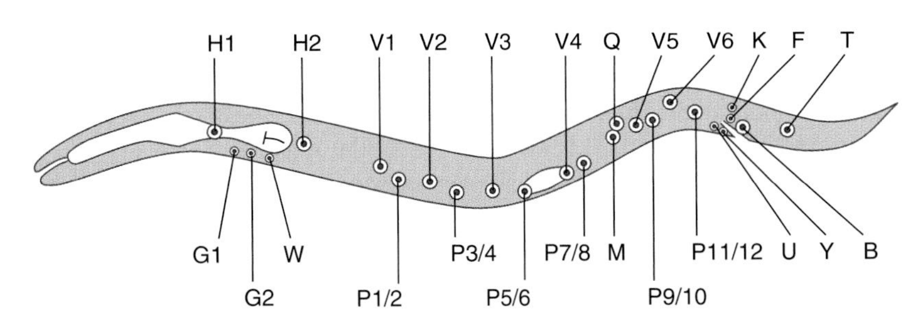

Figure 25.5 Stage L1 larva of *C. elegans* with postembryonic blast cells labeled. Most blast cells are paired, but only the left cell of each pair is shown. G1, G2, B, U, Y, and F are located medially, P1/2 through P11/12 originate bilaterally and then migrate ventrally, at which time they are designated in sequence P1, P2, . . . , P12. (These cells are unrelated to the embryonic cells P_1 through P_4.) B, U, Y, and F are found in both sexes but divide only in males.

positions of postembryonic blast cells in a young L1 larva. The further divisions of these blast cells are as precisely timed and oriented as the embryonic cell divisions. During postembryonic development, about 200 cells fuse into *syncytia* of various sizes, including the hypodermis and certain muscles.

25.2 Localization and Induction during Early Cleavage

The invariance of nematode development allows investigators to analyze how *single cells* are determined. As in other animals, the two major mechanisms of cell determination are cytoplasmic localization and embryonic induction. Both phenomena have been studied in *C. elegans* by experimental as well as genetic and molecular means.

THE FIRST CLEAVAGE GENERATES BLASTOMERES WITH DIFFERENT POTENTIALS

The first cleavage in *C. elegans* generates two blastomeres, AB and P_1. These two cells differ in size (AB is larger), timing of further cleavage (AB divides faster), and *fate* (see Fig. 25.4). AB and P_1 also differ in potency, as shown by isolation experiments in which one of the two blastomeres is destroyed inside the eggshell (Laufer et al., 1980) or removed through a hole made in the eggshell (Priess and Thompson, 1987). The surviving blastomere, which remains inside the eggshell, is then monitored for the types of daughter cells it generates. The cell types produced are identified by their division patterns, by natural markers such as the birefringent ***rhabditin*** granules that are characteristic of gut cells, and/or by *immunostaining* (see Method 4.1). Under these experimental conditions, an isolated P_1 blastomere gives rise to pharynx, body-wall muscle, and gut cells, whereas the descendants of isolated AB blastomeres form hypodermis cells and neurons.

What is the origin of these differences in cellular size, behavior, fate, and potency? Most likely this polarity entails an uneven distribution of cytoplasmic components that are partitioned asymmetrically during first cleavage and act as cytoplasmic determinants (Wood, 1988b; Strome, 1989; Seydoux and Strome, 1999). The nature of these determinants has been actively pursued by investigators using experimental as well as genetic strategies.

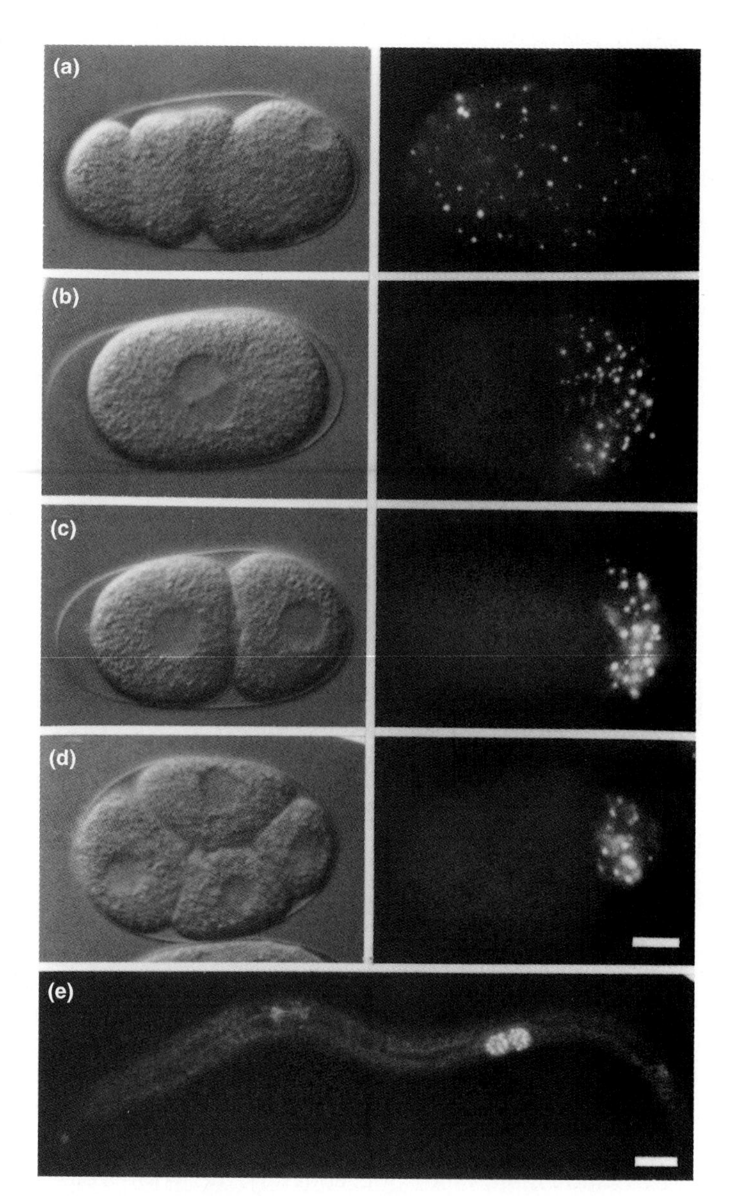

Figure 25.6 Segregation of P granules during early cleavage of *C. elegans.* The left panels show living embryos photographed with differential interference contrast; the right panels and bottom panel show fixed specimens with immunostained P granules (see Method 4.1). Anterior to left, posterior to right. **(a)** In the fertilized egg during pseudocleavage, P granules are dispersed throughout the cytoplasm. **(b)** After pronuclear migration, P granules are localized to the posterior cortex. **(c, d)** P granules are distributed to the P_1 cell of the 2-cell embryo and later to the P_2 cell of the 4-cell embryo. **(e)** P granules are densely packed in the two promordial germ cells of a newly hatched larva.

P GRANULES ARE SEGREGATED INTO GERM LINE CELLS

Much attention has focused on granules resembling the *polar granules* that are associated with germ cell determinants in *Drosophila* and other animals (see Section 8.3). In *C. elegans* embryos, these granules are known as ***P granules*** because they are segregated into the P cells during cleavage (Fig. 25.6). Later during development, P granules are present only in germ line cells except in mature sperm.

Possible mechanisms of P-granule segregation include cytoplasmic transport as well as selective stabilization at the posterior pole. To further investigate this process, Strome and Wood (1983) treated *C. elegans* embryos immediately after fertilization with microtubule inhibitors. These treatments blocked the migration of pronuclei, but the P granules nevertheless became localized near the posterior pole. By contrast, in embryos treated with microfilamemt inhibitors, the P granules coalesced near the center of the embryo instead of accumulating at the posterior pole. Also missing in these embryos were other signs of anteroposterior polarity, such as anterior contractions during pseudocleavage. Moreover, the pronuclei moved together but met in the center instead of posteriorly. All these disturbances occurred only when microfilaments were inhibited during a critical phase of the first cell cycle (Hill and Strome, 1990). These observations suggest that fertilization activates two cytoskeletal systems: a microtubule-based system that moves the pronuclei, and a microfilament-based system for transporting the P granules and generating other asymmetries in the zygote.

Genetic and molecular studies have revealed several P granule components that affect the synthesis and function of mRNA. One component, known as pgl-1, is an RNA-binding protein that is present on P granules at all stages of development. Loss of pgl-1 results in defective P granules and sterility (Kawasaki et al., 1998). Interestingly, pgl-1 function is required for fertility only at elevated temperatures, suggesting that germ line development is inherently sensitive to temperature. Another protein, pie-1, is associated with P granules until the 4-cell stage, after which it becomes prominent in the nuclei of germ line cells, where it inhibits transcription (Batchelder et al., 1999; Seydoux and Strome, 1999). Two other P granule proteins, designated glh-1 and glh-2 for *germ line helicase,* have the molecular characteristics of enzymes that bind to and unwind RNAs (Gruidl et al., 1996). The glh proteins of *C. elegans* are orthologous to the vasa protein, a known helicase and polar granule component of *Drosophila.*

Par GENES AFFECT CYTOPLASMIC LOCALIZATION

Mutagenesis screens for maternal-effect lethal mutations have revealed six genes designated *par-1*$^+$ through *par-6*$^+$ for defective *par*titioning of cytoplasmic components during early cleavage (*S. Guo and K. Kemphues, 1996*). Eggs from homozygous *par* mutant hermaphrodites show abnormalities in cytoplasmic streaming, pseudocleavage, positioning of the first mitotic spindle, cell division timing, and the segregation of P granules. While the phenotypes of *par* mutants are most obvious during the first cell cycle, they also include mislocalizations of other regulatory proteins that affect cell lineage development at subsequent stages (Bowerman et al., 1997).

The par proteins form a complex network of interactions, and several of the par proteins are localized in the early embryo. For example, the par-2 protein is restricted to the posterior cortex of the zygote and becomes partitioned into the P_1 blastomere during the first cleavage (Boyd et al., 1996). Subsequently, the par-2 protein exhibits a similar asymmetrical localization in P_1, P_2, and P_3. Proper localization of the par-2 protein depends on par-3 activity, while in turn the distribution of par-3

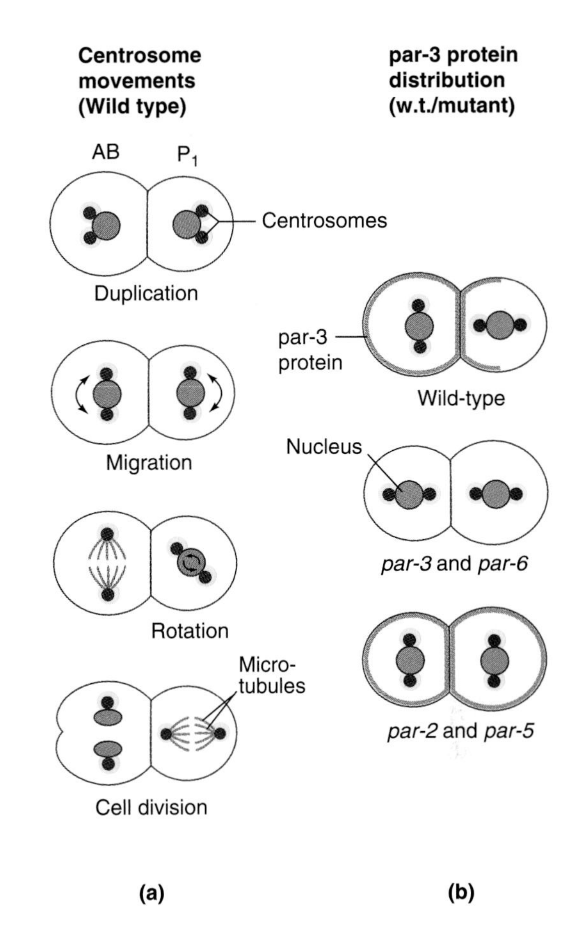

Figure 25.7 Mitotic spindle orientation and par-3 protein distribution in wild-type and mutant *C. elegans* embryos. Anterior is to the left. **(a)** In wild-type embryos, centrosomes duplicate and migrate along the nuclear envelope in both AB and P_1. Only in P_1, the centrosome nuclear complex subsequently rotates 90° to become aligned with the anteroposterior axis. (Astral microtubules are not shown.)
(b) Centrosome position and distribution of par-3 protein in wild-type and various *par* mutant embryos.

depends on par-2. Embryos from hermaphrodites deficient in *par-2* fail to exclude the par-3 protein from the cortex of the posterior half of the P_1 blastomere (Fig. 25.7b). Apparently as a result, a particular rotation of the nucleus with adjacent centrosomes fails to occur in the P_1 blastomere. During the subsequent mitosis of P_1, the mitotic spindle remains perpendicular to the anteroposterior axis, and the P granules are not partitioned into the P_2 blastomere. Thus, the par proteins affect the two key processes in cytoplasmic localization, the uneven distribution of cytoplasmic components and the orientations of cleavage planes.

THREE MATERNAL EFFECT GENES GENERATE A LOCALIZED ACTIVITY THAT PROMOTES EMS IDENTITY

The pharynx of *C. elegans* is formed by descendants of two cell lineages (Fig. 25.8). One lineage, which is derived via ABa from AB and gives rise to the anterior pharynx, will be discussed later. The other lineage,

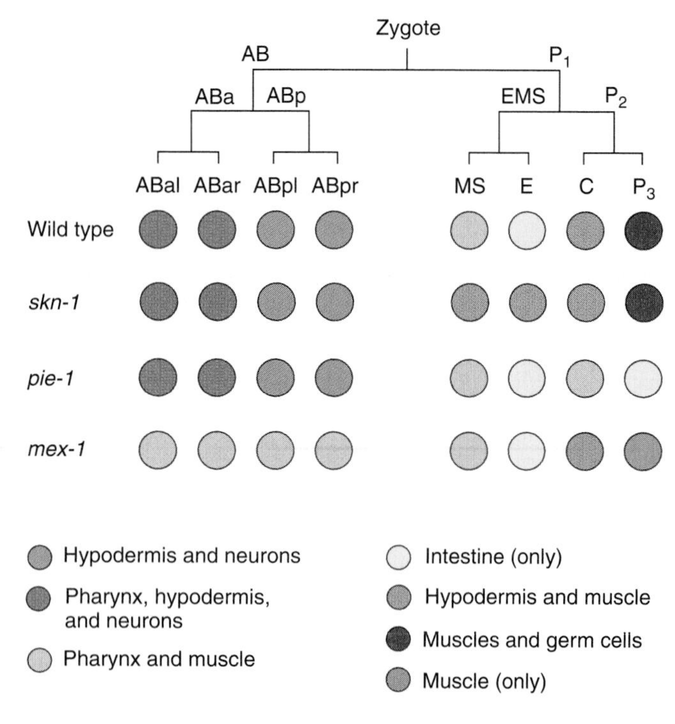

Figure 25.8 Altered blastomere potencies in offspring of hermaphrodites with mutations in the maternal effect genes *pie-1, mex-1,* or *skn-1.* The diagram shows the lineage and names of all *C. elegans* blastomeres up to the 8-cell stage. For instance, ABpr is the right daughter of the posterior daughter of AB. The color-coded circles indicate the types of cells produced by each blastomere after all other blastomeres were killed with a microbeam. In *skn-1* offspring, both EMS descendants develop like additional C founder cells, forming hypodermis and muscle. In *pie-1* offspring, P_2 is transformed into another EMS, so that the E and MS lineages are duplicated in the posterior. In *mex-1* offspring, all four AB granddaughters are transformed into extra MS blastomeres, so that an excess of pharyngeal and muscle cells is formed in the anterior. In addition, P_3 forms only muscle cells.

which forms the posterior pharynx, is derived via MS and EMS from P_1. These cells can give rise to pharyngeal cells in isolation—that is, without inductive interactions. P_1 passes on the ability to form pharynx autonomously to one of its daughter cells, EMS, which in turn passes it on to MS. Thus, some factor(s) sufficient for autonomous pharyngeal cell development must be distributed or activated asymmetrically during the first three cleavage divisions.

A candidate for conferring cell-autonomous pharynx-forming ability is the product of the *skn-1*$^+$ gene, which was discovered in a screen for maternal effect mutants that lack pharyngeal cells (Bowerman et al., 1992). In offspring from *skn-1* mutant hermaphrodites, the EMS descendants behave like C blastomeres, forming hypodermis and body-wall muscle instead of producing pharyngeal, intestinal, and muscle cells as they do in normal embryos (Fig. 25.8). The skn-1 protein has the molecular characteristics of a transcription factor and is therefore likely to control the transcription of embryonic genes that support posterior pharyngeal cell development. For instance, in the E lineage, skn-1 seems to activate the *end-1*$^+$ gene, which is sufficient to determine endoderm cells (J. Zhu et al., 1998).

Immunostaining (see Method 4.1) of wild-type embryos has shown that both P_1 and AB initially contain low levels of nuclear skn-1 protein (Bowerman et al., 1993). Higher levels then accumulate in P_1 and its descendants through the 8-cell stage (Fig. 25.9). Little or no skn-1 is present in ABa or ABp, and the protein is no longer detectable in any nuclei by the 12-cell stage.

The restriction of skn-1 protein to P_1 and its descendants depends on the function of other maternal effect genes, including the *par-1*$^+$ gene mentioned previously and the *mex-1*$^+$ gene. In embryos from *par-1* or *mex-1* mothers, all cell nuclei contain levels of skn-1 protein that are intermediate between the skn-1 levels of the P_1 and AB lineages in wild-type embryos (Fig. 25.9). This observation suggests that the par-1 and mex-1 products are required for the localization of skn-1 mRNA or protein.

Not only is the skn-1 protein necessary for the ability of the EMS descendants to form posterior pharynx and intestine, but it also seems *sufficient* to convert other cells to form pharyngeal and muscle cells. This is suggested by the phenotypes of embryos from hermaphrodites deficient in the maternal effect gene *mex-1*$^+$ (Mello et al., 1992). Such embryos produce too many pharyngeal and muscle cells (*mex* stands for muscle excess). Lineage analysis shows that the extra cells are formed because some or all AB descendants at the 8-cell stage deviate from their fates and develop like MS blastomeres (see Fig. 25.8). The ability of transformed blastomeres to develop like MS cells depends on the presence of the skn-1 gene product. (Offspring from hermaphrodites with both *mex-1* and *skn-1* loss of function do not develop extra pharyngeal muscle.) Because skn-1 is mislocalized to the AB descendants in offspring of *mex-1* hermaphrodites, it appears that the ectopic presence of skn-1 protein in these cells is sufficient to convert them into MS-like cells.

The *skn-1* mutant phenotype is limited to the EMS lineage and does not include any descendants of P_2. The apparent restriction of skn-1 protein activity to EMS requires the function of yet another maternally expressed gene, *pie-1*$^+$, which generally inhibits transcription in germ line blastomeres, as discussed earlier. *Pie* stands for *p*haryngeal and *i*ntestinal *e*xcess, because the loss of *pie-1* function converts P_2 into the likeness of another EMS, giving rise to extra pharyngeal and intestinal cells (see Fig. 25.8). This conversion depends on the maternal skn-1 function, as in the case of the converted AB descendants in mutants lacking *mex-1* function. However, pie-1 does not function to localize skn-1, since skn-1 is present in P_2 in wild type as well as *pie-1* mutant embryos. Thus, pie-1 must prevent skn-1 from promoting EMS fate in P_2 by antagonizing its action.

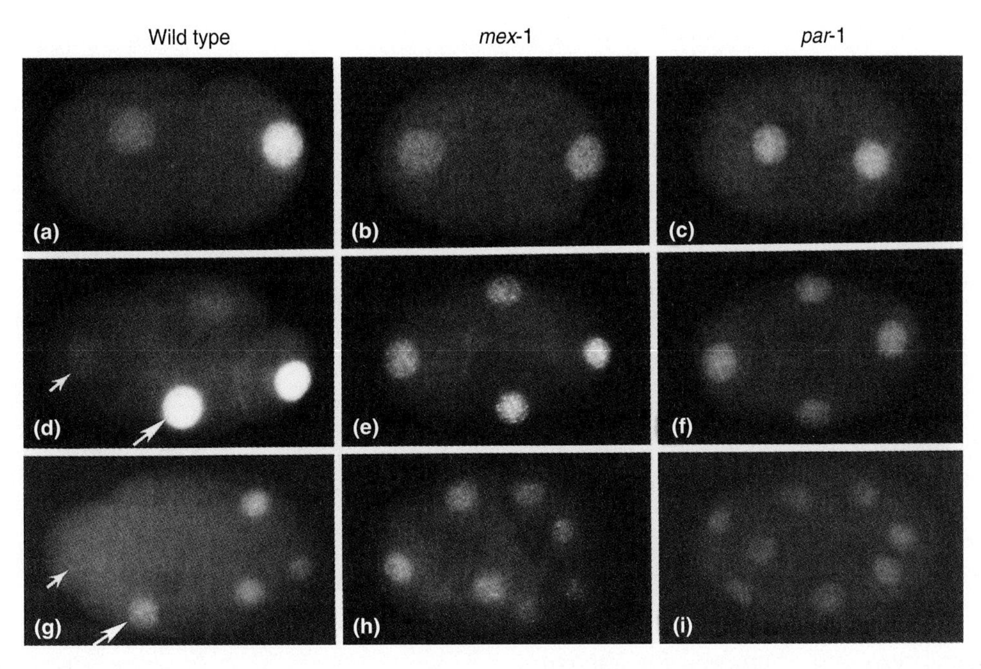

Figure 25.9 Accumulation of skn-1 protein in wild-type and *mex-1* or *par-1* mutant embryos of *C. elegans.* The top, middle, and bottom rows show 2-cell, 4-cell, and 8-cell embryos, respectively. Anterior is to the left. All embryos were immunostained for skn-1 protein (see Method 4.1). **(a–c)** Wild-type embryos. The skn-1 protein accumulates in the nuclei of P_1 and its descendants. At the 4-cell stage, there is a marked difference between skn-1 levels in P_2 and EMS (large arrow) as compared with ABa (small arrow) and ABp (top). **(d–f)** Embryos from *mex-1* mutant mothers and **(g–i)** embryos from *par-1* mutant mothers. In both mutants, all nuclei of an embryo contain about the same level of skn-1 protein.

Taken together, the data indicate that the maternally supplied skn-1 protein is necessary and sufficient to specify the capability of the EMS blastomere to produce posterior pharyngeal cells. In this sense, the skn-1 protein acts like a localized cytoplasmic determinant for EMS cell formation. However, it is really the combined activity of the maternal skn-1, mex-1, and pie-1 proteins that determines EMS.

DETERMINATION OF ANTERIOR PHARYNGEAL MUSCLE CELLS REQUIRES MULTIPLE INDUCTIVE INTERACTIONS

The need for cellular interaction in anterior pharyngeal muscle development was revealed in a set of experiments on the formation of anterior pharyngeal muscle cells (Priess and Thompson, 1987). From lineage analysis it was known that ABa generates anterior pharyngeal muscle, while EMS gives rise to posterior pharyngeal muscle. To learn whether each subset of pharyngeal muscle cells arises independently of the other, the investigators removed specific blastomeres at the 2-cell and 4-cell stages and monitored the development of pharyngeal muscle in the remainder of the embryo. Their results are summarized in Table 25.1.

table 25.1 Development of Pharyngeal Muscle Cells in *C. elegans* Embryos after Removal of Individual Blastomeres at the 2-Cell Stage or the 4-Cell Stage

Removed Cell	Cells Remaining in Eggshell	Development of Pharyngeal Muscle Cells	
		Anterior	Posterior
—	AB, P_1	+	+
P_1	AB	–	–
AB	P_1	–	+
—	ABa, ABp, EMS, P_2	+	+
EMS	ABa, ABp, P_2	–	–
P_2	ABa, ABp, EMs	+	+
ABp	ABa, EMS, P_2	+	+
ABa	ABp, EMS, P_2	+	+

From data of Priess and Thompson (1987). Used with permission.

INDIVIDUAL blastomeres were removed by puncturing the egg with a fine glass needle and applying pressure to extrude the unwanted blastomeres through the opening in the eggshell. After removal of AB at the 2-cell stage, the remaining P_1 blastomere generated pharyngeal muscle, as revealed by immunostaining (see Method 4.1). Conversely, if the P_1 blastomere was extruded, the remaining AB blastomere did not produce pharyngeal muscle. This result showed that P_1 or some of its descendants were necessary for cells in the AB progeny to form anterior pharyngeal muscle. To determine which daughter cell of P_1 fulfilled this role, the researchers removed either P_2 or EMS at the 4-cell stage. Without EMS, the remaining embryo (ABa, ABp, and P_2) did not form pharyngeal muscle. In contrast, after removal of P_2, the remaining embryo (ABa, ABp, and EMS) formed anterior as well as posterior pharyngeal muscle cells. These results showed that EMS, in addition to forming posterior pharyngeal cells itself, induces part of the AB progeny to form anterior pharyngeal cells.

To see whether only ABa—which normally produces anterior pharyngeal muscle cells—is competent to be induced by EMS the investigators removed ABa. In this case, the anterior pharyngeal cells were formed by ABp. Moreover, if ABa and ABp were exchanged inside the eggshell with a blunt needle, the embryos proceeded to develop normally, with the original ABp blastomere (now in the ABa position) giving rise to anterior pharyngeal muscle.

Questions

1. The technique of Priess and Thompson for removing blastomeres is quite laborious. Killing unwanted cells with a laser microbeam is a much quicker and faster operation, but the dead cells stay within the eggshell. Could this present a problem for the interpretation of certain results?
2. In the experiments described above, blastomeres ABa and ABp were equally competent to form pharyngeal cells. What can be concluded from this result, given that only ABa produces pharyngeal cells in normal development?

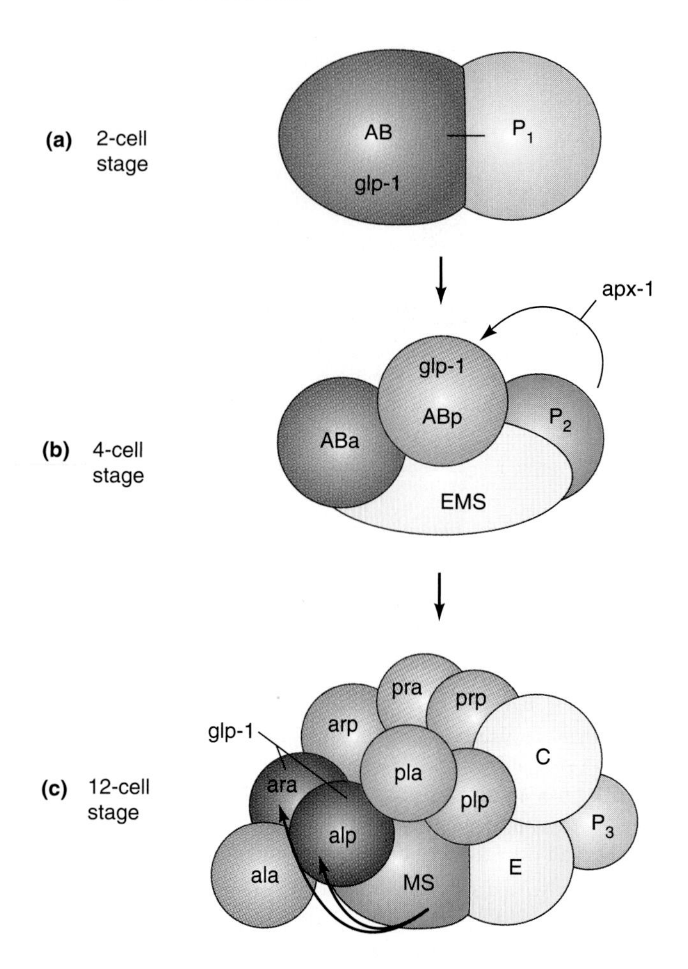

Figure 25.10 Determination of anterior pharyngeal cells in *C. elegans* involves an initial step of cytoplasmic localization followed by inductive interactions. **(a)** At the 2-cell stage, selective translation of maternal glp-1 mRNA in the AB blastomere renders AB, but not P_1, competent to form anterior pharyngeal cells. **(b)** At the 4-cell stage, a non-diffusible signal protein present on the P_2 blastomere and encoded by the *apx-1*$^+$ gene acts on the glp-1 receptor present on the neighboring Abp blastomere. This signal leaves ABp incapable of forming anterior pharyngeal cells. **(c)** At the 12-cell stage, the MS blastomere sends an inducing signal to two granddaughters of ABa. These cells, designated ara for anterior daughter of the right daughter of ABa (AB omitted for brevity) and alp (posterior daughter of the left daughter of ABa) will eventually give rise to anterior pharyngeal cells. The receptor for this signal is again glp-1.

The experiments of Priess and Thompson (1987) indicate that ABa and ABp are born with equivalent potency, so that the restriction of anterior pharyngeal muscle formation to certain descendants of ABa must be ascribed to interactions with other blastomeres, including a positive inductive signal from EMS or some of its descendants. The genetic analysis of these interactions has revealed three initial steps (Fig. 25.10; *Bowerman, 1995*).

The first step in specifying the anterior pharyngeal cell lineage is part of establishing the anteroposterior polarity, which has already been discussed (see Fig. 25.7). This step renders AB, but not P_1, competent to form anterior pharyngeal cells. A key gene in this step, *glp-1*$^+$, was originally discovered in a screen for mutations affecting germ line development (*glp* stands for *g*erm *l*ine *p*roliferation). *glp-1*$^+$ encodes a member of the well-conserved Notch family of transmembrane receptors, which are involved in cell determination in a wide range of metazoans (see Sections 6.4, 13.1, and 14.2). Maternally provided glp-1 mRNA is distributed evenly throughout the early *C. elegans* embryos but is translated into protein only in AB and its descendants (Fig. 25.10a). The asymmetrical translation is mediated by a *cis*-control element in the 3′ UTR of the glp-1 mRNA (Evans et al., 1994). The asymmetry of glp-1 protein synthesis depends on other maternal gene products involved in setting up the anteroposterior polarity, such as par-1 and par-6 (Crittenden, 1997).

The second step in specifying the anterior pharyngeal cell lineage breaks the initial equivalence in potency between ABa and ABp, leaving only ABa capable of forming anterior pharyngeal cells (Fig. 25.10b). Two genes, *glp-1*$^+$ and *apx-1*$^+$, are critically involved. The absence of glp-1 protein converts the ABp lineage into another ABa lineage (Moskowitz et al., 1994). The same effect is observed after removal of P_2 and by preventing P_2 from touching ABp (Mello et al., 1994). Since glp-1 protein is not normally synthesized in P_2 it follows that P_2 is the source of a signal acting on the receptor encoded by *glp-1*$^+$ and deployed in both ABa and ABp (Fig. 25.10b). This signal from P_2 seems to be encoded by a gene named *apx-1*$^+$ for *a*nterior phary*nx* e*x*cess. Embryos lacking *apx-1* function produce an excess of ABa-specific cells at the expense of ABp-specific cells (Mango et al., 1994). The apx-1 protein is similar in sequence to the *Drosophila* protein Delta, which encodes the nondiffusible ligand of the Notch receptor. The activation of the glp-1 receptor by apx-1 determines ABp, rendering it incapable of responding to the later signal from MS.

The third step in specifying the anterior pharyngeal cell lineage is an inducing signal from the MS blastomere to two granddaughters of ABa (Fig. 25.10c). The recipient cells, designated ABara (anterior daughter of the right daughter of ABa) and ABalp (posterior daughter of the left daughter of ABa), which are precursors of the anterior pharyngeal cells. The receptor and signal involved in this induction are again members of the Notch and Delta families, respectively (Hutter and Schnabel, 1995). The same types of receptor and signal are also involved in subsequent interactions among descendants of ABa that provide another demonstration of the common usage of Notch-type signaling in cell determination (Moskowitz and Rothman, 1996).

In summary, the determination of anterior pharyngeal cells in *C. elegans* consists of a series of steps, based on cytoplasmic localization followed by inhibitory as well as activating inductions. The cell constancy and genetic advantages of the worm have made it possible to analyze each of these steps as interactions between single cells. The results obtained with *C. elegans* confirm the concept of cell determination as a stepwise process, which has emerged from earlier studies on other organisms (see Sections 6.5, 12.4, 16.6, and 20.4).

P_2 POLARIZES EMS TO FORM UNEQUAL DAUGHTER CELLS

Another case of cell determination in *C. elegans* is of particular interest because it shows that an inductive interaction may polarize the recipient cell so that it will produce unequal daughter cells during its subsequent division. The EMS blastomere becomes polarized by an interaction with the adjoining P_2 blastomere. During the following division of EMS, the mitotic spindle is oriented perpendicular to the plane of contact between EMS and P_2. The EMS daughter forming next to P_2 becomes the E founder cell, which gives rise exclusively to endoderm. Its sister cell, which is distant from P_2, becomes the MS founder cell, which produces mostly mesoderm (see Fig. 25.4).

To analyze the conditions under which EMS blastomeres would give rise to E founder cells, Goldstein (1992) removed the eggshells from cleavage-stage embryos, separated the blastomeres by forcing embryos through a narrow pipette, and monitored their development in culture. In particular, he assayed the descendants of isolated EMS blastomeres for the ability to form *rhabditin* granules, which are characteristic of gut cells. He found that this ability depended on the time of blastomere separation: EMS blastomeres isolated during the first half of the 4-cell stage never gave rise to gut cells, whereas EMS blastomeres isolated later during the 4-cell stage did (Fig. 25.11a). In cases in which no gut cells formed, both EMS daughters divided in synchrony and with the rhythm characteristic of MS blastomeres, whereas normally the E divides more slowly than MS.

These results could be interpreted two ways. First, the determination of E as the gut founder cell may depend on a previous interaction of EMS with one of its neighbors. Second, the failure of early isolated EMS blastomeres to produce gut cells may reflect a greater sensitivity of younger embryos to damage from handling. To decide which interpretation was correct, Goldstein (1992) recombined early isolated EMS blastomeres with one or two AB daughters or with P_2 (Fig. 25.11b, c). Recombination with P_2, but not with AB daughters, restored the ability of early isolated EMS blastomeres to produce gut cells. This result rules out the sensitivity hypothesis and confirms that the formation of gut cells by isolated EMS cells requires extended contact between EMS and P_2.

In the normal embryo, the P_2 blastomere sits at the posterior end (the "E" end) of EMS. The experiments described above suggest that P_2 polarizes EMS to divide asymmetrically so that the daughter blastomere next to P_2 becomes the gut founder cell E. To test this hypothesis, Goldstein (1993) placed P_2 in random positions on an isolated EMS. He noticed that the two cells stuck together upon contact and did not move relative to each other thereafter. The subsequent cleavage of EMS was oriented so that it produced one daughter cell adjoining to P_2 and the other daughter cell pointing away from P_2. Gut cells consistently differentiated from the EMS daughter that was contacting P_2, whereas the other EMS daughter acquired an MS fate. These results showed that the interaction with P_2 had a dual effect on EMS. First, it oriented the mitotic spindle perpendicular to the plane of contact between EMS and P_2. Second, the contact caused an uneven distribution of extranuclear components along the axis of the mitotic spindle. As part of this polarization, E-determining factors could become segregated into the EMS daughter located next to P_2. Alternatively, E-inhibiting factors might be segregated away from the point of contact with P_2.

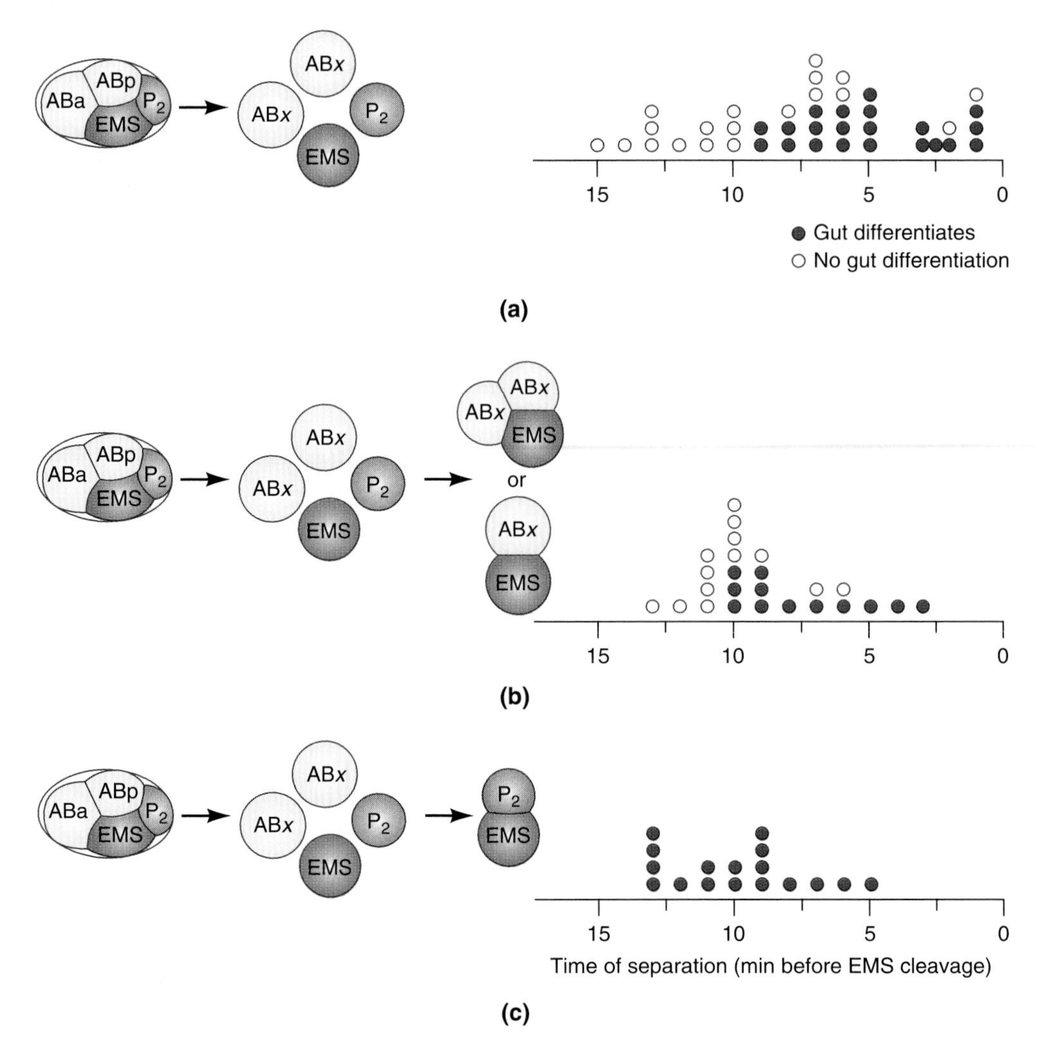

Figure 25.11 Development of gut cells from isolated and recombined EMS blastomeres of *C. elegans.* The experiments are diagrammed to the left, and the results are tallied to the right. Each green circle to the right represents an experiment in which an isolated EMS cell gave rise to gut cells, as indicated by the presence of rhabditin granules. Each open circle represents an experiment in which an isolated EMS cell survived but did not form gut cells. The circles are stacked over a time axis that indicates how much time elapsed between isolation of an EMS cell and its cleavage into E and MS daughter cells. **(a)** Cell isolation only. **(b)** Cell isolation and subsequent recombination (within 1–2 min) of the EMS cell with ABa or ABp or both. **(c)** Cell isolation and subsequent recombination (within 1–2 min) of the EMS cell with P_2. Note that EMS cells isolated early did not form gut cells unless they were recombined with P_2.

These experiments were extended by Schlesinger and coworkers (1999), who confirmed earlier observations of other investigators on wild-type embryos that, after division of the EMS centrosome, the two daughter centrosomes originally align *perpendicular* to the embryo's anteroposterior axis (Fig. 25.12b). However, shortly before mitosis, the EMS nucleus and centrosomes rotate 90°, so that the ensuing mitotic spindle will form *parallel* to the anteroposterior axis. They also confirmed Goldstein's (1995) observation that isolated EMS cells in culture establish their mitotic spindle along the axis defined by the centrosome positions after their initial migration, without subsequent rotation. However, contact with P_2 early in the EMS cell cycle induces rotation of the EMS nucleus and centrosomes, orienting the mitotic spindle perpendicular to the plane of contact between P_2 and EMS (Fig. 25.12d).

Schlesinger and coworkers (1999) also knew from previous studies by other investigators that the unequal division of EMS into E and MS depends on several genes encoding components of the *Wnt signaling pathway* (see Section 9.7). In particular, embryos derived from hermaphrodites lacking the function of any of five *mom*$^+$ genes do not form a gut because both EMS daughters develop as MS cells (Thorpe et al., 1997). Conversely, both EMS daughters develop as E cells in embryos from hermaphrodites lacking a functional *pop-1*$^+$ gene (R. Lin et al., 1995).

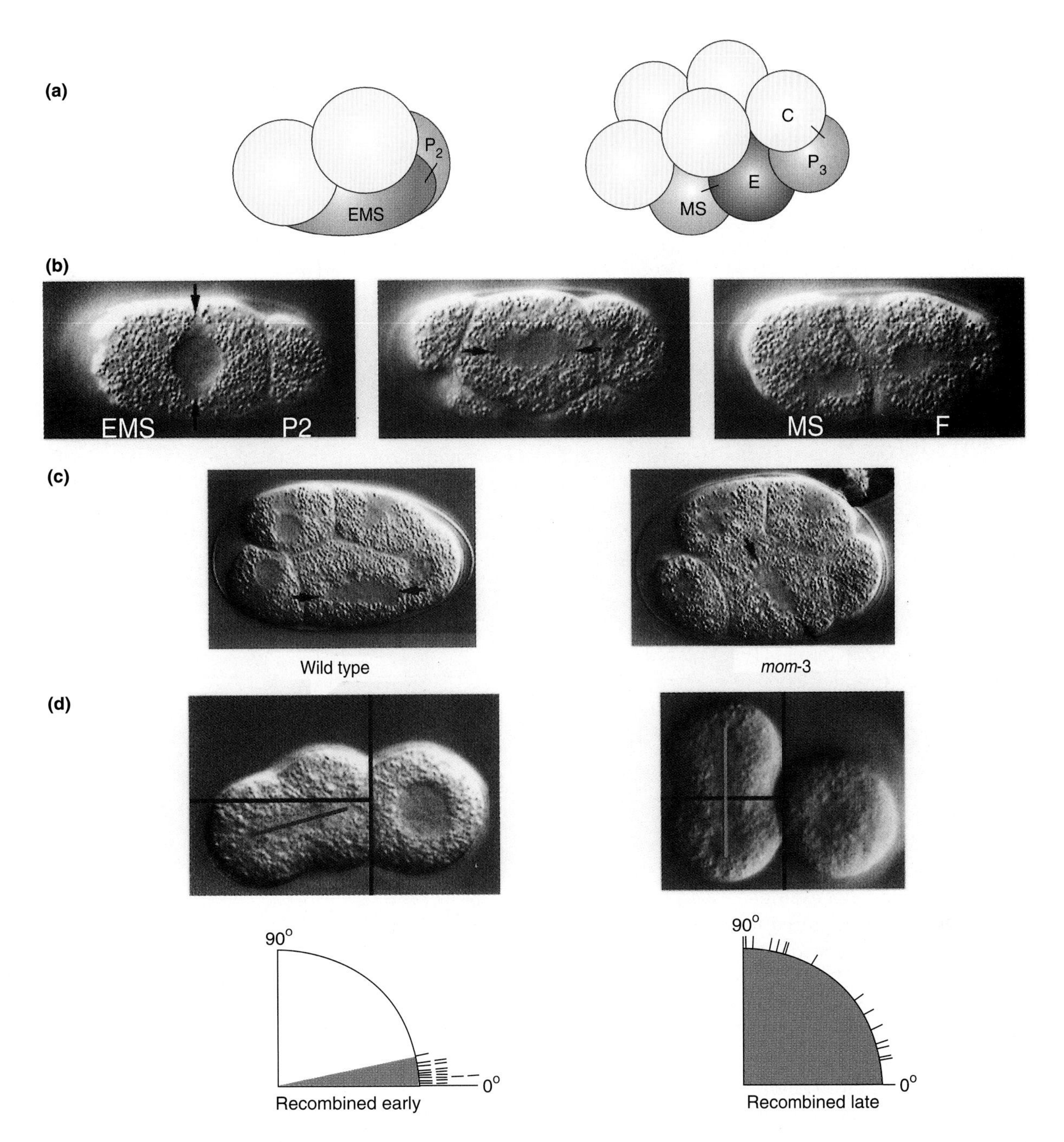

Figure 25.12 Spindle orientation during mitosis of the EMS blastomere in *C. elegans.* Anterior to the left. **(a)** Schematic lateral views of normal embryos at the 4-cell and 8-cell stage. Sister cells connected by short lines. After division of EMS, only the E daughter is in contact with P_2 or its descendants, P_3 and C. **(b)** Ventral views of live wild-type embryos at the 4-cell stage, photographed with differential interference contrast. EMS centrosomes, marked with arrows, are visible due to their exclusion of cytoplasmic yolk granules. (*Left*) The centrosomes are initially oriented along the right-left axis. (*Middle*) After rotation of the centrosome-nucleus complex through 90°, the centrosomes lie along the anteroposterior axis. (*Right*) After cytokinesis, the E cell is located next to P_2. **(c)** Lateral views of live embryos during mitosis of EMS. In the wild-type embryo, the spindle is oriented along the anteroposterior axis. In the embryo derived from a *mom-3* mutant hermaphrodite, the spindle is oriented incorrectly. **(d)** Isolated wild-type blastomeres in culture. EMS and P_2 were separated immediate after division of their mother cell (P_2) and then recombined early or late in the cell cycle of EMS. The angle between the mitotic spindle axis of EMS and a line perpendicular to the plane of contact between EMS and P_2 was measured. After early recombination, the angle was consistently small. After late recombination, the angle varied randomly between 0 and 90°.

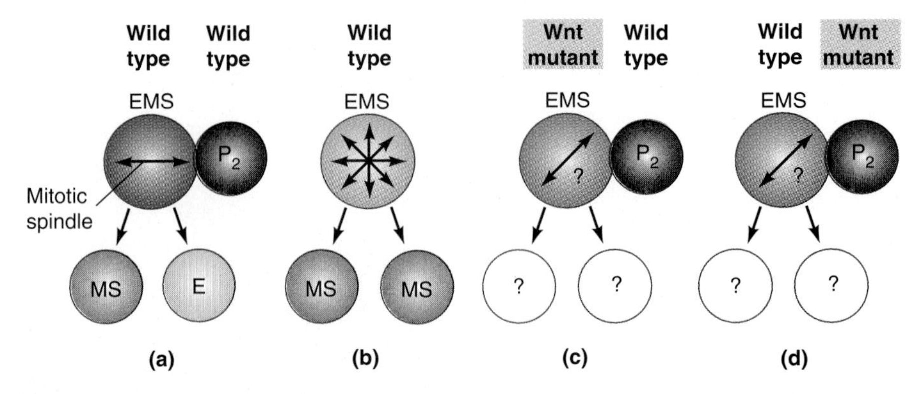

Figure 25.13 Genetic analysis of EMS polarization by its sister cell, P_2, in vitro. **(a)** EMS and P_2 blastomeres are isolated from embryos derived from wild-type hermaphrodites and recombined early during the cell cycle of EMS. The ensuing mitotic spindle of EMS is oriented perpendicular to the plane of contact between EMS and P_2. One daughter cell develops as an E blastomere, forming endoderm, while its sister becomes an MS blastomere and makes mesoderm. **(b)** In an isolated EMS from a wild-type hermaphrodite without an attached P_2, the mitotic spindle is oriented randomly, and both daughter cells become MS cells. **(c, d)** Pairs of blastomeres are formed as described in part a, except that one partner is derived from a hermaphrodite homozygous mutant for a maternally expressed gene encoding a component of the Wnt signaling pathway. The EMS cell is monitored for spindle orientation and the presence of endoderm cells among its descendants. See text.

FOR a genetic analysis of the signal(s) causing EMS polarization, Schlesinger and coworkers (1999) placed together EMS and P_2 cells in culture, with one of the two cells deficient for one of the maternal gene products required for the Wnt signaling pathway (Fig. 25.13). In each combination, the EMS cell was monitored for the orientation of its developing mitotic spindle, and for the fate of its daughter cells. These experiments revealed which gene product was required, in which of the two cells, for correct spindle orientation and/or for the asymmetrical division of EMS into one E cell and one MS cell. For example, *mom-2*$^+$, which encodes the Wnt signal, was required in P_2 but not in EMS. Conversely, *mom-5*$^+$, which encodes the Wnt receptor, was required in EMS but not in P_2. Both these genes were necessary for spindle orientation as well as for asymmetrical EMS division. In contrast, *wrm-1*$^+$, one of the *C. elegans* homologs of β-catenin, was required in EMS and only for asymmetrical division, not for spindle orientation.

The investigators assumed that the "canonical" Wnt signaling pathway, which is well established in *Drosophila* and many vertebrates (Cadigan and Nusse, 1997), applies to *C. elegans*. They also used existing sequence information to establish which of the *C. elegans* genes affecting EMS development were orthologous to canonical Wnt genes. Adding to this body of information, they identified a *C. elegans* ortholog to *gsk3*$^+$, one of the known Wnt signaling genes of other organisms. From this information, they pieced together a branched signaling pathway for the polarizing induction of EMS by P_2 (Fig. 25.14).

Next, Schlesinger and coworkers (1999) performed their standard experiment with wild-type EMS and P_2 cells in the presence of *actinomycin D*, a known inhibitor of transcription. Remarkably, they observed perfectly normal spindle orientation. They obtained the same result after a procedure that selectively inhibits the synthesis of RNA

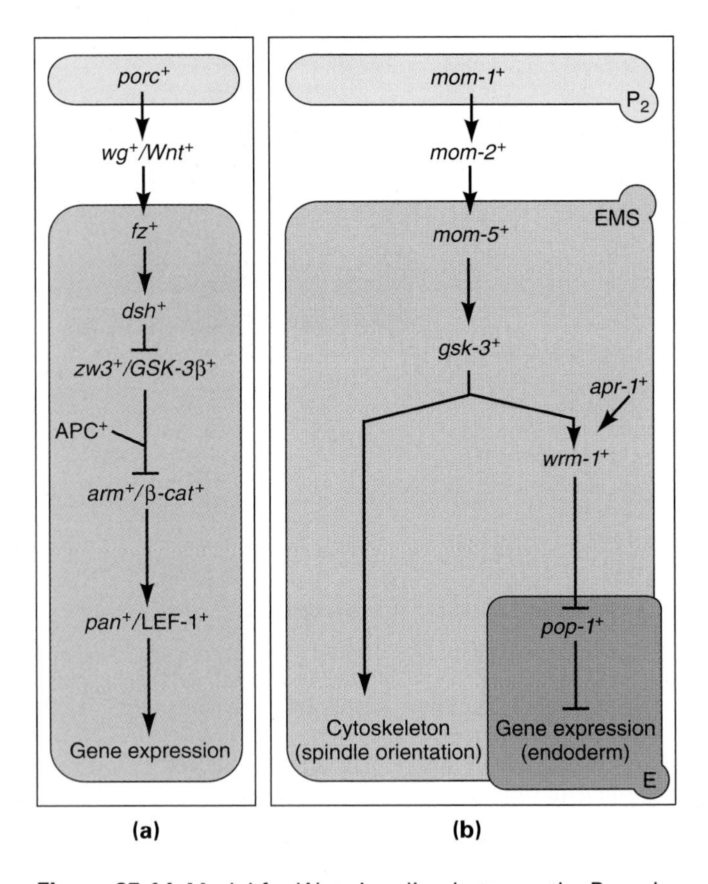

Figure 25.14 Model for Wnt signaling between the P_2 and EMS blastomeres in *C. elegans.* (a) The "canonical" Wnt signaling pathway based on work with mammals and *Drosophila* (Cadigan and Nusse, 1997). (b) Proposed signaling pathway in *C. elegans,* based on the genetic analysis illustrated in Figure 25.13 and on sequence similarities between *C. elegans* genes and their apparent orthologs in other organisms. See text.

polymerase II, the key enzyme of mRNA synthesis. These data show that the branch of the Wnt signaling pathway that controls spindle orientation works without the expression of any embryonic genes.

Questions

1. What did the investigators have to measure in order to quantify the extent to which the mitotic spindle in an EMS cell was oriented by its contact with P_2?
2. What kind of internal control could the investigators have used (and they did) to convince themselves that their methods for inhibiting mRNA synthesis were indeed working? (Hint: It is feasible to introduce any transgene into *C. elegans*.)
3. If an isolated P_2 is recombined with ABa or ABp, no spindle orientation is observed, and no E cells are formed. Thus, it appears that only EMS is competent to respond to the inductive signal from P_2 that generates the E blastomere. What gene product discussed earlier would be a good candidate for conferring this competence to EMS? How could this hypothesis be tested?

Taken together, the experiments described above have shown that the P_2 cell of the *C. elegans* embryo polarizes the EMS cell with regard to mitotic spindle orientation and uneven distribution of extranuclear determinants. Both effects are mediated by the Wnt signaling pathway. Remarkably, the orientation of the spindle can occur without gene transcription, and requires the Wnt pathway components only through gsk-3. This means that this abbreviated pathway can act directly on cytoplasmic components involved in spindle orientation. It will be interesting to find out which cytoplasmic components are the targets of this direct action.

25.3 Heterochronic Genes

Our previous discussions of pattern formation centered on the *spatial* aspect of embryonic development and its genetic control. For example, one of the major questions about segmentation genes and homeotic genes in *Drosophila* was how their expression is limited to certain spatial domains. Here we will focus on the *timing* of development. Timing is critical for the contacts generated during the divisions of neighboring cells, for morphogenetic movements such as gastrulation, and for life cycle events like larval molts or pupation. Changes in the timing of such events are thought to underlie many steps in evolution (Gould, 1977).

In *C. elegans,* each cell division occurs at a particular time in development, and every cell lineage is characterized by the timing as well as the orientation and symmetry of cell divisions. Like other aspects of development, the temporal sequence of events is under genetic control. This is revealed by the existence of *heterochronic mutations,* which cause cells to undergo patterns of cell division and differentiation that are normally observed at a different stage (Ambros and Horvitz, 1984).

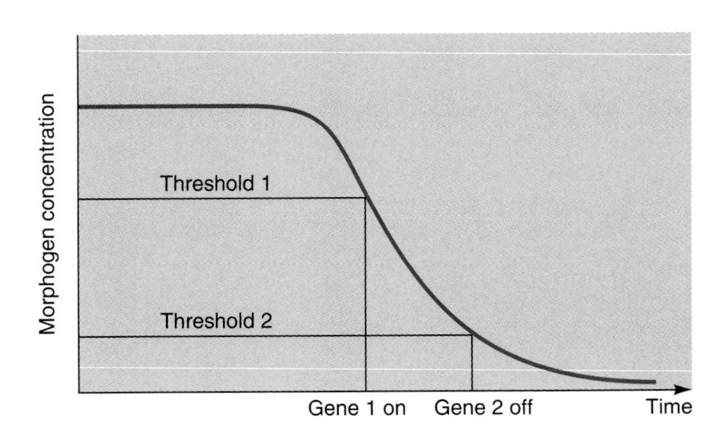

Figure 25.15 Model of a temporal morphogen gradient. The concentration of a morphogenetic signal produced in a given cell or embryonic region varies over time. At certain times during development, the concentration of the signal reaches threshold values at which genes are switched on or off depending on the current signal concentration.

Heterochronic mutations are remarkably similar to homeotic mutations. In our definition of heterochronic mutations, one only needs to replace "heterochronic" with "homeotic" and "stage" with "place" to obtain the correct definition of a homeotic mutation. We can extend this parallel by saying that *morphogen gradients* may exist in time as well as in space. A ***temporal gradient*** of morphogenetic signal would activate different target genes according to its *current* concentration rather than according to its *local* concentration as a spatial gradient does (Fig. 25.15; see Fig. 21.17 for comparison). Thus, the temporal and spatial dimensions of development may be controlled by similar genetic mechanisms.

MUTATIONS IN THE *lin-14*$^+$ GENE ARE HETEROCHRONIC

The *lin-14*$^+$ gene of *C. elegans* plays a central role in the development of hypodermal and intestinal cells as well as neurons (Ruvkun et al., 1991). We focus here on a hypodermal cell lineage that originates from the *postembryonic blast cell T* (see Fig. 25.5). *Lin-14*$^+$ is a heterochronic gene, mutations in which change both the synthesis of lin-14 protein and the division pattern of T (Fig. 25.16). In the wild type, lin-14 protein is synthesized early during the first larval stage (L1), and cell divisions in the T lineage begin during L1 and end during L2. In animals with loss-of-function alleles of *lin-14,* the span of lin-14 protein activity is abbreviated. The same alleles cause precocious execution of cell divisions normally observed in descendant cells at a later stage. Conversely,

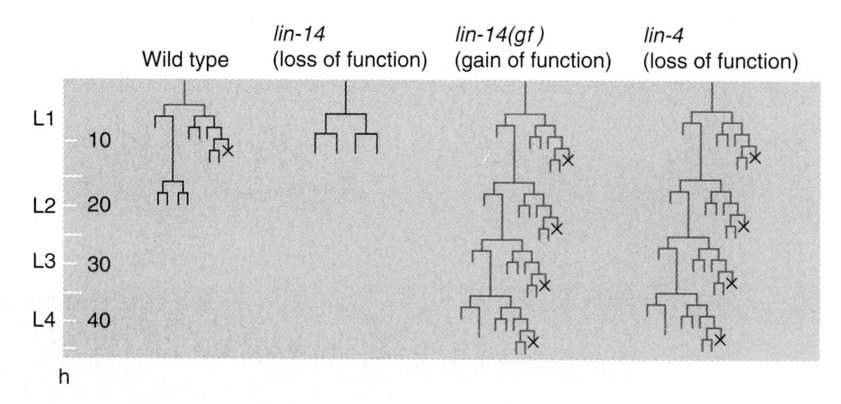

Figure 25.16 The effect of heterochronic mutations on the cell lineage formed by the postembryonic blast cell T (see Fig. 25.5), which contributes to the hypodermis of *C. elegans.* The colored portions of the lineage diagrams represent those cells that express lin-14 protein. The time axis to the left indicates hours after hatching and the four larval stages (L1–L4). In the wild type, T expresses lin-14 protein early during L1 and divides during L1 and L2. A loss-of-function mutant, *lin-14,* lacks any detectable lin-14 protein, and its blast cell T develops a precocious L2-specific lineage at stage L1. A gain-of-function mutant, *lin-14(gf)*, expresses lin-14 protein at all larval stages and reiterates the cell lineage of T characteristic of L1. The same phenotype is produced by a loss-of-function mutation in another gene, *lin-4.*

gain-of-function alleles of *lin-14* affect the T cell lineage in the opposite way: Cell divisions are normal during L1, but later divisions repeat the same division pattern over and over. These alleles also cause an extended span of lin-14 protein synthesis. Thus, abbreviated activity of lin-14 protein causes a precocious "aging" of the cell division pattern, whereas an extended synthesis of lin-14 protein keeps the lineage permanently "young."

Other genes act in the same pathway as *lin-14*$^+$ to control the temporal patterns of postembryonic blast cell divisions. A loss-of-function mutation in *lin-4* causes the same repetitive lineage as *lin-14* gain-of-function mutations (Fig. 25.16). However, the *lin-4* loss-of-function phenotype depends on a functioning *lin-14*$^+$ gene: Mutants defective in both genes show the same phenotype as *lin-14* loss-of-function mutants. These observations indicate that *lin-4*$^+$ normally down-regulates the *lin-14*$^+$ gene after the early L1 stage.

THE *lin-14*$^+$ GENE ENCODES A NUCLEAR PROTEIN THAT FORMS A TEMPORAL CONCENTRATION GRADIENT

The observed relationship between lin-14 protein and the cell division pattern suggests that lin-14 protein may form a *temporal gradient,* with a high concentration early during the L1 stage and a declining concentration thereafter. The gradient may orchestrate the normal sequence of cell divisions in the T lineage.

To test for a temporal gradient, Ruvkun and Giusto (1989) cloned the *lin-14*$^+$ gene and raised antibodies for *immunostaining* the lin-14 protein (see Method 4.1). The antibodies stained specific nuclei in early larvae from wild-type strains but not from strains bearing *lin-14* null alleles, showing that the antibody was directed specifically against lin-14 protein. As expected, lin-14 protein was present in the T cell lineage and in all other postembryonic blast cell lineages that are affected by *lin-14* mutations. The carboxy terminal region of lin-14 contains an unusual expanded nuclear localization domain which is essential for lin-14 function (Hong et al., 2000). These results support the view that lin-14 controls developmental timing in *C. elegans* by regulating gene expression in the nucleus.

In wild-type animals, the staining was most intense in late embryos just before hatching, and in newly hatched L1 larvae. The blast cells containing lin-14 protein underwent their first L1-specific division, but then the lin-14 concentration fell before the next division (Fig. 25.16). By stage L2, the concentration had decreased by a factor of more than 25. In contrast, L3 larvae of a gain-of-function mutant designated *lin-14(gf)* had continuously high concentrations of lin-14 protein in their blast cell nuclei.

THE *lin-4*$^+$ GENE ENCODES SMALL REGULATORY RNAs WITH ANTISENSE COMPLEMENTARITY TO lin-14 mRNA

Two gain-of-function mutations that keep expressing lin-14 protein beyond the time of its normal decline have provided a clue to the mechanism by which *lin-4*$^+$ controls the expression of *lin-14*$^+$ (Wightman et al., 1991). One of these mutations is a large insertion, and the other is a deletion in the same region. Both mutations map to the 3′ UTR of the mRNA.

Since the 3′ UTR of other mRNAs have been shown to interact with proteins that regulate their nucleocytoplasmic transport, half-life, and translation (see Sections 8.4, 8.6, 17.4 , 18.2, and 18.4), it might have been expected that lin-4$^+$ would encode such a regulatory protein. It therefore came as a surprise when it was discovered that *lin-4*$^+$ does not encode a protein (R. C. Lee et al., 1993; Wightman et al., 1993). Instead, *lin-4*$^+$ encodes two small transcripts, about 22 and 61 nucleotides long, with sequences complementary to repeated sequence elements in the 3′ UTR of lin-14 mRNA. The duplexes formed between the lin-4 regulatory RNAs and the lin-14 3′ UTR cause the formation of secondary structures that interfere with translation of the lin-14 mRNA (Ha et al., 1996; Olson and Ambros, 1999).

DOWN-REGULATION OF *lin-14*$^+$ EXPRESSION IS INITIATED BY A DEVELOPMENTAL CUE

In an experiment to investigate what triggers the down-regulation of *lin-14*$^+$ gene expression in normal development, newly hatched larvae were starved. Under these circumstances, lin-14 protein concentration remained high beyond the normal period of 12 h of postembryonic development after which the lin-14 protein disappears from normal larvae. Thus, *lin-14*$^+$ down-regulation is not a function of time but instead depends on some cue that is disturbed by starvation. Furthermore, the continued high level of lin-14 proteins in starved larvae depends on protein synthesis and is associated with a high concentration of lin-14 mRNA. These results suggest that lin-14 protein synthesis continues in starved animals, possibly because the *lin-4*$^+$ gene, which is required for down-regulating lin-14 protein, is activated by a stimulus that depends on feeding.

In summary, the role of the heterochronic gene *lin-14*$^+$ in the temporal regulation of postembryonic cell lineages can be envisioned as follows. A feeding-dependent stimulus activates the *lin-4*$^+$ gene, which encodes small regulatory RNAs that hybridize with the 3′ UTR of lin-14 mRNA with the effect of down-regulating its translation. As the concentration of lin-14 protein falls, it generates a decreasing temporal gradient, which in turn controls a set of target genes that regulate the stage-specific cell division patterns.

25.4 Programmed Cell Death

During most animals' development, cells arise that do not develop further but instead die. Such cell death often occurs in a distinct pattern and plays a role in morphogenesis (Saunders, 1982). For instance, the digits of the limbs of higher vertebrates are sculpted by the programmed death of the cells between them. During the development of the vertebrate nervous system, up to 50% of many types of neurons die after they have already formed synaptic connections with their target cells (Raff et al., 1993; Yamamoto and Henderson, 1999). This massive cell death is ascribed to the failure of the dying neurons to obtain sufficient amounts of growth factors from their target cells. Likewise, hematopoietic cells die rapidly in the absence of appropriate growth factors that they need to divide and differentiate (see Section 20.4). Similar strategies are observed in other cell types that require signals from neighboring cells to survive (G. T. Williams and C. A. Smith, 1993). The survival signals seem to act by suppressing intrinsic cell suicide programs.

The phenomenon of controlled cell death is called ***programmed cell death*** or ***apoptosis*** (Gk. *apo*, "away"; *ptosis*, "drooping, sagging"). The latter term refers to the absence of some signs of accidental cell death, or *necrosis* (Gk. *nekros*, "dead"). In necrosis, cells usually spill their contents, thus eliciting an inflammatory response. In contrast, apoptotic cells condense their chromatin and shrink until they are engulfed by macrophages or other cells. Thus, apoptosis avoids an inflammatory response and appears to be the more economical way of recycling the molecular components of cells.

Of the 1090 somatic cells generated in the hermaphrodite of *C. elegans*, 131 die (Sulston et al., 1983). Similarly, 148 of 1179 somatic cells die in the male. In every individual the same cells die, each in its own time and place. Because of this feature and its favorable genetic properties, *C. elegans* is a particularly useful organism for analyzing apoptosis (*Metzstein et al., 1998;* Vaux and Korsmeyer, 1999). We will first discuss the genetic controls that specify which cells live and which cells die in *C. elegans*, and then briefly explore some orthologous genes found in mammals including humans.

PROGRAMMED CELL DEATH IN *C. ELEGANS* IS CONTROLLED BY A GENETIC PATHWAY

More than a dozen genes in *C. elegans* have specific functions connected with programmed cell death. A key gene is *ced-3*$^+$, which seems to be required for all programmed cell deaths in both embryos as well as larvae (*ced* stands for *ce*ll *d*eath). Loss-of-function mutants in *ced-3* are nearly normal in morphology and behavior, showing that the prevention of programmed cell death is not immediately detrimental to *C. elegans* (H. M. Ellis and H. R. Horvitz, 1986). The extra surviving cells in such mutants do not divide, but some of them differentiate and function. A group of at least six genes, including *ced-1*$^+$, cause the apoptotic cells to be engulfed and degraded by their neighbors. Loss-of-function mutations in these genes allow dead cells to remain present (Fig. 25.17). Yet another gene, *nuc-1*$^+$ (for nuclease-deficient), is necessary for the breakdown of DNA in the engulfed cells (Y. C. Wu et al., 2000).

The *ced-3*$^+$gene, the group represented by *ced-1*$^+$, and the *nuc-1*$^+$ gene define a pathway of cell destruction and removal (Fig. 25.18). However, these genes do not control which cells survive and which die. This selection is made by control genes including *ced-4*$^+$, *ced-9*$^+$, and *egl-1*$^+$. These genes either promote or inhibit cell death in apparently general, nonspecific ways. Other genes, including *ces-1*$^+$ and *ces-2*$^+$ (for *ce*ll death *s*pecification), prevent the programmed death of specific cells without affecting the death of other cells.

THE STRATEGY OF GENETIC MOSAIC ANALYSIS SHOWS THAT *ced-3*$^+$ AND *ced-4*$^+$ ACT CELL-AUTONOMOUSLY

One fundamental question is whether *ced-3*$^+$ acts *cell-autonomously*—that is, whether this gene acts within the cells that die or in other cells that control cell death from outside. A strategy used generally to identify those cells

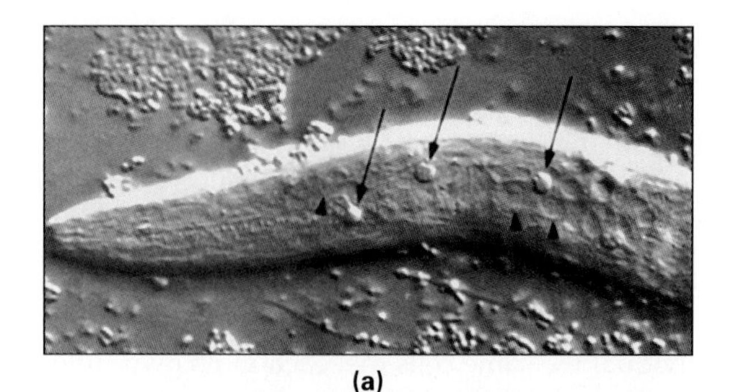

(a)

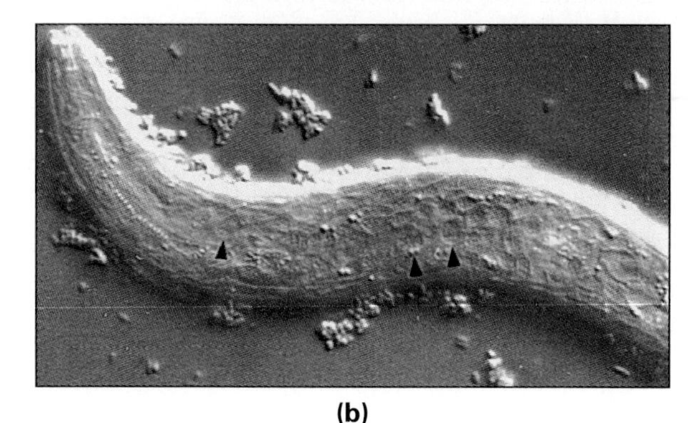

(b)

Figure 25.17 Absence of cell death in *ced-3* mutants of *C. elegans.* **(a)** Photograph of a newly hatched larva that is mutant in the *ced-1* gene. Arrows indicate dying cells, which remain visible because the *ced-1* mutation interferes with the removal of dying cells by other cells, which would occur in the wild type. **(b)** Similar larva mutant in the *ced-1* gene and the *ced-3* gene. No dying cells are detectable, because the *ced-3* mutation interferes with programmed cell death. Arrowheads indicate the same nuclei in both photographs, demonstrating that both are in the same focal plane.

of an organism in which the action of a gene is required is called ***genetic mosaic analysis.*** As suggested by the strategy's name, one needs to generate mosaic (or chimeric) individuals in which a wild-type allele of the gene of interest is present randomly in some cells but not in others. This allele must be associated with a genetic marker that will tell the investigator which cells have a functional allele of the gene of interest. These cells are then tallied in mosaic individuals that show the wild phenotype. Those cells that carry the wild-type allele in *all* of the analyzed mosaic individuals are considered to be the cells where the gene of interest must be active to produce the wild phenotype.

To use this strategy for the analysis of *ced-3* gene in *C. elegans,* Yuan and Horvitz (1990) generated individuals in which some cells were wild-type and others were deficient for *ced-3.* This can be accomplished by a technique using chromosomal fragments that are not well integrated into mitotic spindles and are therefore randomly lost during cleavage (Herman, 1984). The fragments carry marker genes with known cell-specific functions, from which investigators can infer which cell lineages have lost the chromosome fragment and which have kept it.

A suitable fragment carrying the wild-type allele of *ced-3*$^+$ was introduced into individuals homozygous for a loss-of-function allele of *ced-3.* Thus, cell lineages that had lost the fragment were homozygous mutant for *ced-3,* while those retaining the fragment were heterozygous for the wild-type allele. The results showed that programmed cell death requires a wild-type allele of *ced-3*$^+$ in the dying cell itself. This was particularly clear in a pair of neurons, CEMDR and CEMDL, which are physically adjacent in the larva although one originates from ABa and the other from ABp. One of these cells often survived (if its lineage had lost the chromosome fragment), while the adjacent neuron died (if its lineage had kept the chromosome fragment). A corresponding analysis of the *ced-4*$^+$ gene yielded the same results.

The mosaic analysis of Yuan and Horvitz (1990) argues strongly that programmed cell death in *C. elegans* is initiated by the activity of *ced-3*$^+$ and *ced-4*$^+$ in the dying cells themselves and not in neighboring cells. This conclusion implies that the combination of proteins encoded by these genes kills the cells in which they are made.

Cloning and sequencing of the *ced-3*$^+$ gene revealed that it encodes the inactive precursor of an enzyme of the *caspase* (cysteine aspartate–specific protease) family (Yuan et al., 1993; Alnemri, 1997). The targets of the activated enzyme remain to be elucidated. Obviously, apoptotic cells are killed but not immediately destroyed, since their corpses stay around in worms deficient in *ced-1* or other engulfment genes. Possibly, the active ced-3 caspase incapacitates selectively a few proteins that are required for cells to survive.

THE *egl-1, ced-9,* AND *ced-4* GENE PRODUCTS CONTROL THE INITIATION OF APOPTOSIS

In the genetic hierarchy that controls programmed cell death, the gene immediately upstream of *ced-3*$^+$ seems to be *ced-4*$^+$ (Fig. 25.18). Genetic analysis has shown that a loss of function in *ced-4* can be overcome to a large extent by overexpression of *ced-3*$^+$. In contrast, a loss of *ced-3* function can hardly be overcome by overexpression of *ced-4*$^+$. Transcripts from *ced-4*$^+$ accumulate at high levels during embryogenesis, when most programmed cell death occurs (Yuan and Horvitz, 1992). Biochemical data indicate that the ced-4 protein promotes the activation of the ced-3 caspase precursor in a reaction that depends on ATP and other cofactors (Chinnaian et al., 1997a; *Vaux, 1997*).

How are potentially lethal activities like those of *ced-3*$^+$ and *ced-4*$^+$ controlled so that only the appropriate cells are killed? Genetic analysis has shown that a regulatory gene, *ced-9*$^+$, inhibits the activities of *ced-3*$^+$ and *ced-4*$^+$ (Fig. 25.18; Hengartner et al., 1992). Loss-of-function

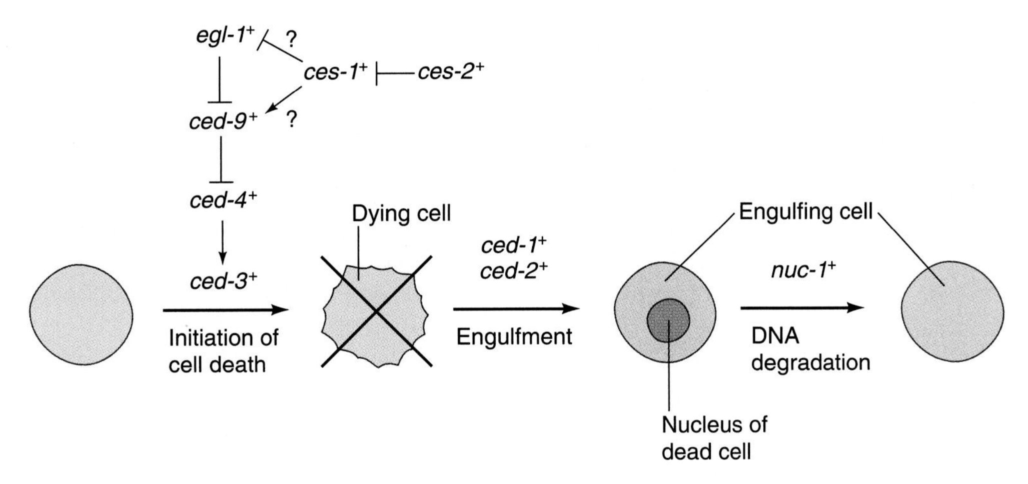

Figure 25.18 Model of a general genetic pathway controlling programmed cell death (apoptosis) in *C. elegans.* In each cell, apoptosis is initiated by the normal function of the *ced-3*$^+$ gene, which encodes a cysteine aspartate-specific protease (caspase). Dying cells are engulfed by neighboring cells, an activity that requires the function of a group of genes represented by *ced-1*$^+$ and *ced-2*$^+$. Degradation of the DNA of the engulfed cell depends on the activity of yet another gene, *nuc-1*$^+$. The initiation of apoptosis by the activation of ced-3 caspase is controlled by a hierarchy of genes. The products of the *egl-1*$^+$, *ced-9*$^+$, and *ced-4*$^+$ genes are made in virtually all cells and interact directly with one another and with the ced-3 precursor protein. In contrast, the *ces-2*$^+$ and *ces-1*$^+$ genes seem to be expressed selectively, interfering with the initiation of apoptosis only in specific cells.

alleles in *ced-9* cause the death of cells that normally survive, including many neurons, eggs in hermaphrodites, and tail cells in males. Direct observation of developing embryos reveals that the missing cells are born by normal cell division but later die. These extra cell deaths do not show up in mutants that are also deficient in *ced-3* or *ced-4.*

Conversely, the cells that normally undergo apoptosis survive in mutants carrying a gain-of-function allele of *ced-9*$^+$, designated *ced-9(gf)*. At least some of the extra cells that survive in *ced-9(gf)* individuals are functional, as they are in individuals carrying loss-of-function alleles of *ced-3* and *ced-4.* These observations suggest that *ced-9*$^+$ inhibits the activities of *ced-3*$^+$ and *ced-4*$^+$. However, in the absence of *ced-4*$^+$, *ced-9*$^+$ fails to protect cells against the destructive effects of *ced-3*$^+$ (Shaham and Horvitz, 1996). Thus, the available genetic data suggest that *ced-9*$^+$ protects cells from programmed death by interfering with *ced-4*$^+$ in a way that prevents the activation of the ced-3 caspase precursor. This interpretation is supported by biochemical data indicating that ced-9 and ced-4 can interact directly (Chinnaian et al., 1997b; *Vaux, 1997*). Moreover, *immunostaining* of ced-9 and ced-4 in wild-type and mutant embryos indicates that ced-4 translocates from mitochondria to the cell nucleus before it activates the ced-3 caspase precursor, and that ced-9 may prevent this translocation (F. Chen et al., 2000).

Finally, ced-9 activity is controlled by the *egl-1*$^+$ gene, which was so designated because hermaphrodites carrying a dominant gain-of-function allele, *egl-1(gf)*, are deficient in *egg laying*. The defect was eventually ascribed to the apoptosis of certain neurons that normally survive. Loss-of-function in *egl-1* prevents virtually all programmed cell death as long as *ced-9*$^+$ is active. However, loss of *egl-1* function does not suppress the extra cell deaths resulting from loss of *ced-9*$^+$ activity (Conradt and Horvitz, 1998). These genetic data indicate that *egl-1*$^+$ activity interferes with *ced-9*$^+$ function (Fig. 25.18). Again, this interpretation is supported by biochemical data indicating that the egl-1 and ced-9 proteins interact directly.

In summary, a hierarchy of gene products, including egl-1, ced-9, and ced-4, controls the conversion of the ced-3 proenzyme into the active caspase that initiates the pathway of apoptosis.

THE GENES *ces-1*$^+$ AND *ces-2*$^+$ CONTROL THE PROGRAMMED DEATH OF SPECIFIC CELLS IN THE PHARYNX

While *ced-9*$^+$ and associated genes prevent death in a wide spectrum of cells, other genes affect the programmed death of specific cells. For instance, *ces-1*$^+$ and *ces-2*$^+$ control the death of specific neurons (R. E. Ellis and H. R. Horvitz, 1991). Both loss-of-function alleles of *ces-2* and a gain-of-function allele of *ces-1,* designed *ces-1(gf)*, affect in particular the sister cells of two neurons designated NSM and I2 (Fig. 25.19). Normally, these sister cells die, and their corpses are clearly visible in larvae that are defective in the genes required for dead cell engulfment. However, in *ces-1(gf)* mutants, the sister cells of NSM and I2 survive and differentiate into neurons, just as the NSM and I2 cells do.

The *ces-1* and *ces-2* mutations prevent the programmed death of specific cells but do not affect the

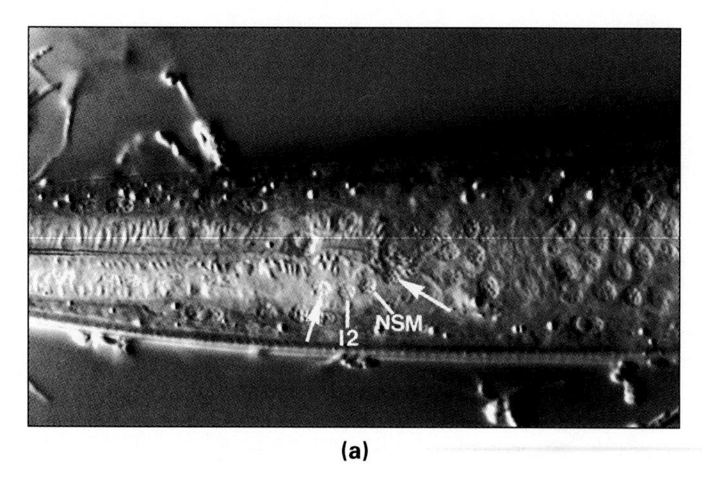

(a)

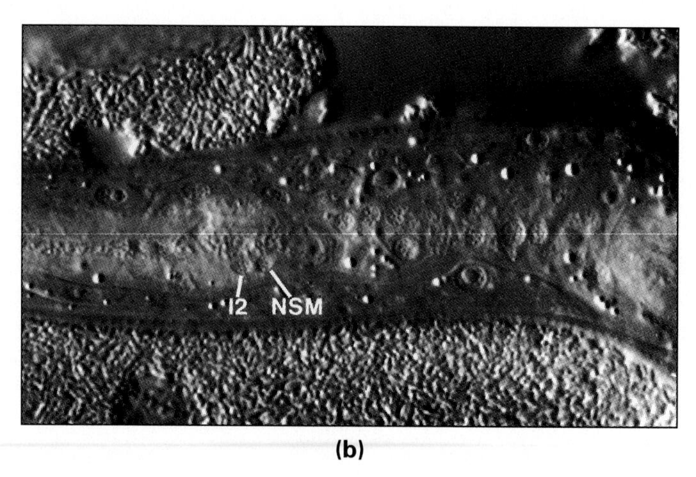

(b)

Figure 25.19 Programmed cell death of specific neurons in third-instar larvae of *C. elegans.* The photographs show **(a)** a *ces-1(gf)* mutant and **(b)** a wild-type animal. The mutant has two extra neurons (arrows) that are not present in the wild type. The extra neurons, which undergo programmed cell death in the wild type, are sister cells of the neurons I2 and NSM.

other cells that die during normal *C. elegans* development. For example, the two mutations do not affect any of the 18 pharyngeal cells that undergo programmed death. Cloning and sequencing of *ces-1*$^{+}$ and *ces-2*$^{+}$ has shown that both genes encode transcription factors, and that the ces-2 protein can bind to a regulatory element of *ces-1*$^{+}$ (Metzstein et al., 1996; Metzstein and Horvitz, 1999). Thus, ces-2 would seem to inhibit the expression of *ces-1*$^{+}$, and ces-1 in turn might inhibit *egl-1*$^{+}$ or activate *ced-9*$^{+}$ (see Fig. 25.18).

THE PRODUCT OF A HUMAN GENE, *bcl-2*$^{+}$, PREVENTS PROGRAMMED CELL DEATH IN *C. ELEGANS*

At least two of the *C. elegans* genes that control programmed cell death, *ced-3*$^{+}$ and *ced-9*$^{+}$, have close functional and molecular counterparts in mammals including humans. The *C. elegans* gene *ced-9*$^{+}$ supports a similar function as the human gene *bcl-2*$^{+}$. In mammals, *bcl-2*$^{+}$ prevents apoptosis in various cell types including embryonic neurons and hematopoietic cells. Indeed, a family of more than 20 similar molecules is present in mammalian cells, most of them antiapoptotic like bcl-2, but others proapoptic, with the ratio of the two apparently controlling the decision whether a cell lives or dies (Hockenberry, 1995; A. Gross et al., 1999).

As a functional test of the similarity between ced-9 and bcl-2 proteins, Vaux and coworkers (1992) transformed *C. elegans* with a cloned human *bcl-2*$^{+}$ gene driven by a *heat shock promoter.* Indeed, they found that the human bcl-2 protein indeed interferes with programmed cell death in *C. elegans.* Subsequent cloning of *ced-9*$^{+}$ revealed considerable sequence similarity with its *bcl-2*$^{+}$ counterpart (Hengartner and Horvitz, 1994). The bcl-2 protein has been detected in mitochondria and also in the nuclear envelope and the endoplasmic reticulum, but the molecular mechanism of its action remains to be elucidated.

Because ced-9 inhibits the activation of ced-3 caspase by ced-4 in *C. elegans*, bcl-2 may antagonize corresponding gene products in mammals. The mammalian gene *Nedd2*$^{+}$ (for neural developmental down-regulation) is strongly expressed in tissues undergoing programmed cell death and encodes a protease that causes apoptosis, just as *ced-3*$^{+}$ does. In order to test whether *bcl-2*$^{+}$ expression inhibits *Nedd2*$^{+}$ activity, Kumar and coworkers (1994) transfected cultured mammalian cells with cDNAs representing these genes. While cells transfected with *Nedd2*$^{+}$ alone underwent apoptosis, cells transfected with both *Nedd2*$^{+}$ and *bcl-2*$^{+}$ survived to a large extent. Similarly, the mammalian protein Apaf-1 has been shown to be orthologous to the ced-4 protein of *C. elegans* (*Metzstein et al., 1998*). Thus, at least some of the key players in the genetic hierarchy that controls apoptosis in *C. elegans* have been conserved in evolution.

25.5 Vulva Development

The vulva of *C. elegans* is a ventral opening that is used for copulation and egg laying. It develops in the hermaphrodite during the larval stages. The formation of the vulva epitomizes the concept of *equivalence groups* of cells in *C. elegans* development. It also illustrates the integration of different signaling pathways during inductive interactions (R. J. Hill and P. W. Sternberg, 1993; *Sundaram and Han, 1996;* Greenwald, 1999).

THE VULVAL PRECURSOR CELLS FORM AN EQUIVALENCE GROUP

The cells giving rise to the vulva belong to the *postembryonic blast cells* that develop during the larval stages to form adult structures later (see Fig. 25.5). Six blast cells, all derived from the ABp blastomere, are lined up

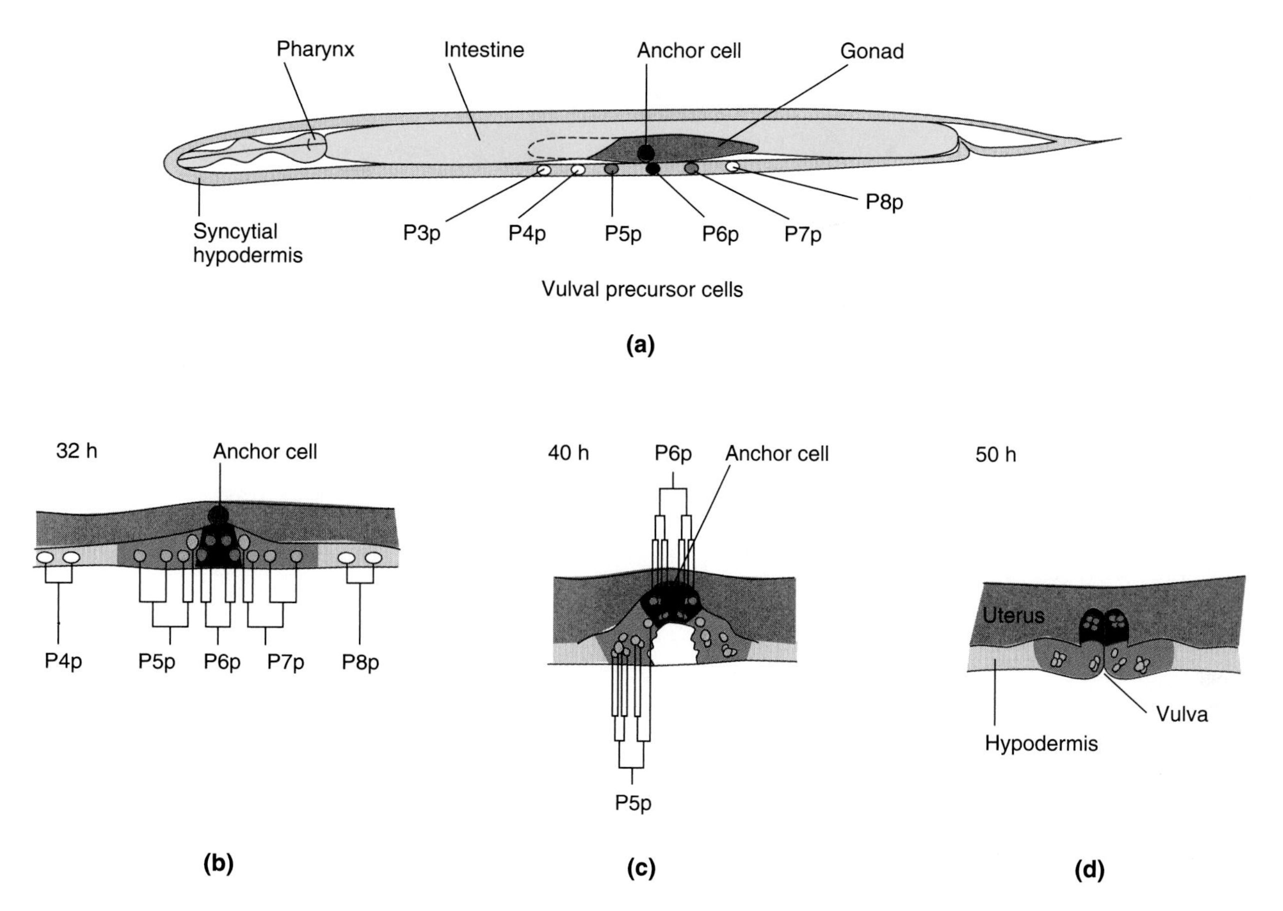

Figure 25.20 Formation of the vulva in *C. elegans.* Six vulval precursor cells (VPCs) have the potency of forming a vulva, but only three normally do. The fate of the VPCs depends on their proximity to the gonadal anchor cell, which will link the vulva to the uterus. The VPC next to the anchor cell forms the inner part of the vulva, and the two neighboring VPCs form the outer vulva. The remaining three VPCs contribute to the hypodermis. **(a)** Early third-stage larva, with the six VPCs, designated P3p to P8p, aligned along the ventral surface of the developing gonad, which contains the anchor cell. **(b, c)** VPC divisions. P6p is normally fated to have eight descendants forming the interior vulva (primary fate). P5p and P7p normally produce seven descendants each, together forming the exterior vulva (secondary fate). The vulva becomes syncytial like the hypodermis. **(d)** Vulval development is complete in the adult.

along the ventral midline of the larva and have the potency to form the vulva of the adult (Fig. 25.20a). These cells are designated P3p, P4p, P5p, P6p, P7p, and P8p, because they are the posterior daughter cells from a ventral series of blast cells labeled P1 through P12. (These cells are unrelated to the germ line cells P_1 through P_4.) All six blast cells, P3p through P8p, have the potency of contributing to the vulva and are therefore called ***vulval precursor cells (VPCs).*** However, only three VPCs are normally fated to form the vulva, while the other three VPCs contribute to the hypodermis. Which VPCs actually form the vulva depends on their proximity to a particular cell in the gonad rudiment, the ***anchor cell,*** which will link the vulva to the uterus.

The VPC closest to the anchor cell undergoes three divisions and forms eight daughter cells in a circular arrangement (Fig. 25.20b–d). This fate is called the ***primary (1°) fate.*** Each of the two neighboring VPCs produces seven descendants in a semicircular arrangement. This fate is referred to as the ***secondary (2°) fate.*** The remaining VPCs undergo only one division, producing two descendants that fuse with other cells forming the hypodermal syncytium. This fate is called the ***tertiary (3°) fate.*** The 22 descendants of the VPCs with primary and secondary fates form a doughnut-shaped syncytium, the center of which forms the vulval opening. The vulva interacts with specific nerve, muscle, and uterus cells during copulation and egg laying.

Although the six VPCs in the *C. elegans* larva express different fates, they have the same potencies, since each of them may adopt any one of the three fates. For such a group of cells that have the same potencies but different fates, scientists studying *C. elegans* have coined the term ***equivalence group.*** The properties of an equivalence group are similar to those of an *embryonic field.* Thus, destroyed cells may be functionally replaced by alternative cells from the same equivalence group. Typically, the cells of an equivalence group are related by lineage but are not clones.

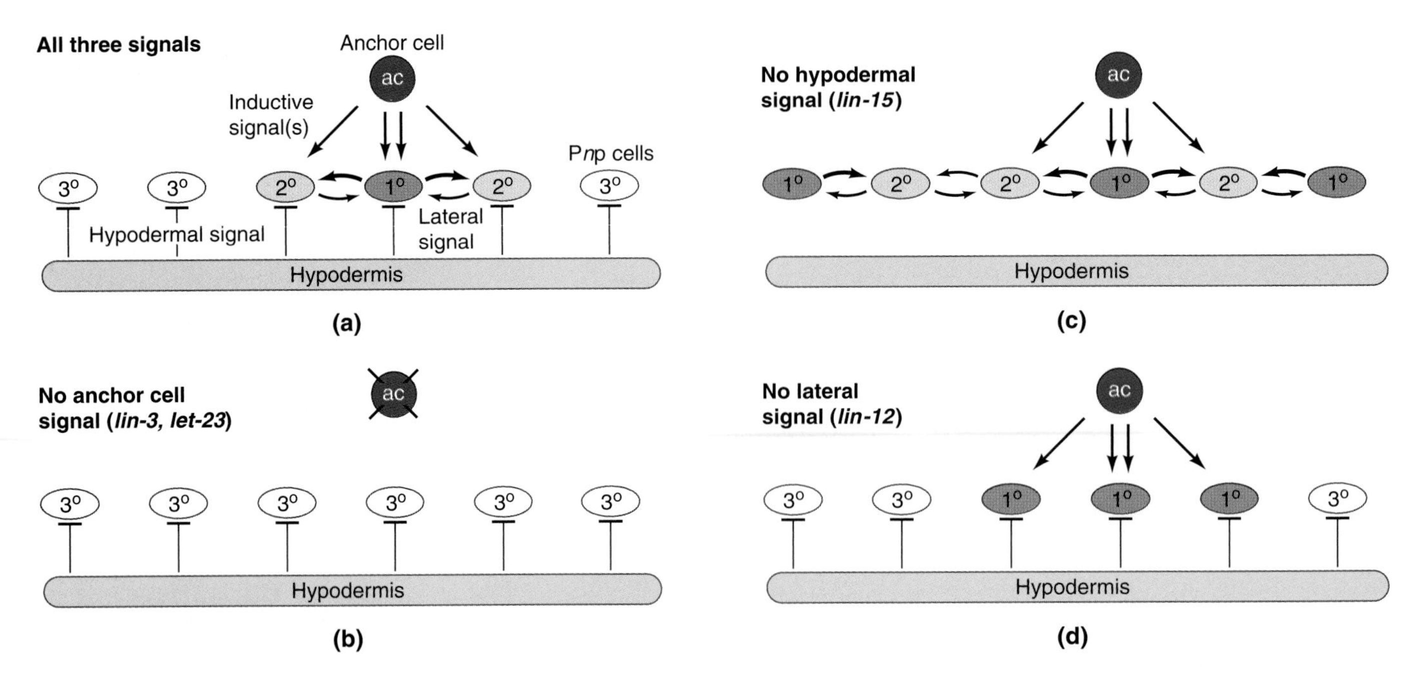

Figure 25.21 Model for the role of three intercellular signals in vulval development. **(a)** General model. First, the anchor cell induces vulva formation in the nearest vulval precursor cell (VPC), and possibly in the two neighboring VPCs as well. Second, the three VPCs next to the anchor cell exchange lateral signals so that a VPC adopting the primary fate prevents its neighbors from doing the same, making them adopt the secondary fate instead. The inductive and lateral signals cooperate so that the primary fate is always expressed next to the gonadal anchor cell (see Fig. 25.24). Third, the hypodermis sends an inhibitory signal that prevents the VPCs from assuming primary and secondary fates except for the three VPCs next to the anchor cell. **(b)** In the absence of an anchor cell, all VPCs assume tertiary fates, because there is no inductive signal to override the inhibitory signal from the hypodermis. The same phenotype is observed in *lin-3* and *let-23* loss-of-function mutants. **(c)** The mutant allele *lin-15* acting in the hypodermis allows the formation of multiple vulvae because the inhibitory signal from the hypodermis is missing. In this case, the pattern of primary and secondary fates among the VPCs varies, but the primary fate is never expressed by two neighboring VPCs. **(d)** In *lin-12* mutants, the three VPCs closest to the anchor cell adopt the primary fate because the mutant is deficient in lateral signaling.

The six VPCs in *C. elegans* were revealed as an equivalence group in the course of laser ablation experiments and through study of certain mutant phenotypes. The available data indicate that normal vulval development requires at least three intercellular signals: an *inductive signal* originating in the gonadal anchor cell, a *lateral signal* exchanged between the VPCs themselves, and an *inhibitory signal* originating in the adjacent hypodermis (Fig. 25.21a).

THE GONADAL ANCHOR CELL INDUCES THE PRIMARY VPC FATE

Vulva development clearly depends on an inductive signal from the gonadal anchor cell. If this cell is destroyed with a laser microbeam, no vulva is formed (Fig. 25.21b). All VPCs then express the tertiary fate, and the resulting animal cannot copulate or lay eggs (Kimble, 1981). Conversely, if all gonadal cells except the anchor cell are destroyed, a vulva still develops. Therefore, the anchor cell is both necessary and sufficient for vulva formation in wild-type *C. elegans*.

Further proof of an inductive signal emanating from the anchor cell comes from certain mutant phenotypes with displaced gonads (J. H. Thomas et al., 1990). If the gonad is displaced anteriorly, so that the anchor cell is next to P5p, this cell instead of P6p adopts the primary fate. Thus, according to the criterion of *heterotopic transplantation*, the gonadal anchor cell *induces* the adjacent VPC to assume the primary fate and the next two VPCs to assume the secondary fate. If the gonad is displaced dorsally, so that the anchor cell is not in direct contact with any VPC, the closest three VPCs nevertheless form a vulva, indicating that the inductive signal is diffusible or can be transmitted through other cells.

The model of vulval induction shown in Figure 25.21 has been confirmed by the genetic and molecular analysis of the genes involved in the process. The genetic analysis started with screens for mutants displaying one of two opposite phenotypes (Fig. 25.22). In animals classified as *vulvaless*, fewer than three VPCs adopt a vulval fate, so that hermaphrodites become defective in egg laying. In the other class of phenotypes, called *multivulva*, more than three VPCs adopt a vulval fate, so that

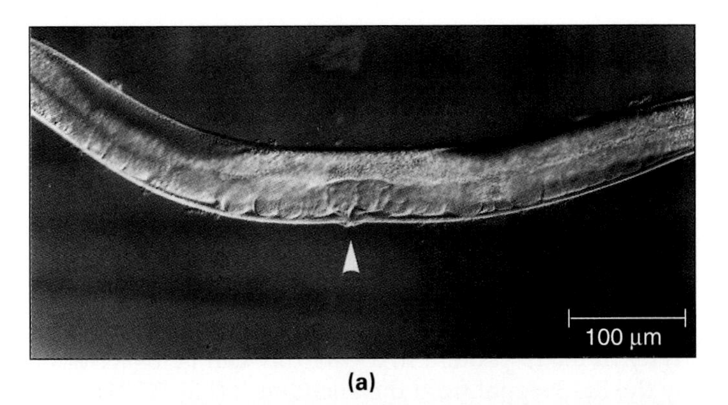

(a)

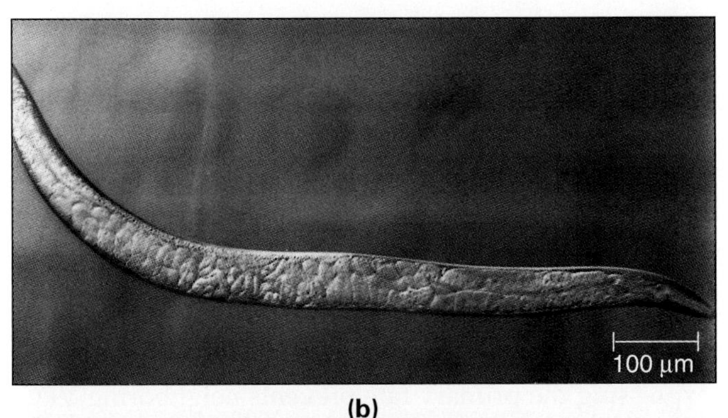

(b)

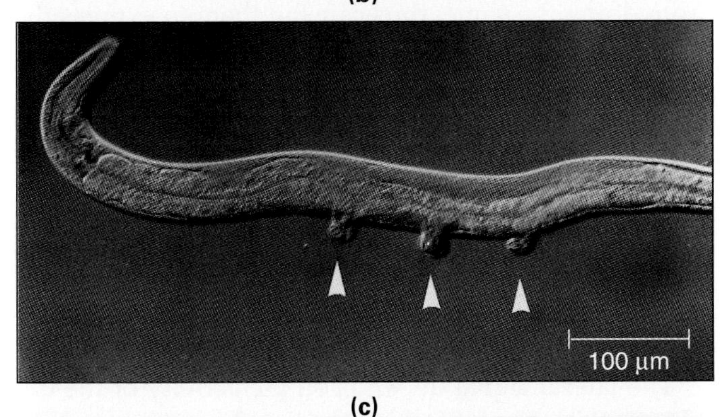

(c)

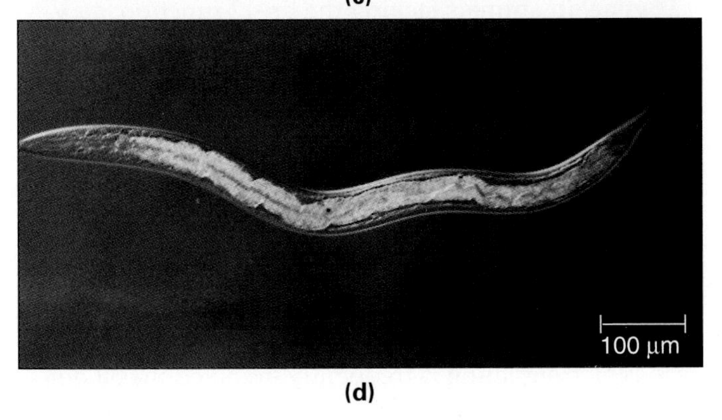

(d)

Figure 25.22 Photomicrographs of *C. elegans lin-3* phenotypes. **(a)** Adult wild-type hermaphrodite. The arrowhead marks the position of the vulva. **(b)** Vulvaless hermaphrodite in which the *lin-3* gene has been interrupted by a transposon. **(c)** Hermaphrodite with multiple copies of a *lin-3*$^+$ transgene. The animal has three vulvalike structures (arrowheads). **(d)** Hermaphrodite with multiple copies of a *lin-3*$^+$ transgene but without a gonad. Most of these animals do not form a vulva.

adult hermaphrodites have ectopic vulval structures (pseudovulvae) protruding from the ventral surface of their body.

One of the mutant alleles resulting in a *vulvaless* phenotype is *lin-3*. Hill and Sternberg (1992) cloned the gene and showed that overexpression of *lin-3*$^+$ from multiple transgenes produced a dominant *multivulva* phenotype. Transgenic animals from which the anchor cell or the entire gonad had been removed were *vulvaless,* indicating that *lin-3*$^+$ and the anchor cell are both required for vulval induction (Fig. 25.22d).

The lin-3 protein predicted from the DNA sequence is a membrane-spanning protein with a consensus sequence, called the EGF repeat, which is also present in *epidermal growth factor* (*EGF*) and related growth factors known primarily from vertebrates (see Section 28.4). To find out where the *lin-3*$^+$ gene is expressed, the investigators inserted the bacterial *lacZ* gene into the first exon of the cloned *lin-3*$^+$ gene. The fusion gene was expected to encode a protein consisting of the extracellular and transmembrane domains of *lin-3*$^+$ followed by the β-galactosidase domain encoded by *lacZ*. Indeed, in transgenic animals this fusion protein still induced the *multivulva* phenotype and produced the galactosidase activity that turns a suitable substrate into a blue stain. The stain was precisely where it should be if *lin-3*$^+$ encodes a vulva-inducing signal: in the gonadal anchor cell (Fig. 25.23). Moreover, the fusion protein began to accumulate when the vulva was induced: as the L2 larva molted into L3. In summary, the biological effects of the lin-3 protein as well as its expression pattern and molecular characteristics are compatible with its possible role as a vulva-inducing signal.

Additional genes with loss-of-function alleles producing the *vulvaless* phenotype include *let-23*$^+$, *sem-5*$^+$, *let-60*$^+$, *lin-45*$^+$, *mek-2*$^+$, and *sur-1*$^+$. Overexpression of the wild-type allele of each of these genes also generates the *multivulva* phenotype. Based on their *epistatic* interactions, these genes form the hierarchy shown in Figure 25.24. Sequence analysis has revealed that these genes encode the components of the *RTK-Ras-MAPK signaling pathway,* which forms the backbone of many induction processes in a wide variety of organisms.

A HYPODERMAL SIGNAL INHIBITS VULVA FORMATION

A signal from the hypodermis inhibits vulva formation from noninduced VPCs. This is indicated by the phenotype of mutants that are deficient in the gene *lin-15* (for cell *lin*eage abnormal). In these mutants, all six VPCs adopt the primary or secondary fate, and multiple vulvalike structures develop (see Fig. 25.21c; Ferguson et al., 1987). Destroying the entire gonad in *lin-15* mutants does not prevent multiple vulva formation. Therefore the mutation does not act by causing overproduction of the gonadal signal.

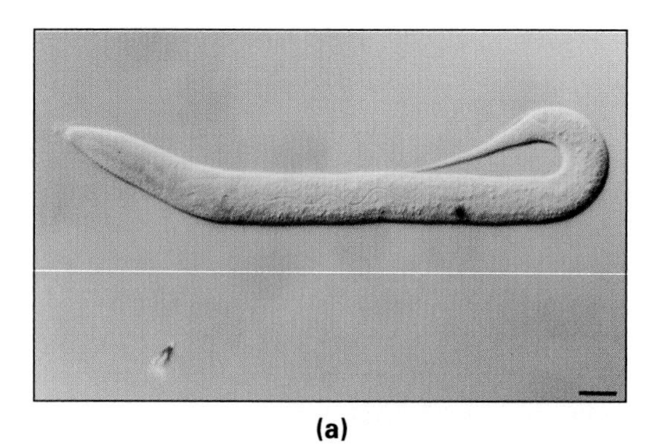

(a)

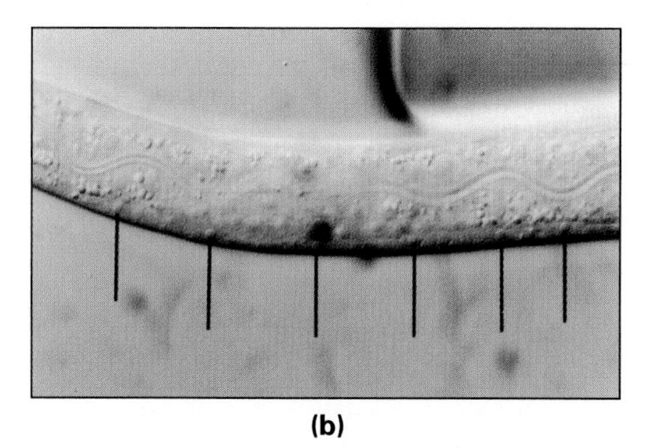

(b)

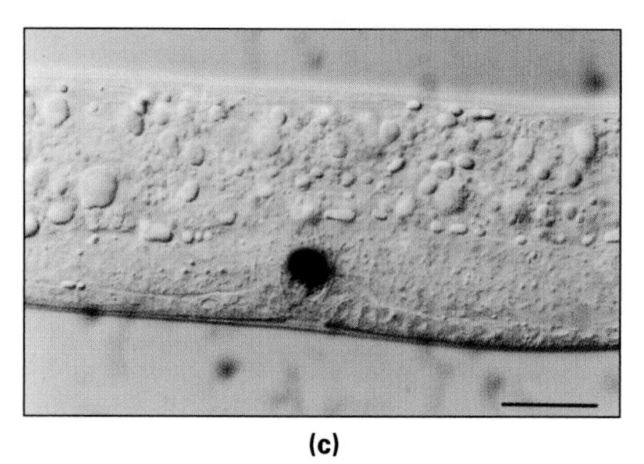

(c)

Figure 25.23 Expression of a *lin-3/lacZ* fusion gene in *Caenorhabditis elegans*. **(a)** Molting L2 larva showing *lin-3/lacZ* expression (blue stain) in the anchor cell of the gonad rudiment. **(b)** Molting L2 larva showing the six vulval precursor cells (marked by vertical lines) and *lin-3/lacZ* expression in the anchor cell. **(c)** L4 larva with the anchor cell on top of the induced vulva. Scale bars 20μm.

To analyze where the *lin-15*$^+$ gene is normally required, Herman and Hedgecock (1990) used the strategy of *genetic mosaic analysis* described in Section 25.4. The investigators engineered hermaphrodites with two X chromosomes, each carrying a recessive loss-of-function allele of *lin-15*$^+$, plus an unstable X chromosome fragment carrying the wild-type allele *lin-15*$^+$. Cell-specific marker genes associated with the different *lin-15* alleles allowed the investigators to deduce from which lineages the X fragment had been lost. The fates adopted by VPCs in various mosaic individuals showed that normal vulva formation requires expression of the *lin-15*$^+$ gene outside the VPCs and the anchor cell, most likely in the hypodermis.

The researchers concluded that a signal dependent on *lin-15*$^+$ expression in the hypodermis inhibits an intrinsic tendency of all VPCs to adopt the primary or the secondary fate and instead promotes their assumption of the tertiary fate. Because the phenotypes caused by *let-23* and *let-60* mutations override those caused by *lin-15* mutations, it appears that lin-15 protein dampens the activity of let-23 protein in a way that can be overridden by the lin-3 signal from the anchor cell (Fig. 25.24).

LATERAL SIGNALS BETWEEN VPCS INFLUENCE THEIR FATES

Lateral signals between VPCs also influence their fates. In *lin-15* mutants with or without an anchor cell, all six VPCs tend to express the primary and secondary fates in an alternating pattern—for example, 2°-1°-2°-1°-2°-1°. Exceptions to this rule are adjacent pairs of secondary fates; adjacent pairs of primary fates are not observed (see Fig. 25.21c). These observations suggest that a VPC expressing the primary fate prevents neighboring VPCs from doing the same, thus causing them to express the secondary fate.

This lateral inhibition hypothesis receives further support from cell removal experiments with *lin-15* mutants (Sternberg, 1988). If the anchor cell and all VPCs except P7p are killed in mutant larvae, then P7p invariably expresses the primary fate. By contrast, if P8p is also left alive, P8p sometimes assumes the primary fate, in which case P7p expresses the secondary fate. These results indicate that the VPCs communicate so that only one of a neighboring pair expresses the primary fate.

The lateral signal depends on the activity of the *lin-12*$^+$ gene: In animals lacking the lin-12 function, *no* VPC expresses the secondary fate; in animals with increased lin-12 function, *all* six VPCs can express the secondary fate (Greenwald et al., 1983). *Mosaic analysis* of *lin-12* revealed that the lin-12 function is required in the VPCs that assume the secondary fate (Seydoux and Greenwald, 1989). Molecular studies showed that *lin-12*$^+$ encodes a membrane receptor of the Notch family, the same receptor type that is also involved in determining the anterior pharyngeal cell lineage (see Section 25.2; Greenwald, 1999). As we have seen in previous chapters, Notch signaling is frequently involved in cell determination by *lateral inhibition*—that is, a situation in which equivalent cells compete for a preferred state of determination and, if they have attained it, prevent their neighbors from doing the same (see Sections 6.4, 13.1, and 14.2). The simplest interpretation of the data reviewed so far is that the lin-12 protein acts as a receptor for a lateral signal sent by VPCs expressing the primary fate. However, the nature of this ligand still needs to be elucidated.

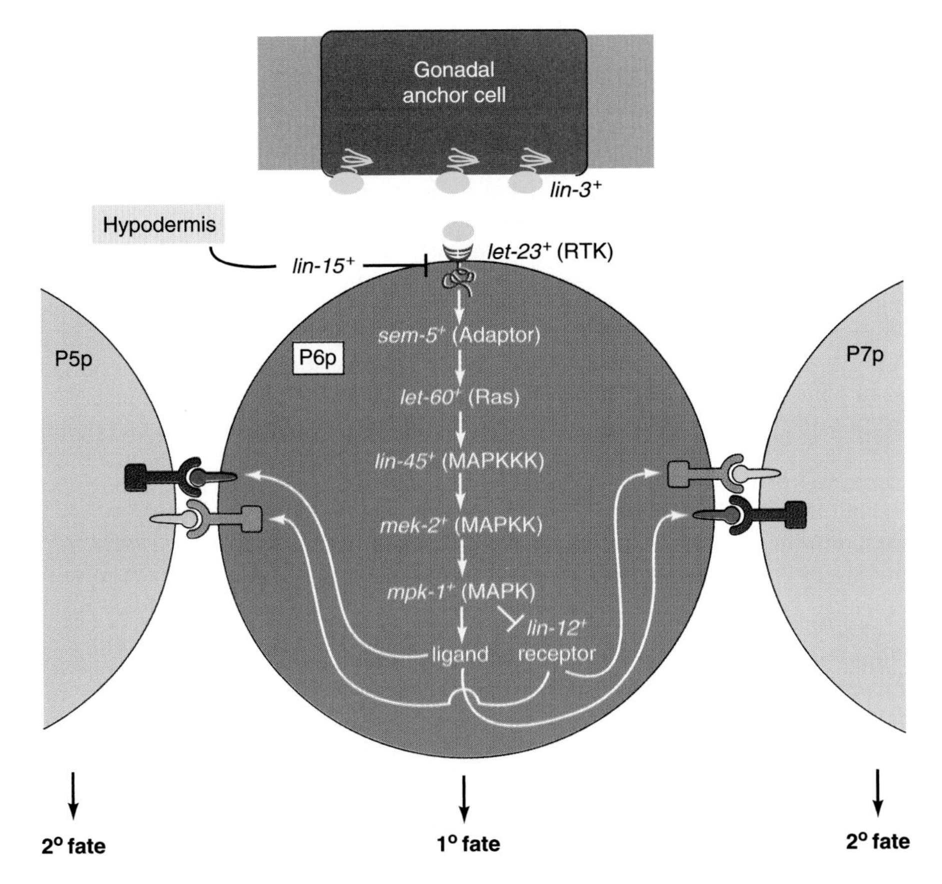

Figure 25.24 Molecular model of vulval induction. The *lin-3*+ gene encodes a signal protein similar to vertebrate epidermal growth factor (EGF). It activates a receptor tyrosine kinase (RTK) encoded by the *let-23*+ gene in the adjacent VPC, usually P6p. The activity of the Ras-MAPK pathway (see Fig. 2.28) biases a lateral inhibition among the VPCs involving a Notch-like receptor, encoded by the *lin-12*+ gene, and its matching ligand. The bias causes an upregulation of the ligand and a downregulation of the receptor. An inhibitory signal, encoded by the *lin-15*+ gene in hypodermal cells, dampens the RTK receptor. This inhibition is overridden by the EGF signal only in the vicinity of the anchor cell. As a result of these interactions, P6p acquires the primary fate, while the two adjacent VPCs acquire the secondary fate in vulva formation.

How do VPCs integrate the inducing signal received from the gonadal anchor cells and the lateral signal exchanged among themselves? Since the anchor cell signal seems to be diffusible, it might act as a *morphogen*, inducing 1° fate at a high concentration and 2° fate at a lower concentration. Alternatively, a diffusible signal from the anchor cells could induce the 1° fate in the one competent cell that receives the signal at the earliest time and/or at the highest concentration. This cell in turn could induce the 2° fate in its next neighbors. Finally, a gradient and a sequential induction mechanism could work in an overlapping or synergistic way. Genetic evidence, while not excluding the gradient model, does support the sequential induction model.

In *lin-12* mutants, the anchor cell signal induces three VPCs to express the primary fate (see Fig. 25.21d). This result shows that the signal from the anchor cell by itself is not sufficient to specify the normal pattern of 2°-1°-2° fates. On the other hand, the fact that in wild-type animals the primary fate is always expressed next to the anchor cell indicates that the anchor cell has a special impact on its nearest neighbor.

Do the VPCs that adopt the 2° fate need the inducing signal from the anchor cell? This question was settled by a *genetic mosaic analysis* for *let-23*, the gene that encodes the receptor for the inducing signal from the anchor cell. VPCs that are deficient for this gene nevertheless acquire the 2° fate as long as they are located next to a VPC that adopts the 1° fate (Simske and Kim, 1995). Thus, a VPC that adopts the 2° fate does not need the inducing signal from the anchor cell provided an adjacent VPC adopts the 1° fate. The simplest interpretation of this result is that the inducing signal from the anchor cell upregulates, through the RTK-Ras-MAPK signaling pathway (see Fig. 2.28), the production of the ligand for the lin-12 receptor in the adjoining VPC (Fig. 25.24; Greenwald, 1999). Upregulation of ligand production in this cell, which normally is P6p, would stimulate the lin-12 receptor in the neighboring VPCs, normally P5p and P7p. Lin-12 activity in these cells would in turn have two effects. First, these cells would adopt the 2° fate. Second, their production of ligand would be down-regulated, thus stabilizing the 1° fate for P6p and the 2° fate for P5p and P7p. The same effects would be reinforced if the RTK-Ras-MAPK pathway would also downregulate the production of lin-12 receptor.

Taken together, the molecular characterizations of the gene products that bring about vulva formation provide striking confirmation of the three-signal model, which was originally based on mutant analysis and cell removal experiments. Further analysis of the intracellular signaling processes within the VPCs that integrate these three signals should greatly enhance our understanding, in molecular terms, of equivalence groups and the integration of different signaling pathways.

SUMMARY

The soil nematode *C. elegans* is very suitable for developmental analysis because of its transparency, simple anatomy, and favorable genetic properties. Moreover, the development of the worm is nearly invariant: The number of cell divisions, their timing and orientation, and the final number of somatic cells are virtually the same for each hermaphrodite and for each male of the species.

During early cleavage, blastomeres are determined by cytoplasmic localization as well as by induction. The segregation of P granules into the germ line cells depends on a microfilament-based cellular transport system during a critical phase of the first cell cycle. The nuclei of the P_1 descendants accumulate the maternally expressed skn-1 protein, which is necessary and sufficient for determining the EMS blastomere. The MS blastomere, a descendant of EMS, induces certain descendants of the AB blastomere to form anterior pharyngeal muscle.

An interaction between the P_2 blastomere and EMS polarizes the latter with regard to mitotic spindle orientation and uneven distribution of extranuclear determinants. As a result, EMS divides unequally into the E blastomere, which produces endoderm, and the MS blastomere, which contributes to mesoderm.

Heterochronic genes in *C. elegans* control the timing of developmental events. Mutations in heterochronic genes cause cells to undergo patterns of cell division and differentiation that are normally observed at a different stage. For instance, the lin-14 protein forms a temporal gradient, with a high concentration during the first larval stage and a declining concentration thereafter. Mutations that shorten the period of lin-14 protein synthesis cause the cell division pattern to age prematurely. Conversely, extended synthesis of lin-14 protein keeps the cell lineage permanently young. The normal down-regulation of lin-14 protein synthesis depends on a regulatory RNA encoded by the *lin-4*$^+$ gene, which is activated by a feeding-dependent stimulus.

Some of the cells that originate by mitosis during the development of *C. elegans* do not divide or differentiate; instead, these cells undergo a process known as programmed cell death or apoptosis. Apoptosis is controlled by a genetic pathway. It is initiated by the *ced-3*$^+$ gene, which encodes a cysteine aspartate–specific protease, or caspase. Whether a cell lives or dies is determined autonomously in each cell and is controlled by regulatory genes that act on *ced-3*$^+$. Some of these regulatory genes act in all cells, while other genes affect only specific cells.

The formation of the vulva in *C. elegans* has been used to study the role of cell interactions during larval development. Six vulval precursor cells (VPCs) located along the ventral midline form an equivalence group: All VPCs have the potency to form different parts of the vulva, but only three of them normally contribute to it. Vulva development requires signals from the anchor cell of the gonad, among the VPCs themselves, and from the hypodermis. The anchor cell produces an inductive signal that is propagated through the RTK-Ras-MAPK signaling pathway in the VPC adjacent to the anchor cells. This signal biases Notch-dependent signaling among the VPCs, which is also dampened by a third signal from the hypodermis.

The nearly invariant cell lineage and other features of *C. elegans* have allowed scientists to analyze the development of single cells with unparalleled precision. Studies of mutants of *C. elegans* have helped to uncover some of the molecular mechanisms involved in such basic developmental processes as cytoplasmic localization, induction, timing, and programmed cell death. At least some of the molecules involved—such as growth factors, their receptors, and regulatory signals for cell death—are shared with vertebrates and presumably other organisms. Thus, many of the data gathered from the tiny roundworm will promote a better understanding of basic molecular mechanisms that underlie the development of many organisms.

Sex Determination

Figure 26.1 Sexual dimorphism. In mammals and many other groups of animals, male and females differ not only in their reproductive organs but also in other physical and behavioral traits. In deer, males are considerably larger and have conspicuous antlers, which they use for fighting. The development of sex-specific traits in mammals is controlled by the SRY^{+} gene on the Y chromosome, which is present in males but absent in females.

THE majority of animal species are ***dioecious*** (Gk. *di-*, "double"; *oikos*, "house"), meaning that individuals exist in the form of two sexes with different gonads. Typically these are ***females*** producing eggs in *ovaries* and ***males*** producing sperm in *testes*. In many dioecious species, the differences between the sexes are not limited to reproductive organs but extend to other characteristics, such as size, ornaments, and weaponry. This situation is called ***sexual dimorphism*** (Gk. *di-; morphe*, "form"). The degree of sexual dimorphism varies: Among vertebrates, mammals (Fig. 26.1) and birds tend to be more sexually dimorphic than fishes and reptiles.

A considerable number of invertebrates are ***monoecious,*** meaning that all individuals of a species look alike and produce both eggs and sperm. Such individuals, which have both testes and ovaries, are called ***hermaphrodites.*** However, some dioecious species have hermaphrodites, too. For instance, roundworms of the species *Caenorhabditis elegans* are either hermaphrodites or males, as described in Section 25.1. Males and hermaphrodites of this species are as distinct as males and females of other types of animals.

The natural event by which an individual of a dioecious species becomes either male or female (hermaphroditic) is called ***sex determination.*** An organism's sex has profound consequences for the course of its development. It determines whether its germ cells will develop into eggs or sperm, and whether its reproductive organs will be male or female . In sexually dimorphic species, sex determination also affects many other organs and behaviors.

Biologists have uncovered various mechanisms of sex determination, which fall into two categories (Table 26.1). In some species, sex is determined *after fertilization* by environmental factors such as temperature or the sex of preexisting adults. This type of sex determination is known as ***environmental sex determination.*** In other species, sex is determined *at fertilization* by the combination of genes that the zygote inherits. This type of sex determination is called ***genotypic sex determination.***

In this chapter, we will first look at examples of environmental sex determination and then focus on genotypic sex determination. The latter has been investigated especially well in three types of animals: the fruit fly *Drosophila,* the roundworm *C. elegans,* and placental mammals, represented by the human and the mouse. A comparison of sex determination in these organisms reveals that certain biological processes are common to all species with genotypic sex determination. However, the genetic and molecular mechanisms that control these processes differ widely and seem to have evolved independently in many taxonomic groups.

26.1 Environmental Sex Determination

A well-known case of environmental sex determination is *Bonellia viridis,* a marine echiuroid worm (Leutert, 1975). Adult females of this species attach to rocks in the ocean. The body of the female measures more than 10 cm, and a long proboscis extends from it for feeding.

table 26.1 Examples of Different Sex Determination Systems

Species	Mechanism	Sexes	
Mouse, human	GSD: dominant Y	XX: female	XY: male
Birds	GSD: dominant W(?)	ZW: female	ZZ: male
Turtles	ESD: temperature	Warm: female	Cool: male
Alligators	ESD: temperature	Cool: female	Warm: male
Insects:			
Drosophila melanogaster	GSD: X:A ratio	XX: female	XY: male
Musca domestica	GSD: dominant M locus	m/m: female	M/m: male
Apis mellifera	GSD: haplodiploidy	Diploid: female	Haploid: male
Nematodes:			
Caenorhabditis elegans	GSD: X:A ratio	XX: hermaphrodite	XO: male
Meloidogyne incognita	ESD: population density	Sparse: female	Crowded: male
Echiuroid worm:			
Bonellia viridis	ESD: female presence	Yes: male	No: female

GSD = genotypic sex determination; ESD = environmental sex determination.

Modified from Hodgkin (1992). Used with permission. © FCSU Press.

The males, which at 1 to 3 mm are tiny by comparison, live as parasites inside the females. The larvae of *Bonellia* live as part of the *plankton*—that is, the small organisms that passively drift in the water. Sex determination occurs when the larvae settle on a substrate and metamorphose. Those that settle in isolation develop into females, whereas those that fall on an adult female become males.

Two hypotheses could explain this single observation: environmental sex determination, or genotypic sex determination coupled with differential mortality. According to the latter hypothesis, genotypically male larvae would survive if they settled on females but die if they settled in isolation, and the converse would happen to genotypically female larvae. To test this hypothesis, Leutert (1975) kept *Bonellia* in laboratory culture. He observed that *more than half* of a known number of larvae developed as males under the influence of females, while *more than half* of the initial number of larvae developed as females in isolation. This result showed that at least some fraction of the larvae had the potential to develop into either sex, with the choice being determined by the environment. This kind of potential is inconsistent with strict genotypic sex determination.

Dependence of sex determination on social context is also observed in several species of fish (Crews, 1994). Orange and white anemone fish are born male and later develop into females. The opposite course is observed in several coral reef fish, which start out female and later become male. The timing of the sex change depends on a social trigger, such as the disappearance of a dominant male or female. Still other fish species, including the butter hamlet, have gonads that produce eggs and sperm simultaneously; individuals may alternate between male and female behavior during successive matings. All types of sex determination that depend on social context seem to maximize the chance of reproduction in small groups of individuals.

An environmental sex determination strategy that maximizes the number of offspring that can be produced under favorable conditions is observed in a nematode, *Meloidogyne incognita,* which is a plant parasite. If nutrients are plentiful and the population density is low, more worms become female, which enhances the reproductive potential of the population.

A puzzling type of environmental sex determination is observed in many reptiles. In all crocodilians, many turtles, and some lizards, sex is determined during embryogenesis by the incubation temperature (J. J. Bull, 1980, 1983; Crews, 1994). These reptiles are egg layers that place their eggs in the ground or in constructed mounds. The temperature of the external environment is the sex-determining factor, and a small temperature difference may cause dramatic changes in the ***sex ratio*** (proportion of males among all recent offspring in a population). For instance, in certain map turtles, the sex ratio drops abruptly from 1.0 (all males) to 0.0 (all females) as the ambient temperature increases from 28 to 30°C (Fig. 26.2a). This pattern of cool temperatures producing males and warm temperatures yielding females is common among turtles. The pattern is reversed in lizards and alligators (Fig. 26.2b). A third pattern of temperature-dependent sex determination, in which females develop at low and high temperature extremes and males develop at intermediate temperatures, is observed in snapping turtles, the leopard gecko, and crocodiles (Fig. 26.2c). In still other reptiles, the sex ratio is not significantly affected by incubation temperature (Fig. 26.2d). The adaptive values of the differing patterns of temperature-dependent sex determination are

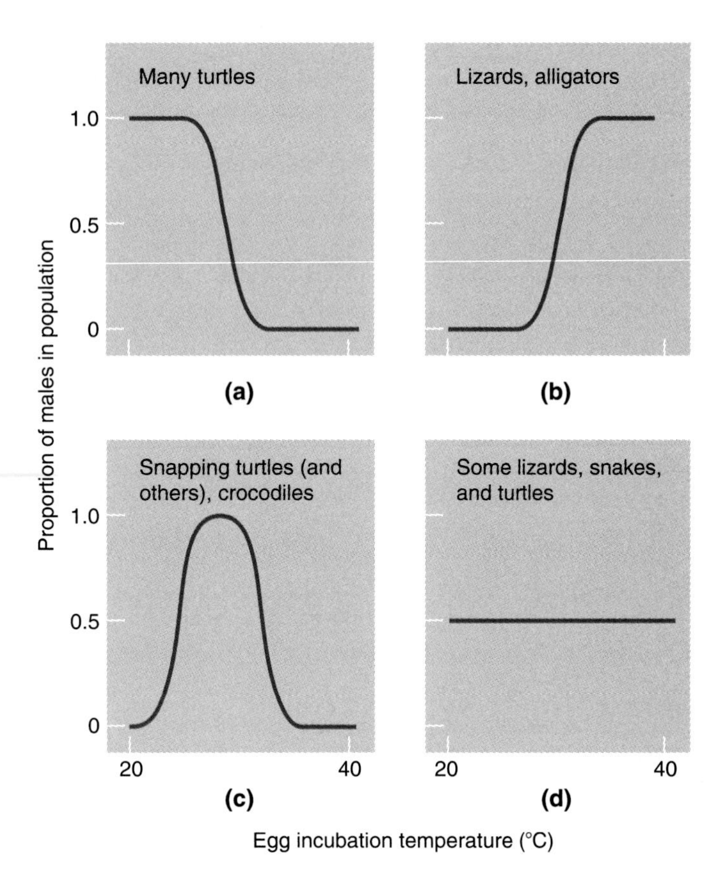

Figure 26.2 Sex ratio (proportion of males in all recent offspring of a population) as a function of egg incubation temperature in different reptiles. Four general patterns are recognized. **(a)** In many turtles, males develop at low temperatures and females at high temperatures. **(b)** The reverse is seen in many lizards and alligators: Females develop at low temperatures, males at high temperatures. **(c)** A pattern of females at low and high temperatures, and males at intermediate temperatures occurs in the snapping turtle, the leopard gecko, and crocodiles. **(d)** In other reptiles, the sex ratio is not significantly influenced by the incubation temperature.

not obvious. Presumably, they allow males to develop at temperatures that benefit males more than females, and vice versa. It will be interesting to find out whether females adaptively modify the sex ratio of their offspring by choosing appropriate nest sites.

In reptiles with temperature-dependent sex determination, the sex is determined by the temperature that prevails midway through embryonic development, when the originally indifferent gonads develop either as ovaries or as testes. After this event, the sex of the animal is fixed for life. This pattern has many similarities with the sexual development of mammals, which also have a sexually indifferent period, which lasts through the sixth week of pregnancy in humans (see Section 27.2). However, the critical event that directs the indifferent gonads to become either testes or ovaries is triggered by temperature in reptiles as opposed to gene activities in mammals.

26.2 Genotypic Sex Determination

Environmental sex determination was the dominant model until the beginning of the twentieth century (Mittwoch, 1985). In particular, human sex determination was ascribed to environmental factors ranging from nutrition to the heat of passion during coitus. However, in humans as well as most other organisms, the sex ratio is close to 0.5 and is remarkably resistant to environmental manipulation. After Mendel's laws gained widespread recognition and chromosomes were discovered as the carriers of genetic information, geneticists began to speculate that a sex ratio of 0.5 could be achieved simply if one sex was homozygous and the other sex heterozygous for some sex-determining gene. Soon thereafter, sexually dimorphic chromosomes (called *sex chromosomes,* for short) were found in several species, including insects (McClung, 1902; E. B. Wilson, 1905) and humans (Painter, 1923).

The well-known mechanism of sex determination in humans and other mammals is diagrammed in Figure 26.3. Here, males have two different sex chromosomes, designated X and Y, whereas females have two X chromosomes. All other chromosomes are indistinguishable between the sexes and are called *autosomes.* A complete haploid set of autosomes is designated A. Males produce two types of sperm with different sets of chromosomes, XA and YA, and are therefore called ***heterogametic.*** Females produce only one chromosomal type of egg, XA, and are called ***homogametic.*** Fertilization by an XA sperm results in an XXAA zygote, which develops into a female. Fertilization by a YA sperm results in an XYAA zygote, which develops into a male.

The heterogametic sex is not necessarily the male. In birds, for instance, females are heterogametic, and their

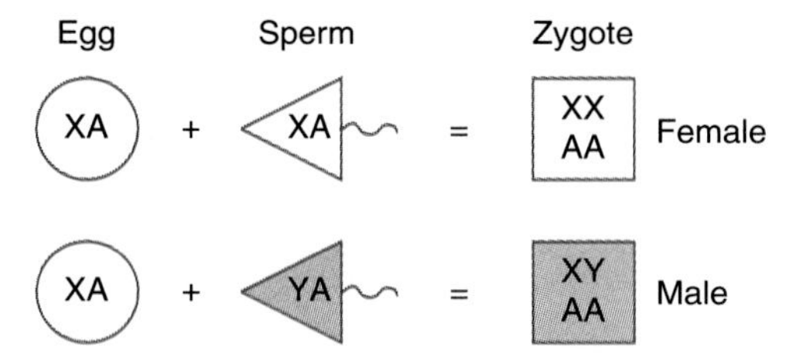

Figure 26.3 Sex determination in mammals. In addition to two sets of autosomes (AA), males have two different sex chromosomes, designated X and Y, whereas females have two X chromosomes. Consequently, males produce two types of sperm with different sets of chromosomes, XA and YA, while females produce only one chromosomal type of egg, XA. Fertilization by an XA sperm results in an XXAA zygote, which develops into a female. Fertilization by a YA sperm results in an XYAA zygote, which develops into a male. This system generates a sex ratio of 0.5, provided that XA and YA sperm have equal chances at fertilization, and that XXAA and XYAA zygotes are equally viable.

sex chromosomes are traditionally designated ZW. Male birds are homogametic, and their sex chromosomes are designated ZZ.

Instead of occurring in two forms, sex chromosomes may also occur in two *numbers*—typically, two chromosomes in one sex and one chromosome in the other sex. The most common system of this type is exemplified by *C. elegans*. Here, hermaphrodites have XXAA diploid cells and produce XA sperm as well as XA eggs, whereas males have XOAA diploid cells and produce sperm that are either XA or OA, with O representing *no* chromosome.

In yet another system of genotypic sex determination, females arise from fertilized eggs whereas males arise from unfertilized eggs. This mechanism, known as ***arrhenotoky*** (Gk. *arrhen*, "male"; *tokos*, "childbirth"), is common among ants, bees, and wasps.

Finally, there are species whose chromosomes are not visibly dimorphic but whose sex determination is nevertheless genotypic. In the simplest case, sex is determined by two alleles of the same gene. One sex is homozygous (*m*/*m*) for one allele and the other sex is heterozygous (*M*/*m*). This system is found in the housefly, *Musca domestica*, which has *m*/*m* females and *M*/*m* males.

Genotypic sex determination systems have changed frequently during evolution (Hodgkin, 1992). A survey of sex-determining systems in amphibians revealed widespread variation in the heterogametic sex and in the extent of sex chromosome dimorphism (Hillis and Green, 1990). Among 63 frog and salamander species, researchers found ZW female heterogamety, ZO female heterogamety, and XY male heterogamety. Phylogenetic analysis suggests that the ancestral state for amphibians was female heterogamety, and that XY male heterogamety has evolved independently at least seven times. In one case, the data suggest that male heterogamety has reversed to female heterogamety.

table 26.2 Chromosome Status and Sexual Phenotype in Three Organisms with XX-XY Sex Determination

Sex Chromosome Status	Human Phenotype	Mouse Phenotype	Drosophila Phenotype
XO	Sterile female*	Fertile female	Sterile male
XX	Normal female	Normal female	Normal female
XXX	Fertile female	Unknown	Sterile female
XY	Normal male	Normal male	Normal male
XXY	Sterile male†	Sterile male	Fertile female
XYY	Fertile male	Semisterile male	Fertile male

* Turner's syndrome.

† Klinefelter's syndrome.

After Sutton (1988). Used with permission. © 1988 Harcourt Brace & Company.

26.3 Sex Determination in Mammals

Sex determination in mammals is genotypic, and the mechanism seems to be fairly well conserved between mice and humans. This conservation has facilitated the genetic and molecular analysis of mammalian sex determination, because data gathered from humans and mice complement each other. On the one hand, abnormal genetic conditions that cause sterility are readily discovered in humans when the afflicted individuals seek medical attention. On the other hand, mice are suitable for experiments that would be unethical with humans.

MALENESS IN MAMMALS DEPENDS ON THE Y CHROMOSOME

The XX/XY system of sex determination is found in mammals and many other taxonomic groups. Yet the genetic mechanisms underlying this system may differ. This is illustrated by the sexual phenotypes of individuals with abnormal combinations of sex chromosomes (Table 26.2). For example, individuals with only one X chromosome (designated XO) are infertile *males* in *Drosophila* but are *females* in humans and mice. Human XO females show *Turner's syndrome*, which includes short stature, ovarian degeneration, and lack of secondary sex characteristics. The chromosome status XXY generates fertile *females* in *Drosophila* but sterile *males* in humans. Such human males show *Klinefelter's syndrome*, which is characterized by long legs, wide hips, narrow shoulders, and lack of spermatogenesis.

From these and other abnormal chromosome configurations, it is clear that development of a male human or mouse depends on the presence of a Y chromosome (Welshons and Russell, 1959). Conversely, the mere absence of a Y chromosome is sufficient for development of a female, which is therefore considered the default program of mammalian sex determination. Genotypic sex determination in many species exemplifies in a striking manner the *principle of default programs*. The presence of a particular genetic signal leads to the development of one sex; in the absence of this signal, the opposite sex will develop by default.

DOSAGE COMPENSATION IN MAMMALS OCCURS BY X CHROMOSOME INACTIVATION

Typical X chromosomes contain a large number of genes that are unrelated to sex determination but nevertheless are not present on the Y chromosome. Consequently, these genes are present in two doses in females and one dose in males, generating a potential imbalance. Studies on losses and duplications of chromosome fragments in otherwise diploid animals show that haploidy or triploidy for more than a few percent of all genes is deleterious or lethal. Apparently, many genes are involved in regulatory networks that are finely tuned and cannot tolerate a major reduction or increase in the number of copies, or ***dosage,*** of a gene. It appears that each group of animals has found some mechanism of ***dosage compensation*** to equalize the expression of non-sex-related genes located on sex chromosomes.

Mammals achieve dosage compensation by inactivating one of the two X chromosomes in nearly all cells of a female during the blastocyst stage (Fig. 26.4; McCarrey and Dilworth, 1992). In the inner cell mass, which gives rise to the embryo proper, the choice of which of the two X chromosomes in any given cell is inactivated seems to be random (Gartler and Riggs, 1983). In female somatic cells, the inactivated X chromosome of a cell will replicate during mitosis but will remain inactive in all the cell's progeny. In the female germ line, inactivated X chromosomes are reactivated when oogonia enter meiotic prophase (Kratzer and Chapman, 1981). Even in the male germ line, the X chromosome is temporarily inactivated between meiosis and fertilization. The inactivated X chromosomes are visible in interphase nuclei as deeply staining bodies of *heterochromatin,* known as *Barr bodies* or ***sex chromatin.*** Observations on Barr bodies in cells from mammals with irregular numbers of chromosomes, such as humans with Turner's or Klinefelter's syndrome, have shown that mammals generally inactivate *all X chromosomes except for one* per diploid set of chromosomes.

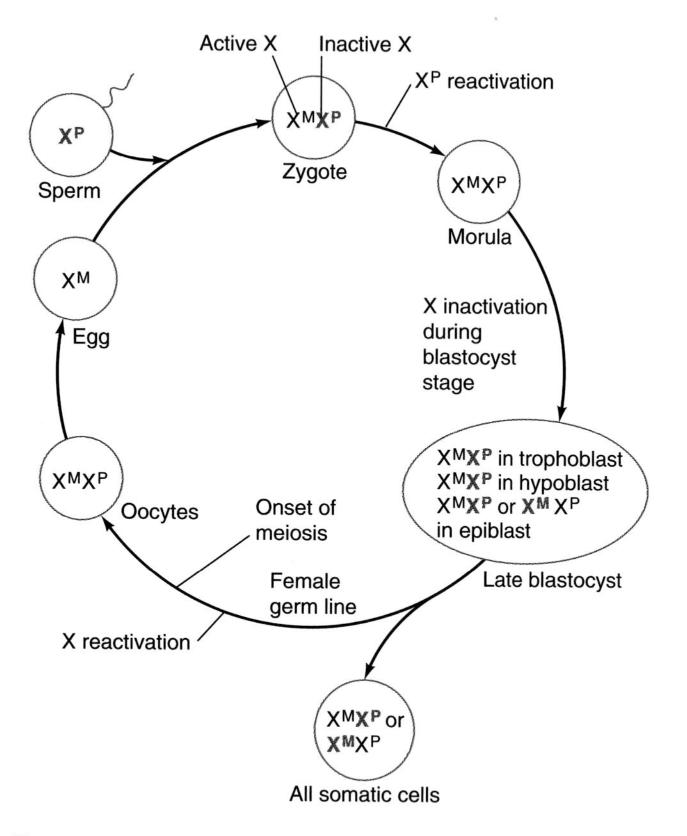

Figure 26.4 Cycle of X chromosome inactivation in the mouse. Both X chromosomes are active in the morula. Inactivation occurs during the blastocyst stage. The paternally inherited X chromosome (X^P) is inactivated selectively in the trophoblast and the hypoblast. In the epiblast, which gives rise to the embryo proper, inactivation occurs randomly in either X^P or the maternally inherited X chromosome (X^M). These inactivations are passed on clonally and are permanent in somatic cells. In the female germ line, the inactivated chromosome is reactivated shortly before meiosis. In the male germ line, the X chromosome is inactivated between meiosis and fertilization.

Nearly all genes in sex chromatin are shut down, but a few genes escape inactivation. These genes are located primarily in the ***pseudoautosomal region,*** which is similar in X and Y chromosomes and therefore behaves like an autosomal region in genetic tests. Presumably, the extra dosage of noninactivated pseudoautosomal genes in XXY males accounts for Klinefelter's syndrome, whereas the reduced dosage of these genes in XO females causes Turner's syndrome.

Random inactivation of all but one X chromosome as a mechanism of dosage compensation was first proposed by Lyon (1961) based on her observation of sex chromatin and on her analysis of coat color patterns in mice. If a female mouse is heterozygous for two alleles (*Pm* and *Pp*) of an X-linked pigmentation gene, then patches expressing *Pm* will alternate in her coat with patches expressing *Pp*. Subsequent studies on cultured cell clones confirmed Lyon's conclusions. Davidson and coworkers (1963) showed that human females heterozygous for genetic alleles encoding two variants of glucose-6-phosphate dehydrogenase (G6PD-A and G6PD-B) have two types of skin cells: One cell type expresses only G6PD-A, and the other type expresses only G6PD-B (Fig. 26.5). Corresponding results were obtained for another human X-linked enzyme, hypoxanthine phosphoribosyltransferase (Migeon, 1971).

The inactivation of an X chromosome is associated with the presence of a specific transcript, designated Xist (for *X-i*nactive-*s*pecific *t*ranscript) in the mouse and XIST in humans (*Kuroda and Meller, 1997;* Panning and Jaenisch, 1998). Xist, which is encoded by an X-linked gene, is necessary and sufficient for chromosome inactivation (Penny et al., 1996; Heard et al., 1999). The Xist transcript shows the characteristics of an mRNA but is not translated. Initially, Xist is transcribed from all X chromosomes present in a cell. However, Xist accumulates selectively along the entire length of those X chromosomes that become inactivated—that is, on all except one X chromosome per diploid cell.

Remarkably, the accumulation of Xist results from its stabilization rather than from an increased rate of

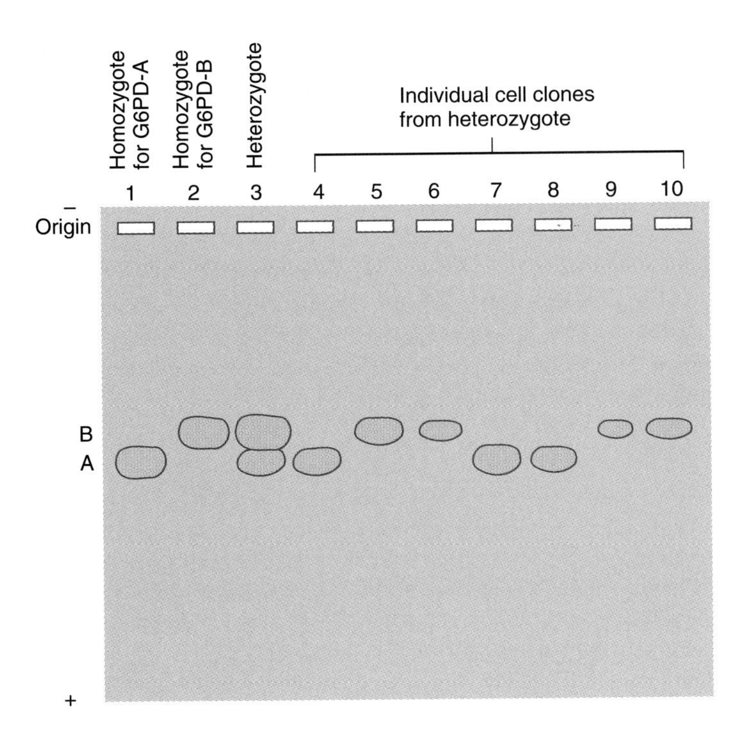

Figure 26.5 Occurrence of two variants of the enzyme glucose-6-phosphate dehydrogenase (G6PD-A and G6PD-B) in human females as revealed by electrophoresis. Enzyme from females homozygous for either variant A or variant B appears in only one band after electrophoresis and specific staining (lanes 1 and 2). Both variants are seen in cultures of mixed skin cells from heterozygous females (lane 3). In contrast, each clone raised from a single skin cell of a heterozygous female (lanes 4 to 10) contains only one variant.

synthesis (Panning et al, 1997; Sheardown et al., 1997). Stable Xist RNA associates with its chromosome in a way that somehow establishes an inactive state, while failure of Xist to accumulate preserves the X chromosome in an active state. This initial decision is subsequently fixed in a step that depends on DNA methylation, after which the transcription of Xist is restricted to inactive X chromosomes (Panning and Jaenisch, 1996). The molecular mechanisms of Xist stabilization and its selective accumulation are still being investigated.

THE FIRST MORPHOLOGICAL SEX DIFFERENCE APPEARS IN THE GONAD

Mammalian development begins with a *sexually indifferent stage,* which in the human lasts through the first 6 weeks of gestation. The first morphological difference in the development of males versus females appears when the indifferent gonad rudiments begin to form either testes or ovaries. The emergence of this difference is referred to as *primary sex differentiation;* it occurs independently of sex hormones but includes the differentiation of those gonadal cells that will produce sex hormones during the following phase, known as *secondary sex differentiation* (see Section 27.2).

The observation that maleness in mammals depends on the presence of a Y chromosome indicates that some signal encoded by the Y chromosome directs the formation of the testes. The part of the Y chromosome that promotes testis development has been termed the ***testis-determining factor,*** abbreviated as ***TDF*** in humans and ***Tdy*** in mice.

When and where in the developing gonad does TDF act? One way of investigating this problem is to identify the gonadal cell type that shows the first visible sexual difference. The gonad develops from two cell lineages: somatic cells and germ line cells. The somatic cells, originating from *intermediate mesoderm,* form a *genital ridge* on the *mesonephros,* whereas the primordial germ cells arise from the posterior yolk sac (see Fig. 3.3). The primordial germ cells migrate to the genital ridge, where they arrive during the 6th week of development in the human and during the 11th or 12th day after mating in the mouse. Shortly before and during the arrival of the primordial germ cells, the coelomic epithelium of the genital ridge proliferates, and columns of epithelial cells penetrate the underlying mesenchyme. Here they surround the germ cells and form irregularly shaped cords called ***primitive sex cords*** (Fig. 26.6a). Through this stage, there is no morphological difference between the gonads of XX and XY individuals, so the gonad at this stage is known as the ***indifferent gonad.***

Among the somatic cells of the gonad, three major lineages are distinguished. During the indifferent stage they are known as the supporting cell lineage, the steroidal cell lineage, and the connective cell lineage. The ***supporting cell lineage*** is derived from the epithelial component of the primitive sex cords. This lineage gives rise to the *Sertoli cells* of the testis and the *follicle cells* of the ovary. The ***steroidal cell lineage*** gives rise to the cells that will produce the gonadal steroid hormones—namely, the *interstitial cells* (*Leydig cells*) of the testis and the *thecal cells* of the ovary. The ***connective cell lineage*** will produce the connective tissues that form the structural framework of the gonads.

The first gonadal cell lineage to show a morphologically apparent sexual difference is the supporting cell lineage. In XX embryos, the primitive sex cords persist only in the cortex (periphery) of the gonad, where the supporting cells form *prefollicle* cells (Fig. 26.6b). Thereafter, the germ cells enter meiosis, an event that begins after 12 weeks of development in the human female but not until the onset of puberty in males (see Fig. 3.6). Later, the cortical cords derived from the primitive sex cords break up into primordial follicles.

In XY embryos, the primitive sex cords persist in the medulla (core) of the gonad, where they form ***testis cords*** (Fig. 26.6c). The supporting cells form *pre-Sertoli cells,* which are the first male-specific cell type to differentiate. The germ cells associated with pre-Sertoli cells

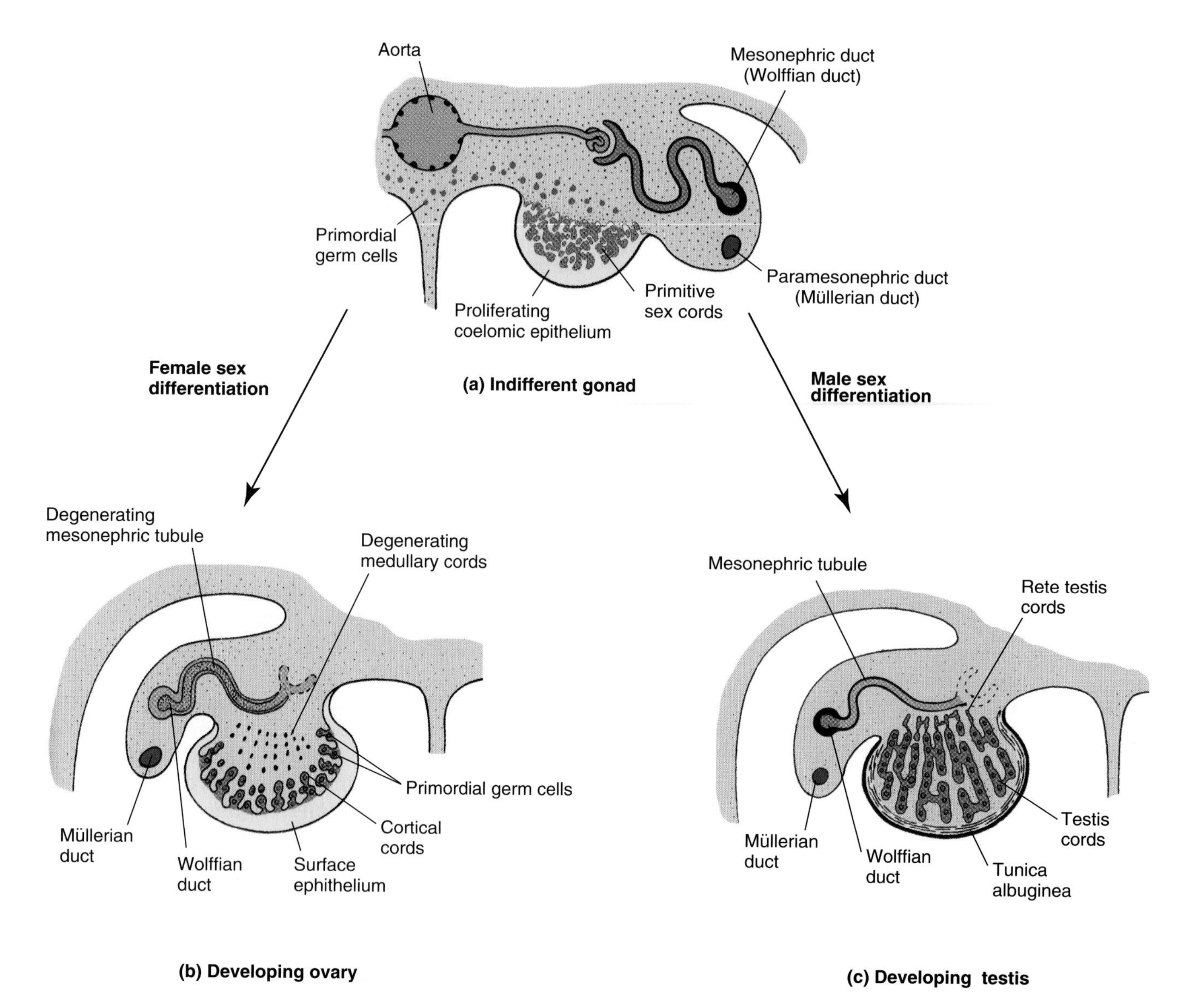

Figure 26.6 Primary sex differentiation in the human. All diagrams show transverse sections. **(a)** Indifferent gonad during the sixth week of gestation. Immigrating primordial germ cells are surrounded by proliferating cells from the coelomic epithelium, which form primitive sex cords traversing the gonad. **(b)** Developing ovary during the seventh week of gestation, showing the persistence of primitive sex cords in the cortex and their degeneration in the medulla (core). **(c)** Testis during the eighth week of development, showing the persistence of the primary sex cords in the medulla, while the cortex develops into a whitish, fibrous coat, the tunica albuginea.

do not enter meiosis—a clear sign that they are not on the path of female development. The testis cords continue to develop into *seminiferous tubules* of the testis.

Supporting cells and germ cells develop in close interaction (McLaren, 1991; Fig. 26.7). If germ cells are absent, as in certain mouse mutants or after experimental interference with germ cell migration, prefollicular cells are formed, but they degenerate later. If germ cells are lost from developing ovaries after meiosis, the follicle cells transdifferentiate into cells that resemble Sertoli cells. Conversely, the formation of pre-Sertoli cells and testis cords inhibits the germ cells from entering meiosis.

The observations described in this section suggest that Tdy may determine the supporting cell lineage of the gonad to develop as pre-Sertoli cells. However, morphological data cannot exclude the possibility that Tdy acts first in a different type of cell without causing a morphological change, and that a signal coming from this type of cell triggers the morphological changes in the supporting cells.

THE TESTIS-DETERMINING FACTOR ACTS PRIMARILY IN PROSPECTIVE SERTOLI CELLS

Conclusive evidence for the supporting cell lineage as the primary site of Tdy action was obtained by *genetic mosaic analysis.* Mice that are genetic mosaics, also known as *chimeric mice,* are routinely obtained by fusing

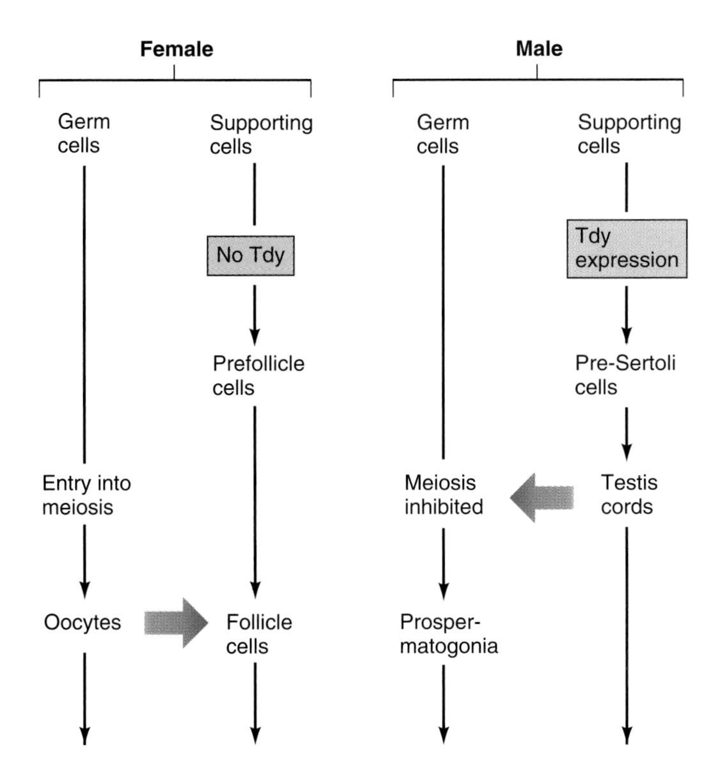

Figure 26.7 Development of supporting cells and germ cells in the fetal mouse gonad. Arrows indicate interactions between the two cell lineages. In males, expression of the testis-determining factor (Tdy) in the supporting cells is followed by the development of pre-Sertoli cells, which inhibit meiosis and induce the germ line cells to form prospermatogonia. In females, the absence of Tdy allows germ cells to enter meiosis and supporting cells to develop into prefollicle cells. The latter depend on the presence of oocytes to develop into follicles.

two embryos from different mouse strains before compaction (see Section 6.4). The chimeras are allowed to develop in foster mothers until they reach the stage of interest for analysis. Chance dictates that 50% of all chimeras will be XX⇔XY chimeras—that is, will be composed of XX and XY cells; these chimeras can be identified by microscopic inspection of their sex chromosomes. If mouse strains with morphologically different Y chromosomes are used, one can also tell which strain contributed the XY cells to a chimera.

Early experiments with mouse chimeras indicated that Tdy does not act in germ line cells: An oocyte in a chimeric mouse was found to have an XY karyotype, indicating that Tdy is not *sufficient* for male germ line development (E. P. Evans et al., 1977). Tdy is also not *necessary* for male germ line initiation, because germ cells without Y chromosomes were seen to develop into spermatogonia, even though they did not proceed with meiosis and spermiogenesis (Burgoyne et al., 1988). Therefore, the *initiation* of male gametogenesis must be an indirect effect of the Y chromosome brought about by creation of a testicular environment, rather than a direct effect of Y chromosome expression in the germ line. Only meiosis and the following stages of spermatogenesis require expression of Y-linked genes in germ line cells, but these genes are different from Tdy (K. Ma et al., 1993).

Once germ cells were excluded as primary sites of Tdy action, the interest focused on the somatic gonadal cell lineages—in particular, the *Sertoli cells* and the *interstitial cells* (Leydig cells), which are the producers of the principal male sex hormones during the secondary phase of sexual differentiation.

TO determine which somatic cell lineage of the testis is the primary site of Tdy action, Burgoyne and coworkers (1988) created XX⇔XY chimeras from two mouse strains that had different electrophoretic variants of the enzyme glucose phosphate isomerase, designated here GPI-A and GPI-B. Chimeras were killed after 12 to 15 days of development, and the testes of developing males were removed for tissue separation. The external layer of connective tissue (tunica albuginea) was removed manually. The remainder of the testes was dissociated into cells by mild digestion with protease and collagenase, and the isolated cells were separated by sedimentation into one fraction containing Sertoli cells and another fraction containing Leydig cells. Proteins from each fraction were separated by gel electrophoresis and stained for GPI activity (Fig. 26.8). The results showed that XX cells contributed randomly to all testicular tissues except the Sertoli cells, to which XX cells never made a significant contribution. These data confirmed the hypothesis that Tdy is necessary and sufficient for Sertoli cell differentiation, but not for the development of other testicular cells.

Questions

1. How could the investigators know whether a given chimeric mouse had XY cells from the GPI-A strain, or the GPI-B strain, or both?
2. How should the result of the electrophoretic test shown in the bottom part of Figure 26.8 look if the XX cells had been contributed by the GPI-B strain?
3. What would you have concluded from a single test result as shown in Figure 26.8 if in the bottom row (XY-GPI-B) there had been no stained band in the middle column (Leydig cells)?

In a follow-up study, Palmer and Burgoyne (1991) used XX⇔XY mouse chimeras in which one donor strain was marked by a globin transgene, which could be detected by *in situ hybridization* (see Method 8.1). This marker allowed the investigators to determine the contributions of XX and XY cells to any testicular cell lineage in microscopic sections. They found that most cell types in fetal XX⇔XY testes were random mixtures of XX and XY cells. Only the Sertoli cells were derived mostly from the XY component of each chimera. This bias became even stronger at later stages of testicular development. However, small proportions of XX Sertoli cells were consistently observed.

The experiments described above show that the supporting cell lineage—which gives rise to the Sertoli cells

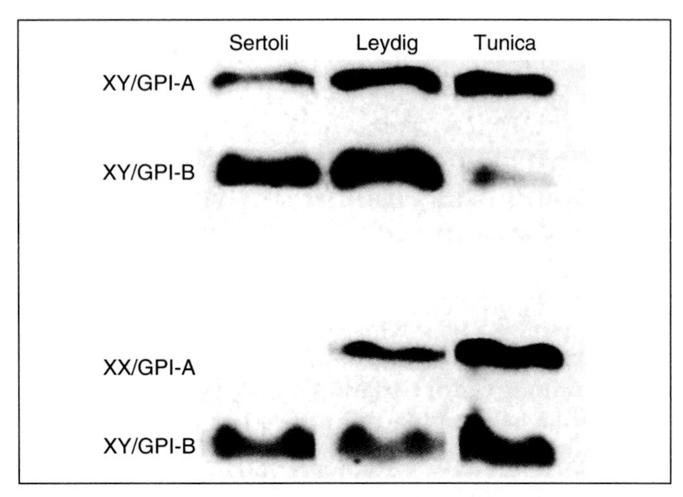

Figure 26.8 Analysis of testicular cell types from chimeric mice for the proportion of cells with Y chromosomes. Proteins from three cell types (top) obtained from individual chimeras were separated by gel electrophoresis and stained for glucose phosphate isomerase (GPI) activity. The two mouse strains used for generating each chimera had different GPI variants, designated GPI-A and GPI-B, and morphologically different sex chromosomes. Upper rows: An XY⇔XY chimera composed of two males had both GPI variants in each testicular tissue. Lower rows: An XX⇔XY chimera of one male and one female had both GPI variants in the tunica albuginea and the Leydig cell fraction. In contrast, the Sertoli cell fraction showed a strong GPI-B band, and a very weak GPI-A band that was detectable only by scanning the gel with a densitometer. Because the GPI-A variant in this chimera was derived from XX cells, it was apparent that very few of these cells contributed to the Sertoli cell fraction.

in males—is the initial site of Tdy function. The Tdy activity in this lineage apparently generates an environment in which other gonadal cell lineages will develop along testicular pathways, regardless of their own sex chromosome status. The small contributions of XX cells to the Sertoli cell lineage in chimeric testes indicate that Sertoli cell determination is subject to the *community effect* discussed in Section 6.3. If there are enough XY cells to get the Tdy-initiated process of Sertoli cell determination under way, a few XX cells can be recruited to join in. This indicates that in the development of a normal male, with all cells having a Y chromosome, the supporting cells in the indifferent gonad promote and stabilize one another's development into Sertoli cells.

26.4 Mapping and Cloning of the Testis-Determining Factor

The discovery that the Y chromosome is necessary for testis development in humans and mice was only the beginning of a long quest to identify the testis-determining factor. Is TDF/Tdy one gene or more? Exactly where on the Y chromosome is it located? What is the nature of the signal encoded by the gene(s)?

TRANSLOCATED Y CHROMOSOME FRAGMENTS CAUSING SEX REVERSAL ARE USED TO MAP TDF

The critical data for mapping TDF on the Y chromosome were provided by human patients whose phenotypic sex did not seem to match their chromosomal constitution. (When discussing these individuals, we will stick to our definition of females as individuals with ovaries, and males as individuals with testes, regardless of their sex chromosomes.) Of interest here are two types of sex-reversed individuals. ***Sex-reversed XX males*** are sterile men carrying what appear to be two X chromosomes and no Y. ***Sex-reversed XY females*** are women with Turner's syndrome carrying one X and one apparent Y chromosome. Most of these cases involve *homologous recombination* between the X and the Y chromosome. Data from such patients were used to map TDF by delineating the smallest Y chromosome fragment that causes sex reversal if translocated to an X chromosome.

The origin of sex-reversed males and females can be explained as the results of rare crossover events between the X and the Y chromosome during male meiosis (Fig. 26.9). However, the postulated exchanges of chromosome fragments are cytologically invisible and genetically difficult to detect because of the paucity of genes on the Y chromosome. Fortunately, cloned DNA probes from the human Y chromosome became available in the 1980s. By applying such probes to *Southern blots* (see Method 15.4) of total genomic DNA from sex-reversed human patients, one can directly demonstrate the postulated exchange of chromosome fragments. The crossover points can then be mapped, and the area containing TDF can be delineated.

Using this method, Palmer and coworkers (1989) mapped the crossover points of four sex-reversed XX males who had sought medical attention for sterility and irregularities in their genitalia (Fig. 26.10; Table 26.3). Each of the four men seemed to have two X chromosomes, but Southern blots of their genomic DNA revealed Y-specific DNA fragments that are not present in genomic DNA of normal females. Clearly, at least one of the chromosomes in each sex-reversal patient contained DNA segments that are normally found on the Y chromosome. In each case, the exchanged DNA included a segment from the pseudoautosomal region, which is homologous in X and Y chromosomes (Figs. 26.10 and 26.11a). None of the exchanged DNA fragments included the ZFY^{+}gene, which researchers had previously thought might be the TDF (Fig. 26.11b). Further probing showed that DNA segments located 10 kb and 20 kb away from the pseudoautosomal region were included in the exchanged DNA, whereas segments located 60 kb or farther away were not part of the exchanged DNA (Figs. 26.10 and 26.11c).

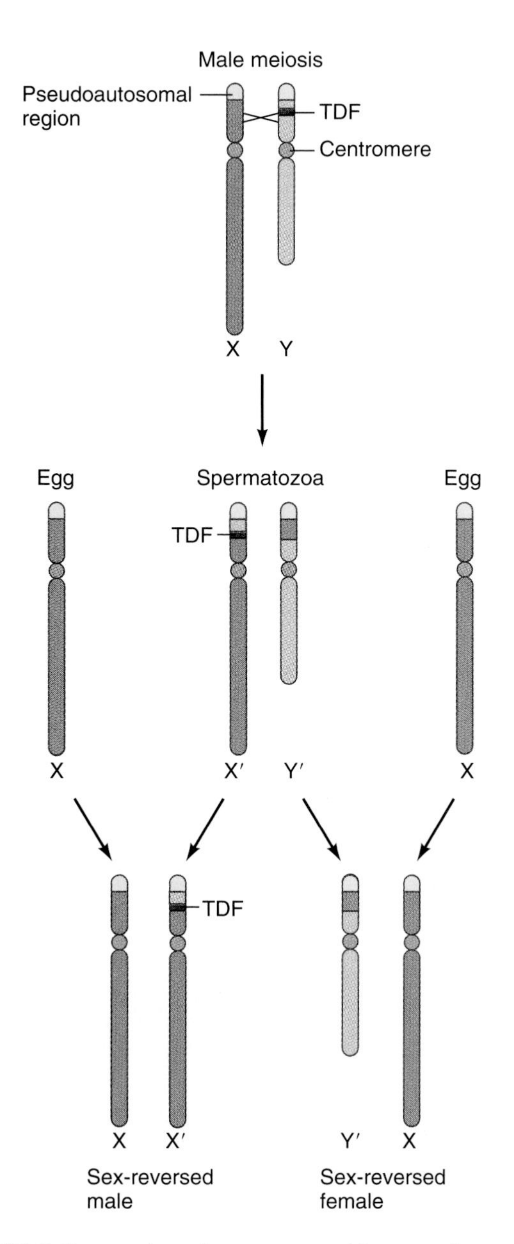

Figure 26.9 Generation of sex-reversed humans by rare X-Y recombination during male meiosis. The top diagram shows one chromatid each from the X and Y chromosomes during meiotic prophase I. The pseudoautosomal region is nearly identical in the two chromosomes. Only this region pairs during meiosis, and recombination events outside this region are correspondingly rare. If they occur, however, and if the exchanged fragment of the Y chromosome contains the testis-determining factor (TDF), then the crossover generates an X-like chromosome (X′) containing TDF and a Y-like chromosome (Y′) lacking TDF. If sperm with such chromosomes fertilize an egg, the resulting zygotes will develop into sex-reversed males and females, respectively.

The data described above allowed the investigators to map the gene(s) acting as TDF within 60 kb of DNA adjacent to the pseudoautosomal region. In a follow-up study using DNA from the same XX males but taking advantage of additional markers, Sinclair and coworkers (1990) further delimited the location of TDF to within 35 kb of DNA. Now the stage was set for identifying any genes in this region and testing them for the properties expected of TDF.

THE *SRY* FUNCTION IS NECESSARY FOR TESTIS DETERMINATION

Within the 35 kb of human DNA delimited as the location of TDF, Sinclair and coworkers (1990) found only one gene, which they termed ***SRY***$^+$ (for sex-determining region Y). It encodes a DNA-binding protein and is conserved among mammals. Gubbay and coworkers (1990) found that an orthologous mouse gene, *Sry*$^+$, was absent in a sex-reversed XY female mouse. These researchers and others (Koopman et al., 1990) then proceeded to test various adult and fetal mouse tissues for transcripts from the *Sry*$^+$ gene. In particular, they analyzed genital ridges from mouse fetuses aged around 11.5 dpc (days postcoitum), because the testis-determining factor (Tdy) is expected to be active at this time and in this tissue in males. Indeed, the researchers found Sry transcripts in male genital ridges from 10.5 through 12.5 dpc but none earlier or later. The transcripts were also absent from female fetuses. Thus, transcription of the *Sry*$^+$ gene was not a general aspect of gonad development but correlated specifically with testis determination.

Moreover, Sry transcripts were found in fetuses homozygous for the *white spotting* allele, which interferes with primordial germ cell migration so that the testes develop without germ cells. The *Sry*$^+$ gene therefore was transcribed in the somatic tissue of the gonad rather than in the germ line. In summary, these observations showed that *Sry*$^+$ was expressed exactly when and where Tdy was expected to be active.

The claim that *SRY*$^+$ (*Sry*$^+$) was the elusive TDF (Tdy) received further support when a sex-reversed human female was found to have a Y chromosome in which *SRY* was not deleted but was *newly mutated*. Berta and coworkers (1990) discovered this patient's point mutation when they were screening sex-reversed patients for alterations in the *SRY* gene. This mutation explained how the patient could have a Y chromosome and still be phenotypically female. Most interestingly, her father—the donor of the mutated Y chromosome—did not carry this mutation and was a normal male, and the same held for her brother; paternity was confirmed with several unrelated DNA probes. Evidently, the *SRY* gene had undergone a point mutation as it was passed on from father to daughter and had in the process lost its testis-determining capacity.

Of course, the simultaneous appearance of the mutation and the sex reversal could have been a coincidence. However, in a similar case involving another sex-reversed female, Jäger and coworkers (1990) found that a new frameshift mutation had taken place in the *SRY* gene. Corresponding observations were made on mice (Lovell-Badge and Robertson, 1990; Gubbay et al., 1992).

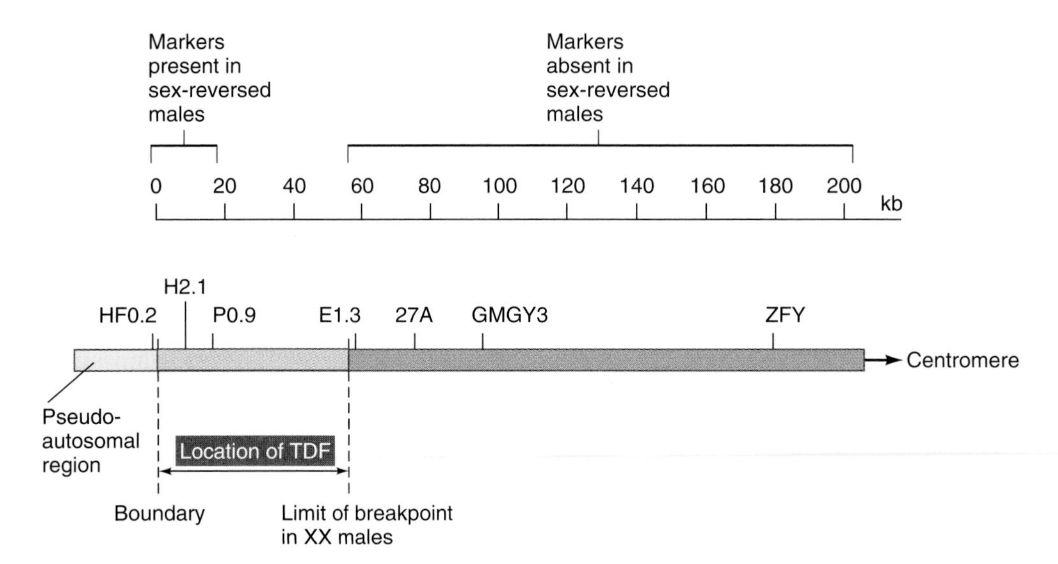

Figure 26.10 Map of the human Y chromosome (colored bar) near the pseudoautosomal region. The scale at the top indicates the distance from the pseudoautosomal boundary in kilobases (kb). The acronyms above the chromosome indicate the positions of cloned DNA segments used as labeled probes on Southern blots of genomic DNA from sex-reversed XX males (see Table 26.3 and Fig. 26.11). The results from those blots indicate that TDF maps to within 60 kb of the pseudoautosomal boundary.

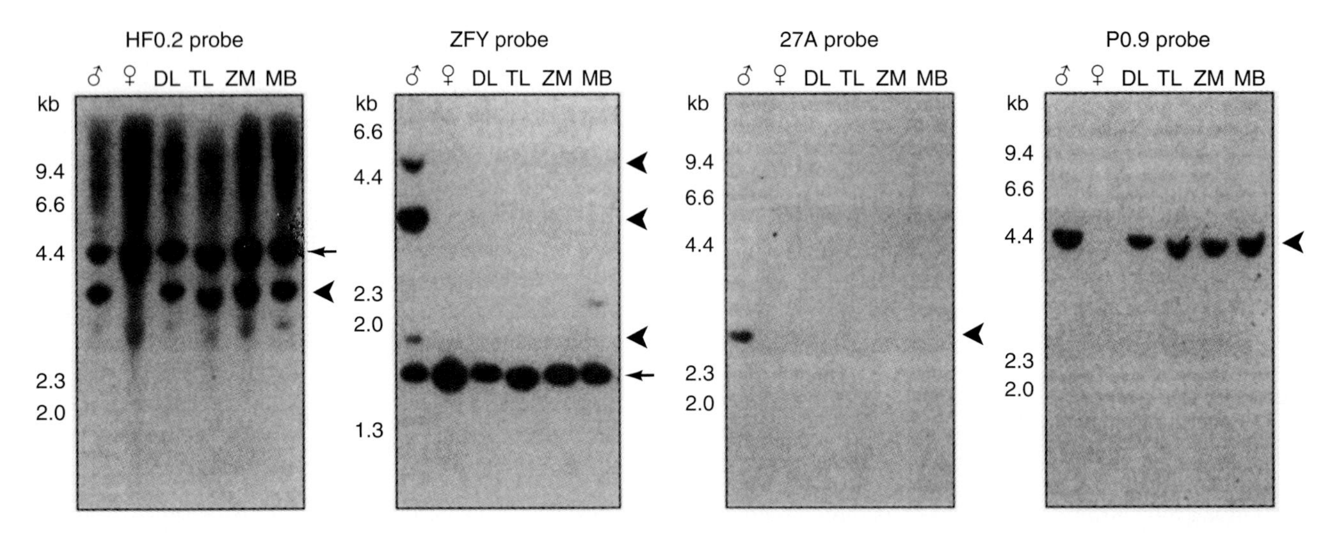

Figure 26.11 Hybridization of sex-specific DNA probes with Southern blots of human genomic DNA (see Method 15.4). The lanes in this experiment contained restriction digests of genomic DNA from a normal human male (♂), a normal human female (♀), and the four sex-reversed XX males (DL, TL, ZM, and MB) also listed in Table 26.3. **(a)** DNA probed with Hf0.2, a pseudoautosomal probe. The probe bound to a 4.5-kb fragment (arrow) of X-specific DNA and a 3.2-kb fragment (arrowhead) of Y-specific DNA reaching across the pseudoautosomal boundary. The latter fragment was present in all sex-reversed patients but missing from normal female genomic DNA. The label at the 3.2-kb level in the female DNA lane is a smear caused by DNA overdigestion, as indicated by the lack of a distinct bulge in the labeling pattern. **(b)** DNA probed with a cDNA clone representing part of the *ZFY*$^{+}$ gene. The probe bound to three Y-specific bands (arrowheads) appearing only in the normal male, indicating that these DNA fragments were not exchanged in the sex-reversed patients. (The weak 2.3-kb band in lane MB was not reproducible and was ascribed to incomplete digestion of the genomic DNA.) There was also a common 1.6-kb band (arrow), confirming that ZFY has a close homolog on the X chromosome. **(c)** DNA probed with clone 27A. Only the normal male was positive, indicating that the corresponding portion of genomic DNA is Y-specific but not exchanged in the sex-reversed patients. **(d)** DNA probed with probe P0.9. The probe bound to DNA from the normal male and the sex-reversed patients, indicating that this DNA is Y-specific and exchanged in the sex-reversed patients. Similar experiments with an E1.3 probe showed that the corresponding portion of genomic DNA is Y-specific and not exchanged in the sex-reversed patients.

table 26.3 Mapping of Genomic DNA Breakpoints in Sex-Reversed XX Males

The data indicate that in each case the breakpoint of the X-Y recombination (see Figs. 25.9 and 25.10) is located between genomic markers P0.9 and E1.3.

Patient	Genitalia	Testis Biopsy	Genomic Markers Present[a]					
			HFO.2	H2.1	P0.9	E1.3	27A	ZFY
ZM	Vaginal pouch; hypospadia[b]	Leydig cells; Sertoli cells; no spermatogonia	+	+	+	–	–	–
MB	Cryptorchidism[c]	Leydig cells; Sertoli cells; no spermatogonia	+	+	+	–	–	–
DL[d]	Hypospadia[b]	Bilateral ovotestis	+	+	+	–	–	–
TL[d]	Uterus; hypospadia[b]	Sertoli cells; no spermatogonia	+	+	+	–	–	–

[a] See map in Figure 26.10.

[b] Urethra opens on lower surface of penis instead of at the tip.

[c] Testis fails to descend into scrotum.

[d] Familial case.

After Palmer et al. (1989). Used with permission. © 1989 Macmillan Magazines Limited.

That a mutation in the *SRY* (*Sry*) gene and the sex reversal should have been coincident in all these independent cases was extremely unlikely. These observations therefore prove beyond reasonable doubt that the SRY/Sry function is *necessary* for testis determination.

THE MOUSE *SRY*+ GENE CAN BE SUFFICIENT FOR TESTIS DETERMINATION

To demonstrate that the SRY^+ (or Sry^+)gene by itself is *sufficient* for testis-determination, it has to be shown that adding SRY^+ or Sry^+ to a normal XX genome causes the development of a male. This test was carried out by Koopman and coworkers (1991) using one of the transgenic mouse techniques described in Section 15.6. The investigators injected mouse eggs with a 14-kb fragment of mouse genomic DNA that contained the Sry^+ gene and, on the basis of sequence analysis, no other genes. The transgenic individuals were tested for their sex chromosome status and for their sexual development.

FERTILIZED eggs were injected with the Sry^+ transgene and allowed to develop in foster mothers. Some of the embryos were removed after 14 days and assayed for their chromosomal status, the integration of the transgene, and their phenotypic sex (Table 26.4). The chromosomal status (karyotype) was assayed by staining for *sex chromatin*,

table 26.4 Analysis of Mouse Embryos Injected with an Sry1 Transgene

The embryos were assayed for the presence of sex chromatin (Barr body), the Sry^+ transgene, the Zfy^+ gene (located on the normal Y chromosome), and the phenotypic sex (testis cords). Among eight chromosomal females (XX karyotype) transgenic for Sry^+, two were phenotypic males, while six were phenotypic females. ND = not determined.

Number of Embryos	Sex Chromatin	Sry^+	Zfy^+	Deduced Karyotype	Transgenic	Phenotypic Sex
63	+	–	–/ND	XX	–	♀
27	–	+	+	XY	ND	♂
58	–	ND	ND	XY	ND	♂
2	–	–	–	XO	–	♀
6	+	+	–	XX	+*	♀
2	+	+	–	XX	+	♂

* In four of these cases, the transgene may not have been present in all cells.

After Koopman et al. (1991). Used with permission. © 1991 Macmillan Magazines Limited.

which indicates the presence of more than one X chromosome and therefore most likely an XX karyotype. In addition, the researchers used Southern blot analysis (see Method 15.4) to test for the *Zfy* gene, which is present on normal Y chromosomes but was not part of the transgene (see Fig. 26.10). Thus, sex chromatin–positive and *Zfy*-negative individuals were classified as XX karyotypes while sex chromatin–negative and *Zfy*-positive individuals were classified as XY karyotypes.

Among seventy-one XX individuals, eight were identified as transgenic by the presence of the Sry^+ gene in Southern blots. Among these eight XX transgenics, two were phenotypically male, as indicated by the formation of testis cords (Fig. 26.12). Histological examination of the transgenics showed that their testis cords were normal and that their testes were indistinguishable from testes of their normal XY siblings. The other six XX transgenics were phenotypically female.

To test the adult phenotype of Sry^+ transgenic mice, some of the injected eggs were allowed to develop to term. Among ninety-three mice born, three were XX transgenics. Of these, two were XX females, and one was an XX male (Fig. 26.13). The XX transgenic male was similar in size to his normal XY littermates. His copulatory behavior was normal, but none of the females with which he mated became pregnant. Internal examination revealed a normal male reproductive tract with no sign of hermaphroditism. The testes were smaller than normal, but again this is normal for sex-reversed XX males. Histologically, the testes of the XX transgenic male contained seminiferous tubules, clearly defined Sertoli cells, Leydig cells, and other testicular cells, but no germ cells undergoing spermatogenesis.

Questions

1. What was the purpose of testing the potentially transgenic mice for the presence of the *Zfy* gene?
2. How do you interpret the fact that the single adult male described here was defective in spermatogenesis?
3. How do you interpret the fact that only 3 out of 11 XX individuals with an Sry^+ transgene developed as gonadal males? Why not all of them?

The experiments of Koopman and others (1991) show that a 14-kb segment of DNA containing Sry^+ can be suf-

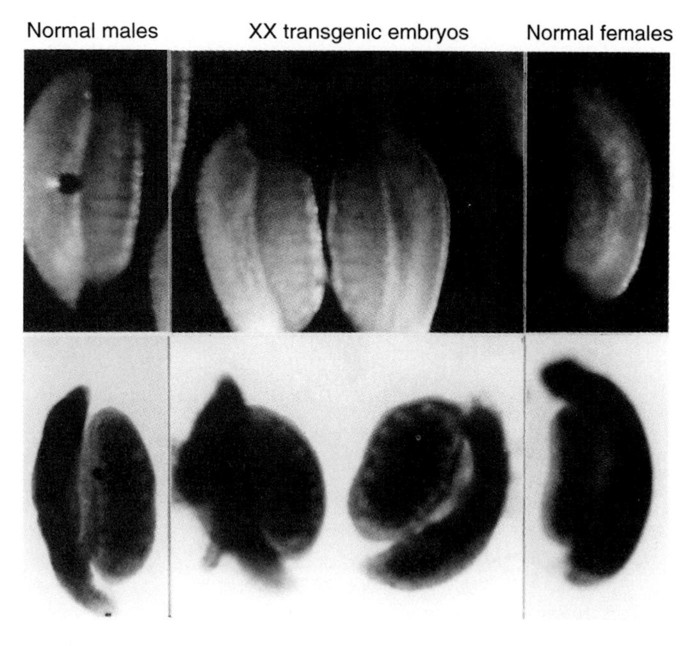

Figure 26.12 Testis development of two mouse embryos with XX karyotype and multiple copies of an Sry^+ transgene (see Table 26.4). The two center panels show pairs of embryonic gonads and adjacent kidneys, excised from the two transgenic XX embryos. The two left panels show single gonads and attached kidneys from normal males. The two panels at the right show single kidneys and gonads from normal females. The top photographs were made with lateral illumination, whereas the bottom photos were made with transmitted light. Note the striated appearance of the male gonad, which is caused by the formation of testis cords (see Figs. 26.6 and 27.4). The gonads of the two XX transgenic embryos have the characteristic stripes associated with testis cord formation.

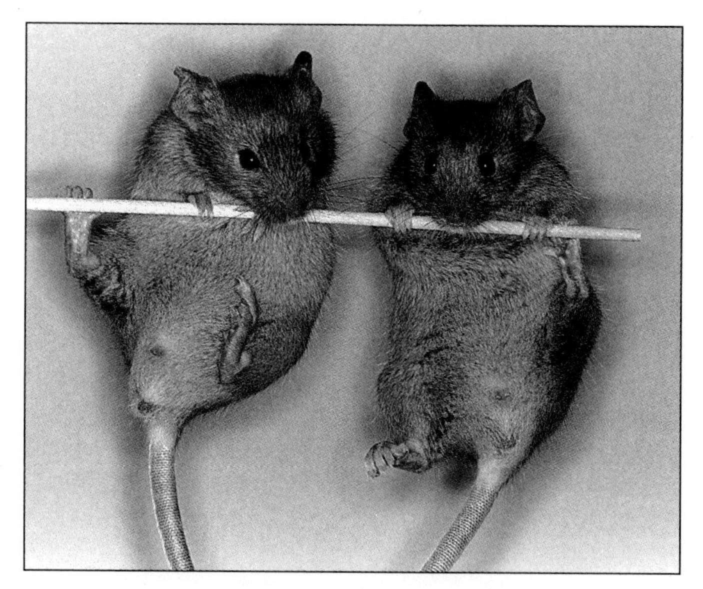

Figure 26.13 Normal and transgenic males. The mouse to the left is a normal male. The mouse to the right is chromosomally female (XX) but made transgenic for the Sry^+ gene, which is normally located on the Y chromosome. The transgenic mouse has the external and internal genitalia of a normal male. He also copulates normally but is sterile. The Sry^+ gene can be sufficient for testis formation and normal male sexual development in an XX individual, except that spermatogenesis requires additional genes located on the Y chromosome.

ficient to determine testis formation and male development. Since no other genes were detected in the DNA segment used as a transgene, it appears that *Sry*$^+$ alone can act as the testis-determining factor in mice. The fact that only a fraction of XX individuals identified as transgenic for *Sry*$^+$ actually developed into males can be ascribed to several limitations of the transgenic mouse technique used (see Section 15.6). Alternatively, complete and consistent male sex determination may require additional Y-linked genes that act in synergistic or overlapping ways with the *SRY*$^+$ gene.

THE *SRY*$^+$ GENE CONTROLS PRIMARY SEX DIFFERENTIATION

The *SRY*$^+$ gene, possibly in concert with one or more other Y-linked genes, controls the first, and essentially genetic, phase of sexual differentiation. At the end of this phase, the indifferent gonad will have developed into a testis or ovary, and the hormones produced in particular by the testis will control the second phase of sexual differentiation. The genetic pathway from *SRY*$^+$ activity to the production of male sex hormones, which essentially defines primary sex differentiation, is just beginning to be explored (*Jiménez and Burgos, 1998; Swain and Lovell-Badge, 1999*).

The *SRY*$^+$ gene encodes a protein of the high mobility group (HMG), a family of DNA-binding proteins. Their DNA-binding domain, the HMG domain, binds to DNA in a sequence-specific manner. The HMG domain may be the only functional domain of the protein, since no activation domain reminiscent of other transcription factors has been found. Also, whereas the HMG domain is highly conserved among marsupials and placental mammals, other parts of the SRY protein are quite divergent even among related species (Whitfield et al., 1993; P. K. Tucker and B. L. Lundrigan, 1993). The HMG domain is thought to considerably distort the bound DNA, perhaps activating its target genes by altering chromatin structure or by bringing together distant DNA sequences (Harley et al., 1992). The restriction of *SRY*$^+$ expression to the gonad during two days of embryonic development implies that there are signals regulating *SRY*$^+$.

A major target of *SRY*$^+$ is the *Dax1*$^+$ gene, an X-linked gene that seems to act essentially as an anti-testis gene in females. It becomes active in the indifferent gonad during the same critical period as *SRY*$^+$, but it does so in both sexes (Swain et al., 1996). As gonad differentiation proceeds, *Dax1*$^+$ is down-regulated in the testis but stays on in the ovary. Genetic evidence indicates that *Dax1*$^+$ is not required for testis development in human XY individuals. However, duplication of the chromosomal region that contains *Dax1*$^+$ can cause the development of sex-reversed XY human females, and overexpression of a *Dax1* transgene in mice may also cause XY female sex reversal (Swain et al., 1998). The simplest interpretation of the available data is that *SRY*$^+$ inhibits *Dax1*$^+$, and that conversely, at least the overexpression of *Dax1*$^+$ interferes with the downstream effects of *SRY*$^+$ (Fig. 26.14).

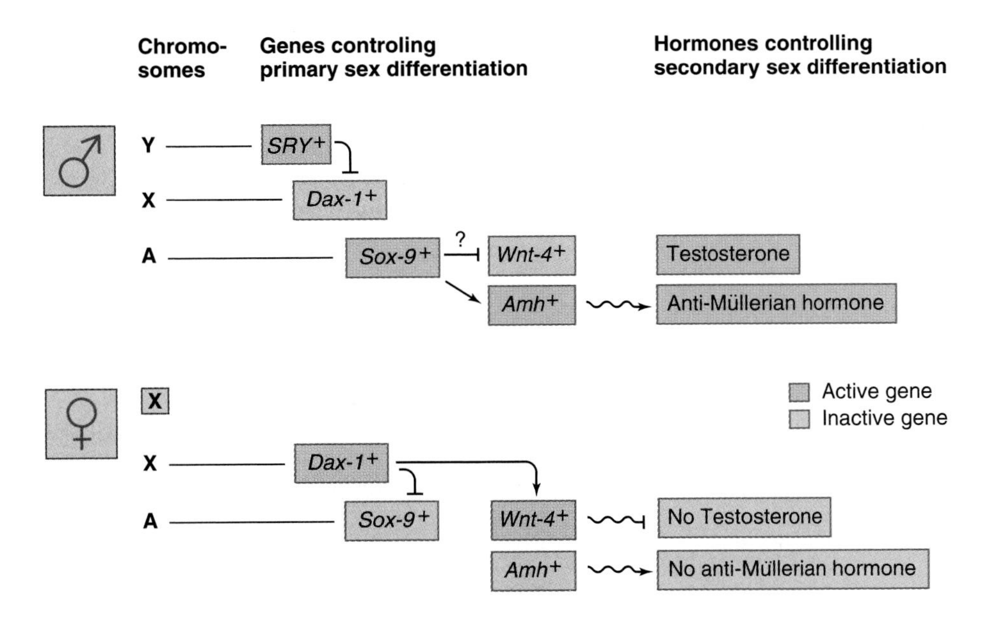

Figure 26.14 Tentative model showing some of the genes involved in primary sex differentiation in mammals. In males, the Y-linked *SRY*$^+$ gene inactivates the *Dax-1*$^+$ gene , whereas in females the *Dax-1*$^+$ gene is expressed from the active X chromosome. The latter gene is thought to control other genes as diagrammed here and explained in the main text. The gene activities occur in the supporting cell lineage as the gonad develops from the indifferent stage to either a testis or an ovary. As a result of primary sex differentiation, the testis—but not the ovary—begins to synthesize anti-Müllerian hormone and testosterone, the hormones that will govern secondary sex differentiation.

A critical target of *Dax1*$^+$ is the *Sox9*$^+$ gene, which basically maintain the testicular development initiated by the short activity of *SRY*$^+$ in males. *Sox9*$^+$ is expressed at low levels in the indifferent gonad of both XX and XY mice. With advancing gonad differentiation, *Sox9*$^+$ is turned off in females but is up-regulated in males, where its expression in Sertoli cells lasts throughout life (Kent et al., 1996). Significantly, human XY individuals with mutations in one copy of *Sox9* show male-to-female sex reversal (Wagner et al., 1994). *Sox9*$^+$ encodes an HMG protein just as *SRY*$^+$ does. The Sox9 protein, however, contains not only a DNA-binding region but also a strong activation domain (Sudbeck et al., 1996). In contrast to SRY, which is restricted to mammals, Sox9 is also highly conserved throughout the vertebrates, despite their different sex determination mechanisms (Morais da Silva et al., 1996). Together, these data indicate that the *Sox9*$^+$ gene is a central element in male primary sex differentiation.

Once the supporting cells of the indifferent gonad are determined to become Sertoli cells, some of the genes characteristic of Sertoli cell function become active. One of these genes is *Amh*$^+$, the gene that codes *anti-Müllerian hormone,* which plays a key role in male secondary sex differentiation (see Section 27.2). *Amh*$^+$ becomes active in Sertoli cells immediately after *Sox9*$^+$, and the regulatory region of *Amh*$^+$ contains a binding site for Sox-9 protein (Arango et al., 1999). Apparently, the *Amh*$^+$ gene is activated cooperatively by the Sox9 protein and SF1, another transcription factor generally involved in gonadal and adrenal gland development (De Santa Barbara et al., 1998).

The genetic control of steroidal cell lineage development involves the *Wnt4*$^+$ gene (Vainio et al., 1999). Male mice homozygous for an apparent null allele of this gene develop normally until birth, whereas such females begin accumulating testosterone in their ovaries and are masculinized in their secondary sex differentiation. Thus, the normal Wnt-4 gene product seems to antagonize testosterone production in females. *Wnt4*$^+$ is expressed in the indifferent gonad before it is down-regulated in the developing testis but remains high in the ovary. Which gonadal cells secrete the Wnt4 signaling protein and which cells respond to it remains to be established.

26.5 Sex Determination and Sex Differentiation in *Drosophila melanogaster*, *C. elegans*, and Mammals

In the last section of this chapter, we will compare sex determination and sex differentiation in the fruit fly *Drosophila melanogaster,* the roundworm *Caenorhabditis elegans,* and placental mammals, represented by the house mouse and the human. We will see that genotypic sex determination controls the same biological functions in all species but that the genetic and molecular control mechanisms differ widely (Parkhurst and Meneely, 1994; Cline and Meyer, 1996).

ONE PRIMARY SIGNAL CONTROLS MULTIPLE ASPECTS OF SEX DIFFERENTIATION

At the most general level, genotypic sex determination involves a primary signal that affects a set of regulatory genes, which in turn control sex differentiation (Table 26.5). The genes that respond to the primary signal form regulatory hierarchies, which diverge into three pathways controlling the three basic aspects of sexual development: *dosage compensation, somatic sex differentiation,* and *germ line sex differentiation.* Typically, each pathway contains a set of regulatory genes, which in turn control a large number of realizator genes. The realizator genes are directly involved in sexually dimorphic functions such as pigmentation, gametogenesis, and mating behavior.

table 26.5 Sex Determination Features in *Drosophila*, in *C. elegans*, and in Humans and Mice

Feature	*Drosophila*	*C. elegans*	Humans and Mice
Sexual dimorphism	Male/female	Male/hermaphrodite	Male/female
Sex chromosomes	XY/XX	XO/XX	XY/XX
Sex-determining signal	X:A ratio	X:A ratio	*SRY*$^+$ activity
Master control gene	*Sex-lethal*$^+$	*xol-1*$^+$	
Dosage compensation	Male hypertranscription	Female hypotranscription	X inactivation
Somatic sex coordination	RNA splicing cascade, cell autonomous	Gene inactivation cascade, induction	Sex hormones
Germ line sex differentiation	*Sex-lethal*$^+$ activity, induction by somatic cells	*tra-1*$^+$ and *fem*$^+$ activity	Induction by somatic cells

THE X:A RATIO IS MEASURED BY NUMERATOR AND DENOMINATOR ELEMENTS

The sex-determining signal in mammals is the *SRY*$^+$ gene product, as discussed earlier. In both *Drosophila* and *C. elegans,* the sex-determining signal is the X:A ratio. This was established for *Drosophila* in a classical study by Bridges (1921), who crossed appropriate fly strains to obtain offspring with different numbers of X chromosomes over two or more sets of autosomes (Fig. 26.15). The results showed that X:A ratios of 0.33 and 0.5 cause male development. A ratio of 0.67 results in intersexes showing mosaics of male and female cells. Ratios between 0.75 and 1.5 support the development of females. Individuals with X:A ratios smaller than 0.33 or greater than 1.5 are not viable. The male/female mosaics observed at an X:A ratio of 0.67 indicate that sex in *Drosophila* is determined cell-autonomously. This ratio seems to be just marginal for engaging the feedback loop that maintains the synthesis of Sex-lethal protein through the female-specific splicing pattern of Sex-lethal pre-mRNA (Cline and Meyer, 1996; see Section 17.2).

Corresponding experiments with *C. elegans* were carried out by Madl and Herman (1979). The results were similar except that individuals with an X:A ratio of 0.67 were males rather than intersexes, and that individuals with X:A ratios between 0.75 and 1.5 were hermaphrodites. The lack of mosaic intersexes in *C. elegans* is explained by a *community effect,* in which cells through local interactions affect the activity of control genes for somatic sex differentiation, as will be discussed later in this section.

How do *Drosophila* and *C. elegans* embryos measure the X:A ratio? The numerator of the X:A ratio is the activity of certain X-linked genes. Obviously, the amounts of products encoded by these so-called ***numerator genes*** increase in proportion to the number of X chromosomes present in a cell. This implies that the numerator genes must escape *dosage compensation,* which equalizes the cellular amount of most X-linked gene products in the two sexes. Either the dosage compensation mechanism does not act on numerator genes, or the X:A ratio is measured and stored before dosage compensation sets in.

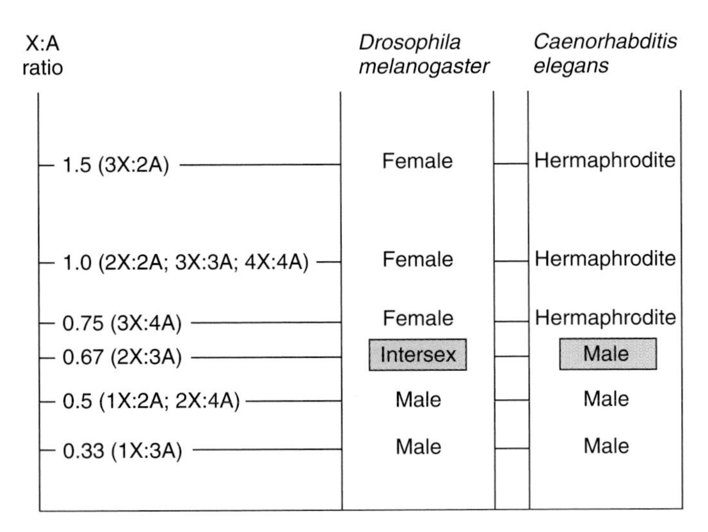

Figure 26.15 Control of the sexual phenotype by the X:A ratio in *Drosophila* and *Caenorhabditis elegans.* Individuals carrying different numbers of X chromosomes and sets of autosomes were obtained by suitable genetic crosses. The columns indicate the X:A ratios that support the development of males, intersexes, or females in *Drosophila,* and males or hermaphrodites in *C. elegans.*

Some of the genes involved in computing the X:A ratio in *Drosophila* have been characterized, including the X-linked numerator genes *sisterless-a*$^+$ and *sisterless-b*$^+$ as well as the autosomal gene *deadpan*$^+$, which may act as a ***denominator gene.*** All three genes encode transcription factors of the *helix-loop-helix* (*HLH*) superfamily, which form homodimers and heterodimers that may enhance or inhibit transcription (see Sections 16.4 and 20.5). Depending on their relative amounts, HLH factors activate, or fail to activate, the *Sex-lethal*$^+$ gene, which is the master regulator of sexual development in *Drosophila.* Activation of *Sex-lethal*$^+$ initiates development of a female, whereas inactivity of *Sex-lethal*$^+$ allows males to develop by default (see Fig. 26.16). Thus, in contrast to the situation in mammals, the female sex is the signal-dependent one in *Drosophila,* whereas males develop by default.

The *sisterless*$^+$ genes owe their name to the fact that loss-of-function alleles are lethal in females but not in males. In XX individuals deficient for *sisterless,* the *Sex-lethal*$^+$ gene is not activated, and the male-specific dosage compensation occurs inappropriately. Such individuals die, while their XY siblings live and develop as males. Conversely, XY individuals carrying additional copies of *sisterless*$^+$ die because *Sex-lethal*$^+$ is activated with the resulting lack of appropriate dosage compensation.

The critical role of the *sisterless-b*$^+$ locus was demonstrated by Erickson and Cline (1991), who rescued females carrying a temperature-sensitive loss-of-function allele of *sisterless-b* by genetic transformation with a cloned genomic DNA segment containing the *sisterless-b*$^+$ allele. The *sisterless-b*$^+$ transgene also rescued *sisterless-a* mutant females, although less efficiently. The sisterless-b protein, also known as T4, turned out to be a transcription factor of the HLH superfamily. Because of the known cooperative properties of HLH factors, the researchers suggested that T4 may act as a numerator in computing the X:A ratio by interacting with other HLH factors in turning on *Sex-lethal*$^+$.

The denominator of the X:A ratio is measured via the nuclear volume, which increases with the number of autosomes and dilutes the numerator elements. Also, transcription factors encoded by embryonically expressed autosomal genes including *deadpan*$^+$ and *groucho*$^+$ specifically interfere with the activation of *Sex-lethal*$^+$ (Younger-Shepherd et al., 1992; Barbash and Cline, 1995; Jimenez et al., 1997). Additional proteins encoded by maternally expressed genes are also involved in communicating the X:A ratio.

DROSOPHILA AND *C. ELEGANS* HAVE MASTER REGULATORY GENES FOR SEX DIFFERENTIATION

The three basic aspects of sexual development—dosage compensation, somatic sex differentiation, and germ line sex differentiation—do not have to be controlled by the same mechanism. Indeed, we have already seen that in mammals the testis-determining factor controls the somatic sex and, indirectly, the gametic sex, whereas dosage compensation is controlled by the X-linked factor *Xist*$^+$. In *Drosophila,* however, all aspects of sexual development are uniformly controlled by the *Sex-lethal*$^+$ gene. A similar master switch in *C. elegans* is the *xol-1*$^+$ gene.

The *Sex-lethal*$^+$ gene of *Drosophila* responds to the relative concentrations of transcription factors that measure the X:A ratio, as discussed previously. A high X:A ratio activates an *establishment promoter* causing the transcription of a pre-mRNA that is spliced and translated into functional Sex-lethal protein (Keyes et al., 1992). The function of the establishment promoter is quickly taken over by transcription from a maintenance promoter of the *Sex-lethal*$^+$ gene. The maintainance promoter becomes active independently of the X:A ratio—that is, in prospective males and females. However, the pre-mRNA transcribed from this promoter is spliced in two alternative ways, yielding either male-specific or female-specific mRNAs (see Fig. 17.10). The male-specific mRNA is translated into a truncated, nonfunctional protein, whereas the female-specific mRNA yields a full-length, functional protein. The male-specific splicing pattern is the default program, whereas the female-specific splicing pattern requires the presence of Sex-lethal protein. Thus, the continued synthesis of Sex-lethal protein depends on a positive feedback loop between Sex-lethal protein and the splicing of female-specific Sex-lethal mRNA. In females, the feedback loop is engaged by a start-up supply of Sex-lethal protein derived from the activity of the establishment promoter (see Section 17.2). In males, no start-up Sex-lethal protein is produced, and all pre-mRNA transcribed from the maintenance promoter is spliced unproductively.

The Sex-lethal protein controls not only the splicing of its own pre-mRNA, but also genetic hierarchies that direct female dosage compensation as well as female somatic development (Fig. 26.16). The latter also leads to the release of an inductive signal in the gonad that promotes oogenesis. Lack of Sex-lethal protein results in the default expression patterns, which lead to male dosage

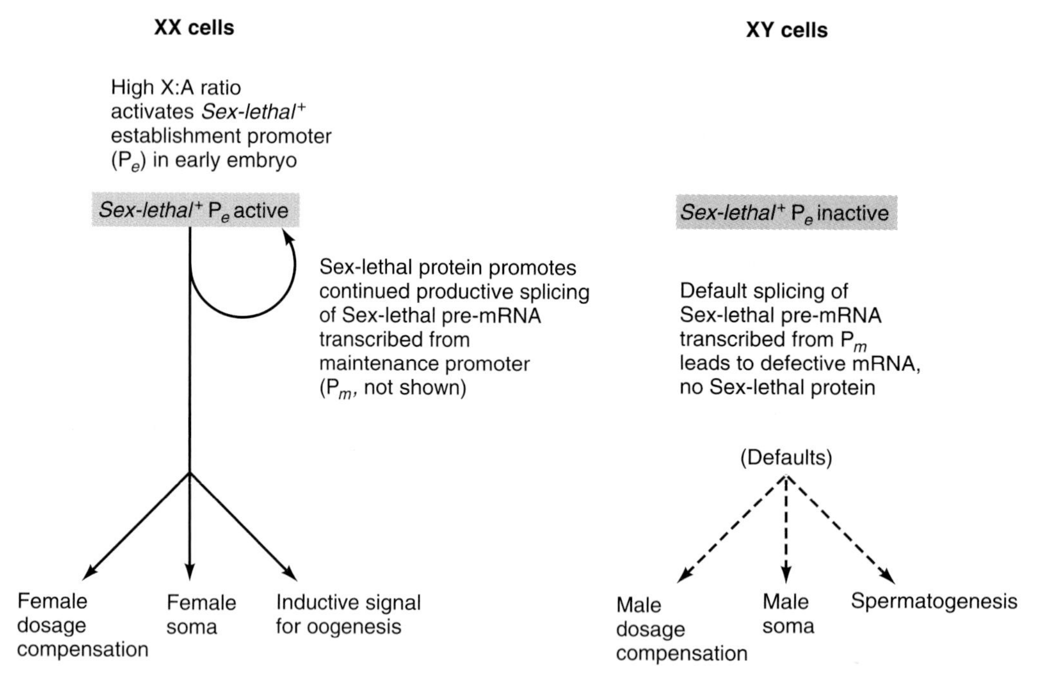

Figure 26.16 Action of the *Drosophila* gene *Sex-lethal*$^+$ as a master regulator of sex determination. At a high X:A ratio, the establishment promoter (P_e) of *Sex-lethal*$^+$ is activated, and the resulting Sex-lethal pre-mRNA is spliced only in one way, which results in functional Sex-lethal protein. When P_e ceases to function and the maintenance promoter (P_m) becomes active, the new pre-mRNA is spliced in the default pattern in males and in a regulated pattern in females (see Fig. 17.10). The female-specific pattern depends on preexisting Sex-lethal protein, which is initially supplied from the P_e transcript. The female-specific splicing of the P_m transcript leads to the synthesis of more Sex-lethal protein, which in turn continues to direct the productive splicing of Sex-lethal pre-mRNA in a positive feedback loop. Sex-lethal protein directs female dosage compensation (see Fig. 26.18) and female somatic development (see Fig. 26. 19). The genetic pathway maintained by Sex-lethal protein also generates an inductive signal for oogenesis (see Fig. 26.22). Lack of functional Sex-lethal protein allows the default programs for male somatic development, male dosage compensation, and spermatogenesis to occur.

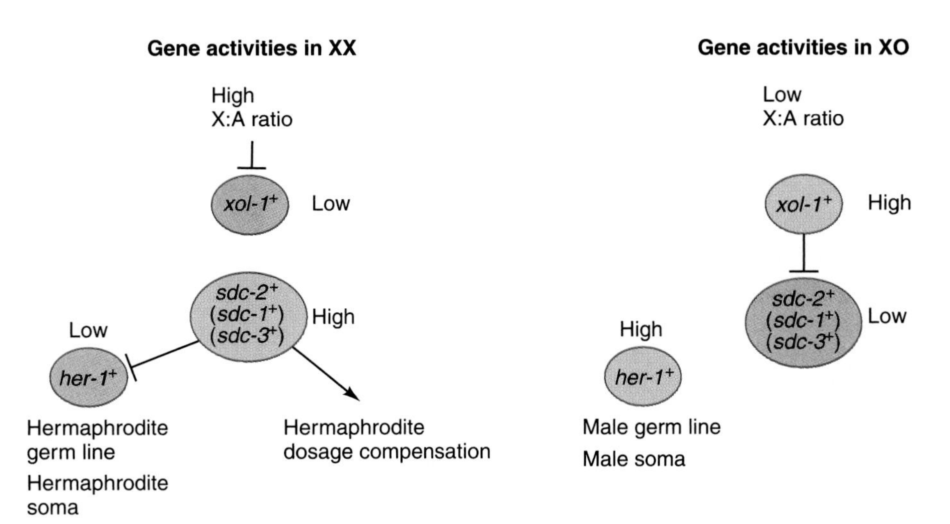

Figure 26.17 Control of sex differentiation and dosage compensation in *C. elegans* by the master regulator gene *xol-1*$^+$. In XX (prospective hermaphrodite) embryos, the high X:A ratio inactivates the *xol-1*$^+$ gene, which permits *sdc-2*$^+$ and the associated *sdc-1*$^+$ and *sdc-3*$^+$ genes to be active. Together, the sdc products activate dosage compensation genes and inhibit the gene *her-1*$^+$ to allow female sex differentiation. In XO (prospective male) embryos, *xol-1*$^+$ becomes active by default and downregulates the *sdc*$^+$ genes, thus permitting high activity of *her-1*$^+$, resulting in male development.

compensation, male somatic development, and spermatogenesis. Some features of these regulatory pathways will be described farther on.

In odd contrast to its role as a master regulator affecting all aspects of sexual development in *Drosophila melanogaster,* the *Sex-lethal*$^+$ gene does not have a sex-determining function in other insects, not even in the house fly *Musca domestica* (Schütt and Nöthiger, 2000). This is a sobering, and intriguing, reminder of the relatively fast pace with which sex-determination mechanisms have evolved.

In *C. elegans,* the master regulator of sex differentiation and dosage compensation is the gene *xol-1*$^+$ (for XO-lethal). Null alleles of *xol-1*$^+$ have no effect in otherwise wild-type XX individuals. However, loss of function in *xol-1* causes XO animals to develop as hermaphrodites and their transcription of X-linked genes to be improperly reduced, with death resulting (L. M. Miller et al., 1988). Similarly to *Sex-lethal*$^+$ in *Drosophila, xol-1*$^+$ is regulated by the X:A ratio at the level of transcription (Cline and Meyer, 1996). Unlike *Sex-lethal*$^+$, however, *xol-1*$^+$ is inhibited by the high X:A ratio of XX individuals—that is, prospective females (Fig. 26.17). The low X:A ratio of XO individuals allows *xol-1*$^+$ to become active by default. Although *xol-1*$^+$ produces three alternatively spliced transcripts, all of them are present in both sexes, and one transcript carries out all *xol-1*$^+$ functions in XO individuals (Rhind et al., 1995). Thus, alternative splicing does not play an essential role in the sex-specific expression of *xol-1*$^+$ in the worm, in contrast to the situation for *Sex-lethal*$^+$ in the fly.

xol-1$^+$ inhibits another gene, *sdc-2*$^+$ (for sex and dosage compensation), which has the converse effects: Loss-of-function mutations in this gene have no effect in XO individuals but cause sex-reversal and overexpression of X-linked genes in XX animals, similar to *Sex-lethal*$^+$ in *Drosophila* (Nussbaum and Meyer, 1989). Because loss of function in the *sdc-2* counteracts the lethality of *xol-1* mutations in XO individuals, *sdc-2*$^+$ must be downstream of *xol-1*$^+$. The effects of the *sdc-2*$^+$ gene are potentiated by *sdc-1*$^+$ and *sdc-3*$^+$, two genes with similar but weaker phenotypes. Together, the *sdc*$^+$ genes regulate separate pathways for somatic sex differentiation and dosage compensation. The *sdc-2*$^+$ gene inhibits, at the level of transcription, the *her-1*$^+$ gene, which controls the somatic sex differentiation pathway (Trent et al., 1991). The dosage compensation pathway is mediated through another set of genes, which will be discussed next.

In summary, the master regulatory genes controlling sexual development in *Drosophila* and *C. elegans* serve the same biological functions but differ in their molecular and regulatory characteristics. These observations indicate that these master regulatory genes are not orthologous but have evolved independently.

DOSAGE COMPENSATION MECHANISMS VARY WIDELY

As discussed earlier, the difference in X-linked gene dosages between the two sexes must be compensated for so that both sexes receive equal amounts of gene products. In mammals, as we have seen, all X chromosomes except one are inactivated, so that both sexes are left with one active X chromosome.

A different mechanism, known as ***male hypertranscription***, achieves the same goal in *Drosophila.* Here, dosage compensation occurs by doubling the transcription rate of X-linked genes in males. This mechanism relies on *cis*-acting enhancer sequences scattered along the X chromosome. Autosomal transgenes inserted into an X chromosome are transcribed at a higher rate in males than in females, and at a higher rate than if they were inserted in an autosome (Lucchesi and Manning, 1987). Male hypertranscription depends on *trans*-acting factors encoded by at least four autosomal genes, known as the MSL set (for *m*ale-*s*pecific *l*ethal). Loss of function in any

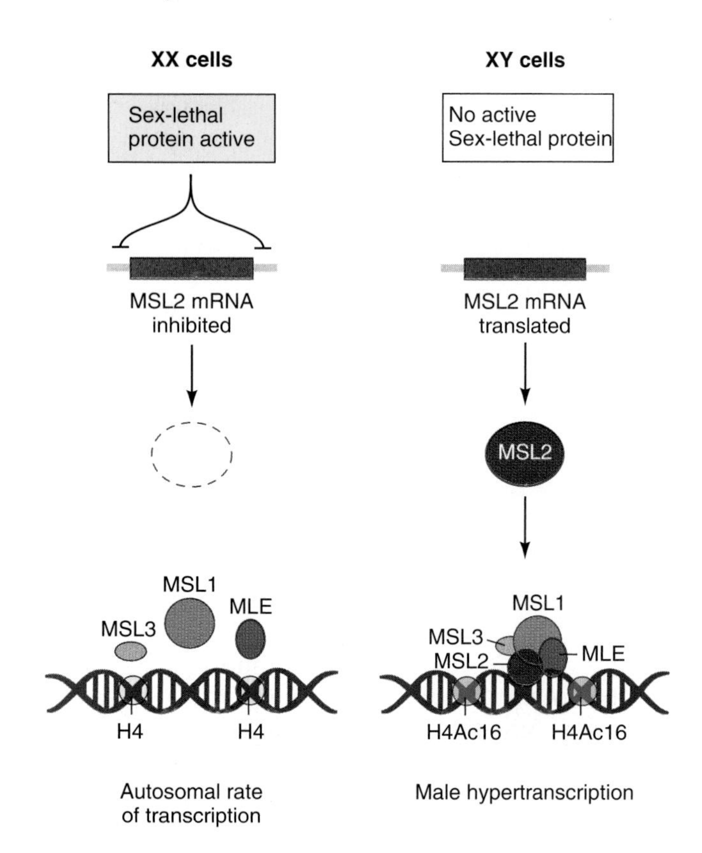

Figure 26.18 Control of *Drosophila* dosage compensation by four proteins of the male-specific lethal (MSL) group and the Sex-lethal protein (Sxl). In XY cells, no active Sxl is present, and MSL2 mRNA is translated into functional MSL2 protein. The latter assembles, along with the three other MSL proteins, at specific sites on the X chromosome. At these sites, lysine-16 of histone H4 is acetylated, resulting in chromosome decondensation and more rapid transcription. In XX cells, Sxl protein inhibits the translation of MSL2 mRNA, and no functional MSL2 protein is made. The MSL proteins then fail to assemble on the X chromosome, histone H4 is not modified, and transcription occurs at the same rate as on autosomes.

of these genes causes the death of XY individuals by insufficient X chromosome transcription. *Immunostaining* (see Method 4.1) of *polytene chromosomes* with antibodies against MSL proteins revealed that the MSL proteins are bound in similar relative amounts to hundreds of sites along the X chromosomes of males but not females (Gorman et al., 1995). The similarity of the pattern of MSL binding sites and the distribution of *cis*-acting enhancer sequences on the X-chromosome strongly suggest that the MSL proteins bind to these sites directly, thereby promoting male hypertranscription (Fig. 26.18).

One member of the MSL group, the maleless protein (MLE), has the molecular characteristics of a *helicase*—that is, an enzyme that unwinds DNA and RNA helices (Gorman et al., 1993). MLE may therefore act to decondense chromosomal DNA. This hypothesis is supported by further observations. First, in polytene chromosomes, the single male X chromosome is more diffuse in appearance than the diploid female X chromosome and the autosomes. Second, a histone H4 isoform acetylated at lysine-16 is found at numerous sites along the male X chromosome but not in male autosomes or any female chromosomes (B. M. Turner et al., 1992). This histone modification is known to locally alter the chromosome structure and to increase the rate of transcription in yeast (L. M. Johnson et al., 1990). Taken together, the available data suggest that the absence of Sex-lethal protein allows the MSL proteins to interact with specific X-linked DNA sequences to decondense the X chromosome, with the result of hypertranscription.

How is male hypertranscription prevented in females? Again, the master gene of sex differentiation in *Drosophila, Sex-lethal*$^{+}$, is involved; loss of Sex-lethal function in XX individuals causes death by hypertranscription of X-linked genes. In this context, however, the critical function of Sex-lethal protein is not in the control of pre-mRNA splicing. Rather, the Sex-lethal protein binds to the mRNA for one of the MSL proteins, male-specific-lethal-2 (Fig. 26.18). By associating with sites in the 5′ UTR and 3′ UTR of this transcript, Sex-lethal protein represses its translation in females (Kelley et al., 1997; Gebauer et al., 1999).

In *C. elegans,* dosage compensation involves at least four genes, collectively known as the DCD set (for *d*osage *c*ompensation *d*umpies, referring to the plump body shape of sublethal alleles). Loss of function in these genes is deleterious to hermaphrodites but not to males. The DCD proteins seem to act by reducing the transcription of X-linked genes in XX individuals but not in XO individuals. Molecular analysis of the protein encoded by the *dumpy-27*$^{+}$ *(dpy-27*$^{+}$*)* gene has provided a clue to its molecular action (Chuang et al., 1994, 1996). The dpy-27 protein is involved in forming protein complexes along the X chromosome of *C. elegans* hermaphrodites. The complex formation seems to be nucleated by the sdc-2 protein, which can localize to the X chromosome without the other components of the dosage compensation complex (Dawes et al., 1999). Thus, a component of the master regulatory switch of sex differentiation in *C. elegans* seems to target the inhibitory complex of dosage compensation proteins directly to both X chromosomes of hermaphrodites (see Fig. 26.17).

In summary, the dosage compensation mechanisms of mammals, flies, and roundworms have in common that they regulate the transcription of X-linked genes. However, beyond this very basic aspect, these three groups of animals use different ways of compensating for the different doses of X-linked genes between the sexes. Mammals rely on (almost) complete inactivation of all X chromosomes except for one. In *Drosophila,* the MSL set of proteins accelerates transcription of the single X chromosome in males. In *C. elegans,* the DCD set slows down the expression of X-linked genes in hermaphrodites.

THE SOMATIC SEXUAL DIFFERENTIATION OCCURS WITH DIFFERENT DEGREES OF CELL AUTONOMY

The most apparent effect of sex determination is the somatic sexual differentiation. In mammals, the first phase of this process is controlled by a genetic hierarchy with the SRY^+ gene at the top (see Fig. 26.14). It sets up the second phase of somatic sex differentiation, which begins when the male embryo produces two sex hormones: *testosterone,* which promotes male secondary sex differentiation, and *anti-Müllerian duct hormone* (*AMH*), which inhibits the development of the female reproductive ducts (see Section 27.2).

In the fruit fly, somatic sex differentiation is controlled by a hierarchy of genes involving, in order, *Sex-lethal*$^+$, *transformer*$^+$, *transformer-2*$^+$, and *doublesex*$^+$ (Fig. 26.19). The regulatory steps in this cascade are all based on differential splicing of pre-mRNAs, as detailed in Section 17.2. The default program at each step results in the development of somatic male characteristics. For instance, XX individuals with loss-of-function alleles of *transformer* develop as somatic males. Conversely, XY individuals carrying a transgenic cDNA for the female-specific transformer mRNA develop as sterile females (McKeown et al., 1988). Differential splicing of doublesex pre-mRNA, in contrast to the other differential splicing steps in this cascade, yields two functional proteins: one male-specific and one female-specific. Both are transcription factors, and they compete with each other for the regulation of their target genes, so that target genes activated by the male doublesex protein are inhited by the female protein and vice versa (Waterbury et al., 1999).

The doublesex proteins control not only morphological features that are sexually dimorphic but also physiological and behavioral traits. Both doublesex proteins, for example, bind directly to the enhancer regions of two genes encoding yolk proteins, with the female-specific protein causing activation and the male-specific protein causing inhibition (Burtis et al., 1991; Coschigano and Wensink, 1993). These genes are expressed in the fat body of females, but not of males. Doublesex proteins also control courtship behavior. Wild-type males inappropriately expressing the female-specific protein are actively courted by other males. This acquisition of feminine sex appeal is likely due to the induction of female pheromone synthesis under the control of the female doublesex protein (Waterbury et al., 1999).

In *C. elegans,* the regulatory cascade for somatic sex differentiation differs greatly from its counterpart in *Drosophila.* It involves more genes, and it does not contain an equivalent of the RNA splicing cascade that forms the centerpiece of somatic sexual differentiation in the fly. In the worm, the genetic pathway for somatic sex differentiation begins with the *her-1*$^+$ gene as discussed earlier (Figs. 26.17 and 26.20). *her-1*$^+$ encodes a secreted protein, which acts as an inhibitory ligand for a receptor

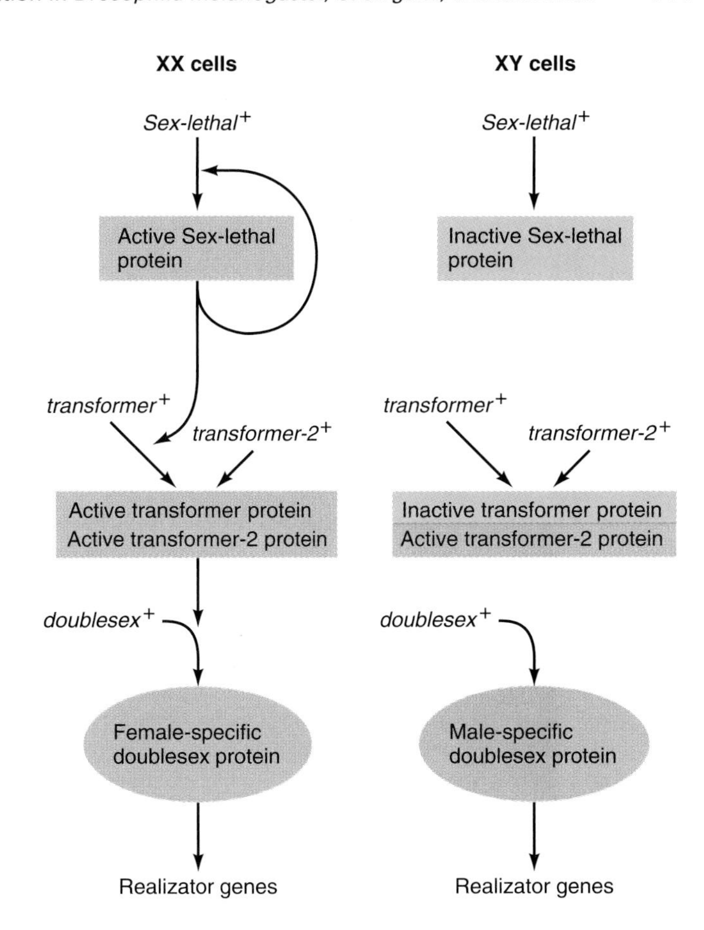

Figure 26.19 Somatic sex differentiation in *Drosophila* by a cascade of differential splicing (see also Fig. 26.16). In XX cells, the presence of Sex-lethal protein regulates the splicing of transformer pre-mRNA so that an active protein will be synthesized. The transformer (tra) protein and the (non-sex-specific) transformer-2 (tra-2) protein together regulate the splicing of doublesex pre-mRNA so that the female-specific doublesex (dsx) protein is produced. In XY cells, the absence of functional Sex-lethal and transformer proteins allows the default splicing of doublesex pre-mRNA to occur. The resulting male-specific mRNA encodes a functional male-specific doublesex protein. Both doublesex proteins regulate sex-specific realizator genes—at least some of them directly.

encoded by *tra-2*$^+$ (Kuwabara and Kimble, 1992; Perry et al., 1993). Tra-2 activity inhibits a group of genes that are known as *fem-1*$^+$, *fem-2*$^+$, and *fem-3*$^+$, because their loss-of-function alleles cause feminization of males. The tra-2 protein seems to antagonize the fem-3 protein by directly interacting with it (Mehra et al., 1999). When the balance between the two proteins favors fem-3, the latter acts along with the other fem proteins to promote male development. The fem genes in turn inhibit *tra-1*$^+$, the final switch gene in the cascade. The tra-1 protein is a rapidly evolving transcription factor (de Bono and Hodgkin, 1996). Its activity causes hermaphrodite development, whereas it absence permits male development, regardless of the activity of the other genes in

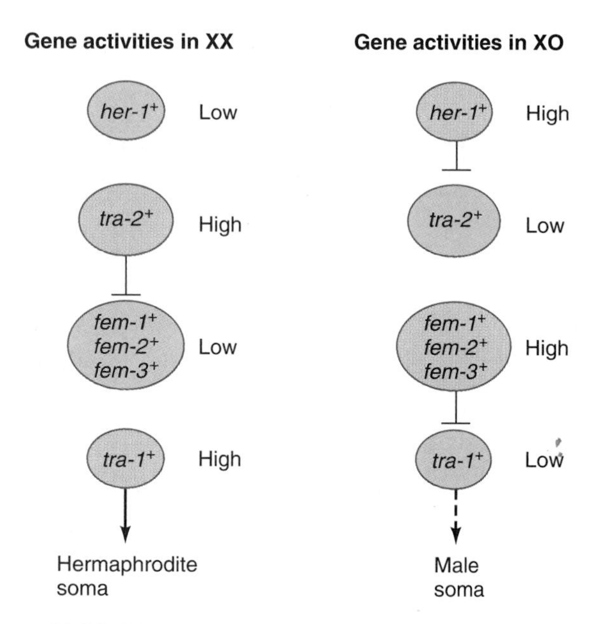

Figure 26.20 Somatic sex differentiation in *C. elegans.* Activity of the *her-1*$^+$ gene is signaled through three more inhibitory steps that eventually control the terminal gene *tra-1*$^+$. High activity of this gene promotes hermaphrodite somatic differentiation, whereas low activity fosters male development.

the cascade. Figure 26.20 outlines the way this cascade functions in determining male and hermaphrodite development in XO and XX individuals, respectively.

The various control mechanisms for somatic sex differentiation confer different degrees of *cell autonomy.* That is, cells of the three types of animals discussed here have different abilities to become male or female independently of their neighbors. In mammals, cell autonomy is limited to the expression of the testis-determining factor in Sertoli cells, and even there a *community effect* is observed, as described earlier. Cell autonomy is completely lost once the sex hormones are produced, as these are carried throughout the body in the blood, and all target cells in a normal individual respond to them. Depending on testosterone level, the target organs of a fetus will become male or female. Even in cases of marginal testosterone level, all target organs develop a similar degree of intersexuality. This occurs, for instance, in some mammalian species when female fetuses are connected by placental blood circulation to male siblings.

In *C. elegans* and *Drosophila,* no sex hormones are carried all over the body as a systemic signal. Instead, the genetic control mechanisms employ signals that act within a single cell or in small groups of neighboring cells. The question then arises just how independent each cell is in its sex determination. Could a single cell develop as a female when all its neighbors are male, or vice versa?

In *C. elegans,* this question has been tested using the strategy of *genetic mosaic analysis* (see Section 25.4). One of the genes tested for cell-autonomous expression was *her-1*$^+$, the first gene in the regulatory cascade with an exclusive role in somatic sexual differentiation (see Fig. 26.20). Expression of *her-1*$^+$ is necessary for development of a male but not of a hermaphrodite. An extra chromosome fragment carrying both *her-1*$^+$ and a cell-autonomous marker gene (*ncl-1*$^+$), if spontaneously lost from an individual chromosomally mutant for these two genes, generates a genetic mosaic for *her-1* and *ncl-1.* Using such mosaics, Hunter and Wood (1992) found that *her-1*$^+$ expression in a sexually dimorphic cell is not necessary for that cell to adopt a male fate. Cells lacking the *her-1*$^+$ gene can be masculinized by neighboring cells expressing the gene. This is in accord with the characterization of the her-1 protein as a secreted signal that inactivates tra-2 receptor. Conversely, cells expressing the *her-1*$^+$ gene can be feminized by neighboring cells lacking this gene activity. Presumably, a cell that expresses *her-1*$^+$ itself but is surrounded by cells failing to express this gene retains enough active tra-2 receptor to undergo hermaphrodite development.

The community effect observed for the expression of *her-1*$^+$ suggests a possible adaptive value for the long cascade of inhibitory gene interactions that control somatic sex development in *C. elegans.* Cells in which the somatic sex differentiation control has been derailed by some unusual environmental effects can be recruited into developmental pathway of their neighbors. Such recruitments should help prevent the formation of sexual chimeras, which tend to be infertile and therefore detrimental to a population.

The highest degree of cell autonomy is observed in the somatic sex differentiation of *Drosophila* and other insects. This is shown dramatically by ***gynandromorphs*** (Gk. *gyne,* "woman"; *andros,* "man"; *morphe,* "form"). Such individuals are sexual chimeras, combining body regions that are fully male with others that are completely female (Fig. 26.21). Gynandromorphs arise from XX embryos during intravitelline cleavage when one X chromosome is not integrated properly into a mitotic spindle and consequently is lost from a nucleus (A. J. Sturtevant, 1929). The progeny of this nucleus are then XO and form male cells, while the other nuclei in the same individual are still XX and form female cells. Recessive mutant alleles for cuticular markers on the persisting X chromosome in male cells can be used to delineate the boundary between XX and XO tissues. In the latter, the mutant alleles are *hemizygous* and therefore expressed, whereas in XX tissues, the same mutant alleles are covered by wild-type alleles on the other X chromosome. If one X chromosome is lost during the first mitosis, about half of the gynandromorph is of the mutant phenotype and the other half, the wild phenotype. If the chromosome loss occurs later, the mutant portion of the gynandromorph is correspondingly smaller. In any case, wherever a body area is sexually dimorphic, the mutant portion has clearly male characteristics, such as abdominal pigment or *sex combs* on the foreleg. The

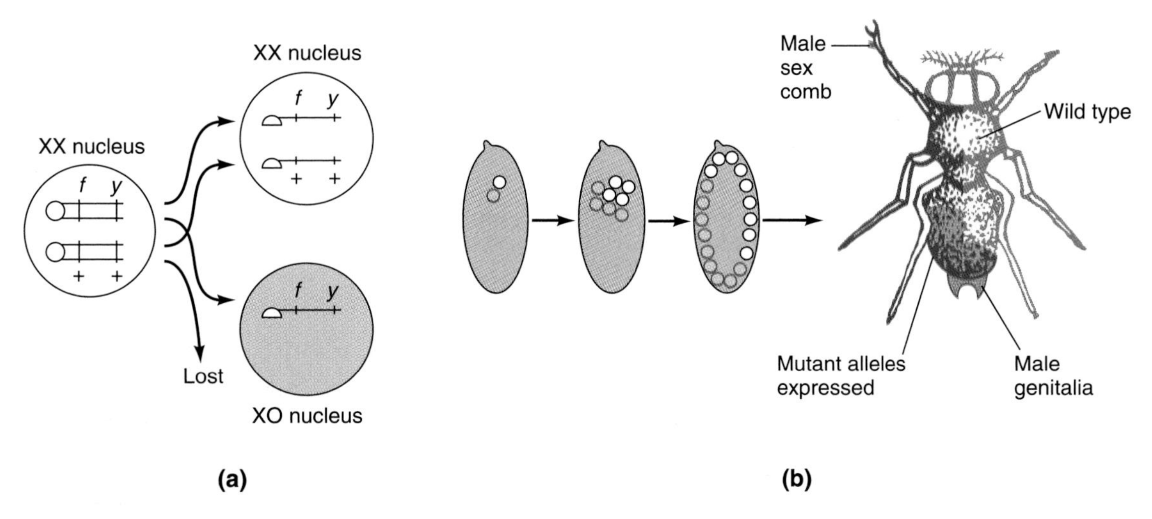

Figure 26.21 Development of a *Drosophila* gynandromorph. **(a)** Early during intravitelline cleavage of an XX individual, a nucleus loses one X chromosome during mitosis, thus generating one XX and one XO daughter nucleus. If the individual was heterozygous for recessive X-linked alleles, such as *forked* (*f*) and *yellow* (*y*), these alleles are now hemizygous in the XO nucleus and its progeny, where they will later be expressed as forked bristles and yellow cuticle. The same alleles are still masked by wild-type alleles in the XX nucleus and its progeny, so that they will not be expressed. **(b)** As the gynandromorph develops into an adult, expression of the X-linked markers clearly delineates XO tissues (color) from XX tissues. Wherever a body area is sexually dimorphic, the XX tissues are fully female, whereas the XO tissues have distinctly male characteristics, such as abdominal pigmentation or sex combs at the forelegs. These observations show that the sex of each somatic cell is determined according to its own X:A ratio.

gynandromorphs therefore confirm that sex determination occurs autonomously in each somatic cell according to its own X:A ratio.

GERM LINE SEX DIFFERENTIATION INVOLVES INTERACTIONS WITH SOMATIC GONADAL CELLS

The third result of sex determination, in addition to dosage compensation and somatic sex differentiation, is germ line sex differentiation. Whether germ cells enter oogenesis or spermatogenesis is in part determined by interactions between germ cells and somatic gonadal cells. The interactions between the supporting cell lineage and the germ line in the mammalian gonad have been discussed earlier in this chapter.

The role of an inductive signal from somatic to germ line cells in *Drosophila* first became apparent after transplantation of *pole cells*—that is, the primordial germ cells. Because the sex of living *Drosophila* embryos is not recognizable to experimenters, pole cell transplantations randomly generate individuals with XX soma and XY pole cells, or vice versa, in addition to correctly matched males and females (Steinmann-Zwicky et al., 1989). In XX somatic gonads, germ cells develop according to their X:A ratio—that is, XX germ cells undergo oogenesis, and XY germ cells take the spermatogenic pathway, although they are arrested as spermatocytes. The results are different in XY gonads. Here, both XX and XY germ cells enter the spermatogenic pathway, although the XX germ cells do not complete it. These results show that the completion of spermatogenesis as well as oogenesis depends on inductive signals from somatic to gonadal cells, and that these signals function appropriately only between XX soma and XX germ line and between XY soma and XY germ line.

A tentative model of female germ line sex differentiation is shown in Figure 26.22. The inductive signal required by XX germ cells to undergo oogenesis depends on the activity of the *transformer*$^+$ and *doublesex*$^+$ genes in the surrounding soma, as indicated by transplantations of XX germ cells into normal and transgenic XY hosts (Steinmann-Zwicky, 1994). Complete oogenesis also depends on the correct expression of certain genes in the germ line, including the *ovarian tumor*$^+$ gene, which responds to the inductive signal, and the *ovo*$^+$ gene, which is regulated independently and encodes a transcription factor that is active only in the female germ line (Oliver et al., 1993; Hinson and Nagoshi, 1999). The activity of the *Sex-lethal*$^+$ gene is also required in the female germ line, and the alternative splicing patterns of Sex-lethal pre-mRNA appear to be the same in germ line and somatic cells (Oliver et al., 1993). The *sans fille*$^+$ gene, which encodes a component of two snRNP particles involved in splicing, has a strong modifying effect on the engagement of the feedback loop that maintains female-specific splicing of Sex-lethal pre-mRNA (Hager and Cline, 1997). Curiously, engagement of this feedback loop is also compatible with spermatogenesis, and it seems that the current picture of germ line differentiation is very preliminary.

In *C. elegans,* the fact that hermaphrodites produce sperm first and then eggs introduces a complication into

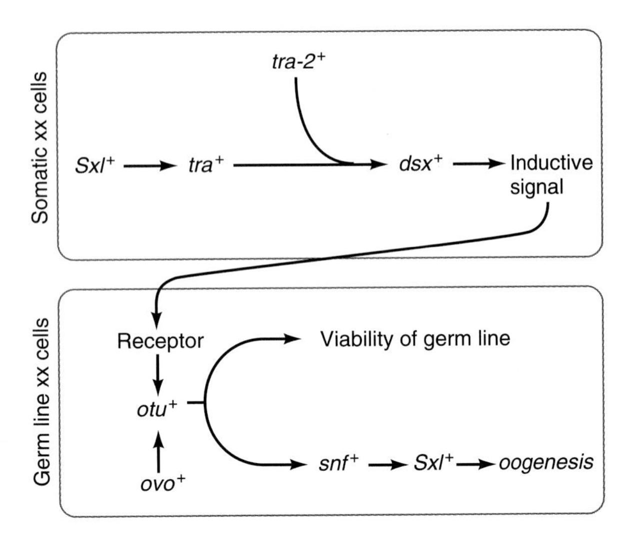

Figure 26.22 A tentative model of female germ line sex differentiation in XX individuals of *Drosophila*. Germ line sex is determined by regulatory hierarchies in both the soma and the germ line. The *Sex-lethal*$^+$ (*Sxl*$^+$) gene is at the top of the hierarchy in the soma but seems to be near the bottom of a different hierarchy in the germ line. Female-specific splicing of Sex-lethal pre-mRNA in the soma starts the cascade of regulation shown in Figure 26.19. This cascade leads to the synthesis of female-specific doublesex (dsx) protein, which in turn helps to promote the synthesis of an inductive signal. The signal binds to a receptor in the germ line cells, which through genes including *ovo*$^+$, *ovarian tumor*$^+$ (*otu*$^+$), and *sans fille*$^+$ (*snf*$^+$) ensures the viability of XX germ line cells and the female-specific splicing of Sex-lethal pre-mRNA.

the genetic control of germ line sex differentiation. A simple model that accommodates this change in gamete production is shown in Figure 26.23. It involves the *fog-2*$^+$ gene, so called because its loss of function causes feminization of the germ line. Its wild-type function is to down-regulate the activity of *tra-2*$^+$ in the germ line of the young adult. This allows the temporary activation of the *fem*$^+$ genes, which inhibit *tra-1*$^+$ so that spermatogenesis can occur. Later in adult life, *fog-2*$^+$ activity ceases, *tra-2*$^+$ inhibits the *fem*$^+$ genes, *tra-1*$^+$ becomes active, and oogenesis ensues. However, the actual genetic pathway of germ line sex differentiation in *C. elegans* appears to be more complex (Cline and Meyer, 1996).

Another regulatory element promoting the switch from spermatogenesis to oogenesis in the *C. elegans* hermaphrodite is known as the *fem-binding factor* (*FBF*), a cytoplasmic protein that inhibits the function of fem-3 mRNA by binding to its 3′ UTR (B. Zhang et al. (1997). Curiously, the RNA-binding domain of FBF is an evolutionarily conserved element that is present in many sequence-specific RNA-binding proteins. These include the pumilio protein, which along with nanos protein binds to the 3′ UTR of hunchback mRNA, thus inhibiting its function in the posterior region of the *Drosophila* embryo (see Section 22.3).

In conclusion, it is apparent that sex determination and sex differentiation are based on a wide variety of genetic controls and molecular mechanisms. This is in contrast to other developmental processes, including cell cycle control, embryonic pattern formation, and programmed cell death, which rely on exceedingly well-conserved genetic and molecular mechanisms.

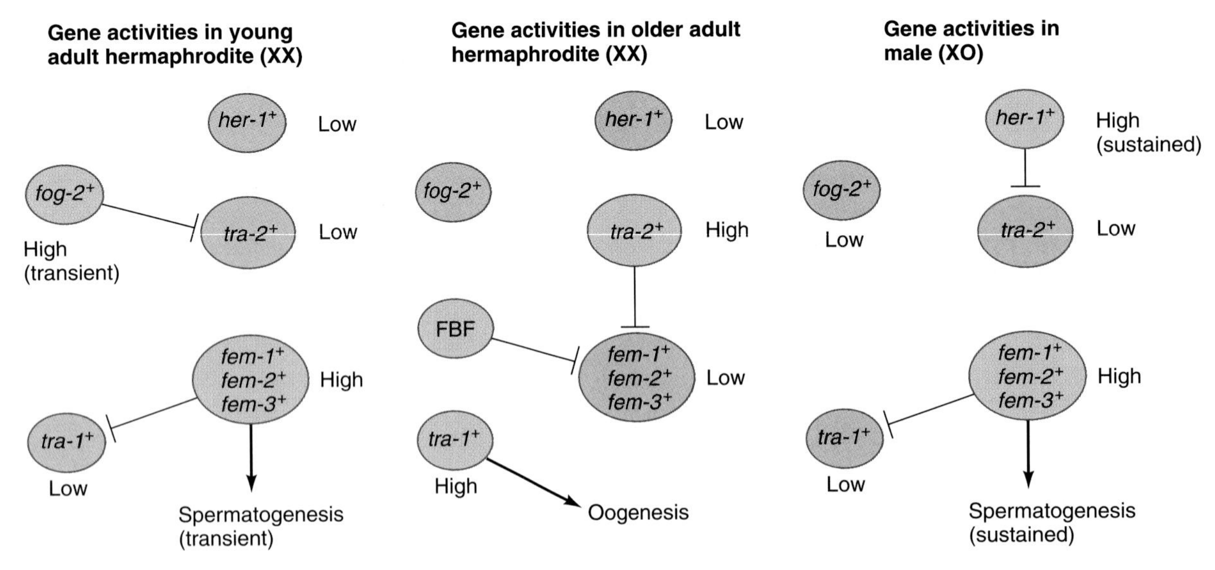

Figure 26.23 Simplified model for the determination of germ line sex in *C. elegans.* Germ line sex differentiation involves many of the genes in the regulatory cascade that directs somatic sex differentiation (see Fig. 26.20). An additional gene, *fog-2*$^+$, inhibits *tra-2*$^+$ in the germ line of the young adult hermaphrodite, so that the *fem*$^+$ genes become temporarily active and promote spermatogenesis, as they do in the male. When *fog-2*$^+$ activity ceases, the *fem*$^+$ genes are repressed and oogenesis can take place. Another regulatory element promoting the switch from spermatogenesis to oogenesis is known as the fem-binding factor (FBF), a cytoplasmic protein that inhibits the function of fem-3 mRNA by binding to its 3′ UTR.

SUMMARY

Most animals are dioecious and more or less sexually dimorphic: they exist as males or females, or in some species as males or hermaphrodites. Sex determination is the natural mechanism by which an animal is determined to become either male or female (or hermaphroditic).

In some species, sex determination is environmentally controlled; sex is determined after fertilization by factors such as temperature or the proximity of sex partners. For most of the species studied, however, sex determination is a function of genotype; sex is determined at fertilization by the combination of genes that the zygote inherits. In the simplest case, one sex is heterozygous for two alleles of the same gene and the other sex is homozygous. This situation easily evolves into a system with entire chromosomes that are sexually dimorphic, such as the X and Y chromosomes of mammals. Sexually dimorphic chromosomes are called sex chromosomes to distinguish them from all other chromosomes, known as autosomes. Instead of having XX females and XY males, other taxonomic groups have XX females (or hermaphrodites) and XO males, or ZW females and ZZ males. The common feature of these different systems is that one sex is heterogametic (producing two types of gametes that differ in their sex-linked genes), while the other sex is homogametic (producing gametes that are identical in their sex-linked genes).

Genotypic sex determination controls three aspects of sexual development: somatic sex differentiation, germ line sex differentiation, and dosage compensation. Somatic sex differentiation brings forth the somatic dichotomy between males and females (or hermaphrodites). Germ line sex differentiation generates either sperm or eggs. Dosage compensation equalizes the expression of non-sex-related genes located on the sex chromosome that is present in two copies in one sex and one copy in the other sex. These three aspects of sexual development are common to mammals, insects, and roundworms. However, the genes and molecular mechanisms that control these pathways of development vary widely.

The critical element of sex determination in mammals is a testis-determining factor (TDF) located on the Y chromosome. This factor can be generated through the activity of a single gene, known as *SRY*$^{+}$. *SRY*$^{+}$ directs a hierarchy of genes including *Dax-1*$^{+}$ and *Sox-9*$^{+}$ in the process of primary sex differentiation. It involves the development of the supporting cells of the indifferent gonad into pre-Sertoli cells, which in turn induce the germ line cells to become spermatogonia. The Sertoli cells also produce anti-Müllerian duct hormone while interstitial (Leydig) cells in the testis produce testosterone. These two hormones will direct the following phase of somatic sex differentiation. Lack of *SRY*$^{+}$ activity allows the supporting cells to become ovarian follicle cells and, indirectly, promotes the development of the germ line cells of the indifferent gonad as oocytes. Dosage compensation in mammals is caused by the virtual inactivation of all X chromosomes in a cell except one.

In the fruit fly *Drosophila melanogaster,* the primary sex-determining signal is the X:A ratio—that is, the number of X chromosomes divided by the number of autosomal sets. An X:A ratio of 1.0 activates a master control gene, *Sex-lethal*$^{+}$. This gene feeds back positively into its own expression and directs all three genetic pathways of sex determination toward female development. An X:A ratio of 0.5 fails to activate *Sex-lethal*$^{+}$ and causes male development by default. Dosage compensation occurs through an increase in the transcription rate of X-linked genes in males.

In the roundworm *Caenorhabditis elegans,* the X:A ratio acts on the *xol-1*$^{+}$ gene, which acts as a master control switch. It controls a set of genes regulating dosage compensation by reducing the transcription rate of X-linked genes in XX individuals. Independently, *xol-1*$^{+}$ acts on a genetic cascade of inhibitory interactions that eventually regulate both somatic and germ line sex differentiation. The latter involves additional control elements that allow the formation of sperm in hermaphrodites before the production of oocytes begins.

Hormonal Control of Development

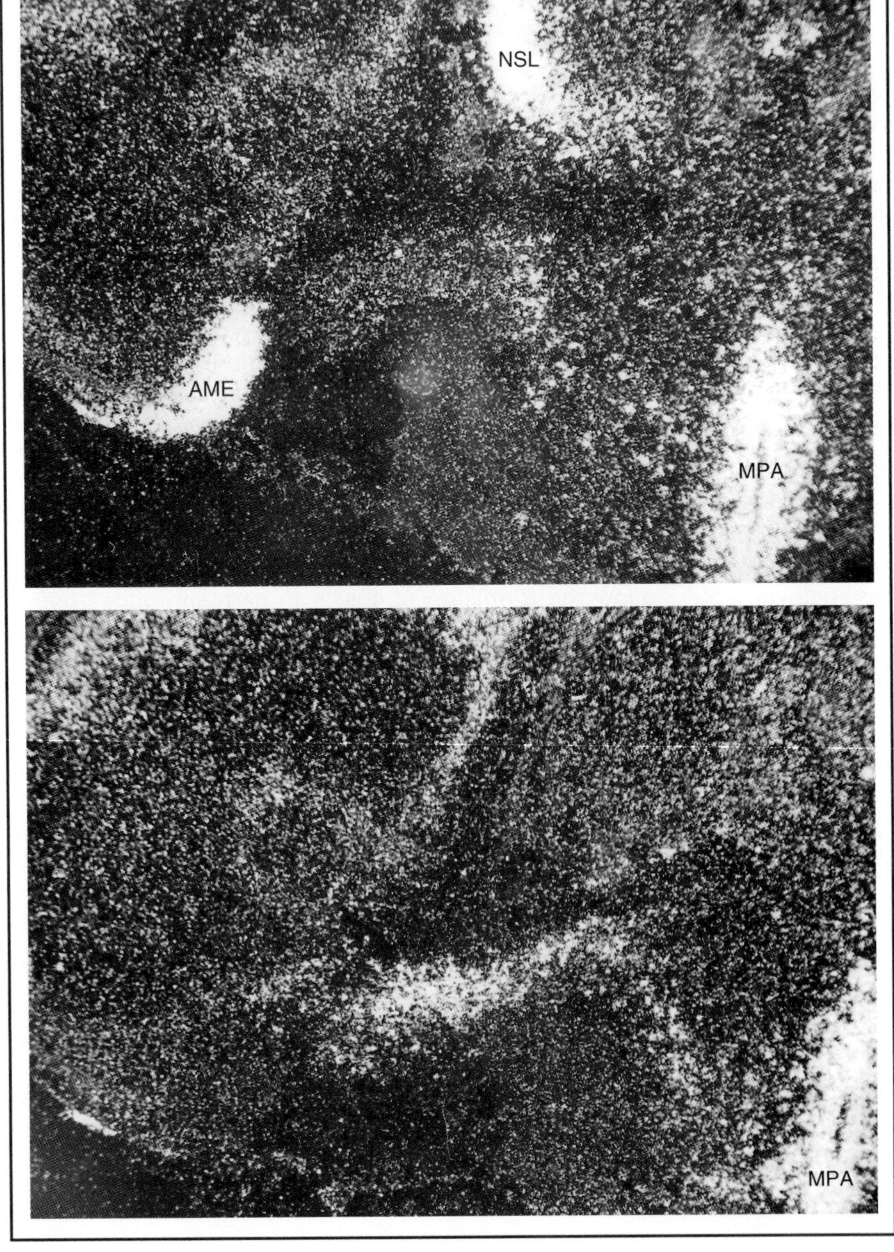

Figure 27.1 In vertebrate brains, some groups of neurons are activated by androgens, hormones circulating primarily in males, whereas other brain areas are affected by the female sex hormones estrogen and progesterone. The prevalence of male or female sex hormones therefore elicits sex-specific combinations of brain activities. This pair of dark-field photographs shows frontal sections of the brain of the whiptail lizard *Cnemidophorus inornatus.* The sections were hybridized in situ with radiolabeled probes that bind either to mRNA for androgen receptor (top) or to mRNA for progesterone receptor (bottom). The white grains reveal the distribution of the mRNA bound by either probe. Some brain areas, including the medial periventricular area (MPA) of the diencephalon, contain the mRNAs for both hormone receptors. The mRNA for androgen receptor is concentrated in two telencephalic areas: the nucleus septalis lateralis (NSL), and the external nucleus of the amygdala (AME). The latter is known to be involved in aggression and male copulatory behavior. The mRNAs for estrogen and progesterone are selectively present in other brain areas associated with the regulation of ovulation and with female sexual receptivity.

CELLS in a developing organism exchange signals that coordinate their actions and allow the organism to adapt to external events. Three principal agents of cell communication have evolved in metazoa: neurons, short-range chemical signals such as inducers and growth factors, and long-range chemical signals called hormones. Neurons send electrical impulses through long extensions, which selectively reach their targets. Inducers and growth factors act locally because they bind rapidly to cell membranes and extracellular materials. In contrast, hormones are chemical signals that are produced by specialized gland cells and then carried all over the organism in the blood or equivalent body fluids.

Like broadcasts, hormonal signals are propagated everywhere but act only on receivers that are appropriately tuned. In other words, hormones elicit a response only from target cells, which have matching hormone receptors and other cofactors that mediate the target cells' responses. Because a hormone accumulates at the receptors to which it binds, the target cells of a hormone can be identified by injection of radiolabeled hormone and subsequent autoradiography. Alternatively, target cells can be revealed by immunostaining of the hormone receptor, or by in situ hybridization with a labeled probe that binds to the mRNA encoding the receptor (Fig. 27.1).

Typically, a hormone coordinates the activities of multiple target tissues or organs. For instance, in the human female, progesterone prepares the inner lining of the uterus (the endometrium) during each menstrual cycle for implantation of an egg; if pregnancy occurs, progesterone also maintains the endometrium, stimulates the growth of the uterine wall, and promotes the formation of mammary ducts. The coordinating action of a hormone often depends on concentration changes over time. Such changes are typically controlled by the rate of synthesis in the hormone-producing glands, which is controlled by feedback loops, other hormones, or nervous system input. The synthesis of progesterone in the ovary, for example, is regulated by luteinizing hormone from the pituitary gland.

In this chapter, we will review some general aspects of hormone function before turning to the hormonal control of sex differentiation in mammals. These discussions will cover the development of the reproductive organs as well as some aspects of brain development and behavior. Next we will explore the hormonal control of insect metamorphosis, which is the subject of exciting new work on the molecular mechanisms of hormone action. Finally, we will examine the role of hormones in amphibian metamorphosis and highlight some of the astounding parallels to hormonal control of insect metamorphosis.

27.1 General Aspects of Hormone Action

In order to act on a target cell, hormones first bind to specific proteins that act as receptors. A bound hormone activates its receptor by conferring on it a specific ability to interact with other molecules. Animal hormones can be divided into two groups: water-soluble and lipid-soluble. The *water-soluble hormones* are mostly peptides, such as the growth hormone released by the anterior pituitary gland. These hormones cannot cross cell plasma membranes; instead they bind to the extracellular domains of receptor proteins. The loaded receptors then act as ion channels or protein kinases, or they interact with other membrane proteins to increase the levels of second messengers such as cAMP or calcium ions (Ca^{2+}); see Section 2.8. The *lipid-soluble hormones* comprise steroids including testosterone, terpenoids such as retinoic acid, and the thyroxines released by the thyroid gland. These hormones diffuse readily through cell plasma membranes. They bind to receptors that are located in the nuclei of target cells, or to receptors that are located in the cytoplasm and move into the nucleus when bound by a ligand. Both types of receptor act as transcription factors (see Section 16.4).

Both water-soluble and lipid-soluble hormones affect the gene expression of their target cells, either by controlling transcription or by modifying the activities of preexisting proteins (see Fig. 2.25). Either way, hormones have powerful effects on cell division, on cell determination and differentiation, and on cell death.

The dependence of hormone action on receptors, and the specific interaction of the receptors with regulatory gene elements or intracellular signaling molecules, confer a remarkable degree of specificity on the action of hormones. This specificity is enhanced when receptors consist of two different polypeptides, or when the ultimate effect of a hormone depends on the presence of additional signaling components. In such cases, the hormone action is limited to those cells in which all of these cofactors are available simultaneously. The receptors and cofactors are deployed in an epigenetic process during development: As an organism becomes more complex, more types of target cells with specific combinations of hormone receptors and assorted cofactors originate.

27.2 Hormonal Control of Sex Differentiation in Mammals

As discussed in the preceding chapter, mammalian sex determination is controlled by the SRY^{+} gene on the Y chromosome. SRY^{+} expression in the supporting-cell lineage causes the indifferent gonad to develop as a testis, whereas lack of SRY^{+} expression allows the gonad to develop as an ovary. We have already discussed this first phase of sexual development, known as *primary sex differentiation*, in Section 26.3. Here, we will focus on the subsequent phase, called *secondary sex differentiation*, which in mammals is governed by hormonal control.

THE SYNTHETIC PATHWAYS FOR MALE AND FEMALE SEX HORMONES ARE INTERCONNECTED

Many vertebrate hormones, including most sex hormones, are steroids. All steroid hormones are derived from cholesterol, as shown in Figure 27.2. Four synthetic steps convert cholesterol into ***progesterone***, a female sex

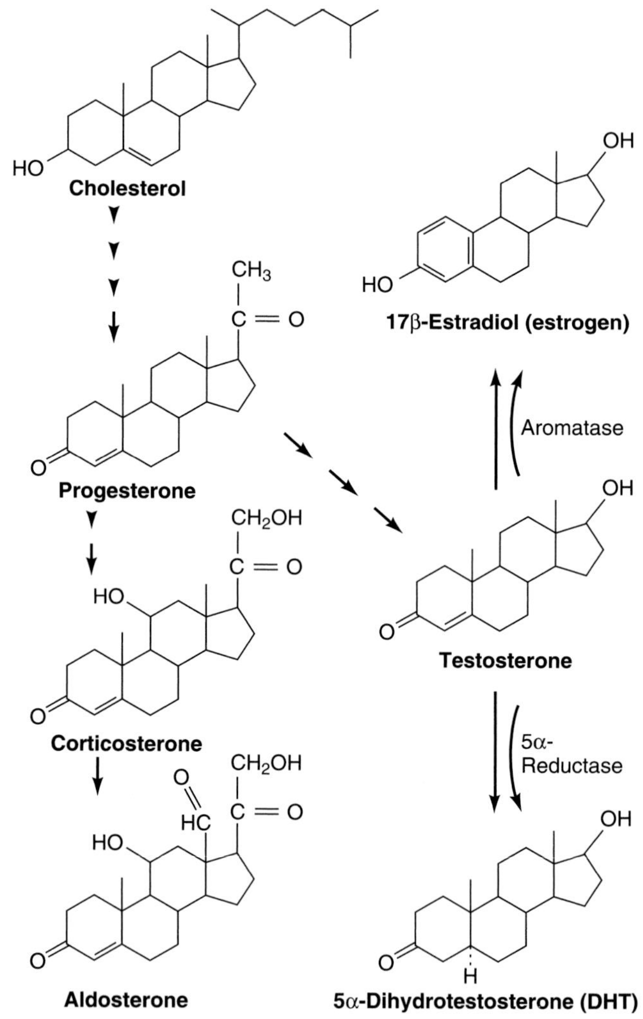

Figure 27.2 Synthetic pathways for vertebrate steroid hormones. All steroids are synthesized from cholesterol. The number of arrows indicates the number of synthetic steps in each pathway. Note that testosterone, the key male sex hormone, is a derivative of the female sex hormone progesterone and the precursor of another female sex hormone, 17β-estradiol (estrogen). The conversion of testosterone into estrogen is catalyzed by the enzyme aromatase. A different enzyme, 5α-reductase, catalyzes the conversion of testosterone into 5α-dihydrotestosterone (DHT).

hormone. Three additional steps transform progesterone into *testosterone,* the key male sex hormone. Testosterone in turn is converted by an enzyme, ***5α-reductase,*** into *5α-**dihydrotestosterone*** **(*DHT*),** another male sex hormone. Collectively, testosterone and DHT are referred to as ***androgens*** (Gk. *andro-*, "man"; *gennan,* "to produce"). Another enzyme, ***aromatase,*** converts testosterone into *17β-estradiol,* a female sex hormone commonly called ***estrogen.*** Progesterone is also a precursor of corticosterone and aldosterone, steroid hormones that are involved in the metabolism of glucose and minerals, respectively.

The principal organs of steroid hormone synthesis are the gonads, but certain synthetic steps also occur in other tissues. Testosterone is synthesized mainly in the interstitial cells, or *Leydig cells,* of the testes. In male mammals, androgens are produced at significant levels during fetal development under the influence of gonadotropic hormones from the placenta. In ovarian follicles, estrogens are synthesized by the *granulosa cells* from the testosterone precursor provided by the adjacent *thecal cells.* In many mammals, including humans, however, the level of estrogen synthesis in female fetuses is low, and female sexual development during this phase occurs mostly by default—that is, in the absence of androgens. With the beginning of adulthood (puberty, in mammals), gonadotropins from the anterior pituitary gland stimulate testes and ovaries to produce testosterone and estrogen, respectively. These hormones control the development and maintenance of features such as breasts in human females and larger muscles in males, which are commonly known as ***secondary sex characteristics.*** (Note that *secondary sex characteristics* are sexually dimorphic features that develop after puberty, whereas *secondary sex differentiation* is a term that covers all hormonally controlled sexual development, beginning during embryogenesis.) In addition to the gonads, the adrenal glands synthesize significant levels of sex hormones.

The pathways of steroid biosynthesis have important biological implications. First, the hormonal differences between males and females are more complex than they appear because the testes have to synthesize progesterone in order to produce testosterone, while the ovaries have to synthesize testosterone as a precursor of estrogen. Second, some critical steps of sex hormone synthesis occur outside the gonads. The pathways that convert testosterone into either DHT or estrogen are especially important. Each conversion depends on a specific enzyme, 5α-reductase or aromatase, respectively. These enzymes are present not only in the gonads but also in target tissues of DHT and estrogen. For instance, significant concentrations of 5α-reductase are present in the indifferent external genitalia of the early embryo. Here testosterone is converted into DHT, which is required for development of the male external genitalia. Later during development, certain areas of the brain in both sexes contain significant amounts of aromatase, the enzyme that converts testosterone into estrogen.

Figure 27.3 provides an overview of sex differentiation and of the role of hormones in the process.

THE GENITAL DUCTS DEVELOP FROM PARALLEL PRECURSORS, THE WOLFFIAN AND MÜLLERIAN DUCTS

As discussed in Section 26.3, the mammalian testis and ovary develop from *complementary* portions of the indifferent gonad: The testis develops from the medulla, whereas the ovary arises from the cortex. As we have seen, this decision is controlled by the *SRY*$^+$ gene. Once the embryonic gonads develop as either testes or ovaries, the following steps of sexual differentiation are controlled entirely by hormones.

The genital ducts that transport and nurture sperm, eggs, and embryos develop from two pairs of embryonic precursors, the *Wolffian ducts* and the *Müllerian ducts* (see Fig. 26.6). Both pairs of ducts are present in parallel at the end of the sexually indifferent stage, which lasts through the sixth week of gestation in the human. Their development is controlled by the presence or absence of two male sex hormones, *testosterone* and ***anti-Müllerian duct hormone* (*AMH*).** The latter is a glycoprotein of the *transforming growth factor*-β family and is produced by the *Sertoli cells* of the testis.

The Wolffian duct is originally part of the mesonephros (embryonic kidney), and it persists in this capacity in lower vertebrates of both sexes (see Fig. 14.24). In mammals and birds, the kidney function of the mesonephros is superseded by the development of the definitive kidney (metanephros). In females, the Wolffian duct subsequently degenerates as its cells undergo *apoptosis.* In males, however, the Wolffian duct becomes part of the reproductive system (Fig. 27.4). Here, the duct gives rise to the *epididymis,* where sperm mature; the *ductus deferens,* which transports the sperm; and the *seminal vesicle,* where most of the seminal fluid is produced. Beyond the seminal vesicle, the ductus deferens is called the *ejaculatory duct.* The Müllerian duct in the male degenerates except for small vestiges, and the degeneration process depends on AMH and its receptor.

The Müllerian duct persists in females, since females produce no AMH. It gives rise to the oviduct, the uterus, and, in mammals, the upper portion of the vagina (Fig. 27.5). The Wolffian duct in the female degenerates except for minor vestiges.

The hormonal control of secondary sex differentiation was first established by observations on cows and rabbits. Lillie (1917) found that female calves were masculinized in their external genitalia, their genital ducts, and even in their gonadal development if they shared the uterine circulatory system with a male twin. Such masculinized female calves were known as *freemartins;*

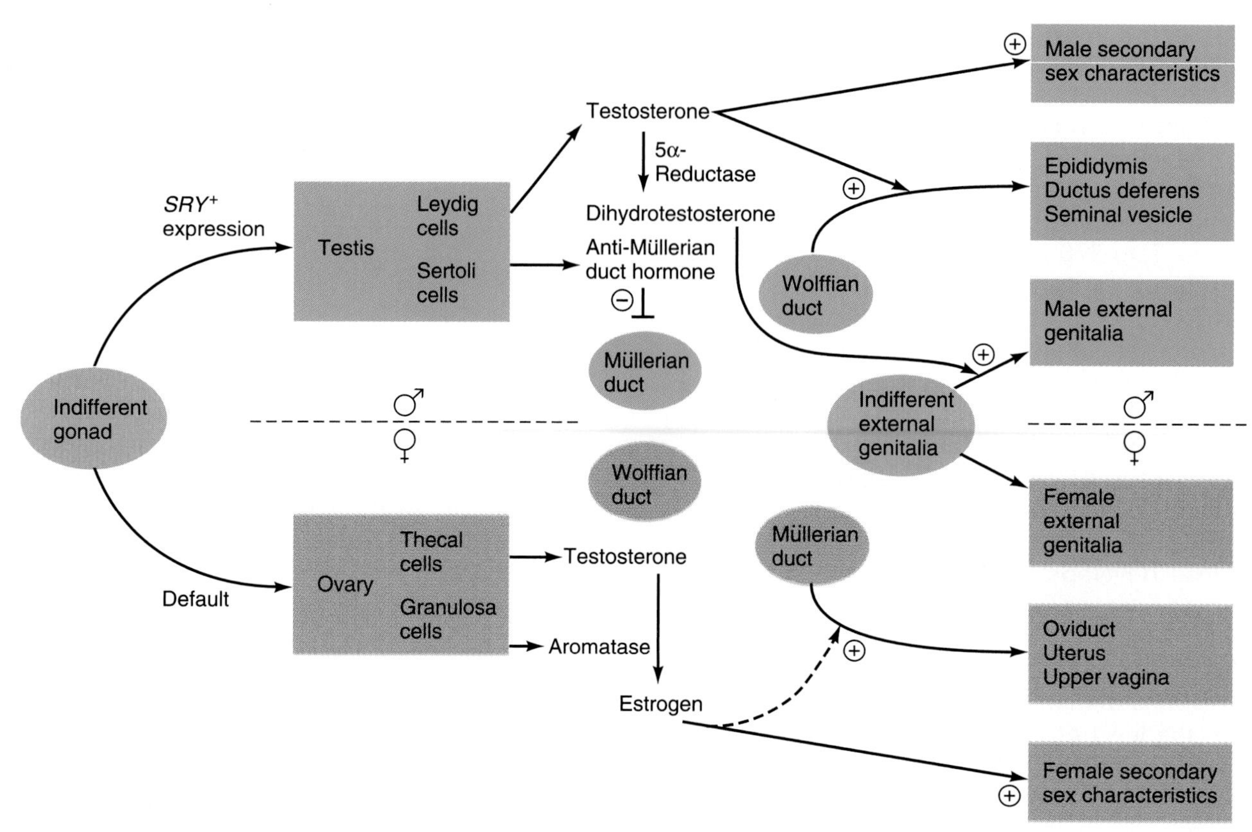

Figure 27.3 Somatic sexual development in mammals. Primary sex differentiation is under genetic control: Expression of the *SRY*$^+$ gene in the indifferent gonad causes testis development, whereas lack of *SRY*$^+$ expression allows ovary development. The following phase of secondary sex differentiation is controlled by hormones. The developing testis produces testosterone in the interstitial, or Leydig, cells and anti-Müllerian duct hormone (AMH) in the Sertoli cells. During fetal development, testosterone causes the Wolffian duct to develop into the epididymis, ductus deferens, and seminal vesicle. Some of the testosterone is converted by 5α-reductase to dihydrotestosterone (DHT). DHT brings about male development of the originally indifferent external genitalia. AMH causes degeneration of the Müllerian duct. Following puberty, testosterone promotes the development of secondary male sex characteristics. In the developing ovary, thecal cells cells within the ovarian follicle produce testosterone, which is converted to estrogen by aromatase in the granulosa cells surrounding the oocyte. However, estrogen levels are low during fetal development, in most species, so that the development of the external female genitalia occurs by default. Likewise, absence of AMH is sufficient for development of the Müllerian duct into the oviduct, uterus, and upper vagina. After puberty, high levels of estrogen promote the development of female secondary sex characteristics.

the conditions of their development suggested that some male factor(s) can pass through the circulatory system and redirect genetically female sex organs toward male development.

In later studies, Jost (reviewed 1965) removed testes or ovaries from rabbit fetuses and monitored the animals' sexual development. In *both* sexes, removal of the gonads was followed by development of the Müllerian ducts into oviducts and a uterus, and by the formation of female external genitalia. These results indicated that the female organs arise by default, whereas the development of male organs requires some testicular factor(s). Thus, the secondary sex differentiation under the control of hormones again demonstrates the *principle of default programs,* as does the primary sex differentiation under control of the SRY gene product (see Section 26.4).

By additional experiments, Jost showed that testes produce at least two such factors. If he grafted a testis next to the ovary of a female rabbit embryo, *two effects* were observed on the operated side: The Wolffian duct developed, and the Müllerian duct degenerated. If instead of a testis he transplanted a crystal of testosterone, only one effect was observed: The Wolffian duct developed, but the Müllerian duct did not degenerate. Subsequent investigations identified the testicular factor that is not replaced by testosterone as AMH (Josso et al., 1976).

More detailed knowledge was obtained by analyzing congenital abnormalities in humans and the phenotypes of knock-out mice. The human ***androgen insensitivity*** syndrome is caused by a mutation in the gene for the androgen receptor—that is, the receptor for testosterone and DHT (Meyer et al., 1975). The gene is X-linked, and the mutant allele is transmitted by heterozygous female carriers. XY individuals who inherit the mutant allele develop testes that produce testosterone and AMH normally. Nevertheless, these individuals develop as

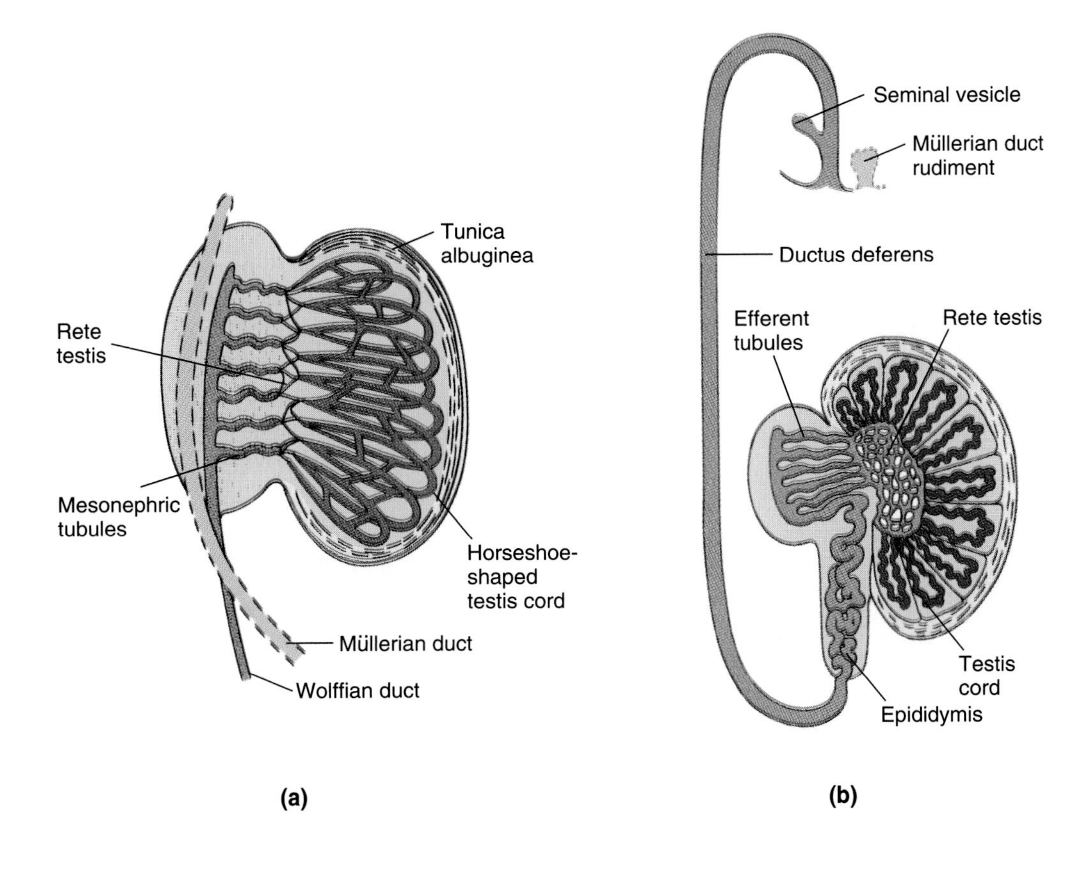

Figure 27.4 Development of the genital ducts in the human male. **(a)** Frontal section of the testis and genital ducts during the fourth month of development. The distal portion of the Wolffian duct is embedded in the mesonephros as the mesonephric tubules connect via the rete testis to the horseshoe-shaped testis cords. The Müllerian duct is still in place but will soon begin to degenerate. **(b)** After the descent of the testis, the mesonephric tubules have become efferent tubules, which will conduct sperm after puberty. Much of the proximal Wolffian duct has been transformed into the convoluted tube of the epididymis, from where the sperm will pass into the ductus deferens (vas deferens). The section of the Wolffian duct distal to the seminal vesicle will form the ejaculatory duct, which connects to the penile urethra.

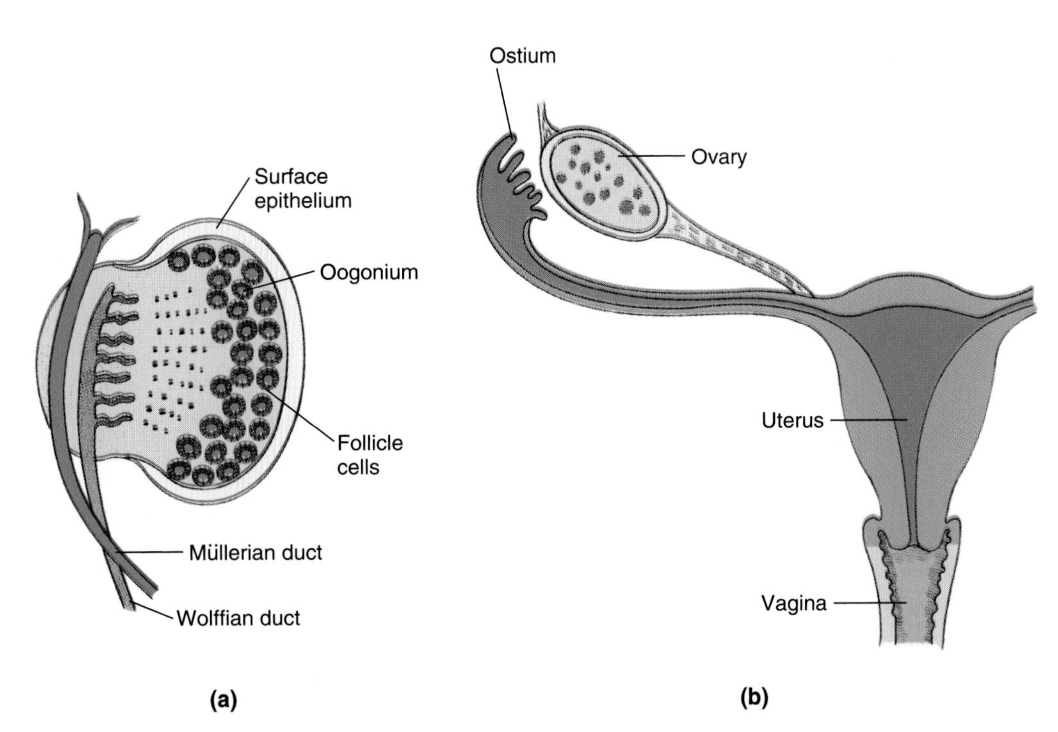

Figure 27.5 Development of the genital ducts in the human female. **(a)** Frontal section of the ovary and genital ducts during the fourth month of development. The cortical sex cords have broken up into individual follicles, each containing an oogonium surrounded by follicle cells. Both the Wolffian and the Müllerian ducts are in place, but the Wolffian duct will soon begin to degenerate. **(b)** In the newborn female, the Wolffian duct has disappeared except for a few vestiges. The proximal part of each Müllerian duct has formed one oviduct with an opening (ostium) into the abdominal cavity, and the distal portions of the two Müllerian ducts have joined to form the uterus and the upper vagina.

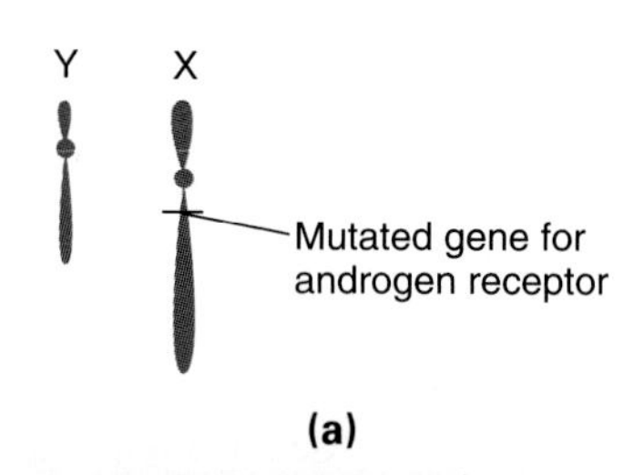

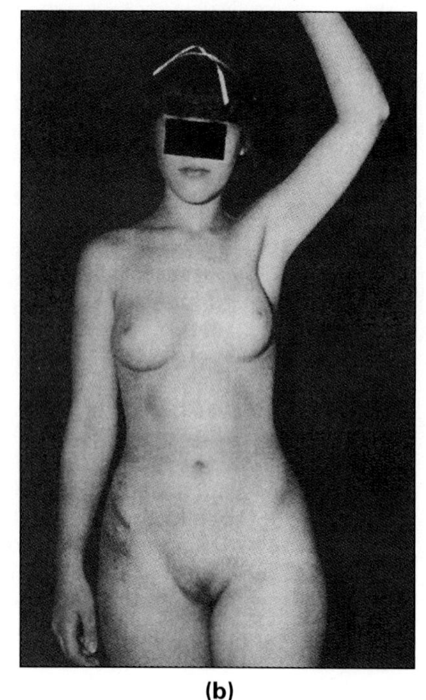

Figure 27.6 Androgen insensitivity in genetic males. **(a)** Such individuals have normal male sex chromosomes (XY) except for a loss-of-function mutation in the X-linked gene that encodes the androgen receptor. **(b)** The individuals have (nondescended) testes and produce androgens, but their cells lack androgen receptors. Instead, they respond to estrogens produced in the adrenal glands. Correspondingly, a distinctly female appearance develops. Internally, however, androgen-insensitive individuals lack the derivatives of the Müllerian duct, because they produce and respond to anti-Müllerian duct hormone.

phenotypic females (Fig. 27.6). Because of the lack of androgen receptors, there is no response to testosterone in the target organs. As a result, the Wolffian duct degenerates, and female external genitalia develop. In the absence of testosterone, the level of estrogen produced in the adrenal glands is sufficient to promote full development of secondary female sex characteristics. This dramatic response indicates that estrogen receptors are present in both males and females. However, the production of AMH in the testes causes the elimination of the Müllerian duct, so that the vagina remains shallow and—due to the lack of ovaries—no menstruations set in at puberty.

The action of AMH in humans is revealed by the *persistent Müllerian duct syndrome* (*PMDS*), which is characterized by the persistence of oviducts and a uterus in otherwise nearly normal men. The condition is usually discovered during surgery for inguinal hernia or cryptorchidism (failure of the testes to descend into the scrotum). In some PMDS cases, the gene encoding AMH is mutated so that no functional AMH is made (Knebelmann et al., 1991). Other PMDS patients may be deficient in the AMH receptor or its downstream signaling components. Male knockout mice deficient for AMH or its receptor possess a complete male reproductive tract as well as a uterus and oviducts (Mishina et al., 1996). Male mice lacking both the androgen receptor and AMH have nondescended testes but develop otherwise female reproductive organs, including uterus and oviducts (Behringer et al., 1994).

MALE AND FEMALE EXTERNAL GENITALIA DEVELOP FROM THE SAME EMBRYONIC PRIMORDIA

Unlike the male and female gonads and genital ducts, which develop from *complementary* embryonic rudiments, the male and female external genitalia develop from the *same* embryonic rudiments.

At the end of the indifferent stage—after 6 weeks, in the human embryo—the *cloaca* has divided into the ***urogenital sinus*** and the ***anus,*** and both openings are closed off by temporary membranes (Fig. 27.7). ***Urethral folds*** on both sides of the urogenital sinus unite anteriorly to form the ***genital tubercle.*** In the male, the urethral folds close ventrally and, together with the genital tubercle, form the ***penis.*** In the female, the genital tubercle forms the ***clitoris,*** while the urethral folds develop into the labia minora. Thus, the urogenital sinus becomes either the ***penile urethra*** in the male or the ***vestibule*** of the vagina in the female. On both sides of the urethral folds at the indifferent stage, a pair of ***genital swellings*** arise. They form either the ***labia majora*** of the female or the ***scrotum*** surrounding the testes in the male.

Hormonal control of external genital differentiation is illustrated by a heritable form of male pseudohermaphroditism that has been observed in human families around the world (Imperato-McGinley, 1997; Nordenskjold and Ivarsson, 1998). The anomaly is caused by a *5α-reductase deficiency,* which renders XY fetuses unable to convert testosterone to DHT. Because testosterone does not activate androgen receptors to the same extent as DHT, such individuals are born with ambiguous external genitalia including a blind vaginal pouch, labia with testes inside, and an enlarged clitoris. In traditional societies, these babies are usually raised as girls until puberty, when the greatly increased level of testosterone causes the penis and scrotum to grow belatedly. As the secondary male sex characteristics also develop, most of these individuals begin to feel like males and adopt male gender roles. However, their urogenital sinus does not close, leaving them with an open urethra on the underside of the penis, a condition known as *hypospadia.*

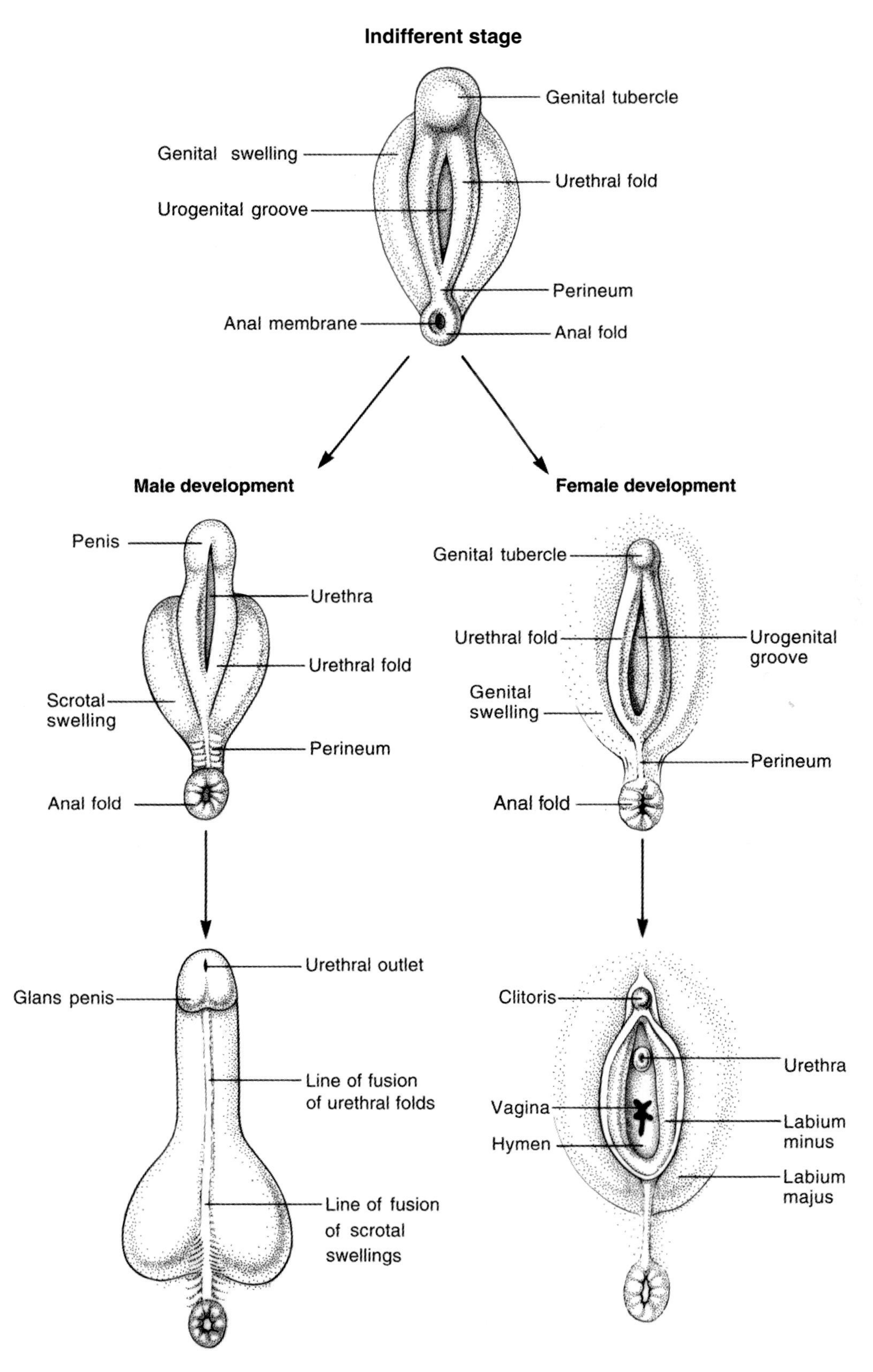

Figure 27.7 Development of the external genitalia in the human male (left) and female (right) from the indifferent stage (top). At the indifferent stage, the urogenital groove becomes separated from the anus by the perineum, and both openings are closed off by temporary membranes. The genital tubercle forms either most of the penis or all of the clitoris. The urogenital groove becomes either the penile urethra or the vestibule of the vagina. The urethral folds form either the part of the penis that surrounds the urethra or the labia minora (sing., labium minus). The genital swellings give rise to either the scrotum or the labia majora (sing., labium majus).

27.3 Hormonal Control of Brain Development and Behavior in Vertebrates

Having seen how sex hormones control the development of the reproductive organs, we will now explore the effects of the same hormones on brain development and on behavior. First, we will explore how hormone receptors—that is, those molecules that account for much of the specificity in hormone action—are distributed in the brain. Then we will examine courtship behavior of songbirds, copulatory behavior in mammals, and the association of these activities with sexually dimorphic brain areas.

ANDROGEN AND ESTROGEN RECEPTORS ARE DISTRIBUTED DIFFERENTLY IN THE BRAIN

Males and females differ in mating and parenting behavior and, more subtly, in some behaviors that are less directly associated with reproduction. Because behaviors are controlled by the central nervous system, and since reproductive behaviors especially are known to depend on hormone levels, researchers have explored how sex hormones interact with the brain, and how these interactions might differ between males and females.

As discussed earlier in this chapter, hormones act through specific receptor proteins, and the distribution of such receptors largely accounts for the selective action of hormones on specific target organs. It was therefore logical for researchers to explore the distribution of estrogen and androgen receptors in the brain. When localizing the receptors themselves turned out to be difficult, they traced the mRNAs encoding these receptors instead (see Fig. 27.1). Other researchers injected radiolabeled estrogen, testosterone, and other hormones into various vertebrates and monitored the accumulation pattern of each hormone in the brain by *autoradiography* (see Method 8.1). As hormones accumulate in cells by binding to their receptors, the observed hormone accumulation patterns reflect hormone receptor distributions. The results obtained with these different methods were the same. Each hormone receptor has its own distinct distribution pattern in the brain. For example, androgen receptor is especially concentrated in brain areas involved in aggression, mounting, and copulation, whereas estrogen and progesterone receptors are more concentrated in brain areas associated with the regulation of ovulation and with female sexual receptivity (Young et al., 1994).

The studies reviewed above have revealed only subtle if any differences between the sexes with regard to the receptor distribution for *any given hormone.* The major differences are not between the sexes but *between the receptors for different hormones.* Therefore, the brain areas stimulated in the presence of androgens differ from the brain areas affected by estrogen and progesterone. On the basis of these results, one would predict that castrated males injected with estrogen and progesterone will show female behaviors and that ovariectomized females injected with androgens will display male behaviors. These predictions are largely confirmed by experimental data, some of which will be examined below. Some conflicting observations, to be discussed subsequently, indicate that certain structural features of the brain itself develop differently in males and females, depending mostly on their prenatal exposure to testosterone.

ANDROGENS CAUSE SEASONAL AND SEXUALLY DIMORPHIC CHANGES IN THE BRAINS OF BIRDS

Songbirds are a large and diverse group of birds that produce complex but stereotypical sounds. Generally, the males sing most actively early in the breeding season, when they attract females by establishing and defending territories. Male songbirds learn their songs from older males. During this early phase of *song acquisition,* the songs of the novice are initially crude and variable, but with continued tutoring and practice they become well defined and invariant in structure. This latter phase of song development is referred to as *song crystallization.*

Like other courtship behaviors, birdsong is strongly influenced by circulating hormone levels (Nottebohm, 1989; Bottjer, 1991; Kelley and Brenowitz, 1992). In male zebra finches, there is a linear correlation between singing and the concentration of testosterone in the blood serum (Pröve, 1978). Castration of a male reduces the rate of singing from about 45 songs to 8 songs per 15-min observation session (Fig. 27.8; Arnold, 1975). If

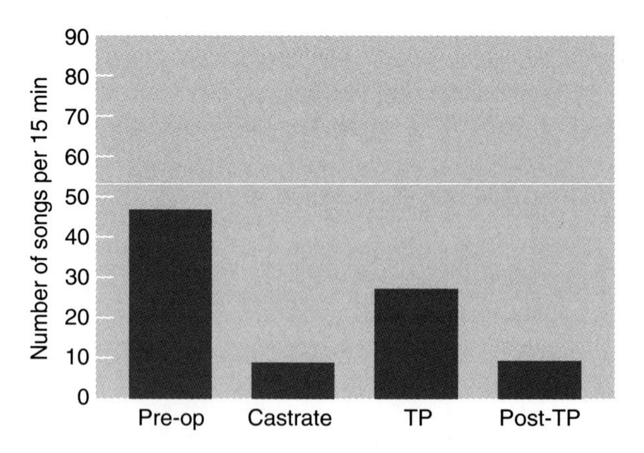

Figure 27.8 Hormonal control of singing in male zebra finches. The rate of song production (number of songs during a 15-min observation session) is shown for normal males (Pre-op), after castration (Castrate), after implantation in castrates of pellets releasing testosterone propionate (TP), and after removal of the implants (Post-TP). The song rate is reduced dramatically after castration but is increased significantly by the hormone implant.

the same male then receives an implant that releases testosterone propionate, his rate of singing increases again to about 27 songs in 15 min. In other songbirds, testosterone injections induce females to sing and cause males to sing out of season (Thorpe, 1958; Nottebohm, 1980). Together, these observations show clearly that the singing of songbirds depends on the level of testosterone or one of its metabolites.

To test which specific hormones promote singing in zebra finches, researchers injected different testosterone metabolites into castrated male zebra finches. Combinations of estrogen and DHT restored singing to its precastration level, whereas either hormone by itself was less effective (Harding, 1983). Similar investigations with sparrows indicate that estrogen facilitates song acquisition, whereas testosterone is required for song crystallization (Marler et al., 1988). Thus, it appears that different testosterone metabolites promote different phases of song development.

There are also anatomical differences between the brains of male and female zebra finches and canaries, especially during the breeding season. Certain forebrain nuclei that are involved in singing are up to six times larger in males than in females (Fig. 27.9; Arnold, 1980). In autoradiographic studies, radiolabel from injected testosterone accumulated selectively in these song control nuclei (Arnold et al., 1976). Three of the nuclei are known as the *higher vocal center* (*HVC*), the *robustus archistriatalis* (*RA*), and the *magnocellular nucleus of the anterior neostriatum* (*MAN*). The RA is innervated by both the HVC and the MAN and depends on these innervations for growth and neuronal survival (F. Johnson and S. W. Bottjer, 1994). In males, these nuclei are much larger in the spring than in other seasons (Nottebohm, 1981); very similar changes in volume occur in females treated with testosterone. Moreover, the proportion of cells in the male HVC and MAN that accumulate radiolabel from injected androgens increases during the time of song crystallization (Bottjer, 1987). These data indicate that testosterone or its metabolites cause a seasonal and sexually dimorphic increase in the volume of the song control nuclei of songbirds. The larger nuclei have more neurons and more synapses, and they produce larger amounts of neuropeptides (Bottjer et al., 1997).

Adult birds continue to generate new neurons from the neuroepithelial layer that lines the brain ventricles (see Section 13.1). DNA labeling with injected radioactive thymidine indicates that neuroblasts born in the neuroepithelial layer move to the higher vocal center, where they become functional neurons before most of them disappear within less than a year. Thus, there seems to be a constant turnover of neurons in the higher vocal center. The larger number of neurons in the male song control nuclei may therefore result from the action of testosterone or its metabolites on neuron generation, migration, differentiation, or death. Among these four parameters, programmed cell death does not seem to play a major role (Burek et al., 1997). In addition to neuron turnover, testosterone and its metabolites may also affect neuronal function in a sexually dimorphic way.

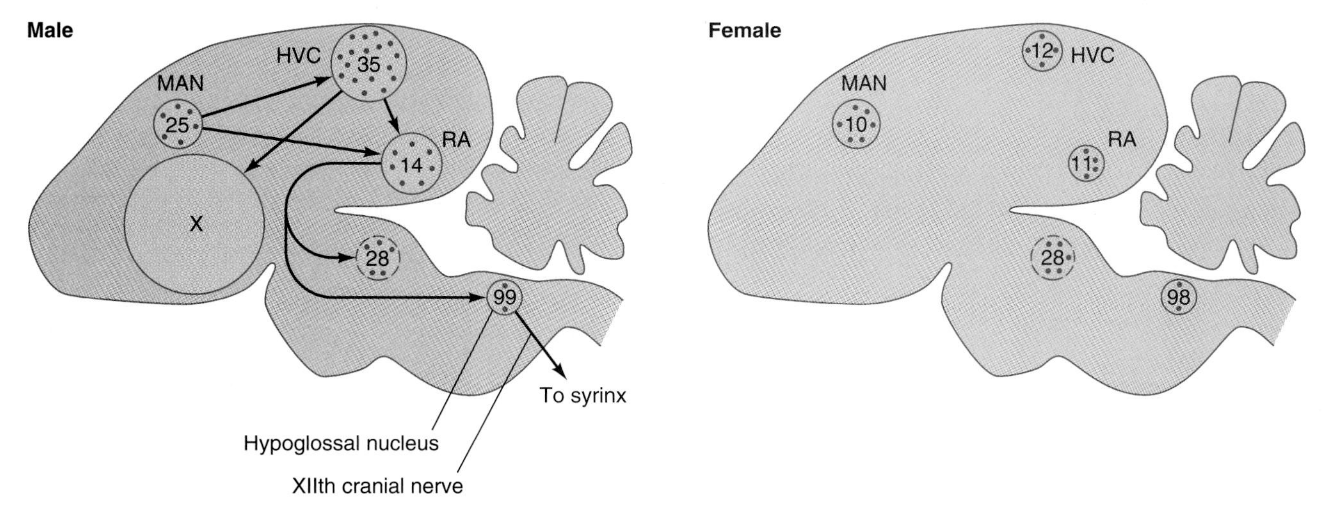

Figure 27.9 Sexual dimorphism of song control nuclei in the songbird brain. The schematic drawings show paramedian sections of male and female songbird brains; anterior to left, dorsal surface up. Circles and arrows represent brain areas and projections thought to be involved in controlling song. The final projection is from the hypoglossal nucleus through the XIIth cranial nerve to the syrinx, the tracheal organ where the birdsong is actually produced. The size of each circle is proportional to the volume of the corresponding brain region in the zebra finch. Circles with dashed lines indicate brain regions whose volumes have not been estimated. Dots mark regions that become labeled after injection of radioactive testosterone. The numbers within each circle indicate the approximate percentage of labeled cells within each region. Note the sexual dimorphism in size and labeling index of the nuclei called the higher vocal center (HVC), the robustus archistriatalis (RA), and the magnocellular nucleus of the anterior neostriatum (MAN). Area X is not distinct in stained sections of brains from female zebra finches.

PRENATAL EXPOSURE TO SEX HORMONES AFFECTS ADULT REPRODUCTIVE BEHAVIOR AND BRAIN ANATOMY IN MAMMALS

The hormonal control of reproductive behavior and brain development has also been studied in mammals (Breedlove, 1992; Crews, 1994). Some of the most stereotypical and easily quantifiable behaviors occur during copulation of rodents. A female rodent arches her back to elevate her rump and head and moves her tail to one side. This posture is called ***lordosis,*** and it is taken as a measure of the female's receptivity to mating. Normally, female rodents exhibit lordosis in response to a mounting male only during a particular phase of the ovulatory cycle. When the ovaries are removed, a female ceases to exhibit lordosis. However, her lordotic behavior can be reactivated by injections of estrogen and progesterone.

When the same hormonal regimen that reactivates lordotic behavior in an ovariectomized female rat is given to a castrated male rat, the male shows almost no lordosis in response to another male's mounting. The fact that castrated male rats and ovariectomized female rats behave differently after the same hormonal treatment indicates that their previous conditioning must be different. Because the central nervous system mediates behavior, the difference presumably lies therein.

One difference between the brains of males and females is the prenatal exposure of the male brain to androgens. If this difference were critical, then female rodents exposed to testosterone prenatally should not show lordosis as adults when given estrogen and progesterone treatment and mounted by a male. This prediction was confirmed in tests with guinea pigs exposed prenatally to testosterone propionate (Phoenix et al., 1959).

The effects of fetal exposure to sex hormones are also seen in mammals that produce litters of multiple young during each pregnancy (vom Saal, 1989). Because the growing fetuses exchange steroid hormones through the placental blood circulation, females developing between two males in the uterus (called 2M females) are exposed to higher levels of testosterone than females developing next to one male (1M females) or no males (0M females). As compared with 0M females, 2M females have masculinized genitalia, take longer to reach puberty, and as adults have shorter and fewer reproductive cycles. 2M females are also less attractive to males and more aggressive to other females. The converse effects are observed for males developing next to females in utero in species where female fetuses synthesize significant amounts of estrogen. In mice, 2F males have smaller seminal vesicles and are less aggressive and sexually more active than 0F males.

These observations inspired the ***organizational hypothesis,*** according to which prenatal exposure to androgens permanently alters the developing brain, causing it to function in male-specific ways later in life. This hypothesis has since been tested for other behaviors and with different mammals and has been confirmed with few exceptions. Interestingly, the sensitive period during which a postnatal behavior is masculinized by prenatal exposure to androgens depends on the behavior and the species studied. Thus, it appears that specific parts of brains have different phases during which they are most sensitive to hormonal conditioning.

Having confirmed the hormonal control of reproductive behavior, researchers looked for sexual dimorphism in those areas of the rodent *brain* that are involved in reproductive behaviors. They identified one such area in rats as the ***sexually dimorphic nucleus of the preoptic area (SDN-POA)*** of the hypothalamus (Raisman and Field, 1973; Gorski et al., 1978). This nucleus is discernible from the 20th day of gestation and becomes sexually dimorphic during the first 10 days after birth. During this period, the volume of the SDN-POA becomes five to six times larger in normal males than in females (Fig. 27.10). In accord with the organizational hypothesis, the SDN-POA of male rats castrated after birth remains small, and the SDN-POA of females treated with androgens shortly before or after birth grows larger. After day 10, hormonal manipulations no longer change the volume of the SDN-POA.

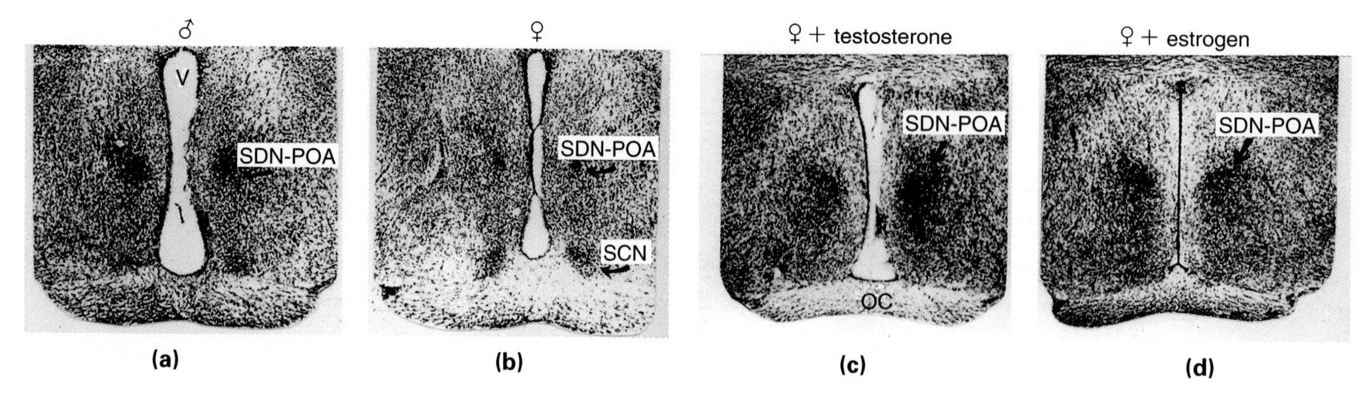

Figure 27.10 The sexually dimorphic nucleus of the preoptic area (SDN-POA) of the rat hypothalamus. The SDN-POA is located on both sides of the third ventricle (V) above the suprachiasmatic nucleus (SCN) on top of the optic chiasm (OC). The volume of the SDN-POA is about five times larger **(a)** in male rats than **(b)** in female rats. Perinatal treatment of female rats with either **(c)** testosterone or **(d)** a synthetic estrogen enlarges (masculinizes) the SDN-POA.

Ironically, the masculinizing effect on the brain of testosterone is exerted by its estrogen metabolite. Male rats with genetic androgen insensitivity, which lack androgen receptors but do have estrogen receptors and aromatase, develop a male-sized SDN-POA that contrasts with their otherwise female appearance (Rosenfeld et al., 1977). Also, treatment of newborn rat females with a synthetic estrogen mimic masculinizes the SDN-POA. Why then do not normal female fetuses, which produce estrogen naturally, develop a male-sized SDN-POA? They escape the masculinizing effect of estrogen on the brain through the action of *α-fetoprotein* (*AFP*), a fetal serum protein that binds to estrogen and prevents its passage through the *blood-brain barrier.* In contrast, testosterone does not bind to AFP and therefore diffuses freely into the brain, where it is converted to estrogen.

In addition to the SDN-POA, several other brain structures are sexually dimorphic. These include the *corpus callosum,* which connects the two cerebral hemispheres (see Fig. 13.19). In rats, the corpus callosum is larger in males than in females (Fitch and Denenberg, 1998). The female corpus callosum is increased by removal of the ovaries, an effect that is compensated by low doses of estrogen administered from day 25 on. Corpora callosa in genetic females are also enlarged by exogenous testosterone administered prior to day 8. In this case, then, the development of a brain structure is affected by exposure to testosterone or its metabolites early during fetal development and by estrogen later.

The human brain also contains sexually dimorphic structures, including the third interstitial nucleus of the anterior hypothalamus, or INAH3 for short (*Swaab and Hofman, 1995*). The homology between the human INAH3 and the rat SDN-POA has not been established. The male INAH3 is only about twice as large, on average, as the female INAH3, and the difference probably would have gone undetected had the rat SDN-POA not been studied first.

Subtle differences between nonreproductive behaviors of men and women are suggested by the different distributions of scores in standardized tests (Hampson and Kimura, 1992; Kimura, 1992). Men tend to score higher on manipulating three-dimensional images, guiding or intercepting projectiles, finding particular shapes in complex figures, and mathematical reasoning. In contrast, women tend to perform better on manual precision tasks, perceptual speed, verbal fluency, and arithmetic tests. These test scores may be biased by different role models and social expectations for boys and girls. Nevertheless, a biological component to these abilities is indicated by the test scores of females affected by a genetic defect called ***congenital adrenal hyperplasia (CAH),*** an adrenal enzyme deficiency that causes excessive androgen levels.

Girls with CAH are born with masculinized external genitalia. This condition can be corrected surgically, and the genetically caused overproduction of androgens can be treated with cortisone. However, the prenatal exposure to androgens seems to have a lasting effect on the behavior of girls with CAH. They grow up more "tomboyish" than their unaffected sisters, according to the girls' own assessments, teachers' accounts, and psychologists' observations of the girls' behavior. In standardized tests, the CAH daughters are superior to their unaffected sisters in spatial manipulations and in discovering hidden figures, both tasks on which males tend to do better. Very similar observations were made on daughters born to mothers who had been given progestin medications during pregnancy to prevent miscarriages. Progestins mimic testosterone in stimulating the androgen receptor. Together, the observations on girls with CAH or progestin exposure suggest that their brains were permanently conditioned by their prenatal hormonal environment.

In summary, data obtained from birds and mammals, including humans, indicate that hormones are powerful regulators of sexual development. Their effects are seen most clearly in physical sexual dimorphisms and adult reproductive behaviors, but they also seem to reach into the development of brain functions that are not directly related to reproduction.

27.4 Hormonal Control of Insect Metamorphosis

While the hormonal control of vertebrate development is of great interest for understanding our own sexuality, the molecular mechanisms of hormone action have been analyzed most successfully for the insect molting hormone *ecdysone.* Studying this hormone has several advantages. First, it is a steroid hormone, so its action on target genes is straightforward, with the hormone-receptor complex acting directly as a transcription factor. Second, the effects of ecdysone on many of its target genes can be seen directly in the puffing patterns of the *polytene chromosomes* of dipteran larvae. Third, the availability of *Drosophila* mutants has facilitated the genetic and molecular analysis of ecdysone action. This has led to the discovery of a control system of regulatory genes and auxiliary transcription factors, which can account for at least some of the puzzling specificity with which hormones act on their target cells (Thummel, 1995, 1997). Before tackling the molecular aspects of ecdysone action, we will briefly survey postembryonic development in insects, in which ecdysone plays a central role.

INSECT MOLTING, PUPATION, AND METAMORPHOSIS ARE CONTROLLED BY HORMONES

Arthropods are covered by a ***cuticle,*** a hardened extracellular shell secreted by the epidermis. Since the cuticle has only a limited capacity for expansion, it needs to be shed periodically to allow growth. The development of insects and other arthropods is punctuated by a series of ***molts***, or *ecdyses*. At each ecdysis, the epidermis withdraws from the old cuticle, expands, and then forms a new, larger cuticle that is initially soft. The newly ecdysed individual moves and exerts pressure with its body fluid to shed the old cuticle and expand the new one.

During molts, insects may change in body form, a process called *metamorphosis*. Primitive insects, such as grasshoppers, have a gradual or *incomplete metamorphosis:* With each molt, the insect comes to look more like an adult. More highly evolved insects such as butterflies, moths, flies, ants, and beetles undergo a dramatic or *complete metamorphosis*. The juvenile stages of these insects are generically termed *larvae* (sing., *larva*); the larvae of individual insect orders have more specific common names, such as caterpillars or maggots. The larva and adult of a species differ widely in their morphology and lifestyle, and they fulfill different functions in the life cycle of the species. Insect larvae are primarily devoted to eating: The caterpillars of large moths increase their body weight several hundredfold. Adult insects typically fly, thus achieving dispersal in addition to reproduction.

Insects with complete metamorphosis are known as ***holometabolous insects;*** they go through a number of larval stages, called ***instars,*** which are separated by molts (Fig. 27.11). After the last larval instar, they molt into an immobile, nonfeeding stage known as the ***pupa,*** during which metamorphosis occurs. During the pupal stage, many larval organs are broken down. Other larval organs, including much of the nervous system and some larval muscles, survive in the adult. While larval tissues are dying, adult structures are being formed. Many parts of the adult epidermis, including wings, legs, antennae, eyes, and genitalia, are made from ***imaginal discs,*** folded epithelial bags that are set aside during larval development and evert during metamorphosis (see Fig. 6.17). Other parts of the epidermis are *serially polymorphic*—that is, they make successively the larval, pupal, and adult cuticles, which often have different pigmentations. At the end of metamorphosis, the insect undergoes a final or adult molt before it *ecloses* from the pupal cuticle. The adult insect is also called the ***imago*** (hence the term *imaginal disc*). The group of flies to which *Drosophila* belongs has a special feature called the ***puparium.*** This is the hardened cuticle of the last larval instar. Instead of being shed during the pupal molt, it forms a barrel-shaped case around the pupa.

The salient feature of metamorphosis in insects (and many other animals) is its tight hormonal control (Nijhout, 1994; L. I. Gilbert et al., 1996). The timing of insect molts as well as their character (larval, pupal, or adult) is regulated by hormone-secreting glands that are controlled in turn by the brain. Figure 27.12 shows a simplified model of this process for a large moth, the tobacco hornworm *Manduca sexta*.

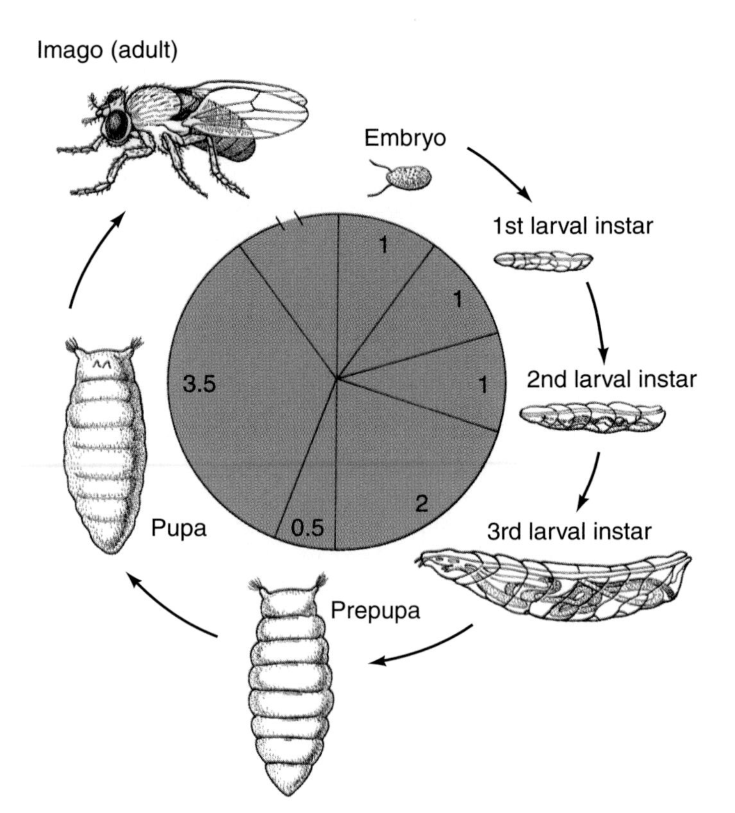

Figure 27.11 Life cycle of *Drosophila*, a holometabolous insect. Seven development stages of the life cycle are shown: the embryo, three larval instars, prepupa, pupa, and adult. The numbers represent the duration of each stage in days. Each molt between successive stages is preceded by an ecdysone pulse.

In response to environmental and endogenous stimuli, *neurosecretory cells* in the brain release a peptide hormone called ***prothoracicotropic hormone (PTTH).*** Through a complex system of *second messengers,* PTTH stimulates the ***prothoracic gland*** to synthesize a steroid hormone, ***ecdysone.*** (Ecdysone is a prohormone that is converted in other tissues to its active form, 20-hydroxyecdysone. For simplicity, we will use the term "ecdysone" for both forms.) Each molt is triggered by a surge of ecdysone in the hemolymph (blood equivalent) of the insect. The character of the molt is determined by the concentration of ***juvenile hormone (JH),*** a hormone produced by a pair of glands called the ***corpora allata*** (sing., *corpus allatum*). The corpora allata are attached to and controlled by the brain. JH essentially inhibits metamorphosis. At high JH concentration, a molt results in another larval instar. JH disappears temporarily during the final larval stage, while ecdydsone rises before the pupal molt. During this molt, JH makes a brief return, which is necessary to prevent precocious formation of adult structures from imaginal discs. During the pupal stage itself, JH disappears permanently, and the subsequent adult molt occurs in response to ecdysone in the absence of JH.

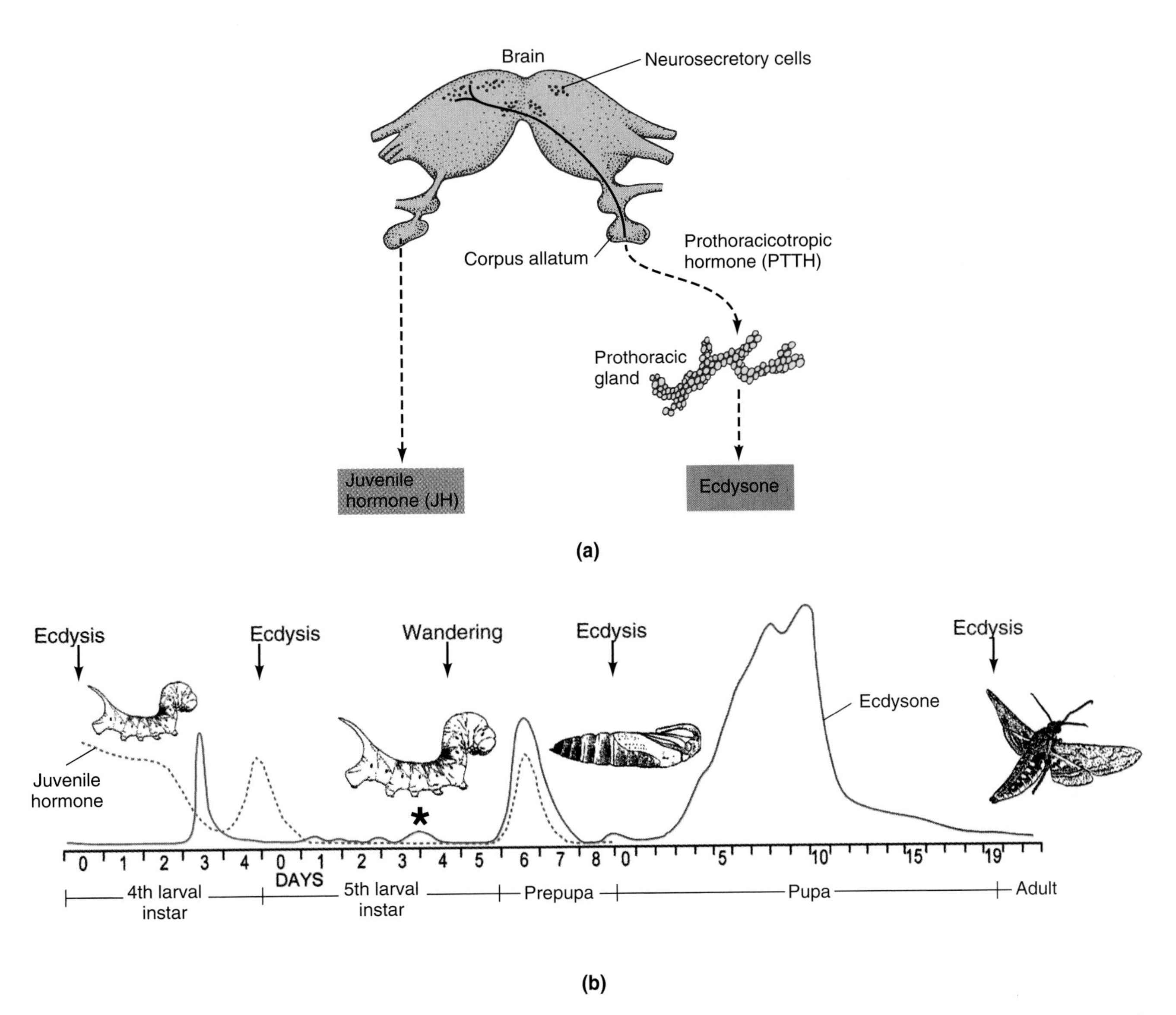

Figure 27.12 Hormonal control of metamorphosis in a moth, the tobacco hornworm *Manduca sexta.* **(a)** All molts seem to be initiated by prothoracicotropic hormone (PTTH), which is synthesized by neurosecretory cells in the brain. They deliver PTTH by axonal transport to the corpora allata (sing., corpus allatum), a pair of glands attached to the brain. PTTH stimulates the prothoracic gland to release the hormone ecdysone. The character of the molt is determined by the concentration of juvenile hormone (JH), which is produced in and released from the corpora allata. **(b)** At high JH levels, molts generate additional larval instars. Metamorphosis is triggered late during the last (fifth) larval instar by an elevated level of ecdysone (asterisk) in the absence of juvenile hormone. This is followed by simultaneous peaks of both ecdysone and JH during the prepupal stage, when the larva no longer feeds. The ecdysone peak triggers the pupal molt, whereas the JH peak prevents the premature formation of adult structures. The final, adult molt is controlled by a large ecdysone peak in the absence of JH.

The hormonal control of insect metamorphosis was first analyzed in the pioneering experiments of Wigglesworth (reviewed 1972). He worked with a bloodsucking bug, *Rhodnius prolixus,* a species with incomplete metamorphosis. The bug goes through five juvenile instars without wings before molting into a winged adult. Wigglesworth used the technique of ***parabiosis***—that is, a connection between two living organisms that allows the exchange of blood or an equivalent body fluid (Fig. 27.13). When a molting fifth-instar juvenile and a first-instar juvenile were joined in parabiosis, they both underwent metamorphosis. Regardless of its minute size, the first instar molted into a precocious adult. In another experiment, Wigglesworth removed the corpora allata from a third-instar juvenile. As a result, the next molt turned this juvenile into a precocious adult. Conversely, when the corpora allata from a fourth-instar juvenile were implanted into a

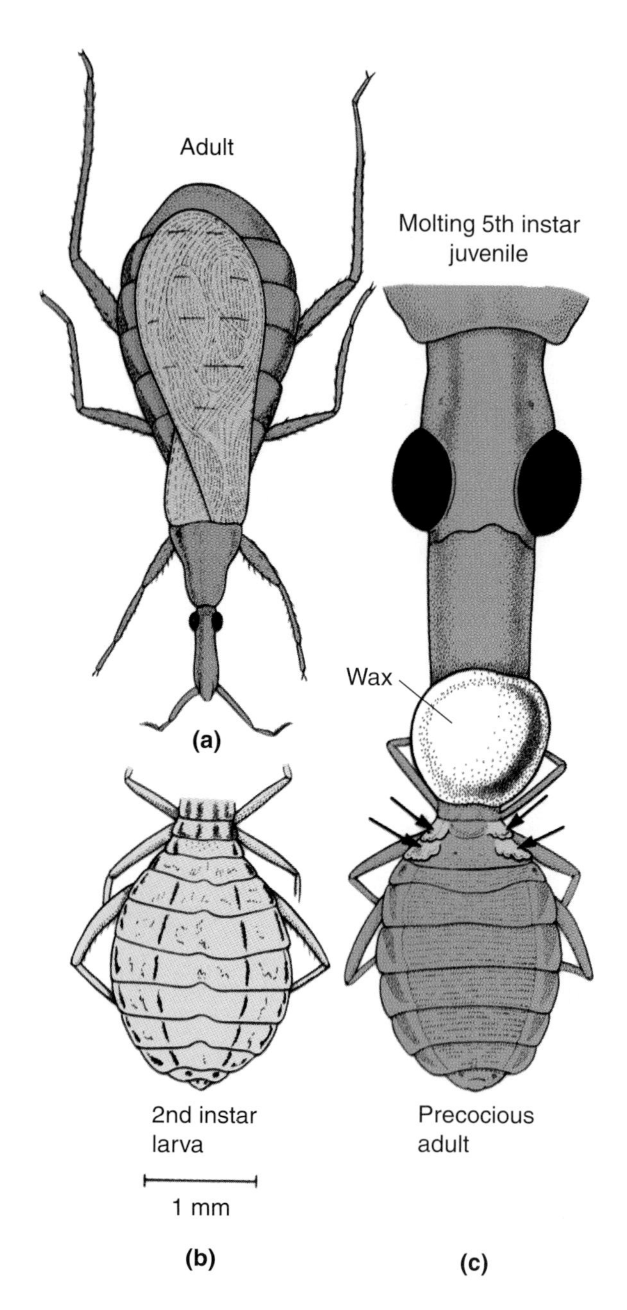

Figure 27.13 Control of metamorphosis by blood-borne signals in the bug *Rhodnius prolixus.* **(a)** Normal adult. **(b)** Thorax and abdomen of a normal second-instar larva at higher magnification. **(c)** Precocious "adult" produced from a first-instar juvenile that was joined with paraffin wax to the head of a molting fifth-instar juvenile. The precocious adult has rudimentary wings (arrows) and the pigmentation pattern of the adult rather than that of the second-instar juvenile.

fifth-instar juvenile, the recipient molted into an oversized "sixth-instar" juvenile instead of becoming an adult.

The mechanisms of JH action are still largely unresolved. However, new work on the molecular basis of ecdysone action, which we will discuss next, has generated much excitement.

PUFFS IN POLYTENE CHROMOSOMES REVEAL GENES RESPONSIVE TO ECDYSONE

Each molt in the life cycle of *Drosophila* is preceded by a pulse of ecdysone; the pulse that is best investigated occurs at the end of the third larval instar, right before puparium formation (Fig. 27.14). A few hours before this pulse, the larva stops feeding and wanders around in search of a place in which to form a puparium. After finding a suitable spot, the larva cements its cuticle onto the substratum with a glue produced in its salivary glands. The larva then assumes a barrel-like shape while its cuticle hardens to form the puparium. Twelve hours later, another ecdysone pulse triggers the eversion of the head, which has been tucked inside the thorax during the larval stages. The stage between puparium formation and head eversion is called the ***prepupa,*** and the subsequent stage, the *pupa.*

During the late third larval instar and the prepupal stage, the polytene chromosomes of the salivary glands are very large and distinct and provide an impressive demonstration of the effects of ecdysone on gene activity. The chromosomal puffs, which signal transcription, can easily be observed at successive stages, because each puff can be assigned a unique position on a cytogenetic map (see Figs. 7.6 and 16.16). The changing pattern of puffs thus directly reveals which genes are transcribed at successive stages.

Repeated observations have shown that each stage of metamorphosis is characterized by a specific combination of puffs (Figs. 27.14 and 27.15). In third-instar larvae only a few ***intermolt puffs*** are visible prior to the wandering stage. As the ecdysone level increases, signaling puparium formation, a small set of fewer than 10 ***early puffs*** appears. These early puffs disappear several hours later, while more than 100 ***late puffs*** appear in a reproducible sequence as ecdysone concentration recedes.

To test whether the complex puffing patterns at the beginning of metamorphosis are all triggered by ecdysone, researchers injected the hormone prematurely. Indeed, the injected ecdysone induced puffs that would normally be seen only later in development (Clever and Karlson, 1960). Extending this work, Ashburner and coworkers (1972, 1974) devised a method for maintaining salivary glands from third-instar larvae in vitro, thus obtaining a degree of control not possible in the intact organism. Again, the addition of ecdysone stimulated the changes in puffing that are normally associated with early metamorphosis (Fig. 27.15).

USING this culture system, Ashburner and coworkers studied the effects of ecdysone addition at various concentrations, and of ecdysone removal. They also examined whether the resulting puffing patterns depended on the ability of the salivary glands to synthesize new proteins. They confirmed that a small number of "early genes"

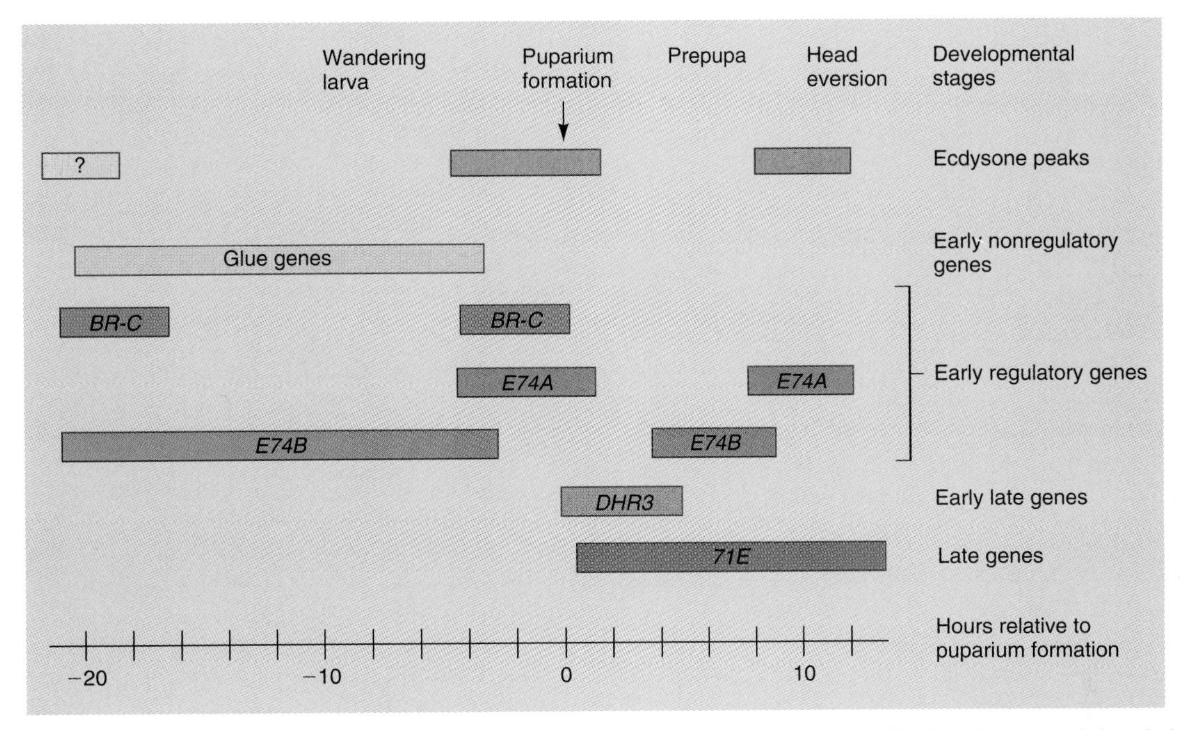

Figure 27.14 Ecdysone peaks and gene activity at the onset of metamorphosis in *Drosophila.* Developmental time is indicated in hours before and after puparium formation. Relative to this time axis, the diagram shows developmental stages, ecdysone peaks, and the activity periods of a few of the genes that form puffs and are transcribed in response to ecdysone.

became puffed within minutes after addition of ecdysone, and they found that inhibition of protein synthesis did not affect the appearance of these puffs. They concluded that the *activation* of the early genes was independent of newly synthesized proteins. However, inhibiting protein synthesis did have an effect on the *regression* of at least some of the early puffs. In the continuous presence of ecdysone, and with protein synthesis available, these puffs reached a maximum size after 4 h and then disappeared. However, if protein synthesis was inhibited, the puffs persisted. This observation suggested that inactivation of the early genes might depend on the proteins they encode or on proteins encoded by their target genes.

In contrast to the small set of early genes, the large group of late genes failed to be activated by ecdysone in the absence of protein synthesis. This result suggested that early-gene proteins might not only limit the activity period of the early genes but also activate the late genes. However, this hypothesis by itself did not explain another remarkable feature of the puffing pattern: The late puffs appeared and regressed in a very strict temporal sequence over a period of about 12 h. A clue to this puzzle was obtained from another experiment: If salivary glands were incubated in ecdysone for a few hours and then washed free of the hormone, many of the late puffs were induced prematurely. It appeared that late puffs were initially repressed in an ecdysone-dependent manner until increasing amounts of early-gene proteins offset the hormone's repressive effects.

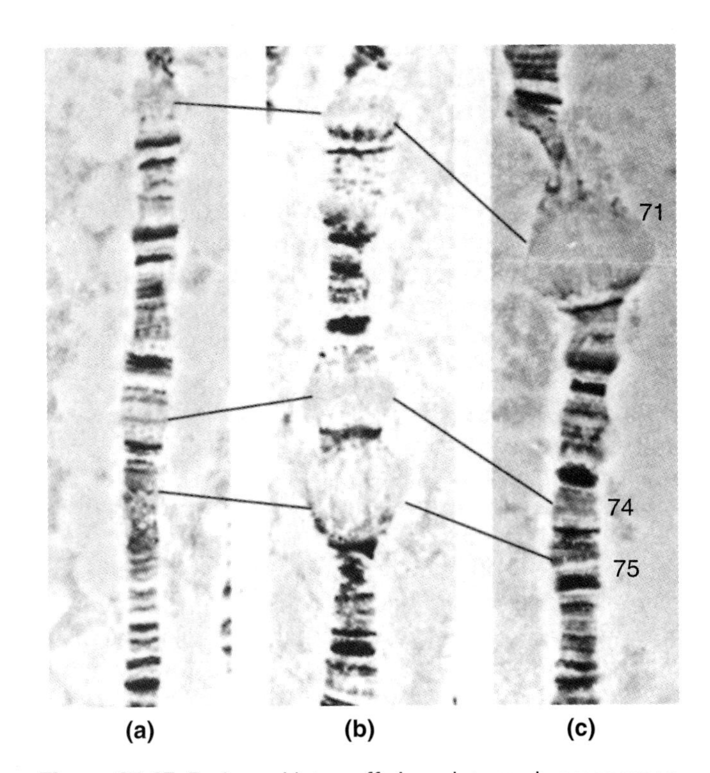

Figure 27.15 Early and late puffs in polytene chromosomes of *Drosophila melanogaster* salivary glands cultured in vitro. The photographs show a segment of the right arm of chromosome III. **(a)** No puffs are visible before ecdysone is added. **(b)** About 5 h after ecdysone addition, two early puffs (74 and 75) are very large, and a late puff (71) is beginning to form. **(c)** About 10 h after ecdysone addition, the early puffs have regressed, and the late puff has grown to full size.

Questions

1. How could Ashburner and coworkers have inhibited protein synthesis, and how could they have tested whether the inhibition was indeed working as intended?
2. What particular type of protein might the investigators have had in mind to block the synthesis of?
3. The investigators explain the failure of late genes to become activated in the presence of inhibitors of protein synthesis by postulating that the activation process itself depends on newly synthesized proteins. Wouldn't it be equally reasonable to assume that (a) puffing depends on ATP or some other donor of chemical energy and (b) inhibiting protein synthesis makes cells "sick" so that their energy production is diminished?

From the experiments described above and follow-up studies, Ashburner (1974, 1990) derived a model for the control of chromosome puffing by ecdysone (Fig. 27.16). Ecdysone binds to a nuclear receptor protein, which then acts as a transcription factor on different groups of target genes. Some of these are *early nonregulatory genes.* For instance, the genes for the glue proteins used by the third-instar larva to cement its cuticle to a substrate are shut off by the ecdysone-receptor complex. Other nonregulatory genes are presumably activated in tissue-specific ways. In addition, the ecdysone-receptor complex activates a small number of *early regulatory genes.* These genes were originally identified as early puffs in salivary gland chromosomes and were later shown to be activated in virtually all other tissues. One early regulatory gene encodes ecdysone receptors. Other genes in this class encode proteins that after a delay cause the regression of the early puffs and the appearance of the late puffs described earlier. Finally, the ecdysone-receptor complex inhibits the *late genes* until the early regulatory proteins overcome the inhibition and trigger transcription of the late genes.

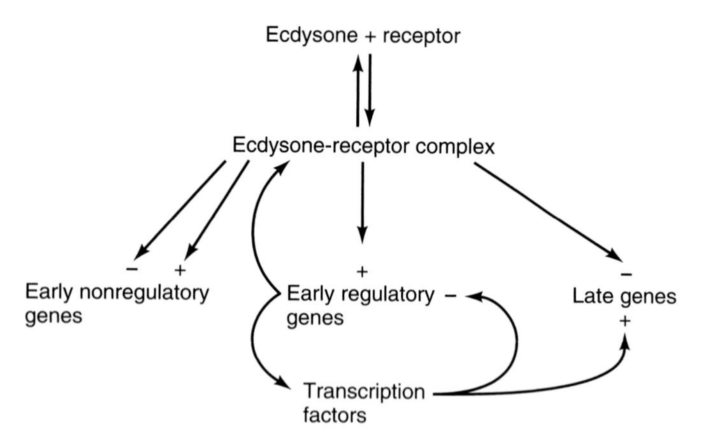

Figure 27.16 Model for gene regulation by ecdysone in *Drosophila* (Ashburner, 1974). Ecdysone forms a complex with a receptor protein, and the ecdysone-receptor complex has multiple effects. First, it acts directly on early nonregulatory genes that have tissue-specific functions. Second, the ecdysone-receptor complex activates a small number of early regulatory genes. After a few hours, the transcription factors encoded by these genes inhibit their own genes and activate a large number of late genes. The products of the early regulatory genes include ecdysone receptor peptides. Finally, the ecdysone-receptor complex inactivates the late genes until this inhibition is overcome by the early regulatory gene proteins.

MULTIPLE ECDYSONE RECEPTORS HAVE TISSUE-SPECIFIC ACTIVITIES

Molecular studies to test the Ashburner model began with the characterization of genes located in early puffs. Of particular interest was the cloning of an early regulatory gene termed *EcR*$^+$ (Koelle et al., 1991). It encodes a group of large peptide isoforms that dimerize with a similar peptide to form ecdysone receptors, according to the following criteria. First, the EcR peptide binds ecdysone and is immunologically indistinguishable from a known ecdysone-binding protein. Second, EcR binds to DNA with high specificity at enhancer motifs known as ecdysone-response elements. Third, EcR is present in ecdysone-responsive cultured cells and absent in nonresponsive cells. The latter can be restored to ecdysone responsiveness by transfection with cDNA encoding EcR. Fourth, EcR is synthesized in vivo at all stages that are ecdysone-responsive.

In accord with the Ashburner model, EcR binds to early puffs and many late puffs. In particular, the ecdysone-EcR complex induces the synthesis of more EcR. This ***autoinduction*** of EcR is indicated by the following data. First, the synthesis of EcR mRNA increases during each ecdysone peak throughout the life cycle of *Drosophila.* Second, EcR mRNA accumulates rapidly after an ecdysone pulse. When cultured larval organs are incubated with ecdysone, EcR mRNA concentration increases after 15 min and reaches a maximum about 1 h after ecdysone addition (Karim and Thummel, 1992). Given a pre-mRNA size of 36 kb and an average transcription rate of about 1.2 kb per min, the transcriptional response certainly appears to be direct. The capability of ecdysone to stimulate the synthesis of EcR mRNA in the absence of protein synthesis was shown directly in *Manduca sexta* (Jindra et al. (1996).

Further analysis of the *EcR*$^+$ gene showed that it includes two overlapping transcription units encoding three peptides, termed EcR-A, EcR-B1, and EcR-B2. All three isoforms have identical DNA-binding and ligand-binding domains but different N-terminal transactivation domains (Fig. 27.17; Talbot et al., 1993). Antibodies to EcR-A and EcR-B1 reveal characteristic expression patterns of these isoforms at the beginning of metamorphosis. For example, the imaginal discs, which give rise to the adult epidermis, show a uniformly high A:B1 ratio of receptor peptides. In contrast, most of the larval

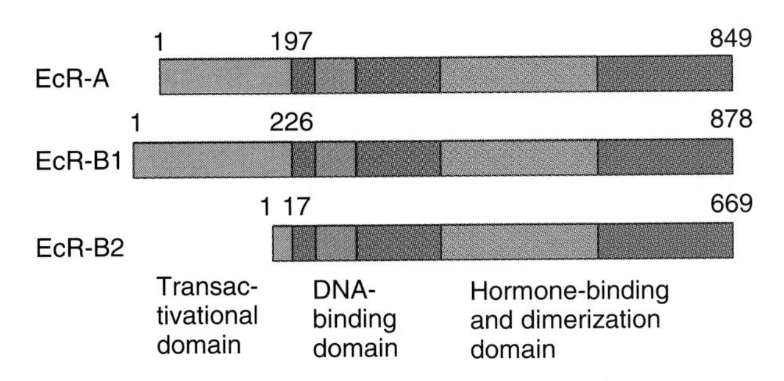

Figure 27.17 Schematic representation of the three peptides encoded by the *EcR*[+] gene of *Drosophila.* All three peptides share a DNA-binding as well as a hormone-binding and dimerization domain. The peptides differ in their N-terminal transactivation domains, which control their activating or inhibitory activities as transcription factors. Numbers indicate amino acid residues, counted from the amino terminus.

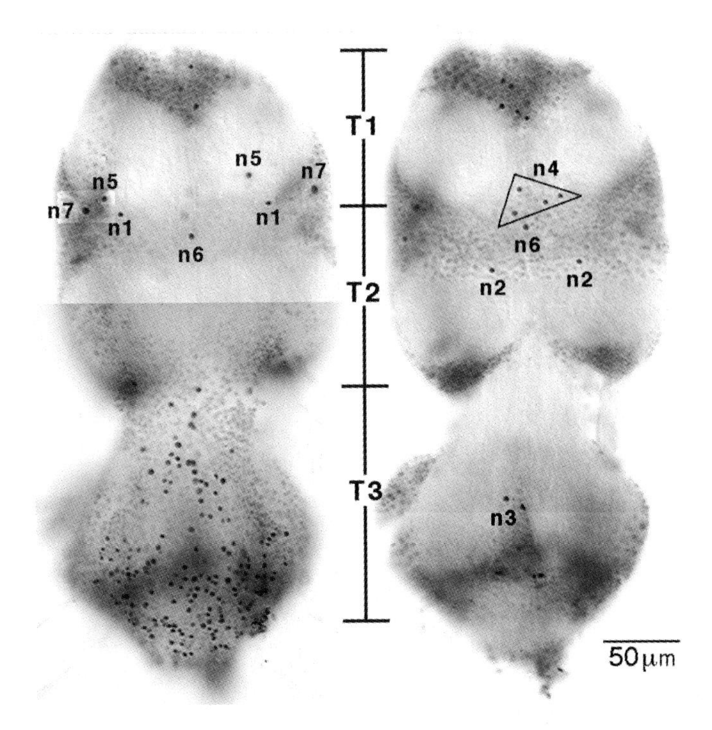

Figure 27.18 Selective accumulation of the ecdysone receptor isoform EcR-A in specific neurons of the central nervous system of a newly eclosed *Drosophila* adult. The brains shown in these photographs were immunostained for the EcR-A polypeptide (see Method 4.1). The deeply stained neurons share the fate of rapid cell death within a few hours after eclosion, thus illustrating a correlation between EcR isoform presence and cell fate. T1, T2, and T3 designate the approximate extents of the CNS portions (neuromeres) that innervate the respective thoracic segments. **(a)** Dorsal view; **(b)** ventral view. The labels designate individual neurons that can be identified reproducibly on the basis of their position.

organs that are broken down during metamorphosis exhibit a low A:B1 ratio of receptor peptides. Such correlations suggest that different combinations of ecdysone receptor peptides may activate specific target genes, which in turn initiate different metamorphic responses, such as disc eversion and organ degradation.

The synthesis of a receptor isoform in a tissue is suggestive of its function but does not prove it. This difficulty has been overcome with the isolation and characterization of mutants that are deficient in the EcR-B1 receptor peptide, while synthesizing the other two EcR isoforms normally. In these mutants, the larval tissues (which synthesize mostly the B1 isoform) behave abnormally, whereas imaginal discs (which produce mostly the A isoform) begin their normal eversion (Bender et al., 1997). Thus, the EcR-B1 receptor peptide is indeed required in those tissues in which it is synthesized.

An intriguing correlation between metamorphic response and the ecdysone receptor EcR-A is observed in the central nervous system of *Drosophila* as well as a moth, *Manduca sexta* (Truman et al., 1994). Generally, most larval neurons produce high levels of EcR-B1 at the start of metamorphosis, while the CNS loses its larval features. Later, during the pupal-adult transformation, they switch to EcR-A synthesis and mature to their adult form. A particular correlation is observed in a set of about 300 neurons that produce a 10-fold higher level of EcR-A than other neurons in *Drosophila* pupae (Fig. 27.18). These neurons, known as *Type II* neurons, share the fate of undergoing *apoptosis* as the adult emerges from the pupal case (Robinow et al., 1993).

The death of *Type II* neurons depends on the function of the genes *grim*[+] and *reaper*[+], which initiate apoptosis in *Drosophila*. Transcripts of these two genes accumulate in *Type II* neurons prior to the onset of apoptosis (Robinow et al., 1997; Draizen et al., 1999). Apoptosis of these neurons also depends on the decline of ecdysone at the end of metamorphosis: Injection of flies with ecdysone at least 3 h before the normal time of cell degeneration prolongs the life of the cells. The same hormone treatment also prevents the accumulation of grim and reaper mRNAs in the cells. Together, these results suggest that the particular response of Type II neurons to ecdysone at the end of metamorphosis involves a high level of EcR-A, which promotes the transcription of *grim*[+] and *reaper*[+], with apoptosis ensuing.

The data discussed so far indicate that the synthesis of different ecdysone receptor isoforms is one way in which the response to ecdysone is made tissue-specific. Another mechanism for "customizing" the response of target cells to ecdysone is based on the fact that steroid receptors bind to their DNA response elements as dimers of two peptides (see Fig. 16.13). All EcR peptides form heterodimers with the peptide encoded by a widely expressed *Drosophila* gene, *ultraspiracle*[+] (Yao et al., 1992). The phenotypes of loss-of-function mutants indicate that the ultraspiracle peptide is required for multiple events in metamorphosis, including eversion of imaginal discs and apoptosis of midgut and salivary gland cells (B. L. Hall and C. S. Thummel, 1998). While EcR peptides bind ecdysone weakly on their own, the heterodimers with ultraspiracle bind the ligand more strongly (Yao et al., 1993). In addition, the EcR-USP dimer seems to bind to the ecdysone response elements of certain genes *without a ligand.* In this ligand-free state, the EcR-USP dimer appears to *inhibit* its target genes, thus preventing metamorphic events from occurring prematurely (Schubiger and Truman, 2000).

The heterodimerization with ultraspiracle provides a basis for tissue-specific responses to ecdysone through competitive inhibition. This mechanism involves so-called ***orphan receptors***—that is, peptides that have the molecular characteristics of nuclear steroid receptors except that they have no known ligand (Thummel, 1995). One orphan receptor, designated DHR38, heterodimerizes with ultraspiracle, thus reducing the formation of functional EcR-ultraspiracle heterodimers. Another orphan receptor, SVP, may form inactive heterodimers with EcR, again reducing the concentration of active EcR-ultraspiracle dimers. Thus, tissues may be able to down-regulate their responsiveness to ecdysone by synthesizing one of these orphan receptors.

The ultraspiracle peptide is *orthologous* to the mammalian retinoid X receptor peptide (RXR), which forms active heterodimers with thyroid hormone receptor and other receptor peptides (Oro et al., 1990; Mangelsdorf and Evans, 1995). In fact, EcR-RXR heterodimers are as effective as EcR-ultraspiracle heterodimers in making cells ecdysone-responsive and in binding to DNA response elements (H. E. Thomas et al., 1993). Thus, the sequence similarity and function of the RXR and ultraspiracle peptides have been conserved in evolution since vertebrates and arthropods diverged more than 500 million years ago. Presumably, such peptides play an important role in diversifying the regulatory effect of hormonal signals on a wide range of target cells.

EARLY REGULATORY GENES DIVERSIFY AND COORDINATE THE RESPONSES OF TARGET CELLS TO ECDYSONE

A third and very powerful mechanism for diversifying the responses of target cells to the ecdysone stimulus, and for proper timing of these diverse responses, is the network of "early regulatory genes" postulated by the Ashburner model (see Fig. 27.16). The genetic and molecular analysis of several of these genes has provided a new understanding of how a series of events as complex as insect metamorphosis can be orchestrated by a few pulses of one hormone (Thummel, 1995, *1997*). While about ten early regulatory genes have been identified, we will focus on two of them, *E74*$^+$ and *DHR3*$^+$.

The *E74*$^+$ gene is located in the 74EF early puff originally observed by Ashburner and colleagues (see Fig. 27.15). *E74*$^+$ is transcribed from two promoters, both of which are activated directly by the ecdysone-receptor complex (Thummel et al., 1990). This leads to the synthesis of two mRNAs, E74A and E74B, which encode two related transcription factors that have unique N-terminal regions but share a common C-terminal region.

The synthesis of E74A and E74B mRNAs is coordinately regulated by each ecdysone pulse. For instance, the pulses that mark the ends of the larval and prepupal stages coincide with high rates of E74 mRNA synthesis (see Fig. 27.14). In each case, however, E74A mRNA synthesis is preceded by a period of E74B mRNA synthesis. While this coordination is explained to a small extent by the greater length of the E74A pre-mRNA, the more important observation is that the E74B promoter responds to a much lower ecdysone concentration. When Karim and Thummel (1991) exposed late third-instar larval organs of *Drosophila* to various concentrations of ecdysone in culture, the two promoters showed different responses as revealed by *Northern blots* (see Method 15.4)

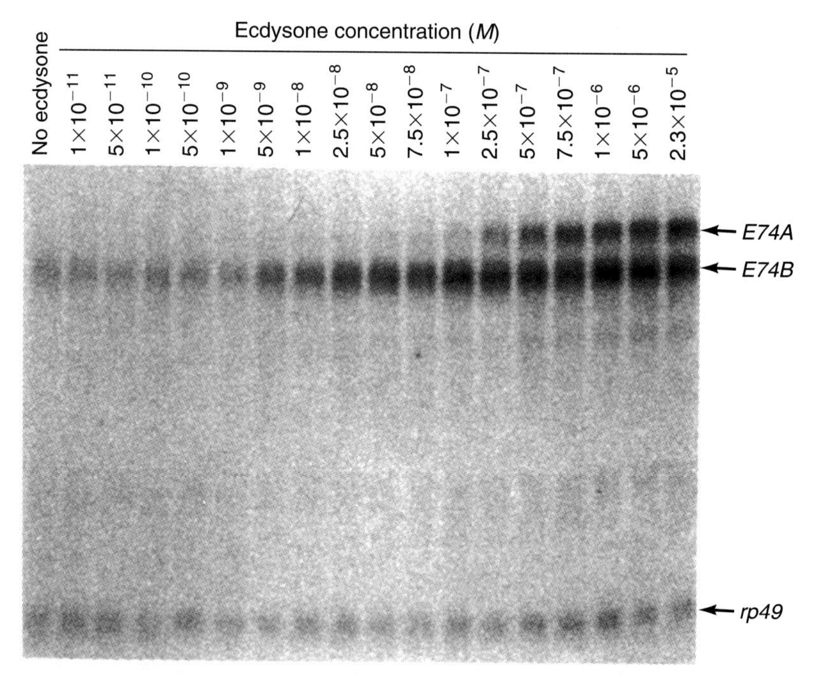

Figure 27.19 Ecdysone dose-response analysis of E74A and E74B mRNA prevalence. Late third-instar larval organs (salivary glands, imaginal discs, gut, and Malpighian tubules) were cultured and treated for 1.5 h with ecdysone at different concentrations. Total RNA was extracted and analyzed by Northern blotting (see Methods 15.4) using a labeled probe representing the common region of both mRNAs. Another labeled probe hybridizing with the mRNA for a ribosomal protein (rp49) was added as a control for equal RNA loading in each lane. The results show that two E74 mRNAs of different sizes accumulate in response to different ecdysone concentrations. The prevalence of E74B mRNA increases above a constant background level as ecdysone concentrations rise between 1×10^{-9} and 7.5×10^{-7} M. In contrast, E74A mRNA has no detectable background level and increases between ecdysone concentrations of 5×10^{-8} and 7.5×10^{-7} M.

of RNA extracted from these organs (Fig. 27.19). Photometric processing of the data indicated that E74B mRNA reaches 50% of its maximum prevalence at an ecdysone concentration of 8×10^{-9} M, whereas E74A mRNA reaches 50% maximum prevalence at 2×10^{-7} M ecdysone. The researchers concluded that the E74A promoter requires a 25-fold higher ecdysone concentration to become activated than the E74B promoter.

In order to assess the functions of E74A and E74B in development, researchers analyzed the effects of two loss-of-function mutations affecting selectively either the E74A mRNA or the E74B mRNA (Fletcher and Thummel, 1995; Fletcher et al. 1995, 1997). They found that E74B mutants are defective in puparium formation and die as prepupae or early pupae, whereas E74A mutants pupariate normally and die at the prepupal or early adult stage. The mutations in *E74*$^+$ also affected the puffing pattern of late genes observed in salivary gland chromosomes of prepupae. Finally, the mutations affected the timing and rate of transcription of many late genes in several tissues.

Together, these data show that the *E74*$^+$ gene acts as a diversifier for the ecdysone signal. Each ecdysone pulse generates a *temporal morphogen gradient* (see Section 25.3), to which *E74*$^+$ responds by generating different combinations of two transcription factors at different ecdysone levels. Another representative of the early regulatory genes postulated by Ashburner (1974), the *Broad-Complex*$^+$ (*BR-C*$^+$), encodes a group of zinc finger transcription factors that are similarly involved in coordinating several temporal and tissue-specific responses to ecdysone at the beginning of metamorphosis (von Kalm et al., 1994). By extension, the total of about ten early regulatory genes generate a much wider variety of DNA-binding proteins that orchestrate the activity of a much larger number of late genes in time. To the extent that the activity of the early genes or the response of late genes is modulated by factors such as the Myo-D proteins or homeodomain proteins, the responses of targets cells to ecdysone will also be diversified in tissue-specific and region-specific ways.

Just as the function of the ecdysone receptors is diversified by nuclear *orphan receptors* as discussed earlier, so is the function of the early regulatory gene products. The orphan receptor encoded by the *DHR3*$^+$ gene may serve as an example. This gene was originally classified as an "early-late gene" because it is a direct target of the ecdysone-receptor complex but appears to require an ecdysone-induced protein for maximum response (Horner et al., 1995). The DHR3 receptor reaches maximum concentration shortly after puparium formation, at the right time to repress early genes, such as *E74A*$^+$ and *BR-C*$^+$ (Fig. 27.20). Also, *immunostaining* (see Method 4.1) of polytene chromosomes indicates that DHR3 protein binds to the puffs containing *E74A*$^+$, *BR-C*$^+$, and *βFTZ-F1*$^+$, suggesting that these genes become associated with DHR3 protein.

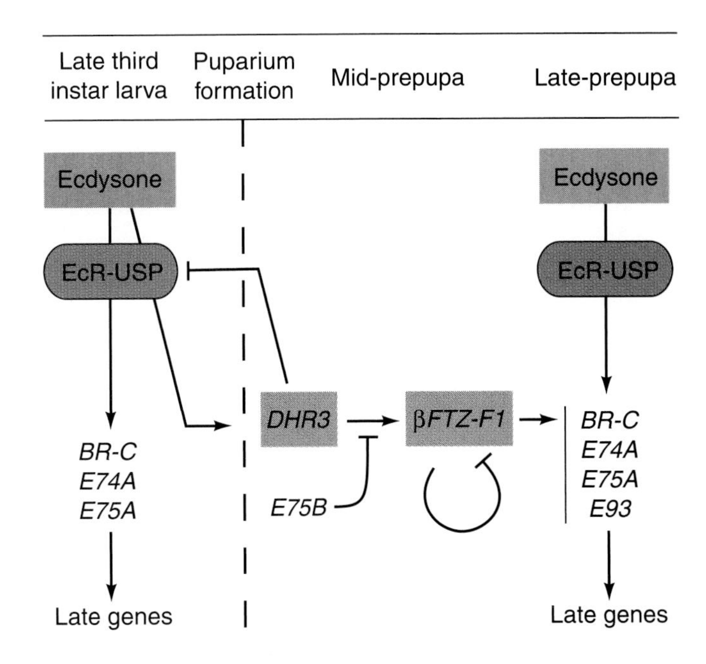

Figure 27.20 Model for early-late gene function during *Drosophila* metamorphosis. Stages of larval and prepupal development are indicated at the top. Ecdysone released at the ends of the larval and prepupal stages binds to the ecdysone receptor/ultraspiracle (EcR/USP) heterodimer and induces early gene (*BR-C*$^+$ and *E74*$^+$) expression. The late larval ecdysone pulse also induces expression of two early-late genes, *DHR3*$^+$ and *E75B*$^+$, both of which encode orphan receptors. DHR3 protein is sufficient to repress early gene expression, apparently through direct interaction with EcR/USP. DHR3 is also sufficient to activate another orphan-receptor-encoding gene, β*FTZ-F1*$^+$, although this activation is temporarily inhibited by heterodimer formation between DHR3 and E75B. βFTZ-F1 in turn seems to promote reactivation of the early genes by the next ecdysone pulse.

IN order to test whether DHR3 is sufficient to inhibit early genes, K. White and coworkers (1997) expressed *DHR3*$^+$ prematurely from a transgene controlled by a *heat shock* promoter. Indeed, after heat-shocking transgenic mid-third-instar larvae, ecdysone-induced synthesis of E74 mRNA was severely reduced, as indicated by *Northern blotting* (see Method 15.4). Furthermore, *βFTZ-F1*$^+$ was transcribed prematurely in these larvae, as early as one day before its normal activity in mid-prepupae.

To determine whether DHR3 protein interacts directly with the ecdysone-receptor complex, the investigators used a yeast *two-hybrid assay,* which makes the transcription of a reporter gene in yeast cells dependent on the binding between the two proteins of interest. Indeed, the investigators found that DHR3 interacts directly with EcR. Deletion mapping of the DHR3 component in this assay indicated that DHR3 binds through its *ligand-binding domain*, which in similar receptors has been shown to be required for dimerization (see Fig. 27.17).

The investigators also showed that DHR3 acts as a transcriptional activator of β*FTZ-F1*$^+$, which in turn seems to facilitate reactivation of the early genes during the subsequent ecdysone pulse. The activation of β*FTZ-F1*$^+$ is temporarily blocked by heterodimerization between DHR3 and another orphan receptor synthesized simultaneously, the E75B protein.

Questions

1. How could the researchers have tested whether DHR3 is not only sufficient but also necessary for inhibiting early genes, such as *E74A*$^+$ and *BR-C*$^+$?
2. If the two-hybrid assay had failed to show binding between DHR3 and EcR, what would have been an alternative mechanism to account for the precocious inhibition of *E74A*$^+$ expression by the heat shock–driven *DHR3*$^+$ transgene?

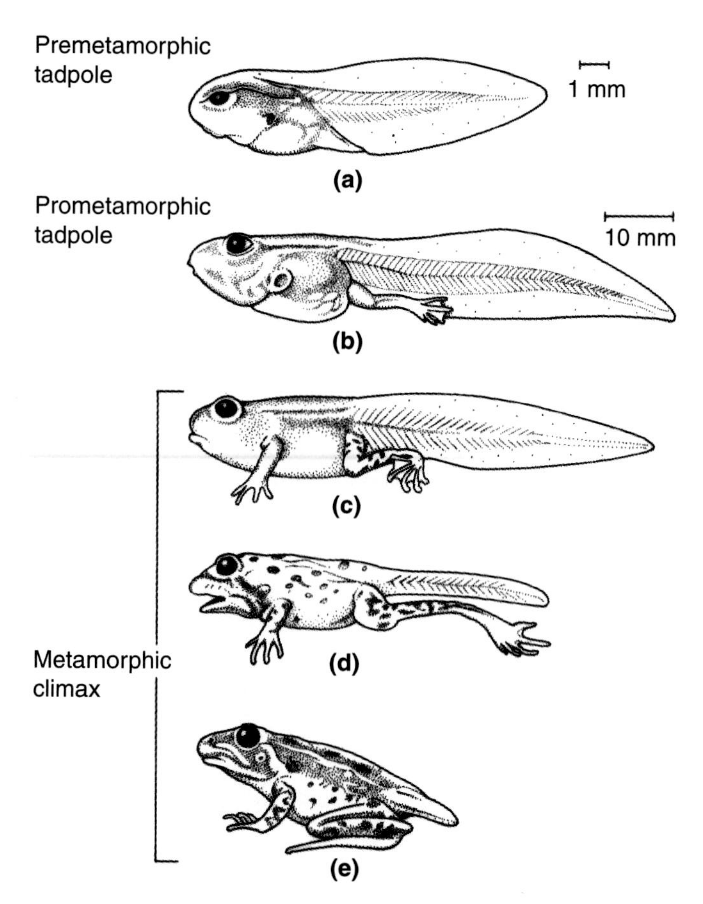

Figure 27.21 Metamorphosis in a typical frog (*Rana pipiens*). **(a)** Premetamorphic tadpole; **(b)** prometamorphic tadpole showing growth of hindlimbs; **(c)** onset of metamorphic climax with eruption of forelimbs; **(d, e)** later climax stages showing the appearance of the adult organization.

The results of White and coworkers (1997) were confirmed by a parallel study of Lam and coworkers (1997), who showed in particular that DHR3 binds to specific elements within the β*FTZ-F1*$^+$ promoter. Both studies have revealed a dual role of the DHR3 orphan receptor in *Drosophila* metamorphosis (Fig. 27.20). One the one hand, DHR3 silences the early regulatory genes after puparium formation. This inhibitory action relies on DHR3 binding to EcR, and therefore presumably, to the EcR-ultraspiracle heterodimer. On the other hand, DHR3 is part of a temporal linkage between the events triggered by successive ecdysone peaks. Here, the role of DHR3 is to activate β*FTZ-F1*$^+$, which in turn seems to promote the reactivation of the early genes by the next ecdysone pulse (Broadus et al., 1999). Both functions of DHR3 illustrate the powerful role of orphan receptors in genetic networks that diversify and orchestrate the responses of targets cells to hormonal stimulus.

The molecular characterization of the early regulatory genes and their targets has validated in an impressive way the Ashburner model (1974), which was based on genetic analysis and observations on chromosome puffs (compare Figs. 27.16 and 27.20). Together, the tissue-specific synthesis of hormone receptors, the competitive inhibitions based on the dimerization of receptor peptides and nuclear orphan receptors, and the complex networks formed by these transcription factors and their target genes, diversify the response to ecdysone in an epigenetic process that coordinates a wide array of cellular responses to a single hormone.

27.5 Hormonal Control of Amphibian Metamorphosis

Amphibians were the first class of vertebrates to conquer the land, and most present-day amphibians still return to the water to reproduce. Consequently, the water is still the habitat of most amphibian larvae. As the larvae make the transition to land-dwelling adults, they undergo metamorphosis.

Anurans (tailless amphibians, including frogs and toads), typically have a dramatic metamorphosis that affects nearly all organ systems and major bodily functions (Tata, 1993). Locomotion changes from swimming with a tail fin to jumping with legs (Fig. 27.21). Gills are replaced with lungs, and blood cells produce a different hemoglobin that binds oxygen less avidly. Mouth, tongue, and gut are greatly modified as the diet changes from herbivorous to carnivorous. The liver and kidney are retooled so that instead of excreting mostly ammonia, the adult excretes mostly urea. The senses of sight and hearing adapt from the physical conditions of water to those of air. Remarkably, all these changes occur while the animal is moving and feeding. There is no quiescent stage, like the insect pupa, devoted exclusively to reconstruction. In urodeles (tailed amphibians, including salamanders), the metamorphic changes are generally less dramatic and vary among species.

The typical frog metamorphosis is divided into three stages (Fig. 27.21). ***Premetamorphosis*** is the period before metamorphosis when the tadpole has no limbs. ***Prometamorphosis*** is the stage when tadpoles have small hindlimbs but no forelimbs. The ***climax of metamorphosis*** begins with the emergence of the forelimbs and extends through the most dramatic changes, including full limb growth and tail resorption. These morphologically defined stages are also characterized by different hormonal and genetic activities.

THE HORMONAL RESPONSE IN AMPHIBIAN METAMORPHOSIS IS ORGAN-SPECIFIC

Early investigators of amphibian metamorphosis took advantage of the feasibility of transplantation experiments with these animals. One of these experiments is of particular interest because it illustrates in a graphic way that the hormonal response varies among target organs.

A given type of tissue may fare quite differently during metamorphosis depending on its location. For instance, bone and skeletal muscle will grow if located in the legs. However, the same tissues will be degraded in the tail. To test whether the response depends on the location of the responding tissue *at the time of metamorphosis,* researchers transplanted a tail from a donor tadpole into the flank of a host tadpole (Fig. 27.22). The new location of the graft was one where there is normally no degradation of bone or muscle. Nevertheless, the grafted tail was degraded and resorbed at about the same pace as the host's own tail (Geigy, 1941). Conversely, when a tadpole eye was transplanted to the tail of a tadpole host, the eye escaped the destruction of the tail and remained attached to the host's rear end at the conclusion of metamorphosis (Schwind, 1933). Thus, the metamorphic response of a tissue is independent of its location during metamorphosis. Rather, the response depends on the *organ* in which the tissue resides, which, of course, is correlated with the location in which the tissue originally developed. In modern terms, these results indicate that the distribution of hormone receptors and other cofactors that control the metamorphic events is organ-specific, apparently as a result of epigenetic processes that have occurred prior to metamorphosis.

AMPHIBIAN METAMORPHOSIS IS CONTROLLED BY THYROID HORMONE

The principal metamorphic hormone in amphibians is ***thyroid hormone (TH),*** which is comparable to insect ecdysone in its actions. It is produced in the *thyroid gland,* located ventrally of the trachea in the neck region. The active components of TH are *thyroxine* (T_4) and *triiodothyronine* (T_3), both iodinated derivatives of the amino acid tyrosine. T_3, which generally seems to be the more active component, is also synthesized from T_4 in tissues other than the thyroid gland. When the thyroid gland is removed from young tadpoles, they grow into giant tadpoles that never metamorphose. Conversely, when TH is applied to young tadpoles with food or by injection, they metamorphose prematurely. TH acts on its target cells through ***thyroid receptors (TRs),*** cytoplasmic receptors that belong to the same superfamily as the ecdysone receptors.

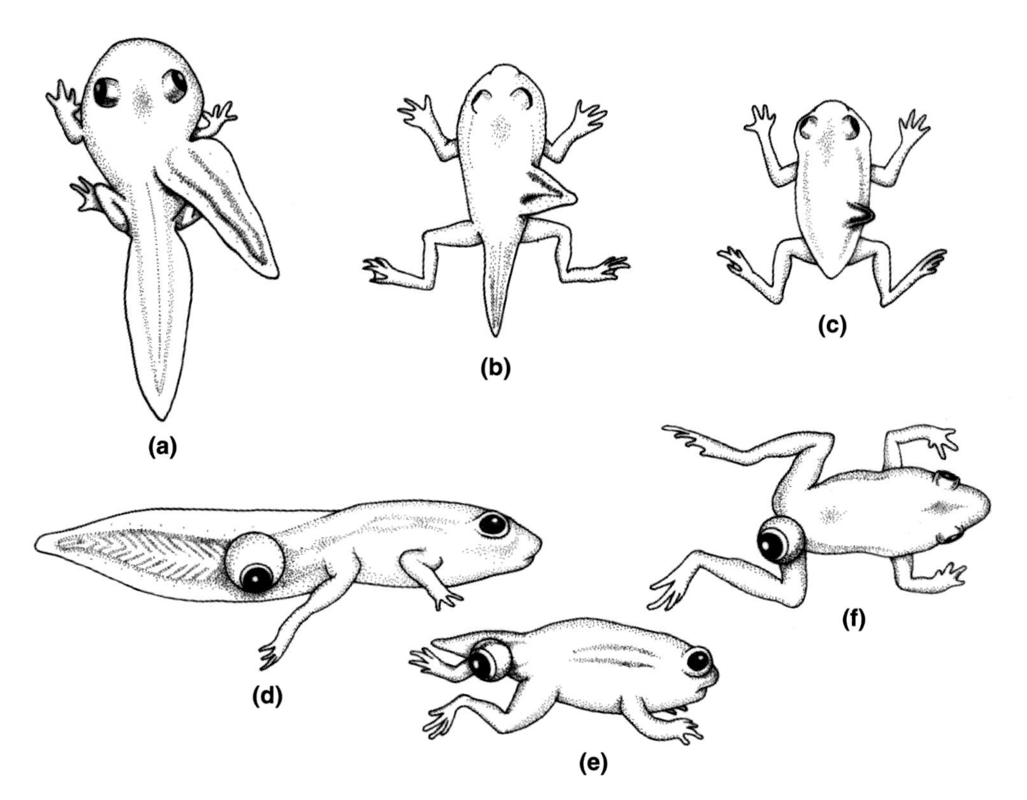

Figure 27.22 Organ specificity of the metamorphic response in frogs. **(a–c)** A tail grafted to the flank is degraded and resorbed along with the host tail. **(d–f)** An eye transplanted to the tail remains unaffected by the degradation of the tail.

Until recently, it was believed that another metamorphic hormone in amphibians, ***prolactin***, is comparable in its effects to insect *juvenile hormone*. Amphibian prolactin is a peptide produced in the anterior pituitary gland. The same hormone in mammals promotes mammary duct development during pregnancy and milk production during nursing. Sheep prolactin, when injected in amphibian tadpoles or added to tadpole organs in culture, acts as an antagonist to TH. However, recent experiments with *Xenopus* prolactin indicate that the latter does not extend tadpole life (Huang and Brown, 2000). Instead, it causes the formation of tailed frogs in which much of the tail muscle has degenerated while the connective tissue remains. Thus, the effect of frog prolactin is much more limited than the action of juvenile hormone in insects, which causes extra larval molts and oversized larvae.

The identification of TH as an amphibian metamorphic hormone has a long history, beginning with the observation that tadpoles metamorphosed prematurely when their food contained sheep thyroid gland (Gudernatsch, 1912). The action of TH and sheep prolactin on tail regression were demonstrated in vitro: Isolated tails kept in organ culture regressed at the same pace seen in vivo when T_3 was added to the culture medium. However, this effect was inhibited when sheep prolactin was present simultaneously (Derby and Etkin, 1968). These observations were confirmed and extended by Tata and coworkers (1991), who studied the modulating effects of sheep prolactin in T_3-induced regression of isolated tails and on the outgrowth of isolated limbs from *Xenopus* tadpoles.

Tails removed from premetamorphic tadpoles retain their size and morphology during 8 days of culture in a serum-free, chemically defined medium (Figs. 27.23 and 27.24). However, the addition of 2×10^{-9} M T_3 causes complete loss of the dorsal and ventral tail fins as well as considerable loss of skin, connective tissue, and musculature. Sheep prolactin added on the same day as T_3 completely prevents the disappearance of the fins and, to a large extent, the loss of muscle and connective tissue. Even if prolactin is added after tail regression has already begun, part of the dorsal fin is retained and the loss in connective tissue and muscle is less severe. A biologically inactive analogue of T_3, 3,5-diiodothyronine, has no effect. Retinoic acid, known as a potent morphogen from other experiments, has little if any effect. Another major event in frog metamorphosis, the outgrowth of limbs, is also promoted by T_3 and inhibited by sheep prolactin in vitro.

The experiments show that T_3 can promote key events of frog metamorphosis in vitro, and they further indicate that these effects are hormone-specific: They are not mimicked by analogues of T_3 . In addition, the experiments show that sheep prolactin inhibits the action of T_3 in culture, especially if administered together with T_3 or shortly thereafter. However, the recent results of Huang and Brown (2000) indicate that frog prolactin

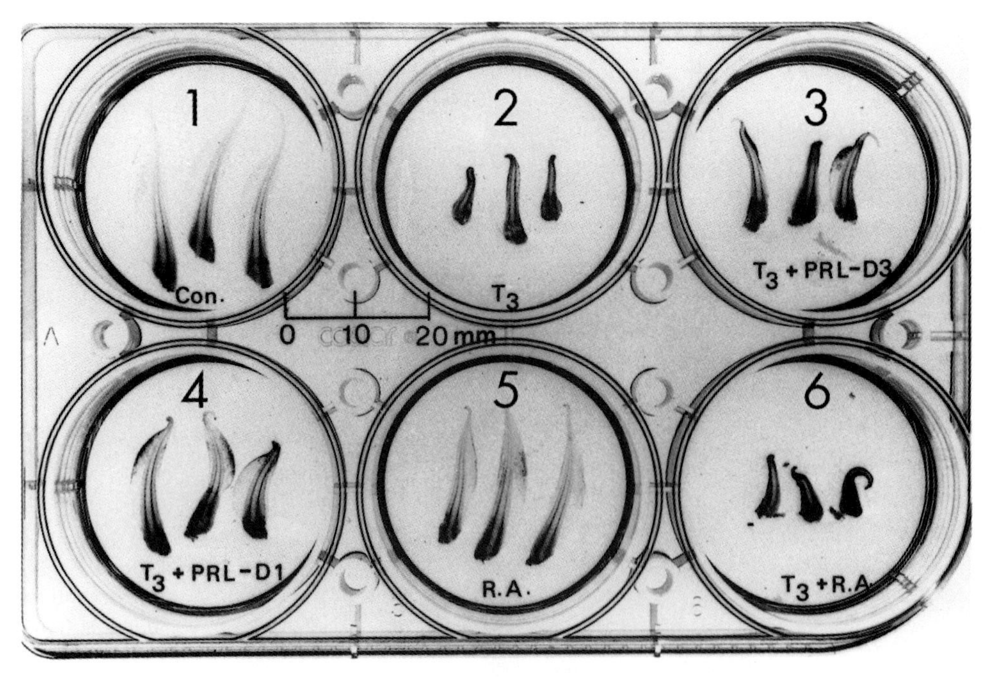

Figure 27.23 Effects of thyroid hormone (T_3), sheep prolactin (PRL), and retinoic acid (RA) on isolated tails from *Xenopus* larvae. Tails from premetamorphic tadpoles were cultured for 1 day in a chemically defined medium before hormones or morphogens were added. The tails were photographed after a total of 8 days in culture. **(1)** Control (no additions). **(2)** T_3 added. **(3)** PRL added 3 days after T_3. **(4)** PRL added 1 day after T_3. **(5)** RA. **(6)** T_3 and RA added simultaneously. Note that the regression caused by T_3 was counteracted effectively by PRL added 1 day after T_3, and less effectively by PRL added 3 days after T_3.

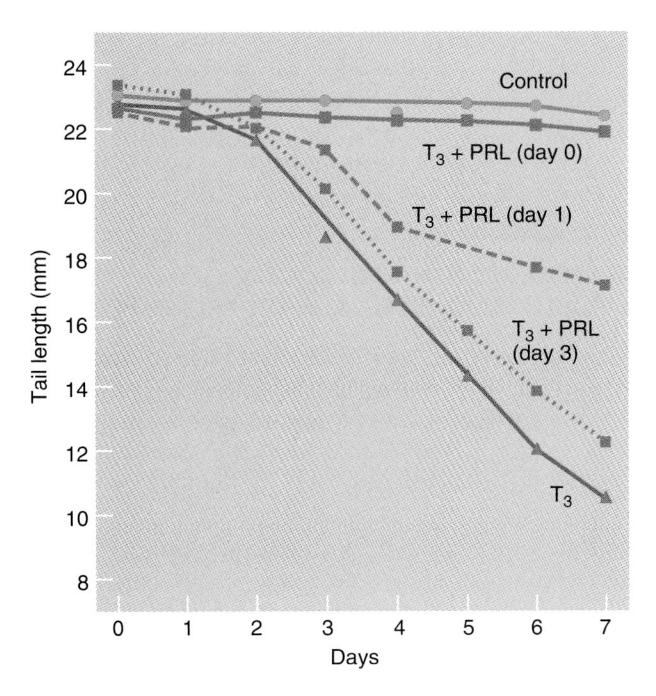

Figure 27.24 Tail regression induced by thyroid hormone (T_3) and its inhibition by sheep prolactin (PRL), in the experiments pictured in Figure 27.23. The tail lengths were measured daily. Relative to the controls kept in medium without hormones, addition of T_3 caused steady tail regression. PRL added simultaneously with T_3 inhibited the regression completely. The inhibitory effect of PRL was diminished when PRL was added 1 day or 3 days after T_3.

interferes with the rèsorption of tail muscle but not connective tissue.

THE PRODUCTION OF THYROID HORMONE IS CONTROLLED BY THE BRAIN

As is the case with insect metamorphosis, amphibian metamorphosis is controlled by the brain (Fig. 27.25). The synthesis of TH in the thyroid gland is initiated by *thyroid-stimulating hormone* (*TSH*), a peptide produced in the anterior pituitary gland. The production of TSH is stimulated by the hypothalamus through ***thyroid-stimulating hormone releasing factor*** (***TSH-RF***). This control hierarchy unfolds as follows (Etkin, 1968).

During premetamorphosis, the signals from the hypothalamus to the pituitary are weak, and the thyroid gland is still in the process of development, so that the TH level is very low. Prometamorphosis begins when the hypothalamus becomes sensitive to a positive TH feedback loop. In response to TH, the hypothalamus then releases TSH-RF, causing an increase in the production of TSH—and consequently TH. The escalating activity of the hypothalamus-pituitary-thyroid axis leads to the climax of metamorphosis, during which the TH concentration reaches its maximum and the metamorphic changes proceed most rapidly. Toward the end of metamorphosis, the synthesis of TH decreases through a negative feedback loop from high TH concentrations on the hypothalamus and through a partial degeneration of the thyroid gland as part of the metamorphic changes.

VARIOUS DEFECTS IN METAMORPHOSIS CAUSE PAEDOMORPHIC DEVELOPMENT IN SALAMANDERS

The hormonal control of metamorphosis in amphibians has provided the basis for an interesting evolutionary phenomenon called ***paedomorphosis*** (Gk. *paido-*, "child"; *morphe,* "form"). Animals of paedomorphic species reach sexual maturity while retaining a juvenile appearance in nonreproductive organ systems. Juvenile characteristics are retained to varying degrees in different salamander species. For example, the axolotl *Ambystoma mexicanum* becomes sexually mature while still living in water and retaining the external gills and large tail fins that are typical of a urodele larva (Fig. 27.26). It also develops some adult characteristics, such as legs,

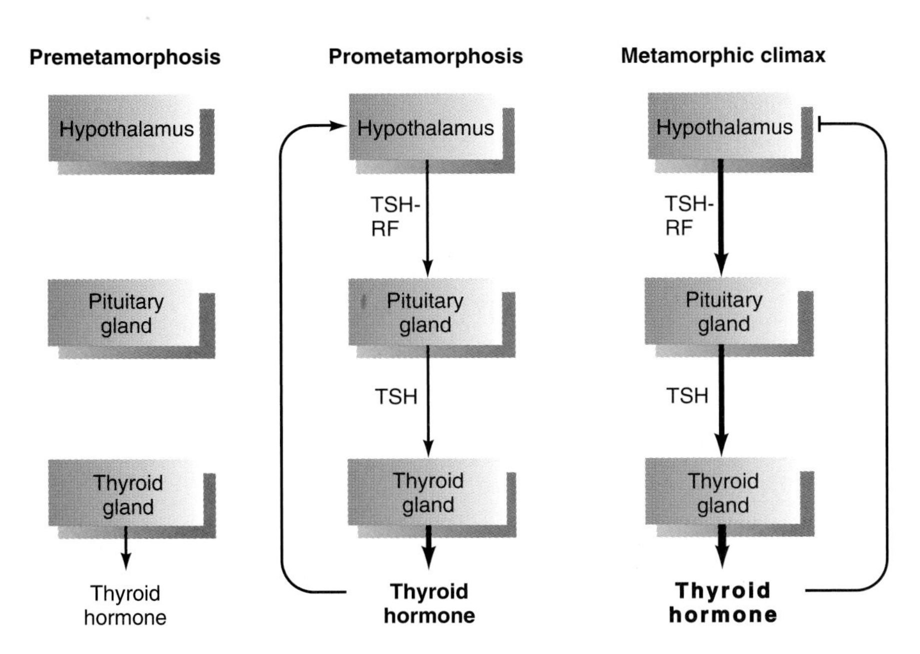

Figure 27.25 Hormonal control of frog metamorphosis. The diagrams show the hypothalamus-pituitary-thyroid axis at consecutive stages. During premetamorphosis, there is no regulatory input from the pituitary gland to the thyroid gland, which then produces minimal amounts of thyroid hormone (TH). During prometamorphosis, the hypothalamus becomes sensitive to a positive feedback loop from TH. The resulting release of thyroid-stimulating hormone releasing factor (TSH-RF) increases the production of thyroid-stimulating hormone (TSH), and thus, TH. The escalating production of TH triggers the metamorphic climax before a reduction in the size of the thyroid gland and a negative feedback loop to the hypothalamus lead to a decrease in TH concentration. See also Figure 27.27.

Figure 27.26 Artificial metamorphosis in the axolotl *Ambystoma mexicanum.* The large individual has been raised normally, is aquatic, and has external gills and a tail fin. The smaller specimen has been treated with thyroxine to induce metamorphosis; individuals like this are usually smaller, lack gills, and are terrestrial.

adult hemoglobin, and excretion of urea, but normally it does not metamorphose completely. Similar examples of paedomorphosis based on partial metamorphosis are observed in other salamander species.

Experimental analysis has shown that in the course of evolution each paedomorphic species has become defective at some point along the signal chain that leads from the hypothalamus via the pituitary and thyroid glands to the target organs of TH (Frieden, 1981). The tiger salamander, *Ambystoma tigrinum,* depends on a certain environmental input to its brain for metamorphosis. In the high and cold parts of the species' range, in the Rocky Mountains, the adults are water-dwelling and paedomorphic; in the lower and warmer parts of its range, the same species undergoes metamorphosis, and the adults become land-living before they mate. Larvae from the high part of the range will metamorphose if they are simply placed in warmer water. It seems that below a certain temperature the hypothalamus of this species does not produce enough TSH-RF to induce metamorphosis. The axolotl *Ambystoma mexicanum* does not normally metamorphose, as noted earlier. However, experimental application of TH or TSH will generate an adult of this species that does not occur in nature (Fig. 27.26). Thus, *A. mexicanum* must be defective in the release of TSH or at an earlier step in the control hierarchy. Finally, some other salamanders, including *Necturus* and *Siren* species, cannot be induced to metamorphose with TH. Presumably, they lack functional TH receptors.

THE RECEPTORS FOR ECDYSONE AND THYROID HORMONE SHARE MANY PROPERTIES

Like insect ecdysone, amphibian TH acts through receptor proteins belonging to the same superfamily of ligand-dependent transcription factors as the receptors of other steroid hormones and of retinoic acid. Besides their general molecular structure, the ecdysone receptors (EcRs) and the TH receptors (TRs) share several properties, including autoinduction, multiple isoforms, heterodimer formation, and tissue-specific distribution.

In *Xenopus laevis,* there are two families of TH receptors, TRα and TRβ (Yaoita and Brown, 1990). The expression of both receptor types begins when the tadpole hatches, but TRα mRNAs are generally more abundant. The prevalence of TRα mRNAs increases during premetamorphosis, is maximal by prometamorphosis, and falls to a lower level after the climax of metamorphosis (Fig. 27.27). In contrast, TRβ mRNAs are barely detectable during premetamorphosis but rise in synchrony with the level of TH in the blood serum. Both TH and TRβ reach a maximum during the metamorphic climax and drop to about 10% of the peak value thereafter. The correlation suggests that TH may induce the synthesis of TRβ mRNAs. Indeed, the expression of TRβ genes can be *autoinduced* precociously in premetamorphic tadpoles if they are kept in water with exogenous T_3 for 2 days.

In an intriguing parallel with the ecdysone receptor discussed earlier, the thyroid hormone-receptor peptide is greatly enhanced in its activity by heterodimer formation with retinoid X receptor (RXR) peptide (Kliewer et al., 1992; Zhang et al., 1992). The TR-RXR heterodimer binds more effectively to its target DNA recognition sequences and enhances transcription more strongly than homodimers of either peptide.

Another striking parallel between insect and amphibian metamorphosis appears to be the diversification of cellular responses by a network of early regulatory genes. Of 17 *Xenopus* genes that were found to be upregulated in response to TH, about half respond rapidly and even in the presence of cycloheximide, an inhibitor of protein synthesis (Z. Wang and D. D. Brown, 1993; D. D. Brown et al., 1996). Four of these early response genes encode transcription factors, as do the early regulatory genes in *Drosophila.* The response of the other upregulated genes is delayed by about a day and dependent on protein synthesis. Some late-responding genes encode proteolytic enzymes and a deiodinase, enzymes expected to be involved in carrying out the tissue

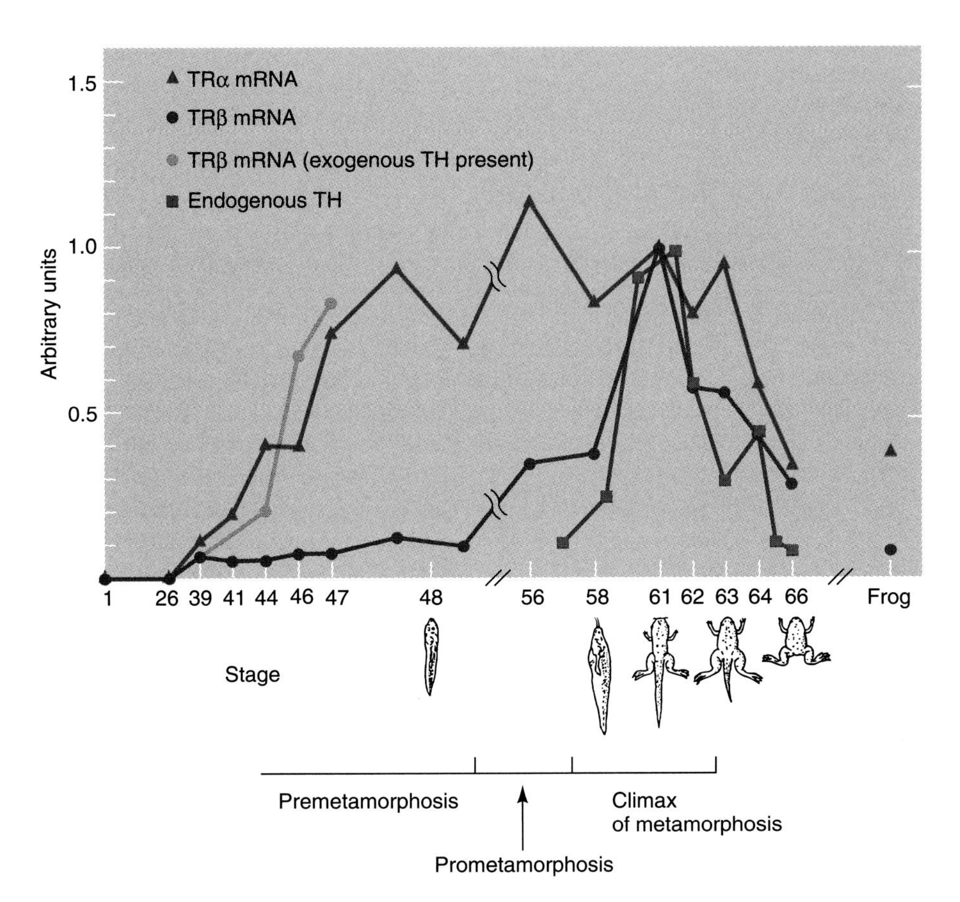

Figure 27.27 Summary of thyroid hormone receptor (TRα and TRβ) gene expression during *Xenopus* development. Developmental stages (abscissa) are represented by diagrams and by metamorphic phases. The prevalence of TRα mRNA, TRβ mRNA, and thyroid hormone is indicated in relative units, with the amount at the peak of metamorphic climax taken as 1.0. Note that the prevalence of TRβ mRNA closely follows the TH concentration. The levels of TRβ mRNA in premetamorphic tadpoles kept in water with exogenous T_3 for 2 days are also shown.

breakdown that forms a major part of metamorphosis and then terminating this process by inactivating TH.

In summary, the hormonal controls of metamorphosis in insects and amphibians share several common features. Are these control mechanisms *homologous*—that is, based on a common ancestral mechanism? On the one hand, the overall regulation by ecdysone and thyroid hormone, respectively, and the control of the glands producing these hormones by input from the brain are striking in their parallelism (compare Figs. 27.12 and 27.25). There are also many shared details in the action of ecdysone receptors and thyroid receptors, as we have seen. On the other hand, features like autoinduction and heterodimer formation are general characteristics of hormone action and transcriptional regulation and are not unique to metamorphosis control. Also, the chemical nature of the metamorphosis hormones is quite different in insects and amphibians and does not support homology. On balance, it seems more likely that the similarities observed are the result of *convergence*—that is, although amphibian and insect metamorphosis evolved separately, they nevertheless acquired analogous features because they were selected to fulfill similar functional requirements.

SUMMARY

Hormones are chemical signals that are produced by specialized glands and transmitted through blood or equivalent body fluids. However, a hormone elicits a response only from those cells that have the appropriate receptors and other cofactors; such cells are called the target cells of the hormone. Water-soluble hormones bind to transmembrane receptors and act through second messengers. Lipid-soluble hormones bind to cytoplasmic or nuclear receptors, which act as transcriptional regulators. Typically, a hormone coordinates the activities of several target tissues or organs.

Secondary sex differentiation in mammals is directed by hormonal control mechanisms. Testicular cells synthesize the principal male sex hormones, testosterone and anti-Müllerian duct hormone. The testosterone derivative, 5α-dihydrotestosterone, directs the male development of originally indifferent external genitalia. Testosterone promotes development of the mesonephric (Wolffian) duct into the epididymis, ductus deferens, and seminal vesicle, while anti-Müllerian duct hormone makes the Müllerian duct undergo apoptosis. In the absence of both hormones, the Wolffian duct disappears while the Müllerian duct gives rise to the oviduct, uterus, and upper vagina. Ovaries produce estrogen, a female sex hormone. Estrogen levels are generally low during fetal development but rise after puberty under the influence of gonadotropic hormones from the pituitary gland.

Hormones also control the sexually dimorphic development of the brain and behavior. The receptors for each sex hormone in the brain have distinct distributions, so that the brain regions stimulated by androgens differ from the brain regions affected by estrogen or progesterone. In songbirds, testosterone or its metabolites promote both male singing and an increase in the number of neurons in certain brain nuclei. According to the organizational hypothesis, prenatal exposure of mammals to testosterone or its metabolites permanently alters the developing brain, causing it to function in male-specific ways later in life. This hypothesis receives its strongest support from the development of reproductive behavior and sexually dimorphic brain areas in rodents. However, the hypothesis seems to hold as well for some nonreproductive human behaviors.

Molecular aspects of hormone action have been studied most extensively in insect metamorphosis, which is controlled by two antagonistic hormones, ecdysone and juvenile hormone. The glands producing these hormones are regulated by the brain. The ecdysone-receptor complex activates a small number of early response genes, which encode transcription factors that regulate a larger number of late response genes. The early genes include, in particular, the gene encoding ecdysone-receptor peptides. The ecdysone receptor is synthesized in three isoforms, which promote different responses in target cells. Each isoform heterodimerizes with the ultraspiracle peptide, an ortholog of the vertebrate retinoid X receptor. The transcriptional effect of these heterodimers is further modulated by orphan receptors and other transcription factors that respond to different ecdysone concentrations. Together, these factors diversify the response to ecdysone in an epigenetic process that choreographs a multitude of cellular responses to a hormonal stimulus.

Amphibian metamorphosis is also controlled by thyroid hormone, the production of which is likewise controlled by the brain. Defects in this control hierarchy have led to the paedomorphic development of several salamander species. Amphibian thyroid hormone receptors share many features with the insect ecdysone receptors, but these similarities seem to be convergent rather than based on homology.

Organismic Growth and Oncogenes

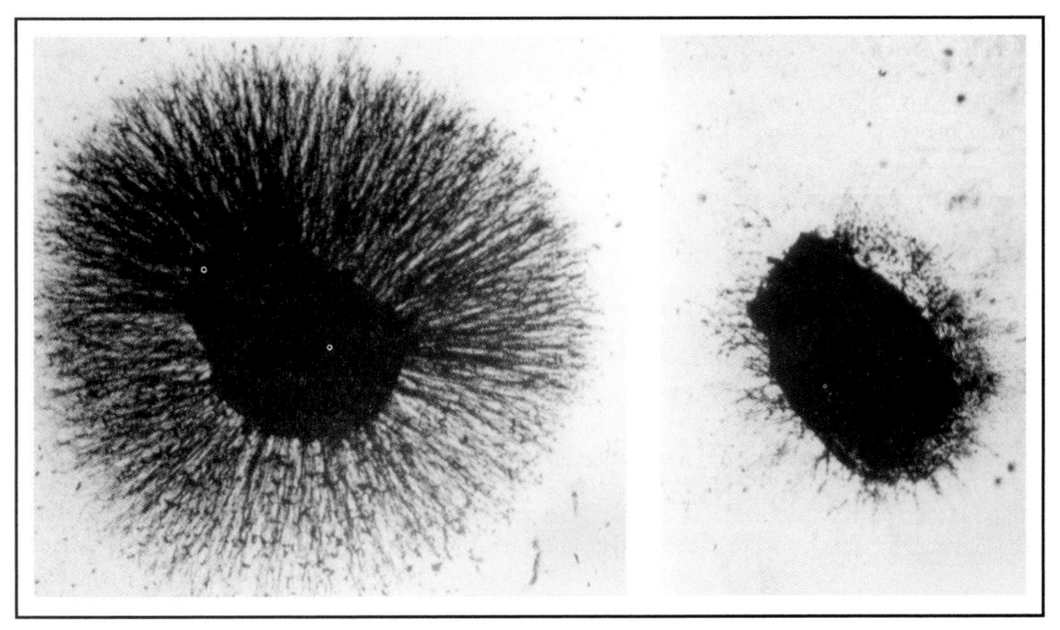

Figure 28.1 Effect of nerve growth factor (NGF) on the growth and differentiation of neurons. The photographs show sensory ganglia removed from chicken embryos and cultured in a semisolid medium. The ganglion to the left was cultured in the presence of purified NGF, a peptide that promotes the survival, growth, and differentiation of nerve cells. The halo around the ganglion consists of extensions that have grown out of the neurons in the ganglion. The ganglion to the right was cultured in the same way but without NGF. NGF was discovered by Rita Levi-Montalcini and Stanley Cohen, who in 1986 received the Nobel prize in Physiology or Medicine for their discovery.

GROWTH is a basic component of development. This is obvious if one compares the size of a human egg, a sphere about 0.1 mm in diameter, with the bulk of a human adult. Growth of an organism is defined as change in mass. The most common form of growth is cell division followed by enlargement of the daughter cells until each has reached the previous size of the mother cell. Other types of positive growth occur when cells increase in mass without cell division, or through the deposition of extracellular matrix.

The classical analysis of growth began with the study of growth curves—that is, coordinate plots of mass versus time. Such measurements allowed quantitative descriptions of how fast an organism has been growing overall, and of the extent to which its component parts have grown at the same rate. By heterospecific transplantation, researchers sought to distinguish factors that affect the growth of the entire organisms from factors that affect specific organs and their component tissues.

A milestone in the physiological analysis of vertebrate growth was the discovery of growth hormone and the control of its synthesis by blood-borne factors released from the brain. While growth hormone is synthesized by specialized cells and carried over the entire body by the circulating blood, other chemical signals known as growth factors are synthesized by various types of cells and act locally. The first growth factor to be discovered was nerve growth factor, a peptide that promotes the growth and differentiation of certain neurons and is released by their natural targets (Fig. 28.1).

The genetic and molecular analysis of growth has led to the discovery of the signal proteins that accelerate or slow down the pace of the cell cycle. The genes encoding these proteins are the same genes that are mutated in cancers. The characterization of the complex networks formed by these genes and their products has led to a deeper understanding of normal growth control and to new ways of cancer therapy.

While growth hormone and growth factors are chemical signals, cells also depend on mechanical input for their ability to survive and to divide. The first clue to this phenomenon has been the anchorage dependence of normal cells—that is, their dependence on contact with a substrate. New research has shown that this effect is mediated by the shape of cells, presumably through gene regulatory pathways that are sensitive to changes in the cytoskeleton.

Another focus of exciting new work is the connection between growth and pattern formation. Cell proliferation in insect imaginal discs is controlled by long-range patterning signals as well as by local cell interactions. Mutations that accelerate or retard cell proliferation tend to be compensated by changes in cell size, while the overall growth and patterning of organs are unaffected. It appears, therefore, that pattern formation and growth are linked by signals that measure overall mass or size rather than number of cells.

28.1 Measurement and Mechanisms of Growth

The scientific exploration of growth began when researchers weighed developing organisms at intervals and plotted the changing weight versus time. Some species, including humans and most other mammals, have *determinate growth,* meaning that after a growth period their mass stays more or less constant. In contrast, plants and many animals have *indeterminate* growth, which means their mass increases throughout their lives, even though the rate of growth tends to slow down with age.

GROWTH IS DEFINED AS CHANGE IN MASS

Growth of an organism is defined as a change in its mass. Growth is usually positive—that is, the mass of living tissue increases—but we also speak of negative growth when the mass decreases. Growth occurs as the net result of an organism's constant replacement of worn-out cells with new ones and the turnover of its extracellular materials. A plot in a coordinate grid of an organism's mass versus time is known as a ***growth curve***. The shape of a growth curve is often sigmoid—that is, it resembles the letter S (Fig. 28.2a).

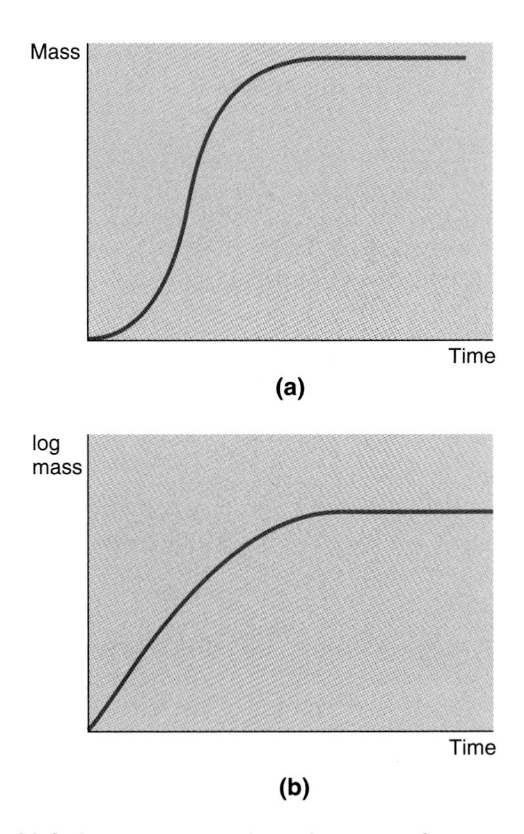

Figure 28.2 Growth curves show the mass of an organism or organ plotted versus time. **(a)** If mass is plotted on a linear scale, the curve is typically sigmoid. **(b)** If mass is plotted on a logarithmic scale, the initial portion of the curve often shows a linear incline.

During the initial phase of a growth curve, the mass increase per unit time is proportional to the mass already present. This relation is expressed by the equation

$$\frac{dM}{dt} = g \cdot M$$

which is equivalent to

$$\frac{1}{M} \cdot \frac{dM}{dt} = g$$

where M represents mass, t is time, and g is a coefficient called the ***relative growth rate.*** As long as the relative growth rate is constant, the growth curve rises exponentially; if the mass is plotted on a logarithmic scale, the exponential part of a growth curve is a straight line (Fig. 28.2b). This type of curve is obtained not only for organisms, but also for freshly seeded cell cultures, and for animal populations that are not limited by predation or by the carrying capacity of the environment. Thus, the exponential part of a growth curve reflects a growth phase during which the cells of an organism divide at constant intervals and restore their original sizes after each division. The exponential phase is followed by a phase of slower growth, which reflects inhibitory interactions or the lack of signals or resources that are required for growth.

GROWTH MAY BE ISOMETRIC OR ALLOMETRIC

Different organs or regions of an animal may grow at the same relative growth rate, a situation described as ***isometric growth.*** More common is ***allometric growth,*** or a process in which some organs grow faster than others. Julian Huxley (1932), who studied allometric growth extensively, used the following equation to describe allometric growth:

$$y = b \cdot x^k$$

which is equivalent to

$$\log y = \log b + k \cdot \log x$$

where y is the mass of one part at a given time, x is the mass of another part (or the entire organism) at the same time, b is a coefficient that gives the mass of y when $x = 1$, and—most important—k is the ***growth ratio***—that is, the relative growth rate of y divided by the relative growth rate of x. If $k = 1$, then x and y grow isometrically; and if $k > 1$, then y is *positively allometric* relative to x. The equation is based on the assumption that the growth ratio itself is constant over extended periods of time, and Huxley found empirically that this is true in many cases. In these cases, k can be determined by plotting y versus x on a double logarithmic coordinate grid and finding the straight line of best fit.

An example of allometric growth is seen in the human fetus, where the head is disproportionately large while the legs are proportionally shorter than in the

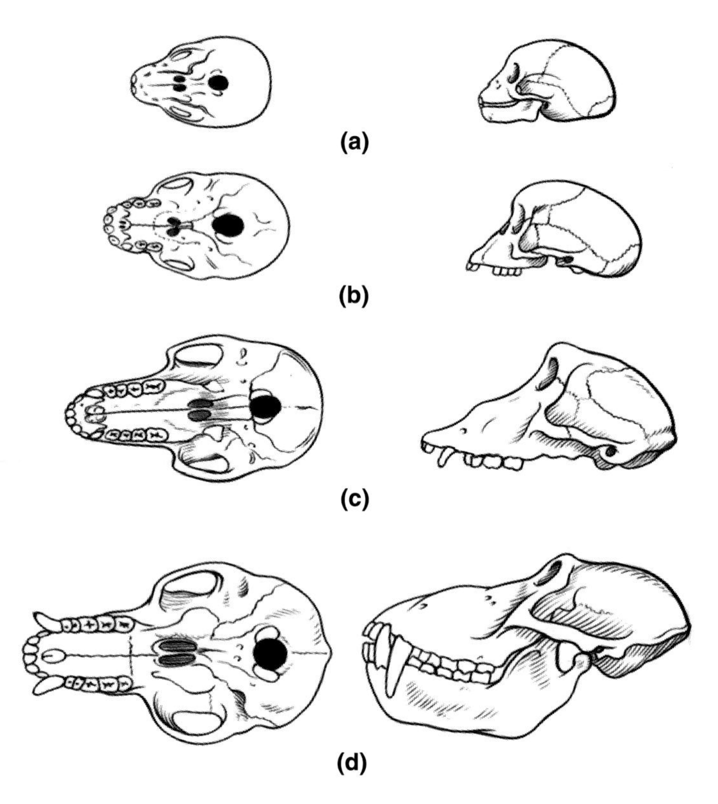

Figure 28.3 Allometric growth of the face relative to the braincase in the baboon: **(a)** newborn; **(b)** juvenile; **(c)** adult female; **(d)** adult male. Note that the face is initially smaller than the braincase but later grows larger, especially in the male.

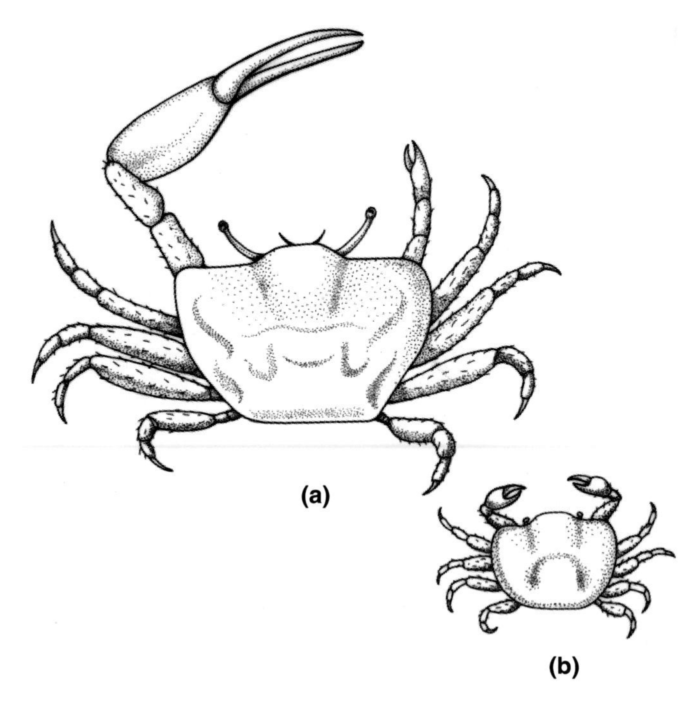

Figure 28.4 Allometric growth of one chela (claw) in the fiddler crab *Uca pugnax.* **(a)** Adult male with disproportionately large left chela; **(b)** juvenile male with chelae of equal size.

adult (see Fig. 1.9). Thus, the legs will grow faster than the head; in other words, their growth ratio relative to the head will be greater than 1. Within the head, the facial parts are positively allometric relative to the braincase. This allometry, which is apparent in the human, is even greater in other mammals and tends to be more pronounced in males (Fig. 28.3).

An extreme example of allometric growth, also sexually dimorphic, is seen in the fiddler crab *Uca pugnax* (Fig. 28.4). The adult male of this species uses one of its chelae (claws) for fighting and display. In females and in young males, both chelae are the same size and relatively small, each representing about 8% of the crab's entire mass. However, as a male approaches adulthood, one chela grows with a growth ratio of 1.62 relative to the entire body, so that eventually this chela accounts for about 38% of the male's total body weight. These observations show that growth ratios and relative growth rates may change in stage-dependent and region-specific ways.

GROWTH OCCURS BY DIFFERENT MECHANISMS

A tissue can grow by several means. The most common mechanism is cell division followed by enlargement of the daughter cells until each has reached the previous size of the mother cell; this type of growth is called ***hyperplasia*** (Gk. *hyper,* "above"; *plassein,* "to form"). Another way for growth to occur is by an increase in mass without cell division, a process called ***hypertrophy*** (Gk. *trophe,* "nourishment"). A third means of growth is the deposition of extracellular matrix, a process called ***accretion*** (Lat. *accrescere,* "to accrue").

Hyperplasia is so common that some authors use "growth" as a synonym for cell division, especially in the context of tumorigenesis. However, it is important to remember that cell division may occur without growth, as in embryonic cleavage, when blastomeres divide but do not subsequently enlarge (see Fig. 5.2). A negative form of hyperplasia is *apoptosis,* or programmed cell death (see Section 25.4).

Hypertrophy is observed in cells that no longer divide. For instance, *adipocytes* (fat cells) store lipids in large cytoplasmic droplets (see Fig. 14.19). If an organism is on a rich diet, these lipid droplets may enlarge enormously without division of the adipocyte. Seasonally positive and negative hypertrophy of adipocytes is common among hibernating animals. A different type of hypertrophy occurs in skeletal muscle fibers. Under conditions of regular exercise and adequate nutrition, skeletal muscle fibers increase in girth through addition of actin and myosin fibrils. In response to muscle tension, so-called satellite cells fuse with existing muscle fibers, thus adding more nuclei to the existing fibers and maintaining a favorable nucleocytoplasmic ratio. The negative form of this type of growth may be caused by a depletion of satellite cells and is known as *dystrophy.*

Growth by accretion is observed mainly in connective tissues such as cartilage and bones. As described in Section 14.3, the extracellular matrix of bones is produced by *osteoblasts* and destroyed by cells of a different type called *osteoclasts*. The dynamic equilibrium between adding and subtracting extracellular matrix, which is under hormonal control, leads to a net positive growth early in life but tends to create a negative balance later, leading to a condition known as *osteoporosis* (Gk. *osteon*, "bone"; *poros*, "pore").

In a growing organ, such as a human leg during puberty, all of the processes discussed here are going on at the same time. Many cells, including the cartilage cells in the *epiphyseal discs* of the long bones, undergo hyperplasia. Elsewhere in bone and other connective tissues, growth occurs by accretion, while muscle fibers increase by hypertrophy.

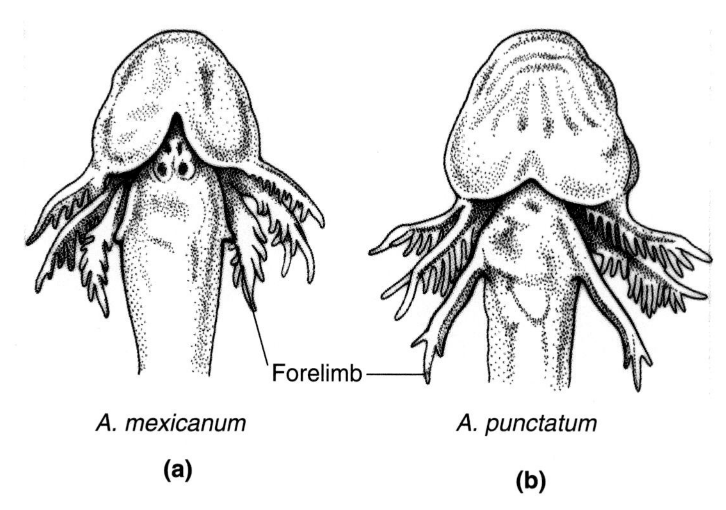

Figure 28.5 Larvae of *Ambystoma mexicanum* **(a)** and *A. punctatum* **(b)** at the beginning of the swimming and feeding stage. Ventral views. Note that *A. punctatum* has well-differentiated forelimbs with digits, whereas *A. mexicanum* has only small nodules at this stage. In this respect, *A. mexicanum* is similar to *A. tigrinum,* which was used in the experiment shown in Figure 28.6.

28.2 Growth Analysis by Heterospecific Transplantation

The relative growth rate of an organ, and its final size, are influenced by a variety of factors, which are partly intrinsic to the organ and partly extrinsic. ***Intrinsic factors*** include the genotype of the organ and its previous developmental history. Intrinsic factors define the maximum rate at which the organ can grow and, in species with determinate growth, its final size under favorable circumstances. ***Extrinsic factors*** include the internal environment provided by the organism for its own cells, such as nutrition and hormones, as well as factors in the external environment, such as temperature. To assess the relative importance of intrinsic and extrinsic factors, researchers have used *heterospecific transplantation*—that is, the replacement of organs from one species with the corresponding organs from another species. These experiments have revealed that intrinsic factors play a major role in growth control, but they also showed that different organ parts may interact with one another in ways that assure the development of a well-proportioned organ (R. G. Harrison, 1935; Weiss, 1939).

THE GROWTH POTENTIAL OF A LIMB IS A PROPERTY OF THE LIMB'S MESODERM

Among salamanders of the genus *Ambystoma,* the forelimbs of certain species differ in their final sizes and in their growth rates at various stages. Two species used in several studies on growth regulation are *Ambystoma punctatum* and *Ambystoma tigrinum. A. tigrinum* is more active and voracious, grows more rapidly, and has an adult length twice that of *A. punctatum.* In *A. punctatum,* the forelimbs appear early in development: They are fully formed, complete with digits, by the time the larva begins to swim and feed (Fig. 28.5). In contrast, the forelimbs of *A. tigrinum* at the corresponding stage are small nodules without visible differentiation. Once they have started to grow, however, they proceed at a much faster rate and eventually grow to a much greater size than *A. punctatum* forelimbs.

TO test whether these specific time characteristics and final sizes are intrinsic to the limbs or to the entire organism, R. G. Harrison (1924) carried out reciprocal heterospecific transplantations. He transplanted forelimb buds before their outgrowth began, during the embryonic period, from the right side of an *A. punctatum* to the right side of an *A. tigrinum* individual and vice versa (Fig. 28.6). The left limb buds remained in place as controls.

The *A. punctatum* limb bud graft on the *A. tigrinum* host developed rapidly during the embryonic period and forged ahead of the normal *A. tigrinum* limb on the opposite side. Its movements were somewhat delayed, but it soon became functional and remained so throughout its life. During the larval period, however, the grafted limb fell behind the normal limb, and at metamorphosis, the grafted limb was about half as long as the control limb on the other side of the host (Fig. 28.6b). This result was observed under a variety of conditions including maximum feeding.

The reciprocal experiment gave corresponding results. The *A. tigrinum* limb bud graft on the *A. punctatum* host embryo grew slowly at first, but once it started, it grew rapidly and eventually became about twice as long as its contralateral control limb (Fig. 28.6e). The grafted limb was also fully functional despite its oversize. Thus, the growth of the transplanted limb buds corresponded unmistakably

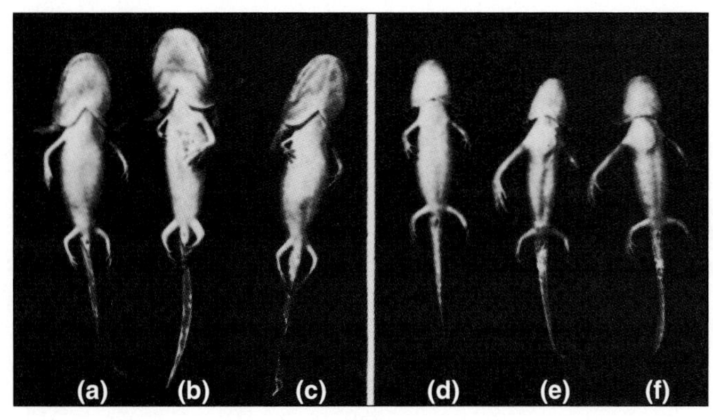

Figure 28.6 Growth of limbs exchanged between two urodele species with different growth characteristics. Limb buds were exchanged reciprocally on the right sides of the embryos (the left sides in the photographs, which are ventral views taken about 3 months later). **(a–c)** *Ambystoma tigrinum;* **(d–f)** *Ambystoma punctatum.* **(a, d)** Ectoderm of the limb bud was replaced with limb bud ectoderm of the other species. All limbs developed according to the size characteristic of the host. **(b, e)** Whole limb buds were exchanged. The grafts developed according to the growth characteristics of the donor. **(c, f)** Mesoderm of the limb buds was exchanged. The grafts developed according to the growth characteristics of the donor.

to the intrinsic time schedule and final size of the donor species.

To further analyze whether the intrinsic growth characteristics resided in a particular part of the grafted limb bud, R. G. Harrison (1935) extended these experiments and grafted only the ectodermal cap or the mesenchymal core of a bud. The effects of transplanting *ectoderm* were only subtle (Fig. 28.6a, d). However, when limb bud *mesoderm* was grafted alone, the outcome was very similar to that obtained by grafting the entire limb bud (Fig. 28.6c, f). Together, these results showed that the intrinsic growth properties of a limb bud reside mostly in its mesenchymal core.

Questions

1. The growth of the limb grafts was analyzed, for the most part, through comparison with the control limb on the opposite side of the *host.* This control limb belonged to a different species and therefore had a different set of intrinsic growth factors, whereas the extrinsic factors were the same for both the transplant and the host control. However, comparisons were also made between the grafted limb and its nontransplanted counterpart in the *donor.* What conclusions could be drawn from the latter comparison?
2. Harrison considered the growth of the *A. tigrinum* grafts in *A. punctatum* hosts to be of particular significance. What could have been his reason?

Harrison's experiments showed that salamander limb buds grow into functional limbs after heterospecific transplantation. The growth potential of the grafted limb is intrinsic to the donor species, and it is fully realized in the host under conditions of maximal feeding. For the most part, these intrinsic growth properties of the limb reside in the mesenchymal core of the limb bud.

THE OPTIC CUP AND THE LENS OF THE EYE ADJUST THEIR GROWTH RATES TO EACH OTHER

Heterospecific transplantations with other organs, including, heart, ear, and eye, yielded results similar to those obtained with limb buds. Eye transplantations are especially interesting because they illustrate that different organ parts may adjust their final sizes to one another.

There are two main constituents of the developing eye: the *optic cup,* which develops from the *optic vesicle;* and the *lens,* which is formed by the overlying *lens epidermis* (see Figs. 1.16 and 13.36). It is possible to transplant the entire eye rudiment or only the lens epidermis or only the optic vesicle.

In separate experiments, R. G. Harrison (1929) and Twitty and Elliott (1934) transplanted entire eye rudiments reciprocally between embryos of *Ambystoma punctatum* and *A. tigrinum.* In both studies, the transplants grew to a final size that was characteristic of the donor species. Thus, the *A. punctatum* host grew to its modest overall size with a large *A. tigrinum* eye, whereas the *A. tigrinum* host grew to its much larger overall size with a small *A. punctatum* eye (Fig. 28.7). The grafted eyes were normally shaped, moved normally, and in many cases seemed to have connected properly to the brain.

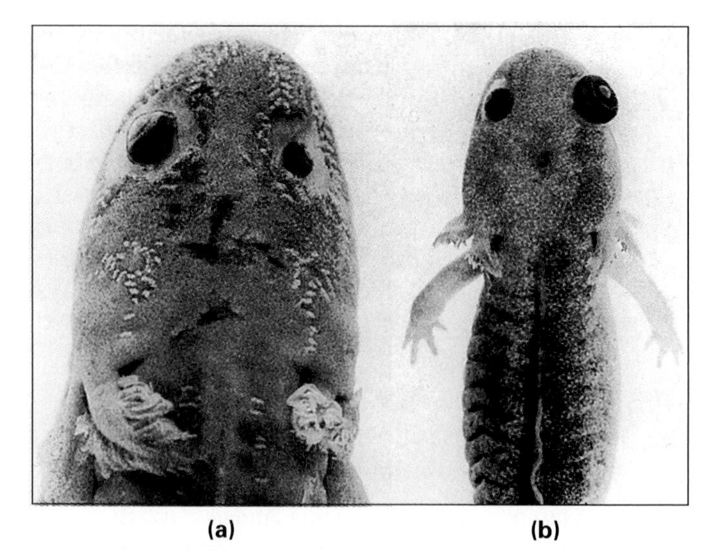

Figure 28.7 Reciprocal transplantation of embryonic eyes between two salamander species of different sizes. The photographs show dorsal views of preserved specimens, with skin removed around the eyes to reveal their sizes. **(a)** *Ambystoma tigrinum* with right eye from *A. punctatum;* **(b)** *A. punctatum* with right eye from *A. tigrinum.* During the months after the operation, *A. tigrinum* grew much faster than *A. punctatum,* but the grafted eyes acquired the sizes characteristic of their donors.

table 28.1 Growth Ratios of Normal, Transplanted, and Chimeric Salamander Eyes

Material Transplanted	Combination	Number of Cases	Number of Survivors	Growth Ratio k	k (P/T) × k (T/P)	Eye Index i Max*	Eye Index i Min	Eye Index i Average	i (P/T) × i (T/P)
Whole eye	P/T	11	5	0.73	0.99	0.62	0.67	0.64	1.15
	T/P	24	12	1.36		2.00	1.63	1.80	
Optic vesicle	P/T	14	2	0.87	0.97	0.77	0.77	0.77	1.01
	T/P	13	2	1.11		1.33	1.28	1.31	
Lens epidermis	P/T	16	7	0.87	1.06	0.79	0.83	0.81	1.03
	T/P	11	5	1.22		1.31	1.21	1.27	

* The terms maximum and minimum are used with respect to the greatest effect—that is, the greatest deviation from unity (1.0).

P/T = *Ambystoma punctatum* graft on *Ambystoma tigrinum*.

T/P = *Ambystoma tigrinum* graft on *Ambystoma punctatum*.

k = relative growth rate of grafted eye (diameter) and normal control on opposite side of the host.

i = ratio of diameter of grafted and host control eyes at close of experiment.

After Harrison (1929, 1935). Used with permission.

To analyze the separate contributions of the eye lens and the optic cup to the overall size of the eye, R. G. Harrison (1929, 1935) carried out similar transplantations except that he grafted only lens epidermis or optic vesicle instead of the entire eye rudiment. As the recipients developed, he periodically measured the sizes of their lenses and optic cups using a microscope with a scale built into one eyepiece. Some specimens were sacrificed for histological sectioning.

From his measurements, Harrison computed the *growth ratio k* of the grafted eye diameter relative to the control eye on the opposite side of the host. He also calculated the eye index *i*—that is, the ratio of the diameters of the grafted and the host control eye at the close of the experiment. The results are compiled in Table 28.1. The number of analyzable cases was small, especially for the optic vesicle transplantations, which were survived by only two individuals in each series. However, the range in the measurements for the eye index in these cases was narrow, suggesting that the average from a larger number of measurements would probably be close to the average from the cases at hand. For both k and i, the greatest departures from unity (1.0)—meaning the greatest differences between the transplanted eye and the host control—were observed after transplantation of entire eye rudiments. Smaller but still significant effects were observed after transplantation of either optic vesicle or lens epidermis. As expected, the products of k values for reciprocal experiments (*A. punctatum* graft on *A. tigrinum* and vice versa) and the products of i values for reciprocal experiments were close to unity.

When *A. tigrinum* lens epidermis was transplanted to an *A. punctatum* embryo, the newly formed lens was initially too large for the host's optic cup (Fig. 28.8). However, the cup nearly caught up with the lens once the host larva began to feed. After 3 to 4 months, the lens had grown less than it would have in a transplant of an entire eye rudiment, whereas the optic cup had grown more than its host control. The resulting eye was intermediate in its overall size between normal *A. tigrinum* and *A. punctatum* eyes, and the lens was nearly in normal proportion to the retina.

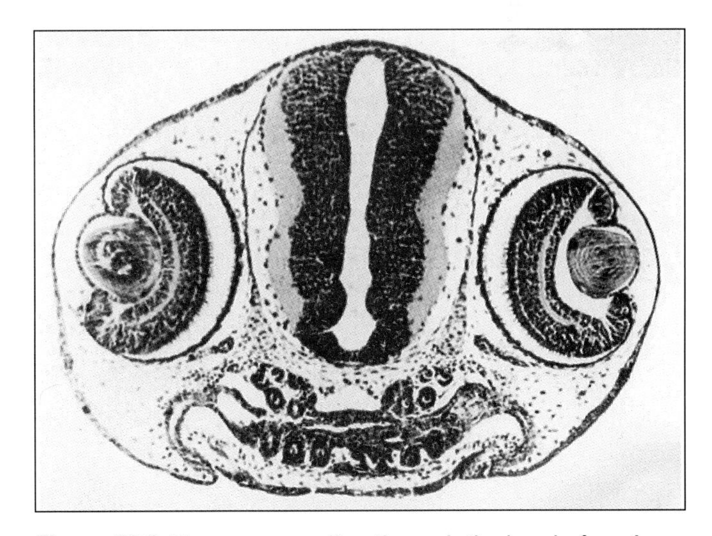

Figure 28.8 Transverse section through the head of an *Ambystoma punctatum* embryo, 8 days after it received a graft of *A. tigrinum* lens epidermis over the right optic vesicle. In the chimeric right eye (to the left in the photograph), the lens was derived from the graft, and the optic cup from the host. In the other eye, both components are derived from the host. Note that the lens is disproportionately large in the chimeric eye.

In the reciprocal experiment, when *A. punctatum* lens epidermis was transplanted to an *A. tigrinum* embryo, the newly formed lens was initially too small for the host optic cup. At the close of the experiment, the lens had grown more than it would have in a transplant of an entire eye rudiment, and the optic cup had grown considerably less than its host control. Again, the final size of the chimeric eye was intermediate between the normal eye sizes of the two species, and the lens was nearly in normal proportion to the eye.

When an *A. tigrinum* optic vesicle was transplanted to an *A. punctatum* embryo, the grafted optic cup initially grew too fast for the lens, which was of host origin. In one case analyzed 19 days after transplantation, the lenses in both eyes were still the same size, indicating that the graft had not yet affected the growth of the lens. After a few months, however, the lenses in the two surviving cases had grown by an average of 13% in each dimension. At the same time, the host lenses retarded the growth of the graft optic cups, which grew only to an average eye index of 1.31 instead of the 1.80 observed after transplantation of an entire eye (Table 28.1). In the reciprocal experiment, the grafted *A. punctatum* optic cup was too small for the lens formed by the *A. tigrinum* host. At the close of the experiment, however, the graft optic cups accelerated in growth, reaching an average eye index of 0.77 instead of the 0.68 value observed after transplantation of an entire eye. At the same time, the *A. tigrinum* lenses in the chimeric eyes were retarded in their growth, so that they reached an average of only 84% of their host control diameter.

The transplantation experiments reviewed above reveal that the optic cup and the lens in chimeric eyes affect each other's growth rate. This is an important result as it indicates that parts of the eye—and presumably other organs as well—exchange signals that can accelerate or slow down the growth of parts so that together they form a well-proportioned organ. This mutual growth regulation between *eye components* is in stark contrast to the lack of growth regulation of the *entire eye*, which, upon transplantation, stubbornly grows according to its intrinsic potential.

Taken together, the transplantation experiments discussed in this section reveal the existence of intrinsic factors that impose species-specific limits on both the growth rate and the final size of organs. However, the establishment and maintenance of these limits may depend on the completeness of the organ and on interactions between its components. The same experiments show that the actual growth of organs also depends on extrinsic factors such as nutrients and other blood-borne components. As an example of such extrinsic factors, growth hormones will be discussed in the following sections.

28.3 Growth Hormones

Exogenous factors affecting vertebrate growth include several hormones. Among these, the one with the strongest and most widespread effects is known as somatotropin, or ***growth hormone* (*GH*)**. The action of GH was demonstrated most clearly in mice with a transgene encoding rat growth hormone. This experiment showed that an elevated level of growth hormone alone was sufficient for mice to grow to twice the mass of their littermates (see Figs. 15.1 and 15.23). At least in mammals, GH has little effect on fetal growth; transgenic mice with enhanced GH levels before birth grow like their normal littermates (Palmiter et al., 1983). However, GH is the single most important stimulus for postnatal growth. GH deficiencies cause dwarfism, whereas GH excess causes gigantism (Fig. 28.9).

Figure 28.9 Gigantism and dwarfism in humans. The photograph shows a normal-sized man (left) walking next to a "giant" (center), whose pituitary gland was overactive, and a "dwarf," whose pituitary was underactive.

GH is produced by the *anterior pituitary gland* under the control of ***growth hormone releasing hormone* (*GHRH*)** from the *hypothalamus* (Fig. 28.10). The action of GHRH is antagonized by *somatostatin*, another signal from the hypothalamus that acts on the anterior pituitary. GH from the anterior pituitary is carried with the blood flow to the heart, from where it is pumped throughout the body. GH acts on a wide range of target tissues including in particular the epiphyseal discs of long bones. Here, GH prompts the division and maturation of chondrocytes, thereby adding more cartilaginous material for bone formation. Muscle, bone marrow, liver, and kidney tissue also respond to GH.

Growth hormone acts both directly and indirectly. The direct effect of GH on its target cells is based on

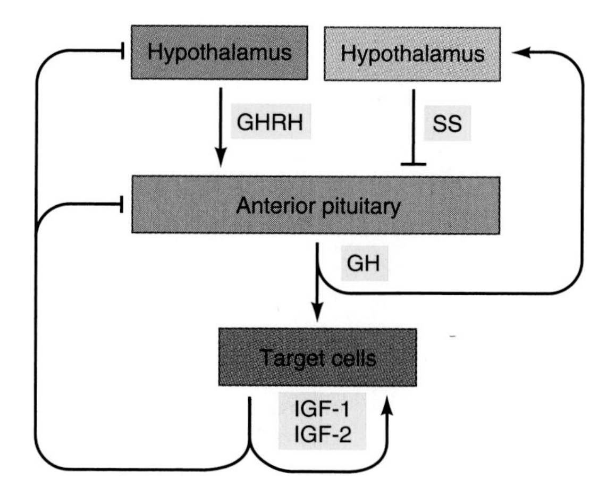

Figure 28.10 Synthesis and function of growth hormone (GH). GH is synthesized in the anterior pituitary gland, under antagonistic control of two hormones, growth hormone releasing hormone (GHRH) and somatostatin (SS). Both are produced in the hypothalamus and carried by a local system of blood capillaries to the anterior pituitary. GH is distributed over the entire body with the bloodstream and acts on a wide range of target cells both directly and indirectly. The direct action is based on steps that boost protein synthesis. The indirect action occurs through the release of two insulin-like growth factors (IGF-1 and IGF-2), which locally stimulate cell division.

promoting steps that are necessary for protein synthesis, such as amino acid uptake and RNA synthesis. The indirect effects of GH are mediated by two signal peptides known as *insulin-like growth factor 1* (*IGF-1*) and *insulin-like growth factor 2* (*IGF-2*). Both peptides stimulate cell division and act as autocrine factors on their producer cells or as paracrine factors on neighboring cells (see Fig. 2.24). IGF-2 is critical for early embryonic growth, whereas IGF-1 is necessary for fetal and postnatal growth. Knock-out mice lacking either of these factors show growth retardation (Tamemoto et al., 1994; M. F. Lopez et al., 1999). IGF-1 can substitute for GH in causing bone growth and weight gain; this was shown when IGF-1 was injected into rats whose pituitary glands had been removed so that their blood contained no endogenous GH (Schoenle et al., 1982). Together, the direct and indirect effects of GH promote the two processes necessary for hyperplasia—namely, cell division and protein synthesis to restore postmitotic cells to the previous size of their mother cell. In addition to GH and IGFs, thyroid hormone and insulin are also important in regulating the postnatal growth of mammals.

28.4 Growth Factors

In addition to ***growth hormone,*** which like other hormones is produced by a specialized endocrine gland, growth is stimulated by a wide range of so-called ***growth factors,*** which are produced by many tissues. Like many hormones, growth factors are signal peptides that interact with specific receptors present in the plasma membranes of target cells. Binding of a growth factor to its receptor starts a cascade of phosphorylations that eventually activate transcription factors (see Fig. 2.28).

While hormones by definition are carried all over the body by blood or equivalent fluids, growth factors typically spread by diffusion and act locally as *paracrine* or *autocrine* signals (see Fig. 2.24). This seemingly crisp distinction—like others in biology—is blurred by exceptions. For instance, the insulin-like growth factors mentioned in the previous section do act locally, but they are also carried by the bloodstream and participate in feedback loops involving the pituitary gland and the hypothalamus (Fig. 28.10). Likewise, the growth factors that stimulate liver regeneration after hepatectomy are carried by the bloodstream (see Section 20.2). Nevertheless, it is useful to think of growth factors as signal peptides that are produced by many types of cells, spread mostly by diffusion, and tend to act locally.

NERVE GROWTH FACTOR PROMOTES THE GROWTH AND DIFFERENTIATION OF CERTAIN NEURONS

The first growth factor to be discovered was called ***nerve growth factor*** (***NGF***) because it is necessary for the survival and differentiation of neurons. NGF was discovered in an exciting set of experiments carried out by Levi-Montalcini, Cohen, and Hamburger, mostly between 1950 and 1960 *(Levi-Montalcini, 1988).* Hamburger, a student of Spemann's, had a long-standing interest in the question of how neurons make selective contact with their appropriate target organs. In particular, he wanted to explore the concept that the survival and differentiation of neurons is controlled by signals emanating from their prospective targets.

In normal development, the neurons of *dorsal root ganglia* send dendrites outward to the periphery of the body and axons inward to the spinal cord (see Fig. 13.20). The rudiments of the segmental ganglia are initially similar in size, but the ganglia that innervate limbs retain a larger volume than those ganglia that innervate segments without limbs. This disparity is brought about mainly by cell death; the more target tissue there is for a ganglion to innervate, the more of its neurons survive. The same effect was observed after *heterotopic transplantation* of extra limb buds to chicken embryos: Near the ectopic limbs, the dorsal root ganglia came to be larger than normal (Hamburger and Levi-Montalcini, 1949). The investigators concluded that all dorsal root ganglia produce an excess of neurons, many of which will die by *default* unless they are rescued by growth-promoting signals spreading from their prospective target organs. The more target issue there is, the more neurons will survive. They termed the hypothetical factor emanating from the neuronal targets *nerve growth factor,* or *NGF.*

In order to determine the nature of NGF, the investigators set up a *bioassay,* in which dorsal root ganglia

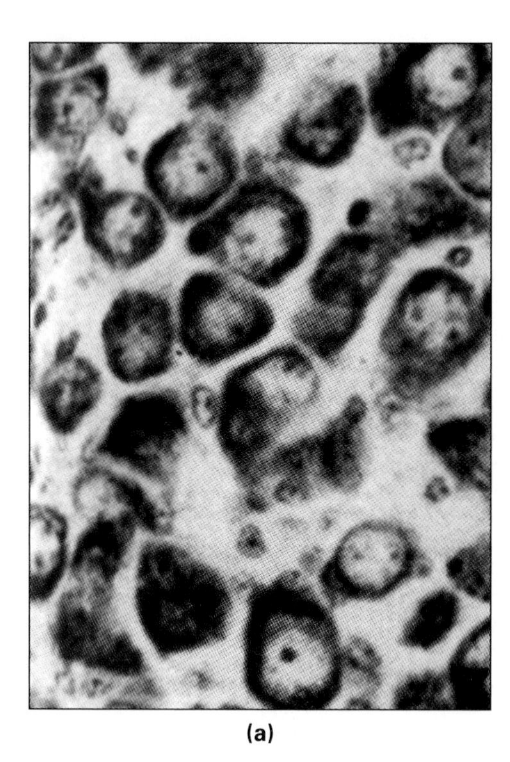

(a)

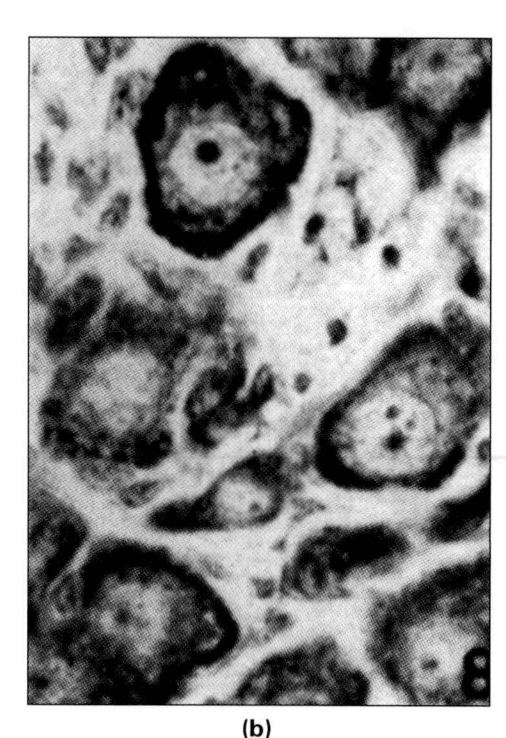

(b)

Figure 28.11 Increased size of neurons in the sympathetic ganglia of mice injected with purified nerve growth factor (NGF). The photographs show histological sections of sympathetic ganglia from two mouse littermates: **(a)** after injection with saline solution only and **(b)** after injection with the same solution containing NGF. Note that the neurons from the NGF-treated mouse are much larger.

were kept in vitro and exposed to biochemical fractions prepared from tissues that showed NGF activity upon heterotopic transplantation in vivo. To make fractionations feasible, the investigators needed tissue that showed high NGF activity and was available in larger quantities than the tiny limb buds of chicken embryos. They found that a particular tumor, a ***sarcoma*** (connective tissue tumor) designated S.180, showed a stunning level of NGF activity upon transplantation into chicken embryos: Nerve fibers from nearby ganglia grew into the tumor instead of innervating adjacent limb buds! Moreover, the ganglia that sent axons into the tumor grew to a larger than normal size. The tumor especially promoted the development of the dorsal root ganglia and the *sympathetic ganglia*—that is, the visceral ganglia forming chains on both sides of the vertebral column (see Figs. 13.20 and 13.27).

As a rapid bioassay, Levi-Montalcini established a technique of maintaining sympathetic and dorsal root ganglia in a semisolid culture medium. This allowed her to test the NGF activity of tissues as well as biochemical fractions by simply adding them to the culture medium. When cultured in the vicinity of S.180 tumor fragments, the ganglia produced a stunning halo of nerve cell processes within 10 h (see Fig. 28.1). Control ganglia cultured in the same way but without S.180 cells displayed only a sparse and irregular outgrowth of nerve fibers. Now the stage was set for testing biochemical fractions prepared from S.180 tissue.

Using the bioassay developed by Levi-Montalcini, Cohen (1960) tested fractions prepared from various tissues for their NGF activity—that is, for their ability to promote the formation of nerve fibers by cultured ganglia. He determined that NGF is a secreted protein that is produced by a wide range of normal tissues and tumors. Levi-Montalcini (1964) injected purified NGF into newborn mice and found that their sympathetic ganglia grew four to six times larger than those of untreated control mice. Cell measurements and cell counts showed that NGF promoted an increase in cell number as well as an increase in the size of individual neurons (Fig. 28.11). Apparently, NGF prevents or reduces the degree of programmed cell death that normally occurs in these ganglia. Injection of NGF did not seem to produce similar changes in *parasympathetic ganglia* or in the central nervous system.

In addition to increasing the number and size of neurons, NGF also has an orienting effect on the outgrowth of neurites (nerve cell extensions) from the sympathetic ganglia. When purified nerve growth factor was injected into the brains of newborn rats, nerve fibers sprouted from the sympathetic ganglia and invaded the brain and spinal cord (Levi-Montalcini and Calissano, 1979). Apparently, the NGF injected into the brain diffused through the motor and sensory roots of the spinal nerves and reached the sympathetic chain ganglia flanking the vertebral column, where the NGF induced the outgrowth of nerve fibers that are not seen in normal rats (Fig. 28.12). This finding implies that the tip of a sprouting sympathetic nerve fiber elongates along a pathway that is determined, at least in part, by a gradient of NGF. The tips then grow in the direction of increasing NGF concentration; this phenomenon is called the ***neurotropic effect*** of NGF (Gk. *tropikos*, "turning"). This effect was confirmed in vitro: In a setup where neurites growing out from cultured neurons

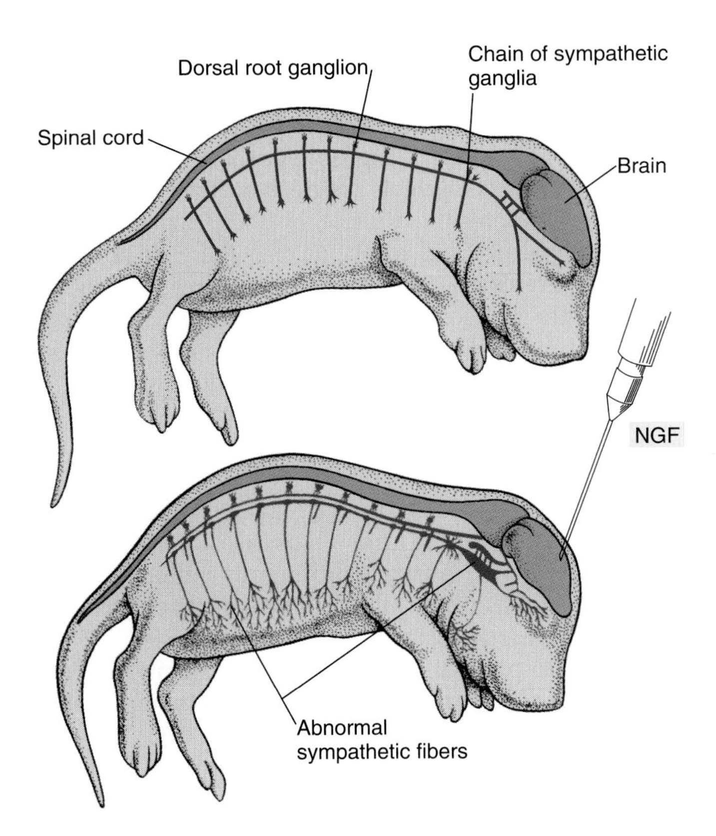

Figure 28.12 Neurotropic effect of nerve growth factor (NGF). NGF purified from mouse salivary glands was injected into the brains of newborn rats; control animals were injected with saline only. The injected fluid diffused into the spinal cord and from there into the spinal nerve roots and the sympathetic ganglia (see Fig. 13.20). In the NGF-treated animals (bottom), the sympathetic ganglia sent out abnormally long and numerous fibers that grew out into the periphery and back through the spinal nerve roots into the spinal cord and as far as the brain. In the saline-injected controls, the sympathetic ganglia sent out their normal complement of fibers into peripheral organs. The results indicate that NGF enhances the growth of sympathetic fibers and causes their orientation toward the highest NGF concentration.

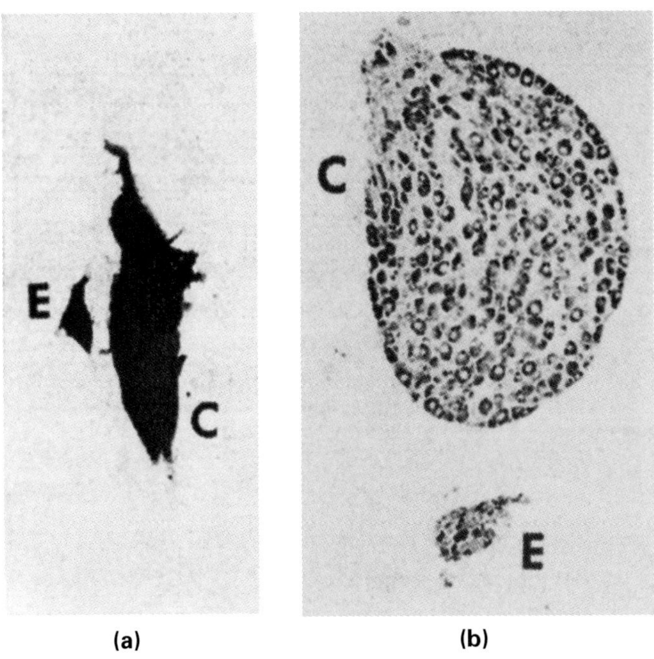

Figure 28.13 Inhibitory effect of antibodies against nerve growth factor on the growth and maintenance of sympathetic ganglia. Experimental (E) mice were injected with antibodies for 3 days after birth, whereas control (C) mice were injected with saline only. **(a)** Wholemounts and **(b)** histological sections of superior cervical ganglia were prepared after 3 months. The ganglia of the experimental mice were much smaller than those of control mice, and their neurons were fewer in number.

could grow into a chamber containing NGF and another chamber without NGF, neurites grew only into the chamber with NGF.

The experiments described so far showed that NGF is *sufficient* to promote the growth and differentiation of certain neurons. To test whether NGF is also *necessary* for this process, Cohen and Levi-Montalcini prepared rabbit anti-NGF antibodies and injected them into newborn mice. A month later, the sympathetic and sensory ganglia in the treated mice were smaller than in the controls. In particular, the sympathetic ganglia were barely visible and contained only a few neurons amounting to 3 to 5% of the normal cell population (Fig. 28.13). Similarly, anti-NGF antibody added to cultured sympathetic ganglion cells caused a massive degeneration of neurons, while the glial and connective tissue cells remained unaffected. The degenerative effect of anti-NGF antibody on neurons was overcome by adding more NGF (Cohen, 1960; Levi-Montalchini and Booker, 1960).

Joining their embryological and biochemical expertise in a brilliant series of experiments, Levi-Montalcini and Cohen showed that various tissues, both tumorous and normal, secrete a nerve growth factor protein that promotes the growth and differentiation of neurons. Nerve growth factor has two general functions. First, it is critical for the survival, growth, and differentiation of neurons in sympathetic and dorsal root ganglia. Second, it has a *tropic* effect that directs the fibers of these neurons to any target that releases it. The importance of these discoveries was acknowledged in 1986 when Cohen and Levi-Montalcini were awarded the Nobel prize in Physiology or Medicine.

Over the following years, Cohen and others characterized NGF as a dimer of two peptides. The effects of NGF described so far are mediated by a receptor with tyrosine kinase A (trkA) activity. *Knock-out mice* lacking the gene for the trkA receptor are lacking sympathetic and dorsal root ganglia, much like mice lacking the NGF gene and mice injected with anti-NGF antibody (Snider, 1994).

Unexpectedly, NGF can also promote programmed cell death, or *apoptosis*, and this effect is mediated by a different receptor, known as $p75^{NTR}$ (Frade and Barde, 1998). This effect of NGF was observed during embryonic development of the retina, when many cells die near the retina's center where axons converge to form the optic nerve. During this time, $p75^{NTR}$ is deployed by all dying cells, whereas the gene encoding the trkA receptor is not expressed (Frade et al., 1996). Thus, the two opposite effects of NGF are mediated by different receptors, and a cell's response to NGF depends on which receptor it deploys on its surface. Similarly distinct, and even paradoxical, effects of one growth factor are observed not only for NGF but also for other growth factors, as we will see next.

OTHER GROWTH FACTORS AFFECT MULTIPLE TARGET CELLS IN A VARIETY OF WAYS

Since the discovery of NGF, many other growth factors have been characterized. Like NGF, they are peptides that are released from cells and act on membrane-bound receptors in the producing cells and/or other target cells. Most growth factors diffuse over short distances, but some are bound to extracellular matrix components, where they can be stored in a localized form for extended periods of time.

Several peptides are termed "growth factors" because they promoted cell division in the studies in which they were first identified, even though further investigations showed that under different conditions the same factors may inhibit cell proliferation. In fact, some growth factors have additional effects that are unrelated to growth, such as embryonic induction (see Sections 9.7, 13.2, 14.2, 14.6, 21.5, and 23.5). Members of the transforming growth factor-β (TGF-β) family can stimulate fibroblasts to synthesize various ECM components and their receptors, including collagens, fibronectins, and integrins (Massagué, 1990). In certain cell types, TGF-β1 can even induce *apoptosis* (G. Fan et al., 1996).

Some growth factors act only on specific cells, while others have a wide range of target cells. NGF is an example of a growth factor with narrow specificity, as are erythropoietin and interleukins, which promote the proliferations of specific blood cells (see Section 20.4). Growth factors with broad specificity include the insulin-like growth factor mentioned in the previous section as well as most of the other growth factors reviewed below. Many of the growth factors with broad specificity are nevertheless named after the cell type that was first identified as a target, despite the fact that later they were found to act on other cell types as well. The following examples will illustrate this somewhat bewildering situation.

While he was still characterizing NGF, Cohen discovered another growth factor (Carpenter and Cohen, 1978). When newborn mice—whose eyes would normally be closed until 10 days after birth—were injected with incompletely purified NGF from mouse salivary glands, they opened their eyes a week ahead of schedule. Cohen found that the precocious eye opening was due to accelerated growth and keratinization of the epidermis of the eyelids. He ascribed this effect to an ***epidermal growth factor* (*EGF*),** which was characterized later. EGF acts on a variety of cell types and may promote or inhibit growth. For instance, EGF promotes the growth of mammary gland ducts, which are derived from epidermis, but also stimulates the division of fibroblasts, which are completely unrelated to epidermal cells. In addition, EGF accelerates the maturation of the pulmonary epithelium and is necessary for the formation of the hard palate. On the other hand, EGF inhibits the proliferation of hair follicle cells and certain cancer cells.

***Fibroblast growth factor* (*FGF*)** stimulates the proliferation of many cell types, including fibroblasts and endothelial cells. Along with EGF, FGF also promotes the division of *myoblasts* before they fuse to form multinuclear *myotubes* (Olwin and Hauschka, 1988). On the other hand, FGF has an antiproliferative effect on some tumor cells. Basic fibroblast growth factor is also involved in mesoderm induction (see Section 9.7). Some FGFs bind to proteoglycans as signals that are frozen in time until a cell contacts them or releases them by enzymatic action.

Yet another growth factor is known as ***platelet-derived growth factor* (*PDGF*),** because platelets release it during blood clotting. In culture, PDGF stimulates the division of two related cell types, fibroblasts and smooth muscle cells. In vivo, this process seems to facilitate wound closure. PDGF also stimulates the proliferation of glioblasts and prevents their premature differentiation. Moreover, EGF and PDGF cooperate with insulin-like growth factors to stimulate the proliferation of fat cells and connective tissue cell precursors.

A large superfamily of more than 20 growth factors is known as the ***transforming growth factor-β* (*TGF-β*)** superfamily (Massagué, 1996). Its first member, TGF-β1, which was found to transform cultured cells from normal to tumorous growth. The mitogenic effect of TGF-β1 and other TGF-β factors seems to be caused indirectly through the stimulation of other growth factors. On the other hand, TGF-β inhibits the growth of various tumors and nontumorous epithelia, and again, it appears that TGF-β exerts this effect by acting upon other molecules that control the cell cycle (Moses et al., 1990).

Some of the most intriguing members of the TGF-β superfamily, including the decapentaplegic, activin, Vg1, nodal, and dorsalin-1 proteins, are involved in pattern formation. In *Drosophila,* decapentaplegic controls the formation of dorsal structures in the embryo, the patterning of the gut, and the growth and patterning of many imaginal discs in the larva (see Section 22.6 and 22.8). In *Xenopus,* Vg1 and activin may be involved in the induction and patterning of mesoderm (see Sections 9.7 and 21.5). In the mouse, nodal seems to be involved in mesoderm formation and axis organization (Zhou et al., 1993).

These examples highlight three striking features of members of the TGF-β superfamily. First, many TGF-β members are involved in setting up polarity axes and specifying embryonic pattern elements. Second, a given family member may play multiple roles in a given organism. Third, TGF-β members show up in these functions throughout the animal kingdom. It appears that these proteins are extremely versatile and effective, and that the genes encoding them have been co-opted—through the addition of new control elements—for a wide spectrum of functions in development.

THE EFFECTS OF A GROWTH FACTOR MAY DEPEND ON THE PRESENCE OF OTHER GROWTH FACTORS

Even in a single cell type, the action of a growth factor may depend on the context set by other growth factors. For example, TGF-β stimulates growth of certain cultured fibroblasts in the presence of PDGF but inhibits their growth if EGF is present (A. B. Roberts et al., 1985). TGF-β can also stimulate replication of *osteoblasts,* but this proliferative effect can be reversed by EGF and FGF (Centrella et al., 1987). Similarly, IL-4 may either enhance or antagonize the growth of an entire group of hematopoietic progenitor cells, depending on the presence of other growth factors, notably interleukin-3 (IL-3) (Rennick et al., 1987). The action of a growth factor may also depend on the differentiated state of a target cell. For instance, TGF-β stimulates cartilage formation early in the development of embryonic mesenchyme and chondroblasts but inhibits it later. In accord with the multiple and context-dependent actions of growth factors, most cells have receptors for several growth factors (Cross and Dexter, 1991).

On the basis of these observations, growth factors seem to be part of a complex cellular signaling language, in which individual growth factors are the equivalents of the letters that compose words. According to this analogy, informational content lies not in an individual growth factor, but in the entire set of growth factors and other signals to which a cell is exposed. The ways in which growth factors exert their combinatorial effects are becoming clearer as the molecular mechanisms of growth factor action are being investigated (*Alevizopoulos and Mermod, 1997*). Growth factors bind to plasma membrane receptors, and the ligand's binding to the receptor's extracellular domain activates the receptor's cytoplasmic domain (Fig. 28.14). The latter is often a tyrosine kinase (see Fig. 2.28) or a serine/threonine kinase. The subsequent chain of signal intermediates eventually controls the activity of transcription factors, which in turn act on groups of target genes. Along these transduction pathways, there are several levels at which signals may branch and reinforce or antagonize one another.

With regard to the classical transplantation experiments described earlier in this chapter, it seems that growth factors and the associated signaling pathways are the best candidates for being those "intrinsic factors" that let limbs grow with species-specific timing and to certain sizes, and allow parts of an organ to adjust their relative growth rates so that a well-proportioned, functional organ results.

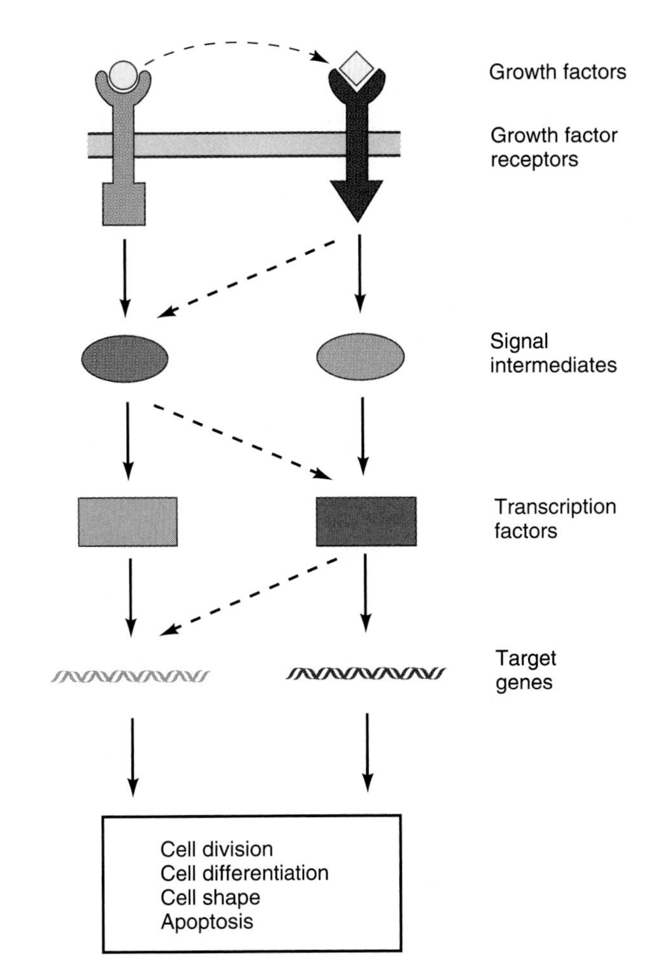

Figure 28.14 Combinatorial action of growth factors. See text.

28.5 Mechanical Control of Cell Survival and Division

While growth hormone and growth factors are chemical signals exchanged between cells, there are also indications that cells depend on mechanical signals for their ability to survive and to divide. Most cells kept in culture show a phenomenon called *anchorage dependence.* They survive only if they adhere to a substratum, such as the bottom of a cell culture dish. If deprived of their anchorage by being kept in suspension, normal cells undergo apoptosis. Anchorage dependence of mammalian cells is mediated, at least in part, by integrins and the activation of the apoptosis-inhibiting gene *bcl-2*$^{+}$ (*Meredith and Schwartz, 1997*). An apparently related property of cultured cells is known as *contact inhibition.* If kept in culture under favorable conditions, they will divide and grow until they begin to touch one another with their dynamic extensions (Fig. 28.15a). At this point, normal

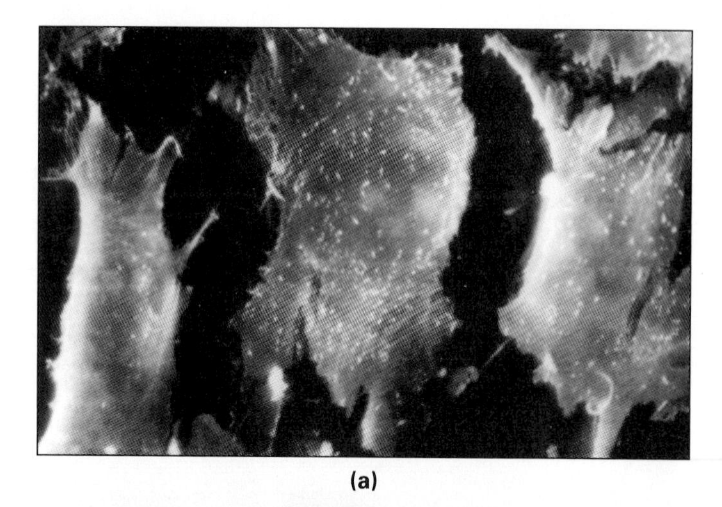

(a)

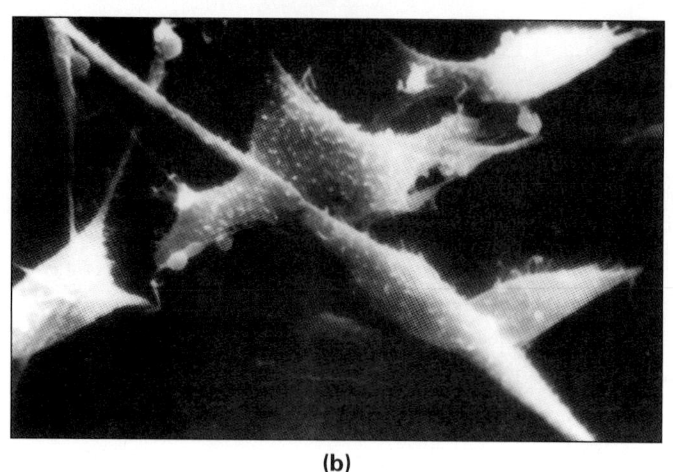

(b)

Figure 28.15 Normal and transformed mouse fibroblasts viewed under the scanning electron microscope. **(a)** Normal cells lie flat on the bottom of the culture dish and stop dividing when they cover most of the bottom as a single cell layer (monolayer). **(b)** After mutations in genes that control the cell cycle, the same cells become independent of contact with a substratum and of the presence of growth factors in the medium. They become round or spindle-shaped and grow on top of each other, and they may divide indefinitely.

cells stop dividing even if sufficient nutrients are available. In contrast, genetically transformed cells may keep dividing and pile on top of one another.

Not only the avoidance of apoptosis, but also the ability of cells to divide depends on anchorage. This was first observed by Folkman and Moscona (1978), who cultured cells on Plexiglas of graded stickiness, on which cells would spread more or less thinly. They found that the degree to which the cells were spread out was closely correlated with their rate of DNA synthesis and cell division. In related experiments, Renshaw and coworkers (1997, 1999) found that activation of MAPK, a key component of the MAP kinase pathway, which transmits the mitogenic (mitosis-stimulating) effect of many growth factors, depends on integrin-mediated cell adhesion.

A problem with the design of the experiments reviewed above was that one could not distinguish between the effects of cell adhesion and cell shape. Were the cells dividing more actively because they had formed a greater area of contact with the substrate or because they assumed a flatter shape? To avoid this ambiguity, C. S. Chen and coworkers (1997) cultured cells on micropatterned surfaces with adhesive islands that were coated with extracellular matrix. By varying the size and spacing of these islands, the investigators could measure the effects of cell contact area (sum area of all adhesion islands covered by a cell) and cell flatness (total surface area covered by a cell if viewed from above). Their results show clearly that the avoidance of apoptosis and the rate of cell division improve with a flat overall cell shape, not with the total cell contact area.

Some ways in which a flat cell shape can promote cell division have been explored. Vascular endothelial cells have ion channels that are activated by mechanical shear stress (Olesen et al., 1988). Mechanical strain on smooth muscle cells release platelet-derived growth factor, which stimulates cell division as an autocrine effect (E. Wilson et al., 1993). These results confirm the general concept of *morphoregulatory cycles,* and specifically the idea that cell shape, presumably via cytoskeletal signals, can feed back on gene activity (see Fig. 11.25).

The critical importance of anchorage dependence and contact inhibition is revealed dramatically by cells that have escaped these controls. The ability to divide and grow in an unrestricted manner in soft agar is one of the hallmarks of tumor cells.

28.6 Cell Cycle Control

How do growth hormones and growth factors control growth? Studies on this subject have focused mostly on *hyperplasia,* and thus on the control of the cell cycle. Such investigations are typically carried out on cultured cell lines that can proliferate indefinitely. Although such cell lines have been selected for survival in vitro, they are widely regarded as models for the behavior of cells in vivo. A commonly used line of cells, known as NIH 3T3 cells, is derived from mouse fibroblasts (Fig. 28.15). These cells proliferate most rapidly when settled on the bottom of a culture dish and covered with a medium containing blood *serum* (the liquid that rises to the top when clotted blood is centrifuged). Serum contains not only nutrients for the cells but also growth factors. The growth factors are present in serum in concentrations of about 10^{-10} M; when they are used up, normal cells stop dividing, whereas tumorous cells keep growing.

As described in Section 2.4, the cell cycle is divided into four major phases known as M (mitosis), G_1 (RNA synthesis), S (DNA synthesis), and G_2 (RNA synthesis). The longest phase is G_1, during which the cell grows in size and synthesizes most of the components necessary for DNA replication and mitosis. When certain conditions, such as nutrients or growth factors, are lacking, cells pause at a so-called ***restriction point*** **(*R*)** located between G_1 and S. Once a cell has moved beyond R, the cell is committed to entering S, G_2, and M. Another restriction point between G_2 and M is important for embryonic cells, but less so in postembryonic cells.

Nondividing cells, such as cultured cells deprived of serum, enter a modified G_1 state known as G_0. If such cells are supplied with fresh serum, they can return from G_0 back to G_1. In multicellular organisms, many cells enter the G_0 state as part of their terminal differentiation, and most of these cells never divide again. The overall growth of an adult tissue by hyperplasia depends on how many of its cells are "parked" in G_0, how many are cycling, and which proportion of them die per unit time.

Whether a cycling cell passes the R point before S phase depends primarily on the assembly of active *cyclin-dependent kinases* (*CDKs*). This requires phosphorylation of certain amino acid residues, and the dephosporylation of others, in these CDKs (see Fig. 2.16). In addition, the CDKs must combine with matching *cyclins*—that is, labile proteins present during specific phases of the cell cycle. We will refer to cyclins, kinases, phosphatases, and other proteins required to assemble active CDKs collectively as ***cell cycle proteins.*** In a wider sense, this term also encompasses proteins required for DNA synthesis, such as thymidine kinase and DNA polymerases.

The regulatory mechanisms for the timely supply of cell cycle proteins are complex and still under investigation (Sherr, 1996). A simplified scheme is shown in Figure 28.16. A critical first step is the binding of one or more growth factors to their matching receptors, which are transmembrane proteins. (In addition, the synthesis of cell cycle proteins also depends on other signals that indicate the availability of nutrients and the successful repair of damage to DNA.) Binding of a growth factor activates its receptor, which in turn may activate other membrane-associated proteins. These proteins are kinases that phosphorylate specific amino acid residues in cytoplasmic signaling proteins such as the RTK-Ras-MAPK pathway (see Fig. 2.28). This cascade of kinases eventually activates transcription factors, which in turn regulate genes encoding cell cycle proteins.

Whereas growth factors and their downstream signaling components have the overall effect of accelerating the cell cycle, other proteins have the opposite effect of slowing down the cell cycle. These proteins include *p53,* which inhibits CDKs, and the *retinoblastoma protein* (*Rb*), which sequesters transcription factors required for the synthesis of cell cycle proteins.

The molecular mechanisms that control the cell cycle are of great importance for a better understanding of normal organismic growth and the design of cancer therapies. One approach taken by investigators has been the study of genes that are mutated in tumors of

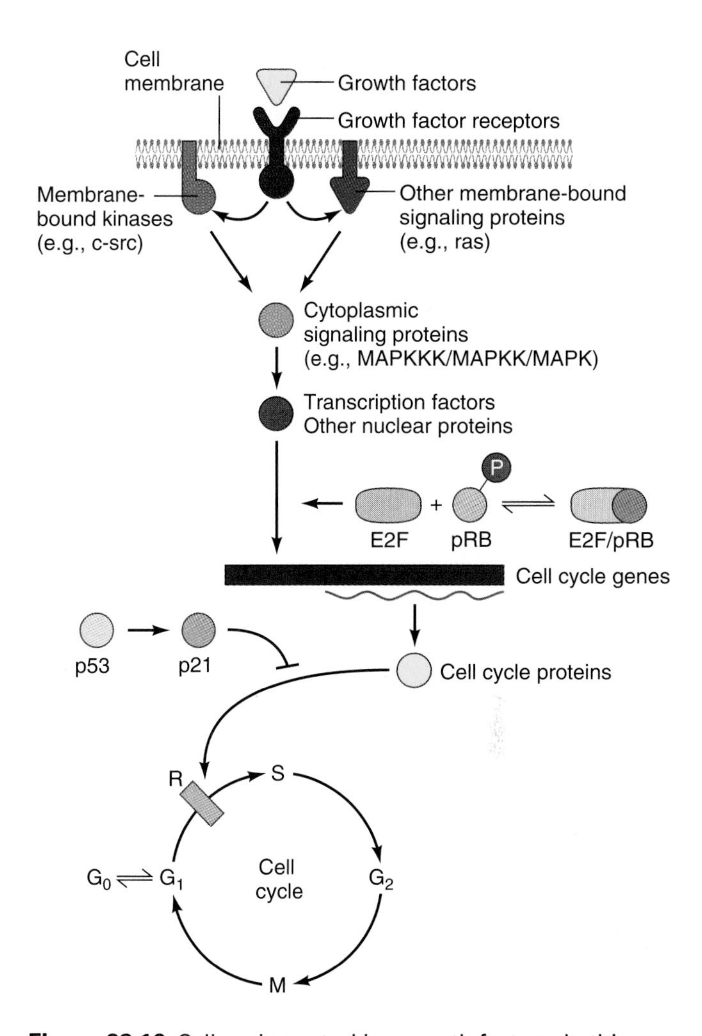

Figure 28.16 Cell cycle control by growth factors. In this simplified scheme, the colored components are those that transfer the signal for the cell to divide (mitogenic signal). Growth factors bind to receptor proteins located in the cell membrane. The bound receptors activate membrane-associated tyrosine kinases, such as the cellular src protein, or other membrane bound signaling proteins, such as ras. These in turn activate signaling proteins located in the cytoplasm, such as the MAPKKK/MAPKK/MAPK cascade (see Fig. 2.28). These proteins then transmit the mitogenic signal to transcription factors and other proteins that promote the expression of cell cycle genes. The products of these genes control the progress of the cell past a restriction point (R) in the cell cycle between G_1 and S phase. They also facilitate the return of cells from the quiescent G_0 state to the cycling G_1 state. The cell cycle proteins are counteracted by tumor suppressor proteins (gray shading). One of these, known as pRB and active in its hypophosphorylated form, sequesters transcription factors (E2F) required for the expression of cell cycle genes. Another tumor suppressor protein, p53, activates the synthesis of another protein, p21, which inhibits cyclin-dependent kinases.

humans and other vertebrates. Indeed, the genes involved in tumorigenesis turn out to be the same genes that also control normal organismic growth.

28.7 Tumor-Related Genes

Cells that escape from their normal growth controls develop into ***tumors*** (Lat. *tumor,* "a swelling"), or ***neoplasms*** (Gk. *neo-,* "new," "recent"; *plasma,* "a thing formed"). Tumors may be *benign* (Lat. "mild") or *malignant* (Lat. "injurious"). Malignant tumors are also called cancers. Besides growing unrestrained, cancers invade other tissues and give off cells that spread via the circulatory system and start additional cancers elsewhere. This chapter will focus on the genetic analysis of tumor formation.

There are two classes of normal eukaryotic genes that can mutate or become translocated so that they promote tumor formation. These two classes are distinguished by the nature of their tumor-causing mutant alleles (Bishop, 1991; T. Hunter, 1991, 1997; *Weinberg, 1996a*). Genes of the first class, known as ***proto-oncogenes***, give rise to dominant *gain-of-function alleles* that cause deregulated growth. These genes generally encode proteins that accelerate the cell cycle (Fig. 28.16). Genes of the second class, called ***tumor suppressor genes***, have tumor-causing *loss-of-function alleles,* which are typically recessive. Thus, both copies of a tumor suppressor gene in a diploid cell must be mutated for the tumorous phenotype to appear. Tumor suppressor genes generally encode proteins that slow down the cell cycle (Fig. 28.16). The examination of tumors often reveals the loss or inactivation of both copies of a tumor suppressor gene and changes in one or more proto-oncogenes.

The term "proto-oncogene" is derived from the term ***oncogene*** (Gk. *oncos,* "mass") and reflects the circuitous way in which proto-oncogenes were discovered. Oncogenes are defined operationally as genes that can transform cultured cells to anchorage independence and uncontrolled growth (Fig. 28.17). The first known oncogenes were transmitted by RNA viruses, and it took an ingenious analysis to show that these oncogenes were *not* intrinsic to their viral carriers. Instead, they were eukaryotic mRNAs that were included erroneously in viral RNA and became mutated or deregulated in the process. Other oncogenes that were identified later are not virus-borne but are simply mutant alleles of normal cellular genes involved in growth control. To emphasize the allelic relationship between oncogenes (virus-borne or not) and their unaltered cellular counterparts, the latter are called proto-oncogenes.

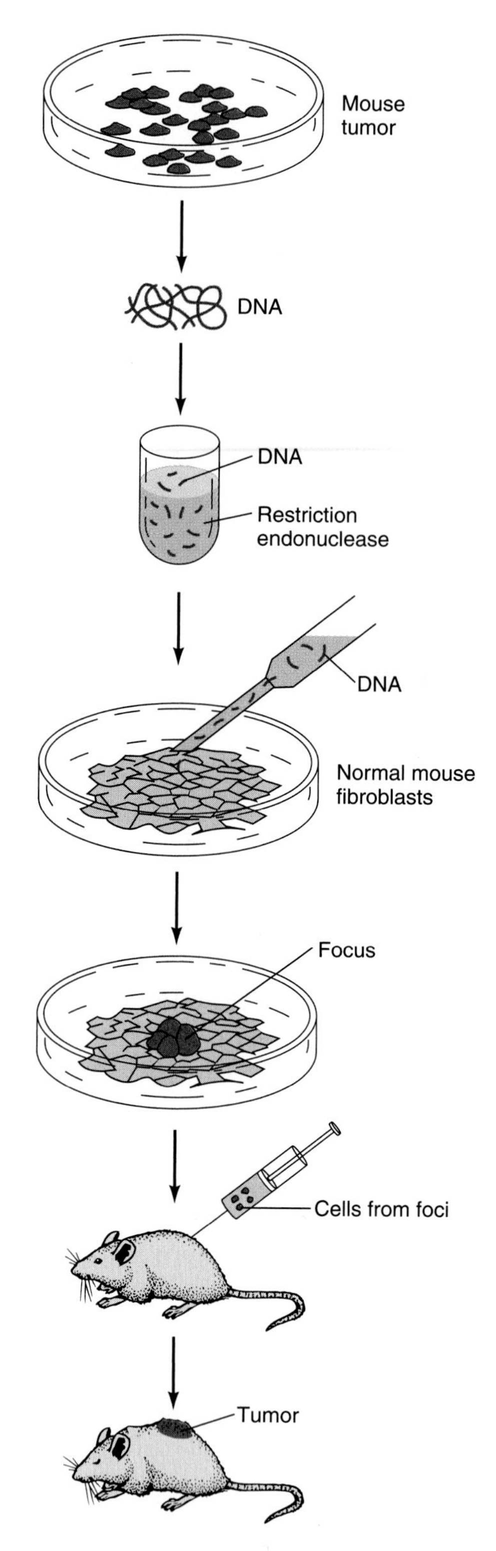

Figure 28.17 Genetic transformation of normal mouse fibroblasts into tumor cells. DNA from a mouse tumor is extracted and cut into large fragments with a restriction endonuclease. These DNA fragments are added to nontumorous mouse fibroblasts, and the cells are subjected to a treatment that promotes the uptake of exogenous DNA. As a result, some of the recipient cells grow into small tumors called foci (sing., focus). Cells from such foci are shown in Figure 28.15b. When transformed cells are transplanted to mice, the cells continue to grow and may kill the host.

ONCOGENES ARE DEREGULATED OR MUTATED PROTO-ONCOGENES

The existence of tumor viruses was first suspected at the start of the twentieth century. A critical discovery was made in 1910 by Francis Peyton Rous, who showed that a cell-free extract from a chicken sarcoma (tumor of connective tissue), when injected into healthy chickens, caused them to develop sarcomas. However, his work was not well received, and because of his peers' disapproval, Rous abandoned this line of work. Decades later, when other researchers established the existence of tumor viruses beyond doubt, Rous was vindicated. In 1966, at the age of 85, he was awarded the Nobel prize. The virus he discovered is now known as the *Rous sarcoma virus* (*Rous SV*).

The Rous SV belongs to the class of ***retroviruses,*** which have an RNA genome. Their name derives from a unique feature of their life cycle: Their genomic RNA must be reverse-transcribed into DNA in order for these viruses to propagate (Varmus, 1982). This unusual step is accomplished by an enzyme, *reverse transcriptase,* which synthesizes single-stranded DNA on a single-stranded RNA template (see Fig. 15.11). Thereafter, the RNA template is replaced with a second DNA strand, and the double-stranded DNA is inserted into the genomic DNA of the host cell. The life cycle of Rous SV, which is a typical retrovirus, is diagrammed in Figure 28.18. If the virus carries only its own genes, the host cell may survive the virus infection, provided that the insertion of the viral DNA into the host genome does not interrupt a critical host gene. However, the retroviral DNA may also contain a wayward segment of the host's genomic DNA or a cDNA that has been reverse-transcribed from an mRNA in the host cytoplasm. In either case, the host gene that has become part of the virus is now transcribed at the same level as the viral genes, and the resulting overexpression may be deleterious to the host. Alternatively, the eukaryotic DNA added erroneously to the viral DNA may become mutated in the process. Either overexpression or mutation may derange the host cell's cycle control, in which case the viral copy of the host gene would be called an oncogene.

The oncogene of the Rous SV was the first oncogene to yield to experimental analysis (Bishop, 1985, 1989). Investigators cut the Rous SV DNA into *restriction fragments* and tested each fragment for the ability to transform normal fibroblasts into cancer cells. This method revealed a gene at the end of the Rous SV DNA, which

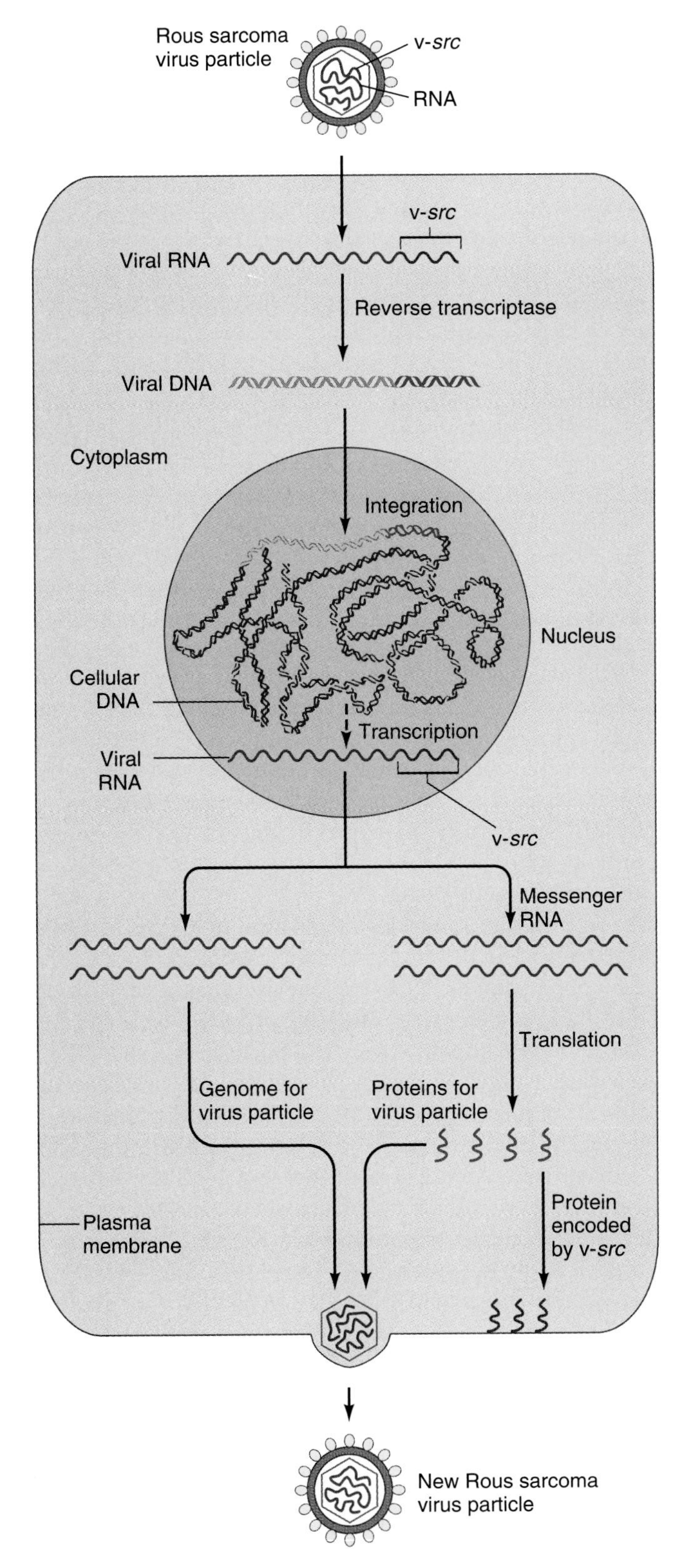

Figure 28.18 Life cycle of the Rous sarcoma virus (Rous SV), which belongs to the class of retroviruses. When a retrovirus infects a eukaryotic host cell, an enzyme known as reverse transcriptase is translated from the single-stranded RNA that serves as the viral genome. The enzyme reverse-transcribes the viral genome into double-stranded DNA. The DNA is integrated into the host's genomic DNA, so that the viral genes are transcribed along with the active host genes. Some of the viral RNA transcripts provide the genomes for a new generation of viruses, while others are translated into viral proteins in the host cell cytoplasm. The viral RNA and proteins are then assembled into new virus particles that are eventually released from the host cell. The Rous SV contains, in addition to its own genes, the v-*src* gene, which is a mutant allele of the eukaryotic gene, c-*src.* While the c-src protein is a ligand-dependent protein kinase, the v-src kinase is active with or without its ligand. This constitutive activity transforms the host cell into a cancer cell.

was named the viral *src* gene, or **v-*src*.** The gene encodes a protein with a plasma membrane-associated *tyrosine kinase,* an enzyme that adds phosphate groups specifically to tyrosine residues of proteins. The greatest reward from the work on the v-*src* gene came with the identification of its eukaryotic counterpart, the src^+ proto-oncogene, or cellular src^+ gene, abbreviated **c-src^+.**

ACCORDING to a hypothesis originally proposed by Huebner and Todaro (1969), oncogenes are part of the normal genome of eukaryotes, and they are harmless unless they are overexpressed or mutated. To test this hypothesis, a research team consisting of Dominique Stehelin, Harold Varmus, J. Michael Bishop, and Peter Vogt (1976) set out to explore whether the genome of healthy chickens contained a gene that was similar to v-*src*. They reasoned that if normal chicken DNA contained a homolog of v-*src*, then single-stranded v-*src* DNA should hybridize to single-stranded genomic DNA from a normal chicken.

To make this test as stringent as possible, they isolated v-*src* so that it would represent exactly the viral DNA that was present in the Rous SV but not in a strain of the same virus that contained only the standard genes of retroviruses and did not cause sarcoma in chicken. To this end, they reverse-transcribed Rous SV RNA in the presence of radiolabeled deoxyribonucleotides to obtain single-stranded cDNA representing the entire genome of the Rous SV (Fig. 28.19). The labeled cDNA was cut into fragments after its RNA template had been hydrolyzed. Next the cDNA fragments were hybridized with an excess of RNA from the nononcogenic virus. Any cDNA fragment that was complementary to a part of this RNA was now present as a cDNA/RNA hybrid. In contrast, any cDNA fragments complementary to v-*src* would find no RNA to hybridize to and would remain single-stranded. Thus, a column chromatography that separates single-stranded from double-stranded nucleic acids was all that was needed to isolate these nonhybridized cDNA fragments—that is, the labeled cDNA fragments representing v-*src*.

Now the investigators were in a position to answer their original question. Would the isolated v-*src* hybridize with single-stranded genomic DNA from healthy chickens? They found that this was the case. However, there were two major differences between the v-*src* gene and its c-src^+ hybridization partner. First, like most eukaryotic genes, the c-src^+ gene contains several introns (Fig. 28.20). Second, in comparison to the combined c-*src* exons, the v-*src* DNA showed mutations that make the encoded v-src protein constitutively active—that is, active without being stimulated by an appropriate cellular signal. In contrast, the c-src protein is folded into a conformation that sequesters its tyrosine kinase domain unless the protein is activated by a loaded growth factor (F. C. Williams et al., 1997). The unregulated activity of v-src protein makes a cell behave as if it were constantly stimulated by a growth factor. This rea-

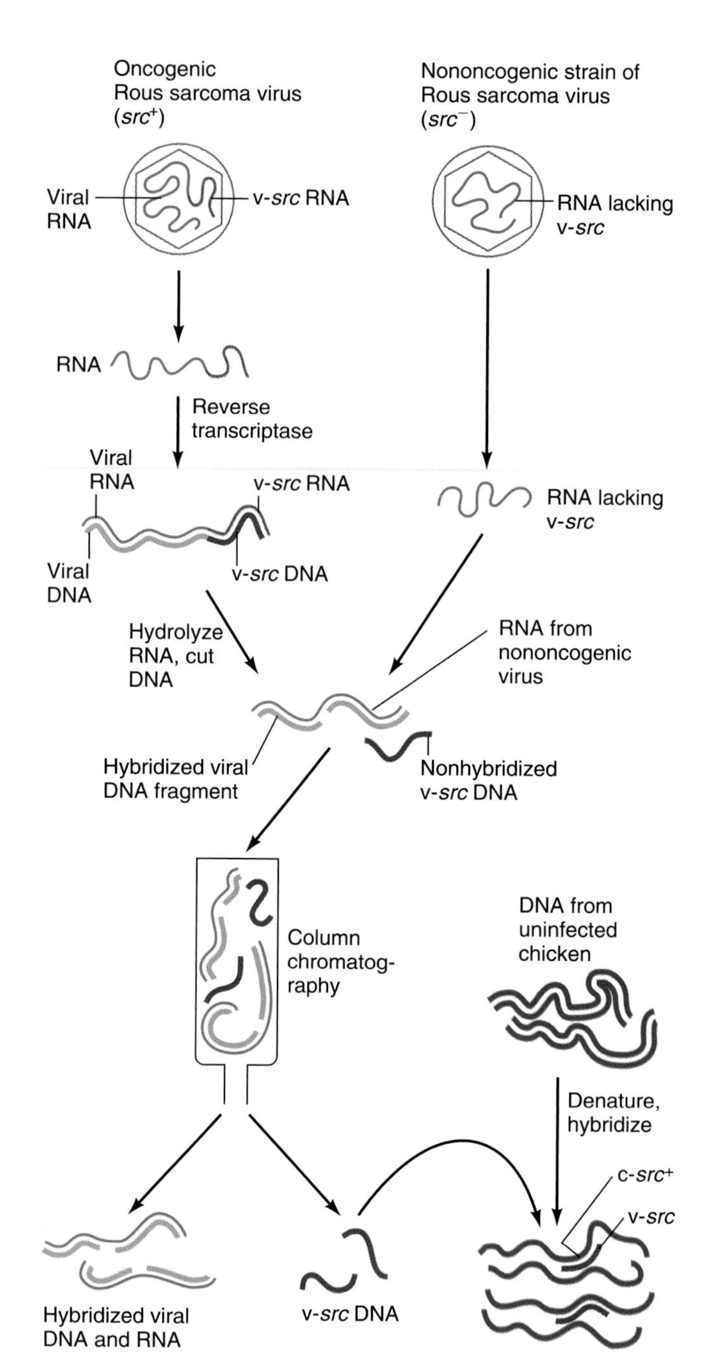

Figure 28.19 Demonstration that the v-*src* gene from the Rous sarcoma virus (Rous SV) has a homolog in the genomic DNA of healthy chickens. RNA from Rous SV carrying the v-*src* gene was reverse-transcribed into single-stranded, labeled DNA. After hydrolysis of the RNA template and fragmentation of the DNA transcript, the labeled DNA fragments were hybridized with RNA extracted from a nononcogenic strain of Rous SV that lacked the v-*src* gene. Most of the DNA hybridized with the deletion-mutant RNA, but the v-*src* DNA fragments found no RNA partners. The hybrids were separated by column chromatography from the single-stranded DNA, which now represented only the v-*src* gene. This probe was mixed with denatured genomic DNA from a normal chicken. The probe hybridized, revealing that DNA from normal chicken cells contained a src^+ proto-oncogene, or cellular src^+ gene (c-src^+).

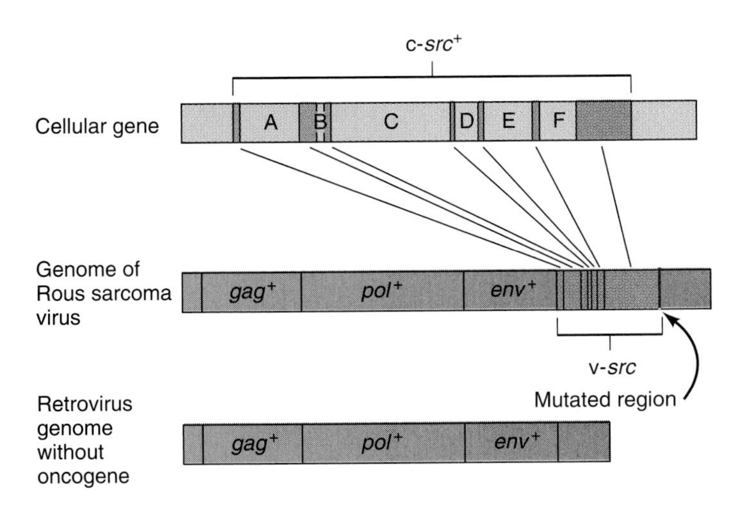

Figure 28.20 Comparison of the cellular *src*$^{+}$ proto-oncogene (c-*src*$^{+}$, shown at the top) with the viral *src* oncogene (v-*src*, shown in the middle). The c-*src*$^{+}$ gene is split by several introns (A to F), whereas its v-*src* homolog is not. The complete genome of the oncogenic Rous sarcoma virus (SV) contains v-*src* and three genes encoding proteins for the viral capsid (*gag*$^{+}$), reverse transcriptase (*pol*$^{+}$), and the spikes of the viral envelope (*env*$^{+}$). The nononcogenic Rous SV strain shown at the bottom does not contain the v-*src* gene. The v-*src* gene is a mutant allele of c-*src*$^{+}$, such that the v-src protein's kinase activity can no longer be regulated. As a result, the v-src protein is constitutively active, transforming its host cell into a cancer cell.

soning explained the oncogenic potential of v-*src*. Additional experiments showed that genomic DNA from other birds as well as fishes and mammals all contained c-*src* homologs that hybridize to the v-*src* probe.

Questions

1. Why did the investigators radiolabel the cDNA reverse-transcribed from Rous SV RNA?
2. DNA can be prepared for electron microscopy so that single strands can be discerned. If this technique were applied to the hybrids of v-*src* cDNA and c-*src* genomic DNA shown in the lower right-hand corner of Figure 28.19, how would they look under the microscope?

The seminal study of Stehelin and coworkers (1976) showed that the v-*src* gene isolated from the Rous SV has close homologs in the normal genomes of all vertebrates. It appears that the Rous SV at some point integrated a c-*src*$^{+}$ gene copy from a host cell. This could have been an actual gene from which all introns were subsequently lost. More likely, the eukaryotic DNA that hitched a ride in the virus was a cDNA transcript made from host mRNA by the viral reverse transcriptase. Subsequent research has revealed cellular proto-oncogenes for almost all retroviral oncogenes.

Establishing the basic relationship between oncogenes and proto-oncogenes was a significant step to a better understanding of both cancer and normal growth control. The transformation of cells from normal to tumorous growth provides a straightforward bioassay for the isolation of oncogenes, which in turn can be used as probes to isolate the corresponding proto-oncogenes. This strategy has revealed many factors in the network of growth control that were previously unknown. For their pioneering work in this area, Bishop and Varmus received the Nobel prize in Physiology or Medicine in 1989.

PROTO-ONCOGENES ENCODE GROWTH FACTORS, GROWTH FACTOR RECEPTORS, SIGNAL PROTEINS, TRANSCRIPTION FACTORS, AND OTHER PROTEINS

The isolation and analysis of proto-oncogenes has confirmed and extended the general scheme of cell cycle control that had previously been based on the behavior of normal fibroblasts (see Fig. 28.16). The proteins encoded by proto-oncogenes generate the ***mitogenic signals*** prompting cells to divide. Mitogenic proteins can be classified as growth factors, growth factor receptors, membrane-bound signaling proteins, cytoplasmic signaling proteins, and nuclear proteins, including transcription factors (Table 28.2).

Growth Factors Several proto-oncogenes encode secreted proteins that are known growth factors or are likely to be growth factors (Cross and Dexter, 1991). They include the *sis*$^{+}$ gene encoding the B chain of platelet-derived growth factor (PDGF) and the *FGF-5*$^{+}$ gene encoding a protein related to fibroblast growth factor (FGF). Proto-oncogene products that are likely growth factors include int-2, which belongs to the fibroblast growth factor family. The *int-2*$^{+}$ gene causes mouse mammary tumors when it is hyperactivated, and thus turned into an oncogene, by the nearby insertion of a virus.

Growth Factor Receptors Other proto-oncogenes encode growth factor receptors (T. Hunter, 1991). Their oncogene homologs encode defective receptors that constantly give off the mitogenic signal that a normal receptor generates only when bound by a ligand. For example, the normal receptor for epidermal growth factor (EGF) has low tyrosine kinase activity when no ligand is bound, and much more tyrosine kinase activity when EGF is bound. The viral oncogene v-*erbB* encodes a truncated version of the EGF receptor. It has a constitutively high tyrosine kinase activity, which stimulates the cell to divide continuously. Another proto-oncogene that encodes a tyrosine kinase receptor is *trk*. The *trk* oncogene was discovered in a human colon cancer, but expression studies in mice showed that transcription of the *trk* proto-oncogene is limited to cranial sensory ganglia and dorsal root ganglia (Martin-Zanca et al., 1990). These ganglia respond to nerve growth factor (NGF),

table 28.2 Functions of Proto-oncogene Products

Proto-oncogene	Product
Class 1: growth factors	
sis$^+$	PDGF B-chain growth factor
int-2$^+$	FGF-related growth factor
hst$^+$ (KS3)	FGF-related growth factor
Class 2: growth factor receptors	
erbB$^+$	EGF receptor protein—tyrosine kinase
neu$^+$	Receptorlike protein—tyrosine kinase
fms$^+$	CSF-1 receptor protein—tyrosine kinase
met$^+$	Receptorlike protein—tyrosine kinase
trk$^+$	NGF receptor
mas$^+$	Angiotensin receptor, not a protein kinase
Class 3: membrane-bound signal proteins	
H-ras$^+$	Membrane-associated GTP-binding/GTPase
K-ras$^+$	Membrane-associated GTP-binding/GTPase
src$^+$	Nonreceptor protein—tyrosine kinase
yes$^+$	Nonreceptor protein—tyrosine kinase
fgr$^+$	Nonreceptor protein—tyrosine kinase
lck$^+$	Nonreceptor protein—tyrosine kinase
Class 4: cytoplasmic signal proteins	
MAPKs$^+$	Mitogen-activated protein kinases
pim-1$^+$	Cytoplasmic protein—serine kinase
mos$^+$	Cytoplasmic protein—serine kinase (cytostatic factor)
cot$^+$	Cytoplasmic protein—serine kinase(?)
Class 5: transcription factors and other nuclear proteins	
myc$^+$	Sequence-specific DNA-binding protein
myb$^+$	Sequence-specific DNA-binding protein
fos$^+$	Combines with c-jun product to form AP-1 transcription factor
jun$^+$	Sequence-specific DNA-binding protein; part of AP-1
erbA$^+$	Thyroxine (T_3) receptor
rel$^+$	NF-κB-related protein
ets$^+$	Sequence-specific DNA-binding protein

Note: The table is incomplete. Some of these proto-oncogenes were originally detected as retroviral oncogenes or tumor oncogenes. Others were identified at the boundaries of chromosomal translocations and at sites of retroviral insertions in tumors, or they were found in the form of amplified genes in tumors.

After Hunter (1991). Used with permission. © Cell Press.

and indeed, binding studies with labeled NGF revealed that the trk protein acts as an NGF receptor (Kaplan et al., 1991).

Membrane-Bound Signal Proteins A large number of proto-oncogenes encode proteins that transmit mitogenic signals from growth factor receptors to the nucleus (Table 28.2; T. Hunter, 1991; Cantley et al., 1991). Many of these signal proteins are membrane-bound tyrosine kinases, including the c-src protein already discussed. Another membrane-associated signal protein is encoded by the c-*ras*$^+$ proto-oncogene, the homolog of the v-*ras* oncogene of two sarcoma viruses. The c-ras protein receives mitogenic signals from activated growth factor receptors (Egan et al., 1993). In its active—that is, signal-transmitting form—c-ras protein is associated with guanosine triphosphate (GTP). This active form of c-ras protein is normally short-lived, because the c-ras protein has a GTPase activity that hydrolyzes GTP to GDP (McCormick, 1989). However, the protein encoded by the v-*ras* oncogene is defective in its GTPase activity, so that the v-ras protein passes on a mitogenic signal continuously.

Cytoplasmic Signal Proteins The membrane-bound signaling proteins pass the mitogenic signal on to cytoplasmic signaling proteins. These include a cascade of mitogen-activated protein kinases (MAPKs), each phosphorylating and thereby activating the protein downstream. Activated ras protein starts this cascade through another kinase, known as the Raf-1 protein (see Fig. 2.28). Oncogenic forms of Raf-1 lack certain N-terminal sequences and are permanently activated. The last signal protein in the MAP kinase pathway activates a transcription factor or another nuclear protein.

Transcription Factors Several proto-oncogene products are components of transcription factors. For instance, the peptides encoded by the c-*fos*$^+$ and c-*jun*$^+$ genes both have *leucine zippers* and form dimers that act as transcription factors (Lewin, 1991). The peptides can form both c-jun homodimers and c-fos/c-jun heterodimers. The heterodimer, also known as AP-1 transcription factor, is more stable. Both genes were discovered as viral oncogenes. The v-jun protein differs from c-jun by a deletion of 27 amino acids near the transcriptional activation domain and three single substitutions. Similarly, v-fos is derived from c-fos by the loss of 50 amino acids near the C-terminus and four substitutions. The selective effect of oncogene expression on a particular tissue may result from a specific combination of the affected transcription factor with other factors present in that tissue; other tissues with different transcription factor combinations may be unaffected.

Cytoskeleton-Related Proteins Proteins involved in the organization of the actin cytoskeleton can also become oncogene products, in accord with the long-standing observations that link the loss of anchorage dependence and the lack of cytoskeletal organization to unrestricted cell proliferation (see Section 28.5). In particular, a family of small of GTPases, known as the Rho family, is often found to be mutated in tumorous cells (Hunter, 1997). This family includes the Rho protein itself, which causes actin stress fiber organization, as well as Rac, which stimulates lamellipodia formation, and Cdc 42, which induces filopodia formation.

ONCOGENES ARISE FROM PROTO-ONCOGENES THROUGH VARIOUS GENETIC EVENTS

Viruses may cause tumors through the overexpression of an oncogene that has become part of the viral genome but encodes a normal or nearly normal gene product. In fact, some viruses turn proto-oncogenes into oncogenes by simply inserting a strong viral promoter next to the proto-oncogene, thus enhancing the proto-oncogene's activity. However, viruses are not the only tumorigenic agents. It has been known for some time that certain types of chemicals and radiation are both mutagenic and carcinogenic. Thus, it was not surprising to find that oncogenes can be derived from the homologous proto-oncogenes through a variety of genetic events.

Many growth factor receptors need to form dimers to acquire their signaling activity, and the dimerization is normally ligand-dependent. Proto-oncogenes that encode such receptors often become oncogenes through mutations that make the dimerization process ligand-independent (T. Hunter, 1997). These receptors then become constitutively active with the result of uncontrolled cell division.

In some cases, a single point mutation is enough to turn a useful proto-oncogene into a deadly oncogene. For example, several human bladder carcinoma cell lines show a *ras* oncogene that differs from its normal proto-oncogene by one base substitution that changes the 12th codon from GGC, which codes for glycine, to GTC, which codes for valine (Reddy et al., 1982; Weinberg, 1983).

Chromosomal rearrangements can also generate oncogenes from proto-oncogenes. For instance, *Burkitt's lymphoma,* a tumor of human B lymphocytes, has been found in association with three independent chromosomal translocations (Fig. 28.21). In each case, the c-*myc*$^+$ proto-oncogene, which encodes a component of a transcription factor that promotes the cell cycle, is translocated next to an immunoglobulin gene (Croce et al., 1983; Croce, 1987; Dalla-Favera et al., 1982; Erikson et al., 1983). The immunoglobulin genes are transcribed most actively in B lymphocytes, as are the inserted c-*myc*$^+$ genes. The forced expression of the cell cycle protein then makes the cells proliferate at a higher than normal rate.

Finally, proto-oncogenes can become oncogenes by selective amplification. As a rule, genomic DNA is replicated evenly, without selective omissions or extra copies (see Section 7.3 and 7.4). However, since this rule is easily broken under selection pressure, the molecular mechanisms for selective replication of certain DNA segments must be present in eukaryotic cells but not normally used. Some proto-oncogenes have a propensity for selective amplification. For instance, the c-*myc*$^+$ gene is selectively amplified in several independent tumors.

TUMOR SUPPRESSOR GENES LIMIT THE FREQUENCY OF CELL DIVISIONS

Whereas oncogenes behave as dominant gain-of-function alleles, other tumor-related genes have been discovered through loss-of-function alleles, which are usually recessive. These genes are referred to as *tumor suppressor genes.* Their normal function is to slow down the cell cycle by inhibiting cell cycle proteins or by interfering with the transcription of cell cycle genes (see Fig. 28.16). The existence of tumor suppressor genes is indicated by several lines of evidence, including cell hybrids and familial cancers (Weinberg, *1996a,* 1996b).

Hybrid cells generated by fusing normal cells with tumor cells do not give rise to tumors unless certain chromosomes are subsequently lost from the hybrids. Thus, the loss of a chromosome or chromosome

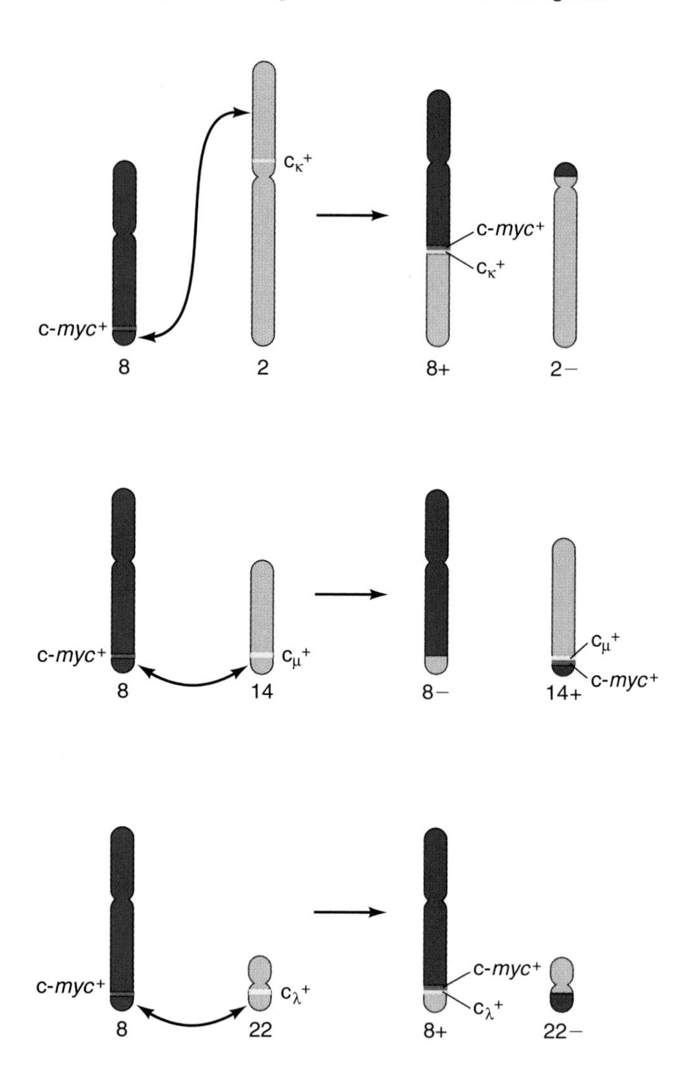

Figure 28.21 Chromosomal translocations giving rise to Burkitt's lymphoma. The diagram depicts human chromosomes; only one of each homologous pair is shown. The c-*myc*$^+$ gene, normally located on human chromosome 8, promotes DNA replication and mitosis. The other chromosomes each carry an immunoglobulin gene, designated $c_\kappa{}^+$, $c_\lambda{}^+$, or $c_\mu{}^+$; these genes are strongly expressed in lymphocytes. If c-*myc*$^+$ is translocated near the promoter of an immunoglobulin gene, c-*myc*$^+$ is expressed more strongly than from its own promoter, and tumorous growth results. Three different translocations of this type have been found between chromosome 8 and chromosome 2, 14, or 22.

fragment can turn a nontumor cell into a tumor cell. This observation indicates that certain genes suppress the tumorigenic potential of cell hybrids and, by extension, of normal cells. By analyzing the effects of chromosome losses from hybrid cells, and through the use of transfer techniques for single chromosomes, it has been possible to identify the chromosomal locations of some of these suppressor genes.

It has long been known that predispositions to certain types of cancers run in families. One of these cancers is *retinoblastoma,* a rare childhood cancer in which immature retina cells proliferate. There are two forms of this cancer: hereditary (familial) and nonhereditary. The

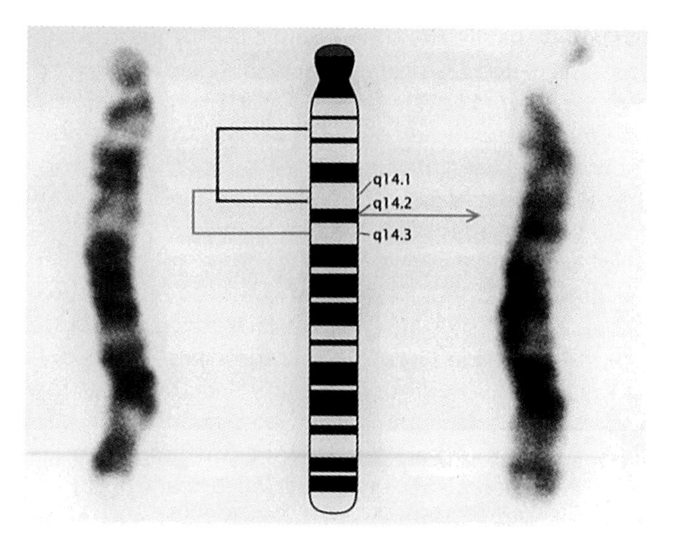

Figure 28.22 Identification of a tumor suppressor gene by means of chromosomal deletions. The photograph to the left shows a normal human chromosome 13, and the drawing in the center is an interpretative diagram. The photograph to the right shows a chromosome 13 from a human retinoblastoma; it lacks the segment that is marked by the narrow bracket in the center diagram. The arrow points to where the segment should be. A different deletion, observed in a chromosome 13 from a different retinoblastoma, is represented by the wide bracket in the diagram. The two deletions overlap in a small segment within band 13.q14.1. The retinoblastoma gene has been cloned from this segment.

hereditary form afflicts both eyes and is often associated with tumors in other organs, whereas the nonhereditary form affects only one eye. Some patients with retinoblastoma have a visibly abnormal karyotype, with a specific deletion on one copy of chromosome 13 (Fig. 28.22). These observations suggest that patients with hereditary retinoblastoma have a predisposition to this and other cancers because they lack one copy of a tumor suppressor gene, designated *Rb*$^+$, in all of their cells. A single mutation in the remaining intact copy of the gene in a developing retina cell should then be sufficient to start a cancer. Indeed, the homologous allele, on the chromosome 13 inherited from the unaffected parent, has been shown to be missing or mutated in retinoblastomas and several other cancers.

Mice in which both copies of the *Rb* gene have been interrupted by *homologous recombination* (see Fig. 15.24) die as embryos, presumably because of irregular mitoses of neurons or blood cells (E. Y.-H. P. Lee et al., 1992). With only one copy of the normal *Rb*$^+$ gene, mice are phenotypically normal at birth but die from pituitary tumors within 10 months. Conversely, overexpression of the human *Rb* gene in transgenic mice results in dwarfism (Bignon et al., 1993). These results indicate that the normal function of tumor suppressor genes is to limit cell growth and that the loss of this function leads to tumor formation.

How does the *Rb*$^+$ gene restrict cell growth? The underlying molecular mechanisms are complex and still being investigated (Dyson, 1998), but a simplified account will suffice here. The protein encoded by *Rb,* designated pRB, accumulates in the nucleus. The phosphorylation of pRB changes with the cell cycle: The protein is underphosphorylated in G_0 phase and during most of G_1; it becomes phosphorylated during late G_1 and remains in this state through S, G_2, and M. The late G_1 phase, when pRB becomes phosphorylated, is the principal restriction point in the adult cell cycle (see Fig. 28.16). Past this point, a cell is committed to divide.

The underphosphorylated form of pRB is the active form, in which it limits the activity of a group of transcription factors, collectively called E2F, which would otherwise promote the transcription of cell cycle genes. Loss of pRB allows these genes to be transcribed at inappropriate times, with the result of tumorous growth. pRB also sequesters cyclin D1, thus putting another brake on the cell cycle (Dowdy et al., 1993). In addition, pRB combines with certain transcription factors to promote cell differentiation. For instance, pRB bound to MyoD promotes myogenesis and inhibits cell proliferation (Gu et al., 1993). Conversely, Id-2, a helix-loop-helix protein that heterodimerizes with MyoD, inhibits myogenesis, and promotes cell proliferation, is bound or sequestered by pRB (Iavarone et al., 1994). These interactions provide another mechanism by which cell proliferation and cell differentiation are regulated antagonistically.

Another tumor suppressor gene, designated *p53*$^+$ for the molecular weight (53 kd) of the encoded protein, is similarly involved in many regulatory pathways that control cell division (Ko and Prives, 1996). The *p53*$^+$ gene is located on the human chromosome 17 in a region that is lost or mutated in a wide range of human cancers, including the common colon, lung, and breast cancers (A. J. Levine, 1997). In most of these cancers there is a loss of both *p53*$^+$ alleles. Introduction of the normal *p53*$^+$ gene into tumor cell lines with little or no endogenous p53 protein suppresses growth. Conversely, like mice without Rb, mice deficient for p53 are born phenotypically normal but are susceptible to spontaneous tumors (Donehower et al., 1992). More dramatically, *Xenopus* embryos develop tumors during early histogenesis if the synthesis of p53 is inhibited by *dominant interference* (Wallingford et al., 1997). In addition to preventing tumors, p53 is also involved in triggering *apoptosis* in cells with DNA damage (Lowe et al., 1993).

Some of the molecular characteristics of p53 protein have been revealed (see Fig. 28.16). In parallel with pRB, p53 is a transcription factor. One of its target genes is the *WAF1*$^+$ gene, also known as *Cip-1*$^+$. Its protein product, named p21 for its molecular mass, has a tumor-suppressing effect (El-Deiry et al., 1993). P21 binds and inactivates several cyclin/CDK complexes, including cyclinD1-CDK4, which is specifically required for cell cycle progression from G_1 to S phase.

Tumor suppressor proteins, the downstream proteins synthesized under their influence, and any drugs that mimic the actions of these proteins are of great medical interest. Further investigation of these agents should greatly enhance our understanding of normal growth control and our ability to prevent cancer formation in people who are genetically susceptible.

28.8 Growth and Pattern Formation

The introduction of genetic and molecular analysis has brought phenomenal advances in understanding the molecular mechanisms of pattern formation, as discussed in Chapters 21 through 25. Likewise, the signaling pathways that control cell cycle have been elucidated in great detail. A persistent puzzle, and the focus of exciting new research efforts, are the connections between cell growth (change in mass), cell division, and pattern formation (Edgar and Lehner, 1996; *Milán, 1998;* Conlon and Raff, 1999; *Weinkove and Leevers, 2000*). While much work needs to be done, progress has been made, especially with *Drosophila.*

HOW DO ORGANISMS MEASURE GROWTH?

How does an organism, or organ, decide when it is getting big enough and slow down its growth? Is the critical parameter the number of cells? In amphibians and other groups, there are related species with different *ploidies* (numbers of genomes per cell). Their cells tend to attain the same *nucleocytoplasmic ratio,* with diploid cells growing twice as big as haploid cells, tetraploid cells growing twice as big as diploid cells, and so forth. However, the overall sizes and body patterns of these related species do not change, which means that similar organisms are built from many small cells or fewer large cells (Fankhauser, 1952).

To study the role of cell division in size control genetically, Weigman and coworkers (1997) used a temperature-sensitive mutant allele of the *Drosophila Cdc2* gene, which encodes a cyclin-dependent kinase that is required for the G_2/M phase transition of the cell cycle. When mutant larvae were shifted to the restrictive temperature, the subsequent rounds of mitosis in imaginal discs were inhibited. However, the genomic DNA continued to be duplicated during S phases, and cells grew much bigger. In other words, the normal cell cycle was replaced with a cycle of DNA replication and cell growth.

Using the *UAS-Gal4* system described in Section 15.6, the investigators partially rescued the mutant flies by expressing a *Cdc2*$^+$ transgene selectively in posterior compartments under the control of the *engrailed* promoter. As a result, they obtained wings in which the posterior compartments were normal in every respect, whereas the anterior compartments were normal in overall size and shape and showed the regular vein pattern but consisted of fewer and bigger cells. Under the same experimental conditions, the expression domains of patterning genes, such as *decapentaplegic*$^+$, *optomotor* *blind*$^+$ and *spalt*$^+$, were nearly normal. Only if the mutants were shifted to the restrictive temperature as early larvae was the development of the imaginal discs disturbed. Under these conditions, both cell number and overall size of the discs were reduced, and expression domains of patterning genes did not resolve properly.

In a related study, Neufeld and coworkers (1998) used mutant alleles of genes encoding cell cycle proteins to accelerate or retard cell proliferation. These manipulations altered cell numbers of wing imaginal discs over a 4- to 5-fold range. Again, these changes were compensated by changes in cell size, so that the overall growth of cell clones and compartments, as well as wing disc morphology, were unaffected. These results indicate that manipulating the cell number through interference with the cell cycle does not necessarily affect the final size of organs or entire organisms, and does not interfere with pattern formation.

Does the rate of protein synthesis limit the final size attained by an organism? A set of *Drosophila* mutants are collectively known as *Minutes* is pertinent here (Morata and Ripoll, 1975). All of these genes seem to be required for protein synthesis, and most of the cloned ones encode ribosomal proteins. Homozygous *Minute* (*M*/*M*) individuals do not develop. However, *heterozygotes* (M^+/M) are viable. They grow slowly but eventually reach normal size and appearance, except that they have smaller bristles. Thus, slowing down the growth process by interfering with the rate of protein synthesis does not necessarily reduce the final size that an organism attains. It therefore came as a surprise when an apparently related *Drosophila* mutant showed a phenotype different from the *Minutes*. In this case, the mutated gene is *ds6k*, for *Drosophila* S6 kinase (Montagne et al., 1999). Lack of ds6k function leaves the overall number of cells unaffected but causes very slow growth, as is the case for the *Minutes*. However, in contrast to the *Minute* phenotype, *ds6k*$^-$ homozygotes also show reduced body size and reduced cell size. The homologous *S6K*$^+$ genes in mammals are known to regulate the activity of ribosomal proteins, but the *Drosophila ds6k*$^+$ gene is part of a signaling pathway that is likely to have multiple functions in cell division, cell survival, protein synthesis, and glucose metabolism (*Weinkove and Leevers, 2000*). Further studies on the role of this pathway in the control of body size should be very rewarding.

PATTERNING SIGNALS CONTROL CELL CYCLING AND GROWTH

The connection between pattern formation and cell proliferation in *Drosophila* imaginal discs is apparent from the striking alterations that result from inhibition or ectopic expression of key patterning signals. For instance, loss of decapentaplegic (dpp) signaling causes an extreme reduction in the size of wings and other appendages, while ectopic expression of dpp result in massive additional growth (see Fig. 22.50). Similarly, inappropriate wingless (wg) signaling early during the development of imaginal discs causes the formation of ectopic dorsoventral *compartment* boundaries, which result in the outgrowth of extra wing and leg parts (Struhl and Basler, 1993; D. Doherty et al., 1996). Curiously, late in wing disc development, wg also induces a zone of nonproliferating cells (ZNC) near the dorsoventral boundary, where cells stop proliferating about 30 h before most other cells. The ZNC includes the cells that give rise to the sensory bristles at the anterior margin of the adult wing.

How do patterning signals like dpp and wg affect cell cycling and growth? Clonal analysis shows that dpp stimulates the proliferation of those cells that express the dpp receptor peptides, which are encoded by the *Punt*$^+$ and *Thickvein*$^+$ genes. Cell clones lacking the activity of either of these genes do not proliferate (Burke and Basler, 1996). However, the rate of cell proliferation does not follow the shape of the proposed double gradient of dpp concentration across the wing (see Fig. 22.48), and no correlation has been reported between the local dpp concentration and any of the known cell cycle protein activities. Thus, it appears that dpp does not simply control the activity of a cell cycle regulator but rather acts in an indirect way.

With regard to wg, the origin of the ZNC has been clarified to an extent (Johnston and Edgar, 1998). In the anterior compartment, the ZNC can be subdivided into three domains (central, dorsal, and ventral), each about four cells wide (Fig. 28.23). Cells in the central domain express the *Notch*$^+$ gene, which represses *Achaete*$^+$ and *scute*$^+$. *Notch*$^+$ also induces the synthesis of wingless protein, which causes cell cycle arrest in G_1 phase, presumably by down-regulating dE2F, a transcription factor required for DNA replication. The same occurs in the entire ZNC of the posterior compartment. In the dorsal and ventral ZNC domains of the anterior compartment, wingless activates *Achaete*$^+$ and *scute*$^+$. These genes in turn inhibit the *string*$^+$, a gene that encodes a phosphatase necessary for entry into G_2 phase (see Fig. 5.34).

In contrast to the cell cycle–arresting functions of wingless in the ZNC, wg promotes cell division and growth in other parts of the wing disc. Here, wingless stimulates the synthesis of Myc protein, which together with other proteins forms heterodimers that act as transcription factors (Johnston et al., 1999). The overall effect of Myc activity in many cells is to promote the G_1 to S transition. It must be concluded that the ways in which patterning molecules such as wg control the cell cycle are context-dependent.

LOCAL CELL INTERACTIONS AFFECT CELL DIVISION AND GROWTH

While long-range patterning signals are clearly important in the control of cell division and growth, there is

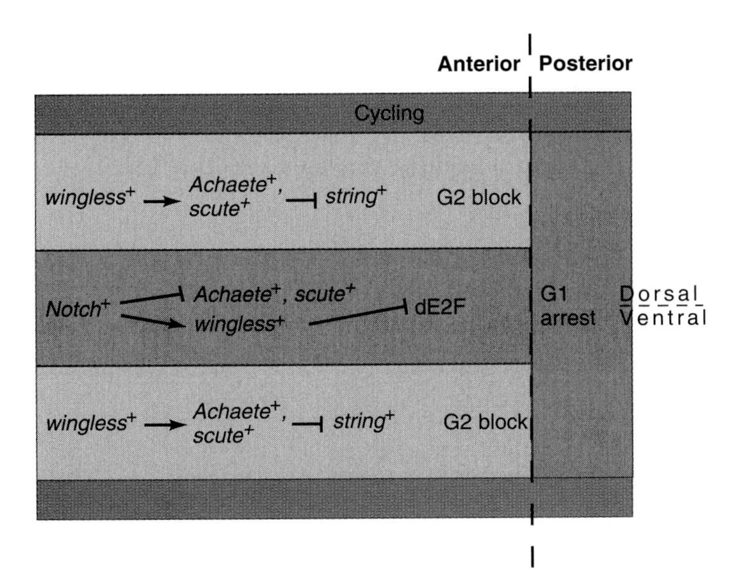

Figure 28.23 Control of cell cycling by the signaling molecules Notch and wingless in the *Drosophila* wing imaginal disc. Straddling the dorsoventral compartment boundary is a zone of nonproliferating cells (ZNC), where cells stop proliferating before most other cells do. In the anterior compartment, the ZNC can be subdivided into three domains (central, dorsal, and ventral). Cells in the central domain express the *Notch*$^+$ gene, which inhibits *Achaete*$^+$ and *scute*$^+$. In addition, *Notch*$^+$ activates *wingless*$^+$, which is also active in the entire ZNC of the posterior compartment. The wingless activity causes cell cycle arrest in G_1 phase, presumably through the inactivation of transcription factor dE2F. In the dorsal and ventral ZNC domains of the anterior compartment, wingless activates *Achaete*$^+$ and *scute*$^+$. These genes inhibit *string*$^+$, which encodes a phosphatase necessary for entry into G_2 phase (see Fig. 5.34).

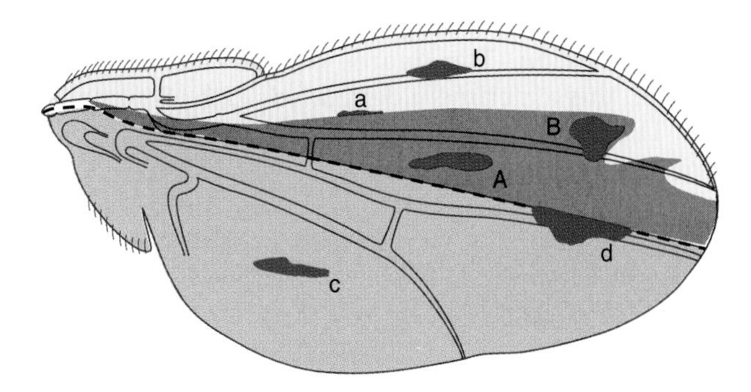

Figure 28.24 Cell competition in the *Drosophila* wing imaginal disc is revealed by analysis of serially generated clones in the adult wing blade. The large M^+/M^+ clone (red) was generated in a M^+/M background by X-ray-induced somatic crossover between homologous chromosomes 2. The small clones (green) were generated later by somatic crossover between homologous chromosomes 3. These clones differ from their surroundings only by expressing a cuticular marker. Comparison of clones located within a large M^+/M^+ clone shows that vicinity to M^+/M territory within in same compartment makes a clone (clone B) grow larger and more rounded than location in the middle of an M^+/M^+ clone or close to a compartment boundary (clone A). Among the cuticular marker clones located within M^+/M territory, those that are adjacent to the M^+/M^+ clone and reside within the same compartment (clone a) tend to become fragmented and stay smaller than clones located away from the M^+/M^+ clone (b, c) and/or in a different compartment (c, d).

also evidence for a role of local cell interactions. Morata and Ripoll (1975) reported some intriguing observations on the *Minute* mutants mentioned earlier. When marked M^+/M^+ homozygous cell clones were generated by X-ray–induced somatic crossover in the wing imaginal discs of M^+/M heterozytes, some of the marked clones grew so large as to almost fill an entire compartment (see Fig. 6.4 and Method 6.1). What happens to the cell clones founded by M^+/M cells, of which there should be 50 or more in a compartment at the time when the M^+/M^+ clone is generated? Are these clones staying small simply because of their intrinsically slow growth, or are they affected in other ways by the fast-growing clone?

In order to answer this question, Simpson and Morata (1981) used a serial irradiation technique. An early irradiation produced large clones and a subsequent irradiation produced smaller clones. The flies used for this treatment were engineered so that somatic crossovers of the second chromosomes produced, in an M^+/M background, large M^+/M^+ clones that also expressed a recessive cuticular marker (marker 1) to make them detectable (red area in Fig. 28.24). Crossovers of the third chromosomes induced in the same flies by a later irradiation generated smaller clones (green areas in Fig. 28.24) that differed from their surroundings only by expressing a second cuticular marker (marker 2). Thus, a marker 2 clone originating in an M^+/M^+ clone retained the M^+/M^+ genotype and was expected to grow fast; likewise, a marker 2 clone originating in an M^+/M clone retained the M^+/M genotype and was expected to grow slowly. Flies with large marker 1 clones and small marker 2 clones in their wing blades were analyzed.

The size and shape of the small marker 2 clones depended on their location. Among clones that originated within a large M^+/M^+ clone, those that were close to M^+/M territory within in same compartment (clone B in Fig. 28.24) grew especially large and more rounded than other clones. Conversely, among clones that originated within M^+/M territory, those that were adjacent to the M^+/M^+ clone and resided within the same compartment (clone A in Fig. 28.24) stayed small and sometimes became fragmented. Cell sizes did not change noticeably, indicating that different clone sizes resulted from differences in mitotic rates rather than cell growth. These observations show that clones within the same compartment interact so that M^+/M^+ clones accelerate their cell cycle, while adjacent M^+/M clones slow down and show signs of cell death. This phenomenon has been termed ***cell competition.*** The degree of cell competition observed in this experiment depended on the *Minute* allele used. The stronger of two alleles employed generated a more acute cell competition.

Both cell competion and the signals that preserve the overall size and pattern of a wing when one cell clone grows dramatically at the expense of others occur in *compartments.* These observations highlight the central role of compartments as functional units in growth and pattern formation. Defining this role in molecular terms is one of the most exciting challenges for the future.

SUMMARY

Growth in an organism is defined as a change in mass. The most common form of growth, called hyperplasia, is cell division followed by enlargement of the daughter cells. Other forms of growth are an increase in cell mass without division, and the deposition of extracellular matrix. The change in mass per unit time relative to the mass already present is called the relative growth rate. Isometric growth occurs when all parts of an organism grow at the same relative rate. More common is allometric growth, or the growth of different regions at different rates.

Heterospecific transplantations indicate that the growth of an organ is controlled by intrinsic and extrinsic factors. Extrinsic factors include the nutrients and growth hormones that circulate in the bloodstream. The single most important growth hormone in vertebrates, called somatotropin or simply growth hormone, is produced in the anterior pituitary gland under the influence of the hypothalamus. GH acts directly by promoting protein synthesis in its target cells and indirectly through signal peptides known as insulin-like growth factors.

Intrinsic factors controlling the growth of organs include growth factors that are released by many tissues and act locally as paracrine or autocrine signals. These factors have growth-related effects and other effects on multiple target cells. The action of a given growth factor may depend on the mix of other growth factors present.

Growth hormones and growth factors act on their target cells through specific receptor proteins in the plasma membrane. The binding of a growth factor receptor to its ligand starts a chain of signals that eventually promotes mitosis. The cytoplasmic domains of the receptors often act as tyrosine kinases, or they activate other membrane-bound tyrosine kinases. From here, mitogenic signals are propagated through cytoplasmic signal proteins to nuclear proteins, including transcription factors. These factors control the assembly of cyclin-dependent kinases and other proteins required for progression through the cell cycle.

The ability of cultured cells to survive and divide depends on their flat shape, which is generated by contact with a substratum. Cells deprived of such contact enter apoptosis. Cell shape, via the cytoskeleton, can change the permeability of certain ion channels and the release of growth factors.

The study of cell cycle control has been greatly enhanced by the isolation and characterization of tumor-related genes. Two classes of these genes are distinguished by their mutant phenotypes. One class, known as proto-oncogenes, gives rise to dominant gain-of-function alleles that cause deregulated growth. The mutant alleles, called oncogenes, are defined by their ability to transform normal cells into tumor cells. Oncogenes are derived from proto-oncogenes by mutations that cause either overexpression of the gene or the synthesis of an altered product. Proto-oncogenes encode all the proteins that transmit mitogenic signals. The second class of tumor-related genes, called tumor suppressor genes, is defined by loss-of-function alleles. The normal function of these genes is to limit the frequency of cell division. Deletions and point mutations in tumor suppressor genes are found in the majority of all human cancers.

Cell proliferation in an organ is controlled by long-range patterning signals as well as by local cell interactions. Mutations that accelerate or retard cell proliferation tend to be compensated by changes in cell size, while the overall growth and patterning of organs are unaffected. Some *Drosophila* mutants that have slow protein synthesis and slow growth also show small cell size and small body size, while others eventually attain normal cell and body size. Further studies on these and other related genes hold great promise for teasing apart the complex relationships that hold together cell size control, organismic growth, and pattern formation.

chapter 29

Senescence

Figure 29.1 If we say a person is "getting old," there may be an ambiguity in this phrase. It could mean that the person is simply adding another birthday. However, there is also a connotation of decreasing vigor, a phenomenon we will call senescence. This photo shows Jeanne Calment of Arles, France, on her 122nd birthday. She died a few months later, on August 4th, 1997, being the oldest human on record.

IN this final chapter, we will return to a basic feature of development—already mentioned in Section 1.2—the fact that all higher organisms exist in *life cycles*. Their development begins with fertilization and proceeds through embryonic and juvenile stages to adulthood. Adults produce gametes, egg or sperm, and their union generates a zygote, the beginning of a new individual. Having reproduced, the adult sooner or later deteriorates (Fig. 29.1). Eventually, the organism will die, leaving a corpse. Therefore, on first view, life cycles seem to be wasteful because they entail the buildup of large bodies that are bound to perish.

A salient feature of typical life cycles is sexual reproduction. In Section 1.2, we have discussed its apparent advantages, including the opportunity to eliminate bad genes and the chance to segregate adaptive mutant alleles from their wild-type counterparts. However, sexual reproduction does not necessarily require that the adults die. Theoretically, adults should be able to reproduce sexually and harness the apparent benefits of this process, but still be potentially immortal.

In common language, "getting old" means either having lived for a long time, or approaching a state of infirmity, or both. For clarity, we will use the term ***aging*** for the simply chronological progress of an individual's life along a time axis. In contrast, we will use the term ***senescence*** (Lat. *senex*, "old man") for the process of deterioration that is associated with aging.

So why do higher organisms senesce and die? This question has prompted many answers, not all of them satisfactory to scientists. One common answer is that old individuals must die to make room for new, more vigorous generations. However, this is a circular argument because it presupposes what it is trying to explain, that older adults are less vigorous than their offspring. The same criticism applies to the argument that older adults are useless to a population because they no longer reproduce. Again, the answer already implies reduced vigor of the older adults, this time in the special form of being no longer fertile. Another common answer, more mechanistic in its nature, is that older adults are no longer capable of repairing the damage that an organisms suffers from the wear and tear of life. But this is more a restatement of the senescence process than an explanation.

For senescence—as well as many other biological processes—one can identify two types of causes, *proximate causes* and *ultimate causes*. The distinction between the two, which goes back to Aristotle, is very useful although eventually they merge into one. A ***proximate cause*** (Lat. *proximus*, "nearest") is a direct mechanical or physiological cause. For instance, many male birds sing at the beginning of the breeding season because their testosterone level is high, and because this hormone acts on specific parts of the brain that drive the singing. If the bird is castrated he no longer sings, and if the testosterone is replaced by injection, the singing resumes (see Section 27.3). This makes testosterone a proximate cause of the bird's singing behavior. An ***ultimate cause*** (Lat. *ultimatus*, "last") refers to the adaptive function that a biological process seems to have—that is, to the way in which it enhances reproduction. The ultimate cause of the male bird's singing is that in doing so he stakes out a territory, which will help him to attract a mate, which is necessary for him to reproduce. Males who are genetically disposed to be good singers are therefore more likely to leave many offspring, who in turn will have inherited the disposition to sing.

In this chapter, we will first define senescence as a statistical phenomenon. Such a definition is appropriate because any physiological definition of senescence would be arbitrary as there are many different ways of measuring vigor, or the lack thereof. The statistical definition of senescence uses a measurement that is independent of any physiological criterion. Using the statistical definition also demonstrates that senescence is a real, and quantitatively major, phenomenon. In the following section, we will explore ultimate causes of senescence. Thereafter, we will consider a few characterized genes that are involved in the control of senescence. Finally, we will survey some proximate causes of senescence that are thought to be of major importance.

29.1 Statistical Definition of Senescence

The process of senescence is apparent from posture, balance, limitations to physical activity, reproductive fitness, slow immune response and wound healing, reduced skill in avoiding accidents and fighting off predators, and many other outward signs. Yet using any one of these manifestations as the defining criterion of senescence would be arbitrary, since an individual senescent by one criterion might not (yet) be senescent by another. A postmenopausal woman may still compete in marathon races, while an arthritic man who cannot walk a mile may still sire a child. To overcome this problem, one might develop a composite parameter, which is sometimes referred to as "biological age" (Arking, 1998). However, this concept is still being developed and will not be considered here.

A definition of senescence that avoids the arbitrariness of choosing one physiological criterion is based on the analysis of groups of individuals. The weakness of this definition is that it disregards variation between individuals, the experience that individuals senesce at different rates.

The primary data for a statistical definition of senescence are survival data. They can be given in the form of *life tables,* which show how many individuals of a *cohort*—for example, all individuals born in Sweden in 1920—are still alive in year 1921, 1922, and so forth. Other useful data are *census data,* indicating how many individuals in a *population* surveyed in a given year are 1 year or older, 2 years or older, and so on. Figure 29.2a shows survivorship curves—that is, plots of the percentage of survivors versus age, based onU.S. census data taken in 1900, 1960, and 1980. The curve for 1900 shows a marked decline in survivorship over the first 2 years, reflecting the susceptibility of human infants to death from congenital abnormalities and infectious diseases. This infant mortality is now much reduced in developed countries, due to improved hygiene and medical care. Past infancy, the survivor curves are approximately S-shaped, with *average life spans* increasing from 58 in 1900 to 81 years in 1980.

Life table data can be used to define a very useful parameter that has been variously called the *age-specific death rate, death rate, force of mortality, age-specific mortality rate, age-specific mortality,* or ***mortality rate.***

$$m(t) = \frac{\text{Number of individuals who die when they are } t \text{ years old.}}{\text{Number of individuals who were alive at the beginning of year } t}$$

The mortality rate, in other words, is the likelihood for an individual at age t to not survive another year. (In this definition, the term "year" may be replaced by any other time interval, such as "month" or "day," as appropriate for the organisms being studied.) Figure 29.2b shows the mortality rate derived from U.S. life tables of 1959–1961—that is, data similar to those used for the 1960 curve shown in Figure 29.2a. The mortality rate is very

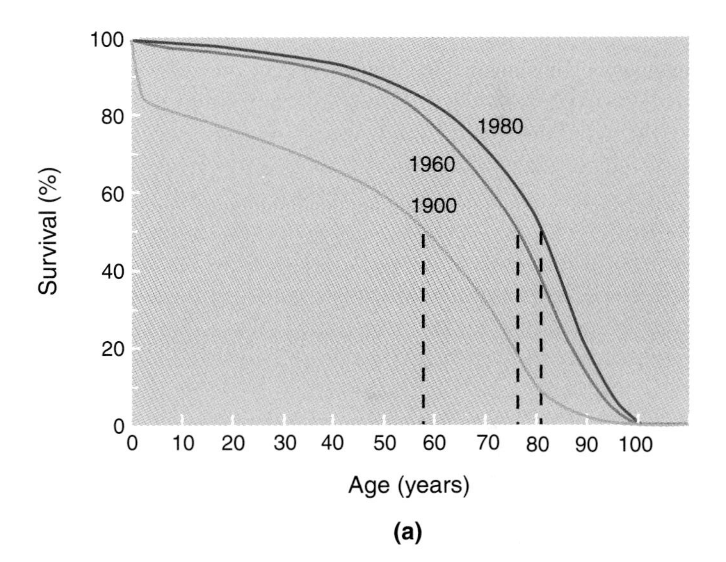

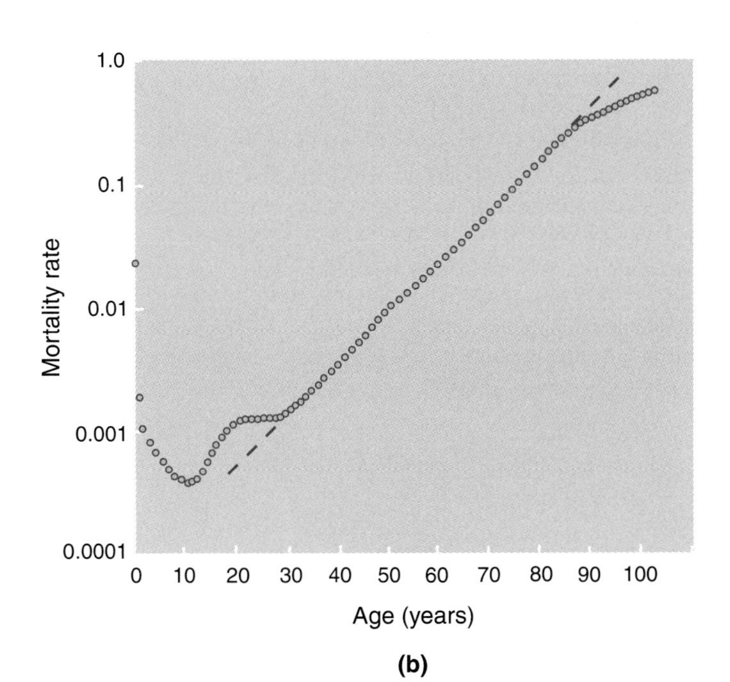

Figure 29.2 Survival and age-specific mortality under conditions of senescence. The primary data for studying aging and senescence are survival curves (or tables). From these, one can compute age-specific mortality rates (see text for definition). **(a)** Survival curves for human females in the United States, based on cross-sectional studies (census data) from 1900, 1960, and 1980. In each census, the percentage of females who had survived for at least 1 year, 2 years, and so forth, was recorded. Each vertical line indicates the age to which 50% of the population had survived. **(b)** Age-specific mortality rate of humans based on U.S. census data 1959–1961. Note the logarithmic scale of the ordinate. Mortality was high during early childhood, mostly from congenital and infectious diseases. Mortality was lowest around age 10, and then increased again for teenagers, mostly due to accidents. Between ages 30 and 85, mortality increased nearly exponentially, resulting in an almost straight line in this semilogarithmic plot. During this interval, the mortality rate doubled approximately every 8 years. The marked increase in mortality rate late in life is the defining criterion of senescence. Past age 90, the mortality rate leveled off for the small segment of the population who had survived that long.

dependent on age. The mortality rate starts high, reflecting infant mortality, and then reaches a minimum around age 10. The mortality rate increases sharply during teenage years, mostly due to accidents, levels off between 20 and 30 years of age, and then increases steadily. Between age 30 and age 80, the mortality curve can be approximated by a straight line in a semilogarithmic plot, which means the mortality rate increases exponentially with time. This part of the mortality curve can therefore be approximated by the following equation.

$$m(t) = A \cdot e^{G \cdot t}$$

which is equivalent to

$$\ln(m(t)) = \ln A + G \cdot t$$

where m = mortality rate, t = time, A is a time-independent adjustment parameter, and G is the ***Gompertz coefficient***, which represents the slope of the straight line in Figure 29.2b. The latter is named after the actuarian Benjamin Gompertz (1825), who first pointed out the exponential increase of the human mortality rate with age. The exponential nature of this increase is an empirical observation that has been made for many species, and at least some of the oxidative damage that is thought to cause senescence seems to increase nearly exponentially (see Section 29.5). However, the early and late portions of the mortality curve for human cohort do not conform to the exponential model of Gompertz. The high mortality rates before age 30 reflect the toll of infant mortality and risk-taking behavior of (especially male) early adults, as discussed earlier. The leveling-off of the human mortality curve beyond age 100 or so is a phenomenon that has come into view only recently as more individuals reach this very old age. Similar observations on declining mortality rates among very old individuals have been made with flies (Carey et al., 1992; Curtsinger et al., 1992). The plateauing of late-life mortality rates may reflect the genetic heterogeneity of populations, with small subgroups predisposed for unusually long life spans (Vaupel et al., 1998), but may also be interpreted differently (Mueller and Rose, 1996).

A parameter that is inversely proportional to the Gompertz coefficient, but easier to picture, is the ***mortality rate doubling time* (*MRDT*)**—that is, the time interval during which the mortality rate doubles. MRDT and G are related by the equation

$$\text{MRDT} = \ln(2) \cdot G^{-1}$$

Based on U.S. census data of 1960, the MRDT for humans between 30 and 80 years of age has been approximately 8 years.

The practical utility of the mortality rate as a human parameter is apparent from its application in term life insurance. The premiums for term life insurance increase steeply with the age of the insured, tracking the mortality rate, which reflects the risk of the insurer to have to pay out the insurance coverage to the designated beneficiary.

To appreciate the definition of senescence based on the increase in the mortality rate, it is helpful to contrast a human survivorship curve with the corresponding curve for a cohort that is *decimated only by random accidents*. A shipment of glass test tubes in a laboratory may serve as an example (Medawar, 1952). Assume a laboratory manager buys a thousand test tubes on a certain date and marks them all in a permanent way before putting them into use. The number of marked test tubes that are still in circulation is recorded thereafter at yearly intervals. This number will decline due to random causes such as breakage, being stolen, or being misplaced. Under these circumstances, the decline will be exponential, as drawn in Figure 29.3a. The mortality rate derived from this curve will be time-independent, and its graph will be a straight horizontal line (Fig. 29.3b).

The difference between the mortality rate function for a cohort of humans (or other living organisms) and a cohort of test tubes defines the phenomenon of senescence. The susceptibility of test tubes to breakage, theft, and so on remains constant (or nearly so) over time,

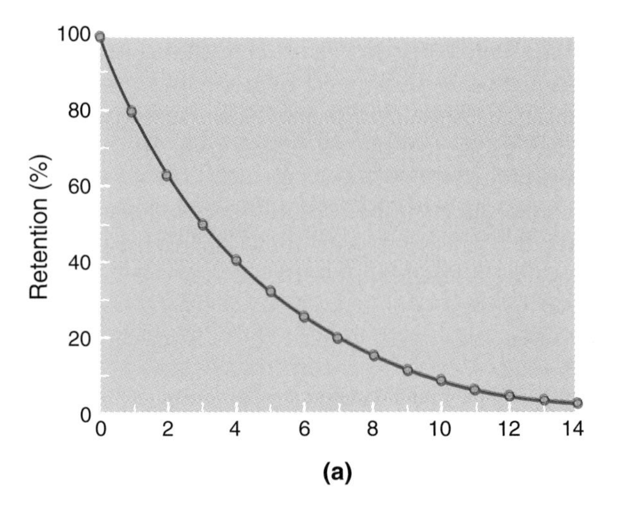

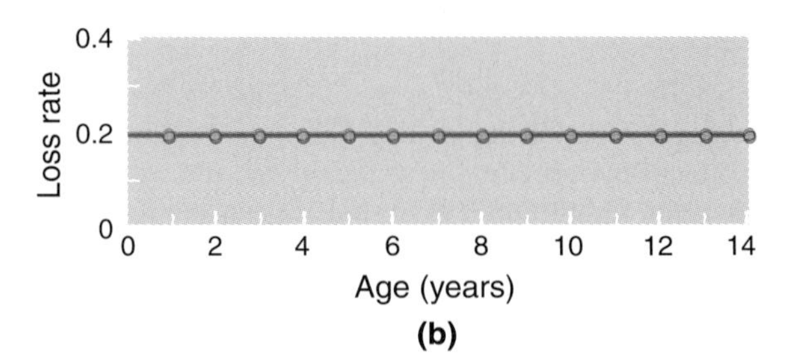

Figure 29.3 Survival and mortality without senescence. **(a)** Retention (or "survival") curve for a cohort of test tubes, which disappear due to causes unrelated to their age, such as misplacement, theft, or accidental breakage. **(b)** The loss (or "mortality") rate is constant over time (compare with Fig. 29.2b).

whereas the susceptibility of humans and other organisms to death from various causes changes with age. *It is the increase of the mortality rate with time that we call senescence.* This definition takes into account all causes for which individuals deteriorate and eventually die, and it avoids the arbitrariness of choosing any particular physiological parameter of declining vigor.

The senescence pattern observed in humans is typical of large animals that produce few young, on which they spend much parental care. Species that produce large numbers of offspring that are hammered incessantly by predation and accidents show survival and mortality curves looking like hybrids of the human and the test tube graphs shown here. The rapidity of senescence—as measured by the Gompertz coefficient or the MRDT—varies widely. Senescence can be gradual, as in humans, or rapid, or negligible (Finch, 1990). Rapid senescence is typical of *semelparous* species—that is, species reproducing only once. Animals such as harlequin flies or Pacific salmon, as well as many annual plants, reproduce once and then fall apart. On the other end of the spectrum, some clam and tree species, and possibly certain fish and reptiles, have life spans extending over decades or even centuries without apparent deterioration.

While senescence is a characteristic of most organisms, it is not limited to the living world. Complex machines such as cars will show senescence (Arking, 1998; Vaupel et al., 1998). Factors that increase the wear and tear on an engine, such as carbon deposits, impeded flow of coolant, and poor maintenance tend to increase with time and to compound each other. As a result, the risk of catastrophic failure will increase with age, which is the statistical definition of senescence. Indeed, the purpose of maintenance and periodic inspections is to slow down senescence by eliminating the risk factors that are easily controlled, like faulty ignition timing, in order to prevent their compounding with factors that are hard to control, like metal fatigue.

29.2 Evolutionary Hypotheses on Senescence

Several hypotheses about ultimate causes of senescence have been proposed. They seek to explain how senescence may have evolved. In this section, we will outline three hypotheses, which are not mutually exclusive. Subsequent sections will deal with the genetic control and proximate causes of senescence.

THE MUTATION ACCUMULATION HYPOTHESIS FOCUSES ON RANDOM DELETERIOUS MUTATIONS

The ***mutation accumulation*** hypothesis proposed by Medawar (1952) is based on two simple observations. First, random mutations occur in all cells, both *germ line* and *somatic,* and most mutations reduce fitness—that is, the ability to reproduce. Mutations in germ line cells may be passed on to offspring unless they are repaired or eliminated by natural selection. Second, in every population, old individuals are less numerous than young individuals. This is due to two causes: senescence and accidental death. Among laboratory animals and in developed human societies, senescence is the major cause. In wild populations, including primitive human societies, most individuals do not live long enough to show senescence. Instead, they die from predation, random accidents, fighting, natural disasters, and other causes that bear little or no relation to the resilience of those who become the casualties. Figure 29.4 shows a survival curve for a cohort of wild lapwing birds in Britain. It decreases almost exponentially, like the "survival" curve for test tubes shown in Figure 29.3a, indicating that senescence was negligible in both cohorts. The same was true for prehistoric humans—their remains hardly ever reveal an age of more than 50 years. Likewise, in the few human societies that still live as hunters and gatherers, individuals older than 50 years are very rare. In contrast, in historic but premodern societies, a minority of humans lived beyond age 50 (Fig. 29.4). However, the overall presence of older individuals, integrated over time, in both the lapwing and the premodern human population amounts only to about 10%. Their contribution to reproduction is zero for postmenopausal females and presumably diminished for older males.

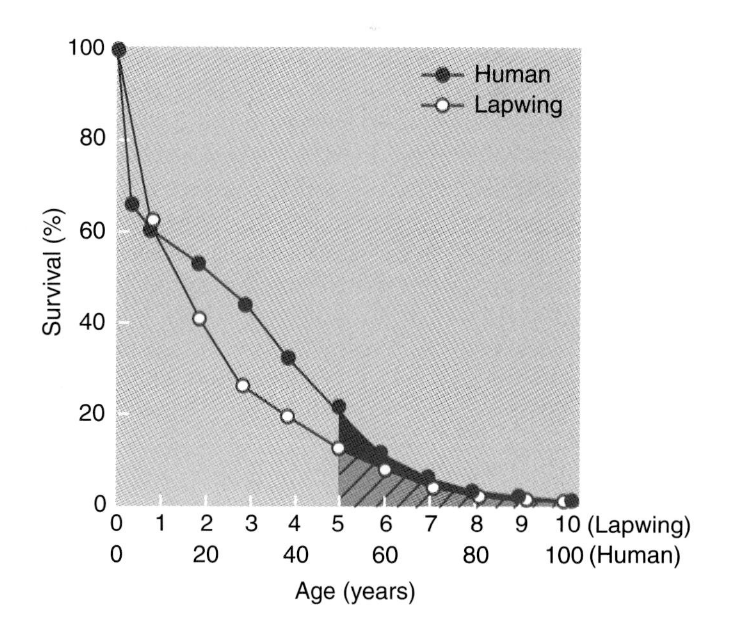

Figure 29.4 Decreasing force of selection with age. The survival curves shown are for a bird (English lapwing) and the human (data for British India, 1921–1930). Individuals who have survived into the second half of their maximum life span (shaded areas) constitute only about 10% of the entire population (total area below the respective curve). The force of natural selection on genetic alleles reducing fitness late in life is correspondingly small.

Thus, with or without senescence, the force of natural selection against random mutations that have deleterious effects *late in life* is minimal. In contrast, mutations acting *early in life* affect a large proportion of the population and are therefore selected against strongly. In summary, the mutation accumulation hypothesis ascribes senescence to the decreasing force of natural selection on the older and therefore smaller segment of a population.

The mutation accumulation hypothesis makes several predictions that have been tested by *comparative observations* on different species and by *forced selection experiments* with laboratory animals (Rose, 1991; Partridge and Barton, 1993). *Semelparous* species, which breed only once in life, should display precipitous senescence following reproduction because there is no natural selection afterward. This pattern is indeed well-known for species such as mayflies, salmon, and many annual plants. Conversely, organisms whose fertility increases with age should show delayed senescence and long life spans. This is in fact observed for large trees, lobsters, and many fish as well as reptilian species.

If the early senescence pattern observed in semelparous species is caused by the accumulation of deleterious mutations due to the lack of natural selection, then this pattern should also evolve in the laboratory, provided the experimentor kills the adults after they have produced their first offspring. This prediction was tested by Sokal (1970), who worked with the flour beetle *Tribolium castaneum*. Sokal bred two strains in which the recently eclosed adults were killed after they had mated and laid their first eggs. Control strains were kept in mass culture, where older adults were allowed to breed while living out their life span. After 40 generations, Sokal observed a decreased median longevity in the experimental strains, as predicted by the mutation accumulation hypothesis. In a parallel study on *Drosophila melanogaster*, Mueller (1987) used a similar design except that a reduction in late female fecundity (number of offspring produced), rather than a shortened life span, was used as a measure of senescence. After 100 generations of forced selection, the females in experimental lines laid significantly fewer eggs late in life, indicating accelerated senescence.

Conversely, the mutation accumulation hypothesis predicts that senescence should be postponed if selection against late-acting mutations is strengthened. A typical experiment of this kind begins with genetically heterogeneous (outbred) cultures from which multiple lines are established. In each line, males and females are allowed to mate normally, but only offspring generated by old adults are used for propagation. Control lines from the same population are propagated normally using offspring from young females. The late reproducing lines are then compared over many generations with the early reproducing lines with regard to longevity, fecundity, performance in certain physiological tests, or biochemical parameters.

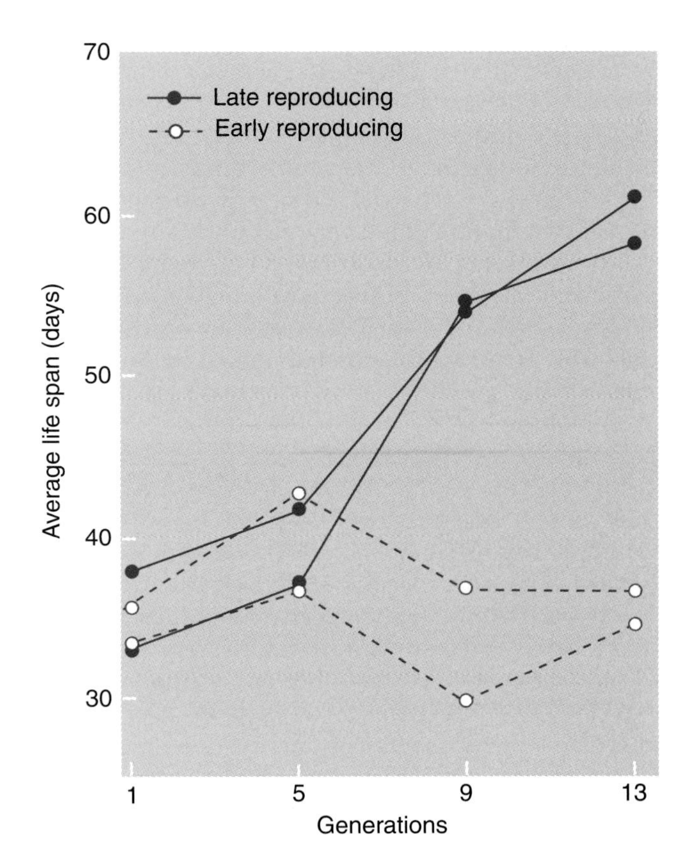

Figure 29.5 Average longevity of late-reproducing versus early-reproducing *Drosophila melanogaster*. In this experiment, two lines from an outbred (genetically heterogeneous) wild-type population were propagated by using only eggs from older females for propagation. Two control lines derived from the same wild-type population were propagated by using only eggs from young females. After only a few generations of breeding, the late reproducing lines showed a much higher average life span than the early reproducing lines

The first experiment of this type was carried out by Wattiaux (1968), who found an extended life span in the late reproducing lines after only a few generations. Similar results were obtained with *Drosophila melanogaster* (Fig. 29.5; Rose and Charlesworth, 1981; Luckinbill et al., 1984). Interestingly, preadult survival and fecundity of young adults declined in the late reproducing lines. We will return to this "trade-off" between early fecundity and late senescence below.

In summary, forced selection experiments with multiple insect species confirmed the predictions derived from the mutation accumulation hypothesis, which ascribes senescence to the deleterious effects of late-acting mutations, against which there is no natural selection.

SENESCENCE CAN BE ASCRIBED TO ANTAGONISTIC PLEIOTROPY

In addition to mutation accumulation, Medawar (1952) and Williams (1957) proposed another ultimate cause of senescence: They postulated the existence of multiple genes, each with two or more alleles that show ***antagonistic pleiotropy*** (Gk. *antagonizesthai,* "to struggle against"; *pleion,* "more"; *tropos,* "direction"). For each of these genes, they suggested there is at least one allele that promotes reproductive success early in life, with the price to be paid later in the form of an accelerated senescence. Other alleles of the same gene would have the opposite effect, limiting the rate of reproduction but affording a longer life span. Natural selection would then favor the combination of alleles that is best suited to the overall life strategy of a population in a given environment.

The production of sex hormones in humans (see Section 27.2) may serve an example. The enhanced production of androgens in men after puberty is clearly beneficial for the development and maintenance of strong bones and muscles as well as for sperm production. However, nearly every man who lives long enough will develop hyperplasia and possibly cancer of the prostate gland, problems that are alleviated by androgen blockage and may be exacerbated by androgen supplementation (Bare and Torti, 1998). Thus, any genetic allele that would enhance androgen production or slow down androgen turnover will have an effect of antagonistic pleiotropy. It will enhance physical and reproductive fitness of its carrier during early and middle adulthood, but it will also contribute to senescence later in life. The compromise that has evolved favors those men who die before they reach old age. Most men did precisely that during almost the entire hominid evolution, as discussed earlier. Thus there was no selective pressure for a better androgen control system to evolve. There is a similar trade-off for estrogen synthesis in women, although this situation is complicated by the phenomenon of menopause and its adaptive role in human life history. The synthesis of other hormones, including growth hormone and insulin-like growth factor, seems to be optimized the same way, enhancing fitness early in life (Lamberts et al., 1997).

Another set of genes that would appear to be prime candidates for antagonistic pleiotropy are those encoding the signals and receptors that set the appetite level. High appetite levels favor the deposition of fat stores, which are adaptive—especially in children and pregnant women—for surviving periods of scarce food. Adverse consequences again appear later in life in the form of atherosclerosis, a disposition to diabetes, and the burdens that overweight places on joints and the cardiopulmonary system. The compromise that evolves in each human population seems to fit the overall levels of physical activity and food supply. Humans are especially prone to overweight problems if they recently descended from families that lived under primitive conditions with much physical activity and scarce food and then transfer to a modern sedentary lifestyle with ample food.

Like the mutation accumulation model, the antagonistic pleiotropy model of senescence is hinged on the decreasing frequency of older individuals in a population. But rather than focusing on the weakening force of selection against *random deleterious* mutations acting late in life, the antagonistic pleiotropy model focuses on genetic alleles that are *potentially adaptive* because they buy increased fitness early in life for the price of an accelerated senescence. Such alleles should be positively selected for if the benefits of the trade-off, which accrue to *all* individuals of a cohort, are substantial and/or if a large fraction of a cohort dies from random accidents before the price of accelerated senescence comes due.

As a particular type of antagonistic pleiotropy, many researchers have noted a correlation between early fecundity and early senescence (Kirkwood and Rose, 1991; Rose, 1991). For instance, the *abnormal abdomen* mutant of *Drosophila mercatorum* greatly enhances early fecundity but also reduces longevity as compared to the wild-type (Templeton et al., 1985). The annual meadow grass *Poa annua* shows the same trade-off between high early reproduction and later survival and reproduction (Law et al., 1977; Law, 1979). When the amphipod *Gammarus lawrencianus* was selected for increased early reproduction, longevity was reduced at the same time (Doyle and Hunte, 1981). Indeed, the propagation of laboratory strains from late-reproducing individuals not only enhances longevity, but also tends to depress early fecundity. This is compatible not only with the mutation accumulation hypothesis, as discussed earlier, but also with antagonistic pleiotropy. The latter interpretation implies that, without forced selection by late reproduction, all or some of the genetic alleles that become prevalent in a population favor early fecundity at the price of early senescence.

IN order to test whether early reproduction is an ultimate cause of early senescence, Sgrò and Partridge (1999) used an *ovo* mutant strain of *Drosophila melanogaster.* The *ovo*$^{+}$ gene encodes a zygotically expressed transcription factor that is required specifically for the development of the female germ line (see Section 26.5; Oliver et al., 1987). The mutant allele (*ovo*^{D1}) used in this study is dominant and halts oogenesis at an early stage. This allele was crossed repeatedly into a wild-type (Dahomey) strain to obtain males that were genetically very similar to Dahomey males except that they were heterozygous for the oogenesis-interrupting allele of *ovo.* Derived from the wild-type Dahomey stock were also two lines obtained under different regimens of forced selection (Fig. 29.6). One line, designated ER for early reproduction, was bred over many generations from young adults, while the other line, designated LR for late reproduction, was bred from old adults. Matings of *ovo*^{D1}/+ males with females from the ER and LR lines produced sterile daughters, which were kept without males, 300 to a cage, to obtain life table data. Genetically equivalent but egg-laying daughters were derived from crossing of +/+ Dahomey males with ER and LR females. Daughters from these matings were caged in the

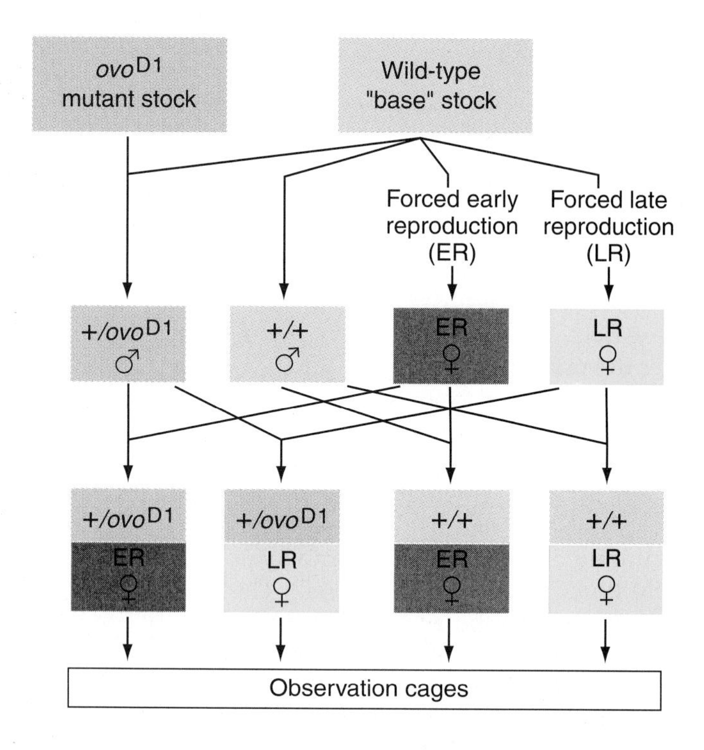

Figure 29.6 Design of an experiment to test the effect of early reproduction on the senescence of *Drosophila melanogaster* females. See main text.

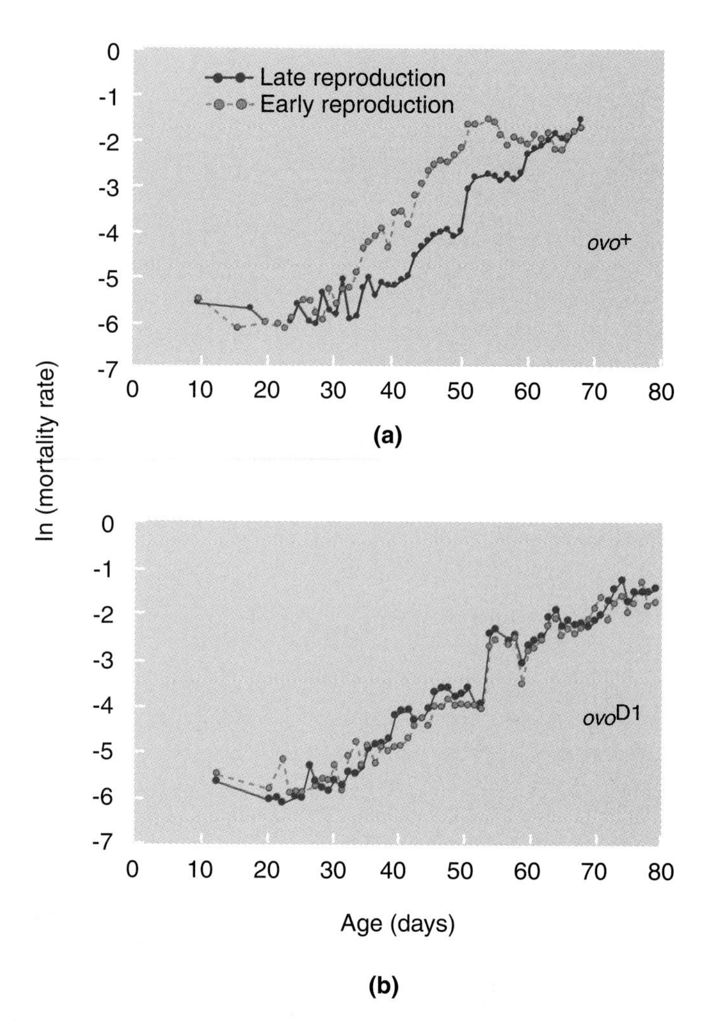

Figure 29.7 Effect of early reproduction on senescence in *Drosophila melanogaster* females (see Fig. 29.6). In the panels shown here, the natural logarithm of the age-specific mortality rate is plotted versus the age of the flies surveyed. The flies' female parents were obtained from laboratory lines selected by either late or early reproduction. **(a)** Data for egg-laying females. During the second half of the flies' life span, mortality was significantly higher for the females derived from the early reproducing line. **(b)** Data for similar females except that they carried the *ovo*^{D1} allele, which blocks oogenesis at an early stage. In the absence of egg production, there was no significant difference in mortality rate between flies from the late reproducing versus the early reproducing line.

same way without males and observed for survival and egg-laying activity. (These eggs, of course, were unfertilized and did not develop.)

Mortality rates for egg-laying daughters (+/+;ER and +/+;LR) are shown in Figure 29.7a. Between 32 and 58 days of age, the LR daughters showed a significantly lower mortality rate than the ER daughters. The LR daughters also had a longer life span (data not shown), whereas the ER daughters had higher early fecundity. These results confirm corresponding data obtained by other investigators and can be ascribed to the different forced selection regimes under which the ER and LR strains were bred. However, when the same comparison was made between *ovo*^{D1}/+;ER and *ovo*^{D1}/+;LR daughters, the difference in mortality rates completely disappeared (Fig. 29.7b). This means the difference in mortality rates between +/+;ER and +/+;LR daughters depends entirely on the absence of the *ovo*^{D1} allele and therefore, most likely, on egg production. When *ovo*^{D1}/+;ER daughters were compared with +/+;ER daughters, the latter again showed significantly higher mortality rates, whereas no such difference was observed between *ovo*^{D1}/+;LR and +/+LR; daughters. This means it is primarily the high rate of egg laying *in early life*, which characterizes the ER line, that accounts for its higher mortality later in life.

Questions

1. Why were all four types of daughters that were observed for survival kept without males?
2. The conclusion of the investigators that early egg laying activity causes early senescence is based on the assumptions that (a) *ovo*^{D1}/+ males are equivalent to +/+ males, except for the presence or absence of one *ovo*^{D1} allele, and that (b) the *ovo*$^{+}$ function is strictly limited to the female germ line. Could the investigators have interfered with egg-laying in a way that avoids the uncertainties associated with these assumptions?
3. Are the results of Sgrò and Partridge (1999) in accord with the mutation accumulation hypothesis?

The experiments described above strongly suggest that early fecundity in *Drosophila* comes at the price of early senescence. A cost associated with early reproduction was also revealed in a study on a bird species, the collared flycatcher, *Ficedula albicollis* (Gustafsson and Pärt, 1990). Females that did not breed during their first year laid larger clutches of eggs during their subsequent years than females who started breeding in their first year. Also, females whose brood sizes were experimentally enlarged or reduced in one year laid smaller or larger, respectively, clutches in subsequent years. In summary, the notion of a trade-off between early reproduction and late senescence has been borne out by several studies with vertebrates as well as invertebrates.

Generally, the predictions derived from the antagonistic pleiotropy hypothesis have been confirmed by comparative observations and experimental tests. Indeed, a wide range of data fits the notion of anatagonistic pleiotropy as a driving force in the evolution of senescence.

ACCORDING TO THE DISPOSABLE SOMA HYPOTHESIS, EVERY INDIVIDUAL IS THE TEMPORARY CARRIER OF HIS/HER GERM LINE

A third hypothesis on the ultimate cause of senescence, not mutually exclusive with the two others discussed so far, has been called the ***disposable soma*** hypothesis of senescence (*Kirkwood, 1996*). It focuses on the separation between *germ line* and *soma* (see Fig. 3.2), taking the perspective that every individual is the temporary carrier of his/her germ line, and that *this carrier function has evolved to be as economical as possible.*

A useful analogy here is car ownership. What car owners try to optimize, if they follow strictly utilitarian principles, is the number of miles driven per dollar spent. Up to a point, this goal is attained by regular maintenance, such as oil changes, air filter replacements, and so on, and by repairing or replacing any damaged parts as the need arises. Initially, the costs of maintaining and repairing a car tend to be low. Over time, however, as the wear and tear of driving take their toll, and as risk factors compound one another in a senescence-like pattern as discussed Section 29.1, there comes a point when future maintenance and repair costs are likely to exceed the payments for a new car. At that point, the economical choice for the car owner is to trade his old car for a new one.

The car ownership analogy suggests that much of the same may happen with soma ownership by the germ line. What evolution has presumably optimized here is the perpetuation and branching of the germ line with a minimum of energy spent on the generation and maintenance of successive carriers—that is, the somatic bodies. Again, the costs of maintaining and repairing a young soma are relatively low. The proteins that do the repairing are at peak performance and work efficiently. Wound healing and recovery from illness are typically fast. Over time, however, the genes that encode repair proteins are compromised by mutations, and recoveries from illness and injury take more time and energy. Eventually, it seems more economical to dispose of the old soma and entrust the germ line to the somatic carrier(s) of the next generation.

Can testable predictions be derived from the disposable soma hypothesis of senescence? One prediction is that organisms with distinct germ lines should show senescence while organisms without a germ line should not. All known vertebrates have a germ line that is separated early in development from the soma, and all land-living vertebrates are also known to senesce. There has been a pervasive notion that some fish and reptilian species would not senesce. Indeed, some fish grow without an apparent upper limit, but there is no clear indication that they do not senesce simultaneously (Comfort, 1979). Likewise, many insects and other arthropods show both a distinct germ line and senescence. Conversely, two annelid worms, *Aelosoma telebrarum* and *Pristina aequiseta,* reproduce by transverse fission and do not seem to senesce (Bell, 1984). The same is reported for some sea anemones, platyhelminthes, and tunicates (Sabbadin, 1979; Bell, 1988). Thus comparative observations on metazoa are compatible with the disposable soma hypothesis. The situations for protozoa and plants is more complicated, but they are not incompatible with the idea that senescence is a characteristic of nonreproductive cells (Rose, 1991).

Another prediction derived from the disposable soma hypothesis is that maintenance and repair mechanisms should operate at the highest levels in the germ line, where cells are potentially immortal. In contrast, somatic cells can be expected to show a less exacting level of repair since they must be maintained just adequately to last until the soma of the next generation can take over. Indeed, mutation rate estimates for germ line cells are much lower than those for somatic cells. The average mutation rate for human germ line cells in vivo is estimated at $4.4 \cdot 10^{-6}$ mutations per locus per generation (Sutton, 1988). Estimates for mutation rates in somatic cells maintained in vitro vary considerably but are generally much higher. Also, visible chromosomal abnormalities in tissue samples taken from mammals become more frequent with age, and accumulate at a rate that would be intolerable in germ line cells (Crowley and Curtis, 1963; Hirsch et al., 1974; Warner and Johnson, 1997). Finally, germ line cells generally divide more slowly than somatic cells, leaving DNA repair mechanisms more time to act before damaged or incorrectly replicated DNA sequences are replicated again.

A third prediction derived from the disposable soma hypothesis is that the cost of maintaining and repairing molecules and cells should increase with age. There seems to be a dearth of experiments that address this question directly. However, the amount of molecular damage in cells, such as carbonyl groups in proteins, does increase with age (see Section 29.5). The damaged proteins are likely to include those that carry out the repairs and replacements of other molecules. As the tools of repair become damaged themselves, they are bound to work less efficiently and to require more energy to maintain a given rate of repair.

Finally, one would predict that the effort spent on soma maintenance is gauged according to the risk of accidental loss. Again the analogy of car ownership is useful. If the risk of accidental loss is low, it is economical to invest in a high-quality car and maintain it well. However, if the risk of accidental loss is high, the better strategy is to buy a cheap car and save on maintenance costs so the down payment for the next car will be on hand quickly. Do living organisms follow the same strategy? One way of seeking an answer to this question is to study species with sexually dimorphic patterns of senescence. In these cases, which include the human, males generally senesce before females do, and their mating behavior typically involves greater risks of death from predation and injury from fighting with other males over access to females.

Another way of testing whether soma maintenance is gauged to the risk of accidental death is to compare different populations of the same species that live in similar environments except for a major difference in one risk factor, such as predation. Such a study has been carried out on two populations of the Virginia opossum, *Didelphis virginiana* (Austad, 1993). One population lives on the mainland in South Carolina, while the other lives on a barrier island off the coast of Georgia. The island population appears to have been genetically isolated for 4000 to 5000 years. The mainland opossums are decimated by birds of prey and by mammalian predators, such as bobcats and wild dogs, whereas the island at 4500 ha is not large enough to support a predator population. The disposable soma hypothesis would therefore predict a lower mortality rate and slower senescence for the island opossums. Indeed, a Gompertz plot of the mortality rates (Fig. 29.8) reveals that the mortality rate at maturity, which is thought to be imposed by the environment rather than caused by senescence, is higher on the mainland. The subsequent increase of mortality with age is also steeper on the mainland, as predicted by the disposable soma hypothesis. Physiological indicators of senescence point in the same direction. Island females were able to better nurse their young during their second reproductive year. In addition, the cross-linking of their collagen fibers, a commonly used biological marker of senescence, increased more slowly with age. The data from this study are also compatible with antagonistic pleiotropy, as the island population raised fewer young (5.66 average) than the mainland population (7.61 average) per year.

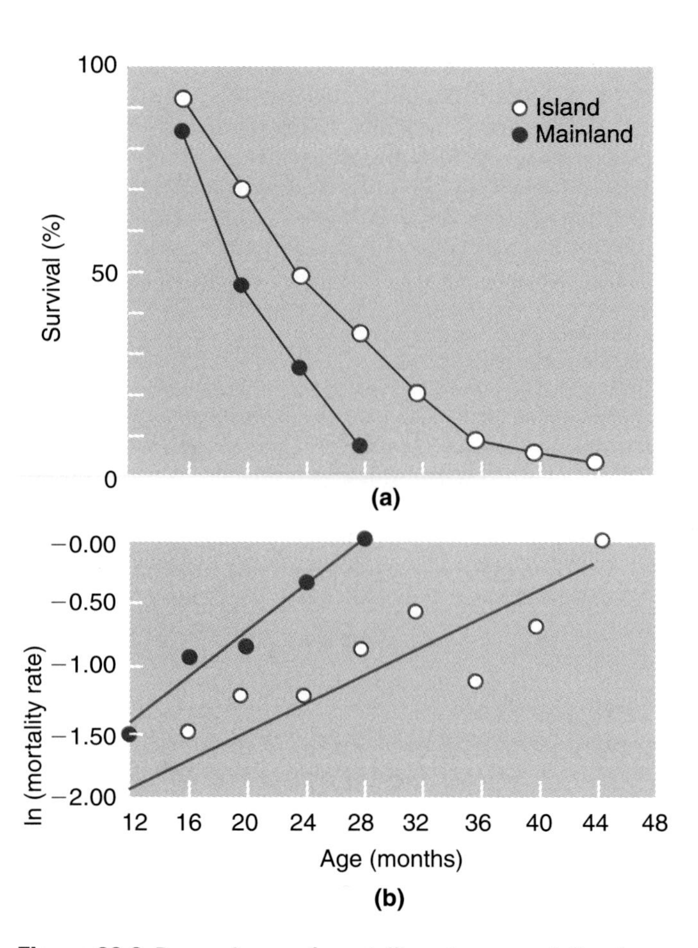

Figure 29.8 Dependence of mortality rate on predation in two populations of the Virginia opossum (*Didelphis virginiana*). One population lives on the mainland in South Carolina, where it is decimated by birds of prey and by mammalian predators. The other opossum population lives off the shore of Georgia on a barrier island, which is too small to support predator populations. On the island, the life span **(a)** is significantly longer, and the mortality rate **(b)** is lower.

In summary, the testable predictions derived from the disposable soma hypothesis of senescence have been confirmed by comparative observations between species and within species. The available data are in accord with the notion that the ultimate cause of senescence is to minimize the costs of maintaining the germ line by a series of disposable somatic carriers.

29.3 Characterized Genes That Affect Animal Life Span

The fact that senescence patterns are subject to change by natural selection and by breeding in the laboratory already shows that senescence is partly under genetic control. This does not mean senescence is controlled by a hierarchy of dedicated genes that has been conserved in evolution, like the cell cycle or apoptosis. To the contrary, the hypotheses on the *ultimate causes* of senescence lead us to expect the involvement of random assortments of genes, which could well differ from one group of organisms to another. Thus we may find genes to play a role in senescence that are otherwise involved in hormonal control, signal transduction, energy metabolism, cellular repair mechanisms, and many other processes.

LIFE SPAN–EXTENDING EFFECTS OF MUTATIONS IN *C. ELEGANS* MAY DEPEND ON THE ENVIRONMENT

In *C. elegans,* several genetic loci have been identified that extend the worm's life span when mutated (*Hekimi et al., 1998;* Wood, 1998). Four of these genes have been cloned and characterized. One of them is known as the *clk-1*$^+$ gene, with *clk* referring to "clock." Loss-of-function alleles of this and three related genes cause a general slowdown of many developmental processes including embryonic cleavage, adult swimming movements, pharyngeal pumping and defecation, as well as senescence (Lakowski and Hekimi, 1996). In accord with his general phenotype, the *clk-1*$^+$ gene encodes a regulatory protein that is required for the synthesis of coenzyme Q, a component of the mitochondrial electron transport chain that carries out *oxydative phosphorylation* (Ewbank et al., 1997). A loss of function in *clk-1* will reduce the supply of coenzyme Q. This will slow down a wide range of metabolic processes, including the production of *oxidants* (see Section 29.5). The extended life span of *clk-1* mutants may therefore result from a combination of two effects: a general slowdown of development and reduced oxidative damage.

The other three cloned genes extending the life span of *C. elegans* are known as *age-1*$^+$, *daf-2*$^+$, and *daf-16*$^+$, with *daf* standing for "dauer formation." These genes are involved in controlling the decision between entering the normal third larval stage (L3) and an alternative *dauer* stage. The latter is an inactive larval stage that allows the worm to survive adverse living conditions (see Section 25.1). Normally, entry into the dauer stage is triggered by a pheromone, which the worms release constitutively, so that high pheromone concentration signals crowding. However, mutants with loss-of-function alleles of *age-1* or *daf-2* enter the dauer stage even in the absence of the pheromone. Conversely, mutants deficient in *daf-16* cannot enter the dauer stage even in the presence of high pheromone concentrations. *Daf-16* is *epistatic* over *age-1* as well as *daf-2,* indicating that *daf-16* acts downstream of *age-1*$^+$ and *daf-2*$^+$. The simplest interpretation of these observations is that *daf-16*$^+$ is normally active at high pheromone concentration, and that its activity is required for entering the dauer stage, whereas *age-1*$^+$ and *daf-2*$^+$ are active at low pheromone concentration and inhibit *daf-16*$^+$ (Fig. 29. 9).

In addition to controlling the entry into the dauer stage, *age-1*$^+$ and *daf-2*$^+$ also affect the life span of the adult (Friedman and Johnson, 1988; Kenyon, 1993). In particular, if *temperature-sensitive* mutants of these genes are raised at a permissive temperature and low density, so that they do not enter the dauer stage, they will develop into adults with a life span about twice as long as that of the wild type. At the same time, the adults become more resistant to high temperature and other stress factors (Lithgow et al., 1994). These aspects of the *age-1* and *daf-2* mutant phenotypes also depend on *daf-16*$^+$.

The *daf-2*$^+$ gene encodes a member of the insulin receptor family (Kimura et al., 1997), and the *age-1*$^+$ gene encodes a subunit of phosphatidylinositol 3 kinase (Morris et al., 1996). The latter acts in signal transduction downstream of membrane receptors such as daf-2. Finally, the *daf-16*$^+$ gene encodes a transcription factor (Lin et al., 1997; Ogg et al., 1997). Thus daf-16 may act on a wide range of target genes promoting entry into the dauer stage as well as stress resistance and an extended adult life span. Meanwhile, age-1 and daf-2 may be part of the pheromone signaling pathway that antagonizes *daf-16*$^+$ function.

THE longer life span and higher stress resistance of *age-1* mutants is not associated with an appreciable loss in fecundity. This seems at odds with the *antagonistic pleiotropy* hypothesis of senescence discussed in Section 29.2. To further investigate this apparent contradiction, Walker and coworkers (2000) did a competitive evolution experiment in the laboratory. They kept wild-type and *age-1* mutant hermaphrodites together on the same agar plates at 20°C, a permissive temperature for the *age-1* allele (*hx546*) used. At intervals, the researchers determined the frequency of the *age-1* allele in the population by removing 100 eggs and shifting them to 27°C. At this restrictive temperature, wild-type worms developed into adults, while *age-1* worms developed into dauer larvae. The percentage of dauer larvae therefore indicated the allele frequency of *age-1.*

As long as the mixed populations were fed constantly with bacteria, there was no appreciable change in *age-1* allele frequency, regardless of whether the frequency at the outset of the experiment had been set at 0.9, or 0.5, or 0.1 (Fig. 29.9b). Thus there was no evidence for a trade-off between longevity and reproductive fitness under these conditions. This situation changed dramatically when the mixed population plates were kept under conditions of cyclical starvation. Under this regimen, the worms were allowed to exhaust their bacterial food and were starved for 4 days before they were fed again. After 6 starvation cycles, with each cycle corresponding to about two generations, the allele frequency of *age-1* had plummeted from its initial value of 0.5 to a mean of 0.06 in five replicate experiments (Fig. 29.9c).

The result of this experiment indicates that the longer life span afforded by the *age-1* mutation can diminish reproductive fitness, depending on food supply. While there is no loss of fitness under conditions of unlimited food supply,

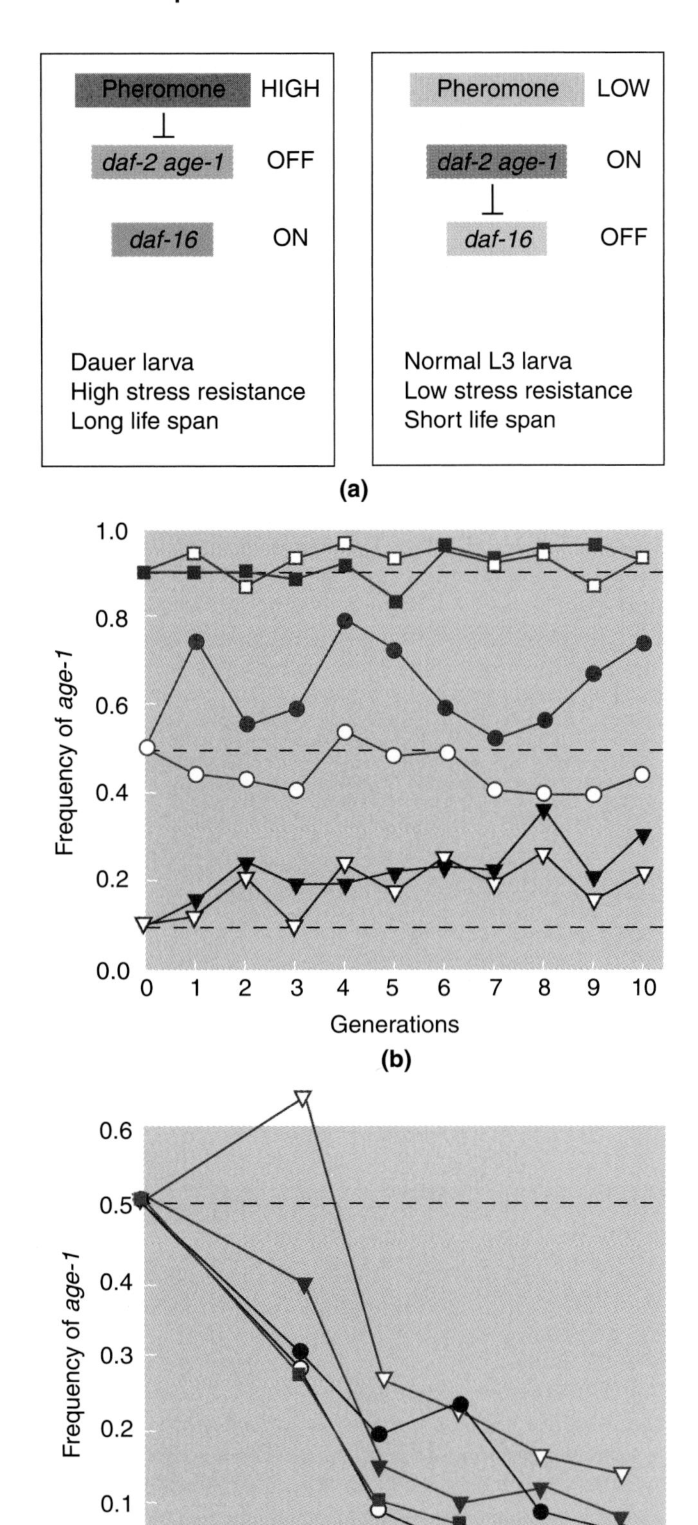

Figure 29.9 Effect of food supply on the outcome of selection experiments in the laboratory. **(a)** The roles of *age-1*$^+$, *daf-2*$^+$, and *daf-16*$^+$ in the formation of dauer larvae, stress resistance, and longevity in the roundworm *C. elegans* (see main text). In particular, temperature-sensitive alleles of *age-1* can considerably extend the worm's life span in the laboratory without significant loss in fecundity. **(b)** Wild-type and *age-1* mutant hermaphrodites were kept together on the same plates with unlimited food supply. Duplicate experiments were done with initial frequencies of the mutant allele set to 0.9, 0.5, and 0.1. The frequency of the mutant allele did not change consistently over 10 generations. **(c)** Similar experiment as shown in part b, except that the worms were starved intermittently, and that 5 replicate experiments were started at an initial *age-1* frequency of 0.5. Under these conditions, the frequency of the mutant allele dropped to a mean of 0.06 after 6 starvation cycles or 12 generations.

the *age-1* mutant displays a markedly reduced fitness under conditions of cyclical starvation. Since cyclical starvation seems to mimic living conditions in the wild more closely than unlimited food supply, there is no conflict between the life history traits of the *age-1* mutant and the antagonistic pleiotropy hypothesis of senescence.

Questions

1. Why did the investigators keep the wild-type and *age-1* mutant worms together on the same plates, instead of keeping the two strains on separate plates?
2. Given that the two strains were kept together, could there have been a problem with heterozygous offspring from matings between individuals from the two strains?

The experiment described above showed that the life history traits of the *age-1* mutant is not at odds with the antagonistic pleiotropy hypothesis of senescence. The results of this experiment also show that environmental conditions can have major effects on the relative fitness of a mutant phenotype relative to the wild type. In particular, the effect of food restriction on senescence can be astounding, as we will discuss further in Section 29.4.

DROSOPHILA MUTANTS REVEAL CORRELATION BETWEEN STRESS RESISTANCE AND LONGEVITY

The *Drosophila melanogaster* laboratory lines that have been selected for extended longevity by late reproduction differ in their physiological characteristics (Jazwinski, 1996). Some lines show improved resistance to oxidative stress, presumably due to enhanced expression of genes encoding antioxidant enzymes (see Section 29.5). Other selected strains show improved resistance to starvation, desiccation, ethanol vapor, and heat. These observations confirm the notion that senescence may be delayed by increased resistance to various stress factors. So far, two genes that affect mortality have been characterized in flies, one in *Drosophila mercatorum* and the other in *Drosophila melanogaster.*

Natural populations of *Drosophila mercatorum* on Hawaii are variable for the number of serially repeated 28S ribosomal genes that bear a 5 kb insert. Females bearing the insert in at least a third of their 28S repeats show the *abnormal abdomen* (*aa*) syndrome. It is characterized by retention of juvenile abdominal cuticle in the adult, a slowdown of larval development, an increase in early female fecundity, and a shortened life span (Templeton et al., 1989). Natural fluctuations in climate affecting the probability that adults will die from desiccation change the frequency of *aa* flies (Templeton et al., 1993). As expected, the *aa* variant is more frequent in drier years. The molecular events through which the mutant genotype causes its pleiotropic effects are still unknown.

A gene that can mutate to extend the laboratory life span of *Drosophila melanogaster* by up to 35% has been named *methusela*$^+$ (Lin et al., 1998). Homozygous (*mth*/*mth*) mutants are also more resistant to starvation, heat, and paraquat, a herbicide that damages cells by generating an aggressive oxidant, *superoxide radical* (see Section 29.5). The original *mth* allele was generated by the insertion of a *P element* into an intron, an event that presumably slows down pre-mRNA splicing but allows some gene function. Two apparent null alleles generated subsequently were homozygous lethal. Cloning and sequencing revealed a predicted mth protein with the characteristics of a G protein–associated transmembrane receptor (see Fig. 2.26). Such proteins are involved in a diverse array of biological activities including neurotransmission, hormonal control, drug responses, and transduction of optical and olfactory stimuli. The data of Lin and coworkers indicate that some function of *mth* is necessary for survival, and that reduced function confers a higher stress resistance and greater longevity than the function of two wild-type alleles.

The two reported mutants affecting longevity in flies confirm the correlation between stress resistance and longevity that was also observed in multiple *C. elegans* mutants. The characterized mutants in both *C. elegans* and *Drosophila* highlight the potential of a variety of genes to affect both stress resistance and senescence.

THE GENE MUTATED IN PERSONS WITH WERNER SYNDROME ENCODES A HELICASE

In humans, a rare heritable disease known as *Werner syndrome* is characterized by the premature appearance of many signs of senescence. Patients appear to be generally normal during childhood but stop growing during their teenage years. While they are still in their twenties, their hair begins to gray, their skin loses its suppleness, and their vision clouds from cataracts (Fig. 29.10). Muscular atrophy, atherosclerosis, osteoporosis, poor wound healing, and a tendency toward diabetes follow. The patients become prone to cancer and other diseases that normally strike at advanced age. Their median age at death is 47 years (Arking, 1998). However, Werner syndrome does not mimic *all* symptoms that are commonly associated with normal human senescence. For instance, patients with Werner syndrome do not develop high blood pressure or Alzheimer disease.

The gene (*WRN*) mutated in persons with Werner syndrome is located on chromosome 8 and has been cloned (Yu et al., 1996). The protein encoded by *WRN*$^+$ has the characteristics of *a helicase*, a type of enzyme that unwinds DNA in preparation for replication, repair, or gene expression. Both copies of *WRN* need to be mutated for Werner syndrome to develop. The mutations found in the *WRN* genes of persons with Werner syndrome indicate that the syndrome results from a complete loss of function of the WRN protein (Yu et al., 1997). WRN seems to interact with *p53*, a tumor-suppressing protein that inhibits the cell cycle (see Section 28.7) and promotes *apoptosis*. The latter step is attenuated in patients with Werner syndrome, which may explain their predisposition for early cancers (Spillare et al., 1999).

It will be interesting to find out exactly which step(s) of DNA replication, repair, or transcription are impaired by the lack of the WRN helicase. In view of the apparent role of DNA damage in senescence (see Section 29.5), it is tempting to speculate that DNA repair is compromised. However, it is conceivable that many aspects of

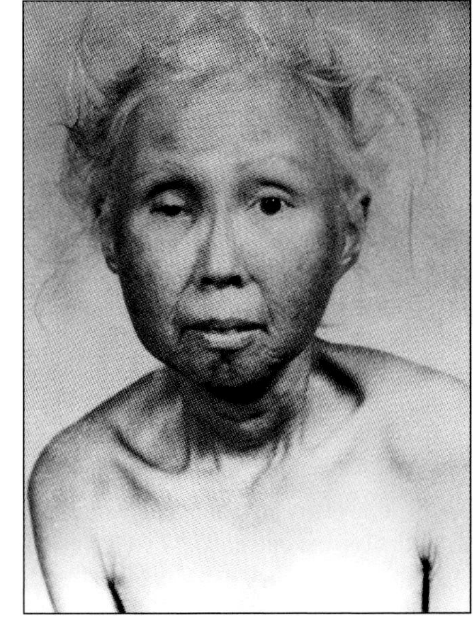

Figure 29.10 As a teenager (left), this Japanese American woman looked normal. At age 48, the accelerated senescence that characterizes Werner syndrome was readily apparent.

senescence result from inappropriate gene expression, which would be promoted by faulty transcription.

The first mutation in a mammal that increases the life span has been reported for mice (Migliaccio et al., 1999). The mutation affects protein p66shc, a cytoplasmic signal transducer involved in the transmission of mitogenic signals from activated receptors to Ras (see Fig. 2.28). Similar to the product of the *methuselah* gene of *Drosophila,* p66shc is part of a common signaling pathway, and perhaps counterintuitively, attenuation of the signal improves longevity and the response to oxidative stress.

In summary, the characterization of a few genes involved in senescence has revealed a wide array of gene functions. So far, there is no indication of a hierarchy of genes that would be dedicated to the control of senescence. However, a common feature of mutant alleles that increase life span is that they also improve resistance to stress factors such as starvation, desiccation, high temperature, or oxidative damage.

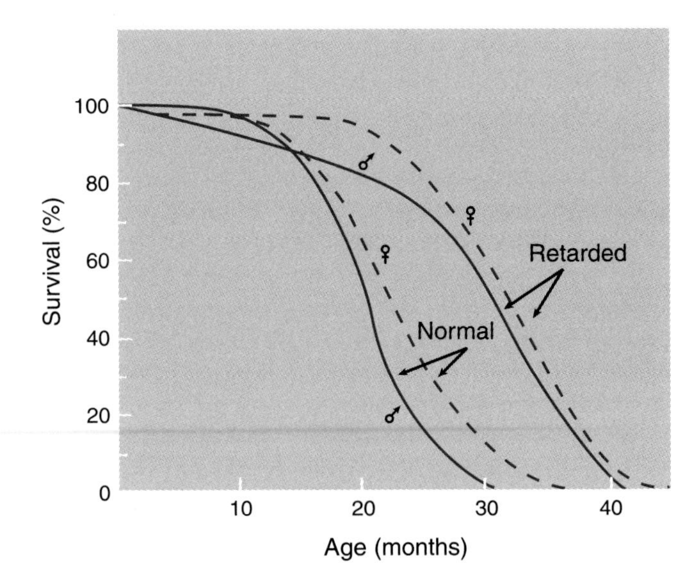

Figure 29.11 Survival rates of normal rats, fed ad libitum, and "retarded" rats, which were kept on a calorie-restricted diet as juveniles and reached sexual maturity only after return to a sufficient diet. The retarded rats remained smaller than their unrestricted counterparts but lived significantly longer.

29.4 Caloric Restriction

Having considered the ultimate causes of senescence and some of the characterized genes involved in the process, we will now turn to proximate causes of senescence. Based on our previous discussions, we expect a multitude of physiological mechanisms to be involved, and this is indeed the case (Finch, 1990; Rose, 1991). We will focus on those mechanisms that have been studied in many laboratories and/or are based on very general features of cells so that they are likely to be important for more than a limited group of species.

The most impressive and easily reproducible way of extending the life span of laboratory mice, rats and hamsters has been ***caloric restriction***—that is, the reduction in calories of an otherwise complete and balanced diet. Experiments now considered classics were carried out by McCay and Crowell (1934) and McCay and coworkers (1943). They fed rats on restricted diets as juveniles, beginning at the time of weaning or 2 weeks later. These rats did not grow beyond the weight they had attained at the time of restriction. The "retarded" animals resumed growth and development when they were given sufficient food at about 2 years of age. However, they remained about 15% smaller than the control animals who had been fed *ad libitum* (as much as they wanted) and had matured rapidly. However, the food-restricted animals lived substantially (about 50%) longer (Fig. 29.11), showing significantly smaller Gompertz coefficients. These experiments have been repeated and extended by numerous investigators. In physiological performance tests, the restricted animals did better than those fed ad libitum (Weindruch and Walford, 1988; Masoro et al., 1989). Food-restricted animals also showed reduced frequencies of atherosclerosis, autoimmune diseases, and many types of cancer. Because the growth-retarding effects of early caloric restriction made this treatment unattractive for human use, the effect of late restriction were also explored. Mice gradually placed on a restricted diet during mid-adulthood responded with a major reduction in body weight followed by a still significantly (about 20%) extended life span (Weindruch and Walford, 1982).

Since the effects of caloric restriction are very robust and potentially relevant to human nutrition, similar experiments are now being carried out with rhesus monkeys. These experiments may be complicated by the difficulties of keeping large mammals in a controlled situation that still provides enough physical activity and social interactions for them to be well. While data on the effect of caloric restriction on the life span of nonhuman primates are not yet available, preliminary results indicate that caloric restriction delays various age-related diseases (Lane et al., 2000).

The life-extending effect of caloric restriction is not limited to mammals. In *C. elegans,* mutations in many genes, called *eat* genes, cause partial starvation by disrupting the function of the pharynx, the feeding organ. Lakowski and Hekimi (1998) found that most *eat* mutations significantly lengthen life span (by up to 50%). In Section 29.3, we have distinguished two types of genes that can mutate to result in extended life spans. One group of genes (including *age-1*$^+$, *daf-2*$^+$, and *daf-16*$^+$) is involved in dauer formation, while another group (including *clk-1*$^+$) slows down development generally by reducing oxidative phosphorylation. Lakowski and Hekimi found that the long life of *eat-2* mutants does not require the activity of *daf-16*$^+$ and that *eat-2; daf-2* double mutants live even longer than the most long-lived *daf-2* mutants. These findings indicate that food restriction lengthens life span by a mechanism distinct from that of dauer-formation mutants. In contrast, food restriction does not further increase the life span of long-lived *clk-1* mutants, suggesting that *clk-1* and caloric restriction affect similar processes.

The physiological mechanism(s) through which caloric restriction delays senescence are still being investigated. In mammals, caloric restriction has a major effect on the endocrine system, causing rapid and major decreases in the blood levels of most hormones, including gonadal hormones (Richardson and Pahlavani, 1994). In human females, caloric restriction and/or extensive physical activity delay menarche and may stop menstrual cycles. The control of the carbohydrate metabolism is changed in such a way that energy requirements can be satisfied efficiently with low blood serum levels of glucose (Feuers et al., 1995). Such a "fuel-efficient" metabolism may reduce the harmful effects of unwanted protein glycosylation. In addition, calorie-restricted animals show, at least in some tissues, higher levels of those enzymes that render oxidants harmless. The central importance of these enzymes will be covered in the following section.

29.5 Oxidative Damage and Organismic Senescence

Except for a few organisms that are specially adapted to life under anaerobic conditions, all plants and animals require oxygen. Animals produce more than 90% of their metabolic energy in their mitochondria, in a process known as ***oxidative phosphorylation.*** It reduces molecular oxygen to water and harnesses much of the liberated energy for the synthesis of adenosine triphosphate (ATP), the cell's major energy currency. Although necessary for the survival of most cells, the reduction of molecular oxygen to water is also a hazardous process that inflicts much damage. According to the oxidative stress hypothesis, oxidative damage to cells is a driving force in organismic senescence.

OXIDATIVE PHOSPHORYLATION GENERATES AGGRESSIVE OXIDANTS AS INTERMEDIATES

During *oxidative phosphorylation,* carrier molecules in the inner mitochondrial membrane transfer electrons to molecular oxygen. By transfer of four electrons (e^-) and capture of four protons (H^+) from the surrounding aqueous medium, an oxygen molecule (O_2) is reduced to two molecules of water (H_2O). As the oxygen reduction occurs in four steps, intermediates are generated as follows (Fig. 29.12; Fridovich, 1978):

- superoxide radical, $\bullet O_2^-$
- hydrogen peroxide, H_2O_2
- hydroxyl radical, $\bullet OH$

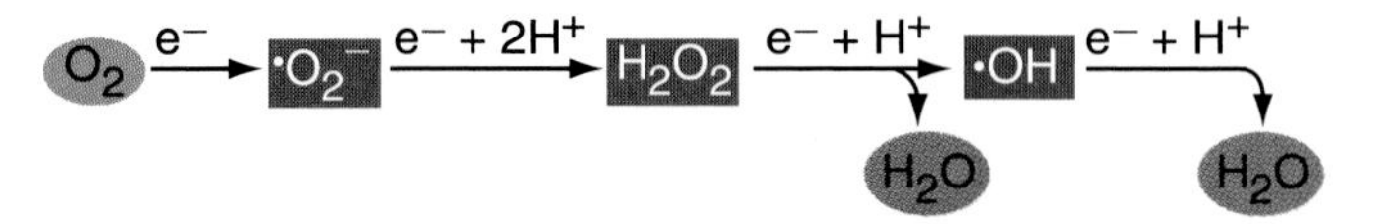

$\bullet O_2^- + \bullet O_2^- + 2H^+ \rightarrow H_2O_2 + O_2$ } Superoxide dismutases

$H_2O_2 + H_2O_2 \rightarrow 2H_2O + O_2$ } Catalases

$H_2O_2 + RH2 \rightarrow 2H_2O + R$ } Peroxidases

Figure 29.12 During normal oxidative phosphorylation, molecular oxygen is reduced to water by the stepwise addition of electrons and hydrogen ions. As by-products, highly reactive oxidants are generated. These primary oxidants are superoxide radical ($\bullet O_2^-$), hydrogen peroxide (H_2O_2), and hydroxyl radical ($\bullet OH$). In eukaryotic cells, several families of enzymes have evolved that convert oxidants into less harmful oxidants and/or harmless molecules.

The intermediates, in particular the two *radicals* (molecules with unpaired electrons), are very reactive. Because they are aggressive captors of electrons, they are also known as ***oxidants.*** By reacting with other molecules, oxidants can generate an array of additional ***reactive oxygen metabolites* (*ROMs*)** and cause extensive damage to biologically important macromolecules.

It is therefore not surprising that a set of enzymes has evolved that convert the oxidants into harmless water and oxygen (Fig. 29.12). In particular, these enzymes metabolize the first two oxidants, superoxide radical and hydrogen peroxide, thus limiting the formation of the third and most damaging oxidant, the hydroxyl radical.

- Superoxide dismutases convert superoxide radical into hydrogen peroxide and oxygen.
- Catalases convert hydrogen peroxide into water and oxygen.
- Peroxidases reduce hydrogen peroxide to water by oxidating other molecules, such as reduced glutathione.

Additional molecules in cells that reduce oxidants to water or other harmless molecules are known as ***antioxidants.*** Natural antioxidants include

- glutathione
- ascorbic acid (vitamin C)
- alpha-tocopherol (vitamin E)
- beta-carotene
- uric acid

OXIDANTS DAMAGE CELLULAR DNA, LIPIDS, AND PROTEINS

Oxidants and ROMs cause substantial damage to major types of cellular molecules, including DNA, lipids, and proteins. Damage to DNA ranges from oxidation products of single nucleotides to larger compounds, all of

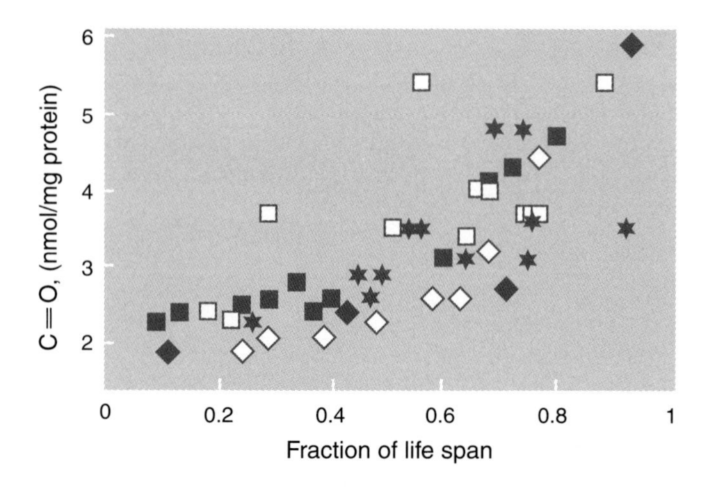

Figure 29.13 Carbonyl (C=O) content of protein from (■) human dermal fibroblasts in culture, (*) human eye lens, (□) human brain obtained at autopsy, (◆) rat liver, and (◇) whole fly. In each case, the carbonyl content increases progressively with age.

which impair DNA replication and transcription, and are thought to increase the chances of tumor development (Randerath et al., 1996). Both mitochondrial and nuclear DNA sustain ongoing oxidative damage, most of which is repaired by enzymatic mechanisms. It is estimated that the steady-state level of oxidized DNA nucleotides per cell is 24,000 in young rats and 66,000 in old rats (Helbock et al., 1998).

Oxidants also cause much damage to proteins by causing conformational changes as well as covalent modifications (Levine and Stadtman, 1996). The latter permanently destroy some or all of a protein's biological activity. A common and easily detected covalent modification is the formation of *carbonyl* (C=O) groups, which most frequently result from oxidation. The carbonyl content of proteins increases dramatically during the second half of the life span of both vertebrates and invertebrates (Fig. 29.13). It is estimated that 40 to 50% of all proteins in old individuals are present in oxidatively damaged form (Stadtman, 1992). Proteins from humans with Werner syndrome show a much much higher carbonyl content than in proteins from age-matched normal individuals. The amount of oxidized proteins in cells reflects the cumulative effects of all oxidant and antioxidant activities and the rate with which oxidized proteins are degraded. Since protein degradation relies on proteases and their regulation, and because oxidative damage to DNA accumulates with age, it seems likely that the protein degradation machinery itself becomes deficient with age.

In addition, oxidants react with lipids to form *lipid peroxides*—that is, lipids containing an –O–O– linkage. Lipid peroxides change the properties of biological membranes. For instance, they increase the uptake of low-density lipoprotein by vascular endothelial cells, and thus hasten the process of atherosclerosis. Lipid peroxide levels in serum are negatively correlated with life span among different mouse strains, and they decrease upon dietary restriction, another effect that may explain part of the life-extending effect of dietary restriction.

OXIDATIVE DAMAGE CAUSES SENESCENCE

According to the oxidative stress hypothesis, much of the deterioration that characterizes senescence is caused by oxidative damage (Sohal and Weindruch, 1996). Three testable predictions have been derived from this hypothesis. (1) Oxidative damage should increase during aging. (2) In comparisons between and within species, long life spans should be correlated with low oxidant and/or high antioxidant activities. (3) Genetic or experimental increase of antioxidant activity should slow down senescence.

The amount of oxidative damage does increase with age in DNA, proteins, and lipids, as discussed above. Indeed, at least some types of damage increase progressively, as does the rate of mortality (see Figs. 29.2b and 29.13). Indeed, the exhalation of ethane and *n*-pentane, both generated in the peroxidation of membrane lipids, increases exponentially (Sagai and Ichinose, 1987). In accord with the second prediction, the life spans of organisms are negatively correlated with peroxide levels and positively correlated with antioxidant levels (Emerit and Chance, 1992). For example, superoxide dismutase (SOD) levels in liver and uric acid levels in blood plasma are positively correlated with life span among rodents and primates. Also, in *age-1* and *daf-2* mutants of *C. elegans*, which have considerably extended life spans, the levels of both SOD and catalase activity are elevated late in life, and there is less oxidative damage to mitochondrial DNA (Lithgow, 1996).

According to the third prediction derived from the oxidative stress hypothesis, senescence should be postponed if antioxidant activity is enhanced by experimental or genetic means.

IN a study on Mongolian gerbils, Carney and coworkers (1991) measured the effects of an antioxidant, *N-tert-butyl-alpha-phenylnitrone* (*PBN*), on brain senescence, in terms of both protein damage and behavior. Male gerbils were injected twice daily with PBN dissolved in saline, while control animals were injected with saline only. In old gerbils, the PBN injections caused a marked decrease in carbonyl residues—the telltale of oxidized proteins—in brain tissue (Fig. 29.14). Indeed, the protein carbonyl content of the PBN-treated old gerbils decreased to the levels of young control males. Other biochemical parameters indicating oxidative damage to brain tissue improved as well. These were the activity levels of two enzymes, glutamine synthetase (GS) and neutral proteases, the latter known to be

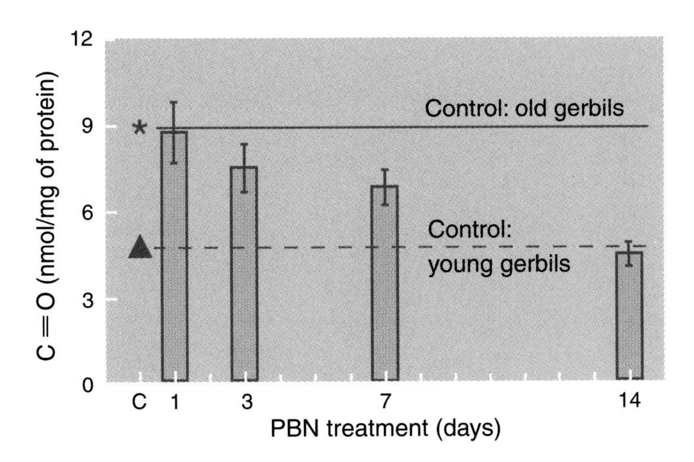

Figure 29.14 Effect of daily treatment with the antioxidant N-tert-butyl-alpha-phenylnitrone (PBN) on the protein carbonyl content of brain from old male gerbils. PBN was injected at 32 mg/kg twice daily. Each column represents the mean and standard error for three brains. The horizontal lines reflect the control values for untreated old gerbils and untreated young gerbils. The PBN treatment reduced the carbonyl content of old gerbil brains to the level found in untreated young gerbils.

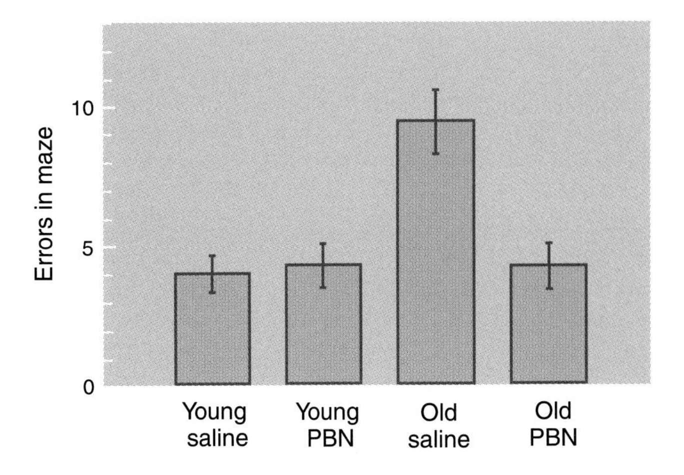

Figure 29.15 Effects of aging and PBN treatment on patrolling behavior of gerbils. Errors are defined here as those arms of a radial maze that are reentered before the gerbil has explored all eight arms of the maze. Old and young gerbils were either injected with PBN (see Fig. 29.14) or control-injected with saline. Each bar represents the mean and standard error for 18 gerbils in each group. PBN treatment greatly reduced the number of errors made by old gerbils but had no significant effect on young gerbils.

involved in the breakdown of defective proteins. The activity levels of both GS and proteases were depressed in old males versus young males, but the PBN injections brought the activity levels in old males back to the levels of young males. In young gerbils, which show less oxidative damage to brain tissue than old gerbils, PBN injections had no significant effect.

Parallel to the biochemical indicators of less oxidative stress, the spatial and temporal memory of old gerbils also improved as indicated by performance in maze tests (Fig. 29.15). Again, the PBN injections had no significant effect on the performance of young gerbils.

Questions

1. What would you expect when the PBN injections were discontinued?
2. How would you test whether the improved memory of the PBN-treated gerbils is based on the antioxidant effect of PBN rather than some other effect of PBN?

The experiments described above indicate that oxidative damage to proteins may limit brain function, and that an antioxidant can reduce the level of protein damage as well as the correlated impairment of brain function in old individuals. The same treatment has no significant effect in young individuals. Thus the antioxidant function counteracts oxidative damage that accumulates late in life.

Because superoxide dismutase (SOD) and catalase play key roles removing oxidants from cells, investigators have tested the effects of increasing the gene dosage for these enzymes (Orr and Sohal, 1994). *Drosophila* flies transgenic either for an additional SOD gene *or* for an additional catalase gene alone showed no consistent increase in longevity. Next, the transgenic lines were crossed to generate 15 lines made transgenic for one additional SOD gene *and* one additional catalase gene. Of these lines, 8 showed an extended life span, while 6 had no change in life span and 1 had a shorter life span. This variability was ascribed to the variable activity and potentially mutagenic effects of transgenes inserted with P-elements (see Section 15.6). The three lines with the longest life spans also showed a lower amount of protein carbonyl content, a delayed loss of walking speed, and a higher rate of oxygen consumption at old age. In similar experiments, Sun and Tower (1999) expressed SOD and/or catalase transgenes using the "flp-out" technique (see Fig. 22.49). This design had the advantage that the control (no transgene expression) and experimental (transgene expressed) flies were genetically equivalent except for the transgene activity. Under these conditions, catalase overexpression significantly increased resistance to hydrogen peroxide but had neutral or slightly negative effects on life span. SOD overexpression extended the mean life span up to 48%. Simultaneous overexpression of catalase with SOD had no added benefit, possibly due to a preexisting excess of catalase. Together, these data confirm that oxidative damage can be a limiting factor for the life span of adult *Drosophila*. However, the life-extending effects of extra SOD or catalase gene copies depend on the preexisting absolute and relative levels of these enzymes.

SUPEROXIDE RADICAL STIMULATES CELL DIVISION AND GROWTH

In addition to its deleterious effects, superoxide radical also seems to act in a signaling pathway that promotes cell division and growth (Irani et al., 1997). Specifically, superoxide radical transmits signal from a membrane-bound receptor tyrosine kinase to transcription factors, in parallel to the MAPK pathway (see Fig. 2.28).

The signaling function of superoxide suggests that low levels of superoxide radical may be necessary for normal cell division and growth. This would explain the protective effect of antioxidants against some forms of cancer. Assuming that superoxide radical has both beneficial and destructive effects, its regulation would appear to be another target for genes with *antagonistic pleiotropy.* Again, one might suspect that the actual level of superoxide that has evolved promotes the reproductive fitness of young individuals while possibly compromising the health of older individuals.

In summary, oxidative phosphorylation generates reactive intermediates, called oxidants, in the reduction of molecular oxygen to water. Oxidants and their metabolites cause unavoidable damage to DNA, proteins, and lipids, which accumulates with age. Several lines of evidence indicate that oxidative damage is a major contributor to organismic senescence.

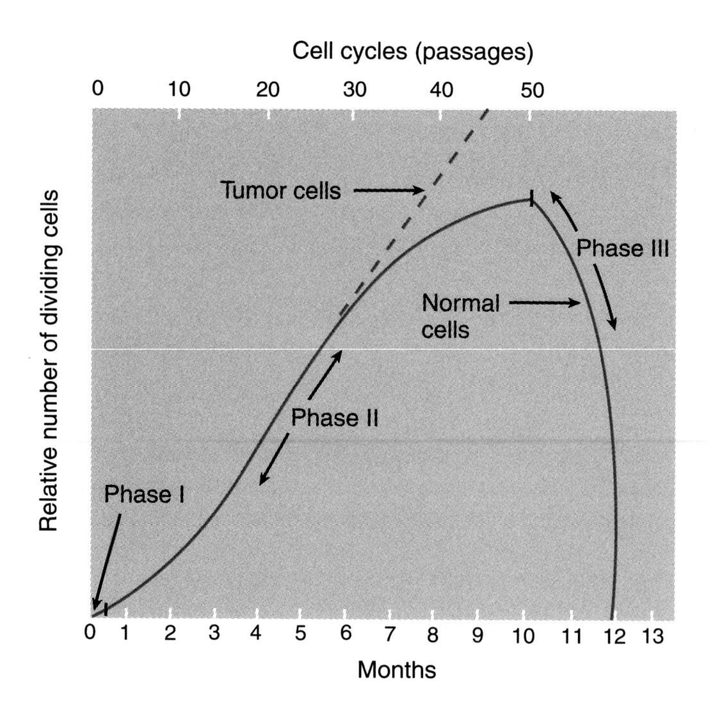

Figure 29.16 Three phases of vertebrate cell proliferation in culture. The ordinate represents the relative number of dividing cells as a measure of the replicative potential of all cells. In phase I, the culture is just establishing itself. During phase II, the cells are proliferating vigorously. The dashed line represents the trajectory of a cell line that has been immortalized by a mutation or the expression of a transgene. Such cells would form a tumor in vivo. In phase III, normal cells cease to grow but do not necessarily die.

29.6 Limited Cell Division and Telomerase

Many eukaryotic cells divide only a limited number of times but then remain viable for a long time without further proliferation. Several lines of evidence suggest that, on the one hand, replicative arrest may contribute to organismic senescence. On the other hand, replicative arrest is a powerful mechanism for suppressing tumors.

SOMATIC MAMMALIAN CELLS SHOW A LIMITED CAPACITY FOR PROLIFERATION

Before 1960, most researchers thought that senescence was a characteristic of organisms rather than cells. It therefore came as a surprise when Hayflick (1965) discovered that cultured mammalian cells show only a limited ability to divide. He distinguished three phases in the development of cells that are cultured in fluid media in glass or plastic dishes or flasks (Fig. 29.16). During phase I, new cells adhere to the surface of the culture vessel and begin to divide. When the cells have completely covered the available surface, usually after a few days, the cells stop dividing, a phenomenon called *contact inhibition* or *confluency* (see Fig. 28.15). To initiate phase II, half of the cells is discarded, and the cells double to reach confluency again. This cycle, called a *passage,* can be repeated many times. Hayflick's fundamental discovery was that there is an end to phase II—that is, an upper limit to the number of passages that the cells will go through. With the beginning of phase III, cell division slows down and eventually ceases. The cells then enter a G_0 state from which they do not return, no matter how accommodating the culture conditions may be. This phenomenon has been observed in cell cultures from all mammals and birds investigated so far. The maximum number of passages that cultured cells will undergo is called *Hayflick's limit* or ***replicative limit.***

The transition from phase II to phase III has also been termed *cellular senescence.* This name suggests that the number of cell divisions is also limited in vivo, and that it is the cause of, or at least contributes to, the process of *organismic senescence.* There is circumstantial evidence that this may be the case. First, the length of phase II—or the number of passages—is somewhat dependent on the age of the donor of the cells. Hayflick (1965) found 35 to 63 passages for fibroblasts taken from fetal human lung, as compared with 14 to 29 passages for fibroblasts taken from adult human lung. An extensive study on human skin fibroblasts showed a decrease in average passage number from 45 for fetal fibroblasts to 28 for fibroblasts from donors aged 70 to 80 years (Martin et al.,

1970). Thus cell divisions completed in vivo seem to count against Hayflick's limit. The same pattern has been found with several other species and cell types sampled (Stanulis-Praeger, 1987). Also, cells taken from persons with *Werner syndrome,* who senesce and die prematurely (see Fig. 29.10), show a greatly reduced capacity for proliferation in culture. Conversely, cells taken from tumors divide without limitation; such cells are called *immortal* or *transformed.*

The replicative limit of cells does not seem to be an artifact of cell culture because serially transplanted tissue shows a similar decline in proliferative capacity in vivo (Krohn, 1962, 1966; Daniel, 1977). Also, the replicative limit changes from species to species and is correlated with the species' maximum life span. However, a major problem with viewing Hayflick's limit as causally related to organismic senescence is that cells taken from senescent donors still undergo many rounds of cell division before they reach their replicative limit.

CHROMOSOMAL ENDS ARE PROTECTED BY TELOMERES

A possible mechanism for triggering the end to cell replication revolves around the ends of eukaryotic chromosomes, known as ***telomeres*** (Fig. 29.17; Gk. *telos,* "end"; *meros,* "part"). Telomeric DNA is characterized by short tandem repeats that contain a block of consecutive G nucleotides. In humans and many other organisms, this repeat sequence is GGGTTA. The length of telomeric DNA varies but generally seems to comprise several thousand base pairs. The telomeric DNA associates with specific proteins, which protect the chromosomal ends from degradation and from attachment to other chromosomes. Thus telomeres protect the ends of chromosomes much like the plastic caps at the ends of shoelaces.

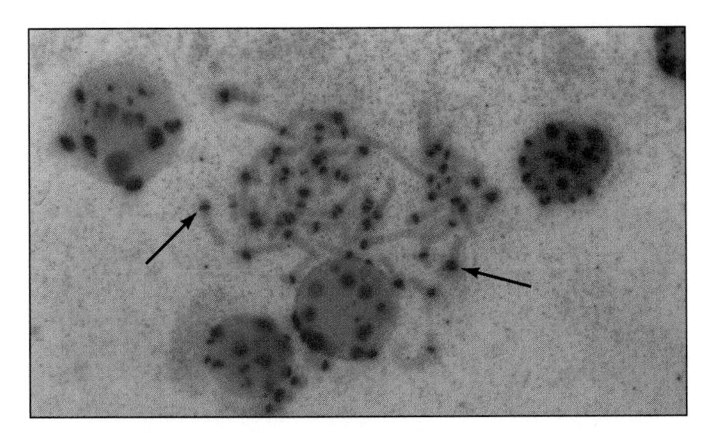

Figure 29.17 Telomeres in mouse chromosomes. Nuclei (blue bodies) from mouse cells in mitosis were broken to release condensed chromosomes. Telomeres (arrows) were hybridized in situ (see Method 8.1) with a DNA probe binding specifically to the telomere repeat sequence.

Telomeres wear off in most cell types because of the so-called *end-replication problem,* which is inherent to the replication of linear (noncircular) DNA, such as the chromosomal DNA of eukaryotes. The core enzyme of DNA replication, DNA polymerase, depends on an RNA primer, which it extends in the 3′ direction. After primer removal, the parental strands are left with 3′ overhangs that cannot be replicated (Fig. 29.18). Therefore, without additional mechanisms, eukaryotic chromosomal DNA should become a little shorter with each round of DNA replication. This has indeed been observed in different cell types in vivo and in vitro (Allsopp et al., 1995). If the DNA lost from the ends contained functional parts of genes, then these genes would be damaged or lost. However, since telomere DNA is noncoding, the losses at the chromosomal ends are inconsequential so long as enough telomeric DNA is left to form a functional telomere.

The average loss of telomeric DNA per cell division in human fibroblasts has been estimated between 50 and 150 base pairs (*Stein and Dulic, 1995;* Martens et al., 2000; Urquidi et al., 2000). Assuming an average of 100 base pairs lost per round of DNA replication, and about 10 kb of DNA per telomere in embryonic cells, a cultured cell would reduce the length of its telomeres to 2 kb, which is thought to be close to minimal, after 80 passages. Based on these numbers, cell replication would seem to stop when telomeres approach minimal length. However, some cells stop dividing when their telomeres are still several kb long. In particular, mice have telomeres that are long enough for multiple lifetimes of shortening, as will be discussed later on. Apparently, these cells have a surveillance mechanism that inhibits mitosis before their telomeres become critically short.

What about germ line cells, which have been dividing for about 1 billion years; tumor cells, most of which seem to be immortal in culture; or even somatic stem cells that might surpass Hayflick's limit during the life time of a single organism? All these cells have an enzyme that rebuilds the telomeric DNA that is lost due to the end-replication problem.

TELOMERASE ACTIVITY ALLOWS UNLIMITED CELL PROLIFERATION

An enzyme known as ***telomerase*** functions to extend the 3′ end—or overhang—of the parental DNA strand (Fig. 29.18). In the absence of complementary DNA, the enzyme uses as a template its own RNA moiety, which includes a sequence complementary to the telomere repeat. After several rounds of extension by telomerase, the 3′ end of the parental DNA strand can be replicated without a net loss of DNA length.

The activity of telomerase in cells is correlated with their proliferative capacity (Blasco et al., 1999). Measurements on different cell types, both in vivo and in vitro,

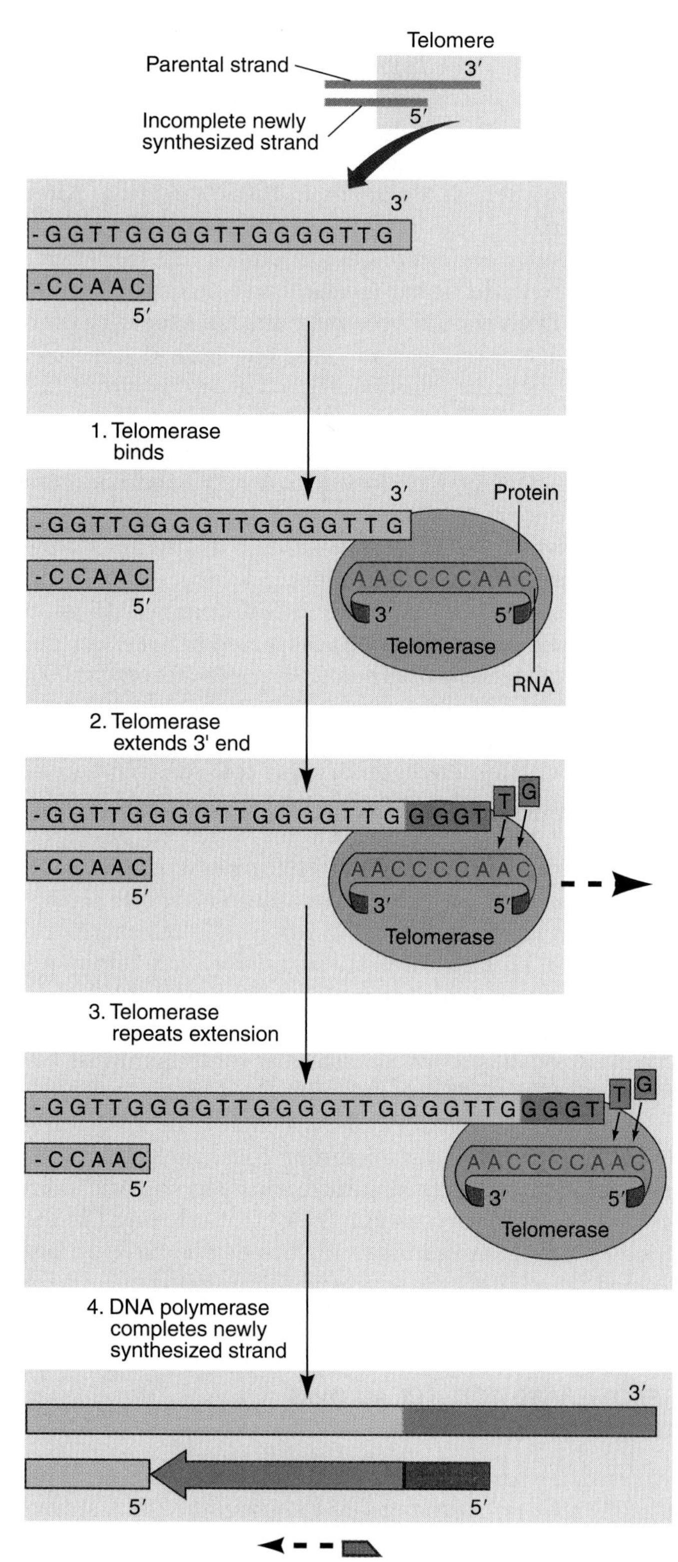

Figure 29.18 Telomeres and telomerase. Telomeres are the ends of eukaryotic chromosomes. Their DNA is characterized by tandem repeats. (Shown here is the GGGTTG repeat found in the protozoon *Tetrahymena.*) At each round of DNA replication, the 3′ end of the parental strand cannot be replicated by DNA polymerase enzymes (end-replication problem). The resulting overhang interacts with telomerase, an enzyme consisting of protein and RNA components. The RNA component contains a 9-nucleotide sequence complementary to 1.5 telomeric repeat units. Part of it hybridizes to the 3′ end of the parental DNA strand so that the remaining part of the telomeric RNA acts as a template for the addition of another telomeric repeat unit. The process is reiterated until the extended 3′ overhang is long enough for regular DNA replication enzymes to extend the newly synthesized DNA strand to the original length of the parental strand.

indicate that telomerase activity is high in most cancer cells and in normal germ line cells, low in somatic stem cells, and nondetectable in other somatic cells (Kim et al., 1994; Allsopp et al., 1995; Colitz et al., 1999).

In order to test whether telomerase can be sufficient to promote cell proliferation beyond Hayflick's limit, investigators have genetically transformed different types of cultured cells with cloned cDNAs encoding telomerase (see Section 15.4). Indeed, such cells maintained telomere length and divided vigorously for many passages beyond Hayflick's limit without displaying other characteristics of cancer cells (Bodnar et al., 1998; Yang et al., 1999). Forced expression of telomerase from a transgene also counteracts the premature senescence of cells from persons with Werner syndrome (Wyllie et al., 2000).

To establish whether telomerase activity is necessary for the normal functions of germ line cells and stem cells, investigators have eliminated the gene encoding the RNA component of telomerase from mice by targeted recombination (see Section 15.5). They found that such mice underwent progressive telomere shortening and could only be bred for a few generations before they expired due to male and female infertility and an increased embryonic lethality associated with failure to close the neural tube. The exact number of generations depended on genetic background and decreased with the original telomere length present in the mouse strain used. The mice of the last possible generation died prematurely, showing reduced proliferative capacity of hematopoietic cells as well as atrophy of the spleen and small intestine (Lee et al., 1998; Herrera et al., 1999). In addition, the knockout mice showed a reduced capacity for wound healing and replacing lost blood (Rudolph et al., 1999). These findings demonstrate a critical role for telomerase in the health of mice, in particular of those organs that depend on a high rate of cell proliferation.

Curiously, the mice lacking telomerase nevertheless showed an increased rate of spontaneous malignancies.

This paradoxical result indicates that some tumor cells have the capacity to divide without telomerase, suggesting that there are alternate ways of overcoming the end-replication problem. The same conclusion must be drawn from the fact that a minority of human tumors maintain telomere length and proliferate without telomerase activity (Bryan et al., 1997).

In summary, the loss of telomerase activity limits the number of cell divisions that somatic cells can undergo. Conversely, the persistance of telomerase activity in germ line cells and somatic stem cells is required for the ongoing ability of these cells to divide.

LOSS OF TELOMERASE ACTIVITY IS A SAFEGUARD AGAINST CANCER

Why have birds and mammals evolved to switch off the telomerase genes in the somatic cells except for stem cells? In terms of the disposable soma hypothesis, is this a trick of the germ line to keep its somatic carriers disposable and maintainance costs low? Could we humans, as the intellectually most advanced somatic carriers, outsmart 100 million years of evolution and decide we don't want to be disposable? Can we drink from the fountain of youth and stay forever young by keeping our telomerase genes on in all our tissues? Probably not.

The lack of telomerase activity in most of our somatic cells not only limits our capability for cell division but also helps us to avoid most tumors and cancers. As discussed in Section 28.7, the development of a tumor requires that a proto-oncogene mutate or be otherwise deregulated to stimulate cell division. In addition, tumor suppressor genes that restrict the cell cycle must be inactivated. By the time both events have occurred, the organism in which this happens may have lived more than half its normal life span. At this point, the replicative limit may be 30 cell divisions away. If so, then the maximum number of descendants that a tumor cell can produce is 2^{30}, or approximately 1 billion, which would amount to a tumor about 1 cm^3 in size.

More than 95% of immortal cell lines show telomerase activity, indicating that most tumors were selected for their ability to overcome the normal inhibition of telomerase gene expression. It seems therefore likely that many more tumors would originate if telomerase genes were routinely expressed in somatic cells. Since the likelihood for the first two steps of tumor development to occur accumulates with age, the elimination of telomerase inhibition would almost certainly cause a dramatic increase in the frequency of tumors in older individuals.

A more promising application of the our recent knowledge about telomerase may be the development of treatments that switch off the inappropriate synthesis of telomerase in cancer cells. This will not cure all types of cancer because some cells seem to find ways of maintaining telomeres without re-expressing telomerase, as mentioned earlier. Another difficulty is that telomerase activity must not be eliminated from healthy stem cells. Nevertheless, in combination with other treatments, anti-telomerase drugs may well become a new weapon in the arsenal against cancer. The fountain of youth, meanwhile, will remain a dream.

SUMMARY

Most higher organisms reproduce sexually by haploid gametes that fuse to generate a zygote, which develops into an adult, which in time will again produce gametes. The cells that produce gametes in each successive generation are an uninterrupted lineage known as the germ line. The other cells that form the developing bodies of each generation are called somatic cells. While germ line cells are potentially immortal, the somatic bodies are bound to die. Each somatic body can therefore be viewed as the temporary carrier of its germ line.

Before somatic bodies die they undergo a process of deterioration called senescence. The ultimate cause of senescence seems to be the propagation of the germ line with minimal costs for the repair and maintenance of the somatic bodies. Senescence may evolve through the accumulation of random deleterious mutations that act late in adult life, when the force of selection against such mutations wanes, as old individuals normally constitute a small and reproductively inactive part of a population. Senescence may also be due to antagonistic pleiotropy of genetic alleles that control trade-offs between enhanced fitness early in life and an accelerated senescence.

The decrease in vigor that characterizes senescence can be measured by a wide array of physiological parameters. However, these parameters do not change in synchrony, so that choosing any one of them as the defining criterion of senescence would be arbitrary. Senescence is therefore defined statistically based on survival data compiled for cohorts of individuals or from census data. The ratio of individuals dying within a time interval divided by the number of individuals alive at the beginning of the interval is called mortality rate. The increase in the mortality rate toward the end of the life span is called senescence.

A few genes have been characterized that affect senescence in *C. elegans,* two *Drosophila* species, the house mouse, and the human. These genes are involved in a wide range of biological processes, most of which affect the resilience of the organism against stress factors such as starvation, desiccation, high temperature, or oxidative damage.

Caloric restriction of mice, rats, and hamsters retards their growth and development but increases their life span substantially, improves their performance in physiological tests, and reduces the frequency of many diseases. The mechanisms by which caloric restriction causes these effects are still under investigation.

Oxidative damage to DNA, proteins, and lipids is caused as an unavoidable side effect of oxidative phosphorylation, which generates reactive intermediates, called oxidants, in the reduction of molecular oxygen to water. Several lines of evidence indicate that oxidative damage is a major contributor to organismic senescence.

Many eukaryotic cells divide only a limited number of times but then remain viable for a long time without further proliferation. Correlative evidence suggest that this replicative arrest may contribute to organismic senescence. A possible mechanism for triggering the end to cell replication revolves around the ends of eukaryotic chromosomes, known as telomeres. A small segment of telomeric DNA is lost after each round of DNA replication, unless it is rebuilt by an enzyme known as telomerase. Telomerase activity is present in germ line cells and tumor cells, allowing these cells to divide indefinitely. The loss of telomerase activity from most somatic cells limits cell proliferation and may contribute to senescence, but it also protects against cancer.

references

Aberle H., Bauer A., Stappert J., Kispert A. and Kemler R. (1997) β-Catenin is a target for the ubiquitin-proteasome pathway. *EMBO J.* **16:** 3797–3804.

Adams C.C. and Workman J.L. (1993) Nucleosome displacement in transcription. *Cell* **72:** 305–308.

Adams J.C. and Watt F.M. (1993) Regulation of development and differentiation by the extracellular matrix. *Development* **117:** 1183–1198.

Adams M.D., Tarng R.S. and Rio D.C. (1997) The alternative splicing factor PSI regulates P-element third intron splicing in vivo. *Genes and Dev.* **11:** 129–138.

Adams M.D., et al. (2000) The genome sequence of *Drosophila melanogaster. Science* **287:** 2185–2195.

Affara N.A. (1991) Sex and the single Y. *BioEssays* **13:** 475–478.

Afzelius B.A. (1976) A human syndrome caused by immotile cilia. *Science* **193:** 317–319.

Afzelius B.A. (1999) Asymmetry of cilia and of mice and men. *Int. J. Dev. Biol.* **43:** 283–286.

Agius E., Oelgeschläger M., Wessely O., Kemp C. and De Robertis E.M. (2000) Endodermal Nodal-related signals and mesoderm induction in *Xenopus. Development* **127:** 1173–1183.

Akam M. (1987) The molecular basis for metameric pattern in the *Drosophila* embryo. *Development* **101:** 1–22.

Akashi K., Traver D., Miyamoto T. and Weissman I. (2000) A clonogenic common myeloid progenitor that gives rise to all myeloid lineages. *Nature* **404:** 193–197.

Alberts B., Bray D., Lewis J., Raff M., Roberts K. and Watson J.D. (1983) *Molecular Biology of the Cell.* New York: Garland Publishing, Inc.

Alberts B., Bray D., Lewis J., Raff M., Roberts K. and Watson J.D. (1989) *Molecular Biology of the Cell.* 2nd ed. New York: Garland Publishing, Inc.

Alberts B., Bray D., Lewis J., Raff M., Roberts K. and Watson J.D. (1994) *Molecular Biology of the Cell.* 3rd ed. New York: Garland Publishing, Inc.

Albrecht-Buehler G. (1977) Daughter 3T3 cells: Are they mirror images of each other? *J. Cell Biol.* **72:** 595–603.

Alessa L. and Kropf D.L. (1999) F-actin marks the rhizoid pole in living *Pelvetia compressa* zygotes. *Development* **126:** 201–209.

Alevizopoulos A. and Mermod N. (1997) Transforming growth factor-beta: The breaking open of a black box. *BioEssays* **19:** 581–591.

Ali I.U. and Hynes R.O. (1977) Effects of cytochalasin B and colchicine on attachment of a major surface protein of fibroblasts. *Biochem. Biophys. Acta* **471:** 16–24.

Allen C.A. and Green D.P. (1997) The mammalian acrosome reaction: Gateway to sperm fusion with the oocyte? *BioEssays* **19:** 241–247.

Allsopp R.C., Chang E., Kashefi-Aazam M., Rogaev E.I., Piatyszek M.A., Shay J.W. and Harley C.B. (1995) Telomere shortening is associated with cell division in vitro and in vivo. *Exp. Cell. Res.* **220:** 194–200.

Alnemri E.S. (1997) Mammalian cell death proteases: A family of highly conserved aspartate specific cysteine proteases. *J. Cell. Biochem.* **64:** 33–42.

Amaya E., Musci T.J. and Kirschner M.W. (1991) Expression of a dominant negative mutant of the FGF receptor disrupts mesoderm formation in *Xenopus embryos. Cell* **66:** 257–270.

Amaya E., Stein P.A., Musci T.J. and Kirschner M.W. (1993) FGF signalling in the early specification of mesoderm in *Xenopus. Development* **118:** 477–487.

Ambros V. and Horvitz H.R. (1984) Heterochronic mutants of the nematode *Caenorhabditis elegans. Science* **226:** 409–416.

Anderson C.L. and Meier S. (1981) The influence of the metameric pattern in the mesoderm on migration of cranial neural crest cells in the chick embryo. *Dev. Biol.* **85:** 383–402.

Anderson D.J., Groves A., Lo L., Ma Q., Rao M., Shah N.M. and Sommer L. (1997) Cell lineage determination and the control of neuronal identity in the neural crest. *Cold Spring Harbor Symp. Quant. Biol.* **LXII:** 493–504.

Anderson K.V. (1998) Pinning down positional information: dorsal-ventral polarity in the *Drosophila* embryo. *Cell* **95:** 439–442.

Anderson K.V. and Nüsslein-Volhard C. (1984) Information for the dorsal-ventral pattern of the *Drosophila* embryo is stored as maternal mRNA. *Nature* **311:**223–227.

Anderson K.V., Bokla L. and Nüsslein-Volhard C. (1985) Establishment of dorsal-ventral polarity in the Drosophila embryo: The induction of polarity by the Toll gene product. *Cell* **42:** 791–798.

Anstrom J.A., Chin J.E., Leaf D.S., Parks A.L. and Raff R.A. (1987) Localization and expression of msp130, a primary mesenchyme lineage-specific cell surface protein of the sea urchin embryo. *Development* **101:** 255–265.

Antequera F. and Bird A. (1993) In: *DNA Methylation: Molecular, Biology and Biological Significance*, J.P. Jost and H.P. Saluz (eds.), 169. Basel: Birkhäuser-Verlag.

Antequera F. and Bird A. (1999) CpG islands as genomic footprints of promoters that are associated with replication origins. *Curr. Biol.* **9:** R661–667.

Arango N.A., Lovell-Badge R. and Behringer R.R. (1999) Targeted mutagenesis of the endogenous mouse Mis gene promoter: In vivo definition of genetic pathways of vertebrate sexual development. *Cell* **99:** 409–419.

Arking R. (1998) *The Biology of Aging.* 2nd ed. Sunderland, MA: Sinauer.

Arnold A.P. (1975) The effects of castration and androgen replacement on song, courtship, and aggression in zebra finches. *J. Exp. Zool.* **191:** 309–326.

Arnold A.P. (1980) Sexual differences in the brain. *Am. Sci.* **68:** 165–173.

Arnold A.P., Nottebohm F. and Pfaff D. (1976) Hormone concentrating cells in vocal control and other areas of the brain of the zebra finch (*Poephila guttata*). *J. Comp. Neurol.* **165:** 487–512.

Arnone M.I. and Davidson E.H. (1997) The hardwiring of development: organization and function of genomic regulatory systems. *Development* **124:** 1851–1864.

Artavanis-Tsakonas S., Matsuno K. and Fortini M.E. (1995) Notch signaling. *Science* **268:** 225–232.

Artavanis-Tsakonas S., Rand M.D. and Lake R.J. (1999) Notch signaling: Cell fate control and signal integration in development. *Science* **284:** 770–776.

Asakura S., Eguchi G. and Iino T. (1966) *Salmonella* flagella: *In vitro* reconstruction and overall shapes of flagellar filaments. *J. Mol. Biol.* **16:** 302–316.

Ashburner M. (1972) Patterns of puffing activity in the salivary gland chromosomes of *Drosophila. Chromosoma* **38:** 255–281.

Ashburner M. (1990) Puffs, genes, and hormones revisited. *Cell* **61:** 1–3.

Ashburner M. and Gelbart W. (1991) *Drosophila* genetic maps. *Drosophila Information Service* **69** (May).

Ashburner M., Chihara C., Meltzer P. and Richards G. (1974) Temporal control of puffing activity in polytene chromosomes. *Cold Spring Harbor Symp. Quant. Biol.* **38:** 655–662.

Ashworth D., Bishop M., Campbell K., Colman A., Kind A., Schnieke A., Blott S., Griffin H., Haley C., McWhir J. and Wilmut I. (1998) DNA microsatellite analysis of Dolly. *Nature* **394:** 329.

Austad S. (1993) Retarded senescence in an insular population of Virginia opossums (*Didelphis virginiana*). *J. Zool.* **229:** 695–708.

Austin C.R. (1952) The "capacitation" of the mammalian spermatozoa. *Nature* **170:** 326.

Austin C.R. (1965) *Fertilization.* Englewood Cliffs, NJ: Prentice Hall.

Avery O.T., MacLeod C.M. and McCarty M. (1944) Studies on the chemical nature of the substance inducing transformation of pneumococcal types: Induction of transformation by a desoxyribonucleic acid fraction isolated from pneumococcus type III. *J. Exp. Med.* **79:** 137–158.

Azpiroz-Leehan R. and Feldmann K.A. (1997) T-DNA insertion mutagenesis in *Arabidopsis:* Going back and forth. *Trends Genet.* **13:** 152–156.

Baker B.S. (1989) Sex in flies: The splice of life. *Nature* **340:** 521–524.

Balinsky B.I. (1975) *An Introduction to Embryology.* Philadelphia: Saunders.

Bandziulis R.J., Swanson M.S. and Dreyfuss G. (1989) RNA-binding proteins as developmental regulators. *Genes and Dev.* **3:** 431–437.

Bank A. (1996) Human somatic gene therapy. *BioEssays* **18:** 999–1007.

Bantock C.R. (1970) Experiments on chromosome elimination in the gall midge, *Mayetiola destructor. J. Embryol. exp. Morphol* **24:** 257–286.

Barbash D.A. and Cline T.W. (1995) Genetic and molecular analysis of the autosomal component of the primary sex determination signal of *Drosophila melanogaster. Genetics* **141:** 1451–1471.

Bare R.L. and Torti F.M. (1998) Endocrine therapy of prostate cancer. *Cancer Treat Res.* **94:** 69–87.

Barker D.D., Wang C., Moore J., Dickinson L.K. and Lehmann R. (1992) Pumilio is essential for function but not for distribution of the *Drosophila* abdominal determinant nanos. *Genes and Dev.* **6:** 2312–2326.

Barkoff A.F., Dickson K.S., Gray N.K. and Wickens M. (2000) Translational control of cyclin B1 mRNA during meiotic maturation: Coordinated repression and cytoplasmic polyadenylation. *Dev. Biol.* **220:** 97–109.

Baroffio A., Dupin E. and Le Douarin N.M. (1991) Common precursors for neural and mesectodermal derivatives in the cephalic neural crest. *Development* **112:** 301–305.

Barth K.A., Kishimoto Y., Rohr K.B., Seydler C., Schulte-Merker S. and Wilson S.W. (1999) Bmp activity establishes a gradient of positional information throughout the entire neural plate. *Development* **126:** 4977–4987.

Basler K. and Struhl G. (1994) Compartment boundaries and the control of the *Drosophila* limb pattern by hedgehog protein. *Nature* **368:** 208–214.

Basson C.T., Huang T., Lin R.C., Bachinsky D.R., Weremowicz S., Vaglio A., Bruzzone R., Quadrelli R., Lerone M., Romeo G., Silengo M., Pereira A., Krieger J., Mesquita S.F., Kamisago M., Morton C.C., Pierpont M.E., Muller C.W., Seidman J.G. and Seidman C.E. (1999) Different TBX5 interactions in heart and limb defined by Holt-Oram syndrome mutations. *Proc. Nat. Acad. Sci. USA* **96:** 2919–2924.

Batchelder C., Dunn M.A., Choy B., Suh Y., Cassie C., Shim E.Y., Shin T.H., Mello C., Seydoux G. and Blackwell T.K. (2000) Transcriptional repression by the *Caenorhabditis elegans* germ-line protein PIE-1. *Genes and Dev.* **13:** 202–212.

Bateson W. (1894) *Materials for the Study of Variation.* London: Macmillan.

Bautzmann H., Holtfreter J., Spemann H. and Mangold O. (1932) Versuche zur Analyse der Induktionsmittel in der Embryonalentwicklung. *Naturwissenschaften* **20:** 971–974.

Bayna E.M., Shaper J.H. and Shur B.D. (1988) Temporally specific involvement of cell surface β-1,4 galactosyltransferase during mouse embryo morula compaction. *Cell* **53:** 145–157.

Baynash A.G., Hosoda K., Giaid A., Richardson J.A., Emoto N., Hammer R.E. and Yanagisawa M. (1994) Interaction of endothelin-3 with endothelin-B receptor is essential for development of epidermal melanocytes and enteric neurons. *Cell* **79:** 1277–1285.

Beachy P.A., Krasnow M.A., Gavis E.R. and Hogness D.S. (1988) An *Ultrabithorax* protein binds sequences near its own and the *Antennapedia* P1 promoters. *Cell* **55:** 1069–1081.

Beall E.L. and Rio D.C. (1997) *Drosophila* P-element transposase is a novel site-specific endonuclease. *Genes and Dev.* **11:** 2137–2151.

Beams H.W. and Kessel R.G. (1974) The problem of germ cell determinants. *Int. Rev. Cytol.* **39:** 413–479.

Beams H.W. and Kessel R.G. (1976) Cytokinesis: A comparative study of cytoplasmic division in animal cells. *Am. Sci.* **64:** 279–290.

Beatty R.A. (1967) Parthenogenesis in vertebrates. In: *Fertilization,* C.B. Metz and A. Monroy (eds.), vol. 1, 413–440. New York: Academic Press.

Becker A.J., McCulloch E.A. and Till J.E. (1963) Cytological demonstration of the clonal nature of spleen cells derived from transplanted mouse marrow cells. *Nature* **197:** 452–454.

Beddington R.S.P. (1994) Induction of a second neural axis by the mouse node. *Development* **120:** 613–620.

Beelman C.A. and Parker R. (1995) Degradation of mRNA in eukaryotes. *Cell* **81:** 179–183.

Beelman C.A., Stevens A., Caponigro G., LaGrandeur T.E., Hatfield L., Fortner D.M. and Parker R. (1996) An essential component of the decapping enzyme required for normal rates of mRNA turnover. *Nature* **382:** 642–646.

Beeman R.W., Stuart J.J., Brown S.J. and Denell R.E. (1993) Structure and function of the homeotic gene complex (HOM-C) in the beetle, *Tribolium castaneum. BioEssays* **15:** 439–444.

Beermann W. (1952) Chromosomes and genes. In: *Developmental Studies on Giant Chromosomes,* W. Beermann (ed.), 1–33. New York: Springer-Verlag.

Behringer R.R., Finegold M.J. and Cate R.L. (1994) Müllerian inhiting substance function during mammalian sexual development. *Cell* **79:** 415–425.

Beisson J. and Sonneborn T.M. (1965) Cytoplasmic inheritance of the organization of the cell cortex in *Paramecium aurelia. Proc. Nat. Acad. Sci. USA* **53:** 275–282.

Bejsovec A. and Wieschaus E. (1993) Segment polarity gene interactions modulate epidermal patterning in *Drosophila* embryos. *Development* **119:** 501–517.

Bel S., Coré N., Djabali M., Kieboom K., Van der Lugt N., Alkema M.J. and Van Lohuizen M. (1998) Genetic Interactions and dosage effects of Polycomb group genes in mice. *Development* **121:** 3543–3551.

Bell A.C. and Felsenfeld G. (1999) Stopped at the border: boundaries and insulators. *Curr. Opin. Genet. Dev.* **9:** 191–198.

Bell G. (1984) Evolutionary and nonevolutionary theories of senescence. *Am. Nat.* **124:** 600–603.

Bell G. (1988) *Sex and Death in Protozoa, the History of an Obsession,* Cambridge, MA: Cambridge University Press.

Bell L.R., Main E.M., Schedl P. and Cline T.W. (1988) *Sex-lethal,* a *Drosophila* sex determination switch gene, exhibits sex-specific RNA splicing and sequence similarity to RNA binding proteins. *Cell* **55:** 1037–1046.

Bell L.R., Horabin J.I., Schedl P. and Cline T.W. (1991) Positive autoregulation of *Sex-lethal* by alternative splicing maintains the female determined state in *Drosophila. Cell* **65:** 229–239.

Bender M., Imam F.B., Talbot W., Ganetzky B. and Hogness D. (1997) *Drosophila* ecdysone receptor mutations reveal functional differences among receptor isoforms. *Cell* **91:** 777–788.

Benezra R., Davis R.L., Lockshon D., Turner D.L. and Weintraub H. (1990) The protein Id: A negative regulator of helix-loop-helix DNA binding proteins. *Cell* **61:** 49–59.

Bennett D. (1975) The T-locus of the mouse. *Cell* **6:** 441–454.

Bentley J.K., Shimomura H. and Garbers D.L. (1986) Retention of a functional resact receptor in isolated sperm plasma membranes. *Cell* **45:** 281–288.

Benya P.D. and Shaffer J.D. (1982) Dedifferentiated chondrocytes reexpress the differentiated collagen phenotype when cultured in agarose gels. *Cell* **30:** 215–224.

Bergsten S.E. and Gavis E.R. (1999) Role for mRNA localization in translational activation but not spatial restriction of nanos RNA. *Development* **126:** 659–669.

Berleth T. and Jürgens G. (1993) The role of the monopteros gene in organising the basal body region of the *Arabidopsis* embryo. *Development* **118:** 575–587.

Bermingham J.R., Jr., Martinez-Arias A., Petitt M.G. and Scott M.P. (1990) Different patterns of transcription from the two *Antennapedia* promoters during *Drosophila* embryogenesis. *Development* **109:** 553–566.

Berridge M.J., Bootman M.D. and Lipp P. (1998) Calcium—a life and death signal. *Nature* **395:** 645–648.

Berta P., Hawkins J.R., Sinclair A.H., Taylor A., Griffiths B.L., Goodfellow P.N. and Fellous M. (1990) Genetic evidence equating *SRY* and the testis-determining factor. *Nature* **348:** 448–449.

Beyer A.L. and Osheim Y.N. (1990) Ultrastructural analysis of the ribonucleoprotein substrate for pre-mRNA processing. In: *The Eukaryotic Nucleus: Molecular Biochemistry and Macromolecular Assemblies,* vol. 2, P.R. Strauss and S.H. Wilson (eds.), 431–445. Caldwell, NJ: Telford Press.

Biehs B., Francois V. and Bier E. (1996) The *Drosophila short gastrulation* gene prevents Dpp from autoactivating and suppressing neurogenesis in the neuroectoderm. *Genes and Dev.* **10:** 2922–2934.

Bier K.H. (1963) Autoradiographische Untersuchungen über die Leistungen des Follikelepithels und der Nährzellen bei der Dotterbildung und Eiweissynthese im Fliegenovar. *Wilhelm Roux's Arch.* **154:** 552–575.

Bier K.H. (1964) Die Kern-Plasma-Relation und das Riesenwachstum der Eizellen. *Zool. Anz.* (Supplement 27): 84–91.

Bignon Y.-J., Chen Y., Chang C.-Y., Riley D.J., Windle J.J., Mellon P.L. and Lee W.-H. (1993) Expression of a retinoblastoma transgene results in dwarf mice. *Genes and Dev.* **7:** 1654–1662.

Binari R.C., Staveley B.E., Johnson W.A., Godavarti R., Saisekharan R. and Manoukian A.S. (1997) Genetic evidence that heparin-like glycosaminoglycans are involved in *wingless* signaling. *Development* **124:** 2623–2632.

Bird A. (1992) The essentials of DNA methylation. *Cell* **70:** 5–8.

Birgbauer E. and Fraser S.E. (1994) Violation of cell lineage restriction compartments in the chick hindbrain. *Development* **120:** 1347–1356.

Bishop J.M. (1985) Viral oncogenes. *Cell* **42:** 23–28.

Bishop J.M. (1989) Viruses, genes and cancer. *Am. Zool.* **29:** 653–666.

Bishop J.M. (1991) Molecular themes in oncogenesis. *Cell* **64:** 235–248.

Bjornson C.R., Rietze R.L., Reynolds B.A., Magli M.C. and Vescovi A.L. (1999) Turning brain into blood: A hematopoietic fate adopted by adult neural stem cells in vivo. *Science* **283:** 534–537.

Blair S.S. (1992) Engrailed expression in the anterior lineage compartment of the developing wing blade of *Drosophila. Development* **115:** 21–33.

Blair S.S. (1995) Compartments and appendage development in *Drosophila. BioEssays* **17:** 299–309.

Blair S.S. and Ralston A. (1997) Smoothened-mediated Hedgehog signalling is required for the maintenance of the anterior-posterior lineage restriction in the developing wing of *Drosophila. Development* **124:** 4053–4063.

Blair S.S., Brower D.L., Thomas J.B. and Zavortink M. (1994) The role of apterous in the control of dorsoventral compartmentalization in PS integrin gene expression in the developing wing of *Drosophila. Development* **120:** 1805–1815.

Blasco M.A., Gasser S.M. and Lingner J. (1999) Telomeres and telomerase. *Genes and Dev.* **13:** 2353–2359.

Blazquez M.A. (1997) Illuminating flowers: Constans induces *leafy* expression. *BioEssays* **19:** 277–279.

Bleil J.D. and Wassarman P.M. (1980) Mammalian sperm and egg interaction: Identification of a glycoprotein in mouse-egg zonae pellucidae possessing receptor activity for sperm. *Cell* **20:** 873–882.

Bleil J.D. and Wassarman P.M. (1986) Autoradiographic visualization of the mouse egg's sperm receptor bound to sperm. *J. Cell Biol.* **102:** 1363–1371.

Bleil J.D. and Wassarman P.M. (1988) Galactose at the nonreducing terminus of O-linked oligosaccharides of mouse egg zona pellucida glycoprotein ZP3 is essential for the glycoprotein's sperm receptor activity. *Proc. Nat. Acad. Sci. USA* **85:** 6778–6782.

Bleil J.D. and Wassarman P.M. (1990) Identification of a ZP3–binding protein on acrosome-intact mouse sperm by photoaffinity crosslinking. *Proc. Nat. Acad. Sci. USA* **87:** 5563–5567.

Bleil J.D., Greve J.M. and Wassarman P.M. (1988) Identification of a secondary sperm receptor in the mouse egg zona pellucida: Role in maintenance of binding of acrosome-reacted sperm to eggs. *Dev. Biol.* **128:** 376–385.

Blondeau J.-P. and Baulieu E.E. (1984) Progesterone receptor characterized by photoaffinity labeling in the plasma membrane of *Xenopus laevis* oocytes. *Biochem. J.* **219:** 785–792.

Bloom W. and Fawcett D.W. (1975) *A Textbook of Histology.* 10th ed. Philadelphia: Saunders.

Blumberg B., Wright C.V.E., De Robertis E.M. and Cho K.W.Y. (1991) Organizer-specific homeobox genes in *Xenopus laevis* embryos. *Science* 253: 194–196.

Bockamp E.-O., McLaughlin F., Murrell A. and Green A.R. (1994) Transcription factors and the regulation of hematopoiesis: Lessons from GATA and SCL proteins. *BioEssays* **16:** 481–488.

Bode H.R., Flick K.M. and Smith G.S. (1976) Regulation of interstitial cell differentiation in *Hydra attenuata.* I. Homeostatic control of interstitial cell population size. *J. Cell Sci.* **20:** 29–46.

Bodemer C.W. (1968) *Modern Embryology.* New York: Holt, Rinehart and Winston.

Bodnar A.G., Ouellette M., Frolkis M., Holt S.E., Chiu C.P., Morin G.B., Harley C.B., Shay J.W., Lichtsteiner S. and Wright W.E. (1998) Extension of life-span by introduction of telomerase into normal human cells. *Science* **279:** 349–352.

Boggs R.T., Gregor P., Idriss S., Belote J.M. and McKeown M. (1987) Regulation of sexual differentiation in *D. melanogaster* via alternative splicing of RNA from the *transformer* gene. *Cell* **50:** 739–747.

Bohn H. (1976) Regeneration of proximal tissues from a more distal amputation level in the insect leg (*Blaberus craniifer, Blattaria*). *Dev. Biol.* **53:** 285–293.

Bomblies K., Dagenais N. and Weigel D. (1999) Redundant enhancers mediate transcriptional repression of *Agamous by Apetala2. Dev. Biol.* **216:** 260–264.

Bonduelle M., Camus M., De Vos A., Staessen C., Tournaye H., Van Assche E., Verheyen G., Devroey P., Liebaers I. and Van Steirteghem A. (1999) Seven years of intracytoplasmic sperm injection and follow-up of 1987 subsequent children. *Hum. Reprod.* **14** (Supplement 1): 243–264.

Borycki A.G., Mendham L. and Emerson C.P. Jr. (1998) Control of somite patterning by Sonic hedgehog and its downstream response genes. *Development* **125:** 777–790.

Bottjer S.W. (1987) Ontogenetic changes in the pattern of androgen accumulation in song-control nuclei of male zebra finches. *J. Neurobiol.* **18:** 125–139.

Bottjer S.W. (1991) Neural and hormonal substrates for song learning in zebra finches. *The Neurosciences* **3:** 481–488.

Bottjer S.W., Roselinsky H. and Tran N.B. (1997) Sex differences in neuropeptide staining of song-control nuclei in zebra finch brains. *Brain Behav. Evol.* **50:** 284–303.

Boucaut J.-C., Darribère T., Boulekbache H. and Thiery J.P. (1984) Prevention of gastrulation but not neurulation by antibodies to fibronectin in amphibian embryos. *Nature* (London) **307:** 364–367.

Boucaut J.-C., Darribère T., Shi D.L., Boulekbache H., Yamada K.M. and Thiery J.P. (1985) Evidence for the role of fibronectin in amphibian gastrulation. *J. Embryol. exp. Morphol.* **89** (Supplement): 211–227.

Boveri T. (1887) Ueber die Differenzierung der Zellkerne während der Furchung des Eies von Ascaris megalocephala. *Anat. Anz.* **2:** 688–693.

Boveri T. (1902) On multipolar mitosis as a means of analysis of the cell nucleus. In: *Foundations of Experimental Embryology,* B.H. Willier and J.M. Oppenheimer (eds.), 74–97. New York: Hafner.

Boveri T. (1907) Zellenstudien VI. Die Entwicklung dispermer Seeigeleier. Ein Beiträg zur Befruchtungslehre und zur Theorie des Kernes. *Jenaer Zeitsch. Naturwiss.* **43:** 1–292.

Bowerman B. (1995) Determinants of blastomere identity in the early *C. elegans* embryo. *BioEssays* **17:** 405–414.

Bowerman B., Eaton B.A. and Priess J.R. (1992) *skn-1*, a maternally expressed gene required to specify the fate of ventral blastomeres in the early *C. elegans* embryo. *Cell* **68:** 1061–1075.

Bowerman B., Draper B.W., Mello C.C. and Priess J.R. (1993) The maternal gene *skn-1* encodes a protein that is distributed unequally in early *C. elegans* embryos. *Cell* **74:** 443–452.

Bowerman B., Ingram M.K. and Hunter C.P. (1997) The maternal par genes and the segregation of cell fate specification activities in early *Caenorhabditis elegans* embryos. *Development* **124:** 3815–3826.

Bowman J.L., Smyth D.R. and Meyerowitz E.M. (1989) Genes directing flower development in *Arabidopsis. Plant Cell* **1:** 37–52.

Bowman J.L., Sakai H., Jack T., Weigel D., Mayer U. and Meyerowitz E.M. (1992) *SUPERMAN,* a regulator of floral homeotic genes in *Arabidopsis. Development* **114:** 599–615.

Bowman J.L., Smyth D.R. and Meyerowitz E.M. (1991) Genetic interactions among floral homeotic genes of *Arabidopsis. Development* **112:** 1–20.

Boyd L., Guo S., Levitan D., Stinchcomb D.T. and Kemphues K.J. (1996) PAR-2 is asymmetrically distributed and promotes association of P granules and PAR-1 with the cortex in *C. elegans* embryos. *Development* **122:** 3075–3084.

Brackenbury R., Thiery J.-P., Rutishauser U. and Edelman G.M. (1977) Adhesion among neural cells of the chick embryo. I. Immunological assay for molecules involved in cell-cell binding. *J. Biol. Chem.* **252:** 6835–6840.

Bradley A. (1987) Production and analysis of chimeric mice. In: *Teratocarcinomas and Embryonic Stem Cells: A Practical Approach,* E.J. Robertson (ed.), 113–151. Oxford: IRL Press.

Brand A.H. and Perrimon N. (1993) Targeted gene expression as a means of altering cell fates and generating dominant phenotypes. *Development* **118:** 401–415.

Branden C. and Tooze J. (1991) *Introduction to Protein Structure.* New York: Garland Publishing, Inc.

Brandner S., Isenman S., Raeber A., Fischer M., Sailer A., Kobayashi Y., Marino S., Weissman C. and Aguzzi A. (1996) Normal host prion protein necessary for scrapie-induced neurotoxicity. *Nature* **379:** 339–343.

Brannon M., Gomperts M., Sumoi L., Moon R.T. and Kimelman D. (1997) A β-catenin/XTcf-3 complex binds to the *siamois* promoter to regulate dorsal axis specification in *Xenopus*. *Genes and Dev.* **11:** 2359–2370.

Braun R.E., Peschon J.J., Behringer R.R., Brinster R.L. and Palmiter R.D. (1989) Protamine 3′-untranslated sequences regulate temporal translational control and subcellular localization of growth hormone in spermatids of transgenic mice. *Genes and Dev.* **3:** 793–802.

Braun T., Rudnicki M.A., Arnold H.-H. and Jaenisch R. (1992) Targeted inactivation of the muscle regulatory gene *Myf-5* results in abnormal rib development and perinatal death. *Cell* **71:** 369–382.

Bray D. and White J.G. (1988) Cortical flow in animal cells. *Science* **239:** 883–888.

Breedlove S.M. (1992) Sexual differentiation of the brain and behavior. In: *Behavioral Endocrinology,* J.B. Becker, S.M. Breedlove and D. Crews (eds.), 39–70. Cambridge, MA: MIT Press.

Breitweiser W., Markussen, F.-H., Horstmann H. and Ephrussi A. (1996) Oskar protein interaction with Vasa represents an essential step in polar granule assembly. *Genes and Dev.* **10:** 2179–2188.

Brennan T.J., Chakraborty T. and Olson E.N. (1991) Mutagenesis of the myogenin basic region identifies an ancient protein motif critical for activation of myogenesis. *Proc. Nat. Acad. Sci. USA* **88:** 5675–5679.

Brenner S. (1974) The genetics of *Caenorhabditis elegans*. *Genetics* **77:** 71–94.

Bretscher M.S. (1996) Moving membrane up to the front of migrating cells. *Cell* **85:** 465–467.

Bridges C.B. (1921) Triploid intersexes in *Drosophila melanogaster*. *Science* **54:** 252–254.

Briegel K., Lim K.-C., Plank C., Beug H., Engel J.D. and Zenke M. (1993) Ectopic expression of a conditional GATA-2/estrogen chimera arrests erythroid differentiation in a hormone-dependent manner. *Genes and Dev.* **7:** 1097–1109.

Briegel K., Bartunek P., Stengl G., Lim K.-C., Beug H., Engel J.D. and Zenke M. (1996) Regulation and function of transcription factor GATA-1 during red blood cell differentiation. *Development* **122:** 3839–3850.

Briggs R. (1977) Genetics of cell type determination. In: *Cell Interactions in Differentiation.* M. Karkinen-Jaäskeläinen, L. Saxen and L. Weiss (eds.), *6th Sigrid Jusélius Foundation Symp.*, 23–43. New York: Academic Press.

Briggs R. and King T.J. (1952) Transplantation of living nuclei from blastula cells into enucleated frogs' eggs. *Proc. Nat. Acad. Sci. USA* **38:** 455–463.

Brinster R.L. (1976) Participation of teratocarcinoma cells in mouse embryo development. *Cancer Res.* **36:** 3412–3414.

Broadus J., McCabe J.R., Endrizzi B., Thummel C.S. and Woodard C.T. (1999) The *Drosophila* beta FTZ-F1 orphan nuclear receptor provides competence for stage-specific responses to the steroid hormone ecdysone. *Mol. Cell* **3:** 143–149.

Bronner-Fraser M. and Fraser S.E. (1991) Cell lineage analysis of the avian neural crest. *Development* (Supplement 2): 17–22.

Bronner-Fraser M. and Stern C. (1991) Effects of mesodermal tissues on avian neural crest cell migration. *Dev. Biol.* **143:** 213–217.

Browder L.W. (1984) *Developmental Biology.* 2nd ed. Philadelphia: Saunders.

Browder, L.W., ed. (1985). *Developmental Biology*, vol. 1, *Oogenesis.* New York: Plenum.

Browder L.W., Erickson C.A. and Jeffery W.A. (1991) *Developmental Biology.* Philadelphia: Saunders.

Brown D.D. and Dawid I.B. (1968) Specific gene amplification in oocytes. *Science* **160:** 272–280.

Brown D.D., Wang Z., Furlow J.D., Kanamori A., Schwartzman R.A., Remo B.F. and Pinder A. (1996) The thyroid hormone-induced tail resorption program during *Xenopus laevis* metamorphosis. *Proc. Nat. Acad. Sci. USA* **93:** 1924–1929.

Brown J.E. and Weiss M.C. (1975) Activation of production of mouse liver enzymes in rat hepatoma-mouse lymphoid hybrids. *Cell* **6:** 481–494.

Brown N.A. and Wolpert L. (1990) The development of handedness in left/right asymmetry. *Development* **109:** 1–9.

Brown N.H. (1993) Integrins hold *Drosophila* together. *BioEssays* **15:** 383–390.

Brueckner M., D'Eustacio P. and Horwich A.L. (1989) Linkage mapping of a mouse gene, *iv,* that controls left-right asymmetry of the heart and viscera. *Proc. Nat. Acad. Sci. USA* **86:** 5035–5038.

Bryan T.M., Englezou A., Dalla-Pozza L., Dunham M.A. and Reddel R.R. (1997) Evidence for an alternative mechanism for maintaining telomere length in human tumors and tumor-derived cell lines. *Nat. Med.* **3:** 1271–1274.

Bryant P.J. and Simpson P. (1984) Intrinsic and extrinsic control of growth in developing organs. *Quart. Rev. Biol.* **59:** 387–415.

Bryant S.V. (1977) Pattern regulation in amphibian limbs. In: *Vertebrate Limb and Somite Morphogenesis,* D.A. Ede, J.R. Hinchliffe and M. Balls (eds.), *3rd Symp. Brit. Soc. Dev. Biol.*, 319–327. Cambridge: Cambridge University Press.

Bryant S.V. and Gardiner D.M. (1992) Retinoic acid, local cell-cell interactions, and pattern formation in vertebrate limbs. *Dev. Biol.* **152:** 1–25.

Bryant S.V. and Iten L.E. (1976) Supernumerary limbs in amphibians: Experimental production in *Notophthalmus viridescens* and a new interpretation of their formation. *Dev. Biol.* **50:** 212–234.

Bryant S.V., French V. and Bryant P.J. (1981) Distal regeneration and symmetry. *Science* **212:** 993–1002.

Büeler H., Fischer M., Lang Y., Bluethmann H., Lipp H.-P., DeArmond S.J., Prusiner S.B., Aguet M. and Weissmann C. (1992) Normal development and behaviour of mice lacking the neuronal cell-surface PrP protein. *Nature* **356:** 577–582.

Büeler H., Aguzzi A., Sailer A., Greiner R.-A., Autenried P., Aguet M. and Weissmann C. (1993) Mice devoid of PrP are resistant to scrapie. *Cell* **73:** 1339–1347.

Bull A. (1966) *Bicaudal,* a genetic factor which affects the polarity of the embryo in *Drosophila melanogaster. J. Exp. Zool.* **161:** 221–241.

Bull J.J. (1980) Sex determination in reptiles. *Quart. Rev. Biol.* **55:** 3–21.

Bull J.J. (1983) *Evolution of Sex Determining Mechanisms.* Menlo Park, CA: Benjamin/Cummings.

Bunch T.A. and Brower D.L. (1993) *Drosophila* cell adhesion molecules. *Curr. Topics Dev. Biol.* **28:** 81–123.

Buratowski S. (1994) The basics of basal transcription by RNA polymerase II. *Cell* **77:** 1–3.

Burek M.J., Nordeen K.W. and Nordeen E.J. (1997) Sexually dimorphic neuron addition to an avian song-control region is not accounted for by sex differences in cell death. *J. Neurobiol.* **33:** 61–71.

Bürglin T.R. (1997) Analysis of TALE superclass homeobox genes (MEIS, PBC, KNOX, Iroquois, TGIF) reveals a novel domain conserved between plants and animals. *Nucleic Acid Res.* **25:** 4173–4180.

Burgoyne P.S., Buehr M., Koopman P., Rossant J. and McLaren A. (1988) Cell-autonomous action of the testis-determining gene: Sertoli cells are exclusively XY in XX↔Y chimaeric mouse testes. *Development* **102:** 443–450.

Burke R. and Basler K. (1996) Dpp receptors are autonomously required for cell proliferation in the entire developing *Drosophila* wing. *Development* **122:** 2261–2269.

Burke T. and Kadonaga J. (1997) The downstream core promoter element, DPE, is conserved from *Drosophila* to humans and is recognized by $TAF_{II}60$ of *Drosophila*. *Genes and Dev.* **11:** 3020–3031.

Burks D.J.; Carballada R., Moore H.D.M. and Saling P.M. (1995) Interaction of a tyrosine kinase from human sperm with the zona pellucida at fertilization. *Science* **269:** 83–86.

Burnside B. (1971) Microtubules and microfilaments in newt neurulation. *Dev. Biol.* **26:** 416–441.

Burnside B. (1973) Microtubules and microfilaments in amphibian neurulation. *Am. Zool.* **88:** 353–372.

Burnside B. and Jacobson A.G. (1968) Analysis of morphogenetic movements in the neural plate of the newt *Taricha torosa*. *Dev. Biol.* **18:** 537–552.

Burtis K.C. and Baker B.S. (1989) *Drosophila doublesex* gene controls somatic sexual differentiation by producing alternatively spliced mRNAs encoding related sex-specific polypeptides. *Cell* **56:** 997–1010.

Burtis K.C., Coschigano K.T., Baker B.S. and Wensink P.C. (1991) The Doublesex proteins of *Drosophila melanogaster* bind directly to a sex-specific yolk protein gene enhancer. *EMBO J.* **10:** 2577–2582.

Busa W.B. and Gimlich R.L. (1989) Lithium-induced teratogenesis in frog embryos prevented by a polyphosphoinositide cycle intermediate or a diacylglycerol analog. *Dev. Biol.* **132:** 315–324.

Busch M.A., Bomblies K. and Weigel D. (1999) Activation of a floral homeotic gene in *Arabidopsis*. *Science* **285:** 585–587.

Buskirk D.R., Thiery J.-P., Rutishauser U. and Edelman G.M. (1980) Antibodies to a neural cell adhesion molecule disrupt histogenesis in cultured chick retinae. *Nature* (London) **285:** 488–489.

Butler E.G. (1955) Regeneration of the urodele forelimb after reversal of its proximo-distal axis. *J. Morphol.* **96:** 265–281.

Butler P.J. and Klug A. (1978) The assembly of a virus. *Scientific American* **239** (November): 62–69.

Cadigan K.M. and Nusse R. (1997) Wnt signaling: A common theme in animal development. *Genes and Dev.* **11:** 3286–3305.

Cambridge S.B., Davis R.L., and Minden J.S. (1997) *Drosophila* mitotic domain boundaries as cell fate boundaries. *Science* **277:** 825–828.

Campbell K.H.S., McWhir J., Ritchie W.A. and Wilmut I. (1996) Sheep cloned by nuclear transfer from a cultured cell line. *Nature* **380:** 64–66.

Campione M., Steinbeisser H., Schweickert A., Deissler K., van Bebber F., Lowe L.A., Nowotschin S., Viebahn C., Haffter P., Kuehn M.R. and Blum M. (1999) The homeobox gene Pitx2: Mediator of asymmetric left-right signaling in vertebrate heart and gut looping. *Development* **126:** 1225–1234.

Cantley L.C., Auger K.R., Carpenter C., Duckworth B., Graziani A., Kapeller R. and Soltoff S. (1991) Oncogenes and signal transduction. *Cell* **64:** 281–302.

Capdevila J., Tabin C. and Johnson R.L. (1998) Control of dorsoventral somite patterning by Wnt-1 and beta-catenin. *Dev. Biol.* **193:** 182–194.

Capdevila J., Vogan K.J., Tabin C. and Izpisua-Belmonte J.C. (2000) Mechanisms of left-right determination in vertebrates. *Cell* **101:** 9–21.

Carlson B.M. (1988) *Patten's Foundations of Embryology*. 5th ed. New York: McGraw-Hill.

Carlson G.A., Hsiao K., Oesch B., Westaway D. and Prusiner S.B. (1991) Genetics of prion infections. *Trends Genet.* **7:** 61–65.

Carmeliet P. et al. (1996) Abnormal blood vessel development and lethality in embryos lacking a single VEGF allele. *Nature* **380:** 435–439.

Carney J.M., Starke-Reed P.E., Oliver C.N., Landum R.W., Cheng M.S., Wu J.F. and Floyd R.A. (1991) Reversal of age-related increase in brain protein oxidation, decrease in enzyme activity, and loss in temporal and spatial memory by chronic administration of the spin-trapping compound N-tert-butyl-alpha-phenylnitrone. *Proc. Nat. Acad. Sci. USA.* **88:** 3633–3636.

Carola R., Harley J.P. and Noback C.R. (1990) *Human Anatomy and Physiology*. New York: McGraw-Hill.

Carpenter G. and Cohen S. (1978) Epidermal growth factors. In: *Biochemical Actions of Hormones*, G. Litwack (ed.), vol. V, 203–247. New York: Academic Press.

Carroll D.J., Ramarao C.S., Mehlmann L.M., Roche S., Terasaki M. and Jaffe L.A. (1997) Calcium release at fertilization in starfish eggs is mediated by phospholipase C-gamma. *J. Cell Biol.* **138:** 1303–1311.

Carroll D.J., Albay D.T., Hoang K.M., O'Neill F.J., Kumano M. and Foltz K.R. (2000) The relationship between calcium, MAP kinase, and DNA synthesis in the sea urchin egg at fertilization. *Dev. Biol.* **217:** 179–191.

Carroll S.B. and Scott M.P. (1986) Zygotically active genes that affect the spatial expression of the *fushi tarazu* segmentation gene during early *Drosophila* embryogenesis. *Cell* **45:** 113–126.

Casanova J. and Struhl G. (1993) The torso receptor localizes as well as transduces the spatial signal specifying terminal body pattern in *Drosophila*. *Nature* **362:** 152–155.

Casanova J., Furriols M., McCormick C.A. and Struhl G. (1995) Similarities between trunk and spätzle, putative extracellular ligands specifying body pattern in *Drosophila*. *Genes and Dev.* **9:** 2539–2544.

Catala M., Teillet M.-A., De Robertis E.M. and Le Douarin N.M. (1996) A spinal cord fate map in the avian embryo: While regressing, Hensen's node lays down the notochord and floor plate thus joining the spinal cord lateral walls. *Development* **122:** 2599–2610.

Cather J.N., Render J.A. and Freeman G. (1986) The relation of time to direction and equality of cleavage in *Ilyanassa* embryos. *Int. J. Invertebrate Reprod. and Dev.* **9:** 179–194.

C. elegans Sequencing Consortium (1998) Genome sequence of the nematode *C. elegans:* A platform for investigating biology. *Science* **282:** 2012–2018.

Centrella M., McCarthy T.L. and Canalis E. (1987) Mitogenesis in fetal rat bone cells simultaneously exposed to type b transforming growth factor and other growth regulators. *FASEB J.* **1:** 312–317.

Chambon P. (1981) Split genes. *Scientific American* **244** (May): 60–77.

Chan A.W.S., Dominko T., Luetjens C.M., Neuber E., Martinovich C., Hewitson L., Simerly C.R. and Schatten G.P. (2000) Clonal propagation of primate offspring by embryo splitting. *Science* **287:** 317–319.

Chandlee J.M. (1991) Analysis of developmentally interesting genes cloned from higher plants by insertional mutagenesis. *Dev. Genet.* **12:** 261–271.

Charité J., de Graff W., Shen S. and Deschamps J. (1994) Ectopic expression of Hoxb-8 causes duplication of the ZPA in the fore limb and homeotic transformation of axial structures. *Cell* **78:** 589–601.

Chellappan S.P., Hiebert S., Mudryj M., Horowitz J.M. and Nevins J.R. (1991) The E2F transcription factor is a cellular target for the RB protein. *Cell* **65:** 1053–1061.

Chen C.M. and Struhl G. (1999) Wingless transduction by the Frizzled and Frizzled2 proteins of *Drosophila. Development* **126:** 5441–5452.

Chen C.S., Mrksich M., Huang S., Whitesides G.M. and Ingber D.E . (1997) Geometric control of cell life and death. *Science* **276:** 1425–1428.

Chen F. and Capecchi M.R. (1997) Targeted mutations in *Hoxa-9* and *Hox-b9* reveal synergistic interactions. *Devel. Biol.* **181:** 186–196.

Chen F., Hersh B.M., Conradt B., Zhou Z., Riemer D., Gruenbaum J. and Horvitz H.R. (2000) Translocation of *C. elegans* CED-4 to nuclear membranes during programmed cell death. *Science* **287:** 1485–1489.

Chen X., Farmer G., Zhu H., Prywes R. and Prives C. (1993) Cooperative DNA binding of p53 with RFIID (TBP): A possible mechanism for transcriptional activation. *Genes and Dev.* **7:** 1837–1849.

Chen Y. and Struhl G. (1996) Dual roles for Patched in sequestering and transducing Hedgehog. *Cell* **87:** 553–563.

Chen Y. and Struhl G. (1998) In vivo evidence that Patched and Smoothened constitute distinct binding and transducing components of a Hedgehog receptor complex. *Development* **125:** 4943–4948.

Chen Y.P., Huang L. and Solursh M. (1994) A concentration gradient of retinoids in the early *Xenopus laevis* embryo. *Dev. Biol.* **161:** 70–76.

Cheng T.-C., Wallace M.C., Merlie J.P. and Olson E.N. (1993) Separable regulatory elements governing *myogenin* transcription in mouse embryogenesis. *Science* **261:** 215–218.

Chia W., Kraut R., Li P., Yang X. and Zavortink M. (1997) On the roles of inscuteable in asymmetric cell divisions in *Drosophila. Cold Spring Harb. Symp. Quant. Biol.* **62:** 79–87.

Chiang C., Litingtung Y., Lee E., Young K.E., Corden J.L., Westphal H. and Beachy P.A. (1996) Cyclopia and defective axial patterning in mice lacking *Sonic hedgehog* gene function. *Nature* **383:** 407–413.

Chiba K., Longo F.J., Kontani K., Katada T. and Hoshi M. (1995) A periodic network of G protein β[gamma] subunit coexisting with cytokeratin filament in starfish oocytes. *Dev. Biol.* **169:** 415–420.

Chinnaiyan A.M., Chaudhary D., O'Rourke K., Koonin E.V. and Dixit V.M. (1997a) Role of CED-4 in the activation of CED-3. *Nature* **388:** 728–729.

Chinnaiyan A.M., O'Rourke K., Lane B.R. and Dixit V.M. (1997b) Interaction of CED-4 with CED-3 and CED-9: A molecular framework for cell death. *Science* **275:** 1122–1126.

Chisaka O. and Capecchi M.R. (1991) Regionally restricted developmental defects resulting from targeted disruption of the mouse homeobox gene *Hox-1.5. Nature* **350:** 473–479.

Chitnis A.J., Henrique D., Lewis J., Ish-Horowitz D. and Kintner C. (1995) Primary neurogenesis in *Xenopus* embryos regulated by a homologue of the *Drosophila* neurogenic gene *Delta. Nature* **375:** 761–766.

Cho C., Bunch D.O., Faure J.E., Goulding E.H., Eddy E.M., Primakoff P. and Myles D.G. (1998) Fertilization defects in sperm from mice lacking fertilin beta. *Science* **281:** 1857–1859.

Cho K.W.Y., Blumberg B., Steinbeisser H. and De Robertis E.M. (1991) Molecular nature of Spemann's organizer: The role of the *Xenopus* homeobox gene *goosecoid. Cell* **67:** 1111–1120.

Choi K., Kennedy M., Kazarov A., Papadimitriou J.C. and Keller G. (1998) A common precursor for hematopietic and endothelial cells. *Development* **125:** 725–732.

Christ B., Jacob H.J. and Jacob M. (1977) Experimental analysis of the origin of the wing musculature in avian embryos. *Anat. Embryol.* **150:** 171–186.

Chuang P.-T., Albertson D.G. and Meyer B.J. (1994) DPY-27, a chromosome condensation protein homolog that regulates *C. elegans* dosage compensation through association with the X chromosome. *Cell* **79:** 459–474.

Chuang P.T., Lieb J.D. and Meyer B.J. (1996) Sex-specific assembly of a dosage compensation complex on the nematode X chromosome. *Science* **274:** 1736–1739.

Cibelli J.B., Stice S.L., Golueke P.J., Kane J.J., Jerry J., Blackwell C., Ponce de Leon F.A. and Robl J.M. (1998) Cloned transgenic calves produced from nonquiescent fetal fibroblasts. *Science* **280:** 1256–1258.

Clack J.A. (1989) Discovery of the earliest known tetrapod stapes. *Nature* **342:** 425–427.

Clark E.A. and Brugge J.S. (1995) Integrins and signal transduction pathways: The road taken. *Science* **268:** 233–239.

Clark I., Giniger E., Ruohola-Baker H., Jan L.Y. and Jan Y.N. (1994) Transient posterior localization of a kinesin fusion protein reflects anteroposterior polarity in the *Drosophila* oocyte. *Curr. Biol.* **4:** 289–300.

Clausi D.A. and Brodland G.W. (1993) Mechanical evaluation of theories of neurulation using computer simulations. *Development* **118:** 1013–1023.

Cleaver O. and Krieg P.A. (1998) VEGF mediates angioblast migration during development of the dorsal aorta in *Xenopus. Development* **125:** 3905–3914.

Clement A.C. (1952) Experimental studies on germinal localization in *Ilyanassa.* I. The role of the polar lobe in determination of the cleavage pattern of its influence in later development. *J. Exp. Zool.* **121:** 593–626.

Clement A.C. (1962) Development of *Ilyanassa* embryo following removal of the D-macromere at successive cleavage stages. *J. Exp. Zool.* **149:** 193–216.

Clements D., Friday R.V. and Woodland H.R. (1999) Mode of action of VegT in mesoderm and endoderm formation. *Development* **126:** 4903–4911.

Clever U. and Karlson P. (1960) Induktion von Puff-Veränderungen in den Speicheldrüsenchromosomen von *Chironomus tentans* durch Ecdyson. *Exp. Cell Res.* **20:** 623–626.

Cline T.W. and Meyer B.J. (1996) VIVE LA DIFFÉRENCE: Males vs females in flies vs worms. *Ann. Rev. Genet.* **30:** 637–702.

Clute P. and Masui Y. (1995) Regulation of the appearance of divison asynchrony and microtubule-dependent chromosome cycles in *Xenopus laevis* embryos. *Dev. Biol.* **171:** 273–285.

Coates M.I. and Cohn M.J. (1998) Fins, limbs, and tails: Outgrowths and axial patterning in vertebrate evolution. *BioEssays* **20:** 371–381.

Coen E.S. and Meyerowitz E.M. (1991) The war of the whorls: Genetic interactions controlling flower development. *Nature* **353:** 31–37.

Coen E.S., Doyle S., Romero J.M., Elliott R., Magrath R. and Carpenter R. (1991) Homeotic genes controlling flower development in *Antirrhinum. Development* (Supplement 1): 149–155.

Cohen A.M. and Konigsberg I.R. (1975) A clonal approach to the problem of neural crest determination. *Dev. Biol.* **46:** 262–282.

Cohen S. (1960) Purification of a nerve growth-promoting protein from the mouse salivary gland and its neurocytotoxic antiserum. *Proc. Nat. Acad. Sci. USA* **46:** 302–311.

Cohen S.M. (1996) Controlling growth of the wing: Vestigial integrates signals from compartment boundaries. *BioEssays* **18:** 855–858.

Cohen-Dayag A., Ralt D., Tur-Kaspa I., Manor M., Makler A., Dor J., Mashiach S. and Eisenbach M. (1994) Sequential acquisition of chemotactic responsiveness by human spermatozoa. *Biol. Reprod.* **50:** 786–90.

Cohn M.J., Izpisua-Belmonte J.C., Abud H., Heath J.K. and Tickle C. (1995) Fibroblast growth factors induce additional limb development from the flank of chick embryos. *Cell* **80:** 739–746.

Cohn M.J., Patel K., Krumlauf R., Wilkinson D.G., Clarke J.D. and Tickle C. (1997) Hox9 genes and vertebrate limb specification. *Nature* **387:** 97–101.

Colbert E.H. (1980) *Evolution of the Vertebrates: A History of the Backboned Animals through Time.* 3rd ed. New York: Wiley.

Colgan D.F. and Manley J.L. (1997) Mechanism and regulation of mRNA polyadenylation. *Genes and Dev.* **11:** 2755–2766.

Colitz C.M., Davidson M.G. and McGahan M.C. (1999) Telomerase activity in lens epithelial cells of normal and cataractous lenses. *Exp. Eye Res.* **69:** 641–649.

Colledge W.H., Carlton M.B., Udy G.B., and Evans M.J. (1994) Disruption of c-mos oncogene causes parthenogenetic development of unfertilized mouse eggs. *Nature* **370:** 65–68.

Coller J.M., Gray N.K. and Wickens M. (1998) mRNA stabilization by poly(A) binding protein is independent of poly(A) and requires translation. *Genes and Dev.* **12:** 3226–3235.

Collignon J., Varlet I. and Robertson E. (1996) Relationship between asymmetric nodal expression and the direction of embryonic turning. *Nature* **381:** 155–158.

Collinge J., Whittington M.A., Sidle K.C.L., Smith C.J., Palmer M.S., Clarke A.R. and Jeffreys J.G.T. (1994) Prion protein is necessary for normal synaptic function. *Nature* **370:** 295–297.

Collinge J., Sidle K.C.L., Meads J., Ironside J. and Hill A.F. (1996) Molecular analysis of prion strain variation and the aetiology of "new variant" CJD. *Nature* **383:** 685–690.

Colwin L.H. and Colwin A.L. (1960) Formation of sperm entry holes in the vitelline membrane of *Hydroides hexagonis* (Annelida) and evidence of their lytic origin. *J. Biophys. Biochem. Cytol.* **7:** 315–320.

Comfort A. (1979) *The Biology of Senescence.* 3rd ed., Edinburgh and London: Churchill Livingstone.

Condie B.G. and Capecchi M.R. (1993) Mice homozygous for a targeted disruption of *Hoxd-3 (Hox-4.1)* exhibit anterior transformations of the first and second cervical vertebrae, the atlas and the axis. *Development* **119:** 579–595.

Condie B.G. and Capecchi M.R. (1994) Mice with targeted disruptions in the paralogous genes *hoxa-3* and *hoxd-3* reveal synergistic interactions. *Nature* **370:** 304–307.

Conklin E.G. (1905) The organization and cell lineage of the ascidian egg. *J. Acad. Nat. Sci. Philadelphia* **13:** 1–119.

Conlon I. and Raff M. (1999) Size Control in Animal Development. *Cell* **96:** 235–244.

Conlon R.A. and Rossant J. (1992) Exogenous retinoic acid rapidly induces anterior ectopic expression of murine *Hox-2* genes in vivo. *Development* **116:** 357–368.

Conlon R.A., Reaume A.G. and Rossant J. (1995) *Notch1* is required for the coordinate segmentation of somites. *Development* **121:** 1533–1545.

Conner S., Leaf D. and Wessel G. (1997) Members of the SNARE hypothesis are associated with cortical granule exocytosisin the sea urchin egg. *Mol. Reprod. Devel.* **48:** 1–13.

Conradt B. and Horvitz H.R. (1998) The *C. elegans* protein EGL-1 is required for programmed cell death and interacts with the Bcl-2–like protein CED-9. *Cell* **93:** 519–529.

Cook P.R. (1999) The organization of replication and transcription. *Science* **284:** 1790–1795.

Copeland N.G., Gilbert D.J., Chao B.C., Donovan P.J., Jenkins N.A., Cosman D., Anderson D., Lyman S.D. and Williams D.E. (1990) Mast cell growth factor maps near the steel locus on mouse chromosome 10 and is deleted in a number of steel alleles. *Cell* **63:** 175–183.

Coschigano K.T. and Wensink P.C. (1993) Sex-specific transcriptional regulation by the male and female doublesex proteins of *Drosophila. Genes and Dev.* **7:** 42–54.

Counce S.J. (1973) The causal analysis of insect embryogenesis. In: *Developmental Systems: Insects,* S.J. Counce and C.H. Waddington (eds.), vol. 2, 1–156. London/New York: Academic Press.

Cowan A.E. and Myles D.G. (1993) Biogenesis of surface domains during spermiogenesis in the guinea pig. *Dev. Biol.* **155:** 124–133.

Craig E.A. (1993) Chaperones: Helpers along the pathways to protein folding. *Science* **260:** 1902–1903.

Crampton H.E. (1894) Reversal of cleavage in a sinistral gastropod. *Ann. New York Acad. Sci.* **8:** 167–170.

Crawford K. and Stocum D.L. (1988) Retinoic acid coordinately proximalizes regenerate pattern and blastema differential affinity in axolotl limbs. *Development* **102:** 687–698.

Crews D. (1994) Animal sexuality. *Scientific American* 270 (January): 108–114.

Crick F.H.C. and Lawrence P.A. (1975) Compartments and polyclones in insect development. *Science* **189:** 340–347.

Crittenden S.L., Rudel D., Binder J., Evans T.C. and Kimble J. (1997) Genes required for GLP-1 asymmetry in the early *C. elegans* embryo. *Dev. Biol.* **181:** 36–46.

Croce C.M. (1987) Role of chromosome translocations in human neoplasia. *Cell* **49:** 155–156.

Croce C.M., Thierfelder W., Erickson J., Nishikura K., Finan J., Lenoir G.M. and Nowell P.C. (1983) Transcriptional activation of an unrearranged and untranslocated c-*myc* oncogene by translocation of a C_λ locus in Burkitt lymphoma cells. *Proc. Nat. Acad. Sci. USA* **80:** 6922–6926.

Cross M. and Dexter T.M. (1991) Growth factors in development, transformation, and tumorigenesis. *Cell* **64:** 271–280.

Cross M.A. and Enver T. (1997) The lineage commitment of haemopoietic progenitor cells. *Curr. Opin. Genet. Dev.* **7:** 609–613.

Crowley C. and Curtis H.J. (1963) The development of somatic mutations in mice with age. *Proc. Natl. Acad. Sci. USA.* **49:** 626–628.

Cubas P., Vincent C. and Coen E. (1999) An epigenetic mutation responsible for natural variation in floral symmetry. *Nature* **401:** 157–161.

Cunningham B.A. (1991) Cell adhesion molecules and the regulation of development. *Am. J. Ob. Gyn.* **164:** 939–948.

Cunningham B.A., Hemperly J.J., Murray B.A., Prediger E.A., Brackenbury R. and Edelman G.M. (1987) Neural cell adhesion molecule: Structure, immunoglobulin-like domains, cell surface modulation, and alternative RNA splicing. *Science* **236:** 799–806.

Curtis D., Lehmann R. and Zamore P. (1995a) Translational regulation in development. *Cell* **81:** 171–178.

Curtis D., Apfeld J. and Lehmann R. (1995b) nanos is an evolutionarily conserved organizer of anterior-posterior polarity. *Development* **121:** 1899–1910.

Curtsinger J.W., Fukui H.H., Townsend D.R. and Vaupel J.W. (1992) Demography of genotypes: Failure of the limited life-span paradigm in *Drosophila melanogaster*. *Science* **258:** 461–463.

Czihak G., Langer H. and Ziegler H. (1976) *Biologie*. Heidelberg: Springer.

Dahanukar, A. and Wharton R.P. (1996) The nanos gradient in *Drosophila* embryos is generated by translational regulation. *Genes and Dev.* **10:** 2610–2620.

Dale L. and Slack J.M.W. (1987a) Fate map for the 32–cell stage of *Xenopus laevis*. *Development* **99:** 527–551.

Dale L. and Slack J.M.W. (1987b) Regional specification within the mesoderm of early embryos of *Xenopus laevis*. *Development* **100:** 279–295.

Dale L., Smith J.C. and Slack J.M.W. (1985) Mesoderm induction in *Xenopus laevis*: A quantitative study using a cell lineage label and tissue-specific antibodies. *J. Embryol. exp. Morphol.* **89:** 289–312.

Dalla-Favera R., Bregni M., Erikson J., Patterson D., Gallo R.C. and Croce C.M. (1982) Human c-*myc onc* gene is located on the region of chromosome 8 that is translocated in Burkitt lymphoma cells. *Proc. Nat. Acad. Sci. USA* **79:** 7824–7827.

Daniel C.W. (1977) Cell longevity: In vivo. In: *Handbook of the Biology of Aging*, C.E. Finch and L. Hayflick (eds.), 122–158. New York: Van Nostrand Reinhold.

Danilchik M.V. and Black S.D. (1988) The first cleavage plane and the embryonic axis are determined by separate mechanisms in *Xenopus laevis*. I. Independence in undisturbed embryos. *Dev. Biol.* **128:** 58–64.

Danilchik M.V. and Denegre J.M. (1991) Deep cytoplasmic rearrangements during early development in *Xenopus laevis*. *Development* **111:** 845–856.

Danos M.C. and Yost H.J. (1995) Linkage of cardiac left-right asymmetry and dorsal-anterior development in *Xenopus*. *Development* **121:** 1467–1474.

Darnell J.E. Jr. (1997) STATS and gene regulation. *Science* **277:** 1630–1635.

Darnell J., Lodish H. and Baltimore D. (1993) *Molecular Cell Biology*. 3rd ed. New York: Scientific American Books.

Darribère T., Yamada K.M., Johnson K.E. and Boucaut J.-C. (1988) The 140 kD fibronectin receptor complex is required for mesodermal cell adhesion during gastrulation in the amphibian *Pleurodeles waltlii*. *Dev. Biol.* **126:** 182–194.

Darwin C. (1868) *The Variation of Animals and Plants under Domestication*. New York: Orange Judd.

Davenport R.W., Dou P., Rehder V. and Kater S.B. (1993) A sensory role for neuronal growth cone filopodia. *Nature* **361:** 721–724.

David C.N. and Murphy S. (1977) Characterization of interstitial stem cells in Hydra by cloning. *Dev. Biol.* **58:** 372–383.

David C.N., Bosch T.C.G., Hobmayer B., Holstein T. and Schmidt T. (1987) Interstitial stem cells in Hydra. In: *Genetic Regulation of Development*, W.F. Loomis (ed.), 289–408. New York: Alan R. Liss.

Davidson E.H. (1986) *Gene Activity in Early Development*. 3rd ed. New York: Academic Press.

Davidson E.H. (1990) How embryos work: A comparative view of diverse models of cell fate specification. *Development* **108:** 365–389.

Davidson L.A., Koehl M.A.R., Keller R. and Oster G. (1995) How do sea urchins invaginate? Using biomechanics to distinguish between mechanisms of primary invagination. *Development* **121:** 2005–2018.

Davidson R.G., Nitowsky H.M. and Childs B. (1963) Demonstration of two populations of cells in the human female heterozygous for glucose-6–phosphate dehydrogenase variants. *Proc. Nat. Acad. Sci. USA* **50:** 481–485.

Davis B.K. (1981) Timing of fertilization in mammals: Sperm cholesterol/phospholipid ratio as determinant of capacitation interval. *Proc. Nat. Acad. Sci. USA* **78:** 7560–7564.

Davis G.S., Phillips H.M. and Steinberg M.S. (1997) Germ-layer surface tensions and "tissue affinities" in *Rana pipiens* gastrulae: Quantitative measurements. *Dev. Biol.* **192:** 630–644.

Davis R.L., Weintraub H. and Lassar A.B. (1987) Expression of a single transfected cDNA converts fibroblasts to myoblasts. *Cell* **51:** 987–1000.

Dawes H.E., Berlin D.S., Lapidus D.M., Nusbaum C., Davis T.L. and Meyer B.J. (1999) Dosage compensation proteins targeted to X chromosomes by a determinant of hermaphrodite fate. *Science* **284:** 1800–1804.

de Bono M. and Hodgkin J. (1996) Evolution of sex determination in *Caenorhabditis:* unusually high divergence of *tra-1* and its functional consequences. *Genetics* **144:** 587–595.

DeLeon D.V., Cox K.H., Angerer L.M. and Angerer R.C. (1983) Most early-variant histone mRNA is contained in the pronucleus of sea urchin eggs. *Dev. Biol.* **100:** 197–206.

Denny P.C. and Tyler A. (1964) Activation of protein synthesis in non-nucleate fragments of sea urchin eggs. *Biochem. Biophys. Res. Commun.* **145:** 245–249.

Derby A. and Etkin W. (1968) Thyroxine induced tail resorption *in vitro* as affected by anterior pituitary hormones. *J. Exp. Zool.* **169:** 1–8.

Derman E., Krauter K., Walling L., Weinberger C., Ray M. and Darnell J.E., Jr. (1981) Transcriptional control in the production of liver-specific mRNAs. *Cell* **23:** 731–739.

De Robertis E.M. and Gurdon J.B. (1977) Gene activation in somatic nuclei after injection into amphibian oocytes. *Proc. Nat. Acad. Sci. USA* **74:** 2470–2474.

De Robertis E.M. and Gurdon J.B. (1979) Gene transplantation and the analysis of development. *Scientific American* 241 (December): 74–82.

De Robertis E.M., Morita E.A. and Cho K.W.Y. (1991) Gradient fields and homeobox genes. *Development* **112:** 669–678.

De Rosa R., Grenier J.K., Andreeva T., Cook C.E., Adoutte A., Akam M., Carrol S.B. and Balavoine G. (1999) Hox genes in brachiopods and priapulids and protostome evolution. *Nature* **399:** 772–776.

De Santa Barbara P., Bonneaud N., Boizet B., Desclozeaux M., Moniot B., Sudbeck P., Scherer G., Poulat F. and Berta P. (1998) Direct interaction of SRY-related protein SOX9 and steroidogenic factor 1 regulates transcription of the human anti-Mullerian hormone gene. *Mol. Cell. Biol.* **18:** 6653–6665.

Deshler J. O., Highett M. I. and Schnapp, B. J. (1997). Localization of *Xenopus* Vg1 mRNA by Vera protein and the endoplasmic reticulum. *Science* **276:** 1128–1131.

Deshpande G., Calhoun G., Yanowitz J.L. and Schedl P.D. (1999) Novel functions of nanos in downregulating mitosis and transcription during the development of the Drosophila germline. *Cell* **99:** 271–281.

Detrick R.J., Dickey D. and Kintner C.R. (1990) The effects of N-cadherin misexpression on morphogenesis in *Xenopus embryos. Neuron* **4:** 493–506.

Dexter T.M. (1982) Stromal cell associated haemo-poiesis. *J. Cell Physiol.* (Supplement 1): 87–94.

Dexter T.M. and Spooncer E. (1987) Growth and differentiation in the hemopoietic system. *Ann. Rev. Cell Biol.* **3:** 423–441.

Diamond M.I., Miner J.N., Yoshinaga S.K. and Yamamoto K.R. (1990) Transcription factor interactions: Selectors of positive or negative regulation from a single DNA element. *Science* **249:** 1266–1272.

Diaz-Benjumea F.J. and Cohen SM (1993) Interaction between dorsal and ventral cells in the imaginal disc directs wing development in *Drosophila. Cell* **75:** 741–752.

Diaz-Benjumea F.J., Cohen B. and Cohen S.M.(1994) Cell interaction between compartments establishes the proximal-distal axis of *Drosophila* legs. *Nature* **372:** 175–179.

DiBerardino M.A. and Hoffner N. (1971) Development and chromosomal constitution of nuclear transplants derived from male germ cells. *J. Exp. Zool.* **176:** 61–72.

DiBerardino M.A. and Hoffner N.J. (1983) Gene reactivation in erythrocytes: Nuclear transplantation in oocytes and eggs of *Rana. Science* **219:** 862–864.

DiBerardino M.A., Hoffner N.J. and Etkin L.D. (1984) Activation of dormant genes in specialized cells. *Science* **224:** 946–952.

Dickerson R.E. and Geis I. (1983) *Hemoglobin.* Menlo Park, CA: Benjamin/Cummings.

Dietrich S. and Gruss P. (1995) undulated phenotypes suggest a role of Pax-1 for the development of vertebral and extravertebral structures. *Dev. Biol.* **167:** 529–548.

Dietrich S., Schubert F.A. and Lumsden A. (1997) Control of dorsoventral pattern in the chick paraxial mesoderm. *Development* **124:** 3895–3908.

Ding D. and Lipshitz H.D. (1993) Localized RNAs and their functions. *BioEssays* **15:** 651–658.

Ding D., Parkhurst S.M. and Lipshitz H.D. (1993) Different genetic requirements for anterior RNA localization revealed by the distribution of Adducin-like transcripts during *Drosophila* oogenesis. *Proc. Nat. Acad. Sci. USA* **90:** 2512–2516.

Dingwall C. and Laskey R. (1992) The nuclear membrane. *Science* **258:** 942–947.

Dixon K.E. (1994) Evolutionary aspects of primordial germ cell formation. In: *Ciba Foundation Symposium* **182:** *Germ Line Development*, J. Marsh and J. Goode (eds.), 92–110. Chichester, England: John Wiley.

Dobrovolskaia-Zavovskaia N. (1927) Sur la mortification spontanée de la queue chez la souris nouveau-née et sur l'existence d'un caractère (facteur) héreditaire "non-viable." *Comp. Rend. Soc. Biol.* **97:** 114–116.

Doherty D., Feger G., Younger-Shephard S., Jan L.Y. and Jan Y.N. (1996) Delta is a ventral to dorsal signal complementary to Serrate, another Notch ligand, in *Drosophila* wing formation. *Genes and Dev.* **10:** 421–434.

Doherty P., Ashton S.V., Moore S.E. and Walsh F.S. (1991) Morphoregulatory activities of NCAM and N-cadherin can be accounted for by G protein-dependent activation of L- and N-type neuronal Ca^{2+} channels. *Cell* **67:** 21–33.

Domingo C. and Keller R. (1995) Induction of notochord cell intercalation behavior and differentiation by progressive signals in the gastrula of *Xenopus laevis. Development* **121:** 3311–3321.

Dominguez I. and Green J.B.A. (2000) Dorsal downregulation of GSK3β by a non-Wnt-like mechanism is an early molecular consequence of cortical rotation in early *Xenopus* embryos. *Development* **127:** 861–868.

Donehower L.A., Harvey M., Slagle B.L., McArthur M.J., Montgomery C.A., Jr., Butel J.S. and Bradley A. (1992) Mice deficient for p53 are developmentally normal but susceptible to spontaneous tumours. *Nature* **356:** 215–221.

Doniach T. (1995) Basic FGF as an inducer of anteroposterior neural pattern. *Cell* **83:** 1067–1070.

Doniach T., Phillips C.R. and Gerhart J.C. (1992) Planar induction of anteroposterior pattern in the developing central nervous system of *Xenopus laevis. Science* **257:** 542–545.

Donne D.G., Viles J.H., Groth D., Mehlhorn I., James T.L., Cohen F.E., Prusiner S.B., Wright P.E. and Dyson H.J. (1997) Structure of the recombinant full-length hamster prion protein PrP(29–231); the N-terminus is highly flexible. *Proc. Natl. Acad. Sci. USA* **94:** 13452–13457.

Dorsett D. (1999) Distant liaisons: Longe range enhancer-promoter interactions in *Drosophila. Curr. Opin. Genet. Dev.* **9:** 505–514.

Dosch R., Gawantka V., Delius H., Blumenstock V. and Niehrs C. (1997) BMP-4 acts as a morphogen in dorsoventral mesoderm patterning in *Xenopus. Development* **124:** 2325–2334.

Dowdy S.F., Hinds P.W., Louie K., Reed S.I., Arnold A. and Weinberg R.A. (1993) Physical interaction of the retinoblastoma protein with human D cyclins. *Cell* **73:** 499–511.

Doyle R.W. and Hunte W. (1981) Demography of an estuarine amphipod (*Gammarus lawrencianus*) experimentally selected for high "r" : A model of the genetic effects of environmental change. *Can. J. Fish. Aquat. Sci.* **38**: 1120–1127.

Draetta G., Brizuela L., Potashkin J. and Beach D. (1987) Identification of p34 and p13, human homologs of the cell cycle regulators of fission yeast encoded by $cdc2^+$ and $suc1^+$. *Cell* **50:** 319–325.

Draetta G., Luca F., Wetendorf J., Brizuela L., Ruder-man J. and Beach D. (1989) cdc2 protein kinase is complexed with both cyclin A and B: Evidence for proteolytic inactivation of MPF. *Cell* **56:** 829–838.

Draizen T.A., Ewer J. and Robinow S. (1999) Genetic and hormonal regulation of the death of peptidergic neurons in the *Drosophila* central nervous system. *J. Neurobiol.* **38:** 455–465.

Drawbridge J., Wolfe A.E., Delgado Y.L. and Steinberg M.S. (1995) The epidermis is a source of directional information for the migrating pronephric duct in *Ambystoma mexicanum* embryos. *Dev. Biol.* **172:** 440–451.

Drews G.N. and Goldberg R.B. (1989) Genetic control of flower development. *Trends Genet.* **5:** 256–261.

Drews G.N., Bowman J.L. and Meyerowitz E.M. (1991a) Negative regulation of the *Arabidopsis* homeotic gene *AGAMOUS* by the *APETALA2* product. *Cell* **65:** 991–1002.

Drews G.N., Weigel D. and Meyerowitz E.M. (1991b) Floral patterning. *Curr. Opin. Genet. Dev.* **1:** 174–179.

Dreyer C., Scholz E. and Hausen P. (1982) The fate of oocyte nuclear proteins during early development of *Xenopus laevis. Wilhelm Roux's Arch.* **191:** 228–233.

Drier E.A., Huang L.H. and Steward R. (1999) Nuclear import of the *Drosophila* Rel protein Dorsal is regulated by phosphorylation. *Genes and Dev.* **13:** 556–568.

Driesch H. (1892) The potency of the first two cleavage cells in echinoderm development. Experimental production of partial and double formations. In: *Foundations of Experimental Embryology,* B.H. Willier and J.M. Oppenheimer (eds.), 38–50. New York: Hafner.

Driesch H. (1894) *Analytische Theorie der organischen Entwicklung.* Leipzig: Engelmann.

Driesch H. (1898) Über rein-mütterliche Charaktere und Bastardlarven von Echiniden. *Wilhelm Roux's Arch.* **7:** 65–102.

Driesch H. (1909) *Philosophie des Organischen.* 2 vols. Leipzig: Engelmann.

Driever W. (1995) Axis formation in zebrafish. *Curr. Opin. Genet. Dev.* **5:** 610–618.

Driever W. and Nüsslein-Volhard C. (1988) A gradient of *bicoid* protein in *Drosophila* embryos. *Cell* **54:** 83–93.

Driever W. and Nüsslein-Volhard C. (1989) The bicoid protein is a positive regulator of *hunchback* transcription in the early *Drosophila* embryo. *Nature* **337:** 138–143.

Driever W., Thoma G. and Nüsslein-Volhard C. (1989a) Determination of spatial domains of zygotic gene expression in the *Drosophila* embryo by the affinity of binding sites for the bicoid morphogen. *Nature* **340:** 363–367.

Driever W., Ma J., Nüsslein-Volhard C. and Ptashne M. (1989b) Rescue of bicoid mutant *Drosophila* embryos by bicoid fusion proteins containing heterologous activating sequences. *Nature* **342:** 149–154.

Driever W., Siegel V. and Nüsslein-Volhard C. (1990) Autonomous determination of anterior structures in the early *Drosophila* embryo by the maternal bicoid morphogen. *Development* **109:** 811–820.

Driever W. et al. (1996) A genetic screen for mutations affecting embryogenesis in zebrafish. *Development* **123:** 37–46.

Dubnicoff T., Valentine S.A., Chen G., Shi T., Lengyel J.A., Paroush Z. and Courey A.J. (1997) Conversion of dorsal from an activator to a repressor by the global corepressor Groucho. *Genes Dev.* **11:** 2952–2957.

Duffy J.B. and Perrimon N. (1994) The torso pathway in *Drosophila:* Lessons on receptor tyrosine kinase signaling and pattern formation. *Dev. Biol.* **166:** 380–395.

Dunphy W.G. and Kumagai A. (1991) The cdc25 protein contains an intrinsic phosphatase activity. *Cell* **67:** 189–196.

Durbin L., Brennan C., Shiomi K., Cooke J., Barrios A., Shanmugalingam S., Guthrie B., Lindberg R., and Holder N. (1998) Eph signaling is required for segmentation and differentiation of the somites. *Genes and Dev.* **12:** 3096–3109.

Durrin L.K., Mann R.K., Kayne P.S. and Grunstein M. (1991) Yeast hisone H4 N-terminal sequence is required for promoter activation in vivo. *Cell* **65:** 1023–1031.

Dutta A., Ruppert J.M., Aster J.C. and Winchester E. (1993) Inhibition of DNA replication factor RPA by p53. *Nature* **365:** 79–82.

Dyson N. (1998) The regulation of E2F by pRB family proteins. *Genes and Dev.* **12:** 2245–2262.

Eddy, E.M. and O'Brien D.A. (1994) The spermatozoon. In: *The Physiology of Reproduction,* E. Knovil and J.D. Neill (eds.), 29–77. New York: Raven Press.

Eddy E.M., Clark J.M., Gong D. and Fenderson B.A. (1981) Origin and migration of primordial germ cells in mammals. *Gamete* **4:** 333–362.

Edelman G.M. (1983) Cell adhesion molecules. *Science* **219:** 450–457.

Edelman G.M. (1984) Cell adhesion molecules: A molecular basis for animal form. *Scientific American* **250** (April): 118–129.

Edelman G.M. (1988) *Topobiology: An Introduction to Molecular Embryology.* New York: Basic Books.

Edgar B.A. and Datar S.A. (1996) Zygotic degradation of two maternal Cdc25 mRNAs terminates *Drosophila's* early cell cycle program. *Genes and Dev.* **10:** 1966–1977.

Edgar B.A. and Lehner C.F. (1996) Developmental control of cell cycle regulators: A fly's perspective. *Science* **274:** 1646–1652.

Edgar B.A. and O'Farrell P.H. (1989) Genetic control of cell division patterns in the *Drosophila* embryo. *Cell* **57:** 177–187.

Edgar B.A. and O'Farrell P.H. (1990) The three postblastoderm cell cycles of *Drosophila* embryogenesis are regulated in G2 by *string. Cell* **62:** 469–480.

Edgar B.A. and Schubiger G. (1986) Parameters controlling transcriptional activation during early *Drosophila* development. Cell **44:** 871–877.

Edgar B.A., Kiehle C.P. and Schubiger G. (1986) Cell cycle control by the nucleo-cytoplasmic ratio in early *Drosophila* development. *Cell* **44:** 365–372.

Edgar B.A., Lehmann D.A., and O'Farrell P.H. (1994a) Transcriptional regulation of *string* (*cdc 25*): a link between developmental programming and the cell cycle. *Development* **120:** 3131–3143.

Edgar B.A., Sprenger F., Duronio R.J., Leopold P. and O'Farrell P.H. (1994b) Distinct molecular mechanisms regulate cell cycle timing at successive stages of *Drosophila* embryogenesis. *Genes and Dev.* **8:** 440–452.

Egan S.E., Giddings B.W., Brooks M.W., Buday L., Sizeland A.M. and Weinberg R.A. (1993) Association of Sos Ras exchange protein with Grb2 is implicated in tyrosine kinase signal transduction and transformation. *Nature* **363:** 45–51.

Eisenbach M. and Ralt D. (1992) Precontact mammalian sperm-egg communication and role in fertilization. *Am. J. Physiol.* **262** (*Cell Physiol.* **31**): C1095–C1101.

Eisenbach M. and Tur-Kaspa I. (1999) Do human eggs attract spermatozoa? *BioEssays* **21:** 203–210.

Ekblom P., Ekblom M., Fecker L., Klein G., Zhang H.-Y., Kadoya Y., Chu M.-L., Mayer U. and Timpl R. (1994) Role of mesenchymal nidogen for epithelial morphogenesis in vitro. *Development* **120:** 2003–2014.

Elbetieha A. and Kalthoff K. (1988) Anterior determinants in embryos of *Chironomus samoensis:* Characterization by rescue bioassay. *Development* **104:** 61–75.

El-Deiry W.S., Tokino T., Velculescu V.E., Levy D.B., Parsons R., Trent J.M., Lin D., Mercer W.E., Kinzler K.W. and Vogelstine B. (1993) WAF1, a potential mediator of p53 tumor suppression. *Cell* **75:** 817–825.

Elinson R.P. (1986) Fertilization in amphibians: The ancestry of the block to polyspermy. *Int. Rev. Cytol.* **101:** 59–100.

Elinson R.P. and Rowning B. (1988) A transient array of parallel microtubules in frog eggs: Potential tracks for a cytoplasmic rotation that specifies the dorso-ventral axis. *Dev. Biol.* **128:** 185–197.

Ellenberger T., Fass D., Arnaud M. and Harrison S.C. (1994) Crystal structure of transcription factor E 47: E-box recognition by basic region helix-loop-helix dimer. *Genes and Dev.* **8:** 970–980.

Ellis H.M. and Horvitz H.R. (1986) Genetic control of programmed cell death in the nematode *C. elegans*. Cell **44:** 817–829.

Emerit I. and Chance B., eds. (1992) *Free Radicals and Aging.* Basel: Birkhaeuser.

Emeson R.B., Hedjran F., Yeakley J.M., Guise J.W. and Rosenfeld M.G. (1990) Alternative production of calcitonin and CGRP mRNA is regulated at the calcitonin-specific splice acceptor. *Nature* **341:** 76–80.

Epel D. (1977) The program of fertilization. *Scientific American* **237** (November): 128–138.

Epel D. (1997) Activation of sperm and egg during fertilization. In: *Handbook of Physiology, Section 14: Cell Physiology,* J.F. Hoffmann and J.J. Jamieson (eds.), 859–884. New York: Oxford University Press.

Epel D., Patton C., Wallace R.W. and Cheung W.Y. (1981) Calmodulin activates NAD kinase of sea urchin eggs: An early response. *Cell* **23:** 543–549.

Ephrussi A. and Lehmann R. (1992) Induction of germ cell formation by *oskar. Nature* **358:** 387–392.

Epperlein H.H. and Löfberg J. (1984) Xanthophores in chromatophore groups of the premigratory neural crest initiate the pigment pattern of the axolotl larva. *Wilhelm Roux's Arch.* **193:** 357–369.

Erickson C.A. and Goins T.L. (1995) Avian neural crest cells can migrate in the dorsolateral path only if they are specified as melanocytes. *Development* **121:** 915–924.

Erickson J.W. and Cline T.W. (1991) Molecular nature of the *Drosophila* sex determination signal and its link to neurogenesis. *Science* **251:** 1071–1074.

Erickson J.W. and Cline T.W. (1993) A bZIP protein, sisterless-a, collaborates with bHLH transcription factors early in *Drosophila* development to determine sex. *Genes and Dev.* **7:** 1688–1702.

Ericson J., Muhr J., Placzek M., Lints T., Jessell, T.M. and Edlund T. (1995) Sonic hedgehog induces the differentiation of ventral forebrain neurons: A common signal for ventral patterning within the neural tube. *Cell* **81:** 747–756.

Ericson J., Morton S., Kawakami A., Roelink H. and Jessel T.M. (1996) Two critical periods of sonic hedgehog signaling required for the specification of motor neuron identity. *Cell* **87:** 661–673.

Ericson J., Rashbass P., Schedl A., Brenner-Morton S., Kawakami A., van Heyningen V. and Jessel T.M. (1997a) *Pax 6* controls progenitor cell identity and neuronal fate in response to graded Shh signaling. *Cell* **90:** 169–180.

Ericson J., Briscoe J., Rashbass P., van Heyningen V. and Jessel T.M. (1997b) Graded Sonic hedgehog signaling and the specification of cell fate in the ventral neural tube. *Cold Spring Harbor Symp. Quant. Biol.* **LXII:** 451–466.

Erikson J., Nishikura K., Ar-Rushdi A., Finan J., Emanuel B., Lenoir G., Nowell P.C. and Croce C.M. (1983) Translocation of an immunoglobulin k locus to a region 3' of an unrearranged c-*myc* oncogene enhances c-*myc* transcription. *Proc. Nat. Acad. Sci. USA* **80:** 7581–7585.

Eriksson P.S., Perfilieva E., Bjork-Eriksson T., Alborn A.M., Nordborg C., Peterson D.A. and Gage F.H. (1998) Neurogenesis in the adult human hippocampus. *Nat. Med.* **4:** 1313–1317.

Etkin L.D. (1997) A new face for the endoplasmic reticulum: RNA localization. *Science* **276:** 1092–1093.

Etkin W. (1968) Hormonal control of amphibian metamorphosis. In: *Metamorphosis,* W. Etkin and L.I. Gilbert (eds.), 313–348. New York: Appleton-Century-Crofts.

Ettensohn C.A. (1985a) Gastrulation in the sea urchin embryo is accompanied by the rearrangement of invaginating epithelial cells. *Dev. Biol.* **112:** 383–390.

Ettensohn C.A. (1985b) Mechanisms of epithelial invagination. *Quart. Rev. Biol.* **60:** 289–307.

Ettensohn C.A. (1992) Cell interactions and mesodermal cell fates in the sea urchin embryo. *Development* (Supplement): 43–51.

Ettensohn C.A. (1999) Cell movements in the sea urchin embryo. *Curr. Opin. Genet. Dev.* **9:** 461–465.

Ettensohn C.A. and Ingersoll E.P. (1992) Morphogenesis of the sea urchin embryo. In: *Morphogenesis,* E.F. Rossomando and S. Alexander (eds.), 189–262. New York: Marcel Dekker.

Ettensohn C.A. and McClay D.R. (1988) Cell lineage conversion in the sea urchin embryo. *Dev. Biol.* **125:** 396–409.

Evans E.P., Ford C.E. and Lyon M.F. (1977) Direct evidence of the capacity of the XY germ cell in the mouse to become an oocyte. *Nature* **267:** 430–431.

Evans T.C., Crittenden S.L., Kodoyianni V. and Kimble J. (1994) Translational control of maternal *glp-1* mRNA establishes an asymmetry in the *C. elegans* embryo. *Cell* **77:** 183–194.

Evrard Y.A., Lun Y., Aulehla A., Gan L. and Johnson R.L. (1998) *lunatic fringe* is an essential mediator of somite segmentation and patterning. *Nature* **394:** 377–381.

Ewbank J.J., Barnes T.M., Lakowski B., Lussier M., Bussey H. and Hekimi S. (1997) Structural and functional conservation of the *Caenorhabditis elegans* timing gene *clk-1*. *Science* **275**: 980–983.

Eyal-Giladi H. (1984) The gradual establishment of cell commitments during the early stages of chick development. *Cell Differentiation* **14:** 245–255.

Fagard R. and London I.M. (1981) Relationship between phosphorylation and activity of hemeregulated eukaryotic initiation factor 2 kinase. *Proc. Nat. Acad. Sci. USA* **78:** 866–870.

Fagotto F. and Gumbiner B.M. (1996) Cell contact-dependent signaling. *Dev. Biol.* **180:** 445–54.

Fagotto F., Funayama N., Glück U. and Gumbiner B.M. (1996) Binding to cadherins anatagonizes the signaling activity of β-catenin during axis formation in *Xenopus. J. Cell Biol.* **132:** 1105–1114.

Fainsod A., Steinbeisser H. and DeRobertis E.M. (1994) On the function of BMP-4 in patterning the marginal zone of the *Xenopus* embryo. *EMBO J.* **13:** 5015–5025.

Fajardo M.A., Haugen H.S., Clegg C.H. and Braun R.H. (1997) Separate elements in the 3' untranslated region of the mouse protamine 1 mRNA regulate translational repression and activation during murine spermatogenesis. *Dev. Biol.* **191:** 42–52.

Fallon J.F., Lopez A., Ros M.A., Savage M.P., Olwin B.B. and Simandl B.K. (1994) FGF-2: Apical ectodermal ridge growth signal for chick limb development. *Science* **264:** 104–107.

Fan G., Ma X., Kren B.T. and Steer C.J. (1996) The retinoblastoma gene product inhibits TGF-β1 induced apoptosis in primary rat hepatocytes and human HuH-7 hepatoma cells. *Oncogene* **12:** 1909–1919.

Fan M.J. and Sokol S.Y. (1997) A role for Siamois in Spemann organizer formation. *Development* **124:** 2581–2589.

Farah M.H., Olson J.M., Sucic H.B., Hume R.I., Tapscott S.J. and Turner D.L. (2000) Generation of neurons by transient expression of neural bHLH proteins in mammalian cells. *Development* **127:** 693–702.

Fawcett D.W. (1975) The mammalian spermatozoon. *Dev. Biol.* **44:** 394–436.

Feldman K.A., Marks M.D., Christianson M.L. and Quatrano R.S. (1989) A dwarf mutant of *Arabidopsis* generated by T-DNA insertion mutagenesis. *Science* **243:** 1351–1354.

Felsenfeld G. (1992) Chromatin as an essential part of the transcriptional mechanism. *Nature* 355: 219–224.

Felsenfeld G. and McGhee J.D. (1986) Structure of the 30 nm chromatin fiber. *Cell* **44:** 375–377.

Feng H., Sandlow J.I., Sparks A.E., and Sandra A. (1999) Development of an immunocontraceptive vaccine. Current status. *Reprod. Med.* **44:** 759–765.

Ferguson E.L. and Anderson K. (1992) Decapentaplegic acts as a morphogen to organize the dorsal-ventral pattern in the *Drosophila* embryo. *Cell* **71:** 451–461.

Ferguson E.L., Sternberg P.W. and Horvitz H.R. (1987) A genetic pathway for the specification of the vulval cell lineages of *Caenorhabditis elegans*. *Nature* **326:** 259–267.

Fernandez-Teran M., Piedra M.E., Simandl B.K., Fallon J.F. and Ros M.A. (1997) Limb initiation and development is normal in the absence of the mesonephros. *Dev. Biol.* **189:** 246–255.

Ferrandon D., Elphick L., Nüsslein-Volhard C. and St Johnston D. (1994) Staufen protein associates with the 3' UTR of bicoid mRNA to form particles that move in a microtubule-dependent manner. *Cell* **79:** 1221–1232.

Ferrandon D., Koch I., Westhof E. and Nüsslein-Volhard C. (1997) RNA-RNA interaction is required for the formation of specific bicoid mRNA 3' UTR-STAUFEN ribonucleoprotein particles. *EMBO J.* **16:** 1751–1758.

Ferrara N., Carver-Moore K., Chen H., Dowd M., Lu L., O'Shea K.S., Powell-Braxton L., Hillan K.J. and Moore M.W. (1996) Heterozygous embryonic lethality induced by targeted inactivation of the VEGF gene. *Nature* **380:** 439–442.

Ferrell J.E. Jr. (1999) *Xenopus* oocyte maturation: New lessons from a good egg. *BioEssays* **21:** 833–842.

Fickett J.W. (1996) Coordinate positioning of MEF2 and myogenin binding sites. *Gene* **172:** 19–32.

Finch C.E. (1990) *Longevity, Senescence, and the Genome*. Chicago, University of Chicago Press.

Fink R.D. and McClay D.R. (1985) Three cell recognition changes accompany the ingression of sea urchin primary mesenchyme cells. *Dev. Biol.* **107:** 66–74.

Fire A., Xu S., Montgomery M.K., Kostas S.A., Driver S.E. and Mello C.C. (1998) Potent and specific genetic interference by double-stranded RNA in *C. elegans*. *Nature* **391:** 806–811.

Fitch R.H. and Denenberg V.H. (1998) A role for ovarian hormones in sexual differentiation of the brain. *Behav. Brain Sci.* **21:** 327–352.

Flanagan J.G., Chan D.C. and Leder P. (1991) Transmembrane form of the *kit* ligand growth factor is determined by alternative splicing and is missing in the Sl^d mutant. *Cell* **64:** 1025–1035.

Fleming T.P. (1987) A quantitative analysis of cell allocation to trophectoderm and inner cell mass in the mouse blastocyst. *Dev. Biol.* **119:** 520–531.

Fleming T.P., Pickering S.J., Qasim F. and Maro B. (1986) The generation of cell surface polarity in mouse 8–cell blastomeres: The role of cortical microfilaments analyzed using cytochalasin D. *J. Embryol. exp. Morphol.* **95:** 169–191.

Fletcher J.C. and Thummel C.S. (1995) The *Drosophila* E74 gene is required for the proper stage- and tissue-specific transcription of ecdysone-regulated genes at the onset of metamorphosis. *Development* **121:** 1411–1421.

Fletcher J.C., Burtis K.C., Hogness D.S. and Thummel C.S. (1995) The *Drosophila* E74 gene is required for metamorphosis and plays a role in the polytene chromosome puffing response to ecdysone. *Development* **121:** 1455–1465.

Fletcher J.C., D'Avino P.P. and Thummel C.S. (1997) A steroid-triggered switch in E74 transcription factor isoforms regulates the timing of secondary-response gene expression. *Proc. Natl. Acad. Sci. USA* **94:** 4582–4586.

Fletcher J.C., Brand U., Running M.P., Simon R. and Meyerowitz E.M. (1999) Signaling of cell fate decisions by clavata3 in *Arabidopsis* shoot meristems. *Science* **283:** 1911–1914.

Florman H.M. and Wassarman P.M. (1985) O-linked oligosaccharides of mouse egg ZP3 account for its sperm receptor activity. *Cell* **41:** 313–324.

Foe V.E. (1989) Mitotic domains reveal early commitment of cells in *Drosophila* embryos. *Development* **107:** 1–22.

Foe V.E. and Alberts B.M. (1983) Studies of nuclear and cytoplasmic behavior during the five mitotic cycles that precede gastrulation in *Drosophila* embryogenesis. *J. Cell Sci.* **61:** 31–70.

Folkman J. and Moscona A. (1978) Role of cell shape in growth control. *Nature* **273:** 345–349.

Follette P.J. and O'Farrell P.H. (1997) Connecting cell behavior to patterning: Lessons from the cell cycle. *Cell* **88:** 309–314.

Foltz K.R. and Lennarz W.J. (1993) The molecular basis of sea urchin gamete interactions at the egg plasma membrane. *Dev. Biol.* **158:** 46–61.

Foltz K.R., Partin J.S. and Lennarz W.J. (1993) Sea urchin egg receptor for sperm: Sequence similarity of binding domain and hsp70. *Science* **259:** 1421–1425.

Forbes A. and Lehmann R. (1998) Nanos and pumilio have critical roles in the development and function of *Drosophila* germline stem cells. *Development* **125:** 679–690.

Foty R.A., Pfleger C.M., Forgacs G. and Steinberg M.S. (1996) Surface tensions of embryonic tissues predict their mutual envelopment behavior. *Dev. Biol.* **122:** 1611–1620.

Fox R.O., Evans P.A. and Dobson C.M. (1986) Multiple conformations of a protein demonstrated by magnetization transfer NMR spectroscopy. *Nature* **320:** 192–194.

Frade J.M. and Barde Y.-A. (1998) Nerve growth factor: two receptors, multiple functions. *BioEssays* **20:** 137–145.

Frade J.M., Rodriguez-Tebar A. and Barde Y.A. (1996) Induction of cell death by endogenous nerve growth factor through its p75 receptor. *Nature* **383:** 166–168.

François V. and Bier E. (1995) *Xenopus* chordin and *Drosophila* short gastrulation genes encode homologous proteins functioning in dorsal-ventral axis formation. *Cell* **80:** 19–20.

Frank E. and Sanes J.R. (1991) Lineage of neurons and glia in chick dorsal root ganglia: Analysis in vivo with a recombinant retrovirus. *Development* **111:** 895–908.

Frankel J. (1989) *Pattern Formation: Ciliate Studies and Models.* New York: Oxford University Press.

Fankhauser G. (1952) Nucleocytoplasmic relations in amphibian development. *Int. Rev. Cytol.* **1:** 165–193.

Fraser S., Keynes R. and Lumsden A. (1990) Segmentation in the chick embryo hindbrain is defined by cell lineage restrictions. *Nature* **344:** 431–435.

Fraser S.E., Carhart M.S., Murray B.A., Chuong C.-M. and Edelman G.M. (1988) Alterations in the *Xenopus* retinotectal projection by antibodies to *Xenopus* N-CAM. *Dev. Biol.* **129:** 217–230.

Freeman G. (1979) The multiple roles which cell division can play in the localization of developmental potential. In: *Determinants of Spatial Organization,* S. Subtelny and I.R. Konigsberg (eds.), *37th Symp. Soc. Dev. Biol.,* 53–76. New York: Academic Press.

Freeman G. (1983) The role of egg organization in the generation of cleavage patterns. In *Time, Space, and Pattern in Embryonic Development,* W.R. Jeffery and R.A. Raff (eds.), 171–196. New York: Alan R. Liss.

Freeman G. (1988) The factors that promote the development of symmetry properties in aggregates from dissociated echinoid embryos. *Wilhelm Roux's Arch.* **197:** 394–405.

Freeman G. (1996) The role of localized cell surface-associated glycoproteins during fertilization in the hydrozoan *Aequorea. Dev. Biol.* **179:** 17–26.

Freeman G. and Lundelius J.W. (1982) The developmental genetics of dextrality and sinistrality in the gastropod *Lymnaea peregra. Wilhelm Roux's Arch.* **191:** 69–83.

Freeman G. and Miller R.L. (1982) Hydrozoan eggs can only be fertilized at the site of polar body formation. *Dev. Biol.* 94: 142–152.

French V., Bryant P.J. and Bryant S.V. (1976) Pattern regulation in epimorphic fields. *Science* **193:** 969–981.

Fridovich I. (1978) The biology of oxygen radicals. *Science* **201:** 875–880.

Frieden E. (1981) The dual role of thyroid hormones in vertebrate development and calorigenesis. In: *Metamorphosis: A Problem in Developmental Biology,* 2nd ed., L.I. Gilbert and E. Frieden (eds.), 545–564. New York: Plenum.

Friedman D.B. and Johnson T.E. (1988) A mutation in the *age-1* gene in *Caenorhabditis elegans* lengthens life and reduces hermaphrodite fertility. *Genetics* **118:** 75–86.

Friedmann, T., ed. (2000) *The Development of Human Gene Therapy.* Cold Spring Harbor, NY. Cold Spring Harbor Laboratory Press.

Friedrich G. and Soriano P. (1991) Promoter traps in embryonic stem cells: A genetic screen to identify and mutate developmental genes in mice. *Genes and Dev.* **5:** 1513–1523.

Frigerio G., Burri M., Bopp D., Baumgartner S. and Noll M. (1986) Structure of the segmentation gene *paired* and the *Drosophila* PRD gene set as part of a gene network. *Cell* **47:** 735–746.

Fristrom D. (1976) The mechanism of evagination of imaginal discs of *Drosophila melanogaster.* III. Evidence for cell rearrangement. *Dev. Biol.* **54:** 163–171.

Fristrom D., Wilcox M. and Fristrom J. (1993) The distribution of PS integrins, laminin A and F-actin during key stages in *Drosophila* wing development. *Development* **117:** 509–523.

Frohnhöfer H.G. and Nüsslein-Volhard C. (1986) Organization of anterior pattern in the *Drosophila* embryo by the maternal gene *bicoid. Nature* **324:** 120–125.

Frohnhöfer H.G. and Nüsslein-Volhard C. (1987) Maternal genes required for the anterior localization of *bicoid* activity in the embryo of *Drosophila. Genes and Dev.* 1: 880–890.

Frydman J., Nimmesgern E., Ohtsuka K. and Hartl F.U. (1994) Folding of nascent polypeptide chains in a high molecular mass assembly with molecular chaperones. *Nature* **370:** 111–117.

Fuchs E. and Cleveland D.W. (1998) A structural scaffolding of intermediate filaments in health and disease. *Science* **279:** 514–519.

Fuchs E. and Segre J.A. (2000) Stem cells: A new lease on life. *Cell* **100:** 143–155.

Fujisue M., Kobayakawa Y. and Yamana K. (1993) Occurrence of dorsal axis-inducing activity around the vegetal pole of an uncleaved *Xenopus* egg and displacement to the equatorial region by cortical rotation. *Development* **118:** 163–170.

Fulka J. Jr., First N.L., Loi P. and Moor R.M. (1999) Cloning by somatic cell nuclear transfer. *BioEssays* **20:** 847–851.

Fullilove S.L. and Jacobson A.G. (1971) Nuclear elongation and cytokinesis in *Drosophila* montana. *Dev. Biol.* **26:** 560–577.

Funayama N., Fagotto F., McCrea P. and Gumbiner B.M. (1995) Embryonic axis induction by the armadillo repeat domain of β-catenin: Evidence for intracellular signaling. *J.Cell Biol.* **128:** 959–968.

Furuno N. et al. (1994) Suppression of DNA replication via Mos function during meiotic divisions in *Xenopus* oocytes. *EMBO J.* **13:** 2399–2410.

Galione A., Jones K.T., Lai F.A. and Swann K. (1997) A cytosolic sperm protein factor mobilizes Ca^{2+} from intracellular stores by activating multiple Ca^{2+} release mechanisms independently of low molecular weight messengers. *J. Biol. Chem.* **272:** 28901–28905.

Gall J.G. (1991) Spliceosomes and snurposomes. *Science* **252:** 1499–1500.

Gallicano G.I., Schwarz S.M., McGaughey R.W. and Capco D.G. (1993) Protein kinase C, a pivotal regulator of hamster egg activation, functions after elevation of intracellular free calcium. *Dev. Biol.* **156:** 94–106.

Gallicano G.I., McGaughey R.W. and Capco D.G. (1997) Activation of protein kinase C after fertilization is required for remodeling the mouse egg into the zygote. *Mol. Reprod. Dev.* **46:** 587–601.

Gallicano G.I., Yousef M.C. and Capco D.G. (1997) PKC—a pivotal regulator of early development. *BioEssays* **19:** 29–36.

Gao Q. and Finkelstein R. (1998) Targeting gene expression to the head: The *Drosophila orthodenticle* gene is a direct target of the Bicoid morphogen. *Development* **125:** 4185–4193.

Garcia-Bellido A. (1975) Genetic control of wing disc development in *Drosophila.* In: *Cell Patterning, Ciba Foundation Symposia* **29:** 161–182. Amsterdam: Associated Scientific Publishers.

Garcia-Bellido A. and Lewis E.B. (1976) Autonomous cellular differentiation of homeotic bithorax mutants of *Drosophila melanogaster*. *Dev. Biol.* **48:** 400–410.

Garcia-Bellido A., Lawrence P.A. and Morata G. (1979) Compartments in animal development. *Scientific American* **241** (July): 102–110.

Garcia-Fernàndez J. and Holland P.W.H. (1994) Archetypal organization of the amphioxus Hox gene cluster. *Nature* **370:** 563–566.

Gard D.L., Cha B.J. and King E. (1997) The organization and animal-vegetal asymmetry of cytokeratin filaments in stage VI *Xenopus* oocytes is dependent upon F-actin and microtubules. *Dev. Biol.* **184:** 95–114.

Garrod A.E. (1909) *Inborn Errors of Metabolism*. Reprinted 1963 with a supplement by Harry Harris. London: Oxford University Press.

Gartler S.M. and Riggs A.D. (1983) Mammalian X-chromosome inactivation. *Ann. Rev. Genet.* **17:** 155–190.

Gasser C.S. and Fraley R.T. (1992) Transgenic crops. *Scientific American* **266** (June): 62–69.

Gasser R.F. (1975) *Atlas of Human Embryos*. New York: Harper and Row.

Gaul U. and Jäckle H. (1990) Role of gap genes in early *Drosophila* development. *Adv. Genet.* **27:** 239–275.

Gautier J., Solomon M.J., Booher R.N., Bazan J.F. and Kirschner M.W. (1991) cdc25 is a specific tyrosine phosphatase that directly activates p34^{cdc2}. *Cell* **87:** 197–211.

Gavalas A., Studer M., Lumsden A., Rijli F.M., Krumlauf R. and Chambon P. (1998) Hoxa1 and Hoxb1 synergize in patterning the hindbrain, cranial nerves and second pharyngeal arch. *Development* **125:** 1123–1136.

Gavis E.R. and Lehmann R. (1994) Translational regulation of nanos by RNA localization. *Nature* **369:** 315–318.

Gebauer F. and Richter J.D. (1997) Synthesis and function of Mos: The control switch of vertebrate oocyte meiosis. *BioEssays* **19:** 23–28.

Gebauer F., Corona D.F., Preiss T., Becker P.B. and Hentze M.W. (1999) Translational control of dosage compensation in *Drosophila* by Sex-lethal: Cooperative silencing via the 5' and 3' UTRs of msl-2 mRNA is independent of the poly(A) tail. *EMBO J.* **18:** 6146–6154.

Gehring W. (1968) The stability of the determined state in cultures of imaginal discs in *Drosophila*. In: *The Stability of the Differentiated State*, H. Ursprung (ed.), 136–154. New York: Springer-Verlag.

Gehring W.J. (1985) Homeotic genes, the homeobox, and the genetic control of development. *Cold Spring Harbor Symp. Quant. Biol.* **50:** 243–251.

Gehring W.J. (1992) The homeobox in perspective. *Trends in Biochemical Sciences* **17:** 277–280.

Gehring W.J., Müller M., Affolter M., Percival-Smith A., Billeter M., Qian Y.Q., Otting G. and Wüthrich K. (1990) The structure of the homeodomain and its functional implications. *Trends Genet.* **6:** 323–329.

Gehring W.J., Qian Y.Q., Billeter M., Furukubo-Tokunaga K., Schier A.F., Resendez-Perez D., Affolter M., Otting G. and Wüthrich K. (1994) Homeodomain-DNA recognition. *Cell* **78:** 211–223.

Geigy R. (1941) Die Metamorphose als Folge gewebsspezifischer Determination. *Revue Suisse de Zoologie* **48:** 483–494.

Geisler R., Bergmann A., Hiromi Y. and Nüsslein-Volhard C. (1992) *Cactus*, a gene involved in dorsoventral pattern formation of *Drosophila*, is related to IκB gene family of vertebrates. *Cell* **71:** 613–621.

Gerhart J. and Keller R. (1986) Region-specific cell activities in amphibian gastrulation. *Ann. Rev. Cell Biol.* **2:** 201–229.

Gerhart J., Danilchik M., Doniach T., Roberts S., Rowning B. and Stewart R. (1989) Cortical rotation of the *Xenopus* egg: Consequences for the anteroposterior pattern of embryonic dorsal development. *Development* 107 (Supplement): 37–51.

Ghiglione C., Perrimon N. and Perkins L.A. (1999) Quantitative variations in the level of MAPK activity control patterning of the embryonic termini in *Drosophila. Dev. Biol.* **205:** 181–193.

Gierer A. (1974) Hydra as a model for the development of biological form. *Scientific American* **231** (December): 44–54.

Gilbert L.I., Tata J.R. and Atkinson B.G. (1996) *Metamorphosis: Post-embryonic Reprogramming of Gene Expression in Amphibian and Insect Cells*. San Diego: Academic Press.

Gilbert S.F. (1988) *Developmental Biology*. 2nd ed. Sunderland, MA: Sinauer Associates.

Gilbert S.F. (1991) *Developmental Biology*. 3rd ed. Sunderland, MA: Sinauer Associates.

Giles R.E., Blanc H., Cann H.M. and Wallace D.C. (1980) Maternal inheritance of human mitochondrial DNA. *Proc. Nat. Acad. Sci. USA* **77:** 6715–6719.

Gillespie L.L., Paterno G.D. and Slack J.M.W. (1989) Analysis of competence: Receptors for fibroblast growth factor in early *Xenopus* embryos. *Development* **106:** 203–208.

Gimlich R.L. (1986) Acquisition of developmental autonomy in the equatorial region of the *Xenopus* embryo. *Dev. Biol.* **115:** 340–352.

Gimlich R.L. and Cook J. (1983) Cell lineage and the induction of nervous systems in amphibian development. *Nature* **306:** 471–473.

Gimlich R.L. and Gerhart J.C. (1984) Early cellular interactions promote embryonic axis formation in *Xenopus* laevis. *Dev. Biol.* **104:** 117–130.

Glabe C.G. and Clark D. (1991) The sequence of the *Arbacia punctulata* bindin cDNA and implications for the structural basis of species-specific sperm adhesion and fertilization. *Dev. Biol.* **143:** 282–288.

Glabe C.G. and Lennarz W.J. (1979) Species-specific sperm adhesion in sea urchins: A quantitative investigation of bindin-mediated egg agglutination. *J. Cell Biol.* **83:** 595–604.

Glass C.K. and Rosenfeld M.G. (2000) The coregulator exchange in transcriptional functions of nuclear receptors. *Genes and Dev.* **14:** 121–141.

Godsave S.F. and Slack J.M.W. (1991) Single cell analysis of mesoderm formation in the *Xenopus* embryo. *Development* **111:** 523–530.

Gold B., Fujimoto H., Kramer J.M., Erickson R.P. and Hecht N.B. (1983) Haploid accumulation and translational control of phosphoglycerate kinase-2 messenger RNA during mouse spermatogenesis. *Dev. Biol.* **98:** 392–399.

Goldberg R.B., de Paiva G. and Yadegari R. (1994) Plant embryogenesis: Zygote to seed. *Science* **266:** 605–614.

Golde D.W. (1991) The stem cell. *Scientific American* 265 (December): 86–93.

Goldschmidt R. (1938) *Physiological Genetics*. New York: McGraw-Hill.

Goldstein B. (1992) Induction of gut in *Caenorhabditis elegans* embryos. *Nature* **357:** 255–257.

Goldstein B. (1993) Establishment of gut fate in the E lineage of *C. elegans*: The role of lineage-dependent mechanisms and cell interactions. *Development* **118:** 1267–1277.

Goldstein B. (1995) Cell contacts orient some cell division axes in the *Caenorhabditis elegans* embryo. *J. Cell Biol.* **129:** 1071–1080.

Goldstein B. and Freeman G. (1997) Axis specification in animal development. *BioEssays* **19:** 105–116.

Goldstein B. and Hird S.N. (1996) Specification of the anterior-posterior axis in *C. elegans. Development* **122:** 1467–1474.

Gompertz B. (1825) On the nature of the function expressive of the law of human mortality and on a new mode of determining the value of life contingencies. *Philos. Trans. Royal Soc. London* **115**: 513–585.

Gong X.H., Dubois D.H., Miller D.J. and Shur B.D. (1995) Activation of a G protein complex by aggregation of β-1,4 galactosyltransferase on the surface of sperm. *Science* **269:** 1718–1721.

Gonzales-Reyes A. and St Johnston D. (1998) Patterning of the follicle cell epithelium along the anterior-posterior axis during *Drosophila* oogenesis. *Development* **125:** 2837–2846.

Gonzales-Reyes A., Elliot H. and St. Johnston D. (1995) Polarization of both major body axes in *Drosophila* by *gurken-torpedo* signaling. *Nature* **375:** 654–658.

Goodrich J.A., Cutler G. and Tijan R. (1996) Contacts in context: Promoter specificity and macromolecular interactions in transcription. *Cell* **84:** 825–830.

Goodrich J., Puangsomlee P., Martin M., Long D., Meyerowitz E.M. and Coupland G. (1997) A Polycomb group gene regulates homeotic gene expression in *Arabidopsis. Nature* **386:** 44–51.

Goossen B., Caughman S.W., Harford J.B., Klausner R.D. and Hentze M.W. (1990) Translational repression by a complex between the iron-responsive element of ferritin messenger RNA and its specific cytoplasmic binding protein is position-dependent *in vivo. EMBO J.* **9:** 4127–4133.

Gordon M.Y., Riley G.P., Watt S.M. and Greaves M.F. (1987) Compartmentalization of a hematopoietic growth factor (GM-CSF) by glycosaminoglycans in the bone marrow microenvironment. *Nature* **326:** 403–405.

Gordon J.W., Scangos G.A., Plotkin D.J., Barbosa J.A. and Ruddle F.H. (1980) Genetic transformation of mouse embryos by microinjection of purified DNA. *Proc. Nat. Acad. Sci. USA* **77:** 7380–7384.

Gordon R. (1985) A review of the theories of vertebrate neurulation and their relationship to the mechanics of neural tube birth defects. *J. Embryol. Exp. Morphol.* **89** (Supplement): 229–255.

Gordon R. and Brodland G.W. (1987) The cytoskeletal mechanics of brain morphogenesis: Cell state splitters cause primary neural induction. *Cell Biophysics* **11:** 177–238.

Gordon R. and Jacobson A.G. (1978) The shaping of tissues in embryos. *Scientific American* **238** (June): 106–113.

Görlich D. and Mattaj I.W. (1996) Nucleocytoplasmic transport. *Science* **271:** 1513–1518.

Gorman M., Franke A. and Baker B.S. (1995) Molecular characterization of the *male-specific-lethal-3* gene and investigation of the regulation of dosage compensation in *Drosophila. Development* **121:** 463–475.

Gorman M., Kuroda M.I. and Baker B.S. (1993) Regulation of the sex-specific binding of the maleless dosage compensation protein to the male X chromosome in *Drosophila. Cell* **72:** 39–49.

Gorski R.A., Gordon J.H., Shryne J.E. and Southam A.M. (1978) Evidence for a morphological sex difference within the medial preoptic area of the rat brain. *Brain Res.* **143:** 333–346.

Gould A., Morrison A., Sproat G., White R.A.H. and Krumlauf R. (1997) Positive cross-regulation and enhancer sharing: Two mechanisms for specifying overlapping Hox expression patterns. *Genes and Dev.* **11:** 900–913.

Gould A.P. and White R.A.H. (1992) Connectin, a target of homeotic gene control in *Drosophila. Development* **116:** 1163–1174.

Gould S.J. (1977) *Ontogeny and Phylogeny.* Cambridge, MA: Belknap/Harvard.

Gould S.J. (1990) An earful of jaw. *Natural History* **3:** 12–23.

Goulding M.D., Lumsden A. and Gruss P. (1993) Signals from the notochord and floor plate regulate the region-specific expression of two Pax genes in the developing spinal cord. *Development* **117:** 1001–1016.

Govind S. and Steward R. (1991) Dorsoventral pattern formation in *Drosophila*: Signal transduction and nuclear targeting. *Trends Genet.* **7:** 119–125.

Graba Y., Aragnol D. and Pradel J. (1997) *Drosophila* Hox complex downstream targets and the function of homeotic genes. *BioEssays* **19:** 379–388.

Grace M., Bagchi M., Ahmad F., Yeager T., Olson C., Chakravarty I., Nasron N., Banerjee A. and Gupta N.K. (1984) Protein synthesis in rabbit reticulocytes: Characteristics of the protein factor RF that reverses inhibition of protein synthesis in heme-deficient reticulocyte lysates. *Proc. Nat. Acad. Sci. USA* **79:** 6517–6521.

Graff J.M. (1997) Embryonic patterning: To BMP or not to BMP, that is the question. *Cell* **89:** 171–174.

Graff J.M., Thies R.S., Song J.J., Celeste A.J. and Melton D.A. (1994) Studies with a *Xenopus* BMP receptor suggest that ventral mesoderm-inducing signals override dorsal signals in vivo. *Cell* **79:** 169–179.

Graham A., Papalopulu N. and Krumlauf R. (1989) The murine and *Drosophila* homeobox gene complexes have common features of organization and expression. *Cell* **57:** 367–378.

Graham C.F. and Wareing P.F. (1976) *The Developmental Biology of Plants and Animals.* Philadelphia: Saunders.

Graham G.J., Wright E.G., Hewick R., Wolpe S.D., Wilkie N.M., Donaldson D., Lorimore S. and Pragnell I.B. (1990) Identification and characterization of an inhibitor of haemopoietic stem cell proliferation. *Nature* **344:** 442–444.

Grainger R.M., Henry J.J. and Henderson R.A. (1988) Reinvestigation of the role of the optic vesicle in embryonic lens induction. *Development* **102:** 517–526.

Grant P. (1978) *Biology of Developing Systems.* New York: Holt, Rinehart and Winston.

Grebe M., Gadea J., Steinmann T., Kientz M., Rahfeld J.U., Salchert K., Koncz C. and Jürgens G. (2000) A conserved domain of the *Arabidopsis* GNOM protein mediates subunit interaction and cyclophilin 5 binding. *Plant Cell* **12:** 343–356.

Green G.R. and Poccia E.L. (1985) Phosphorylation of sea urchin sperm H1 and H2B histones precedes chromatin decondensation and H1 exchange during pronuclear formation. *Dev. Biol.* **108:** 235–245.

Green J.B.A.,New H.V. and Smith J.C. (1992) Responses of embryonic *Xenopus* cells to activin and FGF are separated by multiple dose thresholds and correspond to distinct axes of the mesoderm. *Cell* **71:** 731–739.

Green M.C. (1989) Catalog of mutant genes and polymorphic loci. In: *Genetic Variants and Strains of the Laboratory Mouse,* 2nd ed., M.F. Lyon and A.G. Searle (eds.), 12–403. New York: Oxford University Press.

Greenwald I. (1999) LIN-12/Notch signaling: Lessons from worms and flies. *Genes and Dev.* **12:** 1751–1762.

Greenwald I.S., Sternberg P.W. and Horvitz H.R. (1983) The lin-12 locus specifies cell fates in *Caenorhabditis elegans. Cell* **34:** 435–444.

Greenwood S. and Struhl G. (1997) Different levels of Ras activity can specify distinct transcriptional and morphological consequences in early *Drosophila* embryos. *Development* **124:** 4879–4886.

Grenningloh G., Bieber A.J., Rehm E.J., Snow P.M., Traquina Z.R., Hortsch M., Patel N.H. and Goodman C.S. (1990) Molecular genetics of neuronal recognition in *Drosophila*: Evolution and function of immunoglobulin superfamily cell adhesion molecules. *Cold Spring Harbor Symp. Quant. Biol.* **55:** 327–340.

Griffith F. (1928) The significance of pneumococcal types. *J. Hyg.* (London) **27:** 113–159.

Grimes G.W. (1990) Inheritance of cortical patterns in ciliated protozoa. In: *Cytoplasmic Organization Systems,* G.M. Malacinski (ed.), 23–43. New York: McGraw-Hill.

Grimes G.W. and Aufderheide K.J. (1991) *Cellular Aspects of Pattern Formation: The Problem of Assembly.* Basel/New York: Karger.

Grimes G.W. and Hammersmith R.L. (1980) Analysis of the effects of encystment and excystment on incomplete doublets of *Oxytricha fallax. J. Embryol. exp. Morphol.* **59:** 19–26.

Grimes G.W., McKenna M.E., Goldsmith-Spoegler C.M. and Knaupp E.A. (1980) Patterning and assembly of ciliature are independent processes in hypotrich ciliates. *Science* **209:** 281–283.

Griswold M.D. (1995) Interactions between germ cells and Sertoli cells in the testis. *Biol. Reprod.* **52:** 211–216.

Gross A., McDonnell J.M. and Korsmeyer S.M. (1999) BCL-2 family members and the mitochondria in apoptosis. *Genes and Dev.* **13:** 1899–1911.

Gross P.R. and Cousineau G.H. (1964) Macromolecular synthesis and the influence of actinomycin D on early development. *Exp. Cell Res.* **33:** 368–395.

Grosshans J., Schnorrer F. and Nüsslein-Volhard. C. (1999) Oligomerisation of Tube and Pelle leads to nuclear localisation of dorsal. *Mech. Dev.* **81:** 127–138.

Grosveld F. (1999) Activation by locus control regions. *Curr. Opin. Genet. Dev.* **9:** 152–157.

Gruidl M.E., Smith P.A., Kuznicki K.A., McCrone J.S., Kirchner J., Roussel D.L., Strome S. and Bennett K.L. (1996) Multiple potential germ-line helicases are components of the germline-specific P granules of *Caenorhabditis elegans. Proc. Nat. Acad. Sci. USA* **93:** 13837–13842.

Grunstein M. (1997) Histone acetylation inchromatin structure and transcription. *Nature* **389:** 349–352.

Grunz H. and Tacke L. (1986) The inducing capacity of the presumptive endoderm of *Xenopus laevis* studied by transfilter experiments. *Wilhelm Roux's Arch.* **195:** 467–473.

Gu H., Marth J.D., Orban P.C., Mossmann H. and Rajewski K. (1994) Deletion of a DNA polymerase β gene segment in T cells using cell type-specific gene targeting. *Science* **265:** 103–106.

Gu W., Schneider J.W., Condorelli G., Kaushal S., Mahdavi V. and Nadal-Ginard B. (1993) Interaction of myogenic factors and the retinoblastoma protein mediates muscle cell commitment and differentiation. *Cell* **72:** 309–324.

Gubbay J., Collignon J., Koopman, P., Capel B., Economou A., Münsterberg A., Vivian N., Goodfellow P. and Lovell-Badge R. (1990) A gene mapping to the sex-determining region of the mouse Y chromosome is a member of a novel family of embryonically expressed genes. *Nature* **346:** 245–250.

Gubbay J., Vivian N., Economou A., Jackson D., Goodfellow P. and Lovell-Badge R. (1992) Inverted repeat structure of the *Sry* locus in mice. *Proc. Nat. Acad. Sci. USA* **89:** 7953–7957.

Gudernatsch J.F. (1912) Feeding experiments on tadpoles. I. The influence of specific organs given as food on growth and differentiation. A contribution to the knowledge of organs with internal secretion. *Wilhelm Roux's Arch.* **35:** 457–483.

Guger K.A. and Gumbiner B.M. (1995) β-catenin has wnt-like activity and mimics the Nieuwkoop signaling center in *Xenopus* dorsal-ventral patterning. *Dev. Biol.* **172:** 115–125.

Guillemot F., Lo L.-C., Johnson J.E., Auerbach A., Anderson D.J. and Joyner A.L. (1993) Mammalian *achete-scute* homolog 1 is required for the early development of olfactory and autonomic neurons. *Cell* **75:** 463–476.

Gulyas B.J. (1975) A reexamination of the cleavage patterns in eutherian mammalian eggs: Rotation of the blastomere pairs during second cleavage in the rabbit. *J. Exp. Zool.* **193:** 235–248.

Gumbiner B. (1990) Generation and maintenance of epithelial cell polarity. *Curr. Op. Cell Biol.* **2:** 881–887.

Gumbiner B.M. (2000) Regulation of cadherin adhesive activity. *J. Cell Biol.* **148:** 399–404.

Gumbiner G. (1995) Signal transduction by β-catenin. *Curr. Op. Cell Biol.* **7:** 734–760.

Gumbiner G. (1996) Cell adhesion: The molecular basis of tissue architecture and morphogenesis. *Cell* **84:** 345–357.

Gundersen R.W. (1987) Response of sensory neurites and growth cones to patterned substrata of laminin and fibronectin in vitro. *Dev. Biol.* **121:** 423–431.

Gunkel N., Yano T., Markussen F.H., Olsen L.C. and Ephrussi A. (1998) Localization-dependent translation requires a functional interaction between the 5′ and 3′ ends of oskar mRNA. *Genes and Dev.* **12:** 1652–1664.

Guo M., Jan L.Y. and Jan Y.N. (1996) Control of daughter cell fates during asymmetric division: interaction of Numb and Notch. *Neuron* **17:** 27–41.

Guo S. and Kemphues K.J. (1996) Molecular genetics of asymmetric cleavage in the early *Caenorhabditis* embryo. *Curr. Opin. Genet. Dev.* **6:** 408–415.

Gurdon J.B. (1962) The developmental capacity of nuclei taken from intestinal epithelial cells of feeding tadpoles. *J. Embryol. exp. Morphol.* **10:** 622–640.

Gurdon J.B. (1968) Transplanted nuclei and cell differentiation. *Scientific American* **219** (December): 24–35.

Gurdon J.B., Laskey R.A. and Reeves O.R. (1975) The developmental capacity of nuclei transplanted from keratinized skin cells of adult frogs. *J. Embryol. exp. Morphol.* **34:** 93–112.

Gurdon J.B., Fairman S., Mohun T.J. and Brennan S. (1985) Activation of muscle-specific actin genes in *Xenopus* development by an induction between animal and vegetal cells of a blastula. *Cell* **41:** 913–922.

Gurdon J.B., Lemaire P. and Kato K. (1993) Community effects and related phenomena in development. *Cell* **75:** 831–834.

Gurdon J.B., Harger P., Mitchell A. and Lemaire P. (1994) Activin signalling and response to a morphogen gradient. *Nature* **371:** 487–492.

Gurdon J.B., Mitchell A. and Mahony D. (1995) Direct and continuous assessment by cells of their position in a morphogen gradient. *Nature* **376:** 520–521.

Gustafson-Brown C., Savidge B. and Yanofsky M.F. (1994) Regulation of the arabidopsis floral homeotic gene *APETALA1*. *Cell* **76:** 131–143.

Gustafsson L. and Pärt, T. (1990) Acceleration of senescence in the collared flycatcher *Ficedula albicollis* by reproductive costs. *Nature* **347:** 279–281.

Guthrie S. and Lumsden A. (1991) Formation and regeneration of rhombomere boundaries in the developing chick hindbrain. *Development* **112:** 221–229.

Guthrie S., Muchamore I., Kuroiwa A., Marshall H., Krumlauf R. and Lumsden A. (1992) Neuroectodermal autonomy of *Hox-2.9* expression revealed by rhombomere transpositions. *Nature* **356:** 157–159.

Guthrie S., Prince V. and Lumsden A. (1993) Selective dispersal of avian rhombomere cells in orthotopic and heterotopic grafts. *Development* **118:** 527–538.

Gutzeit H. (1986) The role of microtubules in the differentiation of ovarian follicles during vitellogenesis in *Drosophila. Roux's Arch. Dev. Biol.* **195:** 173–181.

Guyette W.A., Matusik R.J. and Rosen J.M. (1979) Prolactin-mediated transcriptional and post-transcriptional control of casein gene expression. *Cell* **17:** 1013–1023.

Gwatkin R.B.L. (1977) *Fertilization Mechanisms in Man and Mammals*. New York: Plenum.

Ha I., Wightman B. and Ruvkun G. (1996) A bulged lin-4/lin-14 RNA duplex is sufficient for *Caenorhabditis elegans* lin-14 temporal gradient formation. *Genes and Dev.* **10:** 3041–3050.

Hable W.E. and Kropf D.L. (2000) Sperm entry introduces polarity in fucoid zygotes. *Development* **127:** 493–501.

Hadek R. (1969) *Mammalian Fertilization: An Atlas of Ultrastructure*. New York: Academic Press.

Hadorn E. (1948) Genetische und entwicklungsphysiologische Probleme der Insektenontogenese. *Folia Biotheoretica* **3:** 109–126.

Hadorn E. (1963) Differenzierungsleistungen wiederholt fragmentierter Teilstücke männlicher Genitalscheiben von *Drosophila melanogaster* nach Kultur in vivo. *Dev. Biol.* **7(1):** 617–629.

Hadorn E. (1968) Transdetermination in cells. *Scientific American* **219** (November): 110–120.

Haecker V. (1912) Untersuchungen über Elementareigenschaften. *Zeitschr. indukt. Abst. Vererb.* **8:** 36–47.

Hafen E., Levine M. and Gehring W. (1984) Regulation of *Antennapedia* transcript distribution by the *bithorax* complex in *Drosophila. Nature* **307:** 287–289.

Haffter P. et al. (1996) The identification of genes with unique and essential functions in the development of the zebrafish, *Danio rerio. Development* **123:** 1–36.

Hagan K.W., Ruiz-Echervarria M.J., Quan Y. and Peltz S.W. (1995) Characterization of cis-acting sequences and decay intermediates involved in nonsense-mediated mRNA turnover. *Mol. Cell Biol.* **15:** 809–823.

Hagedorn H.H. and Kunkel J.G. (1979) Vitellogenin and vitellin in insects. *Ann. Rev. Entomol.* **24:** 474–505.

Hager J.H. and Cline T.W. (1997) Induction of female Sex-lethal RNA splicing in male germ cells: Implications for *Drosophila* germline sex determination. *Development* **124:** 5033–5048.

Halevy O., Novitch B.G., Spicer D.B., Skapex S.X., Rhee J., Hannon G.J., Beach D. and Lassar A.B. (1995) Correlation of terminal cell cycle arrest of skeletal muscle with induction of p21 by MyoD. *Science* **267:** 1018–1021.

Hall B.K. and Hörstadius S. (1988) *The Neural Crest*. London: Oxford University Press.

Hall B.L. and Thummel C.S. (1998) The RXR homolog Ultraspiracle is an essential components of the *Drosophila* ecdysone receptor. *Development* **125:** 4709–4717.

Hall P.A. and Watt F.M. (1989) Stem cells: The generation and maintenance of cellular diversity. *Development* **106:** 619–633.

Halprin K.M. (1972) Epidermal "turnover time"—A reexamination. *J. Invest. Dermatol.* **86:** 14–19.

Ham R.G. and Veomett M.J. (1980) *Mechanisms of Development*. St. Louis, MO: C.V. Mosby.

Hamaguchi M.S. and Hiramoto Y. (1980) Fertilization process in the heart urchin, *Clypeaster japonicus*, observed with a differential interference microscope. *Dev. Growth Differ.* **22:** 517–530.

Hamaguchi M.S. and Hiramoto Y. (1981) Activation of sea urchin eggs by microinjection of calcium buffers. *Exp. Cell Res.* **134:** 171–179.

Hamann T., Mayer U. and Jürgens G. (1999) The auxin-insensitive *bodenlos* mutation affects primary root formation and apical-basal patterning in the *Arabidopsis* embryo. *Development* **126:** 1387–1395.

Hamburger V. (1960) *A Manual of Experimental Embryology*. Rev. ed. Chicago/London: University of Chicago Press.

Hamburger V. (1988) *The Heritage of Experimental Embryology: Hans Spemann and the Organizer*. New York: Oxford University Press.

Hamburger V. and Hamilton H.L. (1951) A series of normal stages in the development of the chick embryo. *J. Morph.* **88:** 49–92.

Hamburger V. and Levi-Montalcini R. (1949) Proliferation, differentiation and degeneration in the spinal ganglia of the chick embryo under normal and experimental conditions. *J. Exp. Zool.* **111:** 457–501.

Hampson E. and Kimura D. (1992) Sex differences and hormonal influences on cognitive function in humans. In: Behavioral Endocrinology, J.B. Becker, S.M. Breedlove and D. Crews (eds.), 347–400. Cambridge, MA: MIT Press.

Handyside A.H. (1978) Time of commitment of inside cells isolated from preimplantation mouse embryos. *J. Embryol. exp. Morphol.* **45:** 37–53.

Hammerschmidt M., Serbedzija G.N. and McMahon A.P. (1996) Genetic analysis of dorsoventral pattern formation in the zebrafish: Requirement of a BMP-like ventralizing activity and its dorsal repressor. *Genes and Dev.* **10:** 2452–2461.

Hanks M., Wurst W., Anson-Cartwright L., Auerbach A.B. and Joyner A.L. (1995) Rescue of the En-1 mutant phenotype by replacement of En-1 with En-2. *Science* **269:** 679–682.

Hara K. (1977) The cleavage pattern of the axolotl egg studied by cinematography and cell counting. *Wilhelm Roux's Arch.* **181:** 73–87.

Hardin J. (1988) The role of secondary mesenchyme cells during sea urchin gastrulation studied by laser ablation. *Development* **103:** 317–324.

Hardin J. (1989) Local shifts in position and polarized motility drive cell arrangement during sea urchin gastrulation. *Dev. Biol.* **136:** 430–445.

Hardin J. and McClay D.R. (1990) Target recognition by the archenteron during sea urchin gastrulation. *Dev. Biol.* **142:** 86–102.

Hardin J.D. and Cheng L.Y. (1986) The mechanisms and mechanics of archenteron elongation during sea urchin gastrulation. *Dev. Biol.* **115:** 490–501.

Harding C.F. (1983) Hormonal specificity and activation of social behavior in the male zebra finch. In: *Hormones and Behaviour in Higher Vertebrates,* J. Balthazart, E. Prove and R. Gilles (eds.), 275–289. Berlin: Springer-Verlag.

Hardtke C.S. and Berleth T. (1998) The *Arabidopsis* gene *MONOPTEROS* encodes a transcription factor mediating embryo axis formation and vascular development. *EMBO J.* **17:** 1405–1411.

Harley V.R., Jackson D.I., Hextall P.J., Hawkins J.R., Berkovitz G.D., Sockanathan S., Lovell-Badge R. and Goodfellow P.N. (1992) DNA binding activity of recombinant *SRY* from normal males and XY females. *Science* **255:** 453–456.

Harper J.W., Adami G.R., Wei N., Keyomarsi K. and Elledge S.J. (1993) The p21 Cdk-interacting protein Cip1 is a potent inhibitor of G1 cyclin-dependent kinases. *Cell* **75:** 805–816.

Harris H. (1974) *Nucleus and Cytoplasm.* 3rd ed. Oxford: Clarendon Press.

Harrison P.R., Birnie G.D., Hell A., Humphries S., Young R.D. and Paul J. (1974) Kinetic studies of gene frequency. I. Use of a DNA copy of reticulocyte 9 S RNA to estimate globin gene dosage in mouse tissues. *J. Mol. Biol.* **84:** 539–554.

Harrison R.G. (1918) Experiments on the development of the forelimb of *Ambystoma,* a self-differentiating equipotential system. *J. Exp. Zool.* **25:** 413–461.

Harrison R.G. (1924) Some unexpected results of the heteroplastic transplantation of limbs. *Proc. Nat. Acad. Sci. USA* **10:** 69–74.

Harrison R.G. (1929) Correlation in the development and growth of the eye studied by means of heteroplastic transplantation. *Wilhelm Roux's Arch.* **120:** 1–55.

Harrison R.G. (1935) Heteroplastic grafting in embryology. *The Harvey Lectures,* 1933–1934: 116–157.

Hartley R.S., Sible J.C., Lewellyn A.L. and Maller J.L. (1997) A role for cyclinE/Cdk2 in the timing of the midblastula transition in *Xenopus* embryos. *Dev. Biol.* **188:** 312–321.

Harvey E.B. (1940) A comparison of the development of nucleate and non-nucleate eggs of *Arbacia punctulata. Biol. Bull.* **79:** 166–187.

Hashimoto N. et al. (1994) Parthenogenetic activation of oocytes in c-mos-deficient mice. *Nature* **370:** 68–71.

Hasty P., Bradley A., Morris J.H., Edmondson D.G., Venuti J.M., Olson E.N. and Klein W.H. (1993) Muscle deficiency and neonatal death in mice with a targeted mutation in the *myogenin* gene. *Nature* **364:** 501–506.

Hathaway H.J. and Shur B.D. (1992) Cell surface β1,4–galactosyltransferase mediates neural crest cell migration and neurulation *in vivo. J. Cell Biol.* **117:** 369–382.

Hay E.D. (1981) Collagen and embryonic development. In: *Cell Biology of Extracellular Matrix.* E.D. Hay (ed.), 379–409. New York: Plenum.

Hay E.D., ed. (1991) *Cell Biology of Extracellular Matrix.* 2nd ed. New York: Plenum.

Hayflick L. (1965) The limited *in vitro* lifetime of human diploid cell strains. *Exp. Cell Res.* **37**: 614–636.

Hayflick L. (1966) Cell Culture and the aging phenomenon. In: *Topics in the Biology of Aging: A Symposium Held at the Salk Institute for Biological Studies*, Peter L. Krohn (ed.), 83–100. New York and London: Interscience.

Hawley S.H.B., Wünnenberg-Stapleton K., Hashimoto C., Laurent M., Watabe T., Blumberg B.W. and Cho K.W.Y. (1995) Disruption of BMP signals in embryonic *Xenopus* ectoderm leads to direct neural induction. *Genes and Dev.* **9:** 2923–2935.

He H.T., Barbet J., Chaix J.-C. and Goridis C. (1986) Phosphatidyl-inositol is involved in the membrane attachment of N-CAM 120, the smallest component of the neural cell adhesion molecule. *EMBO J.* **5:** 2489–2494.

He X., Saint-Jeannet J.-P., Woodgett J.R., Varmus H.E. and Dawid I. (1995) Glycogen synthase kinase-3 and dorsoventral patterning in *Xenopus* embryos. *Nature* **374:** 617.

Heard E., Mongelard F., Arnaud D., Chureau C., Vourc'h C. and Avner P. (1999) Human XIST yeast artificial chromosome transgenes show partial X inactivation center function in mouse embryonic stem cells. *Proc. Natl. Acad. Sci. USA* **96:** 6841–6846.

Heasman J. (1997) Patterning the *Xenopus* blastula. *Development* **124:** 4179–4191.

Heasman J., Crawford A., Goldstone K., Garner-Hamrick P., Gumbiner B., McCrea P., Kintner C., Yoshida-Noro C. and Wylie C. (1994) Overexpression of cadherins and underexpression of β-catenin inhibit dorsal mesoderm induction in early *Xenopus* embryos. *Cell* **79:** 791–803.

Heasman J., Ginsberg D., Geiger B., Goldstone K., Pratt T., Yoshida-Noro C. and Wylie C. (1994) A functional test for maternally inherited cadherin in *Xenopus* shows its importance in cell adhesion at the blastula stage. *Development* **120**: 49–57.

Heasman J., Quarmby J. and Wylie C.C. (1984) The mitochondrial cloud of *Xenopus* oocytes: The source of germinal granule material. *Dev. Biol.* **105:** 458–469.

Hedgepeth C.M., Conrad L.J., Zhang J., Huang H.-C., Lee V.M.Y. and Klein P.S. (1997) Activation of the Wnt signaling pathway: A molecular mechanism for lithium action. *Dev. Biol.* **185:** 82–91.

Heemskerk J. and DiNardo S. (1994) *Drosophila hedgehog* acts as a morphogen in cellular patterning. *Cell* **76:** 449–460.

Hegner R.W. (1927) *College Zoology.* New York: Macmillan.

Hekimi S., Lakowski B. and Ewbank, J.J. (1998) Molecular genetics of life span in *C. elegans:* How much can it teach us? *Trends Genet.* **14:** 14–20.

Helbock H.J., Beckman K.B., Shigenaga M.K., Walter P.B., Woodall A.A., Yeo H.C. and Ames B.N. (1998) DNA oxidation matters: The HPLC-electrochemical detection assay of 8–oxo-deoxyguanosine and 8–oxo-guanine. *Proc. Nat. Acad. Sci. USA.* **95:** 288–293.

Held L.I., Jr. (1992) *Models for Embryonic Periodicity.* Basel: Karger.

Helde K.A., Wilson E.T., Cretekos C.J. and Grunwald D.J. (1994) Contribution of early cells to the fate map of the zebrafish gastrula. *Science* **265:** 517–520.

Hemmati-Brivanlou A. and Melton D.A. (1992) A truncated activin receptor inhibits mesoderm induction and formation of axial structures in *Xenopus* embryos. *Nature* **359:** 609–614.

Hemmati-Brivanlou A. and Melton D.A. (1997) Vertebrate embryonic cells will become nerve cells unless told otherwise. *Cell* **88:** 13–17.

Hemmingsen S.M., Woodford C., van der Vies S.M., Tilly K., Dennis D.T., Georgopoulos C.P., Hendrix R.W. and Elliz R.J. (1988) Homologous plant and bacterial proteins chaperone oligomeric protein assembly. *Nature* **333:** 330–334.

Hengartner M.O. and Horvitz H.R. (1994) *C. elegans* survival gene *ced-9* encodes a functional homolog of the mammalian proto-oncogene *bcl-2*. *Cell* **8:** 1613–1626.

Hengartner M.O., Ellis R.E. and Horvitz H.R. (1992) *Caenorhabditis elegans* gene *ced-9* protects cells from programmed cell death. *Nature* **356:** 494–499.

Henikoff S. and Meneely P.M. (1993) Unwinding dosage compensation. *Cell* **72:** 1–2.

Hepker J., Wang Q.T., Motzny C.K., Holmgren R. and Orenic T.V. (1997) *Drosophila cubitus interruptus* forms a negative feedback loop with *patched* and regulates expresion of Hedgehog target genes. *Development* **124:** 549–558.

Herman R.K. (1984) Analysis of genetic mosaics of the nematode *Caenorhabditis elegans*. *Genetics* **108:** 165–180.

Herman R.K. and Hedgecock E.M. (1990) Limitation of the size of the vulval primordium of *Caenorhabditis elegans* by *lin-15* expression in surrounding hypodermis. *Nature* **348:** 169–171.

Herrera E., Samper E., Martin-Caballero J., Flores J.M., Lee H.W. and Blasco M.A. (1999) Disease states associated with telomerase deficiency appear earlier in mice with short telomeres. *EMBO J.* **18:** 2950–2960.

Herrmann B.G., Labeit S., Poustka A., King T.R. and Lehrach H. (1990) Cloning of the T gene required in mesoderm formation in the mouse. *Nature* **343:** 617–622.

Hibino T., Nishikata T. and Hishikata H. (1998) Centrosome-attracting body: A novel structure closely related to unequal cleavages in the ascidian embryo. *Devel. Growth Differ.* **40**:85–95.

Hill A.F., Desbruslais M., Joiner S., Sidle K.C.L., Gowland I. and Collinge J. (1997) The same prion strain causes vCJD and BSE. *Nature* **389:** 448–450.

Hill D.P. and Strome S. (1990) Brief cytochalasin-induced disruption of microfilaments during a critical interval in 1–cell *C. elegans* embryos alters the partitioning of developmental instructions to the 2–cell embryo. *Development* **108:** 159–172.

Hill R.J. and Sternberg P.W. (1992) The gene *lin-3* encodes an inductive signal for vulval development in *C. elegans*. *Nature* **358:** 470–476.

Hill R.J. and Sternberg P.W. (1993) Cell fate patterning during *C. elegans* vulval development. *Development* (Supplement): 9–18.

Hillis D.M. and Green D.M. (1990) Evolutionary changes of heterogametic sex in the phylogenetic history of amphibians. *J. Evol. Biol.* **3:** 49–64.

Hinchliffe J.R. and Johnson D.R. (1980) *The Development of the Vertebrate Limb: An Approach through Experiment, Genetics, and Evolution.* Oxford: Clarendon Press.

Hinson S. and Nagoshi R.N. (1999) Regulatory and functional interactions between the somatic sex regulatory gene *transformer* and the germ line genes *ovo* and *ovarian tumor*. *Development* **126:** 861–871.

Hinton H.E. (1970) Insect egg shells. *Scientific American* **223** (August): 84–91.

Hirata J., Nakagoshi H., Nabeshima Y. and Matsuzaki F. (1995) Asymmetric segregation of the homeodomain protein Prospero during *Drosophila* development. *Nature* **377:** 627–630.

Hirokawa N. (1998) Kinesin and dynein superfamily proteins and the mechanism of organelle transport. *Science* **279:** 519–526.

Hiromi Y. and Gehring W.J. (1987) Regulation and function of the *Drosophila* segmentation gene *fushi* tarazu. Cell **50:** 963–974.

Hirsch E., Iglesias A., Potocnik A.J., Hartmann U. and Fässler R. (1996) Impaired migration but not differentiation haematopoietic stem cells in the absence of β_1–integrins. *Nature* **380:** 171–175.

Hirsch H.R., Johnson H.A., Curtis H.J. and Pavelec M. (1974) The influence of temperature on chromosome aberrations in tissue culture: relation to thermal aging. *Exp. Gerontol.* **9:** 221–225.

His W. (1874) *Unsere Körperform und das physiologische Problem ihrer Enstehung: Briefe an einen befreundeten Naturforscher.* Leipzig: F.C.W. Vogel.

Hisatake K., Roeder R.G. and Horikoshi M. (1993) Functional dissection of TFIIB domains required for TFIIB-TFIID-promoter complex formation and basal transcription activity. *Nature* **363:** 744–747.

Ho R. (1992) Axis formation in the embryo of the zebrafish *Brachydanio rerio*. *Sem. Dev. Biol.* **3:** 53–64.

Hoch M., Seifert E. and Jäckle H. (1991) Gene expression mediated by cis-acting sequences of the *Krüppel* gene in response to the *Drosophila* morphogens *bicoid* and *hunchback*. *EMBO J.* **10:** 2267–2278.

Hoch M., Gerwin N., Taubert H. and Jäckle H. (1992) Competition for overlapping sites in the regulatory region of the *Drosophila* gene *Krüppel*. *Science* **256:** 94–97.

Hockenberry D.M. (1995) bcl-2, a novel regulator of cell death. *BioEssays* **17:** 631–638.

Hodgkin J. (1985) Males, hermaphrodites, and females: Sex determination in *Caenorhabditis elegans*. *Trends Genet.* **1:** 85–88.

Hodgkin J. (1992) Genetic sex determination mechanisms and evolution. *BioEssays* **14:** 253–261.

Hofmann T.J. and Cole M.D. (1996) The TAL1/Scl basic helix-loop-helix protein blocks myogenic differentiation and E-box dependent transactivation. *Oncogene* **13:** 617–624.

Hoffman A., Sinn E., Yamamoto T., Wang J., Roy A., Horikoshi M. and Roeder R.G. (1990) Highly conserved core domain and unique N terminus with presumptive regulatory motifs in a human TATA factor (TFIID). *Nature* **346:** 387–390.

Hogan B. (1996) Bone morphogenetic proteins: multifunctional regulators of vertebrate development. *Genes and Devel.* **10:** 1580–1594.

Holder N. and Klein R. (1999) Eph receptors and ephrins: Effectors of morphogenesis. *Development* **126:** 2033–2044.

Holley S.A. and Ferguson E.L. (1997) Fish are like flies are like frogs: conservation of dorsal-ventral patterning mechanisms. *BioEssays* **19:** 281–284.

Holley S.A., Jackson P.D., Sasai Y., Lu B., De Robertis E.M., Hoffmann F.M. and Ferguson E.L. (1995) A conserved system for dorsal-ventral patterning in insects and vertebrates involving sog and chordin. *Nature* **376:** 249–253.

Holley S.A., Neul J.L., Attisano L., Wrana J.L., Sasai Y., O'Connor M.B., De Robertis E.M. and Ferguson E.L. (1996) The *Xenopus* dorsalizing factor noggin ventralizes *Drosophila* embryos by preventing DPP from activating its receptor. *Cell* **86:** 607–617.

Holowacz T. and Elinson R.P. (1993) Cortical cytoplasm, which induces dorsal axis formation in *Xenopus*, is inactivated by UV irradiation of the oocyte. *Development* **119:** 277–285.

Holowacz T. and Elinson R.P. (1995) Properties of dorsal activity found in cortical cytoplasm of *Xenopus* eggs. *Development* **121:** 2789–2798.

Holtfreter J. (1933) Die totale Exogastrulation, eine Selbstablösung des Ektoderms vom Entomesoderm. *Wilhelm Roux's Arch.* **129:** 669–793.

Holtfreter J. (1938) Differenzierungspotenzen isolierter Teile der Anurengastrula. *Wilhelm Roux's Arch.* **138:** 657–738.

Holtfreter J. (1943) A study of the mechanics of gastrulation: Part I. *J. Exp. Zool.* **94:** 261–318.

Holtfreter J. (1944) A study of the mechanics of gastrulation: Part II. *J. Exp. Zool.* **95:** 171–212.

Holtfreter J. (1945) Neuralization and epidermization of gastrula ectoderm. *J. Exp. Zool.* **98:** 161–208.

Holtfreter J. (1946) Structure, motility and locomotion in isolated embryonic amphibian cells. *J. Morph.* **79:** 27–62.

Hong Y., Lee R.C. and Ambros V. (2000) Structure and function analysis of LIN-14, a temporal regulator of postembryonic developmental events in *Caenorhabditis elegans*. *Mol. Cell Biol.* **20:** 2285–2295.

Honig L.S. and Summerbell D. (1985) Maps of strength of positional signalling activity in the developing chick wing bud. *J. Embryol. exp. Morphol.* **87:** 163–174.

Hopper A.F. and Hart N.H. (1985) *Foundations of Animal Development.* 2nd ed. New York: Oxford University Press.

Horabin J.I. and Schedl P. (1993) *Sex-lethal* autoregulation requires multiple cis-acting elements upstream and downstream of the male exon and appears to depend largely on controlling the use of the male exon 5' splice site. *Mol. Cell Biol.* **13:** 7734–7746.

Horabin J.I. and Schedl P. (1996) Splicing of the *Drosophila* Sex-lethal early transcripts involves exon skipping that is independent of Sex-lethal protein. *RNA* **2:** 1–10.

Horder T.J., Witkowski J.A. and Wylie C.C., eds. (1985) *A History of Embryology.* Cambridge: Cambridge University Press.

Horner M., Chen T. and Thummel C.S. (1995) Ecdysone regulation and DNA binding properties of *Drosophila* nuclear hormone receptor family members. *Dev. Biol.* **168:** 490–502.

Hörstadius S. (1973) *Experimental Embryology of Echinoderms.* Oxford: Clarendon Press.

Horvitz H.R. and Sternberg P.W. (1991) Multiple intercellular signalling systems control the development of the *Caenorhabditis elegans* vulva. *Nature* **351:** 535–541.

Hoshijima K., Inoue K., Higuchi I., Sakamoto H. and Shimura Y. (1991) Control of *doublesex* alternative splicing by *transformer* and *transformer-2* in *Drosophila*. *Science* **252:** 833–836.

Houart C., Westerfield M. and Wilson S.W. (1998) A small population of anterior cells patterns the forebrain during zebrafish gastrulation. *Nature* **391:** 788–792.

Houliston E. (1994) Microtubule translocation and polymerization during cortical rotation in *Xenopus* eggs. Development **120:** 1213–1220.

Howard K., Ingham P. and Rushlow C. (1988) Region-specific alleles of the *Drosophila* segmentation gene *hairy*. *Genes and Dev.* **2:** 1037–1046.

Howard K.R. and Struhl G. (1990) Decoding positional information: Regulation of the pair-rule gene hairy. *Development* **110:** 1223–1231.

Hozumi N. and Tonegawa S. (1976) Evidence for somatic rearrangement of immunoglobulin genes coding for variable and constant regions. *Proc. Nat. Acad. Sci. USA* **73:** 3628–3632.

Hu M., Krause D., Greaves M., Sharkis S., Dexter M., Heyworth C. and Enver T. (1997) Multilineage gene expression precedes commitment in the hemopoietic system. *Genes and Dev.* **11:** 774–785.

Huang H. and Brown D.D. (2000) Prolactin is not a juvenile hormone in *Xenopus laevis* metamorphosis. *Proc. Nat. Acad. Sci. USA* **97:** 195–199.

Huber T.L., Zhou Y., Mead P.E. and Zon L.I. (1998) Cooperative effects of growth factors involved in the induction of hematopoietic mesoderm. *Blood* **92:** 4128–4137.

Huebner R.J. and Todaro G.J. (1969) Oncogenes of RNA tumor viruses as determinants of cancer. *Proc. Nat. Acad. Sci. USA* **64:** 1087–1094.

Hülskamp M. and Tautz D. (1991) Gap genes and gradients—The logic behind the gaps. *BioEssays* **13:** 261–268.

Hülskamp M., Schröder C., Pfeifle C., Jäckle H. and Tautz D. (1989) Posterior segmentation of the *Drosophila* embryo in the absence of a maternal posterior organizer gene. *Nature* **338:** 629–632.

Hummel K.P. and Chapman D.B. (1959) Visceral inversion and associated anomalies in the mouse. *J. Hered.* **50:** 9–23.

Hunt P., Guilsano M., Cook M., Sham M., Faiella A., Wilkinson D., Boncinelli E. and Krumlauf R. (1991a) A distinct *Hox* code for the branchial region of the vertebrate head. *Nature* **353:** 861–864.

Hunt P., Whiting J., Muchamore I., Marshall H. and Krumlauf R. (1991b) Homeobox genes and models for patterning the hindbrain and branchial arches. *Development* (Supplement 1): 187–196.

Hunter C.P. and Wood W.B. (1992) Evidence from mosaic analysis of the masculinizing gene her-1 for cell interactions in *C. elegans* sex determination. *Nature* **355:** 551–555.

Hunter T. (1991) Cooperation between oncogenes. *Cell* **64:** 249–270.

Hunter T. (1997) Oncoprotein networks. *Cell* **88:** 333–346.

Hunter T. (2000) Signaling—2000 and beyond. *Cell* **100:** 113–127.

Hunter T. and Karin M. (1992) The regulation of transcription by phosphorylation. *Cell* **70:** 375–387.

Hutter H. and Schnabel R. (1995) Establishment of left-right asymmetry in the *Caenorhabditis elegans* embryo: A multistep process involving a series of inductive events. *Development* **121:** 3417–3424.

Huxley J. and deBeer G.R. (1963) *Elements of Experimental Embryology.* New York: Hafner.

Huxley J.S. (1932) *Problems of Relative Growth.* New York: Dial.

Hynes R.O. (1994) The impact of of molecular biology on models of cell adhesion. *BioEssays* **16:** 663–669.

Hynes R.O. and Destree A.T. (1978) Relationships between fibronectin (LETS protein) and actin. *Cell* **15:** 875–886.

Hynes R.O. and Lander A.D. (1992) Contact and adhesive specificities in the associations, migrations, and targeting of cells and axons. Cell **68:** 303–322.

Iatrou K., Spira A.W. and Dixon G.H. (1978) Protamine messenger RNA: Evidence for early synthesis and accumulation during spermatogenesis in rainbow trout. *Dev. Biol.* **64:** 82–98.

Iavarone A., Garg P., Lasorella A., Hsu J. and Israel M.A. (1994) The helix-loop-helix protein Id-2 enhances cell proliferation and binds to the retino-blastoma protein. *Genes and Dev.* **8:** 1270–1284.

Illmensee K. (1973) The potentialities of transplanted early gastrula nuclei of *Drosophila melanogaster.* Production of their imago descendants by germ-line transplantation. *Wilhelm Roux's Arch.* **171:** 331–343.

Illmensee K. and Mahowald A.P. (1974) Transplantation of posterior polar plasm in *Drosophila*. Induction of germ cells at the anterior pole of the egg. *Proc. Nat. Acad. Sci. USA* **71:** 1016–1020.

Illmensee K., Mahowald A.P. and Loomis M.R. (1976) The ontogeny of germ plasm during oogenesis in *Drosophila. Dev. Biol.* **49:** 40–65.

Imperato-McGinley J. (1997) 5 alpha-reductase-2 deficiency. *Curr. Ther. Endocrinol. Metab.* **6:** 384–387.

Imperato-McGinley J., Guerrero L., Gautier T. and Peterson R.E. (1974) Steroid 5–alpha-reductase deficiency in man: An inherited form of male pseudohermaphroditism. *Science* **186:** 1213–1215.

Ingham P.W. (1988) The molecular genetics of embryonic pattern formation in *Drosophila. Nature* **335:** 25–34.

Ingham P.W. (1990) Genetic control of segmental patterning in the *Drosophila* embryo. In: *Genetics of Pattern Formation and Growth Control,* A.P. Mahowald (ed.), *48th Symp. Soc. Dev. Biol.*, 181–196. New York: Wiley/Alan R. Liss.

Ingham P.W. and Martinez-Arias A. (1992) Boundaries and fields in early embryos. *Cell* **68:** 221–235.

Ingham P.W., Baker N.E. and Martinez-Arias A. (1988) Regulation of segment polarity genes in the *Drosophila* blastoderm by *fushi tarazu* and *even skipped. Nature* **331:** 73–75.

Ingram R., Charrier B., Scollan C. and Meyer P. (1999) Transgenic tobacco plants expressing the *Drosophila* Polycomb (Pc) chromodomain show developmental alterations: Possible role of Pc chromodomain proteins in chromatin-mediated gene regulation in plants. *Plant Cell* **11:** 1047–1060.

Irani K., Xia Y., Zweier J.L., Sollott S.J., Der C.J., Fearon E.R., Sundaresan M., Finkel T. and Goldschmidt-Clermont P.J. (1997) Mitogenic signaling mediated by oxidants in Ras-transformed fibroblasts. *Science* **275:** 1649–1652.

Irish V., Lehmann R. and Akam M. (1989) The *Drosophila* posterior-group gene *nanos* functions by repressing *hunchback* activity. *Nature* **338:** 646–648.

Irish V.F. (1999) Patterning the flower. *Dev. Biol.* **209:** 211–220.

Irvine K.D. and Wieschaus E. (1994) *fringe,* a boundary-specific signaling molecule, mediates interactions between dorsal and ventral cells during *Drosophila* wing development. *Cell* **79:** 595–606.

Irvine K.D., Helfand S.L. and Hogness D.S. (1991) The large upstream control region of the *Drosophila* homeotic gene *Ultrabithorax. Development* **111:** 407–424.

Izpisúa-Belmonte J.C., Tickle C., Dolle P., Wolpert L. and Duboule D. (1991) Expression of the homeobox *Hox-4* genes and the specification of position in chick wing development. *Nature* **350:** 585–589.

Jack T., Brockman L.L. and Meyerowitz E.M. (1992) The homeotic gene *APETALA3* of *Arabidopsis thaliana* encodes a MADS-box and is expressed in petals and stamens. *Cell* **68:** 683–697.

Jack T., Fox G.L. and Meyerowitz E.M. (1994) *Arabidopsis* homeotic gene *APETALA3* ectopic expression: Transcriptional and posttranscriptional regulation determine floral organ identity. *Cell* **76:** 703–716.

Jäckle H. and Eagleson G.W. (1980) Spatial distribution of abundant proteins in oocytes and fertilized eggs of the Mexican axolotl (*Ambystoma mexicanum*). *Dev. Biol.* **75:** 492–499.

Jäckle H., Tautz D., Schuh R., Seifert E. and Lehmann R. (1986) Cross-regulatory interactions among the gap genes of *Drosophila*. Nature **324:** 668–670.

Jackson G.S., Hosszu L.L.P., Power A., Hill A.F., Kenney J., Saibil H., Craven C.J., Waltho J.P., Clarke A.R. and Collinge J. (1999) Reversible conversion of monomeric human prion protein between native and fibrilogenic conformations. *Science* **283:** 1935–1937.

Jackson I.J. (1989) The mouse. In: *Genes and Embryos,* D.M. Glover and B.D. Hames (eds.), 165–221. Oxford: IRL Press.

Jackson R.J. (1993) Cytoplasmic regulation of mRNA function: The importance of the 3' untranslated region. Cell **74:** 9–14.

Jackson R.J. and Wickens M. (1997) Translational control impinging on the 5' untranslated region and initiation factor proteins. *Curr. Opin. Genet. Dev.* **7:** 233–241.

Jacobson A.G. (1966) Inductive processes in embryonic development. *Science* **152:** 25–34.

Jacobson A.G. (1978) Some forces that shape the nervous system. *Zoon* **6:** 13–21.

Jacobson A.G. (1981) Morphogenesis of the neural plate and tube. In: *Morphogenesis and Pattern Formation,* T.G. Connelly, L.L. Brinkley and B.M. Carlson (eds.), 233–263. New York: Raven Press.

Jacobson A.G. (1985) Adhesion and movement of cells may be coupled to produce neurulation. In: *The Cell in Contact: Adhesions and Junctions as Morphogenetic Determinants,* G.M. Edelman and J.P. Thiery (eds.), 49–65. New York: Wiley.

Jacobson A.G. (1988) Somitomeres: Mesodermal segments of vertebrate embryos. *Development* 104 (Supplement): 208–220.

Jacobson A.G. (1991) Experimental analyses of the shaping of the neural plate and tube. *Amer. Zool.* **31:** 628–643.

Jacobson A.G. (1994) Normal neurulation in amphibians. In: *Neural Tube Defects,* G. Beck and J. Marsh (eds.), *Ciba Foundation Symposium* 181, 6–24. Chichester: Wiley.

Jacobson A.G. (1998) Somitomeres. *Principles of Medical Embryology,* vol. 11, *Developmental Biology,* 209–228.

Jacobson A.G. and Gordon R. (1976) Changes in the shape of the developing vertebrate nervous system analyzed experimentally, mathematically and by computer simulation. *J. Exp. Zool.* **197:** 191–246.

Jacobson A.G. and Moury J.D. (1995) Tissue boundaries and cell behavior during neurulation. *Dev. Biol.* **171:** 98–110.

Jacobson A.G. and Sater A.K. (1988) Features of embryonic induction. *Development* **104:** 341–359.

Jacobson A.G., Oster G.F., Odell G.M. and Cheng L.Y. (1986) Neurulation and the cortical tractor model for epithelial folding. *J. Embryol. exp. Morphol.* **96:** 19–49.

Jaffe L.A. (1976) Fast block to polyspermy in sea urchins is electrically mediated. *Nature* **261:** 68–71.

Jaffe L.A. and Gould M. (1985) Polyspermy-preventing mechanisms. In: *Biology of Fertilization,* C.B. Metz and A. Monroy (eds.), vol. 3, 223–249. New York: Academic Press.

Jaffe L.F. (1966) Electrical currents through the developing *Fucus* egg. *Proc. Nat. Acad. Sci. USA* **56:** 1102–1109.

Jäger R.J., Anvret M., Hall K. and Scherer G. (1990) A human XY female with a frame shift mutation in the candidate testis-determining gene *SRY*. *Nature* **348:** 452–454.

Jan Y.N. and Jan L.Y. (1998) Asymmetric cell division. *Nature* **392:** 775–778.

Jan Y.N. and Jan L.Y. (2000) Polarity in cell division: What frames thy fearful asymmety? *Cell* **100:** 599–602.

Jazwinski S.M. (1996) Longevity, genes and aging. *Science* **273:** 55–59.

Jeannottee L., Lemieux M., Charron J., Poirier F. and Robertson E. (1993) Specification of axial identity in the mouse: Role of the *Hoxa-5* (*Hox1.3*) gene. *Genes and Dev.* **7:** 2085–2096.

Jeffery W.R. (1984) Pattern formation by ooplasmic segregation in ascidian eggs. *Biol. Bull.* **166:** 277–298.

Jeffery W.R. (1990) Ultraviolet irradiation during ooplasmic segregation prevents gastrulation, sensory cell induction, and axis formation in the ascidian embryo. *Dev. Biol.* **140:** 388–400.

Jeffery W.R. and Meier S. (1983) A yellow crescent cytoskeletal domain in ascidian eggs and its role in early development. *Dev. Biol.* **96:** 125–143.

Jeffery W.R. and Swalla B.J. (1997) Tunicates. In: *Embryology: Constructing the Organism,* S.F. Gilbert and A.M. Raunio (eds.), 331–364. Sunderland, MA: Sinauer Associates.

Jegla D.E. and Sussex I.M. (1989) Cell lineage patterns in the shoot meristem of the sunflower embryo in the dry seed. *Dev. Biol.* **131:** 215–225.

Jen W.C., Wettstein D., Turner D., Chitnis A. and Kintner C. (1997) The Notch ligand, X-Delta-2, mediates segmentation of the paraxial mesoderm in *Xenopus* embryos. *Development* **124:** 1169–1178.

Jen Y., Weintraub H. and Benezra R. (1992) Overexpression of Id protein inhibits the muscle differentiation program: In vivo association of Id with E2A proteins. *Genes and Dev.* 6: 1466–1479.

Jenkins F.A. (1969) The evolution and development of the dens of the mammalian axis. *Anat. Rec.* **164:** 173–184.

Jentsch S. and Schlenker S. (1995) Selective protein degradation: A journey's end within the proteasome. *Cell* **82:** 881–884.

Jeppesen P. and Turner B.M. (1993) The inactive X chromosome in female mammals is distinguished by a lack of histone H4 acetylation, a cytogenetic marker for gene expression. *Cell* **74:** 281–289.

Jimenez G., Paroush Z. and Ish-Horowicz D. (1997) Groucho acts as a co-repressor for a subset of negative regulators, including Hairy and Engrailed. *Genes and Dev.* **11:** 3072–3082.

Jiménez R. and Burgos M. (1998) Mammalian sex determination: Joining pieces of the genetic puzzle. *BioEssays* **20:** 696–699(11).

Jindra M., Malone F., Hiruma K. and Riddiford L.M. (1996) Developmental profiles and ecdysteroid regulation of the mRNAs for two ecdysone receptor isoforms in the epidermis and wings of the tobacco hornworm, *Manduca sexta. Dev. Biol.* **180:** 258–272.

Johansson C.B., Momma S., Clarke D.L., Risling M., Lendahl U. and Frisen J. (1999) Identification of a neural stem cell in the adult mammalian central nervous system. *Cell* **96:** 25–34.

Johe K.K., Hazel T.G., Muller T., Dugich-Djordjevic M.M. and McKay R.D.G. (1996) Single factors direct the differentiation of stem cells from the fetal and adult central nervous system. *Genes and Dev.* **10:** 3129–3140.

Johnson F. and Bottjer S.W. (1994) Afferent influences on cell death and birth during development of a cortical nucleus necessary for learned vocal behavior in zebra finches. *Development* **120:** 13–24.

Johnson K.E., Nakatsuji N. and Boucaut J.-C. (1990) Extracellular matrix control of cell migration during amphibian gastrulation. In: *Cytoplasmic Organization Systems,* G.M. Malacinski (ed.), 349–374. New York: McGraw-Hill.

Johnson L.M., Kayne P.S., Kahn E.S. and Grunstein M. (1990) Genetic evidence for an interaction between *SIR3* and histone H4 in the repression of the silent mating loci in *Saccharomyces cerevisiae. Proc. Nat. Acad. Sci. USA* **87:** 6286–6290.

Johnson M.H., Maro B. and Takeichi M. (1986) The role of cell adhesion in the synchronization and orientation of polarization in 8–cell mouse blastomeres. *J. Embryol. exp. Morphol.* **93:** 239–255.

Johnson P.T.J., Lunde K.B., Ritchie E.G. and Launer A.E. (1999) The effect of trematode infection on amphibian limb development and survivorship. *Science* **284:** 802–804.

Johnson R.L. and Tabin C.J. (1997) Molecular models for vertebrate limb development. *Cell* **90:** 979–990.

Johnston L.A. and Edgar B.A. (1998) Wingless and Notch regulate cell cycle arrest in the developing *Drosophila* wing. *Nature* **394:** 82–84.

Johnston L.A., Prober D.A., Edgar B.A., Eisenmann R.N. and Gallant P. (1999) *Drosophila* myc regulates cellular growth during development. *Cell* **98:** 779–790 .

Johnston M.C. (1966) A radioautographic study of the migration and fate of cranial neural crest cells in the chick embryo. *Anat. Rec.* **156:** 143–156.

Jones B. and McGinnis W. (1993) The regulation of empty spiracles by abdominal-B mediates an abdominal segment identity function. *Genes and Dev.* **7:** 229–240.

Jones C.M. and Smith J.C. (1998) Establishment of a BMP-4 morphogen gradient by long-range inhibition. *Dev. Biol.* **194:** 12–17.

Jones P.A. (2000) The DNA methylation paradox. *Trends Genet.* **15:** 34–37.

Jones P.H., Harper S. and Watt F.M. (1995) Stem cell patterning and fate in human epidermis. *Cell* **80:** 83–93.

Jongens T.A., Ackerman L.D., Swedlow J.R., Jan L.Y. and Jan Y.N. (1994) *Germ cell-less* encodes a cell-type-specific nuclear pore-associated protein and functions early in the germ-cell specification pathway of *Drosophila. Genes and Dev.* **8:** 2123–2136.

Josso N., Picard M.-Y. and Tran D. (1976) The antimüllerian hormone. *Rec. Prog. Horm. Res.* **33:** 117–167.

Jost A. (1965) Gonadal hormones in the sex determination of the mammalian foetus. In: *Organogenesis,* R.L. DeHann and H. Ursprung (eds.), 611–628. New York: Holt, Rinehart and Winston.

Jürgens G. (1995) Axis Formation in plant embryogenesis. Cues and clues. *Cell* **81:** 467–470.

Jürgens G., Mayer U., Ruiz R.A.T., Berleth T. and Miséra S. (1991) Genetic analysis of pattern formation in the *Arabidopsis* embryo. *Development* (Supplement 1): 27–38.

Kablar B., Krastel K., Ying C., Asakura A., Tapscott S.J. and Rudnicki M.A. (1997) Myo-D and Myf-5 differentially regulate the development of limb versus trunk skeletal muscle. *Development* **124:** 4729–4738.

Kafatos F.C. (1976) Sequential cell polymorphism: A fundamental concept in developmental biology. *Adv. Insect Physiol.* **12:** 1–15.

Kafatos F.C., Regier J.C., Mazur G.D., Nadel M.R., Blau H.M., Petri W.H., Wyman A.R., Gelinas R.E., Moore P.B., Paul M., Efstratiadis A., Vournakis J.N., Goldsmith M.R., Hunsley J.R., Baker B., Nardi J. and Koehler M. (1977) The egg shell of insects: Differentiation-specific proteins and the control of their synthesis and accumulation during development. In: *Results and Problems in Cell Differentiation*, W. Beerman (ed.), vol. 8, 45–143. Berlin/New York: Springer-Verlag.

Kageura H. (1990) Spatial distribution of the capability to initiate a secondary embryo in the 32–cell embryo of *Xenopus laevis*. *Dev. Biol.* **142:** 432–438.

Kageura H. (1997) Activation of dorsal development by contact between the cortical dorsal determinant and the equatorial core cytoplasm in eggs of *Xenopus laevis*. *Development* **124:** 1543–1551.

Kahn C.R., Coyle J.T. and Cohen A.M. (1980) Head and trunk neural crest *in vitro*: Autonomic neuron differentiation. *Dev. Biol.* **77:** 340–348.

Kalthoff K., Rau K.-G. and Edmond J.C. (1982) Modifying effects of ultraviolet irradiation on the development of abnormal body patterns in centrifuged insect embryos (*Smittia* sp., Chironomidae, Diptera). *Dev. Biol.* **91:** 413–422.

Kandel E.R., Schwartz J.H. and Jessell T.M. (1991) *Principles of Neuroscience*. 3rd ed. New York: Elsevier.

Kandler-Singer I. and Kalthoff K. (1976) RNase sensitivity of an anterior morphogenetic determinant in an insect egg (*Smittia* sp., Chironomidae, Diptera). *Proc. Nat. Acad. Sci. USA* **73:** 3739–3743.

Kane D.A and Kimmel C.B. (1993) The zebrafish midblastula transition. *Development* **119:** 447–456.

Kao K.R., Masui Y. and Elinson R.P. (1986) Lithium-induced respecification of pattern in *Xenopus laevis* embryos. *Nature* **322:** 371–373.

Kaplan D.R., Hempstead B.L., Martin-Zanca D., Chao M.V. and Parada L.F. (1991) The *trk* proto-oncogene product: A signal transducing receptor for nerve growth factor. *Science* **252:** 554–558.

Kappen C., Schughart K. and Ruddle F.H. (1989) Two steps in the evolution of Antennapedia-class vertebrate homeobox genes. *Proc. Nat. Acad. Sci. USA* **86:** 5459–5463.

Karim F.D. and Thummel C.S. (1991) Ecdysone coordinates the timing and amounts of E74A and E74B transcription in *Drosophila*. *Genes and Dev.* **5:** 1067–1079.

Karim F.D. and Thummel C.S. (1992) Temporal coordination of regulatory gene expression by the steroid hormone ecdysone. *EMBO J.* **11:** 4083–4093.

Karin M., Castrillo J.-L. and Theill L.E. (1990) Growth hormone gene regulation: A paradigm for cell-type-specific gene activation. *Trends Genet.* **6:** 92–96.

Karp G. and Berrill N.J. (1981) *Development*. 2nd ed. New York: McGraw-Hill.

Kato Y., Tani T., Sotomaru Y., Kurokawa K., Kato J., Doguchi H., Yasue H. and Tsunoda Y. (1998) Eight calves cloned from somatic cells of a single adult. *Science* **282:** 2095–2098.

Kauffman S.A. (1973) Control circuits for determination and transdetermination. *Science* **181:** 310–318.

Kauffman S.A. (1980) Heterotopic transplantation in the syncytial blastoderm of *Drosophila*: Evidence for anterior and posterior nuclear commitments. *Wilhelm Roux's Arch.* **189:** 135–145.

Kaufman T.C., Seeger M.A. and Olsen G. (1990) Molecular and genetic organization of the Antennapedia gene complex of *Drosophila melanogaster*. *Adv. Genet.* **27:** 309–362.

Kawasaki I., Shim Y.H., Kirchner J., Kaminker J., Wood W.B. and Strome S. (1998) PGL-1, a predicted RNA-binding component of germ granules, is essential for fertility in *C. elegans*. *Cell* **94:** 635–645.

Keeton W.T. and Gould J.L. (1986) *Biological Science*. 4th ed. New York: Norton.

Keller G., Paige C., Gilboa E. and Wagner E.R. (1985) Expression of a foreign gene in myeloid and lymphoid cells derived from multipotent hematopoietic precursors. *Nature* **318:** 149–154.

Keller R. (1991) Early embryonic development of *Xenopus laevis*. In: *Methods in Cell Biology*, B.K. Kay and H.B. Peng (eds.), vol. 36, 61–113. New York: Academic Press.

Keller R. and Danilchik M. (1988) Regional expression, pattern and timing of convergence and extension during gastrulation of *Xenopus laevis*. *Development* **103:** 193–209.

Keller R., Shih J. and Domingo C. (1992a) The patterning and functioning of protrusive activity during convergence and extension of the *Xenopus* organiser. *Development* (Supplement): 81–91.

Keller R., Shih J. and Sater A. (1992b) The cellular basis of the convergence and extension of the *Xenopus* neural plate. *Developmental Dynamics* **193:** 199–217.

Keller R.E. (1978) Time-lapse cinemicrographic analysis of superficial cell behavior during and prior to gastrulation in *Xenopus laevis*. *J. Morphol.* **157:** 223–248.

Keller R.E. (1980) The cellular basis of epiboly: An SEM study of deep cell rearrangement during gastrulation of *Xenopus laevis*. *J. Embryol. exp. Morphol.* **60:** 201–243.

Keller R.E. (1981) An experimental analysis of the role of bottle cells and the deep marginal zone in gastrulation of *Xenopus laevis*. *J. Exp. Zool.* **216:** 81–101.

Keller R.E. (1986) The cellular basis of amphibian gastrulation. In: *Developmental Biology: A Comprehensive Synthesis*, vol. 2; *The Cellular Basis of Morphogenesis*, L.W. Browder (ed.), 241–327. New York: Plenum.

Keller R.E. and Jansa S. (1992) *Xenopus* gastrulation without a blastocoel roof. *Developmental Dynamics* **195:** 162–176.

Keller R.E. and Tibbetts P. (1989) Mediolateral cell intercalation in the dorsal, axial mesoderm of *Xenopus laevis*. *Dev. Biol.* **131:** 539–549.

Keller R.E., Danilchik M., Gimlich R. and Shih J. (1985) The function and mechanism of convergent extension during gastrulation of *Xenopus laevis*. *J. Embryol. exp. Morphol.* **89** (Supplement): 185–209.

Keller S.H. and Vacquier V.D. (1994) The isolation of acrosome-reaction-inducing glycoproteins from sea urchin egg jelly. *Dev. Biol.* **162:** 304–312.

Keller W. (1995) No end yet to messenger RNA 3' processing. *Cell* **81:** 829–832.

Kelley D.B. and Brenowitz E. (1992) Hormonal influences on courtship behaviors. In: *Behavioral Endocrinology*, J.B. Becker, S.M. Breedlove and D. Crews (eds.), 187–218. Cambridge, MA: MIT Press.

Kelley R.L., Wang J., Bell L. and Kuroda M.I. (1997) Sex lethal controls dosage compensation in *Drosophila* by a non-splicing mechanism. *Nature* **387:** 195–199.

Kelly R.O. (1981) The developing limb. In: *Morphogenesis and Pattern Formation*, T.G. Connelly, L.L. Brinkley and B.M. Carlson (eds.), 49–85. New York: Raven Press.

Kelly S.J. (1977) Studies of the developmental potential of 4- and 8-cell stage mouse blastomeres. *J. Exp. Zool.* **200:** 365–376.

Kemphues K.J. (1989) *Caenorhabditis*. In: *Genes and Embryos*, D.M. Glover and B.D. Hames (eds.), 95–126. Oxford: IRL Press.

Kent J., Wheatley S.C., Andrews J.E., Sinclair A.H. and Koopman P. (1996) A male-specific role for *Sox9* in vertebrate sex determination. *Development* **122:** 2813–2822.

Kenyon C., Chang J., Gensch E., Rudner A. and Tabtiang R. (1993) A. *C. elegans* mutant that lives twice as long as wild type. *Nature* **366**: 461–464.

Kerstetter R.A., Laudenciachingcuanco D., Smith L.G. and Hake S. (1997) Loss-of-function mutations in the maize homeobox gene Knotted1 are defective in shoot meristem maintenance. *Development* **124:** 3045–3054.

Kessel M. and Gruss P. (1990) Murine development control genes. *Science* **249:** 374–379.

Kessel M. and Gruss P. (1991) Homeotic transformations of murine vertebrae and concomitant alteration of *Hox* codes induced by retinoic acid. *Cell* **67:** 89–104.

Kessel M., Balling R. and Gruss P. (1990) Variations of cervical vertebrae after expression of a *Hox 1.1* transgene in mice. *Cell* **61:** 301–308.

Kessler D.S. and Melton D.A. (1994) Vertebrate embryonic induction: Mesodermal and neural patterning. *Science* **266:** 596–604.

Keyes L.N., Cline T.W. and Schedl P. (1992) The primary sex determination signal of *Drosophila* acts at the level of transcription. *Cell* **68:** 933–943.

Keynes R.J. and Stern C.D. (1984) Segmentation in the vertebrate nervous system. *Nature* **310:** 786–789.

Keys D.N., Lewis D.L., Selegue J.E., Pearson B.J., Goodrich L.V., Johnson R.L., Gates J., Scott M.P. and Carroll S.B. (1999) Recruitment of a *hedgehog* regulatory circuit in butterfly eyespot evolution. *Science* **283:** 532–534.

Kidd S. (1992) Characterization of the *Drosophila* cactus locus and analysis of interactions between cactus and dorsal proteins. *Cell* **71:** 623–635.

Kieny M. and Chevallier A. (1979) Autonomy of tendon development in the embryonic chick wing. *J. Embryol. exp. Morphol.* **49:** 153–165.

Kim N.W., Piatyszek M.A., Prowse K.R., Harley C.B., West M.D., Ho P.L., Coviello G.M., Wright W.E., Weinrich S.L. and Shay J.W. (1994) Specific association of human telomerase activity with immortal cells and cancer. *Science* **266:** 2011–2015.

Kimble J. (1981) Alterations in cell lineage following laser ablation of cells in the somatic gonad of *C. elegans*. *Dev. Biol.* **87:** 286–300.

Kimelman D., Abraham J.A., Haaparanta T., Palsi T.M. and Kirschner M.W. (1988) The presence of fibroblast growth factor in the frog egg: Its role as a natural mesoderm inducer. *Science* **242:** 1053–1056.

Kimelman D. and Maas A. (1992) Induction of dorsal and ventral mesoderm by ectopically expressed *Xenopus* fibroblast growth factor. *Development* **114:** 261–269.

Kimelman D., Christian J.L. and Moon R.T. (1992) Synergistic principles of development: Overlapping patterning systems in *Xenopus* mesoderm induction. *Development* **116:** 1–9.

Kim-Ha J., Kerr K. and Macdonald P.M. (1995) Translational regulation of oskar mRNA by Bruno, an ovarian RNA-binding protein, is essential. *Cell* **81:** 403–412.

Kimura D. (1992) Sex differences in the brain. *Scientific American* **267** (September): 119–125.

Kimura K.D., Tissenbaum H.A., Liu Y. and Ruvkun G. (1997) *daf-2*, an insulin receptor-like gene that regulates longevity and diapause in *Caenorhabditis elegans*. *Science* **277:** 942–946.

King M.L., Zhou Y. and Bubunenko M. (1999) Polarizing genetic information in the egg: RNA localization in the frog oocyte. *BioEssays* **21:** 546–557.

King R.C. (1970) *Ovarian Development in Drosophila melanogaster*. New York: Academic Press.

King T.J. (1966) Nuclear transplantation in amphibia. *Methods Cell Physiol.* **2:** 1–36.

Kingsley D.M. (1994) The TGF-β superfamily: New members, new receptors, and new genetic tests of function in different organisms. *Genes and Dev.* **8:** 133–146.

Kintner C.R. (1992) Regulation of embryonic cell adhesion by the cadherin cytoplasmic domain. *Cell* **69:** 225–236.

Kintner C.R. and Melton D.A. (1987) Expression of *Xenopus* N-CAM RNA in ectoderm is an early response to neural induction. *Development* **99:** 311–325.

Kirby M.L. (1989) Plasticity and predetermination of mesencephalic and trunk neural crest transplanted into the region of cardial neural crest. *Dev. Biol.* **134:** 402–412.

Kirkpatrick M. and Jenkins C.D. (1989) Genetic segregation and the maintenance of sexual reproduction. *Nature* **339:** 300–302.

Kirkwood T.B. (1996) Human senescence. *BioEssays* **18:** 1009–1016.

Kirkwood T.B. and Rose M.R. (1991) Evolution of senescence: Late survival sacrificed for reproduction. *Phil. Trans. R. Soc. London* **B 332:** 15–24.

Kislauskis E.H., Zhu X. and Singer R.H. (1994) Sequences responsible for intracellular localization of β-actin mRNA also affect cell phenotype. *J. Cell Biol.* **127:** 441–451.

Kispert A., Vainio S. and McMahon A.P. (1998) Wnt-4 is a mesenchymal signal for epithelial transformation of metanephric mesenchyme in the developing kidney. *Development* **125:** 4225–4234.

Kleene K.C. (1989) Poly(A) shortening accompanies the activation of five mRNAs during spermiogenesis in the mouse. *Development* **106:** 367–373.

Klein P.S. and Melton D.A. (1996) A molecular mechanism for the effect of lithium on development. *Proc. Nat. Acad. Sci. USA* **93:** 8855–8859.

Kliewer S.A., Umesono K., Mangelsdorf D.J. and Evans R.M. (1992) Retinoid X receptor interacts with nuclear receptors in retinoic acid, thyroid hormone and vitamin D_3 signalling. *Nature* **355:** 446–449.

Kline D., Simoncini L., Mandel G., Maue R.A., Kado R.T. and Jaffe L.A. (1988) Fertilization events induced by neurotransmitters after injection of mRNA in *Xenopus* eggs. *Science* **241:** 464–467.

Klingler M., Soong J., Butler B. and Gergen P. (1996) Disperse versus compact elements for the regulation of *runt* stripes in *Drosophila*. *Dev. Biol.* **177:** 73–84.

Kloc M. and Etkin L. (1994) Delocalization of Vg1 mRNA from the vegetal cortex in *Xenopus* oocytes after destruction of Xlsirt RNA. *Science* **265:** 1101–1103.

Kloc M. and Etkin L. (1995) Two distinct pathways for the localization of RNAs at the vegetal cortex in *Xenopus* oocytes. *Development* **121:** 287–297.

Knebelmann B., Boussin L., Guerrier D., Legeai L., Kahn A., Josso N. and Picard J.-Y. (1991) Anti-Müllerian hormone Bruxelles: A nonsense mutation associated with the persistent Müllerian duct syndrome. *Proc. Nat. Acad. Sci. USA* **88:** 3767–3771.

Knoblich J.A. (1997) Mechanisms of asymmetric cell division during animal development. *Curr. Opin. Cell Biol.* **9:** 833–841.

Knoblich J.A., Jan L.Y. and Jan Y.N. (1995) Asymmetric segregation of Numb and Prospero during cell division. *Nature* **377:** 624–627.

Ko L.J. and Prives C. (1996) p53: puzzle and paradigm. *Genes and Dev.* **10:** 1054–1072.

Kobayashi S. and Okada M. (1989) Restoration of pole cell-forming ability to u.v.-irradiated *Drosophila* embryos by injection of mitochondrial lrRNA. *Development* **107:** 733–742.

Kobayashi S., Amikura R. and Okada M. (1993) Presence of mitochondrial large ribosomal RNA outside mitochondria in germ plasm of *Drosophila melanogaster. Science* **260:** 1521–1524.

Kobayashi S., Yamada M., Asaoka M. and Kitamura T. (1996) Essential role for the posterior morphogen nanos for germ line development in *Drosophila. Nature* **380:** 708–711.

Kochav S. and Eyal-Giladi H. (1971) Bilateral symmetry in chick embryo: Determination by gravity. *Science* **171:** 1027–1029.

Koelle M.R., Talbot W.S., Segraves W.A., Bender M.T., Cherbas P. and Hogness D.S. (1991) The *Drosophila* EcR gene encodes an ecdysone receptor, a new member of the steroid receptor superfamily. *Cell* **67:** 59–77.

Köhler G. and Milstein C. (1975) Continuous cultures of fused cells secreting antibody of predefined specificity. *Nature* **256:** 495–497.

Kondo M., Weissman I.L. and Akashi K. (1997) Identification of clonogenic common lymphoid progenitors in mouse bone marrow. *Cell* **91:** 661–672.

Kondrashov A.S. (1988) Deleterious mutations and the evolution of sexual reproduction. Nature **336:** 435–440.

Konsolaki M. and Schüpbach T. (1998) *Windbeutel,* a gene required for dorsoventral patterning in *Drosophila,* encodes a protein that has homologies to vertebrate proteins of the endoplasmic reticulum. *Genes and Dev.* **12:** 120–131.

Koopman P., Münsterberg A., Capel B., Vivian N. and Lovell-Badge R. (1990) Expression of a candidate sex-determining gene during mouse testis differentiation. *Nature* **348:** 450–452.

Koopman P., Gubbay J., Vivian N., Goodfellow P. and Lovell-Badge R. (1991) Male development of chromosomally female mice transgenic for *Sry. Nature* **351:** 117–121.

Koornneef M., Alonso-Blanco C., Peeters A.J.M. and Soppe W. (1998) Genetic control of flowering time in *Arabidopsis. Plant Physiol. Plant Mol. Biol.* **49:** 345–370.

Kornberg T.B. and Krasnow M.A. (2000) The *Drosophila* genome sequence: Implications for biology and medicine. *Science* **287:** 2218–2220.

Kraatz G. (1876) Deutsche Entomol. *Zeitschr.* **20:** 377.

Krasnow M.A., Saffman E.E., Kornfeld K. and Hogness D.S. (1989) Transcriptional activation and repression by *Ultrabithorax* proteins in cultured *Drosophila* cells. *Cell* **57:** 1031–1043.

Kratzer P.G. and Chapman V.M. (1981) X-chromosome reactivation in oocytes of *Mus caroli. Proc. Nat. Acad. Sci. USA* **78:** 3093–3097.

Kraut R., Chia W., Jan L.Y., Jan Y.N. and Knoblich J.A. (1996) Role of *inscuteable* in orienting asymmetric cell division in *Drosophila. Nature* **383:** 50–55.

Krings M., Stone A., Schmitz R.W., Krainitzki H., Stoneking M. and Pääbo S. (1997) Neanderthal DNA sequences and the origin of modern humans. *Cell* **90:** 19–30.

Krohn P.L. (1962) Heterochronic tranplantation in the study of ageing. *Proc. R. Soc. London* **B157**: 128–147.

Krohn P.L. (1966) Transplantation and aging. In: *Topics in the Biology of Aging: A Symposium Held at the Salk Institute for Biological Studies,* P.L. Krohn (ed.), 125–139. New York: Interscience Publishers.

Kropf D.L. (1992) Establishment and expression of cellular polarity in fucoid zygotes. *Microbiol. Rev.* **56:** 316–336.

Kropf D.L., Kloareg B. and Quatrano R.S. (1988) Cell wall is required for fixation of the embryonic axis in *Fucus* zygotes. *Science* **239:** 187–190.

Krull C.E., Lansford R., Gale N.W., Collazo A., Marcelle C., Yancopoulos G.D., Fraser S.E. and Bronner-Fraser M. (1997) Interactions of Eph-related receptors and ligands confer rostrocaudal pattern to trunk neural crest migration. *Curr. Biol.* **7:** 571–580.

Krumlauf R. (1994) *Hox* genes in vertebrate development. *Cell* **78:** 191–201.

Kühn A. (1961) *Grundriss der allgemeinen Zoologie.* 14th ed. Stuttgart: Georg Thieme.

Kühn A. (1971) *Lectures on Developmental Physiology.* Translated by R. Milkman. Berlin/Heidelberg/New York: Springer.

Kuhn R., Schäfer U. and Schäfer M. (1988) Cis-acting regions sufficient for spermatocyte-specific transcriptional and spermatid-specific translational control of the *Drosophila melanogaster* gene *mst(3)gl-9. EMBO J.* **7:** 447–454.

Kuisk I.R., Li H., Tran D. and Capetanaki Y. (1996) A single MEF2 site governs desmin transcription in both heart and skeletal muscle during mouse embryogenesis. *Dev. Biol.* **174:** 1–13.

Kulesa P. and Fraser S. (2000) In ovo time-lapse analysis of chick hindbrain neural crest cell migration shows cell interactions during migration to the branchial arches. *Development* **127:** 1161–1172.

Kulesa P., Bronner-Fraser M. and Fraser S. (2000) In ovo time-lapse analysis after dorsal neural tube ablation shows rerouting of chick hindbrain neural crest. *Development* **127:** 2843–2852.

Kulessa H., Frampton J. and Graf T. (1995) GATA-1 reprograms avian myelomonocytic cell lines into eosinophils, thromboblasts, and erythroblasts. *Genes and Dev.* **9:** 1250–1262.

Kumar N.M. and Gilula N.B. (1996) The gap junction communication channel. *Cell* **84:** 381–388.

Kumar S., Kinoshita M., Noda M., Copeland N.G. and Jenkins N.A. (1994) Induction of apoptosis by the mouse *Nedd2* gene, which encodes a protein similar to the *C. elegans* cell death gene *ced-3* and the mammalian IL-1β-converting enzyme. *Genes and Dev.* **8:** 1613–1626.

Kuo M.-H. and Allis C.D. (1998) Roles of histone acetyltransferases and deacetylases in gene regulation. *BioEssays* **20:** 615–626.

Kuo M.-H., Zhou J., Jambeck P., Churchill M.A. and Allis C.D. (1998) The histone acetyltransferase activity of yeast Gcn5p is required for the activation of target genes. *Genes and Dev.* **12:** 627–639.

Kuratani S.C. and Eichele G. (1993) Rhombomere transplantation repatterns the segmental organization of cranial nerves and reveals cell-autonomous expression of a homeodomain protein. *Development* **117:** 105–117.

Kuroda M. and Meller V.H. (1997) Transient Xist-ence. *Cell* **91:** 9–11.

Kuwabara P.E. and Kimble J. (1992) Molecular genetics of sex determination in *C. elegans*. *Trends in Genetics* **8:** 164–168.

Kwon Y.K. and Hecht, N.B. (1993) Binding of a phosphoprotein to the 3' untranslated region of the mouse protamine 2 mRNA temporarily represses its translation. *Mol. Cell Biol.* **13:** 6547–6557.

LaBonne C. and Bronner-Fraser M. (1998) Neural crest induction in *Xenopus*: evidence for a two-signal model. *Development* **125:** 2403–2414.

Lacalli T. (1996) Dorsoventral axis inversion: a phylogenetic perspective. *BioEssays* **18:** 251–254.

Lagrange T., Kapanidis A.N., Tang H., Reinberg D. and Ebright R.H. (1998) New core promoter element in RNA polymerase II-dependent transcription: Sequence-specific DNA binding by transcription factor IIB. *Genes and Dev.* **12:** 34–44.

Laidlaw M. and Wessel G.M. (1994) Cortical granule biogenesis is active throughout oogenesis in sea urchins. *Development* **120:** 1325–1333.

Lakowski B. and Hekimi S. (1996) Determination of lifespan in *Caenorhabditis elegans* by four clock genes. *Science* **272:** 1010–1013.

Lakowski B. and Hekimi S. (1998) The genetics of caloric restriction in *Caenorhabditis elegans*. *Proc. Nat. Acad. Sci. USA* **95:** 13091–13096.

Lallier T., Leblanc G., Artinger K.B. and Bronner-Fraser M. (1992) Cranial and trunk neural crest cells use different mechanisms for attachment to extracellular matrices. *Development* **116:** 531–541.

Lam G.T., Jiang C. and Thummel C.S. (1997) Coordination of larval and prepupal gene expression by the DHR3 orphan receptor during *Drosophila* metamorphosis. *Development* **124:** 1757–1769.

Lamberts S.W.J., van den Beld A.W. and van der Lely A.-J. (1997) The endocrinology of aging. *Science* **278:** 419–424.

Lane M.A., Tilmont E.M., De Angelis H., Handy A., Ingram D.K., Kemnitz J.W. and Roth G.S. (2000) Short-term calorie restriction improves disease-related markers in older male rhesus monkeys (*Macaca mulatta*). *Mech. Ageing Dev.* **112:** 185–196.

Lane M.C., Koehl M.A.R., Wilt F. and Keller R.E. (1993) A role for regulated secretion of apical extracellular matrix during epithelial invagination in the sea urchin. *Development* **117:** 1049–1060.

Langeland J.A. and Kimmel C.B. (1997) Fishes. In: *Embryology: Constructing the Organism*. S.F. Gilbert and A.M. Raunio (eds.), 383–407. Sunderland, MA: Sinauer.

Langman J. (1981) *Medical Embryology*. 4th ed. Baltimore: Williams and Wilkins.

Larabell C.A., Rowning B.A., Wells J., Wu M. and Gerhart J. (1996) Confocal microscopy analysis of living *Xenopus* eggs and the mechanism of cortical rotation. *Development* **122:** 1281–1289.

Larabell C.A., Torres M., Rowning B.A., Yost C., Miller J.R., Wu M., Kimelman D. and Moon R.T. (1997). Establishment of the dorso-ventral axis in *Xenopus* embryos is presaged by early asymmetries in β-catenin which are modulated by Wnt signaling. *J. Cell Biol.* **136:** 1123–1136.

Larsen W.J. (1993) *Human Embryology*. New York: Churchill Livingstone.

Laskey R.A. (1974) Biochemical processes in early development. In: *Companion to Biochemistry*, A.T. Bull, J.R. Lagnado, J.O. Thomas and K.F. Tipton (eds.), vol. 2, 137–160. London: Longman.

Laskey R.A. and Gurdon J.B. (1970) Genetic content of adult somatic cells tested by nuclear transplantation from cultured cells. *Nature* **228:** 1332–1334.

Lasky L.A. (1992) Selectins: Interpreters of cell-specific carbohydrate information during inflammation. *Science* **258:** 964–969.

Lassar A.B., Davis R.L., Wright W.E., Kadesch T., Murre C., Voronova A., Baltimore D. and Weintraub H. (1991) Functional activity of myogenic HLH proteins requires hetero-oligomerization with E12/E47–like proteins in vivo. *Cell* **66:** 305–315.

Latham K.E. (1999) Epigenetic modification and imprinting of the mammalian genome during development. *Curr. Top. Dev. Biol.* **43:** 1–49.

Laufer E., Dahn R., Orozco O.E., Yeo C.Y., Pisenti J., Henrique D., Abbott U.K., Fallon J.F. and Tabin C. (1997) Expression of Radical fringe in limb-bud ectoderm regulates apical ectodermal ridge formation. *Nature* **386:** 366–373.

Laufer J.S., Bazzicalupo P. and Wood W.B. (1980) Segregation of developmental potential in early embryos of *Caenorhabditis elegans*. *Cell* **19:** 569–577.

Lauffenburger D.A. and Horwitz R.F. (1996) Cell migration: A physically integrated molecular process. *Cell* **84:** 359–369.

Laurant M.N., Blitz I.L., Hashimoto C., Rothbächer U. and Cho K.W.-Y. (1997) The *Xenopus* homeobox gene *Twin* mediates Wnt induction of *Goosecoid* in establishment of Spemann's organizer. *Development* **124:** 4905–4916.

Laux T. and Jürgens G. (1997) Embryogenesis: A new start in life. *Plant Cell* **9:** 989–1000.

Law R. (1979) The cost of reproduction in annual meadow grass. *Am. Nat.* **113**: 3–16.

Law R., Bradshaw A.D. and Putwain P.D. (1977) Life-history variation in *Poa annua*. *Evolution* **31**: 233–246.

Lawrence J.B. and Singer R.H. (1986) Intracellular localization of mRNAs for cytoskeletal proteins. *Cell* **45:** 407–415.

Lawrence P.A. (1992) *The Making of a Fly: The Genetics of Animal Design*. Cambridge, MA: Blackwell.

Lawrence P.A. and Johnston P. (1989) Pattern formation in the *Drosophila* embryo: Allocation of cells to parasegments by *even-skipped* and *fushi tarazu*. *Development* **105:** 761–767.

Lawrence P.A. and Morata G. (1993) A no-wing situation. *Nature* **366:** 305–306.

Lawrence P.A. and Morata G. (1994) Homeobox genes: Their functions in *Drosophila* segmentation and Pattern formation. *Cell* **78:** 181–189.

Lawrence P.A. and Struhl G. (1996) Morphogens, compartments, and pattern: Lessons from *Drosophila*? *Cell* **85:** 951–961.

Lawson K.A., Dunn N.R., Roelen B.A., Zeinstra L.M., Davis A.M., Wright C.V., Korving J.P. and Hogan B.L. (1999) Bmp4 is required for the generation of primordial germ cells in the mouse embryo. *Genes and Dev.* **13:** 424–436.

Layton W.M., Jr. (1976) Random determination of a developmental process: Reversal of normal visceral asymmetry in the mouse. *J. Heredity* **67:** 336–338.

Lazaris-Karatzas A., Montine K.S. and Sonenberg N. (1990) Malignant transformation by a eukaryotic initiation factor subunit that binds to mRNA 5' cap. *Nature* **345:** 544–547.

Lecuit T. and Cohen S.M. (1997) Proximodistal axis formation in the *Drosophila* leg. *Nature* **388:** 139–145.

Lecuit T., Brook W.J., Ng M., Calleja M., Sun H. and Cohen S.M. (1996) Two distinct mechanisms for long-range patterning by Decapentaplegic in the *Drosophila* wing. *Nature* **381:** 387–393.

Le Douarin N.M. (1969) Particularités du noyau interphasique chez la caille japonaise (*Coturnix coturnix japonica*). Utilisation des ces particularités comme "marquage biologique" dans les recherches sur les interactions tissulaires et les migrations cellulaires au course de l'ontogenèse. *Bull. Biol. Fr. Belg.* **103:** 435–452.

Le Douarin N.M. (1986) Cell line segregation during peripheral nervous system ontogeny. *Science* **231:** 1515–1522.

Le Douarin N.M., Dupin E. and Ziller C. (1994) Genetic and epigenetic control in neural crest development. *Curr. Opin. Genet. Dev.* **4:** 685–695.

Lee E.Y.-H.P., Chang C.-Y., Hu N., Wang Y.-C.J., Lai C.-C., Herrup K., Lee W.-H. and Bradley A. (1992) Mice deficient for Rb are nonviable and show defects in neurogenesis and haematopoiesis. *Nature* **359:** 288–294.

Lee H.W., Blasco M.A., Gottlieb G.J., Horner J.W. 2nd, Greider C.W. and DePinho R.A. (1998) Essential role of mouse telomerase in highly proliferative organs. *Nature* **392:** 569–574.

Lee L., Tirnauer J.S., Li J., Schuyler S.C., Liu J.Y. and Pellman D. (2000) Positioning of the mitotic spindle by a cortical-microtubule capture mechanism. *Science* **287:** 2260–2262.

Lee R.C., Feinbaum R.L. and Ambros V. (1993) The *C. elegans* heterochronic gene *lin-4* encodes small RNAs with antisense complementarity to *lin-14*. *Cell* **75:** 843–854.

Lehmann R. and Nüsslein-Volhard C. (1986) Abdominal segmentation, pole cell formation, and embryonic polarity require the localized activity of *oskar*, a maternal gene in *Drosophila*. *Cell* **47:** 141–152.

Lehmann R. and Nüsslein-Volhard C. (1991) The maternal gene *nanos* has a central role in posterior pattern formation in the *Drosophila* embryo. *Development* **112:** 679–691.

Lehmann R. and Rongo C. (1993) Germ plasm formation and germ cell determination. *Seminars Dev. Biol.* **4:** 149–159.

Lemon B.D. and Freedman L.P. (1999) Nuclear receptor cofactors as chromatin remodelers. *Curr. Opin. Genet. Dev.* **9:** 499–504.

LeMosy E.K., Kemler D. and Hashimoto C. (1998) Role of Nudel protease activation in triggering dorsoventral polarization of the *Drosophila* embryo. *Development* **125:** 4045–4053.

Le Mouellic H., Lallemand Y. and Brulet P. (1990) Targeted replacement of the homeobox gene *Hox-3.1* by the *Escherichia coli lacZ* in mouse chimeric embryos. *Proc. Nat. Acad. Sci. USA* **87:** 4712–4716.

Le Mouellic H., Lallemand Y. and Brulet P. (1992) Homeosis in the mouse induced by a null mutation in the *Hox-3.1* gene. *Cell* **69:** 251–264.

Lenhoff H.M. and Lenhoff S.G. (1988) Trembley's polyps. *Scientific American* **258** (April): 108–113.

Leptin M. and Grunewald B. (1990) Cell shape changes during gastrulation in *Drosophila*. *Development* **110:** 73–84.

Leutert R. (1975) Sex determination in *Bonellia*. In: *Intersexuality in the Animal Kingdom*, R. Reinboth (ed.), 84–90. Berlin/New York: Springer-Verlag.

Levi-Montalcini R. (1964) Growth control of nerve cells by a protein factor and its antiserum. *Science* **143:** 105–110.

Levi-Montalcini R. (1988) *In Praise of Imperfection*. New York: Basic Books.

Levi-Montalcini R. and Booker B. (1960) Destruction of the sympathetic ganglia in mammals by an antiserum to a nerve growth protein. *Proc. Nat. Acad. Sci. USA* **46:** 384–391.

Levi-Montalcini R. and Calissano P. (1979) The nerve-growth factor. *Scientific American* **240** (June): 68–77.

Levin M. (1997) Left-right asymmetry in vertebrate embryogenesis. *BioEssays* **19:** 287–296.

Levin M. and Mercola M. (1998) The compulsion of chirality: Toward an understanding of left-right asymmetry. *Genes and Dev.* **12:** 763–769.

Levin M., Johnson R.L., Stern C.D., Kuehn M. and Tabin C. (1995) A molecular pathway determining left-right asymmetry in chick embryogenesis. *Cell* **82:** 803–814.

Levin M., Pagan S., Roberts D.J., Cooke J., Kuehn M. and Tabin C. (1997) Left/right patterning signals and the independent regulation of different aspects of situs in the chick embryo. *Dev. Biol.* **189:** 57–67.

Levine A., Bashan-Ahrend A., Budai-Hadrian O., Gartenberg D., Menasherow S. and Wides R. (1994) odd Oz: A novel *Drosophila* pair rule gene. *Cell* **77:** 587–598.

Levine A.J. (1997) p53, the cellular gate keeper for growth and division. *Cell* **88:** 323–331.

Levine M., Garen A., Lepesant J.-A. and Lepesant-Kejzlarova J. (1981) Constancy of somatic DNA organization in developmentally regulated regions of the *Drosophila* genome. *Proc. Nat. Acad. Sci. USA* **78:** 2417–2421.

Levine R.L. and Stadtman E.R. (1996) Protein modification with aging. In: *Handbook of the Biology of Aging*, E.L. Schneider and J.W. Rowe (eds.), 184–197. San Diego: Academic Press.

Lewin B. (1980) *Gene Expression*. 2nd ed. vol. 2. New York: Wiley/Interscience.

Lewin B. (1990) *Genes IV*. Oxford: Oxford University Press.

Lewin B. (1991) Oncogenic conversion by regulatory changes in transcription factors. *Cell* **64:** 303–312.

Lewis E.B. (1963) Genes and developmental pathways. *Am. Zool.* **3:** 33–56.

Lewis E.B. (1978) A gene complex controlling segmentation in *Drosophila*. *Nature* **276:** 565–570.

Leyton L., Leguen P., Bunch D. and Saling P.M. (1992) Regulation of mouse gamete interaction by a sperm tyrosine kinase. *Proc. Nat. Acad. Sci. USA* **89:** 11692–11695.

Leyton L., Tomes C. and Saling P.M. (1995) LL95 monoclonal antibody mimics functional effects of ZP3 on mouse sperm: Evidence that the antigen recognized is not hexokinase. *Mol. Reprod. Dev.* **42:** 347–358.

Li E., Bestor T.H. and Jaenisch R. (1992) Targeted mutation of the DNA methytransferase gene results in embryonic lethality. *Cell* **69:** 915–926.

Li E., Beard C. and Jaenisch R. (1993) Role for DNA methylation in genomic imprinting. *Nature* **366:** 362–365.

Li L. and Vaessin H. (2000) Pan-neural Prospero terminates cell proliferation during *Drosophila* neurogenesis. *Genes and Dev.* **14:** 147–151.

Liaw G.-J., Rudolph K.M., Huang J.D., Dubnikoff T., Courey A.J. and Lengyel J.A. (1995) The torso response element binds GAGA and NTF-1/Elf-1, and regulates tailless by relief of repression. *Genes and Dev.* **9:** 3163–3176.

Lieberfarb M.E., Chu T., Wreden C., Theurkauf W., Gergen P.J. and Strickland S. (1996) Mutations that perturb poly(A)-dependent maternal mRNA activation block the initiation of development. *Development* **122:** 579–588.

Liem K.F. Jr., Tremml G., Roelink H. and Jessel T.M. (1995) Dorsal differentiation of neural plate cells induced by BMP-mediated signals from epidermal ectoderm. *Cell* **82:** 969–979.

Liem K.F.Jr., Tremml G. and Jessel T.M. (1997) A role for the roof plate and its resident TGFβ-related proteins in neuronal patterning in the dorsal spinal cord. *Cell* **91:** 127–138.

Li L., Zhou J., James G., Heller-Harrison R., Czech M.P. and Olson E.N. (1992) FGF inactivates myogenic helix-loop-helix proteins through phosphorylation of a conserved protein kinase C site in their DNA-binding domains. *Cell* **71:** 1181–1194.

Liljegren S.J., Gustafson-Brown C., Pinyopich A., Ditta G.S. and Yanowski M.F. (1999) Interactions among *apetala1*, *leafy*, and *terminal flower1* specify meristem fate. *Plant Cell* **11:** 1007–1018.

Lillie F.R. (1917) The freemartin: A study of the action of sex hormones in foetal life of cattle. *J. Exp. Zool.* 23: 371–452.

Lin K., Dorman J.B., Rodan A. and Kenyon C. (1997) *daf-16:* An HNF-3/forkhead family member that can function to double the life-span of *Caenorhabditis elegans*. *Science* **278:** 1319–1322.

Lin R., Thompson S. and Priess J.R. (1995) *pop-1* encodes an HMG box protein required for the specification of a mesoderm precursor in early *C. elegans* embryos. *Cell* **83:** 599–609.

Lin X., Buff E.M., Perrimon N. and Michelson A.M. (1999) Heparan sulfate proteoglycans are essential for FGF receptor signaling during *Drosophila* embryonic development. *Development* **126:** 3715–3723.

Lin Y., Mahan K., Lathrop W.F., Myles D.G. and Primakoff P. (1994) A hyaluronidase activity of the sperm plasma membrane protein PH-20 enables sperm to penetrate the cumulus cell layer surrounding the egg. *J. Cell Biol.* **125:** 1157–1163.

Lin Y.-J., Seroude L. and Benzer S. (1998) Extended life-span and stress resistance in the *Drosophila* mutant *methuselah*. *Science* **282:** 943–946.

Linsenmayer T.F. (1991) Collagen. In: *Cell Biology of Extracellular Matrix*, 2nd ed., E.D. Hay (ed.), 7–44. New York: Plenum.

Lithgow G.J. (1996) Invertebrate gerontology: The age mutations of *Caenorhabditis elegans*. *BioEssays* **18:** 809–815.

Lithgow G.J., White T.M., Hinerfeld D.A. and Johnson T.E. (1994) Thermotolerance of a long-lived mutant of *Caenorhabditis elegans*. *J. Gerontol.* **49:** B270–276.

Littlefield C.L. (1985) Germ cells in Hydra oligactis males. I. Isolation of a subpopulation of interstitial cells that is developmentally restricted to sperm production. *Dev. Biol.* **112:** 185–193.

Littlefield C.L. (1991) Cell lineages in *Hydra*: Isolation and characterization of an interstitial stem cell restricted to egg production in *Hydra oligactis*. *Dev. Biol.* **143:** 378–388.

Littlefield C.L., Dunne J.F. and Bode H.R. (1985) Spermatogenesis in *Hydra oligactis*. I. Morphological description and characterization using a monoclonal antibody specific for cells of the spermatogenic pathway. *Dev. Biol.* **110:** 308–320.

Lodish H.F. and Small B. (1976) Different lifetimes of reticulocyte messenger RNA. *Cell* **7:** 59–65.

Löfberg J., Perris R. and Epperlein H. (1989) Timing in the regulation of neural crest cell migration: Retarded "maturation" of regional extracellular matrix inhibits pigment cell migration in embryos of the white axolotl mutant. *Dev. Biol.* **131:** 168–181.

Logan C., Hornbruch A., Campbell I. and Lumsden A. (1997) The role of Engrailed in establishing the dorsoventral axis of the chicklimb. *Development* **124:** 2317–2324.

Logan M. and Tabin C.J. (1999) Role of Pitx1 upstream of Tbx4 in specification of hindlimb identity. *Science* **283:** 1736–1739.

Lohr J.L., Danos M.C. and Yost H.J. (1997) Left-right asymmetry of a nodal-related gene is regulated by dorsoanterior midline structures during *Xenopus* development. *Development* **124:** 1465–1472.

Lohs-Schardin M. (1982) Dicephalic—A *Drosophila* mutant affecting polarity in follicle organization and embryonic patterning. *Wilhelm Roux's Arch.* **191:** 28–36.

Lohs-Schardin M., Cremer C. and Nüsslein-Volhard C. (1979) A fate map for the larval epidermis of *Drosophila melanogaster*: Localized cuticle defects following irradiation of the blastoderm with an ultraviolet laser microbeam. *Dev. Biol.* **73:** 239–255.

London C., Akers R. and Phillips C. (1988) Expression of Epi 1, an epidermis-specific marker in *Xenopus* laevis embryos, is specified prior to gastrulation. *Dev. Biol.* **129:** 380–389.

Longo F.J. (1989) Egg cortical architecture. In: *The Cell Biology of Fertilization*, H. Schatten and G. Schatten (eds.), 105–138. San Diego: Academic Press.

Lonie A., D'Andrea R., Paro R. and Saint R. (1994) Molecular characterisation of the *Polycomblike* gene of *Drosophila melanogaster*, a *trans*-acting negative regulator of homeotic gene expression. *Development* **120:** 2629–2636.

Lopez L.C., Bayna E.M., Litoff D., Shaper N.L., Shaper J.H. and Shur B.D. (1985) Receptor function of mouse sperm surface galactosyltransferase during fertilization. *J. Cell Biol.* **101:** 1501–1510.

Lopez L.C., Youakim A., Evans S.C. and Shur B.D. (1991) Evidence for a molecular distinction between golgi and cell surface forms of β1,4–galactosyltransferase. *J. Biol. Chem.* **266:** 15984–15991.

Lopez M.F., Dikkes P., Zurakowski D., Villa-Komaroff L. and Majzoub J.A. (1999) Regulation of hepatic glycogen in the insulin-like growth factor II-deficient mouse. *Endocrinology* **140:** 1442–1448.

Lopez-Antunez L. (1971) *Atlas of Human Anatomy*. Philadelphia: Saunders.

Lorca T., Cruzalegui F.H., Fesquet D., Cavadore J.-C., Méry J., Means A. and Dorée M. (1993) Calmodulin-dependent protein kinase II mediates inactivation of MPF and CSF upon fertilization in *Xenopus* eggs. *Nature* **366:** 270–273.

Lovell-Badge R. and Robertson E. (1990) XY female mice resulting from a heritable mutation in the primary testis-determining gene, *Tdy*. *Development* **109:** 635–646.

Lowe L.A., Supp D.M., Sampath K., Yokoyama T., Wright C.V.E., Potter S.S., Overbeek P. and Kuehn M.R. (1996) Conserved left-right asymmetry of nodal expression and alterations in murine situs inversus. *Nature* **381:** 158–161.

Lowe S.W., Schmitt E.M., Smith S.W., Osborne B.A. and Jacks T. (1993) p53 is required for radiation-induced apoptosis in mouse thymocytes. *Nature* **362:** 847–849.

Lucas W.J. (1995) Plasmodesmata: Intercellular channels for macromolecular transport in plants. *Curr. Opin. Cell Biol.* **7:** 673–80.

Lucchesi J. and Manning J.E. (1987) Gene dosage compensation in *Drosophila melanogaster*. *Advances in Genetics* **24:** 371–429.

Luckett W.P. (1978) Origin and differentiation of the yolk sac and extraembryonic mesoderm in presomite human and rhesus monkey embryos. *Am. J. Anat.* **152:** 59–58.

Luckinbill L.S., Arking R., Clare M.J., Cirocco W.C. and Buck S.A.(1984) Selection for delayed senescence in *Drosophila melanogaster*. *Evolution* **38**: 996–1003.

Lüer K. and Technau G.M. (1992) Primary culture of single ectodermal precursors of *Drosophila* reveals dorsoventral prepattern of intrinsic neurogenic and epidermogenic capabilities at the early gastrula stage. *Development* **116:** 377–385.

Lufkin T., Dierich A., LeMeur M., Mark M. and Chambon P. (1991) Disruption of the *Hox-1.6* homeobox gene results in defects in a region corresponding to its rostral domain of expression. *Cell* **66:** 1105–1119.

Lufkin T., Mark M., Hart C.P., Dollé P., LeMeur M. and Chambon P. (1992) Homeotic transformation of the occipital bones of the skull by ectopic expression of a homeobox gene. *Nature* **359:** 835–841.

Lukowitz W., Mayer U. and Jürgens G. (1996) Cytokinesis in the Arabidopsis embryo involves the syntaxin-related KNOLLE gene product. *Cell* **84:** 61–71.

Lumsden A. (1991) Motorizing the spinal cord. Cell **64:** 471–473.

Lund E.J. (1922) Electrical control of polarity in an egg. *Proc. Soc. Exp. Biol.* **20:** 113.

Lundblad V. and Wright W.E. (1996) Telomeres and telomerase: A simple picture becomes complex. *Cell* **87:** 369–375.

Luo D., Carpenter R., Copsey L., Vincent C., Clark J. and Coen E. (1999) Control of organ asymmetry in flowers of *Antirrhinum*. *Cell* **99:** 367–376.

Lutz D.A., Hamaguchi Y. and Inoué S. (1988) Micro-manipulation studies of the asymmetric positioning of the maturation spindle in *Chaetopterus* sp. oocytes. I. Anchorage of the spindle to the cortex and migration of a displaced spindle. *Cell Motility and the Cytoskeleton* **11:** 83–96.

Lyden D. et al. (1999) Id1 and Id3 are required for neurogenesis, angiogenesis and vascularization of tumor xenografts. *Nature* **401:** 670–677.

Lynn J.W., McCulloh D.H. and Chambers E.L. (1988) Voltage clamp studies on fertilization in sea urchin eggs. II. Current patterns in relation to sperm entry, nonentry, and activation. *Dev. Biol.* **128:** 305–323.

Lyon M.F. (1961) Gene action in the X-chromosome of the mouse (*Mus musculus L.*). *Nature* **190:** 372–373.

Ma H. (1994) The unfolding drama of flower development: Recent results from genetic and molecular analyses. *Genes and Dev.* **8:** 745–756.

Ma K., Inglis J.D., Sharkey A., Bickmore W.A., Hill R.E., Prosser E.J., Speed R.M., Thomson E.J., Jobling M., Taylor K., Wolfe J., Cooke H.J., Hargreave T.B. and Chandley A.C. (1993) A Y chromosome gene family with RNA-binding protein homology: Candidates for the azoospermia factor AZF controlling human spermatogenesis. *Cell* **75:** 1287–1295.

Ma Q., Kintner C. and Anderson D.J. (1996) Identification of neurogenin, a vertebrate neuronal determination gene. *Cell* **87:** 43–52.

MacCabe J.A., Errick J. and Saunders J.W. Jr. (1974) Ectodermal control of the dorsoventral axis in the leg bud of the chick embryo. *Dev. Biol.* **39:** 69–82.

Macdonald P.M. (1990) *bicoid* mRNA localization signal: Phylogenetic conservation of function and RNA secondary structure. *Development* **110:** 161–171.

Macdonald P.M., Kerr K. (1997) Redundant RNA recognition events in bicoid mRNA localization. *RNA* **3:** 1413–1420.

Macdonald P.M. and Kerr K. (1998) Mutational Analysis of an RNA recognition element that mediates localization of *bicoid* mRNA. *Mol. Cell Biol.* **18:** 3788–3795.

Macdonald P.M., Kerr K., Smith J.L. and Leask A. (1993) RNA regulatory element BLE1 directs the early steps of bicoid mRNA localization. *Development* **118:** 1233–1243.

MacDougall C., Harbison D. and Bownes M. (1995) The developmental consequences of alternate splicing in sex determination and differentiation in *Drosophila*. *Dev. Biol.* **172:** 353–376.

Maden M. (1993) The homeotic transformation of tails into limbs in *Rana temporaria* by retinoids. *Dev. Biol.* **159:** 379–391.

Maden M. and Holder N. (1992) Retinoic acid and the development of the central nervous system. *BioEssays* **14:** 431–438.

Madl J.E. and Herman R.K. (1979) Polyploids and sex determination in *Caenorhabditis elegans*. *Genetics* **93:** 393–402.

Maekawa M., Ishizaki T., Boku S., Watanabe N., Fujita A., Iwamatsu A., Obinata T., Ohashi K., Mizuno K. and Narumiya S. (1999) Signaling from Rho to the actin cytoskeleton through protein kinases ROCK and LIM-kinase. *Science* **285:** 895–898.

Mahajan-Miklos S. and Cooley L. (1994) Intercellular cytoplasm transport during *Drosophila* oogenesis. *Dev. Biol.* **165:** 336–351.

Mahowald A.P. (1971) Polar granules of *Drosophila*. IV. Cytochemical studies showing loss of RNA from polar granules during early stages of embryogenesis. *J. Exp. Zool.* **176:** 345–352.

Mahowald A.P. (1972) Oogenesis. In: *Developmental Systems: Insects*, S.J. Counce (ed.), 1–47. New York: Academic Press.

Mahowald A.P., Illmensee K. and Turner F.R. (1976) Interspecific transplantation of polar plasm between *Drosophila* embryos. *J. Cell Biol.* **70:** 358–373.

Mahowald A.P., Allis C.D., Karrer K.M., Underwood E.M. and Waring G.L. (1979) Germ plasm and pole cells of *Drosophila*. In: *Determinants of Spatial Organization*, S. Subtelny and I.R. Konigsberg (eds.), *37th Symp. Soc. Dev. Biol.*, 127–146. New York: Academic Press.

Malacinski G.M., Brothers A.J. and Chung H.-M. (1977) Destruction of components of the neural induction system of the amphibian egg with ultraviolet irradiation. *Dev. Biol.* **56:** 24–39.

Malicki J., Schughart K. and McGinnis W. (1990) Mouse *Hox-2.2* specifies thoracic segmental identity in *Drosophila* embryos and larvae. *Cell* **63:** 961–967.

Malinda K.M., Fisher G.W. and Ettensohn C.A. (1995) Four-dimensional microscopic analysis of the filopodial behavior of primary mesenchyme cells during gastrulation in sea urchin embryos. *Dev. Biol.* **172:** 552–566.

Mancilla A. and Mayor R. (1996) Neural crest formation in *Xenopus laevis:* Mechanisms of *Xslug* induction. *Dev. Biol.* **177:** 580–589.

Mangelsdorf D.J. and Evans R.M. (1992) Retinoid receptors as transcription factors. In: *Transcriptional Regulation,* K. Yamamoto and S.L. McKnight (eds.), chapt. 42. Cold Spring Harbor, NY: Cold Spring Harbor Laboratory Press.

Mangelsdorf D.J. and Evans R.M. (1995) The RXR heterodimers and orphan receptors. *Cell* **83:** 841–850.

Mangiarotti G., Zuker C., Chisholm R.L. and Lodish H.F. (1983) Different mRNAs have different nuclear transit times in *Dictyostelium discoideum* aggregates. *Mol. Cell. Biol.* **3:** 1511–1517.

Mango S.E., Thorpe C.J., Martin P.R., Chamberlain S.H. and Bowerman B. (1994) Two maternal genes, *apx-1* and *pie-1,* are required to distinguish the fates of equivalent blastomeres in *C. elegans* embryos. *Development* **120:** 2305–2315.

Mangold O. (1933) Über die Induktionsfähigkeit der verschiedenen Bezirke der Neurula von Urodelen. *Naturwissenschaften* **21:** 761–766.

Maniatis T. (1991) Mechanisms of alternative pre-mRNA splicing. *Science* **251:** 33–34.

Mann R.S. (1995) The specificity of homeotic gene function. *BioEssays* **17:** 855–863.

Mann R.S. (1997) Why are *Hox* genes clustered? *BioEssays* **19:** 661–664.

Mann R.S. and Hogness D.S. (1990) Functional dissection of *Ultrabithorax* proteins in *D. melanogaster*. *Cell* **60:** 597–610.

Mannervik M. (1999) Target genes of homeodomain proteins. *BioEssays* **21:** 267–270.

Mannervik M., Nibu Y., Zhang H. and Levine M. (1999) Transcriptional coregulators in development. *Science* **284:** 606–609.

Marcelle C., Stark M.R. and Bronner-Fraser M. (1997) Coordinate actions of BMPs, Wnts, Shh and noggin mediate patterning of the dorsal somite. *Development* **124:** 3955–3963.

Marcum B.A. and Campbell R.D. (1978) *Development* of *Hydra* lacking nerve and interstitial cells. *J. Cell Sci.* **29:** 17–33.

Markert C.L. and Petters R.M. (1978) Manufactured hexaparental mice show that adults are derived from three embryonic cells. *Science* **202:** 56–58.

Marler P., Peters S., Ball G.F., Dufty A.M., Jr. and Wingfield J.C. (1988) The role of sex steroids in the acquisition and production of birdsong. *Nature* **336:** 770–772.

Maroto M., Reshef R., Münsterberg A.E., Koester S., Goulding M. and Lassar A.B. (1997) Ectopic *Pax-3* activates *Myo-D* and *Myf-5* expression in embryonic mesoderm and neural tissue. *Cell* **89:** 139–148.

Marqués G., Musacchio M., Shimell M.J., Wünnenberg-Stapleton K., Cho K.W.Y. and O'Connor M.B. (1997) Production of a DPP activity gradient in the early *Drosophila* embryo through the opposing actions of the SOG and TLD proteins. *Cell* **91:** 417–426.

Marshall C.J. (1991) Tumor suppressor genes. Cell **64:** 313–326.

Marshall H., Nonchev S., Sham M.H., Muchamore I., Lumsden A. and Krumlauf R. (1992) Retinoic acid alters hindbrain *Hox* code and induces transformation of rhombomeres 2/3 into a 4/5 identity. *Nature* **360:** 737–741.

Martens U.M., Chavez E.A., Poon S.S., Schmoor C. and Lansdorp P.M. (2000) Accumulation of short telomeres in human fibroblasts prior to replicative senescence. *Exp. Cell Res.* **256:** 291–299.

Marti E., Bumcrot D.A., Takada R. and McMahon A.P. (1995) Requirement of 19K form of Sonic hedgehog for induction of distinct ventral cell types in CNS explants. *Nature* **375:** 322–325.

Martin G.M., Sprague C.A. and Epstein C.J. (1970) Replicative life-span of cultivated human cells: Effects of donor's age, tissue and genotype. *Lab. Invest.* **23**: 867–892.

Martin J.-R., Raibaud A. and Ollo R. (1994) Terminal pattern elements in *Drosophila* embryo induced by the torso-like protein. *Nature* **367:** 741–745.

Martinez-Arias A. and Lawrence P.A. (1985) Parasegments and compartments in the *Drosophila* embryo. *Nature* **313:** 639–642.

Martins G.G., Summers R.G. and Morrill J.B. (1998) Cells are added to the archenteron during and following secondary invagination in the sea urchin *Lytechinus variegatus. Dev. Biol.* **198:** 330–342.

Martin-Zanca D., Barbacid M. and Parada L.F. (1990) Expression of the trk proto-oncogene is restricted to the sensory cranial and spinal ganglia of neural crest origin in mouse development. *Genes and Dev.* **4:** 683–694.

Masoro E.J., Katz M.S. and McMahan C.A. (1989) Evidence for the glycation hypothesis of aging from the food-restricted rodent model. *J. Gerontol.* **44**: B20–22.

Massagué J. (1996) The transforming growth factor-β family. *Ann. Rev. Cell Biol.* **6:** 597–641.

Masuda M. and Sato H. (1984) Asynchronization of cell division is concurrently related with ciliogenesis in sea urchin blastulae. *Dev. Growth Differ.* **26:** 281–294.

Masui Y. and Markert C.L. (1971) Cytoplasmic control of nuclear behavior during meiotic maturation of frog oocytes. *J. Exp. Zool.* **19:** 129–146.

Matese J.C., Black S. and McClay D.R. (1997) Regulated exocytosis and sequential construction of the extracellular matrix surrounding the sea urchin zygote. *Dev. Biol.* **186:** 16–26.

Mayer U., Torres Ruiz R.A., Berleth T., Miséra S. and Jürgens G. (1991) Mutations affecting body organization in the *Arabidopsis* embryo. Nature **353:** 402–407.

Mayer U., Büttner G. and Jürgens G. (1993) Apical-basal pattern formation in the *Arabidopsis* embryo: Studies on the role of the *gnom* gene. *Development* **117:** 149–162.

Maynard Smith J. (1989) *Evolutionary Genetics.* New York: Oxford University Press.

McCain E.R. and McClay D.R. (1994) The establishment of bilateral asymmetry in sea urchin embryos. *Development* **120:** 395–404.

McCarrey J.R. and Dilworth D.D. (1992) Expression of *Xist* in mouse germ cells correlates with X-chromosome inactivation. *Nature Genetics* **2:** 200–203.

McCay C.M. and Crowell M.F. (1934) Prolonging the life span. *Sci. Monthly* **39**: 405– 414.

McCay C.M., Sperling G. and Barnes L.L. (1943) Growth, ageing, chronic diseases, and life span in rats. *Arch. Biochem. Biophys.* **2:** 469–479.

McClay D.R., Armstrong N.A. and Hardin J. (1992) Pattern formation during gastrulation in the sea urchin embryo. *Development* (Supplement): 33–41.

McClintock B. (1952) Chromosome organization and genic expression. *Cold Spring Harbor Symp. Quant. Biol.* **16:** 13–47.

McClung C.E. (1902) The accessory chromosome—Sex determinant? *Biol. Bull.* **3:** 43–84.

McCormick F. (1989) ras GTPase activating protein: Signal transmitter and signal terminator. *Cell* **56:** 5–8.

McDaniel C.N. and Poethig R.S. (1988) Cell-lineage patterns in the shoot apical meristem of the germinating maize embryo. *Planta* **175:** 13–22.

McElreavey K., Vilain E., Abbas N., Herskowitz I. and Fellous M. (1993) A regulatory cascade hypothesis for mammalian sex determination: SRY represses a negative regulator of male development. *Proc. Nat. Acad. Sci. USA* **90:** 3368–3372.

McGinnis N., Kuziora M.A. and McGinnis W. (1990) Human *Hox-4.2* and *Drosophila* deformed encode similar regulatory specificities in *Drosophila* embryos and larvae. *Cell* **63:** 969–976.

McGinnis W., Levine M.S., Hafen E., Kuroiwa A. and Gehring W.J. (1984) A conserved DNA sequence in homoeotic genes of the *Drosophila* Antennapedia and bithorax complexes. *Nature* **308:** 428–433.

McGonigle B., Bouhidel K. and Irish V. (1996) Nuclear localization of the *Arabidopsis* apetala3 and pistillata gene products depends on their simultaneous expression. *Genes and Dev.* **10:** 1812–1821.

McGrath J. and Solter D. (1984) Completion of mouse embryogenesis requires both the maternal and the paternal genome. *Cell* **37:** 179–183.

McIntosh J.R. and McDonald K.L. (1989) The mitotic spindle. *Scientific American* (October) **261:** 48–56.

McKay R. (1997) Stem cells in the central nervous system. *Science* **276:** 66–71.

McKeown M., Belote J.M. and Boggs R.T. (1988) Ectopic expression of the female *transformer* gene product leads to female differentiation of chromosomally male *Drosophila*. *Cell* **53:** 887–895.

McKinnell R.G. (1978) *Cloning: Nuclear Transplantation in Amphibia.* Minneapolis: University of Minnesota Press.

McKinney M.L. and McNamara K.J. (1991) *Heterochrony: The Evolution of Ontogeny.* New York: Plenum.

McKnight G.S., Pennequin P. and Schimke R.T. (1975) Induction of ovalbumin mRNA sequences by estrogen and progesterone in chick oviduct as measured by hybridization to complementary DNA. *J. Biol. Chem.* **250:** 8105–8110.

McKnight S.L. (1991) Molecular zippers in gene regulation. *Scientific American* (April) **264:** 54–64.

McLaren A. (1976) *Mammalian Chimeras.* Cambridge: Cambridge University Press.

McLaren A. (1991) Development of the mammalian gonad: The fate of the supporting cell lineage. *BioEssays* **13:** 151–156.

Mead P.E., Kelley C.M., Hahn P.S., Piedad O. and Zon L.I. (1998) SCL specifies hematopoietic mesoderm in *Xenopus* embryos. *Development* **125:** 2611–2620.

Medawar P.B. (1957) *An Unsolved Problem of Biology.* London: H.K. Lewis.

Medvinsky A.L., Samoylina N.L., Müller A.M. and Dzierzak E.D. (1993) An early pre-liver intraembryonic source of CFU-S in the developing mouse. *Nature* **364:** 64–67.

Meehan T., Schlatt S., O'Brian M.K., de Kretzer D.M. and Loveland L. (2000) Regulation of germ cell and Sertoli cell development by activin, follistatin, and FSH. *Dev. Biol.* **220:** 225–237.

Mege R.-M., Matsuzaki F., Gallin W.J., Goldberg J.I., Cunningham B.A. and Edelman G.M. (1988) Construction of epithelioid sheets by transfection of mouse sarcoma cells with cDNAs for chicken cell adhesion molecules. *Proc. Nat. Acad. Sci. USA* **85:** 7274–7278.

Megeney L.A., Kablar B., Garrett K., Anderson J.E. and Rudnicki M.A. (1996) MyoD is required for myogenic stem function in adult skeletal muscle. *Genes and Dev.* **10:** 1173–1183.

Mehra A., Gaudet J., Heck L., Kuwabara P.E. and Spence A.M. (1999) Negative regulation of male development in *Caenorhabditis elegans* by a protein-protein interaction between TRA-2A and FEM-3. *Genes and Dev.* **13:** 1453–1463.

Meier S. (1979) Development of the chick embryo mesoblast. *Dev. Biol.* **73:** 25–45.

Meinke D.W. (1991a) Genetic analysis of plant development. In: *Plant Physiology: A Treatise*, vol. 10, *Growth and Development*, F.C. Steward and R.G.S. Bidwell (eds.), 437–490. New York: Academic Press.

Meinke D.W. (1991b) Perspectives on genetic analysis of plant embryogenesis. *Plant Cell* 3: 857–866.

Mello C.C., Draper B.W., Krause M., Weintraub H. and Priess J.R. (1992) The *pie-1* and *mex-1* genes and maternal control of blastomere identity in early *C. elegans* embryos. *Cell* **70:** 163–176.

Mello C.C., Draper P.W. and Priess J.R. (1994) The maternal genes *apx-1* and *glp-1* and establishment of dorsoventral polarity in the early *C. elegans* embryo. *Cell* **77:** 95–106.

Melton D.A. (1987) Translocation of a localized maternal mRNA to the vegetal pole of *Xenopus* oocytes. *Nature* **328:** 80–82.

Mendez R., Hake L.E., Andresson T., Littlepage L.E., Ruderman J.V. and Richter J.D. (2000) Phosphorylation of CPE binding factor by Eg^2 regulates translation of c-mos mRNA. *Nature* **404:** 302–307.

Meng L., Ely J.J., Stouffer R.L. and Wolf D.P. (1997) Rhesus monkeys produced by nuclear transfer. *Biol. Reprod.* **57:** 454–459.

Meng X. et al. (2000) Regulation of cell fate decision of undifferentiated spermatogonia by GDNF. *Science* **287:** 1489–1493.

Meno C., Saijoh Y., Fujii H., Ikeda M., Yokoyama T., Yokoyama M., Toyoda Y. and Hamada H. (1996) Left-right asymmetric expression of the TGFb-family member lefty in mouse embryos. *Nature* **381:** 151–155.

Meredith J.E. and Schwartz M.A. (1997) Integrins, adhesion, and apoptosis. *Trends Cell Biol.* **7:** 146–150.

Mermod J.J., Schatz G. and Crippa M. (1980) Specific control of messenger translation in *Drosophila* oocytes and embryos. *Dev. Biol.* **75:** 177–186.

Meshcheryakov V. (1978) Orientation of cleavage spindles in pulmonate molluscs. I. Role of blastomere shape in orientation of second cleavage spindles. *Ontogenez* **9:** 559–566.

Metcalf D. (1977) *Hemopoietic Colonies.* Berlin/Heidelberg: Springer-Verlag.

Metcalf D. (1989) The molecular control of cell division, differentiation commitment and maturation in hemopoietic cells. *Nature* **339:** 27–30.

Metcalf D. and Moore M.A.S. (1971) *Hemopoietic Cells.* Amsterdam: North-Holland Publishing Company.

Metzstein M.M. and Horvitz H.R. (1999) The *C. elegans* cell death specification gene ces-1 encodes a snail family zinc finger protein. *Mol. Cell* **4:** 309–319.

Metzstein M.M., Hengartner M.O., Tsung N., Ellis R.E. and Horvitz H.R. (1996) Transcriptional regulator of programmed cell death encoded by *Caenorhabditis elegans* gene *ces-2*. *Nature* **382:** 545–547.

Metzstein M.M., Stanfield G.M. and Horvitz R.H. (1998) Genetics of programmed cell death in *C. elegans:* Past, present and future. *Trends Genet.* **14:** 410–416.

Meyer W.J., III, Migeon B.R. and Migeon C.J. (1975) Locus on human X chromosome for dihydrotestosterone receptor and androgen insensitivity. *Proc. Nat. Acad. Sci. USA* **72:** 1469–1472.

Meyerowitz E.M. (1999) Plants, animals and the logic of development. *Trends Cell Biol.* **9:** M65–68.

Meyerowitz E.M., Bowman J.L., Brockman L.L., Drews G.N., Jack T., Sieburth L.E. and Weigel D. (1991) A genetic and molecular model for flower development in *Arabidopsis thaliana. Development* (Supplement 1): 157–167.

Michalopoulos G.K. and DeFrances M.C. (1997) Liver regeneration. *Science* **276:** 60–66.

Michaud J.L., Lapointe F. and Le Douarin N.M. (1997) The dorsoventral polarity of the presumptive limb is determined by signal produced by the somite and by the lateral somatopleure. *Development* **124:** 1453–1463.

Michod R.E. and Levin B.R., eds. (1987) *The Evolution of Sex: An Examination of Current Ideas.* Sunderland, MA: Sinauer Associates.

Micklem D.R. (1995) mRNA localisation during development. *Dev. Biol.* **172:** 377–395.

Migeon B.R. (1971) Studies of skin fibroblasts from ten families with HGPRT deficiency, with reference to X-chromosomal inactivation. *Am. J. Hum. Genet.* **23:** 199–209.

Migliaccio E., Giorgio M., Mele S., Pelicci G., Reboldi P., Pandolfi P.P., Lanfrancone L. and Pelicci P.G. (1999) The p66shc adaptor protein controls oxidative stress response and life span in mammals. *Nature* **402:** 309–313.

Milán M. (1998) Cell cycle control in the *Drosophila* wing. *BioEssays* **20:** 969–971.

Millar S.E., Chamow S.M., Baur A.W., Oliver C., Robey F. and Dean J. (1989) Vaccination with a synthetic zona pellucida peptide produces long-term contraception in female mice. *Science* **246:** 935–938.

Miller A.D. (1992) Human gene therapy comes of age. *Nature* **357:** 455–460.

Miller D.J., Macek M.B. and Shur B.D. (1992) Complementarity between sperm surface β-1,4-galactosyltransferase and egg-coat ZP3 mediates sperm-egg binding. *Nature* **357:** 589–593.

Miller J., Fraser S.E. and McClay D. (1995) Dynamics of thin filopodia during sea urchin gastrulation. *Development* **121:** 2501–2511.

Miller J.R. and McClay D.R. (1997a) Changes in the pattern of adherens-junction-associated β-catenin accompany morphogenesis in the sea urchin embryo. *Dev. Biol.* **192:** 310–322.

Miller J.R. and McClay D.R. (1997b) Characterization of the role of cadherin in regulating cell adhesion during sea urchin development. *Dev. Biol.* **192:** 323–339.

Miller L.M., Plenefisch J.D., Casson L.P. and Meyer B.J. (1988) *xol-1:* A gene that controls the male modes of both sex determination and dosage compensation in *C. elegans. Cell* **55:** 167–183.

Miller L.M., Gallegos M.E., Morisseau B.A. and Kim S.K. (1993) *lin-31*, a *Caenorhabditis elegans HNF-3/fork head* transcription factor homolog, specifies three alternative cell fates in vulval development. *Genes and Dev.* **7:** 933–947.

Miller R. (1966) Chemotaxis during fertilization in the hydroid *Campanularia. J. Exp. Zool.* **161:** 23–44.

Miller R. (1978) Site-specific agglutination and the timed release of a sperm chemoattractant by the egg of the leptomedusan, *Orthopyxis caliculata. J. Exp. Zool.* **205:** 385–392.

Miller R. (1985) Sperm chemoattraction in the metazoa. In: *Biology of Fertilization*, C.B. Metz Jr. and A. Monroy (eds.), vol. 2, 275–337. Orlando, FL: Academic Press.

Minshull J. (1993) Cyclin synthesis: Who needs it? *BioEssays* **15:** 149–155.

Mintz B. (1965) Experimental genetic mosaicism in the mouse. In: *Preimplantation Stages of Pregnancy*, G.E. Wolstenholme and M. O'Connor (eds.), *Ciba Foundation Symp.*, 194–207. Boston: Little, Brown.

Mintz B. and Illmensee K. (1975) Normal genetically mosaic mice produced from malignant teratocarcinoma cells. *Proc. Nat. Acad. Sci. USA* **72:** 3585–3589.

Mishina Y., Rey R., Finegold M., Matzuk M.M., Joss N., Cate R.L. and Behringer R.R. (1996) Genetic analysis of the Müllerian inhibiting substance signal transduction pathway in mammalian sexual differentiation. *Genes and Dev.* **10:** 2577–2587.

Mitchison T.J. and Cramer L.P. (1996) Actin-based cell motility and cell locomotion. *Cell* **84:** 371–379.

Mittenthal J.E. and Jacobson A.G. (1990) The mechanics of morphogenesis in multicellular embryos. In: *Biomechanics of Active Movement and Deformation of Cells,* NATO ASI Series, vol. H42, N. Akkas (ed.), 320–366. Berlin: Springer-Verlag.

Mittwoch U. (1985) Erroneous theories of sex determination. *J. Med. Genet.* 22: 164–170.

Miyazaki S.-I. (1988) Inositol 1,4,5–trisphosphate-induced calcium release and guanine nucleotide-binding protein-mediated periodic calcium rises in golden hamster eggs. *J. Cell Biol.* **106:** 345–353.

Mizukami Y. and Ma H. (1992) Ectopic expression of the floral homeotic gene *AGAMOUS* in transgenic *Arabidopsis* plants alters floral organ identity. *Cell* **71:** 119–131.

Mochizuki T., Saijoh Y., Tsuchiya K., Shirayoshi Y., Takai S., Taya C., Yonekawa H., Yamada K., Nihei H., Nakatsuji N., Overbeek P.A., Hamada H. and Yokoyama T. (1998) Cloning of inv, a gene that controls left/right asymmetry and kidney development. *Nature* **395:** 177–181.

Molkentin J.D. and Olson N.E. (1996) Defining the regulatory networks for muscle development. *Curr. Opin. Genet. Dev.* **6:** 445–453.

Molkentin J.D., Black B.L., Martin J.F. and Olson N.E. (1995) Cooperative activation of muscle gene expression by MEF2 and myogenic bHLH proteins. *Cell* **83:** 1125–1136.

Montagne J., Stewart M.J., Stocker H., Hafen E., Kozma S.C. and Thomas G. (1999) *Drosophila* S6 kinase: A regulator of cell size. *Science* **285:** 2126–2129.

Moody S.A. (1987) Fates of the blastomeres of the 32–cell-stage *Xenopus* embryo. *Dev. Biol.* **122:** 300–319.

Moon R.T. and Kimelman D. (1998) From cortical rotation to organizer gene expression: toward a molecular explanation of axis specification in *Xenopus. BioEssays* **20:** 536–545.

Mooney D.J. and Mikos A.G. (1999) Growing new organs. *Scientific American* **280** (April): 60–65.

Moore A.R. and Burt A.S. (1939) On the locus and nature of the forces causing gastrulation in the embryos of *Dendraster excentricus. J. Exp. Zool.* **82:** 159–171.

Moore G.D., Kopf G.S. and Schultz R.M. (1993) Complete mouse egg activation in the absence of sperm by stimulation of an exogenous G protein-coupled receptor. *Dev. Biol.* **159:** 669–678.

Moore K.L. (1982) *The Developing Human: Clinically Oriented Embryology.* 3rd ed. Philadelphia: Saunders.

Morais da Silva S., Hacker A., Harley V., Goodfellow P., Swain A. and Lovell-Badge R. (1996) Sox9 expression during gonadal development implies a conserved role for the gene in testis differentiation in mammals and birds. *Nat. Genet.* **14:** 62–68.

Morata G. and Lawrence P.A. (1975) Control of compartment development by the *engrailed* gene in *Drosophila. Nature* **255:** 614–617.

Morata G. and Lawrence P.A. (1978) Cell lineage and homeotic mutants in the development of imaginal discs of *Drosophila.* In: *The Clonal Basis of Development,* S. Subtelny and I.M. Sussex (eds.), 45–60. New York: Academic Press.

Morata G. and Ripoll P. (1975) Minutes: Mutants of *Drosophila* autonomously affecting cell division rate. *Dev. Biol.* **42:** 211–221.

Morgan B.A., Izpisúa-Belmonte J.-C., Duboule D. and Tabin C.J. (1992) Targeted misexpression of *Hox-4.6* in the avian limb bud causes apparent homeotic transformations. *Nature* **358:** 236–239.

Morgan T.H. (1903) The relation between normal and abnormal development of the embryo of the frog, as determined by the effects of lithium chloride in solution. *Wilhelm Roux's Arch.* **16:** 691–712.

Morgan T.H. (1927) *Experimental Embryology.* New York: Columbia University Press.

Morisato D. and Anderson K.V. (1994) The spätzle gene encodes a component of the extracellular signal pathway establishing the dorsal-ventral pattern of the *Drosophila* embryo. *Cell* **76:** 677–688.

Morita Y. and Tilly J.L. (1999) Oocyte apoptosis: Like sand through an hour glass. *Dev. Biol.* **213:** 1–17.

Morrill J.B. (1986) Scanning electron microscopy of embryos. In: *Methods in Cell Biology,* T. Schroeder (ed.), 263–292. New York: Academic Press.

Morrill J.B. and Santos L.L. (1985) A scanning electron micrographical overview of cellular and extracellular patterns during blastulation and gastrulation in the sea urchin, *Lytechinus variegatus.* In: *The Cellular and Molecular Biology of Invertebrate Development,* R.H. Sawyer and R.M. Showman (eds.), 3–33. Columbia: University of South Carolina Press.

Morris J.Z. and Tissenbaum H.A and Ruvkum G. (1996) A phosphatidylinositol-3–OH kinase family member regulating longevity and diapause in *Caenorhabditis elegans. Nature* **382**: 536–539.

Morrison S.J., Shah N.M. and Anderson D.J. (1997) Regulatory mechanisms in stem cell biology. *Cell* **88:** 287–298.

Moskowitz I.P.G. and Rothman J.H. (1996) *lin-12* and *glp-1* are required zygotically for early embryonic cellular interactions and are regulated by maternal GLP-1 signaling in *Caenorhabditis elegans. Development* **122:** 4105–4117.

Moskowitz I.P.G., Gendrau S.B. and Rothman J.H. (1994) Combinatorial specification of blastomere identity by *glp-1*-dependent cellular interactions in the nematode *C. elegans. Development* **120:** 3325–3338.

Moses H.M., Yang E.Y. and Pietenpol J.A. (1990) TGF-β stimulation and inhibition of cell proliferation: New mechanistic insights. *Cell* **63:** 245–247.

Mouches C., Pauplin Y., Agarwal M., Lemieux L., Herzog M., Abadon M., Beyssat-Arnaouty V., Hyrien O., de Saint Vincent B.R., Georghiou G.P. and Pasteur N. (1990) Characterization of amplification core and esterase B1 gene responsible for insecticide resistance in *Culex. Proc. Nat. Acad. Sci. USA* **87:** 2574–2578.

Mountain A. (2000) Gene therapy: The first decade. *Trends Biotechnol.* **18:** 119–128.

Moury J.D. and Jacobson A.G. (1989) Neural fold formation at newly created boundaries between neural plate and epidermis in the axolotl. *Dev. Biol.* **133:** 44–57.

Moury J.D. and Jacobson A.G. (1990) The origins of neural crest cells in the axolotl. *Dev. Biol.* **141:** 243–253.

Mowry K. L. (1996) Complex formation between stage-specific oocyte factors and a *Xenopus* mRNA localization element. *Proc. Nat. Acad. Sci. USA* **93:** 14608–14613.

Mowry K.L. and Melton D.A. (1992) Vegetal messenger RNA localization directed by a 340–nt RNA sequence element in *Xenopus* oocytes. *Science* **255:** 991–994.

Moy G.W. and Vacquier V.D. (1979) Immunoperoxidase localization of bindin during the adhesion of sperm to sea urchin eggs. *Current Topics Dev. Biol.* **13:** 31–44.

Mueller L.D. (1985) Evolution of accelerated senescence in laboratory populations of *Drosophila. Proc. Nat. Acad. Sci.* USA **84**: 1974–1977.

Mueller L.D. and Rose M.R. (1996) Evolutionary theory predicts late-life mortality plateaus. *Proc. Nat. Acad. Sci. USA* **93:** 15249–15253.

Mullen J.R. and DiNardo S. (1995) Establishing parasegments in *Drosophila* embryos: Roles of *odd-skipped* and *naked* genes. *Dev. Biol.* **169:** 295–308.

Müller F., Bernhard W. and Tobler H. (1996) Chromatin diminution in nematodes. *BioEssays* **18:** 133–138.

Müller J., Gaunt S. and Lawrence P.A. (1995) Function of the Polycomb protein is conserved in mice and flies. *Development* **121:** 2847–2852.

Mullins M.C., Hammerschmidt M., Haffter P. and Nüsslein-Volhard C. (1994) Large-scale mutagenesis in the zebrafish: In search of genes controlling development in a vertebrate. *Curr. Biol.* **4:** 189–202.

Muneoka K. and Bryant S.V. (1982) Evidence that patterning mechanisms in developing and regenerating limbs are the same. *Nature* **298:** 369–371.

Munn K. and Steward R. (1995) The anterior-posterior and dorsal ventral axes have a common origin in *Drosophila melanogaster. BioEssays* **17:** 920–922.

Murata Y. and Wharton R.P. (1995) Bonding of pumilio to maternal hunchback mRNA is required for posterior patterning in *Drosophila* embryos. *Cell* **80:** 747–756.

Murray A.W. (1992) Creative blocks: cell-cycle checkpoints and feedback controls. *Nature* **359:** 599–604.

Murray A.W. (1995) Cyclin ubiquitination: The destructive end of mitosis. *Cell* **81:** 149–152.

Murray A.W. and Kirschner M.W. (1989) Cyclin synthesis drives the early embryonic cell cycle. *Nature* **339:** 275–280.

Murtha M.T., Leckman J.F. and Ruddle F.H. (1991) Detection of homeobox genes in development and evolution. *Proc. Nat. Acad. Sci. USA* **88:** 10711–10715.

Myer A., Wagner D.S., Vivian J.L., Olson E.N. and Klein W.H. (1997) Wild-type myoblasts rescue the ability of myogenin-null myoblasts to fuse in vivo. *Dev. Biol.* **185:** 127–138.

Myers G.C. and Manton K.G. (1984) Compression of mortality: Myth or reality? *Gerontologist* **24**: 346–353.

Myles D.G. (1993) Molecular mechanisms of sperm-egg membrane binding and fusion in mammals. *Dev. Biol.* **158:** 35–45.

Nabeshima Y., Hanaoka K., Hayasaka M., Esumi E., Li S., Nonaka I. and Nabeshima Y.-I. (1993) *Myogenin* gene disruption results in perinatal lethality because of severe muscle defect. *Nature* **364:** 532–535.

Nagafuchi A. and Takeichi M. (1988) Cell binding function of E-cadherin is regulated by the cytoplasmic domain. *EMBO J.* **7:** 3679–3684.

Nagafuchi A., Shirayoshi Y., Okazaki K., Yasuda K. and Takeichi M. (1987) Transformation of cell adhesion properties of exogenously introduced E-cadherin cDNA. *Nature* **329:** 341–343.

Nagaraj R., Pickup A.T., Howes R., Moses K., Freeman M. and Banerjee U. (1999) Role of the EGF receptor pathway in growth and patterning of the *Drosophila* wing through the regulation of *vestigial*. *Development* **126:** 975–985.

Nagoshi R.N., McKeown M., Burtis K.C., Belote J.M. and Baker B.S. (1988) The control of alternative splicing at genes regulating sexual differentiation in *D. melanogaster*. *Cell* **53:** 229–236.

Nagy L.M. and Carroll S. (1994) Conservation of *wingless* patterning functions in the short-germ embryos of *Tribolium castaneum*. *Nature* **367:** 460–463.

Nagoshi R.N. and Baker B.S. (1990) Regulation of sex-specific RNA splicing at the *Drosophila doublesex* gene: cis-acting mutations in exon sequences alter sex-specific RNA splicing patterns. *Genes and Dev.* **4:** 89–97.

Nakagawa S. and Takeichi M. (1995) Neural crest cell-cell adhesion controlled by sequential and subpopulation-specific expression of novel cadherins. *Development* **121:** 1321–1332.

Nakamura Y., Ito K. and Isaaksson L.A. (1996) Emerging understanding of translation termination. *Cell* **87:** 147–150.

Nakatsuji N. and Johnson K.E. (1983) Conditioning of a culture substratum by the ectodermal layer promotes attachment and oriented locomotion by amphibian gastrula mesodermal cells. *J. Cell Sci.* **59:** 43–60.

Nakatsuji N., Gould A. and Johnson K.E. (1982) Movement and guidance of migrating mesodermal cells in *Ambystoma maculatum* gastrulae. *J. Cell Sci.* **56:** 207–222.

Nakielny S. and Dreyfuss G. (1997) Nuclear export of proteins and RNAs. *Curr. Opin. Cell Biol.* **9:** 420–429.

Nakielny S. and Dreyfuss G. (1999) Transport of proteins and RNAs in and out of the nucleus. *Cell* **99:** 677–690.

Namba R., Pazdera T.M., Cerrone R.L. and Minden J.S. (1997) *Drosophila* embryonic pattern repair: How embryos respond to *bicoid* dosage alteration. *Development* **124:** 1393–1403.

Namenwirth M. (1974) The inheritance of cell differentiation during limb regeneration in the axolotl. *Dev. Biol.* **41:** 42–56.

Nameroff M. and Munar E. (1976) Inhibition of cellular differentiation by phospholipase C. II. Separation of fusion and recognition among myogenic cells. *Dev. Biol.* **49:** 288–293.

Nan X., Ng H.-H., Johnson C.A., Laherty C.D., Turner B.M., Eisenman R.M. and Bird A. (1998) Transcriptional repression by the methyl-CpG-binding protein MeCP2 involves a histone deacetylase complex. *Nature* **393:** 386–389.

Nayerna K., Reim K., Oberwinkler H. and Engel W. (1994) Diploid expression and translational regulation of rat acrosin gene. *Biochem. Biophys. Res. Commun.* **202:** 88–93.

Needham J. (1959) *A History of Embryology*. London/New York: Abelard-Schuman.

Neer E.J. (1995) Heterotrimeric G proteins: Organizers of transmembrane signals. *Cell* **80:** 249–257.

Nellen D., Burke R., Struhl G. and Basler K. (1996) Direct and long-range action of a DPP morphogen gradient. *Cell* **85:** 357–368.

Neufeld T.P., de la Cruz A.F.A., Johnston L.A. and Edgar B.A. (1998) Coordination of growth and cell division in the *Drosophila* wing. *Cell* **93:** 1183–1193.

Neumann C. and Cohen S. (1997a) Long-range action od Wingless organizes the dorsal-ventral axis of the *Drosophila* wing. *Development* **124:** 871–880.

Neumann C. and Cohen S. (1997b) Morphogens and pattern formation. *BioEssays* **19:** 721–729.

Neuman-Silberberg F.S. and Schüpbach T. (1993) The *Drosophila* dorsoventral patterning gene gurken produces a dorsally localized RNA and encodes a TGFα-like protein. *Cell* **75:** 165–174.

Neuman-Silberberg F.S. and Schupbach T. (1994) Dorsoventral axis formation in *Drosophila* depends on the correct dosage of gene *gurken*. *Development* **120:** 2457–2463.

Newport J.W. and Kirschner M.W. (1982a) A major developmental transition in early *Xenopus* embryos. I. Characterization and timing of cellular changes at midblastula stage. *Cell* **30:** 675–686.

Newport J.W. and Kirschner M.W. (1982b) A major developmental transition in early *Xenopus* embryos. II. Control of the onset of transcription. *Cell* **30:** 687–696.

Ng H.-H. and Bird A. (1999) DNA methylation and chromatin modification. *Curr. Opin. Genet. Dev.* **9:** 158–163.

Nicholls R.D., Saitoh S. and Horsthemke B. (1998) Imprinting in Prader-Willi and Angelman syndromes. *Trends in Genetics* **14:** 194–200.

Nicolet G. (1971) Avian gastrulation. *Adv. Morphog.* **9:** 231–262.

Niehrs C., Keller R., Cho K.W.Y. and De Robertis E.M. (1993) The homeobox gene *goosecoid* controls cell migration in *Xenopus* embryos. *Cell* **72:** 491–503.

Nieto M.A., Sargent M.G., Wilkinson D.G. and Cooke J. (1994) Control of cell behavior during vertebrate development by *Slug*, a zinc finger gene. *Science* **264:** 835–839.

Nieuwkoop P.D. (1952) Activation and organization of the central nervous system in amphibians. III. Synthesis of a new working hypothesis. *J. Exp. Zool.* **120:** 83–108.

Nieuwkoop P.D. (1969a) The formation of the mesoderm in urodelean amphibians. I. Induction by the endoderm. *Wilhelm Roux's Arch.* **162:** 341–373.

Nieuwkoop P.D. (1969b) The formation of the mesoderm in urodelean amphibians. II. The origin of the dorsoventral polarity of the mesoderm. *Wilhelm Roux's Arch.* **163:** 298–315.

Nieuwkoop P.D. and Faber J. (1967) *Normal Table of Xenopus laevis (Daudin)*. 2nd ed. Amsterdam: North-Holland Publishing Company.

Nieuwkoop P.D. and Sutasurya L.A. (1979) *Primordial Germ Cells in the Chordates: Embryognesis and Phylogenesis*, vol. 7. Cambridge: Cambridge University Press.

Nieuwkoop P.D. and Sutasurya L.A. (1981) *Primordial Germ Cells in the Invertebrates: From Epigenesis to Preformation*, vol. 10. Cambridge: Cambridge University Press.

Nigg E.A. (1995) Cyclin-dependant protein kinases: Key regulators of the eukaryotic cell cycle. *BioEssays* **17:** 471–480.

Nigg E.A. (1997) Nucleocytoplasmic transport: Signals, mechanisms, and regulation. *Nature* **386:** 779–787.

Nijhout H.F. (1980) Pattern formation on lepidopteran wings: Determination of an eyespot. *Dev. Biol.* **80:** 267–274.

Nijhout H.F. (1994) *Insect Hormones.* Princeton: Princeton University Press.

Nilson L.A. and Schüpbach T. (1998) Localized requirements for *windbeutel* and *pipe* reveal a dorsoventral prepattern within the follicular epithelium of the *Drosophila* ovary. *Cell* **93:** 253–262.

Nishida H. (1987) Cell lineage analysis in ascidian embryos by intracellular injection of a tracer enzyme. III. Up to the tissue restricted stage. *Dev. Biol.* **121:** 526–541.

Nishida H. (1992) Regionality of egg cytoplasm that promotes muscle differentiation in embryo of the ascidian, *Halocynthia roretzi. Development* **116:** 521–529.

Nishida H. (1994) Localization of determinants for formation of the anteroposterior axis in eggs of the ascidian *Halocynthia roretzi. Development* **120:** 3093–3104.

Nishikata T., Hibino T. and Nishida H. (1999) The centrosome-attracting body, microtubule system, and posterior egg cytoplasm are involved in positioning of cleavage planes in the ascidian embryo. *Dev. Biol.* **209:** 72–85.

Nishimiya-Fujisawa C. and Sugiyama T. (1993) Genetic analysis of developmental mechanisms in *Hydra. Dev. Biol.* **157:** 1–9.

Niswander L., Tickle C., Vogel A., Booth I. and Martin G.R. (1993) FGF-4 replaces the apical ectodermal ridge and directs outgrowth and patterning of the limb. *Cell* **75:** 579–587.

Noden D.M. (1983) The role of the neural crest in patterning of avian cranial, skeletal, connective, and muscle tissues. *Dev. Biol.* **96:** 144–165.

Nohno T., Noji S., Koyama E., Ohyama K., Myokai F., Kuroiwa A., Saito T. and Taniguchi S. (1991) Involvement of the *Chox*-4 chicken homeobox genes in determination of anteroposterior axial polarity during limb development. *Cell* **64:** 1197–1205.

Nomura M. (1973) Assembly of bacterial ribosomes. *Science* **179:** 864–873.

Nonaka S., Tanaka Y., Okada Y., Takeda S., Harada A., Kanai Y., Kido M. and Hirokawa N. (1998) Randomization of left-right asymmetry due to loss of nodal cilia generating left-ward flow of extraembryobnic fluid in mice lacking KIF3B motor protein. *Cell* **95:** 829–837.

Nonet M.L. and Meyer B.J. (1991) Early aspects of *Caenorhabditis elegans* sex determination and dosage compensation are regulated by a zinc-finger protein. *Nature* **351:** 65–68.

Nordenskjold A. and Ivarsson S.A. (1998) Molecular characterization of 5 alpha-reductase type 2 deficiency and fertility in a Swedish family. *J. Clin. Endocrinol. Metab.* **83:** 3236–3238.

Nose A., Nagafuchi A. and Takeichi M. (1988) Expressed recombinant cadherins mediate cell sorting in model systems. *Cell* 54: 993–1001.

Nose A., Tsuji K. and Takeichi M. (1990) Localization of specificity determining sites in cadherin cell adhesion molecules. *Cell* **61:** 147–155.

Nöthiger R. (1972) Larval development of imaginal disks. In: *Results and Problems in Cell Differentiation,* vol. 5, H. Ursprung and R. Nöthiger (eds.), 1–34. Berlin: Springer-Verlag.

Nottebohm F. (1980) Testosterone triggers growth of brain vocal control nuclei in adult female canaries. *Brain Res.* **189:** 429–436.

Nottebohm F. (1981) A brain for all seasons: Cyclical anatomical changes in song control nuclei of the canary brain. *Science* **214:** 1368–1370.

Nottebohm F. (1989) From bird song to neurogenesis. *Scientific American* **260** (February): 74–79.

Nuccitelli R. (1978) Ooplasmic segregation and secretion in the *Pelvetia* eggs is accompanied by a membrane-generated electrical current. *Dev. Biol.* **62:** 13–33.

Nuccitelli R. and Grey R.D. (1984) Controversy over the fast, partial, temporary block to polyspermy in sea urchins: A reevaluation. *Dev. Biol.* **103:** 1–17.

Nuccitelli R. and Jaffe L. (1974) Spontaneous current pulses through developing fucoid eggs. *Proc. Nat. Acad. Sci. USA* **71:** 4855–4859.

Nurse P. (1990) Universal control mechanism regulating onset of M-phase. *Nature* **344:** 503–508.

Nusse R., van Ooyen A., Cox D., Fung Y.K.T. and Varmus H. (1984) Mode of proviral activation of a putative mammary oncogene (*int-1*) on mouse chromosome 15. *Nature* **307:** 131–136.

Nüsslein-Volhard C. (1991) Determination of the embryonic axes of *Drosophila. Development* (Supplement 1): 1–10.

Nüsslein-Volhard C. and Wieschaus E. (1980) Mutations affecting segment number and polarity in *Drosophila. Nature* **287:** 795–801.

Nüsslein-Volhard C., Wieschaus E. and Kluding H. (1984) Mutations affecting the pattern of the larval cuticle in *Drosophila melanogaster. Wilhelm Roux's Arch.* **193:** 267–282.

Nüsslein-Volhard C., Frohnhöfer H.G. and Lehmann R. (1987) Determination of anteroposterior polarity in *Drosophila. Science* **238:** 1675–1681.

Ogg S., Paradis S., Gottlieb S., Patterson G.I., Lee L., Tissenbaum H.A. and Ruvkun G. (1997) The fork head transcription factor DAF-16 transduces insulin-like metabolic and longevity signals in *C. elegans. Nature* **389:** 994–999.

Okada M., Kleinman I.A. and Schneiderman H.A. (1974) Restoration of fertility in sterilized *Drosophila* eggs by transplantation of polar cytoplasm. *Dev. Biol.* **37:** 43–54.

Okada Y., Nonaka S., Tanaka Y., Saijoh Y., Hamada H. and Hirokawa N. (1999) Abnormal nodal flow precedes situs inversus in iv and inv mice. *Mol. Cell* **4:** 459–468.

Olds J.L. et al. (1995) Imaging protein kinase C activation in living sea urchin eggs after fertilization. *Dev. Biol.* **172:** 675–682.

Olesen S.P., Clapham D.E. and Davies P.F. (1988) Haemodynamic shear stress activates a K^+ current in vascular endothelial cells. *Nature* **331:** 168–170.

Oliver B., Perrimon N. and Mahowald A.P. (1987) The *ovo* locus is required for sex-specific germ line maintenance in *Drosophila. Genes and Dev.* **1:** 913–923.

Oliver B., Kim Y.-J. and Baker B.S. (1993) Sex-lethal, master and slave: A hierarchy of germ-line sex determination in *Drosophila. Development* **119:** 897–908.

Olsen P.H, and Ambros V. (1999) The lin-4 regulatory RNA controls developmental timing in *Caenorhabditis elegans* by blocking LIN-14 protein synthesis after the initiation of translation. *Dev. Biol.* **216:** 671–680.

Olson E.N. (1992) Interplay between proliferation and differentiation within the myogenic lineage. *Dev. Biol.* **154:** 261–272.

Olson E.N., Arnold H.-H., Rigby P.W.J. and Wold B.J. (1996) Know your neighbors: Three phenotypes in null mutants of the myogenic bHLH gene *MRF4. Cell* **85:** 1–4.

Olwin B.B. and Hauschka S.D. (1988) Cell surface fibroblast growth factor and epidermal growth factor receptors are permanently lost during skeletal muscle terminal differentiation in culture. *J. Cell Biol.* **107:** 761–769.

O'Neill L.P. and Turner B.M. (1995) Histone H4 acetylation distinguishes coding regions of the human genome from heterochromatin in a differentiation-dependent but transcription-independent manner. *EMBO J.* **14:** 3946–3957.

Orkin S.H. (1996) Development of the hematopoietic system. *Curr. Opin. Genet. Dev.* **6:** 597–602.

Oro A.E., McKeown M. and Evans R.M. (1990) Relationship between the product of the *Drosophila ultraspiracle* locus and vertebrate retinoid X receptor. *Nature* **347:** 298–301.

Orr W.C. and Sohal R.C. (1994) Extension of life-span by over-expression of superoxide dismutase and catalase in *Drosophila melanogaster*. *Science* **263:** 1128–1130.

Oster G.F. (1984) On the crawling of cells. *J. Embryol. exp. Morphol.* **83** (Supplement): 329–364.

Packard D.S., Jr. and Meier S. (1983) An experimental study of the somitomeric organization of the avian segmental plate. *Dev. Biol.* **97:** 191–202.

Painter T.S. (1923) Studies in mammalian spermatogenesis. II. The spermatogenesis of man. *J. Exp. Zool.* 37: 291–335.

Palmeirim I., Henrique D., Ish-Horowicz D. Pourquié O. (1997) Avian *hairy* gene expression identifies a molecular clock linked to vertebrate segmentation and somitogenesis. *Cell* **91:** 639–648.

Palmer M.S., Sinclair A.H., Berta P., Ellis N.A., Goodfellow P.N., Abbas N.E. and Fellous M. (1989) Genetic evidence that *ZFY* is not the testis-determining factor. *Nature* **342:** 937–942.

Palmer M.S., Dryden A.J., Hughes J.T. and Collinge J. (1991) Homozygous prion protein genotype predisposes to sporadic Creutzfeldt-Jakob disease. *Nature* **352:** 340–342.

Palmer S.J. and Burgoyne P.S. (1991) *In situ* analysis of fetal, prepuberal and adult XX↔XY chimaeric mouse testes: Sertoli cells are predominantly, but not exclusively, XY. *Development* **112:** 265–268.

Palmiter R.D., Brinster R.L., Hammer R.E., Trumbauer M.E., Rosenfeld M.G., Birnberg N.C. and Evans R.M. (1982) Dramatic growth of mice that develop from eggs microinjected with metallothionein-growth hormone fusion genes. *Nature* **300:** 611–615.

Pang P.P. and Meyerowitz E.M. (1987) *Arabidopsis thaliana:* A model system for plant molecular biology. *Bio/Technology* **5:** 1177–1181.

Pankratz M.J., Seifert E., Gerwin N., Billi B., Nauber U. and Jäckle H. (1990) Gradients of Krüppel and *knirps* gene products direct pair-rule gene stripe patterning in the posterior region of the *Drosophila* embryo. *Cell* **61:** 309–317.

Pankratz M.J., Busch M., Hoch M., Seifert E. and Jäckle H. (1992) Spatial control of the gap gene *knirps* in the *Drosophila* embryo by posterior morphogen system. Science **255:** 986–989.

Panning B. and Jaenisch R. (1996) DNA hypomethylation can activate Xist expression and silence X-linked genes. *Genes and Dev.* **10:** 1991–2002 (6).

Panning B. and Jaenisch R. (1998) RNA and the epigenetic regulation of X chromosome inactivation. *Cell* **93:** 305–308.

Panning B., Dausman J. and Jaenisch R. (1997) X chromosome inactivation is mediated by Xist RNA stabilization. *Cell* **90:** 907–916(6).

Pardanaut L., Luton D., Prigent M., Bourcheix L.-M., Catala M. and Dieterlein-Lièvre F. (1996) Two distinct endothelial lineages in ontogeny, one of them related to hemopopiesis. *Development* **122:** 1363–1371.

Pardee A.B. (1989) G_1 events and regulation of cell proliferation. *Science* **246:** 603–608.

Parkhurst S.M. and Meneely P.M. (1994) Sex determination and dosage compensation: Lessons from flies and worms. *Science* **264:** 924–932.

Paro R. (1993) Mechanisms of heritable gene repression during development of *Drosophila*. *Curr. Opin. Cell Biol.* **5:** 999–1005.

Parr B.A. and McMahon A.P. (1995) Dorsalizing signal Wnt-7a required for normal polarity of D-V and A-P axes of mouse limb. *Nature* **374:** 350–353.

Partridge L. and Barton N.H. (1993) Optimality, mutation and the evolution of ageing. *Nature* **362**: 305–311.

Parys J.B., McPherson S.M., Mathews L., Campbell K.P. and Longo F.J. (1994) Presence of inositol 1,4,5-trisphosphate receptor, calreticulin, and calsequestrin in eggs of sea urchins and *Xenopus laevis*. *Dev. Biol.* **161:** 466–476.

Passner J.M., Ryoo H.D., Shen L., Mann R.S. and Aggarwaal A.K. (1999) Structure of a DNA-bound Ultrabithorax-Extradenticle homeodomain complex. *Nature* **397:** 714–719.

Pasteels J. (1964) The morphogenetic role of the cortex of the amphibian egg. *Adv. Morphog.* **3:** 363–389.

Patel N.H. (1993) Evolution of insect pattern formation: A molecular analysis of short germband segmentation. In*: Evolutionary Conservation of Developmental Mechanisms,* A.D. Spradling (ed.), 85–110. New York: Wiley/Alan R. Liss.

Patel N.H., Ball E.E. and Goodman C.S. (1992) Changing role of *even-skipped* during the evolution of insect pattern formation. *Nature* **357:** 339–342.

Patel N.H., Condron B.G. and Zinn K. (1994) Pair-rule expression patterns of *even-skipped* are found in both short- and long-germ beetles. *Nature* **367:** 429–434.

Patten B.M. (1964) *Foundations in Embryology.* New York: McGraw-Hill.

Patten B.M. (1971) *Early Embryology of the Chick.* New York: McGraw-Hill.

Pause A., Belsham G.J., Gingras A.-C., Donze O., Lin T.A., Lawrence J.C. Jr. and Sonenberg N. (1994) Insulin-dependent stimulation of protein synthesis by phosphorylation of a regulator of 5'-cap function. *Nature* **371:** 762–767.

Pedersen R.A. (1999) Embryonic stem cells for medicine. *Scientific American* **280** (April): 68–73.

Peifer M. and Bejsovec A. (1992) Knowing your neighbors: Cell interactions determine intrasegmental patterning in *Drosophila*. *Trends Genetics* **8:** 243–249.

Peifer M. and Wieschaus E. (1990) Mutations in the *Drosophila* gene *extradenticle* affect the way specific homeodomain proteins regulate segmental identity. *Genes and Dev.* **4:** 1209–1223.

Peifer M., Karch F. and Bender W. (1987) The bithorax complex: Control of segmental identity. *Genes and Dev.* **1:** 891–898.

Pelling C. (1959) Chromosomal synthesis of ribonucleic acid as shown by incorporation of uridine labeled with tritium. *Nature* **184:** 655–656.

Penny G.D., Kay G.F., Sheardown S.A., Rastan S. and Brockdorff N. (1996) Requirement for *Xist* in X chromosome inactivation. *Nature* **379:** 131–137.

Perbal M.C., Haughn G., Saedler H. and Schwarz-Sommer Z. (1996) Non-cell-autonomous function of the *Antirrhinum* floral homeotic proteins deficiens and globosa is exerted by their polar cell-to-cell trafficking. *Development* **122:** 3433–3441.

Percy J., Kuhn K.L. and Kalthoff K. (1986) Scanning electron microscopic analysis of spontaneous and UV-induced abnormal segment patterns in *Chironomus samoensis* (Diptera, Chironomidae). *Wilhelm Roux's Arch.* **195:** 92–102.

Perez-Terzic C., Jaconi M. and Clapham D.E. (1997) Nuclear calcium and the regulation of the nuclear pore complex. *BioEssays* **19:** 787–792.

Perona R.M. and Wassarman P.M. (1986) Mouse blastocysts hatch in vitro by using a trypsin-like proteinase associated with cells of mural trophectoderm. *Dev. Biol.* **114:** 42–52.

Perris R. (1997) The extracellular matrix in neural crest cell migration. *Trends Neurosci.* **20:** 23–31.

Perris R. and Bronner-Fraser M. (1989) Recent advances in defining the role of the extracellular matrix in neural crest development. *Comm. Dev. Neurobiol.* **1:** 61–83.

Perris R., von Boxberg Y. and Löfberg J. (1988) Local embryonic matrices determine region-specific phenotypes in neural crest cells. *Science* **241:** 86–89.

Perry M.D., Li W., Trent C., Robertson B., Fire A., Hageman J.M. and Wood W.B. (1993) Molecular characterization of the *her-1* gene suggests a direct role in cell signaling during *Caenorhabditis elegans* sex determination. *Genes and Dev.* **7:** 216–228.

Pesce M., Gross M.K. and Schöler H. (1998) In line with our ancestors: Oct-4 and the mammalian germ. *BioEssays* **20:** 722.

Petersen B.E., Bowen W.C., Patrene K.D., Mars W.M., Sullivan A.K., Murase N., Boggs S.S., Greenberger J.S. and Goff J.P. (1999) Bone marrow as a potential source of hepatic oval cells. *Science* **284:** 1168–1170.

Pettway Z., Guillory G. and Bronner-Fraser M. (1990) Molecular mechanisms of avian neural crest cell migration on fibronectin and laminin. *Dev. Biol.* **136:** 335–345.

Pevny L., Simon M.C., Robertson E., Klein W.H., Tsai S.-F., D'Agati V., Orkin S.A. and Constantini F. (1991) Erythroid differentiation in chimaeric mice blocked by a targeted mutation in the gene for transcription factor GATA-1. *Nature* **349:** 257–260.

Peyroche A., Paris S. and Jackson C.L. (1996) Nucleotide exchange on ARF mediated by yeast Gea1 protein. *Nature* **384:** 479–484.

Phelps B.M., Koppel D.E., Primakoff P. and Myles D.G. (1990) Evidence that proteolysis of the surface is an initial step in the mechanism of formation of sperm cell surface domains. *J. Cell Biol.* **111:** 1839–1847.

Phillips H.M. and Steinberg M.S. (1969) Equilibrium measurements of embryonic chick cell adhesiveness. I. Shape equilibrium in centrifugal fields. *Proc. Nat. Acad. Sci. USA* **64:** 121–127.

Phoenix C.H., Goy R.W., Gerall A.A. and Young W.C. (1959) Organizing action of prenatally administered testosterone propionate on the tissues mediating mating behavior in the female guinea pig. *Endocrinology* **65:** 369–382.

Piatigorsky J. (1981) Lens differentiation in vertebrates: A review of cellular and molecular features. *Differentiation* **19:** 134–153.

Piccolo S., Sasai Y., Lu B. and De Robertis E.M. (1996) Dorsoventral patterning in *Xenopus:* Inhibition of ventral signals by direct binding of chordin to BMP-4. *Cell* **88:** 589–598.

Piccolo S., Agius E., Lu B., Goodman S., Dale L. and De Robertis E. (1997) Cleavage of chordin by Xolloid metalloprotease suggests a role for proteolytic processing in the regulation of Spemann organizer activity. *Cell* **91:** 407–416.

Piechaczyk M., Yang J.-Q., Blanchard J.M., Jeanteur P. and Marcu K.B. (1985) Posttranscriptional mechanisms are responsible for accumulation of truncated c-*myc* RNAs in murine plasma cell tumors. *Cell* **42:** 589–597.

Pignoni F., Steingrimsson E. and Lengyel J.A. (1992) *bicoid* and the terminal system activate tailless expression in the early *Drosophila* embryo. *Development* **115:** 239–251.

Piper D.E., Batchelor A.H., Chang C.P., Cleary M.L. and Wolberger C. (1999) Structure of a HoxB1–Pbx1 heterodimer bound to DNA: Role of the hexapeptide and a fourth homeodomain helix in complex formation. *Cell* **96:** 587–597.

Pirotta V. (1998) Polycombing the genome: PcG, TrxG, and chromatin silencing. *Cell* **93:** 333–336.

Pittenger M.F., Mackay A.M., Beck S.C., Jaiswal R.K., Douglas R., Mosca J.D., Moorman M.A., Simonetti D.W., Craig S. and Marshak D.R. (1999) Multilineage potential of adult human mesenchymal stem cells. *Science* **284:** 143–147.

Placzek M., Tessier-Lavigne M., Yamada T., Jessell T. and Dodd J. (1990) Mesodermal control of neural cell identity: Floor plate induction by the notochord. *Science* **250:** 985–988.

Plasterk R.H.A. (1999) The year of the worm. *BioEssays* **21:** 105–109.

Pokrywka N.J. and Stephenson E.C. (1991) Microtubules mediate the localization of *bicoid* RNA during *Drosophila* oogenesis. *Development* **113:** 55–66.

Pons S. and Marti E. (2000) Sonic hedgehog synergizes with the extracellular matrix protein vitronectin to induce spinal motor neuron differentiation. *Development* **127:** 333–342.

Porcher C., Swat W., Rockwell K., Fujiwara Y., Alt F.W. and Orkin S.H. (1996) The T cell leukemia oncoprotein SCL/tal-1 is essential for development of all hematopoietic lineages. *Cell* **86:** 47–57.

Postlethwait J.H. and Schneiderman H.A. (1971) Pattern formation and determination in the antenna of the homeotic mutant *Antennapedia* of *Drosophila melanogaster*. *Dev. Biol.* **25:** 606–640.

Postlethwait J.H. and Talbot W.S. (1997) Zebrafish genomics: From mutants to genes. *Trends Genet.* **13:** 183–190.

Pourquié O. et al. (1996) Lateral and axial signals involved in somite patterning: A role for BMP-4. *Cell* **84:** 461–471.

Preiss T. and Hentze M.W. (1999) From factors to mechanisms: Translation and translational control in eukaryotes. *Curr. Opin. Genet. Dev.* **9:** 515–521.

Prescott D.M. (1988) *Cells.* Boston: Jones and Bartlett.

Priess J.R. and Hirsh E.I. (1986) *Caenorhabditis elegans* morphogenesis: The role of the cytoskeleton in elongation of the embryo. *Dev. Biol.* **117:** 156–173.

Priess J.R. and Thompson J.N. (1987) Cellular interactions in early *C. elegans* embryos. *Cell* **48:** 241–250.

Primakoff P., Lathrop W., Woolman L., Cowan A. and Myles D. (1988) Fully effective contraception in male and female guinea pigs immunized with the sperm protein PH-20. *Nature* **335:** 543–547.

Pröve E. (1978) Courtship and testosterone in male zebra finches. *Zeitschrift für Tierpsychologie* **48:** 47–67.

Prusiner S.B. (1982) Novel proteinaceous infectious particles cause scrapie. *Science* **216:** 136–144.

Prusiner S.B. (1997) Prion diseases and the BSE crisis. *Science* **278:** 245–251.

Prusiner S.B. (1998) Prion protein biology. *Cell* **93:** 337–348.

Prusiner S.B., Groth D., Serban A., Koehler R., Foster D., Torchia M., Burton D., Yang S.-L. and DeArmond S.J. (1993) Ablation of the prion protein (PrP) gene in mice prevents scrapie and facilitates production of anti-PrP antibodies. *Proc. Nat. Acad. Sci. USA* **90:** 10608–10612.

Psychoyos D. and Stern C.D. (1996) Fates and migratory routes of primitive streak cells in the chick embryo. *Development* **122:** 1523–1534.

Puffenberger, E.G., Hosoda K., Washington S.S., Nakao K., deWit D., Yanagisawa M. and Chakravarti A. (1994) A missense mutation of the endothelin-B receptor gene in multigenic Hirschsprung's disease. *Cell* **79:** 1257–1266.

Pursel V.G., Pinkert C.A., Miller K.F., Bolt D.J., Campbell R.G., Palmiter R.D., Brinster R.L. and Hammer R.E. (1989) Genetic engineering of livestock. *Science* **244:** 1281–1288.

Purves D. and Lichtman J.W. (1985) *Principles of Neural Development.* Sunderland, MA: Sinauer Associates.

Püschel A.W., Balling R. and Gruss P. (1991) Separate elements cause lineage restriction and specify boundaries of *Hox-1.1* expression. *Development* **112:** 279–287.

Quatrano R.S. and Shaw S.L. (1997) Role of the cell wall in the determination of cell polarity and the plane of cell division in *Fucus* embryos. *Trends Plant Sci.* **2:** 15–21.

Raff J.W. and Glover D.M. (1989) Centrosomes, and not nuclei, initiate pole cell formation in *Drosophila* embryos. *Cell* **57:** 611–619.

Raff M.C., Barres B.A., Burne J.F., Coles H.S., Ishizaki Y. and Jacobson M.D. (1993) Programmed cell death and the control of cell survival: Lessons from the nervous system. *Science* **262:** 695–700.

Raff R.A. and Kaufman T.C. (1983) *Embryos, Genes, and Evolution: The Developmental-Genetic Basis of Evolutionary Change.* New York: Macmillan.

Raff R.A. and Wray G.A. (1989) Heterochrony: Developmental mechanisms and results. *J. Evol. Biol.* **2:** 409–434.

Raganayaku G., Zhao B., Dokidis A., Molkentin J.D., Olson E.N. and Schulz R.A. (1995) A series of mutations in the D-MEF2 transcription factor reveals multiple functions in larval and adult myogenesis in *Drosophila. Dev. Biol.* **171:** 169–181.

Raisman G. and Field P.M. (1973) Sexual dimorphism in the neuropil of the preoptic area of the rat and its dependence on neonatal androgen. *Brain Res.* **54:** 1–20.

Ralt D., Goldenberg M., Fetterolf P., Thompson D., Dor J., Mashiach S., Garbers D.L. and Eisenbach M. (1991) Sperm attraction to a follicular factor(s) correlates with human egg fertilizability. *Proc. Nat. Acad. Sci. USA* **88:** 2840–2844.

Ramírez-Weber F.A. and Kornberg T. (1999) Cytonemes: Cellular processes that project to the principal signaling center in *Drosophila* imaginal discs. *Cell* **97:** 599–607.

Rancourt D.E., Tsuzuki T. and Capecchi M.R. (1995) Genetic interaction between hoxb-5 and hoxb-6 is revealed by nonallelic noncomplementation. *Genes and Dev.* **9:** 108–122.

Randerath K., Randerath E. and Filburn C. (1996) Genomic and mitochondrial DNA alterations with Aging. In: *Handbook of the Biology of Aging,* E.L. Schneider and J.W. Rowe (eds.), 198–214, San Diego: Academic Press.

Ransick A. and Davidson E.H. (1993) A complete second gut induced by transplanted micromeres in the sea urchin embryo. *Science* **259:** 1134–1138.

Rappaport R. (1967) Cell division: Direct measurement of maximum tension exerted by furrow of echinoderm eggs. *Science* **156:** 1241–1243.

Rappaport R. (1974) Cleavage. In: *Concepts of Development,* J. Lash and J. Whittaker (eds.), 76–98. Stamford, CT: Sinauer Associates.

Rappaport R. and Rappaport B.N. (1994) Cleavage in conical sand dollar eggs. *Dev. Biol.* **164:** 258–266.

Ratcliffe O.J., Amaya I., Vincent C.A., Rothstein S., Carpenter R., Coen E. and Bradley D.J. (1998) Separation of floral and shoot identity in *Arabidopsis. Development* **125:** 1609–1615.

Ratcliffe O.J., Bradley D.J. and Coen E. (1999) Separation of floral and shoot identity in *Arabidopsis. Development* **126:** 1109–1120.

Rau K.-G. and Kalthoff K. (1980) Complete reversal of anteroposterior polarity in a centrifuged insect embryo. *Nature* **287:** 635–637.

Ray R.P. and Schüpbach T. (1996) Intercellular signaling and the polarization of body axes during *Drosophila* oogenesis. *Genes and Dev.* **10:** 1711–1723.

Ray R.P., Arora K., Nüsslein-Volhard C. and Gelbart W.M. (1991) The control of cell fate along the dorsal-ventral axis of the *Drosophila* embryo. *Development* **113:** 35–54.

Razin A. and Shemer R. (1995) DNA methylation in early development. *Hum. Mol. Genet.* **4:** 1751–1755.

Rebagliati M.R., Weeks D.L., Harvey R.P. and Melton D.A. (1985) Identification and cloning of localized maternal RNAs from *Xenopus* eggs. *Cell* **42:** 769–777.

Rebagliati M.R., Toyama R., Fricke C., Haffter P. and Dawid I.B. (1998) Zebrafish nodal-related genes are implicated in axial patterning and establishing left-right asymmetry. *Dev. Biol.* **199:** 261–272.

Reddy E.P., Reynolds R.K., Santos E. and Barbacid M. (1982) A point mutation is responsible for the acquisition of transforming properties by the T24 human bladder cancer oncogene. *Nature* **300:** 149–152.

Reedy M.V., Faraco C.D. and Erickson C.A. (1998) The delayed entry of thoracic neural crest cells into the dorsolateral path is a consequence of the late emigration of melanogenic neural crest cells from the neural tube. *Dev. Biol.* **200:** 234–246.

Reeves O.R. and Laskey R.A. (1975) *In vitro* differentiation of a homogeneous cell population—the epidermis of *Xenopus laevis. J. Embryol. exp. Morphol.* **34:** 75–92.

Regulski M., Dessain S., McGinnis N. and McGinnis W. (1991) High-affinity binding sites for the Deformed protein are required for the function of an autoregulatory enhancer of the *Deformed* gene. *Genes and Dev.* **5:** 278–286.

Reichsman F, Smith L. and Cumberledge S. (1996) Glycosaminoglycans can modulate extracellular localization of the wingless protein and promote signal transduction. *J. Cell Biol.* **135:** 819–827.

Render J. (1983) The second polar lobe of the *Sabellaria cementarium* embryo plays an inhibitory role in apical tuft formation. *Wilhelm Roux's Arch.* **192:** 120–129.

Rennick D., Yang G., Muller-Sieburg C., Smith C., Arai N., Takabe Y. and Gennell L. (1987) Interleukin 4 (B-cell stimulatory factor 1) can enhance or antagonize the factor-dependent growth of hemopoietic progenitor cells. *Proc. Nat. Acad. Sci. USA* **84:** 6889–6893.

Renshaw M.W., Ren X.D. and Schwartz M.A. (1997) Growth factor activation of MAP kinase requires cell adhesion. *EMBO J.* **16:** 5592–5599.

Renshaw M.W., Price L.S. and Schwartz M.A. (1999) Focal adhesion kinase mediates the integrin signaling requirement for growth factor activation of MAP kinase. *J. Cell Biol.* **147:** 611–618.

Reshef R., Maroto M. and Lassar A. (1998) Regulation of dorsal somitic fates: BMPs and noggin control the timing and pattern of myogenic regulator expression. *Genes and Dev.* **12:** 290–303.

Resing K.A., Mansour S.J., Hermann A.S., Johnson R.S., Candia J.M., Fukasawa K., Vande Woude G.F. and Ahn N.G. (1995) Determination of v-Mos-catalyzed phosphorylation sites and autophosphorylation sites on MAP kinase kinase by ESI/MS. *Biochemistry* **34:** 2610–2620.

Reuss C. and Saunders J.W., Jr. (1965) Inductive and axial properties of prospective limb mesoderm in the early chick embryo. *Am. Zool.* **5:** 214.

Reyer R.W. (1954) Regeneration of the lens in the amphibian eye. *Quart. Rev. Biol.* **29:** 1–46.

Rhind N.R., Miller L.M., Kopczynski J.B. and Meyer B.J. (1995) xol-1 acts as an early switch in the *C. elegans* male/hermaphrodite decision. *Cell* **80:** 71–82.

Rhyu M.S. and Knoblich J.A. (1995) Spindle orientation and asymmetric cell fate. *Cell* **82:** 523–526.

Ribbert D. (1979) Chromomeres and puffing in experimentally induced polytene chromosomes of *Calliphora erythrocephala. Chromosoma* **74:** 269–298.

Rice W.R. (1994) Degeneration of a nonrecombining chromosome. *Science* **263:** 230–232.

Richardson A. and Pahlavani M.A. (1994) Thoughts on the evolutionary basis of dietary restriction. In: *Genetics and Evolution of Aging*. M.R. Rose and C.E. Finch (eds.), 226–231, Dordrecht, The Netherlands: Kluwer.

Richardson M.K. (1995) Heterochrony and the phylotypic period. *Dev. Biol.* **172:** 412–421.

Richter J.D. and Smith L.D. (1984) Reversible inhibition of translation by *Xenopus* oocyte-specific proteins. *Nature* **309:** 378–380.

Riddihough G. and Ish-Horowicz D. (1991) Individual stripe regulatory elements in the *Drosophila hairy* promoter respond to maternal, gap, and pair-rule genes. *Genes and Dev.* **5:** 840–854.

Riddle R.D., Johnson R.L., Laufer E. and Tabin C. (1993) *Sonic hedgehog* mediates the polarizing activity of the ZPA. *Cell* **75:** 1401–1416.

Riddle R.D., Ensini M., Nelson C., Tsuchida T., Jessell T.M. and Tabin C. (1995) Induction of the LIM homeobox gene Lmx1 by WNT7a establishes dorsoventral pattern in the vertebrate limb. *Cell* **83:** 631–640.

Riechmann J.L. and Meyerowitz E.M. (1997) Determination of floral organ identity by *Arabidopsis* MADS domain homeotic proteins AP1, AP3, PI, and AG is independent of their DNA-binding specificity. *Mol. Biol. Cell* **8:** 1243–1259.

Riechmann J.L., Krizek B.A. and Meyerowitz E.M. (1996) Dimerization specificity of *Arabidopsis* MADS protein homeotic proteins apetala1, apetala3, pistillata, and agamous. *Proc. Nat. Acad. Sci. USA* **93:** 4793–4798.

Rieger F., Grument M. and Edelman G.M. (1985) N-CAM at the vertebrate neuromuscular junction. *J. Cell Biol.* **101:** 285–293.

Riek R., Hornemann S., Wider G., BilleterM., Glockshuber R. and Wüthrich K. (1996) NMR structure of the mouse prion protein domain PrP(121–231). *Nature* **382:** 180–182.

Rijli F.M., Mark M., Lakkaraju S., Dierich A., Dollé P. and Chambon P. (1993) A homeotic transformation is generated in the rostral branchial region of the head by disruption of *Hoxa-2*, which acts as a selector gene. *Cell* **75:** 1333–1349.

Rijsewijk R., Schuermann M., Wagenaar E., Parren P., Weigel D. and Nusse R. (1987) The *Drosophila* homolog of the mouse mammary oncogene *int-1* is identical to the segment polarity gene *wingless*. *Cell* **50:** 649–657.

Risau W. (1997) Mechanisms of angiogenesis. *Nature* **386:** 671–674.

Ritter W. (1976) Fragmentierungs und Bestrahlungsversuche am Ei von *Smittia* spec. (Diptera, Chironomidae). Master's thesis, Albert Ludwigs Universität, Freiburg.

Robb L. and Begley C.G. (1997) The SCL/TAL1 gene: Roles in normal and malignant hematopoiesis. *BioEssays* **19:** 607–613.

Robbie E.P., Peterson M., Amaya E. and Musci T.J. (1995) Temporal regulation of the *Xenopus* FGF receptor in development: A translation inhibitory element in the 3' untranslated region. *Development* **121:** 1775–1785.

Roberts A.B., Anzano M.A., Wakefield L.M., Roche N.S., Stern D.F. and Sporn M.B. (1985) Type b transforming growth factor: A bifunctional regulator of cellular growth. *Proc. Nat. Acad. Sci. USA* **82:** 119–123.

Roberts R., Gallagher J., Spooncer E., Allen T.D., Bloomfield F. and Dexter T.M. (1988) Heparan sulphate bound growth factors: A mechanism for stromal cell mediated haemopoiesis. *Nature* **332:** 376–378.

Robertson K. and Mason I. (1997) The GDNF-RET signalling partnership. *Trends Genet.* **13:** 1–3.

Robertson S.E., Dockendorff T.C., Leatherman J.L., Faulkner D.L. and Jongens T.A. (1999) *germ cell-less* is required only during the establishment of the germ cell lineage of *Drosophila* and has activities which are dependent and independent of its localization to the nuclear envelope. *Dev. Biol.* **215:** 288–297.

Robinow S., Talbot W.S., Hogness D.S. and Truman J.W. (1993) Programmed cell death in the *Drosophila* DNS is ecdysone-regulated and coupled with a specific ecdysone receptor isoform. *Development* **119:** 1251–1259.

Robinow S., Draizen T.A., Truman J.W. (1997) Genes that induce apoptosis: transcriptional regulation in identified, doomed neurons of the *Drosophila* CNS. *Dev. Biol.* **190:** 206–213.

Robinson K.R. and Jaffe L.F. (1975) Polarizing fucoid eggs drive a calcium current through themselves. *Science* **187:** 70–72.

Robinson W.P. (2000) Mechanisms leading to uniparental disomy and their clinical consequences. *BioEssays* **22:** 452–459.

Rodriguez I. and Basler K. (1997) Control of compartmental affinity boundaries by Hedgehog. *Nature* **389:** 614–618.

Rodriguez-Boulan E. and Nelson W.J. (1989) Morphogenesis of the polarized epithelial cell phenotype. Science **245:** 718–725.

Rodriguez-Esteban C., Schwabe J.W., De La Pena J., Foys B., Eshelman B. and Belmonte J.C. (1997) Radical fringe positions the apical ectodermal ridge at the dorsoventral boundary of the vertebrate limb. *Nature* **386:** 360–366.

Roelink H., Augsburger A., Heemskerk J., Korzh V., Norlin S., Ruiz i Altaba A., Tanabe Y., Placzek M., Edlund T., Jessell T.M. and Dodd J. (1994) Floor plate and motor neuron induction by *vhh-1,* a vertebrate homolog of *hedgehog* expressed by the notochord. Cell **76:** 761–775.

Roelink H., Porter J.A., Chiang C., Tanabe Y., Chang D.T., Beachy P.A. and Jessel T.M. (1995) Floor plate and motor neuron induction by different concentrations of the aminoterminal cleavage product of sonic hedgehog autoproteolysis. *Cell* **81:** 445–455.

Rogers S., Wells R. and Rechsteiner M. (1986) Amino acid sequences common to rapidly degraded proteins: The PEST hypothesis. *Science* **234:** 364–368.

Romanes G.J. (1901) *Darwin and after Darwin.* London: Open Court Publishing.

Romer A.S. (1976) *The Vertebrate Body.* 5th ed. Philadelphia: Saunders.

Ronchi E., Treisman J., Dostatni N., Struhl G. and Desplan C. (1993) Down-regulation of the *Drosophila* morphogen bicoid by the torso receptor-mediated signal transduction cascade. *Cell* **74:** 347–355.

Rong P.M., Teillet M.-A., Ziller C. and Le Douarin N.M. (1992) The neural tube/notochord complex is necessary for vertebral but not limb and body wall striated muscle differentiation. *Development* **115:** 657–672.

Rongo C., Gavis E.R. and Lehmann R. (1995) Lokalization of *oskar* RNA regulates *oskar* translation and requires Oskar protein. *Development* **121:** 2737–2746.

Rorvik D.M. (1978) *In His Image: The Cloning of a Man.* Philadelphia: Lippincott.

Rose M.R. (1991) *Evolutionary Biology of Aging.* New York: University Press.

Rose M.R. and Charlesworth B. (1981) Genetics of life history in *Drosophila melanogaster.* II. Exploratory selection experiments. *Genetics* **97**: 187–96.

Rosenfeld J.M., Daley J.D., Ohno S. and YoungLai E.V. (1977) Central aromatization of testosterone in testicular feminized mice. *Experientia* **33:** 1392–1393.

Rosenquist G.C. (1966) An autoradiographic study of labelled grafts in the chick blastoderm. Development from primitive-streak stages to stage 12. *Contr. Embryol.* **38:** 71–110.

Rosenthal E.T., Hunt T. and Ruderman J.V. (1980) Selective translation of mRNA controls the pattern of protein synthesis during early development of the surf clam, *Spisula solidissima. Cell* **20:** 487–494.

Rosenthal E.T., Tansey T.R. and Ruderman J.V. (1983) Sequence-specific adenylations and deadenylations accompany changes in the translation of maternal messenger RNA after fertilization of *Spisula* oocytes. *J. Mol. Biol.* **166:** 309–327.

Ross J. (1989) The turnover of messenger RNA. *Scientific American* **260** (April): 48–55.

Rossant J. and Hopkins N. (1992) Of fin and fur: Mutational analysis of vertebrate embryonic development. *Genes and Dev.* **6:** 1–13.

Rossant J. and Lis W.T. (1979) Potential of isolated mouse inner cell masses to form trophectoderm derivatives *in vivo. Dev. Biol.* **70:** 255–261.

Rossel M. and Capecchi M.R. (1999). Mice mutant for both Hoxa1 and Hoxb1 show extensive remodeling of the hindbrain and defects in craniofacial development. *Development* **126:** 5027–5040.

Roth S. (1993) Mechanisms of dorsal-ventral axis determination in *Drosophila* embryos revealed by cytoplasmic transplantations. *Development* **117:** 1385–1396.

Roth S., Neumann-Silberberg F.S., Barcelo G. and Schüpbach T. (1995) cornichon and the EGF receptor signaling process are necessary for both anterior-posterior and dorsal-ventral pattern formation in *Drosophila. Cell* **81:** 967–978.

Roth S., Jordan P. and Karess R. (1999) Binuclear *Drosophila* oocytes: Consequences and implications for dorsal-ventral patterning in oogenesis and embryogenesis. *Development* **126:** 927–934.

Rothenberg E.V., Telfer J.C. and Anderson M.K. (1999) Transcriptional regulation of lymphocyte lineage commitment. *BioEssays* **21:** 726–742.

Rothman J.E. (1994) Mechanisms of intracellular protein transport. *Nature* **372:** 55–63.

Roux W. (1885) Beiträge zur Entwicklungsmechanik des Embryo. *Zeitschrift für Biologie* **21:** 411–524.

Rowning B.A., Wells J., Wu M., Gerhart J.C., Moon R.T. and Larabell C.A. (1997) Microtubule-mediated transport of organelles and localization of b-catenin to the future dorsal side of *Xenopus* eggs. *Proc. Nat. Acad. Sci. USA* **94:** 1224–1229.

Rubin G.M. and Spradling A.C. (1982) Genetic transformation of *Drosophila* with transposable element vectors. *Science* **218:** 348–353.

Rudnicki M.A., Braun T., Hinuma S. and Jaenisch R. (1992) Inactivation of *MyoD* in mice leads to up-regulation of the myogenic HLH gene *Myf-5* and results in apparently normal muscle development. *Cell* **71:** 383–390.

Rudnicki M.A., Schnegelsberg P.N.J., Stead R.H., Braun T., Arnold H.-H. and Jaenisch R. (1993) MyoD or *Myf-5* is required for the formation of skeletal muscle. *Cell* **75:** 1351–1359.

Rudolph K.L., Chang S., Lee H.W., Blasco M., Gottlieb G.J., Greider C. and DePinho R.A. (1999) Longevity, stress response, and cancer in aging telomerase-deficient mice. *Cell* **96:** 701–712.

Ruffins S.W. and Ettensohn C.A. (1993) A clonal analysis of secondary mesenchyme cell fates in the sea urchin embryo. *Dev. Biol.* **160:** 285–288.

Ruffins S.W. and Ettensohn C.A. (1996)A fate map of the vegetal plate of the sea urchin (*Lytechinus variegatus*) mesenchyme blastula. *Development* **122:** 253–263.

Ruiz i Altaba A. (1997) Catching a Gli-mpse of Hedgehog. *Cell* **90:** 193–196.

Runft L.L., Watras J. and Jaffe L.A. (1999) Calcium release at fertilization of *Xenopus* eggs requires type I IP(3) receptors, but not SH2 domain-mediated activation of PLC γ or G(q)-mediated activation of PLCβ. *Dev. Biol.* **214:** 399–411.

Ruoslahti E. (1991) Integrins as receptors for extracellular matrix. In: *Cell Biology of Extracellular Matrix,* 2nd ed., E.D. Hay (ed.), 343–363. New York: Plenum.

Rusch J. and Levine M. (1996) Threshold responses to the dorsal regulatory gradient and the subdivision of primary tissue territories in the *Drosophila* embryo. *Curr. Opin. Genet. Dev.* **6:** 416–423.

Rutishauser U., Acheson A., Hall A.K., Mann D.M. and Sunshine J. (1988) The neural cell adhesion molecule (NCAM) as a regulator of cell-cell interactions. *Science* **240:** 53–57.

Ruvinsky I., Oates A.C., Silver L.M. and Ho R.K. (2000) The evolution of paired appendages in vertebrates: T-box genes in the zebrafish. *Dev. Genes Evol.* **210:** 82–91.

Ruvkun G. and Giusto J. (1989) The *Caenorhabditis elegans* heterochronic gene *lin-14* encodes a nuclear protein that forms a temporal developmental switch. *Nature* **338:** 313–319.

Ruvkun G., Wightman B., Bürglin T. and Arasu P. (1991) Dominant gain-of-function mutations that lead to misregulation of the *C. elegans* hetero-chronic gene *lin-14,* and the evolutionary implications of dominant mutations in pattern-formation genes. *Development* (Supplement 1): 47–54.

Ryan A.K., Blumberg B., Rodriguez-Esteban C., Yonei-Tamura S., Tamura K., Tsukui T., de la Pena J., Sabbagh W., Greenwald J., Choe S., Norris D.P., Robertson E.J., Evans R.M., Rosenfeld M.G. and Izpisua Belmonte J.C. (1998) Pitx2 determines left-right asymmetry of internal organs in vertebrates. *Nature* **394:** 545–551.

Sachs A.B. (1993) Messenger RNA degradation in eukaryotes. *Cell* **74:** 413–421.

Sachs A.B., Sarnow P. and Hentze M.W. (1997) Starting at the beginning, middle, and end: Translation initiation in eukaryotes. *Cell* **89:** 831–838.

Sadhu K., Reed S.I., Richardson H. and Russell P. (1990) Human homolog of fission yeast cdc25 mitotic inducer is predominantly expressed in G2. *Proc. Nat. Acad. Sci. USA* **87:** 5139–5143.

Sagai M. and Ichinose T. (1987) Lipid peroxidation and antioxidative protection mechanism in rat lungs upon acute and chronic to nitrogen dioxide. *Environ. Health Perspect.* **73:** 179–189.

Sagata N. (1997) What does Mos do in oocytes and somatic cells? *BioEssays* **19:** 13–21.

Sagata N., Watanabe N., Vande Woude G.F. and Ikawa Y. (1989) The c-*mos* proto-oncogene product is a cytostatic factor responsible for meiotic arrest in vertebrate eggs. *Nature* **342:** 512–518.

Saint-Jeannet J.P., He X., Varmus H.E. and Dawid I.B. (1997) Regulation of dorsal fate in the neuraxis by Wnt-1 and Wnt-3a. *Proc. Nat. Acad. Sci. USA* **94:** 13713–13718.

St. Johnston D. (1995) The intracellular localization of messenger RNAs. *Cell* **81:** 161–170.

St. Johnston D., Driever W., Berleth T., Richstein S. and Nüsslein-Volhard C. (1989) Multiple steps in the localization of bicoid RNA to the anterior pole of the *Drosophila* oocyte. *Development* 107 (Supplement): 13–19.

Sakai H., Medrano L.J. and Meyerowitz E.M. (1995) Role of superman in maintaining *Arabidopsis* floral worl boundaries. *Nature* **378:** 199–203.

Sakai M. (1996) The vegetal determinants required for the Spemann organizer move equatorially during the first cell cycle. *Development* **122:** 2207–2214.

Sallés F.J., Lieberfarb M.E., Wreden C., Gergen P.J. and Strickland S. (1994) Coordinated initiation of *Drosophila* development by regulated polyadenylation of maternal messenger RNAs. *Science* **266:** 1996–1999.

Sander K. (1960) Analyse des ooplasmatischen Reaktionssystems von *Euscelis plebejus* Fall. (Cicadina) durch Isolieren und Kombinieren von Keimteilen. II. Mitteilung: Die Differen zierungsleistungen nach Verlagern von Hinterpolmaterial. *Wilhelm Roux's Arch.* **151:** 660–707.

Sander K. (1975) Pattern specification in the insect embryo. In: *Cell Patterning, Ciba Foundation Symp.* 29 (new series): 241–263. Amsterdam: Elsevier, Excerpta Medica, North-Holland Publishing Company, Associated Scientific Publishers.

Sander K. (1976) Specification of the basic body pattern in insect embryogenesis. *Adv. Insect Physiol.* **12:** 125–238.

Sander K. (1983) The evolution of patterning mechanisms: Gleanings from insect embryogenesis and spermatogenesis. In: *Development and Evolution,* B.C. Goodwin, N. Holder and C.C. Wylie (eds.), 123–159. New York: Cambridge University Press.

Sander K. (1990) Von der Keimplasmatheorie zur synergetischen Musterbildung-Einhundert Jahre entwicklungsbiologischer Ideengeschichte. *Verh. Dtsch. Zool. Ges.* **83:** 133–177.

Sardet C., Speksnijder J., Terasaki M. and Chang P. (1992) Polarity of the ascidian egg cortex before fertilization. *Development* **115:** 221–237.

Sasai Y., Lu B., Steinbeisser H., Geissert D., Gont L.K. and De Robertis E.M. (1994) *Xenopus chordin:* A novel dorsalizing factor activated by organizer-specific homeobox genes. *Cell* **79:** 779–790.

Sater A.K. and Jacobson A.G. (1990a) The role of the dorsal lip in the induction of heart mesoderm in *Xenopus laevis. Development* **108:** 461–470.

Sater A.K. and Jacobson A.G. (1990b) Restriction of the heart morphogenetic field in *Xenopus laevis. Dev. Biol.* **140:** 328–336.

Satoh N. (1994) *Developmental Biology of Ascidians.* New York: Cambridge University Press.

Satoh N. and Jeffery W.R. (1994) Chasing tails in ascidians: Developmental insights into the origin and evolution of chordates. *Trends Genet.* **11:** 354–359.

Sauer F., Hansen S.K. and Tijan R. (1995a) Multiple TAF_{II}s directing synergistic activation of transcription. *Science* **270:** 1783–1788.

Sauer F., Hansen S.K. and Tijan R. (1995b) DNA template and activator-coactivator requirements for transcriptional synergism by *Drosophila* bicoid. *Science* **270:** 1825–1828.

Sauer F., Wassarman D.A., Rubin G.M. and Tijan R. (1996) TAF_{II}s mediate activation of transcription in the *Drosophila* embryo. *Cell* **87:** 1271–1284.

Saunders J.W. Jr. (1948) The proximo-distal sequence of origin of the parts of the chick wing and the role of the ectoderm. *J. Exp. Zool.* **108:** 363–403.

Saunders J.W., Jr. (1970) *Patterns and Principles of Animal Development.* New York: Macmillan.

Saunders J.W., Jr. (1982) *Developmental Biology: Patterns, Problems, and Principles.* New York: Macmillan.

Saunders J.W. Jr. and Gasseling M.T. (1968) Ectoderm-mesenchymal interactions in the origin of wing symmetry. In: *Epithelial-Mesenchymal Interactions,* R. Fleischmajer and R.E. Billingham (eds.), 78–97. Baltimore: Williams and Wilkins.

Savage R. and Phillips C.R. (1989) Signals from the dorsal blastopore lip region during gastrulation bias the ectoderm toward a nonepidermal pathway of differentiation in *Xenopus laevis. Dev. Biol.* **133:** 157–168.

Savagner P., Yamada K.M. and Thiery J.P. (1997) The zinc-finger protein slug causes desmosome dissociation, an initial and necessary step for growth factor-induced epithelial-mesenchymal transition. *J. Cell Biol.* **137:** 1403–1419.

Savant-Bhonsale S. and Montell D.J. (1993) *torso-like* encodes the localized determinant of *Drosophila* terminal pattern formation. *Genes and Dev.* **7:** 2548–2555.

Sawada T. and Schatten G. (1988) Microtubules in ascidian eggs during meiosis, fertilization, and mitosis. *Cell Motility and the Cytoskeleton* **9:** 219–230.

Sawyers C.L., Denny C.T. and Witte O.N. (1991) Leukemia and the disruption of normal hemato-poiesis. *Cell* **64:** 337–350.

Saxén L. (1987) *Organogenesis of the Kidney.* Cambridge: Cambridge University Press.

Saxén L. and Toivonen S. (1962) *Primary Embryonic Induction.* London: Logos Press and Academic Press.

Schäfer M., Börsch D., Hülster A. and Schäfer U. (1993) Expression of a gene duplication encoding conserved sperm tail proteins is translationally regulated in *Drosophila melanogaster. Mol. Cell Biol.* **13:** 1708–1718.

Schäfer M., Nayerna K., Engel W. and Schäfer U. (1995) Translational control in spermatogenesis. *Dev. Biol.* **172:** 344–352.

Schatten G. (1994) The centrosome and its mode of inheritance: The reduction of the centrosome during gametogenesis and its restoration during fertilization. *Dev. Biol.* **165:** 299–335.

Schatten H. and Schatten G., eds. (1989a) *The Cell Biology of Fertilization.* San Diego: Academic Press.

Schatten H. and Schatten G., eds. (1989b) *The Molecular Biology of Fertilization.* San Diego: Academic Press.

Scheer U., Dabauvalle M.-C., Merkert H. and Denavente R. (1988) The nuclear envelope and the organization of the pore complexes. *Cell Biol. Int. Rep.* **12:** 669–689.

Schierenberg E. (1986) Developmental strategies during early embryogenesis of *Caenorhabditis elegans. J. Embryol. exp. Morphol.* 97 (Supplement): 31–44.

Schlesinger A., Shelton C.A., Maloof J.N., Meneghini M. and Bowerman B. (1999) Wnt pathway components orient a mitotic spindle in the early *Caenorhabditis elegans* embryo without requiring gene transcription in the responding cell. *Genes and Dev.* **13:** 2028–2038.

Schlessinger J. (1997) Direct binding and activation of receptor tyrosine kinases by collagen. *Cell* **91:** 869–872.

Schmidt J., Francois V., Bier E. and Kimelman D. (1995a) *Drosophila short gastrulation* induces an ectopic axis in *Xenopus:* evidence for conserved mechanisms of dorsal-ventral patterning. *Development* **121:** 4319–4328.

Schmidt J.E., Suzuki A., Ueno N. and Kimelman D. (1995) Localized BMP-4 mediates dorsal/ventral patterning in the early *Xenopus* embryo. *Dev. Biol.* **169:** 37–50.

Schneider D.S., Hudson K.L., Lin T.-Y. and Anderson K.V. (1991) Dominant and recessive mutations define functional domains of *Toll,* a transmembrane protein required for dorsal-ventral polarity in the *Drosophila* embryo. *Genes and Dev.* **5:** 797–807.

Schneider D.S., Jin Y., Morisato D. and Anderson K.V. (1994) A processed form of Spätzle protein defines dorsal-ventral polarity in the *Drosophila* embryo. *Development* **120:** 1243–1250.

Schneider S., Herrenknecht K., Butz S., Kemler R. and Hausen P. (1993) Catenins in *Xenopus* embryogenesis and their relation to the cadherin-mediated cell-cell adhesion system. *Development* **118:** 629–640.

Schneider S., Steinbeisser H., Warga R.M. and Hausen P. (1996) β-catenin translocation into nuclei demarcates the dorsalizing centers in frog and fish embryos. *Mech. Dev.* **57:** 191–198.

Schneuwly S., Klemenz R. and Gehring W.J. (1987) Redesigning the body plan of *Drosophila* by ectopic expression of the homoeotic gene *Antennapedia. Nature* **325:** 816–818.

Schnieke A.E., Kind A.J., Ritchie W.A., Mycock K., Scott A.R., Ritchie M., Wilmut I., Colman A. and Campbell K.H. (1997) Human factor IX transgenic sheep produced by transfer of nuclei from transfected fetal fibroblasts. *Science* **278:** 2130–2133.

Schoenle E., Zapf J., Humbel R.E. and Froesch E.R. (1982) Insulin-like growth factor I stimulates growth in hypophysectomized rats. *Nature* **296:** 252–253.

Schoenwolf G.C. (1991) Cell movements driving neurulation in avian embryos. *Development* (Supplement 2): 157–168.

Schoenwolf G.C. (1997) Reptiles and birds. In: *Embryology: Constructing the Organism.* S.F. Gilbert and A.M. Raunio (eds.), 437–458. Sunderland, MA: Sinauer.

Schoenwolf G.C. and Smith J.L. (1990) Mechanisms of neurulation: Traditional viewpoint and recent advances. *Development* **109:** 243–270.

Schröder C., Tautz D., Seifert E. and Jäckle H. (1988) Differential regulation of the two transcripts from the *Drosophila* gap segmentation gene hunchback. *EMBO J.* **7:** 2881–2887.

Schröder R. and Sander K. (1993) A comparison of transplantable *bicoid* activity and partial *bicoid* homeobox sequences in several *Drosophila* and blowfly species. *Roux's Arch. Dev. Biol.* **203:** 34–43.

Schroeder T.E. (1972) The contractile ring. II. Determining its brief existence, volumetric changes and vital role in cleaving *Arbacia* eggs. *J. Cell Biol.* **53:** 419–434.

Schroeder T.E. (1981) Development of a "primitive" sea urchin (*Eucidaris tribuloides*): Irregularities in the hyaline layer, micromeres, and primary mesenchyme. *Biol. Bull.* **161:** 141–151.

Schroeder T.E. (1987) Fourth cleavage of sea urchin blastomeres: Microtubule patterns and myosin localization in equal and unequal cell divisions. *Dev. Biol.* **124:** 9–22.

Schubiger M. and Truman J.W. (2000) The RXR ortholog USP suppresses early metamorphic processes in *Drosophila* in the absence of ecdysteroids. *Development* **127:** 1151–1159.

Schüle R., Umesono K., Mangelsdorf D.J., Bolado J., Pike J.W. and Evans R.M. (1990) Jun-Fos and receptors for vitamins A and D recognize a common response element in the human osteocalcin gene. *Cell* **61:** 497–504.

Schüpbach T. (1987) Germ line and soma cooperate during oogenesis to establish the dorsoventral pattern of egg shell and embryo in *Drosophila melanogaster. Cell* **49:** 699–707.

Schüpbach T. and Wieschaus E. (1986a) Maternal-effect mutations altering the anterior-posterior body pattern of the *Drosophila* embryo. *Roux's Arch. Dev. Biol.* **195:** 302–317.

Schüpbach T. and Wieschaus E. (1986b) Germline autonomy of maternal-effect mutations altering the embryonic body pattern of *Drosophila. Dev. Biol.* **113:** 443–448.

Schütt C. and Nöthiger R. (2000) Structure, function and evolution of sex-determining systems in Dipteran insects. *Development* **127:** 667–677.

Schwarz-Sommer Z., Huijser P., Nacken W., Saedler H. and Sommer H. (1990) Genetic control of flower development by homeotic genes in *Antirrhinum majus. Science* **250:** 931–936.

Schweizer G., Ayer-Le Lievre C. and Le Douarin N.M. (1983) Restrictions in developmental capacities in the dorsal root ganglia during the course of development. *Cell Differ.* **13:** 191–200.

Schwind J.L. (1933) Tissue specificity at the time of metamorphosis in frog larvae. *J. Exp. Zool.* **66:** 1–14.

Scott M.P. (1992) Vertebrate homeobox gene nomenclature. *Cell* **71:** 551–553.

Scott M.P. and O'Farrell P.H. (1986) Spatial programming of gene expression in early *Drosophila* embryogenesis. *Ann. Rev. Cell Biol.* **2:** 49–80.

Scott M.P. and Weiner A.J. (1984) Structural relationships among genes that control development: Sequence homology between the *Antennapedia, Ultrabithorax,* and *fushi tarazu* loci of *Drosophila. Proc. Nat. Acad. Sci. USA* **81:** 4115–4119.

Scott M.P., Tamkun J.W. and Hartzell G.W., III (1989) The structure and function of the homeodomain. *Biochem. Biophys. Acta* **989:** 25–48.

Sekine K., Ohuchi H., Fujiwara M., Yamasaki M., Yoshizawa T., Sato T., Yagishita N., Matsui D., Koga Y., Itoh N. and Kato S. (1999) Fgf10 is essential for limb and lung formation. *Nature Genet.* **21:** 138–141.

Selleck M.A.J. and Bronner-Fraser M. (1995) Origins of the avian neural crest: The role of neural plate-epidermal interactions. *Development* **121:** 525–538.

Selleck M.A.J. and Stern C.D. (1991) Fate mapping and cell lineage analysis of Hensen's node in the chick embryo. *Development* **112:** 615–626.

Selleck M.A.J., Scherson T.Y. and Bronner-Fraser M. (1993) Origins of neural crest cell diversity. *Dev. Biol.* **159:** 1–11.

Sen J., Goltz J.S., Stevens L. and Stein D. (1998) Spatially restricted expression of *pipe* in the *Drosophila* egg chamber defines embryonic dorsal-ventral polarity. *Cell* **95:** 471–481.

Service P.M., Hutchinson E.W. and Rose M.R. (1988) Multiple genetic mechanisms for the evolution of senescence in *Drosophila melanogaster. Evolution* **42:** 708–716.

Sessions S.K. and Ruth S.B. (1990) Explanation for naturally occurring supernumerary limbs in amphibians. *J. Exp. Zool.* **254:** 38–47.

Sette C., Bevilacqua A., Bianchini A., Mangia F., Geremia R. and Rossi P. (1997) Parthenogenetic activation of mouse eggs by microinjection of a truncated c-kit tyrosine kinase present in spermatozoa. *Development* **124:** 2267–2274.

Seydoux G. and Greenwald I. (1989) Cell autonomy of *lin-12* function in a cell fate decision in *C. elegans. Cell* **57:** 1237–1245.

Seydoux G. and Strome S. (1999) Launching the germ line in *Caenorhabditis elegans:* Regulation of gene expression in early germ cells. *Development* **126:** 3275–3283.

Sgrò C.M. and Partridge L. (1999) A delayed wave of death from reproduction in *Drosophila. Science* **286:** 2521–2524.

Shah N.M., Groves A.K. and Anderson D.J. (1996) Alternative neural crest cell fates are instructively promoted by TGFβ superfamily members. *Cell* **85:** 331–343.

Shaham S. and Horvitz H.R. (1996) Developing *Caenorhabditis elegans* neurons may contain both cell-death protective and killer activities. *Genes Dev.* **10:** 578–591.

Shalaby F., Rossant J., Yamaguchi T.P., Gertsenstein M., Wu, X.-F., Breitman M.L. and Schuh A.C. (1995) Failure of blood island formation and vasculogenesis in Flk-1–deficient mice. *Nature* **376:** 62–66.

Sham M.H., Vesque C., Nonchev S., Marshall H., Frain M., Das Gupta R., Whiting J., Wilkinson D., Charnay P. and Krumlauf R. (1993) The zinc finger gene *Krox20* regulates *HoxB2* (*Hox2.8*) during hindbrain segmentation. *Cell* **72:** 183–196.

Shaper N.L., Hollis G.F., Douglas J.G., Kirsch I.R. and Shaper J.H. (1988) Characterization of the full length cDNA for murine β-1,4-galactosyltransferase. Novel features at the 5'-end predict two translational start sites at two in-frame AUGs. *J. Biol. Chem.* **263:** 10420–10428.

Shapiro D.J., Blume J.E. and Nielsen D.A. (1987) Regulation of mRNA stability in eukaryotic cells. *BioEssays* **6:** 221–226.

Sharon N. and Lis H. (1993) Carbohydrates in cell recognition: Telltale surface sugars enable cells to identify and interact with one another. New drugs aimed at those carbohydrates could stop infection and inflammation. *Scientific American* **268** (January): 82–89.

Sharp D.J., Yu K.R., Sisson J.C., Sullivan W. and Scholey J.M. (1999) Antagonistic microtubule-sliding motors position mitotic centrosomes in *Drosophila* early embryos. *Nat. Cell Biol.* **1:** 51–54.

Sharpe C.R., Fritz A., De Robertis E.M. and Gurdon J.B. (1987) A homeobox-containing marker of posterior neural differentiation shows the importance of predetermination in neural induction. *Cell* **50:** 749–758.

Sharpe R.M. (1993) Experimental evidence for Sertoli cell-germ cell and Sertoli-Leydig cell interactions. In: *The Sertoli Cell,* L.D. Russell and M.D. Griswold (eds.), 391–418. Clearwater, FL: Cache River Press.

Shaw G. and Kamen R. (1986) A conserved AU sequence from the 3' untranslated region of GM-CSF mRNA mediates selective mRNA degradation. *Cell* **46:** 659–667.

Shaw S.L. and Quatrano R.S. (1996) The role of targeted secretion in the establishment of cell polarity and the orientation of the division plane in *Fucus* zygotes. *Development* **122:** 2623–2630.

Sheardown S.A., Duthie S.M., Johnston C.M., Newall A.E.T., Formstone E.J., Arkell R.M., Nesterova T.B., Alghisi G.-C., Rastan S. and Brockdorff N. (1997) Stabilization of Xist RNA mediates initiation of X chromosome inactivation. *Cell* **91:** 99–107(6).

Sheets M.D., Fox C.A., Hunt T., Vande Woude G. and Wickens M. (1994) The 3'-untranslated regions of c-*mos* and cyclin mRNAs stimulate translation by regulating cytoplasmic polyadenylation. *Genes and Dev.* **8:** 926–938.

Sheets M.D., Wu M. and Wickens M. (1995) Polyadenylation of c-mos mRNA as a control point in *Xenopus* meiotic maturation. *Nature* **374:** 511–516.

Shepard J.F. (1982) The regeneration of potato plants from leaf-cell protoplasts. *Scientific American* **246** (May): 154–165.

Shepherd G.M. (1988) *Neurobiology.* 2nd ed. New York: Oxford University Press.

Shermoen A.W. and O'Farrell P.H. (1991) Progression of the cell cycle through mitosis leads to abortion of nascent transcripts. *Cell* **67:** 303–310.

Sherr C.J. (1996) Cancer cell cycles. *Science* **274:** 1672–1677.

Shi D.-L., Darribère T., Johnson K.E. and Boucaut J.-C. (1989) Initiation of mesodermal cell migration and spreading relative to gastrulation in the urodele amphibian *Pleurodeles waltl. Development* **105:** 351–363.

Shi X., Lu L., Qiu Z., He W. and Frankel J. (1991) Microsurgically generated discontinuities provoke heritable changes in cellular handedness of a ciliate, *Stylonychia mytilus. Development* **111:** 337–356.

Shiels P.G., Kind A.J., Campbell K.H., Waddington D., Wilmut I., Colman A. and Schnieke A.E. (1999) Analysis of telomere lengths in cloned sheep. *Nature* **399:** 316–317.

Shih J. and Fraser S.E. (1995) Distribution of tissue progenitors within the shield region of the zebrafish gastrula. *Development* **121:** 2755–2765.

Shih J. and Keller R. (1992) Patterns of cell motility in the organizer and dorsal mesoderm of *Xenopus laevis*. *Development* **116:** 915–930.

Shilling F.M., Carroll D.J., Muslin A.J., Escobodo J.A., Williams L.T. and Jaffe L.A. (1994) Evidence for both tyrosine kinase and G-protein-coupled pathways leading to starfish egg activation. *Dev. Biol.* **162:** 590–599.

Shilling F.M., Krätzschmar J., Cai H., Weskamp G., Gayko U., Leibow J., Myles D.G., Nuccitelli R., and Blobel C.P. (1997) Identification of metalloprotease/disintegrins in *Xenopus laevis* testis with a potential role in fertilization. *Dev. Biol.* **186:** 155–164.

Shilling F.M., Magie C.R., and Nuccitelli R. (1998) Voltage-dependent activation of frog eggs by a sperm surface disintegrin peptide. *Dev. Biol.* **202:** 113–124.

Shimell M.J., Simon J., Bender W. and O'Connor M.B. (1994) Enhancer point mutation results in a homeotic transformation in *Drosophila*. *Science* **264:** 968–971.

Shin E.K., Shin A., Paulding C., Schaffhausen B. and Yee A.S. (1995) Multiple changes in E2F function and regulation occur upon muscle differentiation. *Mol. Cell Biol.* **15:** 2252–2262.

Shirakawa H. and Miyazaki S. (1999) Spatiotemporal characterization of intracellular Ca^{2+} rise in the acrosome reaction of mammalian spermatozoa induced by zona pellucida. *Dev. Biol.* **208:** 70–78.

Shivdasani R.A., Mayer E.L. and Orkin S.A. (1995) Absence of blood formation in mice lacking the T-cell leukaemia oncoprotein tal-1/SCL. *Nature* **373:** 432–434.

Shulman K. (1974) Defects of the closure of the neural plate. In: *Neurology of Infancy and Childhood*, S. Carter and A.P. Gold (eds.), 20–31. New York: Appleton-Century-Crofts.

Shur B.D. (1989) Galactosyltransferase as a recognition molecule during fertilization and development. In: *The Molecular Biology of Fertilization*, H. Schatten and G. Schatten (eds.), 37–71. San Diego: Academic Press.

Shur B.D. (1991) Cell surface β1,4 galactosyltransferase: Twenty years later. *Glycobiology* **1:** 563–575.

Sieber-Blum M. and Cohen A.M. (1980) Clonal analysis of quail neural crest cells: They are pluripotent and differentiate *in vitro* in the absence of noncrest cells. *Dev. Biol.* **80:** 96–106.

Signer E.N., Dubrova Y.E., Jeffreys A.J., Wilde C., Finch L.M., Wells M., and Peaker M. (1998) DNA fingerprinting Dolly. *Nature* **394:** 329–330.

Silver P.A. (1991) How proteins enter the nucleus. *Cell* **64:** 489–497.

Simcox A.A. and Sang J.H. (1983) When does determination occur in *Drosophila* embryos? *Dev. Biol.* **97:** 212–221.

Simerly P. et al. (1995) The paternal inheritance of the centrosome, the cell's microtubule organizing center, in humans, and the implications for fertility. *Nat. Med.* **1:** 47–52.

Simon M.C., Pevny L., Wiles M.V., Keller G, Constantini F. and Orkin S.A. (1992) Rescue of erythroid development in gene targeted GATA-1 mouse embryonic stem cells. *Nat. Genet.* **1:** 92–98.

Simon M.I., Strathmann M.P. and Gautam N. (1991) Diversity of G proteins in signal transduction. *Science* **252:** 802–808.

Simon R., Ingeño M.I. and Coupland G. (1996) Activation of floral meristem identity genes in *Arabidopsis*. *Nature* **384:** 59–62.

Simpson P. and Morata G. (1981) Differential mitotic rates and patterns of growth in compartments of the *Drosophila* wing. *Dev. Biol.* **85:** 299–308.

Simpson-Brose M., Treisman J. and Desplan C. (1994) Synergy between the hunchback and bicoid morphogens is required for anterior patterning in *Drosophila*. *Cell* **78:** 855–865.

Simske J.S. and Kim S.K. (1995) Sequential signalling during *Caenorhabditis elegans* vulval induction. *Nature* **375:** 142–146.

Sinclair A.H., Berta P., Palmer M.S., Hawkins J.R., Griffiths B.L., Smith M.J., Foster J.W., Frischauf A.-M., Lovell-Badge R. and Goodfellow P.N. (1990) A gene from the human sex-determining region encodes a protein with homology to a conserved DNA-binding motif. *Nature* **346:** 240–244.

Singer M.A., Penton A., Twombly V., Hoffman F.M. and Gelbart W.M. (1997) Signaling through both type 1 dpp receptors is required for anterior-posterior patterning of the entire *Drosophila* wing. *Development* **124:** 79–89.

Sinnot E., Dunn L.C. and Dobzhansky T. (1958) *Principles of Genetics*. 5th ed. New York: McGraw-Hill.

Sive H.L. (1993) The frog princess: A molecular formula for dorsoventral patterning in *Xenopus*. *Genes and Dev.* **7:** 1–12.

Skapek S.X., Rhee J., Spicer D.B. and Lassar A.B. (1995) Inhibition of myogenic differentiation in proliferating myoblasts by cyclin D1–dependent kinase. *Science* **267:** 1022–1024.

Slack J.M.W. (1991) *From Egg to Embryo: Determinative Events in Early Development*. 2nd ed. Cambridge: Cambridge University Press.

Slack J.M.W. and Tannahill D. (1992) Mechanism of anteroposterior axis specification in vertebrates. *Development* **114:** 285–302.

Slack J.M.W., Holland P.W.H. and Graham C.F. (1993) The zootype and the phylotypic stage. *Nature* **361:** 490–492.

Small K.S. and Potter S. (1993) Homeotic transformations and limb defects in Hoxa-11 mutant mice. *Genes and Dev.* **7:** 2318–2328.

Small S., Blair A. and Levine M. (1996) Regulation of two pair-rule stripes by a single enhancer in the *Drosophila* embryo. *Dev. Biol.* **175:** 314–324.

Smibert C.A., Wilson J.E., Kerr K. and Macdonald P.M. (1996) smaug protein represses translation of unlocalized nanos mRNA in the *Drosophila* embryo. *Genes and Dev.* **10:** 2600–2609.

Smith C.L. and DeLotto R. (1994) Ventralizing signal determined by protease activation in *Drosophila* embryogenesis. *Nature* **368:** 548–551.

Smith H.W. (1951) *The Kidney, Structure and Function in Health and Disease*. New York: Oxford University Press.

Smith J.C. (1989) Mesoderm induction and mesoderm-inducing factors in early amphibian development. *Development* **105:** 665–677.

Smith J.C. (1994) *hedgehog*, the floor plate, and the zone of polarizing activity. *Cell* **76:** 193–196.

Smith J.L. and Schoenwolf G.C. (1997) Neurulation: Coming to closure. *Trends Neurosci.* **20:** 510–517.

Smith L.D. (1989) The induction of oocyte maturation: Transmembrane signaling events and regulation of the cell cycle. *Development* **107:** 685–699.

Smith W.C. and Harland R.M. (1992) Expression cloning of noggin, a new dorsalizing factor localized to the Spemann organizer in *Xenopus* embryos. *Cell* **70:** 829–840.

Smith W.C., Knecht A.K., Wu M. and Harland R.M. (1993) Secreted noggin protein mimics the Spemann organizer in dorsalizing *Xenopus* mesoderm. *Nature* **361:** 547–549.

Smith W.C., McKendry R., Ribisi, Jr. S. and Harland R.M. (1995) A *nodal*-related gene defines a physical and functional domain within the Spemann organizer. *Cell* **82:** 37–46.

Snell W.J. and White J.M. (1996) The molecules of mammalian fertilization. *Cell* **85:** 629–637.

Snider W.D. (1994) Functions of the neurotrophins during nervous system development: What the knockouts are teaching us. *Cell* **77:** 627–38.

Sohal R.S. and Weindruch R. (1996) Oxidative stress, caloric restriction, and aging. *Science* **273**: 59–63.

Sokal R.R. (1970) Senescence and genetic load: Evidence from *Tribolium*. *Science* **167**: 1733–1734.

Sokol S. and Melton D.A. (1991) Pre-existent pattern in *Xenopus* animal pole cells revealed by induction with activin. *Nature* **351:** 409–411.

Sokol S.Y. and Melton D.A. (1992) Interaction of Wnt and activin in dorsal mesoderm induction in *Xenopus*. *Dev. Biol.* **154:** 348–355.

Solnica-Krezel L. and Driever W. (1994) Microtubule arrays of the zebrafish yolk cell: Organization and function during epiboly. *Development* **120:** 2443–2455.

Solomon F. (1981) Specification of cell morphology by endogenous determinants. *Cell Biol.* **90:** 547–553.

Sommer R. and Tautz D. (1991) Segmentation gene expression in the housefly *Musca domestica*. *Development* **113:** 419–430.

Sonneborn T.M. (1963) Does preformed cell structure play an essential role in cell heredity? In: *The Nature of Biological Diversity*, J.M. Allen (ed.), 165–221. New York: McGraw-Hill.

Sonneborn T.M. (1970) Gene action in development. *Proc. R. Soc. London B* **176:** 347–366.

Southern E.M. (1975) Detection of specific sequences among DNA fragments separated by gel electrophoresis. *J. Mol. Biol.* **98:** 503–517.

Southworth D. (1996) Gametes and Fertilization in Flowering Plants. *Curr. Top. Dev. Biol.* **34:** 259–279.

Spana E.P. and Doe C.Q. (1995) The prospero transcription factor is asymmetrically localized to the cell cortex during neuroblast mitosis in *Drosophila*. *Development* **121:** 3187–3195.

Spemann H. (1927) Neue Arbeiten über Organisatoren in der tierischen Entwicklung. *Naturwissenschaften* **15:** 946–951.

Spemann H. (1928) Die Entwicklung seitlicher und dorsoventraler Keimhälften bei verzögerter Kernversorgung. *Zeitschr. f. wiss. Zool.* **132:** 105–134.

Spemann H. (1931) Über den Anteil von Implantat und Wirtskeim an der Orientierung und Beschaffenheit der induzierten Embryonalanlage. *Wilhelm Roux's Arch.* **123:** 389–417.

Spemann H. (1938) *Embryonic Development and Induction*. New Haven: Yale University Press. Reprinted 1967. New York: Hafner.

Spemann H. and Mangold H. (1924a) Induction of embryonic primordia by implantation of organizers from a different species. In: *Foundations of Experimental Embryology*, B.H. Willier and J.M. Oppenheimer (eds.), 144–184. New York: Hafner.

Spemann H. and Mangold H. (1924b) Über Induktion von Embryonalanlagen durch Implantation artfremder Organisatoren. *Wilhelm Roux's Arch.* **100:** 599–638.

Spemann H. and Schotté O. (1932) Über xenoplastiche Transplantation als Mittel zur Analyse der embryonalen Induktion. *Naturwissenschaften* **20:** 463–467.

Spillare E.A., Robles A.I., Wang X.W., Shen J.C., Yu C.E., Schellenberg G.D. and Harris CC. (1999) p53–mediated apoptosis is attenuated in Werner syndrome cells. *Genes and Dev.* **13:** 1355–1360.

Sporn M.B. and Roberts A.B. (1988) Peptide growth factors are multifunctional. *Nature* **332:** 217–219.

Spradling A.C. (1981) The organization and amplification of two chromosomal domains containing *Drosophila* chorion genes. *Cell* **27:** 193–201.

Spradling A.C. (1993) Germ line cysts: Communes that work. *Cell* **72:** 649–651.

Spradling A.C. and Mahowald A.P. (1980) Amplification of genes for chorion proteins during oogenesis in *Drosophila melanogaster*. *Proc. Nat. Acad. Sci. USA* **77:** 1096–1100.

Spradling A.C. and Rubin G.M. (1983) The effect of chromosomal position on the expression of the *Drosophila* xanthine dehydrogenase gene. *Cell* **34:** 47–57.

Sprenger F. and Nüsslein-Volhard C. (1992) Torso receptor activity is regulated by a diffusible ligand produced at the extracellular terminal regions of the *Drosophila* egg. *Cell* **71:** 987–1001.

Sprenger F., Stevens L.M. and Nüsslein-Volhard C. (1989) The *Drosophila* gene torso encodes a putative receptor tyrosine kinase. *Nature* **338:** 478–483.

Spritz R.A., Giebel L.B. and Holmes S. (1992) Dominant negative and loss of function mutations of the c-*kit* (mast/stem cell growth factor receptor) proto-oncogene in human piebaldism. *Am. J. Hum. Genet.* **50:** 261–269.

Stadtman E.R. (1992) Protein oxidation and aging. *Science* **257:** 1220–1224.

Stalder J., Groudine M., Dodgson J.B., Engel J.D. and Weintraub H. (1980) Hb switching in chickens. *Cell* **19:** 973–980.

Štanojević D., Hoey T. and Levine M. (1989) Sequence-specific DNA-binding activities of the gap proteins encoded by *hunchback* and *Krüppel* in *Drosophila*. *Nature* **341:** 331–335.

Štanojević D., Small S. and Levine M. (1991) Regulation of a segmentation stripe by overlapping activators and repressors in the *Drosophila* embryo. *Science* **254:** 1385–1387.

Stanulis-Praeger B.M. (1987) Cellular senescence revisited: A review. *Mech. Ageing Dev.* **38**: 1–48.

Starck D. (1965) *Embryologie: Ein Lehrbuch auf allgemein biologischer Grundlage.* Stuttgart: Georg Thieme.

Starr D.B., Matsui W., Thomas J.R. and Yamamoto K.R. (1996) Intracellular receptors use a common mechanism to interpret signaling information at response elements. *Genes and Dev.* **10:** 1271–1283.

Stearns T. and Kirschner M. (1994) In vitro reconstitution of centrosome assembly and function: The central role of t-tubulin. *Cell* **76:** 623–637.

Stebbins-Boaz B., Hake L.E. and Richter J.D. (1996) CPEB controls the cytoplasmic polyadenylation of cyclin, cdk2 and c-mos mRNAs and is necessary for oocyte maturation in *Xenopus*. *EMBO J.* **15:** 2582–2592.

Steeves T.A. and Sussex I.M. (1989) *Patterns in Plant Development.* 2nd ed. Cambridge: Cambridge University Press.

Stehelin D., Varmus H.E., Bishop J.M. and Vogt P. (1976) DNA related to the transforming gene(s) of avian sarcoma viruses is present in normal avian DNA. *Nature* **260:** 170–173.

Stein D. and Nüsslein-Volhard C. (1992) Multiple extracellular activities in *Drosophila* egg perivitelline fluid are required for establishment of embryonic dorsal-ventral polarity. *Cell* **68:** 429–440.

Stein D., Roth S., Vogelsang E. and Nüsslein-Volhard C. (1991) The polarity of the dorsoventral axis in the *Drosophila* embryo is defined by an extracellular signal. *Cell* **65:** 725–735.

Stein G.H. and Dulic V. (1995) Origins of G1 arrest in senescent human fibroblasts. *BioEssays* **17:** 537–543.

Steinberg M.S. (1963) Reconstruction of tissues by dissociated cells. *Science* **141:** 401–408.

Steinberg M.S. (1970) Does differential adhesion govern self-assembly processes in histogenesis? Equilibrium configurations and the emergence of a hierarchy among populations of embryonic cells. *J. Exp. Zool.* **173:** 395–434.

Steinberg M.S. (1998) Goal-directedness in embryonic development. *Integrative Biol.* **1:** 49–59.

Steinberg M.S. and Takeichi M. (1994) Experimental specification of cell sorting, tissue spreading, and specific spatial patterning by quantitative differences in cadherin expression. *Proc. Nat. Acad. Sci. USA* **91:** 206–209.

Steinhardt R.A. and Epel D. (1974) Activation of sea urchin eggs by a calcium ionophore. *Proc. Nat. Acad. Sci. USA.* **71:** 1915–1919.

Steinmann-Zwicky M. (1994) Sex determination of the *Drosophila* germ line: *tra* and *dsx* control somatic inductive signals. *Development* **120:** 707–716.

Steinmann-Zwicky M., Schmid H. and Nöthiger R. (1989) Cell-autonomous and inductive signals can determine the sex of the germ line of *Drosophila* by regulating the gene *Sxl. Cell* **57:** 157–166.

Steinmetz E.J. (1997) Pre-mRNA processing and the CTD of RNA polymerase II: The tail that wags the dog? *Cell* **89:** 491–494.

Steller H. and Pirrotta V. (1985) A transposable P vector that confers selectable G418 resistance to *Drosophila* larvae. *EMBO J.* **4:** 167–171.

Stennard F., Zorn A.M., Ryan K., Garrett N. and Gurdon J.B. (1999) Differential expression of VegT and Antipodean protein isoforms in *Xenopus*. *Mech.Dev.* **86:** 87–98.

Sternberg P.W. (1988) Lateral inhibition during vulval induction in *Caenorhabditis elegans*. *Nature* **335:** 551–554.

Stevens L.M., Frohnhöfer H.G., Klingler M. and Nüsslein-Volhard C. (1990) Localized requirement for *torsolike* expression in follicle cells for the development of terminal anlagen of the *Drosophila* embryo. *Nature* **346:** 660–663.

Steward F.C. (1970) From cultured cells to whole plants: The induction and control of their growth and morphogenesis. *Proc. R. Soc. London B* **175:** 1–30.

Stocum D.L. and Fallon J.F. (1982) Control of pattern formation in urodele limb ontogeny: A review and a hypothesis. *J. Embryol. exp. Morphol.* **69:** 7–36.

Stöger R., Kubicka P., Liu C.-G., Kafri T., Razin A., Cedar H. and Barlow D.P. (1993) Maternal-specific methylation of the imprinted mouse *Igf2r* locus identifies the expressed locus as carrying the imprinting signal. *Cell* **73:** 61–71.

Stopak D. and Harris A.K. (1982) Connective tissue morphogenesis by fibroblast traction. I. Tissue culture observations. *Dev. Biol.* **90:** 383–398.

Stott D., Kispert A. and Herrmann B.G. (1993) Rescue of the tail defect of *Brachyury* mice. *Genes and Dev.* **7:** 197–203.

Strehlow D. and Gilbert W. (1993) A fate map for the first cleavages of the zebra fish. *Nature* **361:** 451–453.

Stricker S.A. (1997) Intracellular injections of a soluble sperm factor trigger calcium oscillations and meiotic maturation in unfertilized oocytes of a marine worm. *Dev. Biol.* **186:** 185–201.

Stricker S.A. (1999) Comparative biology of calcium signaling during fertilization and egg activation in animals. *Dev. Biol.* **211:** 157–176.

Strome S. (1989) Generation of cell diversity during early embryogenesis in the nematode *Caenorhabditis elegans*. *Int. Rev. Cytol.* **114:** 81–123.

Strome S. (1993) Determination of cleavage planes. Cell **72:** 3–6.

Strome S. and Wood W.B. (1983) Generation of asymmetry and segregation of germ-line granules in early *C. elegans* embryos. *Cell* **35:** 15–25.

Struhl G. (1981) A homoeotic mutation transforming leg to antenna in *Drosophila*. *Nature* **292:** 635–638.

Struhl G. (1984) Splitting the bithorax complex of *Drosophila*. *Nature* **308:** 454–457.

Struhl G. (1989) Differing strategies for organizing anterior and posterior body pattern in *Drosophila* embryos. *Nature* **338:** 741–744.

Struhl G. and Basler K. (1993) Organizing activity of wingless protein in *Drosophila*. *Cell* **72:** 527–540.

Struhl G., Johnston P. and Lawrence P.A. (1992) Control of *Drosophila* body pattern by the hunchback morphogen gradient. *Cell* **69:** 237–249.

Struhl K. (1998) Histone acetylation and transcriptional regulatory mechanisms. *Genes and Dev.* **12:** 599–606.

Studer L., Tabar V. and McKay R.D. (1998) Transplantation of expanded mesencephalic precursors leads to recovery in parkinsonian rats. *Nat. Neurosci.* **1:** 290–295.

Studer M., Gavalas A., Marshall H., Ariza-McNaughton L., Rijli F.M., Chambon P. and Krumlauf R. (1998) Genetic interactions between Hoxa1 and Hoxb1 reveal new roles in regulation of early hindbrain patterning. *Development* **125:** 1025–1036.

Sturtevant A.H. (1923) Inheritance of direction of coiling in *Limnaea*. *Science* 58: 269–270.

Sturtevant A.H. (1929) The claret mutant type of *Drosophila simulans*: A study of chromosome elimination and of cell lineage. *Zeitschr. f. wiss. Zool.* **135:** 323–356.

Stüttem I. and Campos-Ortega J.A. (1991) Cell commitment and cell interactions in the ectoderm of *Drosophila melanogaster*. *Development* (Supplement 2): 39–46.

Stutz A., Conne B., Huarte J., Gubler P., Völkel V., Flandin P. and Vasalli J.-D. (1998) Masking, unmasking, and regulated polyadenylation cooperate in the translational control of a dormant mRNA in mouse oocytes. *Genes and Dev.* **12:** 2535–2548.

Stutz F. and Rosbash M. (1998) Nuclear RNA transport. *Genes and Dev.* **12:** 3303–3319.

Sudarwati S. and Nieuwkoop P.D. (1971) Mesoderm formation in the anuran *Xenopus* laevis (Daudin). *Wilhelm Roux's Arch.* **166:** 189–204.

Sudbeck P., Schmitz M.L., Baeuerle P.A. and Scherer G. (1996) Sex-reversal by loss of the C-terminal transactivation domain of human SOX9. *Nat. Genet.* **13:** 230–232.

Sullivan D., Palacios R., Stavnezer J., Taylor J.M., Faras A.J., Diely M.L., Summers N.M., Bishop J.M. and Schimke R.T.

(1973) Synthesis of a deoxyribonucleic acid sequence complementary to ovalbumin messenger ribonucleic acid and quantification of ovalbumin genes. *J. Biol. Chem.* **248:** 7530–7539.

Sulston J.E. and Horvitz H.R. (1977) Post-embryonic cell lineages of the nematode *Caenorhabditis elegans*. *Dev. Biol.* **56:** 110–156.

Sulston J.E. and White J.G. (1980) Regulation and cell autonomy during postembryonic development of *Caenorhabditis elegans*. *Dev. Biol.* **78:** 577–597.

Sulston J.E., Schierenberg E., White J.G. and Thomson J.N. (1983) The embryonic cell lineage of the nematode *Caenorhabditis elegans*. *Dev. Biol.* **100:** 64–119.

Summerbell. (1974) A quantitative analysis of the effect of excision of the AER from the chick limb bud. *J. Embryol. exp. Morphol.* **32:** 651–660.

Summerbell D. and Lewis J.H. (1975) Time, place and positional value in the chick limb bud. *J. Embryol. exp. Morphol.* **33:** 621–643.

Summers R.G., Morrill J.B., Leith A., Marko M., Piston D.W. and Stonebraker A.T. (1993) A stereometric analysis of karyokinesis, cytokinesis and cell arrangements during and following fourth cleavage period in the sea urchin, *Lytechinus variegatus*. *Dev. Growth Differ.* **35:** 41–57.

Sun J. and Tower J. (1999) FLP recombinase-mediated induction of Cu/Zn-superoxide dismutase transgene expression can extend the life span of adult *Drosophila melanogaster* flies. *Mol. Cell Biol.* **19:** 216–228.

Sundaram M. and Han M. (1996) Control and integration of cell signaling pathways during *C. elegans* vulval development. *BioEssays* **18:** 473–480.

Sundin O. and Eichele G. (1990) A homeo domain protein reveals the metameric nature of the developing chick hindbrain. *Genes and Dev.* **4:** 1267–1276.

Sung Z.R., Belachew A., Shunong B. and Bertrand-Garcia R. (1992) EMF, an *Arabidopsis* gene required for vegetative shoot development. *Science* **258:** 1645–1647.

Supp D.M., Witte D.P., Potter S.S. and Brueckner M. (1997) Mutation of an axonemal dynein affects left-right asymmetry in inversus viscerum mice. *Nature* **389:** 963–966.

Supp D.M., Brueckner M., Kuehn M.R., Witte D.P., Lowe L.A., McGrath J., Corrales J. and Potter S.S. (1999) Targeted deletion of the ATP binding domain of left-right dynein confirms its role in specifying development of left-right asymmetries. *Development* **126:** 5495–5504.

Surani M.A. (1998) Imprinting and the initiation of gene silencing in the germ line. *Cell* **93:** 309–312.

Surani M.A., Barton S.C. and Norris M.L. (1986) Nuclear transplantation in the mouse: Heritable differences between the parental genomes after activation of the embryonic genome. *Cell* **45:** 127–136.

Sutasurya L.A. and Nieuwkoop P.D. (1974) The induction of the primordial germ cells in the urodeles. *Wilhelm Roux's Arch. Entw. Org.* **175:** 199–220.

Sutovsky P., Navara C.S. and Schatten G. (1996) Fate of the sperm mitochondria, and the incorporation, conversion, and disassembly of the sperm tail structures during bovine fertilization. *Biol. Reprod.* **55:** 1195–1205.

Sutton H.E. (1988) *An Introduction to Human Genetics*. 4th ed. New York: Harcourt Brace Jovanovich.

Suzuki A., Theis R.S., Yamaji N., Song J.J., Wozney J., Murakami K. and Ueno N. (1994) A truncated BMP receptor affects dorsal-ventral patterning in the early *Xenopus* embryo. *Proc. Nat. Acad. Sci. USA* **91:** 10255–10259.

Swaab D.F. and Hofman M.A. (1995) Sexual differentiation of the human hypothalamus in relation to gender and sexual orientation. *Trends Neurosci.* **18:** 264–270.

Swain A. and Lowell-Badge R. (1999) Mammalian sex determination: a molecular drama. *Genes and Dev.* **13:** 755–767.

Swain A., Zanaria E., Hacker A., Lovell-Badge R. and Camerino G. (1996) Mouse *Dax-1* expression is consistent with a role in sex determination as well as in adrenal and hypothalamus function. *Nat. Genet.* **12:** 404–409.

Swain A., Narvaez V., Burgoyne P., Camerino G. and Lovell-Badge R. (1998) *Dax-1* antagonizes Sry action in mammalian sex determination. *Nature* **391:** 761–767.

Swain J.L., Stewart T.A. and Leder P. (1987) Parental legacy determines methylation and expression of an autosomal transgene: A molecular mechanism for parental imprinting. *Cell* **50:** 719–727.

Swalla B.J. and Jeffery W.R. (1995) A maternal RNA localized in the yellow crescent is segregated to the larval muscle cells during ascidian development. *Dev. Biol.* **170:** 353–364.

Swanson L.W., Simmons D.M., Arriza J., Hammer R., Brinster R.L., Rosenfeld M.G. and Evans R.M. (1985) Novel developmental specificity in the nervous system of transgenic animals expressing growth hormone fusion genes. *Nature* **317:** 363–366.

Sym M., Engebrecht J. and Roeder G.S. (1993) ZIP1 is a synaptonemeal complex protein required for meiotic chromosome synapsis. *Cell* **72:** 365–378.

Szeto D.P., Rodriguez-Esteban C., Ryan A.K., O'Connell S.M., Liu F., Kioussi C., Gleiberman A.S., Izpisua-Belmonte J.C. and Rosenfeld M.G. (1999) Role of the Bicoid-related homeodomain factor Pitx1 in specifying hindlimb morphogenesis and pituitary development. *Genes and Dev.* **13:** 484–494.

Tabin C.J. (1992) Why we have (only) five fingers per hand: Hox genes and the evolution of paired limbs. *Development* **116:** 289–296.

Taipale J. and Keski-Oja J. (1997) Growth factors in the extracellular matrix. *FASEB J.* **11:** 51–59.

Tajbakhsh S. and Spöhrle R. (1998) Somite development: Constructing the vertebrate body. *Cell* **92:** 9–16.

Tajbakhsh S., Rocancourt D., Cossu G. and Buckingham M. (1997) Redefining the genetic hierarchies controlling skeletal myogenesis: *Pax-3* and *Myf-5* act upstrem of *MyoD*. *Cell* **89:** 127–138.

Takeichi M. (1977) Functional correlation between cell adhesive properties and some cell surface proteins. *J. Cell Biol.* **75:** 464–474.

Takeichi M. (1991) Cadherin cell adhesion receptors as a morphogenetic regulator. *Science* **251:** 1451–1455.

Takeichi M. (1995) Morphogenetic roles of classic cadherins. *Curr. Opin. Cell Biol.* **7:** 619–627.

Takeuchi J.K., Koshiba-Takeuchi K., Matsumoto K., Vogel-Hopker A., Naitoh-Matsuo M., Ogura K., Takahashi N., Yasuda K. and Ogura T. (1999) Tbx5 and Tbx4 genes determine the wing/leg identity of limb buds. *Nature* **398:** 810–814.

Talbot W.S., Swyryd E.A. and Hogness D.S. (1993) *Drosophila* tissues with different metamorphic responses to ecdysone express different edcysone receptor isoforms. *Cell* **73:** 1323–1337.

Talwar G.P. and Raghupathy R., eds. (1994) *Birth Control Vaccines.* Boca Raton, FL: CRC Press.

Tam P.P. and Steiner K.A. (1999) Anterior patterning by synergistic activity of the early gastrula organizer and the anterior germ layer tissues of the mouse embryo. *Development* **126:** 5171–5179.

Tam P.P. and Zhou S.X. (1996) The allocation of epiblast cells to ectodermal and germ-line lineages is influenced by the position of the cells in the gastrulating mouse embryo. *Dev. Biol.* **178:** 124–32.

Tam P.P., Goldman D., Camus A. and Schoenwolf G.C. (2000) Early events of somitogenesis in higher vertebrates: allocation of precursor cells during gastrulation and the organization of a meristic pattern in the paraxial mesoderm. *Curr. Top. Dev. Biol.* **47:** 1–32.

Tamemoto H., et al. (1994) Insulin resistance and growth retardation in mice lacking insulin receptor substrate-1. *Nature* **372:** 182–186.

Tan P.B.O. and Kim S.K. (1999) Signaling specificity—the RTK/RAS/MAP kinase pathway in metazoans. *Trends Genet.* **15:** 145–149.

Tanabe Y. and Jessel T.M. (1996) Diversity and pattern in the developing spinal cord. *Science* **274:** 1115–1123.

Tang Y.P., Shimizu E., Dube G.R., Rampon C., Kerchner G.A., Zhuo M., Liu G. and Tsien J.Z. (1999) Genetic enhancement of learning and memory in mice. *Nature* **401:** 63–69.

Tarun S.Z. and Sachs A.B. (1996) Association of the yeast poly(A) tail binding protein with translation initiation factor eIF4G. *EMBO J.* **15:** 7168–7177.

Tata J.R. (1993) Gene expression during metamorphosis: An ideal model for post-embryonic development. *BioEssays* **15:** 239–248.

Tata J.R., Kawahara A. and Baker B.S. (1991) Prolactin inhibits both thyroid hormone-induced morphogenesis and cell death in cultured amphibian larval tissues. *Dev. Biol.* **146:** 72–80.

Tautz D. (1988) Regulation of the *Drosophila* segmentation gene *hunchback* by two maternal morphogenetic centres. *Nature* **332:** 281–284.

Tautz D. (1992) Redundancies, development and the flow of information. *BioEssays* **14:** 263–266.

Taylor M.A. and Smith L.D. (1987) Induction of maturation in small *Xenopus* laevis oocytes. *Dev. Biol.* **121:** 111–118.

Teillet M.A. and Le Douarin N.M. (1983) Consequences of notochord and neural tube excision on the development of the peripheral nervous system in the chick embryo. *Dev. Biol.* **98:** 192–211.

Telfer W.H. (1965) The mechanism and control of yolk formation. *Ann. Rev. Entomol.* **10:** 161–184.

Telfer W.H. (1975) Development and physiology of the oocyte-nurse cell syncytium. *Adv. Insect Physiol.* **11:** 223–319.

Telfer W.H., Huebner E. and Smith D.S. (1982) The cell biology of vitellogenic follicles in *Hyalophora* and *Rhodnius*. In: *Insect Ultrastructure,* R.C. King and H. Akai (eds.), vol. 1, 118–149. New York: Plenum.

Telling G.C., Parchi P., DeArmond S.J., Cortelli P., Montagna P., Gabizon R., Mastrianni J., Lugaresi E., Gambetti P. and Prusiner S.B. (1996) Evidence for the conformation of the pathogenic isoform of the prion protein enciphering and propagating prion diversity. *Science* **274:** 2079–2082.

Templeton A.R., Crease T.J. and Shah F. (1985) The molecular through ecological genetics of *abnormal abdomen* in *Drosphila mercatorum.* I. Basic genetics. *Genetics* **111**: 805–818.

Templeton A.R., Hollocher H., Lawler S. and Johnston J.S. (1989) Natural selection and ribosomal DNA in *Drosophila. Genome* **31:** 296–303.

Templeton A.R., Hollocher H. and Johnston J.S. (1993) The molecular through ecological genetics of abnormal abdomen in *Drosophila mercatorum.* V. Female phenotypic expression on natural genetic backgrounds and in natural environments. *Genetics* **134:** 475–485.

Theriot J.A. and Mitchison T.J. (1991) Actin microfilament dynamics in locomoting cells. *Nature* **352:** 126–132.

Theurkauf W.E. (1994) Microtubules and cytoplasm organization during *Drosophila* oogenesis. *Dev. Biol.* **165:** 352–360.

Theurkauf W.E. and Hazelrigg T.I. (1998) In vivo analyses of cytoplasmic transport and cytoskeletal organiztion during *Drosophila* oogenesis: Characterization of a multi-step anterior localization pathway. *Development* **125:** 3655–3666.

Theurkauf W.E., Alberts B.M., Jan Y.N. and Jongens T.A. (1993) A central role for microtubules in the differentiation of *Drosophila* oocytes. *Development* **118:** 1169–1180.

Thiery J.-P., Brackenbury R., Rutishauser U. and Edelman G.M. (1977) Adhesion among neural cells of the chick embryo. II. Purification and characterization of a cell adhesion molecule from neural retina. *J. Biol. Chem.* **252:** 6841–6845.

Thisse C., Perrin-Schmitt F., Stoetzel C. and Thisse B. (1991) Sequence-specific transactivation of the *Drosophila* twist gene by the *dorsal* gene product. *Cell* **65: 1**191–1201.

Thomas H.E., Stunnenberg H.G. and Stewart A.F. (1993) Heterodimerization of the *Drosophila* ecdysone receptor with retinoid X receptor and *ultraspiracle. Nature* **362:** 471–475.

Thomas J.H., Stern M.J. and Horvitz H.R. (1990) Cell interactions coordinate the development of the *C. elegans* egg-laying system. *Cell* **62:** 1041–1052.

Thomsen G.H. and Melton D.A. (1993) Processed Vg1 protein is an axial mesoderm inducer in *Xenopus. Cell* **74:** 433–441.

Thomson J.A. and Marshall V.S. (1998) Primate embryonic stem cells. *Curr. Top. Dev. Biol.* **38:** 133–165.

Thomson J.A., Itskovitz-Eldor J., Shapiro S.S., Waknitz M.A., Swiergiel J.J., Marshall V.S. and Jones J.M. (1998) Embryonic stem cell lines derived from human blastocysts. *Science* **282:** 1145–1147.

Thorpe C.J., Schlesinger A., Carter J.C. and Bowerman B. (1997) Wnt signaling polarizes an early *C. elegans* blastomere to distinguish endoderm from mesoderm. *Cell* **90:** 695–705.

Thorpe W.H. (1958) The learning of song patterns by birds, with especial reference to the song of the chaffinch, *Fringilla coelebs. Ibis* **100:** 535–570.

Thummel C.S. (1995) From embryogenesis to metamorphosis: The regulation and function of *Drosophila* nuclear receptor superfamily members. *Cell* **83:** 871–877.

Thummel C.S. (1997) Dueling orphans: nuclear receptors coordinate *Drosophila* metamorphosis. *BioEssays* **19:** 669–672.

Thummel C.S., Burtis K.C. and Hogness D.S. (1990) Spatial and temporal patterns of E74 transcription during *Drosophila* development. *Cell* **61:** 101–111.

Tijo J.H. and Puck T.T. (1958) The somatic chromosomes of man. *Proc. Nat. Acad. Sci. USA* **44:** 1229–1237.

Till J.E. and McCulloch E.A. (1961) A direct measurement of the radiation sensitivity of normal mouse bone marrow cells. *Radiat. Res.* **14:** 215–222.

Tjian R. and Maniatis T. (1994) Transcriptional activation: A complex puzzle with few easy pieces. *Cell* **77:** 5–8.

Toivonen S. and Saxén L. (1955) The simultaneous inducing action of liver and bone marrow of the guinea pig in implantation and explantation experiments with embryos of *Triturus. Exp. Cell Res.* (Supplement 3): 346–357.

Tonegawa A., Funuyama N., Ueno N. and Takahashi Y. (1997) Mesodermal subdivision along the mediolateral axis in chicken controlled by different concentrations of BMP-4. *Development* **124:** 1975–1984.

Toole B.P. and Gross J. (1971) The extracellular matrix of the regenerating newt limb: Synthesis and removal of hyaluronate prior to differentiation. *Dev. Biol.* **25:** 57–77.

Torres M., Gomez-Pardo E., Dressler G. and Gruss P. (1995) *Pax-2* controls multiple steps of urogenital development. *Development* **121:** 4057–4065.

Torres-Ruiz, R.A. and Jürgens G. (1994) Mutations in the *fass* gene uncouple pattern formation and morphogenesis in *Arabidopsis* development. *Development* **120:** 2967–2978.

Torres-Ruiz, R.A., Lohner A. and Jürgens G. (1996) The *gurke* gene is required for the normal organization of the apical region in the *Arabidopsis* embryo. *Plant J.* **10:** 1005–1016.

Torrey T.W. and Feduccia A. (1991) *Morphogenesis of the Vertebrates.* 5th ed. New York: Wiley.

Townes P.L. and Holtfreter J. (1955) Directed movements and selective adhesion of embryonic amphibian cells. *J. Exp. Zool.* **128:** 53–120.

Treisman J. and Desplan C. (1989) The products of the *Drosophila* gap genes *hunchback* and *Krüppel* bind to the *hunchback* promoters. *Nature* **341:** 335–337.

Trent C., Purnell B., Gavinski S., Hageman J. and Wood W.B. (1991) Sex-specific transcriptional regulation of the *C. elegans* sex-determining gene *her-1. Mech. Dev.* **34:** 43–56.

Trinkaus J.P. (1963) The cellular basis of Fundulus epiboly. Adhesivity of blastula and gastrula cells in culture. *Dev. Biol.* **7:** 513–532.

Trinkaus J.P. (1984) *Cells into Organs: The Forces That Shape the Embryo.* 2nd ed. Englewood Cliffs, NJ: Prentice Hall.

Trinkaus J.P. (1996) Ingression during early gastrulation of *Fundulus. Dev. Biol.* **177:** 356–370.

Tröbner W., Ramirez L., MotteP., Hue I., Huijser P., Lönning W.E., Saedler H., Sommer H. and Schwarz-Sommer Z. (1992) *globosa,* a homeotic gene which interacts with *deficiens* in the control of *Antirrhinum* floral organogenesis. *EMBO J.* **11:** 4693–4704.

Trotochaud A.E., Hao T., Wu G., Yang Z. and Clark S.E. (1999) The clavata1 receptor-like kinase requires clavata3 for its assembly into a signaling complex that includes KAPP and a Rho-relted protein. *Plant Cell* **11:** 393–405.

Trouche D., Grigoriev M., Lenormand J.-L., Robin P., Leibovitch S.A., Sassone-Corsi P. and Harel-Bellan A. (1993) Repression of c-*fos* promoter by MyoD on muscle cell differentiation. *Nature* **363:** 79–82.

Truman J.W., Talbot W.S., Fahrbach S.E. and Hogness D.S. (1994) Ecdysone receptor expression in the CNS correlates with stage-specific responses to ecdysteroids during *Drosophila* and *Manduca* development. *Development* **120:** 219–234.

Tsai C. and Gergen P. (1995) Pair-rule expression of the *Drosophila fushi tarazu* gene: a nuclear receptor response element mediates the opposing regulatory effects of *runt* and *hairy. Development* **121:** 453–462.

Tsai F.-Y., Keller G., Kuo F.C., Weiss M., Chen J., Rosenblatt M., Alt F.W. and Orkin S.H. (1994) An early hematopoietic defect in mice lacking the transcription factor GATA-2. *Nature* **371:** 221–226.

Tuchmann-Duplessis H., David G. and Haegel P. (1972) *Illustrated Human Embryology,* vol. 1, *Embryogenesis.* New York: Springer-Verlag.

Tucker G.C., Ayoyama H., Lipinski M., Tursz T. and Thiery J.-P. (1984) Identical reactivity of monoclonal antibodies N-1 and NC-1: Conservation in vertebrates on cells derived from neural primordium and on some leukocytes. *Cell Differ.* **14:** 223–230.

Tucker K.L., Beard C., Dausmann J., Jackson-Grusby L., Laird P.W., Lei H., Li E. and Jaenisch R. (1996) Germ-line passage is required for establishment of methylation and expression patterns of imprinted but not of nonimprinted genes. *Genes and Dev.* **10:** 1008–1020.

Tucker P.K. and Lundrigan B.L. (1993) Rapid evolution of the sex determining locus in Old World mice and rats. *Nature* **364:** 715–717.

Turner B.M., Birley A.J. and Lavender J. (1992) Histone H4 isoforms acetylated at specific lysine residues define individual chromosomes and chromatin domains in *Drosophila* polytene nuclei. *Cell* **69:** 375–384.

Turner F.R. and Mahowald A.P. (1976) Scanning electron microscopy of *Drosophila* embryogenesis. I. The structure of the egg envelopes and the formation of the cellular blastoderm. *Dev. Biol.* **50:** 95–108.

Turner P.R. and Jaffe L.A. (1989) G-proteins and the regulation of oocyte maturation and fertilization. In: *The Cell Biology of Fertilization,* H. Schatten and G. Schatten (eds.), 297–318. San Diego: Academic Press.

Twitty V.C. and Elliott H.A. (1934) The relative growth of the amphibian eye, studied by means of transplantation. *J. Exp. Zool.* **68:** 247–291.

Udolph G., Lüer K., Bossing T. and Technau G.M. (1995) Commitment of CNS precursors along the dorsoventral axis of *Drosophila* neuroectoderm. *Science* **269:** 1278–1280.

Urquidi V., Tarin D. and Goodison S. (2000) Role of telomerase in cell senescence and oncogenesis. *Ann. Rev. Med.* **51:** 65–79.

Uzzell T.M. (1964) Relations of the diploid and triploid species of the *Ambystoma jeffersonianum* complex. *Copeia* (1964): 257–300.

Vacquier V.D. (1980) The adhesion of sperm to sea urchin eggs. In: *The Cell Surface: Mediator of Developmental Processes,* S. Subtelny and N.K. Wessels (eds.), 151–168. New York: Academic Press.

Vacquier V.D. (1998) Evolution of gamete recognition proteins. *Science* **281:** 1995–1998.

Vacquier V.D. and Moy G.W. (1977) Isolation of bindin: The protein responsible for the adhesion of sperm to sea urchin eggs. *Proc. Nat. Acad. Sci. USA* **74:** 2456–2460.

Vacquier V.D. and Moy G.W. (1997) The fucose sulfate polymer of egg jelly binds to sperm REJ and is the inducer of the sea urchin sperm acrosome reaction. *Dev. Biol.* **192:** 125–135.

Vainio S., Jalkanen M., Bernfield M. and Saxén L. (1992) Transient expression of syndecan in mesenchymal cell aggregates of the embryonic kidney. *Dev. Biol.* **152:** 221–232.

Vainio S., Heikkilä M., Kispert A., Chin N. and McMahon A.P. (1999) Female development in mammals is regulated by Wnt-4 signalling. *Nature* **397:** 405–409.

Vakaet L. (1984) Early development of birds. In: *Chimeras in Developmental Biology,* N.M. Le Douarin and A. McLaren (eds.), 71–88. London: Academic Press.

Valcárcel J., Singh R., Zamore P.D. and Green M.R. (1993) The protein Sex-lethal antagonizes the splicing factor U2AF to regulate alternative splicing of *transformer* pre-mRNA. *Nature* **362:** 171–175.

Van der Lugt N.M.T., et al. (1994) Posterior transformation, neurological abnormalities, and severe hematopoietic effects in mice with a targeted deletion of the bmi-1 proto-oncogene. *Genes and Dev.* **8:** 757–769.

van Lohuizen, M. (1999) The trithorax-group and Polycomb-group chromatin modifiers: Implications for disease. *Curr. Opin. Genet. Dev.* **9:** 355–361.

Varmus H.E. (1982) Form and function of retroviral proviruses. *Science* **216:** 812–820.

Vasil V. and Hildebrandt A.C. (1965) Differentiation of tobacco plants from single isolated cells in microcultures. *Science* **150:** 889–892.

Vaupel J.D. et al. (1998) Biodemographic trajectories of longevity. *Science* **280:** 855–860.

Vaux D.L. (1997) CED-4, the third horseman of apoptosis. *Cell* **90:** 389–390.

Vaux D.L. and Korsmeyer S.J. (1999) Cell death in development. *Cell* **96:** 245–254.

Vaux D.L., Weissman I.L. and Kim S.K. (1992) Preven-tion of programmed cell death in *Caenorhabditis elegans* by human bcl-2. *Science* **258:** 1955–1957.

Velander W.H., Lubon H. and Drohan W.N. (1997) Transgenic livestock as drug factories. *Scientific American* (January 1997): 70–74.

Venuti J.M., Morris J.S., Vivian J.L., Olson E.N. and Klein W.H. (1995) Myogenin is required for late but not early aspects of myogenesis during mouse development. *J. Cell Biol.* **128:** 563–576.

Verbout A.J. (1985) The development of the vertebral column. *Adv. Anat. Embryol. Cell. Biol.* **90:** 1–120.

Verdonk N.H. and Cather J.N. (1983) Morphogenetic determination and differentiation. In: *The Mollusca,* K.M. Wilbur (ed.), vol 3, 215–252. New York: Academic Press.

Vestweber D., Gossler A., Boller K. and Kemler R. (1987) Expression and distribution of cell adhesion molecule uvomorulin in mouse preimplantation embryos. *Dev. Biol.* **124:** 451–456.

Villee C.A., Solomon E.P., Martin C.E., Martin D.W., Berg L.R. and Davis P.W. (1989) *Biology.* 2nd ed. Philadelphia: Saunders.

Vincent J.-P. and Gerhart J.C. (1987) Subcortical rotation in *Xenopus* eggs: An early step in embryonic axis formation. *Dev. Biol.* **123:** 526–539.

Vincent J.-P. and Lawrence P.A. (1994) *Drosophila wingless* sustains *engrailed* expression only in adjoining cells: Evidence from mosaic embryos. *Cell* **77:** 909–915.

Vincent J.-P. and O'Farrell P.H. (1992) The state of *engrailed* expression is not clonally transmitted during early *Drosophila* development. *Cell* **68:** 923–931.

Vize P.D., Seufert D.W., Carroll T.J. and Wallingford J.B. (1997) Model systems for the study of kidney development: Use of the pronephros in the analysis of organ induction and patterning. *Dev. Biol.* **188:** 189–204.

Vodicka M.A. and Gerhart J.C. (1995) Blastomere derivation and domains of gene expression in the Spemann organizer of *Xenopus laevis. Development* **121:** 3505–3518.

Vogel A., Rodriguez C., Warnken W. and Izpisua Belmonte J.C. (1995) Dorsal cell fate specified by chick Lmx1 during vertebrate limb development. *Nature* **378:** 716–720.

Vogel O. (1978) Pattern formation in the egg of the leafhopper *Euscelis plebejus* fall. (Homoptera): Developmental capacities of fragments isolated from the polar egg regions. *Dev. Biol.* **67:** 357–370.

Vogel O. (1982) Development of complete embryos in drastically deformed leafhopper eggs. *Wilhelm Roux's Arch.* **191:** 134–136.

Vogt W. (1929) Gestaltungsanalyse am Amphibien keim mit örtlicher Vitalfärbung. II. Teil. Gastrulation und Mesodermbildung bei Urodelen und Anuren. *Wilhelm Roux's Arch.* **120:** 384–706.

Vollbrecht E., Veit B., Sinha N. and Hake S. (1991) The developmental gene *Knotted-1* is a member of a maize homeobox gene family. *Nature* **350:** 241–243.

vom Saal F.S. (1989) Sexual differentiation in litter-bearing mammals: Influence of sex of adjacent fetuses *in utero. J. Animal Sci.* **67:** 1824–1840.

Von Allmen, G., Hogga, I., Spierer, A., Karch, F., Bender, W., Gyurkovics, H., Lewis, E. (1996) Splits in fruitfly Hox complexes. *Nature* **380:** 116.

von Baer K.E. (1828) *Entwicklungsgeschichte der Thiere: Beobachtung und Reflexion.* Königsberg: Bornträger.

Von Brunn A. and Kalthoff K. (1983) Photoreversible inhibition by ultraviolet light of germ line development in *Smittia* sp. (Chironomidae, Diptera). *Dev. Biol.* **100:** 426–439.

von Kalm L., Crossgrove K., Von Seggern D., Guild G.M. and Beckendorf S.K. (1994) The Broad-Complex directly controls a tissue-specific response to the steroid hormone ecdysone at the onset of *Drosophila* metamorphosis. *EMBO J.* **13:** 3505–3516.

Wabl M.R., Brun R.B. and DuPasquier L. (1975) Lymphocytes of the toad *Xenopus laevis* have the gene set for promoting tadpole development. *Science* **190:** 1310–1312.

Wagner D., Sablowski R.W.M. and Meyerowitz E. (1999) Transcriptional activation of apetala1 by leafy. *Science* **285:** 582–584.

Wagner E.F. (1990) Mouse genetics meets molecular biology at Cold Spring Harbor. *New Biologist* **2:** 1071–1074.

Wagner T., et al. (1994) Autosomal sex reversal and campomelic dysplasia are caused by mutations in and around the SRY-related gene SOX9. *Cell* **79:** 1111–1120.

Wakahara M. (1990) Cytoplasmic localization and organization of germ cell determinants. In: *Cytoplasmic Organization in Development,* G.M. Malacinski (ed.), vol. 4, 219–242. New York: McGraw-Hill.

Wakayama T. and Yanagimachi R. (1999) Cloning of male mice from adult tail-tip cells. *Nat. Genet.* **22:** 127–128.

Wakayama T., Perry A.C., Zuccotti M., Johnson K.R. and Yanagimachi R. (1998) Full-term development of mice from enucleated oocytes injected with cumulus cell nuclei. *Nature* **394:** 369–374.

Waksman G., Kominos D., Robertson S.C., Pant N., Baltimore D., Birge R.B., Cowburn D., Hanafusa H., Mayer B.J., Overduin M., Resh M.D., Rios C.B., Silverman L. and Kuriyan J. (1992) Crystal structure of the phosphotyrosine recognition domain SH2 of *v-src* complexed with tyrosine-phosphorylated peptides. *Nature* **358:** 646–653.

Walbot V. and Holder N. (1987) *Developmental Biology.* New York: Random House.

Walker D.W., McColl G., Jenkins N.L., Harris J. and Lithgow G.J. (2000) Evolution of lifespan in *C. elegans. Nature* **405:** 296–297.

Wallace H., Maden M. and Wallace B.M. (1974) Participation of cartilage grafts in amphibian limb regeneration. *J. Embryol. exp. Morphol.* **32:** 391–404.

Walldorf U. and Gehring W.J. (1992) *Empty spiracles,* a gap gene containing a homeobox involved in *Drosophila* head development. *EMBO J.* **11:** 2247–2259.

Walldorf U., Fleig R. and Gehring W.J. (1989) Comparison of homeobox-containing genes of the honeybee and *Drosophila. Proc. Nat. Acad. Sci. USA* **86:** 9971–9975.

Wallingford J.B., Seufert D.W., Virta V.C. and Vize P. (1997) p53 activity is essential for normal development in *Xenopus. Curr. Biol.* **7:** 747–757.

Wang C. and Lehmann R. (1991) Nanos is the localized posterior determinant in *Drosophila. Cell* **66:** 637–647.

Wang H.U. and Anderson D.J. (1997) Eph family transmembrane ligands can mediate repulsive guidance of trunk neural crest migration and motor axon outgrowth. *Neuron* **18:** 383–396.

Wang J. and Bell L.R. (1994) The Sex-lethal amino terminus mediates cooperative interactions in RNA binding and is essential for splicing regulation. *Genes and Dev.* **8:** 2072–2085.

Wang N., Butler J.P. and Ingber D.E. (1993) Mechanotransduction across the cell surface and through the cytoskeleton. *Science* **260:** 1124–1127.

Wang Q.T. and Holmgren R.A. (1999) The subcellular localization and activity of *Drosophila cubitus interruptus* are regulated at multiple levels. *Development* **126:** 5097–5106.

Wang S. and Hazelrigg T. (1994) Implications for bcd mRNA localization from spatial distribution of *exu* protein in *Drosophila* oogenesis. *Nature* **369:** 400–403.

Wang Y. and Jaenisch R. (1997) Myogenin can substitute for Myf5 in promoting myogenesis but less efficiently. *Development* **124:** 2507–2513.

Wang Y., Schnegelsberg P.N., Dausman J. and Jaenisch R. (1996) Functional redundancy of the muscle-specific transcription factors Myf5 and myogenin. *Nature* **379:** 823–825.

Wang Z. and Brown D.D. (1993) Thyroid hormone-induced gene expression program for amphibian tail resorption. *J. Biol. Chem.* **268:** 16270–16278.

Ward C.R. and Kopf G.S. (1993) Molecular events mediating sperm activation. *Dev. Biol.* **158:** 9–34.

Ward G.E., Brokaw C.J., Garbers D.L. and Vacquier V.D. (1985) Chemotaxis of *Arbacia punctulata* spermatozoa to resact, a peptide from the egg jelly layer. *J. Cell Biol.* **101:** 2324–2329.

Warner H.R. and Johnson T.E. (1997) Parsing age, mutations and time. *Nat. Genet.* **17:** 368–370.

Wassarman P.M. (1987) The biology and chemistry of fertilization. *Science* **235:** 553–560.

Wassarman P.M. (1990) Profile of a mammalian sperm receptor. *Development* **108:** 1–17.

Wassarman P.M. and Albertini D.F. (1994) The Mammalian Ovum. In: *The Physiology of Reproduction,* E. Knovil and J.D. Neill (eds.), 79–122. New York: Raven Press.

Wassarman P.M. and Litscher E.S. (1990) Sperm-egg recognition mechanisms in mammals. *Curr. Top. Dev. Biol.* **30:** 1–19.

Watanabe T., Kim S., Candia A., Rothbacher U., Hashimoto C., Inoue K. and Cho K.W.Y. (1995) Molecular mechanism of Spemann's organizer formation: Conserved growth factor synergy between *Xenopus* and mouse. *Genes and Dev.* **9:** 3038–3050.

Waterbury J.A., Jackson L.L. and Schedl P. (1999) Analysis of the doublesex female protein in *Drosophila melanogaster:* Role on sexual differentiation and behavior and dependence on intersex. *Genetics* **152:** 1653–1667.

Watson J.D. and Crick F.H.C. (1953) Molecular structure of nucleic acids: A structure for deoxyribose nucleic acid. *Nature* (London) **171:** 737–738.

Watt F.M. and Hogan B.L.M. (2000) Out of Eden: Stem cem cells and their niches. *Science* **287:** 1427–1430.

Wattiaux J.M. (1968) Parental age effects in *Drosophila pseudoobscura. Exp. Gerontol.* **3:** 55–61.

Webster P.J., Liang L., Berg C.A., Lasko P. and Macdonald P. (1997) Translational repressor bruno plays multiple roles in development and is widely conserved. *Genes and Dev.* **11:** 2510–2521.

Wehrle-Haller B. and Weston J.A. (1997) Receptor tyrosine kinase-dependent neural crest migration in response to differentially localized growth factors. *BioEssays* **19:** 337–345.

Weigel D. and Meyerowitz E.M. (1993) Genetic hierarchy controlling flower development. In: *Molecular Basis of Morphogenesis,* M. Bernfield (ed.), 93–107. New York: Alan R. Liss.

Weigel D. and Meyerowitz E.M. (1994) The ABCs of floral homeotic genes. *Cell* **78:** 203–209.

Weigel D., Alvarez J., Smyth D.R., Yanofsky M.F. and Meyerowitz E.M. (1992) *LEAFY* controls floral meristem identity in *Arabidopsis. Cell* **69:** 843–859.

Weigmann K., Cohen S.M. and Lehner C.F. (1997) Cell cycle progression, growth and patterning in imaginal discs despite inhibition of cell division after inactivation of *Drosophila* Cdc2 kinase. *Development* **124:** 3555–3563.

Weinberg R.A. (1983) A molecular basis of cancer. *Scientific American* **249** (November): 126–142.

Weinberg R.A. (1996a) How cancer arises. *Scientific American* (September 1996): 62–70.

Weinberg R.A. (1996b) The retinoblastoma protein and cell cycle control. *Cell* **81:** 323–330.

Weindruch R. and Walford R.L. (1982) Dietary restriction in mice beginning at 1 year of age: Effect on life-span and spontaneous cancer incidence. *Science* **215:** 1415–1418.

Weindruch R. and Walford R.L. (1988) *The Retardation of Aging and Disease by Dietary Restriction.* Springfield, IL: Charles C. Thomas.

Weindruch R., Walford R.L., Fligiel S. and Guthrie D. (1986) The retardation of aging in mice by dietary restriction: Longevity, cancer, immunity and lifetime energy intake. *J. Nutr.* **116:** 641–654.

Weinkove D. and Leevers S.J. (2000) The genetic control of organ growth: insights from *Drosophila. Curr. Opin. Genet. Dev.* **10:** 75–80.

Weintraub H. (1993) The MyoD family and myogenesis: Redundancy, networks, and thresholds. *Cell* **75:** 1241–1244.

Weis L. and Reinberg D. (1992) Transcription by RNA polymerase II: Initiator-directed formation of transcription-competent complexes. *FASEB J.* **6:** 3300–3309.

Weisblat D.A., Zackson S.L., Blair S.S. and Young J.D. (1980) Cell lineage analysis by intracellular injection of fluorescent tracers. *Science* **209:** 1538–1541.

Weiss P. (1939) *Principles of Development: A Text in Experimental Embryology.* New York: Henry Holt and Company.

Weiss P. (1958) Cell contact. *Int. Rev. Cytol.* **7:** 391–423.

Weissman I. (2000) Translating stem and progenitor cell biology to the clinic: Barriers and opportunities. *Science* **287:** 1442–1446.

Weissmann C. (1994) Molecular biology of prion diseases. *Trends Cell Biol.* **4:** 10–14.

Welch M.D. (1999) The world according to Arp: Regulation of actin nucleation by the Arp2/3 complex. *Trends Cell Biol.* **11:** 423–427.

Welch M.D., Mallavarapu A., Rosenblatt J. and Mitchison T.J. (1997) Actin dynamics in vivo. *Curr. Opin. Cell Biol.* **9:** 54–61.

Weliky M., Minsuk S., Keller R. and Oster G. (1991) Notochord morphogenesis in *Xenopus laevis*: Simulation of cell behavior underlying tissue convergence and extension. *Development* **113:** 1231–1244.

Wells D.E., Showman R.M., Klein W.H. and Raff R.A. (1981) Delayed recruitment of maternal histone H3 mRNA in sea urchin embryos. *Nature* **292:** 477–478.

Welshons W.J. and Russell L.B. (1959) The Y chromosome as the bearer of male determining factors in the mouse. *Proc. Nat. Acad. Sci. USA* **45:** 560–566.

Went D.F. and Krause G. (1973) Normal development of mechanically activated, unlaid eggs of an endoparasitic hymenopteran. *Nature* **244:** 454–455.

Westaway D., Zuliani V., Cooper C.M., Da Costa M., Neuman S., Jenny A.L., Detwiler L. and Prusiner S.B. (1994) Homozygosity for prion protein alleles encoding glutamine-171 renders sheep susceptible to natural scrapie. *Genes and Dev.* **8:** 959–969.

Weston J.A. (1963) A radioautographic analysis of the migration and localization of trunk neural crest cells in the chick. *Dev. Biol.* **6:** 279–310.

Wetts R. and Fraser S.E. (1989) Slow intermixing of cells during *Xenopus* embryogenesis contributes to the consistency of blastomere fate map. *Development* **105:** 9–15.

Whalen A.M. and Steward R. (1993) Dissociation of the dorsal-cactus complex and phosphorylation of the dorsal protein correlate with the nuclear localization of *dorsal*. *J. Cell Biol.* **123:** 523–534.

Wharton R.P. and Struhl G. (1991) RNA regulatory elements mediate control of *Drosophila* body pattern by the posterior morphogen nanos. *Cell* **67:** 955–967.

Whetton A.D. and Dexter T.M. (1993) Influence of growth factors and substrates on differentiation of hemopoietic stem cells. *Curr. Opin. Cell Biol.* **5:** 1044–1049.

Whitaker M. and Steinhardt R. (1985) Ion signalling in the sea urchin egg at fertilization. In: *Biology of Fertilization,* C.B. Metz and A. Monroy (eds.), vol. 3, 167–221. Orlando, FL: Academic Press.

Whitaker M. and Swann K. (1993) Lighting the fuse at fertilization. *Development* **117:** 1–12.

White J. and Strome S. (1996) Cleavage plane specification in *C. elegans*: How to divide the spoils. *Cell* **84:** 195–198.

White K., Hurban P., Watanabe T. and Hogness D.S. (1997) Coordination of *Drosophila* metamorphosis by two ecdysone-induced nuclear receptors. *Science* **276:** 114–117.

Whitfield L.S., Lovell-Badge R. and Goodfellow P.N. (1993) Rapid sequence evolution of the mammalian sex-determining gene *SRY*. *Nature* **364:** 713–715.

Whitlock K.E. (1993) Development of *Drosophila* wing sensory neurons in mutants with missing or modified cell surface molecules. *Development* **117:** 1251–1260.

Whittaker J.R. (1973) Segregation during ascidian embryogenesis of egg cytoplasmic information for tissue-specific enzyme development. *Proc. Nat. Acad. Sci. USA* **70:** 2096–2100.

Wickens M., Anderson P. and Jackson R.J. (1997) Life and death in the cytoplasm: Messages from the 3' end. *Curr. Opin. Genet. Dev.* **7:** 220–232.

Wieczorek E., Brand M., Jacq X. and Tora L. (1998) Function of TAF_{II}-containing complex without TBP in transcription by RNA polymerase II. *Nature* **393:** 187–191.

Wieschaus E. and Gehring W. (1976) Clonal analysis of primordial disc cells in the early embryo of *Drosophila melanogaster*. *Dev. Biol.* **50:** 249–263.

Wigglesworth V.B. (1943) The fate of hemoglobin in *Rhodnius prolixus* and other blood-sucking arthropods. *Proc. R. Soc.* London B **131:** 313–339.

Wigglesworth V.B. (1972) *The Principles of Insect Physiology.* 7th ed. London: Chapman and Hall.

Wightman B., Bürglin T.R., Gatto J., Arasu P. and Ruvkun G. (1991) Negative regulatory sequences in the *lin-14* 3'-untranslated region are necessary to generate a temporal switch during *Caenorhabditis elegans* development. *Genes and Dev.* **5:** 1813–1824.

Wightman B., Ha I. and Ruvkun G. (1993) Posttranscriptional regulation of the heterochronic gene *lin-14* by *lin-4* mediates temporal pattern formation in *C. elegans*. *Cell* **75:** 855–862.

Wikramanayake A.H., Uhlinger K., Griffin F.J. and Clark W.H., Jr. (1992) Sperm of the shrimp *Sicyonia ingentis* undergo a bi-phasic capacitation that is accompanied by morphological changes. *Dev. Growth Differ.* **34:** 347–355.

Wikramanayake A.H., Huang L. and Klein W.H. (1998) beta-catenin is essential for patterning the maternally specified animal-vegetal axis in the sea urchin embryo. *Proc. Nat. Acad. Sci. USA* **95:** 9343–9348.

Wilkins A.S. (1986) *Genetic Analysis of Animal Development.* New York: Wiley.

Wilkins A.S. (1993) *Genetic Analysis of Animal Development.* 2nd ed. New York: Wiley.

Wilkinson D.G. (1993) Molecular mechanisms of segmental patterning in the vertebrate hindbrain and neural crest. *BioEssays* **15:** 499–505.

Wilkinson D.G., Bhatt S., Cook M., Boncinelli E. and Krumlauf R. (1989) Segmental expression of *Hox-2* homeobox-containing genes in the developing mouse hindbrain. *Nature* **341:** 405–409.

Wilkinson D.G., Bhatt S. and Hermann B.G. (1990) Expression pattern of the mouse T gene and its role in mesoderm formation. *Nature* **343:** 657–659.

Willadsen S.M. (1986) Nuclear transplantation in sheep embryos. *Nature* **320:** 63–65.

Williams G.C. (1957) Pleiotropy, natural selection and the evolution of senescence. *Evolution* **11**: 398–411.

Williams G.T. and Smith C.A. (1993) Molecular regulation of apoptosis: Genetic controls on cell death. *Cell* **74:** 777–779.

Williams J.C., Weijland A., Gonfloni S., Thompson A., Courtneidge S.A., Superti-Furga G. and Wierenga R.K. (1997) The 2.35 A crystal structure of the inactivated form of chicken Src: A dynamic molecule with multiple regulatory interactions. *J. Mol. Biol.* **274:** 757–775.

Williams R.W. (1998) Plant homeobox genes: Many functions stem from a common motif. *BioEssays* **20:** 280–282).

Willier B.H. and Oppenheimer J.M. (1964) *Foundations of Experimental Embryology.* Englewood Cliffs, NJ: Prentice Hall.

Wilmut I., Schnieke A.E., McWhir J., Kind A.J. and Campbell K.H.S. (1997) Viable offspring derived from fetal and adult mammalian cells. *Nature* **385:** 810–813.

Wilson E., Mai Q., Sudhir K., Weiss R.H. and Ives H.E. (1993) Mechanical strain induces growth of vascular smooth muscle cells via autocrine action of PDGF. *J. Cell Biol.* **123:** 741–747.

Wilson E.B. (1904) Experimental studies on germinal localization. I. The germ regions in the egg of *Dentalium. J. Exp. Zool.* **1:** 1–72.

Wilson E.B. (1905) The chromosomes in relation to the determination of sex in insects. *Science* **22:** 500–502.

Wilson E.B. (1925) *The Cell in Development and Heredity.* 3rd ed. New York: Macmillan.

Wilson H.V. (1907) On some phenomena of coalescence and regeneration in sponges. *J. Exp. Zool.* **5:** 245–258.

Wilson V. and Beddington R.S. (1997) Expression of T protein in the primitive streak is necessary and sufficient for posterior mesoderm movement and somite differentiation. *Dev. Biol.* **192:** 45–58.

Wilson V., Manson L., Skarnes W.C. and Beddington R.S. (1995) The T gene is necessary for normal mesodermal morphogenetic cell movements during gastrulation. *Development* **121:** 887–886.

Winklbauer R. and Keller R.E. (1991) Fibronectin, mesoderm migration, and gastrulation in *Xenopus. Dev. Biol.* **177:** 413–426.

Winklbauer R. and Nagel M. (1991) Directional mesoderm cell migration in the *Xenopus* gastrula. *Dev. Biol.* **148:** 573–589.

Winkler M. (1988) Translational regulation in sea urchin eggs: A complex interaction of biochemical and physiological regulatory mechanisms. *BioEssays* **8:** 157–161.

Witschi E. (1956) *Development of Vertebrates.* Philadelphia: Saunders.

Witthuhn B.A., Quelle F.W., Silvennoinen O., Yi T., Tang B., Miura O. and Ihle J.N. (1993) JAK2 associates with the erythropoietin receptor and is tyrosine phosphorylated and activated following stimulation with erythropoietin. *Cell* **74:** 227–236.

Wodarz A., Ramrath A., Kuchinke U. and Knust E. (1999) Bazooka provides an apical cue for Inscuteable localization in *Drosophila* neuroblasts. *Nature* **402:** 544–547.

Wolberger C., Vershon A.K., Liu B., Johnson A.D. and Pabo C.O. (1991) Crystal structure of a MAT α2 homeodomain-operator complex suggests a general model for homeodomain-DNA interactions. *Cell* **67:** 517–528.

Wolff C., Schröder R., Schulz C., Tautz D. and Klingler M. (1998) Regulation of the *Tribolium* homologues of *caudal* and *hunchback* in *Drosophila:* Evidence for maternal gradient systems in a short germ embryo. *Development* **125:** 3645–3654.

Wolpert L. (1969) Positional information and the spatial pattern of cellular differentiation. *J. Theoret. Biol.* **25:** 1–47.

Wolpert L. (1978) Pattern formation in biological development. *Scientific American* **239** (October): 154–164.

Wolpert L. (1989) Positional information revisited. *Development* **107** (Supplement): 3–12.

Wood W.B. (1980) Bacteriophage T4 morphogenesis as a model for assembly of subcellular structure. *Quart. Rev. Biol.* **55:** 353–367.

Wood W.B., ed. (1988a) *The Nematode Caenorhabditis elegans.* Cold Spring Harbor, NY: Cold Spring Harbor Laboratory Press.

Wood W.B. (1988b) Determination of pattern and fate in early embryos of *Caenorhabditis elegans.* In: *Developmental Biology: A Comprehensive Synthesis,* L.W. Browder (ed.), vol. 5, 57–78. New York: Plenum.

Wood W.B. (1991) Evidence from reversal of handedness in *C. elegans* embryos for early cell interactions determining cell fates. *Nature* **349:** 536–538.

Wood W.B. (1998) Aging of *C. elegans:* Mosaics and mechanisms. *Cell* **95:** 147–150.

Wreden C., Verrotti A.C., Schisa J.A., Lieberfarb M.E. and Strickland S. (1997) Nanos and pumilio establish embryonic polarity in *Drosophila* by promoting posterior deadenylation of hunchback mRNA. *Development* **124:** 3015–3023.

Wu J. and Cohen S.M. (1999) Proximodistal axis formation in the *Drosophila* leg: Subdivision into proximal and distal domains by Homothorax and Distalless. *Development* **126:** 109–117.

Wu X., Vakani R. and Small S. (1998) Two distinct mechanisms for differential positioning of gene expression borders involving the *Drosophila* gap protein giant. *Development* **125:** 3765–3774.

Wu Y.-C., Stanfield G.M. and Horvitz HR. (2000) NUC-1, a *Caenorhabditis elegans* DNase II homolog, functions in an intermediate step of DNA degradation during apoptosis. *Genes and Dev.* **14:** 536–548.

Wyllie F.S., Jones C.J., Skinner J.W., Haughton M.F., Wallis C., Wynford-Thomas D., Faragher R.G. and Kipling D. (2000) Telomerase prevents the accelerated cell ageing of Werner syndrome fibroblasts. *Nat. Genet.* **24:** 16–17.

Xu L., Glass C.K. and Rosenfeld M.G. (1999) Coactivator and corepressor complexes in nuclear receptor function. *Curr. Opin. Genet. Dev.* **9:** 140–147.

Yajima H. (1964) Studies on embryonic determination of the harlequin-fly, *Chironomus dorsalis.* II. Effects of partial irradiation of the egg by ultra-violet light. *J. Embryol. exp. Morphol.* **12:** 59–100.

Yajima H. (1983) Induction of longitudinal double malformations by centrifugation or by partial irradiation of eggs in the chironomid species, *Chironomus samoensis* (Diptera: Chironomidae). *Entomol. Gen.* **8:** 171–191.

Yamada K.M. (1991) Fibronectin and other cell interactive glycoproteins. In: *Cell Biology of Extracellular Matrix,* 2nd ed., E.D. Hay (ed.), 111–146. New York: Plenum.

Yamada T. (1967) Cellular and subcellular events in Wolffian lens regeneration. *Cur. Top. Dev. Biol.* **2:** 247–283.

Yamada T., Placzek M., Tanaka H., Dodd J. and Jessell T.M. (1991) Control of cell pattern in the developing nervous system: Polarizing activity of the floor plate and notochord. *Cell* **64:** 635–647.

Yamamoto Y. and Henderson C.E. (1999) Patterns of programmed cell death in populations of developing spinal motoneurons in chicken, mouse, and rat. *Dev. Biol.* **214:** 60–71.

Yanagimachi R. (1994) Mammalian fertilization. In: *The Physiology of Reproduction,* E. Knobil and J.D. Neill (eds.), 189–317. New York: Raven Press.

Yanagimachi R. and Noda Y.D. (1970) Electron microscope studies of sperm incorporation into the golden hamster egg. *Am. J. Anat.* **128:** 429–462.

Yancopoulos G.D., Klagsbrun M. and Folkman J. (1998) Vasculogenesis, angiogenesis, and growth factors: Ephrins enter the fray at the border. *Cell* **93:** 661–664.

Yandava B.D., Billinghurst L.L. and Snyder E.Y. (1999) "Global" cell replacement is feasible via neural stem cell transplantation: Evidence from the dysmyelinated shiverer mouse brain. *Proc. Nat. Acad. Sci. U S A* **96:** 7029–7034.

Yang J., Chang E., Cherry A.M., Bangs C.D., Oei Y., Bodnar A., Bronstein A., Chiu C.P. and Herron G.S. (1999) Human endothelial cell life extension by telomerase expression. *Biol. Chem.* **274:** 26141–26148.

Yanofsky M.F., Ma H., Bowman J.L., Drews G.N., Feldmann K.A. and Meyerowitz E.M. (1990) The protein encoded by the *Arabidopsis* homeotic gene agamous resembles transcription factors. *Nature* **346:** 35–39.

Yao T.-P., Segraves W.A., Oro A.E., McKeown M. and Evans R.M. (1992) *Drosophila* ultraspiracle modulates ecdysone receptor function via heterodimer formation. *Cell* 71: 63–72.

Yao T.-P., Forman B.M., Jiang Z., Cherbas L., Chen J.D., McKeown M., Cherbas P. and Evans R.M. (1993) Functional ecdysone receptor is the product of EcR and ultraspiracle genes. *Nature* **366:** 476–479.

Yaoita Y. and Brown D.D. (1990) A correlation of thyroid hormone receptor gene expression with amphibian metamorphosis. *Genes and Dev.* **4:** 1917–1924.

Yasuda G.K., Baker J. and Schubiger G. (1991) Temporal regulation of gene expression in the blastoderm *Drosophila* embryo. *Genes and Dev.* **5:** 1800–1812.

Yim D.L., Opresko L.K., Wiley H.S. and Nuccitelli R. (1994) Highly polarized EGF receptor tyrosine kinase activity initiates egg activation in *Xenopus*. *Dev. Biol.* **162:** 41–55.

Yisraeli J.K., Sokol S. and Melton D.A. (1990) A two-step model for the localization of maternal mRNA in *Xenopus* oocytes: Involvement of microtubules and microfilaments in the translocation and anchoring of Vg1 mRNA. *Development* **108:** 289–298.

Yochem J., Weston K. and Greenwald I. (1988) The *Caenorhabditis elegans lin-12* gene encodes a transmembrane protein with overall similarity to *Drosophila Notch. Nature* **335:** 547–551.

Yoder J.A., Soman N.S., Verdine G.L. and Bestor T.H. (1997) DNA (cytosine-5) methyltransferases in mouse cells and tissues. Studies with a mechanism-based probe. *J. Mol. Biol.* **270:** 385–395.

Yokoyama T., Copeland N.G., Jenkins N.A., Mont-gomery C.A., Elder F.F.B. and Overbeek P.A. (1993) Reversal of left-right asymmetry: A situs inversus mutation. *Science* **260:** 679–682.

Yoshida S., Marikawa Y. and Satoh N. (1996) *posterior end mark,* a novel maternal gene encoding a localized factor in the ascidian embryo. *Development* **122:** 2005–2012.

Yost C., Torres M., Miller J.R., Huang E., Kimelman D. and Moon R.T. (1996). The axis-inducing activity, stability, and subcellular distribution of β-catenin is regulated in *Xenopus* embryos by glycogen synthetase kinase 3. *Genes and Dev.* **10:** 1443–1454.

Yost H.J. (1999) Diverse initiation in a conserved left-right pathway? *Curr. Opin. Genet. Dev.* **9:** 422–426.

Young L.J., Lopreato G., Horan K. and Crews D. (1994) Cloning and *in situ* hybridization analysis of estrogen receptor, progesterone receptor and androgen receptor expression in the brain of whiptail lizards (*Cnemidophorus uniparens* and *C. inornatus*). *J. Comp. Neurol.* **347:** 288–300.

Younger-Shepherd S., Vaessin H., Bier E., Jan L.Y. and Jan Y.N. (1992) deadpan, an essential pan-neural gene encoding an HLH protein, acts as a denominator in *Drosophila* sex determination. *Cell* **70:** 911–922.

Yu C.E., Oshima J., Fu Y.H., Wijsman E.M., Hisama F., Alisch R., Matthews S., Nakura J., Miki T., Ouais S., Martin G.M., Mulligan J. and Schellenberg G.D. (1996) Positional cloning of the Werner's syndrome gene. *Science* **272:** 258–262.

Yu C.E., Oshima J. Wijsman E.M., Nakura J., Miki T., Piussan C., Matthews S., Fu Y.H., Mulligan J., Martin G.M. and Schellenberg G.D. (1997) Mutations in the consensus helicase domains of the Werner syndrome gene. *Am. J. Hum. Genet.* **60:** 330–341.

Yuan J. and Horvitz H.R. (1990) The *Caenorhabditis elegans* genes *ced-3* and *ced-4* act cell autonomously to cause programmed cell death. *Dev. Biol.* **138:** 33–41.

Yuan J. and Horvitz H.R. (1992) The *Caenorhabditis elegans* cell death gene *ced-4* encodes a novel protein and is expressed during the period of extensive programmed cell death. *Development* **116:** 309–320.

Yuan J., Shasham S., Ledoux S., Ellis H. and Horvitz H.R. (1993) The *C. elegans* death gene ced-3[1] encodes a protein similar to mammalian interleukin-1β-converting enzyme. *Cell* **75:** 641–652.

Yuge M., Kobayakawa Y., Fujisue M. and Yamana K. (1990) A cytoplasmic determinant for dorsal axis formation in an early embryo of *Xenopus laevis*. *Development* **110:** 1051–1056.

Zachgo S., de Anrade Silva E., MotteP., Tröbner W., Saedler H., Sommer H. and Schwarz-Sommer Z. (1995) Functional analysis of the *Antirrhinum* floral homeotic *deficiens* gene in vivo and in vitro by using a temperature-sensitive mutant. *Development* **121:** 2861–2875.

Zalokar M. and Erk I. (1976) Division and migration of nuclei during early embryogenesis of *Drosophila melanogaster*. *J. Microsc. Biol. Cell.* **25:** 97–106.

Zamore P.D., Tuschl T., Sharp P.D. and Bartel D.P. (2000) RNAi: Double-stranded RNA directs the ATP-dependent cleavage of mRNA at 21 to 23 nucleotide intervals. *Cell* **101:** 25–33.

Zanetti N.C. and Solursh M. (1984) Induction of chondrogenesis in limb mesenchymal cultures by disruption of the actin cytoskeleton. *J. Cell Biol.* **99:** 115–123.

Zecca M., Basler K. and Struhl G. (1995) Sequential organizing activities of engrailed, hedgehog, and decapentaplegic in the *Drosophila* wing. *Development* **121:** 2265–2278.

Zecca M., Basler K. and Struhl G. (1996) Direct and long-range action of a wingless morphogen gradient. *Cell* **87:** 833–844.

Zeller R. and Duboule D. (1997) Dorso-ventral limb polarity and origin of the ridge: on the fringe of independence? *BioEssays* **19:** 541–546.

Zernicka-Goetz M., Pines J., Ryan K., Siemering K.R., Haseloff J., Evans M.J. and Gurdon J.B. (1996) An indelible lineage marker for *Xenopus* using a mutated green fluorescent protein. *Development* **122:** 3719–3724.

Zhang B., Gallegos M., Puoti A., Durkin E., Fields S., Kimble J. and Wickens M.P. (1997) A conserved RNA-binding protein that regulates sexual fates in the *C. elegans* hermaphrodite germ line. *Nature* **390:** 477–484.

Zhang C.-C., Müller J., Hoch M., Jäckle H. and Bienz M. (1991) Target sequences for hunchback in control region conferring *Ultrabithorax* expression boundaries. *Development* **113:** 1171–1179.

Zhang J. and King M.L. (1996) *Xenopus* VegT RNA is localized to the vegetal cortex during oogenesis and encodes a novel T-box transcription factor involved in mesodermal patterning. *Development* **122:** 4119–4129.

Zhang J., Houston D.W., King M.L., Payne C., Wylie C. and Heasman J. (1998) The role of maternal VegT in establishing the primary germ layers in *Xenopus* embryos. *Cell* **94:** 515–524.

Zhang W., Behringer R.R. and Olson E.N. (1995) Inactivation of the myogenic bHLH gene *MRF4* results in up regulation of myogenin and and rib anomalies. *Genes and Dev.* **9:** 1388–1399.

Zhang X.-K., Hoffmann B., Tran P.B.-V., Graupner G. and Pfahl M. (1992) Retinoid X receptor is an auxiliary protein for thyroid hormone and retinoic acid receptors. *Nature* **355:** 441–446.

Zhao J.J., Lazzarini R.A. and Pick L. (1993) The mouse *Hox-1.3* gene is functionally equivalent to the *Drosophila Sex combs reduced* gene. *Genes and Dev.* **7:** 343–354.

Zhou J. and Levine M. (1999) A novel cis-regulatory element, the PTS, mediates an anti-insulator activity in the *Drosophila* embryo. *Cell* **99:** 567–575.

Zhou X., Sasaki H., Lowe L., Hogan B.L.M. and Kuehn M.R. (1993) *Nodal* is a novel TGF-β-like gene expressed in the mouse node during gastrulation. *Nature* **361:** 543–547.

Zhu C., Urano J. and Bell L.R. (1997) The Sex-lethal early splicing pattern uses a default mechanism dependent on the alternative 5' splice sites. *Mol. Cell Biol.* **17:** 1674–1681.

Zhu J., Fukushige T., McGhee J.D. and Rothman J.H. (1998) Reprogramming of early embryonic blastomeres into endodermal progenitors by a *Caenorhabditis elegans* GATA factor. *Genes and Dev.* **12:** 3809–3814.

Zhu Z. and Miller J.B. (1997) MRF4 can substitute for myogenin during early stages of myogenesis. *Dev. Dyn.* **209:** 233–241.

Zimmerman K., Shih J., Bars J., Collazo A. and Anderson D.J. (1993) *XASH-3*, a novel *Xenopus achaete-scute* homolog, provides an early marker of planar neural induction and position along the mediolateral axis of the neural plate. *Development* **119:** 221–232.

Zimmerman L.B., De Jesus-Escobar J.M. and Harland R.M. (1996) The Spemann organizer signal noggin binds and inactivates bone morphogenetic protein 4. *Cell* **86:** 599–606.

Ziomek C.A. and Johnson M.H. (1982) The roles of phenotype and position in guiding the fate of 16–cell mouse blastomeres. *Dev. Biol.* **91:** 440–447.

Zouros E., Freeman K.R., Ball A.O. and Pogson G.H. (1992) Direct evidence for extensive paternal mitochondrial DNA inheritance in the marine mussel *Mytilus. Nature* **359:** 412–414.

Zucker R.S. and Steinhardt R.A. (1978) Prevention of the cortical reaction in fertilized sea urchin eggs by injection of calcium-chelating ligands. *Biochim. Biophys. Acta* **541:** 59–466.

Zusman S., Grinblat Y., Yee G., Kafatos F.C. and Hynes R.O. (1993) Analyses of PS integrin functions during *Drosophila* development. *Development* **118:** 737–750.

Zwilling E. (1955) Ectoderm-mesoderm relationship in the development of the chick embryo limb bud. *J. Exp. Zool.* **128:** 423–441.

Zwilling E. (1956) Interaction between limb bud ectoderm and mesoderm in the chick embryo. I. Axis establishment. *J. Exp. Zool.* **132:** 157–172.

credits for photographs and line art

PHOTOGRAPH CREDITS

Part Openers **I.1:** From Craig M.M. (1992) *Cell* 8(2): Cover. Cell Press. Courtesy of M. M. Craig; **II.1:** From Brinster R.L. and Palmiter R.D. (1986) Introduction of genes into the germ line of animals. *The Harvey Lectures*, Series 80: 1–38. Reprinted by permission of John Wiley & Sons, Inc. Photograph courtesy of R.L. Brinster; **III.1:** From Cheng T.-C., Wallace M.C., Merlie J.P. and Olson E.N. 9 July 1993. *Science*: Cover. ©1993 American Association for the Advancement of Science. Courtesy of T.-C. Cheng and E.N. Olson.

Chapter 1 **1.1:** From Wassarman P.M. (1990) *Development* 108: 1–17. Company of Biologists Ltd. Courtesy of Paul M. Wassarman; **1.2a–i:** From Maro B. Gueth-Hallonet C., Aghion J. and Antony C. (1991) *Development* (Supplement 1): 17–25. Company of Biologists Ltd. Courtesy of Bernard Maro; **1.3:** Reproduced from Blechschmidt E. (1961) *The Stages of Human Development before Birth*. Philadelphia: Saunders; **1.7a–i:** From Johnson K.E., Nakatsuji N. and Boucaut J.-C. (1990) Extracellular matrix control of cell migration during amphibian gastrulation. In: *Cytoplasmic Organization Systems*, G.M. Malacinski (ed.), 349–374. New York: McGraw-Hill. Courtesy of De-Li Shi; **1.10:** Courtesy of Muriel Voter Williams; **1.20a–b:** From Frohnhofer H.G. and Nusslein-Volhard C. (1986) *Nature* 324: 120–125. Courtesy of C. Nusslein-Volhard; **1.21b–c:** Courtesy of E.B. Lewis; **1.22:** Courtesy of Walter Gehring.

Chapter 2 **2.1:** From Rhodin J.A.G. (1975) *An Atlas of Histology*. New York: Oxford University Press. Fig. 2–1, p. 8. Courtesy of Johannes Rhodin; **2.6:** Courtesy of Mary Osborn and Klaus Weber; **2.7b:** Reproduced from Goodenough U.W. and Heuser J.E. (1982) *J. Cell Biol.* 95: 798–815. Rockefeller University Press. Courtesy of U.W. Goodenough; **2.11:** From Inoue, S., and L.G. Tilney (1982) *J. Cell Biol.* 93: 812–819 and Inoue, S. (1981) *J. Cell Biol.* 89: 346–356. Rockefeller University Press. Courtesy of Shinya Inoue, Marine Biological Laboratory, Woods Hole, MA; **2.12:** From Gard D.L., Cha B.J., and King E. (1997) *Devel. Biol.* 184: 95–114. ©1997 by Academic Press, reproduced by permission of the publisher. Courtesy of David L. Gard; **2.15b:** From Beams H.W. and Kessel R.G. (1976) *Am. Sci.* 64: 279–290. Courtesy of R.G. Kessel; **2.21b–c:** From Smith S.J. (1988) Neuronal cytomechanics: the actin-based motility of growth cones. *Science* 242: 709. ©1988 American Association for the Advancement of Science. Courtesy of S.J. Smith.

Chapter 3 **3.1:** Reproduced from Needham J. (1959) *A History of Embryology*. London/New York: Abelard-Schuman; **3.7b:** Provided by D.W. Fawcett; **3.10:** David M. Phillips/Visuals Unlimited; **3.12:** Courtesy of Joseph G. Gall, Carnegie Institution; **3.15a-b:** From K. Bier; courtesy of D. Ribbert; **3.16:** From Telfer W.H., Huebner E. and Smith D.S. (1982) The cell biology of vitellogenic follicles in Hyalophora and Rhodnius. In: *Insect Ultrastructure*, vol. I, 118–149. New York: Plenum. Courtesy of W.H. Telfer; 3.21d: D.W. Fawcett/Visuals Unlimited; **3.22a:** From Bloom W. and Fawcett D.W. (1975) *A Textbook of Histology*, 10th ed. Philadelphia: Saunders. Provided by D.W. Fawcett; **3.22b:** Provided by E.A. Anderson and David Albertini; **3.24b:** Neg#338433. Courtesy Department of Library Services, American Museum of Natural History.

Chapter 4 **4.1:** David M. Phillips/Visuals Unlimited; **4.5a-b:** From Vacquier V.D. (1979) *Current Topics Dev. Biol.* 13: 31–44. Courtesy of V.D. Vacquier, University of California, San Diego; **4.7:** From Epel D. (1977) *Sci. Amer.* 237: 128–138. Provided by Mia Tegner and David Epel; **4.8a–b:** From Epel D. (1977) *Sci. Amer.* 237: 128–138. Courtesy of Gerald Schatten; **4.8c:** Transmission electron micrograph of sperm uteralization through the fertilization core. Courtesy of F.J. Longo; **4.9a:** Reproduced from Hadek R. (1969) *Mammalian Fertilization: An Atlas of Ultrastructure*. New York: Academic Press; **4.14:** From Gilkey J.C., Jaffe L.F., Ridgway E.G. and Reynolds G.T. (1978) *J. Cell Biol.* 76: 448–466. Rockefeller University Press. Courtesy of Lionel Jaffe; **4.15:** From Hamaguchi M.S. and Hiramoto Y. (1980) *Dev. Growth Differ.* 22: 517–530. Courtesy of M.S. Hamaguchi and Y. Hiramoto, Tokyo Institute of Technology; **4.18b–c:** From Schuel H. (1985) *Functions of egg cortical granules. In: Biology of Fertilization*, C.B. Metz and A. Monroy (eds.), vol. 3. New York: Academic Press. Courtesy of Herbert Schuel; **4.18d:** From Vacquier V.D. (1975) *Dev. Biol.* 43: 64–65. ©1975 by Academic Press, reproduced by permission of the publisher. Courtesy of Victor Vacquier; **4.19a–c:** From M. Tegner and D. Epel. 16 February 1973. "Sea urchin sperm-egg interactions studied with the scanning electron microscope." *Science*, 179:685–688. ©1973 American Association for the Advancement of Science. Provided by Mia Tegner and David Epel.

Chapter 5 **5.1:** From Beams H.W. and Kessel R.G. (1976) *Am. Sci.* 64: 279–290. Courtesy of R.G. Kessel; **5.10a–b:** From Fleming T.P., Pickering S.J., Qasim F. and Maro B. (1986) *J. Embryol. exp. Morphol.* 95: 169–191. Company of Biologists Ltd. Courtesy of Tom Fleming; **5.12a-f:** From Bavister B.d. (ed.) (1987) *The Mammalian Preimplantation Embryo: Regulation of Growth and Differention in Vitro*. New York: Plenum. Courtesy of D.E. Boatman and B.D. Bavister; **5.14(all):** From Beams H.W. and Kessel R.G. (1976) *Am. Sci.* 64: 279–290. Courtesy of R.G. Kessel; **5.18(all):** From Turner F.R. and Mahowald A.P. (1976) *Dev. Biol.* 50: 95–108. ©1976 by Academic Press, reproduced by permission of the publisher. Courtesy of F.R. Turner, Indiana University; **5.20a–c:** From Schroeder T.E. (1987) *Dev. Biol.* 124: 9–22. ©1987 by Academic Press, reproduced by permission of the publisher. Courtesy of T.E. Schroeder; **5.22a–d:** From Rappaport R. (1961) *J. Exp. Zool.* 148: 81–89. Photographs courtesy of R. Rappaport. Reprinted by permission of Wiley-Liss, Inc., a subsidiary of John Wiley & Sons, Inc.; **5.23:** From Rappaport R. and Rappaport B.N. (1994) *Devel. Biol.* 164: 258–266. ©1994 by Academic Press, reproduced by permission of the publisher. Courtesy of R. Rappaport; **5.28:** From Hibino T., Nishikata T. and Nishida H. (1998) *Develop. Growth Differ.* 40: p. 88, Fig. 3. Courtesy of H. Nishida; **5.33a,c:** From Edgar B.A. and O'Farrell P.H. (1989) *Cell* 57: 177–187. Cell Press. Courtesy of Bruce Edgar.

Chapter 6 **6.1:** From Spemann H. (1928) *Zeitschrift fur wiss. Zool* 132: 105–134. Print courtesy of Klaus Sander; **6.3a–e:** From Strehlow D. and Gilbert W. (1993) *Nature* 361: 451–453. Courtesy of David Strehlow; **6.12:** Reproduced from Spemann H. (1938) *Embryonic Development and Induction*. New Haven: Yale University Press. Reprinted 1967. New York: Hafner; **6.17b–c:** Courtesy of Dianne Fristrom; **6.23a–e:** From Le Douarin N.L. and McLaren A. (eds.) (1984) *Chimeras in Developmental Biology*. London/Orlando: Academic Press. Courtesy of V.E. Papaioannou; **6.M2a–b:** Reproduced from Garcia-Bellido A., Lawrence P.A. and Morata G. (1979) *Sci. Amer.* 241: 102–110.

Chapter 7 **7.1:** Reproduced from Steward F.C. (1963) *Sci. Amer.* 209: 104–115; **7.4a–h:** From Reyer R.W. (1954) *Quart. Rev. Biol.* 29: 1–46. Courtesy of R.W. Reyer; **7.5a:** John D. Cunningham/Visuals Unlimited; **7.5b:** Courtesy of Veikko Sorsa; **7.6:** Reproduced from Beermann W. (1972) Chromosomes and genes. In: *Developmental Studies on Giant Chromosomes*, W. Beermann (ed.), 1–33. Published with permission from Springer-Verlag, Heidelberg; **7.7:** Preparation and photograph by A. von Brunn. Print supplied by K. Kalthoff; **7.8a:** From Brown D.D. and Dawid I.B. (1968) Specific gene amplification in oocytes. *Science* 160: 272–280. ©1986 American Association for the Advancement of Science. Courtesy of I.B. Dawid; **7.8b:** Courtesy of Ulrich Scheer; **7.15a–b:** Courtesy of John

Gurdon; **7.17b:** From Wilmut I., Schnieke A.E., McWhir J., Kind A.J., and Campbell K.H.S. (1997) *Nature* 385: 810–813. Provided by I. Wilmut; **7.19a-b:** From De Robertis E.M. and Gurdon J.B. (1979) *Sci. Amer.* 241: 74–82. Courtesy of E.M. De Robertis and J.B. Gurdon.

Chapter 8 **8.1a-c:** From Frigerio G., Burri M., Bopp D., Baumgartner S. and Noll M. (1986) *Cell* 47:735–746. Cell Press. Courtesy of Markus Noll; **8.5:** Courtesy of M.R. Dohmen; **8.7d:** From Mahowald A.P. (1971) *J. Exp. Zool.* 176: 329–344. Reprinted by permission of Wiley-Liss, Inc., a subsidiary of John Wiley & Sons, Inc. Courtesy of Anthony P. Mahowald; **8.8b–d:** From Okada M., Kleinman I.A. and Schneiderman H.A. (1974) *Dev. Biol.* 37: 43–54. ©1974 by Academic Press, reproduced by permission of the publisher. Courtesy of M. Okada and S. Kobavashi; **8.10:** Courtesy of Chris Yohn and Ruth Lehmann; **8.12:** From Satoh N., Deno T., Nishida H., Nishikata T. and Makabe K.W. (1990) *Int. Rev. Cytol.* 122: 221–258. Courtesy of N. Satoh; **8.16:** From Jeffery W.R. (1988) *Dev. Biol.* 5: 3–56. ©1988 by Academic Press, reproduced by permission of the publisher. Courtesy of William R. Jeffery; **8.17:** From Kislauskis E.H., Zhu X., and Singer R.H. (1994) *J. Cell Biol.* 127: 441–451. Rockefeller University Press. Courtesy of Edward H. Kislauskis; **8.19a–b:** Courtesy of Klaus Kalthoff; **8.22a–d:** From Rau K.-G. and Kalthoff K. (1980). *Nature* 287: 635–637. Courtesy of Klaus Kalthoff; **8.23:** From Render J. (1983) *Wilhelm Roux's Arch. Dev. Biol.* 192: 120–129. Springer-Verlag. Courtesy of Jo Ann Render; **8.26:** From Melton D.A. (1987) *Nature* 328: 80–82. Courtesy of Douglas Melton.

Chapter 9 **9.1:** Courtesy of Jeffrey Brown and Randall Moon; **9.5a–b:** From Quatrano R.S. and Shaw S.L. (1997) *Trends Plant Sci.* 2: 15–21 Elsevier Science. Courtesy of R. Quatrano; **9.9c:** From Sudarwati S. and Nieuwkoop P.D. (1971) *Wilhelm Roux's Arch. Dev. Biol.* 166: 189–204. Springer-Verlag. Courtesy of Pieter D. Nieuwkoop; **9.10a–c:** From Danilchik M.V. and Denegre J.M. (1991) *Developement* 111: 845–856. Company of Biologists Ltd. Courtesy of Michael V. Danilchik; **9.12a–b:** From Larabell C.A., Rowning B.A., Wells J., Wu M. and Gerhart J. (1996) *Development* 122: 1284, Fig. 3. Company of Biologists Ltd. Courtesy of C. Larabell; **9.14:** From Fujisue M., Kobayakawa Y. and Yamana K. (1993) *Development* 118: 163–170. Company of Biologists Ltd. Courtesy of Megumi Fujisue; **9.22:** From Schneider S., Steinbeisser H., Warga R.M. and Hausen P. (1996) *Mech. Dev.* 57: 191–198. Elsevier Science. Courtesy of P. Hausen; **9.23:** From Lowe L.A., Supp D.M., Sampath K., Yokoyama T., Wright C.V.E., Potter S.S., Overbeek P. and Kuehn M.R. (1996) *Nature* 381: 158–161. Courtesy of M.R. Kuehn.

Chapter 10 **10.1:** Courtesy of Charles Ettensohn; **10.7a–b:** From Ettensohn C.A. and Ingersoll E.P. (1992) Morphogenesis of the sea urchin embryo. In: *Morphogenesis*, E.F. Rossomando and S. Alexander (eds.) 189–262. New York: Marcel Dekker, Inc. N.Y. Courtesy of Charles Ettensohn; **10.9a:** From Hardin J. (1989) *Dev. Biol.* 136: 430–445. ©1989 by Academic Press, reproduced by permission of the publisher. Courtesy of Jeff Hardin; **10.10:** From Miller J., Fraser S.E., and McClay D. (1995) *Development* 121: 2501–2511. Company of Biologists Ltd. Courtesy of D. McClay; **10.18b–c:** From Keller R.E. and Tibbetts P. (1989) *Dev. Biol.* 131: 539–549. ©1989 by Academic Press, reproduced by permission of the publisher. Courtesy of Ray Keller; **10.20a–f:** From Keller R.E. (1980) *J. Embryol. Exp. Morphol.* 60: 201–243. Company of Biologists Ltd. Courtesy of Ray Keller; **10.21a, d:** From Kimmel, C.B., W.W. Ballard, S.R. Kimmel, B. Ullmann, and T.F. Schilling (1995) *Devel. Dyn.* 203: 253–310. Courtesy of C.B. Kimmel and B. Ullmann; **10.25:** Courtesy of J.P. Revel, Caltech; **10.28:** From Catala M., Teillet M.-A., De Robertis E.M., and Le Douarin N.M. (1996) *Development* 122: 2599–2610. Company of Biologists Ltd. Courtesy of N.M. Le Douarin; **10.29a–b:** From Carlson B.M. (1988) *Patten's Foundations of Embryology*, 5th ed. New York: McGraw-Hill. Courtesy of Bruce M. Carlson.

Chapter 11 **11.1:** From Morrill J.B. (1986) Scanning electron microscopy of embryos. In: *Methods in Cell Biology*, T. Schroeder (ed.), 263–292. New York: Academic Press. Courtesy of John B. Morrill; **11.4a–c:** From Steinberg M.S. (1963) Reconstruction of tissues by dissociated cells. *Science* 141: 401–408. ©1963 American Association for the Advancement of Science, Courtesy of M.S. Steinberg; **11.9:** From Kintner C.R. (1992) *Cell* 69: 225–236. Cell Press. Courtesy of Chris Kintner; **11.11:** Courtesy of Kimiko Hayashi and Robert Trelstad; **11.13a:** From Alberts B., Bray D., Lewis J., Raff M. Roberts K. and Watson J.D. (1989) *Molecular Biology of the Cell*. 2nd Ed. New York: Garland Publishing, Inc. Photo by Lawrence Rosenberg; **11.19a:** Courtesy of J.B. Morrill; **11.20:** From Takeichi M. (1988) *Development* 102: 639–655. Company of Biologists Ltd. Courtesy of Masatoshi Takeichi.; **11.21:** Provided by Kurt E. Johnson; **11.22b:** From Shi D.-L., Darribere T., Johnson K.E. and Boucaut J.-C. (1989) *Development* 105: 351–363. Company of Biologists Ltd. Courtesy of De-Li Shi.; **11.23b:** From Boucaut J.-C., Darribere T., Boulekbache H. and Thiery J.P. (1984) *Nature* 307: 364–367. Courtesy of J.C. Boucaut; **11.24b–c:** From Boucaut J.-C., Darribere T., Boulekbache H. and Thiery J.P. (1984) *Nature* 307: 364–367. Courtesy of J.C. Boucaut.

Chapter 12 **12.1a–c:** From Schoenwolf G.C. (1991) *Development* (Supplement 2): 160. Company of Biologists Ltd. Courtesy of Gary Schoenwolf; **12.1d:** Smith J.L. and Schoenwolf G.C. (1997). *Trends in Neurosci.* 20: 510–517. Elsevier Science. Courtesy of G.C. Schoenwolf; **12.12b–d:** Reproduced from Gordon R. and Jacobson A.G. (1978) *Sci. Amer.* 238: 106–113. Courtesy of R. Gordon; **12.19(both):** Courtesy of Lauri Saxen; **12.20b–c:** Reproduced from Bodemer C.W. (1968) *Modern Embryology*. New York: Holt, Rinehart and Winston; **12.23c:** From Keller R. and Danilchik M. (1988) *Development* 103: 193–209. Company of Biologists Ltd. Courtesy of Ray Keller; **12.24a–b:** From Doniach T., Phillips C.R. and Gerhart J.C. (1992) Planar induction of anteroposterior pattern in the developing central nervous system of Xenopus laevis. *Science* 257: 542–545. ©1992 American Association for the Advancement of Science. Courtesy of J.C. Gerhart.

Chapter 13 **13.1:** Reproduced from Palay S.L. and Chan-Palay V. (1974) *Cerebellar Cortex: Cytology and Organization*, Fig. 7, p. 12. Berlin and Heidelberg: Springer-Verlag; **13.6:** H. Webster, D. Fawcett/Visuals Unlimited; **13.9a–b:** From Goulding M.D., Lumsden A. and Gruss P. (1993) *Development* 117: 1001–1016. Company of Biologists Ltd. Courtesy of A. Lumsden; **13.23:** From Tosney K.W. (1978) *Dev. Biol.* 62: 327. ©1978 by Academic Press, reproduced by permission of the publisher. Provided by Kathryn W. Tosney; **13.25b:** From Weston J.A. (1963) *Dev. Biol.* 6:279–310. ©1963 by Academic Press, reproduced by permission of the publisher. Provided by J.A. Weston; **13.26a–c:** From Le Douarin N.M. 28 March 1986. "Cell line segregation during peripheral nervous system ontogeny." *Science*, 231:1515–1522. ©1986 American Association for the Advancement of Science. Courtesy of N. Le Douarin; **13.30a–b:** From Bronner-Fraser M. and Fraser S.E. (1991) *Development* (Supplement 2): 17–22. Company of Biologists Ltd. Print courtesy of Marianne Bronner-Fraser; **13.31a–b:** From Lofberg J., Perris R. and Epperlein H. (1989) *Dev. Biol.* 131: 168–181. ©1989 by Academic Press, reproduced by permission of the publisher. Courtesy of Jan Lofberg; **13.35:** Reproduced from Balinsky B.I. (1975) *An Introduction to Embryology*. Philadelphia: Saunders.

Chapter 14 **14.1:** Courtesy of Chester F. Reather; **14.14b:** From Meier S. (1979) *Dev. Biol.* 73: 25–45. ©1979 by Academic Press, reproduced by permission of the publisher. Photo by Stephen Meier. Courtesy of Antone Jacobson; **14.15:** Courtesy of Chris Kintner; **14.16:** From Takeichi M. (1988) *Development* 102: 639–655. Company of Biologists Ltd. Courtesy of Masatoshi Takeichi; **14.20:** From Fawcett D.W. (1986) *Textbook of Histology*, 11th ed. Philadelphia: Saunders. Provided by D.W. Fawcett; **14.22:** From Rhodin J.A.G. (1974) *Histology: A Text and Atlas*. New York: Oxford University Press. Fig. 10–20, p. 124. Courtesy of Johannes A.G. Rhodin; **14.23a–b:** From Shimada Y., Fischman D.A. and Moscona A.A. (1967) *J. Cell Biol.* 35: 445–453. Rockefeller University Press. Courtesy of Yutaka Shimada; **14.25:** From Poole T. and Steinberg M. (1981)

J. Embryol. Exp. Morphol. 63: 3. Company of Biologists Ltd. Courtesy of Malcolm S. Steinberg; **14.31b–e:** After Cleaver O. and Krieg P.A. (1998) *Development* 125: 3905–3914. Company of Biologists Ltd. Courtesy of Ondine Cleaver; **14.32e:** From Hurle J.M., Icardo J.M. and Ojeda J.L. (1980) *J. Embryol. Exp. Morphol.* 56: 211–223. Company of Biologists Ltd. Courtesy of J.M. Hurle; **14.36b:** D.W. Fawcett/Photo Researchers.

Chapter 15 **15.1:** Courtesy of Ralph Brinster; **15.4:** Courtesy of Trudi Schupbach; **15.5a–b:** From Kaufman, T.C., M.A. Seeger and G. Olsen. 1990. *Advances in Genetics* 27: 309–362. T.R.F. Wright (ed.). San Diego: Academic Press. Courtesy of Thomas C. Kaufman; **15.8:** From Rennebeck G.M., Lader E., Chen Q., Bohm R.A., Caiz. S., Faust C., Magnuson T., Pease L.A. and Artzt K. (1995) *Devel. Biol.* 172: 206–217. ©1995 by Academic Press, reproduced by permission of the publisher. Courtesy of Karen Artzt; **15.9:** From Haffter P. and 16 others (1996) *Development* 123: 1–36. Company of Biologists Ltd. Courtesy of Christiane Nusslein-Volhard; **15.10a–b:** From Meyerowitz E.M., Bowman J.L., Brockman L.L., Drews G.N., Jack T., Sieburth L.E. and Weigel D. (1991) *Development* (Supplement 1): 157–167. Company of Biologists Ltd. Courtesy of Elliot Meyerowitz; **15.12:** From Rubin G.M. and Spradling A.C. (1982) Genetic transformation of *Drosophila* with transposable element vectors. *Science* 218: 348–353. ©1982 American Association for the Advancement of Science. Courtesy of Gerald M. Rubin; **15.13:** Courtesy of Ondine Cleaver and Paul Krieg; **15.15:** From Wang S. and Hazelrigg T. (1994). *Nature* 369: 400–403. Courtesy of T. Hazelrigg; **15.20:** From Rubin G.M. and Spradling A.C. (1982) Genetic transformation of *Drosophila* with transposable element vectors. *Science* 218: 348–353. ©1982 American Association for the Advancement of Science. Courtesy of Gerald M. Rubin.

Chapter 16 **16.1:** From Old R.W., Callan H.G. and Gross K.W. (1977) *J. Cell Sci.* 27: 57–79. Company of Biologists Ltd. Courtesy of R.W. Old; **16.2:** From Hume C.R. and Dodd J. (1993) *Development* 119: 1147–1160. Company of Biologists Ltd. Courtesy of Jane Dodd; **16.6b–c:** From Gehring W.J. (1987) *Cell* 50: 963–974. Cell Press. Courtesy of Walter Gehring; **16.8a–b:** From Nusslein-Volhard C. (1991) *Development* (Supplement 1): 1–10. Company of Biologists Ltd. Courtesy of Christiane Nusslein-Volhard; **16.9a–c:** From Tautz D. (1988) *Nature* 332: 281–284. Courtesy of D. Tautz; **16.11a–d:** From Driever W., Thoma G. and Nusslein-Volhard C. (1989a) *Nature* 340: 363–367. Courtesy of C. Nusslein-Volhard; **16.14:** Reproduced from Littau V.C., Allfrey V.G., Frenster J.H. and Mirsky A.E. (1964) *Proc. Nat. Acad. Sci. USA* 52: 93–100; **16.15a–c:** Reproduced from Moore K.L. (1982) *The Developing Human: Clinically Oriented Embryology.* 3rd ed. Philadelphia: Saunders; **16.16a–b:** From Lezzi M. and Gilbert L.I. (1969) *Proc. Nat. Acad. Sci USA* 64: 498–503. Courtesy of Lawrence I. Gilbert; **16.17a:** Courtesy of Victoria Foe; **16.17b:** From Alberts B., Bray D. Lewis J., Raff M., Roberts K. and Watson J.D. (1983) *Molecular Biology of the Cell*, 388. New York/London: Garland Publishing, Inc. Courtesy of Barbara Hamkalo; **16.21a–d:** From Hafen E., Levine M. and Gehring W. (1984) *Nature* 307: 287–289. Courtesy of E. Hafen.

Chapter 17 **17.1:** From Akey C.W. (1992) The nuclear pore complex: A macromolecular transporter. In: *Nuclear Trafficking*, C.M. Feldherr (ed.), San Diego: Academic Press. Courtesy of Christopher W. Akey; **17.2a:** From Chambon P. (1982) *Sci. Amer.* 244: 60–77. Courtesy of Pierre Chambon; **17.5c:** From Reed R., Griffith J. and Maniatis T. (1988) *Cell* 53: 955. Cell Press. Provided by Jack Griffith; **17.9a–b:** From Wilkins A.S. (1986) *Genetic Analysis of Animal Development*: Reprinted by permission of John Wiley & Sons, Inc. Courtesy of Bruce Baker; **17.9c:** From Wilkins A.S. (1986) *Genetic Analysis of Animal Development*. Reprinted by permission of John Wiley & Sons, Inc. Courtesy of D. Gubb; **17.11:** From Nagoshi R.N., McKeown M., Burtis K.C., Belote J.M. and Baker B.S. (1988) *Cell* 53: 229–236. Cell Press. Courtesy of Michael McKeown; **17.14b:** Adapted from Emeson R.B., Hedjran F., Yeakley J.M., Guise J.W. and Rosenfeld M.G. (1990) *Nature* 341: 76–80. Courtesy of Ronald Emeson; **17.16:** From Scheer U., Dabauvalle M.-C., Merket H. and Denavente R. (1988) *Cell Biol. Int. Rep.* 12: 669–689. Courtesy of Ulrich Scheer; **17.18a–d:** From Angerer R. and Angerer L. (1983) *Dev. Biol.* 100: 197–206. ©1983 by Academic Press, reproduced by permission of the publisher. Courtesy of Lynne M. Angerer; **17.19b–g:** Courtesy of J.T. Bonner.

Chapter 18 **18.1:** From Bloom W. and Fawcett D.W. (1975) *A Textbook of Histology.* Philadelphia: Saunders. Fig. 2–5, p. 41. Provided by D.W. Fawcett; **18.12:** From Rosenthal E.T., Hunt T. and Ruderman J.V. (1980) *Cell* 20: 487–494. Cell Press. Courtesy of Joan Ruderman; **18.13b:** From Rosenthal E.T., Hunt T. and Ruderman J.V. (1980) *Cell* 20: 487–494. Cell Press. Courtesy of Joan Ruderman; **18.13c:** From Rosenthal E.T., Hunt T. and Ruderman J.V. (1980) *Cell* 20: 487–494. Cell Press. Courtesy of Joan Ruderman; **18.15a–b, 18.16:** From Schäfer M., Nayerna K., Engel W. and Schäfer U. (1995) *Devel. Biol.* 172: 344–352. ©1995 by Academic Press, reproduced by permission of the publisher. Courtesy of M. Schäfer; **18.21:** From Marchini M., Morocutti M., Ruggeri A., Koch M.H.J., Bigi A. and Roveri N. (1986) *Connective Tissue Res.* 15: 269–281. Courtesy of Maurizio Marchini.

Chapter 19 **19.1:** From Tamm S.L. (1972) *J. Cell Biol.* 55: 250–255. Rockefeller University Press. Courtesy of Sidney L. Tamm; **19.3a–b:** S.L. Flegler/Visuals Unlimited; **19.3c:** Reproduced from Dykes G., Crepeau R.H. and Edelstein S.J. (1978) *Nature* 272: 509; **19.4a:** From Butler P.J.G. and Klug A. (1978) *Sci. Amer.* 239: 62–69. Provided by P.J.G. Butler; **19.10a–b:** Reproduced from Asakura S., Eguchi G. and Iino T. (1966) *J. Mol. Biol.* 16: 302–316; **19.13b:** Reproduced from Macleod A.G. (1973) *Cytology: The Cell and Its Nucleus.* Kalamazoo, MI: Courtesy, The Upjohn Company; **19.14b:** Reproduced from Ehret C.F. and McArdle E.W. (1974) The structure of Paramecium as viewed from its constituent levels of organization. In: *Paramecium—A Current Survey*, W.J. van Wagtendonk (ed.), 263–338. Amsterdam: Elsevier Scientific Publishing; **19.17:** From Grimes G.W., McKenna M.E., Goldsmith-Spoegler C.M. and Knaupp E.A. (1980) Patterning and assembly of ciliature are independent processes in hypotrich ciliates. *Science* 209: 281–283. ©1980 American Association for the Advancement of Science. Provided by Gary W. Grimes; **19.20a–d:** From Albrecht-Buehler G. (1977) *J. Cell Biol.* 72: 595–603. Rockefeller University Press. Courtesy of Guenter Albrecht-Buehler.

Chapter 20 **20.1:** From Kessel R.G. and Kardon R.H. (1979) *Tissues and Organs: A Text-Atlas of Scanning Electron Microscopy*. 37. San Francisco: W.H. Freeman. Courtesy of Richard G. Kessel; **20.2:** From Bolender R.P. (1974) *J. Cell Biol.* 61: 272. Rockefeller University Press. Courtesy of Robert P. Bolender; **20.3:** From Shih G. and Kessel R. (1982) *Living Images.* Boston, MA: Science Books International. Courtesy of Richard G. Kessel; **20.4:** From Bissel M.J. and Hall H.G. (1987) Form and function in the mammary gland: The role of the extracellular matrix. In: *The Mammary Gland: Development, Regulation, and Function*, M.C. Neville and C.W. Daniel (eds), 128. New York: Plenum. Courtesy of Joanne Emerman; **20.10a–c:** From Littlefield C.L. (1991) *Dev. Biol.* 143: 378–388. ©1991 by Academic Press, reproduced by permission of the publisher. Courtesy of C. Lynne Littlefield; **20.14e:** Reproduced from Keller G., Paige C., Gilboa E. and Wagner E.R. (1985) *Nature* 318: 149–154; **20.21:** From Myer A., Wagner D.S., Vivian J.L., Olson E.N. and Klein W.H. (1997) *Devel. Biol.* 185: 127–138. ©1997 by Academic Press, reproduced by permission of the publisher. Courtesy of A. Myer.

Chapter 21: **21.1:** F. Stuart Westmorland/Photo Researchers; **21.2e–h:** From Ede D.A., Hinchliffe J.R. and Balls M., eds. (1977b) Vertebrate Limb and Somite Morphogenesis. *3rd Symp. Brit. Soc. Dev. Biol.* Cambridge: Cambridge University Press. Courtesy of J.R. Hinchliffe; **21.4:** Provided by Stanley K. Sessions; **21.5a–c:** From De Robertis E.M., Morita E.A. and Cho K.W.Y. (1991) *Development* 112: 669–678. Company of Biologists Ltd. Courtesy of E.M. De Robertis; **21.16e:** From Bryant S. (1977) *3rd Symp. Brit. Soc. Dev. Biol.* Courtesy of Susan Bryant; **21.19:** From Green J.B.A.,New H.V., and

Smith J.C. (1992) *Cell* 71: 731–739. Cell Press. Courtesy of J. Green; **21.21a–c:** From Gurdon J.B., Harger P., Mitchell A. and Lemaire P. (1994) *Nature* 371: 487–492, Fig. 2 a-c. Courtesy of J. Gurdon; **21.25a–b:** From Vogel O. (1982) *Wilhelm Roux's Arch. Dev. Biol.* 191: 134–136. Springer-Verlag. Courtesy of Otto Vogel.

Chapter 22: 22.1: From Gutjahr T., Frei E. and Noll M. (1993) *Development* 117: 609–623. Company of Biologists Ltd. Courtesy of T. Gutjahr and M. Noll; **22.5a–c:** From Carroll S.B., Laughon L., and Thalley B.S. (1988) *Genes and Dev.* 2: 883–890. Cold Spring Harbor Laboratory Press. Courtesy of Stephen Paddock, James Langeland and Sean Carroll; **22.9:** From Gonzales-Reyes A. and St Johnston D. (1998) *Development* 125: 2837–2846. Company of Biologists Ltd. Courtesy of A. Gonzales-Reyes; **22.11a–f:** From Ephrussi A. and Lehmann R. (1992) *Nature* 358: 387–392. Courtesy of Ruth Lehmann; **22.13a:** From Savant-Bhonsale S. and Montell D.J. (1993) *Genes and Dev.* 7: 2548–2555. Cold Spring Harbor Laboratory Press. Courtesy of Denise Montell; **22.15:** Reproduced from Jackle H., Tautz D., Schuh R., Seifert E. and Lehmann R. (1986) *Nature* 324: 668–670; **22.16:** Courtesy of Stephen Paddock, James Langeland and Sean Carroll; **22.17a–e:** Reproduced from Gaul U. and Jackle H. (1990) *Advances in Genetics* 27: 239–275; **22.20:** From Lawrence P.A. (1992) *The Making of a Fly: The Genetics of Animal Design*. Cambridge, MA: Blackwell. Provided by P.A. Lawrence; **22.21a–b:** From Lawrence P.A. and Johnson P. (1989) *Development* 105: 761–767. Company of Biologists Ltd. Provided by P.A. Lawrence; **22.22a–c:** From Howard K., Ingham P. and Rushlow C. (1988) *Genes and Dev.* 2: 1037–1046. Cold Spring Harbor Laboratory Press. Courtesy of P.W. Ingham; **22.24a–d:** Reproduced from Pankratz M.J., Seifert E., Gerwin N., Billi B., Nauber U. and Jackle H. (1990) *Cell* 61:309–317. Cell Press; **22.28:** From Ingham P.W. and Martinez-Arias A. (1992) *Cell* 68: 221–235. Cell Press. Courtesy of P.W. Ingham; **22.31a–b:** From Bermingham J.R., Jr., Martinez-Arias A., Petitt M.G. and Scott M.P. (1990) *Development* 109: 553–566. Company of Biologists Ltd. Courtesy of M.G. Petitt and M.P. Scott; **22.34a-b, 22.35a-c:** From Nusslein-Volhard C. (1991) *Development* (Supplement 1). Company of Biologists Ltd. Courtesy of C. Nusslein-Volhard; **22.39a–f:** From Ray R.P., Arora K., Nusslein-Volhard C. and Gelbart W.M. (1991) *Development* 113: 35–54. Company of Biologists Ltd. Courtesy of Robert P. Ray; **22.40a–c:** From Leptin M. and Grunewald B. (1990) *Development* 110: 73–84. Company of Biologists Ltd. Courtesy of B. Grunewald and Maria Leptin; **22.41** and **22.42:** From Roth S. (1993) *Development* 117: 1385–1396, Company of Biologists Ltd. Courtesy of S. Roth; **22.43b:** From Crick F.H.C. and Lawrence P.A. (1975) Compartments and polyclones in insect development. *Science* 189: 340–347. ©1975 American Association for the Advancement of Science. Provided by P.A. Lawrence; **22.45a,b:** From Blair S.S. (1992) *Development* 115: 21–33. Company of Biologists Ltd. Courtesy of Seth Blair; **22.47a:** From Lawrence P.A. and Morata G. (1993) *Nature* 366: 305–306. Print courtesy of Jim Williams, Stephen Paddock and Sean Carroll; **22.50a–c:** Photographs from Zecca M., Basler K. and Struhl G. (1995) *Development* 121: 2265–2278. Company of Biologists Ltd. Courtesy of K. Basler; **22.51a:** From Neumann and Cohen (1997) *Development* 124: 871–880. Company of Biologists Ltd. Courtesy of S. Cohen; **22.52a-c:** From D. Nellen, R. Burke, G. Struhl, and K. Basler (1996) *Cell* 85: 357–368. Cell Press. Courtesy of K. Basler; **22.53b:** From Ramírez-Weber F.A. and Kornberg T. (1999) *Cell* 97: 599–607. Cell Press. Courtesy of T. Kornberg; **22.54a–g:** From Patel N.H. (1993) Evolution of insect pattern formation: A molecular analysis of short germband segmentation. In: *Evolutionary Conservation of Developmental Mechanisms*, Spradling A.D. (ed.) 85–110. Reprinted by permission of Wiley-Liss, Inc., a subsidiary of John Wiley & Sons, Inc. Photographs courtesy of Nipam Patel.

Chapter 23 23.1: From Friedrich G. and Soriano P. (1991) *Genes and Dev.* 5: 1518. Cold Spring Harbor Laboratory Press. Courtesy of G. Friedrich and P. Soriano; **23.4a:** From Passner J.M., Ryoo H.D., Shen L., Mann R.S. and Aggarwaal A.K. (1999) *Nature* 397: 714–719. Courtesy of A.K. Aggarwaal; **23.4b:** From Piper D.E., Batchelor A.H., Chang C.P., Cleary M.L. and Wolberger C. (1999) *Cell* 96: 587–597. Cell Press. Courtesy of C. Wolberger; **23.8:** From Hunt P. and Krumlauf R. (1991) *Cell* 66: 1075–1078. Cell Press. Courtesy of Robb Krumlauf; **23.9:** From Wilkinson D.G., Bhatt S., Cook M., Boncinelli E. and Krumlauf R. (1989) *Nature* 341: 405–409. Courtesy of Robb Krumlauf; **23.13a–c:** From Le Mouellic H., Lallemand Y. and Brulet P. (1992) *Cell* 69: 251–264. Cell Press. Courtesy of Herve Le Mouellic and Yvan Lallemand; **23.15a–b:** From Kessel M., Balling R. and Gruss P. (1990) *Cell* 61: Cover. Cell Press. Courtesy of Michael Kessel; **23.17a–f:** From Holley S.A., Jackson P.D., Sasai Y., Lu B., De Robertis E.M., Hoffmann F.M. and Ferguson E.L. (1995) *Nature* 376: 249–253. Courtesy of E.L. Ferguson and E.M. De Robertis; **23.18a–i:** From Holley S.A., Jackson P.D., Sasai Y., Lu B., De Robertis E.M., Hoffmann F.M. and Ferguson E.L. (1995) *Nature* 376: 249–253. Courtesy of E.M. De Robertis; **23.19:** From Kelly R.O. (1981) The developing limb. In: *Morphogenesis and Pattern Formation*, T.G. Connelly, L.L. Brinkley and B.M. Carlson (eds.), 49–85. New York: Raven Press. Courtesy of Robert O. Kelly; **23.25(all):** From Riddle R.D., Johnson R.L., Laufer E. and Tabin C. (1993) *Cell* 75: 1401–1416. Cell Press. Courtesy of R.L. Johnson; **23.28a–c:** From Morgan B.A., Izpisua-Belmonte J.-C, Duboule D. and Tabin C.J. (1992) *Nature* 358: 236–239. Courtesy of Bruce A. Morgan.

Chapter 24 24.1: From Bowman J.L., Alvarez J., Weigel D., Meyerowitz E.M. and Smyth D.R. (1993) *Development* 119: 721–743. Company of Biologists Ltd. Courtesy of John L. Bowman; **24.4j–k:** Courtesy of Gerd Jurgens; **24.7b:** Provided by Thomas Eisner, Cornell University; **24.11a–j:** From Mayer U., Torres Ruiz R.A., Berleth T., Misera S. and Jurgens G. (1991) *Nature* 353: 402–407. Courtesy of Gerd Jurgens; **24.12b:** Reproduced from Drews G.N. and Goldberg R.B. (1989) *Trends in Genetics* 5: 256–261. Elsevier Science. Courtesy of G.N. Drews; **24.14a–d, 24.17a-e, 24.19a-d, 24.20:** From Meyerowitz E.M., Bowman J.L., Brockman L.L., Drews G.N., Jack T., Sieburth L.E. and Weigel D. (1991) *Development* (Supplement 1): 157–167. Company of Biologists Ltd. Courtesy of Elliot Meyerowitz; **24.21:** From Bowman J.L., Sakai H., Jack T., Weigel D., Mayer U. and Meyerowitz E.M. (1992) *Development* 114: 599–615. Company of Biologists Ltd. Courtesy of J.L. Bowman; **24.23a–h:** From Coen E.S. and Meyerowitz E.M. (1991) *Nature* 353: 31–37. Courtesy of Elliot Meyerowitz; **24.24:** From Coen E.S. and Meyerowitz E.M. (1991) *Nature* 353: 31–37. Courtesy of Enrico Coen; **24.25a–d:** From Drews G.N., Bowman J.L. and Meyerowitz E.M. (1991a) *Cell* 65: 91–1002. Cell Press. Courtesy of Gary Drews; **24.26:** From Schwarz-Sommer Z., Huijser P., Nacken W., Saedler H. and Sommer H. (1990) Genetic control of flower development by homeotic genes in Antirrhinum majus. *Science* 250: 931–936. ©1990 American Association for the Advancement of Science. Courtesy of Zsuzsanna Schwarz-Sommer.

Chapter 25 25.1: Provided by Dr. Fabio Piano; **25.2a:** Courtesy of Einhard Schierenberg; **25.3a–i:** From Schierenberg E. (1986) *J. Embryol. Exp. Morphol.* 97 (Supplement): 31–44. Company of Biologists Ltd. Courtesy of Einhard Schierenberg; **25.6a–e:** From Strome S. (1989) *Int. Rev. Cytol.* 114: 81–123. Florida: Academic Press. Courtesy of Susan Strome; **25.9a–i:** From Bowerman B., Draper B.W., Mello C.C. and Priess J.R. (1993) *Cell* 74: 443–452. Cell Press. Provided by Bruce Bowerman; **25.12b–d:** From Schlesinger A., Shelton C.A., Maloof J.N. and Bowerman B. (1999) *Genes & Devel.* 13: 2028–2038. Courtesy of A. Schlesinger and B. Bowerman; **25.17a–b:** From Ellis H.M. and Horvitz H.R. (1986) *Cell* 44: 817–829. Cell Press. Courtesy of Hilary Ellis; **25.19a–b:** From Ellis R.E. and Horvitz H.R. (1991) *Development* 112: 591–603. Company of Biologists Ltd. Courtesy of Ronald Ellis; **25.22a–d:** From Hill R.J. and Sternberg P.W. (1992) *Nature* 358: 470–476. Courtesy of Russell J. Hill; **25.23a–c:** From Hill R.J. and Sternberg P.W. (1992) *Nature* 358: 470–476. Courtesy of Russell Hill.

Chapter 26 26.1: Toni Angermayer/Photo Researchers; **26.8:** From Burgoyne P.S., Buehr M., Koopman P., Rossant J. and McLaren A.

(1988) *Development* 102: 443–450. Company of Biologists Ltd. Courtesy of Paul S. Burgoyne; **26.11a–d:** Reproduced from Palmer M.S., Sinclair A.H., Berta P., Ellis N.A., Goodfellow P.N., Abbas N.E. and Fellous M. (1989) *Nature* 342: 937–942. Reprinted. Copyright 1989 Macmillan Magazines Limited; **26.12** and **26.13:** From Koopman P., Gubbay J., Vivian N., Goodfellow P. and Lovell-Badge R. (1991) *Nature* 351: 117–121. Courtesy of R. Lovell-Badge.

Chapter 27 27.1(both): From Young L.J., Lopreato G., Horan K. and Crews D. (1994) *J. Comp. Neurol.* 347: 288–300. Reprinted by permission of Wiley-Liss, Inc., a subsidiary of John Wiley & Sons, Inc. Photographs courtesy of Larry Young and David Crews.; **27.6b:** Courtesy of Leonard Pinsky, McGill University; **27.10a–d:** From Breedlove S.M. (1992) Sexual differentiation of the brain and behavior. In: *Behavioral Endocrinology*, J.B. Becker, S.M. Breedlove and D. Crews (eds.), 29–70. Cambridge, MA: MIT Press. Courtesy of Roger Gorski; **27.15a-c:** Courtesy of Michael Ashburner; **27.18:** From Robinow S., Talbot W.S., Hogness D.S. and Truman J.W. (1993) *Development* 119: 1251–1259. Company of Biologists Ltd. Courtesy of Steven Robinow; **27.19:** From Karim F.D. and Thummel C.S. (1991) *Genes and Dev.* 5: 1067–1079. Cold Spring Harbor Laboratory Press. Courtesy of Felix D. Karim and Carl S. Thummel; **27.23:** From Tata J.R., Kawahara A. and Baker B.S. (1991) *Dev. Biol.* 146: 72–80. ©1991 by Academic Press, reproduced by permission of the publisher. Courtesy of Jamshed R. Tata; **27.26:** Courtesy of S. Duhon and G. Malacinski.

Chapter 28 28.1(both): From Levi-Montalcini R. (1964) Growth control of nerve cells by a protein factor and its antiserum. *Science* 143: 105–110. ©1964 American Association for the Advancement of Science; **28.6:** From Harrison R.G. (1935) Heteroplastic grafting in embryology. *The Harvey Lectures*, 1933–1934: 116–157; **28.7a–b:** Reproduced from Twitty V.C. and Elliot H.A. (1934) *J. Exp. Zool.* 68: 247. Reprinted by permission of Wiley-Liss, Inc., a subsidiary of John Wiley & Sons, Inc.; **28.8:** From Harrison R.G. (1935) Heteroplastic grafting in embryology. *The Harvey Lectures*, 1933–1934: 116–157; **28.9:** Globe Photos; **28.11a–b:** From Levi-Montalcini R. (1964) Growth control of nerve cells by a protein factor and its antiserum. *Science* 143: 105–110. ©1964 American Association for the Advancement of Science; **28.13a–b:** From Levi-Montalcini R. (1964) Growth control of nerve cells by a protein factor and its antiserum. *Science* 143: 105–110. ©1964 American Association for the Advancement of Science; **28.15a–b:** Reproduced from Weinberg R.A. (1983) *Sci. Amer.* 249: 127; **28.22:** Reproduced from Weinberg R.A. (1988) *Sci. Amer.* 259: 46.

Chapter 29 29.1: Figaro Magazine/Liaison Agency; **29.10(both):** Williams & Wilkins; **29.17:** R.J. Baker/Visuals Unlimited.

LINE ART CREDITS

The figures listed below are based on numerical data and/or diagrams of the authors quoted. For each entry, the bibliographical data may be found in the reference list. Credits for interpretive drawings of photos are listed among the photo credits.

Chapter 1 1.4: Duval M. (1889); Patten (1971). **1.5:** Moore, K.L. (1982). **1.6:** Gilbert S.F. (1991). **1.8:** Carlson, B.M. (1988). **1.9:** Huxley, J.S. (1932). **1.11:** Carlson (1988). **1.12:** Sulston, Horvitz (1977) Sulston, J.E., White J.G.(1980). **1.13:** Willier and Oppenheimer (1964). **1.15:** Spemann (1938) and Gilbert (1988). **1.17:** Hamburger, V. (1988).

Chapter 2 2.2: Alberts et al. (1989). **2.3:** Burnside (1971). **2.5:** Alberts et al. (1989). **2.7a:** Albert et al. (1989). **2.14:** McIntosh and McDonald (1989). **2.15:** Beams H.W., Kessel R.G. (1976). **2.16:** Draetta et al. (1989). **2.17:** Alberts et al. (1989). **2.18:** Alberts et al. (1989). **2.20:** Alberts et al. (1989). **2.22:** Hay (1991b). **2.23:** Keeton and Gould (1986). **2.24:** Darnell et al. (1990). **2.25:** Darnell et al. (1990). **2.26:** Darnell et al. (1990). **2.27:** Darnell et al. (1990)

Chapter 3 3.2: Wilson (1925). **3.3:** Langman (1981). **3.6:** Tuchmann-Duplessis et al. (1972). **3.7:** (a, c) Carlson (1988). **3.8:** Bloom and Fawcett (1975). **3.9:** Bloom and Fawcett (1975). **3.13:** King (1970). **3.18:** Austin (1965). **3.19:** Gebauer and Richter (1997). **3.20:** Sagata (1997). **3.23:** Carlson (1988). **3.24a:** Hinton (1970).

Chapter 4. 4.2: Epel (1977). **4.3:** Miller (1966). **4.6:** Glabe and lennarz (1979). **4.9b-d:** Yanagimachi and Noda (1970). **4.10:** Allen and Green (1997). **4.12:** Bleil and Wassarman (1980). **4.16:** Langman (1981). **4.18a:** Gilbert (1988)

Chapter 5 5.3: Czihak, G., Langer, H., and Ziegler, H. (1976). **5.4:** Horstadius, S. (1973); Summers, R.G., Morrill, J.B., Leith, A., Marko, M. Piston D.W., and Stonebraker A.T. (1993). **5.5:** Alberts et al. (1989). **5.6:** Balinsky, B.I. (1975). **5.7:** Grant, P. (1978). **5.8:** Gasser, R.F. (1975). **5.9:** Gulyas, B.J. (1975). **5.11:** Carlson, B.M. (1988). **5.15:** Saunders, J.W., Jr. (1970). **5.17:** Foe and Alberts (1983). **5.19:** Fullilove, S.L., Jacobson, A.G. (1971). **5.22:** Rappaport, R. (1974). **5.23:** Rappaport and Rappaport (1994). **5.24:** Freeman, G. (1983). **5.25:** White and Strome (1996). **5.27:** Hibins et al. (1998). **5.29:** Morgan, T.H. (1927). **5.30:** Sinnot et al. (1958). **5.32:** Foe, V.E. and Alberts, B.M. (1983); Edgar, B.A. and Schubiger, G. (1986). **5.33:** Edgar, B.A. and O'Farrell, P.H. (1989).

Chapter 6 6.2: Kimelman et al. (1992). **6.4:** Garcia-Bellido et al. (1979). **6.m1:** Garcia-Bellido et al. (1979). **6.5:** Lawrence (1992). **6.7:** Lüer and Technan (1992). **6.8:** Lüer and Technan (1992). **6.10:** Balinsky (1975). **6.11:** Saxén and Toivonene (1962). **6.15:** Slack (1991). **6.16:** Slack (1991). **6.18:** Nöthiger (1972). **6.19:** Gehring (1968). **6.20:** Hadorn (1968). **6.21:** Kaufman (1973). **6.22:** Langman (1981).

Chapter 7 7.3: King (1970). **7.5c:** Kühn (1961). **7.10:** Shepard (1982). **7.11:** Spemann (1967). **7.12:** King (1966). **7.13:** McKinnell (1978). **7.14:** Gurdon (1968). **7.16:** Laskey and Gurdon (1970). **7.17a:** Wilmut I., Schnieke A.E., McWhir J. , Kind A.J., Campbell K.H.S. (1997). **7.18:** Gurdon (1968). **7.20:** De Robertis and Gurdon (1979).

Chapter 8 8.3: Bodemer (1968). **8.4:** Saunders (1982). **8.6:** Verdonk and Cather (1983). **8.9:** Illmensee and Mahowald (1974); Mahowald et al. (1979). **8.11:** Driever et al. (1990). **8.13:** Browder et al. (1991). **8.14:** Nishida (1992) and Satoh (1994). **8.15a:** Jeffery and Swalla (1997). **8.15b:** Nishida (1992). **8.16:** (b) Jeffery and Meier (1983). **8.18:** Hirata et al., (1995). **8.24:** Render (1983). **8.25:** Etkin (1997).

Chapter 9 9.3: Walbot and Holder (1987). **9.6:** Quatrano and Shaw (1997). **9.7 :** Zhang et al. (1998). **9.12:** (b) Larabell C.A., Rowning B.A., Wells J., Wu M., Gerhart J. (1996). **9.15:** Gimlich and Gerhart (1984). **9.17:** Dale and Slack (1987b). **9.20:** Dale and Slack (1987b).

Chapter 10 10.5: Hörstadius (1987), Ettensohn (1992). **10.6:** Lane et al. (1993). **10.8:** Fristrom (1976). **10.9:** (b) Hardin (1989). **10.11:** Gerhart and Keller (1986). **10.12:** Holtfreter (1943). **10.13:** Balinsky (1975). **10.14:** Vogt (1929). **10.15:** R.E. Keller (1981). **10.16:** Holtfreter (1933). **10.17:** R.E. Keller et al. (1985). **10.18:** (a) R.E. Keller et al. (1985). **10.19:** R.E. Keller et al. (1985, 1992a). **10.20:** (g) R.E. Keller et al. (1985). **10.21:** Driever (1995), Langeland and Kimmel (1997). **10.22:** Langeland and Kimmel (1997). **10.23:** Patten (1971), Balinksy (1975). **10.24:** Patten (1971), Balinksy (1975). **10.26:** Mittenthal and Jacobson (1990). **10.27:** Patten (1971). **10.30:** Luckett (1978). **10.31:** Langman (1981). **10.32:** Langman (1981).

Chapter 11 11.2: Townes and Holtfreter (1955). **11.3:** Steinberg (1998). **11.5:** Phillips and Steinberg (1969). **11.6:** Cunningham (1991). **11.8:** Pollerberg and Beck-Sickinger (1993). **11.12:** Alberts et al. (1989). **11.13:** (b) Alberts et al. (1989). **11.14:** Alberts et al. (1989). **11.15:** Yamada (1991). **11.16:** Hynes (1992).

11.17: Bernfield and Sanderson (1990). **11.18:** Edleman (1984). **11.19:** (b) Fink and McClay (1985). **11.22:** (a) Shi et al. (1989). **11.23:** (a) Boucaut et al. (1984). **11.24:** (a) Boucaut et al. (1984). **11.28:** Hay (1991b).

Chapter 12 **12.2:** Langman (1981). **12.3**: Hamburger (1960) **12.4:** Lopez-Antunex (1971); More (1982). **12.5:** Burnside (1971); Carlson (1988). **12.6:** Smith and Schoenwolf (1997). **12.7:** Langman (1981). **12.8:** Langeland J.A., Kimmel C.B. (1997). **12.9:** Gordon and Jacobson (1978). **12.10:** Jacobson and Gordon (1976); Gordon and Jacobson (1978). **12.11:** Jacobson (1991) **12.12:** (a) Gordon and Jacobson (1978). **12.14:** Jacobson (1981). **12.15:** Jacobson et al. (1986). **12.17:** Spemann (1938). **12.18:** Carlson (1988). **12.20:** (a) Bodemer (1968). **12.22:** Holtfreter (1943). **12.23:** (a)(b) Keller and Danilchik (1988).

Chapter 13 **13.2:** Carlson (1988). **13.3:** Langman (1981). **13.4:** Carlson (1988). **13.5:** Shepherd (1988). **13.7:** Langman (1981). **13.8:** Lumsden (1991). **13.10:** Tanabe Y. and Jessel T.M. (1996). **13.11:** Ericson J., Briscoe J., Rashbass P., Van Heyningen V., and Jessel T.M. (1997). **13.12:** Langman (1981). **13.13:** Carlson (1988). **13.14:** Langman (1981). **13.15:** Langman (1981). **13.16:** Langman (1981). **13.17:** Langman (1981). **13.18:** Langman (1981). **13.19:** Langman (1981) . **13.20:** Hopper and Hart (1985). **13.21:** Patten (1964). **13.22:** Balinksy (1975). **13.24:** Moury and Jacomson (1989). **13.25:** (a) Weston (1963). **13.27:** Carlson (1988). **13.28:** Le Douarin (1986). **13.29:** Perris and Bronner-Fraser (1989). **13.32:** Perris et al. (1988). **13.33:** Carlson (1988). **13.34:** Langman (1981). **13.36:** Langman (1981). **13.37:** Langman (1981). **13.39:** Langman (1981). **13.40:** Langman (1981). **13.41:** Bloom and Fawcett (1975). **13.42:** Langman (1981). **13.43:** Langman (1981), Carlson (1988).

Chapter 14 **14.2:** Langman (1981). **14.3:** Langman (1981). **14.4:** Langman (1981). **14.5:** Torey and Feduccia (1991). **14.6:** Kühn (1961). **14.7:** (a) Starck (1965). **14.7:** (b) Langman (1981). **14.8:** Langman (1981). **14.9:** Romanes (1901) . **14.10:** Carlson (1988). **14.11:** Langman (1981). **14.12:** Patten (1971). **14.13:** Mittenthal and Jacobson (1990). **14.14:** (a) Meier (1979). **14.17:** Langman (1981). **14.19:** Alberts et al. (1989). **14.21:** Langman (1981). **14.24:** Patten (1988); Vize et al. (1997). **14.26:** Langman (1981). **14.27:** Langman (1981). **14.28:** Langman (1981). **14.29:** Carlson (1988). **14.30:** Langman (1981). **14.31a, f:** Cleaver O. and Krieg P. (1998). **14.32a–d:** Langman (1981). **14.33:** Langman (1981). **14.34:** (a) Romer (1976) (b) Langman (1981). **14.35:** Langman (1981). **14.36:** (a) Bloom and Fawcett (1975). **14.37:** Carlson (1988). **14.38:** Moore (1982). **14.39:** Carlson (1988).

Chapter 15 **15.3:** Wilson E.B. (1925). **15.7:** Chalfie et al. (1981). **15.23:** Palmiter R.D., Brinster R.L., Hammer R.E., Trumbauer M.E., Rosenfeld M.G., Birnberg N.C. and Evans R.M. (1982). **15.m4:** Villee et al. (1989). **15.m5:** Browder et al. (1991).

Chapter 16 **16.5:** Lewin B. (1990). **16.6a:** Hiromi Y. and Gehring W.J. (1987). **16.10:** Driever W. and Nusslein-Volhard C. (1989). **16.11:** Driever W., Thoma G. and Nusslein-Volhard C. (1989a). **16.12:** McKnight G.S., Pennequin P. and Schimke R.T. (1975). **16.13:** Darnell J., Lodish H. and Baltimore D. (1990). **16.20:** After Kaufman, 1973. **16.23:** Alberts B., Bray D., Lewis J., Raff M., Roberts K. and Watson J.D. (1989). **16.24:** After Surani, 1998.

Chapter 17 **17.2b, c:** Chambon (1982). **17.5a, b:** Alberts B., Bray D., Lewis J., Raff M., Roberts K. and Watson J.D., (1989). **17.10:** Baker B.S. (1989). **17.13:** Darnell J., Lodish H. and Baltimore D. (1990). **17.17:** Wells D.E., Showman R.M., Klein W.H. and Raff R.A. (1981). **17.20:** Mangiarotti et al. (1983). **17.21:** Guyette et al. (1979). **17.22:** Alberts et al. (1989).

Chapter 18 **18.2:** Alberts B., Bray D., Lewis J., Raff. M., Roberts K. and Watson J.D. (1989). **18.9:** Sheets M.D., Fox C.A., Hunt T., Vande Woude G. and Wickens M. (1994). **18.10:** Sheets M.D., Fox C.A., Hunt T., Vande Woude G. and Wickens M. (1994). **18.11:** Harvey E.B. (1940). **18.17:** Darnell J., Lodish H. and Baltimore D. (1990). **18.18:** Alberts B., Bray D., Lewis J., Raff. M., Roberts K. and Watson J.D. (1989). **18.19:** Branden C. and Tooze J. (1991). **18.20:** Darnell J., Lodish H. and Baltimore D. (1990).

Chapter 19 **19.2:** Grant P. (1978). **19.4b:** Butler P.J. and Klug A. (1978). **19.5** and **19.7:** Butler P.J. and Klug A. (1978). **19.8:** Grant P. (1978). **19.9:** Wood W.B. (1980). **19.11:** Weissmann C. (1994). **19.12:** Prusiner S.B., Groth D., Serban A., Koehler R., Foster D., Torchia M., Burton D., Yang S.-L. and DeArmond S.J. (1993). **19.13:** (a) Alberts B., Bray D., Lewis J., Raff M., Roberts K., and Watson J.D. (1989). **19.16:** Sonneborn T.M. (1970). **19.18:** Frankel J. (1989). **19.19:** Alberts B., Bray D., Lewis J., Raff M., Roberts K., and Watson J.D. (1989).

Chapter 20 **20.5:** Bloom W. and Fawcett D.W. (1975); Alberts B., Bray D., Lewis J., Raff M., Roberts K., and Watson J.D. (1989). **20.6:** Alberts B., Bray D., Lewis J., Raff M., Roberts K., and Watson J.D. (1989). **20.7:** Alberts B., Bray D., Lewis J., Raff M., Roberts K., and Watson J.D. (1989). **20.8:** Gierer A. (1974). **20.11:** David C.N. and Murphy S. (1977). **20.12:** Golde D.W. (1991). **20.13:** Alberts B., Bray D., Lewis J., Raff M., Roberts K., and Watson J.D. (1989). **20.14:** G. Keller et al. (1985). **20.16:** Alberts B., Bray D., Lewis J., Raff M., Roberts K., and Watson J.D. (1989). **20.17:** Alberts B., Bray D., Lewis J., Raff M., Roberts K., and Watson J.D. (1989). **20.18:** Olson E.N. (1990). **20.21:** From Myer A., Wagner D.S., Vivian J.L., Olson E.N. and Klein W.H. (1997).

Chapter 21 **21.3:** Stocum and Fallon (1982). **21.6:** Graham and Warering (1976); Hans Spemann (1936). **21.7:** Jacobson (1966). **21.8:** Postlethwait and Schneiderman (1971). **21.10:** Walbot and Holder (1987). **21.11:** Saunders (1982). **21.13:** Bryant et al. (1981). **21.15:** Bryant et al (1981). **21.16:** Bryant (1977). **21.19:** Green J.B.A., New H.V. and Smith J.C. (1992). **21.20:** Gurdon J.B., Harger P., Mitchell A., Lemaire P. (1994). **21.22:** Gurdon J.B., Harger P., Mitchell A., Lemaire P. (1994). **21.23:** Gurdon J.B., Mitchell A., Mahony D. (1995). **21.24:** Sander (1975).

Chapter 22 **22.2:** (a-e) Nusslein-Volhard C. (1991), (f) Akam M. (1987), (g) Martinez-Arias A. and Lawrence P.A. (1985). **22.3:** (a-b) Nusslein-Volhard C. (1991). **22.3:** (c) Akam M. (1987). **22.6:** Nusslein-Volhard C., Frohnhofer H.G. and Lehmann R. (1987). **22.7:** Nusslein-Volhard C. (1991). **22.9:** Gonzales et al. (1998). **22.10:** Gao Q. and Finkelstein R. (1998), Driever W., Thoma G. and Nusslein-Volhard C. (1989a); Pignoni F., Steingrimsson E. and Lengyel J.A. (1992); Walldorf U. and Gehring W.J. (1992). **22.12:** Ephrussi A. and Lehmann R. (1992). **22.14:** Scott M.P. and O'Farrell P.H. (1986). **22.18:** Hulskamp M. and Tautz D. (1991). **22.19:** Tautz D. (1988). **22.23:** Pankratz M.J., Seifert E., Gerwin N., Billi B., Nauber U. and Jackle H. (1990); Lawrence P.A. (1992). **22.25:** Pankratz M.J., Seifert E., Gerwin N., Billi B., Nauber U. and Jackle H. (1990). **22.26:** After Akam M. (1987). **22.29:** Lawrence P.A. and Morata G. (1993); Peifer M. and Bejsovec A. (1992). **22.31:** Lawrence and Morata (1994). **22.32:** Lewis E.B. (1963), Peifer et al. (1987). **22.33:** Kaufman T.C., Seeger M.A. and Olsen G. (1990), Gilbert (1991), Slack (1991). **22.43:** (a) Morata G. and Lawrence P.A. (1978). **22.44:** Morata G. and Lawrence P.A. (1975). **22.49:** After Basler and Struhl (1994).

Chapter 23 **23.2:** Gehring W.J., Muller M., Affolter M., Percival-Smith A., Billeter M., Xian Y.Q., Otting G. and Wuthrich K. (1990). **23.3:** Graham A., Papalopulu N. and Krumlauf R. (1989); Mann (1995); De Rosa et al. (1999); Passner et al. (1999). **23.6:** Scott M.P. (1992). **23.10:** Hunt P., Guilsano M., Cook M., Sham M., Faiella A., Wilkinson D., Boncinelli E. and Krumlauf R. (1991a). **23.11:** Fraser S., Jeynes R. and Lumsden A. (1990). **23.12:** Le Mouellic H., Lallemand Y. and Brulet P. (1992). **23.14:** Kessel M. and Gruss P. (1990). **23.16:** After Holley and Ferguson (1997). **23.20:** Saunders J.W., Jr. (1982). **23.21:** After Cohn et al. (1997); Coates and Cohn (1998). **23.22:** Summerbell D. and Lewis J.H. (1975). **23.23:** After Michaud et al. (1997); Zeller and Duboule (1997). **23.24:** Saunders

J.W., Jr. (1982). **23.26:** Tabin C.J. (1991). **23.27:** Morgan B.A., Izpisua-Belmonte J.-C., Duboule D., Tabin C.J. (1920).

Chapter 24 **24.3:** Drews G.N. and Goldberg R.B. (1989). **24.4a–i:** Meinke D.W. (1991b). **24.5:** Meinke D.W. (1991b). **24.6:** Jurgens G., Mayer U., Ruiz R.A.T., Berleth T. and Misera S. (1991). **24.8:** Pang P.P. and Meyerowitz E.M (1987). **24.9:** Jurgens G., Mayer U., Ruiz R.A.T., Berleth T. and Misera S. (1991). **24.10:** Mayer U., Torres Ruiz R.A., Berleth T., Misera S. and Jurgens G. (1991). **24.11:** Bowman et al. (1991). **24.12:** (a) Drews G.N. and Goldberg R.B. (1989). **24.16:** Meyerowitz E.M., Bowman J.L., Brockman L.L., Drews G.N., Jack T., Sieburth L.E. and Weigel D. (1991). **24.22:** Coen E.S. and Meyerowitz E.M. (1991). **24.27:** Jack T., Brockman L.L. and Meyerowitz E.M. (1992).

Chapter 25 **25.2b, c:** Hodgkin J. (1985). **25.4:** Sulston J.E., Schierenberg E., White J.G. and Thomson J.N. (1983). **25.5:** Wilkins A.S. (1986). **25.7:** Guo and Kemphues (1996). **25.8:** Mello C.C., Draper B.W., Krause M., Weintraub H. and Priess J.R. (1992) . **25.10:** Bowerman (1995). **25.11:** Goldstein B. (1992). **25.12a** and **25.14:** Schlesinger A., Shelton C.A., Maloof J.N. and Bowerman B. (1999). **25.16:** Ruvkun G., Wightman B., Burglin T. and Arasu P. (1991). **25.18:** Ellis H.M. and Horvitz H.R. (1986). **25.20** and **25.21:** Horvitz H.R. and Sternberg P.W. (1991).

Chapter 26 **26.2:** Bull J.J. (1983). **26.4:** Gartler S.M. and Riggs A.D. (1983). **26.5:** Davidson R.G., Nitowsky H.M. and Childs B. (1963). **26.6:** Langman (1981). **26.7:** McLaren A. (1991). **26.9:** Affara N.A. (1991). **26.10:** Palmer M.S., Sinclair A.H., Berta P., Ellis N.A., Goodfellow P.N., Abbas N.E. and Fellous M, (1989). **26.15:** Hodgkin J. (1990). **26.16** and **26.17:** Hodgkin J. (1992). **26.18:** Cline T.W. and Meyer B.J. (1996). **26.19** and **26.20:** Hodgkin J. (1992). **26.21:** Walbot V. and Holder N. (1987). **26.22:** Oliver B., Kim Y.-J. and Kajer B.S. (1993). **26.23:** Hodgkin (1992).

Chapter 27 **27.2:** Breedlove (1992). **27.4, 27.5** and **27.7:** Langman (1981). **27.8:** Kelley and Brenowitz (1992). **27.9:** Arnold (1980). **27.11:** Andres and Thummel (1992). **27.12:** Riddiford L.M. (unpublished). **27.13:** Wigglesworth (1972). **27.14:** Andres and Thummel (1992). **27.16:** Ashburner et al. (1974); Andres and Thummel (1992). **27.17:** Talbot et al.(1993). **27.20:** White K.P. et al. (1997). **27.21:** Wistchi (1956). **27.22:** Geigy (1941), Schwind (1933). **27.24:** Tata et al. (1991). **27.27:** Yaoita and Brown (1990).

Chapter 28 **28.3:** Weiss, P. (1939). **28.4:** Morgan T.H. (1927). **28.5:** Harrison R.G. (1935). **28.10:** Karin M., Castrillo J.-L. and Theill L.E. (1990). **28.12:** Levi-Montalcini R. (1979). **29.17:** Weinberg R.A. (1988). **28.18–28.20:** Bishop J.M. (1982). **28.21:** Croce C.M., Thierfelder W., Erickson J., Nishikura K., Finana J., Lenoir G.M.. and Nowell P.C. (1983); Dalla-Favera R., Bregni M., Erikson J., Patterson D., Gallo R.C. and Croce C.M. (1982); Erikson J., Nishikura K., Ar-Rushdi A., Finan J., Emanuel B., Lenoir G., Nowell P.C. and Croce C.M. (1983). **28.23:** After Johnston and Edgar (1998). **28.24:** After Simpson and Morata (1981).

Chapter 29 **29.2a:** Myers and Manton (1984), Arking (1998). **29.2b:** Ham and Veomet (1980), Arking (1998). **29.4:** Arking (1998); Finch (1990). **29.5:** Luckinbill et al. (1984). **29.7:** Sgrò and Partridge (1999). **29.8:** Austad (1993). **29.9:** Walker et al. (2000). **29.11:** McCay et al. (1943), Rose (1991). **29.12:** Fridovich (1978). **29.13:** Levine and Stadtman (1996). **29.14** and **29.15:** Carney et al. (1991). **29.16:** Hayflick (1966), Rose (1991).

subject index

Boldfaced numbers refer to pages where terms are first defined. Page numbers followed by *f* or *t* indicate figures (*f*) or tables (*t*).